THE CEREBRAL CORTEX AND THALAMUS

THE CEREBRAL CORTEX AND THALAMUS

EDITED BY

W. Martin Usrey and S. Murray Sherman

OXFORD
UNIVERSITY PRESS

OXFORD
UNIVERSITY PRESS

Oxford University Press is a department of the University of Oxford. It furthers the University's objective of excellence in research, scholarship, and education by publishing worldwide. Oxford is a registered trade mark of Oxford University Press in the UK and certain other countries.

Published in the United States of America by Oxford University Press
198 Madison Avenue, New York, NY 10016, United States of America.

CIP data is on file at the Library of Congress

ISBN 978–0–19–767615–8

DOI: 10.1093/med/9780197676158.001.0001

Printed by Integrated Books International, United States of America

MIX
Paper
FSC FSC® C183721

CONTENTS

CONTRIBUTORS

Samantha V. Abram, PhD
Department of Psychiatry and Behavioral Sciences
University of California, San Francisco
San Francisco, CA, USA

László Acsády, PhD
Laboratory of Thalamus Research
Institute of Experimental Medicine
Budapest, Hungary

Hillel Adesnik, PhD
Department of Molecular and Cell Biology
University of California, Berkeley
The Helen Wills Neuroscience Institute
University of California, Berkeley
Berkeley, CA, USA

Tim J. Anderson, MD, FRACP
New Zealand Brain Research Institute
Christchurch, New Zealand
Department of Medicine
University of Otago
Christchurch, New Zealand
Department of Neurology
Christchurch Hospital,
Te Whatu Ora Health
Canterbury, New Zealand

Alessandra Angelucci, MD, PhD
Department of Ophthalmology and Visual Science
Moran Eye Institute, University of Utah
Salt Lake City, UT, USA

Victoria M. Bajo, MD, PhD
Department of Physiology,
Anatomy and Genetics
University of Oxford
Oxford, UK

Mary K. L. Baldwin, PhD
Department of Psychology
University of California, Davis
Davis, CA, USA

Sara Bandiera, BSc, MSc
Department of Physiology, Anatomy and Genetics
University of Oxford
Oxford, UK

Annamaria Barczak, PhD
Translational Neuroscience Division
Center for Biomedical Imaging and Neuromodulation
Nathan Kline Institute
Orangeburg, NY, USA

Maxim Bazhenov, PhD
Department of Medicine
University of California, San Diego
La Jolla, CA, USA

Martha Bickford, PhD
Department of Anatomical Sciences and Neurobiology
University of Louisville
Louisville, KY

Martina Bocchetta, MSc, PhD
Dementia Research Centre, Department of
Neurodegenerative Disease
UCL Queen Square Institute of Neurology
University College London
London, UK
Centre for Cognitive and Clinical Neuroscience,
Division of Psychology
Department of Life Sciences, College of Health,
Medicine and Life Sciences
Brunel University London
London, UK

James A. Bourne, PhD
Section on Cellular and
Cognitive Neurodevelopment
National Institute of Mental Health
National Institutes of Health
Bethesda, MD, USA

Farran Briggs, PhD
Ernest J. Del Monte Institute for Neuroscience
University of Rochester
Rochester, NY, USA

Solange P. Brown, MD, PhD
Solomon H. Snyder Department
of Neuroscience
Johns Hopkins University School of Medicine
Baltimore, MD, USA
Kavli Neuroscience Discovery Institute
Johns Hopkins University
Baltimore, MD, USA

Randy M. Bruno, PhD
Department of Physiology, Anatomy and Genetics
University of Oxford
Oxford, UK

Thomas Budde, PhD
Institute of Physiology I
Universität Münster
Münster, Germany

Andreas Burkhalter, PhD
Department of Neuroscience
Washington University School of Medicine
St. Louis, MO, USA

Jianhua Cang, PhD
Department of Biology
Department of Psychology
University of Virginia
Charlottesville, VA, USA

Jérôme Carriot, PhD
Department of Physiology
McGill University
Montréal, Québec, Canada

Francesco Cavarretta, PhD
Department of Biology
Emory University
Atlanta, GA, USA

Maurice J. Chacron, PhD
Department of Physiology
McGill University
Montréal, Québec, Canada

Chinfei Chen, MD, PhD
Department of Neurology
F. M. Kirby Neurobiology Center
Children's Hospital, Boston
Harvard Medical School
Boston, MA, USA

Paul Cisek, PhD
Department of Neuroscience
University of Montréal
Montréal, Québec, Canada

Francisco Clasca, MD, PhD
Department of Anatomy and Neuroscience
Autónoma de Madrid University, Medical School
Madrid, Spain

Gavin J. Clowry, PhD
Newcastle University Biosciences Institute
Newcastle upon Tyne, UK

Emmanuelle Courtiol, PhD
Université Claude Bernard Lyon 1
CNRS, INSERM, CRNL U1028 UMR5292
Bron, France

Kevin P. Cross, PhD
Department of Cell Biology and Physiology,
Neuroscience Center
University of North Carolina at Chapel Hill
Chapel Hill, NC, USA

Vincenzo Crunelli
Neuroscience Division, School of Bioscience
Cardiff University
Cardiff, UK

Kathleen E. Cullen, PhD
Department of Biomedical Engineering
Department of Otolaryngology-Head and Neck Surgery
Department of Neuroscience
Johns Hopkins University School of Medicine
Baltimore, MD, USA
Kavli Neuroscience Discovery Institute
Johns Hopkins University
Baltimore, MD, USA

John C. Dalrymple-Alford, PhD
Te Kura Mahi ā-Hirikapo | School of Psychology, Speech and Hearing
University of Canterbury
Christchurch, New Zealand
New Zealand Brain Research Institute
Christchurch, New Zealand
Department of Medicine
University of Otago
Christchurch, New Zealand

Anita A. Disney, PhD
Department of Neurobiology
Duke University
Durham, NC, USA

Adrian J. Duszkiewicz, PhD
Division of Psychology
Faculty of Natural Sciences
University of Stirling
Stirling, UK

G. Sean Escola, MD, PhD
Department of Psychiatry
Zuckerman Institute
Columbia University
New York, NY, USA

Adrienne L. Fairhall, PhD
Departments of Physiology and Biophysics and Applied Mathematics
and Computational Neuroscience Center
University of Washington
Seattle, WA, USA

Angela Y. Fan, PhD
Section on Cellular and Cognitive Neurodevelopment
National Institute of Mental Health
National Institutes of Health
Bethesda, MD, USA
Australian Regenerative Medicine Institute
Monash University
Clayton, VIC, Australia

Ares Fisher, PhD
Paul G. Allen School of Computer Science and Engineering & Center for Neurotechnology
University of Washington
Seattle, WA, USA

Alfredo Fontanini, MD, PhD
Department of Neurobiology and Behavior
Stony Brook University
Stony Brook, NY, USA

Judith M. Ford, PhD
Mental Health Service
Veterans Affairs San Francisco Healthcare System
San Francisco, CA, USA
Department of Psychiatry and Behavioral Sciences
University of California, San Francisco
San Francisco, CA, USA

Gonzalo Forno, MSc, PhD
School of Psychology
Universidad de los Andes
Santiago, Chile

Karl Friston, MA
Wellcome Trust Centre for Neuroimaging
Institute of Neurology
University College London
London, UK

Veronica C. Galvin, PhD
Department of Neurobiology
Duke University
Durham, NC, USA

Ryan Golden, PhD
Neurosciences Graduate Program
University of California, San Diego
La Jolla, CA, USA

Sophie V. Griswold, BS, MS; PhD Candidate
Department of Biology
Brandeis University
Waltham, MA, USA

Alexander Groh, PhD
Medical Biophysics
Institute of Physiology and Pathophysiology
Ruprecht Karl University of Heidelberg
Heidelberg, Germany

William Guido, PhD
Department of Anatomical Sciences and Neurobiology
University of Louisville, School of Medicine
Louisville, KY, USA

Zengcai V. Guo, PhD
School of Medicine
Tsinghua University
Beijing, China
Tsinghua-Peking Joint Center for Life Sciences
Beijing, China
IDG/McGovern Institute for Brain Research
Tsinghua University
Beijing, China

Michael M. Halassa, MD, PhD
Department of Neuroscience
Tufts University School of Medicine
Boston, MA, USA

Andrew C. Halley, PhD
Center for Neuroscience
University of California, Davis
Davis, CA, USA

Adam W. Hantman, PhD
Edward R. Perl Investigator
Department of Cell Biology and Physiology, Neuroscience Center
University of North Carolina at Chapel Hill
Chapel Hill, NC, USA

Judith A. Hirsch, PhD
Department of Biological Sciences
Dornsife College of Letters, Arts and Sciences,
University of Southern California
Los Angeles, CA, USA

Sonja B. Hofer, PhD
Sainsbury Wellcome Centre for Neural Circuits and Behaviour
University College London
London, UK

Anthony Holtmaat, PhD
Department of Basic Neurosciences and the Center for Neuroscience
Faculty of Medicine
University of Geneva
Geneva, Switzerland

Bryan M. Hooks, PhD
Department of Neurobiology
University of Pittsburgh School of Medicine
Pittsburgh, PA, USA
Office of Naval Research-Reserve Component
Arlington, VA, USA

Michael Hornberger, PhD
Norwich Medical School
University of East Anglia
Norwich, UK

Yujie Hou, PhD
Université Claude Bernard Lyon 1
INSERM, Stem Cell and Brain Research Institute
Bron, France

Kai Hwang, PhD
Department of Psychological and Brain Sciences
Iowa Neuroscience Institute
The University of Iowa
Iowa City, IA, USA

Dieter Jaeger, PhD
Department of Biology
Emory University
Atlanta, GA, USA

Jorge Jaramillo, PhD
Campus Institute for Dynamics of Biological Networks
University of Göttingen
Göttingen, Germany

Mitra Javadzadeh
Sainsbury Wellcome Centre for Neural Circuits and Behaviour
University College London
London, UK

Linxing Preston Jiang
Paul G. Allen School of Computer Science and Engineering & Center for Neurotechnology
University of Washington
Seattle, WA, USA

Igor Kagan, PhD
Decision and Awareness Group, German Primate Center
Leibniz Institute for Primate Research
Goettingen, Germany

Taylor Kahl
Department of Biology
Emory University
Atlanta, GA, USA

Theofanis Karayannis, DPhil
Brain Research Institute
University of Zurich
Zurich, Switzerland

Sabine Kastner
Princeton Neuroscience Institute and Department of Psychology
Princeton University
Princeton, NJ, USA

Henry Kennedy
Université Claude Bernard Lyon 1
INSERM, Stem Cell and Brain Research Institute
Bron, France
Institute of Neuroscience, State Key Laboratory of Neuroscience
Chinese Academy of Sciences (CAS)
Key Laboratory of Primate Neurobiology
Shanghai, China

Andrew J. King, PhD
Department of Physiology, Anatomy and Genetics
University of Oxford
Oxford, UK

Robert T. Knight, MD
Helen Wills Neuroscience Institute
Department of Psychology
University of California, Berkeley
Berkeley, CA, USA

Adam Kohn
Dominick Purpura Department of Neuroscience
Department of Ophthalmology and Visual Sciences
Department of Systems and Computational Biology
Albert Einstein College of Medicine
Bronx, NY, USA

Leah Krubitzer, PhD
Department of Psychology
Center for Neuroscience
University of California, Davis
Davis, CA, USA

Kaushik Lakshminarasimhan, PhD
Zuckerman Institute
Columbia University
New York, NY, USA

Hey-Kyoung Lee, PhD
Professor
Solomon H. Snyder Department of Neuroscience
Zanvyl-Krieger Mind/Brain Institute
Kavli Neuroscience Discovery Institute
Johns Hopkins University
Baltimore, MD, USA

Kok-Sin Lee, MSc
Department of Basic Neurosciences and the Center for Neuroscience
Faculty of Medicine, University of Geneva
Geneva, Switzerland

Nathalie Leresche
Sorbonne Université, CNRS, INSERM
Neurosciences Paris Seine
Institut de Biologie Paris Seine
Paris, France

Jennifer F. Linden, PhD
Ear Institute
Department of Neuroscience, Physiology and Pharmacology
University College London
London, UK

Daniel A. Llano, MD, PhD
Department of Molecular and Integrative Physiology
The Beckman Institute for Advanced Science and Technology
University of Illinois at Urbana-Champaign
Urbana, IL, USA

Laureline Logiaco, PhD
Research Scientist, Simons transition to Independence Fellow
Department of Brain and Cognitive Sciences
Massachusetts Institute of Technology
Cambridge, MA, USA

Arianna Maffei, PhD
Department of Neurobiology and Behavior
Stony Brook University
Stony Brook, NY, USA

Rebecca A. Mease, PhD
Institute of Physiology and Pathophysiology
Ruprecht Karl University of Heidelberg
Heidelberg, Germany

Tracy R. Melzer, PhD
Department of Medicine
University of Otago
Christchurch, New Zealand
New Zealand Brain Research Institute
Christchurch, New Zealand
Te Kura Mahi ā-Hirikapo | School of Psychology
Speech and Hearing
University of Canterbury
Christchurch, New Zealand

Lukas Mesik
Solomon H. Snyder Department of Neuroscience
Zanvyl-Krieger Mind/Brain Institute
Kavli Neuroscience Discovery Institute
Johns Hopkins University
Baltimore, MD, USA

Anna S. Mitchell, PhD
Te Kura Mahi ā- Hirikapo | School of Psychology, Speech, and Hearing
University of Canterbury
Christchurch, New Zealand

Christina Mo, PhD
The Florey Institute of Neuroscience and Mental Health
University of Melbourne
Melbourne, VIC, Australia

Zoltán Molnár
Department of Physiology, Anatomy and Genetics
University of Oxford
Oxford, UK

Christa Müller-Axt, MSc
Technical University of Dresden
Dresden, Germany
Max Planck Institute for Human Cognitive and Brain Sciences
Leipzig, Germany

Allison J. Murphy, PhD
Department of Neuroscience
University of Rochester School of Medicine and Dentistry
Rochester, NY, USA

Ayuno Nakahashi
Department of Neuroscience
University of Montréal
Montréal, Québec, Canada

Marcel Oberlaender, PhD
Postdoctoral Research Fellow
Department of Physiology and Biophysics
University of Colorado
Aurora, CO, USA

Monica N. O'Connell, PhD
Translational Neuroscience Division
Center for Biomedical Imaging and Neuromodulation
Nathan Kline Institute
Orangeburg, NY, USA
Department of Psychiatry
New York University School of Medicine
New York, NY, USA

Ronan O'Shea
Center for Perceptual Systems
Center for Learning and Memory
The University of Texas at Austin
Austin, TX, USA

Lucy M. Palmer
Florey Institute of Neuroscience and Mental Health
University of Melbourne
Melbourne, Victoria, Australia

Leopoldo Petreanu, PhD
Champalimaud Neuroscience Programme
Champalimaud Foundation
Lisbon, Portugal

Adrien Peyrache
Montreal Neurological Institute and Hospital
McGill University
Montreal, QC, Canada

Didier Pinault, PhD
Université de Strasbourg
Strasbourg, France
INSERM U1114, Neuropsychologie cognitive et physiopathologie de la schizophrénie Strasbourg, France
Fédération de Médecine Translationnelle de Strasbourg (FMTS)
Faculté de médecine
Strasbourg, France
Centre de Recherche en Biomédecine de Strasbourg (CRBS)
Strasbourg, France

Anne Lise Pitel, PhD
Normandie Université, UNICAEN, INSERM, U1237
PhIND, Physiopathology and Imaging of Neurological Disorders
NeuroPresage Team, Cyceron
Caen, France

Nicholas J. Priebe
Center for Learning and Memory
The University of Texas at Austin
Austin, TX, USA

Yifat Prut
Edmond & Lily Safra Center for Brain Sciences
The Hebrew University of Jerusalem
Jerusalem, Israel

Rajesh P. N. Rao, PhD
Paul G. Allen School of Computer Science and Engineering & Center for Neurotechnology
University of Washington
Seattle, WA, USA

Alex D. Reyes, PhD
Center for Neural Science
New York University
New York, NY, USA

Adema Ribic
Department of Psychology
University of Virginia
Charlottesville, VA, USA

Yuri B. Saalmann
Department of Psychology
University of Wisconsin-Madison
Madison, WI, USA

Charles E. Schroeder, PhD
Translational Neuroscience Division
Center for Biomedical Imaging and Neuromodulation
Nathan Kline Institute
Orangeburg, NY, USA
Departments of Psychiatry and Neurology
Columbia University College of Physicians and Surgeons
New York, NY, USA

Shailendra Segobin, PhD
Normandie Université, UNICAEN
PSL Université Paris, EPHE, INSERM, U1077
CHU de Caen, GIP Cyceron
Neuropsychologie et Imagerie de la Mémoire Humaine
Caen, France

Mandi C. Severson
Center for Perceptual Systems
Center for Learning and Memory
The University of Texas at Austin
Austin, TX, USA

Eric Shea-Brown
Department of Applied Mathematics and Computational Neuroscience Center
University of Washington
Seattle, WA, USA

S. Murray Sherman, PhD
Department of Neurobiology
University of Chicago
Chicago, IL, USA

Hyeyoung Shin
School of Biological Sciences
Seoul National University
Seoul, Republic of Korea

James M. Shine
Brain and Mind Centre
University of Sydney
Camperdown, NSW, Australia

Stewart Shipp
Institute of Ophthalmology
University College London
London, UK

Samuel G. Solomon, PhD
Institute of Behavioural Neuroscience and Department of Experimental Psychology
University College London
London, UK

Friedrich T. Sommer, PhD
Redwood Center for Theoretical Neuroscience
Helen Wills Neuroscience Institute
University of California, Berkeley
Berkeley, CA, USA
Neuromorphic Computing Laboratory, Intel
Santa Clara, CA, USA

Marc A. Sommer, PhD
Department of Biomedical Engineering
Duke University
Durham, NC, USA

Caleb M. Spiro, MA
Department of Psychology
University of Illinois
Urbana-Champaign
San Francisco, CA, USA

Garrett B. Stanley, PhD
Coulter Department of Biomedical Engineering
Georgia Institute of Technology & Emory University
Atlanta, GA, USA

Catherine M. Sweeney-Reed, MBBS, PhD
Neurocybernetics and Rehabilitation, Department of Neurology, and
Center for Behavioral Brain Sciences (CBBS)
Otto von Guericke University Magdeburg
Magdeburg, Germany

W. Martin Usrey, PhD
Department of Neurology, and the Center for Neuroscience
University of California, Davis
Davis, CA, USA

Stephen D. Van Hooser, PhD
Department of Biology
Brandeis University
Waltham, MA, USA

Carmen Varela, PhD
Department of Psychology
Florida State University
Tallahassee, FL, USA

Julien Vezoli
Université Claude Bernard Lyon 1
INSERM, Stem Cell and Brain Research Institute
Bron, France

Katharina von Kriegstein, MD, Dr.med.
Technical University of Dresden
Dresden, Germany

Melanie Wilke, PhD
Department of Cognitive Neurology
University Medicine Goettingen, Heart-Brain-Center
Goettingen, Germany
Decision and Awareness Group, German Primate Center
Leibniz Institute for Primate Research
Goettingen, Germany

Steven P. Wise, PhD
Olschefskie Institute for the Neurobiology of Knowledge
Bethesda, MD, USA

Mathieu Wolff
Bordeaux Neurocampus
University of Bordeaux
Bordeaux, France

PREFACE

The Cerebral Cortex and Thalamus is a collective effort to bring together, in a single volume, a cohesive account of cortical and thalamic mechanisms for control of behavior, with an emphasis on the importance of interactions between the two structures. There has been a groundswell of new and renewed interest in the thalamus and thalamocortical interactions over the past decade, and this has led to new discoveries demonstrating that the thalamus plays an ongoing and essential role in cortical functioning and, likewise, that the cortex is essential for thalamic functioning. Moreover, it is becoming increasingly clear that the cortex and thalamus are necessary partners for sensory, motor, and cognitive functions. Interactions between the thalamus and cortex are not only essential for proper brain function, they are also sensitive to a range of diseases that can have devastating consequences on individuals and society. Despite the critical roles served by both the cortex and thalamus and their interdependence on each other, most conferences and publications have focused on either the cortex or the thalamus, but not both together. For all of these reasons, the timing felt right for *The Cerebral Cortex and Thalamus.*

The Cerebral Cortex and Thalamus is organized into topical sections with chapters written by more than 100 leading experts in the field. The editors provide opening and closing chapters for the book, and section editors provide an overview for the beginning of each section that sets the stage for the section, discussing the scope of the section and the relationship/synergy of the selected chapters. Because the intended audience for *The Cerebral Cortex and Thalamus* includes graduate students, early career investigators, and investigators looking to shift their research focus, the final chapter concludes with "important and unanswered questions" to inform readers of research directions in need of investigation.

WMU and SMS

ABBREVIATIONS

5-HT	serotonin
3T	3 Tesla
7T	7 Tesla
A1	primary auditory cortex
A2	secondary auditory cortex
aACd	dorsal anterior cingulate cortex
AAV	associated adeno virus
ACh	acetylcholine
AChE	acetylcholine esterase
ACv	ventral anterior cingulate cortex
AD	anterodorsal subnucleus of the ANT
AM	anteromedial subnucleus of the ANT
AMPAR	α-amino-3-hydroxy-5-methyl-4-isoxazolepropionic acid receptor
ANT	anterior nuclei of the thalamus
AV	anteroventral subnucleus of the ANT
BG	basal ganglia
BOLD	blood oxygen level dependent
CalB	calbindin
CalR	calretinin
Cb	cerebellum
cc	corpus callosum
ChR2	adenoassociated virus expressing channelrhodopsin
CMc	caudal cingulate motor area
CMr	rostral cingulate motor area
CP	cortical plate
CT	corticothalamic
DA	dopamine
DBS	deep brain stimulation
DCP	dense cortical plate
DL	dorsolateral visual area
DLc	caudal DL
DLr	rostral DL
DMNT	dorsomedial nuclei of the thalamus
DTB	diencephalic-telencephalic boundary
DTI	diffusion tensor imaging
DWI	diffusion-weighted imaging
E	embryonic day
ECoG	electrocorticographic
EEG	electroencphalographic
EM	electron microscope or microscopy
EPSC	excitatory postsynaptic current
EPSP	excitatory postsynaptic potential
ERh	entorhinal cortex
ERP	event-related potential
FEF	frontal eye field
FGF8	fibroblast growth factor 8
fMRI	functional magnetic resonance imaging
FO	first order
FST	fundus of the superior temporal sulcus
G	gustatory cortex
GABA	gamma-aminobutyric acid
GABAAR	GABAA receptor
GABABR	GABAB receptor
Glu	glutamate
GP	globus pallidus
GPe	globus pallidus, external
GPi	globus pallidus, internal
HFA	high frequency activity
hl	hindlimb representation of M1
HO	higher order
Ia	agranular insular cortex
Iap	posterior Ia
Iav	ventral Ia
IC	inferior colliculus
IC	internal capsule
Id	gusgranular insular cortex
IFL	inner fiber layer
Ig	granular insular cortex
iGluR	ionotropic glutamate receptor
IL	infralimbic cortex
IPSC	inhibitory postsynaptic current
IPSP	inhibitory postsynaptic potential
IT	inferior temporal cortex
ITc	caudal IT
ITi	inferior IT
ITp	polar IT
ITr	rostral IT
IZ	intermediate zone

K	koniocellular
LD	lateral dorsal subnucleus of the ANT
LGE	lateral ganglionic eminence
LGN	lateral geniculate nucleus
LGNd	lateral geniculate nucleus, dorsal division
LGNv	lateral geniculate nucleus, ventral division
LIP	lateral intraparietal area
LO	lateral orbitofrontal cortex
LTD	long-term depression
LTP	long-term potentiation
M	magnocellular
M1	primary motor cortex
M1c	caudal M1
M1r	rostral M1
M2	supplementary motor area (SMA)
MD	mediodorsal nucleus
MDc	central MD
MDcd	caudodorsal MD
MDdl	dorsolateral MD
MDm	medial MD
MDmc	magnocellular MD
MDmf	multiform MD
MDpc	parvocellular MD
MDv	ventral MD
MDvl	ventrolateral MD
MGB	medial geniculate body
MGBd	medial geniculate body, dorsal division
MGBm	medial geniculate body, medial division
MGBv	medial geniculate body, ventral division
MGE	medial ganglionic eminence
mGluR	metabotropic glutamate receptor
MGN	medial geniculate nucleus
MRI	functional magnetic resonance imaging
MST	medial superior temporal area
MT	middle temporal visual area
mya	million years ago
MZ	marginal zone
NA	noradrenaline
NMDAR	N-methyl-D-aspartate receptor
Oc	occipital area
OPt	occipitoparietal transition area
oRG	outer radial glia
OSVZ	outer subventricular zone
OT	olfactory tubercle
P	parvocellular
PAC	phase–amplitude coupling
PCW	postconceptional week
PE-PFG-PG	subdivisions of the PPC
PFC	prefrontal cortex
PFo	granular orbital PFC
Pir	piriform cortex
PL	prelimbic cortex
PMd	dorsal premotor cortex
PMv	ventral premotor cortex
POm	posterior medial nucleus
PPC	posterior parietal cortex
PPc	caudal PPC
PPD	paired-pulse depression
PPF	paired-pulse facilitation
PPr	rostral PPC
PRh	perirhinal cortex
PrV	principal nucleus of Vth nerve
pSP	presubplate
PSPB	pallial subplallial boundary
pTH	prethalamus
Pul	pulvinar
PulM	medial pulvinar
PV	parvalbumin
RMA	rostral motor area
S1	primary somatosensory cortex
S2	secondary somatosensory cortex
SC	superior colliculus
SEF	supplementary eye field
sEEG	stereoencephalographic
sm	stria medularis
SNR	substantia nigra (reticulata)
SP	subplate
SpV	spinal nucleus of Vth nerve
SST	somatostatin
STP	superior temporal polysensory area
SVZ	subventricular zone
TC	thalamocortical
Tc	caudal temporal area
TCA	thalamocortical axon
TE-TEO-TFP-TL-TLO	subdivisions of IT
TRN	thalamic reticular nucleus
V1	primary visual cortex
V2	secondary visual cortex
V3	third visual area
VA	ventral anterior nucleus
V	ventrobasal complex of the thalamus
VGLUT	vesicular glutamate transporter
VIP	ventral intraparietal area
VL	ventral lateral nucleus
VP	ventral posterior nucleus
VPl	ventral posterior lateral nucleus
VPm	ventral posterior medial nucleus
VTel	ventral telencephalon
VZ	ventricular zone
WM	working memory
ZI	zona incerta

SECTION I

INTRODUCTION

1.

INTRODUCTION AND OVERVIEW

W. Martin Usrey and S. Murray Sherman

This book is an attempt to cover two gaps in our appreciation of the critical interplay between the thalamus and the cortex.[1] One is that the tendency in covering these subjects is to treat each in isolation, which overlooks the point that a key to understanding their function is appreciating their essential partnership and interdependence for sensation, action, and cognition. Here, we emphasize the interactions. The other is that the thalamus has been treated as a minor, rather insignificant player in cortical functioning. That is, in the not too distant past, the thalamus was considered a rather boring, machine-like relay of information to cortex. This is consistent with one common view of a relay as an electromagnetic device used only to pass on information rather automatically. But the term "relay" became regularly used in England centuries ago when stagecoaches became a common mode for travel: at the end of a travel stage, a stop was made to change horses and allow travelers rest and refreshment. On overnight stops, the travelers often engaged in various activities, sometimes scandalous, before moving on, or being relayed, the next morning. This provides a better metaphor for our current understanding of thalamic function: the travelers (i.e., messages) are usually or always changed, at least subtly, before being relayed.

This new appreciation of thalamic function and its relationship to the cortex comes with a veritable explosion of research on the subject in the past few decades. This is well documented in the chapters that follow in this book. Our purpose here is to introduce some general features of thalamocortical interactions that will be developed much more thoroughly in later chapters.

SOME BASIC FEATURES OF THALAMOCORTICAL RELATIONSHIPS

Figure 1.1 shows a highly schematic view of the thalamus and its relationship to the cortex.

THALAMUS

The thalamus used to be divided based on developmental criteria into dorsal and ventral parts, with the dorsal part consisting of the nuclear groups containing the relay cells that project to cortex, and the ventral group consisting of the thalamic reticular nucleus (TRN) plus the ventral division of the lateral geniculate nucleus (vLGN), and sometimes the zona incerta (ZI) is also included here. We now have a slightly different appreciation based on modern molecular classification criteria (Puelles et al., 2021; Fratzl and Hofer, 2022); that is, the relay nuclei are now simply called "thalamus," and the TRN is considered "prethalamus" (see **Chapter 2 in Section II**). The thalamus includes relay cells that project to the cortex plus local inhibitory interneurons. The former use glutamate (Glu) as a neurotransmitter, and the latter, gamma-aminobutyric acid (GABA). The relay cell to interneuron ratio across thalamic nuclei and species is roughly 3 to 1 with variation therein. Neither relay nuclei nor relay cells connect directly with one another.

The TRN consists strictly of GABAergic neurons that project only into the thalamus to provide inhibition of relay cells. TRN cells do richly interconnect with one another but rarely make synapses with the local interneurons in the thalamus.

1. Throughout this book, "cortex" refers to the neocortex, which has 6 layers and clear reciprocal relationships with thalamus. Older cortical areas in terms of evolution are the paleocortex, which consists of piriform cortex and secondary olfactory cortex, and the archicortex, which consists of the hippocampal formation. These older areas of cortex are 3-layered, have no consistent clear relationship with the thalamus, and are not much further considered in this book.

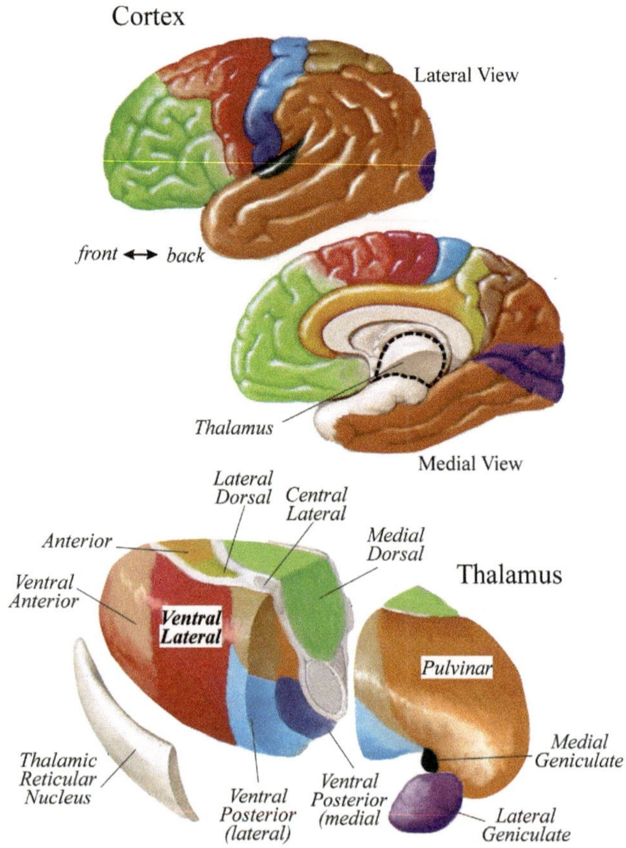

Figure 1.1. Thalamocortical relationships shown schematically. At the top is shown 2 views of cortex, and the dashed black outline shows the location of the thalamus at the center of the brain. Below is an enlarged view of the left thalamus, showing the main relay nuclei and the thalamic reticular nucleus (in white). The thalamic reticular nucleus normally sits tight up against the lateral surface of the relay nuclei, but it is cut off and pulled away to reveal the relay nuclei. The color coding shows the main regions of cortex to which each thalamic nucleus projects. Redrawn from Netter (1977), with updating from Mary Baldwin and Francisco Clasca.

There have been several attempts to subdivide the thalamus into discrete functional types. The most common division is into relay nuclei, as shown in Figure 1.1. Two other attempts stand out. One is a division of thalamic relays into *core* and *matrix* (Jones, 1998). Core projections target cortex only and focus on the middle cortical layers topographically, whereas matrix projections are more diffuse and mainly innervate the upper layers, particularly layer 1. Matrix thalamic cells also provide input beyond cortex, such as to basal ganglia (see **Chapter 11 in Section IV**). There are at least two complications to this classification: (1) whereas some thalamic nuclei are mainly matrix or core, the common feature is that both types of relay are found in most thalamic nuclei; (2) individual thalamocortical cells, via branching, can exhibit a core pattern from one branch and a matrix from another (Rodriguez-Moreno et al., 2020).

The other division is into *first order* (FO) and *higher order* (HO) thalamic relays (Sherman and Guillery, 2013; Sherman, 2016; Usrey and Sherman, 2021). FO relays receive information to relay from a subcortical source (e.g., retinal input to the LGN), whereas HO relays pass on messages from layer 5 of one cortical area to another in transthalamic circuits (e.g., involving the Pul). The same two types of provisos seen in the core/matrix classification apply here: (1) Many if not all thalamic nuclei have both FO and HO components. For instance, some pulvinar (Pul) cells seem to relay inputs from cortical layer 5, making them HO, whereas others relay input from the superior colliculus (SC), a subcortical source, making them FO (Kelly et al., 2003). (2) Some individual thalamic relay cells appear to receive information for relay from both cortex and a subcortical source (Groh et al., 2014; Bickford et al., 2015).

Section II of this volume covers these topics in more detail.

CORTEX

The cortex is a 6-layered structure that covers the brain. In smaller-brained mammals, the cortex fits into the skull as a smooth sheet, but in those with larger brains, packing the cortical surface area into the skull requires extensive folding, leading to sulci (grooves) and gyri (ridges).

Many schemes have been offered to subdivide the cortex. A common one is to recognize the 4 cortical lobes (Figure 1.2A): frontal, parietal, temporal, and occipital. But finer divisions have also been suggested. These generally derive from variations across the cortex in the layering scheme: whereas 6 layers can always be seen in all regions, there is variation in various aspects of the layering. This has led to a veritable cottage industry dating back to the 19th century in using histology to subdivide the cortex. Various schemes produced as few as 5 distinct cortical areas and as many as over 200.

Figure 1.2B shows the scheme with 5 distinct areas. Type 1 has few granule cells associated with layers 2 and 4 and prominently features the pyramidal cells of layers 3 and 5. This is generally associated with motor areas. Type 5 features granule cells of layers 2 and 4 with relatively few layer 3 and 5 pyramidal cells. This is generally associated with primary sensory areas. Intermediary types 2–4 are also identified. Figure 1.2C shows what is probably the most accepted scheme, based on a map made by Brodmann (1909) showing 52 areas. Because this is the most accepted scheme, a few points about it are worth emphasizing. First, there are numbered areas that are commonly referenced: area 17 is primary visual cortex; 1, 2, and 3, primary somatosensory cortex; 41 and 42, primary auditory cortex; and 4, primary motor cortex. Second, beyond these correlations

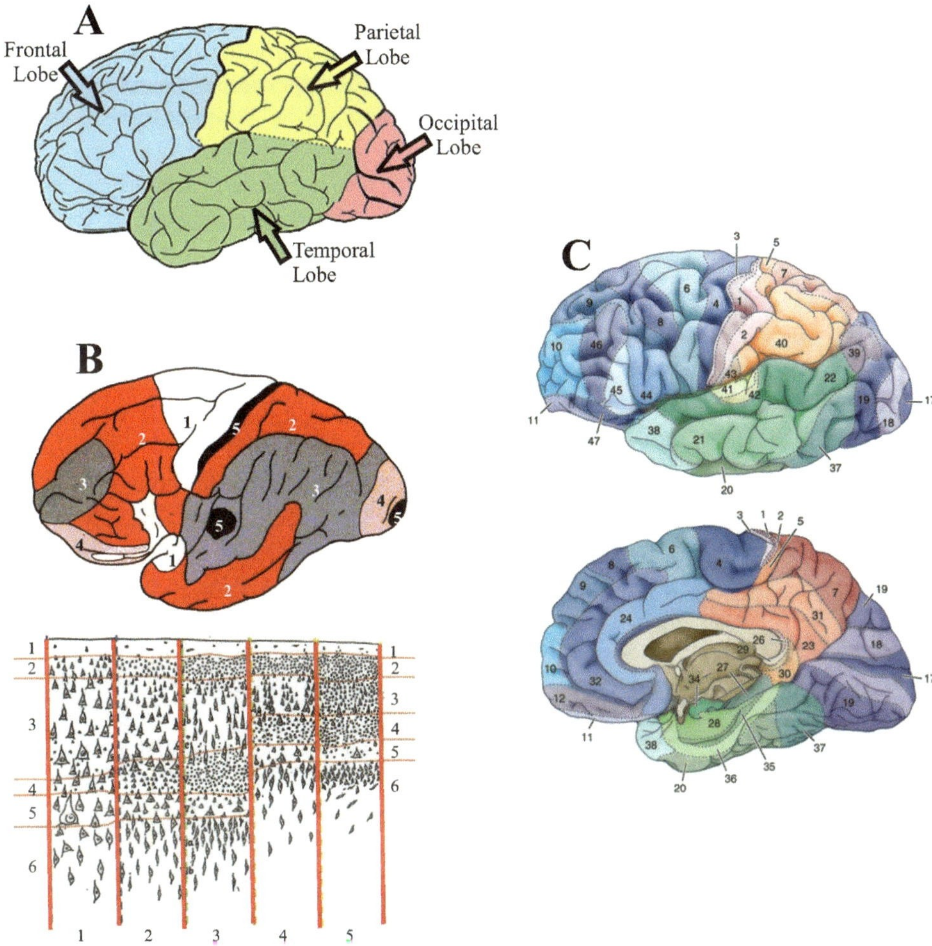

Figure 1.2. Examples of different ways to distinguish cortical areas and types of cortex. (A). Lateral view of the left human cortex with the cortex divided into lobes: frontal, parietal, temporal, and occipital. (B). *Upper panel,* lateral view of monkey cortex with cortical areas distinguished on the basis of "type" of cortex. *Lower panel*, representative sections from 5 types of cortex. Each type has distinct laminar and morphological features (see text for details). Redrawn from Solari and Stoner (2011). (C). Brodmann's parcellation of cortex into 52 areas using different laminar and morphological criteria (upper panel shows lateral view; lower panel shows medial view). Redrawn from Brodmann (1909).

with Brodmann areas, most others have less clear functional significance.

An extensive network of intracortical neuronal circuits, involving multiple cell types, interconnects cortical neurons both within and across the cortical layers. In addition, most of the Glu cortical neurons (generally outside of layer IV) send projections to other cortical areas and/or to subcortical targets. Fundamental to any understanding of the computations performed in the cortex is knowledge of how the cortical intracortical circuits and thalamocortical circuits interact.

Section III covers these topics in more detail.

BETWEEN THALAMUS AND CORTEX

Thalamocortical Relationships

Figure 1.1 shows a very rough approximation of thalamocortical projections. However, details of these projections are much more complex and vary greatly among species. There are, however, several features in common among species for these relationships:

- To the extent that it has been studied, every thalamic nucleus has relay cells that project to the cortex, and typically, each nucleus innervates multiple cortical areas; often single cells do so via branching axons.
- Again, to the extent that it has been studied, every cortical area receives thalamic input, typically from multiple nuclei.
- Virtually all information reaching the cortex passes through the thalamus. This places the thalamus in a key strategic position as a gateway in the information stream headed for the cortex: if the information fails to pass through the gateway, the cortex is not informed thereof, and there is no conscious perception of that information stream.

Corticothalamic Relationships

There is a high level of reciprocity in the connections between the cortex and thalamus. Thus, every cortical area innervates the thalamus, and every thalamic nucleus receives cortical input. Typically, multiple cortical areas innervate each thalamic nucleus.

There are two quite different types of corticothalamic projection: one from layer 6, and the other, layer 5, which are the only sources of subcortical projection from the cortex, and each seems to have a distinct function in thalamocortical circuitry (Sherman and Guillery, 2013; Sherman, 2016; Usrey and Sherman, 2021). Layer 6 projections target all thalamic nuclei and, with rare exceptions involving innervation of the claustrum (Thomson, 2010) or inferior colliculus (Schofield, 2009; Stebbings et al., 2014), target only the thalamus and thalamic reticular nucleus. These inputs to thalamus seem to have a modulatory function rather than providing basic information to be relayed. Layer 5 projections are more selective, targeting HO but not FO nuclei (e.g., Pul but not LGN). Unlike layer 6 projections, these also richly target many extrathalamic regions, often including the spinal cord, and do so by individual axons that branch repeatedly (see below). Thus, most or all layer 5 axons innervating thalamus branch to innervate other subcortical targets as well. Finally, the layer 5 inputs to the thalamus seem to provide basic information to be relayed back to the cortex, thereby creating transthalamic corticocortical pathways.

Section IV covers these topics in more detail.

IMPORTANCE OF EVOLUTIONARY PERSPECTIVE

The brain, which naturally includes the thalamus and cortex, is the result of hundreds of millions of years of evolution. Evolution is messy and complicated, and the end result is a functional organization of neural systems that is more complex than it might be if an intelligent designer had built the whole thing with an end product in mind. One complication is that, in the process of evolution, circuits that arose to efficiently perform functions were generally not discarded as newer (and better?) circuits evolved, and so our brain is now a complex interconnected and interdependent mass of circuits that tend to reflect hierarchical periods of evolutionary history: that is, older circuits are lower in the hierarchy, so that spinal cord is at the bottom, supraspinal brainstem centers higher, and cortex at the top. Put another way, modern engineering, when creating a newer, better gadget to do a job, would put it in place of the older, less efficient one, discarding the latter, keeping the whole with fewer operating pieces and thus producing a simpler design.

Figure 1.3 illustrates this point. A very early vertebrate ancestor is the lamprey, and its brain is shown schematically in Figure 1.3A. There is no cortex yet, but many complex behaviors can be generated and controlled by the brain. This is done largely through central pattern generators found in spinal circuits that are selected and controlled by a brainstem center giving rise to a reticulospinal pathway, and this in turn can be further controlled by telencephalic centers. Because this arrangement works, as evolution proceeds, adding other brainstem control centers and eventually cortex (Figure 1.3B), these older circuits mostly remain and continue to function. Furthermore, when cortex evolved to produce more flexible responses to environmental challenges, there was not evolution of an associated motor plant to which the cortex has special access. Instead, the cortex must operate through and cooperate/coordinate with these older centers. This has real implications for cortical functioning (e.g., Sherman and Usrey, 2021). Furthermore, it should be noted that the only avenue over which the cortex can exert its influence on behavior is from layer 5 subcortical projections; without these, the cortex would be pretty useless.

Sections V–IX cover thalamocortical processing more broadly in the context of these issues, and Section X covers issues of evolution more specifically.

OTHER ISSUES COVERED

Section IX covers aspects of development and plasticity relevant to thalamocortical relationships, Section XI describes more theoretical and computational approaches to these issues, and Section XII describes disease conditions that arise from certain pathologies in thalamocortical circuitry.

SOME CHALLENGES

A major goal of this book is to identify features of the thalamus, cortex, and their relationships that are common to all mammals. For instance, in most if not all cases of study of these topics in mice, one hopes to be studying a feature that is common to many or all mammals, including humans, and not something that is unique to mice. Thus, the search is for features—cells, circuits, nuclei, cortical areas, and so on—that are related across species.

HOMOLOGY VERSUS ANALOGY

Generally speaking, attempts to find such common organizational features across species come down to one of two approaches: *homology*, which implies a close evolutionary

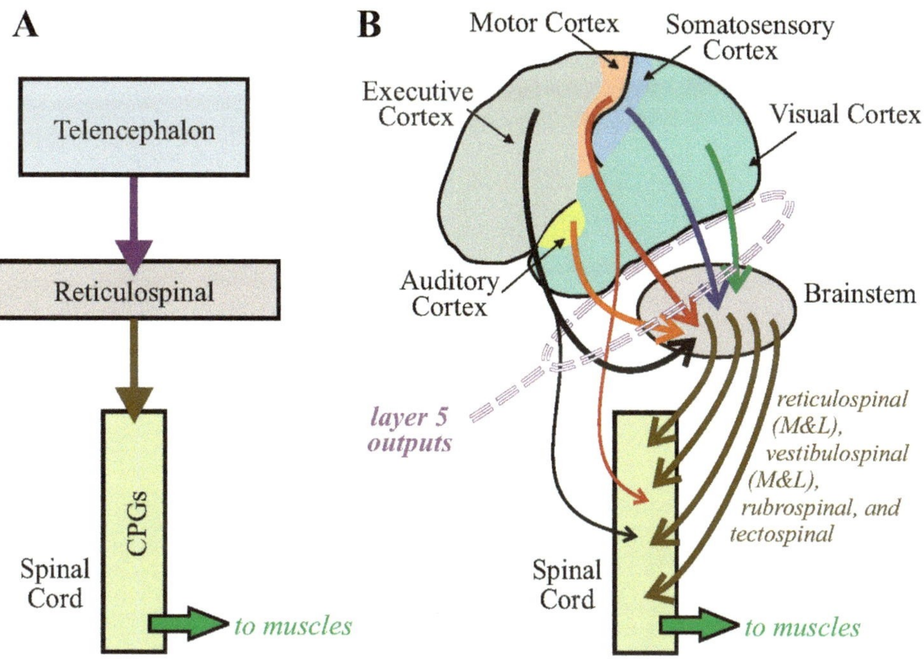

Figure 1.3. Schematic view of 2 evolutionary stages of vertebrate central nervous system. (A). Central nervous system of the lamprey, the most primitive living vertebrate and thus likely representative of early vertebrate species. Here, the central nervous system is dominated largely by limited supraspinal control of central pattern generators (CPGs; for swimming behavior, etc.). Redrawn from Grillner (2003); Stuart (2007); Suryanarayana et al. (2017). (B). Later mammalian evolutionary stage, showing human example. Here, cortical control of lower motor centers has evolved. Note that cortex must operate via layer 5 projections to centers evolved in A, with additional subcortical supraspinal control centers also appearing. Many of the circuits that evolved in (A) remain functional in (B). Redrawn from Sherman and Usrey (2021).

relationship between the compared features, namely, a common ancestor in the recent evolutionary past of the species being compared; or *analogy*, which implies a common function for the compared features.

Figure 1.4 illustrates the difference. Examples of homology include the arms and hands of humans, the wings of birds and bats, and the flippers of whales. Whereas these share a common evolutionary history and occupy similar locations on the body, they look like different structures and perform very different functions: flying versus grasping versus swimming. In contrast, analogy refers to features that are not closely related in evolutionary terms but nevertheless evolved to perform similar functions; this is also known as "convergent evolution." Examples are the fins of fishes and the flippers of whales, structures that did not evolve from a recent common ancestor but did so for the analogous purpose of swimming.

It is important to understand the distinction between homology and analogy because establishing homologous features between species is the "gold standard" for providing meaningful insight into core relationships between biological structures. Establishing homologous relationships across mammalian species for different cortical areas, different thalamic nuclei, different neuronal cell types, and so on, is absolutely crucial for understanding the underlying mammalian design.

For structures such as bones and teeth, the fossil record can be quite useful in establishing homologies. Unfortunately, soft tissues such as brains do not fossilize, and so the fossil record provides limited information for neuroscientists attempting to establish homologies. As a consequence, there has been much controversy and too little certainty in homologies that have been suggested. Thus, the treatment of thalamic and cortical evolution throughout this book (especially in Section X) is often based on comparisons made between currently living species that represent different branches in the evolutionary tree, rather than on fossil evidence, and thus should be considered with healthy skepticism.

TERMINOLOGY

A challenge is applying the same terminology consistently to homologous structures in different species. Of course, part of the problem is the prerequisite of establishing homology across species for, say, a cortical area or thalamic nucleus, before dubbing such with a common name. Regarding thalamus, the naming of nuclei largely occurred over a century ago, before homologies were well established, and so applying the same name to a nucleus in different species may be open to reinterpretation based on better evidence of evolutionary relationships. Establishing homologies between

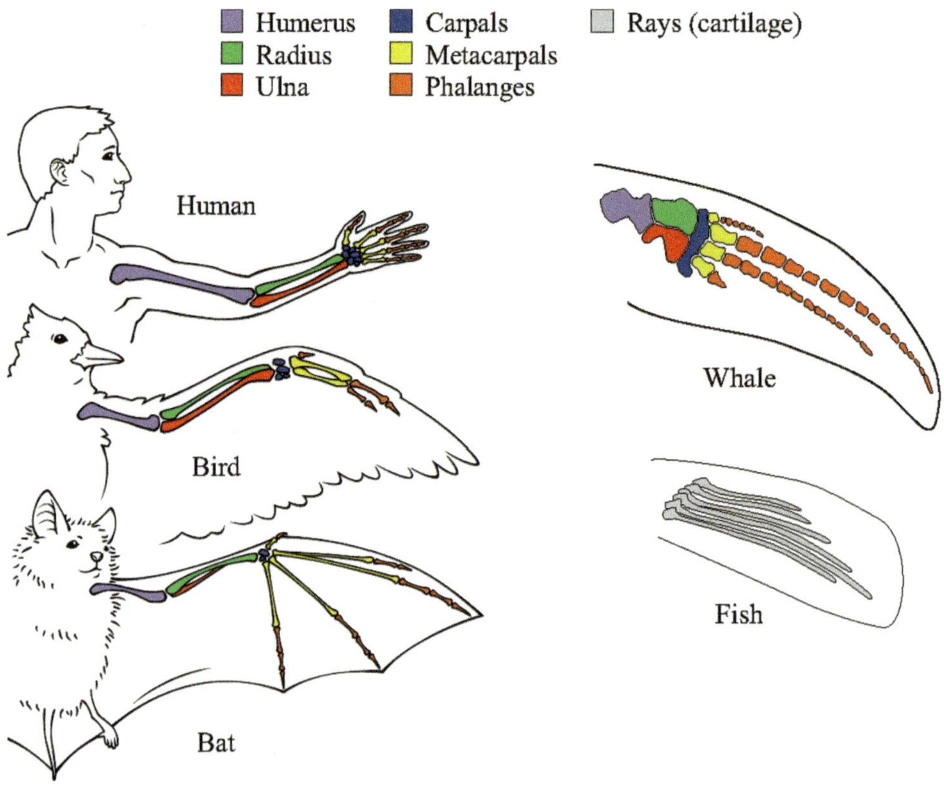

Figure 1.4. The organization of bones in the human arm, bird wing, bat wing, and whale. Although these appendages are used for very different purposes, the tremendous similarity in their organization indicates they are homologous structures. Also shown is the fin of a fish. Although it serves the same function as the whale flipper, its cartilaginous structure shows little resemblance to the bone structure of the whale flipper. Consequently, these structures are considered analogous, rather than homologous. Adapted from https://askabiologist.asu.edu/sites/default/files/resources/coloring_pages/pdf/AAB_bats_coloring_page.pdf.

cortical areas is much more difficult, because identifying these areas across species is limited to rather few species with adequate cortical mapping and not total agreement on how to define areas. Nonetheless, examples of accepted homologies named appropriately are the LGN and V1.

Two examples from thalamus underscore the problem. The Pul is a major thalamic nucleus identified in most mammals, but not in rats or mice. Instead, these rodents have a large thalamic nucleus called "lateral posterior," which is also generally identified in other species. This could lead one to conclude that rats and mice lack a Pul. However, recent evidence suggests that, indeed, the lateral posterior nucleus of rats and mice and Pul in other species are homologous (Zhou et al., 2017). We suggest that these structures in rats and mice be renamed "pulvinar." Another example is the POm, an important thalamic nucleus involved in transthalamic circuitry. This nucleus is recognized in most species but not in primates. The anterior Pul in primates is involved in somatosensory connections much like the POm, and it has been suggested that it should be renamed accordingly (Krubitzer and Kaas, 1989). The limitation here is establishing homologies for thalamic nuclei more generally and then assigning common names to them.

As noted, the problem in doing this with cortical areas is both establishing what constitutes a cortical area and determining potential mammalian homologs for each. Perhaps if this can be done first for thalamic nuclei, one can then use this information in combination with the cortical relationships of each thalamic homolog as an early step in determining cortical homologs.

Such terminological problems are not uncommon in neuroscience, and we have two examples of how these may be resolved. Inconsistent naming of structures in the avian brain troubled that field for decades, but the Avian Brain Nomenclature Forum held at Duke University in July 2002 agreed on consistent terminology (Reiner et al., 2004). Similarly, variable terms applied to insect brains was resolved by the Insect Brain Name Working Group, which arrived at a consensus for such nomenclature (Ito et al., 2014).

Another sometimes confusing terminological issue is whether names are derived from Latin or English. We suggest the latter. Thus, instead of "nucleus reticularis thalami," we suggest "thalamic reticular nucleus," and this also applies to abbreviations thereof, meaning "TRN" instead of "NRT."

UNDERSTANDING THALAMOCORTICAL DIVERSITY: THE CHALLENGE OF CLASSIFICATION

Both thalamocortical and corticothalamic projections display a wide variety of types (Clasca et al., 2012; Economo et al., 2018; Halassa and Sherman, 2019). Clearly, these patterns present a complex system, and an early step in analyzing complex systems is a classification of component elements. For example, early analysis of retinal organization first defined the component neuronal types—receptor, interneuron, and ganglion cell—each of which has been classified into distinct subtypes. In short, we need to know how many distinct motifs and sub-motifs exist for thalamocortical and corticothalamic projections, motifs that are the necessary link between cortex and behavior.

OVERALL GOALS

We emphasize again that the main purpose of this book is twofold. First, the aim is to start to bridge the gap of understanding regarding the interactions and codependence between thalamus and cortex. This is a relatively new way of looking at thalamic and cortical functioning, rather than treating each separately and in isolation, and thus there has been much new research into these relationships. Second, we think it is important to identify key questions that remain unresolved regarding interactions between thalamus and cortex and their cooperative functioning. Understanding the right questions is an important first step in designing future research. The final chapter of the volume is largely devoted to this.

REFERENCES

Bickford ME, Zhou N, Krahe TE, Govindaiah G, Guido W (2015). Retinal and tectal "driver-like" inputs converge in the shell of the mouse dorsal lateral geniculate nucleus. *J Neurosci.* 35:10523–10534.

Brodmann K (1909). *Vergleichende Lokalisationslehre der Grosshirnrinde in ihren Prinzipien dargestellt auf Grund des Zeelenbaues.* Johann Ambrosius Barth Verlag.

Clasca F, Rubio-Garrido P, Jabaudon D (2012). Unveiling the diversity of thalamocortical neuron subtypes. *Eur J Neurosci.* 35:1524–1532.

Economo MN, Viswanathan S, Tasic B, Bas E, Winnubst J, Menon V, Graybuck LT, Nguyen TN, Smith KA, Yao Z, Wang L, Gerfen CR, Chandrashekar J, Zeng H, Looger LL, Svoboda K (2018). Distinct descending motor cortex pathways and their roles in movement. *Nature.* 563:79–84.

Fratzl A, Hofer SB (2022). The caudal prethalamus: Inhibitory switchboard for behavioral control? *Neuron.* 110:2728–2742.

Grillner S (2003). The motor infrastructure: From ion channels to neuronal networks. *Nat Rev Neurosci.* 4:573–586.

Groh A, Bokor H, Mease RA, Plattner VM, Hangya B, Stroh A, Deschenes M, Acsády L (2014). Convergence of cortical and sensory driver inputs on single thalamocortical cells. *Cereb Cortex.* 24:3167–3179.

Halassa MM, Sherman SM (2019). Thalamocortical circuit motifs: A general framework. *Neuron.* 103(5):762–770.

Ito K, Shinomiya K, Ito M, Armstrong JD, Boyan G, Hartenstein V, Harzsch S, Heisenberg M, Homberg U, Jenett A, Keshishian H, Restifo LL, Rässler W, Simpson JH, Strausfeld NJ, Strauss R, Vosshall LB (2014). A systematic nomenclature for the insect brain. *Neuron.* 81:755–765.

Jones EG (1998). A new view of specific and nonspecific thalamocortical connections. *Adv Neurol.* 77:49–71.

Kelly LR, Li J, Carden WB, Bickford ME (2003). Ultrastructure and synaptic targets of tectothalamic terminals in the cat lateral posterior nucleus. *J Comp Neurol.* 464:472–486.

Krubitzer LA, Kaas JH (1989). Cortical integration of parallel pathways in the visual system of primates. *Brain Res.* 478:161–165.

Netter FH (1977). *The CIBA Collection of Medical Illustrations,* Vol. 1: *Nervous System: A Compilation of Paintings on the Normal and Pathologic Anatomy of the Nervous System, with a Supplement on the Hypothalamus.* CIBA.

Puelles L, Diaz C, Stühmer T, Ferran JL, Martínez-de la Torre M, Rubenstein JLR (2021). LacZ-reporter mapping of Dlx5/6 expression and genoarchitectural analysis of the postnatal mouse prethalamus. *J Comp Neurol.* 529:367–420.

Reiner A, Perkel DJ, Mello CV, Jarvis ED (2004). Revised nomenclature for avian telencephalon and some related brainstem nuclei. *J Comp Neurol.* 473:377–414.

Rodriguez-Moreno J, Porrero C, Rollenhagen A, Rubio-Teves M, Casas-Torremocha D, Alonso-Nanclares L, Yakoubi R, Santuy A, Merchan-Perez A, DeFelipe J, Lubke JH, Clasca F (2020). Area-specific synapse structure in branched posterior nucleus axons reveals a new level of complexity in thalamocortical networks. *J Neurosci.* 40:2679.

Schofield BR (2009). Projections to the inferior colliculus from layer VI cells of auditory cortex. *Neurosci.* 159:246–258.

Sherman SM (2016). Thalamus plays a central role in ongoing cortical functioning. *Nat Neurosci.* 19:533–541.

Sherman SM, Guillery RW (2013). *Functional Connections of Cortical Areas: A New View from the Thalamus.* MIT Press.

Sherman SM, Usrey WM (2021). Cortical control of behavior and attention from an evolutionary perspective. *Neuron.* 109:3048–3064.

Solari SV, Stoner R (2011). Cognitive consilience: Primate non-primary neuroanatomical circuits underlying cognition. *Front Neuroanat.* 5:65.

Stebbings KA, Lesicko AM, Llano DA (2014). The auditory corticocollicular system: Molecular and circuit-level considerations. *Hear Res.* 314:51–59.

Stuart DG (2007). Reflections on integrative and comparative movement neuroscience. *Integr Comp Biol.* 47:482–504.

Suryanarayana SM, Robertson B, Wallen P, Grillner S (2017). The lamprey pallium provides a blueprint of the mammalian layered cortex. *Curr Biol.* 27(21):3264–3277.

Thomson AM (2010). Neocortical layer 6, a review. *Front Neuroanat.* 4:13.

Usrey WM, Sherman SM (2021). *Exploring Thalamocortical Interactions: Circuitry for Sensation, Action, and Cognition.* Oxford University Press.

Zhou NA, Maire PS, Masterson SP, Bickford ME (2017). The mouse pulvinar nucleus: Organization of the tectorecipient zones. *Vis Neurosci.* 34:E011.

SECTION II

THALAMIC CELL AND CIRCUIT PROPERTIES

Martha Bickford

The mammalian thalamus was classically divided into an epithalamus, dorsal thalamus, and ventral thalamus. However, modern analyses have firmly established a new, "prosomeric" model of the diencephalon based on embryonic gene expression and cell lineage. Projection neurons of the dorsal thalamus and epithalamus have been shown to derive from precursors in the "prosomere 2" neuropithelial domain of the early diencephalon. In contrast, ventral thalamus neurons derive from precursors in the "prosomere 3" domain, which is separate from and anterior to prosomere 2. Hence, the classic "ventral thalamus" stuctures are actually "prethalamic" in origin.

The dorsal thalamus or thalamus proper is heavily and reciprocally connected with virtually all areas of the cerebral cortex. Consequently, these connections are of critical importance for understanding the brain activity underlying cortical operations such as sensory perception and motor behavior, as well as their contextual modulation. In contrast, prethalamic nuclei do not project to the cortex. Nevertheless, one of these nuclei, the prethalamic reticular nucleus (TRN), is intimately involved in thalamocortical interactions because it receives extensive connections from the thalamus and cerebral cortex, and projects back to the thalamus. Therefore, the chapters in this section discuss the cell and circuit properties of the thalamus as well as the TRN.

As detailed in subsequent sections of this book, each thalamic nucleus is uniquely organized to carry out specialized functions. However, all thalamic nuclei share basic features, and identifying common cellular characteristics and synaptic motifs is a critical first step in providing a framework for understanding thalamic function. Hence, the chapters within this section focus primarily on neuronal architecture and physiological properties that are shared across nuclei. Chapter 2 describes the three basic categories of interconnected neurons that provide the scaffold of the thalamus: glutamatergic thalamocortical (projection cells) that innervate the cortex, TRN, and other forebrain areas; GABAergic intrinsic interneurons that form dendrodendritic synaptic connections with projection cells and interneurons within their nucleus of origin; and GABAergic TRN neurons that are reciprocally connected to projection cells within each thalamic nucleus. Chapter 3 describes the essential ion channels expressed by these three categories of neurons. This chapter explains the functional characteristics of these ion channels which lead to two fundamentally different modes of activity, termed *burst* and *tonic firing*, and also endow the interconnected relay cells, interneurons, and TRN cells with the ability to maintain oscillations which are critical components of the dynamic interplay between the thalamus and the cortex. Chapter 4 reviews the origin, neurotransmitters, and synaptic motifs of a wide variety of thalamic afferents. This chapter describes how heterogeneous afferents interact with the thalamic scaffold neurons to create computational units which can uniquely impact the output of the thalamus. Finally, Chapter 5 reviews the functional properties of thalamic afferents. This chapter describes how thalamic synapses operate, what types of postsynaptic receptors they activate, what intracellular molecular cascades they control, and how synaptic strength is regulated. Together these four chapters outline the fundamental characteristics of thalamic neurons and their afferent connections, providing a framework for the subsequent sections of this book which comprehensively discuss the structure, function, development, and evolution of thalamocortical interactions and diseases that alter them.

2.

THALAMUS AND PRETHALAMUS CELL TYPES AND THEIR INTERCONNECTIONS

Martha Bickford

INTRODUCTION

The cell types that compose the thalamus and prethalamus include neurons, interdigitated astroctyes that surround and support them, oligodendrocytes that produce myelin sheaths that surround axons, endothelial cells and other cell types that maintain and regulate the vasculature, connective tissue cells of the meninges, and microglia and other cells of the immune system. All these cell types are important components of the thalamus/prethalamus, and increasing evidence suggests that glial cells play important and essential roles in maintaining and modulating neuronal activity levels (Khakh and Sofroniew, 2015). Moreover, differences in the proportions of glia cells and neurons vary across the thalamus, potentially influencing activity levels in a nucleus-specific manner (Wei et al., 2011). Blood flow is also a critical factor regulating thalamic function, as blood oxygenation levels can be used as a measure of neuron activity (Hall et al., 2016). However, to set the stage for the subsequent chapters which focus on neuronal architecture and activity patterns in various thalamic nuclei and connected cortical areas, this chapter will focus on the neurons of the thalamus and the interconnected prethalamic reticular nucleus (TRN).

THALAMIC NEURON TYPES

All neurons contain neuron-specific proteins, such as neuronal nuclear protein (NeuN; Kim et al., 2009). Therefore, thalamic neurons can be readily distinguished from surrounding glial cells using antibodies against these proteins (Figure 2.1). Thalamic neurons can then be divided into 2 general types: (1) projection neurons that send their axon outside the thalamus to innervate the cortex and TRN, and often the striatum, amygdala, or other forebrain areas (also referred to as "relay cells"; see **Section IV, Chapter 11**) and (2) interneurons that do not send axons outside their nucleus of origin. The projection neurons develop from neuroepithelial precursors in diencephalic prosomere 2 of the embryonic forebrain (Puelles, 2021), while the interneurons originate in the midbrain and other forebrain areas and migrate into the thalamus at later stages of development (Jager et al., 2021). Thus, these 2 fundamental neuron categories can be distinguished not only by their projection patterns, but also by the expression of genes related to their distinct developmental profiles.

One of the most basic properties that distinguishes thalamic cell types is that projection neurons utilize glutamate as a neurotransmitter, while the interneurons utilize gamma amino butyric acid (GABA) as neurotransmitter. Thus, interneurons can be identified by their GABA content or by the presence of proteins involved in the release of GABA (e.g., glutamic acid decarboxylase, GAD, the GABA synthesizing enzyme; Figure 2.2) while projection cells contain proteins involved in glutamate release (e.g., the type 2 vesicular glutamate transporter, vGLUT2; Nahmani and Erisir, 2005). The proportion of thalamocortical cells versus interneurons depends on the specific thalamic nucleus and the species (Arcelli et al., 1997). However, retrograde tracing studies and immunocytochemical staining for GABA or GAD have demonstrated that in most non-rodent species, GABAergic interneurons make up approximately one-quarter to one-third of the neurons in the thalamus, and the remainder are projection neurons (reviewed in Bickford, 2021). In rodents, such as mice, interneurons are less common, begin virtually absent in some nuclei and constituting only 6% in more densely populated nuclei (Evangelio et al., 2018; Jaeger et al., 2021).

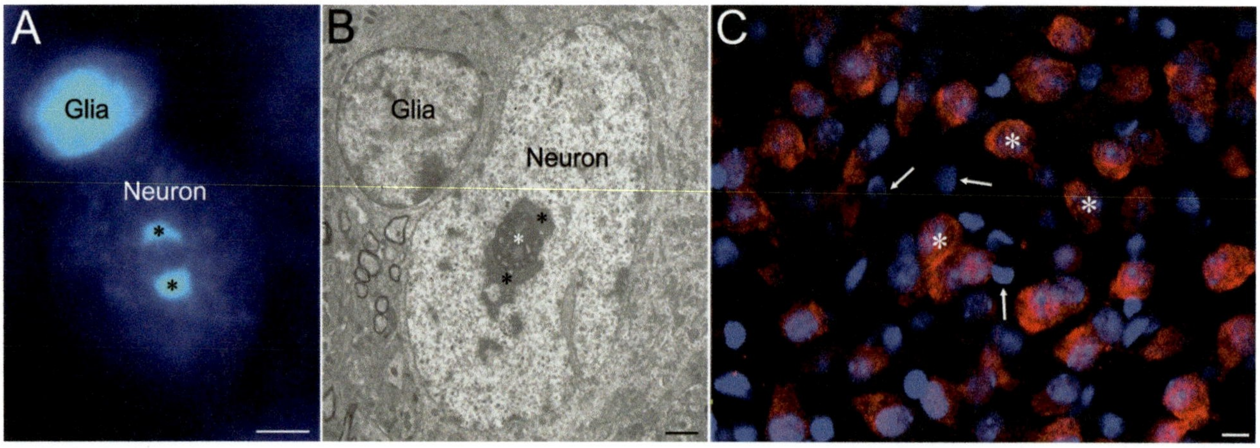

Figure 2.1. Neurons and glia of the thalamus. (A) Confocal image of a neuron and glial cell in the tree shrew dorsal lateral geniculate nucleus (LGNd) stained with 4′,6-diamidino-2-phenylindole (DAPI) which binds to DNA. (B) An electron microscopic image of a neuron and glial cell in the tree shrew LGNd. In neurons, regions of condensed chromatin (A, B, black asterisks) surround nucleoli (B, white asterisk) or are sparsely distributed in the nucleus. In contrast, chromatin is relatively condensed through glial cell nuclei, resulting in intense staining of the entire nucleus with the DAPI stain. (C) Confocal image of the tree shrew LGNd stained with DAPI (blue) and an antibody against neuronal nuclei (NeuN, red). Examples of glia are indicated with white arrows and examples of neurons are indicated with white asterisks. DAPI and NeuN staining in tree shrew tissue revealed glia-to-neuron ratios of 1:1 in the LGNd and 3:1 in the pulvinar nucleus. Scale bars: A = 2 µm, B = 1 µm, C = 10 µm. Figure modified from Wei et al. (2011).

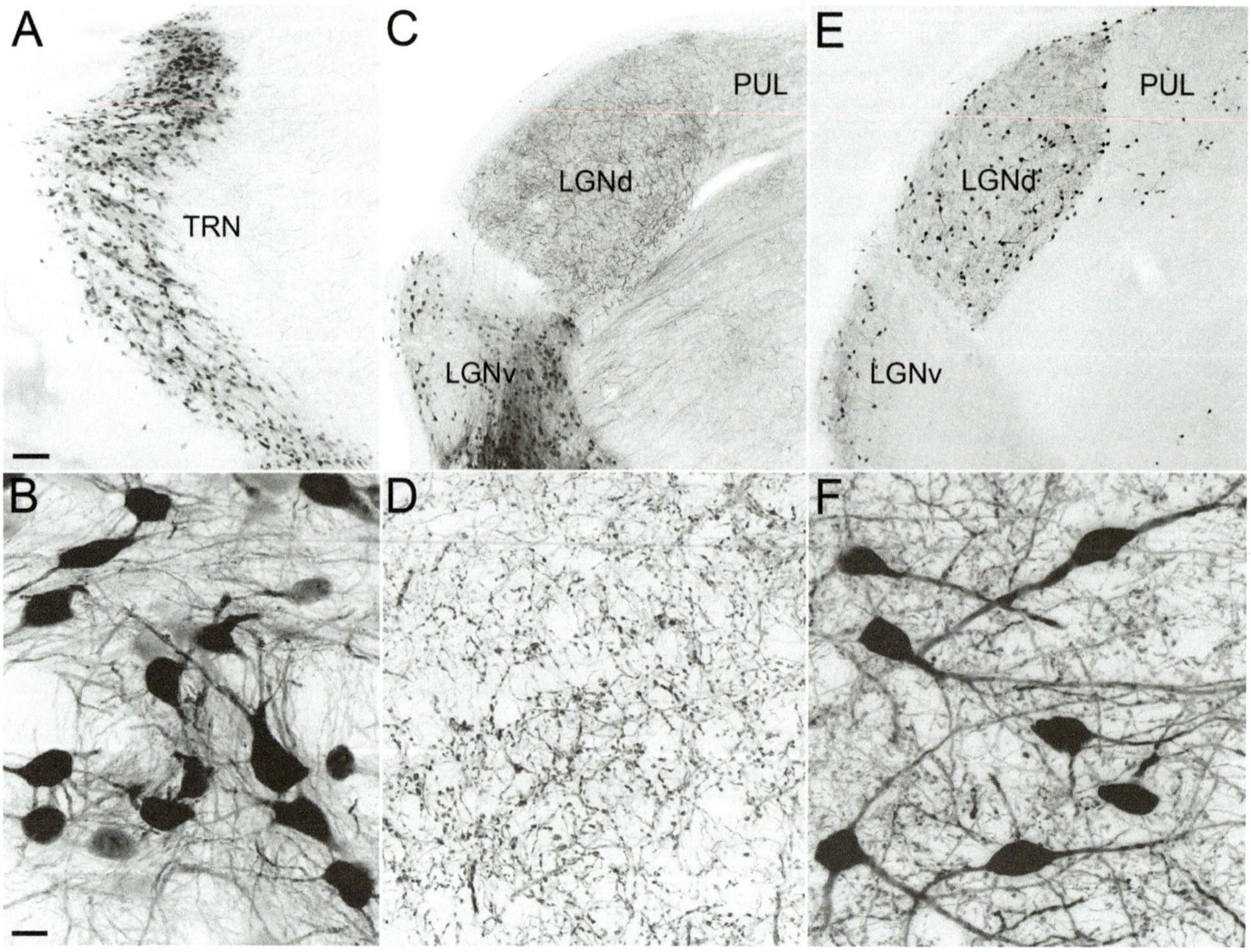

Figure 2.2. GABAergic neurons of the thalamus and prethalamus. Confocal images of coronal sections through the thalamus and prethalamus of GAD65-GFP mice (A–D) illustrate the GABAergic neurons of the prethalamic reticular nucleus (TRN, A, B) and their projections to the dorsal lateral geniculate nucleus (LGNd, C, D). Within the TRN the GABAergic neurons extend thin dendrites, but few puncta are seen (B). Within the LGNd, the axon projections of the TRN cells form a dense population of puncta (D) that innervate projection neurons (Campbell et al., 2020). Confocal images of coronal sections through the thalamus and prethalamus of GAD67GFP mice (E, F) illustrate the GABAergic intrinsic interneurons of the thalamus. In rodents, thalamic intrinsic interneurons are sparse, except in the LGNd, where they extend widespread dendrites and form a dense population of puncta (F) that innervate projection neurons and other interneurons. Scale bars: A = 100 µm, also applies to C and E; B = 10 µm, also applies to D and F.

GLUTAMATERGIC PROJECTION NEURONS

Virtually all thalamic projection neurons cells send axon projections to the cortex, but the distribution and extent of these projections varies widely (see **Section IV, Chapters 11 and 12**). Many thalamocortical axons branch profusely in cortical layer 4, forming highly focused clusters of terminals in this layer. In other cases, thalamocortical axons primarily target other layers, and may be more sparsely and/or widely dispersed. In addition, a subset of thalamic projection neuron axons branch to innervate the striatum, amygdala, or other forebrain areas, and in some cases these subcortical structures may even be their main target (see **Section IV, Chapters 11 and 12**). Nevertheless, the vast majority of projection neurons predominantly target the cortex, and their axons innervate regions of the cortex that are functionally related to their thalamic input (i.e., sensory or motor thalamic nuclei project to corresponding sensory or motor modality-specific areas of the cortex). All thalamocortical neurons also extend collaterals to innervate the TRN as they exit the thalamus, with each thalamic nucleus projecting to restricted sensory or motor modality-specific sectors (considered further in **Section IV, Chapter 16**). Finally, sparse thalamocortical collateral projections within the thalamus have been observed, but these primarily target thalamic interneurons (reviewed in Bickford, 2021). Thus, projection neurons are not directly interconnected. Instead, the activity patterns of projection cells within a given thalamic nucleus may be related by virtue of common synaptic inputs from nonthalamic sources and intrinsic interneurons. The variations in targets and organization of thalamocortical axons form the basis of several systems of thalamic classification and are considered further in **Section IV, Chapters 11 and 12**.

Despite the wide variations in the projection patterns of the thalamic projection neurons, their somatodendritic morphologies are relatively consistent across thalamic nuclei and across species (Figure 2.3A and D). Most projection neurons are multipolar cells that extend dendrites radially from a relatively large soma. However, more detailed examination has revealed variations in the soma sizes of projection neurons, the extent and orientation of dendritic arbors, and the existence of dendritic specializations or appendages (e.g., "claw-, spine-, or grape-like," or most imaginatively, "antlers in velvet"; Majorossy et al., 1965; Guillery, 1966; Figure 2.3A,C). Moreover, in some thalamic nuclei, morphological categories of projection neurons are grouped together to form subregions or lamina. These features have formed the basis of various classification schemes, and in many cases morphological features have been correlated with distinct physiological and other properties. For example, the primate dorsal lateral geniculate nucleus (LGNd) can be divided into distinct lamina based on projection neuron soma sizes (magnocellular, parvocellular, and koniocellular), with each morphological cell type responding to unique visual features and projecting to distinct regions of the striate cortex (Briggs, 2017). Similarly, in the carnivore LGNd, the somatodendritic morphology of projection neuron classes has been related to their physiological functions, and different neuron classes are grouped in different geniculate lamina (see **Section V, Chapter 17**).

A variety of studies have also found that thalamic projection neurons can be divided based on distinct protein expression patterns. For example, the smallest (koniocellular) neurons of the primate LGNd uniquely express the alpha subunit of the type II calmodulin-dependent protein kinase, while the largest (magnocellular) neurons stain with antibodies against the nonphosphorylated form of the high-molecular-weight neurofilament protein (Hendry and Yoshioka, 1994; Chaudhuri et al., 1996). Similar differential staining patterns related to projection neuron soma sizes are also seen in the carnivore LGNd (Hockfield and Sur, 1990; Bickford et al., 1998). More recently, transcriptomic analysis of primate LGNd neurons identified 2 types of koniocellular neurons (Bakken et al., 2021) and gene expression studies of the entire mouse thalamus grouped the projection cells into 11 putative classes based on their transcriptomic profiles. Furthermore, genetic diversity in combination with the axon wiring patterns of projection neurons has been used to group thalamic nuclei into categories with similar gene-expression profiles (Nagalski et al., 2016; Saunders et al., 2018; Roy et al., 2022). Future multimodal approaches for cell classification are likely to provide important insights into the mechanisms by which gene expression contributes to the distinct morphologies, electrophysiological properties, synaptic connections and/or locations of thalamic projection neurons (discussed further in **Section IV, Chapters 11 and 12**).

At the electron microscopic level, projection cells exhibit prominent nuclei and nucleoli (Figure 2.1), abundant rough endoplasmic reticulum, and various cytoskeletal elements. Parallel arrays of microtubules and associated bundles of neurofilaments are prominent within dendritic shafts, while dendritic appendages can contain dense neurofilaments and/or mitochondria, but are mostly devoid of microtubules (Majorossy et al., 1965; Guillery, 1969; Hamos et al., 1985; Figure 2.4). The somata of projection neurons generally receive few synapses, but the remainder of their dendrites are studded with synaptic terminals that originate from various sources (described in **Section II, Chapter 4**). A unique property of projection cells is that their dendrites are anchored to the surrounding neuropil via puncta adherentia, presumed adhesive contacts that may function to limit neurotransmitter diffusion within specialized synaptic zones termed *glomeruli*, or "synaptic islands" (Guillery, 1969; Figure 2.4, asterisks). Within

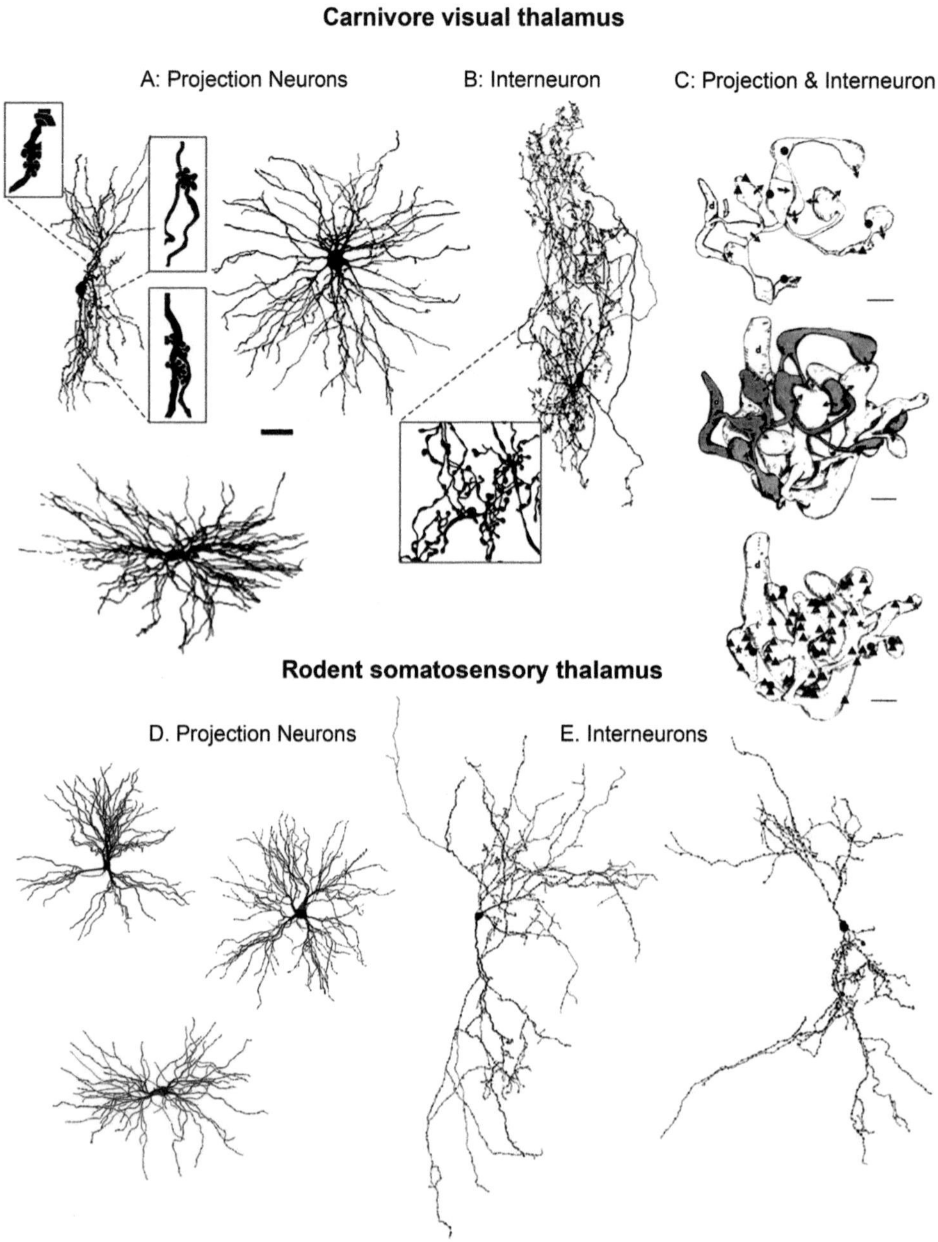

Figure 2.3. Thalamic neuron morphology. Examples of the dendritic morphology of 3 types of projection cells (A) and an interneuron (B) from the cat dorsal lateral geniculate nucleus (LGNd). The insets illustrate examples of projection neuron dendritic appendages and interneuron dendritic terminals. The interconnections of projection cell dendrites and interneuron dendritic terminals in the cat LGNd are illustrated in C; drawings are based on 3-dimensional reconstructions from serial electron micrographs of a labeled interneuron dendrite (top) and its postsynaptic projection neuron dendrite (middle with interneuron, bottom without interneuron). Interneuron synapses are indicated with black arrows. Retinal synapses are indicated by black triangles and other synapses are indicated with black circles. (D) Examples of the dendritic morphology of 3 projection cells in the rat ventroposterior nuclei. (E) Examples of the dendritic morphology of 2 interneurons in the mouse ventroposterior nucleus. Scale bar in A = 50 µm for cells (also applies to B, D, and E) and 10 µm for insets (A and B). Scale bars in C = 1 µm. Panels A and B modified from Sherman (2020). Panel C modified from Hamos et al. (1985). Panel D modified from Iavaron et al. (2019), and Panel E modified from Simko and Markram (2021).

glomeruli, tightly entwined arrangements of synapses occur in close proximity, and no glial processes are seen between them. Instead, glial lamellae can occasionally be seen to surround the outer perimeter of this unique tangle of synaptic connections to form "encapsulated zones" which are somewhat isolated from the surrounding neuropil (reviewed in Bickford, 2021). Between the large, complex, encapsulated zones are the interstitial zones. In these areas, filamentous contacts are rarely seen, and glial lamella can surround individual terminals. In the LGNd, approximately half of the synapses are located in encapsulated zones, associated with the proximal dendrites of projection neurons (closest to the soma), and half are located in the interstitial zones, associated with the distal dendrites of projection cells (farthest

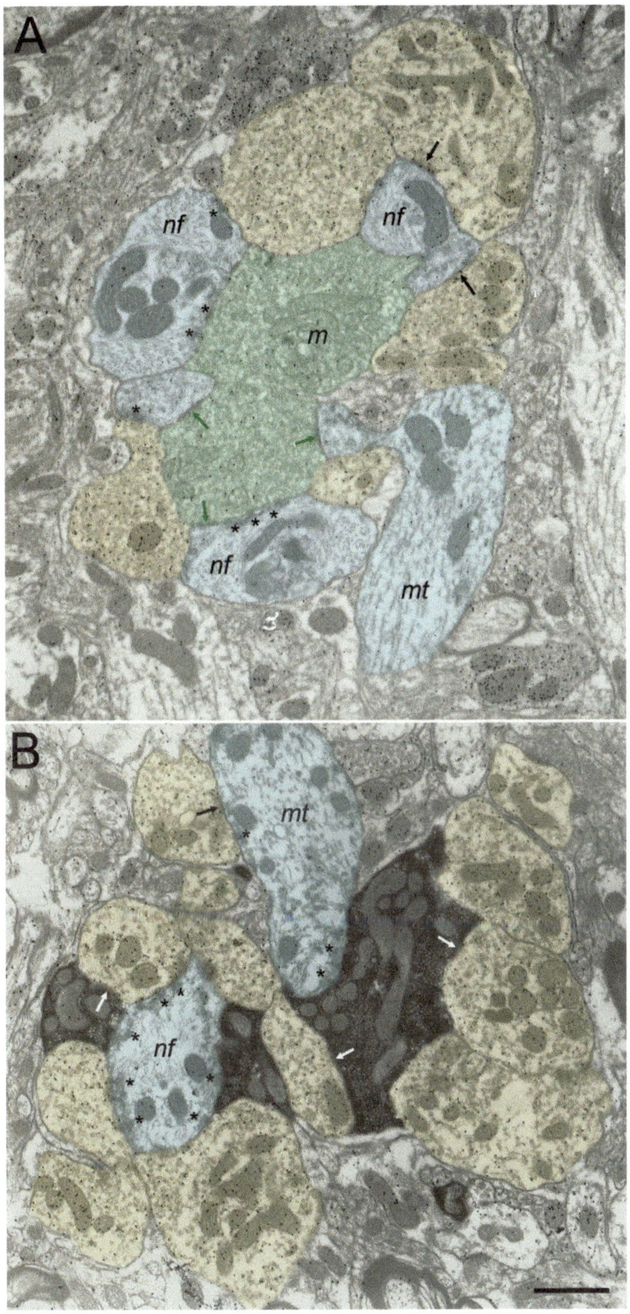

Figure 2.4. Ultrastructure of thalamic glomeruli. Electron micrographs illustrate glomerular synaptic arrangements in the cat dorsal lateral geniculate nucleus (LGNd; A) and pulvinar nucleus (B). The sections were stained with an antibody against GABA tagged with gold particles to identify the GABAergic dendritic terminals of interneurons (pseudo-colored yellow) and the nonGABAergic dendrites of projection neurons (pseudo-colored blue). Projection neurons dendritic shafts contain microtubules (mt) and appendages contain neurofilaments (nf). Projection neurons dendrites are connected to the surrounding neuropil via puncta adherentia (black asterisks). In the LGNd, the glomerulus is innervated by a retinal terminal (pseudo-colored green), identified by its unique pale mitochondria (*m*). Retinogeniculate synapses are indicated by green arrows. In the pulvinar, the glomerulus is innervated by a terminal originating from cortical area 7 (labeled by a tracer injection, dark reaction product). Corticopulvinar synapses are indicated by white arrows. In both the LGNd and pulvinar, synapses formed by the GABAergic dendrites of interneurons (F2 profiles) are indicated by black arrows. Scale bar = 1 μm

from the soma; Guillery, 1969; Wilson et al., 1984). While many of these basic properties are conserved across the thalamus, variations in the complexity and arrangement of encapsulated and interstitial zones are seen across thalamic nuclei and can also contribute to thalamic classification schemes (reviewed in Bickford, 2016, and considered further in **Section II, Chapter 4**).

GABAERGIC THALAMIC INTERNEURONS

In contrast to the thalamic projection neurons, GABAergic interneurons do not project outside their nucleus of origin. Most GABAergic interneurons have relatively small somata and their dendrites can form swellings or appendages that contain synaptic vesicles (Figures 2.3B,C,E, 2.4). Thus, thalamic interneurons are unique in that their dendrites are both pre- and postsynaptic (Hamos et al., 1985; Morgan and Lichtman, 2020). Interneurons are also unique in that the extent and orientation of their dendritic arbors vary greatly. In carnivores, interneurons can display compact dendritic arbors with densely distributed dendritic swellings that are embedded in glomeruli (Figure 2.3B,C) or they can exhibit sparse dendritic arbors that form extraglomerular synapses (Bickford et al., 1999), but the extent of the interneuron dendritic arbors is generally no larger than their projection neuron counterparts. In contrast, in rodents, the dendritic arbors of interneurons are generally much more extensive than those of the adjacent projection cells; relative to projection cells, rodent interneuron dendritic arbors can cover a much greater percentage of the nucleus of origin (Zhu et al., 1999; Iavaron et al., 2019; Morgan and Lichtman, 2020; Simko and Markram, 2021; Figure 2.3E).

When fixed with aldehydes and viewed with an electron microscope, GABAergic terminals in the thalamus are observed to contain flattened or pleomorphic vesicles (Guillery, 1969). Thus, GABAergic synaptic terminals in the thalamus are referred to as "F" profiles, with the interneuron vesicle-filled dendritic swellings referred to as "F2" profiles to distinguish them from "F1" profiles, which are synaptic terminals that contain more densely packed vesicles, stain more intensely with antibodies against GABA, and originate from GABAergic neurons extrinsic to the thalamus (sources and synaptic properties described in **Section II, Chapters 4 and 5**). Interneurons may also release GABA via axon terminals; unlike the dendritic terminals of interneurons which are both pre- and postsynaptic, the interneuron axon terminals are exclusively presynaptic. However, the existing evidence indicates that

Figure 2.4. (continued)

and applies to both panels. Panel A modified from Bickford et al. (2016). Panel B modified from Baldauf et al. (2005).

interneuron axon terminals resemble "F2" profiles in that the vesicles contained within them are relatively loosely packed (Montero, 1987; Morgan and Lichtman, 2020). Finally, some interneuron connections are formed by dendritic shafts where a small collection of vesicles can be seen located adjacent to synapses, while the remainder of the dendrite is devoid of vesicles (Hamos et al., 1985; Bickford et al., 1999; Morgan and Lichtman, 2020).

Thus, thalamic interneurons are multifaceted circuit elements. Synaptic inputs on F2 profiles may cause local GABA release that is independent of activity levels in the rest of the interneuron (Cox et al., 1998). In fact, modeling of interneuron dynamics has suggested that connections involving dendritic terminals might function as local processing units so that each interneuron can participate in multiple independent local circuits (Bloomfield and Sherman, 1989). In contrast, GABA release from interneuron dendritic shafts or axons may mediate more global changes in thalamic activity. Indeed, in vitro studies indicate that active membrane conductances play important roles in regulating GABA release from interneuron dendrites and/or axons to alter thalamic circuits with distinct spatial and temporal dynamics (Acuna-Goycolea et al., 2008; Casale and McCormick, 2011; detailed in **Section II, Chapter 3**).

Genetic fate mapping indicates that GABAergic interneurons migrate into the thalamus from both the midbrain and the forebrain, with neurons originating from each source preferentially occupying select nuclei (Jager et al., 2021). In mice, interneurons that originate from the midbrain (identified by the *Sox14* gene) preferentially occupy caudal nuclei and interneurons that originate from the forebrain (identified by the *Dlx1/2* gene) occupy more rostral thalamic nuclei. In the marmoset thalamus, where interneurons comprise up to 30% of neurons across all nuclei, there is also a differential distribution of midbrain- and presumed forebrain-derived interneurons, although midbrain-derived interneurons are the dominate class (Jager et al., 2021). Further cross-species analysis of interneurons in the LGNd reveal significant differences in the proportions of midbrain- and forebrain-derived interneurons in different species; the mouse LGNd is almost entirely populated by Sox14-positive interneurons while the Sox14-negative/Dlx1/2-positive interneurons represent 10% of interneurons in the marmoset LGNd, 15% in the macaque LGNd, and 40% in the human LGNd (Bakken et al., 2021).

Further genetic analysis of the mouse thalamus indicates that interneurons can be divided into 11 putative subpopulations that express distinct genes (Nagalski et al., 2016; Saunders et al., 2018; Roy et al., 2022). Therefore, it is not surprising that a variety of studies have identified several different classes of GABAergic interneurons based on the expression of different proteins (e.g., calbindin, parvalbumin, or brain nitric oxide synthase), as well as morphological features such as soma size, dendritic or axonal architecture, and the synaptic arrangements they form (reviewed in Bickford, 2021). As for the classification of projection cells, future multimodal approaches that can correlate gene-expression profiles with detailed synaptic features are likely to reveal further diversity in the structure and function of thalamic interneurons both within and across nuclei and species.

INTERCONNECTIONS OF THALAMIC PROJECTION NEURONS AND INTERNEURONS

Thalamic projection neurons and interneurons are extensively interconnected, as are the GABAergic interneurons themselves. A dominant synaptic motif found across many thalamic nuclei is the synaptic triad, which is located within a structure referred to as a glomerulus (described above). Contained within glomeruli are large glutamatergic synaptic terminals (Figure 2.4; sources and other features discussed further in **Section II, Chapters 4 and 5**) that provide input to both the projection neurons and the GABAergic interneurons. Often, these large glutamatergic inputs contact both a projection neuron dendrite and an interneuron dendritic terminal which are themselves interconnected. With this arrangement, activation of the glutamatergic afferent leads to excitation of projection neurons that is rapidly followed by GABA-mediated inhibition via activation of the associated interneuron dendritic terminal (Hirsch et al., 2015; see also **Section II, Chapter 5**).

However, the potential influence of interneurons is extended by the fact that interneurons not only innervate projection neurons but also connect to one another via dendritic terminals (Morgan and Lichtman, 2020; Simko and Markram, 2021). Thus, within each thalamic nucleus the dendritic terminals of interneurons can form connections to locally inhibit projection neuron dendrites, as well as connections with other interneurons, forming a syncytium that may provide more global modulation of thalamic activity. Finally, the occasional intrinsic axon collaterals of projection neurons can innervate interneurons. Where described, these collaterals form large terminals that primarily contact interneurons in spatially restricted zones, potentially serving an additional role in coordinating thalamic activity (Bickford et al., 2008).

NEURONS OF THE PRETHALAMIC RETICULAR NUCLEUS AND THEIR THALAMIC CONNECTIONS

The TRN is a group of GABAergic neurons that surrounds and exclusively innervates the thalamus (Figures 2.2A–D, 2.5A,C). All TRN neurons develop from the prosomere 3 neuroepithelial domain of the embryonic prosencephalon, which is distinct from and anterior to the prosomere 2 domain that becomes the thalamus proper; hence the TRN is currently designated as "prethalamic" (Puelles, 2021). Across species, the majority of TRN neurons express the calcium-binding protein parvalbumin (Li et al., 2020). However, the makeup of the TRN is not homogeneous. For example, in several species, subsets of TRN neurons have been identified that express somatostatin, defining "core" and "shell" regions in some sectors (Conley et al., 1991; Clemente-Perez et al., 2017). Furthermore, mouse studies have revealed "trancriptomic gradients" of gene expression within the TRN that also correspond to "core" and "shell" subdivisions (Li et al., 2020; see **Section IV, Chapter 16**).

As described above, all thalamic projection neurons innervate the TRN via axon collaterals, with each thalamic nucleus projecting to restricted sensory or motor modality-specific sectors as well as "core" and "shell" subdivisions of each sector (Clemente-Perez et al., 2017; MartinezGarcia et al., 2020). Each sector of the TRN in turn feeds back to innervate the modality-specific thalamic areas that innervate it. Thus, the TRN and thalamus are reciprocally connected to elicit opposing effects: the thalamus excites the TRN, while the TRN inhibits the thalamus. Modality-specific sectors of the TRN are also distinguished by their input from corresponding modality-specific cortical areas, and other subcortical nonthalamic sources (discussed further in **Section II, Chapters 4 and 5; Section IV, Chapters 14 and 15**).

Thalamic projection neurons innervate the TRN with large terminals that contain loosely packed vesicles and can be identified by their expression of the type 2 vesicular glutamate transporter (Liu and Jones, 1999; Bickford et al., 2008). In turn, TRN terminals primarily

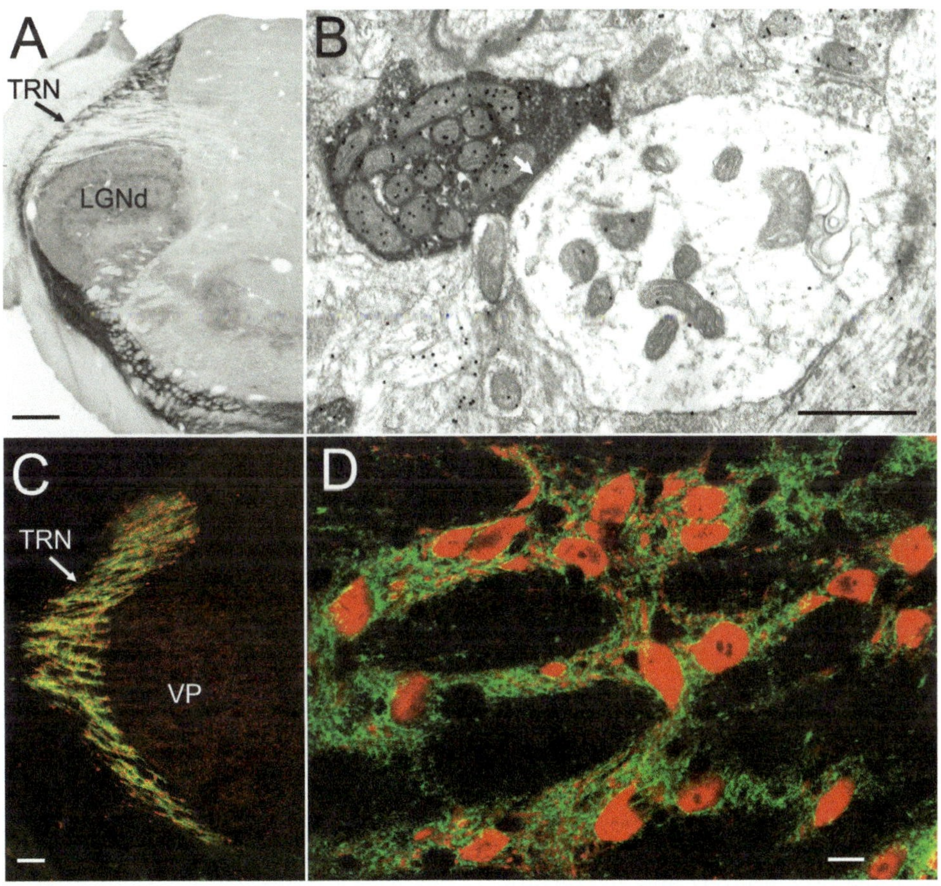

Figure 2.5. Characteristics and connections of TRN neurons. (A) In a section through the cat thalamus, the TRN is revealed by the binding of the lectin wisteria fluoribunda agglutinin (WFA). WFA also binds to regions between the lamina of the dorsal lateral geniculate nucleus (LGNd). (B) The electron micrograph illustrates a synaptic terminal in the cat LGNd, labeled by a tracer injection in the TRN (dark reaction product). The section was stained with an antibody against GABA tagged with gold particles. The high density of gold particles overlying the labeled TRN terminal identify it as GABAergic and it synapses (white arrow) on a projection neuron dendrite (identified as nonGABAergic by the low density of overlying gold particles). (C) In a confocal image of a section through the mouse thalamus, the TRN is identified by the high density of cells labeled with an antibody against parvalbumin (PV; red) and binding of WFA (green). (D) A high magnification confocal image (0.5 um Z axis) of the mouse TRN stained for PV (red) and WFV (green) illustrates the reticular meshwork that surrounds TRN somata and dendrites (perineuronal nets). Axon bundles that travel through the TRN meshwork are unstained. Scale bars: A = 1 mm; B = 1 μm; C = 100 μm; D = 10 μm. Panel A modified from Bickford et al. (2008). Panel B modified from Wang et al. (2001).

innervate projection neuron dendrites with GABAergic terminals that contain densely packed vesicles ("F1" profiles; Figure 2.5B; Wang et al., 2001; Campbell et al., 2020). Thus, TRN-mediated inhibition helps to shape the response properties of projection neurons, including various forms of rhythmic firing states associated with sleep and epilepsy (Huguenard and McCormick, 2007). Coordinated firing across the TRN and thalamus may be further facilitated by interconnections between TRN neurons, which are electrically linked via GAP junction channels (Landisman et al., 2002). TRN neurons are also connected via chemical synapses (Huntsman et al., 1999; Deleuze and Huguenard, 2006), but these connections appear to decline during development so that in the adult fewer TRN-TRN synaptic connections are observed (Hou et al., 2016; see **Section IV, Chapter 16** for further discussion).

TRN firing is also strongly regulated by brain state via cholinergic and other brainstem projections, as well as the unique membrane properties of TRN neurons (McCormick and Prince, 1986; Beierlein, 2014; further discussed in **Section II, Chapters 3, 4, 5, and Section IV, Chapter 16**). Finally, a prominent feature of TRN neurons is that they are surrounded by perineuronal nets, extracellular matrix structures that envelop neurons and potentially regulate synaptic function. TRN perineuronal nets are revealed via the binding of plant lectins such as wisteria fluoribunda agglutinin (WFA; Figure 2.5A,C,D). Increasing evidence suggests that perineuronal nets are plastic structures that vary in density and/or composition during circadian rhythms or other changes in brain states (Cooper et al., 2018; Pantazopoulos et al., 2020). Thus, the perineuronal nets that surround TRN neurons may be integral components of TRN-mediated regulation of thalamic activity.

SUMMARY

The interconnected neurons of the thalamus and prethalamus can be divided into 3 basic categories: (1) glutamatergic projection cells that innervate the cortex, TRN, and other forebrain areas; (2) GABAergic intrinsic interneurons that form dendrodendritic synaptic connections with projection neurons and interneurons within their nucleus of origin; and (3) GABAergic TRN neurons that are reciprocally connected to projection neurons within each thalamic nucleus. These 3 fundamental elements and their canonical connections provide the cellular scaffold of the thalamus. The subsequent chapters of this book build on this scaffold to describe how the diversity of inputs and outputs divide the thalamus into unique functional units that, in partnership with the cortex, mediate sensation, movement, and cognition.

REFERENCES

Acuna-Goycolea C, Brenowitz SD, Regehr WG (2008). Active dendritic conductances dynamically regulate GABA release from thalamic interneurons. *Neuron.* 57:420–431.

Arcelli P, Frassoni C, Regondi MC, Biasi SD, Spreafico R (1997). GABAergic neurons in mammalian thalamus: A marker of thalamic complexity? *Brain Res Bull.* 42:27–37.

Bakken TE, van Velthoven CT, Menon V, et al. (2021). Single-cell and single-nucleus RNA-seq uncovers shared and distinct axes of variation in dorsal LGN neurons in mice, non-human primates, and humans. *eLife.* 10:e64875.

Baldauf ZB, Chomsung RD, Carden WB, May PJ, Bickford ME (2005). Ultrastructural analysis of projections to the pulvinar nucleus of the cat. I: Middle suprasylvian gyrus (areas 5 and 7). *J Comp Neurol.* 485(2):87–107.

Beierlein M (2014). Synaptic mechanisms underlying cholinergic control of thalamic reticular nucleus neurons. *J Physiol.* 592:4137–4145.

Bickford ME (2016). Thalamic circuit diversity: Modulation of the driver/modulator framework. *Front Neural Circuits.* 9:Article 86. doi:10.3389/fncir.2015.00086.

Bickford ME (2021). Synaptic properties of sensory thalamus. In: Murray Sherman, ed., *Oxford Research Encyclopedia of Neuroscience.* Oxford University Press. https://doi.org/10.1093/acrefore/9780190264086.013.352

Bickford ME, Carden WB, Patel NC (1999). Two types of interneurons in the cat visual thalamus are distinguished by morphology, synaptic connections, and nitric oxide synthase content. *J Comp Neurol.* 413:83–100.

Bickford ME, Guido W, Godwin DW (1998). Neurofilament proteins in Y-cells of the cat lateral geniculate nucleus: Normal expression and alteration with visual deprivation. *J Neurosci.* 18:6549–6557.

Bickford ME, Wei H, Eisenback MA, Chomsung RD, Slusarczyk AS, Dankowsi AB (2008). Synaptic organization of thalamocortical axon collaterals in the perigeniculate nucleus and dorsal lateral geniculate nucleus. *J Comp Neurol.* 508:264–285.

Bloomfield SA, Sherman SM (1989). Dendritic current flow in relay cells and interneurons of the cat's lateral geniculate nucleus. *Proc Natl Acad Sci.* 86:3911–3914.

Briggs F (2017). Mammalian visual system organization. In: Murray Sherman, ed., *Oxford Research Encyclopedia of Neuroscience.* Oxford University Press. https://doi.org/10.1093/acrefore/9780190264086.013.66

Campbell PW, Govindaiah G, Masterson SP, Bickford ME, Guido W (2020). Synaptic properties of the feedback connections from the thalamic reticular nucleus to the dorsal lateral geniculate nucleus. *J Neurophysiol.* 124:404–417.

Casale AE, McCormick DA (2011). Active action potential propagation but not initiation in thalamic interneuron dendrites. *J Neurosci.* 31:18289–18302.

Chaudhuri A, Zangenehpour S, Matsubara JA, Cynader MS (1996). Differential expression of neurofilament protein in the visual system of the vervet monkey. *Brain Res.* 709:17–26.

Clemente-Perez A, Makinson SR, Higashikubo B, Brovarney S, Cho FS, Urry A, Holden SS, Wimer M, Dávid C, Fenno LE, Acsády L, Deisseroth K, Paz JT (2017). Distinct thalamic reticular cell types differentially modulate normal and pathological cortical rhythms. *Cell Rep.* 19:2130–2142.

Conley M, Schmechel DE, Diamond IT (1991). Differential distribution of somatostatin-like immunoreactivity in the visual sector of the thalamic reticular nucleus in galago. *Eur J Neurosci.* 3:237–242.

Cooper JM, Halter KA, Prosser RA (2018). Circadian rhythm and sleep-wake systems share the dynamic extracellular synaptic milieu. *Neurobiol Sleep Circadian Rhythms.* 5:15–36.

Cox CL, Zhou Q, Sherman SM (1998). Glutamate locally activates dendritic outputs of thalamic interneurons. *Nature.* 394:478–482.

Deleuze C, Huguenard JR (2006). Distinct electrical and chemical connectivity maps in the thalamic reticular nucleus: Potential roles in synchronization and sensation. *J Neurosci.* 26:8633–8645.

Evangelio M, García-Amado M, Clasca F (2018). Thalamocortical projection neuron and interneuron numbers in the visual thalamic nuclei of the adult C57BL/6 mouse. *Front Neuroanat.* 12:Article 27. http://journal.frontiersin.org/article/10.3389/fnana.2018.00027/full

Guillery RW (1966). A study of Golgi preparations from the dorsal lateral geniculate nucleus of the adult cat. *J Comp Neurol.* 128(1):21–50.

Guillery RW (1969). The organization of synaptic interconnections in the laminae of the dorsal lateral geniculate nucleus of the cat. *Z Zellforsch Mikrosk Anat.* 96:1–38.

Hall CN, Howarth C, Kurth-Nelson Z, Mishra A (2016). Interpreting BOLD: Towards a dialogue between cognitive and cellular neuroscience. *Philos Trans R Soc Lond B Biol Sci.* Oct 5;371(1705):20150348. http://dx.doi.org/10.1098/rstb.2015.0348

Hamos JE, Van Horn SC, Raczkowski D, Uhlrich DJ, Sherman SM (1985). Synaptic connectivity of a local circuit neuron in lateral geniculate nucleus of the cat. *Nature.* 317:618–621.

Hendry SH, Yoshioka T (1994). A neurochemically distinct third channel in the macaque dorsal lateral geniculate nucleus. *Science.* 264:575–577.

Hirsch JA, Wang X, Sommer FT, Martinez LM (2015). How inhibitory circuits in the thalamus serve vision. *Annu Rev Neurosci.* 38:309–329.

Hockfield S, Sur M (1990). Monoclonal antibody Cat-301 identifies Y-cells in the dorsal lateral geniculate nucleus of the cat. *J Comp Neurol.* 300:320–330.

Hou G, Smith AG, Zhang Z-W (2016). Lack of intrinsic GABAergic connections in the thalamic reticular nucleus of the mouse. *J Neurosci.* 36:7246–7252.

Huguenard JR, McCormick DA (2007). Thalamic synchrony and dynamic regulation of global forebrain oscillations. *Trends Neurosci.* 30:350–356.

Huntsman MM, Porcello DM, Homanics GE, DeLorey TM, Huguenard JR (1999). Reciprocal inhibitory connections and network synchrony in the mammalian thalamus. *Science.* 283:541–543.

Iavaron E, Yi J, Shi Y, Zandt B-J, O'Reilly C, Van Geit W, Rossert C, Markram H, Hill SL (2019). Experimentally-constrained biophysical models of tonic and burst firing modes in thalamocortical neurons. *PLoS Comput Biol.* 15(5):e1006753. doi:0.1371/journal.pcbi.1006753.

Jager P, Moore G, Calpin P, Durmishi X, Salgarella I, Menage L, Kita Y, Wang Y, Kim DW, Blackshaw S, Schultz SR, Brickley S, Shimogori T, Delogu A (2021). Dual midbrain and forebrain origins of thalamic inhibitory interneurons. *eLife.* 10:1–29.

Khakh BS, Sofroniew MV (2015). Diversity of astrocyte functions and phenotypes in neural circuits. *Nat Neurosci.* 18:942–952.

Kim KK, Adelstein RS, Kawamoto S (2009). Identification of neuronal nuclei (NeuN) as Fox-3, a new member of the Fox-1 gene family of splicing factors. *J Biol Chem.* 284:31052–31061.

Landisman CE, Long MA, Beierlein M, Deans MR, Paul DL, Connors BW (2002). Electrical synapses in the thalamic reticular nucleus. *J Neurosci.* 22:1002–1009.

Li Y, Lopez-Huerta VG, Adiconis X, Levandowski K, Choi S, Simmons SK, Arias-Garcia MA, Guo B, Yao AY, Blosser TR, Wimmer RD, Aida T, Atamian A, Naik T, Sun X, Bi D, Malhotra D, Hession CC, Shema R, Gomes M, Li T, Hwang E, Krol A, Kowalczyk M, Peca J, Pan G, Halassa MM, Levin JZ, Fu Z, Feng G (2020). Distinct subnetworks of the thalamic reticular nucleus. *Nature.* 583:819–824.

Liu XB, Jones EG (1999). Predominance of corticothalamic synaptic inputs to thalamic reticular nucleus neurons in the rat. *J Comp Neurol.* 414(1):67–79.

Majorossy K, Réthelyi M, Szentágothai J (1965). The large glomerular synapse of the pulvinar. *J Hirnforsch.* 7:415–432.

Martinez-Garcia RI, Voelcker B, Zaltsman JB, Patrick SL, Stevens TR, Connors BW, Cruikshank SJ (2020). Two dynamically distinct circuits drive inhibition in the sensory thalamus. *Nature.* 583:813–818.

McCormick DA, Prince DA (1986). Acetylcholine induces burst firing in thalamic reticular neurones by activating a potassium conductance. *Nature.* 319:402–405.

Montero VM (1987). Ultrastructural identification of synaptic terminals from the axon of type 3 interneurons in the cat lateral geniculate nucleus. *J Comp Neurol.* 264:268–283.

Morgan JL, Lichtman JW (2020). An individual interneuron participates in many kinds of inhibition and innervates much of the mouse visual thalamus. *Neuron.* 106(3):468–481.

Nagalski A, Puelles L, Dabrowski M, Wegierski T, Kuznicki J, Wisniewska MB (2016). Molecular anatomy of the thalamic complex and the underlying transcription factors. *Brain Struct Funct.* 221:2493–2510.

Nahmani M, Erisir A (2005). VGluT2 immunochemistry identifies thalamocortical terminals in layer 4 of adult and developing visual cortex. *J Comp Neurol.* 484:458–473.

Pantazopoulos H, Gisabella B, Rexrode L, Benefield D, Yildiz E, Seltzer P, Valeri J, Chelini G, Reich A, Ardelt M, Berretta S (2020). Circadian rhythms of perineuronal net composition. *eNeuro.* 7:1–21.

Puelles L (2021). Recollections on the origins and development of the prosomeric model. *Front Neuroanat.* 15:Article 787913. https://pubmed.ncbi.nlm.nih.gov/35002639/

Roy DS, Zhang Y, Halassa MM, Feng G (2022). Thalamic subnetworks as units of function. *Nat Neurosci.* 25:140–153.

Saunders A, Macosko EZ, Wysoker A, Goldman M, Krienen FM, de Rivera H, Bien E, Baum M, Bortolin L, Wang S, Goeva A, Nemesh J, Kamitaki N, Brumbaugh S, Kulp D, McCarroll SA (2018). Molecular diversity and specializations among the cells of the adult mouse brain. *Cell.* 174:1015–1030.

Sherman SM (2020). The lateral geniculate nucleus and pulvinar. In: Fritzsch B, ed., *The Senses: A Comprehensive Reference.* 2nd ed. Academic Press:369–391.

Simko J, Markram H (2021). Morphology, physiology and synaptic connectivity of local interneurons in the mouse somatosensory thalamus. *J Physiol.* 599(22):5085–5101.

Wang S, Bickford ME, Van Horn SC, Erisir A, Godwin DW, Sherman SM (2001). Synaptic targets of thalamic reticular nucleus terminals in the visual thalamus of the cat. *J Comp Neurol.* 440:321–341.

Wei H, Bonjean M, Petry HM, Sejnowski TJ, Bickford ME (2011). Thalamic burst firing propensity: A comparison of the dorsal lateral geniculate and pulvinar nuclei in the tree shrew. *J Neurosci.* 31:17287–17299.

Wilson JR, Friedlander MJ, Sherman SM (1984). Fine structural morphology of identified X- and Y-cells in the cat's lateral geniculate nucleus. *Proc R Soc Lond B.* 221:411–436.

Zhu JJ, Uhlrich DJ, Lytton WW (1999). Properties of a hyperpolarization-activated cation current in interneurons in the rat lateral geniculate nucleus. *Neuroscience.* 92(2):4445–4457.

3.

BIOPHYSICAL PROPERTIES OF THALAMIC CELL TYPES

Thomas Budde

INTRODUCTION

In addition to the multitude of anatomically and functionally distinguishable nuclei, the diversity of thalamic function depends on the variance among the different neurons related to their shape, connectivity, and individual electrophysiological properties. In the following, 3 basic cell types will be distinguished: (1) glutamatergic relay neurons that send axons outside a given nucleus to the cortex, TRN, and other brain regions; (2) GABAergic local circuit interneurons (INs) that project within their nucleus of origin; (3) GABAergic neurons of the TRN that surround and innervate the thalamus (see **Chapter 2 in Section II**). In several species, INs are most abundant and best studied in the dorsal part of the lateral geniculate nucleus (dLGN), which represents the sensory relay station of the visual system (see **Chapter 17 in Section V**). The notion of specific intrinsic electrophysiological properties of different neuronal cell types leads to the concept that a neuron of one type cannot functionally replace a neuron of another type and is therefore central for neuronal function (Llinás, 2014). In this respect, an IN cannot replace a TRN neuron, despite releasing the same transmitter. Consequently, neuronal network dynamics are governed not only by the flow of synaptic currents, but also by the intrinsic properties of the involved neurons. Intrinsic electrical properties include both passive and voltage-dependent electrical membrane characteristics that allow neurons to integrate and transfer incoming stimuli and to generate subthreshold oscillations and spontaneous activity in the absence of synaptic input. In addition, nearly all thalamic neurons reveal the peculiar property of bursting, indicating that these neurons belong to a system that is concerned with rhythmic activity such as sleep oscillations (Llinás, 2014).

THE RESTING MEMBRANE POTENTIAL (RMP) AND MEMBRANE INPUT RESISTANCE (R_{IN}) OF THALAMIC NEURONS

The resting membrane potential (RMP) is assumed to be determined by channels active below firing threshold. In thalamic neurons a variety of voltage-dependent ion channels, channels generating a sustained current component at the appropriate membrane voltage, background K^+ channels, and ionic pumps contribute to setting the RMP (Meuth et al., 2006; Amarillo et al., 2014). Beside the background K^+ current (I_{Kb}) generated by members of the two-pore domain K^+ (K_{2P}) channel family, hyperpolarization is provided by steady-state current of I_A (a fast transient A-type K^+ current generated by specific voltage-activated K^+ (K_V) channels) and I_{Kir} (generated by inwardly rectifying K^+ channels). This hyperpolarization is counteracted by depolarizing I_h (hyperpolarization-activated cationic current generated by hyperpolarization-activated cyclic nucleotide-gated [HCN] channels), I_T (low-threshold calcium current generated by Ca_V3 channels), and I_{Na} (sodium current generated by Na_V channels). The 2 latter cation inward currents have the special feature of generating window currents based on the overlap of steady-state activation and inactivation curves, termed $I_{Twindow}$ and I_{NaP} (the persistent Na^+ current may also by generated by specific Na_V channel subtypes). The resulting RMP reveals cell type- and species-specific differences. While rodent relay and TRN neurons reveal RMP around −70 mV (Broicher et al., 2008), INs are more depolarized around -65 mV (Leist et al., 2016). In the cat, a reversed situation was described, with INs as more negative compared to relay neurons (McCormick and Pape, 1988). These differences are likely reflecting different expression profiles of ion channels. In rodent relay neurons, the counterbalancing effects of HCN2 and 2 members of the K_{2P} channel family, namely TWIK-related acid-sensitive K^+ (TASK)3 ($K_{2P}9.1$) and TASK1 ($K_{2P}3.1$) channels were identified as interacting partners (Meuth et al., 2006). It is interesting to note that the coordinated interactions of 2 or more of the ion channels that determine the RMP are also crucial for generation of rhythmic activity (see **Chapter 61 in Section XI**) and resonant behavior on the neuronal network level (Amarillo et al., 2014; Hashimoto, 2020).

Ion channels are considered to be biological conductors in the neuronal membrane with the conductance of a single channel, following Ohm's law (Ceballos et al., 2017). According to the passive cable model of biological membranes, the opening of ion channels increases membrane permeability and decreases R_{in} and the membrane time constant (τ_m). Each neuronal type has a distinct membrane resistance that affects neuronal excitability and integrative properties. Thalamic INs in rodents (around 300–550 MΩ) and in the cat (around 90 MΩ) were found to have higher R_{in} compared to relay neurons (rodent: around 100–150 MΩ; cat around 50 MΩ), which may be the basis of fast spiking found in some IN (McCormick and Pape, 1988; Zhu and Heggelund, 2001; Leist et al., 2016; Zobeiri et al., 2018). K_{2P} channels are considered the classical determinants of R_{in}. Indeed, it was found that the reduction of a K^+ background conductance is associated with an increase in R_{in} in relay neurons (Cox et al., 1998). Conversely, it is known that volatile anesthetics decrease membrane excitability and R_{in} by opening background K^+ channels, especially $K_{2P}9.1/K_{2P}3.1$ (Ries and Puil, 1999).

It should be noted, however, that at least 2 types of voltage-dependent inward currents present in thalamic neurons, namely I_{NaP} and I_T, create regions of negative slope conductance in the corresponding current-voltage curves that are based on a positive feedback between a change of the membrane potential and the activation variables of the channel (Ceballos et al., 2017). This phenomenon relates an increase in current conductance with a paradoxical increase in R_{in} and τ_m, leading to an amplification of excitatory synaptic potentials and neuronal excitability.

THE ELECTROTONIC PROPERTIES OF THALAMIC NEURONS

The term *electrotonic properties of a neuron* describes the manner in which subthreshold voltage changes propagate throughout the axon and the dendritic tree. Each of the 3 main prethalamic/thalamic cell types reveal distinctive electrotonic properties (Halnes et al., 2011; Cox, 2014; Crunelli et al., 2018). In general, soma and dendrites are regarded as electrical elements. Compared with the large "leaky" soma, the thin dendrites are spatially more distributed and are seen as electrically "tight." When a dendritic region receives a local input, a membrane voltage gradient emerges between different locations within the tree. The asymmetric somatic and dendritic conditions further cause local membrane potential changes to attenuate more when spreading from the dendrite toward the soma than in the opposite direction. Based on this effect, most neurons are electrically compact when viewed from the soma. Dendritic patch clamp recordings revealed that this effect is particularly strong in relay and TRN neurons (Crunelli et al., 2018). Although the dendritic trees of these neurons cover large areas within the neuropil (i.e., physical space), their electrotonic space is regarded as small. Thus, relay and TRN neurons behave almost like an isopotential sphere, as if they would not have dendrites as a separate cellular compartment. When the soma is synaptically or experimentally depolarized, the membrane potential in the entire dendritic tree follows with rather little dendritic voltage amplitude attenuation.

Based on the extensive dendritic arborization and small diameters of their dendrites, thalamic local circuit INs reveal strikingly different electrotonic properties (Bloomfield and Sherman, 1989). They appear to be electrotonically much larger than relay neurons; that is, voltage changes evoked at distal dendrites are strongly attenuated at the soma. Consequently, IN may process independent input-output operations in soma and dendrites and even different dendritic locations (see **Chapter 4 in Section II**). This is especially relevant for IN function (Cox, 2014) since these neurons can release γ (gamma)-aminobutyric acid (GABA) via both axon terminals (termed *F1 terminals*) as well as presynaptic, vesicle-containing dendrites (termed *F2 terminals*; see **Chapter 5 in Section II**). Importantly, neurotransmitter-dependent modulation of dendritic ion channels is able to strongly change the compactness of IN (Halnes et al., 2011). In this respect it is known that acetylcholine (ACh) reduces the R_{in} of IN through muscarinic type 2 (M_2) ACh receptor (AChR)-mediated activation of different ion channels, including HCN and $K_{2P}9.1/K_{2P}3.1$ channels (Zhu and Heggelund, 2001; Leist et al., 2017). During certain sleep states, when the cholinergic tone tends to be low, INs are likely to be electronically compact. Under these conditions, INs may provide inhibition within a circumscribed area of the respective thalamic nucleus, as excitatory input at one location in the dendritic tree may result in GABA release throughout the dendritic and axonal arbors. During states of wakefulness, when the cholinergic tone is high, distal dendritic regions may become electronically isolated from each other and from the soma. During these conditions, localized dendritic regions may function as independent processing units. Beside modulatory influences, backpropagating action potentials (bAPs), that is, a transient dendritic depolarization that occurs as a result

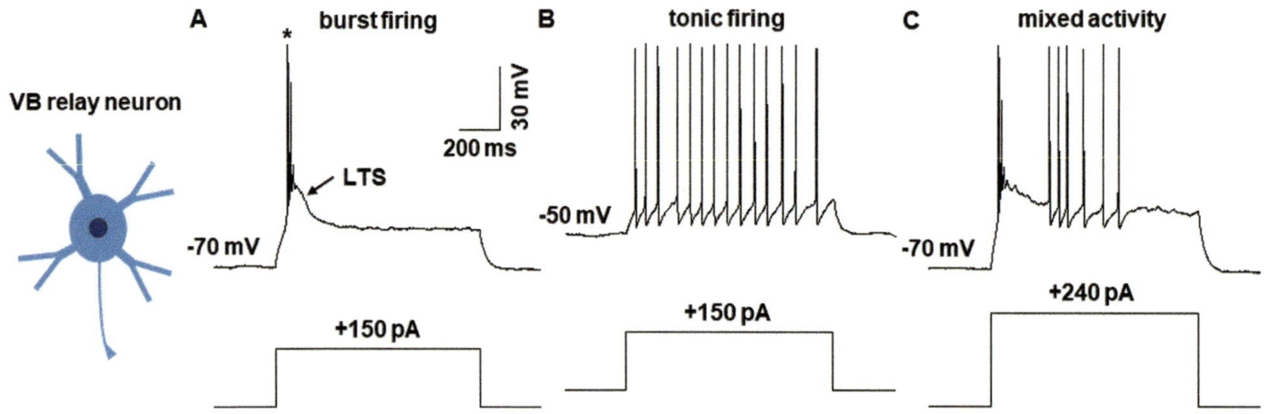

Figure 3.1. Activity modes recorded in a thalamocortical relay neuron. (A) Depolarizing current injections from a hyperpolarized level of the membrane potential triggers burst firing consisting of a low threshold Ca^{2+} spike (LTS; see arrow) crowned by a group of 3 conventional Na^{+}/K^{+} action potentials (see asterisks). (B) Depolarizing current injections from a depolarized level of the membrane potential triggers tonic firing. (C) The separation of burst and tonic firing is not as strict as typically described, as stronger depolarizing current injections from a hyperpolarized level of the membrane potential triggers a mixture of burst and tonic firing.

of the generation of an action potential at the axon hillock, may influence the compactness of thalamic neurons.

MODES OF ACTION POTENTIAL GENERATION IN THALAMIC NEURONS

One of the most intriguing properties of prethalamic/thalamic neurons is their ability to generate action potentials in 2 fundamentally different behavioral state-dependent activity modes. They generate burst activity while hyperpolarized and tonic activity while depolarized (Figure 3.1). States of wakefulness but also rapid eye movement (REM) sleep are characterized by tonic generation of action potentials and high-frequency (γ band frequency), low-amplitude oscillations. States of slow-wave sleep are characterized by highly synchronized oscillatory low-frequency (<15 Hz) and high-amplitude burst activity. While initially thought to mainly occur during states of drowsiness and sleep, bursting can also be found during wakefulness. On the one hand, this may enhance the detectability of sensory stimuli during attentive wakefulness (Crunelli et al., 2018). On the other hand, irregular random bursting and rhythmic low-frequency δ- and θ-band oscillations in the thalamus during wakefulness (collectively termed *thalamocortical dysrhythmia syndromes*, TCD) characterize a number of apparently unrelated neurological and psychiatric conditions, including neurogenic pain, tinnitus, Parkinson's disease, major depressive disorder, and absence epilepsy (Llinás et al., 1999). Especially, the highly synchronized rhythmic burst activity seen during ictal periods of childhood absence epilepsy (CAE) represents an example of the dysfunction (see **Chapter 65 in Section XII**) of the thalamocortical system.

Bursts are grouped action potentials with high intraburst frequencies (100–500 Hz), following sufficiently long periods of membrane hyperpolarization, that are riding on top of a low-threshold Ca^{2+} spike (LTS; Figure 3.1) and are generated by the opening of Ca_V3 channels. The membrane hyperpolarization is necessary to de-inactivate Ca_V3 channels, thereby allowing channel opening with the next depolarization. The cyclic interaction between the membrane currents generated by Ca_V3-, HCN-, and Ca^{2+}-activated K^+ channels renders oscillatory burst activity as an intrinsic property of most prethalamic/thalamic neurons. This activity is synchronized by synaptic mechanisms over the extended thalamocortical network. In this respect, the action potential firing associated with LTS generation in TRN neurons may evoke sufficient $GABA_A$ receptor-associated hyperpolarization in relay neurons, allowing enough removal of Ca_V3 channel inactivation and burst generation by the relay neurons (Crunelli et al., 2014). In turn, LTS-evoked firing of relay neurons excites TRN neurons as part of an interaction that helps to sustain sleep-related oscillations. The generation of an LTS in relay and TRN neurons is based on a unique mechanism that induces depolarization of all cell compartments with simultaneous widespread recruitment of spatially distributed Ca_V3 channels, the global LTS (Crunelli et al., 2018). Initially, INs in cat dLGN were found to lack a typical LTS (McCormick and Pape, 1988). Later, INs in rats were found to generate LTS with relatively long duration and low intraburst action potential frequency (Zhu et al., 1999a). In the mouse dLGN, 2 subtypes of IN were characterized based on the properties of I_h and the expression profile of HCN channels (Leist et al., 2016). Small IN were characterized by a more depolarized RMP, stronger anomalous rectification, generation of rebound bursts, and higher I_h current density compared to large cells.

Heterogeneity of burst firing was also detected among TRN neurons (Lee et al., 2007). Here neurons can be differentiated by their action potential discharge modes and localization within the TRN. Most neurons in the dorsal TRN lack burst discharge or show an atypical burst that consists of an initial action potential, followed by broader small amplitude spikes. In contrast, most neurons in the ventral TRN display a stereotypical burst discharge consisting of a transient, high-frequency discharge of multiple action potentials. Later, TRN neurons were differentiated based on gene-expression profiles. It was found that parvalbumin-, but not somatostatin-expressing neurons exhibit typical burst firing and intrinsic rhythmogenic properties (Clemente-Perez et al., 2017).

It is interesting to note that by adding the K^+ channel blocker 4-AP to non-bursting IN and TRN neurons that reveal a small transient depolarization following release from membrane hyperpolarization, it is possible to unmask a larger amplitude transient depolarization that resembles an LTS (Pape et al., 1994; Lee et al., 2007).

During different behavioral states, bursts appear in characteristic pattern. In defined sleep stages (N1 → N3), bursts occur rhythmically in θ, sleep spindle, and δ frequency bands in relay neurons. Rhythmic bursting during sleep has important functions in long-term memory formation (see **Chapter 54 in Section IX**). During this process, repeated neuronal replay of representations originating from the hippocampus during slow-wave sleep leads to a gradual transformation and integration of representations in neocortical networks (Klinzing et al., 2019).

During phases of relaxed, inattentive wakefulness (closed eyes in humans), the EEG is characterized by α waves and action potential bursts in relay neurons in the same frequency range. During this activity, bursts have a notably lower frequency (50–70 Hz) and are driven by high-threshold Ca^{2+} spikes (HTS) that depend on Ca_V3 and high-voltage-activated (HVA) Ca^{2+} channels of the L-type (Ca_V1) family (Crunelli et al., 2018).

While bursting is less typical for stages of wakefulness, it seems to be relevant for acute and chronic pain perception. Recordings in awake and freely behaving mice confirmed that tonic firing of relay neurons of the ventrobasal thalamic complex (VB) reliably reflects behavioral nociceptive responses, while increased burst firing corresponds to decreased reaction (Huh and Cho, 2013). In addition, an increase in K_V7 channel activity was associated with increased bursting, a higher pain threshold, with fewer VB units responding to noxious stimuli, and increased burst firing in responsive neurons (Cerina et al., 2015). In the context of neuropathic pain, studies performed in humans and rats found that the incidence of burst firing in VB is increased (Lenz, 1992; Jeanmonod et al., 1993; Hains et al., 2006; Wang and Thompson, 2008). Therefore, the role of thalamic burst firing in sensory perception seems to strongly depend on the behavioral context.

Although bursts are widely seen as a stereotyped form of neuronal activity, cell type- and nucleus-specific differences were found (Figure 3.2). In this respect, bursts with different characteristics (e.g., latency to burst, number of action potentials per burst, intraburst frequency) have been described. Beside the general differences between cell types (i.e., relay neurons, IN, TRN neurons), relay neurons from functionally distinct nuclei (sensory nuclei and nuclei involved in cortical-thalamic-cortical interactions; see **Chapter 19 in Section V**; see **Chapter 35 in Section VI**) reveal subtle differences (Broicher et al., 2007b; Broicher et al., 2008; Desai and Varela, 2021). It has been suggested that burst characteristics vary based on thalamocortical hierarchy and sensory modality, with the number of spikes per burst and the latency to burst being essential parameters for thalamocortical transmission. Beside the cell type-specific expression profile of ion channels contributing to burst activity, relative differences in conductance values may account for the distinct differences between burst responses observed in different neurons, suggesting that neurons are individually tuned to optimize network operation under different input conditions (Halnes et al., 2011).

When questioning why the brain seems to extensively generate rhythmic bursting, an energy-intensive form of neuronal activity during rest phases associated with energy preservation, several aspects have to be considered (Crunelli et al., 2018). Compared with a single action potential generated during tonic firing, bursts of action potentials provide several functional advantages (see **Chapter 63 in Section XI**), namely: (1) higher reliability of signal transmission, (2) less sensitivity to noise, (3) more effective triggering of responses in some classes of neocortical neurons, and (4) generation of global somatodendritic Ca^{2+} signals in prethalamic/thalamic neurons that are essential for some forms of cellular plasticity.

Tonic firing occurs when Ca_V3 channels are nearly completely inactivated due to sufficient membrane depolarization (but see below). During attentive wakefulness, where tonic firing is more typical, action potentials are generated independently and proportionally to the incoming stimulus strength (i.e., directly linked to an EPSP), thereby allowing proper linear summation (Sherman, 2001). Thus, larger EPSPs elicit higher action potential rates during tonic firing. Depending on the species under consideration, relay neurons and IN reveal different properties of tonic firing. Tonic firing in relay neurons typically shows a delayed onset of firing during a depolarizing pulse with no pronounced spike-frequency adaptation. While the latter pointed to a lack of the K_V7 channels (that limit repetitive firing in hippocampal and cortical principal neurons) in relay neurons of the cat (McCormick and Pape, 1988),

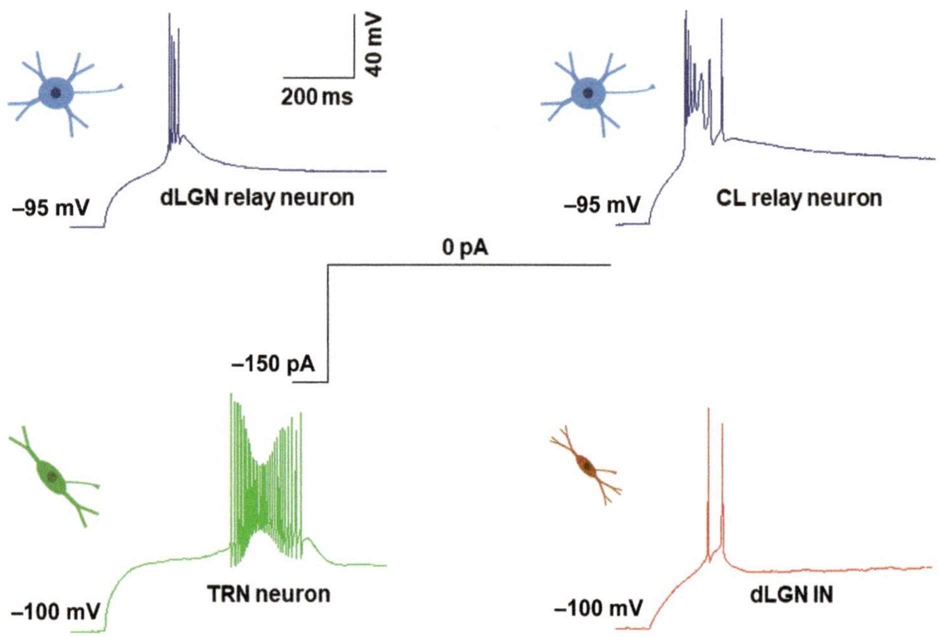

Figure 3.2. Different characteristics of rebound burst responses in different prethalamic/thalamic cell types. Cells are hyperpolarized by current injection. Upon release from membrane hyperpolarization, rebound bursts with distinct properties are generated in a relay neuron from the dorsal part of the lateral geniculate nucleus (dLGN), in a relay neuron from the centrolateral nucleus (CL), in a neuron from the prethalamic reticular nucleus (TRN) and a local circuit interneuron (IN) from the dLGN (as indicated). Please note the differences in LTS onset and duration, as well as in the number and frequency of the generated action potentials.

these channels are functional in rodent relay neurons (Kasten et al., 2007; Cerina et al., 2015). In cats, INs possess some typical characteristics of fast-spiking interneurons found in the brain with short action potentials and a rather linear relationship between firing rate versus injected current (McCormick and Pape, 1988). In rats this firing mode seems to be less pronounced since IN respond to a retinal EPSP with a single action potential when the cell is depolarized (Zhu et al., 1999a). In the murine dLGN, small IN generate higher firing frequency of faster action potentials compared to the large cells (Leist et al., 2016).

DENDRITIC BACKPROPAGATION OF ACTION POTENTIALS

In general, dendritic bAPs have critical roles in spike-timing-dependent synaptic plasticity, the generation of dendritic Ca^{2+} spikes, and triggering dendritic neurotransmitter release in cortical neurons (Crunelli et al., 2018). In prethalamic/thalamic neurons, the membrane capacitance and axial resistance act as low-pass filters, leading to a significant attenuation of bAPs, thereby only influencing dendritic compartments and synapses close to the soma. In relay neurons, the duration of action potentials within an LTS typically increases progressively with increasing spike number (Figure 3.2). Therefore, late actions potentials in a burst propagate more efficiently into the dendritic tree compared with those occurring at burst onset. Consequently, tonic actions potentials that have similar half-widths as late burst spikes undergo similar dendritic attenuation. Therefore, bAPs have a rather limited influence on Ca^{2+}-dependent plasticity compared to the substantial increase in intracellular Ca^{2+} throughout the entire dendritic tree induced by the global LTS. TRN neurons have the intriguing feature to communicate with connected cells electrically via dendrodendritic gap junctions. Therefore, an LTS might modulate electrical synapses throughout the dendritic tree, whereas bAPs can affect only those fairly close to the soma.

CELL-TYPE-SPECIFIC SETS OF ION CHANNELS

The intrinsic electrical properties and the synaptic input-output relationships of neurons are governed by the action of voltage-dependent ion channels. The localization of specific populations of ion channels with distinct functional properties at discrete sites in neurons determine excitability and synaptic transmission (Vacher et al., 2008). Especially the rich membrane behavior of prethalamic/thalamic neurons at subthreshold potentials depends on the diversity and matched properties of ion-conductance mechanisms (Figure 3.3) and is the basis for rhythmic bursting (Amarillo et al., 2014). Especially, 3 families of ion channels that are expressed in differentially combined sets of channel subtypes have been shown to govern the switch between activity modes: K_{2P}, HCN, and Ca_V3 channels.

In the case of K_{2P} channels, one channel subunit consists of 4 transmembrane segments and 2 pore (P)-forming domains (P1, P2) that are arranged as a tandem (Bista et al., 2015a). Two of these subunits comprise a functional channel. In humans, 15 genes encode for K_{2P} channels that are subdivided in 6 subfamilies based on sequence homologies and functional similarities. Several physicochemical stimuli including neurotransmitters, protons, di- and multivalent cations, as well as clinically used anesthetics (bupivacaine; isoflurane), have been shown to modulate K_{2P} channels. In the thalamus the bidirectional modulation of members of the K_{2P} channel family—namely TASK1 ($K_{2P}3.1$) and TASK3 ($K_{2P}9.1$), as well as TWIK-related K^+ (TREK) channel 1 (TREK1; $K_{2P}2.1$)—is of central importance for relay neuron function (Meuth et al., 2003; Bista et al., 2015b). With respect to the GABAergic neurons, $K_{2P}2.1$ and $K_{2P}3.1$ seem to be of importance for the function of TRN neurons and IN, respectively (Bista et al., 2015a; Leist et al., 2017). ACh is an important transmitter released by neurons of the ascending brainstem system and the basal forebrain throughout the thalamocortical system (see **Chapter 5 in Section II**). With respect to modulation of K_{2P} channels by G-protein-coupled receptors, muscarinic ACh receptors (MAChR) and PLCβ play unique roles. While the inhibition of $K_{2P}3.1$ and $K_{2P}9.1$ channels via $M_{1/3}$AChR depolarizes relay neurons and favors the tonic-firing mode, the activation of $K_{2P}3.1$ channels via $M_{2/4}$AChR hyperpolarizes IN. Although muscarinic signaling hyperpolarizes TRN neurons via M_2AChR, effector channels seem to involve inward rectifier K^+ channels rather than K_{2P} channels (Sun et al., 2013). In relay neurons, the enzymatic reaction governed by PLCβ is unique since both substrate breakdown and product synthesis exert important functions (Bista et al., 2015b). While the degradation of PIP_2 leads to the closure of $K_{2P}2.1$ channels, the production of DAG induces the inhibition of $K_{2P}3.1$ and $K_{2P}9.1$ channels.

HCN channels are members of the pore loop cation channel superfamily, with each subunit comprising 6 transmembrane domains and one pore loop. Four of these subunits make up a functional channel. In mammals, the HCN channel family comprises 4 homologous members (HCN1–HCN4) that differ in their voltage-dependency, activation kinetics, and modulation by cyclic nucleotides (Rychlik et al., 2022). While HCN channels allow the passage of a depolarizing mixed Na^+ and K^+ current, these channels structurally resemble K_V channels. However, HCN channels are activated by membrane hyperpolarization, whereas K_V channels are activated by depolarization. With respect to the channel expression profile in the brain, HCN1 and HCN4 are seen as the cortical and thalamic form, respectively. In addition, while HCN2 shows a broad expression throughout the brain, HCN3 is expressed rather specifically in distinct brain areas. Indeed, I_h in relay neurons is mainly based on the expression of HCN2 and HCN4 channels, while HCN3 and HCN1 are less expressed (Ludwig et al., 2003; Zobeiri et al., 2019). Activation of these channels in relay neurons by cAMP production favors the tonic-firing mode. Small INs are characterized by strong expression of HCN4 and HCN3, followed by HCN2 and

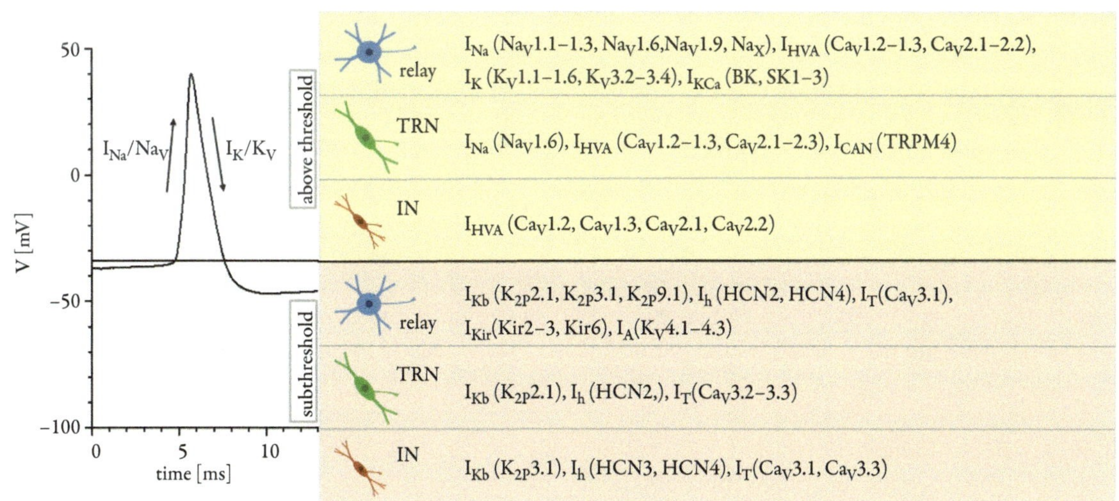

Figure 3.3. Different prethalamic/thalamic cell types express distinct sets of ion channels operating below and above the threshold for generating Na^+/K^+ action potentials. When an external stimulus reaches the threshold potential, all neurons generate action potentials (left part). While the upstroke is determined by the activation of voltage-dependent Na^+ (Na_V) channels, voltage-dependent K^+ (K_V) channels govern the repolarization phase. A selected number of ion channels in relay neurons, neurons of the prethalamic reticular nucleus (TRN), and local circuit interneurons (IN) is shown (right part, as indicated). Corresponding ion channels and resulting membrane currents are grouped (for details, see main text). The sets presented are idealized and mainly deducted from studies on rodents.

HCN1 (Leist et al., 2016). For large IN the abundance of HCN2–HCN4 is generally lower compared to small IN. In TRN neurons, the presence of I_h is only obvious when Ba^{2+}, a blocker of some K_{2P} and inward rectifier channels, is applied (Rateau and Ropert, 2006), suggesting that I_h is present but normally masked by K^+ background currents. When uncovered, I_h in TRN neurons reveals properties similar to those described for the homomeric HCN2 channels. The domination of HCN2 and HCN4 in relay neurons results in an I_h that shows slow activation kinetics and a strong depolarizing shift of the voltage dependency induced by cAMP. The presence of HCN3 in IN renders their I_h sensitive to positive modulation by the membrane phospholipid PIP_2. It is interesting to note that strong expression of HCN3 channels was also found in neurons of the intergeniculate leaflet, another GABAergic thalamic cell type (Ying et al., 2011).

In general, voltage-activated Ca^{2+} channels (Ca_V1–3) are large pore-forming multi-subunit complexes that allow Ca^{2+} influx upon membrane depolarization and control a plethora of cell- and tissue-specific processes, including neurotransmitter release, gene transcription, synaptic plasticity, membrane excitability, and pacemaker activity in neurons (Lanzetti and Di Biase, 2022). The pore forming $Ca_V\alpha1$ subunit is organized in 4 transmembrane domains (I–IV), each containing 6 membrane-spanning helices (S1–S6) and one P-loop. While S1–S4 include the voltage sensor, the S5–S6 segments form the channel pore. Ca_V3 channels activate at low voltages, inactivate rapidly, deactivate slowly, and are characterized by a small single-channel conductance. In prethalamic/thalamic neurons these channels are involved in sleep oscillations that are based on their activation following a period of membrane hyperpolarization leading to the generation of an LTS (Crunelli et al., 2014). For a long time the view on Ca_V3 channel function was restricted to the generation of a stereotypical LTS. However, an intriguing feature of these channels is the generation of a "window current," termed $I_{Twindow}$. This is a small depolarizing tonic current resulting from the fraction of Ca_V3 channels that are open in a narrow membrane potential region confined by the overlap of steady-state activation and inactivation curves. Changing the magnitude of $I_{Twindow}$ by altering the membrane potential within the range of the window allows the reciprocal transformation of slow (<1 Hz) oscillations in individual relay and TRN neurons into delta-frequency oscillations (Crunelli et al., 2018). The generation of $I_{Twindow}$ has another unexpected consequence for tonic action potential firing. Based on the nearly complete inactivation of Ca_V3 at the depolarized membrane potentials found during states of wakefulness, these channels are not considered to modulate tonic action potential firing. However, depolarization associated with sensory AMPA inputs seem to further recruit deinactivated channels, resulting in more I_T that boosts the EPSP and increases the firing probability (Deleuze et al., 2012). Thereby Ca_V3 channels support the reliability of tonic firing in relay neurons. Molecular analyses revealed that different prethalamic/thalamic cell types are characterized by the expression of distinct sets of Ca_V3 channel isoforms and splice variants, resulting in an I_T with different voltage dependency and kinetics (Broicher et al., 2007b; Broicher et al., 2008). Relay neurons and IN of the dLGN show expression of $Ca_V3.1a$ and $Ca_V3.1bc$/$Ca_V3.3$ channels, respectively. Relay neurons of the nonspecific centrolateral nucleus (CL) of the thalamus express $Ca_V3.1a$/bc and $Ca_V3.3$ channels. Prethalamic TRN neurons are characterized by expression of $Ca_V3.2$ and $Ca_V3.3$ channels, resulting in a unique I_T that activates slowly at more positive membrane potentials and shows slower inactivation kinetics compared to other thalamic neurons (Huguenard and Prince, 1992).

Additional ionic conductances contribute to the firing modes of prethalamic/thalamic neurons. Some of them are stated in the following. Although a specific molecular pharmacology was missing for a long time, the Ca^{2+}-activated, nonselective cation current (I_{CAN}) has been suggested to be present in all prethalamic/thalamic cell types based on computer modeling (Bal and McCormick, 1993; Zhu et al., 1999b; Hughes et al., 2002). I_{CAN} provides prolonged membrane depolarization and Ca^{2+} entry, allowing persistent firing, is essential for stabilizing and enhancing the expression of the slow (<1 Hz) oscillations, and triggers Ca^{2+} -dependent mechanisms. In TRN neurons dendritically located Ca^{2+}-dependent small-conductance (SK)-type K^+ channels 2 (SK2) initiate oscillatory bursting (Cueni et al., 2008) that is shaped by Ca^{2+}-induced intracellular Ca^{2+} release (Coulon et al., 2009). Only recently, the molecular nature of I_{CAN} in TRN neurons was identified and shown to be mediated by Ca^{2+}- and voltage-dependent transient receptor potential melastatin 4 (TRPM4) channels (O'Malley et al., 2020).

Another current providing sustained depolarization in relay neurons is I_{NaP}. This current is evoked from potentials around −60 mV and influencing LTS amplitude and latency, thereby shaping the integrative properties of these neurons during both tonic and burst firing (Parri and Crunelli, 1998). Recently the repertoire of ionic mechanisms known to be involved in the generation and control of oscillatory activity in relay neurons was expanded (Tissone et al., 2021). Intrinsic oscillations may be shaped by increased output responses at a preferred frequency, a phenomenon known as intrinsic neuronal resonance. Based on its strong voltage-dependent inward rectification (i.e., a region of negative slope conductance), relay neuron resonance is amplified by the inward rectifier potassium current, I_{Kir}. The molecular nature of this current may be based on members of the Kir2, Kir3 and Kir6 families (Meuth et al., 2003; Thomzig et al., 2005).

The fast transient K^+ current termed I_A has a special role in thalamic neurons by influencing both firing pattern and action potential waveform. Especially (but not exclusively), members of the K_V4 channel family are thought to generate this current based on the interaction with accessory subunits and regulatory proteins. Since these channels activate below action potential threshold, thereby providing a transient hyperpolarzing influence, the onset of tonic firing in response to a depolarization may be delayed (Huguenard et al., 1991; Budde et al., 1992). As their voltage-dependent and kinetic properties are similar to Ca_V3 channels, the functional interplay between K_V4 and Ca_V3 channels controls burst firing and the occurrence and shape of an LTS (Pape et al., 1994). $K_V4.1$–4.3 was found in different thalamic nuclei (Kanyshkova et al., 2011).

A number of additional ion channels operating at or above action potential threshold contribute to tonic firing in relay neurons. Several Na_V ($Na_V1.1$–1.3, $Na_V1.6$, $Na_V1.9$, Na_X) and K_V ($K_V1.1$–1.6, $K_V3.2$–3.4, $K_V7.2$–7.3) channels involved in the generation of action potentials have been detected (Kasten et al., 2007; Decher et al., 2010; Kanyshkova et al., 2011; Makinson et al., 2017). HVA Ca^{2+} channels are thought to shape tonic firing by generating an appropriate Ca^{2+} inward current (I_{HVA}) and thereby activating Ca^{2+}-dependent K^+ channels that generate the membrane outward current, I_{KCa}. In this respect, $Ca_V1.2$–1.3 and $Ca_V2.1$–2.2 were detected in relay neurons, INs, and TRN neurons, while the latter additionally express $Ca_V2.3$ channels (Meuth et al., 2001; Zaman et al., 2011; Kanyshkova et al., 2014). Both, BK (STREX and ZERO splice variants) and SK (SK1–3) channels have been identified in relay neurons, as the molecular substrate for I_{KCa} and SK2 seems to play an important role in TRN neurons (Kasten et al., 2007; Cueni et al., 2009; Ehling et al., 2013).

The distinct sets of ion channel determine RMP, firing pattern, oscillatory activity and resonant properties of prethalamic/thalamic neurons (see **Chapter 61 in Section XI**). One important consequence from this is the preference to intrinsically generate slow (δ frequency band), low threshold oscillations in relay neurons (McCormick and Pape, 1990), faster (5–15 Hz) oscillations regardless of the polarization level in IN (Zhu et al., 1999b) and oscillations at rather depolarized levels in the spindle frequency range (7–14 Hz) in TRN neurons (Bal and McCormick, 1993). In addition, the slow (<1 Hz) oscillations in relay and TRN neurons depend on their distinct ion channel repertoire (Crunelli et al., 2018).

It is interesting to note that some of the ion channels with central importance for the generation of rhythmic oscillatory activity in the thalamocortical system have been suggested as potential drug targets, namely HCN and Ca_V3 channels (Lanzetti and Di Biase, 2022). This is especially relevant for conditions involving mistimed or excess bursting, as found in TCD (Llinás et al., 1999). Although the exact ion channel specificity and concentration range of ethosuximide (ETX) are debated (Leresche et al., 1991), this antiepileptic drug, a Ca_V3 channel blocker, suppresses bursting in animal models of absence epilepsy, central pain syndrome, and cocaine binge abuse (Broicher et al., 2007a; Wang and Thompson, 2008; Bisagno et al., 2010). In addition, the pharmacological suppression of HCN channels in relay neurons prevents absence seizures in rodents (David et al., 2018).

OPEN QUESTIONS

As pointed out above, there is a much higher level of functional diversity in prethalamic/thalamic neurons based on electrophysiological differences than classically thought. This opens up the question of how many cell types exist in these brain areas. Are there many different cell types, or is there a continuum of graded properties in a limited number of cell types? This question can only be answered by multimodal approaches, including the analyses of gene- and protein-expression profiles, cell morphology, neuronal firing pattern, network connectivity, and embedding in functional pathways. Only recently, it was emphasized that the configurations of epigenetic marks seem to be critical for stabilizing cell type-defining characteristics that emerge after the final mitotic division when cells enter their developmental default state (Mukamel and Ngai, 2019). In addition, there seems to be a need for quantitative description of complete single-cell anatomy in cell-type classification (Peng et al., 2021). Beside the standard anatomical, molecular biological, and electrophysiological techniques, this may require the additional use of optogenetics, analysis of neuronal assemblies, computational modeling, whole brain reconstruction, and epigenomic and transcriptomic analyses. Here unique features not found in closely related cells may help to determine the classification. Still, it may be unclear whether differentiated neurons with similar characteristics in different species have indeed the same evolutionarily conserved critical functions (Tosches, 2021).

In terms of ion channel research, the identification of those channels generating the resting Na^+ permeability in thalamic neurons is an important goal. Here NALCN channels (Na$^+$ leak channel, non-selective) that are expressed mainly in the central nervous system may be good candidates (Cochet-Bissuel et al., 2014). This identification may trigger similar scientific advancements compared to that following the characterization of thalamic K_{2P} channels.

Yet it is unclear whether thalamic ion channels indeed represent practical and safe drug targets. For example, the use of specific HCN4 blockers (i.e. drugs directed against the thalamic isoform) may cause side effects on cardiac pacemaking. In addition, it remains to be determined to what extent the sets of ion channel described above are conserved between species.

REFERENCES

Amarillo Y, Zagha E, Mato G, Rudy B, Nadal MS (2014). The interplay of seven subthreshold conductances controls the resting membrane potential and the oscillatory behavior of thalamocortical neurons. *J Neurophysiol.* 112:393–410.

Bal T, McCormick DA (1993). Mechanisms of oscillatory activity in guinea-pig nucleus reticularis thalami in vitro: A mammalian pacemaker. *J Physiol (Lond).* 468:669–691.

Bisagno V, Raineri M, Peskin V, Wikinski SI, Uchitel OD, Llinas RR, Urbano FJ (2010). Effects of T-type calcium channel blockers on cocaine-induced hyperlocomotion and thalamocortical GABAergic abnormalities in mice. *Psychopharmacol.* 212:205–214.

Bista P, Cerina M, Ehling P, Leist M, Pape HC, Meuth SG, Budde T (2015a). The role of two-pore-domain background K^+ (K_{2P}) channels in the thalamus. *Pflügers Arch.* 467:895–905.

Bista P, Pawlowski M, Cerina M, Ehling P, Leist M, Meuth P, Aissaoui A, Borsotto M, Heurteaux C, Decher N, Pape HC, Oliver D, Meuth SG, Budde T (2015b). Differential phospholipase C-dependent modulation of TASK and TREK two-pore domain K(+) channels in rat thalamocortical relay neurons. *J Physiol.* 593:127–144.

Bloomfield SA, Sherman SM (1989). Dendritic current flow in relay cells and interneurons of the cat's lateral geniculate nucleus. *Proc Natl Acad Sci U S A.* 86:3911–3914.

Broicher T, Kanyshkova T, Meuth P, Pape HC, Budde T (2008). Correlation of T-channel coding gene expression, I(T), and the low threshold Ca(2+) spike in the thalamus of a rat model of absence epilepsy. *Mol Cell Neurosci.* 39:384–399.

Broicher T, Seidenbecher T, Meuth P, Munsch T, Meuth SG, Kanyshkova T, Pape HC, Budde T (2007a). T-current related effects of antiepileptic drugs and a Ca(2+) channel antagonist on thalamic relay and local circuit interneurons in a rat model of absence epilepsy. *Neuropharmacol.* 53:431–446.

Broicher T, Kanyshkova T, Landgraf P, Rankovic V, Meuth P, Meuth SG, Pape HC, Budde T (2007b). Specific expression of low-voltage-activated calcium channel isoforms and splice variants in thalamic local circuit interneurons. *Mol Cell Neurosci.* 36:132–145.

Budde T, Mager R, Pape H-C (1992). Different types of potassium outward current in relay neurons acutely isolated from the rat lateral geniculate nucleus. *Eur J Neurosci.* 4:708–722.

Ceballos CC, Roque AC, Leao RM (2017). The role of negative conductances in neuronal subthreshold properties and synaptic integration. *Biophys Rev.* 9:827–834.

Cerina M, Szkudlarek HJ, Coulon P, Meuth P, Kanyshkova T, Nguyen XV, Gobel K, Seidenbecher T, Meuth SG, Pape HC, Budde T (2015). Thalamic Kv 7 channels: pharmacological properties and activity control during noxious signal processing. *Br J Pharmacol.* 172:3126–3140.

Clemente-Perez A, Makinson SR, Higashikubo B, Brovarney S, Cho FS, Urry A, Holden SS, Wimer M, David C, Fenno LE, Acsady L, Deisseroth K, Paz JT (2017). Distinct thalamic reticular cell types differentially modulate normal and pathological cortical rhythms. *Cell Rep.* 19:2130–2142.

Cochet-Bissuel M, Lory P, Monteil A (2014). The sodium leak channel, NALCN, in health and disease. *Front Cell Neurosci.* 8:132.

Coulon P, Herr D, Kanyshkova T, Meuth P, Budde T, Pape HC (2009). Burst discharges in neurons of the thalamic reticular nucleus are shaped by calcium-induced calcium release. *Cell Calcium.* 46:333–346.

Cox CL (2014). Complex regulation of dendritic transmitter release from thalamic interneurons. *Curr Opin Neurobiol.* 29:126–132.

Cox CL, Zhou Q, Sherman SM (1998). Glutamate locally activates dendritic outputs of thalamic interneurons. *Nature.* 394:478–482.

Crunelli V, David F, Leresche N, Lambert RC (2014). Role for T-type Ca^{2+} channels in sleep waves. *Pflügers Arch.* 466:735–745.

Crunelli V, Lorincz ML, Connelly WM, David F, Hughes SW, Lambert RC, Leresche N, Errington AC (2018). Dual function of thalamic low-vigilance state oscillations: rhythm-regulation and plasticity. *Nat Rev Neurosci.* 19:107–118.

Cueni L, Canepari M, Adelman JP, Lüthi A (2009). Ca(2+) signaling by T-type Ca(2+) channels in neurons. *Pflügers Arch.* 457:1161–1172.

Cueni L, Canepari M, Lujan R, Emmenegger Y, Watanabe M, Bond CT, Franken P, Adelman JP, Lüthi A (2008). T-type Ca2+ channels, SK2 channels and SERCAs gate sleep-related oscillations in thalamic dendrites. *Nat Neurosci.* 11:683–692.

David F, Carcak N, Furdan S, Onat F, Gould T, Meszaros A, Di Giovanni G, Hernandez VM, Chan CS, Lorincz ML, Crunelli V (2018). Suppression of hyperpolarization-activated cyclic nucleotide-gated channel function in thalamocortical neurons prevents genetically determined and pharmacologically induced absence seizures. *J Neurosci.* 38:6615–6627.

Decher N, Streit AK, Rapedius M, Netter MF, Marzian S, Ehling P, Schlichthorl G, Craan T, Renigunta V, Kohler A, Dodel RC, Navarro-Polanco RA, Preisig-Müller R, Klebe G, Budde T, Baukrowitz T, Daut J (2010). RNA editing modulates the binding of drugs and highly unsaturated fatty acids to the open pore of Kv potassium channels. *Embo J.* 29:2101–2113.

Deleuze C, David F, Behuret S, Sadoc G, Shin HS, Uebele VN, Renger JJ, Lambert RC, Leresche N, Bal T (2012). T-type calcium channels consolidate tonic action potential output of thalamic neurons to neocortex. *J Neurosci.* 32:12228–12236.

Desai NV, Varela C (2021). Distinct burst properties contribute to the functional diversity of thalamic nuclei. *J Comp Neurol.* 529:3726–3750.

Ehling P, Cerina M, Meuth P, Kanyshkova T, Bista P, Coulon P, Meuth SG, Pape HC, Budde T (2013). Ca(2+)-dependent large conductance K(+) currents in thalamocortical relay neurons of different rat strains. *Pflügers Arch.* 465:469–480.

Hains BC, Saab CY, Waxman SG (2006). Alterations in burst firing of thalamic VPL neurons and reversal by Na(v)1.3 antisense after spinal cord injury. *J Neurophysiol.* 95:3343–3352.

Halnes G, Augustinaite S, Heggelund P, Einevoll GT, Migliore M (2011). A multi-compartment model for interneurons in the dorsal lateral geniculate nucleus. *PLoS Comput Biol.* 7:e1002160.

Hashimoto K (2020). Mechanisms for the resonant property in rodent neurons. *Neurosci Res.* 156:5–13.

Hughes SW, Cope DW, Blethyn KL, Crunelli V (2002). Cellular mechanisms of the slow (<1 Hz) oscillation in thalamocortical neurons in vitro. *Neuron.* 33:947–958.

Huguenard JR, Prince DA (1992) A novel T-type current underlies prolonged Ca2+-dependent burst firing in GABAergic neurons of rat thalamic reticular nucleus. *J Neurosci.* 12:3804–3817.

Huguenard JR, Coulter DA, Prince DA (1991). A fast transient potassium current in thalamic relay neurons: Kinetics of activation and inactivation. *J Neurophysiol.* 66:1304–1315.

Huh Y, Cho J (2013). Discrete pattern of burst stimulation in the ventrobasal thalamus for anti-nociception. *PloS ONE.* 8:e67655.

Jeanmonod D, Magnin M, Morel A (1993). Thalamus and neurogenic pain: physiological, anatomical and clinical data. *Neuroreport.* 4:475–478.

Kanyshkova T, Broicher T, Meuth SG, Pape HC, Budde T (2011). A-type K^+ currents in intralaminar thalamocortical relay neurons. *Pflügers Arch.* 461:545–556.

Kanyshkova T, Ehling P, Cerina M, Meuth P, Zobeiri M, Meuth SG, Pape HC, Budde T (2014). Regionally specific expression of high-voltage-activated calcium channels in thalamic nuclei of epileptic and non-epileptic rats. *Mol Cell Neurosci.* 61:110–122.

Kasten MR, Rudy B, Anderson MP (2007). Differential regulation of action potential firing in adult murine thalamocortical neurons by Kv3.2, Kv1, and SK potassium and N-type calcium channels. *J Physiol.* 584:565–582.

Klinzing JG, Niethard N, Born J (2019). Mechanisms of systems memory consolidation during sleep. *Nat Neurosci.* 22:1598–1610.

Lanzetti S, Di Biase V (2022). Small molecules as modulators of voltage-gated calcium channels in neurological disorders: State of the art and perspectives. *Molecules.* 15:1312.

Lee SH, Govindaiah G, Cox CL (2007). Heterogeneity of firing properties among rat thalamic reticular nucleus neurons. *J Physiol.* 582:195–208.

Leist M, Datunashvili M, Kanyshkova T, Zobeiri M, Aissaoui A, Cerina M, Romanelli MN, Pape HC, Budde T (2016). Two types of interneurons in the mouse lateral geniculate nucleus are characterized by different h-current density. *Sci Rep.* 6:24904.

Leist M, Rinne S, Datunashvili M, Aissaoui A, Pape HC, Decher N, Meuth SG, Budde T (2017). Acetylcholine-dependent upregulation of TASK-1 channels in thalamic interneurons by a smooth muscle-like signalling pathway. *J Physiol.* 595:5875–5893.

Lenz FA (1992). The ventral posterior nucleus of thalamus is involved in the generation of central pain syndromes. *J Pain.* 1:42–51.

Leresche N, Parri HR, Erdemli G, Guyon A, Turner JP, Williams SR, Asprodini E, Crunelli V (1991). On the action of the anti-absence drug ethosuximide in the rat and cat thalamus. *J Neurosci.* 18:4842–4853.

Llinás RR (2014). Intrinsic electrical properties of mammalian neurons and CNS function: A historical perspective. *Front Cell Neurosci.* 8:320.

Llinás RR, Ribary U, Jeanmonod D, Kronberg E, Mitra PP (1999). Thalamocortical dysrhythmia: A neurological and neuropsychiatric syndrome characterized by magnetoencephalography. *Proc Natl Acad Sci U S A.* 96:15222–15227.

Ludwig A, Budde T, Stieber J, Moosmang S, Wahl C, Holthoff K, Langebartels A, Wotjak C, Munsch T, Zong X, Feil S, Feil R, Lancel M, Chien KR, Konnerth A, Pape HC, Biel M, Hofmann F (2003). Absence epilepsy and sinus dysrhythmia in mice lacking the pacemaker channel HCN2. *Embo J.* 22:216–224.

Makinson CD, Tanaka BS, Sorokin JM, Wong JC, Christian CA, Goldin AL, Escayg A, Huguenard JR (2017). Regulation of thalamic and cortical network synchrony by Scn8a. *Neuron.* 93:1165–1179.

McCormick DA, Pape H-C (1988). Acetylcholine inhibits identified interneurons in the cat lateral geniculate nucleus. *Nature.* 334:246–248.

McCormick DA, Pape H-C (1990). Properties of a hyperpolarization-activated cation current and its role in rhythmic oscillation in thalamic relay neurones. *J Physiol (Lond).* 431:291–318.

Meuth SG, Budde T, Kanyshkova T, Broicher T, Munsch T, Pape H-C (2003). Contribution of TWIK-related acid-sensitive K^+ channel 1 (TASK1) and TASK3 channels to the control of activity modes in thalamocortical neurons. *J Neurosci.* 23:6460–6469.

Meuth SG, Budde T, Pape H-C (2001). Differential control of high-voltage activated Ca^{2+} current components by a Ca^{2+}-dependent inactivation mechanism in thalamic relay neurons. *Thalamus & Related Systems.* 1:31–38.

Meuth SG, Kanyshkova T, Meuth P, Landgraf P, Munsch T, Ludwig A, Hofmann F, Pape HC, Budde T (2006). The membrane resting potential of thalamocortical relay neurons is shaped by the interaction among TASK3 and HCN2 channels. *J Neurophysiol.* 96:1517–1529.

Mukamel EA, Ngai J (2019). Perspectives on defining cell types in the brain. *Curr Opin Neurobiol.* 56:61–68.

O'Malley JJ, Seibt F, Chin J, Beierlein M (2020). TRPM4 Conductances in thalamic reticular nucleus neurons generate persistent firing during slow oscillations. *J Neurosci.* 40:4813–4823.

Pape H-C, Budde T, Mager R, Kisvarday Z (1994). Prevention of Ca^{2+}-mediated action potentials in GABAergic local circuit neurons of the thalamus by a transient K^+ current. *J Physiol (Lond).* 478:403–422.

Parri HR, Crunelli V (1998). Sodium current in rat and cat thalamocortical neurons: Role of a non-inactivating component in tonic and burst firing. *J Neurosci.* 18:854–867.

Peng H, Xie P, Liu L, Kuang X, Wang Y, Qu L, Gong H, Jiang S, Li A, Ruan Z, Ding L, Yao Z, Chen C, Chen M, Daigle TL, Dalley R, Ding Z, Duan Y, Feiner A, He P, Hill C, Hirokawa KE, Hong G, Huang L, Kebede S, Kuo HC, Larsen R, Lesnar P, Li L, Li Q, Li X, Li Y, Li Y, Liu A, Lu D, Mok S, Ng L, Nguyen TN, Ouyang Q, Pan J, Shen E, Song Y, Sunkin SM, Tasic B, Veldman MB, Wakeman W, Wan W, Wang P, Wang Q, Wang T, Wang Y, Xiong F, Xiong W, Xu W, Ye M, Yin L, Yu Y, Yuan J, Yuan J, Yun Z, Zeng S, Zhang S, Zhao S, Zhao Z, Zhou Z, Huang ZJ, Esposito L, Hawrylycz MJ, Sorensen SA, Yang XW, Zheng Y, Gu Z, Xie W, Koch C, Luo Q, Harris JA, Wang Y, Zeng H (2021). Morphological diversity of single neurons in molecularly defined cell types. *Nature.* 598:174–181.

Rateau Y, Ropert N (2006). Expression of a functional hyperpolarization-activated current (Ih) in the mouse nucleus reticularis thalami. *J Neurophysiol.* 95:3073–3085.

Ries CR, Puil E (1999). Ionic mechanism of isoflurane's actions on thalamocortical neurons. J *Neurophysiol.* 81:1802–1809.

Rychlik N, Hundehege P, Budde T (2022). Influence of inflammatory processes on thalamocortical activity. *Biol Chem.* 404:303–310.

Sherman SM (2001). Tonic and burst firing: dual modes of thalamocortical relay. *Trends Neurosci.* 24:122–126.

Sun YG, Pita-Almenar JD, Wu CS, Renger JJ, Uebele VN, Lu HC, Beierlein M (2013). Biphasic cholinergic synaptic transmission controls action potential activity in thalamic reticular nucleus neurons. *J Neurosci.* 33:2048–2059.

Thomzig A, Laube P, Prüss H, Veh RW (2005). Pore-forming subunits of K-ATP channels, Kir6.1 and Kir6.2, display prominent differences in regional and cellular distribution in the rat brain. *J Comp Neurol.* 484:313–330.

Tosches MA (2021). From cell types to an integrated understanding of brain evolution: The case of the cerebral cortex. *Annu Rev Cell Dev Biol.* 37:495–517.

Vacher H, Mohapatra DP, Trimmer JS (2008). Localization and targeting of voltage-dependent ion channels in mammalian central neurons. *Physiol Rev.* 88:1407–1447.

Wang G, Thompson SM (2008). Maladaptive homeostatic plasticity in a rodent model of central pain syndrome: Thalamic hyperexcitability after spinothalamic tract lesions. *J Neurosci.* 28:11959–11969.

Ying S, Tibbs GR, Picollo A, Abbas SY, Sanford RL, Accardi A, Hofmann F, Ludwig A, Goldstein PA (2011). PIP2-mediated HCN3 channel gating is crucial for rhythmic burst firing in thalamic intergeniculate leaflet neurons. *J Neurosci,* 31:10412–10423.

Zaman T, Lee K, Park C, Paydar A, Choi JH, Cheong E, Lee CJ, Shin HS (2011). Cav2.3 channels are critical for oscillatory burst discharges in the reticular thalamus and absence epilepsy. *Neuron.* 70:95–108.

Zhu J, Heggelund P (2001). Muscarinic regulation of dendritic and axonal outputs of rat thalamic interneurons: A new cellular mechanism for uncoupling distal dendrites. *J Neurosci.* 21:1148–1159.

Zhu JJ, Uhlrich DJ, Lytton WW (1999a). Burst firing in identified rat geniculate interneurons. *Neurosci.* 91:1445–1460.

Zhu JJ, Lytton WW, Xue JT, Uhlrich DJ (1999b). An intrinsic oscillation in interneurons of the rat lateral geniculate nucleus. *J Neurophysiol.* 81:702–711.

Zobeiri M, Chaudhary R, Blaich A, Rottmann M, Herrmann S, Meuth P, Bista P, Kanyshkova T, Luttjohann A, Narayanan V, Hundehege P, Meuth SG, Romanelli MN, Urbano FJ, Pape HC, Budde T, Ludwig A (2019). The hyperpolarization-activated HCN4 channel is important for proper maintenance of oscillatory activity in the thalamocortical system. *Cereb Cortex.* 29:2291–2304.

Zobeiri M, Chaudhary R, Datunashvili M, Heuermann RJ, Luttjohann A, Narayanan V, Balfanz S, Meuth P, Chetkovich DM, Pape HC, Baumann A, van Luijtelaar G, Budde T (2018). Modulation of thalamocortical oscillations by TRIP8b, an auxiliary subunit for HCN channels. *Brain Struct Funct.* 223:1537–1564.

4.

HETEROGENEITY OF THALAMIC INPUT LANDSCAPES

László Acsády

INTRODUCTION

As the title of this volume suggests, the thalamus and cortex form one, inseparable functional unit. The major target of the thalamic output is the cortex, and the cortex provides by far the most numerous inputs to the thalamus (Jones, 2007a). With a very few exceptions, every thalamic neuron has a cortical projection. In fact cortical projection has been used as a defining feature of "the thalamus" (Jones, 2007b). In return, every cortical region studied so far innervates the thalamus. By now it is clear that this mutual innervation is far more complex than a simple reciprocal connectivity pattern (Deschenes et al., 1998). Rather, multiple, distinctively different pathways link the 2 structures with variable spatial innervation profiles and physiological properties (see **Chapters 11, 14, and 15 in Section IV**; **Chapters 19 and 27 in Section V**; **Chapter 32 in Section VI**). These long-range connections allow rich interactions between the 2 structures and form the neuronal basis of perception and action and everything in between.

Every thalamocortical cell receives inputs from neurons of different origin, conveying information of distinct content. It follows that the core function of the thalamus is to integrate these inputs and to generate a de novo thalamic message for the cortex. Integration can take place by integrating inputs with different properties from the same source (e.g., retinal ganglion cells with different tuning properties) or integrating inputs from different sources (e.g., cortical and subcortical inputs). In the thalamus there are examples of both kinds of integration. As described below in detail, the source, the neurotransmitter, and the structure of these afferents display great heterogeneity in the thalamus. As a consequence, integration of inputs and the generation of the thalamocortical output signal may take many different forms, and thalamocortical neurons display no canonical input integration.

DISTINCT PATTERNS OF INPUT INTEGRATION IN THE SOMATOSENSORY THALAMUS

To gain insight into the heterogeneity of input integration in the thalamus consider first 2 parallel pathways via the somatosensory nuclei, ventral posteromedial nucleus (VPM), and posterior nucleus (PoM) (Figure 4.1) as an example. Afferents of VPM are typical for a classical sensory "relay" nucleus. VPM receives subcortical, excitatory inputs largely from a single origin (principle trigeminal nucleus, PrV) in a highly topographic manner (Pierret et al., 2000). Only few ascending PrV afferents converge on a single VPM neuron (Deschenes et al., 2003). In rodents these all represent a single whisker. In addition, VPM is also contacted by numerous, but weak cortical inputs arising in layer 6 and GABAergic inputs from the thalamic reticular nucleus (TRN) (Figure 4.1A). Earlier it was assumed that integration of few, strong subcortical excitatory afferents with many weaker cortical and TRN inputs by a thalamic neuron applies to other thalamic nuclei as well. However, later it turned out that other nuclei can have different sets of inputs. In the adjacent somatosensory nucleus (POm), beside these 3 important afferents, the neurons are also contacted by other, conceptually different types of cortical and subcortical inputs not present in the VPM (Figure 4.1A). These inputs are able to substantially alter the response properties of POm cells (Lavallée et al., 2005; Groh et al., 2014). In contrast to VPM cells, PoM cells are innervated by powerful excitatory terminals arising in L5 (Groh et al., 2008) of the primary somatosensory cortex and strong inputs from extrathalamic inhibitory (ETI) sources like zona incerta (ZI) (Barthó et al., 2002) and anterior pretectal nucleus (APT) (Bokor et al., 2005). Within POm these inputs display certain areal segregation (Casas-Torremocha et al., 2022), but convergence of the afferents conveying different types of messages can also take place (Figure 4.1B–D) (Lavallée et al., 2005; Groh et al., 2014). As a result, the final output of POm represents a

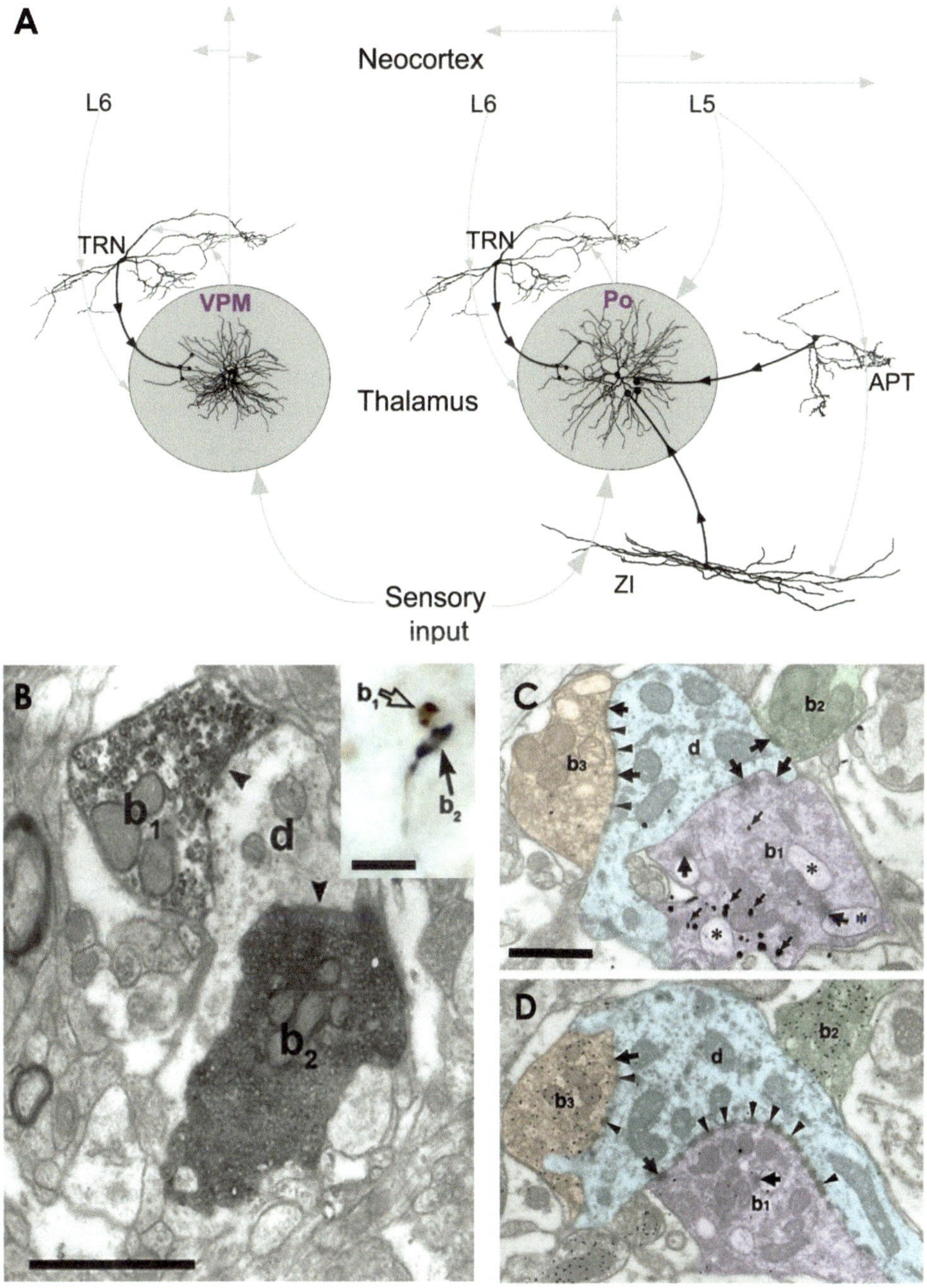

Figure 4.1. Heterogeneity of input integration in 2 somatosensory thalamic nucleus, VPM and Po. (A) Shared input types between the VPM and Po include L6, TRN and the sensory inputs. Note that the exact nuclear origin of sensory inputs differ in VPM and Po (not shown for simplicity) but they represent the same type. Po receives qualitatively different types of inputs from L5, APT, and ZI. Note that these input stations are linked. L5 targets not only Po but also APT and ZI, and the sensory afferents emit a collateral in ZI. (B) Convergence of large, multisynaptic subcortical (b1; brown) and L5 cortical (b2; black) terminals in the dendrite (d) of the same Po cell. Correlated light (inset) and electron microscopic images. Arrowheads indicate synapses. Scalebar: 1 µm. (C–D) Convergence of a large multisynaptic sensory terminal (b1) and a large multisynaptic GABAergic terminal (b3) in the dendrite (d) of the same Po cell. Electron micrographs of adjacent ultrathin sections. In C the anterograde tracer from SpVi is visualized in b1 bouton by silver intensified gold particles (black dots, small arrows). In D the GABA immunoreactivity in b3 is detected by postembedding immunogold method (small black dots). The dendrite (d) is also contacted by a small unisynaptic GABAergic terminal (b2) of probably TRN origin. Large arrows, synapses, arrowheads, puncta adhaerntie, asterisk, dendritic appandages. Scalebar: 0.5 µm.

spatially heterogeneous weighing and integration of various afferents.

Based on the available information, POm is not an exception, and integration of inputs with different properties is common in the thalamus. In fact, transferring information arising in topographically ordered 2D receptor surfaces, as in case of VPM (as well as visual inputs to the lateral geniculate nucleus, dorsal division, LGNd, and auditory inputs to the medial geniculate nucleus ventral division, MGNv), seems to be a special restricted mode of thalamic computation. Apparently the rules of input integration differ in the majority of the thalamus, forming up to 80% of the thalamic volumes in primates (Rovó et al., 2012). But even in the case of sensory information transfer, recent data indicate that thalamocortical neurons can integrate information from various information channels. For example, in the case of vision, individual LGNd neurons integrate inputs from ganglion cells with different receptive field properties (Morgan et al., 2016; Rompani et al., 2017) or from retinal and collicular inputs (Bickford et al., 2015). Unfortunately in the thalamus, unlike in many other brain regions, the study of input integration is in its infancy. Thus, deciphering the rules for generating the thalamic output signal to cortex, and the variability of these rules of integration, awaits further research.

PROPERTIES OF THALAMIC AFFERENTS: AXON TERMINALS WITH ONE VERSUS MANY SYNAPSES

Traditionally, the term *synapse* refers to a single presynaptic active zone and the postsynaptic specialization opposing it. Accordingly, one presynaptic bouton can form one or many synapses. Thalamic inputs display great heterogeneity in the number of synapses established by a single axon terminal. As we shall see below, this heterogeneity has profound functional consequences.

Unisynaptic terminals, as their name suggests, form one synapse with their targets. In case of multisynaptic terminals, a single presynaptic bouton establishes many (2–50) synapses on the same postsynaptic partner (Figure 4.2). These hugely complex synaptic elements (Figure 4.2A,B) in the thalamus were the first to capture the interest of anatomists working with the electron microscope in the 1960s (Figure 4.2C) (see **Chapter 2 in Section II**). Since then it has become clear that in the thalamus, excitatory terminals of both cortical and subcortical sources as well as GABAergic inputs of subcortical origin (Figure 4.2D–F) can form multisynaptic terminals in the thalamus. By now the basic morphological and physiological properties of these complex axon terminals have been elucidated. However, a real understanding of the computations performed by these afferents and their functional consequences is still far away.

Elsewhere in the brain, multisynaptic terminals are exceedingly rare (e.g., mossy terminals in the hippocampus, climbing fibers of the cerebellum, or Calyx of Held in the ascending auditory pathway) (Rollenhagen and Lübke, 2006). In stark contrast, every thalamic nucleus studied so far contains multisynaptic terminals (glutamatergic and/or GABAergic). A rich complexity of uni- and multisynaptic terminals in the thalamus indicates that a core feature of thalamic afferents is the heterogeneous dynamics of synaptic transmission associated with inputs having different number of synapses.

The key variable of multisynaptic terminals that explains the function of these peculiar structures is the close placement of synaptic contacts along the postsynaptic surface. This allows transmitter released from one presynaptic active zone to spill over to the postsynaptic receptors located in other synapses of the same terminal (Telgkamp et al., 2004; Budisantoso et al., 2012). Indeed, modeling studies suggested that in these terminals transmitter can diffuse to several synapses within a millisecond (Telgkamp et al., 2004), resulting in a special case of spillover during which one release site is in a position to influence transmitter action at many other synapses of the same axon terminal.

Surprisingly, in the thalamus the final outcome of this "intra-terminal" spillover is different for multisynaptic excitatory versus inhibitory terminals. Excitatory multisynaptic terminals (Figure 4.2A–C) of both cortical and subcortical origins display robust depression even as a response to relatively low frequency inputs (~10 Hz) (Castro-Alamancos, 2002a; Groh et al., 2008). In contrast, inhibitory multisynaptic terminals are able to maintain the charge transfer without depression at very high presynaptic firing rates (up to 100–200 Hz) (Wanaverbecq et al., 2008; Giber et al., 2015). The mechanisms behind this difference in short-term dynamics have been elucidated at least partially. In the case of excitatory inputs, the high amount of glutamate released initially at one release site rapidly diffuses to many other synapses and quickly saturates their postsynaptic receptors (Budisantoso et al., 2012). A special auxiliary subunit (CKAMP44) has been shown to reduce the rate of recovery of these AMPA receptors (Chen et al., 2018). As a consequence, the glutamate released by the next stimuli will be able to evoke only diminished postsynaptic responses.

In multisynaptic GABAergic terminals (Figure 4.2D–F), transmitter released at one synapse rapidly diffuses to the other synapses of the same terminal as well (Telgkamp et al., 2004). GABA, however, does not saturate the postsynaptic receptors. The key for maintaining inhibitory charge transfer at high presynaptic firing rates is that even if there is a failure of transmitter release at one of the synapses, there is always GABA released at another synaptic contact of the same terminal (Telgkamp et al., 2004). As a result of "intra-terminal" spillover, GABA will be able to diffuse to many other synapses within the synaptic complex and activate sufficient numbers of GABA receptors. As a consequence, the amplitude of inhibitory synaptic barrage will not depress or drop, even at high presynaptic firing rates (Wanaverbecq et al., 2008; Giber et al., 2015).

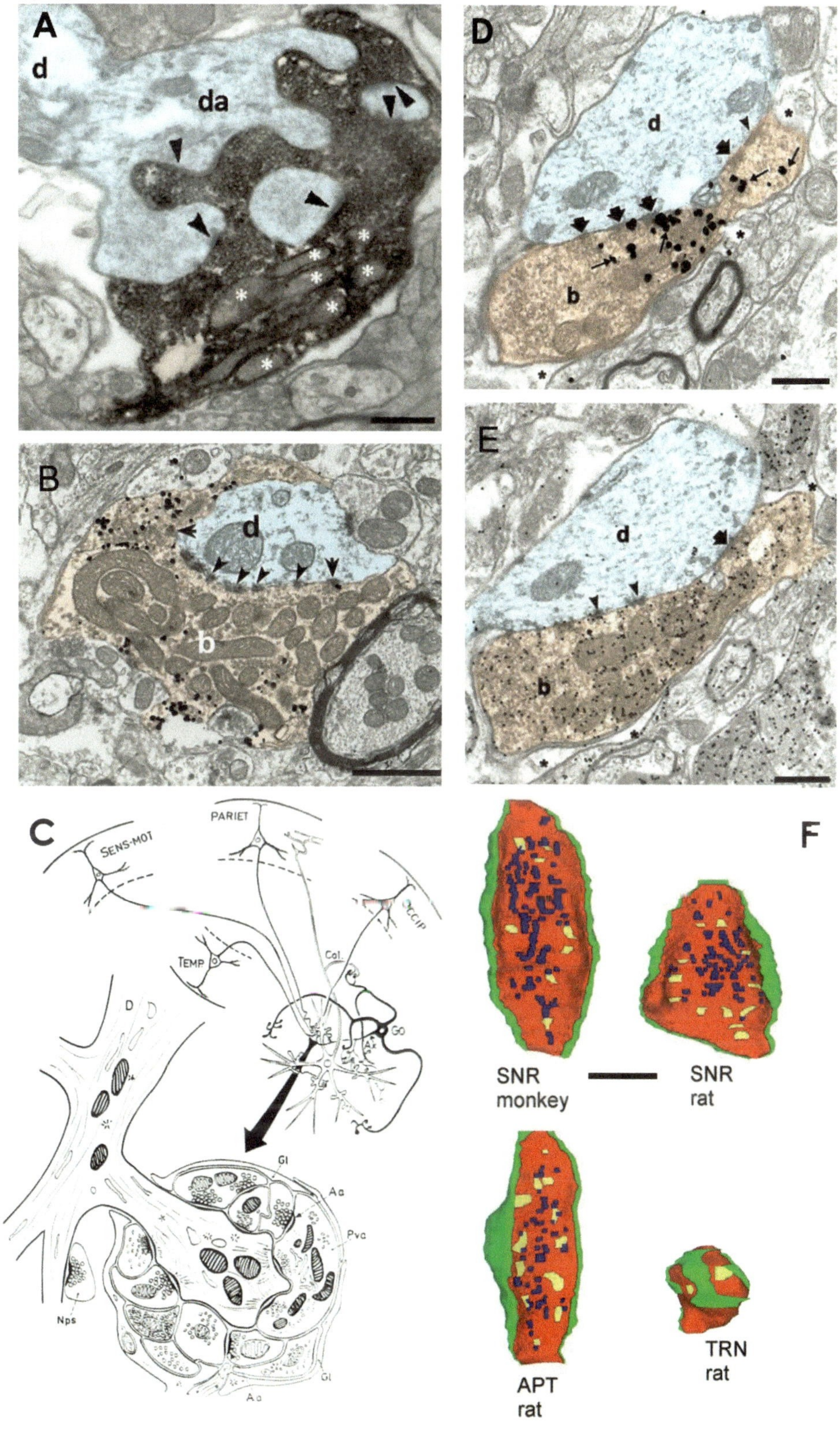

Figure 4.2. Multisynaptic terminal types of the thalamus. (A–B) Large multisynaptic excitatory terminals of L5 origin in the Po of rat (A) and the pulvinar of the macaque (B), labeled by anterograde tracing (DAB chromogene) and vGLUT1 immunostaining (pre-embedding gold), respectively. Scalebars: 0.5µm. d = dendrite; da = dendritic appendages. (C) Early proposal for the integration of large cortical terminals with different cortical origin in the same glomeruli of the pulvinar based on the first electron microscopic observations (Majorossy et al., 1965). This idea has not been formally tested ever since. (D–E) A large multisynaptic GABA terminal of subtantia nigra pars reticulata (SNR) origin contact a proximal dendrited (d) in the ventromedial n. (VM). Adjacent ultrathin sections. In D the anterograde tracer from SNR is visualized in the bouton (b) by silver intensified gold particles (black dots, small arrows). In D the GABA immunoreactivity in is detected by post-embedding immunogold method (small black dots). Thick arrows, synapses; arrowheads, puncta adhaerentia. (F) Extrathalamic GABAergic terminals of various origin in different species share similar morphology. Red = membrane of the terminals; yellow = synapses; blue = puncta adhaerentia; green = glial processes. A TRN terminal is shown for comparison. Note that in this case the TRN terminal have more than 1 synapse (2). These, however, target different dendrites (not shown) and are separated by glia which limits the diffusion of the transmitter.

The functional relevance of these synaptic dynamics is not fully understood. It is unclear why a huge, presynaptic excitatory terminal with a proposed role of "driving" the postsynaptic cell depresses at input firing rates which are well within the normal range of its afferents. It also remains to be determined what the computational advantage is of a constant inhibitory bombardment in selected thalamocortical cells. Several theories have been put forward with some experimental support, but conclusive experiments still need to be performed (Castro-Alamancos, 2002b; Groh et al., 2008; Goldberg et al., 2013; Halassa and Acsády, 2016).

Unisynaptic terminals display short-term plasticity with the opposite dynamics compared to their multisynaptic counterparts. Unisynaptic excitatory terminals (mostly of L6 origin) are facilitatory (Reichova and Sherman, 2004) (see **Chapter 5 in Section II**), whereas unisynaptic inhibitory terminals (mostly of TRN origin) depress at high presynaptic firing rates (Wanaverbecq et al., 2008), although even in a depressed state they can be effective (Campbell et al., 2020). This indicates that the number of synapses established by a single terminal is a key determinant of the synapse dynamics in the thalamus.

CLASSICAL NOMENCLATURE OF THALAMIC AFFERENTS

Traditionally thalamic afferent terminals are divided into large and small terminals (for details, see **Chapter 2 in Section II**). Large, excitatory terminals (often referred to as "drivers") are also call RL terminals since they contain *r*ound vesicles and are *l*arge. These terminals are always multisynaptic. Conversely small, excitatory boutons (or "modulators") are termed RS terminals (S referring to "small") and are almost always unisynaptic. Recently it became clear, though, that in many thalamic regions there are medium-sized excitatory terminals which are in between the RS and RL terminals; hence these were called RM terminals (M for "medium") (Rovó et al., 2012; Bickford et al., 2015; Barsy et al., 2020). RM terminals are multisynaptic but establish fewer active zones than RL terminals and display less short-term depression than true RL terminals. It remains to be determined whether there are indeed 3 classes of terminals (RS, RM, and RL) or a continuum between unisynaptic and multisynaptic terminals, and whether the exact phenotype of the afferent (i.e., facilitating vs. depressing) depends not only on the number of synapses but also on their origin (cortical L6, L5, or subcortical).

In the case of GABAergic terminals, small unisynaptic, inhibitory terminals that derive mainly from the TRN contain flattened vesicles, so they were designated as F1 terminals. F2 terminals refer to the vesicle bearing dendrites of local interneurons. Large multisynaptic GABAergic boutons have been termed F3 terminals (Halassa and Acsády, 2016).

SOURCES OF INPUTS: CORTICAL AFFERENTS

As indicated above, by far the most numerous inputs of the thalamus arise from the cortex (Figure 4.3) (Erişir et al., 1997; Deschenes et al., 1998). With the exception of one recent report of a GABAergic input from the hippocampus (Vetere et al., 2021), all corticothalamic inputs described so far are excitatory. In the case of the classical 6-layered neocortical organization, corticothalamic inputs arise from 3 different cortical cell classes, L6a, L6b, and L5, all with distinct features.

L6a

Numerically, L6a input dominates the thalamic input landscape (Figure 4.3A). These inputs blanket the entire thalamus and can be considered as one of the few canonical inputs which are present in the entire structure (Sherman and Guillery, 2005). They make the thalamus the subcortical structure receiving the densest cortical innervation in the brain. A key feature of classical L6a neurons is that their only subcortical projection involves the thalamus (Bourassa et al., 1995; Kakei et al., 2001). L6a axons form small unisynaptic (RS) terminals that innervate the distal dendrites of thalamocortical neurons (Sherman and Guillery, 2005). To what extent L6a neurons have a direct impact on the instantaneous firing output of thalamocortical neurons is presently unclear and has been detailed elsewhere in this book (**Chapter 14 in Section IV; Chapters 19 and 27 in Section V**). The firing rate of L6 neurons is low (Sirota et al., 2005), they target distal dendrites, and have large, loosely organized axon arbors (Bourassa et al., 1995). This is why many accounts assign a permissive or "modulatory" role to L6a inputs in governing thalamic activity (Sherman and Guillery, 1998).

The L6 action on thalamus is complicated by the collaterals of L6 cells to the TRN. L6 axons target the TRN in a nice topographic manner and arborize in those TRN sectors which have a reciprocal connectivity with the thalamic region they innervate. Indeed, now we have ample evidence that besides the direct excitatory impact on the thalamus, there is a substantial feedforward inhibitory influence on thalamocortical cells mediated by the L6 collaterals in the TRN. Via this feedforward inhibition, L6 inputs to the thalamus are capable, for example, of specifically affecting receptive field properties of the LGNd neurons (**Chapter 14 in Section IV**; **Chapter 19 in Section V**).

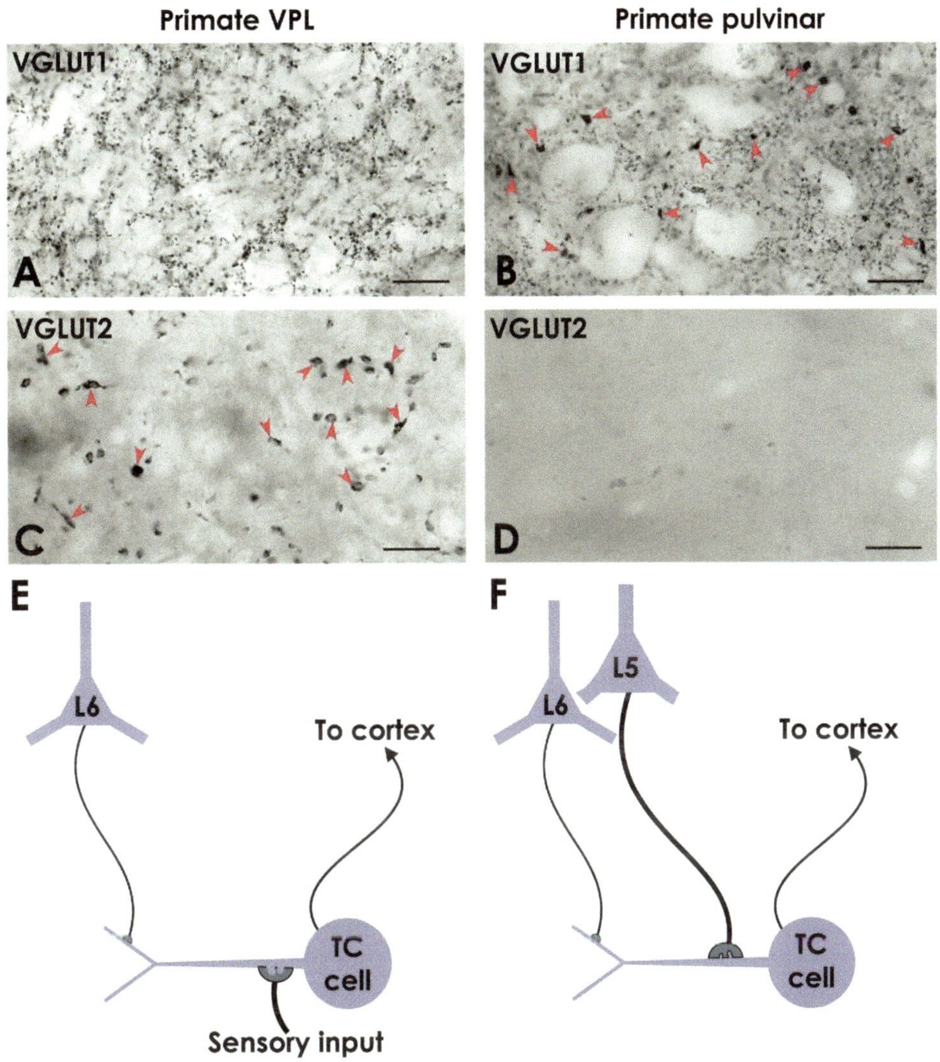

Figure 4.3. Comparison of cortical and subcortical excitatory inputs in the VPL and lateral pulvinar of the macaque monkey. (A–B) High power light microscopic images of vGLUT1-immunostaining which labels cortical inputs. Note that whereas VPL is receiving a huge number of small vGLUT1-positive terminals only, in addition to that in the pulvinar large terminals are also present (arrowheads), consistent with the fact that the former is contacted just by layer 6, whereas the latter is contacted both by layer 5 and 6. (C–D) Adjacent sections immunostained for vGLUT2, which labels subcortical glutamatergic inputs. Note that in the VPL much fewer terminals are labeled compared to vGLUT1; these are, however, all large. In contrast, subcortical excitatory inputs are entirely absent in lateral pulvinar. (E–F) Schematics of the connectivity. VPL represents the classical "relay" nuclei with few large subcortical terminals and many small L6 terminals. In contrast, subcortical drive is absent in the pulvinar; it arises in L5, thus the entire structure is dedicated for cortico-thalamo-cortical communication.

L6b

L6b corticothalamic cells are the least understood and studied cortical cell population. They were just recently described, and very little is known about their activity or role in corticothalamic communication (Bourassa et al., 1995; Hoerder-Suabedissen et al., 2018). Apparently they are the least specific of the 3 communication channels between the cortex and the thalamus. A single thalamic spot will receive convergent inputs from huge cortical L6b populations spanning several cortical regions. They show transient features between L6a and L5 in that they form small unisynaptic terminals like 6a, but innervate only L5 connected (or higher order) thalamic regions (Hoerder-Suabedissen et al., 2018).

L5

One of the most influential results of thalamic research is the discovery of top-down corticothalamic L5 input to the thalamus. This is what turned the traditional view of thalamic function completely upside down and made thalamic experts reconsider the notion that the main function of the thalamus is to relay specific subcortical, sensory information to the cortex. The critical finding was that L5 inputs to the thalamus can be as effective in driving thalamic activity as subcortical excitatory afferents (Ojima, 1994; Bourassa et al., 1995; Sherman and Guillery, 1996). Indeed, the large multisynaptic L5 corticothalamic terminals arising in several cortical regions are indistinguishable from subcortical sensory afferents both in morphological and in physiological features (Hoogland et al., 1991; Vidnyánszky et al., 1996). These L5 boutons form complex, convoluted synaptic complexes with the proximal dendrites of thalamocortical cells, which include vesicle-bearing dendrites (in species having thalamic interneurons), and display a complete glial cover and rows of puncta adhaerentia (Figure 4.1A–C, Figure 4.3B). They evoke fast-rising large amplitude but strongly depressing postsynaptic responses, and upon activation are able to drive thalamocortical cells with short latency and high probability (Groh et al., 2008; Groh et al., 2014). By now it is clear that not all L5 terminals have similar properties. Many frontal cortical regions form small, unisynaptic L5 terminals on their thalamic targets, which can be as effective as their giant counterparts in governing postsynaptic activity (Bokor et al., 2016). The functional consequence of these differences remains to be determined.

Mapping the distribution of excitatory inputs in the primate thalamus revealed that in large thalamic territories, including the lateral and medial pulvinar and the mediodorsal nucleus, large boutons of cortical origin represent the only multisynaptic terminal types (Figure 4.3B). In other words, no subcortical glutamatergic inputs of any type innervate these regions (Figure 4.3C,D) (Rovó et al., 2012). This means that substantial thalamic regions are dedicated entirely to process and return purely cortical information back to the cortex without integrating it with subcortical information. In essence, these territories perform pure cortico-thalamo-cortical operations. In rodent thalamus these territories are smaller and include the ventral and caudal portion of POm (Casas-Torremocha et al., 2022) and to some extent the medial part of the mediodorsal nucleus (MD).

As detailed elsewhere in this book, the output of thalamic territories driven by L5 inputs is highly complex (**Chapter 15 in Section IV**). Thalamocortical cells may return cortical information to the same cortical region that innervates them, forming a classical loop. Several data indicate, however, that the thalamus can route L5 information arising from one cortical area to another cortical area or can have complex axonal arbors, resulting in simultaneous delivery of cortically driven activity to a wide range of cortical regions.

All L5 pyramidal cells having a thalamic collateral innervate usually several subcortical centers, which in return provide a thalamic input (for details, see below) (Guillery and Sherman, 2011). Interestingly, these direct and indirect cortical pathways often target the same thalamic nuclei. Examples include the direct L5 input to VM from M1/M2 cortex (Yamawaki and Shepherd, 2015) paired with the indirect input to this region from the basal ganglia (Bodor et al., 2008), or the direct L5 input to LP from V1 (Blot et al., 2021) and the indirect input via the superior colliculus, or the direct input from S1 to POm (Hoogland et al., 1991; Groh et al., 2008) and the indirect via ZI (Barthó et al., 2002, 2007). Whether the direct and indirect inputs converge on the same neurons or they form parallel streams that return the direct and indirect channels separately to the cortex remains to be determined. For example, recent data from POm indicate that at least in this nucleus the afferents of the direct (L5) and indirect (APT, ZI) pathways can at least partially segregate (Casas-Torremocha et al., 2022).

The first studies described the lack of TRN collaterals from L5 corticothalamic axons originating in S1 and V1 cortices, unlike in the case of L6 (Bourassa et al., 1995). These data fit well with the concept about the similarity between the cortical and the subcortical multisynaptic, excitatory inputs since the latter also reach the thalamus without innervating TRN. In essence, this showed that the L5 input to thalamus is not sculpted by feedforward inhibition via the TRN. Recent data from frontal cortex, however, challenged this idea (Hádinger et al., 2023). According to this, unlike parietal and occipital cortices, L5 afferents from frontal cortical regions provide rich, dense, topographic innervation of the TRN in parallel with L6 TRN collaterals (Hádinger et al., 2023). Moreover, it has been shown that L5-driven TRN regions can innervate thalamic regions not contacted by the direct L5 input to the thalamus, forming the structural background of lateral inhibition. It has to be mentioned, though, that the literature is divided concerning the regional specificity of the L5-TRN input (Carroll et al., 2022).

Finally, it must be acknowledged that our current understanding of the organization of cortical layers is limited to classical 6-layered (or isocortical) regions. There are indications that the rules may be different in allo- or mesocortical regions. For example, the marker NTSR1, which labels L6 corticothalamic cells in neocortical regions, is present in neurons of the mPFC which strongly project to the nucleus accumbens (which a neocortical L6 neuron would never do) and only weakly to the thalamus and TRN (Babiczky and Matyas, 2022). It is also unclear whether the presubicular input to the anterior nuclear group has L6, L5, mixed, or transient features (Vantomme et al., 2020). Future studies should be open to interpretations which differ from our current understanding of the L5–L6 dichotomy.

SOURCES OF INPUTS: TRN AFFERENTS

Beside L6a, the other major input which innervates all thalamic nuclei arises in the TRN. The TRN terminals studied so far establish a single synapse with their postsynaptic partner.

If a single TRN terminal forms more than one synapse, these apparently target different postsynaptic neurons (Bodor et al., 2008). As a consequence, the intra-terminal spillover of the transmitter (described above) will not be effective, so these inputs depress at a high presynaptic firing rate. It should be noted, though, that detailed 3D electron microscopic studies of TRN terminals have only been performed in POm (Bodor et al., 2008); thus, in principle, TRN input may differ from this in other nuclei. TRN neurons target each domain of the thalamocortical neurons (soma, proximal, and distal dendrites) without any preference.

It should be noted that the perisomatic innervation of thalamocortical neurons is much weaker than that of cortical pyramidal cells, and axo-axonic innervation patterns have not been described. This indicates that the perisomatic control of spike output is much weaker in the thalamus, which may indicate a relatively higher importance of dendritic integration in timing thalamic action potentials.

As detailed elsewhere in this book (**Chapter 16 in Section IV**), TRN input is organized in a nice topographic manner. TRN can be divided into sectors and tiers, each targeting well-defined thalamic regions. Most TRN neurons innervate only a single thalamic nucleus. Interestingly, the axons of TRN cells innervated by L5 afferents from the frontal cortex frequently bifurcate and arborize in 2 nuclei (Hádinger et al., 2023).

SOURCES OF INPUTS: SUBCORTICAL AFFERENTS

The thalamus is contacted by a vast number of subcortical centers (from the retina to the spinal cord) which provide GABAergic or glutamatergic inputs (Figure 4.1B–D, Figure 4.3C) (Acsády, 2022). In sharp contrast, the neocortex is almost completely devoid of direct subcortical GABAergic or glutamatergic afferents (except the thalamus, of course) (Acsády, 2022). None of these subcortical inputs can be regarded as canonical in the thalamus since each of these subcortical afferents targets only a well-defined sector within the thalamus. Even if the afferents from a given source target more than one nucleus, most of these inputs are confined to a small proportion of the entire structure. Reflecting the importance of subcortical afferents, most nuclear parcellation of the thalamus follows the boundaries of afferent termination zones arising from distinct subcortical sources.

The GABAergic and glutamatergic subcortical inputs studied so far establish contacts on the thalamocortical cells via large, multisynaptic terminals (F3 and RL or RM types, respectively) (Figure 4.1B–D, Figure 4.2D–F). In the case of the sensory glutamatergic RL afferents, these inputs arise in the spinal cord, and the trigeminal nuclei (somatosensory), inferior colliculus (auditory), nucleus tractus solitary (gustatory), and retina (visual) are responsible for the faithful relay of peripheral sensory information (Sherman and Guillery, 2005). Similarly, multisynaptic RL inputs target the anterior nuclear group from the mammillary body and the verntral lateral nucleus (VL) from the cerebellum, conveying orientation/episodic memory or movement-related information, respectively (Guillery, 1956; Percheron et al., 1996). Both the superficial and deep layers of the superior colliculus provide smaller multisynaptic (RM type) inputs to the LGNd, lateral posterior centrolateral, and supragenicualte/limitans nuclei of the thalamus (Bickford et al., 2015; Barsy et al., 2020). Afferents from the brainstem reticular formation and the hyptothalamus mainly contact the intralaminar and the midline nuclei. Extrathalamic GABAergic inputs arise in the ZI and APT (targeting POm), subtantia nigra pars reticulata and globus pallidus interna (innervating VA and VM), the ventral part of LGN (contacting LP cells), the ventral pallidum (targeting MD), and the rostral part of the reticular formation (innervating the intralaminar nuclei) (Halassa and Acsády, 2016).

SPECIFIC VERSUS NONSPECIFIC INPUTS, FAST VERSUS SLOW TRANSMISSION

Similar to every other brain region, the thalamus also receives diffuse, nonspecific ascending afferents mainly involving cholinergic, serotonergic, noradrenergic, and histaminergic fibers beside specific GABAergic and glutamatergic inputs. All of these inputs are present throughout the structure but display higher density in the midline and in the intralaminar nuclei. In addition, the cholinergic innervation of the DLG is also very dense. Cholinergic terminals have been shown to establish classical synaptic contacts in various thalamic nuclei (Patel et al., 1999; Oda et al., 2003). Noradrenergic terminals have only been studied in the TRN, where they form synapses on TRN cells (Asanuma, 1992). Thus presently it is unclear whether or not noradrenegic, serotonergic, and histminergic terminals establish synapses in the dorsal thalamus. Since the majority of the noradrenergic and serotonergic terminals do not form synapses in the cortex (Umbriaco et al., 1995), this question requires further investigation! Substantial dopaminergic innervation is apparently present only in primates, where it mainly contacts interneurons (García-Cabezas et al., 2009). In rodents, dopaminergic innervation is detectable only in paraventricular thalamic nucleus (García-Cabezas et al., 2009).

These diffuse or nonspecific afferents mostly utilize nonsynaptic neurotransmission and slow metabotropic receptors which are localized both pre-and postsynaptically (Vizi and Lábos, 1991). As a consequence, they play a robust role in modulating state-dependent neurotransmission, or can gate sensory information transfer by tonically acting on

TABLE 4.1 COMPARISON OF THE CORTICAL AND THALAMIC INPUTS

	Cortex	Thalamus
Variability of bouton structure	Minimal	Substantial
Recurrent collaterals	Substantial	Minimal
Major source of inputs	Intrinsic, cortical	Extrinsic (cortical and subcortical)
Diversity of input regions	Minimal	Substantial
Regional heterogeneity of input integration	Minimal	Substantial

various elements of the neuropil (Reggiani et al., 2022), but they certainly do not participate in precise point-to-point information transfer in the millisecond time frame. This latter task, as elsewhere in the brain, is performed mainly by glutamatergic and GABAergic afferents. The only exception to this may be the fast, rapidly adapting nicotinergic acetylcholine neurotransmission. However, how exactly this type of input is influencing thalamic computation is presently unclear.

THE CONCEPT OF INTEGRATING AFFERENTS: THALAMUS VERSUS CORTEX

In order to better understand the basic logic of input integration in the thalamus, it is worth comparing the principles of input organization in the cortex and the thalamus.

In the case of cortical pyramidal cells, the vast majority of GABAergic and glutamatergic afferent originate from other cortical excitatory or inhibitory cells (Harris and Shepherd, 2015). With the exception of thalamocortical inputs, a cortical pyramidal cell receives very few extrinsic GABAergic or glutamatergic inputs from the rest of the brain. Subcortical inputs requiring cortical processing are transferred via the thalamic network. Major exceptions to this rule are the glutamatergic neurons of the olfactory bulb (mitral cells) which directly target pyramidal cells in the pyriform cortex without a thalamic transfer. GABAergic afferent from the basal forebrain to the cortex also bypass the thalamus, albeit these fibers selectively target cortical interneurons, not pyramidal cells (Freund and Meskenaite, 1992).

Local inputs of cortical pyramidal cells originate from the recurrent collaterals of other cortical cell types, arranged in distinct layers which have by now been classified as over 100 different cell types (Zeng, 2022). Most of these axon terminals are unisynaptic and, as a consequence, none of these inputs can dominate the output of the neuron. Thus, cortical input integration is based on the convergence of large number of inputs from different, largely local cell classes with similar strength. If we compare pyramidal cells in various cortical regions, they display no real qualitative difference in the pattern of afferent integration. Apparently the only regionally heterogeneous input which is qualitatively different among cortical areas is provided by the thalamus, since cortical regions differ in the exact set of thalamic nuclei they are innervated by. As a consequence, deciphering the canonical cortical connectivity (or module) in one cortical areas will allow us to extrapolate to the computation performed by individual neurons in the entire structure.

In the thalamus the concept of input organization is different in every respect, which may be the key to understanding thalamic computation (Table 4.1). First, unlike cortical neurons, thalamocortical neurons are not connected to each other via a rich recurrent collateral system. The only indication of occasional intranuclear collaterals have been described in cat LGNd. (Friedlander et al., 1981; Bickford et al., 2008). As a consequence, the afferent space of thalamocortical cell is dominated by extrinsic, not intrinsic inputs. Second, these extrinsic inputs arise from almost the entire neuraxis in the case of the thalamus and are not confined only to few centers, as is the case for the cortex. Third, as described above, beside the classical unisynaptic terminals, every thalamic nucleus studied so far also contains multisynaptic terminals. Fourth, since various thalamic regions differ in the origin and structure of their inputs, the exact set of inputs a thalamic neuron receives is qualitatively different from region to region, so there is no canonical input integration in the thalamus.

To summarize, the cortex is characterized by a highly elaborate computation of the information arising largely from a single afferent source (the thalamus) via very complex local processing. In contrast, the thalamus integrates inputs from many different sources with virtually no interaction between thalamocortical cells. Neocortical computational modules are complex but regionally largely homogeneous (i.e., canonical), whereas thalamic modules are simple (basically consisting of a given set of afferents) but regionally heterogeneous (non-canonical).

LONG-RANGE LOOPS BETWEEN CORTEX AND THALAMUS

According to an influential modern brain theory, the brain can be best understood as a structure which continually generates various forms of spontaneous activity (Buzsáki, 2019). During "action," this intrinsic activity is perturbed by external stimuli in order to generate the appropriate responses to the environmental challenges (the so-called inside-out pattern). Mutually interacting neuronal circuits, forming loops, are critical determinants of spontaneous internal brain activities. There is no doubt that the thalamus and the cortex form such a loop. Spontaneously generated activity of the 2 structures can be best studied during sleep. Indeed, depending on the phase of the sleep, the thalamus and the cortex generate sleep rhythms of various frequencies (Crunelli et al., 2018). Recent data, however, demonstrated that perpetual interactions between the thalamus and the cortex can also be detected during complex awake behaviors and are necessary for proper, adaptive, goal-directed actions (Guo et al., 2017; Schmitt et al., 2017). These thalamocortical interactions involved sequential activation of neuronal ensembles, a hallmark of internally generated activity. Based on these theoretical, morphological, and functional data, the thalamus can be described as an internal pattern generator using the reciprocal connectivity with its cortical partner. This we refer to as a "direct loop" between the cortex and the thalamus (Figure 4.4A).

The cortex, however, not only impacts the thalamic activity directly via the corticothalamic afferents, but also influences it indirectly by controlling the subcortical structures that innerve the thalamus. This we will refer to as the "indirect loop." These 2 types of loops together form "multiple nested loops" (Figure 4.4B) consisting of the cortico-thalamo-cortical loops (the direct loop) and the indirect, long-range loops involving (i) the cortex, (ii) the subcortical input structure of the thalamus, (iii) the thalamus, and finally (iv) the cortex (for the formalization of multiple nested loops, see **Chapter 35 in Section VI** in this book). Direct corticothalamic inputs and the subcortical input of the corresponding indirect loops often converge on the same thalamic regions, indicating that the thalamus is an important integrator of the direct cortical inputs and the indirect cortical information processed by subcortical structures (Figure 4.1A). A classical example of such a multiple nested loop is the interactions between parahippocampal cortical regions and the thalamus. It involves a direct, reciprocal interaction between the cortex and the thalamus (the direct loop), as well as the well-known Papez-circuit (the indirect loop) which includes (i) the subiculum (as a cortical structure), (ii) the mammillary body (as a subcortical input structure of the thalamus), (iii) the anterior nuclear group (as the thalamic station), and (iv) its cortical targets (retrosplenial cortex, pre- and parasubiculum) (Bubb et al., 2017). We know now that this loop works as a functional unit since disrupting the connectivity at any point of the circuit results in similar problems in episodic memory formation. Furthermore, now we have ample evidence at the level of single unit activity that, for example, the anterodorsal thalamic nucleus is an important integrator of the cortical (direct loop) and the subcortical (indirect loop) head direction signals during both wake and sleep states (Peyrache et al., 2015).

It should be kept in mind that not all elements of these long-range loops are excitatory. In several cases the thalamic input stations that convey the cortical message to the thalamus in the indirect loop are GABAergic, and establish giant, multisyanptic inhibitory terminals in the thalamus, as described above (Figure 4.1A). Examples include the basal ganglia, zona incerta, anterior pretectum, or the pontine reticular formation (Halassa and Acsády, 2016). How the excitatory or inhibitory nature of the afferents affects the function of these loops is presently unclear.

In the indirect loops, cortical afferents do not always target directly a thalamic input station. In these cases, cortical information is processed through several subcortical stages before it enters the thalamus (Figure 4.4C). The 2 large motor circuits, the basal ganglia and the cerebellar loops, are classical examples of this organization (for details, see **Chapters 31 and 32 in Section VI**). In the case of the basal ganglia, cortical information is processed in the striatum, globus pallidus, substantia nigra route before it enters the ventromedial/ventral anterior and intralaminar nuclei of the thalamus and returns to the cortex (Goldberg et al., 2013). The cerebro-cerebellar loop utilizes cortical projections to the pontine nuclei to influence cerebellar activity, which via the Purkinje-cells and the deep cerebellar nuclei returns to the ventrolateral and intralaminar nuclei and finally to the motor cortex (Boven et al., 2023).

It should be mentioned that corticofugal members of the direct and indirect loops are frequently the same cells (Guillery and Sherman, 2011). It has been demonstrated that all L5 neurons projecting to the thalamus have collaterals to other subcortical brain structures which will finally serve as a thalamic input station (Guillery and Sherman, 2011). This indicates that the same cortical message is transferred via the direct and indirect loops and can be integrated in the thalamus with the appropriate synaptic delays. This indicates that L5 neurons are especially important elements linking these loops and indeed, as detailed above and elsewhere in this book, L5 inputs are much better suited to impose a precise cortical activity pattern on the thalamus. The basic structure of these multiple nested loops puts the thalamus in a unique position to integrate direct cortical information and the impact of cortical influence on subcortical centers and to transfer this integrated information back to the cortex.

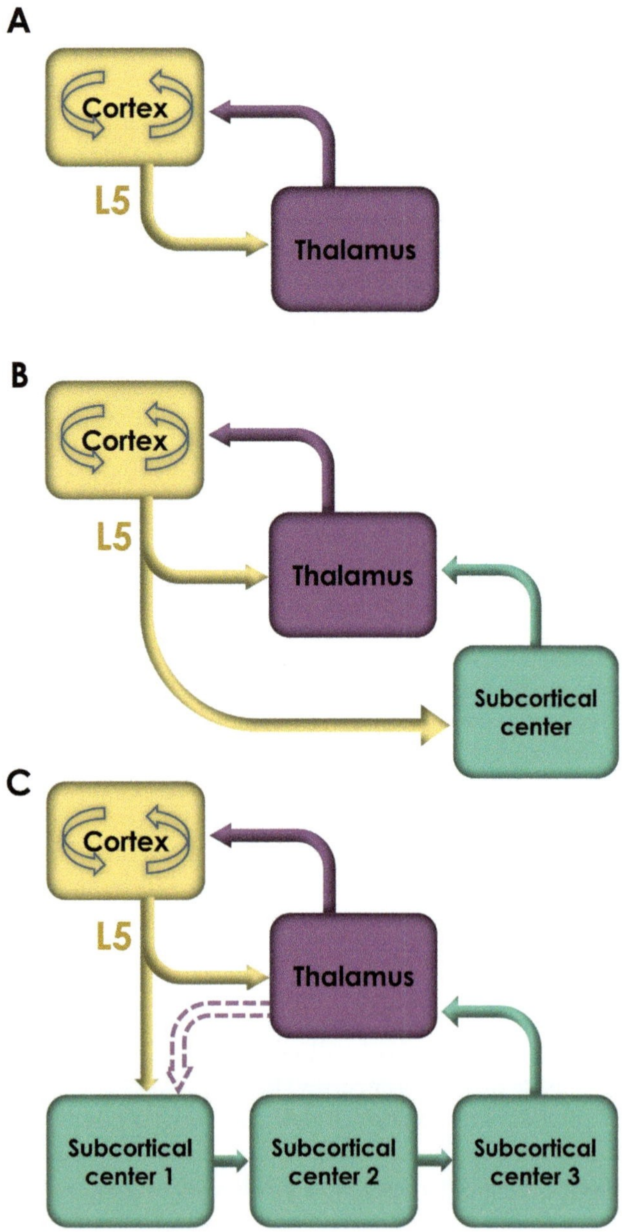

Figure 4.4. Cortex and thalamus display multiple nested loop organization. (A) The inner loop consisting of the corticothalamic and the thalamocortical pathways. Note that the connectivity should not necessarily be reciprocal; open loop organization is also possible. A classic example is the cortico-pulvinar connection (Figure 4.3). To what extent the modulatory L6 pathway can be considered here remains to be determined. Curved arrows in cortex indicate intra-cortical processing. (B) Nested inner and outer loops where the cortical target of the subcortical center has direct thalamic projections. Note that the subcortical center can be glutamatergic (e.g., mammillary body or superior colliculus) or GABAergic (ZI, APT, shown in Figure 4.1). Thalamic afferent from the inner and outer loops frequently target the same nucleus but not necessarily the same thalamic cells. (C) Nested inner and outer loop where the cortical information is processed via several subcortical stages before it re-enters thalamus. Classical examples are the cortico-basal ganglia loop and the cerebro-cerebellar loop (via the pontine nuclei). Striped arrow indicates the thalamo-striatal pathway, which is present only in the basal ganglia loop.

REFERENCES

Acsády L (2022). Organization of thalamic inputs. In: Halassa MM, ed., *The Thalamus*. Cambridge University Press: Chapter 2, pp. 27–44.

Asanuma C (1992). Noradrenergic innervation of the thalamic reticular nucleus: A light and electron microscopic immunohistochemical study in rats. *J Comp Neurol.* 319:299–311.

Babiczky Á, Matyas F (2022). Molecular characteristics and laminar distribution of prefrontal neurons projecting to the mesolimbic system. *eLife*. 11. https://doi.org/10.7554/ELIFE.78813

Barsy B, Kocsis K, Magyar A, Babiczky Á, Szabó M, Veres JM, Hillier D, Ulbert I, Yizhar O, Mátyás F (2020). Associative and plastic thalamic signaling to the lateral amygdala controls fear behavior. *Nat Neurosci.* 23:625–637.

Barthó P, Freund TF, Acsády L (2002). Selective GABAergic innervation of thalamic nuclei from zona incerta. *Eur J Neurosci.* 16:999–1014.

Barthó P, Slézia A, Varga V, Bokor H, Pinault D, Buzsáki G, Acsády L (2007). Cortical control of zona incerta. *J Neurosci.* 27:1670–1681.

Bickford ME, Wei H, Eisenback MA, Chomsung RD, Slusarczyk AS, Dankowsi AB (2008). Synaptic organization of thalamocortical axon collaterals in the perigeniculate nucleus and dorsal lateral geniculate nucleus. *J Comp Neurol.* 508:264–285.

Bickford ME, Zhou N, Krahe TE, Govindaiah G, Guido W (2015). Retinal and tectal "driver-like" inputs converge in the shell of the mouse dorsal lateral geniculate nucleus. *J Neurosci.* 35:10523–10534.

Blot A, Roth MM, Gasler I, Javadzadeh M, Imhof F, Hofer SB (2021). Visual intracortical and transthalamic pathways carry distinct information to cortical areas. *Neuron*. 109:1996–2008.

Bodor ÁL., Giber K, Rovó Z, Ulbert I, Acsády L (2008). Structural correlates of efficient GABAergic transmission in the basal ganglia-thalamus pathway. *J Neurosci.* 28:3090–3102.

Bokor H, Frère SGA, Eyre MD, Slézia A, Ulbert I, Lüthi A, Acsády L (2005). Selective GABAergic control of higher-order thalamic relays. *Neuron*. 45:929–940.

Bokor H, Hadinger N, Acsády L (2016). Efficient cortical control of basal ganglia recipient motor thalamus. *Soc Neurosci Abstr*. 720.18/SS16.

Bourassa J, Pinault D, Deschênes M (1995). Corticothalamic projections from the cortical barrel field to the somatosensory thalamus in rats: A single-fibre study using biocytin as an anterograde tracer. *Eur J Neurosci.* 7:19–30.

Boven E, Pemberton J, Chadderton P, Apps R, Costa RP (2023). Cerebro-cerebellar networks facilitate learning through feedback decoupling. *Nat Commun.* 141(14):1–18.

Bubb EJ, Kinnavane L, Aggleton JP (2017). Hippocampal–diencephalic–cingulate networks for memory and emotion: An anatomical guide. *Brain Neurosci Adv.* 1:1–20. doi:239821281772344.

Budisantoso T, Matsui K, Kamasawa N, Fukazawa Y, Shigemoto R (2012). Mechanisms underlying signal filtering at a multisynapse contact. *J Neurosci.* 32:2357–2376.

Buzsáki G (2019). *The Brain from Inside Out*. Oxford Academic.

Campbell XPW, Govindaiah G, Masterson SP, Bickford XME, Guido W (2020). Synaptic properties of the feedback connections from the thalamic reticular nucleus to the dorsal lateral geniculate nucleus. *J Neurophysiol.* 124:404–417.

Carroll BJ, Sampathkumar V, Kasthuri N, Sherman SM (2022). Layer 5 of cortex innervates the thalamic reticular nucleus in mice. *Proc Natl Acad Sci U S A.* 119(38). https://doi.org/10.1073/PNAS.2205209119

Casas-Torremocha D, Rubio-Teves M, Hoerder-Suabedissen A, Hayashi S, Prensa L, Molnár Z, Porrero C, Clasca F (2022). A combinatorial input landscape in the "higher-order relay" posterior thalamic nucleus. *J Neurosci.* 42:7757–7781.

Castro-Alamancos MA (2002a). Properties of primary sensory (lemniscal) synapses in the ventrobasal thalamus and the relay of high-frequency sensory inputs. *J Neurophysiol.* 87:946–953.

Castro-Alamancos MA (2002b). Different temporal processing of sensory inputs in the rat thalamus during quiescent and information processing states in vivo. *J Physiol.* 539:567–578.

Chen X, Aslam M, Gollisch T, Allen K, Von Engelhardt J (2018). CKAMP44 modulates integration of visual inputs in the lateral geniculate nucleus. *Nat Commun.* 9(1). https://doi.org/10.1038/S41467-017-02415-1

Crunelli V, Lőrincz ML, Connelly WM, David F, Hughes SW, Lambert RC, Leresche N, Errington AC (2018). *Dual function of thalamic low-vigilance state oscillations: rhythm-regulation and plasticity.* Nature Publishing Group.

Deschenes M, Timofeeva E, Lavallee P (2003). The relay of high-frequency sensory signals in the Whisker-to-barreloid pathway. *J Neurosci.* 23:6778–6787.

Deschenes M, Veinante P, Zhang ZW (1998). The organization of corticothalamic projections: reciprocity versus parity. *Brain Res Brain Res Rev.* 28:286–308.

Erişir A, Van Horn SC, Sherman SM (1997). Relative numbers of cortical and brainstem inputs to the lateral geniculate nucleus. *Proc Natl Acad Sci U S A.* 94:1517–1520.

Freund TF, Meskenaite V (1992). Gamma-aminobutyric acid-containing basal forebrain neurons innervate inhibitory interneurons in the neocortex. *Proc Natl Acad Sci U S A.* 89:738–742.

Friedlander MJ, Lin CS, Stanford LR, Sherman SM (1981). Morphology of functionally identified neurons in lateral geniculate nucleus of the cat. *J Neurophysiol.* 46:80–129.

García-Cabezas MÁ, Martínez-Sánchez P, Sánchez-González MÁ, Garzón M, Cavada C (2009). Dopamine innervation in the thalamus: monkey versus rat. *Cereb Cortex.* 19:424–434.

Giber K, Diana MA, Plattner MV, Dugué GP, Bokor H, Rousseau CV, Maglóczky Z, Havas L, Hangya B, Wildner H, Zeilhofer HU, Dieudonné S, Acsády L (2015). A subcortical inhibitory signal for behavioral arrest in the thalamus. *Nat Neurosci.* 18:562–568.

Goldberg JH, Farries MA, Fee MS (2013). Basal ganglia output to the thalamus: Still a paradox. *Trends Neurosci.* 36:695–705.

Groh A, Bokor H, Mease RA, Plattner VM, Hangya B, Stroh A, Deschenes M, Acsády L (2014). Convergence of cortical and sensory driver inputs on single thalamocortical cells. *Cereb Cortex.* 24:3167–3179.

Groh A, de Kock CPJ, Wimmer VC, Sakmann B, Kuner T (2008). Driver or coincidence detector: modal switch of a corticothalamic giant synapse controlled by spontaneous activity and short term depression. *J Neurosci.* 28:9652–9663.

Guillery RW (1956). Degeneration in the post-commissural fornix and the mamillary peduncle of the rat. *J Anat.* 90:350–370.

Guillery RW, Sherman SM (2011). Branched thalamic afferents: what are the messages that they relay to the cortex? *Brain Res Rev.* 66:205–219.

Guo ZV, Inagaki HK, Daie K, Druckmann S, Gerfen CR, Svoboda K (2017). Maintenance of persistent activity in a frontal thalamocortical loop. *Nature.* 545:181–186.

Hádinger N, Bősz E, Tóth B, Vantomme G, Lüthi A, Acsády L (2023). Region-selective control of the thalamic reticular nucleus via cortical layer 5 pyramidal cells. *Nat Neurosci.* 26:116–130.

Halassa MM, Acsády L (2016). Thalamic inhibition: Diverse sources, diverse scales. *Trends Neurosci.* 39(10):680–693.

Harris KD, Shepherd GMG (2015). The neocortical circuit: Themes and variations. *Nat Neurosci.* 18:170–181.

Hoerder-Suabedissen A, Hayashi S, Upton L, Nolan Z, Casas-Torremocha D, Grant E, Viswanathan S, Kanold PO, Clasca F, Kim Y, Molnár Z (2018). Subset of cortical layer 6b neurons selectively innervates higher order thalamic nuclei in mice. *Cereb Cortex.* 28:1882–1897.

Hoogland PV, Wouterlood FG, Welker E, Van der Loos H (1991). Ultrastructure of giant and small thalamic terminals of cortical origin: A study of the projections from the barrel cortex in mice using Phaseolus vulgaris leuco-agglutinin (PHA-L). *Exp Brain Res.* 87:159–172.

Jones EG (2007a). Thalamic neurons, synaptic organization, and functional properties. In: *The Thalamus.* 2nd ed. Cambridge University Press:171–318.

Jones EG (2007b). Principles of thalamic organization. In: *The Thalamus.* 2nd ed. Cambridge University Press:87–171.

Kakei S, Na J, Shinoda Y (2001). Thalamic terminal morphology and distribution of single corticothalamic axons originating from layers 5 and 6 of the cat motor cortex. *J Comp Neurol.* 437:170–185.

Lavallée P, Urbain N, Dufresne C, Bokor H, Acsády L, Deschênes M (2005). Feedforward inhibitory control of sensory information in higher-order thalamic nuclei. *J Neurosci.* 25:7489–7498.

Majorossy K, Réthelyi M, Szentágothai J (1965). The large glomerular synapse of the pulvinar. *J Hirnforsch.* 7:415–432.

Morgan JL, Berger DR, Wetzel AW, Lichtman JW (2016). The fuzzy logic of network connectivity in mouse visual thalamus. *Cell.* 165:192–206.

Oda S, Kuroda M, Kakuta S, Tanihata S, Ishikawa Y, Kishi K (2003). Ultrastructure of ascending cholinergic terminals in the anteroventral thalamic nucleus of the rat: A comparison with the mammillothalamic terminals. *Brain Res Bull.* 59:473–483.

Ojima H (1994). Terminal morphology and distribution of corticothalamic fibers originating from layers 5 and 6 of cat primary auditory cortex. *Cereb Cortex.* 4:646–663.

Patel NC, Breckinridge Carden W, Bickford ME (1999). Synaptic targets of cholinergic terminals in the cat lateral posterior nucleus. *J Comp Neurol.* 410:31–41.

Percheron G, Franqois C, Talbi B, Yelnik J, Fénelon G (1996). The primate motor thalamus. *Brain.* 22:93–181.

Peyrache A, Lacroix MM, Petersen PC, Buzsáki G (2015). Internally organized mechanisms of the head direction sense. *Nat Neurosci.* 18:569–575.

Pierret T, Lavallee P, Deschenes M (2000). Parallel streams for the relay of vibrissal information through thalamic barreloids. *J Neurosci.* 20:7455–7462.

Reggiani JDS, Jiang Q, Barbini M, Lutas A, Liang L, Fernando J, Deng F, Wan J, Li Y, Chen C, Andermann ML (2022). Brainstem serotonin neurons selectively gate retinal information flow to thalamus. *Neuron.* 111:711–726.

Reichova I, Sherman SM (2004). Somatosensory corticothalamic projections: Distinguishing drivers from modulators. *J Neurophysiol.* 92:2185–2197.

Rollenhagen A, Lübke JHR (2006). The morphology of excitatory central synapses: From structure to function. *Cell Tissue Res.* 326:221–237.

Rompani SB, Müllner FE, Wanner A, Zhang C, Roth CN, Yonehara K, Roska B (2017). Different modes of visual integration in the lateral geniculate nucleus revealed by single-cell-initiated transsynaptic tracing. *Neuron.* 93:767–776.

Rovó Z, Ulbert I, Acsády L (2012). Drivers of the primate thalamus. *J Neurosci.* 32:17894–17908.

Schmitt LI, Wimmer RD, Nakajima M, Happ M, Mofakham S, Halassa MM (2017). Thalamic amplification of cortical connectivity sustains attentional control. *Nature.* 545:219–223.

Sherman SM, Guillery RW (1996). Functional organization of thalamocortical relays. *J Neurophysiol.* 76:1367–1395.

Sherman SM, Guillery RW (1998). On the actions that one nerve cell can have on another: distinguishing "drivers" from "modulators." *Proc Natl Acad Sci U S A.* 95:7121–716.

Sherman SM, Guillery RW (2005). The afferent axons to the thalamus: Their structure and connections. In: *Exploring the Thalamus and Its Role in Cortical Functions.* MIT Press:77–137.

Sirota MG, Swadlow HA, Beloozerova IN (2005). Three channels of corticothalamic communication during locomotion. *J Neurosci.* 25:5915–5925.

Telgkamp P, Padgett DE, Ledoux VA, Woolley CS, Raman IM (2004). Maintenance of high-frequency transmission at purkinje to cerebellar nuclear synapses by spillover from boutons with multiple release sites. *Neuron.* 41:113–126.

Umbriaco D, Garcia S, Beaulieu C, Descarries L (1995). Relational features of acetylcholine, noradrenaline, serotonin and GABA axon terminals in the stratum radiatum of adult rat hippocampus (CA1). *Hippocampus.* 5:605–620.

Vantomme G, Rovó Z, Cardis R, Béard E, Katsioudi G, Guadagno A, Perrenoud V, Fernandez LMJ, Lüthi A (2020). A thalamic reticular circuit for head direction cell tuning and spatial navigation. *Cell Rep.* 31(10):107747. doi:10.1016/j.celrep.2020.107747

Vetere G, Xia F, Ramsaran AI, Tran LM, Josselyn SA, Frankland PW (2021). An inhibitory hippocampal–thalamic pathway modulates remote memory retrieval. *Nat Neurosci.* 245(24):685–693.

Vidnyánszky Z, Gorcs T, Negyessy L, Borostyankoi Z, Knopfel T, Hamori J (1996). Immunocytochemical visualization of the mGluR1a metabotropic glutamate receptor at synapses of corticothalamic terminals originating from area 17 of the rat. *Eur J Neurosci.* 8:1061–1071.

Vizi ES, Lábos E (1991). Non-synaptic interactions at presynaptic level. *Prog Neurobiol.* 37:145–163.

Wanaverbecq N, Bodor ÁL, Bokor H, Slézia A, Lüthi A, Acsády L (2008). Contrasting the functional properties of GABAergic axon terminals with single and multiple synapses in the thalamus. *J Neurosci.* 28:11848–11861.

Yamawaki N, Shepherd GMG (2015). Synaptic circuit organization of motor corticothalamic neurons. *J Neurosci.* 35:2293–2307.

Zeng H (2022). What is a cell type and how to define it? *Cell.* 185:2739–2755.

5.

SYNAPTIC PROPERTIES OF THALAMIC AND PRETHALAMIC INPUTS

Carmen Varela

TYPES OF NEUROTRANSMITTER RECEPTORS IN THE THALAMUS

The effect of thalamic inputs on their target neurons is determined by specific pre- and postsynaptic neurotransmitter receptors, the intracellular signaling cascades that they regulate, and the state of the postsynaptic cell (i.e., burst or tonic firing mode). Thalamic synapses contain 2 general types of receptors, ionotropic and metabotropic, that mediate responses with fast and slow temporal dynamics. Within these 2 receptor categories, further subtypes are distinguished based on the receptor sensitivity to ligands and downstream molecular effects. Ionotropic receptors are ligand-gated, transmembrane proteins that form an ion-conducting pore with fast activation dynamics. When open, ions flow rapidly through the channel and the resulting postsynaptic potential allows for rapid effects, in the order of tenths of milliseconds or longer, depending on the receptor (Reiner and Levitz, 2018). Metabotropic receptors are transmembrane proteins that do not have ion-conducting pores in their structure. When a ligand binds to metabotropic receptors, it prompts the activation of intracellular signal transduction cascades that produce a broad range of outcomes, from regulating ion channels to gene transcription. This chapter provides an overview of synaptic effects mediated by neurotransmitter systems in the thalamus (for a review of the presynaptic origin of these systems and additional details on the postsynaptic effect of each neurotransmitter, see McCormick, 1992; Varela, 2014 and **Chapter 4 in Section II**).

IONOTROPIC NEUROTRANSMISSION

Glutamatergic, cholinergic, and GABAergic inputs to the thalamus use ionotropic and metabotropic receptors, and thus these systems control fast neurotransmission in thalamic cells, while also being capable of slower and more complex intracellular effects. Monoaminergic inputs, however, activate only metabotropic receptors, with the possible exception of histamine, which hyperpolarizes thalamic reticular nucleus (TRN) cells by activating a chloride conductance via H2 receptors (Lee et al., 2004).

The presence of specific subunits of ionotropic glutamate receptors (α-amino-3-hydroxy-5-methyl-4-isoxazolepropionic acid receptor [AMPAR]; N-methyl-D-aspartate receptor [NMDAR]; Kainate) in corticothalamic synapses suggests that these inputs evoke stronger NMDAR responses than subcortical glutamatergic inputs. The ionotropic glutamate receptors in thalamic synapses are composed of AMPAR GluR2/3 and NMDAR NR1 subunits (Mineff and Weinberg, 2000; Golshani et al., 2001; Liu, 1997; Wang et al., 2011). However, corticothalamic synapses contain NMDARs with NR2B subunits, which facilitate excitatory postsynaptic potential (EPSP) summation and persistent cell firing under repetitive stimulation of this pathway (Miyata and Imoto, 2006). In addition, corticothalamic synapses in TRN may contain NR2C subunits (Zhang et al., 1999), which lead to a weaker Mg^{2+}-dependent block; this means that NMDARs may contribute to corticothalamic EPSPs at relatively hyperpolarized membrane potentials in TRN (Deleuze and Huguenard, 2016). Corticothalamic synapses in TRN are also more likely to contain AMPAR GluR4 subunits than synapses on thalamocortical cells, a difference that may lead to relatively large corticothalamic EPSPs in TRN compared to the corticothalamic effect in the thalamus (Mineff and Weinberg, 2000; Golshani et al., 2001). The different subunit composition in corticothalamic synapses in TRN and dorsal thalamus has implications for the pathophysiology of absence epilepsy. GluR4 knockout mice—Gria4(-/-) mice—have a selective impairment in corticothalamic-mediated TRN firing (compared to corticothalamic transmission in the thalamus) and present behavioral phenotypes consistent with absence epilepsy (Paz et al., 2011). Mutations in the stargazin protein, which is involved in synaptic targeting of AMPARs in TRN, produce the loss and abnormal localization of AMPARs, disrupt TRN glutamatergic transmission, and lead to epileptic phenotypes.

The main ionotropic receptor for cholinergic neurotransmission in the thalamus is the α4β2 receptor, and cells in the dorsal thalamus of the α4β2 knockout mouse no longer respond to nicotine (Marubio et al., 1999). Using ionotropic receptors, cholinergic input can exert fast control of firing in TRN cells (Beierlein, 2014). In the thalamus, areas that receive cholinergic input from the basal forebrain, such as the mediodorsal and parafascicular nuclei, contain large cholinergic axon boutons (Kuroda and Price, 1991), suggesting that these terminals produce strong EPSPs. Further, a few reports suggest that nicotinic receptors enhance axon conduction in cortical (Kawai et al., 2007) and subcortical (Zhang et al., 2004) glutamatergic inputs, which could promote faster transmission in glutamate systems. Overall, the evidence points to cholinergic inputs as having 2 distinct roles in the thalamus. One effect is specific to nuclei that receive basal forebrain input, which activates thalamic cells on rapid timescales. The other main effect of cholinergic input is common to all areas of the thalamus, which receive brainstem cholinergic input that influences membrane potential at slower timescales, in relation to changes in wakefulness and arousal state (Beierlein, 2014).

GABAergic ionotropic inhibition is mediated by GABAA receptors (GABAARs). Similar to other synaptic inputs, the kinetics of the postsynaptic potentials (PSPs) depend not only on the receptor subunit composition, but also on the location of the receptors in the synaptic membrane (Sanchez-Vives et al., 2021). For instance, GABAARs with δ subunits are found in the extrasynaptic or perisynaptic membrane, where they generate a sizable and tonically active inhibitory current in thalamocortical (but not in TRN) cells that facilitates burst firing (Cope et al., 2005).

METABOTROPIC NEUROTRANSMISSION

The complexity of the molecular cascades controlled by metabotropic receptors is not fully understood, but their mechanisms of action begin with the activation of guanine nucleotide-binding proteins (G proteins). In the thalamus, G proteins engage phospholipase C and adenylyl cyclase enzymatic pathways to control cell excitability through the up- or downregulation of potassium leak (K2P) and inwardly rectifying (GIRK) channels, as well as hyperpolarization-activated cyclic nucleotide-gated (HCN) channels responsible for the I_h current. For instance, group I metabotropic glutamate receptors (mGluRs) (Ngomba and van Luijtelaar, 2018) and muscarinic receptors M1R, M3R, and M5R (Santiago and Abrol, 2019) interact with the Gq/11 subfamily of G proteins, which activates phospholipase C and leads to the production of the second messenger molecules inositol trisphosphate (IP3) and diacylglycerol (DAG). Instead, mGluRs groups II and III, and muscarinic receptors M2R and M4R, activate Gi/o and Gs proteins and regulate adenylyl cyclase activity and the synthesis of cyclic adenosine monophosphate (cAMP).

The Gq/11 cascade is associated with the closing, and the Gi/o cascade with the opening of potassium channels, which respectively increases or decreases input resistance and depolarizes or hyperpolarizes thalamic cells. Likewise, shifting the activation curve of the conductance underlying I_h toward more depolarized (by increasing the production of cAMP) or hyperpolarized (by decreasing cAMP) membrane potentials can respectively depolarize or hyperpolarize thalamic cells. This is because the shift in the activation curve represents a change in the fraction of the conductance active at a particular membrane potential. Ligands such as glutamate (Turner and Salt, 2000), acetylcholine (McCormick and Prince, 1986; Zhu and Uhlrich, 1998; Mooney et al., 2004), noradrenaline (McCormick and Pape, 1990), dopamine (Govindaiah and Cox, 2005), GABA (Cain et al., 2017), and histamine (McCormick and Williamson, 1991) regulate potassium conductances and I_h in thalamic neurons through metabotropic receptors and their associated, G protein–coupled, molecular pathways. The changes in membrane potential (depolarization or hyperpolarization) evoked by metabotropic receptors are relatively slow and long-lasting, in the order of hundreds of milliseconds, a timescale that is well-suited to influence the activation state of the voltage-dependent T-type calcium channels responsible for burst firing (for details on firing mode, see **Chapter 3 in Section II**). Thus, a main function of metabotropic receptors is to switch thalamic cells between tonic and burst firing modes, a hypothesis that has received experimental support (McCormick and von Krosigk, 1992; Bista et al., 2012).

The effect of metabotropic receptors on ion channels that control the resting membrane potential and firing mode (burst, tonic) has been demonstrated in several studies, but effects on other intrinsic properties need further characterization. Serotonin (through 5-HT7 receptors), noradrenaline (via alpha-1 and beta-1 receptors), and dopamine (through D2 receptors) use metabotropic receptors to regulate the action potential after-hyperpolarization and influence spike frequency adaptation in higher order thalamic cells (McCormick and Prince, 1988; Lavin and Grace, 1998; Goaillard and Vincent, 2002). G proteins downregulate voltage-dependent calcium conductances, including those mediated by the T-type calcium channels. However, we do not have a clear idea of how metabotropic receptors may regulate T channels and shape burst firing in thalamus. It is possible that metabotropic receptors modulate additional ion channels to adjust the response properties of thalamic cells in unique ways, depending on the neurotransmitter.

In addition, metabotropic receptors are found in intracellular organelles (Fasciani et al., 2022; Kumar and Puthenveedu, 2022), which raises interesting questions, for example whether thalamic inputs control the function of organelles in the dendrites and soma, and, if that is the case, what transporters mediate the internalization of neurotransmitters to produce such an effect. Activation of metabotropic receptors on the membrane of thalamic cells elicits the release of calcium from intracellular stores through IP3 receptors (Neyer et al., 2016), but it is not known if neurotransmitters released by thalamic inputs reach metabotropic receptors located in the membrane of intracellular organelles.

Lastly, the distinction between ionotropic and metabotropic mediated effects can be blurred when considering that some ionotropic receptors can signal via G protein–coupled mechanisms. Ionotropic receptors that permeate calcium (such as NMDARs) influence G protein–signaling cascades and mGluRs (Alagarsamy et al., 2002), and glutamatergic kainate receptors are ionotropic but can also signal via G proteins (Lerma and Marques, 2013). Furthermore, calcium enters thalamic cells through T- and R-type channels during bursting (Zaman et al., 2011), and burst firing may provide a mechanism to influence intracellular cascades, possibly to promote long-term synaptic plasticity. New tools like the G protein–based sensors to monitor neurotransmitter activity (Sun et al., 2020) may prove key to disentangle the signaling pathways used by different thalamic inputs and to shed light on their interactions.

Metabotropic Glutamate and GABA Receptors in the Thalamus and Prethalamus

Glutamatergic and GABAergic synaptic inputs illustrate how the diversity of metabotropic receptor subtypes and their synaptic location can tune the excitatory and inhibitory balance in thalamic cells. Figure 5.1 shows the location of mGluRs in corticothalamic and TRN-thalamus synapses, which express all mGluRs (1 through 8) except mGluR6 (Ngomba and van Luijtelaar, 2018). Notably, mGluRs contribute to the postsynaptic response evoked by glutamatergic modulator inputs (cortical and subcortical), but not to driver inputs (Reichova and Sherman, 2004; Lee and Sherman, 2010; see **Chapter 4 in Section II** for a description of driver and modulator inputs). Group I mGluRs (mGluR1, mGluR5) mediate the postsynaptic depolarization of thalamic cells and of interneuron F2 terminals (for a review of F2 terminals, see **Chapter 2 in Section II**; Godwin et al., 1996). Instead, group II (mGluR2, mGluR3) and group III receptors (mGluR4, mGluR7–8) are presynaptic and reduce neurotransmitter release (Ngomba and van Luijtelaar, 2018).

GABAB receptors (GABABRs) play an important role in regulating thalamic excitability. These receptors are prominent in extrasynaptic areas and presynaptically, both as auto- and as heteroreceptors in excitatory synapses (Figure 5.2). GABA can diffuse and reach GABABRs following strong synaptic activation, which may occur during synchronous spiking or with burst firing in TRN cells. Presynaptically, GABABRs decrease the release of GABA and glutamate through inhibition of voltage-gated Ca^{2+} channels (Ulrich et al., 2007). Postsynaptically, they evoke a robust hyperpolarization through activation of K^{+} channels (Sanchez-Vives et al., 2021). GABABR also have a critical impact in synchronizing oscillatory activity and can promote abnormal, hypersynchronous discharges (Fogerson and Huguenard, 2016; Sanchez-Vives et al., 2021).

SHORT-TERM SYNAPTIC PLASTICITY

The synaptic properties of thalamic inputs would be of limited value if they were fixed. Short-term changes (in the timescale of milliseconds) in synaptic strength provide the basis to adjust inputs dynamically in a time frame relevant for animal behavior. The short-term plasticity of thalamic inputs has been studied with intracellular recordings in brain slices using paired-pulse or brief train-stimulation protocols and recording the amplitude of evoked PSPs following 2 or more stimulation pulses. The ratio of the amplitude of consecutive PSPs provides a metric to quantify the relative facilitation or depression as a function of the interval between stimulation pulses (stimulation frequency). The main finding from these experiments is that the synaptic strength of drivers and modulator inputs (see **Chapter 4 in Section II** on circuit motifs of thalamic inputs) changes in opposite directions following repetitive stimulation (Li et al., 2003).

Modulator inputs from cortical layer 6 present graded paired-pulse facilitation that is frequency-dependent (Reichova and Sherman, 2004). The facilitation of corticothalamic inputs on thalamocortical cells is maximal with inter-stimulus intervals of 20–30 ms, and continues for at least 400 ms (Granseth et al., 2002; Miyata and Imoto, 2006). Corticothalamic synapses in TRN cells start off evoking a larger EPSP compared to synapses in thalamocortical cells but, upon repetitive stimulation, both pathways reach similar EPSP amplitudes, resulting in stronger short-term facilitation in thalamocortical relative to TRN cells (Cruikshank et al., 2010; Jurgens et al., 2012). Driver inputs show paired-pulse depression with fast dynamics, with the maximal EPSP depression occurring after the second stimulus when drivers are stimulated at 5–10 Hz (Turner and Salt, 1998; Miyata and Imoto, 2006).

The mechanisms underlying short-term plasticity are both pre- and postsynaptic. At the presynaptic level, a major element that determines facilitation or depression is the release probability of synaptic boutons, which is high in the large glutamatergic axon terminals characteristic of drivers and low in small glutamatergic terminals typical of modulators (for details on axon terminal types, see

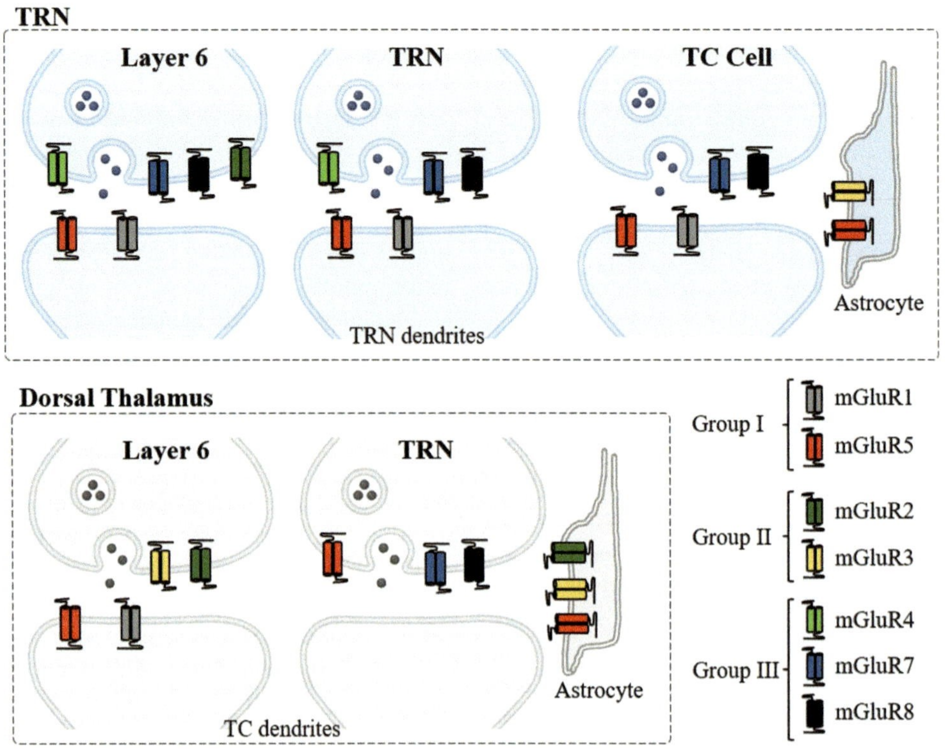

Figure 5.1. Localization of mGluR subtypes in thalamic synapses. Diagram of pre- and postsynaptic terminals within the thalamic reticular nucleus (top) and thalamus (bottom) with the location of mGluRs. Based on Ngomba and van Luijtelaar, Metabotropic glutamate receptors as drug targets for the treatment of absence epilepsy, *Current Opinion in Pharmacology*, 38:43–50, copyright (2018), with permission from Elsevier. Synaptic outlines here and in Figure 5.2 were created with BioRender.com.

Chapter 4 in Section II). Presynaptic calcium concentration and cytoskeletal proteins control transmitter release by mobilizing synaptic vesicles and docking them for release (Zhang and Augustine, 2021). Different variants of the vesicular glutamate transporter (VGLUT) provide another layer to control release probability, with VGLUT1 present in corticothalamic synapses with low-release probability, and VGLUT2 in synapses with high-release probability (Lindström et al., 2020). In general, high-release probability synapses work near saturation, at the upper limit of their response to presynaptic calcium entry, release probability, and postsynaptic receptor sensitivity. These factors dictate a maximal response upon stimulus arrival and cause depression with subsequent fast stimulation. In low-release probability synapses, synaptic release probability increases with repetitive stimulation and enhances the postsynaptic response, but the effect is self-limiting, and as release probability increases it prevents further facilitation. This is suggested by experiments in corticothalamic synapses of somatosensory relay cells, where repeated stimulation augmented the release probability of the synapse and also led to a reduction in the ratio of paired-pulse facilitation after augmentation compared to baseline (Sun and Beierlein, 2011).

The postsynaptic mechanisms of short-term plasticity include receptor desensitization (Kamalova et al., 2020) and receptor saturation (Sun and Beierlein, 2011). These mechanisms introduce different degrees of plasticity in the AMPAR and NMDAR components of the postsynaptic response. For example, with repetitive stimulation, the AMPAR component is more depressed than the NMDAR component in retinogeniculate synapses, where AMPARs contain a Germ cell-specific gene 1–like (GSG1L) auxiliary subunit that stabilizes the desensitized state of the receptor (Kielland and Heggelund, 2002; Kamalova et al., 2020). In corticothalamic synapses, AMPAR desensitization is not a strong contributor to plasticity; instead, AMPAR and NMDAR saturation limit the strength of short-term facilitation (Sun and Beierlein, 2011).

To understand how short-term synaptic plasticity determines the transmission of information through the thalamus, it is crucial to consider the combined effect of plasticity across the synapses of the thalamocortical circuit; that is, corticothalamic synapses (on TRN and thalamocortical cells) present short-term facilitation, but they also control the feedforward inhibition of thalamocortical cells by activating the TRN, and TRN synapses on thalamocortical cells present short-term depression (Campbell et al., 2020). Consequently, the effect of corticothalamic activation varies such that at low frequencies the cortex suppresses spiking in thalamocortical cells, but it increases their firing

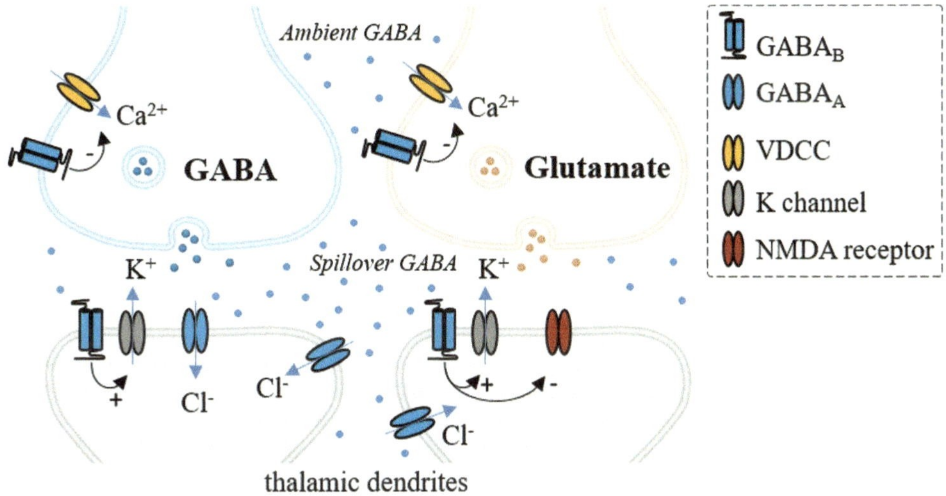

Figure 5.2. GABA receptors mediate interactions between GABAergic and glutamatergic synapses. Extrasynaptic and heterosynaptic GABABR are activated by ambient GABA and by GABA spillover after strong activation of inhibitory terminals. GABABRs reduce neurotransmitter release as autoreceptors (left) and heteroreceptors (right) in presynaptic terminals by downregulating voltage-dependent calcium channels (VDCC). Postsynaptically, GABABRs shunt the effect of glutamatergic inputs through the activation of K^+ leak channels and by promoting the voltage-sensitive Mg^{2+} block of NMDARs.

with repeated activation (Figure 5.3). The reason is that, as the frequency of stimulation increases, the excitatory corticothalamic synapses facilitate, exciting both TRN and thalamocortical cells, but the depression at GABAergic synapses results in overall excitation in the dorsal thalamus (Crandall et al., 2015). These results obtained from intracellular recordings in brain slices are consistent with findings in behaving rodents (Li and Ebner, 2016; Kirchgessner et al., 2020) and emphasize the importance of synaptic dynamics in controlling information flow through the thalamus.

INTERACTIONS BETWEEN THALAMIC INPUTS

Thalamic inputs can interact postsynaptically, through shared intracellular signaling cascades, and presynaptically through heterosynaptic receptors. Although there are still few data, the results are consistent with the idea that heterosynaptic interactions prevent excessive excitation and keep thalamic cells within their normal dynamic range. Retinogeniculate EPSPs are suppressed by presynaptic mGluRs, GABAB, cholinergic, and serotonergic 5-HT1B receptors (Chen and Regehr, 2003; Govindaiah et al., 2012; Lam and Sherman, 2013; Yang et al., 2015). Similar presynaptic mechanisms may be present in corticothalamic synapses, where cholinergic agonists reduce synaptic release probability, leading to a small postsynaptic potential upon initial activation of the corticothalamic input and facilitation with repeated stimuli (Nersisyan et al., 2021). Corticothalamic synapses are also under the control of GABABRs, which inhibit the release of glutamate (Ulrich et al., 2007). Likewise, GABA inputs are modulated presynaptically by nicotinic receptors, which increase neurotransmitter release in the ventral division of the lateral geniculate nucleus (LGNv) and in the medial geniculate nucleus (MGN; Guo et al., 2005; Sottile et al., 2017).

Relatively few studies have looked at crossover effects in other thalamic inputs, and the picture is quite incomplete. An example is the activation of mGluRs, which potentiate cholinergic nicotinic responses in the TRN through postsynaptic mechanisms involving mGluR1 and mGluR5 (Sun et al., 2016). Overall, the experimental data suggest that interactions between neurotransmitter systems provide homeostatic mechanisms to keep thalamic responses within the normal dynamic range, but the rules that define the interplay between thalamic inputs at the synaptic level are far from evident.

CONCLUSION

The literature reviewed in this chapter highlights the diverse and intricate mechanisms implemented at the synaptic level to facilitate the integration and transmission of information through the thalamus. A common theme is that several inputs regulate intrinsic conductances to switch thalamic neurons between burst and tonic firing modes. From that perspective, the inputs to the thalamus may represent independent systems that modulate firing mode under the specific behavioral conditions that activate each input's area of origin. Alternatively, the overlap in signaling cascades may implement synergistic or competitive schemes between thalamic inputs with shared molecular resources. Tools like the

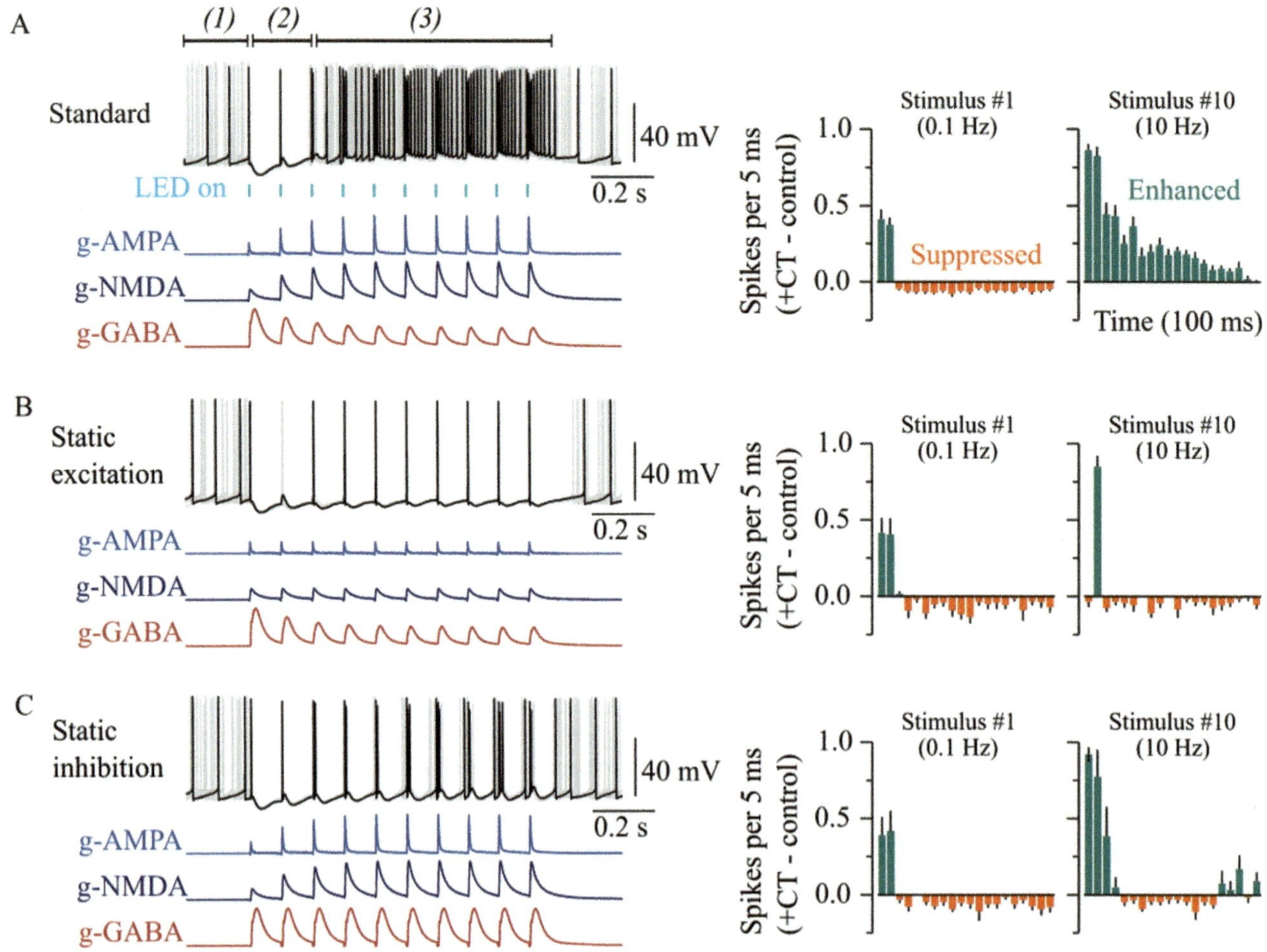

Figure 5.3. Short-term synaptic plasticity determines the suppression or excitation of thalamocortical cells by corticothalamic layer 6 input. In these experiments, neurons in the somatosensory thalamus (ventral posterior medial nucleus, VPm) were recorded intracellularly in brain slices from mice that expressed channelrhodopsin-2 in layer 6 corticothalamic (CT) cells. VPm cells were injected with positive current to produce sustained spiking (control) and to study the effect of repeated light-evoked CT stimuli (10 Hz; once every 10 seconds, 0.1 Hz trains) on firing rate. (A) Left, raster of spikes from a VPm neuron during injection of depolarizing current and CT light activation (overlay of 10 sweeps). Before the light activation starts (1), the cell produces spikes due to injection of positive current. The baseline (control) firing rate is suppressed following CT light activation (2), but the firing rate is enhanced after a few more CT stimulation pulses (3). The *standard* waveforms below show the underlying facilitation of excitatory conductances (g-AMPA, g-NMDA), and the depression of GABA conductances in the cell. To the right, the peri-event histograms show the resultant effect of CT stimulation on the spike output of the neuron minus the control condition (no CT stimulation). The histograms quantify the spike rate in the 100 ms following the first and tenth stimulus, showing that the initial suppression switches to a facilitation by stimulus 10 (note that the facilitation is already apparent in the left, raw traces by stimulus 4). (C) and (D) show the results from follow-up, dynamic clamp experiments used to manipulate the short-term dynamics of AMPAR, NMDAR, and GABAA-mediated conductances and investigate the causal role of short-term plasticity in regulating the spike output of VPm. The results show that, if the short-term plasticity is eliminated from the excitatory (*static excitation*) or from the inhibitory conductances (*static inhibition*), the main effect of the CT input is suppression regardless of the number of stimuli, suggesting a causal link between short-term synaptic plasticity and the dynamic control of thalamic firing output by cortical layer 6. Adapted from Crandall, Cruikshank, and Connors, A corticothalamic switch: Controlling the thalamus with dynamic synapses, *Neuron*, 86(3):768–782, copyright (2015), with permission from Elsevier.

new generation of G protein–based fluorescent sensors to image neurotransmitter activity with subcellular and subsecond resolution are bound to generate essential data to elucidate how the inputs to the thalamus relate to firing mode in behaving animals. Beyond the regulation of firing mode, we need a more complete investigation of how thalamic inputs modulate other intrinsic conductances, as well as how they influence the intracellular environment, a set of experimental questions that will become increasingly tractable with new approaches that allow interrogation of organelle function in vivo.

ACKNOWLEDGMENTS

I thank Dr. Zoe Atherton for helpful comments on an earlier version of this chapter.

REFERENCES

Alagarsamy S, Rouse ST, Junge C, Hubert GW, Gutman D, Smith Y, Conn PJ (2002). NMDA-induced phosphorylation and regulation of mGluR5. *Pharmacol Biochem Behav*. 73:299–306.

Bartlett EL, Smith PH (1999). Anatomic, intrinsic, and synaptic properties of dorsal and ventral division neurons in rat medial geniculate body. *J Neurophysiol*. 81:1999–2016.

Beierlein M (2014). Synaptic mechanisms underlying cholinergic control of thalamic reticular nucleus neurons. *J Physiol*. 592:4137–4145.

Bista P, Meuth SG, Kanyshkova T, Cerina M, Pawlowski M, Ehling P, Landgraf P, Borsotto M, Heurteaux C, Pape H-C, Baukrowitz T, Budde T (2012). Identification of the muscarinic pathway underlying cessation of sleep-related burst activity in rat thalamocortical relay neurons. *Pflugers Arch*. 463:89–102.

Cain SM, Garcia E, Waheed Z, Jones KL, Bushell TJ, Snutch TP (2017). GABAB receptors suppress burst-firing in reticular thalamic neurons. *Channels (Austin)*. 11:574–586.

Campbell PW, Govindaiah G, Masterson SP, Bickford ME, Guido W (2020). Synaptic properties of the feedback connections from the thalamic reticular nucleus to the dorsal lateral geniculate nucleus. *J Neurophysiol*. 124(2):404–417.

Chen C, Regehr WG (2003). Presynaptic modulation of the retinogeniculate synapse. *J Neurosci*. 23:3130–3135.

Cope DW, Hughes SW, Crunelli V (2005). GABAA receptor-mediated tonic inhibition in thalamic neurons. *J Neurosci*. 25:11553–11563.

Crandall SR, Cruikshank SJ, Connors BW (2015). A corticothalamic switch: Controlling the thalamus with dynamic synapses. *Neuron*. 86:768–782.

Cruikshank SJ, Urabe H, Nurmikko AV, Connors BW (2010). Pathway-specific feedforward circuits between thalamus and neocortex revealed by selective optical stimulation of axons. *Neuron*. 65:230–245.

Deleuze C, Huguenard JR (2016). Two classes of excitatory synaptic responses in rat thalamic reticular neurons. *J Neurophysiol*. 116:995–1011.

Fasciani I, Carli M, Petragnano F, Colaianni F, Aloisi G, Maggio R, Scarselli M, Rossi M (2022). GPCRs in intracellular compartments: New targets for drug discovery. *Biomolecules*. 12:1343.

Fogerson PM, Huguenard JR (2016). Tapping the brakes: Cellular and synaptic mechanisms that regulate thalamic oscillations. *Neuron*. 92:687–704.

Goaillard J-M, Vincent P (2002). Serotonin suppresses the slow afterhyperpolarization in rat intralaminar and midline thalamic neurones by activating 5-HT7 receptors. *J Physiol*. 541:453–465.

Godwin DW, Vaughan JW, Sherman SM (1996). Metabotropic glutamate receptors switch visual response mode of lateral geniculate nucleus cells from burst to tonic. *J Neurophysiol*. 76:1800–1816.

Golshani P, Liu XB, Jones EG (2001). Differences in quantal amplitude reflect GluR4- subunit number at corticothalamic synapses on two populations of thalamic neurons. *Proc Natl Acad Sci U S A*. 98:4172–4177.

Govindaiah G, Cox CL (2005). Excitatory actions of dopamine via D1-like receptors in the rat lateral geniculate nucleus. *J Neurophysiol*. 94:3708–3718.

Govindaiah G, Wang T, Gillette MU, Cox CL (2012). Activity-dependent regulation of retinogeniculate signaling by metabotropic glutamate receptors. *J Neurosci*. 32:12820–12831.

Granseth B, Ahlstrand E, Lindström S (2002). Paired pulse facilitation of corticogeniculate EPSCs in the dorsal lateral geniculate nucleus of the rat investigated in vitro. *J Physiol*. 544:477–486.

Guo J-Z, Liu Y, Sorenson EM, Chiappinelli VA (2005). Synaptically released and exogenous ACh Activates different nicotinic receptors to enhance evoked glutamatergic transmission in the lateral geniculate nucleus. *J Neurophysiol*. 94:2549–2560.

Jurgens CWD, Bell KA, McQuiston AR, Guido W (2012). Optogenetic stimulation of the corticothalamic pathway affects relay cells and GABAergic neurons differently in the mouse visual thalamus. *PLoS ONE*. 7:e45717.

Kamalova A, Futai K, Delpire E, Nakagawa T (2020). AMPA receptor auxiliary subunit GSG1L suppresses short-term facilitation in corticothalamic synapses and determines seizure susceptibility. *Cell Rep*. 32:107921.

Kawai H, Lazar R, Metherate R (2007). Nicotinic control of axon excitability regulates thalamocortical transmission. *Nat Neurosci*. 10:1168–1175.

Kielland A, Heggelund P (2002). AMPA and NMDA currents show different short-term depression in the dorsal lateral geniculate nucleus of the rat. *J Physiol*. 542:99–106.

Kirchgessner MA, Franklin AD, Callaway EM (2020). Context-dependent and dynamic functional influence of corticothalamic pathways to first- and higher-order visual thalamus. *Proc Natl Acad Sci U S A*. 117:13066–13077.

Kumar GA, Puthenveedu MA (2022). Diversity and specificity in location-based signaling outputs of neuronal GPCRs. *Curr Opin Neurobiol*. 76:102601.

Kuroda M, Price JL (1991). Ultrastructure and synaptic organization of axon terminals from brainstem structures to the mediodorsal thalamic nucleus of the rat. *J Comp Neurol*. 313:539–552.

Lam Y-W, Sherman SM (2013). Activation of both Group I and Group II metabotropic glutamatergic receptors suppress retinogeniculate transmission. *Neuroscience*. 242:78–84.

Lavin A, Grace AA (1998). Dopamine modulates the responsivity of mediodorsal thalamic cells recorded in vitro. *J Neurosci*. 18:10566–10578.

Lee CC, Sherman SM (2010). Topography and physiology of ascending streams in the auditory tectothalamic pathway. *Proc Natl Acad Sci U S A*. 107:372–377.

Lee KH, Broberger C, Kim U, McCormick DA (2004). Histamine modulates thalamocortical activity by activating a chloride conductance in ferret perigeniculate neurons. *Proc Natl Acad Sci U S A*. 101:6716–6721.

Lerma J, Marques JM (2013). Kainate receptors in health and disease. *Neuron*. 80:292–311.

Li J, Guido W, Bickford ME (2003). Two distinct types of corticothalamic EPSPs and their contribution to short-term synaptic plasticity. *J Neurophysiol*. 90:3429–3440.

Li L, Ebner FF (2016). Cortex dynamically modulates responses of thalamic relay neurons through prolonged circuit-level disinhibition in rat thalamus in vivo. *J Neurophysiol*. 116:2368–2382.

Lindström SH, Sundberg SC, Larsson M, Andersson FK, Broman J, Granseth B (2020). VGluT1 deficiency impairs visual attention and reduces the dynamic range of short-term plasticity at corticothalamic synapses. *Cereb Cortex*. 30:1813–1829.

Liu XB (1997). Subcellular distribution of AMPA and NMDA receptor subunit immunoreactivity in ventral posterior and reticular nuclei of rat and cat thalamus. *J Comp Neurol*. 388:587–602.

Marubio LM, del Mar Arroyo-Jimenez M, Cordero-Erausquin M, Léna C, Le Novère N, de Kerchove d'Exaerde A, Huchet M, Damaj MI, Changeux JP (1999). Reduced antinociception in mice lacking neuronal nicotinic receptor subunits. *Nature*. 398:805–810.

McCormick DA (1992) Neurotransmitter actions in the thalamus and cerebral cortex and their role in neuromodulation of thalamocortical activity. *Prog Neurobiol*. 39:337–388.

McCormick DA, Pape HC (1990). Properties of a hyperpolarization-activated cation current and its role in rhythmic oscillation in thalamic relay neurones. *J Physiol (Lond)*. 431:291–318.

McCormick DA, Prince DA (1986). Acetylcholine induces burst firing in thalamic reticular neurones by activating a potassium conductance. *Nature*. 319:402–405.

McCormick DA, Prince DA (1988). Noradrenergic modulation of firing pattern in guinea pig and cat thalamic neurons, in vitro. *J Neurophysiol.* 59:978–996.
McCormick DA, von Krosigk M (1992). Corticothalamic activation modulates thalamic firing through glutamate "metabotropic" receptors. *Proc Natl Acad Sci U S A.* 89:2774–2778.
McCormick DA, Williamson A (1991). Modulation of neuronal firing mode in cat and guinea pig LGNd by histamine: Possible cellular mechanisms of histaminergic control of arousal. *J Neurosci.* 11:3188–3199.
Mineff EM, Weinberg RJ (2000). Differential synaptic distribution of AMPA receptor subunits in the ventral posterior and reticular thalamic nuclei of the rat. *Neuroscience.* 101:969–982.
Miyata M, Imoto K (2006). Different composition of glutamate receptors in corticothalamic and lemniscal synaptic responses and their roles in the firing responses of ventrobasal thalamic neurons in juvenile mice. *J Physiol.* 575:161–174.
Mooney DM, Zhang L, Basile C, Senatorov VV, Ngsee J, Omar A, Hu B (2004). Distinct forms of cholinergic modulation in parallel thalamic sensory pathways. *Proc Natl Acad Sci.* 101:320–324.
Nersisyan S, Bekisz M, Kublik E, Granseth B, Wróbel A (2021). Cholinergic and noradrenergic modulation of corticothalamic synaptic input from layer 6 to the posteromedial thalamic nucleus in the rat. *Front Neural Circuits.* 15:624381.
Neyer C, Herr D, Kohmann D, Budde T, Pape H-C, Coulon P (2016). mGluR-mediated calcium signalling in the thalamic reticular nucleus. *Cell Calcium.* 59:312–323.
Ngomba RT, van Luijtelaar G (2018). Metabotropic glutamate receptors as drug targets for the treatment of absence epilepsy. *Curr Opin Pharmacol.* 38:43–50.
Paz JT, Bryant AS, Peng K, Fenno L, Yizhar O, Frankel WN, Deisseroth K, Huguenard JR (2011). A new mode of corticothalamic transmission revealed in the Gria4(-/-) model of absence epilepsy. *Nat Neurosci.* 14:1167–1173.
Reichova I, Sherman SM (2004). Somatosensory corticothalamic projections: distinguishing drivers from modulators. *J Neurophysiol.* 92:2185–2197.
Reiner A, Levitz J (2018). Glutamatergic signaling in the central nervous system: Ionotropic and metabotropic receptors in concert. *Neuron.* 98:1080–1098.
Sanchez-Vives MV, Barbero-Castillo A, Perez-Zabalza M, Reig R (2021). GABAB receptors: Modulation of thalamocortical dynamics and synaptic plasticity. *Neuroscience.* 456:131–142.
Santiago LJ, Abrol R (2019). Understanding G protein selectivity of muscarinic acetylcholine receptors using computational methods. *Int J Mol Sci.* 20:5290.
Sottile SY, Hackett TA, Cai R, Ling L, Llano DA, Caspary DM (2017). Presynaptic neuronal nicotinic receptors differentially shape select inputs to auditory thalamus and are negatively impacted by aging. *J Neurosci.* 37:11377–11389.
Sun F, Zhou J, Dai B, Qian T, Zeng J, Li X, Zhuo Y, Zhang Y, Wang Y, Qian C, Tan K, Feng J, Dong H, Lin D, Cui G, Li Y (2020). Next-generation GRAB sensors for monitoring dopaminergic activity in vivo. *Nat Methods.* 17:1156–1166.
Sun Y-G, Beierlein M (2011). Receptor saturation controls short-term synaptic plasticity at corticothalamic synapses. *J Neurophysiol.* 105:2319–2329.
Sun Y-G, Rupprecht V, Zhou L, Dasgupta R, Seibt F, Beierlein M (2016). mGluR1 and mGluR5 synergistically control cholinergic synaptic transmission in the thalamic reticular nucleus. *J Neurosci.* 36:7886–7896.
Turner JP, Salt TE (1998). Characterization of sensory and corticothalamic excitatory inputs to rat thalamocortical neurones in vitro. *J Physiol.* 510:829–843.
Turner JP, Salt TE (2000). Synaptic activation of the group I metabotropic glutamate receptor mGlu1 on the thalamocortical neurons of the rat dorsal lateral geniculate nucleus in vitro. *Neuroscience.* 100: 493–505.
Ulrich D, Besseyrias V, Bettler B (2007). Functional mapping of GABAB-receptor subtypes in the thalamus. *J Neurophysiol.* 98: 3791–3795.
Varela C (2014). Thalamic neuromodulation and its implications for executive networks. *Front Neural Circuits.* 8:69.
Wang H, Liu H, Zhang Z (2011). Elimination of redundant synaptic inputs in the absence of synaptic strengthening. *J Neurosci.* 31:16675–16684.
Yang D, Zhou Q, Labroska V, Qin S, Darbalaei S, Wu Y, Yuliantie E, Xie L, Tao H, Cheng J, Liu Q, Zhao S, Shui W, Jiang Y, Wang M-W (2021). G protein-coupled receptors: Structure- and function-based drug discovery. *Sig Transduct Target Ther.* 6:1–27.
Yang Y-C, Hu C-C, Lai Y-C (2015) Non-additive modulation of synaptic transmission by serotonin, adenosine, and cholinergic modulators in the sensory thalamus. *Front Cell Neurosci.* 9. Accessed December 11, 2022. https://www.frontiersin.org/articles/10.3389/fncel.2015.00060.
Zaman T, Lee K, Park C, Paydar A, Choi JH, Cheong E, Lee CJ, Shin H-S (2011). Cav2.3 channels are critical for oscillatory burst discharges in the reticular thalamus and absence epilepsy. *Neuron.* 70: 95–108.
Zhang C-L, Verbny Y, Malek SA, Stys PK, Chiu SY (2004). Nicotinic acetylcholine receptors in mouse and rat optic nerves. *J Neurophysiol.* 91:1025–1035.
Zhang M, Augustine GJ (2021). Synapsins and the synaptic vesicle reserve pool: Floats or anchors? *Cells.* 10:658.
Zhang Y, Llinas RR, Lisman JE (2009). Inhibition of NMDARs in the nucleus reticularis of the thalamus produces delta frequency bursting. *Front Neural Circuits.* 3:20.
Zhu JJ, Uhlrich DJ (1998). Cellular mechanisms underlying two muscarinic receptor-mediated depolarizing responses in relay cells of the rat lateral geniculate nucleus. *Neuroscience.* 87:767–781.

SECTION III

CORTICAL CELL AND CIRCUIT PROPERTIES

Alex D. Reyes

Much of an animal's ability to volitionally interact with the external world is mediated by cortex. Information about the environment is transduced into electrical activity by sensory organs and eventually reaches the cortex, in most cases via the thalamus. What happens to the signal after entering the cortex is of great interest to the neuroscience community and is the overarching theme of this section.

The simple 6-layer organization of the cortex belies its enormous complexity. Each layer has a unique set of neurons, each of which can have different morphology and intrinsic properties. The same stimulus delivered to each type can evoke profoundly different firing behavior; in some cases, different behaviors can even be elicited by stimulating different cellular compartments. Each cell type does not form willy-nilly connections with each other, but instead targets specific excitatory and inhibitory neurons within and across laminae. The different circuits likely produce spatially and temporally divergent firing patterns that can potentially be used to encode select aspects of the incoming signal. Adding to the complexity is the fact that the cortex does not process only one signal, but likely combines signals from other brain regions. Finally, the input-output properties of individual neurons and networks can be changed dramatically by neuromodulators. The sheer number of often interacting variables that must be considered and controlled makes studying cortical function very difficult.

Nevertheless, a multidimensional approach with techniques borrowed from molecular biology, chemistry, physics, mathematics, and medicine has shed light on some aspects of cortical function. Neuroanatomical methods combined with genetic analyses has helped identify cell types and their short- and long-range connections. In vivo electrophysiological recordings and imaging studies in intact animals have revealed how a large population of neurons respond to natural stimuli. In vitro whole-cell recordings in more reduced preparations have yielded detailed information about the biophysical and synaptic properties of neurons. More recently, the development of optogenetic techniques has provided a means for activating or silencing specific cell types to assess their role in processing information. Finally, theoretical analyses combined with simulations suggest computations that can be performed in the cortex given known experimental variables.

The chapters in this section review key experimental findings that elucidate how signals from the thalamus are represented and processed in the cortex. Chapter 6 discusses the routes that thalamic signals take after entering the cortex. The focus is on morphologically identified pyramidal neurons in each layer, their anatomical connections, and responses during natural stimuli. These determine the flow of signals across laminae and the possible transformations that occur. Chapter 7 examines the processing power of excitatory pyramidal neurons. These neurons have extensive dendrites that span multiple laminae. Moreover, voltage gated ion channels are not uniformly distributed on the neuron, but tend to be denser preferentially in some compartments, potentially allowing them to respond uniquely to different laminar inputs. Chapter 8 discusses the role of inhibitory neurons in shaping the responses of neural networks to stimuli. The heterogeneous population of inhibitory neurons have diverse genetic makeup, morphologies, firing patterns, and connection schemes. Though few in number, the inhibitory neurons can exert powerful effects on the computations performed by networks. Chapter 9 examines the conditions for propagating and transforming signals across laminae successfully. Simulations of feedforward networks are used to elucidate how network parameters affect receptive fields and their transmission through successive layers. Finally, Chapter 10 discusses neuromodulators that affect cortical function. The different types of neuromodulators have unique effects on individual cells and circuits, thereby changing their input-output characteristics. Hopefully, the chapters will provide a good foundation and intuition for understanding cortical function.

6.

ANATOMICAL AND PHYSIOLOGICAL BASIS FOR THE FLOW OF THALAMOCORTICAL INPUTS ACROSS THE LAYERS OF THE PRIMARY SENSORY CORTEX

Marcel Oberlaender

INTRODUCTION

This chapter introduces the basic organization of thalamocortical (TC), intracortical (IC) and corticothalamic (CT) pathways. For this purpose, I will mainly review and reanalyze some of our work from the past decade on the anatomical and physiological organization of the posteromedial subfield in the vibrissae representation of the rat primary somatosensory cortex (S1, i.e., barrel cortex; Woolsey and Van der Loos, 1970). On the example of this well-studied, yet—when compared to other areas and species—uniquely organized primary sensory area, I seek to address the following general questions: *How cell type-specific are the anatomical and physiological properties of thalamorecipient excitatory neurons within and across cortical layers, and how do these properties shape the transformation of thalamocortical input into corticothalamic output?*

The rodent somatosensory system comprises a prominent representation of the facial vibrissae (whiskers). On each side of the rat snout, there are horizontal rows (A- to E-row) and vertical arcs of whiskers (arc-0: α to δ, arc-1: A1 to E1, etc.), which form an orderly array of low-threshold mechanoreceptors (Deschenes et al., 1998). Each peripheral fiber innervating these mechanoreceptors responds to only one vibrissa, and the arrangement of the vibrissal pad is maintained in arrays of cellular aggregates referred to as barrelettes in the brainstem, barreloids in the thalamus, and barrels in layer 4 (L4) of the primary somatosensory cortex (Deschenes et al., 1998). The one-to-one correspondence of somatotopically organized, segregated structures in thalamus and cortex with individual facial whiskers (Welker, 1976) has provided us—like many others—with a precise anatomical reference frame that allows us to investigate repeatedly the same cells and circuits within a thalamocortical system with different experimental approaches.

THALAMOCORTICAL PATHWAYS

The rodent vibrissal system shares many of the common features that characterize sensory systems in mammals. Neurons in different brainstem nuclei that receive vibrissal primary afferents contribute axons to the trigeminothalamic tract (Deschenes et al., 1998). The main streams of these ascending fibers innervate relay cells in the contralateral primary and secondary thalamus, and the axons of these relay cells then innervate the cerebral cortex, including the barrel cortex (Deschenes et al., 1998). Thalamic afferents to the barrel cortex arise mainly from 2 areas (Figure 6.1): the ventral posterior medial nucleus (VPm) and the medial posterior nuclei (POm). Intralaminar nuclei (e.g., the central lateral nucleus), the ventral medial and ventral lateral nuclei, and the medial geniculate body also provide thalamic input to the barrel cortex, but these projections are sparse or involve only small numbers of relay cells, and are hence not considered in this chapter (see Deschenes et al., 1998, for a more comprehensive review).

The VPm is the main route through which sensory input from the vibrissae enters the cerebral cortex (Bruno and Sakmann, 2006). Upon the onset of both passive deflections and active touches during rhythmic back-and-forth movements of the vibrissae (i.e., exploratory whisking), VPm neurons reliably elicit fast action potential responses (Bruno and Sakmann, 2006; de Kock et al., 2007; Moore et al., 2015), and relay the resulting synchronous volley of excitation toward the barrel cortex (Figure 6.1A). Within the barrel cortex, VPm axons establish an extremely precise one-to-one relationship between the barreloids and the barrels. However, single neuron reconstructions revealed that VPm axons generally extend vertically beyond the barrels (Figure 6.1B), and in some cases even reach the pial surface (Oda et al., 2004). Throughout the cortical

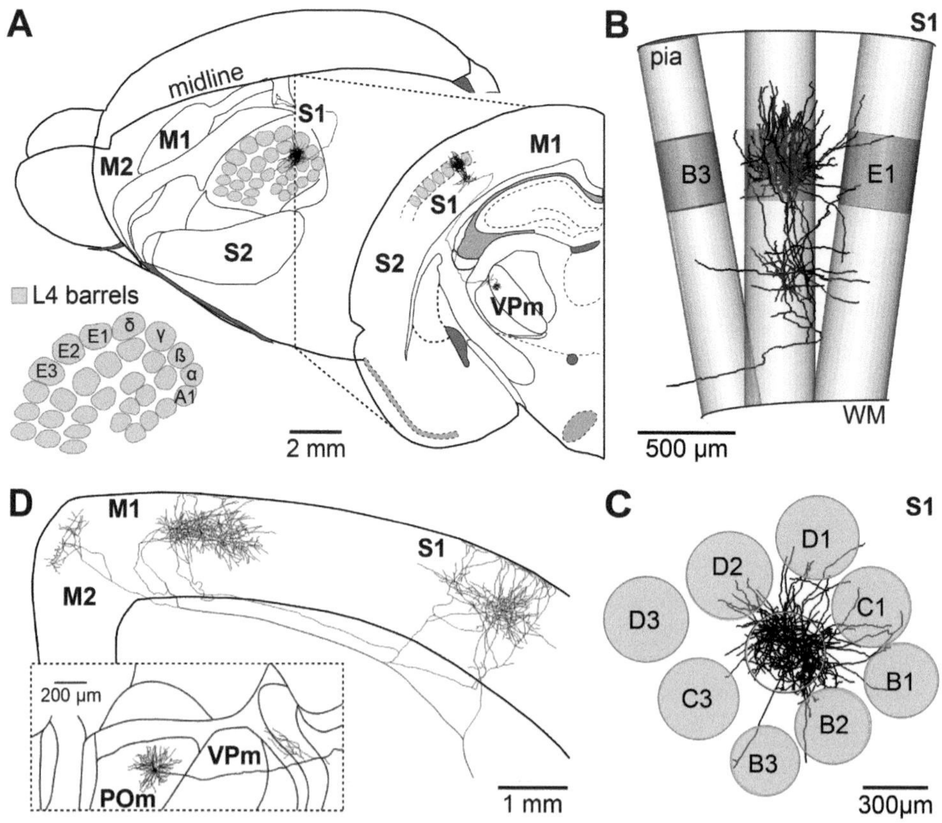

Figure 6.1. Thalamocortical pathways. (A). Tangential view of the rat brain with the barrel field in S1. Coronal view shows the input to S1 from the VPm. (B). Coronal view of single VPm axon in barrel cortex (modified from Oberlaender et al., 2012a). (C). Tangential view of the axon from panel B. (D). Sagittal view of single POm axon in rat cortex (modified from Deschenes et al., 1998).

depth, the horizontal extent of VPm axons remains confined to the dimensions of single barrels (Figure 6.1C). Sensory input from a single vibrissa hence activates an approximately cylindrical cortical volume from the white matter (WM) to the pia—the innervation volume of axons from the same barreloid—which is often referred to as a *cortical barrel column* (Wimmer et al., 2010).

The POm gives rise to the second main route of thalamic projections to the barrel cortex (Yu et al., 2006). Compared to the VPm, the POm is more heterogeneously organized. Axons from different parts of the POm project to different, yet all of the sensorimotor regions of the cortex (Deschenes et al., 1998), including the secondary somatosensory (S2), primary (M1) and secondary motor (M2) cortices (Figure 6.1D). Projections to the barrel cortex originate from scattered neurons throughout the POm (Deschenes et al., 1998). Within the barrel cortex, POm axons extend far beyond the dimensions of single barrel columns, and appear to target the septa between barrels (Deschenes et al., 1998). Some POm neurons show responses to active touches of the vibrissae (Moore et al., 2015; Mease et al., 2016c). However, the organization of responding neurons within the POm, their specific axonal projection patterns, and whether the barrel cortex is among their targets, are unclear. Thus, for the purpose of this chapter, I will not examine inputs from the POm to the barrel cortex. Instead, I will focus on the VPm, and examine how barrel cortex transforms sensory input from this primary thalamocortical pathway into corticothalamic output to the POm.

THALAMORECIPIENT CORTICAL LAYERS

Changes in neuron density along the vertical cortex axis (i.e., from the pia toward the white matter) gave rise to the concept of cytoarchitectonic layers (Brodmann, 1909). The cerebral cortex is typically subdivided into 6 layers (i.e., L1–6), but there are many exceptions to this across cortical areas and species. The subdivision into cortical layers has been widely accepted to provide a first-order criterion to discriminate between neuronal cell types. Countless studies have thus grouped neurons by their laminar soma locations to provide layer-specific analyses, and to describe the signal flow and circuits within and between areas of the cerebral cortex. To precisely identify layer borders in the rat barrel cortex, we reconstructed all somata within the volume that represents the 24 major vibrissae (α–δ, A1–E4). These data revealed the numbers and distributions of all excitatory

(Figure 6.2A) and inhibitory somata (Figure 6.2B) for 24 barrel columns, and the variability of these quantities across rats (Meyer et al., 2013). The resulting layer borders are represented by the following distances from the pial surface (mean ± SD, N = 4 rats): L1–L2: 157 ± 16 µm; L2–L3: 296 ± 30 µm; L3–L4: 575 ± 57 µm; L4–L5: 900 ± 50 µm; L5–L6: 1,411 ± 28 µm; L6–WM: 1,973 ± 44 µm.

These average values conceal that the thickness and depth of each layer vary considerably across the barrel cortex (Figure 6.2A,B). These differences do not reflect a linear scaling with cortical thickness, which varies by more than 500µm across the rat barrel cortex. Instead, different diameters of the barrel columns compensate for cortical thickness differences (Egger et al., 2012). As a result, the volume of barrel columns is similar for vibrissae that are located within the same row, and progressively increases by a factor of 3 from the A-row toward the E-row. While the depths and extents of the layers change in a vibrissa- and row-specific manner, the neuron densities per layer do not (Figure 6.2A,B). Consequently, analogous to the 3-fold increase in volume, also the number of neurons per barrel column—and per layer therein—systematically increases from ~9,000 in the A-row to more than 27,000 in the E-row (Meyer et al., 2013). Notably, this observation disproved the *basic uniformity principle*, a long-standing belief that the number of neurons underneath a square millimeter of the cortical surface is constant across areas and species (Rockel et al., 1980; Carlo and Stevens, 2013).

Similar to the gradient in the barrel cortex, we observed a gradient across the VPm, where the numbers of neurons per barreloid varied between 100 and 400 (Meyer et al., 2013). As a result, the ratio of VPm to S1 neurons is constant for each vibrissa (1:70). Moreover, VPm axons have the same vertical distributions with respect to the laminar organization of each barrel column (Wimmer et al., 2010)—that is, their densest innervation delineates layer 4, and their second densest innervation coincides with the respective depth of the border between layers 5 and 6 (Figure 6.2C). The densest innervations by POm axons delineate layer 1 across the barrel cortex and the respective depth of the border between layers 4 and 5 (Figure 6.2D). Finally, the overall path lengths of individual axons, as well as the densities of axonal boutons (i.e., presynaptic sites), do not differ between VPm neurons depending on which vibrissae they represent (Oberlaender et al., 2012b). Thus, despite the fact that the number of neurons that represents each vibrissa differs substantially in the thalamocortical system, the layer-specific connectivity appears to be well-maintained across the barrel cortex for both of its main thalamic input pathways. For the remainder of this chapter, I will hence restrict my considerations to thalamocortical input from a single VPm barreloid.

THALAMORECIPIENT CORTICAL CELL TYPES

It has become increasingly clear that a cortical layer does not contain a homogeneous neuronal cell type. To classify cell types of excitatory neurons in the rat barrel cortex, we reconstructed the dendrite and axon morphologies for a representative set of randomly sampled in vivo recorded neurons (Narayanan et al., 2015; Rojas-Piloni et al., 2017; Guest et al., 2022). For each neuron, we reconstructed the

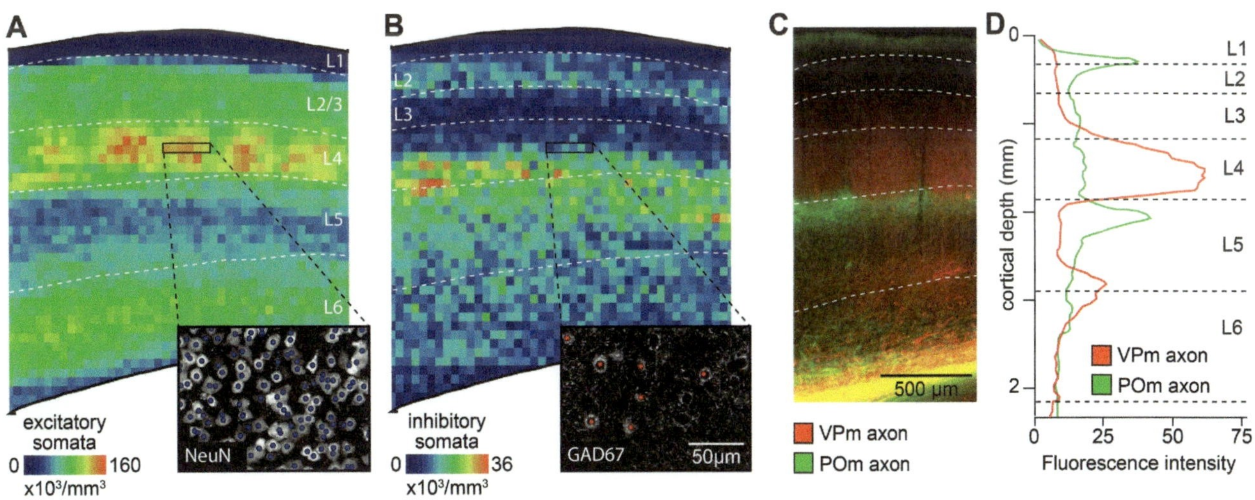

Figure 6.2. Layer-specific thalamocortical pathways. (A). Semicoronal view of the excitatory soma density for the volume that comprises whisker arc-2 of the rat barrel cortex (A2, B2, C2, D2, E2). The 5 barrels in L4 are clearly visible as segregated spots of high soma density: E2 (left) to A2 (right). Densities are derived by marking the center locations of NeuN-labeled somata, as shown in the inset for a small example image from L4. (B). Distribution of inhibitory somata analogous to panel A. Inset shows center locations of GAD67-labeled somata corresponding to the image in panel A. Panels A and B are modified from Meyer et al. (2013). (C). Fluorescent labeling of VPm and POm axons in the same animal. (D). Profiles of VPm and POm bouton fluorescence intensity. Panels C and D are modified from Wimmer et al. (2010).

pial surface, white matter tract, and barrel field to evaluate its soma position, as well as dendrite and axon morphology, with respect to the laminar and columnar organization of the barrel cortex at 50µm precision (Egger et al., 2012). In the following, I will summarize some of the anatomical features that distinguish neurons of the major excitatory cortical cell types (Figure 6.3A,B) (see Harris and Shepherd, 2015; Narayanan et al., 2017 for more comprehensive reviews).

The supragranular layers 2/3 comprise 2 excitatory cell types. Layer 2 pyramids (L2PY) have apical dendrites that lack a prominent trunk, and which terminate with elaborate, often tilted, tufts that extend beyond the barrel column in which the soma is located (i.e., principal column). By contrast, layer 3 pyramids (L3PY) have apical dendrites with trunks that terminate with less elaborate tufts in layer 1. Neurons of both types have elaborate axonal projections within the barrel cortex, and project long-range axons to other cortical areas—that is, they belong to the broad class of intratelencephalic neurons (ITs). Within the barrel cortex, L2PY axons extend far beyond the principal column in layers 1 to 3, and to a lesser degree in layer 5. Compared to L2PY, L3PY axons innervate layers 1 to 3 less densely, but are denser in layer 5. Both L2PY and L3PY can send their long-range axons to either M1 or S2 (Chen et al., 2013). However, because the somata of L2PY and L3PY intermingle, and M1-projecting neurons intermingle with those projecting to S2, it is unclear how neuronal cell type and long-range targets relate to one another.

The granular layer 4 comprises 3 excitatory cell types. Most abundant are spiny-stellates (L4ss). L4ss differ from other excitatory neurons by their lack of an apical dendrite. Their somata cluster around the dense innervation of VPm axons, and their dendrites are often polarized toward the barrel center. L4ss somata and dendrites hence delineate the barrels. Second most abundant are star-pyramids (L4sp). L4sp have short apical dendrites that terminate without tufts in layers 2/3. They are equally abundant in upper layer

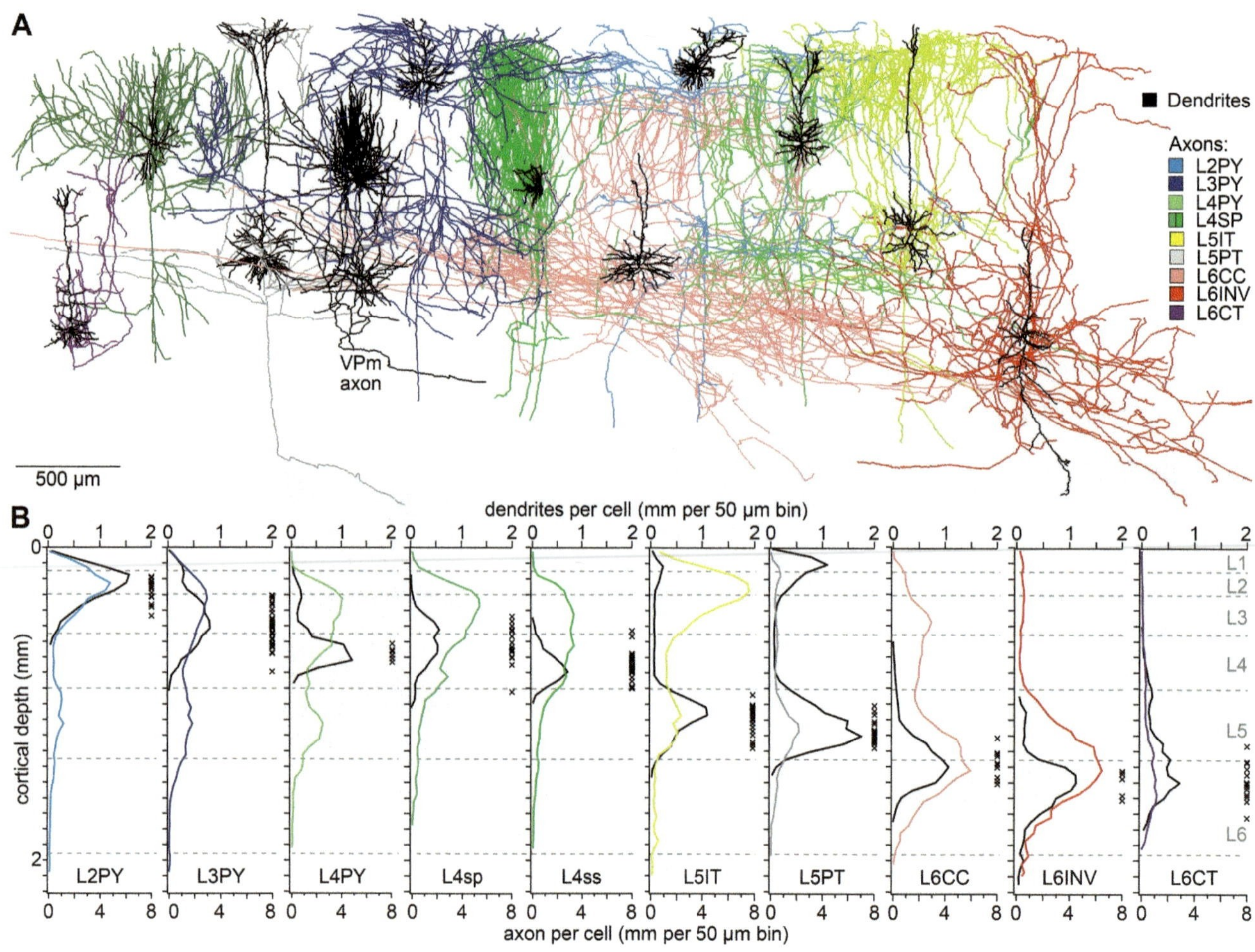

Figure 6.3. Cell type-specific organization of thalamorecipient intracortical pathways. (A). Dendrite (black) and axon morphologies (colored by cell type, intracortical part of one VPm axon in black) of in vivo labeled neurons in rat barrel cortex that are representative examples for each of the 10 major excitatory cell types of the cerebral cortex (modified from Egger et al., 2020). (B). Soma positions (x), and density profiles along the cortical depth of dendrites (black lines) and axons (colored by cell type analogous to panel A) for representative samples of in vivo labeled neurons for each of the 10 major excitatory cell types (modified from Narayanan et al., 2015).

4 and lower layer 3, where they intermingle with L4ss, and L2PY and L3PY, respectively. Least abundant are layer 4 pyramids (L4PY), which are located mostly in the septa between barrels. L4PY have apical dendrites with tufts that terminate in layer 2. Axons of the 3 granular cell types do not innervate layer 1. Apart from this difference, L4PY axons resemble those of L3PY. By contrast, L4sp and L4ss axons are confined to layers 2–4. Like the VPm axons, L4ss axons remain confined to the principal column—which is the exception rather than the norm for excitatory neurons in barrel cortex. L4sp and L4ss have no axons outside the barrel cortex. Whether L4PY project long-range axons to other cortical areas is yet unclear.

The infragranular layers 5/6 comprise 5 excitatory cell types: 2 types of layer 5 pyramids with apical dendrites extending to layer 1, and 3 types of layer 6 pyramids with apical dendrites terminating in layers 4/5. Intratelencephalic neurons (L5IT) are most abundant in upper layer 5 (L5A), pyramidal tract neurons (L5PT) in lower layer 5 (L5B). Apical dendrites of L5IT terminate with narrow tufts, and are hence referred to as *slender-tufted*. L5PT terminate with elaborate apical tufts, and are hence referred to as *thick-tufted*. L5IT have elaborate axons that extend far beyond single columns in layers 1–3, but that remain confined to the principal column in layer 5. Compared to other excitatory cell types, L5PT have sparse axons, indicating that these neurons contribute relatively little to local computations. L5IT project long-range axons to other cortical areas (e.g., M1) and to the striatum (Harris and Shepherd, 2015). In contrast, L5PT provide output to subcortical brain areas via long-range projections to a variety of midbrain, tectum, and brainstem nuclei (Harris and Shepherd, 2015). L5PT are segregated with respect to their primary target areas in the POm, superior colliculus, pons, or spinal trigeminal complex (Rojas-Piloni et al., 2017). L5PT with these target areas intermingle, but form target-related sublayers within layer 5—an organizational principle that generalizes to other cortical areas and species (Economo et al., 2018).

L5PT also intermingle with corticocortical neurons (L6CC), whose somata cluster around the innervation peak of VPm axons at the layer 5/6 border (Egger et al., 2020; see Figure 6.1D). L6CC have apical dendrites that terminate without tufts in layer 4. L6CC have the most elaborate axons of all cell types in barrel cortex. Their axons innervate all layers, but are most elaborate in layers 5/6, where they span horizontally across the entire barrel cortex. Corticocortical neurons deeper in layer 6 have similarly elaborate horizontal axon projections, which are, however, more confined to layers 5/6. These deeper corticocortical neurons have apical dendrites that often extend toward the white matter, and we hence refer to this heterogeneous class of neurons as inverted pyramids (L6INV). Most abundant in the infragranular layers are corticothalamic neurons in layer 6 (L6CT). L6CT have apical dendrites that terminate with narrow tufts in layer 4, and much narrower basal dendrites compared to the other pyramidal types. L6CT have sparse local axons that remain confined to layers 4–6 of the principal column, and send long-range axons to the VPm, or to both VPm and POm (Deschenes et al., 1998).

THALAMOCORTICAL SIGNAL FLOW

The analysis of single neuron morphologies provides 2 major insights that we must consider for describing the flow of thalamocortical input through the cortical circuitry. First, excitatory cortical neurons have elaborate horizontal axon projection patterns, whose extents generally, and by far, exceed those of primary thalamocortical axons. Consequently, even though sensory input from a single vibrissa can only activate neurons within a single barrel column, thalamorecipient neurons within this principal column can monosynaptically activate neurons across the entire barrel cortex—and across all layers therein. Second, somata of excitatory neurons with different dendrite morphologies, axon projection patterns, and/or long-range cortical and subcortical target areas intermingle within and across layers. Thus, layer-specific descriptions of thalamocortical signal flow need to be complemented with descriptions that take all neurons in a cortical area, their cell types, and ideally also their long-range axonal target areas into account.

To address this challenge quantitatively, we generated a digital model of the rat barrel cortex (Udvary et al., 2022), by combining our reconstructions of the in vivo recorded single neurons in VPm (Figure 6.1) and barrel cortex (Figure 6.3), with our reconstructions of the barrel cortex geometry, cytoarchitecture, and cellular organization (Figure 6.2). The neuropil structure and the connectome that the model predicts are remarkably consistent with a rich body of empirical data from several decades of barrel cortex research by many different laboratories. Because it mimics prospective dense electron microscopy reconstructions of thalamocortical and excitatory intracortical pathways throughout the barrel cortex, the model is considered as a first realistic approximation for the anatomical basis that underlies sensory-evoked cortical signal flow at cellular and cell type levels (Egger et al., 2020; Guest et al., 2022).

Table 6.1 provides quantitative estimates from the model for how relay cells in a single barreloid of the VPm synaptically innervate the different cell types of excitatory neurons in the barrel cortex, and for how neurons of these cell types connect to one another within a cortical barrel column. For the purpose of this chapter, I will highlight just 2 important aspects that affect our considerations for the flow of sensory-evoked thalamocortical input. First, VPm

TABLE 6.1 ANATOMICAL BASIS FOR CELL TYPE-SPECIFIC CORTICAL SIGNAL FLOW OF THALAMOCORTICAL INPUTS

Cell type	Number in column	Connection probability (%)									
		L2PY	L3PY	L4PY	L4sp	L4ss	L5IT	L5PT	L6CC	L6INV	L6CT
VPM	350	10	38	52	37	41	32	43	27	21	17
L2PY	1306	43	30	16	10	5	21	38	9	7	6
L3PY	2146	22	22	18	11	8	18	30	11	7	6
L4PY	435	23	31	25	15	10	17	30	10	7	5
L4sp	1354	35	45	39	26	21	22	32	10	7	7
L4ss	2181	31	48	45	32	30	26	35	10	7	8
L5IT	1156	39	30	16	11	6	23	42	12	10	9
L5PT	858	7	7	7	4	3	14	28	16	11	9
L6CC	1104	10	14	11	6	4	11	21	22	24	17
L6INV	639	5	6	4	3	2	18	34	36	37	26
L6CT	3102	<1	2	2	3	2	8	15	16	18	12

≥30% ≥20% ≥15%

Estimates for how relay cells in the VPm that represent the C2 whisker synaptically innervate the different cell types of excitatory neurons in the barrel cortex, and for how neurons of these cell types connect to one another within the C2 barrel column. Connection probabilities are derived statistically from a digital model that mimics dense electron microscopy reconstructions of the rat barrel cortex, and by assuming that axo-dendritic overlap is necessary, but not sufficient, to account for the formation of synaptic connections (Udvary et al., 2022).

axons converge onto excitatory neurons across all layers and cell types, an anatomical prediction that is well-supported by functional connectivity mappings from several in vitro (Petreanu et al., 2009; Rah et al., 2013) and in vivo studies (Bruno and Sakmann, 2006; Constantinople and Bruno, 2013). Second, neurons of all cell types connect to one another across layers. Such translaminar input is particularly strong to the supragranular and infragranular layer 5 cell types. L5IT and L5PT receive inputs from all layers, but particularly the L5PT receive strong convergent input from all cell types. L5PT hence stand out as the major output cell type, but also the major *integrator* cell type of the cerebral cortex. Translaminar input to the cell types in layers 4 and 6 is weak. There, excitatory input originates primarily from neurons within the same layer, and from the VPm—indicating that the layer 4 and 6 cell types *distribute* sensory input to the long-range projecting neurons in layers 2/3 and 5. Taken together, the anatomy of the VPm and barrel cortex suggests that—beyond activation of layer 4—sensory input could activate in parallel a plethora of highly interconnected, cell type-specific local and long-range pathways.

Whether and how neurons respond to thalamocortical input depends, however, on many additional cellular and circuit properties. The dendritic location and strength of synapses, the biophysical properties of the dendrites, the degrees of synchrony, and the balance between feedforward excitation and inhibition are just some of the physiological constraints that—in addition to the anatomy—determine the flow of thalamocortical input across the cortical circuitry. To account for these constraints, we complemented our anatomical data with recording data of sensory-evoked single neuron responses in the VPm and across all layers of the barrel cortex (de Kock et al., 2007; Oberlaender et al., 2012a; Egger et al., 2020). For this purpose, we passively deflected single vibrissa with a piezoelectric bimorph, which enabled us to measure the time (i.e., latency) between the first sensory-evoked action potential and the onset of the sensory stimulus with sub-millisecond precision. For each neuron we measured onset latencies in response to deflections of the somatotopically aligned principal whisker (PW), and to each surrounding whisker (SW) (Figure 6.4A). Moreover, we reconstructed the morphologies of the recorded neurons to analyze their responses in a layer- and cell type-specific manner (Egger et al., 2020).

Shortly after the onset of the synchronous volley of sensory-evoked excitation in the VPm, and consistent with the anatomy, excitatory neurons across all layers respond similarly fast to PW stimuli (Figure 6.4B,C). However, whether and how fast neurons respond differ between cell types, and for PW versus SW stimuli. Within the supragranular and granular layers, L4ss and L4sp respond the fastest to PW stimuli, followed by L3PY. L4ss and L4sp show delayed responses to SW stimuli, whereas L3PY respond exclusively to the PW. The opposite is true for L4PY, which do not respond to the PW, but show delayed responses to SWs. L2PY have no responses to deflections of any of the whiskers. This cell type-specific latency and receptive field

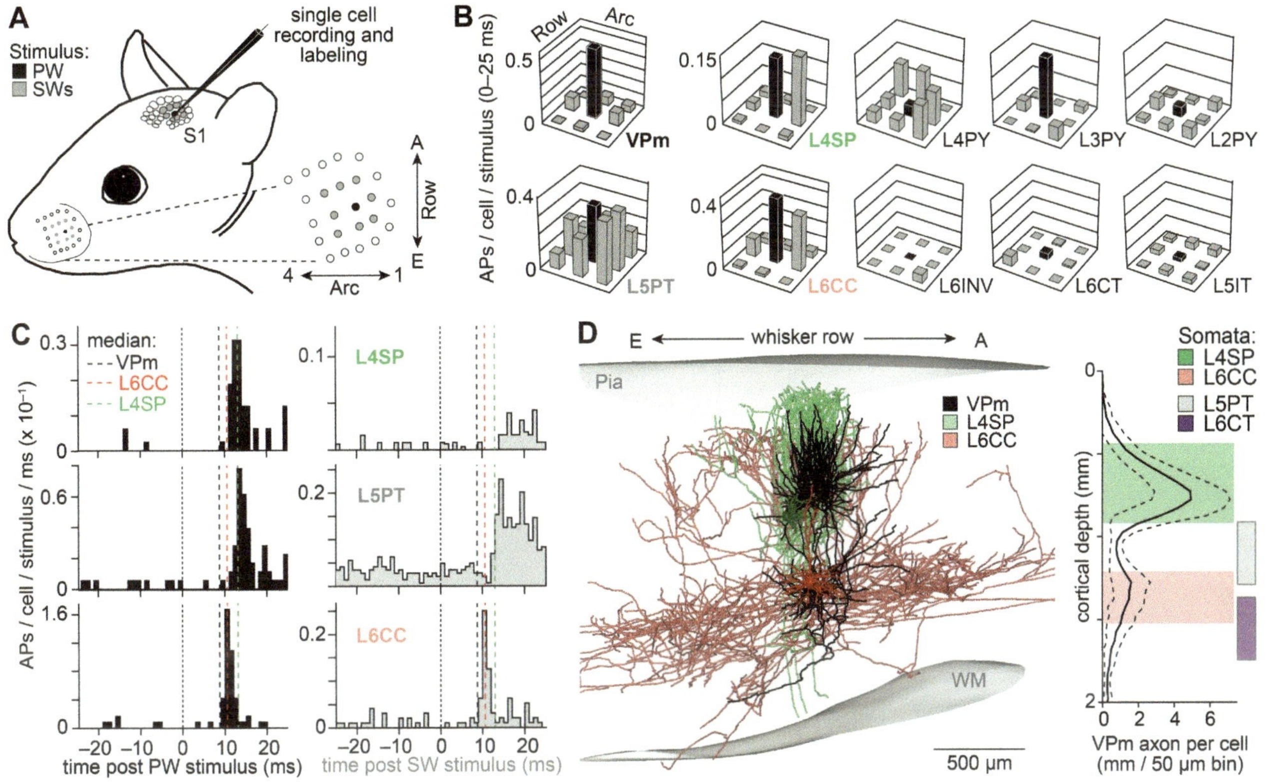

Figure 6.4. Functional basis for cell type-specific signal flow of thalamocortical inputs. (A). Schematic of single neuron recordings during principal whisker (PW) and surround whisker (SW) stimuli. (B). Whisker receptive fields for excitatory cell types. (C). PSTHs of PW- (left) and SW-evoked (right) responses for L4SP, L5PT, and L6CC neurons. Vertical dashed lines demarcate onsets of PW-evoked responses across VPm, L6CC, and L4SP neurons. (D). Examples of in vivo labeled axons in rat barrel cortex. Right panel: locations of somata of L4SP and L6CC coincide with the 2 innervation peaks of VPm axons. All panels are modified from Egger et al. (2020).

data, together with the axonal projection patterns, reveal the following flow of sensory-evoked thalamocortical input through the granular and supragranular layers: (1) the only responses that are driven directly by the VPm are those to PW stimuli of the spiny neurons in layer 4 (L4SP, i.e., L4ss and L4sp); (2) PW input then flows within the principal column from L4SP to L3PY; and (3) then from L3PY to surrounding barrel columns, and to other cortical areas (e.g., M1 or S2). This TC→L4→L2/3→CC pathway has been described across sensory areas and species—and it is hence referred to as the *canonical cortical circuit* (Douglas and Martin, 2004).

According the canonical cortical circuit concept, the flow of thalamocortical input from layer 4 to layers 2/3 should also drive responses in layer 5. While the anatomy supports this hypothesis, the physiology does not (Figure 6.4B,C). L5IT have no responses to deflections of any of the whiskers. L5PT have reliable responses, but their latencies rival those of L4SP, and hence precede those of L3PY. Moreover, L5PT have fast responses to PW and SW stimuli, an observation which neither the receptive fields nor the horizontal axons of L4SP or L3PY can explain. If not the canonical cortical circuit, which other pathways could underlie the responses of L5PT? Compared to all other cell types, L6CC show the fastest, most reliable, and most synchronous responses to PW and SW stimuli. As a result, and because of their elaborate horizontal axons (Figure 6.4D), L6CC relay sensory input from the thalamus across the entire barrel cortex. L6CC thereby drive the fast and broadly tuned responses of L5PT (Egger et al., 2020). Taken together, sensory input drives 2 parallel streams of signal flow in the barrel cortex: (1) the horizontally organized VPm→L6CC pathway results in fast signal flow to subcortical areas via activation of L5PT; (2) the vertically organized VPm→L4SP pathway results in fast signal flow to other cortical areas via activation of L3PY. This concept of a *bistratified thalamocortical circuit* (Figure 6.5A) is supported by manipulations which showed that deactivation of L6CC (Egger et al., 2020), but not of L4SP (Constantinople and Bruno, 2013), abolishes sensory responses in L5PT.

CORTICOTHALAMIC PATHWAYS

The anatomy and physiology of the barrel cortex reveal L5PT as the central element by which cortical circuits transform sensory input from the thalamus

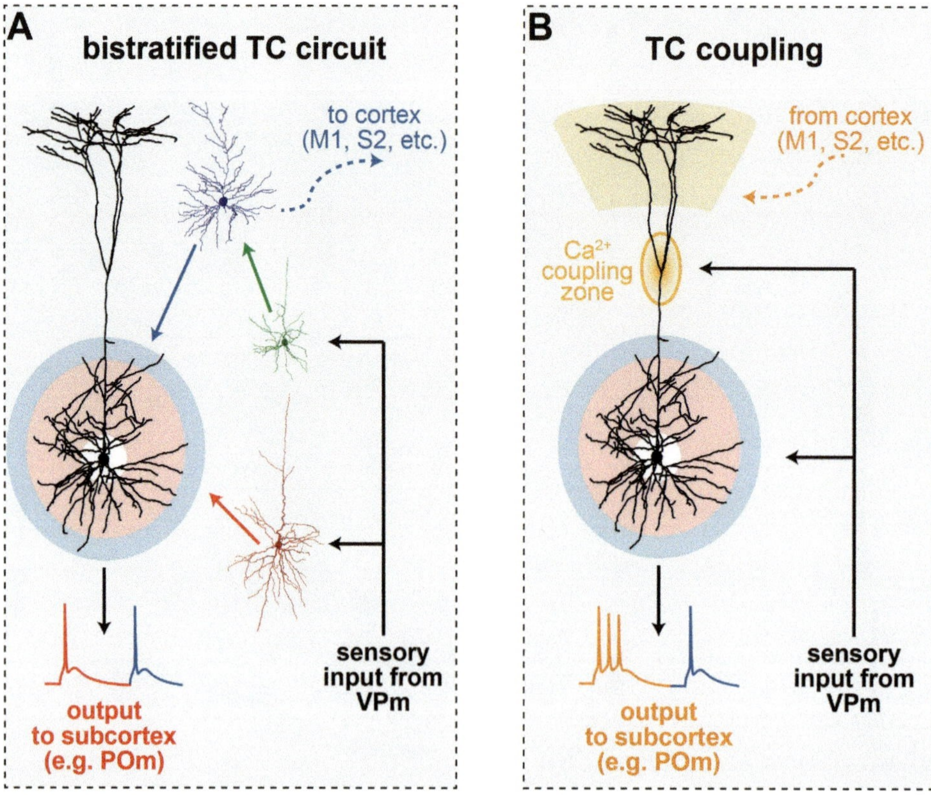

Figure 6.5. Suggested circuit models for how sensory cortex transforms thalamocortical input into corticothalamic output. (A). Sensory-evoked input drives 2 thalamocortical pathways in parallel. First, thalamus drives APs in L6CCs (red), which in turn drive fast onset responses in the subcortically projecting L5PTs. Second, thalamus drives APs in L4SPs (green), which in turn drive responses in corticocortical projecting L2/3PYs (blue). (B). In addition to receiving sensory-evoked thalamocortical input indirectly via the TC→L6CC and TC→L4SP→L2/3PY pathways, L5PTs receive also direct thalamocortical input from VPm neurons along both their proximal and apical dendrites. Depending on the prestimulus depolarization of the apical tuft (by, e.g., top-down pathways), this direct thalamocortical input can drive dendritic calcium APs, which then modulate the sensory-evoked cortical output patterns of L5PTs with high-frequency bursts of APs.

into corticothalamic output. This insight may come as a surprise, as L6CT outnumber the subset of POm-projecting L5PT by at least 1 order of magnitude (Rojas-Piloni et al., 2017). However, sensory-evoked responses are generally absent in L6CT (de Kock et al., 2007; de Kock et al., 2021). Thus, while observations in the primary visual cortex indicate that L6CT process vestibular information (Velez-Fort et al., 2014) and, likely related to this, can sharpen receptive fields (Born et al., 2021), the function of L6CT in the barrel cortex remains unclear. Moreover, L6CT synapses onto thalamic relay cells are weak (Groh et al., 2008). In stark contrast to L6CT, L5PT form strong synaptic connections with POm neurons via large bouton terminals (Groh et al., 2008), and sensory input—even from a single vibrissa—reliably activates L5PT throughout the entire barrel cortex. Within the population of L5PT, those projecting to the POm have even the most reliable responses to the onset of whisker stimuli (Rojas-Piloni et al., 2017). Such sensory-evoked corticothalamic feedback from the barrel cortex via L5PT can facilitate whisker-based perception (Qi et al., 2022).

The ability of L5PT to combine sensory signals with internal information streams is thought to a form a basis for perception and the resulting behavior (Aru et al., 2020; Takahashi et al., 2020). Along their dendrites, L5PT combine information streams arriving at all layers, which include inputs to their apical tufts from several top-down pathways (e.g., M1; Mao et al., 2011), as well as from the POm (Petreanu et al., 2009). In order to contribute to cortical output, these distal inputs must, however, cross a high threshold to evoke dendritic calcium spikes (Schiller et al., 1997), but can then generate bursts of somatic action potentials (Larkum et al., 1999) (see **Chapter 7 in Section III**). Direct evidence that this calcium-dependent spiking mechanism is utilized for perception was reported for L5PT in barrel cortex (Takahashi et al., 2020). Elevated calcium activity in the apical dendrites, in combination with somatic burst firing, correlates with the threshold for perceiving whisker stimuli (Takahashi et al., 2016).

We recently found that direct input from the VPm drives the coupling of sensory signals with top-down information streams in the dendrites of L5PT (Guest et al., 2022).

VPm axons converge strongly onto L5PT (Constantinople and Bruno, 2013), where they form connections with the basal and apical dendrites (Petreanu et al., 2009; Rah et al., 2013). VPm synapses are most densely distributed along the apical trunk, where their density increases from the soma toward the dendritic calcium domain (Guest et al., 2022). Direct VPm input does not generally drive calcium action potentials in L5PT. However, the more inputs impinge onto the apical tufts preceding a stimulus (e.g., by M1, S2, or POm), the higher the probability that the synchronous volley of VPm input drives calcium action potentials in L5PT. Thus, in the absence of prestimulus depolarization of the apical tuft, L5PT broadcast single action potentials that are driven by L6CC to their downstream targets. In the presence of prestimulus depolarization, VPm-driven calcium spikes modulate this output by L5PT into bursts. This *TC coupling* mechanism (Guest et al., 2022) by the TC→L5PT pathway ensures that already the first sensory-evoked action potentials that leave the cerebral cortex can be modulated by nonsensory inputs from top-down pathways (Figure 6.5B). This is important, because L5PT→POm synapses act as low-pass filters when bursts arrive in isolation (Groh et al., 2008). In contrast, multiple bursts that impinge synchronously onto POm neurons facilitate their responses (Mease et al., 2016b; Mease et al., 2017), and hence their thalamocortical feedback to the barrel cortex (Mease et al., 2016a) and other cortical areas. Thus, via TC coupling, and by evoking the signal flow through a bistratified circuit, primary thalamus can facilitate the fast activation of the transthalamic L5PT→POm→cortex pathways.

DISCUSSION

By using the rat barrel cortex as a model, I introduced the basic organization of thalamocortical, intracortical, and corticothalamic circuits in a primary sensory area. The anatomy revealed that axons from the primary thalamus (VPm) relay sensory input in parallel to excitatory neurons of all cell types and across all layers. However, the physiology revealed that VPm input drives sensory-evoked action potentials only in those cell types whose somata cluster around its 2 densest innervation domains. The TC→L4SP pathway evokes vertical signal flow within a barrel column that drives L3PY, which provide output to other cortical areas (e.g., S2). Preceding the signal flow through the canonical TC→L4→L2/3 pathway, the TC→L6CC pathway evokes horizontal signal flow that drives L5PT across the barrel cortex, which provide output to subcortical areas (e.g., POm). This concept of a bistratified thalamocortical circuit is not entirely new. A remarkable review article argued already more than 3 decades ago that the anatomy and physiology of the macaque primary visual cortex suggest parallel activation of layer 4 and horizontally projecting neurons at the layer 5/6 border (Lund, 1988). Bistratified primary thalamocortical axons, as well as horizontally projecting L6CCs, are also observed in the primary visual cortex of cats (Antonini and Stryker, 1993; Karube et al., 2017) and mice (Velez-Fort et al., 2014). Thus, the bistratified thalamocortical circuit is likely the basic "canonical" circuit architecture by which primary sensory areas transform input from the thalamus into cortical—including corticothalamic—output.

The bistratified thalamocortical circuit is embedded into other local and long-range circuits, which I did not address here (e.g., disinhibition via POm; Audette et al., 2019; Williams and Holtmaat, 2019). It is hence important to emphasize that context and behavioral task will generally modulate—for example, via TC coupling—how sensory input flows through the cerebral cortex. Thus, the bistratified thalamocortical circuit, as introduced here, should be considered as the basic configuration by which the cerebral cortex transforms thalamocortical input into corticothalamic output.

REFERENCES

Antonini A, Stryker MP (1993). Rapid remodeling of axonal arbors in the visual cortex. *Science*. 260:1819–1821.

Aru J, Suzuki M, Larkum ME (2020). Cellular mechanisms of conscious processing. *Trends Cogn Sci*. 24:814–825.

Audette NJ, Bernhard SM, Ray A, Stewart LT, Barth AL (2019). Rapid plasticity of higher-order thalamocortical inputs during sensory learning. *Neuron*. 103:277–291 e274.

Bast A, Guest JM, Fruengel R, Narayanan RT, de Kock CPJ, Oberlaender M (2023). Thalamus drives active dendritic computations in cortex. In. www.bioRxiv.org. doi:https://doi.org/10.1101/2021.10.21.465325

Born G, Schneider-Soupiadis FA, Erisken S, Vaiceliunaite A, Lao CL, Mobarhan MH, Spacek MA, Einevoll GT, Busse L (2021). Corticothalamic feedback sculpts visual spatial integration in mouse thalamus. *Nat Neurosci*. 24:1711–1720.

Brodmann K (1909). *Vergleichende Lokalisationslehre der Großhirnrinde—in ihren Prinzipien dargestellt auf Grund des Zellenbaues*. Barth.

Bruno RM, Sakmann B (2006). Cortex is driven by weak but synchronously active thalamocortical synapses. *Science*. 312:1622–1627.

Carlo CN, Stevens CF (2013). Structural uniformity of neocortex, revisited. *Proc Natl Acad Sci U S A*. 110:1488–1493.

Chen JL, Carta S, Soldado-Magraner J, Schneider BL, Helmchen F (2013). Behaviour-dependent recruitment of long-range projection neurons in somatosensory cortex. *Nature*. 499:336–340.

Constantinople CM, Bruno RM (2013). Deep cortical layers are activated directly by thalamus. *Science*. 340:1591–1594.

de Kock CP, Bruno RM, Spors H, Sakmann B (2007). Layer- and cell-type-specific suprathreshold stimulus representation in rat primary somatosensory cortex. *J Physiol*. 581:139–154.

de Kock CPJ, Pie J, Pieneman AW, Mease RA, Bast A, Guest JM, Oberlaender M, Mansvelder HD, Sakmann B (2021). High-frequency burst spiking in layer 5 thick-tufted pyramids of rat primary somatosensory cortex encodes exploratory touch. *Commun Biol*. 4:709.

Deschenes M, Veinante P, Zhang ZW (1998). The organization of corticothalamic projections: reciprocity versus parity. *Brain Res Rev.* 28:286–308.

Douglas RJ, Martin KA (2004). Neuronal circuits of the neocortex. *Annu Rev Neurosci.* 27:419–451.

Economo MN, Viswanathan S, Tasic B, Bas E, Winnubst J, Menon V, Graybuck LT, Nguyen TN, Smith KA, Yao Z, Wang L, Gerfen CR, Chandrashekar J, Zeng H, Looger LL, Svoboda K (2018). Distinct descending motor cortex pathways and their roles in movement. *Nature.* 563:79–84.

Egger R, Narayanan RT, Guest JM, Bast A, Udvary D, Messore LF, Das S, de Kock CPJ, Oberlaender M (2020). Cortical output is gated by horizontally projecting neurons in the deep layers. *Neuron.* 105:122–137 e128.

Egger R, Narayanan RT, Helmstaedter M, de Kock CP, Oberlaender M (2012). 3D reconstruction and standardization of the rat vibrissal cortex for precise registration of single neuron morphology. *PLoS Comput Biol.* 8:e1002837.

Groh A, de Kock CP, Wimmer VC, Sakmann B, Kuner T (2008). Driver or coincidence detector: modal switch of a corticothalamic giant synapse controlled by spontaneous activity and short-term depression. *J Neurosci.* 28:9652–9663.

Harris KD, Shepherd GM (2015). The neocortical circuit: themes and variations. *Nat Neurosci.* 18:170–181.

Karube F, Sari K, Kisvarday ZF (2017). Axon topography of layer 6 spiny cells to orientation map in the primary visual cortex of the cat (area 18). *Brain Struct Funct.* 222:1401–1426.

Larkum ME, Zhu JJ, Sakmann B (1999). A new cellular mechanism for coupling inputs arriving at different cortical layers. *Nature.* 398:338–341.

Lund JS (1988). Anatomical organization of macaque monkey striate visual cortex. *Annu Rev Neurosci.* 11:253–288.

Mao T, Kusefoglu D, Hooks BM, Huber D, Petreanu L, Svoboda K (2011). Long-range neuronal circuits underlying the interaction between sensory and motor cortex. *Neuron.* 72:111–123.

Mease RA, Kuner T, Fairhall AL, Groh A (2017). Multiplexed spike coding and adaptation in the thalamus. *Cell Rep.* 19:1130–1140.

Mease RA, Metz M, Groh A (2016a). Cortical sensory responses are enhanced by the higher-order thalamus. *Cell Rep.* 14:208–215.

Mease RA, Sumser A, Sakmann B, Groh A (2016b). Corticothalamic spike transfer via the L5B-POm pathway in vivo. *Cereb Cortex.* 26:3461–3475.

Mease RA, Sumser A, Sakmann B, Groh A (2016c). Cortical dependence of whisker responses in posterior medial thalamus in vivo. *Cereb Cortex.* 26:3534–3543.

Meyer HS, Egger R, Guest JM, Foerster R, Reissl S, Oberlaender M (2013). Cellular organization of cortical barrel columns is whisker-specific. *Proc Natl Acad Sci U S A.* 110:19113–19118.

Moore JD, Mercer Lindsay N, Deschenes M, Kleinfeld D (2015). Vibrissa self-motion and touch are reliably encoded along the same somatosensory pathway from brainstem through thalamus. *PLoS Biol.* 13:e1002253.

Narayanan RT, Egger R, Johnson AS, Mansvelder HD, Sakmann B, de Kock CP, Oberlaender M (2015). Beyond columnar organization: cell type- and target layer-specific principles of horizontal axon projection patterns in rat vibrissal cortex. *Cereb Cortex.* 25:4450–4468.

Narayanan RT, Udvary D, Oberlaender M (2017). Cell type-specific structural organization of the six layers in rat barrel cortex. *Front Neuroanat.* 11:91.

Oberlaender M, de Kock CP, Bruno RM, Ramirez A, Meyer HS, Dercksen VJ, Helmstaedter M, Sakmann B (2012a). Cell type-specific three-dimensional structure of thalamocortical circuits in a column of rat vibrissal cortex. *Cereb Cortex.* 22:2375–2391.

Oberlaender M, Ramirez A, Bruno RM (2012b). Sensory experience restructures thalamocortical axons during adulthood. *Neuron.* 74:648–655.

Oda S, Kishi K, Yang J, Chen S, Yokofujita J, Igarashi H, Tanihata S, Kuroda M (2004). Thalamocortical projection from the ventral posteromedial nucleus sends its collaterals to layer I of the primary somatosensory cortex in rat. *Neurosci Lett.* 367:394–398.

Petreanu L, Mao T, Sternson SM, Svoboda K (2009). The subcellular organization of neocortical excitatory connections. *Nature.* 457:1142–1145.

Qi J, Ye C, Naskar S, Inacio AR, Lee S (2022). Posteromedial thalamic nucleus activity significantly contributes to perceptual discrimination. *PLoS Biol.* 20:e3001896.

Rah JC, Bas E, Colonell J, Mishchenko Y, Karsh B, Fetter RD, Myers EW, Chklovskii DB, Svoboda K, Harris TD, Isaac JT (2013). Thalamocortical input onto layer 5 pyramidal neurons measured using quantitative large-scale array tomography. *Front Neural Circuits.* 7:177.

Rockel AJ, Hiorns RW, Powell TP (1980). The basic uniformity in structure of the neocortex. *Brain.* 103:221–244.

Rojas-Piloni G, Guest JM, Egger R, Johnson AS, Sakmann B, Oberlaender M (2017). Relationships between structure, in vivo function and long-range axonal target of cortical pyramidal tract neurons. *Nat Commun.* 8(1):870.

Schiller J, Schiller Y, Stuart G, Sakmann B (1997). Calcium action potentials restricted to distal apical dendrites of rat neocortical pyramidal neurons. *J Physiol.* 505 (Pt 3):605–616.

Takahashi N, Ebner C, Sigl-Glockner J, Moberg S, Nierwetberg S, Larkum ME (2020). Active dendritic currents gate descending cortical outputs in perception. *Nat Neurosci.* 23:1277–1285.

Takahashi N, Oertner TG, Hegemann P, Larkum ME (2016). Active cortical dendrites modulate perception. *Science.* 354:1587–1590.

Udvary D, Harth P, Macke JH, Hege HC, de Kock CPJ, Sakmann B, Oberlaender M (2022). The impact of neuron morphology on cortical network architecture. *Cell Rep.* 39(2):110677.

Velez-Fort M, Rousseau CV, Niedworok CJ, Wickersham IR, Rancz EA, Brown AP, Strom M, Margrie TW (2014). The stimulus selectivity and connectivity of layer six principal cells reveals cortical microcircuits underlying visual processing. *Neuron.* 83 :1431–1443.

Welker C (1976). Receptive fields of barrels in the somatosensory neocortex of the rat. *J Comp Neurol.* 166:173–189.

Williams LE, Holtmaat A (2019). Higher-order thalamocortical inputs gate synaptic long-term potentiation via disinhibition. *Neuron.* 101:91–102.

Wimmer VC, Bruno RM, de Kock CP, Kuner T, Sakmann B (2010). Dimensions of a projection column and architecture of VPM and POm axons in rat vibrissal cortex. *Cereb Cortex.* 20:2265–2276.

Woolsey TA, Van der Loos H (1970). The structural organization of layer IV in the somatosensory region (SI) of mouse cerebral cortex: The description of a cortical field composed of discrete cytoarchitectonic units. *Brain Res.* 17:205–242.

Yu C, Derdikman D, Haidarliu S, Ahissar E (2006). Parallel thalamic pathways for whisking and touch signals in the rat. *PLoS Biol.* 4:e124.

7.

BIOPHYSICAL PROPERTIES OF CORTICAL EXCITATORY NEURONS

Lucy M. Palmer

INTRODUCTION

At the heart of cognitive processes is the cortex. Therefore, understanding cognition requires in-depth knowledge of the cortex and the neurons which reside in this brain structure. Within the cortex, there are a vast number of different classes of neurons, which have different input and output patterns, as well as different biophysical properties. These characteristics endow cortical neurons with powerful integrative capabilities which increase overall neural signaling. In this chapter, the biophysical properties of cortical excitatory neurons will be explored, detailing (1) the morphology of cortical excitatory neurons, (2) ion channels and their distribution in cortical neurons, (3) dendritic integration in cortical dendrites (including calcium spike, NMDA spikes, sodium spikes/spikelets), (4) dendritic integration of thalamic input, and (5) firing modes of cortical excitatory neurons. Although this chapter cannot present all known biophysical properties of cortical excitatory neurons, it provides an overview of important processes that contribute to signaling within cortical neurons.

MORPHOLOGY OF CORTICAL EXCITATORY NEURONS

Cortical neurons are endowed with elaborate morphologies which are largely defined by their dendritic tree. However, it is difficult to discuss the morphology of cortical neurons, and their dendrites, without first acknowledging the highly laminar structure of the cortex (see **Chapter 6 in Section III**) which determines the size and location of the neurons which reside in the different layers (Radnikow and Feldmeyer, 2018) (Figure 7.1). Except for layer 1, all of the remaining cortical layers (layers 2–6) contain excitatory neurons. These pyramidal neurons are the principal cell type in the cortex comprising approximately 70%–85% of the entire neuronal population (DeFelipe and Farinas, 1992).

Defined by their dendritic morphology and somatic location within the cortical layers, the different pyramidal neuron subclasses are glutamatergic and have spines for receiving synaptic input along the length of their dendrites. Dendrites can be categorized into distinct domains depending on their cell type and morphological characteristics. Basal dendrites project from the base of the soma, whereas apical dendrites project from the somatic apex, and oblique and tuft dendrites extend from the proximal and distal region of the apical dendrite, respectively (Figure 7.2). Since dendrites within a single neuron are located in spatially distinct regions of the brain structure in which they reside, they typically receive different input streams of information (Figure 7.3). Basal dendrites receive the majority of synaptic inputs (Larkman, 1991) which largely carry feedforward information from the thalamus (Felleman and Van Essen, 1991). In contrast, tuft dendrites largely receive long-range feedback input from recurrent excitatory connections and other cortical areas, the thalamus (including the posterior medial nucleus [POm] of the thalamus) (Rubio-Garrido et al., 2009), the secondary somatosensory cortex (Cauller et al., 1998), and parahippocampal structures (Witter and Groenewegen, 1986). This separation of input pathways which target different locations within a single cortical pyramidal neuron would, in theory, provide individual neurons with a cellular mechanism to simultaneously or independently process different information pathways. By providing multiple independent integrative units, dendritic compartmentalization in cortical neurons may ultimately act to increase the computational power and integrative capabilities of individual neurons (Koch et al., 1982; Beaulieu-Laroche et al., 2018; Godenzini et al., 2022).

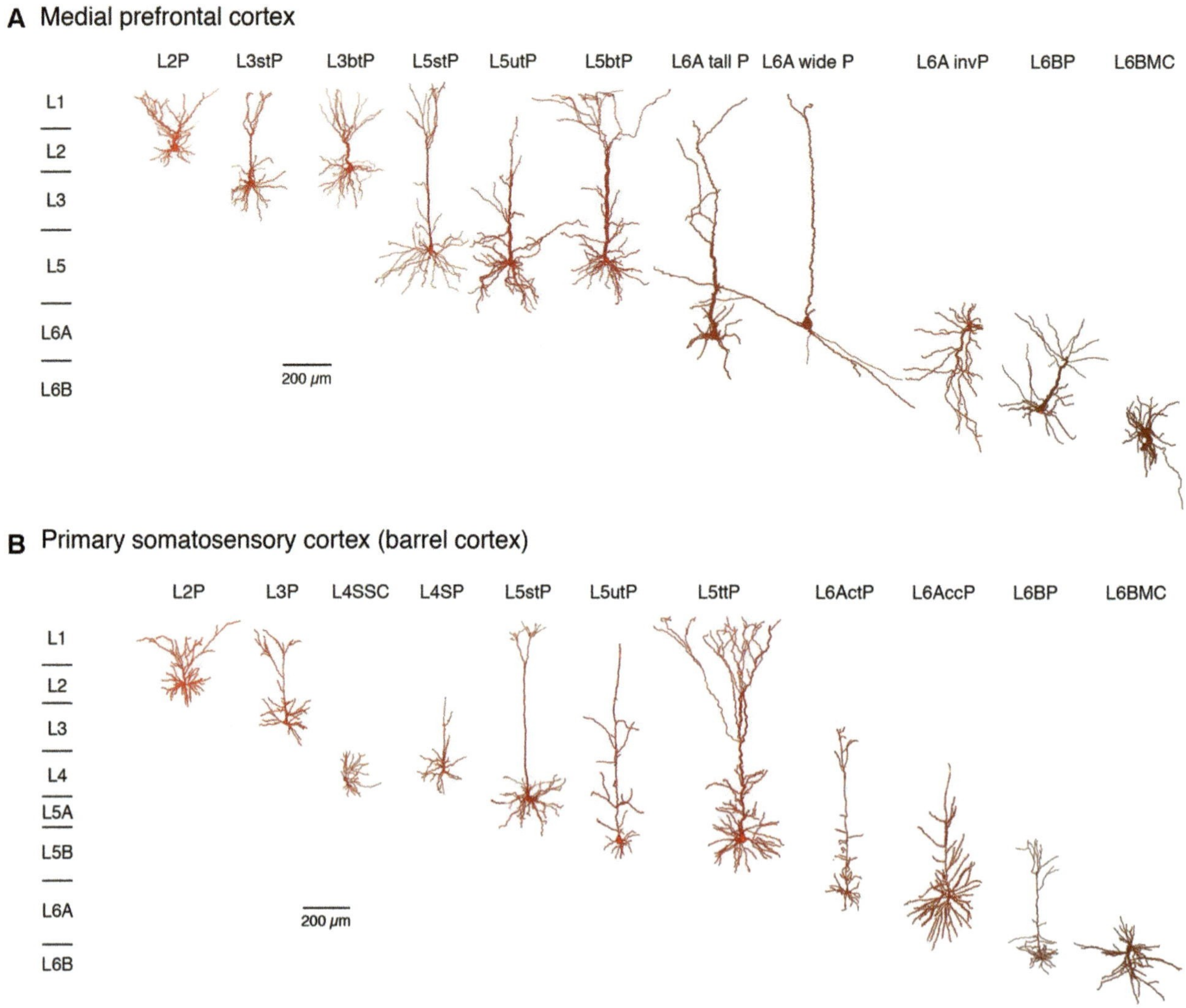

Figure 7.1. Morphology of excitatory cortical pyramidal neurons. The morphology of excitatory neuron types in the (A) medial prefrontal and (B) barrel cortex according to the cortical layer. Figure adapted from Radnikow and Feldmeyer (2018).

ION CHANNELS AND THEIR DISTRIBUTION IN CORTICAL NEURONS

Our understanding of cortical neurons, and how they receive and integrate incoming information, has greatly increased following the characterization of many dendritic voltage-gated ion channels (Migliore and Shepherd, 2002). The main ion channels that contribute to neural excitability, that is, sodium, calcium, potassium, and hyperpolarization-activated cation channels (HCN channels), have been well characterized.

Similar to hippocampal neurons, TTX-sensitive sodium channels have been recorded at a relatively uniform density from the soma to the tuft dendrites (Stuart and Sakmann, 1994). Sodium channels play a crucial role in supporting backpropagating action potentials into the dendrite (Colbert et al., 1997; Stuart et al., 1997), which has been shown to be important for spike-timing dependent plasticity (Letzkus et al., 2006; Sjostrom and Hausser, 2006). Backpropagating action potentials are also influenced by I_A and I_{KD} -type K^+ channels (Harnett et al., 2013) which have been shown to have a density that is either relatively constant throughout the dendritic arbor (Bekkers, 2000a, 2000b) or decreases with increasing distance from the soma (Schaefer et al., 2007). Potassium conductances also act to modulate dendritic branch excitability and synaptic strength (Harnett et al., 2013), making them an important modulator of signaling within cortical neurons. Likewise, Ih conductance mediated by HCN channels have also been shown to influence synaptic integration in cortical neurons, leading to a decrease in the generation and spread of dendritic activity (Stuart and Spruston, 1998; Kole et al., 2006; Harnett et al., 2015), with densities that are elevated into the apical tuft regions (Harnett et al., 2015). Although the density of calcium channels along the dendritic axis of cortical neurons is currently unknown, evidence for the presence of different subtypes of calcium channels (including T-, N-, P/Q-, L-type) have been found across the entire dendritic arbor (Markram et al., 1995; Schiller et al., 1997; Perez-Garci et al., 2013). Taken together, cortical neurons are endowed with different distributions of ion channels, which plays a crucial role in determining the impact of a given synaptic input and shaping neuronal excitability.

DENDRITIC INTEGRATION

Historically, dendrites were often treated as passive structures which collect synaptic signals and relay them to the action potential initiation zone in the axon (Llinás et al., 1968). In seminal work over half a century ago, Wilfrid Rall showed that passive dendritic filtering properties prolong the time window for synaptic summation of distal inputs, enabling greater integration (Rall, 1967). However, it is now well established that dendrites contain active conductances which support the propagation of the somatic action potential back into the dendritic tree (Stuart and Sakmann, 1994a; Johnston et al., 1999) and transform synaptic activity into large nonlinear voltage events. These so-called dendritic spikes or plateau potentials (for a review, see Larkum and Nevian, 2008) influence neuronal computation not only at the level of a single neuron, but also at the level of the neural network. In cortical neurons, dendritic spikes can be mediated by different voltage-gated or ligand-gated channels (Figure 7.2).

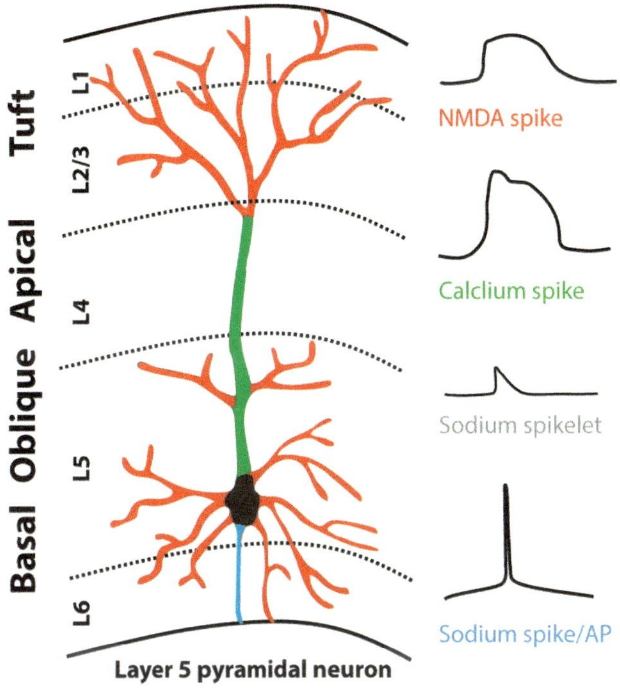

Figure 7.2. The generation of different dendritic spikes in excitatory cortical pyramidal neurons. Schematic of a layer 5 pyramidal neuron illustrating the different dendritic regions (tuft, apical, oblique, basal) and the dendritic location where the different dendritic spikes are generated (red = NMDA spike; green = calcium spike; blue = sodium spike/action potential [AP]). Note, sodium spikelets occur in all dendritic branches.

CALCIUM SPIKE

Local input onto the distal apical dendrite generates all-or-none calcium spikes (Larkum et al., 1999) that are mediated primarily by L-type Ca channels (Perez-Garci et al., 2013). Occurring predominantly in layer 5 pyramidal neurons, and, to a lesser extent, layer 2/3 pyramidal neurons, calcium spikes are generated at a specialized initiation zone, located in the distal apical dendrite, which has a high density of voltage-dependent calcium channels (Schiller et al., 1997). Once evoked, calcium spikes have a considerably longer duration than somatic sodium spikes and have been shown to play an important role in cellular signaling. First, they have a very powerful effect on the firing of the neuron, switching the firing pattern to burst-firing mode (Larkum and Zhu, 2002). Furthermore, since integration with backpropagating action potentials can lower the threshold for their initiation, calcium spikes are thought to act as a coincidence-detection mechanism for inputs arriving at different dendritic locations (Larkum et al., 1999). Since calcium spikes can have a powerful influence on cellular firing patterns, it is reasonable to predict that calcium spikes may play a crucial role in important brain processes. By investigating the occurrence of calcium spikes in the awake behaving animal, studies have shown that calcium spikes within cortical pyramidal neurons play an important role in sensory detection (Xu et al., 2012), perception (Manita et al., 2015; Takahashi et al., 2016), visual encoding (Francioni et al., 2019), navigation (Voigts and Harnett, 2020), decision-making (Kerlin et al., 2019), and learning (Doron et al., 2020).

NMDA SPIKES

The distal tuft dendrite of neocortical pyramidal neurons is electrically remote from the axonal action potential initiation zone and yet receives a large proportion of synaptic inputs. To counteract the strong attenuation of distal input along the apical dendrite, basal and apical tuft can generate so-called NMDA spikes (Polsky et al., 2004). NMDA spikes, as the name implies, are generated by the regenerative activation of N-methyl-D-aspartate (NMDA) receptor channels (after relief from voltage-dependent Mg^{2+} block), resulting in a large nonlinear voltage response to highly localized synaptic input. Although individual NMDA spikes do not have a great impact on the firing at the soma, they increase the electrical coupling of fine dendrites to the soma (Nevian et al., 2007), as well as contributing to reaching threshold for action potential generation in the axon (Larkum et al., 2009). Further, NMDA spikes have been recorded in human pyramidal neurons (Testa-Silva et al., 2022), illustrating their potential importance in cellular signaling in the human brain. Taken together, direct recordings in vitro and in computer simulations have shown that NMDA spikes can, in principle, serve as an effective mechanism for coupling distal synaptic input to neuronal output in vivo. Indeed, NMDA spikes have been shown to occur in cortical pyramidal neurons in vivo, encoding sensory information (Lavzin et al., 2012; Palmer et al., 2014) and cortical map reorganization (Pages et al., 2021). Overall, NMDA spikes act as a local integrative mechanism for transferring information received on thin dendrites to the soma, and therefore may act as a gateway for information within and between neurons.

SODIUM SPIKES/SPIKELETS

The generation of local sodium spikes, also known as fast prepotentials (Spencer and Kandel, 1961), has been shown in most dendritic regions of cortical pyramidal neurons (Nevian et al., 2007; Kalmbach et al., 2017). Similar to the sodium-dependent action potentials generated in the soma, dendritic sodium spikes have a fast rise time and brief duration. However, in contrast to other dendritic spikes, this means that sodium spikelets are strongly attenuated along the dendritic branch (Nevian et al., 2007), and therefore do not influence the generation of action potentials at the soma. Instead, they are believed to play a role in integrating synchronous synaptic input and synaptic plasticity (Frick and Johnston, 2005). During behavior, sodium spikelets have been shown to be important for orientation tuning (Smith et al., 2013) and movement direction (Moore et al., 2017).

OPTIMAL STIMULI FOR ACTIVATING REGENERATIVE PROCESSES

The generation of dendritic spikes requires synaptic input to arrive onto the dendritic branch in both a spatially and temporally restricted manner. For example, to generate NMDA spikes, computational modeling has illustrated that approximately 8 synaptic inputs need to arrive simultaneously onto a 35 μm region of the dendritic branch (Palmer et al., 2014). Furthermore, the generation of multiple NMDA spikes has been shown to potentially lead to the generation of a calcium spike, which then leads to additional action potential generation at the soma (Xu et al., 2012).

DENDRITIC INTEGRATION OF THALAMIC INPUT

The thalamus is regarded as the gateway to the cortex due to its strategic central position and extensive connections with virtually every cortical and subcortical area (Oh et al., 2014). The thalamus comprises a multitude of nuclei that attend to different sensory modalities and higher-order cognition. First-order nuclei (also known as primary relay) transfer modality-specific sensory information directly from the periphery to primary sensory areas in the cortex (Bickford, 2015). This pathway to the cortex through the first-order thalamic nuclei has been termed *feedforward*, or *bottom-up*, input and largely targets the middle cortical layers (Oh et al., 2014). Located next to the first-order nuclei, the higher-order thalamic nuclei are driven primarily by cortical inputs and to a lesser extent by peripheral sensory inputs (see **Chapter 4 in Section II**). This pathway has been termed *feedback*, or *top-down*, input and has been shown to target the upper layers of the cortex (Wimmer et al., 2010). While the goal of primary thalamic nuclei is to faithfully transfer sensory information to the cortex, the higher-order thalamus can act as a coincidence detector of cortical and peripheral inputs (Bickford, 2015).

Broadly speaking, the first-order and higher-order thalamic nuclei specifically target different regions on an individual cortical pyramidal neuron (Figure 7.3). Combining our knowledge about the integration of information in the different dendritic regions (described above) and the cortical layers targeted by the different thalamic nuclei, this would result in the different thalamic input having a different influence on the various integrative processes within a single cortical neuron. For example, the feedforward first-order thalamic input pathway targets the apical, oblique, and basal dendrites of pyramidal neurons. Since these dendritic locations are electrotonically closer to the soma and axon, this thalamic input would have, in theory, greater impact on the generation of axonal action potentials. In contrast, higher-order thalamic nuclei target the upper layers of the cortex where the tuft and distal apical dendrites of excitatory layer 2/3 and layer 5 pyramidal neurons reside. Since tuft and distal apical dendrites generate local NMDA spikes and global calcium spikes respectively, this feedback thalamic input pathway may have a direct influence on these dendritic integrative voltage events. This was illustrated by Gambino et al. (2014), who elegantly illustrated that NMDA-mediated long-lasting depolarizations, akin to dendritic plateau potentials/spikes, in layer 2/3 pyramidal neurons within the barrel cortex, were generated by the higher-order thalamic nuclei subserving the somatosensory cortex, the posteriomedial nucleus of the thalamus (POm). Here, inhibition of POm activity during rhythmic whisker stimulation suppressed the generation of these potentials (Gambino et al., 2014). Furthermore, cortical sensory responses in cortical pyramidal neurons within the barrel cortex were influenced by thalamocortical projections from both higher-order thalamus (Mease et al., 2016) and first-order thalamus (Constantinople and Bruno, 2013).

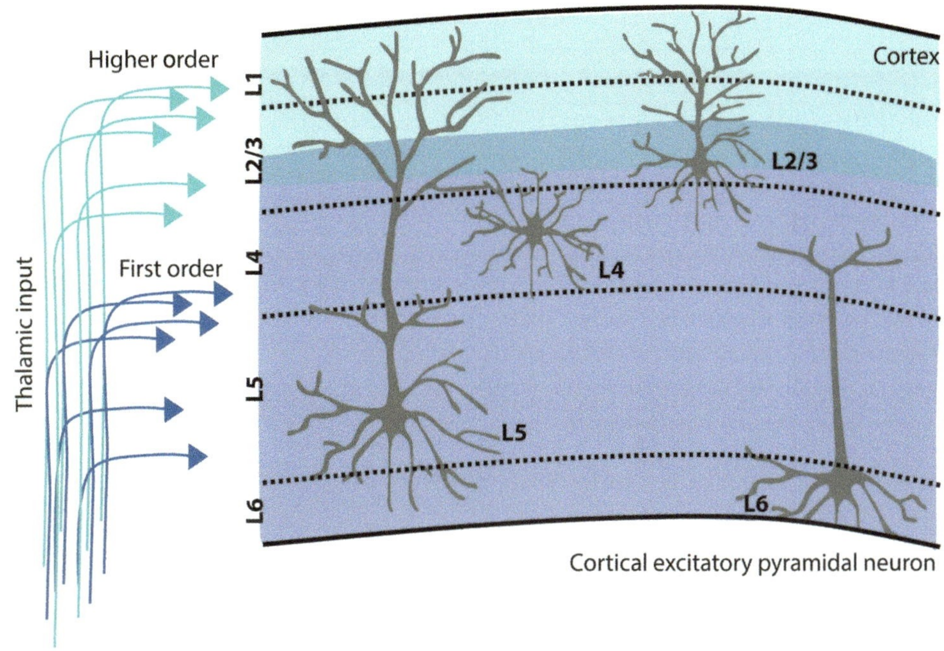

Figure 7.3. Thalamic input to the cortex and the different excitatory pyramidal neurons. First-order thalamic input into the cortex predominantly targets layer 4 (L4), whereas higher-order thalamic input predominantly targets the upper cortical layers. The spatial organization of the different input pathways targets different dendritic branches of excitatory cortical pyramidal neurons.

Thalamic input to cortical excitatory neurons is further complicated not only by different dendritic targets of the direct connectivity, but also by the different microcircuits activated. Higher-order thalamic input has been shown to target various interneuron subtypes within the cortex, which in turn synapse with excitatory neurons (Audette et al., 2018). Taking into the account the different cortical layers and neurons targeted by the different thalamic nuclei, the dendritic integration of thalamic input within cortical networks is difficult to unravel, and is an exciting area of future research.

FIRING MODES OF CORTICAL EXCITATORY NEURONS

At rest, individual excitatory cortical neurons have low firing rates. Typically, cortical neurons are spontaneously active at rates less than 1 Hz (Brecht and Sakmann, 2002). Additionally, in response to sensory input, cortical neurons also have low rates of firing, and are therefore often referred to as sparse encoders. For example, neurons in primary auditory cortex produce, at most, a single spike in response to a sound (DeWeese et al., 2003). These low rates of action potentials, during both spontaneous and evoked activity, are thought to be a mechanism for neurons to conserve energy (Lennie, 2003). However, cortical neurons have also been shown to generate bursts of action potentials which have high fidelity of signaling in synaptically coupled cells leading to learning (Payeur et al., 2021). As discussed above, local input onto the distal apical dendrite of cortical neurons can generate all-or-none calcium spikes, which have a very powerful effect on the firing of the neuron, switching the firing pattern to burst mode (Larkum and Zhu, 2002). Therefore, since they signal a "special" large voltage event being generated in the dendrite, burst firing in cortical neurons is a powerful and meaningful signal. Shown to occur in response to sensory information (Xu et al., 2012; Takahashi et al., 2016), dendritic calcium spikes and subsequent burst firing may act to increase the efficiency of sensory encoding. However, interestingly, burst firing is more prominent in layer 5 than in layer 2/3 pyramidal neurons (Waters et al., 2003) which may reflect different signaling in the different layers.

CONCLUDING REMARKS

The task of understanding how neurons translate input to output is central to explaining brain function. Since the majority of inputs arrive at the dendrites of neurons, it is critical to understand the processing performed by dendritic trees that influences action potential output. This can be achieved by looking at different levels of detail in a single neuron, from the activity in a single spine to an entire dendritic arbor. Historically, even though dendrites of various neurons were shown to have active membranes, dendrites were often treated as non-active structures that collected synaptic signals and relayed them passively to the axonal action potential initiation zone. However, it is now well established that dendrites have active conductances and specialized biophysical characteristics which support various cellular processes and input transformations. We are at the tip of the iceberg in our understanding of cortical computation, and extensive future research is required to unravel how the biophysical properties of individual neurons combine with the various input patterns to generate overall brain function. Much of what we currently know about the biophysical properties of cortical neurons is based on in vitro work. Although insightful, future efforts will need to delve into the intact brain to truly understand the contribution and modulation of cortical computation during cognitive behavior.

REFERENCES

Audette NJ, Urban-Ciecko J, Matsushita M, Barth AL (2018). POm thalamocortical input drives layer-specific microcircuits in somatosensory cortex. *Cereb Cortex*. 28:1312–1328.

Beaulieu-Laroche L, Toloza EHS, van der Goes MS, Lafourcade M, Barnagian D, Williams ZM, Eskandar EN, Frosch MP, Cash SS, Harnett MT (2018). Enhanced dendritic compartmentalization in human cortical neurons. *Cell*. 175:643–651.

Bekkers JM (2000a). Distribution and activation of voltage-gated potassium channels in cell-attached and outside-out patches from large layer 5 cortical pyramidal neurons of the rat. *J Physiol*. 525(Pt 3):611–620.

Bekkers JM (2000b). Properties of voltage-gated potassium currents in nucleated patches from large layer 5 cortical pyramidal neurons of the rat. *J Physiol*. 525(Pt 3):593–609.

Bickford ME (2015). Thalamic circuit diversity: Modulation of the driver/modulator framework. *Front Neural Circuits*. 9:86.

Brecht M, Sakmann B (2002). Dynamic representation of whisker deflection by synaptic potentials in spiny stellate and pyramidal cells in the barrels and septa of layer 4 rat somatosensory cortex. *J Physiol*. 543:49–70.

Cauller LJ, Clancy B, Connors BW (1998). Backward cortical projections to primary somatosensory cortex in rats extend long horizontal axons in layer I. *J Comp Neurol*. 390:297–310.

Colbert CM, Magee JC, Hoffman DA, Johnston D (1997). Slow recovery from inactivation of Na+ channels underlies the activity-dependent attenuation of dendritic action potentials in hippocampal CA1 pyramidal neurons. *J Neurosci*. 17:6512–6521.

Constantinople CM, Bruno RM (2013). Deep cortical layers are activated directly by thalamus. *Science*. 340:1591–1594.

DeFelipe J, Farinas I (1992). The pyramidal neuron of the cerebral cortex: Morphological and chemical characteristics of the synaptic inputs. *Prog Neurobiol*. 39:563–607.

DeWeese MR, Wehr M, Zador AM (2003). Binary spiking in auditory cortex. *J Neurosci*. 23:7940–7949.

Doron G, Shin JN, Takahashi N, Druke M, Bocklisch C, Skenderi S, de Mont L, Toumazou M, Ledderose J, Brecht M, Naud R, Larkum ME (2020). Perirhinal input to neocortical layer 1 controls learning. *Science*. 370:eaaz3136.

Felleman DJ, Van Essen DC (1991). Distributed hierarchical processing in the primate cerebral cortex. *Cereb Cortex*. 1:1–47.

Francioni V, Padamsey Z, Rochefort NL (2019). High and asymmetric somato-dendritic coupling of V1 layer 5 neurons independent of visual stimulation and locomotion. *eLife*. 8:e49145.

Frick A, Johnston D (2005). Plasticity of dendritic excitability. *J Neurobiol*. 64:100–115.

Gambino F, Pages S, Kehayas V, Baptista D, Tatti R, Carleton A, Holtmaat A (2014). Sensory-evoked LTP driven by dendritic plateau potentials in vivo. *Nature*. 515:116–119.

Godenzini L, Shai AS, Palmer LM (2022). Dendritic compartmentalization of learning-related plasticity. *eNeuro*. 9.

Harnett MT, Magee JC, Williams SR (2015). Distribution and function of HCN channels in the apical dendritic tuft of neocortical pyramidal neurons. *J Neurosci*. 35:1024–1037.

Harnett MT, Xu NL, Magee JC, Williams SR (2013). Potassium channels control the interaction between active dendritic integration compartments in layer 5 cortical pyramidal neurons. *Neuron*. 79:516–529.

Kalmbach BE, Gray R, Johnston D, Cook EP (2017). Systems-based analysis of dendritic nonlinearities reveals temporal feature extraction in mouse L5 cortical neurons. *J Neurophysiol*. 117:2188–2208.

Kerlin A, Mohar B, Flickinger D, MacLennan BJ, Dean MB, Davis C, Spruston N, Svoboda K (2019). Functional clustering of dendritic activity during decision-making. *eLife*. 8.

Koch C, Poggio T, Torre V (1982). Retinal ganglion cells: A functional interpretation of dendritic morphology. *Philos Trans R Soc Lond B Biol Sci*. 298:227–263.

Kole MH, Hallermann S, Stuart GJ (2006). Single Ih channels in pyramidal neuron dendrites: Properties, distribution, and impact on action potential output. *J Neurosci*. 26:1677–1687.

Larkman AU (1991). Dendritic morphology of pyramidal neurones of the visual cortex of the rat: I. Branching patterns. *J Comp Neurol*. 306:307–319.

Larkum ME, Nevian T (2008). Synaptic clustering by dendritic signalling mechanisms. *Curr Opin Neurobiol*. 18:321–331.

Larkum ME, Nevian T, Sandler M, Polsky A, Schiller J (2009). Synaptic integration in tuft dendrites of layer 5 pyramidal neurons: a new unifying principle. *Science*. 325:756–760.

Larkum ME, Zhu JJ (2002). Signaling of layer 1 and whisker-evoked Ca2+ and Na+ action potentials in distal and terminal dendrites of rat neocortical pyramidal neurons in vitro and in vivo. *J Neurosci*. 22:6991–7005.

Larkum ME, Zhu JJ, Sakmann B (1999). A new cellular mechanism for coupling inputs arriving at different cortical layers. *Nature*. 398:338–341.

Lavzin M, Rapoport S, Polsky A, Garion L, Schiller J (2012). Nonlinear dendritic processing determines angular tuning of barrel cortex neurons in vivo. *Nature*. 490:397–401.

Lennie P (2003). The cost of cortical computation. *Curr Biol*. 13:493–497.

Letzkus JJ, Kampa BM, Stuart GJ (2006). Learning rules for spike timing-dependent plasticity depend on dendritic synapse location. *J Neurosci*. 26:10420–10429.

Manita S, Suzuki T, Homma C, Matsumoto T, Odagawa M, Yamada K, Ota K, Matsubara C, Inutsuka A, Sato M, Ohkura M, Yamanaka A, Yanagawa Y, Nakai J, Hayashi Y, Larkum ME, Murayama M (2015). A top-down cortical circuit for accurate sensory perception. *Neuron*. 86:1304–1316.

Markram H, Helm PJ, Sakmann B (1995). Dendritic calcium transients evoked by single back-propagating action potentials in rat neocortical pyramidal neurons. *J Physiol*. 485 (Pt 1):1–20.

Mease RA, Metz M, Groh A (2016). Cortical sensory responses are enhanced by the higher-order thalamus. *Cell Rep*. 14:208–215.

Migliore M, Shepherd GM (2002). Emerging rules for the distributions of active dendritic conductances. *Nat Rev Neurosci*. 3:362–370.

Moore JJ, Ravassard PM, Ho D, Acharya L, Kees AL, Vuong C, Mehta MR (2017). Dynamics of cortical dendritic membrane potential and spikes in freely behaving rats. *Science*. 355.

Nevian T, Larkum ME, Polsky A, Schiller J (2007). Properties of basal dendrites of layer 5 pyramidal neurons: A direct patch-clamp recording study. *Nat Neurosci*. 10:206–214.

Oh SW et al. (2014). A mesoscale connectome of the mouse brain. *Nature*. 508:207–214.

Pages S, Chenouard N, Chereau R, Kouskoff V, Gambino F, Holtmaat A (2021). An increase in dendritic plateau potentials is associated with experience-dependent cortical map reorganization. *Proc Natl Acad Sci U S A*. 118.

Palmer LM, Shai AS, Reeve JE, Anderson HL, Paulsen O, Larkum ME (2014). NMDA spikes enhance action potential generation during sensory input. *Nat Neurosci*. 17:383–390.

Payeur A, Guerguiev J, Zenke F, Richards BA, Naud R (2021). Burst-dependent synaptic plasticity can coordinate learning in hierarchical circuits. *Nat Neurosci*. 24:1010–1019.

Perez-Garci E, Larkum ME, Nevian T (2013). Inhibition of dendritic Ca2+ spikes by GABAB receptors in cortical pyramidal neurons is mediated by a direct Gi/o-beta-subunit interaction with Cav1 channels. *J Physiol*. 591:1599–1612.

Polsky A, Mel BW, Schiller J (2004). Computational subunits in thin dendrites of pyramidal cells. *Nat Neurosci*. 7:621–627.

Radnikow G, Feldmeyer D (2018). Layer- and Cell Type-Specific Modulation of Excitatory Neuronal Activity in the Neocortex. *Front Neuroanat*. 12:1.

Rall W (1967). Distinguishing theoretical synaptic potentials computed for different soma-dendritic distributions of synaptic input. *J Neurophysiol*. 30:1138–1168.

Rubio-Garrido P, Perez-de-Manzo F, Porrero C, Galazo MJ, Clasca F (2009). Thalamic input to distal apical dendrites in neocortical layer 1 is massive and highly convergent. *Cereb Cortex*. 19:2380–2395.

Schaefer AT, Helmstaedter M, Schmitt AC, Bar-Yehuda D, Almog M, Ben-Porat H, Sakmann B, Korngreen A (2007). Dendritic voltage-gated K+ conductance gradient in pyramidal neurones of neocortical layer 5B from rats. *J Physiol*. 579:737–752.

Schiller J, Schiller Y, Stuart G, Sakmann B (1997). Calcium action potentials restricted to distal apical dendrites of rat neocortical pyramidal neurons. *J Physiol*. 505 (Pt 3):605–616.

Sjostrom PJ, Hausser M (2006). A cooperative switch determines the sign of synaptic plasticity in distal dendrites of neocortical pyramidal neurons. *Neuron*. 51:227–238.

Smith SL, Smith IT, Branco T, Hausser M (2013). Dendritic spikes enhance stimulus selectivity in cortical neurons in vivo. *Nature*. 503:115–120.

Spencer WA, Kandel ER (1961). Electrophysiology of hippocampal neurons: III. Firing level and time constant. *J Neurophysiol*. 24:260–271.

Stuart G, Spruston N (1998). Determinants of voltage attenuation in neocortical pyramidal neuron dendrites. *J Neurosci*. 18:3501–3510.

Stuart G, Spruston N, Sakmann B, Hausser M (1997). Action potential initiation and backpropagation in neurons of the mammalian CNS. *Trends Neurosci*. 20:125–131.

Stuart GJ, Sakmann B (1994). Active propagation of somatic action potentials into neocortical pyramidal cell dendrites. *Nature*. 367:69–72.

Takahashi N, Oertner TG, Hegemann P, Larkum ME (2016). Active cortical dendrites modulate perception. *Science*. 354:1587–1590.

Testa-Silva G, Rosier M, Honnuraiah S, Guzulaitis R, Megias AM, French C, King J, Drummond K, Palmer LM, Stuart GJ (2022). High synaptic threshold for dendritic NMDA spike generation in human layer 2/3 pyramidal neurons. *Cell Rep*. 41:111787.

Voigts J, Harnett MT (2020) Somatic and dendritic encoding of spatial variables in retrosplenial cortex differs during 2D navigation. *Neuron*. 105:237–245.

Waters J, Larkum M, Sakmann B, Helmchen F (2003). Supralinear $Ca2^{+}$ influx into dendritic tufts of layer 2/3 neocortical pyramidal neurons in vitro and in vivo. *J Neurosci*. 23:8558–8567.

Wimmer VC, Bruno RM, de Kock CP, Kuner T, Sakmann B (2010). Dimensions of a projection column and architecture of VPM and POm axons in rat vibrissal cortex. *Cereb Cortex*. 20:2265–2276.

Witter MP, Groenewegen HJ (1986). Connections of the parahippocampal cortex in the cat. III. Cortical and thalamic efferents. *J Comp Neurol*. 252:1–31.

Xu NL, Harnett MT, Williams SR, Huber D, O'Connor DH, Svoboda K, Magee JC (2012). Nonlinear dendritic integration of sensory and motor input during an active sensing task. *Nature*. 492:247–251.

8.

FUNCTIONAL ROLES OF CORTICAL INHIBITORY INTERNEURONS

Hyeyoung Shin and Hillel Adesnik

Given the detailed cell type classification of a neuron, how well can we predict its spiking behavior in various sensory, cognitive, and motor contexts? Conversely, given its spiking behavior under a diversity of brain states, sensory stimuli, and behavioral tasks, how well can we predict a neuron's cell type?

A key finding in research on cortical inhibitory interneurons (INs) was the discovery that classifications based on intrinsic physiological properties, morphological characteristics, and molecular profiles are remarkably well aligned (Kubota et al., 2016; Tremblay et al., 2016); that is, a neuron's transcriptome predicts its morphological and electrophysiological properties (Tasic et al., 2018; Gouwens et al., 2019). Another cornerstone discovery was the establishment of connectivity rules within and between distinct cell types. Parvalbumin-expressing INs (PVs) target the soma of their postsynaptic partners and receive highly convergent input from feedforward thalamocortical pathways. Somatostatin-expressing INs (SSTs) target the dendrites and are nearly exclusively connected by recurrent cortical sources, often within the same cortical layer. Finally, vasoactive intestinal peptide expressing INs (VIP) mainly target SSTs and drive net disinhibition of excitatory principal cells (PCs). The local and long-range connectivity patterns form the basis for distinct response properties in vivo. For example, computational models that obey cell-type specific connectivity rules recapitulate sensory response properties (Garcia Del Molino et al., 2017; Keller et al., 2020; Veit et al., 2022). Incredibly, the reverse is also possible: by allowing the neural network model to iteratively converge to a solution that fits the experimental data, known connectivity rules can be recovered from experimentally measured sensory response properties (Palmigiano et al., 2020; Mossing et al., 2021).

To understand INs' contributions to cortical dynamics and computation, a large set of studies in the past decade set out to decipher the functional differences between IN types during various sensory, cognitive, and motor contexts, conditioned on arousal level and brain state. While originally there was a growing consensus that single IN types preferentially contribute to specific computations, it has recently become apparent that this thinking is likely flawed. Put simply, the idealized goal of finding computations that are mediated by a single IN type is proving to be elusive. Recent experimental and theoretical work instead supports a model where several IN classes interact during a given computation (Phillips and Hasenstaub, 2016; Garcia Del Molino et al., 2017; Palmigiano et al., 2020; Mossing et al., 2021). There are at least 3 main reasons for the lack of a one-to-one relationship between IN types and function: first, inhibition cannot act independently of each other because IN types are synaptically interconnected; second, substantial heterogeneity exists even between INs of the same class; third, single neurons exhibit variable input-output transformations depending on brain states. If the field is seeking a "unified model" for the role of IN types in cortical computations, we need a more comprehensive, sophisticated, and quantitative framework that incorporates all the recurrent connections between INs and excitatory neurons. In this chapter, we will elaborate on these complex interactions and suggest paths to advance our understanding of the relationship between IN type and function to arrive at a more holistic model of IN contributions to cortical computation.

CLASSIFICATION: MORPHOLOGY, ELECTROPHYSIOLOGY, TRANSCRIPTOMICS, AND CONNECTIVITY

In this chapter, we describe 5 major IN types; PV, SST, VIP, synuclein gamma expressing INs (Sncg), and lysosomal associated membrane protein 5 expressing INs (Lamp5). We detail their morphological, electrophysiological, and transcriptomic characteristics. We elaborate on the connectivity patterns between IN types and PCs. Due to the overwhelming abundance of literature in mice, most of the findings summarized here pertain to the mouse neocortex. Note that homology in IN types has been demonstrated to a limited extent between mice and humans (Hodge et al., 2019).

Ramon y Cajal's pioneering work opened the gateway for morphological classification of cortical neuronal types. Electrophysiological classification of cortical neurons followed decades later, arguably with the seminal finding that excitatory and inhibitory neurons have largely distinct spike waveforms: neurons with narrower spike waveform and fast-spiking behavior were found to be GABAergic inhibitory interneurons (McCormick et al., 1985). The application of molecular biology techniques to brain slices led to the discovery that neocortical inhibitory interneurons are composed of 3 molecular types: parvalbumin (PV), somatostatin (SST), and serotonin receptor 5HT3a (5HT3aR) expressing neurons (Rudy et al., 2011). More recently, large-scale high-throughput transcriptomics revealed distinct clusters that showed good correspondence with previously identified morphological and electrophysiological classifications properties (Tasic et al., 2018; Gouwens et al., 2019). Current consensus divides inhibitory interneurons into 5 major groups; PV, SST, VIP, Sncg, and Lamp5 (Tasic et al., 2018). VIP, Sncg, and Lamp5 were previously grouped as 5HT3aR-expressing INs (Gouwens et al., 2019). The basic characteristics of these 5 IN types are summarized in Table 8.1. Representative morphology for IN types is illustrated in Figure 8.1. In addition, we refer the reader to excellent reviews on neocortical IN types (Kubota et al., 2016; Tremblay et al., 2016).

IN types have consistent characteristics across neocortical areas, providing strong support for the idea of "canonical microcircuits" (Harris and Shepherd, 2015). Interestingly, IN types show higher consistency across neocortical areas than excitatory cells (Tasic et al., 2018). In addition, IN types show characteristic laminar distributions (Tremblay et al., 2016). A neuron's layer is defined based on the soma's location. PVs and SSTs can be found in all layers except layer 1. PVs are most prevalent in layer 4, while SSTs are mostly in other layers. VIPs are predominantly in layers 2/3. Sncg and Lamp5 are mostly supragranular, distributed in layer 1 and layers 2/3 (Tasic et al., 2018). Lamp5 is the most abundant cell type in layer 1. Despite the overall consistency, there are slight variations in INs' distributions across areas. Strikingly, SST/PV cellular ratio increases along the neocortical hierarchy, in the order of motor/somatosensory < audio/visual < medial association < medial prefrontal < lateral network (Wang, 2020).

Perhaps the most important feature of IN types is their stereotypical local connectivity. As described above, IN types show distinct targeting of postsynaptic compartments; PV and Sncg neurons target the perisomatic region, SST

TABLE 8.1 CORTICAL INHIBITORY INTERNEURON TYPES AND THEIR MORPHOLOGICAL AND ELECTROPHYSIOLOGICAL PROPERTIES

Type (%)	Morphology	Electrophysiology	Notes
PV (40%)	Basket (targets soma) Chandelier (targets axon initial segment, i.e., axo-axonic)	Fast-spiking Narrow spike waveforms (<0.4ms between peak and trough of action potential) Little to no spike frequency adaptation Strongly depressing excitatory input Moderately depressing output	Developmental origin: MGE (medial ganglionic eminence) Relatively late development
SST (30%)	Martinotti (axon plexus in layer 1, targets apical tuft dendrites of pyramidal neurons) Non-Martinotti Targets dendrites	Strongly facilitating excitatory inputs Output can be depressing or facilitating	Developmental origin: MGE
VIP (12%)	Bipolar (vertically extending dendrites and axons) Targets dendrites	High input resistance	Developmental origin: CGE (caudal ganglionic eminence) Expresses 5HT3aR (5HT3a receptor)
Sncg	Multipolar Basket Targets soma	Regular or burst spiking	Developmental origin: CGE Expresses 5HT3aR, VIP and CCK (cholectystokinin) Previously grouped with VIP
Lamp5	Neurogliaform (short, radial dendrites and wider, denser spherical axonal plexus, aka spiderweb cells)	Slow, long-lasting IPSPs Volume transmission via extrasynaptic $GABA_B$ receptors The only known IN type to elicit unitary $GABA_B$ responses with a single action potential	Developmental origin: CGE Expresses 5HT3aR, Ndnf (neuron-derived neurotrophic factor)

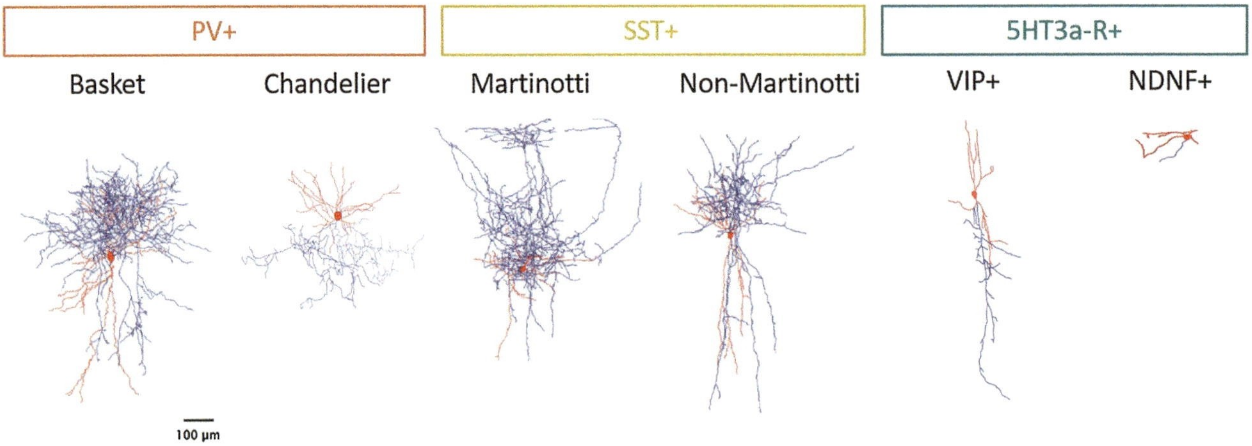

Figure 8.1. Representative morphology of inhibitory interneuron (IN) subtypes. Soma and dendrites are in red, axons are in blue. PVs are subdivided into soma-targeting basket type and axon-targeting chandelier type cells. SSTs are subdivided into Martinotti cells, characterized by dense axonal arborization in layer 1, and non-Martinotti cells. VIPs typically show bipolar morphology, with vertically extending dendrites and axons. Lamp5s, also known as NDNF INs, are neurogliaform cells with short, radial dendrites and wider, denser spherical axonal plexus. Neurogliaform cells were famously described as spiderweb cells by Ramon y Cajal. Reproduced with permission from Studer and Barkat (2022).

and VIP target dendrites, and Lamp5 release GABA extrasynaptically. Remarkably, gap junctions, or electrical synapses, are only shared between interneurons of the same type. Such "homotypic" coupling has been observed within PV and SST neurons (Hestrin and Galarreta, 2005). Less evidence supports homotypic coupling for other IN types, and gap junction specificity within and between IN subtypes (e.g., Martinotti vs. non-Martinotti subtypes of SST) is largely unknown (Tremblay et al., 2016).

Recent advances in microcircuit connectivity mapping have shed light on INs' specificity of postsynaptic cell types (Figure 8.2) (Pfeffer et al., 2013; Jiang et al., 2015; Kubota et al., 2016; Campagnola et al., 2022). VIPs preferentially target SSTs (Pfeffer et al., 2013; Pi et al., 2013). SSTs target nearly all other cell types but largely avoid targeting each other. In contrast, PVs are densely interconnected with each other. PVs and SSTs have similar input and output connection probability with PCs. Note that many of the comprehensive connectivity studies have focused on mouse visual cortex. As such, an interesting future direction would be to survey the similarities and differences in microcircuit connectivity across neocortical areas (see Staiger and Petersen, 2021; Studer and Barkat, 2022 for reviews of connectivity patterns in barrel cortex and auditory cortex, respectively).

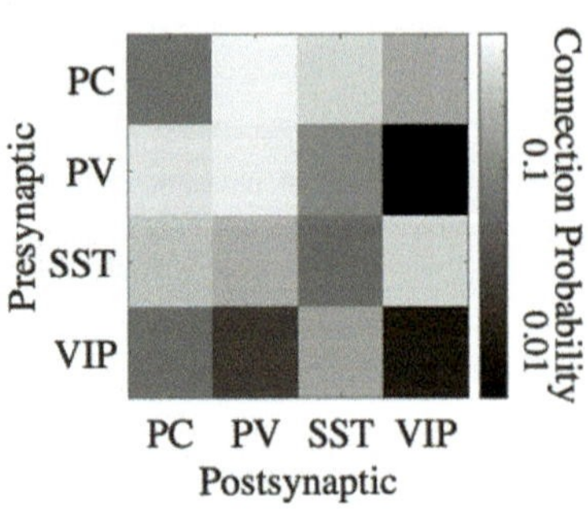

Figure 8.2. Connection probability between INs and excitatory principal cells (PCs). Connection probability is shown in logscale. Of note, PV recurrently inhibits itself whereas SST does not. VIP selectively inhibits SST. Represented data are from mouse V1 layer 2/3 (Campagnola et al., 2022).

NETWORK DYNAMICS: EXCITATION-INHIBITION BALANCE, NEURAL OSCILLATIONS, AND BRAIN STATE

In this section, we discuss how inhibition governs network dynamics. This topic is deeply rooted in theoretical neuroscience. For decades, theoreticians have been analyzing the dynamics of excitatory-inhibitory (EI) networks. An EI network refers to a simple network simulation of excitatory and inhibitory point neurons, discarding details about excitatory and inhibitory cell types and neuronal compartments such as dendrites, soma, and axons. EI network models have been useful for understanding dynamical regimes of the neocortex, including the question of how inhibition balances excitation ("EI balance"). Many theorists have also used EI networks to study neural network oscillations (Tiesinga and Sejnowski, 2009; Buzsáki and Wang, 2012). As our understanding of IN types has progressed, computational models incorporating IN types and their characteristics have gained popularity. These cell-type models of intermediate complexity are particularly useful for interpreting or predicting the effects of cell-type-specific manipulations such as optogenetics (Veit et al., 2017; Veit et al., 2022; Keller et al., 2020; Millman et al., 2020; Mossing et al., 2021). Computational models that are built for explaining experimental results often take a reverse-engineering approach: these models aim to decipher which design principles of the model are necessary and sufficient for recapitulating the experimental observations. The counter approach is to build a highly sophisticated biophysically plausible model incorporating detailed morphology, electrophysiology, biophysics, and connectivity rules (Traub et al., 2005; Billeh et al., 2020). The goal of these highly complex models is to make generative predictions about neural responses in various contexts and to recapitulate data from a battery of experiments. In this section, we will limit our discussion to EI networks and cell type models of intermediate complexity. Our aim is to appreciate how computational modeling has advanced our understanding of INs' role in shaping network dynamics.

EI balance is formally defined as a fixed ratio between excitatory and inhibitory conductances. EI balance can be indirectly measured as the ratio of excitatory versus inhibitory spikes and/or stable baseline firing rate of single neurons in vivo. EI balance is ubiquitously maintained during sensation, cognition, and action, as well as during spontaneous states (Isaacson and Scanziani, 2011). At the most basic level, EI balance prevents runaway excitation, as seen in epilepsy. Beyond this, the computational implications of EI balance have been a topic of intense theoretical investigation (van Vreeswijk and Sompolinsky, 1996; Renart et al., 2010; Rubin et al., 2015; Ahmadian and Miller, 2021). Simulated EI network behavior depends heavily on model parameters such as sparse versus dense connectivity (i.e., low versus high connection probability between E and I population), and the relative magnitude of external versus endogenous drive (Ahmadian and Miller, 2021). In densely connected EI network models, excitatory and inhibitory fluctuations are balanced at fast timescales, and these networks exhibit near-zero spike count correlation (Renart et al., 2010). Meanwhile, in sparsely connected EI networks with strong synapses, EI balance leads to Poisson-like inter-spike interval distributions (van Vreeswijk and Sompolinsky, 1996).

Fairly strong experimental evidence supports the notion that the sensory cortex can be modeled as an inhibition stabilized network (ISN), which refers to a network where strong recurrent excitation is balanced by strong recurrent inhibition (Ahmadian and Miller, 2021; Sadeh and Clopath, 2021). In an ISN, recurrent excitation is strong enough to be unstable by itself, but feedback inhibition prevents runaway excitation. *Feedback inhibition* refers to reciprocal connectivity between PCs and INs (i.e., when INs inhibit the PCs that excite them; Figure 8.3A). A hallmark of ISNs is the paradoxical effect (i.e., when increasing/decreasing the input drive to INs has the opposite effect on their output activity). For example, optogenetic activation of PV neurons not only leads to suppression of the

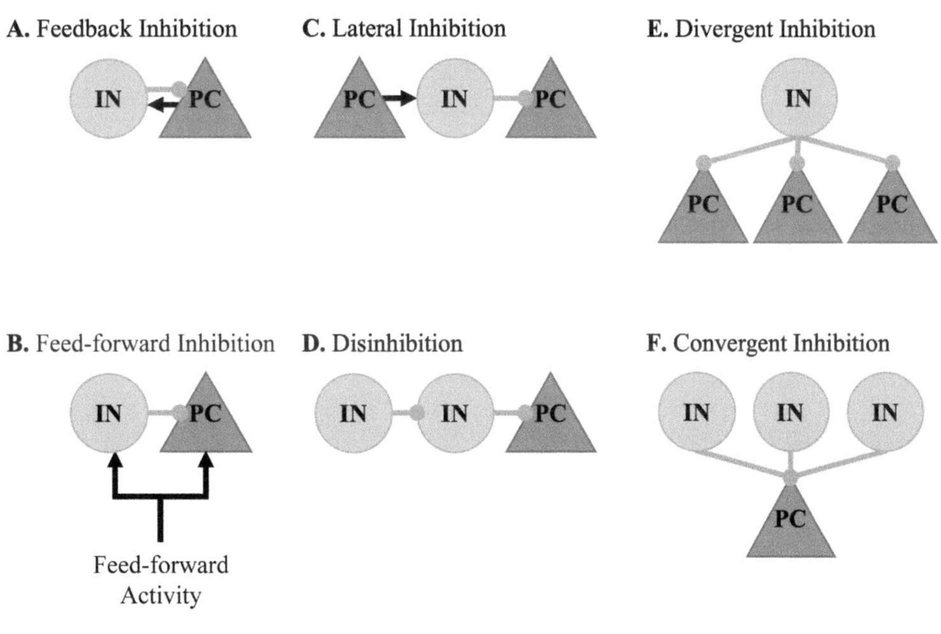

Figure 8.3. Inhibitory connectivity motifs in neocortex. (A). Feedback inhibition occurs when an IN and a PC are bidirectionally connected with one another. This connectivity motif forms the basis of excitatory-inhibitory (EI) network models, and recapitulates network dynamics such as pyramidal-interneuron network gamma oscillations (Tiesinga and Sejnowski, 2009). (B). Feedforward inhibition occurs when PCs get inhibited by an IN that receives a common drive as itself. In layer 4 of primary sensory cortex, this common drive typically originates from the thalamus. Feedforward inhibition is generally mediated by PVs. (C). Lateral inhibition occurs when INat inhibitory output projects to PCs adjacent from those providing excitatory input to the IN. Surround suppression is an instance of lateral inhibition and is mediated by SST. (D). Disinhibition refers to inhibition of inhibition. The most common form of disinhibition occurs when VIP inhibits SST, thereby disinhibiting PC. (E). Divergent inhibition occurs when a single IN projects to many postsynaptic partners. (F). Convergent inhibition occurs when many INs converge onto one postsynaptic cell.

local network, but also can reduce activity of PV neurons themselves (Sanzeni et al., 2020). Conversely, optogenetic silencing of PVs often lead to runaway excitation that also increases PVs' firing rate (Veit et al., 2017). Although paradoxical effects have been observed during optogenetic silencing of both PVs and SSTs (Adesnik, 2017; Kato et al., 2017), PVs are thought to be the main stabilizing force in the ISN (Palmigiano et al., 2020; Sanzeni et al., 2020). We will discuss the functional role of ISN in sensory processing in the following section.

A stabilized supralinear network (SSN) refers to an EI network that shows supralinearity when external inputs are weak, and sublinearity when external inputs are strong (Rubin et al., 2015; Adesnik, 2017; Ahmadian and Miller, 2021; Mossing et al., 2021). The SSN model posits that supralinearity originates from the supralinear transformation between membrane potential and firing rate (Priebe and Ferster, 2008), whereas the stabilizing sublinearity originates from recurrent inhibition (Ahmadian and Miller, 2021; Sadeh and Clopath, 2021). SSN occurs in sparsely connected EI networks where the external input to the network is comparable to the endogenous recurrent excitation and inhibition internally generated in the network (Rubin et al., 2015; Ahmadian and Miller, 2021). In sum, EI balance underlies ISN and SSN, which in turn recapitulates network behavior observed in perturbation experiments.

In addition to EI balance, inhibition has a key role in generating and shaping brain rhythms, also known as neural oscillations. Brain rhythms reflect brain states, and are organized logarithmically in frequency; gamma (30–80Hz), beta (15–30Hz), alpha (8–15Hz), theta (4–8Hz), and delta (0.1–4Hz) oscillations (Buzsáki and Draguhn, 2004). Brain rhythms can be measured intracranially in the local field potentials, or extracranially using magnetoencephalography or electroencephalography. Brain rhythms are thought to arise from synchronous neural activity with a characteristic temporal property. There are 3 possible origins for neural oscillations observed in a particular brain area: they can be intrinsically generated by resonance of the membrane potential (Sciamanna and Wilson, 2011); they can emerge from recurrent network interactions (Whittington et al., 2000; Tiesinga and Sejnowski, 2009); or they can be inherited from upstream areas (Saleem et al., 2017). Due to space constraints, we will limit our discussion to gamma oscillations.

Gamma oscillations are prevalent in the neocortex of awake behaving animals, and their functional relevance has been heavily debated (Fries, 2009; Buzsáki and Wang, 2012; Ray and Maunsell, 2015). Importantly, there appear to be multiple IN circuits that can generate gamma rhythms in different dynamical contexts. For many years, PVs were thought to be the primary mediators of network gamma rhythms owing to their strong reciprocal connectivity with excitatory neurons and their somatically targeted synapses, which are ideally placed to time spike output of PCs (Cardin et al., 2009; Sohal et al., 2009; Buzsáki and Wang, 2012; Chen et al., 2017). However, recent studies have demonstrated that SSTs can also mediate stimulus-induced and optogenetically induced gamma oscillations, albeit at the lower end of the gamma band range (Veit et al., 2017). The functional distinction of PV- and SST-mediated gamma remains to be further investigated.

Behavioral state is interlinked with brain state and potently modulates IN activity in all areas of the neocortex. Locomotion, whisking, and dilated pupils are indicators of high arousal states. The activity of different types and subtypes of INs correlate differently with arousal states (Fu et al., 2014; Pakan et al., 2016; Bugeon et al., 2022). In visual cortex, PVs and SSTs are heterogeneously affected by locomotion (which appears to depend on total scene luminance), whereas Sncg, VIP, and Lamp5 neurons are more homogenously activated by locomotion (Fu et al., 2014; Pakan et al., 2016; Bugeon et al., 2022). How locomotion modulates IN types in other areas of the neocortex is less well known (Studer and Barkat, 2022).

FUNCTIONAL ROLE IN NEOCORTICAL COMPUTATIONS: INHIBITION DURING SENSORY PROCESSING

The advent of optogenetics and the introduction of cell-type-specific driver mouse lines initially gave hope that it was only a matter of time until neuroscientists delineated the functional role of each IN type in cortical computation. In some initial efforts, evidence appeared to support the idea that each IN type supports specific computations; for example, feedforward inhibition by PV (Figure 8.3B) (Gabernet et al., 2005; Cruikshank et al., 2007), lateral inhibition by SST (Figure 8.3C) (Adesnik et al., 2012; Nienborg et al., 2013; Lakunina et al., 2020), and disinhibition by VIP through SST (Figure 8.3D) (Pfeffer et al., 2013; Pi et al., 2013). However, most computations have proven to be complicated to crack. In this section, we first describe each IN type's sensory response properties and their well-known roles in sensory processing. Then, we discuss INs' role in shaping feature selectivity, a question that has been riddled with mixed, and sometimes contradictory, evidence. Due to space limitations, we focus on sensory processing in the healthy brain.

PVs best-known role in sensory processing is *feedforward inhibition*, that is, when INs inhibit PCs that receive a common drive (Figure 8.3B) (Swadlow, 2003; Cruikshank et al., 2007). This common drive is provided by thalamocortical inputs to layer 4 of primary sensory areas, where PV cell density is the highest (Tremblay et al., 2016). Sensory-driven thalamic activity recruits feedforward inhibition with high efficiency, resulting in precise spike timing relative to sensory onset (Gabernet et al., 2005; Cruikshank et al., 2007). Although many fast-spiking neurons, with high correspondence to PVs, are sensory driven (Swadlow, 2003), non-sensory-responsive fast-spiking neurons exist in sensory cortices and may carry out distinct functional roles in coordinating spike synchrony during sensory processing (Shin and Moore, 2019).

SSTs, in contrast, have a prominent role in *lateral inhibition* (Figure 8.3C). Lateral inhibition refers to the inhibition of PCs adjacent in topographic and/or feature space, which manifests as a broader zone of inhibition than excitation (Isaacson and Scanziani, 2011; Adesnik, 2017; Kato et al., 2017). SSTs' role in lateral inhibition is largely due to their intracortical synaptic connectivity. SSTs in layers 2/3 for example, receive nearly all their local excitatory input from neurons in their own layer and almost no input from layer 4 (Adesnik et al., 2012). They also receive substantial synaptic inputs from more distant layers 2/3 PCs within the same cortical area. In turn, SSTs powerfully project back onto excitatory neurons, as well as onto most other types of INs. A sensory correlate of lateral inhibition is surround suppression. In mouse primary sensory cortex, optogenetic suppression of SSTs leads to diminished surround suppression in PCs (Adesnik et al., 2012; Nienborg et al., 2013; Lakunina et al., 2020). These results are well explained by SST response properties: their visually driven activity increases monotonically with visual stimulus size, unlike PCs and PVs that show surround suppression with sizes larger than their receptive field size. When a large high-contrast visual stimulus is present, SSTs inhibit both PCs and PVs via lateral inhibition, leading to an ISN (Ozeki et al., 2009; Adesnik, 2017; Kato et al., 2017).

VIPs mainly target SST cells and therefore primarily drive *disinhibition* of PCs (Figure 8.3D) (Pfeffer et al., 2013; Pi et al., 2013). VIPs' sensory response properties tend to be inversely proportional to SSTs, suggesting that VIPs inhibit SSTs in a divisive manner (Mossing et al., 2021). In V1, VIPs strongly prefer low contrast (Millman et al., 2020; Mossing et al., 2021). VIPs' inhibition of SST puts VIPs in a prime position to modulate surround suppression. Several studies have suggested that VIPs modulate the contrast- and orientation-dependence of surround suppression (Keller et al., 2020; Mossing et al., 2021; Veit et al., 2022).

In contrast to the ample data on the physiology, connectivity, and functional roles of PVs, SSTs, and VIPs, much less is known about Sncg and Lamp5 neurons. The Lamp5 transcriptomic type corresponds to neurogliaform cells, which have been studied with the Ndnf-Cre transgenic line (Cohen-Kashi Malina et al., 2021). Due to their efficacy in driving prolonged $GABA_B$ responses (Oláh et al., 2009), a recent modeling study proposed that neurogliaform cells act as powerful cortical switches that can turn off an entire cortical column (Law et al., 2022). In apparent contradiction, neurogliaform activity is highly correlated with arousal and are driven by sensory stimuli (Cohen-Kashi Malina et al., 2021). Neurogliaform cells inhibit pyramidal cells' apical dendrites and disinhibit their soma by selectively attenuating feedforward inhibition by PVs, suggesting that neurogliaform activity may switch the brain state to preferentially respond to salient and unexpected bottom-up sensory inputs (Cohen-Kashi Malina et al., 2021). Sncg neurons, which express both VIP and CCK, have often been lumped with VIP bipolar neurons. As such, Sncgs' independent characteristics have scarcely been investigated. A recent study found that Sncgs' activity is uniformly suppressed by visual stimuli (Bugeon et al., 2022).

Despite initial hints of success, deciphering the causal role of IN types during a cortical computation turned out not to be a simple matter of probing different IN types with optogenetics for most computations. For example, optogenetic activation and inactivation of IN types can have asymmetric effects on PCs depending on PCs' activity levels (Phillips and Hasenstaub, 2016; Garcia Del Molino et al., 2017; Mossing et al., 2021). In other words, IN's influence on PCs depends not only on the activity of the presynaptic IN, but also on the activity of the postsynaptic PC.

As briefly mentioned above, there are at least 3 reasons why one might fail to find a robust causal relationship between an IN type and a cortical computation. First, inhibition by different INs cannot act independently of one another. INs are interconnected among themselves and with PCs. Even if two microcircuit components are not directly connected, they can influence each other through an intermediary component. This means that no single component in the network is truly independent. Thus, any computation that involves more than one type of connection influences the rest of the network. Second, there are many levels of categorization for interneuron types (i.e., there are IN subtypes and subtypes of subtypes). At the finest level, every neuron is distinct in at least one aspect, whether it be morphology, electrophysiology, transcriptome, or input and output connectivity. This is not meant to undermine the fact that INs fall into discrete categories with distinctive characteristics, but rather to emphasize that there is variability along a continuum within each type. Third, neural networks can exhibit diverse dynamic regimes depending on the brain state. For example, a cortical neuron's response to the same sensory stimulus can vary drastically at different levels of arousal. Further, arousal level modulates different IN types in different ways, as described above (Fu et al., 2014; Bugeon et al., 2022). Accordingly, how a given cortical computation engages IN types may depend on the brain state.

For these reasons, our understanding of certain cortical computations is becoming increasingly complicated. One such question concerns how inhibition shapes PCs' feature selectivity (Priebe and Ferster, 2008; Isaacson and Scanziani, 2011; Harris and Mrsic-Flogel, 2013; Wood et al., 2017). This question has often been addressed in the framework of orientation tuning in V1. Feature selectivity can be shaped in several different ways. First, feature selectivity may be inherited from an upstream area. Second, feedforward inputs from upstream areas may be nonlinearly transformed. Third, feature selectivity may be sharpened locally by lateral inhibition, or by inhibition that is broader than excitation. When inhibition is broader than excitation and is net subtractive, it can sharpen orientation tuning via an "iceberg effect," a floor effect caused by PCs not reaching spike threshold at non-preferred orientations (Harris and Mrsic-Flogel, 2013). On the contrary, divisive inhibition does not cause a change in tuning width (Atallah et al., 2012; Lee et al., 2012, 2014; Wilson et al., 2012; El-Boustani and Sur, 2014).

Whether an IN type causes subtractive versus divisive inhibition of orientation tuning is intensely debated with several seemingly contradictory results (Atallah et al., 2012; Lee et al., 2012; Lee et al., 2014; Wilson et al., 2012; El-Boustani and Sur, 2014). Three pioneering studies aimed to determine whether inhibition by PVs and SSTs are divisive versus subtractive (Atallah et al., 2012; Lee et al., 2012; Wilson et al., 2012). These studies came to contradictory conclusions, which was explained by differences in the temporal pattern of optogenetic stimulation (El-Boustani and Sur, 2014; Lee et al., 2014). As such, inhibitory influence on PCs' feature selectivity depends on INs' spatiotemporal pattern of activity, beyond simply their IN type.

In summary, the dominant focus of cortical IN research in the past decade has been to assign a specific role to a specific IN type. However, recent studies highlight the need for a more holistic view, taking into consideration the interactions between the components of a dynamical system.

IMPORTANT AND UNANSWERED QUESTIONS

Recent advances in molecular techniques and genetic tools, concurrent with cell-type-specific neural activity monitoring and manipulation techniques, have led to countless exciting discoveries regarding the role of neocortical INs in perception, cognition, and action, as well as during various brain states. The congruence of INs' morphology, electrophysiology, and transcriptomic types across cortical areas supports the idea of canonical microcircuits repeated across the neocortex (Harris and Shepherd, 2015; Tasic et al., 2018). Even more strikingly, IN types show distinct connectivity rules that constrain the dynamics and function of interconnected circuit elements (Figure 8.2) (Pfeffer et al., 2013; Jiang et al., 2015; Campagnola et al., 2022).

Despite these and other landmark findings, the IN field is at a stage of growing complexity. In most cases, a given computation can involve various INs in different ways depending on the external context and the internal brain state (Phillips and Hasenstaub, 2016; Garcia Del Molino et al., 2017; Mossing et al., 2021). Moreover, it is becoming clear that a system of interacting components cannot be understood in whole by examining individual components separately. In this vein, it would be most desirable to develop and adopt techniques that allow us to simultaneously record from all neurons and identify the neuronal types in a high-throughput manner; for example, 2-photon imaging of neural activity followed by the identification of each imaged neuron's cell type via post-mortem correlative transcriptional analysis (Bugeon et al., 2022).

A further major area of interest is expanding our understanding of connectivity rules, both in terms of higher-order connectivity and connectivity as a function of in vivo response properties. With regard to higher-order connectivity, of key interest is the degree to which each IN type engages in different connectivity motifs such as feedback inhibition, feedforward inhibition, lateral inhibition, and disinhibition; as well as the divergence and convergence of inputs and outputs (Figure 8.3) (Luo, 2021). With regard to how connectivity patterns depend on feature selectivity, it would be important to determine whether topographically organized features (e.g., receptive fields in V1, tonotopy in A1, whisker columns in S1) are shaped differently from features that have salt-and-pepper topography (e.g., orientation in mouse V1). For addressing these questions, two-photon holographic optogenetics is a highly promising technique (Hage et al., 2022).

Lastly, perhaps the most important question regarding cortical INs is the degree to which findings in mice generalize to other mammalian species, including humans. There is scarce but promising evidence for homology in transcriptomic types with conserved electrophysiological and morphological types (Hodge et al., 2019). However, INs' functional role in the human neocortex is far less understood. A silver lining is that theoretical modeling and computational simulations can reveal dynamical regimes and connectivity patterns necessary and sufficient for a given cortical computation. To this end, computational neuroscientists are building highly detailed large-scale simulations, with the aim of inferring function from transcriptomics, morphology, electrophysiology, and connectivity (Traub et al., 2005; Billeh et al., 2020).

REFERENCES

Adesnik H (2017). Synaptic mechanisms of feature coding in the visual cortex of awake mice. *Neuron*. 95:1147–1159. http://dx.doi.org/10.1016/j.neuron.2017.08.014.

Adesnik H, Bruns W, Taniguchi H, Huang ZJ, Scanziani M (2012). A neural circuit for spatial summation in visual cortex. Nature 490:226–231. http://dx.doi.org/10.1038/nature11526.

Ahmadian Y, Miller KD (2021). What is the dynamical regime of cerebral cortex? *Neuron*. 109:3373–3391. https://doi.org/10.1016/j.neuron.2021.07.031.

Atallah BV, Bruns W, Carandini M, Scanziani M (2012). Parvalbumin-expressing interneurons linearly transform cortical responses to visual stimuli. *Neuron*. 73:159–170. http://www.cell.com/article/S0896627311010944/fulltext.

Billeh YN, Cai B, Gratiy SL, Dai K, Iyer R, Gouwens NW, Abbasi-Asl R, Jia X, Siegle JH, Olsen SR, Koch C, Mihalas S, Arkhipov A (2020). Systematic integration of structural and functional data into multi-scale models of mouse primary visual cortex. *Neuron*. 106:388–403. https://doi.org/10.1016/j.neuron.2020.01.040.

Bugeon S, Duffield J, Dipoppa M, Ritoux A, Prankerd I, Nicoloutsopoulos D, Orme D, Shinn M, Peng H, Forrest H, Viduolyte A, Reddy CB, Isogai Y, Carandini M, Harris KD (2022). A transcriptomic axis predicts state modulation of cortical interneurons. *Nature*. 607:330–338.

Buzsáki G, Draguhn A (2004). Neuronal oscillations in cortical networks. *Science*. 304:1926–1929. http://www.ncbi.nlm.nih.gov/pubmed/15218136.

Buzsáki G, Wang X-J (2012). Mechanisms of gamma oscillations. *Annu Rev Neurosci*. 35:203–225. http://www.ncbi.nlm.nih.gov/pubmed/22443509.

Campagnola L, Seeman SC, Chartrand T, Kim L, Hoggarth A, Gamlin C, Ito S, Trinh J, Davoudian P, Radaelli C, Kim MH (2022) Local connectivity and synaptic dynamics in mouse and human neocortex. *Science*. 375:eabj5861.

Cardin JA, Carlén M, Meletis K, Knoblich U, Zhang F, Deisseroth K, Tsai LH, Moore CI (2009). Driving fast-spiking cells induces gamma rhythm and controls sensory responses. *Nature*. 459:663–667. http://www.pubmedcentral.nih.gov/articlerender.fcgi?artid=3655711&tool=pmcentrez&rendertype=abstract.

Chen G, Zhang Y, Li X, Zhao X, Ye Q, Lin Y, Tao HW, Rasch MJ, Zhang X (2017). Distinct inhibitory circuits orchestrate cortical beta and gamma band oscillations. *Neuron*. 96:1403–1418. https://doi.org/10.1016/j.neuron.2017.11.033.

Cohen-Kashi Malina K, Tsivourakis E, Kushinsky D, Apelblat D, Shtiglitz S, Zohar E, Sokoletsky M, Tasaka G, Mizrahi A, Lampl I, Spiegel I (2021). NDNF interneurons in layer 1 gain-modulate whole cortical columns according to an animal's behavioral state. *Neuron*. 109:2150–2164. https://doi.org/10.1016/j.neuron.2021.05.001.

Cruikshank SJ, Lewis TJ, Connors BW (2007). Synaptic basis for intense thalamocortical activation of feedforward inhibitory cells in neocortex. *Nat Neurosci*. 10:462–468.

El-Boustani S, Sur M (2014). Response-dependent dynamics of cell-specific inhibition in cortical networks in vivo. *Nat Commun*. 5:5689. http://www.nature.com/articles/ncomms6689.

Fries P (2009). Neuronal gamma-band synchronization as a fundamental process in cortical computation. *Annu Rev Neurosci*. 32:209–224. http://www.ncbi.nlm.nih.gov/pubmed/19400723.

Fu Y, Tucciarone JM, Espinosa JS, Sheng N, Darcy DP, Nicoll RA, Huang ZJ, Stryker MP (2014). A cortical circuit for gain control by behavioral state. *Cell*. 156:1139–1152.

Gabernet L, Jadhav SP, Feldman DE, Carandini M, Scanziani M (2005). Somatosensory integration controlled by dynamic thalamocortical feed-forward inhibition. *Neuron*. 48:315–327.

Garcia Del Molino LC, Yang GR, Mejias JF, Wang XJ (2017). Paradoxical response reversal of top- down modulation in cortical circuits with three interneuron types. *eLife*. 6:1–15.

Gouwens NW, Sorensen SA, Berg J, Lee C, Jarsky T, Ting J, Sunkin SM, Feng D, Anastassiou CA, Barkan E, Bickley K (2019) Classification of electrophysiological and morphological neuron types in the mouse visual cortex. *Nat Neurosci*. 22:1182–1195. http://www.nature.com/articles/s41593-019-0417-0.

Hage TA, Bosma-Moody A, Baker CA, Kratz M, Campagnola L, Jarsky T, Zeng H, Murphy GJ (2022). Synaptic connectivity to L2/3 of primary visual cortex measured by two-photon optogenetic stimulation. *eLife*. 11:1–46.

Harris KD, Mrsic-Flogel TD (2013). Cortical connectivity and sensory coding. *Nature*. 503:51–58. http://www.ncbi.nlm.nih.gov/pubmed/24201278.

Harris KD, Shepherd GMGG (2015). The neocortical circuit: Themes and variations. *Nat Neurosci*. 18:170–181. http://dx.doi.org/10.1038/nn.3917.

Hestrin S, Galarreta M (2005). Electrical synapses define networks of neocortical GABAergic neurons. *Trends Neurosci*. 28:304–309.

Hodge RD, Bakken TE, Miller JA, Smith KA, Barkan ER, Graybuck LT, Close JL, Long B, Johansen N, Penn O, Yao Z (2019) Conserved cell types with divergent features in human versus mouse cortex. *Nature*. 573:61–68. http://dx.doi.org/10.1038/s41586-019-1506-7.

Isaacson JS, Scanziani M (2011). How inhibition shapes cortical activity. *Neuron*. 72:231–243. http://www.pubmedcentral.nih.gov/articlerender.fcgi?artid=3236361&tool=pmcentrez&rendertype=abstract.

Jiang X, Shen S, Cadwell CR, Berens P, Sinz F, Ecker AS, Patel S, Tolias AS (2015). Principles of connectivity among morphologically defined cell types in adult neocortex. *Science*. 350:aac9462. http://www.sciencemag.org/cgi/doi/10.1126/science.aac9462.

Kato HK, Asinov SK, Isaacson JS (2017). Network-level control of frequency tuning in auditory cortex. *Neuron*. 95:412–423. http://dx.doi.org/10.1016/j.neuron.2017.06.019.

Keller AJ, Dipoppa M, Roth MM, Caudill MS, Ingrosso A, Miller KD, Scanziani M (2020). A disinhibitory circuit for contextual modulation in primary visual cortex. *Neuron*. 108:1181–1193. https://doi.org/10.1016/j.neuron.2020.11.013.

Kubota Y, Karube F, Nomura M, Kawaguchi Y (2016). The diversity of cortical inhibitory synapses. *Front Neural Circuits*. 10:27.

Lakunina AA, Nardoci MB, Ahmadian Y, Jaramillo S (2020). Somatostatin-expressing interneurons in the auditory cortex mediate sustained suppression by spectral surround. *J Neurosci*. 40:3564–3575.

Law RG, Pugliese S, Shin H, Sliva DD, Lee S, Neymotin S, Moore C, Jones SR (2022). Thalamocortical mechanisms regulating the relationship between transient beta events and human tactile perception. *Cereb Cortex*. 32:668–688.

Lee S-H, Kwan AC, Dan Y (2014). Interneuron subtypes and orientation tuning. *Nature*. 508:E1–E2. http://www.nature.com/articles/nature13130.

Lee S-H, Kwan AC, Zhang S, Phoumthipphavong V, Flannery JG, Masmanidis SC, Taniguchi H, Huang ZJ, Zhang F, Boyden ES, Deisseroth K, Dan Y (2012). Activation of specific interneurons improves V1 feature selectivity and visual perception. *Nature*. 488:379–383. http://dx.doi.org/10.1038/nature11312.

Luo L (2021). Architectures of neuronal circuits. *Science*. 373:eabg7285.

McCormick DA, Connors BW, Lighthall JW, Prince DA (1985). Comparative electrophysiology of pyramidal and sparsely spiny stellate neurons of the neocortex. *J Neurophysiol*. 54:782–806.

Millman DJ, Ocker GK, Caldejon S, Kato I, Larkin JD, Lee EK, Luviano J, Nayan C, Nguyen T V., North K, Seid S, White C, Lecoq J, Reid C, Buice MA, de Vries SEJ (2020). VIP interneurons in mouse primary visual cortex selectively enhance responses to weak but specific stimuli. *eLife*. 9:1–22.

Mossing DP, Veit J, Palmigiano A, Miller KD, Adesnik H (2021). Antagonistic inhibitory subnetworks control cooperation and competition across cortical space. *bioRxiv*. https://doi.org/10.1101/2021.03.31.437953

Nienborg H, Hasenstaub A, Nauhaus I, Taniguchi H, Huang ZJ, Callaway EM (2013). Contrast dependence and differential contributions from somatostatin- and parvalbumin-expressing neurons to spatial integration in mouse V1. *J Neurosci*. 33:11145–11154. http://www.pubmedcentral.nih.gov/articlerender.fcgi?artid=3718383&tool=pmcentrez&rendertype=abstract.

Oláh S, Füle M, Komlósi G, Varga C, Báldi R, Barzó P, Tamás G (2009). Regulation of cortical microcircuits by unitary GABA-mediated volume transmission. *Nature*. 461:1278–1281.

Ozeki H, Finn IM, Schaffer ES, Miller KD, Ferster D (2009). Inhibitory stabilization of the cortical network underlies visual surround suppression. *Neuron*. 62:578–592. http://dx.doi.org/10.1016/j.neuron.2009.03.028.

Pakan JMPP, Lowe SC, Dylda E, Keemink SW, Currie SP, Coutts CA, Rochefort NL (2016). Behavioral-state modulation of inhibition is context-dependent and cell type specific in mouse visual cortex. *eLife*. 5:1–18.

Palmigiano A, Fumarola F, Mossing DP, Kraynyukova N, Adesnik H, Miller KD, Miller KD (2020). Structure and variability of optogenetic responses identify the operating regime of cortex. *bioRxiv*. https://doi.org/10.1101/2020.11.11.378729.

Pfeffer CK, Xue M, He M, Huang ZJ, Scanziani M (2013). Inhibition of inhibition in visual cortex: The logic of connections between molecularly distinct interneurons. *Nat Neurosci*. 16:1068–1076. http://dx.doi.org/10.1038/nn.3446.

Phillips EAK, Hasenstaub AR (2016). Asymmetric effects of activating and inactivating cortical interneurons. *eLife*. 5:1–22.

Pi HJ, Hangya B, Kvitsiani D, Sanders JI, Huang ZJ, Kepecs A (2013). Cortical interneurons that specialize in disinhibitory control. *Nature*. 503:521–524.

Priebe NJ, Ferster D (2008) Inhibition, spike threshold, and stimulus selectivity in primary visual cortex. *Neuron*. 57:482–497.

Ray S, Maunsell JHR (2015). Do gamma oscillations play a role in cerebral cortex? *Trends Cogn Sci*. 19:78–85. http://dx.doi.org/10.1016/j.tics.2014.12.002.

Renart A, De La Rocha J, Bartho P, Hollender L, Parga N, Reyes A, Harris KD (2010). The asynchronous state in cortical circuits. *Science*. 327:587–590. https://www.science.org/doi/10.1126/science.1179850.

Rubin DB, VanHooser SD, Miller KD (2015). The stabilized supralinear network: A unifying circuit motif underlying multi-input integration in sensory cortex. *Neuron*. 85:402–417. http://dx.doi.org/10.1016/j.neuron.2014.12.026.

Rudy B, Fishell G, Lee S, Hjerling-Leffler J (2011). Three groups of interneurons account for nearly 100% of neocortical GABAergic neurons. *Dev Neurobiol*. 71:45–61. http://www.pubmedcentral.nih.gov/articlerender.fcgi?artid=3556905&tool=pmcentrez&rendertype=abstract.

Sadeh S, Clopath C (2021). Inhibitory stabilization and cortical computation. *Nat Rev Neurosci*. 22:21–37. http://dx.doi.org/10.1038/s41583-020-00390-z.

Saleem AB, Lien AD, Krumin M, Haider B, Rosón MR, Ayaz A, Reinhold K, Busse L, Carandini M, Harris KD, Carandini M (2017). Subcortical source and modulation of the narrowband gamma oscillation in mouse visual cortex. *Neuron*. 93:315–322.

Sanzeni A, Akitake B, Goldbach HC, Leedy CE, Brunel N, Histed MH (2020). Inhibition stabilization is a widespread property of cortical networks. *eLife*. 9:1–39.

Sciamanna G, Wilson CJ (2011). The ionic mechanism of gamma resonance in rat striatal fast-spiking neurons. *J Neurophysiol*. 106:2936–2949. http://jn.physiology.org/cgi/doi/10.1152/jn.00280.2011.

Shin H, Moore CI (2019). Persistent gamma spiking in SI nonsensory fast spiking cells predicts perceptual success. *Neuron*. 103:1150–1163. https://doi.org/10.1016/j.neuron.2019.06.014.

Sohal VS, Zhang F, Yizhar O, Deisseroth K (2009). Parvalbumin neurons and gamma rhythms enhance cortical circuit performance. *Nature*. 459:698–702. http://dx.doi.org/10.1038/nature07991.

Staiger JF, Petersen CCH (2021). Neuronal circuits in barrel cortex for whisker sensory perception. *Physiol Rev*. 101:353–415.

Studer F, Barkat TR (2022). Inhibition in the auditory cortex. *Neurosci Biobehav Rev*. 132:61–75. https://doi.org/10.1016/j.neubiorev.2021.11.021.

Swadlow HA (2003). Fast-spike interneurons and feedforward inhibition in awake sensory neocortex. *Cereb Cortex*. 13:25–32. https://academic.oup.com/cercor/article-lookup/doi/10.1093/cercor/13.1.25.

Tasic B, Yao Z, Graybuck LT, Smith KA, Nguyen TN, Bertagnolli D, Goldy J, Garren E, Economo MN, Viswanathan S, Penn O (2018). Shared and distinct transcriptomic cell types across neocortical areas. *Nature*. 563:72–78. http://dx.doi.org/10.1038/s41586-018-0654-5.

Tiesinga P, Sejnowski TJ (2009). Cortical enlightenment: Are attentional gamma oscillations driven by ING or PING? *Neuron*. 63:727–732. http://www.pubmedcentral.nih.gov/articlerender.fcgi?artid=2778762&tool=pmcentrez&rendertype=abstract.

Traub RD, Contreras D, Cunningham MO, Murray H, LeBeau FENN, Roopun A, Bibbig A, Wilent WB, Higley MJ, Whittington MA (2005). Single-column thalamocortical network model exhibiting gamma oscillations, sleep spindles, and epileptogenic bursts. *J Neurophysiol.* 93:2194–2232. http://www.ncbi.nlm.nih.gov/pubmed/15525801.

Tremblay R, Lee S, Rudy B (2016). GABAergic interneurons in the neocortex: From cellular properties to circuits. *Neuron*. 91:260–292. http://dx.doi.org/10.1016/j.neuron.2016.06.033.

van Vreeswijk C, Sompolinsky H (1996). Chaos in neuronal networks with balanced excitatory and inhibitory activity. *Science*. 274:1724–1726.

Veit J, Hakim R, Jadi MP, Sejnowski TJ, Adesnik H (2017). Cortical gamma band synchronization through somatostatin interneurons. *Nat Neurosci*. 20:951–959. http://www.nature.com/doifinder/10.1038/nn.4562.

Veit J, Handy G, Mossing DP, Doiron B, Adesnik H (2022). Cortical VIP neurons locally control the gain but globally control the coherence of gamma band rhythms. *Neuron*. 111:405–417. https://doi.org/10.1016/j.neuron.2022.10.036.

Wang XJ (2020). Macroscopic gradients of synaptic excitation and inhibition in the neocortex. *Nat Rev Neurosci*. 21:169–178. http://dx.doi.org/10.1038/s41583-020-0262-x.

Whittington MA, Traub RD, Kopell N, Ermentrout B, Buhl EH (2000). Inhibition-based rhythms: Experimental and mathematical observations on network dynamics. *Int J Psychophysiol*. 38:315–336.

Wilson NR, Runyan CA, Wang FL, Sur M (2012). Division and subtraction by distinct cortical inhibitory networks in vivo. *Nature*. 488:343–348. http://www.pubmedcentral.nih.gov/articlerender.fcgi?artid=3653570&tool=pmcentrez&rendertype=abstract.

Wood KC, Blackwell JM, Geffen MN (2017). Cortical inhibitory interneurons control sensory processing. *Curr Opin Neurobiol*. 46:200–207. http://dx.doi.org/10.1016/j.conb.2017.08.018.

9.

TRANSFORMATION OF RECEPTIVE FIELDS ACROSS LAMINAE

Alex D. Reyes

INTRODUCTION

A long-standing question in neuroscience is how sensory signals from the thalamus are represented in the cortex. The laminar organization of the cortex suggests that thalamic inputs entering the cortex are processed in successive stages before exiting to other brain regions. Decades of electrophysiological recordings and more recent imaging studies reveal that signals are not simply relayed from one lamina to the next but are transformed. The receptive fields of neurons differ qualitatively and quantitatively, suggesting that the signals from the thalamorecipient laminae are combined and/or refined in subsequent stages. This is perhaps not surprising, given that each lamina is composed of different cell types and network architectures (see **Chapters 6 and 8 in Section III**).

However, understanding the transformation processes is challenging for many reasons. First, signals are not processed in a strictly hierarchical fashion. There are multiple entry points for thalamic inputs, leading to different pathways through the cortex that converge and interact at common sites. There are also inherent and largely unavoidable and often unwanted signal transformations that occur generally in multilayered networks, independent of cell types, synaptic dynamics, and network properties. A further complication is that inputs from other sensory and non-sensory brain regions can modulate the neuronal responses, depending on the state of the animal (Sakata and Harris, 2009; Schneider et al., 2018).

This chapter explores the mechanisms and circuit properties that mediate signal processing within and across cortical laminae. Because it is not possible to cover in great detail the massive dataset accumulated over decades of research in a variety of species, the focus will be on receptive field properties common to topographically organized primary auditory, somatosensory, and visual cortices. Unfortunately, many of the nuanced details of each system are by necessity omitted. The first sections briefly review the anatomical pathways for signals entering the thalamorecipient layers and summarize the associated changes in receptive field properties across laminae. Then, in the final sections, the conditions for propagating and transforming receptive fields across laminae are discussed and explored with simulations.

INTRA- AND INTER-LAMINAR CONNECTIONS

The connections between neurons within and across lamina provide the anatomical substrate for the flow of information through the cortex. An in-depth discussion of the anatomical connections can be found in **Chapter 6 of Section III** of this book, and only a brief summary will be given here. In primary visual, auditory, and somatosensory cortices, the afferents from thalamus terminate with varying degrees in laminae 1–6 (henceforth designated as L1–6) but most densely in L4 (Blasdel and Lund, 1983; Burkhalter, 1989; Hornung and Garey, 1981; Huang and Winer, 2000; Jensen and Killackey, 1987; Ji et al., 2016; Kimura et al., 2003; Larsen and Callaway, 2006; Mitani et al., 1985; Winer, 1984). These findings are important because they indicate that there are multiple entry points for signals from the thalamus.

The spatial extent of axons from the thalamus can be broad, on the order of 100s of microns to millimeters (Blasdel and Lund, 1983; Cetas et al., 1999; Jensen and Killackey, 1987). The densest terminations are clustered over a smaller area (Blasdel and Lund, 1983; Cetas et al., 1999; Jensen and Killackey, 1987). Functionally, this gives rise to an important feature of topographically organized networks: inputs from a single thalamic neuron are not uniformly distributed but tend to be more concentrated over a relatively small population of neurons. Both excitatory (pyramidal and spiny stellate) cells and fast-spiking PV interneurons are innervated, and to a lesser extent somatostatin positive interneurons (Agmon and Connors, 1991; Ji et al., 2016; Schiff and Reyes, 2012; Wehr and Zador,

2003), forming a feedforward inhibitory circuit (see below; see also **Chapter 8 in Section III**).

From lamina 4, the signal propagates toward the upper laminae. Layer 4 excitatory neurons connect to excitatory neurons in lamina 2/3 (Campagnola et al., 2022; Feldmeyer et al., 2002; Feldmeyer and Sakmann, 2000; Levy and Reyes, 2012; Shepherd and Svoboda, 2005; Yabuta and Callaway, 1998; Dantzker and Callaway, 2000). Moreover, the connection appears to be unidirectional; reciprocal connections between L4 and L2/3 are weak at best (Campagnola et al., 2022; Feldmeyer et al., 2002; Kratz and Manis, 2015; Martinez and Alonso, 2001). The L2/3 neurons in turn project to L5 (Campagnola et al., 2022; Reyes and Sakmann, 1999). This, then, comprises the canonical pathway for signals that originate in L4 and propagate to L2/3 and down to L5 (Ahissar et al., 2001; Callaway, 2004; Sakata and Harris, 2009). However, signals can also flow in the opposite direction, starting from thalamorecipient laminae 5 and 6. L5 excitatory neurons send axons to L2/3 (Briggs and Callaway, 2005; Burkhalter, 1989; Campagnola et al., 2022) and L6 neurons to L4 (Kratz and Manis, 2015) and L2/3 (Oviedo et al., 2010). Thus, there is no simple hierarchical flow of signals; instead, the anatomical connections suggest multiple starting points and pathways.

The inter- and intra-layer axonal arbors of cortical neurons are also topographically organized. Though broad (Gilbert, 1977), the terminals are again not uniformly distributed in their target layers. Tract tracing of axons from injection sites in L4 indicates that the areas of densest labeling in L2/3 were comparable to those of the injection site (approximately 100–250μm diameter), becoming very sparse beyond 600μm (Burkhalter, 1989). Similar observations were made from labeled neurons using in vitro slice preparations (Larsen and Callaway, 2006; Wang et al., 2018), though the full extent may be underestimated due to slicing artifacts. It is worth noting that the spatial extents of the axons are, at least in somatosensory cortex, substantially broader than thalamocortical afferents (see **Chapter 6 in Section III**), suggesting that signals from the thalamus may be strongly dispersed across layers.

Signals may also propagate horizontally within a lamina as connections between neurons within a layer support recurrent activity. The properties of synapses between neurons have been studied extensively by performing simultaneous whole-cell recordings from pairs of neurons in an in vitro slice preparation (Campagnola et al., 2022; Feldmeyer et al., 2002; Feldmeyer and Sakmann, 2000; Levy and Reyes, 2012; Oswald and Reyes, 2008; Reyes and Sakmann, 1999). These experiments show that the unitary synaptic potentials evoked by single neurons are small, with amplitudes on the order of a few hundred microvolts. The probability of connection decreases with intersomatic distance, approximating a Gaussian distribution. Notably, the connections between excitatory cells are weak (peak connection probability 0.1), while those between excitatory and inhibitory cells are relatively stronger (peak probability 0.5; Campagnola et al., 2022; Levy and Reyes, 2012; Oswald and Reyes, 2008). The standard deviation of the Gaussian fit ranges from 100 to 130 μm.

RECEPTIVE FIELD PROPERTIES ACROSS LAMINAE

The receptive fields of neurons describe how neurons respond to varying sensory stimuli: typically, a subtle change in the stimulus causes incremental changes in the firing of individual neurons and/or spatial distribution of activated populations of neurons. For example, in the visual system, the firing of neurons can be modulated by the location of the stimuli in the visual field (retinotopy), the orientation and direction of a moving bar, or the stimulus contrast (Hubel and Wiesel, 1962). In auditory cortex, neurons are sensitive to the frequency and intensity of tones (Sugimoto et al., 1997; Wallace and Palmer, 2008). Neurons in somatosensory system can respond to vibration or touch delivered to a particular part of the body (Sur et al., 1985; Chapin, 1986; DiCarlo and Johnson, 2000) or, in the case of the rat barrel cortex, to whisker deflections (Brecht et al., 2003; Brecht and Sakmann, 2002; De Kock et al., 2007; Manns et al., 2004).

A feature of these sensory areas is that neurons are tuned, or respond best, to some aspect(s) of the stimulus. In the visual system, there is an optimal location, bar orientation, length, or direction and velocity of motion. Auditory neurons have a preferred tone frequency, which can elicit a response even at low intensities. Somatosensory neurons are most sensitive to a specific aspect of sensory modality, such as in the rodent barrel cortex where neurons respond best to the direction and/or velocity of whisker movement. Moreover, there is a systematic change in the preferred stimulus of neurons along an axis or axes parallel to brain surface. In the auditory system, neurons along the "tonotopic" axis will exhibit small differences in preferred frequencies such that those at one end of the axis prefer low frequencies and those at the other end prefer high frequencies. Similarly, in the visual system, there are axes for stimulus location and for orientation.

In the axis orthogonal to the brain surface, neurons tend to have similar tuning properties (Hubel and Wiesel, 1962; Chapin, 1986; Sugimoto et al., 1997; Wallace and Palmer, 2008). The next section will discuss differences that provide important clues as to how signals propagate and what transformations may occur across layers.

RESPONSE LATENCIES

The latencies, or delay, of the responses relative to the onset of the stimulus can in principle reveal the site of origin of the signal and the subsequent flow through the laminae. Neurons in thalamorecipient laminae would be expected to become active first, followed by those in downstream laminae after a delay associated with polysynaptic connections and subthreshold integration time constants. Though there are several variables that affect latencies of evoked action potentials, a ballpark estimate of expected delays can be obtained by examining synaptic potentials recorded from connected pairs of pyramidal neurons (Levy and Reyes, 2012; Oswald and Reyes, 2008). The typical latency between pyramidal cells is about 1.5 ms, and the time-to-peak of the synaptic potential is about 10 ms. Thus, under the condition where just enough synaptic potentials arrive synchronously to cross threshold (see Figure 9.2A), the action potentials would be evoked with a latency of about 11.5 ms but could decrease theoretically (but probably not realistically) to 1.5 ms with a very large number of inputs (the rate of rise increases so that threshold crossings occur earlier). This relatively large range, coupled with variability in the latencies of thalamically evoked EPSPs (range: 0.6–3 ms; Mitani and Shimokouchi, 1985), makes it difficult to determine with certainty the sequence of events.

In the primary auditory cortex of guinea pigs, the latency of evoked action potentials is shortest in L5 and 6, about 2 ms ahead of responses evoked in L2–4 (Wallace and Palmer, 2008; but see Sugimoto et al., 1997). In the barrel somatosensory cortex of rodents, action potentials evoked with whisker stimulation have latencies that were similar in L4–6 (De Kock et al., 2007), although other studies show that the shortest latency responses can be evoked in L5 (Constantinople and Bruno, 2013) or in L4 (Ahissar et al., 2001). One study, where simultaneous recordings were made from L2–5, suggests sequential procession of signals: L4 cells are activated first (10 ms delay after stimulus), followed later by cells in L2/3 (15 ms) and 5 (16 ms). In visual cortex, the shortest latency responses were in layers 4 and 6, with the longest latencies tending toward supraganular layers (Maunsell and Gibson, 1992).

In summary, it is very difficult to determine the sequence of events from the latencies alone, as the results are quite variable within and across sensory modalities. This is not entirely unexpected given the multiple entry points from the thalamus and the convergent and interacting pathways (see above).

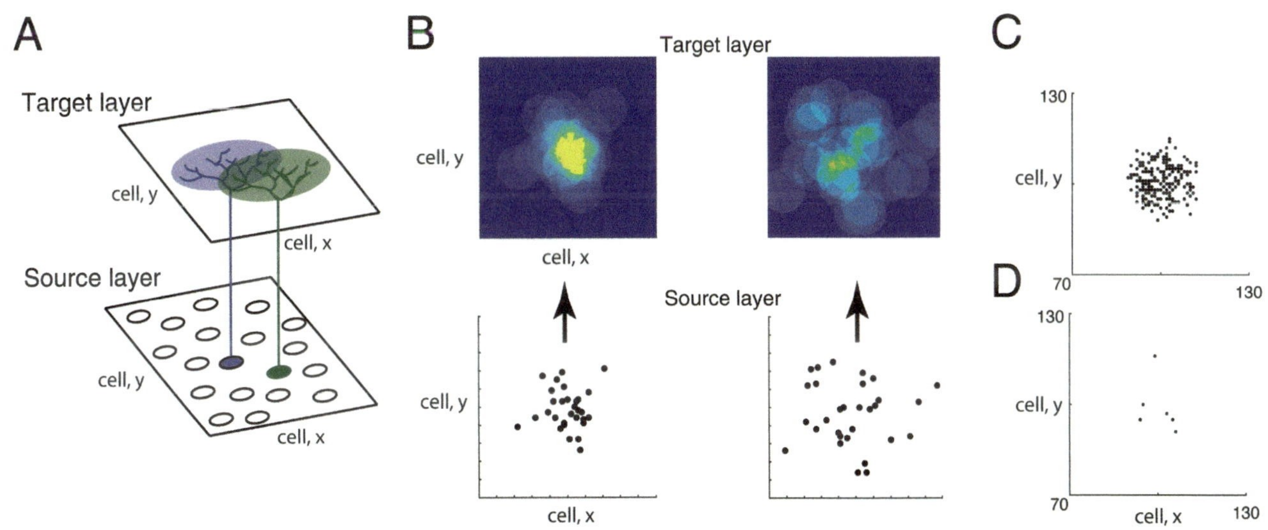

Figure 9.1. Anatomical determinants of signal propagation. (A). Overlapping synaptic fields in target layer from 2 neurons in the source layer. The axons are topographically organized so that the terminal distributions are confined to a small area in the target layer. (B). Distribution of synaptic terminals in neurons in the target layer (top) following firing of a population of neurons in the source layer (black dots, bottom). Extensive overlap occurs near the center (yellow). Right, same as in B except that the active source cells are broadly distributed. The overlap is substantially less. (C). Simulations showing the firing evoked in the 2-dimensional target layer when the active source neurons (not shown) are densely distributed. (D). Same as in C but for sparsely distributed source neurons. Model details: Each layer consists of excitatory pyramidal and inhibitory interneurons (200 × 200 cells; 25% inhibitory), with experimentally measured connections schemes and synaptic potentials. Within a layer, the excitatory neurons form reciprocal connections with each other and with inhibitory neurons. The neurons are adaptive-integrate-and-fire with parameters that match experimentally measured values. Across layers, source neurons synapse onto excitatory and inhibitory cells in the target layer. Connections from source neurons to target follow a Gaussian distributed probability with peak value of 0.5 and a standard deviation σ = 10–35 cell units. Only the active pyramidal cells are shown in the graphs.

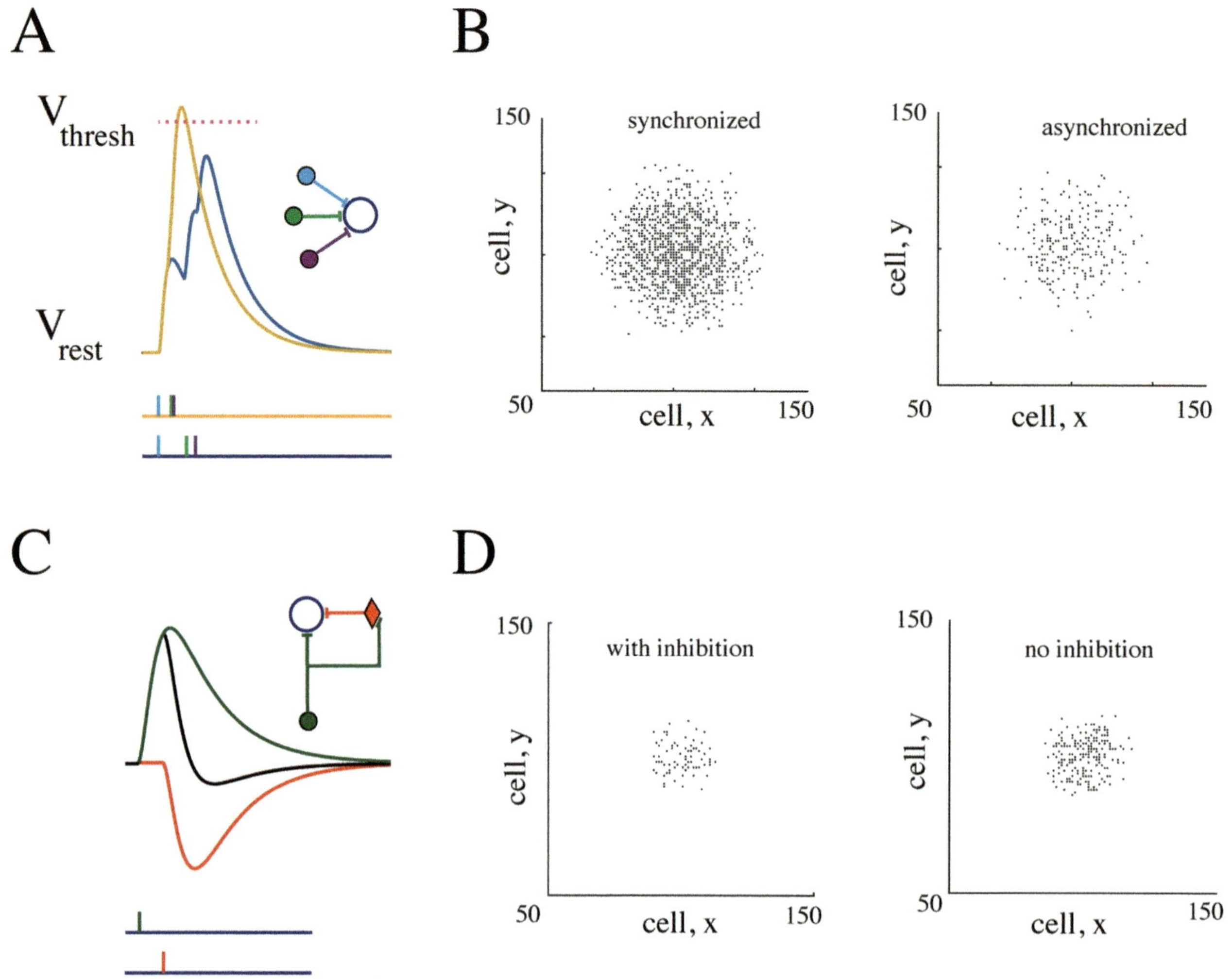

Figure 9.2. Temporal determinants of firing. (A). Composite synaptic potential evoked by presynaptic cells in a postsynaptic cell. Near synchronous spikes (bottom, orange baseline) from 3 presynaptic cells (inset, colored circles) generate a composite synaptic potential in the target cell that exceeds threshold (top, orange trace). In contrast, temporally dispersed firing of presynaptic cells (blue baseline) generates subthreshold synaptic potentials (top, blue trace). (B). Simulations showing action potentials evoked in a cell with synchronous (left) and asynchronous (right) presynaptic firing. (C). Composite synaptic potential evoked by mixed excitatory (green) and inhibitory (red) inputs. The circuit is in a feedforward inhibitory configuration (inset). (D). Simulations showing firing in the target layer with (left) and without (right) inhibition.

INTERLAMINAR DIFFERENCES IN RECEPTIVE FIELDS

There are systematic changes in the receptive field properties across layers in the primary sensory cortices. In many cases, the laminar differences seem to reflect refinement and/or integration of components of the incoming signals from the thalamorecipient layers. While there are differences among the sensory modalities, there appear to be some general trends.

The most consistent observation across the sensory modalities is that the stimulus-evoked responses tend to be stronger in thalamorecipient laminae and weaker and sparser in the supragranular laminae (L1–3). In the rodent barrel somatosensory cortex, the evoked responses are greatest in L5, followed by L4 and then L2/3 (De Kock et al., 2007; Manns et al., 2004). Calcium imaging of large populations indicates that fewer neurons become active in L2/3 (Voelcker et al., 2022). Similar observations were made in the visual (Hirsch et al., 2002; Martinez et al., 2005) and auditory systems (Sakata and Harris, 2009).

Tuning, or sensitivity to changes in stimuli, also differed across laminae. In the auditory system of gerbils and guinea pigs, tuning was sharper in supragranular laminae: the frequency response areas obtained by presenting tones of varying frequencies and intensities were narrower in L2/3 compared to L4 and L5 (Sugimoto et al., 1997; Wallace and Palmer, 2008). In monkey somatosensory cortex, neurons are more sharply tuned to the direction of cutaneous stimuli in laminae 2/3 compared to L4 (DiCarlo and Johnson, 2000). Finally, in the monkey visual cortex, tuning sensitivity to oriented grating stimuli varied across laminae in a complicated manner: neurons were narrowly tuned in upper lamina 4 (subdivision 4A) and lamina 2 and upper lamina 3 (subdivision 3A) and more broadly tuned in lower lamina 3 (subdivision 3B) (Ringach et al., 2002).

There are also changes in the size of the receptive field across laminae. A caveat is that determining the receptive field size can be tricky because the receptive field sizes vary with stimulus parameters (e.g., intensity) (Chapin, 1986); the true receptive field size is predicated on a particular stimulus parameter being sufficiently extensive in order to probe fully the range of responses. In the monkey visual system, the receptive fields are smallest in upper laminae 2/3 (Sceniak et al., 2001). Similarly, in the auditory system, frequency response areas are narrowest in the upper layers (Sugimoto et al., 1997; Wallace and Palmer, 2008). The opposite occurs in somatosensory cortex, where the receptive fields evoked with cutaneous stimuli (touch, vibration, or electrical stimulation) are larger in L2/3 than in L4 (Chapin, 1986; Sur et al., 1985). Similarly, in the rodent barrel cortex, the L2/3 receptive fields (obtained by deflecting the primary and surrounding whiskers and measuring the sub- and suprathreshold responses) are larger than those in thalamorecipient L4 and L5 (Brecht et al., 2003; Brecht and Sakmann, 2002; De Kock et al., 2007; Manns et al., 2004).

Finally, more complex stimuli are needed to drive neurons in L2/3. In many cases, the L2/3 receptive fields appear to be a composition of the simpler receptive fields in thalamorecipient layers. In the visual system, 2 broad classes of receptive fields are "simple" and "complex." Simple cells have bar-shaped receptive fields with distinct "on" and "off" regions where light stimulus can excite and inhibit the cells, respectively (Hubel and Wiesel, 1962; Hirsch and Gilbert, 1991); in contrast, complex cells have no such segregated "on/off" regions (Gilbert, 1977) and respond nonlinearly to stimuli of increasing size (Hubel and Wiesel, 1962; Movshon et al., 1978). It is postulated that the complex cell receptive fields are constructed by combining simple cell receptive fields in L4 (Gilbert, 1977; Martinez et al., 2002; Martinez et al., 2005). This appears to be true in the cat visual cortex, where simple cells are densest in thalamorecipient L4, with complex cells densest below and above L4 (Gilbert, 1977), but may not be true in monkey cortex, where simple and complex cells are equally distributed across laminae (Ringach et al., 2002).

Similarly, in somatosensory cortex, receptive field shapes in nongranular layers are not simple expansion or scaling of L4 receptive fields. The responses to cutaneous stimulation in layers 2/3 of rodents appear to be a combination of L4 receptive fields (Chapin, 1986). Neurons in L2 of barrel cortex are more multimodal than those in L4 in the sense that they are equally responsive to stimulation of nonprimary whiskers (Brecht et al., 2003; Brecht and Sakmann, 2002; De Kock et al., 2007; Manns et al., 2004; Voelcker et al., 2022). Whole-cell recordings reveal that these cells receive significant synaptic inputs from nonprimary whiskers (Brecht et al., 2003; Brecht and Sakmann, 2002; Manns et al., 2004). However, care should be taken in making broad generalizations (see **Chapter 6 in Section III**): for example, L4 pyramidal cells in exact opposite to L4 stellate cells and L3 pyramidal cells respond more to secondary whisker stimulation than to primary whisker stimulation.

SUMMARY

While the receptive fields of neurons vary across laminae, understanding the underlying process is not straightforward owing to multiple entry points for thalamic inputs and hence potentially convergent or parallel pathways. The strongest conclusion that can be drawn from the response latencies is that the signal starts at the thalamorecipient layers and propagates toward supragranular layers; the subsequent sequence of events is less clear. Compared to thalamorecipient neurons in L4 and L5, the responses of L2/3 neurons tend to be weaker and sparser. There were also differences, albeit inconsistent across laminae and species, in the tuning and receptive field size. Finally, there is evidence in some systems that the more complex receptive fields are constructed from simpler receptive fields in the thalamorecipient layers.

In the next section, simulations with a multilayered network will be used to ferret out the possible mechanisms underlying the preceding observations, and to identify the possible sources of inconsistencies. While the model is based experimental data (Levy and Reyes, 2011; Reyes, 2021), it is by no means complete. Specifically, the model lacks the detailed biophysical properties of neurons (see **Chapter 7 in Section III**) and the complex axonal arborizations (see **Chapter 6 in Section III**). However, the main goal is not to replicate specific experimental observations, but rather to develop an intuition for salient phenomena regarding transformations across laminae.

CONDITIONS FOR SIGNAL PROPAGATION

Consider a hypothetical 2-layer network consisting of a "source" layer whose neurons send axons to a "target" layer (Figure 9.1). For a neuron in the target layer to fire, it must receive sufficient excitatory inputs from source neurons to reach threshold. In cortex, the difference between resting potential and threshold for firing range from 10–20 mV. Given that the excitatory synaptic potentials (EPSPs) range from 0.1 to 1 mV (Levy and Reyes, 2012; Oswald and Reyes, 2008), many inputs are needed to evoke firing. This process, however, is much more complicated than just simply counting the number of inputs, as several additional variables come into play. To simplify the following discussion, only the responses to brief stimuli are considered. Brief stimuli—such as light flashes in the visual system, whisker deflection in the somatosensory system, and tone pips in the auditory system—activate primarily feedforward networks and minimize recurrent excitatory activity within layers (Levy and Reyes, 2011).

SPATIAL DISTRIBUTION OF ACTIVE SOURCE NEURONS

If topographically organized, the axons from source neurons terminate within a relatively confined area in the source layer (henceforth termed *synaptic fields*; Figure 9.1A). The axon collaterals from a single source neuron ramify to innervate several target neurons. Conversely, each target neuron receives inputs from multiple source neurons. Intuitively, the number of inputs a target cell receives depends on the extent of overlap of synaptic fields, which increases with the diameter of the synaptic field and decreases when the source cells are far apart from each other (Figure 9.1A). Figure 9.1B, left, shows a population of active source neurons in a 2-dimensional network (bottom) and their associated synaptic fields in the target layer (top). The densely packed set of active source cells generate extensive overlapping synaptic fields in the target layer (yellow); target cells within that area would receive sufficient inputs to fire. In contrast, when the active source cells are spread out (right, bottom), the target cells would receive only subthreshold inputs (top) and will not fire.

These concepts can be illustrated with a simulation of a 2-layer network where each layer consists of excitatory and inhibitory cells whose connection architecture and synaptic properties are based on experimental data (Levy and Reyes, 2011, 2012). Each 2-dimensional layer consists of 200 × 200 grid of cells. Details of the model are described elsewhere (Levy and Reyes, 2011; Reyes, 2021) and briefly summarized in the caption of Figure 9.1. As expected, more activity is generated in the target layer (compare C with D) when the active neurons in the source layer (not shown) are more densely packed.

CONNECTION PROBABILITY BETWEEN SOURCE AND TARGET NEURONS

The number of inputs a target neuron receives from source neurons depends critically on the connection probability. If the connection probability is low, more source neurons need to be active for the target neurons to receive sufficient excitatory drive to fire. If the connection probability is too low, it will be difficult to propagate activity beyond a single layer. On the other hand, a very high connection probability could overexcite the network, leading to uncontrolled, high-frequency firing and spatial spread of activity that will be further amplified in successive layers (see below).

TIMING OF ACTIVE SOURCE NEURONS

The timing of inputs is also an important determinant of neuronal firing. Because unitary synaptic potentials are brief and small, firing is predicated on many inputs arriving within a small time window. The maximum composite excitatory postsynaptic potential (EPSP) can be achieved if the active source neurons are synchronous. Figure 9.2A shows that the EPSP (orange, top trace) reaches threshold (dotted line) when the source neurons (lower trace, orange baseline) are active within a small time window, but not when the source neurons' firing is more dispersed (dark blue traces at top and bottom). The simulations in Figure 9.2B confirm that the activity in the target layer is greater when the source neurons fire synchronously (left) compared when they fire asynchronously (right).

There is evidence that these synchronous events occur under natural conditions. Intracellular recordings in vivo reveal large transient events that are in the range of 10 mV are evoked during sensory stimulation (Hirsch and Gilbert, 1991; Okun and Lampl, 2008; Wehr and Zador, 2003); these are consistent with the near simultaneous arrival of many, small unitary synaptic potentials. In somatosensory cortex, the synchronous firing of thalamic cells is crucial for evoking firing in cortex (Bruno and Sakmann, 2006). Finally, synchronous arrival of inputs is the basis for coincidence detection that underlies spatial localization of sound (Carr and Konishi, 1990).

LOCAL INHIBITORY INTERNEURONS

Though inhibitory neurons make up only 10%–25% of neurons in cortex (depending on brain areas and species), they have diverse morphologies, firing patterns, and connection schemes (Campagnola et al., 2022; Jiang et al., 2015). The basic properties and morphologies of inhibitory neurons are discussed in detail in **Chapter 8 of Section III**, and only a brief summary will be given here. Incorporating these different cell types is not practical or necessary, so the focus will be on the so-called fast spiking (FS) inhibitory neurons. FS neurons receive powerful inputs from the thalamus (Ji et al., 2016; Schiff and Reyes, 2012) and form high probability connections with neighboring pyramidal cells (Campagnola et al., 2022; Levy and Reyes, 2012). Their axons synapse on the soma or axons of pyramidal neurons and so can effectively suppress firing (Tremblay et al., 2016).

Functionally, FS cells are the primary source of feedforward inhibition in cortical circuits (Figure 9.2C; but see **Chapter 8 in Section III**). Brief stimulation either in vitro or in vivo yields the stereotypical sequence of an

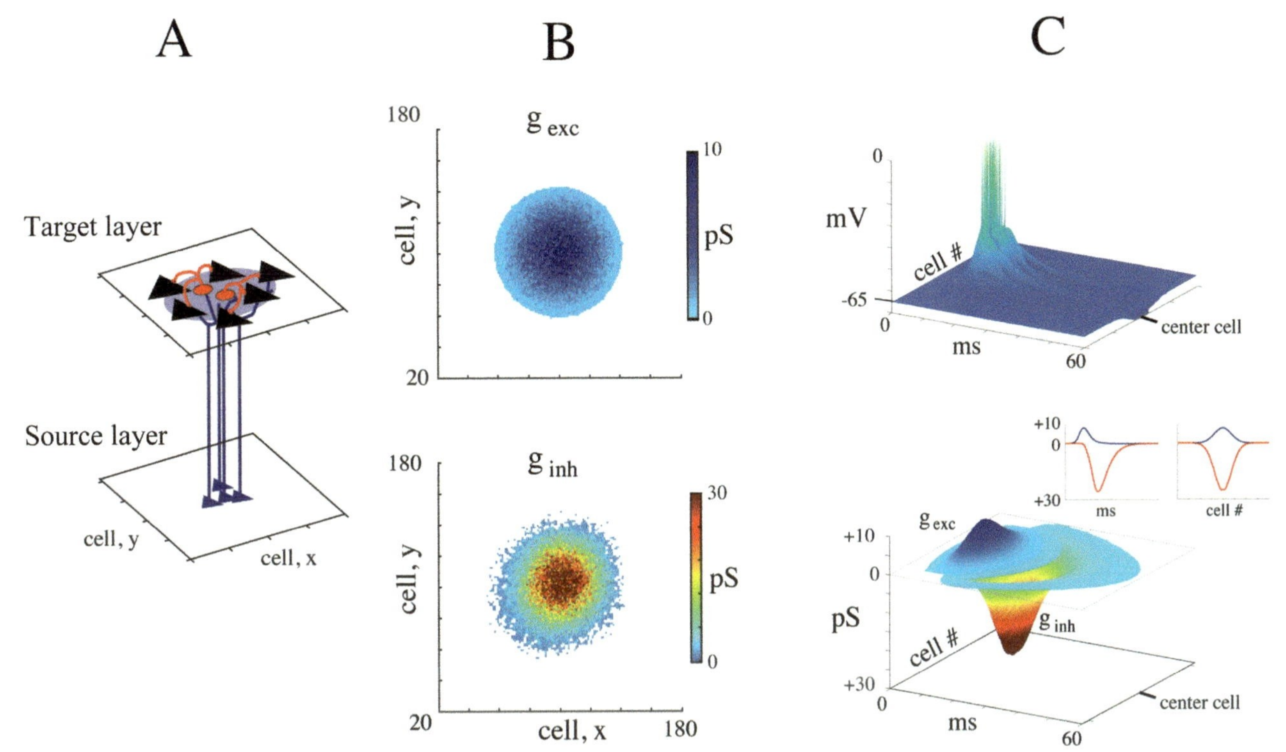

Figure 9.3. Spatiotemporal properties of synaptic inputs in simulations. (A). Schematic of interaction between excitatory (E, black triangles) and inhibitory (I, red circles) in target layer when source layer neurons become active. (B). Top: simulation showing the excitatory conductances (g_{exc}) evoked in neurons of the 2-dimensional target layer; bottom: inhibitory conductances (g_{inh}) evoked in source neurons. (C). Top: membrane potential of neurons in the target layer, arranged with increasing distance from the center of the network. Bottom: spatiotemporal profiles of the excitatory and inhibitory conductance. Note that the polarity of the inhibitory conductance is flipped (has positive values) for illustrative purposes. Insets show temporal (left) and spatial (right) profile of conductances.

EPSP, followed by an inhibitory postsynaptic potential (IPSP) after a short delay (Agmon and Connors, 1991; Gabernet et al., 2005; Hirsch and Gilbert, 1991; Mitani and Shimokouchi, 1985; Moore and Nelson, 1998; Schiff and Reyes, 2012). The delay is important because even if the IPSP is very large, there is a time window where the rising edge of the EPSP can trigger an action potential (Figure 9.2C). Simulations show that inhibition reduces firing significantly (Figure 9.2D, compare left and right). There are cases, however, when the order may be reversed, as with sequential activation of "on/off" regions in the visual cortex with moving bars (Hirsch et al., 2002; Martinez et al., 2002) or with activation of somatosensory neurons using directed cutaneous stimuli (DiCarlo and Johnson, 2000).

There has been some debate whether excitatory-inhibitory circuits are configured as lateral inhibition (aka "Mexican hat" profile, where inhibition has a larger receptive field; Wu et al., 2008) or are co-tuned (where both excitation and inhibition profiles are similar; Wehr and Zador, 2003). Each circuit configuration will have different effects on the receptive fields of neurons. However, modeling studies suggest that either configuration is not hardwired into the circuitry but rather can switch between the two, depending on the spatial extent of the input (Levy and Reyes, 2011).

Active excitatory neurons in the source layer generate synaptic inputs into both excitatory (E) and inhibitory (I) cells in the target layer; the I cells in turn generate inhibitory inputs to surrounding E cells (Figure 9.3A). The simulations in Figure 9.3B show the spatial footprint of excitatory conductance g_{exc} (top) and inhibitory conductance g_{inh} (bottom) evoked in neurons of the target layer. The inputs result in action potentials near the center (Figure 9.3C; voltage traces arranged based on distance of cells from center). The spatiotemporal profile of g_{exc} and g_{inh} (bottom) show that the action potentials are immediately halted by the inhibition in the spatial and temporal dimensions.

MULTILAYER PROPAGATION

The propagation of signals through feedforward networks is conceptually simple but is actually a very complicated process (Diesmann et al., 1999; Reyes, 2003; Vogels and Abbott, 2005). Ideally, for stable and reliable transmission, the signal that enters one end of the network should be qualitatively and quantitatively similar to that exiting at the other end. However, in most cases, strong synchronous firing develops during propagation regardless of input.

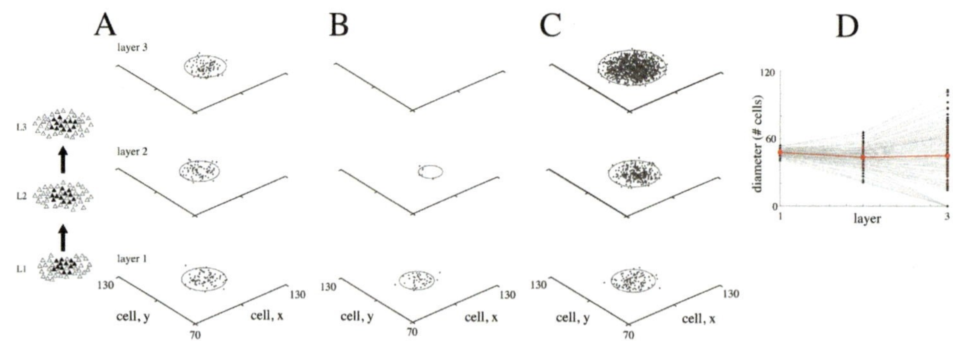

Figure 9.4. Propagation of signals. Simulations were with a 3-layer network (inset). (A). With stable propagation, the number and spatial extent of the active neurons (black dots) were nearly the same in all layers. The outer boundary of the activated area is fitted with a circle. (B). With diminishing propagation, the number of active neurons decreased in successive layers and eventually disappeared in layer 3. (C). With expanding propagation, the number of active neurons and spatial extent increased rapidly in successive layers. (D). Superimposed plots of circle diameter versus layer over many trials with randomized initial conditions in layer 1. Red trace is mean ± SD values.

Indeed, theoretical analyses suggested that there are essentially 2 states: either the signal propagates and synchronizes, or the signal fails to propagate. Importantly, stable propagation was possible only over a very narrow parameter space (Diesmann et al., 1999). Synchrony is remarkably resistant to changes in stimulus conditions and network configurations (Reyes, 2003; Vogels and Abbott, 2005), although recent experiments using a culture preparation suggest that asynchronous propagation can be extended in the presence of NMDA-mediated synapses in combination with recurrent activity (Barral et al., 2019).

Figure 9.4 shows propagation of signals in a 3-layer network. Propagation through these abstract layers may be thought of as propagation through the canonical L4–L2/3–L5 pathway. With stable propagation (Figure 9.4A), the number of active neurons and the spatial extent remain constant through the 3 layers. With diminished propagation (B), the number of active neurons decreases substantially in the next layer and fails to propagate beyond 1–2 layers. With expanding propagation (C), the number of active neurons and area increased rapidly with successive layers. Had there been more layers, nearly all neurons within each layer would have become active.

As might be expected from previous analyses of feedforward networks (Diesmann et al., 1999), it is difficult to produce stable propagation. Figure 9.4D plots the changes in the diameter of a circle enclosing the active neurons (see Reyes, 2021, for details of calculation) versus layer for randomized initial conditions in layer 1. The mean diameter (red) was constant across layers, but there was substantial trial-to-trial variability. In most trials, the diameter either decreased or increased and rarely remained constant.

As mentioned above, successful propagation depends on the connection probability between neurons across layers. For these simulations the connection probability was set to either 0.6 or 0.9. Setting the probability to 0.3 rarely allowed propagation to layer 3, though a more systematic exploration of parameters is needed before arriving at definitive conclusions. This is potentially problematic given that connection probabilities between across layers (L4 pyramidal cells to L2/3 pyramidal cells; L2/3 to L5; L5 to L2/3) appear to be low (<0.2; Campagnola et al., 2022).

Figure 9.5 shows dot rasters (left panels of A and B) and the spatiotemporal profiles (right panels) of the neurons' membrane potentials (blue) and underlying excitatory synaptic conductances (brown). Neurons are arranged according to their distance from the center neuron. With diminished propagation (A), the decrease in the number of evoked action potentials in one layer caused a progressive decrease in excitatory drive and number of action potentials in successive layers. Only subthreshold potentials reached layer 3. The opposite occurred with expanding propagation (B). The uncontrolled increases in activity were in part fueled by the increase in synchrony in successive layers, as indicated by the decrease in standard deviation of initial action potential times (values in insets of dot rasters). This raises the probability of evoking action potentials in a larger population; the activity spreads laterally, resulting in an increase in diameters (Figure 9.4D).

Figure 9.6A shows the computed receptive field of a single neuron at the center of each layer. The 2-dimensional receptive field (middle) may be thought of as the response of a neuron to a stimulus with 2 parameters (i.e., position in a Cartesian plane of visual stimulus, area of skin probed, or direction of whisker deflection). Projecting the receptive field onto one axis (bottom) gives the tuning to a single stimulus parameter (e.g., tone frequency in the auditory system). For diminishing propagation (Figure 9.6B), the receptive field size decreases in successive layers. The opposite occurs for expanding propagation (Figure 9.6C).

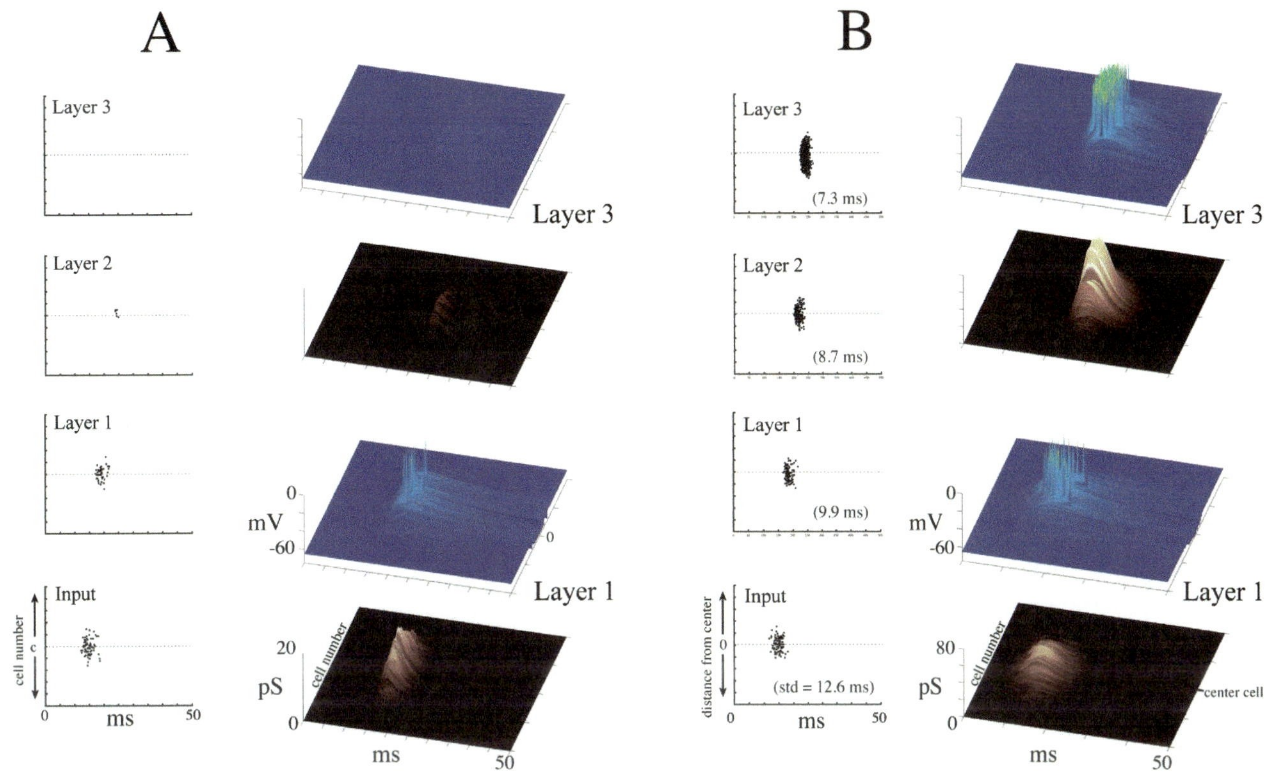

Figure 9.5. Spatiotemporal profile of propagated activity. (A). Left: dot rasters showing action potentials of active cells vs time. Cells arranged according to distance from the center of the network. Right: spatiotemporal profile of membrane potential (blue) and underlying excitatory synaptic conductance (brown) for layers 1 and 3. (B). Same as in A, but for an expanding propagation. Numbers in parentheses give the standard deviations of the action potential onset.

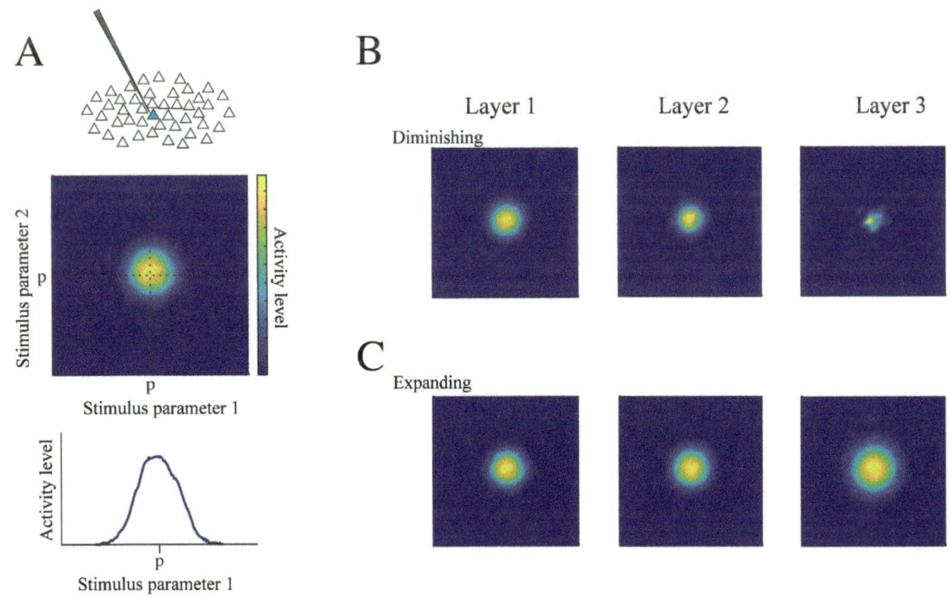

Figure 9.6. Changes in receptive field across laminae. (A). Top: schematic of electrode recording from a single neuron (cyan) in the network. Middle: receptive of the cyan neuron obtained by documenting the changes in its firing after making small changes in the stimulus. Activity is maximum at its preferred (p) stimulus parameter and decreases as it deviates from p. Lower: activity profile along a single axis. (B). Changes in receptive field for a center neuron in 3 layers when propagation diminishes across 3 layers. (C). Same for an expanding propagation.

SUMMARY

It is generally difficult to maintain stable propagation, even for just 3 layers. Expanding propagation could be advantageous for encoding (e.g., activity levels in the first layer) because small differences are amplified, but could produce catastrophic spread of activity to other brain areas due to developing synchrony. Based on the experimental data, diminished propagation appears to be the predominant mode in the cortex. Documented connection probabilities between neurons across layers are low, which favors reduced activity in successive layers. Moreover, in vivo experiments indicate that the thalamorecipient layers 4 and 5 have greater evoked activity than L2/3 (De Kock et al., 2007; Manns et al., 2004; Martinez et al., 2005; Sakata and Harris, 2009; Voelcker et al., 2022). This is actually inconsistent with propagation through the canonical L4 to L2/3 to L5 pathway since the simulations suggest that activities monotonically decrease or increase across layers.

CONSTRUCTION OF COMPLEX RECEPTIVE FIELDS

In the visual (at least in cats) and somatosensory cortices, the receptive field properties of neurons in L2/3 appear to be composed of simpler receptive fields in thalamorecipient layer 4 (Chapin, 1986; Hirsch et al., 2002; Voelcker et al., 2022). The constructed receptive fields are further shaped by local inhibitory circuits within a lamina (De La Rocha et al., 2008; Movshon et al., 1978; Petersen and Crochet, 2013). This conglomeration of receptive fields could partially account for the increased receptive field size observed in somatosensory cortex with cutaneous stimuli (Sur et al., 1985; Chapin, 1986). In this section, some salient features of receptive fields in the target layer obtained by combining receptive fields of neurons in the source layer will be examined with the simulations. The simple simulations are by no means exhaustive, but they do help identify the variables that play a role in shaping the constructed receptive fields. A simple example is the construction of a larger elongated receptive field in the target layer from circular receptive fields in the source layer (Figure 9.7A). In this scenario, a population of source neurons (blue) are activated by a stimulus to generate an excitatory synaptic field in the target layer (blue filled circle). A neuron in the target layer (black triangle) will therefore inherit the response properties of the blue neurons. In addition, local inhibitory interneurons in the target layer will be activated, generating an inhibitory synaptic field (dotted red circles). Thus, the composite synaptic field, in this case, is an excitatory component (Figure 9.7 A *i*, bottom, filled blue circle) surrounded by an inhibitory component (filled red circle on bottom). In the simulation, the single stimulus will not fire the target cell (*i* in Figure 9.7 B); though the neuron receives primarily excitatory (E, blue bars in A *i*, right) with little inhibition (I, red bar), the net input is weak (E + I, black bar).

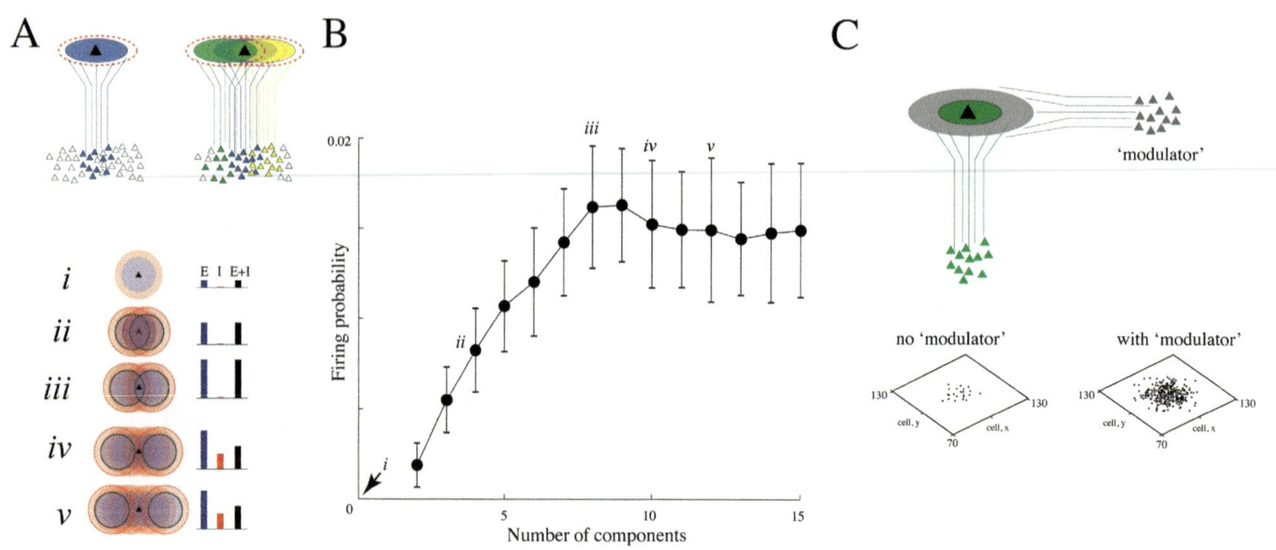

Figure 9.7. Combining receptive fields. (A). Top: hypothetical synaptic field evoked in target layer when a single population of source neurons (left, blue) and 2 others (right, green and yellow) become active. Bottom: *i*–*v* show the excitation (E = blue), inhibition (I = red) and net drive (E + I = black) when additional sets of source cells are added laterally. (B). Plot of neuron firing probability versus number of components. Firing probability increases to a peak and then declines to a steady value. (C). Hypothetical scenario where the response to inputs from a source population (green) is enhanced by an external, subthreshold input (modulator). Bottom: simulations show increase in number of active neurons in presence of modulator signal.

When 2 neighboring sets of source neurons (Figure 9.7A top right, yellow and green neurons) are co-activated, the resultant synaptic fields overlap with those of the blue neurons so that the target neuron now receives inputs from all 3 populations (Figure 9.7A, top right, yellow and green filled circles). The excitation (E, blue bars in Figure 9.7A, *ii*) in the target cell increased while the inhibition (I, red bars) remained low. Consequently, the net synaptic drive (E + I, black bars) increases, causing the firing probability to also increase (*ii* in Figure 9.7B). As more components are added laterally, the net excitation and firing probability continue to increase. At some point, additional components will no longer increase excitability since they would be too far away for their excitatory fields to contribute to the cell's total excitation (*iv*). However, because of lateral inhibition, the inhibitory drive increases, causing a decrease in the net excitability and in firing probability (*iv* in Figure 9.7B). Eventually, additional components will be too far for the inhibitory flanks to overlap, and the firing probability remains at a lower level of firing probability (*v*). This phenomenon is reminiscent of spatial summation of responses in monkey visual cortex (Sceniak et al., 2001).

Another scenario is when the target layer receives external inputs from, for example, other brain regions (Figure 9.7C). In this example, a diffuse subthreshold input from an external source enhances the firing evoked in the target layer, effectively increasing the receptive field of each neuron (lower traces).

CONCLUSIONS

The differences in the receptive field properties across laminae indicate that incoming signals from the thalamus are not only propagated but also transformed through the activities of a wide variety of neurons and unique circuitry in each layer. Taking into account the conditions needed for propagation, the numerous experimental data showing the responses in each layer, and simulation results, it seems likely that diminished propagation is the dominant mode in the cortex. This is advantageous because the potentially catastrophic spread of activity associated with expanding propagation can be avoided. Moreover, the reduced, non-saturated activity within a layer would be conducive to mixing of signals from other sources to produce a signal that incorporates features of other systems. A priori, each lamina—with different sets of neurons, connectivity patterns, and inputs—is designed for transformation of signals; simple propagation does not require multiple stages and can be more easily accomplished with direct connections.

REFERENCES

Agmon A, Connors B (1991). Thalamocortical responses of mouse somatosensory (barrel) cortex in vitro. *Neuroscience.* 41(2–3):365–379.

Ahissar E, Sosnik R, Bagdasarian K, Haidarliu S (2001). Temporal frequency of whisker movement. II. Laminar organization of cortical representations. *J Neurophysiol.* 86(1):354–367.

Barral J, Wang XJ, Reyes AD (2019). Propagation of temporal and rate signals in cultured multilayer networks. *Nature Comm.* 10(1):1–14.

Blasdel GG, Lund JS (1983). Termination of afferent axons in macaque striate cortex. *J Neurosci. 3*(7):1389–1413.

Brecht M, Roth A, Sakmann B (2003). Dynamic receptive fields of reconstructed pyramidal cells in layers 3 and 2 of rat somatosensory barrel cortex. *J Physiol.* 553(1):243–265.

Brecht M, Sakmann B (2002). Dynamic representation of whisker deflection by synaptic potentials in spiny stellate and pyramidal cells in the barrels and septa of layer 4 rat somatosensory cortex. *J Physiol.* 543(1):49–70.

Briggs F, Callaway EM (2005). Laminar patterns of local excitatory input to layer 5 neurons in macaque primary visual cortex. *Cereb Cortex.* 15(5):479–488.

Bruno RM, Sakmann B. (2006). Cortex is driven by weak but synchronously active thalamocortical synapses. *Science.* 312(5780):1622–1627.

Burkhalter A (1989). Intrinsic connections of rat primary visual cortex: Laminar organization of axonal projections. *J Comp Neurol.* 279(2):171–186.

Callaway EM (2004). Feedforward, feedback and inhibitory connections in primate visual cortex. *Neural Netw.* 17(5–6):625–632.

Campagnola L, Seeman SC, Chartrand T, Kim L, Hoggarth A, Gamlin C, Ito S, Trinh J, Davoudian P, Radaelli C, et al. (2022). Local connectivity and synaptic dynamics in mouse and human neocortex. *Science.* 375(6585):eabj5861.

Carr C, Konishi M (1990). A circuit for detection of interaural time differences in the brain stem of the barn owl. *J Neurosci.* 10(10):3227–3246.

Cetas J, De Venecia R, McMullen N (1999). Thalamocortical afferents of Lorente de No: Medial geniculate axons that project to primary auditory cortex have collateral branches to layer I. *Brain Res.* 830(1):203–208.

Chapin J (1986). Laminar differences in sizes, shapes, and response profiles of cutaneous receptive fields in the rat SI cortex. *Exper Brain Res.* 62(3), 549–559.

Constantinople CM, Bruno RM (2013). Deep cortical layers are activated directly by thalamus. *Science.* 340(6140):1591–1594.

Dantzker J, Callaway E (2000). Laminar sources of synaptic input to cortical inhibitory interneu-rons and pyramidal neurons. *Nature Neurosci.* 3(7):701–707.

De Kock C, Bruno RM, Spors H, Sakmann B (2007). Layer-and cell-type-specific suprathreshold stimulus representation in rat primary somatosensory cortex. *J Physiol.* 581(1):139–154.

De La Rocha J, Marchetti C, Schiff M, Reyes AD (2008). Linking the response properties of cells in auditory cortex with network architecture: Cotuning versus lateral inhibition. *J Neurosci. 28*(37):9151–9163.

DiCarlo JJ, Johnson KO (2000). Spatial and temporal structure of receptive fields in primate somatosensory area 3b: Effects of stimulus scanning direction and orientation. *J Neurosci.* 20(1):495–510.

Diesmann M, Gewaltig M-O, Aertsen A (1999). Stable propagation of synchronous spiking in cortical neural networks. *Nature.* 402(6761):529–533.

Feldmeyer D, Lübke J, Silver RA, Sakmann B (2002). Synaptic connections between layer 4 spiny neurone-layer 2/3 pyramidal cell pairs in juvenile rat barrel cortex: Physiology and anatomy

of interlaminar signalling within a cortical column. *J Physiol.* 538(3):803–822.
Feldmeyer D, Sakmann B (2000). Synaptic efficacy and reliability of excitatory connections between the principal neurones of the input (layer 4) and output layer (layer 5) of the neocortex. *J Physiol.* 525(1):31–39.
Gabernet L, Jadhav SP, Feldman DE, Carandini M, Scanziani M (2005). Somatosensory integration controlled by dynamic thalamocortical feed-forward inhibition. *Neuron.* 48(2):315–327.
Gilbert CD (1977). Laminar differences in receptive field properties of cells in cat primary visual cortex. *J Physiol.* 268(2):391–421.
Hirsch JA, Gilbert CD (1991). Synaptic physiology of horizontal connections in the cat's visual cortex. *J Neurosci.* 11(6):1800–1809.
Hirsch JA, Martinez LM, Alonso J-M, Desai K, Pillai C, Pierre C (2002). Synaptic physiology of the flow of information in the cat's visual cortex in vivo. *J Physiol.* 540(1):335–350.
Hornung J, Garey L (1981). The thalamic projection to cat visual cortex: Ultrastructure of neurons identified by golgi impregnation or retrograde horseradish peroxidase transport. *Neuroscience.* 6(6):1053–1068.
Huang CL, Winer JA (2000). Auditory thalamocortical projections in the cat: Laminar and areal patterns of input. *J Comp Neurol.* 427(2):302–331.
Hubel DH, Wiesel TN (1962). Receptive fields, binocular interaction and functional architecture in the cat's visual cortex. *J Physiol.* 160(1):106–154.
Jensen KF, Killackey HP (1987). Terminal arbors of axons projecting to the somatosensory cortex of the adult rat. I. The normal morphology of specific thalamocortical afferents. *J Neurosci.* 7(11):3529–3543.
Ji X-Y, Zingg B, Mesik L, Xiao Z, Zhang LI, Tao HW (2016). Thalamocortical innervation pattern in mouse auditory and visual cortex: Laminar and cell-type specificity. *Cereb Cortex.* 26(6):2612–2625.
Jiang X, Shen S, Cadwell CR, Berens P, Sinz F, Ecker AS, Patel S, Tolias AS (2015). Principles of connectivity among morphologically defined cell types in adult neocortex. *Science.* 350(6264):aac9462.
Kimura A, Donishi T, Sakoda T, Hazama M, Tamai Y (2003). Auditory thalamic nuclei projections to the temporal cortex in the rat. *Neuroscience.* 117(4):1003–1016.
Kratz MB, Manis PB (2015). Spatial organization of excitatory synaptic inputs to layer 4 neurons in mouse primary auditory cortex. *Front Neural Circuits.* 9:17.
Larsen DD, Callaway EM (2006). Development of layer-specific axonal arborizations in mouse primary somatosensory cortex. *J Comp Neurol.* 494(3):398–414.
Levy RB, Reyes AD (2011). Coexistence of lateral and co-tuned inhibitory configurations in cortical networks. *PLoS Comp Biol.* 7(10):e1002161.
Levy RB, Reyes AD (2012). Spatial profile of excitatory and inhibitory synaptic connectivity in mouse primary auditory cortex. *J Neurosci.* 32(16):5609–5619.
Manns ID, Sakmann B, Brecht M (2004). Sub-and suprathreshold receptive field properties of pyramidal neurones in layers 5a and 5b of rat somatosensory barrel cortex. *J Physiol.* 556(2):601–622.
Martinez LM, Alonso J-M (2001). Construction of complex receptive fields in cat primary visual cortex. *Neuron.* 32(3):515–525.
Martinez LM, Alonso J-M, Reid RC, Hirsch JA (2002). Laminar processing of stimulus orientation in cat visual cortex. *J Physiol.* 540(1):321–333.
Martinez LM, Wang Q, Reid RC, Pillai C, Alonso J-M, Sommer FT, Hirsch JA (2005). Receptive field structure varies with layer in the primary visual cortex. *Nature Neurosci.* 8(3):372–379.
Maunsell JH, Gibson JR (1992). Visual response latencies in striate cortex of the macaque monkey. *J Neurophysiol.* 68(4):1332–1344.
Mitani A, Shimokouchi M (1985). Neuronal connections in the primary auditory cortex: An electrophysiological study in the cat. *J Comp Neurol.* 235(4):417–429.
Mitani A, Shimokouchi M, Itoh K, Nomura S, Kudo M, Mizuno N (1985). Morphology and laminar organization of electrophysiologically identified neurons in the primary auditory cortex in the cat. *J Comp Neurol.* 235(4):430–447.
Moore CI, Nelson SB (1998). Spatio-temporal subthreshold receptive fields in the vibrissa representation of rat primary somatosensory cortex. *J Neurophysiol.* 80(6):2882–2892.
Movshon JA, Thompson ID, Tolhurst DJ (1978). Receptive field organization of complex cells in the cat's striate cortex. *J Physiology.* 283(1):79–99.
Okun M, Lampl I (2008). Instantaneous correlation of excitation and inhibition during ongoing and sensory-evoked activities. *Nature Neurosci.* 11(5):535–537.
Oswald A-MM, Reyes AD (2008). Maturation of intrinsic and synaptic properties of layer 2/3 pyramidal neurons in mouse auditory cortex. *J Neurophysiol.* 99(6):2998–3008.
Oviedo HV, Bureau I, Svoboda K, Zador AM (2010). The functional asymmetry of auditory cortex is reflected in the organization of local cortical circuits. *Nature Neurosci.* 13(11):1413–1420.
Petersen CC, Crochet S (2013). Synaptic computation and sensory processing in neocortical layer 2/3. *Neuron.* 78(1):28–48.
Reyes AD. (2003). Synchrony-dependent propagation of firing rate in iteratively constructed networks in vitro. *Nature Neurosci.* 6(6):593–599.
Reyes A, Sakmann B (1999). Developmental switch in the short-term modification of unitary epsps evoked in layer 2/3 and layer 5 pyramidal neurons of rat neocortex. *J Neurosci.* 19(10):3827–3835.
Reyes AD (2021). Mathematical framework for place coding in the auditory system. PLoS Comp *Biol.* 17(8):e1009251.
Ringach DL, Shapley RM, Hawken MJ (2002). Orientation selectivity in macaque v1: Diversity and laminar dependence. *J Neurosci.* 22(13):5639–5651.
Sakata S, Harris KD (2009). Laminar structure of spontaneous and sensory-evoked population activity in auditory cortex. *Neuron.* 64(3):404–418.
Sceniak MP, Hawken MJ, Shapley R (2001). Visual spatial characterization of macaque v1 neurons. *J Neurophysiol.* 85(5):1873–1887.
Schiff ML, Reyes AD (2012). Characterization of thalamocortical responses of regular-spiking and fast-spiking neurons of the mouse auditory cortex in vitro and in silico. *J Neurophysiol.* 107(5):1476–1488.
Schneider DM, Sundararajan J, Mooney R (2018). A cortical filter that learns to suppress the acoustic consequences of movement. *Nature.* 561(7723):391–395.
Shepherd GM, Svoboda K (2005). Laminar and columnar organization of ascending excitatory projections to layer 2/3 pyramidal neurons in rat barrel cortex. *J Neurosci.* 25(24):5670–5679.
Sugimoto S, Sakurada M, Horikawa J, Taniguchi I (1997). The columnar and layer-specific response properties of neurons in the primary auditory cortex of mongolian gerbils. *Hearing Res.* 112(1–2), 175–185.
Sur M, Garraghty PE, Bruce CJ (1985). Somatosensory cortex in macaque monkeys: Laminar differences in receptive field size in areas 3b and 1. *Brain Res.* 342(2):391–395.
Tremblay R, Lee S, Rudy B (2016). Gabaergic interneurons in the neocortex: From cellular properties to circuits. *Neuron.* 91(2):260–292.
Voelcker B, Pancholi R, Peron S (2022). Transformation of primary sensory cortical represen- tations from layer 4 to layer 2. *Nature Comm.* 13(1):1–13.
Vogels TP, Abbott LF (2005). Signal propagation and logic gating in networks of integrate-and-fire neurons. *J Neurosci.* 25(46):10786–10795.
Wallace M, Palmer A (2008). Laminar differences in the response properties of cells in the primary auditory cortex. *Exper Brain Res.* 184(2):179–191.
Wang Y, Ye M, Kuang X, Li Y, Hu S (2018). A simplified morphological classification scheme for pyramidal cells in six layers of primary somatosensory cortex of juvenile rats. *IBRO Reports.* 5:74–90.

Wehr M, Zador AM (2003). Balanced inhibition underlies tuning and sharpens spike timing in auditory cortex. *Nature.* 426(6965):442–446.

Winer J (1984). Identification and structure of neurons in the medial geniculate body projecting to primary auditory cortex (ai) in the cat. *Neuroscience.* 13(2):395–413.

Wu GK, Arbuckle R, Liu B-H, Tao HW, Zhang LI (2008). Lateral sharpening of cortical frequency tuning by approximately balanced inhibition. *Neuron.* 58(1):132–143.

Yabuta NH, Callaway EM (1998). Functional streams and local connections of layer 4c neurons in primary visual cortex of the macaque monkey. *J Neurosci. 18*(22):9489–9499.

10.

NEUROMODULATION IN CORTICAL CIRCUITS

Veronica C. Galvin and Anita A. Disney

WHAT IS NEUROMODULATION, AND WHERE DOES IT COME FROM?

All circuits in the central nervous system are subject to neuromodulation all of the time; cortical circuits are no exception. There is no cortical computation that is not at least constrained—more often it is importantly modified—by the instantaneous neurochemical composition of the extracellular space for the relevant cortical circuit. This is because chemical neuromodulators are state specifying for neural circuits.

In this chapter we focus on 6-layered cortex (also known as neocortex, which we will simply refer to as "cortex") and define neuromodulation as the actions within a circuit of molecules and peptides that dynamically modify synaptic transmission, neural excitability, or both. An interesting extended definition of neuromodulation as "not-neurotransmission" is offered by Katz (1999). Together, all of the neuromodulators active in the cortex create a dynamic neurochemical backdrop, or context, which constrains how cortical circuits process their various inputs, and how and when plastic changes occur based on those inputs. While the focus of this chapter is neuromodulation in cortical circuits, it is interesting to note in the broader context that while the primary subcortical neuromodulatory nuclei provide input to both thalamus and cortex (Figure 10.1), it is not the case that all cortical areas and thalamic nuclei are innervated by the same structures—that is, neuromodulation is by no means a collection of homogeneous signals across the brain. For example, in primates neither the ventral tegmental area (VTA) nor the locus coeruleus innervate the lateral geniculate nucleus of the thalamus, and the VTA does not innervate the primary visual cortex either. The functional relevance of the fact that the visual thalamus and visual cortex draw modulatory influences from different subcortical sources is considered elsewhere (Disney, 2021; Krueger and Disney, 2019).

At least in principle, any neuron with an axon in the cortex can be a source of neuromodulation, as can many non-neuronal cells. Both of the primary cortical neurotransmitters—γ-aminobutyric acid (GABA) and glutamate—can act as neuromodulators, usually when signaling via spillover to receptors outside synapses. Furthermore, many cortical inhibitory (i.e., GABAergic) and excitatory (glutamatergic) neurons can release modulatory amino acids (such as glycine and D-serine) or peptides, in addition to their primary neurotransmitters. Indeed, cortical inhibitory neurons are often classified by their expression of signaling peptides such as somatostatin and vasoactive intestinal peptide. Finally, molecules such as kynurenic acid and nitric oxide that arise from glia and endothelial cells within the central nervous system, and gonadal hormones from outside the central nervous system, are also active in cortical circuits on various timescales. In the end, the neurochemical component of contextual or "state" signals in the cortex probably comprises the joint action of tens, perhaps hundreds, of molecules arising from tissues across the brain and body. Thus, what we cover in this chapter barely scratches the surface of a rich, powerful, and currently poorly understood feature of cortical computation.

Most commonly, when people speak of neuromodulatory effects in cortical circuits, they have in mind the impact of *extrinsic* cortical neuromodulators (i.e., those produced by neurons whose cell bodies lie in subcortical nuclei; see Figure 10.1). The primary extrinsic neuromodulators for cortex are: acetylcholine (ACh, from the basal forebrain), norepinephrine (NE, also known as noradrenaline, from the locus coeruleus), dopamine (DA, from the ventral tegmental area), serotonin (5-HT, from the raphe nuclei), and histamine (HIST, from the tuberomammillary nucleus). These molecules, along with some of the modulatory effects of the neurotransmitters glutamate and GABA, will be the focus of this chapter. Because species differences in the structure and function of neuromodulatory systems are substantial (more on this later), we will generally limit ourselves to considering data from one primate species, macaque monkeys. Where we bring in data from another species, it is noted explicitly.

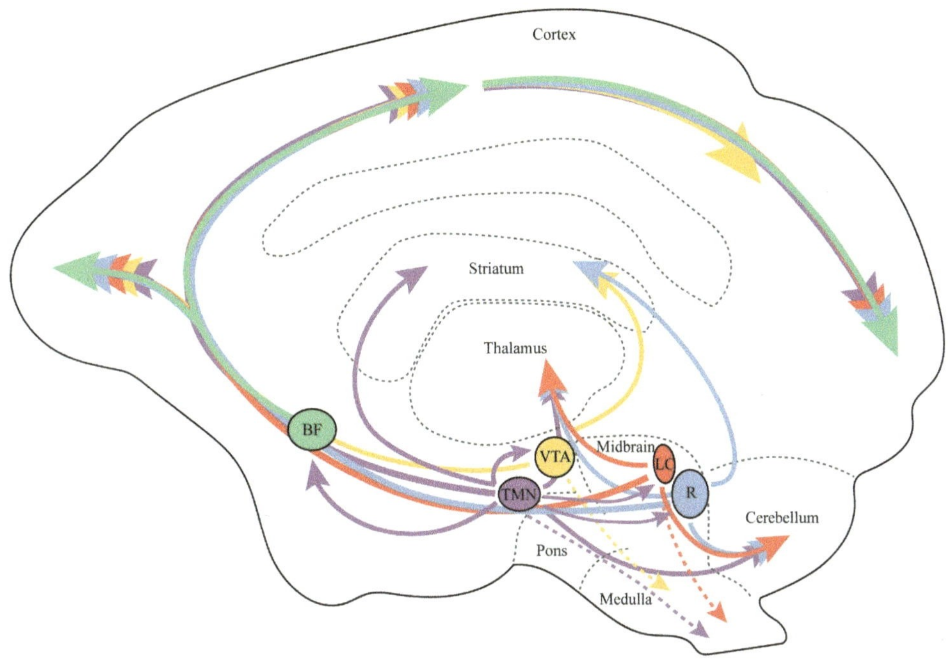

Figure 10.1. Extrinsic sources of neuromodulation for the cortex. The primary subcortical neuromodulatory systems that project to cortex are: the cholinergic system, which in primates primarily arises from the basal forebrain (BF) and releases ACh (green); the noradrenergic system arising from the locus coeruleus (LC) which releases NE (red); the dopaminergic system from the ventral tegmental area (VTA) which releases DA (yellow); the serotonergic system from the raphe nuclei (R) which release 5-HT (blue); and the histaminergic system from the tuberomammillary nucleus (TMN) which releases HIST (purple).

HOW ARE CORTICAL NEUROMODULATORY SIGNALS RECEIVED AND INTEGRATED?

Receiving neuromodulatory signals is critically important for the proper functioning of the cortex. In fact, at least some cortical circuits appear to depend upon the release of ACh in particular for circuit activation to take place at all. For example, around half of the neurons in the *visual cortex* of cats become *unresponsive to visual stimuli* when deprived of ACh (Sato et al., 1987). Similarly, depriving prefrontal cortex (PFC) in macaques of ACh produces behavioral deficits on working memory tasks that are similar to the deficits observed with cortical ablation (Goldman et al., 1971), and at the circuit level, blocking nicotinic (see Table 10.1) ACh receptors in the dorsolateral PFC causes a loss of essential delay-period neural activity in a working memory task (Yang et al., 2013).

Most cortical neuromodulators (but not all; Trimmer, 1999) act via ligand-receptor interactions, mediated by metabotropic, and in some cases ionotropic, receptors. All of the neuromodulators addressed in this chapter have multiple metabotropic receptors (also known as G protein-coupled receptors [GPCRs]; see Table 10.1 for a list of receptors by type) in which a ligand-binding domain is associated with a G protein complex that comprises 3 subunits; Gα, Gβ, and Gγ. Ligand binding to a GPCR triggers one or more of a dazzling array of intracellular signaling cascades, determined by the specific class of Gα subunit involved. In the cortex, the primary Gα subunit types are Gi/o, Gs, and Gq (reviewed by Weis and Kobilka, 2018). While there are common heuristics around the effect of activating particular Gα subtypes, in reality the consequence of activating any GPCR depends importantly on local targets for the relevant downstream cascade(s), and on levels and management of intracellular calcium. For example, altering levels of second messengers downstream of Gα activation in dorsolateral PFC or in primary visual cortex (V1) can result in opposing effects on the firing rates of neurons (Yang et al., 2018). This point can also be emphasized by considering the case of the m1 muscarinic ACh receptor, which is commonly coupled to a potassium current that controls spike-frequency adaptation (Brown and Adams, 1980). This current is, in fact, often called the "m-current" for the association with the m1 receptor. However, in primates, m1 receptors are strongly expressed by almost all fast-spiking inhibitory interneurons that, by definition, do not show spike frequency adaptation and thus have no "m-current." And so, while it is often the case that Gs- and Gq-coupled receptors will tend to have net excitatory effects on neurons, and Gi/o-coupled receptors are commonly associated with net inhibition, care must be taken to determine the coupling and functional effects of GPCR activation in any specific circuit of interest. One rule of thumb that does hold, however, is that all neuromodulatory systems have at least

TABLE 10.1 RECEPTORS EXPRESSED IN CORTEX FOR THE PRIMARY EXTRINSIC NEUROMODULATORS, GABA, AND GLUTAMATE

Ligand	Receptors	
	Ionotropic	**Metabotropic**
Acetylcholine	Nicotinic receptors, cation channels. *High affinity** heteropentameric with diverse subunit composition, primarily α3/α4/β2/β4 *Low affinity** homopentamers of α7 subunits, high Ca^{2+} permeability	Muscarinic receptors *Pirenzepine sensitive* Gq-coupled. m1, m3, m5 subtypes *Pirenzepine insensitive* Gi coupled. m2, m4 subtypes
Dopamine	None	*D1-like* Gs coupled D1, D5 subtypes *D2-like* Gi coupled D2, D3, D4 subtypes
GABA	GABAa receptor, chloride channels with diverse subunit composition	GABAb–Gi coupled
Glutamate	α-amino-3-hydroxy-5-methyl-4-isoxazolepropionic acid (AMPA) receptor, heterotetrameric cation channels comprising various combinations of 4 subunit types designated GluR1-4 Kainate receptor, heterotetrameric cation channels comprising various combinations of 5 subunit types designated GluK1–5 N-methyl-D-aspartate (NMDA) receptor, heterotetrameric calcium permeable cation channels comprising 2 obligatory NR1 subunits usually coupled to 2 NR2 subunits (4 types, NR2A–D) or less commonly NR3 subunits (NR3A,B)	*Group 1* Gq coupled mGluR1, mGluR5 *Group 2* Gi/o coupled mGluR2, mGluR3 *Group 3* Gi/o coupled, pharmacologically distinct from Group 2 mGluR4, mGluR6, mGluR7, mGluR8
Histamine	Possible chloride channel†	H1–Gq coupled H2–Gs coupled H3–Gi coupled
Norepinephrine	None	α1–Gq coupled α2–Gi coupled β–Gs coupled
Serotonin	5-HT3, pentameric cation channels with diverse subunit composition.	5-HT1–Gi coupled 5-HT2–Gq coupled 5-HT4–Gs coupled 5-HT5–Gi coupled 5-HT6–Gs coupled 5-HT7–Gs coupled

* Note that "high" and "low" affinity in the case of nicotinic ACh receptors refers to the affinity for nicotine, not for the endogenous ligand.

† Saras et al. (2008).

one Gi/o-coupled receptor that, when expressed presynaptically (i.e., on an axon), provides autoinhibitory feedback on that axon; thus the so-called autoreceptor for dopamine is expressed by dopaminergic axons and reduces dopamine release. Table 10.1 lists the known Gi/o coupled receptors for each neuromodulator.

In contrast to GPCRs, ionotropic receptors are ligand-gated ion channels in which a complex of protein subunits forms an ion-permeable pore in the neuronal membrane. Ligand binding to an ionotropic receptor thus directly contributes to membrane polarization, and in some cases to the in-flow of calcium, which is itself an important intracellular signal (reviewed by Berridge et al., 2003). There are ionotropic receptors for 5-HT and ACh, and probably for HIST as well. When GABA and glutamate act as neuromodulators, it is generally via GPCRs; signaling by GABA and glutamate through ionotropic receptors is understood to represent neurotransmission (for a discussion of differences between neurotransmission and neuromodulation, see Katz 1999).

After ligand binding to a receptor, modulatory signals can be integrated (Figure 10.2) both biochemically (via calcium inflow or GPCR activation of intracellular signaling cascades) and at the level of membrane permeability and voltage polarization (via ionotropic receptors or GPCRs coupled directly to ion channels). Integration of GPCR-mediated effects yields particularly intriguing possibilities (Fribourg et al., 2011); for example, group 1 metabotropic glutamate receptors (mGluRs) and m1 muscarinic ACh receptors are both Gq-coupled and are co-expressed by hippocampal neurons in rodents. In these neurons, co-activation of muscarinic ACh receptors and group 1 mGluRs has synergistic effects on neuronal activity on short (Park and Spruston, 2012) and long (Moore et al.,

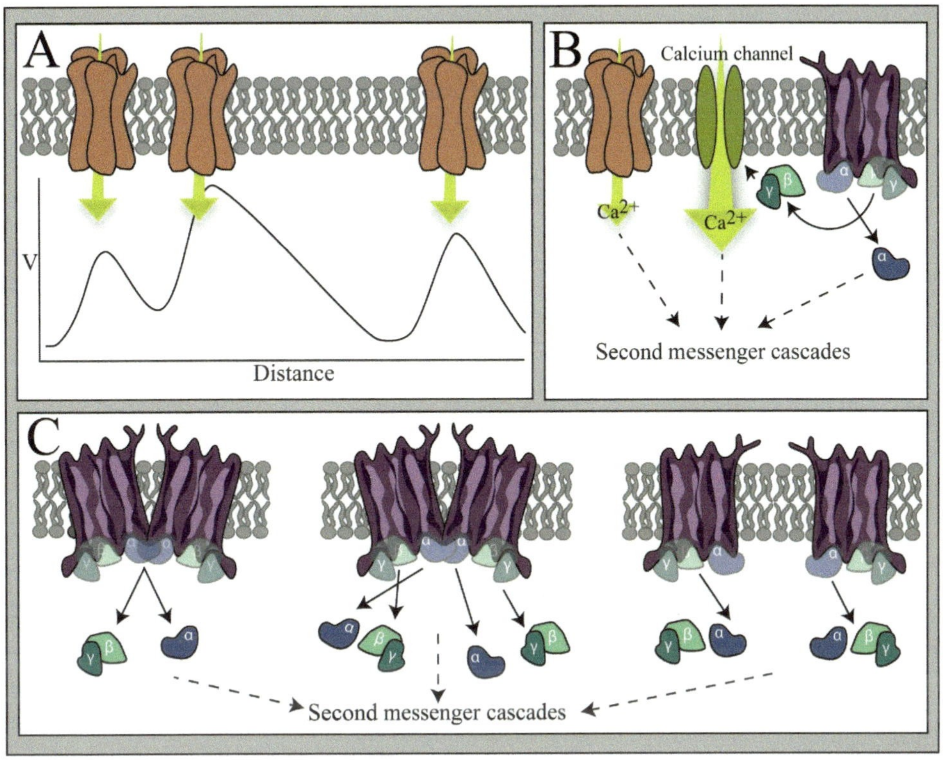

Figure 10.2. Integration of modulatory signals. (A) When neuromodulatory signals are integrated at the level of membrane voltage, standard principles underlying the summation of graded potentials hold such that spatial summation of potentials (left) is a function of proximity on the membrane. Voltage changes (V) from receptors at some distance from each other (right) are not integrated. (B) Some ionotropic receptors (left, orange) pass calcium (Ca^{2+}), and some metabotropic receptors (right, purple) are coupled to calcium-permeable channels (center, green). The resulting intracellular calcium signals interact with second messenger cascades and can be integrated, when allowed by proximity and local calcium sequestration. (C) Signals arising from the activated G protein subunits (α, ß, γ) of a metabotropic receptor can be integrated at the level of a single G protein interacting with 2 receptors (left); by interactions at the level of the G proteins, for complexes anchored together in the membrane (middle); or by interactions within the activation cascades of intracellular second messengers (right).

2009) timescales. Fascinatingly, although these receptors have the same G protein coupling, the short-term effect of co-activation is greater than the effect of a concentration-matched increase in activation of one receptor type alone, indicating integration more complex than a simple linear sum. The interaction between similarly coupled receptors need not, however, be synergistic; in macaque dorsolateral PFC, the α1 NE receptor and the type 2A 5-HT receptor—both of which are Gq-coupled—are co-expressed by pyramidal neurons and have opposing effects on neuronal activity during spatial working memory tasks (Jakab and Goldman-Rakic, 1998; Williams et al., 2002; Datta et al., 2019).

Both voltage and biochemical integration have the important implication that the signal carried by a particular molecule (e.g., a "serotonergic" or "dopaminergic" signal, which are in many ways problematic concepts; see Disney, 2021) need not remain distinct (i.e., from the "point of view" of the neuron, if there is such a thing). That is, when ionotropic or metabotropic receptors are expressed in close proximity on a membrane, they may converge upon the same integrators, and the original molecular identity of a signal (serotonergic or dopaminergic, for example) may be lost. It thus seems likely that the local membrane environment and subcellular compartmentalization around a receptor—such as anchoring proteins that determine proximity to other receptors and calcium-binding proteins that regulate dispersion of intracellular calcium—impact, even define, the way a neuron uses its modulatory inputs. While co-location and anchoring of receptors at synapses has been a focus of study (reviewed by Colledge and Froehner, 1998), the extent to which receptors on membranes outside synapses (extrasynaptic receptors) are co-located is generally not known.

CONSIDERING SPACE AND TIME

The prevalence of extrasynaptic signaling by neuromodulators has at times been a matter of debate, particularly for ACh (Disney and Higley, 2020; Sarter and Lustig, 2020). Neurochemical signaling outside synapses (or volume transmission) has often been presented in a dichotomy with synaptic signaling. Based on this dichotomy, attempts have been made to quantify anatomically identified release sites that participate in synaptic signaling (those with a closely apposed membrane where receptors are expressed; Figure 10.3A) versus sites that contribute to volume transmission (those with no nearby receptive membrane specialization; Figure 10.3B). With the exception of ACh, to which we return in a moment, there is no evidence that quantal release of neuromodulators is so low and/or that reuptake mechanisms are so fast and/or high capacity that modulatory molecules are constrained to remain within synapses (Figure 10.3C). This suggests that, at least anatomically, the synaptic versus volume transmission dichotomy may be uninformative because molecules probably escape synapses. ACh, on the other hand, is broken down by an extraordinarily fast enzyme, acetylcholinesterase (AChE), the ability of which to exert near-perfect spatial control over synaptic signaling is essential to the function of the neuromuscular junction. We have addressed elsewhere (Disney and Higley, 2020) the reasons this extraordinary capacity is probably not applicable to ACh signaling in cortex, and thus why cortical ACh almost certainly participates in extrasynaptic signaling, along with all of the other primary cortical neuromodulators.

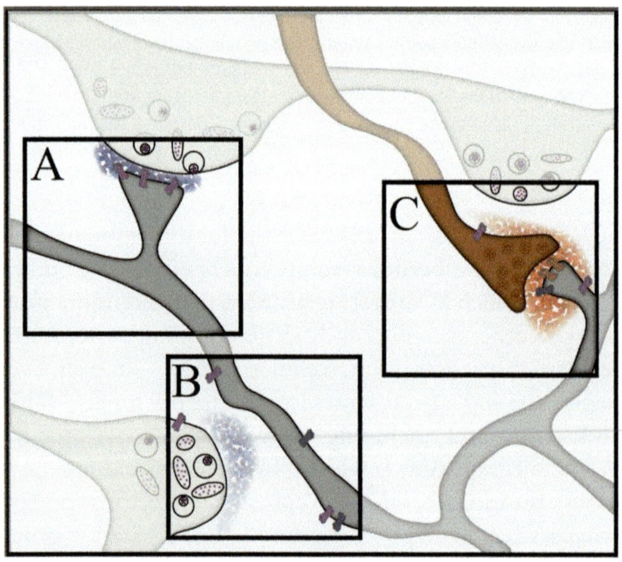

Figure 10.3. Synaptic and extrasynaptic signaling in cortex. (A) Synaptic transmission occurs when the release site (top of inset, with vesicles) and receptive membrane (bottom of inset, expressing purple receptors on the cell surface) are closely apposed, and reuptake mechanisms are sufficient to constrain quantal release (blue cloud) to the synaptic cleft. (B) Purely extrasynaptic transmission occurs when a release site (left of inset) has no nearby receptive membrane. In this case, target receptors (blue and purple) can be expressed on various extrasynaptic cellular compartments, including dendrites (gray) and axons (left of inset). (C) A combination of synaptic and extrasynaptic transmission can occur when release exceeds reuptake capacity and "spillover" (note spread of orange cloud) takes place, in which signaling molecules escape the synaptic cleft to diffuse through the broader extracellular space.

In the end, the prevalence of both ionotropic and metabotropic modulatory receptors on non-synaptic neuronal membranes (Disney et al., 2006; Wang et al., 2007; Yang et al. 2013; Gamo et al., 2015; Jin et al., 2018; Datta et al., 2019; Galvin et al., 2020; and see Vizi et al., 2004, for a review of earlier studies) and the responsiveness to multiple neuromodulators of non-neuronal cells such as glia and endothelial cells that do not make synapses (Liu et al., 2016; Kofuji and Araque, 2021; Iadecola, 2017) strongly argue for extrasynaptic signaling by all neuromodulators. That said, modulatory synapses certainly exist. In our view, discussions of which signaling mode (synaptic or extrasynaptic) is prevalent, "stronger," or somehow more important—whether for individual molecules or neuromodulation in general—are more illuminative of assumptions about brain function than of brain function itself. Most of us are, however, more familiar with synaptic communication and its variations and implications, so we consider here some implications of extrasynaptic communication in more detail, not because it is more important, but because it is less familiar.

The need to reach extrasynaptic receptors which are, by definition, at some distance from release sites means that diffusion through the extracellular space (ECS) is an important feature of neuromodulatory signaling. When diffusion is a factor, concentrations will differ across a circuit. This is a nontrivial observation; receptors differ in their ligand-binding affinities and so, at lower concentrations, fewer receptor types can participate in signaling. Diffusion also means that there is a limit to how fast signaling can be. This does not mean that neuromodulatory signaling is *necessarily* slow; there is no reason to believe that synaptic neuromodulation cannot be as fast as synaptic neurotransmission, particularly when ionotropic receptors are involved. The existence of modulatory synapses *and* extrasynaptic receptors, and of fast ionotropic as well as slower metabotropic receptors, instead means that neuromodulation in cortex operates on timescales that vary by multiple orders of magnitudes—from milliseconds to minutes and hours and, in the case of plasticity, well beyond.

Together, the wide expression of GPCRs in cortex and the fact that for NE and DA there are no known ionotropic receptors mean that mid- to long-timescale modulatory effects (i.e., those on timescales of hundreds of milliseconds or more) are common. Signaling timescales are intimately connected with length-scale when diffusion is involved, and multiple factors contribute to diffusion distances for neuromodulators in cortical circuits. One factor is obviously the distance from a release site to the nearest receptors. How far beyond those nearest receptors a molecule might travel depends in part upon the availability and capacity of signal termination mechanisms,

TABLE 10.2 NEUROMODULATOR SYNTHESIS AND DEGRADATION

Molecule	Synthesis	Termination
Acetylcholine	Choline acetyltransferase	Acetylcholinesterase, followed by choline uptake transporter
Dopamine	Tyrosine hydroxylase	Dopamine or norepinephrine uptake transporters
GABA	Glutamic acid decarboxylase	Uptake transporters
Glutamate	Glutaminase, 2-oxoglutarate	Uptake transporters (expressed on astrocytes)
Histamine	Histidine decarboxylase	Histamine-N-methyltransferase
Norepinephrine	Tyrosine hydroxylase, followed by dopamine β hydroxylase	Norepinephrine or dopamine uptake transporters
Serotonin	Tryptophan hydroxylase, followed by aromatic amino acid decarboxylase	Serotonin uptake transporter

which include degrading enzymes and uptake transporters (see Table 10.2). In the time (and space) between release and receptor binding or termination, molecules traverse the ECS, the physical features of which also contribute to time- and length-scales of diffusion (Syková and Vargová, 2008).

THERE IS NO CANONICAL MODULATORY CIRCUIT FOR CORTEX

While all cortical areas are subject to neuromodulation, the particular mix of modulatory molecules acting in a given cortical area and the laminar patterns of receptor expression and axonal innervation differ between cortical areas and across species. Species differences demonstrated for the modulatory systems range from the subtle—a nicotinic receptor subtype substitution at thalamocortical synapses in the barrel cortex of mice versus rats (Gil et al., 1997)—to the profound: a near-absence of a dopaminergic innervation of the primary visual cortex of primates versus a clear VTA innervation of V1 in rodent (Berger et al., 1991; for more on species differences, see Morrison and Foote, 1986; Foote and Morrison, 1987; McCormick and Williamson, 1991; Wilson and Molliver, 1991a, 1991b; Sánchez-González et al., 2005; Coppola and Disney, 2018). The resulting lack of a single, stereotypical modulatory circuit that can be studied in one "model system" and then applied across cortical areas or species does not, however, mean that there are not common or "canonical" computations or operations being performed by neuromodulatory systems. In the sections that follow, we will consider evidence for a putatively canonical function being performed by ACh across sensory systems and species, often served by a similar circuit-level mechanism. We will also look at evidence that even within a species, the monoamine (NE, DA, and 5-HT) modulatory systems differ widely across the cortex.

NICOTINIC ACH RECEPTORS ENHANCE THE FEEDFORWARD INPUT FROM THE THALAMUS TO THE CORTEX

Across cortical areas and in rats, cats, and monkeys, high-affinity (heteropentameric; see Table 10.1) nicotinic ACh receptors are expressed on axons arriving in the primary sensory cortices from first-order thalamic nuclei serving vision (Prusky et al., 1987; Kimura et al., 1999; Disney et al., 2007), touch (Gil et al., 1997), and hearing (Hsieh et al., 2000). Because they are ligand-gated cation channels, activating these receptors contributes to depolarizing the presynaptic membrane and opening voltage-gated calcium channels; the resulting influx of calcium then goes on to trigger vesicle fusion and neurotransmitter release. In one species (mice) the nicotinic receptor subtype being expressed at this point in the circuit (the homomeric receptor comprising 5 $\alpha7$ subunits; Gil et al., 1997) itself passes calcium with high affinity and can thus contribute to vesicle release in a voltage-independent manner. It has been suggested that the enhanced thalamocortical transmission arising from activation of nicotinic ACh receptors on thalamocortical axons, combined with the suppression of corticocortical interactions that accompanies activation of muscarinic ACh receptors in the cortex (via m2 ACh receptor–mediated suppression of glutamate release or m1 ACh receptor–mediated enhancement of local soma-targeting inhibition, or both), is a canonical computation (Disney and Robert, 2019) that yields preferential processing of sensory inputs, perhaps in support of vigilance or attention (Hasselmo and McGaughy, 2004).

Interestingly, at least in the primary somatosensory cortex of rats, the same corticocortical synapses that are suppressed by muscarinic modulation are subject to (suppressive) modulatory effects of GABA acting through GABAb receptors (Gil et al., 1997), suggesting that pathway-selective modulation may be a motif for modulatory systems more broadly, not just for the cholinergic system. Selective receptor expression can also support differential modulation of pathways within the cortex (not just between the cortex and the thalamus); in macaques, the pathway from area 32 of the anterior cingulate cortex to area 9 of the dorsolateral PFC carries muscarinic ACh receptors, while the pathway from area 46 to area 9 (both within dorsolateral PFC) does not appear to be a target for muscarinic modulation (Medalla and Barbas, 2012).

Finally, there is evidence that nicotinic receptors are also expressed on thalamocortical axons arriving in PFC and parietal cortices, at least in rats (Lavine et al., 1997), and positron emission tomography with radiolabeled nicotinic ligands suggests that, in humans, the mediodorsal thalamic nucleus has the highest expression of nicotinic receptors of all the thalamic nuclei (Garibotto et al., 2020). It thus seems likely that there is nicotinic modulation of non-sensory thalamocortical pathways. Cholinergic modulation of corticothalamic transmission is less well studied, but during development in the frontal lobe of rats, corticothalamic neurons in layer 6 (the cell bodies in the cortex, not the axons in the thalamus) strongly respond to nicotinic modulation (Kassam et al., 2008). Layer 6 neurons in rats generally (i.e., undifferentiated with respect to projection target) continue to be highly responsive to nicotinic modulation into adulthood (Proulx et al., 2014).

Interestingly, a motif in which high levels of ACh yield preferential processing of ascending sensory data was first identified in the piriform cortex of rats, mediated by muscarinic suppression of lateral connectivity without an accompanying nicotinic enhancement of input from the periphery (Hasselmo and Bower, 1991). Thus modulatory motifs can be conserved across species or systems, even when the supporting receptor expression diverges (Disney and Robert, 2019), and there certainly are species differences at the mechanistic level for this motif. As we have already mentioned, the specific nicotinic receptor subtype that underlies the thalamic enhancement is of the high-affinity type in rats, and the low-affinity type in mice (Gil et al., 1997), and there are species differences in the type and localization of muscarinic receptors underlying the suppression as well (Disney and Reynolds, 2014; Disney and Robert, 2019). There is value in a continued search for conserved modulatory motifs both at the mechanistic and algorithmic levels. At the mechanistic level, however, differences across the cortex outnumber the similarities, even for ACh. For example, expression of the terminating enzyme AChE (see Table 10.2) differs so profoundly across brain areas that it is used as an architectonic marker in brain atlases, along with stains for Nissl bodies and myelin.

DIFFERENCES IN MONOAMINERGIC SYSTEMS BETWEEN AND WITHIN SPECIES

One of the earliest descriptions of differences in neuromodulation between cortical areas was for the monoamines in macaques. Brown et al. (1979) measured levels of 5-HT, DA, and NE across all 4 lobes of the macaque cortex and found that each molecule showed a different pattern of concentrations across the cortical mantle. Levels of 5-HT were fairly similar (and quite high) in all areas studied, and there was a rostro-caudal gradient for DA, with high levels in the orbitofrontal and dorsolateral PFC and low levels in the visual cortices. Our recent unpublished data indicate that in macaques, DA release from the VTA into the most posterior cortical area (V1) is below limits of detection for the most sensitive assays. This is unsurprising given that the VTA innervation of V1 in macaques is negligible (Lewis et al., 1987). It is important here to note that while it is common to try to use innervation density as an indicator of neuromodulator concentration, one cannot actually predict DA concentrations across the cortex (Brown et al., 1979) from VTA innervation (Lewis et al., 1987). Furthermore, DA receptor distributions match neither the concentration gradients, nor the patterns of innervation (Froudist-Walsh et al., 2021). It is also important to consider that in studying homogenized tissue, as do Brown et al., one learns something akin to the dynamic range of a modulatory system. That is, if all of the dopamine available in the axonal arbor were released at once, what would be the maximum concentration that can be achieved in the extracellular space (ECS)? It should not be concluded that concentrations reported from studies of homogenized tissue reflect actual, or even relative differences between, true ECS levels.

In the study by Brown et al. (1979), NE concentration had a different pattern again (when compared with 5-HT and DA), with moderate levels in the prefrontal and temporal cortices, low levels in the visual cortices, and high levels in the somatomotor areas around the central sulcus. As is the case for DA, levels of NE cannot be predicted from patterns of cortical innervation (Morrison and Foote, 1986) and patterns of receptor expression differ again (Niu et al., 2020; Rapan et al., 2022). Finally, care might be taken in interpreting the data on 5-HT levels (Brown et al., 1979); the animals in that study were not exsanguinated prior to tissue collection and homogenization, so there was blood in the homogenate. 5-HT levels in blood are high, and likely similar across the cortex. Our unpublished data on visual cortex from an exsanguinated macaque indicates 5-HT levels are both lower and more variable that previously reported (Figure 10.4). Again, patterns of serotonergic innervation differ across the cortex, and do not match patterns of receptor expression (Morrison and Foote, 1986; Watakabe et al., 2009; Palomero-Gallagher et al., 2013; Rapan et al., 2022).

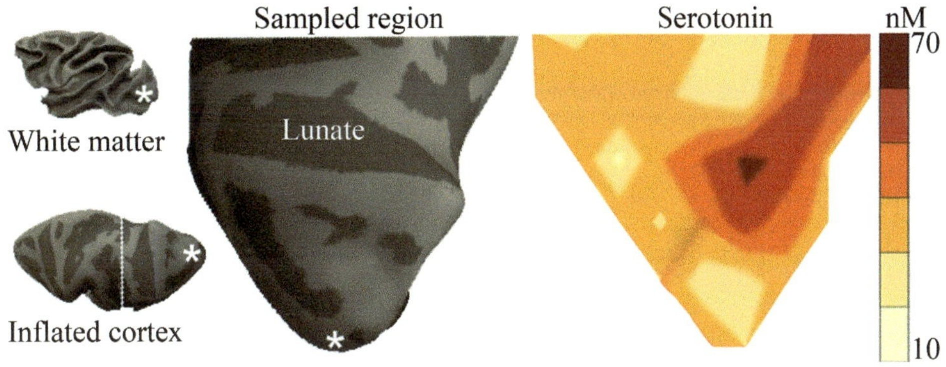

Figure 10.4. Serotonin levels in the posterior cortex of a macaque. 5-HT was quantified for the visual cortex, including the tissue from the prelunate gyrus to the occipital pole (approximate blocking cut indicated by the white dashed line on the inflated cortex, bottom left). Left: white matter and inflated cortex models showing a lateral view of the left (sampled) hemisphere with the occipital pole (*) to the right. Middle: the dorsal surface of the sampled region from the inflated cortex model; the lunate sulcus is indicated, as is the occipital pole (*), at bottom. Right: cubes of cortex were cut, homogenized, and analyzed by liquid chromatography with mass spectrometry. Raw concentration values (range: 10–70 nM) were used to generate an interpolated map of 5-HT concentration, which it can be seen varies considerably over this region of the visual cortices.

Not considered here, in the interests of space, are (1) differences in receptor expression across cortical layers and cell classes, (2) differences in receptor affinities, and (3) the complex patterns of expression for mechanisms of termination (reviewed for early vision by Disney, 2021). Differences in receptor expression, in particular, are considerable and are an important means by which cortical cells and circuits can respond in a differential manner, even when they are exposed to a diffuse signal. Differences in patterns of receptor expression can be observed between adjacent cortical areas in macaques (Disney et al., 2006; Palomero-Gallagher et al., 2013; Niu et al., 2020; Rapan et al., 2022) and are also observed for a given cortical area, when comparisons are made between species (Disney and Reynolds, 2014). We have proposed previously that multivariate data on the anatomy of the cholinergic system, including patterns of receptor expression, axonal innervation, and signal termination, suggest that the cortex comprises a collection of "compartments" (which may or may not line up neatly with architectonically defined cortical areas) between which ACh operates quite differently (Coppola et al., 2016). Based on differences across the cortex in patterns of innervation, receptor expression and affinity, signal termination, and concentration dynamic range, it seems likely that the same will turn out to be true for the monoaminergic systems as well.

CLOSING THOUGHTS

One common question we have not considered in this chapter is what specific neuromodulators might be "for." This is, in our view, a problematic concept that one of us has discussed elsewhere (Disney, 2021). As the first of 2 final thoughts, it is important to recall that the subcortical modulatory nuclei considered in this chapter are highly interconnected such that, for example, the locus coeruleus innervates the cortex *and* other modulatory nuclei as well. So the release of 5-HT into cortex, for example, represents an integrated input that incorporates information about whatever computations are being performed in the raphe nuclei, as well as the computations being performed in the LC. Modulatory signals arriving in cortex are thus not independent of each other. Our final thought is this: It is interesting to consider the functional consequences of the fact that the thalamic nuclei can receive modulatory input from structures different from those innervating their cortical partners (Krueger and Disney, 2019; Disney, 2021). This means that thalamic nuclei likely comprise yet another collection of modulatory compartments, in addition to those in the cortex. And so it is no longer evidence-based to view neuromodulation as diffuse, or global, in character; neuromodulators should be considered complex, dynamic, locally differentiated, and locally regulated participants in cortical state specification and context-setting.

ACKNOWLEDGMENTS

The authors thank Alev Brigande for help with Figure 10.4, including data analysis, and Dr. Juliane Krueger for help with Figure 10.1.

REFERENCES

Berger B, Gaspar P, Verney C (1991). Dopaminergic innervation of the cerebral cortex: unexpected differences between rodents and primates. *Trends Neurosci.* 14:21–27.

Berridge MJ, Bootman MD, Roderick HL (2003). Calcium signalling: dynamics, homeostasis and remodelling. *Nat Rev Mol Cell Biol.* 4:517–529.

Brown RM, Crane AM, Goldman PS (1979). Regional distribution of monoamines in the cerebral cortex and subcortical structures of the rhesus monkey: Concentrations and in vivo synthesis rates. *Brain Res.* 168:133–150.

Brown DA, Adams PR (1980). Muscarinic suppression of a novel voltage-sensitive K+ current in a vertebrate neurone. *Nature.* 283:673–676.

Colledge M, Froehner SC (1998). To muster a cluster: Anchoring neurotransmitter receptors at synapses. *Proc Natl Acad Sci.* 95:3341–3343.

Coppola JJ, Disney AA (2018). Is there a canonical cortical circuit for the cholinergic system? Anatomical differences across common model systems. *Front Neural Circuits.* 12:8.

Coppola JJ, Ward NJ, Jadi MP, Disney AA (2016). Modulatory compartments in cortex and local regulation of cholinergic tone. *J Physiol Paris.* 110:3–9.

Datta D, Yang S-T, Galvin VC, Solder J, Luo F, Morozov YM, Arellano J, Duque A, Rakic P, Arnsten AFT, Wang M (2019). Noradrenergic α1-adrenoceptor actions in the primate dorsolateral prefrontal cortex. *J Neurosci.* 39:2722–2734.

Disney AA (2021). Neuromodulatory control of early visual processing in macaque. *Annu Rev Vis Sci.* 7:181–199.

Disney AA, Aoki C, Hawken MJ (2007). Gain modulation by nicotine in macaque v1. *Neuron.* 56:701–713.

Disney AA, Domakonda KV, Aoki C (2006). Differential expression of muscarinic acetylcholine receptors across excitatory and inhibitory cells in visual cortical areas V1 and V2 of the macaque monkey. *J Comp Neurol.* 499:49–63.

Disney AA, Higley MJ (2020). Diverse spatiotemporal scales of cholinergic signaling in the neocortex. *J Neurosci.* 40:720–725.

Disney AA, Reynolds JH (2014). Expression of m1-type muscarinic acetylcholine receptors by parvalbumin-immunoreactive neurons in the primary visual cortex: A comparative study of rat, guinea pig, ferret, macaque, and human. *J Comp Neurol.* 522:986–1003.

Disney AA, Robert JS (2019). Translational implications of the anatomical nonequivalence of functionally equivalent cholinergic circuit motifs. *Proc Natl Acad Sci U S A.* 116:26181–26186.

Foote SL, Morrison JH (1987). Extrathalamic modulation of cortical function. *Ann Rev Neurosci.* 10:67–95.

Fribourg M, Moreno JL, Holloway T, Provasi D, Baki L, Mahajan R, Park G, Adney SK, Hatcher C, Eltit JM, Ruta JD (2011). Decoding the signaling of a GPCR heteromeric complex reveals a unifying mechanism of action of antipsychotic drugs. *Cell.* 147:1011–1023.

Froudist-Walsh S, Bliss DP, Ding X, Rapan L, Niu M, Knoblauch K, Zilles K, Kennedy H, Palomero-Gallagher N, Wang X-J (2021). A dopamine gradient controls access to distributed working memory in the large-scale monkey cortex. *Neuron.* 109:3500–3520.

Galvin VC, Yang ST, Paspalas CD, Yang Y, Jin LE, Datta D, Morozov YM, Lightbourne TC, Lowet AS, Rakic P, Arnsten AFT, Wang M (2020). Muscarinic m1 receptors modulate working memory performance and activity via KCNQ potassium channels in the primate prefrontal cortex. *Neuron.* 106:649–661.

Gamo NJ, Lur G, Higley MJ, Wang M, Paspalas CD, Vijayraghavan S, Yang Y, Ramos BP, Peng K, Kata A, Boven L, Lin F, Roman L, Lee D, Arnsten AFT (2015). Stress impairs prefrontal cortical function via D1 dopamine receptor interactions with hyperpolarization-activated cyclic nucleotide-gated channels. *Biol Psych.* 78:860–870.

Garibotto V, Wissmeyer M, Giavri Z, Ratib O, Picard F (2020). Nicotinic acetylcholine receptor density in the "higher-order" thalamus projecting to the prefrontal cortex in humans: A PET study. *Mol Imag Biol.* 22:417–424.

Gil Z, Connors BW, Amitai Y (1997). Differential regulation of neocortical synapses by neuromodulators and activity. *Neuron.* 19:679–686.

Goldman PS, Rosvold HE, Vest B, Galkin TW (1971). Analysis of the delayed-alternation deficit produced by dorsolateral prefrontal lesions in the rhesus monkey. *J Comp Physiol Psychol.* 77:212–220.

Hasselmo ME, Bower JM (1991). Selective suppression of afferent but not intrinsic fiber synaptic transmission by 2-amino-4-phosphonobutyric acid (AP4) in piriform cortex. *Brain Res.* 548:248–255.

Hasselmo ME, McGaughy J (2004). High acetylcholine levels set circuit dynamics for attention and encoding and low acetylcholine levels set dynamics for consolidation. *Prog Brain Res.* 145:207–231.

Hsieh CY, Cruikshank SJ, Metherate R (2000). Differential modulation of auditory thalamocortical and intracortical synaptic transmission by cholinergic agonist. *Brain Res.* 880:51–64.

Iadecola C (2017). The neurovascular unit coming of age: A journey through neurovascular coupling in health and disease. *Neuron.* 96:17–42.

Jakab RL, Goldman-Rakic PS (1998). 5-Hydroxytryptamine2A serotonin receptors in the primate cerebral cortex: possible site of action of hallucinogenic and antipsychotic drugs in pyramidal cell apical dendrites. *Proc Natl Acad Sci U S A.* 95:735–740.

Jin LE, Wang M, Galvin VC, Lightbourne TC, Conn PJ, Arnsten AFT, Paspalas CD (2018). mGluR2 versus mGluR3 metabotropic glutamate receptors in primate dorsolateral prefrontal cortex: Postsynaptic mGluR3 strengthen working memory networks. *Cereb Cortex.* 28:974–987

Kassam SM, Herman PM, Goodfellow NM, Alves NC, Lambe EK (2008). Developmental excitation of corticothalamic neurons by nicotinic acetylcholine receptors. *J Neurosci.* 28:8756–8764.

Katz PS (1999). What are we talking about? Modes of neuronal communication. In: Katz P, ed. *Beyond neurotransmission: Neuromodulation and its importance for information processing.* Oxford University Press.

Kimura F, Fukuda M, Tsumoto T (1999). Acetylcholine suppresses the spread of excitation in the visual cortex revealed by optical recording: Possible differential effect depending on the source of input. *Eur J Neurosci.* 11:3597–3609.

Kofuji P, Araque A (2021). G-protein-coupled receptors in astrocyte-neuron communication. *Neuroscience.* 456:71–84.

Krueger J, Disney AA (2019). Structure and function of dual-source cholinergic modulation in early vision. *J Comp Neurol.* 527:738–750.

Lavine N, Reuben M, Clarke PBS (1997). A population of nicotinic receptors is associated with thalamocortical afferents in the adult rat: Laminal and areal analysis. *J Comp Neurol.* 380:175–190.

Lewis DA, Campbell MJ, Foote SL, Goldstein M, Morrison JH (1987). The distribution of tyrosine hydroxylase-immunoreactive fibers in primate neocortex is widespread but regionally specific. *J Neurosci.* 7:279–290.

Liu H, Leak RK, Hu X (2016). Neurotransmitter receptors on microglia. *Stroke Vasc Neurol.* 1:52–58.

McCormick DA, Williamson A (1991). Modulation of neuronal firing mode in cat and guinea pig LGNd by histamine: Possible cellular mechanisms of histaminergic control of arousal. *J Neurosci.* 11:3188–3199.

Medalla M, Barbas H (2012). The anterior cingulate cortex may enhance inhibition of lateral prefrontal cortex via m2 cholinergic receptors at dual synaptic sites. *J Neurosci.* 32:15611–15625.

Moore SJ, Cooper DC, Spruston N (2009). Plasticity of burst firing induced by synergistic activation of metabotropic glutamate and acetylcholine receptors. *Neuron.* 61:287–300.

Morrison JH, Foote SL (1986). Noradrenergic and serotoninergic innervation of cortical, thalamic, and tectal visual structures in Old and New World monkeys. *J Comp Neurol.* 243:117–138.

Niu M, Impieri D, Rapan L, Funck T, Palomero-Gallagher N, Zilles K (2020). Receptor-driven, multimodal mapping of cortical areas in the macaque monkey intraparietal sulcus. *eLife*. 9.

Palomero-Gallagher N, Zilles K, Schleicher A, Vogt BA (2013). Cyto- and receptor architecture of area 32 in human and macaque brains. *J Comp Neurol.* 521:3272–3286.

Park J-Y, Spruston N (2012). Synergistic actions of metabotropic acetylcholine and glutamate receptors on the excitability of hippocampal CA1 pyramidal neurons. *J Neurosci.* 32:6081–6091.

Proulx E, Piva M, Tian MK, Bailey CDC, Lambe EK (2014). Nicotinic acetylcholine receptors in attention circuitry: the role of layer VI neurons of prefrontal cortex. *Cell Mol Life Sci.* 71:1225–1244.

Prusky GT, Shaw C, Cynader MS (1987). Nicotine receptors are located on lateral geniculate nucleus terminals in cat visual cortex. *Brain Res.* 412:131–138.

Rapan L, Niu M, Zhao L, Funck T, Amunts K, Zilles K, Palomero-Gallagher N (2022). Receptor architecture of macaque and human early visual areas: not equal, but comparable. *Brain Struct Funct.* 227:1247–1263.

Sánchez-González MA, García-Cabezas MA, Rico B, Cavada C (2005). The primate thalamus is a key target for brain dopamine. *J Neurosci.* 25:6076–6083.

Saras A, Gisselmann G, Vogt-Eisele AK, Erlkamp KS, Kletke O, Pusch H, Hatt H (2008). Histamine action on vertebrate GABAA receptors: Direct channel gating and potentiation of GABA responses. *J Biol Chem.* 283:10470–10475.

Sarter M, Lustig C (2020). Forebrain cholinergic signaling: Wired and phasic, not tonic, and causing behavior. *J Neurosci.* 40: 712–719.

Sato H, Hata Y, Hagihara K, Tsumoto T (1987). Effects of cholinergic depletion on neuron activities in the cat visual cortex. *J Neurophysiol.* 58:781–794.

Syková E, Vargová L (2008). Extrasynaptic transmission and the diffusion parameters of the extracellular space. *Neurochem Int.* 52 :5–13.

Trimmer BA (1999). The messenger is not the message; or is it? In: Katz P, ed., *Beyond neurotransmission: Neuromodulation and its importance for information processing.* Oxford University Press.

Vizi ES, Kiss JP, Lendvai B (2004). Nonsynaptic communication in the central nervous system. *Neurochem Int.* 45:443–451.

Wang M, Ramos BP, Paspalas CD, Shu Y, Simen A, Duque A, Vijayraghavan S, Brennan A, Dudley A, Nou E, Mazer JA, McCormick DA, Arnsten AFT (2007). alpha2A-adrenoceptors strengthen working memory ne2rks by inhibiting cAMP-HCN channel signaling in prefrontal cortex. *Cell.* 129:397–410.

Watakabe A, Komatsu Y, Sadakane O, Shimegi S, Takahata T, Higo N, Tochitani S, Hashikawa T, Naito T, Osaki H, Sakamoto H, Okamoto M, Ishikawa A, Hara S-I, Akasaki T, Sato H, Yamamori T (2009). Enriched expression of serotonin 1B and 2A receptor genes in macaque visual cortex and their bidirectional modulatory effects on neuronal responses. *Cereb Cortex.* 19:1915–1928.

Weis WI, Kobilka BK (2018). The molecular basis of G protein-coupled receptor activation. *Annu Rev Biochem.* 87:897–919.

Williams GV, Rao SG, Goldman-Rakic PS (2002). The physiological role of 5-HT2A receptors in working memory. *J Neurosci.* 22:2843–2854.

Wilson MA, Molliver ME (1991a). The organization of serotonergic projections to cerebral cortex in primates: regional distribution of axon terminals. *Neuroscience.* 44:537–553.

Wilson MA, Molliver ME (1991b). The organization of serotonergic projections to cerebral cortex in primates: retrograde transport studies. *Neuroscience.* 44:555–570.

Yang Y, Paspalas CD, Jin LE, Picciotto MR, Arnsten AFT, Wang M (2013). Nicotinic α7 receptors enhance NMDA cognitive circuits in dorsolateral prefrontal cortex. *Proc Natl Acad Sci U S A.* 110:12078–12083.

Yang ST, Wang M, Paspalas CD, Crimins JL, Altman MT, Mazer JA, Arnsten AFT (2018). Core differences in synaptic signaling between primary visual and dorsolateral prefrontal cortex. *Cereb Cortex.* 28:1458–1471.

SECTION IV

PATHWAYS BETWEEN THE THALAMUS AND THE CORTEX

Francisco Clasca and Sonja B. Hofer

A massive array of reciprocal axonal pathways binds the cerebral cortex and the thalamus into a complex and highly integrated network. Thalamic inputs powerfully influence cortical activity and function (Sherman 2016; Halassa and Kastner 2017). In turn, computations in all thalamic nuclei are influenced by cortical inputs and, for some of these nuclei, the cerebral cortex is the predominant or exclusive source of input (Acsady 2022). In some sense, therefore, the thalamus can be viewed as the deepest layer of the cortex (Sherman and Guillery 2006). Understanding thalamocortical interactions, and more generally the role of the thalamus within the distributed functional networks of the brain, crucially relies on knowledge about the detailed connectivity patterns and the logic of information flow between the cerebral cortex and the thalamus.

The chapters in this section provide a comprehensive review of the various long-range projection pathways that interconnect the thalamus and the cortex (Figure IV.1). In addition, interactions of these pathways with other forebrain regions, particularly the (pre)thalamic reticular nucleus (TRN), striatum, and amygdala are covered as well.

The organizational principles of thalamic output pathways are reviewed by Clasca at different levels of description in the **Chapters 11 and 12**. In **Chapter 11**, the layout of the thalamus output pathways is examined at the level of thalamic nuclei, as revealed by population-level anatomical tracing methods. Pathways to isocortical and non-isocortical areas of the cerebral cortex, as well as those to subcortical forebrain regions, are reviewed. The data from different thalamic areas and species are briefly summarized and compared. **Chapter 12** then reviews thalamic output pathways on the single-cell level. Powerful methods for labeling individual neurons are beginning to show patterns in the diversity in the projection patterns and circuit architectures of thalamic neurons. Together with functional and transcriptomic data, single-cell

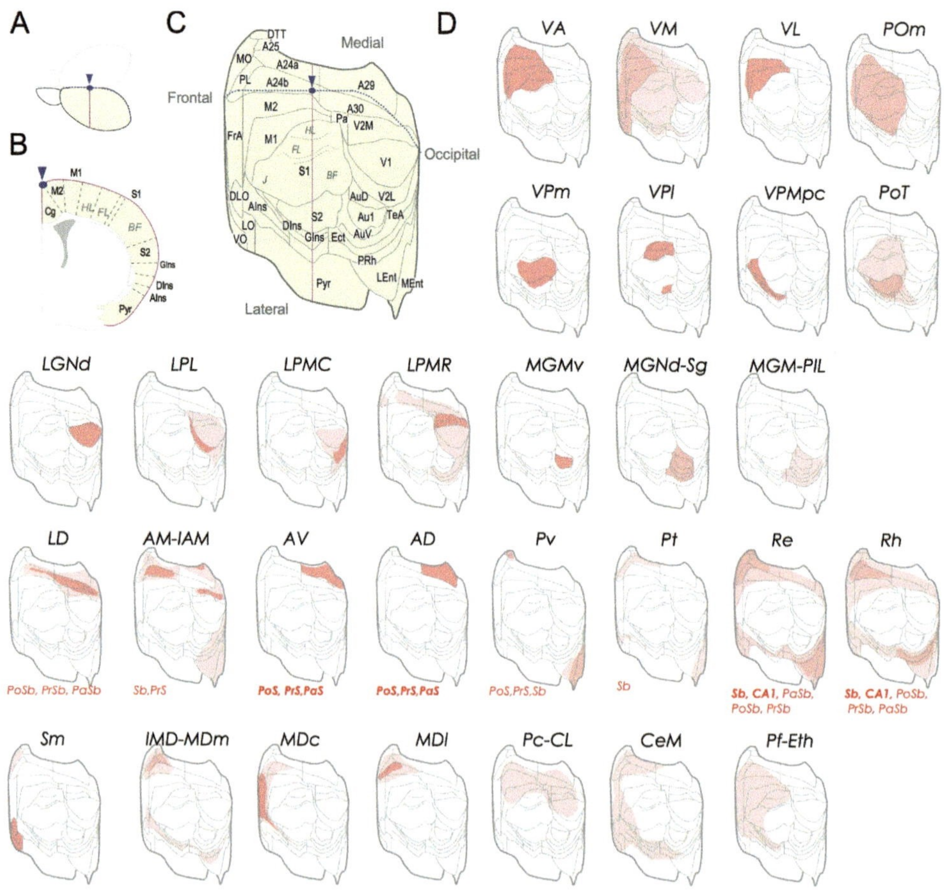

Figure IV.1. Highly simplified cartoon diagram of the circuits linking the thalamus and cerebral cortex and the projection cell classes mediating each of these circuits. Numbers indicate cortical layers. For clarity, projections to subcortical structures other than the TRN are not included.

morphologies provide the basis for a robust classification of thalamic projection neurons into several high-level cell classes.

In **Chapter 13**, Holtmaat and Lee focus on the structure and functional properties of thalamocortical (TC) synapses. After a brief review of the evolution of methods used to investigate these synapses, this chapter summarizes the principles of TC synaptic structure and function in the adult brain. This review emphasizes common features of these synapses across thalamic nuclei, cortical areas, and species, but it also provides evidence for differences between different types of TC axons.

The following two chapters survey the organization and functional properties of the two main projection systems directed from the cortex to the thalamus. In **Chapter 14**, Murphy and Briggs survey the prominent and diverse projections from CT neurons in cortical layer 6. This chapter reviews layer 6 neurons, their axonal wiring, and synaptic contacts onto thalamic projection neurons as well as onto local GABAergic interneurons in the thalamus and TRN cells. Moreover, the authors discuss the possible role of local and long-range TRN inhibition in the dynamic regulation of the excitation/inhibition balance of thalamic neurons. Finally, this chapter covers evidence for the functional impact of layer 6 CT neurons on the tuning, timing, and coding properties of thalamic activity.

Chapter 15, by Mease and Groh, focuses on the input system to the thalamus sustained by collateral branches of cortical layer 5 neurons projecting to the brainstem. The function, evolutionary history, anatomy, and physiology of these layer 5 (L5) CT pathways differ fundamentally from those of the better studied layer 6 CT pathways. Most importantly, L5 CT inputs target HO nuclei but are mostly absent from the FO sensory relay nuclei, in contrast to L6 projections, which target all thalamic nuclei. While many aspects of the L5 CT pathway function remain unknown, the chapter examines the physiological and anatomical properties of these circuits and their roles in the temporally precise and sparse excitatory drive of thalamus cells from the cortex, as well as the integration of cortical signals with other inputs relevant for cognitive processes.

The final chapter in this section, **Chapter 16** by Pinault, reviews the cellular organization of the TRN and the layout of its synaptic circuits with the thalamus and the cortex. The chapter describes recurrent and lateral inhibition in the thalamus mediated by reciprocal TRN-thalamus connections, as well as intrinsic oscillatory properties of TRN cells and communication between them via electrical and chemical synapses that are able to generate synchronized activity and lateral inhibition within TRN. The author describes several TRN sectors, classified based on their connections with specific sets of thalamic and cortical areas, and proposes that focal activity in these sectors may be flexibly combined to allow combinatorial filtering of signals between thalamus and cortex.

REFERENCES

Acsády L (2022). Organization of thalamic inputs. In: Halassa M, ed. *The Thalamus*. Cambridge University Press:27–44.
Halassa MM, Kastner S (2017). Thalamic functions in distributed cognitive control. *Nat Neurosci*. 20(12):1669–1679.
Sherman SM (2016). Thalamus plays a central role in ongoing cortical functioning. *Nat Neurosci*. 19(4):533–541.
Sherman SM, Guillery RW (2006). *Exploring the Thalamus and Its Role in Cortical Function*. 2nd ed. MIT Press.

11.

THALAMIC OUTPUT PATHWAYS

GLOBAL LAYOUT AND NUCLEUS-LEVEL ORGANIZATION

Francisco Clasca

INTRODUCTION

A wealth of studies employing classic (lesion/axonal transport) connection-tracing methods have over several decades extensively mapped and analyzed the input and output pathways from the various thalamic nuclei, in different species. These data are foundational for our current understanding of the thalamus role within the multiregional functional networks of the brain. Each of the nuclei of the thalamus has been found to send projections both to specific regions of the cerebral cortex and to the prethalamic reticular nucleus (TRN). Many nuclei innervate, in addition, one or more subcortical structures such as the striatum, amygdala, and basal forebrain. In the cortex and striatum, the thalamic axon pathways originating from the various thalamic nuclei have been shown to diverge and/or converge in complex spatial patterns that are highly selective and often complementary.

In mammals, axonal pathways coming out of the right or left thalami are directed only to structures situated in their respective side of the brain. As the thalamic axons leave their parent cell bodies, they always extend first away from the midline and toward the lateral or frontal edge of the thalamus, mostly as part of small, myelinated fiber bundles. Subsequently, they traverse the external medullary lamina and TRN. The axons arising from a given nucleus innervate the same sector of the TRN before continuing into the internal capsule, first as a tight and orderly bundle in the pallidal segment of the capsule, and then progressively fanning out in the striatal region of the capsule. Along this segment of their path, some thalamic axons may target subcortical structures such as the striatum or amygdala (Table 11.1). As they enter the pallial white matter, thalamic axons radiate orderly in all directions to reach their target areas of the cerebral cortex. Finally, the axons invade the cortical gray matter of one or several areas extending into the cortex predominantly in radial fashion to finally arborize, often very profusely, into specific layers. In this way, the thalamic axons reach all of the cerebral cortex, except for the dentate gyrus and CA4–CA3 fields (Table 11.1, Figure 11.1).

THALAMOCORTICAL PATHWAYS

PROJECTIONS TO THE CEREBRAL ISOCORTEX

Each isocortical area receives overlapping projections from several thalamic nuclei in specific and highly variable proportions.

Nucleus-Specific Patterns of Divergence and Area-Specific Innervation

The axons arising from each thalamic nucleus target a specific cortical territory (Figure 11.1). This may consist of just one cortical area innervated in a dense and highly topographic manner (e.g., lateral geniculate nucleus [LGNd]), or several areas innervated as multiple foci of variable density (e.g., posterior nucleus [Po], Figure 11.1D, 11.2B), or of large swaths of the cerebral hemisphere, mostly innervated in a roughly continuous and sparse manner (e.g., ventromedial nucleus [VM], Figure 11.1D).

Lamina-Specific Distribution in the Cerebral Isocortex

Axons that originate in individual thalamic nuclei arborize across the cortical laminae in area-specific patterns (Figure 11.2). This organization was first detected by studies carrying out injections of anterograde axonal transport tracers restricted to individual nuclei (Burton and Jones 1976; Ferster and LeVay 1978;

TABLE 11.1 CORTICAL AREAS AND SUBCORTICAL STRUCTURES TARGETED BY THE AXONS ORIGINATED IN EACH NUCLEUS OF THE MOUSE THALAMUS

Thalamic Nuclei	Cortical Areas	Subcortical Areas
Anterodorsal (AD)	**A29**. PoS, PrS	TRN
Anteroventral (AV)	**A29, PoS**, A30, PrS	TRN
Anteromedial and interanteromedial (AM–IAM)	**A24b**, M2, A24a	TRN, CeA, Cla
Lateral dorsal (LD)	**A24b, A30**, A29, PrS, PoS	TRN
Ventral anterior (VA)	**M2, M1**, A24a, S1 (spinal)	TRN, CPu
Ventral medial (VM)	**FrA, PL, IL**, M2, M1, S1, AIns, DLO, VL, LO, MO, V2M, V1	TRN, CPu
Ventral lateral (VL)	**M1, M2**	TRN
Posterior (POm)	**S1, S2, M1**, M2	TRN, CPu, LA
Posterior triangular (PoT)	**S2**, S1, GIns, Ect	TRN, CPu, LA
Ventral posterior (VPm)	**S1** (trigeminal), S2	TRN, CPu
Ventral posterior lateral (VPl)	**S1** (spinal), S2	TRN
Ventral posteromedial parvocellular division (VPMpc)	**DIns**, S1 (tongue)	TRN, LA m
Dorsal lateral geniculate (LGNd)	**V1**	TRN
Lateral posterior medial, rostral division (LPMr)	**V2M**, V1, A30, A24b, TeA, Ect	TRN
Lateral posterior medial, caudal division (LPMc)	**Te**, V2L, V1, Ect	TRN, CPu, LA
Lateral posterior lateral (LPL)	**V2L**, V1, V2M, TeA	TRN,
Medial geniculate, ventral (MGNv)	**Au1**	TRN
Medial geniculate, dorsal (MGNd), and suprageniculate (Sg)	**AuV, AuD** Au1, TeA	TRN
Medial geniculate medial–post-intralaminar (MGM-PiL)	Au1, AuD, AuV, TeA, Ect, Prh, V2L	TRN, CPu, LA
Central lateral (CL)	**M2, M1**, A30, A29, Pa, V2M, V1	TRN, **CPu**, Cla
Paracentral (Pc)	**A30, A29**, M2, M1, Pa, A29	TRN, CPu, BLA,
Central medial (CeM)	**A24a, 24b, VLO, DLO**, M2, AIns, DIns, M1	TRN, **Acb** CPu, BLA, LA, Cla, VPal
Parafascicular (Pf) + Ethmoid (Eth) nuclei	M1, M2, FrA, AIns, DIns, S1, S2, PL, A24b	TRN, **CPu, Acb, Sth**, GPe, SNc, SI, Cla, CeA
Paraventricular (PV)	DTT, IL, AIns, MEnt, Lent, PoSb, PrSb	TRN, **Acb, CeA, MeA, BNST**, AON, BLA, Cla, CPu,
	VPal, DH, OT, LS, SCN, Arc	
Paratenial (Pt)	**PL, IL, MO** AI, A24a, FrA, VO, LO, LEnt, Sb	TRN, BLA, BMA, CeA, Acb, CPu, BNST, LS, AON, Cla
Reuniens (Re)	**CA1, Sb, LEnt A29, PL**, PrSb, PoSb, PaS, Prh, MEnt, IL DTT,	TRN, Cla, SI, ZI, MS
	MO, VO, A24a, A24b, FrA, M2, AIns, APyr, VLO, Pyr	
Rhomboid (Rh)	**CA1, Sb, PrSb, PoSb**, PaS, FrA M2, Prh, LEnt, IL,	TRN, **Acb**, OT, Cla, BLA
	MO, A29, A30	

TABLE 11.1 CONTINUED

Thalamic Nuclei	Cortical Areas	Subcortical Areas
Mediodorsal, medial, and intermediate (MDm–IMD)	**PL**, FrA, A24a, IL, M2, AIns, VO, VLO, LO, MO, Ect	TRN, **CPu, Acb**
Mediodorsal, central	**VO, VLO, LO**, DLO, FrA, M2, PL. A24b, AIns	TRN, CPu
Mediodorsal, lateral	**M2, A24b. A24a, PL**, IL	TRN, **CPu**
Submedius (Sm)	**VLO, VO, LO**, MO	TRN, CPu

MAIN SOURCES: Vertes et al., 2006, 2012; Vertes and Hoover 2008; Van der Werf et al., 2002; Smith et al., 2012, 2019; Kimura et al., 2003.

NOTES: Regions are delineated and named according to Paxinos and Franklin mouse brain atlas (Paxinos G, Franklin KBJ (2019), *The Mouse Brain in Stereotaxic Coordinates*, 5th ed. Academic Press-Elsevier). In each case, boldface is used to indicate the most heavily innervated structures.

ABBREVIATIONS: A24a: cingulate cortex, area 24a; A24b: cingulate cortex, area 24b; A29: cingulate cortex, area 29; A30: cingulate cortex, area 30; Acc: accumbens nucleus; AIns: agranular insular cortex; CeA: central amygdaloid nucleus; AON: anterior olfactory nucleus; APyr: amygdalopiriform transition area; Arc: arcuate hypothalamic nucleus; Au1: primary auditory cortex; AuD: secondary auditory cortex, dorsal area; AuV: secondary auditory cortex, ventral area; BLA: basolateral amygdaloid nucleus; BMA: basomedial amygdaloid nucleus; BNST: bed nucleus of the stria terminalis; CA1: field CA1 of the hippocampus; CeAm: central amygdaloid nucleus; Cla: claustrum; CPu: caudate-putamen (striatum); DHN: dorsal hypothalamic nucleus; DIns: dysgranular insular cortex; DLO: dorsolateral orbital cortex; DMH: dorsomedial hypothalamic nucleus; DTT: dorsal taenia tecta; Ect: ectorhinal cortex; FrA: frontal association cortex; GIns: granular insular cortex; GPe: globus pallidus, external part; IL: infralimbic cortex; LA: lateral amydgdaloid nucleus; LEnt: lateral entorhinal area; LO: lateral orbital area; LS: lateral septal nucleus; M1: primary motor cortex; M2: secondary motor cortex; MeA: medial amygdaloid nucleus; MEnt: medial entorhinal cortex; MO: medial orbital cortex; MS: medial septal nucleus; OT: olfactory tubercle; Pa: parietal area; PaS: parasubiculum; PL: prelimbic cortex; PoS: postsubiculum; Prh: perirhinal cortex; PrS: presubiculum; Pyr: pyriform cortex; S1: primary somatosensory cortex; S2: secondary somatosensory cortex; SCN: suprachiasmatic nucleus; SI: substantia innominate; SNc: substantia nigra, compact part; Sth: subthalamic nucleus; TeA: temporal association cortex; TRN: reticular nucleus (prethalamus); V1: primary visual cortex; V2L: secondary visual cortex, lateral area; V2M: secondary visual cortex, medial area; VLO: ventrolateral orbital cortex; VPal: ventral pallidum; ZI: zona incerta.

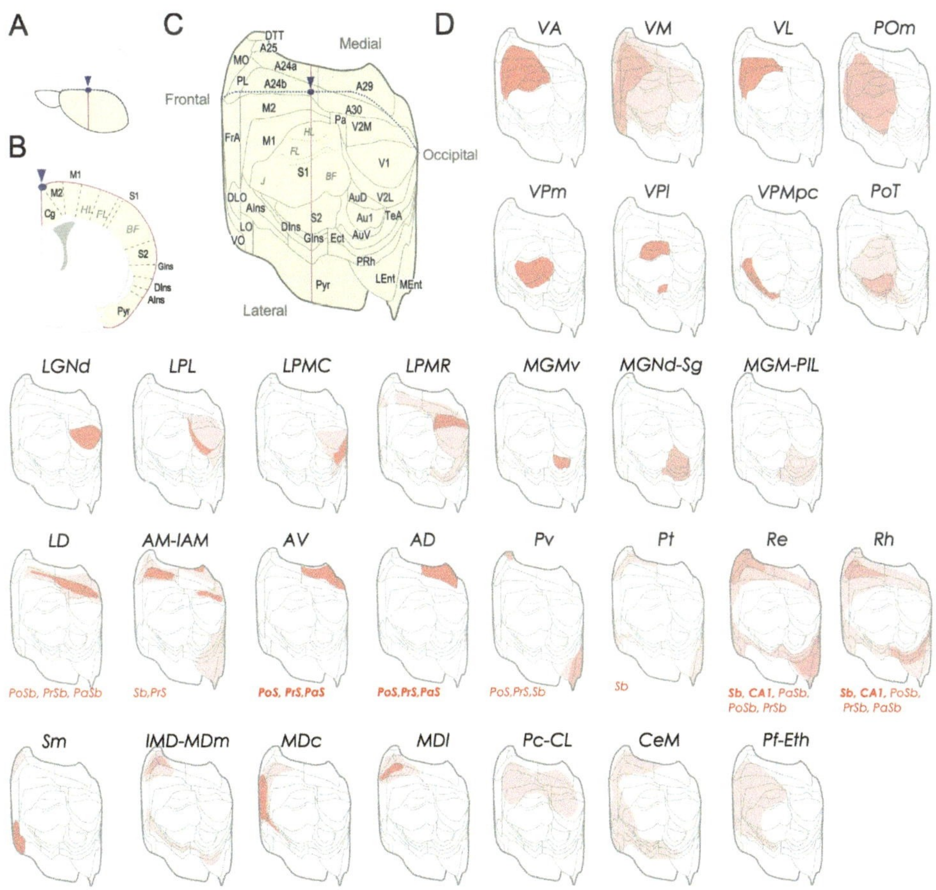

Figure 11.1 Cortical territories innervated by the axons arising from each of the nuclei of the mouse thalamus. (A–C). Two-dimensional map of the cortical areas of the mouse left cerebral hemisphere. The map was generated by aligning along the dorsomedial edge of the hemisphere (dashed blue line in panel C) the pial surface contours from all the coronal section in the Paxinos and Franklin brain atlas, each at its precise anteroposterior distance to Bregma (blue arrowhead). For reference, one such contours is shown as a pink line, in panels A and C. The corresponding coronal section is shown in panel B. (D). Maps of the cortical territory innervated by each nucleus, as indicated on the top of each map. Differences in innervation density are represented, in highly simplified form, by differences in red shading saturation. Parahippocampal and hippocampal areas are not included in the map; when present, thalamic projections to these non-isocortical areas are indicated in red letter case, under the corresponding map. BF = barrelfield region of S1; DP = dorsal peduncular cortex; FL = forelimb region of S1; HL = hindlimb region of S1; J = jaw and lips region of S1; V2MM = secondary visual cortex, mediomedial area; V2ML = secondary visual cortex, mediolateral area. (For other abbreviations, see Table 11.1.)

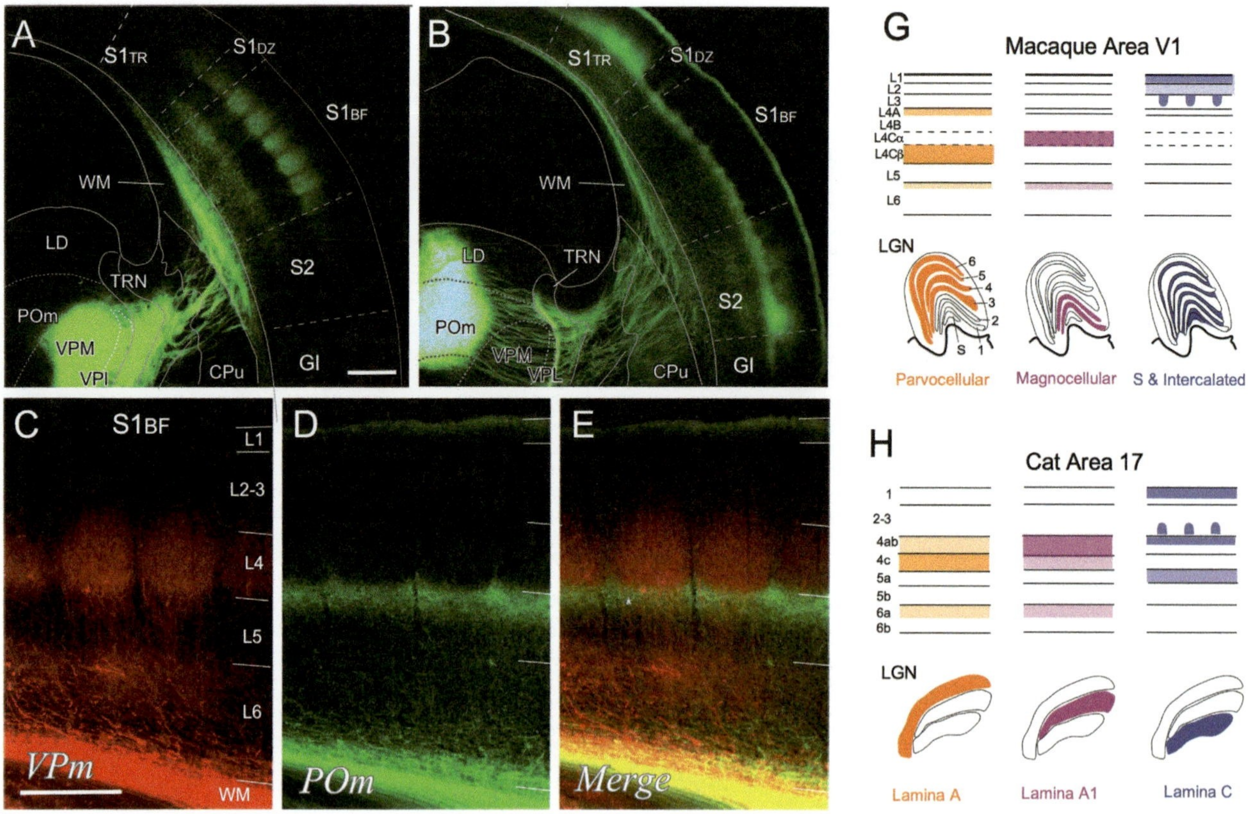

Figure 11.2. Nucleus-, area-, and lamina-specific patterns of thalalamocortical arborization. (A–B). Labeling by adenoviral vector transfection for the enhanced green fluorescent protein in 2 different animals. The mouse shown in A was transfected in VPm and the mouse in B was transfected in POm. Note the compact mass of fluorescently labeled cell bodies and dendrites in the thalamus. Labeled axons can be seen extending across the TRN, internal capsule (IC), and superficial part of the cortical white matter (WM) to terminate in different areas and layers. (C–E). Dual anterograde labeling of VPm and POm axons. Using different viral vectors in the same animal, VPm cells were made to express a red fluorescent protein, and POm cells to express the green fluorescent protein. S1dz = dysgranular zone of the vibrissal S1 cortex. (Other abbreviations as in Table 11.1.) Bar: 500 mm. Modified from Wimmer et al. (2010). (G–H). Schematic representation of the cortical layer distributions in the visual cortex of macaques and cats of axons originated in different subnuclei (laminae) of the dorsal lateral geniculate (LGN) of these 2 species. Modified from Jones (2007).

Herkenham 1980, 1986; Huang and Winer 2000; Vertes and Hoover 2006, 2008; Vertes et al., 2012; Smith et al., 2012; Smith et al., 2019). For example, tracer injections in some nuclei produced dense thalamocortical axon labeling in cortical layers 3, 4, and 6, while injections in other nuclei labeled axons in layers 5 and 1, or in layers 5-6, or even almost exclusively in layer 1. These studies also noted that the axons from nuclei that innervate several areas frequently show diverse laminar distribution pattens in each area (Figure 11.2B).

Connection tracing methods based on lesions or bulk-injected chemical tracers involved, by necessity, the simultaneous labeling of many projection cells. Bulk tracer injections could not individually resolve the wiring of pathways that are different yet spatially overlapped. This was the case of the thalamocortical pathways. Nevertheless, as evidence grew over the years that the populations of axons arising from different nuclei had dissimilar terminal distribution patterns, several classifications were proposed (see Figure 11.3 and Box 11.1).

PROJECTIONS TO ALLOCORTEX

Projections to Hippocampal and Parahippocampal Areas

Some thalamic nuclei are the origin of robust projections to entorhinal, subicular, and hippocampal fields. These projections have been demonstrated in rodents (Herkenham 1978; Wouterlood et al., 1990; Van Groen et al., 1999; Vertes et al., 2006; Vertes and Hoover, 2008) and primates (Amaral and Cowan 1980; Insausti et al., 1987). In all species investigated, the connections originate only from a limited and similar collection of nuclei (Table 11.1, Figure 11.1).

Axons from Re and Rh arborize in the stratum lacunosum-moleculare of the CA1 field at all septotemporal levels of the hippocampus (Figure 11.4A). There is some topographical ordering, as dorsal parts of Re and Rh project to septal/anterior parts of the hippocampus and ventral parts of Re to the temporal/posterior levels. Thalamic inputs to the subiculum originate from Re, Rh, PV, and Pt (Herkenham,

BOX 11.1 A BRIEF HISTORY OF THALAMOCORTICAL PATHWAY AND NUCLEI TAXONOMIES

Early studies inferred the existence of connections from the cell degeneration produced in the thalamus by experimental cortex lesions. These studies noted that while some nuclei degenerated massively following a lesion to a single cortical area, other nuclei degenerated only when several areas were lesioned. This led to a distinction between thalamocortical projections that were *essential* to a given area (focused just on that area) or *sustaining* to the area (spread, in addition, to other areas; Rose & Woolsey, 1949).

At about the same time, a series of electrocorticographic studies explored the cortical wave responses to electrical stimulation in different thalamic nuclei. Pulse trains (5–15Hz) applied to sensory relay nuclei were reported to produce a fast wave that grew in amplitude with repetition in a limited cortical region (*augmenting* response). In contrast, when the same stimuli were applied around the internal medullary lamina, or in dorsomedial thalamus, a more delayed wave reportedly appeared over a wide cortical region (*recruiting* response; Dempsey and Morison 1941). Oddly, these recording data were conflated (Morison and Dempsey 1942; Jasper 1966) with a speculation by Lorente de No (1938) about the possible thalamic origin of 2 kinds of axons he had observed entering the cortex from the white matter, in Golgi-stained samples. Lorente had dubbed "specific" some thick axons that entered into the cortex directly from the white matter to arborize focally in its middle layers and "non-specific" other thinner axons that first branched in the subcortical white matter and then innervated loosely several cortical layers, including layer 1. While the distinction between recruiting and augmenting responses and their relation to particular thalamic nuclei were soon abandoned (Schlag and Villablanca, 1967; Steriade, 2001), the unfounded notion that the intralaminar nuclei of the thalamus innervated the cortex via widely branched and layer 1-directed "non-specific" axons endured for a long time (Jones 2007).

Axonal transport-based methods improved the resolution of long-range pathway tracing. Based on experiments involving radioactive amino acid iontophoretic microinjections and long-exposure autoradiography, Miles Herkenham (1980, 1986) detected clear differences in the laminar position and density in the cortex of the labeled thalamocortical axon arborizations arising from different thalamic nuclei (Figure 11.3A). He proposed a nuclei classification into 3 categories, but explicitly refrained from inferring the existence of specific thalamocortical cell types, due to ambiguity of the autoradiographic data (Figure 11.3B). Herkenham emphasized that different nuclei converging in the same area usually target different laminae (compare Figure 11.3Aa–3Ab, 3Ac–3Ad–3Ae; or 3Af with 3Ag), and that axons from a single nucleus may target different laminae in adjacent areas (compare Figure 3Ab–3Ac, or the shift in labeling between areas V1 and V2 in Figure 3Af).

A cell-specific hypothesis cell-specific hypothesis was proposed by Edward G. Jones (1998, 2007) based on retrograde and anterograde pathway tracing studies in nonhuman primates and data from the literature (Avendaño et al., 1990; Macchi et al., 1996; Figure 11.3C). Jones posited the existence of 2 main thalamocortical cell classes intermingled in variable proportions in each nucleus of the thalamus: (a) a broad class spread as a "matrix" across the thalamus and characterized by having widely spanning axons directed to the superficial layers of the cortex; and (b) a "core" class, present only in some nuclei and characterized by their axons targeting the cortex middle layers of a specific area. In addition, he emphasized that, at least in macaques, these 2 cell classes could be distinguished by their selective expression of one of 2 calcium binding proteins, calbindin or parvalbumin, respectively.

While not related per se to the wiring of the thalamocortical pathways, other classifications based on the *type of inputs that thalamic nuclei receive* are often compounded with the classifications described above and deserve mentioning here. For example, early tract-tracing studies established that only some nuclei of the thalamus are the end point of ascending brainstem and retinal pathways, while other thalamic nuclei receive their connections mainly from cortex. This led to a distinction between "primary" or "extrinsic" versus "association" or "intrinsic" nuclei (Walker, 1938; Rose and Woolsey, 1949). Over several decades, these notions became widely used as framework for the analysis of thalamic nuclei and their connections with the cortex.

The above notions were redefined and extended by Ray Guillery and Murray Sherman based on new synaptic ultrastructure, single-cell tracing, and electrophysiological data (Sherman and Guillery 1996, 1998; Guillery and Sherman, 2006; Figure 11.3C). These authors distinguished between "first-order relay" (FO) nuclei that relay to the cortex ascending signals from the subcortical systems versus "higher-order relay" (HO) nuclei that relay back to cortex signals received from the cortex itself, via collaterals of layer 5 axons that are primarily directed to the brainstem. Their key insight was that the structure and signal transmission properties of cortical L5 synapses in HO nuclei and subcortical synapses in FO nuclei were the same. Both are highly effective ("driver") glutamatergic synapses able to determine the main information content (e.g., receptive field properties) of the postsynaptic cells, in contrast with many other less effective ("modulator") glutamatergic synapses that are only capable of altering the probability of certain aspects of signal transmission (Sherman, 2016).

Despite being based on indirect and clearly incomplete data, the above classifications still frame current analyses of thalamocortical interactions. The diversity revealed by new single-cell methods (see **Chapter 12 in Section IV**), however, goes well beyond what the classic thalamocortical pathway taxonomies envisioned. Likewise, a recently discovered variety in patterns of input convergence upon individual cells implies that thalamic cell computations transcend mere signal relay (Acsady, 2022).

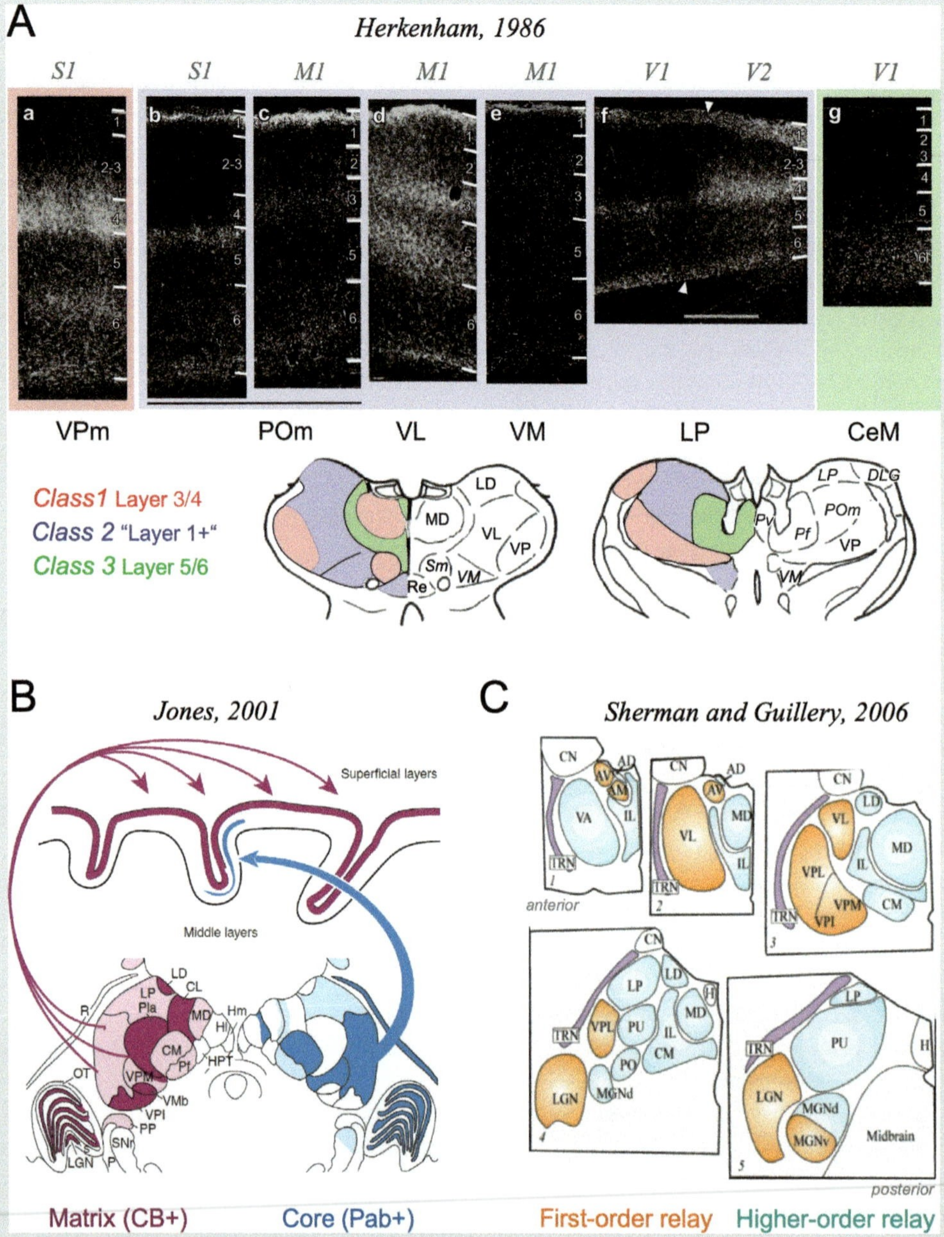

Figure 11.3. Global thalamic nuclei taxonomies based on classic tracing methods. (A–B). Herkenham (1986) tripartite taxonomy of the rat thalamus nuclei based on the laminar patterns of labeling in the cortex. (A): Panels Aa–Ag show darkfield images of the autoradiographic labeling in different cortical areas (indicated at the top) produced by selective tracer microinjections in each of the nuclei indicated under each image. Arrowheads in panel Af indicate the histological border between areas V1 and V2. Bar = 500 μm. A classification of nuclei into 3 main categories is displayed on coronal section diagrams of the rat thalamus. Category 1 nuclei (pale red) aim their projections at the middle (3–4) cortical layers, leaving layer 1 free of labeling. Category 2 nuclei (green) project mainly to layers 5–6, in widespread and sparse manner. Nuclei of category 3 (blue) innervate cortical layer 1 robustly, alone or in addition to other layers (Layer 1+). Adapted from Henkenham (1980, 1986). (B). Dual taxonomy of the thalamocortical projection neurons proposed by Jones (1998, 2007). Nuclei are represented on a coronal diagram of the macaque thalamus; idealized cortical folds are shown at the top. Jones envisioned that two main cell classes were present in the thalamus: (1) a pervasive "matrix" (crimson) of cells distributed unevenly through the thalamus, which express calbindin 28K (CB) and whose axons would provide diffuse input to the superficial layers of the cortex, unconstrained by area borders (multiheaded arrow); and (2) a more limited set of "core" neurons imposed in some nuclei (blue) which mostly express parvalbumin (Pab), whose axons innervate the middle layers of a specific cortical area (blue arrow). Reproduced from Jones EG (2001). (C). Diagram representation of the classification by Sherman and Guillery (1996) of thalamic nuclei in first order relay (orange) and higher order relay (pale blue) nuclei. Idealized coronal sections (1–5) represent the nuclei in a generalized primate thalamus. See original papers for the non-standard abbreviations. Modified from Guillery and Sherman (2006).

1986; Wouterlood et al., 1990; Vertes et al., 2006; Vertes and Hoover, 2008). In contrast, inputs to the presubiculum, postsubiculum, and parasubiculum originate in AV, AD, LD, and, to a lesser extent, in AM (Figure 11.4A–D). The projections target mainly the middle and superficial layers of these areas (Van Groen et al., 1999; Figure 11.4B); weaker projections to layer 1 arise from Re, Rh, and PT (Vertes et al., 2006; Vertes and Hoover; 2008; Figure 11.4D). Other sources of thalamic input to the entorhinal areas arises from CeM (Insausti et al., 1987; Wouterlood et al., 1990; Van Groen et al., 1999; Figure 11.4A, 4C–D). Thalamic innervation of the perirhinal cortex (Ect, in rodents) arises also from PV and Re, and this cortex receives, in addition, robust connections from LPMc and the medial division of the medial geniculate (MGm-PIL). In macaques, entorhinal and perirhinal areas receive a substantial projection from the medial pulvinar (Insausti et al., 1986). Interactions of the thalamus with the hippocampal and parahippocampal areas are involved in spatial and non-spatial navigation, memory encoding, and learning (reviewed in Mathiasen et al., 2020).

Projections to Olfactory Cortices

A few thalamic nuclei send sparse projections to olfactory cortical areas such as the anterior olfactory nucleus, prepyriform cortex, periamygdaloid cortex, and transitional areas of the orbital cortex (DPP in Table 11.1). These have been described in macaques (Dermon and Barbas, 1994) and rats (Vertes and Hoover, 2008). They originate from Pv, Pt, Re, and Rh.

THALAMUS PROJECTIONS TO THE AMYGDALA

In addition to the cortical regions of the forebrain, the thalamus innervates several subcortical structures. Studies in rodents, cats, and primates have shown that several nuclei of the dorsal thalamus innervate different nuclei of the amygdaloid complex. Depending on the parts of the complex they innervate, 2 main groups of nuclei can be distinguished.

A first group situated along the thalamus midline target mainly the central (CeA), basolateral (BLA), and basomedial (BMA) amygdala nuclei. The nuclei at the origin of these projections include PT, PV, Rh, CeM, PIL, SPf, IMD, and IAM Figure 11.4F; Turner and Herkenham, 1991; Russchen et al., 1982; Vertes et al., 2006; Vertes and Hoover, 2008; Hsu and Price, 2009; Vertes et al., 2012). Thalamic axons target the principal amygdala neurons both on their spines and dendritic shafts, but not the local interneurons (Amir et al., 2019). Midline thalamic projections may thus influence BLA and CeA neurons' activity with minimal feedforward inhibition, modulating their responses to emotionally salient internal or external events, in affective and fear-related behaviors (Vertes et al., 2015; Amir et al., 2019).

A second group of thalamic nuclei selectively target the lateral amygdaloid nucleus (LA), which is the major sensory input area of the amygdala, and relays this information to other amygdaloid nuclei for further processing. The heaviest thalamic input to LA arises from the medial portion of the medial geniculate nucleus (MGNm), suprageniculate (Sg), subparafascicular (Spf), and peripeduncular nuclei (Russchen et al., 1982; Ledoux et al., 1990; Turner and Herkerham 1991; Smith et al., 2019). Double retrograde labeling evidence indicates that many of these projections to LA are collaterals of axons simultaneously projecting to the auditory and association temporal cortex (Doron and Ledoux, 2000). Inputs to LA have been shown to be necessary for associative conditioning of fear responses to acoustic stimuli (Rumpel et al., 2005). In addition, LA receives direct projections from LPMc in rodents (Figure 11.4E) or from comparable portions of the pulvinar in primates (Burton and Jones 1976; Day-Brown et al., 2010; Elorette et al., 2018). These thalamo-amygdaloid connections establish synaptic contacts only with principal LA neurons (Day-Brown et al., 2010). They have been postulated to carry fast, low-resolution visual signals to the amygdala that are important for the innate ability of reacting to visual cues related to species-specific predators (Wei et al., 2015).

THALAMOSTRIATAL PATHWAYS

Thalamic projections to the striatum are the most robust and extensively studied of the subcortical thalamic projections, mainly because of their involvement in motor systems disorders. Only a brief overview can be presented here; for an extended review, see Smith et al. (2014).

The thalamostriatal pathways arise from a diverse collection of nuclei across the thalamus, and reach the whole dorsal and ventral striatum, with a preference for its matrix compartment. They consist of 2 anatomically and functionally segregated systems.

The first system originates in the centromedian-parafascicular nucleus (CM-Pf) in primates, or its equivalent in rodents Pf + Eth (Berendse and Groenewegen, 1990; Francois et al., 1991; Sadikot et al., 1992a, 1992b; Deschenes et al., 1995; McFarland and Haber, 2000; Erro et al., 2002; Raju et al., 2006). This nucleus sends heavy and topographically organized projections to specific regions of the dorsal striatum that selectively target the matrix compartment (Sadikot et al., 1990, 1992a; Kita et al., 2016). Characteristically, these pathways terminate always as multiple separated foci of dense lacelike arborizations that display numerous short branches with frequent varicosities and club-like appendages (Deschenes et al., 1995; Figure 11.5D,F).

Electron microscopic studies have shown that the CM-Pf axons target both direct- and indirect-pathway medium spiny stellate neurons (MSN), establishing excitatory

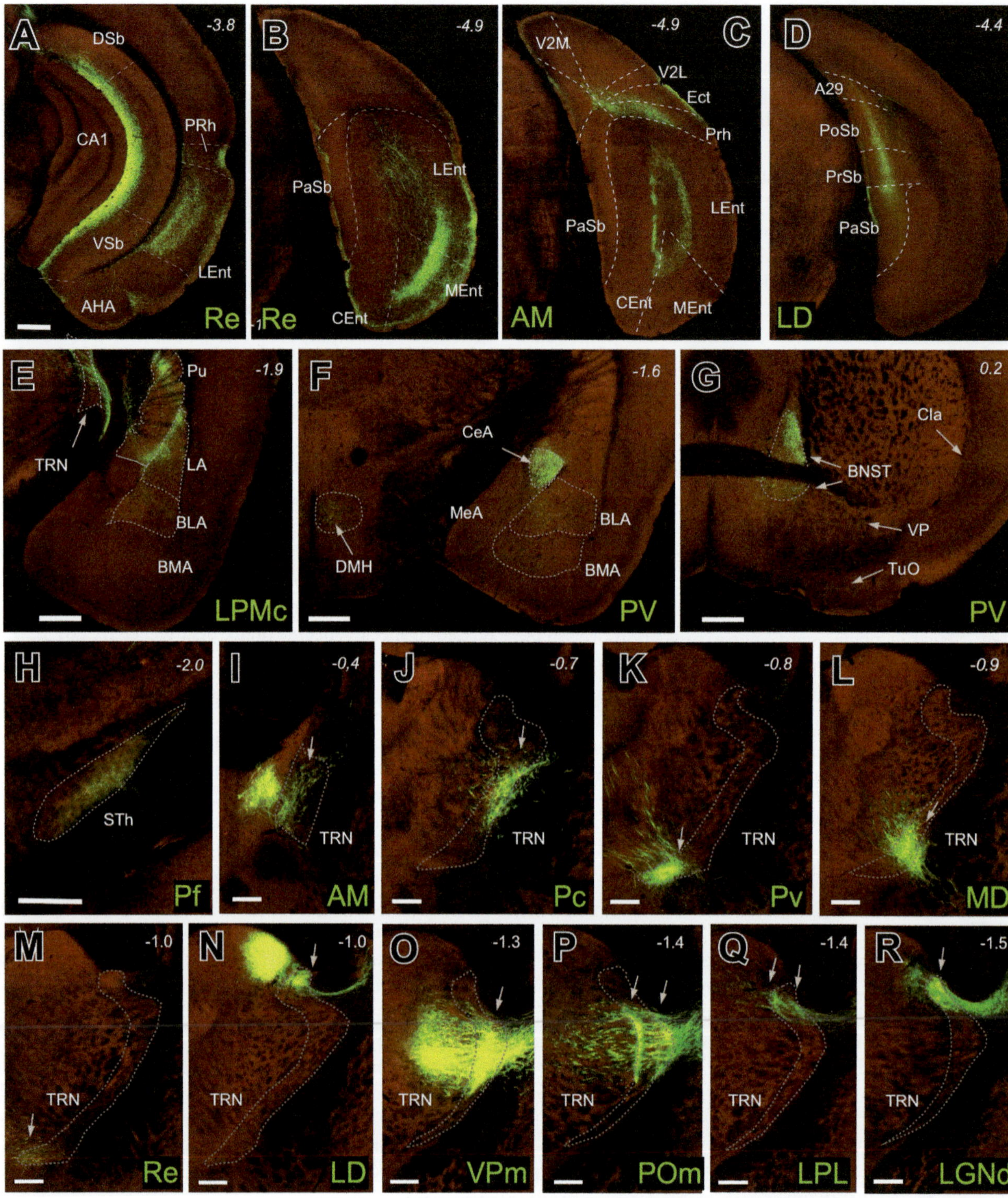

Figure 11.4. Thalamic projections to mouse non-isocortical cerebral cortex areas, amygdala basal forebrain, and reticular prethalamic nucleus. Transfection-labeled thalamocortical axons are shown on coronal brain sections. Conventions as in Figure 11.2. (A–B). Labeled Re axons. Note the band of heavy labeling in the stratum lacunosum-moleculare of CA1 and adjacent dorsal and ventral subicular fields (DSub, VSub). Sparser labeled axons are present in the deep layers of Lent and MEnt, in the same experiment. (C). Labeled AM axons visible in the deep layers of the lateral (LEnt), medial MEnt), and caudal (CEnt) entorhinal areas. Additional labeled axons are present in the perihinal cortex (Prh). (D). Labeled LD axons in the postsubiculum (PoSb), presubiculum (PreSb), and parasubiculum (PaSb). (E). Projection from the caudal division of the lateral posterior nucleus (LPMc) to Pu, LA, and BLA. Note also the labeling in TRN. (F–G). PV projection to BNST and several nuclei of the amygdala (CeA, BLA, BMA), as well as the ventral claustrum (Cla) CeA. Note also in G a projection to the dorsomedial hypothalamic nucleus (DMH). (H). Projection from the parafascicular nucleus (Pf) to the subthalamic nucleus (Sth). (I–R). In different animals, clusters of neurons were transfected in individual thalamic nuclei. Images show the coronal section containing the densest labeling in the TRN (delineated by a dashed line). The nucleus transfected in each case is indicated on the left upper corner of each image. Level relative to bregma is indicated in the lower right corner. Arrows indicate the arborization plexuses visible in TRN of the axonal bundles transfected in each case. Differences in fluorescence intensity mainly reflect the variable size of the transfected cell populations. Images from the Mouse Connectivity Projection datasets, http://connectivity.brain-map.org. A–B: 265286700-Re; C: 266174045-AM; D: 272969333-LD; E: 178489574; F–G: 263106571; H: 272968624; I: 592698832-AM; J: 183071513-PC; K: 263106751-PVT; L, 272970747-MD; M: 496964969-Re; N: 272969333-LD; O: 312240825-VPM; P: 292212456-PO; Q: 183175010-LP; R: 479670988-DLG. Bar A–H = 500 µm, I–R = 250 µm.

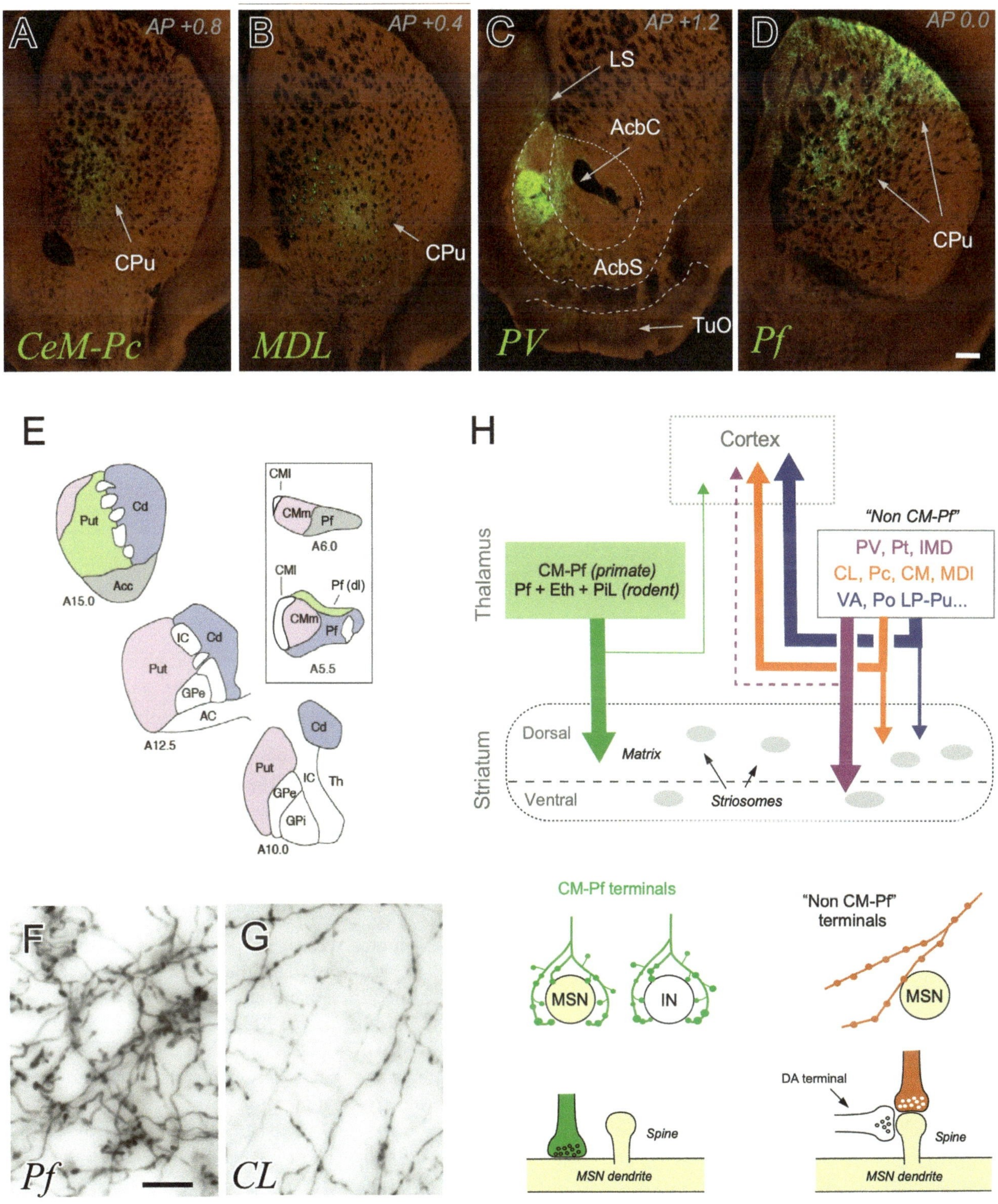

Figure 11.5. The 2 thalamostriatal systems. (A). Axonal arborizations in the caudate-putamen (CPu) from a cluster of cells in the CeM-Pc (panel A) or in MD (panel B) transfected with the green fluorescent protein gene. Note that this axon arborization is fine-grained and evenly distributed. A few small bundlets of brightly fluorescent axon trunks traveling across the striatum toward the cerebral cortex are also visible. (C). Arborizations in the accumbens shell (AcbS) and core (AcbC) from cells transfected in PV. Additional arborizations spread to the lateral septum (LS) and olfactory tubercle (TuO). (D). Dense and discontinuous labeling produced by a cluster of transfected neurons in Pf. Images from the Mouse Connectivity Projection datasets, http://connectivity.brain-map.org. A: experiment 147135107; B and E: 272970747; C: 263106751; D and F: 299732738. Bar A–D = 500 μm. (E). Diagram illustrating the topographic distribution of projections from the centromedian-parafascicular (CM/Pf) nucleus of macaques. Three rostrocaudal coronal section levels (stereotaxic coordinates at the bottom left) of the striatal complex are shown. Two coronal section diagrams of the complex are illustrated in the box. Note that this nucleus alone innervates the whole dorsal and ventral striatum in topographic fashion. Reproduced from Smith et al. (2014). (F–G). High-magnification detail of terminal arborizations labeled with biotinylated dextranamine injected in mouse Pf or CL.

synapses mostly on their dendrite trunks (Raju et al., 2006; Lacey et al., 2007), as well as on most of the striatal interneuron cell classes (Sidibe and Smith, 1999). There is in vitro evidence in rodents that the synapses of this system see predominantly N-methyl D-aspartate (NMDA) receptors on their postsynaptic target (Ellender et al., 2013). Through these projections, the CM/Pf nucleus gains spatially precise access to the striatal complex, thereby making it able to powerfully and selectively modulate both motor and nonmotor basal ganglia functions.

The second thalamostriatal system originates from a disparate group of thalamic nuclei that includes the intralaminar (CL, Pc, CM), midline (PV, IMD, PT), and association or motor nuclei (LP-Pulvinar, MDl, Po, VA, VL) as well (Herkenham, 1986; Smith & Parent, 1986; Berendse and Groenewegen, 1990; Giménez Amaya et al., 1995; McFarland and Haber, 2000; Erro et al., 2002; Vertes and Hoover, 2008; Vertes et al., 2012; Hsu and Price, 2009). Together, these "non-CM/PF" axons cover the whole striatum, each nucleus projecting to a particular region according to the functionally segregated organization of the striatum, so that sensorimotor-, associative-, and limbic-related thalamic nuclei innervate functionally corresponding regions of the dorsal or ventral striatum. Remarkably, despite their widely diverse origins, these axons are wired to the striatal circuits in essentially the same manner. Terminal branches are diffusely spread, elongated, and relatively simple (Deschenes et al., 1995). Most nuclei innervate mainly the matrix compartment, although midline and some CeM innervate the striosomal compartment as well (Kita et al., 2016). The "non-CM/PF" axons target MSN, but not striatal interneurons. On MSN, they form almost exclusively synapses on spines, many of which are co-innervated by dopaminergic synaptic boutons. The synapses are mediated preferentially by non-NMDA receptors. Remarkably, the "non-CM/Pf" synaptic organization is identical to that of the corticostriatal axons (Smith and Bolam, 1990; Ellender et al., 2013).

THALAMIC PROJECTIONS TO OTHER BASAL FOREBRAIN STRUCTURES

The CM-Pf nucleus innervates several structures of the basal ganglia system, such as the subthalamic nucleus (STh), globus pallidus, and substantia nigra. The projection to STh is the most robust, and it is present in rats and monkeys (Sadikot et al., 1992a; Lanciego et al., 2004; Kita et al., 2016). Axon terminals that originate in different portions of CM-Pf distribute within STh in consonance with the functional and connective subdomain organization of this nucleus (review in Alkemade 2013). The projections induce the activation of the last step of the indirect basal ganglia pathway, acting as an inhibitory feedback loop of the basal ganglia circuitry.

Small but consistent projections to the claustrum have been reported from CL, CeM, PV, Pt, and Re (Le Vay and Sherk, 1981; Vertes et al., 2006; Vertes et al., 2012; Vertes and Hoover, 2008) in different species.

The midline nuclei, and particularly PV and IMD, send robust projections to the bed nucleus of the stria terminalis (Figure 11.4G), lateral septum (Figure 11.5C), and substantia innominata. In addition, PV and IMD neurons innervate some nuclei of the hypothalamus, namely the dorsomedial, suprachiasmatic, and arcuate (Figure 11.7B; Vertes et al., 2006; Vertes and Hoover, 2008; Hsu and Price, 2009). These connections have been related to the involvement of PV-IMD as a high-centrality node in the forebrain networks that regulate arousal and homeostatic behavior (Penzo and Gao, 2021).

THALAMIC PROJECTIONS TO THE PRETHALAMIC RETICULAR NUCLEUS

All thalamic nuclei project to the TRN (Figure 11.4I–R). These connections are always collateral branches of the thalamic projection neuron axons that continue toward other forebrain targets (see **Chapters 12 and 16 in Section IV**). Each thalamic nucleus innervates a discrete region of the TRN and receives reciprocal connections from it. The location of this zone is related to the functional modality (visual, auditory, motor, etc.) of the nucleus, as well as to its topographic position within the thalamus. Hence, nuclei related to the same functional modality target overlapping zones of the TRN (Crabtree, 1998; Guillery and Harting, 2003).

Even when 2 nuclei target the same TRN sector, their thalamoreticular arborizations tend to remain separate across the traverse dimension of the nucleus, often referred to as core and shell compartments. As the TRN cells have their dendrites extended as flat discs parallel to the TRN plane, this organization implies that each thalamic nucleus

Figure 11.5. (continued)

Bar = 10 µm. (H). Summary diagram illustrating the key differences between the 2 thalamostriatal systems. The CM-Pf system (left side, in green) innervates the whole striatum via axons that form clustered terminal arborizations (panel F). These axons establish synapses on both medium spiny neurons (MSN) and striatal interneurons (IN, middle row), preferentially on dendritic trunks (bottom row). A second system (non CM/Pf, right side) arises from many other thalamic nuclei (see Table 11.1 and Figure 11.5A–C, E, G). The axonal arborizations of this system cover the whole striatum, in loose topographic fashion. They form relatively simple, linear terminal branches (middle row; see also panel B) that establish synapses on MSN, mainly on dendritic spine heads (bottom row). The same spines are frequently contacted by dopaminergic axons from the substantia nigra compacta ("DA terminal"). Based on Smith et al. (2014).

can selectively interact with different TRN cell populations and establish largely independent subnetworks.

CONCLUSION

In recent decades, a wealth of studies carried out in several model species with population-level tract tracing methods have illuminated the complex circuitry that links the thalamus with the cerebral cortex and other forebrain structures. Along with electrophysiological data, these studies are foundational for our current understanding of the thalamus organization and its interactions with the cerebral cortex, striatum and other central nervous system cell populations. Recently, new single-cell methods have begun to add critical wiring and quantitative precision to the account of the thalamic circuits presented in this chapter (see **Chapter 12 in Section IV**). Along with the population-level information reviewed here, the single-cell data are paving the way to biologically accurate models of the thalamocortical cicuitry and to the development of functional probes for the experimental manipulation of specific thalamic cell populations.

ACKNOWLEDGMENTS

This work was supported by funding from Spain's Ministerio de Ciencia, Innovación y Universidades (PID2020-115780GB-I00).

REFERENCES

Acsády L (2022). Organization of thalamic inputs. In: Halassa M, ed. *The Thalamus*. Cambridge University Press: 27–44.

Alkemade A (2013) Subdivisions and anatomical boundaries of the subthalamic nucleus. *J Neurosci*. 33(22):9233–9234.

Amaral DG, Cowan WM (1980). Subcortical afferents to the hippocampal formation in the monkey. *J Comp Neurol*. 189(4):573–591.

Amir A, Paré JF, Smith Y, Paré D (2019). Midline thalamic inputs to the amygdala: Ultrastructure and synaptic targets. *J Comp Neurol*. 527(5):942–956. doi:10.1002/cne.24557.

Avendaño C, Stepniewska I, Rausell E, Reinoso-Suárez F (1990). Segregation and heterogeneity of thalamic cell populations projecting to superficial layers of posterior parietal cortex: a retrograde tracer study in cat and monkey. *Neuroscience*. 39(3):547–559.

Berendse HW, Groenewegen HJ (1990). Organization of the thalamostriatal projections in the rat, with special emphasis on the ventral striatum. *J Comp Neurol*. 299(2):187–228.

Burton H, Jones EG (1976). The posterior thalamic region and its cortical projection in New World and Old World monkeys. *J Comp Neurol*. 168(2):249–301.

Crabtree JW (1998). Organization in the auditory sector of the cat's thalamic reticular nucleus. *J Comp Neurol*. 390(2):167–182.

Day-Brown JD, Wei H, Chomsung RD, Petry HM, Bickford ME (2010). Pulvinar projections to the striatum and amygdala in the tree shrew. *Front Neuroanat*. 4(143):1–11. doi:10.3389/fnana.2010.00143.

Dempsey, EW, Morison, RS (1942). The production of rhythmically recurrent cortical potentials after localized thalamic stimulation. *Amer J. Physiol*. 135:293–300.

Dermon CR, Barbas H (1994). Contralateral thalamic projections predominantly reach transitional cortices in the rhesus monkey. *J Comp Neurol*. 344(4):508–531.

Deschênes M, Bourassa J, Parent A (1995). Two different types of thalamic fibers innervate the rat striatum. *Brain Res*. 701(1–2):288–292.

Doron NN, Ledoux JE (2000). Cells in the posterior thalamus project to both amygdala and temporal cortex: A quantitative retrograde double-labeling study in the rat. *J Comp Neurol*. 425(2):257–274.

Ellender TJ, Harwood J, Kosillo P, Capogna M, Bolam JP (2013). Heterogeneous properties of central lateral and parafascicular thalamic synapses in the striatum. *J Physiol*. 591(1):257–272.

Elorette C, Forcelli PA, Saunders RC, Malkova L (2018). Colocalization of tectal inputs with amygdala-projecting neurons in the macaque pulvinar. *Front Neural Circuits*. 12(91):1–12. doi:10.3389/fncir.2018.00091.

Erro M, Lanciego JL, Gimenez-Amaya JM (2002). Re-examination of the thalamostriatal projections in the rat with retrograde tracers. *Neurosci Res*. 42(1):45–55.

Ferster D, LeVay S (1978). The axonal arborizations of lateral geniculate neurons in the striate cortex of the cat. *J Comp Neurol*. 182(4):923–944.

François C, Percheron G, Parent A, Sadikot AF, Fenelon G, Yelnik J (1991). Topography of the projection from the central complex of the thalamus to the sensorimotor striatal territory in monkeys. *J Comp Neurol*. 305(1):17–34.

Giménez-Amaya JM, McFarland NR, de las Heras S, Haber SN (1995). Organization of thalamic projections to the ventral striatum in the primate. *J Comp Neurol*. 354(1):127–149.

Guillery RW, Harting JK (2003). Structure and connections of the thalamic reticular nucleus: Advancing views over half a century. *J Comp Neurol*. 463(4):360–371.

Herkenham M (1978). The connections of the nucleus reuniens thalami: Evidence for a direct thalamo-hippocampal pathway in the rat. *J Comp Neurol*. 177:589–610.

Herkenham M (1980). Laminar organization of thalamic projections to the rat neocortex. *Science*. 207:532–535.

Herkenham M (1986). New perspectives on the organization and evolution of nonspecific thalamocortical projections. In: Jones EG, Peters A, eds., *Cerebral Cortex. vol 5. Sensory-Motor Areas and Aspects of Cortical Connectivity*. Plenum/Springer:403–445.

Hsu DT, Price JL (2009). Paraventricular thalamic nucleus: Subcortical connections and innervation by serotonin, orexin, and corticotropin-releasing hormone in macaque monkeys. *J Comp Neurol*. 512(6):825–848.

Huang CL, Winer JA (2000). Auditory thalamocortical projections in the cat: Laminar and areal patterns of input. *J Comp Neurol*. 427(2):302–331.

Insausti R, Amaral DG, Cowan WM (1987). The entorhinal cortex of the monkey: III. Subcortical afferents. *J Comp Neurol*. 264:396–408.

Jasper, H (1966). Unspecific thalamocortical relations. In: Field J, Magoun HW, Hall VE, eds. *Handbook of Physiology*, Section 1: Neurophysiology. American Physiological Association:1307–1231.

Jones EG (1998). Viewpoint: The core and matrix of thalamic organization. *Neuroscience*. 85:331–345.

Jones EG (2001). The thalamic matrix and thalamocortical synchrony. *Trends Neurosci*. 24(10):595–601.

Jones EG (2007). *The Thalamus*. 2nd ed. Cambridge University Press.

Kimura A, Donishi T, Sakoda T, Hazama M, Tamai Y (2003) Auditory thalamic nuclei projections to the temporal cortex in the rat. *Neuroscience*. 117(4):1003–1016.

Kita T, Shigematsu N, Kita H (2016). Intralaminar and tectal projections to the subthalamus in the rat. *Eur J Neurosci.* 44:2899–2908.

Lacey CJ, Bolam JP, Magill PJ (2007). Novel and distinct operational principles of intralaminar thalamic neurons and their striatal projections. *J Neurosci.* 27(16):4374–4384.

Lanciego JL, Gonzalo N, Castle M, Sanchez-Escobar C, Aymerich MS, Obeso JA (2004). Thalamic innervation of striatal and subthalamic neurons projecting to the rat entopeduncular nucleus. *Eur J Neurosci.* 19(5):1267–1277.

LeDoux JE, Farb C, Ruggiero DA (1990). Topographic organization of neurons in the acoustic thalamus that project to the amygdala. *J Neurosci.* 10(4):1043–1054.

LeVay S, Sherk H (1981). The visual claustrum of the cat. I. Structure and connections. *J Neurosci.* 1(9):956–980.

Lorente de No, R (1938). Cerebral cortex: Architecture, intracortical connections, motor projections. In: Fulton J, ed. *Physiology of the Nervous System.* Oxford University Press:291–340.

Macchi G, Bentivoglio M, Minciacchi D, Molinari M (1986). Organisation des connexions thalamiques. *Rev. Neurol. (Paris).* 142:267–282.

Marani E, Heida T, Lakke EA, Usunoff KG (2008). The subthalamic nucleus. Part I: development, cytology, topography and connections. *Adv Anat Embryol Cell Biol.* 198:1–113.

Mathiasen ML, O'Mara SM, Aggleton JP (2020). The anterior thalamic nuclei and nucleus reuniens: So similar but so different. *Neurosci Biobehav Rev.* 119:268–280.

McFarland NR, Haber SN (2000). Convergent inputs from thalamic motor nuclei and frontal cortical areas to the dorsal striatum in the primate. *J Neurosci.* 20(10):3798–3813.

Morison RS, Dempsey EW (1941). A study of thalamo-cortical relations. *Am J Physiol.* 135:281–292.

Penzo MA, Gao C (2021). The paraventricular nucleus of the thalamus: An integrative node underlying homeostatic behavior. *Trends Neurosci.* 44(7):538–549.

Raju DV, Shah DJ, Wright TM, Hall RA, Smith Y (2006). Differential synaptology of vGluT2-containing thalamostriatal afferents between the patch and matrix compartments in rats. *J Comp Neurol.* 499(2):231–243.

Rose JE, Woolsey CN (1949). The relations of thalamic connections, cellular structure and evocable electrical activity in the auditory region of the cat. *J Comp Neurol.* 91:441–466.

Rumpel S, LeDoux J, Zador A, Malinow R (2005). Postsynaptic receptor trafficking underlying a form of associative learning. *Science.* 308:83–88.

Russchen FT (1982). Amygdalopetal projections in the cat. II. Subcortical afferent connections. A study with retrograde tracing techniques. *The.* 207(2):157–176

Sadikot AF, Parent A, François C (1990). The centre médian and parafascicular thalamic nuclei project respectively to the sensorimotor and associative-limbic striatal territories in the squirrel monkey. *Brain Res.* 510(1):161–165.

Sadikot AF, Parent A, François C (1992a). Efferent connections of the centromedian and parafascicular thalamic nuclei in the squirrel monkey: A PHA-L study of subcortical projections. *J Comp Neurol.* 315(2):137–159.

Sadikot AF, Parent A, Smith Y, Bolam JP (1992b). Efferent connections of the centromedian and parafascicular thalamic nuclei in the squirrel monkey: A light and electron microscopic study of the thalamostriatal projection in relation to striatal heterogeneity. *J Comp Neurol.* 320(2):228–242.

Schlag J, Villablanca J (1967). Cortical incremental responses to thalamic stimulation. *Brain Res.* 6(1):119–142.

Sherman SM (2016). Thalamus plays a central role in ongoing cortical functioning. *Nat Neurosci.* 19(4):533–541.

Sherman, SM, Guillery RW (1996). Functional organization of thalamocortical relays. *J Neurophysiol.* 76:1367–1395.

Sherman SM, Guillery RW (1998). On the actions that one nerve cell can have on another: Distinguishing "drivers" from "modulators." *Proc Natl Acad Sci U S A.* 95:7121–7126.

Sherman SM, Guillery RW (2006). *Exploring the thalamus and its role in cortical function.* 2nd ed. MIT Press.

Sidibé M, Smith Y (1999). Thalamic inputs to striatal interneurons in monkeys: Synaptic organization and co-localization of calcium binding proteins. *Neuroscience.* 89(4):1189–1208.

Smith AD, Bolam JP (1990). The neural network of the basal ganglia as revealed by the study of synaptic connections of identified neurones. *Trends Neurosci.* 13(7):259–265.

Smith PH, Uhlrich DJ, Manning KA (2019) Evaluation of medial division of the medial geniculate (MGM) and posterior intralaminar nucleus (PIN) inputs to the rat auditory cortex, amygdala, and striatum. *J Comp Neurol.* 527(9):1478–1494.

Smith PH, Uhlrich DJ, Manning KA, Banks MI (2012). Thalamocortical projections to rat auditory cortex from the ventral and dorsal divisions of the medial geniculate nucleus. *J Comp Neurol.* 520(1):34–51.

Smith Y, Galvan A, Ellender TJ, Doig N, Villalba RM, Huerta-Ocampo I, Wichmann T, Bolam JP (2014). The thalamostriatal system in normal and diseased states. *Front Syst Neurosci.* 8(5):1–18. doi:10.3389/fnsys.2014.00005.

Smith Y, Parent A (1986). Differential connections of caudate nucleus and putamen in the squirrel monkey (*Saimiri sciureus*). *Neuroscience.* 18(2):347–371.

Steriade M (2001). Impact of network activities on neuronal properties in corticothalamic systems. *J Neurophysiol.* 86:1–39.

Turner BH, Herkenham M (1991). Thalamoamygdaloid projections in the rat: A test of the amygdala's role in sensory processing. *J Comp Neurol.* 313(2):295–325.

Van der Werf YD, Witter MP, Groenewegen HJ (2002) The intralaminar and midline nuclei of the thalamus. Anatomical and functional evidence for participation in processes of arousal and awareness. *Brain Res Brain Res Rev.* 39(2-3):107–140.

van Groen T, Kadish I, Wyss JM (1999). Efferent connections of the anteromedial nucleus of the thalamus of the rat. *Brain Res Rev.* 30(1):1–26.

Vertes RP, Hoover WB (2008). Projections of the paraventricular and paratenial nuclei of the dorsal midline thalamus in the rat. *J Comp Neurol.* 508(2), 212–237.

Vertes RP, Hoover WB, Do Valle AC, Sherman A, Rodriguez JJ (2006). Efferent projections of reuniens and rhomboid nuclei of the thalamus in the rat. *J Comp Neurol.* 499(5), 768–796.

Vertes RP, Hoover WB, Rodriguez JJ (2012). Projections of the central medial nucleus of the thalamus in the rat: node in cortical, striatal and limbic forebrain circuitry. *Neuroscience.* 219, 120–136.

Vertes RP, Linley SB, Hoover WB (2015). Limbic circuitry of the midline thalamus. *Neurosci Biobehav Rev.* 54, 89–107.

Walker, AE (1938). *The Primate Thalamus.* University of Chicago Press.

Wei P, Liu N, Zhang Z, Liu X, Tang Y, He X, Wu B, Zhou Z, Liu Y, Li J, Zhang Y, Zhou X, Xu L, Chen L, Bi G, Hu X, Xu F, Wang L (2015). Processing of visually evoked innate fear by a non-canonical thalamic pathway. *Nat Commun.* 6:6756.

Wimmer VC, Bruno RM, de Kock CP, Kuner T, Sakmann B (2010). Dimensions of a projection column and architecture of VPM and POm axons in rat vibrissal cortex. *Cereb Cortex.* 20(10):2265–2276.

Wouterlood FG, Saldana E, Witter MP (1990). Projection from the nucleus reuniens thalami to the hippocampal region: Light and electron microscopic tracing study in the rat with the anterograde tracer Phaseolus vulgaris-leucoagglutinin. *J Comp Neurol.* 296(2):179–203.

12.

THALAMIC OUTPUT PATHWAYS

SINGLE-CELL ORGANIZATION LEVEL

Francisco Clasca

INTRODUCTION

Thalamic projection neurons (TPNs) are the predominant thalamus cell class. They receive direct synapses, in specific combinations, from structures all over the central nervous system, from the spinal cord to the cerebral cortex. These cells lack local recurrent excitatory connectivity; thus, while large-scale inhibitory modulation between nuclei may occur, thalamic computations are essentially determined by the inputs that each TPN receives. In turn, TPNs innervate the cerebral cortex and other forebrain structures though richly branched glutamatergic axons. As a result, each individual TPN acts largely as an integration and broadcasting node within one of the multi-regional sensory, motor, affective, cognitive, or arousal networks of the brain.

Over the past century, studies that used classic (lesion/axonal transport) connection-tracing methods discovered the pathways that link the thalamus with the cerebral cortex, striatum, amygdala, and so on, and analyzed them in different mammal species (see **Chapter 11 in Section IV**). However, the precise wiring of these circuits remained until recently a matter of speculation because the classic methods could not resolve individual axons from cell body to terminal synaptic branches. New methods that enable the direct delineation of complete individual projection neuron morphologies (Figure 12.1; Pinault et al., 1996; Furuta et al., 2001; Economo et al., 2016; Kuramoto, 2019), have opened the era of cell-specific, quantitative connectomics. Applied to date almost exclusively in rodents, these methods have already revealed an unsuspected diversity and specificity in arborization patterns among TPN axons.

A CATALOG OF HIGH-LEVEL ("CANONICAL") THALAMIC PROJECTION NEURON CLASSES

Modern single-cell anatomical, functional, and molecular methods have revealed that while all TPNs share fundamental phenotypic traits (e.g., glutamatergic neurotransmission, general morphology, exclusive targeting of ipsilateral forebrain structures, innervation of the TRN), variation in other traits is sufficiently wide and discontinuous to identify several high-level TPN classes.

Axonal morphology is the trait that varies most widely among TPN, is readily quantifiable, and is relatively stable. Thus, it provides a robust primary axis of variation for the definition of TPN types. Smaller, but consistent, variations in somatodendritic morphology, electrophysiological properties, and transcriptomic profiles largely align with the various axonal motifs.

I recently proposed a catalog of 6 major "canonical" TPN classes for the rodent thalamus primarily based on axonal wiring motifs (Table 12.1; see also Clasca, 2022). Albeit with markedly different saturation, the available data already cover virtually all the thalamus nuclei; thus, the catalog probably captures the full gamut of variation among TPN. At the same time, it is certain that subclasses will be identified within most of these 6 high-level categories in the future, as more single-cell morphologies and single-cell transcriptomic data become available.

In the following sections, variations in axonal architecture ("axonal wiring motifs") are first analyzed in detail. The relationships observed between the wiring motifs and variations in somatodendritic morphology, transcriptomic profiles, and/or membrane properties and conductances are subsequently explored. The last section of this chapter examines the distribution of the various TPN classes across the thalamus and its implications for understanding the functional organization and significance of thalamic nuclei.

SINGLE-CELL LEVEL VARIATION IN AXONAL MORPHOLOGY

An unexpected finding from the application of single-cell labeling methods to the study of thalamofugal projections

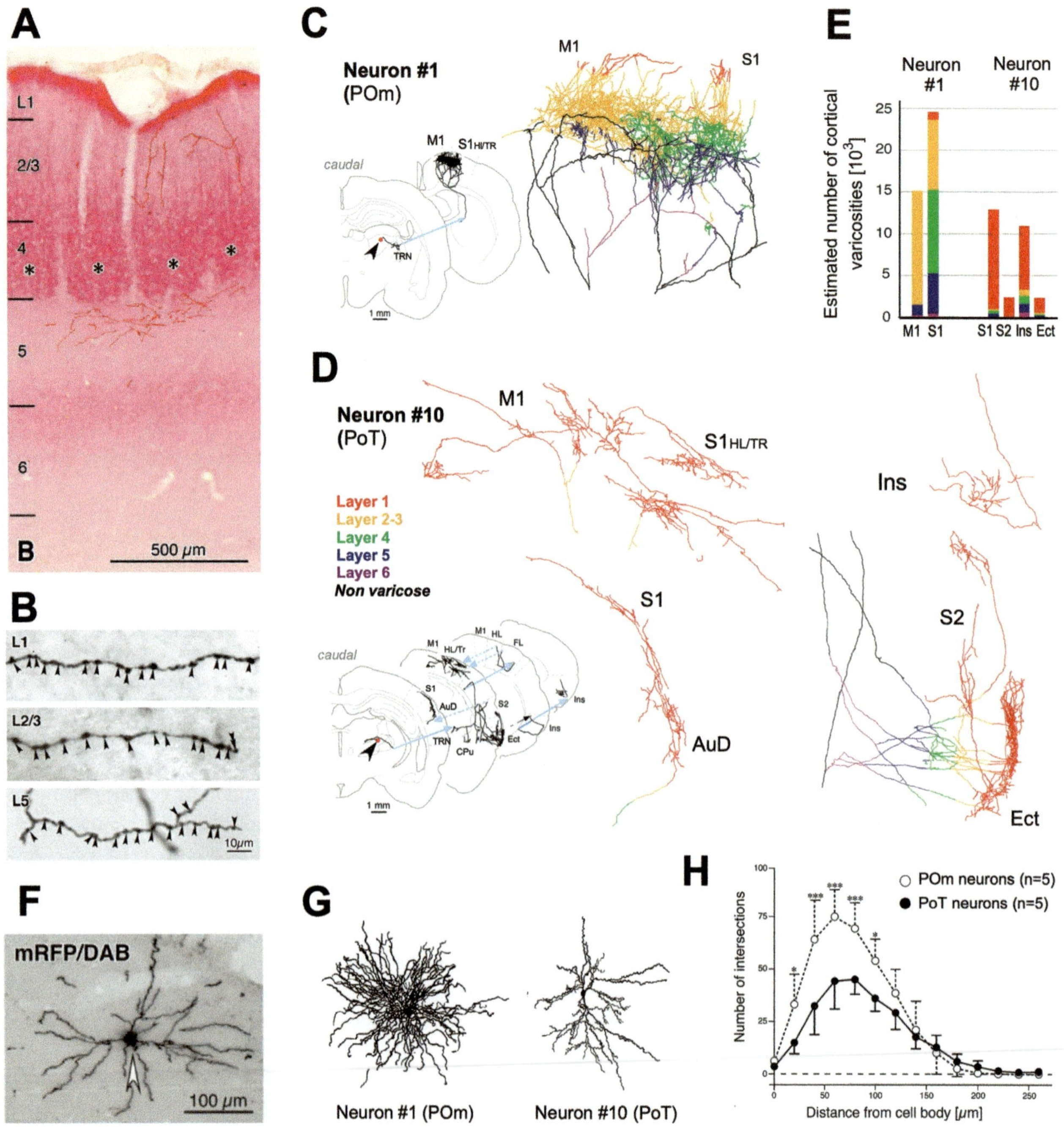

Figure 12.1 Viral vector labeling and analysis of individual TPNs in rat POm and PoT. (A). Coronal section (50-μm thick) sample of S1. Immunostaining with antibodies against the modified red fluorescent protein (mRFP) developed with immunoperoxidase and diaminobenzindine (DAB, brown) histochemistry reveals labeled segments of the axon arborization. Co-staining with an antibody against the vesicular glutamate transporter developed with Tris-aminophenylmethane + p-cresol (pink) labels dense bands of terminal puncta in layer 1, the layer 4 barrels (asterisks) and upper layer 6. Interbarrel septa are indicated with arrows. (B). Detail of the axon varicosities (putative synaptic boutons) in layers (L) 1, 2/3, and 5. (C). Reconstruction of the axon (black line) from a TPN whose soma is in POm ("neuron 1," arrowhead). The axon is represented on 2 serial coronal section contours of the rat brain. A pale blue arrow links the cut ends of axon portions split that are represented on separate coronal diagrams. The image on the right is a magnified complete intracortical arborization of this cell, reconstructed from serial sections. Branches are color-coded according to laminar position. Nonvaricose (probably nonsynaptic) regions are drawn in black. (D). Reconstruction of a TPN ("neuron 10") whose soma (arrowhead) is in the triangular portion of POm (PoT). Note the small branch in the caudateputamen (CPu). (E). The estimated number of axon varicosities per area and layer of the cerebral cortex of single-labeled POm neurons was calculated using the data on the axonal length of the neurons and the average linear density of axon varicosities in each area and layer. The length of axon fibers was estimated by multiplying the length of varicose axons projected onto a frontal plane by $\pi/2$. The S1 cortex included the FL and HL cortices. Neurons 1 and 10 had ~43,000 and ~28.000 estimated cortical varicosities in total, respectively. Neuron 10 had, in addition, a small branch in the striatum not included in the estimation. (F). 50-μm-thick coronal section containing the cell soma and part of the dendritic tree of a transfected POm neuron. (G). Somatodendritic compartment reconstructions of the 2 neurons illustrated in panels C–E. (H). Comparison of the radial dendritic structure (Shollar analysis) of neurons in rostral POm versus caudal POm/PoT, including the 2 cells illustrated in panel G, revealed that dendritic complexity was statistically significant at distances of 20–100 ealed that dendritic complexity was statistiBonferroni multiple comparison test). The open and filled circles are the averaged data for rostral POm and caudal POm/PoT neurons, respectively. Bars indicate SD. Adapted from Ohno et al. (2012). mRFP/DAB = red fluorescent protein immuonstained using DAB as chromogen; for other abbreviations, see Table 11.1 in Chapter 11 in Section IV.

TABLE 12.1 A CATALOG OF 6 TOP-LEVEL ("CANONICAL") PTN CELL CLASSES OF THE RODENT THALAMUS

Class	Shape	Nuclei Where Prevalent	TRN	Main Target	Cortex Areas Pattern Intensity	Cortex Main layers	Striatum	Amyg	Other	Firing	CaBP
Unifocal thalamocortical **(UTC)**	Bushy	VPl, VPm, MGNv, LGNd Sm(d)	+	Neocortex	1 Focal +++	4–3, 6	–	–	–	Fast firing; short latency bursts	PV ++/0 T+++
Multifocal thalamocortical **(MTC)**	Bushy	Po, VL VPMPc, VPc, MD LP, MGNd, Sm(v) [LD, AV, AD, AM]	+	Neocortex [PaHC]	2–5 Variable +++	1, 3–5a [1,3 PH]	Dorsal Str + / 0 Diffuse	LAm	–	Reg. firing; long latency bursts	CB ++ PV + ++
Subpial thalamocortical **(STC)**	Bushy	VA, VM, MGNm, PoT [Re, Rh]	+	Neocortex [PaHC, HC]	≥2 Spread +++	1 [1,5 PH]	+ Dorsal Str Diffuse 0			Reg. firing; long latency bursts	CB +++ CR +++ [Re]
Anterior thalamostriatal **(ATS)**	Bushy	CL, Pc, CeM Some Pf	+	Dorsal striatum Cortex	≥2 Spread ++	3–5a	+++ Dorsal Str Diffuse	BLAm	Cla	Reg. firing; long latency bursts	CB + PV+ N– EFCaBP++
Midline thalamostriatal **(MTS)**	Bushy	PVA, Pt, IMD, Spf	+	Ventral striatum	+ / 0	6, 3, 1	+++ Ventral Str Diffuse	CeAm BLAm MeAm	BNST, SI, AoN, Cla DHN, LS, VP, OT	?	CR ++ CB+++ N– EFCaBP++
Posterior thalamostriatal **(PTS)**	Reticular	Pf, Eth, PiL	+	Dorsal striatum	≥2 Spread +	6, 5b	+++ Dorsal Str Clustered	CeAm	Sth EP Cla, SI GP	High IR; no bursts	CB++ CR +++

NOTES: Each cell class is identified by their axon wiring motif, and linked to other traits such as somatodendritic morphology, membrane properties, and calcium-binding protein-expression patterns. For the latter, just the differential expression of 5 different proteins is indicated; they are chosen as an example because, historically, some of these (CB and PV) were widely used as specific markers for thalamic cell types. Data given in brackets indicate features present only in cells of a particular nucleus within the class.

CaBP: calcium-binding protein genes; CB: calbindin 28K; PV: parvalbumin; CR: calretinin; N-EFCaBP: N-terminal EF-hand calcium-binding protein; PH: parahippocampal áreas; Str: striatum. For other abbreviations, see Table 11.1 in Chapter 11 of Section IV.

Data are derived from Deschênes et al., (1996a,b, 1998); Kuramoto et al. (2009, 2015, 2017a, 2017b); Ohno (2012); Noseda et al. (2011); Nakamura et al. (2015); Unzai et al. (2017); Gheorghita et al. (2006); Phillips et al. (2019); Winnubst et al. (2019); Rodriguez-Moreno et al. (2020); Zhang H. et al. (2021); Peng et al. (2021). Electrophysiological data based on Sherodziya and Timofeev (2014); Jhangiani-Jashanmal et al. (2016); Phillips et al. (2019); Desai and Varela (2021). Gene expression data from Lein et al. (2007) and Phillips et al. (2019).

has been the prevalence and variation among TPN axons of branched morphologies. For example, many thalamic axons have been found to innervate the cortex and striatum and/or amygdala simultaneously, dispelling the notion of strictly separate thalamocortical, thalamostriatal, thalamoamygdaloid (etc.) circuits. Likewise, it has been discovered that different branches of the same axon targeting 2 cortical areas can arborize in different layers and have different synapse structure and postsynaptic targets in each area, dispelling earlier notions notion of simple cell class-defining laminar-projection patterns (Rodriguez Moreno et al., 2020; see **Chapter 11 in Section IV**).

From a functional standpoint, the variety in axon morphologies is relevant because any given morphology powerfully constrains the role that the TPN can play within forebrain networks. For example, axons that establish focal point-to-point circuits, and are able to faithfully preserve topological information within an input matrix, with minimal convergence, while axons that create branched point-to-multipoint circuits may act as backbones for signal broadcasting, synchronization, or cross-correlation between separate forebrain cell populations.

Some TPNs innervate only the cerebral cortex, some few innervate only the striatum, and many others innervate both, in variable proportions. A first general division can be thus made between *thalamocortical* or *thalamostriatal* neurons, each of which have most of their varicose branches (synaptic regions) in the cerebral cortex or in the striatum, respectively. Under each of these 2 primary groupings, 3 main wiring motifs can be further distinguished, up to a total of 6 main classes, as described in the following.

UNIFOCAL THALAMOCORTICAL (UTC) NEURONS

These TPNs have axons that emit subcortical collaterals only to TRN and then arborize in a single area of the isocortex, usually forming one profuse and compact tuft of varicose branches in layers 4 and 3. Synaptic boutons in these branches are relatively large and often multisynaptic (Familtsev et al., 2016; Rodriguez-Moreno et al., 2020). The axons may leave some additional sparser branches in layer 6. While some branch tips may reach layer 1, they typically do not branch extensively into it. Nevertheless, axons from neurons in the peripheral (shell) regions of MGNv or LGNd innervate the middle layers in a looser manner and, unlike other UTC axons, they may give off some branches in layer 1. The unifocal TC cells are the best characterized of all TPNs, as they are prevalent in the intensively studied sensory relay nuclei (VPl, VPm, LGNd, MGNv, VPMpc and dorsal Sm; Kuramoto et al., 2017b).

MULTIFOCAL THALAMOCORTICAL (MTC) NEURONS

The axons of MTC neurons characteristically branch to target at least 2 cortical areas, which can be adjacent or far apart and may include non-isocortical areas as well. They form localized intracortical arborizations whose profusion, laminar distribution, and bouton sizes may vary markedly between branches. In addition, many MTC axons give off branches that arborize diffusely in the striatum. MTC class neurons populate many of the nuclei of the dorsal thalamus, including as the anterior POm, LP-Pulvinar, VL, ventrolateral domain of VPm, MGNd, most of MD, and ventral Sm. Neurons in LD, AM-IAM, AV, and AD also fit the general MTC profile, although it is of note that these cells axons innervate both isocortical and parahippocampal areas.

SUBPIAL THALAMOCORTICAL (STC) NEURONS

The axons of STC neurons target, as a rule, several cortical areas and produce numerous collaterals that cross cortical layers 6–2 with minimal or no arborization, but then arborize extensively as "flat" subpial plexuses within the outer, subpial part of cortical layer 1. Some, but not all, STC axons send branches to the dorsal striatum. MTC neurons constitute several of the nuclei situated in the ventral tier of the thalamus, namely, VA, VM, as well as the caudal portion of POm and the adjacent MGM. Because of their robust and widespread subpial innervation of the cortex, Re and Rh neurons are included in the MTC profile. However, these neurons might well be regarded as a separate class of their own, as their axons not only innervate layer 1 in the isocortex, hippocampal, and parahippocampal areas, but also target layer 5 in some parahippocampal areas. Besides, the gene expression profile of Re cells differs significantly from that of other MTC neurons (see below; Phillips et al., 2019).

ANTERIOR THALAMOSTRIATAL (ATS) NEURONS

The axons of ATS neurons characteristically place most of their varicose branches in the dorsal striatum; these branches are elongated and widely spread (Deschenes et al., 1995). Their synapses target the middle spiny neurons, but not the striatal interneurons (Smith et al., 2014). In addition, ATS axons extend to the isocortex, where they form relatively loose terminal arborizations, mostly in the deep and middle layers. The ATS neurons populate the "anterior intralaminar" nuclei (CL, Pc, CeM). In addition, about a third of Pf neurons in rodents reportedly show ATS features (Deschenes et al., 1996; Beatty et al., 2009; Unzai et al., 2017).

MIDLINE THALAMOSTRIATAL (MTS) NEURONS

The axons of MTS neurons arborize though multiple branches, in ventral striatum BNST, CeA, and substantia innominata. Unlike any other TPN, some of these cells target the hypothalamus (Table 12.1). Their cortical branch is relatively simple and may often be absent; when present, the cortical branch innervates mainly regions transitional between the frontal isocortex and olfactory cortices. Most MTS neurons express calretinin and/or calbindin (Matyas et al., 2018). MTS neurons are localized along the dorsal (PVA, Pt, IMD) and caudomedial (Spf) borders of the dorsal thalamus.

POSTERIOR THALAMOSTRIATAL (PTS) NEURONS

In rodents, the PTS cells constitute a large majority in Pf, Eth, and PIL (Deschenes et al., 1996b; Smith et al., 2006). The main target of PTS cell axons is the matrix compartment of the dorsal striatum, where each axon forms several separate focal plexuses (Figure 12.2). PTS axons target both projection neurons (MSN) and interneurons and establish synapses preferentially on dendritic trunks. Besides, the PTS axons innervate other basal ganglia nuclei, particularly the subthalamic nucleus (Deschenes et al., 1996b; Unzai et al., 2017). While in rodents virtually all PTS axons send a branch to the frontal and parietal isocortex, which arborizes in layers 5 and 6 in a loose fashion, in primates,

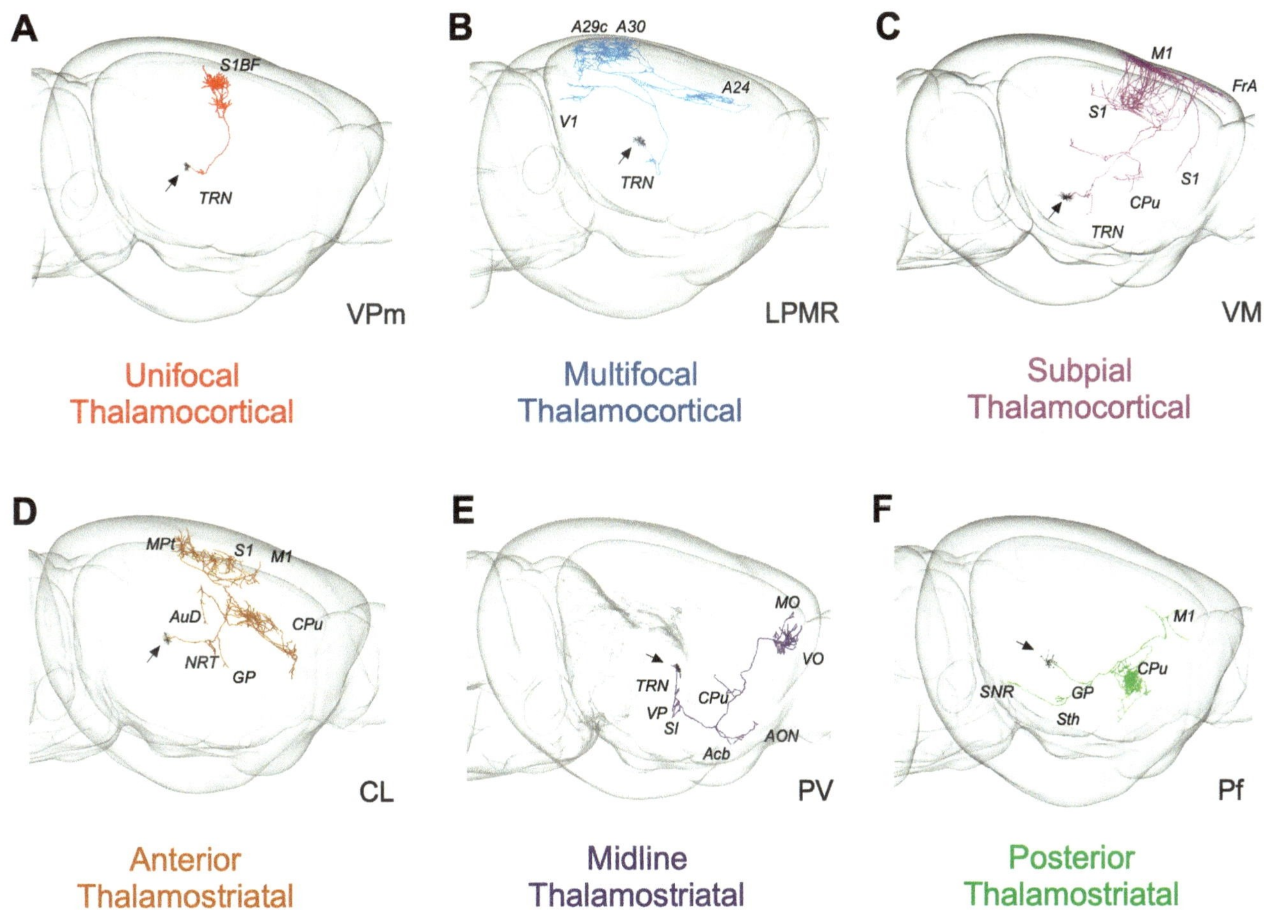

Figure 12.2 Representative examples of fully reconstructed mouse single-cell PTN axonal morphologies. In each panel, a single individual TPN is represented on a "transparent envelope" visualization of the whole mouse brain. One cell is shown within each transparent brain envelope. Cell bodies and dendritic trees are shown in black ink and indicated with an arrowhead. The thalamic nucleus where the cell soma was located is indicated on the lower right corner of each panel. Cortical and subcortical regions targeted by the axon are labeled. These 2-dimensional projection images of 3-dimensional reconstructions in a 5-point (spherical) perspective are presented; because of the spherical distortion, no calibration bar can be provided. VPm = ventral posteromedial nucleus. LPMR = lateral posteromedial rostral nucleus; VM = ventromedial nucleus; CL = central lateral nucleus; PV = paraventricular nucleus; Pf = parafascicular nucleus; for other abbreviations, see Table 11 in Chapter 11 of Section IV. A–D and F modified from the Mouselight database (Winnubst et al., 2020; https://www.janelia.org/project-team/mouselight) A: AA1224; B: AA0341; C: AA0176; D: AA0051; F: AA1439. Panel E modified from the Allen Braintell Brain Image Library (Peng et al., 2021; https://doi.brainimagelibrary.org/doi/10.35077/g.25); experiment 1119749775_17545_4550-X20564-Y25436.

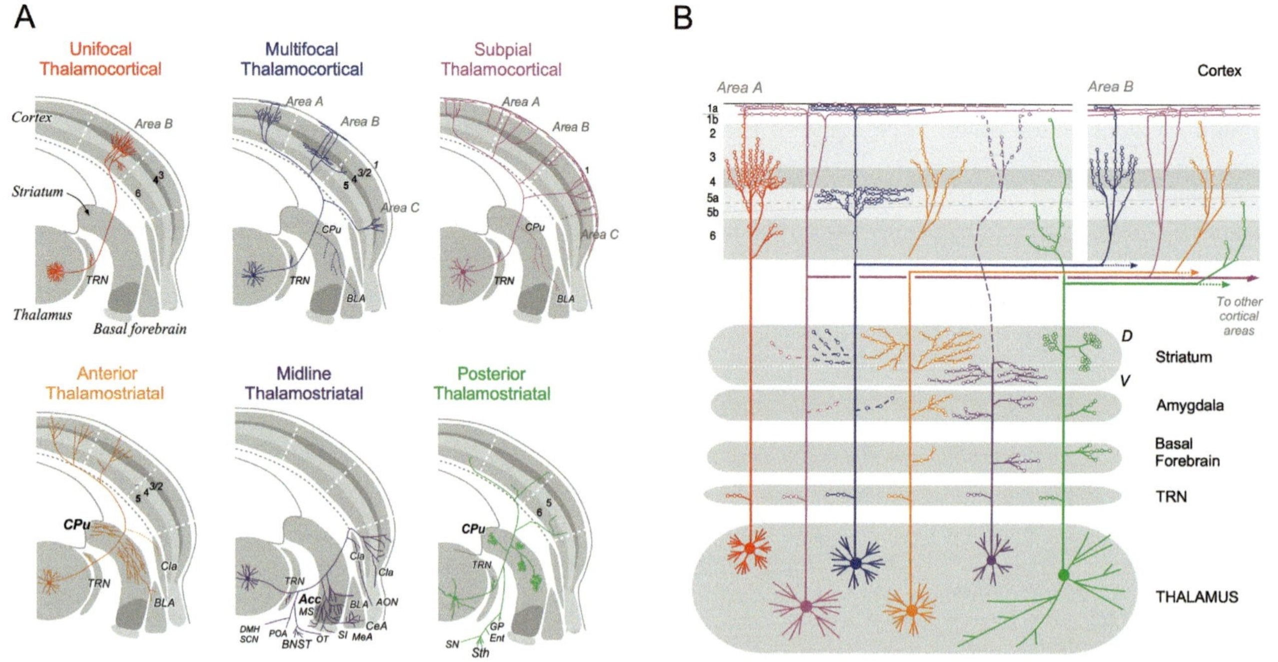

Figure 12.3 Diagram comparison of the main morphological features that characterize the 6 "canonical" cell classes present in the rodent thalamus. (A). Idealized neurons representing each axonal wiring motif are drawn on a cartoon diagram of the forebrain. Dashed lines indicate branches that may often be absent. Structures receiving the heaviest innervation are highlighted in large letter case and boldface. (B). Diagram comparing the variations in axonal branching patterns. Small open circles indicate axon varicosities (putative synaptic regions). Dashed lines indicate branches that may be absent in some cells of a particular class. A white dashed line is used to separate the dorsal and ventral portions of the striatum. Somatodendritic morphologies are represented in highly simplified form. TC = thalamocortical; TS = thalamostriatal (see Table 11.1 in Chapter 11 of Section IV for other abbreviations).

about a third of PTS cells were reported to lack a cortical branch (Parent and Parent, 2005).

Despite their evident diversity, all TPN axons share the same fundamental layout: they extend from a point in the thalamus to a distal point, usually located in the cerebral cortex, giving off some varicose collateral branches to cortical or subcortical structures along the way. Differences stem from the presence or absence of these primary branches and the local elaboration of patterns of the branches (Figure 12.3).

SINGLE-CELL LEVEL VARIATION IN SOMATODENDRITIC MORPHOLOGY

In comparison with the widely different axon architectures just described, variation among TPNs in cell somata and dendritic trees morphology seems modest. As a rule, multiple (5–12) thick primary dendrites sprout from the cell body in multipolar or slightly bi-tufted arrangements (Figure 12.1F–G). After a short (20–50 mm) stretch, each primary dendrite divides again into a tuft of thinner secondary branches that diverge in a radial fashion and that together give TPNs their typical "bushy" or "isodendritic" appearance. The secondary dendrites may have irregularly spaced appendages resembling slender spines or stubby protrusions, which in some TPNs may be clustered together (see **Chapter 2 in Section II**).

The TPN somatodendritic morphologies were first described with Golgi and intracellular recording and labeling methods (review in Jones, 2007). In recent years the single-cell labeling methods have allowed, for the first time, the reliable correlation between somatodendritic and axonal morphologies (Clasca, 2022).

These comparisons have revealed a tendency for primary dendrite branches to be less numerous and secondary branches to be longer in TPNs that have more widely spread axon arbors (Figure 12.1G–H; Kuramoto et al., 2009; Kuramoto et al., 2015; Ohno et al., 2012, Nakamura et al., 2015). For example, UTC neurons tend to have more numerous primary dendrites and richly branched but relatively short secondary dendrites than MTC; likewise, MTC and ATS cells have more numerous primary dendrites and more branched but relatively shorter secondary dendrites than STC.

A clear departure from this typical morphology occurs in the posterior intralaminar nuclei (Pf, Eth, and PIL). Instead of a bushy layout, most cells in these nuclei have few (3–7) primary dendrites that extend divergently over twice as much as other TPN, give off very few secondary branches, and show frequent spines and appendages (Deschenes et al., 1996b; Lacey et al., 2007).

SINGLE-CELL LEVEL VARIATION IN ELECTROPHYSIOLOGICAL PROPERTIES AND GENE EXPRESSION

Cell membrane properties and conductances are widely shared across nuclei (Llinás and Jahnsen, 1982; Jahnsen and Llinás, 1984; Sherodziya and Timofeev, 2014; Desai and Varela, 2021). Conductances include a low-threshold, transient calcium current that operates via T-type Ca^{2+} channels that mediate, in voltage-dependent manner, an inward current that produces a transient depolarization. These channels are inactivated by depolarization, but they become de-inactivated by hyperpolarization; hence, at hyperpolarized voltages they can elicit a burst of sodium-dependent high-frequency action potentials known as "bursts." Interactions between this and other conductances allow TPNs to display different intrinsic rhythms (see **Chapter 2 in Section II**).

Upon such basic similarity, consistent differences between nuclei have recently been detected (Table 12.1). As a rule, the action potentials of UTC are narrower, depolarization thresholds lower, and frequencies of tonic firing higher than other TPNs (Landisman and Connors, 2007; Nakamura et al., 2015; Phillips et al., 2019; Desai and Varela, 2021). Response properties in ATS, MTC and STC neurons are basically the same; however, neurons in CeM show small but consistent differences in their response at hyperpolarized membrane potentials (Jhangiani-Jashanmal et al., 2016). As was the case for the morphological features, the membrane properties of Pf cells depart markedly from the all other TPNs; their resting potentials are more depolarized, their action potentials are larger, and their input resistances are lower (Smith et al., 2006; Lacey et al., 2007; Beatty et al., 2009; Sheroziya and Timofeev, 2014).

Small but consistent differences in the number of spikes fired on bursts and in the time base of membrane repolarization between TPNs in different nuclei have also been reported (Landisman and Connors, 2007; Nakamura et al., 2015; Jhangiani-Jashanmal et al., 2016; Phillips et al., 2019; Desai and Varela 2021). Again an outlier, cells in Pf show a reduction or complete absence of the low-threshold, voltage-sensitive calcium conductance. As a result, the voltage-dependent burst response is minimal or absent in Pf cells (Smith et al., 2006; Lacey et al., 2007; Beatty et al., 2009).

Genomic techniques have greatly expanded the number of genes (to about 4,000) whose expression has been found to significantly vary across the thalamus (Lein et al., 2007; Nagalski et al., 2016; Phillips et al., 2019). These techniques now allow a characterization of putative cell types based on single-cell-level gene-expression data using unsupervised clustering methods. Remarkably, the single-cell-level differences detected are mostly gradual and highly combinatorial in nature. Thus, at least in the mouse thalamus, gene-expression patterns are best viewed as variation across a continuum, rather than specific genes sharply delineating cell classes. More detailed and complete transcriptomic analyses in rodents and other species, including single-cell transcriptomes, are expected to eventually provide key data for the systematization of TPN cell diversity (Winnubst et al., 2020).

A recent unsupervised nearly full-genome analysis of gene profiles was able to detect 3 main groupings among TPN, and they were only loosely aligned with axonal morphotypes (Table 12.1; Phillips et al., 2019; Bakken et al., 2021; Peng et al., 2021). There is mounting evidence that projection neuronal cell groups from within a single transcriptomic cluster can be further separated based on their anatomical projections and connectivity patterns (Kim et al., 2020; Bakken et al., 2021; Zhang Z. et al., 2021). The implication is that connectivity is a key trait for cell-type identification in long-range projection neurons and that, in these cells, gene expression alone cannot be used to fully annotate cell types. In some sense, it is to be expected that the relationship in TPNs between transcriptome and axonal morphology is not simple, as the latter reflects a developmental history, which is not necessarily captured by a snapshot of a cell gene-expression profile (Winnubst et al., 2020). To arrive at meaningful descriptions of TPN types, therefore, gene expression must be linked to structural and functional properties.

A SINGLE-CELL VIEW OF THE THALAMUS AND THALAMIC NUCLEI

CITOARCHITECTONIC NUCLEI AND HIGH-LEVEL TPN CELL CLASSES

A key insight that emerges from the single-cell analysis is that most cytoarchitectonic nuclei are constituted by neurons of a single high-level TPN class. Transitions between cell classes occur mostly at cytoarchitectonic nuclei boundaries and are at least as abrupt as the nuclei borders themselves. Hence, the spatial distribution of the high-level TPN cell classes largely aligns with cytoarchitectonic nuclei borders (Figure 12.4A).

In addition, the nuclei formed by TPNs of the same high-level class show a clear tendency to cluster along "tiers" arching from the diencephalic ventricle wall to the

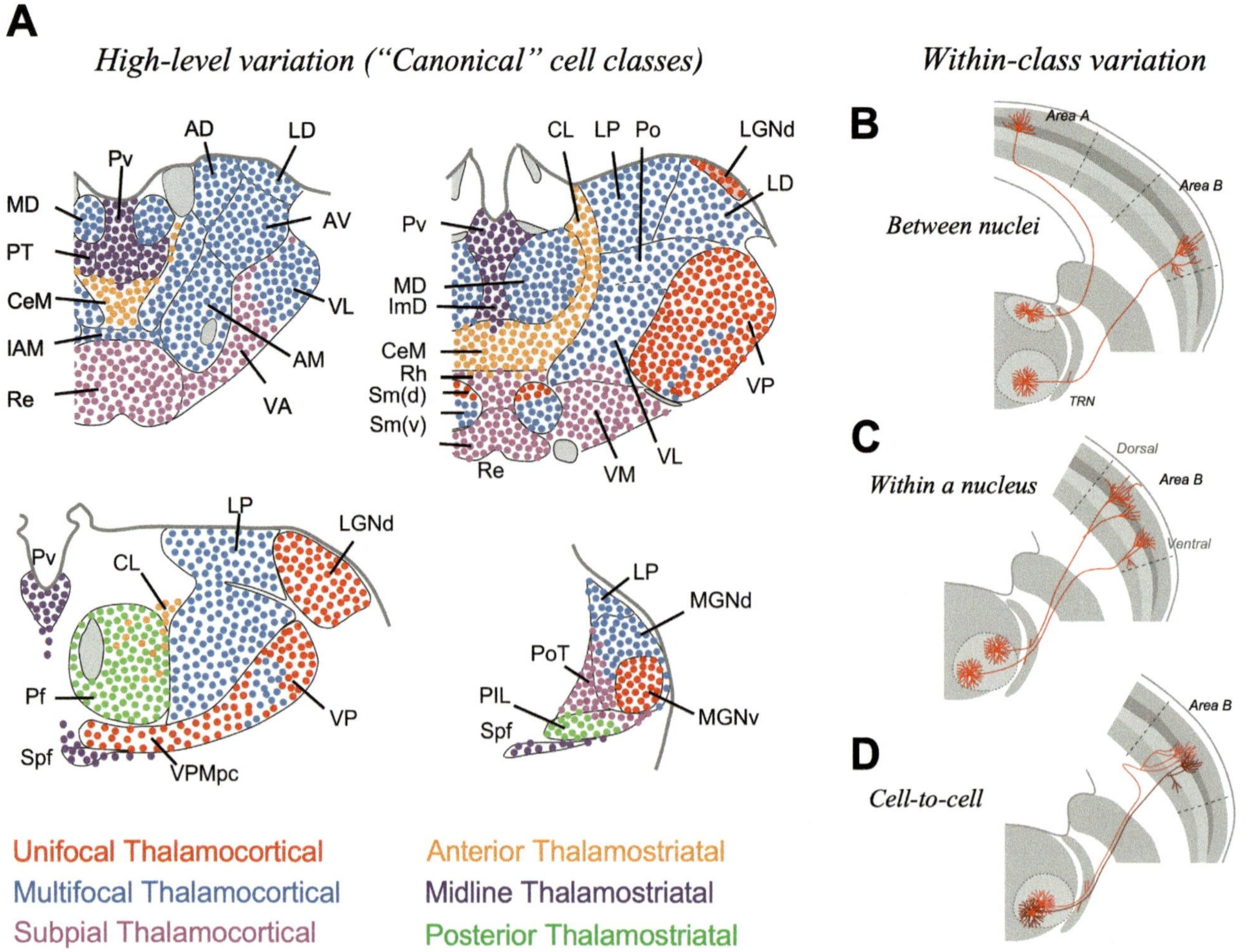

Figure 12.4 Distribution across the mouse thalamus of the 6 high-level ("canonical") PTN classes listed in Table 12.1. (A) Four coronal section levels of the mouse thalamus are shown. Note that these classes largely align with some of the classic cytoarchitectonic nuclei borders. See list for abbreviations. (B-D). Cartoon depiction of the diverse ranges of small-scale variation observed among the TPN of the same canonical cell class.

dorsal pial surface of the thalamus. For example, note in Figure 12.4A in that the nuclei containing UTC cells (in red) form a discontinuous tier that stretches from VPMpc to LGNd. Similar groupings are detectable for most other cell classes. It is of note that these TPN class-specific groupings bear a remarkable resemblance to territories occupied by the thalamic neurons derived from individual neuropithelial precursor clones in the embryonic diencephalon (Shi et al., 2017; Wong et al., 2018), suggesting a possible clonal relationship among cells of the same adult high-level TPN class.

There is evidence, however, that in some few rodent nuclei 2 different top-level TPN classes may be partly intermingled (Figure 12.4A). This is best documented for the rodent Pf; as noted earlier, this nucleus contains a significant population of ATC cells amidst a majority of PTC cells (Lacey et al., 2007; Beatty et al., 2009). There is also evidence that MTC are present in some subnuclei in the ventrolateral domain of the predominantly UTC VPm nucleus (Pierret et al., 2000; Zhang H. et al., 2021). Likewise, anterior, calbindin-poor regions of POm are predominantly MTC, while the caudal and triangular (PoT) calbindin-rich portions of the nucleus show STC morphologies (Figure 12.1; Ohno et al., 2012; el-Boustani et al., 2020). It is yet unclear, however, to which extent these cases represent truly mixed cell populations or, rather, undetected borders between contiguous, but functionally different, nuclear domains.

A practical consequence of cell class homogeneity within most thalamic nuclei is that multicellular probes (tracer injections, optogenetic vectors, etc.) can be applied to small and well-localized nuclear domains to investigate the properties of a single cell class. For decades his fact was unknowingly exploited by connection-tracing studies using bulk-labeling anterograde tracer injections (see **Chapter 11 in Section IV**). At or near nuclei borders, however, multicell probes may yield a mixture of signals from more than one TPN cell class.

SMALL-SCALE VARIATION WITHIN TPN CLASSES

Under each of the above 6 high-level groupings, smaller-scale variation is pervasive across nuclei and within nuclei.

Between nuclei containing cells of the same high-level class, differences involve mainly the specific forebrain area(s) that the axons target (e.g., occipital/visual vs. parietal/somatosensory cortices, or dorsolateral vs. dorsomedial striatum, etc.) (Figure 12.4B).

Within nuclei made of a single high-level cell class, the variation involves mainly the point(s) of the cortex or subcortical structures targeted by the cells' axons within the area(s) innervated by the nucleus. Such variation is the basis, for example, of the orderly spatial gradients of projection observed in most thalamic nuclei. In the first-order relay nuclei, these projection gradients have a high spatial granularity and preserve in the thalamic projection to the cortex the topological relationships of thalamus input pathways. For non-sensory relay nuclei, the functional logic behind variation in axon branching is less well understood.

Within most thalamic nuclei, the gradients of variation are smooth and roughly continuous along 1 or 2 spatial axes. However, in the large and highly specialized first-order relay sensory nuclei of some species (e.g., LGNd of primates and carnivores), spatially ordered cito- or chemoarchitectonic modules ("lamellae," "rods," "barreloids," laminae," and "sublaminae," etc.) are detectable. Such highly ordered cell groupings optimize distances and minimize wiring between neurons.

Other axis of variation among cells of the same nucleus involves the distributions and/or relative weight of the terminal branch plexuses in particular cortical layers or sublayers (Figure 12.4C). These differences are best documented for LGNd and the primary visual cortex. There, TPN subpopulations with specific sublaminar patterns of axon arborization in the visual cortex have been shown to have slightly different somatodendritic morphologies and response properties. Moreover, these LGNd subpopulations receive specific retinal input submodalities, and in highly visual species tend to arrange in cytoarchitectonically detectable layers and sublayers (Boyd and Matsubara, 1996; Jones, 2007; Kerchensteiner and Guido, 2017; see **Chapter 17 in Section V**).

In addition, fine-grain cell-to-cell fluctuations in axonal morphology are always present among the cells of the same nucleus, even between cells that are adjacent to each other. These variations may involve, for example, a different profusion of terminal branches, or even the absence or duplication of a primary branch (Figure 12.5D). They likely represent random end results of the interplay between cell-specific gene-expression programs and non-cell-autonomous spatial and activity-dependent signals during axonal development (Clasca et al., 2016; Martini et al., 2021). Likewise, high-resolution single-cell electrophysiological (Jhangiani-Jashanmal et al., 2016; Desai and Varela, 2021) and transcriptomic (Phillips et al., 2019) studies consistently report a low-level of cell-to cell variation within individual thalamic nuclei.

CONCLUSION AND OUTLOOK

Single-cell labeling methods launched an era of quantitative and cell-specific brain connectomics. As these methods are low-yield and labor-intensive, however, they were until recently applied to a limited set of thalamic nuclei, and only in lab rodents. New industrial-scale 3D morphology pipelines have recently begun to accelerate this process and are expected to saturate the range of cell variationfor the adult mouse brain in the coming years. An important caveat, however, is that morphologies in these large datasets are not yet as thoroughly curated for completeness or accurate spatial registration to tissue compartments as those in peer-reviewed papers, and they do not include critical information about axon varicosity distribution and size. Parallel efforts are underway to elucidate PTN transcriptomic profiles with single-cell resolution. The expectation is that quantitatively robust samples of morphological phenotypic traits and transcriptomic data eventually will be combined to provide a definitive classification of mice TPN cell and circuit diversity (Winnubst et al., 2020), thus allowing the biologically accurate modeling and selective functional manipulation of these key brain circuits.

Except for one published study that investigated Pf-Cm neurons of a New World anthropoid primate (Parent and Parent, 2005), single-cell axon labeling methods have been applied to date only to lab rodents. Interestingly, this study revealed that individual monkey Pf-Cm neurons closely resemble typical rodent PTS neurons. It also detected two PTS subtypes not observed in the rodent Pf nucleus. This finding hints at the tantalizing possibility that new multiregional forebrain circuits may have emerged in evolution simply through the modification of TPN axon-branching motifs. In the thalamus and beyond, comparative analyses of axonal structure at the single-cell level may hold important clues to the evolution of brain function and behavior.

ACKNOWLEDGMENTS

This work was supported by funding from Spain's Ministerio de Ciencia, Innovación y Universidades (PID2020-115780GB-I00) and PID2020-115780GB-I00). Discussions with Drs. César Porrrero, Mario Rubio-Teves and Pablo Rubio-Garrido on some of the ideas presented in this chapter are gratefully acknowledged.

REFERENCES

Bakken TE, van Velthoven CT, Menon V, Hodge RD, Yao Z, Nguyen TN, Graybuck LT, Horwitz GD, Bertagnolli D, Goldy J, Yanny AM, Garren E, Parry S, Casper T, Shehata SI, Barkan ER, Szafer A, Levi BP, Dee N, Smith KA, Sunkin SM, Bernard A, Phillips J, Hawrylycz MJ, Koch C, Murphy GJ, Lein E, Zeng H, Tasic B (2021). Single-cell and single-nucleus RNA-seq uncovers shared and distinct axes of variation in dorsal LGN neurons in mice, non-human primates, and humans. *eLife*. 10:e64875.

Beatty JA, Sylwestrak EL, Cox CL (2009). Two distinct populations of projection neurons in the rat lateral parafascicular thalamic nucleus and their cholinergic responsiveness. *Neuroscience*. 162:155–173.

Boyd JD, Matsubara JA (1996). Laminar and columnar patterns of geniculocortical projections in the cat: Relationship to cytochrome oxidase. *J Comp Neurol*. 365:659–682.

Clasca, F (2022). Thalamic output pathways. In: Halassa MM, ed., *The Thalamus*. Cambridge University Press:45–70.

Clasca F, Porrero C, Galazo M, Rubio-Garrido P, Evangelio M (2016). Anatomy and development of multi-specific thalamocortical axons: Implications for cortical dynamics and evolution. In: Rockland KS, ed. *Axons and Brain Architecture*. Elsevier:69–92.

Desai NV, Varela C (2021). Distinct burst properties contribute to the functional diversity of thalamic nuclei. *J Comp Neurol*. 529:3726–3750.

Deschênes M, Veinante P, Zhang ZW (1998). The organization of corticothalamic projections: reciprocity versus parity. *Brain Res Rev*. 28:286–308.

Deschênes M, Bourassa J, Parent A (1995). Two different types of thalamic fibers innervate the rat striatum. *Brain Res*. 701:288–292.

Deschênes M, Bourassa J, Parent A (1996a). Striatal and cortical projections of single neurons from the central lateral thalamic nucleus in the rat. *Neuroscience*. 72:679–687.

Deschênes M, Bourassa J, Doan VD, Parent A (1996b). A single-cell study of the axonal projections arising from the posterior intralaminar thalamic nuclei in the rat. *Eur J Neurosci*. 8:329–343.

Economo MN, Clack NG, Lavis LD, Gerfen CR, Svoboda K, Myers EW, Chandrashekar J (2016). A platform for brain-wide imaging and reconstruction of individual neurons. *eLife*. 5:e10566.

El-Boustani S, Sermet BS, Foustoukos G, Oram TB, Yizhar O, Petersen CCH (2020). Anatomically and functionally distinct thalamocortical inputs to primary and secondary mouse whisker somatosensory cortices. *Nat Commun*. 11(1):3342.

Familtsev D, Quiggins R. Masterson SP, Dang W, Slusarczyk AS, Petry HM, Bickford ME (2016). Ultrastructure of geniculocortical synaptic connections in the tree shrew striate cortex. *J Comp Neurol*. 524:1292–1306.

Furuta T, Tomioka R, Taki K, Nakamura K, Tamamaki N, Kaneko T (2001). In vivo transduction of central neurons using recombinant Sindbis virus: Golgi-like labeling of dendrites and axons with membrane-targeted fluorescent proteins. *J Histochem Cytochem*. 49:1497–1508.

Gheorghita F, Kraftsik R, Dubois R, Welker E (2006). Structural basis for map formation in the thalamocortical pathway of the barrelless mouse. *J Neurosci*. 26:10057–10067.

Harris JA, Mihalas S, Hirokawa KE, Whitesell JD, Choi H, Bernard A, Bohn P, Caldejon S, Casal L, Cho A, Feiner A, Feng D, Gaudreault N, Gerfen CR, Graddis N, Groblewski PA, Henry AM, Ho A, Howard R, Knox JE, Kuan L, Kuang X, Lecoq J, Lesnar P, Li Y, Luviano J, McConoughey S, Mortrud MT, Naeemi M, Ng L, Oh, SW, Ouellette B, Shen E, Sorensen SA, Wakeman W, Wang Q, Wang Y, Williford A, Phillips JW, Jones AR, Koch C, Zeng H (2019). Hierarchical organization of cortical and thalamic connectivity. *Nature*. 575:195–202.

Jahnsen H, Llinás R (1984). Electrophysiological properties of guinea-pig thalamic neurones: an in vitro study. *J Physiol*. 349:205–226.

Jhangiani-Jashanmal IT, Yamamoto R, Gungor NZ, Paré D (2016). Electroresponsive properties of rat central medial thalamic neurons. *J Neurophysiol*. 115:1533–1541.

Jones EG (2007). Thalamic neurons, synaptic organization, and functional properties. In: *The Thalamus*. 2nd ed. Cambridge University Press:1:171–317.

Kerschensteiner D, Guido W (2017). Organization of the dorsal lateral geniculate nucleus in the mouse. *Vis Neurosci*. 34:E008.

Kim EJ, Zhang,Z, Huang L, Ito-Cole T, Jacobs MW, Juavinett AL, Senturk G, Hu M, Ku M, Ecker JR, Callaway EM (2020). Extraction of distinct neuronal cell types from within a genetically continuous population. *Neuron*. 107:274–282.

Kuramoto E (2019). Method for labeling and reconstruction of single neurons using Sindbis virus vectors. *J Chem Neuroanat*. 100(101648):1–12.

Kuramoto E, Furuta T, Nakamura KC, Unzai T, Hioki H, Kaneko T (2009). Two types of thalamocortical projections from the motor thalamic nuclei of the rat: A single neuron-tracing study using viral vectors. *Cereb Cortex*. 19:2065–2077.

Kuramoto E, Ohno S, Furuta T, Unzai T, Tanaka YR, Hioki H, Kaneko T (2015). Ventral medial nucleus neurons send thalamocortical afferents more widely and more preferentially to layer 1 than neurons of the ventral anterior–ventral lateral nuclear complex in the rat. *Cereb Cortex*. 25:221–235.

Kuramoto E, Pan S, Furuta T, Tanaka YR, Iwai H, Yamanaka A, Ohno S, Kaneko T, Goto T, Hioki H (2017a). Individual mediodorsal thalamic neurons project to multiple areas of the rat prefrontal cortex: A single neuron-tracing study using virus vectors. *J Comp Neurol*. 525:166–185.

Kuramoto E, Iwai H, Yamanaka A, Ohno S, Seki H, Tanaka YR, Furuta T, Hioki H, Goto T (2017b). Dorsal and ventral parts of thalamic nucleus submedius project to different areas of rat orbitofrontal cortex: A single neuron-tracing study using virus vectors. *J Comp Neurol*. 525:3821–3839.

Landisman CE, Connors BW (2007). VPM and PoM nuclei of the rat somatosensory thalamus: Intrinsic neuronal properties and corticothalamic feedback. *Cereb Cortex*. 17:2853–2865.

Lein ES, Hawrylycz MJ, Ao N, Ayres M, Bensinger A, Bernard A, Boe AF, Boguski MS, Brockway KS, Byrnes EJ, Chen L, Chen L, Chen TM, Chin MC, Chong J, Crook BE, Czaplinska A, Dang CN, Datta S, Dee NR, Desaki AL, Desta T, Diep E, Dolbeare TA, Donelan MJ, Dong HW, Dougherty JG, Duncan BJ, Ebbert AJ, Eichele G, Estin LK, Faber C, Facer BA, Fields R, Fischer SR, Fliss TP, Frensley C, Gates SN, Glattfelder KJ, Halverson KR, Hart MR, Hohmann JG, Howell MP, Jeung DP, Johnson RA, Karr PT, Kawal R, Kidney JM, Knapik RH, Kuan CL, Lake JH, Laramee AR, Larsen KD, Lau C, Lemon TA, Liang AJ, Liu Y, Luong LT, Michaels J, Morgan JJ, Morgan RJ, Mortrud MT, Mosqueda NF, Ng LL, Ng R, Orta GJ, Overly CC, Pak TH, Parry SE, Pathak SD, Pearson OC, Puchalski RB, Riley ZL, Rockett HR, Rowland SA, Royall JJ, Ruiz MJ, Sarno NR, Schaffnit K, Shapovalova NV, Sivisay T, Slaughterbeck CR, Smith SC, Smith KA, Smith BI, Sodt AJ, Stewart NN, Stumpf KR, Sunkin SM, Sutram M, Tam A, Teemer CD, Thaller C, Thompson CL, Varnam LR, Visel A, Whitlock RM, Wohnoutka PE, Wolkey CK, Wong VY, Wood M, Yaylaoglu MB, Young RC, Youngstrom BL, Yuan XF, Zhang B, Zwingman TA, Jones AR (2007). Genome-wide atlas of gene expression in the adult mouse brain. *Nature*. 445:168–176.

Llinás RR, Jahnsen H (1982). Electrophysiology of mammalian thalamic neurones in vitro. *Nature*. 297:406–408.

Martini FJ, Guillamón-Vivancos T, Moreno-Juan V, Valdeolmillos M, López-Bendito G (2021). Spontaneous activity in developing thalamic and cortical sensory networks. *Neuron*. 109:2519–2534.

Mátyás F, Komlósi G, Babiczky Á, Kocsis K, Barthó P, Barsy B, Dávid C, Kanti V, Porrero C, Magyar A, Szűcs I, Clasca F, Acsády L (2018). A highly collateralized thalamic cell type with arousal-predicting activity serves as a key hub for graded state transitions in the forebrain. *Nat Neurosci*. 21(11):1551–1562.

Nagalski A, Puelles L, Dabrowski M, Wegierski T, Kuznicki J, Wisniewska MB (2016). Molecular anatomy of the thalamic complex and the underlying transcription factors. *Brain Struct Funct*. 221(5):2493–2510.

Nakamura H, Hioki H, Furuta, Kaneko T (2015). Different cortical projections from three subdivisions of the rat lateral posterior thalamic nucleus: A single-neuron tracing study with viral vectors. *Eur J Neurosci*. 41:1294–1310.

Noseda R, Jakubowski M, Kainz V, Borsook D, Burstein R (2011). Cortical projections of functionally identified thalamic trigeminovascular neurons: Implications for migraine headache and its associated symptoms. *J Neurosci*. 31:14204–14217.

Ohno S, Kuramoto E, Furuta T, Hioki H, Tanaka YR, Fujiyama F, Sonomura T, Uemura M, Sugiyama K, Kaneko T (2012). A morphological analysis of thalamocortical axon fibers of rat posterior thalamic nuclei: a single neuron tracing study with viral vectors. *Cereb Cortex*. 22:2840–2857.

Parent M, Parent A (2005). Single-axon tracing and three-dimensional reconstruction of centre median-parafascicular thalamic neurons in primates. *J Comp Neurol*. 481:127–144.

Peng H, Xie P, Liu L, Kuang X, Wang Y, Qu L, Gong H, Jiang S, Li A, Ruan Z, Ding L, Yao Z, Chen C, Chen M., Daigle TL, Dalley R, Ding Z, Duan Y, Feiner A, He P, Hill C, Hirokawa KE, Hong G, Huang L, Kebede S, Kuo HC, Larsen R, Lesnar P, Li L, Li Q, Li X, Li Y, Li Y, Liu A, Lu D, Mok S, Ng L, Nguyen TN, Ouyang Q, Pan J, Shen E, Song Y, Sunkin SM, Tasic B, Veldman MB, Wakeman W, Wan W, Wang P, Wang Q, Wang T, Wang Y, Xiong F, Xiong W, Xu W, Ye M, Yin L, Yu Y, Yuan J, Yuan J, Yun Z, Zeng S, Zhang S, Zhao S, Zhao Z, Zhou Z, Huang ZJ, Esposito L, Hawrylycz MJ, Sorensen SA, Yang XW, Zheng Y, Gu Z, Xie W, Koch C, Luo Q, Harris JA, Wang Y, Zeng H (2021). Morphological diversity of single neurons in molecularly defined cell types. *Nature*. 598:174–181.

Phillips, JW, Schulmann A, Hara E, Winnubst J, Liu C, Valakh V, Wang L, Shields BC, Korff W, Chandrashekar J, Lemire AL, Mensh B, Dudman JT, Nelson SB, Hantman AW (2019). A repeated molecular architecture across thalamic pathways. *Nat Neurosci*. 22:1925–1935.

Pierret T, Lavallée P, Deschênes M (2000). Parallel streams for the relay of vibrissal information through thalamic barreloids. *J Neurosci*. 20:7455–7462.

Pinault D (1996). A novel single-cell staining procedure performed in vivo under electrophysiological control: Morpho-functional features of juxtacellularly labeled thalamic cells and other central neurons with biocytin or Neurobiotin. *J Neurosci Meth*. 65:113–136.

Rodriguez-Moreno J, Porrero C, Rollenhagen A, Rubio-Teves M, Casas-Torremocha D, Alonso-Nanclares L, Yakoubi R, Santuy A, Merchan-Pérez A, DeFelipe J, Lübke JHR, Clasca F (2020). Area-specific synapse structure in branched posterior nucleus axons reveals a new level of complexity in thalamocortical networks. *J Neurosci*. 40:2663–2679.

Sheroziya M, Timofeev I (2014). Global intracellular slow-wave dynamics of the thalamocortical system. *J Neurosci*. 34:8875–8893.

Shi W, Xianyu A, Han Z, Tang X, Li Z, Zhong H, Mao T, Huang K, Shi SH (2017). Ontogenetic establishment of order-specific nuclear organization in the mammalian thalamus. *Nat Neurosci*. 20:516–528.

Smith PH, Bartlett EL, Kowalkowski A (2006). Unique combination of anatomy and physiology in cells of the rat paralaminar thalamic nuclei adjacent to the medial geniculate body. *J Comp Neurol*. 496:314–334.

Smith Y, Galvan A, Ellender TJ, Doig N, Villalba RM, Huerta-Ocampo I, Wichmann T, Bolam JP (2014). The thalamostriatal system in normal and diseased states. *Front Syst Neurosci*. 8(5):1–18.

Unzai T, Kuramoto E, Kaneko T, Fujiyama F (2017). Quantitative analyses of the projection of individual neurons from the midline thalamic nuclei to the striosome and matrix compartments of the rat striatum. *Cereb Cortex*. 27:1164–1181.

Winnubst J, Spruston N, Harris JA (2020). Linking axon morphology to gene expression: A strategy for neuronal cell-type classification. *Curr Opin Neurobiol*. 65:70–76.

Wong SZH, Scott EP, Mu W, Guo X, Borgenheimer E, Freeman M, Ming GL, Wu QF, Song H, Nakagawa Y (2018). In vivo clonal analysis reveals spatiotemporal regulation of thalamic nucleogenesis. *PLoS Biology*. 16(4):e2005211.

Zhang H, Wang X, Guo W, Li A, Chen R, Huang F, Liu X, Chen Y, Li N, Liu X, Xu T, Xue Z, Zeng S (2021). Cross-streams through the ventral posteromedial thalamic nucleus to convey vibrissal information. *Front Neuroanat*. 15(724861):1–17.

Zhang Z, Zhou J, Tan P, Pang Y, Rivkin AC, Kirchgessner MA, Williams E, Lee CT, Liu H, Franklin AD, Miyazaki PA, Bartlett A, Aldridge AI, Vu M, Boggeman L, Fitzpatrick C, Nery JR, Castanon RG, Rashid M, Jacobs MW, Ito-Cole T, O'Connor C, Pinto-Duartec A, Dominguez B, Smith JB, Niu SY, Lee KF, Jin X, Mukamel EA, Behrens MM, Ecker JR, Callaway EM. (2021). Epigenomic diversity of cortical projection neurons in the mouse brain. *Nature*. 598:167–173.

13.

THALAMOCORTICAL SYNAPSES

Anthony Holtmaat and Kok-Sin Lee

INTRODUCTION

Synapses of thalamocortical (TC) afferents have long been a focus of research (Jones and Powell, 1970). In the early days, synapses in cortex were commonly studied using techniques such as silver staining, histochemistry, and electron microscopy (EM). Later, immunocytochemistry for relatively selective presynaptic marker proteins as well as anterograde tracers allowed for targeted studies of the structure of TC synapses, their distributions, and connectivity. Nowadays, TC projections can be targeted using combinations of transgenesis and viral vector-mediated expression of recombinant proteins. Such vital approaches have allowed combinatorial application of various techniques to study TC synapses, including electrophysiological evaluation; optogenetic interrogation of function and connectivity; and in vivo imaging followed by retrospective 3D EM reconstructions (see Figure 13.1). These techniques have allowed for detailed comparisons with other synapses, thereby providing insights into their contribution to cortical processing.

In this chapter we will summarize some of the gained insights into TC synaptic structure and function in the adult brain. For the sake of simplicity, we will refer to TC synapses as a group, independent of their thalamic origin, position in cortex, or animal species, and only specify these when they institute distinct features. Similarly, we will cautiously distinguish TC projections from first-order (FO) or higher-order (HO) thalamic nuclei, rather than stipulating the exact nucleus and/or sensory modality to which they belong (as discussed in **Chapters 11 and 12 in Section IV**), with the understanding that for some cell groups in the thalamus, it may still be undecided if they harbor FO or HO properties.

STRUCTURAL FEATURES OF TC SYNAPSES

TC axons establish either *en passant* or *terminaux* boutons that synapse mostly on dendritic spines and to a smaller degree on dendritic shafts (White et al., 2004). Synapses in nonvaricose segments of TC axons are rare, but a small number of nonsynaptic *en passant* boutons have been reported (Rodriguez-Moreno et al., 2020).

There are currently no endogenous markers that are unique for TC synapses. Nonetheless, one of the most striking molecular differences between TC and corticocortical (CC) synapses is the divergent expression of vesicular glutamate transporter (VGluT) protein isoforms, responsible for transporting glutamate into synaptic vesicles. TC afferents are predominantly immunoreactive for isoform VGluT2 (Fujiyama et al., 2001), whereas other axons in the cortex express isoform VGluT1. Other, possibly differentiating proteins for thalamic neurons are currently being identified using single-cell genetic profiling, but it is yet unclear if any of those will allow for the selective detection of TC synapses (Phillips et al., 2019).

TC boutons have synapses with pronounced postsynaptic densities (PSDs) and relatively wide synaptic clefts. This gives them an asymmetric appearance in electron micrographs, which is typical for glutamatergic synapses. These features are found in TC synapses across cortical regions and sensory modalities (Kharazia and Weinberg, 1994).

Distinctly selective markers for the postsynaptic elements of TC synapses are lacking. PSDs of TC synapses seem to contain many of the scaffold proteins that are prevalent in excitatory neurons, but none is differentially enriched in TC synapses. Nonetheless, the characterization of PSD composition provides a promising avenue to selectively identify TC synapses (Collins et al., 2006).

TC boutons often have distinct structural features (see Figure 13.2 and Table 13.1). For example, in cortical layer 4 (L4) they are 2–4 times larger than nearby boutons of nonthalamocortical origins (Bopp et al., 2017). Independent observations in noncomparative studies appear to concur with the observed contrast between TC and other boutons in L4 (Rollenhagen et al., 2015; Rodriguez-Moreno et al., 2018; Rodriguez-Moreno et al., 2020). Due

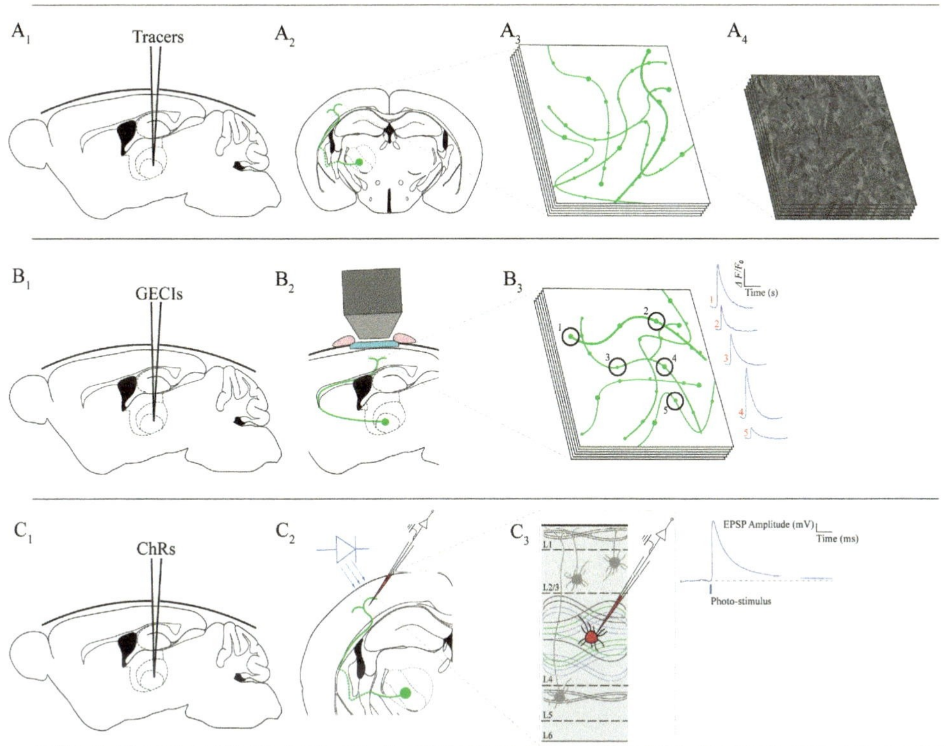

Figure 13.1. Schematic overview of strategies to investigate TC synapses. Rodent TC synapses can be labeled with: (A_1) anterograde tracers such as *Phaseolus vulgaris leucoagglutinin* (PHA-L), biotinylated dextran amine (BDA), or adeno-associated viral vectors that mediate the expression of fluorescent proteins; (B_1) high-performance cation sensors, such as genetically encoded calcium indicators (GECIs), for imaging activity of axons and microcompartments; (C_1) light-gated channelrhodhopsins (ChRs) that allows for photo-stimulation of targeted TC axons enabling opto-interrogation of the associated synaptic circuitry. (A_2). Schematic of a coronal section illustrating the trajectory of labeled TC axons (green) innervating the neocortex. Immunohistochemistry using VGluT2 antibodies may also be employed to positively identify TC axons. (A_3). Schematic of an image stack containing TC axons (green) and interspaced varicosities (green dots) that can be visualized using light microscopy. (A_4). A series of electron micrographs taken from the labeled brain sections allows for the identification and precise reconstruction of TC synapses. (B_2). Schematic illustrating in vivo imaging of GECIs expressed by TC axons. (B_3). Schematic of an image stack containing TC axons (green) with varicosities (green dots). Open circles (black) indicate regions-of-interest (ROIs). The insets on the right are fictive fluorescence transients ($\Delta F/F_0$) representing the calcium responses from the corresponding ROIs. (C_2). Schematic showing ChR-positive TC axons (green) in the neocortex. Photo-stimulation of the labeled axons (blue) and electrophysiology recordings. (C_3). Schematic of an electrophysiological recording in a L4 neuron (red) that receives TC inputs (green) among others on its dendrites. The inset on the right shows a fictive excitatory postsynaptic potential (EPSP) evoked in such a photo-stimulation experiment.

to a paucity in rigorously designed comparative studies between unequivocally identified synapses of both TC and non-TC origins, it is unclear if the observations in L4 can be extrapolated to other TC afferents. For example, even TC boutons from the same thalamic origin can show marked structural variations across different areas or layers (Rodriguez-Moreno et al., 2020).

Many of the TC boutons in L4 (>85%) are found to contain more than 1 mitochondria, often distributed within proximity to the synapses, while only 43% of the general population of synapses in this layer display this feature (Rodriguez-Moreno et al., 2018). The number of vesicles (i.e., the size of its vesicle pool) scales with TC bouton volume (Rodriguez-Moreno et al., 2018), but vesicle density seems comparable to other excitatory synapses (Bopp et al., 2017).

Despite the notion that synaptic TC boutons are large, there is no clear consensus as to whether this also results in larger PSDs as compared to other asymmetric synapses in the surrounding neuropil, although the PSD surface area does seem to modestly increase with bouton volume (Bopp et al., 2017; Rodriguez-Moreno et al., 2018). TC synapses have additional distinctive PSD features, such as non-macular shapes, but their functional role is unclear (see Table 13.1).

Large TC varicosities tend to establish asymmetrical synapses with multiple postsynaptic partners, also termed *multi-synaptic boutons* (MSBs) (see Figure 13.2A and C). This suggests that TC synapses transfer information to a multitude of postsynaptic cells (see below). In fact, more than 50% of TC boutons in L4 were found to be MSBs (Rodriguez-Moreno et al., 2018), while the proportion of MSBs among randomly selected excitatory boutons is much lower (~10%–18%) (Kasthuri et al., 2015). The presence of MSBs has even been used as a criterion for the identification of TC synapses in dense volumetric reconstructions of L4

TABLE 13.1 SELECTED LITERATURE FOR COMPARISON OF ULTRASTRUCTURAL PARAMETERS BETWEEN THALAMOCORTICAL (TC) AND CORTICO-CORTICAL (CC) SYNAPSES

Species	Thalamic Origin or Identity	Cortical Region and Layer	Mean/Median Bouton *vol.* in μm^3 [range]	% Bouton w. >1 *Mito.* {*mito.*:bouton *vol.* ratio} Median Volume in μm^3	% of MSBs {ratio of synapses per bouton} % *sp.* co-innervated by *sym.* synapse	% Non-macular PSDs Mean/Median PSD Area in μm^2 [range]	Reference
				Thalamocortical synapses			
Mice	VBm	S1 L4	*NR*[a]	*NR*[a]	>25% {1.2} *NR*[a]	*NR*[a]	White et al., 2004
Rat	VGLUT2+	MFC L2/3	*NR*[a]	*NR*[a]	*NR*[a] {*NR*[a]} 14.5%	~67% 0.16 ± 0.07[c] *NR*[a]	Kubota et al., 2007
Mice	VGLUT2+	M1 L4	0.17	*NR*[a] {0.20[d]} 0.03	~28% {1.3} *NR*[a]	*NR*[a] 0.06[j]	Bopp et al., 2017
		S1 L4	0.31	*NR*[a] {0.20[d]} 0.06	~67% {2.1} *NR*[a]	*NR*[a] *0.04*[j]	
Mice	VPMdm	S1 L4	0.46 ± 0.27 [0.06–1.59]	92% {0.21[d]} 0.09	53% {1.6} 15%	>50% 0.21 ± 0.11 [0.03–0.47[c]]	Rodriguez-Moreno et al., 2018
Mice	POm	S1 LI	0.37 ± 0.21 [0.01–0.89]	90% {0.15[d]} 0.05	12% {1.1} 3%	57% 0.16 ± 0.10 [0.02–0.51]	Rodriguez-Moreno et al., 2018
		S1 L5a	0.24 ± 0.13 [0.03–0.80]	92% {0.26[d]} 0.06	9% {1.1} 18%	35% 0.11±0.06 [0.01–0.25]	
		MI L3/4	0.39 ± 0.22 [0.09–1.06]	88% {0.20[d]} 0.07	29% {1.4} 9%	59% 0.1 ± 0.05 [0.03 – 0.41]	
Mice	POm	S2 all layers	1.6 µm (all layers)[g] ~2.01 gm (L2/3)	~ 78% (all layers) ~84% (L2/3)	*NR*[a] {1.4 (all layers) ~1.6 (L2/3)} *NR*[a]	*NR*[a]	Sampathkumar et al., 2021
	Unlabeled[f]		1.25 µm (all layers)[g] ~1.33 µm (L5)	~ 37% (all layers) ~44% (L1)	*NR*[a] {1.2 (all layers) ~1.3 (L4)} *NR*[a]		
				Cortico-cortical synapses			
Rat	Intracortical[n]	S1 L4	0.20 ± 0.07 [0.01–1.36]	*NR*[a] {0.20} 0.02	~5% {*NR*[a]} *NR*[a]	~35% 0.18 ± 0.06*P* *NR*[a]	Rollenhegan et al., 2015
Mice	Mixed	S1 L5 and L6	*NR*[a] [0.18–0.37[r]]	~33%[s]	~18 %	*Values in pixel metric*	Kasthuri et al., 2015
Mice	VGLUT−	M1 L4	0.07	*NR*[a] {<<0.11[d, h]} 0.00	Unknown {1.1} *NR* [a]	*NR*[a] 0.056[j]	Bopp et al., 2017
		S1 L4	0.07	*NR*[a] {0.11[d]} 0.008	~19% {1.2) *NR*	*NR*[a] 0.039[j]	
Mice	Mixed	V1 L2/3	SSBs: 0.10 ± 0.01 [0.07–0.11] MSBs: 0.21 ± 0.09 [0.13 – 0–36]	*NR*[a]	~14% {1.0–1.2[q]} 2%–7%	19 – 35% (all areas and species) 0.08 ± 0.01 [0.07– 0.10]	Hsu et al., 2017
		FC L2/3	SSBs: 0.08 ± 0.03 [0.05–0.13] MSBs: 0.14±0.01 [0.12–0.15]	*NR*[a]	~10% {see above[q]} 5%–14%	See above 0.07 ± 0.02 [0.05–0.10]	
Monkey	Mixed	V1 L2/3	SSBs: 0.20 ± 0.04 [0.14–0.25] MSBs: 0.45±0.05 [0.39–0.52]	*NR*[a]	~3% {see above[q]} 2.8 – 3.7%	See above 0.08 ± 0.01 [0.06–0.09]	

TABLE 13.1 CONTINUED

Species	Thalamic Origin or Identity	Cortical Region and Layer	Mean/Median Bouton *vol.* in μm³ [range]	% Bouton w. >1 *Mito.* {*mito.*:bouton *vol.* ratio} Median Volume in μm³	% of MSBs {ratio of synapses per bouton} % *sp.* co-innervated by *sym.* synapse	% Non-macular PSDs Mean/Median PSD Area in μm² [range]	Reference
		LPFC L2/3	SSBs: 0.30 ± 0.01 [0.28–0.31] MSBs: 0.47±0.11 [0.36–0.66]	*NR*[a]	~3% {see above[q]} 1.34%–5%	See above 0.11 ± 0.01 [0.10–0.12]	
Rat	Intracortical	S1 L5b	0.38 ± 0.23 [0.02–1.73]	~97% {0.15} 0.05	*NR*[a]	~35% 0.31 ± 0.21[p] *NR*[a]	Rollenhagen et al., 2015

NOTES: * values reported here are either mean ± standard deviation with range wherever possible or median values reported alone. Wherever appropriate, values are corrected to 2 decimal places for consistency; [a] *NR*: values are not reported; [b] estimated from reported number of varicosities that made multiple synapses and those that made none at all; [c] synapse area of all spines that were T2 or VGLUT2-immunolableled; [d] estimated as the ratio of the median mitochondrial volume per bouton volume against the median bouton volume; [e] mean and range of PSD contour area; [f] axons that were not labeled, thereby do not stain for DAB, within the same sub-volumes in which labeled axons were sampled; [g] mean diameter of labeled or unlabeled terminals; [h] estimated to be less than the last known ratio as median volume of mitochondria was reportedly very small (<0.008); [j] reported median value of PSD area, range was not specified but can be referred from cumulative histogram; [m] inferred from percentage of PSDs with perforated AZs; [n] we will use the terms *intracortical* and *cortico-cortical* interchangeably, preserving the original nomenclature used in the studies wherever possible; [p] reported values for postsynaptic surface area of active zones; [q] estimated from reported values where 87%–99% that were single bouton with single synapse; [r] range values of values that include "unknown" boutons as well as boutons clearly indicated as MSBs; [s] estimated from mitochondria count per synapse in presynaptic axon, possibly an underestimation.

ABBREVIATIONS: CC: cortico-cortical; FC: frontal cortex; POm: posterior medial complex of the thalamus.; LPFC: lateral prefrontal cortex; M1: motor cortex; *mito.*: mitochondria; MSBs: multi-synaptic boutons; S1: primary somatosensory cortex; *sp.*: spines; SSBs: single-synaptic bouton; *sym.*: symmetrical; TC: thalamocortical; V1: primary visual cortex; VBm: ventrobasal of the thalamus; VGLUT2: vesicular glutamate transporter 2; *vol.*: volume; VPMdm: dorsomedial portion of the rodent ventral posteromedial nucleus.

(Motta et al., 2019). However, in other layers of the cortex, TC boutons may establish MSBs at lower rates (Rodriguez-Moreno et al., 2020).

One may speculate that the large TC synapses in L4 are relatively potent since synaptic strength generally scales with bouton size, presynaptic vesicle content, as well as the corresponding PSD surface areas (Holler et al., 2021). An increased presence of presynaptic mitochondria may also be indicative of high synaptic performance (Cserép et al., 2018). We will see later that the unitary TC synaptic response in L4 is indeed relatively strong when measured under isolated conditions.

EM studies have shown that most TC synapses are on spines with morphologies that vary from large stubby to mushroom shapes. Only a few TC-recipient spines are classified as thin or filopodial-like (Rollenhagen et al., 2015; Rodriguez-Moreno et al., 2018). As is common for large spines, many of those targeted by TC afferents contain spine apparatus (see Figure 13.2C), which are specialized forms of the endoplasmic reticulum that store and release calcium and are therefore associated with structural and functional plasticity (Harris and Weinberg, 2012). The functional role of spine apparatus for TC synapses has not been determined directly, but it may equip them with a higher capacity to dynamically regulate postsynaptic calcium concentration and hence synaptic efficacy and plasticity.

Dendritic spines that receive excitatory synaptic input are sometimes also targeted by inhibitory inputs. Ultrastructural studies have shown that spines with excitatory synapses from TC inputs are often co-innervated by a second symmetrical, presumably inhibitory synapse (see Figure 13.2B), which is rare for spines receiving CC inputs (Kubota et al., 2007). This observation has important implications for TC circuit function since near-synaptic inhibition can powerfully modulate or even shunt excitation. Such inhibitory effects can be exerted by inputs directly on dendritic shafts, but may become more specific and effective when compartmentalized within the spine.

TC POSTSYNAPTIC TARGETS

Generally, the density of TC synapses follows the distribution of their afferents; that is, more synapses are found in areas with profuse TC innervation. TC afferents have been found in virtually all cortical regions and layers of isocortex, as well as parts of allocortex (as discussed in **Chapters 11 and 12 in Section IV**). In accordance, monosynaptic inputs have been reported on almost all main neuron types.

L4 and L5 of primary sensory and motor cortices, being the main input layers for FO thalamic nuclei, contain high densities of TC synapses which are found on spiny stellate cells and pyramidal neurons, as well as fast-spiking inhibitory neurons (Cruikshank et al., 2010). Most of the latter are presumably parvalbumin-expressing (PV) neurons (Staiger et al., 1996). On both excitatory and inhibitory neurons in L4, TC synapses are broadly distributed along dendrites, but with subtle biases toward proximal dendrites (Richardson et al., 2009; Schoonover et al., 2014).

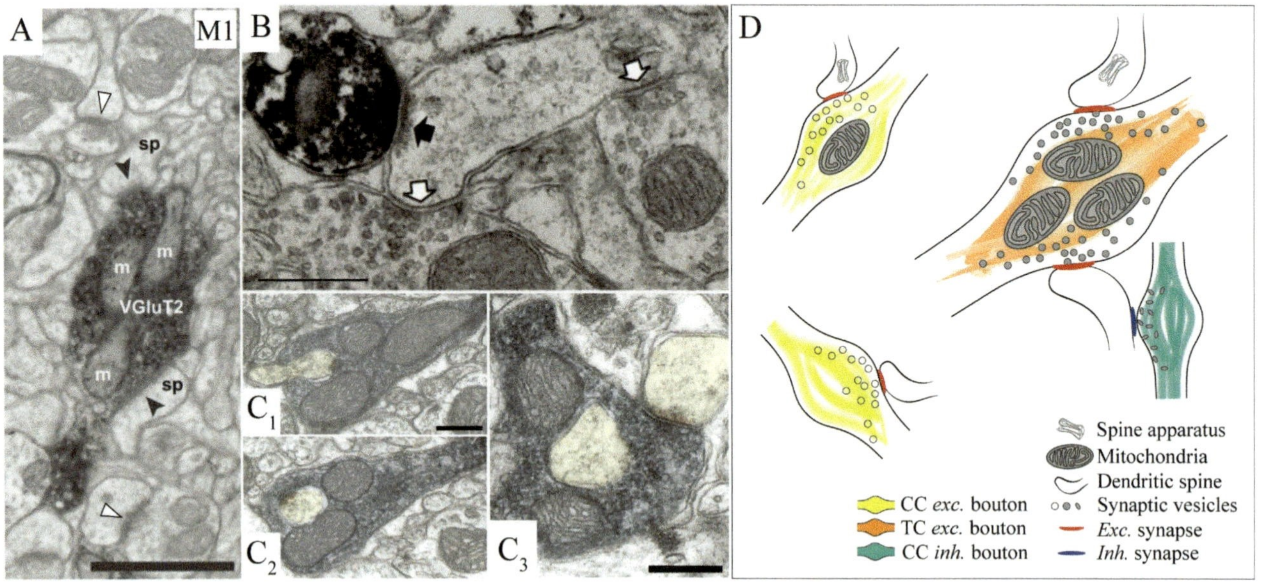

Figure 13.2. Ultrastructure of synapses visualized with electron microscopy. (A and B). Electron micrographs of boutons immuno-reactive for VGluT2, which appears as electron-dense material on synaptic vesicles. (A). A large VGluT2+ bouton in L4 of M1 containing multiple mitochondria (m) and synapses with multiple target spines (*sp*; black arrow). Smaller unlabeled boutons with vesicles, presumably CC, in the surrounding neuropil also form asymmetric synapses with spines (white arrow). (B). A spine head innervated by a VGluT2+ bouton (black arrow). Two additional symmetrical synapses with the same spine (white arrows). (C_{1-3}). Serial TEM images of a TC synaptic bouton in L4 of S1 identified by BDA-tracer. A TC bouton (transparent blue) with multiple mitochondria contacting 2 spines (transparent yellow) with PSDs (brown). One of the spines invaginates the bouton. (D). Comparative schematic of TC and CC structural features in L4 detected using EM. TC boutons (right) are large, have a large mitochondria content, and tend to be multi-synaptic (MSBs). Spines that receive TC synapses are more often co-innervated by a symmetrical synapse as compared to CC synapses (TC vs. CC: ~9% in L4; ~16% in L1; and ~14% in L2/3 vs. <3% in all layers). CC boutons are smaller, rarely form MSBs, and have fewer mitochondria. But when present, the mitochondria:bouton volume ratio is similar to that of TC synapses. Scale bars in A: 1.0 μm, B: 0.5 μm, and in C_1 and C_3: 0.25 μm. Adapted from Kubota et al. (2007); Bopp et al. (2017); Rodriguez-Moreno et al. (2018).

In other layers, TC afferents also target a variety of neurons. Here, the interdigitated FO and HO innervation patterns, combined with the various configurations of apical and basal dendritic arbors, generate a myriad of connectivity motifs. For example, basal dendrites of slender-tufted neurons in L5a may receive a relatively dense innervation from HO afferents, whereas thick-tufted neurons in L5b receive a more FO-biased input on their basal dendrites (Bureau et al., 2006; Petreanu et al., 2009; Audette et al., 2018; Sermet et al., 2019). In general, the extent of axonal and dendritic overlap is a good predictor of the amount of TC synaptic connections (Meyer et al., 2010), although exceptions have been described. For example, L5b pyramidal neurons with extensive tufts in L1 seem to receive lower than expected HO TC input (Petreanu et al., 2009). Interdigitated TC innervation patterns in layers and cell-type specific targeting are also present in secondary sensory cortices (Llano and Sherman, 2008; El-Boustani et al., 2020) and in higher order cortices, such as the medial prefrontal and the retrosplenial cortex (Anastasiades et al., 2021; Lomi et al., 2021).

Many PV and vasoactive intestinal peptide-expressing (VIP) inhibitory interneurons in L2/3 and L5 tend to receive synaptic input from HO TC projections, but somatostatin-expressing (SOM) cells receive little to none (Audette et al., 2018; Williams and Holtmaat, 2018; Sermet et al., 2019). The input patterns of HO TC inputs to interneurons residing in L1 are diverse but remain underexplored. Discrete connectivity patterns of TC afferents to interneurons give rise to functionally important modulatory circuit motifs, such as thalamus-driven feedforward inhibition and disinhibition (see **Chapter 8 in Section III**), as well as presynaptic inhibition (Pardi et al., 2020).

FUNCTIONAL FEATURES OF TC SYNAPSES

The functional properties of TC synapses are analogous to those of CC and other glutamatergic connections in the cortex in that they operate by fast transmission mediated by ionotropic glutamate receptors (iGluRs). Pharmacological studies have indicated that adult TC postsynaptic sites contain mostly α-amino-3-hydroxy-5-methyl-4-isoxazolepropionic acid receptors (AMPARs) and N-methyl-D-aspartate receptors (NMDARs). Some synapses can be devoid of AMPARs, but this pertains largely to early stages of postnatal development (Isaac et al., 1997) (see **Chapter 50 in Section IX**). Depending on the TC-recipient neuron, postsynaptic elements can also variably contain one or more members of the metabotropic glutamate receptor (mGluR) family (Viaene et al., 2011). mGluRs account for additional slower forms of transmission and may even modulate postsynaptic responses or play a role in (short-term) synaptic plasticity. Additionally, modulatory receptors may be present on TC-recipient cells, but their specific role in TC-synaptic transmission is often unclear. Gamma(γ)-aminobutyric acid receptors (GABARs) are probably most noteworthy in this respect, since inhibitory synapses are frequently found on spines in proximity to TC-synapses (see above).

Presynaptic elements of TC projections contain a myriad of both ionotropic and metabotropic receptors, which may modulate transmission (Gil et al., 1997; Porter and Nieves, 2004; Barre et al., 2016; Wang et al., 2019). Agonists for these receptors arrive at the TC pre-synapse by targeted axo-axonal synaptic contacts, extra-synaptic diffusion, or even through large-distance volume transmission. Most of the implicated receptor mechanisms suppress presynaptic function, but some have been reported to enhance synaptic efficacy and possibly gate postsynaptic plasticity processes (Blundon et al., 2011; Barre et al., 2016).

Owing to the differing quantities of pre- and postsynaptic receptors and consequently a manifold of mechanisms for synaptic information transfer, a variety of stereotypical responses have been observed in TC-recipient excitatory neurons upon thalamic stimulation. Sherman and colleagues have carefully characterized many of such responses in brain slices and were able to define several stereotypes. Among these, the 2 most distinctive types are characterized by the presence of paired-pulse depression or facilitation (referred to by many as class 1 and class 2, or driver and modulator-types, respectively) (Lee and Sherman, 2008) (see **Chapter 7 in Section III**). It should be noted that variations exist within these 2 general response types depending on the recipient cell type and cortical layer, and that others have used different taxonomies (e.g., Stratford et al., 1996).

Typically, the activation of FO and HO TC inputs to their respective FO and HO cortical target areas results in similar types of excitatory postsynaptic class 1 responses which include large amplitude excitatory postsynaptic potentials (EPSPs >4 mV) with short onset-latencies (5–7 ms) that are characteristic of glutamatergic monosynaptic responses (Lee and Sherman, 2008). These responses are usually relatively consistent from trial to trial, with low onset jitter (<0.5 ms), and display all-or-none activation profiles, meaning that full-blown responses are either seen or completely absent. When the frequency of stimulation is increased, they tend to show a depression of responses (i.e., a paired-pulse ratio [$EPSP_2/EPSP_1$] <1.0), probably because of decreasing probabilities or quantities of release upon subsequent action potentials in the train, or due to desensitization of postsynaptic receptors (see Figure 13.3A). The depressive nature of these types of synapses stands in contrast with many of the CC synaptic connections, which tend to facilitate at high frequency ranges (Gil et al., 1997).

Class 2 types of TC-mediated responses include those from HO thalamic nuclei to primary sensory cortex. These often-termed feedback circuits are characterized by smaller initial responses (<2 mV) and longer onset-latencies (~8 ms) (Viaene et al., 2011). These TC-evoked responses tend to be graded (i.e., inconsistent) and may facilitate during trains of thalamic stimulation, which is in stark contrast to the paired-pulse depression that can be observed in HO TC synaptic input to secondary sensory cortex. A peculiar additional characteristic of these synaptic responses is the presence of slow components in the excitatory EPSP that cannot be abolished by iGluR antagonists, but only by group I mGluR antagonists, indicative of the involvement of metabotropic mechanisms for prolonged postsynaptic depolarization (see Figure 13.3B).

TC inputs onto fast-spiking, putative inhibitory, neurons also display varying and distinct profiles. They are typically more powerful and show more short-term depression than those onto regular-spiking excitatory neurons (Kloc and Maffei, 2014), but these differences may be less pronounced in vivo (Sedigh-Sarvestani et al., 2017).

TC SYNAPSE-MEDIATED INFORMATION TRANSFER

CORTICAL ACTIVITY

The efficiency of information transfer will highly depend on the release parameters and postsynaptic receptor mechanisms described above. In turn, these factors are likely to depend on whether the parent neuron fires in tonic or burst mode (see **Chapter 3 in Section II**). The exact consequences of the different firing modes on glutamate release by TC synapses have not been investigated, but burst firing generally leads to potent postsynaptic activation (Swadlow and Gusev, 2001). The mode of firing may also have consequences on short-term synaptic adaptations and response dynamics, including paired-pulse plasticity, nonlinear depolarization, and the activation of postsynaptic mGluRs. The latter mechanisms may be particularly sensitive to firing rates, since mGluRs are typically located perisynaptically and may therefore require synaptic spillover of glutamate.

Electrophysiological measurements in vitro of excitatory neurons in TC-recipient layers (e.g., L4 of primary sensory cortical areas) show that TC-evoked unitary synaptic responses are several-fold stronger than those evoked

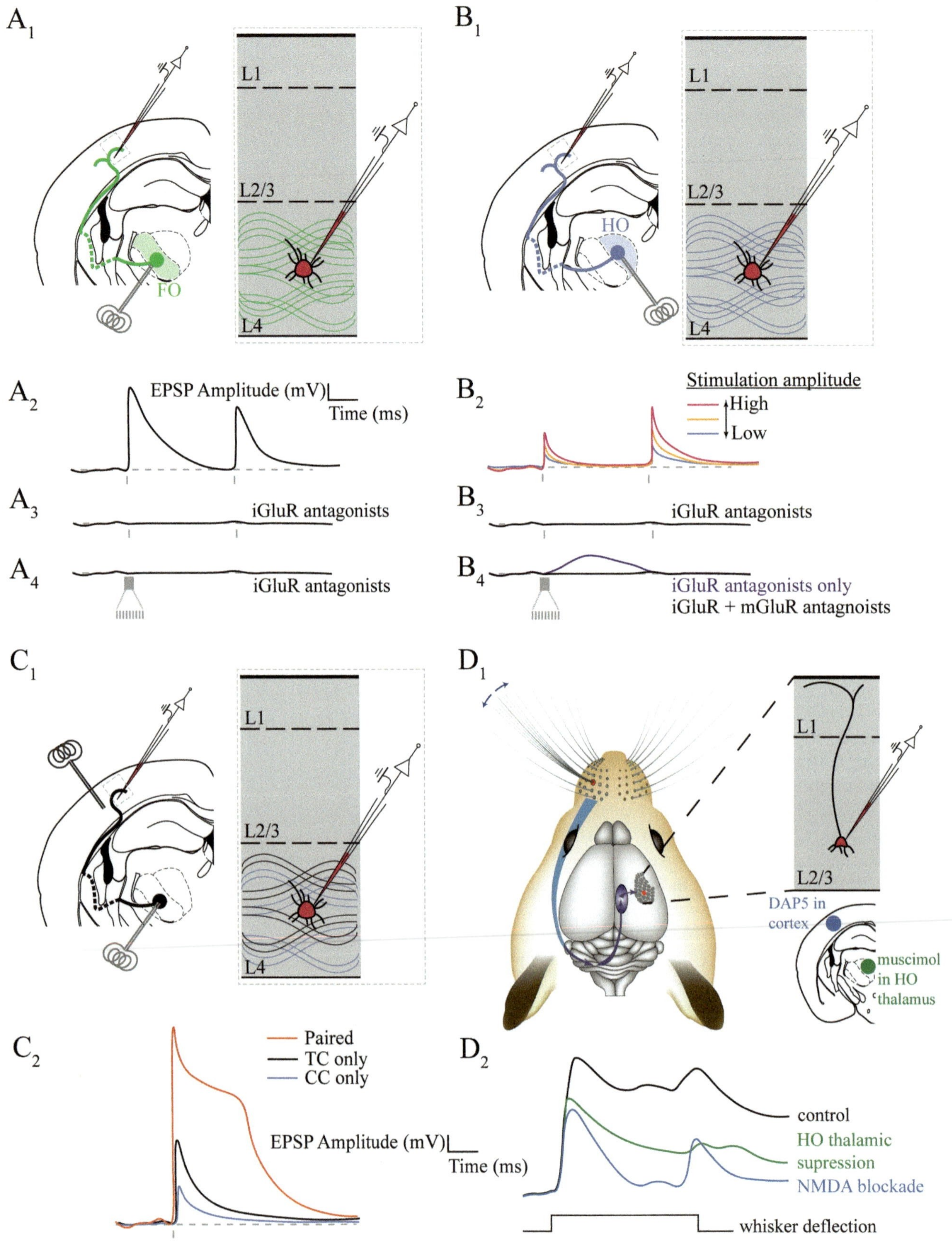

Figure 13.3. Schematic examples of electrophysiological recordings to measure TC-evoked synaptic responses. (A_1). Recording of responses as evoked by first-order (FO) and higher-order (HO) thalamocortical pathways, for example, FO thalamic nuclei to primary cortical areas (VPM→S1) or HO thalamic nuclei to secondary cortical areas (MGd→A2). (A_2). Simplified traces of whole-cell current-clamp recordings in L4 in response to thalamic stimulation. The large initial EPSP depresses upon repeated stimulations. (A_3). The repeated thalamic stimulation fails to evoke responses in the presence of

by CC inputs (Gil et al., 1999). Although the synapses examined exhibited similar quanta size distribution, TC synapses appear to have more release sites with higher release probabilities (Gil et al., 1999). Inhibitory interneurons are often even more powerfully activated by TC inputs than excitatory neurons, despite being largely outnumbered by them (Porter et al., 2001).

How do these findings translate in vivo? The work in brain slices prompted the hypothesis that the influence of TC input on cortical activity will be large. However, several observations may question such an extrapolation. First, TC inputs were found to account for the minority of synapses onto cortical neurons (Peters and Payne, 1993). Second, TC synapses in vivo were often found to be weak and similar in strength to CC synapses (Schoonover et al., 2014; Sedigh-Sarvestani et al., 2017). It has been speculated that since TC synapses are sensitive to paired-pulse depression, high spontaneous activity rates of thalamic afferents may offset the postsynaptic responses to be relatively weak in vivo (Banitt et al., 2007). This in turn would diminish the amplitude difference between TC and CC EPSPs observed in vitro (Chung et al., 2002). Third, robust feedforward inhibition within the cortical circuitry is often seen upon thalamic stimulation, possibly due to the strong TC connectivity to interneurons (see **Chapter 8 in Section III**). On the other hand, inhibitory interneurons are regularly interconnected, thereby mediating modes of inhibitory control known as disinhibition. Altogether, while information transfer between thalamus and cortex is generally efficient, the mechanisms by which this occurs in vivo are compound, and not necessarily linearly scaling with the synaptic efficacy measured in vitro.

SYNAPTIC CONVERGENCE

The findings above led to a cardinal question: How can TC inputs drive activity in cortex so efficiently and reliably if they are outnumbered by other synapses that are similarly efficacious? Three critical features of TC activity and connectivity make this possible. First, sensory stimuli readily activate thalamic neurons, and their onset is impeccably precise and synchronous (Alonso et al., 1996; Bruno and Sakmann, 2006). TC input may thereby produce cortical up-states and synchronously drive EPSPs in similarly tuned recipient neurons (Roy and Alloway, 2001; Poulet et al., 2012; Stanley et al., 2012). Second, TC connection probabilities are reported to be high (~30%) for neurons in L4, with conservative estimates suggesting that excitatory cells are likely to receive connections from as many as 90 thalamic neurons and inhibitory cells may receive even more (from ~150 thalamic neurons) (Bruno and Sakmann, 2006). Thus, TC synapses arising from multiple parent neurons in FO thalamus spatially converge onto recipient neurons in the cortex (L4, L5, and L6) and may activate them synchronously, even if they are individually relatively weak (Constantinople and Bruno, 2013; Schoonover et al., 2014). Some of the above principles may also hold true for HO TC inputs. However, further exploration is needed since HO TC inputs tend to transmit information in a more distributed manner (Ohno et al., 2012; Roth et al., 2016; Rodriguez-Moreno et al., 2020).

As noted above, TC synaptic connectivity to interneurons is robust, and may in addition converge on individual neurons (Bagnall et al., 2011). In fact, interneurons may receive input from many convergent yet dissimilar TC afferents, and thereby influence the local excitatory circuitry through low-latency and broadly tuned feedforward inhibition. (Bereshpolova et al., 2020).

SYNAPTIC COOPERATIVITY

A third feature that may amplify signals from TC synapses and thereby facilitate the transfer of their information to cortical circuits lays in the cooperativity among their synapses, as well as with others from local and long-range inputs.

In basal dendritic compartments of L5 neurons, TC synapses have been observed to be clustered within

Figure 13.3. (continued)

iGluR antagonists (DNQX and AP5). (A_4). High-frequency stimulation (HFS) in the presence of iGluR antagonists does not evoke a mGluR-mediated response. (B_1). Recording of responses as evoked by HO thalamocortical inputs to primary cortical areas (e.g., POm→S1). (B_2) Simplified traces of whole-cell current-clamp recordings in L4 in response to various intensities of thalamic stimulation. The small initial EPSP shows a graded increase in amplitude upon larger stimulation intensities and facilitates in response to repeated stimulations. (B_3). The repeated thalamic stimulation fails to evoke responses in the presence of ionotropic glutamate receptor antagonists (DNQX and AP5). (B_4). HFS in the presence of iGluR antagonists shows presence of a mGluR-mediated response (purple trace) that is abolished with the addition of group I mGluR antagonists (LY367385 and MPEP, black trace). (C_1). Recordings upon dual stimulation of CC (blue) and TC (black) synaptic pathways. (C_2). Simplified traces showing that single pathway stimulation results in relatively modest EPSPs (black and blue), whereas the paired stimulation evokes supralinear responses (i.e., larger than the sum of the single-pathway EPSPs). (D_1). In vivo recording of whisker (i.e., sensory)–evoked responses in a L2/3 pyramidal neuron, and the pharmacological dissection thereof using DAP5 to block NMDARs in cortex, or muscimol (GABAR agonist) to suppress activity in HO thalamus. (D_2). Simplified traces showing that a whisker deflection can drive supralinear long-latency responses. These are blocked either by NMDAR antagonists, or by suppressing HO thalamic activity. Note that suppression of HO activity predominantly impairs the NMDAR-component of the response, indicating that HO TC synapses may play a role in driving supralinear responses in cortical pyramidal neurons. Adapted from work described in Lee and Sherman (2008); Viaene et al. (2011); Lavzin et al. (2012); Gambino et al. (2014).

branches that could enhance their integration (Rah et al., 2013). Similarly, on pyramidal cell apical tufts TC synapses may cluster, either among themselves or with CC types (Kim et al., 2021). While individual TC synapses at large electronic distances from the soma are inadequate to evoke action potentials, robust cooperation of locally clustered synapses may be essential. Indeed, it has been shown that FO and HO TC inputs can be summed supralinearly (see Figure 13.3D) or coupled to the soma through regenerative dendritic propagation (Llinás et al., 2002; Gambino et al., 2014; Suzuki and Larkum, 2020), which in turn may be facilitated by modulatory metabotropic mechanisms (Suzuki and Larkum, 2020). TC-mediated activation of disinhibitory microcircuit motifs may also promote synaptic integration, dendritic propagation, and plasticity (Williams and Holtmaat, 2018; Mukherjee et al., 2021).

TC inputs to L4 cooperate with local CC circuitry by connecting preferentially to locally interconnected networks (e.g., connected pairs of L4 → L2/3 neurons are innervated by the same TC afferents) (Morgenstern et al., 2016). Moreover, co-activation of TC afferents with CC circuits drives supralinear amplification (see Figure 13.3C), which may determine receptive field properties and plasticity (Lavzin et al., 2012; Williams and Holtmaat, 2018).

PLASTICITY

TC projections are thought to be structurally and functionally stable beyond postnatal critical periods. However, with the advent of new labeling and imaging techniques, experiments involving sustained surveyance of identified TC axons in vivo have shown that TC inputs and their synapses can display functional alterations upon changes in sensory experience and learning (e.g., Jaepel et al., 2017; Audette et al., 2019). Furthermore, putative TC axons incidentally protract and retract over small distances in the adult cortex. Their boutons may appear and disappear (~15%) and continue to change in size under baseline conditions, albeit to a lesser extent when compared with CC axons (Paola et al., 2006). These findings were extended by studies showing that TC synapses can undergo substantial structural modifications in response to sensory deprivation, and thereby contribute to plasticity of adult cortical circuits (Oberlaender et al., 2012).

REFERENCES

Alonso J-M, Usrey WM, Reid RC (1996). Precisely correlated firing in cells of the lateral geniculate nucleus. *Nature*. 383:815–819.

Anastasiades PG, Collins DP, Carter AG (2021). Mediodorsal and ventromedial thalamus engage distinct L1 circuits in the prefrontal cortex. *Neuron*. 109:314–330.

Audette NJ, Bernhard SM, Ray A, Stewart LT, Barth AL (2019). Rapid plasticity of higher-order thalamocortical inputs during sensory learning. *Neuron*. 103:277–291.

Audette NJ, Urban-Ciecko J, Matsushita M, Barth AL (2018). POm thalamocortical input drives layer-specific microcircuits in somatosensory cortex. *Cerebral Cortex*. 28:1312–1328.

Bagnall MW, Hull C, Bushong EA, Ellisman MH, Scanziani M (2011). Multiple clusters of release sites formed by individual thalamic afferents onto cortical interneurons ensure reliable transmission. *Neuron*. 71:180–194.

Banitt Y, Martin KAC, Segev I (2007). A biologically realistic model of contrast invariant orientation tuning by thalamocortical synaptic depression. *J Neurosci*. 27:10230–10239.

Barre A, Berthoux C, Bundel DD, Valjent E, Bockaert J, Marin P, Bécamel C (2016). Presynaptic serotonin 2A receptors modulate thalamocortical plasticity and associative learning. *Proc Natl Acad Sci*. 113:E1382–E1391.

Bereshpolova Y, Hei X, Alonso J-M, Swadlow HA (2020). Three rules govern thalamocortical connectivity of fast-spike inhibitory interneurons in the visual cortex. *eLife*. 9:e60102.

Blundon JA, Bayazitov IT, Zakharenko SS (2011). Presynaptic gating of postsynaptically expressed plasticity at mature thalamocortical synapses. *J Neurosci*. 31:16012–16025.

Bopp R, Holler-Rickauer S, Martin KAC, Schuhknecht GFP (2017). An ultrastructural study of the thalamic input to layer 4 of primary motor and primary somatosensory cortex in the mouse. *J Neurosci*. 37:2435–2448.

Bruno RM, Sakmann B (2006). Cortex is driven by weak but synchronously active thalamocortical synapses. *Science*. 312:1622–1627.

Bureau I, Paul F von, Svoboda K (2006). Interdigitated paralemniscal and lemniscal pathways in the mouse barrel cortex. *PLoS Biology*. 4:e382.

Chung S, Li X, Nelson SB (2002). Short-term depression at thalamocortical synapses contributes to rapid adaptation of cortical sensory responses in vivo. *Neuron*. 34:437–446.

Collins MO, Husi H, Yu L, Brandon JM, Anderson CNG, Blackstock WP, Choudhary JS, Grant SGN (2006). Molecular characterization and comparison of the components and multiprotein complexes in the postsynaptic proteome. *J Neurochem*. 97:16–23.

Constantinople CM, Bruno RM (2013). Deep cortical layers are activated directly by thalamus. *Science*. 340:1591–1594.

Cruikshank SJ, Urabe H, Nurmikko AV, Connors BW (2010). Pathway-specific feedforward circuits between thalamus and neocortex revealed by selective optical stimulation of axons. *Neuron*. 65:230–245.

Cserép C, Pósfai B, Schwarcz AD, Dénes Á (2018). Mitochondrial ultrastructure is coupled to synaptic performance at axonal release sites. *Eneuro*. 5:0390–17.2018.

El-Boustani S, Sermet BS, Foustoukos G, Oram TB, Yizhar O, Petersen CCH (2020). Anatomically and functionally distinct thalamocortical inputs to primary and secondary mouse whisker somatosensory cortices. *Nat Commun*. 11:3342.

Fujiyama F, Furuta T, Kaneko T (2001). Immunocytochemical localization of candidates for vesicular glutamate transporters in the rat cerebral cortex. *J Comp Neurol*. 435:379–387.

Gambino F, Pagès S, Kehayas V, Baptista D, Tatti R, Carleton A, Holtmaat A (2014). Sensory-evoked LTP driven by dendritic plateau potentials in vivo. *Nature*. 515:116.

Gil Z, Connors BW, Amitai Y (1997). Differential regulation of neocortical synapses by neuromodulators and activity. *Neuron*. 19:679–686.

Gil Z, Connors BW, Amitai Y (1999). Efficacy of thalamocortical and intracortical synaptic connections quanta, innervation, and reliability. *Neuron*. 23:385–397.

Harris KM, Weinberg RJ (2012). Ultrastructure of synapses in the mammalian brain. *Csh Perspect Biol*. 4:a005587.

Holler S, Köstinger G, Martin KAC, Schuhknecht GFP, Stratford KJ (2021). Structure and function of a neocortical synapse. *Nature.* 591:111–116.

Hsu A, Luebke JI, Medalla M (2017) Comparative ultrastructural features of excitatory synapses in the visual and frontal cortices of the adult mouse and monkey. *J Comp Neurol.* 525:2175–2191.

Isaac JTR, Crair MC, Nicoll RA, Malenka RC (1997). Silent synapses during development of thalamocortical inputs. *Neuron.* 18:269–280.

Jaepel J, Hübener M, Bonhoeffer T, Rose T (2017). Lateral geniculate neurons projecting to primary visual cortex show ocular dominance plasticity in adult mice. *Nat Neurosci.* 20:1708–1714.

Jones EG, Powell TPS (1970). An electron microscopic study of the laminar pattern and mode of termination of afferent fibre pathways in the somatic sensory cortex of the cat. *Phil Trans Roy Soc Lond B.* 257:45–62.

Kasthuri N, Hayworth KJ, Berger DR, Schalek RL, Conchello JA, Knowles-Barley S, Lee D, Vázquez-Reina A, Kaynig V, Jones TR, Roberts M, Morgan JL, Tapia JC, Seung HS, Roncal WG, Vogelstein JT, Burns R, Sussman DL, Priebe CE, Pfister H, Lichtman JW (2015). Saturated reconstruction of a volume of neocortex. *Cell.* 162:648–661.

Kharazia V, Weinberg R (1994). Glutamate in thalamic fibers terminating in layer IV of primary sensory cortex. *J Neurosci.* 14:6021–6032.

Kim N, Bahn S, Choi JH, Kim JS, Rah J-C (2021). Synapses from the motor cortex and a high-order thalamic nucleus are spatially clustered in proximity to each other in the distal tuft dendrites of mouse somatosensory cortex. *Cereb Cortex.* 32:737–754.

Kloc M, Maffei A (2014). Target-specific properties of thalamocortical synapses onto layer 4 of mouse primary visual cortex. *J Neurosci.* 34(46):15455–15465.

Kubota Y, Hatada S, Kondo S, Karube F, Kawaguchi Y (2007). Neocortical inhibitory terminals innervate dendritic spines targeted by thalamocortical afferents. *J Neurosci.* 27:1139–1150.

Lavzin M, Rapoport S, Polsky A, Garion L, Schiller J (2012). Nonlinear dendritic processing determines angular tuning of barrel cortex neurons in vivo. *Nature.* 490:397–401.

Lee CC, Sherman SM (2008). Synaptic properties of thalamic and intracortical inputs to layer 4 of the first- and higher-order cortical areas in the auditory and somatosensory systems. *J Neurophysiol.* 100:317–326.

Llano DA, Sherman SM (2008). Evidence for nonreciprocal organization of the mouse auditory thalamocortical-corticothalamic projection systems. *J Comp Neurol.* 507:1209–1227.

Llinás RR, Leznik E, Urbano FJ (2002). Temporal binding via cortical coincidence detection of specific and nonspecific thalamocortical inputs: A voltage-dependent dye-imaging study in mouse brain slices. *Proc Natl Acad Sci.* 99:449–454.

Lomi E, Mathiasen ML, Cheng HY, Zhang N, Aggleton JP, Mitchell AS, Jeffery KJ (2021). Evidence for two distinct thalamocortical circuits in retrosplenial cortex. *Neurobiol Learn Mem.* 185:107525.

Meyer HS, Wimmer VC, Hemberger M, Bruno RM, Kock CPJ de, Frick A, Sakmann B, Helmstaedter M (2010). Cell type–specific thalamic innervation in a column of rat vibrissal cortex. *Cereb Cortex.* 20:2287–2303.

Morgenstern NA, Bourg J, Petreanu L (2016). Multilaminar networks of cortical neurons integrate common inputs from sensory thalamus. *Nat Neurosci.* 19:1034–1040.

Motta A, Berning M, Boergens KM, Staffler B, Beining M, Loomba S, Hennig P, Wissler H, Helmstaedter M (2019). Dense connectomic reconstruction in layer 4 of the somatosensory cortex. *Science.* 366:eaay3134.

Mukherjee A, Lam NH, Wimmer RD, Halassa MM (2021). Thalamic circuits for independent control of prefrontal signal and noise. *Nature.* 600:100–104.

Oberlaender M, Ramirez A, Bruno RM (2012). Sensory experience restructures thalamocortical axons during adulthood. *Neuron.* 74:648–655.

Ohno S, Kuramoto E, Furuta T, Hioki H, Tanaka YR, Fujiyama F, Sonomura T, Uemura M, Sugiyama K, Kaneko T (2012). A morphological analysis of thalamocortical axon fibers of rat posterior thalamic nuclei: A single neuron tracing study with viral vectors. *Cereb Cortex.* 22:2840–2857.

Paola VD, Holtmaat A, Knott G, Song S, Wilbrecht L, Caroni P, Svoboda K (2006). Cell type-specific structural plasticity of axonal branches and boutons in the adult neocortex. *Neuron.* 49:861–875.

Pardi MB, Vogenstahl J, Dalmay T, Spanò T, Pu D-L, Naumann LB, Kretschmer F, Sprekeler H, Letzkus JJ (2020). A thalamocortical top-down circuit for associative memory. *Science.* 370:844–848.

Peters A, Payne BR (1993). Numerical relationships between geniculocortical afferents and pyramidal cell modules in cat primary visual cortex. *Cereb Cortex.* 3:69–78.

Petreanu L, Mao T, Sternson SM, Svoboda K (2009). The subcellular organization of neocortical excitatory connections. *Nature.* 457:1142–1145.

Phillips JW, Schulmann A, Hara E, Winnubst J, Liu C, Valakh V, Wang L, Shields BC, Korff W, Chandrashekar J, Lemire AL, Mensh B, Dudman JT, Nelson SB, Hantman AW (2019). A repeated molecular architecture across thalamic pathways. *Nat Neurosci.* 22:1925–1935.

Porter JT, Johnson CK, Agmon A (2001). Diverse types of interneurons generate thalamus-evoked feedforward inhibition in the mouse barrel cortex. *J Neurosci.* 21:2699–2710.

Porter JT, Nieves D (2004). Presynaptic GABAB receptors modulate thalamic excitation of inhibitory and excitatory neurons in the mouse barrel cortex. *J Neurophysiol.* 92:2762–2770.

Poulet JFA, Fernandez LMJ, Crochet S, Petersen CCH (2012). Thalamic control of cortical states. *Nat Neurosci.* 15:370–372.

Rah J-C, Bas E, Colonell J, Mishchenko Y, Karsh B, Fetter RD, Myers EW, Chklovskii DB, Svoboda K, Harris TD, Isaac JTR (2013). Thalamocortical input onto layer 5 pyramidal neurons measured using quantitative large-scale array tomography. *Front Neural Circuit.* 7:177.

Richardson RJ, Blundon JA, Bayazitov IT, Zakharenko SS (2009). Connectivity patterns revealed by mapping of active inputs on dendrites of thalamorecipient neurons in the auditory cortex. *J Neurosci.* 29:6406–6417.

Rodriguez-Moreno J, Porrero C, Rollenhagen A, Rubio-Teves M, Casas-Torremocha D, Alonso-Nanclares L, Yakoubi R, Santuy A, Merchan-Pérez A, DeFelipe J, Lübke JH, Clasca F (2020). Area-specific synapse structure in branched posterior nucleus axons reveals a new level of complexity in thalamocortical networks. *J Neurosci.* 40:2886–2919.

Rodriguez-Moreno J, Rollenhagen A, Arlandis J, Santuy A, Merchan-Pérez A, DeFelipe J, Lübke JHR, Clasca F (2018). Quantitative 3D ultrastructure of thalamocortical synapses from the "lemniscal" ventral posteromedial nucleus in mouse barrel cortex. *Cereb Cortex.* 28:3159–3175.

Rollenhagen A, Klook K, Sätzler K, Qi G, Anstötz M, Feldmeyer D, Lübke JHR (2015). Structural determinants underlying the high efficacy of synaptic transmission and plasticity at synaptic boutons in layer 4 of the adult rat "barrel cortex." *Brain Struct Funct.* 220:3185–3209.

Roth MM, Dahmen JC, Muir DR, Imhof F, Martini FJ, Hofer SB (2016). Thalamic nuclei convey diverse contextual information to layer 1 of visual cortex. *Nat Neurosci.* 19:299–307.

Roy SA, Alloway KD (2001). Coincidence detection or temporal integration? What the neurons in somatosensory cortex are doing. *J Neurosci.* 21:2462–2473.

Sampathkumar V, Miller-Hansen A, Sherman SM, Kasthuri N (2021) An ultrastructural connectomic analysis of a higher-order thalamocortical circuit in the mouse. *Eur J Neurosci.* 54:750–762.

Schoonover CE, Tapia J-C, Schilling VC, Wimmer V, Blazeski R, Zhang W, Mason CA, Bruno RM (2014). Comparative strength

and dendritic organization of thalamocortical and corticocortical synapses onto excitatory layer 4 neurons. *J Neurosci.* 34:6746–6758.
Sedigh-Sarvestani M, Vigeland L, Fernandez-Lamo I, Taylor MM, Palmer LA, Contreras D (2017). Intracellular, in vivo, dynamics of thalamocortical synapses in visual cortex. *J Neurosci.* 37:5250–5262.
Sermet BS, Truschow P, Feyerabend M, Mayrhofer JM, Oram TB, Yizhar O, Staiger JF, Petersen CC (2019). Pathway-, layer- and cell-type-specific thalamic input to mouse barrel cortex. *eLife.* 8:e52665.
Staiger JF, Zilles K, Freund TF (1996). Distribution of GABAergic elements postsynaptic to ventroposteromedial thalamic projections in layer IV of rat barrel cortex. *Eur J Neurosci.* 8:2273–2285.
Stanley GB, Jin J, Wang Y, Desbordes G, Wang Q, Black MJ, Alonso J-M (2012). Visual orientation and directional selectivity through thalamic synchrony. *J Neurosci.* 32:9073–9088.
Stratford KJ, Tarczy-Hornoch K, Martin KAC, Bannister NJ, Jack JJB (1996). Excitatory synaptic inputs to spiny stellate cells in cat visual cortex. *Nature.* 382:258–261.
Suzuki M, Larkum ME (2020). General anesthesia decouples cortical pyramidal neurons. Cell. 180:666–676.
Swadlow HA, Gusev AG (2001). The impact of "bursting" thalamic impulses at a neocortical synapse. *Nat Neurosci.* 4:402–408.
Viaene AN, Petrof I, Sherman SM (2011). Properties of the thalamic projection from the posterior medial nucleus to primary and secondary somatosensory cortices in the mouse. *Proc Natl Acad Sci.* 108:18156–18161.
Wang L, Kloc M, Maher E, Erisir A, Maffei A (2019). Presynaptic GABAA receptors modulate thalamocortical inputs in layer 4 of rat V1. *Cereb Cortex.* 29:921–936.
White EL, Weinfeld E, Lev DL (2004). Quantitative analysis of synaptic distribution along thalamocortical axons in adult mouse barrels. *J Comp Neurol.* 479:56–69.
Williams LE, Holtmaat A (2018). Higher-order thalamocortical inputs gate synaptic long-term potentiation via disinhibition. *Neuron.* 101:91–102.

14.

STRUCTURE AND FUNCTION OF CORTICOTHALAMIC PATHWAYS OF LAYER 6 NEURONS

Allison J. Murphy and Farran Briggs

INTRODUCTION

Connections from neurons in the deepest layer of the cortex, layer 6, with various nuclei in the thalamus are ubiquitous in the mammalian brain (Jones, 1985). These layer 6 corticothalamic (L6 CT) neurons have been described in primary visual (Guillery, 1966; Robson, 1983), auditory (Conley et al., 1991), and somatosensory (Bourassa et al., 1995; Killackey and Sherman, 2003) cortices, higher-order sensory cortices (Killackey and Sherman, 2003; Adusei et al., 2021), motor cortex (Kakei et al., 2001), and frontal cortical areas (Schwartz et al., 1991), projecting to their associated first-order (FO) and higher-order (HO) thalamic nuclei (see Figure 14.1). It was long assumed that L6 CT neurons in primary sensory cortices project to FO thalamus, which receives input from the sensory periphery, while L6 CT neurons in nonprimary sensory cortical areas, including motor and frontal cortices, targeted HO thalamic nuclei (i.e., those lacking peripheral input). But this dichotomy may not be accurate. L6 CT neurons targeting FO visual thalamus have been identified throughout extrastriate visual cortex, even though the proportions of these neurons are low compared to L6 CT neurons in primary visual cortex (Adusei et al., 2021). Additionally, some L6 CT neurons in primary sensory cortices project to HO thalamus (Conley and Raczkowski, 1990; Bourassa et al., 1995; Usrey and Fitzpatrick, 1996; Killackey and Sherman, 2003; Whilden et al., 2021). Accordingly, L6 CT neurons are likely present in all cortical areas and are capable of targeting either (or both) FO and HO thalamus (Figure 14.1), providing a vast and diverse network of CT connectivity in the brain. Furthermore, L6 CT neurons are likely quite distinct from separate populations of CT neurons originating in cortical layer 5 that target HO thalamus with more powerful synapses (Sherman and Guillery, 1998; see also **Chapter 15 in Section IV** in this volume).

There is at least one defining feature common to all L6 CT neurons, and that is the structure of their axons in the thalamus, defined as "type 1" axons (Guillery, 1966). L6 CT type 1 axons have diffuse branches that contact multiple thalamic cells on distal dendrites with small terminal boutons containing small, round vesicles (Guillery, 1966; Robson, 1983) (Figure 14.2A,B). Given the structure and physiology of these axons, L6 CT inputs are "modulatory" in their functional influence over thalamic activity, rather than "driving" (Sherman and Guillery, 1998). Interestingly, although their axon terminals are small, L6 CT synapses are numerous in both FO and HO thalamic nuclei, as well as in the thalamic reticular nucleus (TRN) (Erişir et al., 1997; Van Horn and Sherman, 2007; see also **Chapter**

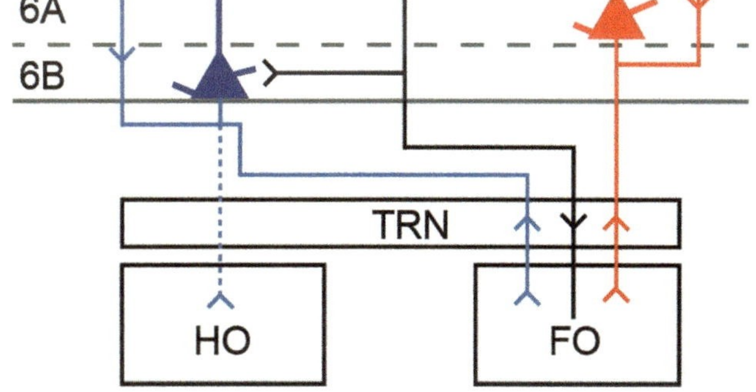

Figure 14.1. Example circuit diagram of upper/layer 6A (red) and lower/layer 6B (blue) L6 CT neurons in primary sensory cortex. L6 CT cells in layer 6A project to first-order (FO) thalamus, while cells in layer 6B can project to both FO and higher-order (HO) thalamus. CT cells also send collaterals to various overlying cortical layers.

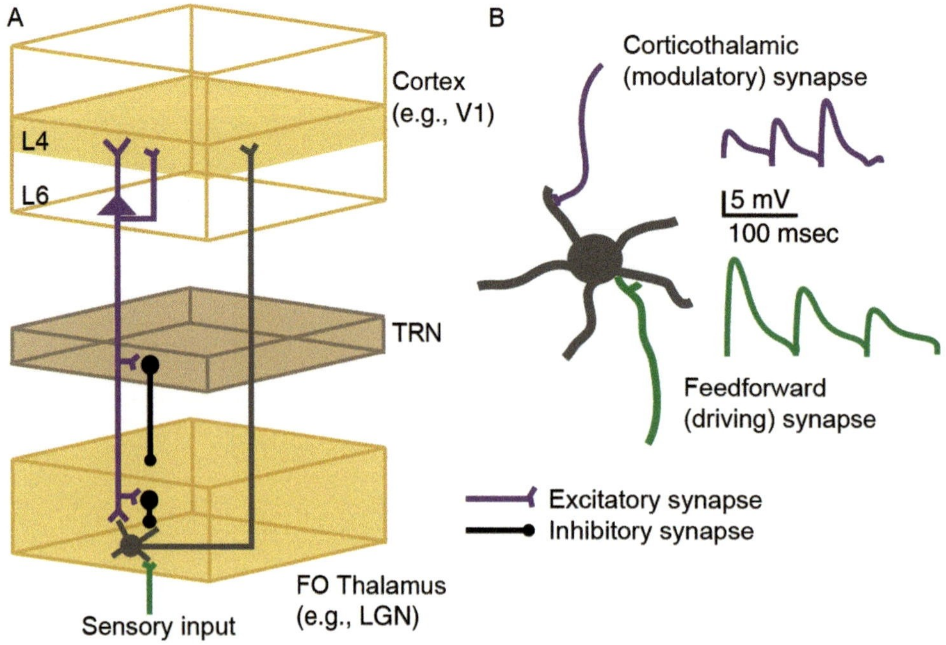

Figure 14.2. (A). Schematic of L6 CT neurons (purple) and their postsynaptic targets in FO thalamus. Note, inhibitory interneurons are not present in all FO thalamic nuclei. (B). Comparison of L6 CT "modulatory" synapses (purple) and feedforward "driving" synapses (green) onto a thalamic relay cell (gray). Traces depict thalamic EPSPs following repeated presynaptic stimulation; small EPSPs from modulators (purple) facilitate while large EPSPs from drivers (green) depress.

16 in Section IV). In FO thalamus, axon terminals from the cortex outnumber driving input from peripheral sensory structures (Erişir et al., 1997). L6 CT axons target thalamocortical relay neurons and inhibitory interneurons when present (Weber et al., 1989). In addition, many CT axons send collaterals into the TRN, where they synapse onto TRN inhibitory neurons (Conley et al., 1991; Bourassa et al., 1995) (Figure 14.2A). In sensory systems with spatial organization, such as retinotopy, somatotopy, or tonotopy, CT projections maintain topography (Claps and Casagrande, 1990; Winer et al., 2001; Ichida and Casagrande, 2002). Thus, L6 CT axons provide a well-organized, anatomically robust yet modulatory input to the thalamus, capable of recruiting both excitatory and inhibitory neuronal networks. This connectivity enables the cortex to dynamically regulate thalamic relay cells through net excitatory and inhibitory states.

DIVERSITY OF L6 CT CELL TYPES: ANATOMY

L6 CT neurons are anatomically, physiologically, and functionally diverse. Across primates and carnivores, there are at least 2 types of CT neurons based on local dendritic morphology. Short pyramidal cells have shorter apical dendrites that rarely reach above layer 4, with arbors mostly in layer 4; and tall pyramidal cells have taller apical dendrites with more extensive arborizations in layer 5 (Briggs, 2010) (Figure 14.3). In primate primary visual cortex (V1), short cells are confined to upper layer 6 and tall cells are mostly in lower layer 6 (Briggs and Callaway, 2001; Briggs et al., 2016). Note that in many species, layer 6 is divided into 2 tiers: 6A and 6B; in primate V1, L6 CT neurons are located in the upper third and lower third tiers of layer 6 and are absent from the middle third. For simplicity, we refer here to L6 CT neurons in L6A or L6B, acknowledging that in primates these should indicate upper and lower thirds of layer 6, respectively. In carnivore V1, distinct short and tall CT cell types are more mixed in layer 6 (Katz, 1987; Hasse et al., 2018). In mouse V1, L6 CT neurons are morphologically distinguishable from other layer 6 cells and show some diversity in apical dendrite height, but represent more of a continuum rather than distinct short and tall cell types (Brumberg et al., 2003; Olsen et al., 2012; Augustinaite and Kuhn, 2020). Beyond short and tall pyramidal CT types, additional, rarer CT cell types are present in primate and carnivore visual cortex, including spiny stellate CT cells with arbors exclusively in layer 6, tall pyramidal CT cells with apical dendritic tufts approaching the pia, and tilted cells (Katz, 1987; Ichida and Casagrande, 2002; Briggs et al., 2016; Hasse et al., 2018; Adusei et al., 2021) (Figure 14.3). L6 CT pyramidal cells in primary auditory and somatosensory cortices in both cats and rodents tend to have apical dendrites that terminate at or below layer 3, with some rodent L6 CT cell dendritic arbors not reaching above layer 5 (Zhang and Deschênes, 1997; Prieto and Winer, 1999; Llano and Sherman, 2009).

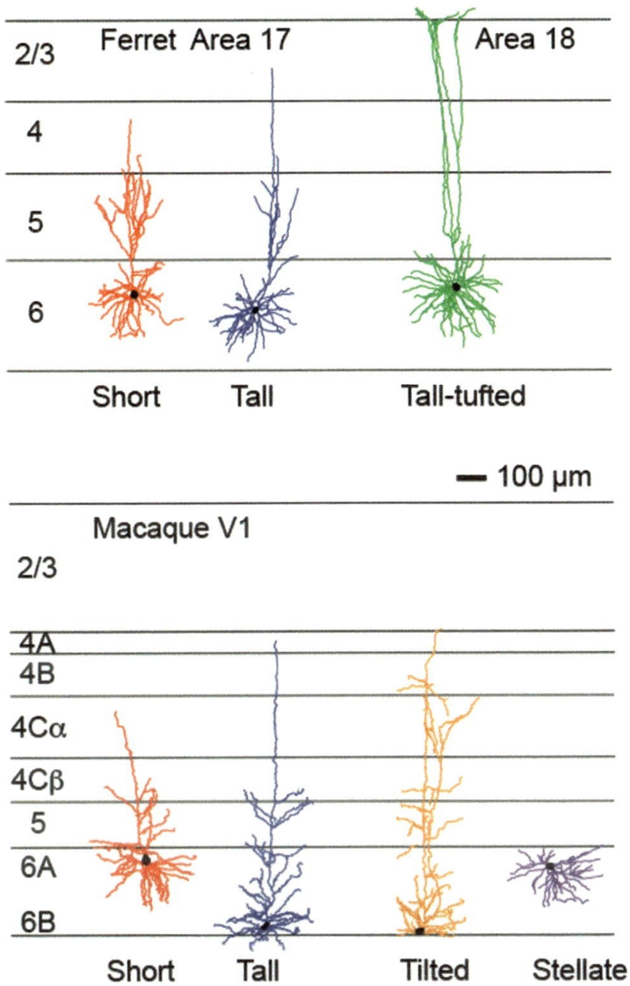

Figure 14.3. Comparison of L6 CT neurons in visual cortex projecting to LGN in carnivore and primate. Dendritic arborizations shown for each cell type; cell bodies are black; scale bar at bottom applies to all. Examples include: short (red), tall (blue), and tall-tufted (green) pyramidal neurons and tilted (orange) and spiny stellate (purple) cells. Adapted from Hasse et al. (2018).

In addition to dendritic morphological diversity, subpopulations of L6 CT neurons show distinct projection patterns in the thalamus. In primary sensory areas, CT cells in L6A tend to project only to FO sensory cortex, while CT cells in L6B project to FO and/or HO thalamic nuclei (Figure 14.1); this has been demonstrated among V1 L6 CT neurons in tree shrews, galagoes, and rodents, targeting lateral geniculate nucleus (LGN) and pulvinar (Conley and Raczkowski, 1990; Bourassa et al., 1995; Usrey and Fitzpatrick, 1996), as well as among primary somatosensory cortex (S1) L6 CT cells in rodents, targeting the ventral posterior medial (VPM) and posterior medial (POm) nuclei (Bourassa et al., 1995; Killackey and Sherman, 2003; Whilden et al., 2021). A subpopulation of CT cells in L6B in mice projects to HO thalamus only, without sending collaterals to the TRN (Hoerder-Suabedissen et al., 2018). Interestingly, L6A versus L6B CT cells in rodents that target FO versus either/both FO and HO thalamus tend to display different dendritic morphologies: L6A CT cells have arborizations through layer 4, while L6B CT cells tend to have dendrites restricted to layers 5 and 6 (Zhang and Deschênes, 1997). Furthermore, there are transcriptional differences between L6A versus L6B CT neurons projecting to FO versus FO and/or HO thalamus in the rodent somatosensory system (Chevée et al., 2018). It is not yet known whether these different projection patterns originating in L6A or L6B are present through association and higher-order sensory cortices or if this is a characteristic unique to primary sensory cortex.

The diversity of morphological types within the broader L6 CT population suggests multiple functional roles for L6 CT cells and/or parallel information-processing streams among CT networks. In primate V1, type Iβ (short) corticogeniculate neurons in L6A project to the parvocellular layers of the LGN, while L6B type IC (tall) corticogeniculate neurons project to the magnocellular LGN layers, with some spread into neighboring koniocellular layers (Briggs et al., 2016; Fitzpatrick et al., 1994; Ichida and Casagrande, 2002; see also **Chapter 19 in Section V**). In addition, type Iβ cells receive local cortical inputs from layer 4Cβ, the parvocellular-recipient zone of layer 4, as well as layers 2/3, while type IC cells receive local inputs from layer 4Cα, the magnocellular-recipient zone of layer 4, and layer 4B (Briggs and Callaway, 2001). This organization supports the notion that separate CT circuits convey distinct qualities of visual information via parallel processing streams.

However, this degree of segregation and specialization of L6 CT circuits may be unique to the primate visual system. In carnivores, tall and short corticogeniculate morphologies are present, but intermixed in layer 6 (Hasse et al., 2018; Katz, 1987). Also, corticogeniculate axons span LGN layers, but can preferentially target the A LGN layer with matching eye preference (Robson, 1983; Claps and Casagrande, 1990; Murphy and Sillito, 1996). Thus, differences in local dendritic morphology, axon thickness, and preference for A or C LGN layers may indicate some segregation of parallel visual CT circuits in carnivores as well (Murphy and Sillito, 1996). Across species, anatomical data suggest multiple types of CT neurons, with at least one major division based on local dendritic morphology (short/L6A, tall/L6B) and/or at least one major division by function (projection to FO or FO and HO thalamus) (Figure 14.1). These neurons are thus well positioned to provide feedback to multiple divisions of the thalamus and facilitate cortico-thalamo-cortical communication in distinct circuits. Further area-specific specializations, similar to the parallel stream-specific CT circuits observed in the primate visual system, may also be revealed with further study.

DIVERSITY OF L6 CT CELL TYPES: PHYSIOLOGY

Early studies aimed at revealing the physiological properties of CT neurons utilized antidromic stimulation to identify CT neurons in vivo. Results demonstrated that CT neurons have a broad range of axonal conduction velocities, from fast transmission comparable with thalamocortical velocities to very slow velocities, likely from small, unmyelinated axons (Stoelzel et al., 2017). Distinct subpopulations of CT neurons based on axon conduction velocity have been observed in the visual cortex of carnivores (Harvey, 1978; Tsumoto and Suda, 1980), rabbits (Swadlow and Weyand, 1987; Stoelzel et al., 2017), and primates (Briggs and Usrey, 2007, 2009) as well in rodent somatosensory cortex (Kwegyir-Afful and Simons, 2009) and cat motor cortex (Sirota et al., 2005). The fact that CT neurons across species and sensory modalities can be distinguished based on differing axon conduction speeds suggests that distinct CT subpopulations operate on different timescales and may represent functionally diverse CT circuits, even parallel processing streams. In many brain areas, this notion is supported by the fact that these CT subtypes also differ in their responses to sensory stimuli. In carnivore V1, CT cells with slower-conducting axons are classified as Simple cells and faster-conducting CT cells are classified as Complex cells (Harvey, 1978; Tsumoto and Suda, 1980; Briggs and Usrey, 2007). The Simple/Complex cell distinction was defined by Hubel and Wiesel (Hubel and Wiesel, 1962) to describe unique spatial receptive field properties of V1 neurons that related to these neurons' position within the cortical processing hierarchy: Simple cells were likely thalamocortical-recipients, while Complex cells likely derived their response preferences from cortical circuits. Thus, Simple and Complex CT subtypes may receive unique thalamocortical and local circuit inputs.

Interestingly, in cat and rabbit V1 and cat motor cortex, the slowest-conducting CT cells are difficult to drive with visual or motor stimulation (Tsumoto and Suda, 1980; Swadlow and Weyand, 1987; Sirota et al., 2005; Stoelzel et al., 2017). Low firing rates among these CT neurons could reflect sub-threshold inputs or net inhibition: juxtacellular current injection in rodent somatosensory CT neurons revealed diverse responses to whisker stimulation (Kwegyir-Afful and Simons, 2009). In macaque V1, layer 6 corticogeniculate neurons are distinguishable based on their axon conduction velocities, Simple/Complex receptive field classification, and visual stimulus tuning properties (Briggs and Usrey, 2009) which also likely correlate with local dendritic morphology (Briggs et al., 2016). Specifically, the tuning preferences of fast-conducting Complex CT cells (likely IC type) are similar to magnocellular LGN neurons, medium-conducting Simple CT cells (likely Iβ type) share tuning properties with parvocellular LGN cells, and slow-conducting Complex cells (likely tilted type) show tuning similar to koniocellular LGN cells, including receiving S-cone input. The fast-conducting CT cells in primate V1 can also receive direct suprathreshold input from LGN relay cells, providing the basis for a fast communication loop between LGN and V1 (Briggs and Usrey, 2007). Thus, in the primate visual system, L6 CT circuits are organized into parallel streams aligned with the feedforward retino-geniculo-cortical parallel processing streams (see also **Chapter 19 in Section V**). Such stereotyped and strict segregation of L6 CT cell types into parallel processing streams may be a special case in primate V1. More "relaxed" versions of parallel processing within L6 CT circuits may exist across species and sensory modalities (see also **Chapters 24 and 27 in Section V**).

In addition to differences in responses to sensory stimulation, subpopulations of L6 CT neurons may also be distinguishable based on modulations by behavioral state. In awake mouse V1, L6 CT neuronal responses can be enhanced, suppressed, or unchanged by visual stimuli and are also modulated by the animal's arousal and locomotive state (Augustinaite and Kuhn, 2020). Similarly, a proportion of L6 CT neurons in barrel cortex of awake mice are modulated by the action of whisking onset (Dash et al., 2022). In rabbit V1, CT neuronal response properties also change based on arousal state (Stoelzel et al., 2017). These behavioral state-dependent modulations of L6 CT responses could have significant implications for the net excitatory or inhibitory impacts of CT feedback on thalamocortical networks. Further studies of L6 CT neurons in awake, behaving animals will help reveal the functional role of distinct CT neuronal subpopulations.

SUBCORTICAL AND LOCAL CORTICAL INPUTS ONTO L6 CT NEURONS

As suggested by their dendritic morphology, L6 CT neurons can receive thalamocortical and/or local cortical inputs from neurons in multiple layers. In mice, most L6 CT neurons have basal dendritic arbors in layer 6, extend apical dendrites through layers 5 and 4 (Brumberg et al., 2003; Olsen et al., 2012), and receive their strongest local cortical inputs from cells in layer 4 and CT and cortico-cortical cells in layer 6 (Tanaka et al., 2011). Local cortical inputs onto L6 CT cell types in primate V1 differ from those in rodents in that both tall and short L6 CT cell types receive local input from superficial layer neurons (Briggs and Callaway, 2001), while rodent layer 6 neurons in general only receive local input from deeper layers (Llano and Sherman, 2009). In primates and rodents, some L6 CT neurons receive monosynaptic thalamic input (Briggs and Usrey, 2007; Constantinople and Bruno, 2013), providing a substrate for rapid communication of sensory signals from the thalamus to the cortex and back again (Figure 14.1).

L6 CT neurons in rodent sensory cortices have been shown to receive a variety of inputs beyond FO thalamic and local cortical inputs. In mouse S1, L6 CT neurons that project to VPM and/or POm can receive inputs from the amygdala, primary motor cortex (M1), and other HO thalamic nuclei (Whilden et al., 2021). Similarly, mouse auditory cortex L6 CT neurons receive motor preparatory input, likely from the globus pallidus, during an active listening task (Clayton et al., 2021). Together, these findings suggest that L6 CT neurons may integrate sensory-motor signals to guide behavior. Whether similar convergence of sensory and motor inputs occurs at L6 CT neurons in non-rodent species remains to be discovered.

When considering subcortical and local cortical inputs onto L6 CT neurons, it is important to also note that these cells have axon collaterals within various local cortical layers. In other words, a copy of the message that L6 CT neurons relay to the thalamus is also communicated locally. Morphologically distinct L6 CT neurons have unique local axonal arborizations, often targeting the same cortical layers that provide their strongest local inputs (Katz, 1987; Briggs and Callaway, 2001; Brumberg et al., 2003; Briggs et al., 2016). In the mouse, L6 CT neurons make robust local connections onto both excitatory and inhibitory neurons in layers 5 and 6. Specifically, L6 CT cells generate strong excitation of pyramidal neurons in layer 5A (Kim et al., 2014). They also form synapses onto a specific subpopulation of inhibitory neurons in L6A that themselves have axons spanning all cortical layers (Bortone et al., 2014; Frandolig et al., 2019). Indeed, it is the L6 CT activation of these layer 6 inhibitory neurons that is thought to underly the substantial net inhibition of the cortex that occurs upon synchronized optogenetic stimulation of L6 CT neurons in mouse sensory cortex (Olsen et al., 2012; Bortone et al., 2014; Kim et al., 2014; Frandolig et al., 2019). It is not clear whether L6 CT neurons in non-rodent species form similar excitatory and inhibitory local connections within the deepest cortical layers. However, the possibility that L6 CT neurons across species produce net excitation or net inhibition locally within the cortex, as they may do in the thalamus, is an intriguing hypothesis that requires further testing.

L6 CT SYNAPSES IN THE THALAMUS

The influence that L6 CT feedback exerts on thalamic neurons is a function of its synaptic connections. L6 CT synapses in the thalamus are classified as modulatory due to their small size and distal placement along thalamic neuronal dendrites. All L6 CT neurons are glutamatergic, producing small excitatory postsynaptic currents (EPSCs) in their postsynaptic target cells, mediated by both ionotropic AMPA and NMDA receptors and metabotropic glutamate receptors (McCormick and von Krosigk, 1992; Li et al., 2003; Landisman and Connors, 2007). Paired-pulse stimulation of L6 CT axons leads to subsequent excitatory postsynaptic potential (EPSP) amplitude facilitation, in contrast to driving inputs onto thalamic cells that produce large EPSPs that depress with repeated stimulation (Von Krosigk et al., 1999; Li et al., 2003) (Figure 14.2B). L6 CT inputs to thalamus are always modulatory in nature, regardless of whether they target FO or HO thalamus (Li et al., 2003; Landisman and Connors, 2007; Kirchgessner et al., 2020). This is in contrast to layer 5 CT neurons that provide consistent driving inputs to HO thalamus such as POm or pulvinar (Sherman and Guillery, 1998). While numerous studies across species have shown that activating L6 CT neurons does not cause large changes in thalamic neuronal activity, consistent with its modulatory role (Denman and Contreras, 2015; Hasse and Briggs, 2017), it is interesting to note that there may be circumstances under which L6 CT feedback could exert a more powerful influence over thalamic neurons. This is due in part to the fact that L6 CT synapses are numerous (Erişir et al., 1997). Computational modeling, supported by slice physiology data, suggests that synchronous activation of many L6 CT inputs could generate a large EPSP mediated by Ca2+ and NMDA channels (Connelly et al., 2016). Thus, although L6 CT axons appear designed for subtle manipulation of their postsynaptic targets, there may be as yet unexplored conditions under which L6 CT circuits could generate more powerful influence.

Within the thalamus, CT axons synapse onto relay cells, inhibitory interneurons (when present), and neurons in the TRN (Figure 14.2A), and the postsynaptic responses to CT inputs vary subtly across these neuronal classes. Inhibitory interneurons produce a slow, gradual depolarization in response to L6 CT stimulation, compared to the faster EPSPs in relay cells (Jurgens et al., 2012). CT synapses on TRN neurons are more powerful than those onto relay cells, causing EPSPs that are larger in magnitude and facilitate to a lesser degree relative to EPSPs in relay neurons (Golshani et al., 2001; Landisman and Connors, 2007; Jurgens et al., 2012). Different postsynaptic responses to CT inputs on TRN neurons may be mediated by a higher proportion of GluR4 AMPA receptor subunits at CT-TRN synapses (Golshani et al., 2001). CT inputs to TRN can also activate local inhibitory networks within the TRN (Zhang and Jones, 2004).

These differential effects of L6 CT neuronal inputs on different thalamic cell classes indicate that L6 CT feedback may control dynamic excitatory/inhibitory interactions within the thalamus. Through connections onto various neurons that provide multiple sources of inhibition for thalamic relay cells, the net effect of L6 CT feedback on a thalamic relay cell can be excitatory or inhibitory (Landisman and Connors, 2007; Jurgens et al., 2012; Denman and Contreras, 2015). The balance between direct, monosynaptic excitation and indirect inhibition of a relay cell depends on the strength and location of CT excitation as well as the frequency with which L6 CT axons are stimulated. With low-frequency L6 CT stimulation, relay cells experience brief excitation followed by inhibition from the TRN. Under high-frequency, sustained L6 CT stimulation, TRN activation overpowers excitation, resulting in strong inhibition of relay cells. Moderate stimulation of L6 CT axons, around 10 Hz, is optimal for net excitation of relay cells (Crandall et al., 2015; Kirchgessner et al., 2020). This pattern of net excitation with medium-frequency stimulation and net inhibition with low- and high-frequency stimulation is present among L6 CT inputs to both FO and HO thalamic nuclei (e.g., LGN and pulvinar; Kirchgessner et al., 2020). Another factor dictating net excitatory/inhibitory influence of L6 CT feedback on thalamic relay cells is topographic alignment. When L6 CT neurons have receptive fields that are aligned with their thalamic targets in terms of spatial pattern or frequency preference, CT activation produces excitation; stimulation of L6 CT neurons whose receptive fields are not aligned with thalamic neurons causes inhibition (Tsumoto et al., 1978; Temereanca and Simons, 2004; Wang et al., 2016). Perhaps through these various mechanisms that influence the balance of excitation and inhibition within thalamic relay cells, L6 CT feedback could also play a role in switching thalamic relay cells between burst and tonic firing modes, which can influence thalamocortical communication (Lu et al., 1992; McCormick and von Krosigk, 1992; Kirchgessner et al., 2020). Taken together, these findings suggest that L6 CT neurons provide context-dependent modulation of the thalamus that can alter information transmission to cortex through adjustments of the balance of excitation and inhibition as well as the firing mode of thalamic relay neurons.

ROLE OF L6 CT FEEDBACK IN SENSORY PROCESSING AND BEHAVIOR

Because the influence of L6 CT circuits on thalamic neurons is both subtle and multifaceted, elucidating their functional contributions remains an enduring puzzle in systems neuroscience. Many manipulations of L6 CT neurons have been conducted across a variety of species, most often focused on L6 CT neurons in primary sensory cortex and their influence over FO thalamic neurons. FO sensory thalamic relay cells inherit their receptive field properties from driving inputs they receive from the sensory periphery (e.g., LGN relay cells' receptive fields reflect the center-surround receptive field organization of retinal ganglion cells). In spite of L6 CT inputs being modulatory, effects of manipulating CT feedback have been observed across sensory systems. Perhaps the most dramatic effects have been observed in the auditory system, where activation of auditory cortex L6 CT neurons causes relay cells in the ventral medial geniculate body (MGNv) to shift their frequency preferences toward those of the CT neurons, a process termed "egocentric selection" (Suga and Ma, 2003). Activating auditory cortex L6 CT cells also enhances gain and sharpens tuning curves among MGN cells with matched frequency tuning (Suga and Ma, 2003), while suppressing gain and broadening tuning among MGN cells with broad tuning (Tang et al., 2012).

In stark contrast to the more robust effects of L6 CT feedback in the auditory system, receptive field properties and tuning among LGN relay cells are largely unaltered by L6 CT feedback. Manipulating V1 L6 CT feedback alters the preferred size of LGN receptive fields, but otherwise minimally impacts visual tuning properties and causes mixed changes in response gain, including increases and decreases in responsivity to visual stimuli (Przybyszewski et al., 2000; Denman and Contreras, 2015; Hasse and Briggs, 2017). Subtle effects reminiscent of egocentric selection have also been observed in visual and somatosensory systems. Specifically, when the spatial receptive fields of the L6 CT neurons are aligned with those of their thalamic targets, enhancing CT feedback sharpens the spatial resolution of their target thalamic neuronal receptive fields and increases suppression arising from the extraclassical surround; when CT and thalamic receptive fields are misaligned, activating CT feedback suppresses thalamic responses (Temereanca and Simons, 2004; Wang et al., 2016; Hasse and Briggs, 2017; Born et al., 2021). Future work is necessary to discover how processes like egocentric selection that require aligned excitation and misaligned inhibition are mediated by CT circuits.

In the visual system, increasing evidence suggests that V1 L6 CT feedback may play a significant role in regulating the timing and precision of LGN responses to visual stimuli (Andolina et al., 2007; Hasse and Briggs, 2017). In particular, V1 L6 CT feedback reduces the variability of LGN responses to visual inputs and increases the information content transmitted to V1 (Andolina et al., 2007; Murphy et al., 2020). Additionally, cooling V1 in awake and behaving primates, which suppresses L6 CT feedback to LGN, significantly decreased the information contained in LGN spike trains (McClurkin et al., 1994). Comparison of results across these studies is particularly relevant because similar effects of V1 L6 CT feedback were observed across species and in both anesthetized and alert and behaving animals, suggesting that reducing LGN response variability is a critical role for L6 CT feedback in the visual system. Furthermore, studies have found that V1 L6 CT feedback may coordinate and synchronize the responses of LGN relay cells (Sillito et al., 1994; Andolina et al., 2007), which also can increase the efficacy and strength of visual signals transmitted to the cortex (Alonso et al., 1996). Accordingly, the first major feedback projection in sensory processing hierarchies may precisely regulate signal-to-noise in sensory inputs to cortex. Whether L6 CT connections to HO thalamic nuclei and originating in non-sensory cortical areas perform similar timing or variability control has yet to be determined.

Although studies in which L6 CT circuits are manipulated during behavioral tasks are limited, there is evidence that CT feedback is important for sensory detection tasks (Homma et al., 2017). Activating L6 CT feedback during behavior may bias thalamic activity to favor either detection or discrimination of sensory stimuli, depending on the timing of CT activation in relation to the stimulus (Guo et al., 2017). These results further reinforce the notion that L6 CT feedback is critical for timing control in the thalamus. In mice, L6 CT neurons in sensory areas such as barrel and auditory cortex can show motor-related activation during behavior (Clayton et al., 2021; Dash et al., 2022), indicating that L6 CT neurons may be integrating task-relevant input. In addition, visual attention modulates neuronal activity in the LGN and TRN (McAlonan et al., 2006), with attention signals most likely relayed from L6 CT neurons to the thalamus. Notably, in modeling how L6 CT feedback reduces LGN response variability, Murphy et al. (2020) discovered a similar reduction in gain variability in anesthetized animals as observed in awake animals when attention was directed toward task-relevant sensory stimuli. Thus, L6 CT circuits are likely involved in a number of behavioral processes including signaling arousal state, task-switching, sensory-motor integration, and attention.

Finally, L6 CT pathways throughout the brain connected to various thalamic nuclei may play a key role in regulating changes in oscillatory rhythms associated with sleep (Steriade, 2003). Both CT-mediated excitation and inhibition through the TRN are necessary for the formation and termination of sleep spindles during non-REM sleep (Steriade, 2003). Stimulation of V1 layer 6 cells is sufficient to induce coherent LGN spiking, and disrupting V1 L6 CT neuronal activity during NREM sleep can impede sensory plasticity that occurs during sleep (Durkin et al., 2017). Furthermore, simply deleting a calcium channel subunit in L6 CT cells in mice was sufficient to induce an absence epilepsy phenotype, indicating that this circuit may be important in pathological rhythmic states (Bomben et al., 2016). Sleep/wake transitions are likely to involve significant CT coordination that involves multiple pathways in addition to L6 CT circuits. However, it is important to consider the possibility that L6 CT circuits could multiplex sensory and arousal information, for example, to help produce a specific brain state and also to facilitate plasticity and learning during sleep.

CONCLUSION

Projections from layer 6 of the cortex to the thalamus are ubiquitous throughout the brain, suggesting that these circuits are critical for brain function. L6 CT cells are among the most diverse neurons in the cortex based on dendritic/axonal morphology, physiological response properties, and perhaps other properties including transcriptional repertoires. Connections of L6 CT neurons in the thalamus are anatomically robust, with large numbers of terminals providing modulatory input to thalamic relay cells, interneurons, and GABAergic neurons in the TRN. These connections enable L6 CT feedback to dynamically regulate the balance of excitation and inhibition in the thalamus, including regulating the firing mode of relay cells, perhaps in a context- or task-relevant manner. A number of open questions remain regarding the structure and function of L6 CT neurons throughout the cortex. Most studies have focused on L6 CT neurons in primary sensory cortex and their projections to FO thalamus. Accordingly, much less is currently known about the L6 CT circuits connecting sensory association cortices, motor cortices, and frontal cortices with various thalamic nuclei. Whether trends observed among L6 CT neurons in primary sensory cortex are recapitulated in higher cortical areas remains to be seen. Additionally, the functional contributions of L6 CT inputs to both FO and HO thalamic nuclei are not resolved. Do L6 CT circuits regulate timing and precision of FO thalamic responses across sensory modalities and/or among HO thalamic cells, as they do in the visual system? Finally, as the majority of L6 CT circuit interrogation has been done in anesthetized animals or in the absence of behavior, critical questions remain about the functional contributions of these pathways to sensory-guided and goal-directed behavior. Fortunately, rapid technological developments now make many of these complicated experiments more tractable.

REFERENCES

Adusei M, Hasse JM, Briggs F (2021). Morphological evidence for multiple distinct channels of corticogeniculate feedback originating in mid-level extrastriate visual areas of the ferret. *Brain Struct Funct.* 226:2777–2791.

Alonso J-M, Usrey WM, Reid RC (1996). Precisely correlated firing in cells of the lateral geniculate nucleus. *Nature.* 383:815–819.

Andolina IM, Jones HE, Wang W, Sillito AM (2007). Corticothalamic feedback enhances stimulus response precision in the visual system. *Proc Natl Acad Sci U S A.* 104:1685–1690.

Augustinaite S, Kuhn B (2020). Complementary Ca2+ activity of sensory activated and suppressed layer 6 corticothalamic neurons reflects behavioral state. *Curr Biol.* 30:3945–3960.

Bomben VC, Aiba I, Qian J, Mark MD, Herlitze S, Noebels JL (2016). Isolated P/Q calcium channel deletion in layer VI corticothalamic neurons generates absence epilepsy. *J Neurosci.* 36:405–418.

Born G, Schneider-Soupiadis FA, Erisken S, Vaiceliunaite A, Lao CL, Mobarhan MH, Spacek MA, Einvoll GT, Busse L (2021). Corticothalamic feedback sculpts visual spatial integration in mouse thalamus. *Nature Neurosci.* 24:1711–1720.

Bortone DS, Olsen SR, Scanziani M (2014). Translaminar inhibitory cells recruited by layer 6 corticothalamic neurons suppress visual cortex. *Neuron.* 82:474–485.

Bourassa J, Pinault D, Deschênes M (1995). Corticothalamic projections from the cortical barrel field to the somatosensory thalamus in rats: A single-fibre study using biocytin as an anterograde tracer. *Eur J Neurosci.* 7:19–30.

Briggs F (010). Organizing principles of cortical layer 6. *Frontiers in Neural Circuits.* 12:1–8.

Briggs F, Callaway EM (2001). Layer-specific input to distinct cell types in layer 6 of monkey primary visual cortex. *J Neurosci.* 21:3600–3608.

Briggs F, Kiley CW, Callaway EM, Usrey WM (2016). Morphological substrates for parallel streams of corticogeniculate feedback originating in both V1 and V2 of the macaque monkey. *Neuron.* 90:388–399.

Briggs F, Usrey WM (2007). A fast, reciprocal pathway between the lateral geniculate nucleus and visual cortex in the macaque monkey. *J Neurosci.* 27:5431–5436.

Briggs F, Usrey WM (2009). Parallel processing in the corticogeniculate pathway of the macaque monkey. *Neuron.* 62:135–146.

Brumberg JC, Hamzei-Sichani F, Yuste R (2003). Morphological and physiological characterization of layer VI corticofugal neurons of mouse primary visual cortex. *J Neurophysiol.* 89:2854–2867,

Chevée M, Robertson JDJ, Cannon GH, Brown SP, Goff LA (2018). Variation in activity state, axonal projection, and position define the transcriptional identity of individual neocortical projection neurons. *Cell Rep.* 22:441.

Claps A, Casagrande VA (1990). The distribution and morphology of corticogeniculate axons in ferrets. *Brain Res.* 530:126–129.

Clayton KK, Williamson RS, Hancock KE, Tasaka G ichi, Mizrahi A, Hackett TA, Polley DB (2021). Auditory corticothalamic neurons are recruited by motor preparatory inputs. *Curr Biol.* 31:310–321.

Conley M, Kupersmith AC, Diamond IT (1991). The organization of projections from subdivisions of the auditory cortex and thalamus to the auditory sector of the thalamic reticular nucleus in galago. *Eur J Neurosci.* 3:1089–1103.

Conley M, Raczkowski D (1990). Sublaminar organization within layer VI of the striate cortex in galago. *J Comp Neurol.* 302:425–436.

Connelly WM, Crunelli V, Errington AC (2016). Passive synaptic normalization and input synchrony-dependent amplification of cortical feedback in thalamocortical neuron dendrites. *J Neurosci.* 36:3735–3754.

Constantinople CM, Bruno RM (2013). Deep cortical layers are activated directly by thalamus. *Science.* 80(340):1591–1594.

Crandall SR, Cruikshank SJ, Connors BW (2015). A corticothalamic switch: Controlling the thalamus with dynamic synapses. *Neuron.* 86:768–782.

Dash S, Autio DM, Crandall SR (2022). State-dependent modulation of activity in distinct layer 6 corticothalamic neurons in barrel cortex of awake mice. *J Neurosci.* 42:6551–6565.

Denman DJ, Contreras D (2015). Complex effects on in vivo visual responses by specific projections from mouse cortical layer 6 to dorsal lateral geniculate nucleus. *J Neurosci.* 35:9265–9280.

Durkin J, Suresh AK, Colbath J, Broussard C, Wu J, Zochowski M, Aton SJ (2017). Cortically coordinated NREM thalamocortical oscillations play an essential, instructive role in visual system plasticity. *Proc Natl Acad Sci U S A.* 114:10485–10490.

Erişir A, Van Horn SC, Sherman SM (1997). Relative numbers of cortical and brainstem inputs to the lateral geniculate nucleus. *Proc Natl Acad Sci U S A.* 94:1517–1520.

Fitzpatrick D, Usrey WM, Schofield BR, Einstein G (1994). The sublaminar organization of corticogeniculate neurons in layer 6 of macaque striate cortex. *Vis Neurosci.* 11:307–315.

Frandolig JE, Matney CJ, Lee K, Kim J, Chevée M, Kim SJ, Bickert AA, Brown SP (2019). The synaptic organization of layer 6 circuits reveals inhibition as a major output of a neocortical sublamina. *Cell Rep.* 28:3131–3143.

Golshani P, Liu XB, Jones EG (2001). Differences in quantal amplitude reflect GLuR4-subunit number at corticothalamic synapses on two populations of thalamic neurons. *Proc Natl Acad Sci U S A.* 98:4172–4177.

Guillery RW (1966). A study of Golgi preparations from the dorsal lateral geniculate nucleus of the adult cat. *J Comp Neurol.* 128:21–49.

Guo W, Clause AR, Barth-Maron A, Polley DB (2017). A corticothalamic circuit for dynamic switching between feature detection and discrimination. *Neuron.* 95:180–194.

Harvey AR (1978). Characteristics of corticothalamic neurons in area 17 of the cat. *Neurosci Lett.* 7:177–181.

Hasse JM, Bragg EM, Murphy AJ, Briggs F (2018). Morphological heterogeneity among corticogeniculate neurons in ferrets: quantification and comparison with a previous report in macaque monkeys. *J Comp Neurol.* 527: 546–557.

Hasse JM, Briggs F (2017). Corticogeniculate feedback sharpens the temporal precision and spatial resolution of visual signals in the ferret. *Proc Natl Acad Sci U S A.* 114:E6222–E6230.

Hoerder-Suabedissen A, Hayashi S, Upton L, Nolan Z, Casas-Torremocha D, Grant E, Viswanathan S, Kanold PO, Clasca F, Kim Y, Molnár Z (2018). Subset of cortical layer 6b neurons selectively innervates higher order thalamic nuclei in mice. *Cereb Cortex (New York, NY).* 28:1882–1897.

Homma NY, Happel MFK, Nodal FR, Ohl FW, King AJ, Bajo VM (2017). A role for auditory corticothalamic feedback in the perception of complex sounds. *J Neurosci.* 37:6149–6161.

Hubel DH, Wiesel TN (1962). Receptive fields, binocular interaction and functional architecture in the cat's visual cortex. *J Physiol.* 160:106–154.

Ichida JM, Casagrande VA (2002). Organization of the feedback pathway from striate cortex (V1) to the lateral geniculate nucleus (LGN) in the owl monkey (aotus trivirgatus). *J Comp Neurol.* 454: 272–283.

Jones EG (1985). *The Thalamus.* Springer.

Jurgens CWD, Bell KA, McQuiston AR, Guido W (2012). Optogenetic stimulation of the corticothalamic pathway affects relay cells and GABAergic neurons differently in the mouse visual thalamus. *PLoS ONE.* 7:e45717.

Kakei S, Na J, Shinoda Y (2001). Thalamic terminal morphology and distribution of single corticothalamic axons originating from layers 5 and 6 of the cat motor cortex. *J Comp Neurol.* 437:170–185.

Katz LC (1987). Local circuitry of identified projection neurons in cat visual cortex brain slices. *J Neurosci.* 7:1223–1249.

Killackey HP, Sherman SM (2003). Corticothalamic projections from the rat primary somatosensory cortex. *J Neurosci.* 23:7381–7384.

Kim J, Matney CJ, Blankenship A, Hestrin S, Brown SP (2014). Layer 6 corticothalamic neurons activate a cortical output layer, layer 5a. *J Neurosci*. 34:9656–9664.

Kirchgessner MA, Franklin AD, Callaway EM (2020). Context-dependent and dynamic functional influence of corticothalamic pathways to first- and higher-order visual thalamus. *Proc Natl Acad Sci U S A*. 117:13066–13077.

Kwegyir-Afful EE, Simons DJ (2009). Subthreshold receptive field properties distinguish different classes of corticothalamic neurons in the somatosensory system. *J Neurosci*. 29:964–972.

Landisman CE, Connors BW (2007). VPM and POm nuclei of the rat somatosensory thalamus: Intrinsic neuronal properties and corticothalamic feedback. *Cereb Cortex*. 17:2853–2865.

Li J, Guido W, Bickford ME (2003). Two distinct types of corticothalamic EPSPs and their contribution to short-term synaptic plasticity. *J Neurophysiol*. 90:3429–3440.

Llano DA, Sherman SM (2009). Differences in intrinsic properties and local network connectivity of identified layer 5 and layer 6 adult mouse auditory corticothalamic neurons support a dual corticothalamic projection hypothesis. *Cereb Cortex*. 19:2810–2826.

Lu SM, Guido W, Sherman SM (1992). Effects of membrane voltage on receptive field properties of lateral geniculate neurons in the cat: Contributions of the low-threshold Ca2+ conductance. *J Neurophysiol*. 68:2185–2198.

McAlonan K, Cavanaugh J, Wurtz RH (2006). Attentional modulation of thalamic reticular neurons. *J Neurosci*. 26:4444–4450.

McClurkin JW, Optican, LM, Richmond BJ (1994). Cortical feedback increases visual information transmitted by monkey parvocellular lateral geniculate nucleus neurons. *Vis Neurosci*. 11:601–617.

McCormick DA, von Krosigk M (1992). Corticothalamic activation modulates thalamic firing through glutamate "metabotropic" receptors. *Proc Natl Acad Sci*. 89:2774–2778.

Murphy AJ, Shaw L, Hasse JM, Goris RLT, Briggs F (2021). Optogenetic activation of corticogeniculate feedback stabilizes response gain and increases information coding in LGN neurons. *J Comput Neurosci*. 49:259–271.

Murphy PC, Sillito AM (1996). Functional morphology of the feedback pathway from area 17 of the cat visual cortex to the lateral geniculate nucleus. *J Neurosci*. 16:1180–1192.

Olsen SR, Bortone DS, Adesnik H, Scanziani M (2012). Gain control by layer six in cortical circuits of vision. *Nature*. 483:47–52.

Prieto JJ, Winer JA (1999). Layer VI in cat primary auditory cortex: Golgi study and sublaminar origins of projection neurons. *J Comp Neurol*. 404:332–358.

Przybyszewski AW, Gaska JP, Foote W, Pollen DA (2000). Striate cortex increases contrast gain of macaque LGN neurons. *Vis Neurosci*. 17:485–494.

Robson JA (1983). The morphology of corticofugal axons to the dorsal lateral geniculate nucleus in the cat. *J Comp Neurol*. 216:89–103.

Schwartz ML, Dekker JJ, Goldman-Rakic PS (1991). Dual mode of corticothalamic synaptic termination in the mediodorsal nucleus of the rhesus monkey. *J Comp Neurol*. 309:289–304.

Sherman SM, Guillery RW (1998). On the actions that one nerve cell can have on another: Distinguishing "drivers" from "modulators." *Neurobiology*. 95:7121–7126.

Sillito AM, Jones HE, Gerstein GL, West DC (1994). Feature-linked synchronization of thalamic relay cell firing induced by feedback from the visual cortex. *Nature*. 369:479–482.

Sirota MG, Swadlow HA, Beloozerova IN (2005). Three channels of corticothalamic communication during locomotion. *J Neurosci*. 25:5915–5925.

Steriade M (2003). The corticothalamic system in sleep. *Front Biosci*. 8:878–899.

Stoelzel CR, Bereshpolova Y, Alonso J-M, Swadlow HA (2017). Axonal conduction delays, brain state, and corticogeniculate communication. *J Neurosci*. 37:6342–6358.

Suga N, Ma X (2003). Multiparametric corticofugal modulation and plasticity in the auditory system. *Nat Rev Neurosci*. 410 4:783–794.

Swadlow HA, Weyand TG (1987). Corticogeniculate neurons, corticotectal neurons, and suspected interneurons in visual cortex of awake rabbits: Receptive-field properties, axonal properties, and effects of EEG arousal. *J Neurophysiol*. 57: 977–1001.

Tanaka YR, Tanaka YH, Konno M, Fujiyama F, Sonomura T, Okamoto-Furuta K, Kameda H, Hioki H, Furuta T, Nakamura KC, Kaneko T (2011). Local connections of excitatory neurons to corticothalamic neurons in the rat barrel cortex. *J Neurosci*. 31:18223–18236.

Tang J, Yang W, Suga N (2012). Modulation of thalamic auditory neurons by the primary auditory cortex. *J Neurophysiol*. 108:935–942.

Temereanca S, Simons DJ (2004). Functional topography of corticothalamic feedback enhances thalamic spatial response tuning in the somatosensory whisker/barrel system. *Neuron*. 41:639–651.

Tsumoto T, Creutzfeldt OD, Legendy CR (1978). Functional organization of the corticofugal system from visual cortex to lateral geniculate nucleus in the cat. *Exp Brain Res*. 32:345–364.

Tsumoto T, Suda K (1980). Three groups of cortico-geniculate neurons and their distribution in binocular and monocular segments of cat striate cortex. *J Comp Neurol*. 193:223–236.

Usrey WM, Fitzpatrick D (1996). Specificity in the axonal connections of layer VI neurons in tree shrew striate cortex: Evidence for distinct granular and supragranular systems. *J Neurosci*. 16:1203–1218.

Van Horn SC, Sherman SM (2007). Fewer driver synapses in higher order than in first order thalamic relays. *Neuroscience*. 146:463–470.

Von Krosigk M, Monckton JE, Reiner PB, McCormick DA (1999). Dynamic properties of corticothalamic excitatory postsynaptic potentials and thalamic reticular inhibitory postsynaptic potentials in thalamocortical neurons of the guinea-pig dorsal lateral geniculate nucleus. *Neurosci*. 91:7–20.

Wang W, Andolina IM, Lu Y, Jones HE, Sillito AM (2016). Focal gain control of thalamic visual receptive fields by layer 6 corticothalamic feedback. *Cereb Cortex*. 28:267–280.

Weber AJ, Kalil RE, Behan M (1989). Synaptic connections between corticogeniculate axons and interneurons in the dorsal lateral geniculate nucleus of the cat. *J Comp Neurol*. 289:156–164.

Whilden CM, Chevée M, An SY, Brown SP (2021). The synaptic inputs and thalamic projections of two classes of layer 6 corticothalamic neurons in primary somatosensory cortex of the mouse. *J Comp Neurol*. 529:3751–3771.

Winer JA, Diehl JJ, Larue DT (2001). Projections of auditory cortex to the medial geniculate body of the cat. *J Comp Neurol*. 430:27–55.

Zhang L, Jones EG (2004). Corticothalamic inhibition in the thalamic reticular nucleus. *J Neurophysiol*. 91:759–766.

Zhang ZW, Deschênes M (1997). Intracortical axonal projections of lamina VI cells of the primary somatosensory cortex in the rat: A single-cell labeling study. *J Neurosci*. 17:6365.

15.

CORTICOTHALAMIC LAYER 5 PATHWAYS AND SYNAPSES

Rebecca A. Mease and Alexander Groh

INTRODUCTION

The thalamus receives profuse "top-down" connectivity from the deepest cortical layers, layers 5 (L5) and 6 (L6). Although first described by Cajal over a century ago, resolving corticothalamic (CT) interactions between well-defined cortical and thalamic populations has required recent advances in cell-type-specific methods for dissecting the structure and function of neuronal circuits.

This chapter focuses on the distinct subset of CT projections from L5 which provide "driver" synapses to the thalamus, in that the spiking output of L5-recipient thalamic neurons is largely contingent upon this CT input. The presence of these L5 driver inputs is the defining criterion for the influential "higher-order" versus "first-order" terminology introduced by Sherman and colleagues: in brief, higher-order and first-order thalamic regions are driven by L5 or peripheral (e.g., brainstem, retina) drivers, respectively. Both categories of nuclei receive modulatory CT input from L6. Although we contrast L5 and L6 CT properties where illustrative, readers should see **Chapter 14 in Section IV** for a full discussion of L6 CT pathways and **Chapters 11 and 12 in Section IV** for higher-order/first-order and driver/modulator concepts.

L5 neurons targeting thalamus have been specified by soma location ("L5B") or dendritic morphology ("L5 thick-tufted") and are termed *pyramidal tract neurons*, to distinguish this population from L5 intratelencephalic corticofugal neurons projecting to striatum and contralateral cortex rather than thalamus (Harris and Shepherd, 2015). Here, we use "L5 CT" to underscore the contrast with the L6 CT pathway. We emphasize physiological and anatomical studies allowing manipulations and labeling with cell-type-specific resolution. In particular, the *Rbp4*, *Npr3*, and *thy-1* transgenic mouse lines have been critical for cell-type-specific targeting of L5 pathways using deep brain or in vitro electrophysiology methods combined with specific CT manipulations and labeling (Groh et al., 2014; Guo et al., 2020; Kirchgessner et al., 2020; Prasad et al., 2020). However, studies in non-human primates—particularly in the visual system and thalamic targets of prefrontal cortex—have contributed much of the foundational anatomical understanding of L5-recipient thalamic regions and appreciation of these circuits' contribution to higher-level sensory processing and attention (Halassa and Kastner, 2017; Rikhye et al., 2018b).

A powerful motivation to study L5-recipient regions of the thalamus is that L5 CT innervation of the thalamus appears to be a generic principle of vertebrate brain organization. First-order nuclei handling primary sensory input lack L5 innervation, yet these regions have served as models of thalamic information processing for decades. However, L5-recipient higher-order thalamic nuclei (see **Chapters 11 and 12 in Section IV**) make up the majority of the thalamus by volume. Seen in this light, most signals transmitted and transformed by the thalamus are actually internally generated by the cortex itself. Intriguingly, L5 CT innervation appears to be a default developmental scheme, as removing peripheral sensory inputs to thalamus triggers genetic programs promoting L5 CT innervation of first-order thalamic nuclei, and the transcriptomes of L5-recipient regions are more similar across modalities than to their modality-specific first-order counterparts (Frangeul et al., 2016).

Diverse L5 CT pathways have been identified (Figure 15.1), with emerging roles in sensation, perception, and cognition, and several circuits implicated in disease (Mease and Gonzalez, 2021). As this chapter will explore, drive from L5 provides a link between cortical regions via powerful, selective transthalamic pathways, as well as recurrent connectivity important for learning and plasticity. The remainder of this chapter introduces L5 CT connectivity motifs and pathways from different cortical regions (Figures 15.1 and 15.2), and outlines key L5 CT anatomical and physiological properties (Figures 15.3 and 15.4).

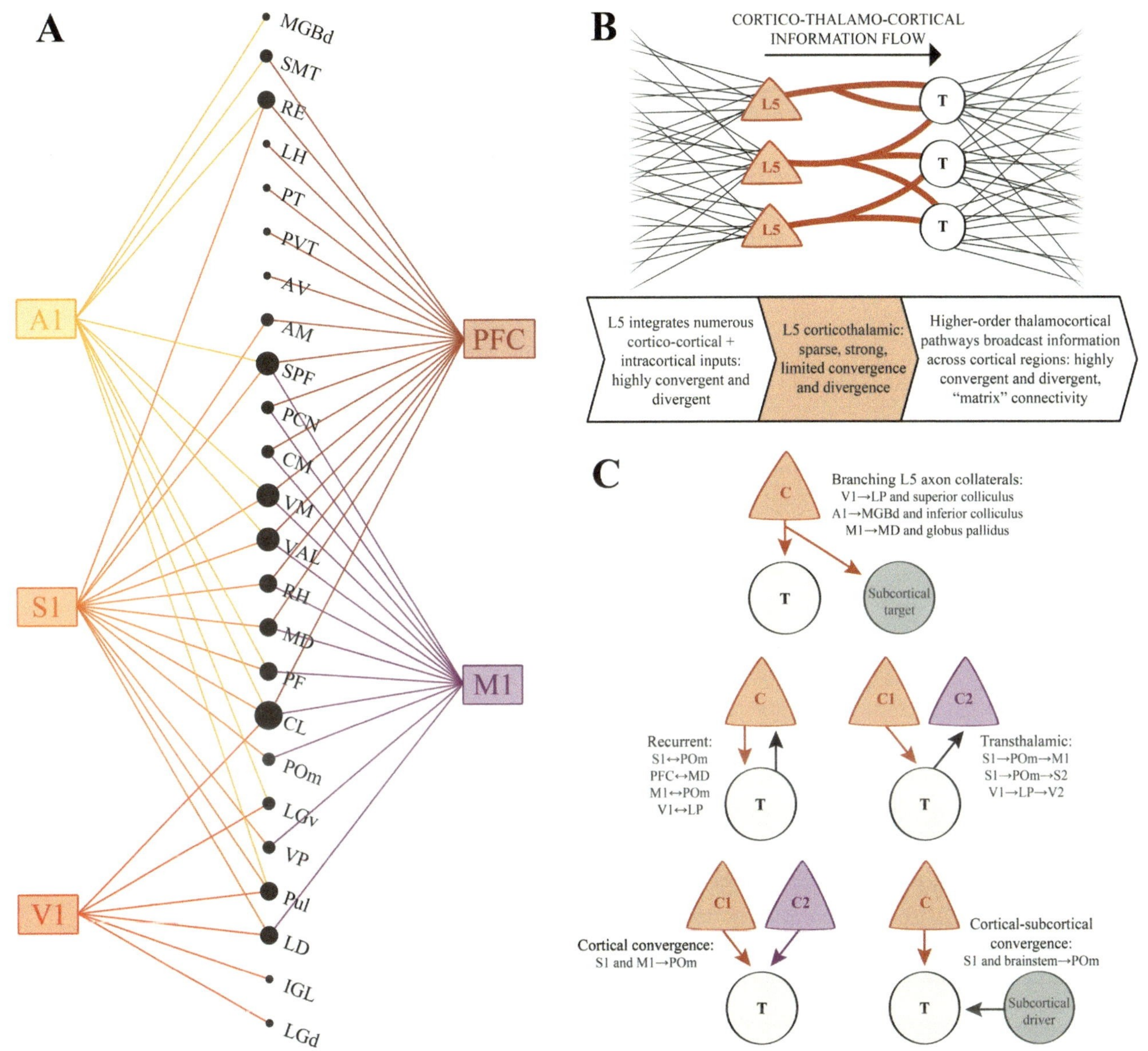

Figure 15.1. L5 corticothalamic pathways for sensation and perception. (A). Simplified overview of L5 corticothalamic projections to higher-order thalamic regions, with connectivity (from Harris et al., 2019; Prasad et al., 2020). Only a minority of pathways have been characterized at the level of synaptic and cellular properties. (B). Cortico-thalamo-cortical cartoon showing (1) convergent intracortical and cortico-cortico input to L5 CT neurons; (2) sparse corticothalamic connectivity with limited convergence and divergence at the level of single higher-order thalamic neurons; and (3) divergent, profuse higher-order thalamocortical projections. (C). Simplified L5 CT connectivity motifs: branching L5 pathways (upper); transthalamic and recurrent input-output motifs (middle); convergence of different L5 CT pathways versus convergence of L5 CT and subcortical drivers (lower).

L5 CORTICOTHALAMIC CIRCUITS AND CONNECTIVITY

Most if not all neocortical regions connect with thalamic targets via L5 CT projections, although detailed structural and functional characterization is uneven so far (Sherman, 2016; Harris et al., 2019; Prasad et al., 2020; Mease and Gonzalez, 2021). As Figure 15.1A illustrates, many thalamic targets are innervated by 2 or more L5 CT sources, but to what extent this implied convergence occurs on the single-neuron level is a topic of ongoing research. Considering how CT connectivity patterns relate to function is particularly important in the context of the thalamus. As there is minimal local processing via recurrent intrathalamic connectivity, the input to a given nucleus—in combination with the intrinsic properties of single relay neurons—will largely dictate the information-processing role of that nucleus.

In the rodent somatosensory system, L5 CT neurons in primary somatosensory cortex (S1) target the posterior medial (POm) nucleus; several recent studies in this system have contributed substantial functional and anatomical data to the current understanding of L5 CT pathways (e.g. Groh et al., 2008; Mo and Sherman, 2019; Casas-Torremocha et al., 2022). In the visual system, the pulvinar, also called lateral posterior nucleus (LP) in rodents, receives L5 CT input from the primary visual cortex (V1) (Bennett et al., 2019; de Souza et al., 2019). In the auditory system, L5 CT neurons in primary auditory cortex (A1) target the dorsal portion of the medial geniculate body (MGBd) (Lee and Sherman, 2011; Sherman, 2017). In primary motor cortex (M1), L5 CT neurons can be categorized by innervation target (Winnubst et al., 2019). One subgroup synapses in the parafascicular (PF) and mediodorsal (MD) thalamic nuclei, and another disjoint subgroup projects to ventral medial nucleus (VM), paracentral nucleus (PCN), and the posterior complex (PO). Yamawaki and Shepherd (2015) and Guo et al. (2020) also report that M1 L5 CT neurons project to the ventrolateral (VL) and PO. Many other L5 CT-recipient thalamic nuclei outside of traditionally sensorimotor regions exist (Figure 15.1), with diverse roles in cognition and pathology. The L5 CT projections from the prefrontal cortex (PFC) are possibly the best-studied to date, targeting "associative" thalamic nuclei—for example, MD, VM, and centrolateral nucleus (CL)—relevant for cognition, decision-making, and attention.

Overall, L5 CT connectivity patterns (Figure 15.1C) vary according to cortical region and subregions of target thalamic nuclei, as well as specific TC projection patterns from higher-order thalamic nuclei (as reviewed in Mease and Gonzalez, 2021; Shepherd and Yamawaki, 2021). L5 CT connectivity appears to be less generalizable than the largely characteristic somatotopic alignment between cortical (pre-) and thalamic (post-) neurons seen in L6 CT connectivity. The following section highlights notable motifs in L5 CT connectivity (Figure 15.1B,C).

SPARSE AND STRONG L5 CORTICOTHALAMIC PROJECTIONS

L5 pyramidal neurons are highly integrative. As the major output conduit for cortical information to subcortical circuits responsible for motor control and behavior (Sherman and Usrey, 2021) (see **Chapter 6 in Section III**), L5 CT neurons' extensive dendrites integrate information from all cortical layers (Figure 15.2A), with output axons that branch minimally within the cortex, and rather funnel collected cortico-cortical and intracortical information to higher-order thalamic nuclei and other subcortical targets via branching axon collaterals. Second, L5 CT pathways connect sparsely but powerfully to thalamic targets, in contrast to diffuse, numerous L6 CT synapses. For example, single S1 L5 CT neurons innervate around 8 postsynaptic POm neurons, and each POm neuron receives fewer than 10 L5 CT inputs, from just 2–3 presynaptic L5 CT neurons (Mease et al., 2016a; Sumser et al., 2017). Similar estimates from pulvinar (Rockland, 2019) support a connectivity scheme of strikingly limited convergence and divergence at the level of single L5 CT neurons and thalamic targets. In the mouse somatosensory whisker system, this scheme preserves the cortical whisker map in POm (Sumser et al., 2017); similarly, in the visual system, V1 CT inputs project a map of visual space onto anterior LP (Bennett et al., 2019).

BRANCHING PATHWAYS FROM L5 TARGET MULTIPLE SUBCORTICAL STRUCTURES

L5 CT neurons form branching collaterals innervating multiple subcortical targets (Figure 15.1C, Figure 15.2B), in contrast to L6 CT projections, which are restricted to the thalamus. This pattern implies that information conveyed to higher-order thalamic nuclei is likely also received by other targets, many of which interact with higher-order

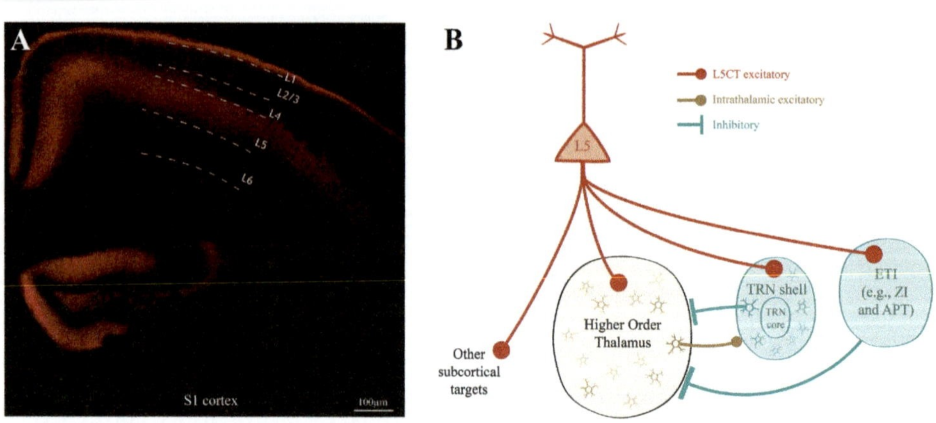

Figure 15.2. Projections from L5 pyramidal neurons target higher-order thalamic nuclei and subcortical inhibitory networks. (A). L5 CT projections from primary sensory cortex (S1) in *Rbp4-cre* mouse line (recolored). (B). Cartoon of L5 CT (red) direct excitatory connectivity to thalamus (beige) and subcortical inhibitory (cyan) networks.

thalamic nuclei and are important for motor functions (e.g., brainstem, superior colliculus). For example, A1 L5 CT populations also innervate the inferior colliculus (IC), but the extent and functional relevance of this branching is not yet known (Asokan et al., 2018; Williamson and Polley, 2019), and subgroups of M1 L5 CT neurons innervate VM and MD thalamic targets as well as nonthalamic globus pallidus and the nucleus of the posterior commissure (Winnubst et al., 2019). Branching L5 CT projections from PFC to MD and VM have been suggested to also coordinate synchronized thalamic spiking across nuclei (Collins et al., 2018). Axonal tracing demonstrates that single L5 CT neurons can innervate many subcortical regions (Veinante et al., 2000; Sherman, 2016; Guo C et al., 2017; Rockland, 2019; Prasad et al., 2020), although there is also evidence that distinct L5 populations favor specific subcortical target combinations (Hattox and Nelson, 2007; Rojas-Piloni et al., 2017). In summary, the same presynaptic L5 message may be sent simultaneously to several subcortical structures, but the local subcortical processing of this information will depend on the specific properties of different L5-subcortical synapses and the intrinsic properties of postsynaptic neurons (Collins et al., 2018).

L5 CORTICO-THALAMO-CORTICAL CONNECTIVITY MOTIFS

As covered in detail in **Chapter 42 in Section VII**, L5 CT pathways are critical legs in *transthalamic* pathways (Figure 15.1C), in which L5 CT information is distributed to the cortex via higher-order thalamic nuclei. These disynaptic L5-thalamus-cortex pathways exist alongside direct cortico-cortical pathways which do not involve the thalamus. Transthalamic signaling via higher-order thalamic nuclei has been demonstrated between primary and secondary cortices in somatosensory, visual, and auditory sensory modalities (Theyel et al., 2010; Lee and Sherman, 2011; Lee, 2015; Sherman, 2017; Bennett et al., 2019; Blot et al., 2021), as well as between S1 and M1 (Mo and Sherman, 2019), and appears to carry information complementary to or distinct from that carried by cortico-cortical pathways (Bennett et al., 2019; Blot et al., 2021). Moreover, some cortical regions which provide L5 CT input to a given thalamic nucleus are in turn innervated by TC projections from that same nucleus. This "closed-loop" structure can support persistent activity (e.g., Guo ZV et al., 2017). Evidence for such *recurrent* connectivity (Figure 15.1C) has been reported in L5 CT pathways from S1, V1, M1, and PFC. For example, Guo et al. (2020) find that in the hindlimb TC system, M1 and S1 target largely segregated, non-overlapping subregions in POm, which are recurrently connected to their cortical partner.

L5 CT DRIVER CONVERGENCE

L5 CT driver convergence is another circuit motif with important functional implications (Figure 15.1C). It is well established that several cortical regions may send L5 CT input to a single higher-order thalamic region (Figure 15.1A), but high-resolution anatomical methods have recently revealed convergence of drivers from different sources onto single thalamic neurons. For example, Sampathkumar et al. (2021) find that half of POm neurons receive convergent driver input from S1 and M1, while M1 input was never observed in isolation. Moreover, subdivisions of higher-order thalamic nuclei can receive driving input from both cortical and non-cortical sources, blurring the strict first- versus higher-order categorization based on driving inputs. In POm, single neurons receive both S1 L5 CT and brainstem driver inputs and can functionally encode temporal lags between cortical and brainstem signals (Groh et al., 2014). In a comprehensive anatomical survey, Casas-Torremocha et al. (2022) show that most of POm receives sensory and/or motor L5 CT input, but innervation by subcortical sources is heterogeneous, resulting in functionally diverse convergence zones. In the motor thalamus, non-canonical drivers from cerebellum may be integrated with L5 CT input (Bickford, 2015).

In sum, several lines of evidence support segregation of L5 CT recipient nuclei into functionally distinct subnetworks defined by different combinations of input driver sources and TC innervation patterns. Overall, driver convergence is an important motif because it suggests that single thalamic neurons can combine previously segregated sources of information and potentially do useful computations on these inputs on single-neuron timescales (i.e., tens of milliseconds).

SYNAPTIC AND CELLULAR PROPERTIES OF L5 CT PATHWAYS

Information from L5 reaches the thalamus via sparse but powerful excitatory synapses that undergo rapid use-dependent depression (Figure 15.3). These properties are largely consistent across sensory modalities and L5 CT circuits, suggesting that L5 drive of thalamic relay neurons is tuned for adaptive and temporally precise transmission of cortical information to the thalamus. This functional characteristic arises from synergy between distinctive synaptic and intrinsic properties, as discussed below.

STRUCTURE

As characteristic drivers, L5 CT synapses are structurally and functionally homologous to synapses carrying information from the sensory periphery. L5 CT synapses are

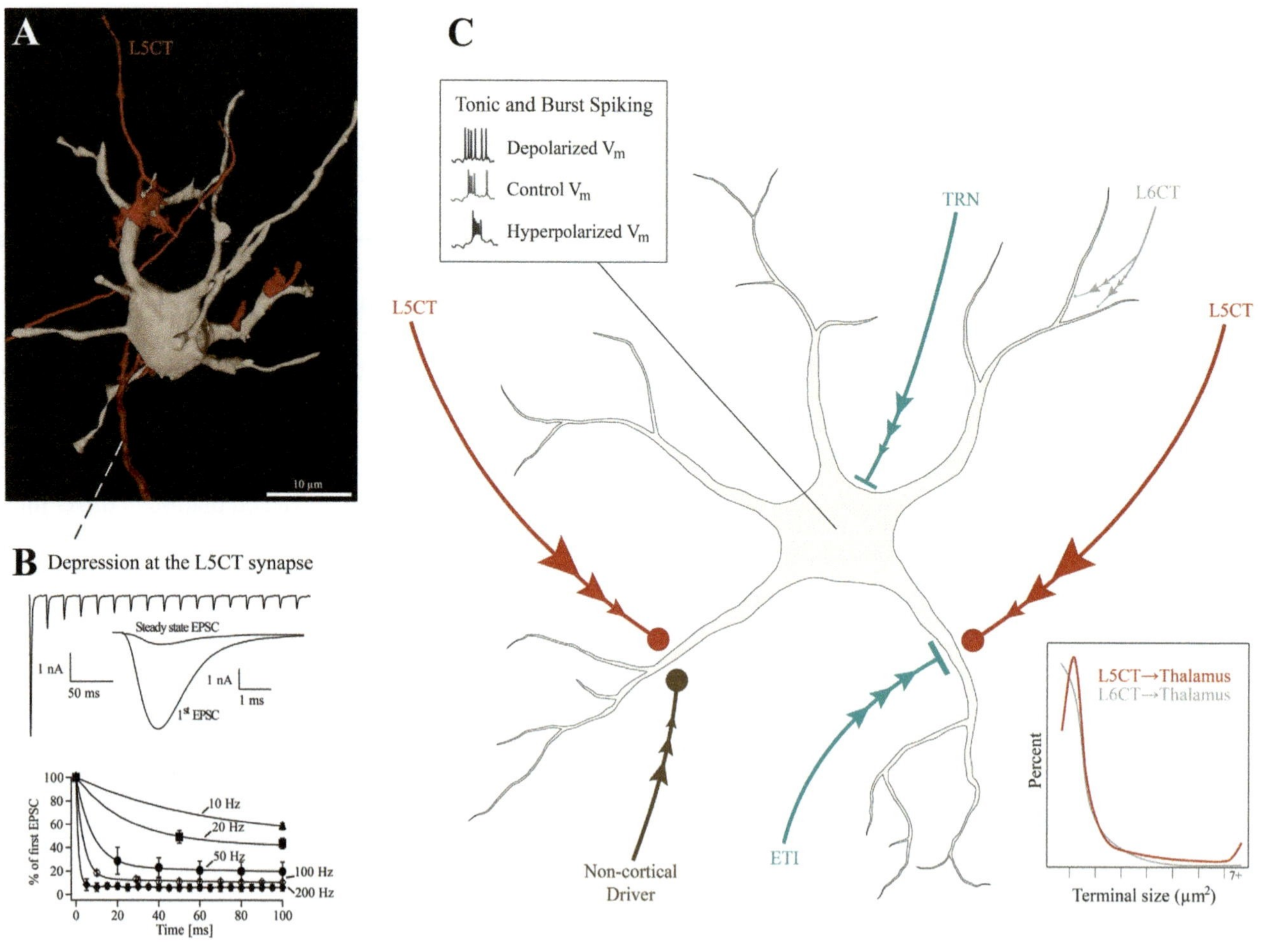

Figure 15.3. Organization and dynamics of L5 CT inputs to higher-order thalamic neurons. (A). Reconstruction of a neuron located in the posteromedial (POm) nucleus: corticothalamic giant terminals (red) formed between L5 CT pyramidal neurons and a POm neuron (beige). Adapted from Groh et al. (2008). (B). Upper: voltage-clamp experimental data showing pronounced synaptic depression from high-frequency single fiber stimulation. Lower: depression as a function of stimulation frequency. Adapted from Groh et al. (2008). (C). Schematic of synaptic inputs to high-order thalamic neurons. Arrows: L5 CT drivers, noncortical drivers, and TRN synapses depress; incertothalamic inhibitory synapses do not, and L6 CT synapses facilitate. Lower box: cartoon of relative size distributions of L5 CT and L6 CT terminals, data replotted and simplified from Prasad et al. (2020). Upper box: burst and tonic spiking modes depend on membrane potential; unpublished simulated data from the authors.

located on proximal dendrites of thalamic neurons (Figure 15.3A,C) and thus electrotonically close to the soma (Hoogland et al., 1991; Bourassa et al., 1995; Deschênes et al., 1998; Rouiller and Welker, 2000), meaning that inputs experience little passive attenuation.

L5 CT synapses are typically much larger and scarcer than L6 CT synapses, but vary significantly in size (0.5–10+ μm; Figure 15.3C). For example, in the mouse, L5 CT synapses from V1 and PFC are smaller than synapses from S1 (Prasad et al., 2020). However, subpopulations of "giant" terminals (size >2 μm) exist in auditory, visual, and somatosensory thalamic nuclei (Llano and Sherman, 2008; Hoerder-Suabedissen et al., 2018; Prasad et al., 2020). Such giant boutons contain multiple synapses (Hoerder-Suabedissen et al. 2018) and develop via activity-dependent presynaptic mechanisms (Hayashi et al., 2021) (see **Chapter 53 in Section IX**).

PHYSIOLOGY

A single L5 CT fiber may provide only a handful of boutons per thalamic neuron, but still elicits powerful excitatory input drive due to a high probability of glutamate release and a high postsynaptic density of AMPA and NMDA receptors (Hoogland et al., 1991; Li et al., 2003a; Reichova and Sherman, 2004; Groh et al., 2008). Indeed, stimulation of a single L5 fiber can evoke unitary excitatory postsynaptic potentials (EPSPs) greater than 10 mV in response to large (~1 nA in mouse) excitatory postsynaptic currents (EPSCs). These large receptor potentials are further amplified by low-threshold transient calcium currents which support thalamic bursting (Li et al., 2003a; Groh et al., 2008). Notably, EPSCs have markedly fast kinetics due to the postsynaptic GLUA4 subunit (reviewed in Mease and Gonzalez, 2021).

However, the driver strength of L5 CT inputs is highly plastic (Figure 15.3B), as fast depletion of releasable vesicles leads to fast and extreme frequency-dependent synaptic depression (Reichova and Sherman, 2004; Groh et al., 2008). The combination of large, temporally precise unitary inputs with synaptic depression means that L5 presynaptic firing patterns can be boosted or attenuated in the thalamus, dependent on previous activity history. While direct physiological measurements are lacking from many pathways, in general, temporally isolated inputs can be magnified into output bursts of thalamic spikes, but input frequencies greater than 2 Hz lead to pronounced depression, with a halving of synaptic strength at frequencies around 20 Hz (reviewed in Mease and Gonzalez, 2021).

L5 ENGAGES SUBCORTICAL INHIBITION OF THALAMUS

L5 CT projections are mediators of both direct excitation and several sources of inhibition to higher-order thalamic nuclei (Figure 15.2B, Figure 15.3C).

Extrathalamic Inhibition

In addition to targeting the thalamus, L5 projections branch to diverse subcortical regions, including extrathalamic inhibitory (ETI) regions (Figure 15.2B) which provide powerful, precise inhibition to higher-order—but not first-order—thalamic nuclei (Halassa and Acsady, 2016) (see **Chapters 11 and 12 in Section IV**). This L5 recruitment of ETI regions such as zona incerta (ZI) and anterior pretectum (APT) means that whatever strong excitation descending L5 drive provides to thalamus will be followed by inhibition in rapid succession. Notably, large GABAergic ETI terminals (Figure 15.3C) collocate closely with giant L5 CT terminals and support stable inhibition even at high frequencies, suggesting tight inhibitory control over descending cortical driving excitation.

L5-TRN Synapses

Until recently, it was thought that L5 neurons did not synapse in the thalamic reticular nucleus (TRN). Indeed, TRN was thought to be recruited solely by thalamic relay neurons and collaterals of L6 CT projections (see **Chapter 16 in Section IV**). However, recent studies (Prasad et al., 2020; Carroll et al., 2022; Hádinger et al., 2022) have revealed that L5 CT neurons in many cortical regions make large synapses in the same TRN region (TRNshell; Figure 15.2B) that inhibits—and is excited by—higher-order thalamic nuclei (Li et al., 2020; Martinez-Garcia et al., 2020) (Figure 15.2) (see **Chapters 11 and 12 in Section IV**). The exact cortical origin of these synapses is still being resolved. Prasad et al. (2020) and Carroll et al. (2022) report that L5 CT pathways from S1, V1, M1, and PFC synapse in TRN, while Hádinger et al. (2022) find that only L5 CT neurons in PFC project to TRN. While the full functional import of this newly identified L5-mediated inhibition of higher-order thalamic nuclei is not fully understood, Hádinger et al. (2022) find that the L5 CT pathway from PFC serves to broadcast cortical synchrony to TRN.

FUNCTIONS OF L5 CT PATHWAYS

L5 CT PATHWAYS LINK SENSORY AND MOTOR SYSTEMS

A potential substrate for sensorimotor interactions is that subset of higher-order thalamic neurons convergent driver input from sensorimotor cortical regions as well as subcortical areas controlling motor output (see **Chapters 11 and 12 in Section IV**). For example, the concept of "super-integration" of information from the cortex, basal ganglia, and cerebellum was highlighted in the motor thalamus (reviewed in Bickford, 2015). However, reconciling reported anatomy and function is an ongoing task—for example, projections from V1 to LP overlap with terminals from SC (Li et al., 2003b; Masterson et al., 2009), but Bennett et al. (2019) suggest that SC- and V1-driven thalamic circuits are largely distinct. Notably, anatomical evidence suggests that POm neurons could integrate motor and somatosensory signals via convergent driver input from S1 and M1 L5 CT neurons (Sampathkumar et al., 2021).

As framed by Sherman (2016) and Sherman and Usrey (2021), this arrangement allows the cortex to influence behavior and motor output by exercising fast control over evolutionarily ancient midbrain and brainstem structures and simultaneously receive an efference copy of these commands via L5 CT projections to higher-order thalamic nuclei. Indeed, one function of L5 may be to extract immediately salient and "actionable" results of cortical computations within an information band compatible with input to subcortical structures, enabling cortical computations to influence behavior within a useful time frame.

L5 CT PATHWAYS DRIVE HIGHER-ORDER THALAMIC NUCLEI

In vivo, spiking in higher-order thalamic nuclei depends strongly on L5 CT input, as shown by cortical activation (Figure 15.4) and inactivation in several modalities (Diamond et al., 1992; Groh et al., 2014; Mease et al., 2016a; Bennett et al., 2019; Kirchgessner et al., 2020; Jung et al., 2022). Functionally, this dependence on L5 means that higher-order thalamic neurons are less tightly tuned

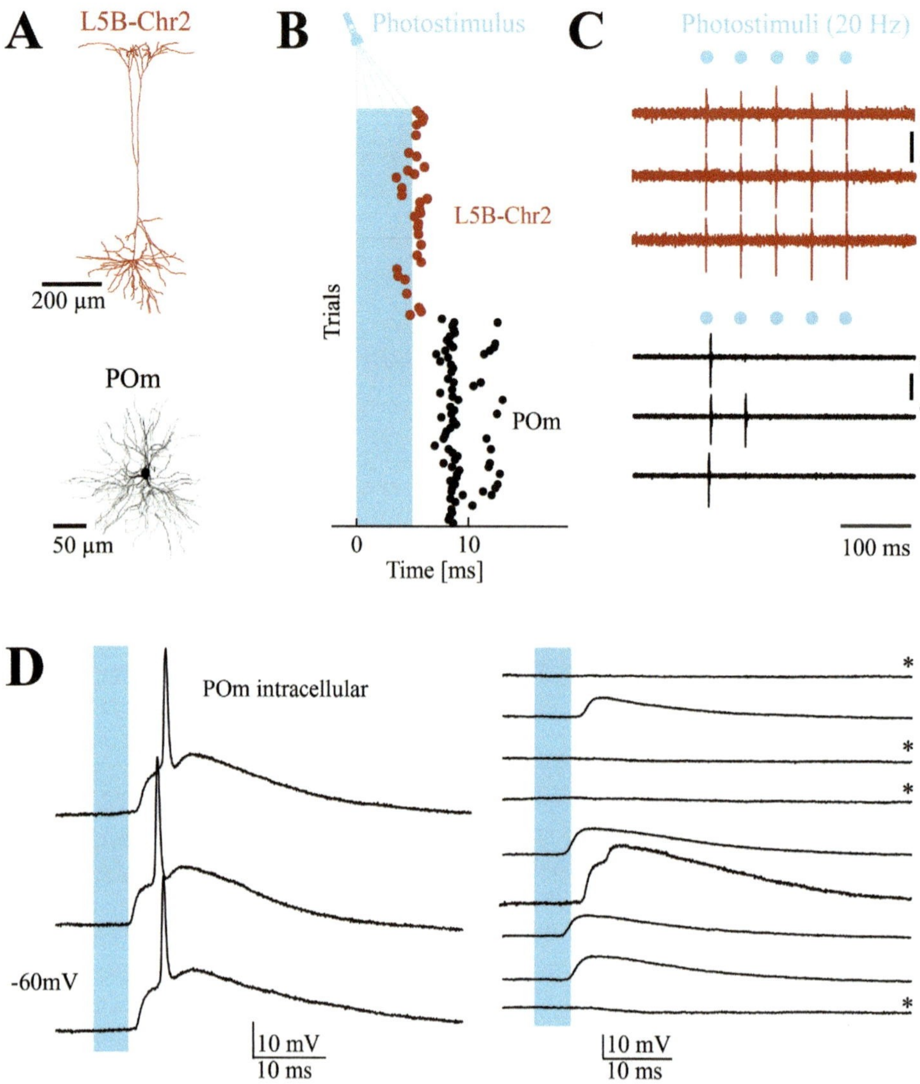

Figure 15.4. Optogenetic stimulation of the L5 CT pathway monosynaptically drives higher-order thalamic nuclei in vivo. (A). Reconstructed L5 CT (barrel cortex) and higher-order thalamic (POm) neurons in mice. (B–D). Electrophysiological recordings in a head-fixed anesthetized mouse expressing ChR2 in L5 CT neurons (thy-1 line). (B). Raster plots of simultaneous L5 CT (upper) and POm (lower) juxtasomal recordings during optogenetic stimulation of L5 CT show short-latency L5 CT responses and POm responses. (C) Simultaneous entrainment of L5 CT neuron to 20 Hz stimulation (upper) and phasic response of POm neuron (lower), high-pass filtered voltage signals from juxtasomal recordings. (D). Whole-cell patch clamp recording from POm neuron during minimal optogenetic stimulation of L5 CT neurons shows suprathreshold (left), and subthreshold EPSPs (right) interspersed with failure trials (*). Adapted from Mease et al. (2016a).

to primary sensory inputs, and thus show broader sensory tuning in comparison to those of first-order nuclei (Moore et al., 2015; Urbain et al., 2015; Mease et al., 2016b; Bennett et al., 2019; Williamson and Polley, 2019; Qi et al., 2022), instead "inheriting" the receptive fields of L5 CT inputs. Given the driver properties of L5 CT synapses and sparsity of this pathway (Sumser et al., 2017; Rockland, 2019), single higher-order thalamic neurons are effectively driven by just a handful of active presynaptic neurons (Groh et al., 2008; Mease et al., 2016a) and can therefore transmit the activity of very small L5 CT populations. This finding may have implications for transthalamic communication between cortical regions, such as propagation of slow-wave activity (Stroh et al., 2013). Notably, in nonhuman primates, the pulvinar has a causal role in transthalamic synchronization of remote cortical regions, rerouting cortical information on-the-fly, based on attentional demands (Halassa and Kastner, 2017). Moreover, in the context of converging driver inputs, higher-order thalamic neurons could act as single-neuron integrators of sparse activity in different cortical and subcortical regions.

As Figure 15.4 illustrates, specific optogenetic activation of L5 CT neurons in vivo leads to reliable short-latency POm spiking via unitary EPSPs, but only phasically, as high-frequency stimulation of L5 only transiently activates POm (Figure 15.4C), suggesting that depression of the L5 CT synapses and intrinsic neuronal adaptation also sculpts CT information flow in vivo. The strong drive of higher-order thalamic nuclei by L5 CT projections, in combination with strong synaptic depression and intrinsic thalamic bursting, prompts the question of exactly what patterns of information present in the complex spike trains of L5 neurons are transmitted and transformed via the L5 CT pathway. In vitro, higher-order thalamic spike trains preferentially encode coincident activation of separate L5 CT fibers (Groh et al., 2008), and can transmit information in separate high- and low-frequency bands (see **Chapter 60 in Section XI**), suggested to correspond to bursting and synchrony in presynaptic L5 CT spike trains (Mease et al., 2017). These encoding properties suggest that higher-order thalamic neurons could potentially route distinct types of information via different TC pathways, but these observations have yet to be tested in vivo.

RELEVANCE OF L5 CT PATHWAYS FOR LEARNING AND COGNITION

The emerging importance of higher-order thalamic nuclei in myriad cognitive tasks (Halassa and Kastner, 2017; Wolff et al., 2020) underscores the importance of L5 CT control of these regions. L5 CT neurons themselves are central players in theories of cortical function and have been suggested as a neuronal substrate for alignment of internal context and novel external information via coincidence detection, and have been linked to conscious perception (Aru et al., 2020). Moreover, higher-order TC synapses are critical loci for learning and plasticity in different sensory systems (reviewed in Mease and Gonzalez, 2021) (see **Chapter 13 in Section IV**). However, studies holistically examining L5 CT and higher-order TC information flow together are sparse. Nonetheless, several key observations highlight L5 CTC circuits as critical to cognitive processes beyond sensory processing.

First, higher-order "associative" thalamic regions have been implicated in cognition, decision-making, and attention (Wolff et al., 2020) (see **Chapters 45, 46, and 48 in Section VIII; Chapter 54 in Section IX**). Such nuclei may orchestrate cortical representations important for timely cognitive tasks by tuning functional connectivity (Halassa and Kastner, 2017; Nakajima and Halassa, 2017). MD is perhaps the best-studied example and receives strong L5 CT input from PFC (Bolkan et al., 2017; Schmitt et al., 2017; Alcaraz et al., 2018; Rikhye et al., 2018a); disruptions in PFC-MD communication may contribute to cognitive and behavioral deficits in schizophrenia, Alzheimer's disease, and pain disorders (reviewed in Mease and Gonzalez, 2021; see also **Section XII of this volume**). Overall, PFC-MD interactions appear to promote cognitive flexibility by supporting switches in cortical ensemble representations of task rules in rodents (Schmitt et al., 2017; Rikhye et al., 2018a) and rule-dependent shifts in attention in humans (Halassa and Kastner, 2017).

Finally, converging evidence from the somatosensory and auditory systems (reviewed in Mease and Gonzalez, 2021) suggests that higher-order TC projections carry signals about stimulus saliency and behavioral context, and also exhibit plasticity during associative learning; TC modulation of dendritic plateau potentials and recruitment of cortical inhibitory networks play key roles in these processes (see **Chapter 13 in Section IV**). Speculatively, the strong link between the cortex and thalamus via L5 means that such learning signals could arise from thalamic computation on L5 spiking patterns. In a direct demonstration of L5 CT pathways' potential contribution to such processes, Qi et al. (2022) found that mouse performance in a texture discrimination task is largely supported by POm, and moreover, this function depends on S1 L5 CT input. Overall, more precise delineation of L5 CT's role in conveying saliency information may help harmonize the various roles attributed to higher-order thalamic nuclei (i.e., in pain, arousal, and attention).

CONCLUSIONS: L5 CT PATHWAYS AS FUTURE TARGETS FOR RESEARCH

Our current understanding of the thalamus has evolved rapidly over the past decade, thanks largely to a resurgence of interest in the thalamus as an embedded participant in cortical dynamics, rather than a passive relay. In this chapter, we have emphasized the viewpoint that knowledge of L5 CT connectivity and functional properties is critical to understand how the majority of the thalamus functions. Thanks to technological advances in targeting specific neuronal pathways and recording from deep brain structures, we have begun to resolve some potentially generalizable rules about L5 CT's participation in thalamocortical dynamics, such as robust L5 CT transfer of cortical information, engagement of strong subcortical inhibition of higher-order thalamic nuclei, and separate modulatory and driving functions for reciprocal and transthalamic innervation of cortex. Nonetheless, many aspects of L5 CT organization and function remain unknown.

Moreover, a general theory of how L5 CT pathways contribute to brain activity dynamics is lacking. This is not surprising, given the diverse nonlinear single-neuron and synaptic properties of these networks, and complex connectivity with subcortical regions, some of which we have discussed here. Overall, the synaptic properties of the L5 CT pathway and the complex, bi-stable excitability of thalamic neurons suggest highly selective information-processing capabilities. Further efforts to gain mechanistic insight into these circuits is necessary for understanding sensory processing and perception in health and disease. Moreover, untangling cortico-thalamo-cortical function at the level of computation is a grand opportunity for neuroscientists to learn from a conserved circuit architecture that has been elaborated on during evolution for different niches. From an evolutionary standpoint, the neocortex—and thus L5 neurons—are relatively late additions to the brain. The anatomical funneling of information via L5 suggests dimensionality reduction in CT information transfer for efficient redistribution back to cortex, and in parallel, a privileged messenger function for L5 to coordinate distributed control of phylogenetically older subcortical circuits.

In closing, we would like to highlight the relatively unexplored yet rich computation performed in L5 CT pathways as a fruitful focus for future research. Fast distribution of cortical signals via L5 and processing in the thalamus appears to be a core function of mammalian brains. Identifying the computational primitives is central to understanding brain function and dysfunction, and could potentially inform biologically inspired artificial information processing.

ACKNOWLEDGMENTS

We thank Heather Lou Mease for figure preparation and Antonio Gonzalez for histology in Figure 15.2. RAM is supported by grants from the Chica-Heinz-Schaller Foundation and Brigitte-Schlieben-Lange Program of Baden-Württemberg Ministry of Science.

REFERENCES

Alcaraz F, Fresno V, Marchand AR, Kremer EJ, Coutureau E, Wolff M (2018). Thalamocortical and corticothalamic pathways differentially contribute to goal-directed behaviors in the rat. *eLife.* 7:e32517.

Aru J, Suzuki M, Larkum ME (2020). Cellular mechanisms of conscious processing. *Trends Cogn Sci.* 24:814–825.

Asokan MM, Williamson RS, Hancock KE, Polley DB (2018). Sensory overamplification in layer 5 auditory corticofugal projection neurons following cochlear nerve synaptic damage. *Nat Commun.* 9:1762–1770.

Bennett C, Gale SD, Garrett ME, Newton ML, Callaway EM, Murphy GJ, Olsen SR (2019). Higher-order thalamic circuits channel parallel streams of visual information in mice. *Neuron.* 102:477–492.

Bickford ME (2015). Thalamic circuit diversity: Modulation of the driver/modulator framework. *Front Neural Circuits.* 9:86.

Blot A, Roth MM, Gasler I, Javadzadeh M, Imhof F, Hofer SB (2021). Visual intracortical and transthalamic pathways carry distinct information to cortical areas. *Neuron.* 109:1996–2008. http://dx.doi.org/10.1016/j.neuron.2021.04.017.

Bolkan SS, Stujenske JM, Parnaudeau S, Spellman TJ, Rauffenbart C, Abbas AI, Harris AZ, Gordon JA, Kellendonk C (2017). Thalamic projections sustain prefrontal activity during working memory maintenance. *Nat Neurosci.* 20:987–996.

Bourassa J, Pinault D, Deschênes M (1995). Corticothalamic projections from the cortical barrel field to the somatosensory thalamus in rats: a single-fibre study using biocytin as an anterograde tracer. *Eur J Neurosci.* 7:19–30.

Carroll BJ, Sampathkumar V, Kasthuri N, Sherman SM (2022). Layer 5 of cortex innervates the thalamic reticular nucleus in mice. *Proc Natl Acad Sci U S A.* 119:e2205209119.

Casas-Torremocha D, Rubio-Teves M, Hoerder-Suabedissen A, Hayashi S, Prensa L, Molnár Z, Porrero C, Clasca F (2022). A combinatorial input landscape in the "higher-order relay" posterior thalamic nucleus. *J Neurosci.* 42:7757–7781. http://dx.doi.org/10.1523/JNEUROSCI.0698-22.2022.

Collins DP, Anastasiades PG, Marlin JJ, Carter AG (2018). Reciprocal circuits linking the prefrontal cortex with dorsal and ventral thalamic nuclei. *Neuron.* 98:366–379.

Deschênes M, Veinante P, Zhang ZW (1998). The organization of corticothalamic projections: Reciprocity versus parity. *Brain Res Rev.* 28:286–308.

de Souza BOF, Cortes N, Casanova C (2019). Pulvinar modulates contrast responses in the visual cortex as a function of cortical hierarchy. *Cereb Cortex.* 30:1068–1086.

Diamond ME, Armstrong-James M, Budway MJ, Ebner FF (1992). Somatic sensory responses in the rostral sector of the posterior group (POm) and in the ventral posterior medial nucleus (VPM) of the rat thalamus: dependence on the barrel field cortex. *J Comp Neurol.* 319:66–84.

Frangeul L, Pouchelon G, Telley L, Lefort S, Luscher C, Jabaudon D (2016). A cross-modal genetic framework for the development and plasticity of sensory pathways. *Nature.* 538:96–98.

Groh A, Bokor H, Mease RA, Plattner VM, Hangya B, Stroh A, Deschenes M, Acsady L (2014). Convergence of cortical and sensory driver inputs on single thalamocortical cells. *Cereb Cortex.* 24:3167–3179. http://dx.doi.org/10.1093/cercor/bht173.

Groh A, de Kock CPJ, Wimmer VC, Sakmann B, Kuner T (2008). Driver or coincidence detector: modal switch of a corticothalamic giant synapse controlled by spontaneous activity and short-term depression. *J Neurosci.* 28:9652–9663.

Guo C, Peng J, Zhang Y, Li A, Li Y, Yuan J, Xu X, Ren M, Gong H, Chen S (2017). Single-axon level morphological analysis of corticofugal projection neurons in mouse barrel field. *Sci Rep.* 7:2846.

Guo K, Yamawaki N, Barrett JM, Tapies M, Shepherd GMG (2020). Cortico-thalamo-cortical circuits of mouse forelimb S1 are organized primarily as recurrent loops. *J Neurosci.* 40:2849–2858.

Guo ZV, Inagaki HK, Daie K, Druckmann S, Gerfen CR, Svoboda K (2017). Maintenance of persistent activity in a frontal thalamocortical loop. *Nature.* 545:181–186. http://dx.doi.org/10.1038/nature22324.

Hádinger N, Bősz E, Tóth B, Vantomme G, Lüthi A, Acsády L (2022). Region-selective control of the thalamic reticular nucleus via cortical layer 5 pyramidal cells. *Nat Neurosci.* 1:116–130. http://dx.doi.org/10.1038/s41593-022-01217-z.

Halassa MM, Acsady L (2016). Thalamic inhibition: Diverse sources, diverse scales. *Trends Neurosci.* 39:680–693.

Halassa MM, Kastner S (2017). Thalamic functions in distributed cognitive control. *Nat Neurosci.* 20:1669–1679.

Harris JA, Mihalas S, Hirokawa KE, Whitesell JD, Choi H, Bernard A, Bohn P, Caldejon S, Casal L, Cho A, Feiner A, Feng D, Gaudreault N, Gerfen CR, Graddis N, Groblewski PA, Henry AM, Ho A, Howard R, Knox JE, Kuan L, Kuang X, Lecoq J, Lesnar P, Li Y, Luviano J, McConoughey S, Mortrud MT, Naeemi M, Ng L, Oh SW, Ouellette B, Shen E, Sorensen SA, Wakeman W, Wang Q, Wang Y, Williford A, Phillips JW, Jones AR, Koch C, Zeng H (2019). Hierarchical organization of cortical and thalamic connectivity. *Nature.* 575:195–202.

Harris KD, Shepherd GM (2015). The neocortical circuit: themes and variations. *Nat Neurosci.* 18:170–181.

Hattox AM, Nelson SB (2007). Layer V neurons in mouse cortex projecting to different targets have distinct physiological properties. *J Neurophysiol.* 98:3330–3340.

Hayashi S, Hoerder-Suabedissen A, Kiyokage E, Maclachlan C, Toida K, Knott G, Molnár Z (2021). Maturation of complex synaptic connections of layer 5 cortical axons in the posterior thalamic nucleus requires SNAP25. *Cereb Cortex*. 31:2625–2638.

Hoerder-Suabedissen A, Hayashi S, Upton L, Nolan Z, Casas-Torremocha D, Grant E, Viswanathan S, Kanold PO, Clasca F, Kim Y, Molnár Z (2018). Subset of cortical layer 6b neurons selectively innervates higher order thalamic nuclei in mice. *Cereb Cortex*. 28:1882–1897.

Hoogland PV, Wouterlood FG, Welker E, Van der Loos H (1991). Ultrastructure of giant and small thalamic terminals of cortical origin: A study of the projections from the barrel cortex in mice using Phaseolus vulgaris leuco-agglutinin (PHA-L). *Exp Brain Res*. 87:159–172.

Jung WB, Jiang H, Lee S, Kim S-G (2022). Dissection of brain-wide resting-state and functional somatosensory circuits by fMRI with optogenetic silencing. *Proc Natl Acad Sci U S A*. 119:e2113313119, http://dx.doi.org/10.1073/pnas.2113313119.

Kirchgessner MA, Franklin AD, Callaway EM (2020). Context-dependent and dynamic functional influence of corticothalamic pathways to first- and higher-order visual thalamus. *Proc Natl Acad Sci U S A*. 117:13066–13077. http://dx.doi.org/10.1073/pnas.200 2080117.

Lee CC (2015). Exploring functions for the non-lemniscal auditory thalamus. *Front Neural Circuits*. 9:69.

Lee CC, Sherman SM (2011). On the classification of pathways in the auditory midbrain, thalamus, and cortex. *Hear Res*. 276:79–87.

Li J, Guido W, Bickford ME (2003a). Two distinct types of corticothalamic EPSPs and their contribution to short-term synaptic plasticity. *J Neurophysiol*. 90:3429–3440.

Li J, Wang S, Bickford ME (2003b). Comparison of the ultrastructure of cortical and retinal terminals in the rat dorsal lateral geniculate and lateral posterior nuclei. *J Comp Neurol*. 460:394–409.

Li Y, Lopez-Huerta VG, Adiconis X, Levandowski K, Choi S, Simmons SK, Arias-Garcia MA, Guo B, Yao AY, Blosser TR, Wimmer RD, Aida T, Atamian A, Naik T, Sun X, Bi D, Malhotra D, Hession CC, Shema R, Gomes M, Li T, Hwang E, Krol A, Kowalczyk M, Peça J, Pan G, Halassa MM, Levin JZ, Fu Z, Feng G (2020). Distinct subnetworks of the thalamic reticular nucleus. *Nature*. 583:819–824. http://dx.doi.org/10.1038/s41586-020-2504-5.

Llano DA, Sherman SM (2008). Evidence for nonreciprocal organization of the mouse auditory thalamocortical-corticothalamic projection systems. *J Comp Neurol*. 507:1209–1227.

Martinez-Garcia RI, Voelcker B, Zaltsman JB, Patrick SL, Stevens TR, Connors BW, Cruikshank SJ (2020). Two dynamically distinct circuits drive inhibition in the sensory thalamus. *Nature*. 583:813–818. http://dx.doi.org/10.1038/s41586-020-2512-5.

Masterson SP, Li J, Bickford ME (2009). Synaptic organization of the tectorecipient zone of the rat lateral posterior nucleus. *J Comp Neurol*. 515:647–663.

Mease RA, Gonzalez AJ (2021). Corticothalamic pathways from layer 5: Emerging roles in computation and pathology. *Front Neural Circuits*. 15:88.

Mease RA, Kuner T, Fairhall AL, Groh A (2017). Multiplexed spike coding and adaptation in the thalamus. *Cell Rep*. 19:1130–1140.

Mease RA, Sumser A, Sakmann B, Groh A (2016a). Corticothalamic spike transfer via the L5B-POm pathway in vivo. *Cereb Cortex*. 26:3461–3475.

Mease RA, Sumser A, Sakmann B, Groh A (2016b). Cortical dependence of whisker responses in posterior medial thalamus in vivo. *Cereb Cortex*. 26:3534–3543.

Mo C, Sherman S (2019). A sensorimotor pathway via higher-order thalamus. *J Neurosci*. 39:692–704.

Moore JD, Mercer Lindsay N, Deschênes M, Kleinfeld D (2015). Vibrissa self-motion and touch are reliably encoded along the same somatosensory pathway from brainstem through thalamus. *PLoS Biol*. 13:e1002253.

Nakajima M, Halassa MM (2017). Thalamic control of functional cortical connectivity. *Curr Opin Neurobiol*. 44:127–131.

Prasad JA, Carroll BJ, Sherman SM (2020). Layer 5 corticofugal projections from diverse cortical areas: variations on a pattern of thalamic and extra-thalamic targets. *J Neurosci*. 40:5785–5796. http://dx.doi.org/10.1523/JNEUROSCI.0529-20.2020.

Qi J, Ye C, Naskar S, Inácio AR, Lee S (2022). Posteromedial thalamic nucleus activity significantly contributes to perceptual discrimination. *PLoS Biol*. 20:e3001896.

Reichova I, Sherman SM (2004). Somatosensory corticothalamic projections: Distinguishing drivers from modulators. *J Neurophysiol*. 92:2185–2197.

Rikhye RV, Gilra A, Halassa MM (2018a). Thalamic regulation of switching between cortical representations enables cognitive flexibility. *Nat Neurosci*. 21:1753–1763.

Rikhye RV, Wimmer RD, Halassa MM (2018b). Toward an integrative theory of thalamic function. *Annu Rev Neurosci*. 41:163–183.

Rockland KS (2019). Corticothalamic axon morphologies and network architecture. *Eur J Neurosci*. 49:969–977.

Rojas-Piloni G, Guest JM, Egger R, Johnson AS, Sakmann B, Oberlaender M (2017). Relationships between structure, in vivo function and long-range axonal target of cortical pyramidal tract neurons. *Nat Commun*. 8:870.

Rouiller EM, Welker E (2000). A comparative analysis of the morphology of corticothalamic projections in mammals. *Brain Res Bull*. 53:727–741.

Sampathkumar V, Miller-Hansen A, Sherman SM, Kasthuri N (2021). Integration of signals from different cortical areas in higher order thalamic neurons. *Proc Natl Acad Sci U S A*. 118:e2104137118. http://dx.doi.org/10.1073/pnas.2104137118.

Schmitt LI, Wimmer RD, Nakajima M, Happ M, Mofakham S, Halassa MM (2017). Thalamic amplification of cortical connectivity sustains attentional control. *Nature*. 545:219–223.

Shepherd GMG, Yamawaki N (2021). Untangling the cortico-thalamo-cortical loop: cellular pieces of a knotty circuit puzzle. *Nat Rev Neurosci*. 22:389–406.http://dx.doi.org/10.1038/s41 583-021-00459-3.

Sherman SM (2016). Thalamus plays a central role in ongoing cortical functioning. *Nat Neurosci*. 19:533–541.

Sherman SM (2017). Functioning of circuits connecting thalamus and cortex. *Compr Physiol*. 7:713–739.

Sherman SM, Usrey WM (2021). Cortical control of behavior and attention from an evolutionary perspective. *Neuron*. 109:3048–3054. http://dx.doi.org/10.1016/j.neuron.2021.06.021.

Stroh A, Adelsberger H, Groh A, Rühlmann C, Fischer S, Schierloh A, Deisseroth K, Konnerth A (2013). Making waves: Initiation and propagation of corticothalamic Ca2+ waves in vivo. *Neuron*. 77:1136–1150.

Sumser A, Mease RA, Sakmann B, Groh A (2017). Organization and somatotopy of corticothalamic projections from L5B in mouse barrel cortex. *Proc Natl Acad Sci U S A*. 114:8853–8858.

Theyel BB, Llano DA, Sherman SM (2010). The corticothalamocortical circuit drives higher-order cortex in the mouse. *Nat Neurosci*. 13:84–88.

Urbain N, Salin PA, Libourel PA, Comte JC, Gentet LJ, Petersen CC (2015). Whisking-related changes in neuronal firing and membrane potential dynamics in the somatosensory thalamus of awake mice. *Cell Rep*. 13:647–656.

Veinante P, Lavallée P, Deschênes M (2000). Corticothalamic projections from layer 5 of the vibrissal barrel cortex in the rat. *J Comp Neurol*. 424:197–204.

Williamson RS, Polley DB (2019). Parallel pathways for sound processing and functional connectivity among layer 5 and 6 auditory corticofugal neurons. *eLife*. 8:e42974. http://dx.doi.org/10.7554/eLife.42974.

Winnubst J, Bas E, Ferreira TA, Wu Z, Economo MN, Edson P, Arthur BJ, Bruns C, Rokicki K, Schauder D, Olbris DJ, Murphy SD,

Ackerman DG, Arshadi C, Baldwin P, Blake R, Elsayed A, Hasan M, Ramirez D, Dos Santos B, Weldon M, Zafar A, Dudman JT, Gerfen CR, Hantman AW, Korff W, Sternson SM, Spruston N, Svoboda K, Chandrashekar J (2019). Reconstruction of 1,000 projection neurons reveals new cell types and organization of long-range connectivity in the mouse brain. *Cell.* 179:268–281.

Wolff M, Morceau S, Folkard R, Martin-Cortecero J, Groh A (2020). A thalamic bridge from sensory perception to cognition. *Neurosci Biobehav Rev.* 120:222–235.

Yamawaki N, Shepherd GMG (2015). Synaptic circuit organization of motor corticothalamic neurons. *J Neurosci.* 35:2293–2307.

16.

THE THALAMIC RETICULAR NUCLEUS

ANATOMO-FUNCTIONAL MECHANISMS AND CONCEPT

Didier Pinault

FOREWORD

The thalamus and the neocortex are functionally linked by the corticothalamic (CT) and thalamocortical (TC) pathways, thereby forming the cortico-thalamo-cortical (CT-TC) networks (see **Section IV**). These are so highly and spatiotemporally distributed that they are implied in almost all functional modalities, including sensory perception, thought, emotions, attention, and consciousness (see **Section VIII**). Any volume unit (e.g., column) of the cerebral cortex is concerned, and it is mono/multi-synaptically connected with many other nearby and remote units, directly (cortico-cortical connections) and via the thalamus (cortico-thalamo-cortical connections) (see **Chapters 40 and 42 in Section VII**). To the thalamus-cortex couple, we must add the thalamic reticular nucleus (TRN), which is reciprocally connected with the dorsal thalamus.

In this chapter, with a holistic and reductionist view, I present a critical survey of the principal anatomical and structural properties of the GABAergic TRN neurons, and the main organizing principles of the thalamus-TRN (or TC-TRN) connectome, the bases of TRN-mediated recurrent and lateral inhibitions. I will not take into consideration the GABAergic interneurons (or local circuit cells) of the dorsal thalamus, which are intrinsic sources of thalamic inhibitions. Our knowledge on this topic is mainly based on data from comprehensive studies performed especially in rodents, whose most of the nuclei of their dorsal thalamus contain few or no GABAergic interneurons (see **Chapter 2 in Section II**). Many studies of the thalamus-TRN connectome conducted in diverse mammals, including nonhuman primates, rodents, and carnivores, have yielded mostly similar findings. In addition, the extrathalamic inhibitory inputs originating from the zona incerta, anterior pretectum, basal ganglia, and pontine reticular formation will not be discussed here (see **Chapters 4 and 5 in Section II**). The present chapter offers supporting material to the unified theory that the functional connectivity between glutamatergic and GABAergic neurons, endowed with state-dependent intrinsic and synaptic properties, generates a wide range of wake/sleep/function-related neural oscillations (see **Chapter 3 in Section II; Chapter 32 in Section VI**).

A SHELL-SHAPED GABAERGIC STRUCTURE AND ITS CONNECTOME

The thalamic reticular nucleus is a curved, shell-shaped structure located, in each hemisphere, at the thalamus–white matter interface, between the internal capsule and the external medullary lamina. The TRN is observed in all mammals (Pritz, 2015). In its embryological origin, the TRN is a nucleus of the prethalamus (formerly known as the ventral thalamus) (Nakagawa, 2019). Ventromedially, it is adjacent to the zona incerta, and caudally, to the ventrolateral geniculate nucleus. It contains only GABAergic neurons, which innervate the dorsal thalamus through the activation of $GABA_A$ and $GABA_B$ receptors (see **Chapter 3 in Section II**). Immunocytochemical stainings have revealed that TRN neurons contain diverse Ca^{++}-binding proteins but, principally, parvalbumin (Pinault, 2004). The peptide hormone somatostatin is also present in the TRN of various species, including in humans (Burgunder and Young, 1992). Thalamic reticular nucleus neurons are surrounded by perineuronal nets, which are organized extracellular matrix structures that envelop neurons to regulate neuronal maturation, synaptic connectivity, and plasticity (see **Chapter 2 in Section II**).

The primary afferents of the TRN are axon collaterals from the CT and TC neurons (Figure 16.1), which are glutamatergic and make asymmetrical synaptic junctions on the TRN dendrites (Ohara and Lieberman, 1985;

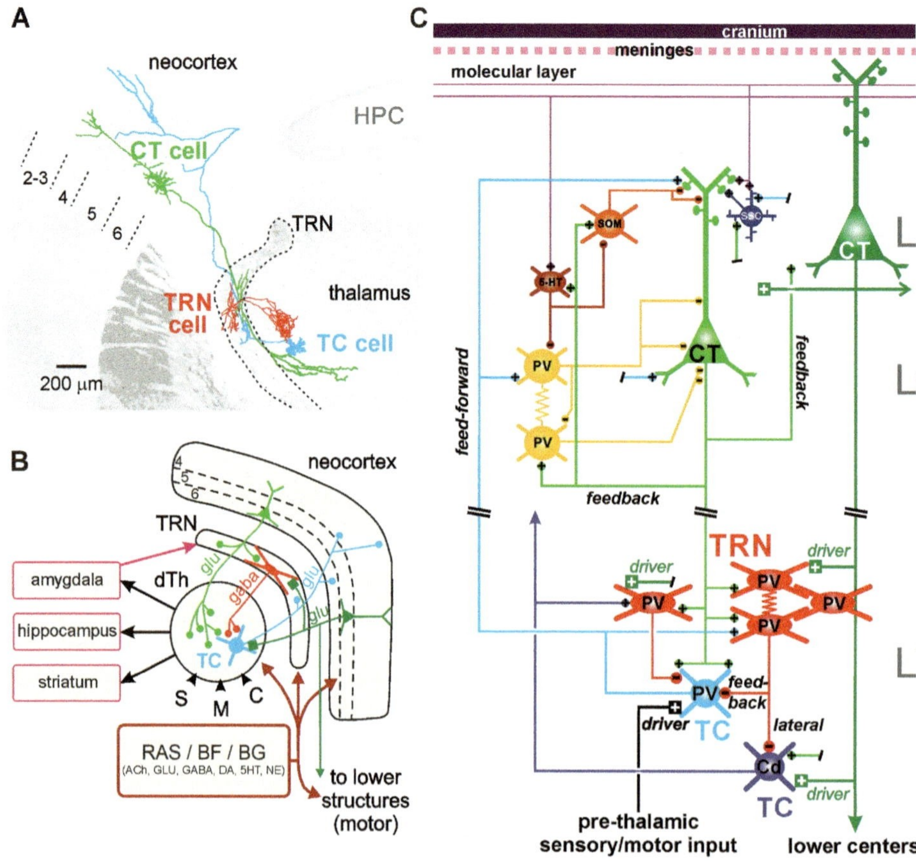

Figure 16.1. The rodent thalamic reticular nucleus connectome. (A) Mounting of reconstructed juxtacellularly labeled neurons of the somatosensory system. Both the layer VI CT (in green) and the TC (in blue) neurons are glutamatergic and their principal axon crosses the TRN, where it gives rise to axon collaterals. This 3-neuron CT-TRN-TC circuit is common to thalamic FO and HO nuclei. HPC, hippocampus. (B) Simplified drawing of the afferets and efferents of the 4-neuron CT-TRN-TC circuit, which includes the 2, layer 5, and 6 CT neuron. The TRN receives sensory (S), motor (M) and cognitive/associative (C) inputs. RSA = reticular activating system; BF = basal forebrain; BG = basal ganglia; dTh = dorsal thalamus. Adapted from Pinault (2017). (C) The thalamus-TRN circuit is hypothesized as the layer 7 of the neocortex. The cortex (layers 1–6) and the thalamus (layer 7) work together, horizontally (cortex←→cortex) and vertically (cortex←→thalamus). The thalamus is also an essential relay between multiple cortical areas. For simplicity, the connectome of layer 5 CT neurons is not illustrated. Like the thalamic afferent sensory inputs, the layer 5 CT neurons are considered as drivers. In contrast, layer 6 CT neurons are modulator. Of importance, the layer 5 CT inputs in the thalamus (TC + TRN) are axon collaterals of corticofugal cells that innervate lower motor centers, including the brainstem. In this model, it is assumed that axodendritic (chemical synapses) and dendrodentritic electrical (via gap junctions) coupling exists between the GABAergic parvalbumin (PV) positive cortical (basket and chandelier), and thalamic (TRN) neurons. CT = corticothalamic; SOM = somatostatin; ssc = spiny stellate cells; TC = thalamocortical; TRN = thalamic reticular nucleus. Adapted from Pinault (2017).

Williamson et al., 1993). The thalamus is reciprocally connected with both the TRN and the cerebral cortex, thereby forming loop CT-TRN-TC circuits (Figure 16.1A). The TRN receives topographically mapped CT inputs, and converging inputs from different cortical areas (Crabtree, 1992a, 1992b). In contrast to the CT neurons, TC neurons do not talk to each other directly. The TRN also receives diffuse neuromodulatory (cholinergic, glutamatergic, GABAergic, dopaminergic, serotoninergic, and noradrenergic) inputs from the brainstem, the basal forebrain, the globus pallidus, and the substantia nigra (see **Chapters 4 and 5 in Section II**). The hodology of the TRN is summed in Figure 16.1B. The loop TC-TRN-CT circuits are the downstream part of the ascending reticular activating system. The TRN of primates and rodents also receives robust excitatory inputs from the amygdala, a key emotional structure involved in attentional processes (Zikopoulos and Barbas, 2012; Aizenberg et al., 2019). Because the thalamus works tightly with the cerebral cortex, Sherman and Guillery (2009) suggested that one could see the thalamus as the layer 7 of the cortex, as illustrated in Figure 16.1C.

DIVISIONS AND SUBDIVISIONS

SECTORS

Neurophysiological and anatomical studies performed in primates, carnivores, and rodents have demonstrated the existence of 7 topographically organized functional sectors: 5 sensory (visual, auditory, gustatory, somatosensory, and visceral), 1 motor, and 1 limbic (Pinault, 2004). The visual sector is located in the dorsocaudal region, the auditory in the ventrocaudal region, the gustatory in the ventromedial-most portion, the somatosensory in the ventrocentral part, the visceral in the rostroventral part, and the motor and limbic sectors in the rostral part of the TRN. In Figure 16.2A, the 7 sectors are illustrated in highly schematic fashion on a simplified 2-D map. The frontiers between the divisions (sectors) and subdivisions (tiers) remain anatomically/histochemically undefined.

Each sector is reciprocally connected with more than one thalamic nucleus, generally one FO and one or more HO nuclei, and receive inputs from more than one thalamic region and more than one cortical area. The corresponding connections generally are topographically organized, thereby forming anatomo-functional maps. Until now, no study has demonstrated the existence of an olfactory sector in the TRN. The somatosensory sector is relatively enlarged. As in the cortex and thalamus, each part of the body is unequally represented (e.g., very enlarged face projections), revealing a deformed figurine (Figure 16.2A). The somatotopic organization was demonstrated in the rat (Shosaku et al., 1984), cat, and monkey (Pollin and Rokyta, 1982). The head and face occupy about two-thirds of the somatosensory sector, distributed in the ventral half and in the dorsocaudal part.

Inter-sectorial differences are observed in and in between species. For instance, in the nonhuman primate visual sector, morphologically heterogeneous subpopulations of TRN cells were observed in the mediolateral axis (Bragg et al., 2017). Neurons in the center of the visual sector transmit global (arousal/cognitive) signals to foveal visual field representations in the dorsal lateral geniculate nucleus, while neurons located more

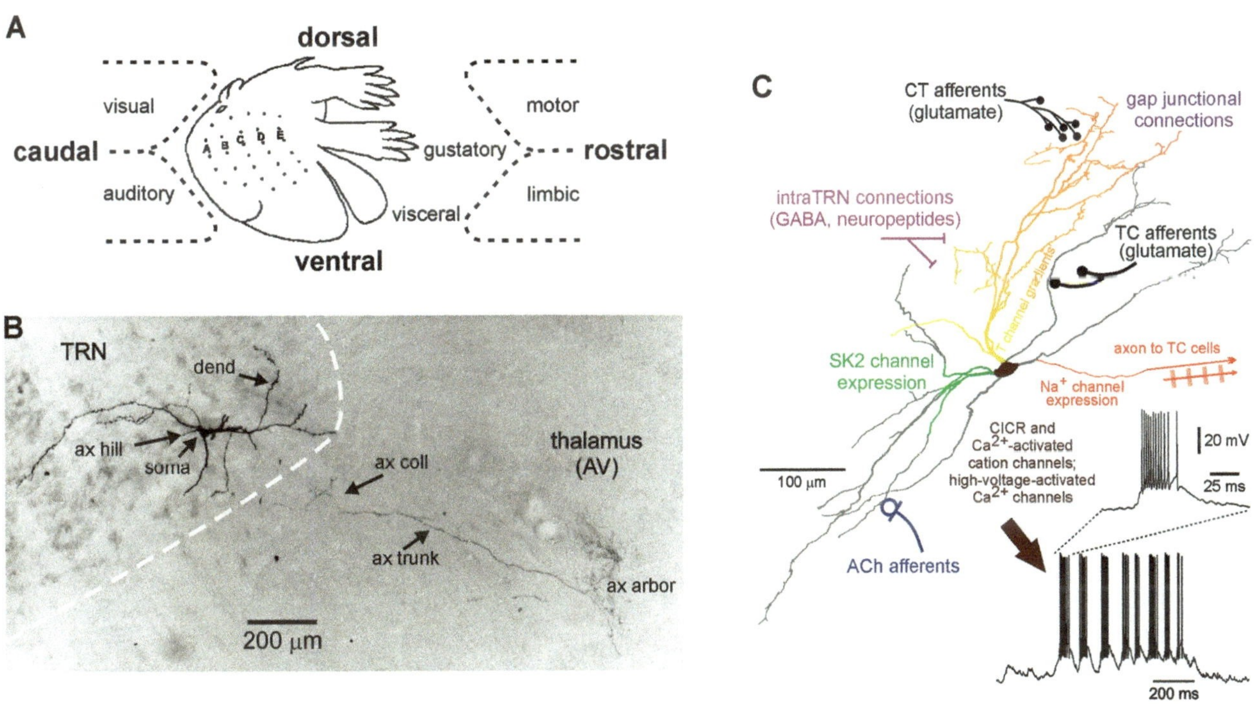

Figure 16.2. Principal global and cellular features of the rodent TRN. (A). Two-dimensional schematic representation of the 7 sectors. The frontiers between sectors are unknown. Adapted from Shosaku et al. (1984). (B). Dendritic and axonal ramifications of a juxtacellularly labeled thalamic reticular nucleus (TRN) cell projecting to the anterior ventral (AV) thalamic nucleus. This cell was located in the rostrodorsal pole of the TRN. This photomicrograph shows parts of somatodendritic and axonal ramifications contained within the same horizontal 100-µm-thick section. One of its 2 thalamic-projecting axonal branches is visible during almost all of its long course. Arrows indicate the initial segment and the cut ends of the 2 main axonal branches. Ax = axon; ax hill = axon hillock; dend = dendrite. Adapted from Pinault and Deschenes (1998). (C). Morphological, ionic, and synaptic properties of a reconstructed rat TRN cell labeled with Neurobiotin® during a juxtacellular recording. Note the multiple thin dendrites emerging from the soma, some of which branch extensively and form hairy appendages. The subcellular distribution is indicated for T channels (orange-brown gradient, with darker colors indicating greater expression levels) and for small-conductance type 2 (SK2) channels (green). SK2 channel expression was demonstrated up to ~100 µm from the soma. Single dendrites were chosen to illustrate T and SK2 channel distribution in color code. Mechanisms of burst control involving high-voltage-activated Ca^{++} channels, Ca^{++}-induced Ca^{++} release (CICR), and Ca++-activated cation channels were described at the somatic level. The axon, emerging from a primary dendrite, expresses high levels of Na^{+} channels (red). Synaptic afferents for which their subcellular localization is known are positioned accordingly. Excitatory afferents are labeled with a circle, inhibitory afferents with a line. Cholinergic afferents mediate both excitation and inhibition. Insets show a single burst and repetitive burst discharge (sleep spindle). The axon transmits only the action potentials, the firing pattern (tonic or burst mode). ACh = acetylcholine; TC = thalamocortical; CT = corticothalamic. Adapted from Luthi (2014).

laterally transmit global signals to peripheral visual field representations in the dorsal lateral geniculate nucleus. These recent observations bring a degree of anatomo-functional complexity in the TRN to the well-known retinotopic organization of both the TRN visual sector and the related dorsal lateral geniculate nucleus.

TIERS

Reticular sectors too are divided into anatomical recipient regions of distinct but functionally related thalamic nuclei and cortical areas. One representative example is the somatosensory sector, which contains, in the TRN thickness, 3 tiers (medial, middle, and lateral), at least in carnivores (Crabtree, 1996) and rodents (Pinault and Deschenes, 1998a; Lam and Sherman, 2011). The lateral and middle tiers are, in an ordered and parallel fashion, connected with the FO thalamic nuclei (the lateral and medial parts of the ventral posterior nucleus, the ventrobasal complex), whereas the medial tier is, with a non-well-defined topography, connected with the posterior thalamic nuclear group, an HO nucleus. Some TRN cells from the lateral and middle tiers project to both FO and HO nuclei.

FIRST-ORDER (FO) AND HIGHER-ORDER (HO) SUBNETWORKS

The multilevel structural organization of each TRN sector is determined by the matching of its outputs with the related thalamic and cortical inputs (Crabtree, 1999). For a long while, we thought that layer 6 but not layer 5 CT neurons innervate the TRN (Bourassa and Deschênes, 1995; Bourassa et al., 1995; Guillery and Sherman, 2002). Recently, Sherman and colleagues demonstrated that, like the dorsal thalamus, the TRN includes FO (core) and HO (shell or edge) subnetworks (Prasad et al., 2020; Carroll et al., 2022). This important finding upsets our current view of the functional connections between the cortex and the thalamus. Layer 5 CT inputs locally innervate the TRN edge, whereas layer 6 CT inputs diffusely innervate both the edge and core of the TRN (see **Chapters 14 and 15 in Section IV**). So, layer 5 and 6 CT axons cross the TRN, where they give off collaterals to innervate mapped subnetworks. Layer 5 CT axon terminations are larger than those originating from layer 6, suggesting that layer 5 CT neurons exert a driver function (potent synaptic effect). However, layer 6 afferents would exert a modulatory-like function as they diffusely target both FO and HO nuclei of the dorsal thalamus (Bourassa and Deschênes, 1995; Bourassa et al., 1995) and both FO and HO subregions of the TRN (Whilden et al., 2021).

Thanks to the combination of single-cell molecular, anatomical, and neurophysiological methods in the mouse, the notion of "core and shell organization" in distinct sectors (visual, auditory, motor, and somatosensory) of the TRN was introduced (Li et al., 2020). More specifically, a majority of TRN cells projecting to thalamic FO nuclei are located in the core, whereas most of the TRN cells located in the shell project to HO nuclei. This core-shell organization is somehow reminiscent of the structure-function organization of the visual sector of primates (Bragg et al., 2017). Neurophysiological findings suggest differential roles for the core and shell TRN cells in regulating sleep-related oscillations (Li et al., 2020). The core is more prone than the shell to generate sustained rhythmic burst firing, like that shown in Figure 16.2C. The notion of "core and shell/edge" in the TRN is somehow reminiscent of the "core-matrix" theory of the dorsal thalamus because of its heterogeneous cyto- and axonal architectures (see **Chapters 11 and 12 in Section IV**).

THE SOMATODENDRITIC FUNCTIONAL ARCHITECTURE

Ohara and Havton (1996) reconstructed in 3 dimensions individual rodent TRN cells retrogradely labeled from the dorsal thalamus with a neuronal tracer. Thereby, they convincingly revealed that neurons from different TRN regions have a similar somatodendritic architecture. This finding was confirmed in the Deschênes's laboratory using the 3-D reconstruction of juxtacellularly recorded (electrophysiological identification) and labeled TRN neurons (Pinault et al., 1995a; Pinault and Deschenes, 1998a). A photomicrograph of a typical juxtacellularly labeled TRN cell is shown in Figure 16.2B. Thalamic reticular nucleus neurons have a round, ovoid, or polygonal perikaryon, from which arise 2–6 smooth or varicose ramifying dendrites bearing appendages of variable size, length, and architecture, but no spines (Figure 16.2C). Each primary dendrite ramifies into secondary, tertiary, and quaternary branches bearing filamentous ramifications with varicosities. The dendritic arbors together form a dendritic domain that fits the geometric constraints of the shell-like form of the nucleus, where pass complex meshes of intertwining branching axons, especially TC and CT axons. The space occupied by each dendritic arbor is variable, whereas that of the dendritic domain, which includes all dendritic arbors, is nearly similar from one TRN cell to the other. The cell body is not systematically located at the geographical center of the parental dendritic domain, suggesting that the latter occupies a space determined by the limits of an anatomo-functional entity or slab (Crabtree, 1992a). Depending on the viewing angle, the dendritic domain is either ellipsoid with dendrites extending principally dorsoventrally (rostral or caudal view) or rostrocaudally (ventral or dorsal view), or multipolar (lateral or medial view) (Ohara and Havton, 1996).

The somatodendritic membranes of TRN cells are characterized by sets of intrinsic and synaptic ionic properties, which allow them to fire, in an all-or-none fashion, either a single action potential, or a high-frequency (up to 600 Hz) burst of 2–15 action potentials. The bursting mode is determined principally by low-threshold T-type Ca^{++} channels (see **Chapter 3 in Section II**). A representative intracellular sleep spindle (repetitive burst firing) is shown in Figure 16.2C.

PARALLEL AND DIVERGENT AXONAL PROJECTIONS TO THE DORSAL THALAMUS

With the development over time of stereotaxic-guided anatomical methods (antero- and retrograde labelings), it has been well demonstrated that, in various species, TRN axons project into all the anterior, dorsal, intralaminar, posterior, and ventral thalamic nuclei according to a loose parallel pattern with a certain degree of divergence. A given TRN locus gives rise to divergent axonal projections to intralaminar and midline nuclei in both rats (Kolmac and Mitrofanis, 1997) and cats (Steriade et al., 1984). In the feline somatosensory system, there is a dual organization of the TRN axonal projections: mapped in the FO nucleus (ventrobasal complex), and non-mapped in the nonspecific nucleus (posterior complex, equivalent to the Puvinar in humans) (Crabtree, 1996).

Using the single-cell juxtacellular recording/labeling technique in anesthetized rats, Pinault and Deschênes revealed the somatodendritic and axonal architectures of more than 100 labeled TRN neurons (Pinault et al., 1995b, 1995a; Pinault and Deschenes, 1998a). Furthermore, the distribution of the labeled cells within the thickness of the TRN seems to obey to a strip-like arrangement. For instance, in the somatosensory system, cells projecting to the posterior complex (HO nucleus) occupy the medial part of the TRN, whereas those projecting to the ventral-posteromedial and ventral-posterolateral nuclei (FO nuclei) are located in, respectively, the middle and lateral tiers of the nucleus (Figure 16.3A). Thereby, it was demonstrated the existence of at least 4 configurations of TRN axonal projections: (1) 2 adjacent TRN cells without overlap of their dendritic trees innervate 2 adjacent districts within a given thalamic nucleus without overlapping axonal arbors; (2) 2 TRN neurons with overlapping dendritic trees have overlapping axonal arbors; (3) a minority (<5%) of TRN neurons can have an axon that splits into 2 branches innervating a FO nucleus and a functionally related HO nucleus (Figure 16.3B); (4) 2 TRN neurons with overlapping dendritic trees can innervate 2 distinct (FO&HO or HO&HO) thalamic nuclei (Figure 16.3C); (5) most of the TRN neurons in the medial tier innervate medial thalamic nuclei, whereas most of those in the lateral tiers innervate more lateral nuclei. Nevertheless, a minority (<5%) of the lateral tier can innervate both lateral (FO, e.g., ventrobasal complex) and more medial (HO, e.g., medial part of the posterior complex) nuclei.

The axonal arborization of most TRN cells is confined within the limits of a single thalamic nucleus. In the FO somatosensory nucleus (ventrobasal complex), termination fields form short rods (diameter approximately 150 μm, length approximately 200–300 μm) densely packed with grape-like boutons and varicosities; termination fields in the related HO thalamic nucleus (medial part of the posterior complex) are larger, much less dense, and they are contained within a horizontal slab of tissue (thickness: ~200 μm; mediolateral width: ~400 μm; rostrocaudal length: ~1 mm) (Figure 16.3A,B).

Structural and morphometric analysis of juxtacellularly stained rats' TRN axon terminal arbors indicated that the average number of boutons within the axonal arbor of a single TRN cell is 4000 ± 1000 (Pinault and Deschenes, 1998a). The clustering of boutons optimally matches the somatodendritic architecture of the corresponding postsynaptic TC neurons. Within the axonal patches, terminals tend to form grape-like clusters or strings according to the tufted or linear architecture of the postsynaptic dendrites, a spatial arrangement to make optimal the synaptically mediated TRN-TC integrative functions. This rule seems to apply to both the FO (e.g., ventrobasal complex) and the HO (e.g., lateral posterior and/or posterior nuclear group) thalamic nuclei, which contain bushy and star-like relay cells, respectively (Figure 16.3D). Three-dimensional observations of the distribution of the TRN axonal boutons revealed that, first, the size of the axonal patches is within the dimensions of the dendritic spread of the target TC neurons; second, the smaller and denser the axonal patches are, the smaller the target neurons are and the larger the amount of their dendritic surface per volume unit of tissue is. In addition, the densest axonal patches contain more grapes and fewer strings of boutons than do the sparse patches. Terminations of TRN axons make Gray type II axodendritic and axosomatic synaptic contacts with TC neurons (Liu et al., 1995).

Taken together, these findings validate early observations (Scheibel and Scheibel, 1966) about a well-organized topographical organization of the TRN axonal projections, which includes overlapping, parallel (mapped in specific thalamic nuclei), and divergent (non-mapped in nonspecific nuclei) beams.

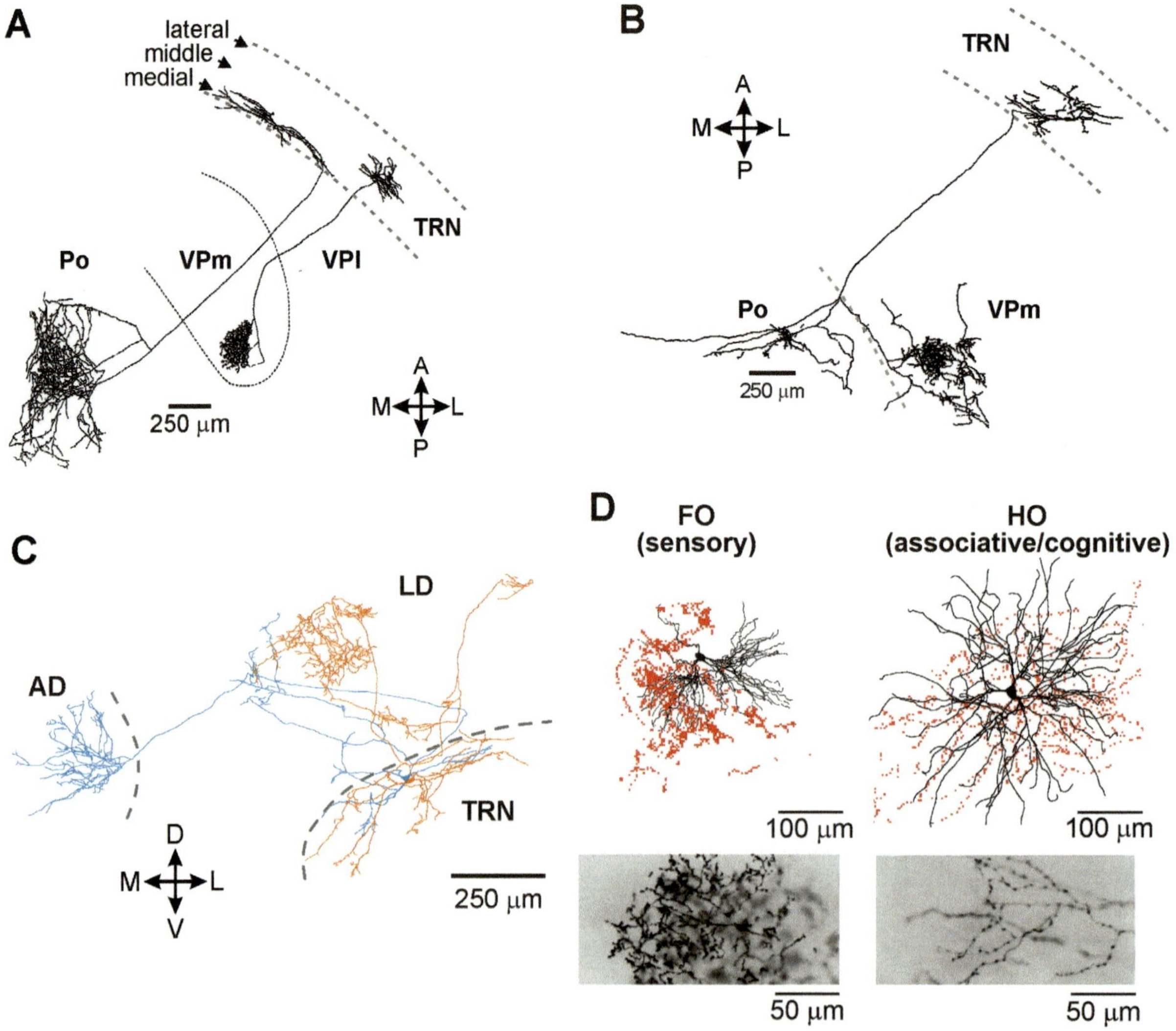

Figure 16.3. Innervation patterns of TRN axons and their endings. (A). Dorsal view of 2 juxtacellularly labeled TRN cells (camera lucida drawings), one projecting to the posterior group (Po) of the thalamus, the other to the medial part of the specific somatosensory (ventral posterior, or VPm) nucleus. The TRN is divided into 3 tiers: medial, middle, and lateral. (B). The axonal trunk of this TRN cell (camera lucida drawing, dorsal view) split into 2 branches, one innervating the VPm, the other the Po. (C). Dorsal view of 2 juxtacellularly labeled TRN cells with overlapping somatodendritic fields (3-D reconstruction) projecting into the anterior dorsal (AD) and lateral dorsal (LD) thalamic nuclei. The AD/LD-projecting cell (in blue) displayed a few axonal collaterals in the LD nucleus and deployed its main arbor just as it entered into the AD nucleus, whereas the LD-projecting cell (in orange) arborizes only within the LD nucleus. A = anterior; D = dorsal; L = lateral; M = medial; P = posterior; V = ventral. (A) and (B) adapted from Pinault et al. (1995a); (C) from Pinault and Deschenes (1998). (D). The density and spatial distribution of the TRN axon terminations match the dendritic architecture of the recipient TC neurons. Left: Posterior view of a 3-D mapping of all the tracer-filled boutons (in red), which belonged to a specific, ventrobasal (VB)-projecting TRN cell. All the boutons are contained in a 200-μm-thick tissue block. Right: Posterior view of a 3-D mapping of all the tracer-filled boutons (in red), which belonged to a nonspecific, Po-projecting TRN cell. All the boutons are contained in a 200-μm-thick tissue block. Below: The photomicrograph shows part of the terminal axonal arbors and endings found in horizontal sections of the VB (left) and Po (right). Adapted from Pinault and Deschenes (1998). The 2 photomicrographs are from Pinault et al. (1995a), and Pinault et al. (1995b).

OPEN-LOOP AND CLOSED-LOOP CONNECTIONS

Anatomical and functional observations demonstrated that the thalamus and the TRN are reciprocally connected (Pinault, 2004). However, the principle of reciprocity seems that it cannot fully apply at the cellular level. Juxtacellular anterograde labeling of a recorded TRN cell—the tip of the recording electrode being very close to its perikaryon—can make it fully visible, including its terminal axonal ramifications in the related nucleus of the dorsal thalamus (Pinault et al., 1995a). In addition, in rare instances (4%), the somatodendritic complex of a TC neuron—simultaneous single-cell retrograde juxta-axonal labeling—is observed at the periphery of the anterogradely stained TRN axonal arbor (Pinault and Deschenes, 1998b) (Figure 16.4A). It is worth knowing that this low percentage of double-cell labeling reflects one of the "pitfalls" (salutary multi-cell labeling) of the single-cell juxtacellular labeling technique. This means that the proportion of open circuits, if they really exist (actual synaptic contact between the 2 anterograde- and retrograde-labeled TRN and TC neurons), is underestimated. Axon collaterals of the retrograde-labeled TC neuron were observed to spread within the somatodendritic field of the anterograde-labeled

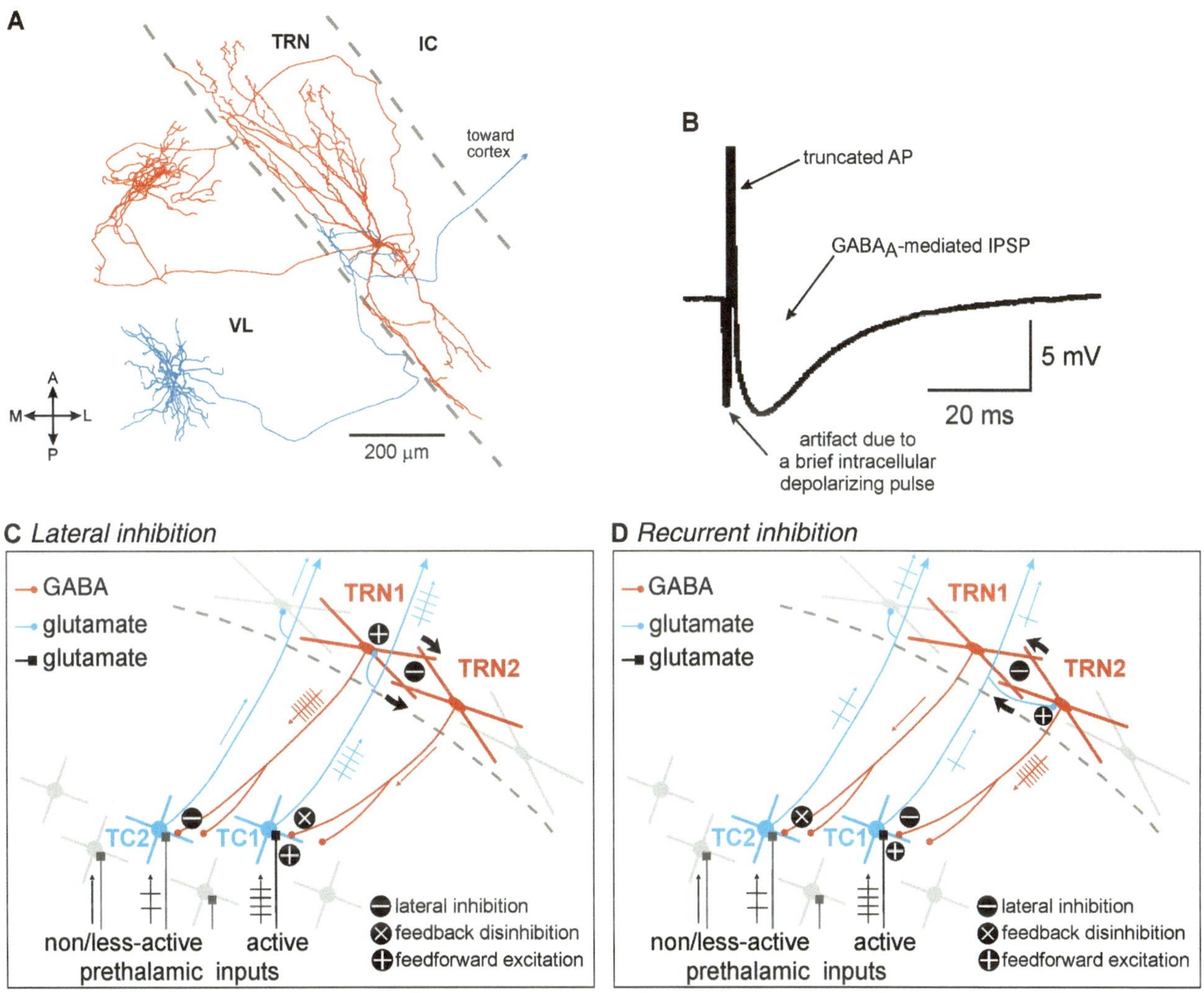

Figure 16.4. Principles of lateral and recurrent inhibitions. The GABAergic TRN (red) and glutamatergic TC (blue) cells are, respectively, inhibitory and excitatory, and they form more open-loop than closed-loop circuits. (A). Dorsal view of a reconstructed juxtacellularly stained pair of TRN-TC neurons. The TRN and TC neurons were simultaneously labeled during, respectively, a simultaneous juxtasomatodendritic and juxta-axonal iontophoresis of biocytin with the same micropipette into the TRN (the recording target). A = anterior; L = lateral; M = medial; P = posterior. (B). Intracellular recording made in an in vitro adult cat thalamic slice. Hyperpolarization in a TC cell of the visual thalamus following an action potential evoked from a brief, depolarizing pulse delivered intracellularly (averaging of 20 individual traces). The hyperpolarization, relatively long-lasting, is in great part mediated by activation of $GABA_A$ receptors since it was reduced in amplitude by the application bicuculline in the bath. (C, D). Schematic drawings illustrating likely synaptic connections between single TC (blue) and TRN (red) neurons. For simplicity, one of the terminal axonal boutons of one TC neuron contacts one TRN cell. Similarly, one of the terminal boutons of one TRN cell makes a synaptic contact with one TC neuron. (C) Lateral inhibition involves the activation of a mono-synaptic open-loop TC1→TRN1. This activation starts by a presynaptic volley from an active input. Here, the presynaptic volley is from a thalamic afferent driver type input (e.g., of sensory nature). One can imagine the presynaptic volley could also be from a driver input of a layer 5 CT neuron. So, activation of the open loop starts with neuron TC1. The cell TC1, via its axon collateral, excites cell TRN1, and that cell then inhibits cell TC2 (lateral or surround inhibition). The neuron TRN1, via inhibitory dendrodendritic synapses, inhibits cell TRN2, and that neuron then disinhibits the neuron TC1 (feedback disinhibition). (D) Recurrent inhibition involves the activation of a di-synaptic closed-loop TC1←→TRN2. This activation starts by a presynaptic volley from an active input. Here, the presynaptic volley is from a thalamic afferent driver type input (e.g., of sensory nature). One can imagine the presynaptic volley could also be from a driver input of a layer V CT neuron. So, activation of the closed-loop starts with neuron TC1. The cell TC1, via its axon collateral, excites cell TRN2, then TRN2 inhibits TC1 (recurrent or feedback inhibition). The neuron TRN2, via inhibitory dendrodendritic synapses, inhibits cell TRN1, and that neuron then disinhibits the neuron TC2 (feedback disinhibition). A and C adapted from Pinault and Deschenes (1998). B adapted from Lo and Sherman (1994).

TRN cell. This is a finding in favor of the hypothesis of a mechanism of TRN-mediated lateral inhibition in the dorsal thalamus (Figure 16.4C).

Intracellular recordings of TC neurons in slices of the cat visual thalamus revealed a post-action potential long-lasting hyperpolarization, mediated by the activation of $GABA_A$ receptors (bicuculline-sensitive), in only a few (6 of 36) geniculate cells (Lo and Sherman, 1994) (Figure 16.4B). This postsynaptic event was interpreted as being the manifestation of either an actual feedback (or recurrent) inhibition (Figure 16.4D) or a lateral (or surround) inhibition (Figure 16.4C) generated by a coincident action potential in another TC neuron.

Other physiological in vivo (Shosaku, 1986; Shosaku et al., 1989) or in vitro (Gentet and Ulrich, 2003) studies revealed findings also supporting the hypothesis of lateral and recurrent inhibitions in the loop TC-TRN circuits. They showed a small proportion of TC-TRN cellular pairs with closed disynaptic loops (<20%). Furthermore, cross-correlation analysis of paired TC-TRN recordings in the somatosensory system of anesthetized rats revealed that TRN and TC neurons with receptive fields on the same vibrissa can interact with each other (excitations and/or inhibitions) (Shosaku, 1986). Conversely, no significant functional interaction was detected between TRN and TC neurons having different vibrissae-related receptive fields. From this comprehensive study, one TRN neuron receives excitatory inputs from approximately 40% of the TC neurons with receptive fields on the same vibrissa and sends inhibitory outputs to approximately 55%. Since these 2 groups of TC neurons were overlapping, the TRN neuron would have reciprocal connections with approximately 20% of the TC neurons with receptive fields on the same vibrissa. Shosaku (1986) concluded that TRN axonal projections' convergence and divergence from a significant number of thalamic neurons with receptive fields on the same vibrissa play a role in channeling ongoing sensory information in the TC system.

STRUCTURAL SUBSTRATES OF INTRANUCLEAR CELL-CELL COMMUNICATIONS

The potential structural substrates of intranuclear cell-cell communications are still the object of investigation. Dendrites of neighboring TRN neurons can run in a parallel fashion over long distances (a few hundreds of μm) and aggregate to form tightly interwoven clusters or bundles (of 2–5 dendrites) (Figure 16.5A,B) embedded in a dense neuropil of presynaptic boutons, most of them arising from the glutamatergic CT and TC axons (Scheibel and Scheibel, 1966). These dendritic bundles are an anatomical signature of potential synaptic (chemical and electrical) and nonsynaptic (via electric fields) cell-cell communications within the TRN.

SYNAPTIC INTERACTIONS

Anatomo-functional and pharmacological studies demonstrated the existence of mutual inhibitory synaptic interactions between TRN cells. In the ferret's perigeniculate nucleus (visual sector of the TRN) maintained in vitro, local application of glutamate evoked $GABA_A$-dependent inhibitory postsynaptic potentials in cells located in the neighborhood of the application (Sanchez-Vives et al., 1997). Dual single-cell recordings of adjacent TRN cells in thalamic slices of young rodents (rats and mice) were, however, unable to reveal the presence of inhibitory chemical synaptic connections (Landisman et al., 2002). Taken together, these findings suggest that the excitation of a pool of TRN cells can trigger remote widespread inhibitions in the TRN. Intra-TRN inhibitions would limit the number and synchrony of TRN cellular bursts (Sohal and Huguenard, 2003).

CHEMICAL AXOAXONIC SYNAPSES

Of more than 100 juxtacellularly labeled TRN neurons of the adult rat, about 10% have poorly ramifying varicose local axon collaterals, which remain indistinguishable from parent distal dendrites (Pinault et al., 1997). Correlated light- and electron-microscope observations of such intrinsic axon-like processes revealed that their varicosities and intervaricose segments are—like the hillock and initial segment of TRN axons—postsynaptic structures, which are contacted by a large number of boutons that, for most of them, form asymmetrical synapses and are non-immunoreactive for GAB_A (Pinault et al., 1997). They, therefore, are the structural substrate of non-GABA-mediated, excitatory, or modulatory axoaxonic synapses. Thereby, these non-GABAergic axoaxonic synapses cannot be directly implicated in intranuclear cell-cell communications. On the other hand, some of the intra-TRN collaterals of TC axons subserve a powerful control on the output of TRN neurons. Because the axon initial segment is the privileged site for action potential initiation, this region is strategically more important than the somatodendritic complex for the control of action potential initiation. Therefore, direct control of this particular region by excitatory inputs might shunt the conventional somatodendritic integration processes. This is an important issue to clarify because thalamic inputs on TRN axons might be a powerful mechanism to generate synchronized oscillations in the TC-TRN system. Assuming that one TC neuron pinpoints the axon hillock and/or the initial segment of several TRN neurons, an action

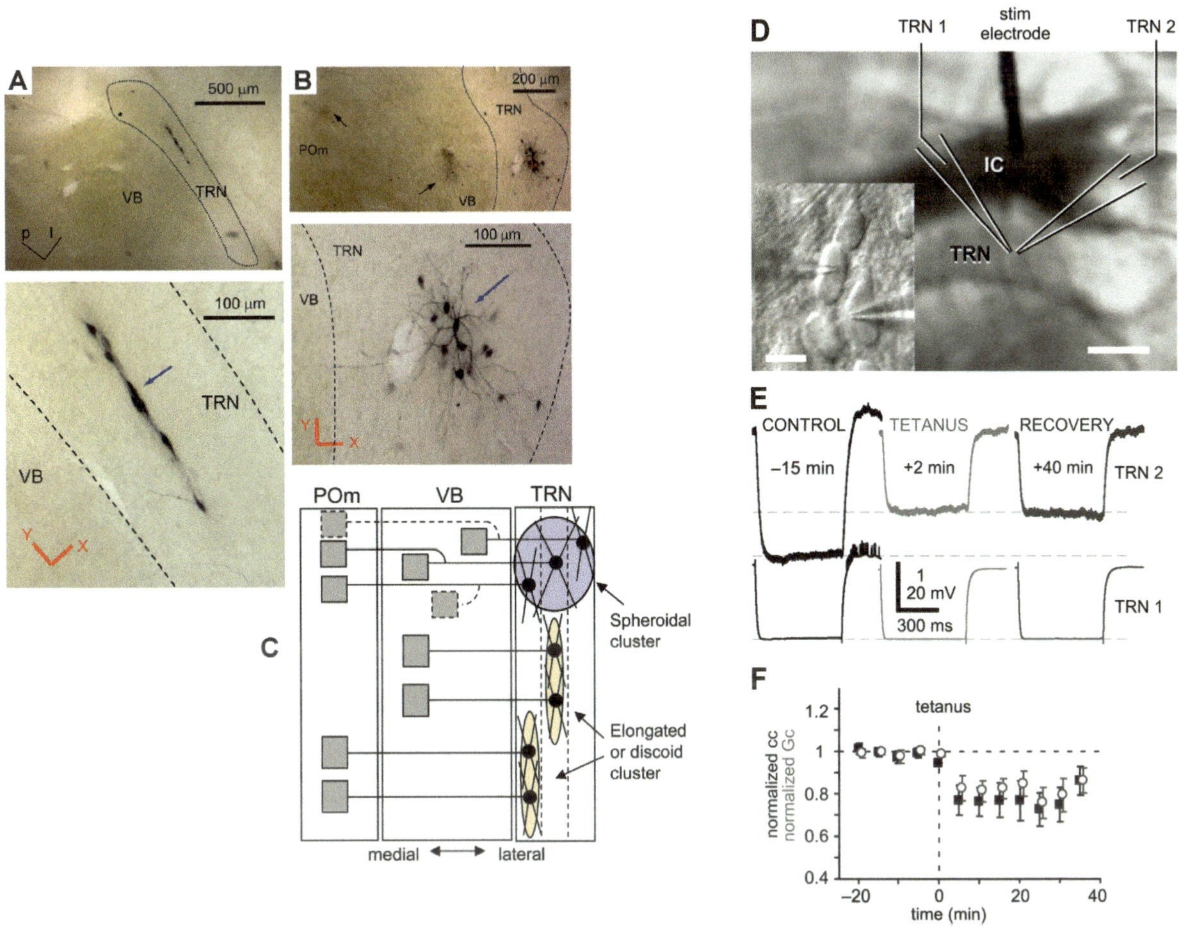

Figure 16.5. Anatomo-functional properties of TRN gap juctions. (A). Elongated dye-coupled clusters from mouse TRN. An example of an elongated cluster from a mouse horizontal slice. Top: low-magnification image of the cluster. p = posterior; l = lateral. Bottom: high-magnification image of the same cluster. The blue arrow indicates the injected neuron. Red lines indicate the directions of the *x* and *y* axes used for analysis. (B). Spherical dye-coupled clusters from mouse TRN. An example of a spherical cluster from a mouse TC slice. Top: low-magnification image of the cluster. Black arrows indicate axon terminals from an injected neuron. dp = dorsoposterior; la = latero-anterior. Bottom: high-magnification image of the same cluster. The blue arrow indicates the injected neuron. Red lines indicate approximate directions of the *x* and *y* axes. Some cells are blurry because they are out of the focal plane. (C). Schematic diagram of distinct gap junction-coupling connectivity patterns. The blue circle indicates the connectivity pattern of spherical clusters, and the yellow ellipses indicate connectivity patterns of elongated or discoid clusters. Black circles and gray boxes indicate somata and axon terminals, respectively. Lines indicate axons or dendrites. Axons with dashed lines in the spherical cluster indicate uncertainty whether neurons in this cluster type are exclusively cells with branched axons or mixed populations with branched and non-branched axons. Adapted from Lee et al. (2014). (D). Effects of tetanic stimulation on gap junction strength. IR-DIC (infrared differential interference contrast) image of the thalamocortical slice with diagram of tetanus paradigm. Scale bars indicate 1 mm and 20 mm (inset). Inset: high-magnification view of recorded neurons in TRN. (E) Injected cell responses (TRN 1) and electrical coupling responses (TRN 2) before tetanus, after tetanus, and after recovery. Control cc equaled 0.11 and 0.07 immediately post-tetanus to the end of recording. (F) Time-course measure of coupling coefficient in a pair of neurons before and after tetanus. Normalized cc (black squares) and Gc (gray open circles) for 9 pairs of cells before and after tetanic stimulation. Y values are normalized to the average of values collected before stimulation for each pair. Each trace is the average of 10 to 30 trials (mean ± SE). Adapted from Landisman and Connors (2005).

potential or burst discharge in the TC cell could make numerous TRN neurons fire or burst simultaneously, which, in turn, could inhibit a remote subset of TRN neurons in a synchronous manner, and so on. The synchronization of bursting TRN neurons during sleep spindles, therefore, could be generated either by cell-cell communication in the TRN and/or by TC inputs on their axon hillock and/or their initial segment.

CHEMICAL DENDRODENDRITIC SYNAPSES

Electron microscope examinations of TRN dendrites of adult cats and rats revealed that they form a local network of symmetrical GABAergic dendrodendritic synapses (Ide, 1982; Montero and Singer, 1984; Deschenes et al., 1985; Williamson et al., 1994; Pinault et al., 1997). The presynaptic dendrites contained pleomorphic vesicles, small cisterna of endoplasmic reticulum, microtubules, and mitochondria, whereas the postsynaptic dendrites were morphologically similar except that they were devoid of synaptic vesicles. Dendrodendritic synapses may play a significant role in the TRN, inhibitions or excitations. The polarity (hyperpolarization vs. depolarization) of membrane potential changes induced by such synapses may depend on the value of the chloride equilibrium potential with respect to the actual membrane potential of the cell. Whether or not TRN dendrodendritic synapses are reciprocal remains an open question.

PUNCTA ADHAERENTIAE

Membrane cell adhesion junctions represent an intriguing singular morphological characteristic of the TRN (Pinault et al., 1997). They may be an important structural substrate for synaptic (electrical and chemical) and nonsynaptic (via electric fields) functions. Currently, we can only speculate as to what the functional significance of such membrane appositions is. For instance, they might increase extracellular resistance, thereby facilitating nonsynaptic (ionic and/or electric) interactions. Also, assuming that the dendritic membranes of TRN cells contain unpaired connexon channels, they would operate in at least 2 ways: (1) they might form transient gap-junctional intercellular channels, which would be involved in calcium-related oscillatory activities (Goodenough and Paul, 2003); and/or (2) such hemichannels might be involved in the control of dendritic Ca^{++} currents, assuming that these connexons are numerous enough to generate efficient current sinks (Kamermans et al., 2001).

ELECTRICAL DENDRODENDRITIC SYNAPSES

In contrast to chemical synapses, electrical synapses are not concerned with a synaptic cleft, as the membranes of the electrically connected cells are juxtaposed. Electrical synapses are composed of clusters of gap junction channels made of proteins called connexins, which allow the direct and bidirectional, low-resistance spread/exchange (of charged ions and small signaling molecules) from one cell to another without delay. Thereby, nearby neurons can mutually interact. Electrical coupling via gap junctions of GABAergic neurons generates a strong inhibitory synaptic drive, which entrains and synchronizes the firing of their postsynaptic neurons (Deans et al., 2001; Traub et al., 2001; Buhl et al., 2003). The morphological substratum of electrical synapses is thought to be gap junctions. Connexin-36 has recently been observed at focal sites on plasma membranes of TRN dendrites and somata in the postnatal mouse brain without any association with identifiable gap junctions (Liu and Jones, 2003). Such close membrane junctional complexes are difficult to see in brain tissue.

Dye coupling could be revealed in thalamic slices of pre-adult mice. In the TRN, 2 types of spatially elongated clusters of electrically coupled neurons (Lee et al., 2014) (Figure 16.5A–C): (1) clusters in the medial or middle tiers, with axons projecting to either a FO or HO nucleus of the dorsal thalamus; (2) clusters occupying the 3 (medial, middle, and lateral) tiers, and innervating both FO and HO nuclei. Dye coupling was reduced among cells of connexin-36 knock-out mice (Lee et al., 2014).

For now, our knowledge of TRN electrical synapses is principally based on comprehensive functional studies performed in TRN slices from pre-adult rodents. Paired-cell recordings (inter-cell distance: up to 35 μm) in thalamic slices of young rodents (rat and mouse, P14–P21) demonstrated the presence of functional electrical cellular couplings in the TRN, with strong low-pass filtering characteristics (Landisman et al., 2002). Such couplings have not been recorded in mice with a null mutation for the connexion-36 gene. They are thought to play a significant role in cellular synchronizations, especially of low-frequency events such as low-threshold Ca^{++} spikes, so acting as low-pass filters. Such electrically generated synchronizations (in the range of 10 Hz) concern small clusters of TRN cells (Long et al., 2004). Intracellular recordings performed in thalamic slices of the adult cat, combined with the activation of metabotropic glutamate receptors, demonstrated the persistence of functional electrical synapses in the TRN (Blethyn et al., 2008).

Like chemical synapses, electrical synapses are endowed with activity-dependent plastic properties, which would play a critical role in the flow of information in the TC system. When 2 electrically coupled TRN neurons are made to fire synchronized bursts, the electrical synapses connecting them undergo long-term depression (Haas et al., 2011). Activation of group I and group II metabotropic glutamate receptors causes long-term depression and long-term potentiation of electrical synapses in the rat TRN (Landisman and Connors, 2005; Wang et al., 2015) (Figure 16.5D–F). The plasticity of electrical synapses is also modulated by $GABA_B$ receptors, suggesting that afferent GABAergic inputs from the basal forebrain, globus pallidus, or the substantia nigra pars reticulata may play a critical role in the intra-TRN inhibitory effects (Wang and Haas, 2021). Together, these findings support the hypothesis that TRN electrical dendrodendritic synapses with their plasticity mechanisms can interact with other neurotransmitter systems (Pereda et al., 2013; Wang et al., 2015; Wang and Haas, 2021).

SEARCHLIGHTS OF ATTENTION DEFINED BY SETS OF DIVERSE INHIBITIONS

As the guardian of the gateway to the cortex, the TRN is expected to direct searchlights of attention simultaneously to functionally relevant thalamic districts in FO and HO nuclei (Crick, 1984). Based on the patterns of innervation, it is tempting to propose that the TRN may perform at least 2 types of control over TC information on its way to the cerebral cortex: (1) one type, a parallel processing, would involve the mapped parallel TRN axonal projections toward FO and HO thalamic nuclei; (2) the other type, a combining processing, would principally involve the divergent axonal projections to both FO and HO thalamic nuclei. Knowing that inputs from different cortical areas may converge into a common TRN sector (Crabtree, 1992a) suggests that divergence of TRN axonal projections from this sector to the thalamus mirrors its converging afferents.

Attention impacts activities of neurons and neural circuits during sensory information processing, which involves target amplification and distractor suppression. For instance, shift in visual attention can influence the activity of visual TC neurons via TRN-mediated inhibitions (McAlonan et al., 2006, 2008). The sensory thalamus-TRN circuits are under the influence of the prefrontal cortex, basal ganglia pathways, and amygdala, which enable selection between sensory modalities and amplification of sensory responses (Aizenberg et al., 2019; Nakajima et al., 2019). As the attention to a sensory modality interferes with the attention to other sensory modalities, the TRN, in concert with the related FO and HO thalamic nuclei, may be a crucial player in cross-modal attentional gating. The cell-to-network mechanisms underlying both cross-modal attention and cross-modal sensory interactions are the object of speculation. Based on the presented material, I suggest that the cross-modal interferences are underpinned by sets of TRN-mediated lateral and recurrent inhibitions, which may include crosstalk inhibitions. So, the TRN would generate, in the dorsal thalamus and via lateral or surround inhibition, thalamic hot spots of synchronized TC activities. The hodology and innervation patterns of the loop CT-TRN-TC circuit strongly indicate that the GABAergic TRN neurons would be implicated in both bottom-up (from environment/sensory receptors to the cerebral cortex) and top-down (from the cortex to subcortical structures/motor commands) processing. This means that TRN-mediated lateral and recurrent inhibitions would target synchronized TC activities induced by afferent inputs (sensory/motor) and layer 5 cortical (cognitive) driver inputs. Combined recurrent and surround inhibitions would lead to an increase in the TC signal-to-noise ratio of relevant information, for instance during sensory processing/perception. The lateral inhibition would attenuate or delete irrelevant or distractive information (distractor suppression), thus ensuring the TC transfer of attended information (target amplification) to the target cortical areas. In other words, lateral inhibition and its counterpart feedback disinhibition would contribute to enhancing the center-surround contrast (Figure 16.4C), for instance during sensory processing of discrimination task (attention/perception). In short, TRN-mediated lateral inhibition would, as a spatial filter, recruit pertinent information in the highly distributed TC systems. As for the role of recurrent inhibition, it would be involved more in temporal than spatial filtering. It would reduce the firing of the TC neurons that triggered the feedback inhibition (Figure 16.4D), thereby leading to synchronized TC activities. Both lateral and recurrent inhibitions may function in isolation or in concert during normal CT-TRN-TC operations. These combined various inhibitions would generate maps of gradients of TC activities underlying all functional modalities.

With its anatomo-functional divisions, subdivisions, and chemical and electrical synaptic properties, the TRN seems endowed with a modular set of filters with versatile functioning during both bottom-up and top-down processing. Also, with its topographically mapped TC and CT inputs, the TRN resembles a "combinatorial matrix" tightly and loosely involved in the TC integration, attention-related filtering, and transfer of specific (sensory and motor) and contextual (cognitive and emotional) information.

Understanding the anatomical and architectural properties of the loop CT-TRN-TC circuits is fundamental to apprehending their physiological properties and functions, and the pathophysiology of neurobiological disorders (see **Chapters 65 and 66 in Section XII**). Lesions of the TRN lead to an attentional deficit (Bucherelli et al., 1993; Weese et al., 1999). Dysfunction of the TRN may cause abnormal attention-related behavior and a decrease in cognitive performances in many neurological and psychiatric disorders, including schizophrenia, autism, and attention deficit hyperactivity disorder (Ferrarelli and Tononi, 2011; Pinault, 2011; Saletin et al., 2017; Krol et al., 2018; Steullet et al., 2018). Advances in understanding the fundamental anatomo-functional properties of neural networks allow us to apprehend their pathophysiological properties better in an attempt to develop innovative therapy (Pinault, 2017).

ACKNOWLEDGMENTS

This chapter was prepared with the financial support of the Université de Strasbourg, and the French Institute of Health and Medical Research (INSERM, Institut National de la Santé et de la Recherche Médicale).

REFERENCES

Aizenberg M, Rolon-Martinez S, Pham T, Rao W, Haas JS, Geffen MN (2019). Projection from the amygdala to the thalamic reticular nucleus amplifies cortical sound responses. *Cell Rep.* 28:605–615.

Blethyn KL, Hughes SW, Crunelli V (2008). Evidence for electrical synapses between neurons of the nucleus reticularis thalami in the adult brain in vitro. *Thalamus Relat Syst.* 4:13–20.

Bourassa J, Deschênes M (1995). Corticothalamic projections from the primary visual cortex in rats: A single fiber study using biocytin as an anterograde tracer. *Neuroscience.* 66:253–263.

Bourassa J, Pinault D, Deschenes M (1995). Corticothalamic projections from the cortical barrel field to the somatosensory thalamus in rats: A single-fibre study using biocytin as an anterograde tracer. *Eur J Neurosci.* 7:19–30.

Bragg EM, Fairless EA, Liu S, Briggs F (2017). Morphology of visual sector thalamic reticular neurons in the macaque monkey suggests retinotopically specialized, parallel stream-mixed input to the lateral geniculate nucleus. *J Comp Neurol.* 525:1273–1290.

Bucherelli C, Tassoni G, Bures J (1993). Differential effect of functional ablation of thalamic reticular nucleus on the acquisition of passive and active avoidance. *Int J Neurosci.* 73:77–84.

Buhl DL, Harris KD, Hormuzdi SG, Monyer H, Buzsaki G (2003). Selective impairment of hippocampal gamma oscillations in connexin-36 knock-out mouse in vivo. *J Neurosci.* 23:1013–1018.

Burgunder JM, Young WS (1992). Expression of cholecystokinin and somatostatin genes in the human thalamus. *J Comp Neurol.* 324:14–22.

Carroll BJ, Sampathkumar V, Kasthuri N, Sherman SM (2022). Layer 5 of cortex innervates the thalamic reticular nucleus in mice. *Proc Natl Acad Sci U S A.* 119 (38):e2205209119.

Crabtree JW (1992a). The somatotopic organization within the cat's thalamic reticular nucleus. *Eur J Neurosci.* 4:1352–1361.

Crabtree JW (1992b). The somatotopic organization within the rabbit's thalamic reticular nucleus. *Eur J Neurosci.* 4:1343–1351.

Crabtree JW (1996). Organization in the somatosensory sector of the cat's thalamic reticular nucleus. *J Comp Neurol.* 366:207–222.

Crabtree JW (1999). Intrathalamic sensory connections mediated by the thalamic reticular nucleus. *Cell Mol Life Sci.* 56:683–700.

Crick F (1984). Function of the thalamic reticular complex: The searchlight hypothesis. *Proc Natl Acad Sci U S A.* 81:4586–4590.

Deans MR, Gibson JR, Sellitto C, Connors BW, Paul DL (2001). Synchronous activity of inhibitory networks in neocortex requires electrical synapses containing connexin36. *Neuron.* 31:477–485.

Deschenes M, Madariaga-Domich A, Steriade M (1985). Dendrodendritic synapses in the cat reticularis thalami nucleus: A structural basis for thalamic spindle synchronization. *Brain Res.* 334:165–168.

Ferrarelli F, Tononi G (2011). The thalamic reticular nucleus and schizophrenia. *Schizophr Bull.* 37:306–315.

Gentet LJ, Ulrich D (2003). Strong, reliable and precise synaptic connections between thalamic relay cells and neurones of the nucleus reticularis in juvenile rats. *J Physiol.* 546:801–811.

Goodenough DA, Paul DL (2003). Beyond the gap: Functions of unpaired connexon channels. *Nat Rev Mol Cell Biol.* 4:285–294.

Guillery RW, Sherman SM (2002). Thalamic relay functions and their role in corticocortical communication: generalizations from the visual system. *Neuron.* 33:163–175.

Haas JS, Zavala B, Landisman CE (2011). Activity-dependent long-term depression of electrical synapses. *Science.* 334:389–393.

Ide LS (1982). The fine structure of the perigeniculate nucleus in the cat. *J Comp Neurol.* 210:317–334.

Kamermans M, Fahrenfort I, Schultz K, Janssen-Bienhold U, Sjoerdsma T, Weiler R (2001). Hemichannel-mediated inhibition in the outer retina. *Science.* 292:1178–1180.

Kolmac CI, Mitrofanis J (1997). Organisation of the reticular thalamic projection to the intralaminar and midline nuclei in rats. *J Comp Neurol.* 377:165–178.

Krol A, Wimmer RD, Halassa MM, Feng G (2018). Thalamic reticular dysfunction as a circuit endophenotype in neurodevelopmental disorders. *Neuron.* 98:282–295.

Lam YW, Sherman SM (2011). Functional organization of the thalamic input to the thalamic reticular nucleus. *J Neurosci.* 31:6791–6799.

Landisman CE, Connors BW (2005). Long-term modulation of electrical synapses in the mammalian thalamus. *Science.* 310:1809–1813.

Landisman CE, Long MA, Beierlein M, Deans MR, Paul DL, Connors BW (2002). Electrical synapses in the thalamic reticular nucleus. *J Neurosci.* 22:1002–1009.

Lee SC, Patrick SL, Richardson KA, Connors BW (2014). Two functionally distinct networks of gap junction-coupled inhibitory neurons in the thalamic reticular nucleus. *J Neurosci.* 34:13170–13182.

Li Y, Lopez-Huerta VG, Adiconis X, Levandowski K, Choi S, Simmons SK, Arias-Garcia MA, Guo B, Yao AY, Blosser TR, Wimmer RD, Aida T, Atamian A, Naik T, Sun X, Bi D, Malhotra D, Hession CC, Shema R, Gomes M, Li T, Hwang E, Krol A, Kowalczyk M, Peca J, Pan G, Halassa MM, Levin JZ, Fu Z, Feng G (2020) Distinct subnetworks of the thalamic reticular nucleus. *Nature.* 583:819–824.

Liu XB, Jones EG (2003). Fine structural localization of connexin-36 immunoreactivity in mouse cerebral cortex and thalamus. *J Comp Neurol.* 466:457–467.

Liu XB, Warren RA, Jones EG (1995). Synaptic distribution of afferents from reticular nucleus in ventroposterior nucleus of cat thalamus. *J Comp Neurol.* 352:187–202.

Lo FS, Sherman SM (1994). Feedback inhibition in the cat's lateral geniculate nucleus. *Exp Brain Res.* 100:365–368.

Long MA, Landisman CE, Connors BW (2004). Small clusters of electrically coupled neurons generate synchronous rhythms in the thalamic reticular nucleus. *J Neurosci.* 24:341–349.

McAlonan K, Cavanaugh J, Wurtz RH (2006). Attentional modulation of thalamic reticular neurons. *J Neurosci.* 26:4444–4450.

McAlonan K, Cavanaugh J, Wurtz RH (2008). Guarding the gateway to cortex with attention in visual thalamus. *Nature.* 456:391–394.

Montero VM, Singer W (1984). Ultrastructure and synaptic relations of neural elements containing glutamic acid decarboxylase (GAD) in the perigeniculate nucleus of the cat. A light and electron microscopic immunocytochemical study. *Exp Brain Res.* 56:115–125.

Nakagawa Y (2019). Development of the thalamus: From early patterning to regulation of cortical functions. *Wiley Interdiscip Rev Dev Biol.* 8:e345.

Nakajima M, Schmitt LI, Halassa MM (2019). Prefrontal cortex regulates sensory filtering through a basal ganglia-to-thalamus pathway. *Neuron.* 103:445–458.

Ohara PT, Lieberman AR (1985). The thalamic reticular nucleus of the adult rat: Experimental anatomical studies. *J Neurocytol.* 14:365–411.

Ohara PT, Havton LA (1996). Dendritic arbors of neurons from different regions of the rat thalamic reticular nucleus share a similar orientation. *Brain Res.* 731:236–240.

Pereda AE, Curti S, Hoge G, Cachope R, Flores CE, Rash JE (2013). Gap junction-mediated electrical transmission: regulatory mechanisms and plasticity. *Biochim Biophys Acta.* 1828:134–146.

Pinault D (2004). The thalamic reticular nucleus: Structure, function and concept. *Brain Res Rev.* 46:1–31.
Pinault D (2011). Dysfunctional thalamus-related networks in schizophrenia. *Schizophr Bull.* 37:238–243.
Pinault D (2017). A neurophysiological perspective on a preventive treatment against schizophrenia using transcranial electric stimulation of the corticothalamic pathway. *Brain Sci.* 7, 34, https://doi.org/10.3390/brainsci7040034.
Pinault D, Deschenes M (1998a). Projection and innervation patterns of individual thalamic reticular axons in the thalamus of the adult rat: A three-dimensional, graphic, and morphometric analysis. *J Comp Neurol.* 391:180–203.
Pinault D, Deschenes M (1998b). Anatomical evidence for a mechanism of lateral inhibition in the rat thalamus. *Eur J Neurosci.* 10:3462–3469.
Pinault D, Bourassa J, Deschenes M (1995a). The axonal arborization of single thalamic reticular neurons in the somatosensory thalamus of the rat. *Eur J Neurosci.* 7:31–40.
Pinault D, Bourassa J, Deschenes M (1995b). Thalamic reticular input to the rat visual thalamus: A single fiber study using biocytin as an anterograde tracer. *Brain Res.* 670:147–152.
Pinault D, Smith Y, Deschenes M (1997). Dendrodendritic and axoaxonic synapses in the thalamic reticular nucleus of the adult rat. *J Neurosci.* 17:3215–3233.
Pollin B, Rokyta R (1982). Somatotopic organization of nucleus reticularis thalami in chronic awake cats and monkeys. *Brain Res.* 250:211–221.
Prasad JA, Carroll BJ, Sherman SM (2020). Layer 5 corticofugal projections from diverse cortical areas: Variations on a pattern of thalamic and extrathalamic targets. *J Neurosci.* 40:5785–5796.
Pritz MB (2015). Crocodilian forebrain: Evolution and development. *Integr Comp Biol.* 55:949–961.
Saletin JM, Coon WG, Carskadon MA (2017). Stage 2 sleep EEG sigma activity and motor learning in childhood ADHD: A pilot study. *J Clin Child Adolesc Psychol.* 46:188–197.
Sanchez-Vives MV, Bal T, McCormick DA (1997). Inhibitory interactions between perigeniculate GABAergic neurons. *J Neurosci.* 17:8894–8908.
Scheibel ME, Scheibel AB (1966). The organization of the ventral anterior nucleus of the thalamus: A Golgi study. *Brain Res.* 1:250–268.
Sherman SM, Guillery RW (2009) Exploring the thalamus and its role in cortical function, 2nd Edition. The MIT Press.
Shosaku A (1986). Cross-correlation analysis of a recurrent inhibitory circuit in the rat thalamus. *J Neurophysiol.* 55:1030–1043.
Shosaku A, Kayama Y, Sumitomo I (1984). Somatotopic organization in the rat thalamic reticular nucleus. *Brain Res.* 311:57–63.
Shosaku A, Kayama Y, Sumitomo I, Sugitani M, Iwama K (1989). Analysis of recurrent inhibitory circuit in rat thalamus: Neurophysiology of the thalamic reticular nucleus. *Prog Neurobiol.* 32:77–102.
Sohal VS, Huguenard JR (2003). Inhibitory interconnections control burst pattern and emergent network synchrony in reticular thalamus. *J Neurosci.* 23:8978–8988.
Steriade M, Parent A, Hada J (1984). Thalamic projections of nucleus reticularis thalami of cat: A study using retrograde transport of horseradish peroxidase and fluorescent tracers. *J Comp Neurol.* 229:531–547.
Steullet P, Cabungcal JH, Bukhari SA, Ardelt MI, Pantazopoulos H, Hamati F, Salt TE, Cuenod M, Do KQ, Berretta S (2018). The thalamic reticular nucleus in schizophrenia and bipolar disorder: Role of parvalbumin-expressing neuron networks and oxidative stress. *Mol Psychiatry.* 23:2057–2065.
Traub RD, Kopell N, Bibbig A, Buhl EH, LeBeau FE, Whittington MA (2001). Gap junctions between interneuron dendrites can enhance synchrony of gamma oscillations in distributed networks. *J Neurosci.* 21:9478–9486.
Wang H, Haas JS (2021). GABABR modulation of electrical synapses and plasticity in the thalamic reticular nucleus. *Int J Mol Sci.* 22(22):12138. doi:10.3390/ijms222212138.
Wang Z, Neely R, Landisman CE (2015). Activation of group I and group II metabotropic glutamate receptors causes LTD and LTP of electrical synapses in the rat thalamic reticular nucleus. *J Neurosci.* 35:7616–7625.
Weese GD, Phillips JM, Brown VJ (1999). Attentional orienting is impaired by unilateral lesions of the thalamic reticular nucleus in the rat. *J Neurosci.* 19:10135–10139.
Whilden CM, Chevee M, An SY, Brown SP (2021). The synaptic inputs and thalamic projections of two classes of layer 6 corticothalamic neurons in primary somatosensory cortex of the mouse. *J Comp Neurol.* 529:3751–3771.
Williamson AM, Ohara PT, Ralston HJ (1993). Electron microscopic evidence that cortical terminals make direct contact onto cells of the thalamic reticular nucleus in the monkey. *Brain Res.* 631:175–179.
Williamson AM, Ohara PT, Ralston DD, Milroy AM, Ralston HJ (1994). Analysis of gamma-aminobutyric acidergic synaptic contacts in the thalamic reticular nucleus of the monkey. *J Comp Neurol.* 349:182–192.
Zikopoulos B, Barbas H (2012). Pathways for emotions and attention converge on the thalamic reticular nucleus in primates. *J Neurosci.* 32:5338–5350.

SECTION V

SENSORY SYSTEMS

Andrew J. King and Judith A. Hirsch

The sensory pathways in the central nervous system (CNS) are often viewed as a series of hierarchically organized structures that transmit information detected by the peripheral receptor cells in the sense organs to the cerebral cortex, where neural responses are associated with perception. Within this simple feedforward framework, the first-order sensory nuclei in the thalamus are often considered to be passive relays that direct modality-specific information to the thalamorecipient layers of the corresponding primary cortical areas. This is an impoverished perspective, however; it overlooks the presence within each sensory system of multiple parallel processing channels, including at the level of both the thalamus and the cortex. Furthermore, the feedforward model of sensory processing does not take account of the massive descending projections that enable the cortical areas to influence the activity of neurons both within the regions of the thalamus from which they receive the bulk of their ascending inputs and in a host of other subcortical structures. Consequently, the corticocentric focus that has dominated our understanding of the neural basis for perception needs to be expanded in light of the growing evidence that sensory processing is continually updated by the bidirectional flow of signals between the thalamus and the cortex.

The chapters in this section examine thalamocortical interactions in the visual, auditory, somatosensory, vestibular, olfactory, and gustatory systems, as well as in the context of multisensory processing. We have chosen to group the chapters by sensory modality, but they address a number of common topics. These include (1) the organization and connections of the major sensory first-order and higher-order thalamic nuclei; (2) how sensory representations change between the first-order thalamic nuclei and primary sensory cortical areas; (3) the organization and functions of corticothalamic feedback from layer 6 neurons, which return massive descending connections to their associated first-order and higher-order thalamic nuclei, and (4) similarities and differences across species and modalities. Some consideration is also given to the role of layer 5 neurons in sensory cortical areas, which deliver driving inputs to higher-order nuclei in the sensory thalamus, thereby transmitting information to those nuclei and providing a possible transthalamic route by which different cortical areas can communicate with each other.

In Chapter 17, Samuel G. Solomon reviews the structure, connections, and response properties of neurons in the dorsal lateral geniculate nucleus (LGN) in different species and considers the functional advantages of organizing this first-order thalamic nucleus into layer-specific parallel pathways that transmit visual motion and form information to the primary visual cortex (V1). In Chapter 18, Ronan O'Shea, Mandi C. Severson, and Nicholas J. Priebe address the origins of the well-known selectivity of V1 neurons for stimulus orientation and direction of movement. They again emphasize the parallel nature of information processing at the level of the cortex and conclude that the spatiotemporal arrangement of LGN inputs constrains the receptive field properties of V1 neurons, which are then shaped by local intracortical circuits. W. Martin Usrey in Chapter 19 examines the massive feedback projections to the LGN from layer 6 neurons in V1, which are also organized into parallel streams that modulate the spatial and temporal properties of LGN neurons. In Chapter 20, Angela Y. Fan and James A. Bourne consider the role of the pulvinar, a higher-order thalamic nucleus, which, unlike the LGN, is connected to most areas of the visual cortex. Focusing primarily on primates, they discuss the possible involvement of the pulvinar in the early maturation of visuomotor behaviors and in blindsight.

We then turn to the auditory system, where the thalamus is again organized into a series of parallel pathways. In this case, however, the medial geniculate body (MGB) is not a single nucleus, but instead comprises three principal subdivisions: the ventral division (MGBv), which is the first-order nucleus that relays information to the primary auditory cortex (A1), and the dorsal (MGBd) and medial subdivisions (MGBm), which are both higher-order nuclei. Consequently, signals are distributed from the MGB to multiple cortical areas, as well as to other brain regions. In Chapter 21, Daniel A. Llano reviews the cellular organization, inputs, and outputs of the MGB, which not only provides the gateway to the auditory cortex, but is also closely linked with brain regions involved in processing the cognitive and emotional aspects of sounds. In Chapter 22, Jennifer F. Linden compares the response properties of neurons in the MGBv and in A1. In contrast to the greater visual feature selectivity of V1 neurons over those in the LGN, evidence for a comparable thalamocortical transformation in the representation of auditory information has not been found, other than differences in temporal fidelity to rapidly changing sounds and in sensitivity to the long-term statistics of sounds. One aspect of auditory processing where a difference between the thalamus and the cortex might be expected is speech perception, where previous research has been dominated by studies on the cortex. However, as Katharina von Kriegstein and Christa Müller-Axt describe in Chapter 23, neuroimaging studies and clinical observations in patients have shown that sensory thalamic areas are also involved in auditory and visual speech recognition, possibly through top-down predictive coding implemented by corticocothalamic interactions. Chapter 24, by Victoria M. Bajo and Andrew J. King, stays with this theme by discussing the organization and potential functions of auditory corticothalamic pathways. As in other sensory systems, descending projections appear to dynamically modulate the responses of MGB neurons, with research on auditory corticothalamic feedback also demonstrating the behavioral importance of these circuits.

The next three chapters address thalamocortical interactions in the somatosensory system. In Chapter 25, Christina Mo describes the organization of the ascending pathways that process somatosensory information in primates and rodents, where, once again, parallel processing streams are a key feature. Because of their use of whisker movements for exploratory behavior and tactile discrimination, rodents are particularly reliant on somatosensation. Using the rodent whisker system, Randy M. Bruno returns in Chapter 26 to the question of the extent to which sensory representations in the cortex are inherited or transformed from their thalamic inputs. He shows that thalamocortical axons from the first-order ventral posterior medial (VPM) nucleus branch to terminate in both the middle and deeper layers of the mouse barrel cortex, and that the response properties of the cortical neurons in those layers are largely determined by their thalamic inputs. Solange Brown then looks at the organization of descending corticothalamic circuits in the somatosensory system, with a focus on barrel cortex, in Chapter 27. She highlights the diversity of layer 6 corticothalamic neurons, subtypes of which differ in both their long-range and intracortical inputs, as well as in the first- (VPM) and higher-order (posterior medial, POm) thalamic nuclei and nuclear subdivisions they target.

We next turn to those sensory systems where there has been much less emphasis on thalamocortical processing. As Kathleen E. Cullen, Jérôme Carriot, and Maurice J. Chacron discuss in Chapter 28, there is no single thalamic region that exclusively processes vestibular signals. Instead, multiple areas of the thalamus receive vestibular inputs. They concentrate on two regions: the ventral posterior lateral nucleus (VPL), which is normally associated with the somatosensory system, but also plays a role in encoding unexpected self-motion, and the anterior dorsal nucleus (ADN), which integrates vestibular and other sensory cues to generate a head direction signal. Chapter 29, by Emmanuelle Courtiol, Arianna Maffei, and Alfredo Fontanini, considers the chemical senses. As is often pointed out, the olfactory system is the exception to the rule that all ascending sensory information must pass through the thalamus in order to reach the cerebral cortex. But while there is no first-order olfactory thalamic nucleus, the higher-order mediodorsal thalamus (MD) does receive olfactory inputs. By contrast, a primary sensory area, the ventral posteromedial nucleus of the thalamus (VPMpc), is involved in encoding taste quality and palatability and is reciprocally connected with the gustatory portion of the insular cortex.

Although it is convenient to consider each sensory system in isolation in the laboratory, an ability to integrate information across the senses is critical for many aspects of perception and behavior. Multisensory convergence is found widely across the cerebral cortex, where it is typically thought to arise from corticocortical connections. In Chapter 30, the final chapter in this section, Annamaria Barczak, Monica N. O'Connell, and Charles E. Schroeder explore the role of the thalamus in multisensory processing. They address the possibility that the thalamus contributes to the modulation of ongoing neuronal oscillations, which has been implicated in the integration of different sensory inputs, and consider whether corticothalamocortical loops enable communication between different sensory cortical areas.

17.

A COMPARATIVE APPROACH TO PARALLEL GENICULOCORTICAL PATHWAYS

Samuel G. Solomon

ANATOMICAL AND FUNCTIONAL EVIDENCE FOR PARALLEL PATHWAYS

Our understanding of parallel pathways has been driven by distinct stages of research, each dominated by different species (particularly domestic cat, nonhuman primate and human, and laboratory mouse), and it is helpful to understand the evolution of understanding by presenting this work in partially chronological order. The primary literature here is extensive, and this short chapter cannot do it all justice, so the reader is encouraged to explore excellent reviews, which are indicated where they are available.

INITIAL STUDIES

Anatomical work had already shown the importance of the dorsal lateral geniculate nucleus (LGN) in connecting retina to primary visual cortex (V1), and clear segregation of nerve cell bodies within LGN, before techniques for recording from single neurons in central brain areas were established. Physiological measurements in the retina and cortex were already providing important insights, and initial measurements in cat LGN therefore aimed to establish if visual responses of LGN resembled their retinal afferents or cortical targets. This work found presence of ON- and OFF-receptive fields with center-surround receptive field organization like that already described for retinal ganglion cells (RGCs). Comparison of response properties between dorsal (A, A1) layers and ventral (C) layers of cat LGN hinted at functional segregation, with larger receptive fields in layer C accompanied by weaker responses and longer visual latency. Contemporaneous measurements from monkey LGN also showed receptive fields with center-surround organization, and additional presence of color-sensitive (cone-opponent) responses. This work also hinted at functional segregation, with stronger cone-opponency in dorsal (parvocellular, P) layers and weaker cone-opponency in ventral (magnocellular, M) layers.

X, Y AND W PATHWAYS IN THE CAT

Techniques for tracing of brain pathways were still in their infancy, so early progress was driven by physiological measurements, particularly from cat RGCs. Several attempts were made to derive classification schemes to explain these functional properties, but the scheme that gained most traction was the identification of 2 dominant subclasses of cat RGC (X and Y) with further subdivision into ON and OFF cells within each class (Enroth-Cugell and Robson, 1966). The functional properties of X- and Y-cells were, respectively, linear or nonlinear spatial summation, smaller or larger receptive fields, sustained or transient light response, and slower or faster conduction velocity. This scheme became particularly attractive when the functional classes were shown to correspond to morphologically distinct classes of RGC (Figure 17.1a; Boycott and Wassle, 1974). X and Y RGCs had different central projections: X-cells projected to layers A/A1 of LGN; Y-cells also projected to A/A1, but had collaterals including LGN layer C and superior colliculus. Other RGCs, with diverse morphology and diverse receptive field properties, were grouped into a third class ("W"; Stone and Hoffmann, 1972), the projections of which included layer C of LGN as well as other central brain areas. Measurements from cat LGN showed that the functional distinctions between X-, Y- and W-like RGCs were largely preserved in thalamocortical cells, and further, that the morphology of X, Y, and W thalamocortical cells was different, even when they were in the same LGN layer (Sherman and Spear, 1982). The emerging picture of parallel retino-geniculo-cortical pathways was reinforced when the axons of X, Y, and W LGN cells were found to project onto different layers of V1 (for simplicity here we refer to area 17 as V1), and by the discovery that while LGN X-cells

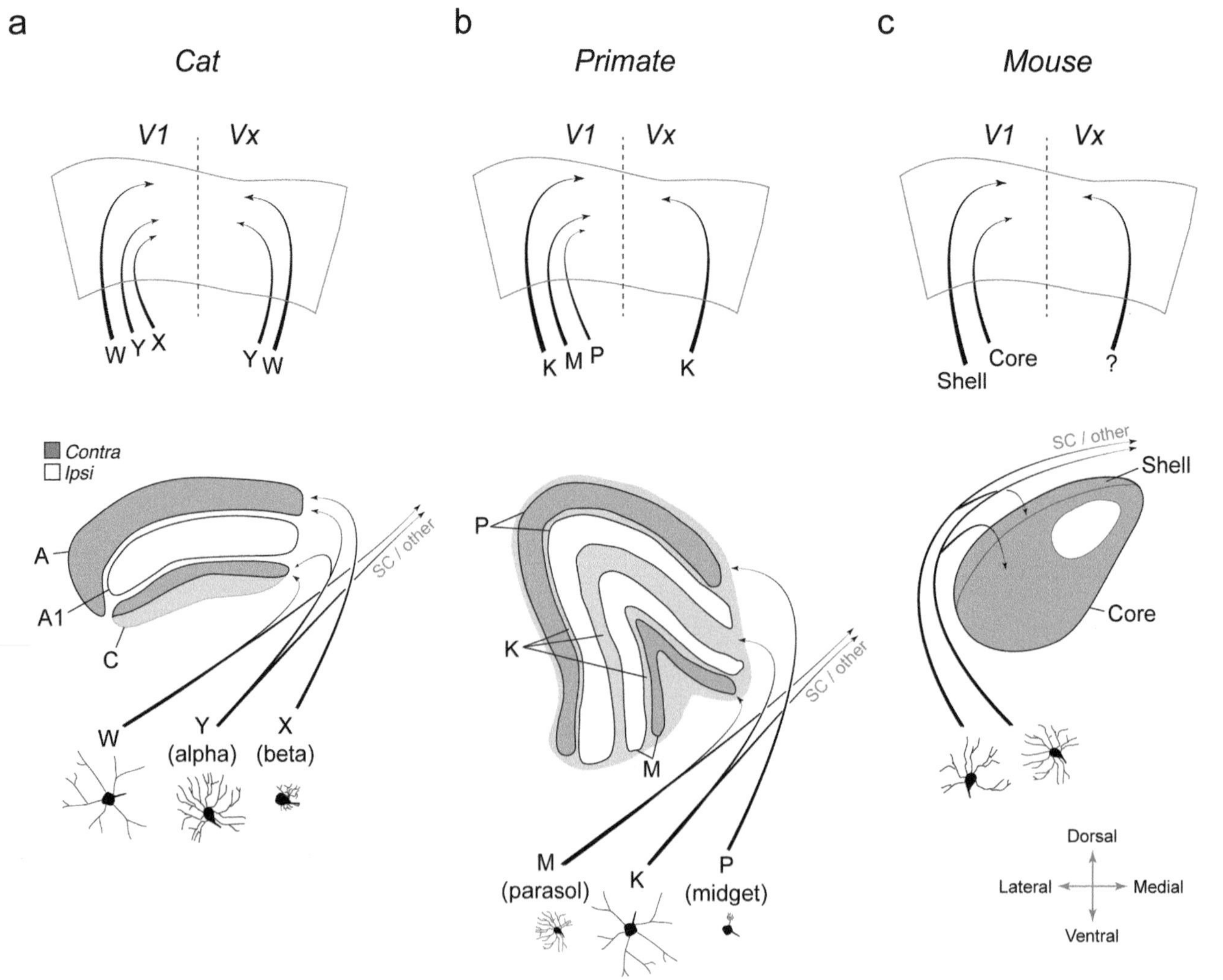

Figure 17.1. Overview of retino-geniculo-cortical pathways in cat, primate, and mouse. Top panels provide schematic overview of dorsal lateral geniculate nucleus (LGN) input to primary visual cortex (V1) and other visual cortical areas (Vx). Bottom panels show the overall structure of the LGN in each case. Layers receiving dominant input from the ipsilateral eye are indicated by unfilled regions; layers receiving dominant input from the contralateral eye are indicated by filled regions, and some of the retinal ganglion cell (RGC) input from the contralateral eye is indicated by the cartoons below. (a). Cat. LGN can be subdivided into A, A1, and C layers. Layers A/A1 receive major input from Y (alpha) RGCs and X (beta) RGCs, while layer C receives mainly W-pathway RGC inputs, comprising diverse classes. Y- and W-pathway RGCs in particular also project to other brain regions, including superior colliculus (SC). (b). Primate. Schematic is based on the slightly simpler organization found in New World (marmoset) monkey. LGN can be subdivided into P, M, and K subdivisions. P layers receive major input from midget RGCs, M layers receive major input from parasol RGCs, K layers from a diverse set of RGCs. M- and K-pathway RGCs often also project to other brain regions. (c). Mouse. LGN can be subdivided into core and shell regions. RGC classes in mouse are less morphologically distinguishable. Most RGCs also project to SC.

projected only to V1, LGN Y- and W-cells had additional cortical targets.

P, M AND K PATHWAYS IN PRIMATES

Rapidly improving techniques for tracing brain pathways showed that P-layers of the monkey LGN were targeted by a small and numerous class of RGC (midget; Figure 17.1b), while M-layers were targeted by a larger and less numerous class (parasol), and that LGN P- and M-cells projected to different sub-laminae of layer 4C in V1. Analysis of visual responses showed that neurons in the P-layers had sustained visual responses, small receptive fields, and stronger cone-opponency, while those in the M-layers had transient responses, larger receptive fields, and little cone-opponency. Further, P- and M-cells had overlapping but distinct responses to achromatic stimuli: M-cells responded better to low-contrast stimuli, and they responded to higher rates of temporal flicker. The high sampling density of P-cells, and the presence of cone-opponency, suggested they were important for high spatial acuity and color vision; the high-contrast sensitivity and fast temporal response of M-cells suggested they were important for motion vision. Subsequent behavioral studies in experimentally lesioned monkeys provided broadly consistent evidence, though the contribution of P- and M-cells to cortical processing and hence perception has been the subject of extensive debate (see, e.g., DeYoe and Van Essen, 1988; Livingstone and Hubel, 1988; Shapley, 1990; Merigan and Maunsell, 1993).

Overall anatomical and functional similarity between the 2 major retino-geniculo-cortical pathways in cat (X/Y) and monkey (P/M) encouraged attempts to assimilate them (Stone and Dreher, 1982). Attempts to align X/P and M/Y were, however, challenged particularly by the observation that only some M-cells showed the nonlinear spatial summation found in all Y-cells (Shapley and Perry, 1986). Whether nonlinear M-cells are a distinct subset or the extreme of a continuous distribution remains unclear. Additional challenges, however, remained in the fact that both X- and Y-cells in cat showed the presence of "contrast gain controls," such that response latency decreased at high contrast, response amplitude saturated at high visual contrast, and response often depended on spatial and temporal context within which a stimulus was viewed. These gain controls were present in M-cells in monkey, but largely absent from P-cells.

Absent from most early work on parallel pathways in monkey was consideration of the other, non-standard receptive fields and their anatomical correlates (Casagrande, 1994). However, the projection to monkey LGN was known to include a variety of morphologically distinct RGCs, RGCs with non-standard receptive field properties were known to be present, and anatomical work had shown that neurons intercalated between the P and M LGN layers ("K cells") expressed distinct neurochemicals that are also expressed by neurons in neighboring inferior pulvinar (e.g., Bakken et al., 2021), showed distinct projection to superficial layers of V1, and included neurons with direct projections to extrastriate visual cortex, bypassing V1. Evidence that these non-standard pathways might be important for vision was provided by recordings from one of the non-P-non-M RGC types that projected to LGN ("small bistratified cells"), which showed a distinct type of cone-opponent receptive field ("blue-ON"; Dacey and Lee, 1994). Subsequent measurements from LGN of New World (Martin et al., 1997) and Old World monkeys (Roy et al., 2009) showed that blue-ON responses could be localized to K-layers, and measurements of thalamocortical activity in V1 (Chatterjee and Callaway, 2003) showed projection of blue-ON responses onto superficial layers, where many K-pathway LGN cells project, distinct from deeper P- and M-like thalamocortical inputs. Other work has shown additional, diverse visual response properties among K-cells in LGN, at least in the New World marmoset monkey, including nonlinear spatial summation, "blue-ON" and "blue-OFF" cone-opponent receptive fields, orientation selectivity, "suppressed-by-contrast" responses, and binocular response. K-cell pathways are therefore functionally and anatomically distinct from those of P- and M-cells. Whether and how this collection of pathways can be likened to the similarly broad collection of "W-cells" in cat remains unclear, though it is useful to note here that "blue-ON" color response is localized to layer C in cat LGN, and therefore is part of the W pathway.

The lamination pattern of human LGN resembles that of nonhuman primates and the organization of retino-geniculo-cortical pathways in the human visual system is likely similar to that of nonhuman primates. The connectivity between RGCs and central brain areas is not established, but morphological classes resembling P- and M-pathway monkey RGCs, as well as other RGC types, are now clear (Grunert and Martin, 2020). Recordings from donated human retina have shown that the visual response of putative P- and M-pathway RGCs are similar to monkey. Functional MRI of LGN in human has shown the presence of voxels that may prefer achromatic or red-green chromatic modulation, and the positioning of these voxels is consistent with M- and P-layers, respectively. Anatomical MRI has also shown white matter pathways linking LGN and V1, and revealed additional pathways linking LGN to extrastriate visual cortex. These data are broadly consistent with the connectivity of P, M, and K pathways in monkey, though the resolution of the techniques is not yet sufficient to specify exact sources and targets.

CORE AND SHELL PATHWAYS IN MOUSE

The availability of expanding genetic toolkits and the resultant ability to image functional activity in genetically defined cell classes increased interest in mouse retino-geniculo-cortical pathways. Mouse LGN appears homogenous in the Nissl stains that reveal cellular layers in cat and primate LGN; however, analysis of projections from the retina and other brain regions reveals "hidden" lamination. Mouse LGN is therefore usually subdivided into a thin dorsal "shell" and larger ventral "core" regions, as in rat (Reese, 1988; Kerschensteiner and Guido, 2017). The shell receives retinal input predominantly from contralateral eye, while the core comprises a larger contralateral projection surrounding a smaller ipsilateral projection. Morphologically, most thalamocortical neurons in shell resemble W-type thalamocortical neurons in cat, while those in core resemble X- or Y-type, with some regional variation (Guido, 2018). Retrograde tracing from V1 suggests thalamocortical neurons in shell project to superficial layers of V1, while those in core project to deeper layers of V1 (Cruz-Martin et al., 2014), but further characterizations are surprisingly spare. Large-scale surveys indicated 2 types of LGN axons in middle-upper layers of V1, and a third type in lower layers (Peng et al., 2021); there was little evidence for thalamocortical projection outside V1, though a small projection from, for example, shell may yet to be discovered.

LGN is a target of about one-third of mouse RGCs, including ~30 of the ~40 RGC classes that can be identified by various quantitative methods (mouse RGCs are less morphologically distinct than cat or primate RGCs). Many or most of these RGCs send collaterals to superior colliculus, the major projection of the mouse retina. LGN core receives relatively strong projections from "alpha-like" RGCs, while LGN shell receives relatively strong projections from direction-selective RGCs (Liang and Chen, 2020; Kerschensteiner, 2022). There is also extensive influence of non-canonical photoreception, derived from intrinsically photosensitive RGCs.

Early recordings from mouse LGN found the presence of sustained and transient responses and center-surround receptive fields. Later work found that these were complemented by orientation selective, direction selective, and "suppressed-by-contrast" responses consistent with retinal projections. Many cells (especially in awake animals) showed nonlinear spatial summation, but it is not clear that they form a discrete class. Center-surround receptive fields are relatively more prevalent in core, while non-standard receptive fields including direction and orientation selectivity are relatively more prevalent in shell. Consistently, imaging of LGN afferents in V1 has found evidence for orientation and direction-selective input to superficial and middle layers. Increased prevalence of orientation tuning in LGN over retina may suggest recombination of direction-selective RGC responses into orientation-selective LGN responses.

HOW PARALLEL ARE PARALLEL PATHWAYS?

A central concern is whether and how processing within LGN alters the signals provided by RGCs. The question here is whether there is convergence of multiple RGC classes onto a single LGN cell, convergence within RGC classes, or convergence of non-retinal inputs, that may change the nature of the signals that parallel pathways convey. The answer to these questions may be species- and pathway-dependent. In cat, there are at least twice as many LGN neurons (~510,000) as retinal afferents (~151,000; Chalupa et al., 1984; Williams et al., 1993). Anatomical measurements have consistently implied convergence and divergence of RGC signals in cat LGN. Several lines of evidence have suggested convergence of a small number (~1–4) of RGC inputs to each LGN cell, including recordings from connected pairs of RGC and LGN neurons (Cleland et al., 1971), simultaneous measurement of slow ("S") potential (synaptic activity of an RGC) and spiking activity within LGN (Kaplan and Shapley, 1984), and intracellular measurements from X-like LGN cells (Martinez et al., 2014). Convergence may be higher within the Y pathway, but the visual preferences of retinal inputs are usually the same as that of their target LGN cell, suggesting limited convergence across RGC classes. Functional measurements from LGN cells of the same class have also shown precisely correlated spiking activity, indicating divergent output from the same RGC. The picture in cat LGN is therefore one of resampling and interpolation across RGC inputs, potentially enhancing signal propagation without substantially disturbing parallel functional organization (Alonso et al., 2006; Martinez et al., 2014).

In primate, LGN neurons (~1.28 M) are about as numerous as RGC afferents, suggesting that neither divergence nor convergence is necessary (Spear et al., 1996). As with cat, the different functional properties of P- and M-cells appear to be built in retina, and the major functional properties of LGN cells are not readily distinguished from RGC afferents (Kaplan and Shapley, 1984). There is less direct evidence, but simultaneous measurement of S-potential and LGN spiking activity has also suggested limited functional convergence. There is limited binocular interactions, but at least in New World monkey a small number of LGN K-cells show strong binocular input (for a review, see Dougherty et al., 2019). Simultaneous recordings have not shown precisely correlated firing in nearby LGN cells, suggesting little divergence of RGC outputs to multiple LGN neurons.

There are fewer LGN neurons (~17,000) than RGC (~45,000) in mouse (Jeon et al., 1998; Seecharan et al., 2003), but what fraction of RGCs project to the LGN is not precisely known, and what convergence there is remains a matter of debate. It appears likely that there are a limited number (~2–5) of strong RGC inputs to each LGN cell, and a larger number of weak inputs. Functional work in shell suggested that some cells receive strong input from a single RGC class, but that most cells probably derived weak input from multiple RGC types. Additional work has shown that most LGN cells are driven by inputs from a single eye, even though anatomical connections can be less discriminating, so the functional consequence of weak connections remain mysterious.

LGN is a target of several non-retinal brain areas. These inputs could dilute pathway segregation, but many appear to be pathway specific. Inputs from superior colliculus primarily target C-layers in cat LGN, K-layers in primate, and shell—with parallel projection to core—in mouse. Similarly, inputs from parabigeminal nucleus, which have quite different organization in the 3 species, nevertheless target the C-, K-, and shell layers, respectively. "Feedback" signals from layer 6 of visual cortex, a major synaptic input to LGN, may also show pathway specificity, at least in primate (Briggs, 2020). (This is considered in more detail in **Chapter 19 in Section V**.)

The P-, M-, and K-layers of primate LGN receive input from distinct groups of corticothalamic neurons in V1, and perhaps V2. Remarkably, these corticothalamic neurons show distinct functional properties that resemble those of the layers they project to. The functional consequences of these pathway-specific non-retinal influences remain unclear, but presumably help explain why infra-slow oscillations are prominent in K-layers but not in P- or M-layers.

IMPORTANCE OF COLOR VISION

Psychophysical work showed long ago that the capacity for color vision in humans relied on receptoral and post-receptoral mechanisms. First, typical human color vision requires 3 types of broadly tuned cone photoreceptor, each sensitive to overlapping but different parts of the visible spectrum (short [S], medium [M], or long [L] wavelengths). Second, color vision requires post-receptoral mechanisms that compare the activity of these receptors ("cone opponency"). As most mammals are dichromatic, with only S- and M-cones, the properties of the cone-opponent pathways have been explored most extensively in trichromatic nonhuman primates where L-cones are also present. The details have been reviewed extensively elsewhere (e.g., Solomon and Lennie, 2007), but some points are particularly relevant to concepts of parallel processing.

Early measurements from trichromatic primates revealed the presence of cone-opponent receptive fields in retina and LGN, but were less certain about how these receptive fields were arranged. This changed when human psychophysical work established stimuli that could isolate the activity of each cone photoreceptor, and specified the importance of 2 cone-opponent mechanisms (MacLeod and Boynton, 1979; Krauskopf et al., 1982). One mechanism compared the activity of L- and M-cones (L-M; "red-green"), and the other compared S-cones to the sum of L- and M-cones (S-[L+M]; "blue-yellow"). Concomitant computational and theoretical work showed that cone-opponent channels with similar structure allowed efficient encoding of chromatic information (Buchsbaum and Gottschalk, 1983). Application to recordings in and around P-layers of monkey LGN showed the presence of 2 distinct functional clusters (Derrington et al., 1984), with patterns of cone-inputs similar to those implied by human psychophysics and theory. In addition, M-cells generally respond as if they summed the signals of L- and M-cone photoreceptors (L+M; "luminance"). These measurements supported a tight link between theory, perception, and physiology, and they also hinted that there might be distinct substrates for blue-yellow signals (components of which were subsequently described, as detailed above). Evolution of our understanding of color vision therefore illustrates the importance of precise control of visual stimulation, and the alignment of theory, perceptual, and neuroscientific work.

While it is now generally accepted that blue-yellow chromatic signals are primarily conveyed to the cortex by K pathways, there remain competing proposals for the analysis of other chromatic dimensions. One debate is whether red-green signals are carried by midget RGCs and LGN P-cells, or by other pathways, yet to be defined, but likely analogous to the blue-yellow K pathways (Rodieck, 1991; Patterson et al., 2019). The second debate is whether the receptive fields of P-cells are specialized for analyzing red-green color information, or for analyzing achromatic spatial information (Kaplan et al., 1990; Lennie et al., 1991). Comparison of dichromatic and trichromatic individuals among New World monkeys has not yet revealed structural differences in retina, LGN, or cortex, and the achromatic visual response properties of P-cells in LGN of dichromatic animals are indistinguishable from those in trichromatic animals. The third debate is which aspects of achromatic spatial vision are supported by P- or M-cells (Shapley, 1990; for a related discussion in cat, see Sherman, 1985). Note that while the receptive fields of P-cells are generally smaller than those of M-cells at the same location in the visual field, this does not imply that only P-cells could signal the location of an object. This is partly because M-cells have greater contrast sensitivity than P-cells, and are therefore better able to distinguish the position of stimuli within their receptive field.

Central to all these debates is the question of whether P-cells "multiplex" signals for both red-green (L-M) and luminance (L+M) vision in trichromatic animals. P-cells certainly respond to appropriate luminance and chromatic stimuli, but unfortunately that observation does not answer the question. This is because the receptive fields of P-cells are very small, particularly those near the center of gaze (i.e., in or near the fovea), where their excitatory receptive field center is dominated by a single cone photoreceptor (i.e., an L-cone or an M-cone). Even if P-cells were not specialized for red-green color vision, the larger receptive field surround would be driven by a mix of L- and M-cones and would have different spectral sensitivity to that of the receptive field center (the "random-wiring" hypothesis; Ingling and Martinez-Uriegas, 1983). The receptive fields of P-cells near the center of gaze must therefore be cone-opponent, even if not specialized to be so. Finely controlled measurements have addressed crucial elements of these debates, but the broader questions remain unresolved, and we are left with the uncomfortable reminder that observing that a class of neurons responds to a particular stimulus is not sufficient to specify its role (cf. "linking hypotheses"; Brindley, 1960).

WHY HAVE PARALLEL PATHWAYS?

The short history above makes clear that there is a great deal of knowledge about the organization and functional properties of parallel retino-geniculo-cortical pathways, but we have much less idea as to *why* they are as they are, *why* the details of their structural and functional organization vary as they do among species, or even *how* these organizations influence subsequent analysis.

There are 2 major perspectives on the purpose of early visual processing. One, anchored by the observation of center-surround, circularly symmetric receptive fields in ON- and OFF-retinal ganglion cells, is that the purpose of RGCs is to efficiently transmit a version of the retinal image to the rest of the brain through the bottleneck of the optic nerve, along channels with limited dynamic range (Barlow, 1961). This perspective is fairly agnostic (and therefore flexible) to the visual preferences of different parallel pathways in a particular species, as long as together they sample the retinal image appropriately. The other, anchored by Lettvin et al.'s (1959) survey of RGCs in the frog, observing responses selective for object motion or curvature, is that the purpose of an RGC is to measure "how much there is in a stimulus of that quality which excites [it] maximally, naming that quality" (Lettvin et al., 1959, p. 1951; see also Barlow, 1972). Different researchers have explored and named different features (a brief list: local edge detector, suppressed-by-contrast, loom-detector, convexity detector, object-motion selective; or even simply orientation- or direction-selective, and sustained or transient).

The retina could therefore either efficiently transmit information about the retinal image to subsequent brain areas, or provide estimates of the probability of features in that image. Said another way, the retina could either defer decisions about the content of the visible world to subsequent stages of processing (akin to making the "least commitment"; Marr, 1982, p. 106), or make decisions and communicate the results of those decisions to guide action. These perspectives have different implications for understanding the organization of parallel pathways.

If the purpose of the retino-geniculo-cortical pathway were to transmit a representation of the retinal image to the cortex, then further parcellation into multiple subpathways should mainly reflect optimizations that improve encoding of the retinal image or extend the range of vision. For example, it might be impossible for a single neural code to support requisite fine spatial resolution and fine temporal resolution. In this case, one pathway may be optimized to transmit highest spatial resolution, and another may be optimized to transmit highest temporal resolution. Many have suggested that this is what is accomplished by the subdivision of the primate visual pathway into P and M pathways, or the subdivision of the cat visual pathway into X and Y classes (Merigan and Maunsell, 1993; Schiller, 1993; Silveira and De Mello Jr., 1998; Casagrande et al., 2009). Similarly, presence of ON and OFF channels within each of the subdivisions, or the presence of parallel "red-green" and "blue-yellow" cone-opponent pathways, like the presence of center-surround receptive fields, may reflect solutions that reduce the redundancy of retinal signals, or overcome the limited dynamic range of real neurons (Attneave, 1954; Barlow, 1961, 1986) and improve transmission of the retinal image to the cortex. This perspective appears to require relatively complete coverage ("tiling") of the retinal image by each representation, and may also require quasi-linear processing, to enable subsequent recombination and extraction of relevant information (Kaplan, 2003).

If the purpose of the retino-geniculo-cortical pathway were instead to indicate the presence of features to guide action, then parcellation into parallel pathways should mainly reflect the nature of the features that it is important for the organism to represent ("we have been tempted, for example, to call the convexity detectors 'bug perceivers'"; Lettvin et al., 1959, p. 1951). The space of potential features (e.g., convexity) is, however, very large, and the fitness they provide is hard to define. We can gain some traction by supposing that if a ganglion cell is attempting to signal the presence of a particular feature, then it might be insensitive to otherwise irrelevant variations in the local retinal image (Barlow, 1972). For example, a neuron whose task is to signal the presence of a small object may respond regardless of the precise location or intensity (brighter or darker than its background) of that small object within its receptive field, much like a Y-cell in the retina, or complex cell in the cortex (Roska and Meister, 2014). Similarly, a neuron whose task is to signal the presence of a "blue" (or more precisely, a short-wavelength reflecting) object may respond regardless of the intensity of light reflected from that object. This approach would have risks—changes in viewpoint or lighting, for example, can dramatically alter the image of an object or feature on the retina. Generating feature selectivity often also requires nonlinear processing, which may be an impediment to subsequent recombination.

Recent computational work exploring deep convolutional networks with structure similar to the visual pathway suggests that these 2 perspectives may be considered as variations (Ocko et al., 2018; Lindsey et al., 2019). Imposing an energy budget produced 2 classes of quasi-linear receptive field (each with ON and OFF subtypes): one with smaller receptive fields and more sustained visual response, another with larger receptive fields and more transient visual response (Ocko et al., 2018). Further, center-surround receptive fields were favored when there was a bottleneck (e.g., the optic nerve), or when complex subsequent processing needed to be supported, because such organization retained more information about the retinal image. Relaxation of the bottleneck or simplification of subsequent processing

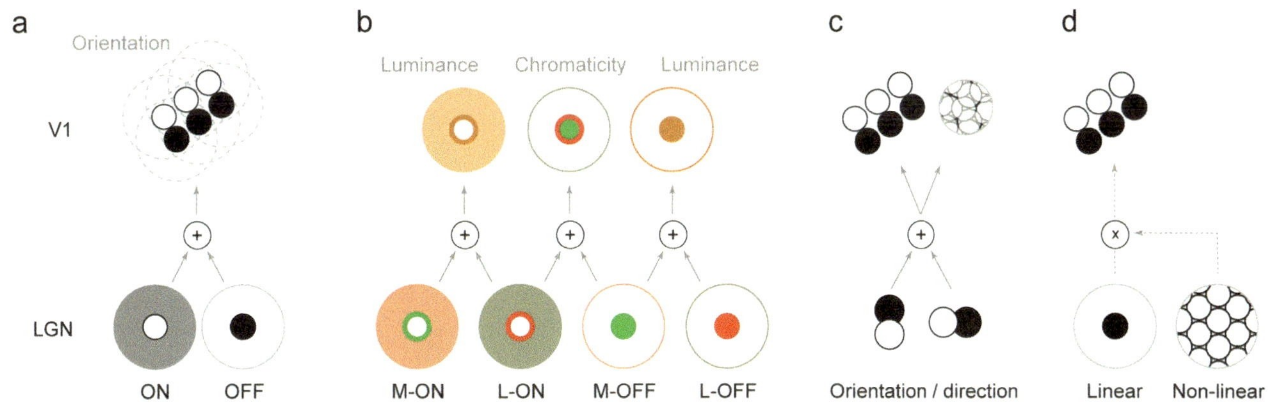

Figure 17.2. Some of the major functional properties of cells in LGN, and potential contribution to cortical receptive fields. (a). Most cells in LGN of cat and primate, and many in mouse, show center-surround receptive field organization. Some are ON-center, others are OFF-center. Orientation selectivity in visual cortex is often explained by summation of LGN afferents with appropriate spatial location and polarity (ON, OFF). (b). Many cells in P-layers of primate LGN show center-surround organization, with one cone photoreceptor (L-cone, or M-cone) dominating the receptive field center. Appropriate summation over these LGN cells can produce cortical receptive fields sensitive to luminance, or chromaticity. (c). Many cells in LGN of cat and primate show receptive fields with weak orientation sensitivity; some LGN cells in primate, and many cells in mouse, show evidence of stronger orientation selectivity. Summation over LGN cells may contribute to cortical orientation selectivity (top, left), but may also result in more complicated receptive field organization (top, right), particularly when strong nonlinearities are required to build orientation sensitivity in LGN receptive fields. (d). Other interactions between LGN pathways are possible. For example, the output of pathways with largely linear receptive fields may be gated by the activity in more nonlinear pathways.

allowed spatially richer, nonlinear retinal receptive fields to emerge (Lindsey et al., 2019). It is useful to note here that there are about 475,000 cells in mouse V1 (Herculano-Houzel et al., 2013), about 32 M in cat (Peters and Yilmaz, 1993), and in monkey estimates vary between 140 M and 416 M (Vanni et al., 2020). Thus, there are between 110 and 325 V1 cells for every LGN cell in monkey, and in mouse there are 30, suggesting that the retino-geniculo-cortical pathway may be a more problematic functional bottleneck for monkey than for mouse. What might drive relatively larger and more involved cortical processing is not known, though species with greater visual cortical expansion usually have more frontally placed eyes, greater binocular overlap, and a fovea or area centralis (Kremkow and Alonso, 2018).

Better understanding of why these parallel retino-geniculo-cortical pathways are as they are requires better understanding of how their signals interact (if they do) in the cortex (Figure 17.2). For example, convergence of quasi-linear signals provided by P-cells and K-cells can explain chromatic responses in monkey cortex; similar operations would also allow de-multiplexing of luminance and chromatic signals from P-cells (Lennie and D'Zmura, 1988). Convergence of more nonlinear signals may be more difficult to achieve, and if instead the signals of parallel pathways were kept largely separate in the cortex, they may play quite distinct roles. For example, the signals of P- and M-cells may make distinct contributions to subsequent cortical processing (Livingstone and Hubel, 1988). Or, M-cell signals may instead be thought of as gating the progression of signals from P-cells, and similarly for Y- and X-cells (Lennie, 1980; Shapley, 1992). Or, because much of the visual information needed for behavior can be extracted without fine spatial sampling of the retinal image, X-cells may only be needed where fine-grained texture is required (Sherman and Spear, 1982; Sherman, 1985). Others have likened the W pathways to "ambient" vision, with X and Y pathways providing "focal" vision (Trevarthan, 1968; Stone et al., 1979); have likened P and M pathways to "epicritic" and "protopathic" (Mollon, 1990); or have proposed that K pathways form a matrix, with similarity to pulvinar and widespread connections to cortical areas, that coordinates and modulates the signals provided by neurons in core P and M pathways (Jones, 2001). There is likely to be some truth in all these explanations—a single explanation for the role of parallel retino-geniculo-cortical pathways in an individual species, or the diversity of organization seen across species, may not be possible or even desirable.

REFERENCES

Alonso JM, Yeh CI, Weng C, Stoelzel C (2006). Retinogeniculate connections: A balancing act between connection specificity and receptive field diversity. *Prog Brain Res*. 154:3–13.

Attneave F (1954) Some informational aspects of visual perception. *Psychol Rev*. 61:183–193.

Bakken TE, van Velthoven CT, Menon V, Hodge RD, Yao Z, Nguyen TN, Graybuck LT, Horwitz GD, Bertagnolli D, Goldy J, Yanny AM, Garren E, Parry S, Casper T, Shehata SI, Barken ER, Szafer A, Levi BP, Dee N, Smith KA, Sunkin SM, Bernard A, Phillips J, Hawrylycz MJ, Koch C, Murphy GJ, Lein E, Zeng H, Tasic B (2021). Single-cell and single-nucleus RNA-seq uncovers shared and distinct axes of variation in dorsal LGN neurons in mice, non-human primates, and humans. *eLife*. 10:e64875.

Barlow HB (1961). Possible principles underlying the transformations of sensory messages. In: Rosenblith WA, ed. *Sensory Communication*. MIT Press: 217–234.
Barlow HB (1972). Single units and sensation: A neuron doctrine for perceptual psychology? *Perception*. 1:371–394.
Barlow HB (1986). Why have multiple cortical areas? *Vision Res*. 26:81–90.
Boycott BB, Wassle H (1974). The morphological types of ganglion cells of the domestic cat's retina. *J Physiol*. 240:397–419.
Briggs F (2020). Role of feedback connections in central visual processing. *Annu Rev Vis Sci*. 6:313–334.
Brindley GS (1960). *Physiology of the Retina and Visual Pathway*. Edward Arnold.
Buchsbaum G, Gottschalk A (1983). Trichromacy, opponent colours coding and optimum colour information transmission in the retina. *Proc R Soc Lond B Biol Sci*. 220:89–113.
Casagrande VA (1994). A third parallel visual pathway to primate area V1. *Trends Neurosci*. 17:305–310.
Casagrande VA, Khaytin I, Boyd J (2009). The evolution of parallel pathways in the brains of primates. In: Kaas JH, ed. *Evolutionary Neuroscience*. Elsevier: 871–892.
Chalupa LM, Williams RW, Henderson Z (1984). Binocular interaction in the fetal cat regulates the size of the ganglion cell population. *Neuroscience*. 12:1039–1046.
Chatterjee S, Callaway EM (2003). Parallel colour-opponent pathways to primary visual cortex. *Nature*. 426:668–671.
Cleland BG, Dubin MW, Levick WR (1971). Simultaneous recording of input and output of lateral geniculate neurones. *Nat New Biol*. 231:191–192.
Cruz-Martin A, El-Danaf RN, Osakada F, Sriram B, Dhande OS, Nguyen PL, Callaway EM, Ghosh A, Huberman AD (2014). A dedicated circuit links direction-selective retinal ganglion cells to the primary visual cortex. *Nature*. 507:358–361.
Dacey DM, Lee BB (1994). The "blue-on" opponent pathway in primate retina originates from a distinct bistratified ganglion cell type. *Nature*. 367:731–735.
Derrington AM, Krauskopf J, Lennie P (1984). Chromatic mechanisms in lateral geniculate nucleus of macaque. *J Physiol*. 357:241–265.
DeYoe EA, Van Essen DC (1988). Concurrent processing streams in monkey visual cortex. *Trends Neurosci*. 11:219–226.
Dougherty K, Schmid MC, Maier A (2019). Binocular response modulation in the lateral geniculate nucleus. *J Comp Neurol*. 527:522–534.
Enroth-Cugell C, Robson JG (1966). The contrast sensitivity of retinal ganglion cells of the cat. *J Physiol*. 187:517–552.
Grunert U, Martin PR (2020). Cell types and cell circuits in human and non-human primate retina. *Prog Retin Eye Res*. 78:100844.
Guido W (2018). Development, form, and function of the mouse visual thalamus. *J Neurophysiol*. 120:211–225.
Herculano-Houzel S, Watson C, Paxinos G (2013). Distribution of neurons in functional areas of the mouse cerebral cortex reveals quantitatively different cortical zones. *Front Neuroanat*. 7:35.
Ingling CR Jr, Martinez-Uriegas E (1983). The relationship between spectral sensitivity and spatial sensitivity for the primate r-g X-channel. *Vision Res*. 23:1495–1500.
Jeon CJ, Strettoi E, Masland RH (1998). The major cell populations of the mouse retina. *J Neurosci*. 18:8936–8946.
Jones EG (2001). The thalamic matrix and thalamocortical synchrony. *Trends Neurosci*. 24:595–601.
Kaplan E (2003). The M, P and K pathways in the primate visual system. In: Chalupa LM, Werner JS, eds. *The Visual Neurosciences*. MIT Press:481–494.
Kaplan E, Shapley R (1984). The origin of the S (slow) potential in the mammalian lateral geniculate nucleus. *Exp Brain Res*. 55:111–116.
Kaplan E, Lee BB, Shapley R (1990). New views of primate retinal function. *Prog Retin Res*. 9:273–336.
Kerschensteiner D (2022). Feature detection by retinal ganglion cells. *Annu Rev Vis Sci*. 8:135–169.
Kerschensteiner D, Guido W (2017). Organization of the dorsal lateral geniculate nucleus in the mouse. *Vis Neurosci*. 34:E008.
Krauskopf J, Williams DR, Heeley DW (1982). Cardinal directions of color space. *Vision Res*. 22:1123–1131.
Kremkow J, Alonso JM (2018). Thalamocortical circuits and functional architecture. *Annu Rev Vis Sci*. 4:263–285.
Lennie P (1980). Parallel visual pathways: A review. *Vision Res*. 20:561–594.
Lennie P, D'Zmura M (1988). Mechanisms of color vision. *Crit Rev Neurobiol*. 3:333–400.
Lennie P, Haake PW, Williams DR (1991). The design of chromatically opponent receptive fields. In: Landy MS, Movshon JA, eds. *Computational Models of Visual Processing*. MIT Press:71–82.
Lettvin JY, Maturana HR, McCulloch WS, Pitts WH (1959). What the frog's eye tells the frog's brain. *Proc IRE*. 47:1940–1951.
Liang L, Chen C (2020) Organization, function, and development of the mouse retinogeniculate synapse. *Annu Rev Vis Sci*. 6:261–285.
Lindsey J, Ocko SA, Ganguli S, Deny S (2019). A unified theory of early visual representations from retina to cortex through anatomically constrained deep CNNs. arXiv:1901.00945v00941.
Livingstone M, Hubel D (1988). Segregation of form, color, movement, and depth: Anatomy, physiology, and perception. *Science*. 240:740–749.
MacLeod DI, Boynton RM (1979). Chromaticity diagram showing cone excitation by stimuli of equal luminance. *J Opt Soc Am*. 69:1183–1186.
Marr D (1982). *Vision*. Freeman.
Martin PR, White AJ, Goodchild AK, Wilder HD, Sefton AE (1997). Evidence that blue-on cells are part of the third geniculocortical pathway in primates. *Eur J Neurosci*. 9:1536–1541.
Martinez LM, Molano-Mazon M, Wang X, Sommer FT, Hirsch JA (2014). Statistical wiring of thalamic receptive fields optimizes spatial sampling of the retinal image. *Neuron*. 81:943–956.
Merigan WH, Maunsell JH (1993). How parallel are the primate visual pathways? *Annu Rev Neurosci*. 16:369–402.
Mollon JD (1990). Neurobiology: The club-sandwich mystery. *Nature*. 343:16–17.
Ocko SA, Lindsey J, Ganguli S, Deny S (2018). The emergence of multiple retinal cell types through efficient coding of natural movies. *Adv Neural Inf Process Syst*. 31:9411–9422.
Patterson SS, Neitz M, Neitz J (2019). Reconciling color vision models with midget ganglion cell receptive fields. *Front Neurosci*. 13:865.
Peng H, Xie P, Liu L, Kuang X, Wang Y, Qu L, Gong H, Jiang S, Li A, Ruan Z, Ding L, Yao Z, Chen C, Chen M, Daigle TL, Dalley R, Ding Z, Duan Y, Feiner A, He P, Hill C, Hirokawa KE, Hong G, Hunag L, Kebede S, Kuo HC, Larsen R, Lesnar P, Li L, Li Q, Li X, Li Y, Li Y, Liu A, Lu D, Mok S, Ng L, Nguyen TN, Ouyang Q, Pan J, Shen E, Song Y, Sunkin SM, Tasic B, Veldman MB, Wakeman W, Wan W, Wang P, Wang Q, Wang T, Wany Y, Xiong F, Xiong W, Xu W, Ye M, Yin L, Yu Y, Yuan J, Yuan J, Yun Z, Zeng S, Zhang S, Zhao S, Zhao Z, Zhou Z, Huang ZJ, Esposito L, Hawrylycz MJ, Sorensen SA, Yang XW, Zheng Y, Gu Z, Xie W, Koch C, Luo Q, Harris JA, Wang Y, Zeng H (2021). Morphological diversity of single neurons in molecularly defined cell types. *Nature*. 598:174–181.
Peters A, Yilmaz E (1993). Neuronal organization in area 17 of cat visual cortex. *Cereb Cortex*. 3:49–68.
Reese BE (1988). "Hidden lamination" in the dorsal lateral geniculate nucleus: The functional organization of this thalamic region in the rat. *Brain Res*. 472:119–137.
Rodieck RW (1991). Which cells code for color? In: Valberg A, Lee BB, eds. *From Pigments to Perception*. Plenum Press:83–93.
Roska B, Meister M (2014) The retina dissects the visual scene into distinct features. In: Chalupa L, Werner JS, eds. *The New Visual Neurosciences*. MIT Press:163–182.
Roy S, Jayakumar J, Martin PR, Dreher B, Saalmann YB, Hu D, Vidyasagar TR (2009). Segregation of short-wavelength-sensitive

(S) cone signals in the macaque dorsal lateral geniculate nucleus. *Eur J Neurosci*. 30:1517–1526.

Schiller PH (1993). Parallel pathways in the visual system. In: Gulyas B, Ottoson D, Roland PE, eds. *Functional Organisation of the Human Visual Cortex*. Pergamon:43–58.

Seecharan DJ, Kulkarni AL, Lu L, Rosen GD, Williams RW (2003). Genetic control of interconnected neuronal populations in the mouse primary visual system. *J Neurosci*. 23:11178–11188.

Shapley R (1990). Visual sensitivity and parallel retinocortical channels. *Annu Rev Psychol*. 41:635–658.

Shapley R (1992). Parallel retinocortical channels: X and Y and P and M. In: Brannan JR, ed. *Applications of Parallel Processing in Vision*. North-Holland:3–36.

Shapley R, Perry VH (1986). Cat and monkey retinal ganglion cells and their visual functional roles. *Trends Neurosci*. 9:229–235.

Sherman SM (1985). Parallel W-, X- and Y-cell pathways in the cat: A model for visual function. In: Rose D, Dobson VG, eds. *Models of the Visual Cortex*. Wiley:71–84.

Sherman SM, Spear PD (1982). Organization of visual pathways in normal and visually deprived cats. *Physiol Rev*. 62:738–855.

Silveira LCL, De Mello HD Jr. (1998). Parallel pathways of primate vision. In: Chalupa L, Finlay B, eds. *Development and Organization of the Retina*. Plenum Press:173–199.

Solomon SG, Lennie P (2007). The machinery of colour vision. *Nat Rev Neurosci*. 8:276–286.

Spear PD, Kim CB, Ahmad A, Tom BW (1996). Relationship between numbers of retinal ganglion cells and lateral geniculate neurons in the rhesus monkey. *Vis Neurosci*. 13:199–203.

Stone J, Dreher B (1982) Parallel processing of information in the visual pathways: A general principle of sensory coding? *Trends Neurosci*. 5:441–446.

Stone J, Dreher B, Leventhal A (1979). Hierarchical and parallel mechanisms in the organization of visual cortex. *Brain Res*. 180:345–394.

Stone J, Hoffmann KP (1972). Very slow-conducting ganglion cells in the cat's retina: A major, new functional type? *Brain Res*. 43:610–616.

Trevarthan CB (1968). Two mechanisms of vision in primates. *Psychologische Forschung*. 31:299–337.

Vanni S, Hokkanen H, Werner F, Angelucci A (2020). Anatomy and physiology of macaque visual cortical areas V1, V2, and V5/MT: Bases for biologically realistic models. *Cereb Cortex*. 30:3483–3517.

Williams RW, Cavada C, Reinoso-Suarez F (1993). Rapid evolution of the visual system: A cellular assay of the retina and dorsal lateral geniculate nucleus of the Spanish wildcat and the domestic cat. *J Neurosci*. 13:208–228.

18.

THALAMIC CONTRIBUTIONS TO VISUAL CORTICAL RECEPTIVE FIELDS

Ronan O'Shea, Mandi C. Severson, and Nicholas J. Priebe

INTRODUCTION

The retina transduces light into electrical signals that are sent to the lateral geniculate nucleus (LGN) of the thalamus and then to primary visual cortex (V1). At each stage of visual processing, the representation of the visual world is transformed. For example, center-surround, circularly symmetric receptive fields (RFs) initially emerge in the retina, whereas spatially elongated RFs, which generate orientation selectivity, emerge in V1. Our understanding of the computations performed by the early visual system is based on these hierarchical changes in representation. In particular, it has been thought that several features of visual processing result from the convergence of geniculate inputs onto visual cortical neurons, including orientation selectivity, motion selectivity, and chromatic sensitivity. However, there is evidence for all of these tuning properties in LGN cells, suggesting that the representation of these features may originate earlier in the visual pathway. In this chapter, we will compare the anatomy and projections of the LGN to V1 across several species of mammals. We will then discuss the geniculate contributions to orientation and motion selectivity and describe mechanisms that could explain how inputs from the LGN converge in V1 to transform visual representations.

THE GENICULOCORTICAL PATHWAY

Across mammalian species, LGN anatomy is characterized by the laminar segregation of neurons receiving input from each eye and from distinct functional classes of retinal ganglion cells. This segregation seeds the canonical parallel-processing streams of visual cortex, which are optimized for the encoding of distinct aspects of visual input. Figure 18.1 summarizes the geniculocortical projection patterns of the three major animal models in vision research: primate, cat, and mouse.

The relationship between laminar segregation and physiology in the geniculocortical pathway has been most clearly delineated in the primate, where the parasol and midget retinal ganglion cells project to the magnocellular (M) and parvocellular (P) layers, respectively, interdigitated with the less clearly understood koniocellular layers (Malpeli and Baker 1975). The M and P parallel streams represent a trade-off in visual processing between temporal and spatial resolution. These streams remain segregated through distinct projection patterns to V1. M afferents have high temporal resolution and project to cortical layer 4Cα, whereas P afferents have high spatial resolution and project to cortical layers 4C and 4A (Kaplan and Shapley, 1982). These segregated projection patterns and distinct response characteristics have implicated the M and P pathways in the perception of motion and form, respectively (Livingstone and Hubel, 1988). The M inputs are well-suited to provide input to the dorsal visual pathway, where the majority of cells in area MT show tuning for visual motion. The P inputs, having higher spatial acuity and sensitivity to chromatic contrast, are well-suited for the identification of visual patterns, such as natural textures, that define and segregate objects in space. Thus, the parvocellular inputs may underlie the ventral visual pathway projecting to inferotemporal cortex, where cells show selectivity for particular object features.

Behavioral studies employing lesions to specific layers of primate LGN have revealed perceptual deficits concomitant with the distinct tuning of the M and P pathways (Schiller et al., 1990). These studies emphasize the parallel nature of these pathways, in that lesions to one pathway leave encoding of features by the other intact. Visual information encoded in the separate layers of the primate LGN remains segregated up to higher brain regions driving behavior.

The potential role of koniocellular projections in this parallel processing scheme has remained mysterious. The

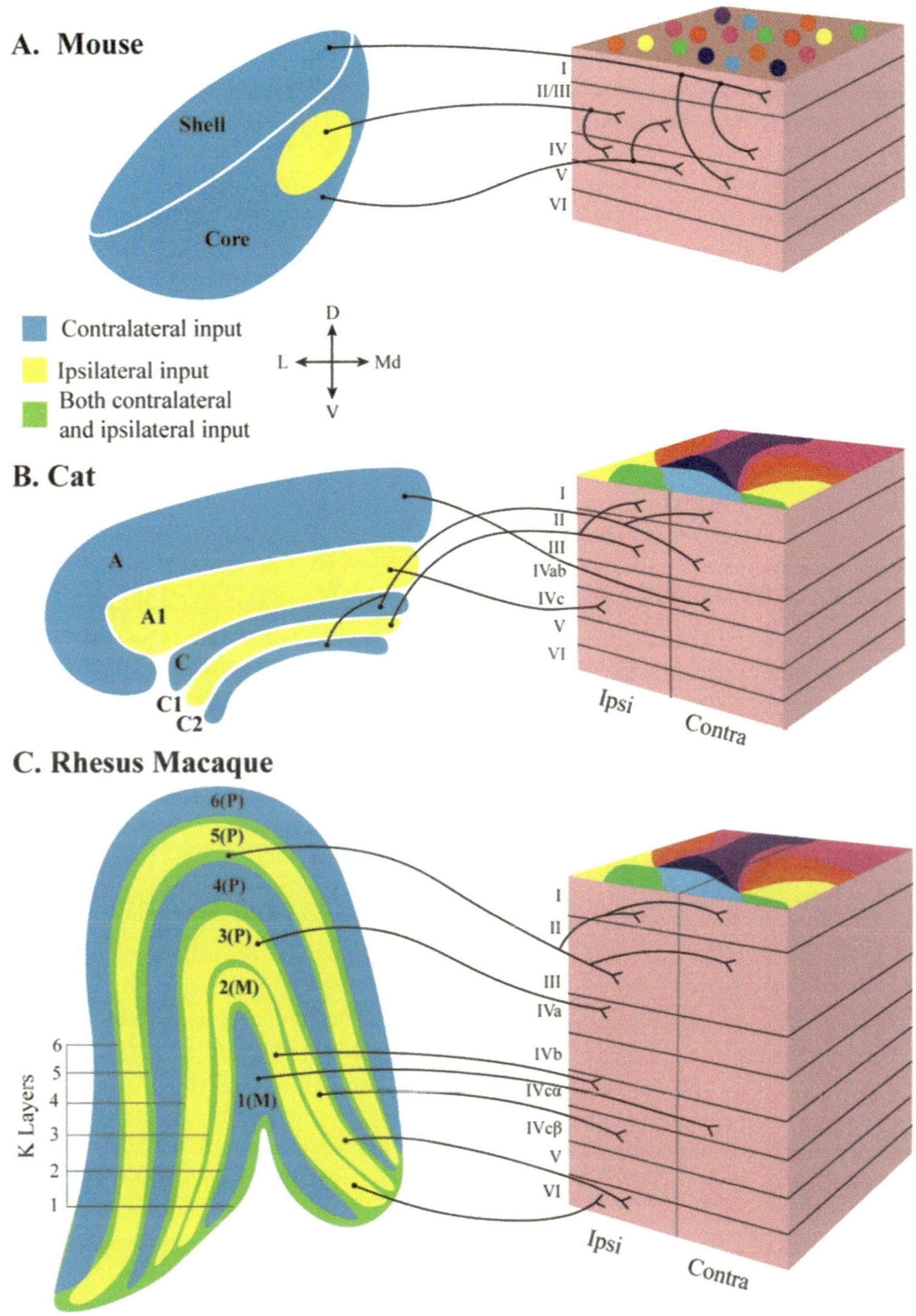

Figure 18.1. Geniculocortical projections. Major geniculocortical projection patterns and orientation tuning maps in: (A) Mouse (Krahe et al., 2011). (B) Cat (Ferster and Levay, 1978; Gilbert and Wiesel, 1979; Humphrey et al., 1985). (C) Rhesus macaque (Winfield et al., 1982; Hubel and Wiesel, 1972); M = magnocellular; P = parvocellular; K = koniocellular; D = dorsal; V = ventral, L = lateral; Md = medial.

heterogeneous physiological properties of cells in these layers have raised the question of whether these cells should be classified together at all (Martin and Solomon, 2018). Unlike their M and P counterparts, koniocellular afferents project directly to supragranular (upper) layers. The unique projection pattern suggests that these cells could play a unique role in visual processing (Casagrande, 1994). A subset of these cells show tuning for chromatic contrast at short wavelengths (Roy et al., 2009). Other cells in the koniocellular layers show high baseline firing rates that are suppressed by contrast (Tailby et al., 2007). Other studies have found that a large proportion of these cells have no visual response (White et al., 2001). A subset of these cells shows strong orientation tuning, in contrast to M and P populations (Cheong et al., 2013). Slow, oscillatory signals are present in koniocellular ensembles, but not in the

M or P layers, suggesting that koniocellular afferents could contribute a modulatory signal to the visual response in V1 carried by the other 2 pathways (Cheong et al., 2011).

As in the primate, distinct innervation patterns generate laminar divisions in the cat and mouse LGN. In the cat, X ganglion cells send eye-specific input to layer A or A1, and Y retinal ganglion cells send their projections to all 3 layers (Fukuda and Stone, 1974). An additional laminar subdivision in the cat LGN, the C layers, receives mixed selectivity inputs, like the primate koniocellular pathway. Although it appears that these pathways in cats and primates segregate information in similar ways, the relationship between X and Y cells in carnivores and the magnocellular and parvocellular cells in primates remains unclear (Kaplan and Shapley, 1982; Shapley and Hugh Perry, 1986).

The mouse LGN exhibits 2 functional subdivisions: the core and the shell. The shell receives direction selective input from the retina, while the core is innervated by a diverse population of center-surround retinal ganglion cells (Piscopo et al., 2013). Both layers receive input from the contralateral eye, while only part of the core gets input from the ipsilateral eye (Krahe et al., 2011).

GENICULOCORTICAL CONVERGENCE AND THE GENERATION OF CORTICAL FEATURE SELECTIVITY

The canonical V1 RF emerges from the convergence of geniculate afferents on a single V1 cell. We will discuss two examples of how this convergence contributes to feature selectivity in V1: orientation selectivity and direction selectivity. When Hubel and Wiesel first described the V1 RF, they proposed a feedforward model in which orientation selectivity is directly linked to the convergence of spatially offset geniculate RFs. A wealth of research in the subsequent decades has sought to describe the contributions of this feedforward mechanism, along with intracortical mechanisms, to the V1 RF.

THE HIERARCHICAL FEEDFORWARD MODEL OF THE SPATIAL RF AND ORIENTATION SELECTIVITY

Retinal ganglion cell and LGN RFs are classically characterized as being circularly symmetric with either an ON or OFF center and a surround of the opposite polarity (Kuffler, 1953). Hubel and Wiesel were the first to characterize RFs in cat V1 and to develop hypotheses about how these RFs might emerge (Hubel and Wiesel, 1962). In contrast to RFs found in the LGN, RFs at the first stage of processing in V1, layer 4, are elongated with segregated ON and OFF subfields. These neurons, termed *simple cells*, responded robustly to bars of the same orientation as the aligned ON and OFF subfields and not at all to bars of the orthogonal orientation (Hubel and Wiesel, 1962). Hubel and Wiesel suggested that this orientation selectivity could emerge if LGN cells with spatially offset RFs along a common axis converge onto a target layer 4 cortical neuron (Hubel and Wiesel, 1962).

Three lines of evidence lend strong support to the feedforward hypothesis of the V1 RF. First, an analysis of LGN neurons monosynaptically connected to a V1 neuron revealed a strong spatial overlap between the RFs of ON and OFF geniculate inputs and the subfields of the same polarity in the cortical RF (Reid and Alonso, 1995). These results demonstrate that input from the LGN significantly contributes to the visual response of these V1 cells. The exact number of LGN inputs that drive a simple cell, however, has been a topic of debate (Box 18.1).

Second, the LGN RFs that innervate a cortical column are spatially offset along the preferred orientation of that cortical column. This was demonstrated by silencing visual cortex and measuring the spatial RFs of individual geniculate afferents projecting to a cortical column with known orientation tuning. The LGN RFs were spatially offset along a common orientation axis, consistent with Hubel and Wiesel's feedforward hypothesis (Chapman and Stryker, 1991).

Third, the orientation selectivity of the aggregate geniculate input onto a V1 cell matches the corresponding orientation selectivity of cortical neurons (Ferster et al., 1996; Chung and Ferster, 1998). Ferster and colleagues used intracellular recording to measure the synaptic drive onto neurons while inactivating the cortical circuit. They found that the convergent LGN input to simple cells was well-tuned for orientation and made up 46% of excitatory input to simple cells (Chung and Ferster, 1998). More recent work recording simultaneously from a set of geniculate afferents projecting to a single cortical column has shown that the aggregate ON/OFF structure of the LGN RFs predicts the orientation tuning in V1 (Jin et al., 2011b).

These studies indicate that convergent input from the LGN is sufficient to produce orientation tuning in V1 simple cells, but the precise structure of this feedforward input is still unclear, as multiple mechanisms are consistent with these results. Four different classes of models of geniculocortical convergence have been proposed to explain the orientation selectivity observed in V1 (Figure 18.2A):

1. *Hubel and Wiesel*: As discussed above, Hubel and Wiesel proposed, based on their experiments in cat and primate V1, that the segregated ON/OFF subfields of V1 simple cell RFs could be constructed by pooling ON and OFF geniculate afferents with RFs along a single axis of elongation in visual space. As a result of this selective pooling, both the ON and OFF subfields

BOX 18.1. HOW MANY GENICULATE AFFERENTS CONVERGE ONTO A V1 CELL?

GENICULOCORTICAL CONVERGENCE GIVES RISE TO SOME RF PROPERTIES OBSERVED IN V1, BUT HOW MANY LGN INPUTS INNERVATE A SINGLE V1 CELL?

Electron microscopy gives an estimate of approximately 200 geniculocortical synapses per cortical cell (Peters and Payne, 1993). This estimate provides the upper bound on convergence since a single LGN cell may make multiple synapses onto a V1 cell and the strength of each synapse is unknown.

To overcome this limitation, physiological studies have sought to estimate the effective number of geniculocortical afferents using paired recordings in LGN and V1 (Lien and Scanziani, 2018). By measuring the excitatory postsynaptic current (EPSC) coincident with a presynaptic geniculate spike, Lien and Scanziani estimated that a single LGN cell contributed 1.25% ± 1.51% of the total geniculate excitation to a V1 cell. Under the assumption that each geniculate afferent contributes a comparable amount of excitation to V1, approximately 80 ± 10 geniculate afferents contribute to the observed feedforward visual response. This value, however, dramatically varied across paired recordings, and was highly skewed. This raises the possibility that across the entire population, a small number of geniculate afferents dominate the responses of a V1 cell. With this in mind, the effective geniculocortical convergence could be relatively sparse, with fewer than 10 geniculate afferents contributing 90% or more of V1 feedforward excitation.

A recent analysis of V1 RFs estimated geniculocortical convergence to be much lower than Lien and Scanziani's estimate (Ringach, 2021). By modeling the RFs of a population of simultaneously imaged mouse V1 cells as a function of a common pool of geniculate RFs, it was revealed that only a few LGN inputs were necessary to account for each neuron's response. One way to reconcile these estimates of convergence is a model in which a few, strong projections, represented in the long tail of Lien and Scanziani's distribution of charge contributions, dominate the input to V1 cells.

Resolving the discrepancy in these estimates of geniculocortical convergence has important implications for models of V1 feature selectivity. In a regime of dense convergence, a single LGN cell has relatively little influence over the V1 response, and any inhomogeneities in space or time between inputs will get averaged out. Some models, such as the "random connectivity" model of orientation tuning, require that selectivity emerges from random spatial inhomogeneities in the distribution of ON and OFF afferents projecting to a V1 cell (Hansel and van Vreeswijk, 2012; Pattadkal et al., 2018). As the number of inputs increases, the inhomogeneities decline, in both the spatial and temporal domain, which impacts whether these inhomogeneities may form the basis of orientation or direction selectivity. Under assumptions of dense convergence, afferents instead must be pooled with a precise structure to maintain tuning in space or time.

of the V1 RF are orientation tuned (monkey: Hubel and Wiesel, 1968; cat: DeAngelis et al., 1993). The combination of these subfields creates the bandpass spatial frequency tuning of the RF.

2. *Spatially offset*: Spatially segregated ON and OFF subfields need not be elongated to produce an orientation-tuned RF. Instead, the spatial offset of ON and OFF geniculate afferents can provide a basis for orientation selectivity. The orientation preference of a V1 RF is given by the angle between the centers of the ON and OFF subfields. In the mouse, studies mapping the ON and OFF geniculate projections to a V1 cell have shown that these subfields appear circularly symmetric and spatially offset from one another, such that their alignment predicts the preferred orientation of the V1 cell (Liu et al., 2010; Lien and Scanziani, 2013).
3. *Random connectivity*: Orientation selectivity may arise from a random pooling of local LGN afferents. Orientation-tuned RFs can emerge from this non-selective pooling due to random inhomogeneities in the distribution of ON and OFF RFs (Ringach, 2004; Hansel and van Vreeswijk, 2012; Pattadkal et al., 2018). Since there is significant overlap between ON and OFF subregions in this RF model, V1 cells will show "complex" responses and orientation tuning that depends on the spatial frequency of the stimulus. This model has recently been used to explain orientation tuning properties observed in mouse V1 (Pattadkal et al., 2018).
4. *Inherited orientation tuning*: In contrast to the classical view of LGN RFs as being circularly symmetric and non-selective for direction of motion, mouse LGN cells have significant orientation and direction tuning (Piscopo et al., 2013; Scholl et al., 2013). These results suggest that V1 orientation tuning could be inherited directly from the tuning in the LGN. Orientation-selective geniculate afferents with similar preference converge onto single V1 neurons, though the prevalence of this orientation-selective convergence may be stronger in layer 2/3 than layer 4 (Kondo and Ohki, 2016; Sun et al., 2016).

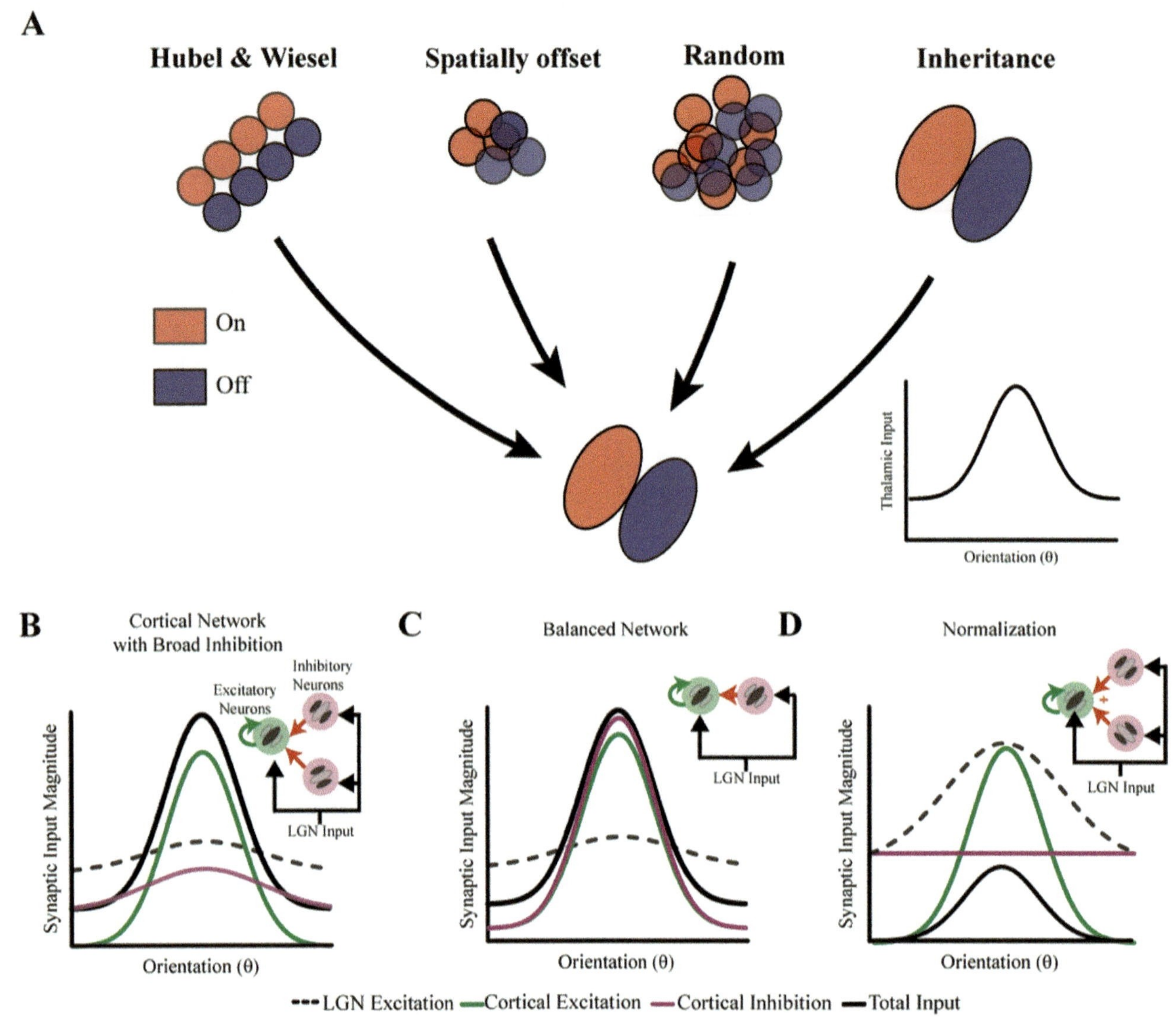

Figure 18.2. Models of orientation selectivity. (A). Mechanisms of feedforward orientation selectivity based on the convergence of ON (red) and OFF (blue) afferents. (B–D). Orientation tuning curves for cortical models of orientation selectivity with B broad inhibition, C balanced excitation-inhibition network, and D normalization.

INTRACORTICAL CONNECTIVITY FOR ORIENTATION SELECTIVITY

A primary feature of the feedforward model is that it is based solely on excitatory, geniculocortical synapses, with no role for cortical excitatory or inhibitory circuitry. There has been extensive debate on the role that intracortical circuitry plays in shaping orientation selectivity.

A natural extension of the feedforward hypothesis is that inhibition could play a role by suppressing responses to non-preferred orientations (Blakemore and Tobin, 1972). This could occur if inhibitory neurons were less selective for orientation or if inhibitory neurons preferentially converged onto excitatory neurons selective for distinct orientations. Indeed, blocking inhibition in visual cortex dramatically alters orientation tuning, either by introducing a gain offset to the tuning function (Katzner et al., 2011) or by broadening tuning itself (Sillito, 1975). Cortical inhibition may play a role in sharpening orientation selectivity (Figure 18.2B) (Douglas et al., 1995), countering excitatory drive ("the balanced hypothesis"; Figure 18.2C) (Ben-Yishai et al., 1995; Hansel and van Vreeswijk, 2012), or adjusting the gain of cortical responses ("the normalization model"; Figure 18.2D) (Albrecht and Geisler, 1991; Heeger, 1992).

There has also been considerable work examining the role of recurrent excitatory connectivity in the emergence of orientation selectivity. Across mammals it is clear that excitatory neurons that share orientation preference selectively interact (Gilbert and Wiesel, 1989; Ko et al., 2011). This specific connectivity is thought to act like an amplifier, enhancing cortical responses to weak inputs from the LGN (Ben-Yishai et al., 1995). The hypotheses about the roles of inhibitory and excitatory circuitry do not necessarily compete with the feedforward hypothesis. Instead, these cortical circuit models build on its foundation to describe how the cortex shapes the input it gets from the LGN.

DIRECTION SELECTIVITY THROUGH GENICULOCORTICAL CONVERGENCE

Similar to orientation tuning, several proposals have been made about how LGN input shapes direction selectivity. These proposals include a dedicated circuit for direction selective retinal signals, temporally precise pooling of non-direction selective geniculate inputs in V1 layer 4, and the de novo computation of direction selectivity in V1 layer 2/3 by spatially imbalanced excitatory and inhibitory inputs to excitatory cells. These mechanisms are likely to act in concert to build a representation of visual motion in the brain.

DEDICATED PATHWAY FOR DIRECTION SELECTIVE RETINAL GANGLION CELLS

Direction selectivity in retinal ganglion cells was first identified in the rabbit retina (Barlow and Hill, 1963), though it was not immediately clear how this direction selectivity is related to that found in V1. Part of the reason for this is that while direction selectivity is commonly observed in the mammalian visual cortex, the presence of retinal directional selectivity varies dramatically across species (Cleland and Levick, 1974; Kim et al., 2022). As in the rabbit, direction-selective retinal ganglion cells are prevalent in the mouse, and there exists a specific projection from these neurons to the shell region of the LGN (Huberman et al., 2009). This specific circuit relays information to layer 1 of V1, providing a dedicated motion pathway from the retina to visual cortex (Cruz-Martín et al., 2014).

Cortical direction selectivity in mice is sensitive to these retinal signals. Using genetic manipulations to eliminate the direction-selective response in the mouse retina alters the distribution of direction selectivity in the V1 population (Hillier et al., 2017), and reduces the direction selectivity of cells in the LGN shell. The effects of this genetic manipulation on cortical selectivity depends on direction preference, disrupting cells tuned for motion in the posterior direction while leaving selectivity for other directions intact. These results indicate that retinal direction selectivity is not solely responsible for generating direction selectivity in mouse cortex, but contributes to tuning for posterior motion in particular.

DIRECTION SELECTIVITY EMERGES AT THE GENICULOCORTICAL SYNAPSE

Alternatively, V1 direction selectivity could emerge from the convergence of non-direction selective geniculate inputs with distinct spatial and temporal properties onto single V1 neurons. This hypothesis extends the spatial feedforward model for orientation selectivity posited by Hubel and Wiesel to the temporal domain (Adelson and Bergen, 1985; Watson and Ahumada, 1985). In this model the geniculate inputs vary in both spatial preference and response latency, such that they are activated simultaneously as an image drifts in the preferred direction, but asynchronously for the null direction. Pooling over these inputs generates a tilted spatiotemporal RF such that a V1 neuron will selectively respond to motion in one direction over the other (Figure 18.3).

The LGN-V1 circuit appears well suited to implement this model, as LGN RFs tile visual space without strong direction tuning, particularly in carnivores and primates. Importantly, geniculate cells show a range of temporal spiking patterns in response to visual input, from fast,

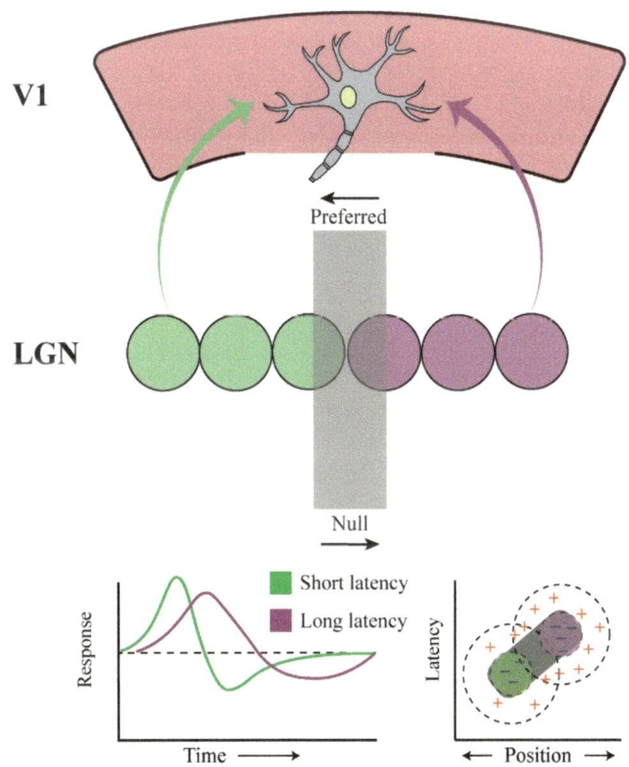

Figure 18.3. Direction selectivity through geniculocortical convergence. Top: a dark gray bar drifts across the visual field, activating short-latency (green) and long-latency (purple) geniculate afferents. Bottom, left: temporal response profile of short- and long-latency geniculate afferents. Bottom, right: these afferents converge onto a single V1 neuron. Convergence generates a spatiotemporal RF.

transient bursts to slow, sustained activity. In the cat, response latencies exist along a continuum from "lagged" to "non-lagged" geniculate populations (Saul and Humphrey, 1990). In the mouse, "transient" versus "sustained" response profiles have been observed (Piscopo et al., 2013). In general, these differences in response properties provide the necessary temporal offsets that, in combination with spatial offset, could generate cortical direction selectivity.

Multiple experiments are consistent with geniculocortical convergence being responsible for cortical direction selectivity. First, in this proposal, the timing and spatial tuning of the geniculate inputs determine direction selectivity, not the direction selectivity of the LGN cells. This idea implies that the direction-selective responses of neurons in V1 should be predicted from the response to stationary signals (Reid et al., 1991; Lien and Scanziani 2018). All of these studies demonstrate that the responses to motion may be recapitulated from the responses to stationary contrast reversing gratings, indicating that inherent geniculate-direction selectivity is not a primary component of cortical-direction selectivity. Second, cortical silencing does not disrupt direction selectivity (Ferster et al., 1996; Lien and Scanziani, 2018).

For this model of feedforward direction selectivity to work, afferents with distinct response latencies must be pooled with a precise spatial offset. It remains to be seen what rules might govern this convergence, or whether random pooling of a local set of geniculate afferents is sufficient to generate the observed tuning. How temporal differences emerge is not known, but appropriate latency differences exist in the LGN across species (Saul and Humphrey, 1990; Priebe and Ferster, 2012; Piscopo et al., 2013). In addition, the lag in ON relative to OFF responses, which has been observed in both the retina and LGN, could contribute to direction selectivity at the geniculocortical synapse (Jin et al., 2011a; Chariker et al., 2021).

INTRACORTICAL ORIGIN OF DIRECTION SELECTIVITY DUE TO SPATIAL IMBALANCE OF EXCITATION AND INHIBITION

While strong evidence indicates a central role for geniculate input in the emergence of direction selectivity, intracortical mechanisms may also play a powerful role. Using viral tracing technology, it has been possible to map the excitatory and inhibitory neurons that converge on single cells. Cortical direction selectivity is related to asymmetries in the spatial pattern of this input. Specifically, inhibitory inputs have offset receptive fields such that delayed inhibition will dampen excitatory responses in the null direction (Rossi et al., 2020). This cortical circuit for direction selectivity is akin to the circuit proposed by Barlow and Levick for motion selectivity in the retina (Barlow and Levick, 1965). This cortical circuit may be working in concert with thalamocortical mechanisms to amplify direction selectivity.

LARGE-SCALE ORGANIZATION OF GENICULATE INPUT

Up to this point we have discussed how the RF of a single V1 cell can be constructed from the geniculate input. The organization of RFs across the V1 population appears to be directly related to the structure of geniculate inputs.

Across species, ON and OFF geniculate afferents cluster in separate patches of V1 (Jin et al., 2008; Tring et al., 2022). As a result, patches of cortex show dominant responses to either increments or decrements in luminance (Figure 18.4A) (ON and OFF domains, respectively).

This orderly arrangement of geniculocortical projections is related to the distribution of orientation tuning across the V1 population. Moving horizontally through V1, the magnitudes of ON and OFF responses are anti-correlated, gradually shifting between patches of cortex dominated by responses to one polarity and patches with balanced ON/OFF responses (Figure 18.4A) (Kremkow et al., 2016). Importantly, the spatial arrangement and relative strength of ON and OFF projections to a point in the cortex is highly predictive of the orientation preference of a V1 neuron located at that point (Jin et al., 2011b). As a result, simple cells with segregated ON/OFF subfields are located at points on the cortex between the peaks of the ON and OFF domains. The retinotopic scatter of ON domains is larger than for OFF domains, such that the continuous procession of orientation preference at a single point in the retinotopy within a V1 hypercolumn can be interpreted as a gradual rotation of the ON subfield around a stable OFF "anchor" (Kremkow et al., 2016). Taken together, these results indicate that the pattern of geniculate projections shape the RF properties of the V1 population (Ringach, 2011).

These results indicate that LGN neurons project to V1 in an organized manner based on response polarity, and that this topographic organization constrains the orientation tuning of V1 neurons. This organization may be a general phenomenon underlying orientation tuning in the cortex, as even mouse V1, which lacks a columnar architecture, has clear ON and OFF domains that predict orientation tuning (Tring et al., 2022). Why then, given these stereotyped projection patterns, do some species have columnar organization for orientation tuning while others do not? The answer may lie in the dramatic differences in the mapping of the visual input from LGN to cortex across species.

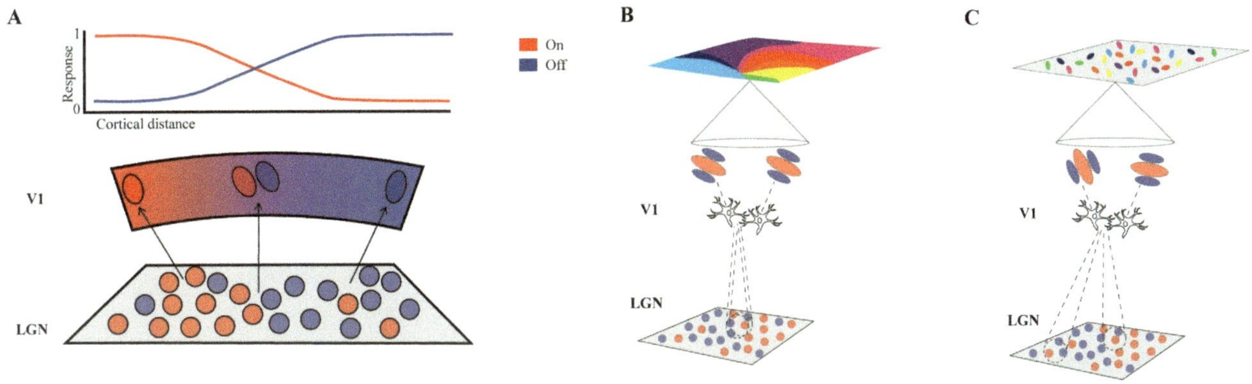

Figure 18.4. ON and OFF domains across the cortical map. (A). Top: the relative magnitude of responses to increments (ON) and decrements (OFF) in luminance varies horizontally across cortex. Bottom: ON (red) and OFF (blue) geniculate afferents project to distinct patches in cortex and constrain V1 orientation tuning. (B). Two neighboring V1 simple cells pool ON and OFF inputs from the LGN. V1 orientation columns emerge in regimes of high sampling density. (C). Same as B, but with low sampling density, such that neighboring simple cells in V1 have uncorrelated orientation tuning, producing a salt-and-pepper orientation map in V1.

Across mammals, the number of V1 neurons increases superlinearly with the number of LGN neurons (Stevens, 2001). One interpretation of this is that increased geniculate afferent density results in more cortical space being devoted to each point in visual space (Najafian et al., 2022). Neighboring V1 neurons in a regime of high geniculate afferent density, such as in the macaque, will receive almost entirely overlapping feedforward inputs (Figure 18.4B). We have seen that the arrangement of ON and OFF input constrains orientation tuning. Thus, neighboring cells in macaque V1 will have the same orientation tuning, with gradual shifts as the balance of geniculate inputs shifts across cortex. In the mouse, however, the geniculate input is sampled at a much coarser scale. This means that neighboring neurons will receive very different inputs, and are therefore unlikely to have similar orientation tuning. This difference in sampling density of the LGN ON/OFF mosaic could explain why mouse V1, which represents the entire visual field in just 4 mm^2 of cortex, has a salt-and-pepper, rather than columnar, orientation map (Figure 18.4C) (Jang et al., 2020).

CONCLUSIONS

We have reviewed evidence that much of the canonical tuning of V1 neurons is seeded in the projections from the LGN. Across species, retinal responses encoding distinct components of the visual scene, broadly classified as motion and form, remain segregated in the LGN. Distinct geniculocortical projection patterns retain this parallel processing which persists all the way to the level of perception. The defining RF properties of V1, selectivity for orientation and direction of motion, can emerge from the convergence of spatially and temporally offset geniculate afferents onto a single V1 cell. Indeed, across the V1 population, RF properties are largely constrained by a local set of geniculate projections. It remains to be seen to what extent cortical computations refine, or perhaps interact with, the computations performed by geniculocortical convergence.

REFERENCES

Adelson EH, Bergen JR (1985). Spatiotemporal energy models for the perception of motion. *J Opt Soc Am A*. 2:284–299.
Albrecht DG, Geisler WS (1991). Motion selectivity and the contrast-response function of simple cells in visual cortex. *Vis Neurosci*. 7: 531–546.
Barlow HB, Hill RM (1963). Selective sensitivity to direction of movement in ganglion cells of the rabbit retina. *Science*. 139:412–414.
Barlow HB, Levick WR (1965). The mechanism of directionally selective units in rabbit's retina. *J Physiol*. 178:477–504.
Ben-Yishai R, Bar-Or RL, Sompolinsky H (1995). Theory of orientation tuning in visual cortex. *Proc Natl Acad Sci U S A*. 92:3844–3848.
Blakemore C, Tobin EA (1972). Lateral inhibition between orientation detectors in the cat's visual cortex. *Exp Brain Res*. 15:439–440.
Casagrande VA (1994). A third parallel visual pathway to primate area V1. *Trends Neurosci*. 17:305–310.
Chapman B, Zahs K, Stryker M (1991). Relation of cortical cell orientation selectivity to alignment of receptive fields of the geniculocortical afferents that arborize within a single orientation column in ferret visual cortex. *J Neurosci*. 11:1347–1358.
Chariker L, Shapley R, Hawken M, Young L-S (2021). A theory of direction selectivity for macaque primary visual cortex. *Proc Natl Acad Sci U S A*. 118(32):e2105062118.
Cheong SK, Tailby C, Martin PR, Levitt JB, Solomon SG (2011) Slow intrinsic rhythm in the koniocellular visual pathway. *PNAS*. 108:14659–14663.
Cheong SK, Tailby C, Solomon SG, Martin PR (2013). Cortical-like receptive fields in the lateral geniculate nucleus of marmoset monkeys. *J Neurosci*. 33(16):6864–6876.
Chung S, Ferster D (1998). Strength and orientation tuning of the thalamic input to simple cells revealed by electrically evoked cortical suppression. *Neuron*. 20:1177–1189.

Cleland BG, Levick WR (1974). Properties of rarely encountered types of ganglion cells in the cat's retina and an overall classification. *J Physiol.* 240:457–492.

Cruz-Martín A, El-Danaf RN, Osakada F, Sriram B, Dhande OS, Nguyen PL, Callaway EM, Ghosh A, Huberman AD (2014). A dedicated circuit links direction-selective retinal ganglion cells to the primary visual cortex. *Nature.* 507:358–361.

DeAngelis GC, Ohzawa I, Freeman RD (1993). Spatiotemporal organization of simple-cell receptive fields in the cat's striate cortex. I. General characteristics and postnatal development. *J Neurophysiol.* 69:1091–1117.

Douglas RJ, Koch C, Mahowald M, Martin KA, Suarez HH (1995). Recurrent excitation in neocortical circuits. *Science.* 269:981–985.

Ferster D, Chung S, Wheat H (1996). Orientation selectivity of thalamic input to simple cells of cat visual cortex. *Nature.* 380:249–252.

Ferster D, Levay S (1978). The axonal arborizations of lateral geniculate neurons in the striate cortex of the cat. *J Comp Neur.* 182:923–944.

Fukuda Y, Stone J (1974). Retinal distribution and central projections of Y-, X-, and W-cells of the cat's retina. *J Neurophysiol.* 37(4): 749–772.

Gilbert CD, Wiesel TN (1979). Morphology and intracortical projections of functionally characterised neurones in the cat visual cortex. *Nature.* 280:120–125.

Gilbert CD, Wiesel TN (1989). Columnar specificity of intrinsic horizontal and corticocortical connections in cat visual cortex. *J Neurosci.* 9:2432–2442.

Hansel D, van Vreeswijk C (2012). The mechanism of orientation selectivity in primary visual cortex without a functional map. *J Neurosci.* 32:4049–4064.

Heeger DJ (1992). Normalization of cell responses in cat striate cortex. *Vis Neurosci.* 9:181–197.

Hillier D, Fiscella M, Drinnenberg A, Trenholm S, Rompani SB, Raics Z, Katona G, Juettner J, Hierlemann A, Rozsa B, Roska B (2017). Causal evidence for retina dependent and independent visual motion computations in mouse cortex. *Nat Neurosci.* 20:960–968.

Hubel DH, Wiesel TN (1962). Receptive fields, binocular interaction and functional architecture in the cat's visual cortex. *J Physiol.* 160:106–154.

Hubel DH, Wiesel TN (1968). Receptive fields and functional architecture of monkey striate cortex. *J Physiol.* 195:215–243.

Hubel DH, Wiesel TN (1972). Laminar and columnar distribution of geniculo-cortical fibers in the macaque monkey. *J Comp Neurol.* 146:421–450.

Humphrey AL, Sur M, Uhlrich DJ, Sherman SM (1985). Projection patterns of individual X- and Y-cell axons from the lateral geniculate nucleus to cortical area 17 in the cat. *J Comp Neurol.* 233:159–189.

Jang J, Song M, Paik S-B (2020). Retino-cortical mapping ratio predicts columnar and salt-and-pepper organization in mammalian visual cortex. *Cell Rep.* 30:3270–3279.

Jin JZ, Weng C, Yeh CI, Gordon JA, Ruthazer ES, Stryker MP, Swadlow HA, Alonso JM (2008). On and off domains of geniculate afferents in cat primary visual cortex. *Nat Neurosci.* 11:88–94.

Jin, J, Wang Y, Lashgari R, Swadlow HA, Alonso JM (2011a). Faster thalamocortical processing for dark than light visual targets. *J Neurosci.* 31:17471–17479.

Jin J, Wang Y, Swadlow HA, Alonso JM (2011b). Population receptive fields of ON and OFF thalamic inputs to an orientation column in visual cortex. *Nat Neurosci.* 14:232–238.

Kaplan E, Shapley RM (1982). X and Y cells in the lateral geniculate nucleus of macaque monkeys. *J Physiol.* 330:125–143.

Katzner S, Busse L, Carandini M (2011). GABAA inhibition controls response gain in visual cortex. *J Neurosci.* 31:5931–5941.

Kim YJ, Peterson BB, Crook JD, Joo HR, Wu J, Puller C, Robinson FR, Gamlin PD, Yau K-W, Viana F, Troy JB, Smith RG, Packer OS, Detwiler PB, Dacey DM (2022). Origins of direction selectivity in the primate retina. *Nat Commun.* 13:2862.

Ko H, Hofer SB, Pichler B, Buchanan KA, Sjöström PJ, Mrsic-Flogel TD (2011). Functional specificity of local synaptic connections in neocortical networks. *Nature.* 473:87–91.

Kondo S, Ohki K (2016). Laminar differences in the orientation selectivity of geniculate afferents in mouse primary visual cortex. *Nat Neurosci.* 19:316–319.

Krahe TE, El-Danaf RN, Dilger EK, Henderson SC, Guido W (2011). Morphologically distinct classes of relay cells exhibit regional preferences in the dorsal lateral geniculate nucleus of the mouse. *J Neurosci.* 31(48):17437–17448.

Kremkow J, Jin J, Wang Y, Alonso JM (2016). Principles underlying sensory map topography in primary visual cortex. *Nature.* 533:52–57.

Kuffler SW (1953). Discharge patterns and functional organization of mammalian retina. *J Neurophysiol.* 16:37–68.

Lien AD, Scanziani M (2013). Tuned thalamic excitation is amplified by visual cortical circuits. *Nat Neurosci.* 16:1315–1323.

Lien AD, Scanziani M (2018). Cortical direction selectivity emerges at convergence of thalamic synapses. *Nature.* 558:80–86.

Liu B, Li P, Sun YJ, Li Y, Zhang LI, Tao HW (2010). Intervening inhibition underlies simple-cell receptive field structure in visual cortex. *Nat Neurosci.* 13:89–96.

Livingstone L, Hubel, D (1988). Segregation of form, color, movement, and depth: Anatomy, physiology, and perception. *Science.* 240: 740–749.

Malpeli JG, Baker FH (1975). The representation of the visual field in the lateral geniculate nucleus of *Macaca mulatta*. *J Comp Neurol.* 161:569–594.

Martin PR, Solomon SG (2018). The koniocellular whiteboard. *J Comp Neur.* 527:505–507.

Najafian S, Koch E, Teh KL, Jin J, Rahimi-Nasrabadi H, Zaidi Q, Kremkow J, Alonso J-M (2022). A theory of cortical map formation in the visual brain. *Nat Commun.* 13:2303.

Pattadkal JJ, Mato G, van Vreeswijk C, Priebe NJ, Hansel D (2018). Emergent orientation selectivity from random networks in mouse visual cortex. *Cell Rep.* 24:2042–2050.

Peters A, Payne BR (1993). Numerical relationships between geniculocortical afferents and pyramidal cell modules in cat primary visual cortex. *Cereb Cortex.* 3:69–78.

Piscopo DM, El-Danaf RN, Huberman AD, Niell CM (2013). Diverse visual features encoded in mouse lateral geniculate nucleus. *J Neurosci.* 33:4642–4656.

Priebe NJ, Ferster D (2012). Mechanisms of neuronal computation in mammalian visual cortex. *Neuron.* 75: 194–208.

Reid RC, Alonso J-M (1995). Specificity of monosynaptic connections from thalamus to visual cortex. *Nature.* 378:281–284.

Reid RC, Soodak RE, Shapley RM (1991). Directional selectivity and spatiotemporal structure of receptive fields of simple cells in cat striate cortex. *J Neurophysiol.* 66:505–529.

Ringach DL (2004). Haphazard wiring of simple receptive fields and orientation columns in visual cortex. *J Neurophysiol.* 92:468–476.

Ringach DL (2011). You get what you get and you don't get upset. *Nat Neurosci.* 14:123–124.

Ringach DL (2021). Sparse thalamocortical convergence. *Curr Biol.* 31:2199–2202.

Rossi LF, Harris KD, Carandini M (2020). Spatial connectivity matches direction selectivity in visual cortex. *Nature.* 588:648–652.

Roy S, Jayakumar J, Martin, PR, Dreher B, Saalmann YB, Daping H, Vidyasagar TR (2009). Segregation of short-wavelength-sensitive (S) cone signals in the macaque dorsal lateral geniculate nucleus. *Eur J Neurosci.* 30:1517–1526.

Saul AB, Humphrey AL (1990). Spatial and temporal response properties of lagged and nonlagged cells in cat lateral geniculate nucleus. *J Neurophysiol.* 64:206–224.

Schiller PH, Logothetis NK, Charles ER (1990). Functions of the colour-opponent and broad-band channels of the visual system. *Nature.* 343:68–70.

Scholl B, Tan AYY, Corey J, Priebe NJ (2013) Emergence of orientation selectivity in the mammalian visual pathway. *J Neurosci.* 33:10616–10624.

Shapley R, Hugh Perry V (1986). Cat and monkey retinal ganglion cells and their visual functional roles. *Trends Neurosci.* 9:229–235.

Sillito AM (1975). The contribution of inhibitory mechanisms to the receptive field properties of neurones in the striate cortex of the cat. *J Phys.* 250:305–329.

Stevens, CF (2001). An evolutionary scaling law for the primate visual system and its basis in cortical function. *Nature.* 6834:193–105.

Sun W, Tan Z, Mensh BD, Ji N (2016). Thalamus provides layer 4 of primary visual cortex with orientation- and direction-tuned inputs. *Nat Neurosci.* 19:308–315.

Tailby C, Solomon SG, Dhruv NT, Majaj NJ, Sokol SH, Lennie P (2007). A new code for contrast in the primate visual pathway. *J Neurosci.* 27:3904–3909.

Tring E, Duan KK, Ringach DL (2022). ON/OFF domains shape receptive field structure in mouse visual cortex. *Nat Commun.* 13:2466.

Watson AB, Ahumada AJ (1985). Model of human visual-motion sensing. *J Opt Soc Am A.* 2:322–341.

White, AJR, Solomon SG, Martin PR (2001). Spatial properties of koniocellular cells in the lateral geniculate nucleus of the marmoset *Callithrix jacchus. J Physiol.* 533:519–535.

Winfield DA, Rivera-Dominguez M, Powell TP (1982). The termination of geniculocortical fibres in area 17 of the visual cortex in the macaque monkey. *Brain Res.* 231(1):19–32.

19.

CORTICOGENICULATE FEEDBACK AND VISUAL PROCESSING

W. Martin Usrey

INTRODUCTION

A reciprocal arrangement of feedforward and feedback projections interconnects the dorsal thalamus and the cerebral cortex. In the visual system, relay cells in the lateral geniculate nucleus (LGN) of the thalamus provide target neurons in primary visual cortex (V1) with feedforward information from the eye. The cortex, in turn, sends extensive feedback projections to the LGN to modulate LGN responses to retinal input. Feedback projections to the LGN arise from a subset of glutamatergic pyramidal neurons in layer 6 of cortex (Gilbert and Kelly, 1975; Lund et al., 1975; Hendrickson et al., 1978). Interestingly, these same neurons, known as corticogeniculate neurons, have branching axons that also provide local synaptic input to the layers of cortex that receive feedforward input from the LGN (Lund and Boothe, 1975; Katz, 1987; Usrey and Fitzpatrick, 1996; Briggs et al., 2016). Thus, based on both their local (cortical) and long-distance (LGN) projections, corticogeniculate neurons are in a strategic position to influence the processing and transmission of signals from the LGN to V1. Within the LGN, corticogeniculate axons provide both monosynaptic excitation and disynaptic inhibition onto relay neurons with rate-dependent interactions (Crandall et al., 2015; Kirchgessner et al., 2020), indicating that the influence of corticogeniculate activity on geniculocortical communication is complex. In the following sections, we explore the organization and function of the corticogeniculate pathway linking V1 with the LGN. (For more information about general features of corticothalamic feedback and feedback in other sensory systems, please see **Chapter 14 in Section IV and Chapters 24 and 27 in Section V**). Finally, because most studies on corticogeniculate feedback have been conducted with monkeys and cats, an emphasis is placed on discussing results from these species, with additional results included from other model organisms to highlight common themes.

BRIEF REVIEW OF FEEDFORWARD PROJECTIONS FROM RETINA TO LGN TO V1

Results from several studies indicate that the corticogeniculate feedback pathway is anchored in the parallel feedforward pathways of information flow from the retina to the LGN to the visual cortex, otherwise known as the retino-geniculo-cortical pathway. Because of this, it is useful to review both the general properties of the feedforward projections and the organization of these projections into parallel pathways before considering the organization and function of corticogeniculate feedback projections. (More detailed descriptions of the organization of the retino-geniculo-cortical pathway and of how feedforward projections from LGN contribute to the receptive field properties of V1 neurons are provided in **Chapters 17 and 18 in Section V**.)

GENERAL PROPERTIES OF RETINAL INPUTS TO THE LGN

It may seem surprising that retinal axons provide a minority (~5%) of the synapses onto individual LGN neurons (reviewed in Usrey and Alitto, 2015). Nevertheless, the retina is the primary driver of LGN activity (Hubel and Wiesel 1961; Cleland et al., 1971), evoking upwards of 95% of the spikes produced by individual LGN neurons (Kaplan and Shapley, 1984; Sincich et al., 2007). Several features of retinogeniculate synapses underlie this strong coupling, including their large size, proximal location to the cell soma, and large amplitude excitatory postsynaptic potentials (EPSPs; reviewed in Usrey and Sherman, 2021). Indeed, retinal evoked EPSPs are often large enough that they can be recorded with extracellular electrodes. These events, called S-potentials, can be measured simultaneously with the geniculate cell's action potentials to compare the response properties of pre- and postsynaptic neurons. It is worth noting that extracellular recordings are only able to distinguish S-potentials coming from the strongest of retinal inputs onto a geniculate neuron. Thus, different recording methods are needed to examine retinogeniculate communication in both weakly and strongly coupled retinal ganglion cells and LGN neurons.

With 2 electrodes, one in the eye and the other in the LGN, it is possible to study retinogeniculate communication for cell pairs with either strong or weak connections, which is useful as individual LGN neurons often receive convergent input from up to 5 retinal ganglion cells, with one of the retinal ganglion cells providing strong input and the remainder providing weaker input (Cleland, 1986; Usrey et al., 1999). One important distinction between the responses of retinal ganglion cells and LGN neurons is that retinal ganglion cells produce many more action potentials than their geniculate target neurons, which provides the opportunity for extraretinal inputs, such as corticogeniculate feedback, to either increase or decrease the successful communication of retinal spikes. The strength of retinogeniculate communication can be quantified using measures of efficacy (i.e., the percentage of retinal spikes that evoke spikes in the postsynaptic geniculate cell) and contribution (i.e., the percentage of geniculate cell spikes evoked by a particular retinal ganglion cell). Using S-potential and dual recording methods, the range of contribution values across weakly and strongly connected cell pairs is 1%–100% (Kaplan and Shapley, 1984; Usrey et al., 1999; Sincich et al., 2007). As might be expected, there is a direct relationship between contribution values and similarity between pre- and postsynaptic receptive fields in terms of their location, size, and sign (on vs. off; Usrey et al., 1999). The similarity between retinal and LGN receptive fields underlies, in part, the early view that the LGN is a straightforward and simple relay of retinal signals en route to the cortex; however, more recent work demonstrates that extraretinal signals (including those from corticogeniculate feedback) modulate retinal efficacy and transform geniculate responses in both the spatial and temporal domains.

PARALLEL PATHWAYS FROM RETINA TO CORTEX

Distinct classes of retinal ganglion cells provide target neurons in the LGN and, subsequently, their target neurons in V1 with different information about specific aspects of the visual scene. In all mammals where it has been studied, there are 2 major pathways to visual cortex, one that targets neurons in cortical layer 4 and the other that targets neurons in the superficial layers. Although the pathway to the superficial layers is considered evolutionarily to be the oldest, much more is known about the pathway to layer 4, which we will discuss first.

In monkeys and cats, the projection to layer 4 contains 2 distinct parallel pathways (Figure 19.1), referred to as the parvocellular and magnocellular pathways (monkeys) and the X and Y pathways (cats). Estimates suggest that monkeys and cats share a common ancestor approximately 80 million years ago. Although their evolutionary paths since then have led to many species-specific differences, there are extensive similarities (both in the retina and the LGN) between monkey parvocellular neurons and cat X cells and between monkey magnocellular neurons and cat Y cells (reviewed in Sherman, 1985; Cleland, 1986; Usrey and Alitto, 2015), indicating that the cell classes are homologous (Usrey and Sherman, 2022; but see Shapley and Perry, 1996). For instance, monkey parvocellular neurons and cat X cells have smaller cell bodies, smaller dendritic fields, smaller receptive fields, slower conducting axons, slower visual responses, and more sustained visual responses than their magnocellular and Y cell counterparts (reviewed in Usrey and Sherman, 2021). Within layer 4 of V1, the axons of monkey parvocellular neurons and cat X cells arborize beneath those of monkey magnocellular neurons and cat Y cells, thereby providing a substrate for the 2 pathways to remain segregated within the early stages of cortical processing (Figure 19.1; Hendrickson et al., 1978; Blasdel and Lund, 1983; Freund et al., 1985; Humphrey et al., 1985). Beyond their anatomical and physiological distinctions, molecular markers that are selective for magnocellular versus parvocellular neurons in the monkey are also selective for X cells versus Y cells in the cat (Hendry et al., 1984; Hockfield and Sur, 1990; Iwai et al., 2013).

The pathway from the LGN to the superficial layers of cortex contains many additional classes of cells. Unfortunately, and perhaps because much less is known about these cells compared to their layer 4 projecting counterparts, the neurons that comprise this diverse group of cells are often grouped together in a single category, referred to as koniocellular neurons in monkeys and W cells in cats (Figure 19.1). As is the case with layer 4 projecting geniculate neurons, the superficial projecting neurons receive retinal input from distinct classes of retinal ganglion cells (Szmajda et al., 2008); however, unlike the layer 4 projecting neurons, superficial projecting neurons also receive input from neurons in the superior colliculus (Fitzpatrick et al., 1980; Bickford et al., 2015), an input consistent with the pathway being evolutionarily older.

GENERAL PROPERTIES OF CORTICOGENICULATE FEEDBACK PROJECTIONS

In contrast to retinal axons that provide a small percentage of the synapses made with LGN neurons (~5%), corticogeniculate axons provide ~40% of all synapses onto LGN neurons (Figure 19.2; Guillery, 1969; Erisir

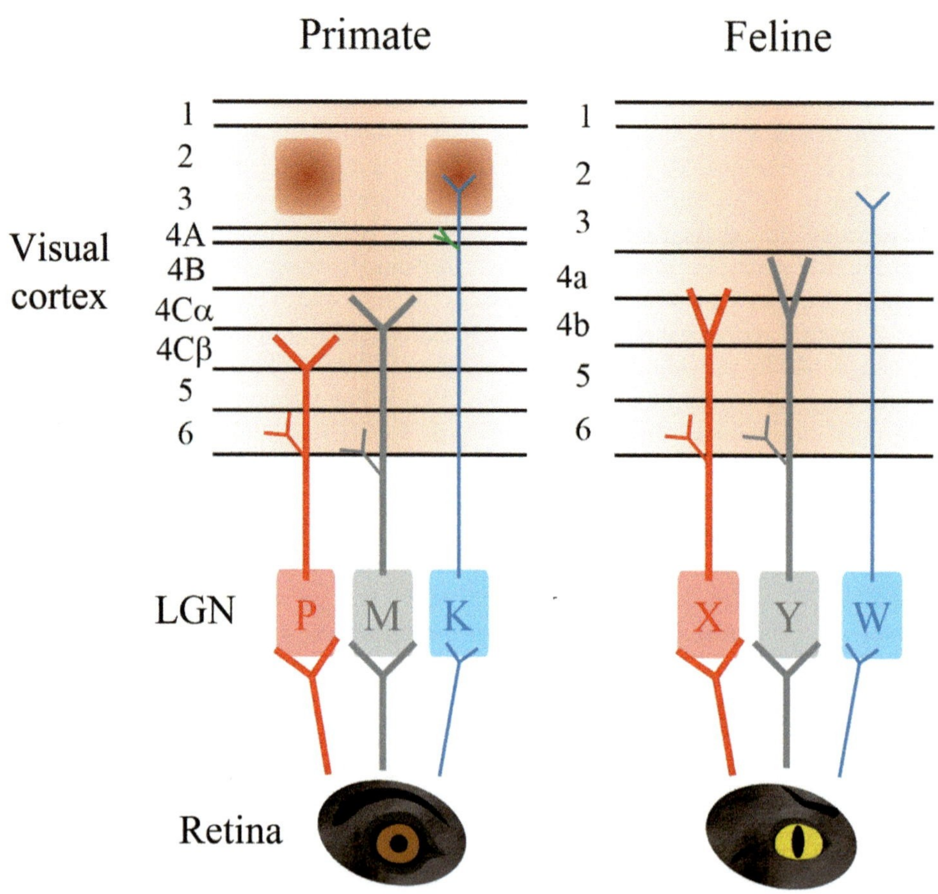

Figure 19.1. Schematic organization of the retino-geniculo-cortical pathway in the primate and carnivore. In all mammals, including these examples, the majority of geniculo-cortical axons terminate in cortical layer 4, a minority of axons terminate in cortical layers 1–3 and 6. There are 3 parallel pathways in the primate: the parvocellular (P) pathway terminates in layer 4Cβ, the magnocellular (M) pathway terminates in layer 4Cα, and the koniocellular (K) terminates in the layer 1–3 blobs. In the feline, there are also 3 major parallel pathways to cortex: the X-cell pathway terminates in layer 4b, the Y-cell pathway terminates in layer 4a, and the W-cell pathway terminates in layers 1–3. Adapted from Rathbun and Usrey (2008).

et al., 1997). Despite these numbers, the corticogeniculate pathway plays a modulatory rather than driving role in evoking geniculate action potentials (Sherman and Guillery, 1998). This is due, in part, to the smaller and more distal location of corticogeniculate synapses compared to retinal synapses (Guillery, 1969; Erisir et al., 1997). Corticogeniculate synapses also have a mixture of metabotropic and ionotropic receptors, whereas retinogeniculate synapses only have ionotropic receptors (Figure 19.2). In addition to making synapses with relay cells in the LGN, corticogeniculate axons also make synapses with local GABAergic interneurons and with GABAergic neurons in the thalamic reticular nucleus (TRN). Importantly, both of these GABAergic targets make synapses, in turn, onto LGN relay cells (Figure 19.2). Thus, corticogeniculate feedback can have a direct excitatory and disynaptic inhibitory influence on LGN relay cell activity (Figure 19.2). Consequently, by modulating the membrane potential of geniculate relay cells toward or away from their spike threshold, corticogeniculate feedback can increase or decrease the successful communication of

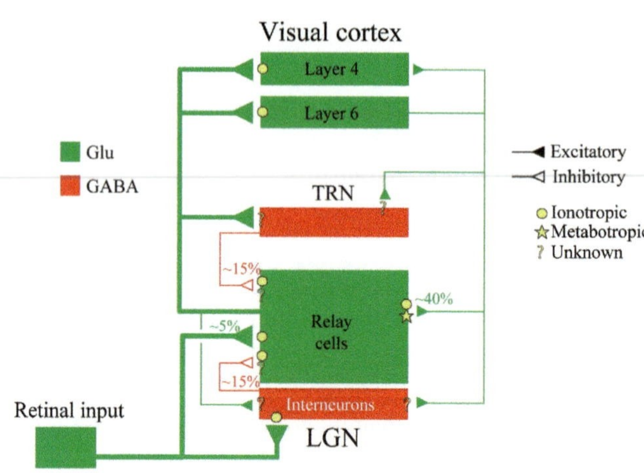

Figure 19.2. Schematic view of the major excitatory and inhibitory circuits involving the LGN (brainstem inputs not indicated). The key to the left indicates the major transmitter systems involved (glutamate vs. GABA), and that to the right, whether the postsynaptic receptors involved are ionotropic or metabotropic or both or unknown and whether the input is excitatory or inhibitory. LGN neurons receive excitatory feedforward input from the retina and corticogeniculate feedback input that is monosynaptic excitatory and disynaptic inhibitory. Adapted from Usrey and Sherman (2021).

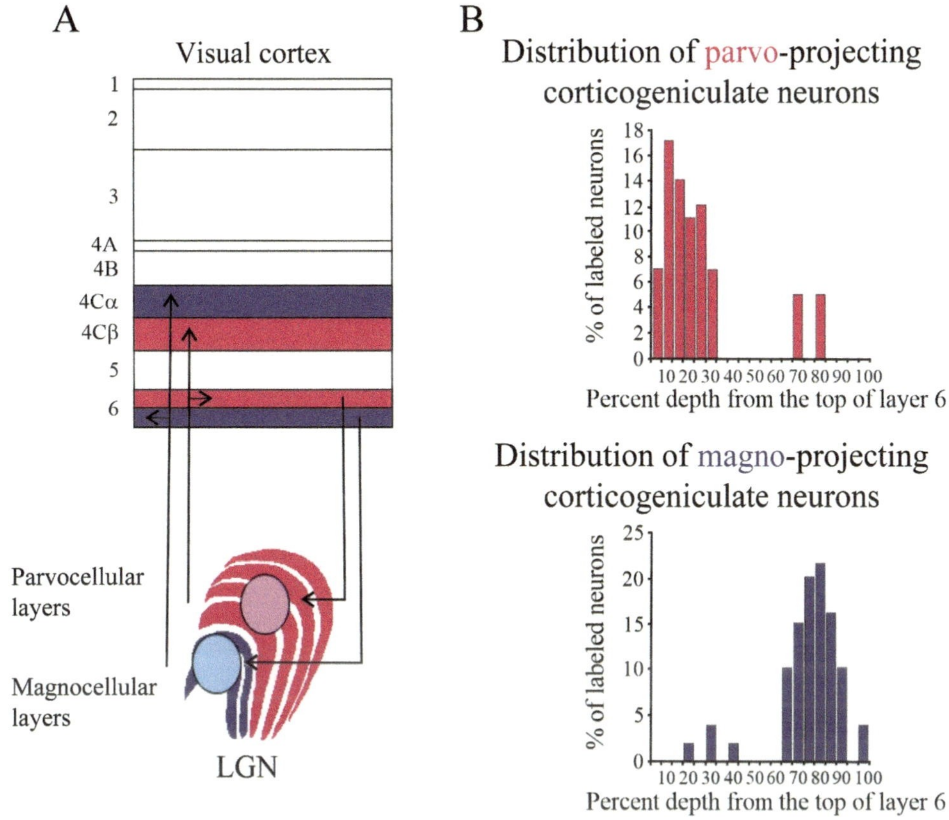

Figure 19.3. Sublaminar organization of corticogeniculate neurons projecting to the parvocellular and magnocellular layers of the LGN. (A). Circuit diagram illustrating the organization of feedforward geniculocortical projections. Axons from magnocellular neurons innervate neurons in cortical layer 4Cα and lower layer 6; axons from parvocellular neurons innervate neurons in cortical layer 4Cβ and upper layer 6. The koniocellular pathway is not shown in this figure. The colored ovals overlapping and restricted to the parvocellular (pink) and magnocellular (lavender) layers represent injection sites of retrograde tracers that labeled corticogeniculate neurons (B) in the upper and lower divisions of layer 6, respectively. Based on data from Fitzpatrick et al. (1994).

retinal action potentials, as well as play a role in determining whether geniculate neurons respond to retinal input with burst or tonic responses (described below).

Compared to the axon arbors of retinal ganglion cells, the axon arbors of corticogeniculate neurons are much larger, approximately 2–3 times larger (Murphy and Sillito, 1996). This difference allows corticogeniculate axons to have access to a larger pool of LGN neurons than do individual retinal axons. It also allows individual LGN neurons to receive corticogeniculate input corresponding to a broader region of visual space than that provided directly from the retina. These properties have important implications when considering the influence of corticogeniculate feedback on visual responses in the LGN.

PARALLEL STREAMS OF CORTICOGENICULATE FEEDBACK

Just as parallel feedforward pathways supply the cortex with distinct information about the visual scene, there is strong evidence for parallel streams of corticogeniculate feedback. In monkeys, neurons in the parvocellular layers of the LGN receive feedback input from corticogeniculate neurons with cell bodies in the upper half of layer 6 (Figure 19.3; Conley and Raczkowski, 1990, Fitzpatrick et al., 1994; Ichida et al., 2014). In contrast, neurons in the magnocellular layers receive feedback input from corticogeniculate neurons in the lower half of layer 6. The division of layer 6 into an upper half and a lower half based on the projection targets of its corticogeniculate neurons is further emphasized when one compares the input layer 6 receives from the LGN. As geniculate axons en route to layer 4 traverse through layer 6, axons targeting the parvocellular division of layer 4 often branch and provide a collateral projection that makes synapses in the upper half of layer 6, whereas those targeting the magnocellular division of layer 4 branch and send collateral projections that make synapses in the bottom of layer 6 (Hendrickson et al., 1978; Blasdel and Lund, 1983).

The physiology of corticogeniculate neurons also supports the idea of parallel streams of corticogeniculate feedback, as feedback neurons supplying the parvocellular and magnocellular layers of the LGN have visual responses that share many features in common with those of their target neurons, including their contrast sensitivity and their temporal frequency tuning (Figure 19.4; Briggs and Usrey, 2009). In addition, corticogeniculate neurons with visual responses that align with the magnocellular pathway have complex receptive fields and fast-conducting axons, whereas corticogeniculate neurons with visual responses that align with the parvocellular pathway have simple receptive fields and slower-conducting axons (Briggs and Usrey, 2009). A third group of corticogeniculate neurons in the monkey, with cell bodies located in the lowest portion of layer 6, have response properties that align well with the koniocellular neurons in the LGN (Briggs and Usrey, 2009; Ichida et al., 2014). These neurons have the slowest conducting axons and receptive fields that include input from the short-wavelength (blue) cones in the retina (Briggs and Usrey, 2009), properties characteristic of koniocellular geniculate neurons.

There is also strong evidence for parallel streams of feedback in the cat, where one group of corticogeniculate neurons has fast-conducting axons and complex receptive fields, and the other has slower-conducting axons and simple receptive fields (Tsumoto and Suda, 1980). Less is known about corticogeniculate neurons in other species; however, existing evidence from tree shrews, ferrets, and rabbits (Usrey and Fitzpatrick, 1996; Briggs and Usrey, 2005; Stoelzel et al., 2017; Hasse et al., 2019) supports the idea that corticogeniculate feedback is organized into parallel streams that can exert stream-specific influences on the feedforward transmission of visual signals to the cortex.

CORTICOGENICULATE EFFECTS ON VISUAL PROCESSING

There are 2 broad categories of identified roles for corticogeniculate feedback in visual processing, and both can be accounted for by modulating the responsiveness of LGN neurons to incoming retinal signals. The first category involves effects on the spatial properties of geniculate receptive fields, and the second involves effects on the temporal properties of geniculate responses.

In the spatial domain, although the receptive fields of LGN neurons and their retinal inputs are generally similar, LGN neurons often have greater extraclassical surround suppression than their retinal counterparts, particularly following a short delay from the initial visual response (Fisher et al., 2017). Extraclassical surround suppression is a receptive field property in which responses to stimuli within a neuron's classical receptive field are suppressed by stimuli that extend into the extraclassical surround (Jones et al., 2000). It is also a receptive field property that is particularly prominent in off-center LGN neurons compared to on-center neurons and in magnocellular neurons compared to parvocellular neurons (Alitto and Usrey, 2008; Archer et al., 2021; Lankow and Usrey, 2022). Furthermore, it is a receptive field property that becomes more pronounced as stimulus contrast increases (Archer et al., 2021), as is the case in the cortex (Sceniak et al., 1999).

Results from studies in monkeys, cats, and mice support the idea that corticogeniculate feedback strengthens extraclassical surround suppression in LGN neurons (Murphy and Sillito, 1987; Sillito and Jones, 2002; Born et al., 2021). Moreover, there is evidence indicating that corticogeniculate feedback has a center/surround influence on geniculate activity, whereby feedback from cortical neurons with retinotopically overlapping receptive fields is net excitatory and feedback from neighboring neurons is net suppressive (Jones et al., 2012; Born et al., 2021). As a consequence, the excitatory (center) and suppressive (surround) effects of feedback, which presumably involve spatially organized monosynaptic excitation and disynaptic inhibition (Figure 19.5), would serve to sharpen the spatial boundaries of LGN receptive fields. This role appears to be generalized to other sensory systems, as cortical feedback in the somatosensory system has been shown to enhance the responses of thalamic neurons with matching receptive field properties and to suppress the responses of neurons with mismatched properties (Temereanca and Simons, 2004; Li and Ebner, 2007).

In the temporal domain, corticogeniculate feedback is reported to influence the transmission of visual information in several ways. For instance, corticogeniculate feedback is reported to increase the gain of LGN responses to weak stimuli, such as low-contrast visual stimuli (Przybyszewski et al., 2000; see also Olsen et al., 2012; Wang et al., 2018). Corticogeniculate feedback has also been shown to decrease the response latency of geniculate neurons to visual stimuli, to enhance the temporal precision of geniculate responses, and to decrease response variability (Andolina et al., 2007; Hasse and Briggs, 2017; Murphy et al., 2021). For small ensembles of geniculate cells excited with a common stimulus, feedback is reported to synchronize the timing of their spikes (Sillito et al., 1994; but see Brody, 1998). For larger ensembles of geniculate cells, modeling efforts indicate that corticothalamic projections can increase thalamic synchrony and oscillations in the cortico-thalamo-cortical loop (Destexhe et al., 1999), a role supported by experimental work showing that corticothalamic stimulation can trigger different patterns of correlated activity within the thalamus that is dependent on the mode of stimulation and inhibitory interactions with the TRN (Blumenfeld and McCormick, 2000; see also Bastos et al., 2014).

Corticothalamic feedback is also reported to influence whether or not geniculate neurons respond to retinal inputs with bursts of spikes or tonic spikes (Andolina et al., 2013; Ortuño et al., 2014; Born et al., 2021). Across thalamic nuclei, relay cells have 2 response modes—burst

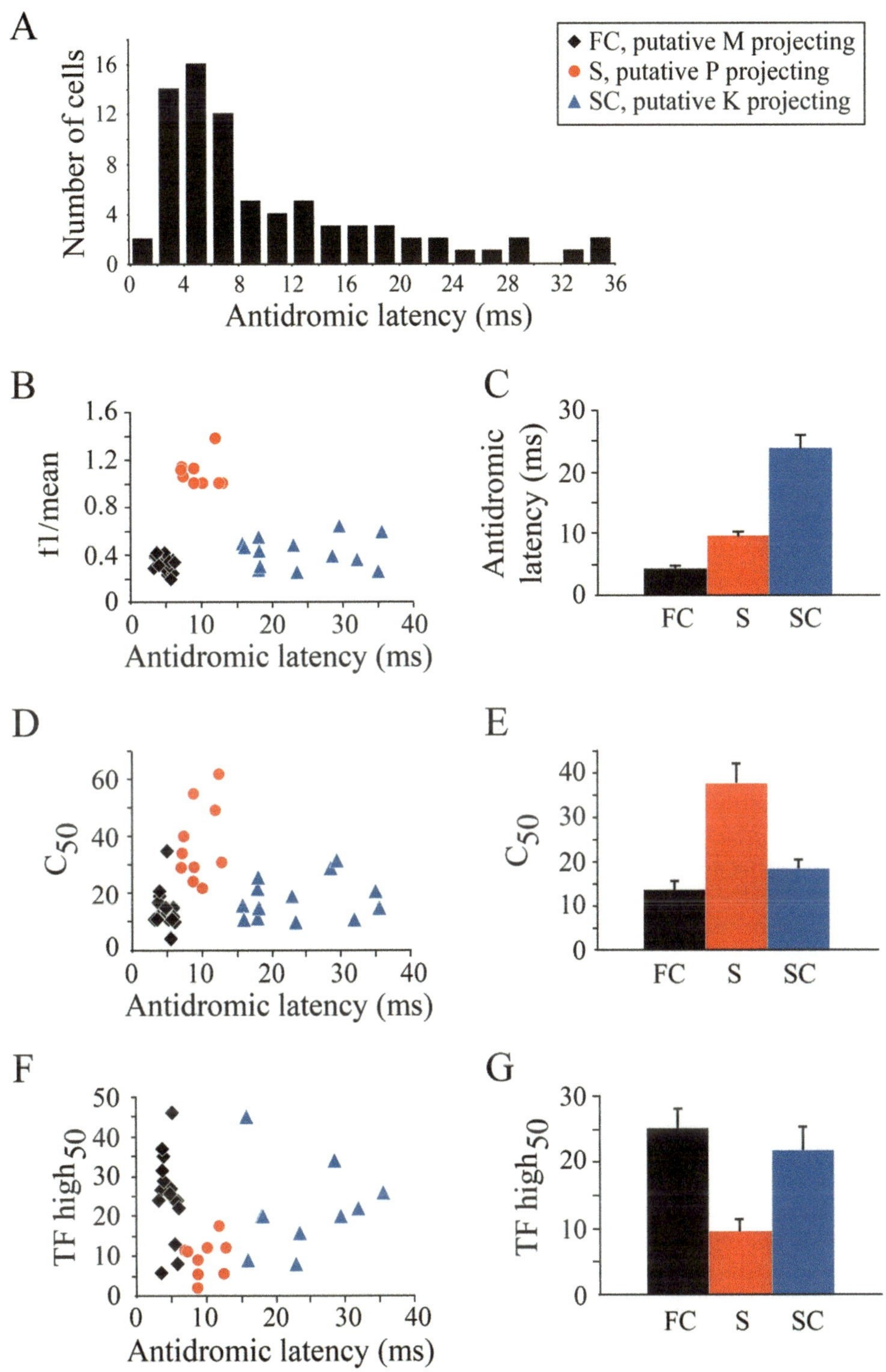

Figure 19.4. Physiological properties of identified corticogeniculate neurons in the alert monkey. (A). Distribution of antidromic response latencies for 78 corticogeniculate neurons. (B). f1 to f0 ratio of visual responses to drifting sinusoidal gratings versus antidromic latency for 40 corticogeniculate neurons. Neurons with ratios <1.0 are simple cells; neurons with ratios >1.0 are complex cells. Three distinct clusters of neurons are evident: fast complex cells (FC, black diamonds) with putative magnocellular projecting axons; simple cells (S, red circles) with putative parvocellular projecting axons; and slow complex cells (SC, blue triangles) with putative koniocellular projecting axons. (C). Average antidromic latencies of the fast complex, simple, and slow complex cells. All 3 classes of corticogeniculate neurons are significantly different from each other in terms of antidromic latency. (D and E). Comparison of the contrast to evoke a half-maximum response (C50), versus antidromic latency for fast complex, simple, and slow complex cells. The fast complex cells (putative magnocellular projecting) are more sensitive to low-contrast stimuli than are the simple cells (putative parvocellular projecting. (F and G). Comparison of the highest temporal frequencies to evoke a half-maximum response (TF high50) versus antidromic latency for fast complex, simple, and slow complex cells. The fast complex cells (putative magnocellular projecting) follow stimuli at higher temporal frequencies than the simple cells (putative parvocellular projecting). For all histograms, error bars represent SEM. Adapted from Briggs and Usrey (2009).

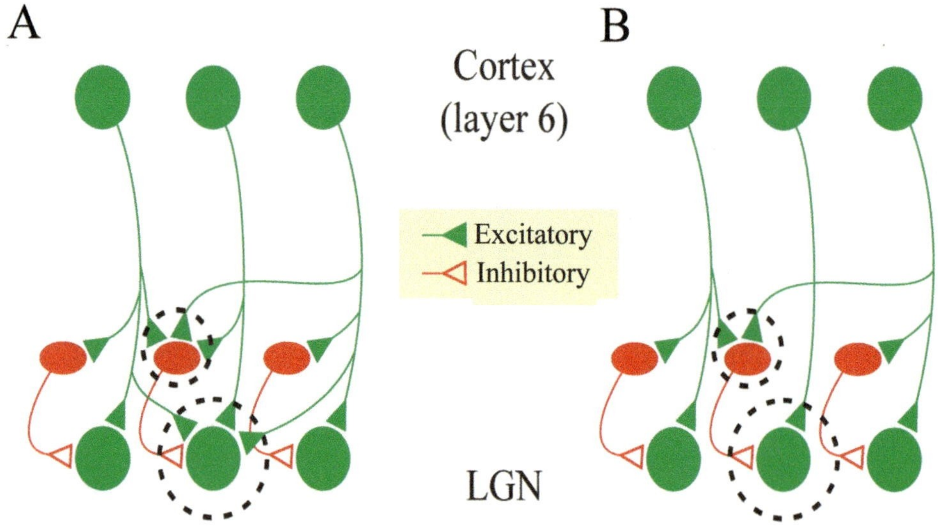

Figure 19.5. Two patterns among others possible for corticogeniculate projections to TRN cells, geniculate interneurons, and geniculate relay cells. (A). The traditional view of feedback circuitry that includes a pattern of monosynaptic excitation and disynaptic inhibition without any center/surround organization. The inhibitory neuron (dashed circle, red cell) provides disynaptic inhibition to a relay cell (dashed circle, green cell) that has a similar spatial scale as the monosynaptic excitation. (B). A more complicated pattern of feedback circuitry, but consistent with physiological results, indicating feedback enhances responses to stimuli overlapping a relay cell's classical receptive field and suppresses responses when stimuli extend beyond the classical receptive field, in which the relay cell receives a center-surround organization of feedback input with monosynaptic excitation (center) and disynaptic inhibition (surround).

mode and tonic mode—that differ substantially in their temporal features (reviewed in Usrey and Sherman, 2021). Ultimately, it is the membrane potential history of a thalamic neuron that determines the response mode, and feedback is one source of input that contributes to that history. Specifically, if a geniculate neuron is hyperpolarized below typical resting levels (e.g., −75 mV vs. −65 mV) for ~100 ms or longer, then T-type Ca^{2+} channels become de-inactivated and subsequent membrane depolarization leads to a low-threshold Ca^{2+} spike that subsequently triggers a burst of Na^{+}-based spikes. In contrast, if a geniculate neuron is less hyperpolarized (for ~100 ms or longer), then the T-type channels are in the inactivated state and unable to open in response to depolarizing inputs. In this state, a geniculate neuron will respond to incoming inputs with regularly spaced tonic spikes. Thus, based on the inactivation state of the T-type channels, LGN neurons operate in either burst mode or tonic mode. Because feedback provides monosynaptic excitation and disynaptic inhibition to LGN relay cells, feedback has the opportunity to shift relay cells between response modes, with tonic mode providing the cortex with higher-fidelity information about the visual stimulus and burst mode being more effective in driving the cortex (Sherman, 2001; Swadlow and Gusev, 2001).

Finally, behavioral state has a strong influence on the temporal features of LGN responses to sensory stimuli, and feedback projections are thought to contribute to this influence. Two extreme examples of behavioral state, sleep and spatial attention, illustrate the range of effects. With sleep, a state that involves many neural circuits including corticothalamic projections, the communication of visual signals to cortex is greatly diminished and disrupted (Livingstone and Hubel, 1981; Steriade, 2005). In contrast, with spatial attention there is an enhancement of geniculate activity (O'Connor et al., 2002; McAlonan et al., 2008) and successful geniculocortical communication (Briggs et al., 2013), which likely involves input from corticogeniculate neurons (Béhuret et al., 2015).

CONCLUDING REMARKS

The LGN and V1 are interconnected with each other via a reciprocal arrangement of feedforward and feedback projections. Projections from the LGN to V1 provide the cortex with visual signals about the outside world. The cortex, in turn, sends feedback projections to the LGN that modulate those signals. Results from a growing number of studies demonstrate that the corticogeniculate feedback pathway is organized into parallel streams that generally reflect the organization of the parallel feedforward pathways. The influence of these projections can be net excitatory or net inhibitory, and the interplay between excitation and inhibition likely underlies the reported effects of feedback on the spatial and temporal features of visual responses in the LGN. Importantly, with feedback the cortex can shape the input it receives about the visual world. Such dynamics underscore the view that the cortex and thalamus cannot be understood when viewed in isolation of each other, but rather are best understood when viewed together as an integrated system.

REFERENCES

Alitto HJ, Usrey WM (2008). Origin and dynamics of extraclassical suppression in the lateral geniculate nucleus of the macaque monkey. *Neuron*. 57:135–146.

Andolina IM, Jones HE, Sillito AM (2013). Effects of cortical feedback on the spatial properties of relay cells in the lateral geniculate nucleus. *J Neurophysiol*. 109:889–899.

Andolina IM, Jones HE, Wang W, Sillito AM (2007). Corticothalamic feedback enhances stimulus response precision in the visual system. *Proc Natl Acad Sci U S A*. 104:1685–1690.

Archer DR, Alitto HJ, Usrey WM (2021). Stimulus contrast affects spatial integration in the lateral geniculate nucleus of macaque monkeys. *J Neurosci*. 41:6246–6256.

Bastos AM, Briggs F, Alitto HJ, Mangun G R, Usrey WM (2014). Simultaneous recordings from the primary visual cortex and lateral geniculate nucleus reveal rhythmic interactions and a cortical source for gamma-band oscillations. *J Neurosci*. 34:7639–7644.

Béhuret S, Deleuze C, Bal T (2015). Corticothalamic synaptic noise as a mechanism for selective attention in thalamic neurons. *Front Neural Circuits*. 9:80.

Bickford ME, Zhou N, Krahe TE, Govindaiah G, Guido W (2015). Retinal and tectal "driver-like" inputs converge in the shell of the mouse dorsal lateral geniculate nucleus. *J Neurosci*. 35:10523–10534.

Blasdel GG, Lund JS (1983). Termination of afferent axons in macaque striate cortex. *J Neurosci*. 3:1389–1413.

Blumenfeld H, McCormick DA (2000). Corticothalamic inputs control the pattern of activity generated in thalamocortical networks. *J Neurosci*. 20:5153–5162.

Born G, Schneider-Soupiadis FA, Erisken S, Vaiceliunaite A, Lao CL, Mobarhan MH, Spacek MA, Einevoll GT, Busse L (2021). Corticothalamic feedback sculpts visual spatial integration in mouse thalamus. *Nat Neurosci*. 24:1711–1720.

Briggs F, Mangun GR, Usrey WM (2013). Attention enhances synaptic efficacy and signal-to-noise in neural circuits. *Nature*. 499:476–480

Briggs F, Usrey WM (2005). Temporal properties of feedforward and feedback pathways between the thalamus and visual cortex in the ferret. *Thalamus Rel Syst*. 3:133–139.

Briggs F, Usrey WM (2009). Parallel processing in the corticogeniculate pathway of the macaque monkey. *Neuron*. 62:135–146.

Briggs F, Kiley CW, Callaway EM, Usrey WM (2016). Morphological substrates for parallel streams of corticogeniculate feedback originating in both V1 and V2 of the macaque monkey. *Neuron*. 90:388–399.

Brody CD (1998). Slow covariations in neuronal resting potentials can lead to artefactually fast cross-correlations in their spike trains. *J Neurophysiol*. 80:3345–3351.

Cleland BG (1986). The dorsal lateral geniculate nucleus of the cat. In: Pettigrew JD, Sanderson KJ, Levick WR, eds. *Visual Neuroscience*. Cambridge University Press:111–120.

Cleland BG, Dubin MW, Levick WR (1971). Simultaneous recording of input and output of lateral geniculate neurones. *Nat New Biol*. 231:191–192.

Conley M, Raczkowski D (1990). Sublaminar organization within layer VI of the striate cortex in Galago. *J Comp Neurol*. 302:425–436.

Crandall SR, Cruikshank SJ, Connors BW (2015). A corticothalamic switch: Controlling the thalamus with dynamic synapses. *Neuron*. 86:768–782.

Destexhe A, Contreras D, Steriade M (1999). Cortically-induced coherence of a thalamic-generated oscillation. *Neuroscience*. 92:427–443.

Erisir A, Van Horn SC, Sherman SM (1997). Relative numbers of cortical and brainstem inputs to the lateral geniculate nucleus. *Proc Natl Acad Sci U S A*. 94:1517–1520.

Fisher TG, Alitto HA, and Usrey WM (2017). Retinal and non-retinal contributions to extraclassical surround suppression in the lateral geniculate nucleus. *J Neurosci*. 37:226–235.

Fitzpatrick D, Carey RG, Diamond IT (1980). The projection of the superior colliculus upon the lateral geniculate body in tupaia glis and galago senegalensis. *Brain Res*. 194:494–499.

Fitzpatrick D, Usrey WM, Schofield BR, Einstein G. 1994. The sublaminar organization of neurons in layer 6 of macaque striate cortex. *Vis Neurosci*. 11:307–315.

Freund TF, Martin KAC, Whitteridge D (1985). Innervation of cat visual areas 17 and 18 by physiologically identified X- and Y-type thalamic afferents. I. Arborization patterns and quantitative distribution of postsynaptic elements. *J Comp Neurol*. 242:263–274.

Gilbert CD, Kelly JP (1975). The projections of cells in different layers of the cat's visual cortex. *J Comp Neurol*. 163:81–106.

Guillery RW (1969). A quantitative study of synaptic interconnections in the dorsal lateral geniculate nucleus of the cat. *Z Zellforsch*. 96:39–48.

Hasse JM, Bragg EM, Murphy AJ, Briggs F (2019). Morphological heterogeneity among corticogeniculate neurons in ferrets: Quantification and comparison with a previous report in macaque monkeys. *J Comp Neurol*. 527:546–557.

Hasse JM, Briggs F (2017). Corticogeniculate feedback sharpens the temporal precision and spatial resolution of visual signals in the ferret. *Proc Natl Acad Sci U S A*. 114(30):E6222–E6230.

Hendrickson AE, Wilson JR, Ogren MP (1978). The neuroanatomical organization of pathways between the dorsal lateral geniculate nucleus and visual cortex in Old World and New World primates. *J Comp Neurol*. 182:123–136.

Hendry SHC, Hockfield S, Jones EG, McKay R (1984). Monoclonal antibody that identifies subsets of neurones in the central visual system of monkey and cat. *Nature*. 307:267–269.

Hockfield S, Sur M (1990). Monoclonal antibody cat-301 identifies Y-cells in the dorsal lateral geniculate nucleus of the cat. *J Comp Neurol*. 300:320–330.

Hubel DH, Wiesel TN (1961). Integrative action in the cat's lateral geniculate body. *J Physiol*. 155:385–398.

Humphrey AL, Sur M, Uhlrich DJ, Sherman SM (1985). Projection patterns of individual X- and Y-cell axons from the lateral geniculate nucleus to cortical area 17 in the cat. *J Comp Neurol*. 233:159–189.

Ichida JM, Mavity-Hudson JA, Casagrande VA (2014). Distinct patterns of corticogeniculate feedback to different layers of the lateral geniculate nucleus. *Eye Brain*. 6(Suppl 1):57–73.

Iwai L, Ohashi Y, van der List D, Usrey WM, Miyashita Y, Kawasaki H (2013). FoxP2 is a parvocellular-specific transcription factor in the visual thalamus of monkeys and ferrets. *Cereb Cortex*. 23:2204–2212.

Jones HE, Andolina IM, Oakely NM, Murphy PC, Sillito AM (2000). Spatial summation in lateral geniculate nucleus and visual cortex. *Exp Brain Res*. 135:279–284.

Jones HE, Andolina IM, Ahmed B, Shipp SD, Clements JT, Grieve KL, Cudeiro J, Salt TE, Sillito AM (2012). Differential feedback modulation of center and surround mechanisms in parvocellular cells in the visual thalamus. *J Neurosci*. 32:15946–15951.

Kaplan E, Shapley R (1984). The origin of the S (slow) potential in the mammalian lateral geniculate nucleus. *Exp Brain Res*. 55:111–116.

Katz LC (1987). Local circuitry of identified projection neurons in cat visual cortex brain slices. *J Neurosci*. 4:1223–1249.

Lankow BS, Usrey WM (2022). Contextual modulation of feedforward inputs to primary visual cortex. *Front Systems Neurosci*. 16:818633. doi:10.3389/fnsys.2022.818633.

Li L, Ebner FF (2007). Cortical modulation of spatial and angular tuning maps in the rat thalamus. *J Neurosci*. 27:167–179.

Livingstone MS, Hubel DH (1981). Effects of sleep and arousal on the processing of visual information in the cat. *Nature*. 291:554–561.

Lund JS, Boothe R (1975). Interlaminar connections and pyramidal neuron organization in the visual cortex, area 17, of the macaque monkey. *J Comp Neurol*. 159:305–334.

Lund JS, Lund RD, Hendrickson AE, Bunt AH, Fuchs AF (1975). The origin of efferent pathways from the primary visual cortex, area 17 of the macaque monkey. *J Comp Neurol*. 164:287–304.

Kirchgessner MA, Franklin AD, Callaway EM (2020). Context-dependent and dynamic functional influence of corticothalamic pathways to first- and higher-order visual thalamus. *Proc Natl Acad Sci U S A*. 117:13066–13077.

McAlonan K, Cavanaugh J, Wurtz RH (2008). Guarding the gateway to cortex with attention in visual thalamus. *Nature*. 456:391–394.

Murphy AJ, Shaw L, Hasse JM, Goris RLT, Briggs F (2021). Optogenetic activation of corticogeniculate feedback stabilizes response gain and increases information coding in LGN neurons. *J Comput Neurosci*. 49:259–271.

Murphy PC, Sillito AM (1996). Functional morphology of the feedback pathway from area 17 of the cat visual cortex to the lateral geniculate nucleus. *J Neurosci*. 16:1180–1192.

O'Connor DH, Fukui MM, Pinsk MA, Kastner S (2002). Attention modulates responses in the human lateral geniculate nucleus. *Nature Neurosci*. 5:1203–1209.

Olsen SR, Bortone DS, Adesnik H, Scanziani M (2012). Gain control by layer six in cortical circuits of vision. *Nature*. 483:47–52

Ortuño T, Grieve KL, Cao R, Cudeiro J, Rivadulla C (2014). Bursting thalamic responses in awake monkey contribute to visual detection and are modulated by corticofugal feedback. *Front Behav Neurosci*. 8:198.

Przybyszewski AW, Gaska JP, Foote W, Pollen DA (2000). Striate cortex increases contrast gain of macaque LGN neurons. *Vis Neurosci*. 17:485–494.

Rathbun DL, Usrey WM (2008). *The geniculo-striate pathway*. In: Encyclopedia of Neuroscience. Eds, Binder MD, Hirokawa N, Windhorst U, and Hirsch MC. Springer-Verlag. Heidelberg, Germany.

Sceniak MP, Ringach DL, Hawken MJ, Shapley R (1999). Contrast's effect on spatial summation by macaque V1 neurons. *Nat Neurosci*. 2:733–739.

Shapley R, Perry VH (1986). Cat and monkey retinal ganglion cells and their visual functional roles. *Trends Neurosci*. 9:229–235.

Sherman SM (1985). Functional organization of the W-, X-, and Y-cell pathways in the cat: A review and hypothesis. In: Sprague JM, Epstein AN, eds. *Progress in Psychobiology and Physiological Psychology*. Academic Press; 11:233–314.

Sherman SM (2001). Tonic and burst firing: Dual modes of thalamocortical relay. *Trends Neurosci*. 24:122–126.

Sherman SM, Guillery RW (1998). On the actions that one nerve cell can have on another: Distinguishing "drivers" from "modulators." *Proc Natl Acad Sci U S A*. 95:7121–7126.

Sillito AM, Jones HE (2002). Corticothalamic interactions in the transfer of visual information. *Philos Trans R Soc Lond B Biol Sci*. 357:1739–1752.

Sillito AM, Jones HE, Gerstein GL, West DC (1994). Feature-linked synchronization of thalamic relay cell firing induced by feedback from the visual cortex. *Nature*. 369:479–482.

Sincich LC, Adams DL, Economides JR, Horton JC (2007). Transmission of spike trains at the retinogeniculate synapse. *J Neurosci*. 27:2683–2692.

Steriade M (2005). Sleep, epilepsy and thalamic reticular inhibitory neurons. *Trends Neurosci*. 28:317–324.

Stoelzel CR, Bereshpolova Y, Alonso J-M, Swadlow HA (2017). Axonal conduction delays, brain state, and corticogeniculate communication. *J Neurosci*. 37:6342–6358.

Swadlow HA, Gusev AG (2001). The impact of "bursting" thalamic impulses at a neocortical synapse. *Nat Neurosci*. 4:402–408.

Szmajda BA, Grünert U, Martin PR (2008). Retinal ganglion cell inputs to the koniocellular pathway. *J Comp Neurol*. 510:251–268.

Temereanca S, Simons DJ (2004). Functional topography of corticothalamic feedback enhances thalamic spatial response tuning in the somatosensory whisker/barrel system. *Neuron*. 41: 639–651.

Tsumoto T, Suda K (1980). Three groups of cortico-geniculate neurons and their distribution in binocular and monocular segments of cat striate cortex. *J Comp Neurol*. 193:223–236.

Usrey WM, Alitto HJ (2015). Visual functions of the thalamus. *Ann Rev Vision Science*. 1:351–371.

Usrey WM, Fitzpatrick D (1996). Specificity in the axonal connections of layer VI neurons in tree shrew striate cortex: Evidence for distinct granular and supragranular systems. *J Neurosci*. 16: 1203–1218.

Usrey WM, Reppas JB, Reid RC (1999). Specificity and strength of retinogeniculate connections. *J Neurophysiol*. 82:3527–3540.

Usrey WM, Sherman SM (2021). *Exploring Thalamocortical Interactions: Circuitry for Sensation, Action, and Cognition*. Oxford University Press.

Usrey WM, Sherman SM (2022). Evolution, homology, cell classification, and parallel processing for vision. In: *Oxford Research Encyclopedia of Neuroscience*. Oxford University Press.

Wang W, Andolina IM, Lu Y, Jones HE, Sillito AM (2018). Focal gain control of thalamic visual receptive fields by layer 6 corticothalamic feedback. *Cereb Cortex*. 28:267–280.

20.

VISUAL PROCESSING IN THE PULVINAR

Angela Y. Fan and James A. Bourne

INTRODUCTION

Humans have evolved to be exceptionally visual animals, with more than half of the cerebral cortex dedicated to processing visual information. Coinciding with the evolutionary expansion of the visual cortex has been a concomitant expansion of the visual diencephalon, specifically, the pulvinar nuclei (Figure 20.1), the thalamus most significant but least understood corpus. This inimitable expansion of thalamic nuclei on such a grand scale is often overlooked in favor of the evolved function of the cortex. However, the pulvinar should be more routinely recognized as a convergence point for the rapid transmission of information widely in the cerebral cortex.

The pulvinar nuclei have evolved to become a more significant constituent of the thalamus during mammalian evolution (Figure 20.1, P/B ratios). Contemporary neuroanatomy studies have settled on defining 3 primary nuclei for primates, including humans: the lateral (PL), inferior (PI), and medial (PM), with PI and PL being considered the "visual pulvinar." Additional pulvinar subdivisions have evolved, especially within the inferior pulvinar nucleus, based on cytoarchitectonics and connectivity (Figure 20.1). This raises the question: Is this pulvinar involved in neural processes that are more complex, elaborate, and sophisticated, and exclusive to the primates in which an increased volumetric presence is observed?

A significant controversy that has plagued the community is the presence of a pulvinar homolog in rodents. Originally defined as the lateral posterior nucleus (LP), this thalamus region is a primitive homolog of the primate pulvinar (Harting et al., 1972), based on the significant and similar connectivity pattern with the superior colliculus. Although the rodent pulvinar does not possess the same degree of complexity as the primate, the model's versatility has undeniably furthered our understanding of the visual pulvinar and associated transthalamic circuits. The LP consists of several subdivisions (Table 20.2; Figure 20.1), divided by response characteristics to visual stimuli and their connectivity. Examining response properties by classic electrophysiological approaches, LPl cells were preferentially tuned for visual stimuli with a higher temporal frequency, preferring higher-speed processing.

Early studies of visual processing in the primate had only defined a circuit from the retina to the primary visual cortex (V1) via the lateral geniculate nucleus (LGN). However, Diamond and Hall (1969) postulated that a second pathway from the retina to the visual cortex might also afford significant visual information. This proposed second pathway included the pulvinar nucleus of the thalamus, which could support visual input to multiple visual cortices. Despite this considerable advance in our understanding, much of vision research has focused on uncovering the complex interconnectivity between visual cortices and their isolated function independent of thalamic input. For the pulvinar nuclei, this was likely because it is considered a prototypic "higher-order" thalamic nucleus, meaning that most of its information is derived from layers 5 and 6 of the cortex. Yet, the function of this circuitry remained elusive, and its role notably recondite. Consequently, there remained little focus on the pulvinar and its afferent and efferent connectivity with the visual system, even though in the 1980s, groups discovered that neurons across the pulvinar nuclei were responsive to various visual stimuli (Benevento and Miller, 1981; Robinson and Petersen, 1985).

The world map we recognize today has evolved significantly since the first paper impressions by the Greek cartographer Anaximander in the 6th century BCE through human perseverance and expanded methodologies. Analogously, our understanding of pulvinar anatomy has evolved through the congruence of opinion from several nonhuman primate pulvinar functional anatomy studies,

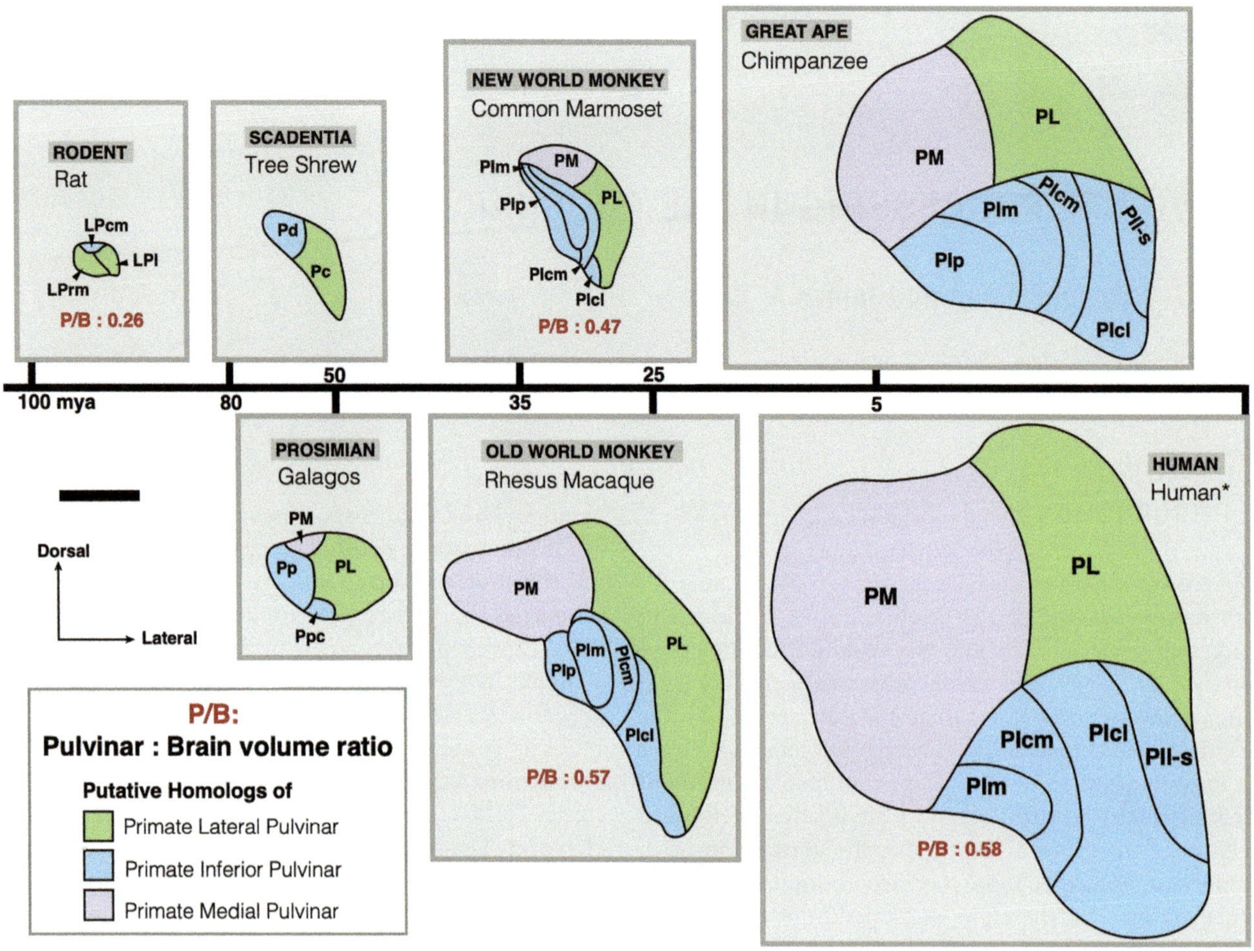

Figure 20.1. A phylogenetic tree of the pulvinar nuclei of notable members of the Euarchontoglires that have been routinely utilized in anatomical and physiological visual neuroscience, revealing the emergence of nuclei and specific subdivisions and an expanded volume compared to the total thalamus. Schematics are based on chemoarchitectural studies: Rat pulvinar adapted from Scholl et al. (2021). Tree shrew; galagos; common marmoset; rhesus macaque pulvinars adapted from Baldwin and Bourne (2020). Chimpanzee and human pulvinars adapted from Cola et al. (1999, 2005); Baldwin et al. (2017). P/B = pulvinar:total brain volume ratio (log[vol(pul)]/log[vol(brain)]) ratio to substantiate the allometric expansion of the pulvinar relative to gross brain size in the primates. Data and calculations from Chalfin et al. (2007). All abbreviations and acronyms can be found in Table 20.2. Scale bar: 2 mm.

*A more recent study mapping the functional connectivity of the human pulvinar can be accessed at Arcaro, Pinsk, and Kastner (2015). However, this study uses MRI imaging, lacking in the resolution that histology provides to demarcate subdivisions of PI.

revealing several general connectivity gradients with the cortex (Shipp, 2003; Kaas and Lyon, 2007). Subdivisions at the ventrolateral portion of the pulvinar are more likely to correlate to the posterior cortex. In contrast, the dorsomedial pulvinar subnuclei are more likely to link to the anterior cortex (Figure 20.2). These trends are also supported by studies based on functional connectivity. The pulvinar is often subdivided in the coronal plane along the dorsal and ventral axis rather than specific nuclei and subdivisions observed cytoarchitectonically. The inferior pulvinar and the ventral portion of the PL (PLvl) form the ventral pulvinar (vPul).

In contrast, above the brachium of the superior colliculus, the dorsal pulvinar (dPul) encompasses the dorsal portion of PL (PLdm) and the medial pulvinar (PM) (see Figure 20.2A and Table 20.2; see also **Chapter 71 in Section XII**, Figure 1A), although a small portion of PI extends above the brachium. While the dPul and vPul respond to various visual stimuli, the vPul possesses more visual pulvinar constituents and connectivity with the primary visual cortex (V1) and extrastriate cortex, and the dPul interacts with the frontoparietal and superior temporal regions. This simple demarcation has been instrumental in reconciling human functional organization and connectivity data and other data derived from imaging studies (Barron et al., 2015) with the cyto-chemoarchitectural framework derived from earlier studies (Morel et al., 1997; Cola et al., 1999), and

TABLE 20.1 GLOSSARY OF TERMS

Anthropoid primates	A subgroup of primates that currently include New World monkeys, Old World monkeys, apes, and humans
Brachium	A white-matter structure that connects the cerebellum to the superior colliculus
Calbindin	A calcium binding protein. In the context of the Core and Matrix theory, it is used to mark the diffusely and widely projecting neurons of the matrix.
Calcium imaging	A biological technique that employs indicators that fluoresce upon calcium binding. Neural activity can be derived from the recording fluorescence.
Cerebral cortex	The outer surface layer of the brain, anatomically segregated by function
Cytoarchitectonics	The study of the cellular makeup of neural tissue
Dorsal and ventral visual streams	Two hypothetical separate anatomical pathways that process perception and action. Both pathways originate at the primary visual cortex; the "what" pathway runs along the ventral surface of the cerebral cortex, while the "where" pathway runs along the dorsal surface.
Dyslexia	A learning disorder that is characterized by difficulty reading
Euarchontoglires	A superorder of mammals, containing 5 subgroups: rodents, lagomorphs, tree shrews, colugos, and primates
GABAergic	Neurons that primarily use GABA as a neurotransmitter. GABAergic interactions are generally inhibitory.
Glutamatergic	Neurons that primarily use glutamate as a neurotransmitter. Glutamatergic interactions are generally excitatory.
Haplorhine	Suborder of primates (*simple nose/dry nose*) including tarsiers, monkeys, apes and humans
Hemianopic/ Hemianopia	The loss of vision in half of the visual field, caused by damage to the brain
Koniocellular cell	A small neuron that resides in the Koniocellular layers of the LGN. They are hypothesized to contributed to eye movement–related signals.
MRI	Magnetic resonance imaging, a medical imaging technique that uses magnetic fields to noninvasively decipher the internal anatomy of bodily organs
Neural synchrony	The correlation of brain activity across 2 or more areas across the function of time. The presence of such rhythmic activity, sometimes referred to as oscillations, suggests a direct connection and functional relationship between areas.
Oculomotor	Related to physical movements of the eye
Optogenetics	A biological technique used to manipulate neuronal activity by opening and closing light-activated channels on neurons via illumination of neural tissue
Parvalbumin	A calcium-binding protein; in this context, used to mark fast-spiking GABAergic inhibitory interneurons
Primary visual cortex	A functionally distinct region of the cerebral cortex, dedicated to the primary processing of visual signals. It is located in the occipital end of the brain, and the terminal of visual information from the primary visual pathway.
Retina	A thin layer of nerve tissue that lines the inside of the eyeball, covered in light-sensitive receptors; transmits visual information through the optic nerve
Retinogeniculatestriate pathway	This pathway is synonymous to the primary visual pathway, describing the neural areas that visual information is transmitted through. Visual information originates at the retina, is relayed to the LGN via optic fibers and undergoes processing in the primary visual cortex (V1).
Retinotopy/Retinotopic	The preservation of cell organization between retinal cells and neurons. This characteristic allows for the mapping of visual fields in substrates of the visual stream.
Saccade	Rapid movement of gaze between 2 points of visual fixation
Salience	The quality, especially of visual stimuli, of being particularly noticeable
Scotoma	A blind spot in the visual field
Simian primates	An infraorder of primates that contain all animals that are otherwise known as monkeys and apes
Stereoacuity	A measure of the acuteness of one's depth perception
Subcortical	Substrates that reside underneath the cerebral cortex. The thalamus is a subcortical region.

(*continued*)

TABLE 20.1 CONTINUED

Superior colliculus	A subnuclei of the thalamus, generally associated with integration of visual stimuli, and generating gaze movements
Supragranular	Generally referring to layers 1, 2, and 3 of the cerebral cortex, which are cytoarchitecturally distinct from the granular layers
Teleology/Teleological	An approach that focuses on the explanation of phenomena by the purpose they serve, rather than the cause from which they arise
Thalamocortical	Connections between neurons of the thalamus and the cerebral cortex
Thalamus	A subnuclei of the diencephalon
Top-down	Used to describe neural activity that is instigated by internal goals, rather than a reaction to sensory information
Visual diencephalon	The diencephalon is a part of the forebrain, containing the epithalamus, hypothalamus, ventral thalamus, and thalamus. The visual diencephalon refers to the subnuclei of this region that contribute to or are associated with visual processing.
Visuomotor	Related to a coordination of visual perception and motor output

TABLE 20.2 VARIANCE IN THE NOMENCLATURE OF PULVINAR NUCLEI AND SUBDIVISIONS ACROSS SPECIES

LPcm	Lateral posterior nucleus, caudomedial	Rat/mouse
LPl	Lateral posterior nucleus, lateral	
LPrm	Lateral posterior nucleus, rostromedial	
Pd	Pulvinar, dorsal	Tree shrew
Pc	Pulvinar, central	
Pp	Pulvinar, posterior	Galagos
Ppc	Pulvinar, posterior central	
PM	Medial pulvinar	Common marmoset, rhesus macaque, chimpanzee, human
PLdm	Lateral pulvinar, dorsomedial	
PLvl	Lateral pulvinar, ventrolateral	
PIp	Inferior pulvinar, posterior	
PIm	Inferior pulvinar, medial	
PIcm	Inferior pulvinar, central medial	
PIcl	Inferior pulvinar, central lateral	
PIl-s	Inferior pulvinar, lateral-shell	Human

enabling a multi-species comparison of the connectivity gradients for the primates.

The ability to better define the architectonic framework of the primate visual pulvinar has led to a resurgence of interest, enabling a more significant interpretation of the visual pulvinar as a retinotopically organized (Figure 20.2B) information-processing center that signals salience. The emergence of non-invasive MRI-associated technologies has confirmed the presence of visual field maps in the human pulvinar (Arcaro et al., 2015). It has enabled the discovery of an expanded pulvinar influence on attentional, emotional, and mnemonic networks (Barron et al., 2015) (which will be discussed in **Chapter 45 in Section VII and Chapter 71 in Section XII**). However, an example of the unique and exquisite function of the visual pulvinar that cannot go without mentioning is the finding from an electrophysiological study in macaque monkeys that the PM and PL respond faster and more robustly to snake stimuli than to monkey faces and hands and geometric shapes (Le et al., 2013). This discovery supports the theory that the threat of snakes has influenced the evolution of the primate brain and that the visual pulvinar is a central neural substrate in more complex visual processing and behavior. Other studies have also hypothesized that the visual pulvinar might represent a critical relay structure, conveying threat-related visual information through the subcortical visual circuit (Bertini et al., 2018).

With the assistance of cellular composition studies, anatomical demarcation has been critical in defining function. This review will unpack the role of the pulvinar in visual processing through known anatomical correlates. An additional focus will be on lesion studies and how they have been instrumental in defining a more complex function for the pulvinar across the life span, especially during develpment.

ANATOMICAL CONSTRUCTS OF CIRCUITS DEFINE PULVINAR FUNCTION

Unlike the primary visual thalamic nucleus, the LGN, the visual pulvinar nuclei have extensive anatomical and functional connectivity with most visual cortices, forming distinct and dynamic feedforward and feedback loops (Figure 20.2A). This has inspired numerous studies to continue dissecting their anatomical and functional role, attempting to unpack a causal relationship between the pulvinar and visual processing. Pulvinar functional anatomy varies significantly across species, reflecting the pulvinar's dramatic evolutionary expansion (Figure 20.1), though the connections with the visual cortex are highly conserved across the simian primates. Through the labyrinth of connectivity emerges a functional construct enabling the nuclei to manipulate cortical activity and shape whole brain dynamics continuously. The pancortical connectivity profile of the primate pulvinar nuclei makes it an exceptionally appropriate candidate for the modulation and gating of visual information. Parsing the pulvinar nuclei and subdivisions through various classical schemas, outlined below, has been an invaluable organizing principle in defining the construct and channels associated with complex functions.

THE CORE AND MATRIX CONSTRUCT

The "core and matrix" theory, postulated initially by Jones (1998), attempts to correlate the cellular makeup of the thalamus with function, with 2 components that serve a dual purpose in the cortex. The pulvinar nuclei are an interesting example of this model, as both elements have emerged in primates, including humans. Each subdivision comprises mixed populations of glutamatergic neurons that innervate the cortex (Jones, 2009; Clasca et al., 2012). These projection neurons express the calcium-binding proteins parvalbumin and calbindin, critical in buffering intracellular calcium and enabling a cell to achieve homeostasis quickly. However, how they regulate neuronal/circuit function remains to be entirely determined. Parvalbumin-expressing "core" cells are usually dense within primary sensory thalamic nuclei (e.g., LGN) and are typically thought to carry external information and "drive" feedforward activity (Jones, 2009), innervating primarily layer 4. In contrast, calbindin-expressing "matrix" cells are more evident in higher-order thalamic nuclei (Jones, 2009), such as the pulvinar, and tend to project to supragranular layers more diffusely.

The extent and location of laminar projections to the cortex are also characteristics often used to delineate thalamic hierarchy (Clasca et al., 2012). However, the importance of this anatomical organizing principle for shaping

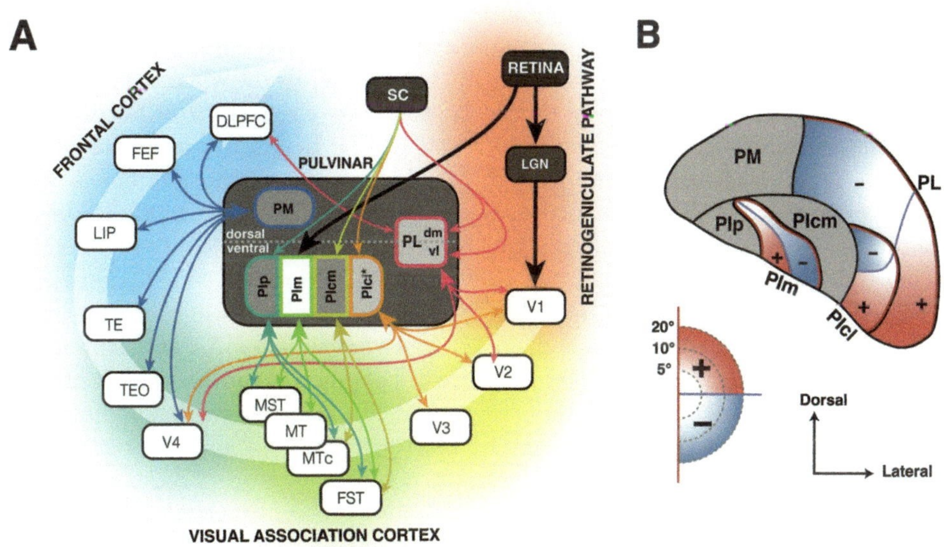

Figure 20.2. (A) Schematic depicting the notable connections between the cortex and subdivisions of the primate pulvinar. Data collated from Grieve et al. (2000); Bridge et al. (2016); Homman-Ludiye et al. (2019). Relative calbindin-positive labeling in specific nuclei and subdivisions of the pulvinar (data from O'Brien et al., 2001; Warner et al., 2010) are shaded in gray. Nearly all pulvinar-cortex connections are reciprocal, depicted by double-headed arrows. The superior colliculus (SC) is a thalamic structure that coordinates eye movements. It receives visual information from both the retina and V1 and is a major input source into subdivisions of the inferior pulvinar. Cortical substrates have been loosely categorized by function and location. The large arrow behind the substrates depicts a general "flow" in visual processing, from early visual area V1 located at the occipital end of the brain to the more functionally specialized extrastriate cortex and onto the frontal cortex. (B) Retinotopic organization of PIm, PIcm, and PL identified in various primate species. Abbreviations not listed in Table 20.1: DLPFC = dorsolateral prefrontal cortex; FEF = frontal eye fields; FST = fundus of the superior temporal visual area; LGN = lateral geniculate nucleus; LIP = lateral intraparietal cortex; MST = medial superior temporal; MT = medial temporal; MTc = medial temporal crescent; SC = superior colliculus; TE = temporal lobe; TEO = temporal lobe occipital; V1 = primary visual cortex; V2 = visual area 2; V3 = visual area 3; V4 = visual area 4.Note: This diagram is not anatomically correct and has been simplified for this review. It does not encapsulate the complete thalamocortical connectivity profile of the pulvinar.

patterns of whole-brain activity and connectivity has been less well characterized.

THE EMERGENCE OF A "CORE" REGION IN THE PRIMATE INFERIOR PULVINAR

Through comparative studies, we have seen that the emergence of an admixture of both calbindin and parvalbumin cell types can be observed within a single thalamic nucleus in primates. The primate LGN is a classic example where calbindin is restricted to cells in the koniocellular (K) or interlaminar regions. However, for the pulvinar, a single subdivision in the inferior pulvinar, the medial subdivision (PIm), is most clearly present and has emerged in the anthropoid primates as the only parvalbumin-rich calbindin-poor region. Often referred to as the calbindin "hole," its cytoarchitecture aligns with the sentiment that PIm is the "core" region of the pulvinar. Further corroborating the evidence that PIm is a "core" driving region of the visual pulvinar is the presence of a direct retinal input (O'Brien et al., 2001; Kwan et al., 2019; Grünert et al., 2021). This is unlike any other region of the pulvinar, including flanking PI, Pip, and PIcm subdivisions, which are the recipients of indirect retinal input via the superficial layers of the superior colliculus and which convey saccadic suppression information (Berman and Wurtz, 2010).

Further, PIm possesses an ascending pathway to the middle temporal (MT) visual cortical area (Lin and Kaas, 1980), a central hub in the dorsal visual stream involved in motion processing (Figure 20.2A). It was revealed that PIm contained substantial retinotopic (first-order; linearly organized) organization of ascending pathway to area MT (Figure 20.2B), further highlighting the precision and potency of the projection (Mundinano et al., 2019b). Within retrogradely labeled cells from MT in this study, all expressed the excitatory marker glutaminase and immunopositivity for parvalbumin. This finding is significant yet not surprising, as there have been years of studies investigating the role of PIm's involvement in establishing early-life vision before the maturation of the retinogeniculate-striate pathway.

The expansion of PI and the prospective teleological origin of a clear PIm in haplorhine primates is corroborated by evidence that MT was once part of a larger region of the temporal cortex and recipient of pulvinar input via the SC, which diminished as area MT became a distinct area in early primates (Kaas et al., 2022). This demonstrates that complex interrelationships between the pulvinar and cortex underpin the anatomy and function of the core and matrix theory. These patterns are likely driven through evolution and help fulfill the functional and developmental bottleneck that occurred across the development of the primates.

AN EXPANSIVE "MODULATORY" INFLUENCE OF THE PULVINAR OVER THE VISUAL CORTEX

The pulvinar is commonly described as a central hub for the modulation and regulation of transcortical communication, especially across visual cortices. Its extensive reciprocal connectivity with substrates of the entire visual processing pathway mimics direct corticocortical connections (Shipp, 2003), a prime substrate for the possession of efference copy signals. Given such an anatomically advantageous connectivity profile, many studies have been curious to understand the influence of the pulvinar over the visual cortex. A common and widely accepted concept is that the pulvinar acts as a global "modulator." What this exactly means, however, varies by interpretation. This section will consolidate opinions from multiple studies.

The term "modulator" originates from Sherman and Guillery's driver and modulator theory (Sherman and Guillery, 1998), categorizing glutaminergic synapses as being either "drivers" that transmit a robust signal capable of influencing the receptive field of the recipient of their input, or "modulators" which modulate signal transmission from cortical area to cortical area, providing global regulation across the wider cortex. (These concepts will be reviewed in **Chapter 72 of Section XII**.) The structural and cellular definitions of a "modulator" overlap significantly with many other prominent theories in neuroscience, such as the matrix from the core and matrix theory. Cortico-pulvinar connectivity has been profiled as mostly modulatory. However, pulvino-cortical circuits, in conjunction with complex entanglements of neural networks, make it much harder to localize and attribute a "modulatory" role to cell populations primarily defined by a set of anatomical criteria. Instead, the pulvinar's influence on the cortex has been inferred by electrophysiological studies demonstrating alterations in oscillatory activity in lesion studies and causality analyses in cortical synchrony.

Although most pulvino-cortical circuits have been described to possess a modulatory function, it has been proposed that there is a reciprocal dependent relationship between the lateral pulvinar (PL) and V1 (Purushothaman et al., 2012). This study revealed that a PL lesion leads to diminished excitability in the supragranular layers (output) of V1, and excitation of the PL leads to increased visual responses in the same V1 receptive fields. Although the magnitude of this manipulation is not necessarily commensurate with biological activity, we can at least substantiate that the pulvinar can exert some degree of influence on the early visual cortex. Further, a PL lesion decreased the strength of synchronous activity between the cat homologs of V1 and V2 (Shumikhina and Molotchnikoff, 1999) and, more recently, V1 and V4 (Cortes et al., 2020). Comparing the effect of a pulvinar lesion on the cortical activity between the cat homologs of V1 and V4, the same group (Souza et al., 2019) proposes that the modulatory role of the pulvinar is more pronounced further along the visual-processing hierarchy.

Although the thalamocortical maps of rodents are not as complex as those observed in primates, it is reassuring that the connectivity observed in mouse LP/pulvinar neurons with anterolateral cortical areas, representative of the higher visual areas (Blot et al., 2021), is similar to the general patterns of connectivity observed in primates. Mouse LP/pulvinar neurons also carry distributed information from visual scenes, such as motion (Roth et al., 2016), reminiscent of the response properties observed in the primate inferior pulvinar, which share connections with extrastriate cortical fields whose neurons are responsive to motion, such as MT.

The necessity of intact pulvinar-cortical interactions in visual processing has been demonstrated by many studies, particularly in relationship with visual area V4, a critical component of the processing pathway in the brain that is primarily associated with object recognition, color, and surface form representation. At this stage of visual processing, much of corticocortical activity is regulated by attention. (This concept will be further expanded in **Chapter 45 of Section VIII** and has been reviewed by the same author [Kastner et al., 2012]). In a visual processing sense, Saalmann and colleagues (2012) revealed that the dorsal pulvinar (Figure 20.2A) could synchronize oscillatory activity between V4 and TEO, 2 cortical areas involved in object recognition. This function is further appreciated when such synchronized oscillatory activity is lost after a reversible lesion to the dorsomedial pulvinar (Zhou et al., 2016; Eradath et al., 2021). However, within the specific cortical areas, there were no changes in total firing rates or local field potential power changes, suggesting that the effect of pulvinar inhibition is purely on modulation across the visual areas. It does not affect an excitatory driving input.

Given the robust connectivity that the pulvinar has with the frontal cortical areas (Figure 20.2A), it is possible that top-down influences can be exerted on the visual cortex from the prefrontal cortex via the pulvinar. During attentive engagement, this is demonstrated through the observed sequential engagement of frontal and intraparietal areas, such as the FEF, LIP, and pulvinar (Fiebelkorn et al., 2019).

Despite the obstinate literature gap surrounding cellular mechanisms of the pulvinar's influence on the cortex, electrophysiological studies have painted a fuller picture of the pulvinar's functional profile—a modulator of visual processing. However, the extent to which this process governs complex behaviors remains to be determined.

THE PULVINAR EMERGES AS A NEURAL SUBSTRATE ENABLING PRECIPITOUS VISUAL CORTICAL AND BEHAVIORAL MATURATION

One incredible feature of the pulvinar nuclei that has yet to be described for any other nucleus of the thalamus is the exuberance of connectivity during development. For anthropoid primates, this has likely enabled the more rapid maturation of specific visual cortical areas and circuitry, and thus the ability to perform tasks necessary for neonatal survival in a brain with an expanded and more complex arrangement of interconnectivity which takes many months or years to be fully realized. In newborn humans, for example, vision is typified by poor shape perception and acuity yet good motion perception (Lewis and Maurer, 2005; Braddick and Atkinson, 2011), while aspects of vision pertaining to object discrimination, such as color, shape, and size, do not emerge until 6 months of age (Siu and Murphy, 2018). While previous studies would suggest that the innate visual circuitry involves the superior colliculus, recent works offer a counter opinion. One seminal study revealed the transient nature of the retina-PIm-MT pathway (Figure 20.2A) without the involvement of the superior colliculus (Warner et al., 2012), which regresses following birth to a sparse connection by adulthood (Cowey et al., 1994; O'Brien et al., 2001; Warner et al., 2010). The switch in dominance from the retinopulvinar-MT pathway to the LGN-V1-MT pathway in adulthood is a major developmental milestone and further highlights the need for an alternative route for visual information to reach the cortex during development which has evolved in anthropoid primates.

In addition to supporting the capacity for survival from an early age, the retinopulvinar-MT pathway likely affords an additional role in steering the organization and maturation of the extrastriate cortex. Maturation of the visual cortex is generally understood to occur hierarchically, like the order in which visual information is processed along the cortex in adulthood (Mundinano et al., 2015). However, one substrate, area MT, stands out as an "early bloomer," which has a maturation profile akin to V1. An explanation for this could be its involvement in the early retinopulvinar-MT pathway, where a visual input from the pulvinar during the early days of development appears to promote experience-dependent maturation. This early maturation of area MT aligns with the observation of good motion perception in babies. The importance of this early-developing circuit was recently demonstrated, where disruption of this pathway in early life leads to permanent anatomical, connectional, and reach-and-grasp behavior deficits (see below).

DISRUPTION ASSISTS IN DEFINING FUNCTION

In the mid-19th century, physician Paul Broca studied a patient's brain who could only produce one syllable verbally. He discovered a highly circumscribed lesion in the inferior frontal gyrus, now named Broca's Area, and attributed it to the role of speech production and articulation.

This was one of the first anatomical proofs that brain function was localized and just one of many monumental lesion studies that have aided in the functional profiling of the human brain. Lesion studies are essential and influential in providing direct evidence for a necessary relationship between an anatomical structure and behavior.

In contemporary neuroscience, lesion studies and imaging, recording, and behavioral methods have been utilized, offering abundant options for testing novel hypotheses and defining functional domains. However, the foundations of lesion studies are rooted in confirming function from damage in human patients.

BLINDSIGHT

An area of research where the contribution of the pulvinar has seen the most controversy is in "blindsight," a rare phenomenon observed in humans following a lesion of V1, thereby disconnecting the primary geniculostriate input (Figure 20.3A) and so-called conscious vision.

Blindsight garnered increasing interest in the 20th century (Weiskrantz, 2004), as its presence challenged 2 well-established notions; that visual function relied on V1, and behavior relied on conscious perception. Individuals with this condition, surprisingly, have the variable capacity to respond to visual stimuli within the associated contralesional blind field, suggesting the presence of alternative pathways that bypass V1, which can facilitate residual vision. Regarding the "stimuli," these are often moving stimuli where the subject (e.g., patient GY) only responds to fast-moving and high-contrast stimuli (Rees, 2008). Others have described a less pronounced effect but have the "feeling" that there was a change within the scotoma rather than being a visual percept per se.

Further, there have been cases where subjects can respond correctly to emotional facial expressions (e.g., happy, sad, angry) in their scotoma, commonly referred to as affective blindsight. In addition, studies revealed that in hemianopic patients (loss of one-half of the vertical visual field; unilateral V1 lesion) without blindsight, unseen fearful faces can be implicitly processed, inducing enhanced visual encoding (Cecere et al., 2014).

There are 3 main contending theories surrounding an explanation for blindsight. First, it was proposed that total obliteration of V1 was highly unlikely, and small islands of tissue were spared enough to drive unconscious visual perception (Radoeva et al., 2008). This study, however, only examines a single patient and ignores the wealth of possibility in subcortical circuitry.

Clinical and experimental studies have reported the residual function in the lateral geniculate nucleus (LGN) following a V1 lesion and suggested its critical role in enabling residual visual capacity. Despite severe retrograde degeneration of the LGN following a visual cortex lesion (Hendrickson et al., 2015), a small population of cells residing in the K layers survived and maintained a projection to MT (Figure 20.3B) (Cowey and Stoerig, 1989; Sincich et al., 2004). This projection was confirmed in human blindsight patients (Ajina and Bridge, 2018), forming the modern-day LGN hypothesis of blindsight.

Many early studies also focused on the superior colliculus (Leh et al., 2010; Kato et al., 2011; Georgy et al., 2016) and, later, its input to PI, a circuit defined initially

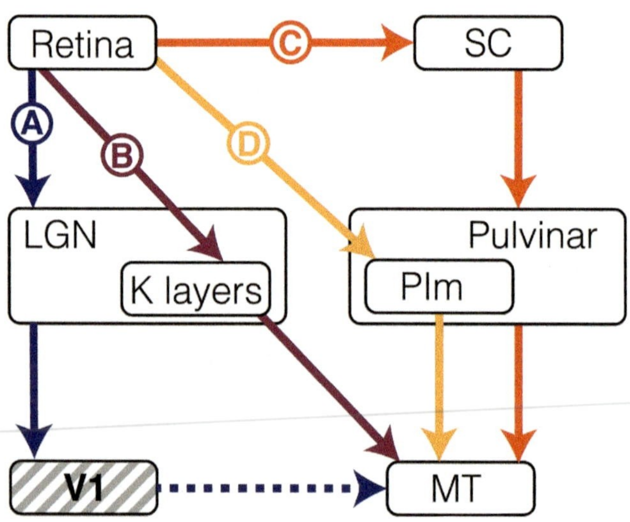

Figure 20.3. Simplified schematic summarizing pathways purportedly underpinning blindsight. The middle temporal (MT) visual cortical area has been a central focus because primate (including human) studies identified activity in the area following a V1 lesion, residual behaviors associated with MT function (e.g., motion), and numerous alternate retinal relays. (A). Degeneration of the retinogeniculate-MT pathway occurs after a V1 lesion, with cell loss observed in both geniculate and retina, and which is more extensive in early life (Hendrickson et al., 2015). (B). Proposed "koniocellular (K) layer hypothesis", whereby a small residual relay from the retina to K cell layers of the LGN, which subsequently relays to MT, survives following V1 damage. (C). Retinal input to SC, which reaches extrastriate cortex via a dense (PI a dense Plm) or topographically organized input (PL) to the extrastriate cortex. (D). Retained tPlm" pathway hypothesized to facilitate extensive residual vision, primarily when damage to V1 occurs during early life, before maturation of the V1-MT pathway.

as being responsible for saccadic suppression (Berman and Wurtz, 2010; Berman et al., 2017) but recently shown to facilitate salience-based visual responses in the absence of V1 (Kinoshita et al., 2019) (Figure 20.3C). However, the additional discovery of retinal input into PIm (Warner et al., 2010; Kwan et al., 2019; Grünert et al., 2021) consolidated the possibility of the pulvinar being a central hub for the relay of visual information to MT (Figure 20.3D). This formed another competing hypothesis against the advocacy for the LGN being the sole neural substrate underpinning blindsight. While these circuits are almost certainly implicated in the capacity to detect visual stimuli within the scotoma, the pulvinar has also been implicated as a neural substrate in the circuitry responsible for affective blindsight. In a study of hemianopic patients who had also received an ipsilateral lesion of the pulvinar, disruption of the implicit visual processing of fearful stimuli occurred (Bertini et al., 2018). This confirms that the pulvinar is a critical neural substrate in detecting visual threats, most likely through the superior colliculus-pulvinar-amygdala route. Still, the exact subcircuit has yet to be defined, but probably invokes the PL and circuit demonstrated in snake detection (Le et al., 2013). In addition, Barleben and colleagues demonstrate the plausibility of a functional pulvinar-MT connection through functional magnetic resonance imaging, robust enough to detect moving stimuli in blind spots (Barleben et al., 2015).

Does the Answer Emerge in Development?

Development has often been an inopportunely overlooked factor in blindsight studies. However, monkey studies (Moore et al., 1996; Gross et al., 2004) reveal that residual visual function is generally better preserved if damage to V1 occurs earlier in life. In terms of the pulvinar, Warner and colleagues (2015) demonstrated that early-life V1 damage to monkeys could decrease the early-life pruning of the retina-PIm-MT pathway (Figure 20.3D) and is robust enough to facilitate vision in the absence of a fully developed V1. Extraordinarily, this was recently corroborated in a case study of a child who received a bilateral occipital lobe injury (including most of V1) before 2 weeks of age yet exhibited significant retention of conscious vision and improved clinical outcomes compared to traditional blindsight patients (Mundinano et al., 2019a). MRI tractography confirmed the more substantial presence of the retina-pulvinar-MT pathway compared to aged-matched controls, akin to that observed in the monkey studies. Together, these findings highlight the importance of the pulvinar in residual visual capacity following a V1 lesion, as well as the need to address the neural underpinnings of blindsight across the function of age.

OTHER VISUAL MANIFESTATIONS ASSOCIATED WITH PULVINAR LESIONS

Consistent with the extensive connectivity of the pulvinar subdivisions across the visual processing hierarchy, lesions in humans manifest behaviorally across a spectrum of functions. These impairments can be exclusively visual, such as a loss of stereoacuity (Takayama et al., 1994), or higher-order, such as difficulty processing fear-associated visual stimuli (Bertini et al., 2018). Snow and colleagues (2009) examined how patients with pulvinar lesions performed on visual discrimination tasks. They discovered that their discriminatory ability is compromised in the presence of a salient distractor. This suggests that the pulvinar is involved in the attentional allocation and distractor filtering and may also play more of a role in maintaining vision rather than actively driving it.

Furthermore, reduced volume and abnormal activity levels in the pulvinar have been shown to correlate with developmental dyslexia (Maisog et al., 2008; Jednoróg et al., 2015), a disorder in reading, one of the most complex oculomotor activities. Yet at the crux, it is a deficit in visual-information processing, but also highly comorbid with visual disabilities such as deficiency in contrast perception and spatial attention. (**Chapter 71 of Section XII** describes thalamic dysfunction-associated disease and pathology in detail.)

CLOSING REMARKS AND FUTURE DIRECTIONS

This chapter has reviewed past and present research and has described mechanisms in which vision relies on support from pulvinar-cortex circuits. More important than the simple conclusion of the pulvinar's involvement in visual processing, pulvinar research in the past few decades has brought to light traditionally overlooked factors. Cross-species evolutionary studies have highlighted areas of the brain that potentially underpin complex behaviors and neural processes exclusive to primates. In contrast, developmental studies have unveiled thalamocortical circuits in the brain involving the pulvinar that has never been considered prior. Together, these findings start new conversations that challenge traditional philosophies. Notable reviews for further reading are listed in Box 20.1.

There is still a considerable distance to travel before fully understanding the pulvinar's role in visual processing. Human studies mostly revolve around noninvasive imaging, lacking the cellular resolution to dissect precise cellular circuits truly. Future research's primary goal should be to bridge the gap between the laboratory and the clinic.

Although electrophysiological and lesion studies in nonhuman primates provide valuable guidance and insight

BOX 20.1 RECOMMENDED FURTHER READING

Bridge H, Leopold DA, Bourne JA (2016). Adaptive pulvinar circuitry supports visual cognition. *Trends Cogn Sci.* 20(2):146–157.

Fox DM, Goodale MA, Bourne JA. (2020). The age-dependent neural substrates of blindsight. *Trends Neurosci.* 43(4):242–252.

Kaas JH, Baldwin MKL (2019). The evolution of the pulvinar complex in primates and its role in the dorsal and ventral streams of cortical processing. *Vision.* 4(1):3.

into global effects, conclusions still rely on inference. These nonspecific approaches generally need the resolution to observe layer-specific functional consequences and cellular mechanisms. With the resurgence of interest in animal models of psychiatry, we can expect a wealth of new technology, such as optogenetic tools and calcium imaging, to fill the gap between the neurons and behavior (Scott and Bourne, 2021). Rodent studies have begun to pave the way for the new generation of visual neuroscience, integrating multiple manipulation and recording techniques with behavioral studies. While a certain amount of validation may still be required for these tools to be fully incorporated into primate visual neuroscience, we can expect a surge of discoveries and knowledge in this realm within the coming years. Through collaboration between systems and cellular neuroscience, the thalamus and cortex, and cross-species studies, we can become closer to establishing causal rather than associative relationships between the pulvinar and visual processing.

REFERENCES

Ajina S, Bridge H (2018). Blindsight relies on a functional connection between hMT+ and the lateral geniculate nucleus, not the pulvinar. *PloS Biol.* 16:e2005769.

Arcaro MJ, Pinsk MA, Kastner S (2015). The anatomical and functional organization of the human visual pulvinar. *J Neurosci.* 35:9848–9871.

Baldwin MKL, Balaram P, Kaas JH (2017). The evolution and functions of nuclei of the visual pulvinar in primates. *J Comp Neurol.* 525:3207–3226.

Baldwin MKL, Bourne JA (2020). The evolution of subcortical pathways to the extrastriate cortex. In *Evolutionary Neuroscience.* Academic Press; 565–587.

Barleben M, Stoppe CM, Kaufmann J, Merkel C, Wecke T, Goertler M, Heinze H-J, Hopf J-M, Schoenfeld MA (2015). Neural correlates of visual motion processing without awareness in patients with striate cortex and pulvinar lesions. *Hum Brain Mapp.* 36(4):1585–1594.

Barron DS, Eickhoff SB, Clos M, Fox PT (2015). Human pulvinar functional organization and connectivity. *Hum Brain Mapp.* 36:2417–2431.

Benevento L, Miller J (1981). Visual responses of single neurons in the caudal lateral pulvinar of the macaque monkey. *J Neurosci.* 1:1268–1278.

Berman RA, Cavanaugh J, McAlonan K, Wurtz RH (2017). A circuit for saccadic suppression in the primate brain. *J Neurophysiol.* 117:1720–1735.

Berman RA, Wurtz RH (2010). Functional identification of a pulvinar path from superior colliculus to cortical area MT. *J Neurosci.* 30:6342–6354.

Bertini C, Pietrelli M, Braghittoni D, Làdavas E (2018). Pulvinar lesions disrupt fear-related implicit visual processing in hemianopic patients. *Front Psychol.* 9:2329.

Blot A, Roth MM, Gasler I, Javadzadeh M, Imhof F, Hofer SB (2021). Visual intracortical and transthalamic pathways carry distinct information to cortical areas. *Neuron.* 109:1996–2008.

Braddick O, Atkinson J (2011). Development of human visual function. *Vision Res.* 51:1588–1609.

Bridge H, Leopold DA, Bourne JA (2016). Adaptive pulvinar circuitry supports visual cognition. *Trends Cogn Sci.* 20:146–157.

Cecere R, Bertini C, Maier ME, Làdavas E (2014). Unseen fearful faces influence face encoding: Evidence from ERPs in hemianopic patients. *J Cogn Neurosci.* 26:2564–2577.

Chalfin BP, Cheung DT, Muniz JAPC, Silveira LC de L, Finlay BL (2007). Scaling of neuron number and volume of the pulvinar complex in new world primates: Comparisons with humans, other primates, and mammals. *J Comp Neurol.* 504:265–274.

Clasca F, Rubio-Garrido P, Jabaudon D (2012). Unveiling the diversity of thalamocortical neuron subtypes. *Eur J Neurosci.* 35:1524–1532.

Cola MG, Gray DN, Seltzer B, Cusick CG (1999). Human thalamus. *Neuroreport.* 10:3733–3738.

Cola MG, Seltzer B, Preuss TM, Cusick CG (2005). Neurochemical organization of chimpanzee inferior pulvinar complex. *J Comp Neurol.* 484:299–312.

Cortes N, Souza BOF de, Casanova C (2020). Pulvinar modulates synchrony across visual cortical areas. *Vision.* 4:22.

Cowey A, Stoerig P (1989). Projection patterns of surviving neurons in the dorsal lateral geniculate nucleus following discrete lesions of striate cortex: Implications for residual vision. *Exp Brain Res.* 75:631–638.

Cowey A, Stoerig P, Bannister M (1994). Retinal ganglion cells labelled from the pulvinar nucleus in macaque monkeys. *Neuroscience.* 61:691–705.

Diamond IT, Hall WC (1969). Evolution of neocortex. *Science.* 164:251–262.

Eradath MK, Pinsk MA, Kastner S (2021). A causal role for the pulvinar in coordinating task-independent cortico-cortical interactions. *J Comp Neurol.* 529(17):3772–3784.

Fiebelkorn IC, Pinsk MA, Kastner S (2019). The mediodorsal pulvinar coordinates the macaque frontoparietal network during rhythmic spatial attention. *Nat Commun.* 10:215.

Georgy L, Celeghin A, Marzi CA, Tamietto M, Ptito A (2016). The superior colliculus is sensitive to gestalt-like stimulus configuration in hemispherectomy patients. *Cortex.* 81:151–161.

Grieve KL, Acuña C, Cudeiro J (2000). The primate pulvinar nuclei: Vision and action. *Trends Neurosci.* 23:35–39.

Gross CG, Moore T, Rodman HR (2004). Visually guided behavior after V1 lesions in young and adult monkeys and its relation to blindsight in humans. *Prog Brain Res.* 144:279–294.

Grünert U, Lee SCS, Kwan WC, Mundinano I-C, Bourne JA, Martin PR (2021). Retinal ganglion cells projecting to superior colliculus and pulvinar in marmoset. *Brain Struct Funct.* 226:2745–2762.

Harting JK, Hall WC, Diamond IT (1972). Evolution of the pulvinar (Part 1 of 2). *Brain Behav Evol.* 6:424–438.

Hendrickson A, Warner CE, Possin D, Huang J, Kwan WC, Bourne JA (2015). Retrograde transneuronal degeneration in the retina and lateral geniculate nucleus of the V1-lesioned marmoset monkey. *Brain Struct Funct*. 220:351–360.

Homman-Ludiye J, Mundinano IC, Kwan WC, Bourne JA (2019). Extensive connectivity between the medial pulvinar and the cortex revealed in the marmoset monkey. *Cereb Cortex*. 30:1797–1812.

Jednoróg K, Marchewka A, Altarelli I, Lopez AKM, Ermingen-Marbach M van, Grande M, Grabowska A, Heim S, Ramus F (2015). How reliable are gray matter disruptions in specific reading disability across multiple countries and languages? Insights from a large-scale voxel-based morphometry study. *Hum Brain Mapp*. 36:1741–1754.

Jones EG (1998). Viewpoint: The core and matrix of thalamic organization. *Neuroscience*. 85:331–345.

Jones EG (2009). Synchrony in the interconnected circuitry of the thalamus and cerebral cortex. *Ann NY Acad Sci*. 1157:10–23.

Kaas JH, Lyon DC (2007). Pulvinar contributions to the dorsal and ventral streams of visual processing in primates. *Brain Res Rev*. 55:285–296.

Kaas JH, Qi H-X, Stepniewska I (2022). Escaping the nocturnal bottleneck, and the evolution of the dorsal and ventral streams of visual processing in primates. *Phil Trans Royal Soc B*. 377(1844):20210293.

Kastner S, Saalmann YB, Schneider KA (2012). Thalamic control of visual attention. *The Neuroscience of Attention: Attentional Control and Selection*. 54–80.

Kato R, Takaura K, Ikeda T, Yoshida M, Isa T (2011). Contribution of the retino-tectal pathway to visually guided saccades after lesion of the primary visual cortex in monkeys. *Eur J Neurosci*. 33:1952–1960.

Kinoshita M, Kato R, Isa K, Kobayashi K, Kobayashi K, Onoe H, Isa T (2019). Dissecting the circuit for blindsight to reveal the critical role of pulvinar and superior colliculus. *Nat Commun*. 10:135.

Kwan WC, Mundinano I, Souza MJ de, Lee SCS, Martin PR, Grünert U, Bourne JA (2019). Unravelling the subcortical and retinal circuitry of the primate inferior pulvinar. *J Comp Neurol*. 527:558–576.

Le QV, Isbell LA, Matsumoto J, Nguyen M, Hori E, Maior RS, Tomaz C, Tran AH, Ono T, Nishijo H (2013). Pulvinar neurons reveal neurobiological evidence of past selection for rapid detection of snakes. *Proc Natl Acad Sci*. 110:19000–19005.

Leh SE, Ptito A, Schönwiesner M, Chakravarty MM, Mullen KT (2010). Blindsight mediated by an S-cone-independent collicular pathway: An fMRI study in hemispherectomized subjects. *J Cogn Neurosci*. 22:670–682.

Lewis TL, Maurer D (2005). Multiple sensitive periods in human visual development: Evidence from visually deprived children. *Dev Psychobiol*. 46:163–183.

Lin C-S, Kaas JH (1980). Projections from the medial nucleus of the inferior pulvinar complex to the middle temporal area of the visual cortex. *Neuroscience*. 5:2219–2228.

Maisog JM, Einbinder ER, Flowers DL, Turkeltaub PE, Eden GF (2008). A meta-analysis of functional neuroimaging studies of dyslexia. *Ann NY Acad Sci*. 1145:237–259.

Moore T, Rodman HR, Repp AB, Gross CG, Mezrich RS (1996). Greater residual vision in monkeys after striate cortex damage in infancy. *J Neurophysiol*. 76:3928–3933.

Morel A, Magnin M, Jeanmonod D (1997). Multiarchitectonic and stereotactic atlas of the human thalamus. *J Comp Neurol*. 387:588–630.

Mundinano I-C, Chen J, Souza M de, Sarossy MG, Joanisse MF, Goodale MA, Bourne JA (2019a). More than blindsight: Case report of a child with extraordinary visual capacity following perinatal bilateral occipital lobe injury. *Neuropsychologia*. 128:178–186.

Mundinano I-C, Kwan WC, Bourne JA (2015.) Mapping the mosaic sequence of primate visual cortical development. *Front Neuroanat*. 9:132.

Mundinano I-C, Kwan WC, Bourne JA (2019b). Retinotopic specializations of cortical and thalamic inputs to area MT. *Proc Natl Acad Sci*. 116:23326–23331.

O'Brien BJ, Abel PL, Olavarria JF (2001). The retinal input to calbindin-D28k-defined subdivisions in macaque inferior pulvinar. *Neurosci Lett*. 312:145–148.

Purushothaman G, Marion R, Li K, Casagrande VA (2012). Gating and control of primary visual cortex by pulvinar. *Nat Neurosci*. 15:905–912.

Radoeva PD, Prasad S, Brainard DH, Aguirre GK (2008). Neural activity within area V1 reflects unconscious visual performance in a case of blindsight. *J Cogn Neurosci*. 20:1927–1939.

Rees G (2008). The anatomy of blindsight. *Brain*. 131:1414–1415.

Robinson DL, Petersen SE (1985). Responses of pulvinar neurons to real and self-induced stimulus movement. *Brain Res*. 338:392–394.

Roth MM, Dahmen JC, Muir DR, Imhof F, Martini FJ, Hofer SB (2016). Thalamic nuclei convey diverse contextual information to layer 1 of visual cortex. *Nat Neurosci*. 19:299–307.

Saalmann YB, Pinsk MA, Wang L, Li X, Kastner S (2012). The pulvinar regulates information transmission between cortical areas based on attention demands. *Science*. 337:753–756.

Scholl LR, Foik AT, Lyon DC (2021). Projections between visual cortex and pulvinar in the rat. *J Comp Neurol*. 529:129–140.

Scott JT, Bourne JA (2021). Modelling behaviors relevant to brain disorders in the nonhuman primate: Are we there yet? *Prog Neurobiol*. 208:102183.

Sherman SM, Guillery RW (1998). On the actions that one nerve cell can have on another: Distinguishing "drivers" from "modulators." *Proc Natl Acad Sci*. 95:7121–7126.

Shipp S (2003). The functional logic of corticopulvinar connections. *Phil Trans Royal Soc Lond Ser B Biol Sci*. 358:1605–1624.

Shumikhina S, Molotchnikoff S (1999). Pulvinar participates in synchronizing neural assemblies in the visual cortex, in cats. *Neurosci Lett*. 272:135–139.

Sincich LC, Park KF, Wohlgemuth MJ, Horton JC (2004). Bypassing V1: A direct geniculate input to area MT. *Nat Neurosci*. 7:1123–1128.

Siu CR, Murphy KM (2018). The development of human visual cortex and clinical implications. *Eye Brain*. 10:25–36.

Snow JC, Allen HA, Rafal RD, Humphreys GW (2009). Impaired attentional selection following lesions to human pulvinar: Evidence for homology between human and monkey. *Proc Natl Acad Sci*. 106:4054–4059.

Souza BOF de, Cortes N, Casanova C (2019). Pulvinar modulates contrast responses in the visual cortex as a function of cortical hierarchy. *Cereb Cortex*. 30:1068–1086.

Takayama Y, Sugishita M, Kido T, Ogawa M, Fukuyama H, Akiguchi I (1994). Impaired stereoacuity due to a lesion in the left pulvinar. *J Neurol Neurosurg Psychiatry*. 57:652.

Warner CE, Goldshmit Y, Bourne JA (2010). Retinal afferents synapse with relay cells targeting the middle temporal area in the pulvinar and lateral geniculate nuclei. *Front Neuroanat*. 4:8.

Warner CE, Kwan WC, Bourne JA (2012). The early maturation of visual cortical area MT is dependent on input from the retinorecipient medial portion of the inferior pulvinar. *J Neurosci*. 32:17073–17085.

Warner CE, Kwan WC, Wright D, Johnston LA, Egan GF, Bourne JA (2015). Preservation of vision by the pulvinar following early-life primary visual cortex lesions. *Curr Biol*. 25(4):424–434. doi:10.1016/j.cub.2014.12.028. Epub 2015 Jan 15.

Weiskrantz L (2004). Roots of blindsight. *Prog Brain Res*. 144:227–241.

Yang F-C, Burwell RD (2020). Neuronal activity in the rat pulvinar correlates with multiple higher-order cognitive functions. *Vision*. 4:15.

Zhou H, Schafer RJ, Desimone R (2016). Pulvinar-cortex interactions in vision and attention. *Neuron*. 89:209–220.

Zhou N, Maire PS, Masterson SP, Bickford ME (2017). The mouse pulvinar nucleus: Organization of the tectorecipient zones. *Visual Neurosci*. 34:E011.

21.

AUDITORY THALAMIC NUCLEI, CELL TYPES, AND PARALLEL PATHWAYS

Daniel A. Llano

INTRODUCTION

The auditory portion of the mammalian thalamus is known as the medial geniculate body (MGB; sometimes referred to as the medial geniculate nucleus, MGN). It is an obligatory relay for acoustic information to reach the acoustically responsive regions of the cerebral cortex. The MGB appears to be an important structure for processing behaviorally relevant acoustic information. For example, in human studies, activity in the MGB is strongly modulated by task demands and acoustic components that are critical for speech perception (von Kriegstein et al., 2008) (see **Chapter 23 in Section V** for more details). Animal studies have confirmed that MGB neurons respond to the cognitive features of auditory learning tasks (Gilad et al., 2020). In addition, disruption of MGB circuits may be found in disease states such as dyslexia and schizophrenia (Galaburda et al., 1994; Horga et al., 2014; Xiao and Llano, 2020). These data highlight the importance of developing a greater understanding of how the MGB is organized.

As will be reviewed in detail below, the main acoustic afferent input to the MGB is from the inferior colliculus (IC), though there is evidence that a smaller projection to the MGB is derived from lower auditory brainstem nuclei, such as the cochlear nuclei and superior olive. Like other thalamic regions, the MGB also receives neuromodulatory inputs from brainstem cholinergic and noradrenergic regions, and serotoninergic centers. A major descending input to the MGB is derived from the auditory cortex, which appears to be at least as dense as inputs coming from lower centers. Another major input is derived from the thalamic reticular nucleus, a GABAergic structure that partially surrounds the thalamus. The thalamic reticular nucleus receives widespread inputs from non-auditory regions of the thalamus and cortex, as well as from the basal ganglia and cholinergic inputs from the basal forebrain, and therefore has the capacity to modulate auditory thalamic function based on ongoing cognitive demands.

The MGB projects to multiple brain regions. Its major output that is thought to be important for acoustic perception is to the auditory cortex. There are multiple auditory cortical fields, and the patterns of inputs from MGB to particular fields and layers are complex and are covered below. The MGB also sends a strong projection to the striatum, thought to be important for motor decision-making. Both the auditory cortex and MGB send projections to the amygdala, which are thought to be important for providing the conditioned stimulus to sound-based fear conditioning. Finally, the MGB has an evolutionarily conserved descending projection to the inferior colliculus. This projection has not yet been well characterized, but has been described in monkeys, cats, and rodents, as well as amphibians, suggesting that it may be an evolutionarily ancient pathway.

Thus, the MGB is strategically situated to link acoustic and cognitive features needed to process complex sounds in varying behavioral contexts. In addition, the MGB appears to be organized as a series of parallel pathways, whereby it receives a diverse array of inputs to its various subdivisions, which then distribute the processed information to multiple other brain regions. Below, we review what is known about the organization of these various pathways and propose open questions to help further our understanding of this important, yet mysterious, structure.

SUBDIVISIONS AND CYTOARCHITECTURE OF THE MGB

Traditional approaches to delineate subdivisions of the MGB have relied on cytoarchitectural approaches with either Nissl or Golgi stains. Such approaches have led to the identification of at least 3 different subdivisions of the MGB. A ventral division (MGBv) contains neurons that are organized roughly into laminae with relatively small and uniformly sized cell bodies. Nissl- and Golgi-based approaches have revealed that most cells in the MGBv have a tufted or bi-tufted appearance (as is seen in other primary thalamic nuclei) and have their dendrites organized into laminae that approximate the colliculothalamic fibers entering the MGBv (McMullen et al., 2005). The MGBv is often broken up into a lateral nucleus (sometimes referred to as *pars lateralis*), where the lamination is more conspicuous, and a more medial nucleus (sometimes referred to as *pars ovoidea*), where cells are arranged concentrically (Clerici and Coleman, 1990). The dorsal division (MGBd) contains multiple neuronal types, including bi-tufted, as well as stellate neurons, with a lower packing density without a discernable laminar organization (Clerici and Coleman, 1990). Based on packing density and cell type, the MGBd can be parcellated into subnuclei, depending on the species and the staining approach. The most commonly defined subnuclei of the MGBd are the superficial dorsal (SD), deep dorsal (DD), and suprageniculate nucleus (SG), which is more commonly grouped with the paralaminar nuclei, described below. The medial division (MGBm) contains neurons with heterogeneous morphology, some with very large somata and radiating dendrites. There is no laminar organization, and the MGBm has the lowest packing density of neurons. Along the medial border of the MGB are nuclei often referred to as *paralaminar nuclei*, within which the MGBm and suprageniculate are sometimes grouped. In addition to the MGBm and SG, these nuclei include the posterior intralaminar nucleus and peripeduncular nucleus, and these regions are distinguished from the MGBv and MGBd by receiving more multisensory inputs and by sending projections more diffusely to the auditory cortex than MGBd and MGBm, and by having subcortical projections to striatum and amygdala (see below). Functionally, the nuclei of the MGB are often categorized as "lemniscal" (MGBv), such that they provide a relatively high-fidelity representation of the incoming sound stimuli and "non-lemniscal" (MGBd and MGBm), whose neuronal responses to sound are more variable and more context-dependent.

Neurochemical approaches have greatly enhanced our ability to discern different subnuclei of the MGB. A common approach that has been used across the thalamus to delineate primary lemniscal from non-lemniscal thalamic nuclei is immunostaining for either parvalbumin or calbindin calcium-binding proteins. These 2 proteins show a generally complementary pattern across thalamic nuclei, with parvalbumin-positive regions generally being considered to be part of the "core" or "first-order" (lemniscal) nuclei, while calbindin-positive regions generally being part of "matrix" or "higher order" (non-lemniscal) nuclei. In the MGB, across multiple mammalian species, calbindin strongly stains neuronal somata in the MGBd, while parvalbumin stains both somata and neuropil in the MGBv (Hashikawa et al., 1991; Molinari et al., 1995; Cruikshank et al., 2001). Neuropil staining in the mouse MGBv is quite strong, making it challenging to discern individual neurons from neuropil. In the paralaminar nuclei, including MGBm, the delineation is less clear. In mouse, MGBm has portions that stain for both parvalbumin and calbindin, whereas the posterior intralaminar and peripeduncular regions stain positively for calbindin (Cruikshank et al., 2001). See Figure 21.1 for an example of the MGB subdivisions in mouse delineated with Nissl staining, parvalbumin immunostaining, and calbindin immunostaining (Cruikshank et al., 2001). In rat and rabbit, MGBm has been described as staining strongly for parvalbumin (de Venecia et al., 1995; Márquez-Legorreta et al., 2016). Calretinin staining roughly approximates calbindin staining, but tends not to stain the SG nucleus (Lu et al., 2009; Márquez-Legorreta et al., 2016).

The proportion of intrinsic GABAergic neurons in the MGB differs across species. In most studied species, the density of GABAergic neurons is roughly 20%–30% of neurons across MGB subdivisions, with a slightly greater proportion being found in the MGBv. For reasons that are unclear, rodents, at least one species of echolocating bat (the mustached bat), and guinea pigs (personal communication, Drs. Jeffrey Mellott and Brett Schofield) have a relative dearth of GABAergic neurons within the MGB (<5% of the total) (Winer and Larue, 1988). In the case of rodents, intrinsic GABAergic neurons are nearly absent across all thalamic nuclei, with the exception of the lateral geniculate nucleus, which is the primary visual thalamic region. For example, in rats and mice, the density of GABAergic neurons is approximately 1% of the total, though the morphological characteristics of this small number of neurons are similar to what has been seen in species with a greater proportion of GABAergic neurons (Winer and Larue, 1988).

ASCENDING AND DESCENDING CONNECTIVITY OF THE MGB

The MGB receives at least as many descending projections as it does ascending connections (Rouiller and De Ribaupierre, 1985). The major source of ascending acoustic input to the MGB is the inferior colliculus, but unlike primary ascending inputs to non-auditory portions of the sensory thalamus, the main auditory input to the MGB contains a substantial proportion of GABAergic neurons derived from all inferior colliculus subdivisions in all

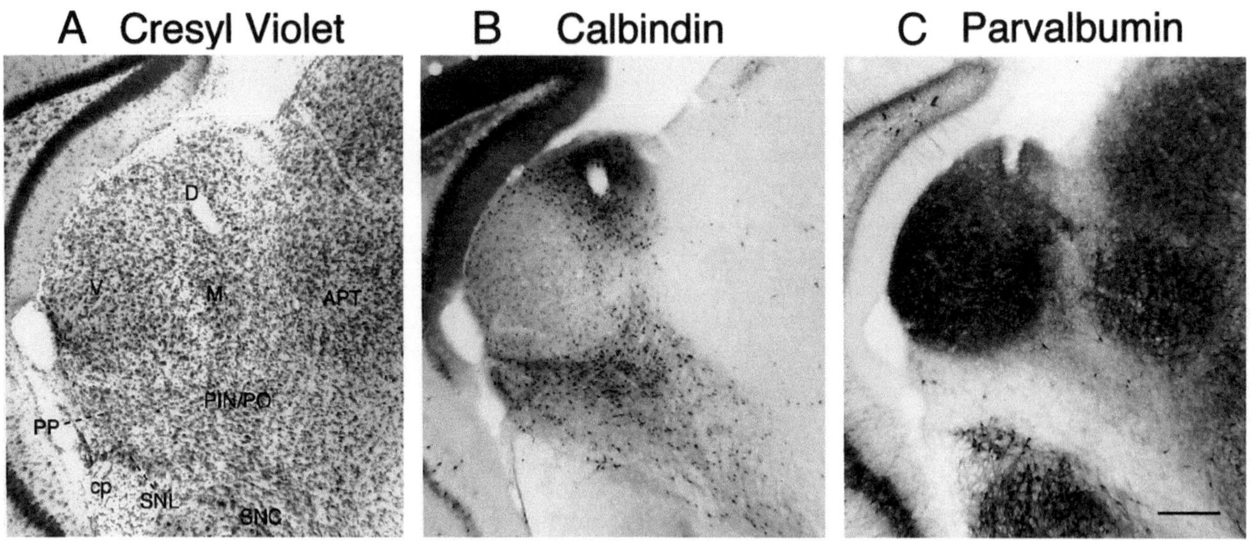

Figure 21.1. Coronal sections of the adult mouse brain stained for Nissl substance (A) or immunostained for calbindin (B) or parvalbumin (C), illustrating the various subdivisions of the MGB. Abbreviations: APT = anterior pretectal nucleus; cp = cerebral peduncle; D = dorsal division of MGB; M = medial division of MGB; PIN/PO = posterior intralaminar nucleus/posterior thalamic nucleus; PP = peripeduncular nucleus; SNC = substantia nigra (compacta); SNL = substantia nigra (lateral); V = ventral division of MGB. Scale bar = 200 µm. Obtained with permission from Cruikshank et al. (2001).

mammalian species that have been studied (cat, mouse, rat, guinea pig) (Winer et al., 1996; Peruzzi et al., 1997; Mellott et al., 2014b; Mellott et al., 2014a). The inhibitory components are derived from multiple different types of GABAergic neurons (GABAergic neurons with or without perineuronal nets and/or VGLUT2 rings) in the inferior colliculus. Ascending GABAergic neurons comprise a substantial proportion of the total ascending input (at least 20%) in animals both with and without a significant proportion of intrinsic GABAergic interneurons in the MGB (Mellott et al., 2014b) and target all subdivisions of the MGB.

Specific subdivisions of MGB receive different sets of inputs from the inferior colliculus and have different projection fields to the auditory cortex. For example, the MGBv, which is the main lemniscal structure of the MGB, receives primary ascending input from the central nucleus of the inferior colliculus (Rouiller and De Ribaupierre, 1985; Mellott et al., 2014b). The colliculothalamic projections to the MGBv end in populations of axodendritic terminals that are larger on average than those to the MGBd (Bartlett et al., 2000). Terminal size may be an important indicator of the ability of a synaptic pathway to influence its postsynaptic targets. The "driver/modulator" dichotomy has been proposed as a way to conceptualize the potential impacts that a neural pathway can have on its target, with large terminals having a greater capacity to transfer sensory information along a hierarchical pathway (Sherman and Guillery, 1998). In the case of the colliculothalamic pathway, the projection from inferior colliculus to MGB, in addition to having larger terminals than similar projections to the MGBd, has synaptic properties with the physiological characteristics of being drivers (large glutamate ionotropic postsynaptic responses that depress with repeated stimulation) compared to those to the MGBd (Lee and Sherman, 2010).

The MGBv canonically sends its major projection to the primary auditory cortex to middle cortical layers, though there is a smaller component to layer 6 as well as branches to layer 1 (Rouiller and De Ribaupierre, 1985; Kimura et al., 2003; Llano and Sherman, 2008). The projection from MGBv to auditory cortex appears to be highly topographic (Hackett et al., 2011). Figure 21.2 illustrates the tight topography of this core projection. As shown, injection of retrograde tracers into adjacent areas of primary auditory cortex label adjacent areas in MGBv. Over recent years, detailed studies have shown that MGBv in fact contains multiple separate populations of neurons that project to different auditory cortex subfields, such as anterior auditory field, insular auditory cortex, an "ultrasonic" auditory field, and areas immediately dorsal and ventral to the primary auditory cortex (Storace et al., 2010; Read et al., 2011). The MGBv also receives a major projection primarily from layer 6 of the auditory cortex, dominated by small terminals (Llano and Sherman, 2008).

The synaptic physiology of thalamocortical synapses between MGBv axons and neurons in the primary auditory cortex has been studied both in vivo and in vitro. In vivo intracellular recordings of auditory cortex neurons have shown that electrical stimulation of MGB produces depressing synaptic responses (Han et al., 2012). Similarly, in brain slice preparations, stimulation of auditory thalamic afferents to layer 4 cortical excitatory neurons produces large, depressing

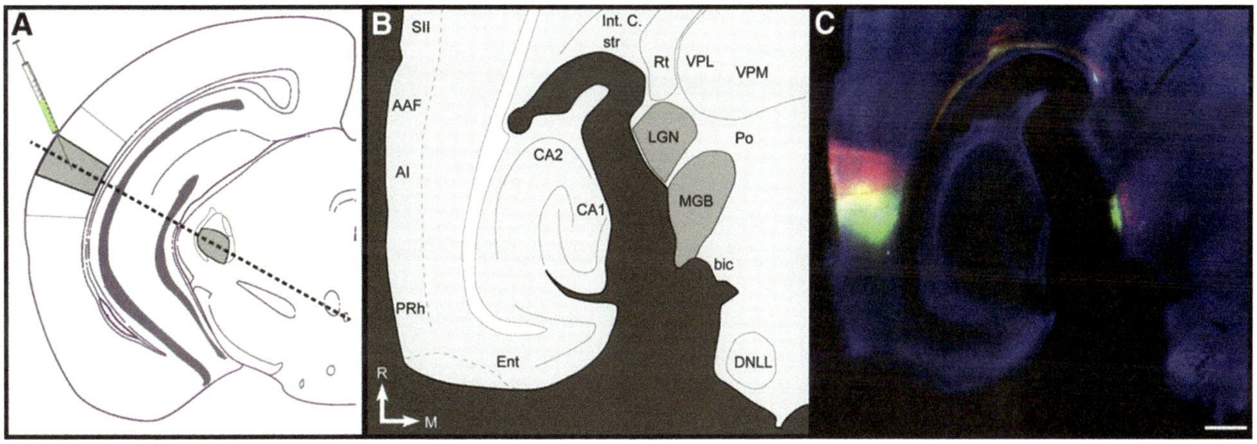

Figure 21.2. Illustration of topography in the MGBv to primary auditory cortex projection. Retrograde tracers were injected into the 7 kHz (green) and 22.5 kHz (red) region of the primary AC. Sections were cut in the plane shown in A. B illustrates a map of nuclei shown in C. C illustrates the topographic mapping of thalamocortical projections such that the thalamocortical cell bodies in MGBv remain segregated based on best frequency and primary auditory cortex injection site. Abbreviations: AAF = anterior auditory field; AI = primary auditory cortex; bic = brachium of the inferior colliculus; DNLL = dorsal nucleus of the lateral lemniscus; Ent = entorhinal cortex; Int C = internal capsule; LGN = lateral geniculate nucleus; M = medial; Po = posterior thalamic complex; PRh = perirhinal cortex; R = rostral; Rt = reticular nucleus; SII = secondary somatosensory cortex; str = superior thalamic radiation; VPL = ventral posterolateral thalamic nucleus; VPM = ventral posteromedial thalamic nucleus. Scale bar = 250 μm. Obtained with permission from Hackett et al. (2011).

postsynaptic potentials without a metabotropic component, suggestive of driver-type synapses. Additional studies reveal that synaptic inputs from MGB to parvalbumin-positive GABAergic interneurons were stronger than those to corresponding excitatory neurons (Ji et al., 2016). These data suggest that the nature of the communicated messages between MGB and auditory cortex show driver-like responses in layer 4 and are distributed to both excitatory and inhibitory cortical neurons.

The MGBd primarily projects to non-primary parts of the auditory cortex. We should note that the distinction between primary and non-primary auditory cortex is not entirely clear in all species, given recent data showing the existence of multiple tonotopic fields across the auditory cortex (Tsukano et al., 2017; Romero et al., 2020). Here, *non-primary auditory cortex* refers to the regions outside of canonical primary auditory cortex and anterior auditory field, which have mirror-image tonotopic maps. Projections from MGBd are found generally in the ventral-most regions of the auditory cortex, either in secondary auditory cortex or areas adjacent, or parabelt regions in the primate, and these projections are more distributed across cortical layers, with a greater proportion to layer 1 than that seen in the MGBv to primary auditory cortex pathway (Rouiller and De Ribaupierre, 1985; Kimura et al., 2003; Llano and Sherman, 2008). Corticothalamic projections to MGBd are derived from multiple areas of the auditory cortex and emanate from cortical layers 5 and 6 (Ojima, 1994; Llano and Sherman, 2008, 2009). A small population of "giant" terminals from auditory cortex have been found in MGBd, and have been speculated to serve as corticothalamic drivers to MGBd (Ojima, 1994; Bartlett et al., 2000; Llano and Sherman, 2008). In the somatosensory system, evidence suggests that corticothalamic drivers from layer 5 serve as the first leg of a cortico-thalamocortical route through the somatosensory hierarchy (Theyel et al., 2010). Whether such a pathway exists in the auditory system has not yet been established. (For more description of the auditory corticothalamic system, see **Chapter 24 in Section V**.)

The MGBm and associated paralaminar nuclei have the most varied pattern of inputs and outputs and are arguably the most enigmatic of the auditory thalamic nuclei. The MGBm receives input from the lateral cortex of the inferior colliculus, as well as major inputs from the deep layers of the superior colliculus, and primary somatosensory inputs from the spinothalamic tract and from midbrain tegmentum (Winer and Morest, 1983; Lesicko and Llano, 2020). These polysensory inputs may be critical for the role that these nuclei putatively play in fear conditioning to painful stimuli and the multisensory responses of MGBm neurons (Bordi and LeDoux, 1994). As such, the MGBm and paralaminar nuclei send a major projection to the amygdala (LeDoux et al., 1991). Their neurons also project to the striatum, and may be important for motor decision-making based on acoustic cues. In terms of their cortical projections, MGBm has widespread projections throughout the temporal cortex, much of which terminates outside of the middle cortical layers, in layers 1 and 6 (Linke, 1999; Kimura et al., 2003) and does not appear to be topographically organized (Hackett et al., 2011). Layer 1 projections are generally diffuse and non-topographic, suggesting that projections from MGBm may play a role in synchronizing activity across large areas of the cortex or entraining oscillatory activity. The MGBm and associated regions also receive corticothalamic projections, some of which contain large terminals (Llano and Sherman, 2008), though these have been less well characterized than those to the other areas of the MGB.

An additional projection from the MGB that does not appear to have a homolog in other sensory systems is the descending projection from MGB to inferior colliculus. These descending projections from the MGB have been the least well characterized of all MGB-based pathways. They are sparse and appear to originate from non-lemniscal portions of the MGB, as well as the posterior intralaminar nuclei, peripeduncular nucleus, and the brachium of the inferior colliculus, and appear to avoid the central nucleus of the inferior colliculus (Kuwabara and Zook, 2000; Winer, 2005; Llano et al., 2014; Patel et al., 2017). These projections have been described across multiple vertebrate species, including amphibians, and are thought to be evolutionarily ancient (Macias and Llano, 2022). It is not yet known if they are part of a larger corticocollicular system and receive corticothalamic descending projections, or if they operate independently of the corticocollicular system.

The MGB also receives inhibitory inputs from several different sources. As outlined above, there is a dense ascending GABAergic projection from the inferior colliculus to the MGB, observed in animals both with and without a substantial supply of intrinsic MGB GABAergic interneurons (Winer et al., 1996; Peruzzi et al., 1997; Mellott et al., 2014b; Mellott et al., 2014a). In addition, the thalamic reticular nucleus, a thin sheet of GABAergic neurons derived from the ventral thalamus that partially surrounds the dorsal thalamus, sends dense projections to the MGB. These projections are strong enough to inhibit ascending activity en route to the auditory cortex and are stimulated by layer 6 of auditory cortex (Ibrahim et al., 2021). Interestingly, the portions of the thalamic reticular nucleus that project to the MGB also receive substantial inputs from non-auditory regions, such as prefrontal cortex, amygdala, and non-auditory regions of the thalamus (Zikopoulos and Barbas, 2006; Kimura et al., 2007; Zikopoulos and Barbas, 2012; Aizenberg et al., 2019), suggesting that the thalamic reticular nucleus may provide a conduit by which non-auditory information influences MGB activity. Finally, the zona incerta sends GABAergic projections to the MGB, mostly to the MGBd and MGBm (Mitrofanis, 2002), though the functional significance of these projections has yet to be established.

In addition to acoustic and other sensory afferents to the MGB, the MGB also receives a rich supply of modulatory influences from brainstem cholinergic, noradrenergic, and serotoninergic nuclei. For example, thalamic regions receive cholinergic inputs from the pedunculopontine and laterodorsal tegmental nuclei (Motts and Schofield, 2010, 2011) and these projections are thought to be important in modulating thalamic responses under different states of arousal. Both muscarinic and nicotinic receptors have been identified in the MGB (Breese et al., 1997; Varela and Sherman, 2007; Hamada et al., 2010; Sottile et al., 2017). Though these projections are mostly excitatory in the MGB, MGBd is more likely than MGBv to be hyperpolarized by muscarinic cholinergic activation (Mooney et al., 2004; Varela and Sherman, 2007), a common theme seen in higher-order sensory thalamic nuclei (Varela and Sherman, 2007) and likely related to differential expression of muscarinic receptor subtypes in these regions. Much less work has been done to examine the impacts of other brainstem-derived neuromodulators on the MGB. There is clear evidence for the presence of norepinephrine and serotonin terminals and/or receptors in the MGB (McCormick and Prince, 1988; Monckton and McCormick, 2002; Varela and Sherman, 2007), but the details of their innervation patterns and functional roles have yet to be examined in detail.

PHYSIOLOGICAL RESPONSE PROPERTIES OF THE AUDITORY THALAMUS

Acoustic and multisensory response properties in the MGB vary based on subdivision. Early work in the anesthetized cat established that MGBv neurons have shorter acoustic response latencies, sharper frequency tuning, and more consistently acoustically driven responses than neurons in the MGBd and MGBm (Calford, 1983). The MGBd and MGBm had a higher proportion of labile response properties. Subsequent work in other species has confirmed and extended these core differences between the subdivisions, though some work in mouse has shown that MGBm response latencies can be short (Anderson and Linden, 2011; Hackett et al., 2011). Work from multiple species has shown that the MGBv is tonotopically organized along the histologically defined laminae with neurons that have primarily V-shaped tuning curves, though it is likely that multiple tonotopic maps exist (Hackett et al., 2011; Horie et al., 2013). The MGBd has acoustic response latencies approximately twice as long as MGBv, broad tuning, and a larger proportion of acoustically unresponsive cells (Calford, 1983; Anderson and Linden, 2011). The long latencies, coupled with the large cortically derived terminals mentioned above (Ojima, 1994; Bartlett et al., 2000; Llano and Sherman, 2008), open the possibility that MGBd responses are driven by cortical, rather than collicular, inputs. The MGBm, reflecting the high degree of cellular heterogeneity in this structure, does not clearly align itself as being lemniscal or non-lemniscal. It has relatively short latencies and appears to be tonotopically organized, but as noted previously, it has labile response properties, broad tuning, and multimodal responses (Bordi and LeDoux, 1994; Anderson and Linden, 2011; Hackett et al., 2011).

Several studies have examined temporal response properties in the MGB. In general, MGB neurons have diminished phase locking compared to neurons in the inferior colliculus, but greater phase locking than that seen in the auditory cortex (Llano and Feng, 1999; Asokan et al., 2021). Less work has examined temporal response properties across the MGB subdivisions. Experiments in the awake marmoset has shown that the MGBv has stronger phase locking than neurons in the other subdivisions (Bartlett and Wang, 2011), consistent with its role as providing a pathway for relatively unprocessed signals from the midbrain to reach the cortex. (A more detailed assessment of how sound representations change between the MGB and auditory cortex is provided in **Chapter 22 in Section V**.)

Cellular properties of MGB neurons resemble those of other thalamic nuclei more broadly. For example, MGB neurons fire tonic spikes when depolarized and display low-threshold bursting when depolarized after a long period of hyperpolarization. One exception is the MGBm, whose neurons tend to not show typical bi-tufted morphology seen in thalamocortical neurons, and have a diminished proportion of low-threshold bursting neurons (Smith et al., 2006). Across MGBd and MGBv, intrinsic electrophysiological properties are mostly similar to each other (Bartlett and Smith, 1999), though some differences in response to synaptic stimulation have been noted. For example, neurons in the MGBd are more likely to burst in response to stimulation of afferents from the inferior colliculus when stimulated at the level of the brachium (Hu et al., 1994). These data are consistent with findings in the sensory thalamus more generally that higher-order nuclei are more likely to burst in vivo than first-order nuclei (Ramcharan et al., 2005). In addition, tufted neurons, which are found more commonly in MGBv and receive primarily excitatory afferent input, tend to show paired-pulse depression when collicular afferents are stimulated. In contrast, stellate cells, which are found more commonly in the MGBd and receive mixed inputs, are more likely to show paired-pulse facilitation with collicular afferent stimulation (Bartlett and Smith, 2002). These data suggest that despite the overall similarity of the intrinsic properties found in the MGB, the different synaptic properties of the colliculothalamic afferents may contribute to some of the differences in response properties between MGBv and MGBd neurons.

SUMMARY AND OPEN QUESTIONS

Given the neurochemical and cytoarchitectonic differences between MGB subdivisions, the segregation of inputs to the subdivisions and differences in the areal outputs of the subdivisions, it is likely that the MGB functions as at least 3 *distinct parallel pathways*. However, cross-talk between the pathways exists, mediated in part by corticothalamic projections and by the broadly projecting paralaminar system comprising the MGBm, suprageniculate, posterior intralaminar nuclei, and the peripeduncular nucleus, as outlined in Figure 21.3. Given the heterogeneity that exists within the subnuclei, particularly the paralaminar nuclei, there may in fact be many more than 3 pathways.

Despite the great deal of detailed information about the MGB that has emerged over the past decades, there is much more to be learned. One particularly compelling set of questions surrounds the highly numerous corticothalamic projections (see also **Chapter 24 in Section V**). These projections are seen across all thalamic nuclei, are derived from layers 5 and 6, are highly branched, and end in a subpopulation of small and large terminals, with the large terminals being derived from layer 5. Multiple roles have been proposed for these projections, including a role for predictive coding, for modulation of thalamic response properties based on the behavioral state of the animal or as a primary information-bearing pathway to transfer information across long-range cortical networks. Indeed, given the heterogeneity in neurochemical, connectional, and physiological properties seen in the MGB, all of these functions could coexist, and could be separable based on subdivision of MGB or at an even finer scale.

In addition, many open questions remain about the MGBm and associated nuclei (suprageniculate, peripeduncular, and posterior intralaminar nucleus). These nuclei have neurons that seem to violate many of the "rules" of thalamic organization: their axons tend to avoid middle cortical layers, their neurons tend to not fire bursts, and their neurons have local collaterals, suggesting that local integration could occur here. In many ways, these neurons function more like intralaminar thalamic neurons, which have been implicated in facilitating arousal and/or gamma oscillations in the cortex via projections outside of the middle cortical layers. In addition, the MGBm appears to be the earliest entry point for auditory information to enter the limbic system. Thus, it may be fruitful to study the impact of combined MGBm/MGBv or MGBm/MGBd stimulation to determine if these pathways interact at the level of the cortex or limbic structures.

Finally, classic work from the thalamic reticular nucleus suggested that this structure is important in producing sleep spindles and generalized spike-and-wave seizures under pathological states. More recent data reviewed above suggest that the thalamic reticular nucleus plays a role in integrating information across sensory, limbic, and/or cognitive networks. How these purported new roles of the thalamic reticular nucleus interface with the parallel pathways through the MGB has scarcely been studied. Thus, given the

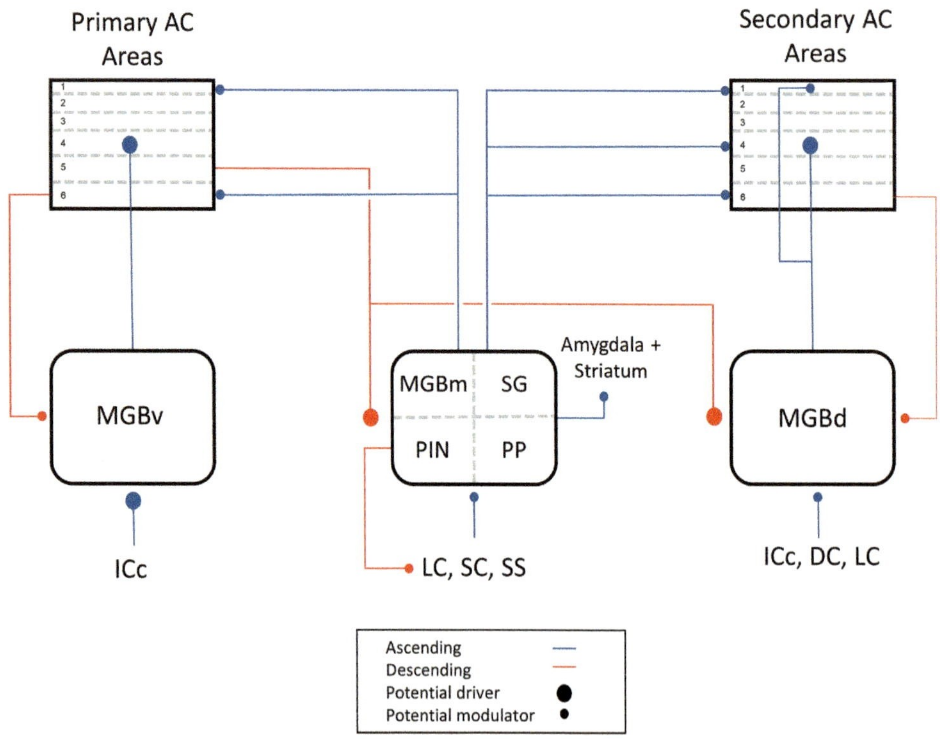

Figure 21.3. Diagram of the major connectivity of the 3 main subdivisions of the MGB. For projections where affirmative anatomical or physiological evidence exists for driver type synapses, a large terminal is shown. All other terminals are assumed to be modulators. For simplicity, projections involving non-auditory portions of the cortex, thalamic reticular nucleus inputs and brainstem neuromodulator inputs are omitted. Abbreviations: DC = dorsal cortex of the inferior colliculus, ICc = central nucleus of the inferior colliculus, LC = lateral cortex of the inferior colliculus, PIN = posterior intralaminar nuclei, PP = peripeduncular nuclei, SC = superior colliculus, SS = somatosensory or pain-related brainstem nuclei.

importance of the MGB in auditory cognition, the many open questions outlined above, and the explosion in new molecular, optical, and physiological approaches available to neuroscientists over the recent decade, the future looks bright for new discoveries to be made and major advances to occur in our understanding of this enigmatic structure.

REFERENCES

Aizenberg M, Rolón-Martínez S, Pham T, Rao W, Haas JS, Geffen MN (2019). Projection from the amygdala to the thalamic reticular nucleus amplifies cortical sound responses. *Cell Reports*. 28:605–615.

Anderson LA, Linden JF (2011). Physiological differences between histologically defined subdivisions in the mouse auditory thalamus. *Hear Res*. 274:48–60.

Asokan MM, Williamson RS, Hancock KE, Polley DB (2021). Inverted central auditory hierarchies for encoding local intervals and global temporal patterns. *Curr Biol*. 31:1762–1770.e1764.

Bartlett E, Stark J, Guillery R, Smith P (2000). Comparison of the fine structure of cortical and collicular terminals in the rat medial geniculate body. *Neuroscience*. 100:811–828.

Bartlett EL, Smith PH (1999). Anatomic, intrinsic, and synaptic properties of dorsal and ventral division neurons in rat medial geniculate body. *J Neurophysiol*. 81:1999–2016.

Bartlett EL, Smith PH (2002). Effects of paired-pulse and repetitive stimulation on neurons in the rat medial geniculate body. *Neuroscience*. 113:957–974.

Bartlett EL, Wang X (2011). Correlation of neural response properties with auditory thalamus subdivisions in the awake marmoset. *J Neurophysiol*. 105:2647–2667.

Bordi F, LeDoux JE (1994). Response properties of single units in areas of rat auditory thalamus that project to the amygdala. II. Cells receiving convergent auditory and somatosensory inputs and cells antidromically activated by amygdala stimulation. *Exp Brain Res*. 98:275–286.

Breese CR, Adams C, Logel J, Drebing C, Rollins Y, Barnhart M, Sullivan B, Demasters BK, Freedman R, Leonard S (1997). Comparison of the regional expression of nicotinic acetylcholine receptor alpha7 mRNA and [125I]-alpha-bungarotoxin binding in human postmortem brain. *J Comp Neurol*. 387:385–398.

Calford MB (1983). The parcellation of the medial geniculate body of the cat defined by the auditory response properties of single units. *J Neurosci*. 3:2350–2364.

Clerici WJ, Coleman JR (1990). Anatomy of the rat medial geniculate body: I. Cytoarchitecture, myeloarchitecture, and neocortical connectivity. *J Comp Neurol*. 297:14–31.

Cruikshank S, Killackey H, Metherate R (2001). Parvalbumin and calbindin are differentially distributed within primary and secondary subregions of the mouse auditory forebrain. *Neuroscience*. 105:553–569.

de Venecia RK, Smelser CB, Lossman SD, McMullen NT (1995). Complementary expression of parvalbumin and calbindin D-28k delineates subdivisions of the rabbit medial geniculate body. *J Comp Neurol*. 359:595–612.

Galaburda AM, Menard MT, Rosen GD (1994). Evidence for aberrant auditory anatomy in developmental dyslexia. *Proc Natl Acad Sci*. 91:8010–8013.

Gilad A, Maor I, Mizrahi A (2020). Learning-related population dynamics in the auditory thalamus. *Elife*. 9:e56307.

Hackett TA, Barkat TR, O'Brien BM, Hensch TK, Polley DB (2011). Linking topography to tonotopy in the mouse auditory thalamocortical circuit. *J Neurosci*. 31:2983–2995.

Hamada S, Houtani T, Trifonov S, Kase M, Maruyama M, Shimizu J, Yamashita T, Tomoda K, Sugimoto T (2010). Histological determination of the areas enriched in cholinergic terminals and M2 and M3 muscarinic receptors in the mouse central auditory system. *Anat Rec (Hoboken)*. 293:1393–1399.

Han L, Zhang Y, Lou Y, Xiong Y (2012). Thalamic activation modulates the responses of neurons in rat primary auditory cortex: An in vivo intracellular recording study. *PLoS ONE*. 7:e34837.

Hashikawa T, Rausell E, Molinari M, Jones E (1991). Parvalbumin- and calbindin-containing neurons in the monkey medial geniculate complex: Differential distribution and cortical layer specific projections. *Brain Res*. 544:335–341.

Horga G, Fernandez-Egea E, Mane A, Font M, Schatz KC, Falcon C, Lomena F, Bernardo M, Parellada E (2014). Brain metabolism during hallucination-like auditory stimulation in schizophrenia. *PloS ONE*. 9:e84987.

Horie M, Tsukano H, Hishida R, Takebayashi H, Shibuki K (2013). Dual compartments of the ventral division of the medial geniculate body projecting to the core region of the auditory cortex in C57BL/6 mice. *Neurosci Res*. 76:207–212.

Hu B, Senatorov V, Mooney D (1994). Lemniscal and non-lemniscal synaptic transmission in rat auditory thalamus. *J Physiol*. 479(Pt 2):217–231.

Ibrahim BA, Murphy CA, Yudintsev G, Shinagawa Y, Banks MI, Llano DA (2021). Corticothalamic gating of population auditory thalamocortical transmission in mouse. *eLife*. 10:e56645.

Ji X-y, Zingg B, Mesik L, Xiao Z, Zhang LI, Tao HW (2016). Thalamocortical innervation pattern in mouse auditory and visual cortex: Laminar and cell-type specificity. *Cereb Cortex*. 26:2612–2625.

Kimura A, Imbe H, Donishi T, Tamai Y (2007). Axonal projections of single auditory neurons in the thalamic reticular nucleus: Implications for tonotopy-related gating function and cross-modal modulation. *Eur J Neurosci*. 26:3524–3535.

Kimura A, Donishi T, Sakoda T, Hazama M, Tamai Y (2003). Auditory thalamic nuclei projections to the temporal cortex in the rat. *Neuroscience*. 117:1003–1016.

Kuwabara N, Zook JM (2000). Geniculo-collicular descending projections in the gerbil. *Brain Res*. 878:79–87.

LeDoux JE, Farb CR, Romanski LM (1991). Overlapping projections to the amygdala and striatum from auditory processing areas of the thalamus and cortex. *Neurosci Lett*. 134:139–144.

Lee CC, Sherman SM (2010). Topography and physiology of ascending streams in the auditory tectothalamic pathway. *Proc Natl Acad Sci*. 107:372–377.

Lesicko AM, Llano DA (2020). Circuit mechanisms underlying the segregation and integration of parallel processing streams in the inferior colliculus. *J Neurosci*. 40:6328–6344.

Linke R (1999). Organization of projections to temporal cortex originating in the thalamic posterior intralaminar nucleus of the rat. *Exper Brain Res*. 127:314–320.

Llano DA, Feng AS (1999). Response characteristics of neurons in the medial geniculate body of the little brown bat to simple and temporally-patterned sounds. *J Comp Physiol A*. 184:371–385.

Llano DA, Sherman SM (2008). Evidence for nonreciprocal organization of the mouse auditory thalamocortical-corticothalamic projection systems. *J Comp Neurol*. 507:1209–1227.

Llano DA, Sherman SM (2009). Differences in intrinsic properties and local network connectivity of identified layer 5 and layer 6 adult mouse auditory corticothalamic neurons support a dual corticothalamic projection hypothesis. *Cereb Cortex*. 19:2810–2826.

Llano DA, Slater BJ, Lesicko AM, Stebbings KA (2014). An auditory colliculothalamocortical brain slice preparation in mouse. *J Neurophysiol*. 111:197–207.

Lu E, Llano D, Sherman S (2009). Different distributions of calbindin and calretinin immunostaining across the medial and dorsal divisions of the mouse medial geniculate body. *Hear Res*. 257:16–23.

Macias S, Llano DA (2022) Descending projections to the auditory midbrain: evolutionary considerations. *J Comp Physiol A*. 209(1):1–13.

Márquez-Legorreta E, Horta-Junior JAC, Berrebi AS, Saldaña E (2016). Organization of the zone of transition between the pretectum and the thalamus, with emphasis on the pretectothalamic lamina. *Front Neuroanat*. 10:82.

McCormick DA, Prince DA (1988). Noradrenergic modulation of firing pattern in guinea pig and cat thalamic neurons, in vitro. *J Neurophysiol*. 59:978–996.

McMullen N, Velenovsky D, Holmes M (2005). Auditory thalamic organization: cellular slabs, dendritic arbors and tectothalamic axons underlying the frequency map. *Neuroscience*. 136:927–943.

Mellott JG, Foster NL, Ohl AP, Schofield BR (2014a). Excitatory and inhibitory projections in parallel pathways from the inferior colliculus to the auditory thalamus. *Front Neuroanat*. 8:124.

Mellott JG, Foster NL, Nakamoto KT, Motts SD, Schofield BR (2014b). Distribution of GABAergic cells in the inferior colliculus that project to the thalamus. *Front Neuroanat*. 8:17.

Mitrofanis J (2002). Evidence for an auditory subsector within the zona incerta of rats. *Anat Embryol*. 205:453–462.

Molinari M, Dell'Anna M, Rausell E, Leggio M, Hashikawa T, Jones E (1995). Auditory thalamocortical pathways defined in monkeys by calcium-binding protein immunoreactivity. *J Comp Neurol*. 362:171–194.

Monckton JE, McCormick DA (2002). Neuromodulatory role of serotonin in the ferret thalamus. *J Neurophysiol*. 87:2124–2136.

Mooney D, Zhang L, Basile C, Senatorov V, Ngsee J, Omar A, Hu B (2004). Distinct forms of cholinergic modulation in parallel thalamic sensory pathways. *Proc Natl Acad Sci*. 101:320–324.

Motts SD, Schofield BR (2010). Cholinergic and non-cholinergic projections from the pedunculopontine and laterodorsal tegmental nuclei to the medial geniculate body in guinea pigs. *Front Neuroanat*. 4:137.

Motts SD, Schofield BR (2011). Cholinergic cells in the tegmentum send branching projections to the inferior colliculus and the medial geniculate body. *Neuroscience*. 179:120–130.

Ojima H (1994). Terminal morphology and distribution of corticothalamic fibers originating from layers 5 and 6 of cat primary auditory cortex. *Cereb Cortex*. 4:646–663.

Patel MB, Sons S, Yudintsev G, Lesicko AM, Yang L, Taha GA, Pierce SM, Llano DA (2017). Anatomical characterization of subcortical descending projections to the inferior colliculus in mouse. *J Comp Neurol*. 525:885–900.

Peruzzi D, Bartlett E, Smith PH, Oliver DL (1997). A monosynaptic GABAergic input from the inferior colliculus to the medial geniculate body in rat. *J Neurosci*. 17:3766–3777.

Ramcharan EJ, Gnadt JW, Sherman SM (2005). Higher-order thalamic relays burst more than first-order relays. *Proc Natl Acad Sci U S A*. 102:12236–12241.

Read HL, Nauen DW, Escabí MA, Miller LM, Schreiner CE, Winer JA (2011). Distinct core thalamocortical pathways to central and dorsal primary auditory cortex. *Hear Res*. 274:95–104.

Romero S, Hight AE, Clayton KK, Resnik J, Williamson RS, Hancock KE, Polley DB (2020). Cellular and widefield imaging of sound frequency organization in primary and higher order fields of the mouse auditory cortex. *Cereb Cortex*. 30:1603–1622.

Rouiller E, De Ribaupierre F (1985). Origin of afferents to physiologically defined regions of the medial geniculate body of the cat: Ventral and dorsal divisions. *Hear Res.* 19:97–114.

Sherman SM, Guillery R (1998). On the actions that one nerve cell can have on another: Distinguishing "drivers" from "modulators." *Proc Natl Acad Sci.* 95:7121–7126.

Smith PH, Bartlett EL, Kowalkowski A (2006). Unique combination of anatomy and physiology in cells of the rat paralaminar thalamic nuclei adjacent to the medial geniculate body. *J Comp Neurol.* 496:314–334.

Sottile SY, Ling L, Cox BC, Caspary DM (2017). Impact of ageing on postsynaptic neuronal nicotinic neurotransmission in auditory thalamus. *J Physiol.* 595:5375–5385.

Storace DA, Higgins NC, Read HL (2010). Thalamic label patterns suggest primary and ventral auditory fields are distinct core regions. *J Comp Neurol.* 518:1630–1646.

Theyel BB, Llano DA, Sherman SM (2010). The corticothalamocortical circuit drives higher-order cortex in the mouse. *Nat Neurosci.* 13:84–88.

Tsukano H, Horie M, Ohga S, Takahashi K, Kubota Y, Hishida R, Takebayashi H, Shibuki K (2017). Reconsidering tonotopic maps in the auditory cortex and lemniscal auditory thalamus in mice. *Front Neural Circ.* 11:14.

Varela C, Sherman SM (2007). Differences in response to muscarinic activation between first and higher order thalamic relays. *J Neurophysiol.* 98:3538–3547.

von Kriegstein K, Patterson RD, Griffiths TD (2008). Task-dependent modulation of medial geniculate body is behaviorally relevant for speech recognition. *Curr Biol.* 18:1855–1859.

Winer JA (2005). Three systems of descending projections to the inferior colliculus. In: Jeffrey Winer, Christoph Schreiner, eds. *The Inferior Colliculus.* Springer; 231–247.

Winer JA, Larue DT (1988). Anatomy of glutamic acid decarboxylase immunoreactive neurons and axons in the rat medial geniculate body. *J Comp Neurol.* 278:47–68.

Winer JA, Morest DK (1983). The medial division of the medial geniculate body of the cat: implications for thalamic organization. *J Neurosci.* 3:2629–2651.

Winer JA, Saint Marie RL, Larue DT, Oliver D (1996). GABAergic feedforward projections from the inferior colliculus to the medial geniculate body. *Proc Natl Acad Sci.* 93:8005–8010.

Xiao G, Llano DA (2020). Hitting the right spot: NMDA receptors in the auditory thalamus may hold the key to understanding schizophrenia. *Int J Neuropsychopharmacol.* 23:578–580.

Zikopoulos B, Barbas H (2006) Prefrontal projections to the thalamic reticular nucleus form a unique circuit for attentional mechanisms. *J Neurosci.* 26:7348–7361.

Zikopoulos B, Barbas H (2012). Pathways for emotions and attention converge on the thalamic reticular nucleus in primates. *J Neurosci.* 32:5338–5350.

22.

SENSORY REPRESENTATIONS IN THE AUDITORY CORTEX AND THALAMUS

Jennifer F. Linden

INTRODUCTION

Over the past 50 years, research on sensory representations in the auditory cortex and thalamus has been motivated in part by a question inspired by research on the visual system. Hubel and Wiesel's discovery of simple and complex orientation-selective cells in visual cortex (Hubel and Wiesel, 1959, 1962) inspired decades of research on the mechanisms by which orientation selectivity and other emergent features of visual cortical representations are generated by thalamocortical and intracortical interactions. In the auditory system, however, there are fewer obvious differences between cortical and thalamic sensory representations in any species, and there is little agreement on what receptive-field properties are "emergent" at the cortical or even thalamocortical level. The most striking and consistent of the differences between auditory cortical and thalamic sound representations tend to be quantitative rather than qualitative; for example, an increase in the timescale of forward suppression and adaptation to repeated stimuli.

Thus, while research on the visual cortex and thalamus was propelled in part by the search for mechanisms underlying receptive-field properties identified as thalamocortically emergent since the 1960s (at least in primates), research on the auditory cortex and thalamus over the same period has been driven more by the search for any thalamocortically emergent receptive-field properties at all. This search has inspired the use of increasingly complex stimulus sets and approaches to analyzing receptive fields, as researchers have sought to determine whether the "right" stimulus set might reveal uniquely cortical or thalamocortical receptive-field properties. While it remains debatable whether any particular features of sound representations are truly emergent at the thalamocortical level in the auditory system, the diversity of stimulus sets used to probe these representations has provided rich insights into more quantitative differences between cortical and thalamic neurons, particularly with regard to temporal response properties.

Here, I use the diversity of stimulus sets as an organizing principle for discussing cortical and thalamic sensory representations. I begin by reviewing key findings regarding representation of relatively simple sounds, such as tones, clicks, and noise bursts. Next, I discuss insights gained from analysis of neuronal responses to complex artificial sounds, such as dynamic ripple stimuli and dynamic random chords. Then, I consider representation of more temporally extended sound sequences, such as those used to examine "stimulus-specific adaptation" and representation of regular or repeating sound patterns. Finally, I discuss cortical and thalamic representations of natural sounds, particularly animal vocalizations and naturalistic noise (see **Chapter 23 in Section V** for discussion of human speech representations).

The chapter focuses primarily on reviewing studies that have directly compared auditory cortical and thalamic sensory representations at the single-neuron and/or neuronal population levels. This constraint means that some features of auditory cortical representations are not discussed, because they have been minimally investigated in the thalamus. Additionally, following the literature, this review focuses mainly on spectral and temporal properties of cortical and thalamic responses, rather than binaural selectivity or other response properties. Topics such as the effects of activation or deactivation of cortex on thalamus; cortical and thalamic contributions to brain oscillations; and the roles of the cortex and thalamus in auditory plasticity and learning are addressed in other chapters.

Several limitations of this comparison of auditory cortical and thalamic sensory representations should be noted at the outset. First, most of the relevant papers have compared neuronal response properties between the core (primary) auditory cortex (A1) and either the auditory thalamus as a whole (medial geniculate body, or MGB) or the lemniscal auditory thalamus (ventral MGB, or vMGB). However, previous work suggests that sensory representations differ substantially between lemniscal and non-lemniscal thalamic subregions (e.g., Anderson et al., 2009; Antunes et al., 2010) (see **Chapter 21 in Section V** for a more detailed description of the organization of the MGB). Second, even within primary auditory cortex, most research has focused on properties of neurons in layers 2–4, but there are known differences in sensory representation between superficial and deep layers (e.g., Christianson et al., 2011) and between deep-layer cell types projecting to different targets (e.g., Williamson and Polley, 2019). Third, similarities and differences between cortical and thalamic sensory representations depend upon brain state; for example, response properties of auditory cortical and thalamic neurons differ between awake and sleep states (for reviews, see Edeline, 2003; Hennevin et al., 2007).

Finally, and perhaps most importantly, sensory representations in the auditory cortex and thalamus are surely somewhat species-specific, optimized for the sensory environment and behavioral goals of the animal. This point is well illustrated by the example of echolocating bats, where cortical and thalamic (as well as midbrain and brainstem) auditory representations are specialized for detection of sound combinations arising from the bat's sonar calls and their echoes (Wenstrup et al., 2012; Suga, 2015). However, the literature on auditory cortical and thalamic sound representations is not extensive enough to permit conclusive analysis of species differences, so instead I have highlighted the issue by specifying the animal species for all key references, and by illustrating how the most common animal models used for auditory cortical and thalamic research have changed over time (Figure 22.1A).

Despite the limitations of the research literature, there has been notable progress toward understanding what the auditory cortex and thalamus are contributing to sound representation in the brain. Historical trends in the use of different stimulus types and response measures (Figure 22.1B) reveal not only diverse approaches with distinct strengths and limitations (Table 22.1), but also a growing appreciation for the role of adaptation to sound statistics. Key experimental insights are summarized in a figure describing differences between auditory cortical and thalamic sensory representations (Figure 22.2).

REPRESENTATION OF SIMPLE SOUNDS

Most investigations of cortical and thalamic responses to simple sounds have found modest or no differences in spectral response properties, with more profound differences in temporal properties and response reliability. Early studies described broader frequency tuning curves in cortex than thalamus, but this observation has not been upheld by later work. Cortical neurons tend to respond more phasically and less reliably to tones than thalamic neurons (Creutzfeldt et al., 1980) and therefore, frequency-intensity response areas are typically noisier and harder to measure in the cortex than in the thalamus. However, frequency tuning is not consistently broader in A1 than MGB, even when thalamic recordings are restricted to the vMGB. For example, Bartlett et al. (2011) found that selectivity of neurons for sound frequency was comparably narrow in the A1 and vMGB of awake marmoset monkeys, and often much narrower than observed in the auditory nerve. Proportions of narrowly tuned neurons were observed to be somewhat lower in A1 than vMGB, but quantitative comparisons are complicated by the fact that tone durations and analysis procedures used for defining the frequency tuning curve borders were different for A1 and vMGB (Bartlett et al., 2011). In anesthetized mice, optogenetic silencing of intracortical activity during intracellular recordings in A1 layer 4 neurons reduced the magnitude of responses to tones but did not affect spectral range or tuning (Li et al., 2013), indicating that tone responses in cortical cells preserve the frequency tuning of their thalamic inputs. Interestingly, responses to tone offsets are also similar in cortex and thalamus; Solyga and Barkat (2021) found that offset responses in the auditory cortex of awake mice are largely inherited from the MGB. Even modulation of tone responses during different sound categorization tasks may be similar in the cortex and the thalamus. In rats performing a flexible high/low sound frequency categorization task, A1 and MGB neurons showed equivalent modulation of responses to the same tone by its category association (Jaramillo et al., 2014). Similar results were obtained in monkeys performing a frequency-step direction-categorization task (Selezneva et al., 2017).

The most striking difference between auditory cortical and thalamic representations of simple sounds is that thalamic neurons respond with higher temporal fidelity to rapidly varying sounds. In an influential early paper on auditory thalamocortical transformations in awake guinea pig, Creutzfeldt et al. (1980) recorded simultaneously from pairs of thalamic and cortical neurons with correlated spiking and showed that thalamic neurons were capable of following amplitude-modulated (or frequency-modulated) sounds at much higher rates than cortical neurons. Later investigations in marmoset monkeys confirmed this finding, and also demonstrated a transformation from more "synchronous" representation of repetition rate in the thalamus to a more predominantly "nonsynchronous" rate-code representation in the auditory cortex (Bartlett and Wang, 2007). More recent studies have also demonstrated increased temporal integration and adaptation from thalamus to cortex in mice, for example in the time course of forward suppression (Xiong et al., 2020). Thalamocortical

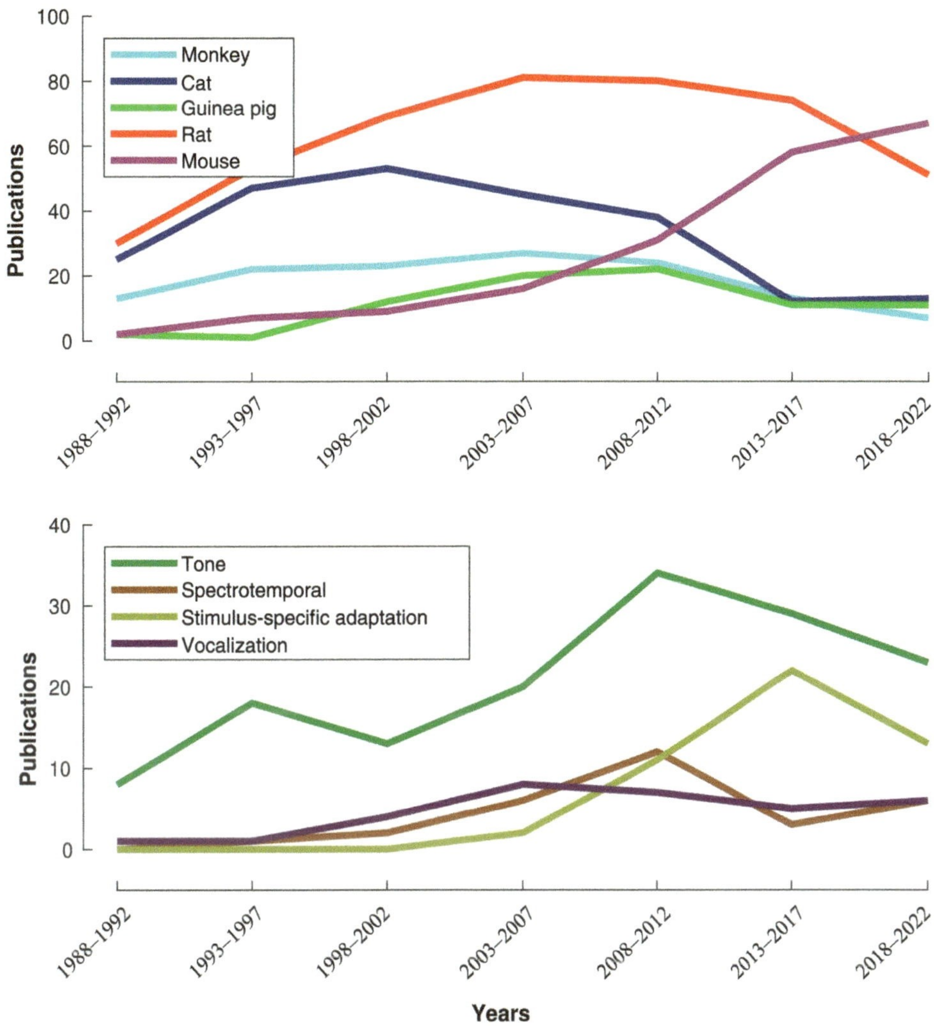

Figure 22.1. Historical trends in research comparing neuronal response properties between the auditory cortex and thalamus. Plots show results from Clarivate Web of Science searches in 5-year intervals for topic (title/abstract/keywords) "auditory AND (cortex OR cortical) AND (thalamus OR thalamic) AND neuron*," plus one of the terms shown in legends. Top: number of publications using the mouse has increased greatly over the past 15 years at the expense of several other model systems with hearing ranges more similar to that of humans. (Ferret, gerbil, bat, and other auditory models excluded for small numbers overall.) This shift follows, of course, the 21st-century development in mice of powerful genetic tools for manipulating neural circuitry, such as optogenetics. Until these genetic tools become easier to use in other species, the reduction in diversity of model systems could increase the risk that species-specific phenomena may be misinterpreted as fundamental characteristics of mammalian thalamocortical transformations. Bottom: number of publications with topic words commonly associated with different stimulus types: "spectrotemporal" for spectrotemporal receptive field estimation; "stimulus-specific adaptation" for sound sequences; and "vocalization" for natural sounds. The "tone" category includes many publications that also appear in the other categories. Note the surge of interest in stimulus-specific adaptation, reflecting the possibility that sensitivity to long-term sound statistics is dependent on cortical activity.

transformation in the auditory system therefore seems to replicate a transformation that also occurs across the entire ascending auditory system, from cochlea to cortex: transition from temporal to rate codes, with increasing rate-based representation of rapidly varying sound signals and temporal representation only of slower modulated signals (Wang et al., 2008).

Another prominent feature of thalamocortical transformation of simple sounds seems to be that the cortical representation is far less reliably stimulus-driven and less robust to major changes in brain state than the thalamic representation. Using 2-photon calcium imaging to record from large ensembles of cortical and thalamic neurons in awake versus anesthetized mice, Filipchuk et al. (2022) recently found that patterns of evoked population activity were easily distinguished from patterns of spontaneous activity in the awake animal in both cortex and thalamus, but in the anesthetized state, distinct differences between patterns of spontaneous and sound-evoked population activity persisted only in the thalamus, not in the cortex. This result echoes conclusions from earlier studies using electrophysiological recording methods in cat (Zurita et al., 1994;

TABLE 22.1 KEY STRENGTHS AND LIMITATIONS OF DIFFERENT STIMULUS TYPES AND RESPONSE MEASURES FOR INVESTIGATING SENSORY REPRESENTATIONS IN THE AUDITORY CORTEX AND THALAMUS

Stimuli	Response Measures	Strengths	Limitations
Simple sounds (tones, noise bursts, clicks, click trains, FM sweeps, AM tones or noise)	Frequency-intensity response areas, modulation transfer functions, synchronization, FM/ AM tuning	Detailed description of stimulus-response function for selected stimulus parameters; computationally simple	Strong assumptions required about what relevant stimulus parameters are; often experimentally inefficient
Complex artificial sounds (dynamic ripples, dynamic random chord stimuli)	Spectrotemporal receptive fields (STRFs), nonlinear stimulus-response functions	Comparatively unbiased and agnostic exploration of stimulus sensitivities; experimentally efficient	Requires assumptions about nature of nonlinearities in stimulus-response function; often computationally intensive
Sound sequences (oddball stimuli, regular versus random patterns)	Stimulus-specific adaptation, regularity sensitivity	Allows analysis of representation and adaptation over a wide range of timescales	Experimentally very time-consuming, so stimulus parameters must be tightly constrained
Natural sounds (vocalizations, ambient environmental sounds)	Vocalization selectivity, categorical perception, noise invariance	Ecologically relevant to animal studied; possibly, also relevant to human language perception	If used for stimulus-response function estimation, higher-order autocorrelations can produce misleading results

Cotillon-Williams and Edeline, 2003) and rat (Cotillon-Williams and Edeline, 2003), which found that different forms of anesthesia could have profoundly different effects on cortical versus thalamic responses to simple sounds.

Thus, to a first approximation, cortical and thalamic representations of simple sounds are remarkably similar, with prominent differences observed mainly in temporal response properties such as forward suppression, and in less stimulus-related properties such as robustness of representations to changes in brain state. A caveat to this conclusion, of course, is that differences between cortical and thalamic representations of simple sounds may be evident in the projections of auditory cortical and thalamic neurons to particular targets, even if they are obscured in the mixed population of projection neurons within the areas themselves. For example, in the mouse, optogenetic inhibition of projections from the auditory cortex to the striatum produces a more frequency-specific suppression than inhibition of projections from the MGB to the striatum (Chen et al., 2019). Differences in amplitude modulation tuning between thalamostriatal and corticostriatal projections have also been observed (Ponvert and Jaramillo, 2019).

REPRESENTATION OF COMPLEX SOUNDS

Spectrotemporally rich sounds, such as "dynamic ripple" or "dynamic random chord" stimuli, can be powerful tools for probing thalamic and cortical sound representations (Meyer et al., 2016). Techniques for estimating stimulus-response functions from neuronal responses to complex sounds (broadly, reverse-correlation techniques and related nonlinear approaches) have many strengths for analysis of sensory representations. Most importantly, these techniques provide a more stimulus-agnostic, less hypothesis-driven answer to the question: What sound features are represented by thalamic and cortical activity? Another strength is that complex artificial sounds are in some sense more "natural" than many simple sounds used in auditory experiments, at least in that the sounds are spectrally rich and temporally varying. A third, more practical, advantage of using complex sounds with stimulus-response function estimation techniques is experimental efficiency; the same stimulus and analysis procedures can be used to estimate response properties of all neurons (including simultaneously recorded neurons) systematically, without the online optimization that is often necessary for efficient evaluation of response properties that depend (for example) on frequency tuning.

However, complex sounds and stimulus-response function estimation techniques also have limitations for analysis of sensory representations, especially in auditory cortex and thalamus. Most importantly, while the stimulus space explored is very large compared to that which can realistically be probed using simple sounds, the stimulus-response function model must be constrained to make estimation of its parameters computationally tractable. The constraint is usually that the stimulus-response function is assumed to be linear in the stimulus spectrogram, or linear with well-defined nonlinearities applied at either the input or the output to the linear stage (Meyer et al., 2016). The approach is therefore relatively agnostic about stimulus sensitivities, but not agnostic about the nature of response nonlinearities—which might ultimately be the most important features of thalamocortical transformations. A related limitation is that the complex

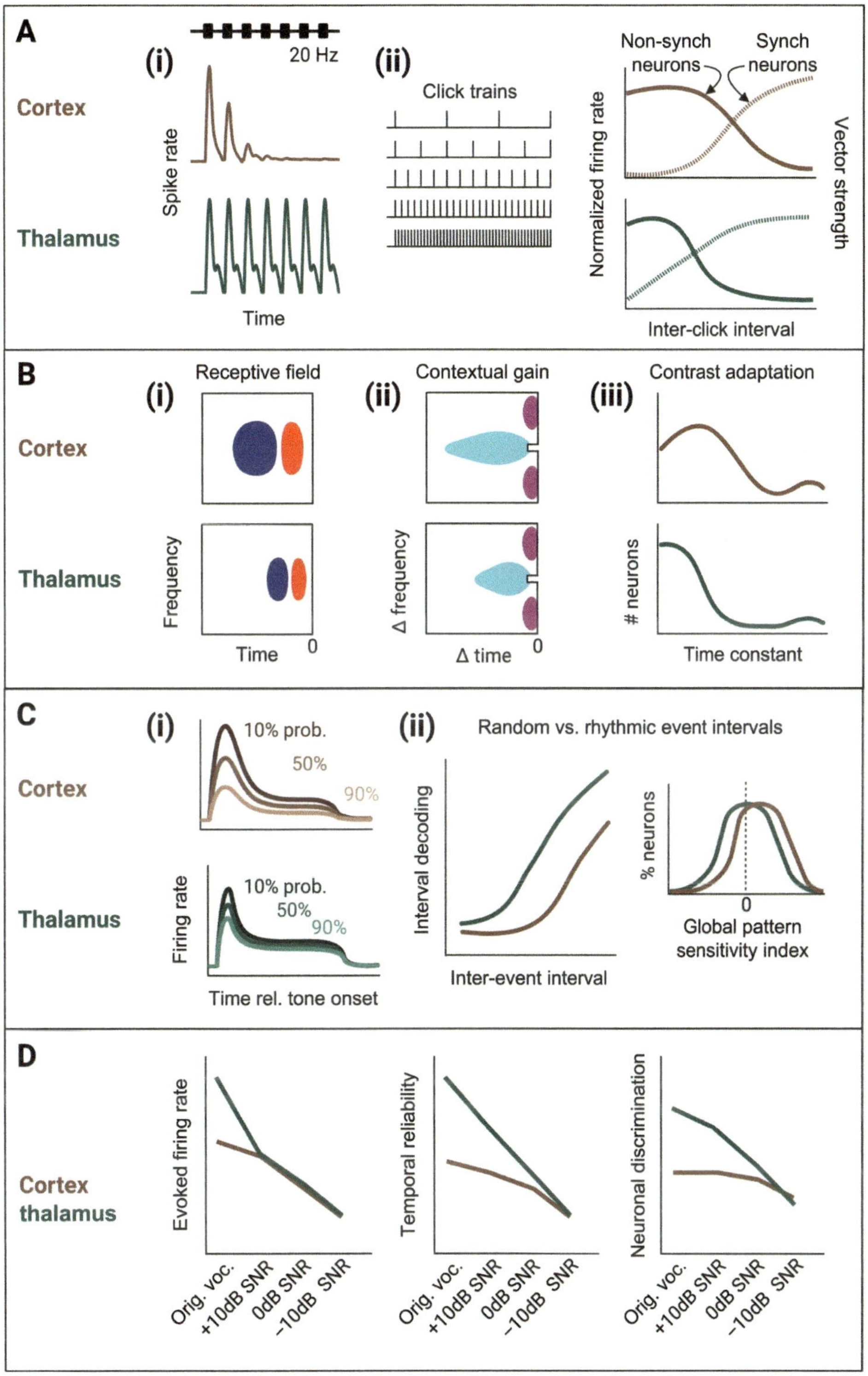

Figure 22.2. Key differences between sensory representations in A1 and MGB. (A). Differences revealed using simple sound stimuli. A(i). Cortical neurons are less capable of following high rates of amplitude modulation than thalamic neurons (Creutzfeldt et al., 1980). A(ii). Cortical and thalamic neurons typically encode inter-click intervals either with synchronized firing (synch neurons) or by increasing firing rate in a nonsynchronized manner as the inter-click interval decreases (nonsynch neurons). The transition between synchronized and rate coding in neuronal population activity occurs at longer inter-click intervals in cortex than thalamus (Bartlett and Wang 2007). (B). Differences revealed using complex artificial stimuli, such as dynamic ripples and dynamic random chords. B(i). Spectrotemporal receptive fields are typically more temporally extended in cortex than thalamus (Miller et al., 2001, 2002). Plots can be interpreted as spectrograms of the typical stimulus preceding a spike. *X*-axis = time to spike; red = excitatory subfield; blue = inhibitory subfield. B(ii). Contextual gain modulation, and specifically delayed suppression, is more temporally extended in cortex than thalamus (Williamson et al., 2016). Notch in plots signifies the target tone, which can be any spectrotemporal element in a complex stimulus; colors indicate modulatory effects of neighboring sounds on the gain of the neuronal response to the target tone. *X*- and *y*-axes = time and frequency

(*continued*)

sounds used for stimulus-response function estimation must not have higher-order autocorrelations in the space used for representing the "receptive field" of the neuron, because the receptive-field estimate can be distorted by artifactual structure arising from interactions between higher-order autocorrelations in the stimulus and unmodeled nonlinear aspects of the neuronal response (Christianson et al., 2008; Meyer et al., 2016). In practice, this means that stimulus-response function estimation techniques cannot be relied upon to produce conservative, unbiased depictions of neuronal receptive-field properties when natural sounds are used as the complex stimuli.

Both the strengths and the limitations of stimulus-response function estimation using complex sounds have helped to advance understanding of cortical and thalamic sound representations over the past 2 decades. In an influential early paper, Miller et al. (2001) used dynamic ripple stimuli to estimate spectrotemporal receptive fields (STRFs) in anesthetized cat, comparing STRFs between pairs of A1 and vMGB neurons with correlated firing. This analysis revealed significant transformation and reconstruction of sensory representations between thalamus and cortex. Cortical STRFs could be inherited from thalamic STRFs, constructed from smaller units, or assembled by cooperative activity (Miller et al., 2001). Key conclusions from previous studies using simple sounds also held for cortical and thalamic responses to complex sounds; STRF analysis revealed slower temporal modulation preferences in cortex than thalamus and similar spectral modulation preferences (Miller et al., 2002). Importantly, analysis of dynamic-ripple modulation transfer functions revealed no differences in spectrotemporal asymmetry between cortical and thalamic STRFs (Miller et al., 2002), suggesting that frequency-modulation (FM) selectivity—an obvious possible auditory analog of visual orientation or direction selectivity—is similar in auditory cortex and thalamus. However, since STRFs are linear in the spectrogram of the stimulus, this work left open the question of how nonlinear properties of responses to complex sounds might differ between cortex and thalamus.

More recent studies have attempted to address this question for response nonlinearities such as spectrotemporal context sensitivity and contrast gain control, using stimulus-response function models with input or output nonlinearities. For example, Williamson et al., 2016 introduced stimulus-response functions with input nonlinearities designed to capture interactions between pairs of spectrotemporal inputs within a complex sound (e.g., nonlinear forward suppression and 2-tone combination sensitivity). Analyzing neuronal responses to dynamic random chord stimuli in anesthetized mice, they found remarkably similar patterns of nonlinear context sensitivity in cortical and thalamic neurons. The only obvious difference was that the time course of nonlinear forward suppression was significantly more elongated in cortex than thalamus (Williamson et al., 2016). A different approach, using stimulus-response functions with an output nonlinearity dependent on spectrotemporal contrast, has revealed contrast gain control at multiple levels of the ascending auditory pathway in both ferrets and mice (Rabinowitz et al., 2013; Lohse et al., 2020). Again, the main difference observed between thalamus and cortex related to temporal aspects of the nonlinearities; time constants for adaptation to spectrotemporal contrast were longer in the cortex than the thalamus (Lohse et al., 2020). Thus, experiments using complex sounds and stimulus-response function estimation techniques indicate that both linear and nonlinear properties of sensory representations in the auditory cortex and thalamus are qualitatively similar (or at least, similarly diverse), with consistent quantitative differences primarily in the temporal domain.

REPRESENTATION OF SOUND SEQUENCES AND STATISTICS

Cortical and thalamic representation of sound sequences and long-term sound statistics accelerated in the early 2000s with the discovery of strong stimulus-specific adaptation (SSA) in the auditory cortex (Ulanovsky et al., 2003; Ulanovsky et al.,

Figure 22.2. **(continued)**

relative to the target tone; magenta = positive gain (facilitation); cyan = negative gain (suppression). B(iii). Time constants for adaptation to changes in spectrotemporal contrast are longer in the cortex than thalamus (Lohse et al., 2020). (C). Differences revealed using sound sequences and patterns. C(i). Stimulus-specific adaptation is stronger in the cortex than thalamus (Ulanovsky et al., 2003; Parras et al., 2017). Plots illustrate typical responses to the same tone when presented with 10% (deviant), 90% (standard), or 50% (control) probability in a sound sequence with a tone of another frequency that evokes similar responses from the neuron under control conditions. C(ii). Analysis of neuronal population responses to sequences of noise burst events with either rhythmic or random inter-event intervals shows that thalamic neurons represent local intervals (time between noise burst events) more accurately than cortical neurons do, while cortical neurons are more sensitive to the existence of global patterns (Asokan et al., 2021). (D). Differences revealed using vocalizations with or without naturalistic noise. Without noise (orig. voc.), thalamic neurons respond more strongly and more reliably to conspecific vocalizations than cortical neurons, and neuronal discrimination of different examples of the same type of vocalization is better in thalamus than cortex. However, as the signal-to-noise decreases (+10 to −10 dB SNR), evoked firing rates, temporal reliability, and neuronal discrimination become more similar in thalamus and cortex (Souffi et al., 2020). Thus, cortical neurons represent vocalizations more poorly than thalamic neurons overall, but are less sensitive to changes in noise level. All illustrations are original cartoons of phenomena shown in data in the cited references. Created with BioRender.com.

2004). A neuron is said to exhibit SSA if the magnitude of its response to a specific stimulus within a sound sequence depends on the probability of that stimulus occurring over many seconds. SSA differs from simple firing-rate adaptation in both its stimulus-specificity and its long timescale. Importantly, SSA can manifest not only as a stimulus-specific suppression of neuronal responses to frequently repeated sounds ("standards"), but also as a stimulus-specific facilitation of neuronal responses to rare sounds ("deviants"). Thus, SSA has been interpreted as a neuronal correlate of auditory expectation violation and prediction error.

Over the past 2 decades, multiple studies have found that SSA is stronger in the auditory cortex than the thalamus. The first 2 papers to describe SSA in the auditory cortex found no evidence for SSA in the auditory thalamus of anesthetized cats (Ulanovsky et al., 2003; Ulanovsky et al., 2004). Later work, mostly in rats and mice, demonstrated that SSA does occur in both the MGB and the inferior colliculus (IC) (Anderson et al., 2009; Antunes et al., 2010; Ayala et al., 2012). However, SSA is strongest in the auditory cortex, and cortical neurons are also more likely than thalamic neurons to show "true deviance detection": greater responsiveness to a stimulus occurring as a deviant in an otherwise predictable sound sequence than to the same stimulus in an unpredictable sound sequence (Parras et al., 2017). Remarkably, cortical neurons exhibit robust SSA even to different instances of multi-frequency tone clouds, while MGB (and IC) neurons do not (Harpaz et al., 2021). These observations indicate that while SSA in thalamic (and midbrain) neurons might arise from relatively simple mechanisms, such as short-term adaptation of narrowly tuned synaptic inputs, such mechanisms cannot fully account for SSA in cortical neurons. More generally, the SSA results demonstrate that cortical neurons are more sensitive to long-term sound statistics than thalamic neurons.

Other studies of cortical and thalamic representation of sound sequences and long-term sound statistics have come to similar conclusions. For example, in anesthetized rats, Asari and Zador (2009) recorded auditory cortical and thalamic responses to sequences of different natural and complex artificial sounds and examined how responses to a particular sound segment depended on the preceding sound context. Time constants for adaptation to a change in sound statistics were approximately 1 second for synaptic inputs to A1 neurons but less than 100 ms for spiking activity in MGB, indicating that the prolonged cortical adaptation timescale arises from thalamocortical or intracortical mechanisms. Interestingly, Cooke et al. (2020) recently reported very late "post-stimulatory" activity in auditory cortical neurons which persisted for hundreds of milliseconds after sound stimulation in awake, passive listening marmosets. This post-stimulatory activity was shorter in duration, longer in latency, and much less commonly observed overall in the MGB than A1, suggesting a thalamocortical, intracortical, or corticocortical origin for prolonged cortical aftereffects of sensory stimulation.

Finally, in one of the most systematic comparisons of cortical and thalamic representational timescales to date, Asokan et al. (2021) recorded simultaneously from neurons in A1, MGB, and IC of awake mice while presenting sequences of noise bursts in random or repeating inter-burst interval patterns. Neuronal ensemble activity in each area was used either to decode local inter-burst interval duration or to classify the global sequence as random versus regular pattern. Decoding accuracy for identification of short intervals deteriorated across the IC-MGB-A1 hierarchy, while classification accuracy for discrimination of random versus regular patterns improved (Asokan et al., 2021). These data clearly demonstrate that the thalamocortical transformation produces a loss in temporal fidelity at short timescales, but a gain in context sensitivity at long timescales.

REPRESENTATION OF NATURAL SOUNDS

Natural sounds such as vocalizations and naturalistic noise are the signals that the auditory system evolved to perceive. In principle, then, differences between cortical and thalamic representations of natural sounds could provide particularly useful information for defining the nature of the auditory thalamocortical transformation. Here I consider this possibility, focusing on studies that have examined neuronal responses to vocalizations or naturalistic noise in both A1 and MGB of animals. (Cortical and thalamic processing of speech sounds in the human brain is discussed in **Chapter 23 in Section V**.)

One of the most well-studied examples of vocalization processing in mammals is the representation of pulsed bio-sonar calls and their echoes in the auditory system of echolocating bat species. Combination-sensitive neurons in the auditory brainstem and midbrain, tuned to bio-sonar signal components and echoes at varying delays, are transformed through collicular and thalamic processing into topographic maps of echo delay in the auditory cortex (Wenstrup et al., 2012; Suga, 2015). In the mustached bat, echo delay tuning is sharper in the thalamus and cortex than in the midbrain (Yan and Suga, 1996). However, thalamic and cortical representations of echo delay are similar (Suga, 2015; Butman and Suga, 2019), and evidence suggests that cortical combination sensitivity is largely inherited from the midbrain and thalamus (Wenstrup et al., 2012). Thus, differences between cortical and thalamic representations have been challenging to identify even in the case of an auditory system highly specialized for a distinctive vocalization-based, sound-guided behavior.

Cortical and thalamic representations of conspecific social vocalizations have been compared most often in guinea pigs, which have a large repertoire of species-specific communication calls and a hearing range similar to that of humans. Wallace et al. (2005) compared responses of cortical and thalamic neurons to the low-frequency "purr" call presented to anesthetized guinea pigs; no differences in neuronal response types or temporal fidelity were observed between A1 and MGB. Huetz et al. (2009) reported similar observations in their investigation of cortical and thalamic vocalization selectivity, studying responses to 4 different types of conspecific vocalizations in both anesthetized and awake guinea pigs. There were no differences between A1 and MGB in the percentage of neurons that were informative about vocalization identity, nor even in the percentage of neurons that carried vocalization-related information primarily in spike timing rather than spike count (Huetz et al., 2009). Suta et al. (2013) also found largely similar A1 and MGB responses to 4 types of vocalizations in anesthetized guinea pigs. Recently, Montes-Lourido et al. (2021) compared cortical and thalamic representations of a wider range of 8 conspecific vocalizations in awake guinea pigs; they also examined differences between the thalamorecipient layer 4 and supragranular layers 2/3 (L4 and L2/3) in auditory cortex. Interestingly, vocalization selectivity was similar in MGB and A1 L4, but substantially higher in A1 L2/3 (Montes-Lourido et al., 2021). Together, these results suggest that representations of conspecific vocalizations in guinea pigs are surprisingly unaffected by thalamocortical transformation, but are refined by intracortical processing.

Perhaps the thalamocortical transformation contributes more significantly to the separation of vocalizations from noisy background sounds? Recordings in cortex and thalamus (and midbrain) of anesthetized ferret have revealed that A1 neurons respond most robustly to naturalistic noise with a 1/f power spectrum, whereas MGB neurons are less selective for natural noise characteristics (Garcia-Lazaro et al., 2011). Vocalizations and other ecologically important signals often have temporal envelopes that are more slowly modulated than natural noise backgrounds. Does greater sensitivity to long-term sound statistics and naturalistic noise characteristics in cortex than thalamus mean that cortical neurons distinguish vocalizations from background noise more effectively than thalamic neurons do?

This question has been addressed in a few studies so far, with intriguing results. One particularly influential paper (Chechik et al., 2006) compared A1, MGB, and IC responses in anesthetized cat to natural bird chirps in noise and to various decompositions of the natural sounds into signal and noise components. For all forms of neural representation examined (spike count, spike timing, or latency codes), informational redundancy between the stimulus representations was highest in the IC and lower in both the MGB and A1 (Chechik et al., 2006). Further work revealed that IC neurons preserved the largest amount of information about both abstract entities (e.g., bird chirp, echoes, ambient noise) and detailed spectrotemporal features, but the ratio of abstract to spectrotemporal information was higher in A1 and MGB (Chechik and Nelken, 2012). Again, however, minimal differences were observed between A1 and MGB; thus, there was no evidence for a special role of the thalamocortical transformation in separating signal and noise components in natural sounds.

More recent experiments have provided evidence that both MGB and IC contribute at least as significantly as A1 does to separation of ecologically relevant signals from noise. Recording in the cochlear nucleus, the central nucleus of the IC, the ventral MGB, A1, or the secondary auditory cortex in anesthetized guinea pigs, Edeline and colleagues examined neuronal discrimination of 4 examples of a guinea pig alarm call masked by frozen noise, either vocalization "chorus" noise or vocalization-shaped stationary noise (Souffi et al., 2020; Souffi et al., 2021; Souffi et al., 2022). The 4 alarm call examples were more effectively distinguished by neurons or neuronal populations in subcortical than cortical structures, even in noise (Souffi et al., 2020). Cortical responses to the vocalizations were more robust than subcortical responses to increasing noise, but even at the highest noise level tested (−10 dB SNR), MGB and IC neurons retained more vocalization-related information than A1 neurons (Souffi et al., 2020; Souffi et al., 2021). Furthermore, the proportion of neurons that were more sensitive to vocalization signals than noise was highest in the MGB and IC, not in the cortex (Souffi et al., 2021).

Thus, separation of vocalizations from background noise appears to be a distributed function of the entire central auditory system, not a specific function of the thalamocortical transformation or the cortex itself. Indeed, Souffi et al. (2022) found that neurons throughout the auditory system (not just in the cortex) maintained good ability to track slow modulations in the stimulus as noise increased. Neuronal discrimination of vocalizations deteriorated with increasing noise, not because of changes or differences in modulation tracking at cortical versus subcortical levels, but because the temporal envelopes of vocalizations degraded by noise were more similar acoustically than temporal envelopes of noise-free vocalizations (Souffi et al., 2022). Notably, different conclusions were obtained from an earlier study in ferrets using variable rather than frozen-noise tokens; cortical representations were found to be more noise-invariant in this context than IC representations (Rabinowitz et al., 2013). The differences may depend critically upon the answer to an almost philosophical question about how to define auditory "noise": Is it an acoustic waveform with particular spectral and temporal properties, or the component of the sound that is most variable?

IMPORTANT AND UNANSWERED QUESTIONS

For decades, much research on sensory representations in the auditory cortex and thalamus has been driven by the question: What thalamocortically emergent property of auditory receptive fields might be analogous to orientation selectivity in the primate visual system? This question still remains unanswered, even after many years of research utilizing diverse stimulus sets, including complex sounds, sound sequences, and natural sounds. To a first approximation, neuronal response properties in the auditory cortex and thalamus have been found to be surprisingly similar, with differences primarily in the extent of temporal integration, temporal adaptation, representational stability, and sensitivity to long-term sound statistics. Temporal processing timescales generally lengthen across the entire central auditory system, from brainstem to cortex (for a recent review, see Kopp-Scheinpflug and Linden, 2020). Thus, differences in temporal processing between cortex and thalamus are not an emergent thalamocortical phenomenon like orientation selectivity in primates, but a continuation of a transformation that occurs throughout the ascending central auditory pathway in all species studied.

Where, then, to go from here? The strongest evidence for a special role for cortical or thalamocortical processing in sound representation has come from studies of sensitivity to sound sequences and global temporal patterns. To the extent that identifying an emergent thalamocortical phenomenon is still the goal, then understanding thalamocortical, intracortical, and corticocortical mechanisms enabling sensitivity to long-term sound statistics is clearly an important focus for future investigation.

However, it is possible that Hubel and Wiesel's seminal discoveries in the visual system have not only inspired but also obscured the most important research questions in the auditory system. Functionally and physiologically (and even histologically), the primary auditory cortex may be more analogous to visual association areas than to primary visual cortex (King and Nelken, 2009). Moreover, as discussed in other chapters in this volume, extensive research on central auditory plasticity already suggests that the most emergent role of the auditory cortex and thalamus relates not to representation of particular sound features, but to *learning* about sound features. Perhaps equivalently from a computational viewpoint, the emergent role of auditory cortex and thalamus may be learning to discount spectrotemporal detail that varies within but not between auditory objects in the real world. Thus, the most important unanswered questions about sensory representations in the auditory cortex and thalamus may not be what those representations are, but how invariances in the representations arise through sensory experience.

REFERENCES

Anderson LA, Christianson GB, Linden JF (2009). Stimulus-specific adaptation occurs in the auditory thalamus. *J Neurosci.* 29:7359–7363.

Antunes FM, Nelken I, Covey E, Malmierca MS (2010). Stimulus-specific adaptation in the auditory thalamus of the anesthetized rat. *PLoS ONE.* 5:e14071.

Asari H, Zador AM (2009). Long-lasting context dependence constrains neural encoding models in rodent auditory cortex. *J Neurophysiol.* 102:2638–2656.

Asokan MM, Williamson RS, Hancock KE, Polley DB (2021). Inverted central auditory hierarchies for encoding local intervals and global temporal patterns. *Curr Biol.* 31:1762–1770.

Ayala YA, Pérez-González D, Duque D, Nelken I, Malmierca MS (2012). Frequency discrimination and stimulus deviance in the inferior colliculus and cochlear nucleus. *Front Neural Circuits.* 6(119), https://doi.org/10.3389/fncir.2012.00119.

Bartlett EL, Sadagopan S, Wang X (2011). Fine frequency tuning in monkey auditory cortex and thalamus. *J Neurophysiol.* 106:849–859.

Bartlett EL, Wang X (2007). Neural representations of temporally modulated signals in the auditory thalamus of awake primates. *J Neurophysiol.* 97:1005–1017.

Butman JA, Suga N (2019). Inhibitory mechanisms shaping delay-tuned combination-sensitivity in the auditory cortex and thalamus of the mustached bat. *Hear Res.* 373:71–84.

Chechik G, Anderson MJ, Bar-Yosef O, Young ED, Tishby N, Nelken I (2006). Reduction of information redundancy in the ascending auditory pathway. *Neuron.* 51:359–368.

Chechik G, Nelken I (2012). Auditory abstraction from spectro-temporal features to coding auditory entities. *Proc Natl Acad Sci U S A.* 109:18968–18973.

Chen L, Wang X, Ge S, Xiong Q (2019). Medial geniculate body and primary auditory cortex differentially contribute to striatal sound representations. *Nat Commun.* 10(418), https://doi.org/10.1038/s41467-019-08350-7.

Christianson GB, Sahani M, Linden JF (2008). The consequences of response nonlinearities for interpretation of spectrotemporal receptive fields. *J Neurosci.* 28:446–455.

Christianson GB, Sahani M, Linden JF (2011). Depth-dependent temporal response properties in core auditory cortex. *J Neurosci.* 31:12837–12848.

Cooke JE, Lee JJ, Bartlett EL, Wang X, Bendor D (2020). Post-stimulatory activity in primate auditory cortex evoked by sensory stimulation during passive listening. *Sci Rep.* 10(13885), https://doi.org/10.1038/s41598-020-70397-0.

Cotillon-Williams N, Edeline J-M (2003). Evoked oscillations in the thalamo-cortical auditory system are present in anesthetized but not in unanesthetized rats. *J Neurophysiol.* 89:1968–1984.

Creutzfeldt O, Hellweg FC, Schreiner C (1980). Thalamocortical transformation of responses to complex auditory stimuli. *Exp Brain Res.* 39:87–104.

Edeline J-M (2003). The thalamo-cortical auditory receptive fields: regulation by the states of vigilance, learning and the neuromodulatory systems. *Exp Brain Res.* 153:554–572.

Filipchuk A, Schwenkgrub J, Destexhe A, Bathellier B (2022). Awake perception is associated with dedicated neuronal assemblies in the cerebral cortex. *Nat Neurosci.* 25:1327–1338.

Garcia-Lazaro JA, Ahmed B, Schnupp JWH (2011). Emergence of tuning to natural stimulus statistics along the central auditory pathway. *PLoS ONE*. 6:e22584.

Harpaz M, Jankowski MM, Khouri L, Nelken I (2021). Emergence of abstract sound representations in the ascending auditory system. *Prog Neurobiol.* 202(102049), https://doi.org/10.1016/j.pneurobio.2021.102049.

Hennevin E, Huetz C, Edeline J-M (2007). Neural representations during sleep: from sensory processing to memory traces. *Neurobiol Learn Mem*. 87:416–440.

Hubel DH, Wiesel TN (1959). Receptive fields of single neurones in the cat's striate cortex. *J Physiol (Lond)*. 148:574–591.

Hubel DH, Wiesel TN (1962). Receptive fields, binocular interaction and functional architecture in the cat's visual cortex. *J Physiol (Lond)*. 160:106–154.

Huetz C, Philibert B, Edeline J-M (2009). A spike-timing code for discriminating conspecific vocalizations in the thalamocortical system of anesthetized and awake guinea pigs. *J Neurosci*. 29:334–350.

Jaramillo S, Borges K, Zador AM (2014). Auditory thalamus and auditory cortex are equally modulated by context during flexible categorization of sounds. *J Neurosci*. 34:5291–5301.

King AJ, Nelken I (2009). Unraveling the principles of auditory cortical processing: Can we learn from the visual system? *Nat Neurosci*. 12:698–701.

Kopp-Scheinpflug C, Linden JF (2020). Coding of Temporal Information. In: Fritzsch B, Grothe B, eds. *The Senses: A Comprehensive Reference*. Academic Press:691–712.

Li L, Li Y, Zhou M, Tao HW, Zhang LI (2013). Intracortical multiplication of thalamocortical signals in mouse auditory cortex. *Nat Neurosci*. 16:1179–1181.

Lohse M, Bajo VM, King AJ, Willmore BDB (2020). Neural circuits underlying auditory contrast gain control and their perceptual implications. *Nat Commun*. 11(324), https://doi.org/10.1038/s41467-019-14163-5.

Meyer AF, Williamson RS, Linden JF, Sahani M (2016.) Models of neuronal stimulus-response functions: Elaboration, estimation, and evaluation. *Front Syst Neurosci*. 10(109), https://doi.org/10.3389/fnsys.2016.00109.

Miller LM, Escabí MA, Read HL, Schreiner CE (2001). Functional convergence of response properties in the auditory thalamocortical system. *Neuron*. 32:151–160.

Miller LM, Escabí MA, Read HL, Schreiner CE (2002). Spectrotemporal receptive fields in the lemniscal auditory thalamus and cortex. *J Neurophysiol*. 87:516–527.

Montes-Lourido P, Kar M, David SV, Sadagopan S (2021). Neuronal selectivity to complex vocalization features emerges in the superficial layers of primary auditory cortex. *PLoS Biol*. 19:e3001299.

Parras GG, Nieto-Diego J, Carbajal GV, Valdés-Baizabal C, Escera C, Malmierca MS (2017). Neurons along the auditory pathway exhibit a hierarchical organization of prediction error. *Nat Commun*. 8(2148), https://doi.org/10.1038/s41467-017-02038-6.

Ponvert ND, Jaramillo S (2019). Auditory thalamostriatal and corticostriatal pathways convey complementary information about sound features. *J Neurosci*. 39:271–280.

Rabinowitz NC, Willmore BDB, King AJ, Schnupp JWH (2013). Constructing noise-invariant representations of sound in the auditory pathway. *PLoS Biol*. 11:e1001710.

Selezneva E, Oshurkova E, Scheich H, Brosch M (2017). Category-specific neuronal activity in left and right auditory cortex and in medial geniculate body of monkeys. *PLoS ONE*. 12:e0186556.

Solyga M, Barkat TR (2021). Emergence and function of cortical offset responses in sound termination detection. *eLife*. 10:e72240.

Souffi S, Lorenzi C, Huetz C, Edeline JM (2021). Robustness to noise in the auditory system: A distributed and predictable property. *eNeuro*. 8(2), https://doi.org/10.1523/ENEURO.0043-21.2021.

Souffi S, Lorenzi C, Varnet L, Huetz C, Edeline J-M (2020). Noise-sensitive but more precise subcortical representations coexist with robust cortical encoding of natural vocalizations. *J Neurosci*. 40:5228–5246.

Souffi S, Varnet L, Zaidi M, Bathellier B, Huetz C, Edeline J-M (2022). Reduction in sound discrimination in noise is related to envelope similarity and not to a decrease in envelope tracking abilities. *J Physiol (Lond)*. 601(1):123–149.

Suga N (2015). Neural processing of auditory signals in the time domain: Delay-tuned coincidence detectors in the mustached bat. *Hear Res*. 324:19–36.

Suta D, Popelář J, Burianová J, Syka J (2013). Cortical representation of species-specific vocalizations in guinea pig. *PLoS ONE*. 8:e65432.

Ulanovsky N, Las L, Farkas D, Nelken I (2004). Multiple time scales of adaptation in auditory cortex neurons. *J Neurosci*. 24:10440–10453.

Ulanovsky N, Las L, Nelken I (2003). Processing of low-probability sounds by cortical neurons. *Nat Neurosci*. 6:391–398.

Wallace MN, Shackleton TM, Anderson LA, Palmer AR (2005). Representation of the purr call in the guinea pig primary auditory cortex. *Hear Res*. 204:115–126.

Wang X, Lu T, Bendor D, Bartlett E (2008). Neural coding of temporal information in auditory thalamus and cortex. *Neuroscience*. 154:294–303.

Wenstrup JJ, Nataraj K, Sanchez JT (2012). Mechanisms of spectral and temporal integration in the mustached bat inferior colliculus. *Front Neural Circuits*. 6(75), https://doi.org/10.3389/fncir.2012.00075.

Williamson RS, Ahrens MB, Linden JF, Sahani M (2016). Input-specific gain modulation by local sensory context shapes cortical and thalamic responses to complex sounds. *Neuron*. 91:467–481.

Williamson RS, Polley DB (2019). Parallel pathways for sound processing and functional connectivity among layer 5 and 6 auditory corticofugal neurons. *eLife*. 8:e42974.

Xiong C, Liu X, Kong L, Yan J (2020). Thalamic gating contributes to forward suppression in the auditory cortex. *PLoS ONE*. 15:e0236760.

Yan J, Suga N (1996). The midbrain creates and the thalamus sharpens echo-delay tuning for the cortical representation of target-distance information in the mustached bat. *Hear Res*. 93:102–110.

Zurita P, Villa AE, de Ribaupierre Y, de Ribaupierre F, Rouiller EM (1994). Changes of single unit activity in the cat's auditory thalamus and cortex associated to different anesthetic conditions. *Neurosci Res*. 19:303–316.

23.

THE ROLE OF THE THALAMUS FOR HUMAN AUDITORY AND VISUAL SPEECH PERCEPTION

Katharina von Kriegstein and Christa Müller-Axt

In this chapter, we start with a definition of speech perception and an brief overview of cerebral cortex regions that are relevant for speech perception. We then review how the thalamus, its subdivisions, and connection to the cerebral cortex contribute to auditory and visual speech perception.

WHAT IS SPEECH (AND LANGUAGE) PERCEPTION?

Humans communicate with each other largely based on speech. Human speech is produced by the human glottal folds and vocal tract (Figure 23.1A). The sound produced by the glottal folds (Figure 23.1B) is filtered by the vocal tract. The movement of the vocal tract articulators, such as the jaws and tongue, produces different speech sounds (Figure 23.1C). Although speech is often conceived of as an auditory phenomenon, it has a visual counterpart—the visible vocal tract and articulator movements. In cognitive neuroscience, speech perception is often studied in the context of language (Price, 2010). In language, speech sounds are connected according to phonotactical rules to form syllables; syllables are connected to form words with meaning; words are organized according to grammatical rules within sentences.

CEREBRAL CORTEX AND SPEECH RECOGNITION

Some of the first insights on how the brain processes speech and language come from classical neuropsychological studies on patients with speech perception or production impairments due to brain lesions (for review, Rutten, 2022). Current prominent neuroscience models of speech and language perception are based on neuropsychological work, as well as neuroimaging results in patients and healthy controls. A review of these models is beyond the focus of the present chapter. A clear common theme, however, in all these models is that for auditory speech perception, auditory primary and association cortices play a large role (Hickok and Poeppel, 2007; Price, 2012; Bhaya-Grossman and Chang, 2022). In particular, the perisylvian regions, including the planum temporale and posterior superior temporal gyrus and sulcus, are known to be important for processing speech sounds, although their specific functional role in speech sound processing is an open question (Griffiths and Warren, 2002; Hickok and Poeppel, 2007). In addition, motor, somatosensory, and visual cortices that represent articulatory movement are also involved in speech sound perception (Stokes et al., 2019; Franken et al., 2022). Such involvement fits with the view that the brain, in order to process and recognize speech sounds, uses internal generative models of the speaker to simulate and predict the incoming sensory information (von Kriegstein et al., 2008b; van Wassenhove, 2013).

In comparison to auditory speech, much less is known about the processing of visual speech in the brain. Visual speech recognition is based on the visible articulator movements. Visual speech can enable comprehension of what is said when the auditory signal is entirely absent (Campbell, 2008). Visual speech is also likely important in face-to-face conversation with high levels of auditory background noise (Ross et al., 2007) and facilitates comprehension in people with auditory impairments (Auer and Bernstein, 2007). A cerebral cortex area that is specialized in visual speech recognition is the so-called temporal visual speech area (TVSA) in posterior superior temporal gyrus and sulcus (Bernstein et al., 2011; Riedel et al., 2015).

In addition to the cerebral cortex, the thalamus is involved in speech processing and recognition, as will be explored in this chapter.

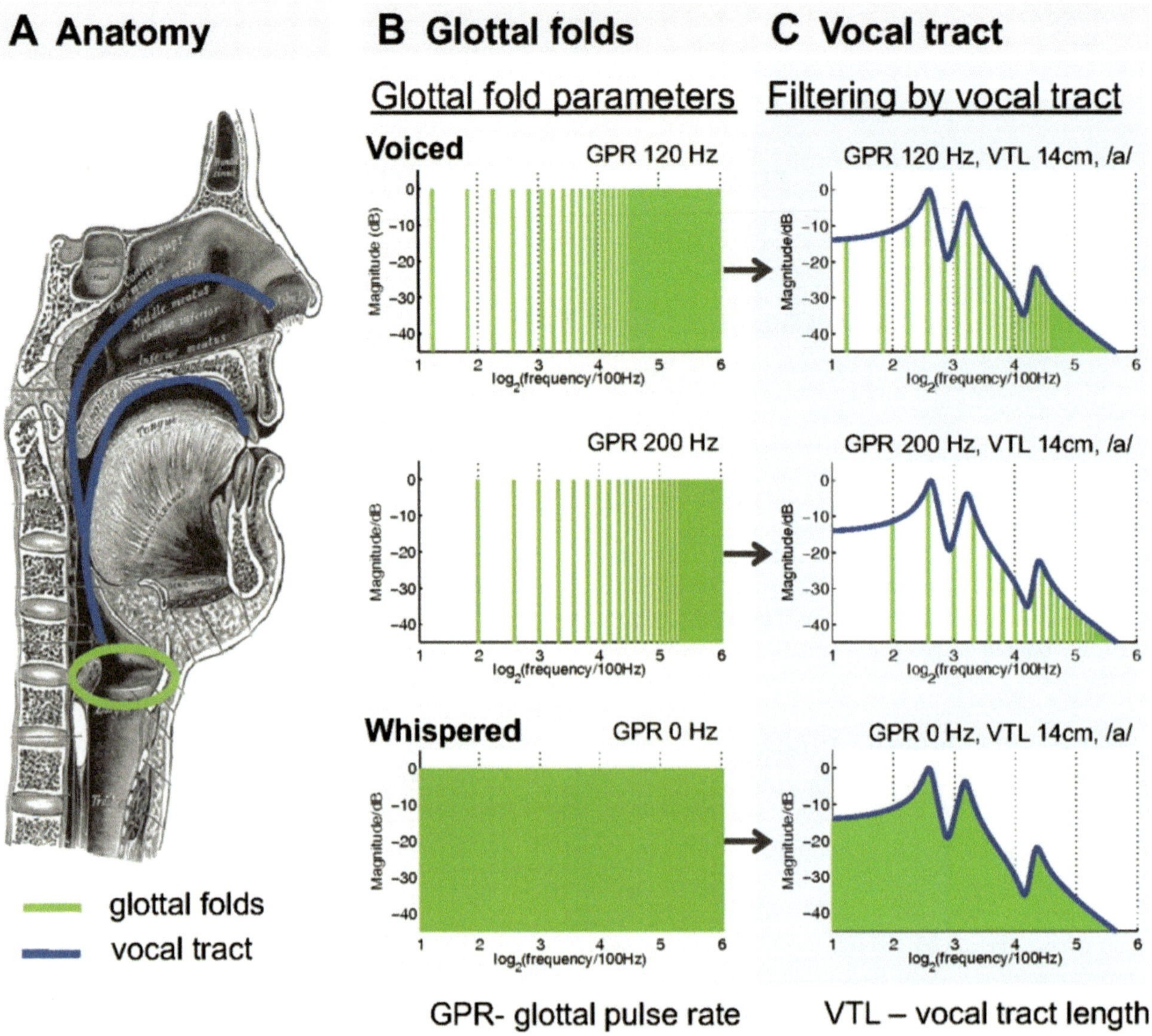

Figure 23.1. (A). Speech is generated by the human glottal folds and vocal tract. The panels show a sagittal section through a human head and neck. Green circle = glottal folds; blue lines = extension of the vocal tract from glottal folds to tip of the nose and lips. (B). The 3 plots show 3 different sounds determined by glottal fold parameters. In voiced speech, the vibration of the glottal folds results in voices with lower (120 Hz GPR; top) or higher fundamental frequency (200 Hz GPR; middle). If glottal folds are constricted, they produce a noise-like sound that is heard as whispered speech (0 Hz GPR; bottom). (C). The vocal tract filters the sound wave coming from the glottal folds, which introduces amplitude peaks at certain frequencies ("formants"; blue lines). The different glottal fold parameters do not influence the formant position. Adapted from von Kriegstein et al. (2010).

THALAMUS AND SPEECH PERCEPTION

THALAMIC APHASIA AND DEEP BRAIN STIMULATION

Aphasia is the difficulty to comprehend or produce language, which can be caused by brain lesions. Aphasia often occurs due to lesions in the cerebral cortex language network. However, focal lesions in the thalamus can also produce aphasia, termed *thalamic aphasia* (Schmahmann, 2003). In a patient sample with focal thalamic lesions (Rangus et al., 2022), 44% received an aphasia diagnosis based on diagnostic criteria requiring impairment in one production and one reception task. Thalamic aphasia symptoms are usually mild, and most patients recover significantly. This could explain why studies involving patients with focal thalamic lesions, but not explicitly testing for aphasia, report lower prevalence of thalamic aphasia (14%; Schaller-Paule et al., 2021). What exact symptoms thalamic aphasia entails is variable across studies (Rangus et al., 2022). For example, Rangus et al. (2022) report that impairments in fluent spontaneous speech and in higher-order language skills, such as complex understanding of speech, are characteristic of focal thalamic lesions. In contrast, Schmahmann (2003) summarizes thalamic aphasia symptoms as consisting of anomia with decreased verbal output and impaired fluency, impairment of comprehension, and fluent paraphasic speech, with preserved repetition. The localization of the focal lesions across studies and patients often differs, and since the diverse thalamic nuclei have a variety of different functions it is likely that thalamic aphasia can present with various symptoms. Thus, it is difficult to come up with a unified account of thalamic aphasia. It would be necessary to associate behavioral symptoms with lesion location in specific nuclei, which has not yet been systematically done. In the case of stroke, symptoms could be related to a specific vascular territory (Schmahmann, 2003). Nevertheless, the existence of thalamic aphasia gives a first insight into the role of thalamic nuclei in speech and language function.

Information gained during the implementation of deep brain stimulation (DBS) also has yielded insight into the involvement of the thalamus in language function (Klostermann and Ehlen, 2013; Krugel et al., 2014; Wang et al., 2022). DBS involves placing electrodes in nuclei located deep inside the brain, such as specific nuclei of the thalamus, to treat neurological or psychiatric disease such as epilepsy. DBS of thalamic targets influences word processing and production (Klostermann and Ehlen, 2013; Krugel et al., 2014; Tiedt et al., 2021; Wang et al., 2022). For example, DBS of the ventral intermediate nucleus, a well-established therapeutic target, leads to simplification of speech, such as the use of simplified syntactical structures or the use of more common words in spontaneous speech (Tiedt et al., 2021). Insights derived from DBS studies on auditory and visual speech perception and recognition, which is the focus of the present chapter, are to our knowledge missing.

NONINVASIVE METHODS FOR INVESTIGATING SPEECH PERCEPTION IN HEALTHY HUMANS IN VIVO

Investigating the human thalamus in healthy humans in vivo comes with considerable technical challenges (Forstmann et al., 2016). However, over the past decade, novel magnetic resonance imaging (MRI) techniques have been developed that now make access to the thalamus in healthy humans in vivo more achievable. In this section we review the difficulties encountered and the recent technological solutions.

There are several reasons for the difficulty of investigating the thalamus in healthy humans in vivo with MRI. First, due to its location at the center of the human brain, there is a great distance of the thalamus from the receiver elements of the radiofrequency coil. The receiver elements of the radiofrequency coil pick up the MR signals. The large distance between thalamus and receiver elements therefore directly translates into a low signal-to-noise ratio (SNR), given that the measurable MR signal falls off with increasing distance from the receiver coil. Second, many thalamic nuclei are small (e.g., the lateral geniculate nucleus comprises about 120 mm^3; Andrews et al., 1997) and tightly packed. This means that the spatial resolution of the MRI has to be maximal to reduce partial volume effects (i.e., receiving a mixed MR signal from different tissue types or nuclei in vicinity to the thalamic nucleus of interest). Third, relative to the cerebral cortex, the thalamus demonstrates a comparatively low contrast-to-noise ratio (CNR) on MR imaging data (Tourdias et al., 2014). This severely compromises the ability to recognize different thalamic nuclei to their full extent and clearly distinguish them from other neighboring intra-thalamic nuclei (Lipin et al., 2021). For example, with conventional MRI it is difficult to clearly dissociate the lateral geniculate nucleus (LGN) from neighboring nuclei, such as the intergeniculate pulvinar and the medial geniculate body (MGB).

Novel high-field MRI methods offer a promising approach to substantially mitigate these problems. This is due to the inherent gain in both signal and image contrast obtained at high field strengths (for review, van der Zwaag et al., 2016). These assets of high-field MRI open up the opportunity of acquiring data with high isotropic image resolutions and enhanced CNR, permitting differentiation between various tissue structures at unprecedented small spatial scales (García-Gomar et al., 2019). Meanwhile, several advances in structural high-field imaging and data processing have made significant contributions to mapping thalamic nuclei in humans in vivo. In terms of acquisition strategies, novel custom-tailored imaging protocols, suppressing or "nulling" signals from gray or white matter, substantially enhance image contrast and the visual detection of individual thalamic nuclei (e.g., Tourdias et al., 2014). In addition (and possibly one of the most exciting acquisition developments), novel quantitative MRI (qMRI) strategies enable direct measurements of the magnetic tissue properties of the entire brain (Ma et al., 2013). Quantitative MRI data permit conclusions to be drawn about the underlying microstructure of the tissue (Edwards et al., 2018) and can be leveraged to generate optimized synthetic MRI contrasts (for review, Ji et al., 2022). These sets of improved contrasts from structural high-field MRI have both facilitated manual segmentations of thalamic nuclei (Tourdias et al., 2014; Müller-Axt et al., 2021) and have paved the way for advanced processing algorithms enabling (semi-)automatic classification of individual thalamic nuclei at high resolution (Liu et al., 2020).

High-field imaging has also proven valuable for functional MRI (fMRI) studies of the human thalamus (e.g., Denison et al., 2014; Moerel et al., 2015; Mihai et al., 2019; Sitek et al., 2019; Tabas et al., 2020). This is due to a higher sensitivity to functional brain responses resulting from a stronger blood oxygen level-dependent (BOLD) contrast at high-field (for review, see van der Zwaag et al., 2016).

AUDITORY SPEECH PERCEPTION AND THE MGB

In animals, invasive methods have led to an advanced understanding of how sounds, including vocalizations, are

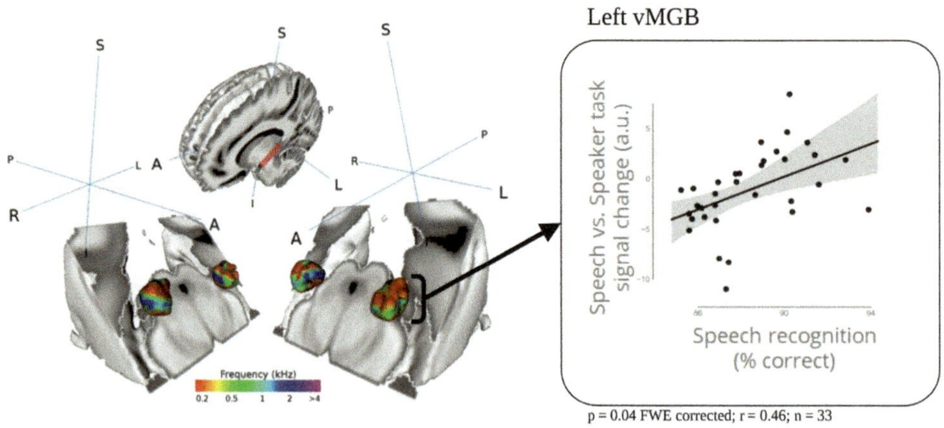

Figure 23.2. Novel developments in high-resolution fMRI reveal a ventrally located tonotopic gradient in the MGB which is attributed to the vMGB. The plot shows that the strength of vMGB responses during a speech recognition task (in contrast to a control task) correlates positively with how well participants recognized speech. Figure adapted from Mihai et al. (2019). R = right; L = left; P = posterior; A = anterior; S = superior; I = inferior; vMGB = ventral division of the medial geniculate body; SVC = small volume correction; FWE = family-wise-error corrected.

represented in the cerebral cortex and sensory pathways, which has also a long history of taking thalamus processing into account (see **Chapter 22 in Section V**). In humans, interest in speech and language research on thalamus is more recent. Neuroimaging of thalamic contribution to speech perception has focused by-and-large on the MGB. As in other species, in humans the MGB consists of 3 divisions (Winer, 1984; Kiwitz et al., 2022): the ventral MGB (MGBv), the dorsal MGB (MGBd), and the medial MGB (MGBm). In analogy to findings in animals, the MGBv in humans is considered to be the primary part of the auditory thalamus based on post-mortem investigations (Winer, 1984). Using the recently developed high-field approaches (see below), functional neuroimaging studies in humans have revealed that the MGB contains 2 tonotopic maps, of which the ventral map is attributed to the MGBv (Moerel et al., 2015; Mihai et al., 2019).

In animals, neuronal representations in the MGB reflect the basic spectrotemporal features of the sound, and there seems to be no special status of calls in contrast to other sounds with spectrotemporal features. There are, however, indications that responses to species-specific calls are often higher than their time-reversed versions, even though the spectrotemporal features are very similar (Suta et al., 2007). (The representation of speech and communication sounds in animals is reviewed in Bartlett, 2013, and in **Chapter 22 in Section V** of this volume.) Most of this knowledge was gained from anesthetized or awake animals with no specific tasks involved. Human studies of speech representation in the MGB are rare. A commonly used method for investigating speech processing in the subcortical auditory pathway involves the recording of auditory brainstem responses, which is predominantly used in experimental paradigms in which participants passively listen to speech. Brainstem recordings mostly reflect processes between the cochlear nucleus and the inferior colliculus (for a recent review, see Chandrasekaran, 2022) and therefore have little relevance for understanding thalamic involvement in speech perception. Nevertheless, wave VI of the brainstem responses seem to reflect processing in the MGB (Hashimoto et al., 1981), but dedicated research on the relation of wave VI to speech processing is, to our knowledge, missing.

fMRI studies in humans, however, show that tasks can have significant influences on the responses in MGB and are relevant particularly for speech processing (von Kriegstein et al., 2008a; Diaz et al., 2012; Mihai et al., 2019). It is well known that there are strong corticofugal projections to the auditory thalamus (Winer, 2006; reviewed in **Chapter 24 in Section V** of this volume), and it has become widely accepted that the MGB is modulated by cognitive demands and is involved in active information regulation. In the case of speech, fMRI studies revealed a task-dependent modulation in the MGB during auditory speech recognition (von Kriegstein et al., 2008a; Diaz et al., 2012). There were 4 key findings from these studies. First, bilateral MGB showed significantly higher responses to an auditory speech-recognition task than to control tasks, independent of attentional load or task-difficulty (von Kriegstein et al., 2008a; Diaz et al., 2012). The control tasks focused on speaker identity or loudness level and were performed on exactly the same stimuli as the speech tasks. Second, the performance level in the auditory speech recognition task was significantly correlated with task-dependent modulation in the MGB of the left hemisphere across participants (Mihai et al., 2019) (Figure 23.2). Third, the task-dependent modulation is not only present in higher-order parts of the MGB, but also in MGBv (Mihai et al., 2019) (Figure 23.2). Fourth, the task-dependent modulation in left MGB is prominent for recognition of speech in background noise (Mihai et al., 2021). Why would the human brain use task-dependent modulation of the sensory thalamus for speech? Before we provide a potential mechanistic explanation for the task-dependent modulation for speech, we review similar findings on task-dependent modulation for speech in the visual sensory thalamus.

VISUAL SPEECH PERCEPTION AND LGN

In the visual thalamus, the LGN, BOLD responses are modulated by attention to visual stimuli, such as flickering checkerboards or moving dots. For example, LGN responses increased when participants attended to moving white dots, in contrast to stationary colored dots, in a dot display (Schneider, 2011). In addition, task-dependent modulation for speech also occurs in the LGN (Díaz et al., 2018). Two fMRI studies showed that recognizing visual speech in contrast to control tasks led to higher responses in the LGN. The modulation seemed to be specific to speech, because a task that recognized finger movements on a mobile phone in contrast to the identity of the phone led to no significant task-dependent modulation. The study had too low spatial resolution to investigate which part of LGN is involved in the task-dependent modulation for speech. Since visual speech is a motion signal, it might particularly activate those regions of the LGN that are specialized for motion processing (i.e., the magnocellular layers of the LGN [LGN-M]; Denison et al., 2014). However, there is to date no published study that shows that this is indeed the case. Although it is currently impossible to resolve different layers of the LGN using conventional MRI, recent developments in MR techniques (see below) make subsection imaging of the LGN feasible both with functional (Denison et al., 2014) as well as structural (Müller-Axt et al., 2021) MR approaches. Denison et al. (2014) mapped the LGN M and P subdivisions in individual human subjects using stimuli that exploit the distinct functional response properties of M and P neurons. Müller-Axt et al. (2021) discovered that M and P LGN subsections can be dissociated based on the myelin-content using qMRI. In contrast to fMRI, structural qMRI (1) is characterized by a higher signal-to-noise ratio, allowing for acquisitions with higher spatial resolutions; (2) generally is less taxing for participants as no lengthy functional design is required; and (3) allows conclusions to be drawn about the underlying microstructure, such as tissue myelination (Müller-Axt et al., 2021). Thus, recent developments have provided techniques with which it is possible to investigate the role of LGN subsections in speech perception in healthy humans in vivo with unprecedented spatial resolution.

MECHANISTIC EXPLANATION OF TASK-DEPENDENT MODULATION IN SPEECH

Findings of corticofugal influences on processing in sensory thalamus have been explained by the view that sensory thalamic nuclei gate sensory information according to attentional demands (for review, see Saalmann and Kastner, 2011). However, the task-dependent modulation for speech in the auditory and visual modalities (see below) is difficult to explain by attentional mechanisms, as attention is required for both tasks (i.e., the speech task and the control tasks). In addition, the task-dependent modulation occurs independently of task-difficulty: Task-dependent modulation for speech has been found in conditions where the control task is more difficult than the speech task and also when the speech task has the same or higher difficulty than the control task. This suggests that attentional modulation due to, for example, higher task demands is very unlikely to explain the task-dependent modulation for speech in the MGB and LGN. Our lab has put forward a qualitative model to explain sensory thalamic involvement in speech perception (von Kriegstein et al., 2008a). We call this model dynamic predictive coding model (DPC-model) of corticothalamic interactions (Figure 23.3). It is to date a qualitative model, but its composition lends itself to translation to a computational model.

The DPC-model has 3 basic ingredients. First, it rests heavily on predictive processing within the Bayesian brain hypothesis (see also **Chapter 41 in Section VII and Chapter 64 in Section XI**). Predictive processing is a leading theoretical framework for sensory processing. In the predictive processing view, the brain constantly generates predictions of the sensory world, and perception emerges from the comparison between these predictions and the actual sensory input. Although predictive processing has been formulated for cerebral cortex pathways, there is good evidence that subcortical sensory pathway structures, including the sensory thalamus, are also involved (Asilador and Llano, 2020; Tabas et al., 2020; see also **Chapter 21 in Section V**). A similar concept was mentioned briefly as "hypothesis-testing" by Sillito et al. (1994). The second ingredient for the DPC model is the brain's hierarchical organization that features different timescales. It is assumed that levels closer to the sensory input encode faster dynamics of the stimulus than levels further away from the sensory input (Kiebel et al., 2008). In accordance with this view, the MGB is tuned to higher frequencies of temporal modulation in relation to auditory cortex (Giraud et al., 2000) (see also **Chapter 22 in Section V**). The MGB and LGN are therefore particularly suited to represent speech stimuli at high temporal fidelity. The third ingredient is based on the findings in nonhuman animals that show that thalamic receptive fields change dynamically in response to inputs from the cerebral cortex (Sillito et al., 1994; Hasse and Briggs, 2017; see also **Chapter 24 in Section V**).

How do these 3 ingredients fit together? For humans, the optimized encoding of the relatively fast dynamics of articulatory movements (i.e., auditory and visual speech)

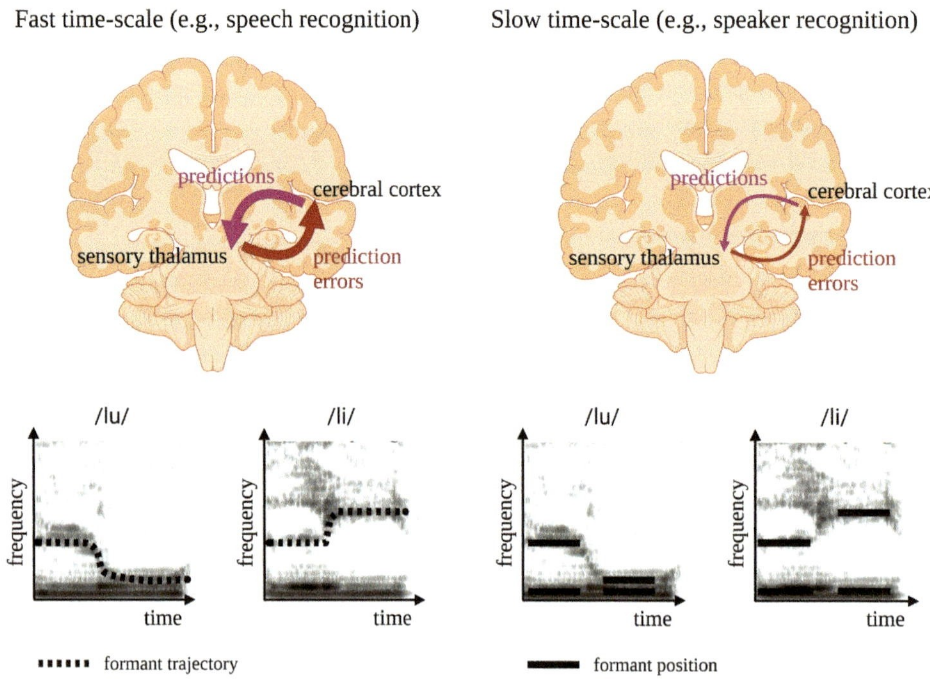

Figure 23.3. Schematic of the dynamic predictive coding (DPC) model for cortico-thalamic interactions in speech perception. In the DPC view, cortico-thalamic interaction can be explained by a predictive coding scheme, where the cortex dynamically modulates the sensory thalamus based on predictions about the incoming sensory information. This means that for fast, dynamically changing, predictable stimuli, such as speech sounds, predictions from the cerebral cortex are used to tune processing in the sensory thalamus in a dynamic iterative fashion to predict the incoming sensory input on a millisecond by millisecond time frame. In contrast, for slower varying information, such as information about the speaker identity, the iterative processes are much slower (indicated by thinner arrows in the figure). Both situations imply that receptive field properties of sensory thalamus neurons changed over time in accordance with the predictions derived from the cerebral cortex. Part of the figure was created using BioRender.com.

are critical for speech recognition and communication (Shannon et al., 1995). In the DPC model, the brain contains an internal generative model of the speech stream with slower dynamics encoded by auditory and visual cortex that provides predictions about input arriving at lower levels of the processing hierarchy (von Kriegstein et al., 2008a). These dynamic predictions modulate the response properties of the first-order sensory thalamus neurons to optimize the early stages of speech recognition. The corticothalamic interactions are thought to occur on a millisecond-by-millisecond timescale in a dynamic, iterative fashion (i.e., in recurrent cortical-subcortical dynamics). In speech recognition, speech sounds need to be represented at high fidelity, in contrast to other tasks required for communication, such as identifying the voice of the speaker. For example, the differences between the 2 syllables presented in Figure 23.3 are largely due to different formant transitions that occur at millisecond timescales. In contrast, recognizing speakers just requires computing summary statistics across the speech signals. Thus, dynamic predictive coding of sensory thalamus responses might have a particularly high frequency when processing such fast-varying and predictable stimuli as speech. We speculate that this mechanism is the reason for the observed task-dependent modulation for speech in the sensory thalamus (see below).

If the DPC model is correct, one would assume that responses in the sensory thalamus can be explained by predictive coding. For the auditory modality, there are good indications that this is indeed the case from experiments in animals (reviewed in Asilador and Llano, 2020; Antunes and Malmierca, 2021) as well as in humans (reviewed in Tabas and von Kriegstein, 2021). Most of the studies investigating predictive coding include simple sounds such as sinusoids. However, more complex sounds, such as frequency sweeps, which are a basic constituent of speech, are also processed in the MGB in a manner that is congruent with a predictive coding mechanism (Tabas et al., 2021). A more direct test for the involvement of predictive coding in speech processing in the sensory thalamus is, however, missing.

Whether the LGN uses a similar predictive coding mechanism is to date relatively unexplored. However, computational models are consistent with a predictive coding architecture between LGN and V1 (Jehee and Ballard, 2009; Lian et al., 2019).

DEVELOPMENTAL DYSLEXIA AS A MODEL OF SENSORY THALAMUS ALTERATIONS

The involvement of the sensory thalamus in speech recognition also has clinical relevance. Developmental dyslexia (DD) is the most common learning disorder, affecting millions of children and adults worldwide (Shaywitz et al., 2008). The reading impairments characteristic of DD are most often diagnosed in the first years of school, but persist into adulthood. Individuals with DD commonly struggle in the educational system and often develop secondary psychological symptoms, such as depression or anxiety (Klassen et al., 2013). One of the key behavioral findings in DD is that most people with DD struggle with phonological tasks (i.e., tasks that involve processing of speech sounds [phonemes]). Similarly, there is evidence that recognizing speech from visual face information is impaired in DD (van Laarhoven et al., 2018).

Most neuroscientific accounts associate DD with dysfunction in those cerebral cortex areas that are unique to human speech processing and writing function (Ozernov-Palchik and Gaab, 2016). This view, however, leaves unexplained why DD is also associated with alterations in sensory thalamic nuclei (Figure 23.4). In the past century, 2 studies on 5 post-mortem brains have shown that DD is associated with histological alterations in the magnocellular part of the LGN and the MGB (Livingstone et al., 1991; Galaburda et al., 1994). Furthermore, in vivo findings in adult dyslexics point to a dysfunction of sensory thalamic nuclei and their connections to cerebral cortex (Diaz et al., 2012; Müller-Axt et al., 2017; Tschentscher et al., 2019). Functional MRI has shown that the task-dependent modulation of speech sounds in the left MGB is reduced in adults with DD in comparison to matched controls (Diaz et al., 2012). In the same participant group, diffusion-weighted MRI revealed that, compared to controls, individuals with DD showed a reduction in white matter connectivity between left MGB and the left motion-sensitive planum temporale—a cortical region that processes auditory motion (Tschentscher et al., 2019). Similarly, there is a reduction of white matter connectivity in DD compared to controls between left LGN and the left V5/MT—a cortical region that processes visual motion. In all 3 studies (Diaz et al., 2012; Müller-Axt et al., 2017; Tschentscher et al., 2019), associations were found with a key dyslexia diagnostic score (i.e., the ability to rapidly name letters and numbers, RANln). To date it is unclear at what stage in development corticothalamic alterations in DD arise. A study testing children at 2 time points (5–6 y and 8–9 y age) did not find alterations in MGB or LGN in DD relative to controls (Kuhl et al., 2020). One reason for this could be that the DD and control groups in that study did not differ in RANln abilities. Another possibility is that sensory thalamus alterations only occur later in development.

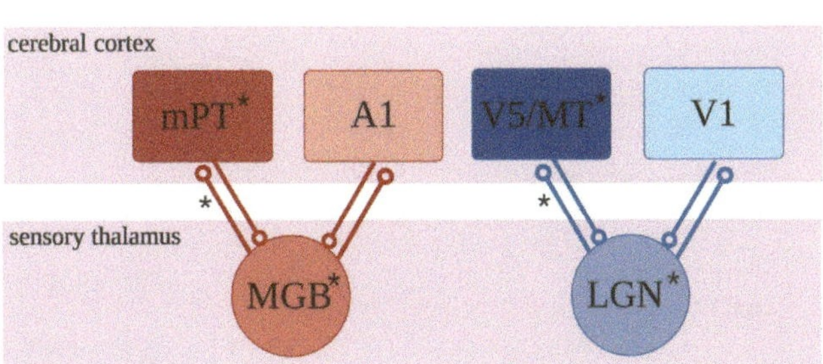

Figure 23.4. Overview of alterations found in developmental dyslexia (DD). mPT = motion-sensitive auditory cortex; A1 = primary auditory cortex; V5/hMT = motion-sensitive visual cortex; MGB = medial geniculate body; LGN = lateral geniculate nucleus. The figure was created using BioRender.com.

At the time of the first reports of histological alterations in DD, the sensory thalamus nuclei were viewed as relay stations, which implies sending sensory information from the periphery to the cerebral cortex in a "bottom-up" fashion. This view is also reflected in an influential theory of DD, the magnocellular theory, which was developed at that time. A key assumption of the magnocellular theory is that DD symptoms can be explained by deficits in processing temporal information in the magnocellular system of the human brain. However, there is now evidence that sensory thalamus alterations in DD are subtler than assumed by the magnocellular theory. For example, alterations in the left MGB in DD were only found when the task emphasized speech recognition, but not for passive listening to speech or for other tasks (Díaz et al., 2012). The DPC model proposes that sensory thalamus alterations in DD can be explained by a "top-down" mechanism: individuals with DD make less use of the predictive information that is required to optimize processing in sensory thalamus for fast-varying, human communication stimuli such as speech. This means that instead of a generic bottom-up deficit, as postulated by the magnocellular theory, dyslexics have a specific "top-down" deficit, such as a reduced precision in the predictions derived from mPT and V5/MT. The DPC model explanation of dyslexia symptoms integrates well with theories of dyslexia that focus on predictive coding or lateralization in the cerebral cortex (Lehongre et al., 2011; Jaffe-Dax et al., 2015).

CONCLUSION

In the past, research on human speech perception has been dominated by work on the cerebral cortex. This has resulted in refined knowledge of the cerebral cortex regions involved in processing auditory and visual speech. Research on how sensory thalamic areas contribute to speech perception in humans is in its infancy. Many general properties of representations of speech sounds in the thalamus will likely be gained by extrapolating from animal work (see **Chapter 22 in Section V**). There are, however, functions that can only be investigated in humans, such as the versatile recognition of sounds and articulatory movements for speech recognition, and the alterations of sensory thalamus representations of speech sounds during language learning. Fortunately, methodological advances make it now possible to investigate sensory thalamic nuclei, their subsections, and their interaction with the cerebral cortex with more reliability and resolution than was previously possible. Most important for this future research are the advances in MRI, including the higher spatial resolution that is now possible, as well as assessment of tissue properties. In addition, recent developments in transcranial focused ultrasound stimulation (tFUS) offer the possibility to selectively target thalamic nuclei, which could be used in the future to assess the role of thalamic nuclei in speech processing (Legon et al., 2018). We hope that these technological developments will generate an in-depth understanding of corticothalamic interactions in speech processing in primary sensory as well as higher-level thalamus in the decades to come. This research will not only be important for our basic science understanding of brain function, but also likely for a mechanistic understanding of clinical disorders, such as developmental dyslexia.

NOTE

The section 'Non-invasive methods for investigating speech perception in healthy humans in-vivo' will also appear in the forthcoming PhD thesis by C.Müller-Axt.

REFERENCES

Andrews TJ, Halpern SD, Purves D (1997). Correlated size variations in human visual cortex, lateral geniculate nucleus, and optic tract. *J Neurosci*. 17:2859–2868.

Antunes FM, Malmierca MS (2021). Corticothalamic pathways in auditory processing: Recent advances and insights from other sensory systems. *Front Neural Circuits*. 15:721186.

Asilador A, Llano DA (2020). Top-down inference in the auditory system: Potential roles for corticofugal projections. *Front Neural Circuits*. 14:615259.

Auer ET, Bernstein LE (2007). Enhanced visual speech perception in individuals with early-onset hearing impairment. *J Speech Lang Hear Res*. 50:1157–1165.

Bartlett EL (2013). The organization and physiology of the auditory thalamus and its role in processing acoustic features important for speech perception. *Brain Lang*. 126:29–48.

Bernstein LE, Jiang J, Pantazis D, Lu ZL, Joshi A (2011). Visual phonetic processing localized using speech and nonspeech face gestures in video and point-light displays. *Hum Brain Mapp*. 32:1660–1676.

Bhaya-Grossman I, Chang EF (2022). Speech computations of the human superior temporal gyrus. *Annu Rev Psychol*. 73:79–102.

Campbell R (2008). The processing of audio-visual speech: Empirical and neural bases. *Philos Trans R Soc Lond B Biol Sci*. 363:1001–1010.

Chandrasekaran BT, Tessmer R, Nike Gnanateja G (2022). Subcortical processing of speech sounds. In: Holt LL, Peele JE, Coffin AB, Popper AN, Fay RR, eds. *Speech Perception*. Springer; 13–44.

Denison RN, Vu AT, Yacoub E, Feinberg DA, Silver MA (2014). Functional mapping of the magnocellular and parvocellular subdivisions of human LGN. *Neuroimage*. 102(Pt 2):358–369.

Díaz B, Blank H, von Kriegstein K (2018). Task-dependent modulation of the visual sensory thalamus assists visual-speech recognition. *Neuroimage*. 178:721–734.

Diaz B, Hintz F, Kiebel SJ, von Kriegstein K (2012). Dysfunction of the auditory thalamus in developmental dyslexia. *Proc Natl Acad Sci U S A*. 109:13841–13846.

Edwards LJ, Kirilina E, Mohammadi S, Weiskopf N (2018). Microstructural imaging of human neocortex in vivo. *Neuroimage*. 182:184–206.

Forstmann BU, de Hollander G, van Maanen L, Alkemade A, Keuken MC (2016). Towards a mechanistic understanding of the human subcortex. *Nat Rev Neurosci*. 18:57–65.

Franken MK, Liu BC, Ostry DJ (2022). Towards a somatosensory theory of speech perception. *J Neurophysiol*. 128:1683–1695.

Galaburda AM, Menard MT, Rosen GD (1994). Evidence for aberrant auditory anatomy in developmental dyslexia. *Proc Natl Acad Sci U S A*. 91:8010–8013.

García-Gomar MG, Strong C, Toschi N, Singh K, Rosen BR, Wald LL, Bianciardi M (2019). In vivo probabilistic structural atlas of the inferior and superior colliculi, medial and lateral geniculate nuclei and superior olivary complex in humans based on 7 Tesla MRI. *Front Neurosci*. 13:764.

Giraud AL, Lorenzi C, Ashburner J, Wable J, Johnsrude I, Frackowiak R, Kleinschmidt A (2000). Representation of the temporal envelope of sounds in the human brain. *J Neurophysiol*. 84:1588–1598.

Griffiths TD, Warren JD (2002). The planum temporale as a computational hub. *Trends Neurosci*. 25:348–353.

Hashimoto I, Ishiyama Y, Yoshimoto T, Nemoto S (1981). Brain-stem auditory-evoked potentials recorded directly from human brainstem and thalamus. *Brain*. 104:841–859.

Hasse JM, Briggs F (2017). A cross-species comparison of corticogeniculate structure and function. *Vis Neurosci*. 34:E016.

Hickok G, Poeppel D (2007). The cortical organization of speech processing. *Nat Rev Neurosci*. 8:393–402.

Jaffe-Dax S, Raviv O, Jacoby N, Loewenstein Y, Ahissar M (2015). A computational model of implicit memory captures dyslexics' perceptual deficits. *J Neurosci*. 35:12116–12126.

Jehee JF, Ballard DH (2009). Predictive feedback can account for biphasic responses in the lateral geniculate nucleus. *PLoS Comput Biol*. 5:e1000373.

Ji S, Yang D, Lee J, Choi SH, Kim H, Kang KM (2022). Synthetic MRI: Technologies and applications in neuroradiology. *JMRI*. 55:1013–1025.

Kiebel SJ, Daunizeau J, Friston KJ (2008). A hierarchy of time-scales and the brain. *PLoS Comput Biol*. 4:e1000209.

Kiwitz K, Brandstetter A, Schiffer C, Bludau S, Mohlberg H, Omidyeganeh M, Massicotte P, Amunts K (2022). Cytoarchitectonic maps of the human metathalamus in 3D space. *Front Neuroanat*. 16:837485.

Klassen RM, Tze VM, Hannok W (2013). Internalizing problems of adults with learning disabilities: A meta-analysis. *J Learn Disabil*. 46:317–327.

Klostermann F, Ehlen F (2013). Functional roles of the thalamus for language capacities. *Front Syst Neurosci*. 7:32.

Krugel LK, Ehlen F, Tiedt HO, Kuhn AA, Klostermann F (2014). Differential impact of thalamic versus subthalamic deep brain stimulation on lexical processing. *Neuropsychologia*. 63:175–184.

Kuhl U, Neef NE, Kraft I, Schaadt G, Dörr L, Brauer J, Czepezauer I, Müller B, Wilcke A, Kirsten H, Emmrich F, Boltze J, Friederici AD, Skeide MA (2020). The emergence of dyslexia in the developing brain. *Neuroimage.* 211:116633.

Legon W, Ai L, Bansal P, Mueller JK (2018). Neuromodulation with single-element transcranial focused ultrasound in human thalamus. *Hum Brain Mapp.* 39:1995–2006.

Lehongre K, Ramus F, Villiermet N, Schwartz D, Giraud AL (2011). Altered low-γ sampling in auditory cortex accounts for the three main facets of dyslexia. *Neuron.* 72:1080–1090.

Lian Y, Grayden DB, Kameneva T, Meffin H, Burkitt AN (2019). Toward a biologically plausible model of LGN-V1 pathways based on efficient coding. *Front Neural Circ.* 13:13.

Lipin M, Bennett J, Ying GS, Yu Y, Ashtari M (2021). Improving the quantification of the lateral geniculate nucleus in magnetic resonance imaging using a novel 3D-edge enhancement technique. *Front Comput Neurosci.* 15:708866.

Liu Y, D'Haese PF, Newton AT, Dawant BM (2020). Generation of human thalamus atlases from 7 T data and application to intrathalamic nuclei segmentation in clinical 3 T T1-weighted images. *Magn Reson Imaging.* 65:114–128.

Livingstone MS, Rosen GD, Drislane FW, Galaburda AM (1991). Physiological and anatomical evidence for a magnocellular defect in developmental dyslexia. *Proc Natl Acad Sci U S A.* 88:7943–7947.

Ma D, Gulani V, Seiberlich N, Liu K, Sunshine JL, Duerk JL, Griswold MA (2013). Magnetic resonance fingerprinting. *Nature.* 495:187–192.

Mihai PG, Moerel M, de Martino F, Trampel R, Kiebel S, von Kriegstein K (2019). Modulation of tonotopic ventral medial geniculate body is behaviorally relevant for speech recognition. *eLife.* 8:e44837.

Mihai PG, Tschentscher N, Von Kriegstein K (2021). Modulation of the primary auditory thalamus when recognizing speech with background noise. *J Neurosci.* 41:7136–7147.

Moerel M, De Martino F, Ugurbil K, Yacoub E, Formisano E (2015). Processing of frequency and location in human subcortical auditory structures. *Sci Rep.* 5:17048.

Müller-Axt C, Anwander A, von Kriegstein K (2017). Altered structural connectivity of the left visual thalamus in developmental dyslexia. *Curr Biol.* 27:3692–3698.

Müller-Axt C, Eichner C, Rusch H, Kauffmann L, Bazin PL, Anwander A, Morawski M, von Kriegstein K (2021). Mapping the human lateral geniculate nucleus and its cytoarchitectonic subdivisions using quantitative MRI. *Neuroimage.* 244:118559.

Ozernov-Palchik O, Gaab N (2016). Tackling the "dyslexia paradox": Reading brain and behavior for early markers of developmental dyslexia. *Wiley Interdiscip Rev Cogn Sci.* 7:156–176.

Price CJ (2010). The anatomy of language: A review of 100 fMRI studies published in 2009. *Ann N Y Acad Sci.* 1191:62–88.

Price CJ (2012). A review and synthesis of the first 20 years of PET and fMRI studies of heard speech, spoken language and reading. *Neuroimage.* 62:816–847.

Rangus I, Fritsch M, Endres M, Udke B, Nolte CH (2022). Frequency and phenotype of thalamic aphasia. *J Neurol.* 269:368–376.

Riedel P, Ragert P, Schelinski S, Kiebel SJ, von Kriegstein K (2015). Visual face-movement sensitive cortex is relevant for auditory-only speech recognition. *Cortex.* 68:86–99.

Ross LA, Saint-Amour D, Leavitt VM, Javitt DC, Foxe JJ (2007). Do you see what i am saying? Exploring visual enhancement of speech comprehension in noisy environments. *Cereb Cortex.* 17:1147–1153.

Rutten G-J (2022). Broca-Wernicke theories: A historical perspective. In: Hillis AE, Fridriksson J, eds. *Handbook of Clinical Neurology.* Elsevier:25–34.

Saalmann YB, Kastner S (2011). Cognitive and perceptual functions of the visual thalamus. *Neuron.* 71:209–223.

Schaller-Paule MA, Oeckel AM, Schüre J-R, Keil F, Hattingen E, Foerch C, Rauch M (2021). Isolated thalamic stroke: Analysis of clinical characteristics and asymmetry of lesion distribution in a retrospective cohort study. *Neurol Res Pract.* 3:49.

Schmahmann JD (2003). Vascular syndromes of the thalamus. *Stroke.* 34:2264–2278.

Schneider KA (2011). Subcortical mechanisms of feature-based attention. *J Neurosci.* 31:8643–8653.

Shannon RV, Zeng FG, Kamath V, Wygonski J, Ekelid M (1995). Speech recognition with primarily temporal cues. *Science.* 270:303–304.

Shaywitz SE, Morris R, Shaywitz BA (2008). The education of dyslexic children from childhood to young adulthood. *Annu Rev Psychol.* 59:451–475.

Sillito AM, Jones HE, Gerstein GL, West DC (1994). Feature-linked synchronization of thalamic relay cell firing induced by feedback from the visual cortex. *Nature.* 369:479–482.

Sitek KR, Gulban OF, Calabrese E, Johnson GA, Lage-Castellanos A, Moerel M, Ghosh SS, De Martino F (2019). Mapping the human subcortical auditory system using histology, postmortem MRI and in vivo MRI at 7T. *eLife.* 8:e48932.

Stokes RC, Venezia JH, Hickok G (2019). The motor system's [modest] contribution to speech perception. *Psychon Bull Rev.* 26:1354–1366.

Suta D, Popelar J, Kvasnak E, Syka J (2007). Representation of species-specific vocalizations in the medial geniculate body of the guinea pig. *Exp Brain Res.* 183:377–388.

Tabas A, Kiebel S, Marxen M, von Kriegstein K (2021). Fast frequency modulation is encoded according to the listener expectations in the human subcortical auditory pathway. https://arxiv.org/abs/2108.02066.

Tabas A, Mihai G, Kiebel S, Trampel R, von Kriegstein K (2020). Abstract rules drive adaptation in the subcortical sensory pathway. *eLife.* 9:1–19.

Tabas A, von Kriegstein K (2021). Adjudicating between local and global architectures of predictive processing in the subcortical auditory pathway. *Front Neural Circ.* 15:644743.

Tiedt HO, Ehlen F, Wyrobnik M, Klostermann F (2021). Thalamic but not subthalamic neuromodulation simplifies word use in spontaneous language. *Front Hum Neurosci.* 15:656188.

Tourdias T, Saranathan M, Levesque IR, Su J, Rutt BK (2014). Visualization of intra-thalamic nuclei with optimized white-matter-nulled MPRAGE at 7T. *Neuroimage.* 84:534–545.

Tschentscher N, Ruisinger A, Blank H, Diaz B, von Kriegstein K (2019). Reduced structural connectivity between left auditory thalamus and the motion-sensitive planum temporale in developmental dyslexia. *J Neurosci.* 39:1720–1732.

van der Zwaag W, Schäfer A, Marques JP, Turner R, Trampel R (2016). Recent applications of UHF-MRI in the study of human brain function and structure: A review. *NMR Biomed.* 29:1274–1288.

van Laarhoven T, Keetels M, Schakel L, Vroomen J (2018). Audio-visual speech in noise perception in dyslexia. *Dev Sci.* 21:e12504.

van Wassenhove V (2013). Speech through ears and eyes: interfacing the senses with the supramodal brain. *Front Psychol.* 4:388.

von Kriegstein K, Patterson RD, Griffiths TD (2008a). Task-dependent modulation of medial geniculate body is behaviorally relevant for speech recognition. *Curr Biol.* 18:1855–1859.

von Kriegstein K, Dogan O, Grüter M, Giraud AL, Kell CA, Grüter T, Kleinschmidt A, Kiebel SJ (2008b). Simulation of talking faces in the human brain improves auditory speech recognition. *Proc Natl Acad Sci U S A.* 105:6747–6752.

von Kriegstein K, Smith DRR, Patterson RD, Kiebel SJ, Griffiths TD (2010). How the human brain recognizes speech in the context of changing speakers. *J Neurosci.* 30:629–638.

Wang D, Lipski WJ, Bush A, Chrabaszcz A, Dastolfo-Hromack CA, Dickey M, Fiez JA, Richardson RM (2022). Lateralized and region-specific thalamic processing of lexical status during reading aloud. *J Neurosci.* 42:3228.

Winer J (1984). The human medial geniculate body. *Hear Res.* 15:225–247.

Winer JA (2006). Decoding the auditory corticofugal systems. *Hear Res.* 212:1–8.

24.

CORTICOTHALAMIC INTERACTIONS IN THE AUDITORY SYSTEM

Victoria M. Bajo and Andrew J. King

INTRODUCTION

Information about the auditory world is transmitted along a series of hierarchically organized structures, beginning with the peripheral receptor cells in the cochlea of the inner ear and passing via nuclei in the brainstem, midbrain, and thalamus to the auditory cortex. As in other sensory systems, descending projections from pyramidal neurons located in layers 5 and 6 enable the cortex to communicate with the thalamus and other subcortical sites. In the auditory system, these descending projections are particularly extensive since they target every level of the ascending pathway. As well as connecting each auditory cortical field with the region of the medial geniculate body (MGB) from which it receives most of its thalamic input, non-reciprocal corticofugal projections reach a number of subcortical auditory nuclei, including other parts of the MGB, the inferior colliculus (IC) in the midbrain, and the superior olivary complex and cochlear nucleus in the brainstem (Winer, 2006). Furthermore, additional descending pathways originate from these subcortical structures, including the olivocochlear bundle, which modulates the activity of the cochlear hair cells and the primary afferent axons in the auditory nerve. This massive network of descending connections therefore has the potential to regulate information processing throughout the auditory system, all the way down to the biomechanical properties of the cochlea, as well as in non-auditory brain regions such as the striatum.

Our understanding of the functions of the various descending pathways that have been defined anatomically in the auditory system is still quite limited, with a wide range of possibilities proposed (Winer, 2006; Antunes and Malmierca, 2021; Souffi et al., 2021). Identifying the specific contributions of auditory corticothalamic pathways is made more challenging by the existence of cortical inputs to earlier levels of auditory hierarchy. This means that in addition to influencing sensory processing in the thalamus via its direct descending inputs, the auditory cortex can potentially shape the ascending inputs to the MGB. Nevertheless, investigation of the anatomical and physiological properties of corticothalamic neurons and of the way they influence the activity and response properties of neurons in the MGB has provided a number of insights into their role in auditory processing. In the course of these studies, it has become increasingly apparent that corticothalamic modulation most likely operates in challenging listening conditions (Souffi et al., 2021), highlighting the importance of investigating these circuits in behaving animals.

ORGANIZATION OF CORTICOTHALAMIC PATHWAYS

The organization of the corticothalamic circuits follows a common pattern in the somatosensory, visual, and auditory systems (Usrey and Sherman, 2019) (see **Chapters 19 and 27 in Section V** for a discussion of visual and somatosensory corticothalamic pathways, respectively). In each modality, the descending axons of layer 6 neurons primarily connect cortical areas with the area of the sensory thalamus from which they receive their ascending inputs, thereby forming reciprocal feedback projections that are thought to have a modulatory influence on thalamic processing. By contrast, cortical layer 5 "driver" neurons make predominantly non-reciprocal connections with higher-order thalamic nuclei, which may facilitate communication between different cortical areas (Usrey and Sherman, 2019). (For a more general description of the properties of these descending cortical projection neurons, readers are referred to **Chapters 14 and 15 in Section IV**.)

In contrast to the lateral geniculate nucleus, which is a first-order thalamic nucleus that relays visual information from the retina mainly to the primary visual cortex, from where it is distributed to higher-level cortical areas for further processing, the MGB comprises both first-order and higher-order nuclei. Consequently, auditory information is conveyed from the different subdivisions of the MGB to the primary and non-primary auditory fields in the temporal cortex via a series of parallel channels. In keeping with the general plan for the organization of corticothalamic pathways, the auditory thalamocortical projections are accompanied by reciprocal feedback connections from layer 6 neurons in these cortical areas to all regions of the MGB and by layer 5 inputs that specifically target the higher-order auditory thalamic nuclei (Ojima, 1994; Winer et al., 2001; Llano and Sherman, 2008). In this chapter, we will focus exclusively on the pathways from auditory (and to some extent other sensory) cortical areas to the MGB, though descending projections from prefrontal or parietal areas may provide additional sources of cortical input.

THE AUDITORY THALAMUS

A full description of the subdivisions, cytoarchitecture, and connections of the MGB is provided in **Chapter 21 in Section V**. Briefly, there are 3 principal regions: the ventral (MGBv), dorsal (MGBd), and medial (MGBm) divisions, which can be distinguished (and further subdivided) anatomically (Morest, 1964) and physiologically (Calford, 1983). The MGB is characterized by a lack of—or at least very limited—local connectivity between the different subdivisions and an absence of commissural connections (Winer and Morest, 1983). The MGBv, MGBd, and MGBm have distinct patterns of ascending and descending connections, indicating that they form separate, parallel processing channels within the auditory thalamus.

The first-order nucleus is the MGBv, which is part of the tonotopically organized core or lemniscal auditory pathway, relaying signals from the central nucleus of the IC almost exclusively to primary cortical areas where thalamic inputs mainly terminate in the middle cortical layers (Lee and Winer, 2008; Hackett et al., 2011; Vasquez-Lopez et al., 2017). Interestingly, the ascending inputs from the IC to the MGBv (and other divisions of the MGB) can be either excitatory or inhibitory (Winer et al., 1996; Bartlett and Smith, 1999; Bartlett et al., 2000). This appears to be a distinctive property of the auditory thalamus that is not shared by other sensory systems, which has implications for the way ascending and descending inputs interact to shape its neuronal response properties. (How auditory signals are transformed between the MGBv and the primary auditory cortex, A1, is considered in **Chapter 22 in Section V**.)

The MGBd and MGBm are higher-order auditory thalamic nuclei that receive most of their ascending auditory input from the IC shell, which surrounds the central nucleus. These are non-lemniscal regions of the auditory pathway in which neuronal responses are often characterized by broader frequency tuning than that exhibited by neurons located in the MGBv, less well defined or nonexistent tonotopic mapping, and, particularly in MGBm, a greater incidence of inputs from other sensory modalities (see **Chapter 30 in Section V** for a discussion of multisensory convergence in the thalamus). The targets of these non-lemniscal thalamic nuclei also differ, since MGBd neurons principally project to non-primary auditory cortical areas, while those in the MGBm mainly innervate layer 1 throughout the auditory cortex (Huang and Winer, 2000; Smith et al., 2019; Vasquez-Lopez et al., 2017). Furthermore, neurons in the MGBm and its medially adjacent areas, the suprageniculate nucleus and the posterior intralaminar nucleus, project to regions outside the auditory pathway, including the basal ganglia and amygdala (Moriizumi and Hattori, 1992; Bordi and LeDoux, 1994; Vasquez-Lopez et al., 2017, Smith et al., 2019), with the latter pathway playing a key role in auditory fear conditioning.

Corticothalamic projections originate from layer 5 and 6 pyramidal neurons located in all auditory cortical areas and target all the MGB divisions (Figure 24.1). Although differences between species have been reported, these projections are generally both divergent, with each cortical area innervating multiple regions of the auditory thalamus, and convergent, with each MGB division receiving inputs from more than one cortical area (Anderson et al., 1980; Ojima, 1994; Bajo et al., 1995; Rouiller and Welker, 2000; Winer et al., 2001; Llano and Sherman, 2008; Budinger et al., 2013). The descending projections arising from cortical layer 6 mirror the organization of the ascending thalamocortical projections and are topographically organized. Consequently, tonotopically organized regions of the auditory thalamus and cortex are reciprocally connected, with MGBv neurons receiving their principal descending input from A1 neurons with matching selectivity for sound frequency. This has important implications for the way the tuning properties of MGBv neurons are affected by activity in A1. Similarly, reciprocal connections involving layer 6 neurons are found between non-lemniscal areas, such as MGBd and secondary auditory cortex. As previously mentioned, however, non-reciprocal connections between the auditory thalamus and cortex also exist. In particular, neurons in the MGBd receive descending inputs from layer 5 neurons located in both non-lemniscal and lemniscal cortical areas, providing a potential substrate for routing information via the thalamus from core to belt regions of the auditory cortex (Llano and Sherman, 2008) (Figure 24.1).

PROPERTIES OF AUDITORY CORTICOTHALAMIC NEURONS

Targeted *in vitro* recordings in mice have shown that layer 6 corticothalamic neurons integrate signals locally within the auditory cortex and produce trains of action potentials in

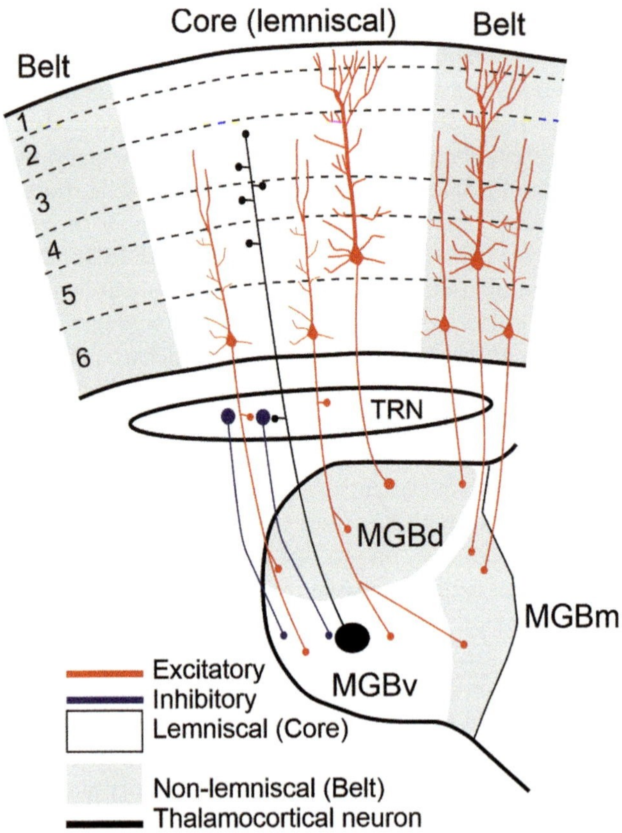

Figure 24.1. Organization of auditory corticothalamic projections. Layer 5 and 6 pyramidal neurons (red) are the origin of the corticothalamic pathways, which form a set of parallel descending channels to the MGB. Layer 6 corticothalamic axons mostly mirror the thalamocortical inputs and have small synaptic terminals, providing modulatory feedback to the same thalamic subdivisions from which they receive their ascending inputs. Layer 6 neurons in core (primary) cortex predominantly target the first-order MGBv, whereas those in belt (non-primary) cortex project to the higher-order MGBd and MGBm. Layer 5 corticothalamic neurons have large axon terminals and form feedforward connections that cross those channels (shown here for a neuron in core auditory cortex projecting to the MGBd), potentially facilitating interactions between different cortical areas via transthalamic pathways. Ascending thalamocortical inputs (black) mainly target the middle layers (particularly layer 4) of the auditory cortex and, for clarity, are illustrated here for the lemniscal projection from MGBv to core auditory cortex only.

response to excitatory inputs (Llano and Sherman, 2009), while *in vivo* recordings have shown that they respond in a sparse and selective manner to sounds (Williamson and Polley, 2019). The axons of layer 6 corticothalamic neurons give rise to profuse terminal fields with small synaptic boutons in the MGB (Ojima, 1994; Rouiller and Welker, 2000; Llano and Sherman, 2008). These are excitatory, glutamatergic projections that predominantly terminate on the dendrites of the MGB neurons (Bartlett and Smith, 1999; Bartlett et al., 2000) and, like layer 6 corticothalamic neurons in other sensory systems, are thought to have a modulatory rather than a driving effect on the transmission of information through the thalamus (Lee and Sherman, 2010).

In addition to providing direct excitatory inputs to thalamocortical relay neurons, the axons of layer 6 neurons have collateral branches that terminate on GABAergic neurons in the neighboring thalamic reticular nucleus (TRN), which provide a source of inhibitory input to the MGB (Crabtree, 1998; Bartlett et al., 2000) (Figure 24.1). This is important because the proportion of intrinsic GABAergic neurons in the MGB varies markedly between species, from approximately 30% in cats (Huang et al., 1999) to less than 1% in rodents (Winer and Larue, 1996). The combination of direct corticothalamic excitation and disynaptic inhibition via the TRN provides a basis for cortical neurons to selectively and dynamically modulate the activity of thalamic relay cells (Guo et al., 2017) and to desynchronize their activity (Ibrahim et al., 2021), thereby controlling the flow of ascending information.

In contrast to those located in layer 6, layer 5 corticofugal neurons integrate information more widely within the cortex and display bursting activity in response to low threshold stimulation, suggesting that they relay signals reliably to their targets (Llano and Sherman, 2009). They also have longer response latencies, show broader stimulus selectivity (Williamson and Polley, 2019), and their axons and terminals are larger than those of layer 6 corticothalamic neurons (Ojima, 1994; Rouiller and Welker, 2000; Llano and Sherman, 2008). Layer 5 neurons are therefore thought to act as drivers that help to define the response characteristics of their subcortical target neurons (Llano and Sherman, 2009; Lee and Sherman, 2010; Sherman and Guillery, 2013). They broadcast signals from the upper cortical layers to widespread downstream targets, which include not only the MGB but also the IC and auditory brainstem nuclei, as well as the striatum and amygdala. In terms of corticothalamic processing, the properties of layer 5 neurons are consistent with their providing a feedforward pathway for relaying information between different cortical areas (Bajo et al., 1995; Llano and Sherman, 2008).

CORTICOFUGAL MODULATION OF SOUND PROCESSING IN THE AUDITORY THALAMUS

A number of studies have shown that the excitability and response properties of neurons in the MGB can be altered by manipulating the activity of neurons in the auditory cortex. Historically, this has been achieved by using pharmacological methods or cooling to deactivate the cortex, or by delivering trains of electrical pulses (sometimes paired with sound presentation) to specific regions of A1 or other cortical areas while recording from neurons in the MGB in anesthetized or passively listening animals. Although these approaches have shown that the auditory cortex does influence MGB response properties, they lack layer specificity and will therefore indiscriminately alter the activity of different types of cortical neuron and potentially affect other downstream targets, including those that provide ascending input to the thalamus. However, the relatively recent availability of methods for targeting and manipulating the activity of specific cortical cell types, particularly in transgenic mice where cell-type and layer-specific markers are available, has provided new insights into the role of auditory corticothalamic (and other corticofugal) pathways.

Studies in which electrophysiological recordings in the MGB of anesthetized animals have been made while deactivating the ipsilateral auditory cortex suggest that corticothalamic neurons have a net excitatory influence on their subcortical targets (Villa et al., 1991; Antunes and Malmierca, 2011; Lohse et al., 2020) (Figure 24.2A,B). In addition to making them less excitable, silencing auditory cortex in anesthetized mice alters the reliability and linearity of the MGBv responses (Lohse et al., 2020). Although changes in the best frequency of some MGB neurons during A1 cooling have been reported in cats (Villa et al., 1991), optogenetic silencing in mice leaves the spectrotemporal tuning properties of MGBv neurons largely unchanged, indicating that these tuning properties arise within the ascending subcortical auditory pathway (Lohse et al., 2020). The ability of MGB neurons to adapt to the statistics of sound stimulation also appears to be independent of the corticothalamic feedback, since both stimulus-specific adaptation (Antunes and Malmierca, 2011) and contrast gain control (Lohse et al., 2020) remain intact when the auditory cortex is deactivated (Figure 24.2C), indicating that these properties are also established at earlier stages in the auditory processing hierarchy.

Because it is possible that activity in corticothalamic circuits may be suppressed under anesthesia, it is important to also carry out these studies in awake animals. In awake mustached bats, pharmacological inactivation of the Doppler-shifted constant-frequency region of A1—which contains an expanded representation of a narrow range of frequencies important for echolocation—also weakens the sound-driven responses in the MGB (Zhang and Suga,

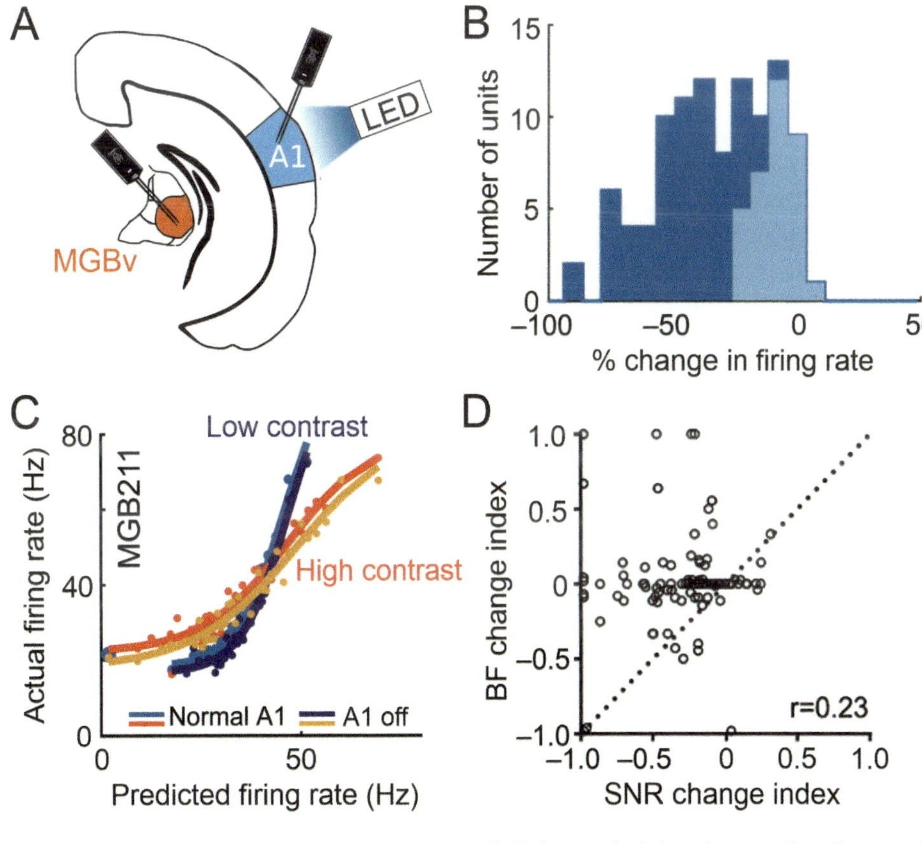

Figure 24.2. Modulation of MGB activity following auditory cortical inactivation. (A). Various methods have been used to silence cortical activity, including application of the $GABA_A$ receptor agonist muscimol, cooling probes, and, as depicted here, photoactivation of GABAergic neurons in which ChR2-YFP is expressed. (B). Transient optogenetic silencing of the auditory cortex in anesthetized mice reduced the mean firing rate of MGBv neurons (dark blue areas represent units that were significantly affected by cortical silencing, $p<0.05$, *t*-test), but left many of their response properties intact, including (C) contrast adaptation, as shown by the difference in the slope of their output nonlinearities when the stimulus contrast was changed. (D). Cooling the auditory cortex in awake marmosets increased spontaneous firing and reduced sound-driven responses, thereby reducing the signal-to-noise ratio (SNR) of MGBv neuronal responses without having a systematic effect on their best frequencies (BF). This suggests that corticothalamic inputs may enhance sound processing in the MGB. A–C adapted from Lohse et al. (2020); D reproduced with permission from Wang et al. (2023).

1997). However, cooling A1 in awake marmosets has been found to produce heterogeneous effects on the spontaneous and stimulus-driven activity of MGB neurons, as well as changes in their temporal firing patterns and tuning for sound frequency, amplitude modulation frequency, and spatial location (Jeschke et al., 2022; Wang et al., 2023). In one study in awake marmosets, A1 inactivation resulted in a frequency-specific reduction in the sound-driven response of most MGB neurons and an increase in spontaneous activity, reducing the signal-to-noise ratio in their responses, which suggests that corticothalamic feedback may increase the gain of MGB neurons in a manner comparable to that observed in the visual and somatosensory systems (Wang et al., 2023) (Figure 24.2D).

The complex and heterogenous effects of deactivating the cortex on the auditory thalamus may reflect differences between species and brain state, inconsistencies in the extent to which the relevant cortical area was silenced, and in the case of studies in which cooling probes were used, possible direct effects on activity in the MGB. Many studies have therefore turned instead to methods for activating the auditory cortex. Although the physiological relevance of the activity patterns generated is questionable, focal stimulation of the cortex has the advantage of allowing the relationship between the tuning properties of neurons at the site of activation and in the MGB to be explored. As with the cortical deactivation experiments, electrical activation of specific sites in auditory cortex has revealed a range of effects on the activity of MGB neurons. Studies in cats (He, 1997) and guinea pigs (He et al., 2002) have shown that A1 activation has a predominantly facilitatory effect on the tone-evoked responses of MGBv neurons, particularly when they are tuned to the same frequency as the cortical neurons at the simulation site.

A more complex set of frequency-dependent changes has been observed in bats following weak focal activation or deactivation of cortical neurons (Figure 24.3). If MGBv neurons are tuned to the same frequency as the cortical neurons recorded at the site of the stimulating electrode ("matched" neurons), their preferred sound frequency remains the same after the cortex is activated, but their responses to this frequency can be enhanced and their tuning sharpened (Zhang and Suga, 2000). By contrast, the responses of "unmatched" MGBv are inhibited at their best frequencies and facilitated at other frequencies, giving rise to shifts in best frequency either away from (Zhang and Suga, 2000) or toward (Tang et al., 2012) that of the activated cortical neurons (Figure 24.3). The opposite pattern of changes was measured in the MGBv when A1 was focally deactivated instead (Zhang et al., 1997). In contrast to these effects on sharply tuned neurons in the MGBv, activation of auditory cortex has been shown to suppress the activity of neurons located in the MGBm (He, 2003; Tang et al., 2012) and to degrade their frequency selectivity (Tang et al., 2012), demonstrating that neurons in lemniscal and non-lemniscal regions of the auditory thalamus differ in the way they respond to cortical activation.

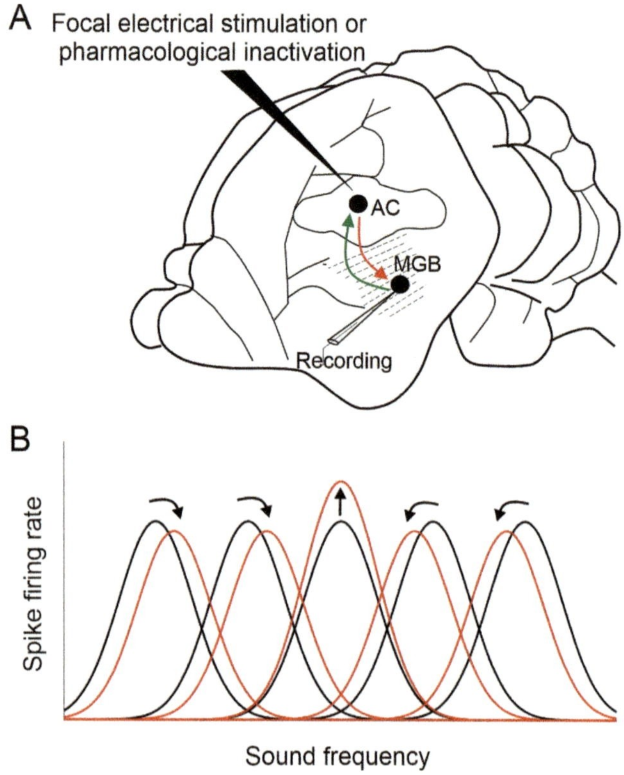

Figure 24.3. Egocentric selection in the bat corticofugal system. (A). Lateral view of the brain of a mustached bat, one of the species used most in cortical stimulation and inactivation experiments, depicting the reciprocal ascending and descending connections between the auditory cortex (AC) and the MGB in the thalamus. (B). Focal electrical stimulation in the cortex results in facilitation of the responses of MGB neurons that have tuning properties matched to those of cortical neurons at the site of the stimulating electrode. The tuning of unmatched MGB neurons may shift toward that of the stimulated cortical neurons (as illustrated here), resulting in an expanded representation of the stimulus feature. Shifts in tuning away from that of the stimulated cortical neurons have also been described, compressing the representation of sound frequency in the auditory thalamus. The direction of the observed shift in tuning properties, which also occurs at other subcortical levels in the auditory system, depends on the species used and the site of cortical activation. Based on studies reviewed in Suga (2012).

Corticofugal inputs can induce related effects at earlier processing levels in the auditory pathway, and in the IC in particular similar changes in neuronal response properties have been observed for a range of stimulus parameters other than sound frequency (Suga, 2012). These findings suggest that one function of the descending pathways is to adjust the representation of particular sound properties in subcortical structures and therefore to regulate and potentially amplify the inputs to the active part of the cortex. This so-called egocentric selection of sensory inputs has been studied most extensively in the auditory system, especially in certain species of bats, but may occur more widely since comparable effects of corticofugal modulation have been described in other sensory systems too (Li and Ebner, 2007; Cudeiro and Sillito, 2006).

An important question is whether layer 6 corticothalamic modulation operates when A1 is driven by sounds rather than by the artificial stimulation used in these studies. In rats, many layer 6 neurons in A1 are under strong local inhibitory control, with sound stimuli suppressing their spontaneous activity (Zhou et al., 2010). This suggests that a gating mechanism may exist, with corticothalamic feedback activated only under certain conditions, such as during active listening or when animals are engaged in auditory learning tasks.

It is also interesting to note that the subcortical changes induced by activating the cortex resemble the frequency-specific plasticity produced by auditory fear conditioning, where combining tones of a particular frequency with a mild electric shock can alter the best frequencies of A1 and MGB neurons (Edeline and Weinberger, 1991). Moreover, combining electrical stimulation of the cholinergic basal forebrain with tone presentation can produce similar shifts in the best frequency of MGBv neurons toward the frequency of the paired tone (Zhang and Yan, 2008). These changes were abolished when the auditory cortex was deactivated. Collectively, these studies suggest that corticothalamic circuits may play a key role in auditory associative learning, a situation where it would clearly be useful to relieve any inhibitory control over the activity of cortical layer 6 neurons.

MULTISENSORY PROCESSING

Convergence of inputs from different sensory modalities is widely found in the thalamus (see **Chapter 30 in Section V** for a comprehensive description). Neurons in the medial part of the auditory thalamus, which includes the MGBm, as well as the suprageniculate nucleus and the posterior intralaminar nucleus, respond to auditory and somatosensory stimuli (Bordi and LeDoux, 1994; Lohse et al., 2021). In rodents, the sound-driven responses of neurons in the lemniscal MGBv can also be suppressed by somatosensory stimulation (Kimura and Imbe, 2018; Lohse et al., 2021), and these effects are then transmitted to A1.

The auditory TRN receives descending cortical inputs from nonprimary visual and somatosensory areas, suggesting that it is likely to be involved in cross-modal thalamic processing (Kimura et al., 2012). However, in mice, the auditory TRN does not appear to contribute to the suppression of auditory responses by whisker stimulation (Lohse et al., 2021). These cross-modal interactions in the MGB are mediated by descending projections from the primary somatosensory cortex (S1) (Lohse et al., 2021). Optogenetic activation of layer 5 neurons in S1 replicates the faciliatory effects of whisker stimulation on neurons in the medial auditory thalamus. Furthermore, S1 neurons innervate a subset of GABAergic neurons in the lateral shell of the IC, which project to and suppress the sound-evoked responses of neurons in the MGBv and MGBd. Cross-modal corticofugal projections to the auditory midbrain and thalamus are therefore involved in integrating multisensory signals and in enabling communication between different sensory cortical areas. These findings also suggest that a functional distinction may exist between the medial and lateral regions of the MGB in terms of the way the neurons found there respond to multisensory stimuli and their connections with other brain areas.

CORTICOTHALAMIC INFLUENCES ON AUDITORY PERCEPTION

Although cortical activation and silencing studies have revealed how corticothalamic pathways can affect processing in the MGB, it remains largely unclear what their role is in active listening. Corticothalamic feedback has been implicated in task-dependent modulation of the MGB in humans performing a speech-recognition task (von Kriegstein et al., 2008; see **Chapter 23 in Section V** for more details). More direct evidence has been obtained from animal studies in which the behavioral consequences of manipulating specific corticofugal circuits have been measured.

Selective lesioning by chromophore-targeted laser photolysis of approximately 60% of A1 layer 6 neurons that project to the MGBv impaired the ability of ferrets to detect a single mistuned frequency in a complex tone comprising 16 harmonics (Homma et al., 2017) (Figure 24.4). The task was used to measure the sensitivity of the animals to the harmonic structure of complex sounds, which, at least in humans, is an important cue for grouping together the different frequency components of a sound so that they can be segregated from other concurrent sounds (de Cheveigné et al., 1997). Consequently, the reduced behavioral sensitivity to "mistuning" suggests that descending layer 6 projections from A1 to the MGBv contribute to sound source segregation through the ability of the animals to perceive the harmonic structure of complex sounds. Computational modeling also supports a role for corticothalamic feedback in identifying sound sources in complex auditory scenes (Otazu and Leibold, 2011). Together, these studies suggest that corticothalamic modulation may play an important role in auditory scene analysis.

Further evidence that layer 6 corticothalamic neurons can influence auditory perception has come from a study in mice in which optogenetic activation of these neurons

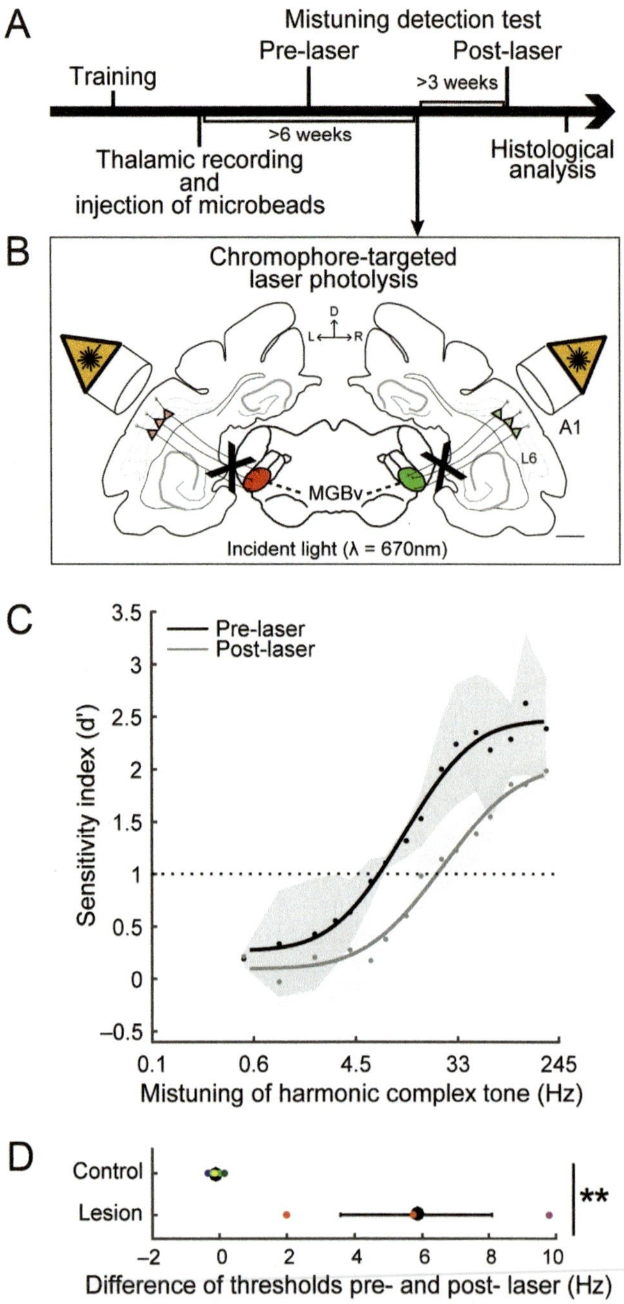

Figure 24.4. Corticothalamic feedback and auditory scene analysis. (A). Ferrets were trained using a go/no-go behavioral paradigm to detect the presence of a mistuned harmonic in a complex tone. Bilateral injections of fluorescent microbeads conjugated with a light-sensitive chromophore were then made into the MGBv (left, red; right, green) under the guidance of electrophysiological recordings. The animals were then tested again on the behavioral task before and after eliminating layer 6 corticothalamic neurons in A1. (B). Apoptosis of the retrogradely labeled layer 6 corticothalamic neurons was induced by illuminating the auditory cortex on both sides with near-infrared laser light. D = dorsal; L = left; R = right; L6 = cortical layer 6. Scale bar = 1 mm. (C). The psychometric curves plot d′ (± SEM) as a function of the degree of mistuning (on a log scale). The shift in the curve toward larger mistuning values after selective elimination of corticothalamic neurons (gray) compared to baseline (black) indicates that mistuning detection performance was impaired. (D). The change in threshold after laser illumination was larger for the ferrets in which layer 6 corticothalamic neurons had been selectively eliminated than in control animals (2-tailed unpaired *t* test, $**p < 0.01$). Data from individual animals are represented by colored dots. Adapted from Homma et al. (2017).

was followed at different times by presentation of an auditory stimulus (Guo et al., 2017). As the interval was increased, sound-driven responses in A1 switched from being suppressed to enhanced as a result of changes in cortical excitability due to inhibitory interneurons resetting the phase of low-frequency cortical rhythms. Suppression of cortical responses was associated with improved discrimination of sound frequencies, but poorer detection thresholds, whereas the increase in cortical gain, which was accompanied by enhanced responses in the MGBv, biased the animals' behavior in the opposite direction so that they now showed improved sound detection at the expense of poorer discrimination (Guo et al., 2017). This is consistent with evidence that dopaminergic modulation results in sustained activation of the auditory cortex through positive recurrent feedback between the cortex and thalamus, which supports better performance on a sound-detection task (Happel et al., 2014).

It remains to be determined what factors drive corticothalamic neurons during active listening, particularly if their activity is behaviorally gated by intracortical circuits (Zhou et al., 2010). However, the recent demonstration that layer 6 corticothalamic neurons in mice receive inputs from motor-related areas, such as the globus pallidus, and begin spiking hundreds of milliseconds prior to orofacial movements that trigger sounds and rewards (Clayton et al., 2021) highlights the likely importance of non-auditory inputs in determining when they respond.

CLINICAL RELEVANCE OF CORTICOTHALAMIC CIRCUITS

Hearing impairment resulting from the loss of cochlear hair cells or of the spiral ganglion neurons that supply their afferent innervation results in substantial compensatory plasticity in the auditory system (Hermann and Butler, 2021). Increased response gain at higher levels of the auditory pathway has been identified as a potential basis for both hyperacusis, an abnormally high sensitivity to sound, and tinnitus, a condition in which phantom sounds are perceived in the absence of acoustic input (Auerbach et al., 2014). Indeed, recent work in mice has provided evidence for a direct link between the activity of auditory thalamocortical neurons and behavioral hypersensitivity (McGill et al., 2022). Gain changes have also been reported in the MGB in response to cochlear deafferentation (Chambers et al., 2016). While the role of corticothalamic feedback has yet to be explored, it seems likely that descending pathways are involved in this plasticity since increases in gain have been observed in the axons of cortical neurons that project to the IC (Asokan et al., 2018). Although changes in MGB circuits have been widely implicated in tinnitus pathology (Caspary and Llano, 2017), there is so far only limited evidence to support an involvement of corticothalamic interactions (Weisz et al., 2007).

Corticothalamic feedback to the MGB can improve the neural coding of weak temporal sound cues, which may help to counteract the impairments in temporal processing that are commonly associated with aging (Kommajosyula et al., 2021). Reducing the modulation depth of sinusoidal amplitude modulated stimuli caused the responses of rat MGB neurons to show repetition enhancement and less adaptation, so that they responded more reliably to the repeating stimulus. This effect was reversed by silencing corticothalamic neurons, raising the intriguing possibility that the cortex may fine-tune thalamic responses to degraded temporal cues so that they become more salient.

Descending corticofugal influences on sound processing at lower levels of the auditory pathway have also been implicated in the beneficial effects of musical experience and training on speech-in-noise perception (Kraus and Chandrasekaran, 2010). In a wider context, changes in auditory corticothalamic interactions are thought to occur in schizophrenia (Tu et al., 2015) and developmental dyslexia (see **Chapter 23 in Section V**). Auditory corticothalamic circuits may also contribute to pain perception since inhibition of a cross-modal glutamatergic projection from layer 6 neurons in the auditory cortex to the posterior nucleus and the ventroposterior nucleus in the thalamus mediates sound-induced analgesia (Zhou et al., 2022).

CONCLUSIONS

The anatomical organization of auditory corticothalamic pathways is broadly consistent with that found in other sensory systems. The first-order MGBv receives reciprocally organized feedback from layer 6 neurons located in core auditory cortical areas, whereas the higher-order MGBd and MGMm receive a combination of reciprocal and nonreciprocal connections that involve both layer 5 and layer 6 neurons. There is growing evidence in the auditory system that different corticothalamic projections have distinct roles: layer 5 neurons are thought to drive downstream neurons and to mediate communication between different cortical areas projections, whereas layer 6 neurons modulate the spiking activity and tuning properties of MGB neurons. Nevertheless, more research is needed on their network and physiological properties in order to better understand how these neurons are activated and what effect they have on their downstream targets.

The results of experiments in which the auditory cortex was either activated or deactivated suggest that the descending projections can induce diverse changes in the activity and response properties of MGB neurons. The interpretation of some of these findings is complicated by differences between species, brain state, and the methods used for modifying cortical activity. Nevertheless, the idea that a key function of corticofugal modulation is to gate the flow of incoming information by selectivity enhancing subcortical responses to specific stimulus properties has gained traction. This may serve to amplify the representation of sound features that become particularly relevant for behavior in a given sensory context, such as during learning.

Key advances in our understanding of the functional role of corticothalamic pathways have recently come from studies in which the activity of specific neuronal populations has been altered in animals performing sound detection or discrimination tasks. In particular, evidence is building that corticothalamic modulation may be particularly important for processing sound signals in challenging listening conditions, such as when multiple sound sources are present or when acoustic information is degraded as a result of aging or hearing loss (Souffi et al., 2021). Another recently popular idea is that corticothalamic modulation may be viewed as a form of predictive coding in which signals from the cortex provide top-down predictions about current sensory inputs, which are then compared with incoming activity and used to update thalamic responses (Antunes and Malmierca, 2021; Asilador and Llano, 2021; Kommajosyula et al., 2021; see also **Chapter 23 in Section V** for a discussion of this in the context of speech perception). In this respect, it is interesting that predictive coding has also been proposed as a possible explanation for tinnitus (Hullfish et al., 2019).

Much remains to be discovered about how corticothalamic neurons affect information processing, both locally in the cortex and in their target neurons in the MGB, under what conditions they are active, and how they contribute to changes in auditory perceptual abilities. However, the availability of new methods for monitoring and manipulating the activity of these neurons in animal models has the potential to provide unprecedented insights into our functional understanding of how the cortex and thalamus interact during active listening.

ACKNOWLEDGMENT

The authors' research is supported by a Wellcome Principal Research Fellowship (WT108369/Z/2015/Z) to AJK.

REFERENCES

Andersen RA, Knight PL, Merzenich MM (1980). The thalamocortical and corticothalamic connections of AI, AII, and the anterior auditory field (AAF) in the cat: evidence for two largely segregated systems of connections. *J Comp Neurol.* 194:663–701.

Antunes FM, Malmierca MS (2011). Effect of auditory cortex deactivation on stimulus-specific adaptation in the medial geniculate body. *J Neurosci.* 31:17306–17316.

Antunes FM, Malmierca MS (2021). Corticothalamic pathways in auditory processing: recent advances and insights from other sensory systems. *Front Neural Circuits.* 15:1–25. Article 721186.

Asilador A, Llano DA (2021). Top-down inference in the auditory system: potential roles for corticofugal projections. *Front Neural Circuits.* 14:1–20. Article 615259.

Asokan MM, Williamson RS, Hancock KE, Polley DB (2018). Sensory overamplification in layer 5 auditory corticofugal projection neurons following cochlear nerve synaptic damage. *Nat Commun.* 9:1–10. Article 2468.

Auerbach BD, Rodrigues PV, Salvi RJ (2014). Central gain control in tinnitus and hyperacusis. *Front Neurol.* 5:1–21. Article 206.

Bajo VM, Rouiller EM, Welker E, Clarke S, Villa AE, de Ribaupierre Y, de Ribaupierre F (1995). Morphology and spatial distribution of corticothalamic terminals originating from the cat auditory cortex. *Hear Res.* 83:161–174.

Bartlett EL, Smith PH (1999). Anatomic, intrinsic, and synaptic properties of dorsal and ventral division neurons in rat medial geniculate body. *J Neurophysiol.* 81:1999–2016.

Bartlett EL, Stark JM, Guillery RW, Smith PH (2000). Comparison of the fine structure of cortical and collicular terminals in the rat medial geniculate body. *Neuroscience.* 100:811–828.

Bordi F, LeDoux JE (1994). Response properties of single units in areas of rat auditory thalamus that project to the amygdala. II. Cells receiving convergent auditory and somatosensory inputs and cells antidromically activated by amygdala stimulation. *Exp Brain Res.* 98:275–286.

Budinger E, Brosch M, Scheich H, Mylius J (2013). The subcortical auditory structures in the Mongolian gerbil: II. Frequency-related topography of the connections with cortical field AI. *J Comp Neurol.* 521:2772–2797.

Calford MB (1983). The parcellation of the medial geniculate body of the cat defined by the auditory response properties of single units. *J Neurosci.* 3:2350–2364.

Caspary DM, Llano DA (2017). Auditory thalamic circuits and $GABA_A$ receptor function: Putative mechanisms in tinnitus pathology. *Hear Res.* 349:197–207.

Chambers AR, Salazar JJ, Polley DB (2016). Persistent thalamic sound processing despite profound cochlear denervation. *Front Neural Circuits.* 10:1–13. Article 72.

Clayton KK, Williamson RS, Hancock KE, Tasaka GI, Mizrahi A, Hackett TA, Polley DB (2021). Auditory corticothalamic neurons are recruited by motor preparatory inputs. *Curr Biol.* 31:310–321.

Crabtree JW (1998). Organization in the auditory sector of the cat's thalamic reticular nucleus. *J Comp Neurol.* 390:167–182.

Cudeiro J, Sillito AM (2006). Looking back: Corticothalamic feedback and early visual processing. *Trends Neurosci.* 29:298–306.

de Cheveigné A, McAdams S, Marin CMH (1997). Concurrent vowel identification. II. Effects of phase, harmonicity, and task. *J Acoust Soc Am.* 101:2848–2856.

Edeline JM, Weinberger NM (1991). Subcortical adaptive filtering in the auditory system: Associative receptive field plasticity in the dorsal medial geniculate body. *Behav Neurosci.* 105:154–175.

Guo W, Clause AR, Barth-Maron A, Polley DB (2017). A corticothalamic circuit for dynamic switching between feature detection and discrimination. *Neuron.* 95:180–194.

Hackett TA, Barkat TR, O'Brien BM, Hensch TK, Polley DB (2011). Linking topography to tonotopy in the mouse auditory thalamocortical circuit. *J Neurosci.* 31:2983–2995.

Happel MF, Deliano M, Handschuh J, Ohl FW (2014). Dopamine-modulated recurrent corticoefferent feedback in primary sensory cortex promotes detection of behaviorally relevant stimuli. *J Neurosci.* 34:1234–1247.

He J (1997). Modulatory effects of regional cortical activation on the onset responses of the cat medial geniculate neurons. *J Neurophysiol.* 77:896–908.

He J (2003). Corticofugal modulation on both ON and OFF responses in the nonlemniscal auditory thalamus of the guinea pig. *J Neurophysiol.* 89:367–381.

He J, Yu YQ, Xiong Y, Hashikawa T, Chan YS (2002). Modulatory effect of cortical activation on the lemniscal auditory thalamus of the guinea pig. *J Neurophysiol.* 88:1040–1050.

Herrmann B, Butler BE (2021). Hearing loss and brain plasticity: The hyperactivity phenomenon. *Brain Struct Funct.* 226:2019–2039.

Homma NY, Happel MFK, Nodal FR, Ohl FW, King AJ, Bajo VM (2017). A role for auditory corticothalamic feedback in the perception of complex sounds. *J Neurosci.* 37:6149–6161.

Huang CL, Larue DT, Winer JA (1999). GABAergic organization of the cat medial geniculate body. *J Comp Neurol.* 415:368–392.

Huang CL, Winer JA (2000). Auditory thalamocortical projections in the cat: Laminar and areal patterns of input. *J Comp Neurol.* 427:302–331.

Hullfish J, Sedley W, Vanneste S (2019). Prediction and perception: Insights for (and from) tinnitus. *Neurosci Biobehav Rev.* 102:1–12.

Ibrahim BA, Murphy CA, Yudintsev G, Shinagawa Y, Banks MI, Llano DA (2021). Corticothalamic gating of population auditory thalamocortical transmission in mouse. *eLife.* 10:e56645.

Jeschke M, Ohl FW, Wang X (2022). Effects of cortical cooling on sound processing in auditory cortex and thalamus of awake marmosets. *Front Neural Circuits.* 15:1–24. Article 786740.

Kimura A, Imbe H (2018). Robust subthreshold cross-modal modulation of auditory response by cutaneous electrical stimulation in first- and higher-order auditory thalamic nuclei. *Neuroscience.* 372:161–180.

Kimura A, Yokoi I, Imbe H, Donishi T, Kaneoke Y (2012). Auditory thalamic reticular nucleus of the rat: Anatomical nodes for modulation of auditory and cross-modal sensory processing in the loop connectivity between the cortex and thalamus. *J Comp Neurol.* 520:1457–1480.

Kommajosyula SP, Bartlett EL, Cai R, Ling L, Caspary DM (2021). Corticothalamic projections deliver enhanced responses to medial geniculate body as a function of the temporal reliability of the stimulus. *J Physiol.* 599:5465–5484.

Kraus N, Chandrasekaran B (2010). Music training for the development of auditory skills. *Nat Rev Neurosci.* 11:599–605.

Lee CC, Sherman SM (2010). Drivers and modulators in the central auditory pathways. *Front Neurosci.* 4:79–86.

Lee CC, Winer JA (2008). Connections of cat auditory cortex: I. Thalamocortical system. *J Comp Neurol.* 507:1879–900.

Li L, Ebner FF (2007). Cortical modulation of spatial and angular tuning maps in the rat thalamus. *J Neurosci.* 27:167–179.

Llano DA, Sherman SM (2008). Evidence for nonreciprocal organization of the mouse auditory thalamocortical-corticothalamic projection systems. *J Comp Neurol.* 507:1209–1227.

Llano DA, Sherman SM (2009). Differences in intrinsic properties and local network connectivity of identified layer 5 and layer 6 adult mouse auditory corticothalamic neurons support a dual corticothalamic projection hypothesis. *Cereb Cortex.* 19:2810–2826.

Lohse M, Bajo VM, King AJ, Willmore BDB (2020). Neural circuits underlying auditory contrast gain control and their perceptual implications. *Nat Commun.* 11:1–13. Article 324.

Lohse M, Dahmen JC, Bajo VM, King AJ (2021). Subcortical circuits mediate communication between primary sensory cortical areas in mice. *Nat Commun.* 12:1–14. Article 3916.

McGill M, Hight AE, Watanabe YL, Parthasarathy A, Cai D, Clayton K, Hancock KE, Takesian A, Kujawa SG, Polley DB (2022). Neural signatures of auditory hypersensitivity following acoustic trauma. *eLife.* 11:e80015.

Morest DK (1964). The neuronal architecture of the medial geniculate body of the cat. *J Anat.* 98:611–630.

Moriizumi T, Hattori T (1992). Ultrastructural morphology of projections from the medial geniculate nucleus and its adjacent region to the basal ganglia. *Brain Res Bull.* 29:193–198.

Ojima H (1994). Terminal morphology and distribution of corticothalamic fibers originating from layers 5 and 6 of cat primary auditory cortex. *Cereb Cortex.* 4:646–663.

Otazu GH, Leibold C (2011). A corticothalamic circuit model for sound identification in complex scenes. *PLoS ONE.* 6:e24270.

Rouiller EM, Welker E (2000). A comparative analysis of the morphology of corticothalamic projections in mammals. *Brain Res Bull.* 53:727–741.

Sherman SM, Guillery RW (2013). *Exploring the Thalamus and Its Role in Cortical Function*. 2nd ed. MIT Press.

Smith PH, Uhlrich DJ, Manning KA (2019). Evaluation of medial division of the medial geniculate (MGM) and posterior intralaminar nucleus (PIN) inputs to the rat auditory cortex, amygdala, and striatum. *J Comp Neurol.* 527:1478–1494.

Souffi S, Nodal FR, Bajo VM, Edeline JM (2021). When and how does the auditory cortex influence subcortical auditory structures? New insights about the roles of descending cortical projections. *Front Neurosci.* 15:1–17. Article 690223.

Suga N (2012). Tuning shifts of the auditory system by corticocortical and corticofugal projections and conditioning. *Neurosci Biobehav Rev.* 36:969–988.

Tang J, Yang W, Suga N (2012). Modulation of thalamic auditory neurons by the primary auditory cortex. *J Neurophysiol.* 108:935–942.

Tu PC, Lee YC, Chen YS, Hsu JW, Li CT, Su TP (2015). Network-specific cortico-thalamic dysconnection in schizophrenia revealed by intrinsic functional connectivity analyses. *Schizophr Res.* 166:137–143.

Usrey WM, Sherman SM (2019). Corticofugal circuits: Communication lines from the cortex to the rest of the brain. *J Comp Neurol.* 527:640–650.

Vasquez-Lopez SA, Weissenberger Y, Lohse M, Keating P, King AJ, Dahmen JC (2017). Thalamic input to auditory cortex is locally heterogeneous but globally tonotopic. *eLife.* 6:e25141.

Villa AE, Rouiller EM, Simm GM, Zurita P, Y de Ribaupierre Y, de Ribaupierre F (1991). Corticofugal modulation of the information processing in the auditory thalamus of the cat. *Exp Brain Res.* 86:506–517.

Von Kriegstein K, Patterson RD, Griffiths TD (2008). Task-dependent modulation of medial geniculate body is behaviorally relevant for speech recognition. *Curr Biol.* 18:1855–1859.

Wang X, Zhang Y, Zhu L, Bai S, Li R, Sun H, Qi R, Cai R, Li M, Jia G, Cao X, Schriver KE, Li X, Gao L (2023). Selective corticofugal modulation on sound processing in auditory thalamus of awake marmosets. *Cereb Cortex.* 33:3372–3386.

Weisz N, Müller S, Schlee W, Dohrmann K, Hartmann T, Elbert T (2007). The neural code of auditory phantom perception. *J Neurosci.* 27:1479–1484.

Williamson RS, Polley DB (2019). Parallel pathways for sound processing and functional connectivity among layer 5 and 6 auditory corticofugal neurons. *eLife.* 8:e42974.

Winer JA (2006). Decoding the auditory corticofugal systems. *Hear Res.* 212:1–8.

Winer JA, Diehl JJ, Larue DT (2001). Projections of auditory cortex to the medial geniculate body of the cat. *J Comp Neurol.* 430:27–55.

Winer JA, Larue DT (1996). Evolution of GABAergic circuitry in the mammalian medial geniculate body. *Proc Natl Acad Sci U S A.* 93:3083–3087.

Winer JA, Morest DK (1983). The medial division of the medial geniculate body of the cat: implications for thalamic organization. *J Neurosci.* 3:2629–2651.

Winer JA, Saint Marie RL, Larue DT, Oliver DL (1996). GABAergic feedforward projections from the inferior colliculus to the medial geniculate body. *Proc Natl Acad Sci U S A.* 93:8005–8010.

Zhang Y, Suga N (1997). Corticofugal amplification of subcortical responses to single tone stimuli in the mustached bat. *J Neurophysiol.* 78:3489–3492.

Zhang Y, Suga N (2000). Modulation of responses and frequency tuning of thalamic and collicular neurons by cortical activation in mustached bats. *J Neurophysiol.* 84:325–333.

Zhang Y, Suga N, Yan J (1997). Corticofugal modulation of frequency processing in bat auditory system. *Nature.* 387:900–903.

Zhang Y, Yan J (2008). Corticothalamic feedback for sound-specific plasticity of auditory thalamic neurons elicited by tones paired with basal forebrain stimulation. *Cereb Cortex.* 18:1521–1528.

Zhou Y, Liu BH, Wu GK, Kim YJ, Xiao Z, Tao HW, Zhang LI (2010). Preceding inhibition silences layer 6 neurons in auditory cortex. *Neuron.* 65:706–717.

Zhou W, Ye C, Wang H, Mao Y, Zhang W, Liu A, Yang CL, Li T, Hayashi L, Zhao W, Chen L, Liu Y, Tao W, Zhang Z (2022). Sound induces analgesia through corticothalamic circuits. *Science.* 377(6602):198–204.

25.

GENERAL ORGANIZATION AND PARALLEL PATHWAYS IN THE SOMATOSENSORY SYSTEM

Christina Mo

INTRODUCTION

Somatosensation is the physical experience of the world through our bodies—a literal sense of self. The loss of sight, hearing, and smell can be overcome, but the loss of somatosensation is disastrous and it is the one sense alone that can support the lived experience (Donaldson, 1890). Somatosensory receptors are located inside and on the entire outside of the body, including receptors of touch, pain, temperature, and proprioception. With such a broad array of inputs, the somatosensory system exemplifies the foundational principle of parallel processing. Different stimuli are transduced by peripheral receptors into a neural code, which is processed separately through discrete anatomical channels up to the cortex. Parallel pathways provide a structure for processing the representations in higher-order cortex. The localization of the stimuli in or on the body is generally maintained in a spatial map at each level through to the cortex, called topography—neurons that innervate adjacent sites on the receptor sheet project to adjacent sites in the brain.

The first step in the system for establishing parallel pathways is its specialized receptors which encode different types of somatosensory stimuli (for review, see Delhaye et al., 2018). Some are mechanoreceptors that respond to physical changes and form parts of nerve endings in the skin, joints, and muscles. They are highly tuned to pressure and vibration. Other receptors respond to temperature changes; in fact, 6 thermoreceptors have distinct temperature-tuning profiles. Chemical-responsive receptors and proprioceptive receptors (spindles) located in muscle, bone, and tendons detect contraction and release. Nociceptors, which detect stimulus extremes, are specialized bare nerve endings responsive to tissue or visceral injury. Neighboring receptors are innervated by sensory afferents, which bundle together to form nerves. Cell bodies of sensory neurons are housed in dorsal root ganglia that enter the spinal cord or the trigeminal ganglia of the head. Unique to the somatosensory system, the sensory neurons are delivered through multiple nerve bundles, perhaps relying on a robust organization more than other sensory systems. Information from the different receptor types is processed in segregated channels via the major ascending pathways to the cortex. Perception ultimately involves integration of the information from these separate representations. This chapter provides a brief overview of the 2 main somatosensory pathways in humans, followed by a more detailed description of the neural circuits at the level of the thalamus and cortex, as understood from monkey and rodent models. The discussion is limited to ascending somatosensory pathways within the central nervous system.

PARALLEL PATHWAYS FROM PERIPHERY TO CORTEX

Somatosensory information in humans is processed through 2 main ascending pathways (Figure 25.1). Fine touch, proprioception, and vibration stimuli are delivered to the brain via a segment of spinal cord called the dorsal columns, forming the *dorsal-column medial lemniscal pathway*. In contrast, crude touch, pain, and temperature information ascends via the ventral spinothalamic tract of the spinal cord to form the *spinothalamic pathway* (Koeppen and Stanton, 2008). The pathways also differ in the anatomical entry points to the spinal cord, the location of decussation, the brainstem nuclei, and the thalamic nuclei involved.

The pathways are similar in their 3-neuron relay structure to reach somatosensory cortex. Peripheral axons innervating sensory receptors form the first-order neurons and their cell bodies are located in the dorsal root ganglion of the spinal nerve. If stimuli are located on the face or head, the first-order neurons are in the trigeminal or cranial nerve

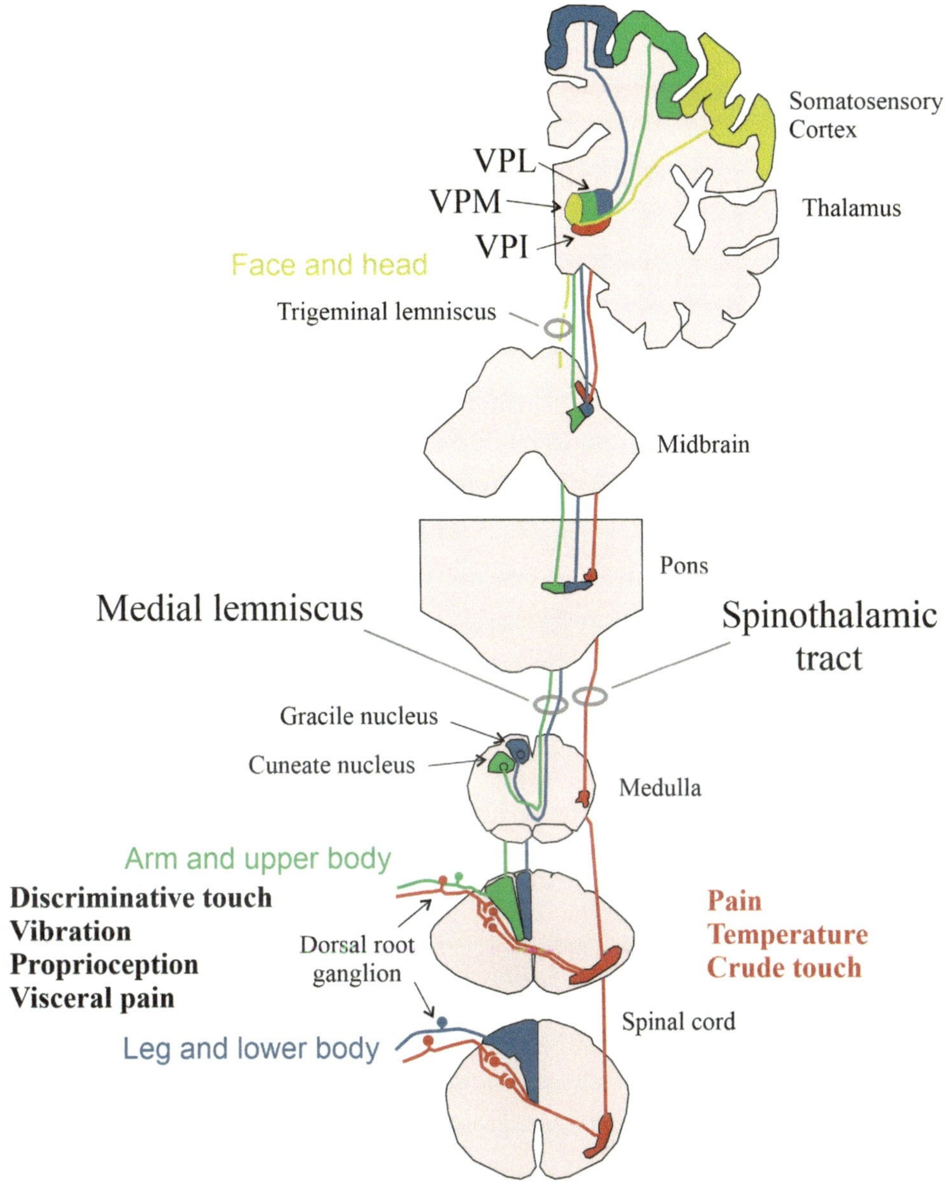

Figure 25.1. The 2 main ascending somatosensory pathways in humans. The dorsal column medial lemniscal pathway carries discriminative touch and conscious proprioceptive input from the body via dorsal root ganglia in the spinal cord. The inputs are relayed via the dorsal columns to synapse in one of the dorsal column nuclei: lower body inputs in the gracile nucleus (blue) and upper body inputs in the cuneate nucleus (green). These second-order neurons then cross the midline and traverse the medial lemniscus tract to synapse in thalamus. The spinothalamic pathway carries pain and temperature sensations from the body via dorsal root ganglia to synapse in the dorsal horn of the spinal cord before crossing the midline to ascend through brainstem to thalamus. Sensations from the head and face do not ascend the spinal cord but enter through the trigeminal ganglion to synapse in the trigeminal nuclei (not shown). Only the ventroposterior complex of the thalamus and a coronal section of primary somatosensory cortex is shown for clarity. Adapted from Koeppen and Stanton (2008).

ganglia. The cell bodies of the second-order neurons are located in either the spinal cord or brainstem and they cross to the contralateral side to continue ascending to the thalamus. The cell bodies of the third-order neurons are located in the thalamus, or cerebellum for unconscious posture processing via the spinocerebellar tract, and project to S1.

The ventroposterior nuclear complex (VP) is the main thalamic target for ascending somatosensory information, which includes the ventral posterior lateral nucleus (VPL), ventral posterior medial nucleus (VPm), the ventroposterior superior nucleus (VPS), and the ventroposterior inferior nucleus (VPI). Information from the body ascends the medial lemniscus tract and mainly projects to VPL. Information from the face and head travels via the trigeminothalamic tract and is the main input to VPM. The spinothalamic pathway also weakly inputs to VPL and VPM, but mainly targets VPI, in addition to the posterior nucleus and central lateral nucleus of the intralaminar complex of the thalamus. The primary somatosensory cortex is located in the postcentral gyrus, which is subdivided into Brodmann areas 3a, 3b, 1, and 2.

MEDIAL LEMNISCUS PATHWAY

Sensations of body posture, conscious proprioception, discrimination, pressure, movement, vibration, and visceral pain are sent via the dorsal columns of the spinal cord to the medulla and ascend via the medial lemniscus tract from the medulla to the brain (Figure 25.1). These signals activate mechano- or thermo- sensitive receptors in the skin, muscle, joints, or internal organs, each with diverse characteristics such as level of myelination, fast- or slow- adapting fibers, nature of the nerve ending, and vibrational tuning.

First-order axons travel up the spinal cord via the dorsal columns (fasciculus gracilis and fasciculus cuneatus) to synapse in one of the brainstem complex dorsal column nuclei. These nuclei are the gracile nucleus, which receives lower body and leg inputs, and the cuneate nucleus, which receives upper body and arm inputs. Touch sensations from the head and face do not ascend the spinal cord but enter through the trigeminal ganglion to synapse in the trigeminal nuclei. Although somatotopy is present, it is the stimulus modality that largely determines the organization, with mechanosensory and proprioceptive afferents arranged in a medial-lateral pattern throughout the entire dorsal column and somatotopic maps within each modality (Niu et al., 2013). The second-order neurons immediately cross the midline and ascend via the medial lemniscus tract on the opposite side, to synapse in the contralateral VPL of the thalamus. Cutaneous inputs dominate areas 3b and 1, while proprioceptive input dominates areas 3a and 2. For head and face input from the trigeminal nuclei, second-order neurons relay to the contralateral VPm via the trigeminal lemniscus tract. From here, third-order neurons project to the somatosensory cortex according to somatotopy.

As a supplement to topography, there is a disproportionate emphasis on detailed input, allowing the creation of a "homunculus," a caricature of proportional cortical representations. In 3 dimensions, the result is a disheveled man with an enlarged face, swollen lips and tongue, with a grotesque advantage in a mitt-less game of baseball. The original rendition was drawn by a medical illustrator commissioned by the neurosurgeon Wilder Penfield to illustrate his findings on cortical stimulation of epileptic patients during brain surgery (Penfield and Rasmussen, 1950). The human homunculus map has since been confirmed by microstimulation and updated by functional magnetic resonance imaging (fMRI) (Muret et al., 2022). Re-drawings over the decades have degraded any chance of accuracy, but its intended purpose was little more than an evocative, artistic expression of functionally important body parts.

SPINOTHALAMIC PATHWAY

Discriminative pain and temperature sensations are delivered to the brain via the spinothalamic tract and other anterolateral pathways. Like the medial lemniscus pathway, these sensory signals also enter the spinal cord via dorsal root ganglia, but synapse immediately in the dorsal horn marginal zone (lamina I) and deeper in the dorsal horn (lamina V). Second-order sensory neurons cross to the contralateral spinal cord to ascend the anterolateral white matter to VPL. Pain and temperature projections synapse in the VPI portion of the ventroposterior complex, the posterior nucleus, and the intralaminar nuclei of the thalamus. From there, third-order neurons project to somatosensory cortex, but also cortical areas that are involved in affective responses, such as the cingulate gyrus and the insula, which have limbic system functions. The equivalent trigeminal pathway is relayed by the spinal trigeminal nucleus and projects to the contralateral VPM and intralaminar nuclei. As part of the anterolateral system, the spinoreticular and spinomesencephalic pathways convey nociceptive information to other thalamic nuclei, the reticular formation, and limbic system.

SPINOCEREBELLAR PATHWAYS

The spinocerebellar pathways process subconscious proprioceptive information required for body equilibrium and posture. These receptors are found in the muscles, tendons, periosteum, and joints. Information from these receptors target the cerebellum and cortex via a 2-neuron relay called the spinocerebellar pathway. First-order neurons with cell bodies in the spinal ganglia terminate in the thoracic nucleus and posterior gray horn of the spinal cord. The second-order neurons then ascend via the anterior or posterior spinocerebellar tracts to synapse on cerebellar nuclei via the cerebellar peduncles.

Conscious proprioceptive information is transmitted by yet another route that partially overlaps with the dorsal column–medial lemniscal pathway. Conscious awareness of body movement (kinesthesia) is thought to be supported by projections to the cortex, which ascend via the dorsal column pathway, synapsing in the cuneate nucleus and then VPL. There are also trigeminocerebellar pathways for conveying tactile and proprioceptive information from the head to the cerebellum.

Human somatosensory research has mainly focused on data obtained from cortical measures through fMRI, magnetoencephalography and electroencephalography. Detailed somatosensory maps can now be reproduced with noninvasive, high-resolution functional imaging, which can be combined with computational modeling to study somatosensory perception, attention, and anticipation signals (Liang et al., 2011). However, to understand the sensory and cognitive processes related to somatosensation, single-cell recording and behavioral experiments performed in monkeys (Roma et al., 2012) are invaluable. Functional connectivity is generally consistent between monkey and human somatosensory cortices (Thomas et al., 2021), supporting the use of nonhuman primates as a model.

PARALLEL PATHWAYS THROUGH THALAMUS AND CORTEX: ANIMAL MODELS

Animal models offer a more detailed understanding of neural processing at the synaptic and circuit levels. The majority of somatosensory research has focused on tactile processing in rodents and monkeys, which, like the pathways discussed herein, have also occurred in parallel with little integrative interpretation (for review, see O'Connor et al., 2021). Researchers have focused on each species' specializations: The hand in monkeys dominates the representation in S1, with the palm and digits in enlarged detail (Nelson et al., 1980), and the representation of the whiskers in mice dominates almost half of S1 (Krubitzer et al., 2011) (Figure 25.2). Hands in a diurnal bipedal animal are used for precise control and manipulation, while whiskers in a nocturnal prey animal are used for navigation, foraging and detection. Thus, despite within-animal advantages of studying salient cortical cartography, species specific behavioral differences lead to different neural wiring and interpretations.

There are profound differences between the somatosensory organization in rodents and monkeys, starting with simply the number of devoted cortical areas (Figure 25.2). Rodents have one primary somatosensory cortex (S1); S2, which is part of the parietal ventral area (PV); and a posterior parietal cortex. In contrast, monkeys have S1 (area 3b) and surrounding somatosensory areas 3a, 1, and 2. In addition to S2 and PV, there are association areas in monkeys, such as the ventral somatosensory area, parietal rostral region, as well as multiple posterior parietal areas relevant to somatosensory processing (Krubitzer and Dooley, 2013). While monkeys clearly have more specialized cortical areas for processing, mice have proportionally more cortical real estate devoted to somatosensation. In mice, M1 is also discussed as part of S1 and S2 signaling loops due to the

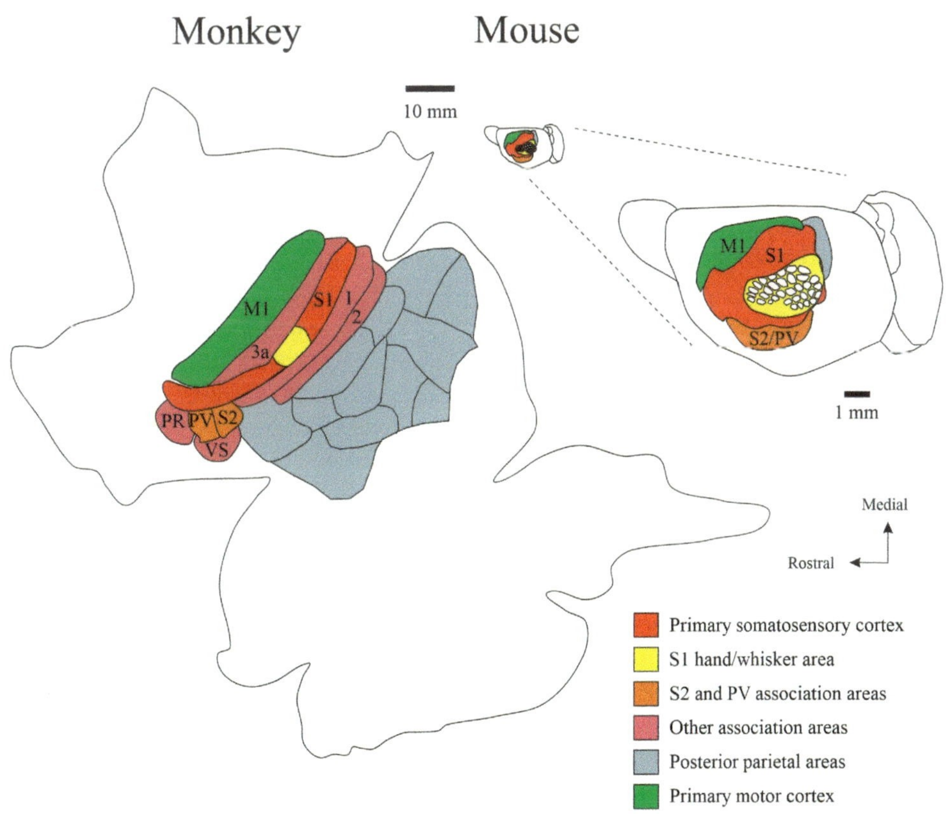

Figure 25.2. Somatosensory cortical areas in the macaque monkey and mouse. Flattened cortex of the macaque (left) compared to mouse (middle) to relative scale , showing relative devotion to somatosensory areas and proportionally large hand and whisker barrel representations (yellow). Enlarged view of mouse cortex is shown on the right. Mice and monkeys have homologous primary somatosensory cortex (red), secondary somatosensory areas (orange), posterior parietal areas (blue), and primary motor cortex (green). Other associated somatosensory areas (pink) are only found in monkeys. Also note the greatly expanded and multiple posterior parietal areas in the monkey compared to the mouse. Adapted from Krubitzer and Dooley (2013); Kirkcaldie (2012).

existence of direct, reciprocal, and somatotopic projections between the areas (Feldmeyer et al., 2013). In contrast, monkey M1 lacks direct connectivity with S1, instead signaling via area 2 and posterior parietal area 5. However, recent imaging shows strong functional association between area 3b and M1 representations in the macaque (Thomas et al., 2021). In monkeys, there are multiple thalamic nuclei involved in somatosensory processing, whereas rodent thalamic nuclei are limited to VP and POm. There are no homologs of many monkey association areas and thalamic nuclei, and thus rats and mice have a much less complex somatosensory network. Such basic species differences make it difficult to establish fundamental organizational principles in mammals.

PRIMATE SOMATOSENSORY PATHWAYS

The monkey somatosensory thalamus comprises the ventroposterior complex which has 3 subdivisions, each somatotopically organized and each receiving different inputs (Padberg et al., 2009): VP, which includes a lateral part (VPL) and medial part (VPm); VPS; and VPI. The thalamocortical circuitry has been mapped in detail, but less focus has been placed on its functional contribution due to its historical casting as an uninteresting passive relay of peripheral signals to the cortex. Cutaneous projections show strong topography in the primary somatosensory cortex, which is generally accepted as Brodmann area 3b. The other somatosensory cortical areas include area 1, which processes cutaneous input; area 3a, which processes proprioceptive input; and area 2, which integrates cutaneous and proprioceptive inputs. Much of the research in nonhuman primates has leveraged access to, and manipulation of, cortical areas. This has led to a deeper understanding of stimulus processing at an integrative level in somatosensory and association cortex (Saal and Bensmaia, 2014).

MEDIAL LEMNISCUS AND SPINOTHALAMIC PATHWAYS IN THE MONKEY

Nonhuman primates also have a medial lemniscus-dorsal columns pathway to process discriminative touch and a spinothalamic pathway to process pain and temperature. Similar to humans, these pathways ascend from the dorsal column nuclei and dorsal horn, respectively to synapse in the thalamus and then the cortex.

Projections from the dorsal column nuclei carry hand and foot cutaneous information to VPL, while projections from the trigeminal nucleus carry face and head information to VPM. The majority of VP cells project to area 3b and area 1. Area 1 contains a higher-order representation of touch just caudal to area 3b. VP neurons also sparsely project to areas 3a, 2, and 5 (Padberg et al., 2009), but not significantly to S2, which is served by VPI. The anterior pulvinar also sends projections to area 3b, 1, 2, and likely 3a, acting to interconnect somatosensory areas with other cortices.

Muscle spindles and other deep body receptors project proprioceptive information via the medial lemniscus to VPS, which in turn projects to areas 2 and 5 (Delhaye et al., 2018). The posterior division of the ventral lateral nucleus (VLp) also has neurons that respond to muscle and joint movement, which projects densely to motor and premotor areas and to area 3a (Padberg et al., 2009). In macaque monkeys, multiple thalamic nuclei beyond VP are associated with the somatosensory system, such as the anterior pulvinar (Padberg et al., 2009). In VPL, the majority of the medial lemniscus tract inputs are modulated by GABA, forming triadic synapses with interneurons, in contrast to inputs of the spinothalamic pathway which do not contact GABAergic appendages (Ralston and Ralston, 1994). This suggests that ascending noxious information is not inhibited through the thalamic relay.

The spinothalamic tract carries noxious and non-noxious stimuli from cutaneous and deep structures and projects via VPI to S2 for thermoreceptive and nociceptive processing. VPI densely projects to higher-order cortical areas and sparsely projects to somatosensory cortex and area 5 (Disbrow et al., 2003). To a lesser extent, there are also inputs to the thalamic posterior nuclear complex, the caudal part of VPL (VPLc), and the centrolateral complex. The posterior part of the ventral medial nucleus (VMpo) also encodes pain and temperature. The major cortical targets of the spinothalamic tract include granular insular cortex, S2, and several cortical areas in the cingulate sulcus.

In the monkey, the medial lemniscus and spinothalamic pathways are kept separate through the thalamus until their convergence in S1 (Saal and Bensmaia, 2014). In particular, pathways carrying noxious and non-noxious stimuli are spatially separated. This is in contrast with the rat, where the lemniscal and spinothalamic projections terminate on the same segment of dendrite in VPL (Ma et al., 1987). About half of S1 neurons show synaptic summation and integration of cutaneous and proprioceptive input from the hand (Kim Sung et al., 2015). The overlapping interconnections of somatosensory cortical regions suggest that processing is integrated once the pathways reach S1 (Disbrow et al., 2003).

S1 has a proportionally large hand representation (Nelson et al., 1980) and single-digit representations are spatially separated by septa. S1 cells are highly tuned to textures, and encode coarse and fine features using a spatial and temporal code (Lieber and Bensmaia, 2019). The higher-order areas S2 and PV show more complex response properties and encode decision variables relevant to the task (Long et al., 2022). The ventral stream, in the lateral parietal cortex, includes S2 and PV. The dorsal stream in the PPC includes areas 5 and 7 and is responsive to motor movements, such as reach and grasp.

RODENT SOMATOSENSORY PATHWAYS

Relative to other senses, rodents depend on their somatosensation to a much larger extent than humans, particularly the sensations from the stiff, functional hairs on the face—the whiskers (or vibrissae, from the Latin word "to vibrate"). Rats and mice are nocturnal and use their whiskers to navigate through dimly lit, preferably unexposed spaces, but also to perform other tasks such as detecting obstacles, sensing air flow, locating and discriminating objects, and for social interaction. Their importance is reflected by the extraordinary amount of S1 space dedicated to their representation (Krubitzer et al., 2011) and in the clustered input to layer 4 creating "barrel-like" cytoarchitecture (Woolsey and Van der Loos, 1970) (Figure 25.2). Individual barrels are separated by cell-poor septa and represent an individual whisker. It is no surprise that researchers have also dedicated a lot of their own cortex to understanding the rodent whisker system. It offers a spectacular example of parallel processing where individual sense organs have a discrete neural representation along the sensory pathway. Despite the provisos for generalization to humans, several synaptic and circuit-level lessons can be learned from studying the whisker pathways in the rodent, which benefits from the technological advances to manipulate and monitor cells. Here, we will focus on the whiskers and thus trigeminal-based touch pathways; however, rodents also have homologous medial lemniscus and spinothalamic pathways to process inputs from the body (Figure 25.3).

Whisker input is carried by the trigeminal nerve, and these afferents synapse in the trigeminal brainstem nuclei; some innervate PrV or SpV, and some bifurcate to innervate both. Each neuron is activated by deflection of only one vibrissa (Zucker and Welker, 1969). Projections from the principal nucleus of the trigeminal nucleus (PrV) or the spinal trigeminal nucleus (SpV) to thalamic nuclei and somatosensory cortical areas form the lemniscal, paralemniscal, and extralemniscal pathways in the rodent (Ahissar et al., 2000; Yu et al., 2006). However, it should be noted that streams of axons weave between these channels, and individual cells can bifurcate and branch to innervate multiple structures.

LEMNISCAL PATHWAY

Similar to the dorsal column medial lemniscus pathway, the lemniscal pathway in rodents is the first-order relay of tactile information from brainstem PrV to VPm in thalamus, to S1. A topographic map of each whisker is found not only in the cortex as barrels, but also as "barrelettes" in the brainstem nuclei and "barreloids" in the first-order VPm of thalamus (Ma and Woolsey, 1984; Van Der Loos, 1976). Thus, each whisker on the whisker pad is individually represented in neural space at each level of processing, and the barrel map in S1 remarkably mimics the whisker layout on the whisker pad.

Stimulation of a single whisker causes focal, precisely timed, short-latency action potentials in dorsomedial VPm (4–8 msec) and S1 (8–12 msec) (Bruno and Sakmann, 2006; Diamond et al., 1992). The fast and robust neural response to the stimulus is mediated via efficient driver synaptic transmission (Sherman and Guillery, 1998) from PrV to VPm, and VPm to S1 (Mo et al., 2017; Viaene et al., 2011a). All layers of S1 receive excitatory monosynaptic input from VPm, but layer 4 barrels receive the highest density of axonal inputs from the lemniscal pathway (El-Boustani et al., 2020). Barreloid and barrel organization reflects cell response preferences to whisker deflection angle, velocity, and direction (Bruno et al., 2003; Kaloti et al., 2016). These properties of the lemniscal projections allow the pathway to convey precise touch information and also encode the phase of the whisk cycle (Yu et al., 2006; Curtis and Kleinfeld, 2009).

PARALEMNISCAL PATHWAY

In contrast to the lemniscal pathway, the paralemniscal pathway for whisker-based sensation is broader in its anatomy and function. It transmits through the rostral part of the brainstem trigeminal nucleus interpolaris (SpVIr) to POm, and then the somatosensory cortical areas (S1, S2) and M1 (Figure 25.3). While PrV of the lemniscal pathway shows a somatotopic whisker map, SpVI of the paralemniscal pathway lacks barrelettes. Accordingly, neural representations of whisker stimuli along this pathway show longer latency and more variable responses in POm compared to VPm (Urbain et al., 2015). This poorer response is partly due to zona incerta inhibition, a regulatory gate found in most higher-order nuclei (Trageser and Keller, 2004). The receptive fields of POm neurons are also larger (6–8 whiskers) and this nucleus responds to multi-whisker movement or stimulation (Ahissar et al., 2000), nociceptive stimuli (Frangeul et al., 2014), and arousal state (Petty et al., 2021).

In alignment with the idea that driver projections are required to support the propagation of receptive field information (Sherman and Guillery, 1998), the synapses along the paralemniscal pathway are modulatory from SpV to POm, and POm to S1 (Mo et al., 2017; Viaene et al., 2011b). An exception is a subset of SpV to POm inputs which are driver and spatially intermixed with modulatory

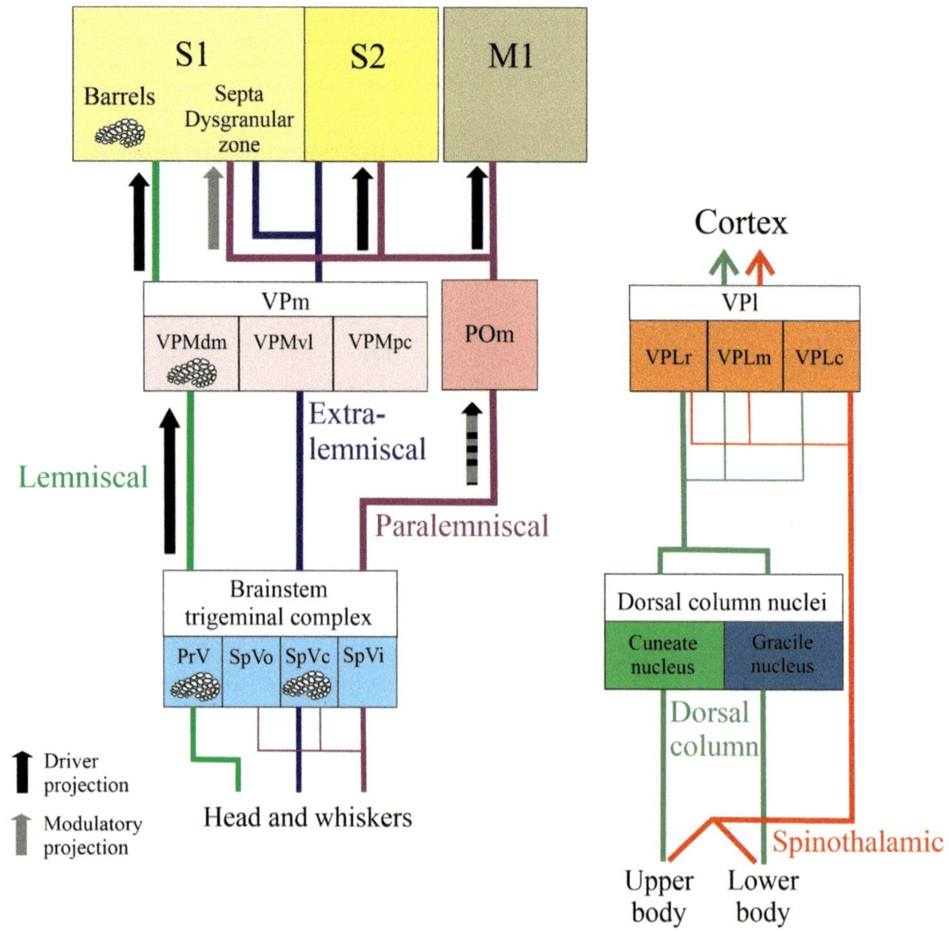

Figure 25.3. Parallel ascending somatosensory pathways in the rodent. Pathways in the rodent from the whiskers (left) and body (right) to brainstem, thalamus, and cortex. The lemniscal pathway transmits sensory input from the whiskers to S1 via driver projections. The paralemniscal pathway from brainstem to POm is a mix of driver and modulator projections and instead driven by layer 5 of S1 (not shown). The output of POm is modulatory to S1 but driver to S2 and M1. Input from the body is delivered via the dorsal column and spinothalamic pathways to thalamus and cortex. Cortical inputs have been omitted for clarity.

inputs (Mo et al., 2017). Driver pathways do not need to be the anatomical majority to exert their strong postsynaptic influence, though it is unknown which of the POm targets are driven or modulated. As discussed in **Chapter 15 in Section IV**, POm is driven by the layer 5 neurons of S1. Instead of transmitting sensory information from the periphery, the paralemniscal pathway may be involved with higher-order cortical or motor processing since POm projections to S2 and M1 are driver synapses to all cortical layers (Miller-Hansen and Sherman, 2022; Mo and Sherman, 2019). There are also other cortical targets, including M2, A1, insular cortex, and frontal cortex (Ohno et al., 2011; Hunnicutt et al., 2016).

Further complicating things is that most individual POm cells branch to project to multiple cortical targets (Ohno et al., 2011), and although the paralemniscal pathway has been traditionally defined as projecting to S1 (Bureau et al., 2006), recent trans-synaptic tracing experiments have revealed that SpVi inputs to anterior POm, which projects to S2 layer 4 (El-Boustani et al., 2020). This suggests that the main paralemniscal pathway targets S2, not S1. The same study also identified a pathway for POm cells which do not receive SpV inputs (i.e., non-paralemniscal), and found that they project to layer 5A and layer 1 of S1 and S2. Considering the involvement of POm in multiple cortical circuits, clusters of POm cells likely participate in cortical pathways and other clusters in ascending peripheral pathways. Through the use of genetic, synaptic, and tracing tools, details of the circuits can be dissected in the mouse (El-Boustani et al., 2020; Elbaz et al., 2022), but currently the function of the paralemniscal pathway remains a debated topic.

The lemniscal input to S1 layer 4 and paralemniscal input to S1 layer 5A may be integrated by a monosynaptic connection between neurons in these 2 layers (Feldmeyer et al., 2005) or within a layer (Mease et al., 2016). However, others have found that the 2 pathways remain segregated at least 2 synapses downstream within

S1, with the lemniscal pathway projecting to layer 4, layer 5B, and layer 6A, and subsequently to layer 3, while the paralemniscal projections mainly synapse in layer 5A and then project to layer 2 (Bureau et al., 2006). If the main paralemniscal route targets S2 rather than S1 (El-Boustani et al., 2020), this would also support discrete initial processing of lemniscal and paralemniscal channels in mouse cortex.

EXTRALEMNISCAL AND OTHER PATHWAYS

The extralemniscal pathway is often considered a third trigemino-thalamo-cortical pathway in the mouse whisker system. It begins in the caudal division of SpV interpolaris (SpVIc), projects to the ventrolateral part of VPm (VPmvl) in the barreloid tail, and projects to the columns of S1 septal area, S1 dysgranular zone, and layers 4 and 6 of S2 (Yu et al., 2006; Pierret et al., 2000). There are multi-whisker, inter-barrelette cells in SpVIc, but the barreloids become indistinguishable in VPmvl compared to VPm of the lemniscal pathway (Pierret et al., 2000). The extralemniscal pathway encodes multi-whisker touch but not whisking alone (Yu et al., 2006) and is also involved in processing pain based on the nociceptive responses reported in the S1 dysgranular zone, an area adjacent to barrel cortex but with a poorly defined layer 4. The dysgranular zone also receives proprioceptive input from deep muscle stimulations or joint rotations (Chapin and Lin, 1984).

Here we have described 3 ascending pathways in rodents, but other whisker pathways have been identified (Ebert et al., 2021), such as from PrV to the "head" of VPm barreloids to S1 septa (Urbain and Deschênes, 2007). In addition, some whisker parallel pathways are functionally described, such as in reference to human visual pathways: S1 to M1 is a location "where" pathway, and S1 to S2 is a feature "what" pathway (Chen et al., 2013).

PROCESSING IN PARALLEL VERSUS SERIAL

We have described parallel streams of the ascending pathways which relay information to brainstem to thalamus to cortex. One discussion point is whether S1 and S2 process information simultaneously or in serial, hierarchical order. Early studies in multiple species report that S2 could be activated via thalamus independent of deactivated S1, supporting a parallel and distributed processing strategy (Garraghty et al., 1991; Zhang et al., 1996). Other work showed the opposite, where S2 activation is serially dependent on S1, and leads to the hypothesis that hierarchical processing is an evolved organization only present in monkeys, and perhaps other higher primates (Pons et al., 1992; Rowe et al., 1996). These results may also be a comment on hierarchical organization in different species in general.

In mice, there is evidence that hierarchical somatosensory processing may be acquired during development (Cai et al., 2022; Gómez et al., 2021), where input to cortex is processed in parallel until postnatal day 12, at which time responses to tactile stimulation are processed in S1 before M1 (Gómez et al., 2021). The paralemniscal pathway to S2 is considered by some as a parallel channel to process tactile information in rodents (Hubatz et al., 2020). However, synaptic characterizations of driver projections support hierarchical processing where the lemniscal pathway comprises driver synapses to S1, which activates corticofugal layer 5 projections from S1 layer 5 neurons to POm, which then synapse on driver projections to S2 (Theyel et al., 2010) (see **Chapter 42 in Section VII**). In contrast to the rodent lemniscal pathway, which serves as a faithful information channel from periphery to cortex, POm of the paralemniscal pathway is a hub for inputs and outputs, which include multiple non-somatosensory cortical areas (Hunnicutt et al., 2016; Casas-Torremocha et al., 2022). It would thus be restrictive to view its role as a parallel relay of the lemniscal pathway.

Human studies are also conflicting in the hierarchical model (Kalberlah et al., 2013; Khoshnejad et al., 2014) versus parallel model (Liang et al., 2011; Cung et al., 2014) of somatosensory processing. To reconcile this, some have suggested that time is a crucial factor in serial versus parallel processing. Based on MRI data, stimuli are processed in parallel in the first 100 msec, but then a serial processing mode takes over (Klingner et al., 2016). Thus, S2 receives input from a new stimulus at the same time as S1, but subsequent processing in S2 is augmented by preprocessed S1 input. Another hypothesis is that modality plays a role, where proprioceptive and nociceptive inputs are processed in parallel, but tactile input is processed in series (Ploner et al., 1999). A limitation of using human neuroimaging data to address this question is that the signal is integrated over many seconds and centimeters of brain tissue, providing an example of when invasive access to non-human neural pathways is a superior approach. Thus, despite activation of S1 and S2 almost simultaneously in response to a stimulus, hierarchical processing can be considered to be the main principle that underlies the processing of tactile stimuli.

CONCLUSIONS

Somatosensation is the body sense and therefore is perhaps the most crucial sense for interacting with the world. It is specialized for functionally important body parts, and for this reason, homology is an important consideration when extrapolating the neural circuit details in animal models to that of humans. Nonetheless, there are segregated channels across humans, monkeys, and rodents at each level of processing, illuminating a conserved logic to parallel pathways in processing somatosensory inputs. Understanding how the complex circuitry integrates the pathways for sensation and perception will require tracing across multiple synapses, combined with functional synaptic assessments. Such experiments have already stimulated a reconsideration of the pathways as peripheral sensory channels, cortical processing channels, or a mix or both.

REFERENCES

Ahissar E, Sosnik R, Haidarliu S. (2000). Transformation from temporal to rate coding in a somatosensory thalamocortical pathway. *Nature.* 406(6793):302–306. doi:10.1038/35018568.

Bruno RM, Khatri V, Land PW, Simons DJ. (2003). Thalamocortical angular tuning domains within individual barrels of rat somatosensory cortex. *J Neurosci.* 23(29):9565–9574. doi:10.1523/jneurosci.23-29-09565.2003.

Bruno RM, Sakmann B. (2006). Cortex is driven by weak but synchronously active thalamocortical synapses. *Science.* 312(5780):1622–1627. doi:10.1126/science.1124593.

Bureau I, von Saint Paul F, Svoboda K. (2006). Interdigitated paralemniscal and lemniscal pathways in the mouse barrel cortex. *PLOS Biol.* 4(12):e382. doi:10.1371/journal.pbio.0040382.

Cai L, Yang JW, Wang CF, Chou SJ, Luhmann HJ, Karayannis T. (2022). Identification of a developmental switch in information transfer between whisker S1 and S2 cortex in mice. *J Neurosci.* 42(22):4435–4448. doi:10.1523/jneurosci.2246-21.2022.

Casas-Torremocha D, Rubio-Teves M, Hoerder-Suabedissen A, Hayashi S, Prensa L, Molnár Z, Porrero C, Clasca F (2022). A combinatorial input landscape in the "higher-order relay" posterior thalamic nucleus. *J Neurosci.* 42(41):7757–7781. doi:10.1523/jneurosci.0698-22.2022.

Chapin JK, Lin CS. (1984). Mapping the body representation in the SI cortex of anesthetized and awake rats. *J Comp Neurol.* 229(2):199–213. doi:10.1002/cne.902290206.

Chen JL, Carta S, Soldado-Magraner J, Schneider BL, Helmchen F. (2013). Behaviour-dependent recruitment of long-range projection neurons in somatosensory cortex. *Nature.* 499(7458):336–340. doi:10.1038/nature12236.

Chung YG, Han SW, Kim H-S, Chung S-C, Park J-Y, Wallraven C, Kim S-P (2014). Intra- and inter-hemispheric effective connectivity in the human somatosensory cortex during pressure stimulation. *BMC Neurosci.* 15(1):43. doi:10.1186/1471-2202-15-43.

Curtis JC, Kleinfeld D. (2009). Phase-to-rate transformations encode touch in cortical neurons of a scanning sensorimotor system. *Nat Neurosci.* 12(4):492–501. doi:10.1038/nn.2283.

Delhaye BP, Long KH, Bensmaia SJ. (2018). Neural basis of touch and proprioception in primate cortex. *Compr Physiol.* 8(4):1575–1602. doi:10.1002/cphy.c170033.

Diamond ME, Armstrong-James M, Ebner FF. (1992). Somatic sensory responses in the rostral sector of the posterior group (POm) and in the ventral posterior medial nucleus (VPM) of the rat thalamus. *J Comp Neurol.* 318(4):462–476. doi:https://doi.org/10.1002/cne.903180410.

Disbrow E, Litinas E, Recanzone G, Padberg J, Krubitzer L. (2003). Thalamocortical connections of the parietal ventral area (PV) and the second somatosensory area (S2) in macaque monkeys. *J Comp Neurol.* 462:382–399. doi:10.1002/cne.10731.

Donaldson HH. (1890). Anatomical observations on the brain and several sense-organs of the blind deaf-mute, Laura Dewey Bridgman. *Am J Psychol.* 3(3):293–342. doi:10.2307/1411697.

Ebert C, Bagdasarian K, Haidarliu S, Ahissar E, Wallach A. (2021). Interactions of Whisking and Touch Signals in the Rat Brainstem. *J Neurosci.* 41(22):4826–4839. doi:10.1523/jneurosci.1410-20.2021.

El-Boustani S, Sermet BS, Foustoukos G, Oram TB, Yizhar O, Petersen CCH. (2020). Anatomically and functionally distinct thalamocortical inputs to primary and secondary mouse whisker somatosensory cortices. *Nat Commun.* 11(1):3342. doi:10.1038/s41467-020-17087-7.

Elbaz M, Callado Perez A, Demers M, Zhao S, Foo C, Kleinfeld D, Deschênes M (2022). A vibrissa pathway that activates the limbic system. *eLife.* 11:e72096. doi:10.7554/eLife.72096.

Feldmeyer D, Brecht M, Helmchen F, Petersen CCH, Poulet JFA, Staiger JF, Luhmann HJ, Schwarz C (2013). Barrel cortex function. *Prog Neurobiol.* 103:3–27. doi:10.1016/j.pneurobio.2012.11.002.

Feldmeyer D, Roth A, Sakmann B. (2005). Monosynaptic connections between pairs of spiny stellate cells in layer 4 and pyramidal cells in layer 5A indicate that lemniscal and paralemniscal afferent pathways converge in the infragranular somatosensory cortex. *J Neurosci.* 25(13):3423–3431. doi:10.1523/jneurosci.5227-04.2005.

Frangeul L, Porrero C, Garcia-Amado M, Maimone B, Maniglier M, Clasca F, Jabaudon D (2014). Specific activation of the paralemniscal pathway during nociception. *Eur J Neurosci.* 39(9):1455–1464. doi:https://doi.org/10.1111/ejn.12524.

Garraghty PE, Florence SL, Tenhula WN, Kaas JH. (1991). Parallel thalamic activation of the first and second somatosensory areas in prosimian primates and tree shrews. *J Comp Neurol.* 311(2):289–299. doi:https://doi.org/10.1002/cne.903110209.

Gómez LJ, Dooley JC, Sokoloff G, Blumberg MS. (2021). Parallel and Serial Sensory Processing in Developing Primary Somatosensory and Motor Cortex. *J Neurosci.* 41(15):3418–3431. doi:10.1523/jneurosci.2614-20.2021.

Hubatz S, Hucher G, Shulz DE, Férézou I. (2020). Spatiotemporal properties of whisker-evoked tactile responses in the mouse secondary somatosensory cortex. *Sci Rep.* 10(1):763. doi:10.1038/s41598-020-57684-6.

Hunnicutt BJ, Jongbloets BC, Birdsong WT, Gertz KJ, Zhong H, Mao T. (2016). A comprehensive excitatory input map of the striatum reveals novel functional organization. *eLife.* 5:e19103. doi:10.7554/eLife.19103

Kalberlah C, Villringer A, Pleger B. (2013). Dynamic causal modeling suggests serial processing of tactile vibratory stimuli in the human somatosensory cortex: An fMRI study. *NeuroImage.* 74:164–171. doi:https://doi.org/10.1016/j.neuroimage.2013.02.018.

Kaloti AS, Johnson EC, Bresee CS, Naufel SN, Perich MG, Jones DL, Hartmann MJZ (2016). Representation of stimulus speed and direction in vibrissal-sensitive regions of the trigeminal nuclei: A comparison of single unit and population responses. *PLoS ONE.* 11(7):e0158399. doi:10.1371/journal.pone.0158399.

Khoshnejad M, Piché M, Saleh S, Duncan G, Rainville P. (2014). Serial processing in primary and secondary somatosensory cortex: A DCM analysis of human fMRI data in response to innocuous and noxious electrical stimulation. *Neurosci Lett.* 577:83–88. doi:https://doi.org/10.1016/j.neulet.2014.06.013.

Kim SS, Gomez-Ramirez M, Thakur PH, Hsiao SS. (2015). Multimodal interactions between proprioceptive and cutaneous signals in primary somatosensory cortex. *Neuron.* 86(2):555–566. doi:https://doi.org/10.1016/j.neuron.2015.03.020.

Kirkcaldie M. (2012). Neocortex. In: Watson CR Paxinos G, Puelles L, eds. *The Mouse Nervous System.* Academic Press:52–111.

Klingner CM, Brodoehl S, Huonker R, Witte OW. (2016). The processing of somatosensory information shifts from an early parallel into a serial processing mode: A combined fMRI/MEG study. Original research. *Front Syst Neurosci.* 10: doi:10.3389/fnsys.2016.00103.

Koeppen B, Stanton B. (2008). The somatosensory system. In: *Berne & Levy Physiology*. 6th ed. Mosby:105–122.

Krubitzer L, Campi KL, Cooke DF. (2011). All rodents are not the same: A modern synthesis of cortical organization. *Brain Behav Evol.* 78(1):51–93. doi:10.1159/000327320.

Krubitzer L, Dooley JC. (2013). Cortical plasticity within and across lifetimes: How can development inform us about phenotypic transformations? *Front Hum Neurosci.* 7:620. doi:10.3389/fnhum.2013.00620.

Liang M, Mouraux A, Iannetti GD. (2011). Parallel processing of nociceptive and non-nociceptive somatosensory information in the human primary and secondary somatosensory cortices: Evidence from dynamic causal modeling of functional magnetic resonance imaging data. *J Neurosci.* 31(24):8976–8985. doi:10.1523/jneurosci.6207-10.2011.

Lieber JD, Bensmaia SJ. (2019). High-dimensional representation of texture in somatosensory cortex of primates. *Proc Natl Acad Sci.* 116(8):3268–3277. doi:doi:10.1073/pnas.1818501116.

Long KH, Greenspon CM, van Driesche A, Lieber JD, Bensmaia SJ. (2022). Texture coding in higher order somatosensory cortices of primates. *bioRxiv*. 504511. doi:10.1101/2022.08.19.504511.

Ma PM, Woolsey TA. (1984). Cytoarchitectonic correlates of the vibrissae in the medullary trigeminal complex of the mouse. *Brain Res.* 306(1):374–379. doi:https://doi.org/10.1016/0006-8993(84)90390-1.

Ma W, Peschanski M, Ralston HJ. (1987). The differential synaptic organization of the spinal and lemniscal projections to the ventrobasal complex of the rat thalamus: Evidence for convergence of the two systems upon single thalamic neurons. *Neuroscience.* 22(3):925–934. doi:https://doi.org/10.1016/0306-4522(87)92970-8.

Mease RA, Metz M, Groh A. (2016). Cortical sensory responses are enhanced by the higher-order thalamus. *Cell Rep.* 14(2):208–215. doi:https://doi.org/10.1016/j.celrep.2015.12.026.

Miller-Hansen AJ, Sherman SM. (2022). Conserved patterns of functional organization between cortex and thalamus in mice. *Proc Natl Acad Sci.* 119(21):e2201481119. doi:doi:10.1073/pnas.2201481119.

Mo C, Petrof I, Viaene AN, Sherman SM. (2017). Synaptic properties of the lemniscal and paralemniscal pathways to the mouse somatosensory thalamus. *Proc Natl Acad Sci.* 114(30):E6212–E6221. doi:10.1073/pnas.1703222114.

Mo C, Sherman SM. (2019). A sensorimotor pathway via higher-order thalamus. *J Neurosci.* 39(4):692–704. doi:10.1523/jneurosci.1467-18.2018.

Muret D, Root V, Kieliba P, Clode D, Makin TR. (2022). Beyond body maps: Information content of specific body parts is distributed across the somatosensory homunculus. *Cell Rep.* 38(11):110523. doi:https://doi.org/10.1016/j.celrep.2022.110523.

Nelson RJ, Sur M, Felleman DJ, Kaas JH. (1980). Representations of the body surface in postcentral parietal cortex of Macaca fascicularis. *J Comp Neurol.* 192(4):611–643. doi:10.1002/cne.901920402

Niu J, Ding L, Li JJ, Kim H, Liu J, Li H, Moberly A, Badea TC, Duncan ID, Son Y-J, Scherer SS, Luo W (2013). Modality-based organization of ascending somatosensory axons in the direct dorsal column pathway. *J Neurosci.* 33(45):17691. doi:10.1523/JNEUROSCI.3429-13.2013.

O'Connor DH, Krubitzer L, Bensmaia S. (2021). Of mice and monkeys: Somatosensory processing in two prominent animal models. *Prog Neurobiol.* 201:102008. doi:10.1016/j.pneurobio.2021.102008.

Ohno S, Kuramoto E, Furuta T, Hioki H, Tanaka YR, Fujiyama F, Sonomura T, Uemura M, Sugiyama K, Kaneko T (2011). A morphological analysis of thalamocortical axon fibers of rat posterior thalamic nuclei: A single neuron tracing study with viral vectors. *Cereb Cortex.* 22(12):2840–2857. doi:10.1093/cercor/bhr356.

Padberg J, Cerkevich C, Engle J, Rajan AT, Recanzone G, Kaas J, Krubitzer L (2009). Thalamocortical connections of parietal somatosensory cortical fields in macaque monkeys are highly divergent and convergent. *Cerebral Cortex.* 19(9):2038–2064. doi:10.1093/cercor/bhn229.

Penfield W, Rasmussen T. (1950). *The Cerebral Cortex of Man: A Clinical Study of Localization of Function*. Macmillan.

Petty GH, Kinnischtzke AK, Hong YK, Bruno RM. (2021). Effects of arousal and movement on secondary somatosensory and visual thalamus. *eLife*. 10:e67611. doi:10.7554/eLife.67611.

Pierret T, Lavallée P, Deschênes M. (2000). Parallel streams for the relay of vibrissal information through thalamic barreloids. *J Neurosci.* 20(19):7455–7462. doi:10.1523/jneurosci.20-19-07455.2000.

Ploner M, Schmitz F, Freund H-J, Schnitzler A. (1999). Parallel activation of primary and secondary somatosensory cortices in human pain processing. *J Neurophysiol.* 81(6):3100–3104. doi:10.1152/jn.1999.81.6.3100.

Pons TP, Garraghty PE, Mishkin M. (1992). Serial and parallel processing of tactual information in somatosensory cortex of rhesus monkeys. *J Neurophysiol.* 68(2):518–527. doi:10.1152/jn.1992.68.2.518.

Ralston HJ, 3rd, Ralston DD. (1994). Medial lemniscal and spinal projections to the macaque thalamus: an electron microscopic study of differing GABAergic circuitry serving thalamic somatosensory mechanisms. *J Neurosci.* 14(5 Pt 1):2485–502. doi:10.1523/jneurosci.14-05-02485.1994.

Romo R, Lemus L, de Lafuente V. (2012). Sense, memory, and decision-making in the somatosensory cortical network. *Curr Opin Neurobiol.* 22(6):914–919. doi:10.1016/j.conb.2012.08.002.

Rowe M, Turman A, Murray G, Zhang H. (1996). Parallel organization of somatosensory cortical areas I and II for tactile processing. *Clin Exper Pharmacol Physiol.* 23(10–11):931–938. doi:https://doi.org/10.1111/j.1440-1681.1996.tb01145.x.

Saal HP, Bensmaia SJ. (2014). Touch is a team effort: Interplay of submodalities in cutaneous sensibility. *Trends Neurosci.* 37(12):689–697. doi:10.1016/j.tins.2014.08.012.

Sherman SM, Guillery RW. (1998). On the actions that one nerve cell can have on another: Distinguishing "drivers" from "modulators." *Proc Natl Acad Sci.* 95(12):7121–7126. doi:10.1073/pnas.95.12.7121.

Theyel BB, Llano DA, Sherman SM. (2010). The corticothalamocortical circuit drives higher-order cortex in the mouse. *Nat Neurosci.* 13(1):84–88. doi:10.1038/nn.2449.

Thomas J, Sharma D, Mohanta S, Jain N. (2021). Resting-state functional networks of different topographic representations in the somatosensory cortex of macaque monkeys and humans. *NeuroImage.* 228:117694. doi:https://doi.org/10.1016/j.neuroimage.2020.117694.

Trageser JC, Keller A. (2004). Reducing the uncertainty: Gating of peripheral inputs by zona incerta. *J Neurosci.* 24(40):8911–8915. doi:10.1523/jneurosci.3218-04.2004.

Urbain N, Deschênes M. (2007) Motor cortex gates vibrissal responses in a thalamocortical projection pathway. *Neuron.* 56(4):714–725. doi:https://doi.org/10.1016/j.neuron.2007.10.023.

Urbain N, Salin PA, Libourel PA, Comte JC, Gentet LJ, Petersen CCH. (2015). Whisking-related changes in neuronal firing and membrane potential dynamics in the somatosensory thalamus of awake mice. *Cell Rep.* 13(4):647–656. doi:10.1016/j.celrep.2015.09.029.

Van Der Loos H. (1976). Barreloids in mouse somatosensory thalamus. *Neurosci Lett.* 2(1):1–6. doi:10.1016/0304-3940(76)90036-7.

Viaene AN, Petrof I, Sherman SM. (2011a). Synaptic properties of thalamic input to the subgranular layers of primary somatosensory and auditory cortices in the mouse. *J Neurosci.* 31(36):12738–12747. doi:10.1523/jneurosci.1565-11.2011.

Viaene AN, Petrof I, Sherman SM. (2011b). Properties of the thalamic projection from the posterior medial nucleus to primary and secondary somatosensory cortices in the mouse. *Proc Natl Acad Sci.* 108(44):18156–18161. doi:10.1073/pnas.1114828108.

Woolsey TA, Van der Loos H. (1970). The structural organization of layer IV in the somatosensory region (SI) of mouse cerebral cortex: The description of a cortical field composed of discrete cytoarchitectonic units. *Brain Res.* 17(2):205–242. doi:10.1016/0006-8993(70)90079-x.

Yu C, Derdikman D, Haidarliu S, Ahissar E. (2006). Parallel thalamic pathways for whisking and touch signals in the rat. *PLOS Biol.* 4(5):e124. doi:10.1371/journal.pbio.0040124

Zhang HQ, Murray GM, Turman AB, Mackie PD, Coleman GT, Rowe MJ. (1996). Parallel processing in cerebral cortex of the marmoset monkey: Effect of reversible SI inactivation on tactile responses in SII. *J Neurophysiol.* 76(6):3633–3655. doi:10.1152/jn.1996.76.6.3633

Zucker E, Welker W. (1969). Coding of somatic sensory input by vibrissae neurons in the rat's trigeminal ganglion. *Brain Res.* 12(1):138–156.

26.

DUPLICATED THALAMOCORTICAL INPUT

LESSONS FROM THE SOMATOSENSORY SYSTEM

Randy M. Bruno

To what extent are the properties of cortex established by its inputs from the thalamus? Whether the cortex—or any structure—inherits or transforms its inputs has been an enduring question in neuroscience. This question has been extensively addressed in multiple sensory systems where sensory stimuli facilitate investigation of the circuitry and its computations. The rodent whisker system has been particularly useful in such studies: each of its several dozen large facial whiskers (vibrissae) corresponds to an anatomically identifiable column in the cortex. This clear mapping of sensors to networks has been beneficial to mapping connectivity and characterizing thalamocortical interactions. This chapter focuses on the somatosensory system, with an eye toward other modalities.

Across the neocortex, thalamocortical synapses are known to be established in every cortical layer and yet are concentrated in particular cortical layers. What does this thalamocortical architecture contribute to cortical activity? This chapter reviews anatomical and physiological results that relate to this question.

DUPLICATED THALAMOCORTICAL INNERVATION

The primary (first-order) thalamic nuclei of the somatosensory system are the ventral posterior medial (VPM), which represents the face, and the ventral posterior lateral (VPL), which represents all other parts of the body. In rodents, VPM has a prominent representation of the large facial whiskers (vibrissae) on which rodents rely so heavily for exploring the world. Bulk labeling of axons shows that rodent VPM arborizes most densely in layer 4 (L4) but has a distinct second zone of sparse collaterals in the deeper cortical layers, around the L5/6 border (Herkenham, 1980; Chmielowska et al., 1989; Senft and Woolsey, 1991; Lu and Lin, 1993; Wimmer et al., 2010). Rather than reflecting 2 separate populations of thalamocortical neurons, these 2 zones of thalamocortical arbors are collaterals of the same cells: Reconstruction of single VPM axons, such as in Figure 26.1A, reveals that each axon typically bifurcates to innervate both L4 and the L5/6 border (Jensen and Killackey, 1987; Pierret et al., 2000; Gheorghita et al., 2006; Oberlaender et al., 2012a). Thus, all the information that VPM makes available to L4 of somatosensory cortex has the potential to be fully duplicated in L5/6.

Dual thalamocortical engagement of L4 and L5/6 appears common to mammalian species and neocortical systems. Cat VPL axons similarly innervate the body representation of somatosensory cortex (Landry and Deschenes, 1981; Landry et al., 1987). One of the earliest single thalamocortical axon studies is of the cat lateral geniculate nucleus (Ferster and LeVay, 1978), which found that almost all of the reconstructed axons target both L4 and the deeper layers (see Figure 26.1B). The more evolved visual cortex of nonhuman primates may receive a greater variety of LGN termination patterns but still contain this basic dual-target pattern (Hendrickson et al., 1978; Blasdel and Lund, 1983). Staining of the thalamocortical marker VGluT2 suggests similar dual targeting in macaque and human primary visual cortex (Garcia-Marin et al., 2013). The pattern has also been observed in cat and rat MGB projections to auditory cortex (Huang and Winer, 2000; Smith et al., 2012), rodent VA/VL projections to motor cortex (Kuramoto et al., 2009), and rodent MD projections to frontal cortex (Kuramoto et al., 2017).

The secondary somatosensory thalamic nucleus of the rodent (the posterior or posterior medial, POm) arborizes in a completely different set of cortical layers, namely L1 and L5A (Herkenham, 1980; Lu and Lin, 1993; Wimmer et al., 2010). Labeling VPM and POm separately with viruses that express different fluorophores reveals distinct segregation of these 2 thalamocortical pathways (see

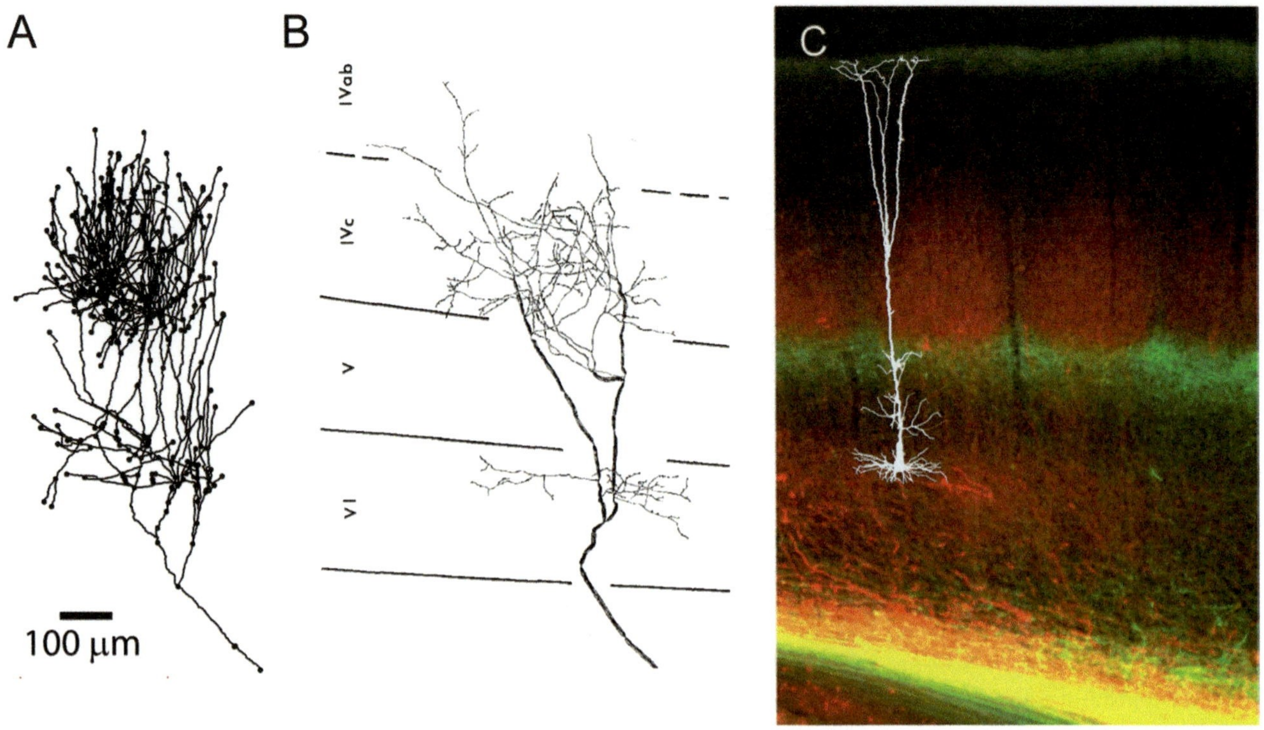

Figure 26.1. Duplicated thalamocortical innervation of L4 and L5/6. (A). Reconstructed VPM axon filled in an adult rat by whole-cell recording. Adapted from Oberlaender et al. (2012a). (B). Reconstructed LGN axon filled in an adult cat by sharp recording. Adapted from Ferster and Levay (1978). (C). Thalamocortical section through barrel cortex showing VPM (red) and POm axons (green) labeled in the same rat with AAV-mediated expression of different fluorophores. Adapted from Wimmer et al. (2010). A reconstruction of a L5B pyramidal neuron from a different animal has been overlaid for illustration purposes.

Figure 26.1C). Few reconstructions have been made of individual POm axons (Ohno et al., 2012), but those that exist suggest that, like for rodent VPM, information from any given POm cell is also duplicated in its 2 target layers. Interestingly, many corticocortical projections to the primary somatosensory cortex, including from motor cortex and secondary somatosensory cortex, also exhibit a dual-target pattern (Veinante and Deschenes, 2003; Zhang and Bruno, 2019). This architectural feature of duplicated innervation in the upper and deeper cortical may be common across cortex-targeting projections.

THE VPM-L4 PATHWAY: AN IDEAL SYNCHRONY DETECTOR

As primary thalamus most densely innervates cortical L4, thalamocortical studies in the somatosensory system have largely focused on the VPM-L4 circuit. One of the most conspicuous features of this circuit is a strong feedforward inhibition (FFI) motif, which has substantial consequences for cortical responses.

This motif is found in diverse brain areas, including but not limited to the neocortex, hippocampus, cerebellum, and amygdala. As shown in Figure 26.2A, the key ingredients of a strong FFI circuit are the following: (1) a group of presynaptic neurons directly excites both the excitatory (spiny, glutamatergic) and inhibitory (aspiny, GABAergic/glycinergic) neurons of a second postsynaptic group, but (2) provides greater synaptic input to the inhibitory neurons than excitatory neurons, and (3) the postsynaptic neurons are interconnected. Simulations show that circuits lacking inhibition simply relay presynaptic activity to postsynaptic neurons (Figure 26.2B,C). In contrast, postsynaptic neurons in FFI circuits are highly sensitive to the relative timing of action potentials among presynaptic neurons (Pinto et al., 1996; Kremkow et al., 2010). Rapid, simultaneous increases in discharges of presynaptic neurons excite the postsynaptic group before excitation is overwhelmed by disynaptic inhibition (Figure 26.2D). In contrast, slow, temporally uncoordinated changes in presynaptic activity elicits fewer or no action potentials among postsynaptic networks (Figure 26.2E). Thus, evoked excitation has a limited "window of opportunity" to escape inhibition (Pinto et al., 1996; Pinto et al., 2000; Gabernet et al., 2005).

While VPM thalamocortical (TC) axons diverge to contact both excitatory and inhibitory L4 neurons, both in vivo and in vitro studies show that inhibitory neurons receive approximately twice as many synapses and with approximately twice the efficacy (Bruno and Simons, 2002;

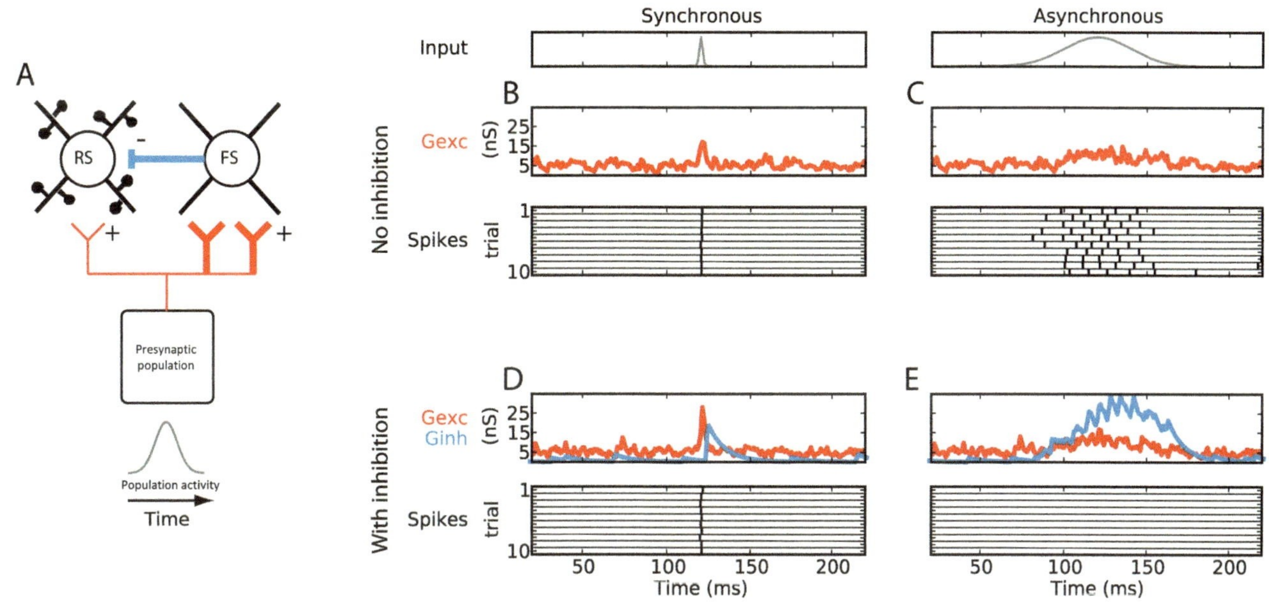

Figure 26.2. Circuits with strong feedforward inhibition (FFI) can selectively gate synchronous over asynchronous thalamic inputs. (A). Minimal FFI model circuit. (B, C). Integration in circuits without FFI and (D, E) with FFI. Shown are the time courses of input population activity (gray) and, for a single postsynaptic excitatory neuron, the total excitatory conductance received (Gexc, red), total inhibitory conductance (Ginh, blue), and spiking output (black). Spike responses are shown for 10 consecutive trials, whereas conductances are for a single trial. In both the absence and presence of inhibition, synchronous input drives postsynaptic output (B, D), but asynchronous inputs do not produce spiking output in the FFI circuit (E). Adapted from Kremkow et al. (2010) with permission from Springer.

Cruikshank et al., 2007). This differential afferent drive is complemented by strong, convergent inhibitory synapses within L4 (Koelbl et al., 2015) and faster membrane time constants among inhibitory neurons (Cruikshank et al., 2007). Together, these 3 mechanisms can ensure that inhibition eventually dominates over excitation.

How can one group of neurons transmit anything to a downstream network through such a suppressive circuit? Individual thalamocortical connections are relatively weak in vivo, depolarizing postsynaptic neurons on average by only ~0.5 mV per thalamic action potential in at least both the rat somatosensory (Bruno and Sakmann, 2006; Schoonover et al., 2014) and cat visual systems (Sedigh-Sarvestani et al., 2017). Furthermore, in awake animals, at least barrel and auditory cortex operate in low-firing rate regimes (Hromadka et al., 2008; O'Connor et al., 2010) with neurons spending little time near threshold (DeWeese and Zador, 2006; Constantinople and Bruno, 2011). Single afferent neurons alone are, therefore, unlikely to play a major role in the coding and transmission of information at this stage.

Thalamic afferents are, however, highly convergent: an excitatory L4 neuron receives synapses from an estimated ~80–90 topographically aligned thalamic neurons (Bruno and Simons, 2002; Bruno and Sakmann, 2006). Thalamocortical neurons spontaneously fire action potentials relatively independently but are tightly time-locked to whisker deflection during sensory stimulation (Pinto et al., 2000). Correlated TC discharges derive from potent common synaptic inputs: Trigeminal brainstem axons diverge to synapse onto many thalamic neurons, individually eliciting depolarizations of up to ~10 mV in vivo (Deschenes et al., 2003). Consequently, effective sensory stimuli induce transient, strong, millisecond-scale synchrony among TC neurons (Bruno and Sakmann, 2006; Temereanca et al., 2008; Wang et al., 2010b) (see also Figure 26.3D, left inset), building feedforward excitation fast enough for L4 neurons to reach threshold before being suppressed by FFI.

Other neocortical sensory circuits share some of this architecture. Assessed functionally, the probability of monosynaptic connections between topographically aligned thalamic and cortical neurons is similar across visual, auditory, and somatosensory systems (Reid and Alonso, 1995; Miller et al., 2001; Bruno and Sakmann, 2006). Anatomically, ~200 spines on a L4 spiny stellate neuron in cat visual cortex receive TC synapses from the lateral geniculate nucleus (LGN) (da Costa and Martin, 2011), similar to the number of TC contacts onto spiny stellate neurons in rat barrel cortex (Meyer et al., 2010; Schoonover et al., 2014). In contrast to VPM projections, LGN connections onto excitatory versus inhibitory neurons of visual cortex appear equivalent in synaptic strength. However, fast-spiking cells still receive a larger proportion of connections, specifically from LGN cells that are not precisely matching cortical cells (Sedigh-Sarvestani et al., 2017). This is consistent with

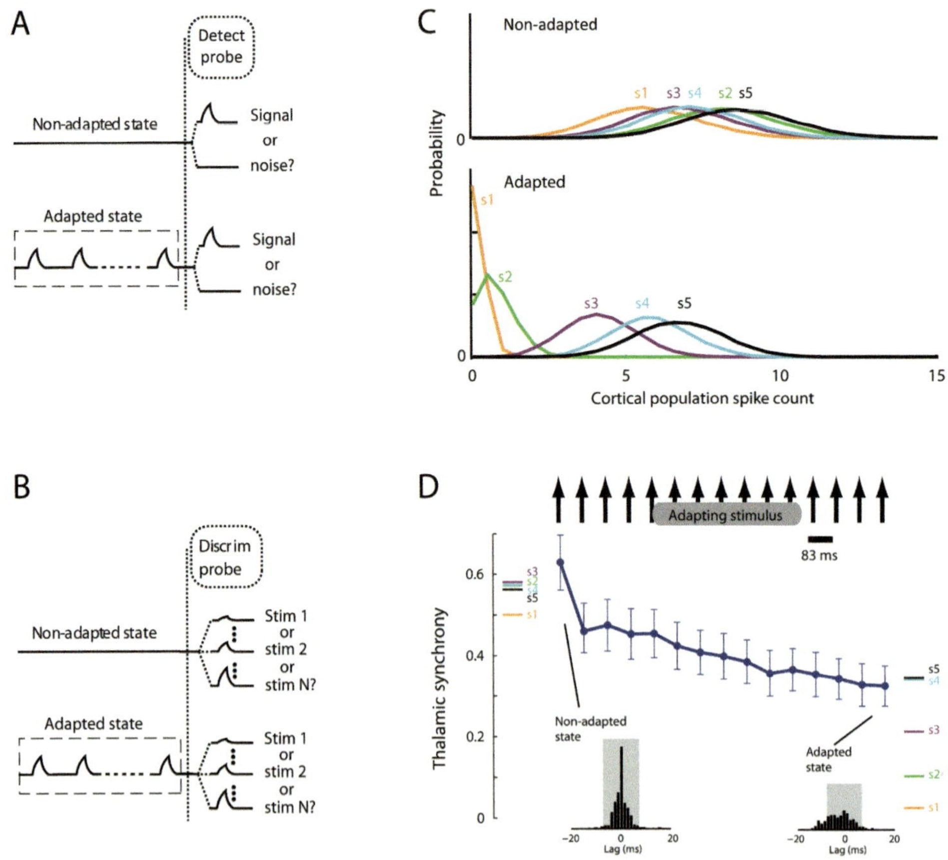

Figure 26.3. Adaptation enhances discrimination and degrades detection via thalamic synchrony. (A). Detection analyses of how well an ideal observer can classify the presence/absence of a stimulus following either no preceding stimuli (top) or a train of adapting stimuli (bottom). (B). Discrimination analyses of how well an ideal observer could identify which of multiple stimuli were presented with and without adaptation. (C). Comparison of cortical population responses to 5 different stimuli with (bottom) and without (top) adaptation. (D). Thalamic synchrony, measured by the sum of the normalized cross-correlograms of thalamic pairs (insets) over lags ±7.5 ms, decreases with adaptation but to different degrees for each stimulus, diversifying subsequent cortical responses. Colored bars at left: synchrony evoked by 5 different stimulus velocities without preceding stimuli. Colored bars at right: synchrony evoked by same stimulus set after a train of adapting stimuli. Line: synchrony evoked by each pulse in an adapting train (adapting stimulus is similar to s4 and s5). Adapted from Wang et al. (2010b) with permission from Macmillan Publishers Ltd.

L4 inhibitory neurons in visual cortex having broadly tuned receptive fields (reviewed in Alonso and Swadlow, 2005). Visual stimuli evoke strong inhibitory conductances in cat L4 neurons only milliseconds after excitatory conductances (Cardin et al., 2010). Thus, L4 of visual cortex, like somatosensory L4 (Wilent and Contreras, 2005), may possesses a temporal window of opportunity that, along with nonlinear integration among excitatory cells, enhances detection of synchronous inputs (Usrey et al., 2000; Cardin et al., 2010; Wang et al., 2010a). Visual thalamus affords such synchronous inputs: LGN cells with overlapping receptive fields exhibit millisecond-scale correlations during visual stimuli (Alonso et al., 1996).

Thalamic synchrony signals more than the mere presence of a sensory stimulus. The degree of synchrony varies with stimulus features. For instance, velocity of whisker deflection affects correlation of extracellularly recorded pairs of thalamic neurons in vivo, even after accounting for overall differences in firing rates (Temereanca et al., 2008). Velocity, which is analogous to contrast in vision, is an important parameter of whisker sensation. As a rodent explores objects, surface friction interacts with the whiskers to produce an irregular pattern of "stick-slip" motions. Whisker slips across textured surfaces are transient high-velocity movements, and their frequency and velocity vary with surface roughness. The relationship of whisker velocity and neuronal synchrony was initially suggested by the temporally precise responses of sequentially recorded single thalamic units (Pinto et al., 2000), which further suggested relationships with the exact whisker stimulated, movement direction, the stimulus amplitude, and contact duration (Pinto et al., 2003). In contrast, free-whisking in air produces less synchronous thalamic activity than touches do, whereby L4 fast-spiking inhibitory cells prevent excitatory neurons from responding (Yu et al., 2019). The FFI circuit therefore largely filters out self-generated motion signals due to whisking, as depicted in the model in Figure 26.2E.

Paired recordings have shown directly that high-frequency repetitive stimuli decrease thalamic synchrony, with important consequences for perception (Temereanca et al., 2008; Wang et al., 2010b). From the perspective of an ideal observer of spiking activity, cortical neurons *detect* the presence of stimuli best on the first stimulus presentation (Figure 26.3A, top). Performance degrades when the stimulus has been presented previously (bottom). However, cortical response magnitudes can *discriminate* among multiple stimuli (e.g., 2 velocities) best after cortical cells are adapted by several presentations (Figure 26.3B). Thus, adaptation degrades detection but enhances discrimination. Prior to adaptation, all stimuli evoke large, easily distinguished cortical spiking responses (Figure 26.3C, top). After adaptation, spike counts decrease but individual responses cover a larger range (bottom). These trends do not arise trivially from changes in the response magnitudes of thalamic neurons, which can be used to discriminate different stimuli equally well before and after adaptation (Wang et al., 2010b).

Cortical output instead reflects adaptation in synchrony. As the system adapts to a train of rapidly presented stimuli, the level of thalamic synchrony is reduced (Figure 26.3D). This degrades detection because stimulus-induced synchrony now falls closer to noise levels and evokes fewer cortical spikes. Although synchrony scales with velocity, the synchrony evoked by even relatively low velocities is robust upon first presentation (Figure 26.3D, colored bars at left), enabling their detection. For repetitive stimuli, however, the synchrony difference between 2 distinct stimulus types is magnified by adaptation even though the absolute synchrony falls (colored bars on right). Via the strong FFI circuit, the wider range of thalamic synchrony produces a more extensive and therefore discriminable, albeit less detectable, range of cortical spikes. A similar pattern also emerges for multiple directions of whisker movement, suggesting that the effects of adaptation on synchrony generalize to many stimulus dimensions.

THE VPM-L5/6 PATHWAY: DUPLICATED SENSORY INPUT

As described above, VPM thalamus clearly innervates 2 separate targets: L4 and the border region between L5 and L6 (Herkenham, 1980; Jensen and Killackey, 1987; Chmielowska et al., 1989; Senft and Woolsey, 1991; Lu and Lin, 1993; Pierret et al., 2000; Gheorghita et al., 2006; Wimmer et al., 2010; Oberlaender et al., 2012a). While the pathway into L5/6 has been known about for some decades, it is often thought to be modulatory due to the denser arborization of axon in L4. However, in vitro slice studies in the somatosensory and auditory cortex clearly demonstrate that L5B and L6 receive thalamocortical synapses with typical driver-like properties (Viaene et al., 2011). Paired recordings in living animals reveal a surprisingly high rate of VPM-L5B connectivity: approximately 50 thalamocortical neurons are estimated to converge onto the typical L5B pyramidal neuron (Constantinople and Bruno, 2013). Even with relatively weak ~0.5 mV connections, the strong synchrony of converging thalamocortical neurons would suggest that these sparse collaterals could be sufficient to elicit action potentials in deep layers.

To what extent, then, is L5/6 activity driven by sparse thalamocortical input versus intracortical input from superficial layers? In anesthetized rats, silencing of barrel cortex by applying muscimol to the cortex followed by a local restoration of L5 activity with bicuculline initially suggested that L5 responses depended on superficial layers (Wright and Fox, 2010). However, simpler, more focal manipulations in awake narcotized rats yield the opposite conclusion. A small bolus of lidocaine in L4 blocks transmission along thalamocortical branches in L4, L4 action potential generation, L4 axons innervating L2/3, and L2/3 axons descending into L5/6, effectively severing the upper layers from the deeper ones. Surprisingly, this pharmacological "cut" does not alter subthreshold synaptic input to L5/6 or suprathreshold L5/6 action potential output during stimulation of the receptive field center (Constantinople and Bruno, 2013). Analogous silencing of L5/6 in a neighboring cortical column with lidocaine (Figure 26.4C) suppresses the surround receptive field corresponding to the targeted column but not the receptive field center (Egger et al., 2020). These 2 studies demonstrate that L5/6 can synthesize its own receptive fields directly from thalamocortical input. Complementary computational modeling of the anatomical circuits indicates that the extensive axons of L6 corticocortical cells are likely a key circuit element in this process. Similar to these pharmacological studies, optogenetic silencing of L4 has little effect on the L5/6 spiking activity in somatosensory and visual cortex of awake running mice experiencing whisker or visual stimuli (Pluta et al., 2015). Thus, on the basis of multiple studies, sparse thalamocortical innervation of the deep layers accounts for nearly all the sensory evoked activity in L5/6.

How can so few axonal branches do so much? Synaptic boutons have relatively uniform density along thalamocortical axonal branches in the cortex, meaning that axonal length is a fair proxy for synaptic density. L5/6 receives only about 20%–25% as much thalamocortical axon as L4 does (Gheorghita et al., 2006; Oberlaender et al., 2012a). This factor 4 difference in innervation would be incorrectly interpreted to be a sign of weak thalamocortical effect. In fact, the average membrane potential of L5/6 pyramidal neurons is only about half as far from spike threshold as that of L4 excitatory neurons (Constantinople and Bruno, 2013). Furthermore, the density of excitatory

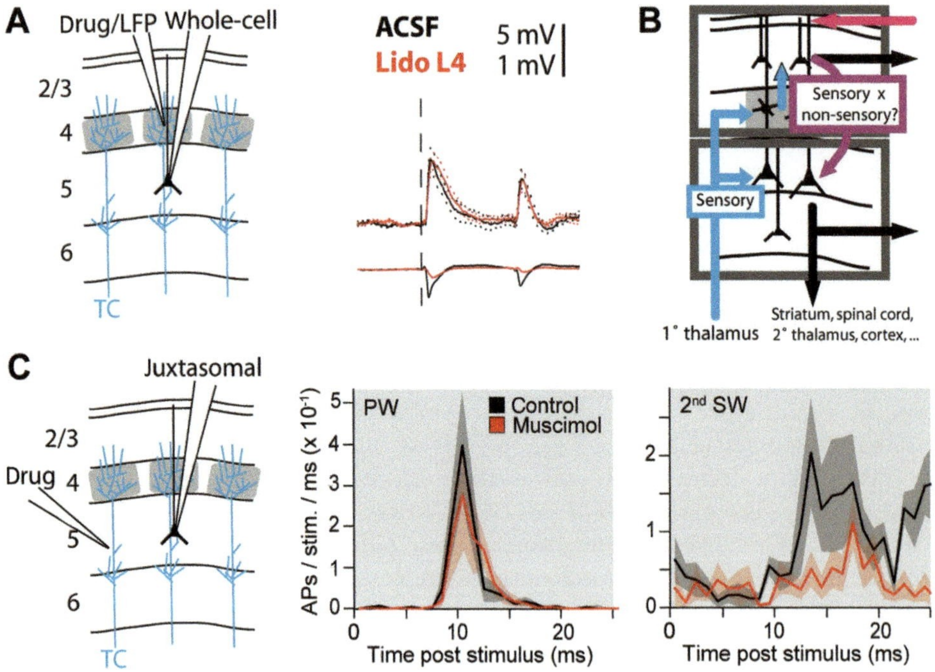

Figure 26.4. L5/6 synthesizes its own response properties directly from thalamocortical inputs. (A). Left: schematic of whole-cell recordings from L5/6 cells during inactivation of the barrel by lidocaine injection. Blue lines schematically depict single TC axons, among hundreds per column. Right: population averages of L5/6 PSP (top) and L4 LFP (bottom) responses to whisker stimulation (dashed line). Dotted lines = SEM. (B). Thalamus copies sensory information to both an upper stratum (L4 and L2/3) and a lower stratum (L5/6), which differ in coding properties and downstream targets. Adapted from Constantinople and Bruno (2013). (C). Left: schematic of juxtasomal recordings from L5 cells during muscimol silencing of L5/6 in the neighboring column. Middle: PSTHs of receptive field center stimulation across L5 cells during control conditions (black) and lidocaine silencing (red). Right: PSTHs of surround stimulation for the same cells in the middle panel. Adapted from Egger et al. (2020).

neurons in L5/6 is half that of L4 (Oberlaender et al., 2012b). Given these differences in intrinsic properties and anatomical density, L5/6 would require only a quarter of L4's thalamocortical input to achieve the same degree of activation.

VPM inputs onto L6 are wired with a similar strong feedforward inhibitory circuit to that found in L4 (Beierlein and Connors, 2002). The strong correspondence of the distributions of parvalbumin-positive inhibitory cells, which mediate feedforward inhibition, to thalamocortical terminals raises the possibility of a similar circuit operating in L5. However, optogenetic stimulation of VPM axons in slices has not revealed strong parvalbumin-mediated inhibition (Sermet et al., 2019). In vivo intracellular recordings demonstrate that, whereas L4 is highly sensitive to synchronous inputs, deep layers are subject to slower, weaker inhibition and may be better suited to integrating inputs over time (Wilent and Contreras, 2004).

THE POM-S1 PATHWAY

POm, which targets L1 and L5A, does not play a simple role in generating cortical sensory responses. Pharmacological blockade of synaptic inputs onto the L1 apical tufts of L5 pyramidal neurons has virtually no effect on whisker-evoked responses recorded intracellularly (Constantinople and Bruno, 2013). Optogenetic stimulation of POm axons rarely elicits L5 action potentials, but does increase and draw out L5 responses to sensory stimuli (Mease et al., 2016a). L2/3 exhibits similar phenomena during POm axon stimulation, apparently due to the triggering of sustained (up to a second long) all-or-none beta-range oscillatory activity in POm (Zhang and Bruno, 2019). POm is also far more effective than feedback connections from other cortical areas, such as secondary somatosensory cortex and primary motor cortex, in modulating primary somatosensory cortex responses to tactile stimuli. These enhancing effects of POm would be consistent with a role in directing attention and/or plasticity (Roelfsema and Holtmaat, 2018).

Physiological characterization of POm supports such a higher functional role. While not insensitive to peripheral stimuli, POm is less responsive than VPM and is largely dependent on somatosensory cortex for activity (Diamond et al., 1992; Mease et al., 2016b). Movements, such as free whisking in air, also modulate POm less than VPM (Moore et al., 2015). Movement modulation is more likely to reflect arousal changes accompanying actions than true motor signals: Movement has the same effect on POm activity even after blockade of reafferent signals or motor efference copy, such as from motor cortex and superior colliculus; and visual LP activity is just as well correlated with whisking even though it lacks direct connections with the somatosensory system (Petty et al., 2021). During detection tasks, actions linked to reward elevate POm activity, and optogenetically suppressing POm during training slows learning (La Terra et al., 2022). Indeed, inhibiting POm impairs perceptual discriminability of textures (Qi et al., 2022). Thus, secondary somatosensory thalamus may exist to alter tactile stimuli processing both in the moment and in the future.

SUMMARY AND FUTURE DIRECTIONS

Thalamocortical axons from primary thalamic nuclei branch to innervate both the middle and deep cortical layers, duplicating information in both targets. Direct thalamocortical engagement of multiple layers is a feature of cortical architecture across many modalities and species. In the somatosensory system, branches in the middle layers are part of a strong feedforward inhibitory circuit, which sensitizes layer 4 to thalamic population synchrony. Thalamic synchrony can code for multiple stimulus dimensions, including space, direction, and velocity, and adaptation can enhance the thalamic synchrony code to improve discrimination. The sparse thalamocortical branches in L5/6 define an entirely independent pathway for information to reach cortex and can account for the generation of L5/6 receptive fields. Thus, sensory response properties of both the upper and deeper layers of cortex may be strongly dictated by inputs from primary thalamic nuclei.

Many major questions, nevertheless, remain. If the sensory properties of cortical layers are largely dictated by primary thalamocortical inputs and local recurrence, what do L2/3 and L5/6 communicate to one another? Do upper and deeper layers perform separate computational functions to support behavior? Why do the TC-L4 and TC-L5/6 input pathways use slightly different inhibitory circuits? What do secondary thalamic nuclei like POm encode, and how do they alter cortical processing in behaviorally useful ways? These and other unknowns ultimately will require considering a cortical area's function in active behavior.

REFERENCES

Alonso JM, Swadlow HA (2005). Thalamocortical specificity and the synthesis of sensory cortical receptive fields. *J Neurophysiol.* 94:26–32.

Alonso JM, Usrey WM, Reid RC (1996). Precisely correlated firing in cells of the lateral geniculate nucleus. *Nature.* 383:815–819.

Beierlein M, Connors BW (2002). Short-term dynamics of thalamocortical and intracortical synapses onto layer 6 neurons in neocortex. *J Neurophysiol.* 88:1924–1932.

Blasdel GG, Lund JS (1983). Termination of afferent axons in macaque striate cortex. *J Neurosci.* 3:1389–1413.

Bruno RM, Sakmann B (2006). Cortex is driven by weak but synchronously active thalamocortical synapses. *Science.* 312:1622–1627.

Bruno RM, Simons DJ (2002). Feedforward mechanisms of excitatory and inhibitory cortical receptive fields. *J Neurosci.* 22:10966–10975.

Cardin JA, Kumbhani RD, Contreras D, Palmer LA (2010). Cellular mechanisms of temporal sensitivity in visual cortex neurons. *J Neurosci.* 30:3652–3662.

Chmielowska J, Carvell GE, Simons DJ (1989). Spatial organization of thalamocortical and corticothalamic projection systems in the rat SmI barrel cortex. *J Comp Neurol.* 285:325–338.

Constantinople CM, Bruno RM (2011). Effects and mechanisms of wakefulness on local cortical networks. *Neuron.* 69:1061–1068.

Constantinople CM, Bruno RM (2013). Deep cortical layers are activated directly by thalamus. *Science.* 340:1591–1594.

Cruikshank SJ, Lewis TJ, Connors BW (2007). Synaptic basis for intense thalamocortical activation of feedforward inhibitory cells in neocortex. *Nat Neurosci.* 10:462–468.

da Costa NM, Martin KA (2011). How thalamus connects to spiny stellate cells in the cat's visual cortex. *J Neurosci.* 31:2925–2937.

Deschenes M, Timofeeva E, Lavallee P (2003). The relay of high-frequency sensory signals in the Whisker-to-barreloid pathway. *J Neurosci.* 23:6778–6787.

DeWeese MR, Zador AM (2006). Non-Gaussian membrane potential dynamics imply sparse, synchronous activity in auditory cortex. *J Neurosci.* 26:12206–12218.

Diamond ME, Armstrong-James M, Budway MJ, Ebner FF (1992). Somatic sensory responses in the rostral sector of the posterior group (POm) and in the ventral posterior medial nucleus (VPM) of the rat thalamus: Dependence on the barrel field cortex. *J Comp Neurol.* 319:66–84.

Egger R, Narayanan RT, Guest JM, Bast A, Udvary D, Messore LF, Das S, de Kock CPJ, Oberlaender M (2020). Cortical output is gated by horizontally projecting neurons in the deep layers. *Neuron.* 105:122–137.

Ferster D, LeVay S (1978). The axonal arborizations of lateral geniculate neurons in the striate cortex of the cat. *J Comp Neurol.* 182:923–944.

Gabernet L, Jadhav SP, Feldman DE, Carandini M, Scanziani M (2005). Somatosensory integration controlled by dynamic thalamocortical feed-forward inhibition. *Neuron.* 48:315–327.

Garcia-Marin V, Ahmed TH, Afzal YC, Hawken MJ (2013). Distribution of vesicular glutamate transporter 2 (VGluT2) in the primary visual cortex of the macaque and human. *J Comp Neurol.* 521:130–151.

Gheorghita F, Kraftsik R, Dubois R, Welker E (2006). Structural basis for map formation in the thalamocortical pathway of the barrelless mouse. *J Neurosci.* 26:10057–10067.

Hendrickson AE, Wilson JR, Ogren MP (1978). The neuroanatomical organization of pathways between the dorsal lateral geniculate nucleus and visual cortex in Old World and New World primates. *J Comp Neurol.* 182:123–136.

Herkenham M (1980). Laminar organization of thalamic projections to the rat neocortex. *Science.* 207:532–535.

Hromadka T, Deweese MR, Zador AM (2008). Sparse representation of sounds in the unanesthetized auditory cortex. *PLoS Biol.* 6:e16.

Huang CL, Winer JA (2000). Auditory thalamocortical projections in the cat: Laminar and areal patterns of input. *J Comp Neurol.* 427:302–331.

Jensen KF, Killackey HP (1987). Terminal arbors of axons projecting to the somatosensory cortex of the adult rat. I. The normal morphology of specific thalamocortical afferents. *J Neurosci.* 7:3529–3543.

Koelbl C, Helmstaedter M, Lubke J, Feldmeyer D (2015). A barrel-related interneuron in layer 4 of rat somatosensory cortex with a high intrabarrel connectivity. *Cereb Cortex.* 25:713–725.

Kremkow J, Perrinet LU, Masson GS, Aertsen A (2010). Functional consequences of correlated excitatory and inhibitory conductances in cortical networks. *J Comput Neurosci*. 28:579–594.

Kuramoto E, Furuta T, Nakamura KC, Unzai T, Hioki H, Kaneko T (2009). Two types of thalamocortical projections from the motor thalamic nuclei of the rat: A single neuron-tracing study using viral vectors. *Cereb Cortex*. 19:2065–2077.

Kuramoto E, Pan S, Furuta T, Tanaka YR, Iwai H, Yamanaka A, Ohno S, Kaneko T, Goto T, Hioki H (2017). Individual mediodorsal thalamic neurons project to multiple areas of the rat prefrontal cortex: A single neuron-tracing study using virus vectors. *J Comp Neurol*. 525:166–185.

La Terra D, Bjerre AS, Rosier M, Masuda R, Ryan TJ, Palmer LM (2022). The role of higher-order thalamus during learning and correct performance in goal-directed behavior. *eLife*. 11:e77177.

Landry P, Deschenes M (1981). Intracortical arborizations and receptive fields of identified ventrobasal thalamocortical afferents to the primary somatic sensory cortex in the cat. *J Comp Neurol*. 199:345–371.

Landry P, Diadori P, Leclerc S, Dykes RW (1987). Morphological and electrophysiological characteristics of somatosensory thalamocortical axons studied with intra-axonal staining and recording in the cat. *Exp Brain Res*. 65:317–330.

Lu SM, Lin RC (1993). Thalamic afferents of the rat barrel cortex: A light- and electron-microscopic study using Phaseolus vulgaris leucoagglutinin as an anterograde tracer. *Somatosens Mot Res*. 10:1–16.

Mease RA, Metz M, Groh A (2016a). Cortical sensory responses are enhanced by the higher-order thalamus. *Cell Rep*. 14:208–215.

Mease RA, Sumser A, Sakmann B, Groh A (2016b). Cortical dependence of whisker responses in posterior medial thalamus in vivo. *Cereb Cortex*. 26:3534–3543.

Meyer HS, Wimmer VC, Hemberger M, Bruno RM, de Kock CP, Frick A, Sakmann B, Helmstaedter M (2010). Cell type-specific thalamic innervation in a column of rat vibrissal cortex. *Cereb Cortex*. 20:2287–2303.

Miller LM, Escabi MA, Read HL, Schreiner CE (2001). Functional convergence of response properties in the auditory thalamocortical system. *Neuron*. 32:151–160.

Moore JD, Deschenes M, Kleinfeld D (2015). Vibrissa self-motion and touch are encoded along the same somatosensory pathway from brainstem through thalamus. *PLoS Biol*. 13:e1002253.

O'Connor DH, Peron SP, Huber D, Svoboda K (2010). Neural activity in barrel cortex underlying vibrissa-based object localization in mice. *Neuron*. 67:1048–1061.

Oberlaender M, Ramirez A, Bruno RM (2012a). Sensory experience restructures thalamocortical axons during adulthood. *Neuron*. 74:648–655.

Oberlaender M, de Kock CP, Bruno RM, Ramirez A, Meyer HS, Dercksen VJ, Helmstaedter M, Sakmann B (2012b). Cell type-specific three-dimensional structure of thalamocortical circuits in a column of rat vibrissal cortex. *Cereb Cortex*. 22:2375–2391.

Ohno S, Kuramoto E, Furuta T, Hioki H, Tanaka YR, Fujiyama F, Sonomura T, Uemura M, Sugiyama K, Kaneko T (2012). A morphological analysis of thalamocortical axon fibers of rat posterior thalamic nuclei: A single neuron tracing study with viral vectors. *Cereb Cortex*. 22:2840–2857.

Petty GH, Kinnischtzke AK, Hong YK, Bruno RM (2021). Effects of arousal and movement on secondary somatosensory and visual thalamus. *eLife*. 10:e67611.

Pierret T, Lavallee P, Deschenes M (2000). Parallel streams for the relay of vibrissal information through thalamic barreloids. *J Neurosci*. 20:7455–7462.

Pinto DJ, Brumberg JC, Simons DJ (2000). Circuit dynamics and coding strategies in rodent somatosensory cortex. *J Neurophysiol*. 83:1158–1166.

Pinto DJ, Brumberg JC, Simons DJ, Ermentrout GB (1996). A quantitative population model of whisker barrels: Re-examining the Wilson-Cowan equations. *J Comput Neurosci*. 3:247–264.

Pinto DJ, Hartings JA, Brumberg JC, Simons DJ (2003). Cortical damping: Analysis of thalamocortical response transformations in rodent barrel cortex. *Cereb Cortex*. 13:33–44.

Pluta S, Naka A, Veit J, Telian G, Yao L, Hakim R, Taylor D, Adesnik H (2015). A direct translaminar inhibitory circuit tunes cortical output. *Nat Neurosci*. 18(11):1631–1640.

Qi J, Ye C, Naskar S, Inácio AR, Lee S (2022). Posteromedial thalamic nucleus activity significantly contributes to perceptual discrimination. *PLoS Biol*. 20(11):e3001896.

Reid RC, Alonso JM (1995). Specificity of monosynaptic connections from thalamus to visual cortex. *Nature*. 378:281–284.

Roelfsema PR, Holtmaat A (2018). Control of synaptic plasticity in deep cortical networks. *Nat Rev Neurosci*. 19:166–180.

Schoonover CE, Tapia JC, Schilling VC, Wimmer V, Blazeski R, Zhang W, Mason CA, Bruno RM (2014). Comparative strength and dendritic organization of thalamocortical and corticocortical synapses onto excitatory layer 4 neurons. *J Neurosci*. 34:6746–6758.

Sedigh-Sarvestani M, Vigeland L, Fernandez-Lamo I, Taylor MM, Palmer LA, Contreras D (2017). Intracellular, in vivo, dynamics of thalamocortical synapses in visual cortex. *J Neurosci*. 37:5250–5262.

Senft SL, Woolsey TA (1991). Growth of thalamic afferents into mouse barrel cortex. *Cereb Cortex*. 1:308–335.

Sermet BS, Truschow P, Feyerabend M, Mayrhofer JM, Oram TB, Yizhar O, Staiger JF, Petersen CC (2019). Pathway-, layer- and cell-type-specific thalamic input to mouse barrel cortex. *eLife*. 8:e52665.

Smith PH, Uhlrich DJ, Manning KA, Banks MI (2012). Thalamocortical projections to rat auditory cortex from the ventral and dorsal divisions of the medial geniculate nucleus. *J Comp Neurol*. 520:34–51.

Temereanca S, Brown EN, Simons DJ (2008). Rapid changes in thalamic firing synchrony during repetitive whisker stimulation. *J Neurosci*. 28:11153–11164.

Usrey WM, Alonso JM, Reid RC (2000). Synaptic interactions between thalamic inputs to simple cells in cat visual cortex. *J Neurosci*. 20:5461–5467.

Veinante P, Deschenes M (2003). Single-cell study of motor cortex projections to the barrel field in rats. *J Comp Neurol*. 464:98–103.

Viaene AN, Petrof I, Sherman SM (2011). Synaptic properties of thalamic input to the subgranular layers of primary somatosensory and auditory cortices in the mouse. *J Neurosci*. 31:12738–12747.

Wang HP, Spencer D, Fellous JM, Sejnowski TJ (2010a). Synchrony of thalamocortical inputs maximizes cortical reliability. *Science*. 328:106–109.

Wang Q, Webber RM, Stanley GB (2010b). Thalamic synchrony and the adaptive gating of information flow to cortex. *Nat Neurosci*. 13:1534–1541.

Wilent WB, Contreras D (2004). Synaptic responses to whisker deflections in rat barrel cortex as a function of cortical layer and stimulus intensity. *J Neurosci*. 24:3985–3998.

Wilent WB, Contreras D (2005). Dynamics of excitation and inhibition underlying stimulus selectivity in rat somatosensory cortex. *Nat Neurosci*. 8:1364–1370.

Wimmer VC, Bruno RM, de Kock CP, Kuner T, Sakmann B (2010). Dimensions of a projection column and architecture of VPM and POm axons in rat vibrissal cortex. *Cereb Cortex*. 20:2265–2276.

Wright N, Fox K (2010). Origins of cortical layer V surround receptive fields in the rat barrel cortex. *J Neurophysiol*. 103:709–724.

Yu J, Hu H, Agmon A, Svoboda K (2019). Recruitment of GABAergic interneurons in the barrel cortex during active tactile behavior. *Neuron*. 104:412–427.

Zhang W, Bruno RM (2019). High-order thalamic inputs to primary somatosensory cortex are stronger and longer lasting than cortical inputs. *eLife*. 8:e44158.

27.

LAYER 6 CORTICOTHALAMIC NEURONS IN THE SOMATOSENSORY SYSTEM

Solange P. Brown

DIVERSITY OF LAYER 6 CORTICOTHALAMIC CELLS

Corticothalamic neurons in primary sensory cortex of mammals have been traditionally divided into 2 types, layer 5 corticothalamic neurons (L5 CThNs), which send axonal projections to the thalamus as well as to other subcortical brain regions, and layer 6 corticothalamic neurons (L6 CThNs), which project exclusively to the thalamus (see **Chapters 14 and 15 in Section IV**). The morphological, electrophysiological, and functional characteristics of these 2 types of corticothalamic neurons are distinct and have been well described (Sherman and Guillery, 1998; Shepherd and Yamawaki, 2021). However, a number of studies of sensory cortex across a range of mammalian systems indicate that L6 CThNs are themselves composed of different subtypes, with distinct patterns of axonal projections into the thalamus and different intracortical synaptic partners. This diversity of L6 CThN subtypes has been one challenge to understanding how L6 CThNs contribute to sensory processing and higher-order cognition.

In the primary somatosensory cortex of rodents, L6 CThNs represent about half of the neurons in layer 6 (L6) (Zhang and Deschênes, 1997; Crandall et al., 2017); the other major class of excitatory neurons in L6 are corticocortical neurons. The axonal inputs of L6 CThNs to the primary somatosensory thalamic nucleus, the ventral posterior complex (VP), are anatomically more numerous than axonal inputs carrying information from the sensory periphery (Liu et al., 1995; Deschênes et al., 1998), as observed in the visual system (see **Chapter 19 in Section V**). Although L6 CThNs are often considered one class of neuron, studies of L6 CThNs using multiple modalities have revealed additional diversity within this population. In the somatosensory cortex representing rodent whiskers, the barrel cortex, where L6 CThNs have been particularly well studied, at least 3 subtypes of L6 CThNs have been described (Bourassa et al., 1995; Zhang and Deschênes, 1997; Kwegyir-Afful and Simons, 2009; Chevée et al., 2018; Hoerder-Suabedissen et al., 2018; Frandolig et al., 2019; Dash et al., 2022). The increasing attention to the different classes of L6 CThNs, particularly in studies of rodent somatosensory cortex, is beginning to suggest potential roles for the diversity of L6 CThNs types, but how this diversity of subtypes contributes to perception and behavior remains unclear. Here, the subtypes of L6 CThNs in the barrel cortex are contrasted and compared.

THALAMIC PROJECTIONS OF LAYER 6 CORTICOTHALAMIC CELLS

L6 CThN axonal projections out of the somatosensory cortex are restricted to the thalamus (Deschênes et al., 1994). Studies of the long-range axonal projections of L6 CThNs demonstrate that there are at least 3 major morphological types in the barrel cortex: VPM-only L6 CThNs, VPM/POm L6 CThNs, and POm-only L6 CThNs (see Figure 27.1; see figure caption for key to these abbreviations) (Bourassa et al., 1995; Zhang and Deschênes, 1997; Killackey and Sherman, 2003; Chevée et al., 2018; Hoerder-Suabedissen et al., 2018; Frandolig et al., 2019). As their name indicates, VPM-only L6 CThNs send axons exclusively into the ventral posterior medial nucleus (VPM) of the thalamus, the primary somatosensory thalamic nucleus for the whisker system (Bourassa et al., 1995; Killackey and Sherman, 2003; Chevée et al., 2018; Whilden et al., 2021). The location of the cell bodies of VPM-only L6 CThNs is biased toward the upper half of L6 (Bourassa et al., 1995; Killackey and Sherman, 2003; Chevée et al., 2018; Whilden et al., 2021), where their spatial organization reflects the tangential organization of the barrels in layer 4 (L4): the cell bodies of VPM-only L6 CThNs form denser patches in infrabarrels under each overlaying L4 barrel (Killackey and Sherman, 2003; Crandall et al., 2017; Frandolig et al., 2019). In contrast, corticocortical neurons are concentrated under the septal regions of L4 barrels, between the L6 infrabarrels (Crandall et al., 2017). Prior work indicates that VPM-only L6 CThNs project in a topographic manner to the thalamic barreloids, with VPM-only L6 CThNs in one barrel projecting in a one-to-one relationship to the corresponding barreloid (Hoogland et al., 1987; Land et al., 1995; Deschênes et al., 1998).

Corticothalamic neurons in the lower half of L6 in the barrel cortex project to the posterior medial nucleus (POm) of the thalamus (Bourassa et al., 1995; Zhang and Deschênes, 1997; Killackey and Sherman, 2003; Chevée et al., 2018; Hoerder-Suabedissen et al., 2018; Frandolig et al., 2019; Ansorge et al., 2020; Casas-Torremocha et al., 2022). The most common type of POm-projecting L6 CThN, VPM/POm L6CThNs, sends axonal projections to both VPM and POm, while a third, seemingly rarer type, POm-only L6 CThNs, projects to POm only. Anatomical studies suggest that VPM/POm L6 CThNs also project topographically into VPM (Hoogland et al., 1987; Land et al., 1995; Deschênes et al., 1998), but in contrast to VPM-only L6 CThNs, they are thought to target their corresponding barreloid as well as neighboring barreloids. POm-only L6 CThNs appear to project to only a subregion of the POm (Hoerder-Suabedissen et al., 2018), but how the projections into POm of VPM/POm and POm-only L6 CThNs differ has not been fully elucidated. The morphological differences among these subtypes of L6 CThNs suggest that they form multiple information streams from L6 of primary somatosensory cortex to thalamic nuclei. These anatomical results demonstrating multiple parallel pathways via L6 CThNs to the thalamus in rodent somatosensory cortex are consistent with studies of other primary sensory areas in other mammals, including primates and carnivores (Fitzpatrick et al., 1994; Bourassa and Deschênes, 1995; Usrey and Fitzpatrick, 1996; Briggs et al., 2016; Hoerder-Suabedissen et al., 2018; Hasse et al., 2019).

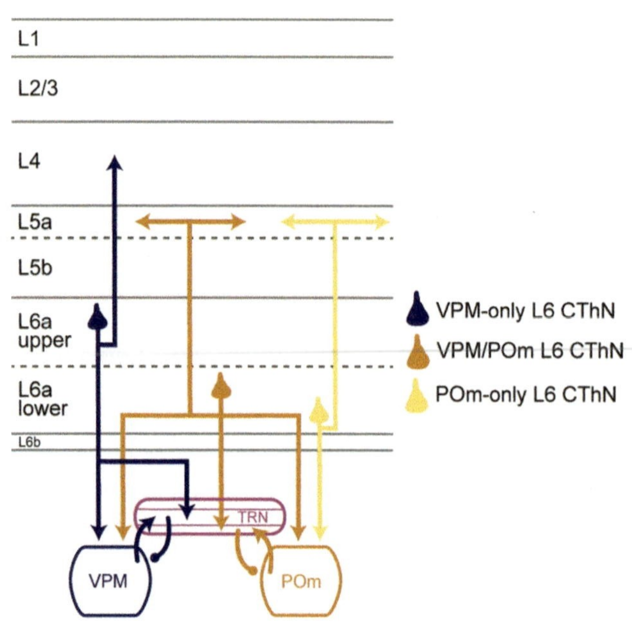

Figure 27.1. Three morphological classes of layer 6 corticothalamic neurons in the rodent barrel cortex: VPM-only L6 CThNs, VPM/POm L6 CThNs, and POm-only L6 CThNs. Whether POm-only L6 CThNs project to the thalamic reticular nucleus remains unclear. Layer 6b is differentiated from layer 6a because it is composed primarily of neurons expressing connective tissue growth factor that do not project to the thalamus. Arrows indicate excitatory connections; closed circles indicate inhibitory connections. VPM = ventral posterior medial nucleus of the thalamus; POm = posterior medial nucleus of the thalamus; TRN = thalamic reticular nucleus.

Studies of L6 CThNs in somatosensory cortex using other experimental approaches support the hypothesis that there are multiple parallel streams of L6 CThN input to the thalamus (Landry and Dykes, 1985; Kwegyir-Afful and Simons, 2009; Chevée et al., 2018; Frandolig et al., 2019; Dash et al., 2022). Recent high-throughput transcriptomic studies using the single-cell transcriptional profiles of cortical neurons to classify neurons into cell types revealed multiple subtypes of L6 CThNs (Chevée et al., 2018; Yao et al., 2021). Furthermore, combining transcriptomic profiling and retrograde labeling identified gene-expression differences that suggest functional differences among the L6 CThN subtypes (Chevée et al., 2018). For example, the gene *Adcyap1*, which encodes the neuropeptide PACAP, and the enzyme required to process this neuropeptide, *Pam*, are expressed at significantly higher levels in VPM-only L6 CThNs than in L6 CThNs that project to POm (Chevée et al., 2018), mirroring a differential distribution of PACAP receptors in the thalamus. In addition, expression of *Grp*, the gene encoding gastrin-releasing peptide, is significantly higher in L6 CThNs projecting to POm than to VPM (Chevée et al., 2018). *Grp* expression also has been identified in approximately half of L6 CThNs in auditory cortex, and *Grp* had been implicated in the regulation of VIP-positive inhibitory interneurons and fear conditioning in auditory cortex (Melzer et al., 2021). However, the relationship between the anatomically defined L6 CThN types and types defined using other features is not always clear. For example, in vivo recordings from CThNs in somatosensory cortex have identified multiple functional types (Landry and Dykes, 1985; Kwegyir-Afful and Simons, 2009; Dash et al., 2022), but whether these types relate to morphological classes of CThNs is not known.

An additional obstacle to understanding the function of L6 CThNs in the whisker-associated somatosensory system is that the major thalamic nuclei involved in somatosensory processing, VPM and POm, are composed of subregions with different complements of synaptic inputs and outputs and, presumably, function (Urbain and Deschenes, 2007; Ohno et al., 2012; Groh et al., 2014; Sumser et al., 2017;

El-Boustani et al., 2020; Guo et al., 2020; Casas-Torremocha et al., 2022). The distribution of the axonal projections of the different types of L6 CThNs into these subregions and their functional influence on their different postsynaptic targets have not yet been fully elucidated. Nonetheless, anatomical studies of the population of L6 CThNs in rodent barrel cortex indicate that, overall, L6 CThNs project topographically into VPM (Hoogland et al., 1987; Deschênes et al., 1994; Land et al., 1995), although there are likely differences in the precise organization of these projections across L6 CThN subtypes. Furthermore, it is unclear whether L6 CThNs in primary somatosensory cortex (S1) project to VPMvl, a subregion of VPM with cells that exhibit large, multi-whisker receptive fields, in contrast to the single-whisker receptive fields typically found in the bodies of the barreloids (Bokor et al., 2008). L6 CThNs in secondary somatosensory cortex (S2) project to VMPvl, but not to the barreloids in VPM (Bokor et al., 2008). Recent studies suggest that the distribution of L6 CThN axonal projections into POm depends upon subtype. VPM/POm L6 CThNs appear to project to a much broader extent of POm than POm-only L6 CThNs which target a subregion of POm (Hoerder-Suabedissen et al., 2018). Determining the postsynaptic targets of the different classes of L6 CThNs and their functional impact on the activity of different classes of thalamic neurons across the subregions of VPM and POm remains an important outstanding goal.

Axonal boutons formed by L6 CThNs in VPM are small relative to those formed by L5 CThNs in POm, contain vesicular glutamate transporter 1 (VGluT1), and target distal dendrites of thalamic neurons (Bourassa et al., 1995; Reichova and Sherman, 2004; Groh et al., 2014). These synaptic connections are mediated by both ionotropic and metabotropic glutamate receptors (Eaton and Salt, 1996; Reichova and Sherman, 2004; Landisman and Connors, 2007). They are initially weak and facilitate during ongoing activity, due in part to presynaptic Synaptotagmin 7 (Golshani et al., 2001; Reichova and Sherman, 2004; Landisman and Connors, 2007; Cruikshank et al., 2010; Crandall et al., 2015; Jackman et al., 2016). The L6 CThN input from the S1-forelimb region forms similar facilitating synapses in VPL (Guo et al., 2020). Determining the properties of S1 L6 CThN synapses onto POm thalamocortical neurons has been more challenging, with some studies suggesting that these synapses are also small, initially weak, and facilitating, while others indicate that they are composed of a mixture of small and large boutons, forming both weak, facilitating synapses and strong, depressing ones (Reichova and Sherman, 2004; Landisman and Connors, 2007; Hoerder-Suabedissen et al., 2018; Ansorge et al., 2020; Guo et al., 2020). Whether these mixed results depend on the L6 CThN subtype, the subregion of POm targeted, and/or the postsynaptic cell type, remains unknown. The synapses of L6 CThNs in somatosensory thalamus also exhibit long-term changes in synaptic strength (Castro-Alamancos and Calcagnotto, 1999; Hsu et al., 2010). Again, these studies have not distinguished among different L6 CThN subtypes, so how these synaptic properties may differ across L6 CThN subtypes and subregions within somatosensory thalamus is not well understood.

The thalamic reticular nucleus (TRN), a thin thalamic nucleus composed of GABAergic neurons, is a major source of inhibitory input to thalamic neurons in VPM and POm. The axons of S1 L6 CThNs project to the TRN, and their TRN projections are topographically aligned with those to VPM (Temereanca and Simons, 2004; Lam and Sherman, 2010). Recent studies indicate that the different subtypes of L6 CThNs target different subregions of the somatosensory TRN. VPM-only L6 CThNs send axonal projections to the core of the TRN, a zone that is reciprocally connected with VPM (Bourassa et al., 1995; Pinault et al., 1995; Lam and Sherman, 2007; Lee et al., 2014; Clemente-Perez et al., 2017; Li et al., 2020; Martinez-Garcia et al., 2020; Whilden et al., 2021). Like VPM-only L6 CThNs, VPM/POm L6 CThNs also project to the TRN, but their axons ramify in the medial TRN, a region of the TRN that is reciprocally connected with POm (Bourassa et al., 1995; Pinault et al., 1995; Lam and Sherman, 2007; Lee et al., 2014; Clemente-Perez et al., 2017; Li et al., 2020; Martinez-Garcia et al., 2020; Whilden et al., 2021). The projection pattern of POm-only L6 CThNs into the TRN is less clear, as morphological reconstructions of POm-only L6 CThNs in rats show axonal branching within the TRN, while in mouse, reconstructed POm-only L6 CThNs bypass the TRN (Bourassa et al., 1995; Hoerder-Suabedissen et al., 2018). However, only a few POm-only L6 CThNs have been fully reconstructed; thus whether this is a species difference or a sign of additional subtypes among POm-only L6 CThNs is not known. Nonetheless, the differences in the axonal projections to the TRN across the different L6 CThN cell types, combined with studies demonstrating functional differences in these TRN regions, indicate that the effects of feedforward inhibition mediated by different L6 CThN subtypes onto thalamic neurons will likely be distinct.

The synaptic strength of L6 CThN synapses onto TRN neurons is greater than onto the dendrites of thalamocortical neurons (Golshani et al., 2001; Cruikshank et al., 2010). Similar to L6 CThN synapses onto thalamocortical neurons in VPM, L6 CThN synapses onto TRN neurons largely exhibit short-term facilitation; synapses from TRN neurons onto thalamic neurons, in contrast, depress (Golshani et al., 2001; Cruikshank et al., 2010; Crandall et al., 2015). Thus, the overall functional effects of L6 CThN input to somatosensory thalamus is complex as, depending on the firing pattern of the L6 CThNs, it can vary from suppressive to excitatory due to differences in the frequency-dependent short-term synaptic plasticity at the different synapse types along these pathways (Mease et al., 2014; Crandall et al., 2015). Corticothalamic activation can also lead to long-term changes in the excitability of TRN neurons (Fernandez et al., 2017). These studies, however, have probed the function of L6 CThNs without differentiating among the different L6 CThN subtypes or the different subregions of the TRN. Thus, there may be additional complexity in the functional impact on thalamic neurons of the circuits formed by S1 L6 CThNs and the TRN depending on the L6 CThN subtype and the TRN subnetwork.

INTRACORTICAL MORPHOLOGY AND SYNAPTIC ORGANIZATION OF LAYER 6 CORTICOTHALAMIC TYPES

In addition to sending axonal projections to the thalamus, S1 L6 CThNs are also integrated into intracortical circuits. The intracortical axonal and dendritic morphologies of different subtypes of L6 CThNs in rodent S1 are distinct (Figure 27.1), indicating that they interact with different intracortical synaptic partners, but their precise morphology differs between rats and mice. The functional implications of these species differences are not known, but variations in the morphology of the intracortical processes of L6 CThNs across animals are likewise seen in visual cortex (Katz, 1987; Usrey and Fitzpatrick, 1996; Olsen et al., 2012; Briggs et al., 2016; Hasse et al., 2019). The axonal and dendritic arbors of VPM-only L6 CThNs in rodents are more restricted to barrel columns than those of POm-projecting L6 CThNs (Zhang and Deschênes, 1997; Frandolig et al., 2019). In rats, the intracortical axonal and dendritic processes of VPM-only L6 CThNs primarily terminate in the main thalamorecipient layer, L4 (Zhang and Deschênes, 1997). In mice, some VPM-only L6 CThNs extend axons or apical dendrites to layer 1 (L1) (Frandolig et al., 2019). In both rats and mice, the apical dendrites of POm-projecting L6 CThNs primarily terminate in layer 5a (L5a), although again, in mice, some extend to L1. The intracortical axons of POm-projecting L6 CThNs primarily terminate in L5a, where they extend horizontally, in contrast to the columnar arrangement of VPM-only L6 CThNs (Zhang and Deschênes, 1997; Frandolig et al., 2019). However, a few reconstructions of POm-only L6 CThNs from mice suggest that some POm-only L6 CThNs lack an intracortical axonal arborization (Hoerder-Suabedissen et al., 2018). The morphologies of the intracortical processes of the different subtypes of L6 CThNs place different constraints on their synaptic partners, but why they differ across species is not understood.

Studies of the synaptic outputs of L6 CThNs using whole-cell recordings from pairs of neurons in acute slices of somatosensory cortex recorded in vitro have revealed that the synaptic relationships of the different L6 CThN subtypes with inhibitory neurons are distinct (Frandolig et al., 2019). Co-localized with VPM-only L6 CThNs in upper L6 is a specialized type of parvalbumin-positive (PV), fast-spiking inhibitory interneuron that sends axonal projections up through the cortical layers toward L1. VPM-only L6 CThNs form frequent synaptic connections onto these interlaminar PV interneurons, while the POm-projecting L6 CThNs do not (Frandolig et al., 2019). These interlaminar PV interneurons have also been described in rat somatosensory cortex (Kumar and Ohana, 2008) and in mouse visual cortex (Bortone et al., 2014) where they strongly modulate cortical activity (Olsen et al., 2012; Bortone et al., 2014). In contrast, POm-projecting L6 CThNs synapse onto local PV interneurons with axons largely confined to L6 (Frandolig et al., 2019). Because of these distinct synaptic relationships with the different types of PV inhibitory interneurons, the activity of a local group of VPM-only L6 CThNs will inhibit the overlying cortical column, while POm-projecting L6 CThNs will inhibit neurons locally within L6.

L6 CThNs also strongly excite PV interneurons in other cortical layers as well as provide input to some cortical excitatory neurons (Kim et al., 2014). Optogenetic activation of L6 CThNs expressing channelrhodopsin-2 (ChR2) in brain slices generates relatively strong excitatory input to pyramidal neurons in L5a (Kim et al., 2014), a layer that is a major target of thalamocortical input from POm (Wimmer et al., 2010) and is enriched for intratelencephalic corticostriatal neurons. Optogenetic activation of L6 CThNs also generates weaker input to excitatory neurons in L4, the primary target of VPM axons. However, paired recordings show that L6 CThNs rarely form synapses onto other pyramidal neurons within L6 (Crandall et al., 2017). These studies, however, did not distinguish among the different L6 CThN subtypes, so it remains possible that local connections onto excitatory neurons vary across subtypes.

Studies of synaptic inputs to L6 CThNs in S1 indicate that they integrate input locally within L6 from neighboring corticocortical neurons (Crandall et al., 2017) and PV interneurons (Frandolig et al., 2019), as well as from other cortical areas, such as primary motor cortex (M1), and subcortical brain regions (Lee et al., 2008; Tanaka et al., 2011; Kim et al., 2014; Yang et al., 2014; Kinnischtzke et al., 2016; Crandall et al., 2017; Frandolig et al., 2019; Ansorge et al., 2020; Whilden et al., 2021; Yamawaki et al., 2021; Yang et al., 2022). VPM-only L6 CThNs receive stronger thalamocortical input from VPM than POm-projecting L6 CThNs (Frandolig et al., 2019), consistent with the VPM axonal arborization found at the L5/6 border (Wimmer et al., 2010), but the VPM input to VPM-only L6 CThNs is weaker than the VPM input onto other neighboring L6 cell types like corticocortical neurons and interlaminar PV neurons (Crandall et al., 2017;

Frandolig et al., 2019). Studies using rabies-based transsynaptic tracing indicate that POm-projecting L6 CThNs, in contrast, receive stronger long-range input from other cortical areas, including secondary somatosensory cortex (S2), motor cortex (M1 and M2), and retrosplenial cortex (RSP), as well as subcortical areas such as the basolateral amygdala (BLA), than do VPM-only L6 CThNs (Whilden et al., 2021). These studies suggest that POm-projecting L6 CThNs integrate more contextual information from these brain regions than VPM-only L6 CThNs.

PERCEPTUAL AND BEHAVIORAL CONTRIBUTIONS OF LAYER 6 CORTICOTHALAMIC TYPES

The contributions of S1 L6 CThNs to perception and behavior remain unclear. L6 CThNs in somatosensory cortex generally exhibit low firing rates (Swadlow, 1989; Lee et al., 2008; Voigts et al., 2020). Studies in which the somatosensory cortex as a whole or L6 CThNs specifically have been acutely or chronically activated or silenced have revealed mixed effects of L6 CThN input on the receptive field properties of neurons in VPM and VPL (Landry and Dykes, 1985; Yuan et al., 1986; Diamond et al., 1992; Ghazanfar et al., 2001; Temereanca and Simons, 2004; Li and Ebner, 2007; Mease et al., 2014; Pauzin and Krieger, 2018; Pauzin et al., 2019). Most of these studies were performed in anesthetized animals, precluding investigations into sensorimotor integration, learning, or higher-order cognitive processing. Furthermore, many of these studies globally inhibited or activated primary sensory cortex, approaches likely to obscure the spatial and temporal contributions of L6 CThNs to receptive field properties. The importance of topographic relationships in understanding the function of L6 CThNs is highlighted by experiments showing that locally activating S1 L6 CThNs enhances the responses in the corresponding barreloid in VPM while suppressing neighboring barreloids (Temereanca and Simons, 2004; Li and Ebner, 2007). Finally, most of these experiments have not tested for functional differences among the different types of L6 CThNs. Nonetheless, these experiments indicate that the receptive field properties of neurons in VPM reflect in part descending input from corticothalamic neurons. Similarly, modulating the activity of L6 CThNs also subtly affects cortical sensory processing, including altering responses to higher-order features of whisker stimuli like change detection (Pauzin and Krieger, 2018; Pauzin et al., 2019; Voigts et al., 2020).

L6 CThNs have been implicated in higher-cognitive processing in studies across sensory systems (Wolff et al., 2021). Although much of this work has focused on the visual and auditory systems, studies in rodent somatosensory cortex are consistent with these findings. For example, L6 CThNs may play a role in sensorimotor integration. In barrel cortex, a subset of L6 CThNs increase their firing activity during whisking (Dash et al., 2022), consistent with studies showing that motor cortex influences L6 CThN activity (Lee et al., 2008; Kinnischtzke et al., 2016). Furthermore, L6 CThNs have long been thought to be the primary projection from S1 to the TRN, which has been increasingly implicated in higher cognitive functions like attentional control and cognitive flexibility (Wolff et al., 2021). L6 CThNs in somatosensory cortex may also play a role in learning. POm synapses onto L5a neurons, the main target of POm axons in S1, rapidly undergo long-term synaptic potentiation during a sensory-association task (Audette et al., 2019). Whether the intracortical synapses of VPM/POm L6 CThNs in L5a (Zhang and Deschênes, 1997; Frandolig et al., 2019), or their long-range synapses in POm or the TRN sector reciprocally connected to POm (Bourassa et al., 1995; Whilden et al., 2021), undergo plasticity and participate in learning-related circuits remains to be tested.

CONCLUSION: OUTSTANDING QUESTIONS

Although there is strong evidence that L6 CThNs generate multiple processing streams within the primary somatosensory cortex, numerous features of their synaptic organization and function remain unaddressed. For example, it is not clear whether all L6 CThN subtypes in primary somatosensory cortex project to the TRN (Bourassa et al., 1995; Hoerder-Suabedissen et al., 2018), a question with implications for how different L6 CThN circuits may contribute to higher cognitive functions like attention. How the spatial organization and temporal activity of different L6 CThN subtypes affect the responses of different cortical and thalamic cell types also requires further study. In addition, which aspects of the organization of somatosensory L6 CThN circuits are shared across sensory cortical areas, in rodents and across other animal species, also requires further study (see **Chapters 19 and 24 in Section V**). The cell types and circuit organization of L6 CThN circuits in visual cortex of the mouse, for example, share many features with those of L6 CThNs in primary somatosensory cortex of rodents, but anatomical and physiological studies suggest there may be additional L6 CThN processing streams with distinct synaptic organization in other animals such as macaques. Cross-regional and cross-species comparisons such as these will reveal basic principles of L6 CThN function. To address these questions, it will also be essential to develop methods for assessing the synaptic organization and functional contributions for each of the different types of L6 CThNs across different model systems.

ACKNOWLEDGMENTS

I thank Seong Yeol An, Su-Jeong Kim, and Alan Gutman-Wei for helpful comments on the chapter and all Brown lab members for insightful discussions. SPB was supported by a Klingenstein-Simons Fellowship in the Neurosciences, the National Institute of Neurological Disorders and Stroke (R01 NS085121), and the National Science Foundation (1656592).

REFERENCES

Ansorge J, Humanes-Valera D, Pauzin FP, Schwarz MK, Krieger P (2020). Cortical layer 6 control of sensory responses in higher-order thalamus. *J Physiol.* 598:3973–4001.

Audette NJ, Bernhard SM, Ray A, Stewart LT, Barth AL (2019). Rapid plasticity of higher-order thalamocortical inputs during sensory learning. *Neuron.* 103:277–291.

Bokor H, Acsady L, Deschenes M (2008). Vibrissal responses of thalamic cells that project to the septal columns of the barrel cortex and to the second somatosensory area. *J Neurosci.* 28:5169–5177.

Bortone DS, Olsen SR, Scanziani M (2014). Translaminar inhibitory cells recruited by layer 6 corticothalamic neurons suppress visual cortex. *Neuron.* 82:474–485.

Bourassa J, Deschênes M (1995). Corticothalamic projections from the primary visual cortex in rats: a single fiber study using biocytin as an anterograde tracer. *Neuroscience.* 66:253–263.

Bourassa J, Pinault D, Deschênes M (1995). Corticothalamic projections from the cortical barrel field to the somatosensory thalamus in rats: A single-fibre study using biocytin as an anterograde tracer. *Eur J Neurosci.* 7:19–30.

Briggs F, Kiley CW, Callaway EM, Usrey WM (2016). Morphological substrates for parallel streams of corticogeniculate feedback originating in both V1 and V2 of the macaque monkey. *Neuron.* 90:388–399.

Casas-Torremocha D, Rubio-Teves M, Hoerder-Suabedissen A, Hayashi S, Prensa L, Molnar Z, Porrero C, Clasca F (2022). A combinatorial input landscape in the "higher-order relay" posterior thalamic nucleus. *J Neurosci.* 42:7757–7781.

Castro-Alamancos MA, Calcagnotto ME (1999). Presynaptic long-term potentiation in corticothalamic synapses. *J Neurosci.* 19:9090–9097.

Chevée M, Robertson JJ, Cannon GH, Brown SP, Goff LA (2018). Variation in activity state, axonal projection, and position define the transcriptional identity of individual neocortical projection neurons. *Cell Rep.* 22:441–455.

Clemente-Perez A, Makinson SR, Higashikubo B, Brovarney S, Cho FS, Urry A, Holden SS, Wimer M, David C, Fenno LE, Acsady L, Deisseroth K, Paz JT (2017). Distinct thalamic reticular cell types differentially modulate normal and pathological cortical rhythms. *Cell Rep.* 19:2130–2142.

Crandall SR, Cruikshank SJ, Connors BW (2015). A corticothalamic switch: controlling the thalamus with dynamic synapses. *Neuron.* 86:768–782.

Crandall SR, Patrick SL, Cruikshank SJ, Connors BW (2017). Infrabarrels are layer 6 circuit modules in the barrel cortex that link long-range inputs and outputs. *Cell Rep.* 21:3065–3078.

Cruikshank SJ, Urabe H, Nurmikko AV, Connors BW (2010). Pathway-specific feedforward circuits between thalamus and neocortex revealed by selective optical stimulation of axons. *Neuron.* 65:230–245.

Dash S, Autio DM, Crandall SR (2022). State-dependent modulation of activity in distinct layer 6 corticothalamic neurons in barrel cortex of awake mice. *J Neurosci.* 42:6551–6565.

Deschênes M, Bourassa J, Pinault D (1994). Corticothalamic projections from layer V cells in rat are collaterals of long-range corticofugal axons. *Brain Res.* 664:215–219.

Deschênes M, Veinante P, Zhang ZW (1998). The organization of corticothalamic projections: reciprocity versus parity. *Brain Res Rev.* 28:286–308.

Diamond ME, Armstrong-James M, Budway MJ, Ebner FF (1992). Somatic sensory responses in the rostral sector of the posterior group (POm) and in the ventral posterior medial nucleus (VPM) of the rat thalamus: Dependence on the barrel field cortex. *J Comp Neurol.* 319:66–84.

Eaton SA, Salt TE (1996) Role of N-methyl-D-aspartate and metabotropic glutamate receptors in corticothalamic excitatory postsynaptic potentials in vivo. *Neuroscience.* 73:1–5.

El-Boustani S, Sermet BS, Foustoukos G, Oram TB, Yizhar O, Petersen CCH (2020). Anatomically and functionally distinct thalamocortical inputs to primary and secondary mouse whisker somatosensory cortices. *Nat Commun.* 11:3342.

Fernandez LMJ, Pellegrini C, Vantomme G, Beard E, Luthi A, Astori S (2017). Cortical afferents onto the nucleus reticularis thalami promote plasticity of low-threshold excitability through GluN2C-NMDARs. *Sci Rep.* 7:12271.

Fitzpatrick D, Usrey WM, Schofield BR, Einstein G (1994). The sublaminar organization of corticogeniculate neurons in layer 6 of macaque striate cortex. *Vis Neurosci.* 11:307–315.

Frandolig JE, Matney CJ, Lee K, Kim J, Chevée M, Kim SJ, Bickert AA, Brown SP (2019). The synaptic organization of layer 6 circuits reveals inhibition as a major output of a neocortical sublamina. *Cell Rep.* 28:3131–3143.

Ghazanfar AA, Krupa DJ, Nicolelis MA (2001). Role of cortical feedback in the receptive field structure and nonlinear response properties of somatosensory thalamic neurons. *Exp Brain Res.* 141:88–100.

Golshani P, Liu XB, Jones EG (2001.) Differences in quantal amplitude reflect GluR4-subunit number at corticothalamic synapses on two populations of thalamic neurons. *Proc Natl Acad Sci U S A.* 98:4172–4177.

Groh A, Bokor H, Mease RA, Plattner VM, Hangya B, Stroh A, Deschenes M, Acsady L (2014). Convergence of cortical and sensory driver inputs on single thalamocortical cells. Cereb Cortex. 24:3167–3179.

Guo K, Yamawaki N, Barrett JM, Tapies M, Shepherd GMG (2020). Cortico-thalamo-cortical circuits of mouse forelimb S1 are organized primarily as recurrent loops. *J Neurosci.* 40:2849–2858.

Hasse JM, Bragg EM, Murphy AJ, Briggs F (2019). Morphological heterogeneity among corticogeniculate neurons in ferrets: Quantification and comparison with a previous report in macaque monkeys. *J Comp Neurol.* 527:546–557.

Hoerder-Suabedissen A, Hayashi S, Upton L, Nolan Z, Casas-Torremocha D, Grant E, Viswanathan S, Kanold PO, Clasca F, Kim Y, Molnar Z (2018). Subset of cortical layer 6b neurons selectively innervates higher order thalamic nuclei in mice. *Cereb Cortex.* 28:1882–1897.

Hoogland PV, Welker E, Van der Loos H (1987). Organization of the projections from barrel cortex to thalamus in mice studied with Phaseolus vulgaris-leucoagglutinin and HRP. *Exp Brain Res.* 68:73–87.

Hsu CL, Yang HW, Yen CT, Min MY (2010). Comparison of synaptic transmission and plasticity between sensory and cortical synapses on relay neurons in the ventrobasal nucleus of the rat thalamus. *J Physiol.* 588:4347–4363.

Jackman SL, Turecek J, Belinsky JE, Regehr WG (2016). The calcium sensor synaptotagmin 7 is required for synaptic facilitation. *Nature.* 529:88–91.

Katz LC (1987). Local circuitry of identified projection neurons in cat visual cortex brain slices. *J Neurosci.* 7:1223–1249.

Killackey HP, Sherman SM (2003). Corticothalamic projections from the rat primary somatosensory cortex. *J Neurosci.* 23:7381–7384.

Kim J, Matney CJ, Blankenship A, Hestrin S, Brown SP (2014). Layer 6 corticothalamic neurons activate a cortical output layer, layer 5a. *J Neurosci.* 34:9656–9664.

Kinnischtzke AK, Fanselow EE, Simons DJ (2016). Target-specific M1 inputs to infragranular S1 pyramidal neurons. *J Neurophysiol.* 116:1261–1274.

Kumar P, Ohana O (2008). Inter- and intralaminar subcircuits of excitatory and inhibitory neurons in layer 6a of the rat barrel cortex. *J Neurophysiol.* 100:1909–1922.

Kwegyir-Afful EE, Simons DJ (2009). Subthreshold receptive field properties distinguish different classes of corticothalamic neurons in the somatosensory system. *J Neurosci.* 29:964–972.

Lam YW, Sherman SM (2007). Different topography of the reticulothalmic inputs to first- and higher-order somatosensory thalamic relays revealed using photostimulation. *J Neurophysiol.* 98:2903–2909.

Lam YW, Sherman SM (2010). Functional organization of the somatosensory cortical layer 6 feedback to the thalamus. *Cereb Cortex.* 20:13–24.

Land PW, Buffer SA, Jr., Yaskosky JD (1995). Barreloids in adult rat thalamus: Three-dimensional architecture and relationship to somatosensory cortical barrels. *J Comp Neurol.* 355:573–588.

Landisman CE, Connors BW (2007). VPM and PoM nuclei of the rat somatosensory thalamus: intrinsic neuronal properties and corticothalamic feedback. *Cereb Cortex.* 17:2853–2865.

Landry P, Dykes RW (1985). Identification of two populations of corticothalamic neurons in cat primary somatosensory cortex. *Exp Brain Res.* 60:289–298.

Lee S, Carvell GE, Simons DJ (2008). Motor modulation of afferent somatosensory circuits. *Nat Neurosci.* 11:1430–1438.

Lee SC, Patrick SL, Richardson KA, Connors BW (2014). Two functionally distinct networks of gap junction-coupled inhibitory neurons in the thalamic reticular nucleus. *J Neurosci.* 34:13170–13182.

Li L, Ebner FF (2007). Cortical modulation of spatial and angular tuning maps in the rat thalamus. *J Neurosci.* 27:167–179.

Li Y, Lopez-Huerta VG, Adiconis X, Levandowski K, Choi S, Simmons SK, Arias-Garcia MA, Guo B, Yao AY, Blosser TR, Wimmer RD, Aida T, Atamian A, Naik T, Sun X, Bi D, Malhotra D, Hession CC, Shema R, Gomes M, Li T, Hwang E, Krol A, Kowalczyk M, Peca J, Pan G, Halassa MM, Levin JZ, Fu Z, Feng G (2020). Distinct subnetworks of the thalamic reticular nucleus. *Nature.* 583:819–824.

Liu XB, Honda CN, Jones EG (1995). Distribution of four types of synapse on physiologically identified relay neurons in the ventral posterior thalamic nucleus of the cat. *J Comp Neurol.* 352:69–91.

Martinez-Garcia RI, Voelcker B, Zaltsman JB, Patrick SL, Stevens TR, Connors BW, Cruikshank SJ (2020). Two dynamically distinct circuits drive inhibition in the sensory thalamus. *Nature.* 583:813–818.

Mease RA, Krieger P, Groh A (2014). Cortical control of adaptation and sensory relay mode in the thalamus. *Proc Natl Acad Sci U S A.* 111:6798–6803.

Melzer S, Newmark ER, Mizuno GO, Hyun M, Philson AC, Quiroli E, Righetti B, Gregory MR, Huang KW, Levasseur J, Tian L, Sabatini BL (2021). Bombesin-like peptide recruits disinhibitory cortical circuits and enhances fear memories. *Cell.* 184:5622–5634.

Ohno S, Kuramoto E, Furuta T, Hioki H, Tanaka YR, Fujiyama F, Sonomura T, Uemura M, Sugiyama K, Kaneko T (2012). A morphological analysis of thalamocortical axon fibers of rat posterior thalamic nuclei: A single neuron tracing study with viral vectors. *Cereb Cortex.* 22:2840–2857.

Olsen SR, Bortone DS, Adesnik H, Scanziani M (2012). Gain control by layer six in cortical circuits of vision. *Nature.* 483:47–52.

Pauzin FP, Krieger P (2018). A corticothalamic circuit for refining tactile encoding. *Cell Rep.* 23:1314–1325.

Pauzin FP, Schwarz N, Krieger P (2019). Activation of corticothalamic layer 6 cells decreases angular tuning in mouse barrel cortex. *Front Neural Circuits.* 13:67.

Pinault D, Bourassa J, Deschênes M (1995). Thalamic reticular input to the rat visual thalamus: A single fiber study using biocytin as an anterograde tracer. *Brain Res.* 670:147–152.

Reichova I, Sherman SM (2004). Somatosensory corticothalamic projections: distinguishing drivers from modulators. *J Neurophysiol.* 92:2185–2197.

Shepherd GMG, Yamawaki N (2021). Untangling the cortico-thalamo-cortical loop: Cellular pieces of a knotty circuit puzzle. *Nat Rev Neurosci.* 22:389–406.

Sherman SM, Guillery RW (1998). On the actions that one nerve cell can have on another: distinguishing "drivers" from "modulators." *Proc Natl Acad Sci U S A.* 95:7121–7126.

Sumser A, Mease RA, Sakmann B, Groh A (2017). Organization and somatotopy of corticothalamic projections from L5B in mouse barrel cortex. *Proc Natl Acad Sci U S A.* 114:8853–8858.

Swadlow HA (1989). Efferent neurons and suspected interneurons in S-1 vibrissa cortex of the awake rabbit: Receptive fields and axonal properties. *J Neurophysiol.* 62:288–308.

Tanaka YR, Tanaka YH, Konno M, Fujiyama F, Sonomura T, Okamoto-Furuta K, Kameda H, Hioki H, Furuta T, Nakamura KC, Kaneko T (2011). Local connections of excitatory neurons to corticothalamic neurons in the rat barrel cortex. *J Neurosci.* 31:18223–18236.

Temereanca S, Simons DJ (2004). Functional topography of corticothalamic feedback enhances thalamic spatial response tuning in the somatosensory whisker/barrel system. *Neuron.* 41:639–651.

Urbain N, Deschenes M (2007). A new thalamic pathway of vibrissal information modulated by the motor cortex. *J Neurosci.* 27:12407–12412.

Usrey WM, Fitzpatrick D (1996). Specificity in the axonal connections of layer VI neurons in tree shrew striate cortex: Evidence for distinct granular and supragranular systems. *J Neurosci.* 16:1203–1218.

Voigts J, Deister CA, Moore CI (2020). Layer 6 ensembles can selectively regulate the behavioral impact and layer-specific representation of sensory deviants. *eLife.* 9:e48957.

Whilden CM, Chevée M, An SY, Brown SP (2021). The synaptic inputs and thalamic projections of two classes of layer 6 corticothalamic neurons in primary somatosensory cortex of the mouse. *J Comp Neurol.* 529:3751–3771.

Wimmer VC, Bruno RM, de Kock CP, Kuner T, Sakmann B (2010). Dimensions of a projection column and architecture of VPM and POm axons in rat vibrissal cortex. *Cereb Cortex.* 20:2265–2276.

Wolff M, Morceau S, Folkard R, Martin-Cortecero J, Groh A (2021). A thalamic bridge from sensory perception to cognition. *Neurosci Biobehav Rev.* 120:222–235.

Yamawaki N, Raineri Tapies MG, Stults A, Smith GA, Shepherd GM (2021). Circuit organization of the excitatory sensorimotor loop through hand/forelimb S1 and M1. *eLife.* 10:e66836.

Yang D, Qi G, Ding C, Feldmeyer D (2022). Layer 6A pyramidal cell subtypes form synaptic microcircuits with distinct functional and structural properties. *Cereb Cortex.* 32:2095–2111.

Yang Q, Chen CC, Ramos RL, Katz E, Keller A, Brumberg JC (2014). Intrinsic properties of and thalamocortical inputs onto identified corticothalamic-VPM neurons. *Somatosens Mot Res.* 31:78–93.

Yao Z, van Velthoven CT, Nguyen TN, Goldy J, Sedeno-Cortes AE, Baftizadeh F, Bertagnolli D, Casper T, Chiang M, Crichton K, Ding SL (2021). A taxonomy of transcriptomic cell types across the isocortex and hippocampal formation. *Cell.* 184:3222–3241.

Yuan B, Morrow TJ, Casey KL (1986). Corticofugal influences of S1 cortex on ventrobasal thalamic neurons in the awake rat. *J Neurosci.* 6:3611–3617.

Zhang ZW, Deschênes M (1997). Intracortical axonal projections of lamina VI cells of the primary somatosensory cortex in the rat: a single-cell labeling study. *J Neurosci.* 17:6365–6379.

28.

THE VESTIBULAR THALAMUS

Kathleen E. Cullen, Jérôme Carriot, and Maurice J. Chacron

INTRODUCTION

Unlike other sensory modalities (e.g., vision, audition), there is no single area of the thalamus that is devoted primarily to processing vestibular information (see Lopez and Blanke, 2011, for review). Specifically, there are two major vestibular thalamic pathways, one posterior and one anterior, that are thought to transmit vestibular information from the sensory periphery to the cortex (for review, see Lopez and Blanke, 2011; Cullen and Taube, 2017; Cullen, 2019; Cullen and Chacron, 2023; see also Figure 28.1).

The posterior pathway mediates self-motion perception and involves the ventroposterior lateral area (VPL) of the thalamus. VPL neurons receive direct monosynaptic input from a subpopulation of vestibular nuclei neurons that are not sensitive to eye movements—referred to as vestibular-only (VO) neurons in the literature (Marlinski and McCrea, 2009; see Figure 28.1, left). In turn, the VPL projects directly to multiple cortical areas that are involved in self-motion processing and perception, including the parietoinsular vestibular cortex (PIVC) and ventral intraparietal cortex, as well as to area 2v of the intraparietal sulcus, and area 3a in the sulcus centralis (reviewed in Lopez and Blanke, 2011; Cullen, 2019; see Figure 28.1, left). In this context, VPL can be considered a first-order area for vestibular processing (i.e., similar to the lateral geniculate nucleus in the visual system; Sherman and Guillery, 2002). Additionally, it is noteworthy that VPL neurons also receive input from other sensory modalities including somatosensory (Poggio and Mountcastle, 1963). The common wisdom has been that VPL neurons merely relay self-motion information from the vestibular nuclei to the cortex (for review, see Lopez and Blanke, 2011; Cullen and Chacron, 2023). However, as detailed below, recent studies have instead shown that VPL neurons play an important role in determining self-motion perception by actively transforming incoming signals from the vestibular nuclei.

In contrast, the anterior pathway is thought to be involved in generating a head-direction signal by integrating both vestibular and extra-vestibular (e.g., visual) cues. This pathway involves the anterior dorsal nucleus (ADN) of the thalamus and is located 4 synapses away from the vestibular nuclei (Cullen and Taube, 2017; Figure 28.1, right). However, it should be noted that neurons within ADN do not primarily respond to vestibular input generated during head motion. Instead, these neurons are thought to encode head direction. The ADN projects to multiple cortical areas such as the dorsal pre-subiculum, the medial entorhinal cortex, and the retrosplenial cortex (see Cullen and Taube, 2017, for review). Thus, while it has been argued that this pathway is similar to other first-order sensory thalamocortical circuits (Petrof and Sherman, 2009), there are some notable differences with the posterior pathway. The common wisdom is that the head-direction signal transmitted by ADN neurons is generated by ring-attractor networks based on head-velocity information transmitted indirectly from the vestibular nuclei. As reviewed below, however, there is considerable evidence that neurons within ascending pathways to the ADN are predominantly sensitive to eye movements, not head movements. Accordingly, the nature of the signal transmitted by ADN neurons remains an open question.

THE POSTERIOR VESTIBULAR THALAMIC PATHWAY

RESPONSES OF VPL NEURONS TO ARTIFICIAL STIMULI

Neurophysiological studies in alert primates have shown that VPL neurons are insensitive to eye movements. Rather, these neurons robustly encode head movements consisting of passive sinusoidal yaw and tilt rotations, as well as translations in the horizontal plane. This is expected, as VPL neurons receive input from VO neurons and not from other eye movement–sensitive neural subpopulations

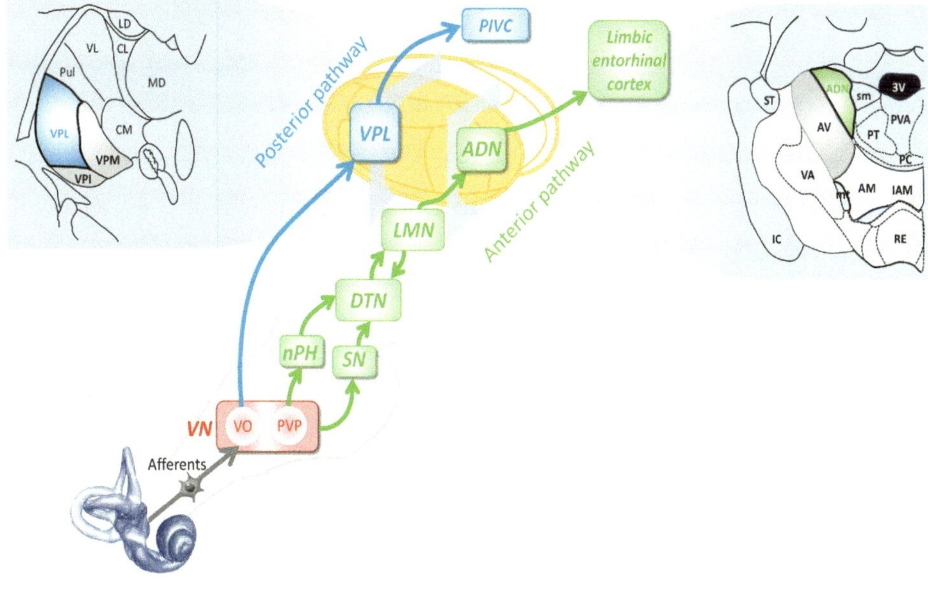

Figure 28.1. Overview highlighting functions and projections of the 2 ascending vestibular pathways. (1). The posterior vestibulothalamic pathway (cyan). From the vestibular nuclei (VN in red), VO neurons project to the ventroposterior lateral nucleus (VPL) before reaching the multiple cortical areas including the parietoinsular vestibular cortex (PIVC). (2). The anterior vestibulothalamic pathway (green). From the vestibular nuclei (VN in red), through the nucleus prepositus hypoglossi (nPH) and supragenual nucleus (SN), information reaches the Head Direction Cells network, composed of the dorsal tegmental nuclei (DTN) and the lateral mammillary nuclei (LMN). Neurons within LMN project to the anterior dorsal nucleus (ADN) in the thalamus and then to the cortex. Left panel: Coronal representation of the posterior thalamus. The ventral part is in cyan. LD = lateral dorsal; VL = ventral lateral; CL = central lateral; MD = medial dorsal; Pul = pulvinar; VPL = ventral posterior lateral; VPM = ventral posterior medial; CM = central medial; VPI = ventral posterior inferior. Right panel: Coronal representation of the anterior thalamus. The anterior dorsal part is in green. ST = stria terminalis; AV = anterior ventral; ADN = anterior dorsal nucleus; sm = stria medullaris; 3V = third ventricle; PT = paratenial nucleus; PVA = paraventricular nucleus; PC = paracentral nucleus; IC = internal capsule; VA = ventral anterior; mt = mammillothalamic track; AM = anteromedial; IAM = interanteromedial; RE = reuniens nucleus.

within the vestibular nuclei (Marlinski and McCrea, 2009). Most prior studies have used artificial self-motion stimuli (e.g., sinusoidal head motion) to investigate the responses of VPL neurons and found remarkable similarity with those of their input VO neurons within the vestibular nuclei (Figure 28.2A,B, top panels). Specifically, in response to sinusoidal head rotations along the yaw axis, VPL neurons displayed high-pass dynamics, with gain increasing as a function of frequency and the firing rate leading (i.e., reaching its maximum before) the head-velocity stimulus (Meng et al., 2007; Meng and Angelaki, 2010; Carriot et al., 2022). Figure 28.2C,D shows the gain (top, dashed curves) and phase (bottom, dashed curves) response profiles of example vestibular nuclei and VPL neurons to artificial yaw rotations, respectively. In response to translations, VPL neurons displayed low-pass dynamics, with gain instead decreasing as a function of frequency and the firing rate lagging the head-acceleration stimulus (Meng et al., 2007). These dynamics are similar to those displayed by a subset of vestibular nuclei neurons (Angelaki and Dickman, 2000).

Based on these results, one might conclude that VPL neurons passively relay self-motion information from the vestibular nuclei to cortical areas with little additional processing. However, it has been shown that VPL neurons display marked response nonlinearities that are not found within the vestibular nuclei. For example, during stimulation, VPL neurons display a form of sensory adaptation termed *contrast gain scaling* (or *contrast gain control*). This is evidenced by the fact that neural sensitivity decreases with increasing stimulus amplitude, such that the firing rate remains within the dynamic range in response to stimuli that would otherwise give rise to rectification and saturation (Marlinski and McCrea, 2008; Dale and Cullen, 2019; Carriot et al., 2021; Carriot et al., 2022). An example of contrast gain scaling display by VPL neurons is shown below in the section "VPL Neurons and Self-Motion Perception." Overall, as further discussed below, the presence of such response nonlinearities suggests that VPL neurons perform additional computations on their input from the vestibular nuclei, as opposed to merely relaying information to cortical areas.

RESPONSES OF VPL NEURONS TO NATURALISTIC STIMULI

As reviewed above, the responses of VPL neurons have been primarily studied using artificial (i.e., sinusoidal) vestibular stimulation. More recently, studies have investigated how

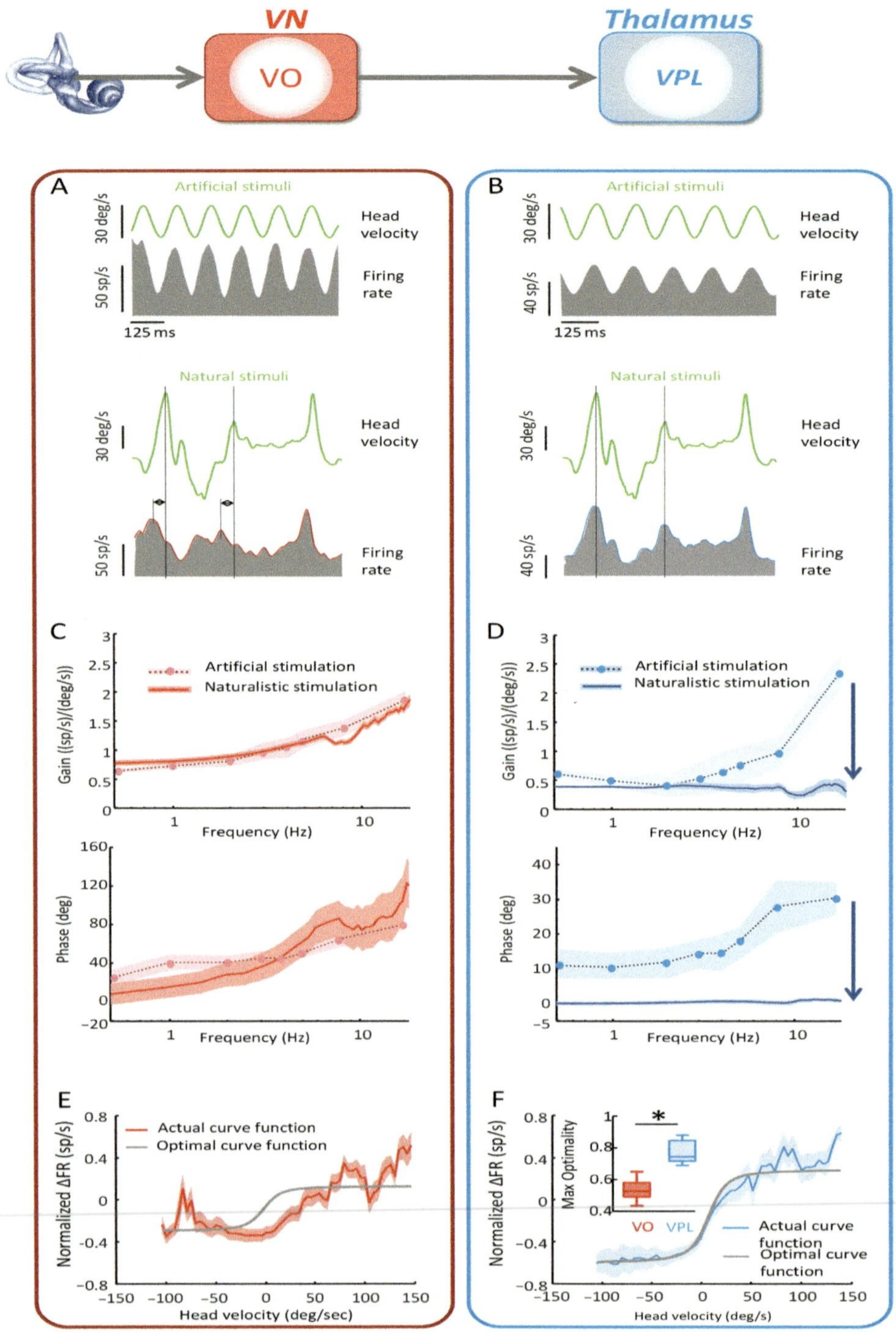

Figure 28.2. Encoding of artificial and naturalistic stimuli within the posterior vestibulothalamic pathway. (Top). Schematic showing the vestibular endorgans projecting to VO neurons within the vestibular nuclei, which in turn project to area VPL of the thalamus. (A). Top: Time series showing an artificial sinusoidal self-motion stimulus (green) with the firing rate response (gray) from a typical VO neuron. Bottom: Same as top, but for a naturalistic self-motion stimulus. (B). Same as A, but for a typical VPL neuron. (C). Top: Population- averaged neural gain as a function of frequency for VN neurons during artificial (dashed lines) and naturalistic (solid lines) stimulation. Note the similar increase with frequency, which is indicative of high- pass tuning. Bottom: Population- averaged phase as a function of frequency for VN neurons during artificial (dashed lines) and naturalistic (solid lines) stimulation. Note the increasing phase lead with frequency. (D). Same as C, but for VPL neurons. Note the important difference (blue arrow) showing that VPL neurons display gains that are constant over the frequency range and no significant phase lead during naturalistic but not artificial stimulation. (E). Actual (red) and optimal (gray) tuning functions for an example VO neuron. (F). Same as E, but for an example VPL neuron. Inset: Population- averaged maximum optimality values were significantly higher for VPL than for VO neurons.

neurons in ascending vestibular pathways respond to naturalistic stimulation, characterized by profiles that more closely resemble those of self-motion experienced during everyday activities (e.g., walking, riding in a vehicle, etc.; Schneider et al., 2015; Mitchell et al., 2018; Mackrous et al., 2020; Carriot et al., 2022). This is an important direction of research because the statistics of natural self-motion stimuli fundamentally differ from those of artificial stimuli typically used in laboratory studies (Carriot et al., 2014, 2017). Notably, natural self-motion stimuli both contain a spectrum of frequencies and can reach much higher amplitudes as compared to artificial stimuli. Natural self-motion stimuli also comprise a mixture of passively applied (e.g., when riding in a vehicle) and voluntary movements (e.g., active head turns, walking) (Carriot et al., 2014, 2017).

A recent study established that VPL neurons respond in a fundamentally different manner to naturalistic versus artificial sinusoidal stimulation (Carriot et al., 2022). As reviewed above in the section "Responses of VPL Neurons to Artificial Stimuli," VPL neurons display characteristic high-pass tuning to artificial rotational stimuli, characterized by increasing gain and phase leads that are similar to those displayed by their input vestibular nuclei neurons (Figure 28.2C,D, dashed curves). Yet, the responses of these same VPL neurons to naturalistic rotational stimuli are characterized by: (1) gains that are independent of frequency, and (2) firing-rate modulations that are approximately in phase with head velocity independent of frequency (Figure 28.2D, solid blue curves). As a result, VPL neural responses faithfully follow the detailed time course of naturalistic self-motion stimuli (Figure 28.2B, compare green and blue in bottom panel). Importantly, this change in response dynamics is not simply inherited from vestibular nuclei neurons whose responses do not faithfully follow the detailed time course of naturalistic self-motion stimuli (Figure 28.2A, compare green and red in bottom panel). Computational modeling further showed that the contrast gain scaling displayed by VPL neurons accounts for their different response dynamics to naturalistic versus artificial stimulation (Carriot et al., 2022). As such, these results demonstrate that VPL neurons do more than simply relay information from vestibular nuclei to cortical areas. Specifically, during naturalistic stimulation, these neurons provide a faithful representation of head velocity that is then transmitted to cortical areas by performing a nonlinear transformation that effectively compensates for the high-pass tuning of their incoming sensory input.

VPL NEURAL RESPONSES ARE ADAPTED SUCH AS TO OPTIMALLY PROCESS NATURAL SELF-MOTION STIMULI

There is growing evidence that the coding strategies used by the brain are adapted to the statistics of the environment that the organism interacts with (for review, see Simoncelli and Olshausen, 2001; Wark et al., 2007; Sharpee et al., 2014). Thus, another reason to study neural responses to natural as opposed to artificial self-motion stimuli is to establish if this is the case in the vestibular system. It is well-known that natural stimuli display redundancy. For example, when inspecting natural images, neighboring spatial locations are strongly correlated in intensity (see Simoncelli and Olshausen, 2001, for review). Such redundancy is also seen in natural self-motion stimuli as their temporal correlation decays with a characteristic timescale (Carriot et al., 2014, 2017). It was initially proposed that sensory systems should take advantage of the redundancy found in natural stimuli in order to maximize information transmission (see Simoncelli and Olshausen, 2001, for review). If one assumes that the firing rate of neurons can only take values within a given interval, then intuitively information is maximized when all values of the firing rate occur with equal probability. As a result, the neural tuning curve is proportional to the cumulative integral of the stimulus probability distribution (Wark et al., 2007).

In the context of the vestibular system, comparison of actual VPL neural curves to those predicted when information is maximized revealed a better match than for vestibular nuclei neurons (Figure 28.2E,F). This finding demonstrates that VPL neurons more optimally encode natural self-motion stimuli. Further, this result holds irrespective of whether self-motion is represented by position, velocity, acceleration, or any combination (Carriot et al., 2022). Specifically, while VPL neural responses best correlate with head velocity under naturalistic stimulation, vestibular nuclei neural responses instead best correlate with a combination of head velocity and acceleration. Thus, encoding of natural self-motion by VPL is more optimal than that of their vestibular nuclei input even if one considers the best representation of naturalistic self-motion for each brain area (Carriot et al., 2022).

The question then arises: Why do vestibular nuclei neurons display high-pass dynamics that best represent a combination of velocity and acceleration, only to have VPL neurons effectively "remove" this transformation in order to recover the head-velocity signal? To answer this question, one must consider that vestibular nuclei VO neurons also project to the spinal cord and thereby mediate vestibulo-spinal reflexes (Goldberg et al., 1987; Sato and Sasaki, 1993). It is thought that their high-pass tuning and phase leads help compensate for the inertia of the head-neck system (Fernandez and Goldberg, 1971; Bilotto et al., 1982; see Cullen, 2019, for review). A second related question is: Why do VPL neural activities faithfully encode head velocity during natural self-motion as opposed to another representation (e.g., acceleration)? To answer this question, one must consider that faithful encoding of head velocity by VPL neurons is in agreement with psychophysical studies showing that head-velocity detection thresholds (i.e., the minimum stimulus amplitude that can be perceived) are approximately constant for frequencies above 0.5 Hz (Grabherr et al., 2008; Valko et al., 2012). Taken together, these findings suggest that the faithful representation of head velocity by VPL neurons plays a fundamental role in determining self-motion perception.

VPL NEURONS AND SELF-MOTION PERCEPTION

How do VPL neurons mediate self-motion perception? Previous psychophysical studies have quantified the ability of human subjects to distinguish between sinusoidal rotation stimuli along the yaw axis with different amplitudes. Overall, the minimum difference in amplitude that can be distinguished (i.e., the just noticeable difference or discrimination threshold) initially increased as a function of the reference stimulus amplitude but then saturated (Mallery et al., 2010; Nesti et al., 2015; Nouri and Karmali, 2018) (Figure 28.3A, solid black line). As such, the dependency of the just noticeable difference on stimulus amplitude deviates from Weber's law, which instead predicts a linear relationship (Weber, 1834).

In general, the neural underpinnings of Weber's law have not been extensively studied. A recent study investigated how VPL neurons contribute to self-motion perception by testing if their responses could account for the observed finding that self-motion perception is better than that predicted from Weber's law (Carriot et al., 2021). Specifically, neural discrimination thresholds were

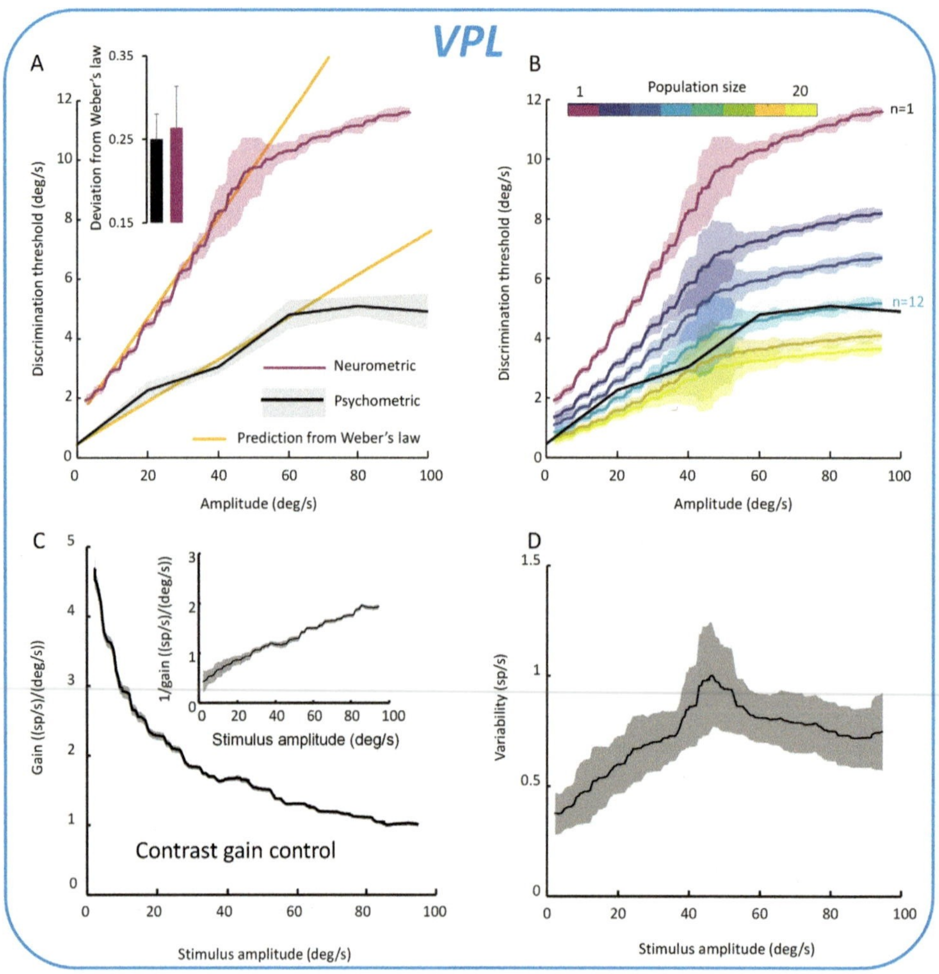

Figure 28.3. The neural basis for violations of Weber's law in self-motion perception. (A). Comparison between VPL neural discrimination thresholds at 0.5 Hz (magenta) and perceptual values also at 0.5 Hz (black) as a function of stimulus amplitude. In both cases, discrimination thresholds increase at a higher rate for low (i.e., < ~50 deg/s) than for high (i.e., > ~50 deg/s) stimulus amplitude, causing a deviation from the Weber's law prediction (compare to orange to magenta and black lines). Inset: quantification of the deviation from Weber's law for neurometric (magenta) and for perception (black) were not significantly different from one another. (B). Neural discrimination thresholds as a function of stimulus amplitude for population sizes of 1 (magenta) to 20 neurons (yellow). Also shown are the perceptual data (black). (C). Population-averaged neural gain as a function of stimulus amplitude. Inset: The reciprocal of gain increases linearly with increasing stimulus amplitude. (D). Variability initially increases but then saturates with increasing stimulus amplitude for VPL neurons. Data adapted from Carriot et al. (2021).

quantified to provide a basis for comparison with previously reported perceptual discrimination thresholds (Mallery et al., 2010). This approach is similar to that used previously to compare neurometric measures with their psychometric counterparts (see, e.g., Sadeghi et al., 2007b; Jamali et al., 2013). Overall, it was found that VPL neural discrimination thresholds displayed a dependency on stimulus amplitude that qualitatively matches that seen for perception (Figure 28.3A). Specifically, while VPL discrimination thresholds were 3-fold higher, they initially increased linearly but then saturated as a function of stimulus amplitude (Figure 28.3A). Furthermore, by pooling the activities of VPL neural populations, it is possible to obtain neural discrimination threshold values that are numerically equal to those obtained for perception (Figure 28.3B). Thus, these results show that VPL neurons provide a substrate as to why the just noticeable difference for high stimulus amplitudes is substantially better than that predicted from Weber's law.

Why exactly do VPL discrimination thresholds initially increase linearly but then saturate with increasing stimulus amplitude? To answer this question, it is useful to consider that neural discrimination thresholds are proportional to the variability-to-gain ratio (Jamali et al., 2013; Carriot et al., 2021). Intuitively this makes sense because "perceptual variability," which is sometimes used to measure the just noticeable difference (Evans and Howarth, 1966), is in part due to the variability displayed by neural activity. Overall, the gain of VPL neurons is inversely proportional to stimulus amplitude, which is a signature of contrast gain scaling (Figure 28.3C). As such, the inverse gain instead increases linearly with stimulus amplitude (Figure 28.3C, inset). If variability were independent of stimulus amplitude, then neural discrimination thresholds should increase linearly with increasing amplitude as they would then be proportional to the inverse gain, consistent with Weber's law. It is for this reason that sensory adaptation in the form of contrast gain scaling is sometimes referred to as Weber adaptation (Donner et al., 1990). Therefore, the deviation from Weber's law observed for neural discrimination thresholds should result from VPL neural variability being dependent on stimulus amplitude. Indeed, VPL variability first increases but is then constant as a function of stimulus amplitude (Figure 28.3D), thereby explaining why neural discrimination thresholds tend to saturate for higher amplitudes. While the nature of the mechanisms responsible for this saturation remain unknown to date, it is possible that these include contrast gain scaling. Indeed, as mentioned above, contrast gain scaling can cause the firing rate response of a neuron to become invariant with increasing amplitude, which would make variability constant.

There are, of course, several caveats when comparing psychophysical results obtained in human subjects to neurophysiological data obtained in nonhuman primates. While there are clear differences between both species, there is reason to believe that there are commonalities in the strategies underlying self-motion perception (Merfeld and Zupan, 2002). It should be noted, moreover, that the pooling of neural activities considered by Carriot et al. (2021) assumed that the neural variabilities were independent of one another. In reality, this is likely not the case, as neurons tend to display correlations between their variabilities, which are referred to as "noise correlations," that can significantly improve or impair encoding by neural populations depending on their relationship with other neural response properties (see Panzeri et al., 2022, for review). Further studies are needed to understand how VPL neural populations encode self-motion and how the responses of individual neurons are integrated to give rise to perception. Finally, we note that the stimuli used in both perceptual (Mallery et al., 2010) and neurophysiological (Carriot et al., 2021) studies varied sinusoidally and were thus artificial. Understanding how the altered dynamics displayed by VPL neurons during naturalistic self-motion impact perception should be investigated in the future.

VPL NEURONS SELECTIVELY ENCODE PASSIVELY APPLIED SELF-MOTION

In the experiments summarized above, both the artificial and naturalistic stimuli were passively applied (i.e., the animal or subject was seated on a motion device that was controlled by a computer). In contrast, the self-motion experienced in our everyday lives comprise a mixture of passively applied and voluntary movements, as mentioned above (Carriot et al., 2014, 2017). Several studies have compared responses of neurons within early vestibular pathways to externally applied self-motion (i.e., passive self-motion) to those resulting from voluntary movements (i.e., active self-motion). In general, the responses of peripheral vestibular afferents are context-independent; they respond similarly to comparable active and passive rotational or translational self-motion (Cullen and Minor, 2002; Sadeghi et al., 2007a; Jamali et al., 2009; Mackrous et al., 2022). Thus, the responses of afferents do not discriminate between active and passive self-motion. However, this is not the case for vestibular neurons in the next stages of processing. Notably, the VO neurons within the vestibular nuclei that project to VPL display substantially attenuated responses to active versus passive self-motion. As a result, these neurons preferentially encode the latter (see Cullen, 2004, 2012, for review). Similarly, neurons within the deep cerebellar nuclei and the anterior vermis demonstrate marked attenuation to active versus passive self-motion (Brooks and Cullen, 2013; Zobeiri and Cullen, 2022).

Importantly, VO and rostral fastigial nucleus neurons further display selectively attenuated responses to active self-motion not only when making a voluntary movement, but also when a voluntary movement occurs concurrently with passively- applied self-motion. The discovery that VO neural responses are suppressed during actively-, but not passively-applied motion has significant ramifications for understanding how self-motion information is gated depending on context. Specifically, the stabilizing commands provided by an intact vestibulo-spinal reflex would be counterproductive during active movements but are essential for regulating posture in reaction to unexpected self-motion (e.g., that experienced while slipping on ice). Accordingly, it is functionally beneficial to reduce the efficacy of vestibulo-spinal reflexes during voluntary self-motion.

Recent studies have shown VPL neurons correspondingly display selectively attenuated responses to active self-motion (Dale and Cullen, 2019) (Figure 28.4A,B). This result is not unexpected, as VO neurons within the vestibular nuclei project directly to VPL. Importantly however, VPL neurons display even stronger response attenuation (~95% vs. ~70%) during active self-motion than do vestibular nuclei VO neurons. This difference implies that response selectivity to passive self-motion is not merely inherited from the vestibular nuclei but is further refined in VPL. Overall, the ventral posterior thalamocortical pathway predominately encodes unexpected self-motion—a feature that likely underlies our perception that the vestibular consequences of active self-motion are the result of our own voluntary behavior, thereby making an essential contribution to our sense of agency.

The question thus arises: What mechanisms mediate selective encoding of passive self-motion? First, the obvious hypothesis that response attenuation involves presynaptic inhibition of the afferent input to VO neurons can be rejected since suppression is not all-or-none. Instead, vestibular nuclei and VPL neurons can continue to respond to the passive component of self-motion when active and passive self-motion occur together, as described above (Figure 28.4C). Second, recent work has shown that attenuation is also not the result of an inhibitory input from an efference copy of the motor command alone (von Holst and Mittelstaedt, 1950). No response (e.g., due to a negative image of the predicted vestibular input resulting from voluntary motion) was seen when animals attempt to make head movements but are prevented from doing so via head restraint (Dale and Cullen, 2019). Rather, response attenuation occurs only when there is a match between the predicted and actual consequences of voluntary motion as determined from the efference copy and proprioceptive signals (Brooks and Cullen, 2014; Cullen and Zobeiri, 2021; Figure 28.4D). Indeed, a cerebellar-based computation is thought to enable the distinction between passively- versus actively-generated vestibular signals at the level of the vestibular and deep cerebellar nuclei (reviewed in Cullen, 2014). We speculate that the cerebellum also plays a key role in mediating the attenuation observed in VPL neurons. Further studies are needed to understand the mechanisms mediating further refinement of response selectivity in VPL neurons beyond that inherited from the vestibular nuclei.

THE ANTERIOR VESTIBULAR THALAMUS

The anterior thalamocortical pathway is thought to encode our static sense of direction during navigation and spatial memory tasks. This thalamocortical pathway consists of direct projections from the anterior dorsal thalamus to the retrosplenial cortex and multisynaptic projections to the entorhinal cortex through pre- and para-subiculum (Cullen and Taube, 2017; Figures 28.1 and 28.5A). Information is transmitted indirectly from the vestibular nuclei to the ADN via a network of structures that include the nucleus prepositus hypoglossi (nPH), dorsal tegmental nucleus, and lateral mammillary nucleus (Figure 28.5A). Neurophysiological recording studies in rodents have demonstrated that individual neurons in the anterior thalamocortical pathway are tuned to a specific directional heading in the horizontal plane that is constant regardless of where the animal is located in space (Cullen and Taube, 2017). Neurons with such tuning are called *head direction* (HD) cells, and the preferred tuning of each HD cell is oriented in a different direction within a given region. Accordingly, each population of HD cells together encode across the full 360° range of directional headings such that the animal's directional orientation is continuously represented.

VESTIBULAR SIGNALS AND THE RING-ATTRACTOR MODEL OF HD CELL GENERATION

The prevailing view is that inputs from early vestibular pathways are essential for the computation of the heading signal encoded by HD cells. Evidence of vestibular input making the primary contribution toward generating the HD signal comes from lesion studies (Stackman and Taube, 1997). Specifically, experiments with horizontal canal-deficient animals and lesions, inactivation, or occlusions of the semicircular canals all result in the lack of head-direction-specific activity when recording from the anterior dorsal nucleus of the thalamus (ADN) or post-subiculum (reviewed in Clark and Taube, 2012; Cullen and Taube, 2017). Similar loss of HD tuning is seen in the anterior thalamus after bilateral neurotoxic lesions of the vestibular labyrinth (Stackman and Taube, 1997). It is commonly assumed that the angular head-velocity information encoded by neurons in the vestibular nuclei is transmitted

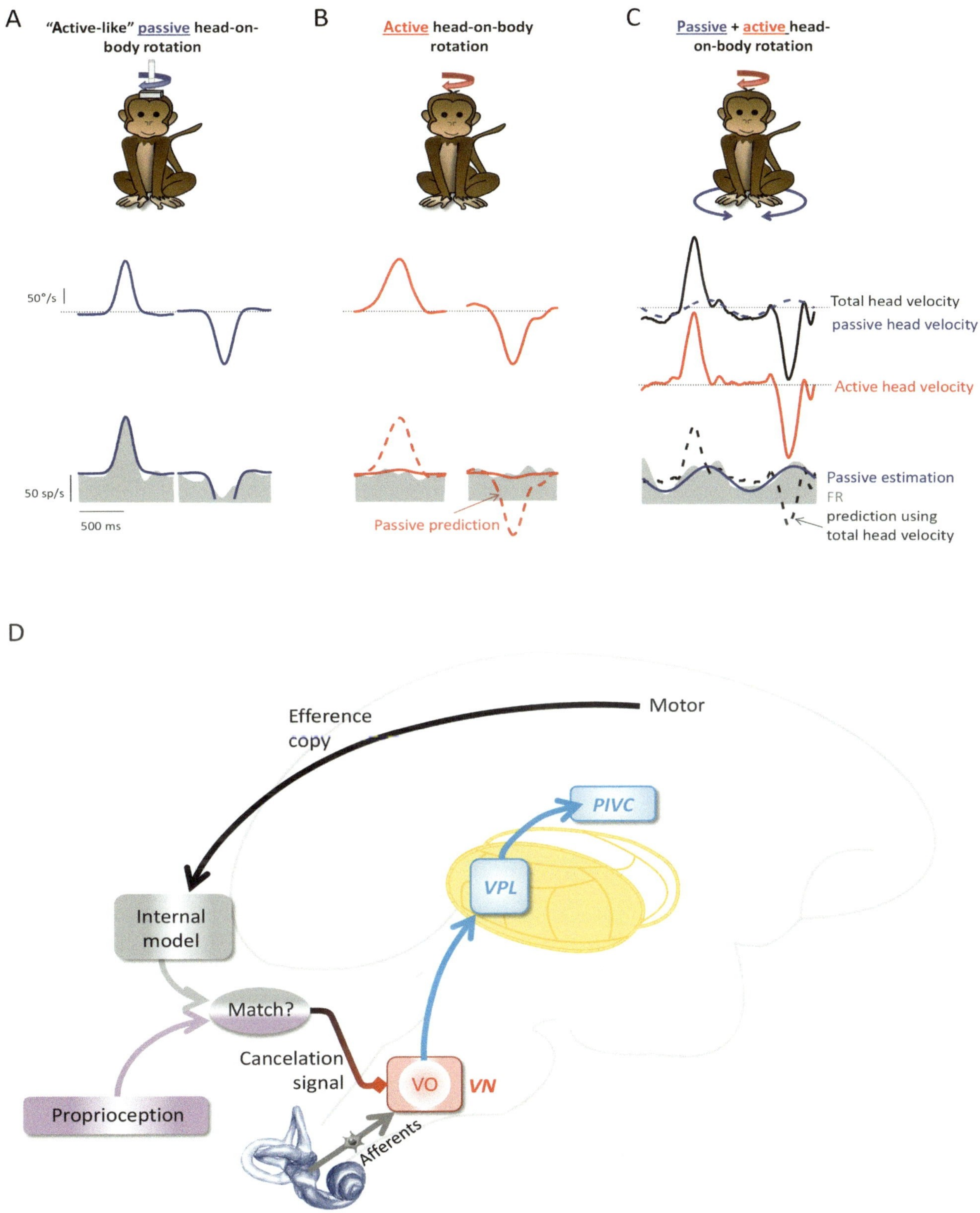

Figure 28.4. Selective encoding of passive self-motion versus self-generated movement by VPL neurons. (A and B). Responses of VPL neurons to passively applied (A, blue) and self-generated (B, red curves) head rotations with comparable trajectories. Firing rate estimations (solid blue and red curves) show robust encoding of the passive stimulation (A, blue) but an attenuation of the response during self-generated movements (B, compare solid and dashed red curves). (C). VPL neurons selectively encode the passive component of concurrent passive and active rotations. The activity of an example neuron (bottom gray shading) in response to self-generated head-on-body rotations (central red curve) during simultaneous passive, whole-body rotations (dashed blue curve) is well fit by the passive component of the motion alone (bottom panel, solid blue curve) and is not well predicted by a response to total head-in-space velocity (black curves). (D). Schematic showing that responses of VO neurons within VN are attenuated during active self-motion because of a cancellation signal that occurs only when there is a match between the efference copy of the motor command and the resulting proprioceptive input. The attenuated responses of VPL neurons during active self-motion are in part inherited from VN. Data adapted from Dale and Cullen (2019).

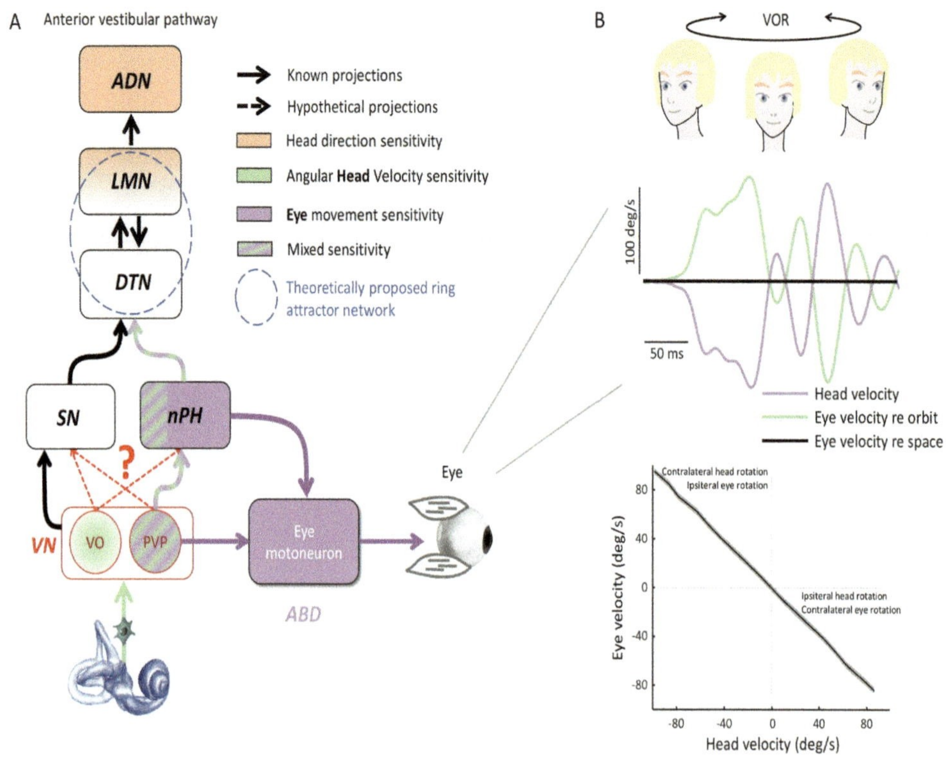

Figure 28.5. How is a head direction signal generated from vestibular input? (A). Schematic showing ascending projections from vestibular pathways toward the ADN. While VO neurons within VN only respond to head velocity (i.e., vestibular input, green), whether these project to higher brain structures such as nPH and SN remains unknown (red dashed lines, question mark). In contrast, PVP neurons within VN project to nPH as well as the abducens nucleus (ABD) whose motoneurons project to eye muscles. However, PVP neurons are sensitive to both head (green) and eye (purple) movements. Overall, 80% of neurons within nPH only respond to eye movements (purple), while the remaining 20% respond to both eye and head movements. (B). Top: Schematic showing that compensatory eye movements maintain gaze direction for different head orientations. Middle: Example head and eye velocity traces obtained during naturalistic stimulation in NHPs during the VOR (top). It is seen that the eye velocity is essentially the opposite of the head velocity. Bottom: Eye velocity as a function of head velocity. There is a very strong negative correlation between both quantities ($R = -0.997$, $p<<10–3$). Data adapted from Mackrous et al. (2020).

to neurons in the nucleus prepositus, dorsal tegmental nucleus, and lateral mammillary nucleus, all of which have been reported to comprise neurons whose activity correlates with angular head velocity (Blair et al., 1998; Stackman and Taube, 1998; Bassett and Taube, 2001). In contrast, ADN neurons are primarily tuned to HD (Taube, 1995) and are, at best, only weakly modulated by angular head velocity (Blair and Sharp, 1995). Thus, current models underlying the computation of the HD typically involve the transformation of vestibular (i.e., head velocity) information into a positionally based heading direction signal.

Ring attractors have been proposed as a model of the HD signal generation. Such attractors are thought to be implemented by the reciprocal connections between the dorsal tegmental nuclei and the lateral mammillary nuclei and transform an input head velocity signal into heading direction (reviewed in Clark and Taube, 2012). Self-motion-based (idiothetic) inputs from angular head-velocity cells are believed to drive the ring-attractor network, while allothetic-based (landmark) inputs such as visual cues are believed to shift the relative orientation of the network so that it is anchored to a new landmark (reviewed in Cullen and Taube, 2017). Indeed, neurons whose activities correlate with HD as well as angular head velocity have been reported in the dorsal tegmental and lateral mammillary nuclei (Blair et al., 1998; Stackman and Taube, 1998; Bassett and Taube, 2001). Specifically, about 75% versus 43% of neurons are tuned to angular head velocity in the dorsal tegmental nuclei versus lateral mammillary nuclei, respectively. Overall, the tuning properties to angular head velocity are very similar between both brain regions.

CONTRIBUTIONS OF VESTIBULAR VERSUS EXTRA-VESTIBULAR INPUTS TOWARD GENERATING THE HD SIGNAL SENT TO THE ADN

Do neurons within the dorsal tegmental and lateral mamillary nuclei receive information about angular head velocity from the vestibular nuclei to compute a HD signal that is then sent to ADN? As reviewed above, neurons in the vestibular nuclei can be divided into 2 basic types based on differences in their sensitivity to passive head motion and eye motion, as well as connections to other brainstem nuclei: VO and eye-movement-sensitive neurons. The latter class makes the key contribution to generating and maintaining the vestibulo-ocular reflex (VOR), an essential reflex which produces compensatory eye movements in the opposite direction to head movements in order to provide stable gaze (Cullen, 2012). These neurons are henceforth referred to as VOR neurons (Figure 28.5A). However, as explained below, neither of these neuron classes are good candidates for providing the head-velocity input to the HD network required in existing models. Below we first consider the signals encoded by VO neurons, and then consider those encoded by the VOR neurons.

VO neurons robustly encode head, but not eye, motion during passive head movements, and contribute to ensuring stable posture and self-motion. There is, however, no evidence to date that these neurons project either directly or indirectly to the dorsal tegmental nucleus via the nPH or to the supragenual nucleus. Furthermore, as mentioned above, the responses of VO neurons are markedly suppressed during active movement, which has significant ramifications for current models of HD signal generation. Specifically, how could an HD signal be computed based on hypothetical input from VO neurons whose responses are strongly attenuated when animals are freely moving within an environment? One potential solution for this apparent conundrum is that the HD network integrates extravestibular self-motion information (e.g., proprioception, motor efference copy) with the residual signals carried by VO neurons during active movements to determine an estimate of direction. Indeed, there is accumulating evidence that HD cell activity is significantly affected by extravestibular self-motion cues (Taube and Burton, 1995; Yoder et al., 2011), which has led to the proposal that these also drive the computation of the HD signal. Recordings from the vestibular nuclei of mice have demonstrated that VO neurons encode a static neck position signal during both active and passive conditions in mice (Medrea and Cullen, 2013). However, VO neurons in rhesus monkeys do not encode a proprioceptively generated and/or neck-motor signal.

Given that VO neurons are not a good candidate for providing head-velocity input to the HD cell network, the question arises: Could an HD signal instead be generated using input from the VOR neurons within the vestibular nuclei? VOR neurons likewise receive direct inputs from the vestibular nerve but also strongly respond to eye movements (see Cullen, 2012, for review). Previous studies have shown that VOR neurons send projections to the nPH, which projects to the HD cell network (Cullen, 2019) (Figure 28.5A). The nPH is also known as the "oculomotor integrator" because it serves an essential function that generates eye-position signals to control all classes of eye movements, including the VOR (see McCrea and Horn, 2006, for review; Figure 28.5A). It is thus not unexpected that nPH neurons predominantly encode eye-movement-related signals as opposed to head velocity during both passive and active movements (Dale and Cullen, 2013).

It thus follows that our existing knowledge of sensorimotor coding by early vestibular pathways and the nPH (Figure 28.5A) is at odds with the prevailing model of HD generation—namely a ring-attractor network that transforms a vestibular-based head-velocity signal into heading direction. Moreover, this greatly complicates the interpretation of prior nPH lesion studies demonstrating impaired HD cell tuning (Butler and Taube, 2015). Given that the nPH serves an essential function in controlling eye movements, lesions to this structure will impair visual stability during movements (Cannon and Robinson, 1987). It is therefore likely that the observed impairment in HD tuning was partly due to altered visual stability, as opposed to a loss of a vestibular information per se, as has been commonly assumed. Taken together, these results show that VOR neurons are also not good candidates for providing head velocity input to the HD cell network.

Another possibility is that information as to angular head velocity originates from higher brain areas. Indeed, several studies have reported the presence of neurons tuned to angular head velocity across multiple cortical areas (visual cortex: Bouvier et al., 2020; dorsal presubiculum: Sharp, 1996; entorhinal cortex: Mallory et al., 2021; retrosplenial cortex: Rancz et al., 2015; Keshavarzi et al., 2022; prefrontal cortex: Rancz et al., 2015). While it is possible that ADN neurons receive information about angular head velocity via such cortical feedback, it should be noted that these prior studies did not explicitly measure eye or head-on-body movements (but see Keshavarzi et al., 2022). As such, the relationship between neural activity and angular head velocity could have been caused by any combination of eye movements, neck proprioception, or motor efference copy.

DO ADN NEURONS ENCODE GAZE INSTEAD OF HEAD DIRECTION?

The above argument challenges the common wisdom that the vestibular-based head-velocity signal that is needed to generate the HD signal via ring-attractor networks comes from the vestibular organs. One possibility is that the so-called HD tuning represents gaze direction as opposed to head direction. Indeed, most studies that have investigated HD tuning across brain areas did not explicitly take into account eye movements (reviewed in Cullen and Taube, 2017). This is important because eye movements can be strongly correlated with head movements (e.g., during the VOR; Figure 28.5B), any relationship between neural activity and head velocity could be due to eye movements. Indeed, it should be noted that eye motoneurons, whose activities determine eye movements, would be classified as being sensitive to angular head velocity by measures used in the field. Thus, the nature of the signals that are received by and conveyed by ADN neurons remains unclear. Further studies that are conducted in more controlled environments (e.g., while passively moving the animal) while monitoring eye movements are needed to better understand how the direction of motion is generated and processed within the brain.

CONCLUSION AND FUTURE DIRECTIONS

Here we have reviewed the current state of knowledge of the function of the anterior and posterior vestibular thalamocortical pathways. Neurons within VPL respond to vestibular (i.e., head angular velocity and/or linear acceleration) input in a nonlinear fashion, thereby enabling optimized and faithful encoding of angular head velocity, as well as selectivity to passively-applied stimulation, as opposed to that resulting from voluntary movements. It is important to note that responses of these neurons to naturalistic translation stimuli have for the most part not been investigated. As such, future work should focus on understanding how neurons within the vestibular thalamus and cortex process translational self-motion. Moreover, studies focusing on how cortical neurons that receive input from VPL (e.g., PIVC, VIP) respond to naturalistic self-motion, as well as to passively-applied versus voluntary motion, are needed. Finally, it is important to note these cortical neurons also respond to signals from other senses such as proprioceptive, somatosensory, and visual. As such, future work is needed to understand how multisensory integration occurs during more natural settings.

The common wisdom is that neurons within ADN receive information about head direction that is generated by the lateral mamillary nucleus–dorsal tegmental nucleus complex by integrating vestibular and visual inputs. ADN neurons then transmit this information to cortical areas such as dorsal presubiculum and retrosplenial. However, there is accumulating evidence that this picture may not be entirely accurate. This is in part because the vast majority of studies used freely moving animals exploring their environment but did not measure eye movements. As such, a number of extra-vestibular cues such as neck proprioception, motor efference copy, and eye movements could be used to obtain information as to head angular velocity without requiring the vestibular apparatus. One possibility is that information as to angular head velocity is obtained from any combination of these cues. Another, likely possibility is that head-direction tuning might reflect gaze-direction tuning, as head direction and gaze direction would be highly correlated during exploration behaviors. Testing these hypotheses should be the focus of future studies.

REFERENCES

Angelaki DE, Dickman JD (2000). Spatiotemporal processing of linear acceleration: Primary afferent and central vestibular neuron responses. *J Neurophysiol.* 84:2113–2132.

Bassett JP, Taube JS (2001). Neural correlates for angular head velocity in the rat dorsal tegmental nucleus. *J Neurosci.* 21:5740–5751.

Bilotto G, Goldberg J, Peterson BW, Wilson VJ (1982). Dynamic properties of vestibular reflexes in the decerebrate cat. Exp Brain Res. 47:343–352.

Blair HT, Sharp PE (1995). Anticipatory head direction signals in anterior thalamus: evidence for a thalamocortical circuit that integrates angular head motion to compute head direction. *J Neurosci.* 15:6260–6270.

Blair HT, Cho J, Sharp PE (1998). Role of the lateral mammillary nucleus in the rat head direction circuit: A combined single unit recording and lesion study. *Neuron.* 21:1387–1397.

Bouvier G, Senzai Y, Scanziani M (2020). Head movements control the activity of primary visual cortex in a luminance-dependent manner. *Neuron.* 108:500–511.

Brooks JX, Cullen KE (2013). The primate cerebellum selectively encodes unexpected self-motion. Curr Biol. 23:947–955.

Brooks JX, Cullen KE (2014). Early vestibular processing does not discriminate active from passive self-motion if there is a discrepancy between predicted and actual proprioceptive feedback. *J Neurophysiol.* 111:2465–2478.

Butler WN, Taube JS (2015). The nucleus prepositus hypoglossi contributes to head direction cell stability in rats. *J Neurosci.* 35:2547–2558.

Cannon SC, Robinson DA (1987). Loss of the neural integrator of the oculomotor system from brain stem lesions in monkey. *J Neurophysiol.* 57:1383–1409.

Carriot J, Cullen KE, Chacron MJ (2021). The neural basis for violations of Weber's law in self-motion perception. *PNAS.* 118:e2025061118.

Carriot J, Jamali M, Chacron MJ, Cullen KE (2014). Statistics of the vestibular input experienced during natural self-motion: Implications for neural processing. *J Neurosci.* 34:8347–8357.

Carriot J, Jamali M, Chacron MJ, Cullen KE (2017). The statistics of the vestibular input experienced during natural self-motion differ between rodents and primates. *J Physiol.* 595:2751–2766.

Carriot J, McAllister G, Hooshangnejad H, Mackrous I, Cullen KE, Chacron MJ (2022). Sensory adaptation mediates efficient and unambiguous encoding of natural stimuli by vestibular thalamocortical pathways. *Nature Comm.* 13:2612.

Clark BJ, Taube JS (2012). Vestibular and attractor network basis of the head direction cell signal in subcortical circuits. *Front Neural Circuits.* 6:7.

Cullen KE (2004). Sensory signals during active versus passive movement. *Curr Opin Neurobiol.* 14:698–706.

Cullen KE (2012). The vestibular system: Multimodal integration and encoding of self-motion for motor control. *Trends Neurosci.* 35:185–196.

Cullen KE (2014). The neural encoding of self-generated and externally applied movement: implications for the perception of self-motion and spatial memory. *Front Integr Neurosci.* 7:108.

Cullen KE (2019). Vestibular processing during natural self-motion: implications for perception and action. *Nat Rev Neurosci.* 20:346–363.

Cullen KE, Chacron MJ (2023). Neural substrates of perception in the vestibular thalamus during natural self-motion: A review. *Curr Res Neurobiol.* 4:100073.

Cullen KE, Minor LB (2002). Semicircular canal afferents similarly encode active and passive head-on-body rotations: Implications for the role of vestibular efference. *J Neurosci.* 22:RC226.

Cullen KE, Taube JS (2017). Our sense of direction: progress, controversies and challenges. *Nat Neurosci.* 20:1465–1473.

Cullen KE, Zobeiri OA (2021). Proprioception and the predictive sensing of active self-motion. *Curr Opin Physiol.* 20:29–38.

Dale A, Cullen KE (2013). The nucleus prepositus predominantly outputs eye movement-related information during passive and active self-motion. *J Neurophysiol.* 109:1900–1911.

Dale A, Cullen KE (2019). The ventral posterior lateral thalamus preferentially encodes externally applied versus active movement: implications for self-motion perception. *Cereb Cortex.* 29:305–318.

Donner K, Copenhagen DR, Reuter T (1990). Weber and noise adaptation in the retina of the toad *Bufo marinus*. *J Gen Physiol.* 95:733–753.

Evans GB, Howarth E (1966). The effect of grip-tension on tactile-kinaesthetic judgement of width. *Q J Exp Psychol.* 18:275–277.

Fernandez C, Goldberg JM (1971). Physiology of peripheral neurons innervating semicircular canals of the squirrel monkey. II. Response to sinusoidal stimulation and dynamics of peripheral vestibular system. *J Neurophysiol.* 34:661–675.

Goldberg JM, Highstein SM, Moschovakis AK, Fernandez C (1987). Inputs from regularly and irregularly discharging vestibular nerve afferents to secondary neurons in the vestibular nuclei of the squirrel monkey. I. An electrophysiological analysis. *J Neurophysiol.* 58:700–718.

Grabherr L, Nicoucar K, Mast FW, Merfeld DM (2008). Vestibular thresholds for yaw rotation about an earth-vertical axis as a function of frequency. *Exp Brain Res.* 186:677–681.

Jamali M, Carriot J, Chacron MJ, Cullen KE (2013). Strong correlations between sensitivity and variability give rise to constant discrimination thresholds across the otolith afferent population. *J Neurosci.* 33:11302–11313.

Jamali M, Sadeghi SG, Cullen KE (2009). Response of vestibular nerve afferents innervating utricle and saccule during passive and active translations. *J Neurophysiol.* 101:141–149.

Keshavarzi S, Bracey EF, Faville RA, Campagner D, Tyson AL, Lenzi SC, Branco T, Margrie TW (2022). Multisensory coding of angular head velocity in the retrosplenial cortex. *Neuron.* 110:532–543 e539.

Lopez C, Blanke O (2011). The thalamocortical vestibular system in animals and humans. *Brain Res Rev.* 67:119–146.

Mackrous I, Carriot J, Cullen KE (2022). Context-independent encoding of passive and active self-motion in vestibular afferent fibers during locomotion in primates. *Nature Comm.* 13:120.

Mackrous I, Carriot J, Cullen KE, Chacron MJ (2020). Neural variability determines coding strategies for natural self-motion in macaque monkeys. *eLife.* 9:e57484.

Mallery RM, Olomu OU, Uchanski RM, Militchin VA, Hullar TE (2010). Human discrimination of rotational velocities. *Exp Brain Res.* 204:11–20.

Mallory CS, Hardcastle K, Campbell MG, Attinger A, Low IIC, Raymond JL, Giocomo LM (2021). Mouse entorhinal cortex encodes a diverse repertoire of self-motion signals. *Nature Comm.* 12:671.

Marlinski V, McCrea RA (2008). Activity of ventroposterior thalamus neurons during rotation and translation in the horizontal plane in the alert squirrel monkey. *J Neurophysiol.* 99:2533–2545.

Marlinski V, McCrea RA (2009). Self-motion signals in vestibular nuclei neurons projecting to the thalamus in the alert squirrel monkey. *J Neurophysiol.* 101:1730–1741.

McCrea RA, Horn AK (2006). Nucleus prepositus. *Prog Brain Res.* 151:205–230.

Medrea I, Cullen KE (2013). Multisensory integration in early vestibular processing in mice: the encoding of passive vs. active motion. *J Neurophysiol.* 110:2704–2717.

Meng H, Angelaki DE (2010) Responses of ventral posterior thalamus neurons to three-dimensional vestibular and optic flow stimulation. *J Neurophysiol.* 103:817–826.

Meng H, May PJ, Dickman JD, Angelaki DE (2007). Vestibular signals in primate thalamus: properties and origins. *J Neurosci.* 27:13590–13602.

Merfeld DM, Zupan LH (2002). Neural processing of gravitoinertial cues in humans. III. Modeling tilt and translation responses. *J Neurophysiol.* 87:819–833.

Mitchell DE, Kwan A, Carriot J, Chacron MJ, Cullen KE (2018). Neuronal variability and tuning are balanced to optimize naturalistic self-motion coding in primate vestibular pathways. *eLife.* 7:e43019.

Nesti A, Beykirch KA, Pretto P, Bulthoff HH (2015). Human discrimination of head-centred visual-inertial yaw rotations. *Exp Brain Res.* 233:3553–3564.

Nouri S, Karmali F (2018). Variability in the vestibulo-ocular reflex and vestibular perception. *Neuroscience.* 393:350–365.

Panzeri S, Moroni M, Safaai H, Harvey CD (2022). The structures and functions of correlations in neural population codes. *Nat Rev Neurosci.* 23:551–567.

Petrof I, Sherman SM (2009). Synaptic properties of the mammillary and cortical afferents to the anterodorsal thalamic nucleus in the mouse. *J Neurosci.* 29:7815–7819.

Poggio GF, Mountcastle VB (1963) The functional properties of ventrobasal thalamic neurons studied in unanesthetized monkeys. *J Neurophysiol.* 26:775–806.

Rancz EA, Moya J, Drawitsch F, Brichta AM, Canals S, Margrie TW (2015). Widespread vestibular activation of the rodent cortex. *J Neurosci.* 35:5926–5934.

Sadeghi SG, Minor LB, Cullen KE (2007a). Response of vestibular-nerve afferents to active and passive rotations under normal conditions and after unilateral labyrinthectomy. *J Neurophysiol.* 97:1503–1514.

Sadeghi SG, Chacron MJ, Taylor MC, Cullen KE (2007b). Neural variability, detection thresholds, and information transmission in the vestibular system. *J Neurosci.* 27:771–781.

Sato F, Sasaki H (1993). Morphological correlations between spontaneously discharging primary vestibular afferents and vestibular nucleus neurons in the cat. *J Comp Neurol.* 333:554–566.

Schneider AD, Jamali M, Carriot J, Chacron MJ, Cullen KE (2015). The increased sensitivity of irregular peripheral canal and otolith vestibular afferents optimizes their encoding of natural stimuli. *J Neurosci.* 35:5522–5536.

Sharp PE (1996). Multiple spatial/behavioral correlates for cells in the rat postsubiculum: Multiple regression analysis and comparison to other hippocampal areas. *Cereb Cortex.* 6:238–259.

Sharpee TO, Calhoun AJ, Chalasani SH (2014). Information theory of adaptation in neurons, behavior, and mood. *Curr Opin Neurobiol.* 25:47–53.

Sherman SM, Guillery RW (2002) The role of the thalamus in the flow of information to the cortex. *Phil Trans Royal Soc Lond B Biol Sci.* 357:1695–1708.

Simoncelli EP, Olshausen BA (2001). Natural image statistics and neural representation. *Annu Rev Neurosci.* 24:1193–1216.

Stackman RW, Taube JS (1997). Firing properties of head direction cells in the rat anterior thalamic nucleus: Dependence on vestibular input. *J Neurosci.* 17:4349–4358.

Stackman RW, Taube JS (1998). Firing properties of rat lateral mammillary single units: head direction, head pitch, and angular head velocity. *J Neurosci.* 18:9020–9037.

Taube JS (1995). Head direction cells recorded in the anterior thalamic nuclei of freely moving rats. *J Neurosci.* 15:70–86.

Taube JS, Burton HL (1995). Head direction cell activity monitored in a novel environment and during a cue conflict situation. *J Neurophysiol.* 74:1953–1971.
Valko Y, Lewis RF, Priesol AJ, Merfeld DM (2012). Vestibular labyrinth contributions to human whole-body motion discrimination. *J Neurosci.* 32:13537–13542.
von Holst E, Mittelstaedt H (1950). Das Reafferenzprinzip. *Naturwissenschaften.* 37:464–476.
Wark B, Lundstrom BN, Fairhall A (2007). Sensory adaptation. *Curr Opin Neurobiol.* 17:423–429.
Weber EH (1834). *De Pulsu, Resorptione, Auditu Et Tactu. Annotationes Anatomicae Et Physiologicae, Auctore.* C. F. Koehler.
Yoder RM, Clark BJ, Brown JE, Lamia MV, Valerio S, Shinder ME, Taube JS (2011). Both visual and idiothetic cues contribute to head direction cell stability during navigation along complex routes. *J Neurophysiol.* 105:2989–3001.
Zobeiri OA, Cullen KE (2022). Distinct representations of body and head motion are dynamically encoded by Purkinje cell populations in the macaque cerebellum. *eLife.* 11:e75018.

29.

THALAMOCORTICAL CONTRIBUTIONS TO CHEMICAL SENSES

Emmanuelle Courtiol, Arianna Maffei, and Alfredo Fontanini

GENERAL INTRODUCTION

Thalamocortical circuits have been historically studied in sensory systems such as audition, somatosensation, and vision. In the context of these systems, thalamic nuclei have been categorized as primary or higher order. This classification is mainly based on the glutamatergic inputs these thalamic nuclei receive, their source, and their properties (i.e. driver/Class I, or modulatory/Class II; Sherman and Guillery, 2006; Sherman, 2012; see **Chapter 1 in Section I**). First-order thalamic nuclei receive and process information from subcortical areas (mainly sensory information) and transmit it to sensory cortices, while higher-order thalamic nuclei receive and process inputs from cortical areas (driver input from layer 5) and transmit them to other cortical areas (Sherman and Guillery, 2006, 2011; Sherman, 2012). The higher-order thalamic nuclei are thought to be critical for the coordination among cortical areas (Saalmann et al., 2012). In the visual system, for example, the lateral geniculate nucleus is the first-order thalamic nucleus and the pulvinar is the higher-order one (Sherman, 2012).

The study of thalamocortical contributions to chemosensation in olfaction and taste has lagged compared to other sensory systems. The reason can be found in the peculiarities of the olfactory and gustatory systems. In the case of the olfactory system, it lacks a primary thalamic nucleus, and instead relies on the olfactory bulb for the processing of peripheral sensory signals. Olfactory signals do recruit a thalamic nucleus, the mediodorsal thalamus (MD), but the importance of this higher-order nucleus in olfaction was acknowledged only recently. As for the gustatory system, it does have a primary thalamic nucleus—the parvicellular portion of the ventral posteromedial nucleus of the thalamus (VPMpc)—however, its function and connectivity are very much understudied compared to inputs from subcortical nuclei such as the basolateral nucleus of the amygdala (BLA).

In this chapter, we will review published work on thalamocortical systems in olfaction and taste, focusing on similarities and differences with other sensory systems and between the 2 chemical senses. We will first present the general connectivity of the system—from sensory epithelia to cortex—and then focus on the main thalamic nucleus subserving each system (MD and VPMpc). Finally, we will introduce additional connections within each system, discuss gaps in our knowledge, and present open questions.

THALAMIC CONTRIBUTION TO OLFACTION

Odorant molecules reach the olfactory epithelium, either ortho- or retro-nasally through respiration, and are detected by olfactory sensory neurons. The olfactory sense has a very distinctive anatomy, as information from olfactory sensory neurons is not processed through a thalamic nucleus before reaching the sensory cortex, like in other sensory modalities. The olfactory sensory neurons project directly to the olfactory bulb. Principal cells of the olfactory bulb then send their information to primary olfactory areas including the piriform cortex (PC), anterior olfactory nucleus, olfactory tubercle, lateral entorhinal cortex, and amygdala. The PC, a 3-layered paleocortex, is the largest area receiving input from the olfactory bulb and has reciprocal connections with the orbitofrontal cortex (OFC; Datiche and Cattarelli, 1996; see Figure 29.1). As a result, the thalamus has been largely ignored in the olfaction literature. However, as mentioned above, a key point when discussing thalamic functions is to distinguish 2 types of thalamic nuclei: first-order and higher-order. If we apply this notion to the olfactory system, there is no first-order/primary sensory thalamic nucleus. Nonetheless, Kay and Sherman (2007) highlighted functional similarities between the olfactory bulb and the lateral geniculate nucleus of the thalamus (Figure 29.1). While there is no conventional first-order/primary olfactory thalamic nucleus, a higher-order olfactory thalamic nucleus has been identified: the MD. The MD is a node where information

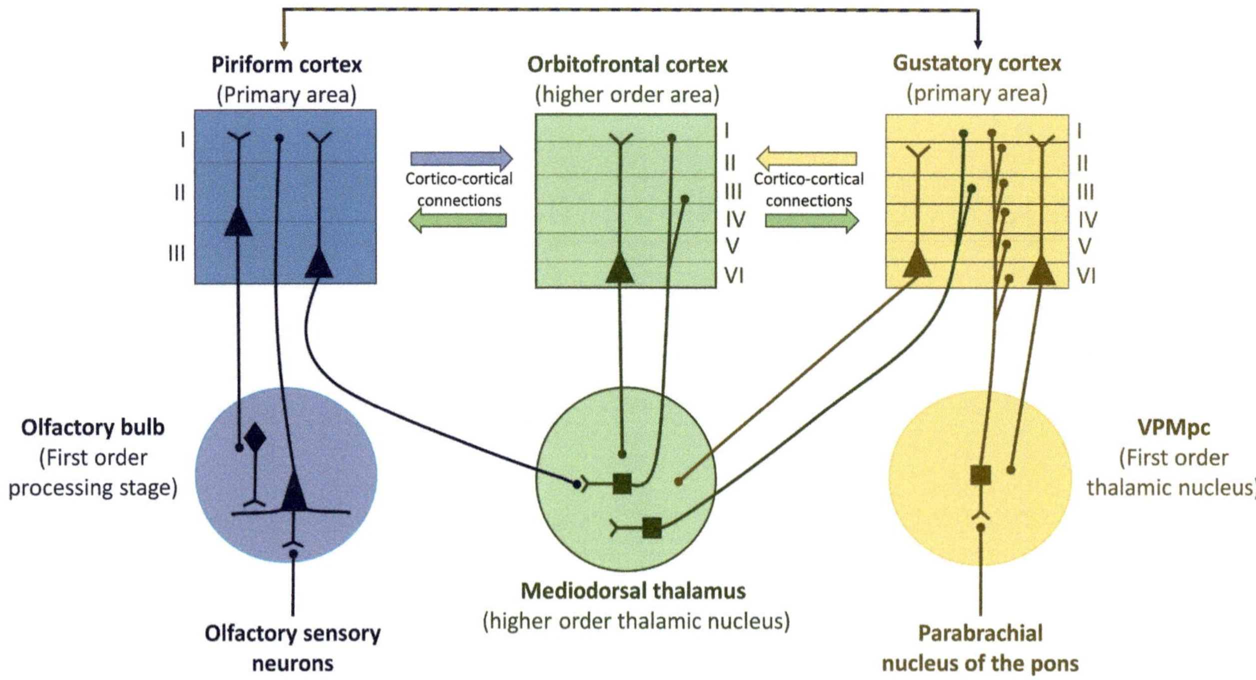

Figure 29.1. The olfactory and gustatory trans- thalamic pathways. The olfactory bulb receives inputs from olfactory sensory neurons. It is thus the first-order processing stage where inhibitory processes occur. Information from the olfactory bulb is sent to the superficial layer of the primary olfactory cortex (piriform cortex, PC). The PC projects back to the olfactory bulb. Cells from the deep layer of the PC project to the mediodorsal nucleus of the thalamus (MD, higher- order thalamic nucleus). The PC also has reciprocal connections with the orbitofrontal cortex (OFC, higher order chemosensory area) and the gustatory cortex (GC). The MD has important reciprocal connections with the OFC. On the gustatory side, the parabrachial nucleus of the pons (which receives inputs from the nucleus of solitary tract) projects to the ventral posteromedial nucleus of the thalamus (VPMpc, first order gustatory thalamic nucleus). VPMpc neurons project to all layers of GC, which in turn sends feedback projection to the VPMpc. Cells from deep layers of GC project to the MD. MD neurons project to the superficial layers of GC forming a cortico- thalamocortical loop. This ensemble of cortico- cortical and cortico- thalamocortical loops highlights the convergence of chemosensory information on the MD- OFC pathway.

converges from multiple brain areas such as the brainstem, lateral hypothalamus, ventral tegmental area, and the thalamic reticular nucleus (TRN) (Cornwall and Phillipson, 1988; Kuroda and Price, 1991). But the most highlighted feature of the MD is its strong reciprocal connection with the prefrontal cortex (Groenewegen, 1988). According to the pattern of afferent and efferent connections, 3 regions of the MD have been defined: the medial, central, and lateral parts (Groenewegen, 1988).

The MD is involved in a wide range of cognitive functions, ranging from working memory processes, adaptive decision-making, goal-directed action, to attentional control, notably because of its dense connectivity with the prefrontal cortex (see **Chapters 45 and 48 in Section VIII and Chapter 54 in Section IX**; Mitchell and Chakraborty, 2013). The MD has also been proposed as a key structure for detecting salient signals and following their behavioral significance (Wolff et al., 2021).

MD AND THE OLFACTORY PATHWAY

Studies of MD in the context of olfaction showed that (1) neural responses can be recorded in the MD of different species following electrical stimulation of the olfactory bulb and lateral olfactory tract or optical stimulation of olfactory sensory neurons (Yarita et al., 1980; Price and Slotnick, 1983; Imamura et al., 1984); (2) single-unit, oscillatory (local field potentials) and functional magnetic resonance imaging (fMRI) blood oxygen level dependent (BOLD) responses to orthonasally or orally sourced odors can be observed in anesthetized and behaving animals (Imamura et al., 1984; Kawagoe et al., 2007; Courtiol and Wilson, 2014; Fredericksen and Samuelsen, 2022; Sanganahalli et al., 2022). Unitary responses can be either excitatory or suppressive (Imamura et al., 1984; Courtiol and Wilson, 2016; Fredericksen and Samuelsen, 2022). In addition, odor-selective neurons do not appear to respond to visual or auditory stimuli (Yarita et al., 1980; Imamura et al., 1984; Kawagoe et al., 2007; Sanganahalli et al., 2022), although some MD neurons are actually chemoselective but multimodal as they do respond to intra-oral odors, tastes, and odor-taste mixtures (Fredericksen and Samuelsen, 2022). These responses to odors originate from the connectivity between the olfactory areas and the MD. Indeed, the PC, olfactory tubercle, amygdala, entorhinal cortex, and anterior olfactory nucleus are all direct recipients of olfactory bulb inputs and project to either the medial or central part of the MD (Cornwall and Phillipson, 1988; Kuroda et al., 1992; Courtiol and Wilson, 2015). The MD also receives projections from the OFC and endopiriform nucleus (Price and Slotnick, 1983). At the synaptic level, projections from the main olfactory cortex, the PC, to the MD are glutamatergic and show 2 types of presynaptic terminals: either small terminals contacting distal dendrites, or large ones mainly opposed to proximal dendrites (Kuroda et al., 1992). In the description by Sherman and Guillery (2006), higher-order thalamic nuclei notably receive Class I driver input from layer 5 of the neocortex. Inputs from PC to MD originate from the deepest layer of the PC (i.e. layer 3, the PC being a paleocortex) (Price and Slotnick, 1983; Figure 29.1). Pelzer et al. (2017) characterized them and demonstrated that some of the large synaptic boutons do display Class I properties. MD also projects back to some structures of the olfactory pathway: the OFC, amygdala, and endopiriform nucleus, but not to the PC (Krettek and Price, 1977).

MD AND OLFACTORY PROCESSES

The few existing studies involving lesions or recordings from MD during olfactory tasks provided some insights into its contributions to olfactory processing (Courtiol and Wilson, 2015). Lesions to the MD do not impact odor detection or simple odor discrimination (Eichenbaum et al., 1980; Sapolsky and Eichenbaum, 1980). However, difficult or novel odor discriminations, odor preference, odor span capacity, and odor reversal learning can be impaired following large damage or inactivation of the MD (Eichenbaum et al., 1980; Sapolsky and Eichenbaum, 1980; Slotnick and Kaneko, 1981; Scott et al., 2020). Odor sampling duration during reversal learning is also impacted when PC to MD projections are inactivated (Courtiol et al., 2019). These different effects seem to be quite specific as they are not observed with visual discrimination or visual reversal learning (Slotnick and Kaneko, 1981). Several studies also suggested a role for MD in olfactory attention in humans (Plailly et al., 2008).

It appears that there is no clearly defined function for MD in olfaction, but rather that MD participates in several olfactory processes when discrimination tasks are complex or when more integrative processes are involved. This is corroborated by electrophysiological recordings in animals performing odor tasks and showing that MD neurons not only respond to odors but also encode stimulus/reward associations (Kawagoe et al., 2007) and spatio-motor variables of the odor task, or display anticipatory activity prior to odorant sampling (Courtiol and Wilson, 2016). Importantly, the activity of MD is also shaped by experience (Fredericksen et al., 2019), as the number of MD neurons responding to odors, assessed through c-fos labeling, is higher when an odor is new. Interestingly, MD has reciprocal connection with the gustatory cortex (Allen et al., 1991b), and the number of MD neurons responding to a novel taste or odor-taste mixture is lower than to a familiar taste or odor-taste mixture, a result opposite to that for odors alone (Fredericksen et al., 2019). This highlights not only the experience-dependent plasticity of the MD, but also a modality-specific activity.

Finally, as higher-order thalamic nuclei appear to mediate coordination between cortical areas, it is interesting to discuss the contribution of MD within the olfactory trans-thalamic pathway (PC-MD-OFC). One of the first reports on this triad was an fMRI study in humans (Plailly et al., 2008). The authors compared the involvement of the direct PC-OFC route versus the indirect one involving the MD during attention to odors. They demonstrated that attention to odors (compared to attention to tone) activated the corticothalamocortical pathway significantly more than the direct PC to OFC route. This is also in line with what was found in rats during simultaneous recordings of MD single-unit activity and local field potentials in PC and OFC during an odor-discrimination task. The PC-MD coupling was stronger during odor sampling, while the MD-OFC coupling was higher during the post-sampling/decision period (Courtiol and Wilson, 2016), demonstrating that the MD contributes to different odor tasks and is differentially engaged together with specific cortical areas depending on the needs of the task.

OPEN QUESTIONS AND SPECULATIONS ON THE MODULATION OF OLFACTORY CORTICO-CORTICAL CIRCUITS BY THE MD

Fundamental questions that remain to be answered include how the MD coordinates with the OFC based on the PC inputs and what the added value is of this parallel trans-thalamic pathway. Several hypotheses can be proposed, although they are neither exhaustive nor exclusive: (1) the MD can prime the OFC to respond to the PC's input, as anticipatory activity was observed in the MD (Courtiol and Wilson, 2016); (2) the MD can amplify the OFC response to relevant stimuli, especially in the context of attention to odors, as was implied by previous results (Plailly et al., 2008; see also **Chapter 46 in Section VIII**); (3) the MD may be useful to compare and update the olfactory representation in the OFC with the actual sensory input coming from PC and other direct recipients of the olfactory bulb. This hypothesis is notably based on recent work on the MD and prefrontal cortex (Griffiths et al., 2022), as well as MRI studies showing that MD response to unexpected odor is higher than to expected odor (Zelano et al., 2011). Future studies will have to address these points to gain an understanding of the contribution of the MD to olfactory processes. For example, experiments using multi-area recordings during different olfactory tasks, which also allow for the investigation of the directionality of the effects or modulation of MD activity while recording PC and OFC, will help addressing the outstanding questions.

THALAMIC CONTRIBUTION TO TASTE

Tastants dissolved into saliva are detected by taste-receptor cells, which, unlike olfactory receptor neurons, are specialized epithelial cells (Stone et al., 1995). Upon binding with specific chemicals, taste-receptor cells release neurotransmitters that activate afferent fibers from 3 cranial nerves (VII, IX, and X) (Roper and Chaudhari, 2017). Gustatory information is then carried into the brain, reaching the brainstem nucleus of solitary tract and continuing to the parabrachial nucleus (PbN) (Norgren and Leonard, 1971; Weiss et al., 2014)—the involvement of the latter in taste has not been demonstrated for primates. From the PbN, gustatory signals bifurcate into 2 distinct pathways: a sensory and a limbic pathway. The sensory pathway goes from the PbN to the parvicellular portion of the ventral posteromedial nucleus of the thalamus (VPMpc) (Cechetto and Saper, 1987; Allen et al., 1991a; Holtz et al., 2015), whereas the limbic pathway involves the basolateral nucleus of the amygdala (BLA), the hypothalamus, the MD, and other subcortical nuclei (Cechetto and Saper, 1987; Allen et al., 1991a). The 2 pathways eventually converge onto the gustatory portion of the insular cortex (GC; Cechetto and Saper, 1987; Stone et al., 2020). At face value, this connectivity reflects the dual function of the gustatory system which processes both the sensory and affective features of taste.

Like the auditory, somatosensory, and visual systems, the gustatory system engages a primary sensory thalamic nucleus, the VPMpc, whose inputs drive cortical neurons. However, the gustatory system presents some distinct features that differentiate it from classical thalamocortical systems. Indeed, the VPMpc is significantly smaller than its auditory, somatosensory, and visual counterparts. Another distinctive feature of the gustatory system is the connectivity of GC with BLA, which can effectively drive GC neurons. VPMpc and BLA inputs converge on a population of GC neurons that could act as an integrator of sensory and limbic information (Cechetto and Saper, 1987; Stone et al., 2020). The gustatory system also shares some similarities with the olfactory system with the involvement of MD and the connectivity of GC with the OB and PC.

VPMpc AND THE SENSORY GUSTATORY PATHWAY

VPMpc neurons receive sensory information via large synaptic inputs from neurons in the gustatory responsive waist region of the PbN. PbN projections are ipsilateral and account for only 2% of all inputs onto VPMpc neurons. While they only contribute a small proportion of synapses onto VPMpc neurons, the size of PbN synaptic terminals is over 5 times larger compared to other inputs. These large terminals extend over an area that almost completely envelops VPMpc relay neurons' proximal dendrites (Holtz et al., 2015), suggesting that PbN inputs have a powerful influence on VPMpc neurons' activity.

The axons of VPMpc relay neurons extend to GC (Cechetto and Saper, 1987; Allen et al., 1991a; Stone et al., 2020) and, differently from other sensory thalamocortical projections, make synapses with neurons in all layers (Haley et al., 2022; Figure 29.1). Laminar specificity of thalamocortical taste projections relies on distinct amplitudes of VPMpc-GC evoked responses. A unique feature of taste thalamocortical inputs is their frequency-dependent short-term dynamics in response to trains of stimuli. VPMpc-GC responses show short-term facilitation at low frequencies and short-term depression at high frequencies of stimulation (Haley et al., 2022). Thus, when interpreted in the context of the definition of driver and modulator proposed for thalamocortical inputs (i.e., drivers are depressing and modulators are facilitating), VPMpc inputs behave as either modulators or drivers depending on the frequency of the incoming pattern of activity. VPMpc afferents in GC also drive feedforward inhibition and, similar to other driver sensory thalamocortical systems (Kloc and Maffei, 2014), they make synapses onto parvalbumin expressing (PV+) inhibitory neurons (Haley et al., 2022). Differently from other sensory areas, the evoked response onto PV+ neurons is comparable in amplitude to that evoked onto pyramidal neurons and shows frequency-dependent short-term dynamics (Haley et al., 2022).

Thalamocortically evoked feedforward inhibitory postsynaptic currents onto GC pyramidal neurons are large, so that the ratio of feedforward excitation to inhibition is low (Haley et al., 2022), indicating the need to overcome powerful inhibitory gating to activate GC. This is consistent with recordings *in vivo* which show relatively low spontaneous firing activity in this region (Bouaichi and Vincis, 2020).

Activation of GC by taste stimuli engages local cortical circuits as well as output circuits including corticothalamic feedback. While the specific neuron groups providing corticothalamic feedback are not as well investigated as in other sensory areas, tracing studies reported the presence of dense corticothalamic inputs from GC to thalamocortical relay neurons in the VPMpc (Holtz et al., 2015), indicating substantial reciprocal connectivity between the 2 regions.

Corticothalamic axons are fine, show varicosities typical of *en passant* synapses and form small bouton terminals onto VPMpc neurons. These small synaptic terminals are asymmetric, typical of glutamatergic inputs. They are located on relay neuron dendrites in proximity with thalamocortically projecting somata, suggesting that thalamocortical neurons are the recipients of corticothalamic feedback, in addition to PbN input (Holtz et al., 2015).

GC sends projections to the taste portion of the TRN (Hanamori, 2003), which, in turn, provides GABAergic input to VPMpc neurons (Stehberg et al., 2001). The specific region of the TRN projecting to VPMpc is in the most ventromedial portion of the TRN, rostral to the anterior portion of the VPMpc (Stehberg et al., 2001). Thus, thalamocortical relay neurons in VPMpc receive feedforward taste information from the PbN, excitatory and feedback input from GC, and inhibitory GABAergic input from neurons in the TRN.

VPMpc AND GUSTATORY PROCESSES

Only a handful of electrophysiological studies have been conducted on the VPMpc. Recordings from primates and rodents show that VPMpc neurons can encode the physicochemical properties of gustatory stimuli, with patterns of activity representing taste qualities (Verhagen et al., 2003; Liu and Fontanini, 2015). In general, VPMpc neurons can be either narrowly or broadly tuned, with most neurons firing in response to stimuli representative of multiple taste qualities in alert rodents (Liu and Fontanini, 2015). Neurons respond to increasing stimulus concentrations with monotonic elevations in firing rates (Verhagen et al., 2003). Thalamic responses are also multimodal, showing activation following somatosensory stimulation. Indeed, oral touch and stimulus temperature are effective in recruiting VPMpc neurons (Verhagen et al., 2003). Finally, it is important to note that VPMpc neurons do not just represent taste quality. In both anesthetized and alert mice, VPMpc neurons encode taste palatability. Stimuli sharing a similar hedonic value (i.e., palatable stimuli such as sucrose and NaCl vs. aversive stimuli such as quinine and citric acid) evoke similar patterns of firing rates (Verhagen et al., 2003; Liu and Fontanini, 2015). Furthermore, the expectation of a taste is very effective in modulating VPMpc activity (Liu and Fontanini, 2015). Altogether these results confirm the primary role of the VPMpc in taste, but also challenge the idea of a clear-cut separation between sensory and affective coding along the sensory and limbic pathways.

Indications on the function of the VPMpc also come from lesion and inactivation studies (Flynn et al., 1991; Samuelsen et al., 2013). Pharmacological silencing of the VPMpc has a dramatic effect on ongoing activity in GC, with a reduction in fast oscillatory activity and an increase in slow oscillations (Samuelsen et al., 2013). Inactivation of VPMpc also affects taste-evoked dynamics and reduces GC neurons' ability to encode taste qualities (Samuelsen et al., 2013). Lesions of the VPMpc also demonstrate an involvement of this nucleus in processes involving expectations, such as anticipatory contrast and autoshaping (Samuelsen et al., 2013). Interestingly, VPMpc lesions do not appear to impact taste discriminations (Flynn et al., 1991) and conditioned taste aversion (Reilly and Pritchard, 1996; Reilly et al., 2003), suggesting that the circuits for chemosensation and learning may be distributed and involve other regions beyond the VPMpc-GC system.

OTHER THALAMIC AND EXTRATHALAMIC CONTRIBUTIONS TO TASTE

In addition to input from the VPMpc, GC receives thalamocortical afferents from the MD and higher-order midline thalamic nuclei such as the nucleus reuniens (RE) (Baylis et al., 1995; Vertes et al., 2006). The organization and properties of RE connections with GC have not been investigated. There is a little more information about MD anatomical interaction with GC. Inputs to MD primarily originate from deep cortical layers (layers 5 and 6) of GC, while MD neurons project back to cortex with terminals in the superficial layers (layers 1 and 3) forming a higher-order cortico-thalamocortical loop (Saper, 1982; Kuroda and Price, 1991; Figure 29.1). The synaptic organization and properties of MD inputs to GC have not been investigated. Evidence from a study of MD synapses onto the anterior cingulate cortex (ACC) suggests that efferents from this nucleus make monosynaptic connections with cortical pyramidal neurons as well as PV+ neurons (Delevich et al., 2015). If the MD projection to GC followed a comparable pattern, it would overlap with that from the primary thalamic nucleus both in spatial distribution of the axonal fields and postsynaptic targets.

From a functional point of view, little is known about the RE and taste. Its connectivity, featuring inputs from many cortical and subcortical regions (including amygdala, hypothalamus, several neuromodulatory nuclei, and PbN) (McKenna and Vertes, 2004), is likely to embody a parallel pathway for taste, sensorimotor, limbic, and visceral information to GC. As for the MD, it is involved in encoding cross-modal signals related to flavor (Fredericksen et al., 2019; Fredericksen and Samuelsen, 2022), expectation, and anticipatory responses to odors predicting immediate arrival of an associated drink (Small et al., 2008; Scott et al., 2020). Thus, the MD likely plays a role in consummatory behaviors (Stein et al., 1990), possibly by mediating cortico-cortical coordination (Mitchell and Chakraborty, 2013). In addition, this nucleus of the thalamus receives a dense projection from the amygdala (Ray et al., 1992), which is likely to contribute to the cortico-amygdalocortical circuit involved in the limbic aspects of chemosensation.

In addition to thalamic nuclei, GC also receives dense projections from the BLA (Saper, 1982; Cechetto and Saper, 1987; Allen et al., 1991a; Haley et al., 2016; Stone et al., 2020)—a nucleus that also has connections with the PC (Majak et al., 2004). The BLA input to GC makes direct, monosynaptic synapses onto a subset of pyramidal neurons, as well as a subset of 2 major groups of GABAergic neurons, those expressing parvalbumin and somatostatin (Haley et al., 2016). BLA inputs show laminar specificity, with densest projections to GC superficial layers (Haley et al., 2016). Furthermore, BLA inputs are distributed across all 3 divisions of GC: granular (gGC), dysgranular (dGC), and agranular (aGC), with a bias for aGC. This pattern is complementary to VPMpc projections, which are distributed in all of GC with a bias for gGC (Stone et al., 2020). A subset of GC neurons receives monosynaptic input from both VPMpc and BLA (Stone et al., 2020), thus integrating thalamocortical and limbic information.

Inactivation studies revealed that silencing BLA, either pharmacologically or optogenetically, affects neural responses in GC (Piette et al., 2012; Samuelsen et al., 2012). Silencing does not affect GC taste quality coding, but rather interferes with the processing of taste hedonics. The contribution of BLA to taste hedonic coding is further confirmed by its involvement in conditioned taste aversion (Levitan et al., 2020).

OPEN QUESTIONS AND SPECULATIONS ON THE ROLE OF VPMPC AND THE INTEGRATION OF THALAMIC AND AMYGDALAR INPUTS IN GC

The study of the gustatory thalamocortical system is in its infancy compared to other sensory systems. Many subject areas that are well understood in audition, somatosensation, and vision remain unexplored in taste. For instance, corticothalamic projections from GC to VPMpc need to be better elucidated, as does the engagement of the TRN, MD, and RE connections to GC. Furthermore, cortico-thalamocortical and cortico-amygdalocortical loops need to be better understood.

From a functional standpoint, it may be tempting to hypothesize a clear separation between sensory and limbic pathways, one relying on VPMpc and one on MD and BLA. However, experimental evidence supports the possibility that signals may be, at least partially, integrated across the 2 pathways and that VPMpc may be doing more than just encoding taste quality. More studies will be needed to clearly establish the differential roles of VPMpc, MD, and BLA. As for the RE, at the moment its role in taste and chemosensation in general is completely unknown.

CONCLUSION

In this chapter we have reviewed the organization of chemosensory thalamocortical systems, highlighting the peculiarities and the many differences with other senses. We treated the olfactory and gustatory systems as independent, dividing the chapter in 2 sections. This choice was motivated by anatomical considerations and by the classical view that portrays olfaction and taste as 2 separate systems. However, olfaction and taste are highly integrated, and their interaction is the basis for taste expectation and for the perception of flavor (Small and Green, 2012; Samuelsen and Vincis, 2021). Odors can relate to taste in 2 ways. When perceived from the external environment through orthonasal smelling, they can announce the imminent availability of taste and hence trigger expectations. When released from food intra-orally, odorants can take a retronasal route and activate the olfactory system. Retronasal olfaction and intra-oral gustation are integrated as a unitary flavor percept. How is flavor encoded in the brain? An old model of flavor processing postulates the convergence of 2 unisensory pathways (i.e., the olfactory and gustatory pathways) into the OFC and attributes the perception of flavor to the activation of OFC neurons (Rolls and Baylis, 1994). A more recent model challenges this hierarchical view and emphasizes the cross-modality of both the olfactory and gustatory system and the distributed representation of flavor (Samuelsen and Fontanini, 2017; Blankenship et al., 2019). This new model lays its foundations on the observation of taste responses in the PC and olfactory responses in GC, as well as on the presence of multiple connections across the 2 systems. MD as well as BLA, both responding to olfactory and gustatory signals and connecting with the PC and GC, are likely to play key roles in this integration. Whether VPMpc neurons can respond to odors remains to be seen, but given their responsiveness to temperature and intraoral tactile stimuli (Verhagen et al., 2003) (additional elements contributing to flavor perception), it cannot be excluded that this thalamic nucleus may also be involved in the integration of taste and olfaction.

In summary, while it is important to continue to understand thalamocortical contributions to olfaction and taste individually—the evidence presented above clearly demonstrates a gap in knowledge compared to other senses—this should be done while taking into account the integration between the chemical senses.

ACKNOWLEDGMENTS

This work was supported by National Institutes of Health grant R01DC019827 to AM; National Institutes of Health grant R01DC013770 to AM and AF; National Institutes of Health grant R01DC018227 to AF; National Institutes of Health grant UF1NS115779 to AF and AM.

REFERENCES

Allen GV, Saper CB, Hurley KM, Cechetto DF (1991a). Organization of visceral and limbic connections in the insular cortex of the rat. *J Comp Neurol.* 311:1–16.

Allen GV, Saper CB, Hurley KM, Cechetto DF (1991b). Organization of visceral and limbic connections in the insular cortex of the rat. *J Comp Neurol.* 311:1–16.

Baylis LL, Rolls ET, Baylis GC (1995). Afferent connections of the caudolateral orbitofrontal cortex taste area of the primate. *Neuroscience.* 64:801–812.

Blankenship ML, Grigorova M, Katz DB, Maier JX (2019). Retronasal odor perception requires taste cortex, but orthonasal does not. *Curr Biol.* 29:62–69.

Bouaichi CG, Vincis R (2020). Cortical processing of chemosensory and hedonic features of taste in active licking mice. *J Neurophysiol.* 123:1995–2009.

Cechetto DF, Saper CB (1987). Evidence for a viscerotopic sensory representation in the cortex and thalamus in the rat. *J Comp Neurol.* 262:27–45.

Cornwall J, Phillipson OT (1988). Afferent projections to the dorsal thalamus of the rat as shown by retrograde lectin transport. I. The mediodorsal nucleus. *Neuroscience.* 24:1035–1049.

Courtiol E, Neiman M, Fleming G, Teixeira CM, Wilson DA (2019). A specific olfactory cortico-thalamic pathway contributing to sampling performance during odor reversal learning. *Brain Struct Funct.* 224:961–971.

Courtiol E, Wilson DA (2014). Thalamic olfaction: Characterizing odor processing in the mediodorsal thalamus of the rat. *J Neurophysiol.* 111:1274–1285.

Courtiol E, Wilson DA (2015). The olfactory thalamus: Unanswered questions about the role of the mediodorsal thalamic nucleus in olfaction. *Front Neural Circuits.* 9:49.

Courtiol E, Wilson DA (2016). Neural representation of odor-guided behavior in the rat olfactory thalamus. *J Neurosci.* 36:5946–5960.

Datiche F, Cattarelli M (1996) Reciprocal and topographic connections between the piriform and prefrontal cortices in the rat: A tracing study using the B subunit of the cholera toxin. *Brain Res Bull.* 41:391–398.

Delevich K, Tucciarone J, Huang ZJ, Li B (2015). The mediodorsal thalamus drives feedforward inhibition in the anterior cingulate cortex via parvalbumin interneurons. *J Neurosci.* 35:5743–5753.

Eichenbaum H, Shedlack KJ, Eckmann KW (1980). Thalamocortical mechanisms in odor-guided behavior. I. Effects of lesions of the mediodorsal thalamic nucleus and frontal cortex on olfactory discrimination in the rat. *Brain Behav Evol.* 17:255–275.

Flynn FW, Grill HJ, Schwartz GJ, Norgren R (1991). Central gustatory lesions: I. Preference and taste reactivity tests. *Behav Neurosci.* 105:933–943.

Fredericksen KE, McQueen KA, Samuelsen CL (2019). Experience-dependent c-fos expression in the mediodorsal thalamus varies with chemosensory modality. *Chem Senses.* 44:41–49.

Fredericksen KE, Samuelsen CL (2022). Neural representation of intraoral olfactory and gustatory signals by the mediodorsal thalamus in alert rats. *J Neurosci.* 42:8136–8153.

Griffiths BJ, Zaehle T, Repplinger S, Schmitt FC, Voges J, Hanslmayr S, Staudigl T (2022). Rhythmic interactions between the mediodorsal thalamus and prefrontal cortex precede human visual perception. *Nat Commun.* 13:3736.

Groenewegen HJ (1988). Organization of the afferent connections of the mediodorsal thalamic nucleus in the rat, related to the mediodorsal-prefrontal topography. *Neuroscience.* 24:379–431.

Haley MS, Fontanini A, Maffei A (2016). Laminar- and target-specific amygdalar inputs in rat primary gustatory cortex. *J Neurosci.* 36:2623–2637.

Haley MS, Fontanini A, Maffei A (2022). Inhibitory gating of thalamocortical inputs onto rat gustatory insular cortex. *bioRxiv.* :2022.2012.2008.519650.

Hanamori T (2003). Chemical stimulation of the thalamic reticular nucleus inhibits the neuronal activity of the posterior insular cortex in rats. *Chem Senses.* 28:717–728.

Holtz SL, Fu A, Loflin W, Corson JA, Erisir A (2015). Morphology and connectivity of parabrachial and cortical inputs to gustatory thalamus in rats. *J Comp Neurol.* 523:139–161.

Imamura K, Onoda N, Takagi SF (1984). Odor response characteristics of thalamic mediodorsal nucleus neurons in the rabbit. *Jpn J Physiol.* 34:55–73.

Kawagoe T, Tamura R, Uwano T, Asahi T, Nishijo H, Eifuku S, Ono T (2007). Neural correlates of stimulus-reward association in the rat mediodorsal thalamus. *Neuroreport.* 18:683–688.

Kay LM, Sherman SM (2007). An argument for an olfactory thalamus. *Trends Neurosci.* 30:47–53.

Kloc M, Maffei A (2014). Target-specific properties of thalamocortical synapses onto layer 4 of mouse primary visual cortex. *J Neurosci.* 34:15455–15465.

Krettek JE, Price JL (1977). The cortical projections of the mediodorsal nucleus and adjacent thalamic nuclei in the rat. *J Comp Neurol.* 171:157–191.

Kuroda M, Murakami K, Kishi K, Price JL (1992). Distribution of the piriform cortical terminals to cells in the central segment of the mediodorsal thalamic nucleus of the rat. *Brain Res.* 595:159–163.

Kuroda M, Price JL (1991) Synaptic organization of projections from basal forebrain structures to the mediodorsal thalamic nucleus of the rat. *J Comp Neurol.* 303:513–533.

Levitan D, Liu C, Yang T, Shima Y, Lin JY, Wachutka J, Marrero Y, Ali Marandi Ghoddousi R, Veiga Beltrame ED, Richter TA, Katz DB, Nelson SB (2020). Deletion of Stk11 and Fos in mouse BLA projection neurons alters intrinsic excitability and impairs formation of long-term aversive memory. *eLife.* 9:e61036.

Liu H, Fontanini A (2015). State dependency of chemosensory coding in the gustatory thalamus (VPMpc) of alert rats. *J Neurosci.* 35:15479–15491.

Majak K, Ronkko S, Kemppainen S, Pitkanen A (2004). Projections from the amygdaloid complex to the piriform cortex: A PHA-L study in the rat. *J Comp Neurol.* 476:414–428.

McKenna JT, Vertes RP (2004). Afferent projections to nucleus reuniens of the thalamus. *J Comp Neurol.* 480:115–142.

Mitchell AS, Chakraborty S (2013). What does the mediodorsal thalamus do? *Front Syst Neurosci.* 7:37.

Norgren R, Leonard CM (1971). Taste pathways in rat brainstem. *Science.* 173:1136–1139.

Pelzer P, Horstmann H, Kuner T (2017). Ultrastructural and functional properties of a giant synapse driving the piriform cortex to mediodorsal thalamus projection. *Front Synaptic Neurosci.* 9:3.

Piette CE, Baez-Santiago MA, Reid EE, Katz DB, Moran A (2012). Inactivation of basolateral amygdala specifically eliminates palatability-related information in cortical sensory responses. *J Neurosci.* 32:9981–9991.

Plailly J, Howard JD, Gitelman DR, Gottfried JA (2008). Attention to odor modulates thalamocortical connectivity in the human brain. *J Neurosci.* 28:5257–5267.

Price JL, Slotnick BM (1983). Dual olfactory representation in the rat thalamus: an anatomical and electrophysiological study. *J Comp Neurol.* 215:63–77.

Ray JP, Russchen FT, Fuller TA, Price JL (1992). Sources of presumptive glutamatergic/aspartatergic afferents to the mediodorsal nucleus of the thalamus in the rat. *J Comp Neurol.* 320:435–456.

Reilly S, Bornovalova M, Dengler C, Trifunovic R (2003). Effects of excitotoxic lesions of the gustatory thalamus on latent inhibition and blocking of conditioned taste aversion in rats. *Brain Res Bull.* 62:117–128.

Reilly S, Pritchard TC (1996). Gustatory thalamus lesions in the rat: II. Aversive and appetitive taste conditioning. *Behav Neurosci.* 110:746–759.

Rolls ET, Baylis LL (1994). Gustatory, olfactory, and visual convergence within the primate orbitofrontal cortex. *J Neurosci.* 14:5437–5452.

Roper SD, Chaudhari N (2017). Taste buds: cells, signals and synapses. *Nat Rev Neurosci.* 18:485–497.

Saalmann YB, Pinsk MA, Wang L, Li X, Kastner S (2012.) The pulvinar regulates information transmission between cortical areas based on attention demands. *Science.* 337:753–756.

Samuelsen CL, Fontanini A (2017). Processing of intraoral olfactory and gustatory signals in the gustatory cortex of awake rats. *J Neurosci.* 37:244–257.

Samuelsen CL, Gardner MP, Fontanini A (2012). Effects of cue-triggered expectation on cortical processing of taste. *Neuron.* 74:410–422.

Samuelsen CL, Gardner MP, Fontanini A (2013). Thalamic contribution to cortical processing of taste and expectation. *J Neurosci.* 33:1815–1827.

Samuelsen CL, Vincis R (2021). Cortical hub for flavor sensation in rodents. *Front Syst Neurosci.* 15:772286.

Sanganahalli BG, Thompson GJ, Parent M, Verhagen JV, Blumenfeld H, Herman P, Hyder F (2022). Thalamic activations in rat brain by fMRI during tactile (forepaw, whisker) and non-tactile (visual, olfactory) sensory stimulations. *PLoS ONE.* 17:e0267916.

Saper CB (1982). Convergence of autonomic and limbic connections in the insular cortex of the rat. *J Comp Neurol.* 210:163–173.

Sapolsky RM, Eichenbaum H (1980). Thalamocortical mechanisms in odor-guided behavior. II. Effects of lesions of the mediodorsal thalamic nucleus and frontal cortex on odor preferences and sexual behavior in the hamster. *Brain Behav Evol.* 17:276–290.

Scott GA, Liu MC, Tahir NB, Zabder NK, Song Y, Greba Q, Howland JG (2020). Roles of the medial prefrontal cortex, mediodorsal thalamus, and their combined circuit for performance of the odor span task in rats: analysis of memory capacity and foraging behavior. *Learn Mem.* 27:67–77.

Sherman SM (2012). Thalamocortical interactions. *Curr Opin Neurobiol.* 22:575–579.

Sherman SM, Guillery RW (2006). *Exploring the Thalamus and Its Role in Cortical Function.* 2nd ed. MIT Press.

Sherman SM, Guillery RW (2011). Distinct functions for direct and transthalamic corticocortical connections. *J Neurophysiol.* 106:1068–1077.

Slotnick BM, Kaneko N (1981). Role of mediodorsal thalamic nucleus in olfactory discrimination learning in rats. *Science.* 214:91–92.

Small DM, Green BG (2012). A proposed model of a flavor modality. In: Murray MM, Wallace MT, eds. *The Neural Bases of Multisensory Processes.* Boca Raton (FL): CRC Press/Taylor & Francis; Chapter 36.

Small DM, Veldhuizen MG, Felsted J, Mak YE, McGlone F (2008). Separable substrates for anticipatory and consummatory food chemosensation. *Neuron.* 57:786–797.

Stehberg J, Acuna-Goycolea C, Ceric F, Torrealba F (2001). The visceral sector of the thalamic reticular nucleus in the rat. *Neuroscience.* 106:745–755.

Stein EA, Carr KD, Simon EJ (1990). Brain stimulation-induced feeding alters regional opioid receptor binding in the rat: an in vivo autoradiographic study. *Brain Res.* 533:213–222.

Stone LM, Finger TE, Tam PP, Tan SS (1995). Taste receptor cells arise from local epithelium, not neurogenic ectoderm. *Proc Natl Acad Sci U S A.* 92:1916–1920.

Stone ME, Fontanini A, Maffei A (2020). Synaptic integration of thalamic and limbic inputs in rodent gustatory cortex. *eNeuro.* 7(1):ENEURO.0199-19.2019.

Verhagen JV, Giza BK, Scott TR (2003). Responses to taste stimulation in the ventroposteromedial nucleus of the thalamus in rats. *J Neurophysiol.* 89:265–275.

Vertes RP, Hoover WB, Do Valle AC, Sherman A, Rodriguez JJ (2006). Efferent projections of reuniens and rhomboid nuclei of the thalamus in the rat. *J Comp Neurol.* 499:768–796.

Weiss MS, Victor JD, Di Lorenzo PM (2014). Taste coding in the parabrachial nucleus of the pons in awake, freely licking rats and comparison with the nucleus of the solitary tract. *J Neurophysiol.* 111:1655–1670.

Wolff M, Morceau S, Folkard R, Martin-Cortecero J, Groh A (2021). A thalamic bridge from sensory perception to cognition. *Neurosci Biobehav Rev.* 120:222–235.

Yarita H, Iino M, Tanabe T, Kogure S, Takagi SF (1980). A transthalamic olfactory pathway to orbitofrontal cortex in the monkey. *J Neurophysiol.* 43:69–85.

Zelano C, Mohanty A, Gottfried JA (2011). Olfactory predictive codes and stimulus templates in piriform cortex. *Neuron.* 72:178–187.

30.

THALAMIC CONTRIBUTIONS TO MULTISENSORY CONVERGENCE AND PROCESSING

Annamaria Barczak, Monica N. O'Connell, and Charles E. Schroeder

INTRODUCTION

In accordance with earlier propositions (Ghazanfar and Schroeder, 2006), the growing body of literature describing evidence for multisensory processing in cortical and thalamic regions supports the idea that the brain is a multisensory structure. In this chapter, we will discuss the role of the thalamus in multisensory processing. First, we will review the relevant anatomical connections of selected major thalamic nuclei. Second, we will present evidence for multisensory/sensorimotor integration occurring within these nuclei. Third, we will discuss other potential mechanisms through which this integration occurs; focusing on (1) the modulation of ongoing oscillations, and (2) the role of cortico-thalamocortical circuits. Finally, we will consider future directions for multisensory research.

In general, we will focus on the auditory, visual, and somatosensory systems as they are related to the thalamic nuclei described below. We will concentrate on information obtained from human and nonhuman primates and will discuss studies in other species only when information specific to primates is lacking.

MAJOR THALAMIC NUCLEI AND THEIR ANATOMICAL CONNECTIONS

We highlight 4 "exemplary" thalamic nuclei and briefly outline their major anatomical connections. The ascending projections provide potential sources of multisensory input/modulation to cortical areas and illustrate key structural underpinnings of multisensory interactions.

MEDIAL GENICULATE BODY (MGB)

The MGB is traditionally considered a major thalamic relay nucleus for auditory processing and is extensively described elsewhere in this book (see **Chapters 21, 22, and 24 in Section V**). Briefly, the MGB is divided into 3 major subdivisions named after their relative positions within the nucleus: ventral (MGBv), dorsal (MGBd), and medial (MGBm). Immunostaining for calcium-binding proteins parvalbumin and calbindin (which demarcate lemniscal and non-lemniscal nuclei, respectively) has revealed that neurons immunoreactive to calbindin are distributed throughout the MGB, forming a matrix to the entire nucleus (Jones, 2001). Across MGB subdivisions, however, the density of calbindin and parvalbumin staining tends to alternate.

MGBv

Nearly all neurons in the primate MGBv are immunoreactive for parvalbumin (Jones, 2001). MGBv neurons also exhibit short latency response onsets to tones, have sharp tuning, and are tonotopically organized. Together, these characteristics indicate that MGBv is part of the auditory lemniscal pathway (Hackett, 2011). The main subcortical inputs to MGBv are from the tonotopically organized central nucleus of the ipsilateral inferior colliculus (CNIC), while projections from MGBv primarily target layers 3 and 4 of primary auditory cortex (A1) (Hackett, 2011).

MGBd

The primate MGBd exhibits strong staining for calbindin and relatively weak staining for parvalbumin (Jones, 2001). Neuronal responses to sounds display a wide range of onset latencies, usually longer than that of MGBv, and exhibit broader frequency tuning. There is a lack of tonotopic organization within the MGBd in any species studied thus far. Subcortical inputs to MGBd come from the dorsal cortex (DCIC) and lateral cortex (LCIC) of the inferior colliculus, which are part of the non-tonotopic or non-lemniscal ascending pathway, and also do not appear to be tonotopically organized (Hackett, 2011). LCIC receives inputs from the ipsilateral superior colliculus (SC), a multisensory structure, and visual cortex (Cooper and Young, 1976). MGBd projections target layer 3b of nonprimary (belt and parabelt) areas of auditory cortex, where tonotopy is weak or absent (Hackett, 2011).

MGBm

The medial subdivision is also known as the magnocellular subdivision because it contains the largest cells of the MGB. Like MGBd, MGBm shows strong immunoreactivity to calbindin but also includes groups of parvalbumin-positive cells dispersed within this matrix of calbindin-positive neurons (Jones, 2001). Neuronal response properties are variable. Some neurons display narrow frequency tuning with short onset latencies like those found in MGBv, while others have broader tuning and longer onset latencies. MGBm receives auditory input from all 3 of the IC divisions (CNIC, LCIC, DCIC), somatosensory input from the spinal-thalamic, dorsal column, and trigeminal pathways, and visual input from the lower layers of the SC (Hackett, 2011). Although projections from MGBm are broadly dispersed to all auditory cortical areas and terminate in layers 1 and 6 (Smiley and Falchier, 2009), tracer studies have shown that only calbindin-positive neurons of MGBm project to layer 1 of auditory cortical areas (Molinari et al., 1995).

LATERAL GENICULATE NUCLEUS (LGN)

The LGN is the main thalamic relay for visual information from the retina to the visual cortex. The LGN is a layered structure that includes 2 magnocellular (M) layers (1 for each eye) and 4 parvocellular (P) layers (2 for each eye). Between and below the M and P layers are 6 koniocellular (K) layers. (For a more detailed anatomical description of the LGN, see Briggs and Usrey, 2011; and **Chapters 17, 18, and 19 in Section V**.)

There are primarily 3 types of LGN thalamocortical neurons. Neurons in P and M layers mainly express parvalbumin, while neurons in the K layers mainly express calbindin. P and M cells project to layers 4c and 6 of V1 (Briggs and Usrey, 2011) and at least 3 different subclasses of K LGN axons in nonhuman primates project to V1 (Casagrande et al., 2007). Koniocellular layers K1 and K2 project to V1 layers 1 and 3a, while LGN axons from layers K3–K6 project to the cytochrome oxidase blobs in layer 3bα of V1. Not all inputs to LGN are relayed to V1. There is a direct connection from LGN to the middle temporal area (MT), most of which is through koniocellular neurons (Sincich et al., 2004). Additionally, there is a population of small LGN neurons that bypass V1 to target V4 neurons and may also be "K" type neurons, but their calcium-binding protein qualities were not reported (Fries, 1981).

Less than 15% of input into the LGN comes directly from the retina. The LGN mostly receives modulatory input from brainstem structures, feedback from different parts of the visual cortex, or inhibitory input from local structures (Sherman and Guillery, 2002). These inputs include the superior colliculus and GABAergic projections from the thalamic reticular nucleus (TRN) (Bickford et al., 2000). In addition to receiving inhibitory projections from TRN, about 35% of cells in the magnocellular layers and 25% of cells in the parvocellular layers within the LGN were also GABAergic, which suggests that inhibition occurs locally as well (Montero and Zempel, 1986).

VENTROPOSTERIOR NUCLEUS (VPN)

The VPN of the thalamus is part of the somatosensory pathway through which sensory information is transmitted from external receptors to the spinal cord, through the medulla, the thalamus, and then to the somatosensory cortex. (For more information, see **Chapters 25 and 26 in Section V**.) The VPN is subdivided into the ventroposterior medial (VPM) and the ventroposterior lateral (VPL) subnuclei.

VPM receives somatosensory information from the face and head through the trigeminal nerve, and includes a matrix of smaller neurons that express calbindin and receive information from the spinal trigeminal complex (Rausell et al., 1992). Also, within the VPM are rod-like clusters of larger neurons that express parvalbumin and receive input from the principal sensory trigeminal nucleus (Jones et al., 1986). These rod-like clusters project to the middle layers of the primary somatosensory cortex (S1), while the matrix neurons project to layers 1 and 2 of S1.

The VPL includes a representation of the rest of the body. Information about touch and proprioception is transmitted through the dorsal column medial lemniscal pathway, while pain and thermal information ascends through the spinothalamic pathway. Neurons along the former pathway tend to express parvalbumin and project to the middle and deep layers of S1, while those in the latter pathway express calbindin and target layer 1 (Rausell et al., 1992).

PULVINAR

The pulvinar is the largest thalamic nucleus and is clearly involved in visual processing; however, it has strong feed-forward and feedback connections with many sensory and higher-order cortices, as well as with the multisensory SC (Cappe et al., 2009a). In addition to displaying overlapping neuronal responses to different sensory stimuli, it is postulated to play a role in cortical multisensory integration. The primate pulvinar consists of 3 primary nuclei:

inferior, lateral, and medial. We will review some key characteristics (but see **Chapter 20 in Section V** for an in-depth discussion).

Inferior Pulvinar (PI)

The degree of calbindin immunoreactivity within the primate PI varies from extremely light (medial PI subdivision) to dense (central PI subdivision) (Gutierrez et al., 1995). PI has reciprocal connections with areas in both the ventral (V1, V2, V4, TEO) and dorsal (MT, MST, FST [a multisensory cortical area]) visual pathways (Kaas and Lyon, 2007). PI projects to extragranular layers 1 and 6 of V1, and layers 1, 3, and 4 of V2 and V3 (Benevento and Rezak, 1976). It also receives direct retinal inputs and is heavily connected with the SC (Warner et al., 2010).

Functional Responses of PI

It has been shown that PI neurons respond to visual stimuli and eye movements in the dark as well as in light (Robinson et al., 1991). Responses to auditory stimuli and hand movements have also been recorded in the PI (Yirmiya and Hocherman, 1987).

Lateral Pulvinar (PL)

Of the 3 nuclei of the pulvinar, PL stains the lightest for calbindin but displays dense staining for parvalbumin throughout (Gutierrez et al., 2000). Like PI, PL is reciprocally connected with visual areas within the dorsal and ventral visual pathways (Adams et al., 2000). PL also receives inputs from the superficial layers of SC. In addition to visual processing, many of the higher-order regions within the dorsal visual stream also play a role in motor control and somatosensation. Additionally, PL is connected with parietal areas PE and PEa whose function is the integration of proprioceptive, somatosensory, and motor signals (Hyvärinen, 1982).

Functional Responses of PL

In addition to being responsive to complex visual stimuli (Nguyen et al., 2013; Van Le et al., 2013), PL neurons also show responses to somatosensory stimuli and arm movements (Yirmiya and Hocherman, 1987). Neurons in ventral PL are also responsive to auditory stimuli (Gattass et al., 1978).

Medial Pulvinar (PM)

The primate PM shows an irregular pattern of light and dark calbindin staining that is complementary to dark and light zones of parvalbumin immunoreactivity (Gutierrez et al., 2000). PM has the most widespread reciprocal connections with the cortex. In fact, a tracer study conducted in macaques showed that projections from the same regions in PM targeted somatosensory, auditory, and premotor cortex, indicating a high degree of divergence (Cappe et al., 2009a). The lateral subdivision of PM (PMl) is reciprocally connected with V1, V2, V4, TEO, and intraparietal area LIP (Shipp, 2003). Subcortically, PMl receives inputs from layer 6 of LGN and the deep layers of SC (Benevento and Standage, 1983). The medial subdivision of PM (PMm) is interconnected with a plethora of auditory-processing regions: belt regions of auditory cortex, the superior temporal gyrus (STG), and the superior temporal sulcus (STS) (Cappe et al., 2009a). PMm has reciprocal connections with somatosensory and proprioceptive posterior parietal areas PE and PEa, inferior parietal cortices, temporal and prefrontal cortices, and premotor cortex (Morel et al., 2005). PMm also receives inputs from the deep layers of SC (Benevento and Standage, 1983).

Functional Responses of PM

Neurons in the PM are responsive to stimuli of many modalities. Primate PM neurons show robust responses to both simple and complex auditory stimuli, somatosensory stimuli, and motor movements (Gattass et al., 1978). Responses related to complex visual stimuli, such as faces and animals, have also been reported in the PM (Van Le et al., 2013).

MECHANISTIC ROLE OF THALAMIC NUCLEI IN MULTISENSORY/MULTIMODAL INTEGRATION

We start by operationally defining some terms whose meanings have varied over the years. *Convergence* describes an anatomical projection of inputs from different sensory modalities onto a target region, and it is a requisite for any further multisensory processes. *Multimodal sensitivity* refers to the capacity of inputs from different sensory and motor systems to influence excitability in neurons of a given region. *Multimodal integration* describes a process wherein the temporally coordinated presentation of stimuli from different sensory modalities or the presentation of sensory stimuli in temporal proximity of a motor act results in a neuronal response that is different from the sum of the sensory and/or motor inputs arriving in isolation.

MULTISENSORY INTEGRATION IN THALAMIC NUCLEI

As previously proposed, it is possible that multisensory effects in cortical areas are inherited from thalamic nuclei or from upstream cortical areas; that is, already integrated multisensory information is quickly passed on to the cortex by feedforward thalamocortical or cortico-cortical projections. Here, we will review evidence of multisensory processing in the major thalamic nuclei.

A functional magnetic resonance imaging (fMRI) study in humans has shown enhanced BOLD (blood oxygen level dependent) responses in both LGN and MGB when a sound co-occurred with a low-intensity visual stimulus, but not when paired with higher-intensity visual stimuli (Figure 30.1A) (Noesselt et al., 2010). Similarly, using fMRI, Baier and colleagues showed modulation of thalamic activity during an audiovisual task; however, this effect occurred in non-modality specific nuclei (Baier et al., 2006).

While, thus far, evidence of suprathreshold responses to extramodal stimuli in primate MGB is lacking, cat and rodent studies have shown that the MGB, especially the medial subdivision (MGBm), contains neurons which display multisensory responses (e.g., photic or vestibular or touch stimulation; Wepsic, 1966; Lohse et al., 2021). Presentation of auditory clicks and cutaneous electrical stimulation of the contralateral paw of cats revealed multimodal neurons in MGBm; clicks resulted in excitation, while somatosensory stimuli resulted in inhibition (Khorevin, 1978). These 2 studies point to the multisensory convergence of inputs on neurons in MGBm. More recent studies have shown evidence for multisensory integration—either response enhancement or suppression—in MGB. For instance, auditory-related single-unit responses were suppressed when paired with cutaneous electrical stimulation of a rat's hindpaw in MGBm and MGBd (Kimura and Imbe, 2018) and, perhaps surprisingly, also in MGBv (Donishi et al., 2011). Similarly, Lohse and colleagues found that auditory-evoked responses in both MGBv and MGBd of the mouse were suppressed by concurrent whisker stimulation (Lohse et al., 2021). They provided evidence that this thalamic suppression was directly passed on to A1 neurons, as thalamocortical axon boutons in layer 1 displayed suppressed responses to auditory stimuli during whisker stimulation too (Figure 30.1B). Concurrent visual and auditory stimuli resulted in an enhancement of auditory responses in the MGBm and MGBd of behaving rats (Komura et al., 2005). Importantly, with the exception of some MGBm neurons, neither cutaneous electrical, whisker, or visual stimulation alone resulted in an evoked response in these studies, indicating that the majority of cross-modal inputs in these cases were strictly modulatory.

There is very little evidence for multisensory processing in LGN. One study in mice found that locomotion slightly enhanced single-unit visual responses in the LGN, but only at unusually high running speeds (Williamson et al., 2015). This study also reported that locomotion suppressed auditory-related responses in MGBv and MGBd, though it was noted that this movement-related modulation of LGN and MGB responses was weaker than that which has been reported in their corresponding primary sensory cortices. Another rodent study found that whisker deflection alone did not alter spontaneous spiking in LGN, nor did it modulate responses to simultaneously presented visual stimuli (Bieler et al., 2018).

Interestingly, Bieler et al. (2018) also found that the bimodal (somatosensory and visual) stimulation did enhance tactile-related responses in the rodent somatosensory relay nucleus, VPM. Visual stimulation alone modulated the firing of approximately 40% of single units sampled in the VPM, by both enhancing and suppressing firing. Similarly, Allen and colleagues (2017) also found evidence of convergence of somatosensory and visual inputs in single units of mouse VPM. These bimodal neurons responded to visual stimulation alone with a robust increase in firing and also exhibited multisensory integration by showing that somatosensory-related responses could be enhanced or suppressed depending on the direction of motion of the visual stimulus. One primate study also found a small subset of VP neurons which were responsive to both vestibular and visual inputs (Meng and Angelaki, 2010). Taken together, these studies point to the convergence and integration of multimodal inputs in the VP complex.

Surprisingly, relatively few studies have directly examined multisensory integration in the pulvinar. A recent study conducted in the PM of behaving macaques found that neurons which responded vigorously to auditory and visual stimuli when presented separately demonstrated suppressive multisensory integration when these stimuli were presented simultaneously (Figure 30.1C) (Vittek et al., 2022). A human imaging study also found multisensory enhancement in the pulvinar during an audiovisual speech-processing task (Ross et al., 2022). Multisensory convergence and integration were similarly demonstrated in neurons recorded from in the cat pulvinar (Avanzini et al., 1980). Single units exhibited suprathreshold responses to separate visual and somatosensory stimuli, but showed an inhibitory interaction during simultaneous visual and somesthetic stimulation.

SENSORIMOTOR INTEGRATION IN THALAMIC NUCLEI

Sensory information can enter the brain passively (e.g., background noise from a freeway as you sit near an open window), or it can be actively sought out (e.g., moving your eyes around a crowded room to find your friend). In the latter scenario, motor-sampling routines are used to actively obtain sensory information from the environment through a process termed *active sensing* (Schroeder et al., 2010). We discuss how the motor system can influence thalamic activity, using the visual system as an example.

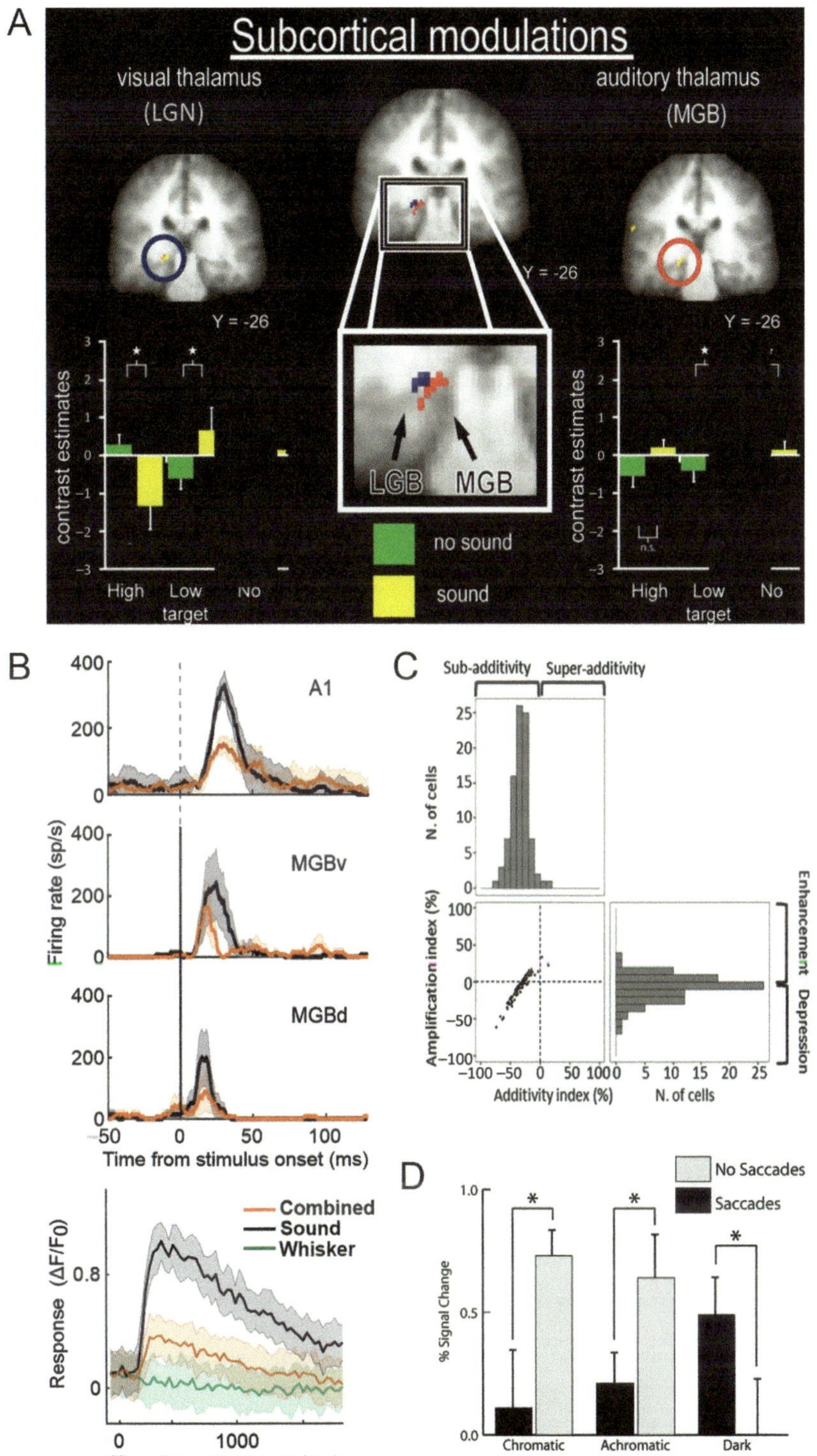

Figure 30.1. Examples of multisensory integration in thalamic nuclei. (A). Brain sections depict blood oxygenation level dependent (BOLD) effects in human LGN (blue) and MGB (red). The bar graphs show BOLD estimates calculated in LGN (left) and MGB (right) during the presentation of a high-intensity (High) or low- (Low) intensity visual target, either paired with (yellow) or without (green) a sound. They show that in both LGN and MGB the co-occurrence of a sound enhanced the BOLD response for the lower-intensity visual target condition more than for the higher-intensity visual target condition. (B). Top 3 line plots are PSTHs demonstrating how responses to best frequency tones recorded in rodent A1, MGBv, and MGBd were suppressed when paired with whisker stimulation (orange) compared to without (black). Bottom line plot shows a response from an MGB thalamocortical axon bouton in layer 1 of A1 in response to an auditory stimulus paired with (orange) and without (black) whisker stimulation, as well as to whisker stimulation alone (green). This indicates that the integration effect of somatosensory stimuli on auditory evoked activity in the thalamus is relayed on to auditory cortex. (C). Plot shows that suppressive and subadditive multisensory integration dominates the multisensory responses of bimodal neurons in medial pulvinar—the neuronal population *Amplification index* distribution, which indicates whether the addition of another modality

(*continued*)

Humans and other primates actively seek visual information using a series of alternating fixations and saccades. These quasi-rhythmic eye movements, occurring at a rate around 3–4 Hz, impose a temporal patterning on the sensory information that enters the brain and have been shown to influence stimulus-response amplitudes even at the level of the thalamus. Using fMRI, activity in human LGN was observed to be strongly modulated by saccades—even in complete darkness (Figure 30.1D) (Sylvester et al., 2005). Royal and colleagues found that 25% of LGN cells showed modulation around saccadic eye movements (Royal et al., 2006). The observed modulation occurred across all 3 cell classes—magnocellular, parvocellular, and koniocellular—and began more than 100 ms before the onset of the eye movement, which suggested to the authors that the modulation was related to motor planning. These findings support the idea that LGN is more than a simple relay nucleus for incoming visual information. Instead, it has been suggested that the LGN is a structure that can influence or be influenced by motor planning (Casagrande et al., 2005). Other thalamic areas that may play an important role in active vision are those that receive input from the superior colliculus (SC). Since the superficial and intermediate layers of the SC respond to visual stimuli and increase activity before saccades, respectively, Wurtz and colleagues (2011) suggested that the thalamic targets of these areas may be potentially involved during active vision. Such targets include (and are not limited to) the LGN, the medial dorsal thalamic nucleus, the thalamic reticular nucleus (TRN), and the inferior and lateral pulvinar. Saccadic corollary signals originating in the SC and projecting through one or more of these diverse circuits (Wurtz et al., 2011) may account for saccadic modulation of activity in visual areas of the thalamus (e.g., Sylvester and Rees, 2006), as well as account for saccadic modulation in brain regions that lack an explicit involvement in vision, such as the anterior thalamic nuclei (Leszczynski et al., 2021).

COMPLEMENTARY ROLES OF CORE AND MATRIX SYSTEMS IN THALAMOCORTICAL PROCESSING

As summarized by Jones and discussed above, a contrasting pattern of calcium-binding protein expression appears to be conserved across the main thalamocortical projections in the visual, auditory, and somatosensory pathways (Jones, 1998, 2001). Jones (1998) termed the parvalbumin-positive projections "core" and the calbindin-positive projections "matrix." Across the 3 nuclear complexes, core neurons carry a precise representation of the sensory receptor surface and project this representation in an orderly fashion to the middle layers of their primary cortical target regions supporting the retinotopic, cochleotopic, and somatotopic maps found therein. In contrast, matrix neurons carry a more complex representation of the receptor surface with large, often poorly circumscribed receptive fields, and properties best exemplified by visual movement and directional selectivity seen in the matrix division of the LGN. Matrix neuron projections are weighted toward supragranular cortical layers, particularly layer 1 (e.g., Figure 30.2A). All of these properties are consistent with the ideas that: (1) core neurons carry rather simple, specific information ("content") and "drive" neurons in a target region to fire action potentials, while (2) matrix neurons carry more diverse and complex bits of information that provide "context" used to modulate the processing of core inputs; that is, modulatory inputs affect the probability that neurons in the target region will fire, given the appropriate driving input. We will elaborate this theme below.

MODULATION OF ONGOING OSCILLATIONS BY NONSPECIFIC THALAMIC INPUTS

In addition to the classical convergence of unimodal signals onto a neuron or area as a mechanism of multisensory integration, there is ample evidence to support the mechanistic role of neuronal oscillations in multisensory processing. Neuronal oscillations reflect the rhythmic fluctuation of neuronal ensembles between high and low excitability states. Thus, their harnessing by the brain during any cognitive process is a powerful tool to augment or suppress any arriving input. In support of their role in multisensory integration, intracranial studies in human and nonhuman primate studies have shown that in cortical regions traditionally considered unisensory, heteromodal stimuli lead to a modulation of the excitability phase (i.e., phase reset) of ongoing neuronal oscillations within specific frequency bands (Lakatos et al., 2007; Lakatos et al., 2009; Mégevand et al., 2020). The Lakatos studies also

Figure 30.1. **(continued)**

increases (response enhancement, index >0) or decreases (response suppression, index <0) the response amplitude relative to the best unisensory response, is shown on the right and is skewed toward depression. The neuronal population *Additivity index* distribution, which categorizes neurons into groups based on the linearity of the multimodal integration: subadditive (index <0), supraadditive (index >0), or additive (index = 0), is shown on the top and is skewed toward subadditivity. (D). Bar graphs show BOLD responses in human LGN during saccade and no saccade conditions in darkness and in the presence of chromatic or achromatic visual stimulation. Saccades significantly increase BOLD responses in LGN compared to a no-saccade condition in the dark. This illustrates that the motor modality can also modulate thalamic processing. Panel A is adapted from Noesselt et al. (2010); panel B is adapted from Lohse et al. (2021); panel C is adapted from Vittek et al. (2022); panel D is adapted from Sylvester et al. (2005).

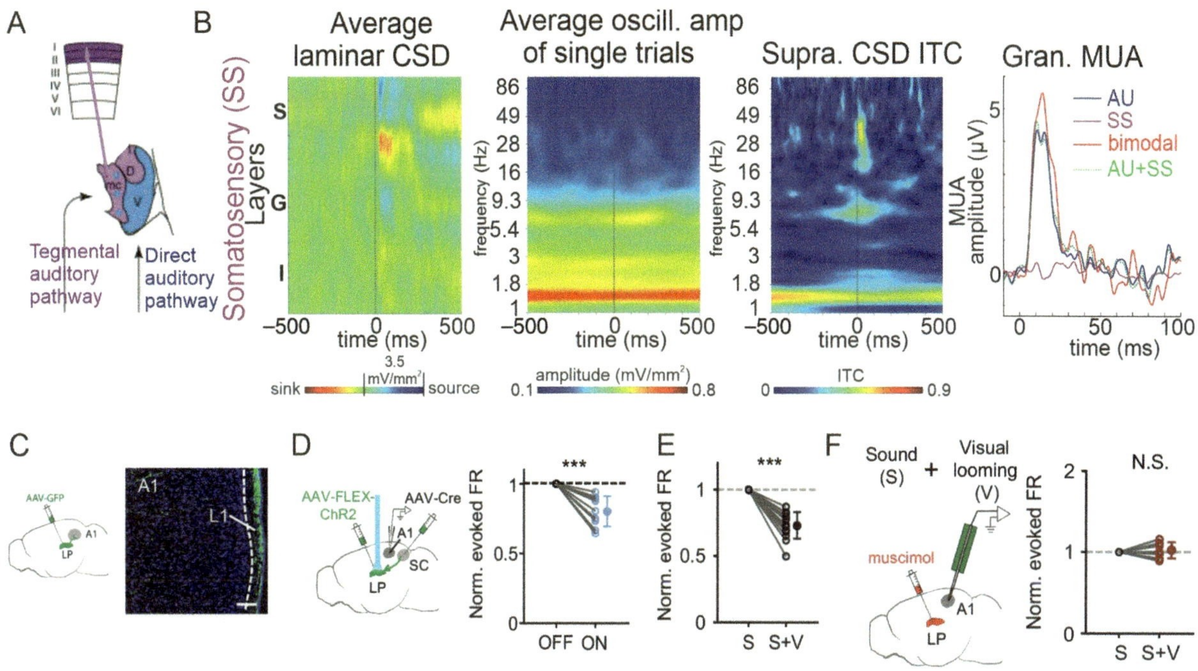

Figure 30.2. Phase reset of neuronal oscillations as a mechanism of multisensory integration and potential pathways. (A). Schematic of MGB showing patterns of staining for calbindin (purple) and parvalbumin (blue), and projection from matrix dominant subdivision to the superficial layers of A1. (B). First color plot shows the laminar profile of somatosensory (SS) related averaged current source density (CSD) responses in A1. Note that the CSD amplitude distribution is heavily biased to the supragranular (S) layers. Middle plot is a time-frequency plot depicting the average oscillatory amplitude of the wavelet transformed single trials, which shows that there is no significant SS event-related power increase in any of the frequency bands. The last time-frequency plot shows phase locking or ITC values of the single trial phases at different times/frequencies. ITC values range between 0 and 1; higher values mean that single trial phases are similar across trials at the same time point, while low values mean that the oscillatory phases are random. SS stimuli cause phase locking only in delta, theta, and gamma frequency bands. Since there is phase locking in discrete frequency bands without a stimulus-related power increase, it can be assumed that pure phase resetting of neuronal oscillations to a specific excitability phase is occurring. Line plots on the extreme right show A1 granular multiunit (MUA) responses to auditory, SS, and bimodal stimuli. Green dotted lines show the arithmetic sum of the unimodal responses. Simultaneous SS and auditory stimulation lead to an enhanced MUA response compared to the unisensory responses or the sum of these. (C). A direct projection from rodent pulvinar (LP) to layer 1 of A1 was confirmed by injection of viral vectors into LP; the right image shows GFP labeled axons concentrated in L1. (D). SC-recipient LP neurons were trans-synaptically labeled using viral vectors injected into both SC and LP. These LP neurons were then optogenetically activated using blue light, leading to reduced evoked firing rate (FR) in A1 layer 2/3 neurons (ON). (E). A similar effect was observed in A1 neurons (i.e., reduced evoked firing rate) when visual (V) and sound (S) stimuli were paired, compared to when S was presented alone. (F). Silencing LP with muscimol negates the effect of visual stimulation on A1 evoked firing rates, indicating the important role of the pulvinar in mediating multisensory modulation of sensory responses. Panel A is adapted from Jones (2001); panel B is adapted from Lakatos et al. (2007); panels C–F are adapted from Chou et al. (2020).

showed that presentation of somatosensory stimuli alone produced reliable reset of ongoing oscillatory activity to a high excitability phase in auditory cortex and, when measured by spiking activity, that concurrent presentation of auditory stimuli (tones) amplified the response to the auditory input. Megevand et al. (2020), working with much more complex audiovisual (speech) stimuli and intracranial recordings in humans, likewise showed a clear contribution of phase reset to the augmentation of an LFP correlate of neuronal spiking activity: broadband high frequency activity (BHA). However, they also demonstrated a clear complementary contribution from evoked responses triggered directly by prominent visual cues (Mégevand et al., 2020).

While the exact source of the input that results in the cross-modal phase reset of cortical oscillations is still debated, it has been proposed that projections from the matrix subdivisions of thalamic nuclei are involved (Jones, 1998; Schroeder and Lakatos, 2009). As noted above, the complementary anatomical and physiological properties of the core and matrix projections strongly align with this idea. The targeting of superficial cortical layers by thalamic matrix cells is of particular interest, as a key signature of cortical phase reset is a low-amplitude response that is earliest in the supragranular layer (Figure 30.2B) (Lakatos et al., 2007; Lakatos et al., 2009). Additionally, as mentioned above, neurons in matrix subdivisions of the thalamus receive cross-modal inputs and show multisensory properties. These properties combined make them an ideal candidate to play a vital role in generating phase reset in cortical areas. In support of this notion, a recent optogenetic study conducted in mice, although not investigating phase reset, showed that a SC → LP (rodent homologue of the primate pulvinar) → A1 layer 1 inhibitory neuron pathway mediated the modulation of auditory processing by sharpening the frequency tuning in A1 during visual stimulation (Figure 30.2C–F) (Chou et al., 2020).

Neuronal oscillations have also been implicated in supporting multisensory integration through their synchronization in different sensory cortical areas, which allows for efficient communication between and integration of the information being processed in these areas (Fries, 2005). Synchrony of neural signals in the alpha and gamma frequency bands across cortical areas have been observed during multisensory tasks (Maier et al., 2008). The primate pulvinar has been shown to synchronize neuronal activity in the alpha and gamma ranges between interconnected visual cortical areas (Saalmann et al., 2012; Zhou et al., 2016), and between frontal and parietal cortical regions (Fiebelkorn et al., 2019) during attention tasks. In contrast, deactivation of the pulvinar has been shown to reduce synchronization of gamma-band oscillatory activity between visual cortical areas in the primate, which highlights the pulvinar's importance in coordinating cortical interactions (Zhou et al., 2016). Since the pulvinar is a site of multisensory convergence and has widespread connections with both sensory and higher-order cortical regions, it is likely to assist in the processing and binding of separate sensory signals emanating from objects. Additionally, the fact that the pulvinar synchronizes activity between interconnected cortical areas in the alpha range is appropriate, as that is the temporal range (~100 ms) which has been shown to be optimal for multisensory integration (Meredith, 2002).

THE ROLE OF CORTICO-THALAMOCORTICAL (CTC) TRANSFER OF INFORMATION IN MULTISENSORY PROCESSING

Another mechanism that would support multisensory integration in the cortex, and which would be compatible with all the other mechanisms mentioned thus far, is the indirect communication of functionally distinct cortical areas through CTC connections (Sherman and Guillery, 2011). Corticothalamic (CT) projections from layer 5 (L5) of one cortical area terminate in higher-order (often multisensory) thalamic nuclei in the form of giant endings, which guarantee rapid synaptic transmission via ionotropic receptors. These L5 projections provide a driving suprathreshold input as opposed to a modulatory-type input, which is conveyed by layer 6 (L6) CT projections. (See **Chapters 14 and 15 in Section IV** for a detailed review of corticothalamic pathways.) The higher-order multisensory nuclei then relay this information on to another cortical area through their widespread connections, hence facilitating multisensory convergence in that sensory cortical area. Additionally, CT projections from different sensory modalities could converge in the same thalamic nucleus (e.g., pulvinar), which then bequeaths this already integrated information to the cortex (Cappe et al., 2009b). In support of this latter process, a recent study showed that somatosensory-related suppression of auditory responses in rodent A1 was due to a corticofugal-thalamocortical pathway: descending projections from S1 targeted the non-lemniscal IC shell, whose projections then inhibited MGBv neurons, which led to suppression of auditory responses in thalamocortical axon boutons and ultimately in A1 neurons (Lohse et al., 2021). Another rodent study demonstrated that a sensorimotor CTC pathway exists between L5 of S1 and primary motor cortex through the posterior medial thalamic nucleus (Mo and Sherman, 2019), further supporting the idea that these CTC pathways can connect functionally distinct cortical areas.

FUTURE DIRECTIONS

Our understanding of the role of the thalamus in multisensory processing has expanded over the last 50 years, but there are still several questions which require additional examination.

DOES THE CORE/MATRIX DICHOTOMY SUPPORTING DRIVER AND MODULATOR FUNCTIONS FOR SENSORY RELAY THALAMOCORTICAL (TC) PROJECTIONS ALSO APPLY TO THOSE OF THE PULVINAR?

As described above for sensory relay nuclei, core parvalbumin-positive projections tend to carry specific sensory information that drive upstream activity, whereas matrix calbindin-positive projections carry more diverse information that modulates activity driven by core projections. An open question is if this arrangement is at work in projections from the pulvinar to the cortex. As with relay nuclei, neuronal staining for parvalbumin and calbindin tends to alternate across the primate pulvinar. However, it is currently unknown whether the projections of the TC calbindin-expressing neurons target layer 1 of their receiving areas, or whether they transmit modulatory information like those of sensory relay matrix neurons. In CTC loops, the pulvinar receives driving input from the cortex via L5 CT projections, but is the quality of the information transmitted in the TC pathway of the loop modulatory, driving, or both? Since the pulvinar has been proposed to support multisensory integration through synchronizing activity in functionally distinct cortical areas via its vast array of anatomical connections, this knowledge would allow us to understand whether this mechanism operates through driving or modulatory pathways.

ARE MULTISENSORY AND SENSORIMOTOR INTEGRATION PROCESSES PRESERVED ACROSS DIFFERENT SPECIES?

It is of paramount importance to consider the possibility that there may be profound differences in the principles governing multisensory and sensorimotor integration across individual species. One reason these mechanistic differences may arise is that homologous structures between species may be anatomically different. For example, the organization and size of the 3 primary nuclei (PI, PL, and PM) of the primate pulvinar described earlier in this chapter are drastically distinct from those found in the mouse pulvinar (Zhou et al., 2017). It stands to reason that the projections to/from this and other homologous structures could vary and provide different substrates for multisensory processing.

Other species-specific differences to consider are the variations in distribution of calbindin- and parvalbumin-positive cells across and within structures. The core and matrix dichotomy that appears in many major primate thalamic nuclei may not apply to all species, as the thalamic nuclei and their targeted projections are often composed of different cell types and occur in varied proportions. These cell types may not express calcium-binding proteins the same way across species, which could indicate that some species do not follow the generalized core and matrix projection patterns described by Jones (1998).

Outside of anatomical variations across species, to explain observed species-specific differences, it would be interesting to quantify if and how the behavioral relevance of a sensory modality influences the level of integration that occurs. For example, primates are vision-oriented and have more robust vision than rodents. Could this sensory-based difference result in greater integration of visual information into other sensory systems for primates when compared to rodents? How does a species' reliance on a specific sensory system influence the degree of multisensory integration observed within and across brain areas?

Multisensory effects can emerge from the convergence of 2 separate inputs from different modalities onto one structure that then responds to the bimodal presentation in a way that differs from a response to either unimodal input individually. These 2 separate inputs can both be driving inputs, which means that both inputs individually elicit responses in the targeted structure. Alternatively, one of the inputs can be driving while the other is a modulatory input. In this case, the former elicits a response by itself while the latter does not. As described earlier, examples of driving-driving convergence have been shown in rodents but are not yet widely observed in NHPs.

Combining electrophysiological recordings with neuromodulatory techniques and/or optogenetics is an ideal route for future studies intending to examine and document substrates for multisensory and sensorimotor integration. By incorporating these methods, some studies using mouse models have shown parvalbumin positive interneurons to form a direct, suppressive connection from superficial and deep layers of medial agranular motor cortex (M2) to auditory cortex. This connection is thought to be part of a circuit through which movement-related corollary discharge signals can be relayed to the auditory cortex (Schneider et al., 2014). Very few such observations have been made in NHPs, in part because the techniques that allow these discoveries in mice cannot yet be reliably applied in NHPs. Ultimately, the continued development and application of complex neuromodulatory techniques in other species, particularly in NHPs, are of incredible value since NHPs are the closest research model for humans. These methods can be used to address many of the open-ended questions posed above.

ACKNOWLEDGMENTS

A.B., M.N.O. and C.E.S. were supported by NIH / NIDCD grant number R01 DC012947 and NIH grant number R01 DC019979.

REFERENCES

Adams MM, Hof PR, Gattass R, Webster MJ, Ungerleider LG (2000). Visual cortical projections and chemoarchitecture of macaque monkey pulvinar. *J Comp Neurol.* 419(3):377–393.

Allen AE, Procyk CA, Brown TM, Lucas RJ (2017). Convergence of visual and whisker responses in the primary somatosensory thalamus (ventral posterior medial region) of the mouse. *J Physiol (Lond).* 595:865–881.

Avanzini G, Broggi G, Franceschetti S, Spreafico R (1980). Multisensory convergence and interaction in the pulvinar-lateralis posterior complex of the cat's thalamus. *Neurosci Lett.* 19:27–32.

Baier B, Kleinschmidt A, Müller NG (2006). Cross-modal processing in early visual and auditory cortices depends on expected statistical relationship of multisensory information. *J Neurosci.* 26:12260–12265.

Benevento LA, Rezak M (1976). The cortical projections of the inferior pulvinar and adjacent lateral pulvinar in the rhesus monkey (*Macaca mulatta*): An autoradiographic study. *Brain Res.* 108:1–24.

Benevento LA, Standage GP (1983). The organization of projections of the retinorecipient and nonretinorecipient nuclei of the pretectal complex and layers of the superior colliculus to the lateral pulvinar and medial pulvinar in the macaque monkey. *J Comp Neurol.* 217:307–336.

Bickford ME, Ramcharan E, Godwin DW, Erişir A, Gnadt J, Sherman SM (2000). Neurotransmitters contained in the subcortical extraretinal inputs to the monkey lateral geniculate nucleus. *J Comp Neurol.* 424:701–717.

Bieler M, Xu X, Marquardt A, Hanganu-Opatz IL (2018). Multisensory integration in rodent tactile but not visual thalamus. *Sci Rep.* 8:15684.

Briggs F, Usrey WM (2011). Corticogeniculate feedback and visual processing in the primate. *J Physiol (Lond).* 589:33–40.

Cappe C, Morel A, Barone P, Rouiller EM (2009a). The thalamocortical projection systems in primate: An anatomical support for multisensory and sensorimotor interplay. *Cereb Cortex.* 19:2025–2037.

Cappe C, Rouiller EM, Barone P (2009b). Multisensory anatomical pathways. *Hear Res.* 258:28–36.

Casagrande VA, Sáry G, Royal D, Ruiz O (2005). On the impact of attention and motor planning on the lateral geniculate nucleus. *Prog Brain Res.* 149:11–29.

Casagrande VA, Yazar F, Jones KD, Ding Y (2007). The morphology of the koniocellular axon pathway in the macaque monkey. *Cereb Cortex.* 17:2334–2345.

Chou X-L, Fang Q, Yan L, Zhong W, Peng B, Li H, Wei J, Tao HW, Zhang LI (2020). Contextual and cross-modality modulation of auditory cortical processing through pulvinar mediated suppression. *eLife.* 9:e54157.

Cooper MH, Young PA (1976). Cortical projections to the inferior colliculus of the cat. *Exp Neurol.* 51:488–502.

Donishi T, Kimura A, Imbe H, Yokoi I, Kaneoke Y (2011). Sub-threshold cross-modal sensory interaction in the thalamus: Lemniscal auditory response in the medial geniculate nucleus is modulated by somatosensory stimulation. *Neuroscience.* 174:200–215.

Fiebelkorn IC, Pinsk MA, Kastner S (2019). The mediodorsal pulvinar coordinates the macaque fronto-parietal network during rhythmic spatial attention. *Nat Commun.* 10:215.

Fries P (2005). A mechanism for cognitive dynamics: neuronal communication through neuronal coherence. *Trends Cogn Sci (Regul Ed).* 9:474–480.

Fries W (1981). The projection from the lateral geniculate nucleus to the prestriate cortex of the macaque monkey. *Proc R Soc Lond B Biol Sci.* 213:73–86.

Gattass R, Sousa AP, Oswaldo-Cruz E (1978). Single unit response types in the pulvinar of the Cebus monkey to multisensory stimulation. *Brain Res.* 158:75–87.

Ghazanfar AA, Schroeder CE (2006). Is neocortex essentially multisensory? *Trends Cogn Sci (Regul Ed).* 10:278–285.

Gutierrez C, Cola MG, Seltzer B, Cusick C (2000). Neurochemical and connectional organization of the dorsal pulvinar complex in monkeys. *J Comp Neurol.* 419:61–86.

Gutierrez C, Yaun A, Cusick CG (1995). Neurochemical subdivisions of the inferior pulvinar in macaque monkeys. *J Comp Neurol.* 363:545–562.

Hackett TA (2011). Information flow in the auditory cortical network. *Hear Res.* 271:133–146.

Hyvärinen J (1982). Posterior parietal lobe of the primate brain. *Physiol Rev.* 62:1060–1129.

Jones EG (1998). Viewpoint: The core and matrix of thalamic organization. *Neuroscience.* 85:331–345.

Jones EG (2001). The thalamic matrix and thalamocortical synchrony. *Trends Neurosci.* 24:595–601.

Jones EG, Hendry SH, Brandon C (1986). Cytochrome oxidase staining reveals functional organization of monkey somatosensory thalamus. *Exp Brain Res.* 62:438–442.

Kaas JH, Lyon DC (2007). Pulvinar contributions to the dorsal and ventral streams of visual processing in primates. *Brain Res Rev.* 55:285–296.

Khorevin VI (1978). Responses of units in the magnocellular part of the medial geniculate body to acoustic and somatosensory stimulation. *Neurophysiology.* 10:89–94.

Kimura A, Imbe H (2018). Robust subthreshold cross-modal modulation of auditory response by cutaneous electrical stimulation in first- and higher-order auditory thalamic nuclei. *Neuroscience.* 372:161–180.

Komura Y, Tamura R, Uwano T, Nishijo H, Ono T (2005). Auditory thalamus integrates visual inputs into behavioral gains. *Nat Neurosci.* 8:1203–1209.

Lakatos P, Chen CM, O'Connell MN, Mills A, Schroeder CE (2007). Neuronal oscillations and multisensory interaction in primary auditory cortex. *Neuron.* 53:279–292.

Lakatos P, O'Connell MN, Barczak A, Mills A, Javitt DC, Schroeder CE (2009). The leading sense: Supramodal control of neurophysiological context by attention. *Neuron.* 64:419–430.

Leszczynski M, Chaieb L, Staudigl T, Enkirch SJ, Fell J, Schroeder CE (2021). Neural activity in the human anterior thalamus during natural vision. *Sci Rep.* 11:17480.

Lohse M, Dahmen JC, Bajo VM, King AJ (2021). Subcortical circuits mediate communication between primary sensory cortical areas in mice. *Nat Commun.* 12:3916.

Maier JX, Chandrasekaran C, Ghazanfar AA (2008). Integration of bimodal looming signals through neuronal coherence in the temporal lobe. *Curr Biol.* 18:963–968.

Mégevand P, Mercier MR, Groppe DM, Zion Golumbic E, Mesgarani N, Beauchamp MS, Schroeder CE, Mehta AD (2020). Crossmodal phase reset and evoked responses provide complementary mechanisms for the influence of visual speech in auditory cortex. *J Neurosci.* 40:8530–8542.

Meng H, Angelaki DE (2010). Responses of ventral posterior thalamus neurons to three-dimensional vestibular and optic flow stimulation. *J Neurophysiol.* 103:817–826.

Meredith MA (2002). On the neuronal basis for multisensory convergence: A brief overview. *Brain Res Cogn Brain Res.* 14:31–40.

Molinari M, Dell'Anna ME, Rausell E, Leggio MG, Hashikawa T, Jones EG (1995). Auditory thalamocortical pathways defined in monkeys by calcium-binding protein immunoreactivity. *J Comp Neurol.* 362:171–194.

Montero VM, Zempel J (1986). The proportion and size of GABA-immunoreactive neurons in the magnocellular and parvocellular layers of the lateral geniculate nucleus of the rhesus monkey. *Exp Brain Res.* 62:215–223.

Morel A, Liu J, Wannier T, Jeanmonod D, Rouiller EM (2005). Divergence and convergence of thalamocortical projections to premotor and supplementary motor cortex: A multiple tracing study in the macaque monkey. *Eur J Neurosci.* 21:1007–1029.

Mo C, Sherman SM (2019). A sensorimotor pathway via higher-order thalamus. *J Neurosci.* 39:692–704.

Nguyen MN, Hori E, Matsumoto J, Tran AH, Ono T, Nishijo H (2013). Neuronal responses to face-like stimuli in the monkey pulvinar. *Eur J Neurosci.* 37:35–51.

Noesselt T, Tyll S, Boehler CN, Budinger E, Heinze H-J, Driver J (2010). Sound-induced enhancement of low-intensity vision: multisensory influences on human sensory-specific cortices and thalamic bodies relate to perceptual enhancement of visual detection sensitivity. *J Neurosci.* 30:13609–13623.

Rausell E, Bae CS, Viñuela A, Huntley GW, Jones EG (1992). Calbindin and parvalbumin cells in monkey VPL thalamic nucleus: Distribution, laminar cortical projections, and relations to spinothalamic terminations. *J Neurosci.* 12:4088–4111.

Robinson DL, McClurkin JW, Kertzman C, Petersen SE (1991). Visual responses of pulvinar and collicular neurons during eye movements of awake, trained macaques. *J Neurophysiol.* 66:485–496.

Ross LA, Molholm S, Butler JS, Bene VAD, Foxe JJ (2022). Neural correlates of multisensory enhancement in audiovisual narrative speech perception: A fMRI investigation. *NeuroImage.* 263:119598.

Royal DW, Sáry G, Schall JD, Casagrande VA (2006). Correlates of motor planning and postsaccadic fixation in the macaque monkey lateral geniculate nucleus. *Exp Brain Res.* 168:62–75.

Saalmann YB, Pinsk MA, Wang L, Li X, Kastner S (2012). The pulvinar regulates information transmission between cortical areas based on attention demands. *Science*. 337:753–756.

Schneider DM, Nelson A, Mooney R (2014). A synaptic and circuit basis for corollary discharge in the auditory cortex. *Nature*. 513:189–194.

Schroeder CE, Lakatos P (2009). Low-frequency neuronal oscillations as instruments of sensory selection. *Trends Neurosci*. 32:9–18.

Schroeder CE, Wilson DA, Radman T, Scharfman H, Lakatos P (2010). Dynamics of active sensing and perceptual selection. *Curr Opin Neurobiol*. 20:172–176.

Sherman SM, Guillery RW (2002). The role of the thalamus in the flow of information to the cortex. *Philos Trans R Soc Lond B Biol Sci*. 357:1695–1708.

Sherman SM, Guillery RW (2011). Distinct functions for direct and transthalamic corticocortical connections. *J Neurophysiol*. 106:1068–1077.

Shipp S (2003). The functional logic of cortico-pulvinar connections. *Philos Trans R Soc Lond B Biol Sci*. 358:1605–1624.

Sincich LC, Park KF, Wohlgemuth MJ, Horton JC (2004). Bypassing V1: A direct geniculate input to area MT. *Nat Neurosci*. 7:1123–1128.

Smiley JF, Falchier A (2009). Multisensory connections of monkey auditory cerebral cortex. *Hear Res*. 258:37–46.

Sylvester R, Haynes J-D, Rees G (2005). Saccades differentially modulate human LGN and V1 responses in the presence and absence of visual stimulation. *Curr Biol*. 15:37–41.

Sylvester R, Rees G (2006). Extraretinal saccadic signals in human LGN and early retinotopic cortex. *NeuroImage*. 30:214–219.

Van Le Q, Isbell LA, Matsumoto J, Nguyen M, Hori E, Maior RS, Tomaz C, Tran AH, Ono T, Nishijo H (2013). Pulvinar neurons reveal neurobiological evidence of past selection for rapid detection of snakes. *Proc Natl Acad Sci U S A*. 110:19000–19005.

Vittek A-L, Juan C, Nowak LG, Girard P, Cappe C (2022). Multisensory integration in neurons of the medial pulvinar of macaque monkey. *Cereb Cortex*. 33(8):4202–4215.

Warner CE, Goldshmit Y, Bourne JA (2010). Retinal afferents synapse with relay cells targeting the middle temporal area in the pulvinar and lateral geniculate nuclei. *Front Neuroanat*. 12(4):8.

Wepsic JG (1966). Multimodal sensory activation of cells in the magnocellular medial geniculate nucleus. *Exp Neurol*. 15:299–318.

Williamson RS, Hancock KE, Shinn-Cunningham BG, Polley DB (2015). Locomotion and task demands differentially modulate thalamic audiovisual processing during active search. *Curr Biol*. 25:1885–1891.

Wurtz RH, McAlonan K, Cavanaugh J, Berman RA (2011). Thalamic pathways for active vision. *Trends Cogn Sci (Regul Ed)*. 15:177–184.

Yirmiya R, Hocherman S (1987). Auditory- and movement-related neural activity interact in the pulvinar of the behaving rhesus monkey. *Brain Res*. 402:93–102.

Zhou H, Schafer RJ, Desimone R (2016). Pulvinar-cortex interactions in vision and attention. *Neuron*. 89:209–220.

Zhou NA, Maire PS, Masterson SP, Bickford ME (2017). The mouse pulvinar nucleus: Organization of the tectorecipient zones. *Vis Neurosci*. 34:E011.

SECTION VI

MOTOR SYSTEMS

Adam W. Hantman and Kevin P. Cross

Mammalian motor systems produce impeccable control of a nearly endless array of actions. Such command arises by integrating bottom-up sensory feedback about the body and the world with top-down processing of desired goals. Examining individual elements of the motor system highlights that actions are not generated by any given region, but rather through numerous nested feedback loops, distributed at every level of the nervous system, including spinal cord, brainstem, and cortex. In this section we focus on the loop between motor cortex and thalamus because of its position in the middle of many of these critical feedback loops. We begin with theories of thalamic and cortical function that serve as frameworks to help maintain consideration of the entire system. Then, we move through the circuits, providing details of motor thalamus and its connections.

What role might the thalamus play, particularly considering its relatively small size compared to the cortex? Chapter 31 tackles this question by exploring models for describing interactions between the thalamus and cortex. First considered are feedforward models which interpret the thalamus as a relay for subcortical information with limited contribution to neural computation. The second class of models considered are feedback models which highlight the bidirectional projections of thalamus and cortex that endows the thalamus with a role in the moment-to-moment generation of the network's dynamic. Such models are exciting new ideas for understanding corticothalamic interactions and shed light on the often overlooked corticothalamic projection.

Motor control requires the integration of bottom-up and top-down inputs. The thalamocortical loop is in a privileged position for such integration since it is a site of convergence for a broad range of inputs from many cortical and subcortical areas. Chapters 32, 33, and 34 cover how sensory-processing areas, including somatosensory cortex and cerebellum, have direct projections to this loop, with the former targeting the motor cortex and the latter targeting the thalamus. Similarly, areas implicated in higher-order processing, including frontal and parietal cortices, also project to this loop by targeting the motor cortex. Chapter 33 highlights some of the key elements of the thalamocortical loop for processing these inputs, including electrophysiological properties of the different cell types in the network, the local connections with each area, and bidirectional projections that couple the motor cortex with the thalamus.

Chapter 34 describes the outputs of the thalamocortical loop. As a prominent driver of motor behaviors, the motor cortex projects to spinal circuits to drive muscle activity along with subcortical structures implicated in low-level control of the body, including the basal ganglia, red nuclei, and reticular nuclei. Also highlighted are projections to other cortical areas, including sensory and higher-order regions, many of which also send reciprocal projections back to motor cortex. Why are projections sent to sensory and higher-order areas? The oculomotor system, as discussed in Chapter 35, might provide a clue. The superior colliculus is a motor structure that sends projections to brainstem nuclei that control saccadic eye movements. However, the superior colliculus also sends projections to various cortical areas through the thalamus. These pathways allow the superior colliculus to modulate and remap the visual responses of cortical neurons depending on upcoming behavior.

Cortical function has recently been considered to be the product of low-dimensional dynamics on which computation emerges. However, it is unclear how thalamus contributes to these cortical dynamics. Chapter 36 highlights how recurrent loops between thalamus and cortex support motor planning of an upcoming motor task. They also describe a pathway from the midbrain to thalamus that initiates the state transition from planning to movement in the thalamocortical loop. Similar observations are highlighted in Chapter 32, which discusses the role of the cerebello-thalamo-cortical pathway in initiating and timing voluntary movements.

While considerable efforts have focused on circuits involving motor thalamus and cortex, much work remains. Motor thalamus is not a simple feedforward transmission, a homogenous set of neurons transferring single modes of information to particular regions of cortex. An appreciation is building that thalamic neurons have a spectrum of intrinsic properties and receive convergent inputs from a diverse set of brain regions. Thus, motor thalamus can be considered as a filter bank engaged in multi-modal integration. Such a conceptualization motivates future efforts to study the computations performed by the input/output transformations of motor thalamus. Modern theories give special consideration to the cortical inputs to motor thalamus since they could provide top-down control. We are excited to see these ideas extended to determine how cortical inputs influence the integration and transmission of bottom-up subcortical information. Population dynamics along these circuits are being revealed by combining simultaneous recordings and perturbations. Such studies force us to redefine the boundaries of relevant dynamical systems, pulling us away from considering brain regions in isolation. Eventually, these efforts might unmoor us from a reductionist approach for understanding motor systems and allow attention to the major modes of brain-wide dynamics in the orchestration of action.

31.

THALAMIC CONTROL OF MOTOR CORTICAL SYNCHRONY

Abdulraheem Nashef and Yifat Prut

The motor thalamus interfaces between major subcortical systems, the cerebellum and the basal ganglia, and cortical areas in the frontal lobe (Kunzle, 1976; Jurgens, 1984; Wiesendanger and Wiesendanger, 1985). Among these subcortical systems, the cerebellum is unique in providing an effective excitatory drive to the motor thalamus, which in turn sends thalamocortical efferents to innervate large areas in the motor cortex. But the motor thalamus is much more than a simple relay station and can modify the transmitted cerebellar signals in a state-dependent manner to affect motor behavior. The extent and nature of this modulation are largely determined by (1) the termination pattern of cerebellar axons within the motor thalamus; (2) the structure and properties of the targeted thalamic circuitry; and (3) the termination pattern of thalamic efferents in the motor cortex. Although many studies have explored the role of the cerebellum in regulating motor control and motor adaptation, the thalamic contribution to this process is not well understood. This chapter reviews current knowledge on the structural organization and functional properties of the cerebellar-thalamocortical system and its role in regulating voluntary movements. It focuses on the ways the cerebellar-recipient motor thalamus modifies the motor cortical integration of cerebellar signals through its effects on the coordinated activity of the motor cortical networks.

STRUCTURAL ORGANIZATION OF THE CEREBELLAR-THALAMOCORTICAL SYSTEM

More than 70% of the neurons in the central nervous system of human and nonhuman primates (NHPs) belong to the cerebellar network (Herculano-Houzel et al., 2007; Azevedo et al., 2009). In humans and NHPs, the major output of this network is its ascending cerebello-thalamo-cortical (CTC) pathway, which is larger than the descending cerebello-spinal system (Armstrong, 1980; Matano and Hirasaki, 1997). The CTC originates from the deep cerebellar nuclei (DCN)—mainly the dentate and interpositus (Orioli and Strick, 1989). Axons from these nuclei form the massive brachium conjunctivum, also known as the superior cerebellar peduncle (SCP), which is further divided into descending and ascending tracts, with the ascending tract reaching specific thalamic nuclei in the posterior part of the ventral lateral thalamus (Ilinsky and Kultas-Ilinsky, 1987; Hirai and Jones, 1989).

The cerebellar receiving thalamic nuclei include several nuclei such as the VPLo, VLc, and area X (Horne and Butler, 1995). Cerebellar inputs are also found in intralaminar thalamic nuclei (Gornati et al., 2018), but in the context of motor control in humans and NHPs, this input is likely of lesser importance (Percheron et al., 1996) and will not be further discussed here. Thalamic nuclei are often divided into "drivers" and "modulators" which have different anatomical and electrophysiological properties (Sherman and Guillery, 1998). Within this scheme, the cerebellar thalamus can be viewed as a "driver," whereas the basal-ganglia receiving thalamus can be considered a modulator (Gornati et al., 2018).

The cerebellar thalamus is somatotopically organized, with an over-representation of proximal (shoulder and elbow) than distal joints (Vitek et al., 1994). Synaptic contacts made by a single DCN axon are broadly distributed across multiple regions in the cerebellar thalamus (Mason et al., 2000) and create structurally large synaptic inputs concentrated around the soma of targeted thalamic cells (Aumann et al., 1994). The efficacy of these synapses is particularly high. Studying these synapses in cats revealed an average excitatory postsynaptic potential (EPSP) amplitude of 1.5–1.7 mv, for axons originating from the dentate or interpositus cerebellar nuclei (Shinoda et al., 1985), thus constituting a highly secured excitatory pathway. From the thalamus, thalamocortical output axons ascend to make excitatory contact with cortical neurons in layers 2/3 and 5 of the frontal lobe (Shinoda and Kakei, 1989). In parallel, the cerebellar thalamus receives feedback inputs from layers 5

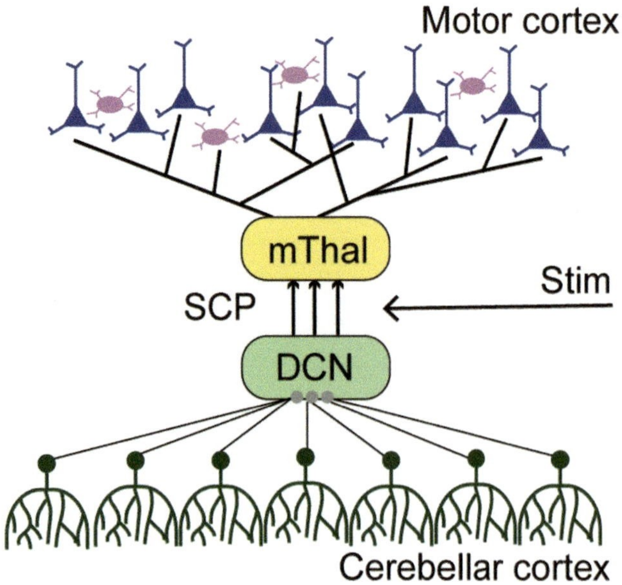

Figure 31.1. Functional organization of information flow in the CTC system. Illustration of the structure of the CTC system. Purkinje cells (cerebellar cortex) converge via inhibitory synapses onto the deep cerebellar nuclei (DCN) at a 40:1 convergence ratio. Axons from the deep cerebellar nuclei send axons via the superior cerebellar peduncle (SCP) to create powerful and extensive excitatory synaptic contacts with neurons in the cerebellar thalamus (mThal). From the motor thalamus, excitatory thalamocortical efferents innervate pyramidal cells and interneurons throughout the motor cortex. The SCP is a target for stimulation (Stim) which activates the thalamocortical portion of the CTC.

and 6 in primary motor cortex (M1) (Shinoda and Kakei, 1989; Kultas-Ilinsky et al., 2003). However, unlike the strong synaptic drive from the DCN, motor cortical feedback to the thalamus mostly terminates on distal dendrites (Galvan et al., 2016). This asymmetric feedforward/feedback organization indicates that the CTC pathway is a primarily feedforward pathway transmitting excitatory signals from the cerebellum to the motor cortex, whereas motor feedback probably modulates information flow through this pathway.

The detailed termination pattern of thalamocortical fibers throughout the motor cortex dictates which motor processes (planning vs. execution) will be the most affected by cerebellar signals and in what way. In rodents (Kuramoto et al., 2015) thalamocortical fibers from the motor thalamus create a set of termination patches along axons which can extend several millimeters along the rostro-caudal axis. This means that while locally the effect of motor TC fibers on cortical cells is probably weaker, this system has a broader effect on remote cortical sites. This change in organizational scheme was interpreted as an indication of the role of thalamic input in coordinating multiple effectors, but it may also coordinate neurons across motor cortical regions and layers. In terms of targeting sites, there is a substantial projection of cerebellar signals to M1 (Hoover and Strick, 1999). However, the majority of the cerebellar output is directed elsewhere, including to the ventral premotor cortex (Orioli and Strick, 1989), SMA (Matelli and Luppino, 1996; Akkal et al., 2007), and other areas where it can affect cognitive functions (Kelly and Strick, 2003). These anatomical findings suggest that unlike the classical view which considered that the cerebellum was mostly involved in motor execution, cerebellar signals can contribute to high-level functions including motor planning and preparation.

One of the most significant structural aspects of the CTC system is the substantial amount of funneling which takes place (Figure 31.1), starting at the vast number of neurons in the cerebellar cortex that converge onto the deep cerebellar nuclei at a ratio of ~40:1 (Person and Raman, 2011) and then go through the SCP, which probably contributes to further funneling (Yamaguchi et al., 1989), to the cerebellar thalamus, where massive convergence onto single thalamic cells has been observed (Rispal-Padel et al., 1987), such that a single thalamic cell can receive input from different deep cerebellar nuclei (Shinoda et al., 1993). The cerebellar receiving thalamus acts as the bottleneck for this funnel, since the thalamocortical innervations of motor cortical sites exhibit an extensive divergence pattern. This organization scheme may also be common to other thalamic motor nuclei such as the mediodorsal thalamus (see **Chapter 46 in Section VIII**). Thus, the motor thalamus may play a key role in integrating and processing the parsimonious information arriving from the cerebellum via the SCP before transmitting it to the motor cortex. The net outcome of this divergent/convergent organization is that multiple cerebellar sites can affect a single cortical region, and that a single cerebellar site may influence remote cortical sites.

EVOLUTIONARY CHANGES IN THE MOTOR THALAMUS AND THE CTC SYSTEM

The CTC system underwent several significant evolutionary changes in terms of its organization pattern. The first was an increase in the ascending part of the SCP that targets the motor thalamus compared to the descending cerebellar pathways (Armstrong, 1980). This evolutionary change was accompanied by a decrease in the size of the magnocellular division of the red nucleus, which is the origin of the rubrospinal system (Nathan and Smith, 1982). Together, these changes highlight an evolutionary shift in which the cerebellar control of motor output has been mediated through its influence on motor cortical activity (Horne and Butler, 1995). Consistent with this transition, we found that stimulating the SCP in NHPs produced broad and very effective activation of motor cortical neurons (in both the primary and premotor cortex), but only produced sparse activation of neurons in the cervical spinal cord (Cohen et al., 2017). The second evolutionary change in the motor thalamocortical system is the segregation between the cerebellar-recipient and basal ganglia–recipient thalamus. Whereas in rodents the terminations of these two systems overlap within what is termed the VA-VL complex, in NHPs the thalamic terminations of these two systems are anatomically segregated (Percheron et al., 1996), and are characterized by distinct histochemical properties (Calzavara et al., 2005). Although anatomical studies showed that cells in these two nuclei may receive synaptic contacts from both systems (Sakai et al., 1996; Bosch-Bouju et al., 2013), the functional correlate of this anatomical overlap appears to be small (Anderson and Turner, 1991), and only a few thalamic neurons have been found to respond to inputs from both structures. The functional implications of the cerebellum-to-basal-ganglia segregation at the subcortical and thalamic levels in the context of voluntary movements are still not fully understood. In addition, recent studies have reported subcortical indirect interactions between the two systems which enable them to share state-related information such as reward (Carta et al., 2019; Baek et al., 2022), thus suggesting that they may act as nodes in a coordinated network (Bostan and Strick, 2018).

PROPERTIES OF THALAMIC AND CORTICAL NEURONS WHICH ARE PART OF THE CTC SYSTEM

To study the CTC system in behaving primates we (Nashef et al., 2018) chronically implanted a stimulating electrode in the SCP of NHPs, ipsilateral to the working hand. This electrode was used to activate the CTC pathway, using a single pulse stimulation protocol, to identify cortical and thalamic sites that are part of the CTC system. In this manner it was possible to identify thalamic and cortical cells as part of the CTC system, and study their task-related properties. Thalamic sites were identified using MRI scans, motor responses to intra-thalamic stimulation bursts and short-latency responses to SCP stimulation. The responses of thalamic neurons to SCP stimulation exhibited short latency (Figure 31.2A) compared to motor cortical cells. During task performance, the thalamic cells had a high background rate, earlier movement-related activity, and less sharp directional tuning compared to motor cortical cells (Figure 31.2B–C). In this respect, these cells are similar to the early onset cells recorded in the motor thalamus of rats (Gaidica et al., 2018). In the motor cortex, CTC cells were shown to be distributed throughout the motor and premotor cortices in similar proportions (Nashef et al., 2018). Here again the oculomotor thalamus behaves similarly projecting to both the frontal eye field and the supplemental eye field (see **Chapter 36 in Section VI**). The typical response pattern involved early excitation followed by inhibition which was long and more pronounced compared to the post-excitation inhibition found in cortico-cortical interactions (Kraskov et al., 2011). The response patterns exhibited certain area-specific properties (Figure 31.3A–C), where the peak time of the premotor response was slightly later and the post-excitation suppression was shorter than the M1 response. This suggests that the SCP response may have triggered a wave of excitation which propagated rostrally from M1 toward the PM. The post-excitation suppression found in M1 had a larger

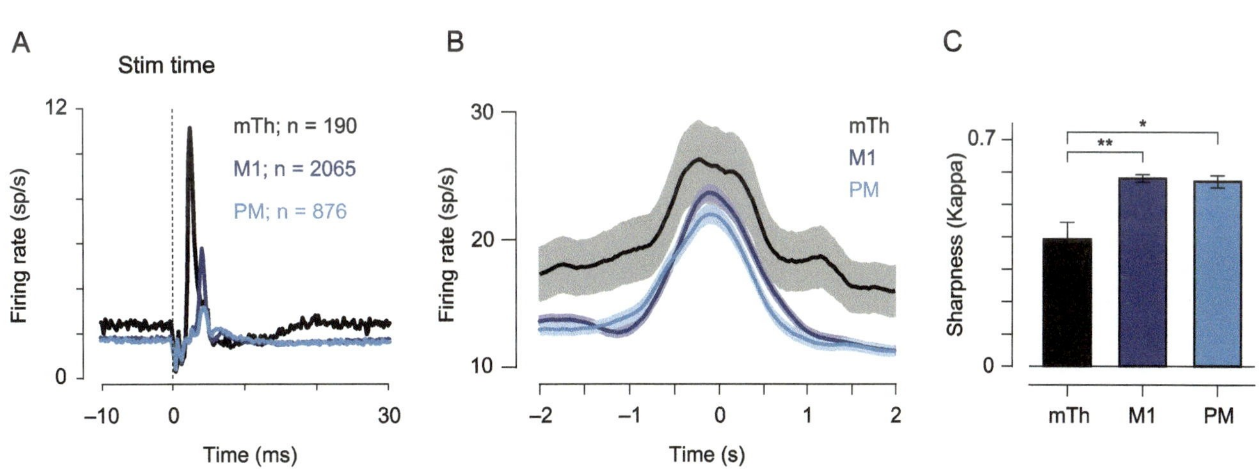

Figure 31.2. Task-related properties of neurons in the motor thalamus and motor cortex. (A). Mean response of thalamic (black), primary motor (M1 = blue) and premotor (PM = light blue) to single-pulse SCP stimulation. Responses are aligned on stimulation time (time = 0). (B). The mean firing rate cells recorded from the motor thalamus (black, *n* = 78 cells), M1 (blue, *n* = 1734 cells) and PM (light blue, *n* = 846 cells), aligned on movement onset. Activity was computed around the preferred target (for tuned cells) or all targets (for untuned cells). (C). The concentration parameter (kappa) which captures the width of the tuning function was calculated from the Von Mises fit, computed for all tuned cells. The kappa was correlated with the width of the tuning curve such that higher kappa values correspond to narrower tuning curves.

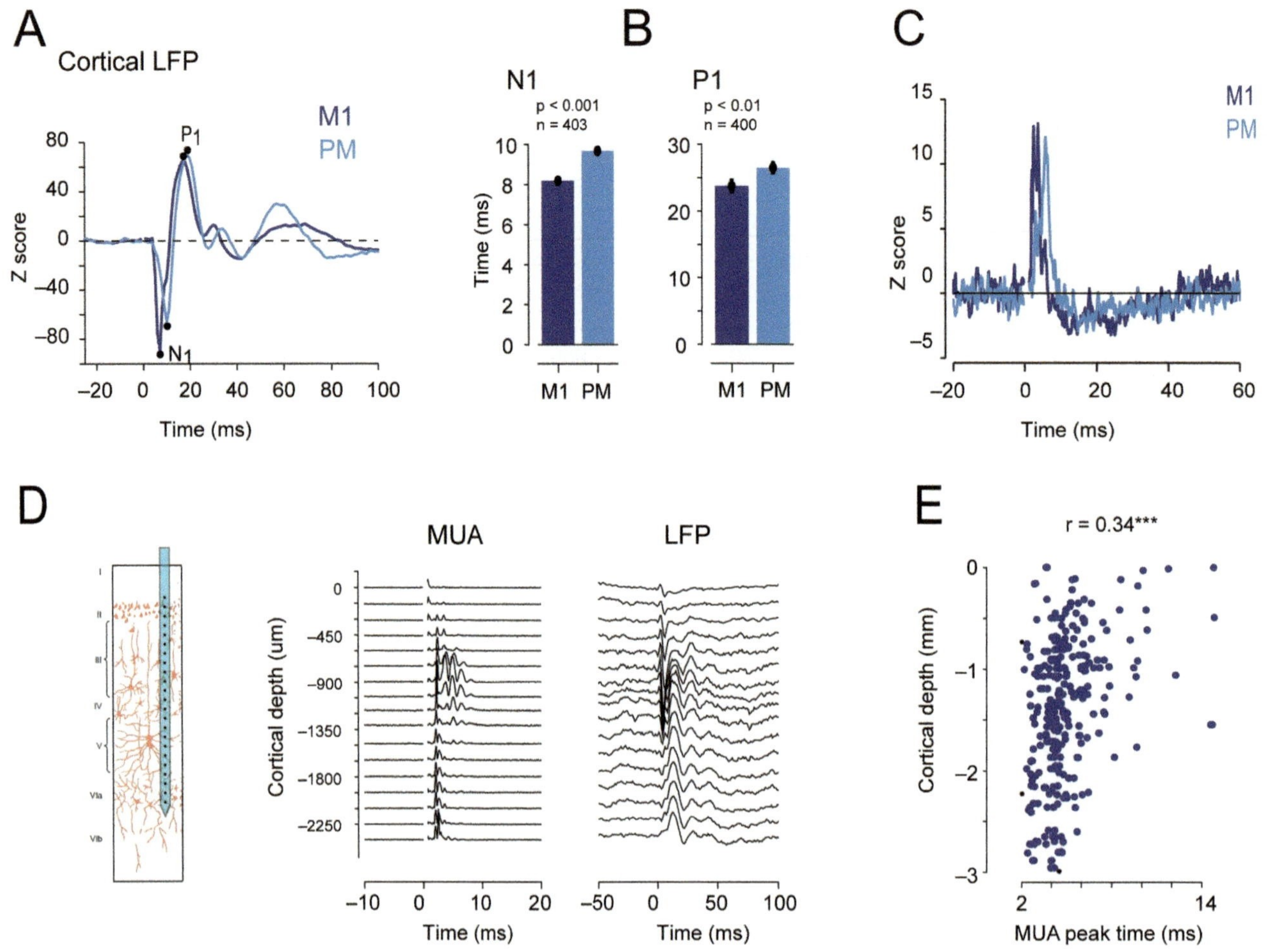

Figure 31.3. Cortical termination pattern of the CTC system. (A). Example of the mean normalized response of cortical LFP recorded simultaneously in M1 and PM sites around the time of SCP stimulation. The evoked response was characterized by early negativity (N1) followed by a positive peak (P1). (B). Comparing the timing of N1 and P1 obtained for M1 and PM sites revealed significant earlier timing in M1 sites (t-test) for both N1 and P1. (C). Example of multi-unit response to SCP stimulation obtained from M1 and PM sites recorded simultaneously. Here again, the M1 response occurred slightly earlier than the PM response. This finding was consistent across M1 and PM sites. (D). Example of a depth-dependent M1 response to SCP stimulation. Recordings were made using a linear probe with 32 sites. Only the bottom sites where spiking activity was found are shown. Mean response of multiunit activity (MUA: left) and local field potential (LFP: right) showed depth-dependent modulation in the timing and pattern of the evoked responses. (E). Depth-dependent response modulation was quantified by measuring the peak time of the multi-unit responses recorded at different cortical depths (here using single electrode recordings). Peak time was correlated with cortical depth such that the responses obtained from deeper sites had an earlier peak than responses recorded from more superficial sites (correlation coefficient $r = .34$, $p < 0.001$).

amplitude and longer duration, which could prevent the stimulus-triggered volley in premotor sites from back-feeding M1 sites. In addition to these area-related modulations, the response of single cells was depth-dependent (Figure 31.3D–E), such that earlier responses occurred in neurons located in deeper layers, which is consistent with previous findings (Strick and Sterling, 1974; Shinoda et al., 1993).

BEHAVIORAL AND NEURAL DEFICITS IN THE ABSENCE OF CEREBELLAR SIGNALS

It has long been suggested that the CTC plays a role in the timing of motor tasks, primarily because pathologies involving the cerebellum, such as cerebellar ataxia, are characterized by deficits in the temporal patterns of motor behavior (Bastian et al., 1996). Computational models, supported by a large amount of anatomical and physiological data, suggest that the cerebellar role primarily consists of producing temporal patterns or temporally precise feed-forward control signals (Horne and Butler, 1995). It was posited that synchronous pauses in cerebellar nuclei may increase the subsequent sensitivity of thalamic cells so that time-related information is forwarded to the motor cortex (Sawyer et al., 1994; Schafer et al., 2021). However, the ways in which these temporal signals actually affect the timing of their target elements (the primary motor cortex and motor effectors) is still unclear. Moreover, the role of the cerebellar thalamus as a processing element in the CTC pathway is often overlooked. Different methods of transient reversible inactivation have been successfully employed to study the specific contribution of different parts of the CTC without the confounding effect of long-term compensatory mechanisms. Early studies showed that cooling the dentate nucleus leads to motor deficits that included increase in the reaction time and the impaired recruitment of muscle synergies (Meyer-Lohmann et al., 1975; Hore and Flament, 1988). These studies also revealed that along with the behavioral deficits, dentate cooling also modified the activity of motor cortical cells, leading to a loss of transient firing at movement onset and a disruption of inhibition in task-related cells and muscles. More recently, optogenetic inactivation techniques have been used to identify the contribution of the CTC system in mice. Silencing the motor thalamus during reach was shown to arrest movements, thus arguing that the motor cortex is driven by input even after movement starts (Sauerbrei et al., 2020). This is a radically different view from previous models which emphasized the role of motor cortical activity in driving movements. In another study, manipulating information flow in the pathway using activation or inactivation was shown to modulate movement initiation (Dacre et al., 2021). Interfering with cerebellar output as early as the preparatory period also interfered with the performance of a delayed response task (Gao et al., 2018). Together, these studies indicate a significant role for cerebellar inputs in planning, initiating, and executing voluntary movements in rodents. The disadvantage of pharmacological, thermal, or optogenetic inactivation techniques is that often they cannot identify which cortical or thalamic cells are part of the CTC system during normal conditions, when the system is intact. In addition, the genetic toolbox developed for rodents is by and large unavailable for NHP studies.

To overcome these obstacles, we took advantage of the chronic SCP electrode to deliver high-frequency stimulation (HFS) while monkeys performed an upper-limb reaching task (Nashef et al., 2019). We found that increasing the frequency of SCP stimulation to 130 Hz resulted in essentially a complete loss of neuronal response to the stimulation pulses (Figure 31.4A–B). The time course of the response suppression was relatively fast (within 3–4 seconds), which allowed for a direct comparison of neural activity and motor behavior before, during, and after the perturbation. The mechanism leading this response suppression is not fully understood. One study reported reduced synaptic transmission during repetitive cerebellar stimulation in neurons in the cerebellar thalamus, consistent with their "driver"-like properties (Schafer et al., 2021). However, this effect may not account for the full reduction in the information flow observed during the long-lasting HFS reported in our study. Applying HFS when monkeys performed an upper-limb reaching task led to substantial motor deficits which were consistent with findings in ataxic patients (Figure 31.4C). These included longer reaction times, reduced movement velocity, increased curvature of the performed trajectories, and the decomposition of multijoint movements, as evidenced by the increase in the relative shoulder-to-elbow onset latencies (Nashef et al., 2019). One common symptom that was not replicated is tremor. This may be related to the fact that in the task design, the monkey's arm was supported, or that tremor is often a progressive outcome of cerebellar ataxia. In addition to the behavioral effects, we found that the HFS had a substantial effect on task-related activity in both motor thalamus and motor cortex (Figure 31.4D). The task-related activity of thalamic cells was almost completely abolished during this perturbation, indicating that the manipulation was effective in blocking the information flow through the CTC. During HFS, motor cortical cells lost their transient firing at movement onset, but their directional tuning remained the same (Figure 31.4E), suggesting that the CTC system plays a role in shaping temporal but not spatial tuning of motor cortical cells.

THE ROLE OF INHIBITION IN THE CTC PATHWAY

The CTC system is strictly excitatory with glutamatergic synaptic contacts (DCN-to-thalamus and thalamus-to-cortex). However, there are several mechanisms through which inhibition can modulate information flow through the pathway (Figure 31.5). First, intra-thalamic interneurons which are found in the thalamic nuclei of humans and NHPs (Arcelli et al., 1997) can affect neighboring relay cells. Another source of inhibitory control of the motor thalamus is the thalamic reticular nucleus (TRN), which receives collaterals from each incoming and outgoing axons of the thalamus (Jones, 2002). The TRN, in turn, inhibits thalamic neurons in a feedforward and feedback way (Sawyer et al., 1994). TRN neurons are interconnected, which may suggest that the TRN operates as an integrated network that modulates thalamic activity in a broad manner leading to a cross-modal interaction of information. Hence, neurons in the cerebellar thalamus can be affected by signals transmitted in the neighboring basal ganglia and/or somatosensory thalamic nuclei.

The interactions between thalamic axons and cortical circuitry constitute another node for the inhibitory modulation of information flow through the CTC system (Okoro et al., 2022). Studies of the sensory TC system in rodents have revealed a feedforward inhibitory pathway (FFI) in which thalamic axons make direct contact with cortical parvalbumin (PV) interneurons (Gabernet et al.,

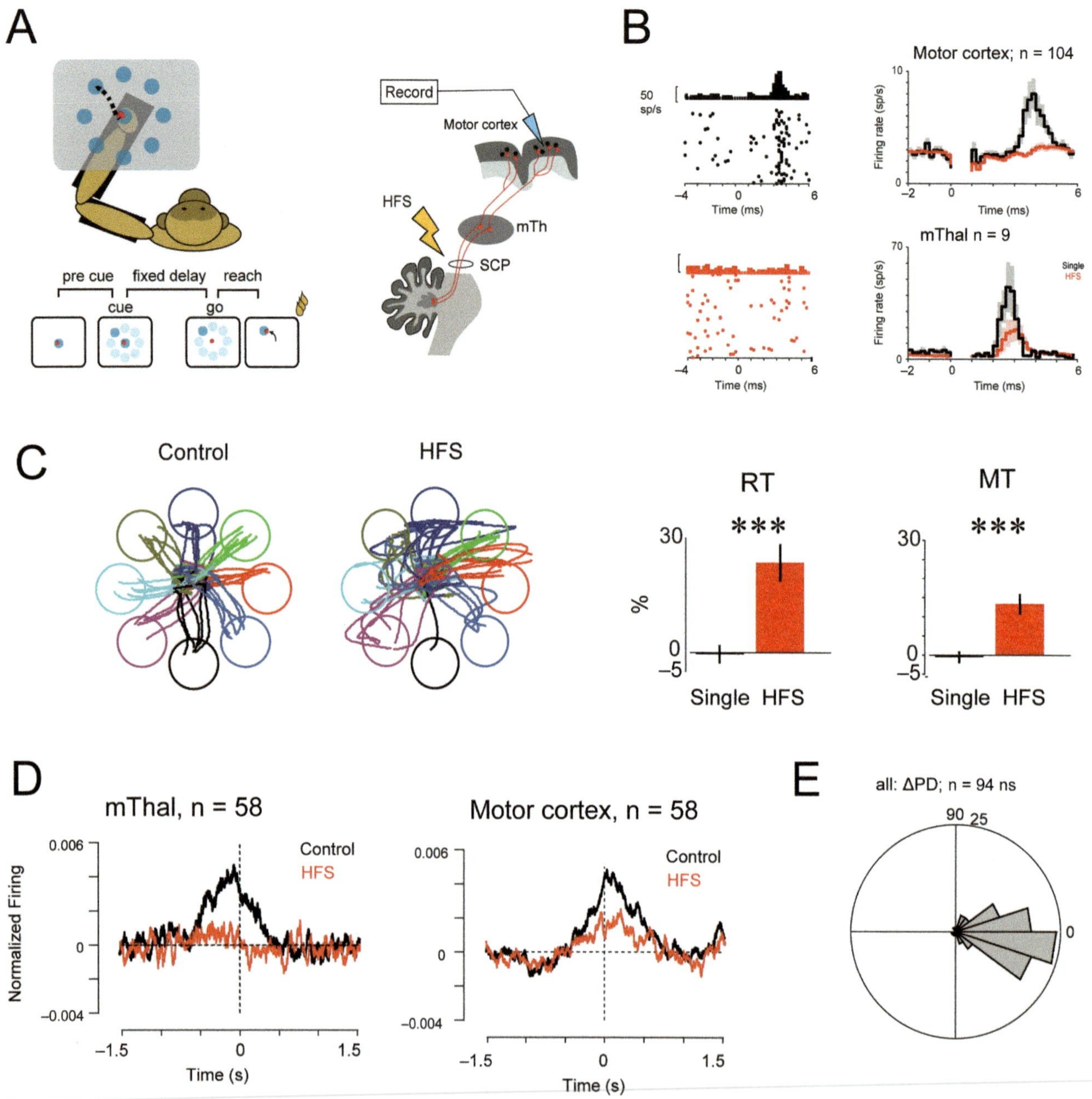

Figure 31.4. Blocking information flow through the CTC using high-frequency stimulation. (A). Experimental paradigm. Monkeys were trained to make center-out reaching movements in a delayed-response paradigm. A stimulating electrode was chronically implanted in the SCP and recordings were made from motor cortical sites. High-frequency stimulation (HFS) was applied through the SCP to interfere with information flow in the CTC. (B). Left panels show an example of a cortical cell that responded to single pulse SCP stimulation applied at 3 Hz (upper raster and PSTH). However, when the same stimulus intensity was applied at high frequency (130 Hz) the cell ceased to response to the stimuli (bottom raster and PSTH). Right panels show the averaged response of 104 cortical cells (upper right) and 9 thalamic cells (lower right) that exhibited a significant response to single pulse stimulation (black curve) but a reduced response when the stimuli were applied at high frequencies (red curve). Shading denotes SEM. Note that under HFS the residual thalamic response was larger than the residual cortical response, probably reflecting the strong cerebellar-thalamic connection. (C). Example of the effect of HFS on motor behavior. Hand trajectory performed by the monkey during control (left panel) and subsequent HFS session revealed a significant change in motor behavior including dysmetric and curved trajectories. In addition, the temporal properties of movement were altered, with a marked increase in reaction time (RT) and movement time (MT). (D). Average task-related activity of 58 pairs of motor cortical (left panel) and thalamic (right panel) cells that were recorded simultaneously during control (black) and HFS trials (red). Data were aligned on movement onset (time = 0). (E) Distribution of single-cell change in preferred direction between HFS and control trials is shown for all cortical cells that were tuned in both conditions. The mean value of the distribution was not significantly different from zero (one sample t-test for circular data with mean angle: $p = 0.59$).

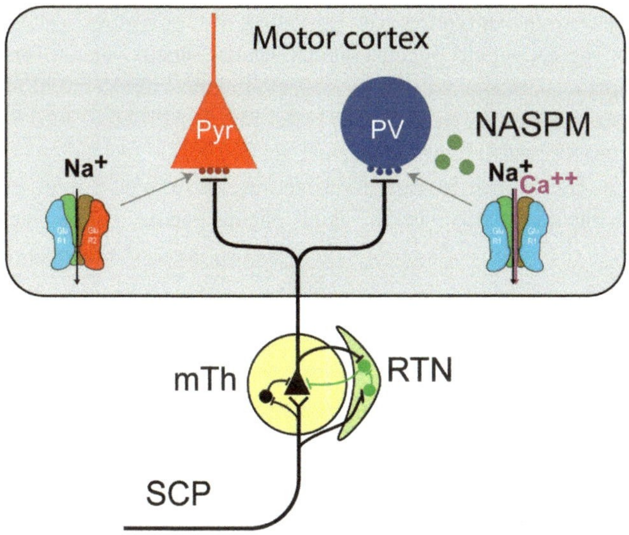

Figure 31.5. Schematic illustration of the inhibitory influence on the CTC pathway. Schematic illustration of the inhibitory influence over the CTC system. In the motor thalamus (mTh) local interneurons (black circles) and reciprocal interactions with the reticular nucleus (RTN) can modulate information arriving from the SCP. In the motor cortex, ascending axons of thalamocortical relay neurons (black triangle) make contact with parvalbumin (PV) and pyramidal (Pyr) neurons. The synapse between the TC fibers and PV neurons is mediated by GluR2-lacking AMPA receptors which are Ca^{+2} permeable and can be blocked using specific drug (NASPM).

2005; Cruikshank et al., 2007). Importantly, these synaptic contacts are mediated by GluR2-lacking AMPA receptors which are unique in terms of their calcium permeability (Hull et al., 2009). The Ca++ permeability makes these synapses more effective (larger synaptic currents) and may also implicate this pathway in plasticity. This termination pattern of thalamic input on cortical areas, where thalamic signals produce excitatory and inhibitory volleys (through feedforward excitation and inhibition), may appear counterintuitive and has led to the hypothesis that it endows the cortical circuitry with a particular sensitivity to thalamic synchrony (Bruno and Sakmann, 2006; Wang et al., 2010). In addition, FFI can enforce a window of opportunity to the cortical processing of incoming sensory information, which could act as a limit on perceptual processes (Gabernet et al., 2005). To date, the functional implications of FFI in sensory systems need to be tested directly.

Unlike the sensory systems, much less is known about the pattern of motor cortical circuitry that integrates thalamic inputs. Motor cortical responses to SCP stimulation are dominated by strong inhibition which appears to truncate the early excitation; however, it is not clear that this suppression is in fact mediated by thalamic-driven FFI. Local iontophoresis of NASPM (a specific GluR2-lacking AMPA receptor antagonist) in the motor cortex of behaving primates reversibly blocked the response of some cortical cells to CTC input (Nashef et al., 2022). In the adult cortex, GluR2-lacking AMPA receptors exist only on interneurons (Kumar et al., 2002). In addition, the set of cells that ceased responding to SCP stimulation in the presence of NASPM were unique in that they all manifested short-duration action potentials, a stronger response to SCP stimulation, and an increased background rate level (Nashef et al., 2022). Taken together, these results support the claim of the existence of FFI in the motor TC system. Using this approach, it was possible to test for differences in task-related activity of motor cortical interneurons and pyramidal cells. The interneurons had a higher background rate and an earlier onset of movement-related activity compared to pyramidal cells, which lagged about 200 ms behind interneurons. A similar temporal order of activity was found in the somatosensory areas of rodents (Yu et al., 2019), though at a shorter time lag, possibly reflecting the slower time course of rate fluctuation in the motor cortex compared to the sensory areas. Finally, interneurons and pyramidal cells were correlated in their firing, particularly around movement onset, possibly due to the thalamic co-drive. This sequence of events, consisting of the leading activity of inhibitory cells and a strong inhibitory-to-excitatory coupling among cortical CTC cells, may amplify the net impact of the TC system on motor cortical activity around movement onset by silencing competing sources of information which may mitigate the impact of cerebellar signals.

THALAMIC CONTROL OF MOTOR CORTICAL SYNCHRONY

The thalamus has long been considered an important orchestrator of cortical states of synchrony. Often these interactions are explored in relation to transitions across different states of consciousness and as a driver for cortical oscillation during sleep (Schreiner et al., 2022). Another example can be found in the whisker system of rodents, where a thalamic drive was shown to decrease cortical oscillations (Poulet et al., 2012) and to mediate the transition between different behavioral states.

However, in addition to regulating the global state of cortical synchrony, the thalamus can also modify local cortical firing when animals are engaged in normal behavior. In terms of the sensory system, it has been claimed that to achieve thalamic control over cortical activity some form of synchrony is necessary to compensate for the relative sparseness of these synapses (Swadlow et al., 2005). In the somatosensory system the cortical response to thalamic input is considered to be dominated by a weak albeit synchronized synaptic drive (Bruno and Sakmann, 2006). Further evidence for the efficacy of well-timed input was reported recently in awake animals, where the switch from tonic to burst firing induced a dynamic filter of sensory input that was based on the timing of incoming inputs without cortical amplification (Borden et al., 2022). Finally, computational models posit that synchronous thalamic input could reliably transmit information in a model of the visual system (Wang et al., 2010). Several studies in the motor system have documented the importance of correlated motor cortical firing to effectively drive downstream elements (Jackson et al., 2003). Although this synchrony could be mediated, at least in part, through local cortical activity, different studies have shown that the thalamic drive to the motor cortex is particularly effective in generating local and global coordinated firing which can participate in shaping motor commands. Stimulating in the deep cerebellar nuclei induced oscillatory muscle responses in monkeys making reaching movements (Aumann and Fetz, 2004). In rats, cerebellar inactivation disrupts the gamma-band coherence of the local field potential between the sensory and motor cortices during whisking, suggesting that the cerebellum participates in behaviorally relevant coordinated activity across cortical regions (Popa et al., 2013).

The ability to identify thalamic and cortical neurons using stimulation of the SCP allowed us to directly assess the role of thalamic input to task-related motor cortical synchrony (Figure 31.6A–B). We found that SCP stimulation can synchronize motor cortical cells (Figure 31.6C), suggesting that this pathway provides a common drive to neighboring motor cortical cells. The role of the motor thalamus in coordinating motor cortical activity was further confirmed when, in the absence of cerebellar signals, the firing of neighboring motor cortical cells became

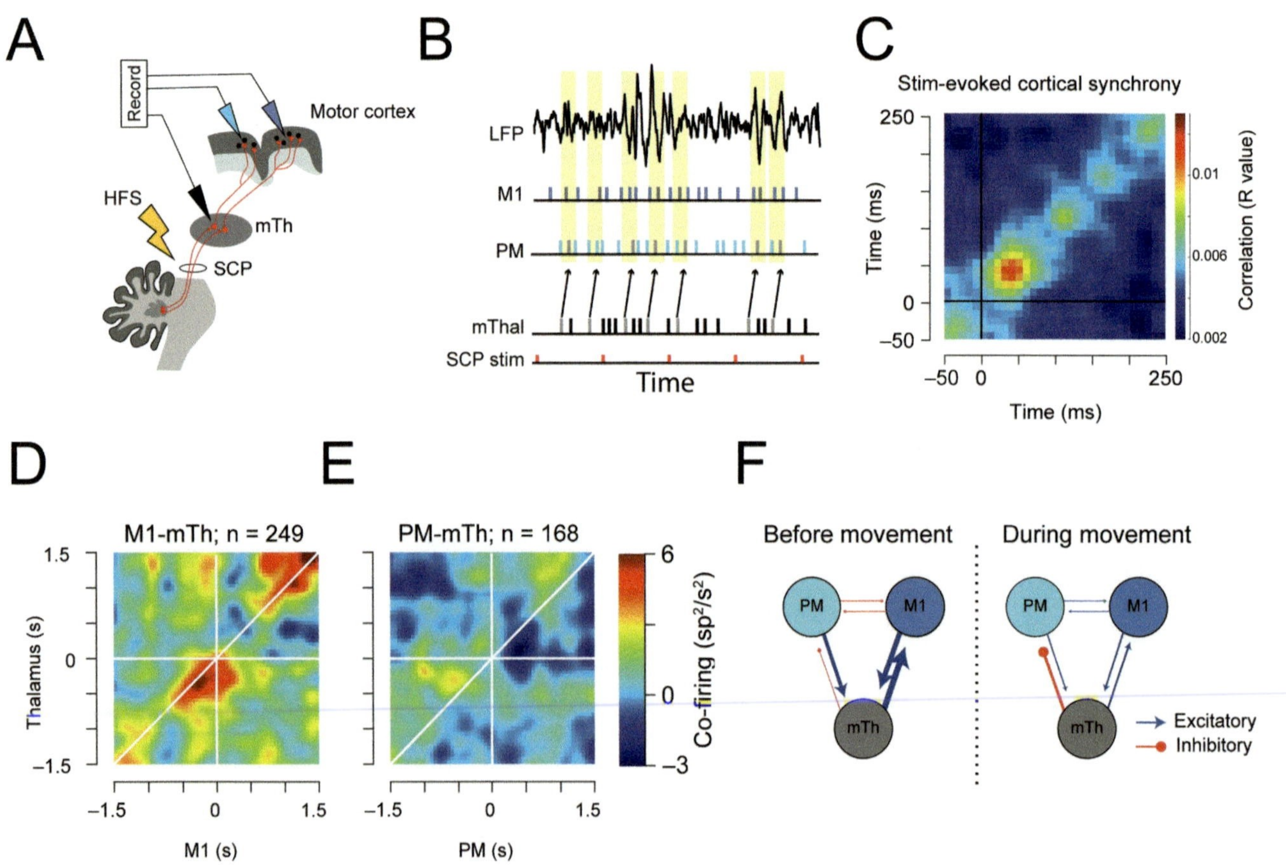

Figure 31.6. Thalamic effect on motor cortical synchrony. (A). Experimental setup. Same as in Figure 31.4A, but in addition, recordings in the motor thalamus. (B). Schematic illustration of the approach used to identify the thalamic regulation of motor cortical synchrony. Single-cell activity was recorded from the motor thalamus (mThal) and motor cortical sites (M1 and PM). The local field potential was used as a global measure of activity. Trigger events were either the SCP stimulation times or the thalamic spikes. The change in firing or correlated firing in relation to these events was computed. (C). Correlated firing of pairs of cortical cells triggered by SCP stimulation. Time zero is the stimulation time and the correlated firing is color-coded such that warm colors indicate increases in correlated firing. (D). Joint Peri-Stimulus Time Histogram computed between the thalamic (*y*-axis) and M1 (*x*-axis) neurons. The plotted matrix depicts the differences between the actual count matrix and the predicted count matrix computed based on the PSTH of the cells. Time zero corresponds to movement onset, and the diagonal depicts the zero-lag inter-unit interactions. The matrix is color-coded by covariance values (sp^2/s^2). Vertical and horizontal white lines correspond to movement onset times. (E). Same as in (D) but for mTh-PM cell pairs. (F). Schematic summary of the directional pairwise net interactions. Connecting lines between sites (circles) correspond to the net correlation between the 2 areas. Arrowheads indicate a positive correlation, and circles indicate a negative correlation. The width of the arrow corresponds to the magnitude of the correlation. The panel on the left was computed during the 500 ms before movement onset, and the panel on the right depicts the interactions computed during the 500 ms after movement onset.

uncorrelated and the stereotypical temporal pattern of activation of joint-specific cortical sites was lost (Nashef et al., 2019). This change in coordinated neural activity subsequently led to uncoordinated reaching movements, consistent with the role of cerebellar signals in coordinated multi-joint movements. Finally, while simultaneously recording the activity of neurons in the cerebellar thalamus and motor or premotor cortex, we found correlated thalamocortical firing around movement onset that exhibited a particular pattern of correlated firing (Figure 31.6D–F). Specifically, just before movement onset, thalamic and M1 cortical cells recorded from somatotopically matching sites were correlated in their firing (Nashef et al., 2021). After movement onset, the thalamic and premotor cells were weakly anti-correlated in their firing. The site-specific temporal pattern of thalamic cells and premotor versus motor cortical cells thus position the cerebellar thalamus in an excellent point to traffic signal from preparation to execution and therefore contribute to the transition from preparation to execution at movement onset.

CONCLUDING REMARKS AND OPEN QUESTIONS

Understanding the role of the cerebellum in controlling motor output poses an enormous challenge to neuroscientists. The complexity of the system and its scope are further complicated by its organization into multiple closed loops. The motor thalamus, which acts as a bottleneck in this divergent/convergent system, may be a good handle for exploring how converging information is relayed to the motor cortex to control motor timing and coordination. New tools are now available in both rodents and NHPs which make it possible to target and perturb the motor thalamus while monitoring the impact of these manipulations on downstream neurons and motor behavior. These techniques allow us to identify the role of subcortical inputs in generating the activity patterns that underlie movements. These studies also underscore the role of cerebellar thalamus in harnessing inhibitory and excitatory cortical systems for controlling temporal properties and coordinated activity across large groups of cortical neurons. Future directions will require (1) using high-density neural recordings to identify the ways that specific thalamic sites control the activity of neurons located in different motor areas and different cortical layers; (2) expanding the study of the intra-thalamic processing of incoming information and its interactions with neighboring systems (the basal-ganglia and somatosensory systems); and (3) using genetic/pharmacological tools to identify cortical neurons that are part of the system to highlight how the thalamic drive contributes to coordinated motor cortical activity when animals perform well-practiced movements and during motor adaptation.

REFERENCES

Akkal D, Dum RP, Strick PL (2007). Supplementary motor area and presupplementary motor area: targets of basal ganglia and cerebellar output. *J Neurosci*. 27:10659–10673.

Anderson ME, Turner RS (1991). Activity of neurons in cerebellar-receiving and pallidal-receiving areas of the thalamus of the behaving monkey. *J Neurophysiol*. 66:879–893.

Arcelli P, Frassoni C, Regondi MC, De Biasi S, Spreafico R (1997). GABAergic neurons in mammalian thalamus: a marker of thalamic complexity? *Brain Res Bull*. 42:27–37.

Armstrong E (1980). A quantitative comparison of the hominoid thalamus. II. Limbic nuclei anterior principalis and lateralis dorsalis. *Am J Phys Anthropol*. 52:43–54.

Aumann TD, Fetz EE (2004). Oscillatory activity in forelimb muscles of behaving monkeys evoked by microstimulation in the cerebellar nuclei. *Neurosci Lett*. 361:106–110.

Aumann TD, Rawson JA, Finkelstein DI, Horne MK (1994). Projections from the lateral and interposed cerebellar nuclei to the thalamus of the rat: A light and electron microscopic study using single and double anterograde labelling. *J Comp Neurol*. 349:165–181.

Azevedo FAC, Carvalho LRB, Grinberg LT, Farfel JM, Ferretti REL, Leite REP, Filho WJ, Lent R, Herculano-Houzel S (2009). Equal numbers of neuronal and nonneuronal cells make the human brain an isometrically scaled-up primate brain. *J Comp Neurol*. 513:532–541.

Baek SJ, Park JS, Kim J, Yamamoto Y, Tanaka-Yamamoto K (2022). VTA-projecting cerebellar neurons mediate stress-dependent depression-like behaviors. *eLife*. 11:e72981.

Bastian AJ, Martin TA, Keating JG, Thach WT (1996). Cerebellar ataxia: Abnormal control of interaction torques across multiple joints. *J Neurophysiol*. 76:492–509.

Borden PY, Wright NC, Morrissette AE, Jaeger D, Haider B, Stanley GB (2022). Thalamic bursting and the role of timing and synchrony in thalamocortical signaling in the awake mouse. *Neuron*. 110:2836–2853.

Bosch-Bouju C, Hyland BI, Parr-Brownlie LC (2013). Motor thalamus integration of cortical, cerebellar and basal ganglia information: implications for normal and parkinsonian conditions. *Front Comput Neurosci*. 7:163.

Bostan AC, Strick PL (2018). The basal ganglia and the cerebellum: nodes in an integrated network. *Nat Rev Neurosci*. 19:338–350.

Bruno RM, Sakmann B (2006). Cortex is driven by weak but synchronously active thalamocortical synapses. *Science*. 312:1622–1627.

Calzavara R, Zappala A, Rozzi S, Matelli M, Luppino G (2005). Neurochemical characterization of the cerebellar-recipient motor thalamic territory in the macaque monkey. *Eur J Neurosci*. 21:1869–1894.

Carta I, Chen CH, Schott AL, Dorizan S, Khodakhah K (2019). Cerebellar modulation of the reward circuitry and social behavior. *Science*. 363:eaav0581.

Cohen O, Harel R, Aumann TD, Israel Z, Prut Y (2017). Parallel processing of internal and external feedback in the spinocerebellar system of primates. *J Neurophysiol*. 118:254–266.

Cruikshank SJ, Lewis TJ, Connors BW (2007). Synaptic basis for intense thalamocortical activation of feedforward inhibitory cells in neocortex. *Nat Neurosci*. 10:462–468.

Dacre J, Colligan M, Clarke T, Ammer JJ, Schiemann J, Chamosa-Pino V, Claudi F, Harston JA, Eleftheriou C, Pakan JMP, Huang CC, Hantman AW, Rochefort NL, Duguid I (2021). A

cerebellar-thalamocortical pathway drives behavioral context-dependent movement initiation. *Neuron*. 109:2326–2338.
Gabernet L, Jadhav SP, Feldman DE, Carandini M, Scanziani M (2005). Somatosensory integration controlled by dynamic thalamocortical feed-forward inhibition. *Neuron*. 48:315–327.
Gaidica M, Hurst A, Cyr C, Leventhal DK (2018). Distinct populations of motor thalamic neurons encode action initiation, action selection, and movement vigor. *J Neurosci*. 38(29):6563–6573. doi:10.1523/JNEUROSCI.0463-18.2018. Epub 2018 Jun 22. PMID: 29934350; PMCID: PMC6052244.
Galvan A, Hu X, Smith Y, Wichmann T (2016). Effects of optogenetic activation of corticothalamic terminals in the motor thalamus of awake monkeys. *J Neurosci*. 36:3519–3530.
Gao Z, Davis C, Thomas AM, Economo MN, Abrego AM, Svoboda K, De Zeeuw CI, Li N (2018). A cortico-cerebellar loop for motor planning. *Nature*. 563:113–116.
Gornati SV, Schafer CB, Eelkman Rooda OHJ, Nigg AL, De Zeeuw CI, Hoebeek FE (2018). Differentiating cerebellar impact on thalamic nuclei. *Cell Rep*. 23:2690–2704.
Herculano-Houzel S, Collins CE, Wong P, Kaas JH (2007). Cellular scaling rules for primate brains. *Proc Natl Acad Sci*. 104:3562–3567.
Hirai T, Jones E (1989). A new parcellation of the human thalamus on the basis of histochemical staining. *Brain Res Rev*. 14:1–34.
Hoover JE, Strick PL (1999). The organization of cerebellar and basal ganglia outputs to primary motor cortex as revealed by retrograde transneuronal transport of herpes simplex virus type 1. *J Neurosci*. 19:1446–1463.
Hore J, Flament D (1988). Changes in motor cortex neural discharge associated with the development of cerebellar limb ataxia. *J Neurophysiol*. 60:1285–1302.
Horne MK, Butler EG (1995). The role of the cerebello-thalamo-cortical pathway in skilled movement. *Prog Neurobiol*. 46:199–213.
Hull C, Isaacson JS, Scanziani M (2009). Postsynaptic mechanisms govern the differential excitation of cortical neurons by thalamic inputs. *J Neurosci*. 29:9127–9136.
Ilinsky I, Kultas-Ilinsky K (1987). Sagittal cytoarchitectonic maps of the Macaca mulatta thalamus with a revised nomenclature of the motor-related nuclei validated by observations on their connectivity. *J Comp Neurol*. 262:331–364.
Jackson A, Gee VJ, Baker SN, Lemon RN (2003). Synchrony between neurons with similar muscle fields in monkey motor cortex. *Neuron*. 38:115–125.
Jones EG (2002). Thalamic circuitry and thalamocortical synchrony. *Philos Trans R Soc Lond B Biol Sci*. 357:1659–1673.
Jurgens U (1984). The efferent and afferent connections of the supplementary motor area. *Brain Res*. 300:63–81.
Kelly RM, Strick PL (2003). Cerebellar loops with motor cortex and prefrontal cortex of a nonhuman primate. *J Neurosci*. 23:8432–8444.
Kraskov A, Prabhu G, Quallo MM, Lemon RN, Brochier T (2011). Ventral premotor-motor cortex interactions in the macaque monkey during grasp: Response of single neurons to intracortical microstimulation. *J Neurosci*. 31:8812–8821.
Kultas-Ilinsky K, Sivan-Loukianova E, Ilinsky IA (2003). Reevaluation of the primary motor cortex connections with the thalamus in primates. *J Comp Neurol*. 457:133–158.
Kumar SS, Bacci A, Kharazia V, Huguenard JR (2002). A developmental switch of AMPA receptor subunits in neocortical pyramidal neurons. *J Neurosci*. 22:3005–3015.
Kunzle H (1976). Thalamic projections from the precentral motor cortex in *Macaca fascicularis*. *Brain Res*. 105:253–267.
Kuramoto E, Ohno S, Furuta T, Unzai T, Tanaka YR, Hioki H, Kaneko T (2015). Ventral medial nucleus neurons send thalamocortical afferents more widely and more preferentially to layer 1 than neurons of the ventral anterior-ventral lateral nuclear complex in the rat. *Cereb Cortex*. 25:221–235.
Mason A, Ilinsky IA, Maldonado S, Kultas-Ilinsky K (2000). Thalamic terminal fields of individual axons from the ventral part of the dentate nucleus of the cerebellum in Macaca mulatta. *J Comp Neurol*. 421:412–428.
Matano S, Hirasaki E (1997). Volumetric comparisons in the cerebellar complex of anthropoids, with special reference to locomotor types. *Am J Phys Anthropol*. 103:173–183.
Matelli M, Luppino G (1996). Thalamic input to mesial and superior area 6 in the macaque monkey. *J Comp Neurol*. 372:59–87.
Meyer-Lohmann J, Conrad B, Matsunami K, Brooks VB (1975). Effects of dentate cooling on precentral unit activity following torque pulse injections into elbow movements. *Brain Res*. 94:237–251.
Nashef A, Cohen O, Harel R, Israel Z, Prut Y (2019). Reversible block of cerebellar outflow reveals cortical circuitry for motor coordination. *Cell Rep*. 27:2608–2619 e2604.
Nashef A, Cohen O, Israel Z, Harel R, Prut Y (2018). Cerebellar shaping of motor cortical firing is correlated with timing of motor actions. *Cell Rep*. 23:1275–1285.
Nashef A, Cohen O, Perlmutter SI, Prut Y (2022). A cerebellar origin of feedforward inhibition to the motor cortex in non-human primates. *Cell Rep*. 39:110803.
Nashef A, Mitelman R, Harel R, Joshua M, Prut Y (2021). Area-specific thalamocortical synchronization underlies the transition from motor planning to execution. *Proc Natl Acad Sci U S A*. 118:e2012658118.
Nathan PW, Smith MC (1982). The rubrospinal and central tegmental tracts in man. *Brain*. 105:223–269.
Okoro SU, Goz RU, Njeri BW, Harish M, Ruff CF, Ross SE, Gerfen C, Hooks BM (2022). Organization of cortical and thalamic input to inhibitory neurons in mouse motor cortex. *J Neurosci*. 42:8095–8112.
Orioli PJ, Strick PL (1989). Cerebellar connections with the motor cortex and the arcuate premotor area: An analysis employing retrograde transneuronal transport of WGA-HRP. *J Comp Neurol*. 288:612–626.
Percheron G, Francois C, Talbi B, Yelnik J, Fenelon G (1996). The primate motor thalamus. *Brain Res Rev*. 22:93–181.
Person AL, Raman IM (2011). Purkinje neuron synchrony elicits time-locked spiking in the cerebellar nuclei. *Nature*. 481:502–505.
Popa D, Spolidoro M, Proville RD, Guyon N, Belliveau L, Lena C (2013). Functional role of the cerebellum in gamma-band synchronization of the sensory and motor cortices. *J Neurosci*. 33:6552–6556.
Poulet JF, Fernandez LM, Crochet S, Petersen CC (2012). Thalamic control of cortical states. *Nat Neurosci*. 15:370–372.
Rispal-Padel L, Troiani D, Harnois C (1987). Converging cerebellofugal inputs to the thalamus. II. Analysis and topography of thalamic EPSPs induced by convergent monosynaptic interpositus and dentate inputs. *Exp Brain Res*. 68:59–72.
Sakai ST, Inase M, Tanji J (1996). Comparison of cerebellothalamic and pallidothalamic projections in the monkey (Macaca fuscata): A double anterograde labeling study. *J Comp Neurol*. 368:215–228.
Sauerbrei BA, Guo JZ, Cohen JD, Mischiati M, Guo W, Kabra M, Verma N, Mensh B, Branson K, Hantman AW (2020). Cortical pattern generation during dexterous movement is input-driven. *Nature*. 577:386–391.
Sawyer SF, Young SJ, Groves PM, Tepper JM (1994). Cerebellar-responsive neurons in the thalamic ventroanterior-ventrolateral complex of rats: In vivo electrophysiology. *Neuroscience*. 63:711–724.
Schafer CB, Gao Z, De Zeeuw CI, Hoebeek FE (2021). Temporal dynamics of the cerebello-cortical convergence in ventro-lateral motor thalamus. *J Physiol*. 599:2055–2073.
Schreiner T, Kaufmann E, Noachtar S, Mehrkens JH, Staudigl T (2022). The human thalamus orchestrates neocortical oscillations during NREM sleep. *Nat Commun*. 13:5231.
Sherman SM, Guillery RW (1998). On the actions that one nerve cell can have on another: distinguishing "drivers" from "modulators." *Proc Natl Acad Sci U S A*. 95:7121–7126.

Shinoda Y, Futami T, Kakei S (1993). Input-output organization of the ventrolateral nucleus of the thalamus. *Stereotact Funct Neurosurg.* 60:17–31.

Shinoda Y, Kakei S (1989). Distribution of terminals of thalamocortical fibers originating from the ventrolateral nucleus of the cat thalamus. *Neurosci Lett.* 96:163–167.

Shinoda Y, Kano M, Futami T (1985). Synaptic organization of the cerebello-thalamo-cerebral pathway in the cat. I. Projection of individual cerebellar nuclei to single pyramidal tract neurons in areas 4 and 6. *Neurosci Res.* 2:133–156.

Strick PL, Sterling P (1974). Synaptic termination of afferents from the ventrolateral nucleus of the thalamus in the cat motor cortex: A light and electron microscopy study. *J Comp Neurol.* 153:77–106.

Swadlow HA, Bezdudnaya T, Gusev AG (2005). Spike timing and synaptic dynamics at the awake thalamocortical synapse. *Prog Brain Res.* 149:91–105.

Vitek JL, Ashe J, DeLong MR, Alexander GE (1994). Physiologic properties and somatotopic organization of the primate motor thalamus. *J Neurophysiol.* 71:1498–1513.

Wang HP, Spencer D, Fellous JM, Sejnowski TJ (2010). Synchrony of thalamocortical inputs maximizes cortical reliability. *Science.* 328:106–109.

Wiesendanger R, Wiesendanger M (1985). The thalamic connections with medial area 6 (supplementary motor cortex) in the monkey (*Macaca fascicularis*). *Exp Brain Res.* 59:91–104.

Yamaguchi K, Goto N, Yamamoto TY (1989). Development of human cerebellar nuclei: Morphometric study. *Acta Anat (Basel).* 136:61–68.

Yu J, Hu H, Agmon A, Svoboda K (2019). Recruitment of GABAergic interneurons in the barrel cortex during active tactile behavior. *Neuron.* 104:412–427.

32.

INTRINSIC PROPERTIES OF NEURONS IN MOTOR THALAMUS AND CONNECTED CORTEX

Francesco Cavarretta, Taylor Kahl, and Dieter Jaeger

INTRODUCTION

The spiking activity of neurons in vivo results from a complex interplay between synaptic inputs and intrinsically generated voltage-gated and/or chemically gated currents (Figure 32.1). In fact, intrinsic currents can form a considerable part of extracellularly measured cortical local field potentials (Ness et al., 2016). Intrinsic currents can delay or inhibit spiking, for example through A-type or M-type potassium channel activation, or cause pacemaking activity, spike burst firing, or rebound bursting. In addition, intrinsic currents can set up resonant properties and dampen or amplify input oscillations in specific frequency bands (Wilson et al., 2018). Further, modulatory inputs acting through G-protein coupled receptors can lead to changing channel densities and kinetics of intrinsic channels. The interplay between synaptic currents and intrinsic currents depends much on the location and strength of the synaptic input and determines the firing response of neurons.

In this chapter we will explore the role of intrinsic properties in motor thalamocortical circuits. As intrinsic current activation depends on the morphology of neurons and spatial configuration of synaptic inputs, we will cover these features as well as ion channel properties, and describe the influence of these properties on spiking activity. Additionally, we will discuss cerebellar and basal ganglia afferent connectivity to motor thalamocortical circuits and implications for motor disorders. What is apparent, however, is that much of the richness of active processing specifically in motor thalamocortical interactions remains yet to be determined. Hence, we will extrapolate to some degree from thalamocortical circuit properties in somatosensory areas. We start with motor thalamic properties, move on to motor cortical areas, and finally briefly discuss ensuing thalamocortical interactions in the motor circuit.

MOTOR THALAMUS

Motor thalamus can be considered a hub of signal integration by which subcortical motor processing and cortical motor processing are linked through multiple interacting feedback networks. In addition, motor thalamus creates the opportunity for trans-thalamic cortico-cortical interactions, which remain poorly understood in motor circuits.

ANATOMY AND CELL TYPES OF MOTOR THALAMIC NUCLEI

Motor thalamus appears conserved across different animal species, though the nomenclature for motor thalamic regions varies across species (Krack et al., 2002). For primates, distinct nomenclatures are used by neurosurgeons and neuroanatomists. Specifically, neuroanatomists distinguish the motor thalamic nuclei by their cytoarchitectural differences, whereas neurosurgeons differentiate motor nuclei by their afferents, as defined by the stereotactic atlas of Schaltenbrand (1977). In this chapter we define motor thalamus as the thalamic nuclei receiving inputs from motor cortical and motor-related subcortical areas, including the cerebellum and basal ganglia (Chevalier and Deniau, 1982; Bodor et al., 2008; Gornati et al., 2018; Guo et al., 2018). Under this definition, motor thalamus encompasses ventro-anterior (VA), ventro-lateral (VL), and ventro-medial (VM) nuclei. In some of the recent literature, motor thalamus also includes the mediodorsal (MD) nuclei, which are, however, connected primarily with the prefrontal rather than motor cortex (see **Chapter 46 in Section VIII**). We will not discuss MD in this chapter.

Thalamocortical (TC) neurons constitute the major neuron population of motor thalamus, whose properties have been extensively analyzed across different nuclei as

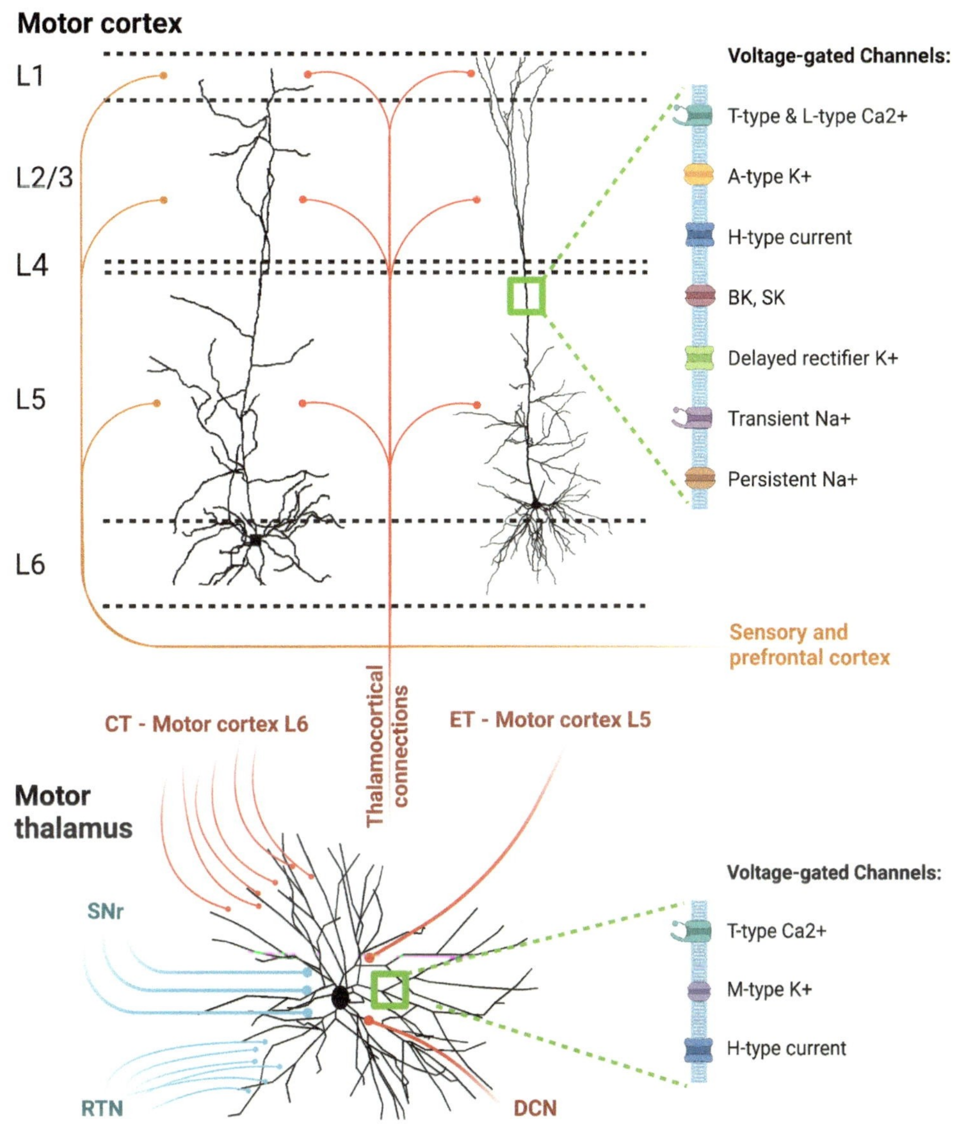

Figure 32.1. Synaptic integration in corticothalamic networks. Motor cortical and motor thalamic neurons have complex dendritic trees that contain multiple voltage- and calcium-gated ion channel types. Synaptic input is integrated in complex nonlinear ways due to the activation of these channels. Hence network dynamics can be expected to be dependent on such nonlinear dendritic integration. Figure created in BioRender.

well as animal species. Multiple morphological subtypes of TC neurons exist, and have distinct densities across motor thalamus (Sawyer et al., 1989; Clasca et al., 2012; Winnubst et al., 2019). Specifically, TC neurons projecting to layer 1 (L1) of cortex have fewer branched dendrites than those projecting to deeper layers (Clasca et al., 2012). Further, distinct groups of TC neurons have specific projection motifs to distinct cortical sub-areas (Winnubst et al., 2019). More recently, TC subtypes have been classified based on genetic transcription data (Phillips et al., 2019). While primary divisions between molecular clusters fall along boundaries between primary, secondary, and tertiary thalamic areas, each individual area also contains multiple clusters of TC neurons with distinct molecular identities. For motor cortex projecting TC neurons, 4 distinct molecular clusters have been isolated (Phillips et al., 2019).

In larger mammals, such as primates and cats, interneurons comprise 30% of motor thalamic cells (Albaugh et al., 2021), while these are absent in the motor thalamus of rodents.

SYNAPTIC AFFERENTS TO MOTOR THALAMUS

A well-known categorization of thalamic afferents delineates 2 classes of synaptic inputs, known as the thalamic "modulators" and "drivers" (see **Chapter 4 in Section II**). The former comprises small round excitatory terminals (RS) that originate from pyramidal cells in cortical layer 6 (L6), which make *en passant* synapses with distal dendrites of TC neurons and display short-term potentiation; the latter comprises large round excitatory terminals (RL) that originate from cortical layer 5 (L5) as well as other subcortical areas, which make multi-synaptic contacts with proximal dendrites, and display short-term depression. Additionally, inhibitory synapses and neuromodulators are typically considered as thalamic "modulators." Further studies have delineated also the class of the "driver-like" inputs, comprising excitatory synapses that share most, but not all, properties of the classic "drivers" and that usually make medium-sized round excitatory terminals (RM) (see **Chapter 4 in Section II**).

While the motor cortex innervates all motor thalamic nuclei, the thalamic targets of basal ganglia (BG) and deep cerebellar nuclei (DCN) delineate 2 different motor thalamic territories (Deniau et al., 1992; Kuramoto et al., 2011), corresponding to distinct distributions of synaptic terminals for "drivers" and "modulators" (Kuramoto et al., 2011; Rovó et al., 2012). Here it is necessary to highlight an anatomical difference between the BG-recipient thalamus of primates and small mammals. In primates, anterior VL receives BG input via the internal globus pallidus (GPi) and VA via the substantia nigra pars reticulata (SNr), while posterior VL receives cerebellar input (Macchi and Jones, 1997). In rodents, VM and rostroventral VA receive BG input via the SNr, while the GPi homolog entopeduncular nucleus (EP) only weakly projects to the thalamus and VL and posterior VA receives cerebellar input (Deniau et al., 1992; Kuramoto et al., 2011). For limb motor control, SNr is the main inhibitory source for motor thalamus in rodents (Chevalier and Deniau, 1982), while it is the GPi in primates. While primary cerebellar and basal ganglia territories are largely separate, the cerebellum also makes smaller terminals in the rodent BG-recipient VM nucleus (Gornati et al., 2018).

In rodents, both SNr and DCN terminals are located in proximity of the soma of TC neurons (Bodor et al., 2008; Gornati et al., 2018). Specifically, SNr terminals make multi-synaptic contacts in perisomatic regions of TC neurons, with each terminal targeting a single neuronal element (Bodor et al., 2008). DCN terminals are of RL-type in VL and the synapses display substantial short-term depression. Moreover, their activation can evoke excitatory postsynaptic currents (EPSCs) with high amplitudes, and a reliable suprathreshold response in thalamocortical neurons (Gornati et al., 2018). In VM, DCN terminals are smaller and display moderate short-term depression. The activation of these synapses evokes EPSCs with smaller amplitudes and no action potential (Gornati et al., 2018). Taken together, these experimental findings suggest that the DCN terminals in VL and VM fall in the class of "driver" and "driver-like" thalamic inputs, respectively. Furthermore, SNr delivers tonic inhibition that can suppress firing activity completely and counterbalance DCN excitation in VM (Chevalier and Deniau, 1982).

Anatomical studies of mammals confirmed that motor thalamus receives corticothalamic projections from both L5 and L6 of primary motor cortex (M1) (Kultas-Ilinsky et al., 2003). Motor cortical projections to thalamus are from axons of corticothalamic (CT) neurons in L6, or from axon collaterals of L5b extratelencephalic (ET) neurons. These axons mostly project ipsilaterally, but may also branch and project contralaterally (Shepherd and Yamawaki, 2021). Corticothalamic projections from L6 to motor thalamus form RS terminals, observed diffusely throughout the entire extent of motor thalamus (Kultas-Ilinsky et al., 2003; Rovó et al., 2012). In contrast, corticothalamic projections from L5 form RL terminals in monkey VA-VL (Kultas-Ilinsky et al., 2003; Rovó et al., 2012). In cats and monkeys, as opposed to the diffuse innervation pattern of L6 projections, axons originating in L5 of M1 make a more compact innervation pattern in motor thalamus (Kultas-Ilinsky et al., 2003), suggesting that these axons have a topographic organization.

The thalamic reticular nucleus (TRN) is a major source of thalamic inhibition that extensively innervates all excitatory thalamic nuclei (see **Chapter 3 in Section II**), and is essential for the maintenance of thalamic rhythms, such as the slow-wave oscillations observed during sleep. TRN terminals make *en passant* synapses with distal dendrites and form extensive patches over several millimeters in VL (Ilinsky and Kultas-Ilinsky, 2002). These findings suggest that SNr and TRN have different scope and function in motor thalamus, and they appear as the inhibitory analogues of the excitatory "drivers" and "modulators." We therefore speculate that the local SNr might mediate reliable and precise inhibitory signals that control spike timing in motor thalamus (Edgerton and Jaeger, 2014), whereas a more global inhibition from TRN might affect the state of the circuit.

Interneurons in primate motor thalamus receive excitatory afferents from motor cortex (Albaugh et al., 2021), dopaminergic inputs from multiple brain sources (Sánchez-González et al., 2005; García-Cabezas et al., 2009), and about 50% of the inhibitory synapses from TRN in motor thalamus (Ilinsky and Kultas-Ilinsky, 2002). Hence TRN input in primate but not rodent motor thalamus may lead to disinhibition of TC neurons. Additionally, interneurons contain synaptic vesicles in the dendrites which provide the substrate for complex dendritic network interactions (Ilinsky and Kultas-Ilinsky, 2002).

NEUROMODULATORY AFFERENTS TO MOTOR THALAMUS

Neuromodulatory afferents fall within the category of thalamic "modulators." Similarly to the glutamatergic inputs from L6 of cortex, neuromodulators are released by RS terminals that originate in multiple brain sources and make synapses *en passant* (see **Chapter 4 in Section II**). Mostly, investigations have been focused on serotonin, noradrenaline, dopamine, and acetylcholine. In rodents and primates, serotonergic afferents to thalamus originate from the dorsal raphe nuclei, yielding a sparse innervation in the ventral thalamus in primates (Moore et al., 1978; Lavoie and Parent, 1991), while no innervation is observable in the VM of rats (Moore et al., 1978). Noradrenergic afferents originate from locus coeruleus, and yield a sparse innervation in VA and VL, where neurons express alpha-1 adrenergic receptors, which have excitatory effects (Pérez-Santos et al., 2021).

Dopamine is another fundamental neuromodulator in mammals, involved in multiple motor functions. Immunochemistry studies have revealed that dopaminergic fibers innervate the VA and VL in primates (García-Cabezas et al., 2009), targeting primarily interneurons (García-Cabezas et al., 2009), while there is far less if any dopaminergic innervation in the motor thalamus of rodents (García-Cabezas et al., 2009), where interneurons are absent.

Cholinergic inputs constitute a major neuromodulatory stream in thalamus that has a global effect (see **Chapter 5 in Section II**) and depolarizes the majority of TC cells during awake states. In rats, cholinergic projections to VL and VM originate from EP and SNr (Kha et al., 2000; Kha et al., 2001), where cholinergic neurons are intermingled with GABAergic and glutamatergic neuron populations (Kha et al., 2000; Kha et al., 2001).

MORPHOLOGICAL AND ELECTRIC PROPERTIES OF MOTOR THALAMIC NEURONS

Physiologically, TC neurons generate the characteristic firing responses (e.g., tonic firing, low-threshold spiking, rebound bursting; Edgerton and Jaeger, 2014; Bichler et al., 2021), which are commonly associated with waking and sleep states (Llinás and Steriade, 2006). These neurons possess the characteristic bushy dendritic tree as observed in other thalamic nuclei (Sawyer et al., 1989; Kuramoto et al., 2009; Gornati et al., 2018), with a dendritic extent up to 200 μm from soma (Sawyer et al., 1989; Kuramoto et al., 2009; Gornati et al., 2018). Modest but significant morphological differences are observed in the VA/VL and VM of rats (Sawyer et al., 1989; Gornati et al., 2018), with the TC neurons in the caudolateral portion of VA/VL having more dendrites than those in the rostromedial portion of VA/VL (receiving predominantly excitatory inputs) (Kuramoto et al., 2009). In addition, ion channel genes show a gradient across VA/VL, leading to narrower action potential waveforms in VL (Phillips et al., 2019). Overall, these findings support the possibility that motor TC neurons may have different electrophysiological properties that finely tune their response to cerebellar and nigral inputs, but further studies are required to characterize these differences. The intrinsic properties and function of primate motor thalamic interneurons remain poorly understood as they cannot be examined easily in brain slice preparations.

INVOLVEMENT OF INTRINSIC PROPERTIES IN DISEASE STATES

Anomalous brain rhythmicity is considered a hallmark of multiple pathological states, such as Parkinson's disease (PD). It has been speculated that entraining thalamocortical loops might increase and underlie the global propagation of these patterns of pathological activity. In motor thalamus, rebound bursting might generate the dysrhythmia underlying motor symptoms of PD, due to the de-inactivation of T-type Ca2+ channels following a strong inhibitory bout. Indeed, inhibitory synapses from SNr are well positioned in motor thalamus to evoke rebound bursting. They make powerful GABAergic synapses with proximal dendrites of thalamocortical cells (Bodor et al., 2008), where T-type Ca2+ channels show the highest densities (Williams and Stuart, 2000). In Parkinsonian states, dopamine depletion alters SNr firing patterns, increasing bursting and synchrony (Willard et al., 2019). Thus, synchronous SNr bursting would be an ideal mechanism for subsequent rebound burst generation, as it would condense the inhibition within ~100 ms, facilitating T-type Ca2+ de-inactivation. Indeed, in vitro experiments showed that direct SNr activation, mimicking tonic firing at 20 Hz or 50 Hz paired pulses, evoked rebound bursting in VM TC neurons in normal conditions (Edgerton and Jaeger, 2014). In addition, whole-cell recordings showed that dopamine depletion induces homeostatic changes in VM TC neurons, such as M-type potassium current suppression (Bichler et al., 2021), enhancing the excitability of TC neurons in VM, assessed as an increase in tonic firing and a prolongation of rebound burst. Therefore, these experimental findings hint at a role of nigral inhibition in evoking rebound bursting in motor thalamus, perhaps exaggerated by dopamine depletion, but further in vivo investigations are needed to test these hypotheses.

Another hallmark of parkinsonian conditions is the exaggeration of beta band oscillations in the local field potential (LFP) of motor thalamus that is phase-locked to VM spiking (Brazhnik et al., 2016). Increased beta coherence was observed between VM and motor cortex as well as VM and SNr in walking animals, achieving higher amplitudes in parkinsonian states (Brazhnik et al., 2016). Blocking VM by muscimol destroyed the beta oscillations, confirming the importance of VM in the transmission of beta oscillations across motor cortex and SNr (Brazhnik et al., 2016). Taking into account that membrane properties can underlie or contribute to the generation of oscillatory activity, the exaggeration in the beta band might be a consequence of homeostatic changes that compensate for increased parkinsonian nigral input, for example, by down-regulating M-type potassium current (Bichler et al., 2021).

MOTOR CORTEX

Motor cortex can be defined by its low stimulation threshold for evoking contralateral movement, a full-body topography, and adjacency with somatosensory cortex which contains a mirrored body topography (Ebbesen and Brecht, 2017). The motor cortices are most active during movement planning and execution.

ANATOMY AND NOMENCLATURE OF MOTOR CORTEX

Motor cortex is divided into 2 major sub-divisions: primary motor cortex (M1 in primates, MOp in rodents) and premotor cortex (also called secondary motor cortex, or M2). Premotor areas are further divided into dorsal and ventral premotor areas and the supplementary motor area. M1 is located in the precentral gyrus in the posterior frontal lobe, and premotor areas are anterior to M1. M1 is distinguished from premotor cortex by its lower threshold for stimulation to elicit contralateral muscle contraction, and by a hierarchy of projections from premotor cortex to M1.

Regions of motor cortex are often defined by the area of the body that they represent. In the rodent literature, the rostral forelimb area (RFA) is considered a premotor area, and the caudal forelimb area (CFA) is considered a primary motor area of cortex that regulates forelimb movements (Hira et al., 2013). The whisker or vibrissal motor area (vM1) controls whisking behavior, and the anterolateral motor cortex (ALM) is involved in orofacial movements and decision-based licking behavior (Guo et al., 2017).

Like all areas of neocortex, motor cortex is divided into 6 layers of cell types with distinct morphologies and synaptic connections. The organization of cortical columns represents a microcircuit that is repeated across cortex, describing excitatory pyramidal neurons (PNs) that project within or outside of cortex, and the local glutamatergic spiny cells and GABAergic interneurons that regulate the activity of PNs (Defelipe et al., 2002). (For a review of the "canonical" cortical microcircuit in somatosensory cortex, see **Chapter 6 in Section III**.)

Motor cortex appears distinct from other areas of neocortex for its expansion of L5, including corticospinal neurons in lower L5b, and a layer 4 (L4) that is significantly smaller compared to sensory cortices. In most mammals, corticospinal neurons from motor cortex synapse onto spinal interneurons that regulate lower motor neurons, which synapse onto skeletal muscle. In primates, some cortical neurons synapse directly onto lower motor neurons, and this specialization is important for controlling distal muscles, and for controlling larynx muscles in humans (Ebbesen and Brecht, 2017).

A recent analysis by the BRAIN Initiative Cell Census Network (BICCN) identified that M1 and its cell types are largely conserved across mice, marmosets, and humans. These cell types include 6 classes of GABAergic interneurons: *Lamp5*, *Sncg*, *Vip*, *SStChodl*, *Sst*, and *Pvalb*, and 5 classes of glutamatergic projection neurons: intratelencephalic (IT), ET, L6 CT, near-projecting (NP), and L6b (BICCN, 2021). GABAergic interneurons are present in all 6 layers of cortex. IT neurons are present in all layers except for L1. ET neurons are located in L5b.

The 2 classes of L5 PNs, ET and IT neurons, are described with different names by different authors and have distinct properties: ET neurons with somas in L5b are known as large, thick-tufted neurons; L5 IT neurons with somas in upper L5 (L5a) are known as smaller, thin-tufted neurons. In a cross-species patch-seq comparison, the BICCN classified primate Betz cells and mouse corticospinal neurons as ET neurons. ET neurons in motor cortex can be divided further into at least 2 classes: brainstem-projecting and non-brainstem-projecting (BICCN, 2021).

ET neurons are sometimes described as pyramidal tract (PT) neurons or subcortically projecting neurons (BICCN, 2021) and there is noted confusion in the field resulting from differences in terminology (Shepherd and Yamawaki, 2021). While "PT" is a common name for the L5 ET neurons, PT neurons reflect only a subset of L5 ET neurons that does not include the entire L5 ET subclass. We refer to PT neurons as ET neurons in this chapter.

In motor cortex, L5 ET neurons are considered the primary output neurons of the cortical columnar microcircuit. These neurons project to the brainstem and spinal cord circuits that include lower motor neurons and interneurons that together control muscle tone and activity. Most L5 ET cells project to multiple targets including the thalamus, the pyramidal tract (to the brainstem and spinal cord), and the midbrain. L5 corticothalamic (CT) neurons are a subtype of L5 ET neurons, and not in the same class as L6 CT neurons (BICCN, 2021). IT neurons project to other neurons within the cortex, and there is a hierarchy of information processing from IT neurons to other IT neurons and from IT to ET neurons, but not from ET to IT neurons (Kiritani et al., 2012).

Corticostriatal projections come from L5 IT and ET neurons and from layer 2/3 (L2/3) IT neurons (Reiner et al., 2010). L2/3 PNs are the primary local excitatory input to L5 PNs (Weiler et al., 2008). L1 lacks excitatory neurons, but contains long-range input to the cortex and local GABAergic interneurons that regulate the excitability of tuft branches of PNs (Schuman et al., 2021). Despite early reports that L4 is not present in motor cortex, recent studies have identified L4-like neurons in mouse MOp and primate M1 with untufted dendrites and axons that project to L2/3 and long-range projections, and receive excitatory input from the thalamus (BICCN, 2021).

The arrangement of cortical layers enables a hierarchy of information processing from neurons in superficial layers to the output neurons in deeper layers of motor cortex. The layer-specific patterns of synaptic innervation interact with the intrinsic properties of neurons in the cortical microcircuit to drive specific output patterns during movement planning and execution.

SYNAPTIC AFFERENTS AND EFFERENTS OF MOTOR CORTEX

Motor cortex receives major inputs from sensory cortex, the BG, and the motor thalamus. PNs in motor cortex project within cortex and to the BG and motor thalamus, and to subcortical targets that implement motor commands. Inputs to motor cortex are layer-specific and may synapse directly onto dendrites of ET or CT neurons or regulate their activity indirectly by synapsing onto IT neurons and GABAergic interneurons.

In rodents, M1 receives thalamic input from the cerebellar-recipient thalamus (VL) and the basal-ganglia-recipient thalamus (VA/VM) and the posterior nucleus (PO) of the thalamus (Shepherd and Yamawaki, 2021). The projections of these neurons are layer-specific: VL axons project to IT and ET neurons around L4 to the L5b/L6 border, and to the dendrites of ET neurons in L1, while VM/VA project to L1 and L5a, and PO projects to IT neurons in L2/3 and L5a, but L6 corticothalamic neurons mostly lack VL or PO input (Shepherd and Yamawaki, 2021). For more detail about the layer-specific inputs and outputs from motor cortex, see **Chapter 33 in Section VI**.

In macaques, VL thalamus projects primarily to caudal motor areas including primary, supplementary, and caudal premotor cortex, while VA projects primarily to rostral premotor areas including cingulate and pre-supplementary areas, and MD projects to dorsolateral and orbital prefrontal cortex (McFarland and Haber, 2002). Thalamic projections were densest in L1, followed by layer 3 and then L5.

In general, motor corticothalamic loops are closed by thalamic projections back to the same cortical column that the thalamic nuclei receive input from. For instance, ALM ET axon collaterals excite ALM-projecting neurons in medial VM, but not the M1-projecting neurons in lateral VM. Vice versa, M1 ET neurons excite M1-projecting neurons in VM, but not the ALM-projecting neurons (Guo et al., 2018). Some divergence exists in this loop as well, however, and VM-projecting ALM neurons additionally excite neurons in MD. For a description of the dynamics of these thalamocortical loops, see **Chapter 35 in Section VI**.

NEUROMODULATORY AFFERENTS TO MOTOR CORTEX

Neural excitability and plasticity are regulated by neuromodulatory inputs to motor cortex. In mice, noradrenergic axons densely innervate all layers of forelimb M1. Blocking noradrenergic receptors in M1 reduced the excitability and firing rate of L5b ET neurons during quiet wakefulness and voluntary movements and blocked the increase in movement-related beta power (Schiemann et al., 2015). Behaviorally, mice were less precise with their contralateral forepaw movements.

Dopamine (DA) regulates cortical activity indirectly via its influence on the BG-thalamocortical loop. DA also regulates motor cortex directly via projections from the ventral tegmental area (VTA) and to a lesser extent from the substantia nigra pars compacta to motor cortex in rats (Hosp et al., 2011). Similar projections exist in primates, and many mammals express D1-like and D2-like receptors in M1 (Cousineau et al., 2022).

A limited number of studies have evaluated the effects of DA on the intrinsic properties of motor cortex, and whether its effects are excitatory or inhibitory remains to be resolved (Cousineau et al., 2022). DA has been shown to act both on PNs in motor cortex and on the GABAergic interneurons that innervate the PNs, which may account for the fast excitatory-inhibitory response observed in M1 after VTA stimulation in rats; VTA terminals co-release glutamate, which could be responsible for the fast excitatory response (Kunori et al., 2014). Multiple studies have demonstrated that DA innervation is necessary for plasticity in motor cortex that enables motor learning, but that DA input is not necessary for the execution of tasks that have already been learned (Cousineau et al., 2022).

Acetylcholinergic input to motor cortex is also important for motor learning. Specifically, cholinergic lesions in motor cortex of rats impaired learning of a skilled forelimb-reaching task (Conner et al., 2010). In control rats, this task increased the area of representation of the forelimb in motor cortex, and this plasticity was blocked by cholinergic depletion. Another study confirmed that ablating cholinergic neurons in mice impaired their performance in learning a lever-pressing task (Ren et al., 2022). During the normal learning process, cholinergic inputs were most active early in learning, and these inputs regulated the activity of VIP and SOM interneurons during learning.

Motor learning is mediated by spine plasticity in motor cortex and depends on protein synthesis; blocking protein synthesis in rodents can inhibit learning a motor task but does not interfere with execution of a learned task, while in primates protein synthesis may also be necessary for the execution of a learned task (Cousineau et al., 2022). M1 seems to be particularly important for learning dexterous movements and may be necessary for executing these movements in primates. At the cellular level, in studies with mice performing a learned forelimb-reaching task or lever-pressing task, dendrites of L5 and L2/3 PNs grow new spines and lose existing spines over the course of learning, and the new spines persist after learning. In separate studies applying D1 or D2 antagonists in M1, the induction of neural plasticity was shown to be dependent on DA input to M1 (Cousineau et al., 2022).

MORPHOLOGICAL AND ELECTRICAL PROPERTIES OF MOTOR CORTICAL NEURONS

L5 ET neurons are large L5 neurons with apical dendrites that reach the pial surface (BICCN, 2021). At least 2 types of L5 ET neurons are present in mouse motor cortex: type 1 neurons have their primary apical bifurcation closer to the soma, with distinct hemi-trees, and type 2 neurons have apical dendrites that bifurcate farther from the soma, with less prominent tuft branching (Otor et al., 2022). L2/3 neurons are smaller, and their tufts also project to L1. Unlike the IT neurons in other layers, L4 IT neurons have minimal or no apical dendrites and lack apical tufts. **Chapter 7 in Section III** further describes the morphologies and other intrinsic properties of cortical excitatory neurons.

The intrinsic properties of L5 neurons have classically been studied from neurons in sensory cortex. L5b ET neurons, compared to L5 IT neurons, have little spike frequency adaptation and strong sag potentials (Dura-Bernal et al., 2022). The L5 IT neurons are called regular-spiking neurons, and L5b ET neurons are called intrinsically bursting. L5 IT neurons have higher input resistance and do not display the characteristic bursts of ET neurons (BICCN, 2021). In motor cortex, there are some differences in ET neurons between species, including low input resistance and sub-threshold responses in primate L5 ET neurons (BICCN, 2021). Inhibitory interneurons are distinguished from PNs by their fast firing rates and narrow spike widths (Ebbesen and Brecht, 2017).

The dendrites of cortical pyramidal neurons have active properties that enable nonlinear summation of synaptic inputs. In sensory cortex, PNs demonstrate multiple classes of spikes including sodium action potentials in the axosomatic region, calcium plateaus in the apical dendrite near its primary bifurcation, and NMDA spikes in thin basal and tuft dendrites (Larkum et al., 2009). Coincident spiking at the soma and in the dendrites is a mechanism of coincidence detection, and activation of L5 apical dendrites in somatosensory cortex is the primary determinant of tactile perception in mice (Takahashi et al., 2020).

In L5 PNs of somatosensory cortex, Ca^{2+} influx accumulates in distinct regions: in an apical band 500 um from soma, or in the basal and proximal dendrites and soma, or in both regions simultaneously (Yuste et al., 1994). The presence of a calcium spike in the apical dendrite, in addition to soma depolarization that alone can generate an AP, when occurring around the same time will elicit bursting (Takahashi et al., 2020). Excitatory inputs to the dendrites increase the excitability of PNs at the soma, making firing more likely. Dendritic excitability is regulated by NMDA and GABAergic inputs to tuft and basal dendrites, with NMDA spikes underlying the transformation from distal synaptic input to dendritic and perisomatic spiking (Larkum et al., 2009).

Cortical PN spiking is regulated by several classes of voltage- or calcium-activated potassium channels, and HCN (hyperpolarization-activated cyclic nucleotide-gated) channels. In sensory cortex, apical calcium spikes are mediated by voltage-gated calcium channels, with a proposed "hot spot" of channels near the primary apical bifurcation.

The analysis of active properties in motor cortex lags far behind that of somatosensory cortex. Similar properties are expected, but need to be confirmed. Heterogeneity also exists within a subpopulation of motor cortical neurons. For instance, HCN expression is high in corticospinal neurons, but low in corticostriatal and corticocortical neurons. HCN channels mediate a hyperpolarization-activated current (Ih) that attenuates glutamate-evoked responses in apical and basal dendrites of corticospinal neurons, regulating synaptic integration (Sheets et al., 2011). H current is important in mice learning a skilled reaching task, as blocking this current reduced accuracy of movements (Boychuk et al., 2017). This mechanism seems to be due to decreased specificity of motor representation in motor cortex.

Also in mouse motor cortex, Kv2 voltage-gated potassium channels regulate burst firing in L5 PNs (Newkirk et al., 2021). Blocking Kv2 converted single spikes to multispike bursts and enhanced fast-afterdepolarization, and these channels are highly expressed perisomatically. While its function in motor cortex is not entirely clear, burst firing in motor cortex may be a critical driver of movement execution and LTP induction in PNs.

INVOLVEMENT OF INTRINSIC PROPERTIES OF MOTOR CORTEX IN MOTOR CONTROL

IT neurons in motor cortex generally have preparatory roles and recurrent connections with the thalamus, while ET neurons have a role in execution and project to the medulla. At least 2 subtypes of ET neurons may also be functionally distinct regarding selectivity for preparation or execution of movement.

In mouse ALM, IT neurons exhibit preparatory activity during a licking task, before ipsi- or contralateral movements, but ET neurons display contralateral selectivity during late movement planning (Li et al., 2015). Also in mouse ALM, there is a distinction between ET neurons located in upper or lower L5b: the upper L5b neurons branch to project to the thalamus and are primarily involved in early preparatory activity before movement initiation (Economo et al., 2018). The lower L5b neurons project to motor centers in the medulla and produce late preparatory activity and motor commands.

In mouse forelimb M1 2 classes of L5 ET neurons have distinct movement-related activity. Type 1 neurons have their primary apical bifurcation closer to the soma than type 2 neurons, and the type 1 neurons preferentially project to the medulla and spinal cord (Otor et al., 2022). Calcium imaging of these cells during walking or reaching showed that the type 1 neurons exhibit more functional compartmentalization than the type 2 neurons, which tend to have synchronous tuft activation. Different branches or hemi-trees within the same type 1 neuron could be selective for distinct motor variables (Otor et al., 2022).

L2/3 and L5 neurons also display distinct patterns of movement-related activity. In mouse vM1 during whisking, in vivo patch-clamp and extracellular recordings demonstrated changes in membrane potential and firing rate during phases of whisking (Sreenivasan et al., 2016). L2/3 PNs were somewhat depolarized with increased firing rates before whisking, followed by hyperpolarization and decreased firing rates at the initiation of whisking. L5 PNs become more depolarized and have increased firing rates during movement initiation and execution. In the same set of experiments, optogenetic stimulation of vibrissal sensory cortex caused excitation, then inhibition, in vM1; following inhibition, rebound excitation in L5 neurons was correlated with the initiation of whisking.

An interesting feature of PNs in motor cortex is that many neurons display reduced firing during movement, which is counterintuitive to the concept that motor cortex is primarily a driver of movement. For example, during in vivo patch-clamp recordings from L5b PNs in mouse forelimb M1, most neurons changed firing rate during voluntary movement, showing either enhanced or suppressed firing (Schiemann et al., 2015). These results suggest that some neurons are selectively active while others are selectively suppressed during movement in order to allow desired movements and inhibit others.

INVOLVEMENT OF INTRINSIC PROPERTIES OF MOTOR CORTEX IN DISEASE STATES

Changes in motor cortex activity are commonly seen in people with PD and in parkinsonian animals. The rate model of PD predicts that following loss of midbrain DA, GABAergic transmission from the BG to the motor thalamus is increased and that glutamatergic transmission from motor thalamus to motor cortex is decreased (Galvan et al., 2015). The rate model has more recently been modified to address that functional consequences of DA depletion are due not only to changes in firing rate, but also changes in firing patterns throughout the BG-thalamocortical loop (Eisinger et al., 2019). For more detail on the symptoms and pathology of PD, see **Chapter 67 in Section XII**.

In parkinsonian animals, corticospinal neurons (but not corticostriatal neurons) are reported to have lower firing rates, or higher burstiness, compared to control animals (Galvan et al., 2015). In monkeys rendered parkinsonian through systemic 1-methyl-4-phenyl-1,2,3,6-tetrahydropyridine (MPTP) injections, neurons recorded in forelimb M1 had no change in firing rate compared to control monkeys, but individual neurons (recorded extracellularly) were more bursty, with long stretches of silence between bursts (Goldberg et al., 2002). In parkinsonian mice, ET neurons had altered intrinsic properties, including an increased firing threshold and widened action potential, due to impaired function of persistent sodium channels and large-conductance calcium-activated potassium channels (Chen et al., 2021).

MOTOR THALAMOCORTICAL LOOPS

Thalamo-cortico-thalamic loops are a common feature across different thalamic regions (Shepherd and Yamawaki, 2021). In the motor system, ALM and VM form an excitatory thalamo-cortico-thalamic loop that supports reverberant activity during motor planning of lick tasks in mice (Guo et al., 2017). Interestingly, a similar closed loop is not observed for cerebellar input recipient VL thalamus in mice (Yamawaki and Shepherd, 2015). Instead M1 strongly excites neurons in other thalamic nuclei (VM, PO, TRN), potentially indicating transthalamic cortical interactions. Whether reverberatory activity is sustained between VM and M1 in mice has not been established. In vivo experiments suggest that basal ganglia input recipient motor thalamus not only drives preparatory activity within the cortico-thalamocortical network, but is also engaged in functions encompassing movement selection, initiation, and vigor encoding (Gaidica et al., 2018; Catanese and Jaeger, 2021; Inagaki et al., 2022).

Findings to date support the hypothesis that reverberation is a feature of basal ganglia–controlled but not cerebellar-controlled thalamocortical interactions. Cerebellum instead may feed predictions about future sensory and cortical states to the cortex in a feedforward manner (Boven et al., 2023).

CONCLUSIONS

It is fair to say that our understanding of how intrinsic properties of motor thalamic and cortical neurons impact motor control in behaving animals remains poor. The scaffold of structural considerations and established intrinsic properties introduced in this chapter can help guide future investigations in this area. Unfortunately, pathway-specific manipulation of intrinsic properties in behaving animals to probe the causality of their function remains underdeveloped compared to our ability to perform optogenetic synaptic manipulations. Nevertheless, observations to date support the notion that intrinsic excitability is important in the control of spike rate and spike timing, as well as in the control of oscillatory network mechanisms. It is also clear that intrinsic properties change in both motor thalamus and in distinct motor cortex cell types during disease states such as PD. In which way such changes are causal in pathological circuit dynamics or reflect compensatory plasticity is still an open question.

REFERENCES

Albaugh DL, Huang C, Ye S, Paré JF, Smith Y (2021). Glutamatergic inputs to GABAergic interneurons in the motor thalamus of control and parkinsonian monkeys. *Eur J Neuroscience*. 53:2049–2060.

BICCN (2021). A multimodal cell census and atlas of the mammalian primary motor cortex. *Nature*. 598:86–102.

Bichler EK, Cavarretta F, Jaeger D (2021). Changes in excitability properties of ventromedial motor thalamic neurons in 6-OHDA lesioned mice. *eNeuro* 8:ENEURO 0436-20.2021:1–18. PMC7920540

Bodor ÁL, Giber K, Rovó Z, Ulbert I, Acsády L (2008). Structural correlates of efficient GABAergic transmission in the basal ganglia–thalamus pathway. *J Neurosci*. 28:3090–3102.

Boven E, Pemberton J, Chadderton P, Apps R, Costa RP (2023). Cerebro-cerebellar networks facilitate learning through feedback decoupling. *Nat Commun*. 14:1–18.

Boychuk JA, Farrell JS, Palmer LA, Singleton AC, Pittman QJ, Teskey GC (2017). HCN channels segregate stimulation-evoked movement responses in neocortex and allow for coordinated forelimb movements in rodents. *J Physiol*. 595:247–263.

Brazhnik E, McCoy AJ, Novikov N, Hatch CE, Walters JR (2016). Ventral medial thalamic nucleus promotes synchronization of increased high beta oscillatory activity in the basal ganglia-thalamocortical network of the hemiparkinsonian rat. *J Neurosci*. 36:4196–4208.

Catanese J, Jaeger D (2021). Premotor ramping of thalamic neuronal activity is modulated by nigral inputs and contributes to control the timing of action release. *J Neurosci*. 41:1878–1891.

Chen L, Daniels S, Kim Y, Chu HY (2021). Cell type-specific decrease of the intrinsic excitability of motor cortical pyramidal neurons in parkinsonism. *J Neurosci*. 41:5553–5565.

Chevalier G, Deniau JM (1982). Inhibitory nigral influence on cerebellar evoked responses in the rat ventromedial thalamic nucleus. *Exp Brain Res*. 48:369–376.

Clasca F, Rubio-Garrido P, Jabaudon D (2012). Unveiling the diversity of thalamocortical neuron subtypes. *Eur J Neurosci*. 35:1524–1532.

Conner JM, Kulczycki M, Tuszynski MH (2010). Unique contributions of distinct cholinergic projections to motor cortical plasticity and learning. *Cereb Cortex*. 20:2739–2748.

Cousineau J, Plateau V, Baufreton J, Le Bon-Jégo M (2022). Dopaminergic modulation of primary motor cortex: From cellular and synaptic mechanisms underlying motor learning to cognitive symptoms in Parkinson's disease. *Neurobiol Disease*. 167 (105674):1–14.

Defelipe J, Alonso-Nanclares L, Arellano J (2002). Microstructure of the neocortex: Comparative aspects. *J Neurocytol*. 31:299–316.

Deniau JM, Kita H, Kitai ST (1992). Patterns of termination of cerebellar and basal ganglia efferents in the rat thalamus: Strictly segregated and partly overlapping projections. *Neurosci Lett*. 144:202–206.

Dura-Bernal S, Neymotin SA, Suter BA, Dacre J, Schiemann J, Duguid I, Shepherd GM, Lytton WW (2022). Multiscale model of primary motor cortex circuits reproduces in vivo cell type-specific dynamics associated with behavior. BioRxiv:1–31. 2022.2002.2003.479040.

Ebbesen CL, Brecht M (2017). Motor cortex—to act or not to act? *Nat Rev Neurosci*. 18:694–705.

Economo MN, Viswanathan S, Tasic B, Bas E, Winnubst J, Menon V, Graybuck LT, Nguyen TN, Smith KA, Yao Z, Wang L, Gerfen CR, Chandrashekar J, Zeng H, Looger LL, Svoboda K (2018). Distinct descending motor cortex pathways and their roles in movement. *Nature*. 563:79–84.

Edgerton JR, Jaeger D (2014) Optogenetic activation of nigral inhibitory inputs to motor thalamus in the mouse reveals classic inhibition with little potential for rebound activation. *Front Cell Neurosci*. 8 (36):1–11.

Eisinger RS, Cernera S, Gittis A, Gunduz A, Okun MS (2019). A review of basal ganglia circuits and physiology: Application to deep brain stimulation. *Parkinsonism Relat Disord*. 59:9–20.

Gaidica M, Hurst A, Cyr C, Leventhal DK (2018). Distinct populations of motor thalamic neurons encode action initiation, action selection, and movement vigor. *J Neurosci*. 38:6563–6573.

Galvan A, Devergnas A, Wichmann T (2015). Alterations in neuronal activity in basal ganglia-thalamocortical circuits in the parkinsonian state. *Front Neuroanat*. 9(5):1–21.

García-Cabezas MA, Martínez-Sánchez P, Sánchez-González MA, Garzón M, Cavada C (2009). Dopamine innervation in the thalamus: Monkey versus rat. *Cereb Cortex*. 19:424–434.

Goldberg JA, Boraud T, Maraton S, Haber SN, Vaadia E, Bergman H (2002). Enhanced synchrony among primary motor cortex neurons in the 1-methyl-4-phenyl-1,2,3,6-tetrahydropyridine primate model of Parkinson's disease. *J Neurosci*. 22:4639–4653.

Gornati SV, Schäfer CB, Rooda OHE, Nigg AL, De Zeeuw CI, Hoebeek FE (2018). Differentiating cerebellar impact on thalamic nuclei. *Cell Rep*. 23:2690–2704.

Guo K, Yamawaki N, Svoboda K, Shepherd GMG (2018). Anterolateral motor cortex connects with a medial subdivision of ventromedial thalamus through cell type-specific circuits, forming an excitatory thalamo-cortico-thalamic loop via layer 1 apical tuft dendrites of layer 5b pyramidal tract type neurons. *J Neurosci*. 38:8787–8797.

Guo ZV, Inagaki HK, Daie K, Druckmann S, Gerfen CR, Svoboda K (2017). Maintenance of persistent activity in a frontal thalamocortical loop. *Nature*. 545:181–186.

Hira R, Ohkubo F, Tanaka YR, Masamizu Y, Augustine GJ, Kasai H, Matsuzaki M (2013). In vivo optogenetic tracing of functional corticocortical connections between motor forelimb areas. *Front Neural Circuits*. 7(55):1–10.

Hosp JA, Pekanovic A, Rioult-Pedotti MS, Luft AR (2011). Dopaminergic projections from midbrain to primary motor cortex mediate motor skill learning. *J Neurosci*. 31:2481–2487.

Ilinsky IA, Kultas-Ilinsky K (2002). Motor thalamic circuits in primates with emphasis on the area targeted in treatment of movement disorders. *Movement Disord*. 17:S9–S14.

Inagaki HK, Chen S, Ridder MC, Sah P, Li N, Yang Z, Hasanbegovic H, Gao Z, Gerfen CR, Svoboda K (2022). A midbrain-thalamus-cortex circuit reorganizes cortical dynamics to initiate movement. *Cell.* 185:1065–1081.

Kha HT, Finkelstein DI, Pow DV, Lawrence AJ, Horne MK (2000). Study of projections from the entopeduncular nucleus to the thalamus of the rat. *J Comp Neurol.* 426:366–377.

Kha HT, Finkelstein DI, Tomas D, Drago J, Pow DV, Horne MK (2001). Projections from the substantia nigra pars reticulata to the motor thalamus of the rat: Single axon reconstructions and immunohistochemical study. *J Comp Neurol.* 440:20–30.

Kiritani T, Wickersham IR, Seung HS, Shepherd GM (2012). Hierarchical connectivity and connection-specific dynamics in the corticospinal-corticostriatal microcircuit in mouse motor cortex. *J Neurosci.* 32:4992–5001.

Krack P, Dostrovsky J, Ilinsky I, Kultas-Ilinsky K, Lenz F, Lozano A, Vitek J (2002). Surgery of the motor thalamus: Problems with the present nomenclatures. *Movement Disord.* 17:S2–S8.

Kultas-Ilinsky K, Sivan-Loukianova E, Ilinsky IA (2003). Reevaluation of the primary motor cortex connections with the thalamus in primates. *J Comp Neurol.* 457:133–158.

Kunori N, Kajiwara R, Takashima I (2014). Voltage-sensitive dye imaging of primary motor cortex activity produced by ventral tegmental area stimulation. *J Neurosci.* 34:8894–8903.

Kuramoto E, Fujiyama F, Nakamura KC, Tanaka Y, Hioki H, Kaneko T (2011). Complementary distribution of glutamatergic cerebellar and GABAergic basal ganglia afferents to the rat motor thalamic nuclei. *Eur J Neurosci.* 33:95–109.

Kuramoto E, Furuta T, Nakamura KC, Unzai T, Hioki H, Kaneko T (2009). Two types of thalamocortical projections from the motor thalamic nuclei of the rat: A single neuron-tracing study using viral vectors. *Cereb Cortex.* 19:2065–2077.

Larkum ME, Nevian T, Sandler M, Polsky A, Schiller J (2009). Synaptic integration in tuft dendrites of layer 5 pyramidal neurons: A new unifying principle. *Science.* 325:756–760.

Lavoie B, Parent A (1991.) Serotoninergic innervation of the thalamus in the primate: An immunohistochemical study. *J Comp Neurol.* 312:1–18.

Li N, Chen TW, Guo ZV, Gerfen CR, Svoboda K (2015). A motor cortex circuit for motor planning and movement. *Nature.* 519:51–56.

Llinás RR, Steriade M (2006). Bursting of thalamic neurons and states of vigilance. *J Neurophysiol.* 95:3297–3308.

Macchi G, Jones EG (1997). Toward an agreement on terminology of nuclear and subnuclear divisions of the motor thalamus. *J Neurosurg.* 86:77–92.

McFarland NR, Haber SN (2002). Thalamic relay nuclei of the basal ganglia form both reciprocal and nonreciprocal cortical connections, linking multiple frontal cortical areas. *J Neurosci.* 22:8117–8132.

Moore RY, Halaris AE, Jones BE (1978). Serotonin neurons of the midbrain raphe: ascending projections. *J Comp Neurol.* 180:417–438.

Ness TV, Remme MWH, Einevoll GT (2016). Active subthreshold dendritic conductances shape the local field potential. *J Physiol.* 594:3809–3825.

Newkirk GS, Guan D, Dembrow N, Armstrong WE, Foehring RC, Spain WJ (2021). Kv2.1 potassium channels regulate repetitive burst firing in extratelencephalic neocortical pyramidal neurons. *Cereb Cortex.* 32(5):1055–1076.

Otor Y, Achvat S, Cermak N, Benisty H, Abboud M, Barak O, Schiller Y, Poleg-Polsky A, Schiller J (2022). Dynamic compartmental computations in tuft dendrites of layer 5 neurons during motor behavior. *Science.* 376:267–275.

Pérez-Santos I, Palomero-Gallagher N, Zilles K, Cavada C (2021). Distribution of the noradrenaline innervation and adrenoceptors in the macaque monkey thalamus. *Cereb Cortex.* 31:4115–4139.

Phillips JW, Schulmann A, Hara E, Winnubst J, Liu C, Valakh V, Wang L, Shields BC, Korff W, Chandrashekar J, Lemire AL, Mensh B, Dudman JT, Nelson SB, Hantman AW (2019). A repeated molecular architecture across thalamic pathways. *Nat Neurosci.* 22:1925–1935.

Reiner A, Hart NM, Lei W, Deng Y (2010). Corticostriatal projection neurons: Dichotomous types and dichotomous functions. *Front Neuroanat.* 4(142):1–15.

Ren C, Peng K, Yang R, Liu W, Liu C, Komiyama T (2022). Global and subtype-specific modulation of cortical inhibitory neurons regulated by acetylcholine during motor learning. *Neuron.* 110:2334–2350.

Rovó Z, Ulbert I, Acsády L (2012). Drivers of the primate thalamus. *J Neurosci.* 32:17894–17908.

Sánchez-González MÁ, García-Cabezas MÁ, Rico B, Cavada C (2005). The primate thalamus is a key target for brain dopamine. *J Neurosci.* 25:6076–6083.

Sawyer SF, Young SJ, Groves PM (1989). Quantitative Golgi study of anatomically identified subdivisions of motor thalamus in the rat. *J Comp Neurol.* 286:1–27.

Schaltenbrand G (1977). *Atlas for Stereotaxy of the Human Brain.* Georg Thieme.

Schiemann J, Puggioni P, Dacre J, Pelko M, Domanski A, van Rossum MC, Duguid I (2015). Cellular mechanisms underlying behavioral state-dependent bidirectional modulation of motor cortex output. *Cell Rep.* 11:1319–1330.

Schuman B, Dellal S, Prönneke A, Machold R, Rudy B (2021). Neocortical layer 1: An elegant solution to top-down and bottom-up integration. *Annu Rev Neurosci.* 44(1):221–252.

Sheets PL, Suter BA, Kiritani T, Chan CS, Surmeier DJ, Shepherd GM (2011). Corticospinal-specific HCN expression in mouse motor cortex: I(h)-dependent synaptic integration as a candidate microcircuit mechanism involved in motor control. *J Neurophysiol.* 106:2216–2231.

Shepherd GMG, Yamawaki N (2021). Untangling the cortico-thalamo-cortical loop: cellular pieces of a knotty circuit puzzle. *Nat Rev Neurosci.* 22:389–406.

Sreenivasan V, Esmaeili V, Kiritani T, Galan K, Crochet S, Petersen CCH (2016). Movement initiation signals in mouse whisker motor cortex. *Neuron.* 92:1368–1382.

Takahashi N, Ebner C, Sigl-Glöckner J, Moberg S, Nierwetberg S, Larkum ME (2020). Active dendritic currents gate descending cortical outputs in perception. *Nat Neurosci.* 23:1277–1285.

Weiler N, Wood L, Yu J, Solla SA, Shepherd GMG (2008). Top-down laminar organization of the excitatory network in motor cortex. *Nat Neurosci.* 11:360–366.

Willard AM, Isett BR, Whalen TC, Mastro KJ, Ki CS, Mao X, Gittis AH (2019). State transitions in the substantia nigra reticulata predict the onset of motor deficits in models of progressive dopamine depletion in mice. *eLife.* 8:e42746:1–27. PMC6402832.

Williams SR, Stuart GJ (2000) Action potential backpropagation and somato-dendritic distribution of ion channels in thalamocortical neurons. *J Neurosci.* 20:1307–1317.

Wilson CJ, Higgs MH, Simmons DV, Morales JC (2018). Oscillations and spike entrainment. F1000Research 7(F1000 Faculty Rev-1960):1–10. PMC6305216.

Winnubst J, Bas E, Ferreira TA, Wu Z, Economo MN, Edson P, Arthur BJ, Bruns C, Rokicki K, Schauder D, Olbris DJ, Murphy SD, Ackerman DG, Arshadi C, Baldwin P, Blake R, Elsayed A, Hasan M, Ramirez D, Dos Santos B, Weldon M, Zafar A, Dudman JT, Gerfen CR, Hantman AW, Korff W, Sternson SM, Spruston N, Svoboda K, Chandrashekar J (2019). Reconstruction of 1,000 projection neurons reveals new cell types and organization of long-range connectivity in the mouse brain. *Cell.* 179:268–281.

Yamawaki N, Shepherd GM (2015). Synaptic circuit organization of motor corticothalamic neurons. *J Neurosci.* 35:2293–2307.

Yuste R, Gutnick MJ, Saar D, Delaney KR, Tank DW (1994). Ca2+ accumulations in dendrites of neocortical pyramidal neurons: An apical band and evidence for two functional compartments. *Neuron.* 13:23–43.

33.

CONNECTIVITY OF MOTOR CORTEX, MOTOR THALAMUS, AND STRIATUM

Bryan M. Hooks

OVERALL THALAMIC AND CORTICAL CIRCUIT ORGANIZATION

The thalamus provides the main source of long-range excitation to the cortex. Although most knowledge of thalamocortical interactions comes from the study of sensory systems, where the thalamus conveys sensory input to primary sensory areas, motor thalamus provides direct input for primary motor cortex and other frontal areas. As with sensory areas, where different nuclei provide layer-specific input to different cortical layers, motor thalamus also includes a range of thalamic nuclei which target the cortex in a layer-specific fashion. However, in contrast to sensory cortex, where primary thalamic input largely targets a granular layer 4 as the input layer, motor cortex is agranular, and thus is targeted differently by its thalamic inputs. To understand where those inputs enter motor cortex and how the thalamus and cortex are connected, this chapter will first describe the laminae, cell types, and local circuit organization of motor cortex. Then it will describe the long-range inputs to motor cortex, addressing both thalamocortical and corticocortical inputs. Next, the long-range targets of motor cortex, including feedback to thalamus, are described. Lastly, the targeting of striatum by both motor cortex and thalamus will be addressed.

LOCAL CIRCUIT ORGANIZATION OF PRIMARY MOTOR CORTEX

CELL TYPES AND LAMINAR ORGANIZATION

Primary motor cortex (M1) in rodents and primates corresponds to Brodmann area 4, a neocortical region whose cytoarchitecture lacks a thick dense granular layer 4 (Brecht et al., 2004; Tennant et al., 2011). In humans, this is just anterior to the central sulcus, where the primary somatosensory cortex (S1; including Brodmann areas 3a, 3b, 1, and 2) is just posterior to it, and is differentiated by a substantial granular layer 4 where the bulk of afferent somatosensory input arrives. Somatosensory and motor cortices are adjacent in other mammals as well. The central sulcus is greatly reduced in size in some primates and absent entirely in rodents, but M1 remains anteriorly adjacent to S1 across species (Figure 33.1).

Motor cortex contains a range of intermingled cell types, which makes disentangling its circuitry a challenge. The major cell types in neocortex are excitatory glutamatergic pyramidal neurons, typically born from radial glia precursors in prenatal development, as well as inhibitory GABAergic interneurons, which tangentially migrate from their embryonic origin in the ganglionic eminences (Network, 2021). These assort into layers, with each layer containing one or more cell types. So, what is a cell type in motor cortex? As in other systems to organize nature, this definition is necessarily imprecise and evokes animated debate among experts. In particular, often there is disagreement over exactly how to differentiate cells into larger or smaller groupings. Although there is not a consensus, I suggest that all of the following must be true of any single cell type in the nervous system:

(a) The cells must share a developmental origin. That is, they must be derived from progenitors which can produce that cell type (though progenitors potentially produce multiple cell types).

(b) The cells must share a molecular phenotype. That is, the collection of proteins they express should be substantially similar, enabling their clustering with other cells of the same type (and to the exclusion of other cell types). Note that it is not strictly necessary that a cell type have a single protein that defines it

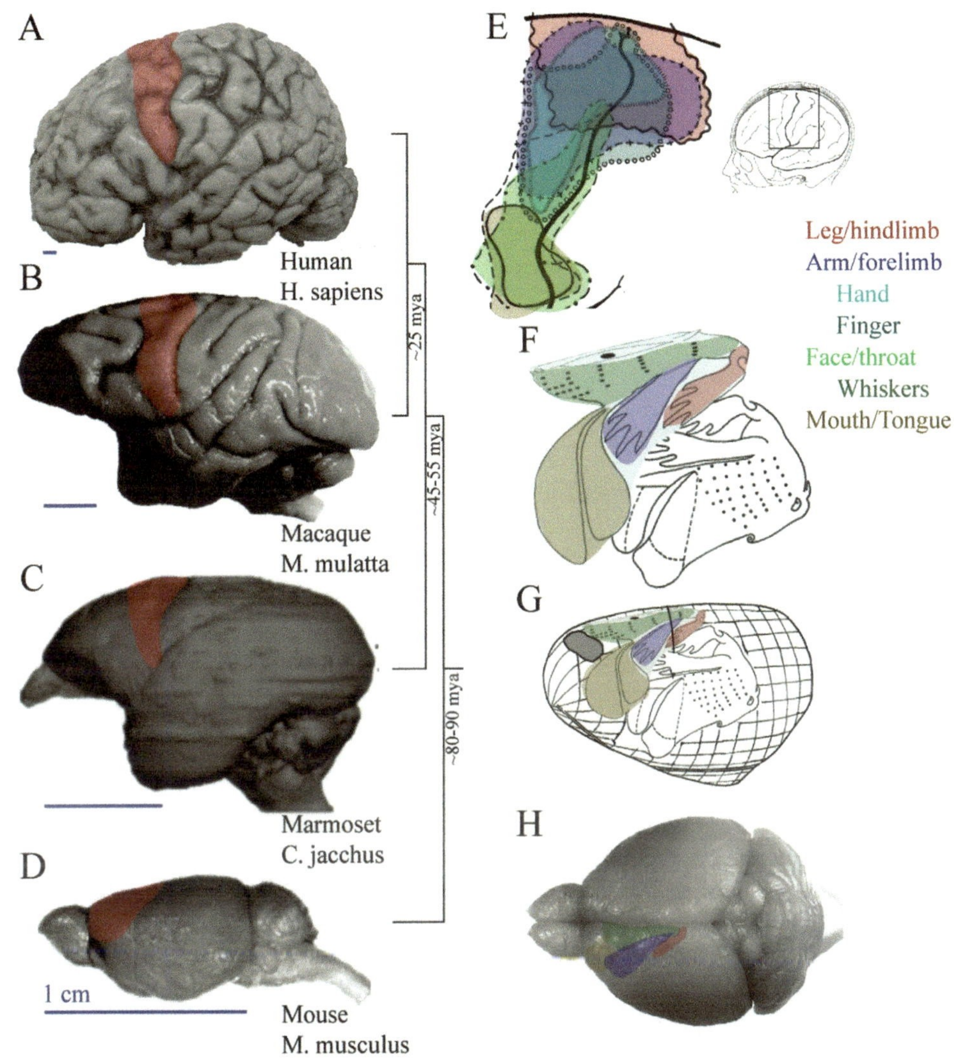

Figure 33.1. Defining primary motor cortex. Primary motor cortex, the lowest threshold cortical location to evoke movement from intracortical microstimulation, is located in frontal cortex of (A) human (Brodmann area 4), (B) macaque, (C) marmoset, and (D) mouse. The map is organized roughly somatotopically. In humans, the map proceeds from the lower limb medially to the trunk and upper limb laterally, with the face most lateral (E, adapted from Schieber (2011)). The primary sensory map is posteriorly adjacent with similar organization. The rodent map (F, adapted from Brecht et al., 2004, *J Comp Neurol*), has a similar somatotopic organization, though the relative sizes of different body parts, such as the whiskers, are of different relative area, roughly proportional to the relevance of whiskers to the animal.

to the exclusion of others (such as hemoglobin or parvalbumin), but that the expression level of a set of proteins it commonly expresses is in a similar ratio to other cells in that cell type.

(c) The cells must share anatomical characteristics, in particular that their dendrites tend to lie in the same brain region, permitting similar input and connectivity, while their axons project to the same targets, with similar degree of axonal elaboration.

(d) The cells must have similar functional properties, in particular shared characteristics of excitable membranes such as firing frequency and action potential shape (which suggest shared channel composition).

(e) The cells must have similar receptive field properties or be active in response to the same stimuli or motor planning. The degree to which this should be subdivided is open for debate (as, for example, whether cells with similar degrees of orientation tuning but in different angles are a distinct type might be disputed).

As it happens, cells generally share most of these characteristics. For each of these parameters, neurons are not uniformly distributed, but instead tend to cluster. Fast-spiking neurons are typically inhibitory interneurons in cortex, and only pyramidal neurons tend to have long-range corticocortical projections, for example. Thus, this definition addresses real categories because neurons are not encountered with randomly selected characteristics, but instead tend to organize into clusters of similar cells. Because of the diversity of cell types, a range of anatomical and functional approaches (corresponding to those in this definition) have been used to quantify what the different cellular components of the cortical circuit exist. Recent efforts (Network, 2021) to make a complete census of the neurons in the motor cortex have used molecular approaches, quantifying gene expression, anatomical assessment, quantifying patterns of axons and dendrites, as well as physiological means (such as recording firing patterns in response to depolarization).

By using similar approaches across species, such as rodent and primate, an approximate number of all types of cortical neurons can be organized hierarchically. For rodents, this identifies ~33 types of excitatory neurons and ~55 types of inhibitory neurons in M1 (Network, 2021). These neural types then will be the building blocks of the circuit diagrams. Due to experimental limits, however, most efforts to map cortical connectivity have been limited to treating these neurons as larger groups, either on the basis of cortical layer or improving upon that with projection pattern (using retrograde label of antidromic stimulation to identify long-range projections of pyramidal neurons), or via data from transgenic approaches that label certain cell types with Cre recombinase or fluorescent proteins. These approaches often lump neurons together, such as the multiple types of parvalbumin (PV) or somatostatin (SST) expressing interneurons, or the multiple classes of intratelencephalic (IT type) or extratelencephalic (ET type; also called pyramidal tract type, PT type) excitatory cells.

EXCITATORY CELL TYPES

Perhaps because excitatory cells make up the majority of cortical neurons, their local circuitry was among the first addressed once it was possible to map excitatory circuits. In addition to anatomical methods, such as single neuron filling and reconstruction, pair recording and other physiological methods have been used to map excitatory connectivity within the local circuit. Cortical pyramidal neurons are divided into subtypes (Harris and Shepherd, 2015), with the range of these described in Figure 33.2. These include neurons whose projections are exclusively within the telencephalon, called *intratelencephalic type* (IT type), which include several types whose somata are in all cortical layers except layer 1 (L1). All IT type neurons are pyramidal in shape, and have an apical dendrite that projects toward L1, as well as extensive basal dendrites. In mouse, the separation of layers 2 and 3 (L2/3) is more difficult than in primates, so these neurons are often grouped together, though the neurons immediately adjacent to L1 have a slightly different morphology than their more typical L2/3 peers since the apical dendrite often emerges bending slightly from the soma. The L2/3 IT type neurons project principally to ipsilateral and contralateral cortex, including the homotypic contralateral motor cortex. Some L2/3 IT type neurons do project to striatum, but less commonly than found in deeper layers. There is not a well-defined layer 4, as previously described, but L4-like pyramidal neurons are found here and connect as might be expected for L4 cells in other primary sensory areas (Yamawaki et al., 2014). Layer 5 has a rich variety of cell types. IT type L5 neurons project extensively to ipsilateral and contralateral striatum and cortex, with some variation in the degree of arborization in each target, as shown by single axon reconstructions. The apical dendrites of these neurons are more likely to be thin tufted, but reach L1 and ramify extensively. Layer 6 IT type neurons do not typically project to striatum. They have an apical dendrite, but it does not reach L1, typically terminating prior to reaching L2/3, and does not have an apical tuft. In some cases, inverted pyramidal neurons are encountered in L6 IT type neurons, with a downward projecting apical dendrite. The variety within layer 6B neuron types is not as well characterized. Their somata lie within white matter, which has made electrophysiological investigation more difficult.

Intermingled with these types are the *extratelencephalic type* (ET type) neurons, also called pyramidal tract type (PT type) layer 5 neurons. These have thick tufted apical dendrites with extensive L1 elaboration. In contrast to L5 IT type cells, ET type neurons do not project across the midline to target cortex and striatum. Instead, they send myelinated projections outside the telencephalon. Their cortical projections within the local circuit and to other cortical targets are attenuated compared to IT type neurons. Their axons do not project to contralateral cortex and striatum, but instead target ipsilateral areas in the cortex, striatum, and thalamus. As these axons pass through striatum, they extend a collateral. The degree of collateralization may differ across species, as the collateral is almost always observed, even dating back to the work of Cajal (1995). Functional connectivity may be of less relevance in primates compared to rodents. The axons passing through striatum seem myelinated and target multiple nuclei in thalamus, including motor thalamic areas (VA/VL), ventromedial thalamus (VM), and midline thalamic nuclei such as MD and PF. Axons continue in the internal capsule and collateralize to targets in brainstem, including superior colliculus, red nucleus, and pontine nuclei. Axons also contribute to the corticospinal tract (for those originating in forelimb and hindlimb regions). These do cross the midline at the pyramidal decussation in medulla. Corresponding pyramidal tract-like axons for facial regions, such as the whiskers, may cross the midline to target the facial motor nucleus.

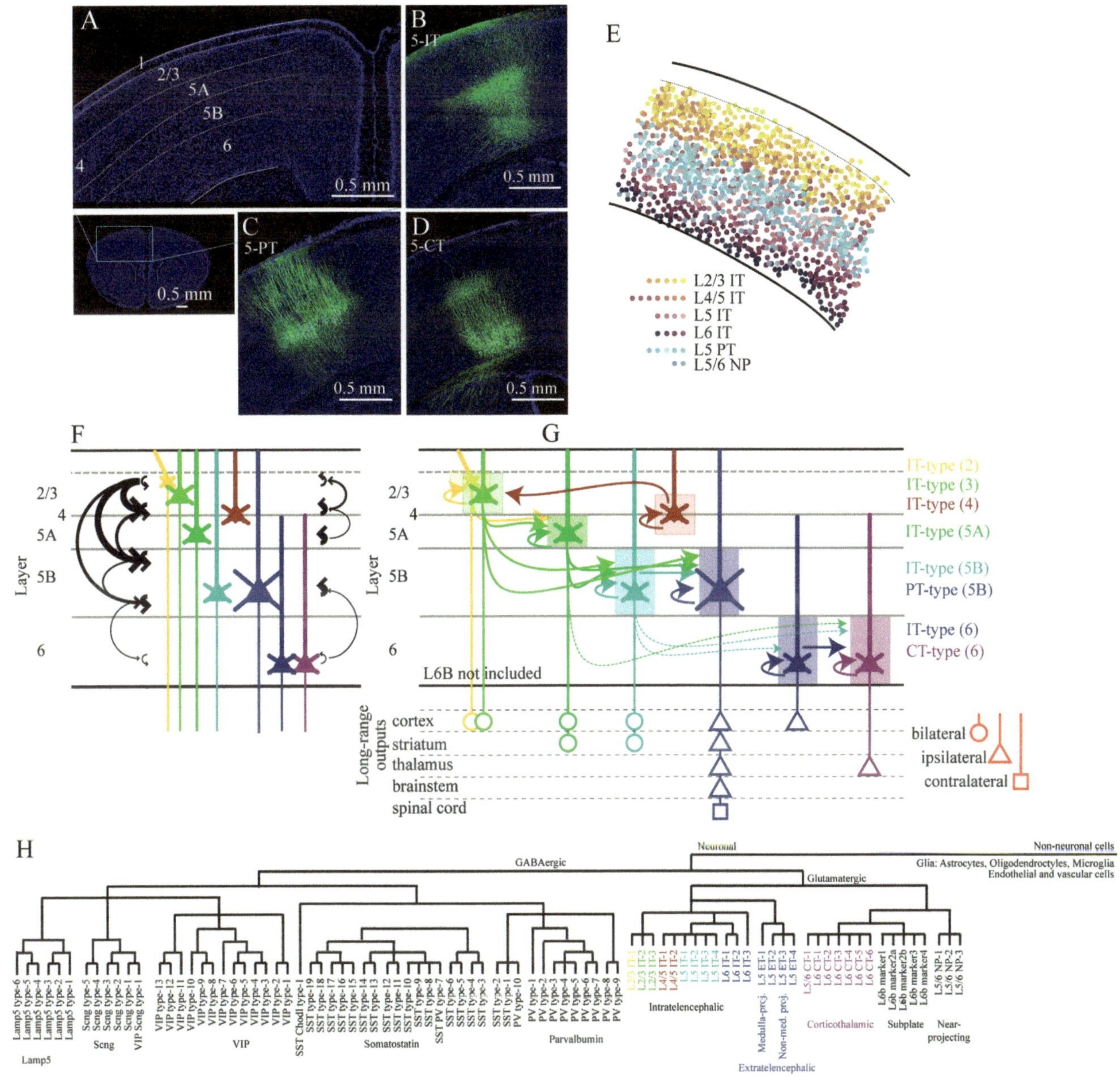

Figure 33.2. Laminar organization of primary motor cortex. Primary motor cortex, shown in coronal section stained in neurotrace blue (A), lacks a granular layer 4 (L4). The coronal section contains some L4 from somatosensory areas. Notably, the relative thickness of cortical layers varies when moving from lateral to medial agranular areas, with L2/3 and L5A becoming thinner, and L5B and L6 becoming relatively wider. (B–D). Transgenic mouse lines exist to label pyramidal neurons in a range of cortical layers, including IT type neurons of L5A and L5B (Tlx3_PL56-Cre, B), PT-type neurons of L5B (Sim1_KJ18-Cre, C), and CT type neurons (Ntsr1_GN220-Cre, D). GFP represents transfection of motor cortex with a Cre-dependent GFP reporter. Somata and apical dendrites can be seen (B–D). (E). A range of molecular subtypes of pyramidal neurons has been defined in motor cortex, which include at least 5 subtypes of IT neurons in L2/3, as well as several IT and PT types in layers 4 and 5 (adapted from BICCN, 2021, *Nature* 598:86–102). (F). These excitatory neurons share excitatory connectivity patterns with other cortical areas, where local connections are dominated by within layer excitatory connectivity and major translaminar projections descending from L2/3 to L5A and L5B (adapted from Weiler et al., 2008, *Nat Neurosci*). (G) Accounting for different excitatory cell types within layers, a characteristic connectivity pattern in motor areas is for IT type neurons in L5 to connect to each other and PT type neurons, but for the long-range output cells (L5 PT type, for example) to connect to each other and their long-range targets but not strongly feed back to colaminar IT type neurons (adapted from Shepherd and Harris, 2015, *Nat Neurosci*). The pattern is the same with L6 for L6 IT type neurons and L6 CT type neurons. (H) The consensus taxonomy of cortical interneurons and pyramidal neurons in primary motor cortex, based on molecular and physiological evidence.

Assessing the number of types of ET type neurons has been of interest, since they potentially collateralize to a large number of targets. However, data from single axon reconstructions, as well as expression data, suggest that there are about 3 subtypes in motor areas in mouse (Economo et al., 2018), including a subset that projects to thalamus but not medulla, and a medulla projecting subset (Winnubst et al., 2019). Primates do include one specialized type of corticospinal neuron, the Betz cell, which is not present in rodents and is marked by a significantly larger neuronal somata (Bakken et al., 2021). It is worth noting that these cells still are only a small minority of all ET type cells in primates (Nudo and Masterton, 1990).

Lastly, layer 6 neurons are varied, including both corticocortical (IT type) and *corticothalamic* neurons (CT type). Though layer 6 of motor cortex has a low firing rate during behavior, it is notable for containing a relatively large number of cells due to its increased layer thickness and cell density. This might enable these neurons to contribute to motor function even with relatively sparse firing. The corticothalamic neurons project to the corresponding motor thalamic nuclei mentioned above. The deepest region of layer 6, called layer 6B, contains excitatory neurons that have received the least attention but potentially have important roles to play, as they seem genetically heterogeneous, including several cell types. These cell bodies are found within denser areas of white matter, which becomes abundant at the bottom of the cortex. Furthermore, they seem to project somewhat more broadly within the cortex compared to excitatory neurons in other layers, as retrograde label in corticocortical injections finds these cells in many cortical areas. In addition to some identified unique primate and human specializations, the relative proportions of these cell types may vary across species. Specifically, L2/3 IT type neurons are likely overrepresented in humans, while L6 CT type neurons are overrepresented in mice (Bakken et al., 2021).

The local connectivity of M1 is now known to replicate some of the main pathways of other neocortical areas. Specifically, within-layer excitatory connectivity is high between pyramidal neurons, as is the case for sensory cortices (Lefort et al., 2009). Major translaminar projections include the descending connections from L2/3 to L5A and the upper half of L5B (Weiler et al., 2008; Hooks et al., 2011). There is also a small ascending projection from L4-like neurons to L2/3 cells (Yamawaki et al., 2014). However, a remarkable additional rule emerges when the subtypes of neurons within each layer are considered. Neurons whose projections are restricted to the local circuit or to the telencephalon (IT type neurons) seem to fall earlier in the pathway than output neurons which project to subcortical targets (ET type/PT type and CT type). Thus, L5 IT type neurons connect to other L5 IT type neurons as well as L5 PT type cells. In contrast, L5 PT type neurons, which do connect to other L5 PT type neurons, do not project to IT type neurons (Anderson et al., 2010). Thus, output seems to be funneled from IT populations to PT type in order to leave the local circuit. Furthermore, this rule also exists within L6 for IT and CT type neurons, where IT type projections connect to L6 CT type cells (but not vice versa) (Yamawaki and Shepherd, 2015). Collectively, the results of local circuit computations seem to be directed to cortical output cells in M1.

INHIBITORY CELL TYPES

The range of morphological phenotypes of interneurons has given rise to the idea that these neurons, though representing only ~10%–20% of cortical neurons in mice, contain many more types than their excitatory counterparts (Tremblay et al., 2016). As in other cortical areas, there are 3 main categories (Figure 33.3) which differ in their developmental origin and laminar distribution, including the parvalbumin-expressing (PV+), somatostatin-expressing (SST+), and a third class marked by a serotonergic receptor (5-HT_{3a}R+). First, PV+ interneurons are present throughout layers 2–6. Though there are ~10 subtypes, the most common and best known is called the basket cell, which is a fast-spiking (high firing rate with narrow spikes) interneuron which targets pyramidal neuron somata (thus the basket shape of the axons). Another well-known subtype is the chandelier cell, which targets the axon initial segment, giving their axons a uniquely identifiable appearance. Although the PV gene has proven a reliable marker for a subset of GABAergic interneurons, PV expression does occur in some pyramidal neurons. Overall, PV+ interneurons comprise about 40% of cortical GABAergic cells. SST+ neurons are also about 40% of inhibitory cells. The range of subtypes is more numerous than PV+ cells, and are found across layers 2–6 as well, with some cells in L1. These are more likely to have a regular spiking phenotype electrophysiologically, in some cases with a lower threshold for spiking. The Martinotti SST+ subtype is well known as for its L1 targeting axon. There is one interneuron type coexpressing PV+ and SST+ . Lastly, the 5-HT_{3a}R+ interneurons include the remainder of GABAergic cells. Their distribution is restricted to layers

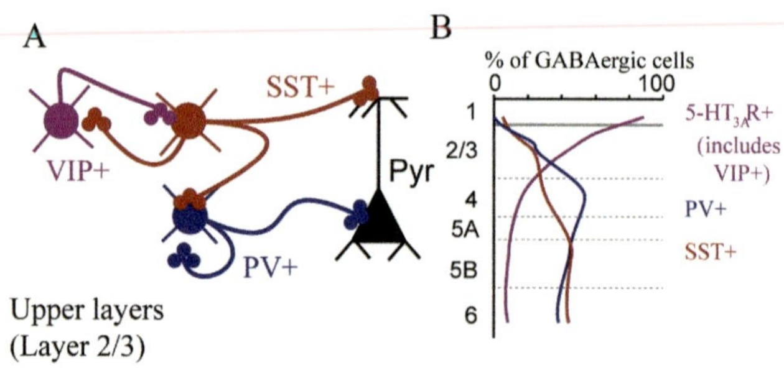

Figure 33.3. Canonical inhibitory connectivity in neocortex. (A). VIP+ interneurons disinhibit cortex by inhibiting SST+ neurons. SST+ cells, which tonically inhibit PV+ neurons and pyramidal (Pyr) neuron apical dendrites, are thus inhibited. Tonic inhibition is relieved by VIP+ neuron activation (Pfeffer et al., 2013). (B). Inhibitory interneurons vary in their laminar distribution across cortex. 5-HT3AR+ neurons are concentrated in L1 and L2/3, including the VIP+ interneurons. PV+ and SST+ interneurons are present across all layers except L1. After Lee et al. (2010).

1, 2/3, and 5A in the cortex. Layer 1 is generally cell sparse, filled mostly with axons contacting distal dendrites; 5-HT_{3a} R+ subtypes are the majority of layer 1 interneurons. One important 5-HT_{3a} R+ which has received substantial attention is the bipolar vasopressin intestinal peptide (VIP+) expressing cells (Lee et al., 2010).

Molecular identification of interneurons across the cortex has enabled the use of tools to study their circuits and identify patterns of connectivity (Hippenmeyer et al., 2005; Taniguchi et al., 2011). Inhibitory connectivity plays an important role in setting cortical gain and preventing runaway excitation (that might result in seizures). Although the degree to which inhibition is specifically integrated into cortical microcircuits is not yet clear, there is strong evidence to suggest that 2 of the major classes of interneurons (PV+ and SST+) connect relatively densely with most pyramidal neurons in their local circuits (Bock et al., 2011; Fino and Yuste, 2011; Packer and Yuste, 2011). The canonical inhibitory microcircuit across cortical areas is as follows: PV+ interneurons mainly act to mediate feedforward inhibition, by inhibiting local pyramidal cell somata, reducing the window for synaptic integration and bursting. SST+ neurons tonically inhibit PV+ neurons and pyramidal neuron apical dendrites in the local circuit. Thus, their activity may show an opposing pattern. This tonic inhibition can be released when VIP+ neurons become active (Pfeffer et al., 2013; Pi et al., 2013). This canonical circuit seems to exist across neocortex, including auditory, somatosensory, and visual sensory cortices, as well as frontal areas. The stimuli that activate VIP+ circuits to trigger local disinhibition vary between regions, however. Furthermore, because the distribution of VIP+ neurons is not uniform across cortical layers, it is not as clear how this circuit operates in deeper cortical layers, like layers 5 and 6. Because different long-range inputs enter the cortical circuit in different layers of motor cortex, this may mean that different rules for recruitment of feedforward inhibition exist across different laminae.

LAMINAR ORGANIZATION OF THALAMOCORTICAL INPUTS

Long-range inputs to primary motor cortex originate from a range of cortical and thalamic areas (Oh et al., 2014; Zingg et al., 2014). The long-range inputs to M1 are summarized in Figure 33.4, showing the general areas providing input. The laminar and cell type specificity of these inputs is shown in Figure 33.5. With few exceptions, such as diffuse neuromodulatory projections from brainstem, virtually all long-range inputs to M1 are excitatory glutamatergic inputs. The origin and targeting of these projections have been dissected using a range of anatomical and physiological techniques, which have given rise to a somewhat orderly picture of the nodes in the local circuit where information from other brain areas enters the motor cortex.

Motor areas of thalamus project directly to M1. The canonical nuclei of motor thalamus are VA (ventroanterior thalamic nucleus) and VL (ventrolateral thalamic nucleus). The motor thalamus is larger and easier to subdivide in humans and nonhuman primates (NHPs) than in rodents (Herkenham, 1980), resulting in the proliferations of names for subdivisions of motor thalamus in primates (Macchi and Jones, 1997). In general, the more anterior area of motor thalamus (VA) receives output from basal ganglia, such as substantia nigra (SNR). The posterior region (VL) reflects the cerebellar output (Hoover and Strick, 1999; Ilinsky and Kultas-Ilinsky, 2002; Phillips et al., 2019). The mouse motor thalamus is called VAL in some annotations, due to the difficulty of separating its components, but it is functionally heterogeneous. In rodents, these areas differ in GAD67 immunoreactivity, with higher GAD67 expression delineating the more anterior and ventral regions (Kuramoto et al., 2009). In mouse, axons from more anterior motor thalamus arborize in layers 1, 2/3, and 5. Axons from more posterior motor thalamus arborize in layers 1 and 2/3 only (Hooks et al., 2013). These also differ in the cells targeted. Posterior components target L2/3 and L5A pyramidal neurons with a pattern similar to S1. This is consistent with data from NHPs, where both S1 and VL target these upper layers, but show differences in plasticity following motor learning (Iriki et al., 1989; Kaneko et al., 1994). However, the laminar pattern of excitation differs for interneurons, with VL and PO axons more likely to target PV+ neurons in L5A and L5B than in L2/3. Thus, these axons are capable of targeting specific neurons in specific layers that vary with postsynaptic identity (Okoro et al., 2022). The more anterior components of VA/VL target excitatory cells in L2/3, L5A, and L5B, including monosynaptic inputs to PT type neurons. In this way, both frontal cortex and some parts of motor thalamus seem to have privileged access to the long-range output neurons of M1. Notably, the posterior nucleus (PO), which is often considered as a higher-order somatosensory nucleus, also projects directly to M1. It is adjacent to VL anatomically, and has a similar pattern of axonal arborization. Indeed, the PO axons can arborize somewhat broadly across S1 and M1 (Ohno et al., 2012).

Ventromedial (VM) thalamus also projects to M1. VM seems to maintain topographic organization such that more lateral regions interconnected with M1 are segregated from more medial areas connected to frontal cortex (such as the anterolateral motor area, ALM; Komiyama et al., 2010). VM axons target layer 1 more specifically (Kuramoto et al.,

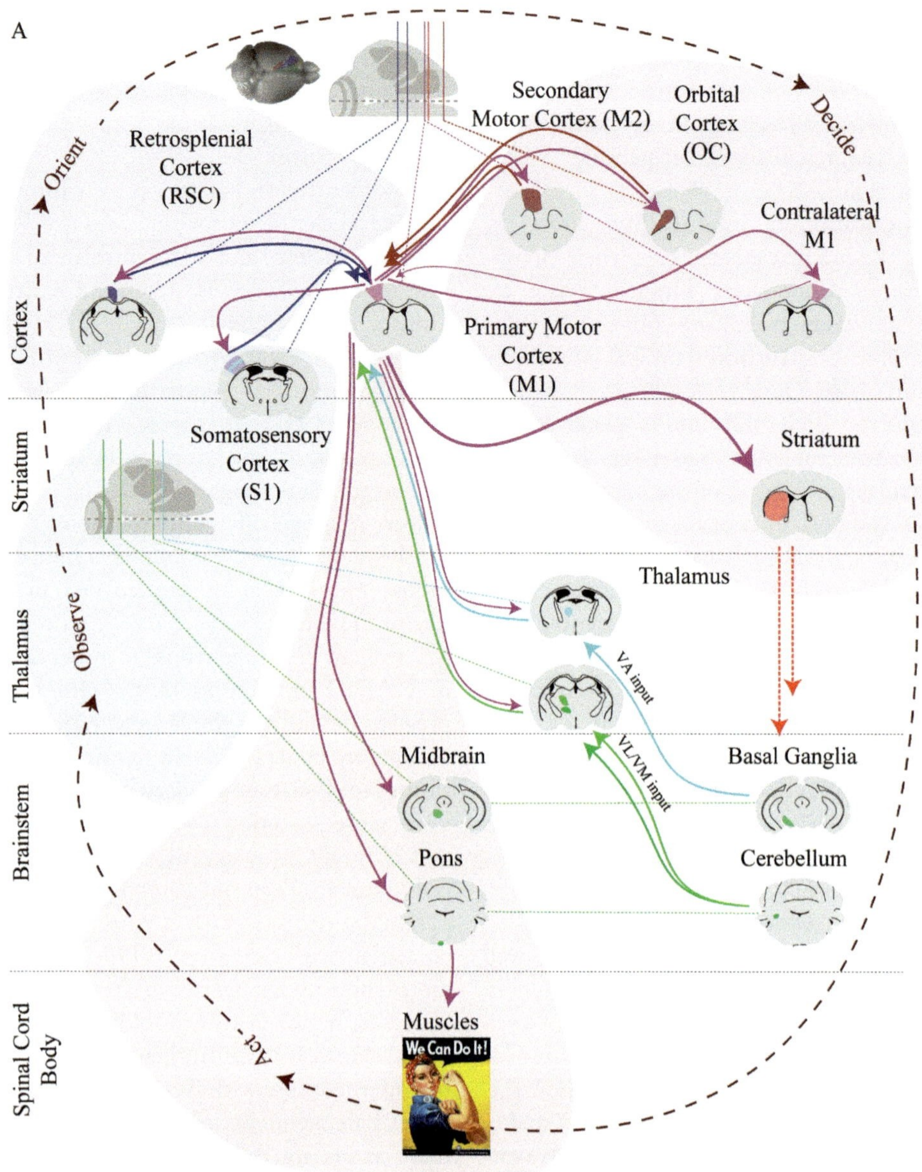

Figure 33.4. Long-range motor cortex pathways. (A). Schematic of the major brain areas connected to motor cortex in rodent. Major brain areas (cortical regions and thalamic nuclei) are presented, without regard for the specific cell types connected. Regions are shown as a single coronal section (many regions, such as striatum, will actually receive connections at many points along the rostrocaudal axis). Regions are hierarchically organized from cortex and thalamus (at top) to brainstem and spinal cord at bottom. This regional organization will be detailed at later points with cell type specificity. Rough depiction of how the motor system fits into the sensory-motor loop (Observe-Orient-Decide-Act) in which organisms sense and respond to their environment.

2015). Based on recordings in frontal areas, it can be inferred that VM axons excite L2/3 IT type neurons and L5B PT type neurons quite strongly, with the strongest connectivity in L1 dendrites (Guo et al., 2018). Notably, input to L6 CT type neurons is very weak, though L6 CT type neurons do directly excite VM neurons. Because VM is implicated in motor planning, it is hypothesized that persistent activity in this circuit plays a role in maintaining a motor plan.

The connectivity of M1 with other thalamic nuclei, such as parafascicular nucleus (PF) and mediodorsal nucleus (MD), is not as well characterized. Since VM and PF seem to be the targets of basal ganglia output via SNR, one pathway by which basal ganglia closes the cortical loop is for the output of VM and PF to return to cortex (Lee et al., 2020). This is consistent with the general theme of M1 circuit organization that output is largely composed of loops, including output to basal ganglia, output to thalamus, and in some cases output to other neocortical areas, that can be fed back to M1. These inputs will be fed through local circuits and in a polysynaptic manner impinge on output neurons in L5B (PT type) and L6 (CT type) that will project to subcortical targets.

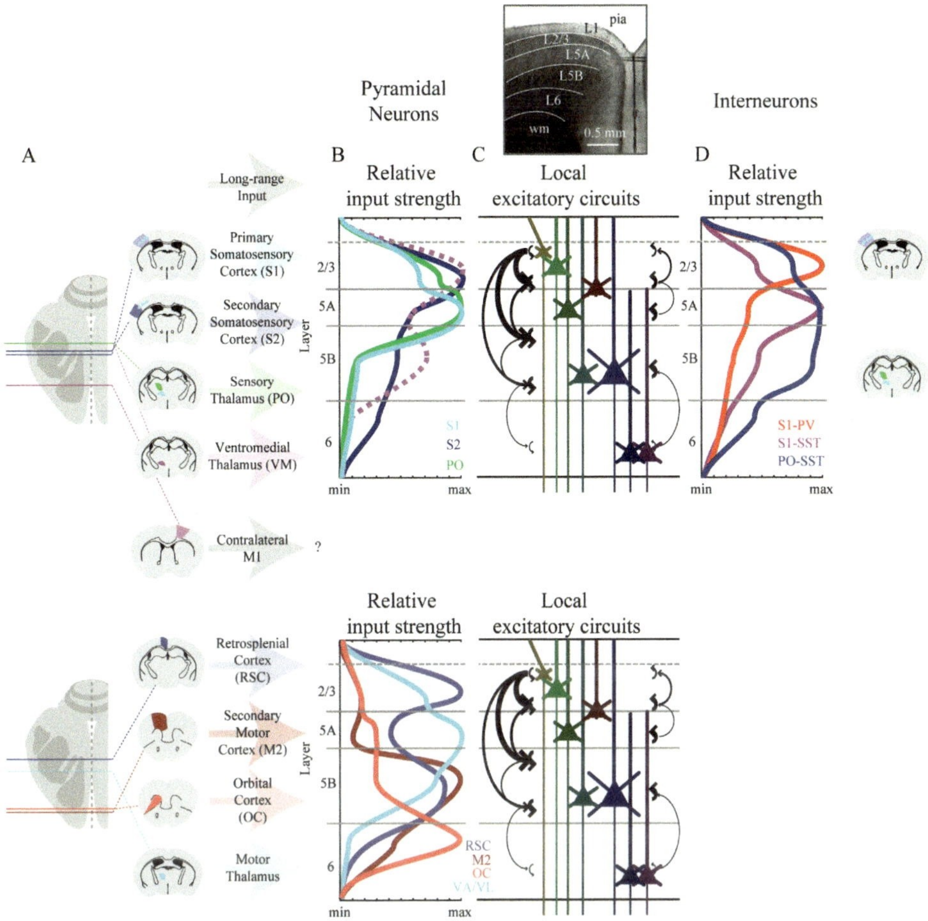

Figure 33.5. Long-range motor cortex inputs. (A). Schematic of the major brain areas providing long-range excitation to mouse motor cortex. Cartoon and sections show anterior/posterior locations. (B). Monosynaptic excitation to pyramidal neurons is shown with relative input strength as a function of cortical layer. Input is divided into 2 rows for clarity. (C). Adjacent cartoon shows typical local connectivity. (D). Long-range inputs to inhibitory interneurons are less well characterized, but differ depending on the interneuron subtype and layer depth. Here, inputs for PV+ and SST+ interneurons are shown.

LAMINAR ORGANIZATION OF CORTICO-CORTICAL INPUTS

Retrograde labels injected into M1 label a range of inputs across ipsilateral and contralateral cortex (Hooks et al., 2013). The strongest ipsilateral cortico-cortical projections to M1 originate in primary somatosensory cortex (S1) in rodents, making this an ideal rodent model to understand cortico-cortical connectivity (Mao et al., 2011). Both S1 and M1 are somatotopically organized, as described above, with the largest component of S1 including the whisker (barrel) cortex, that is reciprocally connected with whisker motor cortex, which lies at the anterior and medial extent of M1. Anatomically, the lateral and medial agranular areas of mouse have been referred to either AGl and AGm or primary and secondary motor cortex, differentiated by the more medial areas having thinner L2/3 and L5A and thicker L5B and L6 closer to the midline flexure. Further, the axons from whisker S1 target whisker M1 slightly laterally within the medial agranular field, such that there might be a slight offset between the microstimulation-defined M1 location for whiskers and the field of somatotopically corresponding M1 axons.

Topographically, the S1 to M1 projection maintains some precision, with more lateral sensory projections targeting more medial motor areas, though the axons arborize over a relatively long anterior/posterior area (the medial/lateral profile is still narrow). In contrast, S1 to M1 projections are not as dense in primates compared to rodents, despite the alignment of the sensory and motor homunculi along the central sulcus. Because in primates the tools needed to dissect these circuits in cell-type-specific detail are lacking, the descriptions below will focus on mouse models. The projections from M1 to S1 originate primarily in L5A, with weaker contributions from L2/3 and L5B and 6. This distinct pattern is shown with retrograde label. The M1 targets follow a pattern that is layer specific as well as cell type specific. Though S1 axons appear to cross all layers, their strongest functional connections target L2/3 and L5A pyramidal neurons, with substantially weaker connectivity to the subcortical output neurons in L5B and 6. It is notable that this laminar pattern is not the same for interneurons. PV+ neurons in L2/3 are the most strongly targeted, with much weaker connections to other layers. The most strongly targeted SST+ neurons reside in L5A and L5B. Thus, as with thalamic input, S1 to M1 axons differently target excitatory and inhibitory neurons in specific layers (Okoro et al., 2022).

Other cortical areas also project to M1. Lateral to S1, secondary somatosensory cortex (S2) also projects directly to M1 (Suter and Shepherd, 2015). Medially, the retrosplenial cortex (RSC) directly projects to medial portions of M1, potentially linking information about memory and spatial navigation to the motor system. This projection does not target lateral M1 areas as densely (more forelimb-associated areas), so it is not clear whether the connectivity rules are generally applicable (Yamawaki et al., 2016). Nevertheless, RSC axons arborize across motor cortex and monosynaptically excite pyramidal neurons in most layers. This excitation seems to peak in L2/3 and also to be strong in deeper areas such as the lower half of L5B, a pattern quite different from S1 input.

Anteriorly, multiple premotor areas connect to M1. These include dorsal premotor areas, which have been called the secondary motor cortex (M2) (Hooks et al., 2013), as well as the rostral forelimb area (Tennant et al., 2011) and the anterior lateral motor area (ALM) (Komiyama et al., 2010). Topographically, more medial premotor regions connect more strongly to the whisker motor areas (also more medial), while the more lateral areas connect more strongly to forelimb areas (Zingg et al., 2014). In sharp contrast to the S1 projection targeting the upper layers of M1, frontal cortex prefers to excite deeper layer cells, including L5B and L6 (Hooks et al., 2013). Indeed, these projections monosynaptically excite both IT type and ET type layer 5 neurons, potentially providing a direct means for frontal regions to modulate the motor output of M1 to the corticospinal tract. Furthermore, another frontal area, the orbital cortex, projects directly to layer 6 in M1 in mouse. The function of this is unknown.

Hierarchies between cortical areas govern their interconnectivity. Motor cortex is no exception. Primary sensory areas are the lowest level of the hierarchy, including S1. When these project to higher areas in the hierarchy, such as M1, they target the upper layers of cortex (L2/3). In contrast, when higher areas project to lower levels, they are likely to excite the deeper layers of cortex (L5B and L6). This is true of M1 output to S1 (Kinnischtzke et al., 2016) as well as frontal cortex (M2 and OC) output to M1 (Hooks et al., 2013). However, the overall picture includes some subtleties (Figure 33.5). RSC is not clearly above or below M1 in the cortical hierarchy, and projects to M1 in a complex pattern, exciting L2/3 but also L5B. Furthermore, the laminar pattern of excitation for any given projection to M1 (for example, S1 to M1) may vary depending on whether the targets are pyramidal neurons or interneurons.

LONG-RANGE TARGETS OF MOTOR CORTEX

Although the roles of motor control in the learning (Kawai et al., 2015) and execution (Guo et al., 2015; Sauerbrei et al., 2020) of motor performance are still debated, it is undeniable that the results of M1 computations must be sent to other parts of the brain and the body to result in movement, as corticomotoneuronal cells are sparse (except for human specializations for dexterous hand movements) (Lemon, 2008). These targets are numerous (Figure 33.6). The corticocortical hierarchy has been discussed briefly above and applies to M1 output to these other areas. Strong targets are typically those providing strong input, so reciprocal connectivity is seen with S1 and frontal cortices (including M2 and OC) (Oh et al., 2014; Zingg et al., 2014). Other areas also targeted by M1 include S2 and RSC. This is not necessarily direct connectivity of projection neurons (2-cell loops), but certainly reciprocal area-to-area loops are present. The laminar nature of these projections is consistent with hierarchical organization: M1 to S1 projections arborize more extensively in infragranular versus supragranular layers (Kinnischtzke et al., 2014), though it is also worth noting that M1 strongly targets layer 1 as well (Lee et al., 2013). M1 projections to more anterior targets (as well as contralateral projections to M1) tend to also have extensive L1 axons, but substantial axon arborization in L2/3, L5A, and L5B.

The major descending projection is from M1 cortex to caudate and putamen (collectively the striatum). This projection is thought to be organized into functionally segregated loops, with M1 sharing targets with a subset of related cortical regions (Alexander et al., 1986). This output mainly originates from L5 neurons, including both IT type and PT type cells projecting to striatum (Hintiryan et al., 2016; Hooks et al., 2018). L2/3 neurons also contribute. Note that the role of PT type collaterals in striatum may be reduced in primates relative to rodents (Turner and DeLong, 2000). While PT axons target exclusively ipsilateral targets, IT type neurons have a range of innervation patterns, including extensive targeting of contralateral cortex (Economo et al., 2016; Winnubst et al., 2019). The range of individual axon morphology includes single IT type cells that arborize in both ipsilateral and contralateral cortex and striatum. But some IT type neurons remain exclusively ipsilateral, while others are mostly contralateral.

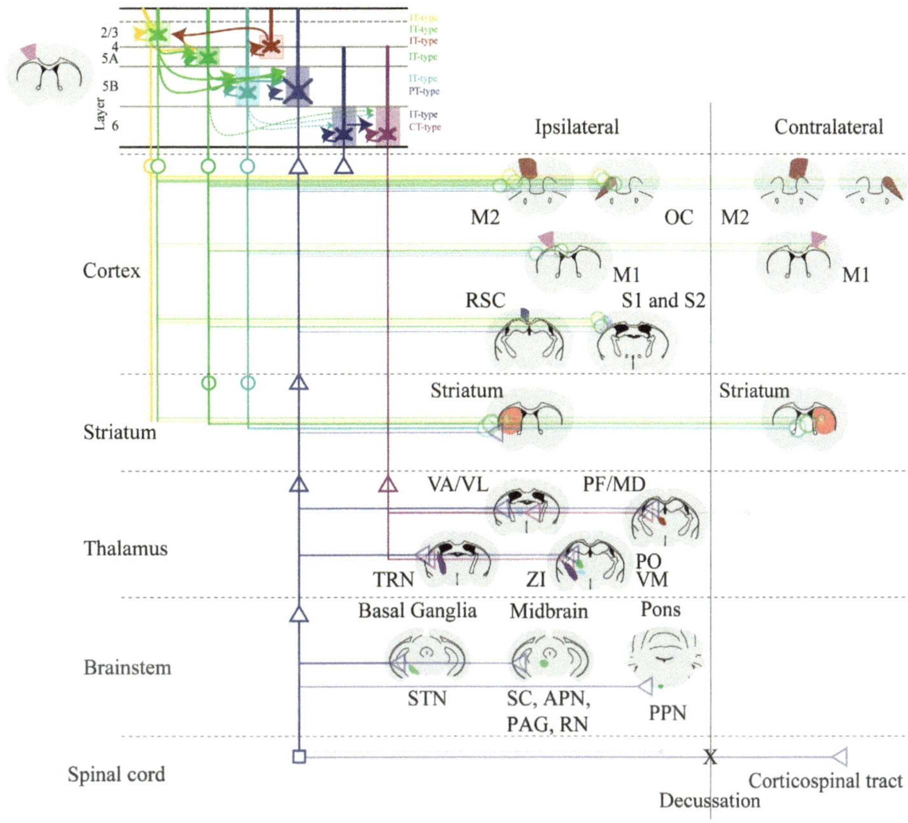

Figure 33.6. Long-range motor cortex projections. Schematic of the major brain areas targeted by motor cortex in rodent. Major brain areas (cortical regions and thalamic nuclei) are presented, without regard for the specific cell types connected. Regions are shown as a single coronal section. Diagram at top shows the local circuit of M1 and the identity of the output neurons (color coded as before). Targets are categorized as being ipsilateral (center column) or contralateral (right column) to the origin of the M1 projection. Thickness of the line is subjectively proportional to projection strength. Abbreviations: APN = anterior pretectal nucleus; CT = corticothalamic type; IT = intratelencephalic type; MD = mediodorsal thalamic nucleus; M1 = primary motor cortex; M2 = secondary motor cortex; OC = orbital cortex; PAG = periaqueductal gray; PF = parafascicular thalamic nucleus; PO = posterior thalamic nucleus; PPN = pedunculopontine nucleus; PT = pyramidal tract type; RN = red nucleus; RSC = retrosplenial cortex; SC = superior colliculus; STN = subthalamic nucleus; S1 = primary somatosensory cortex; S2 = secondary somatosensory cortex; TRN = thalamic reticular nucleus; VA/VL = ventroanterior/ventrolateral thalamic nucleus; VM = ventromedial thalamic nucleus; ZI = zona incerta.

Thalamus is targeted exclusively by PT type and CT type axons, which differ substantially in function and morphology (Sherman and Guillery, 2006). The CT type projection originates from more neurons and targets all the nuclei mentioned previously that project to M1: VA, VL, VM, PF, and MD. Furthermore, the inhibitory nuclei (thalamic reticular nucleus [TRN] and zona incerta [ZI]) are monosynaptically innervated (Urbain and Deschenes, 2007), which may also play a role in gating sensory responses (Lee et al., 2013). Of interest, the M1 projection to thalamus is now known to cross the midline in rodents (Hooks et al., 2013; Szczupak et al., 2021), though the functional significance is not well-known.

M1 projections to midbrain then directly target basal ganglia output in the hyperdirect pathway to subthalamic nucleus (STN) (Gittis et al., 2020; Shepherd, 2013). These axons also target a number of brainstem nuclei, including the red nucleus (RN), the anterior pretectal area (APT), superior colliculus (SC), and basal pontine nuclei such as the pedunculopontine nucleus (PPN). This is certainly true for somatic motor control, such as the legs and arms, though the motor pathways for the facial regions generally terminate in brainstem and may cross the midline to target nuclei such as the facial motor nucleus. Many of the brainstem targets become the sites of origination of separate descending motor control pathways, including the rubrospinal tract and reticulospinal tract (Lemon, 2008). Axons proceed through the pons and enter the medulla (where the tract is known as the medullary pyramids), crossing the midline at the pyramidal decussation to form the lateral corticospinal tract in the lateral portion of spinal white matter. A small, uncrossed portion of this projection is called the anterior corticospinal tract, descending along the midline of spinal cord (they cross at the spinal level they innervate). These pathways then arborize in the laminae of the ventral horn of the spinal cord.

ORGANIZATION OF STRIATAL INPUT

Primary motor cortex's major subcortical target is striatum (caudate and putamen), the input nucleus of the basal ganglia (Figure 33.7). Most of mouse cortex, including M1, sends afferents to striatum. Striatum can generally be divided into dorsolateral, dorsomedial, and ventral areas, with nucleus accumbens part of ventral striatum (Gittis et al., 2020). Each striatal region receives input from specified regions of cortex, with somatosensory and motor inputs generally targeting dorsolateral striatal areas (Hintiryan et al., 2016; Hooks et al., 2018; Hunnicutt et al., 2016). Generalized pathways from the cortex through basal ganglia have long been considered to exist as segregated loops from related cortical areas projecting to overlapping areas of striatum (Alexander et al., 1986). The loop is closed by basal ganglia output from globus pallidus and substantia nigra efferents that target thalamus, which projects back to the associated cortical areas. While cortex provides the major excitatory input, thalamus also provides glutamatergic excitation.

Cortical input to striatum originates from L5, including L5 IT type and PT type neurons, as well as L2/3 IT type neurons to a smaller extent (Gittis et al., 2020). These projections are bilateral for IT type axons, but unilateral for PT type cells. The exact extent to which IT type versus PT type inputs play a role is unclear. That PT type neurons anatomically send a collateral into striatum has been known since Cajal (1995). This is true for both primates and rodents. But the functional import may differ between these species, as recordings in intact macaque motor cortex reveal that IT type neurons in M1 but not PT type neurons can be antidromically stimulated from striatum, suggesting that PT type axons distribute in striatum less extensively than IT type projections (Turner and DeLong, 2000). In

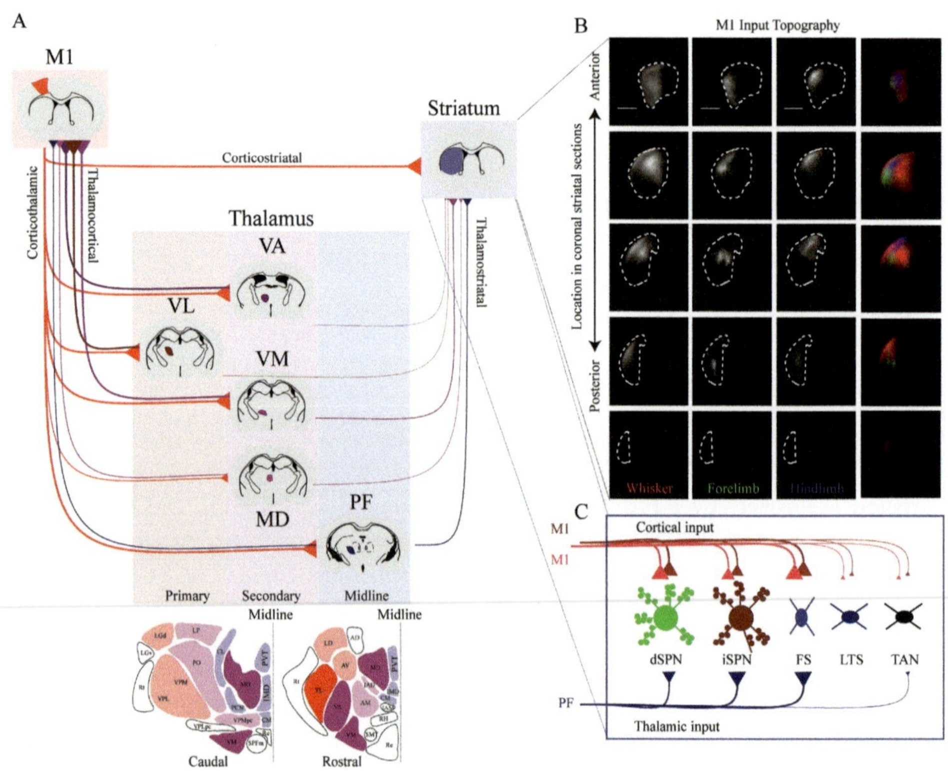

Figure 33.7. Organization of corticostriatal and thalamostriatal inputs. (A). Schematic of corticothalamic, corticostriatal, thalamostriatal, and thalamocortical connectivity for motor cortex and associated thalamic nuclei. Relative line thickness is proportional to input strength. Thalamic nuclei are categorized by their molecular identity as primary (red), secondary (purple), and midline (blue) nuclei (after Phillips et al., 2020). Locations with respect to other nuclei are shown below. (B). Cortical inputs to striatum show somatotopy. In coronal sections of striatum (outlined in white dashed lines), striatal regions targeted by whisker (left), forelimb (center), and hindlimb (right) are shown. These same inputs were color coded to show adjacent targets. Sections span anterior (top) to posterior (bottom) striatum. (C) Cell-type-specific input for corticostriatal and thalamostriatal input to dSPNs, iSPNs, and interneurons including fast-spiking PV+ cells (FS), low-threshold spiking interneurons (LTS), and cholinergic tonically active neurons (TAN). Abbreviations for thalamic nuclei: AD = anterior dorsal; AM = anterior medial; AV = anterior ventral; CL = central lateral; CM = central medial; IAD = interanterodorsal; IAM = interanteromedial; IMD = intermediodorsal; LD = laterodorsal; LGd = lateral geniculate dorsal; LGv = lateral geniculate ventral; LP = lateral posterior; MD = mediodorsal; PCN = paracentral; PF = parafascicular; PO = posterior complex; PVT = paraventricular thalamus; Re = reuniens; RH = rhomboid; RT = reticular thalamic nucleus; SPFm = subparafascicular nucleus magnocellular part; SMT = submedial thalamus; VA = ventral anterior nucleus; VM = ventromedial; VL = ventrolateral; VPM and VPL = ventrobasal complex; VPMpc = ventroposteromedial parvocellular; VPLpc = ventroposterolateral parvocellular.

contrast, in mice, both PT type and IT type axons project extensively within striatum and form functional synapses (Hooks et al., 2018). These inputs are organized, with specific cortical areas targeting defined areas in striatum (Hintiryan et al., 2016;Hunnicutt et al., 2016; Hooks et al., 2018). The motor cortex projects mainly to dorsolateral striatum, covering a relatively long extent along the anterior/posterior axis. Somatosensory projections overlap, with convergence of motor and sensory input from corresponding somatotopic regions (such as forelimb or whisker regions). Frontal areas project extensively and bilaterally in striatum, and target large areas of the dorsomedial and ventral striatum. Visual and auditory sensory areas, in contrast, target a relatively small area in the tail of striatum. It is also worth noting that because each successive stage of the cortico-basal ganglia circuit is smaller, input converges a great deal. In a standard mouse atlas, for example, cortex (25.1%) represents about 5 times as many voxels as striatum (~5.2%) (Oh et al., 2014), with an even steeper convergence in globus pallidus.

Thalamostriatal projections are the other major excitatory input to basal ganglia. As with cortex, these originate from a range of nuclei. The thalamic nuclei most strongly connected to M1 that also project to the corresponding regions of striatum include MD, PF, VM, VA, and VL (Hunnicutt et al., 2016). But these rules are not straightforward. M1 and VA/VL have the strongest reciprocal connectivity, but it seems that PF, MD, and VM send stronger input to striatum than VA and VL. Of these, PF and VM project most strongly to the motor-cortical recipient portions of striatum (Lee et al., 2020). Topography of basal ganglia connectivity is maintained with corticostriatal and thalamostriatal connectivity (Lee et al., 2020), consistent with multiple parallel closed loops in the circuit (Alexander et al., 1986). It is interesting that molecularly, the motor-related thalamic nuclei fall into 3 different clusters by expression pattern, with VL neurons most similar to primary somatosensory and visual nuclei, while VM, VA, and MD are more like secondary nuclei and PF is in a separate midline nuclear category (Phillips et al., 2019), with stronger thalamostriatal feedback from PF and, to a lesser extent, VM.

The striatum contains 2 main opposed classes of striatal projection neurons (SPNs). D1-receptor expressing SPNs (dSPN) project in the direct pathway, bypassing external globus pallidus and directly targeting internal globus pallidus and the SNR. In contrast, D2-receptor expressing SPNs (iSPN) project in the indirect pathway, targeting only external globus pallidus (Gittis et al., 2020). This is thought to be a competitive arrangement, with the direct pathway promoting desired actions or making them more vigorous, while the indirect pathway acts to slow or inhibit undesired actions. Thus, the specificity with which excitatory input targets specific cell types in striatum, especially dSPNs and iSPNs, has long been of interest. Once transgenic lines made tracing inputs to these neurons possible, it was suggested that different cortical areas may target dSPNs and iSPNs with different strength. As suggested by retrograde tracing with monosynaptic rabies, M1 may excite iSPNs more strongly than dSPNs, while other areas, such as S1, may excite dSPNs better (Wall et al., 2013). When recording monosynaptic input to striatal SPNs, cortical as well as thalamic input most strongly excites fast-spiking (PV+) interneurons, with substantial excitation to dSPNs and iSPNs. Other classes of interneuron are only weakly excited. In contrast to retrograde tracing experiments, it is not clear that there is a strong difference in S1 or M1 targeting of direct and indirect pathway SPNs (Johansson and Silberberg, 2020). Thalamic input from PF does excite both classes of SPN, but targets iSPNs more strongly than dSPNs. It remains to be understood how cell-type-specific connectivity contributes to striatal function.

CONCLUSION

M1 is specialized for the planning, execution, and control of motor movements, though the way in which it takes input from other regions and sends outputs to other targets is complex. There are a large number of cell types in M1 cortex (as in all cortical areas), each thought to be connected in a cell-type-specific manner. Each of these receives some subset of local and long-range input, which is as yet incompletely understood, and subsequently sends output to the local circuit as well as to long-range targets in cortical and subcortical pathways. The circuit is complicated because each input targets M1 in a potentially different laminar pattern, and each input also targets a range of cell types, potentially in different ways. The function of M1 output is similarly incompletely understood, as some outputs target regions like basal ganglia and thalamus, which both act in loops to subsequently influence cortical activity. Other output targets, such as brainstem and spinal cord, are closer to the musculature M1 is specialized to control, but the precise way in which M1 encodes movement is not yet fully understood. It is hoped that by understanding the local and long-range connectivity of M1, it may later be understood how this brain region gives rise to the specific output needed for skilled movement. Technical advances have enabled understanding these circuits with greater detail, encompassing a nearly complete census of the circuit elements. This leaves the outstanding, but increasingly approachable, problem of connecting the function of each cellular element to a complete understanding of how the circuit achieves motor control.

REFERENCES

Alexander GE, DeLong MR, Strick PL (1986). Parallel organization of functionally segregated circuits linking basal ganglia and cortex. *Annu Rev Neurosci.* 9:357–381.

Anderson CT, Sheets PL, Kiritani T, Shepherd GM (2010). Sublayer-specific microcircuits of corticospinal and corticostriatal neurons in motor cortex. *Nat Neurosci.* 13:739–744.

Bakken TE, Jorstad NL, Hu Q, Lak BB, Tian W, Kalmbach BE, Crow M, Hodge RD, Krienen FM, Sorensen SA, et al. (2021). Comparative cellular analysis of motor cortex in human, marmoset and mouse. *Nature.* 598:111–119.

Bock DD, Lee WC, Kerlin AM, Andermann ML, Hood G, Wetzel AW, Yurgenson S, Soucy ER, Kim HS, Reid RC (2011). Network anatomy and in vivo physiology of visual cortical neurons. *Nature.* 471:177–182.

Brecht M. Krauss A, Muhammad S, Sinai-Esfahani L, Bellanca S, Margrie TW (2004). Organization of rat vibrissa motor cortex and adjacent areas according to cytoarchitectonics, microstimulation, and intracellular stimulation of identified cells. *J Comp Neurol.* 479:360–373.

Cajal SR y (1995). *Histology of the Nervous System.* Oxford University Press.

Economo MN, Clack NG, Lavis LD, Gerfen CR, Svoboda K, Myers EW, Chandrashekar J (2016). A platform for brain-wide imaging and reconstruction of individual neurons. *eLife.* 5. 10.7554/eLife.10566.

Economo MN, Viswanathan S, Tasic B, Bas E, Winnubst J, Menon V, Graybuck LT, Nguyen TN, Smith KA, Yao Z, Wang L, Gerfen CR, Chandrashekar J, Zeng H, Looger LL, Svoboda K (2018). Distinct descending motor cortex pathways and their roles in movement. *Nature.* 563:79–84.

Fino E, Yuste R (2011). Dense inhibitory connectivity in neocortex. *Neuron.* 69:1188–1203.

Gittis AH, Hooks BM, Gerfen CR (2020). Basal ganglia circuits. In: Rubenstein JLR, Rakic P, eds. *Neural Circuit and Cognitive Development.* Elsevier:221–242.

Guo JZ, Graves AR, Guo WW, Zheng J, Lee A, Rodriguez-Gonzalez J, Li N, Macklin JJ, Phillips JW, Mensh BD, Branson K, Hantman AW (2015). Cortex commands the performance of skilled movement. *eLife.* 4. 10.7554/eLife.10774.

Guo K, Yamawaki N, Svoboda K, Shepherd GMG (2018). Anterolateral motor cortex connects with a medial subdivision of ventromedial thalamus through cell type-specific circuits, forming an excitatory thalamo-cortico-thalamic loop via layer 1 apical tuft dendrites of layer 5b pyramidal tract type neurons. *J Neurosci.* 38:8787–8797.

Harris KD, Shepherd GM (2015). The neocortical circuit: Themes and variations. *Nat Neurosci.* 18:170–181.

Herkenham M (1980). Laminar organization of thalamic projections to the rat neocortex. *Science.* 207:532–535.

Hintiryan H, Foster NN, Bowman I, Bay M, Song MY, Gou L, Yamashita S, Bienkowski MS, Zingg B, Zhu M, Yang XW, Shih JC, Toga AW, Dong HW (2016). The mouse cortico-striatal projectome. *Nat Neurosci.* 19:1100–1114.

Hippenmeyer S, Vrieseling E, Sigrist M, Portmann T, Laengle C, Ladle DR, Arber S (2005). A developmental switch in the response of DRG neurons to ETS transcription factor signaling. *PLoS Biol.* 3:e159.

Hooks BM, Hires SA, Zhang YX, Huber D, Petreanu L, Svoboda K, Shepherd GM (2011). Laminar analysis of excitatory local circuits in vibrissal motor and sensory cortical areas. *PLoS Biol.* 9:e1000572.

Hooks BM, Mao T, Gutnisky DA, Yamawaki N, Svoboda K, Shepherd GM (2013). Organization of cortical and thalamic input to pyramidal neurons in mouse motor cortex. *J Neurosci.* 33:748–760.

Hooks BM, Papale AE, Paletzki RF, Feroze MW, Eastwood BS, Couey JJ, Winnubst J, Chandrashekar J, Gerfen CR (2018). Topographic precision in sensory and motor corticostriatal projections varies across cell type and cortical area. *Nat Commun.* 9:1–16.

Hoover JE, Strick PL (1999). The organization of cerebellar and basal ganglia outputs to primary motor cortex as revealed by retrograde transneuronal transport of herpes simplex virus type 1. *J Neurosci.* 19:1446–1463.

Hunnicutt BJ, Jongbloets BC, Birdsong WT, Gertz KJ, Zhong H, Mao T (2016). A comprehensive excitatory input map of the striatum reveals novel functional organization. *eLife.* 5. 10.7554/eLife.19103.

Ilinsky IA, Kultas-Ilinsky K (2002). Motor thalamic circuits in primates with emphasis on the area targeted in treatment of movement disorders. *Mov Disord.* 17(Suppl 3):S9–S14.

Iriki A, Pavlides C, Keller A, Asanuma H (1989). Long-term potentiation in the motor cortex. *Science.* 245:1385–1387.

Johansson Y, Silberberg G (2020). The functional organization of cortical and thalamic inputs onto five types of striatal neurons is determined by source and target cell identities. *Cell Rep.* 30:1178–1194.

Kaneko T, Caria MA, Asanuma H (1994). Information processing within the motor cortex. II. Intracortical connections between neurons receiving somatosensory cortical input and motor output neurons of the cortex. *J Comp Neurol.* 345:172–184.

Kawai R, Markman T, Poddar R, Ko R, Fantana AL, Dhawale AK, Kampff AR, Olveczky BP (2015). Motor cortex is required for learning but not for executing a motor skill. *Neuron.* 86:800–812.

Kinnischtzke AK, Fanselow EE, Simons DJ (2016). Target-specific M1 inputs to infragranular S1 pyramidal neurons. *J Neurophysiol.* 116:1261–1274.

Kinnischtzke AK, Simons DJ, Fanselow EE (2014). Motor cortex broadly engages excitatory and inhibitory neurons in somatosensory barrel cortex. *Cereb Cortex.* 24:2237–2248.

Komiyama T, Sato TR, O'Connor DH, Zhang YX, Huber D, Hooks BM, Gabitto M, Svoboda K (2010). Learning-related fine-scale specificity imaged in motor cortex circuits of behaving mice. *Nature.* 464:1182–1186.

Kuramoto E, Furuta T, Nakamura KC, Unzai T, Hioki H, Kaneko T (2009). Two types of thalamocortical projections from the motor thalamic nuclei of the rat: A single neuron-tracing study using viral vectors. *Cereb Cortex.* 19:2065–2077.

Kuramoto E, Ohno S, Furuta T, Unzai T, Tanaka YR, Hioki H, Kaneko T (2015). Ventral medial nucleus neurons send thalamocortical afferents more widely and more preferentially to layer 1 than neurons of the ventral anterior-ventral lateral nuclear complex in the rat. *Cereb Cortex.* 25:221–235.

Lee J, Wang W, Sabatini BL (2020). Anatomically segregated basal ganglia pathways allow parallel behavioral modulation. *Nat Neurosci.* 23:1388–1398.

Lee S, Hjerling-Leffler J, Zagha E, Fishell G, Rudy B (2010). The largest group of superficial neocortical GABAergic interneurons expresses ionotropic serotonin receptors. *J Neurosci.* 30:16796–16808.

Lee S, Kruglikov I, Huang ZJ, Fishell G, Rudy B (2013). A disinhibitory circuit mediates motor integration in the somatosensory cortex. *Nat Neurosci.* 16:1662–1670.

Lefort S, Tomm C, Floyd Sarria JC, Petersen CC (2009). The excitatory neuronal network of the C2 barrel column in mouse primary somatosensory cortex. *Neuron.* 61:301–316.

Lemon RN (2008). Descending pathways in motor control. *Annu Rev Neurosci.* 31:195–218.

Macchi G, Jones EG (1997). Toward an agreement on terminology of nuclear and subnuclear divisions of the motor thalamus. *J Neurosurg.* 86:77–92.

Mao T, Kusefoglu D, Hooks BM, Huber D, Petreanu L, Svoboda K (2011). Long-range neuronal circuits underlying the interaction between sensory and motor cortex. *Neuron.* 72:111–123.

Network BICC (2021). A multimodal cell census and atlas of the mammalian primary motor cortex. *Nature.* 598:86–102.

Nudo RJ, Masterton RB (1990). Descending pathways to the spinal cord. III: Sites of origin of the corticospinal tract. *J Comp Neurol.* 296:559–583.

Oh SW, Harris JA, Ng L, Winslow B, Cain N, Mihalas S, Wang Q, Lau C, Kuan L, Henry AM, et al. (2014). A mesoscale connectome of the mouse brain. *Nature.* 508:207–214.

Ohno S, Kuramoto E, Furuta T, Hioki H, Tanaka YR, Fujiyama F, Sonomura T, Uemura M, Sugiyama K, Kaneko T (2012). A morphological analysis of thalamocortical axon fibers of rat posterior thalamic nuclei: a single neuron tracing study with viral vectors. *Cereb Cortex.* 22:2840–2857.

Okoro SU, Goz RU, Njeri BW, Harish M, Ruff CF, Ross SE, Gerfen C, Hooks BM (2022). Organization of Cortical and Thalamic Input to Inhibitory Neurons in Mouse Motor Cortex. *J Neurosci.* 42:8095–8112.

Packer AM, Yuste R (2011). Dense, unspecific connectivity of neocortical parvalbumin-positive interneurons: A canonical microcircuit for inhibition? *J Neurosci.* 31:13260–13271.

Pfeffer CK, Xue M, He M, Huang ZJ, Scanziani M (2013). Inhibition of inhibition in visual cortex: The logic of connections between molecularly distinct interneurons. *Nat Neurosci.* 16:1068–1076.

Phillips JW, Schulmann A, Hara E, Winnubst J, Liu C, Valakh V, Wang L, Shields BC, Korff W, Chandrashekar J, Lemire AL, Mensh B, Dudman JT, Nelson SB, Hantman AW (2019). A repeated molecular architecture across thalamic pathways. *Nat Neurosci.* 22:1925–1935.

Pi HJ, Hangya B, Kvitsiani D, Sanders JI, Huang ZJ, Kepecs A (2013). Cortical interneurons that specialize in disinhibitory control. *Nature.* 503:521–524.

Sauerbrei BA, Guo JZ, Cohen JD, Mischiati M, Guo W, Kabra M, Verma N, Mensh B, Branson K, Hantman, AW (2020). Cortical pattern generation during dexterous movement is input-driven. *Nature.* 577: 386–391.

Schieber MH (2011). Constraints on somatotopic organization in the primary motor cortex. *J Neurphysiol.* 86: 2125–2143.

Shepherd GM (2013). Corticostriatal connectivity and its role in disease. *Nat Rev Neurosci.* 14:278–291.

Sherman SM, Guillery RW (2006). *Exploring the Thalamus and Its Role in Cortical Function.* 2nd ed. MIT Press.

Suter BA, Shepherd GM (2015). Reciprocal interareal connections to corticospinal neurons in mouse M1 and S2. *J Neurosci.* 35:2959–2974.

Szczupak D, Iack PM, Liu C, Consortium, IRC, Tovar-Moll F, Lent R, Silva AC (2021). Direct interhemispheric cortical communication via thalamic commissures: A new white-matter pathway in the rodent brain. *Cereb Cortex.* 31:4642–4651.

Taniguchi H, He M, Wu P, Kim S, Paik R, Sugino K, Kvitsiani D, Fu Y, Lu J, Lin Y, Miyoshi G, Shima Y, Fishell G, Nelson SB, Huang ZJ (2011). A resource of Cre driver lines for genetic targeting of GABAergic neurons in cerebral cortex. *Neuron.* 71: 995–1013.

Tennant KA, Adkins DL, Donlan NA, Asay AL, Thomas N, Kleim JA, Jones TA (2011). The organization of the forelimb representation of the C57BL/6 mouse motor cortex as defined by intracortical microstimulation and cytoarchitecture. *Cereb Cortex.* 21:865–876.

Tremblay R, Lee S. Rudy B (2016). GABAergic interneurons in the neocortex: From cellular properties to circuits. *Neuron.* 91:260–292.

Turner RS, DeLong MR (2000). Corticostriatal activity in primary motor cortex of the macaque. *J Neurosci.* 20:7096–7108.

Urbain N, Deschenes M (2007). Motor cortex gates vibrissal responses in a thalamocortical projection pathway. *Neuron.* 56:714–725.

Wall NR, De La Parra M, Callaway EM, Kreitzer AC (2013). Differential innervation of direct- and indirect-pathway striatal projection neurons. *Neuron.* 79:347–360.

Weiler, N., Wood, L., Yu, J., Solla, S.A., and Shepherd, G.M. (2008). Top-down laminar organization of the excitatory network in motor cortex. *Nat Neurosci.* 11:360–366.

Winnubst J, Bas E, Ferreira TA, Wu Z, Economo MN, Edson P, Arthur BJ, Bruns C, Rokicki K, Schauder D, et al. (2019). Reconstruction of 1,000 projection neurons reveals new cell types and organization of long-range connectivity in the mouse brain. *Cell.* 179:268–281.

Yamawaki N, Borges K, Suter BA, Harris KD, Shepherd GM (2014). A genuine layer 4 in motor cortex with prototypical synaptic circuit connectivity. *eLife.* 3:e05422.

Yamawaki N, Radulovic J, Shepherd GM (2016). A corticocortical circuit directly links retrosplenial cortex to M2 in the mouse. *J Neurosci.* 36:9365–9374.

Yamawaki N, Shepherd GM (2015). Synaptic circuit organization of motor corticothalamic neurons. *J Neurosci.* 35:2293–2307.

Zingg B, Hintiryan H, Gou L, Song MY, Bay M, Bienkowski MS, Foster NN, Yamashita S, Bowman I, Toga AW, Dong HW (2014). Neural networks of the mouse neocortex. *Cell.* 156:1096–1111.

34.

MODELS OF THALAMOCORTICAL INTERACTIONS IN MOTOR CONTROL

G. Sean Escola, Kaushik Lakshminarasimhan, and Laureline Logiaco

INTRODUCTION

Understanding motor control is a fundamental goal of systems neuroscience. Yet a major question remains regarding the role of the interactions between the thalamus and cortex in motor computations. There have been multiple recent models proposed to interrogate this. Thus, it is now timely to review the state of these modeling efforts. Importantly, these recent models of thalamocortical interactions have benefited from the significant expansion of experimental data—in terms of behavior, anatomy, and physiology—and have positioned the role of understanding this circuit centrally in terms of our overall conception of motor control.

We begin by placing models of thalamocortical interactions within the context of motor system modeling writ large, which has greatly expanded our understanding of motor function and control. This body of work has operated at multiple levels of abstraction, from detailed biophysical models of individual neurons (for which there is a vast literature, including in the thalamus; see **Section II** of this volume) to computational and algorithmic models that do not speak directly to implementations within the brain (e.g., Todorov and Jordan, 2002). A class of models that has particular appeal considers the firing rates of individual neurons or populations of neurons, and the correlations between these firing rates across neurons and populations (Gerstner et al., 2014). These network models permit the interrogation of interactions between brain areas and their impacts on behavior. They are thus of particular relevance for the study of thalamocortical interactions and form the focus of our discussion in this chapter.

Firing-rate models themselves can be further subdivided into those that describe neural computation in terms of mean firing rates that are static over some relevant time period and those that consider the moment-by-moment firing-rate dynamics. The former view is inherited from the seminal work of Hubel and Wiesel in visual neuroscience that ascribes to individual cortical cells "receptive fields": static spatiotemporal representations that extract cell-specific statistics from external stimuli and that emerge through several steps of feedforward processing from thalamus to higher-order cortical areas (Hubel and Wiesel, 1962; Yamins et al., 2014). Similarly, motor system models that attribute to individual motor cortical neurons the representation of specific features of movements and that ignore time have been applied to understand neural computation in the motor system (Georgopoulos et al., 1982). However, recent efforts for modeling motor control have followed the alternative perspective in which motor cortical activity results from ongoing continuous dynamics (Shenoy et al., 2013). In this conceptualization, the cortical population as a whole represents instantaneous muscle commands.

As we shall see in detail, despite the fact that the motor cortex and premotor cortex are innervated by multiple motor thalamic nuclei such as the ventromedial (VM), ventroanterior (VA), and ventrolateral (VL) nucleus (see **Chapters 31, 33, and 35 in Section VI**), both static and dynamic models of motor cortical function have often either ignored motor thalamus entirely or have relegated the thalamus to serving as a relay in the flow of information to motor cortex (Doya, 2000). However, a growing body of research explicitly parses the role of thalamocortical interactions and proposes specific computations that this circuit motif enables (Kao et al., 2021; Logiaco et al., 2021; Shine, 2021; Recanatesi et al., 2022). This work considers the importance of corticothalamic feedback in facilitating motor execution and learning (see **Section IV** of this volume).

We propose a taxonomy of models of thalamocortical interactions in the motor system that will permit productive comparison. This taxonomy evaluates models along two dimensions: (1) whether static versus dynamic interpretations of cortical representations are important; and (2) how the anatomy of the circuit—that is, both the feedforward and feedback connections between thalamus and cortex—informs motor function. In view of the recent evidence suggesting that motor cortical dynamics drive motor output in real time for at least some mammalian behaviors (Shenoy et al., 2013; Sauerbrei et al., 2019; Guo et al., 2021), important insights can be gained through the recent theoretical and modeling advances that dissect the contribution of the recurrent cortical and thalamocortical connectivity in shaping rich multi-area dynamics.

HISTORICAL OVERVIEW AND TAXONOMY OF MOTOR THALAMOCORTICAL MODELS

We begin by offering an overview of the history of models of motor control and how they have influenced the development of models of thalamocortical interactions, including the possible influence that earlier results and approaches of sensory neuroscience have had on the field of motor control. This will lead us to define three important classes of models of thalamocortical interactions. In later sections, we dive into the mechanistic details of these classes of models.

POSITIONING THE THALAMUS WITHIN EVOLVING MODELS OF CORTICAL MOTOR CONTROL

One of the earliest lines of research for modeling motor cortical function (Georgopoulos et al., 1982) views motor cortical activity as a distributed representation of relatively static parameters of movements, such as reach angles for arm reaches. As these variables are abstract and slower than the movement timescales, the underlying assumption is that their influence on motor control would be mediated by modulation of a downstream dynamic circuit. This view was influenced by the foundational work of Hubel and Wiesel (1962) in the visual cortex. Similar to the methods used to map receptive fields in sensory areas, this framework involves computing a "tuning curve" that describes the neuron's mean firing rate when an animal performs different behaviors to find the specific movement feature that it "encodes." Anatomically, the thalamus is positioned as a relatively low-dimensional upstream input (see **Sections IV and V** of this volume) with no internal excitatory recurrence (see **Section II**). Thus, in these models, the role of thalamus is a preliminary feedforward processing step, for example, a simple relay (Chevalier and Deniau, 1990). In this scenario—which encompasses an ensemble of models we shall refer to as "Type I"—the thalamus either plays a minor feedforward processing role or is even not explicitly differentiated in the model.

However, two lines of evidence suggest that motor cortical activity should not be viewed as a collection of tuning curves, but rather as a dynamical system whose instantaneous population activity is read out by downstream circuits for online motor control. First, with respect to anatomy, the primary motor cortex is a large recurrent neural network that has embedded corticospinal tract neurons that project to the motor periphery (Friedli et al., 2015). Second, with respect to function, electrophysiology experiments have measured strong correlations between instantaneous activity in the motor cortex and moment-by-moment muscle activity (Shenoy et al., 2013). Given this, some models—which we shall call "Type II"—account for the thalamus as an input to a dynamical cortical system that drives movement.

Finally, additional anatomical data (see **Section IV** of this volume) has established that the thalamocortical circuit is bi-directional, with both feedforward connections from the thalamus and feedback connections from the cortex. Thus, a few recently developed models have considered the loop between the cortex and thalamus (Kao et al., 2021; Logiaco et al., 2021; Lakshminarasimhan et al., 2022; Recanatesi et al., 2022) as essential to shape circuit dynamics to drive downstream movements. We shall call such models "Type III."

It is important to point out that these model classes arise from specific choices made by modelers in order to concentrate on features of the neurobiology that are assumed to be most important. Thus, though Type I and Type II models are simpler, the decision to omit certain details was made to serve the goal of illustrating specific computational hypotheses in a parsimonious manner.

A PROPOSED TAXONOMY OF MOTOR THALAMOCORTICAL MODELS AND ITS LIMITATIONS

Given the historical development of motor thalamocortical models reviewed above, we propose a taxonomy of these according to two axes: (1) whether they assume that motor cortex represents quasi-static high-level parameters of movements or implements complex dynamics that more directly drive muscles; and (2) whether the thalamus acts as a source of external input to the cortex or as a cortico-thalamo-cortical loop that intrinsically participates in the circuit's dynamics (Figure 34.1).

Importantly, it should be noted that these two taxonomic features help conceptualize differences between models that are often of a graded rather than categorical nature. For example, with respect to the static versus dynamic axis, it is possible to design a dynamical model such that a static movement kinematic parameter can be decoded from the population activity at all times (Bachschmid-Romano et al., 2022). In this case, the

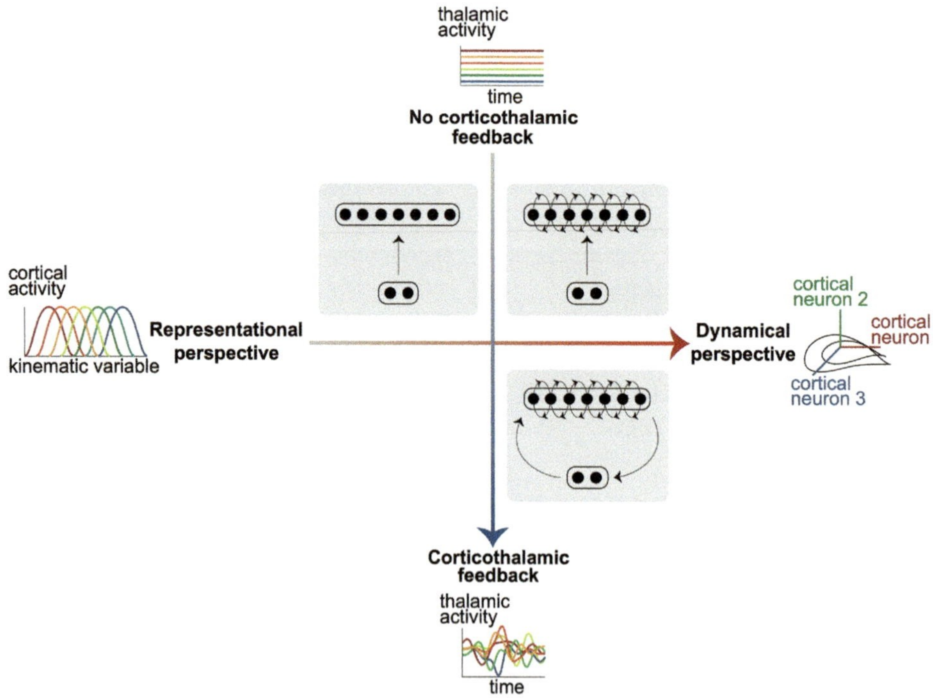

Figure 34.1. A proposed taxonomy of motor thalamocortical models. The horizontal axis delineates models according to the function they attribute to the motor cortex (representational vs. dynamical). The vertical axis identifies models according to the connectivity between the cortex and the thalamus (feedforward vs. loopy).

network dynamics do not contribute to motor control, but rather are included to improve biological realism given the established nature of the motor cortex as a recurrent system and the non-stationarity of motor neural activity seen in recordings. In addition, with respect to the feedforward versus loop axis, the computational consequences of this dichotomy will depend on specific modeling assumptions. For example, the complexity of the nonlinearities in the cortical network can dictate how the cortical dynamics can be modulated, such that even an external input can have a nontrivial impact.

Another important caveat is that different motor thalamocortical subcircuits may fall in different regions along these proposed axes. For instance, it has been suggested that basal ganglia–recipient thalamus may be connected to the motor cortex in a more bidirectional fashion than cerebellum-recipient thalamus (Bosch-Bouju et al., 2013). In addition, while we chose to organize thalamocortical models along these axes, there are other model characteristics worth considering, such as the origin and nature of the extra-cortical signals received by the modeled thalamus (see **Chapter 31 in Section VI**) and the position of the modeled circuit within the premotor to motor hierarchy (see Section VIII). However, despite these considerations, we will argue below that the axes depicted in Figure 34.1 can be helpful guidelines to orient the reader when navigating the space of thalamocortical models of motor function.

TYPE I MODELS: THALAMUS AS A RELAY OF KINEMATIC VARIABLES

As stated in the previous section, early attempts to investigate motor control were heavily influenced by our understanding of sensory processing and, in particular, visual perception. As a result, models of thalamocortical interactions were initially grounded in a perspective in which the motor cortex exerted control over movements by encoding kinematic variables such as the location and/or direction of movements. In this section, we review experimental evidence in favor of representational theories of motor control and the computational role ascribed to thalamocortical interactions by such theories.

BACKGROUND AND MOTIVATION

Inspired by seminal studies of the visual system which identified neurons tuned to static latent variables such as spatial orientation (Hubel and Wiesel, 1962) and the direction of visual motion (Albright, 1984), neural activity in the motor cortex was interpreted in relation to movement parameters such as the joint angle (Evarts, 1968), speed (Inoue et al., 2018), and the direction of movement (Georgopoulos et al., 1982). For example, neural responses in the primary motor cortex (M1) of monkeys performing ballistic reaching movements were shown to depend on the endpoint of the movement. Then, the change (relative to baseline) in the average firing rate of individual cortical neurons h could be captured by a cosine tuning function of the movement direction ϑ (i.e., $h \propto cos\ (\vartheta - \vartheta_0)$ where ϑ_0 denotes the preferred movement direction (Figure 34.2A). The preferred direction of neurons has been shown to vary systematically, constituting a topographic map (Georgopoulos et al., 2007). Therefore, an observer of the neural population in M1 can reliably decode the direction of the monkey's arm movements with high precision (Georgopoulos et al., 1982, 2007). How is this representation of kinematic variables established?

In primary visual cortex, feature tuning is largely a consequence of the precise wiring of feedforward inputs from the retina through the lateral geniculate nucleus of the thalamus, constituting a labeled line code (Hubel and Wiesel, 1962; Chapman and Stryker, 1992; Clay Reid and Alonso, 1995). In contrast, the premotor and motor cortices are innervated by multiple motor thalamic nuclei such as the ventromedial (VM), ventroanterior (VA), and ventrolateral (VL) nucleus, which in turn receive inputs from different regions such as the basal ganglia and the cerebellum. Studies have found no clear evidence for a direction signal in the cerebellum (Trouche and Beaubaton, 1980; Thach et al., 1992; Horne and Butler, 1995). In contrast, the output nuclei of the basal ganglia (GPi/SNr) encode movement direction (Georgopoulos et al., 1983; Turner and Anderson, 1997), as do some neurons in the motor thalamus (Butler et al., 1996). Furthermore, several studies have shown that the activity of neurons in the motor thalamus is triggered by the presentation of a cue and peaks prior to movement onset (Figure 34.2A; Strick, 1976; Butler et al., 1992; Forlano et al., 1993). Lesion studies have also consistently implicated the basal ganglia–thalamocortical pathway in movement initiation (Groenewegen, 2003). Against

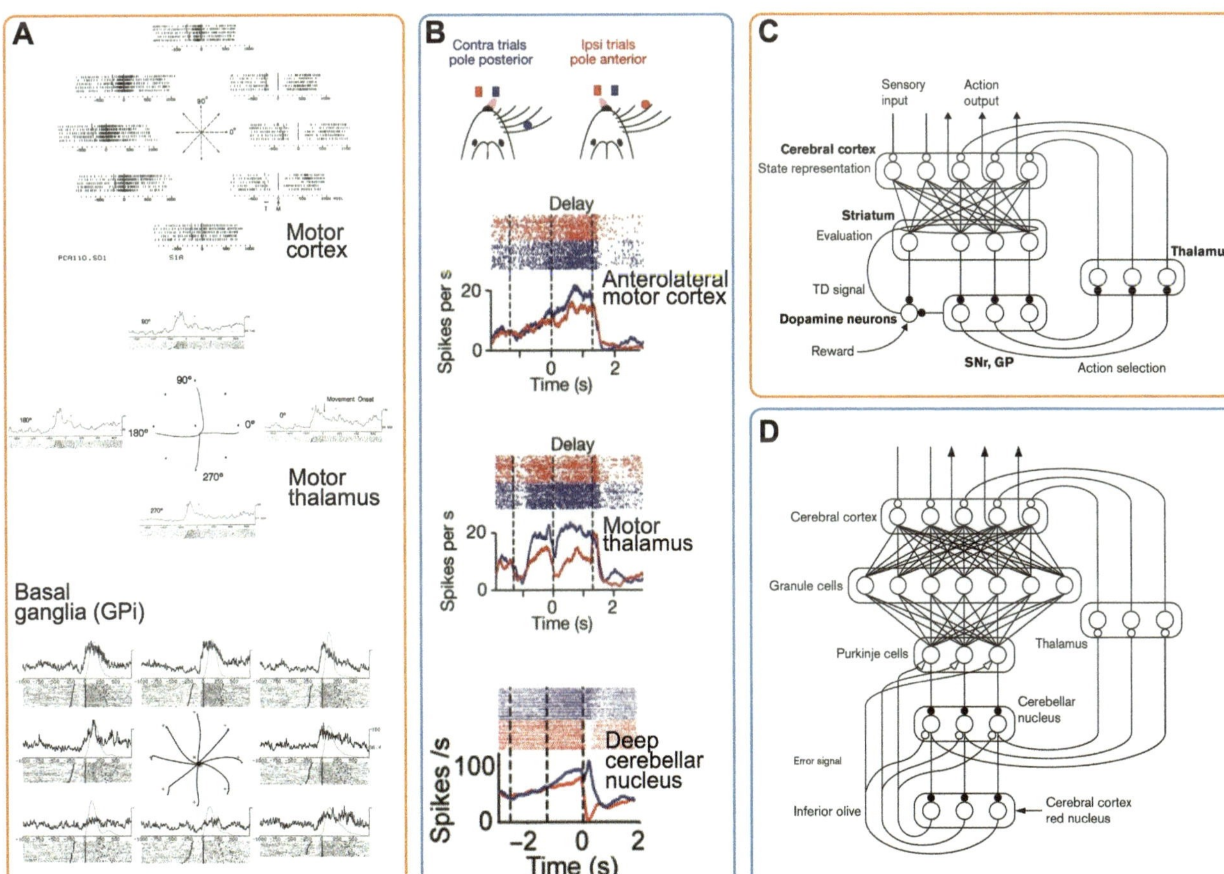

Figure 34.2. Type I models. (A). Responses of example neurons from the primate primary motor cortex (top: from Georgopoulos et al., 1982), motor thalamus (middle: from Inase et al., 1996), and the internal segment of the Globus pallidus (GPi, bottom: from Turner and Anderson, 1997) showing tuning to the direction of arm movement. (B). Activity of neurons in the mouse ALM (top: from Guo et al., 2017), motor thalamus (middle: from (Guo et al., 2017), and deep cerebellar nucleus (bottom: from Gao et al., 2018) showing tuning to the *upcoming* lick direction (delay period activity). (C–D). Standard models of the cortico-basal ganglia-thalamocortical loop (C) and the cortico-cerebello-thalamocortical loop (D) assume that motor thalamus is a relay of subcortical signals to the cortex (from Doya, 2000). Data and the model panels are color coded (orange: basal-ganglia-thalamo-cortical; blue: cerebello-thalamo-cortical).

the backdrop of the models developed to account for the computational role of basal ganglia in movement, a verbal theory without a precise mathematical underpinning has emerged which treats the motor thalamus as a relay of the basal ganglia output to the motor cortex.

Recent studies in mice have used licking behavior to study the role of another prominent pathway through the motor thalamus—the cerebello-thalamocortical pathway—in motor preparation. Similar to the representation of arm movement direction in the motor cortex of monkeys, rodent studies have found a robust representation of the licking direction in the mouse premotor cortex (anterolateral motor cortex, or ALM, to be specific), except these representations were activated *before* movement (Figure 34.2B; Guo et al., 2017). Subsequent work has shown that similar representations are found in the deep cerebellar nuclei (DCN), the output nuclei of the cerebellum (Figure 34.2B). Inactivating the DCN or the VM nucleus of the thalamus (which receives inputs from the DCN) abolishes preparatory activity in the cortex (Gao et al., 2018; Chabrol et al., 2019). Analogous to the role of basal ganglia–recipient motor thalamus, these studies have led to a view of the cerebellum-recipient motor thalamus as a relay of preparatory signals from the cerebellum to the premotor cortex.

MODEL DETAILS

In this section, we review specific examples of Type I models, focusing on basal ganglia and cerebellar-recipient thalamic nuclei. However, motor thalamus also receives inputs from other subcortical structures, such as the superior colliculus and the midbrain (see **Chapter 31 in Section VI**) which could also be modeled along similar lines.

Basal Ganglia–Recipient Thalamic Neurons

Models proposing motor thalamus as a relay of basal ganglia commands to the cortex have been in place since the early 1990s (Albin et al., 1989; Chevalier and Deniau, 1990). Let s and r denote the activity levels in the basal ganglia output nuclei (GPi/SNr) and motor thalamus, respectively. Since neurons in GPi/SNr exhibit high spontaneous firing rates ($s >> 0$) and release GABA into the motor thalamus (Chevalier and Deniau, 1990), these models assume that motor thalamic neurons are chronically inhibited; that is, $r = f(\delta - s) \approx 0$ where f is a rectifying nonlinearity with a small threshold δ. However, when the basal ganglia is activated, the activity of a subset of neurons in GPi/SNr is suppressed ($s < \delta$) and their downstream thalamic targets are transiently disinhibited. In this manner, the thalamus is able to relay commands from the basal ganglia, selectively activating specific representations in the motor cortex (Figure 34.2C; Doya, 2000). The specific mechanism of thalamic disinhibition has since been refined to incorporate a role for the hyperdirect pathway in the basal ganglia. Briefly, the updated view suggests that a delicate interplay between the synaptic latencies and the polarities of parallel cortico-basal ganglia pathways—hyperdirect, direct, and indirect—triggers a sequence of events that lead to inhibition, followed by disinhibition, and then back to inhibition of a subset of neurons in the motor thalamus, resulting in the transient thalamic activation that typically precedes movement initiation (Nambu et al., 2002; Polyakova et al., 2020).

Cerebellar-Recipient Thalamic Neurons

Unlike basal ganglia–recipient thalamic neurons, the cerebellar-recipient thalamus is not under chronic inhibition from the cerebellum. Therefore, the cerebellar outputs (DCN) influence the motor thalamus primarily by excitation rather than disinhibition: that is, $r = f(s - \delta)$. That said, similar to the basal ganglia–thalamocortical pathway above, thalamic neurons targeted by the cerebellothalamic tracts could function as a relay (Figure 34.2D; Doya, 2000). Consistent with this, inhibiting the preparatory activity in the DCN abolishes preparatory activity in the premotor cortex (Gao et al., 2018), which suggests that the motor thalamus allows the cerebellum to mediate transitions between cortical states (Li and Mrsic-Flogel, 2020). An important caveat in interpreting these data is that DCN inactivation induces a mixture of excitation and inhibition in the thalamus and cortex, suggesting that VM thalamus might be more than just a relay.

Thalamocortical Interactions

In the models that consider motor thalamus as a relay, thalamic influence on the cortex can be described as a feedforward drive: $h = f(Ur)$, where h denotes cortical activity, U denotes the thalamocortical connectivity matrix, and f is the nonlinearity of the cortical neurons. An interesting implication of these types of feedforward relay models is that the structure of the thalamocortical weights should dictate the degree to which cortical neurons are tuned. Random thalamocortical interactions would produce a distributed population code resulting in mixed selectivity of preparatory or movement-related parameters in individual cortical neurons. Alternatively, thalamocortical connectivity might be structured by developmental programs or by plasticity-based mechanisms within the cortex (Biane et al., 2016) to ensure high representational fidelity at the level of single neurons. Therefore, a systematic comparison of subcortical and cortical selectivity in single neurons can shed light on the structure of thalamocortical interactions.

LIMITATIONS

A limitation of the representational view assumed by the above models is that they cannot capture how the temporal heterogeneity exhibited by cortical neurons is correlated with moment-by-moment muscle activity and movement kinematics (Shenoy et al., 2013). The significance of considering temporal dynamics in the context of motor control becomes clear when contrasting the nature of computations performed by perceptual and motor systems. In perception, the input to the system is generated externally (i.e., in the world) and the role of the perceptual system is to invert that generative model to infer the latent states responsible for those inputs. In contrast, the goal of motor control is to generate appropriate muscle commands to achieve a desired change of state. Therefore, a purely representational account cannot constitute a causal theory of movement production. A recognition of this limitation has led to the development of models based on dynamical systems theory, which are described below.

TYPE II MODELS: THALAMUS AS AN AFFERENT TO DYNAMICAL CORTICAL MODELS

Motivated to more accurately capture the anatomy and activity features of cortex (Shenoy et al., 2013; Rikhye et al., 2018) or by functional considerations positioning the cortical output as a direct driver of downstream motor effectors (Friedli et al., 2015), several models have moved away from viewing thalamocortical networks as computing static representations and turned instead to considering a dynamical system framework. A subset of these dynamic models consider thalamic activity as an external input to the cortical dynamics. Here we review the main principles behind such models.

BACKGROUND AND MOTIVATION

In the context of motor computations, thalamus is often thought to have a special status that distinguishes it from the many sources of cortical input. Not only does the anatomy indicate robust thalamocortical projections (see **Section IV** of this volume), it also reveals that the signals transmitted by the thalamus are unlike those that could be received through other pathways. Specifically, the thalamus has access to sensorimotor inputs—through the superior colliculus (Crapse and Sommer, 2008), cerebellum (Caligiore et al., 2016), and various midbrain regions (Inagaki et al., 2022)—which can guide (pre)motor cortical dynamics faster than if this information had to be processed by other cortical areas first. This is supported functionally by studies that have revealed direct causal evidence for the importance of thalamic input in dynamic sensorimotor cortical processing (Crapse and Sommer, 2008; Guo et al., 2017; Sauerbrei et al., 2019; Inagaki et al., 2022).

Therefore, several models have been proposed to account for such thalamic input to (pre)motor cortical dynamics. In such models (Figure 34.3A), cortical dynamics are instantiated through a differential equation, which links

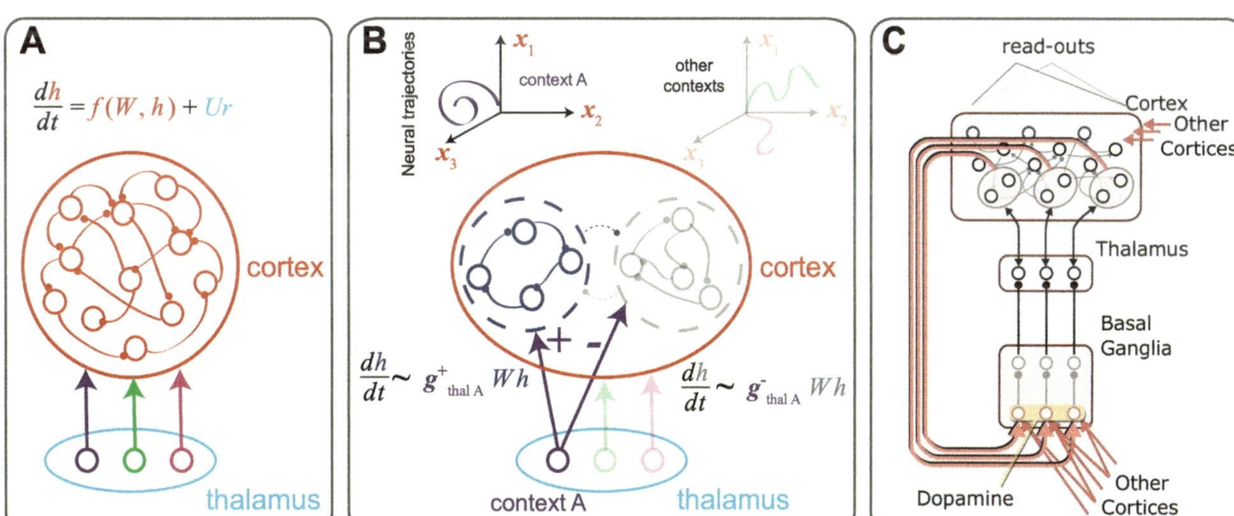

Figure 34.3. Type II models: thalamus as an external feedforward input to cortical dynamics. (A). General framework. (B). An example of an interaction between thalamic input and cortical dynamics (with a mechanism similar to Rikhye et al., 2018). Bottom: The thalamic input gates cortex by acting as a gain factor on the cortical neurons; i.e., the inputs to cortex from the active thalamic neurons operate as multiplicative factors that facilitate (g^+) and suppress (g^-) different cortical subpopulations. Top: If the thalamic input is context-dependent, then this mechanism allows the selection of specific neural trajectories dictated by the cortical neurons that are "gated in." (C). Example of a model adapted from Mannella and Baldassarre (2015), where basal ganglia-to-thalamus signals are proposed to select cortical dynamics.

an infinitesimal change in cortical activation per unit time $\frac{dh}{dt}$ to the sum of the external thalamic input Ur and the cortical recurrent input $f(W, h)$. Therefore, these dynamics read $\frac{dh}{dt} = f(W,h) + Ur$, where h and r are the cortical activities and thalamic rates, W and U are the corticocortical and thalamocortical weights, and f represents some arbitrary function. While the recurrent cortical weights are particularly influential in sculpting complex network dynamics due to their numerous degrees of freedom, the input can also play a non-negligible role, especially if the dynamics include a nonlinearity and the input is particularly strong.

In the following section, we will review the diverse roles that thalamic inputs have been shown to play in various models of Type II.

MODEL DETAILS

Considering neural dynamics—as opposed to a static cortical representation—can substantially alter the mechanisms of action of thalamus on cortex, even in models that remain closest to the representational type. For instance, Bachschmid-Romano et al. (2022) have proposed a model which is built to encode reach direction in a stable manner while constrained to mimic observed cortical activity. In this model, inputs to the cortex—putatively from thalamus—are not only tuned to the characteristics of the movement, as they would be in a Type I model. Rather, this model also includes "untuned" inputs—especially during the movement period—that help shape activity patterns that are shared across movements despite their separate characteristics (Bachschmid-Romano et al., 2022).

The roles that thalamic input can play become even more diverse when moving further away from a fixed representational perspective. For instance, cortical network dynamics are thought to participate in updating representations of the external world to compensate for self-generated movements (Crapse and Sommer, 2008). A feature of this computation is input from thalamus that reflects an "efferent copy"—a broadcast of the motor command. The evidence for such a mechanism has perhaps been strongest for the update of visual representations in response to eye movements, where the mediodorsal (MD) nucleus of the thalamus has been shown (Crapse and Sommer, 2008) to receive saccade-related information directly from the superior colliculus (the putative origin of the motor signal). The MD thalamus then transmits this saccade information to the frontal eye field (Crapse and Sommer, 2008), where it can be integrated with visual information through recurrent dynamics. This mechanism has indeed been successfully implemented in a model of the type illustrated in Figure 34.3A (Rao et al., 2016).

Finally, models have also clarified how feedforward thalamic inputs can act as selectors or gates that can largely impact the activity of cortical circuits, even while their dynamics are largely driven by their internal recurrence. For instance, models have suggested that strong thalamic inputs can leverage the nonlinearity of the transfer function of cortical neurons, therefore modulating their response to cortical recurrent inputs through an effective gain (Figure 34.3B; Rikhye et al., 2018; Stroud et al., 2018). This gain modulation can be pushed to an extreme such that the thalamic input can act like a gate that selects particular cortical subpopulations that will then implement specific dynamics. In such cases, the gain causes relatively large recurrent interactions in some populations while "extinguishing" others through near-zero gain. Such a mechanism was, for instance, proposed in a model of premotor computations, where context-selective thalamic units facilitated cortical context-dependent processing of sensory cues through gating, thereby leading to the correct context-dependent action (Rikhye et al., 2018). Another model where thalamus acted as a gate for cortical subpopulations was implemented in Calderon et al. (2022). Here, in the context of movement sequences, thalamus was used to select the next premotor cortical subpopulation—which indexed a particular movement—in response to basal ganglia activation. Finally, in Mannella and Baldassarre (2015), a basal ganglia–thalamus input pathway was modeled similarly to Rikhye et al. (2018), by gating out irrelevant recurrent subpopulations while selecting and modulating relevant ones that then drive the motor output—putatively through dynamic attractors in the activated cortical subpopulation (Figure 34.3C).

LIMITATIONS OF THE APPROACH AND RELATION TO OTHER MODELS

A limitation of Type II models is that they either fully omit the feedback connections from cortex to thalamus, or do not leverage and explain the specific advantages of the striking thalamic architecture—which is characterized by the absence of local excitation (see **Section II** of this volume) and the presence of long-range bidirectional interactions with the cortex (see **Section IV**). Therefore, in these models, there is no conceptualization of specific computational benefits arising from thalamic processing beyond relaying information from the thalamic afferents.

TYPE III MODELS: THALAMOCORTICAL LOOPS RECONFIGURE CORTICAL DYNAMICS

We now turn to Type III models (Kao et al., 2021; Logiaco et al., 2021; Shine, 2021; Lakshminarasimhan et al., 2022; Recanatesi et al., 2022), whose development was spurred by recent experimental evidence from simultaneous recordings in motor cortex and thalamus and optogenetic perturbation studies (Guo et al., 2017; Sauerbrei et al., 2019). The distinction of these models is that the loop between cortex and thalamus—including the feedback projection from cortex to thalamus in addition to the more well-studied connection from thalamus to cortex—plays a crucial computational role (Kao et al., 2021; Logiaco et al., 2021). Additionally, the synaptic weights in the loop—upon which its computational capability depends—can be set via biologically plausible learning (Lakshminarasimhan et al., 2022). These models demonstrate how thalamocortical loops can remap cortical dynamics to influence behavior. In contrast, for Type I and II models, only the feedforward connection from thalamus to cortex is considered. This difference imbues Type III models with much richer ability to control and sequence movements.

We will expand on this below, but it is important to clarify this crucial distinction immediately. An input to a recurrent neural network (i.e., thalamus to cortex, as in Type I and II models) has a limited capability to impact dynamics. Loops that are bidirectionally coupled to a recurrent network, on the other hand, if their synaptic weights are appropriately tuned, can radically remap network dynamics due to their impact on the recurrent structure itself (Figure 34.4A). As we shall see, this distinction has deep computational implications and thus Type III models make claims about the role of thalamocortical interactions in motor control that differ qualitatively from prior theories. It is important to note that these feedback loops may be polysynaptic (e.g., originating from and returning to different cortical layers; see **Sections II and III** of this

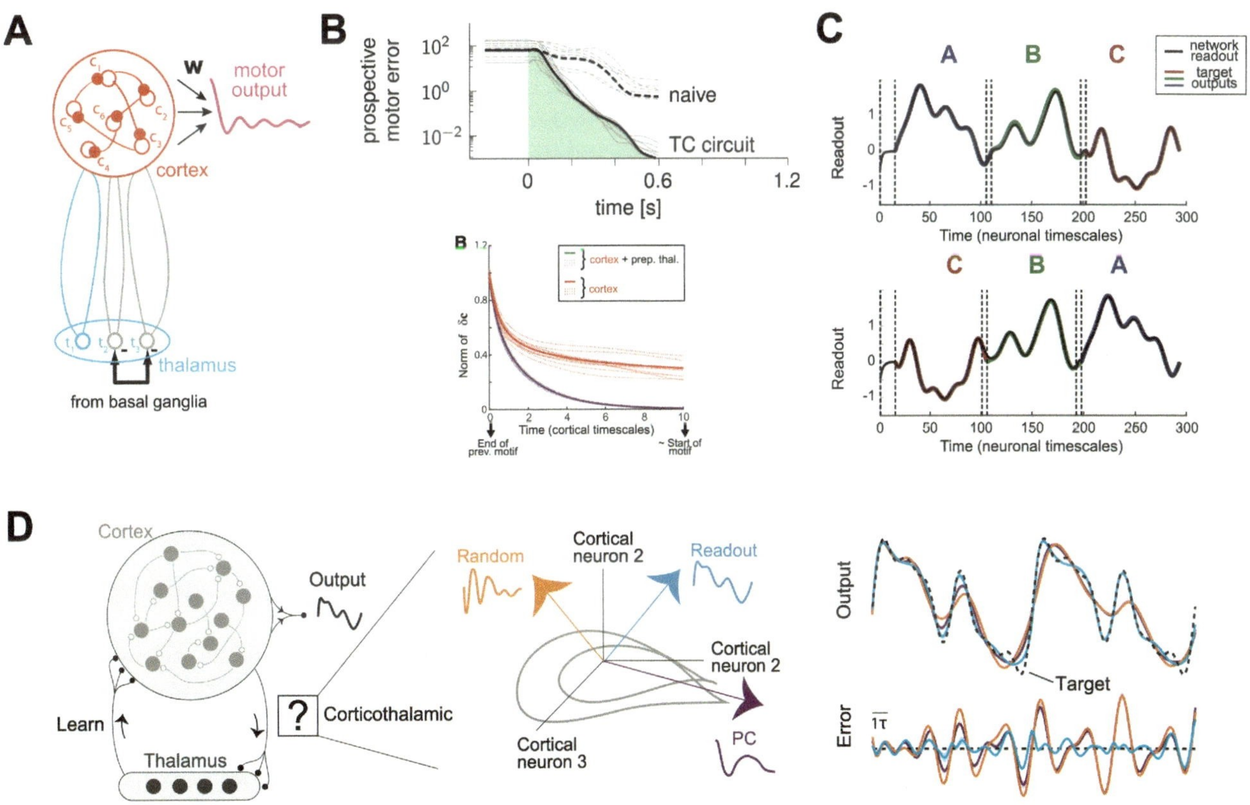

Figure 34.4. Thalamocortical interactions as loops between cortex and thalamus. (A). The basic structure of Type III models in which feedforward connection to cortex and feedback connections to thalamus are both essential. (B). Kao et al. (2021) (top) and Logiaco et al. (2021) (bottom) use thalamocortical loops to instantiate cortical dynamics that permit much more rapid motor preparation than is possible with simple feedforward input. (C). By switching between thalamocortical loops used for motor preparation and motif-specific execution, Logiaco et al. (2021) show that motor sequences can be strung together in arbitrary orders. (D). Lakshminarasimhan et al. (2022) identify conditions under which biologically plausible learning of thalamocortical synaptic weights results in the motor output matching a target function even when the size of the thalamus is small relative to cortex. Successful motor learning occurs if the corticothalamic projection is aligned to the motor readout direction (blue). If the projection is instead aligned to the principal component (PC, purple) directions of the cortical activity, or if the corticothalamic weights are random (orange), then thalamocortical learning is substantially worse. Tau (τ) denotes neuronal time constant. A, B (bottom), and C are adapted from Logiaco et al. (2021); B (top) from Kao et al. (2021); and D from Lakshminarasimhan et al. (2022).

volume), but this detail does not alter the computational characteristics of these models.

BACKGROUND AND MOTIVATION

Experimental Data for the Importance of Thalamocortical Loops

The validity of the critical loop structure in Type III models is motivated by both anatomical and functional findings. With respect to the former, projections to and from the same population of motor cortical neurons can connect to the same populations of thalamic neurons. In other words, cortex and thalamus form a loop (see **Section IV** of this volume). Recent functional studies have demonstrated that these loops are required for many cognitive and motor functions (Guo et al., 2017; Rikhye et al., 2018; Sauerbrei et al., 2019) and that inhibiting higher-order thalamus suppresses cortical activity, while inhibiting cortex impacts thalamic activity (Guo et al., 2017). Together, these results provide strong experimental evidence for the importance of the looped nature of thalamocortical interactions in motor control, rather than a largely feedforward circuit in which the feedback from cortex to thalamus plays at most a limited role, as in Type I and II models.

Theories Conceptualizing the Role of Thalamocortical Loops in Motor Dynamics

Recently, a few theoretical frameworks have emerged that can generate specific hypotheses about how the highly stereotyped thalamocortical loop architecture facilitates motor control. To be analytically tractable and yield robust insight, these frameworks rely on the simplification that the dynamics in the cortex-thalamus system are linear or switching linear (Glaser et al., 2020). While this may seem unrealistic at first blush, in the context of the dynamics of primary motor cortex specifically, this simplification appears to be largely reasonable given that cortical population activity has been shown to be well captured by shared linear dynamical systems for whole classes of movements (Lara et al., 2018; Glaser et al., 2020).

One such framework is optimal control theory, an elegant approach that aims to find the best possible input to a dynamical system—such as a cortical network—so that it evolves according to some desired specification. A power of this framework is that—under some conditions—the optimal input can be identified analytically and, furthermore, can be implemented through feedback loops similar to those between the thalamus and cortex (Kao et al., 2021). The idea that the computational role of thalamocortical interactions in the production of behavior is optimal feedback control is novel and intriguing. A limitation of this framework is that it is unclear how to augment the goal of optimization with other desirable features such as robustness to (unpredictable) variability in this goal or in the dynamics of the system, though steps have been made in this direction outside of standard optimal control (Loeb, 2012; Eysenbach and Levine, 2022). Additionally, with respect to the thalamocortical system per se, integrating the smaller size of the thalamic population relative to the cortex is not straightforward.

Another framework of interest—which proposes a functional role for the relative sizes of a large cortex and a small thalamus—follows from the fact that the cortex-thalamus system can be conceived of as "full-rank" recurrent connectivity in the cortex, modified by "low-rank" connectivity through the thalamus. Concretely, for a population of N cortical neurons, the $N \times N$ synaptic weight matrix W describes the (full-rank) recurrent connectivity. We can then define 2 $N \times K$ matrices U and V which represent the thalamocortical and corticothalamic synaptic connectivity, respectively, with $K \ll N$. The dynamics of the cortex-thalamus system are thus governed by $W + UV^T$, a low-rank modification of W.

This conceptualization connects to a body of recent work that has explored the computational features of low-rank additions to full-rank synaptic weight matrices and their application to neuroscience. These studies have shown that in some settings, desired network computations can be realized even if only the low-rank component is structured (Mastrogiuseppe and Ostojic, 2018). Furthermore, when the weights of the low-rank component UV^T are learned in a way that accounts for the full-rank W, this can substantially reconfigure network dynamics. Relating to the thalamocortical system explicitly, a synergy between cortico-cortical and cortico-thalamo-cortical weights can facilitate solving complex tasks that would otherwise be intractable if these two populations of synapses were uncorrelated (Schuessler et al., 2020a; Logiaco et al., 2021). Moreover, other studies have shown that learning in recurrent neural networks often only induces low-rank changes in recurrent weights (Sussillo and Abbott, 2009; Schuessler et al., 2020a; Schuessler et al., 2020b; Susman et al., 2021; Feulner et al., 2022). While this framework has the advantage of being flexible, it also has specific limitations, including an assumption that the thalamic units respond rapidly relative to cortical dynamics, and its analytical characterization is not as complete as in the case of the optimal control framework.

Note that both of these frameworks emphasize that the computational benefit of thalamocortical loops is achieved through specific tuning of their synaptic weights, which opens up the question of the types of learning rules that are involved in that process.

MODEL DETAILS

Thalamocortical Loops for Motor Preparation

As discussed above, a posited role for the motor cortex is the internal generation of dynamics that participate in driving the musculature during movement. Studies have shown that the evolution of the cortical activity during a ballistic movement (e.g., an arm reach to the left or the right) is highly dependent on the "initial state"—i.e., the neural activity immediately prior to movement onset—and that variability in the initial state is reflected in variability in the behavior (Churchland et al., 2006). These results suggest that an important part of motor control is "motor preparation"—the computation of correctly identifying and setting the initial state.

Recent studies, including Logiaco et al. (2021) and Kao et al. (2021), have proposed Type III models that explore how thalamocortical loops can prepare motor cortex and then trigger movement onset. These models share the mechanism that loops between cortex and thalamus with appropriate synaptic weights can force the cortical activity—despite starting in an arbitrary configuration—to evolve to a desired initial state with much greater rapidity than is possible when driving the cortical network with an isolated external input (i.e., as in Type I and II models; Figure 34.4B). After preparation, movement execution in both models can proceed with negligible error.

In Kao et al. (2021) the authors use optimal control theory to analytically identify thalamocortical loop weights that will drive network activity to configurations that minimize movement errors following preparation. This indirectly results in network dynamics that evolve to correct initial states. Interestingly, they show that only a small number of directions in neural activity space are important for guaranteeing successful motor outputs. Consequently, there is a volume of activity space within which all configurations are valid initial states for a given movement. In alignment with this finding, the solutions the authors derive generate control signals that selectively eliminate errors along the relevant dimensions.

Logiaco et al. (2021), on the other hand, directly finds the synaptic strengths of thalamocortical loops that decay cortical activity as quickly as possible to the initial state that will lead to accurate movement. Of note, in contrast to Kao et al. (2021), this approach does not find solutions that selectively ignore the directions in activity space that are irrelevant for the upcoming movement. On the one hand, this has the disadvantage of performing less optimally when the dynamics during movement execution are shared across movements. However, on the other hand, this leads to thalamocortical loops that can perform robustly across the multiple distinct execution dynamics that are associated with different neural activity subspaces, each supporting specific classes of movements. In addition, while direct optimization loses the connection to the elegance of optimal control theory, advantages include the ability to identify loop weights directly in nonlinear networks and to impose size constraints on the thalamocortical circuit such that the thalamus has many fewer neurons than the cortex (Logiaco and Escola, 2020).

Thalamocortical Loops for Motor Execution and Sequencing

An important problem in motor control is the generation of sequences of varied movement building blocks, or "motifs" (Merel et al., 2019). Important features of this computation include flexibility—stringing motifs together by successfully transitioning between them in any arbitrary and potentially novel order—and extendability—learning new motifs without impacting previously acquired ones. Complicating this problem is the fact that an individual motif may have timescales (i.e., its duration and frequency content) that differ considerably from other motifs. In consequence, a fixed cortical dynamical model may fail to generate all of these motifs without a mechanism to powerfully and flexibly reshape its dynamics. While learning in the motor cortex can potentially allow it to generate motor outputs with certain timescales, this comes at the cost of potentially overwriting prior learning and decreasing the performance for other outputs.

Logiaco et al. (2021) tackled this problem by proposing that—while the cortical connectivity provides a "prior" over the dynamics that may be relevant for the execution of a variety of movements—different thalamocortical loops could be instantiated to support the dynamics needed for specific classes of motifs. It was shown analytically that such loops—which act as low-rank perturbations to the cortical network—can be constructed such that the necessary timescales are guaranteed to exist in the dynamics of the cortex-thalamus system (i.e., $W + UV^T$). Moreover, the thalamocortical loop weights can be tuned to prevent noise accumulation along the direction relevant for movement such that individual motifs can be executed robustly without unrealistic precision in the initial, pre-movement network activity. When combined with the model for fast motor preparation (described above), this permitted the elaboration of arbitrary sequences by switching between thalamocortical loops for motif execution and preparation (Figure 34.4C). It was further shown that the thalamocortical architecture is relatively efficient as it can mitigate the size of the neural networks needed in generating these complex and varied motor sequences.

This study proposed that the basal ganglia could implement the selection of motifs. The basal ganglia provide strong inhibitory input to the thalamus whose pattern changes at the transitions between movements. This suggests that the basal ganglia could "play the thalamic piano" by successively turning on and off different thalamocortical loops to instantiate the different dynamics needed to prepare and execute a sequence of motifs. In this way an abstract sequence code—represented in the basal ganglia output—is converted to a motor pattern that can drive behavior through the dynamics of the cortex-thalamus system.

Another study has specifically explored the computation of sequencing at the premotor level (Recanatesi et al., 2022). In this study, thalamocortical interactions implement context-dependent selection of sequences of cortical states. This model demonstrates that the recurrent cortical connectivity can define different quasi-stable population activity patterns whose activation is sustained through recurrent excitation. Then, a few specific thalamocortical loops acting as a low-rank perturbation to the cortical network can interact with the intrinsic cortical dynamics to drive switching through a sequence of such patterns. The model shows that the properties of the thalamocortical connectivity can shape not only the right sequence order, but also the correct variability in activation time for each cortical population pattern.

Biologically Plausible Learning in Thalamocortical Loops

The models we have just discussed show that it is possible to perform computations by tuning synaptic weights in cortico-thalamo-cortical loops appropriately, but these models have not addressed how the loop synapses can be updated with biologically plausible learning. Several such approaches have been proposed for credit assignment in recurrent networks (Gilra and Gerstner, 2017; Murray, 2019), but a recent study showed that biologically plausible learning is only effective for updating thalamocortical—but not corticothalamic—synapses (Lakshminarasimhan et al., 2022). To resolve this challenge, a model was developed in which structured corticothalamic connectivity—proposed to emerge over evolution and to be independent of any specific motor task—was shown to dramatically enhance learning in thalamocortical synapses. Specifically, the corticothalamic pathway needed to communicate an efferent copy of the motor command (Figure 34.4D).

Supporting this, corticofugal projections are known to send axonal branches to the thalamus (Bourassa et al., 1995; Kita and Kita, 2012; Sherman, 2016) while also targeting lower motor centers (Guillery and Sherman, 2011). Furthermore, a reanalysis of neural data from Sauerbrei et al. (2019) confirmed that the alignment of the activity of thalamic neurons with the cortical motor output was significantly greater than chance during a pellet-grasping task (Lakshminarasimhan et al., 2022).

This study provides important support for the validity of Type III models by demonstrating both that the loop weights which underlie their function can be learned in a biologically plausible way and that aspects of corticothalamic connectivity upon which the model relies are observed anatomically and functionally in extant data.

CONCLUSION

In this chapter, we have reviewed models of thalamocortical interactions in motor control through the lens of their assumptions about the functional relevance of primary motor cortical dynamics during movement, as well as the feedforward versus recurrent aspect of thalamocortical interactions (Figure 34.1).

Given the recent experimental results demonstrating the relevance of fine cortical dynamics to drive at least certain types of movements (Shenoy et al., 2013; Sauerbrei et al., 2019; Guo et al., 2021), we have argued in this chapter that the recent modeling developments showing how recurrent thalamocortical loops can reconfigure cortical dynamics are particularly relevant to shed light on the mechanisms by which thalamocortical interactions can serve motor control. In particular, these models can inspire new analyses and experiments to further test the mechanisms they propose. In a general sense, the Type III models discussed here suggest the importance of bidirectional interactions between thalamus and cortex. While there has been a recent validation of the importance of thalamocortical inputs for motor cortical dynamics (Sauerbrei et al., 2019), to the best of our knowledge more progress remains to be made for testing whether the corticothalamic projections can significantly affect motor thalamus activity. In addition, several models assume that different thalamocortical loops are associated with different types of behaviors or different stages of a task—assumptions that remain to be fully tested. Similarly, as is always the case when considering the importance of structured connectivity, a confirmation of the mechanisms by which thalamocortical interactions may be tuned through experience-dependent learning awaits experiments that are able to monitor these phenomena with the appropriate resolution.

Finally, there are also several open avenues for building further on current thalamocortical models. In particular, there is currently little understanding of the synergistic interactions between different cortical regions and different thalamic nuclei during motor control. Therefore, we expect that this line of research will continue to provide valuable insights to the field of thalamocortical interactions.

REFERENCES

Albin RL, Young AB, Penney JB (1989). The functional anatomy of basal ganglia disorders. *Trends Neurosci.* 12(10):366–375. https://pubmed.ncbi.nlm.nih.gov/2479133/.

Albright TD (1984). Direction and orientation selectivity of neurons in visual area MT of the macaque. *J Neurophysiol.* 52:1106–1130.

Bachschmid-Romano L, Hatsopoulos NG, Brunel N (2022). Interplay between external inputs and recurrent dynamics during movement preparation and execution in a network model of motor cortex. *bioRxiv*. https://www.biorxiv.org/content/biorxiv/early/2022/02/19/2022.02.19.481140.

Biane JS, Takashima Y, Scanziani M, Conner JM, Tuszynski MH (2016). Thalamocortical projections onto behaviorally relevant neurons exhibit plasticity during adult motor learning. *Neuron.* 89:1173–1179.

Bosch-Bouju C, Hyland BI, Parr-Brownlie LC (2013). Motor thalamus integration of cortical, cerebellar and basal ganglia information: implications for normal and parkinsonian conditions. *Front Comput Neurosci.* 7:163. http://dx.doi.org/10.3389/fncom.2013.00163.

Bourassa J, Pinault D, Deschênes M (1995). Corticothalamic projections from the cortical barrel field to the somatosensory thalamus in rats: a single-fibre study using biocytin as an anterograde tracer. *Eur J Neurosci.* 7:19–30.

Butler EG, Finkelstein DI, Harvey MC, Churchward PR, Forlano LM, Horne MK (1996). The relationship between monkey ventrolateral thalamic nucleus activity and kinematic parameters of wrist movement. *Brain Res.* 736:146–159.

Butler EG, Horne MK, Hawkins NJ (1992). The activity of monkey thalamic and motor cortical neurones in a skilled, ballistic movement. *J Physiol.* 445:25.

Calderon CB, Verguts T, Frank MJ (2022). Thunderstruck: The ACDC model of flexible sequences and rhythms in recurrent neural circuits. *PLoS Comput Biol.* 18:e1009854.

Caligiore D, Pezzulo G, Baldassarre G, Bostan AC, Strick PL, Doya K, Helmich RC, Dirkx M, Houk J, Jörntell H, Lago-Rodriguez A, Galea JM, Miall RC, Popa T, Kishore A, Verschure PFMJ, Zucca R, Herreros I (2016). Consensus paper: Towards a systems-level view of cerebellar function: The interplay between cerebellum, basal ganglia, and cortex. *Cerebellum.* 16:203–229.

Chabrol FP, Blot A, Mrsic-Flogel TD (2019). Cerebellar contribution to preparatory activity in motor neocortex. *Neuron.* 103:506–519. https://pubmed.ncbi.nlm.nih.gov/31201123/.

Chapman B, Stryker MP (1992). Origin of orientation tuning in the visual cortex. *Curr Opin Neurobiol.* 2:498–501.

Chevalier G, Deniau JM (1990). Disinhibition as a basic process in the expression of striatal functions. *Trends Neurosci.* 13(7):277–280. https://pubmed.ncbi.nlm.nih.gov/1695403/.

Churchland MM, Afshar A, Shenoy KV (2006). A central source of movement variability. *Neuron.* 52:1085–1096.

Clay Reid R, Alonso J-M (1995). Specificity of monosynaptic connections from thalamus to visual cortex. *Nature.* 378: 281–284.

Crapse TB, Sommer MA (2008). Corollary discharge circuits in the primate brain. *Curr Opin Neurobiol.* 18:552–557.

Doya K (2000). Complementary roles of basal ganglia and cerebellum in learning and motor control. *Curr Opin Neurobiol.* 10(6):732–739. https://pubmed.ncbi.nlm.nih.gov/11240282/.

Evarts EV (1968). Relation of pyramidal tract activity to force exerted during voluntary movement. *J Neurophysiol.* 31:14–27.

Eysenbach B, Levine S (2022). Maximum entropy rl (provably) solves some robust rl problems. arXiv preprint arXiv:2103 06257.

Feulner B, Perich MG, Chowdhury RH, Miller LE, Gallego JA, Clopath C (2022). Small, correlated changes in synaptic connectivity may facilitate rapid motor learning. *Nat Commun.* 13:1–14.

Forlano LM, Horne MK, Butler EG, Finkelstein D (1993). Neural activity in the monkey anterior ventrolateral thalamus during trained, ballistic movements. *J Neurophysiol.* 70:2276–2288.

FriedliL, Rosenzweig ES, Barraud Q, Schubert M, Dominici N, Awai L, Nielson JL, Musienko P, Nout-Lomas Y, Zhong H, Zdunowski S (2015) Pronounced species divergence in corticospinal tract reorganization and functional recovery after lateralized spinal cord injury favors primates. *Sci Transl Med.* 7:302ra134.

Gao Z, Davis C, Thomas AM, Economo MN, Abrego AM, Svoboda K, De Zeeuw CI, Li N (2018). A cortico-cerebellar loop for motor planning. *Nature.* 563(7729):113–116. https://pubmed.ncbi.nlm.nih.gov/30333626/.

Georgopoulos AP, DeLong MR, Crutcher MD (1983). Relations between parameters of step-tracking movements and single cell discharge in the globus pallidus and subthalamic nucleus of the behaving monkey. *J Neurosci.* 3:1586–1598.

Georgopoulos AP, Kalaska JF, Caminiti R, Massey JT (1982). On the relations between the direction of two-dimensional arm movements and cell discharge in primate motor cortex. *J Neurosci.* 2:1527–1537.

Georgopoulos AP, Merchant H, Naselaris T, Amirikian B (2007). Mapping of the preferred direction in the motor cortex. *Proc Natl Acad Sci U S A.* 104:11068–11072.

Gerstner W, Kistler WM, Naud R, Paninski L (2014). *Neuronal Dynamics: From Single Neurons to Networks and Models of Cognition.* Cambridge University Press.

Gilra A, Gerstner W (2017). Predicting non-linear dynamics by stable local learning in a recurrent spiking neural network. *eLife.* 6:e28295. http://dx.doi.org/10.7554/eLife.28295.

Glaser J, Whiteway M, Cunningham JP, Paninski L, Linderman S (2020). Recurrent switching dynamical systems models for multiple interacting neural populations. *Adv Neural Inf Process Syst.* 33:14867–14878.

Groenewegen HJ (2003). The basal ganglia and motor control. *Neural Plast.* 10:107–120.

Guillery RW, Sherman SM (2011). Branched thalamic afferents: What are the messages that they relay to the cortex? *Brain Res Rev.* 66:205–219.

Guo J-Z, Sauerbrei BA, Cohen JD, Mischiati M, Graves AR, Pisanello F, Branson KM, Hantman AW (2021). Disrupting cortico-cerebellar communication impairs dexterity. *eLife.* 10:e65906. https://elifesciences.org/articles/65906.

Guo ZV, Inagaki HK, Daie K, Druckmann S, Gerfen CR, Svoboda K (2017). Maintenance of persistent activity in a frontal thalamocortical loop. *Nature.* 545(7653):181–186. https://pubmed.ncbi.nlm.nih.gov/28467817/.

Horne MK, Butler EG (1995). The role of the cerebello-thalamo-cortical pathway in skilled movement. *Prog Neurobiol.* 46(2-3):199–213. https://pubmed.ncbi.nlm.nih.gov/7568913/ [

Hubel DH, Wiesel TN (1962). Receptive fields, binocular interaction and functional architecture in the cat's visual cortex. *J Physiol.* 160:106–154.

Inagaki HK, Chen S, Ridder MC, Sah P, Li N, Yang Z, Hasanbegovic H, Gao Z, Gerfen CR, Svoboda K (2022). A midbrain-thalamus-cortex circuit reorganizes cortical dynamics to initiate movement. *Cell.* 185:1065–1081.

Inase M, Buford JA, Anderson ME (1996). Changes in the control of arm position, movement, and thalamic discharge during local inactivation in the globus pallidus of the monkey. *J Neurophysiol.* 75(3):1087–1104. https://pubmed.ncbi.nlm.nih.gov/8867120/.

Inoue Y, Mao H, Suway SB, Orellana J, Schwartz AB (2018). Decoding arm speed during reaching. *Nat Commun.* 9:1–14.

Kao T-C, Sadabadi MS, Hennequin G (2021). Optimal anticipatory control as a theory of motor preparation: A thalamo-cortical circuit model. *Neuron.* 109:1567–1581.

Kita T, Kita H (2012). The subthalamic nucleus is one of multiple innervation sites for long-range corticofugal axons: a single-axon tracing study in the rat. *J Neurosci.* 32:5990–5999.

Lakshminarasimhan K, Xie M, Cohen J, Sauerbrei B, Hantman A, Litwin-Kumar A, Escola S (2022). Specific connectivity optimizes learning in thalamocortical loops. bioRxiv:2022.09.27.509618. https://www.biorxiv.org/content/biorxiv/early/2022/09/28/2022.09.27.509618 [

Lara AH, Cunningham JP, Churchland MM (2018). Different population dynamics in the supplementary motor area and motor cortex during reaching. *Nat Commun*. 9:1–16.

Li N, Mrsic-Flogel TD (2020). Cortico-cerebellar interactions during goal-directed behavior. *Curr Opin Neurobiol*. 65:27–37. https://pubmed.ncbi.nlm.nih.gov/32979846/.

Loeb GE (2012). Optimal isn't good enough. *Biol Cybern*. 106:757–765.

Logiaco L, Abbott LF, Escola S (2021). Thalamic control of cortical dynamics in a model of flexible motor sequencing. *Cell Rep*. 35:109090.

Logiaco L, Escola GS (2020). Thalamocortical motor circuit insights for more robust hierarchical control of complex sequences. https://arxiv.org/abs/2006.13332.

Mannella F, Baldassarre G (2015). Selection of cortical dynamics for motor behaviour by the basal ganglia. *Biol Cybern*. 109(6):575–595. https://pubmed.ncbi.nlm.nih.gov/26537483/.

Mastrogiuseppe F, Ostojic S (2018). Linking connectivity, dynamics, and computations in low-rank recurrent neural networks. *Neuron*. 99:609–623.

Merel J, Botvinick M, Wayne G (2019). Hierarchical motor control in mammals and machines. *Nat Commun*. 10:1–12.

Murray JM (2019). Local online learning in recurrent networks with random feedback. *eLife*. 8:e43299. http://dx.doi.org/10.7554/eLife.43299.

Nambu A, Tokuno H, Takada M (2002). Functional significance of the cortico-subthalamo-pallidal "hyperdirect" pathway. *Neurosci Res*. 43(2):111–117. https://pubmed.ncbi.nlm.nih.gov/12067746/.\

Polyakova Z, Chiken S, Hatanaka N, Nambu A (2020). Cortical control of subthalamic neuronal activity through the hyperdirect and indirect pathways in monkeys. *J Neurosci*. 40:7451–7463.

Rao HM, San Juan J, Shen FY, Villa JE, Rafie KS, Sommer MA (2016). Neural network evidence for the coupling of presaccadic visual remapping to predictive eye position updating. *Front Comput Neurosci*. 10:52. http://dx.doi.org/10.3389/fncom.2016.00052.

Recanatesi S, Pereira-Obilinovic U, Murakami M, Mainen Z, Mazzucato L (2022). Metastable attractors explain the variable timing of stable behavioral action sequences. *Neuron*. 110:139–153.

Rikhye RV, Gilra A, Halassa MM (2018). Thalamic regulation of switching between cortical representations enables cognitive flexibility. *Nat Neurosci*. 21:1753–1763.

Sauerbrei BA, Guo J-Z, Cohen JD, Mischiati M, Guo W, Kabra M, Verma N, Mensh B, Branson K, Hantman AW (2019). Cortical pattern generation during dexterous movement is input-driven. Nature. 577:386–391.

Schuessler F, Dubreuil A, Mastrogiuseppe F, Ostojic S, Barak O (2020a). Dynamics of random recurrent networks with correlated low-rank structure. *Phys Rev Res*. 2(1):013111. https://link.aps.org/doi/10.1103/PhysRevResearch.2.013111.

Schuessler F, Mastrogiuseppe F, Dubreuil A, Ostojic S, Barak O (2020b). The interplay between randomness and structure during learning in RNNs. In: Larochelle H, Ranzato M, Hadsell R, Balcan MF, Lin H, eds. *Advances in Neural Information Processing Systems*. Curran Associates:13352–13362.

Shenoy KV, Sahani M, Churchland MM (2013). Cortical control of arm movements: A dynamical systems perspective. *Annu Rev Neurosci*. 36:337–359.

Sherman SM (2016). Thalamus plays a central role in ongoing cortical functioning. *Nat Neurosci*. 19:533–541.

Shine JM (2021). The thalamus integrates the macrosystems of the brain to facilitate complex, adaptive brain network dynamics. *Prog Neurobiol*. 199:101951. https://pubmed.ncbi.nlm.nih.gov/33189781/.

Strick PL (1976). Activity of ventrolateral thalamic neurons during arm movement. *J Neurophysiol*. 39:1032–1044.

Stroud JP, Porter MA, Hennequin G, Vogels TP (2018). Motor primitives in space and time via targeted gain modulation in cortical networks. *Nat Neurosci*. 21:1774–1783.

Susman L, Mastrogiuseppe F, Brenner N, Barak O (2021). Quality of internal representation shapes learning performance in feedback neural networks. *Phys Rev Res*. 3(1):013176. https://link.aps.org/doi/10.1103/PhysRevResearch.3.013176.

Sussillo D, Abbott LF (2009). Generating coherent patterns of activity from chaotic neural networks. *Neuron*. 63:544–557.

Thach WT, Goodkin HP, Keating JG (1992). The cerebellum and the adaptive coordination of movement. *Annu Rev Neurosci*. 15:403–442.

Todorov E, Jordan MI (2002). Optimal feedback control as a theory of motor coordination. *Nat Neurosci*. 5:1226–1235.

Trouche E, Beaubaton D (1980). Initiation of a goal-directed movement in the monkey: Role of the cerebellar dentate nucleus. *Exp Brain Res*. 40:311–321.

Turner RS, Anderson ME (1997). Pallidal discharge related to the kinematics of reaching movements in two dimensions. *J Neurophysiol*. 77:1051–1074.

Yamins DLK, Hong H, Cadieu CF, Solomon EA, Seibert D, DiCarlo JJ (2014). Performance-optimized hierarchical models predict neural responses in higher visual cortex. *Proc Natl Acad Sci U S A*. 111:8619–8624.

35.

THALAMOCORTICAL CONTRIBUTIONS TO NEURAL DYNAMICS AND BEHAVIOR

Jorge Jaramillo and Zengcai V. Guo

INTRODUCTION

Animals exhibit remarkable behaviors. For example, European honeybees use a figure-8 dance to signal the orientation and distance to the location of nectar (Frisch, 1974). Alaskan salmon can swim thousands of miles back to reach the stream of their origin where they will lay eggs. Asian elephants can modify branches to make flyswatters and they can even exhibit mirror self-recognition capabilities. (Hart et al., 2001; Plotnik et al., 2006). Although it is often desirable to study behavior in natural environments, it is in the laboratory that external and internal factors (stimulus drive, emotional states, energy homeostasis, etc.) can be precisely controlled, which greatly facilitates research into neural dynamics underlying behavior. Body restraint paradigms in nonhuman primates, and since the early 21st century in rodents, have enabled precise stimulus control, behavioral monitoring, and neural recording. Extensive research using these paradigms has revealed that cognitive functions, including perception, motor planning, and decision-making, are supported by cortical neural dynamics (Churchland et al., 2012; Inagaki et al., 2019; Mante et al., 2013). The thalamus, sitting in a privileged position to directly influence the cortex, likely plays important roles in shaping cortical neural dynamics and cognitive behaviors.

Most of the knowledge accumulated with respect to thalamic function concerns sensory thalamic nuclei, which receive direct input from the periphery via sense organs (Jones, 2007; Sherman and Guillery, 2013). Interestingly, recent studies have demonstrated contributions from non-sensory thalamic nuclei to cognitive and sensorimotor computations such as attention, working memory, decision-making, and motor planning that had been exclusively ascribed to the cortex (Alcaraz et al., 2018; Bolkan et al., 2017; Guo et al., 2017; McAlonan et al., 2008; Roth et al., 2016; Saalmann et al., 2012; Schmitt et al., 2017). To explain the role of the thalamus in these computations, experimental and computational studies (Jaramillo et al., 2019) have suggested that the activation of higher-order thalamic nuclei may change the dynamical regime of the cortex, for example by enhancing local recurrent connectivity within (Logiaco et al., 2021; Schmitt et al., 2017) and across (Saalmann et al., 2012) cortical areas. Importantly, in addition to reciprocal connections with the cortex, the thalamus constitutes a gateway for subcortical structures to relay information to the cortex (Figure 35.1A). Major subcortical structures, including the basal ganglia, cerebellum, and brainstem nuclei, have extensive projections in the thalamus (see **Chapter 31 in Section VI**). How outputs of these subcortical structures are exactly transformed by thalamocortical projections to regulate cortical dynamics remains unknown.

Here we propose a general computational framework to describe how cortical dynamics relate to subcortical and thalamocortical inputs in the context of a motor-planning task. Specifically, we highlight how thalamocortical anatomy and physiology enable the computations that have been observed during motor planning. We further summarize examples of how subcortical inputs to the thalamus control cortical dynamics within this framework. We conclude with some ideas toward a comprehensive theory of non-sensory thalamic nuclei in the context of anatomy, physiology, and behavior.

COMPUTATIONAL FRAMEWORK FOR REGULATION OF CORTICAL DYNAMICS

Large-scale recording techniques have enabled simultaneous collection of activity from ensembles of neurons (Jun et al., 2017; Stevenson and Kording, 2011). The number of simultaneously recorded neurons is typically much higher than the number of task variables. Thus, in many tasks it has

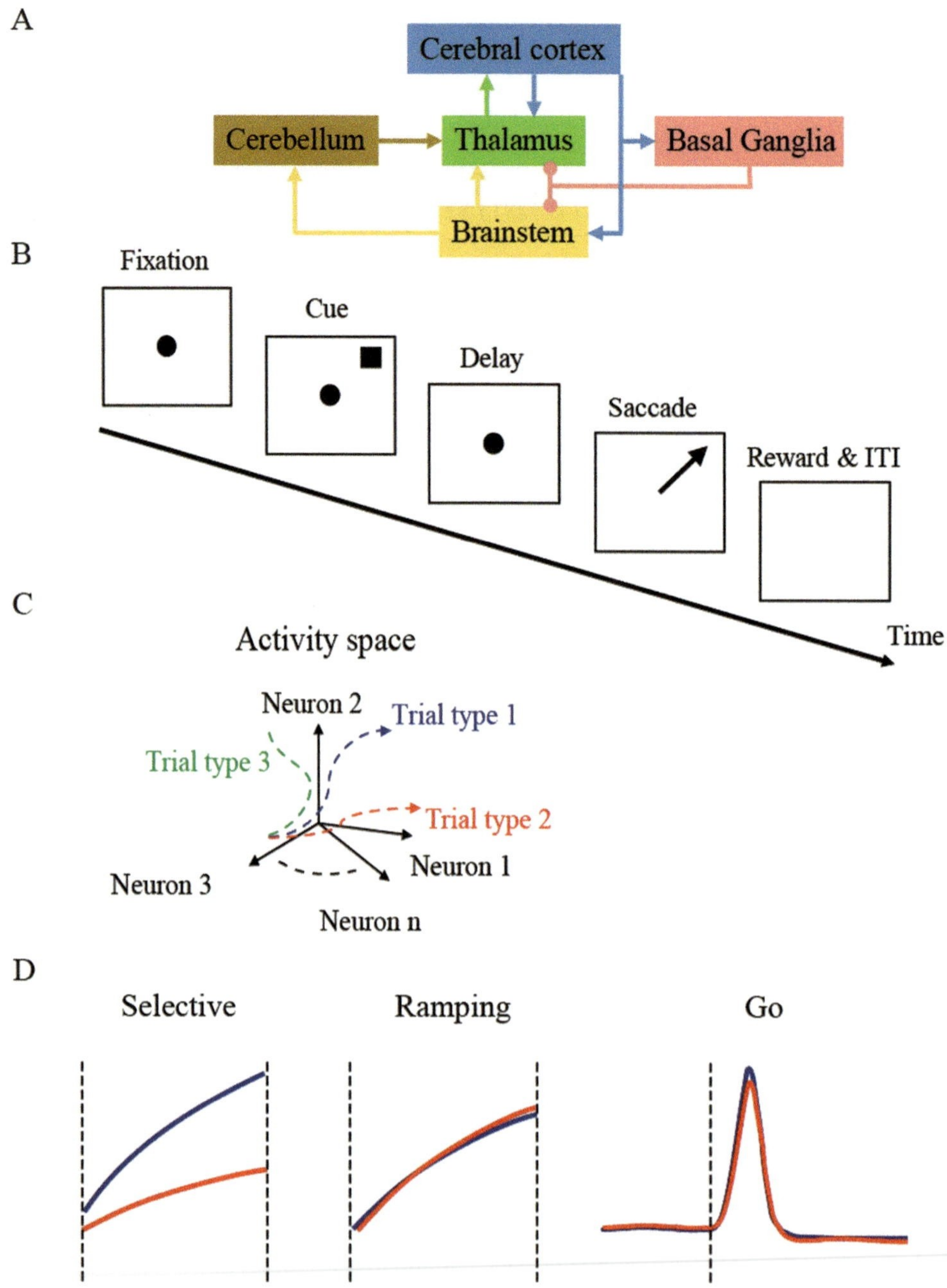

Figure 35.1. Thalamocortical interaction, low dimensional neural dynamics, and behavior. (A). Simplified schematic of thalamocortical interconnections and major subcortical pathways that can influence cortical neural dynamics through the thalamus. Other connections are omitted for clarity. Arrowheads = excitatory connections; circles = inhibitory connections. (B). Trial structure of the oculomotor delayed response (ODR) task. The ODR paradigm is widely used to study neural dynamics underlying working memory. In each trial, there are distinct epochs. (C). Schematic of the activity space with each neuron representing one dimension. Population activity during each trial evolves to form one trajectory. (D). Decomposing neural trajectories into the selective and ramping modes between 2 trial types (blue and red trajectories), and the "go" mode.

been observed that neural dynamics are low dimensional. This means, that only a few "modes" of activity are sufficient to effectively reproduce the complex circuit dynamics for a given task (but see Voitov and Mrsic-Flogel, 2022). We first introduce low-dimensional neural dynamics underlying behavior and then discuss this in the context of thalamocortical projections and in the regulation of cortical dynamical modes.

LOW-DIMENSIONAL NEURAL DYNAMICS AND BEHAVIOR

To understand how information is encoded by populations of neurons, ensemble activity can be analyzed in an activity space, where each dimension corresponds to the activity of one neuron (Figure 35.1C). Population dynamics at different time points trace a trajectory evolving in this high-dimensional space. Several methods based on population-level statistical analysis exist to reduce the dimensionality (reviewed in, e.g., Cunningham and Yu, 2014). Commonly used methods, including principal component analysis (PCA), can effectively reduce the number of dimensions. Although in these methods linear combinations of the original features (e.g., single neuron firing rates) can account for most of the statistical variance, principal components are not as interpretable as the original features.

Activity modes, on the other hand, can be directly extracted based on task-specific considerations. For example, let us suppose that a behavioral task has 2 trial types defined by distinct sensory inputs that guide 2 motor actions—tasks with more sensory inputs and/or motor actions can be mapped onto this basic type. In addition, there are multiple epochs that tile each trial (Figure 35.1B). Population activity within each epoch can be decomposed into 2 dynamical modes, a selective mode that discriminates the 2 trial types and a non-selective mode that reflects a trial-type independent signal that is modulated by the task (Figure 35.1D). At the transition between trial epochs, there is also a third mode that serves as linkage between adjacent epochs. Many memory-guided decision-making paradigms conform to this simple task structure (Figure 35.2A).

CORTICAL MODES DURING MOTOR PLANNING

We illustrate the conceptual framework using a memory-guided response task that contains a motor-planning phase (Guo et al., 2014; Li et al., 2015). In this task, there are 2 trial types: left-lick and right-lick trials based on sensory stimulation and the appropriate motor output. Each trial is separated into 3 epochs, the sample epoch during which sensory stimulus is presented, the delay epoch during which mice maintain a prospective short-term memory (motor planning), and a response epoch during which mice initiate licking as instructed by sensory input (Figure 35.2A). During the transition from the delay to response epoch, an auditory "go" signal triggers the initiation of licking.

Population activity recorded during this task can be projected along a vector in neural activity space that maximally differentiates trial types (Figure 35.2B–D). The vector direction can be obtained using linear discriminant analysis (LDA). LDA requires estimation of a covariance matrix that could be noisy if the number of trials is limited. In practice, the covariance term can be neglected and the vector direction can be obtained as a difference in firing rates for different trial types: av(r_L – r_R), where r_L and r_R denote the population activity vector averaged across correct left- and right-lick trials, respectively, and av denotes a temporal average (e.g., across the delay epoch). By definition, this direction discriminates trial types and is therefore referred to as a selective mode (also coding direction; e.g., Li et al., 2016). The non-selective mode is task specific, and could reflect a timing or urgency signal, that can be defined as av(r_1 – r_0), where r_1 and r_0 denote the activity vector before "go" cue and the baseline activity, respectively (Inagaki et al., 2018). Finally, the "go" mode reflects the acute activity change in the cortex following the "go" trigger signal: av(r_go+ – r_go–), where r_go+ and r_go– denote the activity vector right after and right before the "go" cue, respectively (Inagaki et al., 2022b). These modes are near orthogonal and together account for most of the neural activity variance, thus providing a parsimonious description of cortical dynamics (Figure 35.2E). Importantly, these modes can predict both choice and reaction times (Yang et al., 2022).

The dynamical modes described above can be formulated using a circuit-level computational framework. In this framework, dynamical modes arise from the interaction between excitatory and inhibitory assemblies in the cortex (Figure 35.2F). Moreover, each mode is associated to a specific behaviorally relevant computation. In the memory-guided response task, dynamical modes enable accurate and appropriately timed choice. For example, a selective mode carries information about the stimulus during the stimulus-absent delay period (Li et al., 2016), while the ramp mode may be related to urgency and/or timing (Inagaki et al., 2018; Li et al., 2016; Wang et al., 2021). A phasic mode may be related to transitions between epochs, for example, between motor preparation and motor execution (Inagaki et al., 2022b). Thus, dynamical modes span and uniquely characterize the different epochs of a given behavioral task.

Critically, the cortex does not operate in isolation. It is highly interconnected with the thalamus, which itself is the recipient of subcortical input. *We propose that the thalamus routes subcortical inputs to control dynamical modes for successful behavior*. There are many computations that could be in principle accomplished via subcortical control. For example, a given mode may be selectively amplified or it may be necessary to transition between modes (see also Shine, 2021). Below we review how thalamocortical anatomy and physiology are consistent with the hypothesis of subcortical control, and how different subcortical structures—including the basal ganglia and pedunculopontine nucleus—modulate dynamical modes through thalamocortical projections in the context of motor planning.

THALAMIC REGULATION OF CORTICAL DYNAMICAL MODES

Subcortical control of cortical dynamics depends on the thalamus, which has distinct anatomical and physiological properties. For example, the number of neurons in the

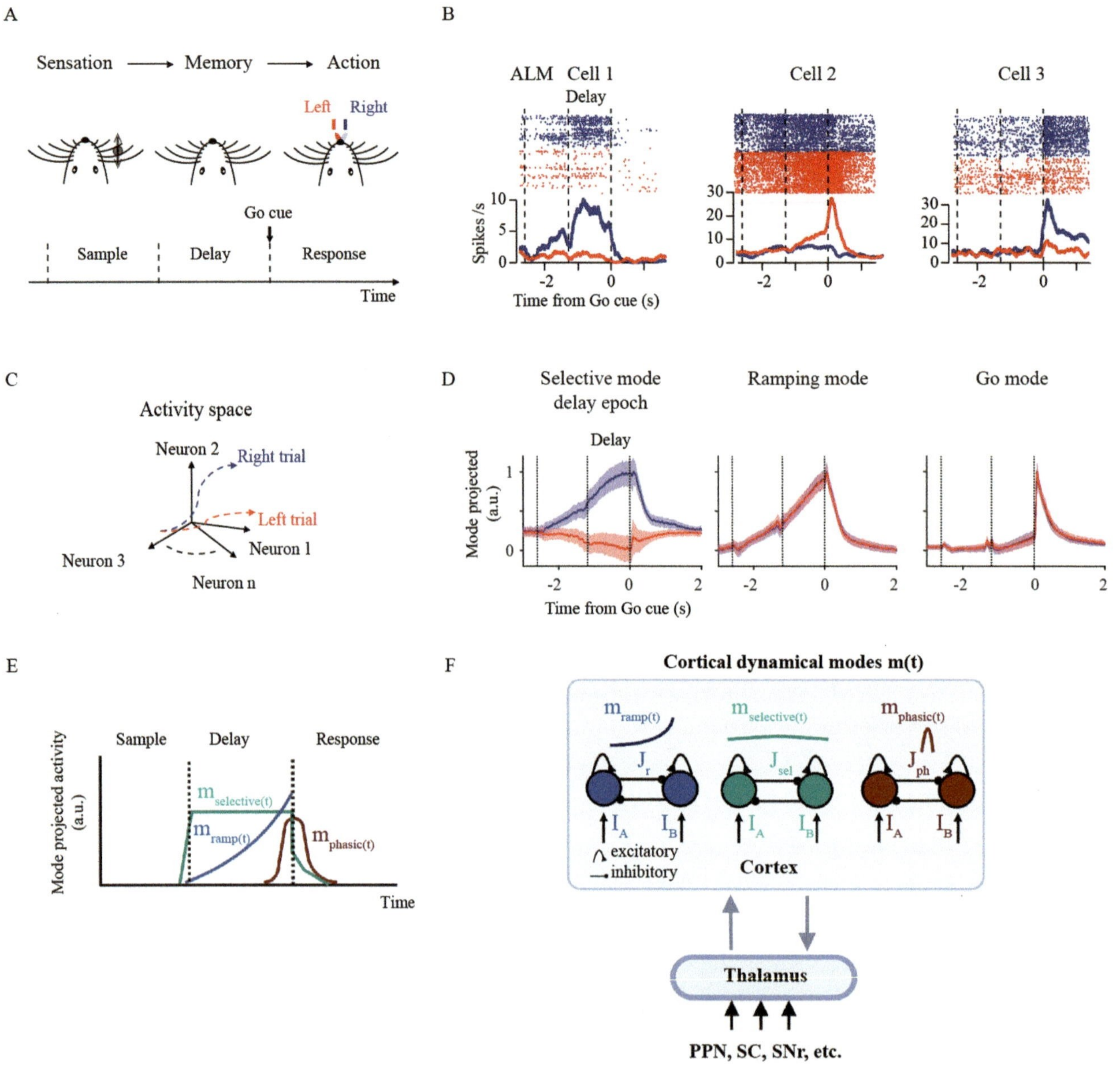

Figure 35.2. Memory-guided delayed response task, dynamical modes and computational implementation. (A). Head-fixed mouse reporting the stimulus strength by directional licking. During the sample epoch, the mouse senses whisker stimulation. During the delay epoch, the animal maintains a memory of past sensory experience and plans for future action. A "go" cue releases the planned action and triggers directional licking. (B). Example ALM neurons showing selective firing during the sample, delay, and/or response epochs. Top: spike raster. Bottom: peri-stimulus time histogram (PSTH). Correct left (red) and right (blue) trials are shown. Dashed lines separate behavioral epochs. Time is aligned to the "go" cue. (C). Schematic of activity space showing neural trajectories in correct left and right trials. (D). Neural trajectories projected along the selective, ramping and "go" directions. (E). Dynamical modes subserve behaviorally relevant computations in the memory-guided delayed response task. (F). Dynamical modes arise from excitatory and inhibitory interactions between selective neural pools in the cortex. Three modes are shown, as in (E). Modes can be controlled (e.g., amplified, suppressed, etc.) by subcortical inputs onto the thalamus. Panels A, B adapted from Guo et al. (2014) with permission; panel D from Inagaki et al. (2022b) with permission.

cortex is higher than that of a typical thalamic nucleus. This allows for the thalamus to control many more cortical cells than the number of available thalamic neurons. Thus, thalamocortical connectivity is suggestive of low-rank perturbations to the cortex (Mastrogiuseppe and Ostojic, 2018). Beyond providing depolarizing input to cortical cells, the thalamus may effectively modify circuits in the cortex (Halassa and Sherman, 2019; Jaramillo et al., 2019; Logiaco et al., 2021), perhaps as a function of task demands.

The motor thalamus, as well as other higher-order thalamic nuclei, are known to form extensive thalamocortical loops (see **Chapter 32 in Section VI**). The details underlying these connectivity patterns may be informative of the ensuing computations. For example, the motor thalamus (ventral medial nucleus, VM, and ventral anterior lateral nucleus, VAL) forms closed excitatory loops with frontal cortex, including the anterior lateral motor cortex (ALM) and the prefrontal cortex (Collins et al., 2018; Guo et al., 2018). Corticothalamic cells from layer 5 project to VM neurons which reciprocally project to the same group of neurons (Collins et al., 2018). Such organization is less evident in layer 6 neurons, the classic source of corticothalamic projections. According to the driver-modulator classification (Sherman and Guillery, 2013), layer 5 synapses onto single thalamic cells are fast and effective and exhibit paired pulse depression. The speed and reliability of these connections are suggestive of the role of the thalamus in sustaining reverberatory activity during the stimulus-absent delay period of motor planning tasks (Guo et al., 2017; Jaramillo et al., 2022). Another thalamocortical circuit feature is transthalamic pathways, which relay cortical information from one cortical area to another (Sherman and Guillery, 2013). A notable thalamic nucleus that conforms to this organization is the pulvinar, which is also involved in motor planning (Wilke et al., 2010).

Below we discuss how projections from motor thalamus to motor cortex participate in the control of cortical dynamical modes and extend this concept to higher-order sensory nuclei (e.g., the pulvinar during visuomotor regulation).

MOTOR THALAMUS-FRONTAL CORTEX INTERCONNECTIONS FOR CORTICAL DYNAMICAL MODES

The motor thalamus has been well known for its role in motor execution and motor learning (Canavan et al., 1989; Goldberg and Fee, 2012; Sauerbrei et al., 2020). Recent studies have ascribed an important role to the motor thalamus in motor planning, in the phase before motor execution takes place (Guo et al., 2017; Yang et al., 2022). Motor planning may involve different computations, including (prospective) short-term memory, decision-making, and motor preparation (Svoboda and Li, 2018). Simultaneous measurements of cortical ensemble activity during thalamic perturbations provide important insights concerning the role of thalamocortical connections (Figure 35.3). Optogenetic inactivation of motor thalamus reduces ALM activity within 5 ms, indicating a direct link between thalamic perturbation and cortical activity reduction. Although ALM receives extensive cortical inputs from the contralateral ALM and motor cortex, disruption of corticocortical communications barely affects ALM membrane potential and firing rates, highlighting the impact of thalamocortical projections on cortical dynamics (Guo et al., 2017).

Strong thalamic inactivation effectively eliminates most of ALM activity, and consequently this also removes the selective and ramping modes (Figure 35.3C–E). Do thalamocortical projections simply provide the excitatory drive required by cortical neurons to fire? Or is the pathway also required for task selective information? Weak inactivation of motor thalamus does not reduce ALM activity in a substantial fraction of neurons. However, this manipulation strongly reduces ALM selectivity (Figure 35.3F). In addition, inactivation also affects dynamical modes coding for stimulus and action, indicating that thalamic activity plays an essential role in sensation, motor planning, and action (Yang et al., 2022). From a circuit-level perspective, activation of higher-order thalamic nuclei may change the dynamical regime of the cortex (Jaramillo et al., 2019; Shine, 2021), for example by enhancing local recurrent connectivity within (Logiaco et al., 2021; Schmitt et al., 2017) and across (Saalmann et al., 2012) cortical areas. From a dynamical systems perspective, one possibility is that the attractor landscape that underlies persistent activity and the selective mode is modulated by thalamic input (Guo et al., 2017; Inagaki et al., 2019; Jaramillo et al., 2022; Schmitt et al., 2017).

OTHER THALAMIC NUCLEI: PULVINAR CONTRIBUTIONS TO VISUOMOTOR BEHAVIOR

The pulvinar is another thalamic nucleus that contributes to motor behaviors. Early studies in the primate pulvinar uncovered its role in attentional processing, as the firing rates of pulvinar neurons are modulated by attentional cues (Robinson and Petersen, 1992). Recent work has expanded pulvinar function to include filtering of distractors (Snow et al., 2009), decision confidence (Komura et al., 2013), visuomotor mismatch (Roth et al., 2016), and visuomotor planning (Wilke et al., 2010). Pulvinar lesions cause pronounced impairments in oculomotor integration and saccadic decision-making (Saalmann et al., 2012). Specifically, pulvinar lesions impair the ability to shift visual attention toward the contralesional hemifield, increasing ipsilesional (ipsilateral to the lesion site) spatial exploration and saccade choice bias (Karnath et al., 2002; Snow et al., 2009; Wilke et al., 2013; Wilke et al., 2010).

To explain the contralateral behavioral effects after unilateral pulvinar lesions, Jaramillo et al. (2019) developed a computational model of pulvino-cortical pathways (Figure 35.4A,B). Such effects are observed because the lesion effectively reduces the excitation toward the contralesional (i.e., affected) visual hemifield, which in turn induces a gain imbalance in the multiregional circuit. The imbalance is amplified by the recurrent circuitry in the cortex and propagates asymmetrically throughout the pulvino-cortical circuit to produce delayed decisions (Figure 35.4C). Importantly, planning and decision behavior in this task is

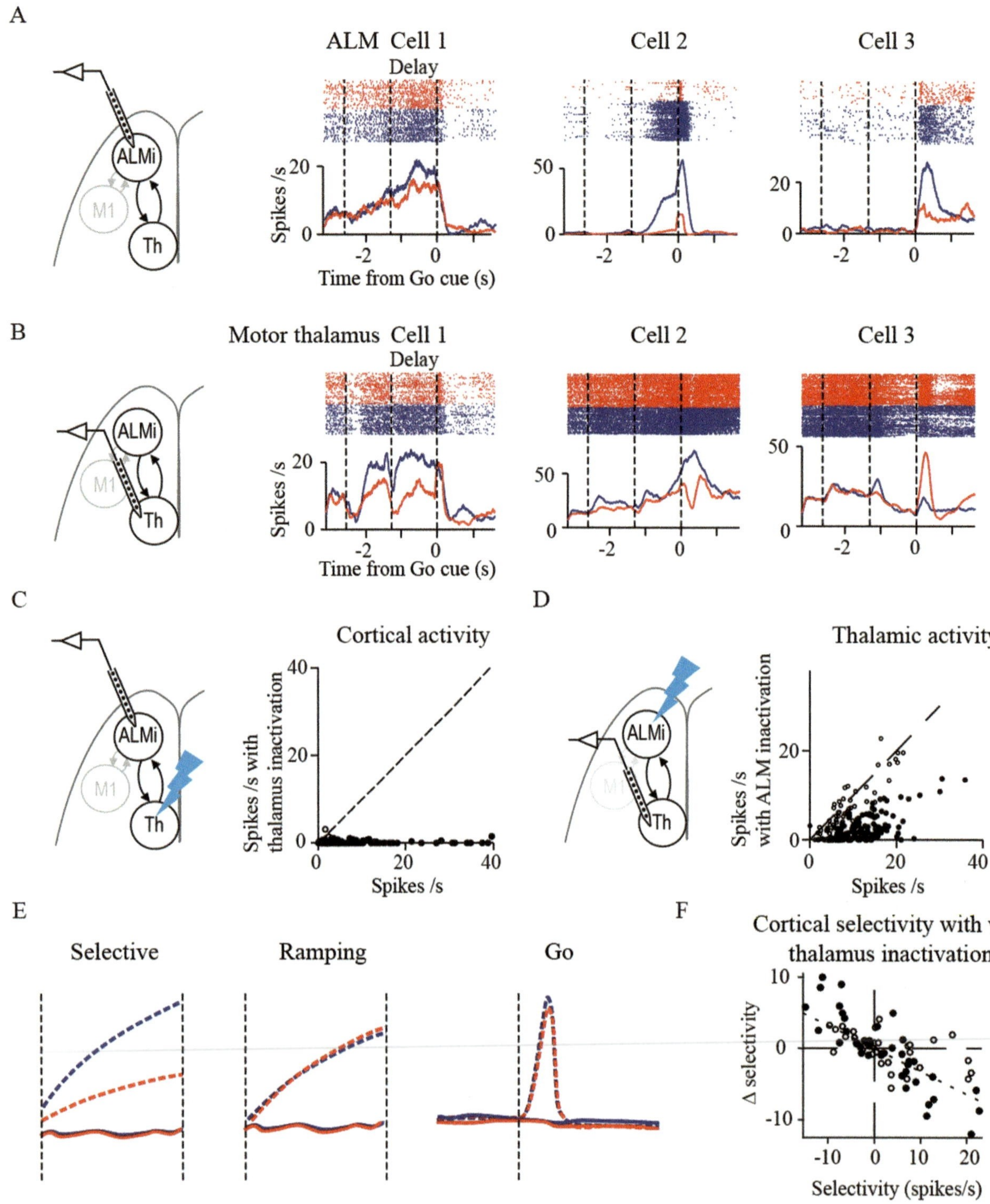

Figure 35.3. Thalamocortical interactions in the memory-guided task. (A, B). Neurons in ALM and motor thalamus show similar response patterns. Left, schematic of silicon probe recordings. Right, example neurons. ALMi, the ipsilateral ALM. Same format as in Figure 35.2B. (C). Inactivation of the motor thalamus abolishes ALM activity. Left: schematic of ALM recording during thalamus inactivation. Right: ALM activity during thalamus inactivation. Firing rate is quantified using the time window of 20–120 ms after photostimulus onset. Filled circles = neurons that are significantly modulated by thalamus inactivation ($p < 0.05$, t-test). Dotted line = unity line. (D). Inactivation of ALM dramatically reduces thalamic activity. Same format as in C. (E). Schematic of selective, ramping, and "go" modes for ALM activity (dotted lines). Removing thalamic activity suppresses each mode (solid). (F). Weak thalamic inactivation abolishes ALM selectivity even in neurons with no mean activity change. Filled circles = neurons with mean firing rate significantly modulated by thalamus inactivation ($p < 0.05$, t-test). Dotted line = linear regression. Panels A, B, C, D, F adapted from Guo et al. (2017) with permission.

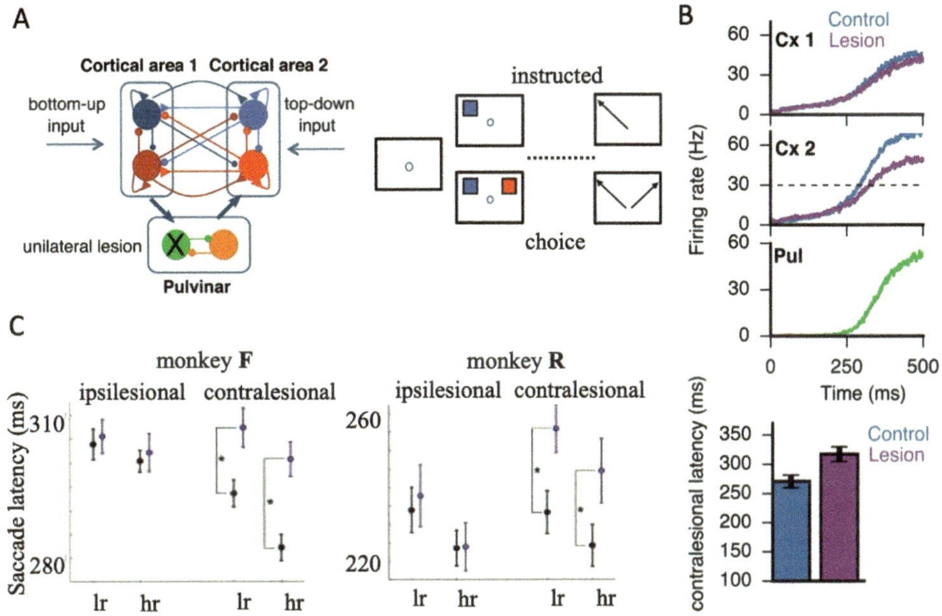

Figure 35.4. Pulvinar and movement planning. (A). Left: schematic of a thalamocortical circuit model composed of 2 cortical areas and the pulvinar. A trans-thalamic pathway connects cortical area 1 to pulvinar and pulvinar to cortical area 2. X marks a unilateral lesion to the pulvinar. Right: instructed and choice tasks, where the subject must make a saccade toward an instructed target or makes a free choice. (B). Numerical simulations of cortical and pulvinar activity in "control" and "lesion" conditions. A ramping mode characterizes the dynamics of the multiregional circuit, predicts reaction times, and is modulated by the pulvinar. Reaction time and choice are recorded after ramping reaches a threshold in cortical area 2 (Cx 2). (C). The ramping mode in B accounts for differences in contralesional latency in the instructed task for low (lr) and high (hr) reward. Panels A and B adapted from Jaramillo et al. (2019); panel C adapted from Wilke et al. (2013).

also characterized by a ramping mode, which upon reaching a predefined threshold, generates an action (Figure 35.4B; see also Figure 35.2).

SUBCORTICAL REGULATION OF CORTICAL DYNAMICAL MODES

While the thalamocortical pathway is able to affect selective and non-selective modes underlying sensation, motor planning, and action, the basal ganglia, cerebellum, and brainstem to thalamus pathways may differentially shape particular dynamical modes. These pathways have been implicated in motor planning and motor execution (Chabrol et al., 2019; Dacre et al., 2021; Gao et al., 2018; Inagaki et al., 2022a; Svoboda and Li, 2018). Here we focus on the basal ganglia (BG)–thalamus and brainstem-thalamus pathways, as these pathways have been shown to specifically regulate the selective and "go" modes during motor planning (Inagaki et al., 2022b; Wang et al., 2021).

BASAL GANGLIA-MOTOR THALAMUS PATHWAY IN SHAPING CORTICAL DYNAMICS

The BG outputs target an array of brainstem nuclei through their GABAergic projections, and these projections also have extensive collaterals innervating major motor and intralaminar thalamic nuclei (McElvain et al., 2021; Mengual et al., 2016). How BG outputs are exactly transformed by these downstream areas to control behavior remains unknown. Neurons in the BG output nuclei , the internal globus pallidus (GPi) and the substantia nigra pars reticulata (SNr), directly inhibit the brainstem motor centers to regulate movements like locomotion, licking, and eye movements (Hikosaka, 2007; Hikosaka et al., 2000; Roseberry et al., 2016; Rossi et al., 2016), whereas the pathway to the thalamus, which in turn projects to the cortex, requires more synaptic steps to reach motor neurons to control specific muscle activity. Do the 2 pathways control behavior using similar mechanisms? Compared with the BG-thalamic pathway, the more direct BG-brainstem pathway, partly due to its privileged position to access motor centers in the brainstem, is relatively better understood. Here we summarize the role of this pathway in gating movements and then contrast it with the more complex BG-thalamic pathway, particularly in the context of motor planning.

The Role of BG-Brainstem Pathway in Gating of Movements

Seminal experiments from Hikosaka and colleagues revealed that the BG-brainstem pathway provides a permissive role in the initiation of saccadic eye movements

(Hikosaka, 2007; Hikosaka et al., 2000). Neurons in the BG output nuclei SNr project to the deep layers of the superior colliculus (SC) whereupon premotor neurons convey a command to initiate saccades (Figure 35.5A). Neurons in SNr exhibit spontaneous activity at a high rate, and this activity tonically inhibits SC to prevent unwanted saccades (Figure 35.5B). In the presence of a rewarded visual target, cortical activation drives a burst of activity in medium spiny neurons (MSNs) in the caudate (CD), the part of the primate striatum involved in movement control. Burst activity in MSNs of the direct pathway suppresses the tonic discharge of SNr neurons. This pause of SNr activity subsequently disinhibits SC neurons, allowing them to generate a burst of action potentials that orient the saccade to the rewarded target. Artificially pausing SNr activity by local infusion of muscimol can induce involuntary saccades to the contralateral visual field (Figure 35.5C), supporting the causal role of this pathway in gating movements. Thus, the BG-brainstem pathway functions as a physiological "gate" that must be opened to allow movements to happen. We refer to this process as the BG gating model.

Studies in licking and locomotion further support this model. SNr neurons send GABAergic projections to the orofacial region of SC (Rossi et al., 2016). During consumption of water, licking related neurons in SNr and SC exhibit antiphase oscillations, and SC neurons discharge action potentials more frequently when SNr neurons fire during troughs of oscillation. Optogenetic activation of SNr projections transiently suppressed activity of SC neurons and reduced licking. SNr also targets the mesencephalic locomotor region (MLR), a brainstem region that is critical for locomotion. Optogenetic stimulation of MSNs in the direct pathway pauses SNr firing, which promotes locomotion by disinhibiting glutamatergic neurons in MLR (Roseberry et al., 2016), just like disinhibiting SC to increase the frequency of saccades (Figure 35.5). Stimulation of MSNs in the indirect pathway has the opposite effect, resulting in suppression of MLR activity and reduced locomotion.

The Role of BG-Motor Thalamus Pathway in Shaping Cortical Dynamical Modes

The gating model is very successful in explaining how the BG-brainstem circuits facilitate actions such as orienting, licking, and locomotion. However, it seems that the gating model cannot be directly applied to the ascending BG-thalamus pathway. If the gating model holds true, we would expect that during movements, BG output neurons pause their activity to disinhibit the motor thalamus. Contrary to this expectation, pallidal and thalamic neurons typically show a similar pattern of activation at movement initiation (Goldberg and Fee, 2012; Inase et al., 1996; Schwab et al., 2020). For example, thalamic neurons and pallidal inputs peaked their firing before song syllable onsets, and decreased their firing rates before syllable offset (Goldberg and Fee, 2012). Similar results were found in macaques performing a reaching task (Schwab et al., 2020). In mice, neurons in the cortex, motor thalamus, and SNr exhibit similar response patterns predicting upcoming choice (Figure 35.6). Thus,

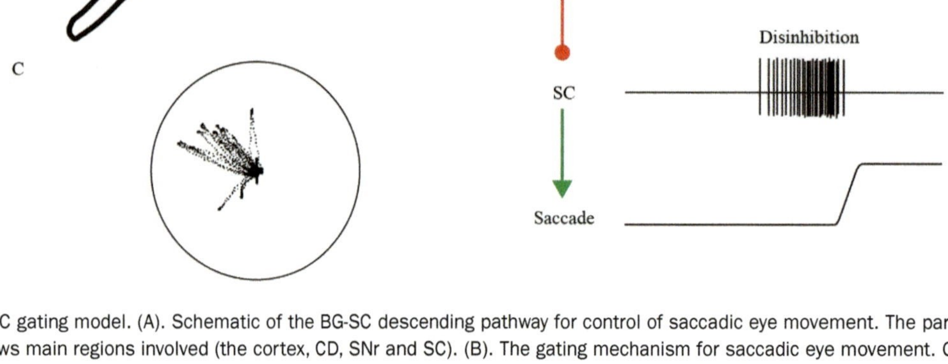

Figure 35.5. The BG-SC gating model. (A). Schematic of the BG-SC descending pathway for control of saccadic eye movement. The parasagittal view of the macaque brain shows main regions involved (the cortex, CD, SNr and SC). (B). The gating mechanism for saccadic eye movement. GABAergic neurons in SNr are tonically active and the SC is normally inhibited by this tonic activity. Projection neurons in the CD are GABAergic, and the phasic activation of the CD interrupts SNr activity. This disinhibition, together with other excitatory inputs, releases a gate for the SC neurons to fire in bursts, leading to a saccadic movement to the contralateral location. (C). Trajectories showing involuntary eye movements of a monkey following muscimol injection into the right SNr. Adapted from Hikosaka O (2007) with permission.

these results consistently indicate that the BG inputs play a modulatory role in shaping thalamic activity, and thalamic neurons are driven by other sources.

What pathway drives activity in higher order thalamic nuclei? The co-activation of BG and motor thalamus suggests that these structures receive a common drive, most likely from the cortex, as the cortex provides extensive excitatory inputs to both the BG and motor thalamus. In support of this view, optogenetic inactivation of mouse motor cortex dramatically reduces thalamic activity (Figure 35.3).

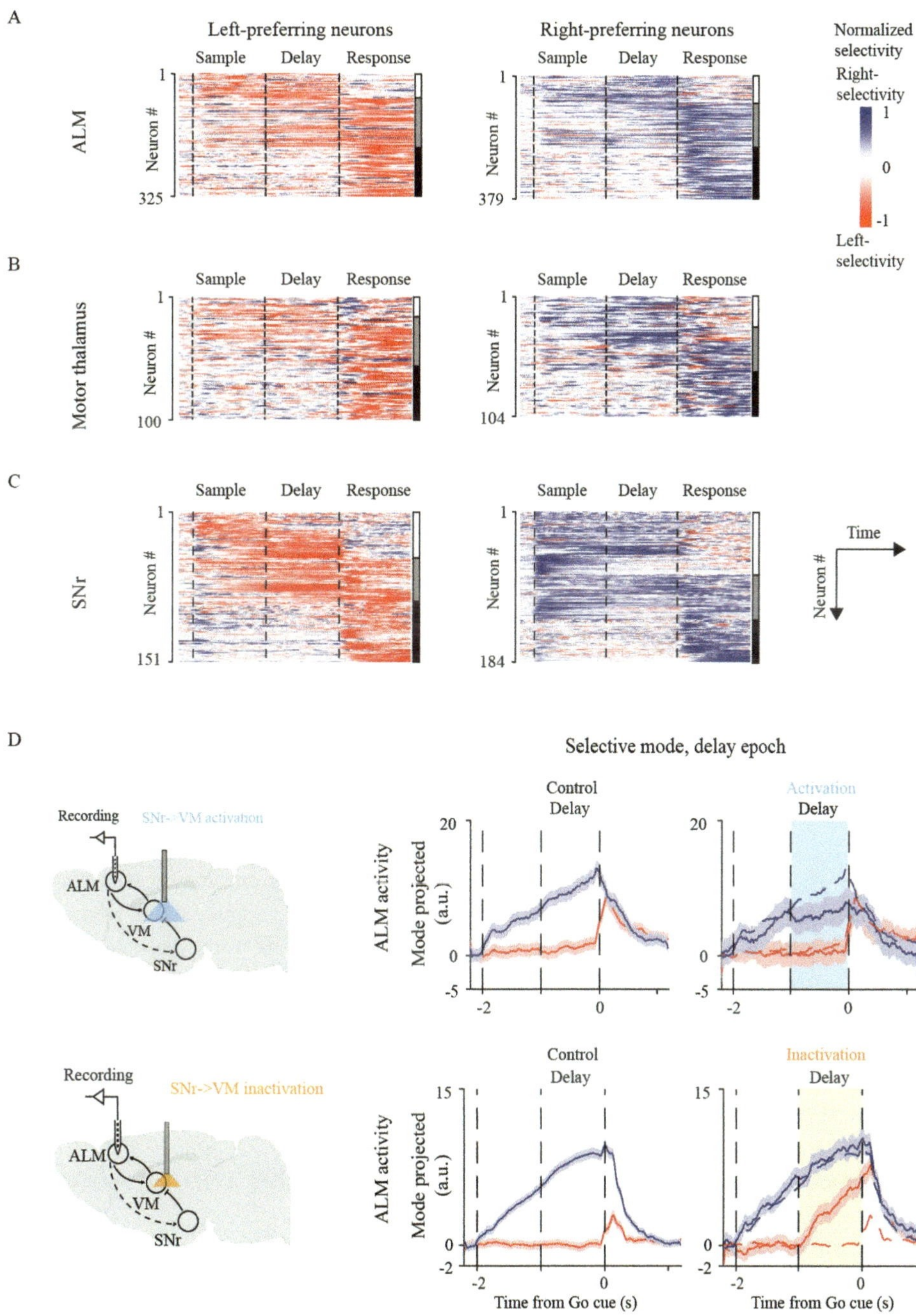

Figure 35.6. ALM, motor thalamus, and SNr population activity and regulation of a cortical selective mode by the BG-thalamus pathway. (A–C). Neurons in ALM, motor thalamus, and SNr show similar response patterns. Left: neurons showing left-trial selectivity. Right: neurons showing right-trial selectivity. Vertical bars on the right: white = neurons with preparatory activity only; gray = neurons with both preparatory activity and peri-movement activity; black = neurons with peri-movement activity only. (D). The BG-thalamus pathway selectively regulates a cortical selective mode. Left: schematic of the cortico-basal ganglia-thalamocortical pathway and optogenetic perturbation of SNr projections in the motor thalamus near VM. Right: neural trajectories with SNr→VM activation (top) or inactivation (bottom). Along the selective mode, activation of SNr terminals specifically affects the right-trial neural trajectory, while inactivation selectively affects the left-trial trajectory. Dotted lines are the control trajectories for comparison. Shading = SEM. Panels A, B adapted from Guo et al. (2017) with permission; panels C, D adapted from Wang et al. (2021) with permission.

Importantly, this is not due to elevated inhibition from SNr, as cortical inactivation slightly reduced, rather than increased, SNr activity (Wang et al., 2021). Consistent with a common drive, physical lesion or pharmacological inactivation of the pallidum did not abolish movement-locked modulations of thalamic activity (Inase et al., 1996). Optogenetic activation or inactivation of SNr afferents to motor thalamus had minimal effects on licking (Wang et al., 2021). However, the effect size seems to depend on the strength of perturbation, and stronger activation induced by higher laser power can still bias licking directions (Catanese and Jaeger, 2021). Thus, multiple evidence supports the notion that cortical inputs drive BG-recipient thalamic activity, while the BG plays a modulatory role.

How does the BG-thalamus pathway modulate cortical dynamical modes? Optogenetic activation and inactivation of BG-thalamic projections bidirectionally bias upcoming choice in the memory-guided task (Wang et al., 2021). Disrupting SNr activity specifically affects cortical trajectories along the selective mode, which points to a role for the SNr in shaping trial-selective activity in the task (Figure 35.6D). This type of specificity is not observed when perturbing another inhibitory source to the thalamus, the thalamic reticular nucleus (TRN). Thus, the BG-thalamic pathway can precisely modulate cortical activity to support choice-specific motor-planning activity (Wang et al., 2021). In terms of the computational framework depicted in Figure 35.2, the BG output targets a selective mode to account for the trial-type specific effects (Jaramillo et al., 2022).

Where is trial-selective information in SNr coming from? ALM provides indirect inputs to SNr through the striatum. Neurons in SNr, like neurons in ALM, exhibit similar response patterns that predict upcoming choice (Figure 35.6). In SNr, neurons with the same task-related selectivity have higher correlated fluctuations, whereas neurons having different preference are anti-correlated (Wang et al., 2021). As there are no recurrent excitations within SNr, the synchrony in firing might be inherited from the cortex. Indeed, neurons in ALM with similar task-related selectivity are preferentially coupled (Daie et al., 2021). Inactivation of ALM to remove task-selectivity also removes SNr selective mode (Wang et al., 2021). Thus, a closed cortico-basal ganglia-thalamo-cortical loop might be specialized in shaping the selective modes.

THE BRAINSTEM-THALAMUS PATHWAY IN TRIGGERING TASK STATE TRANSITIONS

Behavior can be modeled as a sequence of discrete states that are connected by transitions (Figure 35.1B). Transitions can be triggered by the presentation of external cues (Figure 35.2A). External signals induce a dramatic reorganization of neural dynamics in cortical and subcortical structures (Inagaki et al., 2022b; Schmidt et al., 2013). For example, in a stop-signal task, an auditory cue instructs rats to cancel a planned action. Without the "stop" signal, the BG-brainstem pathway functions to release the "gate" to support the planned action. Following the "stop" cue, an excitation from the subthalamic nucleus to SNr races to increase SNr activity, closing the gate to suppress the planned action. While the midbrain descending pathway plays an important role in suppression of planned actions, the ascending pathway can trigger an externally cued action. Neurons in motor thalamus exhibit ramping activity that predicts movement onset (Dacre et al., 2021; Takahashi et al., 2021; Tanaka, 2007). Blocking this activity delays or even abolishes cued action (Dacre et al., 2021; Takahashi et al., 2021).

How does the ascending pathway regulate cortical dynamics to evoke actions? In the memory-guided response task (Figure 35.2A), dispatching an auditory "go" signal triggers a state transition from motor planning to motor execution by evoking short latency signals in midbrain structures, including the pedunculopontine nucleus (PPN) and midbrain reticular nucleus (MRN) (Figure 35.7A). The ascending signal from PPN/MRN drives ALM activity to trigger licking movements indirectly through the thalamus. Interestingly, the "go" cue can be substituted by optogenetic activation of thalamus-projecting PPN/MRN neurons. The stimulation triggers not only correct movements but also the correct evolution of cortical dynamics: the initiation of a "go" mode collapses the selective mode during motor planning and induces the selective mode during licking execution (Figure 35.7B). Moreover, optogenetic inactivation of the same neurons blocks the switching of dynamical modes, indicating that the pathway is necessary for conveying the trigger signal to motor cortex. In terms of the computational framework outlined in Figure 35.2, the PPN/MRN input to thalamus amplifies a selective movement mode during the response period, which inherits its selectivity from a persistent mode during the delay period (Jaramillo et al., 2022).

CONCLUSION AND OUTLOOK

In an era of unprecedented data-acquisition capabilities, we are beginning to unravel how multiregional circuits contribute to cognition. The thalamus is an important and often understudied component of such circuits, particularly in functions that were usually ascribed to the cortex. Here we have identified some of its key anatomical and dynamical properties, including open and closed thalamocortical loops, transthalamic pathways, lack of recurrency as compared to the cortex, and its unique anatomical position

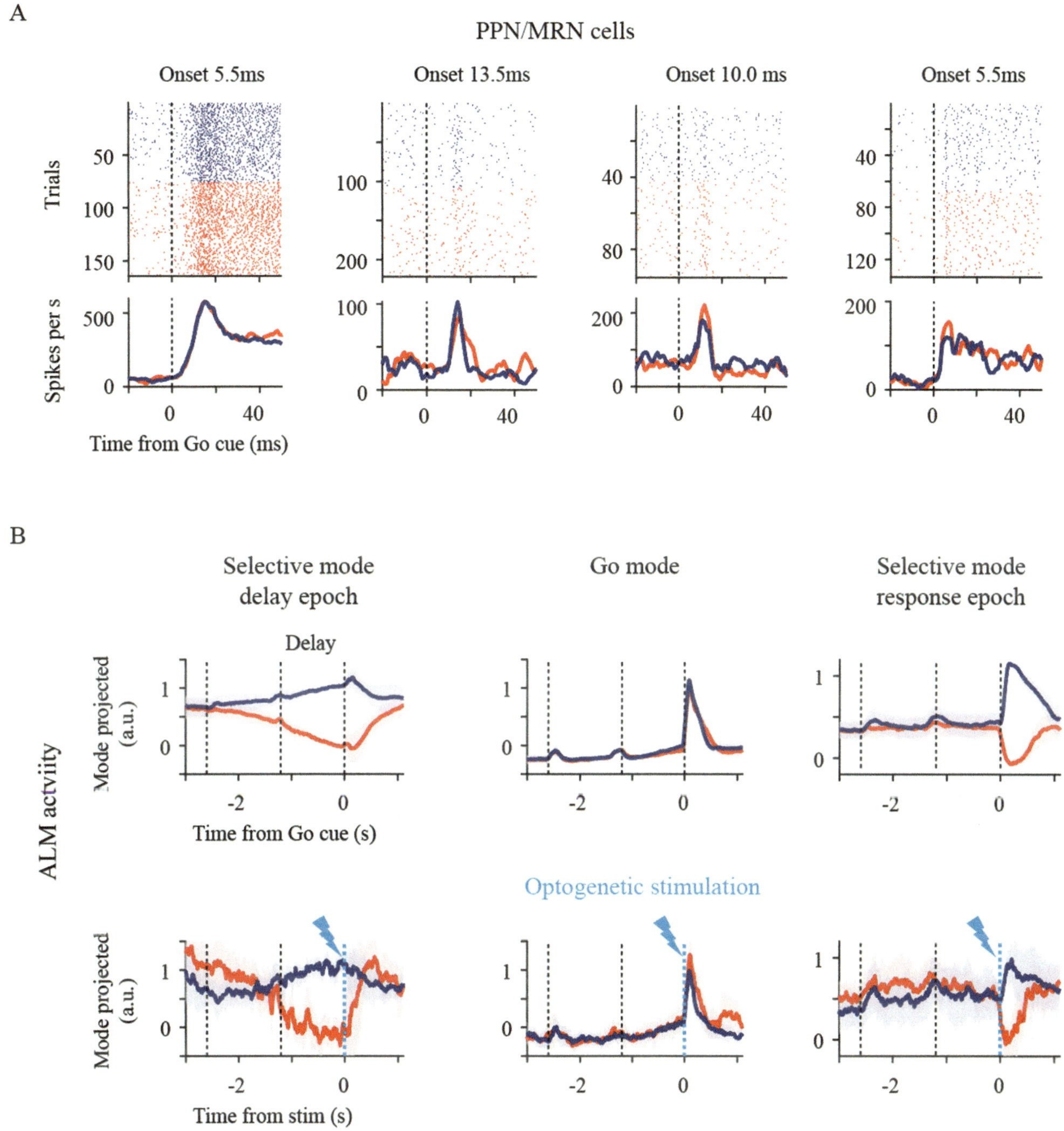

Figure 35.7. The brainstem-thalamus pathway triggers mode switching. (A). Neurons in PPN/MRN show short latency responses to the "go" cue. Same format as in Figure 35.2B. (B). The selective modes before (left) and after (right) the "go" cue. The "go" mode is shown in the middle panels. Photostimulation of PPN/MRN thalamic projections can substitute the "go" cue to induce correct licks. This perturbation also terminates the delay-epoch selective mode to trigger the response-epoch selective mode. The cyan dashed line in each panel signifies time of photostimulation. Shading = SEM. Adapted from Inagaki et al. (2022b) with permission.

that allows other subcortical structures to affect the cortex. We have proposed a computational framework that encapsulates these properties and explains the involvement of the thalamus and other subcortical structures in different computations underlying motor planning (Figure 35.2).

The computational framework developed here has been motivated by studies in mice, for which there are multiple genetic tools to address loss and gain of function, in a transient and reversible manner. Importantly, the framework is general enough as to apply to other thalamic nuclei and species (e.g., macaque in Figure 35.4). We note that there are important differences across species, including in behavioral capabilities that may stem from anatomical and physiological differences in subcortical-cortical circuits. For example, higher-order (non-sensory) thalamic nuclei in mice have few local interneurons, which are otherwise abundant in primates (Evangelio et al., 2018), and there is a more complex organization of the frontal cortex in primates.

It remains to be established what unique computations are conferred by the thalamus to multiregional circuits that are not (or cannot be) implemented by cortico-cortical circuits alone. For example, the motor thalamus and ALM show similar patterns of activity and play similar roles during motor planning, and inactivation of either area profoundly affects activity in the other area (Figures 35.3A–D and 35.6A,B). However, motor thalamus should not be viewed as another cortical layer. One defining characteristic of cortical circuits is excitatory recurrency, which plays a pivotal role in defining receptive fields, plasticity, and in maintaining reverberatory activity. Recurrency is related to neural processing timescales, whereby highly recurrent excitatory circuits are slower than those with less recurrence (Chaudhuri et al., 2015; Murray et al., 2014). Critically, the thalamus does not exhibit such recurrency, as excitatory collaterals are absent (Jones, 2007). This suggests that the thalamus enables fast processing, which is consistent with the relay property of sensory thalamic nuclei. We propose that the "fast relay" property extends to higher-order nuclei (Jaramillo et al., 2019), which form closed and transthalamic loops with the cortex (Sherman and Guillery, 2013) and route subcortical signals to the cortex. Other potential thalamic specializations include synaptic dynamics, synaptic plasticity, and intrinsic channels that allow for different firing modes (see **Chapter 32 in Section VI**). An important difference between the cortex and thalamus is the presence of cortical interhemispheric connections. These connections are important to coordinate cortical dynamics across hemispheres (Chen et al., 2021; Li et al., 2016; Yin et al., 2022). The thalamus, however, lacks such interhemispheric connections, although thalamic nuclei in different hemispheres may be indirectly connected through the TRN. The anatomical and physiological differences between cortical areas and thalamic nuclei and their distinct contributions to neural dynamics and behavior constitute an important point for future research.

The computational framework assigns complementary roles to the cortex, thalamus, and subcortical structures and thus emphasizes a multiregional circuit perspective. The framework, as well as recent experimental studies (Goldberg and Fee, 2012; Schwab et al., 2020; Wang et al., 2021), suggests a revision of the popular gating model of the basal ganglia. The determinants of thalamic neural activity are likely complex and depend not only on the basal ganglia output, but on other structures. In particular, corticothalamic input may activate thalamic neurons despite the presence of inhibition from the SNr. An expanded gating model, for example the computational framework in Figure 35.2, suggests that corticothalamic input may be particularly relevant for cognitive computations (e.g., those occurring during the delay period of a behavioral task where motor execution is not imminent). To fully explore the validity and predictive power of this and other computational frameworks, future experimental studies should consider cell-type-specific physiology and large-scale connectivity.

We focused on subcortical modulation of higher-order thalamus provided by the basal ganglia and brainstem. Importantly, the computational framework we have proposed applies to other thalamic nuclei and subcortical structures. We highlighted the contributions of the motor thalamus and pulvinar to motor planning, but the considerations may apply to first-order thalamic nuclei that are classically regarded as sensory relays. Beyond a relay function, the thalamus may be fundamental in maintaining excitability in the cortex (Purushothaman et al., 2012; Reinhold et al., 2015). However, there are differences between first- and higher-order thalamic nuclei that may be suggestive of their differential involvement in dynamics and behavior (Sherman and Guillery, 2013). One such difference is the propensity of non-sensory thalamic nuclei to form closed and/or transthalamic loops with the cortex, and the diffuse projections across multiple cortical layers. It will be interesting for future work to investigate general thalamocortical functions and circuit dynamical motifs across thalamic nuclei and their cortical targets.

Although we focused on basal ganglia and brainstem inputs onto the thalamus, multiple subcortical structures affect thalamocortical dynamics. For example, the cerebellum participates in movement planning as the activation of a cerebello-thalamic pathway affects trial-selective neural dynamics in the cortex (Gao et al., 2018). We suggest that the output of the cerebellum is an important excitatory input onto the thalamus that modulates cortical activity. Its exact function remains unknown; it may be related to timing or learning. Similarly, the superior colliculus forms part of not only a motor execution circuit, but also a corollary discharge circuit that engages the thalamus (Sommer and Wurtz, 2006). This circuit is another instantiation of our framework, as a corollary discharge can be interpreted as a subcortical signal that is routed to the cortex via the thalamus to control a relevant goal-oriented dynamical mode. In addition, the thalamus is modulated by multiple inhibitory subcortical structures, including SNr, TRN, and zona incerta. The manner in which different sources of feedforward inhibition shape thalamocortical dynamics for different cognitive functions has yet to be elucidated (see Wang et al., 2021). Another interesting question for future theoretical and experimental research is how neuromodulatory centers (e.g., VTA, raphe nuclei, basalis nucleus, etc.) affect either cortex directly or indirectly through the thalamus (see **Chapter 5 in Section II**).

To conclude, groundbreaking advances in experimental neuroscience need accompanying theoretical frameworks. Theories of large-scale brain function are called for, which could explain how computations across cortical (Wang, 2022) and subcortical areas support cognition.

REFERENCES

Alcaraz F, Fresno V, Marchand AR, Kremer EJ, Coutureau E, Wolff M (2018). Thalamocortical and corticothalamic pathways differentially contribute to goal-directed behaviors in the rat. *eLife.* 7:e32517.

Bolkan SS, Stujenske JM, Parnaudeau S, Spellman TJ, Rauffenbart C, Abbas AI, Harris AZ, Gordon JA, Kellendonk C (2017). Thalamic projections sustain prefrontal activity during working memory maintenance. *Nat Neurosci.* 20:987–996.

Canavan AG, Nixon PD, Passingham RE (1989). Motor learning in monkeys (*Macaca fascicularis*) with lesions in motor thalamus. *Exper Brain Res.* 77:113–126.

Catanese J, Jaeger D (2021). Premotor ramping of thalamic neuronal activity is modulated by nigral inputs and contributes to control the timing of action release. *J Neurosci.* 41:1878–1891.

Chabrol FP, Blot A, Mrsic-Flogel TD (2019). Cerebellar contribution to preparatory activity in motor neocortex. *Neuron.* 103:506–519.

Chaudhuri R, Knoblauch K, Gariel MA, Kennedy H, Wang XJ (2015). A large-scale circuit mechanism for hierarchical dynamical processing in the primate cortex. *Neuron.* 88:419–431.

Chen G, Kang B, Lindsey J, Druckmann S, Li N (2021). Modularity and robustness of frontal cortical networks. *Cell.* 184:3717–3730.

Churchland MM, Cunningham JP, Kaufman MT, Foster JD, Nuyujukian P, Ryu SI, Shenoy KV (2012). Neural population dynamics during reaching. *Nature.* 487:51–56.

Collins DP, Anastasiades PG, Marlin JJ, Carter AG (2018). Reciprocal circuits linking the prefrontal cortex with dorsal and ventral thalamic nuclei. *Neuron.* 98:366–379.

Cunningham JP, Yu BM (2014). Dimensionality reduction for large-scale neural recordings. *Nat Neurosci.* 17:1500–1509.

Dacre J, Colligan M, Clarke T, Ammer JJ, Schiemann J, Chamosa-Pino V, Claudi F, Harston JA, Eleftheriou C, Pakan JMP, Huang CC, Hantman AW, Rochefort NL, Duguid I (2021). A cerebellar-thalamocortical pathway drives behavioral context-dependent movement initiation. *Neuron.* 109:2326–2338.

Daie K, Svoboda K, Druckmann S (2021). Targeted photostimulation uncovers circuit motifs supporting short-term memory. *Nat Neurosci.* 24:259–265.

Evangelio M, Garcia-Amado M, Clasca F (2018). Thalamocortical projection neuron and interneuron numbers in the visual thalamic nuclei of the adult C57BL/6 mouse. *Front Neuroanat.* 12:27.

Frisch KV (1974). Decoding the language of the bee. *Science.* 185:663–668.

Gao Z, Davis C, Thomas AM, Economo MN, Abrego AM, Svoboda K, De Zeeuw CI, Li N (2018). A cortico-cerebellar loop for motor planning. *Nature.* 563:113–116.

Goldberg JH, Fee MS (2012). A cortical motor nucleus drives the basal ganglia-recipient thalamus in singing birds. *Nat Neurosci.* 15:620–627.

Guo K, Yamawaki N, Svoboda K, Shepherd GMG (2018). Anterolateral motor cortex connects with a medial subdivision of ventromedial thalamus through cell type-specific circuits, forming an excitatory thalamo-cortico-thalamic loop via layer 1 apical tuft dendrites of layer 5b pyramidal tract type neurons. *J Neurosci.* 38:8787–8797.

Guo ZV, Inagaki HK, Daie K, Druckmann S, Gerfen CR, Svoboda K (2017). Maintenance of persistent activity in a frontal thalamocortical loop. *Nature.* 545:181–186.

Guo ZV, Li N, Huber D, Ophir E, Gutnisky D, Ting JT, Feng G, Svoboda K (2014). Flow of cortical activity underlying a tactile decision in mice. *Neuron.* 81:179–194.

Halassa MM, Sherman SM (2019). Thalamocortical circuit motifs: A general framework. *Neuron.* 103:762–770.

Hart BL, Hart LA, McCoy M, Sarath CR (2001). Cognitive behaviour in Asian elephants: use and modification of branches for fly switching. *Anim Behav.* 62:839–847.

Hikosaka O (2007). Basal ganglia mechanisms of reward-oriented eye movement. *Ann N Y Acad Sci.* 1104:229–249.

Hikosaka O, Takikawa Y, Kawagoe R (2000). Role of the basal ganglia in the control of purposive saccadic eye movements. *Physiol Rev.* 80:953–978.

Inagaki HK, Chen S, Daie K, Finkelstein A, Fontolan L, Romani S, Svoboda K (2022a). Neural algorithms and circuits for motor planning. *Ann Rev Neurosci.* 45:249–271.

Inagaki HK, Chen S, Ridder MC, Sah P, Li N, Yang Z, Hasanbegovic H, Gao Z, Gerfen CR, Svoboda K (2022b). A midbrain-thalamus-cortex circuit reorganizes cortical dynamics to initiate movement. *Cell.* 185:1065–1081.

Inagaki HK, Fontolan L, Romani S, Svoboda K (2019). Discrete attractor dynamics underlies persistent activity in the frontal cortex. *Nature.* 566:212–217.

Inagaki HK, Inagaki M, Romani S, Svoboda K (2018). Low-dimensional and monotonic preparatory activity in mouse anterior lateral motor cortex. *J Neurosci.* 38:4163–4185.

Inase M, Buford JA, Anderson ME (1996). Changes in the control of arm position, movement, and thalamic discharge during local inactivation in the globus pallidus of the monkey. *J Neurophysiol.* 75:1087–1104.

Jaramillo J, Mejias JF, Wang XJ (2019). Engagement of pulvino-cortical feedforward and feedback pathways in cognitive computations. *Neuron.* 101:321–336.

Jaramillo J, Pereira U, Svoboda K, Wang X (2022). Subcortical modulation of cortical dynamics for motor planning: a computational framework. Paper presented at: Cosyne Abstracts 2022.

Jones EG (2007). *The Thalamus.* Cambridge University Press.

Jun JJ, Steinmetz NA, Siegle JH, Denman DJ, Bauza M, Barbarits B, Lee AK, Anastassiou CA, Andrei A, Aydin C, Barbic M, Blanche TJ, Bonin V, Couto J, Dutta B, Gratiy SL, Gutnisky DA, Hausser M, Karsh B, Ledochowitsch P, Lopez CM, Mitelut C, Musa S, Okun M, Pachitariu M, Putzeys J, Rich PD, Rossant C, Sun WL, Svoboda K, Carandini M, Harris KD, Koch C, O'Keefe J, Harris TD (2017). Fully integrated silicon probes for high-density recording of neural activity. *Nature.* 551:232–236.

Karnath HO, Himmelbach M, Rorden C (2002). The subcortical anatomy of human spatial neglect: Putamen, caudate nucleus and pulvinar. *Brain.* 125:350–360.

Komura Y, Nikkuni A, Hirashima N, Uetake T, Miyamoto A (2013). Responses of pulvinar neurons reflect a subject's confidence in visual categorization. *Nat Neurosci.* 16:749–755.

Li N, Chen TW, Guo ZV, Gerfen CR, Svoboda K (2015). A motor cortex circuit for motor planning and movement. *Nature.* 519:51–56.

Li N, Daie K, Svoboda K, Druckmann S (2016). Robust neuronal dynamics in premotor cortex during motor planning. *Nature.* 532:459–464.

Logiaco L, Abbott LF, Escola S (2021). Thalamic control of cortical dynamics in a model of flexible motor sequencing. *Cell Rep.* 35:109090.

Mante V, Sussillo D, Shenoy KV, Newsome WT (2013). Context-dependent computation by recurrent dynamics in prefrontal cortex. *Nature.* 503:78–84.

Mastrogiuseppe F, Ostojic S (2018). Linking connectivity, dynamics, and computations in low-rank recurrent neural networks. *Neuron.* 99:609–623.

McAlonan K, Cavanaugh J, Wurtz RH (2008). Guarding the gateway to cortex with attention in visual thalamus. *Nature.* 456:391–394.

McElvain LE, Chen Y, Moore JD, Brigidi GS, Bloodgood BL, Lim BK, Costa RM, Kleinfeld D (2021). Specific populations of basal ganglia output neurons target distinct brain stem areas while collateralizing throughout the diencephalon. *Neuron*. 109:1721–1738.

Mengual E, Prensa L, Tripathi A, Cebrian C, Mongia S (2016). Comparative analysis of the axonal collateralization patterns of basal ganglia output nuclei in the rat. In: Rockland KS, ed. *Axons and Brain Architecture*. Elsevier Inc.; 47–68.

Murray JD, Bernacchia A, Freedman DJ, Romo R, Wallis JD, Cai X, Padoa-Schioppa C, Pasternak T, Seo H, Lee D, Wang XJ (2014). A hierarchy of intrinsic timescales across primate cortex. *Nat Neurosci*. 17:1661–1663.

Plotnik JM, de Waal FBM, Reiss D (2006). Self-recognition in an Asian elephant. *Proc Natl Acad Sci U S A*. 103:17053–17057.

Purushothaman G, Marion R, Li K, Casagrande VA (2012). Gating and control of primary visual cortex by pulvinar. *Nat Neurosci*. 15:905–912.

Reinhold K, Lien AD, Scanziani M (2015). Distinct recurrent versus afferent dynamics in cortical visual processing. *Nat Neurosci*. 18:1789–1797.

Robinson DL, Petersen SE (1992). The pulvinar and visual salience. *Trends Neurosci*. 15:127–132.

Roseberry TK, Lee AM, Lalive AL, Wilbrecht L, Bonci A, Kreitzer AC (2016). Cell-type-specific control of brainstem locomotor circuits by basal ganglia. *Cell*. 164:526–537.

Rossi MA, Li HE, Lu D, Kim IH, Bartholomew RA, Gaidis E, Barter JW, Kim N, Cai MT, Soderling SH, Yin HH (2016). A GABAergic nigrotectal pathway for coordination of drinking behavior. *Nat Neurosci*. 19:742–748.

Roth MM, Dahmen JC, Muir DR, Imhof F, Martini FJ, Hofer SB (2016). Thalamic nuclei convey diverse contextual information to layer 1 of visual cortex. *Nat Neurosci*. 19:299–307.

Saalmann YB, Pinsk MA, Wang L, Li X, Kastner S (2012). The pulvinar regulates information transmission between cortical areas based on attention demands. *Science*. 337:753–756.

Sauerbrei BA, Guo JZ, Cohen JD, Mischiati M, Guo W, Kabra M, Verma N, Mensh B, Branson K, Hantman AW (2020). Cortical pattern generation during dexterous movement is input-driven. *Nature*. 577:386–391.

Schmidt R, Leventhal DK, Mallet N, Chen F, Berke JD (2013). Canceling actions involves a race between basal ganglia pathways. *Nat Neurosci*. 16:1118–1124.

Schmitt LI, Wimmer RD, Nakajima M, Happ M, Mofakham S, Halassa MM (2017). Thalamic amplification of cortical connectivity sustains attentional control. *Nature*. 545:219–223.

Schwab BC, Kase D, Zimnik A, Rosenbaum R, Codianni MG, Rubin JE, Turner RS (2020). Neural activity during a simple reaching task in macaques is counter to gating and rebound in basal ganglia-thalamic communication. *PLoS Biol*. 18:e3000829.

Sherman SM, Guillery RW (2013). *Functional Connections of Cortical Areas: A New View from the Thalamus*. MIT Press.

Shine JM (2021). The thalamus integrates the macrosystems of the brain to facilitate complex, adaptive brain network dynamics. *Progress Neurobiol*. 199:101951–101962.

Snow JC, Allen HA, Rafal RD, Humphreys GW (2009). Impaired attentional selection following lesions to human pulvinar: Evidence for homology between human and monkey. *Proc Natl Acad Sci U S A*. 106:4054–4059.

Sommer MA, Wurtz RH (2006). Influence of the thalamus on spatial visual processing in frontal cortex. *Nature*. 444:374–377.

Stevenson IH, Kording KP (2011). How advances in neural recording affect data analysis. *Nat Neurosci*. 14:139–142.

Svoboda K, Li N (2018). Neural mechanisms of movement planning: Motor cortex and beyond. *Curr Opin Neurobiol*. 49:33–41.

Takahashi N, Moberg S, Zolnik TA, Catanese J, Sachdev RNS, Larkum ME, Jaeger D (2021). Thalamic input to motor cortex facilitates goal-directed action initiation. *Curr Biol*. 31:4148–4155.

Tanaka M (2007). Cognitive signals in the primate motor thalamus predict saccade timing. *J Neurosci*. 27:12109–12118.

Voitov I, Mrsic-Flogel TD (2022). Cortical feedback loops bind distributed representations of working memory. *Nature*. 608:381–389.

Wang XJ (2022). Theory of the multiregional neocortex: large-scale neural dynamics and distributed cognition. *Ann Rev Neurosci*. 45:533–560.

Wang Y, Yin X, Zhang Z, Li J, Zhao W, Guo ZV (2021). A cortico-basal ganglia-thalamo-cortical channel underlying short-term memory. *Neuron*. 109:3486–3499.

Wilke M, Kagan I, Andersen RA (2013). Effects of pulvinar inactivation on spatial decision-making between equal and asymmetric reward options. *J Cogn Neurosci*. 25:1270–1283.

Wilke M, Turchi J, Smith K, Mishkin M, Leopold DA (2010). Pulvinar inactivation disrupts selection of movement plans. *J Neurosci*. 30:8650–8659.

Yang W, Tipparaju SL, Chen G, Li N (2022). Thalamus-driven functional populations in frontal cortex support decision-making. *Nat Neurosci*. 25:1339–1352.

Yin X, Wang Y, Li J, Guo ZV (2022). Lateralization of short-term memory in the frontal cortex. *Cell Rep*. 40:111190.

36.

OCULOMOTOR THALAMUS AND EFFERENCE COPY

Marc A. Sommer

INTRODUCTION

Neural activity related to eye movements is found across much of the brain. In regions that comprise the central oculomotor generation pathways, including the frontal eye field (FEF) and superior colliculus (SC), the activity can be vigorous and temporally extended, from early preparatory processes to a peak at the initiation of movement. More commonly, eye movement–related activity is revealed as a modulation of visual processing, for example in the primary visual cortex (V1) and the visual-motion selective middle temporal (MT) area. The thalamus is interconnected with all of these areas, and thus it is not surprising that it, too, is permeated with neural signals related to eye movements.

In this chapter, one major type of eye movement will be discussed: the quick, conjugate rotations of the eyes known as *saccades*. Their primary purpose is to look at new parts of a visual scene. In the thalamus, signals related to saccades could have many purposes. We will focus on the two purposes that have been studied the most, *generating* saccades and *monitoring* them. Generation of saccades is the sensorimotor transformation that uses visual (or other sensory) input to calculate and execute the command of where to look next. Monitoring saccades is the process of relaying copies of the saccadic commands within the brain, rather than out toward the muscles. Those signals, known as *efference copies* or *corollary discharge* (we will use the first term), are more modulatory. Efference copies impact the visual system to maintain continuous, stable perception and behavior despite the saccadic eye rotations.

Deciphering saccadic signals in the thalamus presents unique challenges. For the signals to be present, behaving animals must be used. Careful recovery of recording locations is important, because of the closely packed nuclei and their internal regionalizations. Identification of how the thalamic neurons are connected—the sources of their inputs and the targets of their projections—is key to interpretation of function, whether in reference to anatomy or in real time with physiological techniques. Finally, causal studies that include inactivation or stimulation in addition to recording are important for clarifying the functions of the neurons, since a neural signal may in principle be used for many potential purposes. Research during the past 20 years, especially in the nonhuman primate, has made significant progress in combining all of these approaches to advance our understanding of oculomotor thalamus. The work has been complemented by elegant work in other animal models as well, and opportunistic studies of human lesion patients. This chapter will focus on that epoch of research, to extend the summaries of the state of the field provided earlier (e.g., Sommer, 2003).

Defining an "oculomotor thalamus" is as daunting as trying to define an "oculomotor cerebral cortex"; signals related to eye movements can be found across vast, diverse volumes of both structures. For simplification, our focus will be on specific nuclei that have been shown with combinations of behavioral, physiological, and anatomical approaches to play well-defined roles in saccadic functions. These nuclei include the *pulvinar*, which among its other functions relays signals from the SC to MT; the *mediodorsal nucleus*, which connects SC to the FEF; and the *ventrolateral nuclei*, which connect the cerebellum to cerebral cortex.

Other topics will be reviewed where relevant, including thalamic nuclei that are likely to have saccadic functions but have been less studied, nuclei such as the LGN that receive notable saccade-related modulatory input, and oculomotor thalamic nuclei in other, nonprimate animals. Overall, within the constraints of a brief chapter for a general neuroscience audience, the objective is to provide an overview of primate oculomotor thalamus, with references provided throughout for those who want to learn more.

PULVINAR: RELAY OF SACCADIC SUPPRESSION

The pulvinar is a posterior thalamic complex that is generally thought to support sensory processing, especially for vision (for an overview, see **Chapter 20 in Section V**). Early studies of the relation of pulvinar to visuomotor behavior focused on its contributions to visual salience (reviewed by Robinson and Petersen, 1992). Later work using whole-brain functional magnetic resonance imaging (fMRI) in awake, behaving macaques supported the idea that the pulvinar plays a role in a broad network of brain sites involved in making saccades to visual targets (Baker et al., 2006). In terms of its nuclear parcellation and subcortical and cortical connections, however, as well as the logic that ties this anatomy to behavioral functions, the pulvinar is extraordinarily complicated (e.g., Kaas and Lyon, 2007).

One breakthrough in sorting out the saccadic contributions of pulvinar came from the use of identified recordings. The goal of such studies is to find and physiologically study neurons that can be demonstrated to receive input from and/or project to specific brain areas. Berman and Wurtz (2010) accomplished this in behaving monkeys using antidromic and orthodromic stimulation, or both combined. With electrical microstimulation electrodes placed with their tips in the SC, they found numerous pulvinar neurons that could be orthodromically (synaptically) activated. With other stimulating electrodes in visual area MT, they found pulvinar neurons that could be antidromically activated (back-fired). A critical subset of pulvinar neurons could be activated using both methods at once, identifying them as SC-to-MT relay neurons. The relay neurons occupied 2 subnuclei of inferior pulvinar, termed PIm and PIp. This established unequivocally that pathways from the SC ascend via pulvinar to extrastriate cortex.

The main hypothesis about the SC-PI-MT's function was that it helped to suppress visual responses during saccades. This process, known as *saccadic suppression*, helps to reduce signals affected by retinal blur during saccades. Using neural recording and reversible inactivation approaches along the SC-PI-MT pathway, Berman et al. (2017) found saccadic suppression of visual responses at all 3 stages in the circuit, as well as evidence for a causal effect of SC activity on saccadic suppression in area MT. This body of work, the outcome of highly challenging combinations of physiological techniques, therefore provided the first insights into saccade-related roles of ascending pathways through pulvinar in primates (reviewed in more detail by Wurtz et al., 2011).

In other species, knowledge about the role of pulvinar in saccade-related functions is less complete. Limited data from human lesion studies implicated the pulvinar (or posterior thalamus more generally) in ocular fixation and voluntary saccades (Rafal et al., 2004). In mice, inactivation of the pulvinar has confirmed the relay of saccade-related signals to area V1 that is used to distinguish between visual motion due to true stimulus movement in the world from that caused by self-movement of the eyes (Miura and Scanziani, 2022). The latter study suggests that direct efference copy signals (i.e., literal copies of saccade commands) may also course through the pulvinar in primates, but if so, they have not yet been found.

MEDIODORSAL NUCLEUS: RELAY OF EFFERENCE COPY

The main distinguishing feature of the mediodorsal nucleus of the thalamus (MD) is its intricate, reciprocal connections with prefrontal cortex (see **Chapter 54 in Section IX**). Spanning the midline and bounded laterally by a sheet of white matter known as the internal medullary lamina (IML), the MD nucleus occupies a large proportion of thalamus in primates consistent with the expansion of prefrontal cortex through evolution. While much of the medial volume of primate MD interacts with prefrontal areas only tangentially related to saccadic behavior, its lateral aspect, just inside the IML, projects to the FEF and receives input from the SC (Benevento and Fallon, 1975; Goldman-Rakic and Porrino, 1985; Lynch et al., 1994). (This chapter will focus on that portion of MD specified in **Chapter 54 of Section IX** as "MDl.")

The saccade-related and visual properties of MD neurons have been explored extensively with neural recording studies in behaving monkeys (Wyder et al., 2003; Watanabe and Funahashi, 2004; Tanibuchi and Goldman-Rakic, 2005). In general, those investigations aimed at characterizing visuo-oculomotor activity in central thalamus considered broadly, spanning from MD to nuclei within the IML to ventrolateral thalamic zones. Activity patterns are, for the most part, similar across the regions. Much like the frontal cortical regions with which the MD and other thalamic zones interconnect, the central thalamus exhibits an immense diversity of visual- and saccade-related responses, plus activity that links visual and movement responses in delayed oculomotor tasks (termed *delay activity*).

The anatomical evidence for a pathway from the SC, through MD relay neurons, to the FEF (Lynch et al. 1994) was particularly intriguing. Since it was relayed through a narrow zone of lateral MD, however, the prospects of isolating its neurons for behavioral and physiological study seemed low. That changed after unexpected results were found in a study that aimed to examine FEF neurons that project to the SC, using antidromic activation from the SC. In addition to FEF neurons that were antidromically activated, the authors discovered FEF neurons that were orthodromically activated (Sommer and Wurtz, 1998). The short latencies of activation implicated the SC-MD-FEF pathway, prompting follow-up studies that recorded from identified neurons at each stage of the pathway (Sommer and Wurtz, 2004a) and inactivated it at the relay neuron "bottleneck" in MD (Sommer and Wurtz, 2002, 2004b, 2006). The conclusion was that visual-, delay-, and saccade-related neural signals are sent from SC to FEF through the MD relay neurons, with delay activity reduced at the MD stage to emphasize visual and especially prominent pre-saccadic discharge signals in the leg of the circuit that innervated FEF (Sommer and Wurtz, 2004a). Inactivating the pathway had negligible effects on saccade generation, but impaired the monitoring of saccades (Sommer and Wurtz, 2002, 2004b) and the use of saccadic information to dynamically remap visual receptive fields around the time of a saccade in FEF neurons (Sommer and Wurtz, 2006). Overall, the SC-MD-FEF pathway seems to play a role in efference copy for adjusting the visual system to the sudden relocation of retinal images consequent to a saccade, helping to maintain the percept of visual continuity as saccades scan a visual scene (reviewed by Sommer and Wurtz, 2008).

Subsequent studies demonstrated that saccadic signals from the SCs in both hemispheres converge at FEF neurons to provide information about all directions of saccades for efference copy across the visual field (Crapse and Sommer, 2009), and that the dynamics of efference copy-provoked remapping of visual responses may be constructed locally within that cerebral cortical region from the saccadic signals arriving from the SC-MD-FEF pathway (Shin and Sommer, 2012; Rao et al., 2016). Parallel studies mapped out the distribution of putative relay neurons of the pathway throughout MD (Cavanaugh et al., 2020) and showed that presaccadic activity in the FEF is modulated by inactivation of the SC presumably via the pathway (Berman et al., 2009). Of most critical importance in linking the SC-MD-FEF pathway signals to perceptually relevant efference, Cavanaugh et al. (2016) demonstrated that inactivating it at the MD relay stage impaired visual continuity across saccades in monkeys. Similar findings have been reported in a host of human patients with thalamic lesions in and around the MD thalamus, although with some differences from the monkey work (e.g., lateralities of many of the effects) that remain unresolved (Bellebaum et al., 2006; Ostendorf et al., 2010, 2013; Peterburs et al., 2011; Zimmerman et al., 2020).

MD neurons not only project to FEF and other prefrontal areas, they also receive reciprocal projections from them. Studying the signal content of these corticothalamic projections will be an important next step in understanding the source of saccade-related signals in MD, and how they might combine with saccadic activity ascending from the SC. While MD neurons activated orthodromically from the FEF have been found previously (Wurtz et al., 2005), it is ambiguous whether the detected connection is monosynaptic. Newer methods such as optogenetics may help to disambiguate and fully characterize the FEF to MD projection, although the application of such methods to corticothalamic projections in the oculomotor system of monkeys is still developing (Suzuki et al., 2021). Similarly, the array of other possible saccade-related inputs to MD neurons, beyond those from the SC, still need to be examined in detail.

Finally, the demonstration of a pathway for oculomotor efference copy in MD has led to increasingly intense interest in potential impairment of such pathways in psychiatric and motor disorders. Efference copy deficits of many forms have been demonstrated in patients with schizophrenia (e.g., Ford and Mathalon 2019), including deficits in monitoring saccades (e.g., Thakkar and Rolfs 2019).

VENTROLATERAL NUCLEI: RELAY OF SELF-TIMING SIGNALS

For the purpose of brevity, the primate thalamic nuclei within and lateral to the IML that exhibit saccade-related activity and functions will be collectively termed the *ventrolateral nuclei* (but for more information on this large assembly of diverse nuclei, see **Chapters 31 and 32 in Section VI**). The key connectional feature of the ventrolateral complex of nuclei, in terms its relationship with saccadic behavior, is its massive input from the cerebellum (reviewed by Sommer, 2003; Tanaka and Kunimatsu, 2011; Prevosto and Sommer, 2013). In turn, neurons in the ventrolateral nuclei project to saccade-related regions of cerebral cortex, including the FEF, supplementary eye field (SEF), and lateral intraparietal cortex (LIP), among other oculomotor targets (e.g., Bostan et al., 2013; Middleton and Strick, 2001; Prevosto et al., 2010; Prevosto and Sommer, 2013). Studies that examined the general properties of MD neurons during visuo-saccadic tasks typically ventured into and across the IML to the ventrolateral thalamus, and consequently it is known that visual-, delay-, and saccade-related activity is prevalent across that whole swath of MD to ventrolateral thalamus (Wyder et al., 2003; Watanabe and Funahashi, 2004). But is there any particular saccade-related functionality that could be considered a hallmark of the ventrolateral nuclei, as efference copy is in lateral MD?

The first hints of a special role for ventrolateral thalamus in saccadic behavior came from anatomy. Work by Peter Strick's lab made it clear that a major source of likely saccade-related signals to ventrolateral thalamus is the lateral cerebellum and its output node, the dentate nucleus (Middleton and Strick, 2001). Loops are in place between cerebral cortex, cerebellum, and back through ventrolateral thalamus to cerebral cortex, one of which trans-synaptically interconnects the FEF with cerebellum. The first ascending leg of that loop projects from the dentate nucleus to ventrolateral thalamus (see also Lynch et al., 1994). Neural recordings from the zone of dentate nucleus that provides this thalamic projection found neurons that exhibit sustained, often ramping activity in delayed saccade tasks (Ashmore and Sommer, 2013). The hallmark of this dentate neuron activity was its strong relationship to the *timing* of action; it predicted when a saccade would be initiated in delayed, self-timed saccade tasks (Ashmore and Sommer, 2013; Ohmae et al., 2017; Kunimatsu et al., 2018; reviewed by Raghavan et al., 2016, and Tanaka et al., 2021).

The cerebellar-recipient zone of ventrolateral thalamus was studied in detail using this conceptual framework of saccadic action timing in a series of studies by Masaki Tanaka's lab. Reversible inactivation of sites within this ventrolateral thalamic region delayed saccades, particularly when they had to be self-timed (Tanaka, 2006), and neurons in the region exhibited selective delay activity in relation to the timing of their saccades (Tanaka, 2007). Taken together, this body of work spanning the dentate nucleus through its thalamic recipient sites represented a significant breakthrough in our understanding of neural mechanisms for the timing of actions (Raghavan et al., 2016; Tanaka et al., 2021). Moreover, the new findings aligned with prior studies in humans that implicated lateral cerebellum and dentate nucleus in the timing of voluntary behavior (reviewed by Mauk and Buonomano, 2004).

This is not to say that the timing of saccades or other actions is the only oculomotor role of the ventrolateral thalamus. Among its other potential contributions, this part of thalamus may also play a role in efference copy (Bellebaum et al., 2005), saccadic adaptation (Zimmerman et al., 2015), and smooth pursuit (Tanaka, 2005). A useful direction for future work would be to conduct identified pathway studies, as done in monkey pulvinar (Berman and Wurtz, 2010) and MD (Sommer and Wurtz, 2004a), and to examine more closely the relationship between various types of neural activity related to saccades and the fine parcellation of subnuclei in the region (Wyder et al., 2003; Watanabe and Funahashi, 2004).

VENTROANTERIOR NUCLEI: EMERGING VIEWS

The ventroanterior nucleus of the thalamus and an adjacent, anterior subregion of the ventrolateral thalamus (VLo) are typified by inputs from the basal ganglia (Percheron et al., 1996; Macchi and Jones, 1997). These regions, which we will collectively call the *ventroanterior nuclei* for brevity, are largely unexplored in terms of their oculomotor functions in primates. Two lines of evidence suggest that examining the nuclei may be of interest. First, the basal ganglia are well-known to participate in saccade-related behavior (e.g., Hikosaka et al., 2000). Identifying saccade-related neural activity in basal ganglia–recipient thalamus would help to characterize signal processing in the circuit and provide a target for inactivation, to causally determine function. Second, at least one study of neural activity has been conducted during saccadic behavior in a region likely homologous with monkey ventroanterior thalamus, with intriguing results. Leszczynski et al. (2021) recorded extracellular field potentials from the anterior nuclei of surgical patients while they viewed visual stimuli. The end of each saccade (start of fixation) modulated the field potentials in a manner consistent with predictive "resetting" of the system for synchronization of visual processing. As with the other thalamic nuclei considered here, the implication is that the ventroanterior nuclei may be integral to perception and behavior during investigation of a visual scene.

LATERAL GENICULATE NUCLEUS: SACCADIC MODULATION OF VISION

While not part of the "oculomotor thalamus" per se, the LGN is worth reviewing as an example of how saccade-related signals can reach a thalamic region to cause nuanced but potentially important downstream effects. (Sensory functions and organization of the LGN are covered extensively in **Chapters 17, 18, and 19 in Section V**.) The LGN is a visual processor in itself, with an undisputed role in driving the rest of the central visual system. Yet it is subject to extraretinal modulations that in turn pre-saccadically suppress, then post-saccadically enhance, the flow of visual information to V1. The fundamental effects were surmised from cat studies (e.g., Lee and Malpeli, 1998) and psychophysics (reviewed by Ross et al., 1996) and confirmed for primates soon thereafter. Reppas et al. (2002), for example, found a variety of saccadic modulatory effects on the temporal and chromatic properties of visual receptive fields in the primate LGN, and quantified a time course of effect that started distinctly before initiation of a saccade. Royal et al. (2006) confirmed this and argued that the pre-saccadic suppression started so early (100 ms before saccade initiation) that it resembled motor planning more than a motor command, raising questions about the source of the efference copy, or whether it could even be called that (see also Idrees et al., 2020). Moreover, they emphasized that the post-saccadic enhancement dominated the modulation, suggesting that the effects may serve to enhance vision at the start of fixation, more than suppress it during saccades. This is reminiscent of the Leszczynski et al. (2021) data on saccade- and fixation-related phase resets found in human anterior thalamus, as discussed above, suggesting that such oculomotor influences visual processing may be common across thalamic nuclei.

Saccades are used for orientation toward behaviorally relevant visual stimuli, an "overt" form of attention that may be functionally coupled to more "covert," circuit-level shifts of attention in the form of visual response augmentation (reviewed, for example, by Awh et al., 2006). It is therefore of note that visual responses in both the LGN and the adjacent, reciprocally connected thalamic reticular nucleus (TRN) are influenced by spatial attention in behaving monkeys (McAlonan et al., 2008). If one favors a more cognitive perspective, it may be apt to expand one's viewpoint from asking whether thalamic nuclei are "oculomotor" to whether they participate in *active vision*, the integration of retinal information with selective processes internal to the brain. The subregions of the pulvinar and MD nucleus that we have discussed, along with the LGN and TRN and perhaps ventroanterior thalamus, may qualify as participants in such a broader active vision network (Wurtz et al., 2011). In contrast, the ventrolateral thalamus with its action timing functions may occupy a different cognitive and/or motor niche.

OCULOMOTOR THALAMUS IN NON-PRIMATE SPECIES

Although we have focused on primate oculomotor thalamus here, noteworthy work has been performed in other species. As noted above, early findings on saccade-related influences of LGN were accomplished in cats before confirmation in primates. An elegant recent study in mice was mentioned above (Miura and Scanziani, 2022) and readers are encouraged to see that report and its affiliated literature. A particularly interesting study was recently performed in pigeons, thus extending the topic to the avian brain. Yang et al. (2008) studied visual responses in multiple regions of the pigeon brain during spontaneous saccades and the saccade-like fast phases of optokinetic nystagmus. Using a combination of neural recordings and causal studies (activation and inactivation), they demonstrated saccade-related modulations of visual responses in the nucleus opticus principalis thalami (nOPT), considered to be homologous to mammalian LGN. The modulations, an early suppression followed by enhancement of visual responsiveness, paralleled those described above for primate LGN. All told, it seems quite possible that analogs of the variety of saccade-related signals described in this chapter for primates will be found in many other species, although biased by behavioral and environmental demands, as we have described previously in a more general consideration of efference copy signaling used across species (Crapse and Sommer, 2008).

CONCLUSIONS

This objective of this chapter was to summarize the past 20 years of advances in understanding the oculomotor thalamus of primates, and readers are encouraged to see earlier reviews for more historical background (e.g., Sommer 2003; Tanaka and Kunimatsu, 2011). In terms of scientific progress, this recent era has been highly productive. Whereas most of the work on motor thalamus in the 20th century focused on skeletomotor systems, a series of synergistic investigations put oculomotor thalamus more in the spotlight. Physiological characterization of the SC-MD-FEF pathway went hand in hand with other salient findings on efference copy circuits at the same time (e.g., Poulet and Hedwig, 2002). Those advances, combined with elucidation of saccadic suppression mechanism in area MT (Thiele et al. 2002), were followed by functional dissection of the SC-pulvinar-MT pathway (Berman and Wurtz, 2010; Berman et al., 2017). Around the same time, the saccadic self-timing properties of the cerebellar-recipient ventrolateral thalamus were discovered in a series of studies (Tanaka 2006, 2007), in conjunction with supportive findings in the cerebellar source region (Ashmore and Sommer, 2013; Ohmae et al., 2017; Tanaka et al., 2021). Finally, examination of saccade-related influence in LGN helped us to gain a more nuanced view of oculomotor roles in visual processing (Reppas et al., 2002; Royal et al., 2006). Taken together, we are now in a position to think about oculomotor thalamic contributions not only to saccade generation, but also to perception and behavior, including active vision (Wurtz et al., 2011).

Significant findings and viewpoints have been left out of this chapter for reasons of brevity and focus, but should be considered for a more complete consideration of potential mechanisms within, and roles for, the oculomotor thalamus. Throughout this book, important chapters provide context, for example, **Chapter 4 in Section II, Chapters 11, 12, 16, 19, and 20 in Section IV, all of Section VI, and Chapters 45, 46, and 47 in Section VIII**. Beyond this book, readers are urged to review important viewpoints provided by Sherman and Guillery (2002; Sherman 2007).

Despite the flurry of recent progress, we still know much less about oculomotor thalamus than comparable saccade-related regions of cerebral cortex, cerebellum, basal ganglia, and brainstem. Partly this is due to the challenges of studying such an intricately parceled and networked structure. Improved technologies such as cell- and circuit-selective optogenetics, as well as high-density deep neuronal recording probes, are on the horizon for primate brain research. Yet it will remain important to use the new approaches with consideration of strong explanatory concepts such as efference copy and active vision, to gain well-informed and useful data that build on the prior findings described here.

REFERENCES

Ashmore RC, Sommer MA (2013). Delay activity of saccade-related neurons in the caudal dentate nucleus of the macaque cerebellum. *J Neurophysiol.* 109:2129–2144.

Awh E, Armstrong KM, Moore T (2006). Visual and oculomotor selection: Links, causes and implications for spatial attention. *Trends Cogn Sci.* 10:124–130.

Bellebaum C, Hoffmann KP, Koch B, Schwarz M, Daum I (2006). Altered processing of corollary discharge in thalamic lesion patients. *Eur J Neurosci.* 24:2375–2388.

Baker JT, Patel GH, Corbetta M, Snyder LH (2006). Distribution of activity across the monkey cerebral cortical surface, thalamus and midbrain during rapid, visually guided saccades. *Cereb Cortex.* 16:447–459.

Bellebaum C, Daum I, Koch B, Schwarz M, Hoffmann KP (2005). The role of the human thalamus in processing corollary discharge. *Brain.* 128:1139–1154.

Benevento LA, Fallon JH (1975). The ascending projections of the superior colliculus in the rhesus monkey (*Macaca mulatta*). *J Comp Neurol.* 160:339–361.

Berman RA, Cavanaugh J, McAlonan K, Wurtz RH (2017). A circuit for saccadic suppression in the primate brain. *J Neurophysiol.* 117:1720–1735.

Berman RA, Joiner WM, Cavanaugh J, Wurtz RH (2009). Modulation of presaccadic activity in the frontal eye field by the superior colliculus. *J Neurophysiol.* 101:2934–2942.

Berman RA, Wurtz RH (2010). Functional identification of a pulvinar path from superior colliculus to cortical area MT. *J Neurosci.* 30:6342–6354.

Bostan AC, Dum RP, Strick PL (2013). Cerebellar networks with the cerebral cortex and basal ganglia. *Trends Cogn Sci.* 17:241–254.

Cavanaugh J, Berman RA, Joiner WM, Wurtz RH (2016). Saccadic corollary discharge underlies stable visual perception. *J Neurosci.* 36:31–42.

Cavanaugh J, McAlonan K, Wurtz RH (2020). Organization of corollary discharge neurons in monkey medial dorsal thalamus. *J Neurosci.* 40:6367–6378.

Crapse TB, Sommer MA (2008). Corollary discharge across the animal kingdom. *Nat Rev Neurosci.* 9:587–600.

Crapse TB, Sommer MA (2009). Frontal eye field neurons with spatial representations predicted by their subcortical input. *J Neurosci.* 29:5308–5318.

Ford JM, Mathalon DH (2019). Efference copy, corollary discharge, predictive coding, and psychosis. *Biol Psychiatry Cogn Neurosci Neuroimaging.* 4:764–767.

Goldman-Rakic PS, Porrino LJ (1985). The primate mediodorsal (MD) nucleus and its projection to the frontal lobe. *J Comp Neurol.* 242:535–560.

Hikosaka O, Takikawa Y, Kawagoe R (2000). Role of the basal ganglia in the control of purposive saccadic eye movements. *Physiol Rev.* 80:953–978.

Idrees S, Baumann MP, Franke F, Münch TA, Hafed ZM (2020). Perceptual saccadic suppression starts in the retina. *Nat Commun.* 11(1):1977, 1–19.

Kaas JH, Lyon DC (2007). Pulvinar contributions to the dorsal and ventral streams of visual processing in primates. *Brain Res Rev.* 55:285–296.

Kunimatsu J, Suzuki TW, Ohmae S, Tanaka M (2018). Different contributions of preparatory activity in the basal ganglia and cerebellum for self-timing. *eLife.* 7:e35676.

Lee D, Malpeli JG (1998). Effects of saccades on the activity of neurons in the cat lateral geniculate nucleus. *J Neurophysiol.* 79:922–936.

Leszczynski M, Chaieb L, Staudigl T, Enkirch SJ, Fell J, Schroeder CE (2021.) Neural activity in the human anterior thalamus during natural vision. *Sci Rep.* 11:17480. https://www.nature.com/articles/s41598-021-96588-x

Lynch JC, Hoover JE, Strick PL (1994). Input to the primate frontal eye field from the substantia nigra, superior colliculus, and dentate nucleus demonstrated by transneuronal transport. *Exp Brain Res.* 100:181–186.

Macchi G, Jones EG (1997). Toward an agreement on terminology of nuclear and subnuclear divisions of the motor thalamus. *J Neurosurg.* 86:77–92.

Mauk MD, Buonomano DV (2004). The neural basis of temporal processing. *Ann Rev Neurosci.* 27:307–340.

McAlonan K, Cavanaugh J, Wurtz RH (2008). Guarding the gateway to cortex with attention in visual thalamus. *Nature.* 456:391–394.

Middleton FA, Strick PL (2001). Cerebellar projections to the prefrontal cortex of the primate. *J Neurosci.* 21:700–712.

Miura SK, Scanziani M (2022). Distinguishing externally from saccade-induced motion in visual cortex. *Nature.* 610:135–142.

Ohmae S, Kunimatsu J, Tanaka M (2017). Cerebellar roles in self-timing for sub-and supra-second intervals. *J Neurosci.* 37:3511–3522.

Ostendorf F, Liebermann D, Ploner CJ (2010). Human thalamus contributes to perceptual stability across eye movements. *Proc Natl Acad Sci U S A.* 107:1229–1234.

Ostendorf F, Liebermann D, Ploner CJ (2013). A role of the human thalamus in predicting the perceptual consequences of eye movements. *Front Syst Neurosci.* 7:10.

Percheron G, Francois C, Talbi B, Yelnik J, Fenelon G (1996). The primate motor thalamus. *Brain Res Rev.* 22:93–181.

Peterburs J, Pergola G, Koch B, Schwarz M, Hoffmann KP, Daum I, Bellebaum C (2011). Altered error processing following vascular thalamic damage: Evidence from an antisaccade task. *PLoS ONE.* 6:e21517.

Poulet JF, Hedwig B (2002). A corollary discharge maintains auditory sensitivity during sound production. *Nature.* 418:872–876.

Prevosto V, Graf W, Ugolini G (2010). Cerebellar inputs to intraparietal cortex areas LIP and MIP: Functional frameworks for adaptive control of eye movements, reaching, and arm/eye/head movement coordination. *Cereb Cortex.* 20:214–228.

Prevosto V, Sommer MA (2013). Cognitive control of movement via the cerebellar-recipient thalamus. *Front Syst Neurosci.* 7:56.

Rafal R, McGrath M, Machado L, Hindle J (2004). Effects of lesions of the human posterior thalamus on ocular fixation during voluntary and visually triggered saccades. *J Neurol Neurosurg Psychiatry.* 75:1602–1606.

Raghavan RT, Prevosto V, Sommer MA (2016). Contribution of cerebellar loops to action timing. *Curr Opin Behav Sci.* 8:28–34.

Rao HM, Mayo JP, Sommer MA (2016). Circuits for presaccadic visual remapping. *J Neurophysiol.* 116:2624–2636.

Reppas JB, Usrey WM, Reid RC (2002). Saccadic eye movements modulate visual responses in the lateral geniculate nucleus. *Neuron.* 35:961–974.

Robinson DL, Petersen SE (1992). The pulvinar and visual salience. *Trends Neurosci.* 15:127–132.

Ross J, Burr DC, Morrone MC (1996). Suppression of the magnocellular pathways during saccades. *Behav Brain Res.* 80:1–8.

Royal DW, Sary G, Schall JD, Casagrande VA (2006). Correlates of motor planning and postsaccadic fixation in the macaque monkey lateral geniculate nucleus. *Exp Brain Res.* 168:62–75.

Sherman SM (2007). The thalamus is more than just a relay. *Curr Opin Neurobiol.* 17:417–422.

Sherman SM, Guillery RW (2002). The role of the thalamus in the flow of information to the cortex. *Philos Trans R Soc Lond B Biol Sci.* 357:1695–1708.

Shin S, Sommer MA (2012). Division of labor in frontal eye field neurons during presaccadic remapping of visual receptive fields. *J Neurophysiol.* 108:2144–2159.

Sommer MA (2003). The role of the thalamus in motor control. *Curr Opin Neurobiol.* 13:663–670.

Sommer MA, Wurtz RH (1998). Frontal eye field neurons orthodromically activated from the superior colliculus. *J Neurophysiol.* 80:3331–3335.

Sommer MA, Wurtz RH (2002). A pathway in primate brain for internal monitoring of movements. *Science.* 296:1480–1482.

Sommer MA, Wurtz RH (2004a). What the brain stem tells the frontal cortex. I. Oculomotor signals sent from superior colliculus to frontal eye field via mediodorsal thalamus. *J Neurophysiol.* 91: 1381–1402.

Sommer MA, Wurtz RH (2004b). What the brain stem tells the frontal cortex. II. Role of the SC-MD-FEF pathway in corollary discharge. *J Neurophysiol.* 91:1403–1423.

Sommer MA, Wurtz RH (2006). Influence of the thalamus on spatial visual processing in frontal cortex. *Nature.* 444:374–377.

Sommer MA, Wurtz RH (2008). Brain circuits for the internal monitoring of movements. *Ann Rev Neurosci.* 31:317–338.

Suzuki TW, Inoue KI, Takada M, Tanaka M (2021). Effects of optogenetic suppression of cortical input on primate thalamic neuronal activity during goal-directed behavior. *eNeuro.* 8(2). https://doi.org/10.1523/ENEURO.0511-20.2021

Tanaka M (2005). Involvement of the central thalamus in the control of smooth pursuit eye movements. *J Neurosci.* 25:5866–5876.

Tanaka M (2006). Inactivation of the central thalamus delays self-timed saccades. *Nat Neurosci.* 9:20–22.

Tanaka M (2007). Cognitive signals in the primate motor thalamus predict saccade timing. *J Neurosci.* 27:12109–12118.

Tanaka M, Kunimatsu J (2011). Contribution of the central thalamus to the generation of volitional saccades. *Eur J Neurosci.* 33: 2046–2057.

Tanaka M, Kunimatsu J, Suzuki TW, Kameda M, Ohmae S, Uematsu A, Takeya R (2021). Roles of the cerebellum in motor preparation and prediction of timing. *Neuroscience.* 462:220–234.

Tanibuchi I, Goldman-Rakic PS (2005). Comparison of oculomotor neuronal activity in paralaminar and mediodorsal thalamus in the rhesus monkey. *J Neurophysiol.* 93:614–619.

Thakkar KN, Rolfs M (2019). Disrupted corollary discharge in schizophrenia: Evidence from the oculomotor system. *Biol Psychiatry Cogn Neurosci Neuroimaging.* 4:773–781.

Thiele A, Henning P, Kubischik M, Hoffmann KP (2002). Neural mechanisms of saccadic suppression. *Science.* 295:2460–2462.

Watanabe Y, Funahashi S (2004). Neuronal activity throughout the primate mediodorsal nucleus of the thalamus during oculomotor delayed-responses. II. Activity encoding visual versus motor signal. *J Neurophysiol.* 92:1756–1769.

Wurtz RH, McAlonan K, Cavanaugh J, Berman RA (2011). Thalamic pathways for active vision. *Trends Cogn Sci.* 15:177–184.

Wurtz RH, Sommer MA, Cavanaugh J (2005). Drivers from the deep: the contribution of collicular input to thalamocortical processing. *Prog Brain Res.* 149:207–225.

Wyder MT, Massoglia DP, Stanford TR (2003.) Quantitative assessment of the timing and tuning of visual-related, saccade-related, and delay period activity in primate central thalamus. *J Neurophysiol.* 90:2029–2052.

Yang Y, Cao P, Yang Y, Wang SR (2008). Corollary discharge circuits for saccadic modulation of the pigeon visual system. *Nat Neurosci.* 11:595–602.

Zimmermann E, Ghio M, Pergola G, Koch B, Schwarz M, Bellebaum C (2020). Separate and overlapping functional roles for efference copies in the human thalamus. *Neuropsychologia.* 147:107558. https://doi.org/10.1016/j.neuropsychologia.2020.107558

Zimmermann E, Ostendorf F, Ploner C J, Lappe M (2015). Impairment of saccade adaptation in a patient with a focal thalamic lesion. *J Neurophysiol.* 113:2351–2359.

SECTION VII

INTER-AREAL CORTICAL COMMUNICATION

Alessandra Angelucci

In the neocortex, sensory information is processed within multiple hierarchically organized areas, reciprocally connected via feedforward (FF) and feedback (FB) circuits. FF connections carry information from lower to higher level areas, and FB connections send information back from higher to lower areas. Classical models of the visual system postulate that perception is achieved by computations performed at successive hierarchical levels, where neurons at each level acquire increasingly larger and more complex receptive field (RF) properties. In these models, FF connections mediate the complexification of RFs, by integrating converging FF inputs from lower to higher areas, and object recognition occurs largely independently of FB signals (Hubel and Wiesel, 1962; Marr, 1982; DiCarlo et al., 2012). These models fail to account for the existence of an anatomically dense network of FB connections, and of trans-thalamic pathways as an additional route of communication between cortical areas. Moreover, perception is more than a mere representation of sensory stimuli. Object recognition requires extraction of invariant object features, and matching the identified object with internal representations of the sensory world partly hardwired and partly acquired by experience. Moreover, visual perception is affected by spatiotemporal context, the attentional state of the organism, and the organism's motivations and goals, phenomena that cannot be accounted for by purely FF models. This realization has led to a new generation of models in which the bulk of the computational work of the cortex is carried out by information going back and forth over looped FF-FB circuits between pairs of interconnected areas (Carpenter and Grossberg, 1987; Deacon, 1988; Poggio, 1990; Rolls, 1990; Mumford, 1992; Rao and Ballard, 1999; Friston, 2010; Wyatte et al., 2014). These new ideas, in turn, have prompted new experimental studies which have led to our current understanding of the anatomy and physiology of FF and FB systems, the formulation of quantitative models of cortical hierarchy, and hypotheses on the mechanisms of inter-areal communication via both cortico-cortical and cortico-thalamocortical pathways.

The chapters in this section provide an overview of our current understanding of the anatomical and functional organization of cortico-cortical and trans-thalamic pathways for inter-areal cortical communication within the framework of cortical hierarchy. Potential mechanisms and theories for inter-areal communication by these pathways are also presented. The focus is primarily on the visual system of the nonhuman primate and mouse.

Chapter 37 by Vezoli, Yujie, and Kennedy discusses how the concept of cortical hierarchy has evolved from one in which the hierarchy serves to construct increasingly complex RFs, via FF mechanisms, to one in which both FF and FB circuits define cortical hierarchies and counter-stream architectures, and serve complementary, but integrated, roles in visual processing. In Chapter 38, Angelucci and Petreanu review the anatomical and physiological principles governing the organization of FF and FB systems, and how the latter can underlie their differential impact on target neuron responses. These authors depict a high degree of anatomical and functional specificity in the organization of both FF and FB pathways, at the level of areas, functional modules, layers, and cell types. This specificity reaches one of its highest expressions in the modular organization of layer 1 of the mouse primary visual cortex, discussed in Chapter 39 by Burkhalter. Here, the modular expression of the M2 muscarinic receptor provides a scaffold for segregating the thalamocortical and cortico-cortical networks into 3 specialized visual-processing streams.

The following chapters discuss theories of hierarchical computation and mechanisms of inter-areal communication. Chapter 40 by Kohn reviews current understanding of the mechanisms for inter-areal communication and how, within the constraints of a hardwired and fixed anatomical connectivity, this communication can be modulated flexibly and dynamically on a moment-by-moment basis to guide vision-based behaviors. The author reviews evidence for and against various proposed mechanisms for effective and flexible inter-areal communication, including mechanisms based on modulating the synchrony and oscillatory activity in the source network (Fries, 2009; Bastos et al., 2015), the integrative properties of the target network (Hahn et al., 2019), the coordination of activity in the source and target networks, either via *communication through coherence* (Fries, 2015) or via *communication subspace* (Semedo et al., 2019), and the role of the cortico-thalamocortical loops. In Chapter 41, Shipp and Friston provide an in-depth evaluation of the anatomical organization of the primate and mouse FF and FB pathways and of the counter-stream architecture in the context of *predictive coding* theory. This theory views FB as representing a prediction of the external world, based on sensory data and prior experience, which is compared with incoming sensory data. The prediction error, resulting from such comparison, and carried by FF projections, ascends up the cortical hierarchy and refines the higher-level predictions (Mumford, 1992; Rao and Ballard, 1999; Friston, 2010). These authors conclude that the overall organization of cortico-cortical connectivity is largely compliant with the theory of predictive coding.

The role of cortico-thalamocortical loops in inter-areal communication is the focus of Chapter 42 by Javadzadeh and Hofer. These authors describe the fine-scale anatomical organization of three main trans-thalamic circuit motifs (recurrent, FF, and FB) in different neural systems and modalities, and discuss their potential distinct functions. The picture that emerges is one in which communication between cortical areas is not the sole realm of direct intracortical FF and FB pathways, but requires interactions with higher-order thalamic nuclei, via indirect cortico-thalamocortical pathways (Sherman, 2016). These pathways serve many sensory and higher cognitive functions, from maintaining cortical persistent activity during working memory or reward anticipation, to coordinating and gating cortico-cortical communication between distributed cortical areas, to representing efferent copy signals used to predict the sensory consequences of one's motor actions, to attentional control and decision-making. Because of its extensive connectivity with cortical and subcortical structures, the thalamus is ideally positioned to integrate signals from diverse sources, coordinate and consolidate distributed processes across many brain regions and translate them to a common neocortical frame of reference, before sending them back to the cortex (Mumford, 1991).

REFERENCES

Bastos AM, Vezoli J, Bosman CA, Schoffelen JM, Oostenveld R, Dowdall JR, De Weerd P, Kennedy H, Fries P (2015). Visual areas exert feedforward and feedback influences through distinct frequency channels. *Neuron.* 85:390–401.

Carpenter G, Grossberg S (1987). A massively parallel architecture for a self-organizing neural pattern recognition machine. *Comp Vision Graphics Image Proc.* 37:54–115.

Deacon T (1989). Holism and associationism in neuropsychology: An anatomical synthesis. In: Perecman E, ed. *Integrating Theory and Practice in Clinical Neuropsychology.* Erlbaum:1–47.

DiCarlo JJ, Zoccolan D, Rust NC (2012). How does the brain solve visual object recognition? *Neuron.* 73:415–434.

Fries P (2009). Neuronal gamma-band synchronization as a fundamental process in cortical computation. *Annu Rev Neurosci.* 32:209–224.

Fries P (2015). Rhythms for cognition: Communication through coherence. *Neuron.* 88:220–235.

Friston K (2010). The free-energy principle: a unified brain theory? *Nat Rev Neurosci.* 11:127–138.

Hahn G, Ponce-Alvarez A, Deco G, Aertsen A, Kumar A (2019). Portraits of communication in neuronal networks. *Nat Rev Neurosci.* 20:117–127.

Hubel DH, Wiesel TN (1962). Receptive fields, binocular interaction and functional architecture in the cat's visual cortex. *J Physiol (Lond).* 160:106–154.

Marr D (1982). *Vision: A Computational Investigation into the Human Representation and Processing of Visual Information.* Freeman.

Mumford D (1991). On the computational architecture of the neocortex. I. The role of the thalamo-cortical loop. *Biol Cybern.* 65:135–145.

Mumford D (1992). On the computational architecture of the neocortex. II. The role of cortico-cortical loops. *Biol Cybern.* 66:241–251.

Poggio T (1990). A theory of how the brain might work. *Proc Cold Spring Harbor Symp.* 55.

Rao RP, Ballard DH (1999). Predictive coding in the visual cortex: A functional interpretation of some extra-classical receptive-field effects. *Nat Neurosci.* 2:79–87.

Rolls ET (1990). The representation of information in the temporal lobe visual cortical areas of macaques. In: Eckmiller R, ed. *Advanced Neural Computers.* Elsevier:69–78.

Semedo JD, Zandvakili A, Machens CK, Yu BM, Kohn A (2019). Cortical areas interact through a communication subspace. *Neuron.* 102:249–259.

Sherman SM (2016). Thalamus plays a central role in ongoing cortical functioning. *Nat Neurosci.* 19:533–541.

Wyatte D, Jilk DJ, O'Reilly RC (2014). Early recurrent feedback facilitates visual object recognition under challenging conditions. *Front Psychol.* 5:674.

37.

THE EVOLVING CONCEPT OF CORTICAL HIERARCHY

Julien Vezoli, Yujie Hou, and Henry Kennedy

HIERARCHY: THE EARLY YEARS

Early investigations of neural mechanisms underlying perception were largely dominated by the demonstration of the visual receptive field (RF), following Hartline's demonstration of the activation of an optic nerve fiber by light falling on a restricted area of the retina (Hartline, 1938). Later, Hubel and Wiesel (1968) further developed the notion of RF at the cortical level when they explored how the RF evolved across cortical areas. By simultaneously considering the changes in RF organization with respect to laminar and areal location of the recorded neurons, these authors were able to conclude that the response properties that they recorded were the consequence of feedforward processing in a hierarchically organized system. The structural organization of the cortical hierarchy was greatly advanced in a landmark study by Rockland and Pandya (1979). These authors defined feedforward (FF) and feedback (FB) cortico-cortical connections by reference to the interconnectivity of the thalamic relay nucleus of the visual system (lateral geniculate nucleus, LGN) with the primary visual cortex—area V1. Hence, FF LGN neurons target layer (L)4, and FB to the LGN originates from L6. By extension, in cortico-cortical pathways, FF projections originate from supragranular layers and target L4, and FB projections avoid L4 and originate from infragranular layers (Rockland and Pandya, 1979). This was followed by numerous tract-tracing studies that largely confirmed and extended this work, making it possible for Felleman and Van Essen (1991) to construct a collated database and develop a computational model of the cortical hierarchy where FF and FB pathways linked cortical areas across 10 processing levels. The previous tract-tracing studies had greatly increased the numbers of known pathways so that the hierarchy that Felleman and Van Essen described was highly distributed, with numerous parallel ascending and descending pathways.

The notion that the visual RF reflects the anatomical input was challenged by experiments revealing a clear mismatch in the representation of the visual field carried by inputs to a given cortical neuron and the RF size of that neuron. For example, intrinsic connections in visual cortex were such as to suggest that they connected neurons with non-overlapping RFs (Gilbert and Wiesel, 1985). Further, Hubel and Wiesel's demonstration of the increase in RF size as one moves from area V1 to V2 and beyond was confirmed by numerous reports. This implied that neurons in higher-order areas with large RFs project to neurons in lower-order areas should have considerably smaller RFs that will therefore not match the spatial extent the RFs of the higher-order areas' neurons. This indeed was shown to be the case. Visual response mapping experiments coupled with tract tracing indicating that FB projections in cat area V2 representing up to 18 degrees of visual angle converge to an area of V1, encoding visual responses up to 4 degrees of visual field (Salin et al., 1992). Since the investigations of Hartline, researchers had been mapping RFs with discreet visual stimuli; hence the discovery that the RF does not reflect the full spatial extent of inputs to the cortical neuron, and in particular that of the FB input, was unexpected. This introduced the notion of responses beyond the classical RF, which were thought of as being contextual, allowing local-global comparisons (Allman et al., 1985). Subsequently, Angelucci, Bullier, and colleagues were able to quantify the extent of this spatial mismatch, showing that the spatial extent of FB far exceeded that of intrinsic connections, and that the spatial extent increased proportionally with projections arising from progressively higher levels of the cortical hierarchy (Angelucci et al., 2002; Angelucci et al., 2017) (Figure 37.1; see also **Chapter 38 in Section VII** and Figure 38.2B in that chapter). However, the nature of these top-down influences and their overall role in cortical processing remained an unsettled issue, leading Jean Bullier to ask, "What is fed back?" (Bullier, 2006). It turns out, quite a lot, including contextual information (Allman et al., 1985; Angelucci et al., 2017), attention (Luck et al., 1997), and most importantly predictions or expectations (Bastos et al., 2012; Rao and Ballard, 1999). Nevertheless, despite the fact that cortical FB pathways are twice as prevalent as

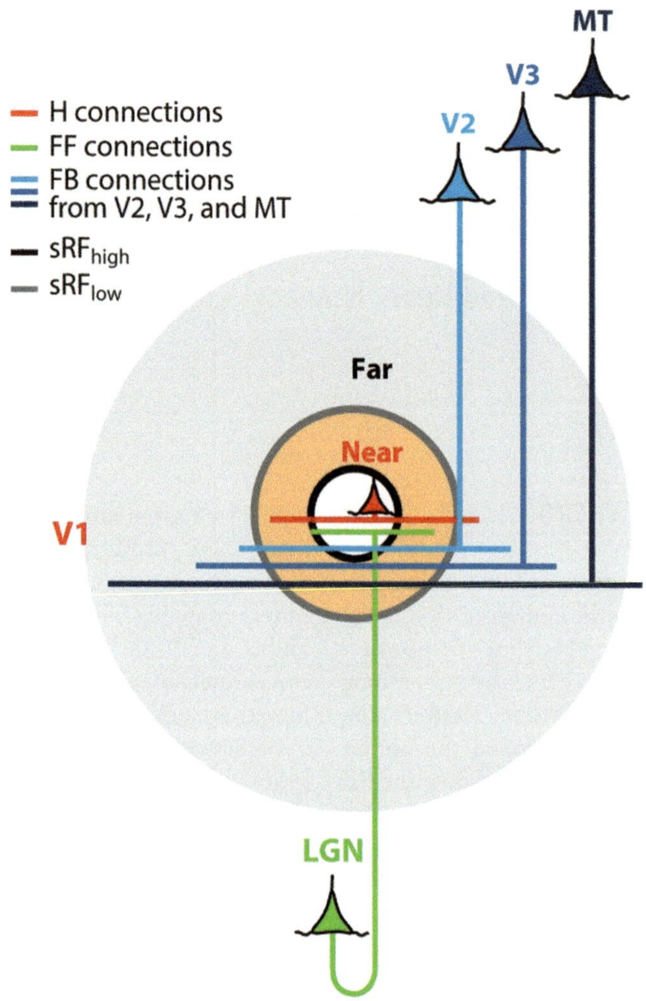

Figure 37.1. Spatial topography of intrinsic, feedback, and feedforward projections to area V1. The visuotopic extent of different connection types is compared to the extent of a V1 neuron's RF (white area), and RF surround, the latter comprising 2 regions, i.e., near (orange annulus) and far (gray annulus), depending on their proximity to the RF. H = horizontal connections intrinsic to area V1; FF = feedforward connections to area V1 from the LGN; FB = feedback connections; $sRF_{high, low}$ = summation RF at high or low contrast, respectively. From Angelucci et al. (2017).

FF pathways (Markov et al., 2014b), the conceptual importance of FB remains highly debated. For example, in recent reviews of the literature, Walsh and colleagues conclude that there is a need for "spurring efforts to recognize the importance of backward connections" (Walsh et al., 2020), while Maric and Domijan (2022) propose "a modest, modulatory role for the FB pathway in cortical processing." This divergence of views results perhaps from the difficulty in comprehending the mechanisms that ensure a synergy of the FF and FB pathways. Here we shall review recent experimental data in human and macaque and to a lesser extent mouse that are beginning to map out a cortical architecture that holds the promise of revealing what such synergistic functions are likely to be.

Sherman and Guillery (1998) introduced the concept of drivers and modulators to explain the 2 types of action of the thalamus on cortical neurons. Given that these authors conceived drivers to transmit the RF properties and the modulators to alter the probability of different aspects of those properties, similar concepts could be relevant to understanding cortico-cortical pathways, where numerous authors have equated FB to a modulatory function (Ekstrom et al., 2009; Marić and Domijan, 2022). However, studies examining the synaptic properties of cortical connections in mouse visual and auditory cortex found "driver" and "modulator" properties that failed to show a correlation with FF and FB anatomical categories (Covic and Sherman, 2011; De Pasquale and Sherman, 2011) in line with modeling findings (Abbott and Chance, 2005). There is some evidence that the supragranular-FB (supra-FB) projections carry activating signals. For instance, inactivation of cat LGN revealed persistent visual responses in the superficial layers of area V1 following activation of area V2 (Mignard and Malpeli, 1991). Moreover, inactivation of area MT in macaque abolishes the response of a small population of cells in the superficial layers of area V2 (Hupe

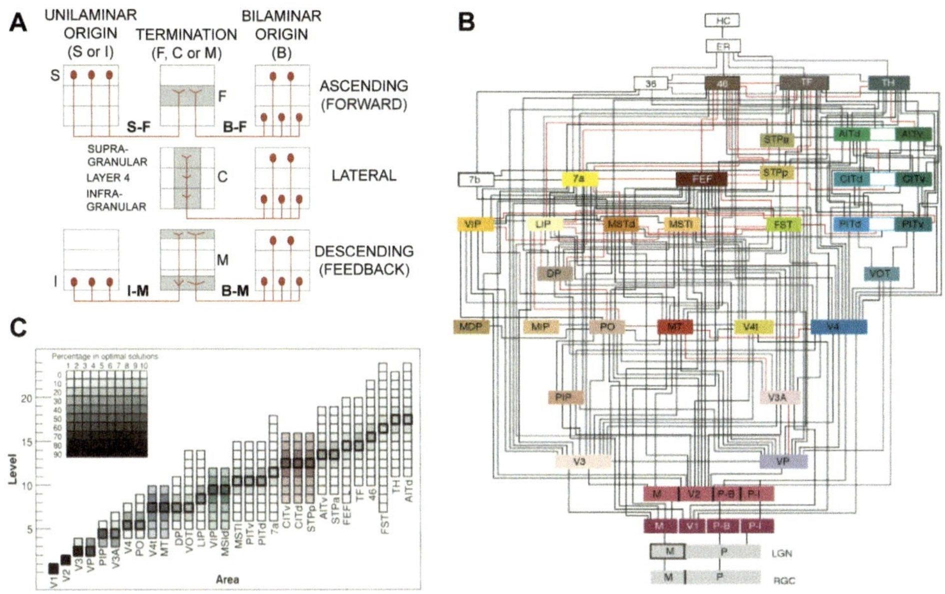

Figure 37.2. The Felleman and Van Essen binary model of cortical hierarchy. (A). Criteria for classifying connections between areas as feedforward (top), lateral (middle), or feedback (bottom). Termination patterns are depicted in the central column; preferentially in layer 4 (F pattern = FF), across all layers (C pattern = lateral), in upper and lower layers avoiding layer 4 (M pattern = FB). Laminar origins from a single layer (left column) either supragranular (S) and therefore FF, or infragranular (I) and therefore FB. Bilaminar (B) origins (right column) terminate in the middle layer (F pattern) and therefore FF, in all layers and therefore lateral, or in upper supra and infragranular layers (M pattern) and therefore FB. (B). The binary hierarchical model. (C). Area frequency distributions for 150,000 optimal orderings. A, B, from Felleman and Van Essen (1991); C from Hilgetag et al. (1996).

et al., 1998). More recently, top-down influences in macaque have been shown to lead to neuron spiking in area V1, even in the absence of visual stimuli (van Kerkoerle et al., 2017) and, when FB connections are selectively inactivated, the attentional increase in the firing of V1 cells primarily located in the supragranular layers is abolished (Debes and Dragoi, 2023).

Further, the conceptual relevance of driver and modulator action in the cortex has been questioned (Vezoli et al., 2021). The terminology of "driving" and "modulatory" implies that it is the nature of the input itself which ensures the neuron response. This needs to be tempered by the fact that the response of a cortical neuron to long-distance input largely reflects the intrinsic property of the local network. All long-distance inputs to cortical neurons, including cortico-cortical inputs, are minute, both anatomically and physiologically, and are thought to undergo amplification by the local recurrent circuit (Douglas et al., 1995; Vezoli et al., 2021). This amplification is highly specific, where highly non-random connectivity ensures the emergence of feature-specific RF properties (Cossell et al., 2015).

HIERARCHY COMES OF AGE

Undoubtedly, the structural cortical hierarchy reflects a very robust set of regularities of the inter-areal network that are observed throughout most of the cortex. It is these regularities that allowed Felleman and Van Essen (1991) to build an influential large-scale model of the cortex, stretching from area V1 in the occipital lobe to area 46 in the prefrontal cortex (Figure 37.2A–B). Much of the interest in cortical hierarchy, up to the end of the 1980s, was that it provided a map of the FF connections across the cortex. These connections were considered to be the main factors in the elaboration of progressively more complex RFs that were being investigated using invasive electrophysiological approaches in animal models and noninvasive imaging approaches in nonhuman primates and humans, and led to influential feedforward modes of cortical function e.g. (Zeki and Shipp 1988). Two influential publications (Mumford, 1992; Rao and Ballard, 1999) brought a more complex notion of hierarchy to the forefront of systems neuroscience, introducing a radical new way of thinking about cortical function that goes back to theories of active inference (Parr et al., 2022; Von Helmholtz, 1867). These authors proposed a hierarchy in which FF and FB played equally important and complementary roles in terms of top-down predictions and bottom-up evidence (Mumford, 1992). These ideas were formalized by Rao and Ballard (1999), focusing on a model that accounted for neuronal responses beyond the classical RF, putting the FB pathway at center stage. Hierarchical predictive coding theory hypothesizes that FB pathways relay top-down expectations or predictions of lower-level neural activities, while the ascending pathways relay the residual error between the prediction and the actual lower-level activities (Rao and

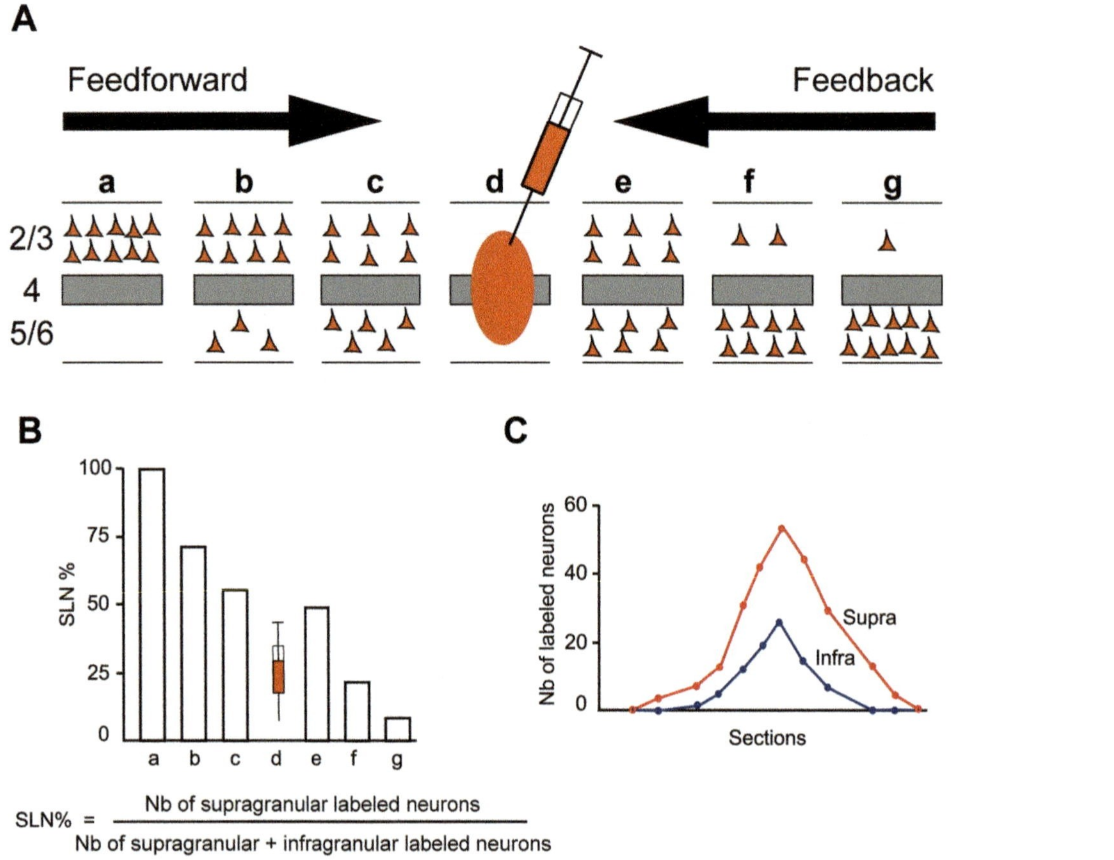

Figure 37.3. Quantitative parameters characterizing the hierarchy. (A). The laminar distribution of parent neurons in each pathway, referred to as SLN (fraction of supragranular layer neurons), is determined by high-frequency sampling and quantitative analysis of labeling. Supra- and infragranular layer neurons contribute to both FF and FB pathways, and their relative proportion is characteristic of each type of pathway. For a given injection there is a gradient of SLN in the labeled areas, from purely FF (SLN = 100%, all the parent neurons are in the supragranular layers) to purely FB (SLN = 0%, all the parent neurons are in the infragranular layers) and a spectrum of intermediate proportions. (B). All labeled areas can then be ordered by decreasing SLN values, and the resulting order is consistent with the hierarchical order according to Felleman and Van Essen. SLN is, thus, used as an indicator of hierarchical distance between areas from the same injection. (C). Reliable estimation of SLN crucially requires sampling labeling throughout the full extent of the projection zone in each area. From Vezoli et al. (2021).

Ballard, 1999) (see also **Chapter 41 in Section VII**). We shall come back to predictive coding later in this chapter, but before we do so, we address how computational theories of cortical function spurred a better understanding of the anatomy of hierarchy and eventually their extension to a functional formulation. Deepening the understanding of cortical hierarchy also required its extension to the human and mouse cortex, the latter being increasingly the main experimental model in neuroscience.

While the Felleman and Van Essen (1991) model of the cortex was extremely powerful, it was nevertheless built on only semi-quantitative data that were collated across numerous studies using different definitions of cortical areas and tracing techniques. One consequence of the nature of the data used in that model is that the hierarchical levels were chosen in an arbitrary fashion. In this way the numerous areas that were not interconnected had unknown hierarchical relations, causing the model to be indeterminate, with several hundred thousand equally likely solutions (Hilgetag et al., 1996) (Figure 37.2C). An alternative approach was to determine levels using a measure of hierarchical distance. Retrograde tracing allows this, because a tracer injection in a target area results in a smooth monotonic distribution of retrograde labeled neurons across the hierarchy (Figure 37.3). The proportion of labeled neurons in supragranular layers (SLN) is higher in lower-order areas, and decreases smoothly approaching the injection site, while in higher-order areas SLN values decrease as one moves away from the injection site (Barone et al., 2000). Tracer injections in just 2 cortical areas provided a determinate hierarchy, where hierarchal rank indicated the operation of a distance rule. In a subsequent study where considerably more pathways were traced (Markov et al., 2014b), the sequence of cortical areas in the determinate model was observed to be very similar to that of the Felleman and Van Essen model, but with 2 importance differences (Figure 37.4). First, in the Felleman and Van Essen model the frontal eye field (area 8) in the prefrontal cortex is at level 8, below area 46, while

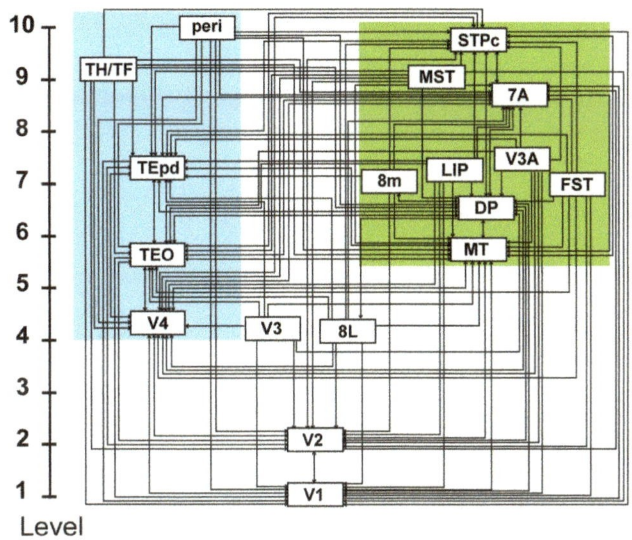

Figure 37.4. The determinate structural hierarchy. Model of hierarchy of visual areas built from the fraction of supragranular neuron (SLN) and the fraction of labeled neuron (FLN) values following retrograde tracer injections in 11 cortical areas. Blue background indicates ventral stream areas; green background, dorsal stream areas. From Markov et al. (2014b).

in the determinate model the frontal eye field is subdivided into areas 8L and 8M; area 8L is located at level 4, on the same level as area V4 (Barone et al., 2000; Markov et al., 2014b), while area 8M is considerably higher in the hierarchy. Second, in the determinate model the dorsal stream is at a significantly higher hierarchical level than the ventral stream, compared to the Felleman and Van Essen model (Markov et al., 2014b) (Figure 37.4).

Graph measures of cortical connectivity in the Markov et al. (2014a) study revealed a graph density of 66%, meaning that two-thirds of all possible connections actually exist. This density is considerably higher than the densities reported in previous collated datasets, which varied between 6% and 34% (Markov et al., 2013). The reported quantitative measures of inter-areal connectivity revealed major mathematical regularities (Figure 37.5) epitomized by a heavy-tailed lognormal distribution of weights spanning 5 orders of magnitude, conforming to a distance rule reflecting exponential decay with inter-areal separation (Ercsey-Ravasz et al., 2013). The physical constraint imposed by the exponential decline of connections strength (Figure 37.5B) and the geometrical constraint of the distance distributions (Figure 37.5B insert), allied with the lognormal distribution of weights, prompted the authors to propose a single-parameter random graph model that predicted numerous features of the cortical network, including core-periphery heterogeneity, local and global communication efficiencies, and community structure (Figure 37.5C–F). The weight-distance relations are the key feature of spatial embedding, which is highly predictable of empirically observed features, which is not the case of models based on binary features (Knoblauch et al., 2016). The spatial embedded model greatly facilitated the development of anatomically realistic dynamical models of the macaque cortex exploiting heterogeneity of local features such as gene expression, receptor density, and microcircuit function, and the quantitatively determined cortical network (Wang et al., 2022). This is exemplified by a study looking at hierarchy in the temporal domain (Chaudhuri et al., 2015). The progressive enlargement of the RF of neurons along the cortical hierarchy is well documented, allowing areas at higher levels to integrate visual signals over a greater spatial extent. In a similar fashion, there is an impact of hierarchy in the temporal domain, with early sensory areas showing shorter time scales and faster dynamics, and higher-order areas showing slower time scales and slower dynamics (Honey et al., 2012). Chaudhuri and colleagues (2015) used the spine count on pyramidal neurons as a proxy for heterogeneity of the local microcircuit (Elston, 2000). Integration of the weighted and directed connectivity with the local heterogeneity enabled these authors to construct a hierarchical model of timescales and to explore the temporal dynamics of the cortex (Chaudhuri et al., 2015). The generality of this approach to dynamic modeling—exploiting local heterogeneity and large-scale structural hierarchy—is further illustrated by dynamic modeling of the spatial cortical gradient of dopamine receptors that effectively determines access to working memory (Froudist-Walsh et al., 2021). Here it was necessary to re-scale weak connections in the cortical matrix, which is justified by the observation that long-distance, weak connections in a functional connectome of the monkey frontal cortex are significantly stronger than their structural counterparts (Hayashi et al., 2022).

While structural models of cortical hierarchy have been highly influential, a legitimate question is whether the brain is actually functionally organized in a hierarchical manner. There are neurophysiological asymmetries that are significant in this respect. First, supragranular layers are the main source of FF structural connections and they exhibit predominant gamma-band synchronization; in contrast, the infragranular layers are the main source of FB connections and exhibit strong beta-band synchronization (Buffalo et al., 2011). Local rhythmic synchronization subserves inter-areal synchronization that has been proposed as a mechanism to facilitate inter-areal communication (Fries, 2005). Hence, the observed laminar frequency channels could exert functional FF and FB influences across the cortical hierarchy. By correlating frequency-directed matrices with anatomical metrics of FF and FB (Markov et al., 2014b), Bastos, Vezoli, and colleagues were able to construct the functional hierarchy of the macaque cortex

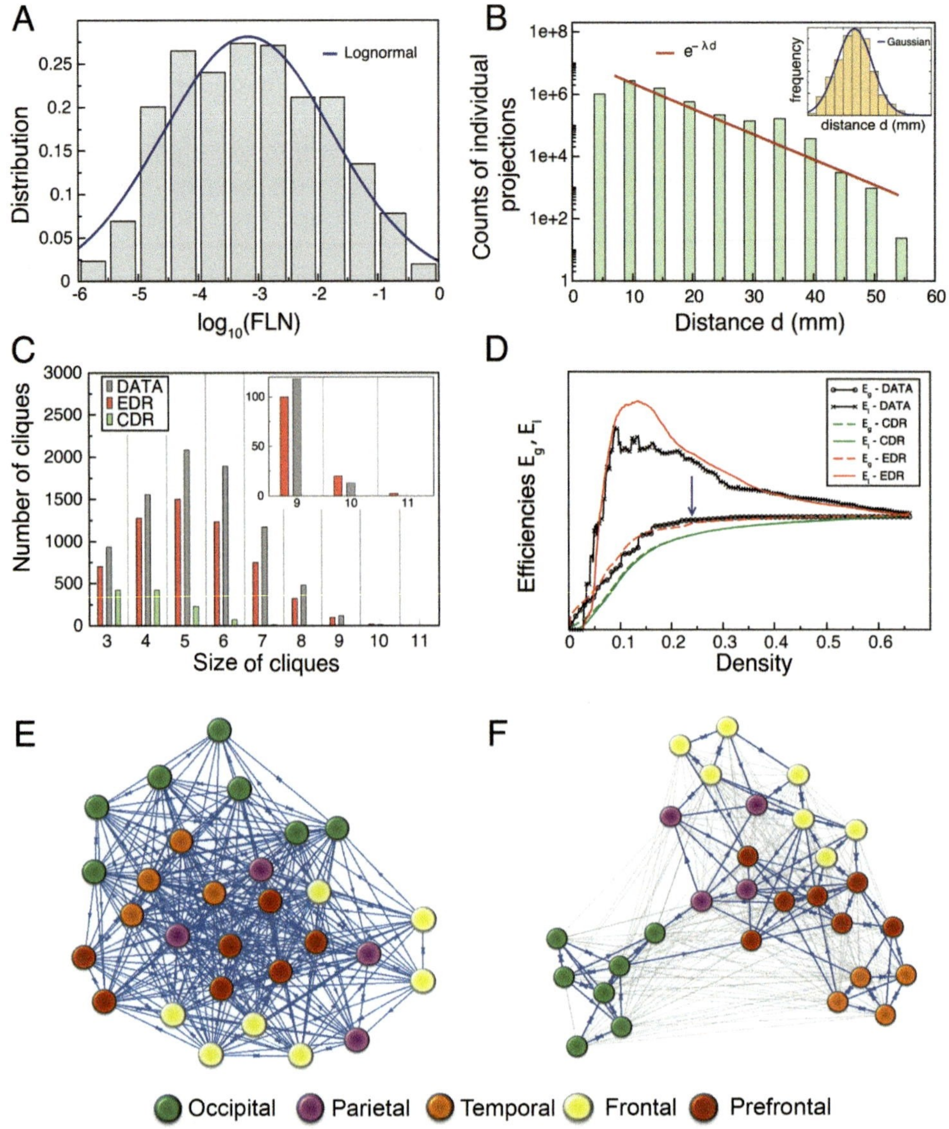

Figure 37.5. A spatial network model of the cortex. (A). Distribution of FLN weights and its approximation by a log-normal distribution. (B). The number of projections as function of projection distance d. Inset: the distribution of interareal distances through the white matter is well approximated by a Gaussian. (C). The distribution of cliques in the G29 × 29 data network is well captured by the Exponential Distance Rule (EDR) model. (D). Global (Eg) and local (El) efficiencies as a function of network density during sequential removal of the weakest link. The EDR model (red) captures both data curves (black) much better than the Constant Distance Rule (CDR, green). (E). The G29 × 29 subgraph using a Kamada-Kawaii force-based layout algorithm with all links considered with unit weight. (F). Same as E but considering the 24% strongest links only (blue arrow in D), with FLN weights. In this case, the areas are clustered into functional regions (see color code). FLN = fraction of labeled neurons, a weight metric. From Markov et al. (2013).

(Bastos et al., 2015). The resulting functional hierarchy was very similar to the anatomical hierarchy, but with one important difference: functional hierarchies display dynamic task-related switching concerning the hierarchical level of area 8L of the frontal eye field with respect to area V4. These results, showing that the input from 8L to area V4 can vary from lateral to FB according to behavioral requirements, are thought to reflect varying degrees of integration of area 8L into the prefrontal network. The functional hierarchy of the macaque cortex has proven to be highly influential for understanding cortical dynamics and has prompted investigation of the functional hierarchy of the human brain. Anatomical metrics of structural hierarchy of visual cortex are not available for the human brain; however, there are well-defined homologies between visual areas in the 2 species. This allowed Michalareas et al. (2016) to use the projection patterns of these areas as defined in the macaque and correlate them with the inter-areal directed influences measured with magnetoencephalography in humans. This showed that FF influences predominate in the gamma

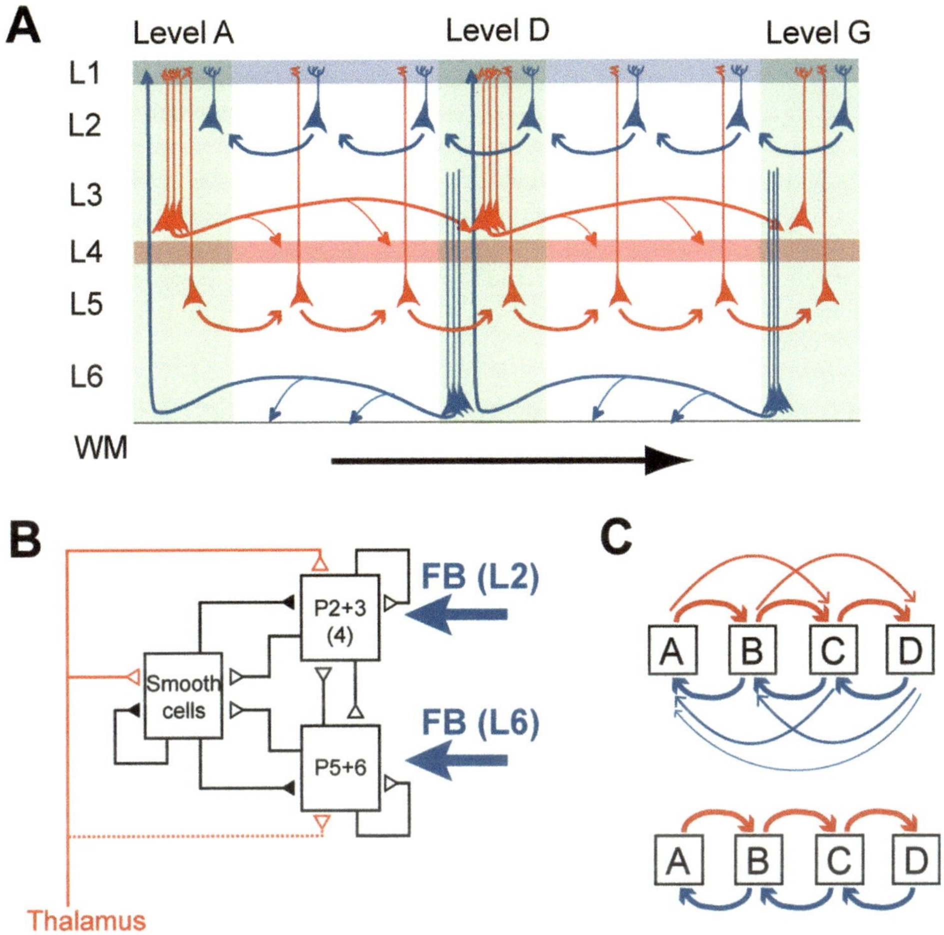

Figure 37.6. Dual counterstream architecture. (A). Neurons in L3 and L5 send FF projections (red) to higher-order areas, which are reciprocated by FB projections (blue) arising from L2 and L6. Simultaneous tracer injections in higher-order and lower-order areas have demonstrated that the upper-layer counterstream shows near 100% segregation of FF and FB neurons; i.e., the FF (FB) neurons do not send axon collaterals to lower (higher)-order areas. However, the evidence that the FF and FB pathways form continuous streams, as depicted here, is indirect; what crucially remains to be elucidated are the laminar details on the connectivity and the cellular targets. Modified from Markov et al. (2014b). (B). The canonical microcircuit showing the 2 FB pathways targeting L2/3 and L5/6, respectively. However, it remains unknown whether the L2-FB and the L6-FB selectively contact the supra- and infragranular layers, respectively, as depicted in the figure, or rather contact both sets of layers in lower areas. P = pyramidal cells; smooth cells: inhibitory cells. Modified from Douglas and Martin (1991). (C). The inter-areal circuitry for serial processing incorrectly assumed in many computational models (lower) is not observed in the cortex, where, instead, each area projects to all upper and lower stream areas (all to all; upper). (However, this difficulty is tackled in Chapter 41 in Section VII). Note, however, that this all-to-all connectivity does not take place at the single neuron level, where, particularly for FF connections in primates, single neurons project predominantly to a single higher area. Modified from Vezoli et al. (2021).

band and FB in the alpha-beta band, and revealed a hierarchy that closely resembled that found in the macaque. However, compared to the macaque, there was a unique feature in the human brain. In the macaque brain FB from the extrastriate cortex influences early visual areas V1 and V2; this contrasts with the human brain, where mid-level ventral stream area V4 appears to receive the bulk of FB influences from higher levels. These observations suggest that the proposed functions of area V4 in the selective enhancement of domain-specific networks, resulting from integration of top-down and bottom-up influences (Roe et al., 2012), are further developed in human compared to macaque.

HIERARCHY AND THE DUAL COUNTERSTREAM ARCHITECTURE

While it is true that FF mainly stem from supragranular layers and FB from infragranular layers, recent findings show that an independent FF and FB pathway also exists in both supra- and infragranular layers. In the macaque supragranular compartment there is an L2 FB pathway immediately below the external granular layer situated immediately below L1, while the L3 FF pathway originates from just above L4 (the internal granular layer) (Figure 37.6A). This constitutes a counterstream consisting of a supra-FB and a supra-FF. Similarly, there is an infragranular L6 FB pathway (infra-FB) and an infragranular L5 FF pathway (infra-FF). This results in a dual counterstream architecture that is found throughout the macaque extrastriate visual cortex (Markov et al., 2014b; Vezoli et al., 2021).

The FF and FB pathways in each laminar compartment share topological features. Compared to the infragranular compartment, the supra-FF and supra-FB pathways show a topographic point-to-point organization. However, the two pathways show important differences too, while the supra-FB covers shorter inter-areal distances, the supra-FF covers relatively longer distances. Hence the diffuse nature of FB pathways is actually characteristic of infra-FB. Compared to the supra-FB, the infra-FB pathway traverse greater inter-areal distance, and this difference is largely responsible for the decrease in SLN with distance in the top-down direction. The longer inter-areal distance of the supra-FF compared to the infra-FF, instead, leads to the augmentation of SLN with increasing bottom-up distances.

The greater inter-areal distance of the infra-FB ensures carrying signals over a considerable portion of the hierarchy and, therefore, highly relevant to its function. Recently, monosynaptic viral tracing in macaque has shown that uniquely in the infragranular pathway, FB neurons form synapses with FB neurons in upstream areas, thereby forming chains of FB connections, which serve to further increase the reach of the infra-FB pathway (Siu et al., 2021). Such FB chains could be very important in human cortex given the suggested shorter inter-areal reach of FB in this species (Michalareas et al., 2016). Overall, the supra-FF is the dominant pathway in the supragranular compartment and the infra-FB in the infragranular compartment. Finally, the spatial segregation of FF and FB in the supragranular counterstream is more pronounced than that of FF and FB in the infragranular counterstream.

While there remain numerous unknowns concerning the dual counterstream architecture (see Figure 37.6), its generality across primate cortex suggests that it could be important for understanding cortical function, particularly with respect to the presence of 2 distinct FB pathways, one in the supragranular compartment and one in the infragranular compartment (Marić and Domijan, 2022; Vezoli et al., 2021). There are numerous reports of important differences in the biology of supra- and infragranular layers in terms of RF properties that could give additional clues to the nature of the information carried in the FF and FB pathways of the 2 compartments (Pettine et al., 2019). More recently, sparse and deep coding have emerged as major differences between the upper and lower compartments of the cortex (reviewed in D'Souza and Burkhalter, 2017; Harris and Mrsic-Flogel, 2013). In sparse coding, only a small subset of neurons is active at any given moment, contrasting with dense coding, where large proportions of neurons are usually active, providing graded responses. It has been hypothesized that the sparse coding observed in the supragranular layer is dependent on the developmental expansion of these layers in primates, made possible by the outer subventricular zone, a specialized developmental germinal zone found in nonhuman primates and humans (Dehay and Kennedy, 2020).

A key feature in the dual counterstream architecture is that there is a structural segregation of FF and FB; that is, that the infra-FB neurons characterized by numerous widespread axon collaterals to distant downstream areas do not also have up-stream axon collaterals (Friston, 2018; Shipp, 2016). Tracer studies have shown that this indeed is the case in both mouse and macaque (Berezovskii et al., 2011; Markov et al., 2014b). While there are hypotheses concerning the role of the supra- and infra-FB pathways in human cognition (see below), we lack a mechanistic understanding of FB pathways in the nonhuman primate cortex. This is not the case in mouse, where FB pathways have been shown to cause disinhibition in their target areas via disynaptic inhibitory circuits involving VIP+ and SOM+ inhibitory cells (Zhang et al., 2014). More recently, Shen and colleagues (2022) examined supra- and infragranular FB pathways in mouse. Their findings show that the projection from LM (the mouse homolog of primate area V2) to area V1 projects to all layers including L4, but has an inhibitory influence on L2/3 pyramids by preferentially contacting SOM+ inhibitory cells in those layers, which in turn inhibit adjacent pyramidal cells. This contrasts with projections from M1 to S1 which had a disinhibitory effect on pyramidal neurons, due to preferentially contacting VIP to SOM+ circuits in L5/6. It is likely that the V2 to V1 supra-FB pathway in macaque has a similar inhibitory effect as the mouse LM to V1 supra-FB pathway given that top-down contextual modulation of area V1 leads to surround suppression in both species (Vangeneugden et al., 2019).

These findings in mouse concerning the supra- and infra-FB pathways are relevant to the theory of predictive coding which postulates that prediction errors ascend the cortical hierarchy and predictions descend the hierarchy (Bastos et al., 2012; Mumford, 1992; Rao and Ballard, 1999). With respect to the dual counterstream architecture, the infra-FB is considered to be suppressive, while the supra-FB has been proposed to be modulatory (Bastos et al., 2012; Friston, 2018; Shipp, 2016) and is associated with precision weighting, thought of as the top-down deployment of attention (Feldman and Friston, 2010) (see also **Chapter 41 in Section VII**). The findings of Shen et al. (2022) fit well with this theory, where the LM to V1 supra-FB, targeting SOM+ and PV+ cells, can provide modulatory action, and L5 pyramids receive inhibitory influences.

The full range of inhibitory motifs of FF and FB pathways in mouse is still not known and remains experimentally challenging. This difficulty is even more acute in primate cortex where genetic models are not available. Further, there are important differences in the cortical connectivity of primates and mouse. For example, while FB projections to L4 have been observed in mouse (Minamisawa et al., 2018; Shen et al., 2022), these non-canonical motifs are absent in primates (Markov et al., 2014b) as well as in rats (Coogan and Burkhalter, 1993). Thus, we cannot assume that the connectivity patterns observed in mouse are necessarily relevant to primate cortex, in line with the notion that mouse biology is highly specialized in many ways (Dehay and Kennedy, 2020).

IS THE CLUE TO THE DUAL COUNTERSTREAM FUNCTION TO BE FOUND IN HIGH RESOLUTION FMRI EXPERIMENTS IN HUMAN SUBJECTS?

The clear separation between the FF and FB pathways in the supragranular counterstream suggests that these pathways have complementary functions, which are fundamentally different from those of the infragranular counterstream for which the segregation between FF and FB projection neurons is less strict. While there are vertical, within-column connections linking the infra- and supragranular compartments, the inter-areal targets of these compartments appear to be largely segregated and restricted to the infra- and supragranular compartments, respectively, suggesting that there might be relatively little direct interaction between the supra- and infragranular pathways (Markov et al., 2014b). Simultaneous recordings in primate areas V1 and V2, and areas V1 and V4, also suggest a high degree of functional segregation (Semedo et al., 2022), providing further support for the independence of FF and FB pathways in the supra- and infragranular counterstream (Marić and Domijan, 2022). These observations are relevant to what is known as cognitive penetration, which is conceived as the modulation and shaping of sensory experience by cognitive experience (Lupyan, 2017). Maric and Domijan argue that cognitive penetration might be graded and depend on hierarchical level. Neuronal responses at the early levels of the visual cortical hierarchy are strongly correlated to visual processing, and as we ascend the hierarchy become increasingly correlated to behavior, in mouse (Siegle et al., 2021), primate (Brincat et al., 2018) and human (Jo et al., 2019), where the sensory-behavior boundary is anterior to V4.

Recent technical development in ultra-high laminar resolution functional magnetic resonance imaging (fMRI) provides a mesoscopic tool that allows probing FF versus FB pathways in human subjects (Jia et al., 2023) provide important clues concerning the differences between the top-down pathways in the supra- and infragranular layers. Using 7T high-resolution fMRI and patterned classification methods in humans presented with a partial occlusion of the visual field, it was found that the signals recorded in the non-stimulated region in V1 supragranular layers could be used to decode the information presented in the surround of the occluded image. The signals in the occluded region were interpreted to represent contextual influences beyond the classical RF (Muckli et al., 2015).

More recently fMRI multivariate decoding analysis in primate area V1 has proved to be a very fruitful approach showing differential laminar representation of errors and predictions (Thomas et al., 2023) in line with predictive coding theory and echoing earlier electrophysiological findings in mouse (Jordan and Keller, 2020). In another study, using 7T high-resolution fMRI, Bergman et al. (2019) compared the activation of the supra- and infragranular layers in human area V1. Their findings suggested that FB influences in response to visual illusions were located in the supragranular layers, consistent with low-level visual illusions being dependent on short-distance contextual signals. In contrast, visual imagery engaged top-down processing in the infragranular layers, presumably originating from a widely distributed network. These authors concluded that at the lower hierarchical levels, supra-FB encodes perceptual-like experiences while infra-FB carries information about inner experience, which is distinct from ongoing perception (Dijkstra and Fleming, 2023).

CONCLUSION AND PERSPECTIVES

David Mumford's Bayesian computational approach to cortical hierarchy was revolutionary, and in a follow-up study he paid particular attention to area V1, which he considered to play a key role where cascading FB priors are echoed in the long-latency responses of its neurons (Lee and Mumford, 2003). Rao and Ballard (1999) proposed that these priors integrate the FB projections into a generative inference process, which enabled them to model the extra-classical receptive field (Figure 37.1). FB neurons are potentially able to carry the computation of the local circuit within which they are embedded, which can include multiple properties such as working memory (Froudist-Walsh et al., 2021), as well as signals integrated over much longer time intervals than found in area V1 (Chaudhuri et al., 2015). Hence, the spatially extended reach of FB connections across the cortical hierarchy (Figure 37.1) could be responsible for a wide range of contextual effects beyond the spatial domain (see also **Chapter 38 in Section VII**). Given the impact of visual field eccentricity on qualitative and quantitative changes in FB projections to area V1, such changing contextual influences likely support mechanisms of active vision (Wang et al., 2022). An important consideration is that the recent discovery of descending FB chains allows the influence of the infra-FB to descend a larger number of hierarchical levels, thereby considerably increasing the potential impact of FB priors (Siu et al., 2021). David Mumford's interpretation of the unique role of area V1 has recently been reinforced by multi-genomic analysis revealing numerous specialized cell types in FF and FB layers in this area (Jorstad et al., 2022).

In macaque cortex there are numerous instances of within-layer intermingling of FF and FB neurons, but vanishingly small instances of neurons that emit axons that both ascend and descend the cortical hierarchy, despite the ubiquity of bifurcation in the primate cortical network (Kennedy and Bullier, 1985; Markov et al., 2014a). Such a segregation could indicate that the function of the FB and FF pathways is an intrinsic feature of the FF and FB projection neuron, thus requiring a dedicated cell type. Testing this hypothesis and developing the tools to manipulate FF and FB cell types are now well within the technical capacity and mission of the BRAIN initiative Cell Atlas Network (BICAN) (Kaiser, 2022). The development of this line of research will have a profound impact on our understanding of the primate cortical hierarchy. These considerations point to area V1 being an ideal model system where parallel studies exploiting invasive investigations in the nonhuman primate and high-resolution fMRI in humans can make important contributions to exploring computational models of the cortex and better defining what makes us human.

Given the differences in the signatures of the top-down signals in supra and infragranular layers in human subjects (Dijkstra and Feming, 2023; Bergmann et al., 2019; Thomas et al., 2023; Muckli et al., 2015), perhaps the outstanding question that needs to be addressed in NHP, is how and where the infra-FB and supra- FB streams interact with the ascending pathways? According to the dual counterstream model the principal targets of the FB streams are concentrated in the compartment corresponding to the layer of origin, although determining the incidence of crossing between supra and infragranular layers is important. Layer 1 of the neocortex has long been considered as a specialized layer for top-down convergence of FB on the ascending pathways (Cauller, 1995; Larkum, 2013) where predominantly supra-FB axons contacting the apical dendritic tufts of L2, L3 and L5 pyramidal cells are known to gate ascending activity (Doron et al., 2020; Takahashi et al., 2020). This will involve carrying out in NHP a quantification of inputs to L1 in a similar manner as in mouse (Ledderose et al., 2023). The main target of infra-FB is L6 (Markov et al., 2014b). On-going experiments with viral tracers and cell type identification completed with electrophysiology will allow us determine the interactions of infra-FB with the ascending projections.

REFERENCES

Abbott LF, Chance FS (2005). Drivers and modulators from push-pull and balanced synaptic input. *Progress Brain Res.* 149:147–155.

Allman J, Miezin R, McGuinness E (1985). Stimulus specific responses from beyond the classical receptive field: Neurophysiological mechanisms for local-global comparisons in visual neurons. *Annu Rev Neurosci.* 8:407–430.

Angelucci A, Bijanzadeh M, Nurminen L, Federer F, Merlin S, Bressloff PC (2017). Circuits and mechanisms for surround modulation in visual cortex. *Annu Rev Neurosci.* 40:425–451.

Angelucci A, Levitt JB, Walton EJ, Hupe JM, Bullier J, Lund JS (2002). Circuits for local and global signal integration in primary visual cortex. *J Neurosci.* 22:8633–8646.

Barone P, Batardiere A, Knoblauch K, Kennedy H (2000). Laminar distribution of neurons in extrastriate areas projecting to visual areas V1 and V4 correlates with the hierarchical rank and indicates the operation of a distance rule. *J Neurosci.* 20:3263–3281.

Bastos AM, Usrey WM, Adams RA, Mangun GR, Fries P, Friston KJ (2012). Canonical microcircuits for predictive coding. *Neuron.* 76:695–711.

Bastos AM, Vezoli J, Bosman CA, Schoffelen JM, Oostenveld R, Dowdall JR, De Weerd P, Kennedy H, Fries P (2015). Visual areas exert feedforward and feedback influences through distinct frequency channels. *Neuron.* 85:390–401.

Berezovskii VK, Nassi JJ, Born RT (2011). Segregation of feedforward and feedback projections in mouse visual cortex. *J Comp Neurol.* 519:3672–3683.

Bergmann J, Morgan AT, Muckli L (2019). Two distinct feedback codes in V1 for 'real' and 'imaginary' internal experiences. bioRxiv. 664870; doi: https://doi.org/10.1101/664870

Brincat SL, Siegel M, von Nicolai C, Miller EK (2018). Gradual progression from sensory to task-related processing in cerebral cortex. *Proc Natl Acad Sci U S A.* 115:e7202–e7211.

Buffalo EA, Fries P, Landman R, Buschman TJ, Desimone R (2011). Laminar differences in gamma and alpha coherence in the ventral stream. *Proc Natl Acad Sci U S A.* 108:11262–11267.

Bullier J (2006). What is fed back? In: van Hemmen JL, Sejnowski TJ, eds. *23 Problems in Systems Neuroscience.* Oxford University Press:103–132.

Cauller L (1995). Layer I of primary sensory neocortex: where top-down converges upon bottom-up. *Behav Brain Res.* 71:163–170.

Chaudhuri R, Knoblauch K, Gariel MA, Kennedy H, Wang XJ (2015). A large-scale circuit mechanism for hierarchical dynamical processing in the primate cortex. *Neuron.* 88:419–431.

Coogan TA, Burkhalter A (1993). Hierarchical organization of areas in rat visual cortex. *J Neurosci.* 13:3749–3772.

Cossell L, Iacaruso MF, Muir DR, Houlton R, Sader EN, Ko H, Hofer SB, Mrsic-Flogel TD (2015). Functional organization of excitatory synaptic strength in primary visual cortex. *Nature.* 518:399–403.

Covic EN, Sherman SM (2011). Synaptic properties of connections between the primary and secondary auditory cortices in mice. *Cereb Cortex.* 21:2425–2441.

D'Souza RD, Burkhalter A (2017). A laminar organization for selective cortico-cortical communication. *Front Neuroanat.* 11:71.

De Pasquale R, Sherman SM (2011). Synaptic properties of corticocortical connections between the primary and secondary visual cortical areas in the mouse. *J Neurosci.* 31:16494–16506.

Debes SR, Dragoi V (2023). Suppressing feedback signals to visual cortex abolishes attentional modulation. *Science.* 379:468–473.

Dehay C, Kennedy H (2020). Evolution of the human brain. Science. 369:506–507.

Dijkstra N, Fleming SM (2023). Subjective signal strength distinguishes reality from imagination. *Nat Commun.* 14:1627.

Doron G, Shin JN, Takahashi N, Druke M, Bocklisch C, Skenderi S, de Mont L, Toumazou M, Ledderose J, Brecht M, Naud R, Larkum ME (2020). Perirhinal input to neocortical layer 1 controls learning. *Science.* 370:eaaz3136.

Douglas RJ, Koch C, Mahowald M, Martin KA, Suarez HH (1995). Recurrent excitation in neocortical circuits. *Science.* 269:981–985.

Douglas RJ, Martin KAC (1991). A functional microcircuit for cat visual cortex. *J Physiol London.* 440:735–769.

Ekstrom LB, Roelfsema PR, Arsenault JT, Kolster H, Vanduffel W (2009). Modulation of the contrast response function by electrical microstimulation of the macaque frontal eye field. *J Neurosci.* 29:10683–10694.

Elston GN (2000). Pyramidal cells of the frontal lobe: All the more spinous to think with. *J Neurosci.* 20:RC95 (1–4).

Ercsey-Ravasz M, Markov NT, Lamy C, Van Essen DC, Knoblauch K, Toroczkai Z, Kennedy H (2013). A predictive network model of cerebral cortical connectivity based on a distance rule. *Neuron.* 80:184–197.

Feldman H, Friston KJ (2010). Attention, uncertainty, and free-energy. *Front Hum Neurosci.* 4:215.

Felleman DJ, Van Essen DC (1991). Distributed hierarchical processing in the primate cerebral cortex. *Cereb Cortex.* 1:1–47.

Fries P (2005). A mechanism for cognitive dynamics: neuronal communication through neuronal coherence. *Trends Cogn Sciences.* 9:474–480.

Friston K (2018). Does predictive coding have a future? *Nat Neurosci.* 21:1019–1021.

Froudist-Walsh S, Bliss DP, Ding X, Rapan L, Niu M, Knoblauch K, Zilles K, Kennedy H, Palomero-Gallagher N, Wang XJ (2021). A dopamine gradient controls access to distributed working memory in the large-scale monkey cortex. *Neuron.* 109:3500–3520 e3513. 10.1016/j.neuron.2021.08.024.

Gilbert CD, Wiesel TN (1985). Intrinsic connectivity and receptive field properties in visual cortex. *Vision Res.* 25:365–374.

Harris KD, Mrsic-Flogel TD (2013). Cortical connectivity and sensory coding. *Nature.* 503:51–58.

Hartline HK (1938). The response of single optic nerve fibers of the vertebrate eye to illumination of the retina. *Am J Physiol.* 121:400–415.

Hayashi T, Kennedy H, Van Essen DC (2022). A spatially embedded cortical connectome reveals complex transformations. *Neuron.* 110:185–187.

Hilgetag CC, O'Neill MA, Young MP (1996). Indeterminate organization of the visual system. *Science.* 271:776–777.

Honey CJ, Thesen T, Donner TH, Silbert LJ, Carlson CE, Devinsky O, Doyle WK, Rubin N, Heeger DJ, Hasson U (2012). Slow cortical dynamics and the accumulation of information over long timescales. *Neuron.* 76:423–434.

Hubel DH, Wiesel TN (1968). Receptive fields and functional architecture of monkey striate cortex. *J Physiol.* 195:215–243.

Hupe JM, James AC, Payne BR, Lomber SG, Girard P, Bullier J (1998). Cortical feedback improves discrimination between figure and background by V1, V2 and V3 neurons. *Nature.* 394:784–787.

Jia K, Goebel R, Kourtzi Z (2023). Ultra-high field imaging of human visual cognition. *Annu Rev Vis Sci.* 9:14.1–14.22.

Jo HG, Ito J, Schulte Holthausen B, Baumann C, Grün S, Habel U, Kellermann T (2019). Task-dependent functional organizations of the visual ventral stream. *Sci Rep.* 9:9316.

Jordan R, Keller GB (2020). Opposing influence of top-down and bottom-up input on excitatory layer 2/3 neurons in mouse primary visual cortex. *Neuron.* 108:1194–1206 e5.

Jorstad NL, Song JH, Exposito-Alonso D, Suresh H, Castro N, Krienen FM, Yanny AM, Close J, Gelfand E, Travaglini KJ (2022). Comparative transcriptomics reveals human-specific cortical features. bioRxiv, 2022.2009. 2019.508480.

Kaiser J (2022). Next goal for U.S. initiative: A human brain atlas. *Science.* 377:1369.

Kennedy H, Bullier J (1985). A double-labeling investigation of the afferent connectivity to cortical areas V1 and V2 of the macaque monkey. *J Neurosci.* 5:2815–2830.

Knoblauch K, Ercsey-Ravasz M, Kennedy H, Toroczkai Z (2016). The brain in space. In: Kennedy H, Van Essen D, Christen Y, eds. *Micro-, Meso- and Macro-Connectomics of the Brain.* Springer:45–74.

Larkum ME (2013). The yin and yang of cortical layer 1. *Nat Neurosci.* 16:114–115.

Ledderose JM, Zolnik TA, Toumazou M, Trimbuch T, Rosenmund C, Eickholt BJ, Jaeger D, Larkum ME, Sachdev RN (2023). Layer 1 of somatosensory cortex: An important site for input to a tiny cortical compartment. *bioRxiv.* 2021.2011.2026.469979.

Lee TS, Mumford D (2003). Hierarchical Bayesian inference in the visual cortex. *J Opt Soc Am A Opt Image Sci Vis.* 20:1434–1448.

Luck SJ, Chelazzi L, Hillyard SA, Desimone R (1997). Neural mechanisms of spatial selective attention in areas V1, V2, and V4 of macaque visual-cortex. *J Neurophysiol.* 77:24–42.

Lupyan G (2017). Changing what you see by changing what you know: The role of attention. *Front Psychol.* 8:553.

Marić M, Domijan D (2022). Dual counterstream architecture may support separation between vision and predictions. *Conscious Cogn.* 103:103375. 10.1016/j.concog.2022.103375.

Markov NT, Ercsey-Ravasz M, Van Essen DC, Knoblauch K, Toroczkai Z, Kennedy H (2013). Cortical high-density counter-stream architectures. *Science.* 342:1238406.

Markov NT, Ercsey-Ravasz MM, Ribeiro Gomes AR, Lamy C, Magrou L, Vezoli J, Misery P, Falchier A, Quilodran R, Gariel MA, Sallet J, Gamanut R, Huissoud C, Clavagnier S, Giroud P, Sappey-Marinier D, Barone P, Dehay C, Toroczkai Z, Knoblauch K, Van Essen DC, Kennedy H (2014a). A weighted and directed interareal connectivity matrix for macaque cerebral cortex. *Cereb Cortex.* 24:17–36.

Markov NT, Vezoli J, Chameau P, Falchier A, Quilodran R, Huissoud C, Lamy C, Misery P, Giroud P, Barone P, Dehay C, Ullman S, Knoblauch K, Kennedy H (2014b). The anatomy of hierarchy: Feedforward and feedback pathways in macaque visual cortex. *J Comp Neurol.* 522:225–259.

Michalareas G, Vezoli J, van Pelt S, Schoffelen JM, Kennedy H, Fries P (2016). Alpha-beta and gamma rhythms subserve feedback and feedforward influences among human visual cortical areas. *Neuron.* 89:384–397.

Mignard M, Malpeli JG (1991). Paths of information-flow through visual cortex. *Science.* 251:1249–1251.

Minamisawa G, Kwon SE, Chevée M, Brown SP, O'Connor DH (2018). A non-canonical feedback circuit for rapid interactions between somatosensory cortices. *Cell Rep.* 23:2718–2731.

Muckli L, De Martino F, Vizioli L, Petro LS, Smith FW, Ugurbil K, Goebel R, Yacoub E (2015). Contextual feedback to superficial layers of V1. *Curr Biol.* 25:2690–2695.

Mumford D (1992). On the computational architecture of the neocortex. II. The role of cortico-cortical loops. *Biol Cybern.* 66:241–251.

Parr T, Pezzulo G, Friston KJ (2022). *Active Inference: The Free Energy Principle in Mind, Brain, and Behavior.* MIT Press.

Pettine WW, Steinmetz NA, Moore T (2019). Laminar segregation of sensory coding and behavioral readout in macaque V4. *Proc Natl Acad Sci U S A. 116*:14749–14754.

Rao RP, Ballard DH (1999). Predictive coding in the visual cortex: A functional interpretation of some extra-classical receptive-field effects. *Nat Neurosci.* 2:79–87.

Rockland KS, Pandya DN (1979). Laminar origins and terminations of cortical connections of the occipital lobe in the rhesus monkey. *Brain Res.* 179:3–20.

Roe AW, Chelazzi L, Connor CE, Conway BR, Fujita I, Gallant JL, Lu H, Vanduffel W (2012). Toward a unified theory of visual area V4. *Neuron.* 74:12–29.

Salin PA, Girard P, Kennedy H, Bullier J (1992). Visuotopic organization of corticocortical connections in the visual system of the cat. *J Comp Neurol.* 320:415–434.

Semedo JD, Jasper AI, Zandvakili A, Krishna A, Aschner A, Machens CK, Kohn A, Yu BM (2022). Feedforward and feedback interactions between visual cortical areas use different population activity patterns. *Nat Commun.* 13:1099.

Shen S, Jiang X, Scala F, Fu J, Fahey P, Kobak D, Tan Z, Zhou N, Reimer J, Sinz F, Tolias AS (2022). Distinct organization of two cortico-cortical feedback pathways. *Nat Commun.* 13:6389.

Sherman SM, Guillery RW (1998). On the actions that one nerve cell can have on another: Distinguishing "drivers" from "modulators." *Proc Natl Acad Sci U S A.* 95:7121–7126.

Shipp S (2016). Neural elements for predictive coding. *Front Psychol.* 7:1792.

Siegle JH, Jia X, Durand S, Gale S, Bennett C, Graddis N, Heller G, Ramirez TK, Choi H, Luviano JA, Groblewski PA, Ahmed R, Arkhipov A, Bernard A, Billeh YN, Brown D, Buice MA, Cain N, Caldejon S, Casal L, Cho A, Chvilicek M, Cox TC, Dai K, Denman DJ, de Vries SEJ, Dietzman R, Esposito L, Farrell C, Feng D, Galbraith J, Garrett M, Gelfand EC, Hancock N, Harris JA, Howard R, Hu B, Hytnen R, Iyer R, Jessett E, Johnson K, Kato I, Kiggins J, Lambert S, Lecoq J, Ledochowitsch P, Lee JH, Leon A, Li Y, Liang E, Long F, Mace K, Melchior J, Millman D, Mollenkopf T, Nayan C, Ng L, Ngo K, Nguyen T, Nicovich PR, North K, Ocker GK, Ollerenshaw D, Oliver M, Pachitariu M, Perkins J, Reding M, Reid D, Robertson M, Ronellenfitch K, Seid S, Slaughterbeck C, Stoecklin M, Sullivan D, Sutton B, Swapp J, Thompson C, Turner K, Wakeman W, Whitesell JD, Williams D, Williford A, Young R, Zeng H, Naylor S, Phillips JW, Reid RC, Mihalas S, Olsen SR, Koch C (2021). Survey of spiking in the mouse visual system reveals functional hierarchy. *Nature.* 592:86–92.

Siu C, Balsor J, Merlin S, Federer F, Angelucci A (2021). A direct interareal feedback-to-feedforward circuit in primate visual cortex. *Nat Commun.* 12:4911.

Takahashi N, Ebner C, Sigl-Glockner J, Moberg S, Nierwetberg S, Larkum ME (2020). Active dendritic currents gate descending cortical outputs in perception. *Nat Neurosci.* 23:1277–1285.

Thomas ER, Haarsma J, Nicholson J, Yon D, Kok P, Press C (2023). Predictions and errors are distinctly represented across V1 layers. *bioRxiv.* 2023.2007.2011.548408.

van Kerkoerle T, Self MW, Roelfsema PR (2017). Layer-specificity in the effects of attention and working memory on activity in primary visual cortex. *Nat Commun.* 8:13804.

Vangeneugden J, van Beest EH, Cohen MX, Lorteije JAM, Mukherjee S, Kirchberger L, Montijn JS, Thamizharasu P, Camillo D, Levelt CN, Roelfsema PR, Self MW, Heimel JA (2019). Activity in lateral visual areas contributes to surround suppression in awake mouse V1. *Curr Biol.* 29:4268–4275.

Vezoli J, Magrou L, Goebel R, Wang XJ, Knoblauch K, Vinck M, Kennedy H (2021). Cortical hierarchy, dual counterstream architecture and the importance of top-down generative networks. *NeuroImage.* 225:117479.

Von Helmholtz H (1867). *Handbuch der physiologischen Optic.* Voss.

Walsh KS, McGovern DP, Clark A, O'Connell RG (2020). Evaluating the neurophysiological evidence for predictive processing as a model of perception. *Ann N Y Acad Sci.* 1464:242–268.

Wang M, Hou Y, Magrou L, Autio JA, Misery P, Coalson T, Reid E, Xu Y, Lamy C, Falchier A, Zhang Q, Poo MM, Dehay C, Glasser MF, Hayashi T, Knoblauch K, Vand Essen D, Shen Z, Kennedy H (2022). Retinotopic organization of feedback projections in primate early visual cortex: implications for active vision. *bioRxiv.* 10.1101/2022.04.27.489651.

Zeki S, Shipp S (1988). The functional logic of cortical connections. *Nature.* 335:311–317.

Zhang S, Xu M, Kamigaki T, Hoang Do JP, Chang WC, Jenvay S, Miyamichi K, Luo L,Dan Y (2014). Selective attention: Long-range and local circuits for top-down modulation of visual cortex processing. *Science.* 345:660–665.

38.

FEEDFORWARD AND FEEDBACK CONNECTIONS

FUNCTIONAL CONNECTIVITY, SYNAPTIC PHYSIOLOGY, AND FUNCTION

Alessandra Angelucci and Leopoldo Petreanu

ANATOMICAL ORGANIZATION OF FEEDFORWARD AND FEEDBACK CONNECTIONS

LAMINAR SPECIFICITY AND CORTICAL HIERARCHY

The long-held proposal that the cortex is organized into a hierarchy of areas was first based on anatomical studies of the primate visual cortex. Rockland and Pandya (1979), by labeling reciprocal connections between macaque visual cortical areas, using the bidirectional tracer horseradish peroxidase, proposed that the differences in the laminar origin and termination between inter-areal connections could indicate the flow of information processing between areas. Feedforward (FF) connections, defined as arising predominantly from superficial layers and terminating in layer (L)4 and lower L3, carry information from lower-order areas to high-order areas. Feedback (FB) connections, instead, defined as originating predominantly in deep layers and terminating outside L4, carry top-down information. Subsequently, Felleman and Van Essen (1991) classified most connections throughout the primate visual system as either FF or FB, and used this qualitative classification to arrange visual areas in hierarchical order. The primary visual cortex (V1 or area 17) was at the lowest hierarchical level, and extrastriate cortical areas (>32 areas) were arranged at different higher levels (up to 10). Inter-areal connections in the cat (Symonds and Rosenquist, 1984) and rat (Coogan and Burkhalter, 1993) visual system, and in other sensory systems (Galaburda and Pandya, 1983; Friedman et al., 1986), were similarly classified as FF or FB. Subsequently, more sensitive neuroanatomical tracing methods (fluorescent tracers, PHA-L, dextrans, and viral vectors carrying the genes for reporter fluorescent proteins) revealed that, in fact, most FF and FB connections arise from both superficial and deep layers, but in different proportions depending on both the information flow (FF vs. FB) and the hierarchical distance between areas. The proportion of neurons of origin in superficial layers is higher in lower-order areas and decreases monotonically in higher-order areas (see Figure 37.3 in **Chapter 37 of Section VII**). This has been used as a quantitative metrics of hierarchical distance between areas (Markov et al., 2014). None of these previous studies investigated the differential contribution of superficial-versus deep-layer FF and FB projections to superficial- and deep-layer terminations; thus it remains unknown whether FF and FB systems are heterogeneous with respect to laminar connectivity and function. Recent studies in mouse are beginning to tackle this question using genetic approaches.

Mapping of the mouse inter-areal connections using anterograde tracers has revealed that the mouse visual cortex is composed of over 10 areas, each having a topographic representation of visual space (Wang and Burkhalter, 2007). These areas comprehensively interconnect, more so than in primates, forming an ultra-dense network, but the strength varies considerably across connections (Gamanut et al., 2018). Inter-areal projections in mouse originate in all layers, except L1 (Young et al., 2021; Yao et al., 2022). Their classification as FF or FB, thus, has been based mainly on laminar termination: FF projections terminate in all layers but more sparsely in L1, while FB projections terminate in L1, L5/6, and weakly in L2 and L4 (D'Souza et al., 2022; Harris et al., 2019). Thus, unlike in higher mammals, L4 in mouse directly participates in inter-areal communication. Interestingly, the laminar termination patterns differ across genetically defined cell types, identifying 6 FF and 3 FB projection types, each possibly subserving different roles in inter-areal communication (Harris et al., 2019). Several metrics have been used to classify inter-areal connections as FF or FB, and to organize the mouse visual areas into a multilevel hierarchy. The ratio of the afferent axons' density in L1 relative to that in L2–4 provides a graded hierarchical index, resulting in 2 hierarchical streams, similar to the primate dorsal and ventral visual pathways (see below), with V1 located at the lower level, and some areas at similar hierarchical levels (D'Souza et al., 2022). A similar hierarchical organization has also been derived using as a metric the ratio of projections originating in superficial layers (L2–4) versus L5 (Yao et al., 2022), as used in primates. All of these metrics indicate that the hierarchy of mouse visual areas is shallower than in primates and cats and comprises fewer levels. (For an in-depth discussion of cortical hierarchies, see **Chapter 37 in Section VII**.)

AREA SPECIFICITY

Most inter-areal projection neurons are either FF or FB, but not both. In monkey visual cortex, the cells of origin of FF and FB projections are largely segregated, with FF projection neurons occupying preferentially L3B and L5, and FB neurons L2/3A and L6. Moreover, injections of distinct retrograde tracers in a pair of cortical areas, one receiving FF, the other FB projections from the same source area, result in few double-labeled neurons (1%–5%) in that source area, in both macaque (Markov et al., 2014) and mouse (Berezovskii et al., 2011). (The separation of FF and FB connections is central to *predictive coding* theories of vision, and is discussed in detail in **Chapter 41 of Section VII**.)

In primates and cats, dual-area retrograde tracer injections have demonstrated that FF projections are highly area specific; that is, individual neurons typically project to a single higher cortical area and only rarely to multiple areas (Bullier et al., 1984). For example, although primate V1 projects to both V2 and V5/MT, only a minute fraction of V1 cells projects to both areas (Sincich and Horton, 2003). Similarly, single V2 neurons project to V4 or area TEO, but rarely both (Nakamura et al., 1993).

Compared to primates, in mouse there is comparatively stronger evidence, based on single axon arbor reconstructions (Winnubst et al., 2019) and dual-area retrograde tracer injections (Kim et al., 2020; Young et al., 2021), that a substantial fraction of FF-projecting V1 neurons innervate multiple areas (see Figure 41.4 in **Chapter 41 of Section VII**). Moreover, high-throughput DNA sequencing of genetically barcoded neurons (MAPseq) has revealed that approximately half of V1 neurons in mouse target multiple higher visual areas (Han et al., 2018); however, the combinations of areas innervated by individual neurons are non-random, suggesting that subclasses of projection neurons broadcast signals to defined groups of areas. To what extent FF connections are significantly more area-specific in primates than in mouse will need to be re-investigated using high-throughput methods such as MAPseq, which, unlike tracer injections, probe connections to a much larger number of target areas. (For a discussion of this point, see also **Chapter 41 in Section VII**).

Whether FB connections in higher mammals show similar area specificity as FF connections is less clear. On the one hand, reconstruction of single FB neurons and dual-area retrograde tracer injections have demonstrated a substantial amount (up to 20%–30%) of FB neuron bifurcations in both cat (Bullier et al., 1984) and primate visual cortex (Kennedy and Bullier, 1985; Rockland and Knutson, 2000), particularly for deep-layer FB neurons. On the other hand, divergent FB connections to multiple areas do not imply that individual V1 neurons receive converging FB inputs from multiple areas. Rabies virus–mediated labeling of brain-wide monosynaptic inputs to V1 neurons projecting to V2 has revealed that these specific neurons receive direct FB inputs exclusively from V2, but not from other extrastriate areas that also send FB to V1 (Siu et al., 2021) (Figure 38.1A). Interestingly, these monosynaptic FB-to-FF neuron contacts arise from V2 L2/3 and L5, but not L6, suggesting that while L5-FB neurons directly and selectively contact V1 neurons sending FF projections to their home area, L6-FB neurons likely contact local, rather than FF-projecting, V1 neurons. These results suggest that FB is a heterogenous entity, consisting of subsystems with distinct functions.

In mice, as in monkeys, a fraction of FB projections innervates other visual areas besides V1 (Han et al., 2018; Winnubst et al., 2019; Young et al., 2021), and V1 projection neurons receive direct FB inputs predominantly (but not exclusively, as in monkey), from the higher-visual areas to which they project (Kim et al., 2020; Young et al., 2021) (Figure 38.1B).

CELLS OF ORIGIN AND SYNAPTIC TARGETS

Inter-areal projections originate from pyramidal neurons, except for small subpopulations originating from long-range GABAergic interneurons and, in cat and monkey V1, spiny stellate cells in subdivisions of L4. Despite mainly consisting of glutamatergic afferents, FF and FB inputs can evoke both inhibition and excitation in postsynaptic areas depending on the cell types they innervate. In rodent and primate visual cortex, 80%–85% of contacts made by most inter-areal FF and FB inputs are with excitatory neurons and 15%–20% with inhibitory neurons (Gonchar and Burkhalter, 2003; Anderson and Martin, 2006, 2009; Yang et al., 2013). EM studies have further shown that synaptic type, size, and morphology are similar for FF and FB projections. Therefore, what may account for the known asymmetry in their physiological effects are differences in laminar termination and the location of their contacts on the postsynaptic neurons (soma and proximal dendrites vs. apical dendrites) (Petreanu et al., 2009). In mouse, V1 FF afferents to pyramidal neurons in L2/3 mainly target perisomatic dendrites, while FB connections mainly target

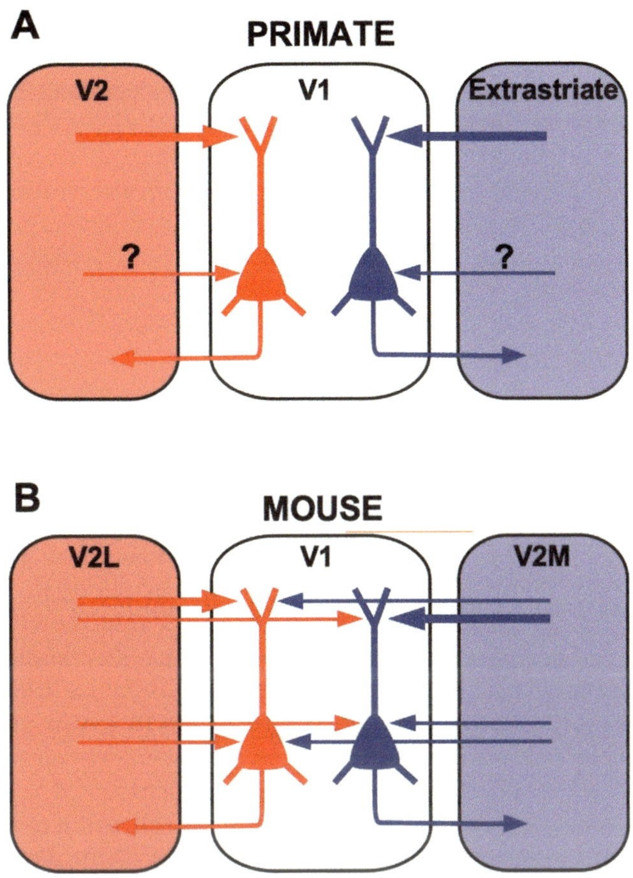

Figure 38.1. Area-specificity of FF and FB pathways. Schematics of V1 FF and FB circuit motifs in primate (A) and mouse (B). Triangles = pyramidal neuron somata; arrows = axonal projections. Arrow thickness indicates projection magnitude. (A). In the primate, most V1 FF neurons project to a single area. FB neurons make monosynaptic contacts with V1 neurons sending reciprocal FF projections to their home area, but not with other V1 neurons projecting to other extrastriate visual areas (Based on data from Siu et al., *2021*). "?": it is unclear whether FB onto the perisomatic dendrites is weaker than to apical dendrites, but this could depend on the layer of termination, as FB projections are as dense in L1 as they are in L5/6 (Federer et al., 2021), but sparser in L2/3 and 4B. (B). In mouse, neurons that project to a single area receive stronger FB inputs from the area to which they project, and weaker inputs from other areas (Kim et al., 2020; Young et al., 2021). FB innervates more strongly the apical dendritic tufts of reciprocating L5 FF neurons; FB inputs onto the perisomatic dendrites are of similar magnitude on reciprocating vs. non-reciprocating V1 neurons. V2L = lateral visual areas; V2M = medial visual areas. Based on data from: Kim et al. (2020), and Young et al. (2021).

apical dendrites in L1 (Figure 38.1B) (Young et al., 2021). Both FF and FB pathways innervate multiple dendritic compartments of L5 pyramids. FF inputs innervate the perisomatic dendrites and trunk of apical dendrites of L5 pyramids, while FB inputs innervate the perisomatic and distal apical dendrites in L1 (Yang et al., 2013; Young et al., 2021) (Figure 38.1B). These multiple innervation sites suggest that FF and FB inputs might be initially integrated independently by different neuronal compartments.

An additional difference between FF and FB that may account for their differential physiological effects is the inhibitory neuron subtypes that they target. In mouse visual and somatosensory cortices, both FF and FB connections innervate all 3 major classes of cortical inhibitory neurons, the parvalbumin- (PV), vasoactive intestinal peptide- (VIP) and somatostatin-positive (SOM) neurons. However, the relative strength of cortico-cortical inputs on these interneuron classes varies across FF and FB pathways, FF inputs activating preferentially PV neurons and FB inputs VIP neurons, a motif that generalizes across areas at various hierarchical stages (Ma et al., 2021). As PV cells mainly contact the somata and proximal dendrites of pyramidal and other PV cells, whereas VIP neurons disinhibit pyramidal neurons via contacts with SOM cells, this connectivity motif suggests that FF connections sharpen sensory responses via feedforward inhibition, while FB connections enhance them via a disinhibitory mechanism. However, the inhibitory neuron types preferentially contacted by inter-areal connections are also area- and layer dependent. For example, in mouse V1, FB inputs from the cingulate cortex target preferentially VIP cells (Zhang et al., 2014; Ma et al., 2021), but those from the lateromedial visual area (LM) target preferentially SOM cells (Shen et al., 2022), and in mouse somatosensory cortex, FB inputs to VIP and SOM cells arise predominantly from deep and superficial layers, respectively (Wall et al., 2016). This heterogeneity suggests that FB inputs from different sources might be differentially integrated in V1 and uniquely affect V1 responses.

Inter-areal projections also target a wide array of excitatory cell types. While FF and FB projections synapse onto most pyramidal cell types, they selectively innervate neurons with specific projection patterns in visual cortex (Kim et al., 2015; Zhang et al., 2016; Young et al., 2021). For example, FF and FB inputs between areas V1 and LM are stronger onto neurons projecting to other cortical areas or the striatum (intratelencephalic neurons), than onto subcortically-projecting neurons in the same layer (Young et al., 2021). Moreover, as discussed above and illustrated in Figure 38.1, FB connections from higher visual areas contact preferentially (mouse; Kim et al., 2020; Young et al., 2021) or exclusively (monkey; Siu et al., 2021) V1 neurons sending reciprocating FF projections to their home area. Similarly, V1 FF inputs to higher areas are stronger onto reciprocating FB neurons than onto neurons within the same layer that project to other areas (Young et al., 2021). Thus, inter-areal connections form monosynaptic FF-FB loops between interconnected areas. Brain-wide monosynaptic inputs to different inhibitory and excitatory cell types were recently mapped for 10 mouse visual cortical areas using rabies viruses (Yao et al., 2022). This study showed that, while the areal presynaptic input patterns are partly determined by cell type, target area and target layer had comparatively more influence.

SYNAPTIC PHYSIOLOGY

The differential impact, efficacy, and synaptic properties of FF and FB inputs has been studied primarily in brain slice preparations of rodent visual cortex using electrical stimulation, glutamate uncaging or, more recently, optogenetic photostimulation of axons. Stimulation of FF and FB projections evokes AMPA- and NMDA-mediated depolarizations in pyramidal neurons, indicating they are excitatory (Dong et al., 2004). Repeated electrical stimulation of excitatory inputs from V1 to higher visual areas evokes facilitating excitatory postsynaptic potentials (EPSPs) in some neurons and depressing ones in others in the same layer (De Pasquale and Sherman, 2011). This diversity in synaptic properties depends on both the laminar position of the FF inputs and the postsynaptic cell type. For example, repeated stimulation of V1 FF inputs to anterior cingulate and retrosplenial cortices evokes depressing EPSPs in pyramidal, PV, and SOM neurons, but facilitating EPSPs in VIP interneurons. Conversely, FB inputs from these areas are facilitating on SOM V1 neurons and depressing on all other cell types (Ma et al., 2021). Thus, in visual cortex, it is a combination of targeted cell type and the specific short-term plasticity properties of the input type that determines the differential impact of FF and FB inputs. In contrast, in somatosensory cortex, FF- and FB-evoked responses of postsynaptic neurons generally show depression upon repeated stimulation, but their effect on SOM neurons can also be facilitating (Naskar et al., 2021). However, a recent study found cortico-cortical inputs to be facilitating onto both excitatory and all types of inhibitory neurons in distant areas, but depressing within the local circuits (Martinetti et al., 2022). Thus, it remains unclear to what extent the short-term plasticity properties of FF and FB inputs across areas and projection types follow common or specialized rules. Further investigation is required to determine general rules of synaptic physiology for FF and FB connections.

The differential impact of FF and FB connections onto the postsynaptic cell also depends on the cellular compartments they differentially innervate (see also preceding section). FB contacts onto the apical dendrites of pyramidal cells in L1 are attenuated and temporally filtered, and are much less efficient at driving postsynaptic neurons than those terminating onto proximal dendrites, such as FF inputs (Williams and Stuart, 2002). Optogenetic stimulation of FB inputs in brain slices can drive interneurons, but rarely pyramidal neurons (Shen et al., 2022), consistent with a modulatory role. Rather than directly driving neurons, FB inputs to L1 can modify how proximal inputs are integrated. This can result in distal FB inputs regulating the gain (Larkum, 2013) or temporal integration window (Shen et al., 2022) of FF excitation. In addition, by inducing NMDA receptor-mediated entrance of calcium and disinhibiting distal dendrites trough the activation of VIP interneurons, FB can induce plasticity in postsynaptic neurons. This might allow optimization of synaptic weight in lower-order areas based on the activity in the higher-order area.

Several studies have investigated the overall impact of FF and FB connections onto the inhibition/excitation balance of their target areas. Stimulation of V1 FF connections evokes EPSPs followed by strong IPSPs (disynaptic inhibition) in most (90%) cells in rat higher-visual areas. In contrast, stimulation of FB pathways evokes purely depolarizing monosynaptic responses in most (80%) V1 cells (Shao and Burkhalter, 1996). Similarly, in L2/3 of mouse LM, FF inputs from V1 evoke stronger disynaptic inhibition than the reciprocal FB inputs (Dong et al., 2004). Thus, the relative balance of excitation and inhibition induced by the different circuits is somewhat different, with FB modulating V1 responses via synaptic mechanisms in which weak inhibition is strongly dominated by excitation. Subcellular ChR2-assisted circuit mapping further demonstrated that the stronger inhibition in the FF versus FB pathway is due to stronger FF than FB inputs onto PV interneurons versus excitatory neurons (Yang et al., 2013). D'Souza et al. (2016) found that the relative recruitment of excitation and PV-mediated inhibition by inter-areal connections depends not only on the flow of information (FF vs. FB), but also on the hierarchical distance between source and target areas, as well as the laminar location of the target neurons. Specifically, in upper, but not lower, layers, inter-areal inputs to PV neurons are stronger than to excitatory pyramidal neurons; this inhibitory/excitatory asymmetry is greater in FF than FB pathways and increases (decreases) with hierarchical distance for the FF (FB) pathway.

Overall, the synaptic physiology and connectivity suggest that FF connections drive high-order neurons while also eliciting feedforward inhibition. On the other hand, FB inputs, which cannot directly drive postsynaptic neurons, may modulate and drive plastic changes in their postsynaptic partners.

FUNCTIONAL ORGANIZATION OF FEEDFORWARD AND FEEDBACK CONNECTIONS

Understanding how inter-areal connections are functionally organized is necessary to understand how they carry out specific computations. In this section we review studies that have examined the spatial extent and patterning of inter-areal connections in relation to the cortical functional maps.

SPATIAL ORGANIZATION

A fundamental organizational feature of the mammalian sensory cortex, particularly of areas at lower hierarchical levels, is the orderly representation of the sensory periphery, so that each area contains a "topographic map." Studies of the visual system have taken advantage of this topographic organization of cortex to determine the extent of visual field communicated by a small population of FF and FB neurons to a target area (the divergence region), or received by this small neuronal population from other areas (the convergence region). In cat and monkey visual cortex, the convergence and divergence regions of both FF and FB connections both span several millimeters (Salin and Bullier, 1995). Moreover, the cells of origin of FF projections (the convergence region) and the axon terminals of the FB projections (the divergence region) labeled by a small bidirectional tracer injection in a higher area are of the same extent and occupy the same cortical region in the lower area. This spatial reciprocity of FF and FB connections exists in the visual system of several species, including cat (Salin et al., 1995), rat (Johnson and Burkhalter, 1997), and monkey (Angelucci et al., 2002), and suggests that FB may be able to control directly and rapidly the FF neurons in the target region that provide their excitatory drive. This hypothesis is further supported by the recent discovery of area-specific monosynaptic FB-to-FF contacts (see above and Figure 38.1).

In cat and monkey, combining tracer injections with electrophysiological RF mapping at the injected site and in the cortical region of expected tracer transport revealed that both FF and FB connections are visuotopically organized, but FB is more divergent and convergent (Salin and Bullier, 1995; Angelucci et al., 2002). While both FF and FB connections connect group of neurons in 2 areas looking collectively at the same region of visual space (Figure 38.2A–B, left panels), compared to FF connections, FB connections provide their target neurons with a much larger region of visual field than the target neuron's RF size (Figure 38.2A–B). Angelucci et al. (2002) further demonstrated in macaque visual cortex that the widespread FB connections from higher cortical areas have similar spatial extent as the RF surround of V1 neurons (Figure 38.2B, right panel; see also Figure 37.1 in **Chapter 37 of Section VII**). In visual cortex, visual stimulation of the RF surround modulates (often suppressing) a neuron's response to a stimulus in its RF, a phenomenon called "surround suppression" (Angelucci and Shushruth, 2013). By providing V1 cells with a large integration region extending beyond their RF, FB connections may, thus, provide an anatomical substrate for surround suppression (Figure 38.2B) (Angelucci et al., 2002; Angelucci and Bressloff, 2006). In contrast, the visuotopically organized FF projections were hypothesized to represent the anatomical substrate for the increasing RF size of neurons along the cortical hierarchy (Figure 38.2A, right panel). This spatial organization of FF and FB circuits likely represents a general principle of inter-areal connectivity, as similar results were reported for macaque somatosensory cortex (Palfi et al., 2018).

In mouse, V1 and most higher visual areas also contain a topographic representation of visual space, and both inter-areal FF and FB projections are topographically organized (Wang and Burkhalter, 2007). Two-photon calcium imaging of area LM FB axon terminals within V1 has shown that while, on average, FB inputs are retinotopically matched with their target V1 neurons, they also show considerable scatter (Marques et al., 2018). About half of the FB inputs to L1 relay information from visual locations over 20° away from the RF center of V1 neurons (Figure 38.2B, left panel). As in monkey, this large FB integration region could generate surround suppression in mouse V1.

PATTERNING OF INTER-AREAL CONNECTIONS AND RELATION TO THE CORTICAL FUNCTIONAL ARCHITECTURE

Feedforward Connections and Parallel Processing

In cat and primates, small injections of retrograde tracers in a cortical area label patches of somata in lower areas. Vice versa, injections of anterograde tracers label patches of axon terminals in higher areas (Shipp and Zeki, 1985; Livingstone and Hubel, 1988; Salin and Bullier, 1995). Surprisingly few studies to date have examined the spatial relation of these patches to the cortical functional architecture. In higher mammals, cortical neurons sharing similar RF properties are grouped into domains (often spanning all layers; thus, also termed "columns"), and such properties are orderly mapped across the tangential domain of cortex. For example, in cat and monkey V1, there is an orderly representation of orientation preference or "orientation map" (Blasdel and Salama, 1986). Other stimulus features (e.g., eye preference, spatial frequency, hue, etc.) are similarly mapped in V1 and other visual areas. Gilbert and Wiesel (1989) demonstrated that FF projections from cat area 17 to area 18 interconnect neurons having similar orientation preference. Like-to-like functional connectivity in the projections from V1 to V2 has also been recently demonstrated in macaque, using focal electrical stimulation of single V1 functional domains (color or orientation) identified by intrinsic signal optical imaging of functional maps in V1 and V2 (Hu and Roe, 2022). This like-to-like organization could generate the orientation preference of V2 neuronal RFs. However, it is unknown whether this

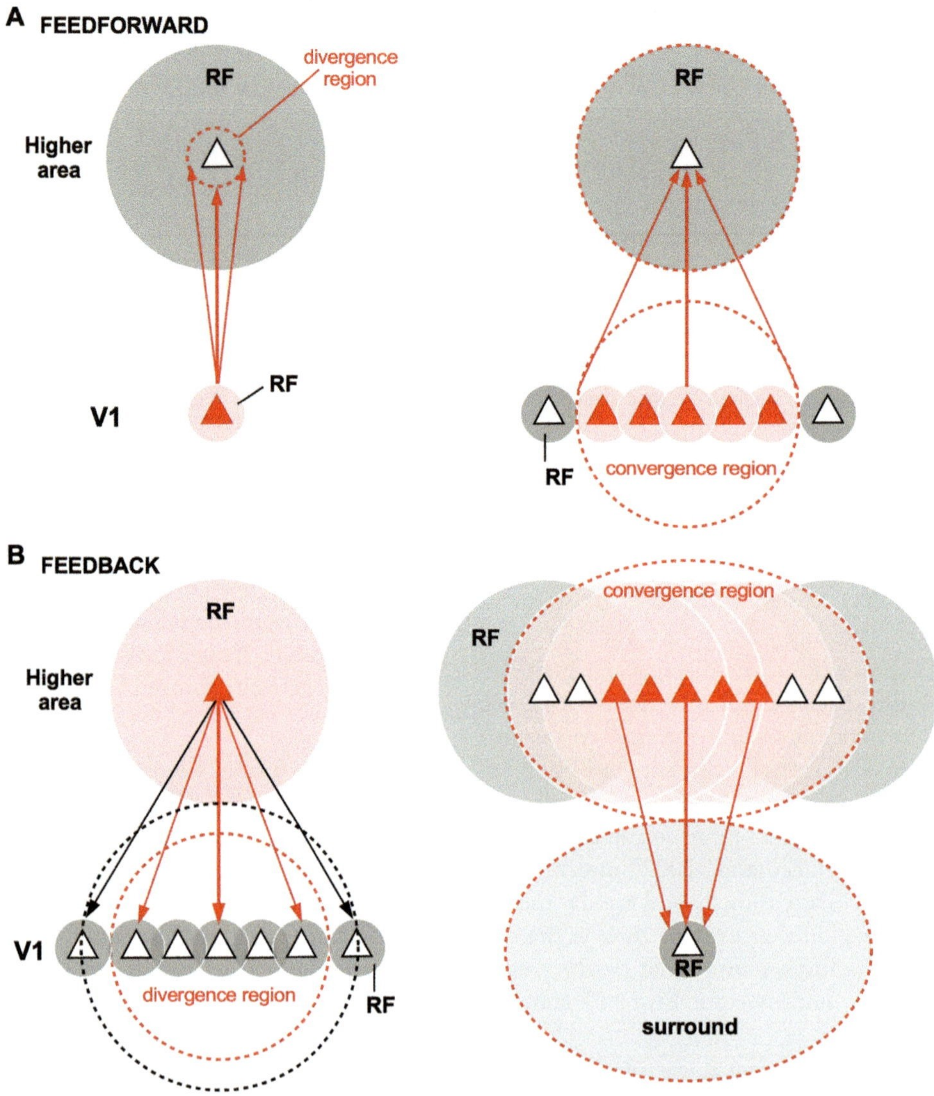

Figure 38.2. Visuotopic organization of FF and FB connections. Visuotopic extent here is defined as the distance between the RF centers of neurons on the opposite margins of the labeled fields. (A). FF connections. Left: schematics illustrating the divergence of FF connections from V1 to higher areas in monkey. FF connections are retinotopic; they diverge to a region in higher areas (red dashed circle) that is equal in visuotopic extent to the RF size of the source FF neurons (pink area), but is smaller than their target neurons' RF (gray area). Right: convergence of FF connections in monkey. The convergence region of FF connections from V1 is equal in extent to the RF size of their target cells and provides the anatomical substrate for these neurons' RF size. (B). FB connections. Left: FB connections are retinotopic in monkey; i.e., they diverge to a region in V1 (red dashed circle) that is equal in visuotopic extent to the RF size of the source FB neurons (pink area), but much larger than the individual V1 neurons' RF size (gray area). In mouse, FB connections are less retinotopic, as a fraction of cells projects to a V1 region (black dashed circle) that is larger than the RF size of the source FB neurons. Right: convergence of FB connections in monkey. The convergence region in V1 of FB projections from higher areas is equal in size to the RF surround of the target V1 cells, and provides the anatomical substrate for these neurons' RF surround.

organization represents a general principle of FF connectivity across functional maps, cortical areas, and species.

In addition to feature maps, primate V1 and V2 contain cytochrome oxidase (CO) maps. These are characteristic patterns obtained when staining V1 and V2 for CO (a mitochondrial enzyme which reveals the level of metabolic activity): an array of dark CO patches (the "blobs") separated by CO-pale zones (the "interblobs"), in V1, and a repeating pattern of dark thick and thin stripes separated by pale stripes, in V2. These CO compartments form the scaffold around which connections between V1 and V2 are organized. Livingstone and Hubel (1988) first proposed 3 anatomically segregated V1-to-V2 pathways, specialized in color, form, and motion/depth information processing. Injections of more sensitive retrograde tracers targeted to different V2 stripe types in macaque (Sincich and Horton, 2002; Federer et al., 2013) and marmoset (Federer et al., 2009) led to a revision of this model, and to the proposal of 4 parallel streams of information processing between V1 and V2 (Figure 38.3A). According to this model, V1

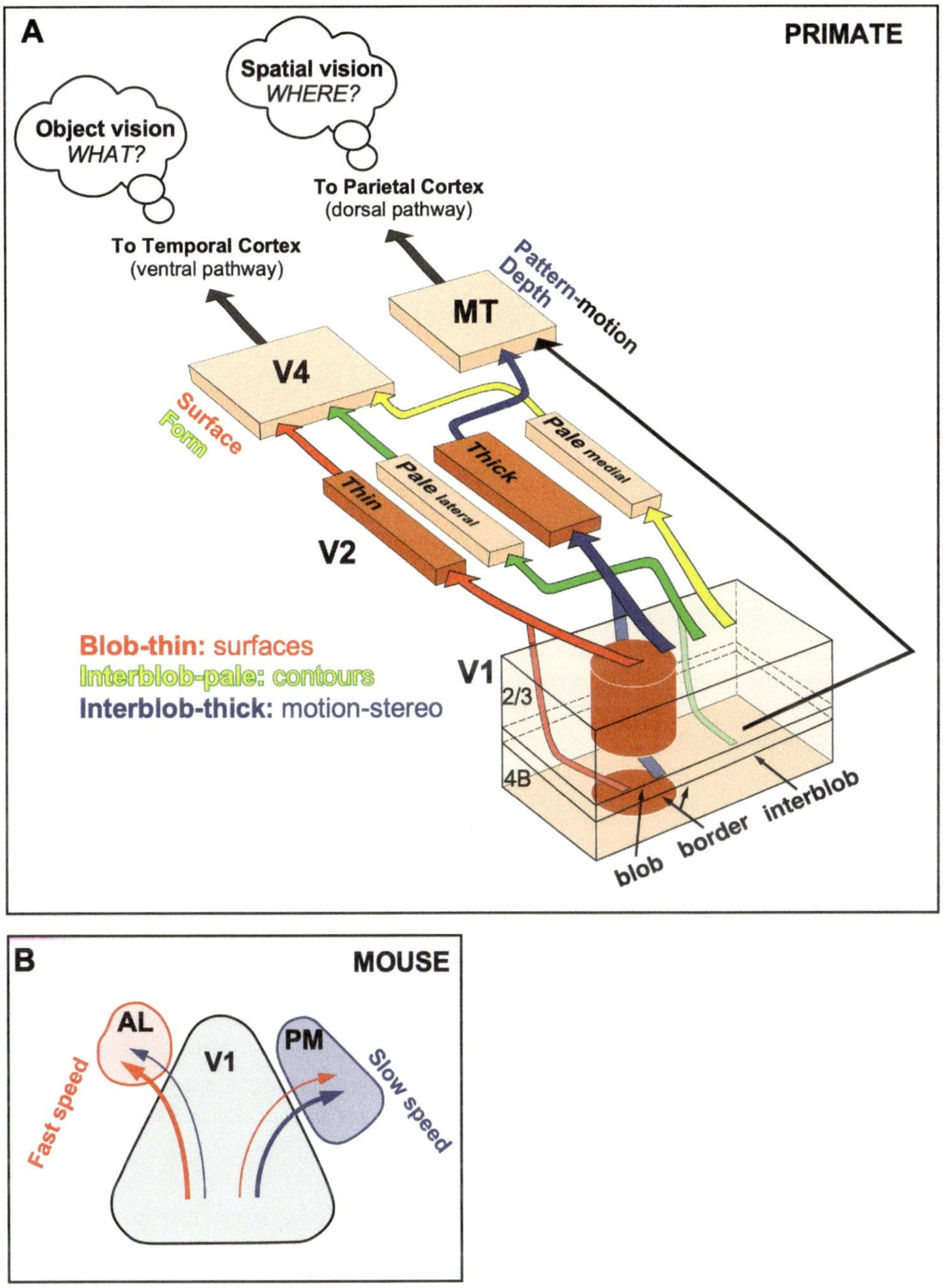

Figure 38.3. Parallel FF pathways from V1 to higher visual areas carry distinct visual information. (A). Primate. Schematics of 4 parallel FF pathways from V1 to V2. Each pathway is represented by arrows of different colors with thickness indicating projection density. Lighter colors = projections from V1 L4B to V2. Sparser projections from V1 L4A and L5/6 to V2 are omitted. CO blob columns (predominantly L2/3 and less so 4A-B and 5/6) project to thin stripes (red). Interblob (in macaque) or blob/interblob border (in marmoset) columns (predominantly from L2/3 and 4B, and less so from 4A and 5/6) project to thick stripes (blue). Two additional pathways project from interblobs to either pale-lateral or pale-medial stripes, with projections to pale-lateral stripes arising predominantly from L2/3 and less so from 4A–B and 5/6 (green), and those to pale-medial stripes arising exclusively (marmoset) or overwhelmingly (macaque) from L2/3 (yellow). Thin and pale stripes send information to temporal cortex via projections to V4 (ventral stream), while thick stripes send information to parietal cortex via projections to MT (dorsal stream). These pathways show functional clustering of RF tuning properties and specialized functional maps. There is also a direct projection from V1 L4B to MT (black arrow), possibly involved in motion processing. Modified from Federer et al. (2009, 2013). (B) Mouse. Schematics of 2 specialized parallel pathways from V1 to higher areas. V1 cells preferring gratings moving at faster speed (red) project predominantly to area AL, while those preferring slower speeds (blue) project predominantly to area PM. The visual preferences of the V1 afferents match the average speed preference of their postsynaptic neurons in AL and PM. Modified from Glickfeld et al. (2013).

CO blobs project predominantly to V2 thin stripes, and interblobs to both thick and pale stripes. All stripe types, except for one set of pale stripes (termed *pale-medial*, which receive projections almost exclusively from L2/3), receive predominant projections from both L2/3 and L4B and sparse projections from L4A and L5/6. Dual-stripe retrograde tracer injections further demonstrated, in both macaque and marmoset, that the interblob projections to thick, pale-lateral and pale-medial stripes arise from distinct neuronal populations (only 1%–8% of cells send projections to more than one stripe type), which, in the marmoset, are additionally spatially segregated within the interblob region (as depicted in Figure 38.3A). Numerous electrophysiological and optical imaging studies have additionally demonstrated at least some degree of functional specificity in the different V1 and V2 CO compartments. The CO blobs and thin stripes contain maps of hue and luminance (Xiao et al., 2003; Lu and Roe, 2007, 2008); the interblobs and pale stripes contain maps of orientation and higher-order orientation, respectively (Ramsden et al., 2001, 2014); and the thick stripes contain maps of retinal disparity and motion (Chen et al., 2008; Lu et al., 2010). The hypothesis that has emerged is that these parallel FF pathways specialize in processing surface properties (blob–thin stripes), object contours (interblob–pale stripes), and stereoscopic depth and motion (interblob–thick stripes) (Figure 38.3A). Importantly, anatomical and functional segregation in the V1 and V2 CO compartments is not absolute, as all V1 compartments contribute at least some small projections to all V2 stripes, and mixed RF properties are found in all V1 and V2 compartments. Perhaps the predominant V1 input to a particular V2 stripe reflects the need to maintain clustering within V2 of stimulus features that are also clustered within V1 compartments (e.g., hue in the blobs and thin stripes). In contrast, inputs from multiple V1 compartments to a single stripe type may serve to map within that stripe type stimulus features that are distributed across V1 compartments (e.g., binocular retinal disparity).

Anatomical and functional segregation is maintained beyond V2, with thin and pale stripes projecting to V4, an area specialized in processing color and form, and thick stripes to MT, an area specialized in processing motion and stereopsis (Shipp and Zeki, 1985). Information from V4 is then relayed to the temporal cortex via the ventral pathway, which specializes in object vision, while information from MT is relayed to the posterior parietal cortex, via the dorsal pathway, which is involved in spatial vision and visually guided action (Ungerleider and Mishkin, 1982) (Figure 38.3A).

There is increasing evidence that in mouse visual cortex FF connections also show like-to-like connectivity. The temporal tuning of V1 neurons in L2/3 and L5 projecting to the anterolateral (AL) and posteromedial (PM) areas differ from each other, and resemble the average tuning of neurons in their respective target areas (Glickfeld et al., 2013) (Figure 38.3B). A similar like-to-like organization exists for V1-to-AL and V1-to-LM projections with respect to spatial frequency, orientation, and direction tuning (Matsui and Ohki, 2013). This suggests that direct V1 inputs contribute to the tuning properties of neurons in these higher-order areas. However, in area AL the spatiotemporal tuning properties of neurons are more correlated with afferents from the pulvinar than from V1 (Blot et al., 2021), suggesting that inputs from this high-order thalamic nucleus also play an important role in shaping the functional properties of visual areas (see also **Chapters 39 and 42 in Section VII**).

In summary, FF connections are topographically organized and highly area-, compartment-, and functionally specific; this specificity serves to parse sensory information into specialized parallel-processing streams, and likely to generate the larger and more complex RFs of neurons along the cortical hierarchy.

Feedback Connections: Patterned and Functionally Specific

It has long been debated whether FB connections are anatomically diffuse and, therefore, unspecific with respect to the CO compartments and/or functional domains that they target (Rockland and Pandya, 1979; Stettler et al., 2002), or whether they are clustered and functionally specific (Angelucci et al., 2002; Shmuel et al., 2005). This controversy is primarily attributable to the lack, in early studies, of sensitive anterograde neuroanatomical tracers capable of labeling FB axons fully and selectively, without also labeling retrogradely the reciprocal FF projections (Angelucci and Bressloff, 2006). Recent advances in neuroanatomical labeling methods, based on viral-mediated delivery of genes for fluorescent proteins to specific projection neurons, have made it possible to resolve this controversy. Selective viral-mediated labeling of FB projections to V1 from specific V2 CO stripes demonstrated that V2-to-V1 FB projections are anatomically clustered and organized into multiple parallel streams that resemble the reciprocal V1-to-V2 FF pathways: V2 thin stripes send FB projections predominantly to CO blobs, while thick and pale stripes project predominantly to interblobs (Federer et al., 2021) (Figure 38.4A). Given the clustering of specific RF properties and functional maps in the different CO compartments (see above), this anatomical segregation of the V2-to-V1 pathways suggests that FB pathways maintain the functional specificity established by the reciprocal FF pathways. Notably, the segregation of FB pathways within CO compartments is not strict, as less dense FB projections occur between the densest labeled

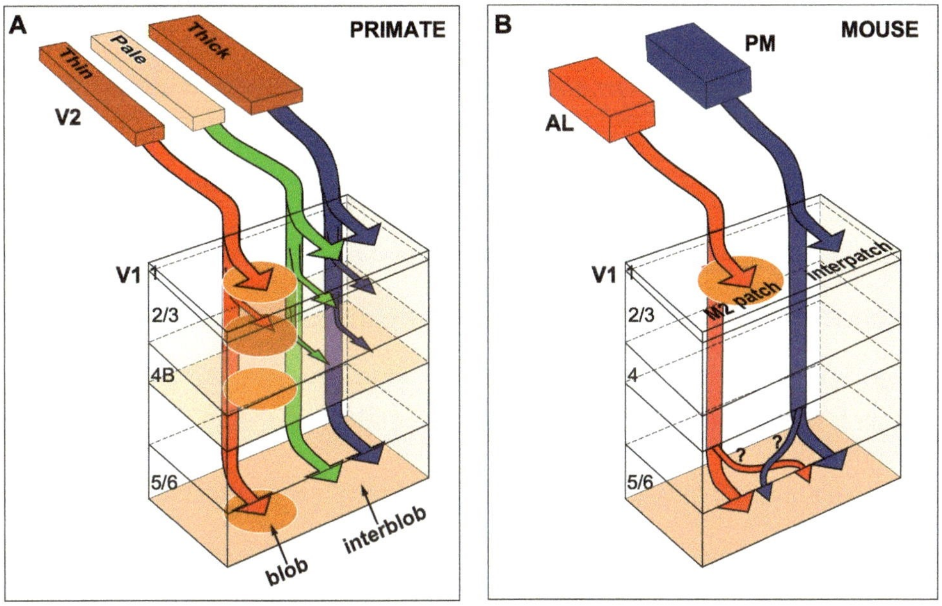

Figure 38.4. Parallel FB pathways to V1. (A) Primate. Schematics of the FB parallel pathways between the V2 CO stripes and the V1 CO compartments in macaque. Arrow thickness indicates projection density. For all stripe types, L1/upper2 and 5/6 receive the densest FB projections, while lower L2 and L3 receive sparser projections. Both thick and pale, but not thin, stripes send FB projections also to L4B. FB projections from thin stripes (red) project predominantly to CO blobs, while FB from thick (blue) and pale (green) stripes project predominantly to interblobs. Modified from Federer et al. (2021). (B) Mouse. FB pathways from higher visual areas to V1 L1 and L5/6. FB to L1 segregates into parallel pathways: FB from area AL clusters into M2 muscarinic acetylcholine receptor-rich patches, while FB from area PM projects to interpatches. M2 receptors are uniformly distributed in L5/6, but it is unknown ("?") whether the FB projections from AL and PM in L5/6 align with the L1 patches and interpatches or intermingle. Based on data from D'Souza et al. (2019). For simplicity of illustration, in both the primate and mouse schematics, the FB projections to different layers are depicted as branches from a single axon, but in both species, it is unknown whether they arise from the same or different FB neurons.

FB terminal patches, suggesting a smaller population of FB neurons projects to multiple CO compartments (Federer et al., 2021). This population of FB neurons may serve to integrate information across streams, or to carry information about a specific stimulus feature to multiple CO compartments representing that feature (e.g., orientation-tuned cells are present in both blobs and interblobs).

Whether FB projections link neurons having similar RF properties is unknown in primates. Two previous studies combining anterograde labeling of V2-to-V1 FB axons with optical imaging of orientation maps reached different conclusions. One study, using a first-generation GFP-expressing adenovirus, which failed to fill long-distant axons completely, reported diffuse and non-orientation-specific V2-to-V1 FB terminations (Stettler et al., 2002). The other study, using the bidirectional tracer BDA, found clustered FB connections linking cortical territories of similar orientation preference along an axis in visual space parallel to the orientation preference of the FB neurons (Shmuel et al., 2005). Due to the bidirectional labeling of both FF and FB axons in the latter study, it was unclear whether the axonal clusters in V1 were FB axon terminals or the axon collaterals of the FF neurons, the latter known to form clusters (see above). A recent study in tree shrew, by combining calcium imaging of V2 FB axon terminals with intrinsic signal optical imaging of V1 orientation maps, found that the FB axon terminals form an organized orientation map which aligns with the V1 orientation map (Zhang et al., 2018). This alignment, however, was less precise than that observed for intra-V1 horizontal connections, suggesting that FB connections may serve to facilitate integration of more elongated contours beyond the RF of V1 neurons.

Studies in mouse visual cortex are consistent with those in primates, with mounting evidence indicating that FB connections in this species are highly organized depending on the area of origin of the projections and the tuning of individual neurons. Viral-mediated anterograde tracing of FB projections from areas AL and PM resulted in interleaved clusters of AL and PM FB-axon terminations in V1 L1 (Ji et al., 2015; D'Souza et al., 2019), which aligned with M2 muscarinic acetylcholine receptor-rich and poor patches, respectively; the latter patches, in turn, correspond to V1 domains specialized in processing distinct spatiotemporal features (Ji et al., 2015; D'Souza et al., 2019) (Figure 38.4B; see also **Chapter 39 in Section VII**). Interestingly, M2-positive patches are also found in macaque V1 L1, where they align with the CO-interblobs (Ji et al., 2015). This anatomical arrangement suggests that FB projections from different areas might target functionally distinct V1 neurons. Consistent with this hypothesis, selective silencing of FB neurons from areas AL and PM reduces visual responses in

V1 neurons having similar spatiotemporal tuning properties as the silenced FB neurons, suggesting like-to-like connectivity (Huh et al., 2018). Notably, as in monkeys, these FB channels match the functional segregation seen in the reciprocal FF pathways from V1 to AL and PM (Glickfeld et al., 2013) (see above).

In addition to this area-specific selectivity, there is also evidence in mouse that, within the same FB projection, the orientation selectivity of the afferent axons determines their V1 targets. Marques et al. (2018) used two-photon imaging of LM FB axon terminals in V1 L1 to address how the organization of FB axon terminals relates to the FB neuron's orientation tuning. They found that in V1, FB terminals extended beyond retinotopically matched locations, thus also contacting V1 neurons with retinotopically non-matching RFs (see also above and Figure 38.2B). Importantly, these contacts at non-matched retinotopic locations were displaced along an axis in visual space perpendicular to the FB axons' preferred orientation, an arrangement that may serve to enhance the representation of visual contours. Further studies are necessary to demonstrate whether this result generalizes across different FB orientations, different cortical areas, and species. In the study of Marques et al. (2018) the orientation tuning of V1 cells postsynaptic to the LM-FB axons was not examined; thus it remains unknown whether FB selectively connects neurons with similar orientation preference in the mouse.

In summary, FB connections are not random or diffuse. They are spatially extensive but topographically organized, and convey to the smaller RFs of lower-area neurons information from beyond their RF. FB is also area-, compartment-, and functional-module-specific, suggesting functionally specialized parallel FB streams. Thus, at least at the lower hierarchical levels, FB does not primarily integrate information across parallel streams, but rather modulates V1 responses to the specific stimulus attributes processed within their home stream, an organization that may support, for example, stream-selective gating by attention, stream-selective surround or contextual modulations, and/or plasticity in stream-specific lower-order neurons.

FUNCTION OF FF AND FB CONNECTIONS: CAUSAL MANIPULATIONS

Our understanding of the function of inter-areal circuits has been hampered by the inability to causally manipulate the activity of FF and FB neurons selectively. Inactivation of cortico-cortical systems has typically been performed by cooling, pharmacological agents, lesions, or optogenetic activation of intra-areal PV interneurons. These approaches lack spatiotemporal precision and cell-type specificity, and affect activity in an entire cortical area, thus disrupting the entire network of connections linking directly or indirectly the 2 areas under study. With this caveat, these studies have generally revealed that FF connections drive sensory responses and contribute to the RF size and tuning properties of neurons in target areas, while FB connections have a modulatory influence. These studies are briefly reviewed below.

In both primates and rodents, cooling or inactivating V1 silences or reduces, respectively, visually evoked activity at retinotopically matched loci in several higher visual areas (Molotchnikoff and Hubert, 1990; Girard et al., 1991; Javadzadeh and Hofer, 2022).

Studies in cats and primates investigating the effects of cortical inactivation on the RF tuning of neurons in downstream areas whose activity was not abolished by this manipulation have been inconclusive, with reports of either decreased or unaffected RF tuning (for a review, see Salin and Bullier, 1995). The discrepant results across these studies may depend on species, manipulated cortical area/pathway, stimulus tuning property, methods of analysis, and, most importantly, the non-selective causal manipulation methods used. Moreover, in both primates and rodents, extrageniculate (colliculo-pulvinar-cortical) pathways also contribute to the RF tuning of neurons in some higher areas, such as direction tuning in primate MT (Girard et al., 1992) or velocity tuning in rodent areas LM, AL, AM, and RL (Tohmi et al., 2014).

It has also been difficult to pinpoint consistent effects of inactivating higher cortical areas on the visual responses of neurons in lower areas to stimulation of their RF, but there is general agreement that these effects are subtle. The most consistently reported effects of higher-area inactivation across studies and species have been reduced gain of neural responses without concomitant changes in tuning (reviewed in Angelucci and Bressloff, 2006; Angelucci et al., 2017). Studies in mouse have additionally shown that the decrease in V1 sensory-evoked responses after silencing higher visual areas is area- and stimulus-dependent, for example, being more marked at the neuron's preferred orientation after silencing area LM (Pafundo et al., 2016), but at the anti-preferred stimulus direction after silencing dorsal-stream areas AL or PM (Oude Lohuis et al., 2022). Moreover, the effects of silencing higher visual areas can be variable and dynamic, with both increases and decreases in sensory-evoked activity occurring in a context-dependent manner (Javadzadeh and Hofer, 2022).

FB connections also provide neurons with a large integration region extending beyond their RFs and coextensive with the RF surround of V1 neurons (Figure 38.2B) (Angelucci et al., 2002), and thus are hypothesized to serve surround suppression (Angelucci and Bressloff, 2006; Angelucci et al., 2017) (see above). However, inactivation studies have provided contrasting results regarding the role of FB in surround suppression. Some studies in primates, cats, and mice observed some reduction in surround suppression after cooling or optogenetically silencing a higher visual area (Hupé et al., 1998; Bardy et al., 2009; Nassi et al., 2013; Vangeneugden et al., 2019). Other studies, instead, found no change in surround suppression after pharmacologically silencing primate V2 (Hupé et al., 2001), or cooling cat postero-temporal visual cortex (Wang et al., 2010). These discrepancies could depend on differences across studies in the magnitude and extent of FB inactivation, the specific cortical area inactivated, or methods of analysis. Most importantly, all these previous studies inactivated entire cortical areas, rather than FB connections selectively, with potentially broader and variable effects.

Only a handful of studies have selectively manipulated the activity of FB neurons. In marmoset, selective optogenetic inactivation of V2 FB axon terminals within V1 resulted in increased RF size, decreased visual responses to stimuli inside the RF, and reduced surround suppression (Nurminen et al., 2018). The magnitude of these effects depended on the degree of FB inactivation, with stronger reduction of V2 FB activity leading to greater increase in RF size, and a progressive overall decrease in response amplitude to all stimulus sizes. Thus, selective and titrated manipulation of FB neuron activity revealed nuanced effects that were missed with coarser inactivation of entire areas. Optogenetic inactivation of FB neurons in mouse areas AL or PM selectively reduced the responses of V1 neurons tuned to spatial frequencies similar to those of the inactivated FB cells (Huh et al., 2018), thus demonstrating that the effects of FB connections are functionally specific. However, as the FB neurons of origin, rather than their terminals, were inactivated in this study, it remains unclear whether the effects of FB inactivation on V1 cells were direct or indirect via these FB-neuron projections to other cortical or subcortical targets.

So far, emphasis has been placed on visual functions for FB, but FB has been implicated in many other, not strictly visual functions, including top-down attention, reward, learning, and predictive coding (Gilbert and Li, 2013). Whether FB connections multiplex these various forms of information is unknown, but evidence for parallel FB channels (see above and Figure 38.4) would suggest that different FB systems perform different functions. However, direct demonstrations of FB in any of these top-down phenomena have been lacking, with the exception of a single recent study in which selective inactivation of FB axons from V4 abolished the attentional modulation of V1 responses observed while the monkey performed a visual attention task (Debes and Dragoi, 2023).

CONCLUSIONS

The distinct organizational and functional properties of FF and FB connections reviewed in this chapter support the long-held view that these 2 pathways play distinct and complementary roles in cortical computation. The conserved organizational and functional properties of these connections across cortical areas and species suggest conserved, canonical functional roles. Although historically it has been challenging to experimentally assess their roles in cortical information processing, recent technological advances, based on optical manipulation of neural circuits, have opened up the possibility to interrogate FF and FB circuits with greater specificity than previously possible.

ACKNOWLEDGMENTS

The authors would like to thank their students and postdoctoral fellows who have contributed to some of the work reviewed in this chapter: S. Shushruth, Frederick Federer, Lars Schwabe, Maryam Bijanzadeh, Semi Ta'afua, Sam Merlin, Lauri Nurminen, Mahlega Hassanpour, Caitlin Siu, Hedi Young, and Tiago Marques. A special thank goes to the 2 people who have inspired and supported my (AA's) work on feedback connections: Jenny S. Lund and Jean Bullier. We would also like to thank our funding agencies and donors: NIH (Grants: R01EY026812, R01 EY019743, R01 EY031959, BRAIN U01 NS099702, to AA), NSF (IOS 1755431 to AA), the Mary Boesche Endowment (to AA), Fundação para a Ciência e a Tecnologia (PTDC/MED-NEU/6645/2020, to LP), and La "Caixa" Foundation (HR22-00778, to LP).

REFERENCES

Anderson JC, Martin KA (2006). Synaptic connection from cortical area V4 to V2 in macaque monkey. *J Comp Neurol.* 495:709–721.

Anderson JC, Martin KAC (2009). The synaptic connections between cortical areas V1 and V2 in macaque monkey. *J Neurosci.* 29:11283–11293.

Angelucci A, Bijanzadeh M, Nurminen L, Federer F, Merlin S, Bressloff PC (2017) Circuits and mechanisms for surround modulation in visual cortex. *Ann Rev Neurosci.* 40:425–451.

Angelucci A, Bressloff PC (2006). The contribution of feedforward, lateral and feedback connections to the classical receptive field center and extra-classical receptive field surround of primate V1 neurons. *Prog Brain Res.* 154:93–121.

Angelucci A, Levitt JB, Walton E, Hupé JM, Bullier J, Lund JS (2002). Circuits for local and global signal integration in primary visual cortex. *J Neurosci.* 22:8633–8646.

Angelucci A, Shushruth S (2013) Beyond the classical receptive field: Surround modulation in primary visual cortex. In: Chalupa LM, Werner JS, eds. *The New Visual Neurosciences*. MIT Press:425–444.

Bardy C, Huang JY, Wang C, Fitzgibbon T, Dreher B (2009). "Top-down" influences of ispilateral or contralateral postero-temporal visual cortices on the extra-classical receptive fields of neurons in cat's striate cortex. *Neurosci.* 158:951–968.

Berezovskii VK, Nassi JJ, Born RT (2011). Segregation of feedforward and feedback projections in mouse visual cortex. *J Comp Neurol.* 519:3672–3683.

Blasdel GG, Salama G (1986). Voltage-sensitive dyes reveal a modular organization in monkey striate cortex. *Nature.* 321:579–585.

Blot A, Roth MM, Gasler I, Javadzadeh M, Imhof F, Hofer SB (2021). Visual intracortical and transthalamic pathways carry distinct information to cortical areas. *Neuron.* 109:1996–2008.

Bullier J, Kennedy H, Salinger W (1984). Branching and laminar origin of projections between visual cortical areas in the cat. *J Comp Neurol.* 228:329–341.

Chen G, Lu HD, Roe AW (2008). A map for horizontal disparity in monkey V2. *Neuron.* 58:442–450.

Coogan TA, Burkhalter A (1993) Hierarchical organization of areas in rat visual cortex. *J Neurosci.* 13:3749–3772.

D'Souza RD, Bista P, Meier AM, Ji W, Burkhalter A (2019). Spatial clustering of inhibition in mouse primary visual cortex. *Neuron.* 104:588–600.

D'Souza RD, Meier AM, Bista P, Wang Q, Burkhalter A (2016). Recruitment of inhibition and excitation across mouse visual cortex depends on the hierarchy of interconnecting areas. *eLife.* 5:e19332.

D'Souza RD, Wang Q, Ji W, Meier AM, Kennedy H, Knoblauch K, Burkhalter A (2022) Hierarchical and nonhierarchical features of the mouse visual cortical network. *Nat Commun.* 13:503.

De Pasquale R, Sherman SM (2011). Synaptic properties of corticocortical connections between the primary and secondary visual cortical areas in the mouse. *J Neurosci.* 31:16494–16506.

Debes SR, Dragoi V (2023). Suppressing feedback signals to visual cortex abolishes attentional modulation. *Science.* 379:468–473.

Dong H, Shao Z, Nerbonne JM, Burkhalter A (2004). Differential depression of inhibitory synaptic responses in feedforward and feedback circuits between different areas of mouse visual cortex. *J Comp Neurol.* 475:361–373.

Federer F, Ichida JM, Jeffs J, Schiessl I, McLoughlin N, Angelucci A (2009). Four projections streams from primate V1 to the cytochrome oxidase stripes of V2. *J Neurosci.* 29:15455–15471.

Federer F, Williams D, Ichida JM, Merlin S, Angelucci A (2013). Two projection streams from macaque V1 to the pale cytochrome oxidase stripes of V2. *J Neurosci.* 33:11530–11539.

Federer F, Ta'afua S F, Merlin S, Hassanpour MS, Angelucci A (2021). Stream-specific feedback inputs to the primate primary visual cortex. *Nature Comm.* 12:228.

Felleman DJ, Van Essen DC (1991). Distributed hierarchical processing in the primate cerebral cortex. *Cereb Cortex.* 1:1–47.

Friedman DP, Murray EA, O'Neill JB, Mishkin M (1986). Cortical connections of the somatosensory fields of the lateral sulcus of macaques: Evidence for a corticolimbic pathway for touch. *J Comp Neurol.* 252:323–347.

Galaburda AM, Pandya DN (1983). The intrinsic architectonic and connectional organization of the superior temporal region of the rhesus monkey. *J Comp Neurol.* 221:169–184.

Gamanut R, Kennedy H, Toroczkai Z, Ercsey-Ravasz M, Van Essen DC, Knoblauch K, Burkhalter A (2018). The mouse cortical connectome, characterized by an ultra-dense cortical graph, maintains specificity by distinct connectivity profiles. *Neuron.* 97:698–715.

Gilbert CD, Li W (2013). Top-down influences on visual processing. *Nat Rev Neurosci.* 14:350–363.

Gilbert CD, Wiesel TN (1989). Columnar specificity of intrinsic horizontal and corticocortical connections in cat visual cortex. *J Neurosci.* 9:2432–2442.

Girard P, Salin PA, Bullier J (1991). Visual activity in areas V3a and V3 during reversible inactivation of area V1 in the macaque monkey. *J Neurophysiol.* 66:1493–1503.

Girard P, Salin PA, Bullier J (1992). Response selectivity of neurons in area MT of the macaque monkey during reversible inactivation of area V1. *J Neurophysiol.* 67:1437–1446.

Glickfeld LL, Andermann ML, Bonin V, Reid RC (2013). Cortico-cortical projections in mouse visual cortex are functionally target specific. *Nat Neurosci.* 16:219–226.

Gonchar Y, Burkhalter A (2003). Distinct GABAergic targets of feedforward and feedback connections between lower and higher areas of rat visual cortex. *J Neurosci.* 23:10904–10912.

Han Y, Kebschull JM, Campbell RAA, Cowan D, Imhof F, Zador AM, Mrsic-Flogel TD (2018). The logic of single-cell projections from visual cortex. *Nature.* 556:51–56.

Harris JA, Mihalas S, Hirokawa KE, Whitesell JD, Choi H, Bernard A, Bohn P, Caldejon S, Casal L, Cho A, Feiner A, Feng D, Gaudreault N, Gerfen CR, Graddis N, Groblewski PA, Henry AM, Ho A, Howard R, Knox JE, Kuan L, Kuang X, Lecoq J, Lesnar P, Li Y, Luviano J, McConoughey S, Mortrud MT, Naeemi M, Ng L, Oh SW, Ouellette B, Shen E, Sorensen SA, Wakeman W, Wang Q, Wang Y, Williford A, Phillips JW, Jones AR, Koch C, Zeng H (2019). Hierarchical organization of cortical and thalamic connectivity. *Nature.* 575:195–202.

Hu JM, Roe AW (2022). Functionally specific and sparse domain-based micro-networks in monkey V1 and V2. *Curr Biol.* 32:2797–2809.

Huh CYL, Peach JP, Bennett C, Vega RM, Hestrin S (2018). Feature-specific organization of feedback pathways in mouse visual cortex. *Curr Biol.* 28:114–120.

Hupé JM, James AC, Girard P, Bullier J (2001). Response modulations by static texture surround in area V1 of the macaque monkey do not depend on feedback connections from V2. *J Neurophysiol.* 85:146–163.

Hupé JM, James AC, Payne BR, Lomber SG, Girard P, Bullier J (1998). Cortical feedback improves discrimination between figure and background by V1, V2 and V3 neurons. *Nature.* 394:784–787.

Javadzadeh M, Hofer SB (2022). Dynamic causal communication channels between neocortical areas. *Neuron.* 110:2470–2483.

Ji W, Gamanut R, Bista P, D'Souza RD, Wang Q, Burkhalter A (2015). Modularity in the organization of mouse primary visual cortex. *Neuron.* 87:632–643.

Johnson RR, Burkhalter A (1997). A polysynaptic feedback circuit in rat visual cortex. *J Neurosci.* 17:7129–7140.

Kennedy H, Bullier J (1985). A double-labeling investigation of the afferent connectivity to cortical area V1 and V2 of the macaque monkey. *J Neurosci.* 5:2815–2830.

Kim EJ, Juavinett AL, Kyubwa EM, Jacobs MW, Callaway EM (2015). Three types of cortical layer 5 neurons that differ in brain-wide connectivity and function. *Neuron.* 88:1253–1267.

Kim EJ, Zhang Z, Huang L, Ito-Cole T, Jacobs MW, Juavinett AL, Senturk G, Hu M, Ku M, Ecker JR, Callaway EM (2020) Extraction of distinct neuronal cell types from within a genetically continuous population. *Neuron.* 107:1–9.

Larkum M (2013). A cellular mechanism for cortical associations: An organizing principle for the cerebral cortex. *Trends Neurosci.* 36:141–151.

Livingstone MS, Hubel DH (1988). Segregation of form, color, movement, and depth: anatomy, physiology, and perception. *Science.* 240:740–749.

Lu HD, Chen G, Tanigawa H, Roe AW (2010). A motion direction map in macaque V2. *Neuron.* 68:1002–1013.

Lu HD, Roe AW (2007). Optical imaging of contrast response in macaque monkey V1 and V2. *Cereb Cortex.* 17:2675–2695.

Lu HD, Roe AW (2008). Functional organization of color domains in V1 and V2 of macaque monkey revealed by optical imaging. *Cereb Cortex.* 18:516–533.

Ma G, Liu Y, Wang L, Xiao Z, Song K, Wang Y, Peng W, Liu X, Wang Z, Jin S, Tao Z, Li CT, Xu T, Xu F, Xu M, Zhang S (2021). Hierarchy in sensory processing reflected by innervation balance on cortical interneurons. *Sci Adv*. 7:eabf5676.

Markov NT, Vezoli J, Chameau P, Falchier A, Quilodran R, Huissoud C, Lamy C, Misery P, Giroud P, Ullman S, Barone P, Dehay C, Knoblauch K, Kennedy H (2014). Anatomy of hierarchy: feedforward and feedback pathways in macaque visual cortex. *J Comp Neurol*. 522:225–259.

Marques T, Nguyen J, Fioreze G, Petreanu L (2018). The functional organization of cortical feedback inputs to primary visual cortex. *Nat Neurosci*. 21:757–764.

Martinetti LE, Bonekamp KE, Autio DM, Kim HH, Crandall SR (2022). Short-term facilitation of long-range corticocortical synapses revealed by selective optical stimulation. *Cereb Cortex*. 32:1932–1949.

Matsui T, Ohki K (2013). Target dependence of orientation and direction selectivity of corticocortical projection neurons in the mouse V1. *Front Neural Circuits*. 7:143.

Molotchnikoff S, Hubert F (1990). Susceptibility of neurons in area 18a to blockade of area 17 in rats. *Brain Res*. 510:223–228.

Nakamura H, Gattass R, Desimone R, Ungerleider LG (1993). The modular organization of projections from areas V1 and V2 to areas V4 and TEO in macaques. *J Neurosci*. 13:3681–3691.

Naskar S, Qi J, Pereira F, Gerfen CR, Lee S (2021). Cell-type-specific recruitment of GABAergic interneurons in the primary somatosensory cortex by long-range inputs. *Cell Rep*. 34:108774.

Nassi JJ, Lomber SG, Born RT (2013). Corticocortical feedback contributes to surround suppression in V1 of the alert primate. *J Neurosci*. 33:8504–8517.

Nurminen L, Merlin S, Bijanzadeh M, Federer F, Angelucci A (2018). Top-down feedback controls spatial summation and response amplitude in primate visual cortex. *Nat Commun*. 9:2281.

Oude Lohuis MN, Canton AC, Pennartz CMA, Olcese U (2022). Higher order visual areas enhance stimulus responsiveness in mouse primary visual cortex. *Cereb Cortex*. 32:3269–3288.

Pafundo DE, Nicholas MA, Zhang R, Kuhlman SJ (2016). Top-down-mediated facilitation in the visual cortex is gated by subcortical neuromodulation. *J Neurosci*. 36:2904–2914.

Palfi E, Zalanyi L, Ashaber M, Palmer C, Kantor O, Roe AW, Friedman RM, Negyessy L (2018). Connectivity of neuronal populations within and between areas of primate somatosensory cortex. *Brain Struct Funct*. 223:2949–2971.

Petreanu L, Mao T, Sternson SM, Svoboda K (2009). The subcellular organization of neocortical excitatory connections. *Nature*. 457:1142–1145.

Ramsden BM, Hung CP, Roe AW (2001). Real and illusory contour processing in area V1 of the primate: a cortical balancing act. *Cereb Cortex*. 11:648–665.

Ramsden BM, Hung CP, Roe AW (2014). Orientation domain diversity in macaque area V2. *Eye Brain*. 6:97–112.

Rockland KS, Knutson T (2000). Feedback connections from area MT of the squirrel monkey to areas V1 and V2. *J Comp Neurol*. 425:345–368.

Rockland KS, Pandya DN (1979). Laminar origins and terminations of cortical connections of the occipital lobe in the Rhesus monkey. *Brain Res*. 179:3–20.

Salin PA, Bullier J (1995). Corticocortical connections in the visual system: Structure and function. *Physiol Rev*. 75:107–154.

Salin PA, Kennedy H, Bullier J (1995). Spatial reciprocity of connections between areas 17 and 18 in the cat. *Can J Physiol Pharmacol*. 73:1339–1347.

Shao Z, Burkhalter A (1996). Different balance of excitation and inhibition in forward and feedback circuits of rat visual cortex. *J Neurosci*. 16:7353–7365.

Shen S, Jiang X, Scala F, Fu J, Fahey P, Kobak D, Tan Z, Zhou N, Reimer J, Sinz F, Tolias AS (2022). Distinct organization of two cortico-cortical feedback pathways. *Nat Commun*. 13:6389.

Shipp S, Zeki S (1985). Segregation of pathways leading from area V2 to areas V4 and V5 of macaque monkey visual cortex. *Nature*. 315:322–325.

Shmuel A, Korman M, Sterkin A, Harel M, Ullman S, Malach R, Grinvald A (2005) Retinotopic axis specificity and selective clustering of feedback projections from V2 to V1 in the owl monkey. *J Neurosci*. 25:2117–2131.

Sincich LC, Horton JC (2002). Divided by cytochrome oxidase: a map of the projections from V1 to V2 in macaques. *Science*. 295:1734–1737.

Sincich LC, Horton JC (2003). Independent projection streams from macaque striate cortex to the second visual area and middle temporal area. *J Neurosci*. 23:5684–5692.

Siu C, Balsor J, Merlin S, Federer F, Angelucci A (2021). A direct interareal feedback-to-feedforward circuit in primate visual cortex. *Nat Commun*. 12:4911.

Stettler DD, Das A, Bennett J, Gilbert CD (2002). Lateral connectivity and contextual interactions in macaque primary visual cortex. *Neuron*. 36:739–750.

Symonds LL, Rosenquist AC (1984). Laminar origins of visual corticocortical connections in the cat. *J Comp Neurol*. 229:39–47.

Tohmi M, Meguro R, Tsukano H, Hishida R, Shibuki K (2014). The extrageniculate visual pathway generates distinct response properties in the higher visual areas of mice. *Curr Biol*. 24:587–597.

Ungerleider LG, Mishkin M (1982). Two cortical visual systems. In: Ingle DJ, Mansfield JW, Goodale MA, eds. *Advances in the Analysis of Visual Behavior*. MIT Press:549–586.

Vangeneugden J, van Beest EH, Cohen MX, Lorteije JAM, Mukherjee S, Kirchberger L, Montijn JS, Thamizharasu P, Camillo D, Levelt CN, Roelfsema PR, Self MW, Heimel JA (2019). Activity in lateral visual areas contributes to surround suppression in awake mouse V1. *Curr Biol*. 29:4268–4275.

Wall NR, De La Parra M, Sorokin JM, Taniguchi H, Huang ZJ, Callaway EM (2016). Brain-wide maps of synaptic input to cortical interneurons. *J Neurosci*. 36:4000–4009.

Wang C, Huang JY, Bardy C, FitzGibbon T, Dreher B (2010). Influence of "feedback" signals on spatial integration in receptive fields of cat area 17 neurons. *Brain Res*. 1328:34–48.

Wang Q, Burkhalter A (2007). Area map of mouse visual cortex. *J Comp Neurol*. 502:339–357.

Williams SR, Stuart GJ (2002). Dependence of EPSP efficacy on synapse location in neocortical pyramidal neurons. *Science*. 295:1907–1910.

Winnubst J, Bas E, Ferreira TA, Wu Z, Economo MN, Edson P, Arthur BJ, Bruns C, Rokicki K, Schauder D, Olbris DJ, Murphy SD, Ackerman DG, Arshadi C, Baldwin P, Blake R, Elsayed A, Hasan M, Ramirez D, Dos Santos B, Weldon M, Zafar A, Dudman JT, Gerfen CR, Hantman AW, Korff W, Sternson SM, Spruston N, Svoboda K, Chandrashekar J (2019). Reconstruction of 1,000 projection neurons reveals new cell types and organization of long-range connectivity in the mouse brain. *Cell*. 179:268–281.

Xiao Y, Wang YI, Felleman DJ (2003). A spatially organized representation of color in macaque cortical area V2. *Nature*. 421:535–439.

Yang W, Carrasquillo Y, Hooks BM, Nerbonne JM, Burkhalter A (2013). Distinct balance of excitation and inhibition in an interareal feedforward and feedback circuit of mouse visual cortex. *J Neurosci*. 33:17373–17384.

Yao S, Wang Q, Hirokawa KE, Ouellette B, Ahmed R, Bomben J, Brouner K, Casal L, Caldejon S, Cho A, Dotson NI, Daigle TL, Egdorf T, Enstrom R, Gary A, Gelfand E, Gorham M, Griffin F, Gu H, Hancock N, Howard R, Kuan L, Lambert S, Lee EK, Luviano J, Mace K, Maxwell M, Mortrud MT, Naeemi M, Nayan C, Ngo NK, Nguyen T, North K, Ransford S, Ruiz A, Seid S, Swapp J, Taormina MJ, Wakeman W, Zhou T, Nicovich PR, Williford A, Potekhina L, McGraw M, Ng L, Groblewski PA, Tasic B, Mihalas S, Harris

JA, Cetin A, Zeng H (2022) A whole-brain monosynaptic input connectome to neuron classes in mouse visual cortex. *Nat Neurosci.* 26:350–364. doi:10.1038/s41593-022-01219-x

Young H, Belbut B, Baeta M, Petreanu L (2021). Laminar-specific cortico-cortical loops in mouse visual cortex. *eLife.* 10:e59551.

Zhang Q, Li H, Chen M, Guo A, Wen Y, Poo M (2018). Functional organization of intrinsic and feedback presynaptic inputs in the primary visual cortex. *Proc Natl Acad Sci U S A.* 115:E5174–E5182.

Zhang S, Xu M, Chang WC, Ma C, Hoang Do JP, Jeong D, Lei T, Fan JL, Dan Y (2016). Organization of long-range inputs and outputs of frontal cortex for top-down control. *Nat Neurosci.* 19:1733–1742.

Zhang S, Xu M, Kamigaki T, Hoang Do JP, Chang WC, Jenvay S, Miyamichi K, Luo L, Dan Y (2014). Long-range and local circuits for top-down modulation of visual cortex processing. *Science.* 345:660–665.

39.

MODULAR ARCHITECTURE OF LAYER 1 AS A TEMPLATE FOR THE ORGANIZATION OF THE MOUSE DISTRIBUTED HIERARCHICAL VISUAL NETWORK

Andreas Burkhalter

FRAMEWORK FOR MODULAR PROCESSING

Vision starts in the retina, where local neuronal networks generate selective responses to image elements of different luminosities, colors, textures, and motions elicited from small regions of the visual field (Roelfsema, 2006). Information from such receptive fields (RF) are encoded by spike trains, which are carried into the brain by the optic nerve for further processing in many different subcortical nuclei and cortical areas along the visual pathway. Neurons in earlier targets are selective for simple stimulus features, and become more highly selective for conjunctive properties in deeper parts of the brain (Roelfsema, 2006). The purpose of such feedforward (FF) processing is to group image elements, segregate object fragments from the background, and create representations in which different attributes of the same object are distributed across multiple areas of the hierarchical network (Felleman and Van Essen, 1991; Roelfsema, 2006).

Constructing representations by pre-attentive FF processing based on saliency maps, however, is just one aspect of vision. Identification of specific objects requires discrimination of different activity patterns and figuring out which discharge patterns are associated with specific objects seen at different locations, in various orientations and sizes. Extracting invariant object features is enabled by attentive processing (Olshausen et al.,1993; Roelfsema, 2006). Once selected by attracting attention, an object can be interpreted by matching external input features with internally represented top-down information acquired by previous visual experience. This generative process depends on recurrent inputs from neighboring neurons with similar RFs, as well as interactions with neurons distributed across the cortical hierarchy, which send gating signals to lower areas and modulate the strength and functional content of responses of groups of neurons that map the same point in space (Lee et al., 2016).

In primary visual cortex (V1), information is represented in a retinotopic map. Embedded in this global map, information is also represented at a finer scale in periodic maps of object features. It has been proposed, based on a computational modeling, that shifting and scaling responses to features of the same object are governed by circuits underlying attention (Olshausen et al., 1993). Such circuits are implemented most economically by clustered connections, which target groups of neurons specialized for encoding of low-level cues (Olshausen et al., 1993). Clustered connections in FF pathways have long been known from retrograde tracings of inter-areal pathways in monkey visual cortex (DeYoe and Van Essen, 1985), indicating that retinotopic maps exhibit modular sub-architectures of functionally specialized columns.

Spatial clustering was considered a unique feature of FF pathways, which was lacking in uniformly distributed and functionally nonspecific feedback (FB) projections (Stettler et al., 2002). This view was challenged by recent findings (reviewed in **Chapter 38 of Section VII**) showing that reciprocal loops between functionally distinct compartments of areas V1 and V2 in monkey visual cortex are topographically precise (Federer et al., 2021). Specifically, FB connections from thin cytochrome oxidase (CO)-rich stripes and those from thick and interstripes in V1 were precisely targeted to layers (L) 1, 2/3, 4A, and 5/6 of CO-reactive blobs and CO-poor interblobs of V1, respectively. Notably, FB inputs to L1 may contact apical dendrites of pyramidal neurons whose cell bodies reside in the layers below (Karimi et al., 2020). Coincident synaptic activation by FB inputs to distal dendrites and FF input to basal dendrites of the same cell may provide nonlinear coupling between top-down and bottom-up inputs and gate responses in functionally distinct cell clusters (Larkum et al., 1999; Fisek et al., 2023).

The picture that emerges from this brief overview of the primate visual network indicates that intracortical FF and FB connections are reciprocal and organized in modular like-to-like fashion. Although the basic structure of the network has been known for more than 30 years (Felleman and Van Essen, 1991), only recently have we learned that the connections between areas and columns are topographically precise (Federer et al., 2021; Angelucci and Petreanu, 2023). The purpose of this modular network architecture may be to spatially segregate the functionally distinct "where," "what," and "how" visual-processing streams (Ungerleider and Mishkin, 1982). Recent discoveries of spatially clustered hierarchical network architectures of thalamocortical and intracortical circuits in mice (Wang et al., 2012; Ji et al., 2015; D'Souza et al., 2019; D'Souza et al., 2022) suggest, based on observations in primates, that the organization may allow stream-selective gating by attention. Conceivably, pre-attentive bottom-up input may gate processing in the dorsal "where" stream, while the ventral "what" stream may be more tightly controlled by selective attention through top-down FB connections (Roth et al., 2016; Wang and Krauzlis, 2018). The proposal is derived from the similarities of the modular network architectures of mouse and primate visual systems. Specifically, it is supported by the module-selective routing of thalamocortical inputs from the dLGN and pulvinar (lateral posterior nucleus), as well as FF and FB connections between V1 and higher cortical areas (Ji et al, 2015; D'Souza et al., 2019; D'Souza et al., 2021; Meier et al., 2021).

MODULES IN MOUSE VISUAL CORTEX

It has long been held, based on the lack of orientation columns (Ohki et al., 2005), that columnar units are absent in rodent V1. While orientation columns in mice may be too thin or the streamlines too curved to be readily observed (Kondo et al., 2016; Maruoka et al., 2017), computational modeling suggests that in mice and rats the sampling ratio between the retina and V1 is too low for the formation of columns (Jang et al., 2020), whose significance for encoding specific types of visual information has been debated (Horton and Adams, 2005). However, there is general agreement from studies in carnivores and primates that cell clusters play a role in the routing of information into different processing streams (Horton and Adams, 2005). Neuronal clustering is known to facilitate correlated activity, which, even when spontaneously generated in the retina, shapes the development of dorsal and ventral visual cortical processing streams and their retinotopic interconnections (Murakami et al, 2022). Thus, it is reasonable to speculate that similar mechanisms underlie the development of the fine-scale connectivity between distinct cell clusters in different areas of primate visual cortex (Federer et al., 2021). However, in mouse barrel cortex, synaptic connectivity in cell clusters, defined by bundling of apical dendrites, is equally dense inside and outside of clusters. While this is consistent with a salt-and-pepper distribution of orientation-tuned cells in mouse V1 (Ohki et al, 2005), it suggests that additional intrinsic molecular mechanisms (Dong et al., 2015) may underlie the nonuniform dendritic architecture in L1 of V1 (Ji et al., 2015; D'Souza et al., 2019).

Despite the evidence against cell clustering in rodents, early studies have shown clustered intra-areal projections in superficial layers of flat-mounted rat V1 (Burkhalter and Charles, 1990). Many years later, in a search for molecular fingerprints of cortical areas (Gamanut et al., 2018), we found in tangential sections through V1 a strikingly periodic pattern of 60–80-μm-wide patches of M2 muscarinic acetylcholine receptor expression in L1 (Ji et al., 2015). Labeling in L2–6 was stratified but tended to be more uniform in the tangential plane (Ji et al., 2015) (Figure 39.1). Similar M2+ patches were present in L1 of rat and monkey V1, indicating that the nonuniform pattern is evolutionarily conserved (Ji et al., 2015) (Figure 39.2). M2 interpatches stood out by their parvalbumin (PV)-rich neuropil (D'Souza et al., 2019) (Figure 39.2), resembling the honeycomb-like mosaic found at the border of L1 and 2 in rat V1 (Ichinohe et al., 2003).

L1 is cell-sparse and contains axons from subcortical and cortical sources (Ji et al 2015; Harris et al., 2019; D'Souza et al., 2019; Meier et al., 2021; D'Souza et al., 2022), including dendritic tufts of pyramidal cells (PCs) whose cell bodies lie in the layers below. Most M2 receptors are membrane bound, and localized presynaptically to axons of inhibitory and excitatory neurons, as well as to cholinergic fibers and apical dendrites of PCs (Mrzljak et al., 1996; Disney et al., 2006). Since cholinergic fibers in L1 are uniformly distributed in the tangential plane (D'Souza et al., 2019), acetylcholine must act in spatially selective receptor-dependent fashion on distinct subsets of axon terminals and PC dendrites in M2+ patches and M2− interpatches.

M2+ AND M2− OUTPUT UNITS

The periodic M2 pattern in L1 resembles the vertical bundling of apical dendrites from PCs located in L2–6 (Peters and Kara, 1987). These 30–70-μm-wide microcolumns of dendrites and axons are smaller than the 200–500-μm-wide ocular dominance, orientation, and barrel columns (Meyer et al., 2013). They originate from vertical chains of cell bodies, known as minicolumns, which are

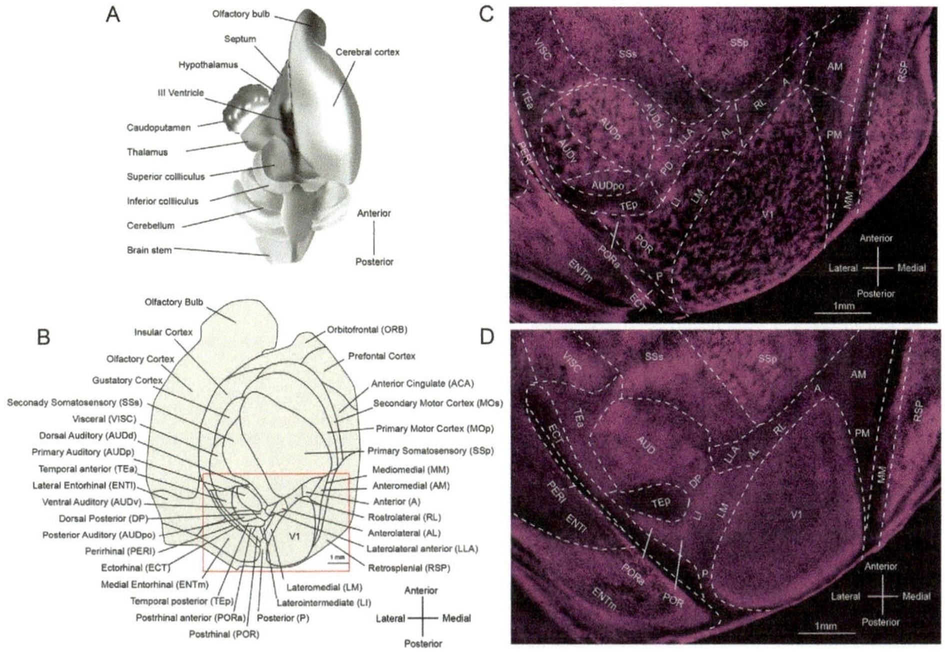

Figure 39.1. Flat-mount of mouse cerebral cortex. (A). Dissection of left mouse cerebral cortex. (B). Diagram of areal partitioning scheme based on the expression patterns of immunostaining for M2 muscarinic acetylcholine receptor and vesicular glutamate transporter 2, cytochrome oxidase histochemistry, and tdTomato fluorescence in parvalbumin tdT transgenic reporter mice (adapted from Gamanut et al., 2018). (C). Tangential section through L1 of flat-mounted mouse visual cortex. Immunostaining with an antibody against the M2 acetylcholine receptor (M2) shows patchy expression in V1 and the surrounding higher visual areas LM, LI, POR, and PORa. The imaged region corresponds to the red boxed part shown in panel A. (D). Tangential section through L4 shows uniform expression of M2 in V1. The images shown in panels C and D are adapted from Meier et al. (2021).

thought to represent elementary units of cortical processing. While roughly similar in width, minicolumns are dedicated to intracolumnar processing, whereas microcolumns exchange dendritic branches between neighboring apical bundles, which makes parent PCs part of an intercolumnar network (Figure 39.3). Estimated from the diameter of dendritic tufts in L1, the tangential diameter of this network is at least 100–350 μm wide (Gouwens et al., 2019). This is larger than the width of dendritic bundles in L2–6, and larger than the 60–80-μm-wide M2+ patches we have found in L1 (Ji et al., 2015). What then is the relationship between M2 patches and microcolumns? One possibility is that M2 is limited to proximal dendritic branches and is not expressed in the tips of the wider distal tufts. Alternatively, the dendritic canopy of individual PCs may be tangentially nonuniform with clusters in M2 patches or M2- interpatches, resembling an umbrella with holes (Figure 39.3). If so, groups of M2+ patches and M2− interpatches may belong to distinct assemblies of microcolumns, whose parent cell bodies are distributed across a surface area with a diameter of at least the size of a single dendritic tuft. Such assemblies of dendritic bundles have been shown, by retrograde tracing of their parent cell bodies, to be connected to the same cortical or subcortical target, and are referred to as "output units" (Innocenti and Vercelli, 2010) (Figure 39.3). Retinotopic mapping in mouse V1 has shown that a total of 6–10 M2+ patches and M2− interpatches have overlapping RFs contained within 250 × 360 μm of surface area (Ji et al., 2015), roughly the spread of apical dendrites in L1 (Gouwens et al., 2019). The robust micro-organization within this space coexists with the more scattered spatial organization of neurons with similar functions (Ohki and Reid, 2007). The organization suggests that M2+ and M2− output units representing the same point in visual space are differentially influenced by cholinergic inputs.

THALAMOCORTICAL CONNECTIONS OF M2+ AND M2− OUTPUT UNITS IN V1

Tracings of the visual pathway in mice have shown that contralateral and ipsilateral projections from one eye terminate in L1–5 of V1 (Antonini et al., 1999). The densest input from the dLGN core was found in L3–4 (Cruz-Martin et al., 2014). Unlike in cats and monkeys, the inputs were uniform and lacked ocular dominance bands (Antonini et al., 1999). Consistent with these observations, individual afferent dLGN axons in L3–5 were shown to be spatially uniform (Antonini et al., 1999). They covered a cortical surface larger than the 250 × 360-μm-wide region of approximately 10 M2+ and M2− modules, dedicated to encode visual information from the 9°–15°-wide classical RF (Antonini et al., 1999; Ji et al., 2015). Thus,

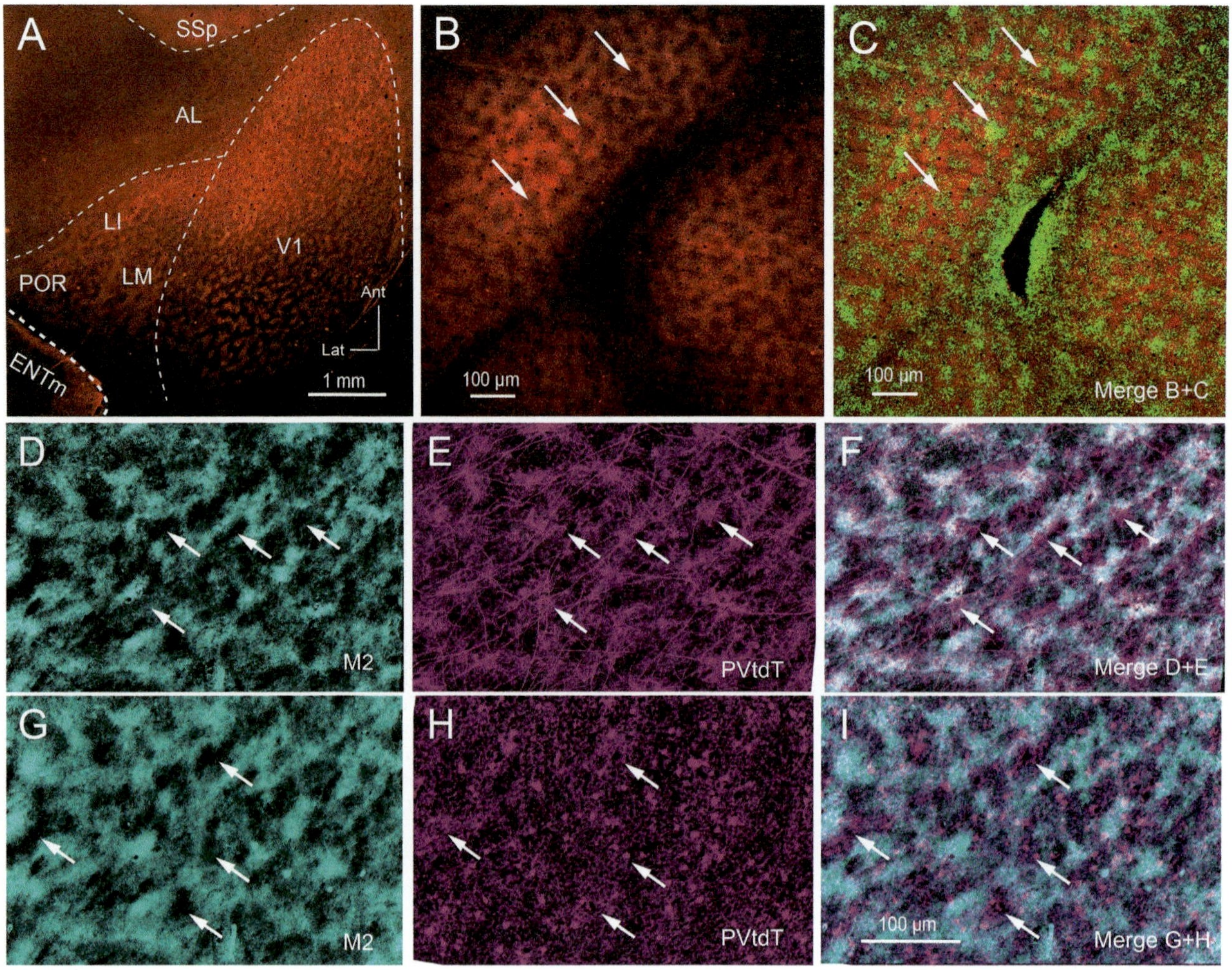

Figure 39.2. Patchy expression of M2 is evolutionarily conserved. (A). Patchy expression of M2 immunostaining in tangential section though L1 of rat visual cortex. (B). Patchy expression of M2 immunostaining in tangential section through L1 of macaque monkey primary visual cortex (V1). Arrows indicate M2− interpatches. (C). Complementary patches of M2 immunostaining (red) in L1 and cytochrome oxidase (CO, green) in L2/3 of macaque V1. Arrows indicate CO-reactive interpatches. (D). M2 immunoreactive patches in tangential section through L1 of mouse V1. Arrows indicate M2− interpatches. (E). Same section as in panel D, showing complementary expression of tdTomato in parvalbumin (PV) processes (arrows) of PVtdT transgenic reporter mice. (F). Overlay of panels D and E. (G–I). Complementary M2 patterns of M2− interpatches in L1, shown in panel G, and PV cell bodies (arrows) in L2/3 shown in panel H. (I). Overlay of panels G and H. Panels A–C adapted from Ji et al. (2015); D–I adapted from D'Souza et al. (2019).

the wide tangential extent of dLGN core inputs to L3–5 easily reaches the basal dendrites of PCs (Gouwens et al., 2019) in M2+ and M2− compartments. In sharp contrast, inputs to L1, which originate from the dLGN shell (Cruz-Martin et al., 2014), are clustered in M2+ patches and avoid M2− interpatches (Ji et al., 2015). Notably, while dLGN core inputs to basal dendrites lack modular specificity, dLGN shell projections to L1 select for a subset of apical PC dendrites, which overlap with M2+ patches (Figure 39.3). A prominent feature of PCs with similar cortical or subcortical connections is that their dendrites are bundled in clusters of approximately 50-µm-wide microcolumns (Innocenti and Vercelli, 2010). In L1 microcolumns expand to 80-120 um-wide clusters, suggesting that PCs contained within an M2+ output unit receive dLGN shell inputs to dendritic tufts in M2+ patches (Figure 39.3).

Notably, a complementary set of output units receive L1 input from the higher-order anterior lateral posterior thalamic nucleus (aLP), which targets M2− interpatches (D'Souza et al., 2019) (Figure 39.3). Although both of the interdigitating M2+ and M2− output units receive dLGN core input to basal dendrites from the same point of visual space, inputs to their apical dendrites in L1 derive from dLGN shell and aLP, respectively (Figure 39.3). Inputs at this distal location play little role in driving stimulus-selective responses (Park et al., 2019). Rather, studies in rodent somatosensory cortex have shown that L1 inputs gate the frequency, pattern, and variability of firing of different types of output neurons (Egger et al., 2015; Doron et al.,

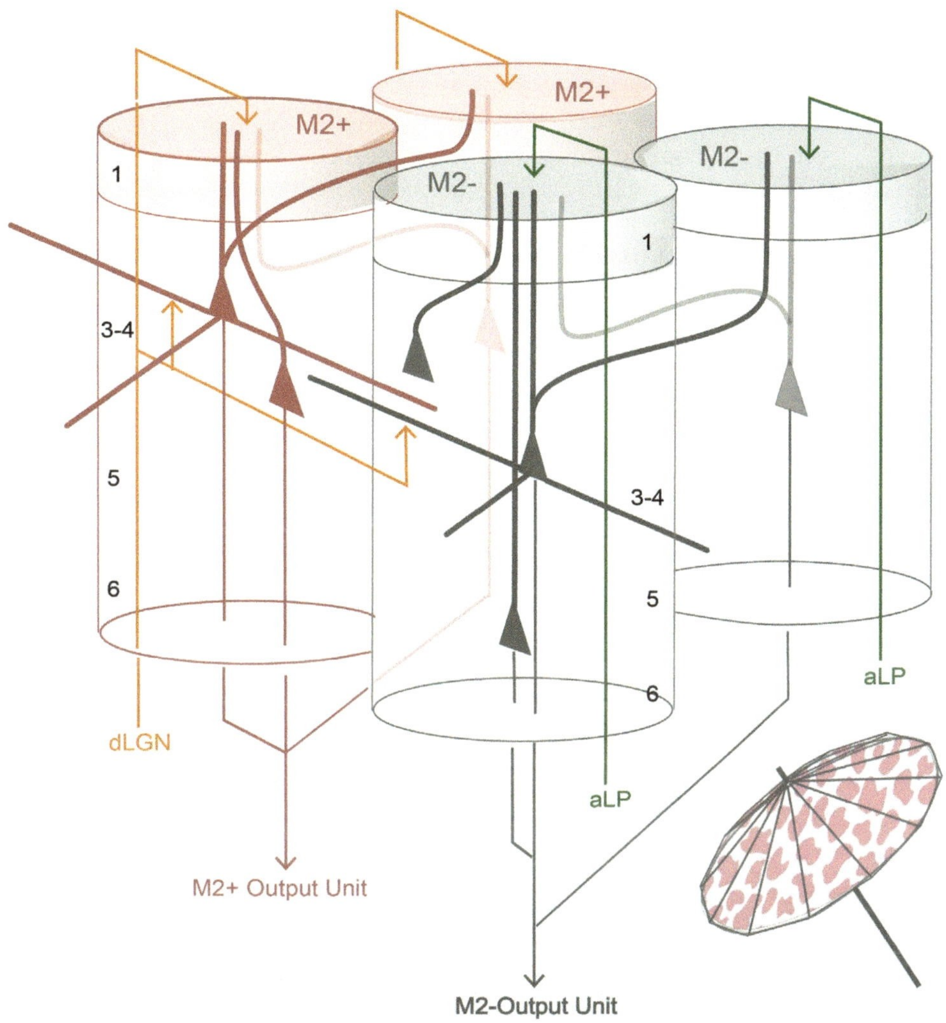

Figure 39.3. Simplified diagram of the networks of M2+ and M2– output units in mouse V1. The overall organization is derived from the non-uniform tangential distribution of M2+ patches (red) and M2– interpatches (gray) in L1. Notice, in situ M2+ patches and M2- interpatches are alternating. For simplicity they are drawn as adjacent M2+ and M2- pairs. The columns across L2–6 are conceived as modules aligned with the patch/interpatch system in L1. Each module contains pyramidal cells (triangles) with apical and basal dendrites (thick lines). Axons are indicated by thin lines. Non-pyramidal cells are omitted for simplicity. Apical dendrites are bundled in microcolumns and terminate in tufts, which overlap with M2+ patches or M2– interpatches in L1. This may resemble an umbrella with red and gray patches, as shown at lower right corner. Each microcolumn sends and receives dendritic branches from 6–10 neighboring modules of the same type. Patch- and interpatch-preferring assemblies of dendrites cover 250×360 µm of surface area, which approximates the overall diameter of the distal dendritic arbor of a pyramidal cell (Scala et al., 2019), and includes neurons in L2–6 representing the same point in visual space (Ji et al., 2015). The parent cell bodies of such assemblies have been shown to project to the same target, and were designated "output units" (Innocenti and Vercelli, 2010). Hence, V1 contains M2+ and M2– output units. M2+ and M2– output units receive inputs from the dLGN core (yellow lines) to basal dendrites in L3/4. Inputs from the dLGN shell selectively target M2+ patches in L1. In contrast, input from the higher-order aLP thalamus (green lines) selectively innervate M2– interpatches in L1. Thus, we propose that V1 contains interdigitating M2+ and M2– output units with distinct inputs from the thalamus, which encode signals from the same retinotopic location.

2020). Calcium recordings from axon terminals in mouse V1 have shown that thalamocortical inputs are an important source of inputs to L1 (Roth et al., 2016). Excitation of dLGN terminals was strongest when externally and self-generated visual motion was matched. In contrast, LP inputs were most active when externally and self-generated visual motion deviated, a condition that may draw attention to unexpected events (Roth et al., 2016). Since the pathways are strikingly patchy (D'Souza et al., 2019), it seems likely that inputs from the dLGN shell and aLP provide distinct behavioral contexts to M2+ and M2– output units for differential downstream processing.

INTRACORTICAL CONNECTIONS OF M2+ AND M2– OUTPUT UNITS

Mouse visual cortex contains at least 10 distinct retinotopically organized areas (Gamanut et al., 2018; Meier et al., 2021), which project to 39 and receive input from 46 intracortical targets (Wang et al., 2012; Gamanut et al., 2018). The rich innervation by a large number of functionally specialized cortical areas (Andermann et al., 2011; Marshel et al., 2011; Siegle et al., 2021; Han et al., 2022) suggests that V1 processes cross-modal sensory and motor signals. In spite of this, it was unexpected to find that outputs from V1 flow into distinct dorsal and ventral processing streams (Wang et al., 2012). Graph theoretical analyses showed that the ventral network includes areas LM (lateromedial), LI (laterointermediate), P (posterior), and POR (postrhinal), while the dorsal network consists of areas AL (anterolateral), RL (rostrolateral), A (anterior), AM (anteromedial), and PM (posteromedial) (Figure 39.4). Wide-field calcium recordings of responses to rich visual noise stimuli in each of these areas showed a continuous distribution of spatiotemporal tuning (Han et al., 2022). However, the properties showed different distributions with preferences for high spatial frequencies, low temporal frequencies, and low speeds in areas of the ventral stream, while areas with opposite biases were found in the dorsal stream (Han et al., 2022).

One of the most consistent features of intracortical networks is that neurons with different functional

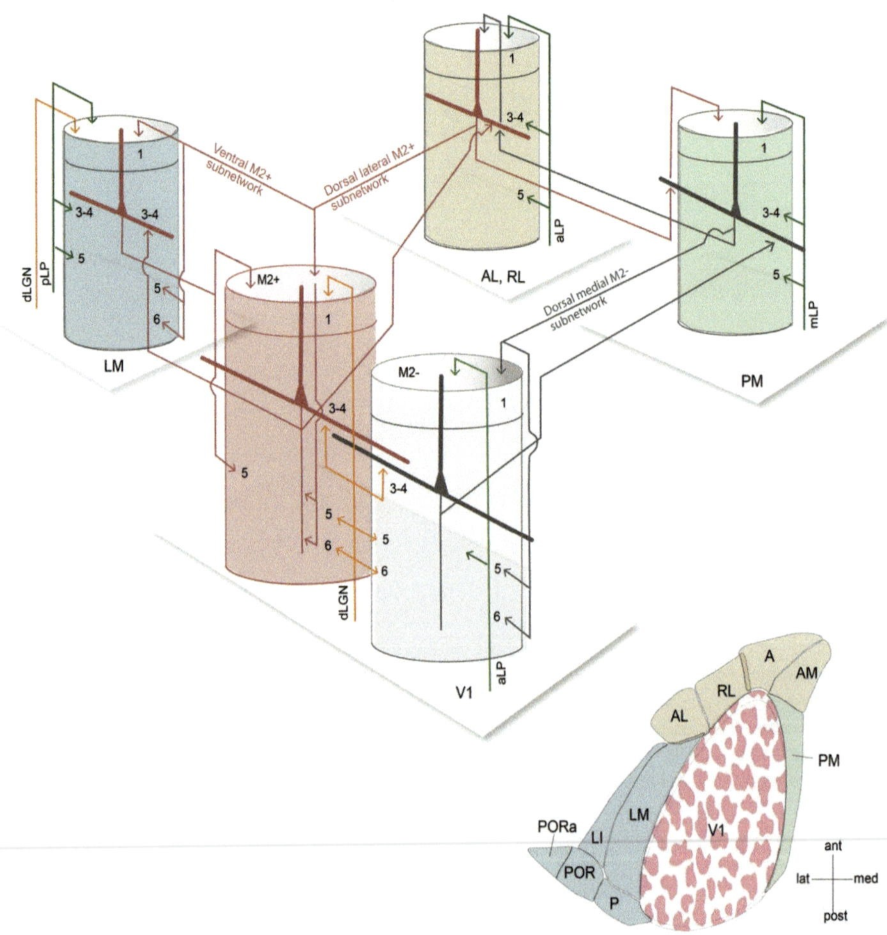

Figure 39.4. Tripartite intracortical processing streams. Simplified diagram of ventral M2+ , lateral dorsal M2+ and medial dorsal M2− cortical networks, interconnecting different areas of mouse visual cortex. A map of 11 visual areas is shown in the lower right corner. The patchy pattern in V1 indicates M2+ patches (red) and M2− interpatches (light gray) in L1. Ventral stream areas LM, LI, P, POR, and PORa (blue). Lateral and anterior dorsal areas AL, RL, A, and AM (tan). Medial dorsal area PM (green). The scheme in the center shows the connectivity of M2+ (red) and M2− (gray) output units composed of red and gray pyramidal cells (triangles) with dendrites (thick lines) and axons (thin lines). It is evident that V1 M2+ and M2− modules show distinct connectivities. The inter- areal connections are reciprocal, and are like- to- like; i.e., outputs from cells with apical dendrites in M2+ patches or M2− interpatches receive inputs to patches or interpatches in L1. M2− modules belong to the M2− medial dorsal stream. They receive input from the dLGN core to L3/4 (yellow lines) and aLP thalamus to L1 (green lines), form like- to- like FF- FB loops (gray lines) with PM (D'Souza et al., 2022), which itself is reciprocally connected by lateral connections (gray and red lines) with AL and RL (D'Souza et al., 2022), and receives input from the mLP (green lines). The M2+ network receives input from the dLGN core to L3/4 and dLGN shell to L1 (yellow lines). The ventral M2+ branch forms a like- to- like FF- FB loop with LM, which itself receives input from the dLGN core (yellow lines) and the pLP (green lines). The lateral dorsal M2+ subnetwork makes looped like- to- like FF- FB connections with AL and RL, which themselves receive input from aLP (green lines) and PM. Thus, in the M2+ lateral dorsal subnetwork, information from the posterior parietal cortex and the medial branch feeds back to M2+ modules in V1, where signals from both dorsal subnetworks may be jointly analyzed and flow downstream to ventral areas. The proposed tripartite structural scheme is aligned with distinct stream- specific visual response properties found by wide- field calcium imaging (Han et al., 2022), and refines the binary layout proposed previously (Wang et al., 2012).

specializations form clusters (Kravitz et al., 2013). How does the M2 modularity of mouse V1 fit into this organization? Reciprocal connectivity is a characteristic feature of primate inter-areal networks (Sui et al., 2021). This is also the case in mouse visual cortex, which shows looped connections between V1 ↔ LM and V1↔PM (Young et al., 2021). We have found that these loops are not uniform, but show a modular organization (Ji et al., 2015; D'Souza et al., 2019; D'Souza et al., 2021). Notably, FB projections from LM→V1 terminate in M2+ patches of L1, where they overlap precisely with clustered dendritic tufts of FF-projecting neurons (D'Souza et al., 2019; D'Souza et al., 2021) (Figure 39.4). Similar like-to-like connections were found in the V1↔PM loop, except that FB axons and apical dendrites of FF-projecting neurons overlap in M2− interpatches (D'Souza et al., 2019; D'Souza et al., 2021). At first sight, this fingerprint suggests that ventral and dorsal streams are spatially segregated in V1. A closer look, however, shows a more multiplexed organization in which M2+ loops of the ventral area, LM, overlap with like-to-like M2+ loops of the dorsal areas AL and RL (D'Souza et al., 2019; D'Souza et al., 2021) (Figure 39.4). Thus, it appears that M2+ patches group the dorsal areas AL and RL of the lateral posterior parietal cortex together with the ventral area, LM. In contrast, the medial dorsal area, PM, is linked to M2− output units of V1. The proposed tripartite structural model of visual processing at early stages of the cortical hierarchy (Harris et al., 2019; D'Souza et al., 2022) aligns well with the 3 distinct functional clusters observed by wide-field calcium imaging (Han et al., 2022). In addition, the proposal is consistent with the framework in primates, which shows that the dorsal stream contains multiple branches (Kravitz et al., 2011).

The mesoscale modularity challenges the long-held belief that FB connections are randomly distributed (Stettler et al., 2002). In fact, inputs from larger RFs of the hierarchically higher areas LM, AL, and RL (Harris et al., 2019; Siegle et al., 2021; D'Souza et al., 2022) not only provide V1 with distinct types of contextual information (Andermann et al., 2011; Marshel et al., 2011; Han et al., 2022), but also selectively combine FB input from ventral, lateral, and medial dorsal areas in M2+ and M2− modules of V1. A simplified diagram of inputs to V1 to apical dendrites in L1 from LM, AL, RL, and PM, a tiny fraction of the 46 feedback-projecting areas (Gamanut et al., 2018), is shown in Figure 39.4. The M2− subnetwork is simple compared to the M2+ subnetwork. It starts with module-nonselective input from the dLGN core to L3–4 onto basal dendrites of PCs in M2− and M2+ modules. Thalamic inputs to L1 are strikingly module-selective, with projections from the dLGN shell to M2+ patches, and inputs from aLP to M2− interpatches (Ji et al., 2015; Bennett et al., 2019; D'Souza et al., 2019). V1 M2− interpatches in addition receive FB input from area PM of the dorsal medial stream (D'Souza et al., 2019), which itself receives input from mLP, AL, and RL to L1, L4, and L5 (Bennett et al., 2019; D'Souza et al., 2022). The V1 M2+ patch network differs in connectivity not only with the thalamus, but also with higher cortical areas. For example, FB inputs from AL and RL converge onto apical dendrites of M2+ patches, avoiding M2− interpatches. Thus, in the simplified diagram shown in Figure 39.4, the M2− and M2+ networks are largely independent. Notably however, FB inputs from LM→V1, AL→V1, and RL→V1 converge on M2+ patches of L1 (D'Souza et al., 2021). This indicates that information from ventral and lateral dorsal areas, which receive distinct thalamocortical input (i.e., LM from dLGN and pLP; AL and RL from aLP; Bennett et al., 2019; Meier et al., 2021) and intracortical inputs (Gamanut et al., 2018) are integrating information in M2+ module-specific output units, that together with complementary M2− units represent the same location in visual space (Figure 39.4).

M2+ AND M2− SUBNETWORKS ARE FUNCTIONALLY DISTINCT

Electrophysiological recordings in mouse V1 have shown that L2/3 neurons aligned with M2+ patches are sensitive to high spatial frequency gratings, while M2− interpatch neurons are more sensitive to the direction and speed of moving dot patterns (Ji et al., 2015). From these results, one expects that LM, the main downstream ventral stream target of M2+ PCs in V1, would be biased for static object features. However, LM encodes visual motion as well (Marshel et al., 2011; Han et al., 2022). Some of these features may be inherited from direction selective (DS) ON-OFF retinal signals, carried by the dLGN shell→V1 pathway (Cruz-Martin et al., 2014) to LM-projecting PCs, whose dendrites reach to M2+ patches in L1. In addition, LM may acquire motion sensitivity from pLP inputs to L4 and L1 (Bennett et al., 2019), inputs which were shown to make POR neurons susceptible to small moving objects (Beltramo and Scanziani, 2019). Thus, M2+ output units in V1 that project to LM and onward to downstream areas of the ventral stream (Wang et al., 2012) may endow neurons with high spatial resolution of slowly moving objects, properties that are well suited for the identification and pursuit of moving prey.

The modular architecture of the dorsal stream can be most effectively linked to visual motion processing, which underlies the perception of optic flow (Marques et al., 2018). There are multiple sources of motion signals. Retina-derived DS is carried to M2+ patches in V1 (Cruz-Martin et al., 2014; Ji et al., 2015; Roth et al., 2016), where it is acquired by RL-projecting neurons in L2/3 of the same module (Rasmussen et al., 2020; D'Souza et al., 2021) (Figure 39.4). Cortex-derived DS in PM-projecting PCs contained in neighboring M2− interpatch modules may emerge from computations within cortical circuits (Lien and Scanziani, 2018). Alternatively DS may be provided by aLP inputs, which are tuned to global motion (Bennett et al., 2019). RL and PM are both tuned to coherent motion of random dot kinematograms (Sit and Goard, 2020). While the sensory representations in RL and PM may be similar, they depend on inputs from M2+ or M2− output units in V1, respectively. The modular architecture of V1 suggests that RL responses may be more strongly affected by eye/head movements (Meyer et al., 2020), whereas PM responses may be driven by movements of the visual scene during self-motion of the body. RL and PM also make like-to-like FB connections to V1 (Figure 39.4). In this way M2+ neurons in V1 may combine inputs from the dLGN shell with aLP-derived FB from RL (Figure 39.4), encode conjunctive eye-centered and body-centered reference frames, and extract heading information that is not available from dLGN input alone (Diamanti et al., 2021). In contrast, M2− V1 PCs integrate FF input from aLP and dorsal medial FB from PM to predict optic flow based in active motor output during spatial navigation that is known to depend on areas of the posterior parietal cortex (Minderer et al., 2019).

The modular architecture of streams also intersects with the transthalamic loop via the LP (Sherman, 2016). To date the data are limited to AL of the M2+ dorsal lateral subnetwork and PM of the M2− dorsal medial substream (Blot et al., 2021). Tracings of trans-thalamic connections have shown that AL and PM are reciprocally connected with distinct types of LP neurons and receive input from largely overlapping sets of visual and nonvisual areas (Blot et al., 2021). Retina-derived DS in V1 may be carried forward from M2+ to AL, a projection that is purely visual and is unaffected by the animal's running speed (Blot et al., 2021). In parallel with this intracortical stream, a trans-thalamic loop from M2− interpatches projects via aLP to AL. Notably, unlike the intracortical pathway, this trans-thalamic loop was shown to carry optic flow signals, which are suppressed by running speed (Blot et al., 2021). This suggests that the trans-thalamic loop may be visually most active when the animal is stationary and is specialized for the processing of visual flow relative to self-motion (Blot et al., 2021). In contrast, trans-thalamic projections to PM carry a wide range of visuomotor signals, which are strongly modulated by locomotion, suggesting that they are relevant the processing of spatial detail during landmark navigation (Andermann et al., 2011; Marshel et al., 2011; Blot et al., 2021; Han et al., 2022).

PERSPECTIVE

Accumulating evidence suggests that the hierarchical framework we have used may be overly simplistic. In fact, Vezoli, Hou, and Kennedy (see **Chapter 37 in Section VII**) make a strong case for the existence of FF and FB pathways in supragranular and infragranular layers, a layout they call "counterstream architecture" (Vezoli et al., 2021). Physiological and anatomical evidence suggests that a similar architecture may exist in mouse V1 (Barzegaran and Plomp, 2022; Shen et al., 2022). Thus, it is possible that the modular architecture we have described here applies preferentially to the supragranular network. This view is supported by the precise topography we have found in the connectivity of M2+ and M2− modules in L1–3. Further, the perspective is consistent with preliminary evidence from our lab, showing that the connectivity in deep layers is less modular and tends to be more uniform.

Another important question is how the activation of M2 receptors by acetylcholine controls the visual network. Recordings in monkey V1 have shown that spatial attention improves the gain of unitary responses, and that attentional modulation is reduced by the muscarinic cholinergic antagonist, scopolamine, while blocking nicotinic receptors has no effect (Herrero and Thiele, 2008). However, acetylcholine (ACh) does more than changing the response gain, it also desynchronizes the activity and makes responses in rat V1 more reliable by signaling through muscarinic ACh receptor (Goard and Dan, 2009). Optogenetic activation of cholinergic input from the basal forebrain has recently been shown to improve encoding of visual information in neuronal populations of mouse V1 (Minces et al., 2017). Such improvements come about by increasing the signal-to-noise ratio in a manner in which stimulus and noise correlations have opposite signs. It has been suggested that activation of M2 may reduce the correlation of intracortical signals while simultaneously enhancing the efficacy of thalamocortical inputs (van Kempen et al., 2017). M2 receptors also play a role in encoding predictive representations of the patterns and temporal order of recurrent stimulus sequences in mouse V1 (Sakar et al., 2022), involving mechanisms of associative reinforcement learning. Such mechanisms have been linked to cortical L1, where the sensitivity of tuft dendrites is enhanced by contextual input (Doron et al., 2020).

ACKNOWLEDGMENTS

I would like to thank Pawan Bista, Will Cater, Rinaldo D. D'Souza, Razavan Gamanut, Weiqing Ji, Henry Kennedy, Ken Knoblauch, Andrew M. Meier, Olaf Sporns, Katia Valkova, David C. Van Essen, and Quanxin Wang; without their contributions this article would not have been possible. This work was supported by the National Institutes of Health grants RO1EY0161184, RO1EY1020523, RO1EY022090, RO1EY027383, and R21EY027946.

REFERENCES

Andermann M, Kerlin AM, Roumis DK, Glickfeld LL, Reid RC (2011). Functional specialization of mouse higher cortical areas. Neuron. 72:1025–1039.

Antonini A, Fagolini M, Stryker MP (1999). Anatomical correlates of functional plasticity in mouse visual cortex. *J Neurosci.* 19: 4388–4406.

Barzegaran E, Plomp G (2022). Four concurrent feedforward and feedback networks with different roles in the visual cortical hierarchy. *PLoS Biol.* 20(2):e3001534. doi:10.1371/journal.pbio.3001535.

Beltramo R, Scanziani M (2019). A collicular visual cortex: Neocortical space for an ancient midbrain visual structure. *Science.* 363:64–69.

Bennett C, Gale SD, Garrett ME, Newton ML, Callaway EM, Murphey GJ, Olsen SR (2019). Higher order thalamic circuits channel parallel streams of visual information in mice. *Neuron.* 102:477–492.

Blot A, Roth MM, Gasler I, Javadzadeh M, Imhof F, Hofer SB (2021). Visual intracortical and transthalamic pathways carry distinct information to cortical areas. *Neuron.* 109:1996–2008.

Burkhalter A, Charles V (1990). Organization of local axon collaterals of efferent projection neurons in rat visual cortex. *J Comp Neurol.* 32:920–934.

Cruz-Martin A, El-Danaf RN, Osaka F, Sriram B, Dhande OS, Nguyen PL, Callaway EM, Gosh A, Huberman AD (2014). A dedicated circuit links direction-selective retinal ganglion cells to the primary visual cortex. *Nature.* 507:358–361.

D'Souza RD, Bista P, Meier AM, Ji W, Burkhalter A (2019). Spatial clustering of inhibition in mouse primary visual cortex. *Neuron.* 104:588–600.

D'Souza RD, Ji W, Burkhalter A (2021). A modular organization of looped interareal and thalamocortical pathways in mouse visual cortex. *Annu Mtg Soc Neurosci.* Abstract P463.09.

D'Souza RD, Wang Q, Ji W, Meier AM, Kennedy H, Knoblauch, Burkhalter A (2022). Hierarchical and nonhierarchical features of the mouse visual cortical network. *Nat Commn.* 13(1):503. doi:10.1038/s41467-022-28035-y.

DeYoe EA, Van Essen DC (1985). Segregation of efferent connections and receptive field properties in visual area V2 of the macaque. *Nature.* 317:58–61.

Diamanti EM, Reddy CB, Schröder S, Muzzu T, Harris KD, Saleem AB, Carandini M (2021). Spatial modulation of visual responses arises in cortex with active navigation. *eLife.* 10:e63705. doi:10.7554/eLife.63705.

Disney AA, Domakonda KV, Aoki C (2006). Differential expression of muscarinic acetylcholine receptors across excitatory and inhibitory cells in visual cortical areas V1 and V2 of the macaque. *J Comp Neurol.* 499:49–63.

Dong X, Shen K, Bülow HE (2015). Intrinsic and extrinsic mechanisms of dendritic morphogenesis. *Annu Rev Neurosci.* 77:271–300.

Doron G, Shin JN, Takahashi N, Drüke M, Bocklisch C, Skenderi S, de Mont L, Toumazou J, Ledderose J, Brecht M, Naud R, Larkum ME (2020). Perirhinal input to layer 1 controls learning. *Science.* 370(6523):eaaz3136. doi:10.1126/science.aaz3136.

Egger R, Schmitt AC, Wallace DJ, Sakmann B, Oberlaender M, Kerr JND (2015). Robustness of sensory-evoked excitation is increased by inhibitory inputs to distal apical tuft dendrites. *Proc Natl Acad Sci U S A.* 132:14072–14077.

Federer F, Ta'afua S, Merlin S, Hasanpour MS, Angelucci A (2021). Stream-specific inputs to the primate primary visual cortex. *Nat Commun.* 12(1):228. doi:10.1038/s41467-020-20505-5.

Felleman DJ, Van Essen DC (1991). Distributed hierarchical processing in the primate. *Cereb Cortex.* 1:1–47.

Fisek M, Herrmann D, Egea-Weiss A, Cloves M, Bauer L, Lee T-Y, Russell LE, Häusser M (2023) Cortico-cortical feedback engages active dendrites in visual cortex. *Nature.* 617(7962):779–776. doi:10.1038/s41586-023-06007-6.

Gamanut R, Kennedy H, Toroczkai Z, Ercsey-Ravasz M, Van Essen DC, Knoblauch K, Burkhalter A (2018). The mouse cortical connectome, characterized by an ultra-dense cortical graph, maintains specificity by distinct connectivity profiles. *Neuron.* 97:698–715.

Goard M, Dan Y (2009) Basal forebrain activation enhances cortical coding of natural scenes. *Nat Neurosci.* 1:1440–1445.

Gouwens NW, Sorensen SA, Berg J, Lee C, Jarsky T, Ting J,Sunkin SM, Feng D, Anaatassiou CA, Barkan E, Bickley K, Blesie N, Braun T, Brouner K, Budzillo A, Caldejon S, Casper T, Castelli D, Chong P, Crichton A, Cuhaciyan C, Daigle TL, Dalley R, Dee N, Desta T, Ding S-L, Dingman S, Doperalski A, Dotson N, Egdorf T, Fisher M, de Frates RA, Garren E, Garwood M, Gary A, Gauderault N, Godfrey K, Gorham M, Gu H, Habel C, Hadley K, Harrington J, Harris JA, Henry A, Hill D, Josephsen S,Kebede S, Kim L, Kroll M, Lee B, Lemon T, Link KE, Liu X, Long B, Mann R, McGraw M, Mihalas S, Mukora A, Murphey GJ, Ng L, Ngo K, Nguyen TN, Nicovich PR, Oldre A, Park D, Parry S, Perkins J, Potekhina L, Reid D, Robertson M, Sandman D, Schroedter M, Slaughterbeck C, Soler-Llavina G, Sulc J, Szafer A, Tasic B, Taskin N, Teeter C, Thatra N,Tung H, Wakeman W, Williams G, Toung R, Zhou Z, Farrel C, Peng H, Hawrylycz MJ, Lein E, Ng L, Arkhipov A, Bernard A, Phillips JW, Zeng H, Koch C (2019). Classification of electrophysiological and morphological neuron types in the mouse visual cortex. *Nat Neurosci.* 22:1182–1195.

Han X, Vermaercke B, Bonin V (2022). Diversity of spatiotemporal coding reveals specialized visual processing streams in mouse visual cortex. *Nat Commn.* 13(1):3249. doi:10.1038/s4167-022-29656-z.

Harris JA, Mihalas S, Hirokawa KE, Whitsell JD, Choi H, Bernard A Bohn P, Caldejon S, Casal L, Cho A, Feiner A, Feng D, et al. (2019). Hierarchical organization of cortical and thalamic connectivity. *Nature.* 575:195–202.

Herrero JL, Roberts MJ, Delicato LS, Gieselmann MA, Dayan P, Thiele A (2008). Acetylcholine contributes through muscarinic receptors to attentional modulation in V1. *Nature.* 454(7208):1110–1114. doi:10.1038/nature07141.

Horton JC, Adams DL (2005). The cortical column: a structure without a function. *Phil Trans R Soc B.* 360:837–862.

Ichinohe N, Fujiyama F, Kaneko T, Rockland KS (2003). Honeycomb-like mosaic at the border of layers 1 and 2 in the cerebral cortex. *J Neurosci.* 23:1372–1382.

Innocenti GM, Vercelli A (2010). Dendritic bundle, minicolumns, columns and cortical output units. *Front Neuroanat.* 4:11. doi:10.3389/neuro.05.011.2010.eCollection 2010.

Jang J, Song M, Paik S-B (2020). Retino-cortical mapping ratio predicts columnar and salt-and-pepper organization in mammalian visual cortex. *Cell Report.* 30:3270–3279.

Ji W, Gamanut R, Bista P, D'Souza RD, Wang Q, Burkhalter A (2015). Modularity in the organization of mouse visual cortex. *Neuron.* 87:632–643.

Karimi A, Odenthal J, Drawitsch F, Boergens KM, Helmstaedter M (2020). Cell-type specific innervation of cortical pyramidal cells at their apical dendrites. *eLife.* 9:e46876. doi:10.7554/eLife.46876.

Kondo S, Yoshida T, Ohki K (2016). Mixed functional microarchitecture for orientation selectivity in the mouse primary visual cortex. *Nat Commun.* 7:13210. doi:10.1038/ncomms13210.

Kravitz DJ, Saleem KS, Baker I, Mishkin M (2011). A new framework for visuospatial processing. *Nat Rev Neurosci.* 12:217–230.

Kravitz DJ, Saleem KS, Baker CI, Ungerleider LG, Mishkin M (2013). The ventral pathway: An expanded neural framework for the processing of object quality. *Trend Cogn Sci.* 17:26–49.

Larkum ME, Zhu JJ, Sakmann B (1999). A new cellular mechanism for coupling inputs arriving at different cortical layers. *Nature.* 398:338–341.

Lee W-C A, Bonin V, Reed M, Graham BJ, Hood G, Glattfelder K, Reid RC (2016). Anatomy and function of an excitatory network in the visual cortex. *Nature.* 532:370–374.

Lien AD, Scanziani M (2018). Cortical direction selectivity emerges at convergence of thalamic synapses. *Nature*. 558:80–86.

Marshel JH, Garrett ME, Nauhaus I, Callaway EM (2011) Functional specialization of seven mouse visual cortical areas. *Neuron*. 72:1040–1054.

Marques T, Summers MT, Fioreze G, Fridman M, Dias RF, Feller MB, Petreanu L (2018). A role for mouse primary visual cortex in motion perception. *Curr Biol*. 28:1703–1713.

Maruoka H, Nakagawa N, Tsuruno S, Sakai S, Yoneda T, Hosoya T (2017). Lattice system of functionally distinct cell types in the neocortex. *Science*. 358:610–615.

Meier AM, Wang Q, Ji W, Ganachaud J, Burkhalter A (2021). Modular network between postrhinal visual cortex, amygdala, and entorhinal cortex. *J Neurosci*. 41:4809–4825.

Meyer AF, O'Keefe J, Poort J (2020). Two distinct types of eye-head coupling in freely moving mice. *Curr Biol*. 30:2116–2130.e6.

Meyer HS, Egger R, Guest JM, Foerster R, Reissl S, Oberlaender M (2013). Cellular organization of cortical barrel columns is whisker-specific. *Proc Natl Aca Sci U S A*. 110:19113–19118.

Minces V, Pinto L, Dan Y, Chiba AA (2017). Cholinergic shaping of neural correlations. *Proc Natl Acad Sci U S A*. 114:5725–5730.

Minderer M, Brown KD, Harvey CD (2019). The spatial structure of neural encoding in mouse posterior parietal cortex during navigation. *Neuron*. 102:232–248.

Mrzljak L, Levey AI, Rakic P (1996). Selective expression of m2 muscarinic receptor in the parvocellular channel of primate visual cortex. *Proc Natl Acad Sci U S* A. 93:7337–7340.

Murakami T, Matsui T, Uemura M, Ohki K (2022). Modular strategy of development of the hierarchical visual network in mice. *Nature*. 608:578–585.

Ohki K, Chung S, Ch'ng Y, Kara P, Reid RC (2005). Functional imaging with cellular resolution reveals precise microarchitecture in visual cortex. *Nature*. 433:597–603.

Ohki K, Reid RC (2007). Specificity and randomness in the visual cortex. *Curr Opin Neurobiol*. 17:401–407.

Olshausen BA, Anderson CH, Van Essen DC (1993). A neurobiological model of visual attention and invariant pattern recognition based on dynamic routing of information. *J Neurosci*. 13:4700–4719.

Park J, Papoutsi A, Ash RT, Marin MA, Poirazi P, Smirnakis SM (2019). Contribution of apical and basal dendrites to orientation encoding in mouse L2/3 pyramidal neurons. *Nat Commun*. 10(1):5372. doi:41467-019-13029-0.

Peters A, Kara DA (1987). The neuronal composition of area 17 of rat visual cortex. IV. Organization of pyramidal cells. *J Comp Neurol*. 260:573–590.

Rasmussen R, Matsumoto A, Sietam MD, Yonehara K (2020). A segregated cortical stream for retinal direction selectivity. *Nat Commun*. 11(1):831. doi:10.1038/s41467-020-14643-z.

Roelfsema P (2006). Cortical algorithms for perceptual grouping. *Annu Rev Neurosci*. 29:203–227.

Roth MM, Dahmen JC, Muir DR, Imhof F, Martini FJ, Hofer SB (2016). Thalamic nuclei convey diverse contextual information to layer 1 of visual cortex. *Nat Neurosci*. 19:299–307.

Sakar S, Reyes CM, Jensen CM, Gavornik JP (2022). M2 receptors are requited for spatiotemporal sequence learning in mouse primary visual cortex. *bioRxiv*. doi:10.1101/2022.02.09.479792. 1–23.

Scala F, Kobak D, Shan S, Bernaerts Y, Laturnus S, Cadwell CR, Hartmanis L, Froudarakis E, Castro JR, Tan ZH, Papadopoulos S, Patel SS, Sandberg R, Berens P, Jiang X, Tolias AS (2019). Layer 4 of mouse neocortex differs in cell types and circuit organization between sensory areas. *Nat Commun*. 10:4174. doi:10.1038/s41467-019-12059-z.

Shen S, Jiang X, Scala F, Fu J, Fahay P, Kobak D, Tan S, Zhou N, Reimer J, Sinz F, Tolias AS (2022). Distinct organization of two cortico-cortical feedback pathways. *Nat Commun*. 13(1):6389. doi:10.1038/s41467-022-33883-9.

Sherman SM (2016). Thalamus plays a central role in ongoing cortical functioning. *Nat Neurosci*. 19:533–541.

Siegle JH, Jia X, Durand S, Gale S, Bennett C, Graddis N, Heller G, Ramirez TK, Choi H, Luviano JA, Groblewski PA, Ahmed R, Arkhipov A, Bernard A, Billeh YN, Brown D, Buice MA, Cain N, Caldejon S, Casal L, Cho A, Chvilicek M, Cox CT, Dai K, Denman DJ, deVries E, Dietzman R, Esposito L, Farrell, Feng D, Galbraith J, Garrett M, Gelfand EC, Hancock N, Harris JA, Howard R, Hu B, Hytnen R, Iyer R, Jessett E, Johnson K, Kato I, Kiggins J, Lambert S, Lecoq J, Ledochowitsch P, Lee JH, Leon A, Li Y, Liang E, Long F, Mace K, Melchior J, Millman D, Mollenkopf T, Nayan C, Ng L, Ngo K, Nguyen T, Nicovich PR, North K, Ocker GK, Ollereshaw D, Oliver M, Pachitariu M, Perkins J, Reding M, Reid D, Robertson M, Ronellenfitch K, Seid S, Slaughterbeck C, Stoecklin M, Sullivan D, Sutton B, Swapp J, Thompon C, Turner K, Wakeman W, Whitesell JD, Willams D, Williford A, Young R, Zeng H, Naylor S, Phillips JW, Reid RC, Mihalas S, Olsen S, Koch C (2021). Survey of spiking in the mouse visual system reveals functional hierarchy. *Nature*. 592(7852):86–92. doi:10.1038/s41586-020-03171-x.

Sit KK, Goard MJ (2020). Distributed and retinotopically asymmetric processing of coherent motion in mouse visual cortex. *Nat Commun*. 11(1):3565. doi:10.1038/s41467-020-17283-5.

Siu C, Balsor J, Merlin S, Federer F, Angelucci A (2021). A direct interareal feedback-to-feedforward circuit in primate visual cortex. *Nat Commn*. 12(1):4911. doi:10.1038/s41467-021-24928-6.

Stettler DD, Das A, Bennett J, Gilbert CD (2002). Lateral connectivity and contextual interactions in macaque primary visual cortex. *Neuron*. 36:739–750.

Ungerleider LG, Mishkin M (1982). Two cortical visual systems. In: Ingle DJ, Goodale M, Mansfield RJW, eds. *Analysis of Visual Behavior*. MIT Press:549–586.

Van Kempen J, Panzeri S, Thiele A (2017). Cholinergic control of information coding. *Trends Neurosci*. 40:522–524.

Vezoli J, Magrou L, Goebel R, Wang XJ, Knoblauch K, Vinck M, Kennedy H (2021). Cortical hierarchy, dual counterstream architecture and the importance of top-down generative networks. *Neuroimage*. 225:117479. doi10.1016/.neuroimage.2020.117479.

Wang L, Krauzlis RJ (2018). Visual selective attention in mice. *Curr Biol*. 28:676–685.

Wang Q, Sporns O, Burkhalter A (2012). Network analysis of corticocortical connections reveals ventral and dorsal processing streams in mouse visual cortex. *J Neurosci*. 32:4386–4399.

Young H, Belbut B, Baeta M, Petreanu L (2021). Laminar-specific cortico-cortical loops in mouse visual cortex. *eLife*. 10:e59551. doi:10.7554/eLife.59551.

40.

MECHANISMS OF CORTICO-CORTICAL COMMUNICATION

Adam Kohn

INTRODUCTION

Inter-areal communication is fundamental to most brain functions since most functions are distributed across multiple anatomical areas. For instance, the visual system of the macaque monkey involves processing that occurs in dozens of cortical areas and several subcortical areas (Felleman and Van Essen, 1990). Multiple nodes of this distributed network interact with other brains areas involved in executive, memory, and motor functions to guide vision-based behaviors.

Clearly, inter-areal communication requires anatomical connectivity. Direct communication between 2 brain areas can only occur if projection neurons in a source area send axons to the target area, to relay activity there. However, knowing that a connection between areas exists tells us little about when activity is relayed. Nor does it define what information is routed or the efficacy of communication—whether signals arriving from a source area alter the activity in the target in a meaningful way.

Understanding inter-areal communication thus requires functional measurements, in addition to defining the anatomical substrate. Functionally, communication between areas is usually assessed by quantifying the statistical dependency of activity in 2 areas. These dependencies might be measured using the activity of a neuron in each area, a group of neurons, or imaged voxels. Statistical dependencies between these signals may involve measures of correlation, coherence, or other forms of predictability (Granger regression, transfer entropy, etc.). Stronger dependencies (e.g., greater coherence or stronger correlations) are interpreted as a stronger interaction or more effective communication.

Empirically, we know that interactions between brain areas vary from moment to moment. For instance, resting-state functional magnetic resonance imaging (fMRI) experiments have shown that temporal correlations between brain areas can change drastically on brief timescales (Hutchison et al., 2013; Lurie et al., 2020). Electrophysiological recordings also indicate that interactions are flexible: they have shown, for instance, that neuronal activity is more correlated between areas with the engagement of attention, evident both in spike-spike correlations or the coherence of local field potentials across areas (Buschman and Miller, 2007; Bosman et al., 2012). Interactions are thus dynamic and selective—they change from moment to moment, such that activity in a given brain area can be correlated with different partner areas at different times.

A central hypothesis in systems and cognitive neuroscience is that flexible behavior—a critical aspect of cognition—arises from these dynamic or flexible inter-areal interactions. For instance, our ability to attend to certain features or locations in our environment might arise from the preferential relaying of activity between relevant areas. Similarly, our ability to react differently to the same sensory input might require differential routing of those signals to decision, action, or motor areas, depending on context.

How then is flexible and selective inter-areal communication achieved? Here I provide a brief overview of mechanisms proposed to modulate inter-areal signaling, some of the evidence in their favor (or against), and some open questions. I focus on signaling in the macaque visual system but borrow insights from other systems when relevant. Inter-areal communication has long been a subject of study. As a result, many of the issues discussed here have been addressed in prior more focused reviews, which are cited for the interested reader.

THE ANATOMICAL PERSPECTIVE

One way to achieve flexible and selective communication is to use different sets of projection neurons in a source area to communicate with different downstream targets. That is, area A could communicate with area B using one set of neurons, and communicate with area C using another set. By shifting activity between these 2 groups of neurons, area

A could selectively and dynamically modulate communication with each downstream target. There is experimental evidence for selective inter-areal pathways. For instance, in mouse V1, distinct groups of neurons project to areas AL and LM (Glickfeld et al., 2013). In macaque cortex, different neurons appear to project to MT and to V2 (Sincich et al., 2003). In the mouse motor cortex, there appear to be at least 2 distinct sets of L5 projection neurons, one projecting to the brainstem and another projecting to thalamic motor nuclei (Economo et al., 2018).

However, the approach of modulating inter-areal signaling by relying on different subsets of projection neurons has clear limitations. First, emerging evidence suggests that the specificity of inter-areal connections (e.g., that a neuron projects to only one target area) may have been overestimated in prior work. Clearly, any limitations in the sensitivity of anatomical labeling will bias connections to appear more selective than they truly are. Modern, more sensitive labeling techniques suggest that projection neurons often have bifurcating axons, so that they provide synaptic input to multiple downstream areas (Han et al., 2018). If neurons project to multiple downstream targets, it becomes untenable to use different neurons to signal selectively to different downstream targets.

A second objection to the proposal that selective communication relies on using specific subsets of projection neurons is the representational explosion needed to encode each signal in the source area, in each subset of projection neurons. Modern labeling methods suggest that roughly 65% of possible inter-areal connections exist (Markov et al., 2014). If each area has many downstream partners and signaling to each area is performed by different projection neurons—and if each of those subpopulations needs to represent the same information (i.e., that which is computed in the source area)—this creates a tremendous redundancy in representation in the source area. In many ways, instantiating selective communication through distinct subsets of neurons would be much like duplicating each area multiple times, once for each of its downstream targets. A seemingly more efficient solution would be to have more permissive or dense inter-areal connectivity, but with mechanisms in place to select a subset of the available connections in a particular situation.

MECHANISMS OF COMMUNICATION

If selective and dynamic communication is not achieved by recruiting different subsets of projection neurons, what mechanism might instead govern the efficacy with which a source area interacts with a target network? There are several existing proposals for how signaling efficacy might be regulated, and these can be roughly divided into mechanisms operating in the source area (or "sender"), the target area (or "receiver"), both areas, or through the properties of the circuitry linking the 2 (Kohn et al., 2020).

A long-standing proposal is that inter-areal signaling efficacy can be regulated by controlling the degree of synchronous spiking activity in the source network (Diesmann et al., 1999; Salinas and Sejnowski, 2001; Kumar et al., 2010). Because the temporal integration of synaptic input by postsynaptic neurons is limited, synchronous synaptic input should be more effective in eliciting downstream activity. However, the degree to which synchrony will influence the efficacy of inter-areal signaling depends on several factors, including the strength and precision of synchrony in the source population, the balance between the inhibition and excitation recruited in the target area, the degree of anatomical convergence and divergence between areas, and the integrative properties of downstream neurons (Diesmann et al., 1999; Salinas and Sejnowski, 2001; Kumar et al., 2010; Zandvakili and Kohn, 2015). Modeling studies have explored these various influences and have shown not only regimes in which synchronous events are potent drivers of downstream activity, but also regimes in which these events are ineffective (Figure 40.1A; Shadlen and Newsome, 1998; Diesmann et al., 1999).

Given the equivocal conclusions from modeling, direct evaluation of the efficacy of synchrony in inter-areal signaling through physiological recordings becomes particularly important. Early work on communication between the lateral geniculate nucleus (LGN) of the thalamus and primary visual cortex (V1) indicated that synchronous LGN spikes were more effective in driving V1 responses (Alonso et al., 1996). Changes in thalamic synchrony with adaptation have also been used to explain concomitant changes in downstream responses in primary sensory cortex (Wang et al., 2010). For communication between cortical areas, synchronous activity within a population of V1 neurons was shown to be associated with a higher probability of spiking in the input layers of V2 (Zandvakili and Kohn, 2015). Thus, there is some evidence that inter-areal signaling can be modulated by the degree of synchrony in a source population.

One way to achieve more synchronous or coordinating spiking activity in a source area is to entrain activity to an ongoing network rhythm or oscillation (Buzsaki, 2006). A widely cited proposal of this type is that inter-areal signaling between visual cortical areas is modulated by gamma rhythms (Fries, 2009; Palmigiano et al., 2017). Gamma oscillations in a source area have been proposed to "bind" the responses of neurons respresenting different visual features (Gray et al., 1989), or to coordinate the activity of neurons encoding an attended feature so that representation of that feature is propagated more effectively downstream (Fries et al., 2001).

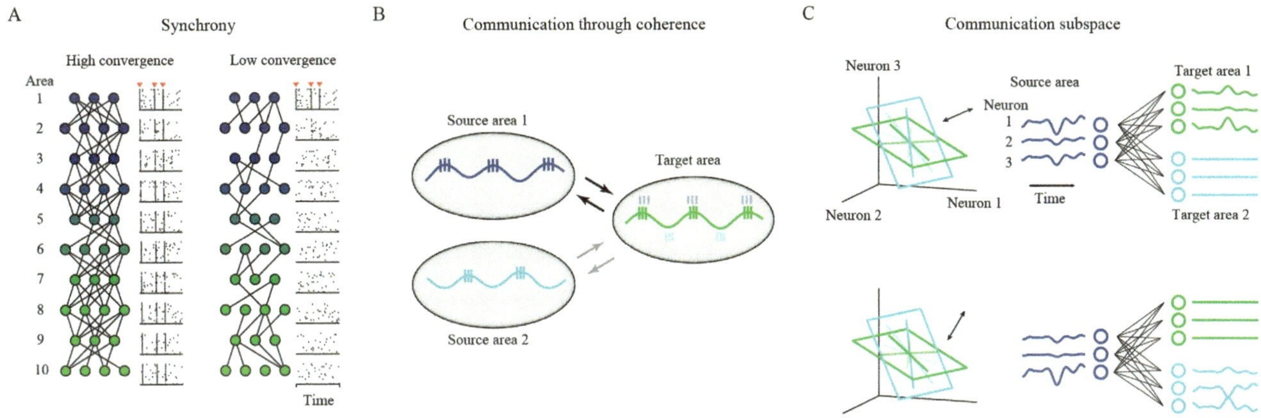

Figure 40.1. (A). Role of synchrony in inter-areal signaling. In the network on the left, there is dense connectivity between areas (layers) in a feedforward network. In this network, synchronous spiking activity in area 1 (top row) propagates effectively through the network. In the network on the right, the same pattern of synchronous activity in the first area fails to propagate effectively. Adapted from Zandvakili and Kohn (2015). (B), Communication through coherence. Spiking activity in 2 source areas is clustered in time, due to local gamma rhythms. Communication between source area 1 and the target area is effective (black arrows) because activity from source area 1 arrives in the target area at times in which that network is more excitable. In contrast, the input from source area 2 arrives when the target network is inhibited, so communication is compromised (gray arrows). Adapted from Fries (2009). (C). Communication subspaces. Top row: interactions between a source area (blue) and 2 target areas (green and cyan) is defined by 2 distinct communication subspaces. These are indicated by 2 distinct planes in the source population activity space (left). Fluctuations in the source area that have fallen within the communication subspace for target area 1 are communicated to that area. Bottom row: a different activity pattern in the source area results in communication to the target area 2. Adapted from Semedo et al. (2019).

A recent variant of the proposal that inter-areal interactions are modulated by oscillations in a source network is the suggestion that feedforward signaling is associated with gamma oscillations, whereas feedback signaling is mediated by alpha or beta rhythms (Bastos et al., 2015; van Kerkoele et al., 2014; but see Gieselmann and Thiele, 2022). This suggestion is consistent with gamma being most prominent in the superficial layers (the site from which feedforward cortico-cortical connections originate; Smith et al., 2013; Gieselmann and Thiele, 2022) and with lower-frequency rhythms like alpha and beta being more prominent in the deep layers (the source of some feedback connections; Maier et al., 2011; Smith et al., 2013). However, many feedback connections originate in the superficial layers as well (Markov et al., 2014), so there is only a partial correspondence with the anatomy. Moreover, both gamma and alpha/beta rhythms can be coherent over large distances (within an area), which may limit their ability to modulate inter-areal signaling selectively (i.e., some projection neurons and not others; Jia et al., 2011).

The proposals discussed thus far have focused on manipulating the source population to alter inter-areal signaling. An alternative is to change the properties of the target population, so that it becomes more receptive to inputs from the source area (or from one source area rather than another). For instance, changing the integrative properties of downstream neurons can alter their reliance on synchronous input to reach threshold for firing (Hahn et al., 2019). These integrative properties could be modulated by controlling background synaptic activity, which determines the membrane time constant (Destexhe et al., 2003; Kuhn et al., 2004), or by neuromodulators. More sophisticated schemes have shown that the dynamics of target networks can be tuned to selectively amplify some inputs over others (Mante et al., 2013).

The efficacy of inter-areal signaling might also be modulated by the coordination or alignment of activity in the source and target networks. A proposal of this type is "communication through coherence" (Figure 40.1B; Fries, 2009; Fries, 2015), which argues that selective and flexible communication is achieved by the phase alignment of gamma rhythms in the source and target networks (but see Ray and Maunsell, 2015). Just as gamma rhythms in the source network can cluster activity in brief windows, gamma in the target network defines epochs in which target neurons are in a more excitable state. This is because gamma oscillations involve rhythmic fluctuations in the strength of inhibition (Buzsaki and Wang, 2012). By adjusting the phase of rhythms in a source-target pair, the signaling efficacy between networks can be modulated: signaling is most potent if source activity arrives when the target network is in a less inhibited state, and least effective if it arrives at the opposite phase. Numerous studies have provided experimental support for the communication through coherence proposal, usually through showing a change in the coherence of gamma rhythms between areas with the allocation of attention (Buschman and Miller, 2007; Gregoriou et al., 2009; Bosman et al., 2012; but see Jia et al., 2013).

An alternative "alignment-based" mechanism for modulating inter-areal signaling is the recently proposed concept of the communication subspace (Figure 40.1C; Semedo et al., 2019). The communication subspace defines which patterns of population-spiking activity in the source area are relayed to the downstream network and which remain private to the source area. The communication subspace was first observed for interactions between primary visual cortex (V1) and area V2 of the macaque monkey, where it was shown that the relevant (communicated) population activity patterns were not those that were most dominant in the source area (Semedo et al., 2019). Subsequent work identified communications subspaces between areas V1 and V4 (Semedo et al., 2019; Semedo et al., 2022), and between area MT and the superior colliculus (Srinath et al., 2021). Related, low-dimensional inter-areal mappings have also been observed in mice (Ebrahami et al., 2022; Javadzadeh and Hofer, 2022). Recent evidence suggests that feedforward and feedback signaling between cortical areas may use distinct communication subspaces (Semedo et al., 2022; Javadzadeh and Hofer, 2022).

The communication subspace between 2 areas can be defined by the pattern of connections linking them. A trivial communication subspace arises with sparse inter-areal connectivity: in this case, only activity among the population of projection neurons is relevant for communication. But a communication subspace does not require sparse connectivity: it can arise in a network with dense connections, if the set of synaptic weights linking 2 areas is low rank, meaning those weights can be described by linear combinations of a set of canonical weights (Semedo et al., 2019). Dense, low-rank inter-areal connectivity provides for both selective communication (only population activity patterns in the source area that match the relevant communication subspace are effectively propagated) and for dynamic modulation of communication (by changing source population activity patterns). Importantly, these features can be achieved without relying on different neurons to communicate with different downstream networks.

Note that the communication subspace relies on properties of the inter-areal circuit that are distinct from those involved in synchrony- or coherence-based schemes. Synchrony requires a limited temporal window for synaptic integration; in the concept of the communication subspace, there is no requirement or consideration of temporal integration (though this could be introduced and might shape the nature of the communication subspace between areas). Communication through coherence involves time-varying windows of excitability to allow for windows of communication. The communication subspace does not require oscillations; it regulates communication by the pattern of activity across neurons rather than across time (see Kohn et al., 2020, for further discussion).

There are several advantages to the communication subspace framework for understanding inter-areal signaling. First, it considers the communication issue through the lens of population activity patterns, which are widely believed to be the relevant scale of measurement for understanding cortical computation and representation (Cunningham and Yu, 2014; Kohn et al., 2016; Ebitz and Hayden, 2021). Second, the concept of low-dimensional population activity has provided important insights into cortical function, including but not limited to motor cortex function (Kaufman et al., 2012), the neural substrate of cognitive functions (Ruff and Cohen, 2019; Panichello and Buschman, 2021), and the constraints on learning (Sadtler et al., 2014). Thus, consideration of inter-areal signaling through the lens of low-dimensional activity patterns may dovetail with understanding of computations and representations in the source and target areas.

THE ROLE OF THE PULVINAR IN MEDIATING INTER-AREAL SIGNALING

Cortico-cortical signaling can occur through at least 2 distinct pathways. The first is direct projections between cortical areas: projection neurons in the superficial layers have long-distance axons that terminate in the input layers of downstream networks, the basis of feedforward signaling; neurons in the deep (and, in some cases, the superficial) layers project back to lower areas, instantiating a feedback pathway (Felleman and Van Essen, 1991; Salin and Bullier, 1995). In addition to these direct connections, cortical areas may communicate with each other through the pulvinar nuclei of the thalamus; that is, through a cortico-pulvinocortical pathway (Sherman, 2016; Halassa and Kastner, 2017; Halassa and Sherman, 2019; Jaramillo et al., 2019). Corticopulvinar projections originate in layers (L)5 and 6; the projections from the pulvinar back to cortex project to L1 and L5 or L2/3 and L4, depending on the target area (Sherman, 2016). Like cortico-cortical connections, cortico-pulvinocortical pathways may relay signals both from lower to higher (feedforward) as well as higher to lower cortex (feedback; see **Chapter 42 of Section VII**).

The cortico-pulvinocortical loop in fact consists of at least 2 distinct loops. The first begins in a source cortical area, extends to the pulvinar, and ends back in the same source area—a feedback loop through the thalamus. This pathway may be important for providing recurrent excitation back to cortex (Guo et al., 2017), possibly functioning as a gating mechanism to control inter-areal communication via direct cortico-cortical projections (Olshausen et al., 1993; Purushothaman et al., 2012).

The second loop involves pulvinocortical projections onto a cortical area different from the one from which inputs derive, a potential basis for inter-areal signaling. Relatively little is known about how signaling via the cortico-pulvinocortical loop complements or interacts with direct cortico-cortical circuitry. Previous work has shown that the cortico-pulvinocortical loop is sufficiently robust, anatomically and functionally, to mediate the successful propagation of signals from lower to higher sensory areas, even in the absence of cortico-cortical connectivity (Theyel et al., 2010). Prior work has also shown that pulvinar inactivation results in disrupted inter-areal coherence (Zhou et al., 2016; Erakath et al., 2021), suggesting some role in mediating cortico-cortical communication. Relatedly, the allocation of attention—which alters cortico-cortical interactions, as described above—results in increased coherence between the pulvinar and various visual areas (Saalman et al., 2012; Zhou et al., 2016).

Whether the signals relayed between areas via the pulvinar are distinct from those relayed via the direct cortico-cortical pathway is unclear. The cortical column, a fundamental organizational motif in the cortex, is based in part on the observation that vertically aligned superficial and deep neurons have similar functional properties (Horton and Adams, 2005). This similarity suggests that the information relayed via the 2 pathways may be similar. However, others have suggested that the signals relayed by the 2 pathways may be distinct. Since L 5 neurons (the origin of corticopulvinar pathway) may provide signals used to guide eye movements, their inputs to the pulvinar might provide an efference copy for higher areas (Sherman, 2016; see Roth et al., 2016, and Blot et al., 2021, for other suggestions). In addition, because the pulvinar receives input from a broad range of cortical and subcortical sources, it is well placed to modulate the signals it receives from one cortical area before passing them on to another area (see **Chapter 42 of Section VII**).

SUMMARY

The study of inter-areal signaling is in its infancy. Previous work has documented that inter-areal signaling can change from moment to moment to meet task demands. Existing proposals for how inter-areal communication is regulated include proposals that focus on properties of the source network, the target network, or the circuitry linking them. Each of these is supported by data, but it remains far from clear which (if any) proposals have merit or whether entirely different explanations are needed.

Beyond debates about the relative merits of existing proposals, there are many fundamental open questions about inter-areal signaling more generally. These include:

(1) How selective are the anatomical pathways linking different cortical areas? Do source neurons project to one or many targets? As reviewed here, emerging evidence suggests denser and less selective projections between areas than previously understood, but this is still an active area of investigation.

(2) Are similar principles used to communicate signals via feedforward and feedback pathways? Or is signaling in these pathways accomplished using distinct mechanisms?

(3) How is flexible signaling guided to accomplish task goals? Most of the proposed signaling modes seemingly require an instructive signal to arrange communication in the desired manner. That is, the proposals are mechanisms for modulating communication, but leave unclear how the desired neurons or patterns of activity are preferentially recruited. Since there is no puppet master, communication must instead self-organize. How this is done is unknown.

(4) How do different mechanisms for modulating signaling affect computation in the source and target network? Patterns of neural activity are passed through circuitry in order to perform computation. Yet, most work on inter-areal signaling focuses solely on getting activity from one network to the next. Considering how signaling mechanisms interact and influence the type of computations that can be accomplished may provide powerful constraints both on proposals for inter-areal signaling and on cortical computation more generally. Alternatively, we may find that the concept of within-area computations or representations become subordinate to inter-areal representations defined on the basis of coordination (Wong et al., 2016; Jia et al., 2022). In this view, the dominant functional motif would be a multi-area neural circuit, rather than separate, discrete circuits that are defined by anatomical (e.g., cytoarchitectonic) boundaries.

Addressing the shortcomings in our understanding will not be straightforward. A great deal remains to be done. But we are positioned to gain significant traction on these issues, with the advent of powerful new recording methods that provide simultaneous access to many thousands of neurons (Machado et al., 2022), distributed across cortical areas, as well as new analytic tools for understanding and interpreting the data (Semedo et al., 2020; Keeley et al., 2020; Perich et al., 2020; Gokcen et al., 2022) and their relationship to behavior (Sani et al., 2021).

REFERENCES

Alonso JM, Usrey WM, Reid RC (1996). Precisely correlated firing in cells of the lateral geniculate nucleus. *Nature*. 383:815–819.

Bastos AM, Vezoli J, Bosman CA, Schoffelen JM, Oostenveld R, Dowdall JR, De Weerd P, Kennedy H, Fries P (2015). Visual areas exert feedforward and feedback influences through distinct frequency channels. *Neuron.* 85:390–401.

Blot A, Roth MM, Gasler I, Javadzadeh M, Imhof F, Hofer SB (2021). Visual intracortical and transthalamic pathways carry distinct information to cortical areas. *Neuron.* 109:1996–2008.

Bosman CA, Schoffelen JM, Brunet N, Oostenveld R, Bastos AM, Womelsdorf T, Rubehn B, Stieglitz T, De Weerd P, Fries P (2012). Attentional stimulus selection through selective synchronization between monkey visual areas. *Neuron.* 75:875–888.

Buschman TJ, Miller EK (2007). Top-down versus bottom-up control of attention in the prefrontal and posterior parietal cortices. *Science.* 315:1860–1862.

Buzsaki G (2006). *Rhythms of the Brain.* Oxford University Press.

Buzsáki G, Wang XJ (2012). Mechanisms of gamma oscillations. *Annu Rev Neurosci.* 35:203–225.

Cunningham JP, Yu BM (2014). Dimensionality reduction for large-scale neural recordings. *Nat Neurosci.* 17:1500–1509.

Destexhe A, Rudolph M, Paré D (2003). The high-conductance state of neocortical neurons in vivo. *Nat Rev Neurosci.* 4:739–751.

Diesmann M, Gewaltig MO, Aertsen A (1999). Stable propagation of synchronous spiking in cortical neural networks. *Nature.* 402:529–533.

Ebitz RB, Hayden BY (2021). The population doctrine in cognitive neuroscience. *Neuron.* 109:3055–3068.

Ebrahimi S, Lecoq J, Rumyantsev O, Tasci T, Zhang Y, Irimia C, Li J, Ganguli S, Schnitzer MJ (2022). Emergent reliability in sensory cortical coding and inter-area communication. *Nature.* 605:713–721.

Economo MN, Viswanathan S, Tasic B, Bas E, Winnubst J, Menon V, Graybuck LT, Nguyen TN, Smith KA, Yao Z, Wang L, Gerfen CR, Chandrashekar J, Zeng H, Looger LL, Svoboda K (2018). Distinct descending motor cortex pathways and their roles in movement. *Nature.* 563:79–84.

Eradath MK, Pinsk MA, Kastner S (2021). A causal role for the pulvinar in coordinating task-independent cortico-cortical interactions. *J Comp Neurol.* 529:3772–3784.

Felleman DJ, Van Essen DC (1991). Distributed hierarchical processing in the primate cerebral cortex. *Cereb Cortex.* 1:1–47.

Fries P (2009). Neuronal gamma-band synchronization as a fundamental process in cortical computation. *Ann Rev Neurosci.* 32:209–224.

Fries P (2015). Rhythms for cognition: Communication through coherence. *Neuron.* 88:220–235.

Fries P, Reynolds JH, Rorie AE, Desimone R (2001). Modulation of oscillatory neuronal synchronization by selective visual attention. *Science.* 291:1560–1563.

Gieselmann MA, Thiele A (2022). Stimulus dependence of directed information exchange between cortical layers in macaque V1. *eLife.* 11:e62949.

Glickfeld LL, Andermann ML, Bonin V, Reid RC (2013). Cortico-cortical projections in mouse visual cortex are functionally target specific. *Nat Neurosci.* 16:219–226.

Gokcen E, Jasper AI, Semedo JD, Zandvakili A, Kohn A, Machens CK, Yu BM (2022). Disentangling the flow of signals between populations of neurons. *Nat Comp Science.* 2:512–525.

Gray CM, König P, Engel AK, Singer W (1989). Oscillatory responses in cat visual cortex exhibit inter-columnar synchronization which reflects global stimulus properties. *Nature.* 338:334–337.

Gregoriou GG, Gotts SJ, Zhou H, Desimone R (2009). High-frequency, long-range coupling between prefrontal and visual cortex during attention. *Science.* 324:1207–1210.

Guo ZV, Inagaki HK, Daie K, Druckmann S, Gerfen CR, Svoboda K (2017). Maintenance of persistent activity in a frontal thalamocortical loop. *Nature.* 545:181–186.

Hahn G, Ponce-Alvarez A, Deco G, Aertsen A, Kumar A (2019). Portraits of communication in neuronal networks. *Nat Rev Neurosci.* 20:117–127.

Halassa MM, Kastner S (2017). Thalamic functions in distributed cognitive control. *Nat Neurosci.* 20:1669–1679.

Halassa MM, Sherman SM (2019). Thalamocortical circuit motifs: A general framework. *Neuron.* 103:762–770.

Han Y, Kebschull JM, Campbell RAA, Cowan D, Imhof F, Zador AM, Mrsic-Flogel TD (2018). The logic of single-cell projections from visual cortex. *Nature.* 556: 51–56.

Horton JC, Adams DL (2005). The cortical column: A structure without a function. *Philos Trans R Soc B Biol Sci.* 360:837–862.

Hutchison RM, Womelsdorf T, Allen EA, Bandettini PA, Calhoun VD, Corbetta M, Della Penna S, Duyn JH, Glover GH, Gonzalez-Castillo J, Handwerker DA, Keilholz S, Kiviniemi V, Leopold DA, de Pasquale F, Sporns O, Walter M, Chang C (2013). Dynamic functional connectivity: promise, issues, and interpretations. *NeuroImage.* 80:360–378.

Jaramillo J, Mejias JF, Wang XJ (2019). Engagement of pulvino-cortical feedforward and feedback pathways in cognitive computations. *Neuron.* 101:321–336.

Javadzadeh M, Hofer SB (2022). Dynamic causal communication channels between neocortical areas. *Neuron.* 110:2470–2483.

Jia X, Siegle JH, Durand S, Heller G, Ramirez TK, Koch C, Olsen SR (2022). Multi-regional module-based signal transmission in mouse visual cortex. *Neuron.* 110:1585–1598.

Jia X, Smith MA, Kohn A (2011). Stimulus selectivity and spatial coherence of gamma components of the local field potential. *J Neurosci.* 31:9390–9403.

Jia X, Tanabe S, Kohn A (2013). Gamma and the coordination of spiking activity in early visual cortex. *Neuron.* 77:762–774.

Kaufman MT, Churchland MM, Ryu SI, Shenoy KV (2012). Cortical activity in the null space: permitting preparation without movement. *Nat Neurosci.* 17(3):440–448.

Keeley SL, Zoltowski DM, Aoi MC, Pillow JW (2020). Modeling statistical dependencies in multi-region spike train data. *Curr Opin Neurobiol.* 65:194–202.

Kohn A, Coen-Cagli R, Kanitscheider I, Pouget A (2016). Correlations and neuronal population information. *Annu Rev Neurosci.* 39:237–256.

Kohn A, Jasper AI, Semedo JD, Gokcen E, Machens CK, Yu BM (2020). Principles of corticocortical communication: Proposed schemes and design considerations. *Trends Neurosci.* 43:725–737.

Kuhn A, Aertsen A, Rotter S (2004). Neuronal integration of synaptic input in the fluctuation- driven regime. *J Neurosci.* 24:2345–2356.

Kumar A, Rotter S, Aertsen A (2010). Spiking activity propagation in neuronal networks: reconciling different perspectives on neural coding. *Nat Rev Neurosci.* 11:615–627.

Lurie DJ, Kessler D, Bassett DS, Betzel RF, Breakspear M, Kheilholz S, Kucyi A, Liégeois R, Lindquist MA, McIntosh AR, Poldrack RA, Shine JM, Thompson WH, Bielczyk NZ, Douw L, Kraft D, Miller RL, Muthuraman M, Pasquini L, Razi A, Vidaurre D, Xie H, Calhoun VD (2020). Questions and controversies in the study of time-varying functional connectivity in resting fMRI. *Netw Neurosci.* 4:30–69.

Machado TA, Kauvar IV, Deisseroth K (2022). Multiregion neuronal activity: The forest and the trees. *Nat Rev Neurosci.* 23:683–704.

Maier A, Aura CJ, Leopold DA (2011). Infragranular sources of sustained local field potential responses in macaque primary visual cortex. *J Neurosci.* 31:1971–1980.

Mante V, Sussillo D, Shenoy KV, Newsome WT (2013). Context-dependent computation by recurrent dynamics in prefrontal cortex. *Nature.* 503:78–84.

Markov NT, Ercsey-Ravasz MM, Ribeiro Gomes AR, Lamy C, Magrou L, Vezoli J, Misery P, Falchier A, Quilodran R, Gariel MA, Sallet J, Gamanut R, Huissoud C, Clavagnier S, Giroud P, Sappey-Marinier D, Barone P, Dehay C, Toroczkai Z, Knoblauch K, Van Essen DC, Kennedy H (2014). A weighted and directed interareal connectivity matrix for macaque cerebral cortex. *Cereb Cortex.* 24:17–36.

Olshausen BA, Anderson CH, Van Essen DC (1993). A neurobiological model of visual attention and invariant pattern recognition based on dynamic routing of information. *J Neurosci.* 13:4700–4719.
Palmigiano A, Geisel T, Wolf F, Battaglia D (2017). Flexible information routing by transient synchrony. *Nat Neurosci.* 20:1014–1022.
Panichello MF, Buschman TJ (2021). Shared mechanisms underlie the control of working memory and attention. *Nature.* 592(7855):601–605.
Purushothaman G, Marion R, Li K, Casagrande VA (2012). Gating and control of primary visual cortex by pulvinar. *Nat Neurosci.* 15:905–912.
Perich MG, Rajan K (2020). Rethinking brain-wide interactions through multi-region "network of networks" models. *Curr Opin Neurobiol.* 65:146–151.
Ray S, Maunsell JH (2015). Do gamma oscillations play a role in cerebral cortex? *Trends Cogn Sci.* 19:78–85.
Roth MM, Dahmen JC, Muir DR, Imhof F, Martini FJ, Hofer SB (2016). Thalamic nuclei convey diverse contextual information to layer 1 of visual cortex. *Nat Neurosci.* 19:299–307.
Ruff DA, Cohen MR (2019). Simultaneous multi-area recordings suggest that attention improves performance by reshaping stimulus representations. *Nat Neurosci.* 22:1669–1676.
Saalmann YB, Pinsk MA, Wang L, Li X, Kastner S (2012). The pulvinar regulates information transmission between cortical areas based on attention demands. *Science.* 337:753–756.
Sadtler PT, Quick KM, Golub MD, Chase SM, Ryu SI, Tyler-Kabara EC, Yu BM, Batista AP (2014). Neural constraints on learning. *Nature.* 512:423–426.
Salin PA, Bullier J (1995). Corticocortical connections in the visual system: structure and function. *Physiol Rev.* 75:107–154.
Salinas E, Sejnowski TJ (2001). Correlated neuronal activity and the flow of neural information. *Nat Rev Neurosci.* 2:539–550.
Sani OG, Abbaspourazad H, Wong YT, Pesaran B, Shanechi MM (2021). Modeling behaviorally relevant neural dynamics enabled by preferential subspace identification. *Nat Neurosci.* 24:140–149.
Semedo JD, Gokcen E, Machens CK, Kohn A, Yu BM (2020). Statistical methods for dissecting interactions between brain areas. *Curr Opin Neurobiol.* 65:59–69.
Semedo JD, Jasper AI, Zandvakili A, Krishna A, Aschner A, Machens CK, Kohn A, Yu BM. (2022). Feedforward and feedback interactions between visual cortical areas use different population activity patterns. *Nat Commun.* 13(1):1099.
Semedo JD, Zandvakili A, Machens CK, Yu BM, Kohn A (2019). Cortical areas interact through a communication subspace. *Neuron.* 102:249–259.
Shadlen MN, Newsome WT (1998). The variable discharge of cortical neurons: Implications for connectivity, computation, and information coding. *J Neurosci.* 18:3870–3896.
Sherman SM (2016). Thalamus plays a central role in ongoing cortical functioning. *Nat Neurosci.* 19:533–541.
Sincich LC, Horton JC (2003). Independent projection streams from macaque striate cortex to the second visual area and middle temporal area. *J Neurosci.* 23:5684–5692.
Smith MA, Jia X, Zandvakili A, Kohn A (2013). Laminar dependence of neuronal correlations in visual cortex. *J Neurophysiol.* 109:940–947.
Srinath R, Ruff DA, Cohen MR (2021). Attention improves information flow between neuronal populations without changing the communication subspace. *Curr Biol.* 31:5299–5313.
Theyel BB, Llano DA, Sherman SM (2010). The corticothalamocortical circuit drives higher-order cortex in the mouse. *Nat Neurosci.* 13:84–88.
van Kerkoerle T, Self MW, Dagnino B, Gariel-Mathis MA, Poort J, van der Togt C, Roelfsema PR (2014). Alpha and gamma oscillations characterize feedback and feedforward processing in monkey visual cortex. *Proc Natl Acad Sci U S A.* 111:14332–14341.
Wang Q, Webber RM, Stanley GB (2010). Thalamic synchrony and the adaptive gating of information flow to cortex. *Nat Neurosci.* 13:1534–1541.
Wong YT, Fabiszak MM, Novikov Y, Daw ND, Pesaran B (2016). Coherent neuronal ensembles are rapidly recruited when making a look-reach decision. *Nat Neurosci.* 19:327–334.
Zandvakili A, Kohn A (2015). Coordinated neuronal activity enhances corticocortical communication. *Neuron.* 87:827–839.
Zhou H, Schafer RJ, Desimone R (2016). Pulvinar-cortex interactions in vision and attention. *Neuron.* 89:209–220.

41.

PREDICTIVE CODING

FORWARD AND BACKWARD CONNECTIVITY

Stewart Shipp and Karl Friston

Predictive coding is what the name suggests: that what the brain encodes are predictions—that might equally be termed *perceptual beliefs* or *motor intentions* (Adams et al., 2013), depending on the brain systems under discussion. For the purposes of this chapter we focus upon the visual system, where predictive coding theory (PCT) has a notable antecedent in Helmholtz's insight that "[w]e always think we see such objects before us as would have to be present in order to bring about the same retinal images" (Helmholtz, 1860/1962). In other words, what we see is effectively a best guess—one that is constrained by sensory input, but additionally reliant upon prior experience—owing to an inadequacy of the retinal image to specify precisely what is, actually, before us.

The mathematical formulation of PCT derives from statistics, and aspires to Bayesian principles of optimal inference. In extended and hierarchical formulations, it constitutes a serial computation (i.e., data assimilation) with reciprocal exchange of "messages" between successive levels, each with a potentially modular aspect: where each module tries to estimate the (unobservable, conditionally independent) states causing (observable) data. Immediately, this lends itself as a model for hierarchical processing in the visual cortex, which is known to build an increasingly abstract representation of sensory content at higher levels (Rao, 1999; Friston, 2008). The theory indicates that backward-going signals from the representation at one level should serve as predictions of neuronal representations at the preceding level. Comparison of predicted and actual activity yields forward-going prediction-error signals, which act to modify the predictions enacting the next iteration of the computation, progressively minimizing the magnitude of error.

Though they may have a temporal dimension, the predictions of PCT are not temporal forecasts per se, so much as ongoing descriptions (or, in terms of probability distributions, "expectations") of scenarios unfolding in the outside world. They come in 2 classes: *causes* (categorical variables such as colors, sizes, familiar items, gestures) and *states* (dynamic relationships between *causes*). While both require reciprocal exchange of prediction and prediction error, it is only for the former, *causes*, that this exchange takes place between adjacent stages; by contrast, the computation of *states* is intrinsic to each stage, or level of abstraction within the serial computation (Kanai et al., 2015) (see Figure 41.1 for a schematic of the message passing implied by this formulation of PCT). There is more theoretical exposition to come, but this is already sufficient to reflect upon the hierarchical nature of cortical anatomy.

THE RELATIONSHIP OF HIERARCHY TO LAMINAR PATTERNS OF CORTICAL CONNECTIVITY

The assignment of cortical areas to specific tiers in a formal hierarchy is contingent upon the distinction of feedforward (FF) and feedbackward (FB) patterns of connection between them, characterizing the laminar distribution of both axon terminals and their neurons of origin (see **Chapters 37, 38, and 39 in Section VII**). Axioms implicit in this exercise are (a) that an area constitutes a computational unit (i.e. that it serves a single stage of a serial hierarchical computation, though likely conducting several in parallel), and (b) that the laminar organization of the computational process is conserved across areas, so generating the characteristic laminar patterns of connection. So far so good, from the perspective of PCT.

The anatomical analysis of hierarchy in macaque visual cortex has evolved from the initial qualitative classification

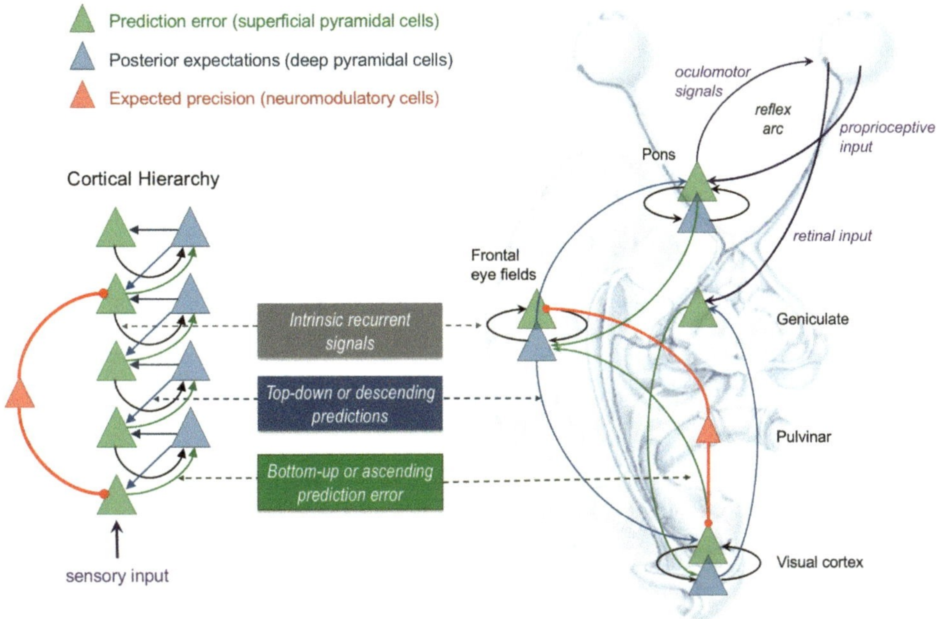

Figure 41.1. Hierarchical predictive coding. This schematic summarizes the hierarchical message passing implied by PCT. Neuronal activity encodes expectations about the *causes* of sensory input, where expectations minimize prediction error. Prediction error is the difference between (ascending) sensory input and (descending) predictions of that input. This minimization rests upon recurrent neuronal message passing among different levels of a cortical hierarchy. For simplicity, we have omitted the encoding of expected *states* within each level, where the *causes* are some nonlinear mixture of *states* that evolve in a predictable way over time. In this schematic, certain (e.g., superficial pyramidal) cells (green triangles) compare the expectations—at each level—with top-down predictions from (e.g., deep pyramidal) cells (blue triangles) of higher levels. Left panel: a simple cortical hierarchy with ascending prediction errors and descending predictions. We have included neuromodulatory gating or gain control (red triangles)—of cells reporting prediction errors—that determines their relative influence on cells encoding expectations. In PCT, this influence rests upon the expected or predicted precision of prediction errors. Right panel: an illustration of global visual processing, showing the putative cells of origin of ascending or feedforward (FF) connections that convey prediction errors (ERR) and descending or feedbackward (FB) connections that construct predictions. The prediction errors are weighted by their expected precision, mediated in part by projections from the pulvinar. In this example, the frontal eye field (FEF) sends predictions to visual cortex, which in turn projects back to the lateral geniculate. However, the FEF also sends proprioceptive predictions to pontine nuclei, which are passed to the oculomotor system to elicit eye movement through classical reflexes (that resolve proprioceptive prediction errors). Note that every top-down prediction is reciprocated with a bottom-up prediction error to ensure predictions are constrained by sensory information. Adapted from Kanai et al. (2015) under a Creative Commons Attribution License (CC-BY-4.0) http://creativecommons.org/licenses/by/4.0/.

of FF versus FB connections (Felleman and Van Essen, 1991), to the development of a quantitative metric of hierarchical distance, the SLN (% supragranular layer neurons of origin) (Barone et al., 2000; Markov et al., 2014a). And, whereas the former was simply a compilation of pairwise area relationships—allocating each area to one of several discrete tiers in an ordinal scale—the latter employs a global statistical analysis of the SLN metric pertaining to all network connections, assigning each area a precise hierarchical level on an interval scale (Markov et al., 2014a; Vezoli et al., 2021). Application of this process to mouse visual cortex is foiled by the fact that, in this species, FF and FB connections differ little in their laminar origins. However, a substitute metric, based on variations in the optical density of labeled axon terminals across layers (L)1 to 3 has been used to achieve the same end, namely, an optimized hierarchy of mouse visual cortical areas (D'Souza et al., 2022).

The systematics of this mouse hierarchy are not, however, without anomalies. If FF and FB connections are re-designated according to the rank of the areas they link, they constitute substantially overlapping ranges of the axon-terminal metric values (D'Souza et al., 2022). Paradoxically, this emergent property challenges the founding axiom (b), above. Relatedly, connections between areas that do not differ significantly in their hierarchical level are deemed to belong to a third category, "lateral," and are found to comprise a range of metric values intermediate between the FF and FB classes of connection. Functional interpretation of these anatomical characteristics posits a complementary role for cryptic "non-hierarchical" network interactions (D'Souza et al., 2022; Gămănuț and Shimaoka, 2022). Analogous anomalies have also arisen in a separate, global hierarchy derived for mouse cortex, encompassing all known areas (Harris et al., 2019). The strategy here identified several distinct pan-laminar terminal connection patterns, whose classification as FF or FB was algorithmically optimized to maximize consistency within a serial ranking of areas. Yet, even the optimal hierarchy displayed instances where the laminar class of a connection, FF or FB, conflicted with its hierarchical direction. Recognizing that the presence of these anomalies acts to compress the number of levels, both studies (Harris et al., 2019; D'Souza et al., 2022) refer to the emergent mouse hierarchy as "shallow"; this is by reference to the systematics of the macaque where such anomalies are indeed less frequent, but not entirely absent (Vezoli et al., 2021).

We suggest that PCT offers a resolution of these paradoxical network properties. Reciprocal FB and FF connections are conceived as the exchange of prediction and prediction-error and recall that, in the case of *states*, this exchange is intrinsic to the computations operating at a single stage or level of abstraction. Rendered between 2 areas at the same level, and conforming with axiom (b), the exchange specific to one particular *state* will yield an FF pattern in one area and an FB pattern in the other. But supposing each area to perform parallel computation of multiple *states*, there is no stipulation that all exchanges should share the same polarity; hence the direction of transmission of prediction for one *state* might be the direction of transmission of error for another (Figure 41.2C). Standard anatomical tracing methodology will likely engage bidirectional connections pertaining to multiple *state* computations. The resulting admixture of FF and FB patterns within a single area could be segregated, or amalgamated to yield what is commonly termed the "lateral" (Felleman and Van Essen, 1991) (or intermediate, pan-laminar) pattern of both terminals and cells of origin, equivalent to a superimposition of the canonical FF and FB patterns. Exactly such variegated laminar patterning has been observed (Ungerleider and Desimone, 1986), potentially a signature of the processing of *states* as envisaged by PCT.

Areas are equally likely to perform parallel computations of multiple *causes*. Extending the logic above, there is no theoretical stricture that the serial ordering of computational stages across areas should be identical for each and every *cause* or, by the same token, that it should invariably match the specific serial order prescribed by an optimized hierarchy. Any departure from such uniformity in the processing of *causes* would yield either aberrant FF or FB connections in the "wrong" direction, or dilate the range of metric values (cellular-SLN, or axon-terminal) associated with forward or backward steps, dependent on the hierarchical systematics employed.

In summary, we propose that FF and FB patterns of connection (in their raw anatomical sense) express a conserved laminar apparatus for computations relating to the exchange of prediction and prediction error, conforming to the structural axioms of hierarchical processing and capable of accounting for emergent anomalies in specific optimized hierarchical orderings of areas.

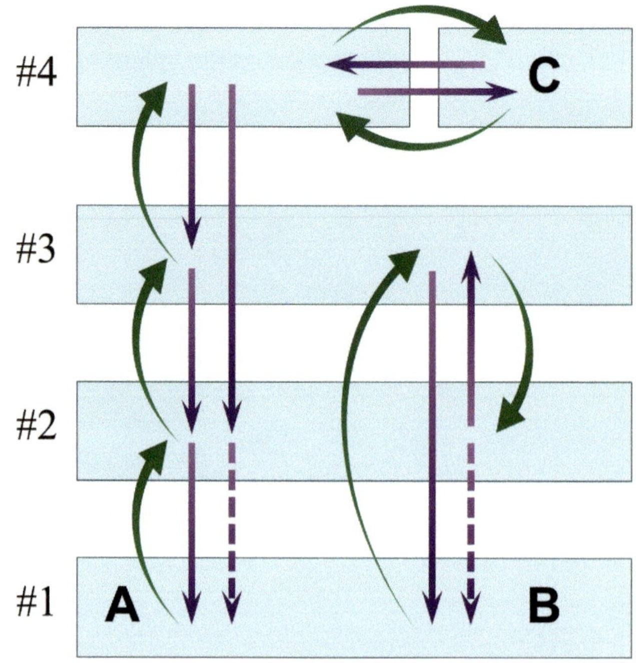

Figure 41.2. How hierarchical stages of computation might align with a cortical area hierarchy. (A). A standard serial computation for *causes* aligning with tiers #1 to #4 of an area hierarchy, with "one to many" FB connectivity (e.g., macaque V1-V2-V4-TEO, where a proportion of FB axons extend collaterals to both V2 and V1; Rockland, 2013). (B). An "anomalous" serial computation for *causes* misaligning with tiers #1–#3 (e.g., macaque V1-V2-V4, where a tiny minority of deep V2 neurons collateralize to both V1 and V4; Markov et al., 2014a). (C). A reciprocal pair of computations for *states* between areas sharing the same tier (e.g., macaque V4–V5, where V4 displays a mix of FF and FB patterns; Ungerleider and Desimone, 1986). Green arrows = FF connection; purple arrows = FB connection; dashed arrows symbolize axon collaterals of the co-axial solid arrow.

THE ANATOMICAL IDENTITY OF FF AND FB NEURONS OF ORIGIN

PCT is adamant that FF and FB neurons must be separate classes (i.e., that a single neuron cannot combine both roles), and therefore should not issue both forward- and backward-going signals through a bifurcating axonal projection. This is one instance where an indication of PCT arguably anticipates the definitive experimental evidence: namely, 2 studies, 1 in mouse (Berezovskii et al., 2011), 1 in macaque (Markov et al., 2014a), published some 20 years after the initial canonization of FF and FB patterns (Felleman and Van Essen, 1991). Both adopt the same strategy, placing dual distinct retrograde tracers in areas A and C of an A-B-C hierarchy (area A being V1 in both cases), and examining area B for dual-labeled neurons. In the macaque, FF- and FB-labeled neurons (respectively projecting to areas V4 and V1) show partial laminar segregation as expected, more so superficially (FF in L3B and FB in L3A and L2) than in the deep layers (FF distributed evenly across L5 and L6, FB denser in L6).

In the mouse, FF- and FB-labeled neurons are co-distributed in a single band centered upon L3, diminishing in density through L4 and deeper layers.

Dual-labeled neurons were infrequent, affirming PCT, but were not totally absent: about 5% in the mouse, and 1%–2% in the macaque, predominantly in the deep layers. Can the presence of a small minority of neurons with both FF and FB projections be dismissed as "mis-wiring"? As we explore below, aspects of neural circuitry that might appear inconsistent with PCT are not uncommon, and our aim here is to expand conceptions of the neural implementation of the theory in potential reconciliation of such findings. In that spirit, we offer the following rationale for the "aberrant" connectivity. Simply put, and as mooted in the preceding section, this could be a case where the serial order of computation does not align with the optimized hierarchy (Figure 41.2B): area B may be forward of both area A (V1) *and* area C for the specific serial computation engaging these dual-labeled neurons. In that case, they would emit FB signals to both areas; as discussed below, FB connections do frequently have multiple targets.

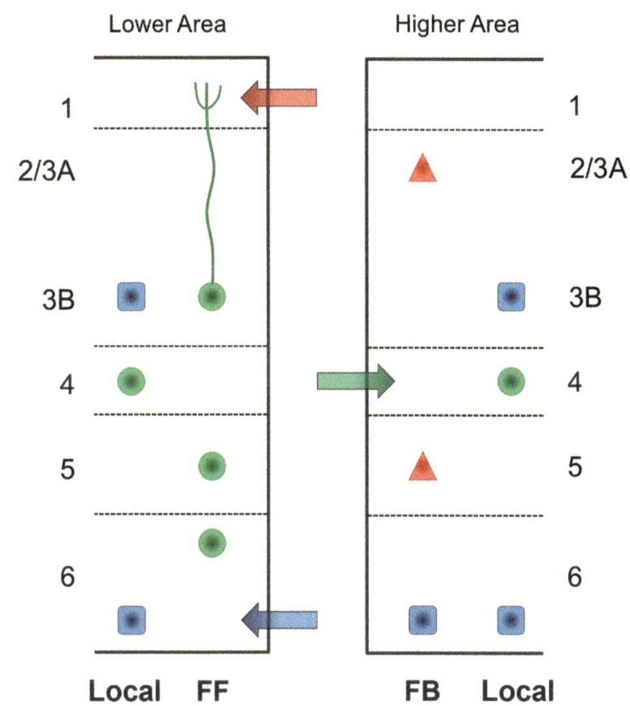

Figure 41.3. Pyramidal neurons and computational units (modeled on primate laminar architecture). Neurons are schematized by layer, and by local, FF, or FB axonal connectivity (shown in columns). Circles and green arrow denote ERR (prediction-error) units and output target; squares and blue arrow denote EXP (expectation—or coding) units and output target; triangles and red arrow denote PRC (precision) units and output target. Each computational unit symbolizes a local population of neurons sharing that functionality.

NEURAL ELEMENTS IMPLEMENTING PREDICTIVE CODING

Confirmation of the separate identity of FF and FB neurons establishes the principle of direct correspondence between specific neural subtypes and the classes of computational unit envisaged by PCT (Bastos et al., 2012; Shipp, 2016). Possessed of an extrinsic (inter-areal) axon, these are excitatory neurons, generally with pyramidal morphology; other pyramidal neurons possess axons that make only local connections intrinsic to an area. Figure 41.3 presents a summary schematic allocating neurons (indexed by laminar location and connectivity) to 3 classes of computational unit: FF and local ERR (error) units; FB and local EXP (expectation) units—equivalent to "coding" or "internal representation" neurons, as termed in other accounts (Keller and Mrsic-Flogel, 2018); and FB precision (PRC) units.

Precision (Rao, 1999; Feldman and Friston, 2010) is a PCT quantity that regulates the impact of error signals (i.e., adjusts their capacity to influence the representation they are fed forward to). Precision signals can have diverse computational origins and might be transmitted by a variety of means, including diffuse neuromodulatory systems. Our schematic illustrates a form of precision that is estimated from the magnitude of received error signals and transmitted backward by the superficial tier of FB neurons in L2/3A, and probably also by L5. The focus of the backward projection upon L1 is known to contact the apical dendritic tufts of FF neurons, both superficial and deep, and local inhibitory interneurons (Schuman et al., 2021); this can modulate (amplify) FF neural activity (Phillips, 2017) and coordinate spike timing (Shen et al., 2022), effectively regulating the gain and downstream influence of error signals. Such loop circuitry (i.e., FB contact upon identified source neurons of a reciprocal FF connection) has been verified in all studies possessing this technical capability (Johnson and Burkhalter, 1997; Kim et al., 2020; Siu et al., 2021; Young et al., 2021); 2 of these (Kim et al., 2020; Siu et al., 2021) are reviewed below.

Our schematic allocates to deep (L6) EXP units the role of prediction-proper (i.e., generating the signal that is ultimately fed to ERR units at the lower level for the purpose of comparison to local EXP signals). Under PCT, this predictive signal is not simply a copy of the higher-level EXP representation, but a nonlinear function of it—a transformation from a higher to a lower level of abstraction or, more figuratively, a translation into a code that the lower area can understand. Anatomically, a deep-to-deep FB pathway will be transmitted to a greater degree through contact upon perisomatic dendrites of deep local pyramidal neurons, matching a functional requirement for a driving influence to relay the predictive information through local processing to the error units sited more superficially.

STRICT RECIPROCITY IN THE PREDICTION/ PREDICTION-ERROR LOOP

The computational apparatus depicted in Figure 41.3 represents a pair of translaminar neural ensembles ("columns" or "modules") dedicated to processing a particular feature at a particular level of abstraction. Evidently, a prediction-error generated in respect of a particular prediction should be conveyed back to the source of that prediction, and to no other destination. PCT therefore indicates that inter-areal connections should indeed be reciprocal, and specific to the columns/modules in question. Cortical areas are, of course, known to form multiple reciprocal connections with each other, probed by techniques that are of varied specificity in determining precisely what connects with what at a cellular level. To what degree is known extrinsic cortical circuitry compliant with PCT?

The first point of note is that an EXP representation active at one level can be predictive of multiple featural qualities at subordinate levels of processing. Hence an output EXP unit might make multiple FB contacts across more than one area (Figure 41.2A). Anatomically, this is known to be the case. Either by tracing the courses of bifurcating axon collaterals, or strategic placement of dual distinct retrograde tracers, such multiplicity of targeting by individual FB neurons has been amply confirmed (Kennedy and Bullier, 1985; Rockland, 2013). Studies to date likely reveal only a subtotal of this projectome—establishing the full distribution of a single FB neuron's output remains a worthwhile challenge. All of the sites receiving this distributed signal should reciprocate with convergent FF error signals targeting the relevant source ensemble (i.e., not restricted to the immediate source FB neurons). Such convergent inter-areal FF projections are well documented in general, but not at a column-to-column or cell-to-cell level of specificity sufficient to arbitrate the issue.

An important concern here is that the nature of what is predicted should be germane to each recipient prediction-error apparatus, even if the identical signal is broadcast over the full arborization of an FB axon. To reconcile the nature of FB circuitry with PCT, we therefore suppose that some component of the nonlinear transformation from expectation to prediction, noted above, takes place within each recipient area—a final adaptation to the nature of the local featural representation. Thus (in contrast to precision signals) such a consideration weighs against direct contact of predictive FB signals upon FF ERR units, but does mandate translaminar pre-processing through deeper pyramids (and interneurons), consistent with the schematic in Figure 41.3.

In summary, therefore, the initial prescription arising from PCT is a configuration of "one-to-many" in FB connections, reciprocated by "many-to-one" in FF connections. While much of the literature is consistent with this picture, there are known exceptions: some are long-standing, while others are the product of the latest viral technology (and likely to proliferate with some rapidity). We recount these exceptions, and develop (one) potential accommodation within the framework of PCT.

THE "ONE-TO-MANY" PATTERN IN FF CONNECTIONS

The conventional wisdom has been that FF projection axons generally conform to a "one-to-one" pattern and target a single area. Examples of FF "manifold" neurons, with multiple targets, are known—but only a handful of studies have used the requisite experimental strategy, placing distinct retrograde tracers in separate, higher-order areas. In the visual system, all such work has detected only small fractions of FF manifold neurons projecting to 2 targets: for example, in L4B of macaque V1, one study (Sincich and Horton, 2003) counted 6 cells projecting to both area V2 and area V5/MT, amidst 891 projecting solely to the former, and 72 solely to the latter. This amounts to 7.7% manifold neurons, expressed as a fraction of the minority target population (here V5). Similarly quantified, some 18% of cells in macaque V2 projecting to area TEO also project to area V4 (Nakamura et al., 1993), and between 5% and 20% of cells in mouse V1 are found to project to various pairs of the higher visual areas bordering V1 in that species (Kim et al., 2020; Young et al., 2021). Technique-wise, all these fractions would be expected to edge upward if the full axonal distribution can be accessed—as reported, so far uniquely, for mouse V1 (Han et al., 2018). Indeed, some 44% of all superficial (L2/3) neurons with output to at least one higher visual area were detected to project to either 1, 2, or 3 additional areas (among 6 tested), while a still larger proportion of a small sub-sample (77%, 23/30 neurons traced by global visualization of their axonal arborizations) projected to multiple sites within the entire ipsilateral cerebrum (Han et al., 2018).

Cortical connections in the mouse are generally more liberal than those in the monkey (Gămănuț et al., 2018), but such a frequency of multiple FF projections prompts further consideration: while much of the patterning of FF connectivity conforms to a first reading of PCT, this cannot be the whole story.

THE "MANY-TO-ONE" PATTERN IN FB CONNECTIONS

Cortical areas are known to receive FB input from multiple sources, but the operative question here is whether multiple

FB inputs ever converge upon the same unit of prediction-error apparatus. If they were to do that, neither the generation of a "one-to-one" nor a "one-to-many" reciprocal FF error signal could properly fulfill the PCT role that we have so far prescribed: the former omits error correction for some sources of prediction, and the latter broadcast would be inapposite for the varying states of prediction across all of those sources.

The development of so-called viral intersection tracing techniques (Callaway and Luo, 2015) has provided the means to tackle anatomical questions of this nature. These methods allow determination of all the sources of input to a restricted subpopulation of neurons—"starter-cells"—for instance, the subset of neurons in area A with output to area B. At time of writing, studies of this type are limited in number, but 2 are particularly significant, both detailing sources of FB input to area V1, 1 in macaque (Siu et al., 2021) and 1 in the mouse (Kim et al., 2020). The prognoses they yield for PCT are rather different.

The study in macaque revealed the global source of input to the subpopulation of V1 neurons with FF projections to area V2 (Siu et al., 2021) (Figure 41.4A). As V1 is the base layer in the hierarchy any extrinsic cortical source of this input is automatically classed as FB, and it was found to derive from no area other than V2. So, this indicates a closed loop circuit between the 2 areas and is precisely the result anticipated by PCT. Incidentally, the target FF neurons in V1 included those in L4B, a few of which (as noted above) have output to area V5 as well as V2; no corresponding FB input from area V5 was detected, though this could possibly indicate a limitation to the sensitivity of the methodology, given that these manifold neurons form a tiny minority of the total starter-cell population.

As discussed above, PCT would lean toward classifying these FB inputs to V1 as precision signals, because they directly contact the FF elements of the prediction-error apparatus. Notably, FB neurons in V2 were equally distributed between the superficial tier (L2/3A) and L5, with relatively few in L6 (Siu et al., 2021). By contrast, standard tracing methods, that detect sources of FB input to *all* cells in V1, reveal a far greater frequency of FB neurons in L6 than L5 of V2 (Markov et al., 2014a). Hence L6 FB neurons in V2 are inferred not to make direct contact with FF neurons in V1; this is consistent with the schematic of Figure 41.3, in which they convey a prediction-proper signal primarily to deep local neurons.

The study in mouse adopted a similar strategy, extending it to examine the global source of FB inputs to 3 separate groups of V1 starter-cells, respectively identified by FF projections to either area PM, or AL, or POR (Kim et al., 2020) (Figure 41.4B–D). In contrast to the organization described above for the macaque, these FB sources were not restricted to a single visual area, nor even to visual cortex per se; some were located in distant nonvisual areas. Less than

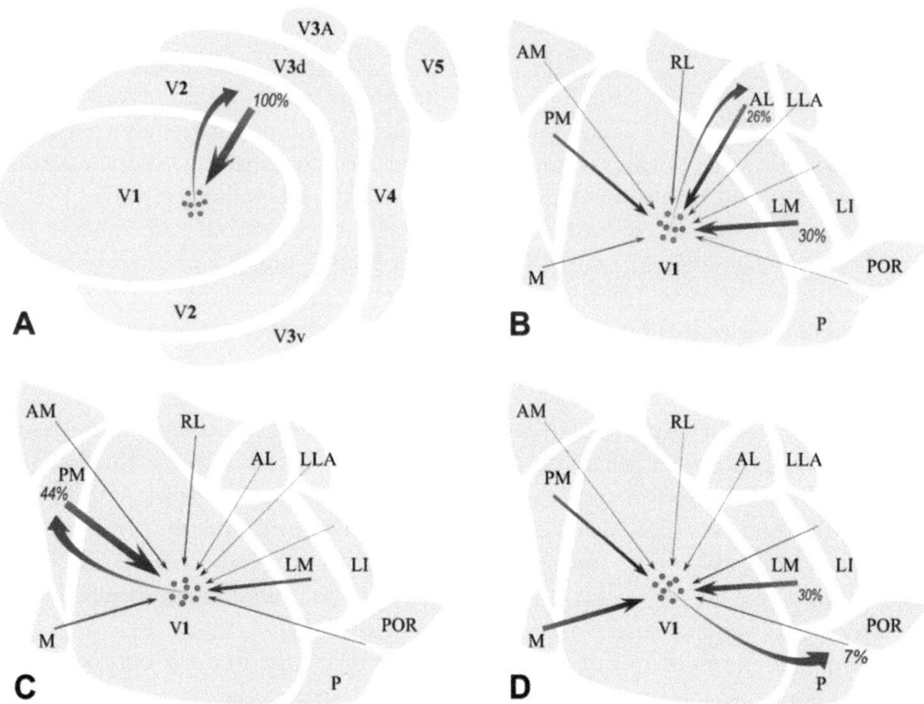

Figure 41.4. Sources of FB input to specific subpopulations of FF neurons with identified outputs. (A). Macaque: single FB source for FF neurons in all layers of V1 with output to area V2. (B–D). Mouse: multiple FB sources for FF neurons in all layers of V1 with output to area AL, or area PM, or area P/POR. Green arrows = FF connection; purple arrows = FB connection; relative width of purple arrows symbolizes relative weight of connections; % scores refer to the principal, and reciprocal FB sources among higher visual areas.

50% were located in the principal source area so as to form a reciprocal loop. In all cases a major contingent arose from visual area LM (hierarchically equivalent to macaque V2; D'Souza et al., 2022), an area associated with AL and PM by virtue of dual projections from some V1 FF manifold neurons (i.e., to LM and AL, or to LM and PM) (Kim et al., 2020). The distribution of FB origins was also broader for input to FF cells in L5, than to FF cells in L2/3. Area LM was actually the principal source of FB input to V1 L5 cells identified to project to area AL, at about 40%, versus 30% from area AL itself. Consistent with this, V1 cells projecting to area P/POR are concentrated in L5 (Kim et al., 2020) and showed a still greater diversification of FB input sources, with P/POR itself responsible for no more than 7%.

Such disjuncture between monkey and mouse—here, a measure of reciprocity varying from 100% to 7% in FB connections to V1—is far from an isolated example; we have already mentioned differences in the laminar architecture of FF and FB connections. Variation in the systematics of cortical network organization is rooted in such species-specific factors as brain size, cortical laminar ontogeny, and behavioral repertoire, that can obscure or disguise conserved principles of cortical (and subcortical) computation. In prospecting for elements of anatomy that demonstrate those conserved principles, we alight upon more abstract commonalities in structure-function relationships, as discussed below.

RE-ENTRANT AND DISTRIBUTIVE PROCESSING OF POOLED-ERROR SIGNALS

By *pooled-error* we refer to convergence of prediction-error signals of varying feature specificity but sharing a common retinotopic location. In outline, this may be implemented in cortical projections to the superior colliculus (SC), undergoing subsequent re-entry through a tecto-thalamo-cortical loop subserving spatial awareness. This subcortical system may be emblematic of broader distributive networks processing such pooled prediction-error signals.

Direct evidence for the transmission of error signals to SC obtains from a particular class of large pyramids located at the L5/6 border in primate V1, often known as Meynert cells. Notably, the axons of these neurons branch to project both cortically and subcortically, specifically to area V5/MT and the SC (Fries et al., 1985; vogt Weisenhorn et al., 1995; Nhan and Callaway, 2012). If the FF signal to V5 is classed as an error signal, so must be the output to SC. A corresponding finding in the mouse pertains to SC-projecting neurons in anterior cingulate cortex (ACC) that also project to prefrontal cortex (Zhang et al., 2016); a separate population of ACC neurons provides FB output to V1 and other visual areas (consistent with separate origins for FF and FB signals). Macaque SC neurons fail to inherit selectivity for features such as orientation, color, and direction of motion from their cortical inputs (Marrocco and Li, 1977; Davidson and Bender, 1991; White et al., 2009)—presumably through pooling inputs of varied selectivity—but retain good retinotopic order; in both species the SC superficial layers are considered to act as a salience map (Phongphanphanee et al., 2014; White et al., 2017).

Across all areas, the majority of SC-projecting cells in L5 do not also project to cortex, though their axons may branch to thalamus as well as SC (Rockland, 1998). Thus we lack direct inference for universal transmission of an error signal by subcortical projections from L5, except to observe that ERR unit activity is a logical basis for a salience map, by virtue of signaling unexpected features (Rao and Ballard, 1999; Spratling, 2012). The SC projects back broadly to visual cortex via the pulvinar nucleus of the thalamus (Shipp, 2004); in mice, this pathway specifically processes local motion salience (Bennett et al., 2019). We have previously discussed how the action of the pulvinar upon the cortex can be considered a form of precision input, regulating cortical activity by middle or superficial layer-dependent mechanisms (Kanai et al., 2015). As noted above, precision can be estimated from the magnitude of error signals, as might be computed over the tecto-pulvinar pathway; equally, PRC units in L5 (or L6), may transmit directly to pulvinar. Either way, from a behavioral perspective, a system based on local salience can be interpreted as a mechanism for exogenous attention (see Figure 41.5).

The pooling process evident in SC may be initiated in L5 itself. Featural information encoding in L5 is typically diminished with respect to more superficial layers (Pettine et al., 2019). One early study (Finlay et al., 1976) specifically examined macaque corticotectal neurons in V1 L5 and described complex-type receptive fields with broader tuning for orientation than similarly classified neurons in other layers—possibly effected by some degree of pooling in the intrinsic transmission from superficial to deep layers. To the extent that the feature specificity of ERR units is eroded in L5, their signals can become relevant to a broader range of higher-level processing directed not so much at elucidating the featural characteristics or identity of a visual object, but its attention-worthiness. The regulation of attention is known to operate through a highly distributed system, involving various forms of salience (or priority) maps across occipital, parietal, and frontal cortices. Below, we elaborate this idea with specific reference to the frontal eye field (FEF).

Revisiting the findings outlined in preceding sections, we can interpret the broader sources of FB input to FF neurons in L5 of mouse V1 as feature-unspecific precision signals regulating the gain of pooled (or partially pooled) ERR units; or, more generally, as interactions among units

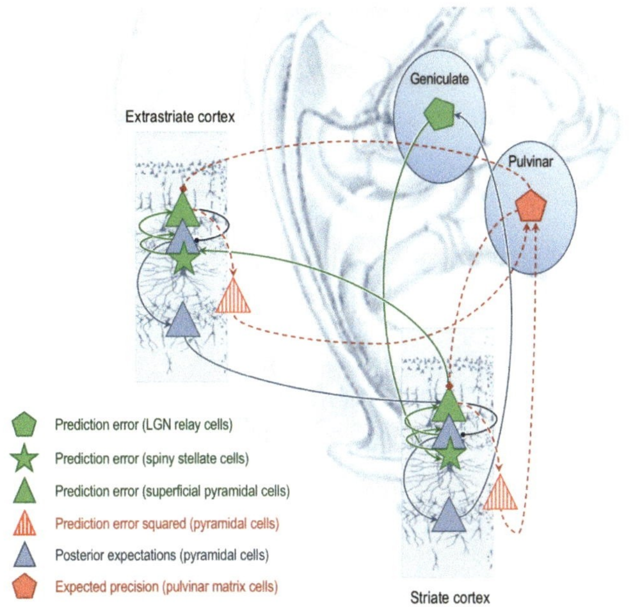

Figure 41.5. Precision and PCT. A schematic illustrating putative laminar-specific connections that are consistent with the precision-based PCT. This architecture (Bastos et al., 2012; Adams et al., 2013) models the neuroanatomy and physiology of canonical microcircuits and laminar specificity of extrinsic connections. Here, deep pyramidal cells—encoding the amplitude of prediction error (squared) inform posterior expectations about precision in the (matrix cells) of the pulvinar. Re-entrant projections from these pulvinar cells act to modulate the gain of superficial pyramidal cells in cortex. FF/ascending connection-stream = green; FB/descending connection-stream = blue. Precision-related streams = dashed red. Arrow terminals = excitatory contacts; ball terminals = indirect inhibitory contacts; diamond terminals = indirect or modulatory contacts. Adapted from Kanai et al. (2015) under a Creative Commons Attribution License (CC-BY-4.0) http://creativecommons.org/licenses/by/4.0/.

implementing a distributed, brain-wide computation of spatial attentional priority. Individual areas may participate in this computation to a greater or lesser degree. Mouse area POR, notably, is heavily influenced by input from the SC (Beltramo and Scanziani, 2019) and involved in spatial awareness; the FF input it receives from V1 is therefore likely specialized toward such a role, and it was the FF V1 cells projecting to POR that received FB input from the most broadly distributed sources (Kim et al., 2020).

While in mouse V1, FF manifold neurons were most frequent in L5 (Kim et al., 2020; Young et al., 2021), this was not the case with the examples cited above for areas V1 and V2 in macaque (Nakamura et al., 1993; Sincich and Horton, 2003). Here (as also in cat area 17; Bullier et al., 1984; Ferrer et al., 1992), manifold neurons were most frequent where FF neurons are most frequent, that is, in the superficial layers, a step or two removed from corticotectal output. Yet, if the rationale we have advanced so far leaves these examples in abeyance, it does lead us to anticipate the presence of FF manifold neurons at higher levels, as explained below.

THE ROLE OF AREA FEF AND ITS HIERARCHICAL STATUS

The FEF is generally considered to function as a "priority" map (Bisley and Mirpour, 2019), a node in the control of spatial attention that integrates physical stimulus salience with endogenous factors such as prior experience, behavioral relevance, scene context, and the like. Visual input to FEF spans the hierarchy, from V1 upward, but with greater weight from mid-tier areas such as V4, TEO, and V5 (Markov et al., 2014b). The laminar origins of these connections conform to the typical FF pattern (Schall et al., 1995). Like SC, FEF is agnostic to object qualities such as motion, color, or shape, presumed to be discarded by featurally convergent patterns of projection upon FEF neurons (Schall et al., 1995). Hence, as with SC, we interpret this as a pooling of prediction-error signals. Their role is to update FEF as to the attention-worthiness of a location, as opposed to modify the state of featural representation in a higher-level visual area, to which they might equally project. The possibility of dual projection to FEF and higher visual areas by neurons in areas such as V4 and V5 has not been investigated, but such FF manifold neurons would not be discordant with PCT.

In a similar vein, the role of FB input to visual areas from FEF cannot be to predict featural qualities, but it is observed to modulate activity according to behavioral priorities (Noudoost and Moore, 2011; Gregoriou et al., 2012); this is equivalent to a precision function in PCT terms. As such, the scheme outlined above (Figures 41.3 and 41.5) anticipates that the laminar source of FB output from FEF to visual areas such as V4 will be skewed toward the superficial layers—as is well documented (Barone et al., 2000; Pouget et al., 2009; Ninomiya et al., 2012). In consequence, under the SLN metric, FEF adopts an anomalously low hierarchical level (Markov et al., 2014a) as these FB projections appear to have FF character. Relatedly, perhaps, a functional index of hierarchy (Bastos et al., 2015) (based upon measures of reciprocal Granger causality between areas at gamma versus beta/alpha oscillatory bands) determined that FEF had an ambivalent hierarchical level; uniquely, among the areas recorded, its level shifted several tiers contingent upon the epoch (the behavioral phase of the monkey's task) at which neural activity was analyzed. While respecting the hierarchical principles of computation envisaged by PCT, these observations question the utility of global schemes of area hierarchy; the broader the construction—across sensory modalities, across behavioral networks, or universal across cortex—the greater the likelihood of conflating diverse and distinct structure-function relationships.

MANIFOLD PROCESSING OF EXP UNIT SIGNALS?

If manifold processing of ERR unit output is plausible under PCT, as we suggest, what about EXP units? To recap, EXP unit activity encodes Bayes-optimal representations of visual scene content and, as depicted in Figure 41.3 and previously (Bastos et al., 2012; Shipp, 2016), we envisage 2 subclasses: deep FB EXP units, that transmit predictions to subordinate levels, and local or intrinsic EXP units that assist the generation of FF ERR signals. Local neurons are rarely described in the anatomical literature, as their positive identification requires axonal reconstructions of sufficient quality to determine that no branch passes from gray to white matter. Functionally speaking, however, the subclassification of some EXP units as "local" is specifically in the context of sensory perception: hierarchical processing yielding progressive abstraction of featural quality. Several brain systems must draw upon current best estimates of feature and object identity for other purposes, and these should be served by export of EXP, not ERR signals: for instance, to the striatum and basal ganglia for learning the sensory context of rewarded actions, or to the hippocampus for memory encoding. Such considerations apply, for example, to the finding that virtually all (97%) of neurons in L2/3 of mouse V1 make extrinsic contacts external to V1, and not only to higher areas of visual cortex; other targets included striatum and amygdala (Han et al., 2018). The paucity of local neurons implies that both ERR and EXP units export signals from mouse V1.

Such long-range connections are characteristic of the mouse brain (Gămănuţ et al., 2018), and are less frequent in large-brained primates; macaque V1, by contrast, does possess identified local neurons in its superficial layers (Callaway and Wiser, 1996; Sawatari and Callaway, 2000) and is not known to project to either the striatum or amygdala. But higher visual areas, such as V4, do project to both structures (Saint-Cyr et al., 1990; Gattass et al., 2014) and—inverting the logic—this might afford a means of identifying putative EXP neurons: suspect EXP neurons and ERR neurons should form separate populations in a dual retrograde tracer experiment with tracer placed in, say, striatum and area TEO.

CONCLUSION

We have taken a close look at the neural circuitry implied by hierarchical predictive coding, focusing upon anatomical studies of the visual system. The requisite neuronal message-passing furnishes specific constraints on laminar and population-specific connections with which the mainstream organization of connectivity is largely compliant. Judicious expansion of the theoretical framework can accommodate some anomalous (e.g., heterarchical) aspects of connectivity, and predict elements yet to be investigated experimentally—especially those pertaining to predictions of precision or predictability that may underwrite a more nuanced kind of predictive coding in the visual brain.

REFERENCES

Adams RA, Shipp S, Friston KJ (2013). Predictions not commands: Active inference in the motor system. *Brain Struct Funct.* 218:611–643.

Barone P, Batardiere A, Knoblauch K, Kennedy H (2000). Laminar distribution of neurons in extrastriate areas projecting to visual areas V1 and V4 correlates with the hierarchical rank and indicates the operation of a distance rule. *J Neurosci.* 20:3263–32681.

Bastos AM, Usrey WM, Adams RA, Mangun GR, Fries P, Friston KJ (2012). Canonical microcircuits for predictive coding. *Neuron.* 76:695–711.

Bastos AM, Vezoli J, Bosman CA, Schoffelen JM, Oostenveld R, Dowdall JR, De Weerd P, Kennedy H, Fries P (2015). Visual areas exert feedforward and feedback influences through distinct frequency channels. *Neuron.* 85:390–401.

Beltramo R, Scanziani M (2019). A collicular visual cortex: Neocortical space for an ancient midbrain visual structure. *Science.* 363:64–69.

Bennett C, Gale SD, Garrett ME, Newton ML, Callaway EM, Murphy GJ, Olsen SR (2019). Higher-order thalamic circuits channel parallel streams of visual information in mice. Neuron. 102:477–+.

Berezovskii VK, Nassi JJ, Born RT (2011). Segregation of feedforward and feedback projections in mouse visual cortex. *J Comp Neurol.* 519:3672–3683.

Bisley JW, Mirpour K (2019). The neural instantiation of a priority map. *Curr Opin Psychol.* 29:108–112.

Bullier J, Kennedy H, Salinger W (1984). Branching and laminar origin of projections between visual cortical areas in the cat. *J Comp Neurol.* 228:329–341.

Callaway EM, Luo L (2015). Monosynaptic circuit tracing with glycoprotein-deleted rabies viruses. *J Neurosci.* 35:8979–8985.

Callaway EM, Wiser AK (1996). Contributions of individual layer 2–5 spiny neurons to local circuits in macaque primary visual cortex. *Vis Neurosci.* 13:907–922.

D'Souza RD, Wang QX, Ji WQ, Meier AM, Kennedy H, Knoblauch K, Burkhalter A (2022) Hierarchical and nonhierarchical features of the mouse visual cortical network. *Nat Commun.* 13:14.

Davidson RM, Bender DB (1991). Selectivity for relative motion in the monkey superior colliculus. *J Neurophysiol.* 65:1115–1133.

Feldman H, Friston KJ (2010). Attention, uncertainty, and free-energy. *Front Hum Neurosci.* 4:215.

Felleman DJ, Van Essen DC (1991). Distributed hierarchical processing in the primate cerebral cortex. *Cereb Cortex.* 1:1–47.

Ferrer JM, Kato N, Price DJ (1992). Organization of association projections from area 17 to areas 18 and 19 and to suprasylvian areas in the cat's visual cortex. *J Comp Neurol.* 316:261–278.

Finlay BL, Schiller PH, Volman SF (1976). Quantitative studies of single-cell properties in monkey striate cortex. IV. Corticotectal cells. *J Neurophysiol.* 39:1352–1361.

Fries W, Keizer K, Kuypers HG (1985). Large layer VI cells in macaque striate cortex (Meynert cells) project to both superior colliculus and prestriate visual area V5. *Exp Brain Res.* 58:613–616.

Friston K (2008). Hierarchical models in the brain. *PLoS Comp* Biol. 4:e1000211.

Gămănuţ R, Shimaoka D (2022). Anatomical and functional connectomes underlying hierarchical visual processing in mouse visual system. *Brain Struct Funct.* 227:1297–1315.

Gămănuț R, Kennedy H, Toroczkai Z, Ercsey-Ravasz M, Van Essen DC, Knoblauch K, Burkhalter A (2018). The mouse cortical connectome, characterized by an ultra-dense cortical graph, maintains specificity by distinct connectivity profiles. *Neuron.* 97:698–715.e610.

Gattass R, Galkin TW, Desimone R, Ungerleider LG (2014). Subcortical connections of area V4 in the macaque. *J Comp Neurol.* 522:1941–1965.

Gregoriou GG, Gotts SJ, Desimone R (2012). Cell-type-specific synchronization of neural activity in FEF with V4 during attention. *Neuron.* 73:581–594.

Han Y, Kebschull JM, Campbell RAA, Cowan D, Imhof F, Zador AM, Mrsic-Flogel TD (2018). The logic of single-cell projections from visual cortex. *Nature.* 556:51–56.

Harris JA, Mihalas S, Hirokawa KE, Whitesell JD, Choi H, Bernard A, Bohn P, Caldejon S, Casal L, Cho A, Feiner A (2019) Hierarchical organization of cortical and thalamic connectivity. Nature 575:195–202.

Helmholtz H (1860/1962). *Handbuch der Physiologischen Optik*, Vol 3. (English trans.: Southall JPC, ed). Dover.

Johnson RR, Burkhalter A (1997). A polysynaptic feedback circuit in rat visual cortex. *J Neurosci.* 17:7129–7140.

Kanai R, Komura Y, Shipp S, Friston K (2015). Cerebral hierarchies: predictive processing, precision and the pulvinar. *Philos Trans R Soc Lond B Biol Sci.* 370:20140169.

Keller GB, Mrsic-Flogel TD (2018). Predictive processing: A Canonical cortical computation. *Neuron.* 100:424–435.

Kennedy H, Bullier J (1985). A double-labeling investigation of the afferent connectivity to cortical areas V1 and V2 of the macaque monkey. *J Neurosci.* 5:2815–2830.

Kim EJ, Zhang Z, Huang L, Ito-Cole T, Jacobs MW, Juavinett AL, Senturk G, Hu M, Ku M, Ecker JR, Callaway EM (2020). Extraction of distinct neuronal cell types from within a genetically continuous population. *Neuron.* 107:274–282.

Markov NT, Vezoli J, Chameau P, Falchier A, Quilodran R, Huissoud C, Lamy C, Misery P, Giroud P, Ullman S, Barone P, Dehay C, Knoblauch K, Kennedy H (2014a). Anatomy of hierarchy: Feedforward and feedback pathways in macaque visual cortex. *J Comp Neurol.* 522:225–259.

Markov NT, Ercsey-Ravasz MM, Ribeiro Gomes AR, Lamy C, Magrou L, Vezoli J, Misery P, Falchier A, Quilodran R, Gariel MA, Sallet J (2014b). A weighted and directed interareal connectivity matrix for macaque cerebral cortex. *Cereb Cortex.* 24:17–36.

Marrocco RT, Li RH (1977). Monkey superior colliculus: Properties of single cells and their afferent inputs. *J Neurophysiol.* 40:844–860.

Nakamura H, Gattass R, Desimone R, Ungerleider LG (1993). The modular organization of projections from areas V1 and V2 to areas V4 and TEO in macaques. *J Neurosci.* 13:3681–3691.

Nhan HL, Callaway EM (2012). Morphology of superior colliculus- and middle temporal area-projecting neurons in primate primary visual cortex. *J Comp Neurol.* 520:52–80.

Ninomiya T, Sawamura H, Inoue K, Takada M (2012). Segregated pathways carrying frontally derived top-down signals to visual areas MT and V4 in macaques. *J Neurosci.* 32:6851–6858.

Noudoost B, Moore T (2011). Control of visual cortical signals by prefrontal dopamine. *Nature.* 474:372–375.

Pettine WW, Steinmetz NA, Moore T (2019). Laminar segregation of sensory coding and behavioral readout in macaque V4. *Proc Natl Acad Sci U S A.* 116:14749–14754.

Phillips WA (2017). Cognitive functions of intracellular mechanisms for contextual amplification. *Brain Cogn.* 112:39–53.

Phongphanphanee P, Marino RA, Kaneda K, Yanagawa Y, Munoz DP, Isa T (2014). Distinct local circuit properties of the superficial and intermediate layers of the rodent superior colliculus. *Eur J Neurosci.* 40:2329–2343.

Pouget P, Stepniewska I, Crowder EA, Leslie MW, Emeric EE, Nelson MJ, Schall JD (2009). Visual and motor connectivity and the distribution of calcium-binding proteins in macaque frontal eye field: implications for saccade target selection. *Front Neuroanat.* 3:2.

Rao RP (1999). An optimal estimation approach to visual perception and learning. *Vision Res.* 39:1963–1989.

Rao RP, Ballard DH (1999). Predictive coding in the visual cortex: A functional interpretation of some extra-classical receptive-field effects. *Nat Neurosci.* 2:79–87.

Rockland KS (1998). Convergence and branching patterns of round, type 2 corticopulvinar axons. *J Comp Neurol.* 390:515–536.

Rockland KS (2013). Collateral branching of long-distance cortical projections in monkey. *J Comp Neurol.* 521:4112–4123.

Saint-Cyr JA, Ungerleider LG, Desimone R (1990). Organization of visual cortical inputs to the striatum and subsequent outputs to the pallidal-nigral complex in the monkey. *J Comp Neurol.* 298:129–156.

Sawatari A, Callaway EM (2000). Diversity and cell type specificity of local excitatory connections to neurons in layer 3B of monkey primary visual cortex. *Neuron.* 25:459–471.

Schall JD, Morel A, King DJ, Bullier J (1995). Topography of visual cortex connections with frontal eye field in macaque: Convergence and segregation of processing streams. *J Neurosci.* 15:4464–4487.

Schuman B, Dellal S, Prönneke A, Machold R, Rudy B (2021). Neocortical layer 1: An elegant solution to top-down and bottom-up integration. *Annu Rev Neurosci.* 44:221–252.

Shen S, Jiang X, Scala F, Fu J, Fahey P, Kobak D, Tan Z, Zhou N, Reimer J, Sinz F, Tolias AS (2022). Distinct organization of two cortico-cortical feedback pathways. *Nat Commun.* 13:6389.

Shipp S (2004). The brain circuitry of attention. Trends Cogn Sci. 8:223–230.

Shipp S (2016). Neural elements for predictive coding. *Front Psychol.* 7:1792.

Sincich LC, Horton JC (2003). Independent projection streams from macaque striate cortex to the second visual area and middle temporal area. *J Neurosci.* 23:5684–5692.

Siu C, Balsor J, Merlin S, Federer F, Angelucci A (2021). A direct interareal feedback-to-feedforward circuit in primate visual cortex. *Nat Commun.* 12:4911.

Spratling MW (2012). Predictive coding as a model of the V1 saliency map hypothesis. *Neural Networks.* 26:7–28.

Ungerleider LG, Desimone R (1986). Cortical connections of visual area MT in the macaque. *J Comp Neurol.* 248:190–222.

Vezoli J, Magrou L, Goebel R, Wang XJ, Knoblauch K, Vinck M, Kennedy H (2021). Cortical hierarchy, dual counterstream architecture and the importance of top-down generative networks. *NeuroImage.* 225:117479.

vogt Weisenhorn DM, Illing RB, Spatz WB (1995). Morphology and connections of neurons in area 17 projecting to the extrastriate areas MT and 19DM and to the superior colliculus in the monkey *Callithrix jacchus*. *J Comp Neurol.* 362:233–255.

White BJ, Berg DJ, Kan JY, Marino RA, Itti L, Munoz DP (2017). Superior colliculus neurons encode a visual saliency map during free viewing of natural dynamic video. *Nat Commun.* 8:14263.

White BJ, Boehnke SE, Marino RA, Itti L, Munoz DP (2009). Color-related signals in the primate superior colliculus. *J Neurosci.* 29:12159–12166.

Young H, Belbut B, Baeta M, Petreanu L (2021). Laminar-specific cortico-cortical loops in mouse visual cortex. *eLife.* 10:e59551.

Zhang S, Xu M, Chang W-C, Ma C, Hoang Do JP, Jeong D, Lei T, Fan JL, Dan Y (2016). Organization of long-range inputs and outputs of frontal cortex for top-down control. *Nat Neurosci.* 19:1733–1742.

42.

CORTICO-CORTICAL COMMUNICATION THROUGH THE THALAMUS

TRANS-THALAMIC PATHWAYS

Mitra Javadzadeh and Sonja B. Hofer

Thalamus and cerebral cortex are extensively and intricately interconnected. These projection patterns provide pathways through which the cortex can receive information from subcortical sources, but they may also allow cortical areas to indirectly communicate with each other or with themselves through the thalamus (Figure 42.1A; Halassa and Sherman, 2019; Shepherd and Yamawaki, 2021). To understand the function of these trans-thalamic pathways, it is crucial to uncover their fine-scale organization, including the input and output connectivity patterns of individual neurons and how information flows through thalamocortical circuits. Several motifs of cortico-thalamocortical (i.e., trans-thalamic) connectivity have been proposed that would impose specific directions of trans-thalamic information flow. In this chapter we will introduce these motifs and evidence for them in the literature, as well as their potential role in different brain functions.

CONNECTIVITY MOTIFS

Trans-thalamic pathways can be classified into three motifs based on their connectivity: recurrent, feedforward, and feedback circuits. Recurrent trans-thalamic circuits connect cortical areas with themselves (Figure 42.1B), while trans-thalamic feedforward and feedback pathways connect hierarchically lower with higher cortical areas, such as primary (V1) and secondary visual cortex (V2), from to lower to higher (feedforward, Figure 42.1C), or vice versa (feedback, Figure 42.1D). While these motifs often coexist, they can be seen as the building blocks of the complex array of cortico-thalamocortical (CTC) interactions.

RECURRENT THALAMOCORTICAL LOOPS

Cortical and thalamic areas are often reciprocally connected, forming recurrent loops. These loops connect diverse cortical areas with both primary and higher-order thalamic nuclei (Rockland, 2021). Projections from primary sensory thalamic nuclei to layer (L)4 of primary cortical areas provide the main channel for sensory information to the cortex. In turn, corticothalamic (CT) neurons in L6 project back from primary cortical to primary thalamic areas. This recurrent motif can be found in visual, auditory and somatosensory pathways (Figure 42.1B, left). The L6 CT feedback projection to first-order thalamus has a mostly modulatory effect on thalamic activity (Hasse and Briggs, 2017; Born et al., 2021); nevertheless, connections from L6 CT neurons make up a substantial fraction (~30%–40%) of synapses onto thalamic neurons (Erişir et al., 1997). (For more detail on how cortical L6 projections affect thalamic neurons, see **Chapter 14 in Section IV**.) L6 CT projections also send collaterals to the thalamic reticular nucleus (TRN), which inhibits primary thalamus. Therefore, activation of L6 CT neurons can both facilitate or, through disynaptic inhibition, suppress activity in primary thalamus (Usrey and Sherman, 2019; Born et al., 2021), and the effect of corticothalamic feedback on thalamic responses can differ depending on the temporal pattern of L6 CT activity (Crandall et al., 2015; Kirchgessner et al., 2020). Such timing-dependent modulation could support a versatile role for these pathways, for instance in shaping temporal dynamics, tuning, and adaptation of thalamic sensory responses (Mease et al., 2014; Guo et al., 2017a). (A detailed review of the potential function of the L6 feedback pathway from V1 to dorsolateral geniculate nucleus can be found in **Chapter 19 in Section V**.) While first-order thalamic projections strongly target L4 of the cortex, they can also innervate other layers (Blasdel and Lund, 1983; Constantinople and Bruno, 2013). In particular, thalamic innervation of L6 suggests a possible direct monosynaptic loop between cortex and thalamus. However, there is little evidence that L6 CT cells receive significant direct thalamic input. Primary thalamocortical loops are thus most likely to involve multiple synapses in the cortex, via intratelencephalic (IT) neurons, from which CT neurons receive most of their input (Shepherd and Yamawaki, 2021).

Cortical areas also form recurrent loops with higher-order thalamus (Figure 42.1B, middle). Higher-order thalamic structures include the pulvinar, posterior medial nucleus (POm) and the dorsal and medial divisions of the medial geniculate nucleus in the visual, somatosensory, and auditory system, respectively. These thalamic nuclei are defined as not being primarily driven by the sensory periphery, but receiving prominent input from various cortical areas. The evidence for recurrent connectivity between cortex and higher-order thalamus comes from various studies showing that CTC connectivity is organized in a reciprocal manner; cortical areas receiving projections from a higher thalamic nucleus are also more likely to project back to it (Rockland, 2021). For instance, in the mouse visual system, axon density from pulvinar projections in cortical areas is proportional to the amount of input it receives from each area (Bennett et al., 2019). Importantly, the cortical input onto higher-order thalamic nuclei not only stems from L6 CT neurons, but also from cortical L5 neurons which exert a driving influence on thalamus (Sherman and Guillery, 2011). These thalamus-projecting L5 neurons are a subtype of the so-called pyramidal tract (PT) neurons that do not project to intracortical target areas (in contrast to IT neurons) (Shepherd and Yamawaki, 2021; for more details on L5 input to thalamus, see **Chapter 15 in Section IV**). Higher-order sensory thalamic nuclei form at least 2 types of recurrent thalamocortical circuits—with primary and with higher-order sensory cortical areas (Figure 42.1B, middle and right). Projections from higher-order thalamic nuclei mainly terminate in L1 and L5a in primary sensory cortical areas where they have a modulatory effect on sensory responses (Rubio-Garrido et al., 2009; Roth et al., 2016). Both L5 and L6b neurons in V1 have been shown to directly connect to pulvinar neurons projecting back to V1 in rodents, and the same recurrent circuit exists between primary somatosensory cortex (S1) and POm (Miller-Hansen and Sherman, 2022), providing feedback loops from thalamus to primary sensory cortices.

In contrast, higher-order thalamic projections prominently innervate L4 and L2/3 in higher-order sensory cortical areas and are therefore thought to exert a strong, driving influence on these cortical circuits (Zhou et al., 2016; Beltramo and Scanziani, 2019). Pulvinar neurons projecting to a specific higher visual cortical area have been found to also receive input from the same area in mice, demonstrating the existence of recurrent higher-order thalamocortical loops (Bennett et al., 2019; Blot et al., 2021). Interestingly, this reciprocal input to thalamus comes mainly from L6 neurons in the target area, indicating that these higher-order recurrent thalamocortical loops resemble the recurrent circuits between primary thalamic and cortical areas. However, these recurrent higher-order thalamocortical loops do not appear to be a dominant feature of thalamocortical interactions, as pulvinar neurons projecting to a specific cortical area do not preferentially receive connections from this area, but equally numerous input from other visual cortical areas, at least when quantified using anatomical monosynaptic rabies-tracing methods (Blot et al., 2021). Nevertheless, cortical inputs can significantly influence the activity of pulvinar neurons projecting back to the same area (Blot et al., 2021).

Much more selective recurrent thalamocortical loops have been described in other modalities. For instance, thalamic nucleus POm is reciprocally interconnected with both S1 and primary motor cortex (M1), and S1-projecting POm neurons receive much stronger input from both L6 CT and L5 PT neurons in S1 than in M1, while M1-projecting POm neurons receive stronger input from M1 (Guo et al., 2020). A thalamocortical loop with similar specificity has been described between anterolateral motor cortex (ALM) and thalamic ventromedial (VM) motor area (Guo et al., 2017b). ALM and motor thalamus have been shown to form a surprisingly powerful recurrent circuit, since optogenetic silencing of motor thalamus strongly disrupts activity in ALM, and vice versa (Guo et al., 2017b). In contrast, M1 and the ventrolateral (VL) nucleus, although prominently connected with each other, do not appear to form a recurrent circuit (Yamawaki and Shepherd, 2015). Recurrent CTC motifs are not limited to sensory and motor areas. Prefrontal cortex (PFC) is reciprocally connected with both MD and VM, as MD and VM neurons projecting to PFC receive input from PT and CT neurons in PFC (Collins et al., 2018).

In general, higher-order thalamic projections preferentially target upper-layer and L5 IT neurons in neocortex, but also—albeit usually to a much lesser degree—thalamus-projecting L5 PT neurons (but not L5 CT neurons) (Shepherd and Yamawaki, 2021). This creates the potential for both polysynaptic as well as weaker monosynaptic thalamocortical loops that may be interconnected. Moreover, thalamic projections also prominently connect to various types of inhibitory interneurons in cortical target areas which have the potential to regulate the excitatory thalamic influence on cortex, and in some thalamocortical pathways can mediate suppression of cortical activity upon activation of higher-order thalamic nuclei (Fang et al., 2020; Mukherjee et al., 2020).

TRANS-THALAMIC FEEDFORWARD PATHWAYS

Higher-order thalamic areas not only project back to the cortical areas they receive input from, but also often innervate many other cortical areas. These projections are highly structured. Different subdivisions of the pulvinar are topographically organized and project to distinct cortical areas (Shipp, 2003; Bennett et al., 2019), Moreover, in some circuits, visual response properties of pulvinar projection neurons match those of the targeted cortical areas (Shipp, 2003; Roth et al., 2016; Bennett et al., 2019; Blot et al., 2021). Importantly, higher-order sensory thalamic areas receive driving input from L5 cortical neurons of primary sensory areas, and innervate L4 of higher cortical areas where they can exert a strong influence on neural activity (Zhou et al., 2016; Beltramo and Scanziani, 2019). Therefore, a long-standing hypothesis, in particular formulated by Murray Sherman and Roy Guillery, proposed that thalamus connects primary cortical areas to higher areas, via indirect, feedforward trans-thalamic pathways that parallel direct cortical connections (Guillery and Sherman, 2002; Figure 42.1C).

In support of this theory, the density of axons directly connecting two cortical areas in the mouse visual system has been shown to be proportional to the density of axons connecting them indirectly through the pulvinar (Bennett et al., 2019). However, to directly demonstrate the existence of feedforward trans-thalamic pathways connecting two cortical areas through the thalamus, thalamic neurons projecting to a higher cortical area have to be shown to themselves receive driving synaptic input from a hierarchically lower cortical area, since excitatory thalamic neurons do not form local connections within an area. Such direct evidence of fine-scale synaptic connectivity in thalamocortical circuits has emerged only recently. Studies in mice using monosynaptic rabies virus tracing techniques showed that pulvinar neurons projecting to a specific higher visual cortical area receive direct synaptic input from L5 neurons in V1, while synaptic input from other visual cortical areas stems mainly from L6 neurons. These findings indicate that the feedforward trans-thalamic connectivity motif is a main feature of visual thalamocortical interactions, because L5 cells form large, driving synapses on the proximal dendrites of thalamic neurons, whereas L6 neurons form weaker synapses thought to be modulatory (Sherman and Guillery, 2011). Consistent with this anatomical evidence, V1 exerts strong influence on the visual responses measured in thalamic projections to higher visual areas (Blot et al., 2021).

A similar trans-thalamic feedforward pathway has been demonstrated in the somatosensory system in brain slices: even in the absence of a direct cortical connection between S1 and S2, S1 can strongly activate neurons in S2 via a trans-thalamic pathway through POm. This effect was shown to be mediated by L5, not L6 CT projections neurons in S1 (Theyel et al., 2010). Moreover, this projection pathway from S1 to POm has been shown to be important for discriminating textures (Qi et al., 2022). Trans-thalamic pathways are not limited to connecting sensory cortical areas. Using viral-tracing techniques and optogenetics, Mo and Sherman described a feedforward sensory-motor pathway connecting L5 neurons in S1 with primary motor cortex (M1) via the POm (Mo and Sherman, 2019).

TRANS-THALAMIC FEEDBACK PATHWAYS

Higher-order thalamic nuclei project back to primary sensory areas with modulatory synapses (Miller-Hansen and Sherman, 2022), and receive input from many higher-order cortical areas. For instance, the pulvinar receives input not only from V1 but from all visual cortical areas (Rovó et al., 2012; Bennett et al., 2019; Blot et al., 2021). These connections could provide the building blocks for trans-thalamic feedback pathways from hierarchically higher to lower sensory cortical areas, paralleling direct intracortical feedback projections (Figure 42.1D). And indeed, pulvinar neurons projecting to V1 receive input from higher visual cortical areas, and similarly, POm neurons projecting to S1 receive input from S2 (Miller-Hansen and Sherman, 2022). But these feedback pathways not only connect sensory areas. Pulvinar and POm also receive input from associative and motor areas, and a trans-thalamic pathway from PFC to V1 through the pulvinar has been shown to enhance visual responses in V1 (Hu et al., 2019). Feedback projections from higher-order thalamic areas to primary sensory cortices possess an interesting feature that may distinguish them from intracortical feedback: brief activation of POm projections to S1 in vivo has a long-lasting effect on S1 activity for up to 1 second, and this effect is not observed when stimulating intracortical feedback inputs from S2 or M1 to S1 (Zhang and Bruno, 2019). This persistent cortical response to higher-order thalamic input is mediated by POm-S1 thalamocortical loops, providing further evidence for the importance of higher-order thalamocortical interactions for cortical function.

COEXISTENCE OF TRANS-THALAMIC MOTIFS

Recurrent, feedforward, and feedback motifs in thalamocortical interactions have been covered separately above, but they rarely exist in isolation. Neurons in

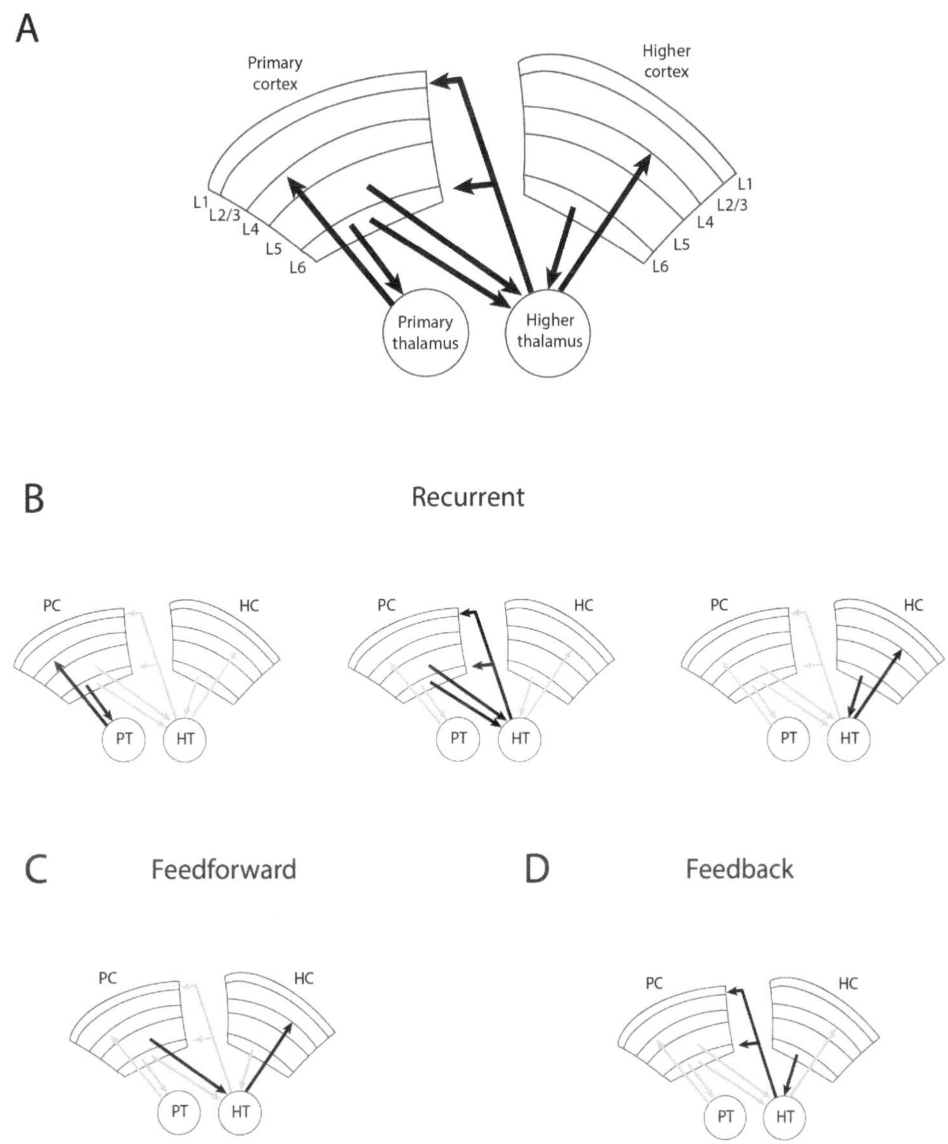

Figure 42.1. Trans-thalamic circuit motifs. (A). Projections between primary and higher-order thalamus and cortical areas with cortical input and output layers. (B). Recurrent trans-thalamic loops between primary cortical and thalamic areas (left), between primary cortical areas and higher-order thalamus (middle), and between higher-order cortical and thalamic areas (right). PC = primary cortex; HC = higher cortex; PT = primary thalamus; HT = higher-order thalamus. (C) Trans-thalamic feedforward pathways, connecting primary to higher cortical areas via higher-order thalamus. (D) Trans-thalamic feedback pathways, connecting higher to primary cortical areas via higher-order thalamus.

higher-order thalamus receive input from the cortical areas they project to, as well as hierarchically lower and higher cortical areas (Bennett et al., 2019; Shepherd and Yamawaki, 2021). Feedforward, recurrent, and feedback information may therefore converge and intermix (Figure 42.1A). Direct evidence for such convergence on the single-cell level comes from electron microscopy studies demonstrating that individual POm neurons receive L5 inputs from both S1 and M1 (Sampathkumar et al., 2021). Convergence of diverse cortical inputs onto thalamic neurons appears to be less pronounced in the motor thalamus, where thalamocortical connectivity motifs have been shown to be more specific (see above, section on recurrent thalamocortical loops; Guo et al., 2017b, 2020).

Higher-order thalamic nuclei are driven not only by projections from the neocortex, but also by input from various subcortical areas. The population of pulvinar neurons projecting to a specific higher visual area, for instance, receives input from cortical visual areas and the superior colliculus, and in POm, driving input from both the sensory brainstem and from S1 can converge onto single neurons (Groh et al., 2014). Convergence patterns of cortical and subcortical inputs are less clear outside sensory thalamus; however, motor areas in the thalamus receive prominent projections from cortical motor areas, cerebellum, and basal ganglia that overlap in some thalamic subregions (Hintzen et al., 2018). The diversity of convergent pathways onto thalamic nuclei indicates that CTC circuits have the capacity to integrate diverse information from cortical and subcortical sources in order to direct and regulate cortical activity. The next section provides a brief overview of potential functions of trans-thalamic pathways.

FUNCTION OF TRANS-THALAMIC CIRCUITS

MAINTENANCE OF ACTIVITY

Neocortical circuits can exhibit long-lasting neural activity patterns that persist in the absence of sensory input. Such persistent patterns have been linked to working memory, motor preparation, or reward anticipation (Romo et al., 1999; Churchland et al., 2010). Maintenance of activity has been proposed to be mediated by reverberations of neural activity in recurrently connected circuits. The mammalian neocortex has a strongly recurrent structure, as the majority of synapses onto cortical neurons come from other excitatory cortical cells (Douglas et al., 1995). Intracortical networks are thus thought to underlie persistent activity patterns (Douglas et al., 1995). However, recent studies demonstrate that cortical circuits cannot maintain neural activity in the absence of thalamic input. Persistent activity in visual cortical areas after presentation of a visual stimulus is eliminated with a very fast time constant of 10 ms after optogenetic silencing of thalamus (Reinhold et al., 2015). A similar effect has been observed in thalamocortical motor circuits: neurons in the mouse anterior lateral motor cortex (ALM) exhibit persistent neural activity during a post-stimulus delay period in which a motor action is remembered and prepared. This persistent cortical activity rapidly ceases when motor thalamus is silenced (Guo et al., 2017b). These studies show that neural activity cannot be sustained by cortical circuits alone, but is instead maintained via cortico-thalamocortical loops.

REGULATING INTERACTIONS BETWEEN CORTICAL AREAS

All brain functions necessitate interactions between distributed networks of brain areas. Information routing between cortical areas is thought to be context-dependent and dynamic (Gonzalez-Castillo and Bandettini, 2018; Javadzadeh and Hofer, 2022), enabling flexible behavioral output and cognitive function. The mechanisms for regulating cortico-cortical interactions are not fully understood, but a widely studied hypothesis suggests changes in synchronization of activity across cortical areas as a mechanism for regulating their influence on each other (Womelsdorf et al., 2007; Fries, 2015). Trans-thalamic pathways between cortical areas could provide a substrate for coordinating activity in distributed cortical areas and for regulating interactions between them. Evidence supporting this theory comes mainly from studies on the role of trans-thalamic pathways through the pulvinar for attentional modulation. The pulvinar appears to coordinate cortical interactions during states of high attention (Saalmann et al., 2012; Fiebelkorn and Kastner, 2019), and inactivation of pulvinar can weaken functional interactions between cortical areas (Eradath et al., 2021). These studies suggest a role for trans-thalamic pathways in regulating communication between networks of cortical areas, for instance by changing the synchrony of activity patterns in different areas, or by changing the excitability of neurons in the target area to make them more or less susceptible to different inputs (Halassa and Kastner, 2017; Boshra and Kastner, 2022).

PARALLEL INFORMATION-PROCESSING PATHWAYS

Increasing evidence suggests that trans-thalamic pathways not only regulate cortical activity and information flow through cortical pathways, but also can function as information-processing pathways themselves, paralleling intracortical feedforward pathways. Trans-thalamic pathways in the visual system have been shown to carry specific and distinct visual signals to different visual cortical areas (Blot et al., 2021). Interestingly, these signals are different from those conveyed by direct intra-cortical projections from V1 to the same visual cortical area. This indicates that cortical and trans-thalamic pathways constitute parallel but specialized sensory-processing streams. Sensory trans-thalamic pathways are likely to be highly integrative, since they receive not only cortical feedforward input (i.e., from V1), but also input from many other cortical areas, as well as subcortical structures such as the superior colliculus (SC), the pretectum, and other brainstem nuclei (Zhou et al., 2016; Blot et al., 2021). Therefore, the thalamus is ideally positioned to integrate signals from diverse sources, before sending them back to the cortex (and the striatum).

For instance, pulvinar projections to higher visual areas in mice exhibit a distinct preference for fast-moving visual stimuli, thought to be at least partly inherited from SC projections to the pulvinar (Tohmi et al., 2014; Blot et al., 2021). Neural activity in the post-rhinal region of visual cortex in mice (POR) has even been shown to be independent of V1, but driven mainly by superior colliculus through the pulvinar (Beltramo and Scanziani, 2019). In motor thalamus, cortical signals can be combined with information from the cerebellum and basal ganglia (Li and Mrsic-Flogel, 2020; Matsuyama and Tanaka, 2021). This convergence of multiple pathways in the thalamus, as well as its diverging outputs to multiple target areas, may allow the thalamus to coordinate and consolidate distributed processes across many brain regions, and translate them to a common (neocortical) frame of reference. Because the brain is organized in functional modules, information is often processed in parallel in multiple distributed networks. These processes need to be combined and unified, across different modalities and cortical and subcortical circuits. Thalamus has been proposed to act as an "active blackboard," continuously reconciling information—different "hypotheses"—received from distributed networks. A reconciled view could then be broadcast back to these networks, ensuring convergence into one common "hypothesis" (Mumford, 1991).

PREDICTIVE CODING

Pulvinar inputs to visual cortical areas in rodents convey not only visual signals, but also information about the context in which the visual signals were received, such as the animal's own movements at the time (Roth et al., 2016; Blot et al., 2021; Miura and Scanziani, 2022). This nonvisual information could represent efference copy signals, used by the brain to predict the sensory consequences (i.e., sensory feedback) of one's own motor actions. Such predictive signals are crucial for interpreting sensory input, and in the predictive coding framework, they are hypothesized to cancel out predicted (i.e., self-induced) sensory input, and to calculate sensory prediction errors—deviations of incoming sensory information from internally generated predictions (Lee and Mumford, 2003; Friston, 2008; Olshausen, 2013). Higher-order thalamic pathways may carry such internal predictions (Sherman and Guillery, 2011; Miura and Scanziani, 2022); however, they have also been shown to convey prediction-error-like signals to cortical areas (Roth et al., 2016; Blot et al., 2021). This input may increase the salience of unpredicted, and therefore particularly relevant, visual information (Hu et al., 2019). Theoretical work has also suggested another role for trans-thalamic pathways in predictive processing. Friston and colleagues proposed that while direct, intracortical pathways conduct perceptual inference (combining predictions and incoming sensory information), trans-thalamic connections may contextualize these processes by adjusting the gain of ascending prediction errors depending on their precision and the confidence level of descending predictions (Komura et al., 2013; Kanai et al., 2015).

HIGHER COGNITIVE FUNCTIONS

Trans-thalamic pathways have been shown to be important for diverse cognitive functions. The thalamus has been strongly linked to working memory and motor preparation, because the persistent cortical activity that is deemed necessary for these processes is dependent on maintenance of activity through trans-thalamic loops (see above) (Bolkan et al., 2017; Guo et al., 2017b). The thalamus is also thought to be important for attentional processes. Trans-thalamic pathways have been proposed to synchronize and connect functional networks across segregated cortical areas, according to attentional demands, to prioritize cortical representations of behaviorally relevant sensory input features (Halassa and Kastner, 2017). Therefore, the proposed role of trans-thalamic pathways in regulating interactions between cortical areas closely links them to attentional control (see Jaramillo et al., 2019, for a mechanistic account based on multi-regional circuit modeling). For instance, synchrony in pulvino-cortical network activity has been suggested to depend on neural activity in pulvinar, which could be modulated by spatial attention (Saalmann et al., 2012). Consistent with this view, pulvinar lesions in humans and monkeys cause attentional deficits (Petersen et al., 1987; Snow et al., 2009). Higher-order thalamocortical interactions, in particular loops between PFC and mediodorsal (MD) nucleus, have also been shown to be important for decision-making: MD input to PFC is important for cognitive flexibility, namely switching behavioral strategies in different learned contexts, and facilitates decision-making under uncertainty (Wimmer et al., 2015; Schmitt et al., 2017). Finally, in humans, the thalamus has been shown to play a central role in language formation, with thalamic lesions leading to aphasia. Specifically, the loops connecting the basal ganglia to the frontal cortical areas via the thalamus have been proposed to support grammatical operations (Crosson, 2021). (These diverse cognitive roles of trans-thalamic pathways will be covered in detail in **Section VIII**.)

CONCLUSIONS

The communication between cerebral cortex and thalamus unfolds through diverse circuit motifs, including recurrent thalamocortical loops, trans-thalamic feedforward and feedback pathways, and these motifs are likely to support different functions of thalamocortical interactions. It is important to corroborate the existence of these motifs and gain a better understanding of how information flows through these pathways. However, many thalamic areas show convergence and divergence of diverse inputs and outputs. Looking at specific trans-thalamic motifs in isolation is thus likely not sufficient for understanding the function of thalamocortical networks and how they process information.

For instance, we still do not have a good understanding of how different pathways through the thalamus overlap on the single-cell level. Do individual thalamic neurons participate in multiple trans-thalamic pathways? Or do these pathways remain segregated? And to what degree do single thalamic neurons integrate cortical and subcortical signals? A fine-scale understanding of trans-thalamic pathways could also shed light on the interplay between the direct and indirect, trans-thalamic flow of information between cortical areas. How do cortical neurons integrate signals from direct cortical projections and trans-thalamic pathways? With the current large-scale efforts to optimize genetic tools to dissect, record from, and manipulate specific circuit elements and pathways, answering these questions and unraveling the fine-scale organization and function of trans-thalamic pathways should be achievable in the near future.

REFERENCES

Beltramo R, Scanziani M (2019). A collicular visual cortex: Neocortical space for an ancient midbrain visual structure. *Science*. 363:64–69.

Bennett C, Gale SD, Garrett ME, Newton ML, Callaway EM, Murphy GJ, Olsen SR (2019). Higher-order thalamic circuits channel parallel streams of visual information in mice. *Neuron*. 102:477–492.

Blasdel GG, Lund JS (1983). Termination of afferent axons in macaque striate cortex. *J Neurosci*. 3:1389–1413.

Blot A, Roth MM, Gasler I, Javadzadeh M, Imhof F, Hofer SB (2021). Visual intracortical and transthalamic pathways carry distinct information to cortical areas. *Neuron*. 109(12):1996–2008.e6. Accessed May 17, 2021. https://www.sciencedirect.com/science/article/pii/S089662732100283X.

Bolkan SS, Stujenske JM, Parnaudeau S, Spellman TJ, Rauffenbart C, Abbas AI, Harris AZ, Gordon JA, Kellendonk C (2017). Thalamic projections sustain prefrontal activity during working memory maintenance. *Nat Neurosci*. 20:987–996.

Born G, Schneider-Soupiadis FA, Erisken S, Vaiceliunaite A, Lao CL, Mobarhan MH, Spacek MA, Einevoll GT, Busse L (2021). Corticothalamic feedback sculpts visual spatial integration in mouse thalamus. *Nat Neurosci*. 24:1711–1720.

Boshra R, Kastner S (2022). Attention control in the primate brain. *Curr Opin Neurobiol*. 76:102605.

Churchland MM, Cunningham JP, Kaufman MT, Ryu SI, Shenoy KV (2010). Cortical preparatory activity: representation of movement or first cog in a dynamical machine? *Neuron*. 68:387–400.

Collins DP, Anastasiades PG, Marlin JJ, Carter AG (2018). Reciprocal circuits linking the prefrontal cortex with dorsal and ventral thalamic nuclei. *Neuron*. 98:366–379.

Constantinople CM, Bruno RM (2013). Deep cortical layers are activated directly by thalamus. *Science*. 340:1591–1594.

Crandall SR, Cruikshank SJ, Connors BW (2015). A corticothalamic switch: Controlling the thalamus with dynamic synapses. *Neuron*. 86:768–782.

Crosson B (2021). The role of the thalamus in declarative and procedural linguistic memory processes. *Front Psychol*. 12:682199.

Douglas RJ, Koch C, Mahowald M, Martin KA, Suarez HH (1995). Recurrent excitation in neocortical circuits. *Science*. 269:981–985.

Eradath MK, Pinsk MA, Kastner S (2021). A causal role for the pulvinar in coordinating task-independent cortico-cortical interactions. *J Comp Neurol*. 529:3772–3784.

Erişir A, Van Horn SC, Sherman SM (1997). Relative numbers of cortical and brainstem inputs to the lateral geniculate nucleus. *Proc Natl Acad Sci*. 94:1517–1520.

Fang Q, Chou X, Peng B, Zhong W, Zhang LI, Tao HW (2020). A differential circuit via retino-colliculo-pulvinar pathway enhances feature selectivity in visual cortex through surround suppression. *Neuron*. 105:355–369.

Fiebelkorn IC, Kastner S (2019). A rhythmic theory of attention. *Trends Cogn Sci*. 23:87–101.

Fries P (2015). Rhythms for cognition: Communication through coherence. *Neuron*. 88:220–235.

Friston K (2008). Hierarchical models in the brain. *PLoS Comp Biol*. 4:e1000211.

Gonzalez-Castillo J, Bandettini PA (2018). Task-based dynamic functional connectivity: Recent findings and open questions. *NeuroImage*. 180:526–533.

Groh A, Bokor H, Mease RA, Plattner VM, Hangya B, Stroh A, Deschenes M, Acsády L (2014). Convergence of cortical and sensory driver inputs on single thalamocortical cells. *Cereb Cortex*. 24:3167–3179.

Guillery RW, Sherman SM (2002). Thalamic relay functions and their role in corticocortical communication: generalizations from the visual system. *Neuron*. 33:163–175.

Guo K, Yamawaki N, Barrett JM, Tapies M, Shepherd GMG (2020). Cortico-thalamo-cortical circuits of mouse forelimb S1 are organized primarily as recurrent loops. *J Neurosci*. 40:2849–2858.

Guo W, Clause AR, Barth-Maron A, Polley DB (2017a). A corticothalamic circuit for dynamic switching between feature detection and discrimination. *Neuron*. 95:180–194.

Guo ZV, Inagaki HK, Daie K, Druckmann S, Gerfen CR, Svoboda K (2017b). Maintenance of persistent activity in a frontal thalamocortical loop. *Nature*. 545:181–186.

Halassa MM, Kastner S (2017). Thalamic functions in distributed cognitive control. *Nat Neurosci*. 20:1669–1679.

Halassa MM, Sherman SM (2019). Thalamocortical Circuit Motifs: A General Framework. *Neuron*. 103:762–770.

Hasse JM, Briggs F (2017). Corticogeniculate feedback sharpens the temporal precision and spatial resolution of visual signals in the ferret. *Proc Natl Acad Sci*. 114:E6222–E6230.

Hintzen A, Pelzer EA, Tittgemeyer M (2018). Thalamic interactions of cerebellum and basal ganglia. *Brain Struct Funct*. 223:569–587.

Hu F, Kamigaki T, Zhang Z, Zhang S, Dan U, Dan Y (2019). Prefrontal corticotectal neurons enhance visual processing through the superior colliculus and pulvinar thalamus. *Neuron*. 104:1141–1152.

Jaramillo J, Mejias JF, Wang X-J (2019). Engagement of pulvino-cortical feedforward and feedback pathways in cognitive computations. *Neuron*. 101:321–336.

Javadzadeh M, Hofer SB (2022). Dynamic causal communication channels between neocortical areas. *Neuron*. 110:2470–2483.

Kanai R, Komura Y, Shipp S, Friston K (2015). Cerebral hierarchies: Predictive processing, precision and the pulvinar. *Philos Trans R Soc Lond B Biol Sci*. 370:20140169.

Kirchgessner MA, Franklin AD, Callaway EM (2020). Context-dependent and dynamic functional influence of corticothalamic pathways to first- and higher-order visual thalamus. *Proc Natl Acad Sci*. 117:13066–13077.

Komura Y, Nikkuni A, Hirashima N, Uetake T, Miyamoto A (2013). Responses of pulvinar neurons reflect a subject's confidence in visual categorization. *Nat Neurosci.* 16:749–755.

Lee TS, Mumford D (2003). Hierarchical Bayesian inference in the visual cortex. *J Opt Soc Am A Opt Image Sci Vis.* 20:1434–1448.

Li N, Mrsic-Flogel TD (2020). Cortico-cerebellar interactions during goal-directed behavior. *Curr Opin Neurobiol.* 65:27–37.

Matsuyama K, Tanaka M (2021). Temporal prediction signals for periodic sensory events in the primate central thalamus. *J Neurosci.* 41(9):1917–1927. Accessed December 12, 2022. https://www.jneurosci.org/content/early/2021/01/12/JNEUROSCI.2151-20.2021.

Mease RA, Krieger P, Groh A (2014). Cortical control of adaptation and sensory relay mode in the thalamus. *Proc Natl Acad Sci U S A.* 111:6798–6803.

Miller-Hansen AJ, Sherman SM (2022). Conserved patterns of functional organization between cortex and thalamus in mice. *Proc Natl Acad Sci U S A.* 119:e2201481119.

Miura SK, Scanziani M (2022). Distinguishing externally from saccade-induced motion in visual cortex. *Nature.* 610:135–142.

Mo C, Sherman SM (2019). A sensorimotor pathway via higher-order thalamus. *J Neurosci.* 39:692–704.

Mukherjee A, Bajwa N, Lam NH, Porrero C, Clasca F, Halassa MM (2020). Variation of connectivity across exemplar sensory and associative thalamocortical loops in the mouse. *eLife.* 9:e62554.

Mumford D (1991). On the computational architecture of the neocortex. *Biol Cybern.* 65:135–145.

Olshausen BA, Mangun GR, Gazzaniga MS (2014). 27 Perception as an inference problem. *Cogn Neurosci.* 295.

Petersen SE, Robinson DL, Morris JD (1987). Contributions of the pulvinar to visual spatial attention. *Neuropsychologia.* 25:97–105.

Qi J, Ye C, Naskar S, Inácio AR, Lee S (2022). Posteromedial thalamic nucleus activity significantly contributes to perceptual discrimination. *PLoS Biol.* 20:e3001896.

Reinhold K, Lien AD, Scanziani M (2015). Distinct recurrent versus afferent dynamics in cortical visual processing. *Nat Neurosci.* 18:1789–1797.

Rockland KS (2021). A closer look at corticothalamic "loops." *Front Neural Circuits.* 15:632668.

Romo R, Brody CD, Hernández A, Lemus L (1999). Neuronal correlates of parametric working memory in the prefrontal cortex. *Nature.* 399:470–473.

Roth MM, Dahmen JC, Muir DR, Imhof F, Martini FJ, Hofer SB (2016). Thalamic nuclei convey diverse contextual information to layer 1 of visual cortex. *Nat Neurosci.* 19:299–307.

Rovó Z, Ulbert I, Acsády L (2012). Drivers of the primate thalamus. *J Neurosci.* 32:17894–17908.

Rubio-Garrido P, Pérez-de-Manzo F, Porrero C, Galazo MJ, Clasca F (2009). Thalamic input to distal apical dendrites in neocortical layer 1 is massive and highly convergent. *Cereb Cortex.* 19:2380–2395.

Saalmann YB, Pinsk MA, Wang L, Li X, Kastner S (2012). The pulvinar regulates information transmission between cortical areas based on attention demands. *Science.* 337:753–756.

Sampathkumar V, Miller-Hansen A, Sherman SM, Kasthuri N (2021). Integration of signals from different cortical areas in higher order thalamic neurons. *Proc Natl Acad Sci U S A.* 118:e2104137118.

Schmitt LI, Wimmer RD, Nakajima M, Happ M, Mofakham S, Halassa MM (2017). Thalamic amplification of cortical connectivity sustains attentional control. *Nature.* 545:219–223.

Shepherd GMG, Yamawaki N (2021) Untangling the cortico-thalamo-cortical loop: cellular pieces of a knotty circuit puzzle. *Nat Rev Neurosci.* 22:389–406.

Sherman SM, Guillery RW (2011). Distinct functions for direct and transthalamic corticocortical connections. *J Neurophysiol.* 106:1068–1077.

Shipp S (2003). The functional logic of cortico-pulvinar connections. *Philos Trans R Soc Lond B Biol Sci.* 358:1605–1624.

Snow JC, Allen HA, Rafal RD, Humphreys GW (2009). Impaired attentional selection following lesions to human pulvinar: Evidence for homology between human and monkey. *Proc Natl Acad Sci U S A.* 106:4054–4059.

Theyel BB, Llano DA, Sherman SM (2010). The corticothalamocortical circuit drives higher-order cortex in the mouse. *Nat Neurosci.* 13:84–88.

Tohmi M, Meguro R, Tsukano H, Hishida R, Shibuki K (2014). The extrageniculate visual pathway generates distinct response properties in the higher visual areas of mice. *Curr Biol.* 24:587–597.

Usrey WM, Sherman SM (2019). Corticofugal circuits: Communication lines from the cortex to the rest of the brain. *J Comp Neurol.* 527:640–650.

Wimmer RD, Schmitt LI, Davidson TJ, Nakajima M, Deisseroth K, Halassa MM (2015). Thalamic control of sensory selection in divided attention. *Nature.* 526:705–709.

Womelsdorf T, Schoffelen J-M, Oostenveld R, Singer W, Desimone R, Engel AK, Fries P (2007). Modulation of neuronal interactions through neuronal synchronization. *Science.* 316:1609–1612.

Yamawaki N, Shepherd GMG (2015). Synaptic circuit organization of motor corticothalamic neurons. *J Neurosci.* 35:2293–2307.

Zhang W, Bruno RM (2019). High-order thalamic inputs to primary somatosensory cortex are stronger and longer lasting than cortical inputs. *eLife.* 8:e44158.

Zhou H, Schafer RJ, Desimone R (2016). Pulvinar-cortex interactions in vision and attention. *Neuron.* 89:209–220.

SECTION VIII

COGNITIVE PROPERTIES

Michael M. Halassa and Sabine Kastner

The thalamus has traditionally been viewed as a relay to the cortex, an idea primarily based on how thalamic structures contribute to early sensory processing. However, only a subset of thalamic nuclei fulfills the criteria of receiving peripheral sensory inputs and projecting to primary sensory areas. In fact, the largest thalamic nuclei in primates, the mediodorsal thalamus and pulvinar, are largely connected to associative cortical areas. Although older theories have suggested that these areas play critical roles in cognitive processing, there has been a recent reinvigoration of such ideas based on contemporary studies that span lesion and intracranial recordings in humans to mechanistic functional circuit dissection in rodents.

In this section, we bring together a number of chapters that build on the idea that the thalamus plays important roles in cognitive processing. We begin by focusing on the role of the thalamus in memory, with Chapter 43 by Sweeney-Reed and Knight highlighting the role of the anterior thalamus in human memory. Chapter 44 by Peyrache and Duszkiewicz then provides mechanistic insights into such process based on rodent studies. Next, we transition to the role of the thalamus in selective attention, with Chapter 45 by Kastner and Usrey highlighting the role of the geniculate and pulvinar in basic computational mechanisms related to sensory prioritization and attention. This is primarily focused on studies from nonhuman primates, but supportive evidence from both humans and rodents is also discussed. This leads to Chapter 46 on the mediodorsal thalamus, by Halassa and Saalmann, and its role in decision-making, again drawing on data across rodents and primates. Chapter 47 by Hwang and Shine utilizes both lesion and fMRI studies in the human brain, connecting thalamic subregions to both cognitive processing on one end and to large-scale cortical network organization on another.

43.

MEMORY AND THE HUMAN ANTERIOR THALAMUS

Catherine M. Sweeney-Reed and Robert T. Knight

INTRODUCTION

Memories define us as individuals and form the basis for the diverse range of activities that comprise daily life. The cerebral cortex and hippocampus are well-established as key brain structures involved in orchestrating the many aspects of memory processing. The thalamus, on the other hand, was long considered simply to relay, rather than actively process, cortical information. In recent years, however, the role of the thalamus, and the anterior thalamus in particular, has gained increasing attention in theories and studies regarding how memories are formed, maintained, and retrieved. The thalamus comprises a collection of nuclei, which have extensive connectivity with different, functionally specialized areas of the cerebral cortex (Vertes et al., 2001; Aggleton et al., 2010). Their reciprocal anatomical connectivity with the hippocampus (Aggleton et al., 2010), as well as their oscillatory activity in the theta frequency range, the dominant hippocampal rhythm (Vertes et al., 2001), have led to consideration of the anterior nuclei of the thalamus (ANT) as a part of an extended hippocampal system (Aggleton et al., 2010). Indeed, the ANT are part of the Papez circuit (Papez, 1937), which supports memory processing (Aggleton and Brown, 1999; Vertes et al., 2001). Widespread connectivity with neocortical networks places the ANT in an ideal neuroanatomical position to coordinate the diverse features represented in an epxisodic memory trace (Figure 43.1).

Due to their deep brain location and small size, most of our knowledge regarding the thalamic nuclei has been derived from lesion, post-mortem, and imaging studies (Aggleton and Brown, 1999; Harding et al., 2000; van der Werf et al., 2003; Aggleton et al., 2010; Pergola et al., 2013). Since the introduction of deep brain stimulation (DBS) of the ANT as a potential treatment for pharmacoresistant epilepsy, both causal information, through altering memory processes by stimulation (Fisher et al., 2010; Oh et al., 2012; Liu et al., 2021), as well as real-time information, gained from the recording of ANT electrophysiological signals during memory processing (Fitzgerald et al., 2013; Sweeney-Reed et al., 2014; Štillová et al., 2015), have enabled new insights into the role of the ANT in mnemonic function. Growing evidence from these studies supports the notion that the ANT play a critical, active role in the information processing underlying mnemonic function, beyond simply relaying hippocampal and neocortical signals (Aggleton et al., 2010; Sweeney-Reed et al., 2021). Simultaneous scalp electroencephalographic (EEG) recordings of neocortical activity have, moreover, provided a window on real-time interaction between the anterior thalamus and cerebral cortex during memory processing (Sweeney-Reed et al., 2014; Sweeney-Reed et al., 2015; Sweeney-Reed et al., 2016). The active role postulated for the ANT lies in the coordination of widespread neocortical oscillatory electrical brain activity (Ketz et al., 2014; Sweeney-Reed et al., 2014), selecting information and facilitating switching between cognitive states (de Luca et al., 2006; Sweeney-Reed et al., 2017; Sweeney-Reed et al., 2021).

We first consider the nomenclature, which varies across the research fields investigating the role of the anterior thalamus in memory. Terminology differs between human and animal literature (see **Chapter 44 in Section VIII**) and between basic neuroscience and clinical (neurology; neurosurgery) communities. We then briefly examine the component processes into which memory is commonly subdivided, including encoding, maintenance, and retrieval, as well as different memory types, including recollection and familiarity, as different hypotheses have been proposed regarding the role of the ANT in these stages and types of memory. Finally, we review findings from the range of data recording modalities used to investigate the role of the anterior thalamus and its interactions with cerebral cortex to support memory processing, including the insights provided by real-time human electrophysiological recordings simultaneously from the ANT and neocortex, which have revealed interactions between the ANT and frontal and parietal neocortex during memory processing.

ANTERIOR THALAMIC SUBNUCLEI: NOMENCLATURE

The collection of nuclei comprising the anterior thalamus are defined through their chemo- and cytoarchitecture, and their anatomical connectivity (Lambert et al., 2017). Their connectivity is widespread, involving reciprocal connections with the cerebral cortex, cerebellum, basal ganglia, brainstem, and spinal cord (Lambert et al., 2017). The ANT are located between the limbs of the internal medullary lamina, and together with the lateral dorsal and medial dorsal nuclei, they comprise the limbic thalamus. Human memory studies with the spatial and temporal resolution required to evaluate the roles of specific thalamic nuclei and subnuclei, made possible by advances in high resolution neuroimaging, as well as intracranial electrophysiological recordings from the thalamus from clinical populations, have cast new light on the roles of these separate structures during real-time memory processing.

Inter-species anatomical differences, including anatomical connectivity, have been observed between the human, monkey, and rat ANT (Bubb et al., 2017). Another important difference between human and animal literature reflects spatial resolution. ANT (or ATN: anterior thalamic nucleus) is frequently used in the singular in clinical literature. Many human neuroimaging studies, for example, consider the anterior thalamus as a single structure, due to spatial resolution both in functional neuroimaging and the structural imaging used to confirm postoperative electrode location in intracranial recordings. However, animal and in vitro studies differentiating between subnuclei refer to the ANT in the plural (Aggleton et al., 2010). Diffusion-weighted imaging (DWI) has recently enabled differentiation between thalamic nuclei in human studies, revealing differing connectivity patterns of the separate subnuclei (Grodd et al., 2020). The four subnuclei comprising the ANT include the anteroventral (AV), anteromedial (AM), anterodorsal (AD), and lateral dorsal (LD) nuclei. These ANT subnuclei have different anatomical connectivity patterns with other brain regions and differing functional roles. Oscillatory neuronal firing in the low theta frequency range, centered around ~3 Hz in human hippocampal recordings (Jacobs, 2014), has a well-recognized association with memory processing (Crespo-Garcia et al., 2010; Watrous et al., 2013). Rhythmical neuronal firing in the theta frequency range is also observed in the AV, AM, and AD (Vertes et al., 2001), with the greatest coherence with the hippocampal theta rhythm shown in AV recordings (Albo et al., 2003).

MEMORY SYSTEMS ENGAGING THE ANTERIOR THALAMUS

Memory encoding, maintenance/storage, consolidation, and retrieval have specific neural correlates (Tulving, 2002; Diekelmann and Born, 2010). Retrieval is also divided into episodic recall and familiarity-based recognition (Carlesimo et al., 2015). While brain structures involved in memory overlap, with hippocampally mediated reactivation of neocortical activity patterns stored at encoding, on subsequent retrieval (Rugg et al., 2008), the differing anatomical connectivity of the ANT compared with the dorsomedial nucleus of the thalamus (DMNT), with prominent prefrontal projections, suggests differential engagement, and different theories ascribe separate roles.

Specific thalamic nuclei engage in particular functional networks (van der Werf et al., 2003; Aggleton et al., 2010; Sweeney-Reed et al., 2014; Sweeney-Reed et al., 2016; Carlesimo et al., 2015). Anatomical and functional evidence suggests ANT involvement in information selection for encoding (Sweeney-Reed et al., 2014; Sweeney-Reed et al., 2015), whereas the DMNT coordinate and select information-retrieval strategies (van der Werf et al., 2003; Aggleton et al., 2010; Staudigl et al., 2012). Dual-process models of memory retrieval have been proposed, suggesting that recall and familiarity depend on separate neuroanatomical systems (Eichenbaum et al., 2008). The ANT have also been proposed to support episodic recollection, as part of an extended hippocampal system (Aggleton et al., 2010), while the DMNT, with prominent frontal neocortex projections, engage in a separate recognition network, in which retrieval involves familiarity, without recall of contextual details (Aggleton et al., 2010; Carlesimo et al., 2015).

MODALITIES FOR INVESTIGATION OF THE HUMAN ANTERIOR THALAMUS

The cerebral cortex has been extensively investigated in memory processing. Access to real-time electrophysiological cerebral cortical activity through scalp EEG has enabled substantial exploration of functionally specialized cerebral cortical regions during a range of cognitive paradigms addressing inter-regional interactions underpinning memory processing. In recent years, intracranial recording including electrocorticographic (ECoG) and stereoencephalographic (sEEG) recordings from patients undergoing investigation for treatment of pharmacoresistant epilepsy have provided greater spatial resolution and also access to high frequency band information associated with

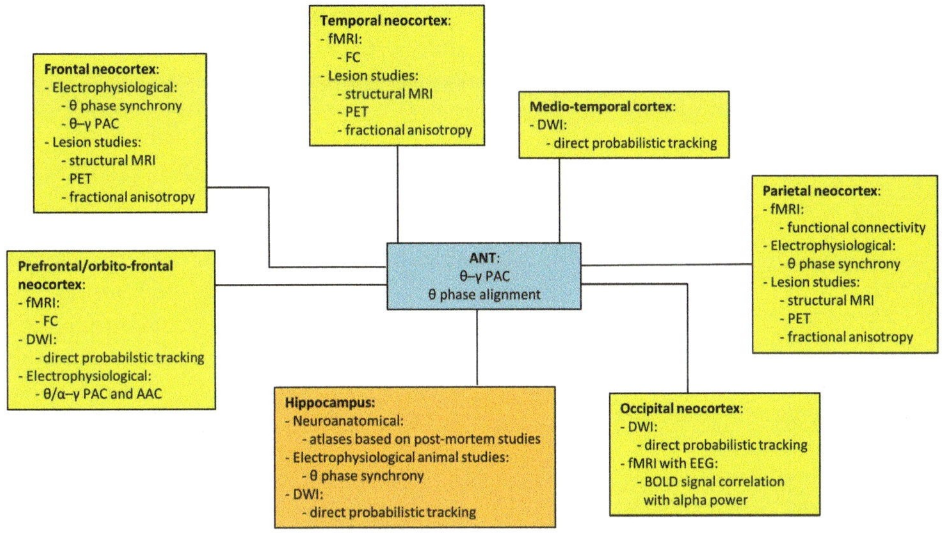

Figure 43.1. Anatomical and functional connectivity (FC) findings between the human anterior nuclei of the thalamus (ANT) and the cerebral cortex and hippocampus (Stenset et al., 2007; Fitzgerald et al., 2013; Pergola et al., 2013; Sweeney-Reed et al., 2014, 2021; Chen et al., 2016; Zotev et al., 2018; Grodd et al., 2020). fMRI = functional magnetic resonance imaging; DWI = diffusion weighted imaging; PET = positron emission tomography; PAC = phase–amplitude coupling; AAC = amplitude–amplitude coupling.

local information processing (Canolty et al., 2006; Canolty and Knight, 2010) (see **Section VII of this volume**). Structural and functional studies also indicate a memory processing role for the anterior thalamus, including post-mortem anatomical studies of patients with anterior thalamic lesions, as well as functional evaluation of memory performance in clinical populations known to have anterior thalamic lesions. Structural neuroimaging confirms lesion location, while functional neuroimaging provides a window on networks that include the ANT. High resolution imaging now allows differentiation between thalamic nuclei engaged in different aspects of memory processing. Finally, the advent of ANT-DBS to treat pharmacoresistant epilepsy has enabled generation of causal evidence for a role for the ANT in memory processing and has led to examination of real-time interactions between ANT and neocortex during memory processing. The findings from these different approaches provide evidence for anatomical connectivity between the human ANT and cerebral cortex, as well as functional connectivity between these brain regions in memory processing (Figure 43.1).

LESIONS STUDIES IN CLINICAL POPULATIONS

The anterior thalamus is affected in certain pathological conditions such as stroke, vascular malformations, and neurodegeneration. Memory deficits associated with ANT pathology confirm a role for the ANT in memory processing and provide causal evidence for that role. Specifically, ANT lesions are associated with anterograde amnesia, an inability to encode episodic and semantic information (Ghika-Schmid and Bogousslavsky, 2000; Harding et al., 2000; van der Werf et al., 2003).

Korsakoff's syndrome results in pathological degeneration in the anterior thalamus and mammillary bodies due to Wernicke's encephalopathy, resulting from thiamine deficiency in alcoholics or other malnourishment conditions. These patients suffer from marked anterograde memory deficits (Kopelman et al., 2009). Interestingly, the memory impairment is specific to declarative memory, leaving implicit memory intact (Harding et al., 2000; Kopelman et al., 2009), fitting with the proposal that the ANT are a part of an extended hippocampal system (Aggleton et al., 2010). Korsakoff patients also show degeneration of the DMNT and evidence of additional prefrontal behavioral deficits in addition to anterograde amnesia. The well-known case of HM, who underwent surgical removal of bilateral parts of the medial temporal lobe, including the hippocampus, also had declarative memory deficits, sparing implicit memory function. Studies of HM led to the conclusion that declarative and implicit memory depend on different brain structures and networks (Spiers et al., 2001). Indeed, these networks involve different thalamic nuclei (Sweeney-Reed et al., 2014; Terzic et al., 2022).

Lacunar infarction can affect specific thalamic nuclei, enabling differentiation of their functional roles through neuropsychological evaluation (van der Werf et al., 2003). Memory deficits are seen in infarction affecting the anterior thalamus (Ghika-Schmid and Bogousslavsky, 2000). They also arise in lacunar stroke patients in whom disconnections in multiple circuits that include the ANT have been identified through tractography (Nishio et al., 2014). Patients with Alzheimer's disease also show ANT pathology, arising at an earlier stage than the well-recognized medial temporal lobe atrophy (Aggleton et al., 2010). Post-mortem findings have revealed deposition of neurofibrillary tangles in the ANT of patients with Alzheimer's disease consistently across cases, particularly affecting the AD (Bernstein et al., 2021). AV volume on structural magnetic resonance imaging (MRI) in Alzheimer's disease was smaller in patients with more marked cognitive, and specifically memory, impairment (Bernstein et al., 2021). ANT lesions in multiple sclerosis patients have also been associated with memory deficits, particularly visuospatial memory, which was more closely linked with ANT than whole thalamus volume (Bergsland et al., 2021) (see also **Section XII of this volume**).

DIFFUSION-WEIGHTED IMAGING

DWI has provided detailed information regarding the anatomical connectivity that enables orchestration of anterior thalamic and cerebral cortical activity during memory processing (Grodd et al., 2020). All four ANT show hippocampal—diencephalic and parahippocampal—retrosplenial networks, engaging in memory and spatial orientation (Child and Benarroch, 2013; Peyrache et al., 2019), but differences between the connectivity of these subnuclei have also been observed. For example, the AM shows thalamocortical connectivity extending further into orbitofrontal, temporal, and occipital cortex than the AD, and despite similarities in AV and AM connectivity, frontal projections from the AM extend more into orbitofrontal cortex (Grodd et al., 2020). The LD, in contrast with the other three subnuclei, projects to the orbitofrontal cortex only on the right side and is the only subnucleus projecting to the left parietal cortex, although connectivity between the LD and the cingulate cortex is similar for the AV (Grodd et al., 2020).

Diffusion tensor imaging (DTI) studies provide a window on potential interactions between remote brain regions by revealing anatomical connectivity, and connectivity alterations have been associated with altered memory function. Increased ANT–neocortical connectivity was seen in patients with hydrocephalus-associated memory deficits and was interpreted as reflecting compensatory processes involving the ANT to ameliorate the memory impairment (Chen et al., 2016). DTI was also employed in a case study of a patient with episodic, semantic, and working memory (WM) deficits (Stenset et al., 2007). Structural MRI documented a left anteromedial thalamus lesion, fluordesoxyglucose positron emission tomography revealed reduced uptake in the left frontal, parietal, and temporal lobes, and fractional anisotropy showed reduced connectivity in the left compared with the right hemisphere. These findings underline the importance of the connectivity between the anterior thalamus and cerebral cortex in memory processing.

FUNCTIONAL IMAGING

Functional imaging indicates a role for the ANT in human memory processing and also for functional connectivity between the ANT and widespread cerebral cortical regions underpinning memory formation. During an associative learning task designed to examine neural correlates of recognition and recall, 3-Tesla (3T) functional magnetic resonance imaging (fMRI) recordings showed specific activation of the ANT during retrieval of items that were recognized and previously encoded associations with those items recalled (Pergola et al., 2013). Greater ANT activation corresponded with higher recall rates. DMNT and associated frontal activations were seen during successful encoding and retrieval of the associations, providing evidence for differential roles for the ANT and DMNT in memory processing.

In another 3T fMRI study, real-time feedback of the blood oxygenation level dependent (BOLD) response in the ANT and DMNT was provided to participants during autobiographical memory retrieval (Zotev et al., 2018). An increased BOLD response was elicited in the ANT and DMNT during episodic memory retrieval when BOLD response neurofeedback was provided, and the BOLD response correlated with the alpha rhythm in EEG recordings over occipital cortex. These findings underline the engagement of these nuclei in memory processing and also provide evidence for thalamocortical networks supporting memory.

Structural 7T MRI has been used to distinguish between thalamic substructures (Calamante et al., 2013), but 7T fMRI has not yet been implemented to differentiate between these nuclei during memory processing. Such an approach would be expected to provide not only a separation between different thalamic nuclei, but also between the subnuclei (e.g., AV/AM/AD/LD) within the ANT.

DEEP BRAIN STIMULATION IN CLINICAL POPULATIONS

The ANT have been selected as the target for high frequency electrical DBS to treat pharmacoresistant epilepsy, as their connectivity with widespread cortical regions, including the temporal lobe, is hypothesized to enable disruption of pharmacoresistant seizure activity with multiple foci not amenable to surgical treatment (Fisher et al., 2010). DBS, which is applied with high success in movement disorders, is thought to induce a "functional lesion" within pathological networks. This potential mechanism of action renders DBS also a technique with which to generate causal evidence regarding the role of the target structure in particular cognitive processes.

Two lines of evidence from ANT-DBS support a role for the ANT in memory function: clinical reports of memory impairment associated with ANT-DBS and formal studies in which patients participated in a memory-based cognitive paradigm with ANT-DBS switched on compared with off. A large study investigating the clinical effects of ANT-DBS in focal epilepsy, the SANTE study (Stimulation of the Anterior Nucleus of the Thalamus for Epilepsy), incorporated subjective evaluation of an impact on memory function at the end of a three-month period, during which patients were blinded as to whether the stimulation was on or off (Fisher et al., 2010). Memory impairment was reported by 13% of patients with stimulation on, compared with only 1.8% by patients not receiving active stimulation. At 5-year follow-up, a quarter of the patients reported memory impairment, although half of these patients had had preexisting memory impairment before DBS commencement (Salanova et al., 2015), and memory (16%) and mood (26%) changes are the most commonly reported reasons for adjusting ANT-DBS parameters (Fasano et al., 2021). However, unilateral, on-demand ANT-DBS resulted in no reports of memory disturbance (Elder et al., 2018), and formal memory assessment in a seven-year follow-up showed no memory deterioration compared with a baseline four weeks prior to electrode implantation (Tröster et al., 2017). Interestingly, improvements in delayed verbal memory performance, without changes in general cognitive abilities or executive function, have been reported (Oh et al., 2012). Moreover, formal evaluation of the effects of ANT-DBS on WM revealed an improvement in performance (Liu et al., 2021). Intracranial electrophysiological recordings from the hippocampus revealed an associated reduction in interictal epileptiform discharges when ANT-DBS was on. Reduction of pathological epileptiform activity that impairs the hippocampal–ANT network could account for the WM improvements. An analogous interpretation was proposed when motor sequence learning was specifically improved on DBS to the ventro-intermediate nucleus of the thalamus in patients treated for essential tremor. The improvement was proposed to result from disruption of pathological processes in the tremor network, which comprises the same brain structures as those engaged in motor sequence learning (Terzic et al., 2022) (see **Section VI of this volume**). Such an effect would depend on whether the stimulated brain structure is engaged in pathological network activity impairing healthy learning or memory function. Comparing ANT-DBS with DMNT-DBS during WM, only DMNT-DBS resulted in WM modulation, with an increase in errors during DMNT-DBS (Peräkylä et al., 2017).

INTRACRANIAL ELECTROPHYSIOLOGICAL RECORDINGS

Brain electrical activity oscillates in multiple frequency ranges (Buzsáki and Draguhn, 2004). Task-relevant large-scale functional networks and neural assemblies form and dissolve, through brief temporal relationships between activity of relevant brain structures, either between the same or different frequency ranges. Long-range phase–phase and phase–amplitude synchrony of neural oscillations are dynamically modulated during different cognitive tasks, enabling long-range communication and providing the timing necessary to facilitate synaptic plasticity and learning (Buzsáki and Draguhn, 2004). Prominent oscillations are seen in the neocortex (Buzsáki and Draguhn, 2004; Canolty et al., 2006) (Figure 43.2A) and hippocampus (Colgin et al., 2009; Fell and Axmacher, 2011) (Figure 43.2B) and have also been observed in the ANT (Fitzgerald et al., 2013; Sweeney-Reed et al., 2014; Sweeney-Reed et al., 2015; Sweeney-Reed et al., 2016; Sweeney-Reed et al., 2017) (Figure 43.2C).

Phase Synchrony

Memory processing is associated with modulation of oscillatory activity in the theta frequency range (Fell and Axmacher, 2011). During memory formation, long-range theta synchrony arises between frontal cortex and hippocampus (Johnson et al., 2018) and between medial temporal lobe (MTL) and parietal cortex (Crespo-Garcia et al., 2010), and during memory retrieval between MTL and frontal and parietal neocortex (Watrous et al., 2013).

Long-range theta (5–6 Hz) phase synchrony has been observed between recordings from the ANT and frontal and parietal neocortex around 1 s post-stimulus during successful memory encoding, suggesting that the ANT are engaged in memory processes, providing real-time evidence for functional connectivity in this part of the proposed circuit underpinning memory processing (Sweeney-Reed et al., 2014) (Figure 43.2C). The timing is in keeping with a range of neural correlates of memory formation previously identified (Osipova et al., 2006; Sweeney-Reed et al., 2014). DMNT signals are often also acquired with ANT recordings, as the electrode probe goes through the ANT targeting the distally located DMNT (Buentjen et al., 2014; Peräkylä et al., 2017). The frontal theta rhythm appears to drive the theta rhythm in the ANT, as revealed by Granger causality during episodic memory formation (Sweeney-Reed et al., 2014), as well as in the MTL during WM (Johnson et al., 2018). The absence of theta synchrony between frontal cortex and the ANT when encoding was not successful, as well as its absence between frontal cortex

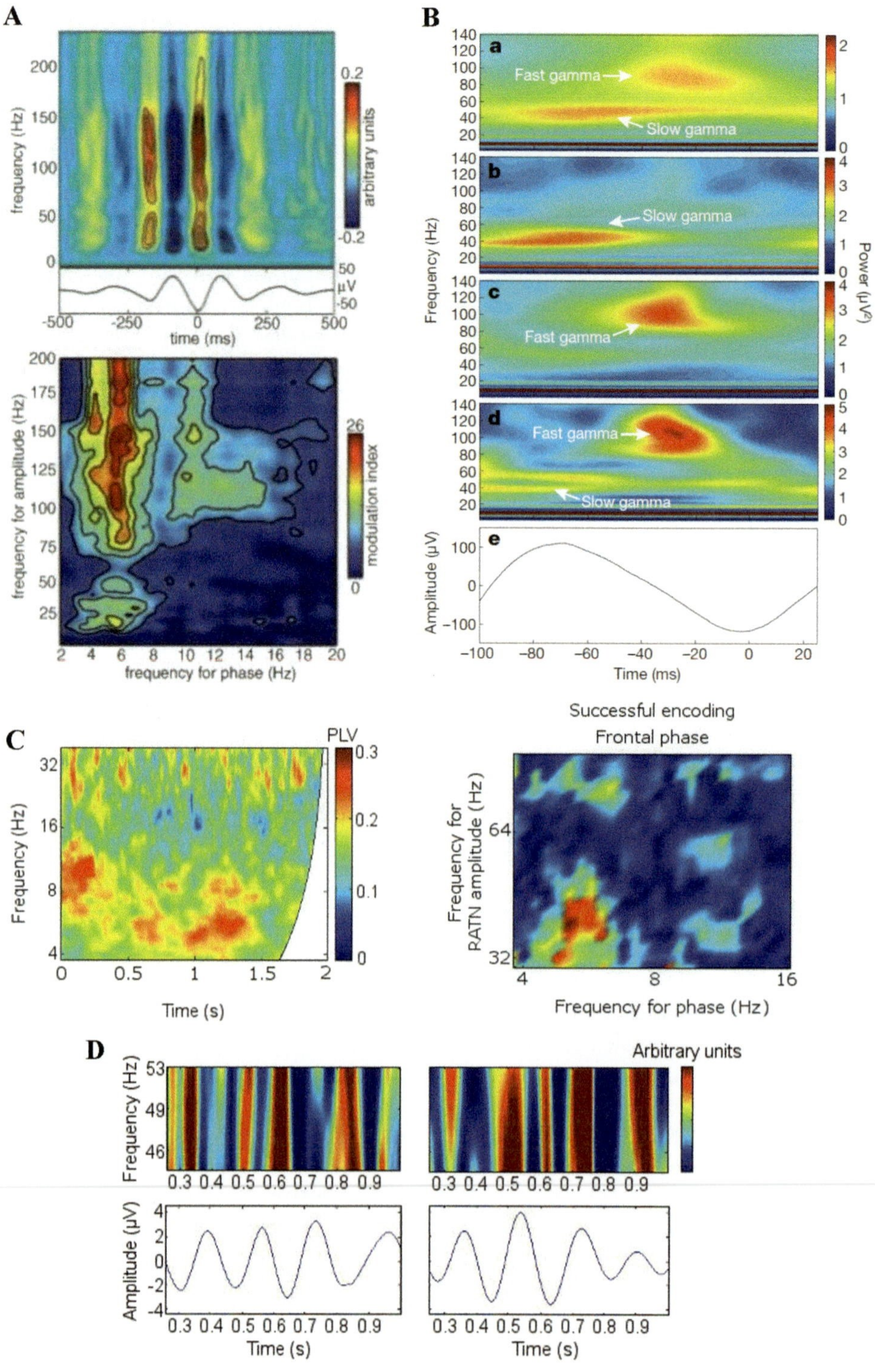

Figure 43.2. Phase–amplitude coupling (PAC) between theta phase and gamma power. (A). Upper panels: neocortical gamma activity (above) coupled with theta phase troughs (below) in electrocorticographic signal from the anterior part of the middle frontal gyrus. Lower panel: modulation index of the theta–gamma PAC above. (Reproduced with permission from Canolty et al., 2006). (B). PAC in the human hippocampus. (a). Mean gamma high frequency activity (HFA) over theta cycles. (b). Theta cycles with slower HFA. (c). Theta cycles with faster HFA. (d). Some theta cycles coupled with slower and faster HFA. €. Mean, unfiltered theta cycles. (Reproduced with permission from Colgin et al., 2009). (C). Theta phase synchrony between frontal neocortex and anterior nuclei of the thalamus (ANT) during successful memory formation (Reproduced from Sweeney-Reed et al., 2014, in accordance with the CC-BY license). (D). Theta–gamma PAC between frontal neocortex and ANT during successful memory formation illustrated for an individual patient. Upper panels: gamma activity amplitude. Lower panels: left: frontal theta phase, right: gamma theta phase. Left: trough-to-peak coupling. Right: peak-to-peak coupling. (Reproduced from Sweeney-Reed et al., 2014, in accordance with the CC-BY license).

and DMNT during successful encoding, suggest a specific network role for the ANT in memory formation.

Phase–Amplitude Coupling

Synchrony also arises between slow (e.g., theta) oscillations and high frequency activity (low gamma: 30–50 Hz; high frequency band activity, HFA: 60–200 Hz) (Canolty et al., 2006; Canolty and Knight, 2010; Sweeney-Reed et al., 2014). HFA indexes local neuronal activity, including single unit and dendritic potentials (Leszczyński et al., 2020). In particular, a relationship between the phase of the slower rhythm and the amplitude of fast activity has been detected in multiple brain structures and linked to function. Such phase–amplitude coupling (PAC) has been reported during memory processing in the neocortex (Canolty et al., 2006) (Figure 43.2A) and hippocampus (Colgin et al., 2009; Axmacher et al., 2010) (Figure 43.2B), as well as in the ANT (Fitzgerald et al., 2013; Sweeney-Reed et al., 2014; Sweeney-Reed et al., 2017) (Figure 43.2C,D). The phase of the slower rhythm is modulated by external environmental and endogenous cognitive processes, reflecting local neuronal excitability and selection of relevant neural assemblies for particular tasks (Canolty et al., 2006; Schroeder and Lakatos, 2009; Axmacher et al., 2010; Canolty and Knight, 2010; Sweeney-Reed et al., 2014), while HFA activity reflects local information processing e.g., object representation (Canolty et al., 2006; Jensen and Colgin, 2007). Timing of the slower rhythm is thought to underpin information integration across multiple spatiotemporal scales, supporting long-range communication and facilitation of neural plasticity (Canolty et al., 2006; Axmacher et al., 2010; Canolty and Knight, 2010).

PAC between the frontal neocortex theta rhythm and ANT gamma amplitude during successful but not unsuccessful memory formation provides further evidence for functional communication between cerebral cortex and ANT during memory encoding (Sweeney-Reed et al., 2014) (Figure 43.2C). Granger causality analysis indicated that the frontal theta rhythm most strongly predicted the ANT theta rhythm approximately one third of a theta cycle prior to the ANT. Trough-to-peak PAC arose between the frontal theta and high frequency ANT activity, while peak-to-peak PAC occurred within the ANT, fitting with this delay (Sweeney-Reed et al., 2014) (Figure 43.2D). The ANT gamma activity coupled with the frontal theta rhythm was narrower in bandwidth during successful than unsuccessful encoding, which was interpreted as reflecting selection, with neuronal populations firing that were relevant to the memory trace being entrained and selected through the frontal theta rhythm (Sweeney-Reed et al., 2014).

Functional neuroimaging studies suggest the thalamus modulates neocortico–neocortical interactions associated with switching between an internally-focused resting state and states involving an external attentional focus (de Luca et al., 2006). ECoG recordings show distinct, task-specific spatial patterns of theta–HFB PAC (Canolty et al., 2006). Widespread neocortex–ANT functional and anatomical connectivity (Aggleton et al., 2010; Fitzgerald et al., 2013; Sweeney-Reed et al., 2014) places the ANT in a position to flexibly establish functional connectivity with task-relevant regions of neocortex engaged in particular cognitive states and tasks (Sweeney-Reed et al., 2021). The "synchronized gating hypothesis" proposes that PAC enables flexible coordination of cortical and subcortical hubs in dynamic networks (Florin and Baillet, 2015). ANT-PAC may enable the ANT to act as a hub, coordinating and switching between task-related networks (Sweeney-Reed et al., 2017; Sweeney-Reed et al., 2021). The differential coupling of HFA with the theta rhythm during continuously recorded ANT activity between rest and memory and novelty oddball tasks is consistent with such a role (Sweeney-Reed et al., 2017).

Timing differences between event-related potentials (ERPs) recorded from the ANT and the hippocampus led to the proposal of a gating role for the ANT (Štillová et al., 2015). Given neocortex–ANT PAC during memory formation (Sweeney-Reed et al., 2014), inter-task differences in PAC in the ANT (Sweeney-Reed et al., 2017), the well-established prefrontal cortex (PFC) role in memory retrieval (Szczepanski and Knight, 2014; Helfrich and Knight, 2016), and known ANT–PFC anatomical connectivity (Aggleton et al., 2010), as well as the proposal that the thalamus facilitates subcortical integration (Wolff and Vann, 2019), we postulated that the ANT could provide a subcortical–PFC gate (Sweeney-Reed et al., 2021), analogously to the proposal that the DMNT regulate prefrontal networks (Pergola et al., 2018).

Phase Alignment

Precise alignment of oscillatory phase to stimulus onset is a prerequisite for synaptic plasticity, as synaptic modification requires pre- and postsynaptic potentials to occur within a narrow time window (Paulsen and Sejnowski, 2000; Buzsáki and Draguhn, 2004; Düzel et al., 2005; Fell and Axmacher, 2011). Theta oscillations are an ongoing phenomenon, but their precise timing, both in frontal neocortex and ANT, relative to stimuli onset, determines whether stimuli are encoded (Sweeney-Reed et al., 2015). Successful memory formation depended on enhanced late post-stimulus theta alignment in the ANT and also on a correlation between theta phase alignment, early post-stimulus in frontal neocortex, later post-stimulus in the ANT, and ANT theta–gamma PAC, suggesting that ANT-specific synaptic modification depends on timing of the frontal theta rhythm input (Sweeney-Reed et al., 2015) (Figure 43.3). Early frontal neocortical phase

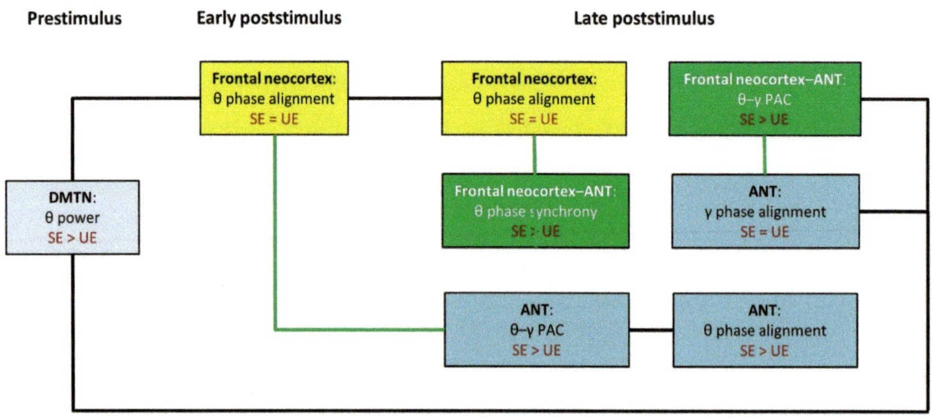

Figure 43.3. Correlations between electrophysiological features in the frontal neocortex and the anterior nuclei of the thalamus (ANT) associated with human memory formation (Sweeney-Reed et al., 2014, 2015, 2016, 2021). Correlations are indicated by black lines. ANT–cortical correlations are shown in green. DMTN = dorsomedial thalamic nuclei; SE = successful encoding; UE = unsuccessful encoding.

alignment is consistent with a narrow time window in which a rapid, early perceptual response occurs, which is a requirement, along with the correlation with later ANT theta–gamma PAC, for subsequent successful stimulus encoding. The correlation with ANT PAC, which was only observed during successful memory formation, suggests a frontal–ANT network supporting memory encoding. These findings underpin the proposal that cerebral cortex and the anterior thalamus are both key brain structures engaged in the neural networks underpinning memory encoding. DMNT phase alignment was not associated with encoding success (Sweeney-Reed et al., 2015). Pre-stimulus DMNT theta power predicted subsequent memory formation as well as several ANT post-stimulus correlates of successful encoding, suggesting that these nuclei have separate, complementary roles in memory encoding (Sweeney-Reed et al., 2016).

CONCLUSIONS

Human lesion, neuroimaging, DBS, and electrophysiological studies provide support for a critical role for the ANT in memory processing, integrating information from widespread neocortical sources (Sweeney-Reed et al., 2021). The ANT is neuroanatomically well-placed to switch between cognitive states and to select local neural assemblies relevant to particular memory traces, and memory formation is associated with real-time modulation of functional connectivity between the cerebral cortex and the ANT.

OUTLOOK

Further investigations for enhancing our understanding of how the ANT and cerebral cortex orchestrate human memory processing will include new approaches to electrophysiological analyses, as well as combining advances in high resolution imaging. Dominant brain oscillatory frequencies vary with brain structure, cognitive task, and between individuals (Klimesch, 1999; Sweeney-Reed and Nasuto, 2009). For example, theta frequencies reported in hippocampal and cortical recordings range between 2 and 14 Hz and serve specific functions, with higher frequencies involved in spatial cognition and lower frequencies in memory encoding (Lega et al., 2012; Goyal et al., 2020). Approaches based on adaptive, data-driven time–frequency decomposition, such as empirical mode decomposition, have the potential to provide new insights into frequency-modulated oscillatory processes, with inter-individual and temporal variation of the oscillatory frequencies mediating functional connectivity in cortical–ANT–hippocampal networks (Sweeney-Reed and Nasuto, 2007; Sweeney-Reed et al., 2018). Combination of DWI with high-resolution functional imaging differentiating between thalamic nuclei and subnuclei (Grodd et al., 2020) will expand our understanding of the structural and functional networks underpinning human memory processing. These advances will also provide greater precision in postoperative determination of intrathalamic electrode location, extending the interpretations possible from intrathalamic recordings from clinical populations.

REFERENCES

Aggleton JP, Brown MW (1999). Episodic memory, amnesia, and the hippocampal-anterior thalamic axis. *Behav Brain Sci.* 22:425–444.

Aggleton JP, O'Mara SM, Vann SD, Wright NF, Tsanov M, Erichsen JT (2010) Hippocampal-anterior thalamic pathways for memory: Uncovering a network of direct and indirect actions. *Eur J Neurosci.* 31:2292–2307.

Albo Z, di Prisco GV, Vertes RP (2003). Anterior thalamic unit discharge profiles and coherence with hippocampal theta rhythm. *Thalamus Relat Syst.* 2:133–144.

Axmacher N, Henseler MM, Jensen O, Weinreich I, Elger CE, Fell J (2010). Cross-frequency coupling supports multi-item working memory in the human hippocampus. *PNAS*. 107:3228–3233.

Bergsland N, Benedict RHB, Dwyer MG, Fuchs TA, Jakimovski D, Schweser F, Tavazzi E, Weinstock-Guttman B, Zivadinov R (2021). Thalamic nuclei volumes and their relationships to neuroperformance in multiple sclerosis: A cross-sectional structural MRI study. *J Magn Res Imaging*. 53:731–739.

Bernstein AS, Rapcsak SZ, Hornberger M, Saranathan M (2021). Structural changes in thalamic nuclei across prodromal and clinical Alzheimer's disease. *J Alzheimer's Dis*. 82:361–371.

Bubb EJ, Kinnavane L, Aggleton JP (2017). Hippocampal–diencephalic–cingulate networks for memory and emotion: An anatomical guide. *Brain Neurosci Adv*. 1:1–20.

Buentjen L, Kopitzki K, Schmitt FC, Voges J, Tempelmann C, Kaufmann J, Kanowski M (2014). Direct targeting of the thalamic anteroventral nucleus for deep brain stimulation by T1-weighted magnetic resonance imaging at 3 T. *Stereotact Funct Neurosurg*. 92:25–30.

Buzsáki G, Draguhn A (2004) Neuronal oscillations in cortical networks. *Science* 304:1926–1929.

Calamante F, Oh SH, Tournier JD, Park SY, Son YD, Chung JY, Chi JG, Jackson GD, Park CW, Kim YB, Connelly A, Cho ZH (2013). Super-resolution track-density imaging of thalamic substructures: Comparison with high-resolution anatomical magnetic resonance imaging at 7.0T. *Hum Brain Mapp*. 34:2538–2548.

Canolty RT, Edwards E, Dalal SS, Soltani M, Nagarajan SS, Kirsch HE, Berger MS, Barbaro NM, Knight RT (2006). High gamma power is phase-locked to theta oscillations in human neocortex. *Science*. 313:1626–1628.

Canolty RT, Knight RT (2010). The functional role of cross-frequency coupling. *Trends Cogn Sci*. 14:506–515.

Carlesimo GA, Lombardi MG, Caltagirone C, Barban F (2015). Recollection and familiarity in the human thalamus. *Neurosci Biobehav R*. 54:18–28.

Chen YC, Chiang SW, Chi CH, Liou M, Kuo DP, Kao HW, Chung HW, Ma HI, Peng GS, Wu YT, Chen CY (2016). Early idiopathic normal pressure hydrocephalus patients with neuropsychological impairment are associated with increased fractional anisotropy in the anterior thalamic nucleus. *Medicine (US)*. 95:1–9.

Child ND, Benarroch EE (2013). Anterior nucleus of the thalamus: Functional organization and clinical implications. *Neurology*. 81:1869–1876.

Colgin LL, Denninger T, Fyhn M, Hafting T, Bonnevie T, Jensen O, Moser M-B, Moser EI (2009). Frequency of gamma oscillations routes flow of information in the hippocampus. *Nature*. 462:353–357.

Crespo-Garcia M, Cantero JL, Pomyalov A, Boccaletti S, Atienza M (2010). Functional neural networks underlying semantic encoding of associative memories. *NeuroImage*. 50:1258–1270.

de Luca M, Beckmann CF, de Stefano N, Matthews PM, Smith SM (2006). fMRI resting state networks define distinct modes of long-distance interactions in the human brain. *NeuroImage*. 29:1359–1367.

Diekelmann S, Born J (2010). The memory function of sleep. *Nat Rev Neurosci*. 11:114–126.

Düzel E, Richardson-Klavehn A, Neufang M, Schott BH, Scholz M, Heinze H-J (2005). Early, partly anticipatory, neural oscillations during identification set the stage for priming. *NeuroImage*. 25:690–700.

Eichenbaum H, Sauvage MM, Fortin NJ, Yonelinas AP (2008). ROCs in rats? Response to Wixted and Squire. *Learn Mem*. 15:691–693.

Elder C, Friedman D, Devinsky O, Doyle W, Dugan P (2018). Responsive neurostimulation targeting the anterior nucleus of the thalamus in 3 patients with treatment-resistant multifocal epilepsy. *Epilepsia Open*. 4:187–192.

Fasano A, Eliashiv D, Herman ST, Lundstrom BN, Polnerow D, Henderson JM, Fisher RS (2021). Experience and consensus on stimulation of the anterior nucleus of thalamus for epilepsy. *Epilepsia*. 62:2883–2898.

Fell J, Axmacher N (2011). The role of phase synchronization in memory processes. *Nat Rev Neurosci*. 12:105–118.

Fisher R, Salanova V, Witt T, Worth R, Henry T, Gross R, Oommen K, Osorio I, Nazzaro J, Labar D, Kaplitt M, Sperling M, Sandok E, Neal J, Handforth A, Stern J, DeSalles A, Chung S, Shetter A, Bergen D, Bakay R, Henderson J, French J, Baltuch G, Rosenfeld W, Youkilis A, Marks W, Garcia P, Barbaro N, Fountain N, Bazil C, Goodman R, McKhann G, Krishnamurthy KB, Papavassiliou S, Epstein C, Pollard J, Tonder L, Grebin J, Coffey R, Graves N, and the SANTE Study Group (2010). Electrical stimulation of the anterior nucleus of thalamus for treatment of refractory epilepsy. Epilepsia 51:899–908.

Fitzgerald THB, Valentin A, Selway R, Richardson MP (2013). Cross-frequency coupling within and between the human thalamus and neocortex. *Front Hum Neurosci*. 7:84:1–13.

Florin E, Baillet S (2015). The brain's resting-state activity is shaped by synchronized cross-frequency coupling of neural oscillations. *NeuroImage*. 111:26–35.

Ghika-Schmid F, Bogousslavsky J (2000). The acute behavioral syndrome of anterior thalamic infarction: A prospective study of 12 cases. *Ann Neurol*. 48:220–227.

Goyal A, Miller J, Qasim SE, Watrous AJ, Zhang H, Stein JM, Inman CS, Gross RE, Willie JT, Lega B, Lin JJ, Sharan A, Wu C, Sperling MR, Sheth SA, McKhann GM, Smith EH, Schevon C, Jacobs J (2020). Functionally distinct high and low theta oscillations in the human hippocampus. *Nat Commun*. 11:1–10.

Grodd W, Kumar VJ, Schüz A, Lindig T, Scheffler K (2020). The anterior and medial thalamic nuclei and the human limbic system: Tracing the structural connectivity using diffusion-weighted imaging. *Sci Rep*. 10:1–25.

Harding A, Halliday G, Caine D, Kril J (2000). Degeneration of anterior thalamic nuclei differentiates alcoholics with amnesia. *Brain*. 123:141–154.

Helfrich RF, Knight RT (2016). Oscillatory dynamics of prefrontal cognitive control. *Trends Cogn Sci*. 20:916–930.

Jacobs J (2014). Hippocampal theta oscillations are slower in humans than in rodents: Implications for models of spatial navigation and memory. *Phil Trans Roy Soc B Biol Sci*. 369:1–9.

Jensen O, Colgin LL (2007). Cross-frequency coupling between neuronal oscillations. *Trends Cogn Sci*. 11:267–269.

Johnson EL, Adams JN, Solbakk AK, Endestad T, Larsson PG, Ivanovic J, Meling TR, Lin JJ, Knight RT (2018). Dynamic frontotemporal systems process space and time in working memory. *PLoS Biol*. 16:1–22.

Ketz NA, Jensen O, O'Reilly RC (2014). Thalamic pathways underlying prefrontal cortex-medial temporal lobe oscillatory interactions. *Trends Neurosci*. 38:3–12.

Klimesch W (1999). EEG alpha and theta oscillations reflect cognitive and memory performance: a review and analysis. *Brain Res Rev*. 29:169–195.

Kopelman MD, Thomson AD, Guerrini I, Marshall EJ (2009). The Korsakoff syndrome: Clinical aspects, psychology and treatment. *Alcohol Alcoholism*. 44:148–154.

Lambert C, Simon H, Colman J, Barrick TR (2017). Defining thalamic nuclei and topographic connectivity gradients in vivo. *NeuroImage*. 158:466–479.

Lega BC, Jacobs J, Kahana M (2012). Human hippocampal theta oscillations and the formation of episodic memories. *Hippocampus*. 22:748–761.

Leszczyński M, Barczak A, Kajikawa Y, Ulbert I, Falchier AY, Tal I, Haegens S, Melloni L, Knight RT, Schroeder CE (2020). Dissociation of broadband high-frequency activity and neuronal firing in the neocortex. *Sci Adv*. 6:1–12.

Liu J, Yu T, Wu J, Pan Y, Tan Z, Liu R, Wang X, Ren L, Wang L (2021). Anterior thalamic stimulation improves working memory precision judgments. *Brain Stimul.* 14:1073–1080.

Nishio Y, Hashimoto M, Ishii K, Ito D, Mugikura S, Takahashi S, Mori E (2014). Multiple thalamo-cortical disconnections in anterior thalamic infarction: Implications for thalamic mechanisms of memory and language. *Neuropsychologia.* 53:264–273.

Oh YS, Kim HJ, Lee KJ, Kim YI, Lim SC, Shon YM (2012). Cognitive improvement after long-term electrical stimulation of bilateral anterior thalamic nucleus in refractory epilepsy patients. *Seizure.* 21:183–187.

Osipova D, Takashima A, Oostenveld R, Fernández G, Maris E, Jensen O (2006). Theta and gamma oscillations predict encoding and retrieval of declarative memory. *J Neurosci.* 26:7523–7531.

Papez JW (1937). A proposed mechanism of emotions. *Arch Neurol Psychiatry.* 38:725–743.

Paulsen O, Sejnowski TJ (2000). Natural patterns of activity and long-term synaptic plasticity. *Curr Opin Neurobiol.* 10:172–179.

Peräkylä J, Sun L, Lehtimäki K, Peltola J, Öhman J, Möttönen T, Ogawa KH, Hartikainen KM (2017). Causal evidence from humans for the role of mediodorsal nucleus of the thalamus in working memory. *J Cogn Neurosci.* 29:2090–2102.

Pergola G, Danet L, Pitel AL, Carlesimo GA, Segobin S, Pariente J, Suchan B, Mitchell AS, Barbeau EJ (2018). The regulatory role of the human mediodorsal thalamus. *Trends Cogn Sci.* 22:1011–1025.

Pergola G, Ranft A, Mathias K, Suchan B (2013). The role of the thalamic nuclei in recognition memory accompanied by recall during encoding and retrieval: An fMRI study. *NeuroImage.* 74:195–208.

Peyrache A, Duszkiewicz AJ, Viejo G, Angeles-Duran S (2019). Thalamocortical processing of the head-direction sense. *Prog Neurobiol.* 183:1–28.

Rugg M, Johnson J, Park H, Uncapher M (2008). Encoding-retrieval overlap in human episodic memory: a functional neuroimaging perspective. *Prog Brain Res.* 169:339–352.

Salanova V, Witt T, Worth R, Henry TR, Gross RE, Nazzaro JM, Labar D, Sperling MR, Sharan A, Sandok E, Handforth A, Stern JM, Chung S, Henderson JM, French J, Baltuch G, Rosenfeld WE; Garcia P, Barbaro NM, Fountain NB, Elias WJ, Goodman RR, Pollard JR, Tröster AI, Irwin CP, Lambrecht K, Graves N, Fisher R, For the SANTE Study (2015). Long-term efficacy and safety of thalamic stimulation for drug-resistant partial epilepsy. *Neurology.* 84:1017–1025.

Schroeder CE, Lakatos P (2009). Low-frequency neuronal oscillations as instruments of sensory selection. *Trends Neurosci.* 32:9–18.

Spiers HJ, Maguire EA, Burgess N (2001). Hippocampal amnesia. *Neurocase.* 7:357–382.

Staudigl T, Zaehle T, Hanslmayr S, Esslinger C, Hinrichs H, Schmitt FC, Heinze H, Richardson-Klavehn A (2012). Memory signals from the thalamus: Early thalamocortical phase synchronization entrains gamma oscillations during long-term memory retrieval. *Neuropsychologia.* 50:3519–3527.

Stenset V, Grambaite R, Reinvang I, Hessen E, Cappelen T, Bjørnerud A, Gjerstad L, Fladby T (2007). Diaschisis after thalamic stroke: A comparison of metabolic and structural changes in a patient with amnesic syndrome. *Acta Neurol Scand.* 115:68–71.

Štillová K, Jurák P, Chládek J, Chrastina J, Halámek J, Bočková M, Goldemundová S, Říha I, Rektor I (2015). The role of anterior nuclei of the thalamus: A subcortical gate in memory processing: an intracerebral recording study. *PLoS ONE.* 10:e0140778, 1–13.

Sweeney-Reed CM, Zaehle T, Voges J, Schmitt FC, Buentjen L, Borchardt V, Walter M, Hinrichs H, Heinze H-J, Rugg MD, Knight RT (2017). Anterior thalamic high frequency band activity is coupled with theta oscillations at rest. *Front Hum Neurosci.* 11:1–13.

Sweeney-Reed CM, Zaehle T, Voges J, Schmitt FC, Buentjen L, Kopitzki K, Hinrichs H, Heinze H-J, Rugg MD, Knight RT, Richardson-Klavehn A (2015). Thalamic theta phase alignment predicts human memory formation and anterior thalamic cross-frequency coupling. *eLife.* e07578:1–9.

Sweeney-Reed CM, Zaehle T, Voges J, Schmitt FC, Buentjen L, Kopitzki K, Richardson-Klavehn A, Hinrichs H, Heinze H-J, Knight RT, Rugg MD (2016). Pre-stimulus thalamic theta power predicts human memory formation. *NeuroImage.* 138:100–108.

Sweeney-Reed CM, Buentjen L, Voges J, Schmitt FC, Zaehle T, Kam JWY, Kaufmann J, Heinze HJ, Hinrichs H, Knight RT, Rugg MD (2021). The role of the anterior nuclei of the thalamus in human memory processing. *Neurosci Biobehav Rev.* 126:146–158.

Sweeney-Reed CM, Nasuto SJ (2007). A novel approach to the detection of synchronisation in EEG based on empirical mode decomposition. *J Comput Neurosci.* 23:79–111.

Sweeney-Reed CM, Nasuto SJ (2009). Detection of neural correlates of self-paced motor activity using empirical mode decomposition phase locking analysis. *J Neurosci Methods.* 184:54–70.

Sweeney-Reed CM, Nasuto SJ, Vieira MF, Andrade AO (2018). Empirical mode decomposition and its extensions applied to EEG analysis: a review. *Adv Data Sci Adapt Anal.* 10:1840001, 1–35.

Sweeney-Reed CM, Zaehle T, Voges J, Schmitt FC, Buentjen L, Kopitzki K, Esslinger C, Hinrichs H, Heinze HJ, Knight RT, Richardson-Klavehn A (2014). Corticothalamic phase synchrony and cross-frequency coupling predict human memory formation. *eLife.* 3:e05352.

Szczepanski SM, Knight RT (2014). Insights into human behavior from lesions to the prefrontal cortex. *Neuron.* 83:1002–1018.

Terzic L, Voegtle A, Farahat A, Hartong N, Galazky I, Nasuto SJ, Andrade A de O, Knight RT, Ivry RB, Voges J, Buentjen L, Sweeney-Reed CM (2022). Deep brain stimulation of the ventrointermediate nucleus of the thalamus to treat essential tremor improves motor sequence learning. *Hum Brain Mapp.* 43:4791–4799.

Tröster AI, Meador KJ, Irwin CP, Fisher RS (2017). Memory and mood outcomes after anterior thalamic stimulation for refractory partial epilepsy. *Seizure Eur J Epil.* 45:133–141.

Tulving E (2002). Episodic memory: from mind to brain. *Annu Rev Psychol.* 53:1–25.

van der Werf YD, Jolles J, Witter MP, Uylings HBM (2003). Contributions of thalamic nuclei to declarative memory functioning. *Cortex.* 39:1047–1062.

Vertes RP, Albo Z, Viana di Prisco G (2001). Theta-rhythmically firing neurons in the anterior thalamus: implications for mnemonic functions of Papez's circuit. *Neuroscience.* 104:619–625.

Watrous AJ, Tandon N, Connor C, Pieters T, Ekstrom AD (2013). Frequency–specific network connectivity increases underlie accurate spatiotemporal memory retrieval. *Nat Neurosci.* 16:349–356.

Wolff M, Vann SD (2019). The cognitive thalamus as a gateway to mental representations. *J Neurosci.* 39:3–14.

Zotev V, Misaki M, Phillips R, Ki C, Bodurka J (2018). Real-time fMRI neurofeedback of the mediodorsal and anterior thalamus enhances correlation between thalamic BOLD activity and alpha EEG rhythm. *Hum Brain Mapp.* 39:1024–1042.

44.

MEMORY AND THE RODENT ANTERIOR THALAMUS

Adrien Peyrache and Adrian J. Duszkiewicz

ANATOMY OF A MEMORY: THE PAPEZ NETWORK AND BEYOND

In 1786, Felix Vicq d'Azyr, using novel brain dissection techniques, described the mammillothalamic tract, a bundle of fibers that originate from the mammillary bodies (MBs) and act as the primary subcortical input to the anterior thalamus (AT). This pathway forms the basis of a larger brain network, the role of which in memory formation would later be gradually uncovered. One hundred fifty years later, James Papez, reviewing the current understanding of anatomy, proposed the existence of a network organized in loops. Specifically, these circuits successively connect the mammillary bodies, anterior thalamus, medial cortex, and subicular complex (Papez, 1937). In turn, the subicular complex projects back to the MBs, closing the loop of the so-called Papez circuit. This network is highly interconnected with the hippocampus and the entorhinal cortex, two structures that play a critical role in memory formation (Scoville and Milner, 1957). This circuit is now understood to have a more complex connectivity than simple, parallel loops (Aggleton and O'Mara, 2022). Therefore, it will be referred to as the *Papez network* in this chapter.

The AT can be divided into three cytoarchitectonically defined sub-regions: the anterodorsal (AD), anteromedial (AM), and anteroventral (AV) nuclei. While traditionally all of these nuclei are classified as part of AT, a recent comprehensive RNAseq assay established that each of them shows a different profile of gene expression, in the case of AM and AV more similar to other neighboring, non-AT nuclei than to each other, and in case of AD unique across all of the thalamus (Phillips et al., 2019) (Figure 44.1A). Nevertheless, despite different genetic identities, what unifies these nuclei are reciprocal connections with the medial cortex and the parahippocampal structures (Figure 44.1B,C).

Interestingly, none of the AT nuclei project directly to the hippocampus proper—the only thalamic nucleus that is known to send axons to the hippocampus is the nucleus reuniens (RE) (Herkenham, 1978), another thalamic nucleus with a unique gene-expression pattern (Phillips et al., 2019). A growing body of evidence suggests that the RE plays a crucial role in modulating hippocampal activity (Ito et al., 2015; Mei et al., 2018). Furthermore, the laterodorsal (LD) nucleus, anatomically positioned dorsal and posterior to AV/AM and classified as part of the lateral thalamus, also forms reciprocal connections with the medial cortex and the parahippocampal areas (Mathiasen et al., 2017). What is more, LD nucleus expresses similar genes to the neighboring AV (Phillips et al., 2019). For these reasons, both RE and LD will also be discussed in this chapter as part of the Papez network.

THE ANTERIOR THALAMUS IS HEAVILY INTERCONNECTED WITH MNEMONIC NETWORKS

The Papez network is anatomically organized into partially segregated pathways. The MBs are divided into two cytoarchitecturally distinct areas which form separated circuits (Figure 44.1C): the lateral mammillary nucleus (LMN) projects to AD, while the medial mammillary nucleus (MMN) targets AM and AV (Shibata, 1992). The MMN can itself be subdivided in different areas projecting to different sub-regions of the AT, in particular the *pars lateralis* (lMMN) and *pars medialis* (mMMN) that connect to the AV and AM, respectively. Each nucleus of the MBs receives inputs from different brainstem nuclei. Thus, the mammillothalamic tract can be considered a collection of independent, non-divergent fibers, connecting distinct

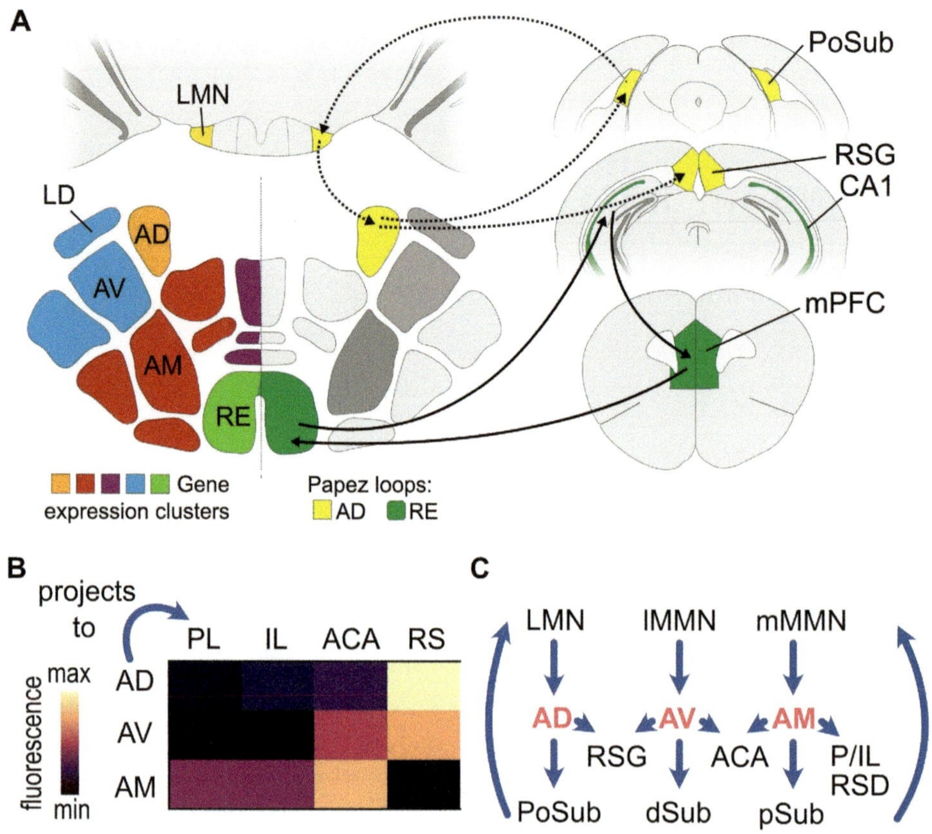

Figure 44.1. Anatomy of the Papez network. (A). Anatomy of the AT, color-coded by different gene expression profiles (Phillips et al., 2019). Two distinct loops are illustrated: the LMN-AD-PoSub and the Re-CA1-PFC circuits. (B). Projection density from AT nuclei to the medial cortex. Adapted from Peyrache (2022b). (C). Segregation and overlap in the anatomical organization of the Papez network. ACA = cingulate cortex; I/PL = infra/prelimbic cortex; d/pSub = distal/proximal subiculum.

sub-regions of MBs and the AT. This anatomical arrangement also reflects the differences in MB-projecting zones in the brainstem.

All AT nuclei project to the parahippocampal areas, in particular the subicular complex and the entorhinal cortex (Van Groen and Wyss, 1995). AT projections to the subicular complex remain segregated, with AD, AV, and AM projecting mainly to the postsubiculum (PoSub), the proximal subiculum (near CA1), and the distal subiculum, respectively. The subiculum-to-MB feedback projections are also anatomically organized, respecting the segregation of AT inputs to the subicular complex (Wright et al., 2010). The individual loops within the Papez network are thus believed to be segregated and to serve distinct functions. An example of this is the LMN-AD-PoSub-LMN loop, which is well established anatomically and is central in conveying and processing the head-direction (HD) signal to the cortical navigation system (Peyrache et al., 2019; Taube, 2007) (Figure 44.2). The projection from AD to PoSub is unique among the AT nuclei in that, in contrast to other thalamocortical projections originating in the AT, it seems to be the main input driving the patterned activity within its cortical target (Goodridge and Taube, 1997; Peyrache et al., 2015). Thus, the AD-PoSub projection in some ways resembles a first-order thalamocortical pathway, with LD possibly playing the role of the second-order modulatory thalamic input to PoSub (Balsamo et al., 2022). Still, LD and AD axon terminals in PoSub layer 1 seem to be anatomically segregated, which suggests that there may be two distinct thalamocortical sub-networks embedded within that cortical area (Balsamo et al., 2022).

In addition to directly contacting the subicular complex, each AT nucleus projects to a different subset of medial cortical targets, including pre- and infralimbic cortex, the cingulate cortex, and the retrosplenial cortex. However, unlike the segregated organization of subcortical limbic pathways, AT projections to the medial cortex overlap (Wright et al., 2013) (Figure 44.1B). AM is the nucleus with the highest rostro-caudal projection span across the medial cortex, with axons reaching rostrally toward anterior cingulate (ACC) and medial prefrontal (mPFC) cortices, and caudally to

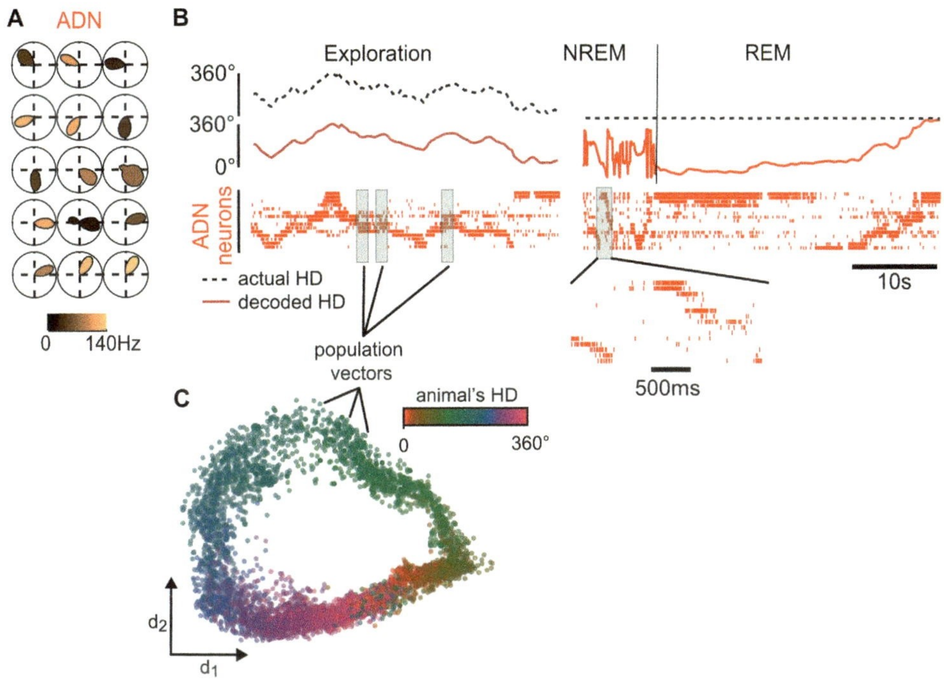

Figure 44.2. Organized activity in a population of head-direction cells in the AD across brain states. (A). Example tuning curves of simultaneously recorded HD neurons in AD. Color coding indicate peak firing rates. (B). Raster plot of neuronal activity (bottom), animal's actual head-direction (top, dotted line) and head-direction decoded from population activity (top, plain red line), during exploration (left) and sleep (right). Activity is still organized during sleep, but while it is virtually indistinguishable from wake during REM, the internal HD signal drifts at a much faster speed during NREM. (C). Low-dimensional projection of neuronal population vectors onto 2 dimensions (d1 and d2) reveals the ring topology of the ensemble activity, compatible with the hypothesis of an attractor network. Adapted from Peyrache et al. (2015); Viejo and Peyrache (2020).

the retrosplenial cortex (RS) as well as the primary and secondary visual cortices (van Groen et al., 1999). LD has a more limited medial reach, with projections terminating in ACC and RS. Finally, AV and AD axons terminate exclusively in RS. The vast majority of the medial cortical projections are reciprocal.

The retrosplenial cortex (RS) is of particular interest, as recent evidence points to its central role in spatial memory (for review, see Alexander et al., 2023). It can be divided into two anatomically distinct but nonetheless heavily interconnected sub-regions, the granular (RSG) and dysgranular (RSD) subregion, that differ in their input/output profiles. Placed medially within the cerebral fissure, RSG receives inputs from the extended hippocampal system, mainly subiculum and PoSub (Wyss and Groen, 1992). RSD, on the other hand, receives strong inputs from the visual cortices in addition to PoSub. These differential connectivity patterns of RSG and RSD are also evident in their connections with the AT. The AM nucleus projects mainly to RSD, while AD projects to RSG (Shibata, 1993; Van Groen and Wyss, 1995).

Finally, RE receives strong inputs from all subdivisions of the mPFC (Herkenham, 1978) and is thought to be the main connection hub between mPFC and the hippocampus (Vertes et al., 2007). The ventral part of CA1 in turn projects back to mPFC, thus completing the loop (Jay et al., 1989).

In summary, the AT, RE, and LD occupy a central position in the organization of a wide network responsible for memory processing (Figure 44.1). Overall, the picture emerging from this complex connectivity is that of a gradient in medial cortical connectivity across the AT, with RE and AM at one end, associated mainly with the frontal areas, and AD/AV at the other, connected with the limbic cortex.

THE ANTERIOR THALAMUS IS NECESSARY FOR MEMORY FORMATION

The Papez network has been demonstrated to play a crucial role in navigation and memory. The bulk of the evidence gathered so far comes from the research investigating the cognitive impairments resulting from lesions at various stages of this pathway. Specifically, both declarative (Squire, 2004) and working memory (Baddeley, 2010) are affected by lesions or manipulations of this network (Aggleton and O'Mara, 2022).

In humans, clinical cases of strokes and alcohol abuse point to a role of the AT and RE in explicit memory (Aggleton and O'Mara, 2022), suggesting that they are necessary for normal functioning of the hippocampal-dependent memory processes. In rodents, hippocampal lesions are best characterized by their effect on spatial navigation deficits. As expected, lesions of the AT lead to severe deficits in a wide range of navigation tasks, similar to the effect of hippocampal lesions (Aggleton and O'Mara, 2022). However, in most cases, the deficits could not be attributed to one specific AT nucleus, but were instead most severe when the entire AT was lesioned (Aggleton and O'Mara, 2022). This is the case, for example, in the Morris Water Maze in which animals learn to find a hidden platform below the water surface, and in which only extended lesions of the AT produce severe learning deficits, in contrast to AD or AV lesions alone (van Groen et al., 2002). The ability to remember previously visited locations defined in an allocentric (map-based) reference frame, but not egocentric one (e.g., "this place on my left"), is impaired in animals with extensive AT lesions, particularly after a delay (Aggleton et al., 1996). Interestingly, lesions of the AT usually lead to more severe cognitive deficits than lesions of the MBs (Aggleton and O'Mara, 2022). Transient inactivation of the RE also disrupts spatial working memory (Maisson et al., 2018).

Thalamic lesions or transient inactivation certainly deprive cortical targets of a major excitatory input, leading to potential confounds in the interpretation of such manipulations. The specific role of AT in spatial cognition is confirmed by the demonstration that immediate early genes (e.g. cFos, a proxy indicator for increased neuronal activity) are expressed at high levels in the AT after animals explore a radial arm maze (Vann et al., 2000). Furthermore, the integrity of the reciprocal interaction between the AT and the subicular complex is required for allocentric orientation, as the selective disruption of these thalamocortical or corticothalamic pathways impairs behavioral performance in a forced alternation task in which the animal has to rely on distal visual cues (Nelson et al., 2020).

COGNITIVE CIRCUITS BETWEEN THE ANTERIOR THALAMUS AND THE CORTEX

One central hypothesis is that different sub-circuits within the Papez network process independent signals and thus play different roles in memory and spatial navigation. However, anatomy alone cannot address the question of the nature of the signals conveyed in each loop and their role in cognitive processes. While nucleus-specific lesion studies are challenging to achieve due to the small size of the individual nuclei and their close proximity to each other, sensory and behavioral correlates of neuronal activity within each nucleus provide important insights into the role of specific brain regions and the nature of the local computations performed by the circuits.

Activity of individual cells in the hippocampus is often bound to a particular point in space (O'Keefe and Dostrovsky, 1971). Soon after the discovery of these so called 'place cells' in the hippocampus, it was postulated that these spatial correlates must result from the integration of lower-level components, including signals related to external sensory inputs and self-movement (O'Keefe and Nadel, 1978). One key property of place cells is that they are preserved in darkness, suggesting that the system can be updated in the absence of visual information and that vestibular inputs must thus play a central role in their generation. The AT receives direct inputs from the tegmento-mamillary system, which is itself targeted by brainstem vestibular nuclei. Hence, the AT is believed to be part of the circuit conveying information related to how the animal moves relative to its environment. In the AT, the head-direction signal is by far the most studied and understood behavioral correlate of the activity of individual neurons (Peyrache et al., 2019; Taube, 2007). In particular, cells in the AD show robust tuning to the animal's current heading, independently of its location and ongoing behavior. From a computational perspective, the HD system is a canonical example of an "attractor" network, that is, a system in which internal circuits and dynamics constrain neuronal population activity into a small number of possible states. For the HD system, these states are continuously distributed along a functional ring, corresponding to low energy levels in the system. One experimental evidence supporting this particular organization of the HD network is the observation that during sleep, HD cell population activity is still organized, pointing toward a "virtual" and seemingly random direction at each moment in time (Peyrache et al., 2015; Viejo and Peyrache, 2020) (Figure 44.2).

The HD signal is broadcast to the cortical navigation system from the brainstem tegmentum via the LMN-AD-PoSub loop, with cells showing strong HD tuning present at each level of this circuit (Taube, 2007). The importance of the loop circuitry is evidenced by the fact that stable HD tuning in AD requires cortical feedback (Goodridge and Taube, 1997). While AD has by far the highest density of HD cells among the AT nuclei, HD-tuned neurons have also been reported in other AT nuclei (Jankowski et al., 2014; Tsanov et al., 2011), likely reflecting their HD-tuned cortical inputs. The effects of LMN-AD-PoSub lesions on spatial navigation are mixed (see Weiss and Derdikman, 2018 for review). Yet, recent studies using optogenetic inactivation point to the important role of AD in recall of recent but not remote memories (Vetere et al., 2021) and, via its projection to RS, in memory encoding (Roy et al., 2021). Moreover, AD seems to be inhibited by a direct projection from CA3 inhibitory neurons during recent memory recall (Vetere et al., 2021). Overall, these recent studies point to a more general role of this nucleus in mnemonic processes beyond the representation of the animal's current direction.

One hallmark of the coupling of a network to the extended hippocampal system is the presence of hippocampal theta-modulated cell activity (Buzsáki, 2002). The theta rhythm is a large-amplitude 4–12 Hz oscillation that orchestrates hippocampal activity and information processing (Buzsáki, 2002). Indeed, MMN neurons (Bland et al., 1995) as well as neurons in most AT nuclei (Tsanov et al., 2011; Vertes et al., 2001; Viejo and Peyrache, 2020) show considerable coupling to the hippocampal theta. The degree of theta-coupling in AT nuclei is likely inherited from their inputs. Thus, among the AT nuclei showing theta-modulated activity, AV seems to be the most strongly theta-coupled (Tsanov et al., 2011) and also receives the strongest input from the theta-modulated MMN, indicating strong association with the hippocampal system. This functional coupling is supported by the postulated role of AV in controlling the specificity of memory retrieval (Roy et al., 2021), perhaps by gating the contextual information stored in the HPC. In contrast, there is a near-complete absence of theta modulation in AD (Viejo and Peyrache, 2020), as well as a relatively small proportion of theta-modulated cells in the upstream LMN, especially among HD cells (Blair et al., 1998).

Next to AD, RE is perhaps the most studied rostral thalamic nucleus involved in higher cognitive functions. It has for a long time been of interest to memory researchers because its connectivity profile puts it in the ideal position to relay information between mPFC and HPC—the same RE neurons that receive inputs from the former also project to the latter (Vertes et al., 2007). While communication in the HPC-to-mPFC direction is thought to be critical for offline memory processes (Peyrache et al., 2009), communication in the opposite direction is thought to contribute to online processes like working memory (Griffin, 2021). When animals are engaged in a spatial working memory task, hippocampal place cells in CA1 (but importantly not in CA3) fire at different rates depending on the future trajectory (Grieves et al., 2016; Wood et al., 2000). RE neurons are spatially tuned (Jankowski et al., 2015) and also show such trajectory-dependent activity; inactivation of RE abolishes such differential activity in CA1 but notably not in CA3, to which RE does not project (Ito et al., 2015). Furthermore, enhanced synchrony in the theta band between mPFC, RE, and CA1 is present while animals are making a working memory–based spatial decision in the same task (Ito et al., 2018). Thus, RE is a thalamic nucleus with a central role in the mPFC-HPC interactions, at least during online behaviors.

In summary, the role of the AT in conveying the HD signal, especially via AD, is now well established. Similarly, there is strong evidence supporting the central role of RE in mediating the online information transfer between mPFC and the hippocampus during working memory tasks. Furthermore, other neurons in the AT, especially in AV, show strong modulation by the hippocampal theta rhythms, but the behavioral and cognitive correlates of these neurons remain unclear.

NEURONAL DYNAMICS SUPPORTING MEMORY FORMATION DURING SLEEP

The formation of explicit memories depends on a cascade of molecular, neuronal as well as circuit-level processes (Dudai, 2004), and sleep is believed to be crucial for this multi-level reorganization (Diekelmann and Born, 2010). During sleep, thalamic neurons change their firing dynamics and certainly contribute to memory processing. In addition, thalamic neuronal population activity codes for "virtual" spatial signals during sleep. This offline activity updates downstream cortical structures as if the animal were awake and may favor the replay of recent memories to support long-term memory formation.

CHANGES IN THALAMIC NEURONAL DYNAMICS DURING SLEEP: BURSTS AND SPINDLES

Neuronal activity in the thalamus shifts dramatically across brain states, from regular spiking during wakefulness to burst firing regimes during Non-Rapid Eye Movement (NREM) sleep (Steriade et al., 1993). One of the hallmarks of NREM sleep is the spindle oscillation in thalamocortical networks. Spindles arise from the coordination of burst firing across the population of thalamic neurons, resulting in intermittent network oscillations at 10–15 Hz and lasting for 0.5–3s that rhythmically modulate their cortical targets (Fernandez and Lüthi, 2019). While the role of spindles remains elusive, it has been hypothesized that they are related to learning and memory processes (see Peyrache and Seibt, 2020, for review).

The existence and significance of bursts and spindles in the anterior thalamus is a topic of discussion (Bandarabadi et al., 2020; Pare et al., 1987; Viejo and Peyrache, 2020). Recent studies have clarified this question. Some neurons in the AT shift to burst mode during NREM, depending on the neuron's functional integration within the Papez natwork. Specifically, bursting neurons are found in all nuclei, with the remarkable exception of the AD, which is entirely devoid of bursts during NREM (Viejo and Peyrache, 2020). In the other AT nuclei, bursting neurons are not anatomically segregated, and two neighboring neurons can show drastically different dynamics during NREM. However, the dynamics of neuronal spike trains are remarkably correlated with neuronal responses to hippocampal population activity, suggesting a link between dynamics and function (or at least, functional integration) within the Papez network (Viejo and Peyrache, 2020).

Interestingly, differences in burstiness across AT nuclei do not seem to correspond to differences in spindle generation. AT nuclei are reciprocally connected with the thalamic reticular nucleus (TRN), a network of inhibitory cells surrounding the thalamus proper which play a key role in burst firing and spindle generation (Fernandez and Lüthi, 2019). The TRN preferentially targets the AD in the AT (Duszkiewicz et al., 2022; Gonzalo-Ruiz and Lieberman, 1995; Vantomme et al., 2020), and activation of TRN inputs to the AD triggers spindles (Bandarabadi et al., 2020). The AD-generated spindles are thus somehow different from other thalamocortical systems where spindles are associated with bursts in thalamocortical neurons. Whether the other AT nuclei generate spindles despite lighter inputs from the TRN is still unknown, yet spindles are detected in most (if not all) medial cortical areas, even the ones that are not primarily targeted by AD (Siapas and Wilson, 1998).

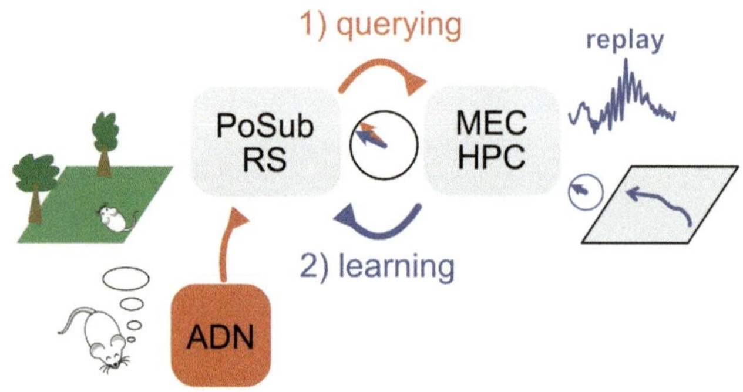

Figure 44.3. A hypothetical mechanism for thalamic querying of downstream cortical areas for memory formation during sleep. During NREM sleep, the HD signal encoded by AD neurons selects a random direction. This direction in turn constrains the ensemble activity in the cortex ensemble activity in the cortex. In the hippocampus, a spatial trajectory is replayed in a previously visited environment—not necessarily corresponding to a previously experienced path. By learning to decode a direction from these replayed trajectories on specific maps; i.e., minimizing the difference between queried (orange) and decoded (blue) directions, the RS, and perhaps at some level the PoSub, generalize the relationship between position and direction across different viewpoints. A spatial memory independent of specific previous experiences thus gradually forms and becomes independent of the hippocampus. Following this query, AD and other AT nuclei generate spindles to favor plasticity in the cortex and stabilize the replayed information (not shown).

THE ROLE OF THE ANTERIOR THALAMUS IN COORDINATING REPLAY AND MEMORY FORMATION IN THE HIPPOCAMPUS AND ASSOCIATED STRUCTURES

During NREM sleep, hippocampal activity is dominated by sharp wave-ripples (SWRs), intermittent bouts of synchronized activity (Buzsáki, 2015). A large body of literature has demonstrated that SWRs during NREM sleep are instrumental in long-term explicit memory formation (Buzsáki, 2015; Joo and Frank, 2018). Specifically, SWRs are hypothesized as the physiological processes through which the hippocampus, which is endowed with rapid learning capacities, takes part in training the cortex, a slow-learning structure (McClelland et al., 1995). During SWRs, neuronal patterns that correspond to recent wake experiences are replayed. Specifically, ensembles of place cells reinstate previously experienced spatial trajectories and can even be generative, forming patterns corresponding to unvisited routes (Swanson et al., 2020). These replay events are believed to convey the information that helps the cortex form long-term memories.

Simultaneous recordings of the hippocampus and medial cortical structures, especially the medial prefrontal cortex, have revealed a complex coordination between spindles and hippocampal SWRs (Siapas and Wilson, 1998). At fine timescale, SWRs which occur during spindles tend to be phase-locked to spindle oscillations. At larger timescales, SWRs tend to precede spindles recorded in the medial cortex—they occur generally more often at the onset of spindles than during spindles. Considering the potential role of spindles in plasticity processes, the SWR-spindle relationship suggests a dynamical coordination between replay and memory formation, the former preceding the latter (Peyrache and Seibt, 2020).

During NREM sleep, the role of the AD nucleus in coordinating medial cortical and parahippocampal activity is not restricted to spindles (Gent et al., 2018). As discussed earlier, the HD signal in AD remains coherent during sleep and randomly drifts along the two possible directions on the ring (Peyrache et al., 2015). This drift stops immediately before hippocampal SWRs and the activity of AD neurons now represents a precise direction during the entire SWR (Viejo and Peyrache, 2020). The origin of this spontaneous activity in AD is still unclear, but it is possible that it is locally generated during NREM, irrespective of the inputs from the MBs. The AD thus influences large cortical networks with an organized and meaningful signal.

The temporal relationship between AD activity and hippocampal SWRs suggests that the spontaneous activity in AD can act as a query for the hippocampal formation, constraining hippocampal neuronal assemblies along specific directions to generate coherent replay (Peyrache, 2022a) (Figure 44.3). The same query could evoke viewpoint-specific neuronal patterns in the RS, aligning RS and hippocampal activity during SWRs. This would favor generalization of spatial knowledge from previously unvisited viewpoints, thus forming the basis for a spatial memory in the RS. The role of other AT neurons is less clear. Many AT neurons (outside AD) discharge more strongly after SWRs (Viejo and Peyrache, 2020), and this is typically what is expected from a higher-order pathway that is conveying messages between cortical regions and enabling the broadcasting of replayed information across large cortical networks.

The RE is a key relay between regions of the medial cortex and parahippocampal structures and it thus may play an important role during offline memory consolidation. The hippocampus projects directly to the medial prefrontal cortex (Jay et al., 1989), as evidenced by the strong modulation of mPFC neurons by hippocampal rhythms (Siapas et al., 2005). Neuronal populations in the mPFC replay previous wake experience–associated activity patterns during NREM (Euston et al., 2007; Peyrache et al., 2009), especially after learning, at the time of SWRs (Peyrache et al., 2009). In turn, the relay of medial prefrontal signals to the hippocampus by the RE may be particularly important for memory processes (Mei et al., 2018; Xu and Südhof, 2013). Hippocampal replay is strengthened during spindles and, during SWRs, RE neurons show high level of burstiness (Varela and Wilson, 2020). RE inputs may thus play a role in a dynamic control of replay in the hippocampo-cortical network, possibly modulating the gain and the routing of replayed information.

CONCLUSION

In this chapter, we have presented the circuitry, dynamics, and behavioral correlates of neurons of the anterior thalamus and have outlined how these relate to memory formation. The emerging picture is that of considerable complexity and heterogeneity, with distinct gene expression profiles but overlapping input/output connectivity. While the functional roles of the two nuclei with oddball gene expression signatures, AD and RE, are relatively well mapped out using novel circuit interrogation techniques, the specific functions of AV, AM, and LD, and especially their individual projections, are still veiled in mystery. Functional mapping of individual AT nuclei is complicated by their relatively small size, irregular shape, and overlapping connectivity, but novel combinatorial approaches targeting multiple inputs/outputs (Phillips et al., 2019) offer a promising avenue toward uncovering their unique contributions to memory, navigation and other cognitive functions.

REFERENCES

Aggleton JP, Hunt PR, Nagle S, Neave N (1996). The effects of selective lesions within the anterior thalamic nuclei on spatial memory in the rat. *Behav Brain Res*. 81:189–198.

Aggleton JP, O'Mara SM (2022). The anterior thalamic nuclei: Core components of a tripartite episodic memory system. *Nat Rev Neurosci*. 23:505–516.

Alexander AS, Place R, Starrett MJ, Chrastil ER, Nitz DA (2023). Rethinking retrosplenial cortex: Perspectives and predictions. *Neuron*. 111:150–175.

Baddeley A (2010). Working memory. *Curr Biol*. 20:R136–R140.

Balsamo G, Blanco-Hernández E, Liang F, Naumann RK, Coletta S, Burgalossi A, Preston-Ferrer P (2022). Modular microcircuit organization of the presubicular head-direction map. *Cell Rep*. 39:110684.

Bandarabadi M, Herrera CG, Gent TC, Bassetti C, Schindler K, Adamantidis AR (2020). A role for spindles in the onset of rapid eye movement sleep. *Nat Commun*. 11:5247.

Blair HT, Cho J, Sharp PE (1998). Role of the lateral mammillary nucleus in the rat head direction circuit: a combined single unit recording and lesion study. Neuron. 21:1387–1397.

Bland BH, Konopacki J, Kirk IJ, Oddie SD, Dickson CT (1995). Discharge patterns of hippocampal theta-related cells in the caudal diencephalon of the urethan-anesthetized rat. *J Neurophysiol*. 74:322–333.

Buzsáki G (2002). Theta oscillations in the hippocampus. *Neuron*. 33:325–340.

Buzsáki G (2015). Hippocampal sharp wave-ripple: A cognitive biomarker for episodic memory and planning. *Hippocampus*. 25:1073–1188.

Diekelmann S, Born J (2010). The memory function of sleep. *Nat Rev Neurosci*. 1:114–126.

Dudai Y (2004). The neurobiology of consolidations, or, how stable is the engram? *Annu Rev Psychol*. 55:51–86.

Duszkiewicz AJ, Carrasco SS, Orhan P, Brown E, Owczarek E, Vite GR, Wood ER, Peyrache A (2022). Reciprocal feature encoding by cortical excitatory and inhibitory neurons. BioRxiv. https://doi.org/10.1101/2022.03.14.484357

Euston DR, Tatsuno M, McNaughton BL (2007). Fast-forward playback of recent memory sequences in prefrontal cortex during sleep. Science. 318:1147–1150. https://doi.org/10.1126/science.1148979

Franklin KBJ, Paxinos G (2007). *The Mouse Brain in Stereotaxic Coordinates*. Academic Press.

Fernandez LMJ, Lüthi A (2019). Sleep spindles: Mechanisms and functions. *Physiol Rev*. 100:805–868.

Gent TC, Bandarabadi M, Herrera CG, Adamantidis AR (2018). Thalamic dual control of sleep and wakefulness. *Nat Neurosci*. 21:974–984.

Gonzalo-Ruiz A, Lieberman AR (1995). Topographic organization of projections from the thalamic reticular nucleus to the anterior thalamic nuclei in the rat. *Brain Res Bull*. 37:17–35.

Goodridge JP, Taube JS (1997). Interaction between the postsubiculum and anterior thalamus in the generation of head direction cell activity. *J Neurosci*. 17:9315–9330.

Grieves RM, Wood ER, Dudchenko PA (2016). Place cells on a maze encode routes rather than destinations. *eLife*. 5:e15986.

Griffin AL (2021). The nucleus reuniens orchestrates prefrontal-hippocampal synchrony during spatial working memory. *Neurosci Biobehav Rev*. 128:415–420.

Herkenham M (1978). The connections of the nucleus reuniens thalami: Evidence for a direct thalamo-hippocampal pathway in the rat. *J Comp Neurol*. 177:589–609.

Ito HT, Moser EI, Moser M-B (2018). Supramammillary nucleus modulates spike-time coordination in the prefrontal-thalamo-hippocampal circuit during navigation. *Neuron.* 99:576–587.

Ito HT, Zhang S-J, Witter MP, Moser EI, Moser M-B (2015). A prefrontal–thalamo–hippocampal circuit for goal-directed spatial navigation. *Nature.* 522:50–55.

Jankowski MM, Islam MN, Wright NF, Vann SD, Erichsen JT, Aggleton JP, O'Mara SM (2014). Nucleus reuniens of the thalamus contains head direction cells. *eLife.* 3:e03075.

Jankowski MM, Passecker J, Islam MN, Vann S, Erichsen JT, Aggleton JP, O'Mara SM (2015). Evidence for spatially-responsive neurons in the rostral thalamus. *Front Behav Neurosci.* 9. https://www.frontiersin.org/articles/10.3389/fnbeh.2015.00256/full?utm_source=twitterfeed&utm_medium=twitter

Jay, T.M., Glowinski, J., Thierry, A.-M., 1989. Selectivity of the hippocampal projection to the prelimbic area of the prefrontal cortex in the rat. *Brain Res.* 505:337–340.

Joo HR, Frank LM (2018). The hippocampal sharp wave–ripple in memory retrieval for immediate use and consolidation. *Nat Rev Neurosci.* 19:744–757.

Maisson DJ-N, Gemzik ZM, Griffin AL (2018). Optogenetic suppression of the nucleus reuniens selectively impairs encoding during spatial working memory. *Neurobiol Learn Mem.* 155:78–85.

Mathiasen ML, Dillingham CM, Kinnavane L, Powell AL, Aggleton JP (2017). Asymmetric cross-hemispheric connections link the rat anterior thalamic nuclei with the cortex and hippocampal formation. *Neuroscience.* 349:128–143.

McClelland JL, McNaughton BL, O'Reilly RC (1995). Why there are complementary learning systems in the hippocampus and neocortex: insights from the successes and failures of connectionist models of learning and memory. *Psychol Rev.* 102:419–457.

Mei H, Logothetis NK, Eschenko O (2018). The activity of thalamic nucleus reuniens is critical for memory retrieval, but not essential for the early phase of "off-line" consolidation. *Learn Mem.* 25:129–137.

Nelson AJD, Kinnavane L, Amin E, O'Mara SM, Aggleton JP (2020). Deconstructing the direct reciprocal hippocampal-anterior thalamic pathways for spatial learning. *J Neurosci.* 40:6978–6990.

O'Keefe J, Dostrovsky J (1971). The hippocampus as a spatial map: Preliminary evidence from unit activity in the freely-moving rat. *Brain Res.* 34:171–175.

O'Keefe J, Nadel L (1978). *The Hippocampus as a Cognitive Map.* Clarendon Press.

Papez JW (1937). A proposed mechanism of emotion. *Arch Neurol Psychiatry.* 38:725–743.

Pare D, Steriade M, Deschenes M, Oakson G (1987). Physiological characteristics of anterior thalamic nuclei, a group devoid of inputs from reticular thalamic nucleus. *J Neurophysiol.* 57:1669–1685.

Peyrache A (2022a). Querying hippocampal replay with subcortical inputs. *Curr Opin Neurobiol.* 77:102645. https://www.sciencedirect.com/science/article/pii/S0959438822001398

Peyrache A (2022b). The thalamus in navigation. In: Halassa MM, ed. *The Thalamus.* Cambridge University Press:340–360.

Peyrache A, Duszkiewicz AJ, Viejo G, Angeles-Duran S (2019). Thalamocortical processing of the head-direction sense. *Prog Neurobiol.* 183:101693.

Peyrache A, Khamassi M, Benchenane K, Wiener SI, Battaglia FP (2009). Replay of rule-learning related neural patterns in the prefrontal cortex during sleep. *Nat Neurosci.* 12:919–926.

Peyrache A, Lacroix MM, Petersen PC, Buzsáki G (2015). Internally organized mechanisms of the head direction sense. *Nat Neurosci.* 18:569–575.

Peyrache A, Seibt J (2020). A mechanism for learning with sleep spindles. *Philos Trans R Soc B Biol Sci.* 375:20190230. https://royalsocietypublishing.org/doi/full/10.1098/rstb.2019.0230

Phillips JW, Schulmann A, Hara E, Winnubst J, Liu C, Valakh V, Wang L, Shields BC, Korff W, Chandrashekar J, Lemire AL, Mensh B, Dudman JT, Nelson SB, Hantman AW (2019). A repeated molecular architecture across thalamic pathways. *Nat Neurosci.* 22:1925–1935.

Roy DS, Zhang Y, Aida T, Choi S, Chen Q, Hou Y, Lea NE, Skaggs KM, Quay JC, Liew M, Maisano H, Le V, Jones C, Xu J, Kong D, Sullivan HA, Saunders A, McCarroll SA, Wickersham IR, Feng G (2021). Anterior thalamic dysfunction underlies cognitive deficits in a subset of neuropsychiatric disease models. *Neuron.* 109:2590–2603.

Scoville WB, Milner B (1957). Loss of recent memory after bilateral hippocampal lesions. *J Neurol Neurosurg Psychiatry.* 20:11–21.

Shibata, H. (1992). Topographic organization of subcortical projections to the anterior thalamic nuclei in the rat. *J Comp Neurol.* 323:117–127.

Shibata H (1993). Direct projections from the anterior thalamic nuclei to the retrohippocampal region in the rat. *J Comp Neurol.* 337:431–445.

Siapas AG, Lubenov EV, Wilson MA (2005). Prefrontal phase locking to hippocampal theta oscillations. *Neuron.* 46:141–151.

Siapas AG, Wilson MA (1998). Coordinated interactions between hippocampal ripples and cortical spindles during slow-wave sleep. *Neuron.* 21:1123–1128.

Squire LR (2004). Memory systems of the brain: A brief history and current perspective. *Neurobiol Learn Mem.* 82:171–177.

Steriade M, McCormick DA, Sejnowski TJ (1993). Thalamocortical oscillations in the sleeping and aroused brain. *Science.* 262:679–685.

Swanson RA, Levenstein D, McClain K, Tingley D, Buzsáki G (2020). Variable specificity of memory trace reactivation during hippocampal sharp wave ripples. *Curr Opin Behav Sci.* 32:126–135.

Taube JS (2007). The head direction signal: origins and sensory-motor integration. *Annu Rev Neurosci.* 30:181–207.

Tsanov M, Chah E, Vann SD, Reilly RB, Erichsen JT, Aggleton JP, O'Mara SM (2011). Theta-modulated head direction cells in the rat anterior thalamus. *J Neurosci.* 31:9489–9502.

van Groen T, Kadish I, Wyss JM (1999). Efferent connections of the anteromedial nucleus of the thalamus of the rat. *Brain Res Rev.* 30:1–26.

van Groen T, Kadish I, Wyss JM (2002). Role of the anterodorsal and anteroventral nuclei of the thalamus in spatial memory in the rat. *Behav Brain Res.* 132:19–28.

van Groen T, Wyss JM (1995). Projections from the anterodorsal and anteroventral nucleus of the thalamus to the limbic cortex in the rat. *J Comp Neurol.* 358:584–604.

Vann SD, Brown MW, Aggleton JP (2000). Fos expression in the rostral thalamic nuclei and associated cortical regions in response to different spatial memory tests. *Neuroscience.* 101:983–991.

Vantomme G, Rovó Z, Cardis R, Béard E, Katsioudi G, Guadagno A, Perrenoud V, Fernandez LMJ, Lüthi A (2020). A thalamic reticular circuit for head direction cell tuning and spatial navigation. *Cell Rep.* 31:107747. https://www.sciencedirect.com/journal/cell-reports/vol/31/issue/10

Varela C, Wilson MA (2020). mPFC spindle cycles organize sparse thalamic activation and recently active CA1 cells during non-REM sleep. *eLife.* 9:e48881.

Vertes RP, Albo Z, Viana Di Prisco G (2001). Theta-rhythmically firing neurons in the anterior thalamus: implications for mnemonic functions of Papez's circuit. *Neuroscience.* 104:619–625.

Vertes RP, Hoover WB, Szigeti-Buck K, Leranth C (2007). Nucleus reuniens of the midline thalamus: Link between the medial prefrontal cortex and the hippocampus. *Brain Res.* Bull. 71:601–609.

Vetere G, Xia F, Ramsaran AI, Tran LM, Josselyn SA, Frankland PW (2021). An inhibitory hippocampal–thalamic pathway modulates remote memory retrieval. *Nat Neurosci.* 24:685–693.

Viejo G, Peyrache A (2020). Precise coupling of the thalamic head-direction system to hippocampal ripples. *Nat Commun.* 11:2524. https://www.nature.com/articles/s41467-020-15842-4

Weiss S, Derdikman D (2018). Role of the head-direction signal in spatial tasks: when and how does it guide behavior? *J Neurophysiol.* 120:78–87.

Wood ER, Dudchenko PA, Robitsek RJ, Eichenbaum H (2000). Hippocampal neurons encode information about different types of memory episodes occurring in the same location. *Neuron.* 27:623–633.

Wright NF, Erichsen JT, Vann SD, O'Mara SM, Aggleton JP (2010). Parallel but separate inputs from limbic cortices to the mammillary bodies and anterior thalamic nuclei in the rat. J. Comp. Neurol. 518:2334–2354.

Wright NF, Vann SD, Erichsen JT, O'Mara SM, Aggleton JP (2013). Segregation of parallel inputs to the anteromedial and anteroventral thalamic nuclei of the rat. *J Comp Neurol.* 521:2966–2986.

Wyss JM, Groen TV (1992). Connections between the retrosplenial cortex and the hippocampal formation in the rat: A review. Hippocampus. 2:1–11.

Xu W, Südhof TC (2013). A neural circuit for memory specificity and generalization. *Science.* 339:1290–1295.

45.

ATTENTION AND THALAMOCORTICAL INTERACTIONS

Sabine Kastner and W. Martin Usrey

INTRODUCTION

Selective attention, the ability to direct attention to specific stimulus features, objects, or locations without moving the eyes, is a cognitive activity known to improve the detection and discrimination of visual stimuli and to decrease the reaction time for behavioral responses (Nobre and Kastner, 2014). Although most studies of selective attention have focused solely on the cerebral cortex, results from an increasing number of experiments studying the visual system of primates indicate the involvement of thalamocortical network interactions in attention. In the visual system of primates, there are more than 20 visual cortical areas, and each area receives input from and projects to the thalamus. Two prominent thalamic nuclei, the dorsal lateral geniculate nucleus (LGN) and the pulvinar, process visual signals and communicate reciprocally with the cortex. Although the LGN and pulvinar are both involved in vision, they have different circuitry and serve different functions. The LGN receives visual signals directly from the retina and relays these signals almost exclusively to primary visual cortex. In contrast, the pulvinar has many distinct divisions that collectively receive feedforward input from every visual cortical area and project, in turn, to the cortex, presumably to influence and/or mediate cortico-cortical communication (Sherman and Guillery, 2013). Based on their sources of feedforward input, retina versus cortex, the LGN and pulvinar are classified as first-order and higher-order thalamic nuclei, respectively (see **Chapter 42 in Section VII** for additional details about thalamic classification). Here, we focus on the cells and circuits of the LGN and pulvinar and review recent work examining the effects of covert spatial attention on neuronal activity and thalamocortical network interactions in primates.

STUDYING SPATIAL ATTENTION IN THE LABORATORY

We use spatial attention throughout our daily lives for all sorts of tasks. With covert spatial attention involving vision, our eyes are directed at one location while our attention is directed elsewhere. The "no-look" pass made by skilled NBA point guards is one example[1] of covert spatial attention (hereafter just referred to as spatial attention). Often used during a fast break as players run full speed from one end of the court to the other, the point guard with the ball will fool the opposing team by looking at a player in one direction while making a precise pass to a different player in a different location. Importantly, the point guard's attention is directed to a target other than where the eyes are looking. In a similar fashion, monkeys spend much of their time focusing their attention on higher-ranking individuals without making direct eye contact, which could be viewed as a threat and have undesired consequences. Working with nonhuman primates in the laboratory, we take advantage of their ability to fixate to (1) present visual stimuli at precise locations relative to the center of gaze, including locations overlapping with the receptive fields of recorded neurons, and (2) present stimuli at various locations for a sufficient time to measure the effects of attention on neural activity and behavior. Accordingly, in a typical experiment, the eyes fixate on a point in the center of a computer screen while multiple stimuli are shown at various pre-selected spatial locations. Based on a cue, the participant is instructed about which stimulus location they should covertly direct their attention toward in anticipation of a change in the stimulus, which they report with an eye movement or lever press to receive a reward. In a minority of trials, the cue is misleading and will direct attention away from the stimulus that will change. For both humans and monkeys, response accuracy and reaction times are better and faster, respectively,

1. Another example for those not familiar with the NBA is what soccer (football) players do with penalty kicks. Here the kicker will often direct their eyes to the opposite side of the goal box from where they covertly attend and kick the ball, leading the goalie to jump away from the ball.

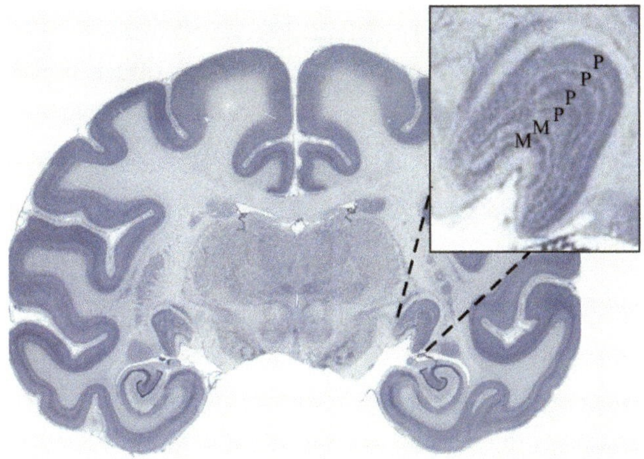

Figure 45.1. Nissl-stained coronal section of macaque brain with the LGN shown at 2 magnifications. The magnocellular (M) and parvocellular (P) layers are indicated; the koniocellular layers (not indicated) are located beneath each of the magnocellular and parvocellular layers.

when the cue is valid compared to when it is not valid. For electrophysiological studies, one of the stimuli is typically positioned over the receptive field of recorded neurons so that neural activity and network interactions can be compared when the attention is directed toward and away from the neurons' receptive fields.

THE LGN AND EFFECTS OF SPATIAL ATTENTION ON VISUAL RESPONSES AND GENICULOCORTICAL NETWORK INTERACTIONS

BRIEF OVERVIEW OF THE ANATOMICAL AND FUNCTIONAL ORGANIZATION OF THE LGN

The LGN is the primary source of visual information being relayed to the cortex. This information is communicated via anatomically and functionally distinct parallel processing streams that originate from distinct classes of retinal ganglion cells and project to different layers/modules in V1. Briefly (see **Chapter 17 in Section V** for additional details), the LGN in Old World monkeys and humans contains 2 magnocellular layers, 4 parvocellular layers, and 6 koniocellular layers (Figure 45.1). The response properties of neurons in each of these layers are generally similar to those of their retinal inputs (reviewed in Usrey and Alitto, 2015). Compared to parvocellular neurons, magnocellular neurons have larger receptive fields, are more sensitive to low-contrast stimuli, respond better to stimuli at high temporal frequencies, and are more suppressed by stimuli that extend beyond the classical receptive field. Magnocellular neurons have little selectivity for the chromatic properties of a stimulus, whereas parvocellular neurons show selectivity within the middle and long wavelengths (green vs. red). Less is known about the response properties of koniocellular neurons; however, some are sensitive to short-wavelength (blue) stimuli and, unlike magnocellular or parvocellular neurons which respond exclusively to one eye, some koniocellular neurons have binocular responses (Cheong et al., 2013). While the focus of past work and this chapter is on spatial attention, it is worth noting that parallel processing streams through the LGN provide a substrate for attentional mechanisms to selectively modulate the processing of specific features within the visual scene (e.g., motion, color, texture, form).

Given the similarity in receptive field properties between the retina and LGN, the question arises as to what purpose the LGN serves. One answer to this question involves the diversity of extra-retinal inputs to LGN neurons that can modulate the gain of LGN responses to incoming retinal signals (Sherman and Guillery, 2013; Usrey and Alitto, 2015). Nonvisual, extra-retinal sources of input to LGN neurons include cholinergic input from the parabrachial nucleus, noradrenergic input from the reticular formation, and serotonergic input from the dorsal raphe nucleus. Although these extra-retinal inputs do not directly evoke LGN responses, they do play an important role in adjusting LGN activity levels as a function of alertness and the sleep-wake cycle (Livingstone and Hubel, 1981; Steriade, 2004; Bereshpolova et al., 2011). In addition to these nonvisual inputs, LGN neurons also receive visually evoked, extra-retinal glutamatergic feedback input from layer (L)6 corticogeniculate neurons and GABAergic input from the thalamic reticular nucleus (TRN), a neighboring nucleus with neurons that integrate feedback input from cortex and feedforward input from the LGN. If, as discussed in the next section, LGN activity is modulated by attention, then it seems likely that the effects of attention include the involvement of the corticogeniculate feedback pathway and/or the TRN.

ATTENTION AUGMENTS VISUAL RESPONSES IN THE LGN

Visual attention has been studied most extensively in the cerebral cortex, where attention typically increases neuronal responses to visual stimuli at attended locations (reviewed in Maunsell, 2015) and increases coherence between single-unit activity and the local field potential (LFP) in specific frequency bands (reviewed in Fries, 2015; Buschman and Kastner, 2015). Although the effects of spatial attention are usually strongest in extrastriate cortical areas (e.g., V4, MT, VIP), attention can also influence neuronal activity at the earliest levels of processing beyond the eye; namely, in V1 and in the LGN (McAdams and Reid, 2005; Hembrook-Short et al., 2019; Briggs et al., 2013; McAlonan et al., 2006, 2008; O'Connor et al., 2002; Schneider and Kastner, 2009).

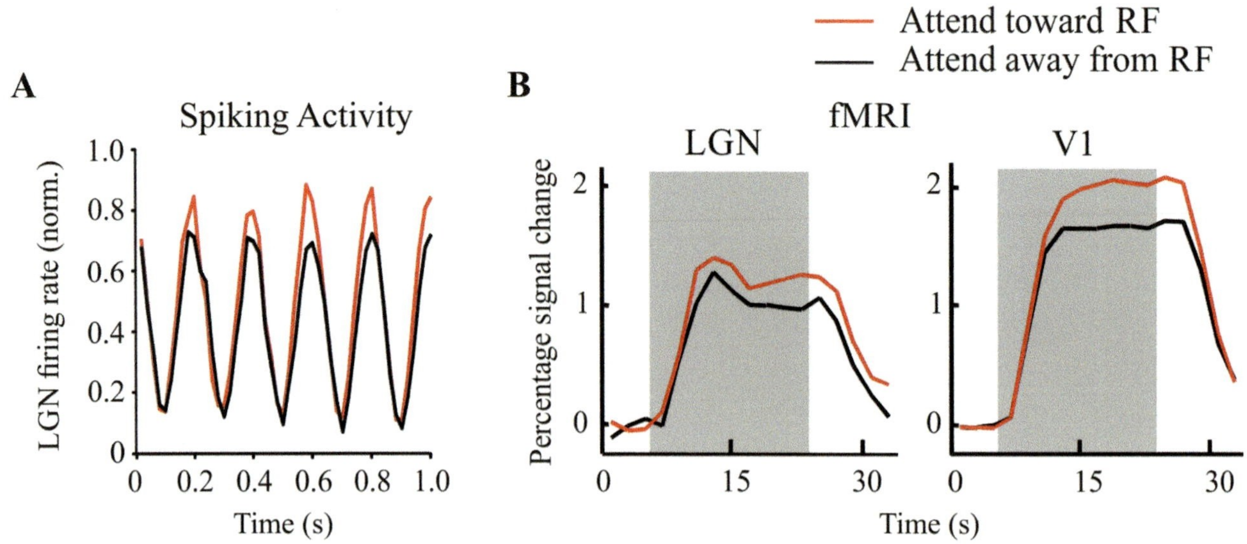

Figure 45.2. Influence of attention on the spiking activity and fMRI signal in the monkey and human LGN, respectively. (A). Plot showing average firing rate of 95 neurons in the monkey LGN when attention is directed toward (red) and away (black) from a drifting sinusoidal grating stimulus positioned over the cells' receptive field. The firing rate is greater when attention is directed towards the cells' receptive fields than when attention is directed away (adapted from Usrey and Alitto, 2015). (B). Plots showing the percentage of fMRI signal change in the human LGN (left) and V1 (right) in response to visual stimuli (shaded area) when individuals attended toward (red) and away (black) from the visual stimulus. For both the LGN and V1, the percentage signal change is greater when individuals attend toward the visual stimulus (adapted from O'Connor et al., 2002).

Within the LGN, spatial attention increases the single-unit activity of individual neurons in macaque monkeys and the blood-oxygenation-level-dependent (BOLD) responses measured during functional magnetic resonance imaging (fMRI) experiments in humans (Figure 45.2; O'Connor et al., 2002; McAlonan et al., 2008; Schneider and Kastner, 2009). Consistent with the view that the mechanisms contributing to the effects of attention in the LGN include the release of inhibition from the TRN, attention has been shown to decrease the activity of TRN neurons (McAlonan et al., 2000, 2006; see also Wimmer et al., 2015). Additional mechanisms likely contribute to attentional effects in the LGN, since the influence of attention on TRN neurons is more transient than it is for neurons in the LGN. Excitatory feedback from cortical L6 neurons is one likely candidate for the extended effects of attention on lateral geniculate neurons. Effects from cortical feedback could provide additional specificity to the effects of attention, as corticogeniculate feedback is composed of parallel pathways (see **Chapter 19 in Section V**) with distinct populations of L6 neurons selectively innervating neurons in the magnocellular, parvocellular, and koniocellular layers of the LGN (Fitzpatrick et al., 1994; Ichida et al., 2014).

ATTENTION AUGMENTS GENICULOCORTICAL COMMUNICATION

In addition to augmenting visual responses in the LGN, spatial attention also increases the strength of geniculocortical communication. This effect was revealed by pairing electrical stimulation in the LGN with recordings from synaptically coupled V1 target neurons in monkeys performing a visual task involving spatial attention. With this preparation, the successful communication of electrically evoked spikes to cortical target neurons could be compared when animals attended toward and away from a particular visual stimulus. As shown in Figure 45.3, attention toward a visual stimulus significantly increases the ability of thalamic neurons to drive postsynaptic responses in V1 (Briggs et al., 2013). Taken together with previous results, attention not only increases the firing rate of LGN neurons, but also increases the success of geniculate spikes in evoking cortical responses.

Rhythmic (also called oscillatory) activity patterns are common throughout the brain and have been proposed to play a role in dynamically facilitating neuronal communication between brain regions that oscillate in phase with each other (Fries, 2015). With regard to thalamocortical interactions, it is interesting to note that oscillatory phase synchronization between the local field potentials (LFPs) measured in the LGN and V1 has been reported for the alpha (8–14 Hz) and beta (15–30 Hz) frequency bands (Bastos et al., 2014). Moreover, an analysis of directed connectivity reveals that beta-band interactions are mediated by geniculocortical feedforward processing, whereas alpha-band interactions are mediated by corticogeniculate feedback processing (Bastos et al., 2014). These oscillatory interactions have been shown to provide a substrate for attentional modulation (in pulvinar; see next section). Future

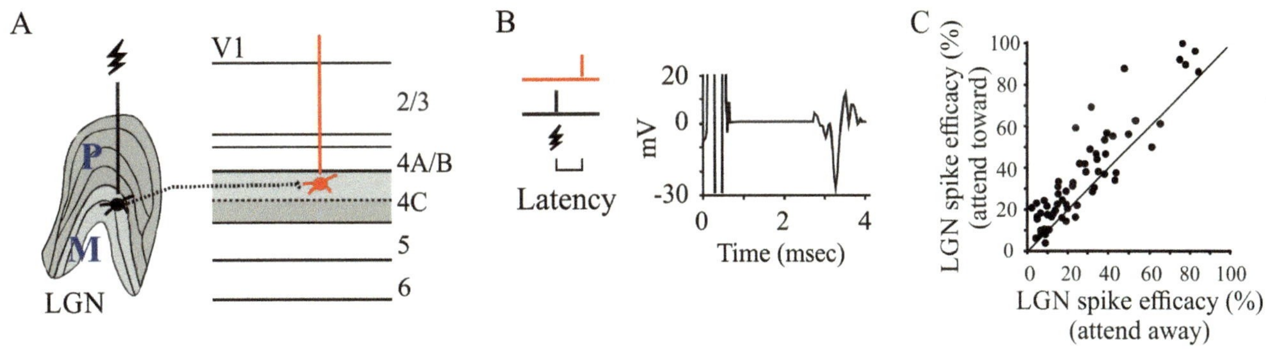

Figure 45.3. Synaptic communication between LGN neurons and target neurons in layer 4C of macaque V1 is enhanced with spatial attention. (A). Diagram illustrating the experimental configuration: a stimulating electrode and recording electrode are placed in retinotopic register in the LGN and V1, respectively. (B). While animals perform a visual spatial attention task, electrical stimulation is reduced such that only a subset of shocks to the LGN evoke a postsynaptic response in the cortex. (C). The efficacy of successful communication of electrically evoked geniculate spikes is increased when attention is directed toward the cells' receptive fields compared to when attention is directed away from the receptive fields (adapted from Briggs et al., 2013).

research will be able to address whether they serve a similar function in corticogeniculate interactions.

EFFECTS OF SPATIAL ATTENTION ON VISUAL RESPONSES IN THE PULVINAR AND PULVINOCORTICAL NETWORK INTERACTIONS

BRIEF OVERVIEW OF THE ANATOMICAL AND FUNCTIONAL ORGANIZATION OF THE PULVINAR

The pulvinar is the largest nucleus in the primate thalamus and is considered a higher-order thalamic nucleus because it forms input-output loops almost exclusively with the cortex. The pulvinar has undergone a significant expansion during evolution, which is on the order of that observed in prefrontal cortex. This in itself suggests that corticopulvinar interactions may play an important role in the increasingly flexible mechanisms underlying perception, action, and cognition that parallel this evolutionary expansion of brain structures.

The pulvinar is primarily interconnected with the extended visual cortical system, which forms the basis for characterizing its main subdivisions: the inferior pulvinar (PI) is interconnected with early visual areas, the lateral pulvinar (PL) with ventral and dorsal extrastriate areas, and the medio-dorsal pulvinar (PM) with higher-order frontal and parietal areas (Figure 45.4A; Kaas and Lyon, 2007). In nonhuman primates, PI and PL both contain a detailed retinotopic map informed by projections from early visual cortex and have accordingly a considerable proportion of visually responsive neurons. These characteristics have been utilized, using MRI, to establish a quite homologous functional organization with the human pulvinar (Figure 45.4C; Arcaro, 2015). The only region that receives input from the sensory periphery—either directly from the retina (Warner, 2012) or via the superior colliculus (Stepniewska, 2000)—and projects to dorsal extrastriate cortex is located in PI, medial to those regions that are interconnected with early visual cortex. In primates, this retino-collicular-pulvinar pathway appears to be particularly important during early development (Bridge, 2016) and for visual plasticity after cortical lesions (Bourne, 2017). Whether this pathway plays a significant functional role in the neurotypical adult brain is less clear. In the context of cognition and specifically selective attention, PM is a particularly critical part of the pulvinar. This subdivision appears to be primate-specific, reflecting the vast expansion of parietal association and frontal cortex in primates relative to other species (Romanski, 1997; Darian-Smith, 1999; Homman-Ludiye, 2019). Since multiple areas in parietal and frontal cortex constitute an attention-control network, it is of critical importance to understand the functional interactions of PM with cortex in the context of attention behaviors.

There are 2 well established corticopulvinar pathways that parallel the canonical input-output relationships constituting the visual processing hierarchy (Marion, 2013; Markov, 2014): (1) a trans-thalamic feedforward pathway (the "green" pathway in Figure 45.4B) that originates from L5 of a "lower" area (e.g., area V2 in Figure 45.4B), loops through a dedicated projection zone in the pulvinar, and feeds into L4 of a "higher" area (e.g., area V4 in Figure 45.4B); and (2) a feedback pathway (the "red" pathway in Figure 45.4B) that originates from L6 of a cortical area and projects to its pulvinar projection zone, also passing through the thalamic reticular nucleus (TRN) (Figure 45.4B). An important characteristic of the trans-thalamic pathway, known as the *replication principle*, is that it indirectly connects 2 cortical areas that share a direct cortico-cortical projection (such as V2 and V4 in Figure 45.4B; Adams, 2000; Shipp, 2003). In this respect, the pulvinar

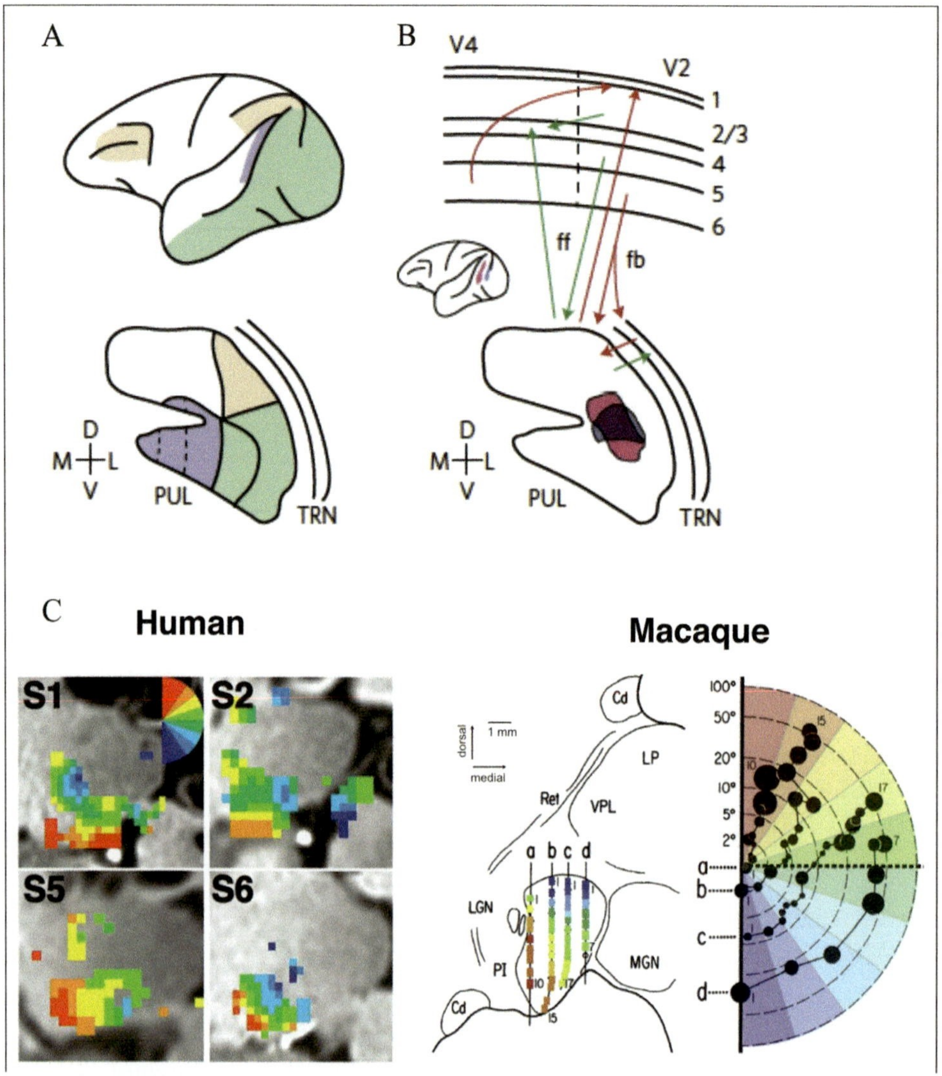

Figure 45.4. Anatomical and functional organization of pulvinar.
(A). Fronto-parietal (pink), medio-temporal (violet) and infero-temporal and occipital (green) cortex are interconnected with different subregions of the pulvinar (PUL). (B). Direct cortico-cortical connections and trans-thalamic feedforward (denoted by green arrows) and feedback (denoted by red arrows) pathways exemplified by V2-pulvino-V4 circuitry. Tracer injections into V2 (blue) and V4 (pink; inset) showed overlapping (purple) projection zones in the pulvinar, illustrating the replication principle of the indirect trans-thalamic pathways (adapted with permission from Figure 8 in Adams et al., 2000).
(C). Comparison of the topographic organization within the ventro-lateral pulvinar between humans and macaques. FMRI-defined polar angle maps (on left) from 4 human subjects (S1, S2, S5, S6, adapted from Arcaro et al., 2015) are compared to the electrophysiologically defined topographic organization of the macaque pulvinar (on right). Electrode penetrations a–d are color-coded relative to receptive field location (adapted with permission from Bender, 1981). PUL = pulvinar; TRN = thalamic reticular nucleus; V2 = visual area 2; V4 = visual area 4; Cd = caudate; LGN = lateral geniculate nucleus; MGN = medial geniculate nucleus; PI = inferior pulvinar; LP = lateral pulvinar.

can be thought of as a mosaic of projection zones that mirrors the cortico-cortical connectivity. The replication principle may be a fundamental feature of corticothalamic organization that extends to all higher-order thalamic nuclei (Phillips, 2019). It is important to note that the input-output loops of the "canonical trans-thalamic pathways," as described above, certainly present a simplification of the rich thalamocortical connectivity that exists (see Rovo, 2012). For example, pulvinocortical projections can have spatially segregated targets in cortex (Rockland, 1998; 1999), and corticopulvinar projections exhibit convergence from multiple sources, particularly in PM (Homman-Ludiye, 2019). However, this framework presents a starting point to explore functions of the trans-thalamic pathways, which we will discuss in the next sections.

REGULATION OF INTER-AREAL CORTICAL INTERACTIONS BY PULVINAR

In a typical visual attention experiment, a stimulus presented to the periphery of the visual field serves as a cue to direct attention to a specific location. Attention is then to be sustained at that location until a target is presented and the animal responds with a manual response or an eye movement. Such a simple task engages a vast network of areas distributed across all major lobes and includes the thalamus (Buschman and Kastner, 2015; Fiebelkorn and Kastner, 2020). The allocation of attention leads to an enhancement of the neural representation (e.g., of the spatial location in the above example) that can occur in several different ways: (1) by enhancing response rates (both on baseline and visually evoked responses; Luck, 1997); (2) by decreasing noise correlations (Cohen, 2009; Mitchell, 2009); and (3) by increasing synchronous firing in local neural populations (Fries, 2015). These modulatory effects of attention have been demonstrated in many parts of the attention network, including the pulvinar. Indeed, the response modulation observed in the pulvinar appears to resemble that of the interconnected cortical areas (Figure 45.5A; Saalmann, 2012; Zhou, 2016; Fiebelkorn, 2019). This could be interpreted as evidence for a function of the trans-thalamic feedforward pathway in routing information through the thalamus, thereby indirectly relaying information from one cortical area to another (Sherman and Guillery, 1998; Theyel, 2010). However, such a relay function is not compatible with a number of observations. For example, effect sizes of response modulation in the cortex and pulvinar can differ substantially (Zhou, 2016; Fiebelkorn, 2019). Also, pulvinar neurons typically show less feature selectivity and larger receptive fields than their cortical inputs (Chalupa, 1989; Purushothaman, 2012). These differences between pulvinar neurons and their cortical inputs suggest convergent inputs from additional sources.

Given that the influence of the cortex on the pulvinar is quite unclear, what is known about the reverse influence of the pulvinar on the cortex? By recording simultaneously from 2 cortical areas in the monkey ventral extrastriate cortex (i.e., areas V4 and TEO) that share both direct cortico-cortical connectivity and a common pulvinar projection zone, Saalmann et al. (2012) probed functions of the trans-thalamic feedforward pathway and found evidence for a thalamic role in coordinating attention-related functional interactions across cortical areas. In addition to attentional modulation of spike rates (Figure 45.5A), attention enhanced coordinated spiking activity within the pulvinar and between the pulvinar and interconnected cortical areas. Such temporal coordination can be measured as a correlation of spiking activity relative to local population activity obtained from the field potential (i.e., spike-field coherence), either within or across areas. Saalmann et al. (2012) measured these functional interactions during the task period following the cue and before target presentation (i.e., the delay period), when the wider cortical attention network is set up for target selection. During the delay period, neurons in all 3 areas showed enhanced coordinated spiking in the alpha and low beta ranges (<20 Hz). Strikingly, the pulvinar appeared to have the strongest influence on coordinated activity in V4 and TEO in a frequency band that matched that of the local attentional modulation (Figure 45.5B). In contrast, either cortical area had little influence on these interactions, suggesting that the pulvinar controlled the temporal coordination of functional interactions between cortical areas. These data are therefore consistent with the pulvinar, particularly the trans-thalamic feedforward pathway, having a primary role as a temporal coordinator, organizing and optimizing temporal dynamics of functional interactions across cortical networks to enhance the efficiency of signal processing between nodes. Fiebelkorn et al. (2019) has since provided evidence that this role of the pulvinar in coordinating cortical interactions is not limited to sensory cortical regions, but also extends to functional interactions between higher-order cortical regions. Zhou et al. (2016) found evidence that the interactions between the pulvinar and the cortex reverse during visual processing. For example, latencies of pulvinar neurons lagged behind those of cortical neurons, and the cortex exerted strong influence on the pulvinar in establishing functional interactions. However, since this study targeted primarily the corticopulvino feedback pathway, these results may reflect fundamental differences between functions of the trans-thalamic feedforward and feedback pathways.

CAUSAL STUDIES ON PULVINAR FUNCTION

Compelling evidence for a functional role of the pulvinar in selective attention has been provided by lesion studies in humans and monkeys demonstrating behavioral impairments. Functional or structural lesions of the pulvinar can lead to (1) thalamic neglect, a deficit in directing attention to contralesional space, as demonstrated in monkeys (Wilke, 2010), (2) spatial coding deficits, characterized by slowing of responses in the affected visual space (Danziger, 2002; Rafal, 1987), and (3) errors in binding featural information, both demonstrated in humans. A particularly striking deficit following pulvinar lesions is the inability to filter distracting stimuli that compete with a target stimulus for neural representation (Danziger, 2004; Snow, 2009), thereby impairing visually guided behavior more broadly, similar to attention deficits observed after lesions of human posterior parietal cortex (Friedman-Hill, 2003).

Studies using causal manipulations have also corroborated the idea that pulvinar regulates cortico-cortical interactions by showing that functional connectivity between cortical areas is weakened as a consequence of pulvinar inactivation (Zhou, 2016; Eradath, 2021). These studies suggest that cortico-cortical information transmission is greatly compromised without the pulvinar's influences. During pulvinar inactivation, neural

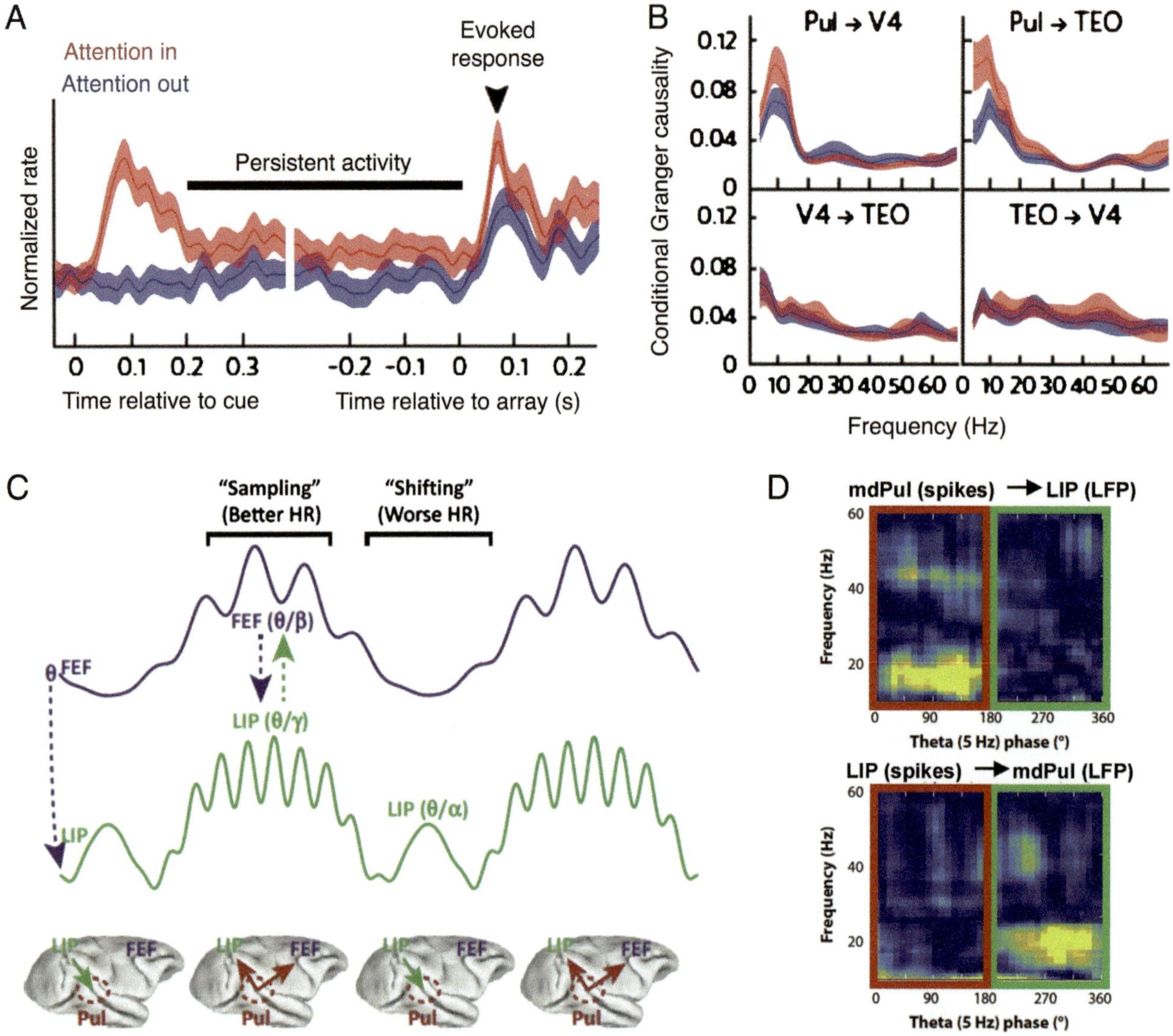

Figure 45.5. Pulvino-cortical interactions during spatial attention. (A). Attentional modulation of pulvinar responses when spatial attention is allocated at the receptive field that is recorded from ("Attention in") versus elsewhere in the visual field ("Attention out") (adapted with permission from Saalmann et al., 2012). (B). Conditional granger causality analysis suggests that pulvinar increases coherence in alpha frequencies in ventral extrastriate areas V4 and TEO during the delay period. In contrast, V4 and TEO do not seem to exert functional influences onto each other or the pulvinar (adapted with permission from Saalmann et al., 2012). (C). Spatial attention is characterized by alternating attentional states that promote either sampling at the attended location, resulting in better behavioral performance (i.e., higher hit rate, HR), or a higher likelihood of shifting attentional resources to another location, associated with worse behavioral outcome (i.e., worse HR). These theta-rhythmic, alternating attentional states are associated with different oscillatory frequencies (e.g., beta and gamma oscillations) and different patterns of functional connectivity (C adapted from Fiebelkorn and Kastner, 2019, with the pulvinar coordinating cortical activity specifically during periods of sensory sampling (D, red box denotes sampling phase, and green box shifting phase; adapted from Fiebelkorn et al., 2019). One theta cycle is about 200–250ms. Pul = pulvinar; V4 = visual area 4; FEF = frontal eye fields; LIP = lateral intraparietal area; mcPul = mediodorsal pulvinar.

activity—spiking as well as gamma activity—was largely reduced in visual cortex (Zhou, 2016; Purushothaman, 2012), indicating a profound loss of responsiveness to visual stimuli. Pulvinar inputs to the cortex may therefore be instrumental in controlling the excitability of neurons in superficial and/or granular layers, allowing them to respond normally to incoming visual information from downstream areas. Such control of excitability could occur via excitatory inputs on pyramidal neurons, or inhibitory control of interneurons, as suggested by increases in baseline firing during pulvinar inactivation (Zhou, 2016). Sensory gain control of cortical neurons by the pulvinar appears to follow a response gain model, in which neural responses are scaled multiplicatively or additively (deSouza, 2019). Thus, another major functional role of the trans-thalamic pathways appears to be that of an "enabler" of cortical function (Lakatos, 2016). In the next section, we will briefly discuss the highly dynamic nature of these pulvinocortical interactions and their associated functions.

TEMPORAL DYNAMICS OF PULVINOCORTICAL INTERACTIONS

Recent work has shown that attentional deployment is highly dynamic, characterized by alternating periods of relatively enhanced or diminished visual processing at the presently attended location (Fiebelkorn and Kastner, 2019, TICS). Even though these alternations in the sensitivity of visual processing are not noted subjectively, they can be readily observed in behavioral data and occur about 4–6 times per second (i.e., at a frequency in the theta range) (VanRullen, 2007; Landau, 2012; Fiebelkorn, 2013; Song, 2014). Importantly, these theta-dependent rhythmic alternations of visual processing have been observed in both humans (Fiebelkorn, 2013; Helfrich, 2018) and nonhuman primates (Fiebelkorn, 2018), thus constituting a fundamental property of attention function that has been preserved for more than 20 million years.

At the neural level, theta-dependent attentional rhythms appear to coordinate 2 alternating states: a "sampling" state that emphasizes the preferential processing of visual information from the attended location, and a "shifting" state when there is an increased likelihood of attentional shifting through either covert attentional shifts (i.e., in the absence of eye movements) or through overt attentional shifts (i.e., with eye movements). These states are characterized by complex dynamics across a fronto-parieto-pulvinar attention network. Specifically, the sampling state is characterized by increased coherence between spiking activity of visual neurons and gamma activity (>35 Hz) measured in the local field potential (LFP) in the lateral intraparietal area (LIP; Figure 45.5C; Fiebelkorn, 2018), as a signature of enhanced visual processing (Fries, 2015), as well as increased coherence of spiking activity of visuomotor neurons with LFP beta activity (~15–30Hz) in the frontal eye fields (FEF; Figure 45.5C; Fiebelkorn, 2018), as a signature of motor action suppression (e.g., eye movements, attention shifts; Pogosyan, 2009; Zhang, 2008). The primate-specific mediodorsal pulvinar (or PM) coordinates these cortico-cortical interactions across the attention network (Fiebelkorn, 2019), similar to what has been described in the previous section. However, it does so exclusively during sampling periods (Figure 45.5D) and not during shifting periods, further underlining the role of the trans-thalamic pathway in sensory gating.

In contrast, shifting periods—associated with relatively worse visual-target detection at the attended location—are associated not only with a release of suppression in the motor system (Fiebelkorn, 2018), but also with greater synchronization of alpha-band activity (~9–14 Hz), specifically within parietal cortex and between parietal cortex and the mediodorsal pulvinar (Figure 45.5D; Fiebelkorn, 2019). Parieto-occipital alpha-band activity is frequently associated with the suppression of sensory processing in human studies (Jensen, 2010; Foxe, 2011), as well as the suppression of cortical neuronal firing (Haegens, 2011). We propose that parietal-driven synchronization in the pulvinar temporarily shuts down the trans-thalamic pathway and its sensory gating function, thereby providing a neural substrate for the frequently observed visual suppression following alpha synchronization in parietal cortex. This pattern of results is consistent with a previous proposal based on human lesion studies, which suggested that the pulvinar was associated with the engagement of attentional resources at a behaviorally relevant location, while parietal cortex was associated with a disengagement of attentional resources (Buschman and Kastner, 2015).

AN EVOLUTIONARY PERSPECTIVE ON ATTENTION

Attention enhances our cognitive abilities to attended objects. While this has countless benefits and applications to our daily lives and has important advantages for survival, it does come at the cost of reducing our cognitive abilities to unattended objects. The most common explanation for this trade-off is that attention is a limited resource. But why is this the case? Given the size, complexity, and computational power of the cortex and thalamocortical networks, why should attention have limitations? Perhaps the answer can be found by considering attention from an evolutionary perspective (see Sherman and Usrey, 2021).

The cerebral cortex evolved subsequently to and then alongside the evolution of lower motor circuits to which it has unique access. Moreover, the cortex can only influence behavior by projections from L5 neurons that activate the older motor centers in the brainstem and spinal cord. These older motor centers may be seen as a bottleneck through which the cortex must operate. This presents a problem. If every cortical area were to operate at maximum capacity to turn its inputs into L5 motor commands, these would all compete for control through subcortical intermediaries, and chaos would likely ensue. Thus, there needs to be some selective process that ensures that only the cortical areas engaged in analyzing environmental objects that are most important, or most crucial to survival, are permitted to control subcortical motor centers. This is where "attention" comes in. Via a variety of mechanisms that includes thalamocortical processes, the appropriate region or regions of cortex are engaged and their L5 corticofugal projections are allowed to dominate subcortical motor regions; other areas of cortex (and their L5 outputs) dealing with less critical environmental events are suppressed. From this perspective, the limitations of attention are a consequence of the subcortical bottlenecks for behavior.

SUMMARY AND CONCLUDING REMARKS

With attention, the cognitive resources of the brain are directed to specific stimulus attributes, such as location, color, shape, and motion. The ability of the brain to selectively focus attention has a direct impact on an organism's success and survival. Although attention has traditionally been viewed as a cortical phenomenon, increasing evidence indicates the thalamus and thalamocortical interactions contribute to attentional processes. Studies primarily with nonhuman primates have provided new insights into the role of the thalamus in visual attention. As with cortical neurons, neurons in the LGN and the pulvinar increase their activity with attention. This effect is stronger in the pulvinar than in the LGN, and mechanisms likely include increased excitation from the cortex, along with disinhibition from the TRN. The pulvinar may also play a more direct role in mediating the effects of attention, by synchronizing rhythmic activity between cortical areas, presumably to enhance communication between areas. Clearly there is much that remains to be learned about how the thalamus and thalamocortical interactions contribute to attention.

REFERENCES

Adams MM, Hof PR, Gattass R, Webster MJ, Ungerleider LG (2000). Visual cortical projections and chemoarchitecture of macaque monkey pulvinar. *J Comp Neurol.* 419:377–393.

Arcaro MJ, Pinsk MA, Kastner S (2015). The anatomical and functional organization of the human visual pulvinar. *J Neurosci.* 35:9848–9871.

Bastos AM, Briggs F, Alitto HJ, Mangun GR, Usrey WM (2014). Simultaneous recordings from the primary visual cortex and lateral geniculate nucleus reveal rhythmic interactions and a cortical source for gamma-band oscillations. *J Neurosci.* 34:7639–7644.

Bender DB (1981). Retinotopic organization of macaque pulvinar. *J Neurophysiol.* 46(3):672–693.

Bereshpolova Y, Stoelzel CR, Zhuang J, Amitai Y, Alonso J-M, Swadlow HA (2011). Getting drowsy? Alert/nonalert transitions and visual thalamocortical network dynamics. *J Neurosci.* 31:17480–17487.

Bourne JA, Morrone MC (2017). Plasticity of visual pathways and function in the developing brain: Is the pulvinar a crucial player? *Front Syst Neurosci.* 11:3.

Bridge H, Leopold DA, Bourne JA (2016). Adaptive pulvinar circuitry supports visual cognition. *Trends Cogn Sci.* 20(2):146–157.

Briggs F, Mangun GR, Usrey WM (2013). Attention enhances synaptic efficacy and the signal-to-noise ratio in neural circuits. *Nature.* 499:476–480.

Buschman TJ, Kastner S (2015). From behavior to neural dynamics: An integrated theory of attention. *Neuron.* 88(1):127–144.

Chalupa LM, Abramson BP (1989). Visual receptive fields in the striate-recipient zone of the lateral posterior-pulvinar complex. *J Neurosci.* 9(1):347–357.

Cheong SK, Tailby C, Solomon SG, Martin PR (2013). Cortical-like receptive fields in the lateral geniculate nucleus of marmoset monkeys. *J Neurosci.* 33:6864–6876.

Cohen MR, Maunsell JH (2009). Attention improves performance primarily by reducing interneuronal correlations. *Nat Neurosci.* 12:1594–1600.

Danziger S, Ward R, Owen V, Rafal R (2002). The effects of unilateral pulvinar damage in humans on reflexive orienting and filtering of irrelevant information. *Behav Neurol.* 13(3–4):95–104.

Danziger S, Ward R, Owen V, Rafal R (2004). Contributions of the human pulvinar to linking vision and action. *CABN.* 4(1):89–99.

Darian-Smith C, Tan A, Edwards S (1999). Comparing thalamocortical and corticothalamic microstructure and spatial reciprocity in the macaque ventral posterolateral nucleus (VPLc) and medial pulvinar. *J Comp Neurol.* 410(2):211–234.

de Souza BOF, Cortes N, Casanova C (2019). Pulvinar modulates contrast response in the visual cortex as a function of cortical hierarchy. *Cereb Cortex.* 30(3):1068–1086.

Eradath MK, Pinsk MA, Kastner S (2021). A causal role for the pulvinar in coordinating task-independent cortico-cortical interactions. *J Comp Neurol.* 529(17):3772–3784.

Fiebelkorn IC, Kastner S (2019). A rhythmic theory of attention. *Trends Cogn Sci.* 23:87–101.

Fiebelkorn IC, Kastner S (2020). Functional specialization in the attention network. *Annu Rev Psychol.* 71(1):221–249.

Fiebelkorn IC, Pinsk MA, Kastner S (2018). A dynamic interplay within the frontoparietal network underlies rhythmic spatial attention. *Neuron.* 99(4):842–853.

Fiebelkorn IC, Pinsk MA, Kastner S (2019). The mediodorsal pulvinar coordinates the macaque fronto-parietal network during rhythmic spatial attention. *Nat Commun.* 10(215):1–15.

Fiebelkorn IC, Saalmann YB, Kastner S (2013). Rhythmic sampling within and between objects despite sustained attention at a cued location. *Curr Biol.* 23(24):2553–2558.

Fitzpatrick D, Usrey WM, Schofield BR, Einstein G (1994). The sublaminar organization of corticogeniculate neurons in layer 6 of macaque striate cortex. *Visual Neurosci.* 11:307–315.

Foxe J, Snyder A (2011). The role of alpha-band brain oscillations as a sensory suppression mechanism during selective attention. *Front Psychol.* 2:154.

Friedman-Hill SR, Robertson LC, Desimone R, Ungerleider LG (2003). Posterior parietal cortex and the filtering of distractors. Proc Natl Acad Sci U S A. 100(7):4263–4268.

Fries P (2015). Rhythms for cognition: communication through coherence. *Neuron.* 88(1):220–235.

Haegens S, Nácher V, Luna R, Romo R, Jensen O (2011). α-Oscillations in the monkey sensorimotor network influence discrimination performance by rhythmical inhibition of neuronal spiking. *Proc Natl Acad Sci U S A.* 108(48):19377–19382.

Helfrich RF, Fiebelkorn IC, Szczepanski SM, Lin JJ, Parvizi J, Knight RT, Kastner S (2018). Neural mechanisms of sustained attention are rhythmic. *Neuron.* 99(4):854–865.

Hembrook-Short JR, Mock VL, Usrey WM, Briggs F (2019). Attention enhances the efficacy of communication in V1 local circuits. *J Neurosci.* 39:1066–1076.

Homman-Ludiye J, Bourne JA (2019). The medial pulvinar: Function, origin and association with neurodevelopmental disorders. *J Anat.* 235(3):507–520.

Ichida JM, Mavity-Hudson JA, Casagrande VA (2014). Distinct patterns of corticogeniculate feedback to different layers of the lateral geniculate nucleus. *Eye Brain.* 2014(6 Suppl 1):57–73.

Jensen O, Mazaheri A (2010). Shaping functional architecture by oscillatory alpha activity: Gating by inhibition. *Front Hum Neurosci.* 4:186.

Kaas JH, Lyon DC (2007). Pulvinar contributions to the dorsal and ventral streams of visual processing in primates. *Brain Res Rev.* 55:285–296.

Lakatos P, O'Connell MN, Barczak A (2016). Pondering the pulvinar. *Neuron.* 89(1):5–7.

Landau AN, Fries P (2012). Attention samples stimuli rhythmically. *Curr Biol.* 22(11):1000–1004.

Livingstone MS, Hubel DH (1981). Effects of sleep and arousal on the processing of visual information in the cat. *Nature.* 291:554–561.

Luck SJ, Chelazzi L, Hillyard SA, Desimone R (1997). Neural mechanisms of spatial selective attention in areas V1, V2, and V4 of macaque visual cortex. *J Neurophysiol.* 77(1):24–42.

Marion RT (2013). *Exploring the Role of the Pulvinar-Cortical Interactions in Vision: A Tale of Maps, Loops and Gates.* PhD thesis, Vanderbilt University, Nashville, TN.

Markov NT, Vezoli J, Chameau P, Falchier A, Quilodran R, Huissoud C, Lamy C, Misery P, Giroud P, Ullman S, Baron P, Dehay C, Knoblauch K, Kennedy H (2014). Anatomy of hierarchy: Feedforward and feedback pathways in macaque visual cortex. *J Comp Neurol.* 522(1):225–259.

Maunsell JHR (2015). Neuronal mechanisms of visual attention. *Annu Rev Vis Sci.* 1:373–391.

McAdams CJ, Reid RC (2005). Attention modulates the responses of simple cells in monkey primary visual cortex. *J Neurosci.* 25:11023–11033.

McAlonan K, Brown VJ, Bowman EM (2000). Thalamic reticular nucleus activation reflects attentional gating during classical conditioning. *J Neurosci.* 20:8897–8901.

McAlonan K, Cavanaugh J, Wurtz RH (2006). Attentional modulation of thalamic reticular neurons. *J Neurosci.* 26:4444–4450.

McAlonan K, Cavanaugh J, Wurtz RH (2008). Guarding the gateway to cortex with attention in visual thalamus. *Nature.* 456: 391–394.

Mitchell JF, Sundberg KA, Reynolds JH (2009). Spatial attention decorrelates intrinsic activity fluctuations in macaque area V4. *Neuron.* 63(6):879–888.

Nobre K, Nobre A, Kastner S (2014). *The Oxford Handbook of Attention.* Oxford University Press.

O'Connor DH, Fukui MM, Pinsk MA, Kastner S (2002). Attention modulates responses in the human lateral geniculate nucleus. *Nat Neurosci.* 5(11):1203–1209.

Phillips JM, Fish LR, Kambi NA, Redinbaugh MJ, Mohanta S, Kecskemeti SR, Saalman YB (2019). Topographic organization of connections between prefrontal cortex and mediodorsal thalamus: Evidence for a general principle of indirect thalamic pathways between directly connected cortical areas. *NeuroImage.* 189: 832–846.

Pogosyan A, Gaynor LD, Eusebio A, Brown P (2009). Boosting cortical activity at beta-band frequencies slows movement in humans. *Curr Biol.* 19(19):1637–1641.

Purushothaman G, Marion R, Li K (2012). Gating and control of primary visual cortex by pulvinar. *Nat Neurosci.* 15(6):905–912.

Rafal RD, Posner MI (1987). Deficits in human visual spatial attention following thalamic lesions. *Proc Natl Acad Sci U S A.* 84(20):7349–7353.

Rockland KS (1998). Convergence and branching patterns of round, type 2 corticopulvinar axons. *J Comp Neurol.* 290(4):515–536.

Rockland KS, Andresen J, Cowie RJ, Robinson DL (1999). Single axon analysis of pulvinocortical connections to several visual areas in the macaque. *J Comp Neurol.* 406(2):221–250.

Romanski LM, Giguere M, Bates JF (1997). Topographic organization of medial pulvinar connections with the prefrontal cortex in the rhesus monkey. *J Comp Neurol.* 379(3):313–332.

Rovo Z, Ulbert I, Acsády L (2012). Drivers of the primate thalamus. *J Neurosci.* 32(49):17894–17908.

Saalmann YB, Pinsk MA, Wang L, Li X, Kastner S (2012). The pulvinar regulates information transmission between cortical areas based on attention demands. *Science.* 337:753–756.

Schneider KA, Kastner S (2009). Effects of sustained spatial attention in the human lateral geniculate nucleus and superior colliculus. *J Neurosci.* 29:1784–1795.

Sherman SM, Guillery RW (2013). *Functional Connections of Cortical Areas: A New View from the Thalamus.* MIT Press.

Sherman SM, Guillery RW (1998). On the actions that one nerve cell can have on another: Distinguishing "drivers" from "modulators." *Proc Natl Acad Sci U S A.* 95(12):7121–7126.

Sherman SM, Usrey WM (2021). Cortical control of behavior and attention from an evolutionary perspective. *Neuron.* 109:3048–3054.

Shipp S (2003). The functional logic of cortico-pulvinar connections. *Phil Trans R Soc B.* 358(1438):1605–1624.

Snow JC, Allen HA, Rafal RD, Humphreys GW (2009). Impaired attentional selection following lesions to human pulvinar: evidence for homology between human and monkey. *Proc Natl Acad Sci U S A.* 106(10):4054–4059.

Song K, Meng M, Chen L, Zhou K, Luo H (2014). Behavioral oscillations in attention: rhythmic α pulses mediated through θ band. *J Neurosci.* 34(14):4837–4844.

Stepniewska I, Qi HX, Kaas JH (2000). Projections of the superior colliculus to subdivisions of the inferior pulvinar in New World and Old World monkeys. *Vis Neurosci.* 17(4):529–549.

Steriade M (2004). Acetylcholine systems and rhythmic activities during the waking–sleep cycle. *Prog. Brain Res.* 145:179–196.

Theyel BB, Llano DA, Sherman SM (2010). The corticothalamocortical circuit drives higher-order cortex in the mouse. *Nat Neurosci.* 13(1):84–88.

Usrey WM, Alitto HJ (2015). Visual functions of the thalamus. *Annu Rev Vis Sci.* 1:351–371.

VanRullen R, Carlson T, Cavanagh P (2007). The blinking spotlight of attention. *Proc Natl Acad Sci U S A.* 104(49):19204–19209.

Warner CE, Kwan WC, Bourne JA (2012). The early maturation of visual cortical area MT is dependent on input from the retinorecipient medial portion of the inferior pulvinar. *J Neurosci.* 32(48):17073–17085.

Wilke M, Turchi J, Smith K, Mishkin M, Leopold DA (2010). Pulvinar inactivation disrupts selection of movement plans. *J Neurosci.* 30(25):8650–8659.

Wimmer RD, Schmitt LI, Davidson TJ, Nakajima M, Deisseroth K, Halassa MM (2015). Thalamic control of sensory selection in divided attention. *Nature.* 526:705–709.

Zhang Y, Chen Y, Bressler SL, Ding M (2008). Response preparation and inhibition: The role of the cortical sensorimotor beta rhythm. *Neuroscience.* 156(1):238–246.

Zhou H, Schafer RJ, Desimone R (2016). Pulvinar-cortex interactions in vision and attention. *Neuron.* 89:209–220.

46.

THE MEDIODORSAL THALAMUS AND DECISION-MAKING

Michael M. Halassa and Yuri B. Saalmann

A BRIEF OVERVIEW OF DECISION-MAKING RESEARCH IN NEUROSCIENCE

A decision is a commitment to a proposition or action plan on the basis of evidence for and value of associated outcomes. Given its centrality to human affairs, the study of decision-making is shared across many disciplines, including neuroscience, psychology, economics, and philosophy. In neuroscience, this endeavor is uniquely capable of revealing the neural mechanisms underlying decision components such as deliberation and commitment, which may generalize to the study of cognition writ large. In addition to providing a mechanistic window to cognition, decision neuroscience is poised to identify targets of intervention in neurological and psychiatric disorders with substantial cognitive components (Mukherjee and Halassa, 2022).

Over the last several decades, the field of decision neuroscience has occupied a number of distinct domains. Broadly speaking, these include the study of perceptual decision-making, value-based decision-making, social decision-making, and memory-guided decision-making. Each of these areas of inquiry have provided invaluable insight not only to the details of each decision-making mode, but also to the common principles of the decision process that transcend individual domains. For example, recent evidence suggests that the idea of sequential sampling in perceptual decisions (serial accumulation of sensory evidence) may be a neural strategy in mnemonic decision as well (Shadlen and Shohamy, 2016), with the "aha" moment associated with recalling a memory being analogous to a commitment to a proposition in the perceptual domain. Similarly, evidence accumulation of the value of different behavioral options, and of cost-benefit trade-offs, contribute to value-based decision-making (Shenhav et al., 2018).

As a general note, the overwhelming majority of decision neuroscience experiments have made some concessions regarding the complexity of the decision process under study. As such, each experiment involves designing a laboratory task that monitors a subject's choice behavior (out of 2 or more possible actions) in response to a form of instruction (e.g., sensory cue) which varies across individual trials and/or blocks of trials. Admittedly, such sacrifices on complexity always leave researchers with some concern about generalizability to real-world decisions. However, the gains made in understanding the neural substrates of these well-controlled cases have been remarkable (Gold and Shadlen, 2007) and are setting the foundation for interpreting neural data from more complex settings applicable to real-world decision-making (Yoo et al., 2021).

PERCEPTUAL DECISION-MAKING

Studying perceptual processes is a long and highly successful tradition in neuroscience. Vision has occupied a special place given its prominence in human perception (Trenholm and Krishnaswamy, 2020). Therefore, we have much insight into the neural mechanisms underlying early visual processing, attentional selection, and visual-guided decision-making.

The random dot motion task (RDM), pioneered by Britten, Movshon, Newsome, and colleagues, has been instrumental to tying decision processes to signal-detection theory and sequential-sampling theory (Gold and Shadlen, 2007). In its basic form, the task requires subjects (humans or macaque monkeys) to perform an action (e.g., left vs. right saccade) in response to moving dots with varying levels of motion coherence (Britten et al., 1992). Signal-detection theory was helpful in revealing that single neurons in visual area MT show neural responses that are highly similar to the subject's choice probability on single trials (Britten et al., 1992). Sequential-sampling theory revealed that area LIP, on the other hand, may be a readout area for area MT, with LIP neurons sequentially integrating bits of visual motion evidence to form a neural correlate of a decision variable. Interestingly, the neural correlates of evidence integration in LIP have given rise to a variety of computational models that have been applied to other neuroscience domains, such as the study of working memory (Mejías and Wang, 2022).

These studies reinforce the notion that the basic computational building blocks of such dynamic motifs may be general across associative cortical areas, including the prefrontal cortex (PFC).

Beyond the RDM, perceptual decision-making has had success in other visual tasks (Okazawa et al., 2021) and also beyond vision (Tsunada et al., 2019). For example, the work by Romo and colleagues on interval discrimination tasks has shown a role for the PFC in guiding the decision-making process in a modality-invariant manner (Romo et al., 1999). It has also shown that the PFC exhibits activity patterns that bridge the stimulus-free time that separates the 2 sensory inputs being compared, consistent with its role in freeing the subject from the immediacy of the sensorimotor context, allowing for deliberation and abstract knowledge to guide future actions (Friedman and Robbins, 2022).

VALUE- AND MEMORY-GUIDED DECISION-MAKING

Many decisions in real life are not about the veracity of corrupt sensory evidence (e.g., the presence/absence or identity of a sound source), but rather about choosing between 2 or more seemingly similar alternatives. Under these conditions, the subjective value drives choice, with areas of the frontal cortex engaging in various elements of the decision process (Lee et al., 2012). For example, this line of research has given rise to the notion that the orbitofrontal cortex harbors a map that encodes the subjective value of propositions and is essential in driving choice behavior in such contexts (Wilson et al., 2014).

Such value-based decisions are inexplicably linked to memory processing, as associative learning is essential to providing subjective values to different objects. Intriguingly, work by Ratcliff and colleagues on lexical decisions has shown that the speed-accuracy trade-off in the memory retrieval component of such tasks resembles bounded evidence accumulation in perceptual decisions. In fact, this work developed the drift-diffusion model of decision-making before it became popular in the perceptual domain (Ratcliff et al., 2016). This classical work, along with more recent studies revealing the role for the hippocampus and ventromedial prefrontal cortex in value-guided decision-making (Bakkour et al., 2019), hint at a number of potential general principles in decision-making: first, and as mentioned, that value-based decision-making is inexplicably linked to mnemonic processing; second, that the notion of sequential sampling of sensory evidence may be analogous to sequential sampling from memory in value-based decision-making, explaining reaction times and speed-accuracy trade-offs under such conditions (Biderman et al., 2020); finally, that the neural correlates underlying these processes involve basic computational primitives of assembling raw pieces of evidence, evidence accumulation, and appropriate readout.

LINKING DECISION-MAKING TO PREFRONTAL CORTICAL FUNCTION

The PFC has occupied a central role in the study of higher-level cognition (Miller and Cohen, 2001). Therefore, it is no surprise that it has also been extensively studied in the context of decision-making. There is evidence to link PFC function to both perceptual and value-based decision-making. However, the evidence indicates that distinct areas of the PFC engage in different components of these processes.

The PFC has extensive connectivity to both cortical and subcortical areas, including those involved in sensory processing. Therefore, areas of the PFC had been known to engage in perception by exerting executive control over sensory processing in extrastriate visual regions, as well as subcortical structures such as the sensory thalamus. Inactivation of PFC is known to eliminate the "top-down" bias in certain attention tasks (Rossi et al., 2007).
However, the precise engagement of PFC areas in the various stages of perceptual decision-making has only been recently explored through appropriately designed tasks. One study to highlight in this context is that by D'Esposito and colleagues (Rahnev et al., 2016), who used functional magnetic resonance imaging (fMRI) and transcranial magnetic stimulation in human subjects performing a visual decision-making task involving speed-accuracy trade-off. This study showed that an anterior-to-posterior gradient of areas in the PFC progressively control later stages of the perceptual decision-making process. Specifically, the dorsolateral PFC (dlPFC) is associated with stimulus judgment, the frontal eye field (FEF) is associated with motor preparation, and the anterior PFC (aPFC) is associated with post-feedback metacognitive judgment. These findings are consistent with neural recordings from macaque monkeys from dlPFC and FEF, and reinforce the notion that the dlPFC operates as a flexible input-output circuit for transforming sensory inputs into abstract rules. This work is also consistent with recent rodent work on the prelimbic cortex, the rodent analog of the primate dlPFC (Hoover and Vertes, 2007), which indicates similar neural machinery underlying the control of perceptual decision-making and attentional processing (Mukherjee et al., 2021).

In value-guided decision-making, the idea that distinct subregions of the PFC play specialized roles has also garnered ample experimental support. Specifically, these studies can be summarized in an influential 2-stream hypothesis put forth by Koechlin and colleagues (Domenech and Koechlin, 2015). This idea contrasts the role of the lateral stream of the PFC (which includes dlPFC) that is involved in mapping of cues onto rules when the dominant decision-making strategy is exploitative (in the broader exploration-exploitation trade-off), and that of the medial stream (including the orbitofrontal cortex, ventromedial PFC, and cingulate cortex) which is involved in mapping the value of objects and determining whether to switch away from exploitation to exploration based on deviation from the expected reward. This framework is powerful in interpreting a variety of results involving lesion studies and is also consistent with the idea of temporal gradients across prefrontal areas (Murray et al., 2014). More specifically, recent macaque experiments have shown that the anterior cingulate cortex, which has the longest temporal timescale across cortical areas, is involved in accumulating errors across trials in a macaque reversal task, and that the accumulated signal drives task switching (Sarafyazd and Jazayeri, 2019).

Given that the evidence supports the notion that individual PFC areas engage in distinct computations across decision-making domains, one key question is how these processes are optimally coordinated. A key candidate region for this coordination is the mediodorsal (MD) thalamus, given its reciprocal connectivity to all parts of the PFC. Below we review the evidence for this proposition and speculate on its role in decision-making and cognitive processing.

MEDIODORSAL THALAMUS

Individual PFC neurons can respond to multiple different task variables, a characteristic known as "mixed selectivity." At the neural population level, such mixed selectivity can be exploited to encode for relevant information across a wide range of tasks and situations (Rigotti et al., 2013). These high-dimensional representations in PFC have the advantage of allowing separation of similar inputs and relatively simple readout (Fusi et al., 2016), supporting a large repertoire of potential actions. There are also situations where lower-dimensional representations can be advantageous, supporting greater generalization in new circumstances and robustness to noise (Badre et al., 2021). When categorizing information, for instance, it is useful to identify the task-relevant features and discard other, irrelevant features. After performing this signal compression, the relevant features can be further processed to guide desired actions. Growing evidence suggests that MD plays a vital role in both signal compression and flexibly transmitting task-relevant information to guide action. The anatomical connectivity between the PFC and MD is characterized by convergent inputs from PFC to MD, which are well-suited to reduce dimensionality. Further, projections from MD to all hierarchical levels of PFC allow the transformed signals to be routed across PFC toward action pathways. We elaborate on these processes below.

GROSS ANATOMY

The MD is the largest medial thalamic nucleus, and it is extensively interconnected with the PFC (Figure 46.1A). Based on cytoarchitectonics, the primate MD is broadly divided into an anteromedial magnocellular zone (MDmc) with large evenly spaced cells, a central parvocellular zone (MDpc) with greater cell-size variability and dense myelination, and a lateral paralamellar zone that can be subdivided into an anterior multiform (MDmf) and caudal densocellular (MDdc) subzone (Olszewski, 1952; Walker, 1940). Although the rodent MD has more homogenous cytoarchitecture, there is a rough tripartite correspondence based on anatomical location. Rodent MD is commonly divided into a fiber-poor medial zone (MDm), a fiber-rich central zone (MDc), and a lateral zone (MDl) near the internal medullary lamina (Van Eden, 1986). Each of these 3 zones, in primates and rodents, has a particular pattern of connectivity with PFC and subcortical regions. Broadly speaking, in primates (macaques), MDmc preferentially connects with orbitofrontal cortex, MDpc connects with the dorsolateral convexity of PFC, and the lateral paralamellar zone of MD connects with the posterior arcuate cortex (Goldman-Rakic and Porrino, 1985; Siwek and Pandya, 1991). In rodents, there is also mapping between distinct divisions of the MD and PFC with a mediolateral thalamic gradient mapping onto a dorsoventral cortical one. Specifically, the MDm connects with vmPFC (including the infralimbic cortex), whereas the MDl connects with the dmPFC (including prelimbic and cingulate regions). The MDc, on the other hand, connects with the OFC (Mukherjee et al., 2021; Schmitt et al., 2017). There are molecular, cellular, and connectivity similarities and differences between the rodent and primate MD, which offer valuable insights into MD-PFC interactions and their contribution to decision-making.

CONVERGENT CONNECTIVITY AND SIGNAL COMPRESSION

MD receives major input from cortical layers 5 and 6 (Figure 46.1B). Approximately 20% of cortical inputs to MD originate in layer (L)5, and 80% originate in L6 (Xiao et al., 2009). These L5 and L6 corticothalamic cells provide 2 different modes of innervation of thalamocortical cells in MD. L5 corticothalamic cells sample their cortical inputs relatively widely due to their extensive basal dendrites (>2 mm in primates), but these cells show low convergence of their terminations in MD, usually less than 8 converging axons

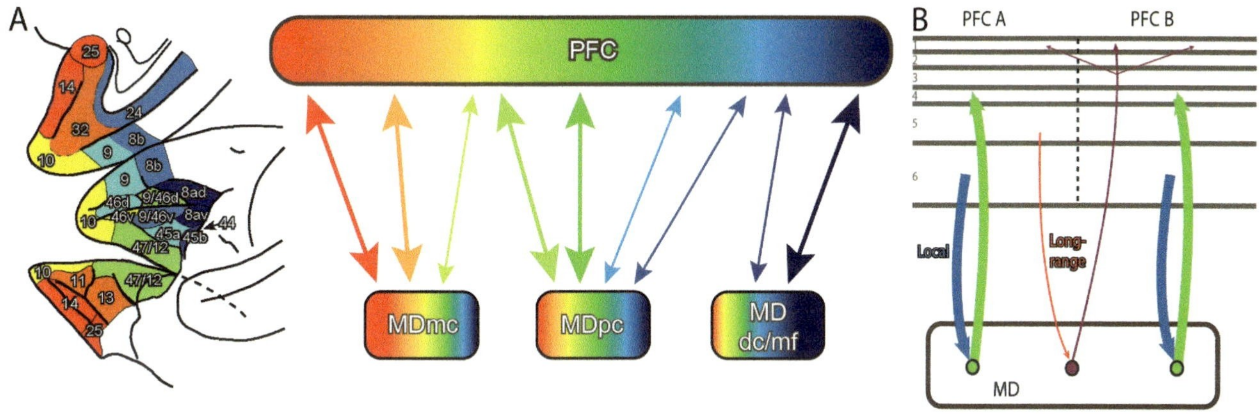

Figure 46.1. Anatomical connectivity between PFC and MD. (A). Left: ventromedial to posterolateral PFC areas (macaque) color-coded according to their topographic connections along a gradient (red-blue) from anteromedial (red) to posterolateral (blue) MD. Right: schematic of connections between PFC areas and MD zones. Color gradients (see left) approximate preferential PFC connections with MD zones. Arrow thickness approximates density of projections. (B). Blue arrow represents modulatory projection from PFC L6 to MD; and green arrow represents driving MD projection to the middle layers of the same PFC area. Together, they form a local, reciprocal cortico-thalamocortical pathway. In contrast, red arrow represents driving projection from PFC L5 to MD; and purple arrow represents modulatory MD projection to superficial cortical layers in a different PFC area. Together, they form a long-range, trans-thalamic cortico-thalamocortical pathway. Arrow thickness represents the proportion of projection density for corticothalamic and thalamocortical pathways: thinnest and thickest arrows approximately 20% and 80% projection density, respectively. Vertical dashed line represents border between 2 PFC areas. Adapted from Phillips et al. (2021).

(Rockland, 2019). L5 cells provide driving input, which have high conduction velocity, terminate in giant boutons, and act via ionotropic glutamate receptors (Rouiller and Welker, 2000). These driving inputs can form part of a triadic synaptic arrangement, in which they are presynaptic to the proximal dendrites of a thalamocortical cell and an interneuron, which inhibits the same thalamocortical cell. Thus, the thalamocortical cell is rapidly excited by the corticothalamic input, followed by inhibition. Because the balance between excitation and inhibition depends on the driving input (Steriade and Deschenes, 1984), the triad offers context-dependent gain control.

In contrast, L6 corticothalamic cells have more restricted sampling of their cortical inputs due to less extensive basal dendrites (e.g., <0.5 mm in primates), but these cells show high convergence of their terminations in MD (Rockland, 2019). L6 cells provide modulatory input, which terminate in smaller boutons, and act via AMPA, NMDA, and metabotropic glutamate receptors (Rouiller and Welker, 2000). These modulatory inputs synapse with the distal dendrites of thalamocortical cells and do not form triads. Overall, MD receives significant convergence of inputs from PFC, enabling signal integration and compression.

MD sensitivity to context, statistical regularity, and level of uncertainty seemingly reflect such dimensionality reduction. In mice trained to use different sets of cues to guide attention, MD cells were not cue-selective, instead responding to the statistical regularity of a cue set across trials. Convergent input from multiple cue-selective PFC cells appears to give rise to this context-sensitivity in MD (Rikhye et al., 2018). When the cues provided different levels of conflicting information, MD cells responded to the level of uncertainty associated with the cued information (Mukherjee et al., 2021). Such a summary statistic about the cues could once again be derived from convergent PFC input. Similar MD representations of uncertainty have been reported in humans. When categorizing stimuli which vary in how close they are to categorical boundaries, the categorization uncertainty modulated MD blood oxygenation level dependent (BOLD) activation (Grinband et al., 2006). This suggests that MD influences the evidence accumulation process in decision-making. In a dot motion task with varying numbers of relevant features, participants' choices were best captured by a drift-diffusion model in which the drift rate varied with uncertainty about feature relevance. MD BOLD activity was associated with the uncertainty-dependent changes in drift rate (Kosciessa et al., 2021). Taken together, such MD representations of context and level of uncertainty help guide decision-making under uncertain conditions.

FLEXIBLE ROUTING OF TASK-RELEVANT INFORMATION IN THE PFC HIERARCHY

The transformed signals in MD need to be routed across PFC toward action pathways, according to behavioral demands. Thalamocortical cells target specific layers in the cortex. A subpopulation of MD cells (containing the calcium-binding protein parvalbumin in MDmc and possibly calretinin in MDpc) (Phillips et al., 2021) influence the middle cortical layers (deep L3 and L4 in primates), and these MD cells provide driving input to PFC neurons

(Sherman and Guillery, 1998). Another subpopulation of MD cells (containing the calcium-binding protein calbindin) influence more superficial cortical layers (L1), and these MD cells provide modulatory input to PFC neurons (Jones, 1998; Sherman and Guillery, 1998). The thalamocortical axons in superficial layers can branch over substantial distances (Jones, 1998), thereby modulating an extended cortical region.

The corticothalamic projections described above and the thalamocortical projections form extensive cortico-thalamo-cortical pathways. These pathways are either reciprocal, originating and terminating in the same cortical site, or trans-thalamic/unreciprocated, originating in one cortical site and terminating in another. L6 corticothalamic cells and thalamocortical cells targeting middle layers constitute reciprocal pathways, whereas L5 corticothalamic cells and thalamocortical cells targeting superficial layers constitute trans-thalamic pathways (McFarland and Haber, 2002). The reciprocal pathway seems well suited to supporting more local working memory processes (Bolkan et al., 2017; Schmitt et al., 2017), and the trans-thalamic pathway suited to coordinating processes across PFC (Saalmann, 2014).

Functional MRI and clinical lesions in humans suggest that PFC is hierarchically organized along the anterior-to-posterior axis of lateral PFC, based on the level of representational abstraction (Koechlin et al., 2003), although the precise details of this organization are debated (Badre and Nee, 2018). A similar organization may be present in macaques as well (Boschin et al., 2015). There is additional evidence for a hierarchical organization of orbital (Kringelbach and Rolls, 2004) and medial (Zarr and Brown, 2016) PFC. The PFC areas across the proposed hierarchies are connected by both cortico-cortical and cortico-thalamo-cortical pathways. Further, the topography of connections between MD and PFC is consistent with the hierarchical nature of information flow across PFC. Across the horizontal axis of MD, there is an ordered topographic gradient of PFC connectivity. Focusing on primates, ventromedial to posterolateral PFC areas are mapped onto anteromedial to posterolateral MD (Goldman-Rakic and Porrino, 1985; Siwek and Pandya, 1991). This gradient in MD is characterized by substantial overlap of connections with cortical areas which are nearby on the cortical surface and connected via cortico-cortical pathways (Phillips et al., 2019). Thus, the cortico-thalamo-cortical pathways appear to follow the proposed hierarchical organizations of functional pathways across PFC.

The cortico-thalamo-cortical pathways position MD to control functional connectivity across PFC. MD has been shown to regulate local connectivity among an ensemble of PFC neurons during working memory. MD enabled sequential firing of PFC cells, and optogenetic deactivation of MD disrupted this sequence and the maintenance of PFC task representations (Schmitt et al., 2017). MD regulation of functional connectivity between distal PFC areas could be supported by longer-range trans-thalamic pathways. In rats, there is also evidence for individual MD neurons projecting to multiple PFC areas (Kuramoto et al., 2017), which could coordinate distributed processing. Although simultaneous intracranial neural recordings from multiple PFC areas and MD in behaving subjects have not yet been published, another thalamic nucleus, the pulvinar, has been shown to control functional connectivity between different cortical areas (Saalmann et al., 2012). Because of similar connectivity principles, it suggests a similar role for MD. In traumatic brain injury patients, damaged MD projections to PFC resulted in simple cortical dynamics derived from electroencephalography (EEG), with only a limited number of cortical states manifesting (Mofakham et al., 2022). *In silico* modeling suggested that MD regulation of local- and long-range functional connectivity across PFC is required for more complex dynamics to emerge (Mofakham et al., 2021). In summary, the available evidence supports the view that MD influences cortical functional connectivity across different scales, to coordinate the distributed computations necessary for flexible decision-making.

CONTRIBUTIONS TO DIFFERENT DECISION-MAKING PROCESSES

Decision-making involves extracting context, working memory, accumulating evidence, valuation, and action selection. There are differential contributions of MD zones to these processes. MDmc contributes to valuation, including the formation of object-reward associations, and episodic-like memory, to guide action selection. MDmc lesions perturb devaluation following reward satiation, suggesting key MDmc roles in adjusting object value (Izquierdo and Murray, 2010) and reward-based action selection (Wicker et al., 2018). Converging inputs from the orbitofrontal cortex, amygdala, and basal forebrain give rise to such value monitoring in MDmc (Murray and Rudebeck, 2013). MDmc also plays a key role in processing new memories and learning new associations toward optimal decision-making, but seemingly not in their retention (Mitchell and Gaffan, 2008). MDmc lesions resulted in impaired learning of object-in-place discriminations, which probe episodic-like memory, but had little effect on information learned prior to the intervention (Izquierdo and Murray, 2010). Inputs from the medial temporal cortex and orbitofrontal cortex enable MDmc to integrate such object, reward, and response information (Mitchell and Chakraborty, 2013).

Evidence suggests that MDpc contributes to working memory and cognitive control more generally, including rule-guided action and behavioral inhibition. Lesions involving MDpc impaired working memory functions in humans and macaques (Zoppelt et al., 2003). In macaques performing spatial working-memory tasks, MDpc neurons showed cue, delay, and action-related activity. The delay period activity was directionally selective, spanned the delay duration, and was reduced on error trials (Fuster and Alexander, 1971, 1973; Takeda and Funahashi, 2002). In mice performing working-memory tasks, MDl sustained activity of ensembles of PFC neurons (Bolkan et al., 2017; Schmitt et al., 2017), and functional interactions between MD and PFC operated in the beta frequency range (13–30Hz). Moreover, optogenetic deactivation of MD impaired PFC delay activity and task performance (Bolkan et al., 2017; Schmitt et al., 2017). MD also can play a role in rule-guided action. MDpc receives basal ganglia input from the substantia nigra pars reticulata (SNr). The pathway from the SNr to MDpc to ventral principal sulcus has been reported to carry rule-related, go/no-go information to ventrolateral PFC (Tanibuchi et al., 2009). During an anti-saccade task, MDpc neurons showed selectivity for pro- and anti-saccades close to response onset (Kunimatsu and Tanaka, 2010). Because anti-saccades require withholding of the pro-saccade to the peripheral rule cue (in addition to saccade generation to its mirror position), it suggests that MDpc may also contribute to behavioral inhibition. The underlying mechanism may involve MD suppressing the irrelevant PFC representation (Rikhye et al., 2018) (Figure 46.2).

Although the task-relevant information is represented across the PFC-MD network, lateral PFC and MDpc make differential contributions. It has been reported that,

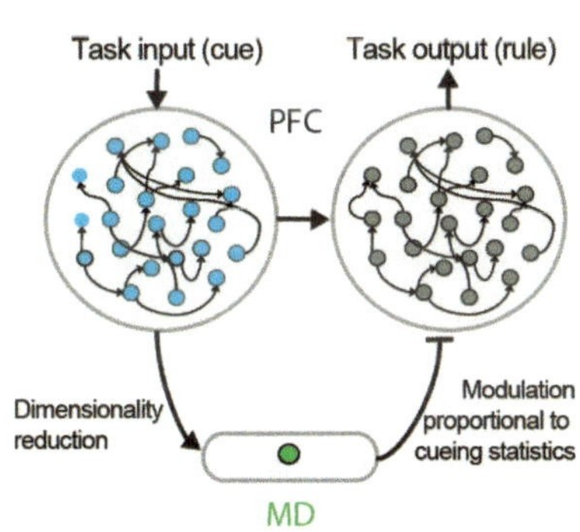

Figure 46.2. Putative interactions between PFC and MD. Cartoon summarizing a subset of interactions between the PFC and MD, highlighting 2 critical features: (1) Convergence of PFC inputs onto MD neurons resulting in a dimensionality reduction of task-relevant variables (and the generation of summary statistics representations such as context and sensory ambiguity). (2) MD output to PFC modulating circuit dynamics in a manner that is proportional to such statistical summaries, such as suppression of context-irrelevant representations (Rikhye et al., 2018) and slowing down accumulation of evidence dynamics when confronted with ambiguous sensory input (Mukherjee et al., 2021). The microcircuits depicted may be within a region or across PFC regions as shown in Figure 46.1.

during working memory tasks, PFC neurons carried more retrospective information about cue locations, while MD neurons carried more prospective information about saccade directions (Takeda and Funahashi, 2002; Watanabe et al., 2009). Similarly, during context processing measured using the dot pattern expectancy task, lateral PFC neurons carried more information about the environmental state, whereas MD neurons carried more information about the forthcoming action. Interaction between lateral PFC and MD was strongest around the time of decision-making (DeNicola et al., 2020). These results are consistent with a high-dimensional PFC representation of the environmental state, and a lower-dimensional MD representation to guide action.

CONCLUSIONS AND FUTURE DIRECTIONS

Intracranial electrophysiology, fMRI, EEG, and clinical lesions have highlighted the role of PFC in decision-making. The relatively fewer studies of MD have nonetheless shown vital and complementary MD roles, consistent with the extensive corticothalamic and thalamocortical anatomical connectivity. These MD roles include signaling context, statistical regularity of behavioral cues, and the level of uncertainty about relevant information, as well as working memory and prospective processes—all key components of decision-making. Mechanistically, this involves MD integrating and compressing the convergent signals from PFC—this dimensionality reduction affords generalization to novel situations and robustness to noise—then flexible transmission of the relevant information across frontal cortex to guide behavior.

Looking forward, simultaneous recordings from PFC and MD, as well as causal manipulations of PFC-MD circuits, are much needed to advance our understanding of the neural computations enabling decision-making. Such past experiments have helped shed light on interactions between MD and PFC (MDpc and lateral PFC in primates, and MDc/l and medial PFC in rodents) for working memory and evidence accumulation during trials of cognitive control tasks. Because primate lesion work suggests that MDmc is important for value and memory-guided decision-making, future simultaneous recordings from medial MD and connected PFC (e.g., orbitofrontal cortex and anterior cingulate cortex in primates) will help us understand the underlying thalamocortical mechanisms in these forms of decision-making, as well as how MD contributes to evidence accumulation over longer timescales (i.e., across trials).

Such work is a necessary step toward a better understanding of neuropsychiatric disorders and the development of new treatment strategies. Impaired decision-making and cognitive control more broadly characterize disorders such as schizophrenia and depression. Evidence suggests that perturbed MD processing and PFC-MD functional connectivity contribute to these disorders (Mukherjee and Halassa, 2022). Thus, PFC-MD circuits represent therapeutic targets for clinical interventions such as deep brain stimulation, transcranial magnetic stimulation, or focused ultrasonic stimulation (Mukherjee and Halassa, 2022; Saalmann et al., 2023).

REFERENCES

Badre D, Bhandari A, Keglovits H, Kikumoto A (2021). The dimensionality of neural representations for control. *Curr Opin Behav Sci.* 38:20–28.

Badre D, Nee DE (2018). Frontal cortex and the hierarchical control of behavior. *Trends Cogn Sci. 22*(2), 170–188.

Bakkour A, Palombo DJ, Zylberberg A, Kang YH, Reid A, Verfaellie M, Shadlen MN, Shohamy D (2019). The hippocampus supports deliberation during value-based decisions. *Elife.* 8:e46080. https://doi.org/10.7554/eLife.46080

Biderman N, Bakkour A, Shohamy D (2020). What are memories for? The hippocampus bridges past experience with future decisions. *Trends Cogn Sci.* 24(7):542–556.

Bolkan SS, Stujenske JM, Parnaudeau S, Spellman TJ, Rauffenbart C, Abbas AI, Harris AZ, Gordon JA, Kellendonk C (2017). Thalamic projections sustain prefrontal activity during working memory maintenance. *Nat Neurosci.* 20(7):987–996.

Boschin EA, Piekema C, Buckley MJ (2015). Essential functions of primate frontopolar cortex in cognition. *Proc Natl Acad Sci U S A.* 112(9):E1020–1027.

Britten KH, Shadlen MN, Newsome WT, Movshon JA (1992). The analysis of visual motion: A comparison of neuronal and psychophysical performance. *J Neurosci.* 12(12):4745–4765.

DeNicola AL, Park MY, Crowe DA, MacDonald AW, Chafee MV (2020). Differential roles of mediodorsal nucleus of the thalamus and prefrontal cortex in decision-making and state representation in a cognitive control task measuring deficits in schizophrenia. *J Neurosci.* 40(8):1650–1667.

Domenech P, Koechlin E (2015). Executive control and decision-making in the prefrontal cortex. *Curr Opin Behav Sci.* 1:101–106.

Friedman NP, Robbins TW (2022). The role of prefrontal cortex in cognitive control and executive function. *Neuropsychopharmacology.* 47(1):72–89.

Fusi S, Miller EK, Rigotti M (2016). Why neurons mix: High dimensionality for higher cognition. *Curr Opin Neurobiol.* 37:66–74.

Fuster JM, Alexander GE (1971). Neuron activity related to short-term memory. *Science.* 173(3997):652–654.

Fuster JM, Alexander GE (1973). Firing changes in cells of the nucleus medialis dorsalis associated with delayed response behavior. *Brain Res.* 61:79–91.

Gold JI, Shadlen MN (2007). The neural basis of decision making. *Annu Rev Neurosci.* 30:535–574.

Goldman-Rakic PS, Porrino LJ (1985). The primate mediodorsal (MD) nucleus and its projection to the frontal lobe. *J Comp Neurol. 242*(4):535–560.

Grinband J, Hirsch J, Ferrera VP (2006). A neural representation of categorization uncertainty in the human brain. *Neuron.* 49(5):757–763.

Hoover WB, Vertes RP (2007). Anatomical analysis of afferent projections to the medial prefrontal cortex in the rat. *Brain Struct Funct.* 212(2):149–179.

Izquierdo A, Murray EA (2010). Functional interaction of medial mediodorsal thalamic nucleus but not nucleus accumbens with amygdala and orbital prefrontal cortex is essential for adaptive response selection after reinforcer devaluation. *J Neurosci.* 30(2):661–669.

Jones EG (1998). A new view of specific and nonspecific thalamocortical connections. *Adv Neurol.* 77:49–71.

Koechlin E, Ody C, Kouneiher F (2003). The architecture of cognitive control in the human prefrontal cortex. *Science.* 302(5648):1181–1185.

Kosciessa JQ, Lindenberger U, Garrett DD (2021). Thalamocortical excitability modulation guides human perception under uncertainty. *Nat Commun.* 12(1):2430.

Kringelbach ML, Rolls ET (2004). The functional neuroanatomy of the human orbitofrontal cortex: evidence from neuroimaging and neuropsychology. *Prog Neurobiol.* 72(5):341–372.

Kunimatsu J, Tanaka M (2010). Roles of the primate motor thalamus in the generation of antisaccades. *J Neurosci.* 30(14):5108–5117.

Kuramoto E, Pan S, Furuta T, Tanaka YR, Iwai H, Yamanaka A, Ohno S, Kaneko T, Goto T, Hioki H (2017). Individual mediodorsal thalamic neurons project to multiple areas of the rat prefrontal cortex: A single neuron-tracing study using virus vectors. *J Comp Neurol.* 525(1):166–185.

Lee D, Seo H, Jung MW (2012). Neural basis of reinforcement learning and decision making. *Annu Rev Neurosci.* 35:287–308.

McFarland NR, Haber SN (2002). Thalamic relay nuclei of the basal ganglia form both reciprocal and nonreciprocal cortical connections, linking multiple frontal cortical areas. *J Neurosci.* 22(18):8117–8132.

Mejías JF, Wang XJ (2022). Mechanisms of distributed working memory in a large-scale network of macaque neocortex. *eLife.* 11:e72136. https://doi.org/10.7554/eLife.72136

Miller EK, Cohen JD (2001). An integrative theory of prefrontal cortex function. *Annu Rev Neurosci.* 24:167–202.

Mitchell AS, Chakraborty S (2013). What does the mediodorsal thalamus do? *Front Syst Neurosci.* 7:37.

Mitchell AS, Gaffan D (2008). The magnocellular mediodorsal thalamus is necessary for memory acquisition, but not retrieval. *J Neurosci.* 28(1):258–263.

Mofakham S, Fry A, Adachi J, Stefancin PL, Duong TQ, Saadon JR, Winans NJ, Sharma H, Feng G, Djuric PM, Mikell CB (2021). Electrocorticography reveals thalamic control of cortical dynamics following traumatic brain injury. *Commun Biol.* 4(1):1210.

Mofakham S, Liu Y, Hensley A, Saadon JR, Gammel T, Cosgrove ME, Adachi J, Mohammad S, Huang C, Djurić PM, Mikell C B. (2022). Injury to thalamocortical projections following traumatic brain injury results in attractor dynamics for cortical networks. *Prog Neurobiol.* 210:102215.

Mukherjee A, Halassa MM (2022). The associative thalamus: A switchboard for cortical operations and a promising target for schizophrenia. *Neuroscientist. 0*(0):10738584221112861. https://doi.org/10.1177/10738584221112861

Mukherjee A, Lam NH, Wimmer RD, Halassa MM (2021). Thalamic circuits for independent control of prefrontal signal and noise. *Nature.* 600(7887):100–104.

Murray EA, Rudebeck PH (2013). The drive to strive: goal generation based on current needs. *Front Neurosci.* 7:112.

Murray JD, Bernacchia A, Freedman DJ, Romo R, Wallis JD, Cai X, Padoa-Schioppa C, Pasternak P, Seo H, Lee D, Wang XJ (2014). A hierarchy of intrinsic timescales across primate cortex. *Nat Neurosci.* 17(12):1661–1663.

Okazawa G, Hatch CE, Mancoo A, Machens CK, Kiani R (2021). Representational geometry of perceptual decisions in the monkey parietal cortex. *Cell.* 184(14):3748–3761.

Olszewski J (1952). *The Thalamus of the Macaca Mulatta: An Atlas for Use with the Stereotaxic Instrument.* Karger.

Phillips JM, Fish LR, Kambi NA, Redinbaugh MJ, Mohanta S, Kecskemeti SR, Saalmann YB (2019). Topographic organization of connections between prefrontal cortex and mediodorsal thalamus: Evidence for a general principle of indirect thalamic pathways between directly connected cortical areas. *NeuroImage*. 189:832–846.
Phillips JM, Kambi NA, Redinbaugh MJ, Mohanta S, Saalmann YB (2021). Disentangling the influences of multiple thalamic nuclei on prefrontal cortex and cognitive control. *Neurosci Biobehav Rev*. 128:487–510.
Rahnev D, Nee DE, Riddle J, Larson AS, D'Esposito M (2016). Causal evidence for frontal cortex organization for perceptual decision making. *Proc Natl Acad Sci U S A*. 113(21):6059–6064.
Ratcliff R, Smith PL, Brown SD, McKoon G (2016). Diffusion decision model: Current issues and history. *Trends Cogn Sci*. 20(4):260–281.
Rigotti M, Barak O, Warden MR, Wang XJ, Daw ND, Miller EK, Fusi S (2013). The importance of mixed selectivity in complex cognitive tasks. *Nature*. 497(7451):585–590.
Rikhye RV, Gilra A, Halassa MM (2018). Thalamic regulation of switching between cortical representations enables cognitive flexibility. *Nat Neurosci*. 21(12):1753–1763.
Rockland KS (2019). Corticothalamic axon morphologies and network architecture. *Eur J Neurosci*. 49(8):969–977.
Romo R, Brody CD, Hernández A, Lemus L (1999). Neuronal correlates of parametric working memory in the prefrontal cortex. *Nature*. 399(6735):470–473.
Rossi AF, Bichot NP, Desimone R, Ungerleider LG (2007). Top down attentional deficits in macaques with lesions of lateral prefrontal cortex. *J Neurosci*. 27(42):11306–11314.
Rouiller EM, Welker E (2000). A comparative analysis of the morphology of corticothalamic projections in mammals. *Brain Res Bull*. 53(6):727–741.
Saalmann YB (2014). Intralaminar and medial thalamic influence on cortical synchrony, information transmission and cognition. *Front Syst Neurosci*. 8:83.
Saalmann YB, Mofakham S, Mikell CB, Djuric PM (2023). Microscale multicircuit brain stimulation: Achieving real-time brain state control for novel applications. *Curr Res Neurobiol*. 4:100071.
Saalmann YB, Pinsk MA, Wang L, Li X, Kastner S (2012). The pulvinar regulates information transmission between cortical areas based on attention demands. *Science*. 337(6095);753–756.
Sarafyazd M, Jazayeri M (2019). Hierarchical reasoning by neural circuits in the frontal cortex. *Science*. 364(6441):eaav8911. https://doi.org/10.1126/science.aav8911
Schmitt LI, Wimmer RD, Nakajima M, Happ M, Mofakham S, Halassa MM (2017). Thalamic amplification of cortical connectivity sustains attentional control. *Nature*. 545(7653):219–223.
Shadlen MN, Shohamy D (2016). Decision making and sequential sampling from memory. *Neuron*. 90(5):927–939.
Shenhav A, Straccia MA, Musslick S, Cohen JD, Botvinick MM (2018). Dissociable neural mechanisms track evidence accumulation for selection of attention versus action. *Nat Commun*. 9(1):2485.
Sherman SM, Guillery RW (1998). On the actions that one nerve cell can have on another: distinguishing "drivers" from "modulators." *Proc Natl Acad Sci U S A*. 95(12):7121–7126.
Siwek DF, Pandya DN (1991). Prefrontal projections to the mediodorsal nucleus of the thalamus in the rhesus monkey. *J Comp Neurol*. 312(4):509–524.
Steriade M, Deschenes M (1984). The thalamus as a neuronal oscillator. *Brain Res*. 320(1):1–63.
Takeda K, Funahashi S (2002). Prefrontal task-related activity representing visual cue location or saccade direction in spatial working memory tasks. *J Neurophysiol*. 87(1):567–588.
Tanibuchi I, Kitano H, Jinnai K (2009). Substantia nigra output to prefrontal cortex via thalamus in monkeys. II. Activity of thalamic relay neurons in delayed conditional go/no-go discrimination task. *J Neurophysiol*. 102(5):2946–2954.
Trenholm S, Krishnaswamy A (2020). An annotated journey through modern visual neuroscience. *J Neurosci*. 40(1):44–53.
Tsunada J, Cohen Y, Gold JI (2019). Post-decision processing in primate prefrontal cortex influences subsequent choices on an auditory decision-making task. *eLife*. 8:e46770. https://doi.org/10.7554/eLife.46770
Van Eden CG (1986). Development of connections between the mediodorsal nucleus of the thalamus and the prefrontal cortex in the rat. *J Comp Neurol*. 244(3):349–359.
Walker AE (1940). A cytoarchitectural study of the prefrontal area of the macaque monkey. *J Comp Neurol*. 73(1):59–86.
Watanabe Y, Takeda K, Funahashi S (2009). Population vector analysis of primate mediodorsal thalamic activity during oculomotor delayed-response performance. *Cereb Cortex*. 19(6):1313–1321.
Wicker E, Turchi J, Malkova L, Forcelli PA (2018). Mediodorsal thalamus is required for discrete phases of goal-directed behavior in macaques. *eLife*. 7:e37325. https://doi.org/10.7554/eLife.37325
Wilson RC, Takahashi YK, Schoenbaum G, Niv Y (2014). Orbitofrontal cortex as a cognitive map of task space. *Neuron*. 81(2):267–279.
Xiao D, Zikopoulos B, Barbas H (2009). Laminar and modular organization of prefrontal projections to multiple thalamic nuclei. *Neuroscience*. 161(4):1067–1081.
Yoo SBM, Hayden BY, Pearson JM (2021). Continuous decisions. *Philos Trans R Soc Lond B Biol Sci*. 376(1819):20190664.
Zarr N, Brown JW (2016). Hierarchical error representation in medial prefrontal cortex. *NeuroImage*. 124(Pt A):238–247.
Zoppelt D, Koch B, Schwarz M, Daum I (2003). Involvement of the mediodorsal thalamic nucleus in mediating recollection and familiarity. *Neuropsychologia*. 41(9):1160–1170.

47.

THALAMIC CONTRIBUTION TO CORTICAL NETWORK FUNCTIONS

Kai Hwang and James M. Shine

INTRODUCTION

The human brain expresses dynamic patterns of neural activity across spatial scales, transitioning from local evoked responses to large-scale, inter-regional network interactions. Localized neural activity and brain network interactions together support a broad range of perceptual, motor, and cognitive functions. For instance, in the visual domain, evoked responses in occipital regions encode elementary stimuli features and retinotopic maps (Wandell et al., 2007; Yacoub et al., 2008), whereases regions in the occipitotemporal cortices selectively encode different categories of visual stimuli (McCandliss et al., 2003; Kanwisher and Yovel, 2006). Beyond the perceptual domain, activity in premotor and motor cortices can instantiate complex motoric representations (Yokoi and Diedrichsen, 2019), and medial, lateral frontal, and posterior parietal regions can encode abstract information such as cognitive states (Schuck et al., 2016) and working memory content (D'Esposito and Postle, 2015). At the network level, integration processes dynamically couple different informational patterns to complete the perceptual-action cycle (Tononi et al., 1994), generating goal-directed behaviors that shape our environment. The need of network integration necessitates a specific modular organizational scheme that balances local processing and large-scale interactions (Sporns and Betzel, 2016). Accordingly, distributed brain regions have been found to form several large-scale cortical networks (Yeo et al., 2011), whose interactions have been shown to subserve a wide range of behavioral and cognitive functions (Cole et al., 2013; Shine et al., 2019a). Indeed, discovering how local and network-level brain network dynamics function is critical for elucidating the neural basis of human cognition.

Most studies that have investigated the structure and properties of human brain networks have focused on the cerebral cortex. However, there is compelling evidence from behavioral, anatomy, and connectivity studies indicating that many features of cortical network functions are in fact inextricably linked to the thalamus. This chapter reviews some of these findings. First, to establish the behavioral significance of the human thalamus, we review studies that demonstrated how the human thalamus is necessary for a diverse range of complex cognitive functions beyond the sensory and motor domains. Second, to explain how thalamic function may be related to cortical functions, we summarize the organization of thalamocortical connectivity. Third, to illustrate how the thalamocortical system can influence whole-brain network dynamics, we summarize the operational principles of thalamocortical interactions derived from animal models and human functional neuroimaging studies. In conclusion, we present a framework that describes how thalamocortical interactions implement different components of processes influencing brain network functions necessary for human cognition. Altogether, this chapter highlights that thalamocortical interaction is a critical substrate of human cognition.

NEUROPSYCHOLOGY STUDIES OF THALAMIC CONTRIBUTION TO HUMAN COGNITION

The thalamus is not a unified structure—instead it can be described by its various subdivisions or nuclei based on distinct anatomical and functional properties. One influential framework classifies thalamic nuclei into first-order and higher-order thalamic nuclei based on its extrinsic connectivity (Sherman and Guillery, 2006). First-order thalamic nuclei, such as the lateral geniculate nucleus (LGN), receive inputs from ascending sensory pathways or other subcortical brain regions, and have been extensively studied in both animal models and human neuroimaging studies as a model system for understanding thalamocortical circuits (O'Connor et al., 2002; Schneider et al., 2004; Bastos et al., 2014). In contrast, higher-order thalamic nuclei, for example the mediodorsal (MD) and pulvinar nucleus, receive inputs primarily from the cortex and take up most of the volume of the thalamus. Several of these nuclei have reciprocal connectivity with associative frontal and parietal association cortices (Goldman-Rakic and Porrino, 1985; Giguere and Goldman-Rakic, 1988; Selemon and Goldman-Rakic, 1988; Zikopoulos and Barbas, 2006), and their precise functions are not as well understood. Their connectivity profiles nevertheless suggest a potential contribution to complex cognitive functions that are traditionally thought to be mediated by associative cortical regions (i.e., working memory, language, and executive function; Goldman-Rakic, 1988; Fuster, 2015).

Brain functions are constrained by the connectivity patterns of their associated networks (Passingham et al., 2002), and given the known anatomical connections of the thalamus, the thalamus should be a critical substrate of cognitive circuitries. There is a long tradition of studying human brain-behavior relationship by examining the effects of focal brain lesions on neuropsychological outcomes (Lezak et al., 2012). Functional neuroimaging, while also compelling in mapping neural substrates for human cognition, is nevertheless correlational, and thus the lesion approach remains the gold standard for establishing whether a human brain region is *necessary* for complex cognitive behavior (Sutterer and Tranel, 2017). While human patients with focal lesions within the thalamus are relatively rare, accumulated case reports and recent experimental neuropsychology studies do provide naturalistic and strong causal evidence that has demonstrated the behavioral significance of the human thalamus.

Indeed, studies have demonstrated that focal lesions to the human thalamus are associated with symptoms and behavioral impairments across many functional domains, including pain (Kim et al., 2022), visual neglect (Watson and Heilman, 1979), distraction and attention deficits (Danziger et al., 1999; Arend et al., 2008; Snow et al., 2009), aphasia (Graff-Radford et al., 1984; Crosson et al., 1986; Hwang et al., 2021), amnesia (von Cramon et al., 1985; Graff-Radford et al., 1990; Danet et al., 2015), working memory deficits (Dagenbach et al., 2001; de Bourbon-Teles et al., 2014), and executive dysfunction (Van der Werf et al., 2003; Liebermann et al., 2013; Hwang et al., 2020). These deficits involved cognitive processes beyond basic sensorimotor processes, and in several studies careful control tests were carried out to isolate cognitive deficits independent of basic visuomotor processes (Hwang et al., 2020; Hwang et al., 2021).

Lesions to thalamic subregions with anatomical projections to specific cortical regions can produce behavioral impairments similar to lesions to the connected cortical regions. For example, thalamic executive dysfunction is associated with lesions to the mediodorsal and anterior thalamus, which has strong anatomical and functional connectivity to medial and lateral frontal regions (Hwang et al., 2020), where frontal lesions can produce similar executive function deficits (Szczepanski and Knight, 2014; Fellows, 2017). Similarly, attention-related deficits are associated with pulvinar lesions (Danziger et al., 1999; Arend et al., 2008; Snow et al., 2009), which projects to both parietal and occipitotemporal regions (Arcaro et al., 2018). This pattern suggests that thalamic lesions have distal effects that disrupt the interconnected thalamocortical circuit, impacting not only localized thalamic function but potentially distal cortical processes as well, a phenomenon known as *diaschsis* (Carrera and Tononi, 2014). Broadly speaking, cortical functions likely depend on thalamocortical inputs and interaction with the thalamus, a point that we will revisit later in this chapter.

Lesion findings provide further insight into how cognitive functions are organized in the thalamus (Figure 47.1A). Some lesion findings indicate the thalamic circuits have segregated and specialized functions (Figure 47.1A, left panel), because in some patients a focal lesion only had circumscribed impairment in a selective domain and spared other functions, also known as "single dissociation" (Hwang et al., 2020). This is consistent with the parallel and segregated thalamocortical circuits described in non-human primate anatomical studies (Alexander et al., 1986). Intriguingly, it appears that in some cases a very focal lesion can have broad behavioral impairments across functional domains (Hwang et al., 2021; Figure 47.1A, right panel). For example, a circumscribed lesion to the anterior-mediodorsal thalamus can produce behavioral impairments in language, memory, and executive function domains (Figure 47.1B). This finding suggests that, at least in some scenarios, thalamocortical circuits may have a convergence organization, where multiple systems converge onto a local thalamic region (Figure 47.1C); hence when the circuitry is disrupted, the behavioral impact is more diffuse and is not constrained to a specific cognitive task.

THE ORGANIZATION OF THALAMOCORTICAL CONNECTIVITY

The clinical human neuropsychology findings reviewed thus far suggest 2 organizational properties of thalamic function: segregation and convergence. How is the thalamocortical system organized to give rise to these properties? Compared to the lesion method, functional magnetic resonance imaging (fMRI) and diffusion imaging methods have better spatial resolution and unbiased coverage, making them well-suited to the investigation of thalamic functional organization. Diffusion imaging studies that tracked white matter pathways between the thalamus and the cerebral cortex have found that every cortical lobule has strong structural connectivity with a specific thalamic sub-region (Behrens et al., 2003; Johansen-Berg et al., 2004). This result suggests that

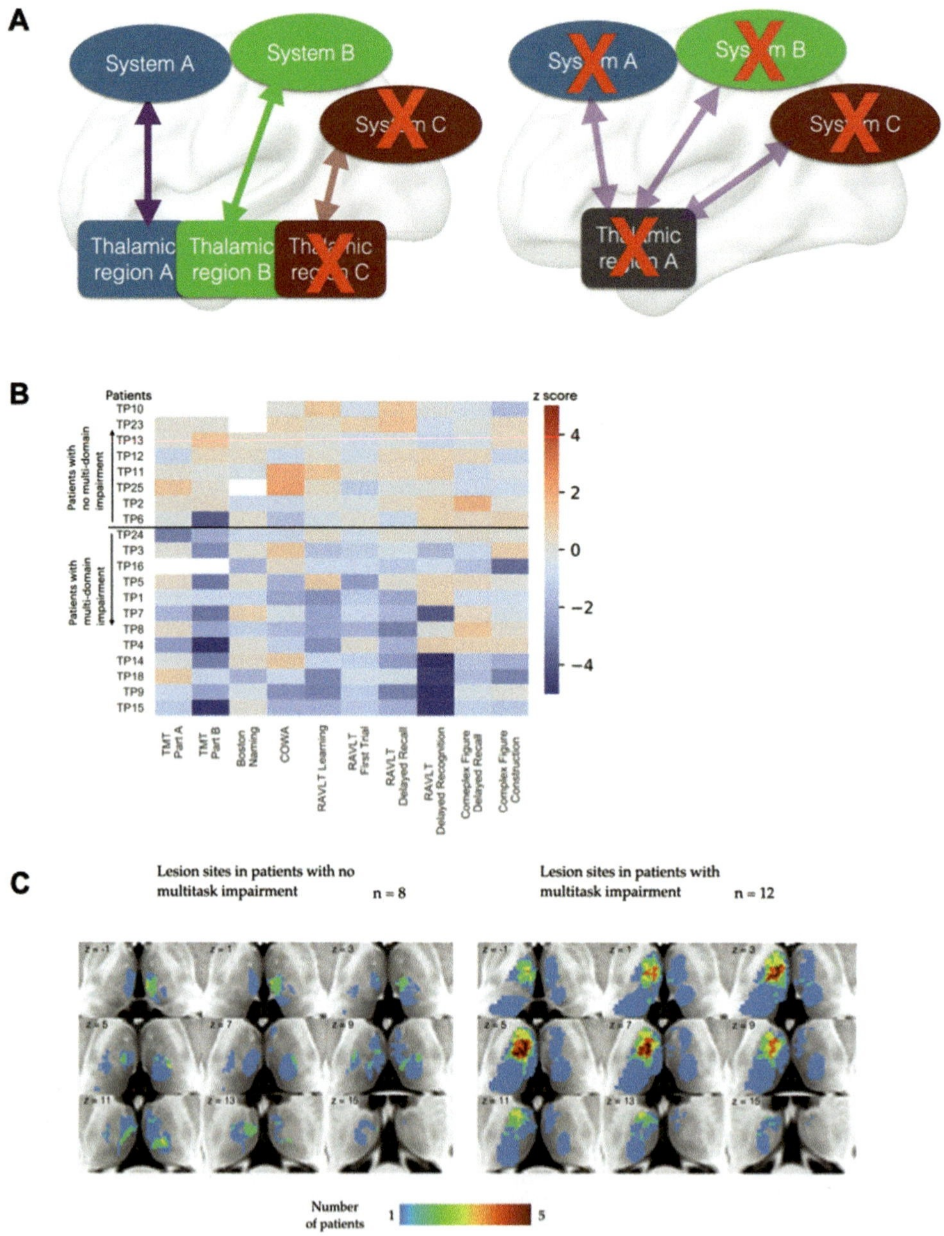

Figure 47.1. Neuropsychology studies of thalamic contribution to human cognition. (A). Schematic of organization of thalamic cognitive circuitries. The left panel depicts parallel thalamocortical circuitries, where a focal lesion can potentially disrupt selective cortical functions. The right panel depicts convergence thalamocortical circuitries, where a focal lesion may disrupt multiple cortical functions. (B). Neuropsychological test scores from patients with thalamic lesions, all test scores transformed to z-score using published population norm. TMT: = Trail Making Test; COWA = Controlled oral word association test; RAVLT = Rey Auditory Verbal Learning Test. (For detailed list of tests, see Hwang et al., 2021.) (C). Patients with a focal lesion in the left anterior-medial-dorsal thalamic showed significant impairment in multiple functional domains. Figure adapted from Hwang et al. (2021) with permission.

the human thalamus may have a complete representation of cortical systems that is interlinked via thalamocortical connectivity. Consistent with this notion, more detailed, systems-level organizations can be interrogated via studies of functional connectivity. Functional connectivity infers patterns of information communication between brain regions, with the assumption that regions that share similar functions will exhibit strong signal correlations among them (Buckner et al., 2013; Cole et al., 2016). One popular technique for estimating the functional connectivity is to measure correlations of spontaneous fluctuations in blood oxygenation level dependent (BOLD) signals between regions, also known as resting-state functional connectivity.

When studies combined resting-state functional connectivity with formal network analyses, key insights on the architecture and function of cortical brain networks

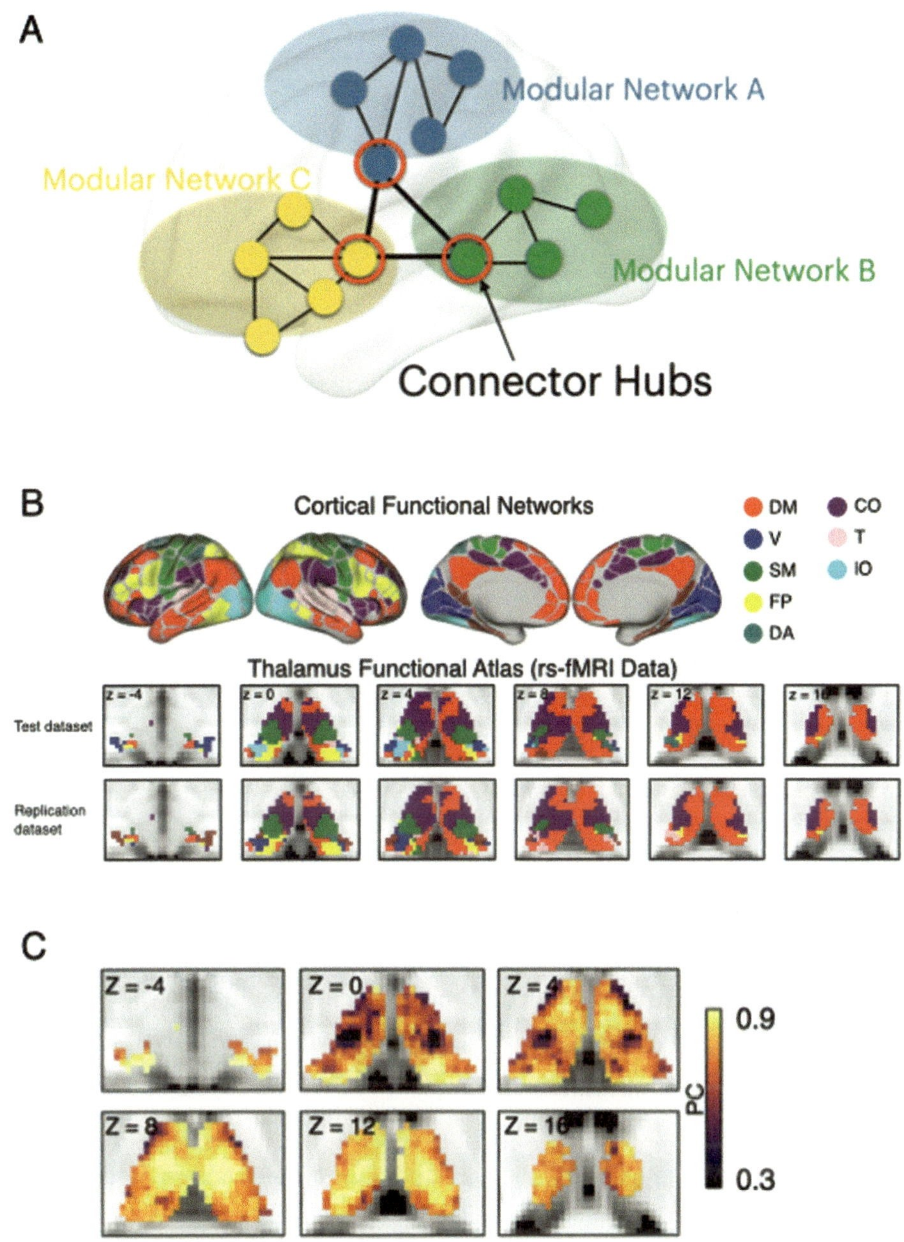

Figure 47.2. Network analysis of thalamocortical connectivity. (A). Brain networks can be represented as graphs, where brain regions are modeled as nodes, and functional connectivity among regions are modeled as edges. Nodes that share strong connectivity can be grouped as modular networks. Nodes with many between-network connections are connector hubs. (B). The thalamus contains a complete representation of cortical functional networks. Cortical functional networks: DM = default mode; V = visual; SM = somatomotor; FP = frontoparietal; DA = dorsal attention; CO = cingulo opercular; T = temporal; IO = lateral occipital. (C). Distribution of thalamic connector hubs. PC (participation coefficient) is a measure of connector hub property. Figure adapted from Hwang et al. (2017), with permission.

were revealed (Bullmore and Sporns, 2009; Figure 47.2A). First, a modular architecture organizes cortical networks: cortical regions within the same module have strong functional connectivity with other cortical regions in the same module, and weaker functional connectivity with regions outside the module. For example, sensory and motor cortices form visual, auditory, and somato-motor networks that are localized and confined to unimodal cortices (Yeo et al., 2011). In contrast, frontoparietal association cortices form several distributed networks including the dorsal-attention, ventral-attention, operculo-cingular, and frontal-parietal networks (Power et al., 2011; Yeo et al., 2011). This cortical network organization is consistent with how task-evoked activities are organized based on the putative cognitive processes engaged, where tasks that engage similar cognitive functions tend to engage brain regions that belong to the same modular network (Bertolero et al., 2015; Yeo et al., 2015).

To connect different modular systems, regions with strong functional connectivity with different networks can be defined as "connector hubs," with a hypothetical function to mediate interactions between functional networks. Studies have found that frontal and parietal cortices have a high proportion of connector hubs (Power et al., 2013; Bertolero et al., 2015). Activity in frontoparietal connector hub regions have been found to exhibit "task-flexible" properties, indicating that they are not tuned to specific cognitive functions but respond to many different cognitive tasks (Duncan, 2001; Cole et al., 2013; Yeo et al., 2015). Thus, hub regions may be important for integrative functions. This modular organization of segregated networks interconnected by hub regions can support both specialized and integrated brain network functions (Shine et al., 2016; Shine et al., 2019a).

Critically, this balanced segregation and integration organization has also been found in the human thalamus. The human thalamus contains a complete representation of cortical functional networks (Figure 47.2B), where every cortical functional network has been found to have a corresponding subdivision in the human thalamus with which it shares strong functional connectivity (Hwang et al., 2017; Greene et al., 2020). For example, the anterior, medial, and posterior subdivisions of the thalamus are most strongly associated with the cingulo-opercular, frontoparietal, and the default mode networks. This finding is consistent with the segregation organization reported in neuropsychology observations, indicating that there is parallel thalamocortical connectivity engaging distinct cortical functional networks.

However, more detailed analyses of thalamocortical function have also reported converging thalamocortical functional connectivity patterns, where *multiple* cortical networks simultaneously exhibited strong functional connectivity with targeted thalamic subregions (Figure 47.2C). These thalamic "connector hubs" were found to be strongest in the anterior, mediodorsal, and posterior part of the thalamus and were found to be even stronger than frontoparietal connector hubs (Hwang et al., 2017). While the precise functions of these thalamic connector hubs are not well understood, they could have similar functions and properties to cortical connector hubs, mediating interactions between cortical function networks (Hwang et al., 2017), and broadly responding to multiple cognitive tasks (Hwang et al., 2022). This converging thalamocortical functional connectivity may also explain some neuropsychology findings, where lesions to thalamic connector hubs may simultaneously disrupt multiple cortical systems, producing cross-domain behavioral impairments (Hwang et al., 2021).

THALAMOCORTICAL INTERACTIONS THAT SUPPORT LOCAL AND NETWORK-LEVEL BRAIN DYNAMICS

The preceding section described how thalamocortical connectivity reflects the organization of thalamic functions. In this section, we describe how thalamocortical interactions can influence brain network activity for human cognition. As described in the beginning of this chapter, brain networks express task-related, dynamic changes at both the local and network levels. There is compelling evidence suggesting that thalamocortical interactions can influence local cortical evoked responses, as well as the modular network organization that mediates interactions between distributed systems. These thalamocortical operations may in turn support the functionally specialized and integrated cognitive functions.

At the local level, thalamocortical projections have been shown to be capable of modulating evoked activity in localized cortical regions that are associated with several cognitive processes, including working memory and selective attention. For example, pharmacological or optogenetic inhibition of the pulvinar or the reticular nucleus during a visual selective attention task reduced attentional modulation of evoked responses recorded in the visual cortex (Wimmer et al., 2015; Zhou et al., 2016). Specifically, the attentional enhancement in evoked response amplitude for the attended target was reduced, suggesting that thalamocortical interactions are necessary for the amplification of task-relevant evoked responses. Parallel findings have been found for working memory as well. During working memory tasks, neural activity occurring after the presentation of a sensory input and before the contingent response (known as "delay activity") is thought to support the active maintenance of working memory content (Sreenivasan and D'Esposito, 2019). In rodents, inhibiting activity in ventrolateral nucleus or MD diminished delay activity in premotor (Guo et al., 2018) and prefrontal cortex (Bolkan et al., 2017). From human fMRI data, it has also been shown that task-evoked responses in the thalamus can predict cortex-wide activity patterns during working memory (Chen et al., 2022) and perceptual, motor, and language-related tasks (Hwang et al., 2022). These results indicate that cortical evoked activity is tightly associated with thalamic responses, and 2 more points can be summarized. First, task-related modulation of local evoked activity is at least partially dependent on thalamocortical inputs. Second, thalamocortical modulation of cortical function is likely organized by parallel thalamocortical circuitries (e.g., pulvinar-occipital, MD-prefrontal, VL-premotor), in which the specific cortical regions that require modulation can engage its corresponding thalamocortical system.

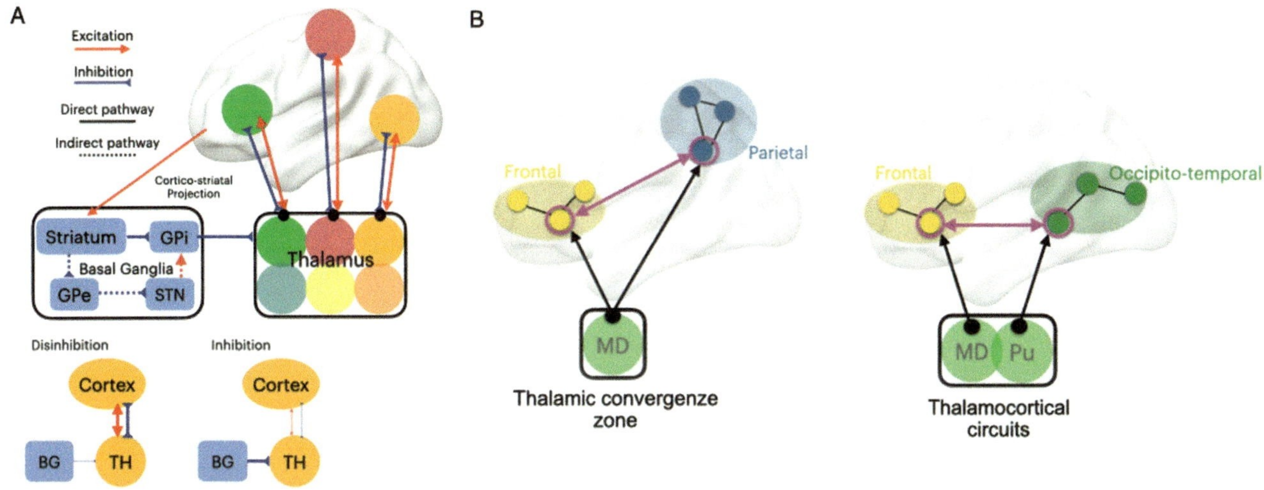

Figure 47.3. Model of thalamocortical influence on local and network level cortical processes. (A). Model of BG-thalamocortical influence on cortical responses. The striatum disinhibits or inhibits the thalamocortical system to increase or decrease cortical responses via the indirect and direct pathway of the basal ganglia circuitry. (B). Two hypothetical models of how the thalamus regulates inter-network interaction. (A). A single thalamic convergenze zone, in this example the mediodorsal thalamus (MD), can regulate cortico-cortical interaction between frontal and parietal regions that it projects to. (B). Multiple different thalamocortical circuits may simultaneously modulate distributed cortical neural activity to facilitate cortico-cortical connectivity. For example, both the mediodorsal nucleus and the pulvinar (Pu) may modulate lateral frontal and occipito-temporal regions, respectively. Figure adapted from Hwang and D'Esposito (2022), with permission.

Computational models have been developed to describe how the thalamocortical system can adaptively influence local evoked responses of different cortical regions (Figure 47.3A). Mechanistically, this process is hypothesized to be mediated by the striatal-thalamic circuit (Frank et al., 2001; Hazy et al., 2006). Specifically, the striatum can disinhibit or inhibit the thalamocortical system to increase or inhibit the evoked activity in cortical circuits. In addition to the basal ganglia, the thalamus also directly receives corticothalamic projections from the cortex (Theyel et al., 2010) as well as inhibitory outputs from the reticular thalamic nucleus (Wimmer et al., 2015) that may modulate its influence on cortical responses. Critically, this model may leverage the segregated, parallel organization of thalamocortical connectivity, to increase or decrease activity at targeted cortical regions, which may in turn influence the specialized local computation of different cortical systems.

At the network level, brain functions also depend on interaction between task-related brain regions. Though direct evidence from human subjects is lacking, findings from animal models suggest that the thalamus may influence cortico-cortical interaction via a cortico-thalamocortical connectivity motif (Guillery and Sherman, 2002). For example, the pulvinar is connected to both V4 and occipitotemporal cortices, and exhibited increased neural synchrony to modulate the level of synchronized oscillation between these regions (Kamiński et al., 2001; Saalmann et al., 2012). Beyond selective cortical regions, interactions between cortical systems are likely supported by a modular organization that balances local information processing and inter-regional communication (Sporns and Betzel, 2016). This organization is known to be influenced by the thalamocortical system, as lesions in the human thalamus have been shown to disrupt the modular organization of cortical networks (Hwang et al., 2017). Other studies have also shown that the modular structure of cortical networks is modulated by task demands, particularly during cognitively demanding tasks that require strong interaction between multiple systems (Cohen and D'Esposito, 2016; Shine et al., 2016). Critically, the temporal and dynamic changes in cortical modularity are tightly linked to similar temporal patterns in both medial and posterior thalamic groups (Shine et al., 2019b). These same thalamic regions also show coordinated fluctuations that align with selective cortical network regions (Müller et al., 2020).

Given the known thalamocortical connectivity organization that we have reviewed, the thalamus can influence cortical network interactions via 2 potential mechanisms (Figure 47.3B). One possibility is that a single thalamic convergence zone area could regulate cortico-cortical interaction between multiple cortical regions to which it projects. Alternatively, multiple segregated thalamocortical circuits may simultaneously modulate distributed cortical regions and impact their interactive patterns. For example, if the lateral prefrontal cortex needs to effectively communicate biasing signals to the early visual cortex through increases in neural interactions between these respective regions, both the mediodorsal nucleus and the pulvinar may modulate lateral frontal and occipital neural activity patterns, respectively. Together, these studies highlight the close correspondence between thalamic activity and the interaction patterns of cortical networks.

CONCLUSION

In this chapter, we have summarized several organization schemes of how thalamocortical interactions influence brain network dynamics that subserve cognitive functions. First, human neuropsychology studies demonstrated the necessity of the human thalamus for a variety of cognitive functions, as lesions to several thalamic subregions impair working memory, executive-, language-, and attention-related functions. The organization of thalamic functions and thalamocortical connectivity adheres to both segregation and convergence principles, as different thalamic subregions are strongly interacting with targeted cortical systems, but also demonstrate convergence properties that may be capable of influencing multiple systems simultaneously. Thalamocortical interaction can modulate the amplitude of local cortical evoked responses, to excite or inhibit cortical activity via dedicated circuitries, but also mediate large-scale, modular interactions between cortical systems. Through these operations, the thalamocortical interactions can influence complex brain network functions that instantiate cognitive processes. The diversity of the thalamocortical organization is therefore crucial for a myriad of adaptive human behaviors.

ACKNOWLEDGMENTS

KH is supported by National Institute of Mental Health grant R01MH122613. JMS is supported by National Health and Medical Research Council grant GNT1193857.

REFERENCES

Alexander GE, DeLong MR, Strick PL (1986). Parallel organization of functionally segregated circuits linking basal ganglia and cortex. *Annu Rev Neurosci.* 9:357–381.

Arcaro MJ, Pinsk MA, Chen J, Kastner S (2018). Organizing principles of pulvino-cortical functional coupling in humans. *Nat Commun.* 9:5382.

Arend I, Rafal R, Ward R (2008). Spatial and temporal deficits are regionally dissociable in patients with pulvinar lesions. *Brain.* 131:2140–2152.

Bastos AM, Briggs F, Alitto HJ, Mangun GR, Martin Usrey W (2014). Simultaneous recordings from the primary visual cortex and lateral geniculate nucleus reveal rhythmic interactions and a cortical source for gamma-band oscillations. *J Neurosci.* 34:7639–7644.

Behrens TEJ, Johansen-Berg H, Woolrich MW, Smith SM, Wheeler-Kingshott CAM, Boulby PA, Barker GJ, Sillery EL, Sheehan K, Ciccarelli O, Thompson AJ, Brady JM, Matthews PM (2003). Non-invasive mapping of connections between human thalamus and cortex using diffusion imaging. *Nat Neurosci.* 6:750–757.

Bertolero MA, Yeo BTT, D'Esposito M (2015). The modular and integrative functional architecture of the human brain. *Proc Natl Acad Sci U S A.* 112:E6798–807.

Bolkan SS, Stujenske JM, Parnaudeau S, Spellman TJ, Rauffenbart C, Abbas AI, Harris AZ, Gordon JA, Kellendonk C (2017). Thalamic projections sustain prefrontal activity during working memory maintenance. *Nat Neurosci.* 20:987–996.

Buckner RL, Krienen FM, Yeo BTT (2013). Opportunities and limitations of intrinsic functional connectivity MRI. *Nat Neurosci.* 16:832–837.

Bullmore E, Sporns O (2009). Complex brain networks: Graph theoretical analysis of structural and functional systems. *Nat Rev Neurosci.* 10:186–198.

Carrera E, Tononi G (2014). Diaschisis: Past, present, future. *Brain.* 137:2408–2422.

Chen X, Sorenson E, Hwang K (2022). Thalamocortical contributions to working memory processes during the n-back task. *Neurobiol Learn Mem.* 197:107701.

Cohen JR, D'Esposito M (2016). The segregation and integration of distinct brain networks and their relationship to cognition. *J Neurosci.* 36:12083–12094.

Cole MW, Ito T, Bassett DS, Schultz DH (2016). Activity flow over resting-state networks shapes cognitive task activations. *Nat Neurosci.* 19:1718–1726.

Cole MW, Reynolds JR, Power JD, Repovs G, Anticevic A, Braver TS (2013). Multi-task connectivity reveals flexible hubs for adaptive task control. *Nat Neurosci.* 16:1348–1355.

Crosson B, Parker JC, Kim AK, Warren RL, Kepes JJ, Tully R (1986). A case of thalamic aphasia with postmortem verification. *Brain Lang.* 29:301–314.

Dagenbach D, Kubat-Silman AK, Absher JR (2001). Human verbal working memory impairments associated with thalamic damage. *Int J Neurosci.* 111:67–87.

Danet L, Barbeau EJ, Eustache P, Planton M, Raposo N, Sibon I, Albucher JF, Bonneville F, Peran P, Pariente J (2015). Thalamic amnesia after infarct: The role of the mammillothalamic tract and mediodorsal nucleus. *Neurology.* 85(24):2107–2115.

Danziger S, Ward R, Owen V, Rafal R (1999). The effects of unilateral pulvinar damage in humans on reflexive orienting and filtering of irrelevant information. *Behav Neurol.* 13:95–104.

de Bourbon-Teles J, Bentley P, Koshino S, Shah K, Dutta A, Malhotra P, Egner T, Husain M, Soto D (2014). Thalamic control of human attention driven by memory and learning. *Curr Biol.* 24:993–999.

D'Esposito M, Postle BR (2015). The cognitive neuroscience of working memory. *Annu Rev Psychol.* 66:115–142.

Duncan J (2001). An adaptive coding model of neural function in prefrontal cortex. *Nat Rev Neurosci.* 2:820–829.

Fellows LK (2017). Cognitive control in the injured brain. In: Egner T, ed. *The Wiley Handbook of Cognitive Control.* John Wiley & Sons:513–538.

Frank MJ, Loughry B, O'Reilly RC (2001). Interactions between frontal cortex and basal ganglia in working memory: a computational model. *Cogn Affect Behav Neurosci.* 1:137–160.

Fuster JM (2015). *The Prefrontal Cortex.* Academic Press.

Giguere M, Goldman-Rakic PS (1988). Mediodorsal nucleus: Areal, laminar, and tangential distribution of afferents and efferents in the frontal lobe of rhesus monkeys. *J Comp Neurol.* 277:195–213.

Goldman-Rakic PS (1988). Topography of cognition: Parallel distributed networks in primate association cortex. *Annu Rev Neurosci.* 11:137–156.

Goldman-Rakic PS, Porrino LJ (1985). The primate mediodorsal (MD) nucleus and its projection to the frontal lobe. *J Comp Neurol.* 242:535–560.

Graff-Radford NR, Eslinger PJ, Damasio AR, Yamada T (1984). Nonhemorrhagic infarction of the thalamus: behavioral, anatomic, and physiologic correlates. *Neurology*. 34:14–23.

Graff-Radford NR, Tranel D, Van Hoesen GW, Brandt JP (1990). Diencephalic amnesia. *Brain*. 113(Pt 1):1–25.

Greene DJ, Marek S, Gordon EM, Siegel JS, Gratton C, Laumann TO, Gilmore AW, Berg JJ, Nguyen AL, Dierker D, Van AN (2020). Integrative and network-specific connectivity of the basal ganglia and thalamus defined in individuals. *Neuron*. 105:742–758.

Guillery RW, Sherman SM (2002). Thalamic relay functions and their role in corticocortical communication: generalizations from the visual system. *Neuron*. 33:163–175.

Guo KH, Yamawaki N, Svoboda K, Shepherd GM (2018). Anterolateral motor cortex connects with a medial subdivision of ventromedial thalamus through cell type-specific circuits, forming an excitatory thalamo-cortico-thalamic loop via layer 1 apical tuft dendrites of layer 5B pyramidal tract type neurons. *J Neurosci*. 38(41):8787–8797.

Hazy TE, Frank MJ, O'Reilly RC (2006). Banishing the homunculus: making working memory work. *Neuroscience*. 139:105–118.

Hwang K, Bertolero MA, Liu WB, D'esposito M (2017). The human thalamus is an integrative hub for functional brain networks. *J Neurosci*. 37:5594–5607.

Hwang K, Bruss J, Tranel D, Boes AD (2020). Network localization of executive function deficits in patients with focal thalamic lesions. *J Cogn Neurosci*. 32:2303–2319.

Hwang K, Shine JM, Bruss J, Tranel D, Boes A (2021). Neuropsychological evidence of multi-domain network hubs in the human thalamus. *eLife*. 10:e69480. http://dx.doi.org/10.7554/eLife.69480.

Hwang K, Shine JM, Cole MW, Sorenson E (2022). Thalamocortical contributions to cognitive task activity. *eLife*. 11:e81282.

Johansen-Berg H, Behrens TEJ, Sillery E, Ciccarelli O, Thompson AJ, Smith SM, Matthews PM (2004). Functional–anatomical validation and individual variation of diffusion tractography-based segmentation of the human thalamus. *Cereb Cortex*. 15:31–39.

Kamiński M, Ding M, Truccolo WA, Bressler SL (2001). Evaluating causal relations in neural systems: granger causality, directed transfer function and statistical assessment of significance. *Biol Cybern*. 85:145–157.

Kanwisher N, Yovel G (2006). The fusiform face area: a cortical region specialized for the perception of faces. *Philos Trans R Soc Lond B Biol Sci*. 361:2109–2128.

Kim NY, Taylor JJ, Kim YW, Borsook D, Joutsa J, Li J, Quesada C, Peyron R, Fox MD (2022). Network effects of brain lesions causing central poststroke pain. *Ann Neurol*. 92:834–845.

Lezak MD, Howieson DB, Bigler ED, Tranel D (2012). *Neuropsychological Assessment*. 5th ed. Oxford University Press, USA.

Liebermann D, Ploner CJ, Kraft A, Kopp UA, Ostendorf F (2013). A dysexecutive syndrome of the medial thalamus. Cortex. 49:40–49.

McCandliss BD, Cohen L, Dehaene S (2003). The visual word form area: expertise for reading in the fusiform gyrus. *Trends Cogn Sci*. 7:293–299.

Müller EJ, Munn B, Hearne LJ, Smith JB, Fulcher B, Arnatkevičiūtė A, Lurie DJ, Cocchi L, Shine JM (2020). Core and matrix thalamic sub-populations relate to spatio-temporal cortical connectivity gradients. *Neuroimage*. 222:117224.

O'Connor DH, Fukui MM, Pinsk MA, Kastner S (2002). Attention modulates responses in the human lateral geniculate nucleus. *Nat Neurosci*. 5:1203–1209.

Passingham RE, Stephan KE, Kötter R (2002). The anatomical basis of functional localization in the cortex. *Nat Rev Neurosci*. 3:606–616.

Power JD, Cohen AL, Nelson SM, Wig GS, Barnes KA, Church JA, Vogel AC, Laumann TO, Miezin FM, Schlaggar BL, Petersen SE (2011). Functional network organization of the human brain. *Neuron*. 72:665–678.

Power JD, Schlaggar BL, Lessov-Schlaggar CN, Petersen SE (2013). Evidence for hubs in human functional brain networks. *Neuron*. 79:798–813.

Saalmann YB, Pinsk MA, Wang L, Li X, Kastner S (2012). The pulvinar regulates information transmission between cortical areas based on attention demands. *Science*. 337:753–756.

Schneider KA, Richter MC, Kastner S (2004). Retinotopic organization and functional subdivisions of the human lateral geniculate nucleus: A high-resolution functional magnetic resonance imaging study. *J Neurosci*. 24:8975–8985.

Schuck NW, Cai MB, Wilson RC, Niv Y (2016). Human orbitofrontal cortex represents a cognitive map of state space. *Neuron*. 91:1402–1412.

Selemon LD, Goldman-Rakic PS (1988). Common cortical and subcortical targets of the dorsolateral prefrontal and posterior parietal cortices in the rhesus monkey: Evidence for a distributed neural network subserving spatially guided behavior. *J Neurosci*. 8:4049–4068.

Sherman SM, Guillery RW (2006). *Exploring the Thalamus and Its Role in Cortical Function*. 2nd ed. MIT Press;2:484.

Shine JM, Bissett PG, Bell PT, Koyejo O, Balsters JH, Gorgolewski KJ, Moodie CA, Poldrack RA (2016). The dynamics of functional brain networks: integrated network states during cognitive task performance. *Neuron*. 92:544–554.

Shine JM, Breakspear M, Bell PT, Ehgoetz Martens KA, Shine R, Koyejo O, Sporns O, Poldrack RA (2019a). Human cognition involves the dynamic integration of neural activity and neuromodulatory systems. *Nat Neurosci*. 22:289–296.

Shine JM, Hearne LJ, Breakspear M, Hwang K, Müller EJ, Sporns O, Poldrack RA, Mattingley JB, Cocchi L (2019b) The low-dimensional neural architecture of cognitive complexity is related to activity in medial thalamic nuclei. *Neuron*. 104:849–855.

Snow JC, Allen HA, Rafal RD, Humphreys GW (2009). Impaired attentional selection following lesions to human pulvinar: Evidence for homology between human and monkey. *Proc Natl Acad Sci U S A*. 106:4054–4059.

Sporns O, Betzel RF (2016). Modular brain networks. *Annu Rev Psychol*. 67:613–640.

Sreenivasan KK, D'Esposito M (2019). The what, where and how of delay activity. *Nat Rev Neurosci*. 20:466–481.

Sutterer MJ, Tranel D (2017). Neuropsychology and cognitive neuroscience in the fMRI era: A recapitulation of localizationist and connectionist views. *Neuropsychology*. 31:972–980.

Szczepanski SM, Knight RT (2014). Insights into human behavior from lesions to the prefrontal cortex. *Neuron*. 83:1002–1018.

Theyel BB, Llano DA, Sherman SM (2010). The corticothalamocortical circuit drives higher-order cortex in the mouse. *Nat Neurosci*. 13:84–88.

Tononi G, Sporns O, Edelman GM (1994). A measure for brain complexity: Relating functional segregation and integration in the nervous system. *Proc Natl Acad Sci U S A*. 91:5033–5037.

Van der Werf YD, Scheltens P, Lindeboom J, Witter MP, Uylings HBM, Jolles J (2003). Deficits of memory, executive functioning and attention following infarction in the thalamus: A study of 22 cases with localised lesions. *Neuropsychologia*. 41:1330–1344.

von Cramon DY, Hebel N, Schuri U (1985). A contribution to the anatomical basis of thalamic amnesia. *Brain*. 108(Pt 4):993–1008.

Wandell BA, Dumoulin SO, Brewer AA (2007). Visual field maps in human cortex. *Neuron*. 56:366–383.

Watson RT, Heilman KM (1979). Thalamic neglect. *Neurology*. 29:690–694.

Wimmer RD, Schmitt LI, Davidson TJ, Nakajima M, Deisseroth K, Halassa MM (2015). Thalamic control of sensory selection in divided attention. *Nature*. 526:705–709.

Yacoub E, Harel N, Ugurbil K (2008). High-field fMRI unveils orientation columns in humans. *Proc Natl Acad Sci U S A*. 105:10607–10612.

Yeo BTT, Krienen FM, Eickhoff SB, Yaakub SN, Fox PT, Buckner RL, Asplund CL, Chee MWL (2015). Functional specialization and flexibility in human association cortex. *Cereb Cortex*. 25:3654–3672.
Yeo BTT, Krienen FM, Sepulcre J, Sabuncu MR, Lashkari D, Hollinshead M, Roffman JL, Smoller JW, Zöllei L, Polimeni JR, Fischl B, Liu H, Buckner RL (2011). The organization of the human cerebral cortex estimated by intrinsic functional connectivity. *J Neurophysiol.* 106:1125–1165.
Yokoi A, Diedrichsen J (2019.) Neural organization of hierarchical motor sequence: Representations in the human neocortex. *Neuron*. 103:1178–1190.
Zhou H, Schafer RJ, Desimone R (2016). Pulvinar-cortex interactions in vision and attention. *Neuron*. 89:209–220.
Zikopoulos B, Barbas H (2006). Prefrontal projections to the thalamic reticular nucleus form a unique circuit for attentional mechanisms. *J Neurosci*. 26:7348–7361.

48.

THALAMIC BASES OF GOAL-DIRECTED BEHAVIORS

Mathieu Wolff

INTRODUCTION

Thalamus research has developed at a growing rate over the past few years and now increasingly implicates thalamic nuclei in cognitive functions, ranging from learning and memory to more executive abilities such as attentional control or decision-making (Wolff et al., 2015; Halassa and Kastner, 2017). While recent technical advances provide us with powerful tools to finely interfere with neural processes, accurately and unambiguously identifying the cognitive events associated with these neural processes remains largely dependent on how behavior is analyzed and interpreted (Balleine, 2019). A critical appraisal of the cognitive processes that are recruited by any behavioral task is thus fundamental. Studies in the field of experimental psychology provide a rich behavioral repertoire to assess the elementary psychological processes upon which complex cognitive events may be built. Several behavioral strategies for controlling ongoing actions or choices exist, and there is a major dichotomy between behaviors motivated by the onset of certain environmental stimuli and those directed toward a specific goal outcome (O'Doherty et al., 2017). The latter class is often referred to as *goal-directed*, to emphasize the intention to achieve a specific outcome (de Wit and Dickinson, 2009). Importantly, these goal-directed behaviors are characterized by a flexible behavioral output to adapt to any change that may occur in the environment. Relatively simple associative processes are currently thought to be largely conserved from rodent to nonhuman and human primate (Balleine and O'Doherty, 2010). Determining their neural bases has therefore considerable translational value. The aim of this chapter is to clarify the thalamic bases of adaptive decisions by focusing on several associative elementary processes linking actions, stimuli, and their outcomes, which all contribute to goal-directed behaviors.

GOAL-DIRECTED BEHAVIORS

The study of decision-making has been a popular pursuit across many scientific fields, and the increasing number of studies aimed at documenting their neural bases certainly calls for a better definition of the concepts that define the so-called goal-directed behaviors. These behaviors are meant to be highly reflective and flexible, as they are guided by explicit knowledge regarding the current value of different behavioral outcomes as well as the causal link between actions and their outcomes. Classically, this has been viewed as fulfilling both a *desire* and a *belief* criteria (de Wit and Dickinson, 2009; Turner and Parkes, 2020) (Figure 48.1A). But how are these concepts assessed in the laboratory?

Typically, goal-directed behaviors are assessed by proceeding first to an instrumental learning phase during which animals learn that certain actions are associated with specific outcomes (Figure 48.1B). For instance, pressing a particular lever earns access to a grain-based pellet, while pressing an alternate lever earns a sucrose-based pellet; that is, animals can learn at this occasion specific action-outcome associations. After that initial phase, the *desire* and the *belief* criteria can be questioned in a relatively independent manner. For the former, manipulating current outcome value is an effective method to track if animals can represent the current value of the action associated with that outcome. For instance, reducing the incentive value of a specific outcome (i.e., outcome "devaluation") can be achieved by using a sensory-specific satiety procedure (Figure 48.1C). To this end, after initial instrumental training, animals are provided with only 1 of the 2 outcomes (grain or sugar pellets) and are free to consume it at will for a fixed duration (typically 1 hour). As a result, when given a choice between the 2 outcomes, the relative value of the recently consumed outcome should decrease. Immediately

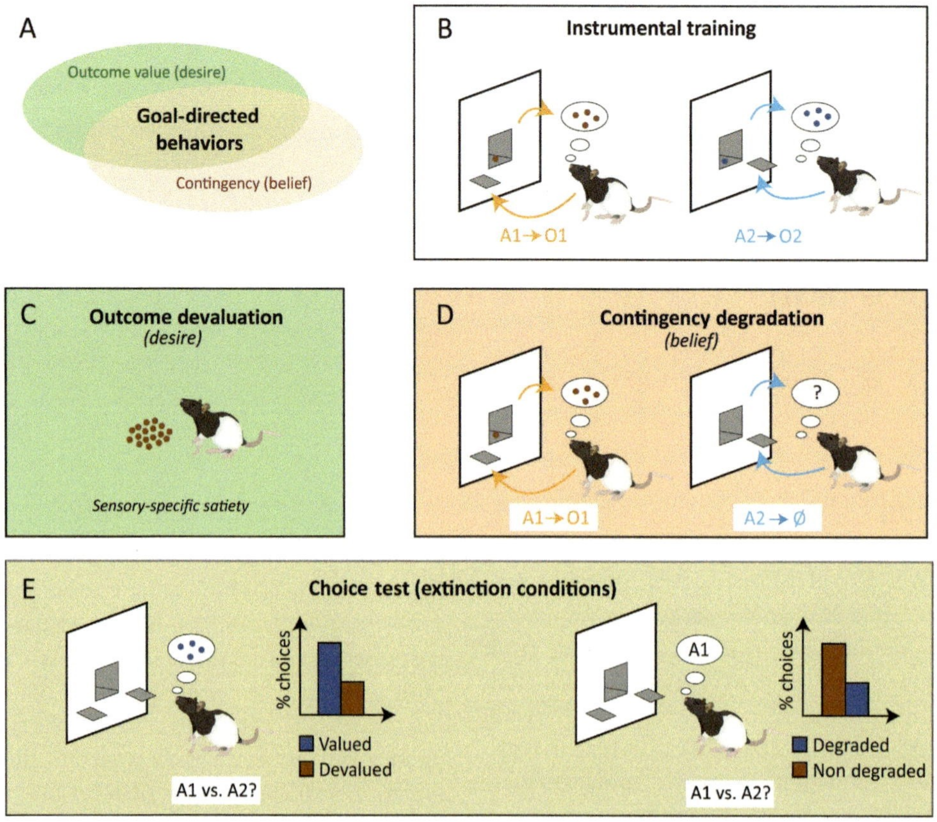

Figure 48.1. Goal-directed behaviors refer to behavior controlled by both current outcome value and current action-outcome contingency. (A). They can be assessed in a relatively simple manner in the behaving rodent by relying on instrumental learning: rats can learn that pressing 2 distinct levers will earn 2 distinct food outcomes, enabling them to form 2 distinct A1→O1 and A2→O2 associations. (B). From there it is possible to assess both the *desire* and the *belief* criteria by manipulating either current outcome value (C) or current action-outcome contingency (D). The outcome of these specific behavioral manipulations is best assessed during a choice test conducted under extinction conditions (E) to ensure that behavior can only be guided by current mental representation. Failure to express a bias for either the still valued outcome or for the non-degraded action is considered evidence for impaired goal-directed behavior if subsequent control assays are not impaired (e.g., a free consumption test for the devaluation outcome effect; see main text).

after outcome devaluation, animals are thus placed back in conditioning chambers with the 2 levers available for the choice test. Importantly, this test is conducted under extinction conditions (no outcomes are provided) to ensure that animal's choices are guided only by their representation of action-outcome associations and especially current outcome value (Figure 48.1E, left). During this test, animals behaving in a goal-directed manner would thus typically bias choice toward the action leading to the non-devalued outcome. Importantly, this procedure translates well to humans (Morris et al., 2018; Wang and Kahnt, 2021), thus providing a framework to identify the neural bases of adaptive decision-making based on current goal value.

An advantage of sensory-specific satiety is that this manipulation of goal value is reversible. When given supplemental instrumental training in the subsequent days, animals again respond in a similar way for both outcomes, opening the possibility to probe the *belief* criterion. This simply corresponds to the causal link between the action and the outcome: animals learn that performing a specific action is needed to access a specific outcome. This causal link can also be specifically manipulated through an instrumental degradation procedure, during which one of the outcomes is provided freely and it is no longer necessary to perform the action to receive this outcome (Figure 48.1D). This degrades the causal link between this specific action and its associated outcome. As a result, in a choice test conducted again under extinction conditions, animals would decrease their performance on this degraded lever and will instead bias their choice toward the action that is consistently needed to access to its associated outcome (Figure 48.1E, right). Taken together, these procedures allow one to pinpoint the nature of the associations that guide behavior and how they are supported by the neural circuits of interest. The next section will focus on studies that have relied on this framework.

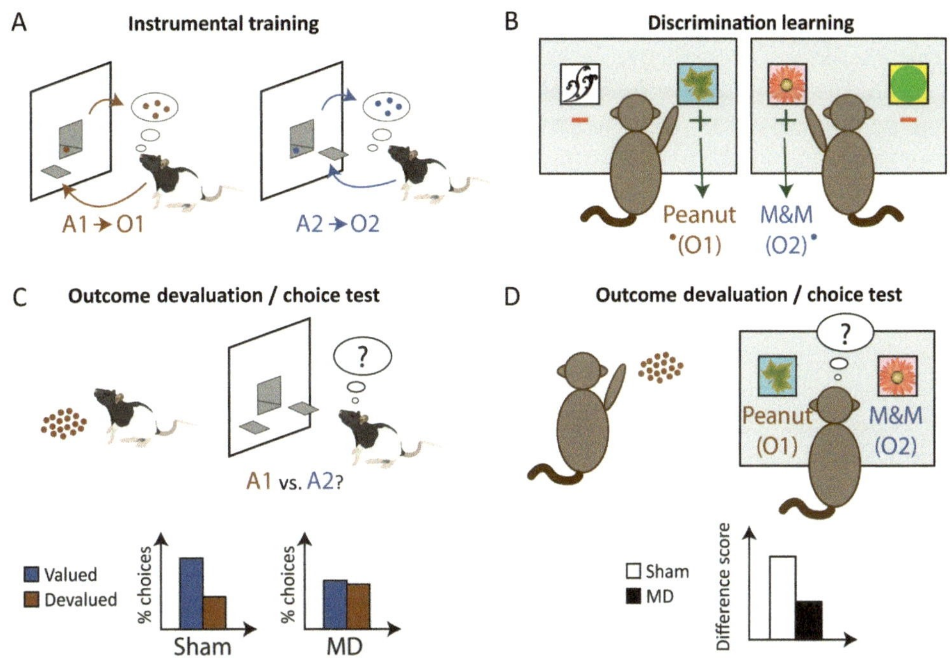

Figure 48.2. Outcome devaluation in rodent and nonhuman primate. By comparison to the rodent standard (A, see also Figure 48.1), an outcome devaluation assay suitable for monkey starts by a more complex discrimination learning phase during which animals learn by trials and errors which visual stimulus (or object) among a pair or 2 is reinforced by a food reward. Importantly, stimuli are reinforced with 2 distinct food rewards (i.e., some items are consistently reinforced with a peanut, some others with M&Ms). (B). Outcome devaluation is similar on both species, with one of the outcomes only being available to the animal (C, D, top). For the choice test conducted in monkey, the animals are now provided with a choice between 2 items previously rewarded so that only the current goal value should guide their choice (D). MD lesion or dysfunction typically reduce differential responding both in monkey (Mitchell et al., 2007; Wicker et al., 2018; Chakraborty et al., 2019) and in rodent (Corbit et al., 2003; Alcaraz et al., 2016b, 2018; Fisher et al., 2020), highlighting a key role for this region in goal-directed behaviors.

THALAMOCORTICAL CIRCUITS AND THE MEDIODORSAL THALAMUS

A large literature has documented the role of prefrontal areas in goal-directed behaviors (for recent reviews, see Coutureau and Parkes, 2018; Turner and Parkes, 2020). These prefrontal areas are densely interconnected with thalamic nuclei, which are now also recognized to critically support such behaviors (Wolff et al., 2015; Pergola et al., 2018; Wolff and Vann, 2019). In particular, the mediodorsal thalamus (MD) has emerged as one of the key thalamic regions due to its extensive reciprocal connections with both the medial prefrontal cortex (mPFC) and the orbitofrontal cortex (OFC; Alcaraz et al., 2016a; Murphy and Deutch, 2018).

ACTION SELECTION

Lesions studies have clearly pointed toward an involvement of the MD in supporting choices based on current goal value. While initial acquisition of an instrumental task is generally not affected by MD lesions, animals with MD lesions are often impaired in the ability to retrieve correct associations after selective outcome devaluation: they tend to generalize re-evaluation to both outcomes with reduced responding for both options (Corbit et al., 2003; Bradfield et al., 2013b; Alcaraz et al., 2016b), while normal animals typically bias their choice toward the action earning the non-devalued outcome. Reduction of overall motivation cannot account for this result, as responding is typically unaffected during rewarded tests (Corbit et al., 2003), as is consumption of the food reward (Corbit et al., 2003; Alcaraz et al., 2016b). Rather, this pattern of results points toward a deficit in the ability to selectively update action-outcome associations. Similar findings have been reported in primates, for which a slightly more complex design is generally used with an initial discrimination phase between arbitrarily selected visual stimuli before proceeding to the devaluation phase. While this initial phase appears to be totally spared by MD lesions (animals with MD lesions can discriminate between the visual stimuli), specific flexible responding after devaluation is clearly impaired (Mitchell et al., 2007; Baxter, 2013; Chakraborty et al., 2019). Thus, lesion data from both rodent and primate studies point toward an inability to respond in a selective manner after outcome devaluation (Figure 48.2). The devaluation procedure does not affect animals' motivation for the still valued outcome (reinforced tests show normal performance, e.g., Corbit et al., 2003; and free consumption assays show a preference for the still valued outcome, e.g., Alcaraz et al., 2018) showing that the deficit is associative in nature. Temporary interventions, which are less susceptible to concerns regarding the specificity of the deficits, have led to similar conclusions and consistently show that action selection is impaired after outcome devaluation (Parnaudeau et al., 2014; in primate: Wicker et al., 2018; in rodent: Fisher et al., 2020; Parnaudeau et al., 2014).

While the above-mentioned studies all focused on the MD as a single brain region, there is clearly interest in better identifying its specific contribution at the circuit level. There is indeed a large agreement in the literature that functional interactions between the MD and prefrontal areas are critical to support flexible responding, from rodent to primate (Browning et al., 2015; Alcaraz et al., 2018; Rikhye et al., 2018; Fresno et al., 2019). More precisely, these interactions are required for choices based on current goal values. Since the MD and prefrontal areas are reciprocally connected, identifying the respective functional contribution of thalamocortical versus corticothalamic pathways is of particular interest. Classic disconnection approaches with contralateral unilateral lesions of thalamic and prefrontal regions have proven to be valuable and to confirm that communication between the MD and the mPFC is necessary to support action selection after outcome devaluation (Bradfield et al., 2013b), but this method cannot capture the specific contribution of these pathways. Relatively few studies have actually attempted to explicitly address this question in the context of decision-making. The availability of viral strategies to select projection-defined neurons now makes it possible to specifically target either thalamocortical or corticothalamic pathways (Morceau et al., 2019). Chemogenetic inhibition of thalamocortical and corticothalamic pathways connecting the MD and the mPFC was found to differentially affect goal-directed behavior (Figure 48.3).

Inhibition of thalamocortical pathway produced a systematic impairment by interfering with both the ability to choose based on current outcome value (*desire*) or on current action-outcome contingency (*belief*). By contrast, inhibiting the thalamocortical pathway only affected choice based on current goal value (*desire*) without affecting choice based on current action-outcome contingency (*belief*) (Alcaraz et al., 2018). The 2 main attributes of goal-directed behavior (value and action-outcome contingency/*desire* and *belief*) thus appear to be differently

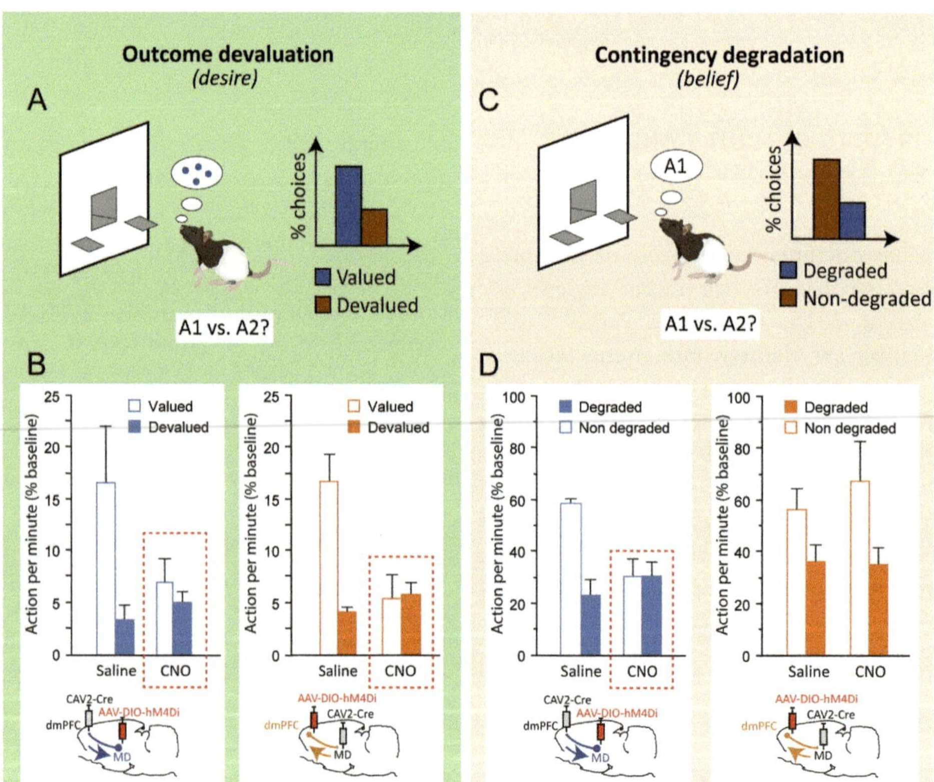

Figure 48.3. Thalamocortical and corticothalamic pathways connecting the mPFC and the MD diffrentially support goal-directed behaviors in the rat. Chemogenetic inhibition of either pathways disrupts outcome devaluation (A, B) but only chemogenetic inhibition of the thalamocortical pathway also removes the ability to adapt to action-outcome contingency degradation in extinction conditions (C, D). Projections-defined thalamic or cortical neurons were targeted by using a cre-dependent viral vector carrying an inhibitory DREADD in one region and the retrograde CAV2-Cre viral vector in its downstream target region. Dataset from Alcaraz et al. (2018).

supported by reciprocal connections between the mPFC and the MD, at least to some extent (Figure 48.3). This suggests a differential functional contribution of the reciprocal pathways connecting prefrontal and thalamic areas, which has also been reported when assessing attentional control (Schmitt et al., 2017) or working memory (Bolkan et al., 2017). The latter 2 studies also pointed toward a general role for thalamocortical pathways in exerting a sustainable influence on cortical activity, which has been confirmed in primates with MD activity reported to be critical for appropriate response selection (DeNicola et al., 2020). In general, thalamocortical projections are largely outnumbered by corticothalamic projections, with as many as 10 descending corticothalamic projections for 1 ascending thalamocortical afferent (Jones, 2002). In frontal systems, further work is required to have a better estimate of the respective proportion, but corticothalamic projections also appear to be abundant (Gabbott et al., 2005) and to dominate over thalamocortical projections (Haber and Calzavara, 2009; Alcaraz et al., 2018). Further work is now required to systematically clarify the contribution of corticothalamic versus thalamocortical pathways, not only at the behavioral but also at anatomical and physiological levels. Provisionally, it appears that the direction of information flows within thalamocortical circuits may constitute one of their distinctive functional features (Wolff and Vann, 2019).

CUE SELECTION

Environmental signals can heavily bias choice between different actions. As such, there is a clear interest in examining how thalamic nuclei may process contextual cues to attend to ongoing behavioral challenges. Classically, a rich body of data using Pavlovian paradigms showed that multiple thalamic nuclei, both first- and higher-order, are necessary to assign a predictive value to discrete cues (Wolff et al., 2021). Beyond Pavlovian conditioning per se, action selection is also influenced by environmental cues, and thalamic nuclei contribute to this influence at sensory and cognitive levels.

Pavlovian-to-instrumental transfer (PIT) provides a standardized and convenient method to examine the influence of environmental cues on action selection. PIT experiments comprise 3 stages. During the first stage, Pavlovian conditioning is used to establish an associative link between a predictive cue and its outcome, typically the delivery of a food reward. Next, the animal learns to press a lever to obtain the same reward (instrumental conditioning). Finally, during the critical PIT test, the predictive cue is now presented while the lever is available to assess the influence of this cue on action selection. Importantly, the lever and the predictive cues are presented together for the first time so that there is no prior direct associative link between them. But because both are associated with the same outcome, instrumental responses are increased when the Pavlovian predictive cue is present, thus defining the transfer of control from the Pavlovian cue to instrumental performance. This rich behavioral approach has been extensively used in the literature to disentangle the factors controlling choice (Holmes et al., 2010).

Using this task, MD lesions have been reported to prevent rats' ability to select the appropriate action based on Pavlovian cues (Ostlund and Balleine, 2008). Important control tests performed at this occasion demonstrated that MD damage spared the ability to choose between actions based on the value or discriminative stimulus properties of the reward, but impaired the expression of stimulus-outcome associations. Pavlovian contingency degradation is another possibility to question how current cue-outcome associations can guide behavior. In essence it is similar to the instrumental degradation task depicted in Figure 48.1, but with distinct cues instead of distinct levers. Classic and more recent evidence has consistently shown that MD lesions typically affect adaptive responding after updating of stimulus-outcome associations (Ostlund and Balleine, 2008; Morceau et al., 2022). This is also the case after lesion of the submedius thalamic nucleus, which is also reciprocally connected with the OFC, but not the mPFC (Alcaraz et al., 2015). Connections with the OFC rather than with the mPFC are thus likely to account for this thalamic contribution on guiding choice based on stimulus-outcome associations (Browning et al., 2015; Alcaraz et al., 2016a). This is consistent with the data indicating that interactions with the mPFC (Alcaraz et al., 2018) but not the OFC (Fresno et al., 2019) support flexible instrumental behaviors. The MD thus contributes to distinct thalamocortical circuits (i.e., including the mPFC or the OFC) depending on current behavioral needs, making this region a key integrative area. However, acknowledging the role of thalamic subcortical afferents may also be necessary to have a better understanding of thalamic functions at the circuit level (Barbas, 2015; Wolff and Vann, 2019). Afferents from the medial ventral pallidum to the MD appear to influence action selection. A temporary disconnection of the MD from this afferent produces a specific deficit during the PIT test, indicating that rats are no longer able to bias response selection as a function of the specific predictive information provided by the conditional stimulus (Leung and Balleine, 2015). This effect appeared to be specific, contrasting with a more general impairment in overall rate of responding that appeared when disconnecting the MD from its ventral tegmental area input. This cognitive deficit is highly reminiscent of that produced by bilateral MD lesions (Ostlund and Balleine, 2008). Altogether, this suggests that the MD may integrate information in widely distributed cortical-subcortical circuits to support flexible behaviors and attend to task-relevant parameters.

Identifying and switching between different rules thus appear as major assets for this thalamic region. An essential role for the MD in sustaining appropriate rule representation in prefrontal areas has actually been demonstrated in a 2-alternative forced choice task (2AFC) which requires mice to switch between 2 rules (attend to visual versus auditory cues) showing that this thalamic region can control functional cortical connectivity (Schmitt et al., 2017). Specifically, thalamic neurons did not directly encode current rules, but rather sustained these rules which were represented at the cortical level. Thalamic activity also become prominent when greater challenges are encountered, for example by rendering the environment more noisy (Mukherjee et al., 2021). Consistent with these highly integrative functions, MD neurons are contacting a large proportion of prefrontal parvalbumin interneurons (Delevich et al., 2015; Mukherjee et al., 2020) and are thus in an ideal position to finely tune the activity of prefrontal microcircuits. By comparison, prototypical first-order nuclei typically contact cortical pyramidal cells to elicit robust neuronal responses, leaving little room for fine modulatory regulation (Mukherjee et al., 2020). An important feature of higher-order nuclei is their cellular heterogeneity. While a subset of MD cells can sustain context-relevant prefrontal representations, other MD cells can actually suppress prefrontal representations that are currently irrelevant. Importantly, the latter are not erased and appear to be preserved for potential future use depending on the behavioral context (Rikhye et al., 2018). To summarize, MD activity as a whole appears to effectively promote cognitive flexibility, possibly due to the complementary functional contributions of distinct cell populations. Interestingly, both computational (Rikhye et al., 2018; Halassa and Sherman, 2019) and conceptual (Wolff et al., 2015; Wolff and Vann, 2019) views on thalamic functions suggest that the MD contributes to cortical functions by focusing attention on currently relevant task parameters, an "attention to learn" function that has also been suggested for other thalamic nuclei (Wright et al., 2015).

CORTICOSTRIATAL CIRCUITS AND THALAMOSTRIATAL PROJECTIONS

Early conceptions on the integration of the thalamus with other cortical and subcortical areas have highlighted this structure as essentially an output station within parallel cortico-basal ganglia-thalamic loops (Alexander and Crutcher, 1990), acting as the final stage before projecting back to the cortex. Previously believed to primarily control motor functions, a more general role for basal ganglia in motivation and cognition has now been acknowledged (Haber and Calzavara, 2009). According to this updated view, core aspects of goal-directed actions and habits are governed by the striatum (Balleine and O'Doherty, 2010).

While the thalamus appears to be the main recipient of inhibitory projections from the basal ganglia before reaching back to cortical areas, it is important to note that some thalamic inputs also directly innervate the striatum. Projections from the posterior intralaminar group, mostly the centromedian and the parafascicular nuclei (Pf), do provide substantial innervation to the dorsal striatum, and these inputs are now considered important for adaptive decision-making and action selection (Girard et al., 2020). Projections from the Pf are indeed particularly dense and well organized at anatomical and physiological levels (Mandelbaum et al., 2019) suggesting that they can contribute to highly specific functions.

ACTION CONTROL

Classic studies have established that the posterior dorsomedial striatum (DMS) is essential for the acquisition and expression of goal-directed actions (Yin et al., 2005; Corbit and Janak, 2010), while the dorsolateral striatum (DLS) appears to support habit formation (Yin et al., 2005; O'Hare et al., 2016). Interestingly, reward-prediction error signaling has also been found in the DMS (Stalnaker et al., 2012), consistent with its role in guiding value-based choices. Consistent with this, flexible responding, as assessed by place reversal, is impaired in a dose-dependent manner by pharmacological inhibition of the Pf (Brown et al., 2010), suggesting that Pf inputs to DMS support flexible responding. Recently, it has been shown that Pf afferents actually directly contact cholinergic interneurons (CINs) in the DMS (Bradfield et al., 2013a). Removal of this Pf-to-CINs pathway by asymmetrical thalamic and striatal lesions did not affect instrumental learning or choice based on current goal value, but rats were no longer able to maintain goal-directed responding after action-outcome reversal. This specific deficit was not apparent in rats sustaining ipsilateral lesions, showing that unilateral lesions alone cannot account for this impairment. Remarkably, this disconnection recapitulated the full extent of the deficit produced by complete removal of the Pf, and the deficit was also fully replicated by a more advanced approach that combined unilateral thalamic lesions with a pharmacological reduction of cholinergic activity in the pDMS in the opposite hemisphere (Bradfield et al., 2013a). This latter finding clearly indicates that the Pf-to-DMS pathway directly contacts CINs to support successful adaptation. This was further confirmed in a subsequent study showing that initial instrumental learning is not affected by thalamostriatal disconnection, but that the same manipulation prevents animals from responding in a goal-directed manner after action-outcome reversal (Bradfield and Balleine, 2017).

Besides the Pf, the posterior intralaminar thalamic group also includes the centromedian nucleus (CM), which contains many neurons that are activated when the reward obtained by performing a specific action is smaller than expected (Minamimoto et al., 2005), suggesting a role for this region in monitoring actual versus expected outcomes, a recurring theme when it comes to thalamic functions (Ullsperger et al., 2014) and an essential ability to maintain goal-directed responding. This activity is suggested to oppose a "reward-based pre-action bias" implemented by basal ganglia as a consequence of the reward history of a specific action (Minamimoto et al., 2009) and is expected to be critical for attention in unexpected situations. Relying on a go/no-go setup in the monkey, recent results suggest that the CM may indeed signal the discrepancy between internal pre-action bias and the experienced outcome, thus enabling the resetting of behavioral bias to engage in adaptive processes (Minamimoto et al., 2014).

Altogether, it thus seems relatively clear that the whole posterior intralaminar group provides a critical input to striatal activity at the service of adaptive, goal-directed behaviors. The contribution of this thalamostriatal input differs from that of the thalamocortical circuits discussed in the previous section in that initial goal-directed actions can be performed without this input. Instead, the thalamostriatal contribution becomes critical when action-outcome contingencies are modified after initial learning to force the updating of instrumental rules.

CUE SELECTION

The involvement of the posterior intralaminar group has also been examined in tasks during which visual cues govern action selection. The activity of thalamic cells in the CM-Pf complex has been shown to transiently vary when instruction cues or target stimuli are presented, and inhibiting this activity has a detrimental effect on reaction time in monkeys (Minamimoto and Kimura, 2002). In rodents, eliminating Pf neurons projecting to the dorsal striatum disrupts action selection in a visual cue-dependent discrimination task even if the intervention is conducted after acquisition of the task (Kato et al., 2011). Importantly, this manipulation does not impact motor skills, suggesting that thalamostriatal inputs may not only support flexible actions, but also convey critical environmental information. Relying on the same behavioral approach, the role of the centrolateral thalamic nucleus (CL), which is part of the rostral intralaminar group, has been more recently examined. Interestingly, removal or chemogenetic inhibition of striatum-projecting CL neurons has no effect on task acquisition. However, when behavioral flexibility was assessed either by reversals or by introducing attentional set-shifting, during which place rather than visual cues were now relevant, a clear deficit developed in experimental animals (Kato et al., 2018). Thus, again, this thalamic contribution appears to become prominent when elements of flexibility are needed to address current challenges.

Interestingly, thalamostriatal projections originating from the CL and the Pf differentially innervate the striatum; they both make synapses on medium spiny neurons, but the Pf targets the DLS whereas the CL targets the more medial portion of the dorsal striatum. This may account for the differential role of thalamostriatal projections at the behavioral level: Pf afferents critically support expression of stimulus-action associations (Kato et al., 2011), while CL afferents are essential for flexible responding (Kato et al., 2018). Together, these data suggest that thalamic inputs may provide essential information regarding environmental cues or task-relevant elements to assist basal ganglia in correct action selection. It also underscores the need to collect additional data at the anatomical and neurophysiological levels and to link this with behavioral data in order to have a clearer understanding on the role of these thalamostriatal inputs at the circuit level.

CONCLUSIONS AND PERSPECTIVES

A long-lasting conceptual effort to capture global thalamic functions by relying on behavioral data has led to the influential proposal that thalamic nuclei may bind together information processed in the frontal and the temporal lobes (Warrington and Weiskrantz, 1982; Wolff et al., 2015). The idea that the thalamus may connect distal brain areas to support cognition has been refined through the years and is supported by the fact that multiple mental pathologies are currently conceptualized as brain-wide connectivity disorders with thalamic dysfunction at the core (Wolff et al., 2015; Wolff and Vann, 2019). This approach complements the ongoing effort to document the functional organization of thalamocortical motifs (Halassa and Sherman, 2019). Both question the functional meaning of the main features of thalamocortical architecture, including the convergence, divergence, and reciprocity of projections between cortical areas and thalamic nuclei (Halassa and Sherman, 2019; Wolff and Vann, 2019). Having multiple accounts of thalamic functions is thus a strength, as a multidisciplinary approach is warranted to tackle a difficult problem. Thalamic nuclei appear to be important contributors to cortical functioning, and this contribution may explain their complex role at the behavioral level.

MD-based thalamocortical circuits thus appear to support acquisition and expression of goal-directed behaviors, while thalamostriatal inputs from the Pf may be required to update that knowledge (Bradfield et al., 2013b). The contribution of thalamic nuclei to decision-making can thus be conceived at multiple circuit levels that act synergistically depending on current behavioral needs. Of course, other thalamic nuclei add to this complexity and are open for other functional contribution. For example, the reticular thalamic nucleus may act as an effective thalamic searchlight (Crick, 1984) by inhibiting the activity of thalamic nuclei conveying currently irrelevant information (Pinault, 2004; Halassa and Acsády, 2016). Transient modulation of this thalamic region is thus observed when attention needs to be focused, from rodents to primates (McAlonan et al., 2006; Wimmer et al., 2015), showing that thalamic activity may therefore be the primary drive to capture environmental features that are currently important for coping with ongoing behavioral challenges. Interestingly, this attention-to-learn function has been proposed for multiple thalamic nuclei (Wolff et al., 2015; Wright et al., 2015), suggesting that it may be a fundamental feature of thalamic functions.

The behavioral data reviewed in this chapter clearly show that thalamic nuclei make a specific and important contribution to cognitive processes supporting decision-making. The cognitive processes considered here are indeed vital for making appropriate decisions, especially in dynamic environments wherein identifying and representing the currently relevant information are critical. The thalamus as a whole continuously supports the chains of processes at the sensory, cognitive, and motor levels that give rise to adaptive actions. Understanding more accurately the nature of the interactions with the cortical and the striatal stages is a clear challenge for the coming years. To address this, a multidisciplinary approach is necessary by joining specific functional interventions, carefully controlled behavioral assays, and computational approaches, possibly in multiple animal models, to primarily focus on highly conserved functional principles. However, as stressed in previous sections, choosing well-established behavioral paradigms is essential to pinpoint unambiguously the processes at play. In many respects, behavioral paradigms can be considered causal interventions that may constrain neural activity and should thus be considered with greatest care. In the framework, it is worthwhile to stress the importance of primate research to increase the translational value of thalamus research (Roberts and Clarke, 2019). Bringing together experimental psychology, advanced neuroscience techniques, and computational approaches appears as the only course to reveal the neural underpinnings of high-order cognition.

ACKNOWLEDGMENTS

I would like to thank Shauna Parkes, Etienne Coutureau, Mike Halassa, and Anna Mitchell for helpful discussions and constructive comments. Mathieu Wolff was supported by grants from the ANR THALACOR (ANR-22-CE37-0015-01) and THALAGRAM as well as the Region Nouvelle-Aquitaine and the FRC.

REFERENCES

Alcaraz F, Fresno V, Marchand AR, Kremer EJ, Coutureau E, Wolff M (2018). Thalamocortical and corticothalamic pathways differentially contribute to goal-directed behaviors in the rat. *eLife*. 7:e32517.

Alcaraz F, Marchand AR, Courtand G, Coutureau E, Wolff M (2016a). Parallel inputs from the mediodorsal thalamus to the prefrontal cortex in the rat. *Eur J Neurosci*. 44:1972–1986.

Alcaraz F, Marchand AR, Vidal E, Guillou A, Faugere A, Coutureau E, Wolff M (2015). Flexible use of predictive cues beyond the orbitofrontal cortex: Role of the submedius thalamic nucleus. *J Neurosci*. 35:13183–13193.

Alcaraz F, Naneix F, Desfosses E, Marchand AR, Wolff M, Coutureau E (2016b). Dissociable effects of anterior and mediodorsal thalamic lesions on spatial goal-directed behavior. *Brain Struct Funct*. 221:79–89.

Alexander GE, Crutcher MD (1990). Functional architecture of basal ganglia circuits: neural substrates of parallel processing. *Trends Neurosci*. 13:266–271.

Balleine BW (2019). The meaning of behavior: Discriminating reflex and volition in the brain. *Neuron*. 104:47–62.

Balleine BW, O'Doherty JP (2010). Human and rodent homologies in action control: Corticostriatal determinants of goal-directed and habitual action. *Neuropsychopharmacology*. 35:48–69.

Barbas H (2015). General cortical and special prefrontal connections: Principles from structure to function. *Annu Rev Neurosci*. 38:269–289.

Baxter MG (2013). Mediodorsal thalamus and cognition in non-human primates. *Front Syst Neurosci*. 7:38.

Bolkan SS, Stujenske JM, Parnaudeau S, Spellman TJ, Rauffenbart C, Abbas AI, Harris AZ, Gordon JA, Kellendonk C (2017). Thalamic projections sustain prefrontal activity during working memory maintenance. *Nat Neurosci*. 20:987–996.

Bradfield LA, Balleine BW (2017). Thalamic control of dorsomedial striatum regulates internal state to guide goal-directed action selection. *J Neurosci*. 37:3721–3733.

Bradfield LA, Bertran-Gonzalez J, Chieng B, Balleine BW (2013a). The thalamostriatal pathway and cholinergic control of goal-directed action: Interlacing new with existing learning in the striatum. *Neuron*. 79:153–166.

Bradfield LA, Hart G, Balleine BW (2013b). The role of the anterior, mediodorsal, and parafascicular thalamus in instrumental conditioning. *Front Syst Neurosci*. 7:51.

Brown HD, Baker PM, Ragozzino ME (2010). The parafascicular thalamic nucleus concomitantly influences behavioral flexibility and dorsomedial striatal acetylcholine output in rats. *J Neurosci*. 30:14390–14398.

Browning PG, Chakraborty S, Mitchell AS (2015). Evidence for mediodorsal thalamus and prefrontal cortex interactions during cognition in macaques. *Cereb Cortex*. 25:4519–4534.

Chakraborty S, Ouhaz Z, Mason S, Mitchell AS (2019). Macaque parvocellular mediodorsal thalamus: Dissociable contributions to learning and adaptive decision-making. *Eur J Neurosci*. 49:1041–1054.

Corbit LH, Janak PH (2010). Posterior dorsomedial striatum is critical for both selective instrumental and Pavlovian reward learning. *Eur J Neurosci*. 31:1312–1321.

Corbit LH, Muir JL, Balleine BW (2003). Lesions of mediodorsal thalamus and anterior thalamic nuclei produce dissociable effects on instrumental conditioning in rats. *Eur J Neurosci*. 18:1286–1294.

Coutureau E, Parkes SL (2018). Cortical determinants of goal-directed behavior. In: Morris R, Bornstein A, Shenhav A, eds. *Goal-Directed Decision Making*. Academic Press:179–197.

Crick F (1984). Function of the thalamic reticular complex: The searchlight hypothesis. *Proc Natl Acad Sci U S A*. 81:4586–4590.

de Wit S, Dickinson A (2009). Associative theories of goal-directed behaviour: A case for animal-human translational models. *Psychol Res.* 73:463–476.

Delevich K, Tucciarone J, Huang ZJ, Li B (2015). The mediodorsal thalamus drives feedforward inhibition in the anterior cingulate cortex via parvalbumin interneurons. *J Neurosci.* 35:5743–5753.

DeNicola AL, Park M-Y, Crowe DA, MacDonald AW, Chafee MV (2020). Differential roles of mediodorsal nucleus of the thalamus and prefrontal cortex in decision-making and state representation in a cognitive control task measuring deficits in schizophrenia. *J Neurosci.* 40:1650–1667.

Fisher H, Pajser A, Pickens CL (2020). Pre-training inactivation of basolateral amygdala and mediodorsal thalamus, but not orbitofrontal cortex or prelimbic cortex, impairs devaluation in a multiple-response/multiple-reinforcer cued operant task. *Behav Brain Res.* 378:112159.

Fresno V, Parkes SL, Faugère A, Coutureau E, Wolff M (2019). A thalamocortical circuit for updating action-outcome associations. *eLife.* 8:e46187.

Gabbott PLA, Warner TA, Jays PRL, Salway P, Busby SJ (2005). Prefrontal cortex in the rat: Projections to subcortical autonomic, motor, and limbic centers. *J Comp Neurol.* 492:145–177.

Girard B, Lienard J, Gutierrez CE, Delord B, Doya K (2021). A biologically constrained spiking neural network model of the primate basal ganglia with overlapping pathways exhibits action selection. *Eur J Neurosci.* 53(7):2254–2277.

Haber SN, Calzavara R (2009). The cortico-basal ganglia integrative network: The role of the thalamus. *Brain Res Bull.* 78:69–74.

Halassa MM, Acsády L (2016). Thalamic inhibition: Diverse sources, diverse scales. *Trends Neurosci.* 39:680–693.

Halassa MM, Kastner S (2017). Thalamic functions in distributed cognitive control. *Nat Neurosci.* 20:1669–1679.

Halassa MM, Sherman SM (2019). Thalamocortical circuit motifs: A general framework. *Neuron.* 103:762–770.

Holmes NM, Marchand AR, Coutureau E (2010). Pavlovian to instrumental transfer: A neurobehavioural perspective. *Neurosci Biobehav Rev.* 34:1277–1295.

Jones EG (2002). Thalamic organization and function after Cajal. *Prog Brain Res.* 136:333–357.

Kato S, Fukabori R, Nishizawa K, Okada K, Yoshioka N, Sugawara M, Maejima Y, Shimomura K, Okamoto M, Eifuku S, Kobayashi K (2018). Action selection and flexible switching controlled by the intralaminar thalamic neurons. *Cell Rep.* 22:2370–2382.

Kato S, Kuramochi M, Kobayashi K, Fukabori R, Okada K, Uchigashima M, Watanabe M, Tsutsui Y, Kobayashi K (2011). Selective neural pathway targeting reveals key roles of thalamostriatal projection in the control of visual discrimination. *J Neurosci.* 31:17169–17179.

Leung BK, Balleine BW (2015). Ventral pallidal projections to mediodorsal thalamus and ventral tegmental area play distinct roles in outcome-specific Pavlovian-instrumental transfer. *J Neurosci.* 35:4953–4964.

Mandelbaum G, Taranda J, Haynes TM, Hochbaum DR, Huang KW, Hyun M, Umadevi Venkataraju K, Straub C, Wang W, Robertson K, Osten P, Sabatini BL (2019). Distinct cortical-thalamic-striatal circuits through the parafascicular nucleus. *Neuron.* 102:636–652.

McAlonan K, Cavanaugh J, Wurtz RH (2006). Attentional modulation of thalamic reticular neurons. *J Neurosci.* 26:4444–4450.

Minamimoto T, Hori Y, Kimura M (2005). Complementary process to response bias in the centromedian nucleus of the thalamus. *Science.* 308:1798–1801.

Minamimoto T, Hori Y, Kimura M (2009). Roles of the thalamic CM-PF complex-basal ganglia circuit in externally driven rebias of action. *Brain Res Bull.* 78:75–79.

Minamimoto T, Hori Y, Yamanaka K, Kimura M (2014). Neural signal for counteracting pre-action bias in the centromedian thalamic nucleus. *Front Syst Neurosci.* 8. Accessed September 7, 2020. https://www.ncbi.nlm.nih.gov/pmc/articles/PMC3904122/.

Minamimoto T, Kimura M (2002). Participation of the thalamic CM-Pf complex in attentional orienting. *J Neurophysiol.* 87:3090–3101.

Mitchell AS, Browning PG, Baxter MG (2007). Neurotoxic lesions of the medial mediodorsal nucleus of the thalamus disrupt reinforcer devaluation effects in rhesus monkeys. *J Neurosci.* 27:11289–11295.

Morceau S, Faugère A, Coutureau E, Wolff M (2022). The mediodorsal thalamus supports adaptive responding based on stimulus-outcome associations. *Curr Res Neurobiol.* 3:100057.

Morceau S, Piquet R, Wolff M, Parkes SL (2019). Targeting reciprocally connected brain regions through CAV-2 mediated interventions. *Front Mol Neurosci.* 12:303.

Morris RW, Cyrzon C, Green MJ, Le Pelley ME, Balleine BW (2018). Impairments in action-outcome learning in schizophrenia. *Transl Psychiatry.* 8:54.

Mukherjee A, Bajwa N, Lam NH, Porrero C, Clasca F, Halassa MM (2020), Variation of connectivity across exemplar sensory and associative thalamocortical loops in the mouse. *eLife.* 9:e62554.

Mukherjee A, Lam NH, Wimmer RD, Halassa MM (2021). Thalamic circuits for independent control of prefrontal signal and noise. *Nature.* 600(7887):100–104.

Murphy MJM, Deutch AY (2018). Organization of afferents to the orbitofrontal cortex in the rat. *J Comp Neurol.* 526:1498–1526.

O'Doherty JP, Cockburn J, Pauli WM (2017). Learning, reward, and decision making. *Annu Rev Psychol.* 68:73–100.

O'Hare JK, Ade KK, Sukharnikova T, Van Hooser SD, Palmeri ML, Yin HH, Calakos N (2016). Pathway-specific striatal substrates for habitual behavior. *Neuron.* 89:472–479.

Ostlund SB, Balleine BW (2008). Differential involvement of the basolateral amygdala and mediodorsal thalamus in instrumental action selection. *J Neurosci.* 28:4398–4405.

Parnaudeau S, Taylor K, Bolkan SS, Ward RD, Balsam PD, Kellendonk C (2015). Mediodorsal thalamus hypofunction impairs flexible goal-directed behavior. *Biol Psychiatry.* 77(5):445–453.

Pergola G, Danet L, Pitel AL, Carlesimo GA, Segobin S, Pariente J, Suchan B, Mitchell AS, Barbeau EJ (2018). The regulatory role of the human mediodorsal thalamus. *Trends Cogn Sci.* 22:1011–1025.

Pinault D (2004). The thalamic reticular nucleus: Structure, function and concept. *Brain Res Rev.* 46:1–31.

Rikhye RV, Gilra A, Halassa MM (2018). Thalamic regulation of switching between cortical representations enables cognitive flexibility. *Nat Neurosci.* 21:1753–1763.

Roberts AC, Clarke HF (2019) Why we need nonhuman primates to study the role of ventromedial prefrontal cortex in the regulation of threat- and reward-elicited responses. *Proc Natl Acad Sci.* 116:26297–26304.

Schmitt LI, Wimmer RD, Nakajima M, Happ M, Mofakham S, Halassa MM (2017). Thalamic amplification of cortical connectivity sustains attentional control. *Nature.* 545:219–223.

Stalnaker TA, Calhoon GG, Ogawa M, Roesch MR, Schoenbaum G (2012). Reward prediction error signaling in posterior dorsomedial striatum is action specific. *J Neurosci.* 32:10296–10305.

Turner KM, Parkes SL (2020). Prefrontal regulation of behavioural control: Evidence from learning theory and translational approaches in rodents. *Neurosci Biobehav Rev.* 118:27–41.

Ullsperger M, Danielmeier C, Jocham G (2014). Neurophysiology of performance monitoring and adaptive behavior. *Physiol Rev.* 94:35–79.

Wang F, Kahnt T (2021). Neural circuits for inference-based decision-making. *Curr Opin Behav Sci.* 41:10–14.

Warrington EK, Weiskrantz L (1982). Amnesia: A disconnection syndrome? *Neuropsychologia.* 20:233–248.

Wicker E, Turchi J, Malkova L, Forcelli PA (2018). Mediodorsal thalamus is required for discrete phases of goal-directed behavior in macaques. *eLife.* 7:e37325.

Wimmer RD, Schmitt LI, Davidson TJ, Nakajima M, Deisseroth K, Halassa MM (2015). Thalamic control of sensory selection in divided attention. *Nature.* 526:705–709.

Wolff M, Alcaraz F, Marchand AR, Coutureau E (2015). Functional heterogeneity of the limbic thalamus: From hippocampal to cortical functions. *Neurosci Biobehav Rev*. 54:120–130.

Wolff M, Morceau S, Folkard R, Martin-Cortecero J, Groh A (2021). A thalamic bridge from sensory perception to cognition. *Neurosci Biobehav Rev*. 120:222–235.

Wolff M, Vann SD (2019). The cognitive thalamus as a gateway to mental representations. *J Neurosci*. 39:3–14.

Wright NF, Vann SD, Aggleton JP, Nelson AJ (2015). A critical role for the anterior thalamus in directing attention to task-relevant stimuli. *J Neurosci*. 35:5480–5488.

Yin HH, Ostlund SB, Knowlton BJ, Balleine BW (2005). The role of the dorsomedial striatum in instrumental conditioning. *Eur J Neurosci*. 22:513–523.

SECTION IX

DEVELOPMENT AND PLASTICITY

Chinfei Chen

Development and plasticity of thalamocortical circuits are critical for sensory and motor function, emotional responses, and normal cognition. In this section, we will focus on these processes for the major components of this circuit: the thalamocortical and corticothalamic projections that form reciprocal connections between circuits in the cortex and those in the thalamus. Here we define plasticity as the ability to change in response to stimuli (e.g., molecular cues and/or activity). Activity is often categorized in 2 forms: sensory- or experience-dependent activity refers to action potential firing that is driven by external stimuli, whereas spontaneous activity is firing in the absence of an obvious trigger. Plasticity is most prominent during development, and disruption of these processes can lead to neurodevelopmental and psychiatric disorders.

Development of thalamic circuits is heavily influenced by the innervation and refinement of the afferent input carrying primary information from the environment or from other regions of the brain. In Chapter 49, Chen summarizes the current understanding of development of a representative afferent input in the visual system, the retinogeniculate synapse. Early maturation of this input depends on spontaneous activity in the form of retinal waves. Additionally, recent studies in mice, supported by previous reports in cat and ferret, demonstrate a later phase of refinement that is driven by experience.

The development of the synaptic circuits between thalamus and cortex plays an important role in setting up later experience-dependent plasticity. (A comprehensive review by Molnar on embryonic development of the thalamocortical projection can be found in Chapter 58 in Section X of this volume.) In Chapter 50, Karayannis extends thalamocortical development to postnatal ages in the rodent somatosensory system, highlighting how early, transient synaptic connections between thalamic inputs and cortical neurons are used as a scaffold for constructing the mature thalamocortical circuit.

Perhaps the best-studied form of experience-dependent developmental plasticity is that of cortical ocular dominance in the visual system. *Ocular dominance* refers to the relative response of binocular neurons to input from one eye versus the other. The pioneering work of Hubel and Wiesel showed that during development, deprivation of one eye for a brief period of time leads to a long-lasting shift in ocular dominance toward the open eye in binocular visual cortex. This shift is most robust during a critical period of development; similar deprivation in the adult does not cause as dramatic a change in response. These findings inspired decades of research aimed at understanding the underlying mechanisms of this plasticity. In Chapter 51, Cang and Ribic summarize recent insights gained in this area of study, detailing the circuit mechanisms that contribute to binocular vision.

Binocularity is not the only sensory feature that exhibits developmental experience-dependent plasticity. In the visual system, subsets of neurons exhibit feature selectivity—that is, they change their firing response to only specific types of visual stimuli. Neurons that respond to motion (such as moving bars in one direction) are called *direction-selective cells*; others that respond to a particular orientation of a bar are referred to as *orientation selective*. Neurons in other sensory and higher-order regions of the brain exhibit selectivity to more complex features. Chapter 52 by Griswold and Van Hooser focuses on the development and plasticity of feature selectivity.

Perhaps the largest component of the thalamocortical circuit, and the last to mature, is the corticothalamic projection. The role of this massive projection is still debated in the adult (see Chapter 14 in Section IV; Chapter 19 in Section V). In Chapter 53, Guido reviews the growing evidence from recent work highlighting evidence supporting an important role for cortical feedback in specifying development of other subcortical regions.

Our understanding of development and plasticity of the thalamocortical circuitry comes largely from studying sensory systems, for the obvious reasons that the circuits can be easily manipulated. While many principles gained from these studies can be generalized to other modalities, some principles are likely unique for a specific thalamocortical circuit. In Chapter 54, Mitchell describes the components and plasticity of an important higher-order thalamocortical circuit, the connections between limbic thalamus and prefrontal cortex, which underlie executive function and the gating of cognitive and emotional processing.

While plasticity of the thalamocortical circuit is most robust in development, studies have shown that it can be enhanced or reactivated in the adult with specific stimuli or in response to disease or injury. Interest in adult plasticity is very high with the hope that, ultimately, we can harness mechanisms underlying this plasticity to repair the mammalian brain. One particular form, cross-modal plasticity, is relevant in this context as it shares mechanisms that are common to development, including those that dictate axon targeting and synaptic remodeling. In the final chapter of this section (Chapter 55), Mesik and Lee describe studies revealing how loss of a sensory modality can trigger cross-modal plasticity of different thalamocortical circuits.

This section is not intended to be exhaustive, but rather to give the reader a flavor of the richness of thalamocortical development. The plasticity described here in one sensory system has often been described in other modalities. While there is a heavy emphasis on the mouse experimental model, efforts are made to compare and contrast plasticity across species. With the development of future tools to study neuronal circuits in other species, we anticipate that our understanding of how developmental and plasticity mechanisms generalize will grow with time. Nevertheless, several common themes emerge from this collection of chapters. First, each component of the thalamocortical circuit develops over overlapping time periods rather than sequentially. Second, distinct patterns of activity, whether spontaneous or stimulus-evoked, seem to drive different developmental sequences. Third, perturbation of one component of the circuit will unquestionably affect other components because of the interdependence of the development of these projections. Thus, development and plasticity of cortical or subcortical regions must be considered together as interacting and synergistic components of a larger circuit.

49.

MECHANISMS UNDERLYING THE REFINEMENT OF THE RETINOGENICULATE SYNAPSE

Chinfei Chen

INTRODUCTION

Studies across many species have contributed to our current understanding of development of the retinogeniculate synapse. This process can be divided into different overlapping phases (Figure 49.1). The initial phase involves the mapping of retinal ganglion cell axons onto the dorsal lateral geniculate nucleus (dLGN), followed by coarse rearrangement of axon arbors into eye-specific layers. A third phase follows that spans eye opening, during which some retinal ganglion cell (RGC) inputs onto a thalamocortical (TC) neuron are eliminated, while others are strengthened. In some species, a late phase of synapse remodeling has been described during which visual experience can alter retinogeniculate connectivity. Together, these 4 phases of synaptic circuit development contribute to the maturation of the receptive field properties of thalamocortical neurons (Figure 49.1).

The retinogeniculate synapse offers several advantages as an experimental model. First, the axons of presynaptic RGCs can be easily visualized in the dLGN by injection of label into the eyes since they are the only neurons that project out of the eye. This allows for easy tracking of retinal inputs from the different eyes. Second, the activity of RGCs can be manipulated by altering visual experience or by using pharmacological or genetic tools. Third, the dLGN synaptic circuit is relatively simple when compared to circuits in other areas of the brain (reviewed in Litvina and Chen, 2017; Guido, 2018; Liang and Chen, 2020; see also **Section II of this volume**). In particular, there is no excitatory feedback projection from the thalamus to the retina, allowing ease of interpretation of functional changes in the connection. Finally, over normal development, the organization of axon terminals in the dLGN exhibit increasingly precise retinotopy, whereby neighboring RGC neurons target neighboring thalamocortical neurons. Additionally, axons from the 2 eyes remodel to occupy distinct dLGN regions, referred to as eye-specific laminae or layers. The number of layers varies across species, but these eye-specific maps serve as a useful experimental assay. Changes in this topographical organization can be monitored functionally and anatomically to assess postnatal refinement of the circuit. Our current understanding of this developmental process is summarized below.

AXON MAPPING

A number of molecular cues regulate the targeting of RGC axons to their proper dLGN regions early in development. One family of cues identified in the mouse are Eph receptors, which are transmembrane receptor tyrosine kinases expressed in a gradient across both the retina and dLGN. The Eph family of receptors bind to their ligands, Ephrins—this interaction enables bidirectional signaling important for axon targeting (Flanagan and Vanderhaeghen, 1998). These receptors are found across species including in mice, ferrets, and primates (Cang and Feldheim, 2013). At the mouse retinogeniculate synapse, Eph/Ephrins A1 and 2 are the prominent family members, with EphA levels expressed highest in RGCs located in the temporal retina and lowest in those located in the nasal retina. In the dLGN, the EphA gradient is high in the dorsal and low in the ventral region of the nucleus. Ephrins are expressed in a complementary gradient to Ephs in both retina and dLGN. The ligand-receptor interaction is repulsive, such that RGCs from the temporal retina target the dorsal dLGN, and those from the nasal retina terminate in the ventral nucleus (see Figure 49.2).

The majority of the mature mouse dLGN is innervated by RGC axons from the contralateral eye, with the exception of one small ipsilateral layer located in the anterior aspect of the nucleus (Figure 49.3A). Double and triple knockouts of multiple Ephrin A ligands (EphrinA2/A5 and A2/A3/

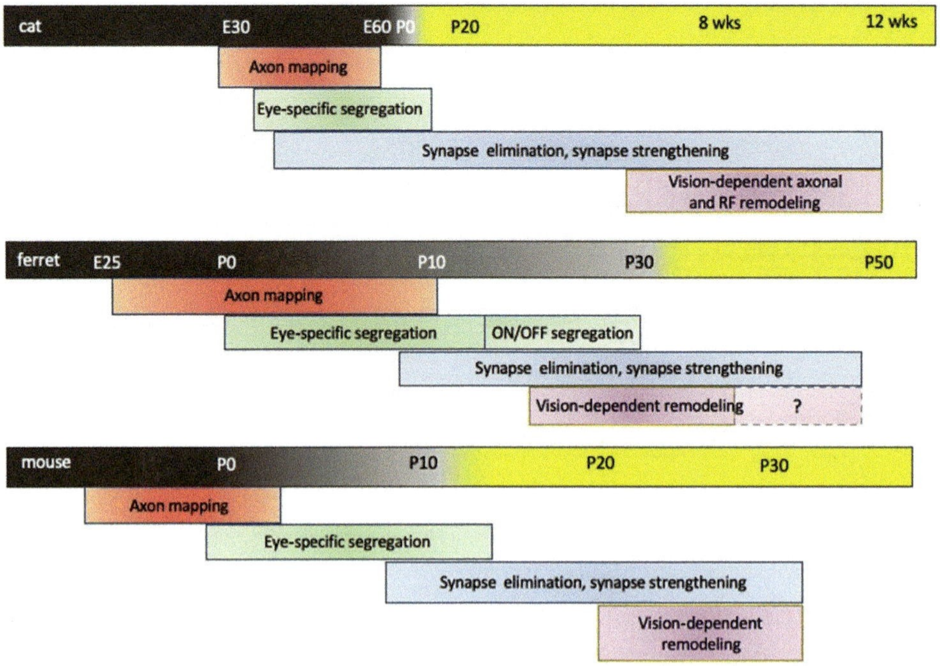

Figure 49.1. Approximate timeline for retinogeniculate development across cat, ferret, and mouse. The duration for each phase varies for each species, as does the age of eye opening. Vision-dependent phase of refinement for cat and ferret is extrapolated from studies that report changes in axon arbors or receptive field properties to visual deprivation (Sherman and Spear, 1982; Sur et al., 1982; Akerman et al., 2002).

A5, respectively) result in disruption of proper targeting of RGC axons in the dLGN and the mislocation of the ipsilateral layer (Pfeiffenberger et al., 2005). In contrast to the single stereotypic location of the ipsilateral layer in wildtype mice, multiple small ipsilateral patches are found throughout the dLGN in the mutants. Yet, despite their improper location, the segregation of ipsi- and contralateral axons still occurs. These observations demonstrate that the mechanisms for axon targeting are distinct from those underlying eye-specific segregation (see below). They also emphasize the important role of Eph/Ephrin protein gradients in specifying the dorsoventral dLGN map (Cang and Feldheim, 2013).

Less is known about the molecular cues that specify the mediolateral dLGN map. Several molecules have been proposed, including Ephrin Bs, wnt-Ryk, and Ten-m3, that exhibit dorsal-ventral gradients in the retina. Other molecules that are important for the interactions of RGC axons in the optic chiasm may also play a role in the initial targeting of inputs from the 2 eyes (Cang and Feldheim, 2013).

THE EMERGENCE OF EYE-SPECIFIC LAYERS OVER DEVELOPMENT

After axon guidance cues help target the RGC growth cone to the correct region of the dLGN, synaptic connections are formed and further refinement occurs. In many species, gross axonal development can be visualized by labeling RGC axons from the 2 eyes with different colors (such as with fluorescent labels linked to cholera toxin subunit β, CTB). For example, early in development, retinal axons from the 2 eyes labeled with red and green colors, respectively, yield yellow territories in the thalamus where eye-specific axons overlap (Figure 49.3A). With age, the overlapping label decreases, leading to distinct red and green layers of innervation (Rakic, 1976; Linden et al., 1981; Shatz and Stryker, 1988; Jaubert-Miazza et al., 2005).

Axon pruning and elaboration underlie the emergence of eye-specific layers. Classic studies in the cat, where individual retinal axons were filled with horseradish peroxidase, revealed a developmental sequence in dLGN where initial inputs with "stick-like" morphology exhibit secondary elaboration in the form of small axon "twigs." With age, these "twigs" become less obvious in the inappropriate region of the dLGN, while the axon arbor in the appropriate region elaborates extensively. This process has been described across species in cat, ferret, and mice (Sretavan and Shatz, 1986; Hahm et al., 1991; Dhande et al., 2011; Hong et al., 2014). Interestingly, however, in monkey, the immature twigs in inappropriate layers are not obvious. Instead, the elaboration and increased complexity of arbors in the appropriate region dominate (Snider et al., 1999). This observation emphasizes that while axon elimination has received the predominant attention in studies of the emergence of eye-specific segregation, increased complexity and elaboration of the axon arbor in the correct eye layer also contribute substantially to the

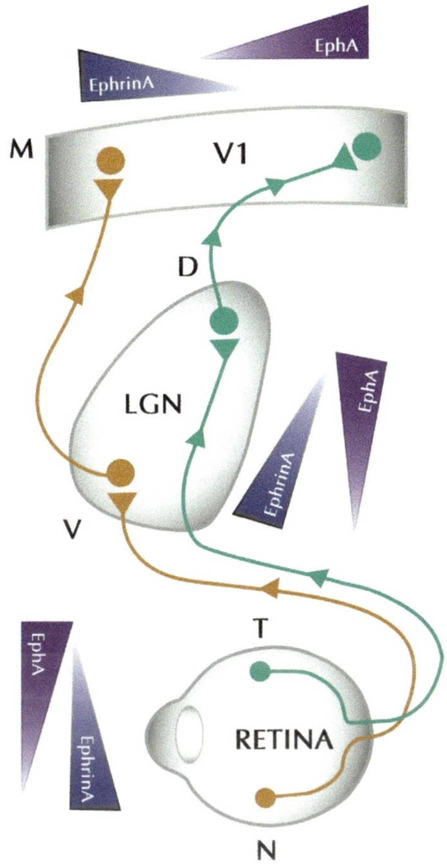

Figure 49.2. Retinal axon targeting is determined by Eph and Ephrin countergradients in retina, dorsal lateral geniculate nucleus (LGN), and visual cortex (V1). V = ventral; D = dorsal; M = medial; L = lateral. Adapted from Cang and Feldheim (2013).

process (Sretavan and Shatz, 1986). Indeed, recent studies in mice quantifying the morphology of individually labeled RGC arbors revealed that increases in arbor complexity are more dominant than axon pruning over development (Dhande et al., 2011; Hong et al., 2014). The size of single axon arbors grow to be quite large, often spanning the entire lamina (Sur et al., 1984; Hamos et al., 1987a; Sur et al., 1987; Conley and Fitzpatrick, 1989; Hahm et al., 1991; Dhande et al., 2011). With standard labeling assays commonly used to assess eye-specific segregation, the enlarged and elaborated arbors would cover a larger dLGN territory and contribute much greater labeled signal when compared to that of immature, stick-like axons. Taken together, eye-specific segregation reflects both the addition and elimination of axon arbors.

MECHANISMS UNDERLYING AXON ARBOR REFINEMENT

Retinal Waves

Classic studies in cat involving perinatal infusion of tetrodotoxin (TTX), a sodium channel inhibitor, or in ferret infused with NMDA receptor antagonists demonstrated the importance of activity in regulating eye-specific segregation (Shatz and Stryker, 1988; Sretavan et al., 1988; Hahm et al., 1991). This regulation of axon remodeling is active, rather than passive, as pharmacologically enhancing activity in one eye led to an increase in its layer in the dLGN (Stellwagen and Shatz, 2002). In fetal rats, correlated spontaneous activity between neighboring neurons was detected even before vision (Maffei and Galli-Resta, 1990). Retinal neurons in newborn ferrets and fetal cat fired in synchronous bursts lasting a few seconds, with inter-burst interval of 1–2 minutes (Meister et al., 1991). Importantly, calcium imaging studies in retinal explants illustrated that these bursts of activity exhibited a striking spatial structure in the form of waves which propagate across the retina with a frequency of 90–120 seconds depending on the stage of development (Wong et al., 1993; Feller et al., 1996; Bansal et al., 2000). Retinal waves have been described in ferret, monkeys, and mice as well as other species. These waves change with development and can be divided into 3 stages: stage I waves occur prenatally and involve cholinergic and gap junction circuits; stage II waves are found postnatally and are driven by nicotinic cholinergic circuits; stage III waves follow a few weeks later and arise from glutamatergic networks (see reviews: Wong, 1999; Huberman et al., 2008).

The discovery of retinal waves afforded strong support to a model where correlated spontaneous activity, combined with Hebbian synaptic plasticity, imparts a higher likelihood that RGC terminals from the same eye target neighboring postsynaptic neurons, rather than those from the opposite eye (Mooney et al., 1993). Further support of this model came from manipulations of retinal waves. Stage II retinal waves correspond to the period of eye-specific formation and arise from acetylcholine (ACh) release from starburst amacrine cells (Wong, 1999). Pharmacologically blocking nicotinic ACh receptors in the retina or knockout of the beta (β)2 subunit of the nicotinic AChR (in β2 KO mice) disrupted correlated retinal activity, the propagation of retinal waves, and eye-specific segregation (Feller et al., 1996; Bansal et al., 2000; Muir-Robinson et al., 2002). This model was further validated with in vivo optogenetic experiments in mice that directly and independently manipulated retinal activity of the 2 eyes. Synchronous activation of both eyes altered segregation in the dLGN, whereas asynchronous activation did not (Zhang et al., 2012).

Interactions between Spontaneous Activity and Molecular Cues

So far, we have considered molecular cues and spontaneous activity as independent entities. However, they exist contemporaneously and interact synergistically to drive proper eye-specific segregation. To understand the

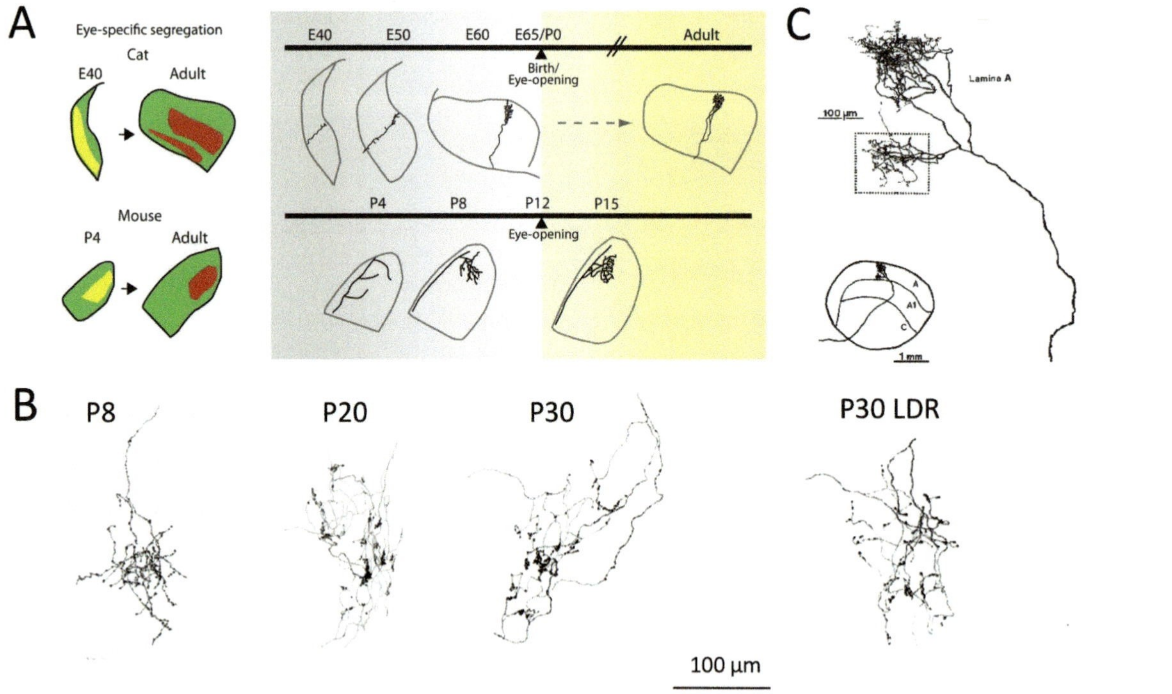

Figure 49.3. Axon arbor remodeling underlie eye-specific segregation. (A). Left: eye-specific segregation in cat and mouse over development. Ipsilateral and contralateral eye projection in red and green, respectively, overlapping projections in yellow. Right: schematic of the changes in RGC axon arbors in cat (top) and mice (bottom). Background color gradient from gray to yellow represents the change in light with eye opening. (Panel adapted from Hong and Chen, 2011). (B). Reconstruction of single mouse axon arbors from a specific RGC type (ON-OFF direction-selective RGC) over development. Axon arbors increase in size and complexity over the developmental phase when functional retinogeniculate convergence decreases. However, over this period, boutons increase in size and become spatially clustered. Visual deprivation from P20 to P30 reduces bouton clustering, but not arbor size or complexity (adapted from Hong et al., 2014). (C). Tracing of a single labeled RGC axon arbor in cat. The region within the dashed box was used for EM reconstruction. This branch of the arbor covered the territory of 40 different dLGN cells, but only made synapses on 4 neurons (adapted from Hamos et al., 1987a).

relationship between these mechanisms, Feldheim and colleagues took advantage of the genetic power of mice to generate triple knockout of Ephrins (A2/3/5) and crossed them with the β2 KO mouse (Pfeiffenberger et al., 2006). In addition to assessing eye-specific segregation, they examined the dLGN for terminal zones of axons labeled by focal injection of DiI (a lipophilic fluorescent label) into the retina. As described before, the triple Ephrin KO mice exhibit abnormally located terminal zones and ipsilateral layer. Yet axons from the 2 eyes segregate normally between the layers. On the other hand, β2 KO mice have terminal zones located in the correct dLGN region that are abnormally large, and eye-specific segregation is disrupted. In genetically altered mice lacking both Ephrins A2/3/5 and the nicotinic receptor β subunit, the phenotype is distinct and more dramatic. A larger region of the dLGN receives overlapping innervation from both eyes, and a distinct ipsilateral layer is lacking when compared to the Ephrin or β2 KO mutations alone. These studies demonstrate that molecular and activity-dependent mechanisms mediate development through independent and synergistic processes.

The Role of Waves in Other Areas of the Mammalian Brain

Spontaneous correlated activity is now understood to be a generalized mechanism for the formation of sensory maps (see also **Chapter 52 in Section IX**). Studies in mice point to a role for retinal waves in the patterning of maps not only in dLGN, but also in the visual cortex and superior colliculus (SC) (Weliky and Katz, 1999; Huberman et al., 2006; Triplett et al., 2009). Additionally, a recent study showed that the waves also play a role in the refinement of feature selectivity in mice. Interestingly, spontaneous activity within the same eye is not entirely random. In the mouse retina, the direction of wave propagation is biased toward the nasal-temporal axis, which corresponds to the direction of self-motion (optic flow, motion forward) (Stafford et al., 2009; Ge et al., 2021). This bias is transiently evident before eye opening during early development in the superior colliculus, where RGC axonal activity propagates in waves anteriorly to posteriorly. Disruption of this propagation bias in the retina results in aberrant direction selectivity in SC neurons of mice after eye opening (Ge et al., 2021).

SYNAPSE REFINEMENT: THE STRENGTHENING AND ELIMINATION OF CONNECTIONS

Concurrent to changes in axon arbor structure are changes in functional connectivity between the retina and neurons of the dLGN. During early postnatal development, connectivity is weak and extracellular single unit recordings from fetal cat dLGN brain slices demonstrated that over 90% of thalamic neurons could be driven by stimulation of either contralateral or ipsilateral optic tract stumps. This proportion decreased to about 50% in the neonate (Shatz and Kirkwood, 1984). This study supported the idea that convergent retinal synapses from both the ipsilateral and contralateral eyes are functional and that there is significant synapse elimination over development. Consistent with this idea is evidence that synapses in the appropriate versus inappropriate lamina from RGCs from the same eye demonstrated little distinction in their morphology save for a difference in the number of synaptic vesicles (Campbell and Shatz, 1992). These observations initially described in cat have been shown to be true in other species, including rodents and ferrets (Mooney et al., 1993; Chen and Regehr, 2000; Lo et al., 2002; Jaubert-Miazza et al., 2005; Guido, 2018). Competition between RGC axons for dLGN neuron connections is not restricted to those with inputs spanning eye-specific layers of the nucleus, but also occurs between RGCs that innervate exclusively within the monocular region. In mice, dLGN slice recordings from the monocular region revealed that changes in connectivity do not end with the completion of eye-specific segregation, the bulk of which occurs before postnatal day (P)10. Recordings around this age revealed that many weak RGC inputs still innervate a given TC neuron. Over the course of 3 postnatal weeks spanning eye opening, the average number of convergent inputs decreases by more than 50% while the strength of a subset of these inputs increases more than 20-fold. These robust changes in connectivity are dependent on spontaneous activity, but not by visual experience (Figure 49.4) (Hooks and Chen, 2006). Synaptic refinement during this period corresponds temporally to the developmental of receptive field properties of dLGN neurons (Tschetter et al., 2018). dLGN receptive fields are initially large, irregularly shaped, and temporally imprecise, and then refine to be smaller, sharper, and more precise over development across all species (cat: Wiesel and Hubel, 1963; Daniels et al., 1978; ferret: Tavazoie and Reid, 2000; primate: Blakemore and Vital-Durand, 1986; mice: Tschetter et al., 2018).

DEVELOPMENTAL CHANGES IN SYNAPTIC RECEPTORS

The strengthening of RGC inputs involves an increase in AMPA receptors (AMPAR) relative to NMDA receptors (NMDAR) such that the ratio of peak AMPAR to NMDAR currents increases with development, presumably by insertion of these receptors into postsynaptic densities. Additionally, the subunit composition of these glutamate receptors changes with age, with a developmental switch of NMDAR subunits NR2B, C, and D to NR2A, and AMPAR subunits GluA2 to GluA1 (reviewed in Litvina and Chen, 2017). Unlike AMPARs, the total NMDAR synaptic response from a given TC neuron does not change markedly over development after eye-specific segregation (Hooks and Chen, 2006), consistent with the idea that synaptic refinement spanning eye opening involves competition between different presynaptic RGC inputs for innervation of a relative stable number of postsynaptic densities or "synaptic slots." Competition between different afferent inputs during the spontaneous and experience-dependent phase of development has been shown to involve both Hebbian and homeostatic mechanisms (Litvina and Chen, 2017).

MORPHOLOGICAL CORRELATES OF SYNAPTIC REFINEMENT

Cellular correlates to the refinement of synaptic connections after eye-specific segregation also include changes in axonal morphology. Reconstruction in mice of single RGC axon arbors showed that rather than retraction of the axon skeleton, the axon arbor size and average bouton size increase after eye-specific segregation, similar to what has been described in other species (Sur et al., 1984; Sretavan and Shatz, 1986; Hamos et al., 1987a; Conley and Fitzpatrick, 1989; Dhande et al., 2011; Hong et al., 2014). Notably, during the period of robust synapse refinement in mice (P10–20), boutons that are initially distributed evenly across the axon arbor become clustered (Hong et al., 2014), reminiscent of an electron microscopy study in adult cat where one labeled RGC input makes multiple synaptic contacts onto the proximal dendrite of one TC neuron (Hamos et al., 1987a). Remarkably, in both species, the large retinal axon spans the territory of tens of cells that it does not make synaptic contacts to. These morphological studies highlight the specificity of the connections made over development.

EXPERIENCE-DEPENDENT PLASTICITY

DEVELOPMENTAL SENSITIVE PERIODS

It was long thought that neuronal circuits stabilized in the dLGN once eye-specific segregation was complete, leaving experience-dependent plasticity to the visual

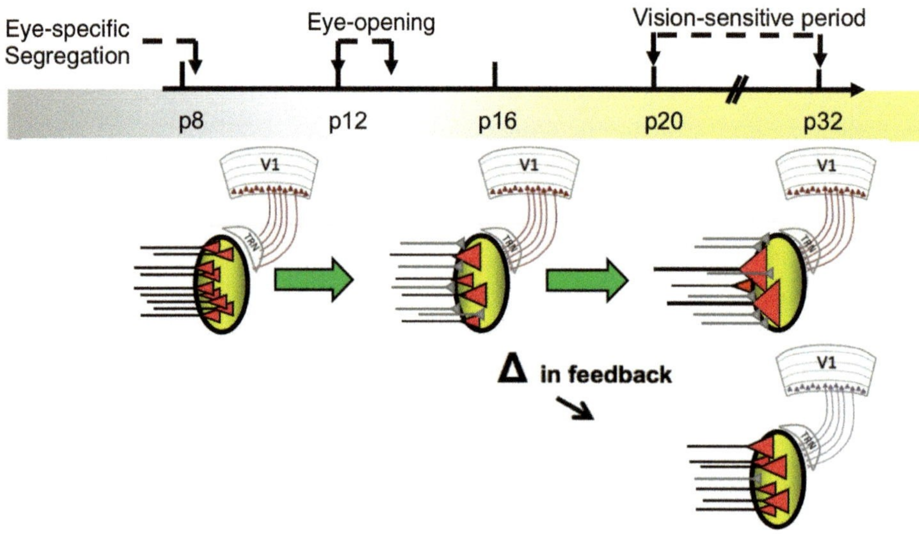

Figure 49.4. Model of retinogeniculate refinement during development in the mouse. Feedforward connections between the retina and dorsal lateral geniculate nucleus (dLGN) are formed at birth and undergo synapse elimination and strengthening between P8–P20. Thalamocortical (TC) neurons (yellow oval) initially receive innervation from multiple weak retinal ganglion cell (RGC) inputs (relative synaptic strength is depicted by the size of the red triangles). With age, some inputs are strengthened, while other inputs are eliminated or weakened until they are not functionally relevant (gray triangles). Cortical feedback connections made from the visual cortex (V1) to the dLGN with disynaptic inhibitory connections through the thalamic reticular nucleus (TRN) develop later and are not complete until P14. From P20–P30, changes in visual experience or in the activity of cortical feedback increase the number of RGC inputs to relay neurons (reprinted with permission from Liang and Chen, 2020).

cortex. Classic studies by Hubel and Wiesel revealed no change in receptive-field properties of dLGN neurons in monocularly deprived cats despite observations that cells in layers receiving inputs from the occluded eye were significantly smaller in size (Wiesel and Hubel, 1963). Yet, some investigators did report alterations in spatial receptive field properties of dLGN neurons and overall reduced activity in the deprived layers when compared to the non-deprived layers. Other studies described changes in the morphology of certain types of RGC axon arbors in visually deprived cats (see review, Sherman and Spear, 1982). More recent studies have revived the idea that the cortex is not the only region that exhibits plasticity to visual experience during development, as well as in the adult.

Refinement in the monocular region of the mouse dLGN is initially dependent on spontaneous activity. About a week after eye opening (which occurs around P12–14), the retinogeniculate synapse becomes sensitive to visual experience (P20–30). Dark rearing beginning at P20 for more than 4 days results in an increase in the number of convergent retinal inputs and a decrease in the average strength of a single RGC input (Hooks and Chen, 2006). This plasticity is limited to a discrete window of time; deprivation beginning 5 days earlier or later did not elicit the same plasticity (Hooks and Chen, 2008). A similar remodeling of retinogeniculate connectivity is observed when cortical feedback activity is disrupted between P20 and P30 (Figure 49.4) (Narushima et al., 2016; Thompson et al., 2016). These observations have led to the hypothesis that experience drives changes in thalamic circuits in order to align feedforward and feedback components of thalamocortical loops around the developmental time when these circuits mature and consolidate (see **Chapter 53 in Section IX**; Thompson et al., 2016).

Morphological correlates to this vision-sensitive phase of refinement were assessed in one type of RGC in mice. Single axon reconstruction revealed de-clustering of boutons without significant changes in the axon arbor in response to visual deprivation (Hong et al., 2014). Only after this late refinement is complete do the RGC axons reduce slightly in size and complexity (between P30 and P60) (Hong et al., 2014). Interestingly, during a developmental time period after eye-specific segregation in cat and ferret, sublaminar segregation is associated with growth and/or retraction of RGC axon arbors, and this process is experience-dependent (Sur et al., 1982; Akerman et al., 2002). These studies also suggest that axon arbor remodeling and synapse refinement may vary depending on the feature selectivity of the RGC.

OCULAR DOMINANCE PLASTICITY

Experience-dependent plasticity has also been described with respect to ocular dominance (OD) in the dLGN.

Based on axon labeling, it was previously assumed that the binocular region of the dLGN exists in early development and disappears with the completion of eye-specific segregation. However, recent studies have described binocularly responding single units in the anterior medial dorsal region of the adult mouse dLGN and the koniocellular region of the adult marmoset (Howarth et al., 2014; Zeater et al., 2015; Jaepel et al., 2017). In the mouse, the binocular neurons become predominantly responsive to the unoccluded eye after monocular deprivation during late development. Whether this dLGN plasticity involves changes in the synaptic weights of convergent RGC inputs from the 2 eyes and/or loss of responsiveness of dLGN neurons is still not clear (Bauer et al., 2021). Evidence suggests that cortical OD plasticity is not directly inherited from the dLGN, but changes in thalamic inhibitory circuits may play a role in the cortical response (Sommeijer et al., 2017). Indeed, unlike cortical ocular dominance plasticity, some dLGN plasticity persists in the adult, as assessed by in vivo calcium imaging of thalamocortical inputs in the cortex (Jaepel et al., 2017). More studies are needed to further elucidate the relationship between thalamic and cortical plasticity.

MOLECULAR MECHANISMS UNDERLYING CIRCUIT DEVELOPMENT

Studies from many labs have used the murine visual system as a model for identifying molecular mechanisms important in the development of visual maps and synaptic circuits. In addition to the aforementioned axon guidance cues, a number of other molecules have been identified as playing a role in axon pruning, synapse refinement, and experience-dependent remodeling. Table 49.1 categorizes many of the identified molecules by their known roles in the different phases of developmental synapse remodeling. Below, a few examples for these different categories are highlighted.

Molecules that affect neuronal activity during development, whether pre- or postsynaptically, were found to disrupt eye-specific segregation without altering the location of the ipsilateral layer (i.e., axon targeting). For example, *Nob* mice, which carry a mutation in the proteoglycan nyctalopin (nyx), exhibit enhanced retinal wave activity after eye opening. Accordingly, initial eye-specific segregation proceeds normally before eye opening but then these layers de-segregate as spontaneous activity increases and presumably obscures the distinction in correlated activity between the 2 eyes (Demas et al., 2006). This study supports the idea that maintenance of segregation also requires proper RGC activity.

One of the most surprising themes that has emerged from developmental studies of the retinogeniculate synapse is the identification of molecules with known roles in the immune system. For example, the class I major histocompatibility complex (MHCI) molecules were identified from a differential activity-dependent gene-expression screen in the fetal cat dLGN. Disruption of the expression of these proteins leads to aberrant eye-specific layer formation and synapse refinement (see reviews: Shatz, 2009; Nagappan-Chettiar et al., 2023). Immune cells have also been implicated in developmental pruning by phagocytosis of inappropriate synapses. Molecules of the complement system, such as C1q or C3, are thought to tag synapses for elimination ("eat me" signals). Other molecules, such as CD47 and SIRPα, are thought to be protective signals ("don't eat me" signals) that prevent proper synapses from being eliminated. Other signaling systems between microglia and neurons are also implicated, with knockout of the Fn14-Tweak signaling disrupting the formation of stable synapses and experience-dependent synaptic plasticity. In addition to microglia, astrocytes can also mediate synapse elimination through the activation of the phagocytic receptors *Megf10* and *Mertk*. Knockout of these recently identified molecules borrowed from the immune system results in varying disruption of retinogeniculate convergence and the formation of eye-specific segregation (see Table 49.1; for review, see Nagappan-Chettiar et al., 2023).

Other identified mechanisms that affect developmental refinement include genes that regulate synaptic function such as *Stargazin* or *Mecp2* (see Table 49.1; Nagappan-Chettiar et al., 2023). Knockout of these genes affects the later vision-dependent phase but not eye-specific segregation, even though the genes are active throughout both developmental phases. One possible explanation for the selective role of these genes in experience-dependent refinement is that altered synaptic function may disrupt precise spike patterns more than correlated activity. Collectively, the results from mechanistic studies have supported the idea that the different phases of retinogeniculate refinement described above involve distinct molecular mechanisms.

ADULT PLASTICITY

There are growing reports across species that suggest that changes in retinogeniculate connectivity may be possible in the adult. Tuning properties of dLGN neurons in cat were shown to change from ON-center responses to OFF-center responses within 10 minutes of pharmacologic blockade of ON-center RGC activity (Moore et al., 2011). Acute suppression of cortical feedback

TABLE 49.1 MOLECULAR UNDERLYING DEVELOPMENTAL REFINEMENT

Gene/Molecule	Known Function	Axon Mapping	ESS	Synapse Refinement	Vision-Dependent Remodeling
Eph/Ephrinsaxon[a,b]	Axon targeting repulsive cues	+	+	+	NT
Ten-m3[a,b]	A transmembrane glycoprotein expressed in a dorsal-ventral gradient in the retina	NT	+	NT	NT
nyx[b]	Nyctalopin,GPI-anchored proteoglycan. Mutation disrupts stage III retinal waves	–	+#	NT	NT
Class I MHC proteins[c]	Cell-mediated immune recognition	+	+	+	NT
C1q and C3 and receptor CR3[d]	Classic complement cascade proteins in immune system	–	+	+	NT
Pentraxins: NP1 and NP2[d]	Immune system proteins known to bind to C1q	–	+*	+*	NT
Fn14 and TWEAK[d]	Fibroblast growth factor-inducible 14: a cell surface pro-inflammatory cytokine receptor	NT	NT	–	+
Jak2[d]	Protein tyrosine kinase implicated in cytokine signaling, spontaneous activity-dependent synapse elimination	–	+	NT	NT
CD47-SIRPα[d]	Immune system proteins, "don't eat me" signal	–	+$^{\phi}$	+$^{\zeta}$	NT
Megf10, MERTK[d]	Phagocytic receptors expressed in astrocytes, recognizes "eat me" signals	–	+	+	NT
SRPX2[d]	Sushi domain protein, complement inhibitor binds to C1q and blocks its activity	–	+$^{\phi}$	+	NT
mGluR1[e]	Glutamate receptor expressed on corticothalamic inputs	NT	NT	–	+
stargazin[d]	AMPA receptor chaperone protein	NT	NT	–	+
Mecp2[d]	Methyl CpG binding protein 2, transcriptional regulator	–	+#	–	+

Summary of molecules known to play a role in different phases of refinement including: (1) axon mapping, assessed by terminal zone analysis or location of ipsilateral patch; (2) eye-specific segregation (ESS); (3) synapse refinement assessed by electrophysiology; (4) vision-dependent remodeling assessed by visual deprivation. Role in specific developmental phase assessed analyzing mutant mice or manipulating gene expression.

+: yes there is a phenotype; –: no phenotype; NT: not tested. #: initially normal, but desegregates with age; *: delayed segregation and synaptic refinement; ϕ: accelerated segregation; ζ: excess synapse pruning. Details about these proteins are reviewed in: *a*: (Cang and Feldheim, 2013); *b*: (Huberman et al., 2008); *c*, (Shatz, 2009); *d*: (Nagappan-Chettiar et al., 2023); *e*: (Narushima et al., 2016).

has also been shown to shift the receptive field position of some TC neurons in awake monkeys (Aguila et al., 2017). More studies are needed to exclude the possibility that these examples of thalamic plasticity are secondary to changes in the retinal or cortical circuitry, but substrates for the changes in thalamic circuits are clearly present. As previously mentioned, even after the period of robust retinogeniculate refinement, retinal axons remain large and broad (see above). In the adult, only focal regions of an axon arbor make connections with thalamocortical regions (Hamos et al., 1987b; Hong et al., 2014). The unused portion of the axon arbor may reflect the past connectivity between RGCs and dLGN neurons, but it also offers the potential for the reactivation of or formation of new synaptic connections, even in the adult.

CONCLUSIONS

We have learned a great deal from the retinogeniculate synapse of how circuits in sensory systems are constructed. The fact that the formation of these circuits involves multiple phases, where connections are made and then remade, is quite telling and suggests that there must be an advantage for activity and experience to sculpt the circuit. This extensive process endows the circuit with adaptability to the external environment. It is likely that the principles learned from this simple synapse can be generalizable to other sensory systems and other thalamocortical loops. Indeed, evidence for both activity and experience-dependent synapse refinement have also been reported in the lemniscal input innervating the somatosensory thalamus, as well as in the auditory system (Lu and Trussell,

2007; Wang and Zhang, 2008). Importantly, the somatosensory thalamic plasticity can be reactivated with injury in the adult (Takeuchi et al., 2012). Thus, understanding the mechanisms by which synaptic circuits are refined is relevant in situations where reactivating plasticity is advantageous for repair of injury or disease. There is a great deal more to be learned from studying this simple synapse. Emerging tools that allow specific labeling, activating, silencing, and monitoring of neuronal populations across species will provide insight into which features of development and plasticity at the retinogeniculate synapse are generalizable to other thalamocortical systems and other species.

ACKNOWLEDGMENTS

I thank members of the Chen laboratory, especially Tracy Yang, for comments on the figures and manuscript and Dr. Eefei Chen for adapting Figure 49.2. This work was supported by National Institutes of Health grants NIH NEI EY013613, NIH P50HD105351 and a HMS-Tan Yang Autism Center grant 362281.

REFERENCES

Aguila J, Cudeiro FJ, Rivadulla C (2017). Suppression of V1 feedback produces a shift in the topographic representation of receptive fields of LGN cells by unmasking latent retinal drives. *Cereb Cortex*. 27:3331–3345.

Akerman CJ, Smyth D, Thompson ID (2002). Visual experience before eye-opening and the development of the retinogeniculate pathway. *Neuron*. 36:869–879.

Bansal A, Singer JH, Hwang BJ, Xu W, Beaudet A, Feller MB (2000). Mice lacking specific nicotinic acetylcholine receptor subunits exhibit dramatically altered spontaneous activity patterns and reveal a limited role for retinal waves in forming ON and OFF circuits in the inner retina. *J Neurosci*. 20:7672–7681.

Bauer J, Weiler S, Fernholz MHP, Laubender D, Scheuss V, Hübener M, Bonhoeffer T, Rose T (2021). Limited functional convergence of eye-specific inputs in the retinogeniculate pathway of the mouse. *Neuron*. 109:2457–2468.e12.

Blakemore C, Vital-Durand F (1986). Organization and post-natal development of the monkey's lateral geniculate nucleus. *J Physiol (Lond)*. 380:453–491.

Campbell G, Shatz CJ (1992). Synapses formed by identified retinogeniculate axons during the segregation of eye input. *J Neurosci*. 12:1847–1858.

Cang J, Feldheim DA (2013). Developmental mechanisms of topographic map formation and alignment. *Annu Rev Neurosci*. 36:51–77.

Chen C, Regehr WG (2000). Developmental remodeling of the retinogeniculate synapse. *Neuron*. 28:955–966.

Conley M, Fitzpatrick D (1989). Morphology of retinogeniculate axons in the macaque. *Vis Neurosci*. 2:287–296.

Daniels JD, Pettigrew JD, Norman JL (1978). Development of single-neuron responses in kitten's lateral geniculate nucleus. *J Neurophysiol*. 41:1373–1393.

Demas J, Sagdullaev BT, Green E, Jaubert-Miazza L, McCall MA, Gregg RG, Wong RO, Guido W (2006). Failure to maintain eye-specific segregation in Nob, a mutant with abnormally patterned retinal activity. *Neuron*. 50:247–259.

Dhande OS, Hua EW, Guh E, Yeh J, Bhatt S, Zhang Y, Ruthazer ES, Feller MB, Crair MC (2011). Development of single retinofugal axon arbors in normal and beta2 knock-out mice. *J Neurosci*. 31:3384–3399.

Feller MB, Wellis DP, Stellwagen D, Werblin FS, Shatz CJ (1996). Requirement for cholinergic synaptic transmission in the propagation of spontaneous retinal waves. *Science*. 272:1182–1187.

Flanagan JG, Vanderhaeghen P (1998). The ephrins and Eph receptors in neural development. *Annu Rev Neurosci*. 21:309–345.

Ge X, Zhang K, Gribizis A, Hamodi AS, Sabino AM, Crair MC (2021). Retinal waves prime visual motion detection by simulating future optic flow. *Science*. 373.

Guido W (2018). Development, form, and function of the mouse visual thalamus. *J Neurophysiol*. 120:211–225.

Hahm JO, Langdon RB, Sur M (1991). Disruption of retinogeniculate afferent segregation by antagonists to NMDA receptors. *Nature*. 351:568–570.

Hamos JE, Van Horn SC, Raczkowski D, Sherman SM (1987a). Synaptic circuits involving an individual retinogeniculate axon in the cat. *J Comp Neurol*. 259:165–192.

Hamos JE, Van Horn SC, Raczkowski D, Sherman SM (1987b). Synaptic circuits involving an individual retinogeniculate axon in the cat. *J Comp Neurol*. 259:165–192.

Hong YK, Chen C (2011). Wiring and rewiring of the retinogeniculate synapse. *Curr Opin Neurobiol*. 21:228–237.

Hong YK, Park S, Litvina EY, Morales J, Sanes JR, Chen C (2014). Refinement of the retinogeniculate synapse by bouton clustering. *Neuron*. 84:332–339.

Hooks BM, Chen C (2006). Distinct roles for spontaneous and visual activity in remodeling of the retinogeniculate synapse. *Neuron*. 52:281–291.

Hooks BM, Chen C (2008). Vision triggers an experience-dependent sensitive period at the retinogeniculate synapse. *J Neurosci*. 28:4807–4817.

Howarth M, Walmsley L, Brown TM (2014). Binocular integration in the mouse lateral geniculate nuclei. *Curr Biol*. 24:1241–1247.

Huberman AD, Feller MB, Chapman B (2008). Mechanisms underlying development of visual maps and receptive fields. *Annu Rev Neurosci*. 31:479–509.

Huberman AD, Speer CM, Chapman B (2006). Spontaneous retinal activity mediates development of ocular dominance columns and binocular receptive fields in v1. *Neuron*. 52:247–254.

Jaepel J, Hübener M, Bonhoeffer T, Rose T (2017). Lateral geniculate neurons projecting to primary visual cortex show ocular dominance plasticity in adult mice. *Nat Pub Group*. 20:1–11.

Jaubert-Miazza L, Green E, Lo FS, Bui K, Mills J, Guido W (2005). Structural and functional composition of the developing retinogeniculate pathway in the mouse. *Vis Neurosci*. 22:661–676.

Liang L, Chen C (2020). Organization, function, and development of the mouse retinogeniculate synapse. *Annu Rev Vis Sci*. 6:261–285.

Linden DC, Guillery RW, Cucchiaro J (1981). The dorsal lateral geniculate nucleus of the normal ferret and its postnatal development. *J Comp Neurol*. 203:189–211.

Litvina EY, Chen C (2017). An evolving view of retinogeniculate transmission. *Vis Neurosci*. 34:E013.

Lo FS, Ziburkus J, Guido W (2002). Synaptic mechanisms regulating the activation of a Ca(2+)-mediated plateau potential in developing relay cells of the LGN. *J Neurophysiol*. 87:1175–1185.

Lu T, Trussell LO (2007). Development and elimination of endbulb synapses in the chick cochlear nucleus. *J Neurosci*. 27:808–817.

Maffei L, Galli-Resta L (1990). Correlation in the discharges of neighboring rat retinal ganglion cells during prenatal life. *Proc Natl Acad Sci U S A*. 87:2861–2864.

Meister M, Wong RO, Baylor DA, Shatz CJ (1991). Synchronous bursts of action potentials in ganglion cells of the developing mammalian retina. *Science.* 252:939–943.
Mooney R, Madison DV, Shatz CJ (1993). Enhancement of transmission at the developing retinogeniculate synapse. *Neuron.* 10:815–825.
Moore BD, Kiley CW, Sun C, Usrey WM (2011). Rapid plasticity of visual responses in the adult lateral geniculate nucleus. *Neuron.* 71:812–819.
Muir-Robinson G, Hwang BJ, Feller MB (2002). Retinogeniculate axons undergo eye-specific segregation in the absence of eye-specific layers. *J Neurosci.* 22:5259–5264.
Nagappan-Chettiar S, Burbridge TJ, Umemori H (2023). Activity-dependent synapse refinement: From mechanisms to molecules. *Neuroscientist.* 4:10738584231170167.
Narushima M, Uchigashima M, Yagasaki Y, Harada T, Nagumo Y, Uesaka N, Hashimoto K, Aiba A, Watanabe M, Miyata M, Kano M (2016). The metabotropic glutamate receptor subtype 1 mediates experience-dependent maintenance of mature synaptic connectivity in the visual thalamus. *Neuron.* 91(5):1097–1109.
Pfeiffenberger C, Cutforth T, Woods G, Yamada J, Renteria RC, Copenhagen DR, Flanagan JG, Feldheim DA (2005). Ephrin-As and neural activity are required for eye-specific patterning during retinogeniculate mapping. *Nat Neurosci.* 8:1022–1027.
Pfeiffenberger C, Yamada J, Feldheim DA (2006). Ephrin-As and patterned retinal activity act together in the development of topographic maps in the primary visual system. *J Neurosci.* 26:12873–12884.
Rakic P (1976). Prenatal genesis of connections subserving ocular dominance in the rhesus monkey. *Nature.* 261:467–471.
Shatz CJ (2009). MHC class I: An unexpected role in neuronal plasticity. *Neuron.* 64:40–45.
Shatz CJ, Kirkwood PA (1984). Prenatal development of functional connections in the cat's retinogeniculate pathway. *J Neurosci.* 4:1378–1397.
Shatz CJ, Stryker MP (1988). Prenatal tetrodotoxin infusion blocks segregation of retinogeniculate afferents. *Science.* 242:87–89.
Sherman SM, Spear PD (1982). Organization of visual pathways in normal and visually deprived cats. *Physiol Rev.* 62:738–855.
Snider CJ, Dehay C, Berland M, Kennedy H, Chalupa LM (1999). Prenatal development of retinogeniculate axons in the macaque monkey during segregation of binocular inputs. *J Neurosci.* 19:220–228.
Sommeijer J-P, Ahmadlou M, Saiepour MH, Seignette K, Min R, Heimel JA, Levelt CN (2017). Thalamic inhibition regulates critical-period plasticity in visual cortex and thalamus. *Nat Pub Group.* 20:1715–1721.
Sretavan DW, Shatz CJ (1986.) Prenatal development of retinal ganglion cell axons: segregation into eye-specific layers within the cat's lateral geniculate nucleus. *J Neurosci.* 6:234–251.
Sretavan DW, Shatz CJ, Stryker MP (1988). Modification of retinal ganglion cell axon morphology by prenatal infusion of tetrodotoxin. *Nature.* 336:468–471.
Stafford BK, Sher A, Litke AM, Feldheim DA (2009). Spatial-temporal patterns of retinal waves underlying activity-dependent refinement of retinofugal projections. *Neuron.* 64:200–212.
Stellwagen D, Shatz CJ (2002). An instructive role for retinal waves in the development of retinogeniculate connectivity. *Neuron.* 33:357–367.
Sur M, Esguerra M, Garraghty PE, Kritzer MF, Sherman SM (1987). Morphology of physiologically identified retinogeniculate X- and Y-axons in the cat. *J Neurophysiol.* 58:1–32.
Sur M, Humphrey AL, Sherman SM (1982). Monocular deprivation affects X- and Y-cell retinogeniculate terminations in cats. *Nature.* 300:183–185.
Sur M, Weller RE, Sherman SM (1984). Development of X- and Y-cell retinogeniculate terminations in kittens. *Nature.* 310: 246–249.
Takeuchi Y, Yamasaki M, Nagumo Y, Imoto K, Watanabe M, Miyata M (2012). Rewiring of afferent fibers in the somatosensory thalamus of mice caused by peripheral sensory nerve transection. *J Neurosci.* 32:6917–6930.
Tavazoie SF, Reid RC (2000). Diverse receptive fields in the lateral geniculate nucleus during thlamocortical development. *Nat Neurosci.* 3:608–616.
Thompson AD, Picard N, Min L, Fagiolini M, Chen C (2016.) Cortical feedback regulates feedforward retinogeniculate refinement. *Neuron.* 91:1021–1033.
Triplett JW, Owens MT, Yamada J, Lemke G, Cang J, Stryker MP, Feldheim DA (2009). Retinal input instructs alignment of visual topographic maps. *Cell.* 139:175–185.
Tschetter WW, Govindaiah G, Etherington IM, Guido W, Niell CM (2018). Refinement of spatial receptive fields in the developing mouse lateral geniculate nucleus is coordinated with excitatory and inhibitory remodeling. *J Neurosci.* 38:4531–4542.
Wang H, Zhang ZW (2008). A critical window for experience-dependent plasticity at whisker sensory relay synapse in the thalamus. *J Neurosci.* 28:13621–13628.
Weliky M, Katz LC (1999). Correlational structure of spontaneous neuronal activity in the developing lateral geniculate nucleus in vivo. *Science (NY).* 285:599–604.
Wiesel TN, Hubel DH (1963). Effects of visual deprivation on morphology and physiology of cells in the cats lateral geniculate body. *J Neurophysiol.* 26:978–993.
Wong RO (1999). Retinal waves and visual system development. *Annu Rev Neurosci.* 22:29–47.
Wong RO, Meister M, Shatz CJ (1993). Transient period of correlated bursting activity during development of the mammalian retina. *Neuron.* 11:923–938.
Zeater N, Cheong SK, Solomon SG, Dreher B, Martin PR (2015). Binocular visual responses in the primate lateral geniculate nucleus. *Curr Biol.* 25:3190–3195.
Zhang J, Ackman JB, Xu H-P, Crair MC (2012). Visual map development depends on the temporal pattern of binocular activity in mice. *Nat Pub Group.* 15:298–307.

50.

POSTNATAL DEVELOPMENT OF THE THALAMOCORTICAL CIRCUIT AND OF SENSORY-DRIVEN CORTICAL ACTIVITY

Theofanis Karayannis

THALAMO-SUBPLATE DEVELOPMENTAL INTERACTIONS

The thalamus establishes connections with the cortex during embryogenesis. As has been found for peripheral sensory organs such as the retina, the thalamus also displays spontaneous activity during the early developmental stages, which propagates to the cortex, as shown by wide-field calcium imaging (Antón-Bolaños et al., 2019). Developmentally, with the arrival of the radially migrating excitatory neurons, the preplate, the first structure to be created in the neocortex, is split into the subplate (SP) on the ventral side and the marginal zone (MZ) on the dorsal side (future layer [L]1) (Bystron et al., 2008). These 2 structures are thought to form a "scaffold" between which the 6-layered (L1–6) neocortex is created. Interestingly, while thalamic fibers reaching the neocortex first target the SP, a population of thalamic fibers also targets the MZ, thereby bringing thalamic activity into the 2 boundaries of the developing cortical plate (Figure 50.1A).

A number of studies have focused on the developmental trajectory of the thalamocortical fibers in the ventral side of the neocortex and their initial contact with the early born population of neurons of the subplate. These contacts are not mere structural interactions, since early thalamic input activation elicits fast excitatory glutamatergic synaptic inputs on SP neurons (Friauf et al., 1990; Hanganu et al., 2002). These early structural and functional interactions are crucial for the maturation of neocortical circuits (for a comprehensive review on subplate neurons, see Kanold and Luhmann, 2010). Elimination of SP neurons using kainic acid injections in vivo at early postnatal days prevents the structural (barrel formation in rodent whisker somatosensory cortex or ocular dominance columns in cats) and the functional maturation of thalamocortical projections onto the main adult thalamorecipient layer (L4) (Ghosh and Shatz 1992; Tolner et al. 2012). At the same time, ablating SP neurons using p75-immunotoxin prevents the developmental upregulation of genes involved in mature fast GABAergic transmission in cortical L4. These genes include mature $GABA_A$ receptor subunits and the transporter KCC2 (Blaesse et al., 2009), potentially resulting in GABA acting as an excitatory neurotransmitter at later stages of development upon SP neuron ablation (Kanold and Shatz, 2006; Kanold and Luhmann, 2010). The sequela of these gene expression changes are unknown, but could lead to the persistence of giant depolarizing potentials (GDPs), observed early in the developing cortex (Ben-Ari et al., 1989).

Even though the SP was originally thought to comprise only of typical glutamatergic excitatory cells born in the dorsal telencephalon, it is now appreciated that it is composed of a variety of cell types with distinct morphological characteristics, some upright, others inverted or bipolar (Kostovic and Rakic, 1980; Hanganu et al., 2002). Moreover, not all of these neurons are glutamatergic, but many are GABAergic neurons that have tangentially migrated from the ventral telencephalon (ganglionic eminences-GE) through the so-called subventricular (SVZ)-intermediate zone (IZ) stream (Anderson et al., 1997; Lavdas et al., 1999). The somatic and dendritic morphological heterogeneity of SP neurons is matched by the diversity of their axonal targeting. While some SP cells have locally ramifying axons, others project to the thalamus (McConnell et al., 1989). Further projection areas include the MZ and the developing cortical plate, as well as distant cortical areas through long-range projections (Friauf et al., 1990; Xue et al., 2022). In accordance with their local projections, some SP cells create a local circuit through electrical and chemical synapses (Dupont et al., 2006). Transcriptomic analysis of SP cells using RNA sequencing has revealed a number of molecular markers that label distinct populations of SP neurons, one of which expresses *Lpar1*. Using transgenic mice that express GFP under the endogenous *Lpar1* gene, recent research in early postnatal cortex identified 2 distinct SP neurons, 1 displaying local axonal elaboration, the other projecting to L1. The 2 types of neurons were shown to also differ in the developmental progression of their inputs (Ghezzi et al., 2021).

Despite its developmental importance, the subplate is a largely transient structure since many of the SP cells disappear during the first stages of postnatal life through developmental cell death. Nevertheless, a subpopulation of SP neurons remains deep in the future L6 of the cortex (as well as interstitial cells in the white matter), where a number of excitatory cells are found to face toward the white matter, tangentially, or are polymorphic (Karayannis et al., 2007), similar to some SP cells during development.

THALAMO-MARGINAL ZONE DEVELOPMENTAL INTERACTIONS

During the birth of deep-layer pyramidal cells and the arrival of thalamic fibers into the subplate, L1 also starts being targeted by the axons of calbindin+ neurons located in several thalamic nuclei (Figure 50.1A) (Gesuita and Karayannis, 2021). Some thalamocortical projections display an extensive subpial arbor, which is exclusive to L1. Other axonal projections of thalamocortical neurons also branch in other cortical layers and even display collaterals in subcortical regions. In rodents, the growth cones of both types of axons extend through the cortical plate all the way to the MZ, with virtually no branching in other layers. Similarly to the subplate-targeting axons, once thalamic fibers reach the upper border of the CP at the end of embryogenesis, they pause for around 2 days before taking a sharp turn and extending tangentially within L1a. Even though much less studied compared to their ventral counterparts, the MZ thalamic axons seem to connect to the resident L1 GABAergic neurons during the first 10 postnatal days and to carry sensory signals that are essential for the development of the morphology and connectivity of L1 reelin+ interneurons (De Marco García et al., 2011; De Marco García et al., 2015). It may well be that some of these inputs are transient in nature, present early only to facilitate the development of some L1 neurons (Che et al., 2018). Nevertheless, adult L1 INs of the prefrontal cortex (PC) receive thalamic inputs which are able to drive their activation in vitro and hence the generation of feedforward (FF) inhibition onto PC dendrites (Cruikshank et al., 2012).

Before the arrival of GABAergic cells in L1, this layer is occupied by another very early-born population of neurons,

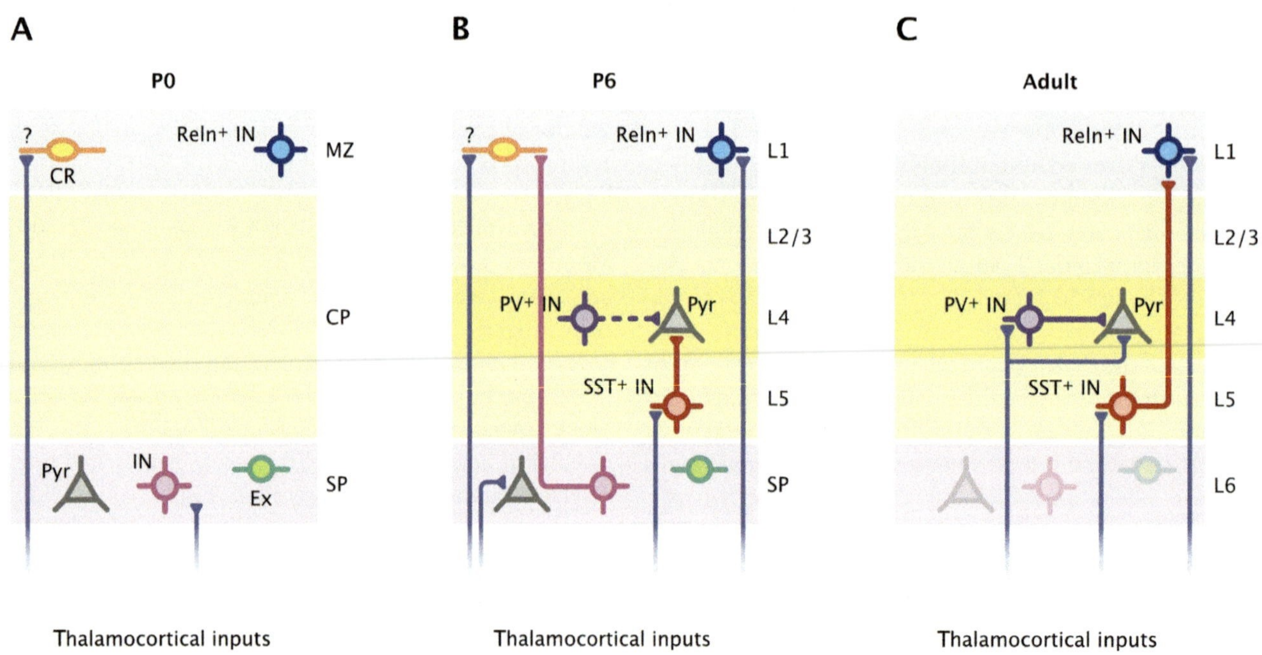

Figure 50.1. The postnatal development of thalamocortical and intracortical connections. (A). At birth (postnatal day 0; P0) thalamic axons target the subplate (SP) and marginal zone (MZ). They synapse onto SP cells and potentially (?) onto Cajal-Retzius cells (CR). The subplate contains a variety of different cell types: pyramidal (Pyr), inhibitory (IN), and other excitatory (Ex) cells. Whether there is a preference for thalamic fibers innervating specific SP types is not clear. (B). By P6 subplate INs project to Layer 1 CR cells. At the same time, thalamic fibers contact L5b somastostatin-expressing (SST) cells, which in turn synapse onto L4 pyramidal cells (Pyr). Parvalbumin-expressing (PV) cells of L4 begin to contact L4 pyramidal cells. Thalamic axons also target L1 reelin-expressing interneurons (Rln) and drive their circuit integration. (C). In the adult, CR have largely disappeared and SP cells are reduced in numbers (faint color). The characteristic feedforward circuit onto L4 Pyr by thalamic input and PV cells has matured, and SST cells connect to L1 interneurons.

Cajal-Retzius (CR) cells. These neurons originate from multiple embryonic progenitor domains that include the hem, the pallial-subpallial boundary, and the septum and migrate to L1 tangentially during embryogenesis (Bielle et al., 2005). Despite their rise to fame as the cells that express and release reelin, they are also connected to the resident circuit. Their inputs seem to be almost exclusively from presynaptic GABAergic cells, some of which originate from deep cortical neurons, including the SP. The initial data for this observation come mainly from extracellular stimulation of deeper layers and recordings of CR cells in vitro (Kirmse et al., 2007). Subsequent research utilizing pseudo-typed rabies virus to trace direct presynaptic cells onto hem-derived CR cells identified SST-expressing (SST^{+}) and calbindin-expressing (CB^{+}) neurons sitting in superficial, but also deep cortical layers (Cocas et al., 2016). Therefore neurons in the deep cortical layers of sensory cortices, including the SP, which receive early thalamic inputs, seem to interact perinatally with MZ cells (Xue et al., 2022) (Figure 50.1B), the function of which is currently not understood. What has nevertheless been uncovered is a role for prenatal spontaneous thalamic activity and NMDA receptors in CR cells in regulating the neonatal density of these cells, but not their cell death (Genescu et al., 2022). These findings suggest that somehow CR cells receive whisker-related information from the thalamus through glutamatergic pathways (Figure 50.1A).

THALAMO-CORTICAL INNERVATION AND STRUCTURAL DEVELOPMENT

After the "waiting period" of thalamic fibers in the SP and the subsequent gradual loss of SP cells, the fibers progress to L4 to target the local cells. During the first few postnatal days, thalamocortical axons refine their targeting and segregate into the characteristic "barrel" structures that correspond somatotopically to the mystaceal vibrissae. This process is largely complete by the end of the first postnatal week, with the creation of "barrels" and the spaces between them called "septa." The formation of the barrels necessitates the engagement of a number of genetic and activity-dependent programs, including sensory-driven activity, which is indispensable. In this regard, it has been shown that removal of the whiskers or severance of the trigeminal nerve that carries the sensory information from birth until P3 leads to a lack of barrel formation, even though thalamocortical fibers are present. In contrast, manipulation of sensory-driven activity after P3–5 does not disrupt the creation of barrels, but has effects on how certain cell types develop (see below). The thalamic fibers that project to L4 and form the barrels originate from the primary whisker thalamic nucleus called the ventro-posterior nucleus (VPN) (Lee and Sherman, 2010). The same nucleus also projects to L5b and L1 of the barrel cortex. In contrast, the secondary higher-order nucleus, the posterior nucleus (POm), sends projections to L5a and L2/3, as well as a different part of L1. Fluorescent labeling of the POm shows the projection fibers going through the septa to reach L1, hence creating a 2-orthogonal-axes segregation between the VPN and POm cortical projections (El-Boustani et al., 2020).

The barrels are an entity pertaining not only to the presynaptic thalamocortical fibers, but also to the organization of the postsynaptic excitatory cells of L4, most of which are of the glutamatergic stellate cell type. L4 stellate cells also segregate into clusters that correspond to the incoming thalamic axons and preferentially orient their dendrites toward the center of the barrel to receive the incoming information. Genetic ablation studies that reduce thalamocortical synaptic transmission have unveiled its importance for the proper formation of the barrels (Narboux-Nême et al., 2012; Li et al., 2013). Further studies found that the presence of NMDA receptors in stellate cells is also key for formation of the barrels and the inward orientation of their dendrites. Deletion of the necessary NMDA receptor subunit NR1 from all excitatory cortical cells disrupts the formation of the barrels (Iwasato et al., 2000), and sparse cell autonomous deletion of the NR2B subunit in stellate cells leads to an increase in the number of dendrites that span multiple barrels (Espinosa et al., 2009). Altering other components of the glutamatergic pathway in stellate cells, such as metabotropic glutamate receptor 5 (mGluR5) and PLCβ (phospholipase C-β), can also lead to defects in the segregation of the cortex into barrels (Erzurumlu and Gaspar, 2012; Ballester-Rosado et al., 2016). In addition, BTBD3, FGF/FGFR, Lhx2, and RORα add to the repertoire of key genes in stellate cells for the refinement of their dendritic tree in the barrels (Wang et al., 2017; Iwasato and Erzurumlu, 2018; Vitalis et al., 2018).

Besides sensory activity and intrinsic genetic programs, other systems also regulate thalamocortical axon development. Serotonergic fibers, as well as those from other neuromodulatory cells, innervate the developing cortex during embryogenesis. There are multiple potential sites of interaction between these ascending axonal systems and thalamocortical fibers—one of the most studied of which being the serotonergic system (Erzurumlu and Gaspar, 2012). Interestingly, the serotonergic transporter (SERT) is found on the membrane of developing thalamocortical terminals, and its removal or that of monoamine oxidase enzyme (MAO) leads to defects in the patterning of the characteristic barrel structure in L4 of the somatosensory whisker system (Lebrand et al., 1996; Cases et al., 1996). These data suggest that there is an early developmental cross-talk between the serotonergic and thalamocortical fibers in building the barrel cortex, a process which may be shared with other neuromodulatory systems.

THALAMO-CORTICAL INNERVATION AND FUNCTIONAL DEVELOPMENT OF CIRCUITS

As mentioned above, the subplate is crucial for proper thalamocortical innervation of L4, which is the considered the primary thalamorecipient layer in adult sensory cortices. After the initial innervation of the subplate by thalamic fibers, the latter transition to L4 excitatory cells, as well as to GABAergic cells of the same layer. This dual innervation drives a classical FF circuit in adult cortex leading to a biphasic synaptic response of excitation followed by inhibition, providing a window for information transfer in the brain within and across regions (Pouille and Scanziani, 2001) (Figure 50.1C). In the developing cortical plate, as is the case in the adult cortex, this FF thalamocortical circuit has been shown to involve L4 parvalbumin-expressing (PV^+) GABAergic cells (Figure 50.1B). These interneurons are functionally innervated by thalamic fibers at the end of the first postnatal week, which is later than excitatory cells (Daw et al. 2007). Even though there does not appear to be FF inhibition onto L4 excitatory cells at postnatal day (P)4/5, GABA signaling is active, and the activation of $GABA_A$ receptors on stellate cells leads to membrane depolarization due to the absence of the transporter KCC2 and hence an inverse concentration gradient of chloride (Daw et al., 2007). Further research revealed that this FF circuit is not hardwired, but some of its components are set up via sensory-dependent mechanisms driving the development of presynaptic inputs onto L4 PV^+ neurons and postsynaptic NMDA receptor components onto stellate cells (Chittajallu et al. 2010). Interestingly, sensory-driven activity, impinging directly or indirectly onto L4 excitatory cells, was also reported as a key component for the assembly of interhemispheric connections provided by these cells. Even though the adult L4 excitatory neurons do not project to the contralateral cortex, in the first 2 postnatal weeks they profoundly do so, and whisker-driven activity has been shown to be important for the proper developmental elimination of these connections over time (De León Reyes et al., 2019).

As already mentioned, L4 is not the only thalamorecipient layer of the cortex receiving inputs from VPN or POm. Research shows that GABAergic cells of deeper and superficial layers also receive inputs from thalamic fibers during the first 2 postnatal weeks. Somatostatin-expressing (SST^+) GABAergic cells of L5b receive thalamic input before PV^+ cells of L4 provide their output to L4 stellate cells. By manipulating SST^+ cells, it was shown that initial thalamocortical connection with these interneurons is important for the proper development of L4 FF inhibition on stellate cells mediated by PV^+ neurons (Figure 50.1B) (Marques-Smith et al., 2016; Tuncdemir et al., 2016). This sequential thalamic innervation of SST^+ cells followed by PV^+ cells aligns with their developmental generation, since the former are the earliest-born GABAergic cells and the first to integrate and mature into the nascent cortical circuits (Miyoshi et al., 2007; Wang et al., 2019). Interestingly, SST^+ cells have been reported to also have an axon that extends toward the pia surface, as well as toward the SP (Butt et al., 2017). This divergent and extensive innervation pattern in the neocortex is in line with their role as "hub" neurons in the hippocampus, orchestrating perinatal circuit activity via a depolarizing action of GABA and the generation of GDPs (Bonifazi et al., 2009). The presence of such a phenomenon in the neocortex has been recently postulated to also exist, with SST^+ cells displaying higher synaptic connectivity than other GABAergic cells at early postnatal stages, thereby disproportionally influencing spontaneous cortical activity (Wang et al., 2019). Furthermore, perinatal activation of SST^+ cells is also important for the development of pyramidal cell dendrites and the formation of spines and glutamatergic synapses onto them (Oh et al., 2016). Taken together, these data suggest that early innervation of SST^+ cells by thalamic inputs orchestrates sensory-driven neonatal activity patterns in the cortex, important for driving the assembly of both glutamatergic and GABAergic circuit components.

The distribution of GABAergic cells is not uniform across all layers, with SST^+ and PV^+ cells being more abundant in the deep layers (4–6) and reelin only-expressing (Rln^+) and vasoactive intestinal peptide–expressing cells (VIP^+) in higher numbers in the superficial layers (1–3). Utilizing a pseudo-typed rabies viral approach that traces neurons providing monosynaptic inputs to the superficial Rln^+ cells, it was uncovered that these neurons receive thalamic input from both the VPN and POm nuclei during the first 2 postnatal weeks (De Marco García et al., 2015). Importantly, these thalamic inputs, as well as whisker-driven activity, but not intracortical inputs, were key in driving not only the axo-dendritic development of these neurons, but also their proper integration into cortical circuits (De Marco García et al., 2015). The importance of these bottom-up inputs in organizing the maturation of intracortical inputs was at the level of the local, as well as top-down inputs coming from the frontal areas (Ibrahim et al. 2021). Further work reported that the thalamocortical inputs innervating these L1 Rln^+ cells also assist through a yet unknown mechanism in the fine-grain formation of the barrels in L4, thereby affecting mouse whisker-dependent sensory-discrimination abilities (Che et al., 2018).

Beyond Rln^+, superficial SST^+ and VIP^+ cells also receive thalamic inputs during development that change over time. Before the end of the second postnatal week and the beginning of mouse whisking (i.e., active tactile exploration through the whiskers), both of these GABAergic interneuron populations receive abundant inputs from thalamic cells residing in the VPN and POm. This connectivity changes drastically after the second postnatal week, when VIP⁺ neurons maintain some connections to the sensory thalamic nuclei, albeit fewer, whereas SST⁺ cells switch their connectivity to receive more from motor-related thalamic nuclei. These nuclei include the ventromedial (VM) and ventrolateral (VL), which in the adult are known to have dense axonal ramifications in L1 of a large part of the neocortex, where the dendrites of superficial SST⁺ can also be found (Kastli et al., 2020). These significant changes in thalamocortical innervation, together with the distinct intracortical inputs onto these 2 GABAergic cells, determine their developmental activation to a multi- or single-whisker stimulus (see also the following **Section of this Chapter** and Kastli et al., 2020). Similar to the importance of thalamic inputs in regulating aspects of the development of Rln⁺ neurons, a sub-population of VIP⁺ neurons necessitates sensory-driven activity to acquire the characteristic bipolar extension of their dendrites (De Marco García et al., 2011; De Marco García et al., 2015). Nevertheless, this is not a generalizable phenomenon, as another sub-population of VIP⁺ neurons that show a multipolar morphology is unaffected by sensory manipulations.

One should take note that even though research on thalamocortical interactions has almost exclusively focused on neurons, they are not the only cells receiving thalamic innervation during development. NG2-expressing oligodendrocyte progenitor cells (NG2⁺ cells) of the barrel cortex receive inputs from thalamic axons at the time when the barrel-septa domains are being formed. The thalamic inputs not only drive the distribution of NG2⁺ cells to the septal regions surrounding the developing barrels, but also regulate their proliferation (Mangin et al., 2012). To what extent similar processes are in place for other non-neuronal cells of the cortex remains to be determined.

THE DEVELOPMENT OF SENSORY-DRIVEN ACTIVITY IN THE SOMATOSENSORY BARREL CORTEX

The somatotopic organization of the brain has been revealed by pioneering studies in both the human cortex and the cortex of other species. The whisker somatosensory system of the mouse is also somatotopically organized along the whole information path and ultimately also in the primary cortical region called the barrel cortex, due to the segregated thalamocortical fiber ramification in L4 (Woolsey and Van der Loos, 1970). Therefore, unilateral whisker stimulation evokes spiking activity in the contralateral barrel cortex in a whisker-specific and somatotopically organized manner (Welker, 1976). In accordance with the connectivity organization of the incoming thalamic input and intracortical information flow, whisker deflection first leads to spiking in L4 and then L2/3 and L5 in the adult rodent. The inter- and intra-layer connections that provide the highways for intra-barrel information flow develop over the first postnatal weeks with L4 excitatory cells projecting and innervating L2/3 cells during the first week, after which L2/3 neurons project to L5 cortical output cells via the septal regions (Sehara et al., 2010) and at the same time begin connecting to each other (Bureau et al., 2004). Based on the literature, the cellular long-term plasticity (long-term potentiation and long-term depression) follows along the same developmental sequence of events (Fox, 2009).

The cortex is electrically active at birth and a number of distinct spontaneously arising patterns have been described in vivo that have different spatial and temporal scales (Luhmann and Khazipov, 2018). Even though the thalamocortical fibers are not yet segregated into barrels and septa, whisker stimulation elicits activity in the barrel cortex of rodents right after birth (Quairiaux et al., 2011; van der Bourg et al., 2019; Cai et al., 2022). This activity is already somewhat organized in a columnar and somatotopic fashion within the first couple of postnatal days, and further refines in subsequent days at the structural and functional level (van der Bourg et al., 2019) (Figure 50.2A). This early vertical organization of activity is dependent on the subplate, which is itself driven strongly by whisker stimuli perinatally (Kanold and Luhmann, 2010). In accordance with the delayed FFI microcircuit formation of L4, the initial whisker-dependent activation of the subplate and L4 do not seem to depend on inhibition. Instead, whisker deflections in the middle of the first postnatal week, or electrical stimulation of the thalamus, lead to a repetitive oscillatory excitation of the VPN cells at the gamma frequency range that is subsequently transferred to L4 stellate cells as a barrage of excitatory inputs at the same frequency, as seen when recording these neurons intracellularly in vivo (Minlebaev et al., 2011; Yang et al., 2013). A few days later, this excitatory barrage is also accompanied by a similar input structure for inhibitory synaptic events (Minlebaev et al., 2011). This perhaps leads to L2/3 cells displaying relatively mature receptive fields, responding stronger to the principal whisker already at the beginning of the second postnatal week (P8) (Bureau et al., 2004). Further recordings from unidentified L2–L5 cells of mice in the second postnatal week showed that whisker-driven excitatory synaptic responses were smaller in amplitude than the ones recorded in the third postnatal week. Nevertheless, the responses recorded in the second week displayed short-term facilitation that could lead to spiking of the recorded cell, in contrast to the later time point, when responses showed short-term depression instead (Borgdorff et al., 2007).

The contribution of the developing inhibition in sensory-driven activation of the cortex has not received as much attention. Using a mouse genetics approach to abolish

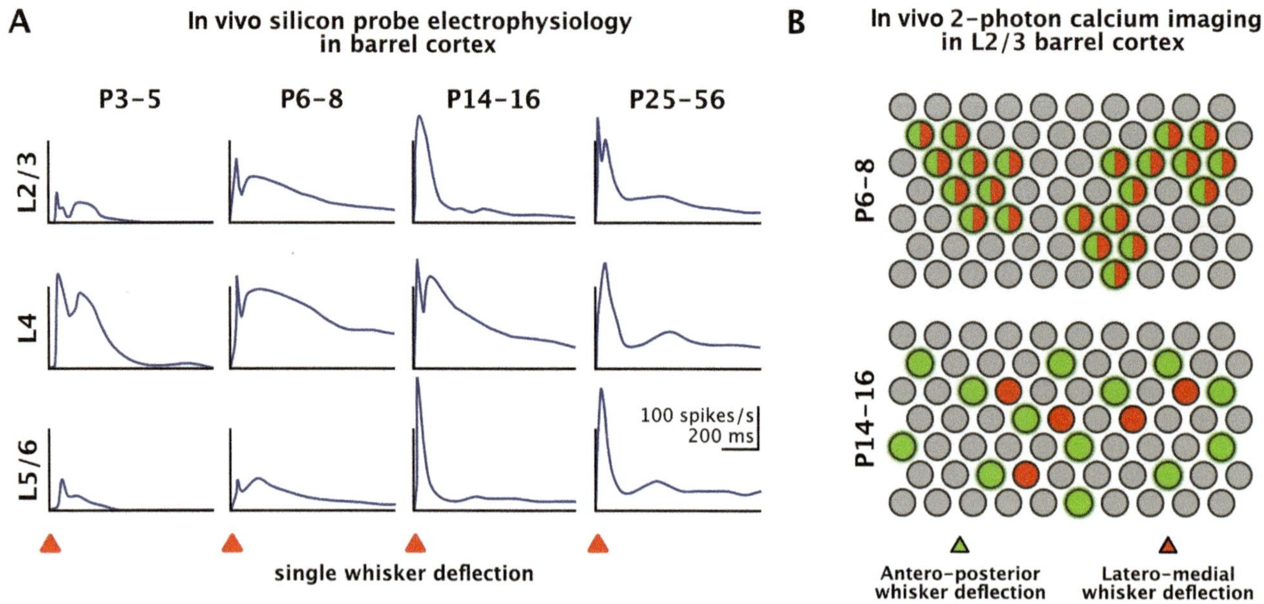

Figure 50.2. The postnatal whisker-driven activation of the somatosensory cortex in vivo. (A). A single whisker deflection leads to very distinct spiking patterns across layers (L) of the barrel cortex at postnatal day (P) 3–5, 6–8, 14–16, and P24–56 as recorded with a multisite silicon probe. At P3–5, activity is stronger in the L4 and much weaker and delayed in L2/3 and L5/6. At P6–8, L2/3 is much more activated compared to the previous time point. Spiking activity in all layers, and especially in L2/3 and L4, is very much prolonged. At P6–8, spiking activity is sharper especially in L2/3 and L5, whereas it remains prolonged in L4. By P25–56, all layers show a narrower spiking profile that displays 2 peaks. (B). A single whisker deflection in different directions leads to distinct activity patterns in L2/3 at P6–8 versus P14–16 as recorded with two-photon calcium imaging. At P6–8, a large population of L2/3 neurons of the barrel cortex is activated whether a whisker deflection is in the antero-posterior (represented by the green color) or latero-medial direction (represented by the red color). In contrast, at P14–16, a smaller fraction of L2/3 neurons is activated with one or the other deflection direction, gaining specificity.

neurotransmitter release by developmentally expressing tetanus toxin, the role of SST$^+$ and VIP$^+$ GABAergic populations in sensory-driven activity was assessed early postnatally. Silencing SST$^+$ interneuron output led to a reduction in early spontaneous activity and a change in the sensory adaptation upon repetitive whisker stimuli. In contrast, VIP$^+$ cell silencing only started having an effect on sensory-driven activity closer to the onset of whisking (Baruchin et al., 2022). Recent work has examined the cellular mechanisms behind the impact of SST$^+$ cells in sensory adaptation. SST$^+$ cells are known to receive excitatory inputs that display strong short-term facilitation. Genetically reducing this synaptic facilitation significantly contributes to the delayed recruitment of SST$^+$ cells and the subsequent effect of the latter in the spiking adaptation of cortical pyramidal cells upon repetitive whisker deflection. Based on the developmental timecourse of the appearance of this short-term facilitation onto SST$^+$ cells in different layers, it was revealed that these interneurons first developmentally contribute to L4 spiking adaptation and at a later time point to L2/3 (Stachniak T, personal communication).

In the adult mice it has been shown that movement of the whiskers, even in the absence of a touch event, leads to increased barrel cortex activity. This whisker-dependent activity change in this sensory area is also brought about by body movement and locomotion. The locomotion effect does not seem to be unique to the barrel cortex, but rather to act as a mechanism for creating an overall increased baseline activity level in different parts of the cortex. A similar phenomenon has been shown to take place in neonatal mouse pups prior to P14, even though they only display unilateral whisker pad twitches and not active bilateral whisking (Dooley et al., 2020). These results confirm the importance of movement-based sensory feedback in driving activity in sensorimotor cortex (Khazipov et al., 2004).

Beyond the primary sensory area (barrel cortex) it has been revealed that passive or active whisker deflection under anesthesia or in awake behaving adult mice respectively also leads to the activation of the secondary whisker cortical area, motor cortex, and higher-order processing areas in the parietal and frontal parts (Ferezou et al., 2007). Recent work showed that the barrel and secondary whisker-related area are activated by passive whisker deflections right after birth, with the 2 areas not showing any functional interaction until the end of the first postnatal week. At this time point the primary area provides functional excitation to the secondary, switching to a more inhibitory action after the end of the second postnatal week (Cai et al., 2022). Other work has reported that the primary whisker motor area is activated together with the barrel cortex upon whisker deflection at the end of the first and beginning of the second postnatal week (Gómez et al., 2021). This parallel bottom-up thalamic activation changes to include a serial interaction of the 2 cortical areas toward the end of the second postnatal week. Overall, these results indicate a parallel processing of whisker-related information by the primary, secondary, and related motor cortical areas until around the end of the first postnatal week, after which sequential cortical processing begins from the barrel cortex to other cortical areas and likely vice versa.

The in vivo extracellular and intracellular electrophysiological recordings performed in developing rodents have been complemented by in vivo single- and two-photon calcium imaging experiments on the barrel cortex or a larger cortical area. Allowing for a spatial and cell type-specific assessment of cortical activity compared to the electrophysiological studies described above, two-photon calcium imaging has uncovered multiple transitions of sensory-evoked and spontaneous cortical activity at the end of the second postnatal week. Whereas prior to P14 many L2/3 cells are synchronously activated, after P14 cell activity is more sparse and desynchronized (Golshani et al., 2009), a switch that does not seem to be dependent on sensory inputs (Figure 50.2B). In addition, around the same "passive" to "active" sensing transition time point, pyramidal neurons of L2/3 were found to acquire specification as to which whisker deflection direction they will mostly respond to (van der Bourg et al., 2017) (Figure 50.2B). Along the same lines and in accordance with structural rearrangements (see above), there is significant change in the activation of L2/3 SST^+ and VIP^+ cells in terms of their "preferred" stimulus, with the former retaining and enhancing preference to multi-whisker deflection after P14, whereas the latter losing it (Kastli et al., 2020).

In conclusion, many decades of research in developmental neuroscience have led to significant progress in mapping out some of the key dynamic events in the assembly of thalamocortical and intracortical circuits. Based on its current trajectory, this line of research promises to bring many more insights into the development of sensory processing in the next few decades.

ACKNOWLEDGMENTS

I would like to thank Lorenzo Gesuita for help with the figures. Research in the Karayannis lab has been supported by the Swiss National Foundation (SNF), the European Research Council, and the University of Zurich. I would like to apologize to all the researchers who have produced relevant work to this chapter but were not cited herein due to space limitations.

REFERENCES

Anderson SA, Eisenstat DD, Shi L, Rubenstein JL (1997). Interneuron migration from basal forebrain to neocortex: dependence on Dlx genes. *Science.* 278:474–476.

Antón-Bolaños N, Sempere-Ferràndez A, Guillamón-Vivancos T, Martini FJ, Pérez-Saiz L, Gezelius H, Filipchuk A, Valdeolmillos M, López-Bendito G (2019). Prenatal activity from thalamic neurons governs the emergence of functional cortical maps in mice. *Science.* 364:987–990.

Ballester-Rosado CJ, Sun H, Huang J-Y, Lu H-C (2016). mGluR5 exerts cell-autonomous influences on the functional and anatomical development of layer IV cortical neurons in the mouse primary somatosensory cortex. *J Neurosci.* 36:8802–8814.

Baruchin LJ, Ghezzi F, Kohl MM, Butt SJB (2022). Contribution of interneuron subtype-specific GABAergic signaling to emergent sensory processing in mouse somatosensory whisker barrel cortex. *Cereb Cortex.* 32:2538–2554.

Ben-Ari Y, Cherubini E, Corradetti R, Gaiarsa JL (1989). Giant synaptic potentials in immature rat CA3 hippocampal neurones. *J Physiol.* 416:303–325.

Bielle F, Griveau A, Narboux-Nême N, Vigneau S, Sigrist M, Arber S, Wassef M, Pierani A (2005). Multiple origins of Cajal-Retzius cells at the borders of the developing pallium. *Nat Neurosci.* 8:1002–1012.

Blaesse P, Airaksinen MS, Rivera C, Kaila K (2009). Cation-chloride cotransporters and neuronal function. *Neuron.* 61:820–838.

Bonifazi P, Goldin M, Picardo MA, Jorquera I, Cattani A, Bianconi G, Represa A, Ben-Ari Y, Cossart R (2009). GABAergic hub neurons orchestrate synchrony in developing hippocampal networks. *Science.* 326:1419–1424.

Borgdorff AJ, Poulet JFA, Petersen CCH (2007). Facilitating sensory responses in developing mouse somatosensory barrel cortex. *J Neurophysiol.* 97:2992–3003.

Bureau I, Shepherd GMG, Svoboda K (2004). Precise development of functional and anatomical columns in the neocortex. *Neuron.* 42:789–801.

Butt SJ, Stacey JA, Teramoto Y, Vagnoni C (2017). A role for GABAergic interneuron diversity in circuit development and plasticity of the neonatal cerebral cortex. *Curr Opin Neurobiol.* 43:149–155.

Bystron I, Blakemore C, Rakic P (2008). Development of the human cerebral cortex: Boulder Committee revisited. *Nat Rev Neurosci.* 9:110–122.

Cai L, Yang J-W, Wang C-F, Chou S-J, Luhmann HJ, Karayannis T (2022). Identification of a developmental switch in information transfer between whisker S1 and S2 cortex in mice. *J Neurosci.* 42:4435–4448.

Cases O, Vitalis T, Seif I, De Maeyer E, Sotelo C, Gaspar P (1996). Lack of barrels in the somatosensory cortex of monoamine oxidase A-deficient mice: role of a serotonin excess during the critical period. *Neuron.* 16:297–307.

Che A, Babij R, Iannone AF, Fetcho RN, Ferrer M, Liston C, Fishell G, De Marco García N V (2018). Layer I interneurons sharpen sensory maps during neonatal development. *Neuron.* 99:98–116.

Chittajallu R, Isaac JTR (2010). Emergence of cortical inhibition by coordinated sensory-driven plasticity at distinct synaptic loci. *Nat Neurosci.* 13:1240–1248.

Cocas LA, Fernandez G, Barch M, Doll J, Zamora Diaz I, Pleasure SJ (2016). Cell type-specific circuit mapping reveals the presynaptic connectivity of developing cortical circuits. *J Neurosci.* 36:3378–3390.

Cruikshank SJ, Ahmed OJ, Stevens TR, Patrick SL, Gonzalez AN, Elmaleh M, Connors BW (2012). Thalamic control of layer 1 circuits in prefrontal cortex. *J Neurosci.* 32:17813–17823.

Daw MI, Ashby MC, Isaac JTR (2007). Coordinated developmental recruitment of latent fast spiking interneurons in layer IV barrel cortex. *Nat Neurosci.* 10:453–461.

De León Reyes NS, Mederos S, Varela I, Weiss LA, Perea G, Galazo MJ, Nieto M (2019). Transient callosal projections of L4 neurons are eliminated for the acquisition of local connectivity. *Nat Commun.* 10:4549.

De Marco García NV, Karayannis T, Fishell G (2011). Neuronal activity is required for the development of specific cortical interneuron subtypes. *Nature.* 472:351–355.

De Marco García NV, Priya R, Tuncdemir SN, Fishell G, Karayannis T (2015). Sensory inputs control the integration of neurogliaform interneurons into cortical circuits. *Nat Neurosci.* 18:393–401.

Dooley JC, Glanz RM, Sokoloff G, Blumberg MS (2020). Self-generated whisker movements drive state-dependent sensory input to developing barrel cortex. *Curr Biol.* 30:2404–2410.e4.

Dupont E, Hanganu IL, Kilb W, Hirsch S, Luhmann HJ (2006). Rapid developmental switch in the mechanisms driving early cortical columnar networks. *Nature.* 439:79–83.

El-Boustani S, Sermet BS, Foustoukos G, Oram TB, Yizhar O, Petersen CCH (2020). Anatomically and functionally distinct thalamocortical inputs to primary and secondary mouse whisker somatosensory cortices. *Nat Commun.* 11:3342:1–12.

Erzurumlu RS, Gaspar P (2012). Development and critical period plasticity of the barrel cortex. *Eur J Neurosci.* 35:1540–1553.

Espinosa JS, Wheeler DG, Tsien RW, Luo L (2009). Uncoupling dendrite growth and patterning: single-cell knockout analysis of NMDA receptor 2B. *Neuron.* 62:205–217.

Ferezou I, Haiss F, Gentet LJ, Aronoff R, Weber B, Petersen CCH (2007). Spatiotemporal dynamics of cortical sensorimotor integration in behaving mice. *Neuron.* 56:907–923.

Fox K (2009). Experience-dependent plasticity mechanisms for neural rehabilitation in somatosensory cortex. Philos Trans R Soc London Ser B Biol Sci. 364:369–381.

Friauf E, McConnell SK, Shatz CJ (1990). Functional synaptic circuits in the subplate during fetal and early postnatal development of cat visual cortex. *J Neurosci.* 10:2601–2613.

Genescu I, Aníbal-Martínez M, Kouskoff V, Chenouard N, Mailhes-Hamon C, Cartonnet H, Lokmane L, Rijli FM, López-Bendito G, Gambino F, Garel S (2022). Dynamic interplay between thalamic activity and Cajal-Retzius cells regulates the wiring of cortical layer 1. *Cell Rep.* 39:110667:1–14.

Gesuita L, Karayannis T (2021). A "marginal" tale: The development of the neocortical layer 1. *Curr Opin Neurobiol.* 66:37–47.

Ghezzi F, Marques-Smith A, Anastasiades PG, Lyngholm D, Vagnoni C, Rowett A, Parameswaran G, Hoerder-Suabedissen A, Nakagawa Y, Molnar Z, Butt SJ (2021). Non-canonical role for Lpar1-EGFP subplate neurons in early postnatal mouse somatosensory cortex. *eLife.* 10:e60810.

Ghosh A, Shatz CJ (1992). Involvement of subplate neurons in the formation of ocular dominance columns. *Science.* 255:1441–1443.

Golshani P, Gonçalves JT, Khoshkhoo S, Mostany R, Smirnakis S, Portera-Cailliau C (2009). Internally mediated developmental desynchronization of neocortical network activity. *J Neurosci.* 29:10890–10899.

Gómez LJ, Dooley JC, Sokoloff G, Blumberg MS (2021). Parallel and serial sensory processing in developing primary somatosensory and motor cortex. *J Neurosci.* 41:3418–3431.

Hanganu IL, Kilb W, Luhmann HJ (2002). Functional synaptic projections onto subplate neurons in neonatal rat somatosensory cortex. J Neurosci. 22:7165–7176.

Ibrahim LA, Huang S, Fernandez-Otero M, Sherer M, Qiu Y, Vemuri S, Xu Q, Machold R, Pouchelon G, Rudy B, Fishell G (2021). Bottom-up inputs are required for establishment of top-down connectivity onto cortical layer 1 neurogliaform cells. *Neuron.* 109(21):3473–3485.

Iwasato T, Datwani A, Wolf AM, Nishiyama H, Taguchi Y, Tonegawa S, Knöpfel T, Erzurumlu RS, Itohara S (2000). Cortex-restricted disruption of NMDAR1 impairs neuronal patterns in the barrel cortex. *Nature.* 406:726–731.

Iwasato T, Erzurumlu RS (2018). Development of tactile sensory circuits in the CNS. *Curr Opin Neurobiol.* 53:66–75.

Kanold PO, Luhmann HJ (2010). The subplate and early cortical circuits. *Annu Rev Neurosci.* 33:23–48.

Kanold PO, Shatz CJ (2006). Subplate neurons regulate maturation of cortical inhibition and outcome of ocular dominance plasticity. *Neuron.* 51:627–638.

Karayannis T, Huerta-Ocampo I, Capogna M (2007). GABAergic and pyramidal neurons of deep cortical layers directly receive and differently integrate callosal input. *Cereb Cortex.* 17:1213–1226.

Kastli R, Vighagen R, van der Bourg A, Argunsah AÖ, Iqbal A, Voigt FF, Kirschenbaum D, Aguzzi A, Helmchen F, Karayannis T (2020). Developmental divergence of sensory stimulus representation in cortical interneurons. *Nat Commun.* 11:5729:1–14.

Khazipov R, Sirota A, Leinekugel X, Holmes GL, Ben-Ari Y, Buzsáki G (2004). Early motor activity drives spindle bursts in the developing somatosensory cortex. *Nature.* 432:758–761.

Kirmse K, Dvorzhak A, Henneberger C, Grantyn R, Kirischuk S (2007). Cajal Retzius cells in the mouse neocortex receive two types of pre- and postsynaptically distinct GABAergic inputs. *J Physiol.* 585:881–895.

Kostovic I, Rakic P (1980). Cytology and time of origin of interstitial neurons in the white matter in infant and adult human and monkey telencephalon. *J Neurocytol.* 9:219–242.

Lavdas AA, Grigoriou M, Pachnis V, Parnavelas JG (1999). The medial ganglionic eminence gives rise to a population of early neurons in the developing cerebral cortex. *J Neurosci.* 19:7881–7888.

Lebrand C, Cases O, Adelbrecht C, Doye A, Alvarez C, El Mestikawy S, Seif I, Gaspar P (1996). Transient uptake and storage of serotonin in developing thalamic neurons. *Neuron.* 17:823–835.

Lee CC, Sherman SM (2010). Drivers and modulators in the central auditory pathways. *Front Neurosci.* 4:79–86.

Li H, Fertuzinhos S, Mohns E, Hnasko TS, Verhage M, Edwards R, Sestan N, Crair MC (2013). Laminar and columnar development of barrel cortex relies on thalamocortical neurotransmission. *Neuron.* 79:970–986.

Luhmann HJ, Khazipov R (2018). Neuronal activity patterns in the developing barrel cortex. *Neuroscience.* 368:256–267.

Mangin J-M, Li P, Scafidi J, Gallo V (2012). Experience-dependent regulation of NG2 progenitors in the developing barrel cortex. *Nat Neurosci.* 15:1192–1194.

Marques-Smith A, Lyngholm D, Kaufmann A-K, Stacey JA, Hoerder-Suabedissen A, Becker EBE, Wilson MC, Molnár Z, Butt SJB (2016). A transient translaminar GABAergic interneuron circuit connects thalamocortical recipient layers in neonatal somatosensory cortex. *Neuron.* 89:536–549.

McConnell SK, Ghosh A, Shatz CJ (1989). Subplate neurons pioneer the first axon pathway from the cerebral cortex. *Science.* 245:978–982.

Minlebaev M, Colonnese M, Tsintsadze T, Sirota A, Khazipov R (2011). Early γ oscillations synchronize developing thalamus and cortex. *Science.* 334:226–229.

Miyoshi G, Butt SJB, Takebayashi H, Fishell G (2007). Physiologically distinct temporal cohorts of cortical interneurons arise from telencephalic Olig2-expressing precursors. *J Neurosci.* 27:7786–7798.

Narboux-Nême N, Evrard A, Ferezou I, Erzurumlu RS, Kaeser PS, Lainé J, Rossier J, Ropert N, Südhof TC, Gaspar P (2012). Neurotransmitter release at the thalamocortical synapse instructs barrel formation but not axon patterning in the somatosensory cortex. *J Neurosci.* 32:6183–6196.

Oh WC, Lutzu S, Castillo PE, Kwon H-B (2016). De novo synaptogenesis induced by GABA in the developing mouse cortex. *Science.* 353:1037–1040.

Pouille F, Scanziani M (2001). Enforcement of temporal fidelity in pyramidal cells by somatic feed-forward inhibition. *Science.* 293:1159–1163.

Quairiaux C, Mégevand P, Kiss JZ, Michel CM (2011). Functional development of large-scale sensorimotor cortical networks in the brain. *J Neurosci.* 31:9574–9584.

Sehara K, Toda T, Iwai L, Wakimoto M, Tanno K, Matsubayashi Y, Kawasaki H (2010). Whisker-related axonal patterns and plasticity of layer 2/3 neurons in the mouse barrel cortex. *J Neurosci.* 30:3082–3092.

Stachniak TJ, Argunsah AÖ, Yang J-W, Cai L, Karayannis T (2023). Presynaptic kainate receptors onto somatostatin interneurons are recruited by activity throughout development and contribute to cortical sensory adaptation. *J Neurosci.* 10.1523:1461–22.

Tolner EA, Sheikh A, Yukin AY, Kaila K, Kanold PO (2012). Subplate neurons promote spindle bursts and thalamocortical patterning in the neonatal rat somatosensory cortex. *J Neurosci.* 32:692–702.

Tuncdemir SN, Wamsley B, Stam FJ, Osakada F, Goulding M, Callaway EM, Rudy B, Fishell G (2016). Early somatostatin interneuron connectivity mediates the maturation of deep layer cortical circuits. *Neuron.* 89:521–535.

van der Bourg A, Yang J-W, Reyes-Puerta V, Laurenczy B, Wieckhorst M, Stüttgen MC, Luhmann HJ, Helmchen F (2017). Layer-specific refinement of sensory coding in developing mouse barrel cortex. *Cereb Cortex.* 27:4835–4850.

van der Bourg A, Yang J-W, Stüttgen MC, Reyes-Puerta V, Helmchen F, Luhmann HJ (2019). Temporal refinement of sensory-evoked activity across layers in developing mouse barrel cortex. *Eur J Neurosci.* 50:2955–2969.

Vitalis T, Dauphinot L, Gressens P, Potier M-C, Mariani J, Gaspar P (2018). RORα coordinates thalamic and cortical maturation to instruct barrel cortex development. *Cereb Cortex.* 28:3994–4007.

Wang C-F, Hsing H-W, Zhuang Z-H, Wen M-H, Chang W-J, Briz CG, Nieto M, Shyu BC, Chou S-J (2017). Lhx2 expression in postmitotic cortical neurons initiates assembly of the thalamocortical somatosensory circuit. *Cell Rep.* 18:849–856.

Wang C-Z, Ma J, Xu Y-Q, Jiang S-N, Chen T-Q, Yuan Z-L, Mao X-Y, Zhang S-Q, Liu L-Y, Fu Y, Yu Y-C (2019). Early-generated interneurons regulate neuronal circuit formation during early postnatal development. *eLife.* May 23;8:e44649.

Welker C (1976). Receptive fields of barrels in the somatosensory neocortex of the rat. *J Comp Neurol.* 166:173–189.

Woolsey TA, Van der Loos H (1970). The structural organization of layer IV in the somatosensory region (SI) of mouse cerebral cortex: The description of a cortical field composed of discrete cytoarchitectonic units. *Brain Res.* 17:205–242.

Xue B, Meng X, Xu Y, Kao JPY, Kanold PO (2022). Transient coupling between infragranular and subplate layers to layer 1 neurons before ear opening and throughout the critical period depends on peripheral activity. *J Neurosci.* 42:1702–1718.

Yang J-W, An S, Sun J-J, Reyes-Puerta V, Kindler J, Berger T, Kilb W, Luhmann HJ (2013). Thalamic network oscillations synchronize ontogenetic columns in the newborn rat barrel cortex. *Cereb Cortex.* 23:1299–1316.

51.

CRITICAL PERIOD PLASTICITY AND DEVELOPMENT OF BINOCULAR VISION

Jianhua Cang and Adema Ribic

INTRODUCTION

When we refer to ensembles of interconnected neurons in the brain as neural circuits, we make an implicit analogy with circuits in electronic devices. In Chinese, for example, computers are also called "electric brains," making the analogy explicit. This analogy is a useful one, as it reminds us in neuroscience studies to determine what direction the information flows in a neural circuit and what computations it performs. However, it should be no surprise to any of us that fundamental differences exist between neural circuits in the brain and electronic circuits in a computer. One of the differences is that, whereas electronic circuits are hardwired, connections in neural circuits can be changed by the activity patterns they experience. Such experience-dependent processes are especially prominent during development, when they are required to establish, refine, or sculpt the connections of many circuits in the central and peripheral nervous systems. Several examples of experience-dependent neural development are reviewed in this book (see **Chapters 49, 52, and 55 in Section IX**), and none has been studied longer (or arguably has been better understood) than binocular development in the visual system.

A major benefit of binocular vision is that it gives us stereoscopic depth and distance perception, based on the small differences between the retinal images when the same object is viewed by the 2 eyes (i.e., binocular disparity). In humans, stereoscopic ability starts at 4–6 months old and continues to improve for years before reaching the adult level (Kiorpes, 2015). During this prolonged period, the growth in facial structure, eye positions, and interocular distance leads to continuous physical changes in binocular disparity, thus demanding corresponding physiological changes in the brain circuits that process binocular information. In other words, experience-dependent processes are needed to allow the visual system to adapt to the changing inputs and individual differences, as well as to differences in the external environment. The importance of visual experience in binocular development can be seen in individuals who had abnormal visual inputs during early childhood. For example, children whose eyes are improperly aligned or imbalanced, if not corrected in time, will grow up with no or severely compromised stereoscopic depth perception (von Noorden and Campos, 2002). Along this line, clinical observations and manipulations in animal studies have identified "critical periods" in early life, when the impacts of visual deprivations on binocular development are the most severe. In other words, experience-dependent changes in binocular circuits are most prominent during these critical periods, and the neural plasticity then declines with age. What mechanisms control the opening and closing of the critical periods, and what mechanisms mediate the experience-dependent changes during these time windows, are important questions in visual development and the topics of this chapter.

THREE MEASURES OF BINOCULAR VISION

Binocular development has been mostly studied in the primary visual cortex (V1), which is traditionally considered to be the first stage of binocular integration (Figure 51.1A). In the V1 input-recipient layer (L4 in carnivores and L4c in primates), the dLGN axonal terminals representing the 2 eyes are anatomically segregated and form alternating bands of eye-specific innervations. Within these so-called ocular dominance columns, most neurons respond only to 1 eye (i.e., monocular neurons). Neurons outside the input-recipient layer receive converging inputs from both eye columns, making them binocularly responsive (Hubel and Wiesel, 1962). In species that do not have ocular dominance columns (e.g., mice), binocular convergence already takes place in L4, with binocular neurons present in all layers of V1.

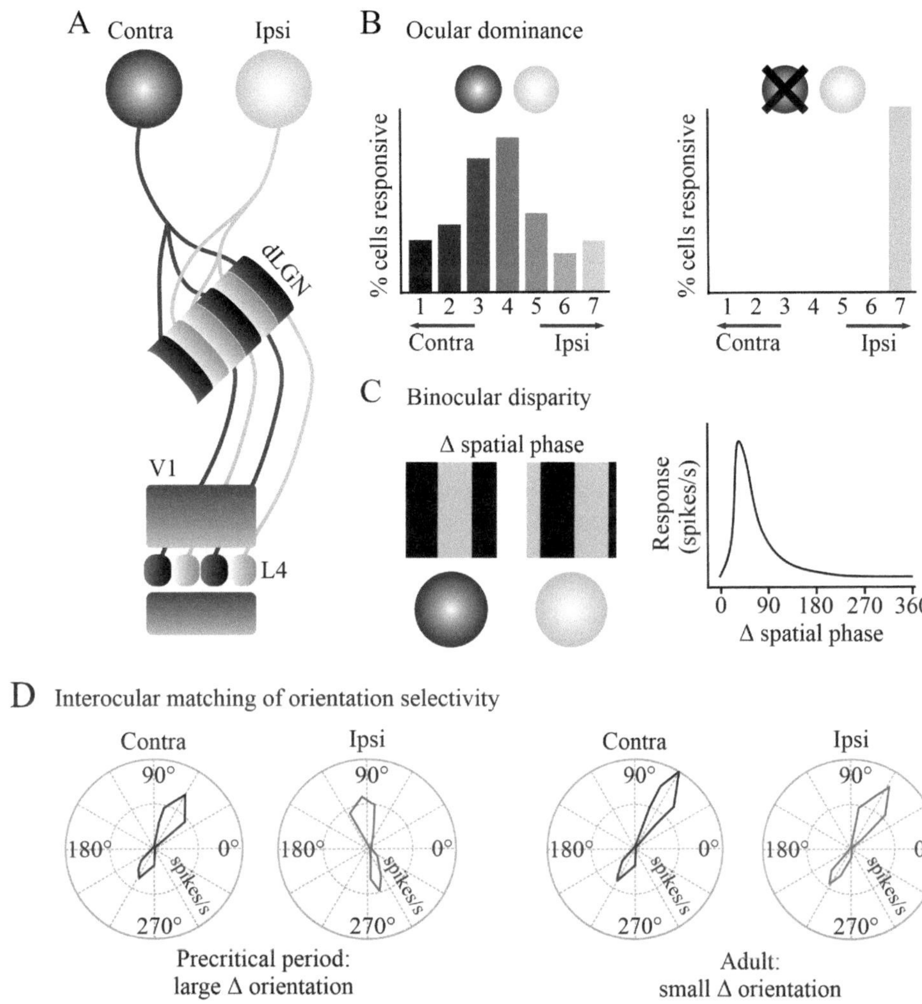

Figure 51.1. Anatomical and physiological basis of binocularity. (A) Schematics of primate visual system. Inputs from the 2 eyes terminate in eye-specific regions in the dorsal lateral geniculate nucleus (dLGN). dLGN neurons project to V1 layer 4, which maintains the segregation of inputs through eye-specific ocular dominance columns. (B) Ocular dominance is assessed through the responsivity of individual neurons to the stimulation of the contralateral and ipsilateral eyes and has been traditionally studied using a 7-point scale (Hubel and Wiesel, 1962), with purely monocular neurons in groups 1 and 7, and neurons with different degrees of ocular dominance in intermediate groups. The distribution shifts away from the deprived eye after monocular deprivation during a critical period in early life, which in monkeys and cats results in a complete loss of binocularity. (C) Binocular disparity selectivity measures the response selectivity of V1 neurons for small differences between the 2 retinal images. Disparity selectivity is essential for depth perception (stereopsis) and requires binocular visual experience to develop. (D) Interocular matching of orientation preference quantifies the similarity of individual V1 neurons' tuning through the 2 eyes. Orientation matching is established during a critical period and requires normal visual experience.

The binocularity of V1 neurons can be described by their relative responsiveness to monocular stimulations, namely ocular dominance, which ranges from responding only to contralateral eye stimulation to responding only to ipsilateral eye, with different levels of binocular neurons in between (Figure 51.1B). Binocular neurons can also be characterized by the similarity in their response selectivity. V1 neurons are well known for their orientation selectivity, and binocular neurons are tuned to similar orientations through the 2 eyes (Hubel and Wiesel, 1962) (Figure 51.1D). The degree of this similarity in tuning is an indicator of binocular visual development. The preferred orientations are often mismatched interocularly for individual V1 neurons in young animals, and the binocular matching process takes place in a critical period and requires visual experience, as first discovered in mice and later confirmed in ferrets (Wang et al., 2010; Chang et al., 2020). Finally, many V1 neurons are tuned to particular binocular disparities (Figure 51.1C), and their activity encodes the signals required for stereoscopic depth perception (Cumming and DeAngelis, 2001). Studies in developing cats and monkeys have observed adult-like disparity selectivity in quite young animals, near the beginning of the critical period (Freeman and Ohzawa, 1992; Chino et al., 1997). Binocular disparity selectivity is weakened by monocular deprivation during the critical period in mice

(Scholl et al., 2017), confirming its dependence on visual experience.

Compared to interocular matching and disparity selectivity, the significance of ocular dominance in binocular vision is not as clear (Cang et al., 2023). In fact, studies in several species have reported a lack of correlation between individual neurons' ocular dominance and the strength of their disparity tuning. Additionally, the population level of ocular dominance does not change during normal critical period development (Sato and Stryker, 2008), suggesting that experience-dependent circuit changes are not required for setting up normal ocular dominance. Despite these important considerations, ocular dominance plasticity—the phenomenon of massive ocular dominance shifts following monocular deprivation during the critical period—has been a productive experimental paradigm. Studies of ocular dominance plasticity have generated a rich body of information regarding the mechanisms underlying experience-dependent neural development and regulation of critical period timing. Here, we review general concepts and conclusions obtained from these studies, as well as current understandings of interocular matching and disparity selectivity. For all 3 measures of binocular vision, we also highlight significant knowledge gaps and unresolved controversies.

OCULAR DOMINANCE PLASTICITY AND CRITICAL PERIODS

In cats and monkeys, most V1 neurons are binocularly responsive, with a small fraction driven exclusively by the contralateral or ipsilateral eye (Hubel and Wiesel, 1962, 1968). Monocular eye closure during postnatal development degrades the cortical responses to the closed eye and increases the proportion of V1 neurons driven by the open eye, a phenomenon Hubel and Wiesel called "ocular dominance plasticity" to signify the functional and territorial expansion ("dominance") of the open eye in the visual cortex (Wiesel and Hubel, 1963). Monocular occlusion was later found to rapidly shrink the thalamocortical arbors serving the closed eye, providing the anatomical substrate for ocular dominance plasticity (Espinosa and Stryker, 2012). While structural deterioration after monocular deprivation was also seen in the lateral geniculate nucleus, robust functional effects appear restricted to the V1. Finally, ocular dominance shift of cortical responses is induced only if monocular deprivation is performed during a brief postnatal window, cannot recover spontanously, and would lead to permanent visual deficits, thus confirming the concept of a critical period.

In addition to monocular occlusion, surgical procedures that misalign the 2 eyes also caused a dramatic shift in ocular dominance, when performed within the critical period (Hubel and Wiesel, 1965). In this case, the vast majority of V1 neurons became monocular, responding to only one eye or the other, but not to both (Hubel and Wiesel, 1965; Kiorpes et al., 1998). A similar shift was also observed following alternating monocular occlusion during the critical period (Hubel and Wiesel, 1965). Together, these findings highlight the vital role of *balanced* visual input in the maturation of binocular vision. These manipulations have been used to model unilateral amblyopia and strabismus in animal studies, which have greatly contributed to the understanding and treatment of these disorders.

Early pharmacological experiments that attempted to identify the synaptic mechanisms of ocular dominance plasticity provided partial insight into the roles of neuronal activity in the effects of monocular deprivation (Hata et al., 1999; Hensch and Stryker, 2004). However, it was the research in mice that identified synaptic and cellular loci of critical period plasticity and its timing. Mice have laterally placed eyes, small binocular field of view, and lack clearly segregated ocular dominance columns. V1 neurons in mice are strongly driven by the contralateral eye even in the binocular zone, resulting in a skewed distribution of ocular dominance (Gordon and Stryker, 1996). Importantly, the ocular dominance distribution can be shifted by monocular deprivation. Similar to cats and monkeys, monocular deprivation in mice is most effective during a brief postanal window and results in degraded responses to the deprived eye (Gordon and Stryker, 1996). Studies in mice have demonstrated that critical period opening requires a certain level of inhibitory neurotransmission, and that critical period onset can be advanced or postponed through manipulation of cortical inhibition (Espinosa and Stryker, 2012). Levels of cortical inhibition slowly rise postnatally and are regulated by visual experience and brain derived neurotrophic factor (BDNF). GABA agonists and overexpression of BDNF can advance critical period onset, while critical period in mice lacking GABA-synthetizing enzyme glutamic acid decarboxylase 65 (GAD65) does not commence unless a GABA agonist is administered to the cortex, illustrating the central role of inhibition in critical period onset (Espinosa and Stryker, 2012).

Cortical inhibition is mostly mediated by 3 classes of interneurons: fast-spiking, parvalbumin positive interneurons (PV+), regular-spiking somatostatin positive (SST+) and vasoactive intestinal peptide positive (VIP+) interneurons (Ribic, 2020). While early studies suggested that PV+ interneurons mediate the effects of GABA agonists on critical period onset, more recent studies demonstrated that local excitatory drive onto PV+ interneurons is necessary for critical period opening (Gu et al., 2013). Monocular deprivation during the critical period results in rapid V1 disinhibition through loss of excitatory drive onto PV+ interneurons and permits subsequent shifts in cortical responses (Kuhlman et al., 2013; Severin et al., 2021).

In addition to disinhibition, residual retinal activity in the closed eye is necessary for the depression of its responses, suggesting that decreased cortical responsiveness to the closed eye proceeds through classical, Hebbian plasticity (Frenkel and Bear, 2004). In mice, 3–4 days of deprivation during the critical period is sufficent to depress the closed eye responses (Gordon and Stryker, 1996; Frenkel and Bear, 2004). Longer deprivation (≥5 days) results in a potentiation of open eye responses that occurs even after the closed eye is inactivated with tetrodotoxin (TTX) (Frenkel and Bear, 2004). As closed eye responses slightly recover after 5 days of deprivation, these findings point toward the involvement of homeostatic plasticity in the effects of prolonged deprivation (Frenkel and Bear, 2004; Mrsic-Flogel et al., 2007). Indeed, mice that lack tumor necrosis factor α (TNFα) show reduced homeostatic scaling of synaptic transmission in response to activity blockade and fail to show open eye potentiation after prolonged deprivation (Espinosa and Stryker, 2012). Deprivation-induced homeostatic plasticity is influenced by behavioral state and is most prominent during the animal's active cycles (Hengen et al., 2016), illustrating the importance of multifaceted in vivo approaches to studying the effects of monocular deprivation in mice.

As previously mentioned, functional effects of monocular deprivation are largely restricted to the visual cortex. However, recent mouse studies have demonstrated deprivation-induced functional plasticity in the dorsolateral geniculate nucleus (dLGN). Unlike structurally segregated layers of cells in the primate dLGN, mouse dLGN has roughly delineated contralateral and ipsilateral eye-specific patches (Liang and Chen, 2020). Mouse dLGN contains cells that can be driven by both eyes (Howarth et al., 2014), and recent studies demonstrate that prolonged monocular deprivation disrupts binocular responses and results in lasting functional changes in mouse dLGN (Rose and Bonhoeffer, 2018). It is unclear if short-term deprivation commonly used to induce ocular dominance plasticity in the V1 is sufficient to induce dLGN plasticity, or if the plasticity in the dLGN takes forms different from the V1, or if this is a mouse-specific feature. Future studies will undoubtedly address this, as well as the relationship between critical period timing in the V1 and plasticity in dLGN. This is particularly important given that V1 layer 4 PV+ interneurons receive the bulk of their excitatory drive from the thalamus (Ribic, 2020).

INTEROCULAR MATCHING OF ORIENTATION PREFERENCE

Compared to ocular dominance plasticity, interocular matching of orientation preference is a more direct demonstration of functional changes that normally take place during development. In mouse and ferret V1, individual neurons are often tuned to different orientations through the 2 eyes in young animals. The mismatch in orientation preference decreases in the following days (~P20–30 in mice; and ~P29–36 in ferrets) and reaches the adult level. Visual deprivation during this period, but not in adulthood, blocks the matching process (Wang et al., 2010; Wang et al., 2013; Chang et al., 2020). This experience-dependent matching of orientation preference is also evident at the population level in ferret V1, where similarly tuned neurons are organized into orientation columns. The monocular orientation maps in P30 ferrets are already adult-like, but they are not matched until 1 week later under normal visual experience (Chang et al., 2020).

Although both ocular dominance plasticity and interocular matching take place during similar periods in early life and are both driven by visual experience, there are important conceptual differences between the 2 phenomena. Ocular dominance plasticity is induced by monocular deprivation in pathological or experimental conditions, where cortical neurons lose responses to the deprived eye. In contrast, interocular matching is driven by normal vision and leads to a beneficial outcome, where cortical cells match their orientation preference to the 2 eyes. Mechanistically, interocular matching and ocular dominance plasticity are also different. Ocular dominance plasticity is a competitive process where eye-specific inputs "fight" for postsynaptic targets based on their activity levels. Interocular matching, on the other hand, is mediated by correlation-based processes where synaptic changes are under the precise and delicate control of the synchronized binocular inputs evoked by normal visual experience.

The receptive field basis of interocular matching of orientation preference has been revealed in mice. The 2 monocular receptive fields of individual V1 neurons show a significant overlap between the same sign subregions (e.g., ON-ON and OFF-OFF). To establish receptive field subregion correspondence requires normal visual experience, although experience-independent processes also seem to contribute to its development (Sarnaik et al., 2014). This result is consistent with the feedforward model of orientation selectivity (Hubel and Wiesel, 1962), which posits that the orientation selectivity of V1 "simple cells" is determined by the layout of the elongated ON and OFF subregions in their receptive fields, resulting from specific arrangement of precise geniculate inputs (Priebe and Ferster, 2012) (see also **Chapter 18 in Section V**). Also consistent with this model, it was shown that the thalamic and cortical inputs serving the same eyes are tuned to similar orientations in adult mice (Gu and Cang, 2016). Both inputs are matched interocularly in adult mice, but not in mice before the critical period. Interestingly, the matching process seems to follow a feedforward sequence, with thalamic inputs initiating their interocular matching before intracortical circuits (Gu and Cang, 2016).

Recent advances in 2-photon calcium imaging have allowed researchers to track the same cells during the matching process, thus revealing how V1 neurons change their monocular orientation preferences to match between the 2 eyes. One study demonstrated that Hebbian mechanisms may mediate binocular matching recovery after visual deprivation, where the weaker eye tended to change its orientation preference to align with that of the dominant eye (Levine et al., 2017). In ferrets, where the authors studied individual tunings and orientation maps in response to binocular stimulation, it is the binocular response that acts as a template during the matching process, and the monocular tuning becomes more similar to the binocular one (Chang et al., 2020). The results from both studies are consistent with a correlation-based matching process (Xu et al., 2020). Finally, imaging studies in young mice reported a substantial turnover between monocular and binocular populations during the period of binocular matching, which seems to depend on the selectivity of individual V1 neurons (Tan et al., 2020; Tan et al., 2021; Tan et al., 2022). These studies also reported a much smaller binocular population in mouse V1 than other studies. Future investigations are needed to resolve this discrepancy and provide a more complete understanding of the mechanisms underlying interocular matching (Cang et al., 2023).

BINOCULAR DISPARITY SELECTIVITY

Encoding binocular disparity is a critical neural computation in stereoscopic vision. Experimentally, disparity selectivity can be studied by stimulating the 2 eyes independently with controlled visual stimuli, such as sinusoidal gratings with varying interocular phase or position differences. Although they do not correspond to specific stereoscopic depths because of the repeated patterns in gratings, dichoptic gratings are quite useful in revealing individual neurons' monocular receptive fields and their binocular relationship. Ohzawa, Freeman, and colleagues used mostly this type of stimuli and performed a series of foundational studies to investigate how cat V1 neurons transform monocular inputs into disparity-selective responses (Ohzawa et al., 1997; Cang et al., 2023). Their studies led to an elegant model that remains the state-of-the-art understanding of disparity computation. This model postulates that the monocular receptive fields of individual binocular V1 neurons are spatially offset (position and/or phase, i.e., ON/OFF subfield composition); and the monocular inputs from the 2 receptive fields are linearly summated before going through an output nonlinearity (e.g., firing threshold). Consequently, only specific disparities that are precisely matched with the spatial offset would cause the neuron to discharge, thereby generating disparity selective responses.

Despite its success in explaining many experimental observations in monkeys (Cumming and Parker, 1997; Prince et al., 2002; Tsao et al., 2003), this model has not been directly tested at the level of circuit connectivity. This is likely due to the still limited circuit-dissection techniques in cats and monkeys. Furthermore, the disparity energy model is purely feedforward and only requires excitatory connections, while ignoring recurrent and inhibitory circuits. Hoping to fill these important gaps, studies have started to use additional model organisms, including mice and, more recently, tree shrews. Tanabe et al. (2022) directly compared disparity selectivity in these 2 species using random dots stereograms. Compared to dichoptic gratings, random dots stereograms are of broad band (i.e., including all spatial frequencies and orientations) and correspond to specific stereoscopic depths. Tree shrew V1 neurons display a high degree of disparity selectivity to random dots stereograms, at a level similar to what has been reported in primates (Tanabe et al., 2022). This is not entirely surprising given the importance of vision in the tree shrew's natural behaviors and the phylogenetic closeness between tree shrews and primates (Petry and Bickford, 2019). In contrast, mouse V1 neurons show a much lower disparity selectivity to random dots stereograms than tree shrew V1 neurons (but see other mouse studies: La Chioma et al., 2019; Samonds et al., 2019). Surprisingly, both mouse and tree shrew V1 neurons show similarly strong phase disparity selectivity to dichoptic gratings. This stimulus-dependent dissociation is inconsistent with the purely feedforward model, but can be explained by a network model including inhibitory and recurrent connections. Intriguingly, this network model suggests that orientation columns, which exists in tree shrews but not in mice, may provide the anatomical substrate needed for the precise connectivity patterns that produced tree shrew–like disparity selectivity (Tanabe et al., 2022). Whether this is indeed the case will have to be tested in future studies.

It is clear from the above discussion that our understanding of disparity computation is still rather rudimentary. Consequently, much less is known about their developmental processes, let alone mechanisms. In monkey V1, adult-like disparity selectivity was seen at postnatal day 6 (Chino et al., 1997), near the beginning of the critical period for ocular dominance plasticity. Similarly, in cat V1, it was reported that disparity selectivity had mostly developed before 3 weeks of age (Freeman and Ohzawa, 1992), again before the peak of the critical period for ocular dominance plasticity. These findings are consistent with the notion that much of the binocular disparity circuits are already in place soon after vision onset, and these circuits are then constantly modified and rewired by the changing binocular inputs caused by the overall growth. The exact developmental process and its underlying circuit and synaptic mechanisms remain to be revealed.

Excitingly, both tree shrews and mice will facilitate these future mechanistic and developmental studies. The sharp disparity selectivity, together with the modern viral, imaging, and behavioral techniques being developed and used in tree shrews, makes them a promising model to study circuit mechanisms of disparity selectivity. Furthermore, tree shrews have a short reproductive cycle, ~45 days in gestation and ~4 months from birth to sexual maturity (Fuchs and Corbach-Söhle, 1999), making them also a viable choice for developmental studies. On the other hand, despite their low selectivity to random dots stereograms, mouse V1 neurons still show clear binocular integration, as indicated by their phase disparity selectivity. The knowledge of cell types in mice and the ability to achieve cell-type specific recording and manipulation are unparalleled. Finally, as mentioned earlier, the long history of using mice in ocular dominance plasticity and critical period studies has accumulated a tremendous amount of information that can guide future disparity selectivity studies.

REOPENING THE CRITICAL PERIODS

After the critical period closes, elevated plasticity characteristic of development tapers off, and short bouts of monocular deprivation that induce robust plasticity in young animals have no effect (Hensch and Quinlan, 2018). In other words, adult cortex is not a permissive environment for large-scale experience-driven changes, impeding recovery of function after injury (Ribic, 2020). One of the main goals of studies on critical period plasticity is therefore to identify ways to "reopen" the critical period in the adult brain. Plasticity in adults is actively restricted through molecular "brakes," such as inhibitory neurotransmission. Reducing cortical inhibition or PV+ neuron activity after the critical period pharmacologically, through loss of excitatory drive, chemogenetic suppression, or through disinhibition by activating SST+ interneurons (Ribic, 2020; Sadahiro et al., 2020), can extend ocular dominance plasticity into adulthood, showing that PV-mediated cortical inhibition restricts plasticity after the critical period. Interestingly, transplantation of PV+ and SST+ progenitors into adult V1 is capable of reopening a critical period for ocular dominance plasticity in adult mice (Tang et al., 2014; Davis et al., 2015), suggesting that the maturation state of inhibitory networks, not the overal level of inhibition, may gate plasticity.

Gain-of-function approaches and environmental manipulations that reduce intracortical inhibition, such as dark exposure and environmental enrichment, also result in heightened plasticity in the adult brain and promote recovery from deprivation-induced amblyopia (He et al., 2006; He et al., 2007; Sale et al., 2007). Multisensory environmental enrichment facilitates the recovery of acuity and ocular dominance in adult rats that have undergone monocular deprivation (Bonaccorsi et al., 2014). Environmental enrichment in the form of voluntary physical exercise is also capable of increasing plasticity and promoting recovery in the adult brain, confirming the therapeutic effects of physical activity on brain plasticity (Kalogeraki et al., 2014; Kaneko and Stryker, 2014). In humans, enrichment in the form of perceptual learning can promote recovery of function in amblyopes, but the mechanism is unclear (Bonaccorsi et al., 2014).

In the adult brain, heightened inhibition is maintained through additional plasticity brakes, such as extracellular matrix (ECM). Experience-dependent expression of ECM forms a highly glycosylated sheath around cells and synapses, and is particularly dense around PV interneurons (Ribic, 2020). Removal of ECM can reopen or extend the critical period and promotes recovery from amblyopia (Bonaccorsi et al., 2014). Reintroduction of light after dark exposure in adult mice activates ECM-regulating molecules, providing a link between dark-exposure induced plasticity and molecular brakes (Murase et al., 2017). In addition to ECM and inhibition, cholinergic modulation, myelin, and synapse-organizing molecules actively restrict plasticity in the adult brain (Espinosa and Stryker, 2012; Ribic et al., 2019). All of the aforementioned plasticity brake mechanisms can impact PV+ interneuron activity (Gu et al., 2016; Ribic et al., 2019; Sadahiro et al., 2020; Stephany et al., 2016), highlighting their central role in regulating plasticity levels in adult brain. On the other hand, it was shown that the maturation of silent glutamatergic synapses onto excitatory neurons closes the critical period for ocular dominance plasticity, largely independent of PV interneurons (Huang et al., 2015). Whether these different processes work through a common mechanism remains unknown.

Despite the many successes in reopening the critical period, it is important to emphasize the limitations of using mice and rats in such studies to promote recovery following amblyopia. Due to the strong contralateral bias in rodent V1, monocular deprivation of the contralateral eye actually shifts the ocular dominance to a more balanced distribution, making V1 neurons more "binocular." This is in drastic contrast to carnivores and primates, where V1 neurons completely lose their responses to the deprived eye and become monocular. Therefore, the starting condition for recovery is rather different between rodents and more binocular animals, where the recovery may be much more difficult due to the initial lack of anatomical and physiological substrates to the deprived eye. Given this important difference, one must carefully interpret the findings from studies in mice. In addition, the fact that many developmental processes take place to restrict plasticity suggests that closing the critical period is a normal step in the maturation of visual functions (Ribic, 2020; Chan et al., 2021). We should therefore be mindful of unintended consequences with manipulations to reopen the critical period.

CONCLUSIONS AND PERSPECTIVES

Heightened plasticity is a hallmark of the developing brain and enables robust and rapid circuit remodeling in response to experience. Developmental plasticity facilitates the maturation of sensory, social, and cognitive functions during temporally restricted sensitive and critical periods. While the original report of critical periods comes from a study of social behaviors, they are most extensively studied at the neuronal level in the visual system, where multiple critical periods for the development of different visual response properties exist (Figure 51.2). These studies, especially those of ocular dominance plasticity, have revealed experience-dependent synaptic changes and have identified approaches to reopening windows of plasticity in adulthood (Hubener and Bonhoeffer, 2014; Hooks and Chen, 2020; Ribic, 2020). However, the bulk of studies on adult ocular dominance plasticity referenced here has unfortunately only tested whether increased plasticity promotes changes of response magnitude and recovery of acuity. To ensure that they are indeed relevant for binocular vision, future studies need to test whether physiological and behavioral recovery of interocular matching and disparity selectivity occur after manipulations to increase plasticity. Along this line, it must be emphasized that every response property of visual cortical neurons likely has a sensitive or critical period for maturation and development, even the properties present before the onset of

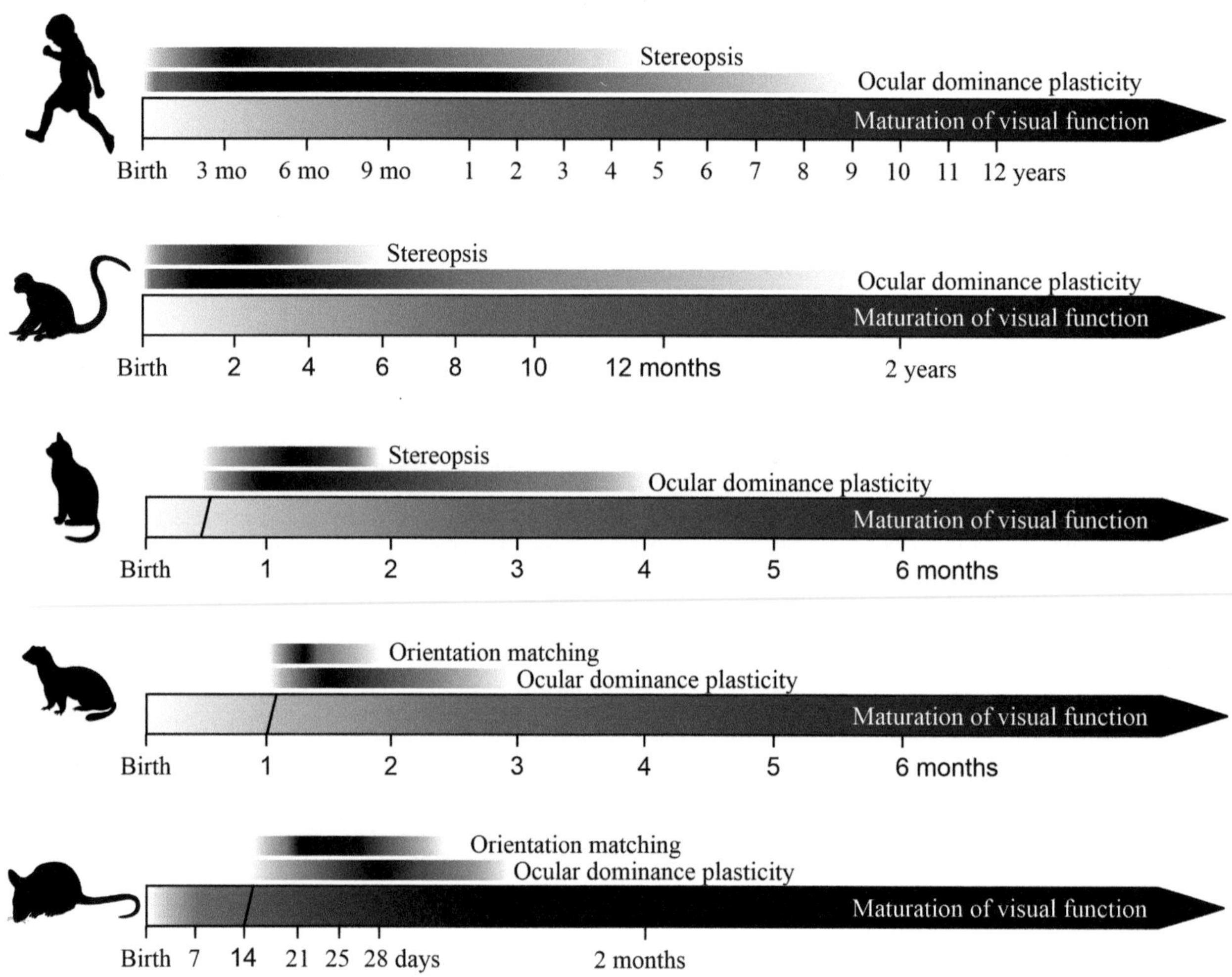

Figure 51.2. Critical periods for binocular development in mammals. Timing of critical periods for ocular dominance plasticity, disparity selectivity development, and interocular matching in humans and common model animals (monkeys, cats, ferrets, and mice). Eye opening in cats, ferrets, and mice is indicated by the oblique line in the timeline. Not all 3 properties are studied in all species. (The following studies were referenced to construct the timelines: Hubel and Wiesel, 1970; Timney, 1983; Gordon and Stryker, 1996; Horton and Hocking, 1997; Issa et al., 1999; Kumagami et al., 2000; Fawcett et al., 2005; Daw, 2006; Wang et al., 2013; Chang et al., 2020.)

external visual experience, such as orientation selectivity. Future studies will need to reveal the precise sequence of events during the normal devleopment of visual functions (Figure 51.2). Such knowledge is necessary for designing optimal approaches for recovery in amblyopia and other visual disorders. Studies of binocular propeties in addition to ocular dominance, and inclusion of additional model organisms, will undoubtedly yield new insights into the mechanisms of visual development and plasticity.

ACKNOWLEDGMENTS

We thank members of the Cang lab and Ribic lab for discussion. The work in the Cang lab is supported by US National Institutes of Health (NIH) grants (R01EY026286, R01EY020950, and UF1NS122040) and Jefferson Scholars Foundation. The work in the Ribic lab is supported by the University of Virginia start-up funds and the National Center for Advancing Translational Science of the National Institutes of Health Award UL1TR003015/KL2TR003016.

REFERENCES

Bonaccorsi J, Berardi N, Sale A (2014). Treatment of amblyopia in the adult: Insights from a new rodent model of visual perceptual learning. *Front Neural Circuits*. 8:82.

Cang J, Fu J, Tanabe S (2023). Neural circuits for binocular vision: Ocular dominance, interocular matching, and disparity selectivity. *Front Neural Circuits*. 17:1084027.

Chan J, Hao X, Liu Q, Cang J, Gu Y (2021). Closing the critical period is required for the maturation of binocular integration in mouse primary visual cortex. *Front Cell Neurosci*. 15:749265.

Chang JT, Whitney D, Fitzpatrick D (2020). Experience-dependent reorganization drives development of a binocularly unified cortical representation of orientation. *Neuron*. 107:338–350.

Chino YM, Smith EL, 3rd, Hatta S, Cheng H (1997). Postnatal development of binocular disparity sensitivity in neurons of the primate visual cortex. *J Neurosci*. 17:296–307.

Cumming BG, DeAngelis GC (2001) The physiology of stereopsis. *Annu Rev Neurosci*. 24:203–238.

Cumming BG, Parker AJ (1997). Responses of primary visual cortical neurons to binocular disparity without depth perception. *Nature*. 389:280–283.

Davis MF, Figueroa Velez DX, Guevarra RP, Yang MC, Habeeb M, Carathedathu MC, Gandhi SP (2015). Inhibitory neuron transplantation into adult visual cortex creates a new critical period that rescues impaired vision. *Neuron*. 86:1055–1066.

Daw NW (2006). *Visual Development*. Springer.

Espinosa JS, Stryker MP (2012). Development and plasticity of the primary visual cortex. *Neuron*. 75:230–249.

Fawcett SL, Wang YZ, Birch EE (2005). The critical period for susceptibility of human stereopsis. *Invest Ophthalmol Vis Sci*. 46:521–525.

Freeman RD, Ohzawa I (1992). Development of binocular vision in the kitten's striate cortex. *J Neurosci*. 12:4721–4736.

Frenkel MY, Bear MF (2004). How monocular deprivation shifts ocular dominance in visual cortex of young mice. *Neuron*. 44:917–923.

Fuchs E, Corbach-Söhle S (1999). Tree shrews. In: Poole T, ed. *UFAW Handbook on the Care and Management of Laboratory Animals*, Vol. 1: *Terrestrial Vertebrates*. 7th ed. Oxford: Blackwell:235–245.

Gordon JA, Stryker MP (1996). Experience-dependent plasticity of binocular responses in the primary visual cortex of the mouse. *J Neurosci*. 16:3274–3286.

Gu Y, Cang J (2016). Binocular matching of thalamocortical and intracortical circuits in the mouse visual cortex. *eLife*. 5:e22032.

Gu Y, Huang S, Chang MC, Worley P, Kirkwood A, Quinlan EM (2013). Obligatory role for the immediate early gene NARP in critical period plasticity. *Neuron*. 79:335–346.

Hata Y, Tsumoto T, Stryker MP (1999). Selective pruning of more active afferents when cat visual cortex is pharmacologically inhibited. *Neuron*. 22:375–381.

He HY, Hodos W, Quinlan EM (2006). Visual deprivation reactivates rapid ocular dominance plasticity in adult visual cortex. *J Neurosci*. 26:2951–2955.

He HY, Ray B, Dennis K, Quinlan EM (2007). Experience-dependent recovery of vision following chronic deprivation amblyopia. *Nat Neurosci*. 10:1134–1136.

Hengen KB, Torrado Pacheco A, McGregor JN, Van Hooser SD, Turrigiano GG (2016). Neuronal firing rate homeostasis is inhibited by sleep and promoted by wake. *Cell*. 165:180–191.

Hensch TK, Quinlan EM (2018). Critical periods in amblyopia. *Vis Neurosci*. 35:E014.

Hensch TK, Stryker MP (2004). Columnar architecture sculpted by GABA circuits in developing cat visual cortex. *Science*. 303:1678–1681.

Hooks BM, Chen C (2020). Circuitry underlying experience-dependent plasticity in the mouse visual system. *Neuron*. 106:21–36.

Horton JC, Hocking DR (1997). Timing of the critical period for plasticity of ocular dominance columns in macaque striate cortex. *J Neurosci*. 17:3684–3709.

Howarth M, Walmsley L, Brown TM (2014). Binocular integration in the mouse lateral geniculate nuclei. *Curr Biol*. 24:1241–1247.

Huang X, Stodieck SK, Goetze B, Cui L, Wong MH, Wenzel C, Hosang L, Dong Y, Lowel S, Schluter OM (2015). Progressive maturation of silent synapses governs the duration of a critical period. *Proc Natl Acad Sci U S A*. 112:E3131–3140.

Hubel DH, Wiesel TN (1962). Receptive fields, binocular interaction and functional architecture in the cat's visual cortex. *J Physiol*. 160:106–154.

Hubel DH, Wiesel TN (1965). Binocular interaction in striate cortex of kittens reared with artificial squint. *J Neurophysiol*. 28:1041–1059.

Hubel DH, Wiesel TN (1968). Receptive fields and functional architecture of monkey striate cortex. *J Physiol*. 195:215–243.

Hubel DH, Wiesel TN (1970). The period of susceptibility to the physiological effects of unilateral eye closure in kittens. *J Physiol*. 206:419–436.

Hubener M, Bonhoeffer T (2014). Neuronal plasticity: Beyond the critical period. *Cell*. 159:727–737.

Issa NP, Trachtenberg JT, Chapman B, Zahs KR, Stryker MP (1999). The critical period for ocular dominance plasticity in the Ferret's visual cortex. *J Neurosci*. 19:6965–6978.

Kalogeraki E, Greifzu F, Haack F, Lowel S (2014). Voluntary physical exercise promotes ocular dominance plasticity in adult mouse primary visual cortex. *J Neurosci*. 34:15476–15481.

Kaneko M, Stryker MP (2014). Sensory experience during locomotion promotes recovery of function in adult visual cortex. *eLife*. 3:e02798.

Kiorpes L (2015). Visual development in primates: Neural mechanisms and critical periods. *Dev Neurobiol*. 75:1080–1090.

Kiorpes L, Kiper DC, O'Keefe LP, Cavanaugh JR, Movshon JA (1998). Neuronal correlates of amblyopia in the visual cortex of macaque monkeys with experimental strabismus and anisometropia. *J Neurosci*. 18:6411–6424.

Kuhlman SJ, Olivas ND, Tring E, Ikrar T, Xu X, Trachtenberg JT (2013). A disinhibitory microcircuit initiates critical-period plasticity in the visual cortex. *Nature*. 501:543–546.

Kumagami T, Zhang B, Smith EL, 3rd, Chino YM (2000). Effect of onset age of strabismus on the binocular responses of neurons in the monkey visual cortex. *Invest Ophthalmol Vis Sci*. 41:948–954.

La Chioma A, Bonhoeffer T, Hubener M (2019). Area-specific mapping of binocular disparity across mouse visual cortex. *Curr Biol*. 29:2954–2960.

Levine JN, Chen H, Gu Y, Cang J (2017). Environmental enrichment rescues binocular matching of orientation preference in the mouse visual cortex. *J Neurosci*. 37:5822–5833.

Liang L, Chen C (2020) Organization, function, and development of the mouse retinogeniculate synapse. *Ann Rev Vis Sci*. 6:261–285.

Mrsic-Flogel TD, Hofer SB, Ohki K, Reid RC, Bonhoeffer T, Hubener M (2007). Homeostatic regulation of eye-specific responses in visual cortex during ocular dominance plasticity. *Neuron*. 54:961–972.

Murase S, Lantz CL, Quinlan EM (2017). Light reintroduction after dark exposure reactivates plasticity in adults via perisynaptic activation of MMP-9. *eLife*. 6:e27345.

Ohzawa I, DeAngelis GC, Freeman RD (1997). The neural coding of stereoscopic depth. *Neuroreport*. 8:iii–xii.

Petry HM, Bickford ME (2019). The second visual system of the tree shrew. *J Comp Neurol*. 527:679–693.

Priebe NJ, Ferster D (2012). Mechanisms of neuronal computation in mammalian visual cortex. *Neuron*. 75:194–208.

Prince SJ, Cumming BG, Parker AJ (2002). Range and mechanism of encoding of horizontal disparity in macaque V1. *J Neurophysiol*. 87:209–221.

Ribic A (2020). Stability in the face of change: Lifelong experience-dependent plasticity in the sensory cortex. *Front Cell Neurosci*. 14:76.

Ribic A, Crair MC, Biederer T (2019). Synapse-selective control of cortical maturation and plasticity by parvalbumin-autonomous action of synCAM 1. *Cell Rep*. 26:381–393 e386.

Rose T, Bonhoeffer T (2018). Experience-dependent plasticity in the lateral geniculate nucleus. *Curr Opin Neurobiol*. 53:22–28.

Sadahiro M, Demars MP, Burman P, Yevoo P, Zimmer A, Morishita H (2020.) Activation of somatostatin interneurons by nicotinic modulator lypd6 enhances plasticity and functional recovery in the adult mouse visual cortex. *J Neurosci*. 40:5214–5227.

Sale A, Maya Vetencourt JF, Medini P, Cenni MC, Baroncelli L, De Pasquale R, Maffei L (2007). Environmental enrichment in adulthood promotes amblyopia recovery through a reduction of intracortical inhibition. *Nat Neurosci*. 10:679–681.

Samonds JM, Choi V, Priebe NJ (2019). Mice discriminate stereoscopic surfaces without fixating in depth. *J Neurosci*. 39:8024–8037.

Sarnaik R, Wang BS, Cang J (2014). Experience-dependent and independent binocular correspondence of receptive field subregions in mouse visual cortex. *Cereb Cortex*. 24:1658–1670.

Sato M, Stryker MP (2008). Distinctive features of adult ocular dominance plasticity. *J Neurosci*. 28:10278–10286.

Scholl B, Pattadkal JJ, Priebe NJ (2017). Binocular disparity selectivity weakened after monocular deprivation in mouse V1. *J Neurosci*. 37:6517–6526.

Severin D, Hong SZ, Roh SE, Huang S, Zhou J, Bridi MCD, Hong I, Murase S, Robertson S, Haberman RP, Huganir RL, Gallagher M, Quinlan EM, Worley P, Kirkwood A (2021). All-or-none disconnection of pyramidal inputs onto parvalbumin-positive interneurons gates ocular dominance plasticity. *Proc Natl Acad Sci U S A*. 118:e2105388118.

Stephany CÉ, Ikrar T, Nguyen C, Xu X, McGee AW (2016). Nogo Receptor 1 Confines a Disinhibitory Microcircuit to the Critical Period in Visual Cortex. *J Neurosci*. 36:11006–11012.

Tan L, Ringach DL, Trachtenberg JT (2022). The development of receptive field tuning properties in mouse binocular primary visual cortex. *J Neurosci*. 42:3546–3556.

Tan L, Ringach DL, Zipursky SL, Trachtenberg JT (2021). Vision is required for the formation of binocular neurons prior to the classical critical period. *Curr Biol*. 31:4305–4313.

Tan L, Tring E, Ringach DL, Zipursky SL, Trachtenberg JT (2020). Vision changes the cellular composition of binocular circuitry during the critical period. *Neuron*. 108:735–747 e736.

Tanabe S, Fu J, Cang J (2022). Strong tuning for stereoscopic depth indicates orientation-specific recurrent circuitry in tree shrew V1. *Curr Biol*. 32:5274–5284.

Tang Y, Stryker MP, Alvarez-Buylla A, Espinosa JS (2014). Cortical plasticity induced by transplantation of embryonic somatostatin or parvalbumin interneurons. *Proc Natl Acad Sci U S A*. 111:18339–18344.

Timney B (1983). The effects of early and late monocular deprivation on binocular depth perception in cats. *Brain Res*. 283:235–243.

Tsao DY, Conway BR, Livingstone MS (2003). Receptive fields of disparity-tuned simple cells in macaque V1. *Neuron*. 38:103–114.

von Noorden GK, Campos EC (2002). *Binocular Vision and Ocular Motility: Theory and Management of Strabismus*. Mosby.

Wang BS, Sarnaik R, Cang J (2010). Critical period plasticity matches binocular orientation preference in the visual cortex. *Neuron*. 65:246–256.

Wang BS, Feng L, Liu M, Liu X, Cang J (2013). Environmental enrichment rescues binocular matching of orientation preference in mice that have a precocious critical period. *Neuron*. 80:198–209.

Wiesel TN, Hubel DH (1963). Single-cell responses in striate cortex of kittens deprived of vision in one eye. *J Neurophysiol*. 26:1003–1017.

Xu X, Cang J, Riecke H (2020). Development and binocular matching of orientation selectivity in visual cortex: a computational model. *J Neurophysiol*. 123:1305–1319.

52.

DEVELOPMENT AND PLASTICITY OF FEATURE SELECTIVITY

Sophie V. Griswold and Stephen D. Van Hooser

OVERVIEW OF THE VISUAL SYSTEM

GROSS DEVELOPMENT OF LGN AND CORTICAL CONNECTIONS

Much of the gross configuration of thalamic and cortical circuits in the mammalian brain occurs prior to the onset of vision, either in utero or before eye opening. The feedforward pathway from the retina, to the lateral geniculate nucleus (LGN), to the primary visual cortex is configured via a combination of molecular guidance cues and spontaneous activity. Prior to the onset of vision, retinotopic arrangement of retinal ganglion cells (RGCs) onto their targets in the LGN is established by way of axon guidance molecules expressed in the retina and LGN (Flanagan, 2006). In many mammals, the LGN begins to organize into distinct, functional layers following the arrival of RGCs in the thalamus. In primates (Kaas et al., 1972) and carnivores, these layers receive eye-specific inputs, with primate LGN divided into 6 layers further differentiated by their magnocellular or parvocellular inputs, and carnivores divided into A, A1, and C layers. These layers segregate early in development, beginning and ending between E49– P12 and P0–21 in cat and ferret, respectively (Issa et al., 1999), and in macaque by E90 (Robinson and Dreher, 1990). By contrast, rat and mouse LGN is organized into just 2 sections, the shell and core, which receive inputs from both eyes (Reese, 1988). Only a small population of ipsilateral RGCs innervate the core, however, and lamination is finished by P9 and P8 in the mouse and rat, respectively (Robinson and Dreher, 1990).

The mammalian thalamocortical synapse is established early in development, with thalamic axons arriving in the subplate just before the birth of layer (L)4 cortical cells (Kanold and Luhmann, 2010), and forming synapses in cortex by E55 and P26 in ferret and cat, respectively (Issa et al., 1999). Mature LGN cells primarily project to L4 of cortex, and to a lesser extent L6 (Callaway, 1998), which sends corticothalamic afferents back to LGN (Robinson and Dreher, 1990). In macaque, the only areas of V1 that do not receive input from LGN are L4B and L5, and regions between cytochrome oxidase blobs in L2/3 (Callaway, 1998). The corticothalamic feedback pathway only completes innervation following eye opening across mammals (Shatz and Rakic, 1981).

SPONTANEOUS ACTIVITY OF THE EARLY RETINA, LGN, AND CORTEX

Even before visual experience shapes the developing visual system, spontaneous activity plays a vital role in configuring cortical and thalamic circuits. Retinal and cortical waves—spontaneous bursts of action potentials that propagate across the developing retina and cortex—typify such activity, and are vital for the refinement of retinotopy and receptive field properties across the developing visual system.

Though originating in the retina, retinal waves drive activity throughout the developing visual circuit, propagating to LGN and on to visual cortex (Torborg and Feller, 2005). Retinal waves have been observed across mammals, occurring either before birth or eye opening, although some species such as xenopus lack retinal waves. Instead, development of the visual system is likely facilitated by waves of activity driven by visual experience in the tadpole stage (Pratt et al., 2016). Retinal waves occur in 3 distinct stages mediated by different forms of neurotransmission. Stage I waves are produced by a combination of cholinergic circuit activity and gap junctions, stage II waves are cholinergic, and stage III waves are glutamatergic (Torborg and Feller, 2005).

Retinal waves ensure that the activity of RGCs that are close together in the retina is correlative, while the activity of distant RGCs is not. Such correlated activity is necessary for retinotopic organization; mice lacking β2-containing nAChRs exhibit spatially uncorrelated retinal waves, and develop less retinotopic refinement than wild-type mice (Torborg and Feller, 2005).

Disrupting stage II and III waves has been demonstrated to have deleterious effects on development in mouse and ferret models. Mice in which stage II waves are abolished demonstrate atypically large, diffuse axonal arborizations in LGN, causing LGN neurons to have abnormally coarse receptive field maps (Grubb et al., 2003). Further, these mice also have imprecise, defective cortical retinotopic maps (Cang et al., 2005). Likewise, RGCs in ferrets in which stage II retinal waves are abolished during the period of eye-specific segregation from P1 to P10 fail to achieve eye-specific segregation in LGN (Torborg and Feller, 2005).

The visual cortex plays an important role in propagating and amplifying spontaneous retinal waves. Before the onset of vision, corticothalamic afferents provide excitatory feedback amplification to LGN. Not merely relaying spontaneous activity originating in the retina back to thalamus, the impact of the corticothalamic loop on retinal waves is multiplicative, accounting for 80% of the drive in spontaneous activity in LGN (Murata and Colonnese, 2016).

Beyond retinal waves, the developing cortex is capable of producing spontaneous, long-range cortical activity without input from LGN or retina (Chiu and Weliky, 2001). Binocular enucleation of ferrets at P21 only slightly impacts the development of clustered horizontal connections, while silencing the cortex abolishes their development (Weliky and Katz, 1999). Similarly, silencing activity in ferret cortex for 23–26 days beginning between P21 and P23 completely abolishes the development of orientation tuning in V1 (Chapman and Stryker, 1993). Clustered horizontal connections exhibit modular specificity, linking regions of the cortex with similar tuning properties to one another (Van Hooser, 2007), suggesting that spontaneous activity may function to entrain modular features of the adult cortex in the developing brain. Further, tightly coupled excitatory and inhibitory functional networks exist in ferret cortex before eye opening, and short-range connections between these networks produce large-scale circuit activity that may subserve later functional organization in visual cortex (Mulholland et al., 2021). Indeed, longitudinal imaging of developing ferret V1 reveals that spontaneous, long-range correlated activity predicts mature network structure (Smith et al., 2018). L2/3 neurons participate in such long-range modular networks, and their correlated patterns of activity are remarkably stable over early development (Smith et al., 2018). Blockade of feedforward retinal and thalamic drive fails to abolish this activity, suggesting that it may originate in cortex (Smith et al., 2018), and pointing to a strong role for spontaneous cortical activity in the development of modular networks in cortex.

LATERAL GENICULATE NUCLEUS

DEVELOPMENT OF CENTER-SURROUND RECEPTIVE FIELD AND DIRECTION-SELECTIVE CELLS IN RETINA/LGN AND INFLUENCE OF EXPERIENCE

As in the visual cortex and LGN, some receptive field features of neurons in the retina develop before the onset of vision, while others undergo vision-dependent refinement following birth or eye opening. ON/OFF responses of LGN neurons are inherited from retinal ganglion cells which demonstrate cell-type segregation in the inner plexiform layer of the retina according to receptive field properties. Retinal ganglion cells first stratify throughout the inner plexiform layer of the retina before becoming restricted to sublamina a, which contains circuitry mediating OFF responses, or sublamina b, which mediates ON responses. This process occurs between P3 and P5 in the cat (Maslim and Stone, 1988) and by P8 in the rat (Gunhan-Agar et al., 2000). Evidence from rat and monkey suggests that the OFF response pathway develops slightly earlier than the ON pathway (Gunhan-Agar et al., 2000) in a process mediated by endogenous activity.

Many retinal ganglion cells exhibit direction selectivity in rodents and lagomorphs (Tiriac and Feller, 2022), and these cells send monosynaptic inputs to LGN (Huberman et al., 2009). Signals passed on to LGN by these monosynaptic inputs are in turn passed on to primary visual cortex (Huberman et al., 2009). This suggests that, in mice, direction selectivity in LGN and visual cortex is inherited from retina instead of computed de novo, although convergence of other non-direction-selective channels such as sustained and transient cells may also contribute to cortical direction selectivity in mice (Lien and Scanziani, 2018). Such direction selective RGCs are absent or very rare in carnivores (Cleland and Levick, 1974) and primates (Kim et al., 2022). Direction-selective RGCs are present at eye opening in mice following dark-rearing, and after abolishing cholinergic retinal waves, suggesting that direction selectivity emerges in retina in an experience-independent manner as a result of molecular cues (Tiriac and Feller, 2022). Dark-rearing experiments in mice suggest that retinal direction selectivity maps also develop in an experience-independent manner, and are well established at eye opening (Tiriac and Feller, 2022).

LGN neurons acquire ON or OFF responses before eye opening in ferret (Akerman et al., 2002) and mouse (Tschetter et al., 2018). In ferret, dark-rearing before eye opening leads to abnormal segregation of ON-OFF layers in LGN, suggesting that visual experience through the closed eyelids may be necessary for normal development (Akerman et al., 2002). Unlike expression of ON- and

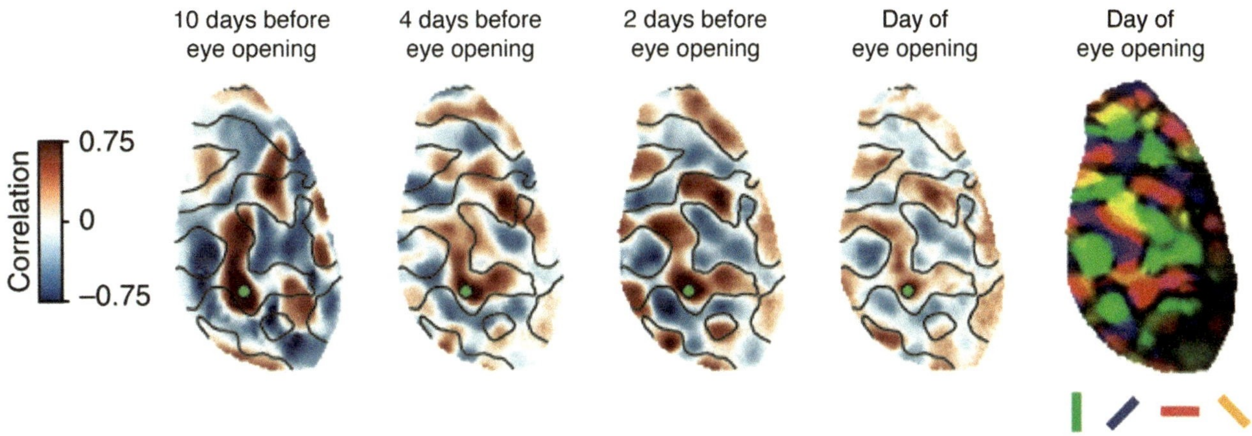

Figure 52.1. Spontaneous activity correlation patterns early in development are strongly related to future orientation maps. Chronic imaging of developing ferret V1 well before eye opening reveals that early spontaneous correlation patterns are highly related to the structure of the future orientation maps. Figure from Smith et al. (2018).

OFF-center RFs, the refinement of receptive field size and precision occurs after eye opening and requires visual experience to proceed. Following the onset of visual experience, LGN receptive fields become smaller, more narrowly tuned, temporally sharper, and exhibit shorter latencies. This transformation has been observed in primates (Blakemore and Vital-Durand, 1986), mouse (Tschetter et al., 2018), cat (Wiesel and Hubel, 1963), and ferret (Akerman et al., 2002). Along with a decrease in receptive field size, LGN cells also demonstrate an increase in peak spatial frequency and stimulus size suppression over this period, paralleled by an overall increase in visual acuity. Visual deprivation of cats in the period following eye opening results in decreased visual acuity of LGN X cells, further demonstrating that typical visual experience is necessary for normal development to proceed (Mower et al., 1982).

CORTEX

DEVELOPMENT OF ORIENTATION SELECTIVITY AND ORIENTATION MAPS AND THE INFLUENCE OF ACTIVITY AND EXPERIENCE

The tendency of cells in the primary visual cortex to fire preferentially when presented with visual stimuli of specific orientations is an important first-order response property in early visual cortex found in all mammals studied. Since its description in the cat by Hubel and Wiesel, the development and mapping of orientation selectivity in cortex has been well described across mammals. The organization of orientation-selective cells in cortex differs between rodents, and carnivores and primates. In rodents there is a disorganized "salt and pepper" distribution of cells according to orientation preference such that neighboring cells generally exhibit different orientation preferences (Rochefort et al., 2011). By contrast, primates and carnivores possess highly organized maps of orientation columns, in which cells preferring similar orientations are grouped together in circular "pinwheel" centers and vary continuously across the surface of the cortex (Chapman et al., 1996). In ferret, modular spontaneous activity patterns are evident in the week before eye opening, and these modules, while refined by experience, become the eventual orientation column modules (Figure 52.1) in mature animals. Orientation maps are weakly present at eye opening (Figure 52.2A1) and become stronger and more defined in the week after eye opening (Chapman et al., 1996; Li et al., 2006). These maps mature over the course of several days, transforming into the patchy, highly defined maps typical of adult animals, and demonstrate strong stability over time (Chapman et al., 1996). The development of orientation maps in kitten follows a similar pattern, becoming detectable around P17, increasing in definition and strength until P30, and exhibiting pronounced durability in map geometry (Godecke et al., 1997).

After the establishment of orientation selectivity, there is a matching and refinement of the orientation-selective inputs from the 2 eyes. (See **Chapter 51 in Section IX** for a description of this process.)

The development of orientation selectivity has been extensively studied in carnivores. This substantial body of work has revealed that orientation tuning emerges prior to the onset of visual experience (or eye opening) in an activity-dependent manner (Figure 52.3A). While this response property is present at eye opening in cat and ferret, only about 25% of cells are sharply tuned for orientation at this point in development in the ferret (Chapman and Stryker, 1993), and 11% in the cat (Albus and Wolf, 1984). Orientation tuning matures in the ferret by postnatal week

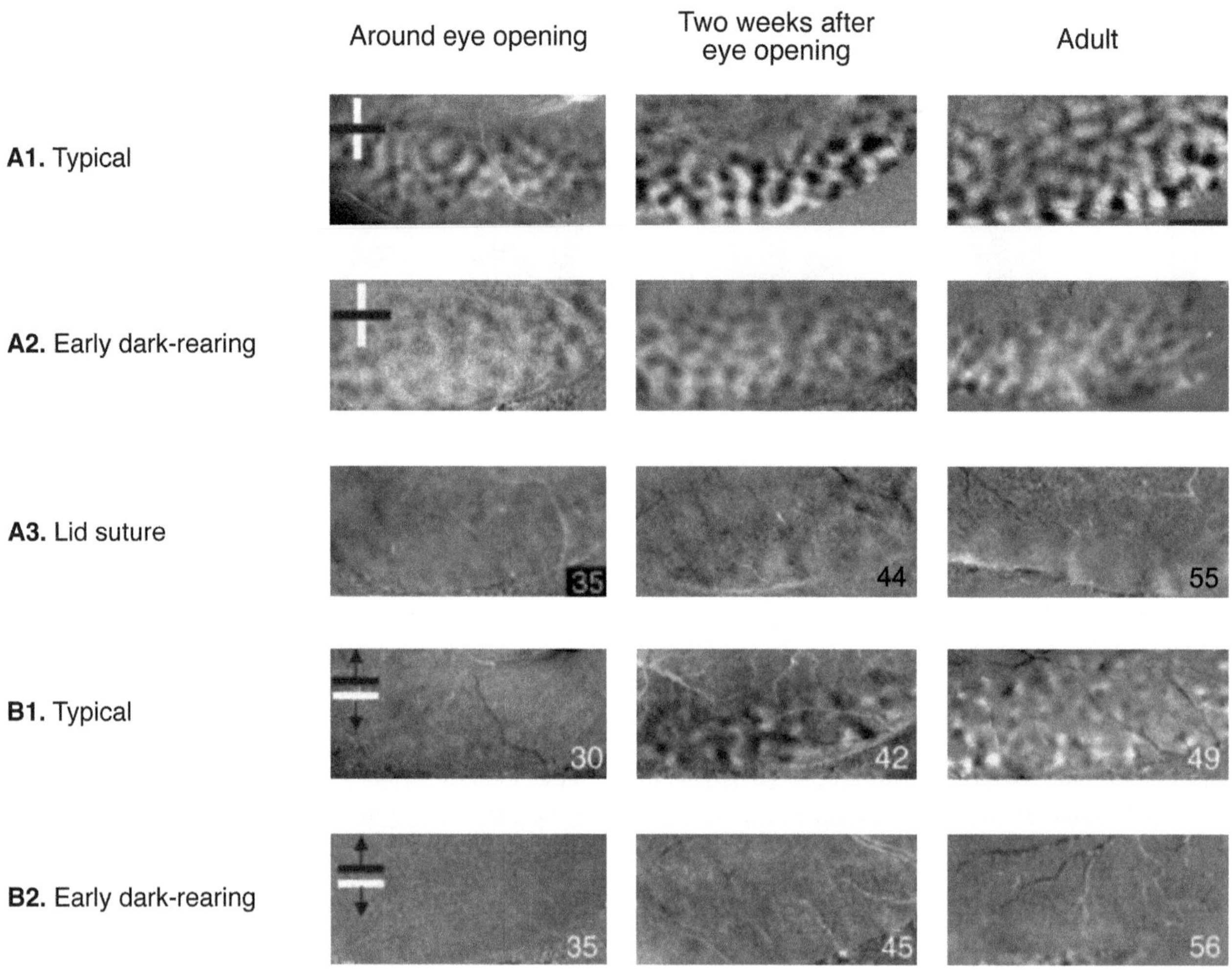

Figure 52.2. Intrinsic signal imaging of ferret V1 reveals the development of orientation and direction maps. (A1). Orientation maps appear before eye opening and gradually mature by postnatal week 7. Images from Li et al. (2006). Left image from a P30 ferret, middle image from a P42 ferret, and right image from a P49 ferret. (A2). Orientation maps emerge in animals that have undergone 2–3 weeks of dark-rearing followed by introduction of normal vision. Images from Li et al. (2006). Left image from a P35 ferret, middle image from a P45 ferret, and right image from a P77 ferret. (A3). Orientation maps fail to develop in ferrets subject to binocular lid suture beginning before eye opening and continuing up to postnatal week 10. Images from White et al. (2001). Left image from a P35 ferret, middle image from a P44 ferret, and right image from a P55 ferret. (B1). Direction maps develop in the weeks following eye opening in an experience dependent manner. Images from Li et al. (2006). Left image from a P30 ferret, middle image from a P42 ferret, and right image from a P49 ferret. (B2). Maturation of direction maps requires early visual experience. Ferrets dark-reared from before the onset of vision do not develop direction maps. Images from Li et al. (2006). Left image from a P35 ferret, middle image from a P45 ferret, and right image from a P56 ferret.

7, with 75% of cortical cells demonstrating strong orientation selectivity (Chapman and Stryker, 1993), and postnatal week 4 in the cat, with 72% of cells demonstrating the mature response property (Albus and Wolf, 1984). Normal development of orientation selectivity requires cortical activity before eye opening and visual experience after eye opening; activity blockade in ferret visual cortex in the weeks before eye opening prevents the maturation of orientation selectivity, as does visual deprivation by binocular lid suture, albeit to a lesser degree (Chapman and Stryker, 1993). Dark-rearing ferrets (Li et al., 2006) and kittens (Buisseret and Imbert, 1976) in the weeks after natural eye opening also inhibits full maturation of strong orientation selectivity in V1 (Figures 52.2A2, 52.3A). This suggests that typical visual experience is needed for the development of sharply tuned orientation selectivity, and that only spontaneous activity, not visual experience, is needed to build initial orientation selectivity. However, abnormal visual experience, in the form of deprivation by lid suture, can greatly degrade orientation selectivity and orientation maps (Crair et al., 1998) (Figures 52.2A3, 52.3A). Similarly, visually deprived rodents demonstrate normal development of orientation selectivity in the first postnatal month (Iwai et al., 2003) which degrades following prolonged deprivation by lid suture (Benevento et al., 1992) (Figure 52.3B). Taken together, these findings suggest

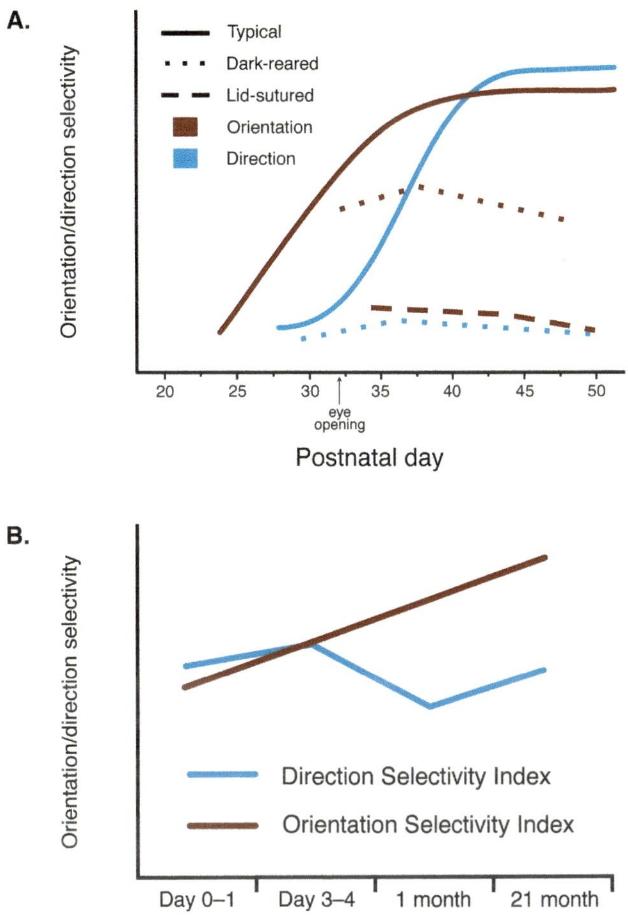

Figure 52.3. Timeline of orientation and direction selectivity in mouse and ferret. (A). Ferret V1 cells are orientation selective and only weakly direction selective at eye opening (Li et al., 2006). Dark rearing prevents the maturation of orientation and direction tuning in V1, although prolonged visual deprivation is necessary to aberrantly impact orientation tuning (White et al., 2001). Binocular lid suture from the time of eye opening destroys orientation tuning in V1 (White et al., 2001). (B). Orientation and direction selectivity are present at eye opening in mice and go on to mature in the following weeks. Data in B from Rochefort et al. (2011).

that the development of orientation selectivity is driven by both intrinsic mechanisms and visual inputs at different timepoints.

Development of orientation selectivity is differentially influenced by experience and intrinsic mechanisms according to cell type in the ferret. Studies in ferret visual cortex have revealed that excitatory and inhibitory cortical neurons possess distinct developmental trajectories (Chang and Fitzpatrick, 2022). While orientation-selective excitatory networks develop before the onset of visual experience (Chiu and Weliky, 2001; Smith et al., 2018), such modular organization develops in inhibitory networks following eye opening in an experience-dependent manner (Chang and Fitzpatrick, 2022). In mice, inhibitory interneurons are sharply tuned for orientation in the days following eye opening, and subsequently broaden their tuning curves with experience (Kuhlman et al., 2011). This broadening has been proposed to open the early critical period for binocular plasticity, as it corresponds with its onset (Kuhlman et al., 2011). While this may be the case in mice, inhibitory interneurons are strengthening their tuning at this point in development in ferret, so experience-dependent broadening of orientation tuning in inhibitory interneurons is not a conserved mechanism gating critical-period plasticity across species (Chang and Fitzpatrick, 2022).

Orientation selectivity in visual cortex is vulnerable to plasticity in mouse and cat when reared in an environment in which a single orientation is presented, deemed *stripe rearing*. Juvenile mice fitted with goggles that afforded only visual stimuli of a single orientation at P25 demonstrate an overrepresentation of the experienced orientation in L2/3 neurons of primary visual cortex (Kreile et al., 2011). Similarly, cats exposed to a single orientation from 2 weeks to about 5 months of age show a profound cortical bias for the trained stimulus (Godecke et al., 1997), and work employing intrinsic optical signaling demonstrates that stripe-reared kittens devote up to twice as much area of cortex to the trained orientation than to the orthogonal. It remains unclear whether these results support an instructive role for the establishment of orientation preferences, or whether the lack of experience with particular orientations causes a degradation of the cells that would typically represent the orientation angles that were filtered out, or whether such cells simply fail to develop the selectivity and responsiveness of cells that prefer the angles that were stimulated (Sengpiel et al., 1999).

DEVELOPMENT OF DIRECTION SELECTIVITY AND COMPONENT MOTION SELECTIVITY IN HIGHER AREAS AND THE INFLUENCE OF EXPERIENCE

Many cells in mammalian V1 demonstrate selectivity for visual stimuli moving in a particular direction. Like orientation selectivity, the development of cortical direction tuning in primates, carnivores, and rodents has been extensively studied. V1 neurons are only weakly tuned for direction of motion at the onset of vision in carnivores (Li et al., 2006) (Figures 52.2B1, 52.3A) and primates (Hatta et al., 1998), with direction tuning maturing in normal cats and ferrets around P42 (Buisseret and Imbert, 1976; Li et al., 2006), and over the first 4 weeks of life in macaques (Hatta et al., 1998). While weakly tuned for direction at eye opening, ferret V1 has been shown to possess a weak direction map before eye opening, suggesting that direction columns are pre-seeded in primary visual cortex in a vision-independent manner (Van Hooser et al., 2012). The maturation of direction selectivity in carnivore V1 requires visual experience, however. Dark-reared ferrets (Li et al.,

2006) and cats (Buisseret and Imbert, 1976) do not demonstrate maturation of direction tuning in V1 following eye opening (Figure 52.2B2). In the ferret, the critical period for direction selectivity comprises the 2 weeks following eye opening (Li et al., 2006); visual experience that is delayed beyond 2 weeks cannot drive the development of direction selectivity. Further, multiple studies in ferret have demonstrated that moving stimuli drive the rapid emergence of direction selectivity in V1 just after eye opening (Van Hooser et al., 2012; Roy et al., 2020), thus elucidating the vital role of visual experience in the development of direction tuning.

One may ask whether the parameters of visual stimuli necessary for the development of direction selectivity are critical for determining the angle preferences of direction selective neurons. Interestingly, experience with only a single direction of motion causes cells that are predisposed to that direction to increase their direction selectivity, but cells that are predisposed to prefer the opposite direction do not become selective (Van Hooser et al., 2012). In addition, providing experience with specific spatiotemporal patterns after the normal onset of visual experience does not direct the spatiotemporal selectivity of cortical neurons, suggesting that spatiotemporal receptive fields are not being built instructively (Roy et al., 2020). Further, nonspecific activation of the naive cortex by optogenetic input also causes an increase in direction selectivity without any information to direct the preference angle of the emerging selectivity (Roy et al., 2016). Taken together, these experiments provide evidence that the parameters of the tuning that emerges were already specified before the onset of visual experience, and that experience was only necessary to finish the development, perhaps by setting response gains.

While no studies of the impact of dark-rearing on the development of direction selectivity have taken place in primates to date, work in humans has shown that individuals with impoverished early visual experience demonstrate a higher threshold of motion detection in comparison with individuals that have experienced normal vision through both eyes, suggesting that visual experience may play an important role in the maturation of direction tuning in primates (Ellemberg et al., 2002).

Neurons in higher motion areas, such as area PSS in the ferret, exhibit selectivity for global motion of a pattern that goes beyond the local features detected in V1 receptive fields (Lempel and Nielsen, 2021). Global motion selectivity develops postnatally, about a week after the development of direction selectivity in area V1 (Lempel and Nielsen, 2021).

In contrast to primates and carnivores, direction selectivity develops independently of visual experience in mouse primary visual cortex (Figure 52.3B); mice born and reared in total darkness demonstrate robust direction tuning at eye opening (Rochefort et al., 2011). Intriguingly, while orientation selective cells are present at eye opening in ferret V1 and subsequently develop direction tuning (Li et al., 2006), this developmental sequence is reversed in mice. In mouse V1, direction-selective cells present at eye opening exhibit strong biases for dorsal, anterior, or posterior directions. With experience, the distribution of direction preferences that are represented broadens, mirroring direction selectivity in the retina (Elstrott et al., 2008), and some neurons lose their direction selectivity and become orientation-selective (Rochefort et al., 2011).

DEVELOPMENT OF NORMALIZATION SUCH AS SURROUND SUPPRESSION, HIGH-LEVEL OBJECT DETECTION, AND FACE DETECTION

While the development of direction selectivity in carnivore V1 requires visual experience, this does not appear to be the case for some forms of normalization, such as cross-orientation or surround suppression, a receptive field property vital for motion perception. Cross-orientation suppression and surround suppression are present in dark- and normally reared ferrets as early as P40 and do not vary significantly with experience (Popovic et al., 2018). Work in primates has yielded different findings in different areas, with surround suppression detectable in macaque V1 by P14, but not until postnatal weeks 4–8 in higher visual areas such as V2 (Danka Mohammed and Khalil, 2020).

Studies in human infants suggest that surround suppression mediated by cortical middle temporal area (MT), an important higher visual region necessary for higher-order motion processing in primates, does not emerge until around 7 months of age (Lin and Tadin, 2019). Studies in young macaques have clarified the developmental timeline of many aspects of motion processing in MT. While MT neurons in infant macaques are as well tuned for speed and direction as those of adults (Kiorpes and Movshon, 2004), their response latencies are longer and their spike rates lower (Movshon et al., 2003). Other response properties are immature in infant macaques as well; pattern direction selective cells, which signal the direction of motion of complex patterns such as plaids, were half as plentiful in area MT of 16-week-old macaques, and even fewer in young animals (Movshon et al., 2003). Evidence that early visual experience plays a role in the development of complex motion detection is mixed. While a model of early strabismus in macaques failed to induce significant alterations in response properties in MT (Kiorpes et al., 1996), functional magnetic resonance imaging (fMRI) studies of amblyopia in human have reported deficits in area MT/V5 (Ho and

Giaschi, 2009), suggesting that abnormal early experience may drive aberrant development in this brain region.

The functional maturation of visual areas responsible for higher-order visual processes appears to proceed more slowly than development of V1 (Ho and Giaschi, 2009), with circuits representing local aspects of visual scenes emerging before those subserving global processing (El-Shamayleh et al., 2010; Danka Mohammed and Khalil, 2020). This is the case for higher form perception arising in extrastriate areas such as contour processing, a receptive field property implicated in the integration of local features of the visual field into coherent, global forms. Evidence suggests that contour processing emerges in V2 (Taylor et al., 2014). Monkeys develop the ability to perform contour integration at 5–6 months (Kiorpes and Bassin, 2003), and humans at 36 months (Pennefather et al., 1999), with performance continuing to improve into the teenage years (Kovacs, 2000). The delayed postnatal development of feedback connections from V2 to V1 have been implicated in the slow emergence of contour processing (Kovacs, 2000). While experimental evidence is lacking, deficits in contour processing in individuals with amblyopia and/or strabismus (Norcia et al., 2005) suggest that the maturation of this receptive field property depends in part on normal binocular input during an early sensitive period in development (Norcia et al., 2005).

In primates, V4 and higher extrastriate areas subserve texture segregation, the division of a visual stimulus into discrete sections by spatial gradients of local feature properties (El-Shamayleh et al., 2010). Macaques are capable of texture discrimination at 6 weeks, with improvements occurring up to 40 weeks (El-Shamayleh et al., 2010). Weak texture segregation is present in human infants at 4 months and matures very slowly, only reaching adult levels at 7–8 years of age (Sireteanu and Rieth, 1992). This response property shows similarly late development in kittens, which only demonstrate texture segregation after P80 (Wilkinson and Crotogino, 1995). While experimental studies examining the impact of visual experience on the development of texture segregation have not been carried out, a study of amblyopic children has identified robust deficits in texture segregation. Interestingly, this deficit was identified in both the amblyopic and unaffected fellow eye, suggesting that the deficit may arise from aberrant development of binocular cells in V4 subserving this RF property (Wang et al., 2007).

This slower developmental trajectory is also true of inferotemporal cortex (IT), one of the higher levels of the primate ventral stream subdivided into functional domains selective for image categories such as faces, objects, and colors. Cells in macaque IT gradually become responsive to complex object stimuli in the months following birth; only 10% of cells in area IT respond to visual stimuli in monkeys younger than 4 months, rising to 50% between 4 and 7 months, and reaching 80% in maturity (Rodman et al., 1993). Selectivity for faces in IT emerges at around 200 days of age in the macaque, before which selectivity for faces is present but weak (Arcaro et al., 2019). Much like orientation columns, patches of cells selective for faces over other complex object stimuli are very stable over time (Arcaro et al., 2019).

Experimental evidence suggests that both innate processes and experience influence domain formation in area IT in human and macaque (Arcaro et al., 2019). IT domains represent both biologically salient categories, such as faces, and culturally entrained categories, such as buildings and text. Interestingly, while these domains are subject to the influence of objects encountered in early development, the locations of specialized domains in area IT are consistent among individuals (Arcaro et al., 2019). Juvenile monkeys intensively trained to recognize 3 sets of objects, consisting of text and digits, tetris blocks, and cartoon faces, developed specialized domains in area IT selective for each category. In spite of the training order of each being different, the location of each domain in area IT was similar across monkeys (Arcaro et al., 2019). Such category entrainment has been demonstrated in juvenile macaques, but fails in adults, suggesting that there is an early critical period for the development of IT domains, and that the object selectivity of domains is malleable during this period (Arcaro et al., 2019). While it was once believed that face recognition was innate in primates, recent work suggests an instructive role for visual stimuli in the development of IT domains. Macaques raised without exposure to faces for the first year of life develop IT domains selective for hands, bodies, and scenes, but not faces, while control monkeys developed domains selective for all of these categories (Arcaro et al., 2019). Further, while control monkeys looked disproportionately at faces in free viewing experiments, face-deprived monkeys did not look disproportionately at faces, but at hands (Arcaro et al., 2019). Taken together, these findings suggest that formation of IT domains is instructively driven by experience with environmentally salient cues, experienced in early development.

While the early face-viewing bias observed in human infants has been held up as evidence that face processing is innate, the experience-dependent development of this bias after birth is consistent with an experience-driven model of face domains. Infants' preference for their mother's face within hours of birth correlates with the amount of visual experience the infant has had, and face preferences continue to refine in the months after birth (Arcaro et al., 2019). While infants have no bias for race at birth, 3-month-old infants demonstrate a decreased ability to recognize other-race faces, a decrease that is inversely correlated with the degree of ethnic diversity in its community and thus related to the extent of exposure to faces of different races (Arcaro et al., 2019). Likewise, infants develop a bias for faces that share the gender of their primary caregiver (Arcaro et al., 2019).

Thus, the development of IT domains is of a kind with receptive field properties at lower levels of visual processing, such as ocular dominance and orientation selectivity: influenced by both innate mechanisms of circuit development, and by exposure to visual stimuli early in life. Much as a mouse reared with contact lenses limiting visual experience with only one angle will develop an overrepresentation of the experienced orientation in visual cortex, young monkeys extensively trained on particular sets of objects will develop IT domains for those stimuli. Nonetheless, many questions remain regarding the development of response properties in IT, and whether the developmental organizing principles of IT are shared across visual cortex, other sensory systems, and even across species.

REFERENCES

Akerman CJ, Smyth D, Thompson ID (2002). Visual experience before eye-opening and the development of the retinogeniculate pathway. *Neuron*. 36:869–879.

Albus K, Wolf W (1984). Early post-natal development of neuronal function in the kitten's visual cortex: A laminar analysis. *J Physiol.* 348:153–185.

Arcaro MJ, Schade PF, Livingstone MS (2019). Universal mechanisms and the development of the face network: What you see is what you get. *Annu Rev Vis Sci.* 5:341–372.

Benevento LA, Bakkum BW, Port JD, Cohen RS (1992). The effects of dark-rearing on the electrophysiology of the rat visual cortex. *Brain Res.* 572:198–207.

Blakemore C, Vital-Durand F (1986). Organization and post-natal development of the monkey's lateral geniculate nucleus. *J Physiol.* 380:453–491.

Buisseret P, Imbert M (1976). Visual cortical cells: Their developmental properties in normal and dark reared kittens. *J Physiol.* 255:511–525.

Callaway EM (1998). Local circuits in primary visual cortex of the macaque monkey. *Annu Rev Neurosci.* 21:47–74.

Cang J, Renteria RC, Kaneko M, Liu X, Copenhagen DR, Stryker MP (2005). Development of precise maps in visual cortex requires patterned spontaneous activity in the retina. *Neuron*. 48:797–809.

Chang JT, Fitzpatrick D (2022). Development of visual response selectivity in cortical GABAergic interneurons. *Nat Commun*. 13:3791.

Chapman B, Stryker MP (1993). Development of orientation selectivity in ferret visual cortex and effects of deprivation. *J Neurosci.* 13:5251–5262.

Chapman B, Stryker MP, Bonhoeffer T (1996). Development of orientation preference maps in ferret primary visual cortex. *J Neurosci.* 16:6443–6453.

Chiu C, Weliky M (2001). Spontaneous activity in developing ferret visual cortex in vivo. *J Neurosci.* 21:8906–8914.

Cleland BG, Levick WR (1974). Properties of rarely encountered types of ganglion cells in the cat's retina and an overall classification. *J Physiol.* 240:457–492.

Crair MC, Gillespie DC, Stryker MP (1998). The role of visual experience in the development of columns in cat visual cortex. *Science*. 279:566–570.

Danka Mohammed CP, Khalil R (2020). Postnatal development of visual cortical function in the mammalian brain. *Front Syst Neurosci.* 14:29.

El-Shamayleh Y, Movshon JA, Kiorpes L (2010). Development of sensitivity to visual texture modulation in macaque monkeys. *J Vis.* 10:11.

Ellemberg D, Lewis TL, Maurer D, Brar S, Brent HP (2002). Better perception of global motion after monocular than after binocular deprivation. *Vision Res.* 42:169–179.

Elstrott J, Anishchenko A, Greschner M, Sher A, Litke AM, Chichilnisky EJ, Feller MB (2008). Direction selectivity in the retina is established independent of visual experience and cholinergic retinal waves. *Neuron*. 58:499–506.

Flanagan JG (2006). Neural map specification by gradients. *Curr Opin Neurobiol.* 16:59–66.

Godecke I, Kim DS, Bonhoeffer T, Singer W (1997). Development of orientation preference maps in area 18 of kitten visual cortex. *Eur J Neurosci.* 9:1754–1762.

Grubb MS, Rossi FM, Changeux JP, Thompson ID (2003). Abnormal functional organization in the dorsal lateral geniculate nucleus of mice lacking the beta 2 subunit of the nicotinic acetylcholine receptor. *Neuron*. 40:1161–1172.

Gunhan-Agar E, Kahn D, Chalupa LM (2000). Segregation of on and off bipolar cell axonal arbors in the absence of retinal ganglion cells. *J Neurosci.* 20:306–314.

Hatta S, Kumagami T, Qian J, Thornton M, Smith EL, 3rd, Chino YM (1998). Nasotemporal directional bias of V1 neurons in young infant monkeys. *Invest Ophthalmol Vis Sci.* 39:2259–2267.

Ho CS, Giaschi DE (2009). Low- and high-level motion perception deficits in anisometropic and strabismic amblyopia: evidence from fMRI. *Vision Res.* 49:2891–2901.

Huberman AD, Wei W, Elstrott J, Stafford BK, Feller MB, Barres BA (2009). Genetic identification of an On-Off direction-selective retinal ganglion cell subtype reveals a layer-specific subcortical map of posterior motion. *Neuron*. 62:327–334.

Issa NP, Trachtenberg JT, Chapman B, Zahs KR, Stryker MP (1999). The critical period for ocular dominance plasticity in the Ferret's visual cortex. *J Neurosci.* 19:6965–6978.

Iwai Y, Fagiolini M, Obata K, Hensch TK (2003). Rapid critical period induction by tonic inhibition in visual cortex. *J Neurosci.* 23:6695–6702.

Kaas JH, Guillery RW, Allman JM (1972). Some principles of organization in the dorsal lateral geniculate nucleus. *Brain Behav Evol.* 6:253–299.

Kanold PO, Luhmann HJ (2010). The subplate and early cortical circuits. *Annu Rev Neurosci.* 33:23–48.

Kim YJ, Peterson BB, Crook JD, Joo HR, Wu J, Puller C, Robinson FR, Gamlin PD, Yau KW, Viana F, Troy JB, Smith RG, Packer OS, Detwiler PB, Dacey DM (2022). Origins of direction selectivity in the primate retina. *Nat Commun*. 13:2862.

Kiorpes L, Bassin SA (2003). Development of contour integration in macaque monkeys. *Vis Neurosci.* 20:567–575.

Kiorpes L, Movshon JA (2004). Development of sensitivity to visual motion in macaque monkeys. *Vis Neurosci.* 21:851–859.

Kiorpes L, Walton PJ, O'Keefe LP, Movshon JA, Lisberger SG (1996). Effects of early-onset artificial strabismus on pursuit eye movements and on neuronal responses in area MT of macaque monkeys. *J Neurosci.* 16:6537–6553.

Kovacs I (2000). Human development of perceptual organization. *Vision Res.* 40:1301–1310.

Kreile AK, Bonhoeffer T, Hubener M (2011). Altered visual experience induces instructive changes of orientation preference in mouse visual cortex. *J Neurosci.* 31:13911–13920.

Kuhlman SJ, Tring E, Trachtenberg JT (2011). Fast-spiking interneurons have an initial orientation bias that is lost with vision. *Nat Neurosci.* 14:1121–1123.

Lempel AA, Nielsen KJ (2021). Development of visual motion integration involves coordination of multiple cortical stages. *eLife*. 10:e59798.

Li Y, Fitzpatrick D, White LE (2006). The development of direction selectivity in ferret visual cortex requires early visual experience. *Nat Neurosci*. 9:676–681.

Lien AD, Scanziani M (2018). Cortical direction selectivity emerges at convergence of thalamic synapses. *Nature*. 558:80–86.

Lin Y, Tadin D (2019). Motion perception: Slow development of center-surround suppression. *Curr Biol*. 29:R878–R880.

Maslim J, Stone J (1988). Time course of stratification of the dendritic fields of ganglion cells in the retina of the cat. *Brain Res Dev*. 44:87–93.

Movshon JA, Albright TD, Stoner GR, Majaj NJ, Smith MA. (2003). Cortical responses to visual motion in alert and anesthetized monkeys. *Nat Neurosci*. 6(1):3.

Mower GD, Burchfiel JL, Duffy FH (1982). Animal models of strabismic amblyopia: physiological studies of visual cortex and the lateral geniculate nucleus. *Brain Res*. 281:311–327.

Mulholland HN, Hein B, Kaschube M, Smith GB (2021). Tightly coupled inhibitory and excitatory functional networks in the developing primary visual cortex. *eLife*. 10:e72456.

Murata Y, Colonnese MT (2016). An excitatory cortical feedback loop gates retinal wave transmission in rodent thalamus. *eLife*. 5:e18816.

Norcia AM, Sampath V, Hou C, Pettet MW (2005). Experience-expectant development of contour integration mechanisms in human visual cortex. *J Vis*. 5:116–130.

Pennefather PM, Chandna A, Kovacs I, Polat U, Norcia AM (1999). Contour detection threshold: repeatability and learning with "contour cards." *Spat Vis*. 12:257–266.

Popovic M, Stacy AK, Kang M, Nanu R, Oettgen CE, Wise DL, Fiser J, Van Hooser SD (2018). Development of cross-orientation suppression and size tuning and the role of experience. *J Neurosci*. 38:2656–2670.

Pratt KG, Hiramoto M, Cline HT (2016). An evolutionarily conserved mechanism for activity-dependent visual circuit development. *Front Neural Circuits*. 10:79.

Reese BE (1988). "Hidden lamination" in the dorsal lateral geniculate nucleus: the functional organization of this thalamic region in the rat. *Brain Res*. 472:119–137.

Robinson SR, Dreher B (1990). The visual pathways of eutherian mammals and marsupials develop according to a common timetable. *Brain Behav Evol*. 36:177–195.

Rochefort NL, Narushima M, Grienberger C, Marandi N, Hill DN, Konnerth A (2011). Development of direction selectivity in mouse cortical neurons. *Neuron*. 71:425–432.

Rodman HR, Scalaidhe SP, Gross CG (1993). Response properties of neurons in temporal cortical visual areas of infant monkeys. *J Neurophysiol*. 70:1115–1136.

Roy A, Wang S, Meschede-Krasa B, Breffle J, Van Hooser SD (2020). An early phase of instructive plasticity before the typical onset of sensory experience. *Nat Commun*. 11:11.

Roy A, Osik JJ, Ritter NJ, Wang S, Shaw JT, Fiser J, Van Hooser SD (2016). Optogenetic spatial and temporal control of cortical circuits on a columnar scale. *J Neurophysiol*. 115:1043–1062.

Sengpiel F, Stawinski P, Bonhoeffer T (1999). Influence of experience on orientation maps in cat visual cortex. *Nat Neurosci*. 2:727–732.

Shatz CJ, Rakic P (1981). The genesis of efferent connections from the visual cortex of the fetal rhesus monkey. *J Comp Neurol*. 196:287–307.

Sireteanu R, Rieth C (1992). Texture segregation in infants and children. *Behav Brain Res*. 49:133–139.

Smith GB, Hein B, Whitney DE, Fitzpatrick D, Kaschube M (2018). Distributed network interactions and their emergence in developing neocortex. *Nat Neurosci*. 21:1600–1608.

Taylor G, Hipp D, Moser A, Dickerson K, Gerhardstein P (2014). The development of contour processing: Evidence from physiology and psychophysics. *Front Psychol*. 5:719.

Tiriac A, Feller MB (2022). Roles of visually evoked and spontaneous activity in the development of retinal direction selectivity maps. *Trends Neurosci*. 45:529–538.

Torborg CL, Feller MB (2005). Spontaneous patterned retinal activity and the refinement of retinal projections. *Prog Neurobiol*. 76:213–235.

Tschetter WW, Govindaiah G, Etherington IM, Guido W, Niell CM (2018). Refinement of spatial receptive fields in the developing mouse lateral geniculate nucleus is coordinated with excitatory and inhibitory remodeling. *J Neurosci*. 38:4531–4542.

Van Hooser SD (2007). Similarity and diversity in visual cortex: Is there a unifying theory of cortical computation? *Neuroscientist*. 13:639–656.

Van Hooser SD, Li Y, Christensson M, Smith GB, White LE, Fitzpatrick D (2012). Initial neighborhood biases and the quality of motion stimulation jointly influence the rapid emergence of direction preference in visual cortex. *J Neurosci*. 32:7258–7266.

Wang J, Ho CS, Giaschi DE (2007). Deficient motion-defined and texture-defined figure-ground segregation in amblyopic children. *J Pediatr Ophthalmol Strabismus*. 44:363–371.

Weliky M, Katz LC (1999). Correlational structure of spontaneous neuronal activity in the developing lateral geniculate nucleus in vivo. *Science*. 285:599–604.

White LE, Coppola DM, Fitzpatrick D (2001). The contribution of sensory experience to the maturation of orientation selectivity in ferret visual cortex. *Nature*. 411:1049–1052.

Wiesel TN, Hubel DH (1963). Effects of visual deprivation on morphology and physiology of cells in the cats lateral geniculate body. *J Neurophysiol*. 26:978–993.

Wilkinson F, Crotogino J (1995). The late onset of visual texture segmentation in kittens. *Behav Brain Res*. 68:201–217.

53.

DEVELOPMENT OF CORTICOTHALAMIC PROJECTIONS

William Guido

INTRODUCTION

The cortex provides massive feedback projections to the thalamus. These inputs arise from subpopulations of cortical projection neurons of layers (L)5 and 6. Descending projections from these deep cortical layers dominate many aspects of thalamic activity, modulating incoming sensory information from the periphery and also serving as a critical node for cortico-thalamocortical signaling (Sherman 2016; Usrey and Sherman, 2019). Much has been written about the functional organization of corticothalamic projections, but comparatively little attention has been given to how, when, and where these pathways originate, grow, and eventually innervate specific thalamic nuclei (but see Grant et al., 2012; Leyva-Diaz and Lopez-Bendito, 2013). Such information is vital to understand adult function and also to provide insight as to the basis of many neurodevelopmental disorders that have at their core a defect in thalamic circuitry.

The purpose of this chapter is to focus on the development of corticothalamic projections and to highlight our current understanding of the origin, cell fate, axon guidance, and layer-specific targeting to first-order and higher-order thalamic nuclei. Importantly, this chapter will also review how the corticothalamic pathway influences the development of other major tracts, including the thalamocortical and retinothalamic pathways.

Finally, it is important to note the recent shift in the choice of animal models used to study corticothalamic development. Although much of our foundational knowledge comes from early work done in carnivores and primates, the mouse as a model organism has moved to the forefront as a tractable experimental platform to examine many of the genetic and molecular mechanisms that control cell type specification, axon guidance, and the lamina specificity of thalamic innervation. As a result, this chapter is mouse-centric, but hopefully the reader will appreciate the utility of adopting the mouse as a model to advance a set of conserved principles that explain corticothalamic development.

OVERVIEW OF CORTICOTHALAMIC PROJECTIONS

A key feature of the cerebral cortex is the presence of several long-distance pathways that originate from excitatory cortical projection neurons. One of the largest groups of cortical projection neurons are corticofugal, with somata that originate in cortical L5 and L6 and with axons that project to subcortical targets. Corticofugal neurons give rise to 2 descending pathways. One is typically referred to as a subcerebral projection, with neurons that are confined to L5 and axonal projections that innervate subcortical structures such as the basal ganglia, midbrain, brainstem, and spinal cord. The other, the subject of this chapter, is the corticothalamic projection, with neurons located in L6 and with a smaller population residing in L5, and with axons that innervate thalamic nuclei.

The form and function of corticothalamic projections are detailed elsewhere in this book (see **Chapters 14 and 15 in Section IV; Chapter 19 in Section V**) and in numerous reviews by Sherman and colleagues (e.g., Sherman and Guillery 2002; Sherman and Guillery, 2011; Sherman 2016; Usrey and Sherman, 2019). Thus, they are only briefly outlined here and in Figure 53.1. L6 corticothalamic neurons innervate all thalamic nuclei. Most prominent and perhaps best characterized are the projections to first-order thalamic nuclei such as the dorsal lateral geniculate nucleus (dLGN), ventrobasal complex (VB), the dorsal division of the medial geniculate nucleus (MGNd), the ventrolateral nucleus (VL), and the anterior thalamic group. L6 projections are modality specific and topographically aligned with their ascending thalamocortical counterparts. They terminate diffusely and form numerous small excitatory synapses on distal regions of thalamocortical neurons. Corticothalamic input from L6 is considered modulatory in nature, providing signals that alter the gain of thalamocortical transmission from the periphery in a state-dependent fashion. By contrast, L5 corticothalamic projections are confined to higher-order thalamic nuclei, which include the pulvinar

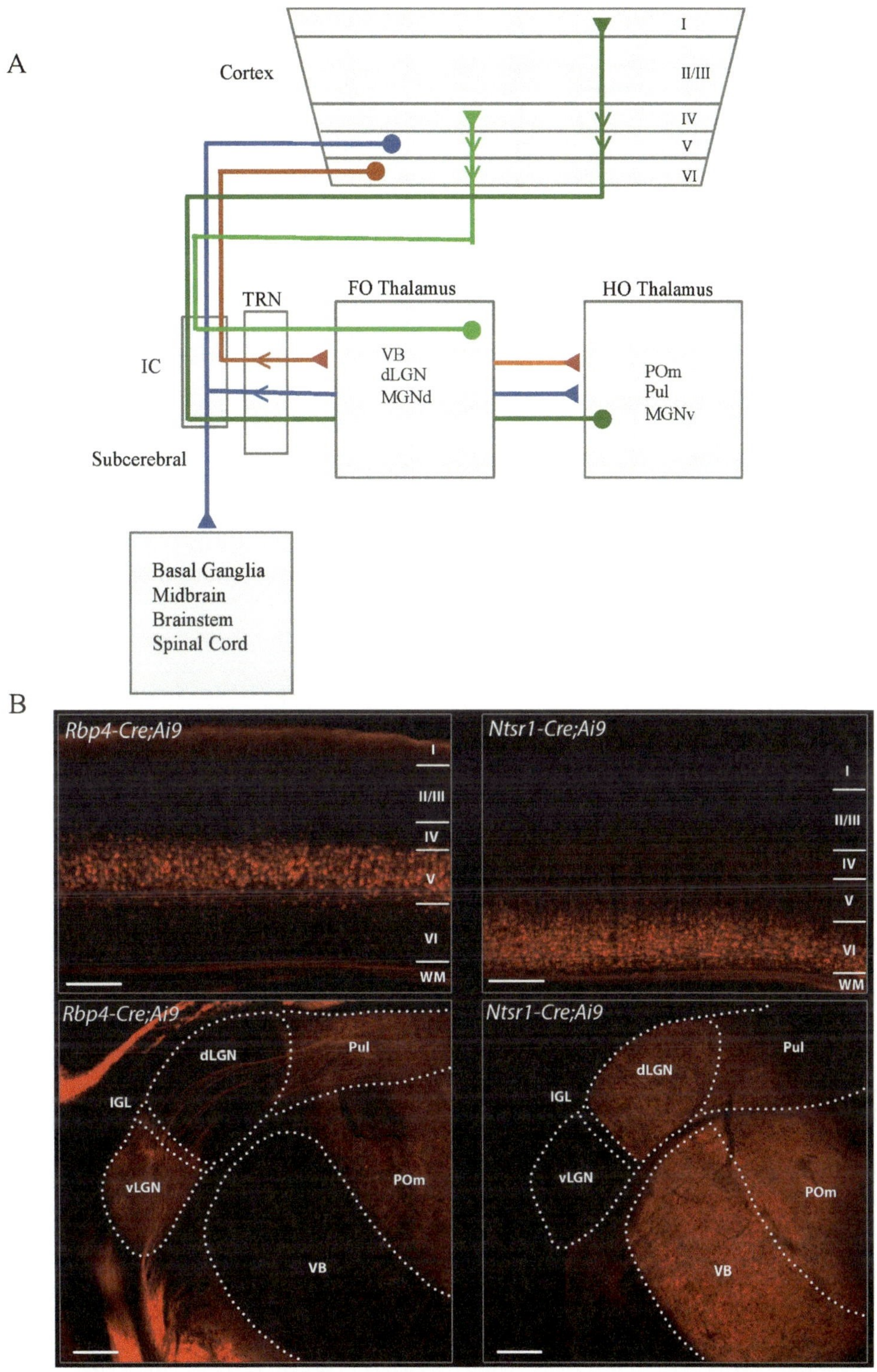

Figure 53.1. Organization of corticothalamic projections. (A). Wiring diagram showing the route and pattern of subcerebral and corticothalamic projections from L5 and L6 of cortex. For clarity, only corticothalamic projections to sensory nuclei are shown. IC = internal capsule; TRN = thalamic reticular nucleus; FO = first order; HO = higher order; VB = ventrobasal complex; dLGN = dorsal lateral geniculate nucleus; MGN = medial geniculate nucleus; POm = posterior medial nucleus; Pul = pulvinar. (B). Visualization of L5 and L6 neurons along with their pattern of thalamic innervation using Cre-driver lines crossed on a td-Tomato (Ai9) reporter. The Rbp4-Cre line (left) expresses Cre-recombinase in L5 cortical neurons, while the Ntsr1-Cre (right) expresses Cre-recombinase in L6 cortical neurons. Both sets of neurons become labeled when these mice are crossed with the tdTomato Ai9 reporter strain. Left: Rbp4-Cre;Ai9 labels L5 neurons (top) and their corticothalamic projections (bottom) as they pass through first-order nuclei such as dLGN and VB and terminate in the higher-order Pul and POm. Note the innervation of ventral lateral geniculate nucleus (vLGN). Right: Ntsr1-Cre;Ai9 labels L6 neurons (top) and their corticothalamic projections (bottom) as they innervate first-order nuclei such as dLGN and VB and higher-order Pul and POm. IGL = intrageniculate leaflet; vLGN = ventral lateral geniculate nucleus. Scale bars: 200 μm.

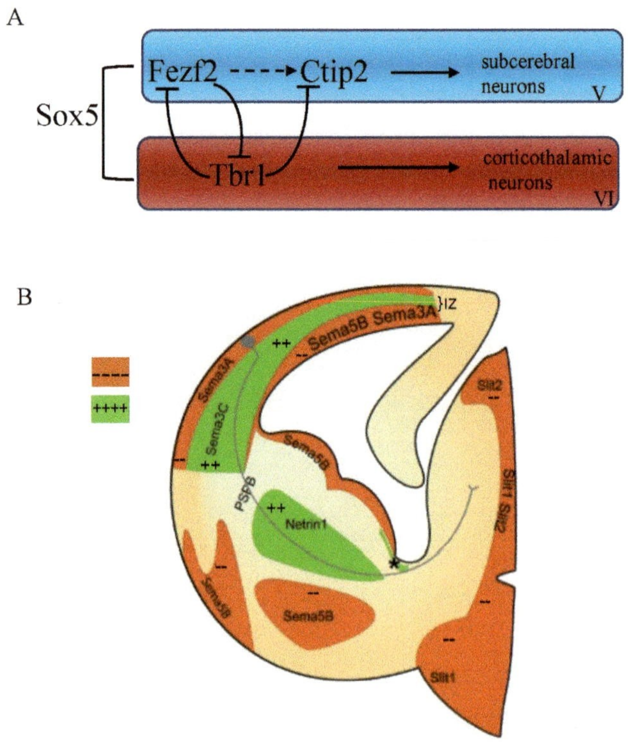

Figure 53.2. Specification of corticothalamic neurons and their guidance cues. (A). Schematic depicting the interactions of key transcription factors that determine corticofugal neuron fate for L5 (blue) and L6 (red). The dashed line between Fezf2 and Ctip2 poses an upstream regulation. Solid, bar-end lines reflect inhibition/repression. Adapted from McKenna et al. (2011) and Leyva and Lopez-Bendito (2013). (B). Schematic from Grant et al. (2012) illustrating the path of corticothalamic axon growth in relation to key guidance molecules that lead axons toward thalamus. Chemoattractant (++) cues are outlined in green and chemorepulsive (–) ones in orange. Asterisk depicts the internal capsule. The bracket shows the approximate location of the intermediate zone (IZ).

(Pul; lateral posterior [LP] nucleus in rodent), the posterior medial nucleus (POm), the ventral division the medial geniculate nucleus (MGNv), and the mediodorsal thalamic group. They have large terminals that synapse on proximal regions of dendrites and provide the primary excitatory drive for thalamocortical neurons. These higher-order thalamic neurons participate in a cortico-thalamocortical circuit that enables different cortical areas to communicate with each other.

SPECIFICATION OF CORTICOTHALAMIC NEURONS AND EARLY AXON GUIDANCE

Early progenitor cells in the ventricular zone give rise to deep cortical layer neurons that express a variety of transcription factors that lead to the specification of subcortical projecting neurons (Molyneaux et al., 2007; Leone et al., 2008; Grant et al., 2012; Greig et al., 2013; Leyva-Diaz and Lopez-Bendito, 2013). Most notable for this chapter are SRY-box gene *Sox5* which controls the differentiation of corticofugal neuron subtypes, and the zinc finger transcription factors Fezf2 and Ctip2, which together regulate the identity of L5 subcerebral and L6 corticothalamic projection neurons (Arlotta et al., 2005; Chen et al., 2005; McKenna et al., 2011; Figure 53.2A). *Sox5* is expressed at high levels in developing corticofugal neurons, and in its absence, neurons destined for L5 and L6 fail to migrate and segregate into distinct laminae (Lai et al., 2008). Fezf2 upregulates Ctip2 expression, and together they work in concert to promote the fate and differentiation of subcerebral projection neurons (Arlotta et al., 2005; Chen et al., 2005; Molyneaux et al., 2005). Additionally, Tbr1, a t-box transcription factor which becomes highly enriched in L6 neurons, acts to reduce Ctip2 and Fezf2 expression, thereby repressing subcerebral fates and promoting the identity and trajectory of corticothalamic neurons (McKenna et al., 2013). Studies in mutant mice further reveal how these factors interact to regulate cell fate divergence of corticothalamic L6 and subcerebral L5 projections. For example, in the absence of Tbr1, L6 neurons continue to express Fezf2 and Ctip2, thereby adopting a L5 cell fate with axons that follow a subcerebral route (McKenna et al., 2011). By contrast, in mice lacking Fezf2, L5 neurons exhibit an increased expression of Tbr1, and a cell fate that promotes L6 neurons with corticothalamic projections (Han et al., 2011; McKenna et al., 2011).

AXON GUIDANCE OF CORTICOTHALAMIC PROJECTIONS

The growth and trajectory of corticothalamic projections are controlled by well-established families of guidance molecules (see Grant et al., 2012; Figure 53.2B). In rodent, newly born deep-layer cortical neurons located in the intermediate zone (a region that sits between the germinal zone and the cortical preplate) begin to grow neurites as early as embryonic day (E)10. Members of the semaphorin family are expressed in the cortical plate and intermediate zone to provide the initial guidance of corticofugal axons. They are repelled from entering the cortical plate by Sema3A and are attracted toward the intermediate zone by Sema3C (Skaliora et al., 1998; Bagnard et al., 2000). The intermediate zone forms a lane between the cortical plate and the ventricular zone where the expression of Sema3A and Sema5B prevent cortical axons from overshooting into the cortical germinal zone (Bagnard et al. 2000; Lett et al., 2009). Within the intermediate zone, Sema3C is arranged in a lateral to medial gradient, attracting growing axons and steering them toward the lateral regions of cortex. Between E13 and E15, axons extend through the intermediate zone until they reach the lateral boundary of the internal capsule, where they pause to await the arrival of more dorsally located axons (Molnar and Cordery, 1999). Soon after, they continue their extension crossing the pallial-subpallial boundary (a region that separates the developing cortex from the striatum) and enter the internal capsule. Here, corticothalamic axons encounter the chemoattractant molecule Netrin-1, which is expressed in the internal capsule and ventral telencephalon (Oeschger et al., 2012). The complementary expression of Netrin-1 in the ventral telencephalon and the chemorepellent expression of Sema5B located in the lateral cortex, which flanks the internal capsule, appear to be responsible for redirecting corticothalamic axons toward the ventral telencephalon (Lett et al., 2009). As corticothalamic axons approach the diencephalon/telencephalon boundary, members of the chemorepellent Slit family direct axons dorsally toward the thalamus, preventing them from crossing midline. These inhibitory cell surface molecules are expressed in large overlapping domains that include the ganglionic eminences, prethalamus, hypothalamus, and the germinal zone of dorsal thalamus (Braisted et al., 2000).

The guidance cues that direct cortical axons to specific thalamic nuclei have yet to be defined. However, once they reach their appropriate targets, complementary gradients of EphA7 receptors and ephrin-A5 ligands play a major role in the topographic mapping of corticothalamic projections within thalamic sensory nuclei (Torii and Levitt, 2005; Torii et al., 2012). For example, using in utero electroporation gene transfer in mice to alter the graded pattern of EphA7 in somatosensory cortex leads to predictable shifts in corticothalamic topography in somatosensory thalamic nuclei (VPM and POm), while leaving their trajectory to the thalamus unperturbed and also sparing the topography of ascending thalamocortical mapping (Torii and Levitt, 2005).

INTERACTIONS BETWEEN CORTICOTHALAMIC AND THALAMOCORTICAL PATHWAYS

During embryogenesis, pioneer axons from the cortex and thalamus first meet in the subpallium, an intermediate telencephalic region that gives rise to the internal capsule. The subpallium serves as an important nexus for the early navigation of corticothalamic and thalamocortical axons as they proceed in opposing directions to innervate their respective targets. Their close proximity to each other at this juncture has led to the formulation of the "handshake hypothesis" as a way to explain how these reciprocal projections work together to assist each other in their respective guidance and targeting (Blakemore and Molnar, 1990; Molnar and Blakemore, 1991, 1995). For many years, how such mutual dependence occurred and the underlying molecular mechanisms that governed it remained poorly defined. On a molecular level, several transcription factors, including Tbr1, Emx2, Gbx2, and Pax6, have been implicated (Leyva Diaz and Bendito-Lopez, 2013). Mutant mice lacking these transcription factor genes lead to pathfinding errors for both corticothalamic and thalamocortical axons (Hevner et al., 2002; Lopez-Bendito et al., 2002). However, questions about the source of transcription factor expression and the type of signaling they engage in have yet to be elucidated. Some clarity about their mutual dependence has emerged from studies that examine where and when corticothalamic and thalamocortical axons meet at the subpallium and pallium-subpallial border. One study that underscores the importance of location and a supporting role for corticothalamic axons reveals that in the absence of differentiated corticothalamic neurons, their axons still progress through the subpallium but fail to cross the pallium-subpallium border and enter the developing cortex (Chen et al., 2012). By using an in vitro co-culture preparation, the authors also show that the introduction of normal cortex was able to rescue cortical entry. A case for a complimentary role, whereby thalamocortical axons support corticothalamic axon pathfinding, features timing as a defining factor (Deck et al., 2013). Early corticofugal axons seem to arrive at ~E14, a day earlier than the arrival of ascending thalamocortical axons, then pause their trajectory until thalamocortical axons appear. However, the genetic ablation of thalamocortical axons leads to a pathfinding error whereby corticothalamic axons fail to wait. Consequently, instead of projecting to thalamus, they remain on a subcerebral trajectory coursing through the globus pallidus and reaching the cerebral peduncle. Thus, it appears that the arrival of thalamocortical axons is needed to steer corticothalamic ones away from a subcerebral course and toward thalamus. The pause in corticothalamic progression is believed to be mediated by Sema3E/Plexin signaling among early pioneer axons. This interaction acts as a brake to slow progression and coordinate the timing of corticothalamic and thalamocortical axon navigation and to ultimately prevent the premature subpallial progression of corticothalamic axons en route to the thalamus (Deck et al., 2013). Together, these studies provide a deeper understanding of the handshake hypothesis and underscore the importance of location and timing as defining features that regulate the mutual interactions between corticothalamic and thalamocortical axons.

TEMPORAL AND SPATIAL REGULATION OF CORTICOTHALAMIC TARGETING

The timing and route of corticothalamic projections have been described in a number of mammals, including rodents, carnivores, and primates (Shatz and Rakic 1981; Miller et al., 1993; Clasca et al., 1995; Molnar and Cordery, 1999; Jacobs et al., 2007). These studies show that corticothalamic projections from the subplate, L5, and L6 grow toward the thalamus with distinct temporal patterns (Grant et al., 2012). The axons of early born subplate neurons are the first to reach the internal capsule and appear to provide structural support and subsequent guidance for the navigation of descending L5 and L6 axons, as well as for ascending thalamocortical axons en route to the cortex. The exact timing of thalamic innervation for L5 and L6 axons in carnivores and primates has been difficult to track with conventional tracing techniques and has been the subject of some debate (Miller et al., 1993, vs. Clasca et al., 1995). However, the advent of cell-type-specific transgenic (e.g., golli-τ-eGFP) and Cre-driver lines (e.g., Rbp4 L5, NTSR1 L6) in the mouse has made it possible to follow independently their perinatal trajectory and innervation of first-order and higher-order nuclei (Jacobs et al., 2007; Seabrook et al., 2013; Grant et al., 2016; Figures 53.1B, 53.3A–B). Together these lines reveal that between E16.5 and E17.5 corticothalamic fibers grow into the internal capsule and then sweep toward the diencephalon to approach the lateral border of the ventral thalamus. At E18.5-P0, corticothalamic axons begin to pass through the thalamic reticular nucleus, cross the external medullary lamina, and enter the dorsal thalamus. The layer-specific innervation of higher-order and first-order nuclei emerge as early as postnatal day (P)0–2, but takes several days to complete, fully innervating thalamic nuclei by P12. During these postnatal ages, L5 axons initially pass through the first-order nuclei of dLGN and VPM to innervate and arborize throughout the adjacent higher-order nuclei such as the Pul and POm. L6 corticothalamic axons follow a similar time course for the innervation of first-order nuclei, with more medial nuclei (e.g., VPM) receiving input before lateral ones, such as dLGN and MGN.

Studies that focus on L6 innervation of dLGN have received the greatest attention and highlight the mechanisms that regulate the timing of innervation, especially in relation to the arrival of retinofugal projections (Figure 53.3C). Using the golli-τ-eGFP transgenic line to visualize L6 innervation of dLGN in conjunction with anterograde labeling of retinogeniculate projections with CTB reveals that CT fibers appear to "wait" at the ventral border of dLGN until retinogeniculate fibers from both eyes innervate dLGN. In fact, corticogeniculate fibers do not fully innervate dLGN until the age of natural eye opening (P12), well after the time when retinal inputs from the 2 eyes innervate, then segregate to form non-overlapping eye specific domains. At this time, structural and functional corticogeniculate synapses emerge in dLGN (Bickford et al., 2010; Jurgens et al., 2012; Seabrook et al., 2013). Between P10 and P12, optogenetic stimulation of corticogeniculate terminals in dLGN evokes weak, somewhat immature excitatory postsynaptic current (EPSC) activity in relay neurons. By 2 weeks of age, dLGN neurons exhibit large-amplitude excitatory activity that increases in amplitude with each successive pulse of a stimulus train (Figure 53.3D). Such facilitation is a hallmark feature of modulator input (Sherman and Guillery, 2011).

It is notable that other mammals including the cat, ferret, and macaque seem to follow a similar sequence (Anker, 1977; Shatz and Rakic, 1981; Clasca et al., 1995), suggesting that the coordination of retinal and cortical inputs is part of a conserved developmental plan in which the retinogeniculate pathway is established before the innervation and formation of corticothalamic connections. Whether such an arrangement, in which ascending sensory afferents from the periphery arrive before descending ones, applies to other first-order sensory thalamic nuclei remains to be tested.

THE ROLE OF SENSORY INPUT AND NEURAL ACTIVITY IN REGULATING THE TIMING AND TARGETING OF CORTICOTHALAMIC PROJECTIONS

The coordinated progression of retinal and cortical innervation of dLGN suggests that retinal signaling somehow regulates the timing of corticogeniculate innervation. Support for this comes from loss of function studies showing that the early elimination of retinal signaling accelerates innervation (Seabrook et al., 2013; Brooks et al., 2013; Grant et al., 2016; Moreno-Juan et al., 2022). Interestingly, the loss of retinal input does not appear to affect the rate of L6 innervation of higher-order visual thalamic nuclei, such as LP (Grant et al., 2016). In the case of dLGN, the premature entry of corticogeniculate fibers is best illustrated in a mutant mouse that lacks Math5, a transcription factor necessary for retinal ganglion progenitor cell differentiation, resulting in a >95% loss of retinal ganglion cells and a brain that is devoid of retinofugal projections (Brown et al., 1998; Wang et al., 2001; El-Danaf et al., 2015). In Math5 null mice, corticogeniculate axons fail to wait at the ventral border of dLGN, and instead rapidly advance to fully innervate the nucleus by the end of the first postnatal week (Seabrook et al., 2013; Brooks et al., 2013). The mechanism underlying this dysregulation involves aggrecan, a repulsive extracellular matrix molecule that shows an early enriched, but transient expression pattern throughout dLGN. The pattern of transient aggrecan expression coincides in a complimentary manner with the progression of corticogeniculate innervation (Brooks et al., 2013; Figure 53.3E). However, in Math5 null mice this repulsive molecule is nearly absent at perinatal ages due to the higher-than-normal levels of degradative proteases that cleave

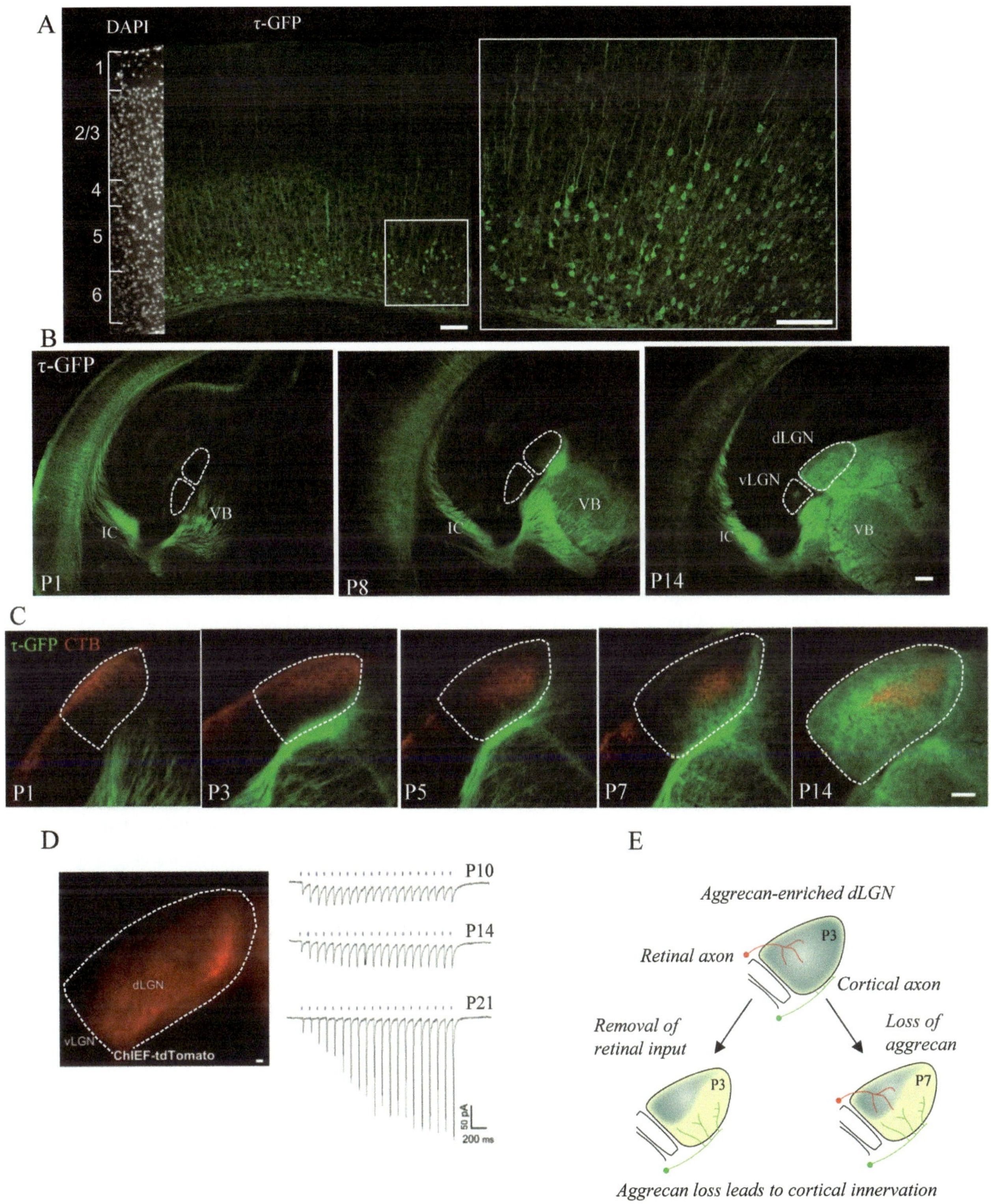

Figure 53.3. Development of L6 corticothalamic projections. (A). Coronal section of visual cortex at P14 showing the distribution of L6 neurons expressing GFP (green) in the golli-τ-eGFP mouse (Jacobs et al., 2007). This transgenic mouse uses a 1.3 kb golli promoter element of the myelin basic protein gene to target expression of a tau-green fluorescent protein (GFP) in L6 cortical neurons. Left inset: DAPI stain depicting the boundaries of cortical layers. Right inset: high-power image for the outlined area. Scale bars: 100μm. (B). The pattern of L6 cortical innervation of the thalamus shown in coronal sections at P1, P8, and P14 in golli-τ-GFP mice. The boundaries of the dorsal lateral geniculate nucleus (dLGN) and the ventral lateral geniculate nucleus (vLGN) are outlined with dashed lines. IC = internal capsule; VB = ventrobasal complex. Scale bar: 200 μm. (C). Innervation of dLGN by cortical and uncrossed retinal projections in coronal sections of golli-τ-GFP mice at different postnatal ages. Shown are merged images of corticogeniculate (green) and uncrossed retinogeniculate (red) projections. Corticogeniculate projections are visualized with GFP and retinogeniculate projections were labeled by making eye injections of CTB conjugated to Alexa Fluor 594 in one eye. The boundaries of dLGN are outlined with dashed

aggrecan, thus allowing corticogeniculate axons to invade dLGN prematurely. These effects are likely due to the disruption of trophic signaling between retinogeniculate fibers and thalamocortical neurons, events that seem regulated by neural activity. Support for the latter comes from studies showing that the disruption of early spontaneous retinal waves leads to an acceleration of corticogeniculate innervation (Grant et al., 2016). Recently, intrinsic patterns of spontaneous thalamic activity have also been implicated. When thalamic activity is silenced, it leads to a delay in corticogeniculate innervation (Moreno-Juan et al., 2022). Together, these studies suggest that different sources of early activity regulate the balance between aggrecan expression and the proteases that cleave it.

Finally, it should be noted that the elimination of retinal input affects the progression and patterning of some descending L5 projections. In the absence of retinal input, L5 axons belonging to the subcerebral, corticotectal pathway prematurely innervate the retino-recipient layers of the superior colliculus (Grant et al., 2016). Interestingly, the distribution of aggrecan expression in the superior colliculus is unaffected by such a loss. In fact, L5 terminations are observed throughout aggrecan-rich regions. Thus, the molecular cues regulating the timing of cortical input to subcortical structures are diverse, largely unknown, and likely target-specific.

The loss of retinal input also affects the targeting of developing L5 corticothalamic projections (Grant et al., 2016; Frangeul et al., 2016). Normally, L5 axons pass through dLGN en route to innervate the higher-order visual thalamic pulvinar nucleus (Figure 53.1B). However, in the absence of retinal input, L5 fibers coursing through now branch in dLGN (Grant et al., 2016). These aberrant processes stain for VGluT1 and synaptophysin, persist into adulthood, and when stimulated optogenetically evoke postsynaptic excitatory activity in dLGN neurons (Grant et al., 2016; Frangeul et al., 2016; Giasafaki et al., 2022). Such "cross-hierarchical plasticity" represents a novel form of rewiring in which L5 input innervates a first-order thalamic nucleus that has lost its sensory-derived, driver input (Grant et al., 2016). Transcriptional analysis comparing the gene-expression patterns of higher-order and first-order nuclei further reveals that in response to neonatal sensory afferent ablation, deafferented first-order nuclei (e.g., dLGN and VPM) take on a higher-order genetic identity (Frangeul et al., 2016). This has led to the idea that hierarchical order, not sensory modality, determines thalamic identity. Moreover, it raises the possibility that thalamic neurons by default possess a higher-order identity, only to take on first-order features as modality-specific input arrives (Frangeul et al., 2016; Giasafaki et al., 2022).

ROLE OF CORTICOTHALAMIC PROJECTIONS IN RETINOGENICULATE TARGETING AND REMODELING

As discussed above, the close apposition of corticothalamic and thalamocortical axons during development provide ample opportunity for mutual interactions. Recent studies also reveal that developing corticothalamic projections can have a long-range impact on the trajectory of sensory pathways from the periphery. In mutant mice that lack a large part of their cortex (Tra2β nulls), retinogeniculate projections fail to innervate dLGN but continue to innervate other retinorecipient targets such as the superior colliculus and ventral LGN (Shanks et al., 2016; Diao et al., 2018). Similar defects are found in Tbr1 null mice where L6 neurons fail to differentiate, as well as in Fezf2 conditional knock-outs which disrupt retinogeniculate projections at embryonic ages (Hevner et al., 2002; Diao et al., 2018). While these results suggest that L6 innervation of dLGN is required for proper retinogeniculate innervation, the underlying mechanism remains unclear, since cortical innervation of dLGN does not occur until well after the retinal fibers from both eyes innervate dLGN (Seabrook et al., 2013). These effects seem even more puzzling in light of evidence demonstrating how the loss of retinal input greatly affects the timing and innervation of corticothalamic axons (Seabrook et al., 2013; Brooks et al., 2013; Grant et al., 2016). To reconcile these conflicting results, it has been suggested that the pioneering corticogeniculate fibers which arrive at the medial boundaries of the thalamus at perinatal ages secrete a growth-promoting factor (albeit unknown) that promotes axonal branching from the optic tract to then innervate dLGN (Shanks et al., 2016). Others have suggested that the absence of cortical input leads to the premature apoptosis of thalamocortical neurons, thereby making them "unsuitable targets" for retinal axons (Diao et al., 2018).

Corticogeniculate input has also been implicated in regulating the refinement of retinogeniculate connections.

Figure 53.3 Continued

lines. Scale bar: 100 μm. (D). Left: coronal section of dLGN at P10 showing the expression of the opsin ChIEF-tdTomato after an injection of an AAV construct containing ChIEF-tdTomato into the deep layers of the visual cortex. Right: whole-cell voltage clamp recordings showing optically evoked corticothalamic responses in relay neurons to a 10 Hz train (20 pulses, blue bars) at P10, P14, and P21. (E). Schematic illustrating the regulation of cortical innervation in dLGN by aggrecan in the presence and absence of retinal innervation. A–C from Seabrook et al. (2013); E from Jurgens et al. (2012).

Soon after corticogeniculate innervation and functional synapse formation with thalamocortical neurons, feedback activity from L6 proves to be essential for the final phase of retinogeniculate refinement. During this period (P20–P30) retinogeniculate connections with thalamic relay neurons are still labile, exhibiting a capacity to weaken or strengthen, depending on nature of visual experience (Liang and Chen, 2020). Indeed, late dark-rearing during this period of synapse stabilization leads to the additional recruitment of weak retinal inputs (Hooks and Chen 2006, 2008). Remarkably, pharmacological or chemogenetic suppression of L6 activity between P20 and P27 leads to a similar increase in retinal convergence (Thompson et al., 2016). In fact, an elevation of L6 activity has similar effects, suggesting that patterned feedback, and not activity levels per se, is needed. Taken together, the studies of Chen and colleagues illustrate that bidirectional exchanges between retinally mediated feedforward and cortically mediated feedback activity are needed for synapse stabilization of retinogeniculate connections. How the interplay between cortical feedback and retinal feedforward signaling work to achieve such remodeling remains an open question.

CONCLUDING COMMENTS

Clearly, studies in the mouse, which provide the capacity to visualize and manipulate developing neurons and circuits in a cell-type-specific manner have greatly advanced our understanding of the mechanisms underlying the development of corticothalamic projections. However, questions still remain. For example, the transcriptional regulation of corticofugal neuron determination is now well established. However, can such knowledge be used to initiate or reprogram neuron fate? If so, could such plasticity lead to axon outgrowth and target appropriate innervation? Additionally, while the molecular mechanisms underlying early axon guidance are known, it is unclear how corticothalamic projections from L5 and L6 choose their thalamic targets. What are the key guidance cues involved in the layer-specific targeting of thalamic nuclei? What is the defining feature, cell type, thalamic order, or modality? While molecules such as aggrecan play a role in the timing of innervation, the ones that govern axon target selection remain largely unknown. A related question pertains to the role of activity. Loss of function studies reveal that neural activity, from either the periphery or the target, plays an important role in axon target interactions as well as the remodeling of newly formed connections. How patterned activity, as well as its interactions with intrinsic genetic programs, works to guide development continues to be an enduring question worthy of investigation.

REFERENCES

Anker RL (1977). The prenatal development of some of the visual pathways in the cat. *J Comp Neurol.* 173(1):185–204.

Arlotta P, Molyneaux BJ, Chen J, Inoue J, Kominami R, Macklis JD (2005). Neuronal subtype-specific genes that control corticospinal motor neuron development in vivo. *Neuron.* 45(2):207–221.

Bagnard D, Thomasset N, Lohrum M, Püschel AW, Bolz J (2000). Spatial distributions of guidance molecules regulate chemorepulsion and chemoattraction of growth cones. *J Neurosci.* 20(3):1030–1035.

Bickford ME, Slusarczyk A, Dilger EK, Krahe TE, Kucuk C, Guido W (2010). Synaptic development of the mouse dorsal lateral geniculate nucleus. *J Comp Neurol.* 518(5):622–635

Blakemore C, Molnár Z (1990). Factors involved in the establishment of specific interconnections between thalamus and cerebral cortex. *Cold Spring Harb Symp Quant Biol.* 55: 491–504.

Braisted, JE, Catalano SM, Stimac, R, Kennedy TE, Tessier-Lavigne M, Shatz CJ, O'Leary DD (2000). Netrin-1 promotes thalamic axon growth and is required for proper development of the thalamocortical projection. *J Neurosci.* 20(15):5792–5801.

Brooks JM, Su J, Levy C, Wang JS, Seabrook TA, Guido W, Fox MA (2013). A molecular mechanism regulating the timing of corticogeniculate innervation. *Cell Rep.* 5(3):573–581.

Brown NL, Kanekar S, Vetter ML, Tucker PK, Gemza DL, Glaser T (1998). Math5 encodes a murine basic helix-loop-helix transcription factor expressed during early stages of retinal neurogenesis. *Development.* 125(23):4821–4833.

Chen B, Schaevitz LR, McConnell SK (2005). Fezl regulates the differentiation and axon targeting of layer 5 subcortical projection neurons in cerebral cortex. *Proc Natl Acad Sci U S A.* 102(47):17184–17189.

Chen Y, Magnani D, Theil T, Pratt T, Price DJ (2012). Evidence that descending cortical axons are essential for thalamocortical axons to cross the pallial-subpallial boundary in the embryonic forebrain. *PLoS ONE.* 7(3):e33105.

Clasca F, Angelucci A, Sur M (1995). Layer-specific programs of development in neocortical projection neurons. *Proc Natl Acad Sci U S A.* 92(24):11145–11149.

Deck M, Lokmane L, Chauvet S, Mailhes C, Keita M, Niquille M, Yoshida M, Yoshida Y, Lebrand C, Mann F, Grove EA, Garel S (2013). Pathfinding of corticothalamic axons relies on a rendezvous with thalamic projections. *Neuron.* 77(3):472–484.

Diao Y, Cui L, Chen Y, Burbridge TJ, Han W, Wirth B, Sestan N, Crair MC, Zhang J (2018). Reciprocal connections between cortex and thalamus contribute to retinal axon targeting to dorsal lateral geniculate nucleus. *Cereb Cortex.* 28(4):1168–1182.

El-Danaf RN, Krahe TE, Dilger EK, Bickford ME, Fox MA, Guido W (2015). Development remodeling of relay cells in the dorsal lateral geniculate nucleus in the absence of retinal input. *Neural Dev.* 10:19.

Frangeul L, Pouchelon G, Telley L, Lefort S, Luscher C, Jabaudon D (2016). A cross-modal genetic framework for the development and plasticity of sensory pathways. *Nature.* 538(7623):96–98.

Giasafaki C, Grant E, Hoerder-Suabedissen A, Hayashi S, Lee S, Molnár Z (2022). Cross-hierarchical plasticity of corticofugal projections to dLGN after neonatal monocular enucleation. *J Comp Neurol.* 530(7):978–997.

Grant E, Hoerder-Suabedissen A, Molnár Z (2012). Development of the corticothalamic projections. *Front Neurosci.* 6:53.

Grant E, Hoerder-Suabedissen A, Molnár Z (2016). The regulation of corticofugal fiber targeting by retinal inputs. *Cereb Cortex.* 26(3):1336–1348.

Greig LC, Woodworth MB, Galazo MJ, Padmanabhan H, Macklis JD (2013). Molecular logic of neocortical projection neuron specification, development and diversity. *Nat Rev Neurosci.* 14(11):755–769.

Han W, Kwan KY, Shim S, Lam MM, Shin Y, Xu X, Zhu Y, Li M, Sestan N (2011). TBR1 directly represses Fezf2 to control the laminar

origin and development of the corticospinal tract. *Proc Natl Acad Sci U S A*. 108(7):3041–3046.
Hevner RF, Miyashita-Lin E, Rubenstein JJ (2002). Cortical and thalamic axon pathfinding defects in Tbr1, Gbx2, and Pax6 mutant mice: Evidence that cortical and thalamic axons interact and guide each other. *J Comp Neurol*. 447(1):8–17.
Hooks BM, Chen C (2006). Distinct roles for spontaneous and visual activity in remodeling of the retinogeniculate synapse. *Neuron*. 52(2):281–291.
Hooks BM, Chen C (2008). Vision triggers an experience-dependent sensitive period at the retinogeniculate synapse. *J Neurosci*. 28(18):4807–4817.
Jacobs EC, Campagnoni C, Kampf K, Reyes SD, Kalra V, Handley V, Xie YY, Hong-Hu Y, Spreur V, Fisher RS, Campagnoni AT (2007). Visualization of corticofugal projections during early cortical development in a tau-GFP-transgenic mouse. *Eur J Neurosci*. 25(1):17–30.
Jurgens CW, Bell KA, McQuiston AR, Guido W (2012). Optogenetic stimulation of the corticothalamic pathway affects relay cells and GABAergic neurons differently in the mouse visual thalamus. *PLoS ONE*. 7(9):e45717.
Lai T, Jabaudon D, Molyneaux BJ, Azim E, Arlotta P, Menezes JR, Macklis JD (2008). SOX5 controls the sequential generation of distinct corticofugal neuron subtypes. *Neuron*. 57(2):232–247.
Leone DP, Srinivasan K, Chen B, Alcamo E, McConnell SK (2008). The determination of projection neuron identity in the developing cerebral cortex. *Curr Opin Neurobiol*. 18(1):28–35.
Lett RL, Wang W, O'Connor TP (2009). Semaphorin 5B is a novel inhibitory cue for corticofugal axons. *Cereb Cortex*. 19(6):1408–1421.
Leyva-Díaz E, López-Bendito G (2013). In and out from the cortex: development of major forebrain connections. *Neuroscience*. 254:26–44.
Liang L, Chen L (2020). Organization, function, and development of the mouse retinogeniculate. *Annu Rev Vis Sci*. 6:261–285.
López-Bendito G, Chan CH, Mallamaci A, Parnavelas J, Molnár Z (2002). Role of Emx2 in the development of the reciprocal connectivity between cortex and thalamus. *J Comp Neurol*. 451(2):153–169.
McKenna WL, Betancourt J, Larkin KA, Abrams B, Guo C, Rubenstein JL, Chen B (2011). Tbr1 and Fezf2 regulate alternate corticofungal neuronal identities during neocortical development. *J Neurosci*. 31(2):549–564.
Miller B, Chou L, Finlay BL (1993). The early development of thalamocortical and corticothalamic projections. *J Comp Neurol*. 335(1):16–41.
Molnár Z, Blakemore C (1991). Lack of regional specificity for connections formed between thalamus and cortex in coculture. *Nature*. 351(6326):475–477.
Molnár Z, Blakemore C (1995). How do thalamic axons find their way to the cortex? *Trends Neurosci*. 18(9):389–397.
Molnár Z, Cordery P (1999). Connections between cells of the internal capsule, thalamus, and cerebral cortex in embryonic rat. *J Comp Neurol*. 413(1):1–25.
Molyneaux BJ, Arlotta P, Hibi M, Macklis JD (2005). Fezl is required for the birth and specification of corticospinal motor neurons. *Neuron*. 47(6):817–831.
Molyneaux BJ, Arlotta P, Menezes JR, Macklis JD (2007). Neuronal subtype specification in the cerebral cortex. *Nat Rev Neurosci*. 8(6):427–437.
Moreno-Juan V, Anibal-Martinez M, Herrero-Navarro Á, Valdeolmillos M, Martini FJ, López-Bendito G (2023). Spontaneous thalamic activity modulates the cortical innervation of the primary visual nucleus of the thalamus. *Neuroscience*. 508:87–97.
Oeschger FM, Wang WZ, Lee S. Garcia-Moreno F, Goffinet AM, Arbonés ML, Rakic S, Molnár Z (2012). Gene expression analysis of the embryonic subplate. *Cereb Cortex*. 22(6):1343–1359.
Seabrook TA, El-Danaf RN, Krahe TE, Fox MA, Guido W (2013). Retinal input regulates the timing of corticogeniculate innervation. *J Neurosci*. 33(24):10085–10097.
Shanks JA, Ito S, Schaevitz L, Yamada J, Chen B, Litke A, Feldheim DA (2016). Corticothalamic axons are essential for retinal ganglion cell axon targeting to the mouse dorsal lateral geniculate nucleus. *J Neurosci*. 36(19):5252–5263.
Shatz CJ, Rakic P (1981). The genesis of efferent connections from the visual cortex of the fetal rhesus monkey. *J Comp Neurol*. 196(2):287–307.
Sherman SM (2016). Thalamus plays a central role in ongoing cortical functioning. *Nat Neurosci*. 19(4):533–541.
Sherman SM, Guillery RW (2002). The role of the thalamus in the flow of information to the cortex. *Philos Trans R Soc Lond B Biol Sci*. 357(1428):1695–1708.
Sherman SW, Guillery RW (2011). Distinct functions for direct and transthalamic corticocortical connections. *J Neurophysiol*. 106(3):1068–1077.
Skaliora I, Singer W, Betz H, Püschel AW (1998). Differential patterns of semaphorin expression in the developing rat brain. *Eur J Neurosci*. 10(4):1215–1229.
Thompson AD, Picard N, Min L, Fagiolini M, Chen C (2016). Cortical feedback regulates feedforward retinogeniculate refinement. *Neuron*. 91(5):1021–1033.
Torii M, Levitt P (2005). Dissociation of corticothalamic and thalamocortical axon targeting by an EphA7-mediated mechanism. *Neuron*. 48(4):563–575.
Torii M, Rakic P, Levitt P (2013). Role of EphA/ephrin--a signaling in the development of topographic maps in mouse corticothalamic projections. *J Comp Neurol*. 521(3):626–637.
Usrey WM, Sherman SM (2019). Corticofugal circuits: Communication lines from the cortex to the rest of the brain. *J Comp Neurol*. 527(3):640–650.
Wang SW, Kim BS, Ding K, Wang H, Sun D, Johnson RL, Klein WH, Gan L (2001). Requirement for math5 in the development of retinal ganglion cells. *Genes Dev*. 15(1):24–29.

54.

THE CONNECTIONS OF THE PREFRONTAL CORTEX AND THALAMUS AND PLASTICITY WITHIN THIS CIRCUITRY

Anna S. Mitchell

INTRODUCTION

One of the main goals in the study of neuroscience is to understand the workings of the human brain in cognitive functions and dysfunction. The prefrontal cortex (PFC) coordinates many cognitive abilities, including decision-making, learning, working memory, planning, and cognitive control, which may be defined as the ability to suppress undesirable actions and remain attentive to relevant stimuli. However, the PFC does not operate in isolation to govern these cognitive processes. Rather, extensive interconnectivity exists between the PFC and other cortical and subcortical structures that provide a plethora of information, including about the immediate environment, the current physiology of the individual, and the relevant behavioral task(s) being performed, so that the PFC may coordinate actions according to current goal(s) (Fuster, 2001). The PFC also governs cognitive processes even without the influence of external inputs (Goldman-Rakic, 1995). How the PFC coordinates this information to direct our actions to fulfill our goals is one of the key questions for neuroscientists (and others) to answer.

Thalamocortical and corticothalamic interactions are the essence of this book. As evidenced from animal neuroscience research, the mediodorsal thalamus (MD) is particularly influential in PFC cognitive functions (see **Chapters 46 and 48 of this volume**). The PFC is robustly interconnected to the MD, and behavioral and cognitive neuroscience research conducted in monkeys and rodents indicates that the MD and PFC interact in ongoing and dynamic ways for appropriate task-relevant actions, in the form of choices, or responses to be produced (e.g., Alcaraz et al., 2018; Browning et al., 2015; Ferguson and Gao, 2018; Parnaudeau et al., 2013; Sommer and Wurtz, 2006). This chapter will focus on PFC cortico-thalamocortical connectivity of some of the thalamic nuclei (i.e., the MD, the anterior thalamic nuclei [ATN] and the nucleus reuniens [NRe]) that connect with the PFC based on neuroanatomical studies in monkeys and rodents so that the complexity of the information received and processed in the PFC for cognition may be considered. These specific interactions are subcircuits in wider neural networks that include structures in the frontal, temporal, parietal, and occipital lobes, the basal ganglia, midbrain, cerebellum, and brainstem. It is not possible to cover all the interconnectivity in this chapter.

Communication between brain structures occurs due to network oscillations at different frequencies (e.g., alpha, beta, theta, and gamma) that produce coordinated interactions across neural networks (Fujisawa and Buzsáki, 2011). It is proposed that cortico-thalamocortical communication occurs in a similar fashion (Floresco and Grace, 2003; Jones, 2009; Steriade and Llinás, 1988). Specifically, for the PFC, these interactions are proposed to help sustain or maintain persistent neural activity within the circuitry for the duration of the trial, to help align cortical neural signals for appropriate behavioral choices or responses in the form of measurable actions to occur (see **Chapter 46** of this volume; Bolkan et al., 2017; Guo et al., 2017).

Additionally, this chapter will consider neuroplasticity of PFC cortico-thalamocortical connectivity and disrupted network communication (altered neural oscillations) in relation to the extended periods of PFC neurodevelopment. In mammals, the PFC develops slowly to reach its maximal capacity, which occurs when axons are fully myelinated. For humans, PFC neurodevelopment continues throughout the teenage years until reaching its maximum developmental capacity during the late teens and early 20s in years of age, while in chimpanzees, neurodevelopment of the cortex coincides with sexual maturity (Miller et al., 2012). In rats, neurodevelopment of the medial PFC peaks at 24 days and the orbital surfaces peak around 30 days (Van Eden and Uylings, 1985). Given that some neurodevelopmental disorders (e.g., schizophrenia) affect the PFC, researchers have investigated the impact of changes to typical PFC neurodevelopment which lead to altered neural oscillations and can disrupt cognitive processes (for reviews, see Ouhaz et al., 2018; Parnaudeau et al., 2018; Uhlhaas and Singer, 2011).

CONNECTIVITY BETWEEN THE PREFRONTAL CORTEX AND THALAMUS IN MONKEYS AND RODENTS

In the primate brain, the granular part of the PFC, which is so defined because it comprises a clearly identifiable internal granular layer of cell bodies, called layer (L)4, in the anterior two-thirds of the frontal lobes. The granular PFC extends from the frontal pole (area 10) downward onto the orbital surface (areas 11, rostral 13, and rostral 14), around the ventral surface, called the inferior frontal gyrus in humans (areas 45 and 47) or the inferior convexity in monkeys (areas 45, 12/47), and onto the lateral surface, called the superior frontal gyrus (area 9) and the middle frontal gyrus in humans (area 46) or principal sulcus in monkeys (area 46). The caudal part of the dorsolateral surface of the PFC (area 9/46) is called the caudal middle frontal gyrus in humans or the caudal principal sulcus in monkeys. The caudal PFC, area 8, is labeled caudal lateral PFC in humans or pre-arcuate in monkeys. At the midline, the dorsomedial PFC includes dorsal area 9 (Fuster, 2001; Goldman-Rakic, 1995; Passingham and Wise, 2012; see **Chapter 57** of this volume).

The agranular PFC, in which the cells bodies in L4 are less conspicuous or nonexistent, is composed of 2 areas: the medial and the lateral agranular PFC. The medial agranular PFC includes the infralimbic (area 25), prelimbic (area 32), and anterior cingulate cortex (ACC: area 24), while the lateral agranular PFC is in the caudal orbital surface (caudal areas 13 and 14) and extends into the agranular insular cortex (Fuster, 2001; Goldman-Rakic, 1995; Passingham and Wise, 2012; see **Chapter 57** of this volume).

Anatomical studies in rodents have confirmed similarities in connectivity, although the PFC in rodents does not have the granular part. Instead, the PFC is described as two spatially separated areas: the rostral quarter of the dorsal bank of the rhinal sulcus, which is comparable to the agranular ventral surface in primates, and the rostral half of the medial hemispheric wall, which is comparable to the dorsomedial and caudal agranular PFC and medial agranular PFC (Uylings et al., 2003).

In mammals, the PFC was defined by Rose and Woolsey in 1948 as the projection area of the MD (Uylings et al., 2003). However, many additional thalamic nuclei, including, but not limited to, the ATN, intralaminar and midline thalamic nuclei, the motor thalamus, the NRe, and the pulvinar also interconnect with the PFC. It is not possible to cover the details of the known neuroanatomical connections of each of these thalamic nuclei in this chapter, so the reader is encouraged to see **Chapter 20**; **Chapter 33**; **Chapter 57 of this volume**; and others for specific details of the cortico-thalamocortical connectivity of these nuclei (e.g., Anastasiades et al., 2021; Collins et al., 2018; Fresno et al., 2019; Guo et al., 2017; Hsu and Price, 2007; Shibata and Naito, 2005; Timbie et al., 2020; Vertes et al., 2023; Xiao et al., 2009; Yu et al., 2022). In addition to focusing on the PFC-MD cortico-thalamocortical connectivity, as already mentioned, this chapter will also explain the ATN and NRe cortico-thalamocortical connections as these thalamic nuclei have PFC connectivity and are known to contribute to cognitive processes.

THE MEDIODORSAL THALAMUS (MD) CORTICO-THALAMOCORTICAL CONNECTIVITY IN MONKEYS AND RODENTS

The MD is a nodal structure within several frontal-striatal-thalamocortical neural circuits. In monkeys, there are 4 MD subdivisions: the magnocellular (MDmc), meaning large cells, is the most medial subdivision located adjacent to the massa intermedia; the parvocellular (MDpc) sits adjacent to the MDmc; the densocellular (MDdc) area is adjacent and dorsal to the MDpc; the pars multiform (MDpm) subdivision is the most lateral part of the MD.

Each primate MD subdivision has dense and reciprocal connections to selective regions of granular and agranular PFC that in combination span the entire frontal lobes (Figure 54.1). Each subdivision contributes to separate frontal-striatal-thalamocortical circuits, with unique afferent and efferent connections that help characterize their influences on cortical cognitive functions. This review of the monkey neuroanatomy will focus on two of these subdivisions: the MDmc (Figure 54.1A) and the MDpc (Figure 54.1B), and will combine the two other subdivisions (MDdc and MDpm) into the MDl (Figure 54.1C) because little research has distinguished either of these latter two subdivisions separately. Neuroanatomical evidence from monkeys is documented in Figure 54.1.

In rodents, there are three MD subdivisions: the medial (MDM) is like the primate MDmc; the central (MDC) is like the primate MDpc; and the lateral (MDL). However, there are some differences in the connectivity given that the rodent PFC is only agranular, so these connections will be documented separately in Figure 54.2.

For primates, as shown in Figure 54.1A, the MDmc is part of a frontal-temporal-basal ganglia circuit connecting preferentially with more ventral and medial surfaces of the PFC, including the orbital surface and ventromedial and ventrolateral PFC of the inferior frontal gyrus in humans or inferior convexity in monkeys, as well as

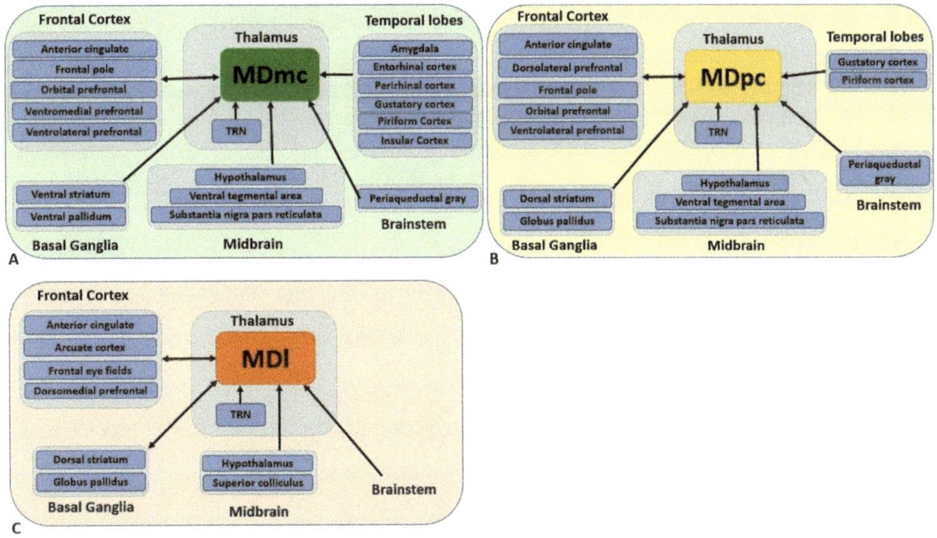

Figure 54.1. Schematic diagrams outlining the main connections of primate mediodorsal thalamic (MD) nucleus subdivisions: (A) the magnocellular (MDmc), (B) the parvocellular (MDpc), and (C) the lateral (MDl) from studies in monkeys. All 3 subdivisions share dense reciprocal connections to the prefrontal cortex (PFC). However, major differences are apparent between each of the subdivisions indicating heterogeneity of their PFC connections. For example, the MDmc receives multiple inputs from structures in the temporal lobes (including visual inputs from the temporal cortex and chemosensory inputs from the piriform and gustatory cortex), while the MDpc receives chemosensory inputs from the piriform and gustatory cortex, but the lateral MD receives no temporal lobes inputs. The differences in the connections of each subdivision highlight how each are likely to functionally influence cognitive processes within the PFC.
Arrowheads indicate the direction of information flow, with double-headed arrows showing reciprocal connections between structures. The gray boxes group each structure into the broader category of brain region it belongs to, e.g., frontal cortex. Additional connections also exist between cortical structures, the midbrain, and brainstem, but these are not depicted here. TRN = thalamic reticular nucleus.

the anterior cingulate cortex (ACC) and the frontal pole. The MDmc also has some L6 connections with dorsolateral and dorsomedial PFC. The amygdala, the perirhinal and entorhinal cortex, and the gustatory, piriform, and insular cortex in the temporal lobes all project to the MDmc, indicating that the MDmc is receiving multiple sensory signals from the temporal cortex. The MDmc also receives substantial GABAergic inputs from the ventral striatum via the ventral pallidum (Barbas et al., 1991; Giguere and Goldman-Rakic, 1988; Goldman-Rakic and Porrino, 1985; Porrino et al., 1981; Ray and Price, 1993; Saunders et al., 2005; Xiao et al., 2009).

As shown in Figure 54.1B, the MDpc forms a frontal–basal ganglia circuit connecting primarily to dorsolateral PFC, called the superior frontal gyrus and middle frontal gyrus in humans or the principal sulcus in monkeys, as well as the ACC and the frontal pole. The MDpc also has some connections with L6 of the dorsomedial, ventrolateral, and ventromedial PFC. The gustatory and piriform cortex also project to the MDpc. In contrast, the lateral MD primarily connects to the dorsomedial PFC, the frontal eye fields (pre-arcuate cortex in monkeys), and dorsal ACC. As shown in Figure 54.1C, the lateral MD does not connect to structures in the temporal lobes. The dorsal striatum and globus pallidus send GABAergic projections to the MDpc and lateral MD, with the lateral MD also projecting back to the striatum (Barbas et al., 1991; Giguere and Goldman-Rakic, 1988; Goldman-Rakic and Porrino, 1985; Ray and Price, 1993).

For rodents, the MDM, as shown in Figure 54.2A, also forms part of a frontal-temporal–basal ganglia circuit connecting the ventral and lateral orbital surface, the agranular insular cortex, and the cingulate cortex. The perirhinal cortex and the gustatory and piriform cortex also project to the MDM, although the perirhinal cortex input is not particularly substantial (Furtak et al., 2007; Kuroda et al., 1992). The MDM also receives substantial GABAergic inputs from the ventral striatum via the ventral pallidum (Krettek and Price, 1977; Kuroda and Price, 1991a; Kuroda and Price, 1991b; Kuroda et al., 1992; Mátyás et al., 2014; Vertes et al., 2023).

As shown in Figure 54.1B, the MDC forms a frontal–basal ganglia circuit connecting primarily to medial PFC, cingulate cortex, and the gustatory and piriform cortex. The MDL primarily connects to the cingulate and premotor cortex (Figure 54.2C). The dorsal striatum and globus pallidus send GABAergic projections to the MDC and MDL (Krettek and Price, 1977; Kuroda and Price, 1991a; Kuroda and Price, 1991b; Kuroda et al., 1992; Mátyás et al., 2014; Vertes et al., 2023).

In primates, there is some degree of overlap for MDmc and MDpc cortico-thalamocortical connections, especially within L6, with both subdivisions interacting with distinct parts of frontal polar cortex (area 10), orbitofrontal

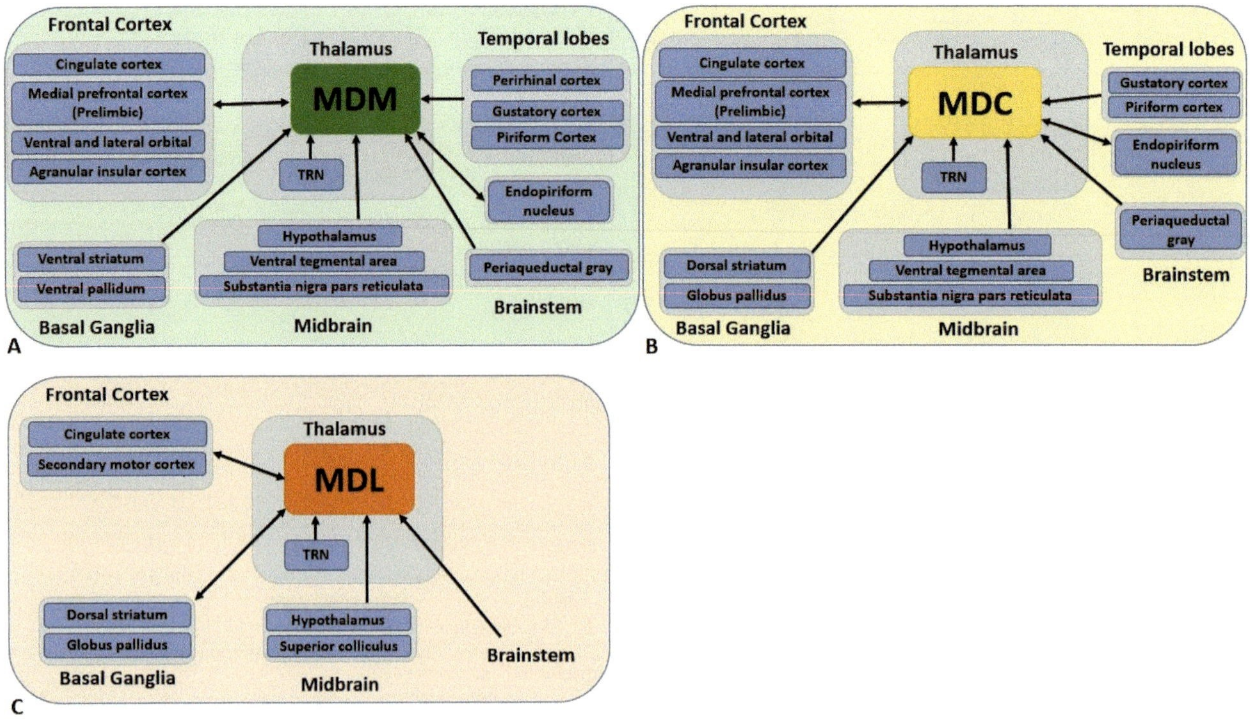

Figure 54.2. Schematic diagrams outlining the main connections of mediodorsal thalamic (MD) nucleus subdivisions of the rodent: (A) the medial (MDM), (B) the central (MDC), and (C) the lateral (MDL) subdivisions. While all 3 subdivisions are reciprocally connected to the agranular prefrontal cortex (PFC), major differences are apparent, indicating the heterogeneity of their PFC connections. The MDM and MDC subdivisions each have unique, reciprocal connections with different parts of the cingulate, medial prefrontal, orbital, and agranular insular cortex, while the MDL connects to the cingulate and secondary motor cortex. Additionally, the MDM and MDC receive differing inputs from the gustatory and piriform cortex, while the MDM also receives signals from the perirhinal cortex, although the inputs are not robust. In contrast, the MDL receives no inputs from the temporal lobes. The differences in the connections of each subdivision highlight how they will each contribute to functionally influence cognitive processes within the PFC. Arrowheads indicate the direction of information flow, with double-headed arrows showing reciprocal connections between structures. The gray boxes group each structure into the broader category of brain region it belongs to, e.g., frontal cortex. Additional connections also exist between cortical structures, the midbrain, and brainstem, but these are not depicted here. TRN = thalamic reticular nucleus.

cortex (areas 11, 12, and 13), dorsolateral and ventrolateral PFC, and ACC (Schwartz et al., 1991; Xiao et al., 2009). As observed in primates and rodents, L6 neurons also collateralize to terminate in the thalamic reticular nucleus (TRN) and other thalamic nuclei (e.g., Collins et al., 2018; Lévesque and Parent, 1998; Xiao et al., 2009; Yanakieva et al., 2022). Additionally, in primates, differences between the MDmc and MDpc emerge from L5 corticothalamic projecting neurons (Schwartz et al., 1991; Timbie and Barbas, 2015; Xiao et al., 2009) with these corticothalamic projections being one of the properties that identifies these MD subdivisions as higher order thalamic relays (see **Chapter 15 of this volume**) and see below, section on higher-order thalamic nuclei and subcortical inputs. L5 cortical neurons whose axons terminate in the MD are collateral branches of major corticofugal axons that project to motor output structures (Rockland, 2019; Sherman and Guillery, 2013; Usrey and Sherman, 2019). In primates, L5 corticothalamic projections to the MDmc originate in the orbitofrontal cortex (area 13; Timbie and Barbas, 2015), while for the MDpc, the driver input from L5 cortical neurons originates in dorsolateral PFC (Schwartz et al., 1991; Guillery, 1995). Further studies are needed to identify L5 corticothalamic projections to different thalamic nuclei in humans, in primates, and in rodents.

Finally, across species within each MD subdivision, a subset of single neurons project to multiple cortical areas (Kuramoto et al., 2017) because of the axonal collateralization of these single neurons. This axonal collateralization allows for the possibility that cortical or thalamic neurons may communicate the same information to multiple areas simultaneously, potentially allowing for network oscillations to coordinate communication across interconnected neural structures and networks (Jones, 2009; Rockland, 2019).

OTHER THALAMIC NUCLEI INTERACTING WITH THE PFC

As mentioned, other thalamic nuclei also interconnect with the PFC. Within the PFC, the overlap in corticothalamocortical connectivity occurs within the orbital, lateral, and medial parts of the PFC and the ACC. The cortico-thalamocortical connectivity of the ATN and NRe will be covered in this section.

The Anterior Thalamic Nuclei (ATN)

The ATN, located more rostral in the dorsal thalamus than the MD, comprise 3 distinct nuclei: the anteromedial nucleus (AM; Figure 54.2A), the anteroventral nucleus (AV; Figure 54.2B), and the anterior dorsal nucleus (AD; Figure 54.2C). The AM has extensive reciprocal connectivity with the PFC, including the orbital surface, dorsomedial PFC (medial prefrontal [prelimbic] and secondary motor cortex in rodents), and ACC (cingulate cortex in rodents). In contrast, the AV receives inputs from the ACC and medial prefrontal (prelimbic) cortex and is interconnected to secondary motor cortex, while the AD only receives inputs from the medial prefrontal (prelimbic) cortex (Bubb et al., 2017; Shibata and Naito, 2005; van Groen et al., 1999; Vertes et al., 2023). Each nucleus of the ATN is also interconnected to different subregions of the retrosplenial cortex, the subicular complex, and the temporal lobes, particularly the entorhinal cortex, with these cortical regions all interconnected and communicating directly with the PFC (Bubb et al., 2017; Lomi et al., 2021; Nassar et al., 2018; Perry and Mitchell, 2019; Perry et al., 2021; Saunders et al., 2005; van Groen et al., 1999). Interestingly, in rodents some AM neurons have axons that collateralize to terminate in the medial PFC and retrosplenial cortex or medial PFC and dorsal subiculum (Yanakieva et al., 2022), suggesting that AM neurons may help coordinate communication between these interconnected cortical structures. Causal evidence from behavioral and cognitive studies and neurophysiology recordings conducted in animal models and humans demonstrates that the ATN support spatial navigation, learning, and memory recall (see **Chapters 43 and 44 of this volume**; Aggleton et al., 2016; Jankowski et al., 2015; Lomi et al., 2023; Perry et al., 2021).

The Nucleus Reuniens (NRe)

The NRe, in the midline thalamus above the third ventricle, extends throughout the entire rostro-caudal axis of the thalamus. The NRe (Figure 54.2D) is reciprocally connected to the medial agranular PFC, specifically with the infralimbic, prelimbic, and cingulate cortex in monkeys and rodents (Hsu and Price, 2007; Vertes et al., 2006). The NRe also projects to the ventral hippocampus (Wouterlood et al., 1990; Vertes et al., 2006). Some NRe neurons have axons that collateralize sending the same information to both the ventral hippocampus and medial PFC (Hoover and Vertes, 2012; Varela et al., 2014), potentially positioning the NRe to coordinate ventral hippocampus and medial PFC neural interactions. In rodents, neurophysiology recordings show spatially responsive neurons in the NRe, while damage to the NRe disrupts learning and memory consolidation (Cassel et al., 2013; Jankowski et al., 2015).

HIGHER-ORDER THALAMIC RELAYS AND SUBCORTICAL INPUTS

As detailed in Chapter 1 of this volume, some thalamic nuclei are proposed to be higher-order thalamic relays (Sherman and Guillery, 2013; Usrey and Sherman, 2019). Higher-order thalamic relays are characterized as having the primary driving inputs originating from branches (collateral axons) of the main corticofugal axons of L5 cortical neurons that are projecting to subcortical output structures. These driver inputs to these higher-order relays make them distinct from first-order thalamic relays, whose driver inputs originate in peripheral or sensory structures (e.g., the lateral geniculate thalamic nucleus sends inputs to L4 of the occipital lobes; Guillery, 1995; Sherman and Guillery, 2013). In relation to thalamic nuclei mentioned in this chapter, higher-order relays are proposed to include subdivisions in the MD and the AM (Guillery, 1995; Mitchell, 2015; Perry and Mitchell, 2019; Sherman and Guillery, 2013). However, only the MDmc and MDpc subdivisions receive a driver input originating from PFC L5 collateral axons (Schwartz et al., 1991; Timbie and Barbas, 2015).

All thalamic nuclei receive unique afferents from other subcortical structures, including the midbrain and brainstem, that will also influence their functioning, in addition to receiving this driver input from L5 neurons (Figures 54.1, 54.2, and 54.3). These additional subcortical inputs to higher-order thalamic structures will help regulate the levels of excitability in these thalamic nuclei, leading to differential downstream influences on behavior. For example, the MDmc receives unidirectional glutamatergic inputs from the amygdala (Porrino et al., 1981; Timbie et al., 2020); the MDmc (MDM) and MDpc (MDC) receive inputs from the hypothalamus and periaqueductal gray (Kuroda et al., 1991a; Kuroda et al., 1991b; Timbie and Barbas, 2015); the AM, AV, and AD each receive unidirectional glutamatergic innervations from distinct parts of the mammillary bodies and brainstem (Guillery, 1955); and the NRe receives inputs from the amygdala, locus coeruleus, and midbrain (Vertes et al., 2006). This combined neuroanatomical evidence of extensive and diverse cortico-thalamocortical interactions and glutamatergic subcortical inputs suggests that rather than being a passive relay of sensory information, the influence of these thalamic nuclei within cortical functioning is ongoing and dynamic (Sherman, 2016).

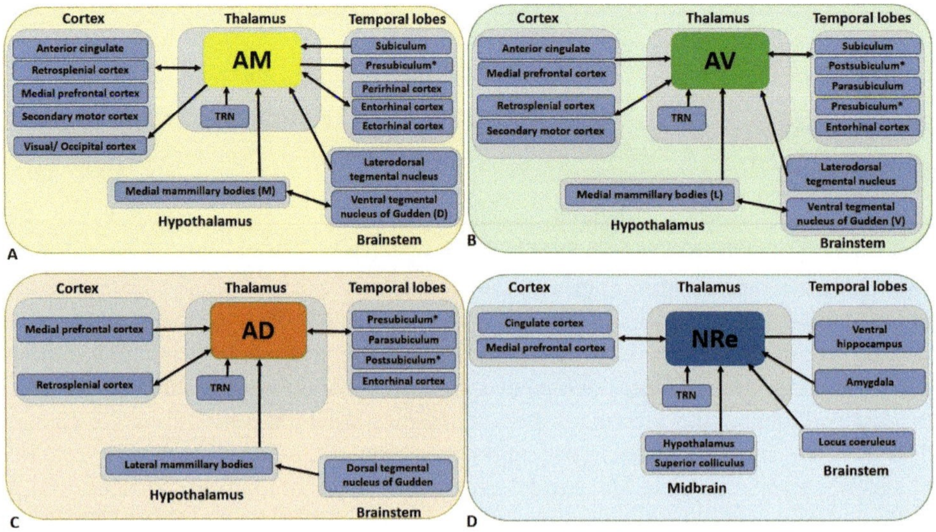

Figure 54.3. Schematic diagrams outlining the main connections of (A) the anteromedial (AM), (B) the anteroventral (AV), and (C) the anterodorsal (AD) subnuclei of the anterior thalamic nuclei, and (D) the nucleus reuniens (NRe) from studies in rats and monkeys. These nuclei share connections with the prefrontal cortex (PFC). However, major differences are apparent between each of the 3 anterior thalamic nuclei and the NRe indicating the heterogeneity of their PFC connections. Additionally, the differences in the connections of each thalamic nucleus highlight how these structural differences in their connectivity are likely to direct their functional influence within the PFC. For example, the AM is broadly connected to many cortical sites, including the PFC and temporal lobes, and the anterior cingulate and visual cortex. In contrast, the AD only receives unidirectional signals from the medial PFC, while the AV and NRe also each have unique PFC connectivity. Arrowheads indicate the direction of information flow, with double-headed arrows showing reciprocal connections between structures. The gray boxes group each structure into the broader category of the brain region it belongs to, e.g., cortex. Additional connections also exist between cortical structures, the hippocampal formation, midbrain, and brainstem, but these are not depicted here. * The presubiculum and postsubiculum are included as separate structures, however, the dorsal part of the presubiculum is commonly known as the postsubiculum. (D) is the dorsal part, and (V) is the ventral part of the ventral tegmental nucleus of Gudden; (L) is the lateral part, and (MD) is the medial part of the medial mammillary bodies; TRN = thalamic reticular nucleus.

SPECIFICITY OF TERMINAL SITE INNERVATIONS

Several studies are investigating differences in thalamocortical innervation within the different layers of mouse medial PFC (e.g., Anastasiades et al., 2021; Collins et al., 2018; Guo et al., 2017). In the mouse brain, evidence shows MD neurons that terminate in L1 of medial PFC innervate the dendrites from L5 pyramidal neurons as well as interneurons (Anastasiades et al., 2021; Collins et al., 2018). Anastasiades and colleagues (2021) showed using retrograde rabies viral tracing that within medial PFC, both L1 neuron-derived-neurotropic-factor (NDNF+) and vasoactive intestinal peptide (VIP+) interneurons receive monosynaptic inputs from MD, indicating that this thalamic nucleus can target both pyramidal neuron dendrites and interneurons located within L1.

Additionally, Collins and colleagues (2018), using current-clamp recordings, tracers, and optogenetics, reported that most medial PFC to MD corticothalamic projecting neurons were in L6 with axons that collateralized to the TRN and the ventromedial thalamic nucleus. With repetitive stimulation, these L6 inputs have a faciliatory effect confirming their modulatory role in this circuitry. In contrast, repetitive stimulation of L5 inputs caused depressing effects on MD thalamic neurons consistent with the driver input proposal of these axon collaterals. Furthermore, thalamocortical neurons from MD to medial PFC strongly activate L2/3 cortico-cortical neurons with the strength of the MD-evoked excitatory postsynaptic potentials (EPSPs) able to cause action potentials, indicating that the MD may provide some driver inputs to medial PFC (Collins et al., 2018).

Many more studies across different species are required to document the neural circuitry of cortico-thalamocortical interactions. These studies could also elucidate the extent of clusters or overlap of terminal sites in the cortex from MD, ATN, NRe, motor thalamus, intralaminar, and pulvinar neurons. This work will help our understanding of how these thalamic nuclei influence cortical cognitive functions and dysfunction.

NEUROPLASTICITY OF PFC-THALAMOCORTICAL NEURAL NETWORKS

As indicated at the beginning of the chapter, in mammals, the PFC has an extended neurodevelopmental maturation period, with humans showing the most extended maturation period (Hibar et al., 2015; Miller et al., 2012; van Eden and Uylings, 1985). Genetic risk factors contribute to the manifestation of many neurological diseases, syndromes, and disorders (e.g., schizophrenia; see van de Leemput et al., 2016) with exposure to environmental factors proposed to interact with some genetic factors through processes described as neuroplasticity (see below). Alterations in neurodevelopment of the PFC and neural synchrony across brain networks because of genetic and/or environmental risk factors are suggested to provide some of the underlying mechanisms that lead to the pathophysiology in schizophrenia and other neurodevelopmental disorders (e.g., Ouhaz et al., 2018; Parnaudeau et al., 2018; Uhlhass and Singer, 2011; van de Leemput et al., 2016).

Neuroplasticity refers to the brain's ability to change and modify its structure and function in response to experiences or injury. While the numbers of neurons in the healthy adult brain are relatively stable, structural changes or remodeling of dendrites occurs throughout the life span, for example, as new memories are formed (Bailey and Kandel, 1993). This remodeling of dendrites then remains largely stable for lifelong memories in mice (Yang et al., 2009). Nowadays, evidence indicates there is an ongoing interplay between genetic and environmental factors that can influence the extent of neuroplasticity or "dysplasticity" that occurs in the brain. Recent reviews (e.g., Vinogradow et al., 2023; Voss et al., 2017) have discussed current neuroscience understanding about neuroplasticity and have highlighted causal evidence established in animal models that has enabled investigations into dysplasticity, which occurs when neural synchrony and typical connectivity go awry because of decreases in oscillatory activity within neural networks.

Additionally, Vinogradow and colleagues (Vinogradow et al., 2023) identify many other factors, in addition to age, critical periods, and genetics, that also contribute to brain neuroplasticity or dysplasticity, including (1) opportunities for an individual to experience "unsupervised learning"—according to the principles of Hebbian learning, this would allow several neurons in a neural network to be altered during learning; (2) opportunities to experience learning associated with receiving "reward" or "punishment," or "novelty" or "success"; (3) an individual's baseline levels of neuromodulators that are linked to learning and attention (e.g., acetylcholine, dopamine, norepinephrine); and (4) spike-timing-dependent-plasticity that leads to combined activation in the presynaptic and postsynaptic neurons. The reader is encouraged to consult this detailed review for current understanding about the contributions of these additional factors (Vinogradow et al., 2023).

Thalamic nuclei, including the MD and pulvinar, and thalamocortical connectivity are altered in neurodevelopmental disorders (e.g., schizophrenia; see **Chapter 66 of this volume**). These alterations in higher-order thalamic nuclei and their interconnections with the cortex could cause cortical dysregulation (altered neural oscillations) and, in turn, dysplasticity, potentially providing part of an underlying explanation for some of the deficits in visual attention and cognitive functions governed by the PFC in this neurodevelopmental disorder (Parnaudeau et al., 2018).

The modifiability of the brain because of neuroplasticity suggests that exposure to a variety of environmental factors and experiences could also help to account for the variability in some of the symptoms that are experienced by individuals in many other neurological diagnoses.

In relation to cortico-thalamocortical connectivity discussed in this chapter, Figures 54.1, 54.2, and 54.3 highlight that each thalamic nucleus receives extensive neuromodulator inputs (including acetylcholine to the ATN, dopamine to the MD, and norepinephrine to ATN, MD, and NRe) from midline and brainstem structures. Certainly, differences in baseline levels of these neuromodulators will impact the ability of these thalamic nuclei to respond to the ongoing neural signals received from interconnected structures. Further studies are required to establish this evidence. However, and potentially, the brain's ability to modify itself because of experiences and environmental factors also provides potential avenues for therapeutic interventions and treatment options. For example, lifestyle alterations may encourage some structural brain changes and improve quality of life, while gene therapies and other novel strategies may provide future effective treatments options.

SUMMARY AND FURTHER DIRECTIONS

In summary, this review highlights how MD subdivisions of the thalamus interact with the PFC as key nodes receiving cortical information via collateral axonal branches from L5 neurons that send corticofugal axons; this information is then transmitted back to cortical areas via MDmc and MDpc thalamocortical inputs. This transmission can be managed in the thalamus in an ongoing, dynamic way via further inputs from other cortical and subcortical structures. Modulatory signals are transmitted from L6 of the PFC and may be altered by inhibitory inputs received from the basal ganglia, and/or excitatory signals from the hypothalamus, midbrain, and temporal lobes. Ongoing transmission within this cortico-thalamocortical connectivity will also depend on the balance of neuromodulation within the MD from midbrain and brainstem inputs. Thus, in addition to the PFC and the basal ganglia being major neural structures involved in cognition, it is timely that the MD also be considered critical in supporting cognition through ongoing partnerships with the frontal lobes. Further to this, the additional brief mention of cortico-thalamocortical connectivity between the PFC and other thalamic nuclei indicate that for the brain to function optimally in higher cognitive processes, the MD, the ATN, and the NRe support the transmission of brain signals across interconnected neural networks. Clearly, there is a great deal more research required to understand how this cortico-thalamocortical neural connectivity changes and modifies after genetic and/or environmental influences.

This cortico-thalamocortical influence, in addition to direct cortico-cortical cognitive processing, proposes that higher-order thalamic relays are integrally involved in distributing copies of neural signals via collateral axonal branches stemming from corticofugal axons that are providing these neural signals to other relevant brain structures (Guillery and Sherman, 2002; Sherman and Guillery, 2013; Sherman, 2016). These thalamic routes of communication may also contribute in an ongoing and rapid manner to update widespread regions of interconnected PFC about our upcoming responses and our current internal physiological states (Mitchell, 2015). If there was damage to a thalamic nucleus, then disrupted transmission of signals across these neural networks could lead to a mismatch in information transfer, and as a result, "noise" may occur within the PFC neural networks that are governing our ongoing cognitive functioning, which could lead to less optimal performance that may manifest as errors in behavior, or unusual or inappropriate decisions.

Given that cortico-thalamocortical neural networks themselves are disrupted in neurodevelopmental disorders (see **Chapter 66 of this volume**), that pathological changes in the dorsal thalamus may be part of altered processes that are taking place throughout the entire network across the life span (see **Chapters 67, Chapter 68, and Chapter 70 of this volume**), and critically, that behavioral and cognitive performance is altered as a consequence of subtle manipulations to the MD or other thalamic nuclei in animals and humans (see **Chapters 46 and 48 of this volume**), it is now, more than ever, relevant to re-evaluate the roles of different subdivisions of the dorsal thalamus in cognitive functions, cognitive control, and neurological disorders and diseases. This chapter provides ongoing causal evidence that continuing to investigate cortico-thalamocortical interactions will prove enlightening for researchers and clinicians studying and seeking to treat neurodevelopmental and neurodegenerative disorders and diseases.

REFERENCES

Aggleton JP, Pralus A, Nelson AJD, Hornberger M (2016). Thalamic pathology and memory loss in early Alzheimer's disease: Moving the focus from the medial temporal lobe to Papez circuit. *Brain.* 139:1877–1890. doi:10.1093/brain/aww083.

Alcaraz F, Fresno V, Marchand AR, Kremer EJ, Coutureau E, Wolff M (2018). Thalamocortical and corticothalamic pathways differentially contribute to goal-directed behaviors in the rat. *eLife.* 7:e32517. doi:10.7554/eLife.32517.

Anastasiades PG, Collins DP, Carter AG (2021). Mediodorsal and ventromedial thalamus engage distinct L1 circuits in the prefrontal cortex. *Neuron.* 109:314–330. doi:10.1016/j.neuron.2020.10.031.

Bailey CH, Kandel ER (1993). Structural changes accompanying memory storage. *Ann Rev Physiol.* 55:397–426.

Barbas H, Henion TH, Dermon CR (1991). Diverse thalamic projections to the prefrontal cortex in the rhesus monkey. *J Comp Neurol.* 313:65–94. doi:10.1002/cne.903130106.

Bolkan SS, Stujenske JM, Parnaudeau S, Spellman TJ, Rauffenbart C, Abbas AI, Harris AZ, Gordon JA, Kellendonk C (2017). Thalamic projections sustain prefrontal activity during working memory maintenance. *Nat Neurosci.* 20:987–996. doi:10.1038/nn.4568.

Browning PG, Chakraborty S, Mitchell AS (2015). Evidence for mediodorsal thalamus and prefrontal cortex interactions during cognition in macaques. *Cereb Cortex.* 25:4519–4534. doi:10.1093/cercor/bhv093.

Bubb EJ, Kinnavane L, Aggleton JP (2017). Hippocampal-diencephalic-cingulate networks for memory and emotion: An anatomical guide. *Brain Neurosci Adv.* 1:2398212817723443. doi:10.1177/2398212817723443.

Cassel JC, Pereira de Vasconcelos A, Loureiro M, Cholvin T, Dalrymple-Alford JC, Vertes RP (2013). The reuniens and rhomboid nuclei: neuroanatomy, electrophysiological characteristics and behavioral implications. *Prog Neurobiol.* 111:34–52. doi:10.1016/j.pneurobio.2013.08.006.

Collins DP, Anastasiades PG, Marlin JJ, Carter AG (2018). Reciprocal circuits linking the prefrontal cortex with dorsal and ventral thalamic nuclei. *Neuron.* 98:366–379. doi:10.1016/j.neuron.2018.03.024.

Ferguson BR, Gao WJ (2018). Thalamic control of cognition and social behavior via regulation of gamma-aminobutyric acidergic signaling and excitation/inhibition balance in the medial prefrontal cortex. *Biol Psychiatry.* 83:657–669. doi:10.1016/j.biopsych.2017.11.033.

Floresco SB, Grace AA (2003). Gating of hippocampal-evoked activity in prefrontal cortical neurons by inputs from the mediodorsal thalamus and ventral tegmental area. *J Neurosci.* 23:3930–3943. doi:10.1523/jneurosci.23-09-03930.2003.

Fresno V, Parkes SL, Faugère A, Coutureau E, Wolff M (2019). A thalamocortical circuit for updating action-outcome associations. *eLife.* 8:e46187. doi:10.7554/eLife.46187.

Fujisawa S, Buzsáki G (2011). A 4 Hz oscillation adaptively synchronizes prefrontal, VTA and hippocampal activities. *Neuron.* 72:153–165. doi:10.1016/j.neuron.2011.08.018.

Furtak SC, Wei SM, Agster KL, Burwell RD (2007). Functional neuroanatomy of the parahippocampal region in the rat: The perirhinal and postrhinal cortices. *Hippocampus.* 17:709–722. doi:10.1002/hipo.20314.

Fuster JM (2001) The prefrontal cortex—an update: Time is of the essence. *Neuron.* 30:319–333. doi:10.1016/s0896-6273(01)00285-9.

Giguere M, Goldman-Rakic PS (1988). Mediodorsal nucleus: areal, laminar, and tangential distribution of afferents and efferents in the frontal lobe of rhesus monkeys. *J Comp Neurol.* 277:195–213. doi:10.1002/cne.902770204.

Goldman-Rakic PS (1995). Architecture of the prefrontal cortex and the central executive. *Ann N Y Acad Sci.* 769:71–83. doi:10.1111/j.1749-6632.1995.tb38132.x.

Goldman-Rakic PS, Porrino LJ (1985). The primate mediodorsal (MD) nucleus and its projection to the frontal lobe. *J Comp Neurol.* 242:535–560. doi:10.1002/cne.902420406.

Guillery RW (1955). A quantitative study of the mamillary bodies and their connexions. *J Anat.* 89:19–32.

Guillery RW (1995). Anatomical evidence concerning the role of the thalamus in corticocortical communication: A brief review. *J Anat.* 187:583–592.

Guillery RW, Sherman SM (2002). Thalamic relay functions and their role in corticocortical communication: generalizations from the visual system. *Neuron.* 33:163–175. doi:10.1016/S0896-6273(01)00582-7.

Guo ZV, Inagaki HK, Daie K, Druckmann S, Gerfen CR, Svoboda K (2017). Maintenance of persistent activity in a frontal thalamocortical loop. *Nature.* 545:181–186. doi:10.1038/nature22324.

Hibar D, Stein J, Renteria M, Arias-Vasquez A, Desrivières S, Jahanshad N, Toro R, Wittfeld K, Abramovic L, Andersson M, Aribisala BS, Armstrong NJ, Bernard M, Bohlken MM, Boks MP, Bralten J, Brown AA, Chakravarty MM, Chen Q, Ching CR, Cuellar-Partida G, den Braber A, Giddaluru S, Goldman AL, Grimm O, Guadalupe T, Hass J, et al. (2015) Common genetic variants influence human subcortical brain structures. *Nature.* 520:224–229. doi:10.1038/nature14101.

Hoover WB, Vertes RP (2012). Collateral projections from nucleus reuniens of thalamus to hippocampus and medial prefrontal cortex in the rat: A single and double retrograde fluorescent labeling study. *Brain Struct Funct.* 217:191–209. doi:10.1007/s00429-011-0345-6.

Hsu DT, Price JL (2007). Midline and intralaminar thalamic connections with the orbital and medial prefrontal networks in macaque monkeys. *J Comp Neurol.* 504:89–111. doi:10.1002/cne.21440.

Jankowski MM, Passecker J, Islam MN, Vann S, Erichsen JT, Aggleton JP, O'Mara SM (2015). Evidence for spatially responsive neurons in the rostral thalamus. *Front Behav Neurosci.* 9:256. doi:10.3389/fnbeh.2015.00256.

Jones EG (2009). Synchrony in the interconnected circuitry of the thalamus and cerebral cortex. *Ann N Y Acad Sci.* 1157:10–23. doi:10.1111/j.1749-6632.2009.04534.x.

Krettek JE, Price JL (1977). The cortical projections of the mediodorsal nucleus and adjacent thalamic nuclei in the rat. *J Comp Neurol.* 171:157–191. doi:10.1002/cne.901710204.

Kuramoto E, Pan S, Furuta T, Tanaka YR, Iwai H, Yamanaka A, Ohno S, Kaneko T, Goto T, Hioki H (2017). Individual mediodorsal thalamic neurons project to multiple areas of the rat prefrontal cortex: A single neuron-tracing study using virus vectors. *J Comp Neurol.* 525:166–185. doi:10.1002/cne.24054.

Kuroda M, Murakami K, Kishi K, Price JL (1992). Distribution of the piriform cortical terminals to cells in the central segment of the mediodorsal thalamic nucleus of the rat. *Brain Res.* 595:159–163. doi:10.1016/0006-8993(92)91468-t.

Kuroda M, Price JL (1991a). Synaptic organization of projections from basal forebrain structures to the mediodorsal thalamic nucleus of the rat. *J Comp Neurol.* 303:513–533. doi:10.1002/cne.903030402.

Kuroda M, Price JL (1991b). Ultrastructure and synaptic organization of axon terminals from brainstem structures to the mediodorsal thalamic nucleus of the rat. *J Comp Neurol.* 313:539–552. doi:10.1002/cne.903130313.

Lévesque M, Parent A (1998). Axonal arborization of corticostriatal and corticothalamic fibers arising from prelimbic cortex in the rat. *Cereb Cortex.* 8:602–613. doi:10.1093/cercor/8.7.602.

Lomi E, Jeffery KJ, Mitchell AS (2023). Convergence of location, direction, and theta in the rat anteroventral thalamic nucleus. *iScience.* 26(7):106993. doi:10.1016/j.isci.2023.106993.

Lomi E, Mathiasen ML, Cheng HY, Zhang N, Aggleton JP, Mitchell AS, Jeffery KJ (2021). Evidence for two distinct thalamocortical circuits in retrosplenial cortex. *Neurobiol Learn Mem.* 185:107525. doi:10.1016/j.nlm.2021.107525.

Mátyás F, Lee J, Shin HS, Acsády L (2014). The fear circuit of the mouse forebrain: connections between the mediodorsal thalamus, frontal cortices and basolateral amygdala. *Eur J Neurosci.* 39:1810–1823. doi:10.1111/ejn.12610.

Miller DJ, Duka T, Stimpson CD, Schapiro SJ, Baze WB, McArthur MJ, Fobbs AJ, Sousa AM, Sestan N, Wildman DE, Lipovich L, Kuzawa CW, Hof PR, Sherwood CC (2012). Prolonged myelination in human neocortical evolution. *Proc Natl Acad Sci U S A.* 109:16480–16485. doi:10.1073/pnas.1117943109.

Mitchell AS (2015). The mediodorsal thalamus as a higher order thalamic relay nucleus important for learning and decision-making. *Neurosci BioBehav Rev.* 54:76–88. doi:10.1016/j.neubiorev.2015.03.001.

Nassar M, Simonnet J, Huang L-W, Mathon B, Cohen I, Bendels MHK, Beraneck M, Miles R, Fricker D (2018). Anterior thalamic excitation and feedforward inhibition of presubicular neurons projecting to medial entorhinal cortex. *J Neurosci.* 38:6411–6425. doi:10.1523/JNEUROSCI.0014-18.2018.

Ouhaz Z, Fleming H, Mitchell AS (2018). Cognitive functions and neurodevelopmental disorders involving the prefrontal cortex and mediodorsal thalamus. *Front Neurosci.* 12:33. doi:10.3389/fnins.2018.00033.

Parnaudeau S, Bolkan SS, Kellendonk C (2018). The mediodorsal thalamus: An essential partner of the prefrontal cortex for cognition. *Biol Psychiatry.* 83:648–656. doi:10.1016/j.biopsych.2017.11.008.

Parnaudeau S, O'Neill PK, Bolkan SS, Ward RD, Abbas AI, Roth BL, Balsam PD, Gordon JA, Kellendonk C (2013). Inhibition of mediodorsal thalamus disrupts thalamofrontal connectivity and cognition. *Neuron.* 77:1151–1162. doi:10.1016/j.neuron.2013.01.038.

Passingham RE, Wise SP (2012). *The Neurobiology of the Prefrontal Cortex: Anatomy, Evolution and the Origin of Insight.* Oxford University Press.

Perry BAL, Lomi E, Mitchell AS (2021). Thalamocortical interactions in cognition and disease: The mediodorsal and anterior thalamic nuclei. *Neurosci Biobehav Rev.* 130:162–177. doi:10.1016/j.neubiorev.2021.05.032.

Perry BAL, Mitchell AS (2019). Considering the evidence for anterior and laterodorsal thalamic nuclei as higher order relays to cortex. *Front Mol Neurosci.* 12:167. doi:10.3389/fnmol.2019.00167.

Porrino LJ, Crane AM, Goldman-Rakic PS (1981). Direct and indirect pathways from the amygdala to the frontal lobe in rhesus monkeys. *J Comp Neurol.* 198:121–136. doi:10.1002/cne.901980111.

Ray JP, Price JL (1993). The organization of projections from the mediodorsal nucleus of the thalamus to orbital and medial prefrontal cortex in macaque monkeys. *J Comp Neurol.* 337:1–31. doi:10.1002/cne.903370102.

Rockland KS (2019). Corticothalamic axon morphologies and network architecture. *Eur J Neurosci.* 49:969–977. doi:10.1111/ejn.13910.

Saunders RC, Mishkin M, Aggleton JP (2005). Projections from the entorhinal cortex, perirhinal cortex, presubiculum, and parasubiculum to the medial thalamus in macaque monkeys: identifying different pathways using disconnection techniques. *Exp Brain Res.* 167:1–16. doi:10.1007/s00221-005-2361-3.

Schwartz ML, Dekker JJ, Goldman-Rakic PS (1991). Dual mode of corticothalamic synaptic termination in the mediodorsal nucleus of the rhesus monkey. *J Comp Neurol.* 309:289–304. doi:10.1002/cne.903090302.

Sherman SM (2016). Thalamus plays a central role in ongoing cortical functioning. *Nat Neurosci.* 19:533–541.

Sherman SM, Guillery RW (2013). *Functional Connections of Cortical Areas: A New View from the Thalamus.* MIT Press.

Shibata H, Naito J (2005). Organization of anterior cingulate and frontal cortical projections to the anterior and laterodorsal thalamic nuclei in the rat. *Brain Res.* 1059:93–103.

Sommer MA, Wurtz RH (2006). Influence of the thalamus on spatial visual processing in frontal cortex. *Nature.* 444:374–377. doi:10.1038/nature05279.

Steriade M, Llinás RR (1988). The functional states of the thalamus and the associated neuronal interplay. *Physiol Rev.* 68:649–742. doi:10.1152/physrev.1988.68.3.649.

Timbie C, Barbas H (2015). Pathways for emotions: specializations in the amygdalar, mediodorsal thalamic, and posterior orbitofrontal network. *J Neurosci.* 35:11976–11987. doi:10.1523/JNEUROSCI.2157-15.2015.

Timbie C, García-Cabezas MÁ, Zikopoulos B, Barbas H (2020). Organization of primate amygdalar-thalamic pathways for emotions. *PLoS Biol.* 18:e3000639. doi:10.1371/journal.pbio.3000639.

Uhlhaas PJ, Singer W (2011). The development of neural synchrony and large-scale cortical networks during adolescence: Relevance for the pathophysiology of schizophrenia and neurodevelopmental

hypothesis. *Schizophr Bull.* 37:514–523. doi:10.1093/schbul/sbr034.
Usrey WM, Sherman SM (2019). Corticofugal circuits: Communication lines from the cortex to the rest of the brain. *J Comp Neurol.* 527:640–650. doi:10.1002/cne.24423.
Uylings HB, Groenewegen HJ, Kolb B (2003). Do rats have a prefrontal cortex? *Behav Brain Res.* 146:3–17. doi:10.1016/j.bbr.2003.09.028.
van de Leemput J, Hess JL, Glatt SJ, Tsuang MT (2016). Genetics of schizophrenia: Historical insights and prevailing evidence. *Adv Genet.* 96:99–141. doi:10.1016/bs.adgen.2016.08.001.
Van Eden CG, Uylings HB (1985). Postnatal volumetric development of the prefrontal cortex in the rat. *J Comp Neurol.* 241:268–274. doi:10.1002/cne.902410303.
van Groen T, Kadish I, Wyss JM (1999). Efferent connections of the anteromedial nucleus of the thalamus of the rat. *Brain Res Rev.* 30:1–26.
Varela C, Kumar S, Yang JY, Wilson MA (2014). Anatomical substrates for direct interactions between hippocampus, medial prefrontal cortex, and the thalamic nucleus reuniens. *Brain Struct Funct.* 219:911–929. doi:10.1007/s00429-013-0543-5.
Vertes RP, Hoover WB, Do Valle AC, Sherman A, Rodriguez JJ (2006). Efferent projections of reuniens and rhomboid nuclei of the thalamus in the rat. *J Comp Neurol.* 499:768–796. doi:10.1002/cne.21135.
Vertes RP, Hoover WB, Witter MP, Yanik MF, Rojas AKP, Linley SB (2023). Projections from the five divisions of the orbital cortex to the thalamus in the rat. *J Comp Neurol.* 531:217–237. doi:10.1002/cne.25419.
Vinogradov S, Chafee MV, Lee E, Morishita H (2023). Psychosis spectrum illnesses as disorders of prefrontal critical period plasticity. *Neuropsychopharmacology.* 48:168–185. doi:10.1038/s41386-022-01451-w.
Voss P, Thomas ME, Cisneros-Franco JM, de Villers-Sidani É (2017). Dynamic brains and the changing rules of neuroplasticity: Implications for learning and recovery. *Front Psychol.* 8:1657. doi:10.3389/fpsyg.2017.01657.
Wouterlood FG, Saldana E, Witter MP (1990). Projection from the nucleus reuniens thalami to the hippocampal region: Light and electron microscopic tracing study in the rat with the anterograde tracer Phaseolus vulgaris-leucoagglutinin. *J Comp Neurol.* 296:179–203. doi:10.1002/cne.902960202.
Xiao D, Zikopoulos B, Barbas H (2009). Laminar and modular organization of prefrontal projections to multiple thalamic nuclei. *Neuroscience.* 161:1067–1081. doi:10.1016/j.neuroscience.2009.04.034.
Yanakieva S, Mathiasen ML, Amin E, Nelson AJD, O'Mara SM, Aggleton JP (2022). Collateral rostral thalamic projections to prelimbic, infralimbic, anterior cingulate and retrosplenial cortices in the rat brain. *Eur J Neurosci.* 56:5869–5887. doi:10.1111/ejn.15819.
Yang G, Pan F, Gan WB (2009). Stably maintained dendritic spines are associated with lifelong memories. *Nature.* 462:920–924. doi:10.1038/nature08577.

55.

CROSS-MODAL PLASTICITY OF CORTICAL AND THALAMIC CIRCUITS

Lukas Mesik and Hey-Kyoung Lee

INTRODUCTION

It is well established that changes in sensory experience can trigger long-lasting changes in the brain. This is especially noted during early postnatal development, where sensory experience is essential for proper functional wiring and maturation of brain circuits. One prominent example is the development of the visual system, where visual experience initiated by eye-opening, along with spontaneous retinal activity, leads to proper development and segregation of eye-specific inputs to the visual thalamus, refinement of thalamocortical connections in the primary visual cortex (V1), and maturation of feature selectivity, such as ocular dominance and orientation selectivity, of V1 neurons (Hooks and Chen, 2020). At a cellular level, visual experience-dependent maturation of the visual circuit is driven by activity-dependent plasticity mechanisms that strengthen and weaken synaptic connections dependent on coincident patterned activity (Cooke and Bear, 2014). Therefore, many forms of vision loss early in development can lead to widespread consequences to normal development of the visual circuits. The role of sensory experience is general across all the sensory cortices (Fox and Wong, 2005; Froemke and Jones, 2011). The critical role of sensory experience in sculpting the sensory circuit is not just restricted to the early postnatal developmental period, but extends through a lifetime to support perceptual learning (Karmarkar and Dan, 2006) and long-term storage of memories (Ji and Wilson, 2007).

While the role of sensory experience in shaping the functional connectivity within its own modality is expected, it is also known that each sensory modality can influence the development and function of the other sensory systems through cross-modal plasticity (Merabet and Pascual-Leone, 2010; Ewall et al., 2021). In humans, early-onset blind subjects often show enhancement of the remaining senses, as seen by demonstrations of better auditory pitch discrimination (Gougoux et al., 2004), sound localization (Collignon et al., 2009), enhanced ability to follow speed-speech (Dietrich et al., 2013), and better tactile functions (Goldreich and Kanics, 2003), including Braille reading (Cohen et al., 1997). Such enhancement in the processing of the remaining senses in early-blind subjects has been implicated to involve 2 distinct brain adaptation mechanisms: one that is termed cross-modal recruitment and another that is referred to as compensatory plasticity (Ewall et al., 2021) (Figure 55.1). Cross-modal recruitment in early-blind subjects has been seen using various functional imaging methods, such as activation of V1 by Braille reading (Sadato et al., 1996) and speech (Roder et al., 2002). Such recruitment of V1 by the remaining senses is thought to add capacity for processing. Compensatory plasticity, on the other hand, refers to adaptation within the spared sensory areas that allows better and/or more refined processing of the remaining senses (Rauschecker, 1995; Huber et al., 2019; Ewall et al., 2021).

It is now evident that cross-modal plasticity is not limited to early sensory loss. Late-onset blind subjects, for example, display enhanced functionality of the remaining senses under certain circumstances, such as better sound localization (Voss et al., 2004) and speech perception (Dietrich et al., 2013). Furthermore, there is evidence that cross-modal recruitment may happen quite rapidly in adult humans, as demonstrated by the emergence of V1 activity with Braille reading after a few days of training on Braille while being blindfolded (Merabet et al., 2008). The rapid recruitment of V1 for Braille reading with short-term visual deprivation of adult subjects suggests that the plasticity is likely mediated by unmasking of functional connectivity that preexists in the adult brain. In support of this idea, subjects that are temporarily blindfolded display tactile-related activity in V1 (Merabet et al., 2007). There is also evidence that the enhanced performance in the spared sensory modality of blind subjects may be due to perceptual learning (e.g., Grant et al., 2000), which suggests that

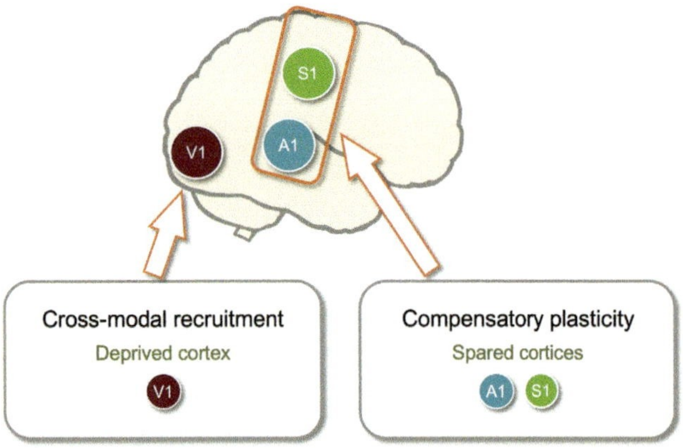

Figure 55.1. Cross-modal plasticity with vision loss. Vision loss leads to functional changes in the deprived sensory cortex (including V1) and in the spared sensory cortices (including A1 and S1) in human subjects. Examples of cross-modal recruitment of V1 are activation of V1 in blind individuals upon Braille reading (Sadato et al., 1996) and speech (Roder et al., 2002). Cross-modal recruitment of V1 by Braille reading can be observed quite rapidly in adults (Merabet et al., 2008). Vision loss also induces compensatory plasticity in the spared sensory cortices. Examples include refinement of tonotopy in A1 of blind individuals (Huber et al., 2019), evidence of more efficient representation of sound in A1 (Stevens and Weaver, 2009), and reduced contralateral bias of A1 and auditory thalamus responses to sound (Coullon et al., 2015). In addition to cortical changes, there are changes to activity in the thalamic and subcortical structures in blind human subjects (Coullon et al., 2015). Both cross-modal recruitment and compensatory plasticity are thought responsible for sensory compensation in the blind. A recent study suggests that concomitant cross-modal recruitment and compensatory plasticity may reallocate computational resources to the deprived sensory cortex (Mattioni et al., 2022).

experience-dependent plasticity mechanisms may mediate the functional gain seen with cross-modal plasticity.

The neural basis of cross-modal plasticity is studied using various animal models. While cross-modal plasticity induced by early-onset sensory loss can lead to anatomical changes, especially in the thalamocortical circuits and cortico-cortical connections, cross-modal adaptation to losing a sensory modality later in life mainly involves functional plasticity in the cortex and the thalamus. In this chapter, we will review the neural mechanisms of cross-modal plasticity, mainly focusing on thalamic and cortical plasticity. Most examples will be from studies on visual deprivation, which is a widely used model of sensory deprivation to examine cross-modal plasticity. In addition, we will provide information on potential synaptic plasticity mechanisms that could mediate cross-modal plasticity.

CROSS-MODAL PLASTICITY WITH EARLY SENSORY LOSS

Whether cross-modal plasticity can lead to anatomical rewiring of the brain has been of interest. At least in models of early sensory loss, there is evidence that anatomical rewiring can happen within the circuits mediating the lost sense, especially when it happens during embryonic or early neonatal development following loss or developmental abnormalities of sensory organs themselves (Lopez-Bendito et al., 2022) (Table 55.1). Typically, cross-modal anatomical changes are less pronounced when sensory loss happens after the initiation of sensory experience. During postnatal development, normal sensory experience is critical for refinement of anatomical connections and proper maturation of the circuits within the primary sensory thalamus and cortex serving that sense. Even earlier in development, prior to sensory experience, proper development of sensory circuits relies on spontaneous activity arising from the sensory organs. For example, in the visual system, normal development of eye-specific input segregation in the visual thalamus (dorsal lateral geniculate nucleus, dLGN) depends on spontaneously generated waves of activity in the retina (Thompson et al., 2017) in addition to guidance cues (Huberman et al., 2008; Cang and Feldheim, 2013). Even diffuse light entering through closed eyelids before eye opening is needed for proper segregation of eye-specific inputs in dLGN (Tiriac et al., 2018). Therefore, congenital blindness caused by improper eye development or retinal loss is expected to cause the most severe defect of retinogeniculate projections in the dLGN.

In carnivores, the anatomical segregation of eye-specific inputs in the thalamorecipient layer (L)4 of the primary visual cortex (V1) is dependent on activity from the retina (Stryker and Harris, 1986). In rodents, the establishment of precise geniculocortical map development in V1 has been shown dependent on both molecular guidance cues and spontaneous retinal waves during early postnatal development (Cang and Feldheim, 2013). Moreover, balanced activity arising from the 2 eyes is critical for the development of normal binocular vision, as exemplified by the loss of binocular neurons in V1 following various monocular deprivation paradigms (Hooks and Chen, 2020). Such a shift in ocular dominance following monocular vision loss is prominent during early postnatal development during a time window defined as the critical period for ocular dominance plasticity (Levelt and Hubener, 2012). In contrast to monocular deprivation, removing vision from both eyes by dark-rearing animals from birth leads to reduced responsiveness of V1 neurons without altering the distribution of ocular dominance across the population (Mower et al., 1981). This result is consistent with the notion that binocular competition drives ocular dominance plasticity. While dark rearing does not shift ocular dominance, early bilateral enucleation alters the anatomical connections in the visual system, causing degeneration of the retinogeniculate axons and reduced size of dLGN in animal models (see Table 55.1). A similar loss of projections from the retina through dLGN to V1 is also observed in early-blind humans (Bock

TABLE 55.1 CROSS-MODAL PLASTICITY WITH EARLY ONSET VISION LOSS OBSERVED IN ANIMAL MODELS

	Species	Mode of Vision Loss	Age of Vision Loss	Age of Observation	Summary of Findings
Anatomical	Mouse	Enucleation	Embryonic (E14.5)	Early postnatal (P8)	Decrease in the size of V1 and dLGN Increase in the size of S1 barrel field, including increase in size of individual barrels
			Neonatal (P0)	Early postnatal (P1 and P4)	Decrease in dLGN size as early as 1 day after enucleation
			Neonatal	Adult (P60)	Decrease in V1 and dLGN size
			Neonatal	Adult	Large increase in V1 innervation by LD
			Postnatal (first week)	Adult (≥P60)	Increase in the size of S1 barrels
		Anophthalmic	Embryonic	Adult (P60)	Decrease in V1 and dLGN size, increase in A1 volume
			Embryonic	Adult	Increased density of extrageniculate (LP and LD) inputs into V1
			Embryonic	Adult	dLGN became partially innervated by somatosensory inputs from the brain stem.
	Rat	Enucleation	Neonatal	Adult	Decrease of V1 innervation by dLGN and increase of innervation by LP
	Hamster	Enucleation	Neonatal	Adult	dLGN enucleated hamsters received inputs from inferior colliculus. V1 received inputs from LD.
	Short-tailed opossum	Enucleation	P4 (before thalamocortical axon innervate to V1)	Adult	Decreased size of dLGN and V1, increased size of S1. S1 received stronger inputs from S2, multimodal cortex, piriform cortex, and motor thalamic nuclei.
		Enucleation	P4 (before thalamocortical axon innervate to V1)	Adult	V1 became innervated by several non-visual thalamic nuclei (including VP, AV, and VL) and developed stronger connectivity with multiple cortical regions (including A1, S1 and multimodal cortex).
	Crab-eating macaque	Enucleation	Embryonic (E51–E109)	Late embryonic- P0	The earlier the enucleation, the smaller the dLGN, inferior pulvinar and the striate cortex. Reduction of striate cortex was accompanied by corresponding expansion of extrastriate cortex.
Functional	Mouse	Enucleation	Neonatal	Adult (~P100)	c-Fos staining showed increased density of auditory-responsive neurons in both medial and lateral higher visual areas.
		Anophthalmic	Embryonic	Adult (~P100)	c-Fos staining showed increased density of auditory-responsive neurons in dLGN, V1, and medial higher visual areas.
	Rat	Enucleation	Neonatal	Adult	About 1/3 of V1 cells responded to auditory stimuli. These cells had higher intensity threshold and wider bandwidth compared to A1 neurons. Sound responses also increased in higher visual areas.
		Enucleation	Neonatal	Adult (≥P80)	Whisker barrels in neonatally enucleated rats responded to more whiskers.
	Hamster	Enucleation	Neonatal	Adult	Many V1 neurons responded to auditory stimuli, although with longer latency and higher intensity threshold compared to A1 neurons.
		Enucleation	Neonatal	Adult	Large increase in somatosensory-responsive neurons in LP

(continued)

TABLE 55.1 CONTINUED

	Species	Mode of Vision Loss	Age of Vision Loss	Age of Observation	Summary of Findings
	Short-tailed opossum	Enucleation	P4 (before thalamocortical axon innervate to V1)	Adult (≥P120)	S1 neurons in early enucleated possums responded to fewer whiskers (sparser coding).
		Enucleation	P4 (before thalamocortical axon innervate to V1)	Adult	Many V1 neurons were found to respond to auditory and/or somatosensory stimuli. In addition, A1 and S1 became more bimodal, often responding to both types of stimuli.
	Cat	Lid suture	Around eye opening (~P10)	Adult (≥2 years)	Neurons in the anterior auditory field of blind cats were more sharply tuned for spatial location of the sound
Behavioral	Rat	Enucleation	Neonatal	Adult (≥P80)	Neonatally enucleated rats learned to solve a maze faster than sighted controls.
	Short-tailed opossum	Enucleation	P4 (before thalamocortical axon innervate to V1)	Adult (≥P120)	Early enucleated opossums were able to discriminate finer sandpaper texture differences.

Abbreviations: AV = anterior ventral; E = embryonic; LD = lateral dorsal; LP = lateral posterior; P = postnatal; S2 = secondary somatosensory cortex; VL = ventrolateral; VP = ventral posterior.

and Fine, 2014). As will be discussed later, there is also evidence that early bilateral enucleation prevents pruning of multisensory inputs to dLGN and V1. These results indicate that not all early vision loss will cause the same types of anatomical changes. This highlights the need to consider the exact nature of sensory loss and how early during development it occurred to determine whether there would be anatomical rewiring during cross-modal plasticity.

One of the interesting findings related to whether there could be widespread anatomical rewiring with cross-modal plasticity has come from a study where early developmental intervention allowed rewiring of visual inputs to the primary auditory cortex (A1) (von Melchner et al., 2000). This study used immature ferret kits where the main inputs from the inferior colliculus to the primary auditory thalamus (medial geniculate nucleus, MGN) and the superior colliculus (SC), which is one of the targets of the retinogeniculate axons, were surgically ablated to enable retinogeniculate axons to rewire to the auditory thalamus. These rewired ferrets were able to use A1 to detect light to guide behavior. This study is interesting in many aspects. First, it suggests that cross-modal anatomical rewiring can happen if normal axon guidance is affected during early development. Second, it suggests that A1 can process rewired visual inputs, which supports the idea that cortical circuits may be equivalent across brain regions.

Multisensory connections, both thalamocortical and corticocortical, arise during postnatal development. Some of these early connections are transient and are pruned by normal sensory experience. For example, in Mongolian gerbils, exuberant multisensory thalamocortical connectivity exists transiently during normal postnatal development, but early sensory deprivation in the form of sciatic nerve transection, ototoxic deafening, or enucleation increases multisensory thalamocortical connectivity by sprouting of axons originating from non-lemniscal thalamic nuclei (Henschke et al., 2018). In the same model system, the normal development of multisensory corticocortical connectivity occurs a bit delayed in comparison to the multisensory thalamocortical connectivity, but early sensory loss leads to an increased proportion of these multisensory corticocortical connections (Henschke et al., 2018). Similar deficits in developmental pruning of multisensory inputs to dLGN and V1 have been observed across different species following early sensory loss (see Lopez-Bendito et al., 2022, for a recent review). These results exemplify that early sensory loss increases anatomical connectivity that contributes multisensory inputs to the deprived primary sensory thalamus and cortex. The increase in multisensory inputs does not constitute large-scale rewiring as observed in the von Melchner et al. (2000) study using early surgical ablations for rerouting the developing retinogeniculate axons. But the results from these studies suggest that cross-modal plasticity following early sensory loss occurs either by preventing the normal pruning of multisensory connections or promoting the sprouting of existing axonal and dendritic elements in the deprived primary sensory cortex. Such adaptation would serve as an anatomical substrate for functional recruitment of the deprived cortex for processing the spared sensory modality, as exemplified by activation of V1 by Braille reading in early-blind subjects (Sadato et al., 1996).

Even in normal primary sensory cortices, cross-modal functional responses are observed, which often appear with a relatively long delay (≥100 ms) after stimulus onset (Teder-Salejarvi et al., 2002; Talsma and Woldorff, 2005). This suggests that routing of cross-modal signals to primary cortices may involve higher-order multisensory and associational areas. But some report short latency responses in the primary sensory cortices with cross-modal sensory stimuli (Lakatos et al., 2007; Iurilli et al., 2012; Sieben et al., 2013), which has been interpreted as being mediated via thalamic or direct connections between primary sensory cortices. In either case, cross-modal multisensory influences on the primary sensory cortical neurons under normal conditions are largely subthreshold and modulatory in nature (Lakatos et al., 2007; Iurilli et al., 2012).

Several studies of early vision loss, either by embryonic or neonatal enucleation, report activation of V1 by auditory or tactile stimuli (Table 55.1). Such cross-modal recruitment of V1 by other sensory inputs is thought to arise from improper pruning of multisensory inputs at the level of the thalamus and/or cortex in the absence of the eyes. There is also evidence that higher-order thalamus, such as lateral dorsal (LD), lateral posterior (LP), or posterior medial (POm) nuclei, may be involved in cross-modal activation of primary sensory cortices (Lopez-Bendito et al., 2022). Regardless of the exact anatomical pathway(s) that bring in cross-modal sensory inputs, activation of V1 by cross-modal recruitment is rather delayed (≥30 or 50 msec after the onset of cross-modal auditory stimulus) compared to that seen in the respective primary sensory cortex (~10 msec) (Izraeli et al., 2002). In addition, cross-modal activation of V1 requires higher intensity stimuli (Izraeli et al., 2002) and is less organized as seen by a lack of tonotopy in V1 with auditory activation (Piche et al., 2007). Collectively, these early enucleation studies highlight that the nature of processing within the cross-modally recruited deprived cortex may not be equivalent to that occurring in the spared sensory cortices.

In addition to cross-modal recruitment, early sensory loss leads to compensatory plasticity in the spared sensory cortices, which is thought to underlie sensory compensation (Rauschecker, 1995). Compensatory plasticity of the spared sensory cortices may occur by anatomical and/or functional changes of the spared sensory cortices. For example, there is an expansion in the size of the barrel cortex in early-blinded mice, and auditory cortical neurons in early-blinded cats display increased auditory spatial tuning (see Table 55.1). Such changes could form the basis of improved discrimination of features in the spared modalities, such as enhanced tactile discrimination (Wong et al., 2011) or sound localization (Roder et al., 1999) in early-blind subjects.

CROSS-MODAL PLASTICITY FOLLOWING LATE-ONSET SENSORY LOSS

Plasticity mechanisms become more restrained in the mature brain, which is well documented across sensory areas. For example, experience-dependent thalamic plasticity largely ends before the critical period of plasticity for the primary sensory cortices (Hooks and Chen, 2020). Across the primary sensory cortices, there is a well-defined critical period for inducing Hebbian plasticity of thalamocortical synapses (Crair and Malenka, 1995; Jiang et al., 2007). In the juvenile brain, some degree of experience-dependent plasticity can be induced in the primary sensory thalamus, which is mediated by corticothalamic feedback (Thompson et al., 2016) or changes in intrinsic thalamic inhibition (Sommeijer et al., 2017). However, even in the adult primary sensory cortices, synaptic plasticity is not completely absent and is quite robust, especially in the superficial layers (Goel et al., 2006; Jiang et al., 2007). Also, sensory deprivation or denervation can enable plasticity in the adult primary sensory cortices. For example, a period of visual deprivation enables ocular dominance plasticity in V1 of adult rats (He et al., 2006), focal scotoma causes rapid remapping of V1 receptive fields in adult cats (Gilbert and Wiesel, 1992; Giannikopoulos and Eysel, 2006), partial cochlear lesion remaps tonotopy in the primary auditory cortex of adult cats (Rajan et al., 1993), and digit amputation or nerve transection leads to remapping of tactile receptive fields in the primary somatosensory cortex of adult monkeys (Merzenich et al., 1983). These studies not only suggest that the adult cortex retains plasticity, but also indicate that ongoing sensory experience may be restricting large-scale reorganization of the sensory cortices. In addition to plasticity within the same modality, sensory deprivation is also quite remarkable at inducing cross-modal plasticity in adults, which involves widespread plasticity across the deprived and the spared primary sensory cortices (Ewall et al., 2021; Lee, 2023).

Cross-modal plasticity induced in adults with late-onset sensory loss is thought to mainly occur through functional changes of preexisting circuits. For example, deafening adult ferrets causes A1 neurons to respond to somatosensory stimulation without gross changes in anatomical connections to A1 (Allman et al., 2009). While A1 neurons respond to touch to the skin, the cross-modal activation differs from tactile responses in S1 because it does not display refined somatotopy, and activation occurs with bilateral touch across the whole head (Allman et al., 2009). This is similar to the cross-modal recruitment observed with early sensory loss as discussed above, which suggests that cross-modal sensory processing differs from the more refined nature of unimodal processing regardless of the onset of sensory loss. The lack of gross anatomical changes and the rather rapid nature of cross-modal recruitment (typically observed after a few days of sensory loss) caused by adult-onset sensory loss suggests that it is either due to unmasking preexisting functional connections or due to plasticity of previously subthreshold functional connections.

Recent studies using mice provide evidence of large-scale functional plasticity taking place in both the deprived and the spared thalamic nuclei and primary sensory cortices following late-onset sensory loss (Figure 55.2). A few days of visual deprivation in the form of dark exposure produces potentiation of the majority of excitatory synaptic inputs onto L2/3 pyramidal neurons (Goel et al., 2006; He et al., 2012; Petrus et al., 2015; Chokshi et al., 2019; Rodriguez et al., 2019). But the potentiation is rather specific to the lateral inputs to L2/3 of V1 without changes in the strength of feedforward synapses, including thalamocortical synapses to L4 or L4 inputs to L2/3 (Petrus et al., 2014; Petrus et al., 2015; Chokshi et al., 2019). In contrast, visual deprivation causes potentiation of feedforward synapses in the spared sensory cortices, including potentiation of thalamocortical synapses and L4 to L2/3 synapses (Jitsuki et al., 2011; Petrus et al., 2014; Petrus et al., 2015). There is evidence that thalamocortical plasticity is likely enabled by specific disinhibition of the spared primary thalamic nucleus (Whitt et al., 2022). The potentiation of thalamocortical synapses in A1 L4 correlated with a lowered neuronal threshold for sound detection (Petrus et al., 2014), suggesting that it may improve auditory function. In addition, there is evidence that there is a refinement of functional circuits within the spared sensory cortex (Jitsuki et al., 2011; Meng et al., 2015, 2017), which may lead to more efficient sensory processing in S1 and A1 L2/3 following late-onset visual deprivation (Jitsuki et al., 2011; Solarana et al., 2019). A recent study reported that late-onset vision loss leads to increased dendritic spine turnover and an overall decrease in spine density in A1 (Han et al., 2022), which is consistent with the functional refinement of circuitry. Compensatory plasticity in the spared sensory cortex is likely a general principle of cortical plasticity because thalamocortical synaptic potentiation is also observed in V1 following late-onset auditory deprivation (Rodriguez et al., 2018). Collectively, these results suggest that late-onset sensory loss can lead to widespread functional plasticity across primary sensory cortices. As with early sensory loss, the mode of sensory loss seems to matter because changes in V1 seem to require a complete loss of visual inputs, as with dark-exposure and enucleation (He et al., 2012), but compensatory plasticity in the spared sensory cortex can be triggered by a milder loss of vision in the form of bilateral eyelid suture (Jitsuki et al., 2011; He et al., 2012). These findings suggest that compensatory plasticity in the spared sensory cortices may occur when vision is degraded to make it behaviorally less useful, while degraded visual inputs may still prevent cross-modal recruitment of V1.

At a mechanistic level, cross-modal synaptic plasticity observed across the primary sensory cortices seems to be mediated by NMDA receptor-dependent Hebbian plasticity that is regulated by metaplasticity triggered upon sensory loss. In particular, the sliding threshold model of metaplasticity is consistent with most of the experimental evidence (Lee, 2023). It is proposed that loss of sensory inputs will slide down the synaptic modification threshold in the deprived sensory cortex to promote long-term potentiation (LTP) at the expense of long-term depression (LTD). As discussed previously, under normal cases, sensory cortices receive subthreshold inputs from other sensory modalities, which modulate and provide context to the main sensory input. It is postulated that lowering the LTP induction threshold with loss of the major sensory input will now allow these subthreshold inputs to potentiate. This in essence could allow the deprived sensory cortex to be recruited by the spared sensory modalities. In the spared sensory cortices, based on the evidence of potentiated feedforward inputs, it is speculated that this may drive the synaptic modification threshold to increase. This type of metaplasticity will cause the activity of less efficient inputs to now fall below the synaptic modification threshold to become weaker via LTD. Hence, in essence such adaptation will increase the feature selectivity of the spared cortical neurons.

CONCLUDING REMARKS

As discussed here, sensory loss leads to robust plasticity of sensory circuits that can allow an organism to navigate and interact with the world more effectively with the remaining senses. One note worthy of emphasis is that how and when a sensory loss occurs need to be considered to fully understand the nature of plasticity and what type of functional compensation can result from it. An early sensory loss that occurs before the full development of the thalamocortical circuit has the most impact on anatomical plasticity, especially with respect to cross-modal recruitment of the deprived sensory thalamus and cortex (Table 55.1). But cross-modal plasticity is not restricted to the early developmental period that is often defined as the critical period. Sensory loss later in life is quite effective at producing functional plasticity in the deprived sensory cortex and produces perhaps more dramatic changes in the spared sensory thalamus and cortex (Figure 55.2). This highlights the capacity of the thalamic and cortical circuits to dynamically adapt to the changes in the sensory environment.

REFERENCES

Allman BL, Keniston LP, Meredith MA (2009). Adult deafness induces somatosensory conversion of ferret auditory cortex. *Proc Natl Acad Sci U S A*. 106:5925–5930.

Bock AS, Fine I (2014). Anatomical and functional plasticity in early blind individuals and the mixture of experts architecture. *Front Human Neurosci*. 8:971.

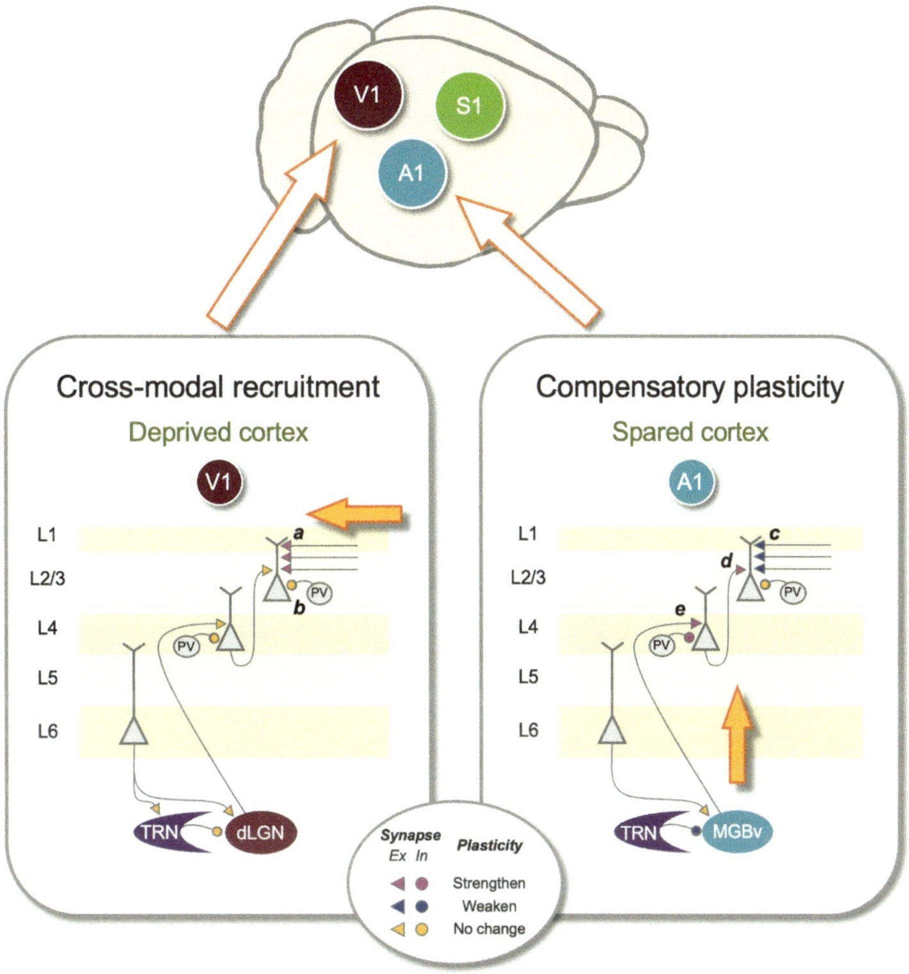

Figure 55.2. Cross-modal synaptic plasticity in adult mice following late-onset vision loss. The deprived sensory cortex (V1) undergoes specific plasticity of functional circuits to enhance the strength of lateral inputs arriving at L2/3 neurons, while the spared sensory cortex (A1) potentiates the feedforward synapses. In addition to the plasticity of the cortical circuits, late-onset visual deprivation leads to plasticity of the thalamic circuit of the spared sensory modality. The figure summarizes data compiled from studies using short-duration dark exposure in adult mice (Petrus et al., 2014; Petrus et al., 2015; Chokshi et al., 2019; Whitt et al., 2022). Plasticity of excitatory (*Ex* = triangle symbols) and inhibitory (*In* = circle symbols: mainly referring to inhibition from parvalbumin [PV]-positive interneurons) synapses are color coded, as shown in the inset. Some of the results are also observed with lid suture or enucleation and in adult rats, as detailed below.

Cross-modal recruitment: in the deprived V1, there is no change in the strength of thalamocortical synapses from dLGN (Petrus et al., 2014) or L4 to L2/3 (Petrus et al., 2015), but there is potentiation of lateral inputs to L2/3 pyramidal neurons (Petrus et al., 2015; Chokshi et al., 2019). Potentiation of lateral inputs to V1 L2/3 (labeled as *a*) is likely reflected in the observed increase in the amplitude of miniature excitatory postsynaptic currents (mEPSCs) recorded in these neurons following visual deprivation initiated in mature mice (Goel and Lee, 2007; He et al., 2012; Petrus et al., 2015; Chokshi et al., 2019) and rats (Goel et al., 2006). Potentiation of V1 L2/3 mEPSCs is also observed with bilateral enucleation, but not with bilateral lid suture (He et al., 2012), which suggests that complete loss of vision is necessary to drive plasticity. Inhibitory synapses onto V1 L2/3 (labeled as *b*) do not change the strength of action potential evoked inhibitory postsynaptic current (eIPSC) amplitude (Petrus et al., 2015; Gao et al., 2017), but there is a reduction in the frequency of miniature inhibitory postsynaptic currents (mIPSCs) (Gao et al., 2017). The selective adaptation of mIPSCs has been suggested to allow homeostatic adaptation of neuronal activity without compromising the temporal profile of excitatory input activity (Gao et al., 2017). In the visual circuit, there is no observed plasticity of the corticothalamic feedback onto dLGN or TRN (Whitt et al., 2022). There was also no change in the strength of TRN inhibition onto dLGN (Whitt et al., 2022). Collectively these results suggest that unimodal sensory deprivation does not produce thalamocortical, corticothalamic, or thalamic plasticity. The main adaptation is observed as the potentiation of the lateral inputs into L2/3 (orange arrow). Long-range direct cortico-cortical inputs or indirect cortico-thalamocortical inputs, as well as inputs from higher-order thalamus, most of which carry multisensory information, are likely reflected in the lateral inputs (Ewall et al., 2021).

Compensatory plasticity: in the spared A1, there is strengthening of thalamocortical inputs to L4 (labeled as *e*) (Petrus et al., 2014) and L4 to L2/3 (labeled as *d*) (Petrus et al., 2015), which may improve feedforward processing of auditory information (orange arrow). Potentiation of L4 to L2/3 (labeled as *d*) is also observed in barrel cortex of adult rats following a few days of bilateral lid suture (Jitsuki et al., 2011). While L4 inputs to L2/3 potentiate, lateral inputs in L2/3 (labeled as *c*) weaken in strength, which correlates with the observed decrease in the amplitude of mEPSCs in mice (He et al., 2012; Petrus et al., 2015) and rats (Goel et al., 2006). The fact that bilateral lid suture can alter synapses in A1 and S1 barrel cortex (Jitsuki et al., 2011; He et al., 2012) suggests that a milder form of vision loss is sufficient to produce compensatory plasticity. Potentiation of thalamocortical synapses (labeled as *e*) observed in A1 following visual deprivation is likely a general mode of compensatory plasticity in the adult brain, because it is also observed in adult V1 after a few days of deafening and recovers thalamocortical LTP, but not LTD (Rodriguez et al., 2018). This suggests that cross-modal sensory deprivation may be an effective way to restore thalamocortical LTP in adults. While visual deprivation does not alter corticothalamic feedback from A1 L6 to MGBv, it decreases the strength of inhibition from TRN to MGBv (Whitt et al., 2022). A decrease in the strength of TRN inhibition to MGBv occurs via postsynaptic weakening and a reduction in short-term depression that is dependent on the activation of endocannabinoid receptors (CB1-Rs) (Whitt et al., 2022). In addition to synaptic strength changes, visual deprivation causes refinement of spatial activity spread of inputs in A1 L4 and L2/3 (Meng et al., 2015, 2017), which may be responsible for the sparsification of tone representation in A1 L2/3 (Solarana et al., 2019).

Cang J, Feldheim DA (2013). Developmental mechanisms of topographic map formation and alignment. *Annu Rev Neurosci.* 36:51–77.

Chokshi V, Gao M, Grier BD, Owens A, Wang H, Worley PF, Lee HK (2019). Input-specific metaplasticity in the visual cortex requires homer1a-mediated mGluR5 signaling. *Neuron.* 104:736–748 e736.

Cohen LG, Celnik P, Pascual-Leone A, Corwell B, Falz L, Dambrosia J, Honda M, Sadato N, Gerloff C, Catala MD, Hallett M (1997). Functional relevance of cross-modal plasticity in blind humans. *Nature.* 389:180–183.

Collignon O, Voss P, Lassonde M, Lepore F (2009). Cross-modal plasticity for the spatial processing of sounds in visually deprived subjects. *Exper Brain Res.* 192:343–358.

Cooke SF, Bear MF (2014). How the mechanisms of long-term synaptic potentiation and depression serve experience-dependent plasticity in primary visual cortex. *Phil Trans R Soc Lond Ser B, Biol Sci.* 369:20130284.

Coullon GS, Jiang F, Fine I, Watkins KE, Bridge H (2015). Subcortical functional reorganization due to early blindness. *J Neurophysiol.* 113:2889–2899.

Crair MC, Malenka RC (1995). A critical period for long-term potentiation at thalamocortical synapses. *Nature.* 375:325–328.

Dietrich S, Hertrich I, Ackermann H (2013). Ultra-fast speech comprehension in blind subjects engages primary visual cortex, fusiform gyrus, and pulvinar: A functional magnetic resonance imaging (fMRI) study. *BMC Neurosci.* 14:74.

Ewall G, Parkins S, Lin A, Jaoui Y, Lee H-K (2021). Cortical and subcortical circuits for cross-modal plasticity induced by loss of vision. *Front Neural Circuits.* 15:665009.

Fox K, Wong RO (2005). A comparison of experience-dependent plasticity in the visual and somatosensory systems. *Neuron.* 48:465–477.

Froemke RC, Jones BJ (2011). Development of auditory cortical synaptic receptive fields. *Neurosci Biobehav Rev.* 35:2105–2113.

Gao M, Whitt JL, Huang S, Lee A, Mihalas S, Kirkwood A, Lee HK (2017). Experience-dependent homeostasis of "noise" at inhibitory synapses preserves information coding in adult visual cortex. *Phil Trans R Soc Lond Ser B, Biol Sci.* 372:20160156.

Giannikopoulos DV, Eysel UT (2006). Dynamics and specificity of cortical map reorganization after retinal lesions. *Proc Natl Acad Sci U S A.* 103:10805–10810.

Gilbert CD, Wiesel TN (1992). Receptive field dynamics in adult primary visual cortex. *Nature.* 356:150–152.

Goel A, Jiang B, Xu LW, Song L, Kirkwood A, Lee HK (2006) Cross-modal regulation of synaptic AMPA receptors in primary sensory cortices by visual experience. *Nat Neurosci.* 9:1001–1003.

Goel A, Lee HK (2007). Persistence of experience-induced homeostatic synaptic plasticity through adulthood in superficial layers of mouse visual cortex. *J Neurosci.* 27:6692–6700.

Goldreich D, Kanics IM (2003) Tactile acuity is enhanced in blindness. *J Neurosci* 23:3439–3445.

Gougoux F, Lepore F, Lassonde M, Voss P, Zatorre RJ, Belin P (2004). Neuropsychology: Pitch discrimination in the early blind. *Nature.* 430:309.

Grant AC, Thiagarajah MC, Sathian K (2000). Tactile perception in blind Braille readers: A psychophysical study of acuity and hyperacuity using gratings and dot patterns. *Percept Psychophys.* 62:301–312.

Han JJ, Noh TS, Suh MW, Kim SH, Kim DH, Kim SJ, Oh SH (2022). Synaptic remodeling of the auditory cortex following bilateral blindness: evidence of cross-modal plasticity. *Exp Neurobiol.* 31:299–306.

He HY, Hodos W, Quinlan EM (2006). Visual deprivation reactivates rapid ocular dominance plasticity in adult visual cortex. *J Neurosci.* 26:2951–2955.

He K, Petrus E, Gammon N, Lee HK (2012). Distinct sensory requirements for unimodal and cross-modal homeostatic synaptic plasticity. *J Neurosci.* 32:8469–8474.

Henschke JU, Oelschlegel AM, Angenstein F, Ohl FW, Goldschmidt J, Kanold PO, Budinger E (2018). Early sensory experience influences the development of multisensory thalamocortical and intracortical connections of primary sensory cortices. *Brain Struct Function.* 223:1165–1190.

Hooks BM, Chen C (2020). Circuitry underlying experience-dependent plasticity in the mouse visual system. *Neuron.* 106:21–36.

Huber E, Chang K, Alvarez I, Hundle A, Bridge H, Fine I (2019). Early blindness shapes cortical representations of auditory frequency within auditory cortex. *J Neurosci.* 39:5143–5152.

Huberman AD, Feller MB, Chapman B (2008). Mechanisms underlying development of visual maps and receptive fields. *Annu Rev Neurosci.* 31:479–509.

Iurilli G, Ghezzi D, Olcese U, Lassi G, Nazzaro C, Tonini R, Tucci V, Benfenati F, Medini P (2012). Sound-driven synaptic inhibition in primary visual cortex. *Neuron.* 73:814–828.

Izraeli R, Koay G, Lamish M, Heicklen-Klein AJ, Heffner HE, Heffner RS, Wollberg Z (2002). Cross-modal neuroplasticity in neonatally enucleated hamsters: Structure, electrophysiology and behaviour. *Eur J. Neurosci.* 15:693–712.

Ji D, Wilson MA (2007). Coordinated memory replay in the visual cortex and hippocampus during sleep. *Nat Neurosci.* 10:100–107.

Jiang B, Trevino M, Kirkwood A (2007). Sequential development of long-term potentiation and depression in different layers of the mouse visual cortex. *J Neurosci.* 27:9648–9652.

Jitsuki S, Takemoto K, Kawasaki T, Tada H, Takahashi A, Becamel C, Sano A, Yuzaki M, Zukin RS, Ziff EB, Kessels HW, Takahashi T (2011). Serotonin mediates cross-modal reorganization of cortical circuits. *Neuron.* 69:780–792.

Karmarkar UR, Dan Y (2006). Experience-dependent plasticity in adult visual cortex. *Neuron.* 52:577–585.

Lakatos P, Chen CM, O'Connell MN, Mills A, Schroeder CE (2007). Neuronal oscillations and multisensory interaction in primary auditory cortex. *Neuron.* 53:279–292.

Lee H-K (2023). Metaplasticity framework for cross-modal synaptic plasticity in adults. *Front Synaptic Neurosci.* 14:1087042.

Levelt CN, Hubener M (2012). Critical-period plasticity in the visual cortex. *Annu Rev Neurosci.* 35:309–330.

Lopez-Bendito G, Anibal-Martinez M, Martini FJ (2022). Cross-modal plasticity in brains deprived of visual input before vision. *Annu Rev Neurosci.* 45:471–489.

Mattioni S, Rezk M, Battal C, Vadlamudi J, Collignon O (2022). Impact of blindness onset on the representation of sound categories in occipital and temporal cortices. *eLife.* 11:e79370.

Meng X, Kao JP, Lee HK, Kanold PO (2015). Visual deprivation causes refinement of intracortical circuits in the auditory cortex. *Cell Rep.* 12:955–964.

Meng X, Kao JP, Lee HK, Kanold PO (2017). Intracortical circuits in thalamorecipient layers of auditory cortex refine after visual deprivation. *eNeuro.* 4.

Merabet LB, Hamilton R, Schlaug G, Swisher JD, Kiriakopoulos ET, Pitskel NB, Kauffman T, Pascual-Leone A (2008). Rapid and reversible recruitment of early visual cortex for touch. *PloS ONE.* 3:e3046.

Merabet LB, Pascual-Leone A (2010). Neural reorganization following sensory loss: the opportunity of change. *Nat Rev Neurosci.* 11:44–52.

Merabet LB, Swisher JD, McMains SA, Halko MA, Amedi A, Pascual-Leone A, Somers DC (2007). Combined activation and deactivation of visual cortex during tactile sensory processing. *J Neurophysiol.* 97:1633–1641.

Merzenich MM, Kaas JH, Wall J, Nelson RJ, Sur M, Felleman D (1983). Topographic reorganization of somatosensory cortical areas 3b and 1 in adult monkeys following restricted deafferentation. *Neuroscience.* 8:33–55.

Mower GD, Berry D, Burchfiel JL, Duffy FH (1981). Comparison of the effects of dark rearing and binocular suture on development and plasticity of cat visual cortex. *Brain Res.* 220:255–267.

Petrus E, Rodriguez G, Patterson R, Connor B, Kanold PO, Lee HK (2015). Vision loss shifts the balance of feedforward and intracortical circuits in opposite directions in mouse primary auditory and visual cortices. *J Neurosci* 35:8790–8801.
Petrus E, Isaiah A, Jones AP, Li D, Wang H, Lee HK, Kanold PO (2014). Crossmodal induction of thalamocortical potentiation leads to enhanced information processing in the auditory cortex. *Neuron*. 81:664–673.
Piche M, Chabot N, Bronchti G, Miceli D, Lepore F, Guillemot JP (2007). Auditory responses in the visual cortex of neonatally enucleated rats. *Neuroscience*. 145:1144–1156.
Rajan R, Irvine DR, Wise LZ, Heil P (1993). Effect of unilateral partial cochlear lesions in adult cats on the representation of lesioned and unlesioned cochleas in primary auditory cortex. *J Comp Neurol*. 338:17–49.
Rauschecker JP (1995). Compensatory plasticity and sensory substitution in the cerebral cortex. *Trends Neurosci*. 18:36–43.
Roder B, Stock O, Bien S, Neville H, Rosler F (2002). Speech processing activates visual cortex in congenitally blind humans. *Eur J. Neurosci.* 16:930–936.
Roder B, Teder-Salejarvi W, Sterr A, Rosler F, Hillyard SA, Neville HJ (1999). Improved auditory spatial tuning in blind humans. *Nature*. 400:162–166.
Rodriguez G, Mesik L, Gao M, Parkins S, Saha R, Lee HK (2019). Disruption of NMDA receptor function prevents normal experience-dependent homeostatic synaptic plasticity in mouse primary visual cortex. *J Neurosci*. 39(39):7664–7673.
Rodriguez G, Chakraborty D, Schrode KM, Saha R, Uribe I, Lauer AM, Lee HK (2018). Cross-modal reinstatement of thalamocortical plasticity accelerates ocular dominance plasticity in adult mice. *Cell Rep*. 24:3433–3440.
Sadato N, Pascual-Leone A, Grafman J, Ibanez V, Deiber MP, Dold G, Hallett M (1996). Activation of the primary visual cortex by Braille reading in blind subjects. *Nature*. 380:526–528.
Sieben K, Roder B, Hanganu-Opatz IL (2013) Oscillatory entrainment of primary somatosensory cortex encodes visual control of tactile processing. *J Neurosci*. 33:5736–5749.
Solarana K, Liu J, Bowen Z, Lee HK, Kanold PO (2019). Temporary visual deprivation causes decorrelation of spatiotemporal population responses in adult mouse auditory cortex. *eNeuro*. 6.
Sommeijer JP, Ahmadlou M, Saiepour MH, Seignette K, Min R, Heimel JA, Levelt CN (2017). Thalamic inhibition regulates critical-period plasticity in visual cortex and thalamus. *Nat Neurosci*. 20:1715–1721.
Stevens AA, Weaver KE (2009). Functional characteristics of auditory cortex in the blind. *Behav Brain Res*. 196:134–138.
Stryker MP, Harris WA (1986). Binocular impulse blockade prevents the formation of ocular dominance columns in cat visual cortex. *J Neurosci*. 6:2117–2133.
Talsma D, Woldorff MG (2005). Selective attention and multisensory integration: multiple phases of effects on the evoked brain activity. *J Cogn Neurosci*. 17:1098–1114.
Teder-Salejarvi WA, McDonald JJ, Di Russo F, Hillyard SA (2002) An analysis of audio-visual crossmodal integration by means of event-related potential (ERP) recordings. *Brain Res Cogn Brain Res*. 14:106–114.
Thompson A, Gribizis A, Chen C, Crair MC (2017). Activity-dependent development of visual receptive fields. *Curr Opin Neurobiol*. 42:136–143.
Thompson AD, Picard N, Min L, Fagiolini M, Chen C (2016). Cortical feedback regulates feedforward retinogeniculate refinement. *Neuron*. 91:1021–1033.
Tiriac A, Smith BE, Feller MB (2018). Light prior to eye opening promotes retinal waves and eye-specific segregation. *Neuron*. 100:1059–1065 e1054.
von Melchner L, Pallas SL, Sur M (2000). Visual behaviour mediated by retinal projections directed to the auditory pathway. *Nature*. 404:871–876.
Voss P, Lassonde M, Gougoux F, Fortin M, Guillemot JP, Lepore F (2004). Early- and late-onset blind individuals show supra-normal auditory abilities in far-space. *Curr Biol*. 14:1734–1738.
Whitt JL, Ewall G, Chakraborty D, Adegbesan A, Lee R, Kanold PO, Lee HK (2022). Visual deprivation selectively reduces thalamic reticular nucleus-mediated inhibition of the auditory thalamus in adults. *J Neurosci*. 42:7921–7930.
Wong M, Gnanakumaran V, Goldreich D (2011). Tactile spatial acuity enhancement in blindness: evidence for experience-dependent mechanisms. *J Neurosci*. 31:7028–7037.

SECTION X

EVOLUTION

Andrew C. Halley and Leah Krubitzer

In this section of the book, our authors have considered the relationship between the neocortex and dorsal thalamus in light of evolution. As neuroscientists, we often focus on a single portion of the brain or specific circuits that comprise it. Here our authors delve into our evolutionary history and examine the adaptive behaviors which drive the evolution of the cortex and thalamus. In Chapter 56, Halley and Krubitzer discuss the coevolution of the neocortex and dorsal thalamus, focusing on the scaling relationships between these structures and within these structures. In general, as the neocortex and dorsal thalamus become larger, particularly in primates, less space is dedicated to regions directly processing sensory input, and more space to regions involved in sensory integration and other higher-level functions via the trans-thalamic pathway. Thus, small brains with a relatively small neocortex are more sensory bound, while larger brains with a relatively larger neocortex expend more space and energy on within structure and across structure communication. In Chapter 57, Baldwin and Wise provide compelling theories for how granular prefrontal and premotor cortex, the medial pulvinar, and the medial dorsal nucleus of the thalamus coevolved in primates, and propose that behaviors associated with food selection, reaching, and leaping in tree canopies drove the evolution of these regions of the brain. They speculate on how these cortical areas and thalamic nuclei, and their connections, evolved to improve foraging choices in the dense rainforests that early primates inhabited.

In Chapter 58, Molnár, Bandiera and Clowry consider the evolution and development of thalamocortical connections. They discuss the interactions between developing corticofugal projections and thalamocortical projections that occur at the pallial-subpallial boundary, which has been altered significantly over the course of evolution. They also examine the impact of thalamocortical input on cortical development. While much of cortical development can occur independent of subcortical input, some aspects of cortical development rely on this input. In macaque monkeys and humans, thalamic projections arrive at the cortex very early, before cortical areas have even formed, and interact with subplate neurons as well as with the cortical progenitor populations. These early thalamocortical and corticothalamic interactions help create cortical areas and their circuitry and allow sensory input to shape them.

Cisek and Nakahashi in Chapter 59 take us further back in our evolutionary history and consider striatal circuits in early vertebrates, tetrapods, and amniotes and the constraints imposed on the evolution of thalamocortical and corticostriatal circuits in mammals. These authors focus on how an animal's adaptive behavioral repertoire, such as searching, ingesting, handling, and defensive actions, shaped representational maps in the neocortex. Specifically, they propose that cortex has ethological action maps, each specialized for particular aspects of an animal's behavioral repertoire. The basal ganglia controls the animal's selection for broad categories of behavior (e.g., walking versus reaching), while an animal's selection for a specific behavior (e.g., reaching for a particular item) occurs through competition within cortical action maps. Aspects of this dual system are highly conserved, but some have been elaborated independently to accommodate the unique peripheral morphology of different animals, and the niche in which they behave.

56.

THE COEVOLUTION OF THE NEOCORTEX AND DORSAL THALAMUS IN MAMMALS

SCALING RELATIONSHIPS BETWEEN AND WITHIN STRUCTURES

Andrew C. Halley and Leah Krubitzer

INTRODUCTION

The neocortex is one of the defining features of mammals, and it differs dramatically in its size and internal organization across species, as does its major source of sensory input, the dorsal thalamus. For example, compared to the smallest present-day primate brain, the mouse lemur, the human brain is over 700 times larger (*Homo*: 1330 cm^3; *Microcebus*: 1.78 cm^3), the human thalamus is 230 times larger (*Homo*: 18.2 cm^3; *Microcebus*: 0.078 cm^3), and the human neocortex is over 1300 times larger (*Homo*: 1006 cm^3; *Microcebus*: 0.74 cm^3) (Stephan et al., 1981). In terms of internal organization, it has been proposed that mammals with a relatively large neocortex have more than one hundred cortical fields, while those with a relatively small neocortex have on the order of 20 cortical fields (Van Essen et al., 2011, 2019; Kaas, 2013). Similarly, mammals with a relatively small dorsal thalamus appear to have substantially fewer nuclei than mammals with a relatively large dorsal thalamus (Jones, 2007).

Despite the dramatic differences in size and internal organization of the neocortex and dorsal thalamus across mammals, it is still possible to identify similar or homologous features of organization in these structures. For instance, the dorsal thalamus is composed of several first-order nuclei—universally shared among extant mammals—that process sensory input from the periphery and ultimately relay that input to the neocortex. These include nuclei associated with visual, somatosensory, and auditory processing; the dorsolateral geniculate nucleus (dLGN), the ventral posterior nucleus (VPN), and the medial geniculate nucleus (MGN), respectively. Like the dorsal thalamus, a number of neocortical fields are highly conserved across species, including (but not limited to) the primary visual (V1), somatosensory (S1), and auditory (A1) areas, which receive inputs from the first-order nuclei noted above. In addition to its inputs from the first-order nuclei of the thalamus, cortical fields other than the primary fields described above have a rich number of connections with higher-order nuclei of the thalamus that both modulate incoming sensory information and drive cortico-thalamocortical circuits (i.e., the trans-thalamic pathway; Sherman, 2016). The ubiquity of some aspects of the functional organization of the neocortex and dorsal thalamus across mammals is remarkable, given how dramatically mammals differ in their absolute brain size and behavioral specializations. However, beyond a constellation of primary sensory cortical fields and first-order thalamic nuclei that have been defined as homologous across all groups of mammals, less is known about the evolution of the higher-order cortical fields and thalamic nuclei, or about their scaling relationships. This is in part due to the difficulties in defining equivalencies across different species.

DEFINING HOMOLOGIES IN THE CORTEX AND THALAMUS

One of the largest problems in making valid comparisons across species is determining the criteria that should be used to determine homologous structures. While many laboratories have traditionally utilized architectonic criteria combined with functional organization and often connections (e.g., Kaas, 1982), modern studies utilize a variety of molecular techniques to define homologous features of the cortex and thalamus. For example, Molnár and colleagues (**Chapter 58 in Section X** of this volume) compare patterns of gene expression across a wide range of vertebrates such as reptiles, birds, and mammals to better understand the origins of the dorsal thalamus and neocortex and corresponding regions within these structures.

These types of studies circumvent the problems that arise when homologous structures take diverse forms (e.g., differ in appearance or relative location) in different lineages, or have different functions and patterns of connectivity.

Despite the difficulties in identifying homologous nuclei of the thalamus and areas of the neocortex across species, decades of comparative studies have shown that mammals share a basic plan of gross brain organization, and that some "new" brain areas may be elaborations of ancestral structures that have been modified, rather than added de novo (Striedter, 2005). Because mammals have been evolving independently for tens of millions to hundreds of millions of years, and have adapted numerous neural, morphological, and behavioral specializations, a central challenge in evolutionary neurobiology is to understand how thalamocortical systems evolve to meet these demands within the constraints imposed by development.

Comparing first-order nuclei of the thalamus and primary cortical fields (e.g., dLGN and V1) can be done with relative ease across wide-ranging groups of mammals, including those with derivations, because their cytoarchitectonic appearance, patterns of connectivity, and functional organization are ubiquitous. Within the thalamus, first-order nuclei ultimately receive their inputs from the skin, muscles, and joints (VPN), the cochlea (MGN), and the retina (dLGN). These homologous nuclei are the easiest to compare across species, as they are defined by a conserved pattern of cytoarchitecture, inputs from the periphery, outputs to the neocortex, and to some extent, neural response properties. By contrast, it is more difficult to determine homologies of "higher-order" nuclei of the thalamus across species for several reasons. They are often less cyotoarchitectonically distinct, their inputs come from a combination of subcortical and cortical sources (the latter of which has changed dramatically in different lineages), their cortical targets are more diverse and diffuse, and in many cases their functional roles remain poorly understood.

The pulvinar nucleus is a perfect case in point (Baldwin et al., 2017; Kaas and Baldwin, 2020). Originally thought to be a thalamic structure that was unique to primates (e.g., Le Gros Clark, 1930), the pulvinar was ultimately shown to have a homolog in non-primate mammals, the product of converging evidence from comparative studies of connectivity, electrophysiology, and histochemistry. In rodents, this nucleus is called the lateral posterior nucleus (LP), which led to confusion because this same terminology had already been applied to an unrelated nucleus in the primate thalamus (e.g., see Schmahmann and Pandya, 1990). Finally, as with many brain structures, more detailed studies have shown that the primate "pulvinar" is not a single nucleus in either function, cytoarchitecture, or connectivity. The inferior and lateral portions of the pulvinar play distinct roles in visual processing (Baldwin et al., 2017; Kaas and Baldwin, 2020), the anterior pulvinar appears to be involved in somatosensation (e.g., Pons et al., 1985; Padberg et al., 2009), and the medial pulvinar is involved in diverse functions such as integrating polymodal information, orienting selective attention, and working memory (see **Chapter 57 in Section X** of this volume). With these complications in mind, in this chapter we use the single term "Pul/LP" to refer to parts of the pulvinar and lateral posterior (LP) nucleus that are thought to be homologous across species, and which have been combined in studies from the literature (e.g., Chalfin et al., 2007, discussed below).

As in the thalamus, establishing homology in the neocortex has been difficult because only a small constellation of cortical fields is unambiguously shared across mammals, including the primary areas (e.g., V1, S1, A1) and a few other areas involved in sensory processing (e.g., second somatosensory area, S2; parietal ventral area, PV; second visual area; V2) or volitional motor control (e.g., primary motor cortex, M1; Krubitzer, 2009; Kaas, 2013). These fields have a unique architectonic appearance, share patterns of cortical and thalamocortical connections, and often share functional features such as topographic organization and magnification of behaviorally relevant sensory receptor arrays.

Comparative data on the relative size of thalamic nuclei and cortical fields are relatively rare, and the majority of studies have focused on structures that compose the visual system. The most widely studied thalamic nucleus is the dorsal lateral geniculate nucleus (dLGN), which can be readily identified in all mammalian species, though it differs in its relative size, the number of layers that compose it, its general morphology, and orientation within the thalamus (Figure 56.1). Figure 56.1 shows matched coronal sections stained for cytochrome oxidase (CO) through the thalamus of 13 mammals, representing 4 mammalian orders (primates, rodents, carnivores, and bats). Even from this limited sample, a number of features are apparent. In anthropoid primates (New and Old World monkeys), the dLGN has a characteristic U-shape, with magnocellular layers located along the ventral aspect of this nucleus (Jones, 2007). In some prosimians such as galagos, the dLGN is ventral with a slight lateral bend, and in some such as mouse lemurs it occupies a more dorsal location (also see McDonald et al., 1993). Within carnivores, the orientation of the dLGN is more variable across species; in the grizzly bear and cat it is laterally oriented, and in ferrets it is lateral to the optic tract. Within rodents, the location and orientation of the dLGN is similar, but there are differences in lamination across species.

Additional studies have measured elements of the pulvinar/LP complex (Pul/LP) and the primary visual cortex (V1) of the neocortex across species, and these will be described below. Scaling data on other regions of the

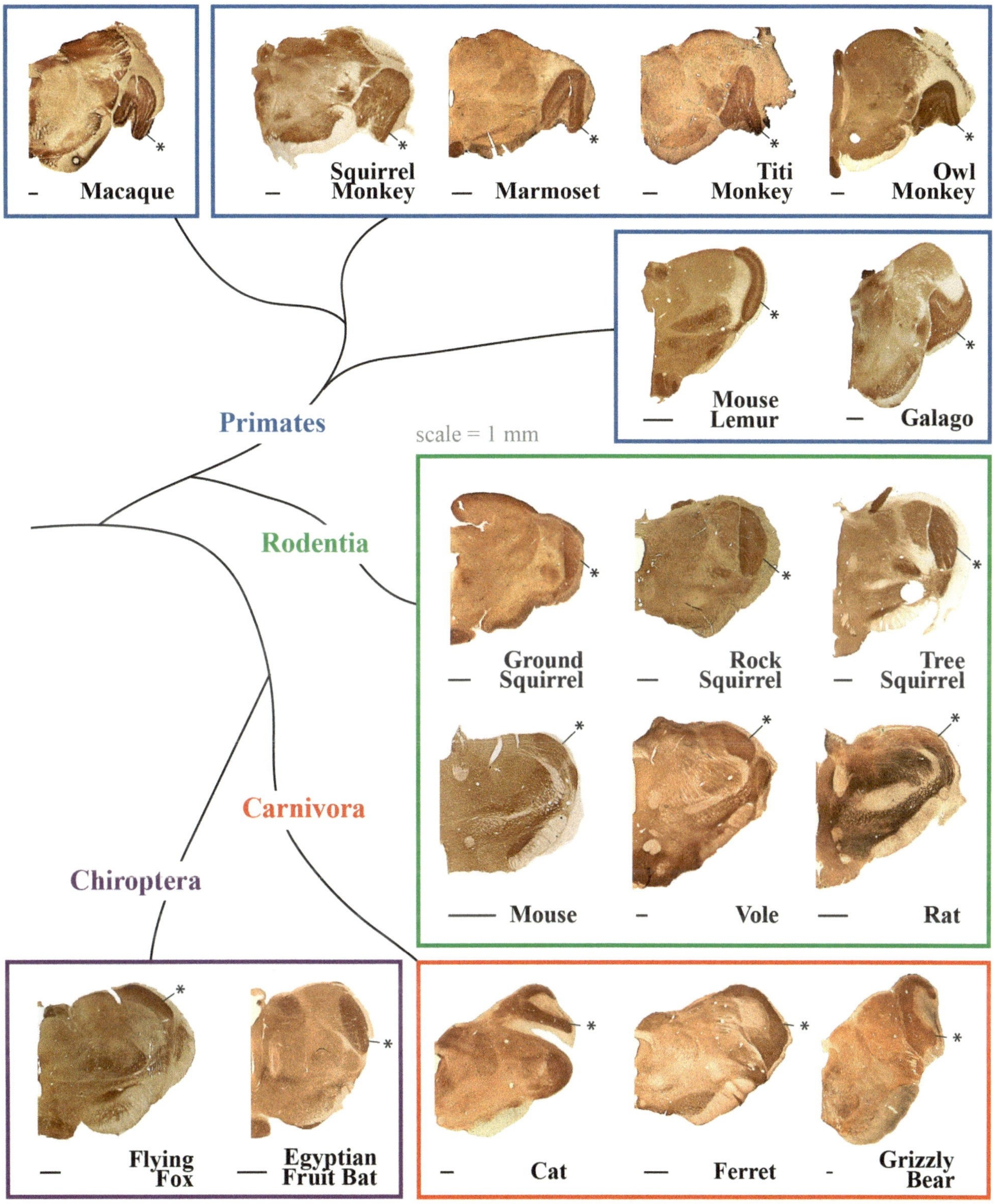

Figure 56.1. Coronal sections stained for cytochrome oxidase (CO) through the dorsal thalamus at the level of the dorsal lateral geniculate nucleus (dLGN), indicated by a star, in 13 mammalian species, representing Old World monkeys, New World monkeys, prosimians, rodents, carnivores, and bats. Similarities in shape and location within orders are apparent (e.g., rodents), as are some striking differences (e.g., compare mouse lemur to the galago). Despite differences within and across orders, the dorsal lateral geniculate can be readily identified across all mammals. Images of the mouse lemur, galago, and tree squirrel are from unpublished materials provided by the Kaas lab; image of the grizzly bear is from Krubitzer et al. (2018); all other images are from unpublished materials in the Krubitzer lab. Scale bars indicate 1 mm.

dorsal thalamus and neocortex are sparse but will be noted in relation to general trends discovered in the visual system.

CONCERTED AND MOSAIC EVOLUTION

Given the diversity in body morphology, behavioral specializations, locomotor styles, and sensorimotor adaptations, mammalian brains have clearly evolved to meet the unique demands of these diverse lifestyles, but it remains unclear how exactly this coevolution of the brain, body, and behavior occurs. First, we might expect that brain structures involved in processing inputs from specialized sensory systems would increase in size relative to other parts of the brain (e.g., more visually specialized species might exhibit enlarged visual structures (e.g., retina, dLGN, and V1)). Theories of *mosaic evolution* propose that specializations in a given sensory system (e.g., visual, somatosensory) are accompanied by changes in the size of brain areas that underlie these adaptations. A classic example in mammals of mosaic brain evolution associated with the overall size of a structure is the expansion of the neocortex in primates (especially monkeys and apes) relative to mammals such as rodents (Barton and Harvey, 2000; Halley and Deacon, 2017). This "neocorticalization" in primates has been proposed to partially explain why primates have more cortical fields than other mammals (Halley and Krubitzer, 2019), and will be discussed below when we compare the size of the neocortex to the size of the structure that provides its primary source of sensory input, the dorsal thalamus (Figure 56.2).

Mosaic evolution of the internal organization of a structure has been observed in a wide range of mammals and includes the magnification of the representation of behaviorally relevant sensory receptor arrays in both the neocortex and dorsal thalamus. This magnification or relative enlargement of sensory representations is the

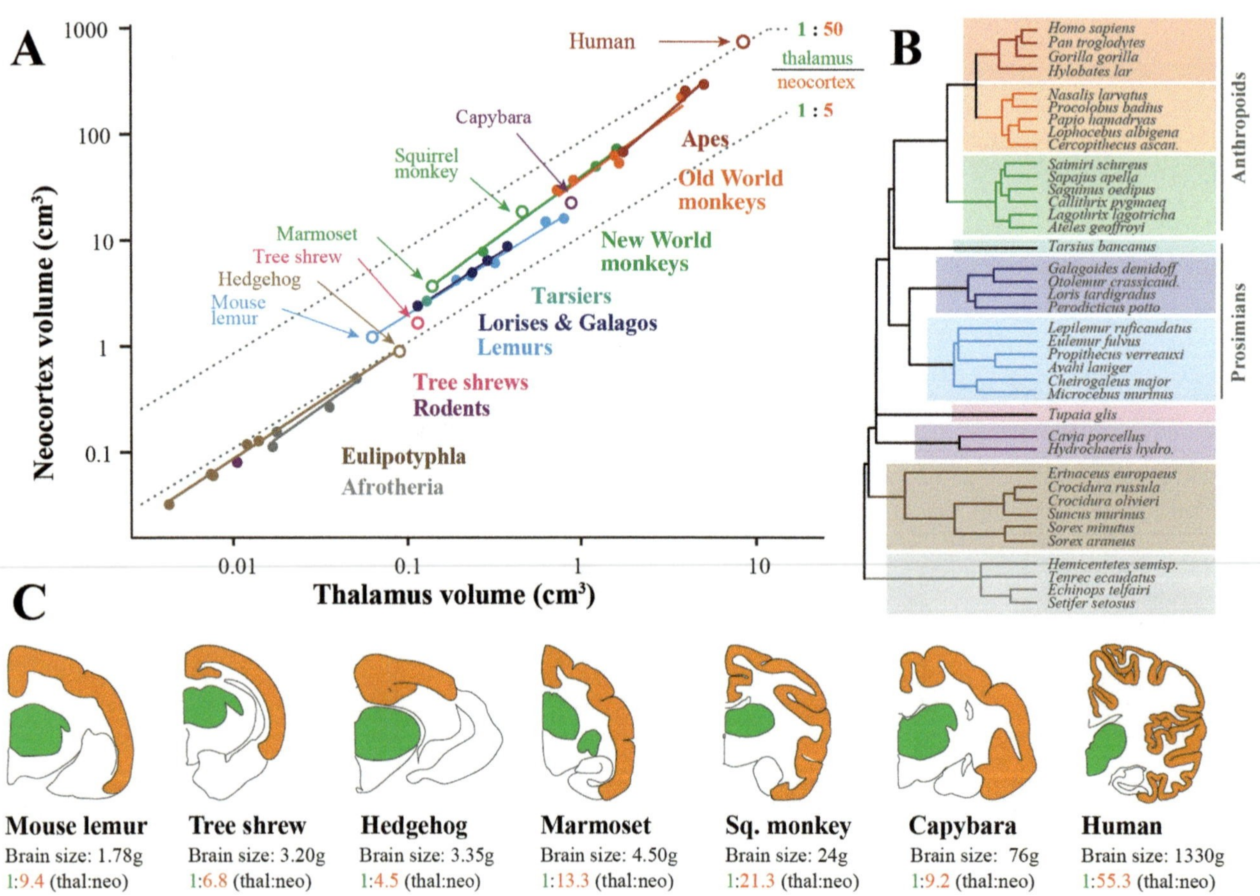

Figure 56.2. Species differ widely in the size of the neocortex relative to the dorsal thalamus, shown here in a comparative dataset. (A). The size of the thalamus plotted against the size of the neocortex across 39 mammalian species. Differences exist both between evolutionary lineages and according to size. Primates are uniquely "neocorticalized" relative to the dorsal thalamus, exhibited by the intercept shift that distinguishes all primates from rodents and shrews. Open circles indicate species whose brains are outlined in coronal section below. (B). The evolutionary relationship between the species included in this plot. (C). Coronal sections through the brain of 7 representative species. An outline of the thalamus is shown in green; an outline of the neocortex is shown in orange. Adapted from Halley and Krubitzer (2019); details in that paper.

manifestation of multiple factors, including specialized morphology with unique arrays of sensory receptors, behaviors associated with these morphologies, and the unique environmental contexts with different affordances within which mammals live and behave. The best examples are found in mammals with extreme specializations, occupying different niches (e.g., burrowing and aquatic), such as the star nosed mole (Catania and Kaas, 1995) and the duck-billed platypus (Krubitzer et al., 1995). The platypus has a distinctive bill with rows of mechanosensory and electrosensory receptors and uses this bill when engaging in ethologically relevant behaviors such as mating, feeding and navigating (Pettigrew, 1999). Somatosensory cortex in the duck-billed platypus is dominated by the representation of the bill; approximately 90% of S1 is devoted to processing inputs from the bill, and about 60% percent of the entire cortical sheet is dominated by the representation of the bill (see Englund and Krubitzer, 2022, for review). Other examples of cortical magnification in S1 include the representation of the nose rays in the star nosed mole (Catania and Kaas, 1995), the representation of the incisors in naked mole rats (Catania and Remple, 2002), and the representations of the glabrous hands, and supralaryngeal tract and oral structures in humans (Penfield, 1937; also see Gandhoke et al., 2019). This type of mosaic evolution of cortical fields has been observed to a greater or lesser extent in every mammal examined for all sensory systems.

The magnification of important sensory surfaces is not limited to the cortex, but is also observed in the thalamus. For example, in the platypus, the dorsal thalamus is dominated by the ventral posterior nucleus (VPN) to such an extent that it is difficult to identify other major first-order nuclei (i.e. dLGN, MGN; Ashwell, 2012; Mikula et al., 2008). In the blind mole rat, the dLGN is extremely small and undifferentiated (Rehkämper et al., 1994), and VPN is enlarged. Interestingly, the relative magnification of the pulvinar complex has independently evolved in different mammals that rely heavily on vision, such as primates and carnivores, which both exhibit a relatively larger pulvinar complex compared to rodents (Chalfin et al., 2007).

While there are clear examples of mosaic evolution in the neocortex and dorsal thalamus, much of the variation in the size of brain structures (not necessarily their internal organization) is a reflection of overall brain size. Large comparative datasets measuring the size of brain areas indicate that the absolute size of a given brain is highly predictive of the size of different structures that compose it. This basic size-associated (allometric) effect is called *concerted evolution* and has been described both across species (Stephan et al., 1981) and within species (Charvet et al., 2013). Even within a given brain structure, size has predictable effects on cell density (i.e., larger brains have lower cell density in general across all brain structures; Herculano-Houzel et al., 2015).

In reality, the variation in mammalian brains is the product of both mosaic and concerted evolution. The relative contribution of each form of evolution is largely debated in comparative neurobiology (see Striedter, 2005, for a review), but both are essential for interpreting neuroembryonic data on the expansion of the neocortex (Halley and Krubitzer, 2019). A central proposal to explain concerted evolutionary effects (i.e., the predictability of brain structure size from overall brain size) is that a conserved pattern of neurogenesis during embryonic development—a ventral-to-dorsal and caudal-to-rostral pattern of neurogenic onset from progenitor pools—creates regularities in the size of adult brain areas (Finlay and Darlington, 1995; Finlay et al., 2001). Even though different species develop their brains over variable time periods, they are highly conservative in both the brain growth rate (Halley, 2017) and the schedule of neurodevelopment during this period (Workman et al., 2013), leading to predictable brain structure sizes in adults.

The following sections of this chapter use published data from the literature to shed new light on general scaling principles of the neocortex and dorsal thalamus. Data on the size of the thalamus versus neocortex (Figure 56.2) are taken from Stephan et al. (1981; reproduced from Halley and Krubitzer, 2019). Data on the scaling of thalamic nuclei in apes (Figure 56.3) are taken from a series of papers by Armstrong (1979, 1980a, 1980b, 1981). The datasets on the size of the dLGN, Pul/LP (Figure 56.4) are taken from Stephan et al. (1981); Barton (1998); Chalfin et al. (2007); Finlay et al. (2014); and Armstrong (1979, 1980a, 1980b, 1981). Data on the size of the dLGN, whole thalamus, V1, and neocortex (Figure 56.5A,C) are taken from Stephan et al. (1981); thalamic reconstructions (Figure 56.5B) were made from unpublished coronal sections digitized from the Kaas lab (mouse lemur) and Krubitzer lab (macaque), while neocortical diagrams (Figure 56.5D) were redrawn from published studies (mouse lemur: Saraf et al., 2019; macaque: Krubitzer and Dooley, 2013). Phylogenetic trees across all figures (Figures 56.2, 56.4, 56.5) were made from a mammalian supertree (Bininda-Emonds et al., 2007).

Below, we draw on these large comparative datasets to better understand how the neocortex and thalamus coevolve in mammals. The next section examines the relative size of the dorsal thalamus with respect to the neocortex, followed by a section that examines the size of individual thalamic nuclei relative to the entire dorsal thalamus and the size of individual neocortical fields relative to the whole neocortex; the subsequent section focuses on the visual system, specifically the relative size of the dLGN and Pul/LP relative to the dorsal thalamus, as well as the size of V1 relative to the entire neocortex.

HOW DOES THE THALAMUS SCALE WITH CHANGES IN THE SIZE OF THE CORTICAL SHEET?: ACROSS-STRUCTURE COMPARISONS

Broadly speaking, the relative size of individual brain structures is associated with absolute brain size—for example, larger brains tend to have a relatively smaller brainstem, and a relatively larger neocortex. But because some mammalian lineages have smaller or larger brains in general, brain structure proportions are also associated with phylogenetic lineages. An important question is: How large is the neocortex relative to the source of its primary sensory inputs, the dorsal thalamus? Figure 56.2 shows the volume of the neocortex compared to the size of the dorsal thalamus in a sample of 39 mammalian species (reproduced from Halley and Krubitzer, 2019). In this overall sample, we can see that the ratio of the thalamus to neocortex varies considerably; from less than 1:5 in the smallest species (the pygmy shrew *Sorex minutus*) to over 1:50 in the largest (humans, *Homo sapiens*). While some of this effect is caused by the expansion of the neocortex in primates specifically (mosaic evolution, e.g., Halley and Deacon, 2017), the general trend of larger mammals having a larger neocortex can be observed within *each* mammalian lineage independent of phylogeny (concerted evolution; see PGLS analysis in Halley and Krubitzer, 2019). Simply put, in larger mammals with larger brains, the neocortex becomes exceptionally large relative to the size of the dorsal thalamus (Figure 56.2).

SCALING OF THALAMIC NUCLEI AND NEOCORTICAL FIELDS: WITHIN-STRUCTURE COMPARISONS

Is there a relationship between the overall size of the dorsal thalamus and the size of the individual nuclei that compose it? Is this relationship the same for the neocortex? Data on the relative size of thalamic nuclei compared to the entire dorsal thalamus are limited, and as noted above, are primarily derived from studies of the size of nuclei in the visual system: the dLGN and Pul/LP. Comparative data on the other nuclei of the thalamus are relatively rare; one exception is a unique study on the size of thalamic nuclei from studies on apes (Armstrong, 1979, 1980a, 1980b, 1981). While limited in its species diversity, this unique dataset offers a window into how individual thalamic nuclei scale relative to one another, and to the overall size of the dorsal thalamus. In this section, we will first examine the trends observed in this ape dataset, and then discuss the limited dataset that examines the size of cortical fields relative to the entire cortical sheet.

The size of 7 thalamic nuclei relative to the entire dorsal thalamus in a sample of 5 ape species (gibbon, chimpanzee, gorilla, orangutan, and human) are plotted in Figure 56.3A. To make this raw data easier to compare, we have shown the regression lines for each nucleus aligned at a common origin, with isometry (dotted line) for reference, in Figure 56.3B. This figure shows that all first-order nuclei of the thalamus scale at a low allometric coefficient relative to the overall thalamus, meaning that in larger ape brains, the dLGN,

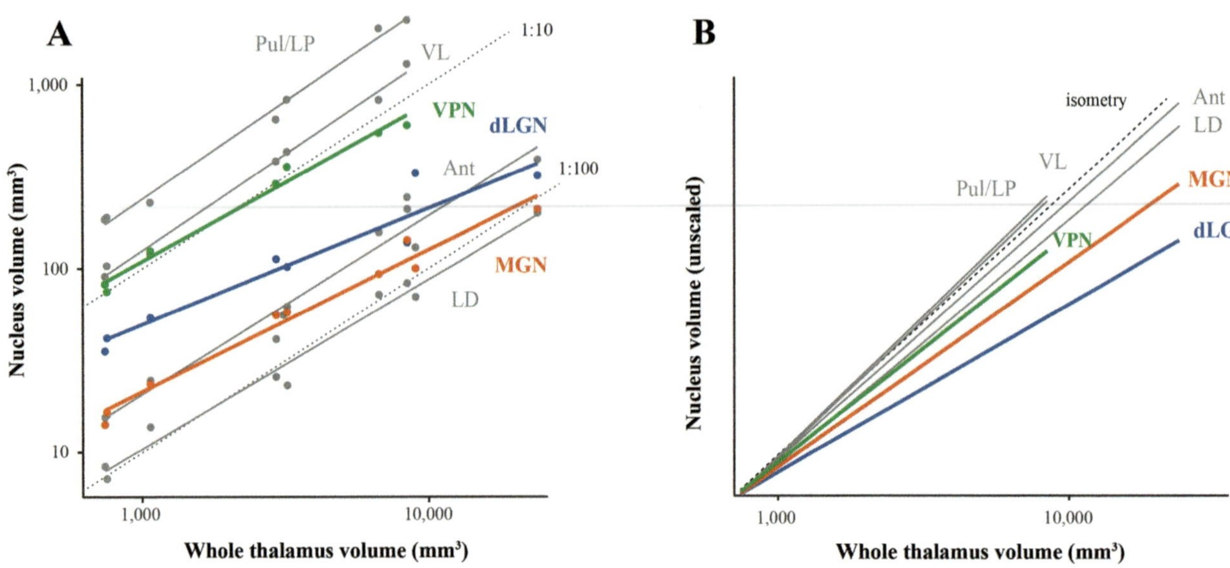

Figure 56.3. The relative size of thalamic nuclei in a sample of apes (gibbon, chimpanzee, gorilla, orangutan, and human) reanalyzed from Armstrong's dataset. (A). The volume of each nucleus is plotted against the total volume of the thalamus across the sample. First-order nuclei are highlighted in color (blue = dLGN; orange = MGN; green = VPN) while higher-order nuclei are collectively shown in gray. (B). The trend lines for this same dataset are aligned at the origin to highlight the different scaling relationships, relative to isometry (black dotted line). The *y*-axis is necessarily unscaled, as the trend lines have been aligned for easier comparison. As ape brains increase in size, the first-order nuclei of the thalamus (VPN, MGN, dLGN) occupy a smaller proportion of the overall thalamus, while higher-order nuclei increase in relative proportions. Data from Armstrong (1979, 1980a, 1980b, 1981).

MGN, and VPN become proportionally the smallest nuclei in the thalamus. By contrast, higher-order nuclei either scale with positive allometry (e.g., Pul/LP; the ventrolateral nucleus, VL), or at higher rates than first-order nuclei (anterior nuclei, Ant; the lateral dorsal nucleus, LD), thereby occupying a larger proportion of the thalamus. Thus, as the thalamus becomes larger, more space is occupied by higher-order nuclei and less space is occupied by first-order nuclei (e.g., Armstrong, 1979, 1980a, 1980b, 1981; Figure 56.3B).

This small dataset, composed of a handful of specimens in 5 species of a highly derived primate lineage—the apes—makes it difficult to draw conclusions about mammals in general. Nevertheless, the Armstrong studies represent the only comparative dataset that measures the relative size of most of the thalamic nuclei. Further, as noted below, this trend is observed in a larger dataset encompassing more species for visual nuclei of the thalamus. Expanding this type of analysis to include more and varied species for more nuclei of the thalamus would greatly inform our understanding of the evolution of the dorsal thalamus, how it changes with alterations in peripheral inputs from the different sensory systems, and how it scales with the neocortex.

Quantitative data on scaling of cortical fields within the neocortex is limited (e.g., Kaskan et al., 2005). However, the basic scaling relationship exhibited in the dorsal thalamus has also been shown for the neocortex. As the cortical sheet increases in size, the primary cortical fields (S1, A1, and V1) become relatively smaller, and a larger proportion of the cortical sheet is devoted to "higher-order" cortical fields involved in multimodal integration and "associative" functions. In mammals with small brains, primary cortical fields and motor cortex occupy a large proportion of the cortical sheet, while in mammals with larger brains—and especially in primates with expanded neocortices—the neocortex is composed of numerous "higher-order" cortical fields not observed in other lineages (Kaas, 2013; Krubitzer et al., 2011; Seelke et al., 2016; Van Essen et al., 2019; Chaplin et al 2013). There is evidence that these higher-order regions in frontal, posterior parietal, and temporal cortex have differentially expanded in primate evolution (e.g., Chalpin et al., 2013; Van Essen and Dieker, 2007; Goldring & Krubitzer, 2017; Reardon et al., 2018) and that they develop at slower rates in humans compared to primary areas (Reardon et al., 2018), suggesting that they may be more influenced by environmental context, and therefore built over the course of an extended postnatal development.

If we combine the scaling relationships observed in the thalamus and neocortex, a central theme emerges: As brains increase in size, the relative size of first-order thalamic nuclei and the cortical fields they innervate become relatively smaller, the thalamus becomes dominated by higher-order nuclei, and the neocortex becomes dominated by higher-order cortical fields. Thus, larger brains appear to dedicate less space to regions directly processing sensory input, and more space to regions involved in sensory integration and other higher-level functions via cortico-cortical connections, or by the expansion of the trans-thalamic pathway. Thus, small brains with a relatively small neocortex are more sensory bound, while larger brains with a relatively larger neocortex expend more space and energy on within-structure and across-structure communication.

EXAMPLES FROM THE VISUAL SYSTEM

The studies above describe general trends in scaling across structures and within structures, but are limited in their sample sizes and the species represented. Much more is known about the evolution of visual regions of the thalamus and neocortex, which have been studied extensively both in primates and other mammalian lineages. Here we re-examine larger comparative datasets on the size of dLGN (Stephan et al., 1981; Barton, 1998; Finlay et al., 2014); Pul/LP (Chalfin et al., 2007); dorsal thalamus, V1, and neocortex (Stephan et al., 1981). Importantly, these data on the scaling of visual structures reinforce several findings described above regarding the general scaling principles of the dorsal thalamus and neocortex.

One interesting comparison from the visual system is the size of dLGN relative to the size of Pul/LP. Figure 56.4 shows the volume of the dLGN plotted against Pul/LP in 31 species, including multiple primates, carnivores, rodents, and ungulates. In the smallest mammals, dLGN and Pul/LP are similar in size (see 1:1 isometric line), while in larger species, Pul/LP is many times larger than the dLGN (approaching the 1:10 isometric line). For example, the human Pul/LP is 3560 mm^3, while the dLGN is 406 mm^3. As a lineage, primates appear to have a larger Pul/LP volume, relative to the dLGN, than non-primate mammals—a "grade shift" noted by Chalfin et al. (2007). One potential complication in the Pul/LP measures is that only the inferior (PI) and lateral (PL) portions of the primate pulvinar are predominantly involved in visual processing (Baldwin et al., 2017; Baldwin and Kaas, 2020). If the medial pulvinar and anterior pulvinar were included in the measurements by Chalfin and colleagues, corresponding to nonhomologous structures in their non-primate sample, it would artificially inflate the size of the primate pulvinar to a degree that might affect analysis.

Within the visual system, one of the more surprising findings is that the size of visual structures (dLGN, Pul/LP, V1) are not always different in nocturnal and diurnal species, despite differences in the distribution of rods and cones in the retina (Wikler and Rakic, 1990; van der Merwe et al., 2018; Finlay et al., 2008) and visual acuity (Ordy and

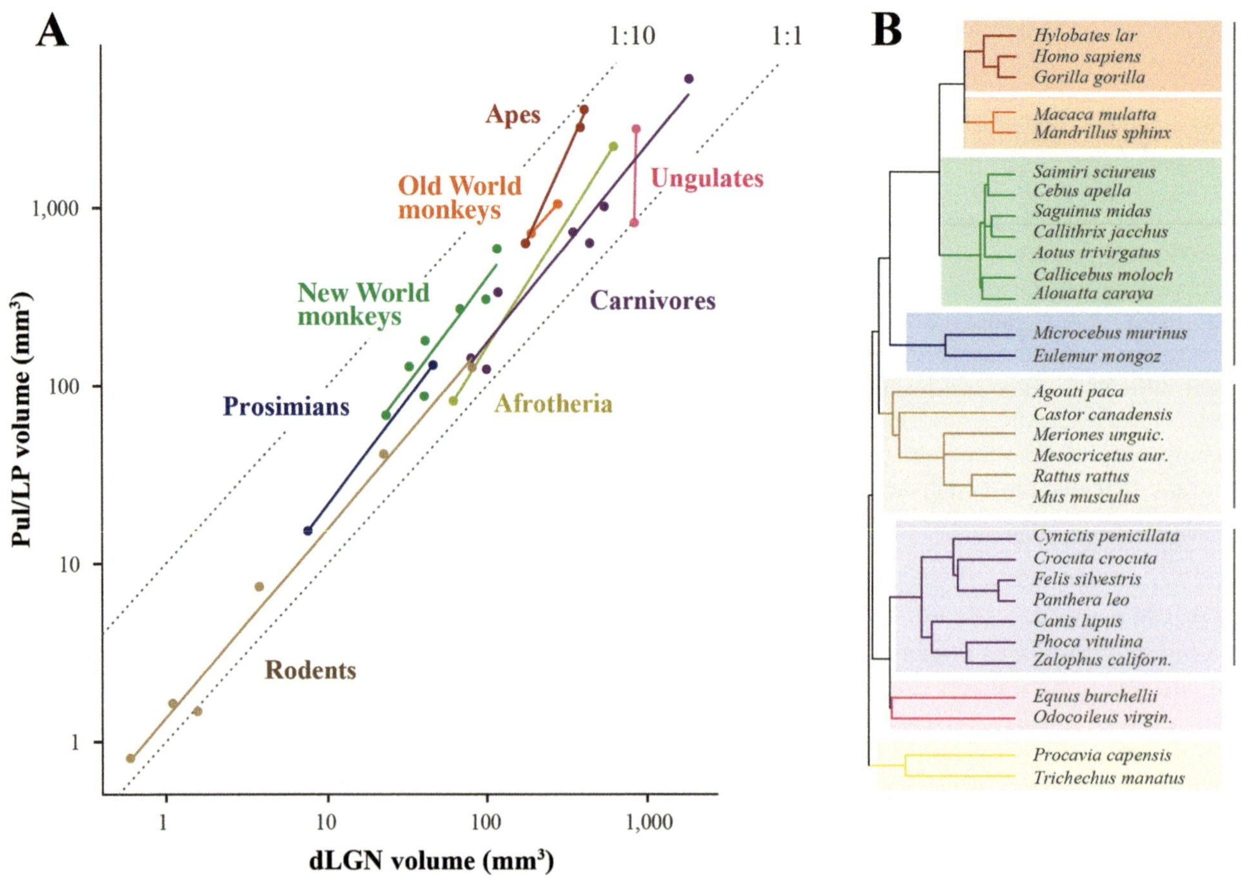

Figure 56.4. The relative size of the pulvinar complex (Pul/LP) and lateral geniculate nucleus (dLGN) in a diverse sample of 31 mammals. In the smallest mammalian brains, the pulvinar and dLGN are relatively similar in size, while in the largest primate brains (e.g., humans and gorillas), the pulvinar can approach 10 times the size of the dLGN. This suggests that in larger brains with larger thalami, first-order nuclei such as the dLGN occupy less space proportionally, while higher-order nuclei such as the pulvinar become exceptionally large. Data are adapted from Chalfin et al. (2007) and Finlay et al. (2014).

Smaorajski, 1968). The size of the pulvinar (Chalfin et al., 2007) and of V1 (Krubitzer and Kaas, 1990; Kaskan et al., 2005) do not appear to be associated with nocturnal/diurnal niches. Owl monkeys, which are the only nocturnal anthropoid primate, have a dLGN that is relatively "normal" in size for a New World monkey of its brain size. This lack of variation in visual structure size across diel niche may reflect the fact that within the geniculocortical visual stream, a small number of active neurons can encode sensory information (i.e., sparse coding; Olshausen and Field 1996), making it unnecessary to increase cell numbers or structure size (Finlay et al., 2014). It also seems possible that while nocturnal and diurnal vision involve different *kinds* of processing demands, the information involved in each vision type still requires a similar number of neurons to encode this information.

Another way of looking at scaling within the visual system is to compare the size of dLGN as a proportion of the thalamus, and V1 as a proportion of the neocortex. Importantly, these comparisons (using different datasets) show the same thalamocortical relationships described in sections above: larger brains devote less space to first-order thalamic nuclei and primary neocortical fields. As noted above for first-order nuclei of the thalamus in apes and in mammals in general, the dLGN scales with negative allometry as the whole thalamus increases in size, composing a smaller proportion of the overall thalamus in larger brains (Figure 56.5A). Note that the overall negative allometric trend is also observed in each mammalian lineage, suggesting an allometric trend that is independent of lineage. Figure 56.5B highlights this trend by showing 3-D reconstructions of the thalamus (dLGN highlighted in blue) in a small- and large-brained primate (*Microcebus* vs. *Macaca*). Similarly, Figure 56.5C shows that V1 scales with negative allometry as the whole neocortex increases in size, composing a smaller proportion of the neocortex in larger brains.

CONCLUSIONS

All mammals—from the tiniest rodents to blue whales—have a neocortex that receives its sensory inputs from the first-order nuclei of the dorsal thalamus. Basic features of this thalamocortical system are shared across mammals, irrespective of brain size, sensorimotor specializations, or ecological niche. The thalamus always contains first-order nuclei for processing somatosensory (VPN), auditory (MGN), and

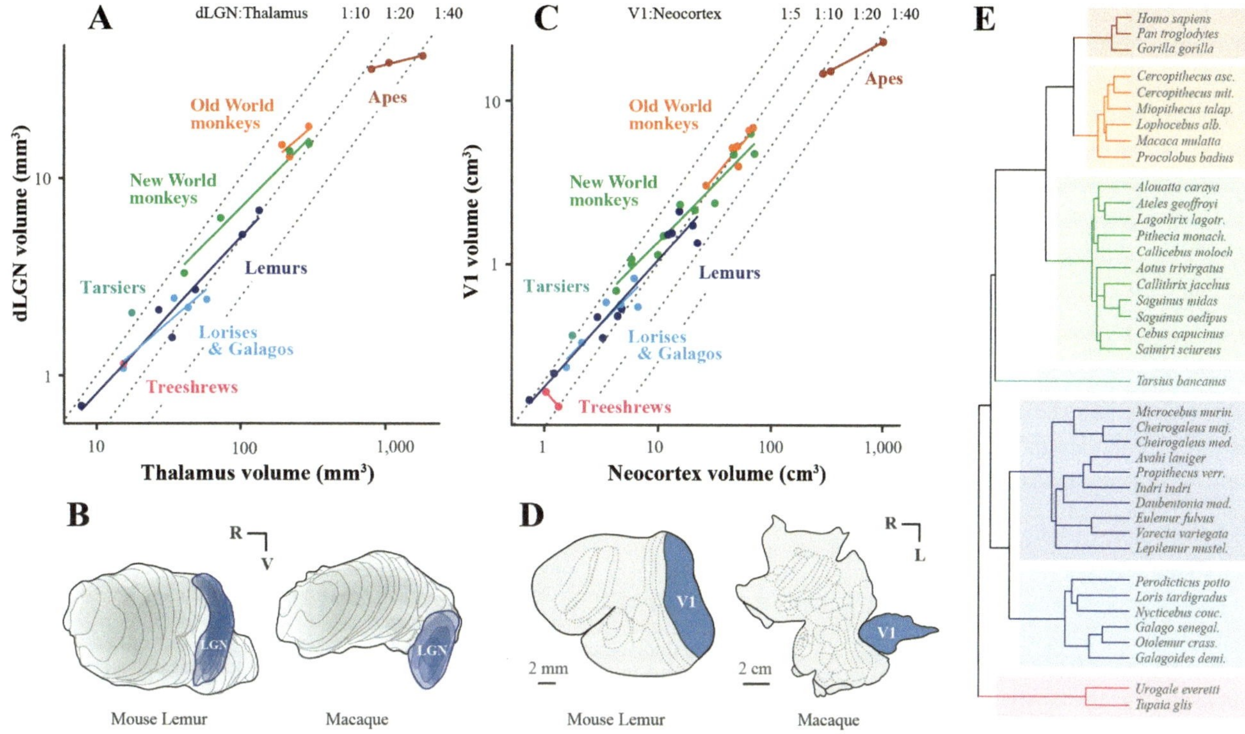

Figure 56.5. Overall brain size has a major influence on the relative size of both thalamic nuclei and cortical fields, demonstrated here by examining first-order visual structures in a sample of primates and closely related tree shrews. (A). In small primates and tree shrews, the lateral geniculate nucleus (dLGN) is relatively large and constitutes between 5% and 10% of the overall thalamus volume. In large-brained primates (i.e., apes) the dLGN is relatively smaller—between 2% and 5% of the thalamus. (B). 3-D reconstructions of the whole thalamus (gray) and the dLGN (blue) in both the mouse lemur (*Microcebus murinus*) and rhesus macaque (*Macaca mulatta*). The dLGN makes up a larger proportion of the thalamus in the smaller-brained mouse lemur. (C). Similarly, smaller primates have a V1 that occupies a relatively large proportion of the neocortex (10%–20%), while the largest primate brains (i.e., apes) have a relatively smaller V1 (2%–5% of total neocortex). (D). This effect can be seen by comparing the relative size of V1 in flattened neocortex. V1 occupies a larger proportion of the neocortex in the mouse lemur when compared with the rhesus macaque. (E). A phylogenetic tree of the species in (A) and (C). The size-associated effects described above are observed within each primate lineage. Data in (A) and (C) are from a comparative dataset detailed in Stephan et al. (1981). Reconstructions of the thalamus in (B) were created from regularly spaced coronal sections of the forebrain stained for cytochrome oxidase (CO) (Mouse lemur: Kaas Lab, Vanderbilt University; Macaque: Krubitzer Lab, UC Davis). Flattened cortex diagrams were adapted from published studies (Mouse lemur: Saraf et al., 2019; Krubitzer and Dooley, 2013).

visual (dLGN) inputs, and these nuclei always project to primary sensory fields of the neocortex (S1, A1, V1, respectively). Comparative work on the size of brain structures suggests that in larger brains, less space is devoted to first-order thalamic nuclei and to the primary cortical fields they innervate, and more space is devoted to higher-order cortical fields involved in sensory integration, and to the thalamic nuclei that provide input to these cortical fields. Essentially, brain size has a major influence on the size of primary versus higher-order structures (i.e., concerted evolution).

Where do these size-associated effects come from? One proposal is that the order of neurogenesis in the thalamus—with first-order nuclei undergoing neurogenesis earlier than higher-order nuclei—generates the allometric patterns we observe in adult brains, much as it is thought to underlie broader allometric patterns across the entire brain (Finlay et al., 2001). First-order nuclei like the dLGN, VPN, and MGN could initiate neurogenesis first, cutting off progenitor proliferation earlier than in other nuclei, and thereby producing relatively smaller nuclei.

While some version of this developmental logic likely affects the evolution of neocortex (e.g., cell density across layers; Charvet et al., 2015), it is clear that the interactions between the developing cortex and thalamus are much more complex (see **Chapter 58 in Section X** of this volume). Further, cortical fields appear to be more variable and plastic in their size than subcortical structures are. As noted above, mammals are famously diverse in the relative size and internal organization of somatosensory, visual, and auditory neocortex. This variability may reflect the fact that cortical field specification occurs at the latest stages of neurodevelopment, when arealization is more heavily influenced by activity-dependent mechanisms than by genes intrinsic to the cortex and thalamus associated with specification of these structures. Accordingly, cortical fields can be greatly impacted by manipulations that occur early in development, such as enucleation or congenital deprivation of visual input. Experimental studies have shown that early loss of visual input has profound effects on the size of cortical fields, thalamocortical and corticocortical connectivity, and functional organization (e.g., Chabot et al., 2007;

Charbonneau et al., 2012; Dooley and Krubitzer, 2018; Karlen et al., 2006; Ramamurthy et al., 2021; Englund et al., 2022). These trends are observed in the dorsal thalamus, but are less pronounced (Englund and Krubitzer, 2022). This effect of late-development plasticity should be most extreme in large-brained species with extended neurodevelopmental schedules (Workman et al., 2013), like primates. This may be particularly true for cortical regions that have differentially expanded in primates over the course of evolution, and that have extended periods of development such as frontal, posterior parietal, and inferotemporal cortex (Reardon, 2018). These higher-order regions of the neocortex appear to be built during development based on activity-dependent mechanisms and environmental context, allowing for the generation of adaptive behavior in an ever-changing sensory milieu.

REFERENCES

Armstrong E (1979). A quantitative comparison of the hominoid thalamus. I. Specific sensory relay nuclei. *Am J Phys Anthropol.* 51:365–382.

Armstrong E (1980a). A quantitative comparison of the hominoid thalamus. II. Limbic nuclei anterior principalis and lateralis dorsalis. *Am J Phys Anthropol.* 52:43–54.

Armstrong E (1980b). A quantitative comparison of the hominoid thalamus. III. A motor substrate: The ventrolateral complex. *Am J Phys Anthropol.* 52:405–419.

Armstrong E (1981). A quantitative comparison of the hominoid thalamus. IV. Posterior association nuclei: The pulvinar and lateral posterior nucleus. *Am J Phys Anthropol.* 55:369–383.

Ashwell KWS (2012). Development of the dorsal and ventral thalamus in platypus (*Ornithorhynchus anatinus*) and short-beaked echidna (*Tachyglossus aculeatus*). *Brain Struct Function.* 217:577–589.

Baldwin MKL, Balaram P, Kaas JH (2017). The evolution and functions of nuclei of the visual pulvinar in primates. *J Comp Neurol.* 525:3207–3226.

Barton RA (1998). Visual specialization and brain evolution in primates. *Proc R Soc Lond B Biol Sci.* 265:1933–1937.

Barton RA, Harvey PH (2000). Mosaic evolution of brain structures in mammals. *Nature.* 405:1055–1058.

Bininda-Emonds ORP, Cardillo M, Jones KE, MacPhee RDE, Beck RMD, Grenyer R, Price SA, Vox RA, Gittleman JL, Purvis A (2007). The delayed rise of present-day mammals. *Nature.* 446:507–512.

Catania KC, Kaas JH (1995). Organization of the somatosensory cortex of the star-nosed mole. *J Comp Neurol.* 351:549–567.

Catania, KC, Remple, MS (2002). Somatosensory cortex dominated by the representation of teeth in the naked mole-rat brain. *Proc Natl Acad Sci.* 99:5692–5697.

Chabot N, Robert S, Tremblay R, Miceli D, Boire D, Bronchti G (2007). Audition differently activates the visual system in neonatally enucleated mice compared with anophthalmic mutants. *Eur J Neurosci.* 26:2334–2348.

Chalfin BP, Cheung DT, Muniz, JAPC, De Lima Silveira, LC, Finlay, BL (2007). Scaling of neuron number and volume of the pulvinar complex in New World primates: Comparisons with humans, other primates, and mammals. *J Comp Neurol.* 504:265–274.

Chaplin TA, Yu H-H, Soares JGM, Gattass R, Rosa MGP (2013). A conserved pattern of differential expansion of cortical areas in simian primates. *J Neurosci.* 33:15120–15125.

Charbonneau V, Laramée M-E, Boucher V, Bronchti G, Boire D (2012). Cortical and subcortical projections to primary visual cortex in anophthalmic, enucleated and sighted mice. *Eur J Neurosci.* 36:2949–2963.

Charvet CJ, Cahalane DJ, Finlay BL (2015). Systematic, cross-cortex variation in neuron numbers in rodents and primates. *Cereb Cortex.* 25:147–160.

Charvet CJ, Darlington RB, Finlay BL (2013). Variation in human brains may facilitate evolutionary change toward a limited range of phenotypes. *Brain Behav Evol.* 81:74–85.

Clark, WE (1930). The thalamus of tarsius. *J Anatomy.* 64:371–414.

Dooley JC, Krubitzer L (2018). Alterations in cortical and thalamic connections of somatosensory cortex following early loss of vision. *J Comp Neurol.* 527:1675–1688.

Englund M, Krubitzer L (2022). Phenotypic alteration in cortical organization and connectivity across different time scales. *Brain Behav Evol.* 97:108–120.

Finlay BL, Charvet CJ, Bastille I, Cheung DT, Muniz JAPC, Silveira LCdL (2014). Scaling the primate lateral geniculate nucleus: Niche and neurodevelopment in the regulation of magnocellular and parvocellular cell number and nucleus volume. *J Comp Neurol.* 522:1839–1857.

Finlay BL, Darlington RB (1995). Linked regularities in the development and evolution of mammalian brains. *Science.* 268:1578–1584.

Finlay BL, Darlington RB, Nicastro N (2001). Developmental structure and evolution. *Behav Brain Sci.* 24:263–308.

Finlay BL, Franco ECS, Yamada ES, Crowley JC, Parsons M, Muniz JAPC, Silveira LCL (2008). Number and topography of cones, rodes and optic nerve axons in New and Old World primates. *Vis Neurosci.* 25:289–299.

Gandhoke GS, Belykh E, Zhao X, Leblanc R, Preul MC (2019). Edwin Boldrey and Wilder Penfield's homunculus: A life given by Mrs. Cantlie (in and out of realism). *World Neurosurg.*132:377–388.

Goldring A, Krubitzer L (2017). Evolution of parietal cortex in mammals: From manipulation to tool use. In: Kaas J, Krubitzer L, eds. *Evolution of Nervous Systems*, Vol. 3. 2nd ed. Elsevier:259–286.

Halley AC (2017). Minimal variation in eutherian brain growth rates during fetal neurogenesis. *Proc R Soc Lond B Biol Sci.* 284:20170219.

Halley AC, Deacon TW (2017). The developmental basis of evolutionary trends in primate encephalization. In: Kaas J, Krubitzer L, eds. *Evolution of Nervous Systems*, Vol. 3. 2nd ed. Elsevier:149–162.

Halley AC, Krubitzer LA (2019). Not all cortical expansions are the same. *Curr Opin Neurobiol.* 56:78–86.

Herculano-Houzel S, Catania K, Manger PR, Kaas JH (2015). Mammalian brains are made of these: A dataset of the numbers and densities of neuronal and nonneuronal cells in the brains of Glires, Primates, Scandentia, Eulipotyphlans, Afrotherians and Artiodachtyls, and their relationship with body mass. *Brain Behav Evol.* 86:145–163.

Jones EG (2007). *The Thalamus.* Cambridge University Press.

Kaas JH (1982). The segregation of function in the nervous system: Why do the sensory systems have so many subdivisions? *Contrib Sensory Physiol.* 7 201–240.

Kaas JH (2012). The evolution of neocortex in primates. *Prog Brain Res.* 195:91–102.

Kaas JH (2013). The evolution of brains from early mammals to humans. *Rev Cogn Neurosci.* 4:33–45.

Kaas JH, Baldwin MKL (2020). The evolution of the pulvinar complex in primates and its role in the dorsal and ventral streams of cortical processing. *Vision.* 4:3.

Karlen SJ, Kahn DM, Krubitzer L (2006). Early blindness results in abnormal corticocortical and thalamocortical connections. *Neuroscience.* 142:843–858.

Kaskan PM, Franco ECS, Yamada ES, Silbeira LCL, Darlington RB, Finlay BL (2005). Peripheral variability and central constancy in mammalian visual system evolution. *Proc R Soc Lond B Biol Sci.* 272:91–100.

Krubitzer LA, Kaas JH (1990). Cortical connections of MT in four species of primates: Areal, modular, and retinotopic patterns. *Vis Neurosci.* 5:165–204.

Krubitzer L (2009). In search of a unifying theory of complex brain evolution. *Ann N Y Acad Sci.* 1156:44–67.

Krubitzer L, Campi KL, Cooke DF (2011). All rodents are not the same: A modern synthesis of cortical organization. *Brain Behav Evol.* 78:51–93.

Krubitzer L, Dooley JC (2013). Cortical plasticity within and across lifetimes: How can development inform us about phenotypic transformation? *Front Human Neurosci.* 7:620.

Krubitzer L, Manger P, Pettigrew JD, and Calford MB. (1995). The organization of neocortex in monotremes: In search of the prototypical plan. *J Comp Neurol.* 351:261–306.

McDonald CT, McGuinness ER, Allman JM (1993). Laminar organization of acetylcholinesterase and cytochrome oxidase in the lateral geniculate nucleus of prosimians. *Neuroscience.* 54:1091–1101.

Mikula S, Manger P, Jones E (2008). The thalamus of the monotremes: Cyto- and myeloarchitecture and chemical neuroanatomy. *Phil Trans R Soc Lond B Biol Sci.* 363:2415–2440.

Olshausen BA, Field DJ (2004). Sparse coding of sensory inputs. *Curr Opin Neurobiol.* 14:481–487.

Ordy JM, Samorajski T (1968). Visual acuity and ERG-CFF in relation to the morphologic organization of the retina among diurnal and nocturnal primates. *Vis Res.* 8:1205–1225.

Padberg J, Cerkevich C, Engle J, Rajan AT, Recanzone G, Kaas J, Krubitzer L (2009). Thalamocortical connections of parietal somatosensory cortical fields in macaque monkeys are highly divergent and convergent. *Cereb Cortex.* 19:2038–2064.

Penfield W, Boldrey E (1937). Somatic motor and sensory representation in the cerebral cortex of man as studied by electrical stimulation. *Brain.* 60:389–443.

Pettigrew, JD (1999). Electroreception in monotremes. *J Exper Biol.* 202:1447–1454.

Pons TP, Kaas JH (1985). Connections of area 2 of somatosensory cortex with the anterior pulvinar and subdivisions of the ventroposterior complex in macaque monkeys. *J Comp Neurol.* 240:16–36.

Ramamurthy DL, Dodson HK, Krubitzer LA (2021). Developmental plasticity of texture discrimination following early vision loss in the marsupial Monodelphis domestica. *J Exper Biol.* 224: jeb236646.

Reardon PK, Seidlitz J, Vandekar S, Liu S, Patel R, Park MTM, Alexander-Bloch A, Clasen LS, Blumenthal JD, Lalonde FM, et al. (2018). Normative brain size variation and brain shape diversity in humans. *Science.* 360:1222–1227.

Rehkämper G, Necker R, Nevo E (1994). Functional anatomy of the thalamus in the blind mole rat Spalax ehrenbergi: An architectonic and electrophysiologically controlled tracing study. *J Comp Neurol.* 347:570–584.

Saraf MP, Balaram P, Pifferi F, Gamanut R, Kennedty H, Kaas JH (2019). Architectonic features and relative locations of primary sensory and related areas of neocortex in mouse lemurs. *J Comp Neurol.* 527:625–639.

Schmahmann JD, Pandya DN (1990). Anatomical investigation of projections from thalamus to posterior parietal cortex in the rhesus monkey: A WGA-HRP and fluorescent tracer study. *J Comp Neurol.* 295:299–326.

Seelke AMH, Yuan, S-M, Perkeybile AM, Krubitzer L, Bales KL (2016). Early experiences can alter the size of cortical fields in prairie voles (*Microtus ochrogaster*). *Environ Epigenetics.* 2(3):1–9.

Sherman SM (2016). Thalamus plays a central role in ongoing cortical functioning. *Nat Neurosci.* 19:533–541.

Stephan H, Frahm H, Baron G (1981). New and revised data on volumes of brain structures in insectivores and primates. *Folia Primatologica.* 35:1–29.

Striedter GF (2005). *Principles of Brain Evolution.* Sinauer Associates.

Van der Merwe I, Lukats A, Blahova V, Oosthuizen MK, Bennett NC, Němec P (2018). The topography of rods, cones and intrinsically photosensitive retinal ganglion cells in the retinas of a nocturnal (*Micaelamys namaquensis*) and a diurnal (*Rhabdomys pumilio*) rodent. *PLoS ONE.* 13:e0202106.

Van Essen DC, Dierker DL (2007). Surface-based and probabilistic atlases of primate cerebral cortex. *Neuron.* 56:209–225.

Van Essen DC, Donahue CJ, Coalson TS, Glasser ME (2019). Cerebral cortical folding, parcellation, and connectivity in humans, nonhuman primates, and mice. *Proc Natl Acad Sci U S A.* 116:26173–26180.

Van Essen DC, Glasser MF, Dierker DL, Harwell J, Coalson T (2011). Parcellations and hemispheric asymmetries of human cerebral cortex analyzed on surface-based atlases. *Cereb Cortex.* 22:2241–2262.

Wikler KC, Rakic P (1990). Distribution of photoreceptor subtypes in the retina of diurnal and nocturnal primates. *J Neurosci.* 10:3390–3401.

Workman AD, Charvet CJ, Clancy B, Darlington RB, Finlay BL (2013). Modeling transformations of neurodevelopmental sequences across mammalian species. *J Neurosci.* 33:7368–7383.

57.

EVOLUTION OF FRONTAL CORTEX AND THALAMUS IN PRIMATES

Mary K. L. Baldwin and Steven P. Wise

INTRODUCTION

In this chapter, we consider the evolution of frontal cortex, especially the prefrontal cortex (PFC) and 2 thalamic nuclei important to its function: the mediodorsal nucleus (MD) and the pulvinar (Pul), specifically the medial pulvinar (PulM). Figure 57.1 illustrates geological terms used in the discussion, as well as phylogenetic relationships among primates and their closest relatives. Understanding these relationships helps identify primate specializations. For instance, when tree shrews, rodents, and primates share a given cortical area, it probably has been inherited from a common ancestor and is therefore homologous in these groups. In contrast, when primates have an area, but tree shrews and rodents do not, it is likely a primate specialization that has no homologs in nonprimates.

EVOLUTION OF FRONTAL CORTEX

EOCENE AND EARLIER

From their origins until the late Eocene, primate brains resembled those of closely related placental mammals: in shape; in volume relative to body mass (encephalization); and in the percentage of brain occupied by neocortex (corticalization). Then, during the late Eocene, primate brains increased in both encephalization and corticalization (top row of asterisks in Figure 57.1), independently in the major primate lineages (Wise, 2023).

In addition to cortical expansion, new parts of the PFC emerged during primate evolution: some in early primates, others in anthropoids (Preuss and Goldman-Rakic, 1991; Preuss, 1995, 2007a, 2007b). Understanding PFC evolution and homologies across species remains a source of confusion, mostly because of inconsistencies in the use of the term "prefrontal" and some misconceptions about homology. Here we provide a brief history, but interested readers should consult Preuss and Wise (2021) for details.

Early comparative neuroanatomists concluded that PFC was unique to primates because of a clear-cut anatomical feature. This feature is so fundamental to how early anatomists parcellated the cortex that it was (and still is) used to define the 4 basic types of cortex: granular, homotypical, dysgranular, and agranular. It is also one of the most reliable markers for defining borders between cortical areas, which often match physiologically defined ones. So, what is this feature? Unlike the frontal cortex of most other mammals, the majority of the frontal cortex in primates has a conspicuous layer of small cells—an inner granular layer, layer (L)4—in Nissl-stained brain sections. Accordingly, the rostral part of PFC is called granular PFC in primates. Unfortunately, as another source of confusion, this term is a misnomer. These areas are homotypical or dysgranular in cytoarchitecture, not genuinely granular, which is the type of cortex found in primary sensory areas.

In the 1940s, a different idea about PFC arose: that it corresponds to areas with MD inputs. By that definition, all placental mammals have a PFC. Researchers of that era thought that PFC was homologous across mammals and that its granular cytoarchitecture emerged during primate evolution, which implies that PFC changed in type: from agranular to granular. For comparison, lamination in the primary visual cortex (V1) differs significantly across mammals and especially among primates (Balaram and Kaas, 2014), yet V1 is clearly homologous in all mammals. But V1 does not change in type, as posited for PFC. It is always granular in cytoarchitecture (koniocortex), and, as far as we know, there are no convincing examples of a transition of the sort suggested for PFC.

It is now known that MD's cortical connections are much more widespread than appreciated at the time, involving many areas beyond the frontal cortex, let alone the PFC. We illustrate these connections later, but for now the key point is that MD input is not necessarily a defining feature of PFC in any species. Accordingly, a revised view

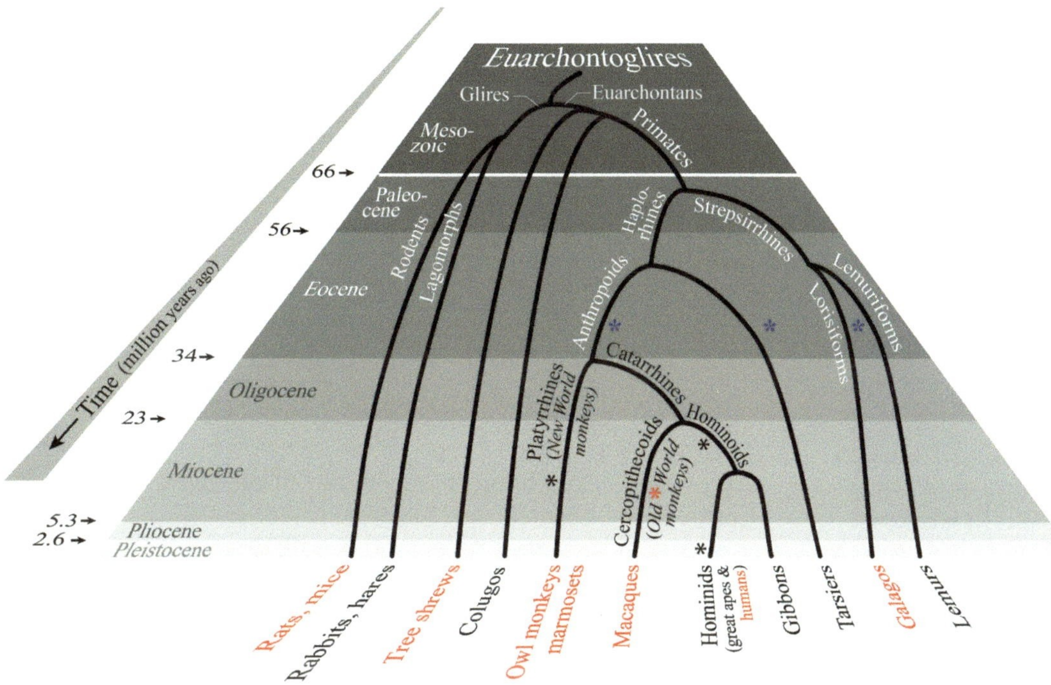

Figure 57.1. Geological epochs and evolutionary tree of Euarchontoglires. Groups in red type are the most intensively studied species in neuroscience. Asterisks mark approximate times of increases in encephalization and corticalization.

of the PFC emerged during the 1990s: that it has a granular part unique to primates, along with an agranular part that all mammals share (Preuss and Goldman-Rakic, 1991; Preuss, 1995, 2007a, 2007b).

Cortical maps in rats, tree shrews, galagos, and a variety of anthropoids indicate that new, granular parts of PFC emerged at various times during primate evolution. Moreover, they evolved in conjunction with a suite of premotor and visual areas that are also unique to primates. Figure 57.2 compares cortical maps in a selection of Euarchontoglires. The blue shading marks primate-specific parts of PFC (with one complication explained below), yellow indicates new and/or expanded visual areas, and orange highlights new motor areas. The hindlimb representation in the primary motor cortex (M1) deserves special consideration. Comparisons of somatotopically organized motor representations in rats (Halley et al., 2020), tree shrews (Remple et al., 2006; Baldwin et al., 2017a), galagos (Wu et al., 2000), and anthropoids suggest that it—or at least most of it—is a primate specialization. The absence of a clear M1 hindlimb representation in tree shrews is the key finding in this regard because, among studied species, they are most closely related to primates. The hindlimb representation in M1 is either absent entirely or very small (Remple et al., 2006; Baldwin et al., 2017a), so a tree shrew–galago comparison suggests that the latter's hindlimb representation is a specialization of M1 in primates. This finding contrasts with the primary somatosensory cortex (S1), which is homologous with area 3b in anthropoids. It is easy to demonstrate a sizable hindlimb representation in the S1 cortex of tree shrews with both tactile receptive-field mapping (Sur et al., 1980) and intracortical microstimulation (Baldwin et al., 2017a). An alternative idea is that M1 and S1 share an overlapping, amalgamated hindlimb representation, but Halley et al. (2020) have argued convincingly against that view.

The selective factors that led to cortical enlargement in Eocene primates involved adaptations for exploiting the terminal branches of angiosperm trees in the rainforests that thrived throughout the Northern Hemisphere at the time. All primates were small, arboreal mammals then, with forward-facing eyes, relatively long fingers and toes, finger- and toenails instead of claws, modifications of Meissner's corpuscles that enhanced tactile sensitivity, a leaping-grasping form of locomotion dominated by elongated hindlimbs, and a distinctive gait suited to movements along an unstable and discontinuous substrate of fine branches. The rainforest canopy, under which primates first evolved, necessitated foraging in dim light, even in daytime. In addition to retinal specializations, new visual and premotor areas guided reaching, leaping, and grasping. Earlier, we mentioned that primate brains expanded during the late Eocene. Most likely, this expansion involved primate-specific cortical areas, which drove an increase in encephalization and corticalization sometime after ~40 million years ago (mya) (Kirk et al., 2014). New parts of posterior parietal cortex (PPC) provided the premotor cortex with information about visual metrics: locations, speeds, directions, and so on, which guided movements. A new (or vastly enlarged) M1 hindlimb representation contributed

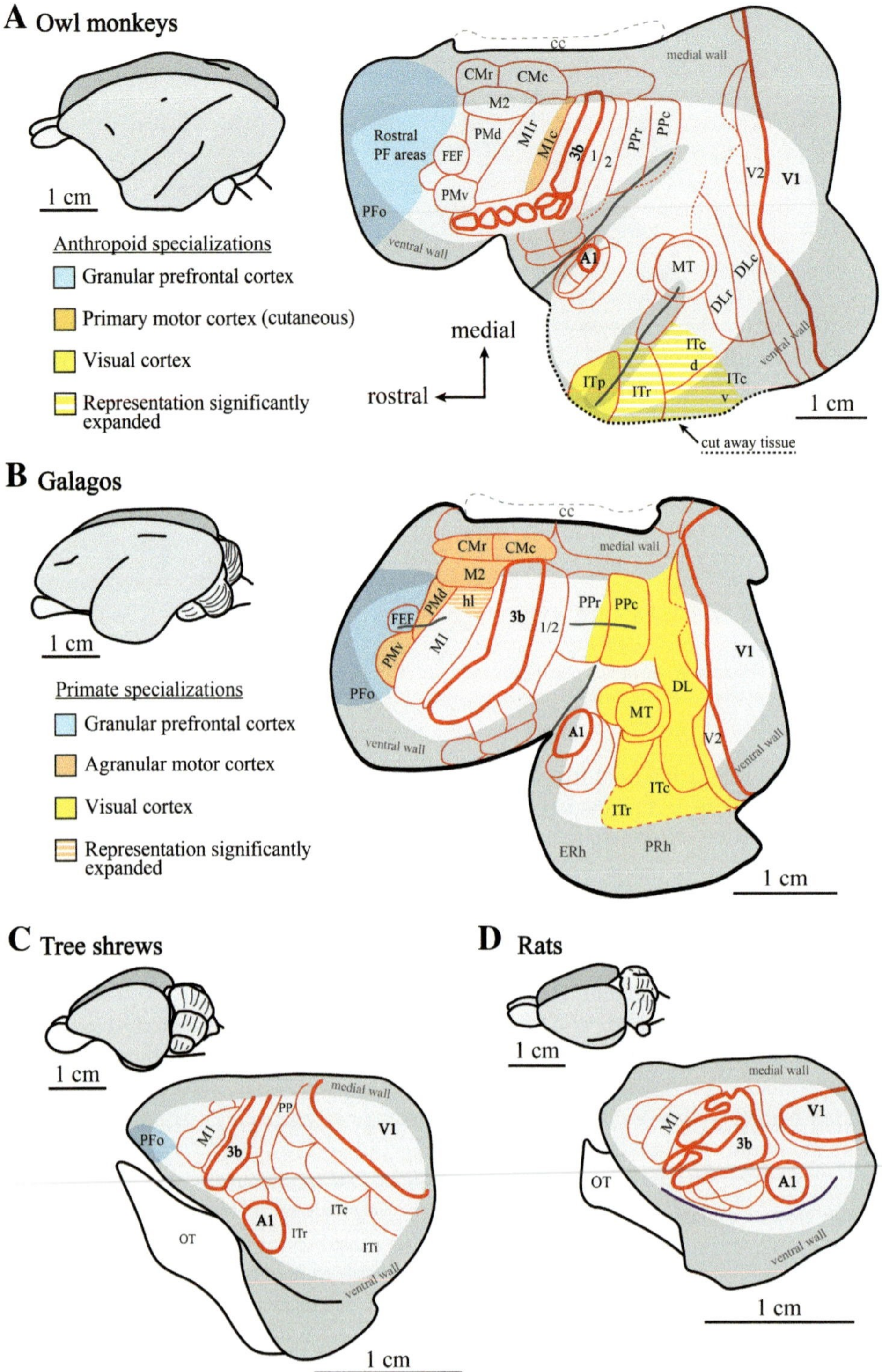

Figure 57.2. Cortical maps in owl monkeys (A), galagos (B), tree shrews (C), and rats (D). Areas with colored shading in panel A are anthropoid specializations; those in panel B are primate specializations. The hatching in panel B recognizes the possibility that a small hindlimb M1 representation might have been present before the emergence of primates. Yellow hatching in panel A indicates that although these visual fields may have been present before the emergence of anthropoids, they have expanded considerably. Note that most of the ventral wall of temporal cortex, including PRh and ERh, are cut away in panel A. Panels C and D contrast with panels A and B in several respects, most notably in the size and complexity of the PFC, premotor cortex, and visual areas. Adapted from Kaas (2008). For a key to abbreviations that appear in the figure, see Abbreviations in the front matter.

to pedal grasping, which aided predator avoidance, leaping-grasping locomotion more generally, and postural stability during foraging. New, primate-specific parts of the PFC, including granular components of the orbital prefrontal (PFo) cortex and the frontal eye field (FEF), improved vision-based foraging in dim illumination. PFo cortex represented valuations of foraging options among hidden or poorly seen food items, while FEF tracked desirable items (and associated visual patterns) via connections with both the dorsal and ventral visual streams. These cortical functions provided crucial advantages in foraging amidst a dense array of items—some nutritious, but many worthless—in the terminal-branch milieu.

We return to the PFo cortex later, but note here its likely emergence in Euarchontans that predate primates, as reflected in the cortical map of tree shrews (Figure 57.2C). Although Wong and Kaas (2009) described a small granular region in the orbitofrontal cortex of tree shrews, which suggests that part of PFo cortex emerged in ancestral Euarchontans, the PFo cortex is larger and more complex in primates than in tree shrews, which indicates that most of it is a primate specialization.

MIOCENE

Anthropoids had evolved by the early Eocene, but they remained small (less than 1 kg) throughout the Eocene and much of the Oligocene, with brains in a range of encephalization and corticalization typical of modern prosimians (or smaller). For example, the Miocene cercopithecoid *Victoriapithecus*, which lived ~15 mya, had a prosimian-size brain (relative to body size), if that. Later, during the middle-to-late Miocene, the cortex expanded independently in both platyrrhines (New World monkeys) and cercopithecoids (Old World monkeys) (Gonzales et al., 2015). In both lineages, the cortex enlarged long after their other characteristic traits evolved, including the fovea, 1–10 kg bodies, and quadrupedal locomotion. At about the same time as the cortex enlarged, a faster mode of locomotion also evolved independently in platyrrhines and cercopithecoids (Almécija et al., 2019). Both developments probably occurred as rainforests contracted during a period of global cooling ~14–9 mya, which decreased moisture in the atmosphere and soil and thereby stressed rainforest habitats.

In hominoids, cortical enlargement (red asterisk in Figure 57.1) occurred somewhat earlier than in platyrrhines and cercopithecoids. Encephalization and corticalization increased as hominoids encountered new, large, and speedy predators during the early Miocene. About 20 mya, these anthropoids became larger (1–10 kg) (Cachel, 2015), and by ~18 mya the Miocene hominoid *Proconsul* had a brain in the modern hominoid size range (Gonzales et al., 2015). At about the same time, hominoids evolved a novel, more upright form of locomotion, which promoted rapid movements over long distances (Almécija et al., 2019). Then, from ~17–15 mya, their descendants diversified and migrated from Africa to Asia. These species resembled contemporary cercopithecoids in size and agility but had a much larger cortex. Despite their massive cortex, most hominoid species died out during the global cooling of ~14–9 mya (Cachel, 2015). Survivors of the late Miocene extinctions included the large hominids (great apes and humans) we know today, which weigh 10s to 100s of kg. But it was in the earlier, smaller, more agile, cercopithecoid-like hominoids that the cortex expanded, most likely in an ancestor shared by gibbons, great apes, and humans.

Several new parts of the granular PFC evolved in anthropoids (Preuss and Goldman-Rakic, 1991; Preuss and Wise, 2021), and the frontal lobe enlarged. Fossil evidence shows that frontal-lobe expansion occurred independently in hominoids, cercopithecoids, and platyrrhines (Gonzales et al., 2015). As the PFC expanded, the olfactory bulbs contracted, also independently, in these 3 anthropoid groups. Thus, there was a size trade-off between the PFC and the olfactory bulbs, indicating an increased reliance on PFC functions at the expense of olfaction (Wise, 2023). This trade-off probably resulted from the role of new anthropoid-specific parts of the PFC in guiding long-distance foraging choices based largely on vision. These areas represent behavioral goals and the means to achieve them, and in Miocene anthropoids they reduced the number of unproductive foraging journeys by speeding learning about where and when to forage (Wise, 2017). Because of predation risks, any decrease in poor foraging choices enhanced fitness.

In addition to new anthropoid-specific parts of the PFC, the frontal cortex changed in other ways. Premotor areas functioned in association with new or expanded parts of the PPC to represent visual affordances: the actions available based on what an individual can see. The evolution of these action maps is discussed elsewhere in this volume (see **Chapter 59 in Section X**). Another change involved the caudal part of M1 (M1c), which developed 2 novel features: a dominance of cutaneous inputs (Tanji and Wise, 1981; Strick and Preston, 1982) and direct monosynaptic connections with spinal motor neurons (Rathelot and Strick, 2009; see also Bortoff and Strick, 1993). These specializations contributed to the control of fine finger movements, which anthropoids use to manipulate fruit in order to evaluate ripeness, among other functions.

PLEISTOCENE

Although hominins split from the chimpanzee-bonobo lineage ~7–5 mya, increases in encephalization and corticalization remained modest for several million years. In early hominins, neither these measures nor sulcal patterns differed appreciably from the ancestral condition or from the brains of modern chimpanzees (Gunz et al., 2020). Then, from ~3–2 mya, encephalization accelerated in several *Homo* species, possibly in association with tool manufacture or social complexity (Püschel et al., 2021). During early *Homo sapiens* evolution, ~315,000 years ago, the frontal squama (forehead) bulged upward and forward, giving the anterior braincase its rounded, bulbous shape. However, the brain remained more elongated overall, and only in later *Homo sapiens*, ~130,000 years ago, did it reach its modern, globular shape and size (Bruner et al., 2013; Scerri et al., 2018). Enlargement of homotypical association areas drove this expansion, including the PFC, PPC, and much of the temporal cortex (Donahue et al., 2018). PFC expansion exceeded that of the PPC and temporal homotypical areas by a small but significant amount (Smaers et al., 2017), as the granular PFC, and especially the lateral frontal pole cortex (area 10), became more prominent (Semendeferi et al., 2001). Eventually, the granular PFC cortex came to occupy ~80% of the human frontal lobe (Elston et al., 2006).

As new homotypical association areas emerged in anthropoids and expanded in hominins, they contributed to several transcortical networks. Among them, the default-mode network includes the medial PFC and the hippocampus. The latter contracted during most of anthropoid evolution, but during human evolution this phylogenetic trend reversed and it expanded dramatically (Schilder et al., 2019; Vanier et al., 2019), which indicates that the hippocampus became more important to humans than to our anthropoid ancestors or to our closest relatives, including other hominids (great apes). Likely contributors to this expansion include species-specific representations of oneself and conspecifics, which originate in the medial PFC and contribute to social cognition and to the sense of participation in events that characterizes human episodic memory (Murray et al., 2017, 2020). Other transcortical networks also changed during human evolution. Of special significance, there were augmentations of PFC-temporal lobe connections as the arcuate fascicle extended to encompass more of the temporal lobe, possibly in relation to language (Roumazeilles et al., 2020). A forthcoming book (Wise, 2023) documents and elaborates on the points made thus far.

THE SEARCH FOR THE FRONTAL EYE FIELD IN TREE SHREWS

Studies of the FEF are especially instructive for understanding PFC evolution. Earlier, we noted that the FEF is a primate specialization, which might surprise readers who know about similar areas in rodents or cats. However, similarities can arise independently in evolution, through convergent evolution or other forms of homoplasy.

As noted earlier, most of the primate-specific parts of PFC are characterized by a distinctive inner granular layer, L4, and in galagos, macaques, and several platyrrhine species the FEF has this characteristic. Furthermore, careful investigation of the frontal cortex using intracortical microstimulation reveals that tree shrews lack a functionally defined FEF, that is an area in frontal cortex where eye movements can be elicited (Remple et al., 2006; Baldwin et al., 2017a). If the FEF were homologous in rats, cats, and primates, then it should exist in tree shrews, which are more closely related to primates than either rats or cats are. The absence of a functional FEF in tree shrews has 1 of 2 implications: (1) either tree shrews lost the FEF during evolution, or (2) the functionally defined FEF-like areas in rats and cats evolved convergently. The first possibility is unlikely because the FEF is a visual area and tree shrews have a well-developed visual system, including a large and complex set of visual areas in the cortex (Lyon et al., 1998). Therefore, what is called the FEF in rats, cats, and primates probably evolved convergently, although a broader comparative analysis is needed to confirm this conclusion.

THALAMIC EVOLUTION RELATED TO PREFRONTAL CORTEX

The discussion to this point has relied on both fossil evidence and comparative studies of extant species. Unfortunately, the former reveals almost nothing about internal brain structures, such as the thalamus. Accordingly, our understanding of thalamic evolution depends almost entirely on comparisons among modern species. Two nuclei are of special significance to the PFC: the mediodorsal nucleus (MD) and the pulvinar (Pul), which are the largest thalamic nuclei in humans. Both enlarge as brains increase in size, but the trajectory of the pulvinar's expansion is steepest in hominoids (Armstrong, 1982; see **Chapter 56 in Section X** of this volume). In addition, new subdivisions of both nuclei have emerged during primate evolution.

All studied mammals have a homolog of the inferior and lateral pulvinar, which are associated with vision (see Baldwin et al., 2017b, for review), but clear evidence for a homolog of the medial pulvinar (PulM) is lacking in nonprimate species (Preuss, 2007a, 2007b; Homman-Ludiye and Bourne, 2019). These findings indicate that PulM is a primate specialization, which augmented the thalamus in an ancestor common to galagos and anthropoids. (Later, we discuss some possible differences between galagos and anthropoids in PulM's frontal connections.)

Understanding the evolution of MD presents some unique challenges. This nucleus, which all placental mammals (and probably all mammals) share, is usually assumed to be mostly conserved across species. Other than the variation in cell composition and architecture that is typically observed in brains of different size, there seem to be few changes in its basic cytoarchitectonic appearance (Jones, 2007). In most species, MD has 2 main divisions based on cell size and packing density: a medial magnocellular part (MDmc) and a central and lateral aspect. In primates, the latter includes the parvocellular MD (MDpc) and, at its lateral extreme, the multiform MD (MDmf). In rodents, such as rats, MD is often divided into medial, central, and lateral subdivisions, with the latter sometimes split into dorsal and ventral parts. Further subdivisions have been described in both rodents and primates (for reviews, see Jones, 2007; Mitchell, 2015; Georgescu et al., 2020), but the cytoarchitectonic differences among MD's subnuclei are less distinct than their labels imply, and localizing borders is inconsistent among studies, even when using similar methods (Armstrong, 1982; Jones, 2007).

The cross-species consistency of MD's architecture contrasts with other thalamic nuclei, such as the lateral geniculate nucleus (LGN) and the visual pulvinar (i.e., the inferior and lateral pulvinar). Across species, the LGN and the visual pulvinar have obvious differences in orientation, cytoarchitectonics, cell types, physiological characteristics, and (for LGN) lamination patterns. These and other differences are evident even among closely related species (Kaas et al., 1978; Kahn and Krubitzer, 2002; Baldwin et al., 2017b), making it clear that these nuclei have changed during evolution. In contrast, MD has much the same cytoarchitectonic appearance in a wide diversity of mammals. However, this architectonic consistency masks extensive remodeling during primate evolution, as subsequent sections explain.

SUBCORTICAL AND CORTICO-THALAMOCORTICAL PATHWAYS

The thalamus influences cortical processing by providing subcortical inputs from the sensory periphery or other brain structures (see **Section IV** of this volume), a transthalamic route for interaction among cortical areas (see **Section VII**), or both. MDmc receives dense subcortical inputs: from the amygdala, globus pallidus, and deep cerebellar nuclei; and it also receives sparse inputs from the locus coeruleus, ventral tegmental area, substantia nigra, dorsal raphé nucleus, ventral pallidum, and substantia innominata (Russchen et al., 1987). In contrast, both PulM and MDpc receive only sparse subcortical inputs: PulM from the amygdala, inferior and superior colliculus, and nucleus of the optic tract (Benevento and Standage, 1983; Homman-Ludiye and Bourne, 2019); MDpc from the zona inserta, superior colliculus, periaqueductal gray, locus coeruleus, ventral tegmental area, and substantia nigra (Benevento and Fallon, 1975; Erickson, 2004; Mitchell, 2015; Georgescu et al., 2020).

The predominance of subcortical inputs to MDmc is evident in studies of VGLUT2 (vesicular glutamate transporter type 2) immunoreactivity: a hallmark of subcortical excitatory input. In macaques, the medial and central aspects of MD, including MDmc, receive large VGLUT2-positive terminal boutons, like sensory and cerebellar relay nuclei (Rovo et al., 2012). MDmf, which has reciprocal connections with the FEF, also receives VGLUT2-positive subcortical inputs, mostly from the superior colliculus. In contrast, the main body of MDpc and PulM have few, if any, VGLUT2-positive terminals. Thus, while MDmc and MDmf mostly relay subcortical signals to the cortex, both MDpc and PulM are more likely to influence cortical activity via cross-area transthalamic routes.

Very little is known about the influence of subcortical inputs to PulM or lateral MD on their cortical targets. Their thalamocortical projections terminate mainly in L1, L3, and L4, with MD also projecting to L5 (Giguere and Goldman-Rakic, 1988; Romanski et al., 1997; McFarland and Haber, 2002; Homman-Ludiye and Bourne, 2019). Some evidence suggests that the FEF receives inputs about pending saccades from the superior colliculus via MDmf (Sommer and Wurtz, 2004), but little is known about other parts of the PFC. In contrast, there is a long history of studies of subcortical inputs to MDmc and its influence on cortical targets (see the section below entitled "Mediodorsal Nucleus").

Figure 57.3 presents a synthesis of studies in both marmosets and macaques. The figure caption explains the pros and cons of combining data from 2 distantly related anthropoids, but this approach provides a compact, cortex-wide view of corticothalamic projections to PulM (Figure 57.3A) and MD (Figure 57.3B). Both nuclei receive signals from several cortical fields within the frontal, parietal, and temporal cortex, many of which are primate-specific, and some of which are anthropoid-specific. Not all the illustrated connections are reciprocal; more areas project to PulM and MD than receive input from them. Overall, this architecture suggests a role in integrating information across cortical areas. Perry et al. (2023) have suggested that MDpc mediates interactions between frontal and temporal cortex, whereas PulM performs a similar function for the frontal-parietal integration, although the anatomy illustrated in Figure 57.3 suggests more widespread and complex influences.

The corticothalamic projections to MDpc and PulM seem similar when described in terms of general connectivity, but an analysis of VGLUT1 immunoreactivity reveals differences. In macaques, much of PulM contains small VGLUT1-positive terminals (Rovo et al., 2012), which are typical of inputs originating from corticothalamic cells in L6. In accord with this finding, retrograde axonal

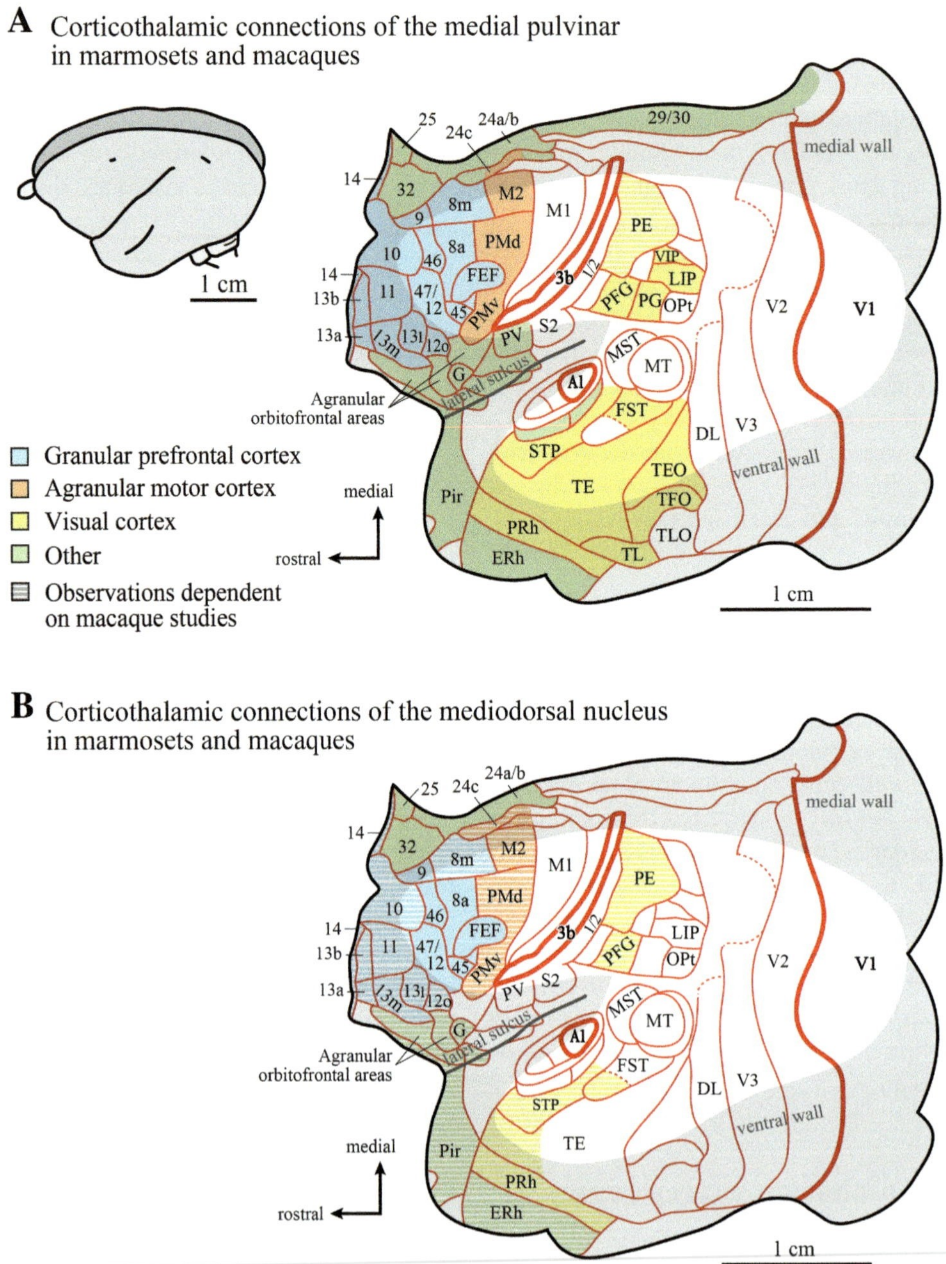

Figure 57.3. Correspondence of primate- and anthropoid-specific cortical areas with PulM and MD connections. Areas with colored shading project to PulM (A) or MD (B), with the color of the shading matching the cortical maps in Figure 57.2A and B. We have plotted data from both marmosets and macaques onto a standard cortical map of marmosets and have used hatched shading to acknowledge that: (1) homology has not been fully established across many of the similarly named cortical fields in marmosets and macaques; and (2) not all the illustrated connections have been described in both. We recognize the downside of combining data from 2 distantly related primates, which have been evolving separately for more than 30 million years and have their own specializations. However, the figure provides a compact summary of the basic pattern of connections. Note that the connections indicated by hatched fills have only been reported for macaques. This convention indicates a lack of knowledge about such connections in marmosets, not necessarily an absence of these connections. Adapted from Homman-Ludiye et al. (2020).

tracing methods in both macaques and marmosets indicate that most cortical projections to PulM originate from L6 (Romanski et al., 1997; Homman-Ludiye et al., 2020), with some temporal areas sending additional projections from L5 (Homman-Ludiye et al., 2020). According to current thinking, corticothalamic inputs from L6 modulate activity within the thalamus, whereas those from L5 drive neuronal activity, much like subcortical inputs do in sensory and cerebellar relay nuclei (Sherman and Guillery, 2002). However, because of their predominance in PulM, L6 corticothalamic inputs likely control its neuronal activity despite the small size of their glutamatergic boutons and notwithstanding the general principle that L6 inputs have modulatory functions.

In contrast to PulM, which has small VGLUT1-positive boutons, MDpc contains large ones (Rovo et al., 2012), which are usually characteristic of L5 corticothalamic inputs. However, only 20% of the corticothalamic cells projecting to MDpc originate from L5, with 80% arising from L6 (Xiao et al., 2009). The preponderance of large VGLUT1-positive boutons in MDpc supports the idea that corticothalamic inputs drive its neuronal activity, but it seems unlikely that they all come from L5 neurons. We can only speculate about this discrepancy, but perhaps large boutons arise from L6 in this instance.

It seems clear that MDpc and PulM emerged in Euarchontans after their divergence from Glires, but more precise conclusions require future research. Unfortunately, our knowledge of MD and PulM anatomy depends almost entirely on studies of rodents and anthropoids. Such comparisons often lead to erroneous evolutionary inferences because of the large gap between these 2 groups. Rodents and primates have been evolving independently since the Mesozoic, probably for more than 70 million years (Figure 57.1), and it would be valuable to know more about species that diverged from anthropoids more recently. New and improved anatomical knowledge that "bridges" the rodent-anthropoid gap, which future studies of galagos and tree shrews can provide, would strengthen the understanding of thalamic evolution significantly.

MEDIAL PULVINAR

As mentioned earlier, Figure 57.3A summarizes connectional data from marmosets and macaques. This part of the figure is based in large part on a recent, comprehensive study of cortical inputs to PulM in common marmosets (Homman-Ludiye et al., 2020). In this species, PulM receives connections from the PFC and other parts of the cortex that have also emerged and expanded in primates or anthropoids, such as the PPC and temporal cortex. Few comprehensive studies have been conducted on PulM connections in primates other than marmosets, but studies of frontal, parietal, and temporal cortex in macaques, owl monkeys, and squirrel monkeys show similar patterns of connectivity (Romanski et al., 1997; Gutierrez et al., 2000; Morel et al., 2005; Gharbawie et al., 2010).

In galagos, connections of PulM with FEF have been reported (Stepniewska et al., 2018), in accord with the idea that the primate FEF emerged early in primate evolution. However, its connections with the PPC and many parts of frontal cortex are weak, if present at all (Fang et al., 2006; Wang et al., 2023). These findings suggest that PulM's connections and functions differ in anthropoids and prosimians, especially in relation to parietofrontal networks. However, for firmer conclusions, future studies are needed, especially in galagos or other strepsirrhines.

Because PulM shares connections with cortical fields involved in visuomotor and auditory processing, as well as executive functions, the suggested roles of PulM have been diverse, including integrating polymodal information, selective attention, working memory, and other cognitive functions (see Homman-Ludiye and Bourne, 2019, for review). However, PulM's mechanisms and precise functions remain poorly understood.

Part of the difficulty in understanding PulM's function—and exactly how PulM influences the cortex—may be due to a lack of knowledge about its internal organization. The architecture of the medial pulvinar is mostly homogeneous. Some reports in macaques have indicated differences in myelin, acetylcholine esterase (AChE), and parvalbumin staining patterns between the medial and lateral aspects of PulM, with such variations correlating with differences in patterns of cortical afferents (Romanski et al., 1997; Gutierrez et al., 2000) and physiological properties (see **Chapter 71 in Section XII**). However, neither the connections nor the histochemical markers have clear-cut borders. Instead, there are gradual transitions from medial to lateral PulM, which are inconsistent among anthropoids. For instance, mediolateral differences observed in macaques are not as clear in marmosets (Homman-Ludiye et al., 2020). These inconsistencies might be a byproduct of independent pulvinar and cortical expansion in different anthropoid lineages. The fact that the frontal lobes expanded independently in hominoids, platyrrhines, and cercopithecoids increases the likelihood of differences among these 3 anthropoid groups in frontal cortex-thalamus interactions. Another proposal about the internal structure of PulM is that it has a mosaic organization, with each microunit receiving input from multiple cortical sources (Cappe et al., 2009; Homman-Ludiye et al., 2020). Cortical projections to PulM are not focused in a way that fosters the identification of topographic organization. Instead, they have a diffuse projection profile within the nucleus. Some terminal fields of single corticothalamic cells in macaques span 1–3 mm (Rockland, 1996), and there are large interneurons unique to PulM, which have been proposed to connect these microunits (Imura and Rockland, 2006). These aspects of PulM anatomy make topographical organization within PulM difficult to discern, but do not exclude it. For instance, there is evidence that auditory areas in the superior temporal sulcus project topographically to PulM, with a medial-to-lateral distribution (Hackett et al., 1998, 2007).

Given that subcortical inputs to PulM are relatively sparse—and that cortical inputs arise from many areas yet consist of small excitatory boutons—it seems most probable that activation of PulM cells requires a convergence of inputs from multiple cortical sources. Cortical areas that project to PulM typically have corticocortical connections with each other as well, which should mutually support corticothalamic activations of PulM neurons. One intriguing finding is that PulM sends a projection to the lateral nucleus of the amygdala (Jones and Burton, 1976). Given the architecture of corticothalamic inputs to PulM, it seems likely that it requires concerted activity in several, widely distributed cortical areas to activate this thalamo-amygdala pathway.

MEDIODORSAL NUCLEUS

For primates, it has long been known that MDmc is connected with both granular and agranular orbitofrontal areas, and MDpc is connected with most of the remainder of granular PFC. Here we propose a reinterpretation of the comparative analysis of Ray and Price (1993), which Figure 57.4 illustrates.

In rats, Ray and Price found that the lateral part of MD projects to a part of the cortex that goes by many names, including RMA (rostral motor area), PrCm (medial "precentral" area), and AGm (the medial agranular area).

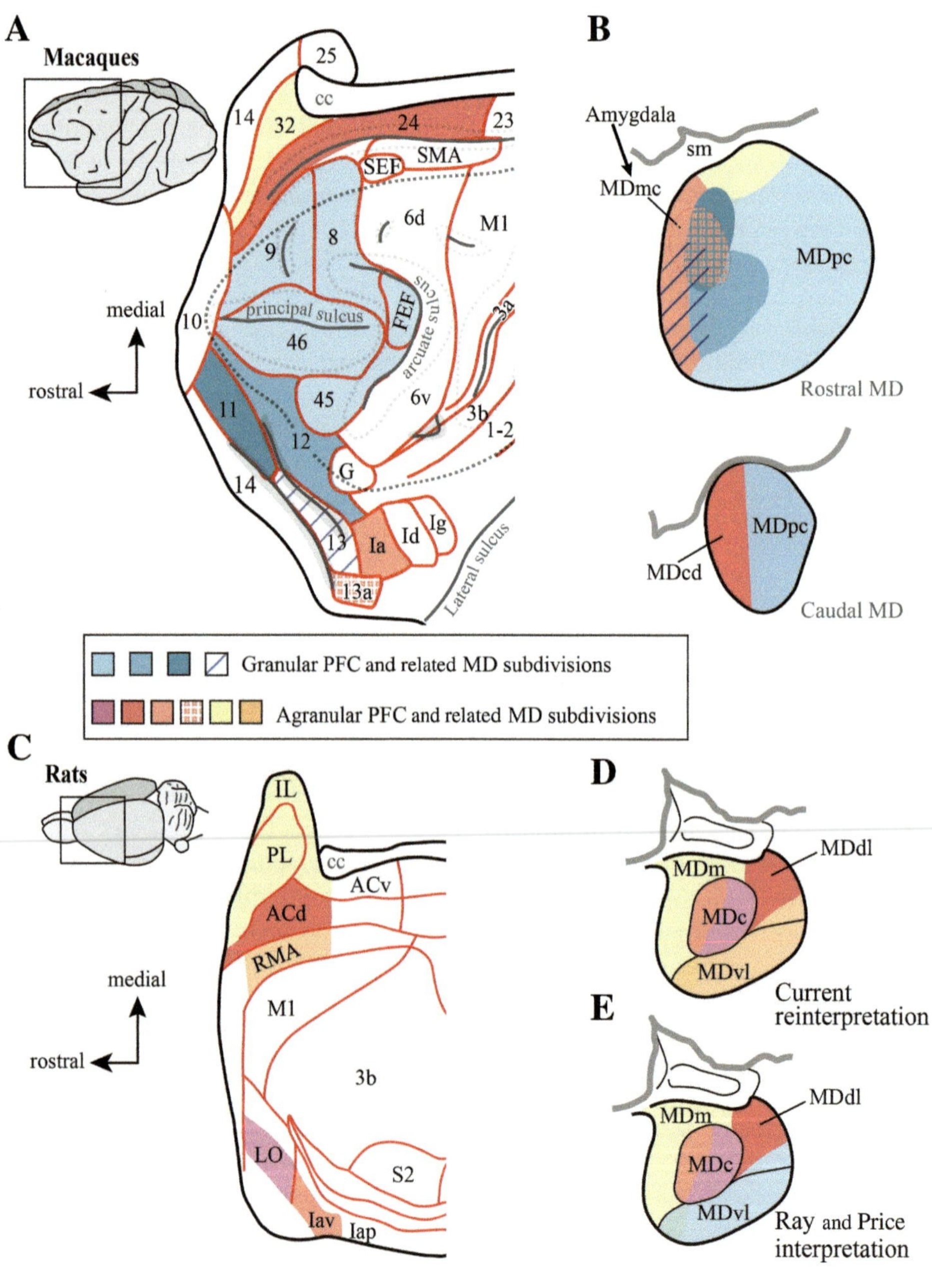

Figure 57.4. A reinterpretation of Ray and Price (1993) for MD's thalamocortical connections in rats and macaques. Primate-specific parts of the PFC and the connected subdivisions of MD are shaded in cool colors (plus an area indicated by blue hatching, for reasons explained in the text). Warm colors mark parts of the thalamus and cortex shared by rats and macaques. In panels B, D, and E, medial is to the left and dorsal is up.

They considered the lateral part of MD in rats (MDvl in Figure 57.4E) to be homologous with MDpc in macaques (Figure 57.4B). Likewise, they considered RMA to be homologous with the granular PFC in macaques (Figure 57.4A).

However, the well-documented corticospinal projections that emanate from RMA show that it is not a prefrontal area (Wise and Donoghue, 1986). No part of the PFC in anthropoids projects directly to the spinal cord. Accordingly, what Ray and Price called MDvl in rats is not homologous to MDpc in macaques. Instead, MDpc is now understood to be a primate specialization, like the granular PFC to which it is connected. (The same goes for MDmf, although it is not illustrated in Figure 57.4B.) The dorsomedial and medial parts of MDmc in macaques (warm colors in Figure 57.4B) have similar cortical connections to the whole MD in rats (Figure 57.4D): mainly with agranular parts of PFC, such as the anterior cingulate cortex (part of area 24 in macaques), prelimbic cortex (area 32), infralimbic cortex (area 25), and agranular insular areas. Medial parts of MD also have connections with the granular PFC, but it is their connections with agranular areas that distinguish them from MDpc and MDmf.

While it is clear that MD expanded laterally during primate evolution (Giarrocco and Averbeck, 2023), there were also significant changes in the parts of MD that are homologous in primates and other mammals: MDmc in primates. Amygdala nuclei that project to MDmc in primates and its homolog (MDm) in rats, such as the basolateral and accessory basal nuclei, are more than 30-fold larger in macaques than in rats, with 10–15-fold more neurons. In contrast, other parts of the amygdala, the central and medial nuclei, are less than 10-fold larger, with only 2–4-fold more neurons (Chareyron et al., 2011).

In summary, the major differences in MD between rats and macaques reflect: (1) the addition of MDpc (or, at least, most of it) and MDmf during primate evolution (Figure 57.4B); and (2) the development of connections between medial parts of MD and the granular PFC.

A listing of possible MD functions is available in reviews by Mitchell and her colleagues (Mitchell, 2015; Perry et al., 2023), but here we focus on a specific example in macaques. Izquierdo and Murray (2010) showed that MDmc lesions in one hemisphere, combined with lesions of the amygdala and granular orbitofrontal (PFo) cortex in the other hemisphere, blocked the updating of a specific type of valuation. In these experiments, monkeys first learn that given object is associated with a specific kind of food. Then, in testing, they choose between 2 objects that block any view of the food items, and they receive the food associated with the object that they choose. After bilateral lesions of MDmc (Mitchell et al., 2007) or the MDmc lesions that Izquierdo and Murray made, monkeys fail to update the valuation of an object-food pair as they consume that kind of food to satiety. So, when "foraging" for hidden food items via object associations, they make poor choices and receive the sated (and therefore undesirable) food. However, when choosing between foods they can see, they make good choices. As explained above, MDmc receives glutamatergic input from the basolateral amygdala and projects to the granular PFo cortex. The impairments caused by MDmc lesions show that an amygdala→MDmc→granular PFo pathway is necessary for updating the appropriate valuations. Other pathways between the amygdala and PFo cortex—such as the direct amygdalocortical projections and routes via other cortical areas—are insufficient to support the updating function. Follow-up experiments show that area 13 (marked by hatched lines in Figure 57.4A, lower left) is the key area within the granular PFo cortex (Murray et al., 2015).

CONCLUSIONS

PulM and most of the lateral MD (MDpc and MDmf) are primate specializations, as are the granular prefrontal areas with which they are connected. When the granular PFC emerged during early primate evolution and new parts of PFC evolved in anthropoids, MDpc and PulM expanded in parallel. Additional evolutionary changes occurred elsewhere in the frontal cortex; most of the premotor areas and most (if not all) of the hindlimb representation in M1 also emerged during primate evolution, most likely in early primates.

MD and PulM function both in relaying information from subcortex to cortex and in transcortical integration. These modes of organization coexist, and natural selection operates on both. MDmc relays information from a subcortical structure, the amygdala, to primate-specific parts of PFo cortex, which is a phylogenetically new output from medial MD. PulM and MDpc function differently. Their thalamocortical cells are driven by inputs from a large suite of cortical areas—many of which are primate- or anthropoid-specific—which also have connections with each other. Although the precise functions of PulM and MDpc remain unknown, they probably play a role in integrating information across sensory modalities and cognitive domains, and their expansion during human evolution points to selection for this mode of organization and the cognitive functions it supports.

REFERENCES

Almécija S, Tallman M, Sallam HM, Fleagle JG, Hammond AS, Seiffert ER (2019). Early anthropoid femora reveal divergent adaptive trajectories in catarrhine hindlimb evolution. *Nat Commun.* 10:4778.

Armstrong E (1982). Mosaic evolution in the primate brain: Differences and similarities in the hominoid thalamus. In: Armstrong E, Falk D, eds. *Primate Brain Evolution*. Springer:131–161.

Balaram P, Kaas JH (2014). Towards a unified scheme of cortical lamination for primary visual cortex across primates: Insights from NeuN and VGLUT2 immunoreactivity. *Front Neuroanat.* 8:81.

Baldwin MKL, Cooke DF, Krubitzer L (2017a). Intracortical microstimulation maps of motor, somatosensory, and posterior parietal cortex in tree shrews (*Tupaia belangeri*) reveal complex movement representations. *Cereb Cortex.* 27:1439–1456.

Baldwin MKL, Balaram P, Kaas JH (2017b). The evolution and functions of nuclei of the visual pulvinar in primates. *J Comp Neurol.* 525:3207–3226.

Benevento LA, Fallon JH (1975). The ascending projections of the superior colliculus in the rhesus monkey (*Macaca mulatta*). *J Comp Neurol.* 160:339–361.

Benevento LA, Standage GP (1983). The organization of projections of the retinorecipient and non-retinorecipient nuclei of the pretectal complex and layers of the superior colliculus to the lateral pulvinar and medial pulvinar in the macaque monkey. *J Comp Neurol.* 217:307–336.

Bortoff GA, Strick PL (1993). Corticospinal terminations in two new-world primates: Further evidence that corticomotoneuronal connections provide part of the neural substrate for manual dexterity. *J Neurosci.* 13:5105–5118.

Bruner E, Athreya S, de la Cuetara JM, Marks T (2013). Geometric variation of the frontal squama in the genus *Homo*: Frontal bulging and the origin of modern human morphology. *Am J Phys Anthropol.* 150:313–323.

Cachel S (2015). *Fossil Primates*. Cambridge University Press.

Cappe C, Morel A, Barone P, Rouiller EM (2009). The thalamocortical projection systems in primate: An anatomical support for multisensory and sensorimotor interplay. *Cereb Cortex.* 19:2025–2037.

Chareyron LJ, Banta Lavenex P, Amaral DG, Lavenex P (2011). Stereological analysis of the rat and monkey amygdala. *J Comp Neurol.* 519:3218–3239.

Donahue CJ, Glasser MF, Preuss TM, Rilling JK, Van Essen DC (2018). Quantitative assessment of prefrontal cortex in humans relative to nonhuman primates. *Proc Natl Acad Sci U S A.* 115:E5183–E5192.

Elston GN, Benavides-Piccione R, Elston A, Zietsch B, DeFelipe J, Manger P, Casagrande V, Kaas JH (2006). Specializations of the granular prefrontal cortex of primates: Implications for cognitive processing. *Anat Rec A Discov Mol Cell Evol Biol.* 288:26–35.

Erickson SL (2004). Subcortical afferents to the lateral mediodorsal thalamus in cynomolgus monkeys. *Neuroscience.* 129:675–690.

Fang PC, Stepniewska I, Kaas JH (2006). The thalamic connections of motor, premotor, and prefrontal areas of cortex in a prosimian primate *(Otolemur garnetti)*. *Neuroscience.* 143:987–1020.

Georgescu IA, Popa D, Zagrean L (2020). The anatomical and functional heterogeneity of the mediodorsal thalamus. *Brain Sci.* 10:624.

Gharbawie OA, Stepniewska I, Burish MJ, Kaas JH (2010). Thalamocortical connections of functional zones in posterior parietal cortex and frontal cortex motor regions in New World monkeys. *Cereb Cortex.* 20:2391–2410.

Giarrocco F, Averbeck BB (2023). Anatomical organization of forebrain circuits in the primate. *Brain Struct Funct.* 228:393–411.

Giguere M, Goldman-Rakic PS (1988). Mediodorsal nucleus: Areal, laminar, and tangential distribution of afferents and efferents in the frontal lobe of rhesus monkeys. *J Comp Neurol.* 277:195–213.

Gonzales LA, Benefit BR, McCrossin ML, Spoor F (2015). Cerebral complexity preceded enlarged brain size and reduced olfactory bulbs in Old World monkeys. *Nat Commun.* 6:7580.

Gunz P, Neubauer S, Falk D, Tafforeau P, Le Cabek A, Smith TM, Kimbel WH, Spoor F, Alemseged Z (2020). *Australopithecus afarensis* endocasts suggest ape-like brain organization and prolonged brain growth. *Sci Adv.* 6:eaaz4729.

Gutierrez C, Cola MG, Seltzer B, Cusick C (2000). Neurochemical and connectional organization of the dorsal pulvinar complex in monkeys. *J Comp Neurol.* 419:61–86.

Hackett TA, De La Mothe LA, Ulbert I, Karmos G, Smiley J, Schroeder CE (2007). Multisensory convergence in auditory cortex. II. Thalamocortical connections of the caudal superior temporal plane. *J Comp Neurol.* 502:924–952.

Hackett TA, Stepniewska I, Kaas JH (1998). Thalamocortical connections of the parabelt auditory cortex in macaque monkeys. *J Comp Neurol.* 400:271–286.

Halley AC, Baldwin MKL, Cooke DF, Englund M, Krubitzer L (2020). Distributed motor control of limb movements in rat motor and somatosensory cortex: The sensorimotor amalgam revisited. *Cereb Cortex.* 30:6296–6312.

Homman-Ludiye J, Bourne JA (2019). The medial pulvinar: Function, origin and association with neurodevelopmental disorders. *J Anat.* 235:507–520.

Homman-Ludiye J, Mundinano IC, Kwan WC, Bourne JA (2020). Extensive connectivity between the medial pulvinar and the cortex revealed in the marmoset monkey. *Cereb Cortex.* 30:1797–1812.

Imura K, Rockland KS (2006). Long-range interneurons within the medial pulvinar nucleus of macaque monkeys. *J Comp Neurol.* 498:649–666.

Izquierdo A, Murray EA (2010). Functional interaction of medial mediodorsal thalamic nucleus but not nucleus accumbens with amygdala and orbital prefrontal cortex is essential for adaptive response selection after reinforcer devaluation. *J Neurosci.* 30:661–669.

Jones EG (2007). *The Thalamus*. 2nd ed. Cambridge University Press.

Jones EG, Burton H (1976). A projection from the medial pulvinar to the amygdala in primates. *Brain Res.* 104:142–147.

Kaas JH (2008). The evolution of the complex sensory and motor systems of the human brain. *Brain Res Bull.* 75:384–390.

Kaas JH, Huerta MF, Weber JT, Harting JK (1978). Patterns of retinal terminations and laminar organization of the lateral geniculate nucleus of primates. *J Comp Neurol.* 182:517–553.

Kahn DM, Krubitzer L (2002). Retinofugal projections in the short-tailed opossum (*Monodelphis domestica*). *J Comp Neurol.* 447:114–127.

Kirk EC, Daghighi P, Macrini TE, Bhullar BA, Rowe TB (2014). Cranial anatomy of the Duchesnean primate *Rooneyia viejaensis*: New insights from high resolution computed tomography. *J Hum Evol.* 74:82–95.

Lyon DC, Jain N, Kaas JH (1998). Cortical connections of striate and extrastriate visual areas in tree shrews. *J Comp Neurol.* 401:109–128.

McFarland NR, Haber SN (2002). Thalamic relay nuclei of the basal ganglia form both reciprocal and nonreciprocal cortical connections, linking multiple frontal cortical areas. *J Neurosci.* 22: 8117–8132.

Mitchell AS (2015). The mediodorsal thalamus as a higher order thalamic relay nucleus important for learning and decision-making. *Neurosci Biobehav Rev.* 54:76–88.

Mitchell AS, Browning PG, Baxter MG (2007). Neurotoxic lesions of the medial mediodorsal nucleus of the thalamus disrupt reinforcer devaluation effects in rhesus monkeys. *J Neurosci.* 27:11289–11295.

Morel A, Liu J, Wannier T, Jeanmonod D, Rouiller EM (2005). Divergence and convergence of thalamocortical projections to premotor and supplementary motor cortex: A multiple tracing study in the macaque monkey. *Eur J Neurosci.* 21:1007–1029.

Murray EA, Moylan EJ, Saleem KS, Basile BM, Turchi J (2015). Specialized areas for value updating and goal selection in the primate orbitofrontal cortex. *eLife.* 4:e11695.

Murray EA, Wise SP, Baldwin MKL, Graham KS (2020). *The Evolutionary Road to Human Memory*. Oxford University Press.

Murray EA, Wise SP, Graham KS (2017). *The Evolution of Memory Systems: Ancestors, Anatomy, and Adaptations*. Oxford University Press.

Perry BAL, Mendez JC, Mitchell AS (2023). Cortico-thalamocortical interactions for learning, memory and decision-making. *J Physiol.* 601:25–35.

Preuss TM (1995). Do rats have prefrontal cortex? The Rose-Woolsey-Akert program reconsidered. *J Cog Neurosci.* 7:1–24.

Preuss TM (2007a). Primate brain evolution in phylogenetic context. In: Kaas JH, Preuss TM, eds. *Evolution of Nervous Systems.* Elsevier, Academic Press:4:2–34.

Preuss TM (2007b) Evolutionary specializations of primate brain systems. In: Ravosa MJ, Dagasto M, eds. *Primate Origins: Adaptations and Evolution.* Springer:625–675.

Preuss TM, Goldman-Rakic PS (1991). Myelo- and cytoarchitecture of the granular frontal cortex and surrounding regions in the strepsirhine primate *Galago* and the anthropoid primate *Macaca. J Comp Neurol.* 310:429–474.

Preuss TM, Wise SP (2021). The evolution of prefrontal cortex. *Neuropsychopharmacol.* 47: 3–19.

Püschel HP, Bertrand OC, O'Reilly JE, Bobe R, Püschel TA (2021). Divergence-time estimates for hominins provide insight into encephalization and body mass trends in human evolution. *Nat Ecol Evol.* 5:808–819.

Rathelot JA, Strick PL (2009). Subdivisions of primary motor cortex based on cortico-motoneuronal cells. *Proc Natl Acad Sci U S A.* 106:918–923.

Ray JP, Price JL (1993). The organization of projections from the mediodorsal nucleus of the thalamus to orbital and medial prefrontal cortex in macaque monkeys. *J Comp Neurol.* 337:1–31.

Remple MS, Reed JL, Stepniewska I, Kaas JH (2006). Organization of frontoparietal cortex in the tree shrew *(Tupaia belangeri).* I. Architecture, microelectrode maps, and corticospinal connections. *J Comp Neurol.* 497:133–154.

Rockland KS (1996). Two types of corticopulvinar terminations: Round (type 2) and elongate (type 1). *J Comp Neurol.* 368:57–87.

Romanski LM, Giguere M, Bates JF, Goldman-Rakic PS (1997). Topographic organization of medial pulvinar connections with the prefrontal cortex in the rhesus monkey. *J Comp Neurol.* 379:313–332.

Roumazeilles L, Eichert N, Bryant KL, Folloni D, Sallet J, Vijayakumar S, Foxley S, Tendler BC, Jbabdi S, Reveley C, Verhagen L, Dershowitz LB, Guthrie M, Flach E, Miller KL, Mars RB (2020). Longitudinal connections and the organization of the temporal cortex in macaques, great apes, and humans. *PLoS Biol.* 18:e3000810.

Rovo Z, Ulbert I, Acsady L (2012) Drivers of the primate thalamus. *J Neurosci.* 32:17894–17908.

Russchen FT, Amaral DG, Price JL (1987). The afferent input to the magnocellular division of the mediodorsal thalamic nucleus in the monkey, *Macaca fascicularis. J Comp Neurol.* 256:175–210.

Scerri EML, Thomas MG, Manica A, Gunz P, Stock JT, Stringer C, Grove M, Huw SG, Timmermann A, Rightmire GP, d'Errico F, Tryon CA, Drake NA, Brooks AS, Dennell RW, Durbin R, Henn BM, Lee-Thorp J, deMenocal P, Petraglia MD, Thompson JC, Scally A, Chikhi L (2018). Did our species evolve in subdivided populations across Africa, and why does it matter? *Trends Ecol Evol.* 33:582–594.

Schilder BM, Petry HM, Hof PR (2019). Evolutionary shifts dramatically reorganized the human hippocampal complex. *J Comp Neurol.* 528:3143–3170.

Semendeferi K, Armstrong E, Schleicher A, Zilles K, Van Hoesen GW (2001). Prefrontal cortex in humans and apes: A comparative study of area 10. *Am J Phys Anthropol.* 114:224–241.

Sherman SM, Guillery RW (2002). The role of the thalamus in the flow of information to the cortex. *Philos Trans R Soc Lond B Biol Sci.* 357:1695–1708.

Smaers JB, Gomez-Robles A, Parks AN, Sherwood CC (2017). Exceptional evolutionary expansion of prefrontal cortex in great apes and humans. *Curr Biol.* 27:714–720.

Sommer MA, Wurtz RH (2004). What the brain stem tells the frontal cortex. II. Role of the SC-MD-FEF pathway in corollary discharge. *J Neurophysiol.* 91:1403–1423.

Stepniewska I, Pouget P, Kaas JH (2018). Frontal eye field in prosimian galagos: Intracortical microstimulation and tracing studies. *J Comp Neurol.* 526:626–652.

Strick PL, Preston JB (1982). Two representations of the hand in area 4 of a primate. II. Somatosensory input organization. *J Neurophysiol.* 48:150–159.

Sur M, Weller RE, Kaas JH (1980). Representation of the body surface in somatosensory area I of tree shrews, *Tupaia glis. J Comp Neurol.* 194:71–95.

Tanji J, Wise SP (1981). Submodality distribution in sensorimotor cortex of the unanesthetized monkey. *J Neurophysiol.* 45:467–481.

Vanier DR, Sherwood CC, Smaers JB (2019). Distinct patterns of hippocampal and neocortical evolution in primates. *Brain Behav Evol.* 93:171–181.

Wang Q, Stepniewska I, Liao CC, Kaas JH (2023). Thalamocortical and corticothalamic connections of multiple functional domains in posterior parietal cortex of galagos. *J Comp Neurol.* 531:25–47.

Wise SP (2017). The evolution of the prefrontal cortex in early primates and anthropoids. In: Kaas JH, Krubitzer L, eds. *The Evolution of Nervous Systems.* 2nd ed. Elsevier:3:387–422.

Wise SP (2023). *Cortical Evolution in Primates: What Primates Are, What Primates Were, and Why the Cortex Changed.* Oxford University Press.

Wise SP, Donoghue JP (1986). The motor cortex of rodents. In: Peters A, Jones EG, eds. *Cerebral Cortex.* Plenum:243–270.

Wong P, Kaas JH (2009). Architectonic subdivisions of neocortex in the tree shrew (*Tupaia belangeri*). *Anat Rec (Hoboken).* 292: 994–1027.

Wu CWH, Bichot NP, Kaas JH (2000). Converging evidence from microstimulation, architecture, and connections for multiple motor areas in the frontal and cingulate cortex of prosimian primates. *J Comp Neurol.* 423:140–177.

Xiao D, Zikopoulos B, Barbas H (2009). Laminar and modular organization of prefrontal projections to multiple thalamic nuclei. *Neuroscience.* 161:1067–1081.

58.

EVOLUTION AND DEVELOPMENT OF THALAMOCORTICAL RELATIONSHIPS

Zoltán Molnár, Sara Bandiera, and Gavin J. Clowry

Thalamus and cortex form functional circuits that process sensory input, regulate brain state, and perform various higher cognitive functions. The development and evolution of the thalamus and cortex are intimately linked. Early anatomical tracing experiments from the cortex revealed that all cortical areas receive thalamocortical projections and all cortical areas communicate via thalamo-corticothalamic circuits (Jones, 2007). Almost all sensory information reaches the cortex through the thalamus (Sherman and Guillery, 2005).

THE GUIDANCE OF EARLY THALAMIC PROJECTIONS THROUGH THE INTERNAL CAPSULE AND CROSSING THE PALLIAL-SUBPALLIAL BOUNDARY

The thalamic projections arrive to the cortex at an early stage, even before the cortical neurogenesis and cortical neuronal migration are complete, and they accumulate among the early generated neurons in the subplate zone (Molnár et al., 2020). In the mouse, the thalamic and cortical neurons are generated around the same embryonic periods (thalamus embryonic (E) day 12–15; cortex E11–E17; Shi et al., 2017). In primate, thalamic neurons are generated relatively earlier (macaque: thalamus E30–E45, Spadory et al., 2022; visual cortex E45–E102, Rakic 1974).

The guidance of early thalamocortical connections relies on multiple mechanisms. The thalamic projections descend through the prethalamus, exit the diencephalon, extend along the primitive internal capsule, and cross the boundaries between the pallidum and ventral pallium and the pallial-subpallial boundary (PSPB) to grow through the subplate zone to approach the cortex (Molnár et al., 2012). Early connectivity, as well as migrating transient cell populations, shapes the trajectory of this axonal tract and assists the crossing of several embryonic developmental boundaries. The initial topography is guided by several factors, some establishing gradients in the ventral telencephalon (VTel). Early tracing studies revealed cell populations in the prethalamus, thalamic reticular nucleus, and perireticular nucleus, with precautious projections to the dorsal thalamus prior to the early outgrowth of the thalamocortical projections (Métin and Godement, 1996; DeCarlos and O'Leary, 1992; Molnár et al., 1998a, 1998b; Molnár and Cordery, 1999; Tuttle et al., 1999). It has been proposed that these early scaffolds assist the thalamocortical afferent outgrowth and entry to the telencephalon, while tangentially migrating neurons have been implicated for the guidance through the internal capsule (López-Bendito et al., 2006; Hanashima et al., 2006; Figure 58.1).

The crossing of the PSPB presents a challenge to the thalamocortical and corticofugal projections. Tracing studies in various species have demonstrated that the early corticofugal projections reach the PSPB at an early stage and they accumulate along these boundaries. Thalamocortical projections arrive at this boundary around the same developmental stage in mouse and rat (Figure 58.1). It has been described that the 2 sets of fibers co-fasciculate to cross this region in mouse and rat (Molnár et al., 1998a, 1998b). According to the "handshake hypothesis," the 2 fiber systems guide each other through this region of the embryonic forebrain, and this explains how ascending thalamic axons start to navigate to their appropriate cortical targets with help from reciprocal descending cortical axons (Molnár and Blakemore, 1995). The hypothesis was based on in vivo observations in mouse and rat that demonstrated an intimate anatomical relationship between developing thalamic and early cortical axons (DeCarlos and O'Leary, 1992; Molnár et al., 1998a, 1998b) and on in vitro findings in co-cultures of thalamus and cortex in which cortical explants from different regions accept innervations from any region of the thalamus and form layer-specific connections

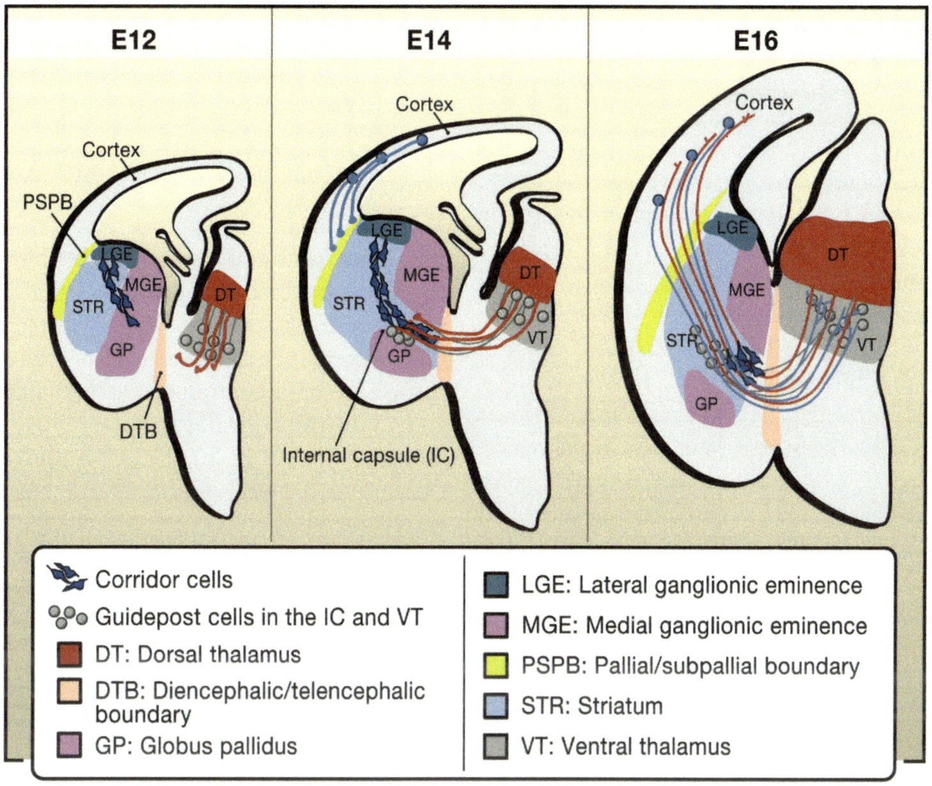

Figure 58.1. Three stages of the early establishment of reciprocal connections between dorsal thalamus and cortex in mouse. The prethalamus (formerly called ventral thalamus) and internal capsule contain guidepost cells (gray) that already have developed projections to the dorsal thalamus and guide the thalamic projections toward the internal capsule through the diencephalic and telencephalic boundary (DTB). Corridor cells (dark blue) originate from the lateral ganglionic eminence (LGE) at embryonic day 12 (E12) and migrate tangentially toward the diencephalon, where they form a permissive "corridor" for the thalamic projections (red) to navigate through the internal capsule. Subsequently the early corticothalamic projections guide the thalamic projections through the pallial-subpallial boundary (PSPB), as originally proposed in the handshake hypothesis (Molnár & Blakemore, 1995). Reproduced from Hanashima et al. (2006).

(Molnár and Blakemore, 1991, 1999). It has been suggested that guidance from descending axons, which is present in vivo but disrupted in explant culture, might be necessary to achieve area-specific patterns of thalamocortical connectivity (Molnár and Blakemore, 1991, 1999). The original formulation of the handshake hypothesis stated that "the descending and ascending axons each pioneer the pathway through their specific segment of the brain and, after a 'handshake' near the internal capsule, each may guide the growth of the other over the PSPB and subsequently through the distal part of its trajectory" (Molnár and Blakemore, 1991, pp. 475–477).

The PSPB, which thalamocortical axons must cross to reach the cortex, is first established as a gene expression boundary. By the age at which thalamic axons approach the pallium (E13 for mouse), the PSPB has developed a striking radial glial fascicle that extends across the trajectory of thalamocortical axons (TCAs) and has a high density of cells, including those of the lateral cortical stream, which migrate across the path of TCAs (Chapouton et al., 2001; Carney et al., 2006, 2009). This is also a feature of the human PSPB (González-Arnay et al., 2017). It has been suggested that these features make this region relatively non-permissive to the passage of thalamic axons and that descending corticofugal axons from the cortex interact with ascending thalamic axons and assist them across this region (Molnár et al., 1998a,b; Molnár and Butler, 2002). The non-permissive environment causes the fibers to fasciculate on each other in tight bundles, rather than spreading out to extend as individual fibers (Molnár et al., 1998a, 1998b). After entering the subplate at the PSPB, the fiber bundles break up into individual fibers, suggesting that the environment in this compartment is more permissive. In rodents, the pioneer corticofugal axons arrive at the PSPB before the thalamic axons, and tracing studies suggest that at least the earliest cohort of these fibers can cross the PSPB without thalamic axons. It is possible that a breakdown of this interaction explains TCA defects in some strains of mutant mice (Hevner et al., 2002; Jones et al., 2002; López-Bendito et al., 2002; López-Bendito and Molnár, 2003; Dwyer et al., 2011; Doyle et al., 2021).

The special properties of the PSPB can only be understood in light of comparative neurology and evolutionary developmental biology. The greatest divergence occurred along these boundaries between sauropsids (birds and reptiles) and mammals (Molnár and Butler, 2002; Bruguier et al., 2020). In sauropsids the dorsal ventricular ridge (DVR) protrudes into the lateral ventricle. In mammals, there is no DVR, but some of the homologous cell groups migrated ventrolaterally to form the claustrum, endopyriform nucleus, and lateral amygdala along the PSPB. This ventrolateral migration adds to the complexity of the PSPB during mammalian development. Relative timing is different for crossing the PSPB in various species.

Butler and Molnár proposed a *collopallial field hypothesis*, which examined the differences in the claustroamygdalar formation between birds, reptiles, and mammals (Molnár and Butler, 2002). The hypothesis held that in mammals, the thalamorecipient collopallium differentiates into deep (claustroamygdalar) and superficial (neocortical) components, whereas in sauropsids, and perhaps in platypus, this split may not occur. The general presence of the claustrum in all extant mammalian taxa (excluding the uncertainty in monotremes), together with shared histochemical, molecular, cytoarchitectural, and hodological features, suggests that the claustrum was already present in the now extinct pan-mammalian common ancestor (Montiel et al., 2011; Puelles, 2017; Suárez et al., 2018). We hypothesize that early during the independent evolution of mammals, some genetic changes enabled pallial tangential migrations. These changes most likely occurred as a byproduct of earlier developing divergences such as early pallial patterning and signaling or proliferation dynamics (García-Moreno and Molnár, 2020). As a main consequence, the corticogenic instructions provided in mammals by Cajal-Retzius cells and subplate neurons are missing in birds (García-Moreno et al., 2018). This very early difference might be the reason behind the diverging strategies of evolution in mammals that developed laminar and columnar organization in neocortex from avian and reptilian pallia which adopted nuclear arrangement. In reptiles, the homolog of the claustrum is located at the most anterior region of the dorsal ventricular ridge, a position consistent with its lateral pallial developmental origin (Norimoto et al., 2020). This suggests a deep conservation of the radially exclusive development of lateral ventral pallia in amniotes. The mammalian novelty of these early and later migratory patterns of glutamatergic neurons is in contrast with the highly conserved tangential migratory patterns of GABAergic neurons in sauropsids (Carrera et al., 2008; Cobos et al., 2001; 2002; Métin et al., 2007; Moreno et al., 2008; Garcia-Moreno and Molnár, 2020). These changes are reflected in the PSPB at the time of thalamocortical development. The extensive rearrangements within this region explain why so many developmental defects are associated with abnormalities and thalamocortical and corticothalamic guidance defects (López-Bendito and Molnár, 2003).

AXON GUIDANCE DEFECTS AT THE PALLIAL-SUBPALLIAL BOUNDARY: THE IMPORTANCE OF EARLY SUBPLATE PROJECTIONS FROM THE CORTEX FOR GUIDING THALAMOCORTICAL CONNECTIONS

The handshake hypothesis explained some of the thalamocortical and cortico-thalamic fiber abnormalities in various mutants (López-Bendito and Molnár, 2003). This hypothesis has been directly tested with conditional mutagenesis to assess the effects of blocking early corticofugal axonal development without disrupting the thalamus, subpallium, or PSPB in the *Emx1Cre;APCloxP/loxP* mutants (Chen et al., 2012). This study found that, whereas thalamic axons still traversed the subpallium in topographic order, they did not cross the PSPB. Normal cortex and mutant cortex stimulate the growth of axons from the thalamus by equal amounts in culture experiments. This suggests that the inability of thalamic axons to cross the PSPB in *Emx1Cre;APCloxP/loxP* mutants is unlikely to be explained by long-range chemorepulsion by the mutant cortex. By providing evidence against alternative explanations and by showing that replacement of mutant cortex with control cortex restored corticofugal efferents and allowed thalamic axons from conditional mutants to cross the PSPB, Chen et al. (2012) provided strong evidence for the notion that cortical efferents are required for the guidance of TCAs across the PSPB. These studies were also aided by our better understanding of embryonic subplate and thalamus gene expression patterns (Osheroff and Hatten, 2009; Oeschger et al., 2012; Hoerder-Suabedissen et al., 2013).

The most compelling evidence to date for the handshake hypothesis that cortical efferents are required to guide TCAs across the PSPB comes from the experiments by Doyle et al. (2021). This study provides careful genetic dissection of the cellular and molecular interactions and demonstrates that subplate neurons at the junction of PSPB play key roles in circuit wiring. Doyle et al. (2021) discovered an unexpectedly selective role for ubiquitous chromatin remodeler *Arid1a* in subplate neuron identity and circuit wiring. Doyle and colleagues described that using *Emx1-Cre* for selective cortical deletion of *Arid1a* leads to extensive mistargeting of intracortical axon tracts, but not corticofugal axon tracts. Such a deletion led to callosal agenesis. Tracing the early cortical projections following *Arid1a* deletion demonstrated that the intracortical axons are markedly misrouted into aberrant bundles. *Arid1a* thus plays non-cell autonomous roles in callosal axon guidance. Doyle et al. (2021) used in utero electroporation to elicit sparse *Arid1a* deletion and observed no callosal axon misrouting. Using RNA-seq to analyze transcriptomic changes, Doyle and colleagues found selective loss of the gene expression in subplate neurons following *Arid1a* deletion. They found that following *Arid1a* deletion, subplate neurons are impaired in their interactions ("handshake") and fasciculation with the thalamocortical projections and their pathfinding of ascending axons into the cortex.

This study suggested that *Arid1a* function was selective and specific to subplate neurons since other cortical neurons could not substitute subplate guidance of TCAs. Subplate neuron expression of *Arid1a* (without expression in other cortical neurons) is sufficient for TCA tract development. Doyle and colleagues (2021) showed that cortical plate specific deletion of *Arid1a* (using *hGFAP-Cre*) has no pathfinding defects, since subplate neurons were spared in this mutant. When *Arid1a* deletion is spared from subplate neurons, a normal "handshake" with TCAs crossing the PSPB and the cortex was observed. This suggests that cortical plate and subplate projections have different roles in early thalamocortical guidance. The authors described similar findings for the corpus callosum. Axons of thalamocortical neurons, which are unaffected by *Arid1a* conditional deletion, have severe guidance defects, and they are misrouted in their trajectories into the barrel cortex. The authors concluded that *Arid1a* plays a non-cell autonomous role in TCA guidance and that subplate neurons have a special role in guiding the thalamocortical projections through the PSPB, as was originally suggested in the handshake hypothesis (Molnár and Blakemore, 1995).

THE PATTERN OF MURINE EARLY THALAMOCORTICAL RELATIONS AT THE TIME OF THE ARRIVAL AND AT THE TIME OF ENTRY TO THE CORTICAL PLATE

Thalamic organization changes during the process of normal development, and it can be altered through (1) manipulations of the early guidance mechanisms in the subpallium; (2) manipulations of genes involved in early cortical regionalization; or (3) changing the flow of sensory input from the sense organs.

The initial relationship between the thalamus and the cortex through the thalamocortical projections is established during very early stages of development in most species explored. The best-studied model system is murine thalamocortical development. Tracing from the embryonic cortex at the time of the arrival of the thalamic projections to the cortex revealed a simple relationship of a 2D cortical sheet with a 3D thalamic structure (Figure 58.2). An anteroposterior sequence of movement of tracer injection sites along the cortex resulted in a mediolateral movement of back-labeled thalamic groups of neurons in the thalamus; whereas a mediodorsal movement of tracer injections along the cortex corresponded to an anteroposterior transition along an anteroposterior line in the thalamus (Molnár and Blakemore, 1991; Molnár et al., 1998a; Figure 58.2).

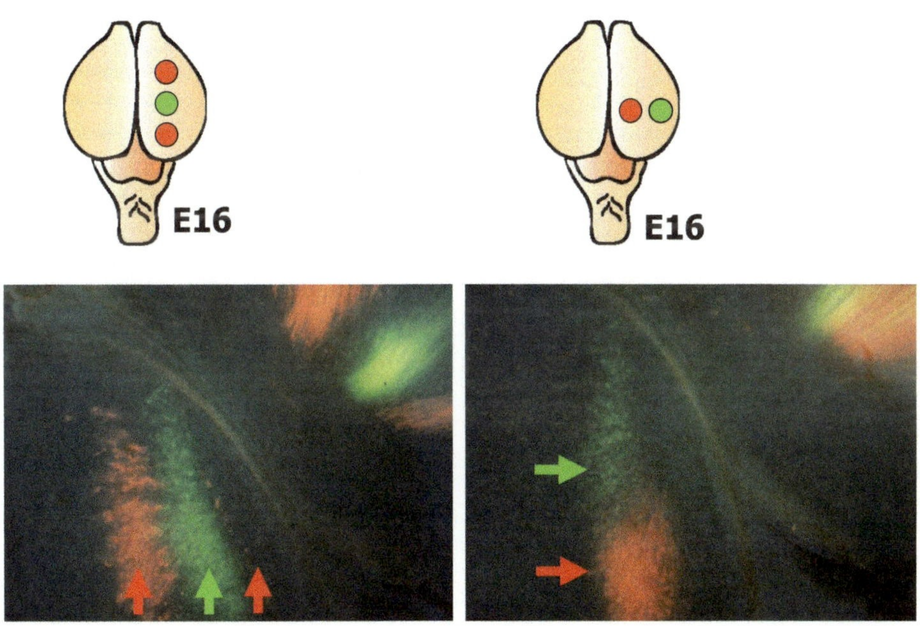

Figure 58.2. The early relationship of thalamocortical projections in an embryonic 16-day-old mouse brain; upper panels depict the relative carbocyanine crystal placement sites into the right hemispheres, and lower panels are horizontal sections through the thalamus (anterior up, lateral right, arrows point to the back-labeled cell groups in the thalamus). Tracing from the embryonic cortex at the time of the arrival of the thalamic projections to the cortex revealed a simple relationship of a 2D cortical sheet with a 3D thalamic structure. An anteroposterior sequence of movement of tracer injection sites along the cortex resulted in a mediolateral movement of back-labeled thalamic groups of neurons in the thalamus (indicated by the 3 arrows, red, green, and red); whereas a mediodorsal placement of 2 crystal placements to the cortex corresponded to an anteroposterior transition along a sagittal line in the thalamus (orange and red arrows in right panel). Figure adopted from Molnár et al. (1998a) and Molnár et al. (2012).

At these early stages, the thalamocortical projections have not entered the cortical plate; they accumulate in the subplate zone below the cortical plate, which is composed of migrating cortical neurons. As some of the cortical neurons complete migration, layers (L)6 and 5 start to form, while the neurons destined for the upper layers are still in the dense cortical plate. As the thalamic projections start to enter L6 and L5 of the developing cortex, this initial simple registration between cortex and thalamus is changing and the fiber-targeting topography is altered. The deployment and initial entry of thalamocortical projections to the subplate zone are considerably modified as the TCAs enter the cortical plate in cortical areas, such as V1 of rodents (Naegele et al., 1988; Krug et al., 1998; Ravary et al., 2003). There are areal differences in the density, topographic precision, and maturity of thalamocortical projections. Whereas TCAs undergo significant rearrangements in the cortex after entry into V1 of rodents, in the rodent S1 the topography is essentially established immediately after entry (Agmon et al., 1993; Agmon et al., 1995). Using retrograde labeling from pairs of carbocyanine dye placements from the putative somatosensory cortex and from the putative visual cortex, regional differences in the density and topography of thalamocortical innervations were revealed (Molnár, 1998). Pairs of injections in the somatosensory cortex demonstrated precise topography within the ventrobasal complex (VB), but similarly distanced pairs of injections to the visual cortex revealed more overlapping thalamocortical projections and loosely arranged back-labeled thalamic cells in the dorsal lateral geniculate nucleus (LGNd). These findings suggest that the initial layout of the thalamocortical projections is subsequently changing in some cortical areas, such as the primary visual area, but not to the same extent in the primary somatosensory cortex, as the thalamic projections arrive in the cortical plate. Such changes might be more substantial in the visual cortex compared to the somatosensory cortex not only in the murine cortex, but also in the ferret (Krug et al., 1998). There are several examples of substantial rearrangements in this region before the final ingrowth into the cortical plate, caused through either manipulation of the early cortical regionalization or changes in the flow of sensory input from the sensory periphery. Indeed, Nelson and LeVay (1985) described a substantial transformation of the visual representation 200 μm away from the visual cortex in the white matter, in an area that could have corresponded to the subplate zone during development.

The initial deployment of thalamic projections and their accumulation in the subplate zone below the cerebral cortex is orchestrated by molecular gradients and early activity patterns in the subplate (Molnár et al., 2002; Molnár et al., 2006; Molnár et al., 2012). The earliest synapses are formed in the subplate and marginal zone (Figure 58.3). Subplate neurons have diverse origins (Hoerder-Suabedissen and Molnár, 2015). Thalamocortical projections accumulate and "wait" in the subplate. The length of this waiting period is different in different species (Catalano et al., 1991). In marsupials the thalamic projections advanced fast and deep to the cortex compared to mice and rat, and the existence of subplate zone and the thalamocortical waiting period in subplate has been initially questioned. Tracing studies in *Monodelphis domestica* demonstrated that the ingrowth of thalamocortical projections follows a faster timecourse, but it is all compatible with the relatively more superficial position of subplate neurons and relative timing of the events of cortical development in marsupials (Molnár et al., 1999). The ingrowth of the thalamocortical projections target the displaced subplate neurons in the shaking rat Kawasaki at the surface of the cortex (Higashi et al., 2005).

Although the subpallium governs the early guidance of TCAs, cortical regionalization, which is controlled by the morphogen FGF8 and gradients of TFs (Pax6, COUP-TFI, Emx2, and Sp8), is sufficient to reorient the thalamocortical map within the neocortex (Garel et al., 2003; Rash and Grove, 2006; O'Leary and Sahara, 2008). The period during which the thalamocortical projections can be rearranged after sensory manipulations may vary considerably. Altering early cortical gene expression patterns of FGF8 imposes shifts, or even 2 opposing cortical gradients of transcription factors, results in corresponding shifts and duplications of thalamocortical projections (Shimogori and Grove, 2005), whereas changes in Pax6 gradients fail to elicit substantial changes in thalamocortical topography (Piñon et al., 2008). FGF8 gradient alterations can lead to duplication of the thalamic input from the same ventrobasal (VB) nucleus into multiple areas. The thalamocortical projections develop additional branches within the white matter, a region that corresponds to the location of subplate neurons (Shimogori and Grove, 2005). The earliest thalamocortical interactions and the eventual thalamocortical entry into the cortical plate are orchestrated by the subplate (Allendoerfer and Shatz, 1994; Kanold and Luhmann, 2010). The recognition of the ultimate target neurons within L4 of the cerebral cortex and the maturation of these connections relies on multiple cellular and molecular mechanisms, including a layer-specific stop signal that is not dependent on the regulated transmitter release from the thalamocortical projections (Molnár and Blakemore, 1991; Blakey et al., 2012; Molnár et al., 2012; Yamamoto and López-Bendito, 2012). The transient circuits between subplate neurons, thalamic afferents, and L4 neurons are now widely recognized as constituting a key mechanism for early circuit formation (Kanold and Luhmann, 2010; Molnár et al., 2020). The subplate neurons integrate into the cortical circuits in an age-specific and area-specific dynamic fashion (Piñon et al., 2009; Hoerder-Suabedissen and Molnár, 2012; Tolner et al., 2012; Viswanathan et al., 2012).

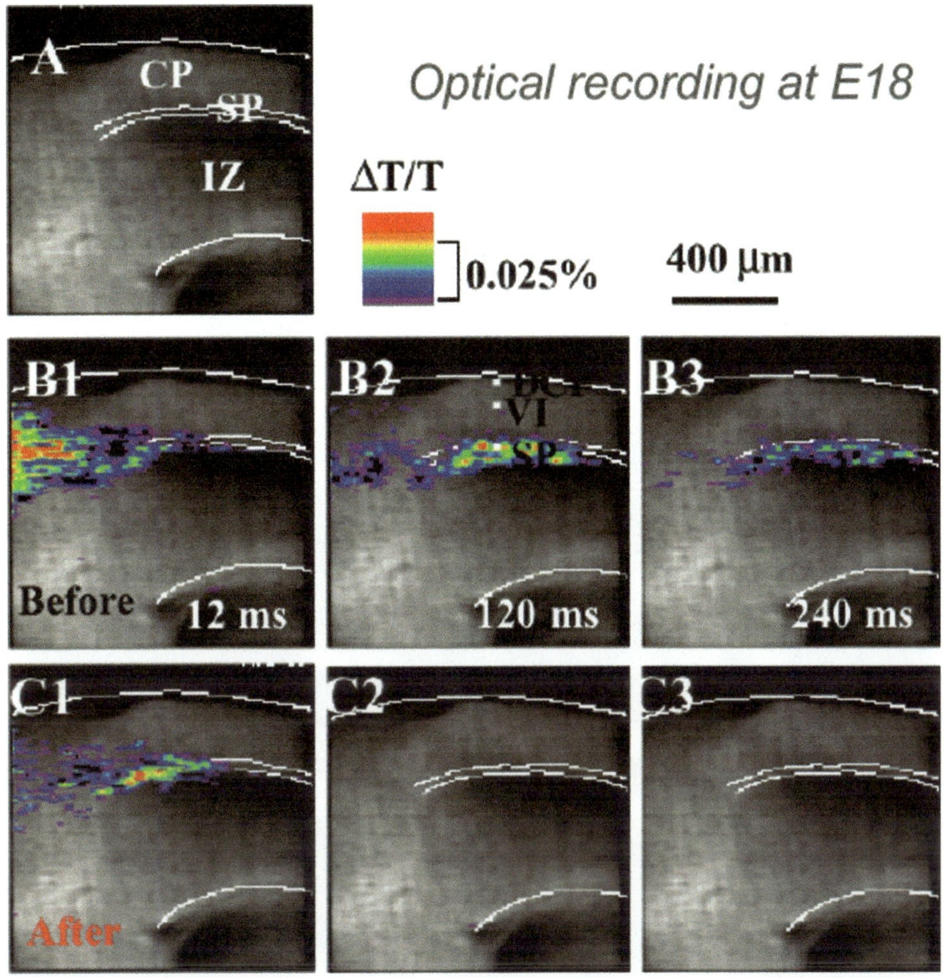

Figure 58.3. Optical recording reveals activation of subplate after thalamic stimulation in the developing neocortex in mice. Thalamocortical slices with thalamus, segment of the putative somatosensory cortex, and their interlinking pathways were prepared at E17–E18, and the thalamus has been stimulated while recording the activity of a selected field of the cortex. (A). Reference image of the thalamocortical slice with cortical plate (CP), subplate (SP), and intermediate zone (IZ). (B1–B3). A typical example of an E17–E18 optical response at 3 selected timepoints (12 ms, 120 ms, 240 ms) that is largely restricted to subplate. Pseudocolor responses superimposed on the reference images show relative changes in transmitted light intensity ΔT/T (where T is the intensity of transmitted light) with lower cutoff at the standard deviation from a mean value of B1. The responses to 16 single 100-μs stimuli of the thalamus (applied at 2-min intervals) from a bipolar stimulating electrode were averaged. (C). The same field is imaged after the application of a cocktail of glutamate receptor blockers (APV and DNQX). There is a fast spike-like response in the internal capsule which reaches the subplate at 12 ms (C1), but the response is abolished at 120 ms and 240 ms (C2 and C3, respectively). Figure adopted from Higashi et al. (2002) and Molnár et al. (2003).

EARLY THALAMOCORTICAL RELATIONS IN THE HUMAN BRAIN

In the mouse, it is well established that TCAs cross the diencephalic/telencephalic boundary (DTB) at embryonic day E12, the PSPB by E14.5, and innervate the subplate by E15.5, where they make synaptic connections with subplate neurons (Miyashita-Lin et al., 1999; Auladell et al., 2000; López-Bendito and Molnár, 2003; Gezelius and López-Bendito, 2017; Antón-Bolaños et al., 2018). Corticothalamic neurons are born prior to the arrival of thalamic afferents (at E11–E13), and this is key to our current understanding of thalamocortical pathfinding to the cortex. Corticothalamic axons cross the PSPB first at E11 ("handshake hypothesis") and are required to guide the TCAs across the PSPB (Molnár et al., 1998a, 1998b; Chen et al., 2012; Molnár et al., 2012). Original studies employing histological approaches, principally acetylcholinesterase histochemistry, and imaging (Krsnik et al., 2017; Išasegi et al., 2018) suggested that there are similar timings for the ingrowth of TCAs to the cortex for human development relative to neurogenesis of L6 and outgrowth of corticothalamic fibers. However, not all TCAs in human express acetylcholinesterase (Kostovic and Rakic, 1984), and our recent study, using different and specific molecular markers, found evidence that as early as 7 postconceptional week (PCW) in human, TCAs have traversed the DTB and reached the PSPB by 7.5 PCW (Figure 58.4; Alzu'bi et al., 2019).

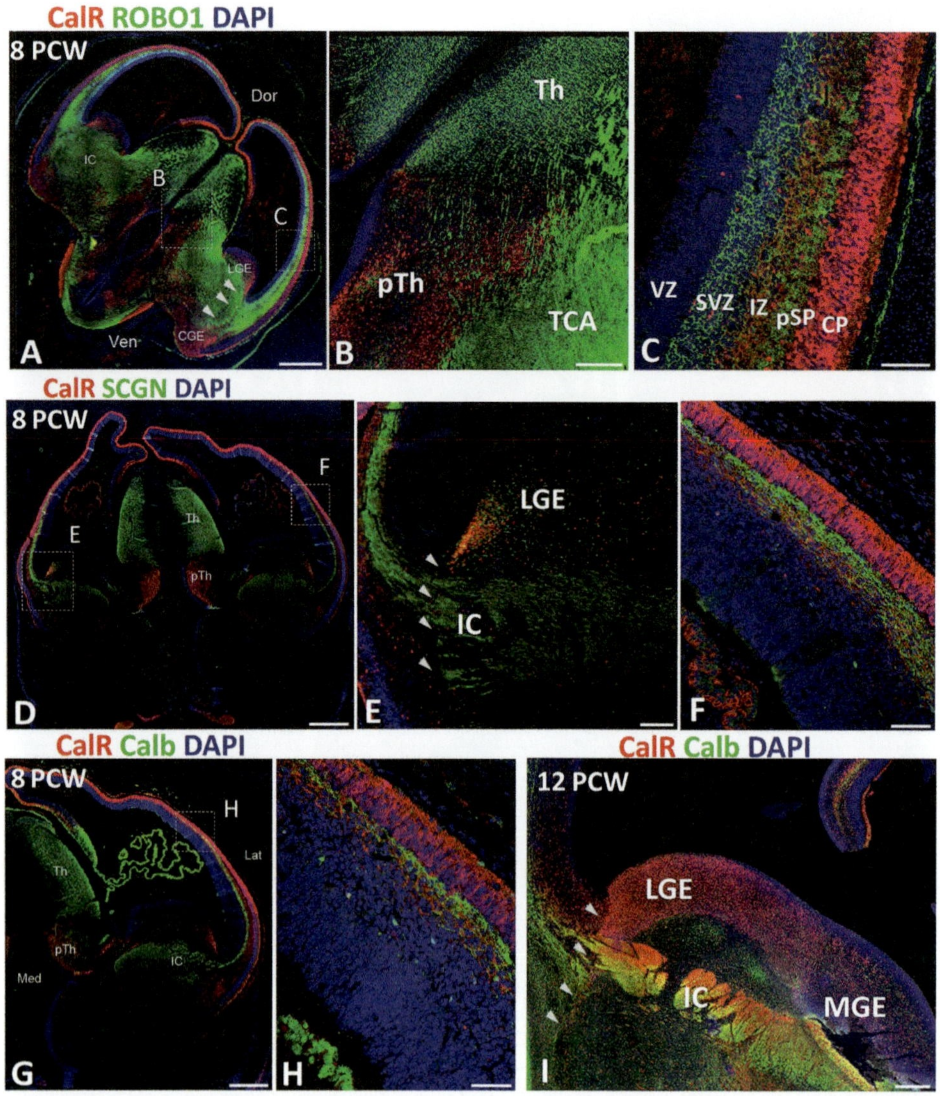

Figure 58.4. Double immunohistochemical labeling of corticothalamic (calretinin) and thalamocortical (ROBO1, secretagogin, calbindin) fibers in human fetal forebrain at 8 postconceptional weeks (PCW). In a coronal section (A–C) strong ROBO1 immunoreactivity (green) extensively labels neurons in the thalamus and thalamic axons in the prethalamus (pTh) internal capsule (IC) lateral cortical intermediate zone (IZ) and presubplate (pSP). Calretinin (CalR) immunoreactivity (red) is present in neuronal cell bodies in the cortical plate (CP) and prethalamus (pTh), is weakly expressed in the IZ, and is not present in the IC or thalamus. Arrows in A indicate the pallial-subpallial boundary (PSPB). D–F illustrate CalR (red) secretagogin (SCGN; green) co-labeling at 8 PCW and further emphasize the lack of CalR+ fibers crossing the PSPB to the ventral telencephalon at this stage (E). Similarly, for CalR (red) and calbindin (CalB; green) co-labeling (G, H). Presumptive CalB+ TCA invade the pSP but CalR+ fibers are not evident in the IC or thalamus. However, by 12 PCW (I) both CalR and CalB immunoreactive fibers are evident in the IC, with CalR+ fibers tending toward dorsal locations, and CalB+ fibers localizing more ventrally. Taken from Alzu'bi et al. (2019) under a creative commons licence. Scale bars: 1 mm in A, D, G; 200 μm in B; 100 μm in C, E, F, H, I.

The immunohistochemical studies on the early fibers in primates (human and marmoset) suggest a differential relationship between thalamocortical and corticothalamic projections from the pattern observed in rodents and carnivores. Thalamocortical projections seem to arrive to the cortex at an earlier stage. L6 formation in human is estimated to take at least 18 days, beginning around 7 PCW (http://translatingtime.org; Workman et al., 2013). Even if corticothalamic neurons start extending axons while still migrating to the cortical plate, TCA reach the PSPB before a substantial majority of corticothalamic neurons are even born, let alone extend axons across the PSPB. Furthermore, even potential CTA from calretinin-positive (CalR+) preplate pioneer neurons (De Carlos and O'Leary, 1992) do not appear to project beyond the PSPB by 8 PCW (Figure 58.4). Similarly, in marmoset brains, TCAs also cross the PSPB before CalR+ corticofugal axons leave the cortex (Alzu'bi et al., 2019). If these immunohistochemical markers are expressed in the earliest corticofugal projections, then this suggests that primate brains follow a

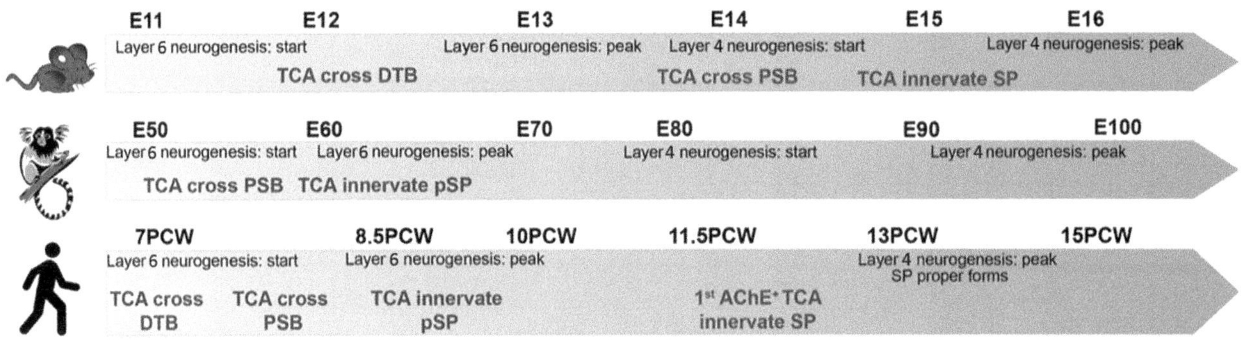

Figure 58.5. Comparison of timelimes of development of TCAs and neocortex in mouse, marmoset, and human. In mouse, L6 neurogenesis precedes the entry of thalamocortical afferents (TCAs) into the cortex by 3 days, during which time L6 formation is almost completed. In human thalamic, fibers cross from the DTB to the PSPB in 4/5 days, compared with 2 days in mouse, and invade the presubplate almost as soon as it is formed. TCA outgrowth in the marmoset brain more closely resembles the human condition than the mouse. Adapted from Alzu'bi et al. (2019) under a creative commons licence.

different pattern of thalamocortical development to that of rodents or other mammalian species (Figure 58.5). A more recent study in humans did find evidence of corticofugal axons, identified by DiI labeling and anti-neurofilament immunofluorescence (Qin et al., 2020) entering the ventral pallium by 8PCW and interacting with ISL1+ guidepost cells as previously described in rodents (Feng et al., 2016). However, currently it is not yet clear what the origin of the NF+ axons is, or whether or not the DiI labeling was both retrograde as well anterograde. Therefore, the "handshake" which is required for corticothalamic afferents (CTA) to guide TCA across the PSPB in rodents may occur at an earlier time point in primates and depend upon a population of very early born pioneer neurons that are not the CalR+ usually associated with this role in other mammals studied. Alternatively, the handshake may not occur at the PSPB at all but may occur at a different meeting point in both marmoset and human brains, perhaps having a different role such as guiding TCA to different regions of the cortex. A recent study in human has demonstrated a specific population of transient deep subplate neurons, restricted in their location and expressing specific cell guidance molecules, proposed to have a role in segregating and directing geniculocortical and pulvinocortical axons to areas 17 and 18 of the visual cortex respectively (Išasegi et al., 2022).

We have observed that early in human development (8-10 PCW) TCA and CTA segregate into 2 different compartments within the IZ of the cortical wall, the TCA preferring to be close to the cortical plate, whereas corticofugal axons (GAP43+ and sometimes CalR+) cluster nearer the subventricular zone (SVZ). Similarly, in the internal capsule putative CalR+ corticofugal fibers segregate dorsally (nearer the SVZ) and TCA ventrally by 12 PCW (Figure 58.4; Alzu'bi et al., 2019). Thus, TCA target the presubplate at a time at which the earliest synapses are forming, as demonstrated by electron microscopy and immunohistochemistry (Kostović and Rakic, 1990; Bayatti et al., 2008; Harkin et al., 2017). Therefore, spontaneous activity in thalamic neurons, transmitted via the presubplate (pSP) to the cortical plate, could influence the expression of genes governing cortical arealization in human cortical neurons, as has been described in mouse from E16 onward (Moreno-Juan et al., 2017), but at a much earlier stage of development. Given the enormous expansion of the human neocortex to include hundreds of cortical fields compared to the relatively small number of cortical fields in mice, and the independent evolution of these lineages of about 80 million years, it is not surprising that differences in developmental programs have evolved, and we observe much greater complexity of cortical arealization in human and monkey (Buckner and Kreinen, 2013; Mundinado et al., 2015; Clowry et al., 2018; Krubitzer and Prescott, 2018), which may require that this process begins at an earlier developmental timepoint.

CORTICOTHALAMIC DEVELOPMENT AND PLASTICITY

Thalamic nuclei can be divided into 2 major categories according to their connectivity. Only first-order thalamic nuclei receive direct and indirect inputs from the sensory periphery. Higher-order thalamic nuclei receive input from the cortex and relay this back to other cortical areas, providing trans-thalamic-cortico-thalamocortical pathways (Figure 58.6B,C). One large difference between rodents and primates, including humans, is the proportion of the cerebral cortex that does not receive inputs from the first-order thalamic nuclei. The expansion of the higher-order thalamic nuclei and the trans-thalamic pathway is one

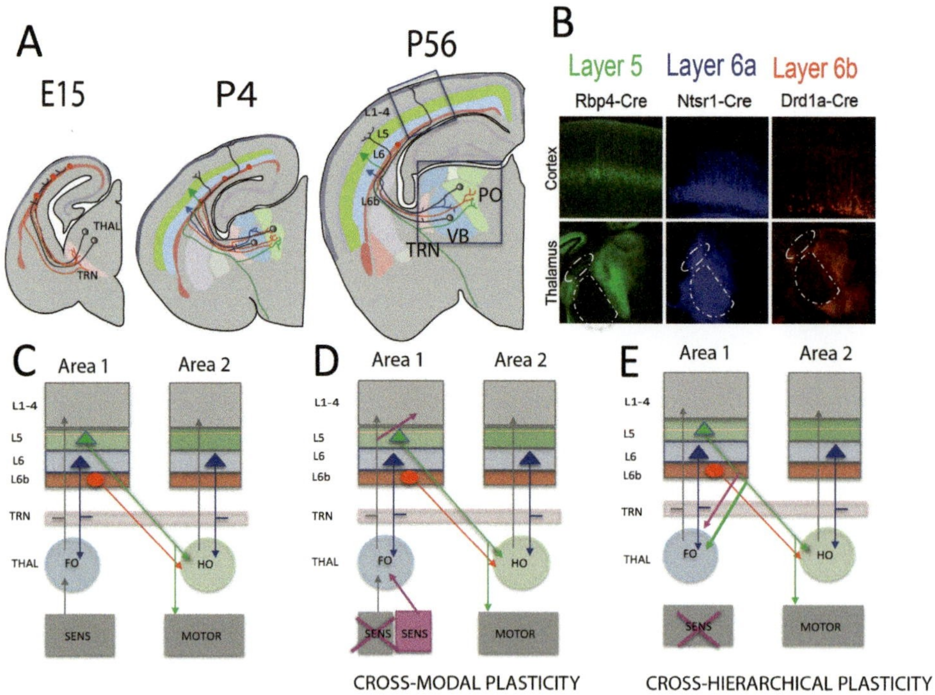

Figure 58.6. Schematic demonstrating the specificity of the connections between the cortex and thalamus using the mouse somatosensory system as an example (based on and modified from Molnár, 2019; Molnár et al., 2020).

(A). Schematic summary of the development of the complexities of the thalamocortical connectivity in the developing (E15, P4) and adult (P56) mouse on an idealized section of mouse brain containing somatosensory cortical connections. The first-order ventrobasal thalamic nucleus (VB) receives somatosensory peripheral input (not indicated). The VB then projects axons (black) to L4 of the primary somatosensory cortex. L6 "modulator" neurons (dark blue) in S1 project back to the VB. L5 neurons (light green) in S1 project to subcerebral structures and make a collateral branch to a higher-order thalamic nucleus, for example, posterior thalamic nucleus (Po). L5 projections give side-branches selectively to higher-order thalamic nuclei. L6 projections innervate both nuclei, whereas some subgroups of L6b selectively innervate higher-order thalamic nuclei (Grant et al., 2016, Hoerder-Suabedissen et al., 2018).

(B). Transgenic mouse lines with selective Cre recombinase expression in subpopulations of cortical projection neurons enabled the study of selective innervation of first- and higher-order thalamic nuclei. White circles illustrate the outline of LGNd and VB in the thalamus. L5 (Rbp4-Cre) and L6b (Drd1a-Cre) projections selectively innervate higher order thalamic nuclei, whereas L6a (Ntsr1-Cre) lack such preference (from Grant et al, 2016, Hoerder-Suabedissen et al., 2018, Hayashi et al., 2020).

(C). Schematic illustration of the possible functional circuits generated by this open-loop connectivity and how persistent subplate in L6b can regulate the trans-thalamic cortico–cortical communication through their projections (red) to higher-order thalamic nuclei (Molnár, 2019). Sensory information is relayed through the first-order thalamic nucleus to the cortex (gray). This cortical area then projects from L6 reciprocally back to the first-order nucleus (blue). Each area is also non–reciprocally connected to a higher-order thalamic nucleus. The L5 input to the thalamus (dark green) is an "efferent copy" of the L5 output to the motor system in the brainstem and spinal cord. This copy is forwarded to a higher cortical area (HO, green). Some persistent subplate (SP) neurons in L6b selectively project to higher-order thalamic nuclei and can regulate the trans-thalamic cortico–cortical communication. Direct cortico–cortical connections are also depicted between cortical layers and cortical areas (pale gray lines). These circuits enable cortical areas to act with other cortical areas and motor apparatus in a coordinated manner.

(D). After loss of a particular sensory input, another sensory input can colonize the affected part of the thalamus, and this modality will be transmitted to a different cortical area. This form of plasticity has been termed *cross-modal plasticity*.

€. Early sensory deprivation can substantially change the relationship of the cortico-thalamocortical projections originating from L6 and L5 (Grant et al., 2016; Frangeul et al., 2016; Giasafaki et al., 2022). The LGNd normally only receives corticothalamic projections from L6, but after neonatal enucleation, L5 projections innervated LGNd and formed synapses (Grant et al., 2016) that were maintained to adulthood (Giasafaki et al., 2022). The form of plasticity when first-order thalamic nuclei that no longer receive sensory input from the periphery are innervated by L5 projections has been termed *cross-hierarchical plasticity* (Grant et al., 2016). CP = cerebral peduncle; FO = first-order thalamic nuclei; GP = globus pallidus; HO = higher-order thalamic nuclei; ic = internal capsule; Po = posterior thalamic nuclei; RTN = reticular thalamic nuclei; SP = subplate; Str = striatum; S1 = primary somatosensory cortex; S2 = secondary association somatosensory cortex; VB = ventrobasal thalamic nuclei; wm = white matter. Figure is modified from Grant et al. (2012) that was originally inspired by Sherman and Guillery (2005) and modified based on the results of Hoerder-Suabedissen et al. (2018).

of the defining features of primate/human. The thalamic projections from first-order thalamic nuclei and higher-order thalamic nuclei target different cortical layers, and in return, the corticofugal projections have different relationships with them. In rodent, neurons from 3 cortical layers contribute to the corticothalamic projections, from L5, L6a, and remnants of the subplate neurons in L6b in rodent. The early tracing experiments in mouse used simple retrograde and anterograde tracing methods that labeled connections regardless of cortical neuronal cell types or laminar origin. Subsequent research has used more selective methods to reveal specific projections from subtypes of cortical projection neurons (Figure 58.6B). In mouse L5 and some populations of L6b, cortical projection neurons selectively target higher-order thalamic nuclei, whereas L6a and some populations of L6b target both first- and higher-order thalamic nuclei (Grant et al., 2012; Hoerder-Suabedissen et al., 2018). The choreography of corticothalamic innervation of the thalamus follows a specific algorithm and is sensitive to alterations in the sensory input. These allocations of L6 and L5 projections can change after early sensory deprivations (Frangeul et al., 2016; Grant et al., 2016; Giasafaki et al., 2022).

The corticofugal projections arrive in the thalamus at different developmental stages. The first corticofugal projections that leave the cortex originate from subplate neurons, followed by L6 and L5 (Grant et al., 2012; Figure 58.6A). The different classes of corticothalamic projections advance with different speed and have different relations to the dorsal thalamus during development. It has been discovered that the first corticofugal projections that enter the thalamus originate from L5 projection neurons (Clasca et al., 1995). The early corticofugal projections from the subplate and L6 reach the proximity of the thalamus early, but entry to the thalamus requires a longer period. Just as the thalamocortical projections accumulate close to the cortex within the subplate zone, the corticofugal projections from L6a and L6b accumulate outside the thalamus. There is a similar "waiting" period for the corticothalamic projections as for the thalamocortical projections. Corticothalamic projections from L6 "wait" outside the thalamus (perhaps TRN) before entering the dorsal thalamus (Grant et al., 2012). While L5 projections enter the dorsal thalamus early, the other projections from L6a and L6b do not enter some thalamic nuclei until after the first postnatal week in mouse. For instance, L6 projections only enter LGNd around P10 in mouse (Grant et al., 2012, 2016). Little is known about the time course of corticothalamic innervation in human, due to lack of specific markers for histology, and the difficulties in separating CTA from TCA in structural magnetic resonance imaging (MRI) studies. However, it does appear that TC-CT development is prolonged in humans, extending to term and beyond, and lasting much longer than the development of corpus callosum pathways (Wang et al., 2017).

That the entry of corticothalamic projections to the dorsal thalamus is dependent on sensory input has been demonstrated for the LGNd in the mouse. Neonatal enucleation accelerates ingrowth of these corticothalamic projections to LGNd, and reducing early activity in the retina has a similar effect (Grant et al., 2016). It has been demonstrated that early sensory deprivation can substantially change the gene expression in LGNd, and this can also change the relationship of the cortico-thalamocortical projections originating from L6 and L5 (Frangeul et al., 2016; Grant et al., 2016; Giasafaki et al., 2022). The LGNd normally only receives corticothalamic projections from L6, but after neonatal enucleation, L5 projections innervated LGNd and formed synapses (Grant et al., 2016) that were maintained to adulthood (Giasafaki et al., 2022). The form of plasticity when first-order thalamic nuclei that no longer receive sensory input from the periphery are innervated by L5 projections, which would normally only innervate higher-order thalamic nuclei, has been termed *cross-hierarchical plasticity* (Grant et al., 2016; Figure 58.6E). This is a different form of plasticity from the cross-modal plasticity (Figure 58.6D) that has been originally described after early loss of sensory input or blocking the activity of the peripheral sensory organs, although cross-hierarchical plasticity can also contribute to the rearrangements of the circuits observed in cross-modal plasticity. The cross-hierarchical plasticity is also associated with gene expression changes that make the deafferented, originally first-order thalamic nuclei more like the higher-order thalamic nuclei (Frangeul et al., 2016; Giasafaki et al., 2022). Both cross-modal and cross-hierarchical plasticity have an impact on the information processing of the thalamo-corticothalamic networks. However, the exact role of the rearranged circuits in cross-hierarchical plasticity is still not understood.

THE EARLY THALAMOCORTICAL INTERACTIONS MEDIATE THE FIRST SIGNALS THAT CONTRIBUTE TO THE AREA-SPECIFIC CORTICAL CIRCUIT FORMATION

The major source of extrinsic modulation to the area-specific development of the cerebral cortex is provided by the early thalamocortical afferents (O'Leary, 1989). Thalamocortical projections arrive in the developing cerebral cortex at a time when neurogenesis, neuronal migration, and the formation of the connections are still in progress. There are variations in the relative maturity of the cortex in various species, but these interactions are present across all mammals. Therefore, thalamic projections can exert influence on area-specific neurogenesis, neuronal migration, differentiation, cell death, and somatodendritic integration of cortical neurons in various cortical areas. There is evidence that neurogenesis, neuronal migration, and area-specific early gene-expression differences can proceed without any thalamocortical input or even without any connectivity between cortex and subcortical structures (Miyashita-Lin et al., 1999). It has been demonstrated that cortical neurons are generated in a similar pattern in a cortex that develops without any input or output (*cerveau isole*) and in matched controls, and some of the early region-specific gene-expression patterns are indistinguishable between them. However, there is convincing evidence for the role of thalamocortical projections for imposing cytoarchitectonic differentiation program onto the cortex that they innervate. Barrels do not form in the *cerveau isole* mice where thalamocortical projections do not innervate the cortex (Zhou et al., 2010), and early enucleation experiments demonstrated the changes in the thalamus and in cytoarchitecture and neuron numbers in the putative visual cortex (Rakic and Williams, 1986; Rakic, 1988; Rakic et al., 1991; Dehay et al., 1996). Early loss of the visual sensory input has a profound effect on the thalamocortical interactions that result in changes in cortical arealization.

Early sensory lesions can have profound effect on the gross morphological appearance (folding) of the brain and demonstrate that the borders between cortical areas are not fixed (Rakic 1988; Dehay et al., 1996). Early enucleations in macaque drastically changed the extent of the primary visual cortex that can be determined from the extent of the Stria Gennari that corresponds to the thick L4 extending within primary visual cortex. There is evidence for changes in progenitor division, cell death, neural migration, neural differentiation, and formation of connections in the developing cerebral cortex prompted by thalamocortical projections on the developing cortex (Valverde, 1967; Polleux et al., 1997; Caric and Price, 1999; Cooke et al., 1999; Lotto et al., 1999; Dehay et al., 2001; Edgar and Price, 2001; Sur and Leamey, 2001).

THALAMIC PROJECTIONS INTERACT WITH CORTICAL PROGENITORS IN THE GERMINAL ZONE IN PRIMATES

The interaction between thalamic axons and cortical progenitors has been explored in mice. Several molecular mechanisms have been implicated in these interactions. Thalamic axons modulate the behavior of cortical precursor cells (i.e., cell-cell direct contacts by ligand-receptor bound; paracrine factors). Gerstmann and colleagues demonstrated that thalamic projections express ephrin A5 ligand. It has been also demonstrated that developing thalamic afferents are in direct contact with the cortical progenitor cells that express the receptor EphA4 in the mouse (Gerstmann et al., 2015). Moreover, neuroactive molecules such as VGF (non-acronymic) are also released by the thalamic axonal terminals and influence the intermediate progenitor cells in the rodent SVZ of the sensory cortices by paracrine mechanisms (Monko et al., 2022). We are interested in such early thalamocortical interactions because they might be also present in larger brains with more prolonged development and their influence on cortical arealization might be more prominent.

Thalamocortical projections extend into the proximity of outer subventricular zone (OSVZ) in carnivores (Shatz and Rakic, 1981). In the human, Išasegi and colleagues (2018) described how the inner sagittal strata (ISS) corresponding to the thalamic axons, intermingled with cells of the OSVZ, while the external sagittal strata (ESS), corresponding to the basal forebrain axons, intermingled with outer/external transient layer of cells. This study defined a multilaminar axonal-cell compartment (MACC) where fibers and cellular layers are closely related. Our recent carbocyanine dye-tracing results also demonstrated that the TCAs of the intermediate zone and the dividing progenitor cells in the germinal zone are in close contact with one another (Bandiera and Molnár, 2022b; Figure 58.7E).

The compartmentalization and proportions of the cortical progenitor populations have undergone evolutionary changes (Smart et al., 2002; Reillo et al., 2011; García-Moreno et al., 2012; Nowakowski et al., 2016; Toda et al., 2016). The anatomical arrangements suggest a larger and more compartmentalized germinal zone with closer interaction between thalamic axons and these progenitors in the larger gyrencephalic brains. Early binocular enucleation causes alterations in the thalamic afferents in the ferret (Reillo et al., 2011) and the monkey (Smart et al., 2002), and this leads to a significant reduction of the primary visual area in the adult animal. The effect of early enucleation on cerebral cortical development has been interpreted differently by different investigators. In the ferret brain, no apparent change was observed in the boundary between area 17 and the adjacent area 18, but thalamic axons seem to influence a non-neurogenic population of OSVZ progenitors that serve as a physical scaffold for neuronal tangential migration underlying the formation of cortical folds. On the other hand, in the primate brain, the reduction of thalamic axons after enucleation seems to directly affect the mitotic rate of OSVZ progenitor cells, and subsequently the proliferation gradient necessary for the proper formation of the boundary between those adjacent areas.

Further evidence for close interaction between thalamic axons and the precursor cells of the primate cortex has been provided by Lukaszewicz and colleagues (2005). They showed that at E78 cortical progenitors in the superficial part of OSVZ (where the TCAs are located) have a shorter S-phase than their counterparts in the deep portion of the OSVZ, further supporting the idea that thalamic axons influence the cell cycle parameters of progenitor cells in their surroundings (Dehay and Kennedy, 2007). These studies in carnivores and primates suggest that extrinsic modulation of the cortical progenitor populations not only alters the neuronal output as described in rodents, but also underlies the acquisition of areal-specific features in the primate developing cortex.

In the human brain (Alzu'bi et al., 2019) the thalamic axons may exert an even more pronounced and widespread

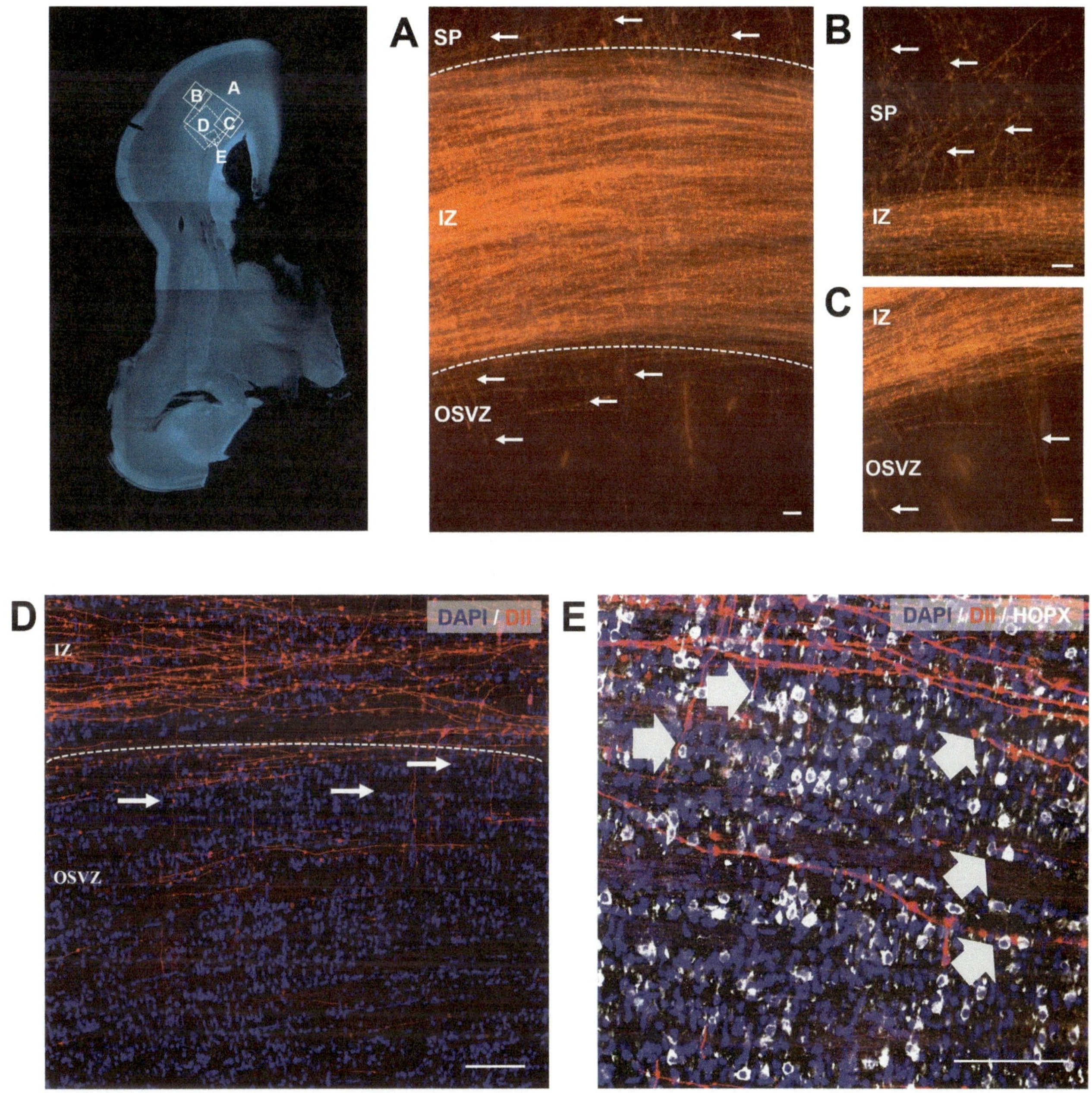

Figure 58.7. Thalamocortical axons interact with the subplate (A, B) and progenitor cells of the cortical outer subventricular zone (A, C, D, E) in the human developing cerebral cortex at 17 postconception week (PCW) from Bandiera and Molnár (2022b). The locations of the high-power images are indicated with white boxes on the low-power image of the entire coronal section of the human brain slice counterstained by DAPI (blue). Epifluorescence (A–C) and laser scanning confocal (D–E) microscopic images showing thalamocortical afferents traced by carbocyanine dye (DiI) and reaching not only the subplate (highlighted in A and B; thin white arrows), but also projecting in the direction of the outer subventricular zone (OSVZ) (highlighted in A, C and D; thin white arrows below intermediate zone, IZ) in the human 17 postconception week (PCW) fetal cortex. Confocal microscopic reconstruction (E) shows the DiI-labeled TCA (red) intermingling with the HOPX-immunoreactive outer radial glial cells (white). Overlapping thalamic fibers and oRGC cellular bodies are indicated by thick white arrows. Cellular nuclei are counterstained with DAPI (blue) in D and E. Scale bars = 100 μm.

effect on cortical neurogenesis (Dehay and Kennedy, 2007; Molnár 2011; Molnár et al., 2011). Tracing early thalamocortical projections with carbocyanine dye (DiI) in fixed post-mortem human brains at midgestational stages (17 PCW) revealed that TCAs project into the underlying OSVZ. Here, HOPX-immunoreactive outer radial glial cells (oRGC) are observed in close relationship with the DiI-labeled axons (Figure 58.7E; Bandiera and Molnár, 2022b). Strikingly, a similar pattern of innervation of the OSVZ by thalamic axons has been documented also in the macaque brain at similar gestation stages (Carney, 2004). These interactions are present at the peak of cortical neurogenesis, but the cellular and molecular aspects have not been elucidated.

Previous literature suggests VGF as a candidate thalamic factor that might account for the simultaneous interaction with both progenitor cells of the OSVZ (Monko et al., 2022) and post-mitotic cortical neurons of the subplate (Sato et al., 2022). Available bulk and single-cell transcriptomic datasets of the human brain (Brainspan.org and Nemoanalytics.org, respectively), along with immunohistochemical and molecular biological approaches applied to post-mortem human material support the notion that VGF is a molecular candidate to mediate thalamocortical interactions in human (Bandiera and Molnár, 2022b). These early interactions might have evolutionary relevance since both the subplate and the OSVZ are massively expanded in the fetal human brain, and they are associated with its increased size and complexity. Therefore, a prolonged and finely tuned modulation of cortical neurogenesis by thalamocortical afferents might explain some unique aspects of the human cerebral cortex.

CONCLUSIONS

Thalamus and cortex and their interlinking circuits via the thalamo-cortico-thalamo-cortical connections evolved in tandem. During development, thalamocortical projections start to interact with the cortex at the time when most of its neurons have not been produced or are migrating into position. Thalamic projections start to arrive in the cortex by crossing through regions of the forebrain that had very substantial modifications during evolution. The developmental changes at the PSPB reflects the diverging mammalian evolutionary path from the ones that sauropsids adopted. The thalamic projections cross the PSBP by associating to transient early corticofugal projections and a transient cell population in the internal capsule. The basic cortical neurogenetic and migration program is autonomous of subcortical input and can largely occur independently from external input. However, thalamic projections arrive in the cortex at an early stage, and they start to mediate some of the first signals that contribute to some of the area-specific cortical circuit formation. The timing and nature of these early interactions can differ in different taxa, but they are present in all mammals. Observations in human suggests that the thalamic projections arrive in the cortex relatively early in comparison to other species and interact not only with subplate neurons, but also with some of the cortical progenitor populations in the germinal zone. The corticothalamic projections also interact with the developing thalamus and have an impact on the formation of cortico-thalamocortical circuits. The choreography of corticothalamic innervation of the thalamus follows a specific algorithm and is sensitive to alterations in the sensory input. These allocations of L6 and L5 projections can change after early sensory deprivations. The development and plasticity of these elaborate thalamo-corticothalamic circuits provide exceptional flexibility to provide a high level of adaptability depending on the evolutionary demands on the individual species.

ACKNOWLEDGMENTS

Some of the ideas and concepts reviewed in this chapter are the updated views articulated in our previous publications: Molnár and Blakemore, 1995; Molnár 1998; López-Bendito and Molnár, 2003; Molnár et al., 2012; Hoerder-Suabedissen and Molnár, 2015; Grant et al., 2016; Alzu'bi et al., 2019; Molnár et al., 2020; Garcia-Moreno and Molnár, 2020; Bandiera and Molnár, 2022a, 2022b). We are grateful to all previous and current laboratory members for their comments and discussions. The work in the laboratory of Zoltán Molnár has been funded by MRC, BBSRC, Royal Society, Wellcome Trust, Oxford Martin School, Anatomical Society, St John's College. ZM is an Einstein Visiting Fellow at Charité—Universitätsmedizin Berlin, Cluster of Excellence NeuroCure, and Institute of Biochemistry (2020–2024). SB is supported by the MRC DTP fellowship, MSGS Studentship from the Department of Physiology Anatomy and Genetics, and the Ramage Scholarship from Balliol College. Work on human tissue has been made possible by the Human Developmental Biology Resource funded by the Wellcome Trust and MRC.

REFERENCES

Agmon A, Yang LT, Jones EG, O'Dowd DK (1995). Topological precision in the thalamic projection to neonatal mouse barrel cortex. *J Neurosci.* 15(1 Pt 2):549–561.

Agmon A, Yang LT, O'Dowd DK, Jones EG (1993). Organized growth of thalamocortical axons from the deep tier of terminations into layer IV of developing mouse barrel cortex. *J Neurosci.* 13:5365–5382.

Allendoerfer KL, Shatz CJ (1994). The subplate, a transient neocortical structure: its role in the development of connections between thalamus and cortex. *Annu Rev Neurosci.* 17:185–218.

Alzu'bi A, Homann-Ludiye J, Bourne JA, Clowry GJ (2019). Thalamocortical fibres invade the subplate much earlier in primates than in rodents. *Cereb Cortex.* 29:1706–1718.

Antón-Bolaños N, Espinosa A, López-Bendito G (2018). Developmental interactions between thalamus and cortex: A true love story. *Curr Opin Neurobiol.* 52:33–41.

Auladell C, Pérez-Sust P, Supèr H, Soriano E (2000). The early development of thalamocortical and corticothalamic projections in the mouse. *Anat Embryol.* 201:169–179.

Bandiera S, Molnár Z (2022a). Development of the thalamocortical systems. In: Halassa M, ed. *The Thalamus* (Chapter 7: 139–162). Cambridge University Press.

Bandiera S, Molnár Z (2022b). Characterisation of the earliest thalamocortical interactions in the human fetal brain. *J Anatomy.* 240(4):788–788.

Bayatti N, Moss JA, Sun L, Ambrose P, Ward JFH, Lindsay S, Clowry GJ (2008). A molecular neuroanatomical study of the developing human neocortex from 8 to 17 post conceptional weeks revealing the early differentiation of the subplate and subventricular zone. *Cereb Cortex.* 18:1536–1548.

Blakey D, Wilson MC, Molnár Z (2012). Termination and initial branch formation of SNAP-25 deficient thalamocortical fibres in heterochronic organotypic co-cultures. *Eur J Neurosci.* 35:1586–1594.

Bruguier H, Suarez R, Manger P, Hoerder-Suabedissen A, Shelton AM, Oliver DK, Packer AM, Ferran JL, García-Moreno F, Puelles L, Molnár Z (2020). In search of common developmental and evolutionary origin of the claustrum and subplate. *J Comp Neurol.* 528(17):2956–2977.

Buckner RL, Krienen FM (2013). The evolution of distributed association networks in the human brain. *Trends Cogn Sci.* 17:648–665.

Caric D, Price DJ (1999). Evidence that the lateral geniculate nucleus regulates the normal development of visual corticocortical projections in the cat. *Exp Neurol.* 156(2):353–362.

Carney RSE (2004). *Thalamocortical development and cell proliferation in fetal primate and rodent cortex*, D. Phil Thesis, Department of Human Anatomy and Genetics, University of Oxford, UK.

Carney RS, Alfonso TB, Cohen D, Dai H, Nery S, Stoica B, Slotkin J, Bregman BS, Fishell G, Corbin JG (2006). Cell migration along the lateral cortical stream to the developing basal telencephalic limbic system. *J Neurosci.* 26:11562–11574.

Carney RS, Cocas LA, Hirata T, Mansfield K, Corbin JG (2009). Differential regulation of telencephalic pallial–subpallial boundary patterning by Pax6 and Gsh2. *Cereb Cortex.* 19:745–759.

Carrera I, Ferreiro-Galve S, Sueiro C, Anadón R, Rodríguez-Moldes I (2008). Tangentially migrating GABAergic cells of subpallial origin invade massively the pallium in developing sharks. *Brain Res Bull.* 75:405–409.

Catalano SM, Robertson RT, Killackey HP (1991). Early ingrowth of thalamocortical afferents to the neocortex of the prenatal rat. *Proc Natl Acad Sci.* 88(8):2999–3003.

Chapouton P, Schuurmans C, Guillemot F, Gotz M (2001). The transcription factor neurogenin 2 restricts cell migration from the cortex to the striatum. *Development.* 128:5149–5159.

Chen Y, Magnani D, Theil T, Pratt T, Price DJ (2012). Evidence that descending cortical axons are essential for thalamocortical axons to cross the pallial–subpallial boundary in the embryonic forebrain. *PLoS ONE.* 7:e33105.

Clasca F, Angelucci A, Sur M (1995). Layer-specific programs of development in neocortical projection neurons. *Proc Natl Acad Sci.* 92(24):11145–11149.

Clowry GJ, Alzu'bi A, Harkin LF, Sarma S, Kerwin J, Lindsay S (2018). Charting the protomap of the human telencephalon. *Semin Cell Dev Biol.* 76:3–14.

Cobos I, Shimamura K, Rubenstein JLR, Salvador Martínez S, Puelles L (2001). Fate map of the Avian anterior forebrain at the four-somite stage, based on the analysis of Quail–Chick Chimeras. *Dev Biol.* 239:46–67.

Cobos I, Puelles L, Cobos I (2002). The avian telencephalic subpallium originates inhibitory neurons that invade tangentially the pallium (dorsal ventricular ridge and cortical areas). *Dev Biol.* 239:30–45.

Cooke S, Grant G, McLauchlan C, Lotto RB, Price DJ (1999). Basic fibroblast growth factor promotes subplate cell survival in explant cultures of embryonic mouse cortex. *Neurosci Lett.* 271(3):143–146.

De Carlos JA, O'Leary DD (1992). Growth and targeting of subplate axons and establishment of major cortical pathways. *J Neurosci.* 12(4):1194–1211.

Dehay C, Giroud P, Berland M, Killackey H, Kennedy H (1996). Contribution of thalamic input to the specification of cytoarchitectonic cortical fields in the primate: Effects of bilateral enucleation in the fetal monkey on the boundaries, dimensions, and gyrification of striate and extrastriate cortex. *J Comp Neurol.* 367:70–89.

Dehay C, Kennedy H (2007). Cell-cycle control and cortical development. *Nat Rev Neurosci.* 8:438–450.

Dehay C, Savatier P, Cortay V, Kennedy H (2001). Cell-cycle kinetics of neocortical precursors are influenced by embryonic thalamic axons. *J Neurosci.* 21(1):201–214.

Doyle DZ, Lam MM, Qalieh A, Qalieh Y, Sorel A, Funk OH, Kwan KY (2021). Chromatin remodeler Arid1a regulates subplate neuron identity and wiring of cortical connectivity. *Proc Natl Acad Sci.* 118:21. e2100686118.

Dwyer ND, Manning DK, Moran JL, Mudbhary R, Fleming MS, Favero CB, Vock VM, O'Leary DD, Walsh CA, Beier DR (2011). A forward genetic screen with a thalamocortical axon reporter mouse yields novel neurodevelopment mutants and a distinct emx2 mutant phenotype. *Neural Dev.* 6:3.

Edgar JM, Price DJ (2001). Radial migration in the cerebral cortex is enhanced by signals from thalamus. *Eur J Neurosci.* 13(9):1745–1754.

Feng J, Xian Q, Guan T, Hu J, Wang M, Huang Y, So KF, Evans SM, Chai G, Goffinet AM, Qu Y, Zhou L (2016). Celsr3 and Fzd3 organize a pioneer neuron scaffold to steer growing thalamocortical axons. *Cereb Cortex.* 26(7):3323–3334.

Frangeul L, Pouchelon G, Telley L, Lefort S, Luscher C, Jabaudon D (2016). A cross-modal genetic framework for the development and plasticity of sensory pathways. *Nature.* 538:96–98.

García-Moreno F, Anderton E, Jankowska M, Begbie J, Encinas JM, Irimia M, Molnár Z (2018). Absence of tangentially migrating glutamatergic neurons in the developing avian brain. *Cell Rep.* 22:96–109.

Garcia-Moreno F, Molnár Z (2020). Variations of telencephalic development that paved the way for neocortical evolution. *Progr Neurol.* 194:101865.

García-Moreno F, Vasistha NA, Trevia N, Bourne JA, Molnár Z (2012). Compartmentalization of cerebral cortical germinal zones in a lissencephalic primate and gyrencephalic rodent. *Cereb Cortex.* 22(2):482–492.

Garel S, Huffman KJ, Rubenstein JL (2003) Molecular regionalization of the neocortex is disrupted in Fgf8 hypomorphic mutants. *Development.* 130:1903–1914.

Gerstmann K, Pensold D, Symmank J, Khundadze M, Hubner CA, Bolz J, Zimmer G (2015). Thalamic afferents influence cortical progenitors via ephrin A5-EphA4 interactions. *Development.* 142(1):140–150.

Gezelius H, López-Bendito G. 2017. Thalamic neuronal specification and early circuit formation. *Dev Neurobiol.* 77:830–843.

Giasafaki C, Eleanor G, Hoerder-Suabedissen A, Hayashi S, Lee S, Molnár Z (2022). Cross-hierarchical plasticity of corticofugal projections to dLGN after neonatal monocular enucleation. *J Comp Neurol.* 530(7):978–997.

González-Arnay E, Gonzáález-Gómez M, Meyer G (2017). A radial glia fascicle leads principal neurons from the pallial-subpallial boundary into the developing human insula. *Front Neuroanat.* 11:111.

Grant E, Hoerder-Suabedissen A, Molnár Z (2012). Development of the corticothalamic projections. *Front Neurosci.* 6:05.

Grant E, Hoerder-Suabedissen A, Molnár Z (2016). The regulation of corticofugal fibre targeting by retinal inputs. *Cereb Cortex.* 26(3):1336–1348.

Hanashima C, Molnár Z, Fishell G (2006). Building bridges to the cortex. *Cell.* 125:24–27.

Harkin LF, Lindsay SJ, Xu Y, Alzu'bi A, Ferrara A, Gullon EA, James OG, Clowry GJ (2017). Neurexins 1–3 each have a distinct pattern of expression in the early developing human cerebral cortex. *Cereb Cortex.* 27:216–232.

Hayashi S, Hoerder-Suabedissen A, Kiyokage E, Maclachlan C, Toida K, Knott GW, Molnár Z (2020). Maturation of complex synaptic connections of layer 5 cortical axons in the posterior thalamic nucleus requires SNAP25. *Cereb Cortex.* 31(5):2625–2638.

Hevner RF, Miyashita-Lin E, Rubenstein JL (2002). Cortical and thalamic axon pathfinding defects in Tbr1, Gbx2, and Pax6 mutant mice: Evidence that cortical and thalamic axons interact and guide each other. *J Comp Neurol.* 447:8–17.

Higashi S, Hioki K, Kurotani T, Molnár Z (2005). Functional thalamocortical synapse reorganization from subplate to layer IV during postnatal development in the reeler-like mutant rat (shaking rat Kawasaki): An optical recording study. *J Neurosci.* 25(6):1395–406.

Higashi S, Molnár Z, Kurotani T, Inokawa H, Toyama K (2002). Functional thalamocortical connections develop during embryonic period in the rat: an optical recording study. *Neuroscience.* 115:1231–1246. doi: 10.1016/s0306-4522(02)00418-9. PMID: 12453494.

Hoerder-Suabedissen A, Hayashi S, Upton L, Nolan Z, Casas-Torremocha D, Grant E, Viswanathan S, Kanold PO, Clasca F, Kim Y, Molnár Z (2018). Subset of cortical layer 6b neurons selectively innervates higher order thalamic nuclei in mice. *Cereb Cortex.* 28(5):1882–1897.

Hoerder-Suabedissen A, Molnár Z (2012). Morphology of mouse subplate cells with identified projection targets changes with age. *J Comp Neurol.* 520:174–185.

Hoerder-Suabedissen A, Molnár Z (2015). Development, evolution and pathology of cortical subplate neurons. *Nat Rev Neurosci.* 16(3):133–146.

Hoerder-Suabedissen A, Oeschger FM, Krishnan ML, Belgard TG, Wang WZ, Lee S, Webber C, Petretto E, Edwards AD, Molnár Z (2013). Expression profiling of mouse subplate reveals a dynamic gene network and disease association with autism and schizophrenia. *Proc Natl Acad Sci U S A.* 110(9):3555–3560.

Išasegi IŽ, Janja Kopić J, Smilović D, Krsnik Ž, Kostović I (2022). Transient subplate sublayer forms unique corridor for differential ingrowth of associative pulvinar and primary visual projection in the prospective visual cortical areas of the human fetal occipital lobe. *Cereb Cortex.* 32(1):110–122.

Išasegi IŽ, Radoš M, Krsnik Ž, Radoš M, Benjak V, Kostović I (2018). Interactive histogenesis of axonal strata and proliferative zones in the human fetal cerebral wall. *Brain Struct Funct.* 223:3919–3943.

Qin J, Wang M, Zhao T, Xiao X, Li X, Yang J, Yi L, Goffinet AM, Qu Y, Zhou L (2020). Early forebrain neurons and scaffold fibers in human embryos. *Cereb Cortex.* 30(3):913–928.

Jones EG (2007). *The Thalamus.* Cambridge University Press.

Jones L, López-Bendito G, Gruss P, Stoykova A, Molnár Z (2002). Pax6 is required for the normal development of the forebrain axonal connections. *Development.* 129:5041–5052.

Kanold PO, Luhmann HJ (2010). The subplate and early cortical circuits. *Annu Rev Neurosci.* 33:23–48.

Kostović I, Rakic P (1984). Development of prestriate visual projections in the monkey and human fetal cerebrum revealed by transient cholinesterase staining. *J Neurosci.* 4:25–42.

Kostović I, Rakic P (1990). Developmental history of the transient subplate zone in the visual and somatosensory cortex of the macaque monkey and human brain. *J Comp Neurol.* 297:441–470.

Krsnik Ž, Majić V, Vasung L, Huang H, Kostović I (2017). Growth of thalamocortical fibers to the somatosensory cortex in the human fetal brain. *Front Neurosci.* 11:233.

Krubitzer LA, Prescott TJ (2018). The combinatorial creature: Cortical phenotypes within and across lifetimes. *Trends Neurosci.* 41:744–762.

Krug K, Smith AL, Thompson ID (1998). The development of topography in the hamster geniculo-cortical projection. *J Neurosci.* 18:5766–5776.

López-Bendito G, Cautinat A, Sánchez JA, Bielle F, Flames N, Garratt AN, Talmage DA, Role LW, Charnay P, Marín O, Garel S (2006). Tangential neuronal migration controls axon guidance: A role for neuregulin-1 in thalamocortical axon navigation. *Cell.* 125:127–142.

López-Bendito G, Chan C-H, Mallamaci A, Parnavelas J, Molnár Z (2002). The role of Emx2 in the development of the reciprocal connectivity between cortex and thalamus. J. Comp. Neurol. 451:153–169.

López-Bendito G, Molnár Z (2003). Thalamocortical development: How are we going to get there? *Nat Rev Neurosci.* 4: 276–289.

Lotto RB, Aitkenhead A, Price DJ (1999). Effects of the thalamus on the development of cerebral cortical efferents in vitro. *J Neurobiol.* 39(2):186–196.

Lukaszewicz A, Savatier P, Cortay V, Giroud P, Huissoud C, Berland M, Kennedy H, Dehay C (2005). G1 phase regulation, area-specific cell cycle control, and cytoarchitectonics in the primate cortex. *Neuron.* 47(3):353–364.

Metin C, Alvarez C, Moudoux D, Vitalis T, Pieau C, Molnar Z (2007). Conserved pattern of tangential neuronal migration during forebrain development. *Development.* 134:2815–2827.

Métin C, Godement P (1996). The ganglionic eminence may be an intermediate target for corticofugal and thalamocortical axons. *J Neurosci.* 16:3219–3235.

Miyashita-Lin EM, Hevner R, Wassarman KM, Martinez S, Rubenstein JL (1999). Early neocortical regionalization in the absence of thalamic innervation. *Science.* 285:906–909.

Molnár Z (1998). *Development of Thalamocortical Connections.* Springer.

Molnár Z (2019). Cortical layer with no known function. *Eur J Neurosci.* 49(7):957–963.

Molnár Z (2011). Evolution of cerebral cortical development. *Brain Behav Evol.* 78(1):94–107.

Molnár Z, Adams R, Blakemore C (1998a). Mechanisms underlying the early establishment of thalamocortical connections in the rat. *J Neurosci.* 18:5723–5745.

Molnár Z, Adams R, Goffinet AM, Blakemore C (1998b). The role of the first postmitotic cortical cells in the development of thalamocortical innervation in the reeler mouse. *J Neurosci.* 18:5746–5765.

Molnár Z, Blakemore C (1991). Lack of regional specificity for connections formed between thalamus and cortex in coculture. *Nature.* 351:475–477.

Molnár Z, Blakemore C (1995). How do thalamic axons find their way to the cortex? *Trends Neurosci.* 18:389–397.

Molnár Z, Blakemore C (1999). Development of signals influencing the growth and termination of thalamocortical axons in organotypic culture. *Exp Neurol.* 156:363–393.

Molnár Z, Butler AB (2002). The corticostriatal junction: a crucial region for forebrain development and evolution. *BioEssays.* 24:530–541.

Molnár Z, Cordery P (1999). Connections between cells of the internal capsule, thalamus, and cerebral cortex in embryonic rat. *J Comp Neurol.* 413:1–25.

Molnár Z, Garel S, López-Bendito G, Maness P, Price DJ (2012). Mechanisms controlling the guidance of thalamocortical axons through the embryonic forebrain. *Eur J Neurosci.* 35(10):1573–1585.

Molnár Z, Higashi S, López-Bendito G (2003). Choreography of early thalamocortical development. *Cereb Cortex.* 13(6):661–669.

Molnár Z, Knott GW, Blakemore C, Saunders NR (1999). Development of thalamocortical projections in the South American gray short-tailed opossum (*Monodelphis domestica*). *J Comp Neurol.* 398:491–514.

Molnár Z, López-Bendito G, Blakey D, Thompson A, Higashi S (2006). Chapter 4. *Earliest Thalamocortical Interactions.* In: Erzurumlu R, Guido W, Molnár Z, eds.: *Development and Plasticity in Sensory Thalamus and Cortex.* Springer:54–78.

Molnár Z, López-Bendito G, Small J, Partridge LD, Blakemore C, Wilson MC (2002). Normal development of embryonic thalamocortical connectivity in the absence of evoked synaptic activity. *J Neurosci.* 22:10313–10323.

Molnár Z, Luhmann HJ, Kanold PO (2020). Transient local and global circuits match spontaneous and sensory driven activity patterns during cortical development. *Science.* 370(6514):eabb2153.

Molnár Z, Vasistha NA, Garcia-Moreno F (2011). Hanging by the tail: Progenitor populations proliferate. *Nat Neurosci.* 14(5):538–540.

Monko T, Rebertus J, Stolley J, Salton SR, Nakagawa Y (2022). Thalamocortical axons regulate neurogenesis and laminar fates in the early sensory cortex. *Proc Natl Acad Sci U S A.* 119(22):e2201355119.

Montiel J, Wang W, Oeschger F, Hoerder-Suabedissen A, Tung W, García-Moreno F, Holm IE, Villalón A, Molnár Z (2011). Hypothesis on the dual origin of the mammalian subplate. *Front Neuroanat.* 5:25.

Moreno N, González A, Rétaux S (2008). Evidences for tangential migrations in Xenopus telencephalon: developmental patterns and cell tracking experiments. *Dev Neurobiol.* 68:504–520.

Moreno-Juan V, Filipchuk A, Antón-Bolanos N, Mezzera C, Gezelius H, Andrés B, Rodríguez-Malmierca L, Susín R, Schaad O, Iwasato T, Schüle R, Rutlin M, Nelson S, Ducret S, Valdeolmillos M, Rijli FM, López-Bendito G (2017). Prenatal thalamic waves regulate cortical area size prior to sensory processing. *Nat Commun.* 8:14172.

Mundinando I-C, Kwan WC, Bourne JA (2015). Mapping the mosaic sequence of primate visual cortical development. *Front Neuroanat.* 9:132.

Naegele JR, Jhaveri S, Schneider GE (1988). Sharpening of topographical projections and maturation of geniculocortical axon arbors in the hamster. *J Comp Neurol.* 277:593–607.

Nelson SB, Le Vay S (1985). Topographic organization of the optic radiation of the cat. *J Comp Neurol.* 240:322–330.

Norimoto H, Fenk LA, Li HH, Tosches MA, Gallego-Flores T, Hain D, Reiter S, Kobayashi R, Macias A, Arends A, Klinkmann M, Laurent G (2020). A claustrum in reptiles and its role in slow-wave sleep. *Nature.* 578(7795):413–418.

Nowakowski TJ, Pollen AA, Sandoval-Espinosa CS, Kriegstein AR (2016). Transformation of the radial glia scaffold demarcates two stages of human cerebral cortex development. *Neuron.* 91(6):1219–1227.

O'Leary DDM (1989). Do cortical areas emerge from a protocortex? *Trends Neurosci.* 12:400–406.

O'Leary DD, Sahara S (2008). Genetic regulation of arealization of the neocortex. *Curr Opin Neurobiol.* 18:90–100.

Oeschger FM, Wang WZ, Lee S, García-Moreno F, Goffinet AM, Arbonés ML, Rakic S, Molnár Z (2012). Gene expression analysis of the embryonic subplate. *Cereb Cortex.* 22(6):1343–1359.

Osheroff H, Hatten ME (2009). Gene expression profiling of preplate neurons destined for the subplate: genes involved in transcription, axon extension, neurotransmitter regulation, steroid hormone signaling, and neuronal survival. *Cereb Cortex.* 19(Suppl 1):126–134.

Piñon MC, Jethwa A, Jacobs E, Campagnoni A, Molnár Z (2009). Dynamic integration of subplate neurons into the cortical barrel field circuitry during postnatal development in the Golli-tau-eGFP (GTE) mouse. *J Physiol.* 587(Pt 9):1903–1915.

Piñon MC, Tuoc TC, Ashery-Padan R, Molnár Z, Stoykova A (2008). Altered molecular regionalization and normal thalamocortical connections in cortex-specific Pax6 knock-out mice. *J Neurosci.* 28:8724–8734.

Polleux F, Dehay C, Moraillon B, Kennedy H (1997). Regulation of neuroblast cell-cycle kinetics plays a crucial role in the generation of unique features of neocortical areas. *J Neurosci.* 17(20):7763–7783.

Puelles L (2017). Comments on the updated tetrapartite pallium model in the mouse and chick, featuring a homologous claustro-insular complex. *Brain Behav Evol.* 90(2):171–189.

Rakic P (1974). Neurons in rhesus monkey visual cortex: Systematic relation between time of origin and eventual disposition. *Science.* 183:425–427.

Rakic P (1988). Specification of cerebral cortical areas. *Science.* 241:170–176.

Rakic P, Suner I, Williams RW (1991). A novel cytoarchitectonic area induced experimentally within the primate visual cortex. *Proc Natl Acad Sci U S A.* 88:2083–2087.

Rakic P, Williams RW (1986). Thalamic regulation of cortical parcellation: An experimental perturbation of the striate cortex in rhesus monkey. *Soc Neurosci Abstr.* 12:1149.

Rash G, Grove EA (2006). Area and layer patterning in the developing cerebral cortex. *Curr Opin Neurobiol.* 16(1):25–34.

Ravary A, Muzerelle A, Hervé D, Pascoli V, Ba-Charvet KN, Girault JA, Welker E, Gaspar P (2003). Adenylate cyclase 1 as a key factor in the refinement of retinal projection maps. *J Neurosci.* 23:2228–2238.

Reillo I, de Juan Romero C, Garcia-Cabezas MA, Borrell V (2011). A role for intermediate radial glia in the tangential expansion of the mammalian cerebral cortex. *Cereb Cortex.* 21(7):1674–1694.

Sato H, Hatakeyama J, Iwasato T, Araki K, Yamamoto N, Shimamura K (2022). Thalamocortical axons control the cytoarchitecture of neocortical layers by area-specific supply of VGF. *eLife.* 11: e67549.

Shatz CJ, Rakic P (1981). The genesis of efferent connections from the visual cortex of the fetal rhesus monkey. *J Comp Neurol.* 196(2):287–307.

Sherman SM, Guillery RW (2005). *Exploring the Thalamus and Its Role in Cortical Function.* MIT Press.

Shi W, Xianyu A, Han Z, Tang X, Li Z, Zhong H, Mao T, Huang K, Shi SH (2017). Ontogenetic establishment of order-specific nuclear organization in the mammalian thalamus. *Nat Neurosci.* 20(4):516–528.

Shimogori T, Grove EA (2005) Fibroblast growth factor 8 regulates neocortical guidance of area-specific thalamic innervation. *J Neurosci.* 25:6550–6560.

Smart IH, Dehay C, Giroud P, Berland M, Kennedy H (2002). Unique morphological features of the proliferative zones and postmitotic compartments of the neural epithelium giving rise to striate and extrastriate cortex in the monkey. *Cereb Cortex.* 12(1):37–53.

Spadory T, Duque A, Selemon LD (2022). Spatial–temporal topography in neurogenesis of the macaque thalamus. *Brain Struct Funct.* 227:1673–1682.

Suárez R, Paolino A, Fenlon LR, Morcom LR, Kozulin P, Kurniawan ND, Richards LJ (2018). A pan-mammalian map of interhemispheric brain connections predates the evolution of the corpus callosum. *Proc Natl Acad Sci U S A.* 115(38):9622–9627.

Sur M, Leamey CA (2001). Development and plasticity of cortical areas and networks. *Nat Rev Neurosci.* 2(4):251–262.

Toda T, Shinmyo Y, Duong TAD, Masuda K, Kawasaki H (2016) An essential role of SVZ progenitors in cortical folding in gyrencephalic mammals. *Sci Rep.* 6:29578.

Tolner, E.A., Sheikh, A., Yukin, A.Y, Kaila K, Kanold PO (2012). Subplate neurons promote spindle bursts and thalamocortical patterning in the neonatal rat somatosensory cortex. *J Neurosci.* 32:692–702.

Tuttle R, Nakagawa Y, Johnson JE, O'Leary DD (1999). Defects in thalamocortical axon pathfinding correlate with altered cell domains in Mash-1-deficient mice. *Development.* 126:1903–1916.

Valverde F (1967). Apical dendritic spines of the visual cortex and light deprivation in the mouse. *Exp Brain Res.* 3(4):337–352.

Viswanathan S, Bandyopadhyay S, Kao JP, Kanold PO (2012). Changing microcircuits in the subplate of the developing cortex. *J Neurosci.* 32:1589–1601.

Wang R, Wilkinson M, Kane T, Takahashi E (2017). Convergence of cortical, thalamocortical, and callosal pathways during human fetal development revealed by diffusion MRI Tractography. *Front Neurosci.* 11:576. doi:10.3389/fnins.2017.00576. PMID: 29163000; PMCID: PMC5671991.

Workman AD, Charvet CJ, Clancy B, Darlington RB, Finlay BL (2013). Modeling transformations of neurodevelopmental sequences across mammalian species. *J Neurosci.* 33:7368–7383.

Yamamoto N, López-Bendito G (2012) Shaping brain connections through spontaneous neural activity. *Eur J Neurosci.* 35: 1595–1604.

Zhou L, Gall D, Qu Y, Prigogine C, Cheron G, Tissir F, Schiffmann SN, Goffinet AM (2010). Maturation of "neocortex isole" in vivo in mice. *J Neurosci.* 30:7928–7939.

59.

EVOLUTION AND ELABORATION OF THALAMOCORTICAL ACTION MAPS

Paul Cisek and Ayuno Nakahashi

INTRODUCTION

The cerebral cortex is widely considered as the seat of human intelligence, responsible for our perceptual, intellectual, and manipulative skills. Its role in neuroscience is so prominent, in fact, that much of the rest of the forebrain is often referred to as "subcortical," below the cortex anatomically but also, one could say, in terms of theoretical prominence. Of course, as neuroscientific knowledge grows, we increasingly appreciate the distributed nature of neural processing, recognizing that functional capacities arise not from modules contained within specific cortical regions of interest, but rather from an interplay between cortical and subcortical systems. Indeed, a given patch of cerebral cortex can perhaps best be understood as a component of converging thalamocortical, cortico-striatal, and cortico-cerebellar circuits.

This distributed picture is most strongly apparent when one examines the growing literature on brain evolution (Figure 59.1). Evolutionary data suggest that the first recognizable structure of the nervous system of our ancestors was the hypothalamus (Arendt et al., 2016), whose elements appeared even before there was any organized brain to speak of, and whose basic function of feedback regulation is the foundation of all behavior (Powers, 1973; Cisek, 2019). The spinal cord and midbrain formed next in mobile chordates (Lacalli, 2008; Albuixech-Crespo et al., 2017), followed by the telencephalon in early vertebrates (Benito-Gutiérrez et al., 2018), extending feedback control toward increasingly sophisticated means of interaction with the world. That early telencephalon already included something like the thalamocortical and cortico-striatal system (Suryanarayana et al., 2022), though without much of the complexity found in mammals or birds. The cerebellum appeared next, in jawed fish, conferring predictive control and completing the basic topological outline (the "Bauplan") of all modern vertebrate brains (Nieuwenhuys and Puelles, 2016). The mammalian cerebral cortex is really just a modification of part of that plan—expanding and elaborating part of the forebrain (the "dorsal pallium"). But the story of evolution is not one of new structures being added upon old ones, like the long-abandoned theory of a mammalian brain added atop a reptilian one. Instead, it is a story of how the circuits (thalamocortical, cortico-striatal, and cortico-cerebellar) expanded and diversified along with the expanding behavioral repertoire.

The goal of the present chapter is to summarize part of that story—the transition from amniotes to early mammals—and to propose some hypotheses on the functional significance of the changes that occurred in thalamocortical and cortico-striatal circuits as the cerebral cortex took on a more central role in controlling interactive behavior. We will propose that it filled that role through specialized action circuits (Stepniewska et al., 2009; Graziano, 2016; Stepniewska et al., 2020) that controlled different kinds of species-typical behavior (e.g., searching, handling, feeding, defense, etc.). Each of these circuits was responsible for specifying and selecting among potential actions of a given type (Cisek, 2007), and their activation at different moments in time was orchestrated by the basal ganglia (Redgrave et al., 1999). It must be noted that this proposal is just a hypothesis, constrained on one side by inferences about basal vertebrates and amniotes, and on the other by data from modern mammals (particularly rodents and primates), which extrapolates putative mechanisms that might have existed in the last common ancestors of mammals over 220 million years ago (mya). While formulating such a hypothesis is clearly a precarious exercise, it nevertheless yields some viable predictions that can be tested, and we hope will be tested, in future experimental work with a wide range of animals.

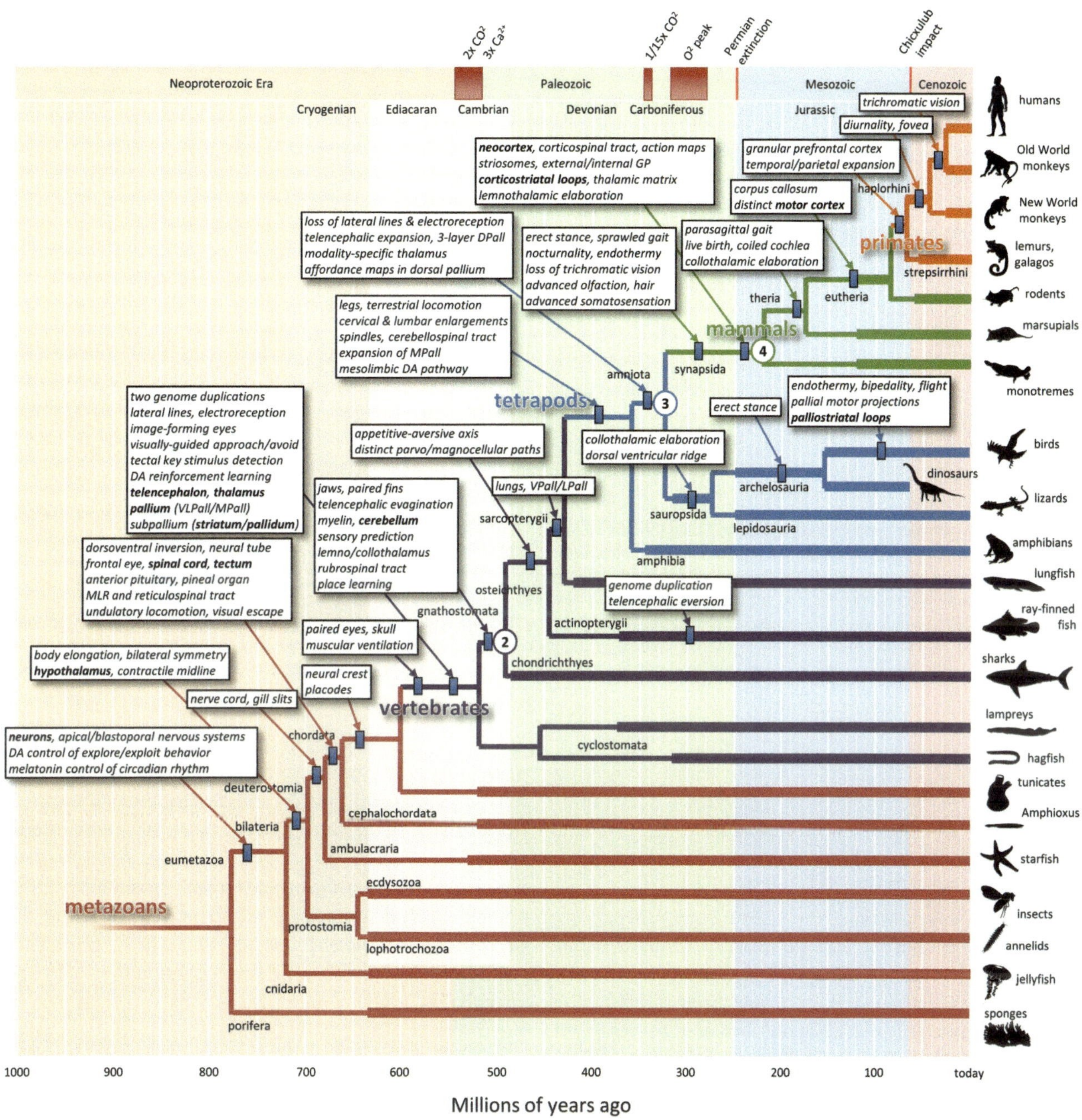

Figure 59.1. Phylogenetic tree of animals, emphasizing the lineage that produced humans, based on Cisek (2022). Small rectangles indicate the latest estimated timing of the innovations listed in the boxes. Circled numbers refer to figures below. The silhouettes of example species are from http://phylopic.org.

EARLY VERTEBRATES, TETRAPODS, AND AMNIOTES

While a thorough discussion of early vertebrate brains is beyond the scope of this chapter, the basic outline is useful for defining the context within which the mammalian thalamocortical and cortico-striatal systems evolved. What follows is a brief description of that outline, starting with early filter-feeding chordates and finishing with the land-dwelling amniote ancestors of modern mammals.

From a topological perspective, all chordate nervous systems are organized as a tube consisting of segments arranged rostrocaudally along the dorsal midline of the body (Nieuwenhuys and Puelles, 2016), as shown in Figure 59.2A. Studies comparing cephalochordates, tunicates, and vertebrates suggest that in the earliest chordates, the most rostral segment was homologous to the hypothalamus, which controlled more caudal segments, homologous to the spinal cord, through descending projections that governed slow swimming (Wicht and Lacalli, 2005).

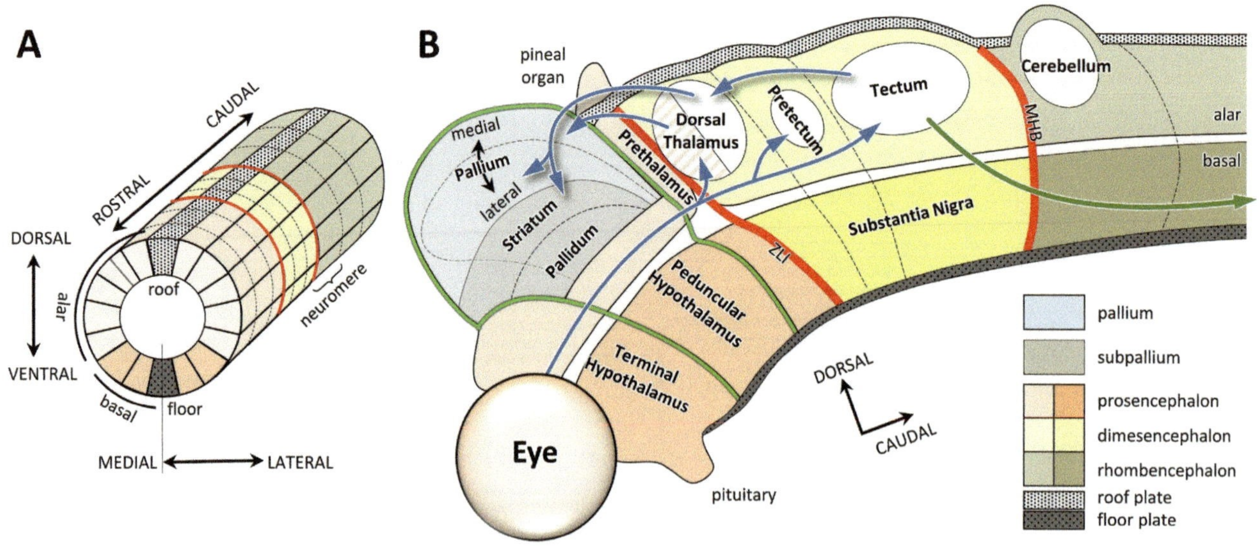

Figure 59.2. Topological organization of the neural tube of vertebrates. (A). The basic structure of the tube, consisting of rostrocaudal segments called neuromeres, each with alar and basal sectors surrounding the ventricular canal. (B). Sagittal view of the rostral tube of vertebrates. The thick red lines indicate 2 gene expression boundaries, the zona limitans intrathalamica (ZLI) and the midbrain-hindbrain boundary (MHB), which delineate the major subdivisions. The first 3 segments are the prosencephalon, which includes the hypothalamus and the prethalamus. The telencephalon (pallium and subpallium) is an alar expansion of the second hypothalamic segment, outlined in green. The next 3 segments are the dimesencephalon, which includes the dorsal thalamus and midbrain. The remaining segments form the hindbrain and spinal cord. Blue arrows show the 2 main routes of visual information, one to the anterior dorsal thalamus (lemnothalamus, striped) and to medial pallium, and one to the tectum and then through the posterior dorsal thalamus (collothalamus, solid) to lateral pallium and striatum. The green arrow shows the tectospinal tract.

In addition, early chordates possessed a rapid escape circuit, which involved a single patch of photoreceptors projecting to a structure in what would later become the dorsal midbrain, likely homologous to the vertebrate tectum (Wicht and Lacalli, 2005), that then projected to the spinal cord. Thus, if a shadow fell on the photoreceptors, the animal would quickly swim away.

In early vertebrates, the photosensitive patch split into two that migrated to the sides of the head (Butler, 2000), forming the lateral eyes, whose projections to the tectum were primarily contralateral. Ipsilateral tectospinal projections mediated oriented escape, while contralateral tectospinal projections mediated approach actions (Saitoh et al., 2007), an organization found in all modern vertebrates (Comoli et al., 2012; Helmbrecht et al., 2018). In addition, part of the hypothalamus enlarged into what would become the telencephalon, including both an external sector called the pallium and an internal sector called the subpallium. The subpallium is homologous to structures including the striatum and pallidum of the basal ganglia, whose dopaminergic reinforcement and descending modulatory control of tectal sensorimotor mechanisms has been conserved for at least 500 million years (Suryanarayana et al., 2022). The pallium is homologous to much of the rest of the forebrain. Figure 59.2B shows some of the major segments shared by all jawed vertebrates.

A recent study by Albuixech-Crespo et al. (2017) revealed where the thalamus belongs in this general scheme. Surprisingly, it does not fit within the classical concept of a *diencephalon*, but is instead composed of 2 parts with somewhat different evolutionary and developmental origins. The dorsal thalamus, including the nuclei that receive sensory input, derives from a segment called the *dimesencephalon*, which also includes the optic tectum (mammalian superior colliculus) and pretectum, all structures receiving direct projections from the retina. In contrast, the prethalamus, including the thalamic reticular nucleus, is part of the prosencephalon, together with the hypothalamus and the rest of the forebrain. In all jawed vertebrates, the dorsal thalamus can be further subdivided into 2 parts (Butler, 2008): an anterior sector called the *lemnothalamus*, which receives visual input from the retina; and a posterior sector called the *collothalamus*, which receives visual input indirectly, via the tectum. In general, the lemnothalamus projects primarily to the medial part of the pallium, while the collothalamus projects to more lateral parts of the pallium as well as the striatum—though this situation will become considerably more complicated in later stages of evolution.

In ancient fish, direct control of sensorimotor behavior was primarily the role of the midbrain, with telencephalic contributions limited to modulatory projections through the basal ganglia (Striedter and Northcutt, 2019). When our ancestors first came out on land, this architecture was largely retained, though many of its parts expanded and elaborated, while others regressed. For example, given the dramatic increase of visual range (MacIver et al., 2017), the

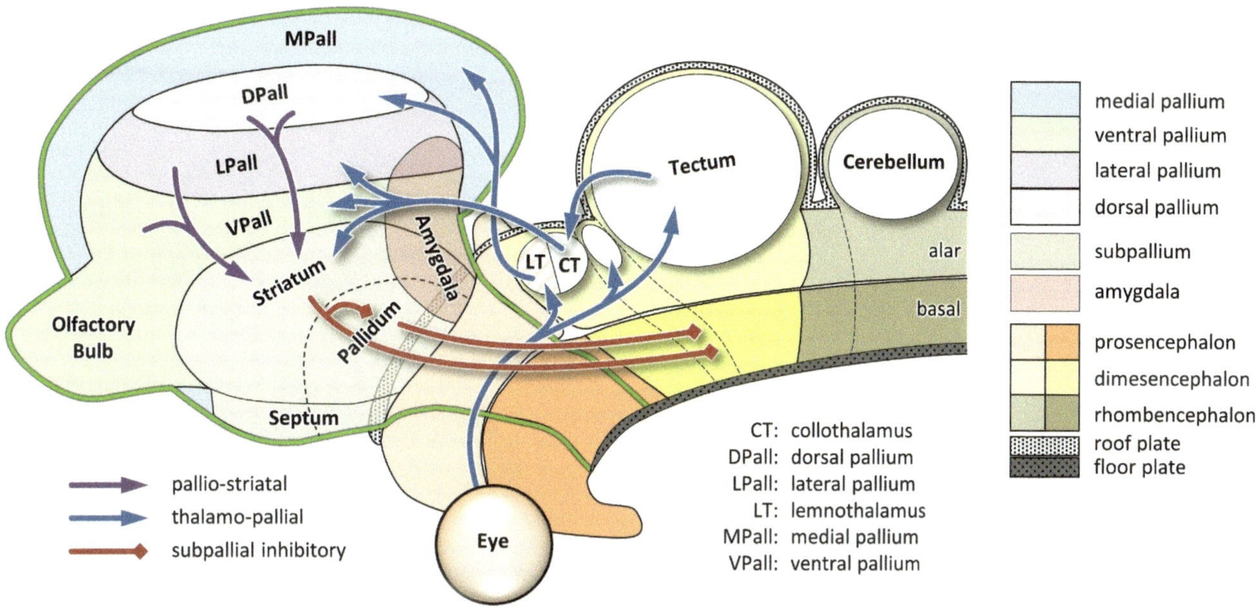

Figure 59.3. Expansion of the telencephalon in amniotes. Here, the 4 subdivisions of the pallium are shown (ventral, lateral, dorsal, and medial), as well as pallio-striatal projections (purple) and downstream projections from the subpallium to the basal midbrain (red).

retino-tecto-spinal mechanisms became even more elaborate, as did lemnothalamic projections to the expanded medial pallium, whose role in navigation could now rely on very distant landmarks. Meanwhile, the lateral line system regressed in significance, and was lost in terrestrial adults.

As amniotes became fully land-dwelling, their behavioral repertoire expanded to exploit the richness of the terrestrial world. This was made possible not only by further expansion of the tectum, but also by an expansion of the telencephalic pallium and its efferent and afferent projections. In particular, while the dorsal thalamus of extant amphibians is largely multisensory, in amniotes there are segregated pathways dedicated to visual, auditory, and somatosensory input (Desfilis et al., 1998; Medina and Reiner, 2000), which target several distinct subregions within the pallium.

Cisek (2022) suggested that the more sophisticated thalamo-pallial circuitry of amniotes allowed more elaborate processing of sensory information useful for specific distinct purposes (Figure 59.3). The first, in ventral and lateral pallium, involved detecting the "key stimuli" (Hinde, 1966) that indicate cues for motivating particular types of actions, such as approaching and ingesting food versus engaging the escape circuits of the midbrain. The second, in dorsal pallium, involved detecting the "affordances" (Gibson, 1979) for interacting with the world, such as finding shelter or a place to rest. The third, in medial pallium, involved guiding navigation using landmarks and odor cues (Jacobs, 2012), elaborating mechanisms already present in anamniotes. However, despite their elaboration, these pallial systems did not themselves control the details of actions, but rather just selectively activated downstream systems, such as the tectospinal circuits, as befitted the current behavioral context.

In reptiles, which dominated the diurnal world, the major sites of elaboration were in the tectum and the tecto-recipient regions of the ventral and lateral pallium, which form the dorsal ventricular ridge. Birds took these elaborations even further, but a discussion of their specific innovations is beyond the scope of this chapter. By contrast, mammals retreated into a nocturnal niche, relying less on vision and more on olfaction, audition, and somatosensation. This led to the expansion of the dorsal pallium, which became the major target of modality-specific projections and gradually took on the primary role of guiding interactive behavior.

THE BRAINS OF EARLY MAMMALS

Figure 59.4 illustrates the topological organization of the hypothetical last common ancestor of monotremes and therian mammals, which lived about 220 mya during the late Triassic period. Already by this time, the dorsal pallium had become the neocortex, with its unusual inside-out developmental sequence resulting in a characteristic 6-layer architecture where, unlike in non-mammalian species, thalamic afferents arrive in a columnar fashion. This allowed the neocortex to expand dramatically without incurring excess connectivity costs. The mammalian pallium is organized as a concentric ring architecture with allocortical ventral

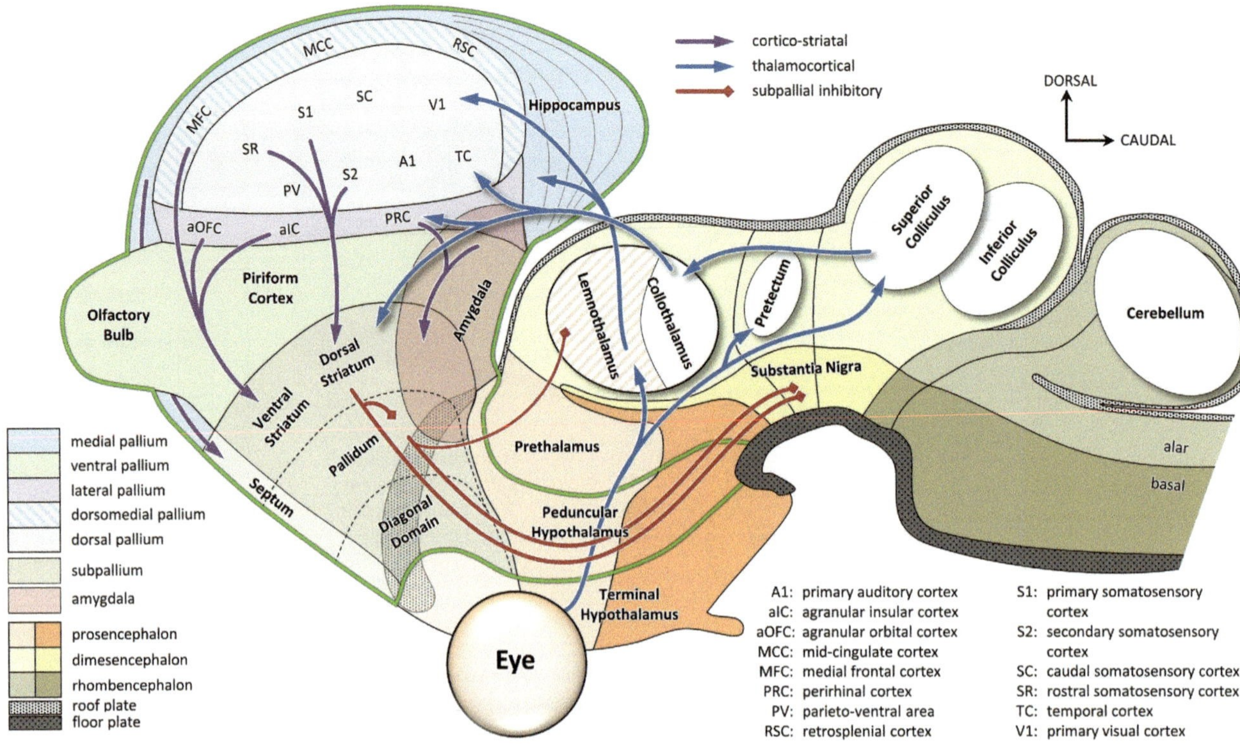

Figure 59.4. Topological organization of a hypothetical ancestral mammalian brain (based on Kaas, 2013; Puelles et al., 2013; Puelles et al., 2019). Blue arrows indicate the 2 major visual pathways, as in Figure 59.2, and purple arrows indicate projections from the pallium to the subpallium, including those of the septo-hippocampal, ventral striatal, dorsal striatal, and amygdalar circuits. Red arrows show the subpallial projections for the dorsal striatal circuit.

and medial pallial regions (3-layer piriform cortex and hippocampus, respectively) surrounding agranular regions of lateral pallium (orbital, insular, and perirhinal cortex) and a medial portion of the dorsal pallium (medial frontal, mid-cingulate, and retrosplenial cortex), themselves surrounding a central island of 6-layered granular neocortex (Puelles et al., 2019). According to the "structural model" of Barbas and colleagues (García-Cabezas et al., 2019), the main bottom-up feedforward flow of sensory information is from that central island outward to the dysgranular and agranular rings, with top-down feedback returning in the opposite direction (see **Chapter 38 in Section VII**).

When thalamic afferents shifted dorsally in mammalian ancestors (Aboitiz et al., 2002), they created several distinct modality-specific regions. These include visual and auditory sensory fields in caudal cortex, as well as 5 somatosensory fields in more rostral regions (Krubitzer et al., 1995; Beck et al., 1996; Huffman et al., 1999; Halley et al., 2020; Kaas et al., 2020). Of these, the *primary somatosensory area* (S1, probably homologous to Brodmann's area 3b in humans) included a complete somatotopic map of the body, with the hindlimbs represented medially and the nose and snout laterally. Just anterior to it lay a *rostral somatosensory area* (SR, putatively homologous to area 3a and primary motor cortex in humans), which mostly responded to tactile inputs on the mouth, vibrissae, and forelimb. Just posterior lay a *caudal somatosensory area* (SC, putatively homologous to areas 1 and 2 and anterior area 5), dominated by the forelimbs and hindlimbs. More laterally there were 2 other regions: the *secondary somatosensory area* (S2) and the *parietal ventral area* (PV), which both included almost a full somatotopy but overrepresenting the mouth and vibrissae. Note that in these early mammals, as in modern monotremes and marsupials, all of these regions played both sensory and motor roles and only differentiated into primary motor versus primary somatosensory cortex in placental mammals (Lende, 1963; Halley et al., 2020).

In addition to the changes in the microarchitecture of the dorsal pallium and its innervation by thalamocortical afferents, a major topological change in the brains of early mammals involved an elaboration of descending projections to the dimesencephalon, particularly to the ventral midbrain (including the red nucleus), optic tectum (now called the superior colliculus), and dorsal thalamus. With a larger number of projections, the particular pattern of neural activity in a source structure can more finely modulate the neural activity of the target. Thus, we suggest that whereas the dorsal pallium of pre-mammalian amniotes was mostly involved in detecting affordances to activate the visuomotor guidance circuits of the midbrain, the expanded neocortex could exert direct control over that guidance (Cisek, 2022). For example, rather than just turning on midbrain locomotion circuits when the animal is oriented toward a hiding spot, it could now precisely sculpt the activity in downstream circuits to adapt locomotion to the spatial details of that hiding spot. Thus, regions in the dorsal pallium specialized for detecting different affordances (a place to hide, a surface to rest upon, a thing to satisfy thirst, etc.) gradually turned into a set of specialized action circuits for controlling different species-typical actions (Graziano, 2016).

NEOCORTICAL ACTION MAPS

Graziano (2016) proposed that the functions of sensorimotor neocortex, particularly the motor, premotor, and parietal cortices, can be understood not just in terms of maps of the body, but as a collection of control networks adapted to the species-typical repertoire of ethologically relevant actions. The initial evidence for this concept emerged with the use of long trains (e.g., 500ms) of electrical microstimulation of the macaque motor cortex, which produced a variety of complex actions such as reaching movements, hand-to-mouth movements, as well as defensive postures and facial expressions (Graziano et al., 2002; Cooke and Graziano, 2004). Chemical manipulation of cortical areas that invoked defensive actions resulted in exaggerated or attenuated responses to an air puff on the face, further suggesting its role in producing ethologically meaningful responses to unpleasant stimuli (Cooke and Graziano, 2004). Similar effects were later found through microstimulation in a wide range of species (reviewed by Graziano, 2016; Kaas et al., 2020). For example, Stepniewska et al. (2014) demonstrated interactions between these different domains in New World monkeys and galagos: when a specific action domain in the primary motor cortex was chemically inactivated, microstimulation of the matching domain in the premotor and posterior parietal cortex resulted in diminished movement, while that of non-matching domains was preserved. Furthermore, simultaneous stimulation of action domains in primary motor cortex and posterior parietal cortex resulted in movements that suggest domain-dependent interactions, such that stimulation of matching domains produced more pronounced movements, whereas that of non-matching domains produced partial or unitary movements (Stepniewska et al., 2020). The concept of "action maps" was thus proposed to assimilate the fragmented body map obtained from traditional, short trains of electrical stimulation, and various ethologically meaningful movements elicited by long stimulation, to understand the function of precentral gyrus as an ensemble of motor systems. Computational modeling studies using self-organizing feature maps demonstrated that the basic arrangement of that fragmented organization could emerge simply through use-dependent convergence of projections to body parts involved in different types of behavior (Aflalo and Graziano, 2006). Thus, while the concept of action maps has initially been quite controversial, it has now been replicated in many species by many independent laboratories, and in our opinion the data now strongly suggest that it is fundamental to understanding neocortical organization from early mammals to primates (Cisek, 2022).

Importantly, different types of actions require very different kinds of action maps, with different relationships between sensory information and motor activity. For example, if you're a mouse exploring a kitchen when someone suddenly enters, then a policy of maintaining contact with a wall on one side of the body is a good strategy for finding a hole in that wall. However, if you're a short-tailed opossum trying to eat some insects, then you want to guide your snout to their nest and flick out your tongue in the right direction. Therefore, one might expect to find within the neocortex different action maps, each specialized for different aspects of an animal's behavioral repertoire. These include several kinds of foraging activity, including searching for food and ingesting it, as well as defensive behaviors such as flight toward shelter and freezing to escape detection. Figure 59.5 shows a preliminary proposal on how these may have mapped to the neocortex of early mammals.

Searching involves a variety of locomotor patterns involving the trunk, forelimbs, and hindlimbs. It is motivated by odors but specified by tactile as well as visual information, and probably relies heavily on input from the navigation system in the medial pallium—the hippocampus. We propose that it involved medial agranular areas lying adjacent to the medial pallium, including retrosplenial cortex (RSC), as well as somatosensory parts of mid-cingulate cortex (MCC). Retrosplenial cortex is interconnected with the hippocampal subiculum, and is strongly implicated in navigation using visual landmarks (Mitchell et al., 2018). It contains cells sensitive to head direction and spatial layout (Jacob et al., 2017), including some specifically involved in keeping track of the direction toward relevant locations (Campagner et al., 2022), and its chemical inactivation impairs place-related contextual memory retrieval (Tsai et al., 2022). The mid-cingulate cortex includes areas involved in the initiation and termination of locomotion (Sachuriga et al., 2021), and work in cats suggests that they are sensitive to the spatial layout of the environment and may mediate between visual cortical regions and the hippocampus (Olson and Musil, 1992). In prosimians, these regions are interconnected with a medial part of parietal cortex that appears specialized for climbing and running (Wang et al., 2021). In macaques, homologous regions appear to have become highly differentiated, and include the supplementary motor area (SMA) and the rostral,

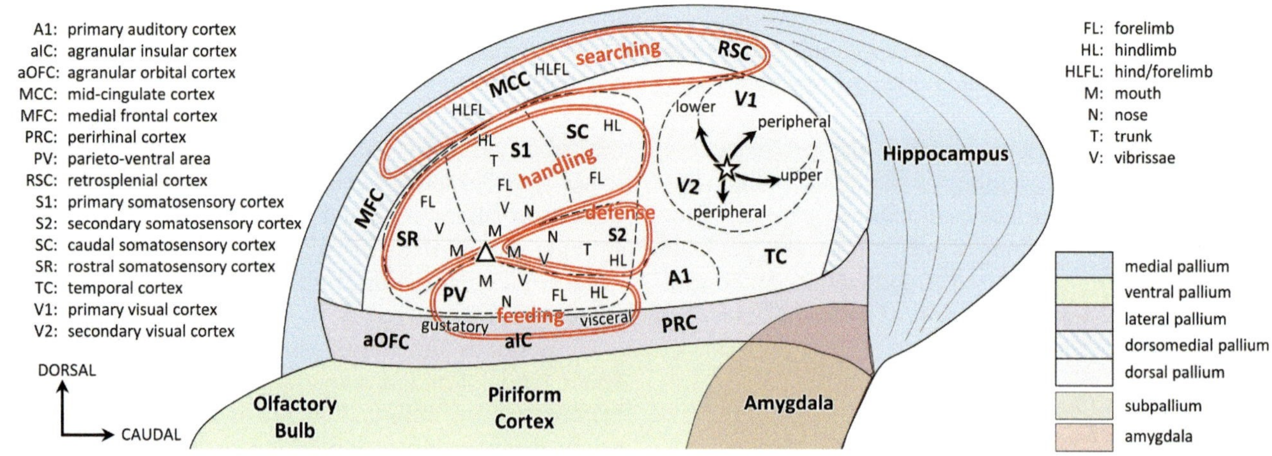

Figure 59.5. Simplified schematic of the proposed organization of the cerebral cortex in early mammals. The star indicates the convergence of primary and secondary visual areas at a region sensitive to central space, and black arrows indicate regions sensitive to the upper, lower, and contralateral peripheral visual field. Small black letters indicate the body parts to which a given cortical site is sensitive, and the triangle indicates the convergence of 4 of the 5 somatosensory maps at a region sensitive to the mouth. Red polygons outline 4 different hypothetical "action maps" involved in searching, handling, feeding, and defensive behaviors.

dorsal, and ventral cingulate motor areas (CMAr, CMAd, CMAv), all of which project to the spinal cord and, when stimulated, produce forelimb and hindlimb movements (Picard and Strick, 1996). The mid-cingulate cortex also includes a region sensitive to vestibular input and optic flow signals, but only those that indicate self-motion (Smith, 2021). In general, the medial edge of the neocortex appears to possess the right properties for the guidance of locomotor behavior: sensitivity to sensory input related to self-motion, projections to effectors relevant for locomotion, and proximity and interconnections with the navigation circuits of the hippocampus.

Feeding behavior, on the other hand, is very different. It strongly involves the mouth and nose, sometimes the forelimbs, and is motivated by local cues from the vibrissae, nose, and tongue, as well as olfactory and visual information on potential food items, modulated by hunger. We suggest that it was governed by lateral regions of the neocortex, including the dysgranular parieto-ventral (PV) map and gustatory cortex, as well as the anterior parts of the lateral pallium, which will ultimately become the agranular parts of orbitofrontal and insular cortex.

In addition, between the activities of searching and feeding, there is another class of activities that ethologists call *handling*, which involves various species-typical behaviors that make food ingestible. Examples include burrowing and digging out a root or a nest of insects, or biting down on a nut to crack its outer shell. Like ingesting, handling relies heavily not only on the mouth and nose, but also on other parts of the body, and involves specifying potential actions using a mixture of olfactory, tactile, and visual information. We suggest that this was governed by the cortical regions lying between the cingulate and dysgranular insular areas, corresponding to the primary somatosensory cortex (S1) as well as the adjacent rostral and caudal somatomotor fields (SR and SC).

A comparison of handling behaviors across different mammalian species supports the notion that these regions are critically involved in handling. For example, in the striped opossum, which uses its elongated fourth digit to extract insects from the bark of trees, the representation of that digit is highly enlarged in S1, but not in the other fields (Huffman et al., 1999). A very different organization is seen in the northern quoll, which captures prey with its mouth under guidance of snout, chin, cheek, and lip vibrissae, and which devotes about 40% of S1 to those sensors. In the platypus, S1 and SR are both dominated by input from its sensitive bill (Krubitzer et al., 1995). This kind of nonuniform allocation of cortical territory is of course also well-known in many mammals. For example, the forelimb representation in S1 (3b) and SR (3a) is enlarged in animals that use them to manipulate their food, such as raccoons, squirrels, and primates (Krubitzer, 2007). Admittedly, all of this could just be due to use-dependent scaling, but it is not seen in the other fields, like S2 or PV. In other words, it appears that the cortical territory devoted to a body part in S1, SR, and SC depends not just on how that body part is used in general, but how it is used for *handling* behaviors.

Finally, defensive behaviors like fleeing to a hiding spot or freezing in place to avoid detection require information from audition and vision (particularly from the upper and peripheral visual field), as well as somatosensory guidance from vibrissae and body hair. Integrating that information could have been the role of cortical regions overlapping the secondary somatosensory field (S2), which contains a large central area devoted to the vibrissae surrounded by regions sensitive to touch on the face and limbs (Beck et al., 1996). It is interconnected with the perirhinal cortex (Furtak et al., 2007), which has been implicated in visual identification of threats (Kent and Brown, 2012), as well as auditory cortex, which can provide cues about approaching predators or other dangers.

In summary, we propose that the mammalian version of the dorsal pallium became far more than the rudimentary "affordance detector" of our amniote ancestors. As its descending projections became more elaborate, it came to directly control a variety of species-typical actions, each guided by a particular subset of the neocortical sheet containing cells sensitive to the kinds of information most useful for that type of action. Searching behavior and general locomotor behavior were guided by a medial part of cortex that received information from the hippocampus and visual cortex sensitive to the lower visual field, and projected to downstream centers controlling the limbs. Feeding behavior was guided by a lateral part that received olfactory, gustatory, and interoceptive signals, as well as somatosensory input from the mouth, and projected to the head and mouth. Defensive behavior was guided by a region close to auditory input and visual cortex sensitive to the peripheral and upper visual field. It is probably incorrect to assume that these regions were ever strictly distinct, especially since they all presumably evolved through diversification of a less specialized ancestral system, but the different requirements of these diverse behaviors demanded a certain degree of specialization. Thus, over time, cells that brought together the right constellation of sensory signals and downstream motor projections for specific aspects of an animal's behavioral repertoire converged (Aflalo and Graziano, 2006; Krubitzer, 2007) to form specialized behavioral subsystems.

However, an increasing diversification of behavioral subsystems comes with a price—these systems cannot all act at once. You can't dig for food while escaping a predator. In principle, one could imagine that cells belonging to different systems could compete against each other through mutual inhibition among cortical regions, but in practice that could not get far because as the cortical sheet expands, the number of necessary inhibitory interconnections would undergo a combinatorial explosion. Instead, we propose that action selection takes place in at least two hierarchical levels: *between* the different action systems, and *within* a given system. More specifically, we propose that competition between different action maps was made possible by the basal ganglia circuits (see Redgrave et al., 1999; Suryanarayana et al., 2022), and that once a given action map was activated, then selection between alternative movements took place within that map in an idiosyncratic representation of the space of potential actions of the corresponding type (Cisek, 2007; Cisek and Pastor-Bernier, 2014).

SELECTION BETWEEN ACTION SYSTEMS

The basal ganglia played a major role in behavioral selection already in early vertebrates (Suryanarayana et al., 2022). This basic architecture was further elaborated in tetrapods and especially amniotes, whereby the basal ganglia became a major arbiter between different types of behavior such as seeking, pursuing or ingesting food, or finding shelter or a warm place to rest. This primarily involved modulation of downstream systems on the basis of state estimates in the forebrain, with the basal ganglia acting as the go-between. However, in mammals these circuits gained a new target—the pallium itself—through a recurrent connection via the thalamic matrix (Jones, 1998; Striedter and Northcutt, 2019). Thus, the basal ganglia could now selectively activate not just downstream systems, but also the very same upstream regions from which they received input. In other words, the basal ganglia could now modulate their own sources of input, creating a positive feedback loop.

The basic topology of the cortico-striatal system is not limited to just that part of the forebrain involving the neocortex. It involves all parts of the pallium, and all parts of the extended striatum, pallidum, and amygdala (Haber et al., 2000; Swanson, 2000; Floresco, 2015; Heilbronner et al., 2016; Murray et al., 2016). Each of these begins with a region of the pallium that projects to a region of the striatum, which projects to the pallidum, which projects to a thalamic region, which then forms a reciprocal excitatory connection with the original pallial area. Figure 59.4 showed the pallio-striatal projections for four types of these circuits: the septo-hippocampal loop, the ventral striatal loop, the dorsal striatal loop, and the amygdalar loop. Figure 59.6 illustrates this in some more detail, particularly for the familiar basal ganglia loop through the dorsal striatum, for which 3 separate circuits of the "direct pathway" are shown. Importantly, while the cortico-striatal and striato-pallidal projections maintain an ordered topography (Alexander et al., 1991), there is a convergence of input from the cortex to the pallidum of a factor of about 30,000 to 1. There is also a significant divergence of the projections from the thalamus back to the neocortex, which have a very particular pattern of innervation. They involve calbindin-expressing neurons in the thalamic matrix (Jones, 1998), which project collaterals to the prethalamus and striatum before arborizing across wide regions of cortex, primarily in superficial layers (Clasca et al., 2016; Shine, 2021). Notably, reconstructions of the axons of individual thalamocortical neurons in mice show that they innervate very large patches of cortex, often extending rostrocaudally over several millimeters across somatomotor, cingulate, or orbital regions (Kuramoto et al., 2015) (see **Chapters 11 and 12 in Section IV**). Consequently, each of the parallel loops shown in Figure 59.6 actually involves a convergence of

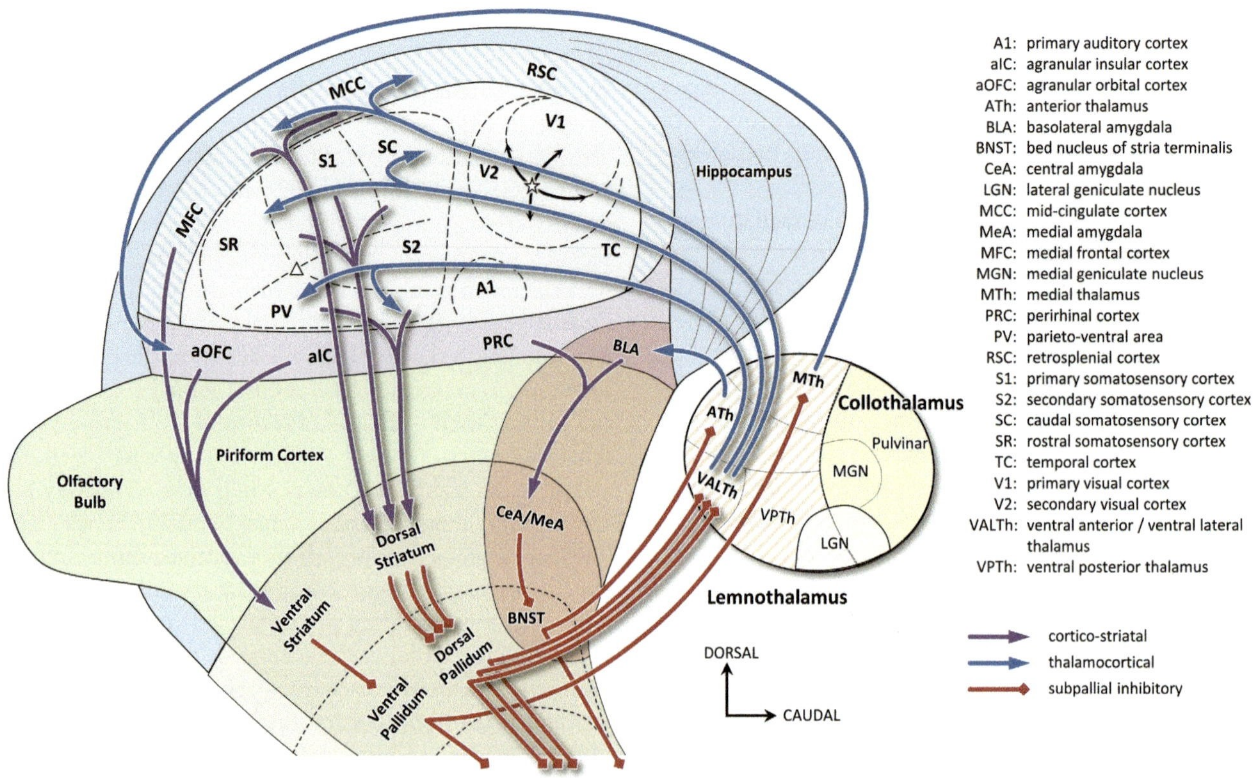

Figure 59.6. Organization of cortico-striatal-thalamocortical loops in a hypothetical ancestral mammalian brain, showing only the "direct pathway." For the loop through the dorsal striatum, 3 separate circuits are shown, corresponding to mechanisms for selecting between the action maps for searching, handling, and feeding. Higher-level loops through the ventral striatum and amygdala are also shown.

excitation from a large patch of cortex onto a specific basal ganglia loop, which competes against other loops before projecting back to excite many of the neurons in that patch.

We propose that this architecture implements a mechanism for selecting between different types of actions (e.g., feeding vs. searching), but does not contribute to selecting among the specific actions of that type (e.g., ingesting one particular item among many). It begins with information about the availability of different types of potential actions, converging in the striatum with information about the behavioral relevance of performing such actions in the context of current needs, resulting in the selection of a particular class of behavior in which to engage at a given moment in time. That class of behavior is then selectively activated via nonspecific thalamocortical fibers projecting from calbindin-expressing neurons in the matrix (Clasca et al., 2016) to widely innervate the superficial layers of a rostrocaudal strip of cortex (Kuramoto et al., 2015) that corresponds to a given action map (Cisek, 2022). Thus, the basal ganglia act as a central arbiter selecting among different kinds of actions (Redgrave et al., 1999) by invigorating different action maps as appropriate for a given behavioral context.

Note that there are two action maps for each kind of behavior, one in each cortical hemisphere. It is possible that selection between these two maps takes place through transcallosal projections (which appeared in placental mammals), but it is also possible that it involves the basal ganglia. In other words, the basal ganglia may be implicated in selecting the arm with which a monkey makes a reaching movement, or the left-versus-right direction in space in which a mouse may run, even if they are not directly involved in selecting the specific details of the action (e.g., the target for the reach).

Part of the information for making this type of high-level selection comes from the ring of dysgranular and agranular cortical regions that project through the ventral striatum, which from the perspective of the "structural model" (García-Cabezas et al., 2019) are concerned with high-level aspects of behavioral control. Much of this may have to do with the regulation of foraging (Hayden et al., 2011; Rushworth et al., 2012; Murray et al., 2016), where lateral frontal cortex (including agranular orbital and insular areas) favors taking advantage of currently available food sources, while the medial frontal cortex (including infralimbic, prelimbic, and anterior cingulate cortex) favors expending energy to obtain something better. This is consistent with anatomical connectivity, whereby medial frontal regions (e.g., anterior cingulate) project more strongly to medial neocortex and whole-body sensorimotor fields, while lateral pallial regions (e.g., orbital cortex) project more to lateral neocortex and head- and mouth-dominated fields. The high-level influences of frontal regions could also act through the basal ganglia themselves, via an "ascending spiral" of overlaps between cortico-striatal and striato-nigral circuits that implement a gradient of connectivity of ventral to dorsal striatum and frontal to somatomotor cortex (Haber et al., 2000).

SELECTION WITHIN ACTION SYSTEMS

Selecting the type of activity in which to engage is only part of the challenge of behavior—one must also specify the details of the particular movements that will be performed. It is one thing to decide to flick one's tongue at a nest of insects, but quite another to aim it at a particular target with adequate precision. The basal ganglia circuits are not ideally suited for this, simply because of their limited resolution in comparison with the complexity of potential actions that mammals can perform. For that, the cortical action maps themselves are a much better candidate. Each map contains millions of cells that cover the range of the idiosyncratic parameters of a given type of action, and the activity pattern across those cells can specify something like a "desirability density function" across the space of those actions (Cisek and Pastor-Bernier, 2014), potentially including multiple peaks that specify several potential actions in parallel (Cisek and Kalaska, 2005). Since multiple actions with the same effector set cannot be performed at the same time, different peaks of activity must compete against each other, presumably biased by information favoring each given choice over others (Cisek, 2007).

For example, in the frontoparietal regions of the primate reaching system, neural activity reflects a competition between simultaneously specified potential reach directions (Cisek and Kalaska, 2005), which is biased by a continuous flow of information on the likelihood or reward value (Klaes et al., 2011; Pastor-Bernier and Cisek, 2011; Thura et al., 2022) of the associated actions. Importantly, this process reflects the geometric contingencies of choosing between reaching movements, exhibiting stronger competition between movements that strongly differ in direction, as opposed to between movements that are similar (Pastor-Bernier and Cisek, 2011). In contrast, neural activity in the basal ganglia regions projecting to the reach-related cortical system does not reflect the competition between actions until it is largely already resolved at the cortical level (Thura et al., 2022). Furthermore, inactivation of the basal ganglia nuclei projecting back to cortex does not impair the selection of reach targets and instead merely reduces the vigor of reaching movements (Desmurget and Turner, 2010). Such findings have been used to argue against the role of basal ganglia in action selection, but in these experiments the selection was always within a given type of action, not between different types. Thus, they are compatible with our proposal that the role of basal ganglia is in selecting between, but not within, a given class of actions.

The proposals made above must be tempered with a few caveats. First, it is probably incorrect to assume that these 2 hierarchical levels of selection (between versus within systems) are strictly independently implemented by the basal ganglia and neocortical action maps, respectively. The reality is almost certainly subtler. For example, it may be that the basal ganglia selectively activate a particular region of cortex to initially favor a given action map, but may then increasingly focus on a particular subset of that map as the movement begins to unfold—thus participating in the process of committing to a specific action (Shine, 2021; Thura et al., 2022). Second, the borders between the cortical maps are not likely to be precisely definable. To implement a given type of behavior, the brain can make use of any relevant constellation of effector-related patches of cortex, which need not be contiguous. Patches that are reliably used in concert may be brought together during evolution or during development and learning (see Aflalo and Graziano, 2006, for a model of how that might occur) but many meaningful actions can be performed using cell activity across a fractured set of circuits. Finally, it is important to emphasize that much of the data upon which this proposal is based comes from studies in modern rodents and primates, so it is admittedly precarious to suggest it as the basic architecture of the early mammalian cortico-striatal and thalamocortical systems. Nevertheless, we conjecture that the basic mechanisms have been conserved among mammals, even as the actual topography of the specific action maps shifted and differentiated in a great variety of ways along different mammalian branches (Krubitzer, 2007).

THE ROAD TO PRIMATES

The above text focuses on early mammals, and proposes a functional interpretation of how the cerebral cortex and basal ganglia together orchestrated their behavior. Much of this has presumably been retained in all mammalian branches, albeit with particular specializations for the highly varied lifestyles to which mammals have adapted. One of those lifestyles is exemplified by primates, which found shelter and food in the branches of trees and, in some lineages, eventually returned to diurnal life and highly visually dominant behavior. (An account that does justice to the associated changes is beyond the scope of this chapter, and we refer the reader instead to **Chapter 57** in this section.) Here, we note just a few of the most relevant changes that took place along the lineage that eventually produced humans.

Anthropoid monkeys succeeded in arboreal life thanks to their grasping and leaping form of locomotion, their large frontally facing eyes, and significant expansions of the neocortex and associated thalamocortical, cortico-striatal, and cortico-cerebellar circuits. Particularly dramatic was the expansion of temporal, frontoparietal, and prefrontal regions, causing the cerebral cortex to bulge out so much that it curved around the insula into the familiar C-shape of primate brains, bringing the hippocampus and amygdala around to the medial wall of the temporal sulcus. Expansion of the temporal cortex elaborated visual processing beyond the key stimulus detection in ancestral perirhinal regions and toward full-fledged object recognition and categorization. Expansion of the frontoparietal network led to a diversification of action maps, particularly those related to handling, into a variety of systems governing the many ways that primates interact with their world (Cisek, 2022). Prefrontal expansion generalized action maps to higher levels of abstraction (Fine and Hayden, 2022), allowing more long-term planning without overt action. For example, instead of approaching food items to evaluate them through touch or smell, primates could use their eyes to foveate objects from far away, appraise them using sophisticated visual processing, and decide whether acquiring them is worth leaving the safety of one's tree branch.

Importantly, these cortical expansions did not take place on their own. In each case, there were concomitant expansions in thalamocortical (Baldwin et al., 2017), cortico-striatal (Heilbronner et al., 2016), and cortico-cerebellar circuits (Balsters et al., 2010). Pezzulo and Cisek (2016) suggested that the latter made it possible for primates to not only select between the affordances that are directly available at a given moment, but to predict how specific actions will make new affordances available, to learn their value in a given behavioral context, and to use this information to bias behavior. Thus, instead of being dominated by the prepotent motivation to exploit a current opportunity, primates could look ahead to future opportunities and select actions with increasingly distant payoffs.

CONCLUDING REMARKS

The human brain is a product of a long and convoluted evolutionary history. That history consisted of long periods of gradual adaptation to specific ecological niches, punctuated by dramatic transitions between niches, including the transition from sessile to mobile life, filter-feeding to visually guided predation, aquatic to terrestrial environments, a retreat into nocturnality, the return to diurnal arboreal life, and the establishment of large social groups. Each of these periods of adaptation and transition was accompanied by significant changes in the brain and body, and in the behavioral repertoire to which both were adapted, but the changes could only occur as elaborations and specializations that took place within the context of the ancestral state. Consequently, we believe it is not possible to understand the organization of the brain by limiting oneself to looking only at its modern abilities and properties without appreciating the convoluted journey that it took (Cisek, 2019).

Here, we propose a hypothesis on the organization of the forebrain of early mammals, constrained on one side by inferences about basal vertebrates and amniotes, and on the other by data from modern mammals, particularly rodents and primates. We focus primarily on the thalamocortical and cortico-striatal systems, and on the organization of sensorimotor areas in the neocortex and surrounding pallial regions, interpreting them in the context of real-time selection of concrete actions. We propose that these cortical areas implement a set of ethological action maps (Graziano, 2016), each specialized toward the needs of particular aspects of an animal's behavioral repertoire. Selection between these systems is governed by the basal ganglia (Redgrave et al., 1999), while selection of movements within a given class of actions occurs through competition within each map (Cisek, 2007). During natural behavior, these circuits work together in unison to implement a hierarchical control system that has been both conserved and highly elaborated toward the specific niches of the great diversity of mammalian species, including primates.

While these ideas are still rather general, they do make some specific testable predictions. For example, we've proposed that the medial edge of the neocortex governs searching behavior, the lateral edge governs feeding, and the territory in-between governs species-specific handling behaviors. All of these are aspects of foraging, which can be readily observed in freely behaving mice. Our hypothesis predicts that activity in these regions should reflect when a mouse engages in these different aspects of foraging, with stronger activity in mid-cingulate cortex while searching, and stronger activity in S1 while reaching and grasping food, even though both include the forelimb. Furthermore, the balance of activity in these regions should vary as the animal depletes a given site and predict when it shifts from handling in one site to searching for another (as shown by Hayden et al., 2011, in frontal regions during a foraging-like task in monkeys). Finally, targeted stimulation of the striatal circuits that ultimately project to medial versus lateral cortical maps should causally influence the animal's trade-off between searching versus handling. For example, stimulation of medium spiny neurons on the direct path in the medial circuit should influence the animal to abandon a given foraging site in favor of searching for another earlier than in conditions without stimulation.

In summary, we propose that a consideration of the evolutionary history of thalamocortical and cortico-striatal circuits in the context of the behavioral repertoire of ancestral species may provide insights into their functions in modern mammalian brains. Here, we've proposed a particular hypothesis on the functional roles of the basal ganglia and neocortex in terms of a hierarchical control system with specialized subcircuits for species-typical actions. This hypothesis may be wrong, but it is testable, and we hope that future research will shed light on how our understanding of these systems must itself evolve.

ACKNOWLEDGMENTS

The authors thank Leah Krubitzer for helpful comments on an earlier version of this chapter. We apologize in advance to all of the investigators whose relevant work could not be cited due to space constraints. This work was supported by grants from the Canadian Institutes of Health Research (PJT-166014) and the Natural Sciences and Engineering Research Council of Canada (RGPIN-05345) to PC and scholarships from the Groupe de Recherche sur le Système Nerveux Central and Fonds de Recherche Québec Nature et Technologies to AN.

REFERENCES

Aboitiz F, Montiel J, Morales D, Concha M (2002). Evolutionary divergence of the reptilian and the mammalian brains: Considerations on connectivity and development. *Brain Res Rev.* 39:141–153.

Aflalo TN, Graziano MS (2006). Possible origins of the complex topographic organization of motor cortex: Reduction of a multidimensional space onto a two-dimensional array. *J Neurosci.* 26:6288–6297.

Albuixech-Crespo B, Lopez-Blanch L, Burguera D, Maeso I, Sanchez-Arrones L, Moreno-Bravo JA, Somorjai I, Pascual-Anaya J, Puelles E, Bovolenta P, Garcia-Fernandez J, Puelles L, Irimia M, Ferran JL (2017). Molecular regionalization of the developing amphioxus neural tube challenges major partitions of the vertebrate brain. *PLoS Biol.* 15:e2001573.

Alexander GE, Crutcher MD, DeLong MR (1991). Basal ganglia-thalamocortical circuits: parallel substrates for motor, oculomotor, "prefrontal" and "limbic" functions. *Progress Brain Res.* 85:119–146.

Arendt D, Tosches MA, Marlow H (2016). From nerve net to nerve ring, nerve cord and brain—evolution of the nervous system. *Nat Rev Neurosci.* 17:61–72.

Baldwin MK, Balaram P, Kaas JH (2017). The evolution and functions of nuclei of the visual pulvinar in primates. *J Comp Neurol.* 525:3207–3226.

Balsters JH, Cussans E, Diedrichsen J, Phillips KA, Preuss TM, Rilling JK, Ramnani N (2010). Evolution of the cerebellar cortex: The selective expansion of prefrontal-projecting cerebellar lobules. *NeuroImage.* 49:2045–2052.

Beck PD, Pospichal MW, Kaas JH (1996). Topography, architecture, and connections of somatosensory cortex in opossums: evidence for five somatosensory areas. *J Comp Neurol.* 366:109–133.

Benito-Gutiérrez È, Gattoni G, Stemmer M, Rohr SD, Schuhmacher LN, Tang J, Marconi A, Jékely G, Arendt D (2018). The dorsoanterior brain of adult amphioxus shares similarities in expression profile and neuronal composition with the vertebrate telencephalon. *BMC Biol.* 19:1–19.

Butler AB (2000). Chordate evolution and the origin of craniates: An old brain in a new head. *Anat Rec.* 261:111–125.

Butler AB (2008). Evolution of the thalamus: A morphological and functional review. *Thalamus Rel Syst.* 4:35–58.

Campagner D, Vale R, Tan YL, Iordanidou P, Pavón Arocas O, Claudi F, Stempel AV, Keshavarzi S, Petersen RS, Margrie TW, Branco T (2022). A cortico-collicular circuit for orienting to shelter during escape. *Nature.* 613(7942):1–9.

Cisek P (2007). Cortical mechanisms of action selection: the affordance competition hypothesis. *Phil Trans R Soc B.* 362:1585–1599.

Cisek P (2019). Resynthesizing behavior through phylogenetic refinement. *Atten Percep Psychophys.* 81:2265–2287.

Cisek P (2022). Evolution of behavioural control from chordates to primates. *Phil Trans R Soc B.* 377:20200522.

Cisek P, Kalaska JF (2005). Neural correlates of reaching decisions in dorsal premotor cortex: Specification of multiple direction choices and final selection of action. *Neuron.* 45:801–814.

Cisek P, Pastor-Bernier A (2014). On the challenges and mechanisms of embodied decision-making. *Phil Trans R Soc B.* 369.

Clasca F, Porrero C, Galazo MJ, Rubio-Garrido P, Evangelio M (2016). Anatomy and development of multispecific thalamocortical axons: Implications for cortical dynamics and evolution. In: Rockland, S ed., *Axons and Brain Architecture.* Elsevier: 69–92.

Comoli E, Das Neves Favaro P, Vautrelle N, Leriche M, Overton PG, Redgrave P (2012). Segregated anatomical input to sub-regions of the rodent superior colliculus associated with approach and defense. *Front Neuroanat.* 6:9.

Cooke DF, Graziano MS (2004). Super-flinchers and nerves of steel: Defensive movements altered by chemical manipulation of a cortical motor area. *Neuron.* 43:585–593.

Desfilis E, Font E, García-Verdugo JM (1998). Trigeminal projections to the dorsal thalamus in a lacertid lizard, *Podarcis hispanica. Brain Behav Evol.* 52:99–110.

Desmurget M, Turner RS (2010). Motor sequences and the basal ganglia: Kinematics, not habits. *J Neurosci.* 30:7685–7690.

Fine JM, Hayden BY (2022). The whole prefrontal cortex is premotor cortex. *Phil Trans R Soc B.* 377:20200524.

Floresco SB (2015). The nucleus accumbens: An interface between cognition, emotion, and action. *Annu Rev Psychol.* 66:25–52.

Furtak SC, Wei SM, Agster KL, Burwell RD (2007). Functional neuroanatomy of the parahippocampal region in the rat: the perirhinal and postrhinal cortices. *Hippocampus.* 17:709–722.

García-Cabezas MÁ, Zikopoulos B, Barbas H (2019). The structural model: A theory linking connections, plasticity, pathology, development and evolution of the cerebral cortex. *Brain Struct Funct.* 224:985–1008.

Gibson JJ (1979). *The Ecological Approach to Visual Perception.* Houghton Mifflin.

Graziano MS (2016)/ Ethological action maps: A paradigm shift for the motor cortex. *Trends Cogn Sci.* 20:121–132.

Graziano MSA, Taylor CSR, Moore T (2002). Complex movements evoked by microstimulation of precentral cortex. *Neuron.* 34:841–851.

Haber SN, Fudge JL, McFarland NR (2000). Striatonigrostriatal pathways in primates form an ascending spiral from the shell to the dorsolateral striatum. *J Neurosci.* 20:2369–2382.

Halley AC, Baldwin MK, Cooke DF, Englund M, Krubitzer L (2020). Distributed motor control of limb movements in rat motor and somatosensory cortex: The sensorimotor amalgam revisited. *Cereb Cortex.* 30:6296–6312.

Hayden BY, Pearson JM, Platt ML (2011). Neuronal basis of sequential foraging decisions in a patchy environment. *Nat Neurosci.* 14:933–939.

Heilbronner SR, Rodriguez-Romaguera J, Quirk GJ, Groenewegen HJ, Haber SN (2016). Circuit-based corticostriatal homologies between rat and primate. *Biol Psychiatry*. 80:509–521.

Helmbrecht TO, Dal Maschio M, Donovan JC, Koutsouli S, Baier H (2018). Topography of a visuomotor transformation. *Neuron*. 100:1429–1445.

Hinde RA (1966). *Animal Behaviour: A Synthesis of Ethology and Comparative Psychology*. McGraw-Hill.

Huffman KJ, Nelson J, Clarey J, Krubitzer L (1999). Organization of somatosensory cortex in three species of marsupials, *Dasyurus hallucatus, Dactylopsila trivirgata*, and *Monodelphis domestica*: Neural correlates of morphological specializations. *J Comp Neurol.* 403:5–32.

Jacob P-Y, Casali G, Spieser L, Page H, Overington D, Jeffery K (2017). An independent, landmark-dominated head-direction signal in dysgranular retrosplenial cortex. *Nat Neurosci*. 20:173–175.

Jacobs LF (2012). From chemotaxis to the cognitive map: the function of olfaction. *Proc Natl Acad Sci U S A*. 109(Suppl 1):10693–10700.

Jones E (1998). The core and matrix of thalamic organization. *Neuroscience*. 85:331–345.

Kaas J, Qi H-X, Stepniewska I (2020). Evolution of parietal-frontal networks in primates. In: Kass J, ed. *Evolutionary Neuroscience*. Elsevier:657–667.

Kaas JH (2013). The evolution of brains from early mammals to humans. *Cogn Sci*. 4:33–45.

Kent BA, Brown TH (2012). Dual functions of perirhinal cortex in fear conditioning. *Hippocampus*. 22:2068–2079.

Klaes C, Westendorff S, Chakrabarti S, Gail A (2011). Choosing goals, not rules: Deciding among rule-based action plans. *Neuron*. 70:536–548.

Krubitzer L (2007). The magnificent compromise: Cortical field evolution in mammals. *Neuron*. 56:201–208.

Krubitzer L, Manger P, Pettigrew J, Calford M (1995). Organization of somatosensory cortex in monotremes: in search of the prototypical plan. *J Comp Neurol*. 351:261–306.

Kuramoto E, Ohno S, Furuta T, Unzai T, Tanaka YR, Hioki H, Kaneko T (2015). Ventral medial nucleus neurons send thalamocortical afferents more widely and more preferentially to layer 1 than neurons of the ventral anterior–ventral lateral nuclear complex in the rat. *Cereb Cortex*. 25:221–235.

Lacalli TC (2008). Basic features of the ancestral chordate brain: a protochordate perspective. *Brain Res Bull*. 75:319–323.

Lende RA (1963). Cerebral cortex: A sensorimotor amalgam in the Marsupialia. *Science*. 141:730–732.

MacIver MA, Schmitz L, Mugan U, Murphey TD, Mobley CD (2017). Massive increase in visual range preceded the origin of terrestrial vertebrates. *Proc Natl Acad Sci*. 114:E2375–E2384.

Medina L, Reiner A (2000). Do birds possess homologues of mammalian primary visual, somatosensory and motor cortices? *Trends Neurosci*. 23:1–12.

Mitchell AS, Czajkowski R, Zhang N, Jeffery K, Nelson AJ (2018). Retrosplenial cortex and its role in spatial cognition. *Brain Neurosci Adv*. 2:2398212818757098.

Murray EA, Wise SP, Graham KS (2016). *The Evolution of Memory Systems: Ancestors, Anatomy, and Adaptations*. Oxford University Press.

Nieuwenhuys R, Puelles L (2016). *Towards a New Neuromorphology*. Springer.

Olson CR, Musil S (1992). Posterior cingulate cortex: Sensory and oculomotor properties of single neurons in behaving cat. *Cereb Cortex*. 2:485–502.

Pastor-Bernier A, Cisek P (2011). Neural correlates of biased competition in premotor cortex. *J Neurosci*. 31:7083–7088.

Pezzulo G, Cisek P (2016). Navigating the affordance landscape: Feedback control as a process model of behavior and cognition. *Trends Cogn Sci*. 20:414–424.

Picard N, Strick PL (1996). Motor areas of the medial wall: A review of their location and functional activation. *Cereb Cortex*. 6:342–353.

Powers WT (1973). *Behavior: The Control of Perception*. Aldine.

Puelles L, Alonso A, García–Calero E, Martínez-de-la-Torre M (2019). Concentric ring topology of mammalian cortical sectors and relevance for patterning studies. *J Comp Neurol*. 527:1731–1752.

Puelles L, Harrison M, Paxinos G, Watson C (2013). A developmental ontology for the mammalian brain based on the prosomeric model. *Trends Neurosci*. 36:570–578.

Redgrave P, Prescott TJ, Gurney K (1999). The basal ganglia: A vertebrate solution to the selection problem? *Neuroscience*. 89:1009–1023.

Rushworth MF, Kolling N, Sallet J, Mars RB (2012). Valuation and decision-making in frontal cortex: One or many serial or parallel systems? *Curr Opin Neurobiol*. 22:946–955.

Sachuriga, Nishimaru H, Takamura Y, Matsumoto J, Ferreira Pereira de Araújo M, Ono T, Nishijo H (2021). Neuronal representation of locomotion during motivated behavior in the mouse anterior cingulate cortex. *Front Syst Neurosci*. 15:655110.

Saitoh K, Menard A, Grillner S (2007). Tectal control of locomotion, steering, and eye movements in lamprey. *J Neurophysiol*. 97:3093–3108.

Shine JM (2021). The thalamus integrates the macrosystems of the brain to facilitate complex, adaptive brain network dynamics. *Progress Neurobiol*. 199:101951.

Smith AT (2021). Cortical visual area CSv as a cingulate motor area: a sensorimotor interface for the control of locomotion. *Brain Struct Funct*. 226:2931–2950.

Stepniewska I, Fang PC, Kaas JH (2009). Organization of the posterior parietal cortex in galagos: I. Functional zones identified by microstimulation. *J Comp Neurol*. 517:765–782.

Stepniewska I, Friedman RM, Miller DJ, Kaas JH (2020). Interactions within and between parallel parietal-frontal networks involved in complex motor behaviors in prosimian galagos and a squirrel monkey. *J Neurophysiol*. 123:34–56.

Stepniewska I, Gharbawie OA, Burish MJ, Kaas JH (2014). Effects of muscimol inactivations of functional domains in motor, premotor, and posterior parietal cortex on complex movements evoked by electrical stimulation. *J Neurophysiol*. 111:1100–1119.

Striedter GF, Northcutt RG (2019). *Brains Through Time: A Natural History of Vertebrates*. Oxford University Press.

Suryanarayana SM, Robertson B, Grillner S (2022). The neural bases of vertebrate motor behaviour through the lens of evolution. *Phil Trans R Soc B*. 377:20200521.

Swanson LW (2000). Cerebral hemisphere regulation of motivated behavior. *Brain Res*. 886:113–164.

Thura D, Cabana J-F, Feghaly A, Cisek P (2022). Integrated neural dynamics of sensorimotor decisions and actions. *PLoS Biol*. 20:e3001861.

Tsai T-C, Yu T-H, Hung Y-C, Fong L-I, Hsu K-S (2022). Distinct contribution of granular and agranular subdivisions of the retrosplenial cortex to remote contextual fear memory retrieval. *J Neurosci*. 42:877–893.

Wang Q, Liao CC, Stepniewska I, Gabi M, Kaas JH (2021). Cortical connections of the functional domain for climbing or running in posterior parietal cortex of galagos. *J Comp Neurol*. 529:2789–2812.

Wicht H, Lacalli TC (2005). The nervous system of amphioxus: Structure, development, and evolutionary significance. *Can J Zool*. 83:122–150.

SECTION XI

COMPUTATIONAL APPROACHES TO THALAMOCORTICAL DYNAMICS, ENCODING, AND ALGORITHMS

Eric Shea-Brown

Theoretical and computational neuroscience uses mathematical tools to distill key mechanisms that generate brain dynamics and to explain how these dynamics create neural representations of the environment that, in turn, drive behavior. These approaches interact with conceptual frameworks and models engaged each day across many fields of neuroscience. They are both richly inspired by and—sometimes immediately and sometimes after rounds of refinement—richly inspiring to experimental studies.

Classically, as today, theoretical and computational neuroscience spans multiple styles and categories of goals and techniques. First is the mechanistic approach, exemplified by perhaps the most famous computational neuroscience result of all, the Hodgkin-Huxley formalism for membrane dynamics. The resulting model was not just brilliant biophysically, but brilliant in terms of mechanism and theory, in establishing that the "parts list" of identifiable membrane currents can indeed be assembled to produce neural spike trains. The next step, and a hugely active modern research area, is to understand how these membrane dynamics interact with connectivity structures to create dynamical patterns across networks of neurons. Complementary to such mechanistic approaches to neural dynamics are efforts to understand how these dynamics encode information. While the fact that neural spike rates can encode stimuli was developed a century ago (e.g., in the work of Lord Adrian), and advanced to the cortex by Hubel and Wiesel's famous studies, this is the tip of the iceberg in terms of possible neural codes. Beyond rates, patterns of spikes across time can serve as additional channels of information. The richness of possibilities explodes further still when we consider patterns of spikes occurring across multiple neurons. Quantifying these patterns, while determining which of their features actually carry information about stimuli (or internal brain processes), is a focus of deeply computational work in neural coding. Perhaps even more exciting and challenging opportunities arise at the interface of neural coding and mechanistic modeling approaches. Finally, normative or "top-down" approaches describe statistically optimal neural codes and, in turn, algorithms for computation. This includes neural network learning—or "training"—approaches, which are undergoing a tremendous resurgence rooted in their long and important history in neuroscience (e.g., in the work of the "connectionists" and the Parallel Distributed Processing Group).

The chapters in this section traverse this range of approaches, highlighting how they can shed light on thalamocortical interactions. Chapter 60, by Mease, Shea-Brown, and Fairhall, provides a general introduction and broad illustrations regarding how neural dynamics support signal transformation and computation across several timescales. Golden and Bazhenov then survey, in Chapter 61, how single-cell mechanisms and network connectivity combine to produce synchronized, oscillatory firing patterns, and describe how these patterns can regulate and participate in a range of cognitive and behavioral processes. Next, in Chapter 62, Stanley continues the theme of temporal aspects of the neural code, with a specific focus on mechanisms that control the flow of encoded information from thalamus to cortex. Here, the synchronous timing of thalamic spikes is critical, due to a specific circuit motif that creates a narrow "window of opportunity" in which multiple thalamic neurons conspire to drive cortical responses. In Chapter 63, Sommer and Hirsch highlight the information theoretical approach to neural coding, focusing on how the thalamus dramatically transforms visual information arriving from retina, often making each spike more informative about stimuli. This includes spatial transformations, framed in terms of receptive fields, as well as temporal transformations driven by cellular-level mechanisms that lead thalamic neurons to produce spike bursts versus more continuous "tonic" spike sequences. Finally, in Chapter 64, Rao, Jiang, and Fisher describe a highly influential top-down approach to understanding thalamocortical processing. In this "predictive coding" approach, each level in the thalamocortical hierarchy combines incoming sensory information with prior expectations, to form successively more abstract (and arguably more behaviorally useful) representations of the sensory world. Overall, we hope that this section celebrates past progress while enticing more scholars of the thalamus and cortex to engage with computational methods and computational neuroscientists in the future, as we all engage with the fascinating open questions ahead.

60.

CONNECTING NEURAL DYNAMICS AND COMPUTATION ACROSS SPATIAL AND TEMPORAL SCALES

Rebecca A. Mease, Eric Shea-Brown, and Adrienne L. Fairhall

FROM CONDUCTANCES TO COMPUTATION

Biophysical modeling, starting from the pioneering work of Hodgkin and Huxley (1952), aims to derive differential equations that describe the dynamics of voltage variations in response to current inputs. These equations incorporate the dynamics of activation and inactivation of the multiple types of ion channels that a neuron contains; in the relatively simple case of the Hodgkin-Huxley model, this includes activation of potassium channels and activation and inactivation of sodium channels. While this powerful modeling approach has helped to delineate quantitatively how different ion channel types contribute to the behavior of neurons, the transformation from current input to voltage or spiking output is expressed in a very complex and often nonintuitive way through the nonlinear interactions of a potentially large number of dynamical variables. It is further often notoriously difficult to precisely fit such a model to data. Computational approaches have addressed these issues in 2 ways. First, one can develop simpler dynamical models with fewer degrees of freedom—or lower dimensionality—that capture the essence of the behavior of the full system. Second, one can use data-driven approaches to extract statistical models for how the system transforms its inputs to outputs (Aljadeff et al., 2016; Kass et al., 2018). This latter type of model captures the coding properties of the neuron by explicitly expressing the features of the input that are associated with neural firing. Ideally, one hopes to connect such a coding model to the underlying biophysical mechanisms.

DETERMINING MODELS FOR COMPUTATION

A key ingredient of a model that expresses the effective computation of a neuron or circuit is its *filter*. To describe this, consider the system receiving an arbitrary time-varying input, $I(t)$, and producing an output as a series of action potentials. If a neural system were purely linear, one would be able to express the resulting voltage as the convolution of this input with a filter, or kernel, $k(t)$:

$$V(t) = V(0) + \int_0^t k(\tau) I(t-\tau)\, d\tau. \tag{1}$$

Spiking neurons are clearly not linear, as they produce action potentials. However, one can still make use of the linear filtering concept by characterizing the stimulus that drives the system toward an action potential in terms of an effective kernel. In this case, the firing rate $r(t)$ can be expressed as a function of the input analogously to (2):

$$r(t) \approx r_0 + \bar{r} \int_0^t k(\tau) I(t-\tau)\, d\tau, \tag{2}$$

where $\bar{r}$ sets the scale of the stimulus-driven variations. The kernel or filter $k(t)$ in this expression can be found using a method called reverse correlation. If the system is driven by a randomly varying input that samples a wide variety of possible inputs (typically Gaussian white noise), one can compute the *spike-triggered average* by averaging the stimulus segments preceding a spike over trials, resulting in a time-varying function $k(t)$, Figure 60.1A. In general, the linear approximation of Equation (2) is insufficient to provide a good approximation to the firing rate, as it cannot capture neural dynamics such as rectification (firing rates cannot be negative) and saturation (firing rates cannot increase arbitrarily with the size of the input). As a fix, one can incorporate a nonlinearity into this description:

$$r(t) \approx \bar{r} g(k * I(t)), \tag{3}$$

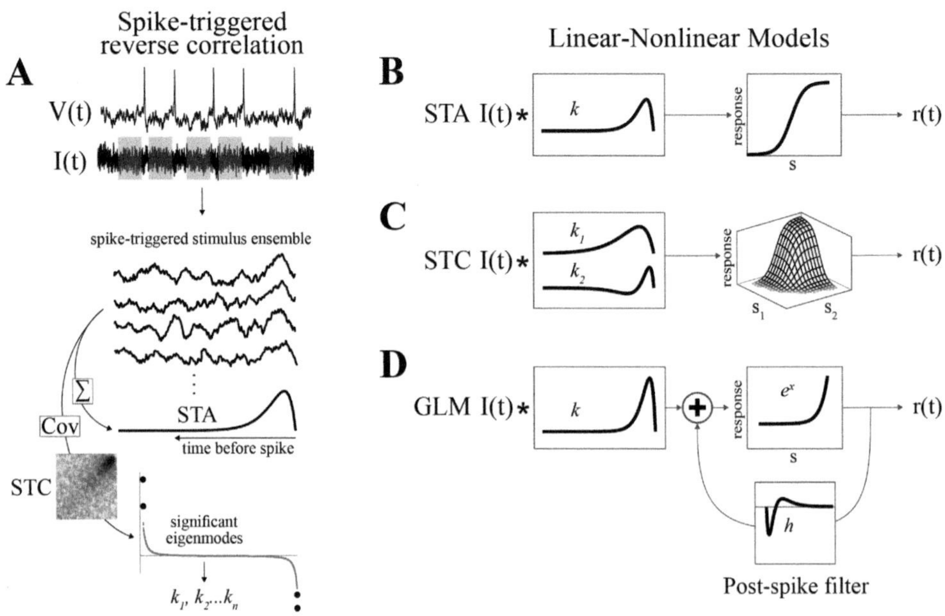

Figure 60.1. Cascade models for capturing neuronal computation and intrinsic encoding properties. (A). Reverse correlation methods: the spike-triggered ensemble comprises the stimuli that precede spikes. From this ensemble one can determine the spike-triggered average (STA) or spike-triggered covariance (STC). (B–D). Cascade models for capturing neuronal dynamics and intrinsic encoding properties. (B). One-dimensional STA model, with a single linear filter k and a nonlinear function relating the filtered stimulus s = k*I to the firing output. (C). n-dimensional model using multiple filters derived from covariance analysis. (D). GLM point process model in which the filtered stimulus is combined with the filtered spike history and then passed through a nonlinearity.

where we will use the following shorthand for convolution:

$$k * I(t) = \int_0^t k(\tau) I(t-\tau) d\tau.$$

Here $g(\cdot)$ is a nonlinear function, the mean rate $\overline{r}$ again sets an overall scale, and we from now on consider $r(t)$ relative to the baseline rate r_0 for simplicity. In Equation (3), one can regard the filter k as a "feature" of the input to which the system is sensitive, extracting some important component of the input, s, by convolution: $s(t) = k * I(t)$. The function g then acts as a "threshold" or more general input/output function that mediates the sensitivity of firing to that stimulus component.

This well-known formulation is known as a "linear/nonlinear" (LN) model, Figure 60.1B,C, and has been used to characterize many neural systems. While very powerful and general, the model of Equation (3) has several shortcomings that limit the accuracy with which it can represent the behavior of neurons. The first is that while one observes spiking through voltage fluctuations, this is just one observable of a high-dimensional set of dynamical conductance variables, as we discussed for the Hodgkin-Huxley system. Thus one might expect that the linear kernel k that governs the approach of the system to spiking threshold needs to take into account the contributions of those unobserved variables in bringing a neuron to threshold. This can be done by allowing the kernel to have more than one dimension. A convenient way to estimate these additional driving factors of the input is to characterize the set, or ensemble, of stimuli that triggers spikes in more detail. So far we have used only the average; if we also consider the covariance of these stimulus samples, we can derive a set of features from the principal components of the set of spike-triggering stimuli that are most predictive of spikes, Figure 60.1A,C. This is known as *spike-triggered covariance* (STC) analysis, where one computes the covariance matrix of the set of spike-triggering stimuli. The eigenvectors of this matrix whose corresponding variance is most different from that of non-spike-triggering stimuli define a set of filters, k_1 to k_n, where the number n of relevant dimensions is hopefully small, –typically ranging from 2 to 8 in published studies, e.g. (Agüera y Arcas et al., 2003; Slee et al., 2005; Rust et al., 2005; Fairhall et al., 2006; Mease et al., 2014b) a model of the form—giving a model of the form

$$r(t) \approx \overline{r} g(k_1 * I, k_s * I, \ldots, k_n * I), \tag{4}$$

a straightforward generalization of Equation (3), Figure 60.1C.

The beauty of this approach is that one can now examine the form of the filters k_i and interpret them in terms of the computation that the circuit properties and biophysics of

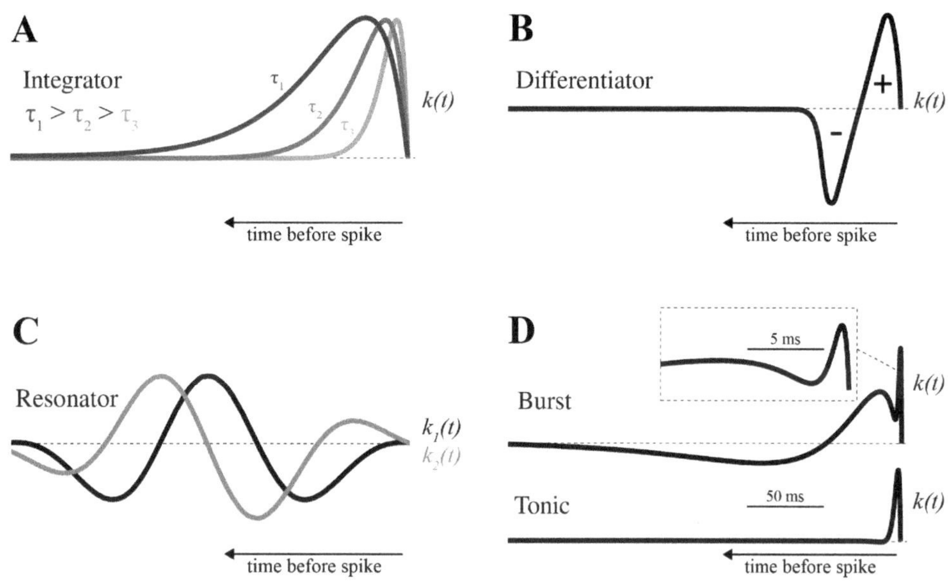

Figure 60.2. Prototypical stimulus filters for single-neuron feature selectivity. (A). Integrators with different time constants. (B). A differentiator is a bimodal filter. (C). A resonator is sensitive to a certain frequency. Such filters often appear in conjugate pairs (sine and cosine). (D). Burst (upper) and tonic (lower) filters for thalamic relay neurons. Modified from Mease et al. (2017).

the neural system generates. For single neurons, 3 typical types of filters emerge. The most typical is a *leaky integrator* (Figure 60.2A). Such a filter sums over recent inputs, but with a weight that decreases as one looks further into the past. This is the filter one would expect from a simple passive membrane resistor-capacitor (RC) circuit, the dynamics that form the basis for the leaky integrate-and-fire neuron model. Such a filter has a decay timescale τ that defines the window of integration of a neural system. To first approximation, cortical neurons are often well fit by such filters, with time constants on order of tens of milliseconds. Other specialized neuron types have very different time constants. For example, neurons in brainstem auditory sound localization circuits need to be extremely temporally precise in order to deliver temporally synchronous information arising from the 2 ears; such neurons have filters with time constants on the order of 1 or 2 milliseconds (Slee et al., 2005).

Another typical type of filter is a *differentiator* (Figure 60.2B). Such a filter takes inputs from the recent past, smoothed in a window of a certain size, and subtracts the weighted inputs from further back in time. Such a filter is therefore bimodal, with positive and negative lobes. It turns out that the Hodgkin-Huxley neuron is best described as a differentiator (Agüera y Arcas et al., 2003), meaning that *changes* in the input most reliably drive spiking.

A third type of filter is a *resonator* (Izhikevich, 2007) (Figure 60.2C). Such a filter contains a fluctuation at a certain frequency; thus by computing the convolution between this filter and the input, one can find moments in time at which that frequency dominates. An example of this occurs in rat barrel cortex, where cortical neurons are sensitive to a certain frequency (Maravall et al., 2007). In this case, covariance analysis shows that there are 2 primary filters, essentially a sine wave and a cosine wave at the same frequency. These 2 filters are then nonlinearly combined to predict that firing occurs at an arbitrary phase in this cycle.

While we have discussed this approach primarily in terms of temporal variations in the stimuli driving neurons (which might be injected current, or whisker movement, or light intensity), these methods are easily generalized to other stimulus dimensions such as spatial structure for visual neurons (Rust et al., 2005) or frequency content (Theunissen et al., 2000), leading to spatial, spatiotemporal, or spectrotemporal filters.

HISTORY DEPENDENCE

While the filter-based models described above are powerful, synaptic inputs or applied stimuli are not the only determinants of spiking, as neuronal excitability at a given time is also critically dependent on spiking history. At a minimum, this is due to refractoriness arising from the (de)inactivation and (de)activation dynamics of voltage-gated sodium and potassium channels. Functionally, the neuron cannot spike or will spike with lower probability immediately after an action potential, thus limiting the firing rate (absolute refractory period) and rendering the neuron less sensitive to additional inputs (relative refractory period). Longer timescale activity dependence arises from longer timescale channel dynamics, including the calcium-dependent activation of potassium channels controlling afterhyperpolarization (AHP) or slow inactivation of potassium and sodium channels. Inputs can also trigger persistent increases in excitability, for example through calcium dynamics that support intrinsic bursting or NMDA plateau potentials (see, e.g., **Chapter 3 in Section II** and **Chapter 7 in Section III**).

Such history-dependent behavior cannot be captured in the purely feedforward, stimulus-selective models of Equations (1–3). Instead, the instantaneous probability of spiking given presentation of an input stimulus must be modulated by a feedback term reflecting the neuron's internal state. Without such a representation, feedforward models predict many "false positive" spikes and erroneously high output firing rates. Furthermore, neglecting interspike interactions influences the estimation of stimulus features (Agüera y Arcas et al., 2003; Mease et al., 2014b).

This issue can be addressed using the *generalized linear model* (GLM; Figure 60.1D), a widely used statistical model that incorporates multiple inputs using linear filters and combines them through a single nonlinear function. GLMs are often autoregressive, that is, one can include a filter that acts on the previous output of the system. In our context, one can define a spike history filter, $h(t)$, which filters the history of the system's activity and alters the probability of subsequent spiking accordingly. The history and stimulus filters are fit simultaneously in order to best capture both stimulus encoding and interspike interactions. Unlike the models described above, GLMs use a prespecified nonlinearity and are fit using maximum likelihood methods. An alternative method to capture history dependence is to include it in an LN model as an additional factor entering the nonlinearity that is calculated directly from the interspike interval distribution (Mease et al., 2014b). In this case, the probability of spiking at any given time is a function of both the stimulus components $s_1,\ldots,s_n$ and the time to last spike.

DIVERSE COMPUTATIONS ACROSS CELL TYPES

Neurons across the brain differ considerably in terms of the synaptic receptor types with which they are driven and the ion channels that they express, leading to distinct dynamics of excitation and post-stimulus refractoriness. While one could model these differences using complete sets of differential equations describing channel dynamics in the Hodgkin-Huxley formalism, one can also use the methods, as we have discussed, to analyze the effective computations of neurons, and to attempt to link these computational differences to underlying biophysical and circuit mechanisms.

GLMs incorporating stimulus and spike-history selectivity can capture a diverse range of dynamical behaviors realized by neurons. Izhikevich (2007) showed that one can simulate a wide range of firing properties using dynamical models consisting of just 2 variables, an equation for the voltage incorporating an intrinsic nonlinearity combined with a threshold to approximate the generation of a spike, along with a feedback variable that provides history dependence:

$$\begin{aligned}\frac{dV}{dt} &= -aV + bV^2 - cW + I(t),\\ \frac{dW}{dt} &= dV - eW,\end{aligned} \tag{5}$$

along with a reset condition on V and W once V exceeds some threshold. Different settings for the parameters a–e produce many firing behaviors, including tonic, phasic, and burst spiking, and also bi-stable spiking initiated and terminated by brief inputs, resonance, and type I and II f-I curves. GLMs are able to capture all of these regimes (Weber and Pillow, 2017). Moreover, simple changes to either $k(t)$ or $h(t)$ can shift neuronal dynamics between distinct dynamical regimes. The GLM fits are also able to capture the extremes of both highly precise spike timing and highly variable (i.e., "super-Poisson") spike trains seen in vivo.

Closely related methods have recently been applied to the Allen Institute's large-scale database of physiological responses sampled from many neuron types in mouse primary visual cortex, to extract cellular dynamics and cluster or categorize them (Teeter et al., 2018). This analysis found functional categories that overlapped with multiple genetically defined cell types. Thus inspired, statistical methods have recently been designed that are tailored to the problem of identifying cell types based on stimulus filters and spike-history filters, and to test hypotheses that there are indeed distinct clusters of these filters' properties. These combine GLM model fitting with hierarchical clustering into a single "simultaneous" modeling step (Zdeblick et al., 2023).

THALAMIC COMPUTATION

Thalamic relay neurons are a fascinating example of how complex intrinsic encoding can be revealed in the LN framework. Due to the presence of both fast spike-generating currents as well as depolarizing currents with slow dynamics—the transient low-threshold calcium current I_T and the hyperpolarization activated cation current (HCN)—thalamic relay neurons show bistable behavior controlled by resting membrane potential (see **Chapter 3 in Section II**). In brief, the voltage-dependent (in)activation of I_T leads to bursting, a well-characterized hallmark of thalamic excitability. At mean hyperpolarized membrane potentials, depolarizing input evokes transient, relatively stereotyped calcium plateaus or "low threshold spikes" that in turn are sufficient to bring the membrane potential near or above the voltage threshold for sodium channels, resulting in high-frequency bursts of sodium action potentials. However, after this brief period of depolarization, I_T inactivates, ending the burst of APs until the neuron remains at hyperpolarized membrane potential sufficient to deinactivate I_T. In contrast, during prolonged depolarization, I_T is largely unavailable and more regular, lower-frequency "tonic" spiking occurs. Within the LN framework, this complex burst-tonic behavior manifests in dramatic changes in filter shape, reflecting the kinetics and voltage-dependencies of these currents (Samengo et al., 2013; Elijah et al., 2015; Mease et al., 2017; Zeldenrust et al., 2018).

In the case of neurons from the higher-order somatosensory thalamus in rodents, a particularly complex picture emerges (Mease et al., 2017). As Figure 60.2D illustrates, in tonic mode, the stimulus filter is narrow (~10 ms) with a largely integrating shape, reminiscent of filters measured for cortical neurons. However, in burst mode, the filter is elaborated into a compound waveform comprising (1) a much broader (~200 ms) biphasic, differentiator waveform, summed with (2) a very narrow (~3 ms) additional filter. The broad filter shape reflects the requirement for I_T to first be deinactivated (initial hyperpolarizing phase) in order to then activate with relatively slow kinetics (subsequent depolarizing phase); the relative contribution of each lobe depends on the mean membrane potential determined by the DC input. The similarity of slowly changing stimuli to this filter sets the size of the evoked calcium spike and, concomitantly, the duration (~spike count) of each burst event. Most strikingly, the presence of the very narrow feature means that during burst mode, high-frequency stimuli can control the exact timing of spikes within a burst.

In sum, this complex arrangement of intrinsic properties allows higher-order thalamic neurons to simultaneously encode information with distinct timescales in burst onset time, burst duration, and interspike intervals within a burst. The driving excitatory inputs to these neurons are largely strong, precise projections from cortical layer (L)5 neurons suggested to carry information in a particularly dense code (see **Chapter 7 in Section III**; **Chapter 15 in Section IV**), integrated with several sources of inhibition and excitation on widely varying timescales (see **Chapter 4 in Section II**). How intrinsic thalamic processing further sculpts and distributes these representations is not yet understood, but the multiple feature selectivity could allow information from distinct presynaptic sources to be combined at the single neuron level.

COMPUTATION ON MULTIPLE TIMESCALES

As the discussion above has indicated, ion channel dynamics at the single neuron level, as well as short and long-term synaptic plasticity, imbue individual neurons and neural circuits with diverse computational properties that can span multiple timescales. One aspect of this is reflected in alternate ways of describing a neuron's output: on one hand, neurons are often characterized in terms of their firing rate versus current input, or f-I curves, which quantify responses to the mean input; on the other hand, one may characterize the details of fluctuations that tend to trigger individual spikes, as discussed above. In both cases, these descriptions depend on the statistics of the stimuli used to sample the model. The shape of f-I curves can be altered by the variance of noise added to the mean input (Chance et al., 2002; Higgs et al., 2006), while LN models are also influenced by the properties of the stimulus. In general, the computation of a neural system can be considered as nonstationary, depending intimately on the potentially time-varying statistics of the input and the history of its own recent behavior. This nonstationarity is often called *adaptation*, a term that incorporates a wide variety of phenomenology (Whitmire and Stanley, 2016; Weber and Fairhall, 2019; Weber et al., 2019).

A classic form of adaptation is a change in firing rate in response to a stimulus that is stepped to a new value and held fixed; generally the firing rate will show a rapid response to the change in the stimulus, then gradually relax toward a resting value. While this decay is regarded as an adaptation to a steady-state input, a similar adaptation of firing rate also occurs in response to a change in the variance or contrast of an input. As the previous discussion will have made clear, such behavior is just one of several possibilities for systems in which the computation can be described through the temporal filters of Figures 60.1 and 60.2, in this case a differentiator, whereby a change in the stimulus drives a large response that then decays away.

One can also evaluate the effect of a change in stimulus condition by fitting an LN model of the system, which, as discussed above, describes how the instantaneous firing rate depends on the recent stimulus input. In many systems, the properties of the LN model, both the filter(s) and the nonlinearity, depend on the statistical properties of the input—its mean, variance/contrast, or even higher-order properties of the stimulus such as correlation structure (Hosoya et al., 2005; Sharpee et al., 2006) (Figure 60.3A,B). These changes may be almost instantaneous, suggesting that they arise from intrinsic nonlinearity, or may take time to develop, suggesting changes in circuit wiring.

GAIN MODULATION

Gain modulation refers to changes in the slope of an input/output function, which can be influenced by other properties of the input (Ferguson and Cardin, 2020). For

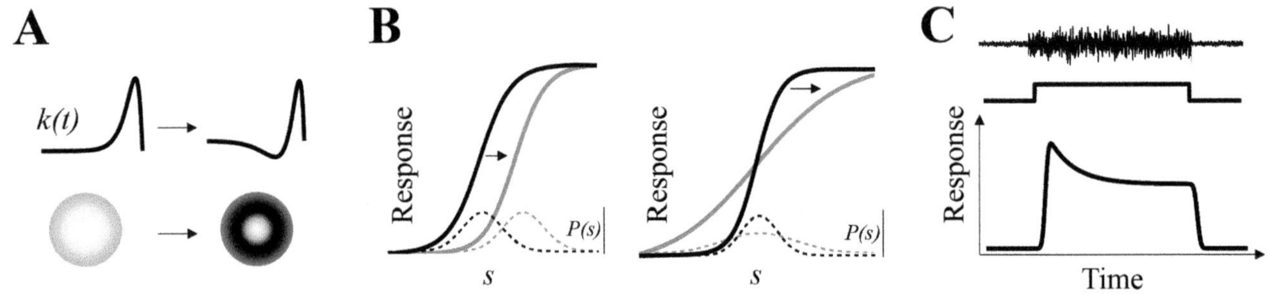

Figure 60.3. Multiple forms of adaptation. (A). Adaptation in cascade models: filters may be changed by the statistics of the input. This can occur for temporal and spatial aspects of the filter. In this example, the filter transforms from integrating to differentiating. (B). Adaptation in cascade models: the slope of the input/output relation between stimulus and response may change in offset, or in slope, i.e., undergo gain modulation. This often occurs in response to changes in the stimulus distribution, shown in dotted lines, where the stimulus distribution may change in mean (left) or variance (right). Such adaptation is a form of efficient coding whereby the system acts to maintain the same rate of information transmission about the new stimulus distribution. (C). Spike-frequency adaptation. The firing rate in response to a step change in input often relaxes following the change, both to steps in mean input and in stimulus variance. In some systems, the changes in firing rate can be modeled as fractional differentiation, which acts as a filter on the envelope of the input.

example, as mentioned, the slope of an f-I curve may be changed by the variance of added noise (Chance et al., 2002; Higgs et al., 2006). In the LN model picture, the slope or offset of the nonlinear function g in Equation (3) may depend on stimulus properties. Gain-scaling adaptation is a special case in which the nonlinearity g adapts to maintain the same dynamic range independent of the overall fluctuation range, σ, of the stimulus (Brenner et al., 2000) (Figure 60.3B), which in general is time-varying (Fairhall et al., 2001). In this case, Equation (3) above becomes

$$r(t) \approx \bar{r}g(s(t)/\sigma(t)), \qquad (6)$$

where the filtered stimulus $s(t)$ is normalized by its local amplitude or envelope $\sigma(t)$. Such a coding scheme maintains the same information encoded per spiking event, regardless of the stimulus amplitude, preventing the system from saturating when the overall stimulus range is large, or from failing to resolve small fluctuations when the input "goes quiet."

This property has been observed in several systems, including the fly H1 neuron (Brenner et al., 2000; Fairhall et al., 2001), the retina (Baccus and Meister, 2002), and barrel cortex (Maravall et al., 2007; Diaz-Quesada and Maravall, 2008; Mease et al., 2013). While it can arise as a circuit property, it can also emerge from quite simple mechanisms at the level of single neurons, according to the balance of voltage-gated sodium (Na_v) and potassium (K_v) channels. For example, in the Hodgkin-Huxley model, increasing the Na_v/Ka_v maximal conductance ratio both shifts selectivity from differentiation to integration (Lundstrom et al., 2008b) and supports gain-scaling behavior (Hong et al., 2008). In developing neurons in the mouse sensorimotor cortex, gain scaling emerges as a function of the Na_v/K_v conductance ratio that increases with age until reaching a stable value: "young" neurons with low Na_v/K_v with limited spiking do not gain scale, while comparatively "mature" neurons with higher Na_v/K_v show gain scaling (Mease et al., 2013). This behavior can be reproduced using both conductance-based and exponential leaky integrate and fire models (Mease et al., 2013).

In the example of thalamic relay neurons described in the previous section, using reverse correlation to parse stimulus selectivity into coding channels with distinct timescales also reveals distinct adaptation motifs for each channel. While the slow encoding channel driving calcium low-threshold spikes is sensitive to global stimulus variance, the fast encoding channel driving timing of sodium APs within bursts shows gain-scaling behavior (i.e., sensitivity to high-frequency inputs adjusts according to overall stimulus gain; Mease et al., 2017). Thus, in burst mode, fluctuations matching the selectivity of the slow channel unlock a transient window of gain invariant encoding of fast inputs (see **Chapter 15 in Section IV**).

LONG TIMESCALES OF ADAPTATION

While the model above demonstrates that one can capture time-varying instantaneous coding properties with an LN model with time-varying components, the mean firing rate also tracks the variations in stimulus conditions on long timescales, that is, $\bar{r}(t)$ is some function F of the slow-varying mean or variance, for example, $\bar{r}(t) = F(\sigma(t))$. Spike-frequency adaptation is an example of this (Figure 60.3C). It has long been observed that some neural systems adapt to step changes in input mean or variance with a multiple time constant or even power-law decay (Drew and Abbott, 2006; Pozzorini et al., 2013). Power law decay can be a reflection of another type of computation:

fractional differentiation, observed in both fly visual neurons (Fairhall et al., 2000) and cortical L5 sensorimotor neurons (Lundstrom et al., 2008a). This intriguing computation is a generalization of differentiation. While a differentiator responds only to change in the input, a fractional differentiator is a "soft" differentiator, giving an enhanced response to change, but settling to a steady state in response to a maintained input that has a unique value. The same operation also leads to a frequency-independent phase shift in response to a sinusoidal input. While first demonstrated in vitro, such a phase-shifted response was observed in vivo both in thalamic ventral posteromedial (VPM) nucleus and in L5 barrel cortical neurons responding to vibrissa motion (Maravall et al., 2007; Lundstrom et al., 2010). Cortical neurons showed a much larger phase shift, indicating that the vibrissa input may have been cascaded through multiple fractional differentiations in its passage through the neural circuitry.

MULTIPLEXING

Thalamic neurons have been demonstrated to convey information in spike patterns (Reinagel and Reid, 2000). In particular, the 2 firing modes, tonic firing and bursting, shaped by the unique channel dynamics of thalamic neurons, have long been proposed to act as distinct channels of information transmission (Sherman and Guillery, 2002). As we have discussed, these 2 modes of firing are associated with different stimulus features (Figure 60.2D), and show distinct adaptation properties (Mease et al., 2017; Zeldenrust et al., 2018). The ability of neurons to communicate through multiple symbol types is a form of multiplexing, whereby the same channel can convey distinct types of information (Oswald et al., 2004; Alenda et al., 2010; Estebanez et al., 2012; Naud and Sprekeler, 2018). The concept of temporal multiplexing can also be applied to the adapting neural codes as described above. For example, one may have a neural response to a fast-varying signal modulated by an overall amplitude σ given by

$$r(t) = \bar{r}(\sigma(t))g(s/\sigma(t)), \tag{7}$$

whereby the slowly varying rate $\bar{r}$ encodes the envelope σ on slow timescales, while fast variations drive individual spikes according to the normalized LN model, $g(s/\sigma)$, conveying information about different stimulus properties on different timescales (Alenda et al., 2010; Lundstrom et al., 2008a).

The sensitivity of thalamic firing properties and thus feature selectivity to a shift in the mean input, as discussed previously, provides a mechanism by which corticothalamic feedback (see, e.g., **Chapter 27 in Section V**) and subsequent feedforward inhibition from thalamic reticular nucleus (**Chapter 16 in Section IV**) can dynamically modulate feedforward processing. Such feedback can control thalamic tonic/burst mode and adaptation to peripheral sensory inputs in a frequency-dependent manner (Mease et al., 2014a; Crandall et al., 2015), sharpen thalamic spatial and temporal receptive fields (**Chapter 14 in Section IV**), and gate information transmission to cortex by controlling thalamic synchrony. Moreover, naturalistic patterns of corticothalamic feedback increase the dynamic range of input encoding by promoting interleaved tonic and bursting modes (Wolfart et al., 2005).

More generally, this intrinsic ability to sculpt feature selectivity based on mean input could contribute to network-level and behavioral effects linked to changes in thalamic excitability. For example, the enhancement of cortical sensory response precision by thalamic hyperpolarization which increases bursting (Borden et al., 2022), and conversely, locus coeruleus's neuromodulatory boosting of thalamic tonic spiking, which enhances sensory coding and discrimination (Rodenkirch et al., 2019).

The generation of spike patterns in response to inputs is a combination of channel, synaptic, and circuit mechanisms that span milliseconds to seconds. While it is sometimes possible to separate the timescales of short fluctuations and slower timescale variations in statistical properties such as mean or variance in order to write down models such as Equation (7), more generally one might consider adaptive dynamics to simply encode stimulus variations on slower timescales, as in the case of fractional differentiation, which can modulate fast encoding (Weber and Fairhall, 2019). One approach to capture encoding on multiple timescales is to fit filters that span a wide range of temporal scales. This analysis has been performed using intracellular recordings from cortical neurons from 3 different sensory areas: visual, somatosensory, and auditory cortices (Latimer et al., 2019). By using a random sequence of punctate stimuli, responses were fit using multidimensional GLMs in which filter outputs were combined nonlinearly. This method extracted a set of filters that spanned an order of magnitude in time, from short filters describing local features to filters integrating over almost a second of stimulus history, making explicit how long timescale stimulus statistics modulate spike rates. These methods could be generalized to model how combinations of feedforward and feedback inputs dynamically influence information processing.

COMPUTATION FROM DYNAMICS

To relate neurons' computational properties to the underlying biophysical mechanisms, one would like to be able to derive filters directly from the underlying dynamical equations. In the case in which the underlying equations are themselves linear, as for example resistor-capacitor circuit models of dendritic trees or cell bodies, this process is direct: the equations can be solved explicitly in terms of a filter that exactly describes responses to inputs (Gabbiani and Cox, 2010). For nonlinear, spiking neurons, the observed voltage is only one of many underlying dynamical variables that interact nonlinearly to contribute to excitation. Thus, the filter of a linear/nonlinear model arises as a linearized approximation to the nonlinear dynamics that drive the voltage to threshold (Agüera y Arcas et al., 2003). The filter is therefore modulatable according to the regime in which the system is operating (Famulare and Fairhall, 2010)—for example, as a DC input shifts the mean subthreshold voltage, as shown above for thalamic neurons (Figure 60.2D), or an increased input variance explores a different range of the subthreshold dynamics (Hong et al., 2007). It should also not be surprising that the probability of firing can often be better accounted for by contributions from multiple stimulus filters that help to capture the influence of unobserved activation and inactivation variables (Agüera y Arcas et al., 2003; Hong et al., 2007). For some simplified model neurons, analytical expressions for the mean firing properties can be derived which allow one to directly link the filtering properties of the model neuron with the parameters of its dynamical equations (Hong et al., 2007; Famulare and Fairhall, 2010; Brunel et al., 2001; Lindner et al., 2005; Trousdale et al., 2012).

FROM COMPUTATION IN NEURONS TO COMPUTATION IN NETWORKS

How do the dynamical properties of single cells scale up to describe how signals are processed across neural networks? Here, we review modeling and theoretical approaches to this problem. These reveal how distinct dynamics, expressed in terms of input-output filters of single cells, drive distinct dynamics at the network level—and how connectivity can systematically reshape neuronal dynamics to tune signal processing at the network level.

Many modeling studies have addressed the question of information propagation in feedforward networks (Figure 60.4A), in which neurons in the first layer are driven with a signal which then flows through the network. The nature of single-neuron coding in individual layers strongly influences the network output, including the synchrony of firing in subsequent layers (Vogels and Abbott, 2005; Ratté et al., 2013), and the ability of the network to transmit different timescales of input drive through multiple layers (van Rossum et al., 2002; Lundstrom et al., 2010; Gjorgjieva et al., 2014). For example, without the gain-scaling properties described above, single neurons are relatively insensitive to the rapid fluctuations local to each neuron and tend to fire synchronously to common large-amplitude inputs (van Rossum et al., 2002; Lundstrom et al., 2010; Gjorgjieva et al., 2014). With gain scaling, each neuron becomes sensitive to fast input fluctuations, efficiently conveying local information but losing the ability to transmit slowly varying population-wide input variations as the signal propagates through the network (Figure 60.4B).

Next, we describe a general approach to understanding signal processing in recurrent neural networks. Once again, we begin with the most basic approximation of individual neurons above, in which they are described by linear (or linearized) input-output filters. These individual neurons may produce outputs described as either continuous rates, or as spikes emitted stochastically according to these rates (Sejnowski, 1976; Lindner et al., 2005; Ostojic and Brunel, 2011; Trousdale et al., 2012). Specifically, as illustrated in Figure 60.4C, we ask: When such a network receives an input signal $I(t)$, how do its single neuron properties—the filters $k(t)$—*combine* with recurrent reverberations of neural outputs according to a connectivity matrix W to produce the network wide response $R(t)$?

The general answer to this question may be expressed in an exact mathematical form (Sejnowski, 1976; Lindner et al., 2005; Ostojic and Brunel, 2011; Trousdale et al., 2012; Ocker et al., 2017). However, the range of possibilities at first appears very complicated to parse, because there are so many possible connectivities W (Pernice et al., 2011; Trousdale et al., 2012; Hu et al., 2014; Ocker et al., 2017). However, it turns out that the impact of a general connectivity matrix on a network's signal processing can often be accurately summarized through a surprisingly small and simple set of connectivity features, or "motifs." These features, illustrated in Figure 60.4D, begin with the probability that any cell pair is connected. The next motif is the "excess" probability that any chain of 3 cells is connected, over and above what we would estimate from the connection probability of cell pairs. The next is the excess probability that any chain of 4 cells is connected, and so forth. These chain motifs define pathways for reverberations of various lengths through the network, and, in the simplest settings, give a complete description of network signal processing through a set of reverberant loops. as schematized in Figure 60.4E (Hu et al., 2018). The strengths of these loops, in turn, determine whether a network might produce a more versus less oscillatory response to a network input—or, equivalently, might be selective for a more versus less oscillatory form of this input (Figure 60.4F).

While this motif-based approach is attractive in how it simplifies analysis of a vast set of possible network configurations, and how it applies to many other metrics of network response beyond that shown above (Pernice et al., 2011; Roxin, 2011; Hu et al., 2014; Martí et al.,

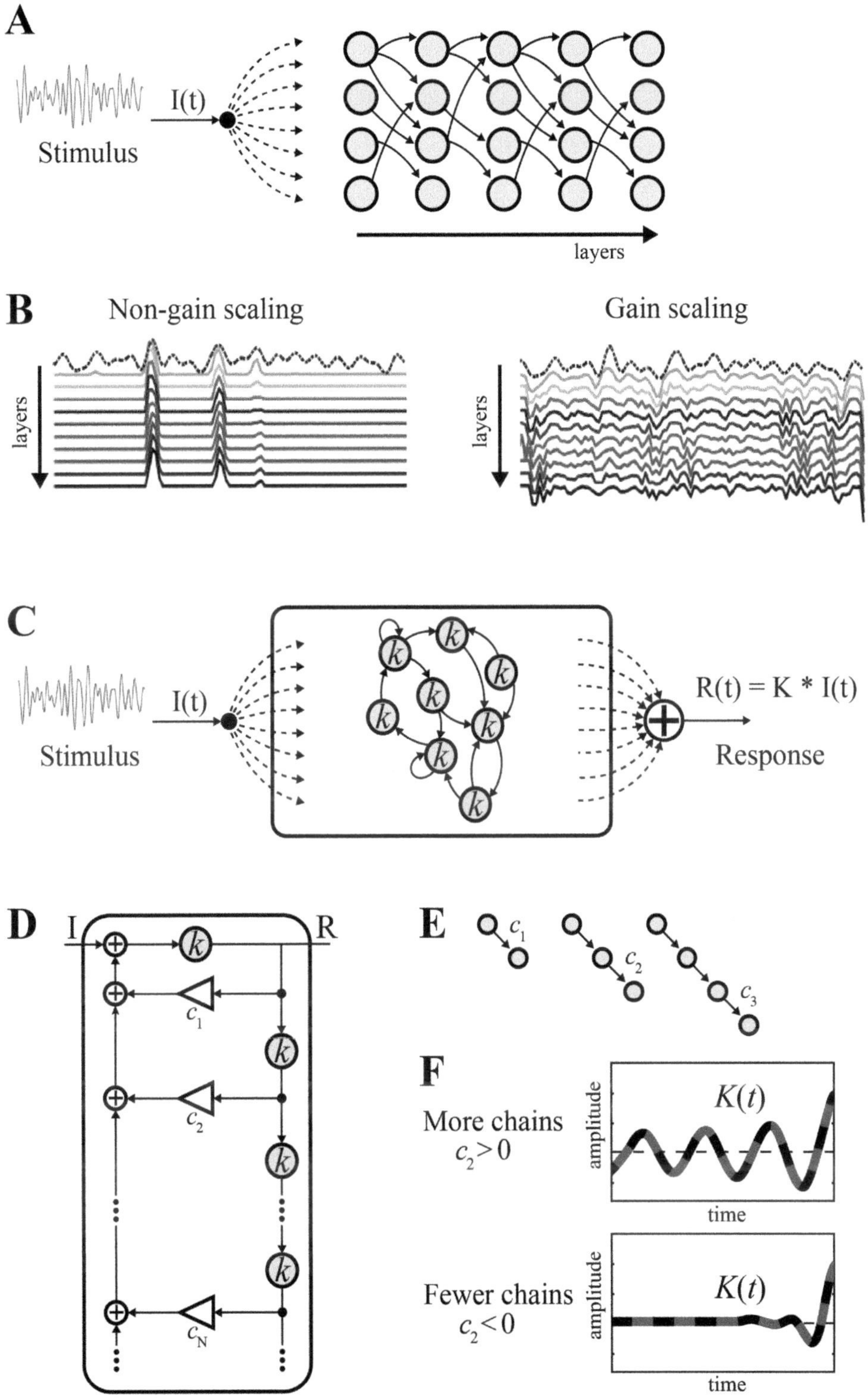

Figure 60.4. Network transformations. (A). Schematic of a feedforward network receiving a stimulus I(t). (B). Propagation of signals through a multi-layer feedforward network. A time-varying input is injected into each neuron in the initial layer (dotted trace) but with independent noise added to each neuron. Depending on whether individual cells have gain-scaling or non-gain-scaling properties, the network either reliably transmits the peaks of the original common input in a wave-like fashion, or loses the common signal in favor of faithfully propagating the noise fluctuations (or local signals). Adapted from Gjorgieva et al. (2014). (C). Schematic of a recurrent network receiving a stimulus *I(t)*; here, individual cells have a stimulus filter *k*, and the question is how the network as a whole reshapes these filters to produce a network-wide filter *K:* the response of the network as a whole is described by the convolution *R(t) = I * K(t).* (D). Network motifs: single connections among pairs of cells, with probability $c1$, length-2 chains of connections with excess probability $c2$, and length-3 chains of connections with excess probability $c3$. These excess probabilities are referred to as motif cumulants in the literature (Ocker et al., 2017). (E). These motifs determine the strength of reverberant loops in the network-wide filter connecting stimulus and response. (F). In networks with the same overall number (probability), more vs. fewer chains can lead to more vs. fewer oscillatory network-wide stimulus filters *K(t)*. Panels C–F modified from Hu et al. (2018).

2018; Recanatesi et al., 2019), it is part of a much wider landscape of analytical and computational techniques for network dynamics. We highlight in particular the textbook *Mathematical Foundations in Neuroscience* (Bard Ermentrout and Terman, 2010) which, among many others, treats methods that describe the emergence of oscillations using nonlinear methods which profoundly extend the assumptions above.

Recent large-scale realistic modeling efforts incorporating detailed morphological and physiological properties from thalamocortical circuits (e.g., from the Human Brain Project and as described in **Chapter 6 in Section III**), suggest exciting ways to test and expand upon these principles.

CONCLUSION

Here we have surveyed how computational methods allow us to determine how biophysical properties of neurons influence the fundamental ways in which they respond to their inputs. These methods reveal a range of response types: integrators, which sum inputs over time; differentiators, which identify changes in these inputs; and resonators, which pick out oscillatory features in these inputs. Each of these subserves different computations that are important for thalamocortical processing. Integrators accumulate signals over long timescales. Differentiators can enhance selectivity to coincident, punctate inputs, over and above steadier background inputs (a major focus of **Chapter 62 in Section XI**). Resonances in single cells interact with feedback delays and synaptic dynamics in networks to produce network-wide oscillations (as developed in detail in **Chapter 61 in Section XI**). The complex thalamic filters act to differentiate slow-varying inputs, but also allow fast fluctuations to precisely control the timing of spikes within a burst (as discussed in **Chapter 15 in Section IV**). The overall shape of stimulus filters, and how this is matched to the temporal statistics of input stimuli, determines whether and when neurons reformat these inputs to spike rates that maximize transmitted information (as described in **Chapter 63 in Section XI**).

Importantly, the mapping of stimuli to responses, even in individual neurons, should not be viewed as a fixed process, but rather as a continuously evolving one, changing in response to the sensory environment as well as local circuit conditions. Above, we have described one fundamental way in which this occurs, via adaptation to stimulus statistics in the single neurons themselves, and we have discussed how stimulus variations on multiple timescales contribute to the system's probability of response. These factors are, once again, just the tip of the iceberg. Other chapters in this volume detail important impacts of long-term kinetics in synapses, which reshape inputs *en route* to neurons, and underlie important aspects of adaptation across thalamocortical networks. Additionally, modulatory inputs can combine with "driving" stimulus inputs to reshape overall input statistics and hence neural response processing. Neuromodulators broadcast throughout neural networks drive further changes in signal processing in the underlying neurons and networks. As many of the chapters in this section detail, computational methods have already illuminated many ways that these mechanisms can combine to shape thalamocortical processing, and as such have cast light on an exciting road ahead, as many more interactions remain to be discovered.

ACKNOWLEDGMENTS

ALF is grateful for support from the Simons Collaboration for the Global Brain and NIH U19NS107609. ALF and ETSB gratefully acknowledge support from NIH UF1NS126485, and ETSB from NIH RF1DA055669 and NSF NCS-FO 2024364. RAM is supported by grants from the Chica-Heinz-Schaller Foundation and the Brigitte-Schlieben-Lange Program (Baden-Württemberg Ministry of Science). ALF, ETSB, and RAM give many thanks to Heather Mease for expert assistance with figures.

REFERENCES

Agüera y Arcas B, Fairhall AL, Bialek W (2003). Computation in a single neuron: Hodgkin and Huxley revisited. *Neural Comput.* 15:1715–1749.

Alenda A, Molano-Mazón M, Panzeri S, Maravall M (2010). Sensory input drives multiple intracellular information streams in somatosensory cortex. *J Neurosci.* 30:10872–10884.

Aljadeff J, Lansdell BJ, Fairhall AL, Kleinfeld D (2016). Analysis of neuronal spike trains, deconstructed. *Neuron.* 91:221–259.

Baccus SA, Meister M (2002). Fast and slow contrast adaptation in retinal circuitry. *Neuron.* 36:909–919.

Bard Ermentrout G, Terman DH (2010). *Mathematical Foundations of Neuroscience.* Springer.

Borden PY, Wright NC, Morrissette AE, Jaeger D, Haider B, Stanley GB (2022). Thalamic bursting and the role of timing and synchrony in thalamocortical signaling in the awake mouse. *Neuron.* 110:2836–2853.

Brenner N, Bialek W, de Ruyter van Steveninck RR (2000). Adaptive rescaling maximizes information transmission. *Neuron.* 26:695–702.

Brunel N, Chance FS, Fourcaud N, Abbott LF (2001). Effects of synaptic noise and filtering on the frequency response of spiking neurons. *Phys Rev Lett.* 86:2186–2189.

Chance FS, Abbott LF, Reyes AD (2002). Gain modulation from background synaptic input. *Neuron.* 35:773–782.

Crandall SR, Cruikshank SJ, Connors BW (2015). A corticothalamic switch: Controlling the thalamus with dynamic synapses. *Neuron.* 86:768–782.

Diaz-Quesada M, Maravall M (2008). Intrinsic mechanisms for adaptive gain rescaling in barrel cortex. *J Neurosci.* 28:696–710.

Drew PJ, Abbott LF (2006). Models and properties of power-law adaptation in neural systems. *J Neurophysiol.* 96:826–833.

Elijah DH, Samengo I, Montemurro MA (2015). Thalamic neuron models encode stimulus information by burst-size modulation. *Front Comput Neurosci.* 9:113.

Estebanez L, El Boustani S, Destexhe A, Shulz DE (2012). Correlated input reveals coexisting coding schemes in a sensory cortex. *Nat Neurosci.* 15:1691–1699.

Fairhall AL, Andrew Burlingame C, Narasimhan R, Harris RA, Puchalla JL, Berry MJ (2006). Selectivity for multiple stimulus features in retinal ganglion cells. J Neurophysiol. 96:2724–2738.

Fairhall AL, Lewen GD, Bialek W, de Ruyter Van Steveninck RR (2001). Efficiency and ambiguity in an adaptive neural code. *Nature.* 412:787–792.

Fairhall A, Lewen G, Bialek W, van Steveninck R (2000). Multiple timescales of adaptation in a neural code. *Adv Neural Inf Process Syst.* 13. Accessed March 1, 2023. https://proceedings.neurips.cc/paper/2000/hash/645098b086d2f9e1e0e939c27f9f2d6f-Abstract.html.

Famulare M, Fairhall A (2010). Feature selection in simple neurons: How coding depends on spiking dynamics. *Neural Comput.* 22:581–598.

Ferguson KA, Cardin JA (2020). Mechanisms underlying gain modulation in the cortex. *Nat Rev Neurosci.* 21:80–92.

Gabbiani F, Cox SJ (2010). *Mathematics for Neuroscientists.* Academic Press.

Gjorgjieva J, Mease RA, Moody WJ, Fairhall AL (2014). Intrinsic neuronal properties switch the mode of information transmission in networks. *PLoS Comput Biol.* 10:e1003962.

Higgs MH, Slee SJ, Spain WJ (2006). Diversity of gain modulation by noise in neocortical neurons: regulation by the slow afterhyperpolarization conductance. *J Neurosci.* 26:8787–8799.

Hodgkin AL, Huxley AF (1952). A quantitative description of membrane current and its application to conduction and excitation in nerve. *J Physiol.* 117:500–544.

Hong S, Agüera y Arcas B, Fairhall AL (2007). Single neuron computation: From dynamical system to feature detector. *Neural Comput.* 19:3133–3172.

Hong S, Lundstrom BN, Fairhall AL (2008). Intrinsic gain modulation and adaptive neural coding. *PLoS Comput Biol.* 4:e1000119.

Hosoya T, Baccus SA, Meister M (2005). Dynamic predictive coding by the retina. *Nature.* 436:71–77.

Hu Y, Brunton SL, Cain N, Mihalas S, Kutz JN, Shea-Brown E (2018). Feedback through graph motifs relates structure and function in complex networks. *Phys Rev E.* 98:062312.

Hu Y, Trousdale J, Josić K, Shea-Brown E (2014). Local paths to global coherence: cutting networks down to size. *Phys Rev E Stat Nonlin Soft Matter Phys.* 89:032802.

Izhikevich EM (2007). *Dynamical Systems in Neuroscience.* MIT Press.

Kass RE, Amari SI, Arai K, Brown EN, Diekman CO, Diesmann M, Doiron B, Eden UT, Fairhall AL, Fiddyment GM, Fukai T, Grün S, Harrison MT, Helias M, Nakahara H, Teramae JN, Thomas PJ, Reimers M, Rodu J, Rotstein HG, Shea-Brown E, Shimazaki H, Shinomoto S, Yu BM, Kramer MA (2018). Computational neuroscience: Mathematical and statistical perspectives. Ann *Rev Stat Applic.* 5:183–214.

Latimer KW, Barbera D, Sokoletsky M, Awwad B, Katz Y, Nelken I, Lampl I, Fairhall AL, Priebe NJ (2019). Multiple timescales account for adaptive responses across sensory cortices. *J Neurosci.* 39:10019–10033.

Lindner B, Doiron B, Longtin A (2005). Theory of oscillatory firing induced by spatially correlated noise and delayed inhibitory feedback. *Phys Rev E Stat Nonlin Soft Matter Phys.* 72:061919.

Lundstrom BN, Fairhall AL, Maravall M (2010). Multiple timescale encoding of slowly varying whisker stimulus envelope in cortical and thalamic neurons in vivo. *J Neurosci.* 30:5071–5077.

Lundstrom BN, Higgs MH, Spain WJ, Fairhall AL (2008a). Fractional differentiation by neocortical pyramidal neurons. *Nat Neurosci.* 11:1335–1342.

Lundstrom BN, Hong S, Higgs MH, Fairhall AL (2008b). Two computational regimes of a single-compartment neuron separated by a planar boundary in conductance space. *Neural Comput.* 20:1239–1260.

Maravall M, Petersen RS, Fairhall AL, Arabzadeh E, Diamond ME (2007). Shifts in coding properties and maintenance of information transmission during adaptation in barrel cortex. *PLoS Biol.* 5:e19.

Martí D, Brunel N, Ostojic S (2018). Correlations between synapses in pairs of neurons slow down dynamics in randomly connected neural networks. *Phys Rev E.* 97:062314.

Mease RA, Famulare M, Gjorgjieva J, Moody WJ, Fairhall AL (2013). Emergence of adaptive computation by single neurons in the developing cortex. *J Neurosci.* 33:12154–12170.

Mease RA, Krieger P, Groh A (2014a). Cortical control of adaptation and sensory relay mode in the thalamus. *Proc Natl Acad Sci U S A.* 111:6798–6803.

Mease RA, Kuner T, Fairhall AL, Groh A (2017). Multiplexed spike coding and adaptation in the thalamus. *Cell Rep.* 19:1130–1140.

Mease RA, Lee S, Moritz AT, Powers RK, Binder MD, Fairhall AL (2014b). Context-dependent coding in single neurons. *J Comput Neurosci.* 37:459–480.

Naud R, Sprekeler H (2018). Sparse bursts optimize information transmission in a multiplexed neural code. *Proc Natl Acad Sci U S A.* 115:E6329–E6338.

Ocker GK, Hu Y, Buice MA, Doiron B, Josić K, Rosenbaum R, Shea-Brown E (2017). From the statistics of connectivity to the statistics of spike times in neuronal networks. *Curr Opin Neurobiol.* 46:109–119.

Ostojic S, Brunel N (2011). From spiking neuron models to linear-nonlinear models. *PLoS Comput Biol.* 7:e1001056.

Oswald A-MM, Chacron MJ, Doiron B, Bastian J, Maler L (2004). Parallel processing of sensory input by bursts and isolated spikes. *J Neurosci.* 24:4351–4362.

Pernice V, Staude B, Cardanobile S, Rotter S (2011). How structure determines correlations in neuronal networks. *PLoS Comput Biol.* 7:e1002059.

Pozzorini C, Naud R, Mensi S, Gerstner W (2013). Temporal whitening by power-law adaptation in neocortical neurons. *Nat Neurosci.* 16:942–948.

Ratté S, Hong S, De Schutter E, Prescott SA (2013). Impact of neuronal properties on network coding: roles of spike initiation dynamics and robust synchrony transfer. *Neuron.* 78:758–772.

Recanatesi S, Ocker GK, Buice MA, Shea-Brown E (2019). Dimensionality in recurrent spiking networks: global trends in activity and local origins in connectivity. *PLoS Comput Biol.* 15:e1006446.

Reinagel P, Reid RC (2000). Temporal coding of visual information in the thalamus. *J Neurosci.* 20:5392–5400.

Rodenkirch C, Liu Y, Schriver BJ, Wang Q (2019). Locus coeruleus activation enhances thalamic feature selectivity via norepinephrine regulation of intrathalamic circuit dynamics. *Nat Neurosci.* 22:120–133.

Roxin A (2011). The role of degree distribution in shaping the dynamics in networks of sparsely connected spiking neurons. *Front Comput Neurosci.* 5:8.

Rust NC, Schwartz O, Movshon JA, Simoncelli EP (2005). Spatiotemporal elements of macaque v1 receptive fields. *Neuron.* 46:945–956.

Samengo I, Mato G, Elijah DH, Schreiber S, Montemurro MA (2013). Linking dynamical and functional properties of intrinsically bursting neurons. *J Comput Neurosci.* 35:213–230.

Sejnowski TJ (1976). On the stochastic dynamics of neuronal interaction. *Biol Cybern.* 22:203–211.

Sharpee TO, Sugihara H, Kurgansky AV, Rebrik SP, Stryker MP, Miller KD (2006). Adaptive filtering enhances information transmission in visual cortex. *Nature.* 439:936–942.

Sherman SM, Guillery RW (2002). The role of the thalamus in the flow of information to the cortex. *Philos Trans R Soc Lond B Biol Sci.* 357:1695–1708.

Slee SJ, Higgs MH, Fairhall AL, Spain WJ (2005). Two-dimensional time coding in the auditory brainstem. *J Neurosci.* 25:9978–9988.

Teeter C, Iyer R, Menon V, Gouwens N, Feng D, Berg J, Szafer A, Cain N, Zeng H, Hawrylycz M, Koch C, Mihalas S (2018). Generalized leaky integrate-and-fire models classify multiple neuron types. *Nat Commun.* 9:709.

Theunissen FE, Sen K, Doupe AJ (2000). Spectral-temporal receptive fields of nonlinear auditory neurons obtained using natural sounds. *J Neurosci.* 20:2315–2331.

Trousdale J, Hu Y, Shea-Brown E, Josić K (2012). Impact of network structure and cellular response on spike time correlations. *PLoS Comput Biol.* 8:e1002408.

van Rossum MCW, Turrigiano GG, Nelson SB (2002). Fast propagation of firing rates through layered networks of noisy neurons. *J Neurosci.* 22:1956–1966.

Vogels TP, Abbott LF (2005). Signal propagation and logic gating in networks of integrate-and-fire neurons. *J Neurosci.* 25:10786–10795.

Weber AI, Fairhall AL (2019). The role of adaptation in neural coding. *Curr Opin Neurobiol.* 58:135–140.

Weber AI, Krishnamurthy K, Fairhall AL (2019). Coding principles in adaptation. *Annu Rev Vis Sci.* 5:427–449.

Weber AI, Pillow JW (2017). Capturing the dynamical repertoire of single neurons with generalized linear models. *Neural Comput.* 29:3260–3289.

Whitmire CJ, Stanley GB (2016). Rapid sensory adaptation redux: A circuit perspective. *Neuron.* 92:298–315.

Wolfart J, Debay D, Le Masson G, Destexhe A, Bal T (2005). Synaptic background activity controls spike transfer from thalamus to cortex. *Nat Neurosci.* 8:1760–1767.

Zdeblick D, Shea-Brown E, Witten D, Buice M (2023). Modeling functional cell types in spike train data. BioRxiv. doi: 10.1101/2023.02.28.530327.

Zeldenrust F, Chameau P, Wadman WJ (2018). Spike and burst coding in thalamocortical relay cells. *PLoS Comput Biol.* 14:e1005960.

61.

MODELING THALAMOCORTICAL OSCILLATIONS

MECHANISMS AND IMPACT

Ryan Golden and Maxim Bazhenov

INTRODUCTION

The thalamus and cortex are some of the most diverse structures in the brain, each housing a wide variety of anatomically and physiologically differentiated sub-regions that serve a range of sensory, motor, and cognitive functions. Despite this diversity of function, much of the communication within and between these structures appears to be organized by a set of common, periodic oscillatory activities. The dominant frequencies and spatiotemporal dynamics of these oscillations are determined by mechanisms of neuronal synchronization that operate simultaneously at different levels of functioning and further depend on the functional state of the brain as a whole. These oscillations have been associated with a wide variety of functions, including level of arousal and vigilance, working memory, cellular plasticity, and memory formation, recall, and consolidation. How these oscillatory modes are generated, coexist, and support neuronal communication has been the subject of intense investigation throughout the last century of neuroscience.

OSCILLATIONS GENERATED WITHIN THE THALAMOCORTICAL SYSTEM

The thalamocortical system supports a variety of oscillatory activities, with the dominant frequencies ranging from slow (0.01–15 Hz) to fast (20–600 Hz). While slow activities are predominantly present during rest, sleep, or under anesthesia, the faster activities can be found across various levels of vigilance and arousal, frequently coupled to the slower ones. At a mechanistic level, these oscillations can be divided into 2 primary types: intrinsic—generated by individual neurons due to the interplay between particular intrinsic currents; and extrinsic—network phenomena which rely on excitatory and inhibitory interactions within neuronal populations. In what follows, we will provide an overview of the mechanisms involved in generating, sustaining, and propagating some of the most prominent oscillations found within the thalamocortical system, as well as the systems-level and cognitive functions they have been found to support.

INFRA-SLOW OSCILLATION

Infra-slow oscillations (0.01–0.2 Hz) have been observed during the resting-state or in spontaneous background fluctuations across a variety of techniques, including functional magnetic resonance imaging (fMRI), electroencephalograph (EEG), and local field potential (LFP) recordings (Fukunaga et al., 2006; Pan et al., 2013). These oscillations were found to be correlated across remote brain regions and are thought to reflect neuronal activity fluctuations between functionally connected areas of the brain.

Mechanism

Although the mechanisms underlying the infra-slow oscillation are still being highly debated, several key factors have been implicated in shaping the functional connectivity which results from this spontaneous activity. Perhaps unsurprisingly, both experimental and computational studies have highlighted the role that the underlying structural connectivity plays in shaping these dynamics (Honey et al., 2007; Deco et al., 2009). Modeling studies have suggested roles for intrinsic noise, coupling strengths, and conduction velocities in the spatiotemporal dynamics of these fluctuations (Honey et al., 2007; Ghosh et al., 2008; Deco et al., 2009).

Many computational studies of infra-slow oscillation have relied on mean-field models, which attempt to simplify the system under study by taking a many-neuron system and averaging across the individual interactions to get a single effective-interaction term which is applied to all spatial components of the system. In practice, greater biological realism can be attained by considering multiple mean-field models at once, according to brain region, or by first dividing interactions into excitatory and inhibitory. It was proposed that the timescale of infra-slow fluctuations could be a result of transient bouts of synchrony between clusters of nodes, and that the functional connectivity arising in the network strongly reflected the underlying structural architecture (Honey et al., 2007; Deco et al., 2009). Phase locking between distant network sites through long-range connections was reported in the mean-field-type models (Honey et al., 2007; Ghosh et al., 2008).

Regarding the biophysical origin of these oscillations, a recent modeling study which used Hodgkin-Huxley neurons suggests that infra-slow oscillations can arise from fluctuations of the Na^+ and K^+ ion concentrations mediated by neuronal and glial activity (Figure 61.1) (Krishnan et al., 2018). Increase of the extracellular K^+ concentration, $[K^+]_o$, results in higher excitability of neurons which may then trigger further elevation of the $[K^+]_o$, leading to a positive feedback loop. Importantly, increase of $[K^+]_o$ can be arbitrarily slow, being determined by the balance of the inward and outward K^+ flux. Increase in the intracellular Na^+ concentration, $[Na^+]_i$, terminates the positive feedback loop, initiating a phase of progressive decrease of the ion concentrations and firing rate. Transitions between these phases depend on the Na^+/K^+ pump. This mechanism, when coupled with realistic long-range structural connectivity, can explain both the timescale of the infra-slow

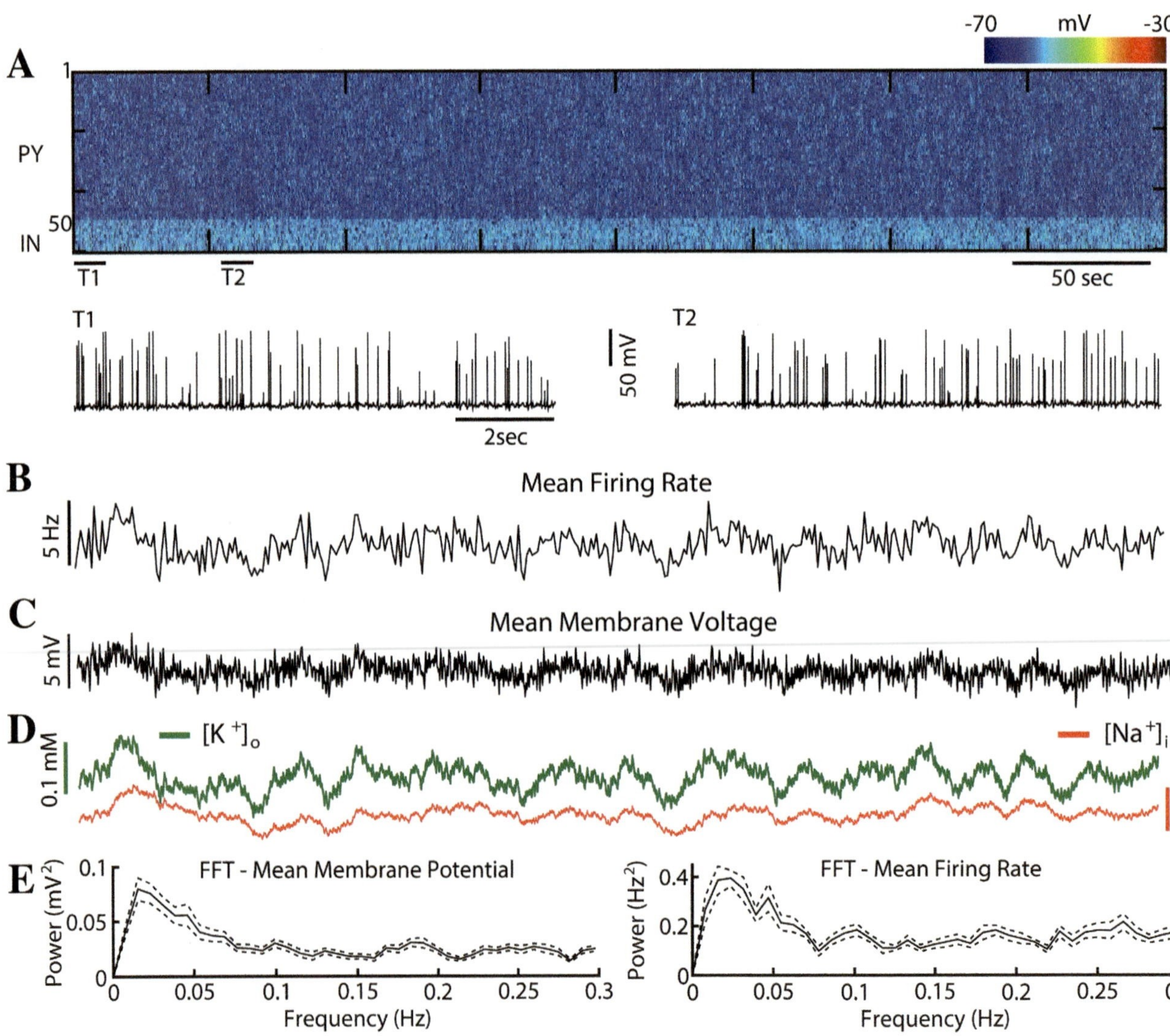

Figure 61.1. Ionic basis of infra-slow oscillations. (A). Upper: spontaneous activity in the network of excitatory pyramidal cells (PY) and inhibitory interneurons (IN) during infra-slow resting state oscillation. Lower: T1 and T2 indicate times expanded below in single CX cell traces. (B–D). Mean CX firing rate, mean filtered (0.001–0.1 Hz) membrane voltage, $[K^+]_o$, and $[Na^+]_i$, respectively. (E). Power spectrums of the mean membrane voltage (left) and mean firing rate (right) in CX cells. Note peak at around 0.02 Hz. Modified from Krishnan et al. (2018).

fluctuations, as well as the local and long-range synchronization and correlations captured by functional connectivity measures on resting-state activity (Krishnan et al., 2018).

Functional Role

The infra-slow oscillation has most prominently been used as a proxy for determining functional connectivity between brain regions (Fukunaga et al., 2006; Zhang and Raichle, 2010), and to delineate functional networks, such as the default-mode network, which vary with brain state and task demands (Fukunaga et al., 2006). Moreover, infra-slow fluctuations in LFP gamma power and neuronal firing rate have been shown to correlate with resting-state fMRI blood-oxygen-level-dependent (BOLD) fluctuations (Pan et al., 2013), which may allow for a more neuron-centric interpretation of BOLD signals. Finally, alterations in the temporal coherence of infra-slow oscillations, as well as the resultant functional connectivity, have been correlated with certain neurological and psychiatric disorders, such as epilepsy and schizophrenia (Zhang and Raichle, 2010).

Resting-state fluctuations increase in specific brain areas during rest compared to task periods. Early observations identified the default mode network (DMN) as a subnetwork that has higher activation and functional connectivity during rest periods compared to task periods. While specific mechanisms of this neuromodulation are poorly understood, a recent modeling study suggested that cholinergic input may influence resting-state activity and its functional connectivity through cellular and synaptic modulation (Sanda et al., 2022). Specifically, selective cholinergic modulation of DMN closely captured transitions between the baseline resting state and states with suppressed DMN fluctuations associated with attention to external tasks.

SLOW OSCILLATION

The slow oscillation (0.3–1 Hz) is a hallmark of deep NREM sleep (stage 3, also called slow wave sleep, or SWS), dominating cortical activity during early periods of sleep and under some types of anesthesia (Steriade et al., 1993b). During a slow oscillation cycle, the entire cortical network alternates between silent (hyperpolarized or "Down") and active (depolarized or "Up") states, each lasting 0.2–1s (Figure 61.2A). Silent periods are periods of disfacilitation (i.e., absence of synaptic activity), while active periods have intensive synaptic activity leading to the generation of fast oscillations within the thalamocortical system.

Mechanism

The slow oscillation is essentially a cortical rhythm. This conclusion is based on 3 basic observations: (1) The slow oscillation survives extensive thalamic lesions (Steriade et al., 1993a); (2) the slow oscillation can be recorded in neocortical slices (Sanchez-Vives and McCormick, 2000); and (3) the slow oscillation is absent in the thalamus of decorticated cats (Timofeev and Steriade, 1996).

Several mechanisms have been proposed to explain sleep slow oscillations. The first depends on spontaneous miniature postsynaptic potentials (mPSPs) caused by the action potential-independent release of transmitter vesicles and regulated at the level of single synapses. Occasional summation of the miniature excitatory postsynaptic potentials (EPSPs) during the hyperpolarized (silent) phase of slow-sleep oscillations activates the persistent sodium current and depolarizes the membrane of cortical pyramidal cells, which is sufficient for spike generation (Bazhenov et al., 2002; Timofeev et al., 2000). This triggers the active phase of the slow oscillation, which is maintained by synaptic activities and the persistent sodium current (Figure 61.2B). Modeling experiments suggest that short-term synaptic depression and the activation of Ca^{2+}-dependent K^+ currents

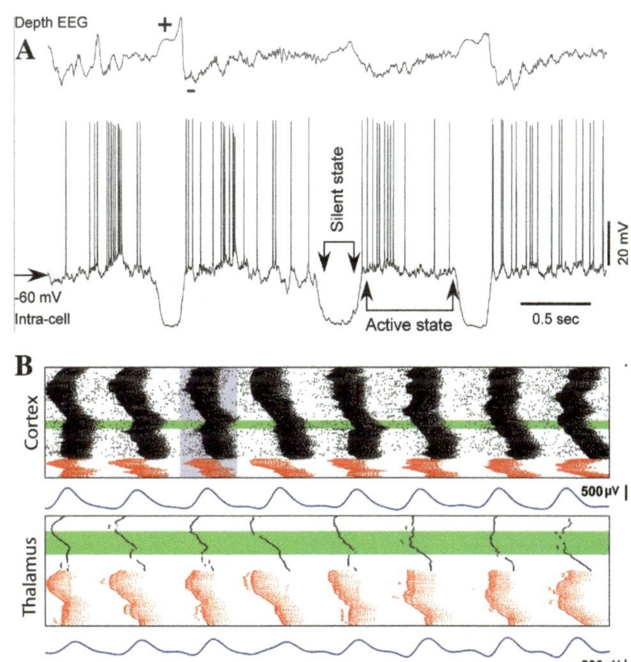

Figure 61.2. Cortical sleep slow oscillation. (A). In vivo, at EEG level the slow oscillation appears as periodic alterations of positive and negative waves (indicated by + and – signs). During EEG depth-positivity, cortical neurons remain in hyperpolarized, silent state. During EEG depth-negativity, cortical neurons move to active states, reveal barrages of synaptic events, and fire action potentials. (B). Model of thalamocortical slow oscillations showing 10 seconds of network activity as spike rastergrams for the cortex (excitatory pyramidal cells: black; inhibitory interneurons: red) and thalamus (excitatory thalamocortical neurons: black; inhibitory reticular interneurons: red). Average LFPs from 100 neurons (green areas) are shown below each rastergram, filtered from 0.5 to 2 Hz. Panel A modified from Timofeev and Bazhenov (2005); panel B modified from Sanda et al. (2021).

were sufficient to terminate the active state (Bazhenov et al., 2002). While the thalamus was not necessary to maintain slow oscillations in this model, it was later found, using a combination of in vivo electrophysiology and modeling, that the normal pattern of the neocortical slow oscillation requires thalamic inputs to synchronize activity across large cortical areas (Lemieux et al., 2014). Inactivation of specific thalamic regions led to the loss of synchrony and periodicity of the slow oscillation in the target cortical cell populations.

Another group of mechanisms relies on spontaneous activity in some populations of cortical neurons. Computer models were proposed where transitions from the Down (silent) to the Up (active) state were initiated by spontaneous spike discharges in a random group of neurons (Compte et al., 2003) or progressive activation of persistent Na^+ currents (Hill and Tononi, 2005). Once started, Up states were maintained by strong recurrent excitation, and the transitions to the Down state were due to a slow Na^+-dependent K^+ and/or depolarization-activated K^+ currents. In vitro studies indicate that a group of neurons, which could initiate active periods, could be either layer L5 intrinsically bursting neurons (Sanchez-Vives and McCormick, 2000) or spatially structured neuronal ensembles (Cossart et al., 2003). In another model, the spontaneous activity arose from recurrent connections between cortical neurons within a given layer (presumably L5), leading to self-sustained irregular activity states within that layer (Destexhe, 2009). Bursting in thalamocortical neurons was proposed as an alternative mechanism triggering Down-to-Up-state transitions in cortical populations (Crunelli and Hughes, 2010).

While the above studies assume that there is an active process that leads to the Down-to-Up-state transition, a recent study (Levenstein et al., 2019) proposed a model of slow oscillation generation where a stable active state is interrupted by transient inactive states. The model predicted how the level of drive and the relative strength of recurrent excitation and adaptation may create a spectrum of dynamical regimes with transitions between Up and Down states. Importantly, this model produced the characteristic asymmetric distributions of Up- and Down-state durations found in vivo.

Transitions from the active state to silence are often better synchronized than the transitions from silence to the active state and show no latency bias for any location or cell type (Sheroziya and Timofeev, 2014). This in vivo result could not be explained by earlier models (Bazhenov et al., 2002; Hill and Tononi, 2005) and suggested that some larger-scale biophysical mechanisms may be involved in manipulating the downward transition during slow oscillation. One possible mechanism is active cortical inhibition. Using biophysical models of SO, it was shown that synaptic inhibition controls the duration and the synchrony of active state termination (Chen et al., 2012). The reduction of interneuron-mediated inhibition led to asynchronous downward transition across the cortical network and broke the regular slow oscillation pattern.

Functional Role

The presence of the slow oscillation during sleep has been reported to be involved in memory consolidation and formation (Rasch and Born, 2013). It has been suggested that the preferred directions of traveling slow waves could shape the sequential firing of neurons involved in reactivation or replay of waking activity patterns during sleep. This hypothesis is known as sleep-dependent consolidation (Rasch and Born, 2013). An alternatively proposed function for the slow oscillation during sleep is the synaptic homeostasis hypothesis (SHY) (Tononi and Cirelli, 2014). The main ideas of SHY are that the waking state is associated with progressive strengthening of synapses mediating learning, while slow-wave sleep down-regulates synapses to facilitate subsequent learning in the following wake cycle. Recent modeling studies have indicated that reactivation resulting from slow oscillations can enhance recent memories (Wei et al., 2016) (Figure 61.3), reduce prospective interference, and rescue retroactive interference (Gonzalez et al., 2020), and may underly relational associative memory formation (Tadros and Bazhenov, 2022).

The reactivation episodes in medial prefrontal cortex (mPFC) are coordinated with those of the hippocampus and with the occurrence of sharp wave-ripples during sleep (Peyrache et al., 2009). Such a relationship has recently been demonstrated in a model of hippocampal-thalamocortical loop (Sanda et al., 2021), while other modeling efforts have shown that input resembling hippocampal replay can bias activity and consolidation during slow waves (Wei et al., 2016). Indeed, the connectivity between hippocampus and prefrontal cortex has been found to increase after a task in which cues presented during SWS were increasing subject performance in memorizing cue-items associations (van Dongen et al., 2012).

DELTA OSCILLATION

The field potential recordings from the neocortex in both humans and animals during sleep reveal the presence of delta oscillation with frequencies 1–4 Hz. Delta and the slow oscillation likely represent 2 distinct phenomena, with differences in the dynamics between the slow and the delta oscillations, as the latter declines in activity from the first to the second NREM sleep episode, whereas the former does not. The delta oscillation likely has 2 different components, one of which originates in the neocortex and another in the thalamus.

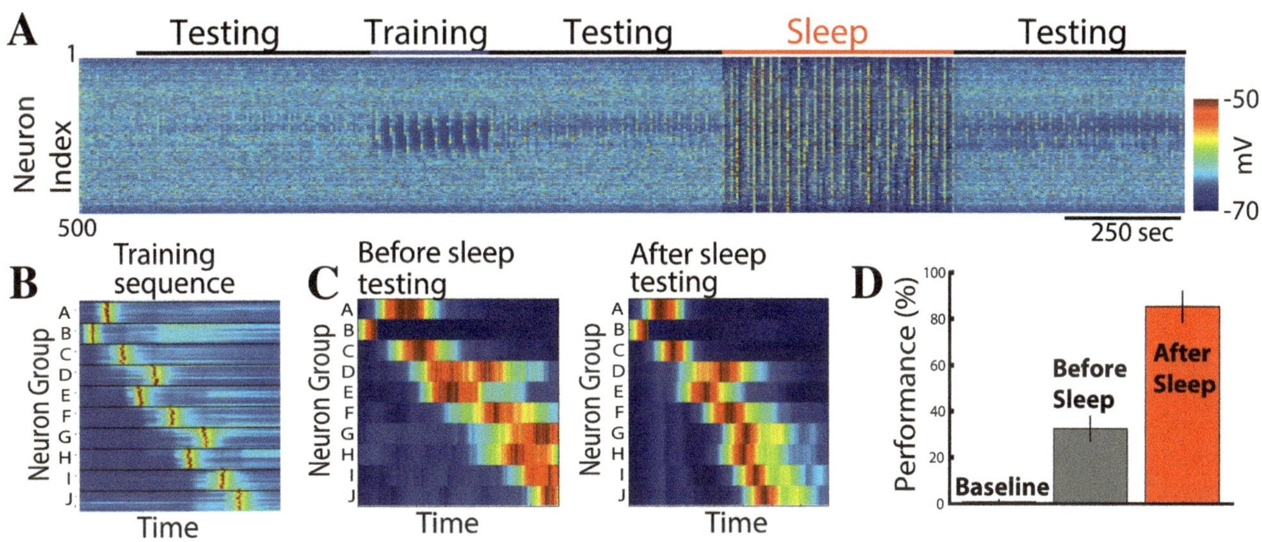

Figure 61.3. Sleep replay improves recall performance for memory sequences. (A). Network activity during periods of training and testing of a memory sequence, and slow-wave sleep. Heat maps of membrane potentials of cortical pyramidal cells are shown. (B). Example of the training protocol used for training a long nonlinear sequence—BACEDFHGIJ. Note that each "letter" in a sequence is represented by a group of 10 neurons. (C). Normalized average group responses during testing periods before (left) and after (right) sleep. Sleep leads to an improvement and tuning of the responses such that testing after sleep results in the correct ordering of group activations and faster completion of the sequence. (D). Performance of the sequence completion at baseline, after training but before sleep, and after sleep (red). Modified from Gonzalez et al. (2020).

Mechanism

Thalamic delta (1–4 Hz) oscillation is a well-known example of rhythmic activity generated intrinsically in thalamic relay neurons. These oscillations arise as an interplay of low-threshold Ca^{2+} current (I_T) and hyperpolarization activated cation current (I_h) and may be observed during deep sleep when thalamocortical (TC) neurons are sufficiently hyperpolarized to deinactivate I_T (McCormick and Pape 1990). Sufficiently long and deep hyperpolarization of a TC neuron removes I_T inactivation and makes possible rebound burst generation, triggered by a depolarized input. The interplay of I_T and I_h during delta oscillation was first described in vitro (McCormick and Pape, 1990) and was later studied with computational models (Lytton et al., 1996). Being dependent on the interplay of intrinsic currents, delta oscillations can be generated in a single thalamic neuron. (For more information on the biophysical properties of thalamic cells, see **Chapter 3 in Section II.**)

Cortical mechanisms of delta are poorly understood. It is possible that slow waves generated during cortical slow oscillations and during delta share the same mechanisms. The differences in frequency of the 2 rhythms may be explained by activities of neuromodulatory systems. Even slight changes of the neuromodulator content dramatically affect the properties of slow waves (Krishnan et al., 2016), which may explain the frequency shift toward the delta band.

Functional Role

Given the potential mechanistic similarities to the slow oscillation and difficulties in properly differentiating the slow and delta oscillations, not much is known about the particular role played by delta. However, one recent study has gained support for the idea that delta waves may weaken memories, while slow oscillations (particularly when paired with spindles) promote memory consolidation (Kim et al., 2019).

SLEEP SPINDLE OSCILLATION

Slow oscillations are commonly coupled with another type of characteristic activity during sleep: sleep spindles, consisting of waxing-and-waning field potentials of 9–15 Hz lasting 1–3 sec. Spindles are found in isolation during stage 2 sleep and appear at Down-to-Up-state transitions during slow-wave (stage 3) sleep.

Mechanism

Spindle-like activity was found in thalamic lateral geniculate nucleus (LGN) slice preparations of ferrets with preserved interconnections with perigeniculate nucleus (Kim et al., 1995), which suggested thalamic origin of sleep spindles. In vivo, in vitro, and modeling studies suggest that the minimal substrate contributing to the generation of spindle oscillations is the thalamus, and the spindles are generated as a result of interaction between thalamic reticular (RE) neurons and thalamocortical (TC) cells. According to this hypothesis, the RE inhibitory neurons fire a spike-burst that elicits inhibitory postsynaptic potentials (IPSPs) in TC neurons; the RE-mediated IPSPs enhance TC cell hyperpolarization, thus increasing deinactivation of the low-threshold Ca^{2+} T-channels and making possible the rebound burst in TC neurons; the TC-mediated EPSPs trigger next burst in RE cells (Destexhe et al., 1993). TC neurons do not fire every cycle of oscillations but intermittently burst with intervening subthreshold oscillations (Kim et al., 1995). An important caveat to this RE-TC hypothesis is that spindles have been reported in the reticular nucleus of the thalamus following deafferentation from all other thalamic nuclei (Steriade et al., 1987).

A more complex spindle-generation model suggests the presence of at least 3 phases with different underlying mechanisms that contribute to spindle generation (Figure 61.4A). The waxing phase of spindles is associated with recruitment of neurons from the dorsal thalamic and reticular nuclei (Steriade et al., 1993a). In vivo recordings and computational models of RE networks found that the reversed IPSPs between the RE neurons can directly trigger low-threshold spike bursts at membrane potentials close to those observed during natural sleep (Bazhenov et al., 1999); this can lead to self-sustained bursting activity within the reticular nuclei. In this model, each sequence of spindle oscillations is followed by waves of activity that persist in the RE network during inter-spindle lulls and initiate new spindle sequences (Bazhenov et al., 2000) (Figure 61.4B,C). The second phase of spindles primarily develops as result of interactions between RE and TC neurons as described above, but cortical firing contributes to the spindle synchronization, where the corticothalamic projections impose the simultaneous excitation of RE and TC neurons. The waning phase occurs as a result of Ca^{2+} induced cAMP upregulation of hyperpolarization activated cation current, I_h, in TC cells, as well as network desynchronization (Bonjean et al., 2011). The effect of I_h up-regulation on spindle termination was first predicted in the models (Destexhe et al., 1993) and later confirmed by in vitro experiments (Luthi and McCormick, 1998). When this up-regulation is prevented, spindles in slices do not wax-wane anymore but remain sustained.

Human sleep spindles exhibit different synchronization properties when measured by EEG or magnetoencephalography (MEG) during simultaneous EEG/MEG recordings, whereby the EEG signal shows strong cross-pair coherence, unlike the MEG signal (Dehghani et al., 2010). Current source-density profiles of cortical depth electrode recordings suggested that spatially coherent spindles could be generated by upper cortical layers, while spatially incoherent spindles could be generated by middle cortical layers (Bonjean et al., 2012). This hypothesis was tested using biologically plausible computational models (Bonjean et al., 2012) using 2 parallel thalamocortical pathways (*core* and *matrix*). In this model, when the thalamocortical and corticothalamic connections fan-outs of the matrix significantly increased while maintaining constant focal connectivity in the core, the cortical spindle activity appeared more synchronous in the matrix compared to the spindle activity in the core. The study concluded that the difference between core and matrix pathways in thalamocortical fan-out explains best the discrepancies in spindle synchronization between scalp EEG and MEG (Bonjean et al., 2012). (See **Chapter 12 in Section IV** for more about the core and matrix pathways.)

The core and matrix pathways could also help to explain a potential mechanistic distinction between slow (9–12 Hz) and fast (12–15 Hz) spindles which dominate in centroparietal and frontal regions, respectively (Molle et al., 2011). However, recent research points to a growing inconsistency between the physiological mechanisms of slow and fast spindles which cannot be completely explained by these 2 pathways, and may suggest a greater intra-cortical role for the generation of slow spindles (Timofeev and Chauvette, 2013).

While the majority of the computational models of sleep are focused on one specific type of oscillation, a few studies have the simulated effects of neuromodulators to explain transitions between different sleep and wake activities. An increase in potassium leak conductances was shown to be sufficient to trigger the transition from wakefulness to sleep with slow oscillations (Hill and Tononi, 2005). A neural mass model implementing the effects of a few neuromodulators was developed in Costa et al. (2016) and revealed coexistence of K-complexes, sleep spindles, and slow oscillations during NREM. Neural mass models are similar to mean-field models in that they rely on statistical averaging of interactions. However, whereas mean-field models evolve over both space and time, neural mass models evolve solely over time, reducing the population to a single point which oscillates according to the average firing rate of the population. A recent study developed a biophysical model of the thalamocortical network that revealed a minimal set of changes in the neurochemical systems that are sufficient for transitions between wake and different sleep stages - stages 2 (N2) and 3 (N3) of NREM and REM sleep (Krishnan et al., 2016) (Figure 61.5). Importantly, the model revealed characteristic stage-specific activities found in humans: spindles in N2, slow oscillations in N3, alpha bursts during REM sleep. (For more detail about neuromodulation in the corticothalamic system, see **Chapter 10 in Section III.**)

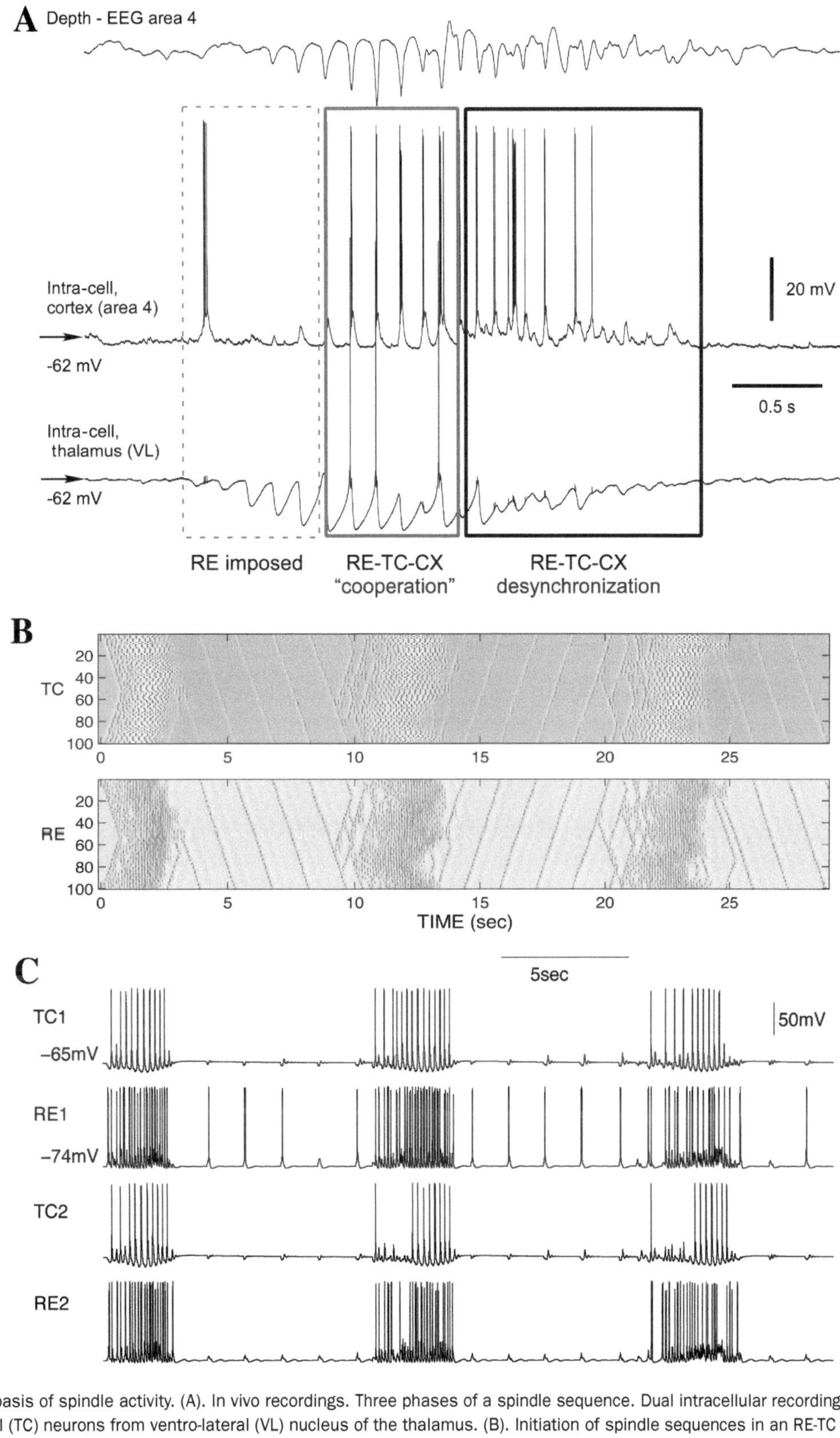

Figure 61.4. Cellular basis of spindle activity. (A). In vivo recordings. Three phases of a spindle sequence. Dual intracellular recording of cortical (area 4) and thalamocortical (TC) neurons from ventro-lateral (VL) nucleus of the thalamus. (B). Initiation of spindle sequences in an RE-TC network with periodic boundary conditions. The sequence of spindle oscillations was initiated at $t = 0$ s by the local stimulation of TC cell #1 (left upper corner of the panel) and then propagated in both directions with constant velocity. After 2–4 s, the spindle sequence was terminated because of the Ca^{2+} up-regulation of I_h current and was followed by localized patterns traveling through the RE nuclei. This activity triggered a new spindle sequence at about t = 10 s. (C). Membrane potentials for two RE-TC. pairs. Panel A modified from Timofeev and Bazhenov (2005); panels B, C modified from Bazhenov et al. (2000).

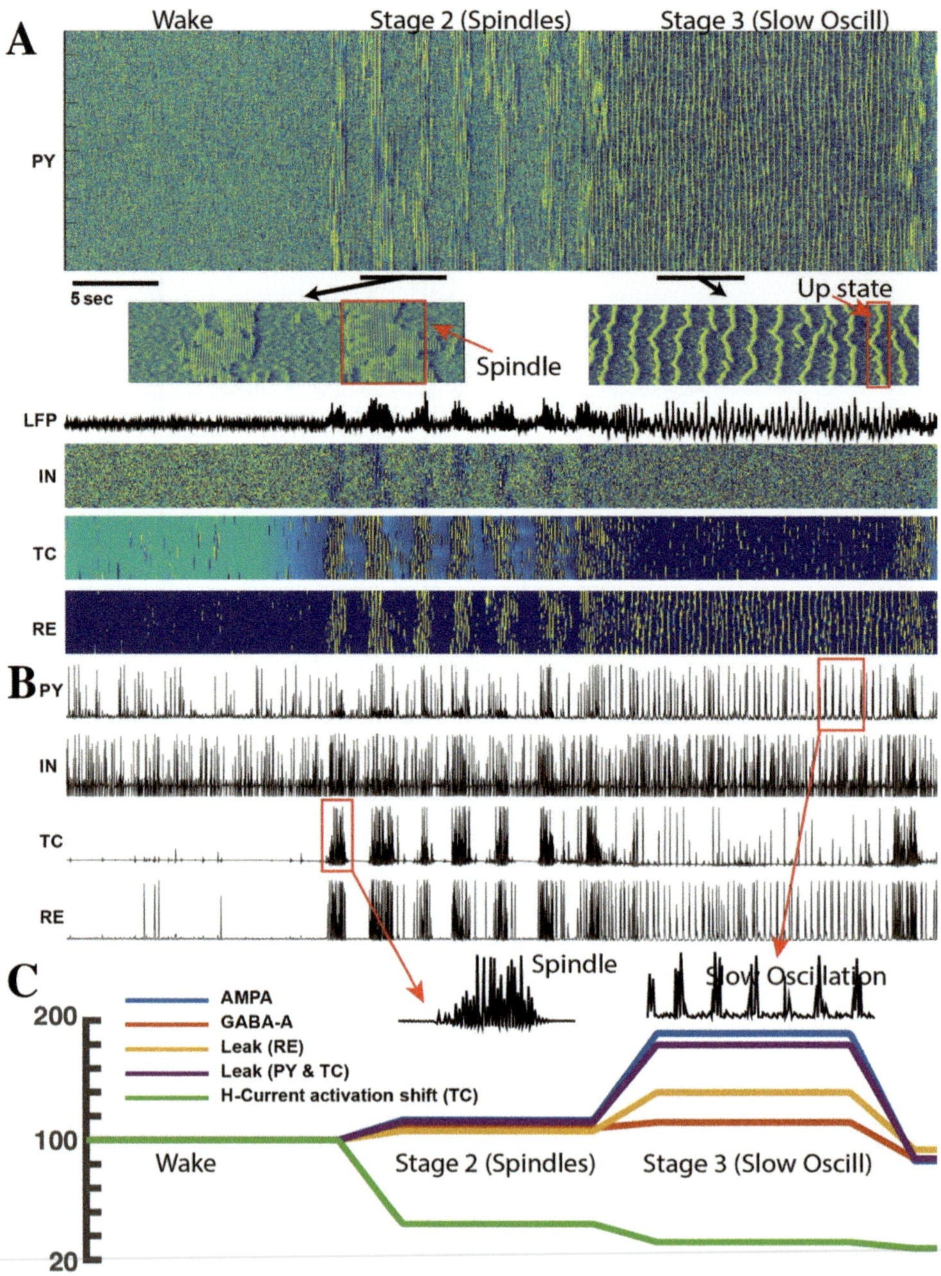

Figure 61.5. Neuromodulatory basis of sleep stages and dominant oscillations. (A). Activity of neurons in the thalamocortical network model with 500 pyramidal cells (PY), 100 interneurons (IN), 100 reticular interneurons (RE), and 100 thalamocortical cells (TC) across periods of waking, stage 2 (spindles), and stage 3 (slow oscillation) NREM sleep. *X*-axis is time and *Y*-axis is neuron index. Colors indicate membrane voltage. The pair of insets below the PY activity show zoomed-in activity during stage 2 (left) and stage 3 (right) sleep, with red boxes highlighting spindle and Up state activity, respectively. Inset just above IN activity shows LFP calculated from PY network (mean voltage of all 500 PY neurons). Note spindles and slow oscillations during N2 and N3, respectively. (B). Membrane voltages of the representative selected neurons of each type from the network. Red boxes highlight and point to insets displaying single cell voltage traces from a TC cell during a spindle (left) and a PY cell during slow oscillations (right). (C). Percent change in conductance of relevant intrinsic and synaptic currents across the 2 stages of sleep referenced to the awake state. Modified from Krishnan et al. (2016).

Functional Role

Studies have shown that spindle oscillations are essential for the consolidation of recent memories (Rasch and Born, 2013) and demonstrated the presence of efficient spindle-dependent short- and middle-term synaptic plasticity (Steriade and Timofeev, 2003). It was suggested that the rhythmic spike-trains or spike-bursts fired by cortical and thalamic neurons during sleep spindles may promote synaptic changes (Steriade and Timofeev, 2003) by massive Ca^{2+} entry in cortical pyramidal neurons (Sejnowski and Destexhe, 2000).

While spindles which are nested within slow oscillations have been reported to correlate strongly with memory consolidation (Molle et al., 2011; Kim et al., 2019), spindles can and often do appear in isolation of slow waves. A recent modeling study (Wei et al., 2018) suggested complementary roles for sleep spindles and slow oscillations in the consolidation of multiple interfering memories. Spindles resulted in local replay, which allowed for relatively independent consolidation of memory traces encoded by overlapping or spatially close populations of neurons. By contrast, slow oscillations elicited more global reactivation, promoting competition that could lead to extinction of weak memory traces in the presence of stronger ones.

BETA/GAMMA OSCILLATIONS

The waking state of the brain is characterized by a low correlation of spike discharges across neighboring neurons and the appearance of transient synchronization events in the beta (15–30 Hz) and gamma (30–60 Hz) frequency ranges (Bressler, 1990). The synchronization in these frequency bands was also found during deep anesthesia, natural slow-wave and REM sleep, when consciousness is either suspended or in a bizarre state. During slow-wave sleep the fast rhythms follow the onset of depth-negative EEG waves. Beta/gamma oscillations may propagate in space leading to beta- and gamma-waves.

Mechanism

Beta/gamma oscillations, ubiquitous throughout the neocortex, have been hypothesized to result from 2 distinct, but non-exclusive mechanisms. The first suggests that transient, feedforward synchronization to high-frequency oscillations in the periphery may impose this high-frequency activity onto the thalamocortical system (Timofeev and Steriade, 1997). The second mechanism of gamma activity generation was described both in vitro and computational models. According to this mechanism, the activity of inhibitory interneurons controls synchronization and is essential to obtain oscillations in the gamma range (Lytton and Sejnowski, 1991). (See **Chapter 8 in Section III** for more about how cortical interneurons shape oscillations through excitation-inhibition balance.)

Modeling studies focusing on the role of inhibition in generating beta/gamma oscillations are of 2 primary types: pyramidal-interneuron gamma (PING) (Borgers and Kopell, 2003), and interneuronal gamma (ING) (Whittington et al., 2000). In the case of PING, intrinsic spiking of excitatory pyramidal cells is thought to drive and progressively synchronize the activity of inhibitory, fast-spiking interneurons through AMPA-receptor mediated synaptic interactions. These interneurons subsequently synchronize the pyramidal cell activity through $GABA_A$-receptor mediated synaptic interactions. Coherent gamma-range oscillations may persist or may appear through occasional increases in spiking synchrony within local groups of cortical neurons. In contrast, ING-based models rely on a reciprocally connected network of inhibitory, fast-spiking interneurons. Activity in this population builds up to temporarily silence the network, which leads to a coordinated release from inhibition across the population, ultimately resulting in more synchronized spiking when firing resumes (Whittington et al., 2000).

Synchrony in these models can be made more robust by including gap junctions (Traub et al., 1999), and transitions between gamma and beta oscillations have been simulated by altering excitatory coupling between pyramidal neurons and by changes in K^+-conductances (Kopell et al., 2000). Interestingly, it has been reported that the network geometry may impact long-range gamma synchronization in the cortex (Bazhenov et al., 2008). Long-range synchronization was found to emerge in 2-dimensional, but not 1-dimensional, networks of cortical neurons. This effect was dependent on the presence of many independent synaptic pathways between any remote sites in the more realistic 2-dimensional system (Figure 61.6).

Functional Role

Given the ubiquity of oscillations in the beta and gamma ranges, these oscillations have come to be associated with a variety of functions, including attentiveness (Rougeul-Buser et al., 1975) and sensory perception (Gray et al., 1989). Moreover, it has been proposed that synchronization in the gamma range may contribute to cognitive processing and temporal binding (Joliot et al., 1994). It has recently been suggested that beta and gamma activity co-modulate one another to accomplish working memory tasks, with gamma activity carrying the relevant remembered information and beta bursts leading to suppression of gamma and gating access to working memory (Lundqvist et al., 2016). In this way, the balance between beta/gamma activity can control the information load present in working memory.

RIPPLE OSCILLATION

Fast oscillations (80-100 Hz in humans, 140-200 Hz in rodents), termed *ripples*, were first described in CA1 hippocampal area and perirhinal cortex, where they were associated with bursts of sharp potentials during anesthesia, behavioral immobility, and natural sleep (Buzsaki, 2015). Later, ripples were found during behavioral states (Jadhav et al., 2012). Recently, precise long-range synchrony of cortical ripples was reported in the human neocortex during both sleep and awake states (Dickey et al., 2022). Although

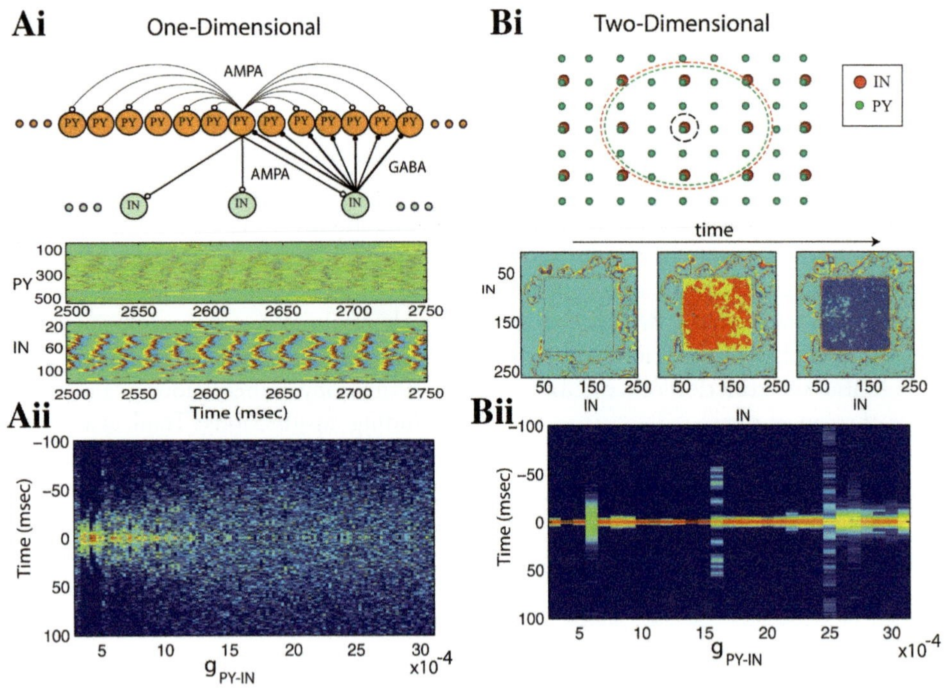

Figure 61.6. Global synchrony of gamma oscillations in 2D but not in 1D network model. Group of PY neurons was entrained to gamma by DC input. (A). 1D model: 512 PY–128 INs. (Ai). Top: connectivity structure. Bottom: spatiotemporal patterns of activity. Spikes = red; dark blue = hyperpolarization. (Aii). Time lags to the main peak of the cross-correlation function between LFPs in different spatial locations plotted for various strengths of PY-IN coupling. Color indicates number of time lags within 1 ms bins (dark blue = no lags). (B). 2D model: 512 × 512; PY: 256 × 256 INs. (Bi). Top: connectivity. Red and green circuits indicate radii of PY-IN and IN-PY connectivity for a single PY and IN neuron. Bottom: snapshots of activity in the IN population during global coherent gamma. (Bii). Same as in (Aii), but for a 2D network. Note zero-lag phase locking between all network sites in 2D. Modified from Bazhenov et al. (2008).

models of cortical ripples are still lacking, the mechanisms behind hippocampal ripples may provide some insight, and are described below.

Mechanism

A recent minimal model of the ripple-generation mechanism in CA1 of the hippocampus (Malerba et al., 2016) indicates that synchronous, large-amplitude input from CA3 to CA1 pyramidal cells and interneurons initiates and sets the frequency of the ripple oscillation observed in CA1. However, it is the local inhibitory spiking within CA1 which sets the duration of the ripple: spike synchrony of inhibitory interneurons progressively decreases over the course of the ripple until the oscillation can no longer be maintained. While this model has some superficial similarities to the ING model for gamma activity (Whittington et al., 2000), it relies on weaker coupling between inhibitory neurons which ensures that the network's baseline state is asynchronous and disorganized. In this regime, the interneuron spiking can be driven by synchronous CA3 inputs without being slowed down by local inhibition within CA1.

Alternative models of ripples–and high-frequency oscillations more generally–have been proposed (Wilson and McNaughton, 1994; Traub and Bibbig, 2000) which view the ripple as a continuation of the event which triggers it, such as CA3 sharp-waves. These models rely on supralinear dendritic summation or gap junctions to achieve ripple-like properties.

Functional Role

In the hippocampus, ripples have been associated with memory formation, recall, and consolidation (Buzsaki, 2015). It is likely that cortical ripples play a similar role. Indeed, recent experimental work has demonstrated coupling between hippocampal and cortical ripples which increases with learning (Khodagholy et al., 2017). Distinct from the function of hippocampal ripples, recent work in humans suggests that widespread co-rippling between cortical sites might act a substrate for binding disparate memory components into a single coherent gestalt (Dickey et al., 2022). It should be noted that cortical ripples in humans have a characteristic frequency of around 90 Hz, and as such may be supported by mechanisms similar to gamma activity, as models have shown such frequencies are at the upper end of the supported frequency range (Bazhenov et al., 2008).

A recent closed loop thalamo-cortico-hippocampal model simulating interaction of hippocampal ripples and cortical slow waves during NREM sleep (Sanda et al.,

2021) explained the experimentally observed distribution of ripple-generation times relative to the Down-to-Up transition of cortical slow waves. The study revealed complex bidirectional interaction of ripples and slow oscillations in which early hippocampal ripples influence transitions to the Up state, while cortical Up states control the occurrence of later ripples, which in turn influence the transition to the Down state. This bidirectional dynamic may prove to be important to the larger functional role of cortical ripples given recent experimental evidence that hippocampal ripples can propagate to the neocortex (Nitzan et al., 2020).

CONCLUSION

Oscillatory activities are a ubiquitous property of brain function, found across all spatial and temporal scales of investigation. The research reviewed in this chapter points to the growing understanding of how these processes are generated, sustained, and propagated throughout the thalamocortical system, as well as the functions they serve at the system and cognitive levels. In particular, we have attempted to highlight the role that computational models of brain circuits have had in this endeavor. While there is certainly more to learn about the biological substrates and functions of each of the oscillations described above, we believe that the next major step for the field lies in investigating how the brain manages to support diverse combinations of these oscillations simultaneously and harmoniously. A recent advance in this direction was taken in a modeling study which demonstrated how interactions between locally generated gamma oscillations embedded in a realistic connectome structure can underpin the metastable oscillatory modes observed in collective brain rhythms (Cabral et al., 2022). Understanding the mechanisms and dynamics underlying these neuronal harmonies will be necessary to understand the macroscopic organization of brain functioning.

ACKNOWLEDGMENTS

Supported by grants from NIH (1R01NS104368, 1R01NS109553, 1R01MH125557).

REFERENCES

Bazhenov M, Rulkov NF, Timofeev I (2008). Effect of synaptic connectivity for long-range synchronization of fast cortical oscillations. *J Neurophysiol.* 100(3):1562–1575.

Bazhenov M, Timofeev I, Steriade M, Sejnowski TJ (1999). Self-sustained rhythmic activity in the thalamic reticular nucleus mediated by depolarizing GABAA receptor potentials. *Nat Neurosci.* 2(2):168–174.

Bazhenov M, Timofeev I, Steriade M, Sejnowski T (2000). Spiking-bursting activity in the thalamic reticular nucleus initiates sequences of spindle oscillations in thalamic networks. *J Neurophysiol.* 84(2):1076–1087.

Bazhenov M, Timofeev I, Steriade M, Sejnowski TJ (2002). Model of thalamocortical slow-wave sleep oscillations and transitions to activated states. *J Neurosci.* 22(19):8691–8704.

Bonjean M, Baker T, Bazhenov M, Cash S, Halgren E, Sejnowski T (2012). Interactions between core and matrix thalamocortical projections in human sleep spindle synchronization. *J Neurosci.* 32(15):5250–5263.

Bonjean M, Baker T, Lemieux M, Timofeev I, Sejnowski T, Bazhenov M (2011). Corticothalamic feedback controls sleep spindle duration in vivo. *J Neurosci* 31(25):9124–9134.

Borgers C, Kopell N (2003). Synchronization in networks of excitatory and inhibitory neurons with sparse, random connectivity. *Neural Comput.* 15(3):509–538.

Bressler SL (1990). The gamma wave: A cortical information carrier? *Trends Neurosci.* 13(5):161–162.

Buzsaki G (2015). Hippocampal sharp wave-ripple: A cognitive biomarker for episodic memory and planning. *Hippocampus.* 25(10):1073–1188.

Cabral J, Castaldo F, Vohryzek J, Litvak V, Bick C, Lambiotte R, Friston K, Kringelbach ML, Deco G (2022). Metastable oscillatory modes emerge from synchronization in the brain spacetime connectome. *Commun Physics.* 5:184. https://doi.org/10.1038/s42005-022-00950-y

Chen JY, Chauvette S, Skorheim S, Timofeev I, Bazhenov M (2012). Interneuron-mediated inhibition synchronizes neuronal activity during slow oscillation. *J Physiol.* 590(Pt 16):3987–4010.

Compte A, Sanchez-Vives MV, McCormick DA, Wang X-J (2003). Cellular and network mechanisms of slow oscillatory activity (<1 Hz) and wave propagations in a cortical network model. *J Neurophysiol.* 89(5):2707–2725.

Cossart R, Aronov D, Yuste R (2003). Attractor dynamics of network UP states in the neocortex. *Nature.* 423(6937):283–288.

Costa MS, Born J, Claussen JC, Martinetz T (2016). Modeling the effect of sleep regulation on a neural mass model. *J Comput Neurosci.* 41(1):15–28.

Crunelli V, Hughes SW (2010). The slow (<1 Hz) rhythm of non-REM sleep: a dialogue between three cardinal oscillators. *Nat Neurosci.* 13(1):9–17.

Deco G, Jirsa V, McIntosh AR, Sporns O, Kotter R (2009). Key role of coupling, delay, and noise in resting brain fluctuations. *Proc Natl Acad Sci U S A.* 106(25):10302–10307.

Dehghani N, Cash SS, Chen CC, Hagler DJ, Jr, Huang M, Dale AM, Halgren E (2010). Divergent cortical generators of MEG and EEG during human sleep spindles suggested by distributed source modeling. *PLoS One.* 5(7):e11454.

Destexhe A (2009). Self-sustained asynchronous irregular states and Up-Down states in thalamic, cortical and thalamocortical networks of nonlinear integrate-and-fire neurons. *J Comput Neurosci.* 27(3):493–506.

Destexhe A, Babloyantz A, Sejnowski TJ (1993). Ionic mechanisms for intrinsic slow oscillations in thalamic relay neurons. *Biophys J.* 65(4):1538–1552.

Dickey CW, Verzhbinsky IA, Jiang X, Rosen BQ, Kajfez S, Stedelin B, Shih JJ, Ben-Haim S, Raslan AM, Eskandar EN, Gonzalez-Martinez J, Cash SS, Halgren E (2022). Widespread ripples synchronize human cortical activity during sleep, waking, and memory recall. *Proc Natl Acad Sci* 119(28):e2107797119.

Fukunaga M, Horovitz SG, van Gelderen P, de Zwart JA, Jansma JM, Ikonomidou VN, Chu R, Deckers RH, Leopold DA, Duyn JH (2006). Large-amplitude, spatially correlated fluctuations in BOLD fMRI signals during extended rest and early sleep stages. *Magn Reson Imaging.* 24(8):979–992.

Ghosh A, Rho Y, McIntosh AR, Kotter R, Jirsa VK (2008). Noise during rest enables the exploration of the brain's dynamic repertoire. *PLoS Comput Biol.* 4(10):e1000196.

Gonzalez OC, Sokolov Y, Krishnan GP, Delanois JE, Bazhenov M (2020). Can sleep protect memories from catastrophic forgetting? *eLife.* 9:e51005.

Gray CM, Konig P, Engel AK, Singer W (1989). Oscillatory responses in cat visual cortex exhibit inter-columnar synchronization which reflects global stimulus properties. *Nature.* 338(6213):334–337.

Hill S, Tononi G (2005). Modeling sleep and wakefulness in the thalamocortical system. *J Neurophysiol.* 93(3):1671–1698.

Honey CJ, Kotter R, Breakspear M, Sporns O (2007). Network structure of cerebral cortex shapes functional connectivity on multiple time scales. *Proc Natl Acad Sci U S A.* 104(24):10240–10245.

Jadhav SP, Kemere C, German PW, Frank LM (2012). Awake hippocampal sharp-wave ripples support spatial memory. *Science.* 336(6087):1454–1458.

Joliot M, Ribary U, Llinas R (1994). Human oscillatory brain activity near 40 Hz coexists with cognitive temporal binding. *Proc Natl Acad Sci U S A.* 91(24):11748–11751.

Khodagholy D, Gelinas JN, Buzsaki G (2017). Learning-enhanced coupling between ripple oscillations in association cortices and hippocampus. *Science.* 358(6361):369–372.

Kim J, Gulati T, Ganguly K (2019). Competing roles of slow oscillations and delta waves in memory consolidation versus forgetting. *Cell.* 179(2):514–526.

Kim U, Bal T, McCormick DA (1995). Spindle waves are propagating synchronized oscillations in the ferret LGNd in vitro. *J Neurophysiol.* 74(3):1301–1323.

Kopell N, Ermentrout GN, Whittington MA, Traub RD (2000). Gamma rhythms and beta rhythms have different synchronization properties. *Proc Natl Acad Sci U S A.* 97(4):1867–1872.

Krishnan GP, Chauvette S, Shamie I, Soltani S, Timofeev I, Cash SS, Halgren E, Bazhenov M (2016). Cellular and neurochemical basis of sleep stages in the thalamocortical network. *eLife.* 5:e18607.

Krishnan GP, Gonzalez OC, Bazhenov M (2018). Origin of slow spontaneous resting-state neuronal fluctuations in brain networks. *Proc Natl Acad Sci U S A.* 115(26):6858–6863.

Lemieux M, Chen J-Y, Lonjers P, Bazhenov M, Timofeev I (2014). The impact of cortical deafferentation on the neocortical slow oscillation. *J Neurosci.* **34**(16):5689–5703.

Levenstein D, Buzsáki G, Rinzel J (2019). NREM sleep in the rodent neocortex and hippocampus reflects excitable dynamics. *Nat Commun.* 10, 2478.

Lundqvist M, Rose J, Herman P, Brincat SL, Buschman TJ, Miller EK (2016). Gamma and beta bursts underlie working memory. *Neuron.* 90(1):152–164.

Luthi A, McCormick DA (1998). Periodicity of thalamic synchronized oscillations: The role of Ca2+- mediated upregulation of Ih. *Neuron.* 20(3):553–563.

Lytton WW, Destexhe A, Sejnowski TJ (1996). Control of slow oscillations in the thalamocortical neuron: a computer model. *Neuroscience.* 70(3):673–684.

Lytton WW, Sejnowski TJ (1991). Simulations of cortical pyramidal neurons synchronized by inhibitory interneurons. *J Neurophysiol.* 66(3):1059–1079.

Malerba P, Krishnan GP, Fellous JM, Bazhenov M (2016). Hippocampal CA1 ripples as inhibitory transients. *PLoS Comput Biol.* 12(4): e1004880.

McCormick DA, Pape HC (1990). Properties of a hyperpolarization-activated cation current and its role in rhythmic oscillation in thalamic relay neurons. *J Physiol.* 431:291–318.

Molle M, Bergmann TO, Marshall L, Born J (2011). Fast and slow spindles during the sleep slow oscillation: Disparate coalescence and engagement in memory processing. *Sleep.* 34(10):1411–1421.

Nitzan N, McKenzie S, Beed P, English DF, Oldani S, Tukker JJ, Buzsaki G, Schmitz D (2020). Propagation of hippocampal ripples to the neocortex by way of a subiculum-retrosplenial pathway. *Nat Commun.* 11:1947.

Pan WJ, Thompson GJ, Magnuson ME, Jaeger D, Keilholz S (2013). Infraslow LFP correlates to resting-state fMRI BOLD signals. *NeuroImage.* 74:288–297.

Peyrache A, Khamassi M, Benchenane K, Wiener SI, Battaglia FP (2009). Replay of rule-learning related neural patterns in the prefrontal cortex during sleep. *Nat Neurosci.* 12(7):919–926.

Rasch B, Born J (2013). About sleep's role in memory. *Physiol Rev.* 93(2):681–766.

Rougeul-Buser A, Bouyer JJ, Buser P (1975). From attentiveness to sleep: A topographical analysis of localized "synchronized" activities on the cortex of normal cat and monkey. *Acta Neurobiol Exp (Warsz).* 35(5–6):805–819.

Sanchez-Vives MV, McCormick DA (2000). Cellular and network mechanisms of rhythmic recurrent activity in neocortex. *Nat Neurosci.* 3(10):1027–1034.

Sanda P, Hlinka J, van den Berg M, Bazhenov M, Keliris G, Krishnan G (2022). Cholinergic modulation supports DMN suppression during resting state. *bioRxiv*: doi.org/10.1101/2022.11.01.514686.

Sanda P, Malerba P, Jiang X, Krishnan P, Gonzalez-Martinez J, Halgren E, Bazhenov M (2021). Bidirectional interaction of hippocampal ripples and cortical slow waves leads to coordinated spiking activity during NREM sleep. *Cereb Cortex.* 31(1):324–340.

Sejnowski TJ, Destexhe A (2000). Why do we sleep? *Brain Res.* 886(1–2):208–223.

Sheroziya M, Timofeev I (2014). Global intracellular slow-wave dynamics of the thalamocortical system. *J Neurosci.* 34(26):8875–8893.

Steriade M, Domich L, Oakson G, Deschenes M (1987). The deafferented reticular thalamic nucleus generates spindle rhythmicity. *J Neurophysiol.* 57(1):260–273.

Steriade M, Nuñez A, Amzica F (1993a). Intracellular analysis of relations between the slow (<1 Hz) neocortical oscillations and other sleep rhythms of electroencephalogram. *J Neurosci.* 13:3266–3283.

Steriade M, Nuñez A, Amzica F (1993b). A novel slow (<1 Hz) oscillation of neocortical neurons in vivo: depolarizing and hyperpolarizing components. *J Neurosci.* 13:3252–3265.

Steriade M, Timofeev I (2003). Neuronal plasticity in thalamocortical networks during sleep and waking oscillations. *Neuron.* 37(4):563–576.

Tadros T, Bazhenov M (2022). Role of sleep in formation of relational associative memory. *J Neurosci.* 42(27):5330–5345.

Timofeev I, Bazhenov M (2005). Mechanisms and biological role of thalamocortical oscillations. In Columbus F, ed., Trends in Chronobiology Research. Nova Science:1–47.

Timofeev I, Chauvette S (2013). The spindles: Are they still thalamic? *Sleep.* 36(6):825–826.

Timofeev I, Grenier F, Bazhenov M, Sejnowski TJ, Steriade M (2000). Origin of slow cortical oscillations in deafferented cortical slabs. *Cereb Cortex.* 10(12):1185–1199. doi:10.1093/cercor/10.12.1185

Timofeev I, Steriade M (1996). Low-frequency rhythms in the thalamus of intact-cortex and decorticated cats. *J Neurophysiol.* 76(6):4152–4168.

Timofeev I, Steriade M (1997). Fast (mainly 30–100 Hz) oscillations in the cat cerebellothalamic pathway and their synchronization with cortical potentials. *J Physiol.* 504(1):153–168.

Tononi G, Cirelli C (2014). Sleep and the price of plasticity: From synaptic and cellular homeostasis to memory consolidation and integration. *Neuron.* 81(1):12–34.

Traub RD, Bibbig A (2000). A model of high-frequency ripples in the hippocampus based on synaptic coupling plus axon-axon gap junctions between pyramidal neurons. *J Neurosci.* 20(6):2086–2093.

Traub RD, Schmitz D, Jefferys JG, Draguhn A (1999). High-frequency population oscillations are predicted to occur in hippocampal pyramidal neuronal networks interconnected by axoaxonal gap junctions. *Neuroscience.* 92(2):407–426.

van Dongen EV, Takashima A, Barth M, Zapp J, Schad LR, Paller KA, Fernandez G (2012). Memory stabilization with targeted reactivation during human slow-wave sleep. *Proc Natl Acad Sci U S A.* 109(26):10575–10580.

Wei Y, Krishnan G, Bazhenov M (2016). Synaptic mechanisms of memory consolidation during sleep slow oscillations. *J Neurosci.* 36(15):4231–4247.

Wei Y, Krishnan GP, Komarov M, Bazhenov M (2018). Differential roles of sleep spindles and sleep slow oscillations in memory consolidation. *PLoS Comput Biol.* 14(7):e1006322.

Whittington MA, Traub RD, Kopell N, Ermentrout B, Buhl EH (2000). Inhibition-based rhythms: experimental and mathematical observations on network dynamics. *Int J Psychophysiol.* 38(3):315–336.

Wilson MA, McNaughton BL (1994). Reactivation of hippocampal ensemble memories during sleep. *Science.* 265(5172): 676–679.

Zhang D, Raichle ME (2010). Disease and the brain's dark energy. *Nat Rev Neurol.* 6(1):15–28.

62.

CONTROL OF THALAMOCORTICAL SIGNALING THROUGH SPIKE TIMING AND SYNCHRONY

Garrett B. Stanley

INTRODUCTION

We extract information about the outside sensory environment through our sensory pathways. Sensory inputs such as sound, light, and touch are transduced by peripheral sensors, which send sensory afferents through early stages of processing, to sensory thalamus, on the way to cortical sensory areas that ultimately underlie sensory perception (see **Section V** of this volume). The notion of the sensory thalamus as a simple relay, connecting signals from peripheral sensors to higher cortical regions, has long been an outdated simplification. It is now widely accepted that the sensory thalamus plays a more nuanced role than this, serving to dynamically gate afferent signaling to the cortex, controlling information flow and ultimately dictating what does and does not get perceived. However, the "how" of this signal processing strategy has only more recently been investigated in detail. The emergence of tools for precisely measuring and manipulating elements of this circuit has made it clear that the sensitivity of the cortex to the precise timing of thalamic inputs, combined with the modulation of timing of thalamic activity, serves as a fundamental gating mechanism for signaling and computation in the pathway. A range of properties likely regulate and influence the timing sensitivity of this circuit. For example, this includes intrinsic properties of the thalamic neurons themselves, such as the membrane time constants, spike thresholds, and spike-frequency adaptation, among other properties (see **Chapter 3 in Section II**). Also important are the properties of the thalamocortical axonal projections, such as the conduction velocity, the diversity of synaptic target locations, and short-term dynamics of thalamocortical synapses (depression, facilitation) (see **Chapters 12 and 13 in Section IV**). And finally, critical are elements of the cortical dynamics, such as the kinetics of postsynaptic conductances and the voltage changes they generate, and interactions among the feedforward interneurons, among other properties (see **Chapters 7 and 8 in Section III**). However, one particularly prominent mechanism that is thought to strongly shape timing sensitivity is the circuit motif of feedforward inhibition at the thalamocortical junction, making the throughput of sensory signaling to the cortex highly dependent upon the timing and synchronization of thalamic inputs. Dynamic modulation of thalamic synchronization thus serves as a fundamental mechanism by which the flow of information, and thus perception, is ultimately controlled in sensory pathways.

THE CIRCUIT

THE MOTIF OF FEEDFORWARD INHIBITION

Although the thalamocortical circuit is complex anatomically and functionally, here we focus on the relatively simple motif of feedforward inhibition as a central and important element of controlling information flow in the pathway. As projections from the primary sensory thalamus synapse onto neurons in layer (L)4 of primary sensory cortex, synaptic connections are made onto both excitatory and inhibitory sub-populations of neurons in a highly conserved, topographic manner (see **Chapter 13 in Section IV**). L4 excitatory neurons are of particular interest, as they provide the primary drive to other cortical layers, and thus represent the information that is ultimately conveyed through the cortex. The L4 inhibitory neurons, however, while acting locally, play a pivotal role in the signaling through feedforward inhibition. L4 excitatory neurons receive direct feedforward excitation from thalamic projections, as well as delayed inhibition through the L4 inhibitory interneurons. This circuit motif is often referred to as "feedforward inhibition" or "disynaptic inhibition," as the L4 excitatory neurons receive feedforward inhibition traveling through 2

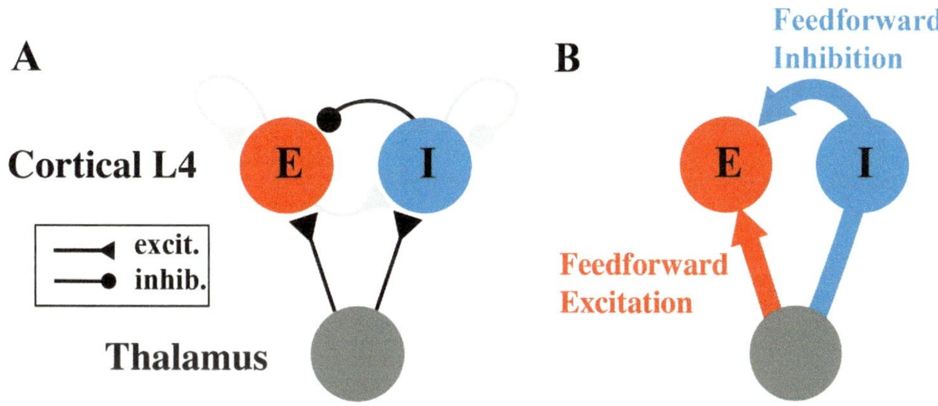

Figure 62.1. Thalamocortical connectivity establishes disynaptic feedforward inhibition. (A). A simplified circuit diagram illustrates the excitatory synaptic inputs from primary sensory thalamus to the excitatory (E, red) and inhibitory (I, blue) neurons in cortical L4. Numerous intracortical connections are present, but highlighted are the prominent inhibitory connections from inhibitory cortical neurons to excitatory. (B). Excitatory neurons in cortical L4 receive direct excitatory input from thalamus (red), followed by delayed inhibition (blue).

synapses. A conceptual illustration of the circuit is shown in Figure 62.1; the full circuit is shown in Figure 62.1A, and net excitatory and inhibitory inputs to the L4 excitatory neurons are highlighted in Figure 62.1B.

TIMING SENSITIVITY AND THE WINDOW OF OPPORTUNITY

The dynamic nature of feedforward inhibition at the thalamocortical junction has been elucidated in several key studies that employed in vivo recording of the transmembrane voltage of cortical neurons. Across numerous studies, it has been shown that there is a clear time delay between the initial sensory-driven excitatory input and the subsequent inhibitory input (Gabernet et al., 2005; Wehr and Zador, 2005; Wilent and Contreras, 2005). For example, in vivo intracellular recordings in the whisker pathway of the anesthetized rat (Figure 62.2A) reveal that the total conductance for sensory-driven excitatory cortical neurons is composed of a quick transient excitatory conductance, followed ~1 ms later by the onset of an increase in inhibitory conductance, as shown in Figure 62.2B (Wilent and Contreras, 2005). Similar recordings in the auditory pathway of the anesthetized rat, in response to an auditory stimulus, reveal a pattern of short-latency excitation followed by a slightly delayed inhibition (Wehr and Zador, 2005), with a stereotyped delay between the excitatory and inhibitory components by ~1–4 ms. The combination of the direct, feedforward excitation and the delayed inhibition established by the indirect disynaptic input has been widely recognized to establish an extreme sensitivity to the timing of thalamic inputs to cortex. This establishes a "window of opportunity" for the feedforward thalamic input, as these excitatory inputs can potentially drive the postsynaptic cortical neuron to spike, but only during a relatively short interval, before the delayed inhibition comes to "shut the door" and blunt the excitatory effect of any subsequent arrivals of thalamic input (Pinto et al., 2000). It is important to note that although the latency between the

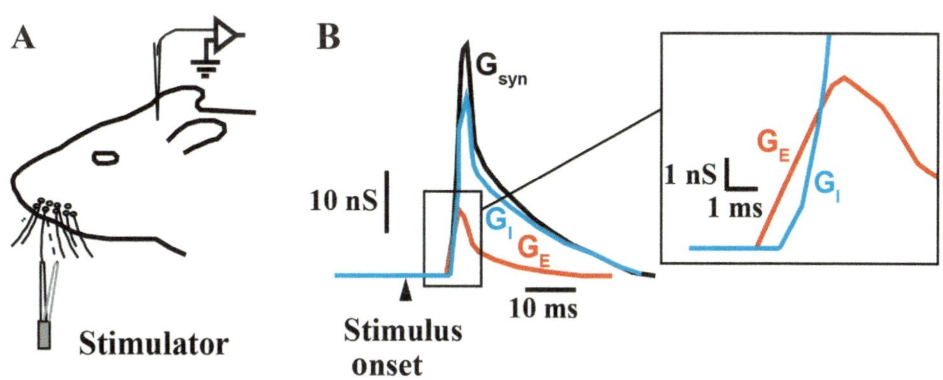

Figure 62.2. Intracellular effects of disynaptic inhibition. (A). In vivo intracellular recordings in S1 cortex of anesthetized rat, in response to controlled whisker stimulation. (B). Conductance measures from an L4 neuron, for the total synaptic conductance (G_{syn}, black), the excitatory component (G_E, red), and the inhibitory component (G_I, blue). Inset shows the delayed inhibition. Adapted from Wilent and Contreras (2005).

initiation of excitation and inhibition is on the order of 1–4 ms and perhaps even at the low end of this range (Yu et al., 2016), the subsequent evolution of the excitatory and inhibitory dynamics, coupled with the aggregate properties of the population, establishes an overall timescale of ~10–15 ms for the window in the face of ongoing activity (Gabernet et al., 2005), though there is some evidence that this may vary across sensory modalities and experimental conditions.

CONVERGENCE OF PRIMARY THALAMIC RELAY NEURONS

The thalamus provides a strong excitatory drive to cortical L4, relaying sensory signals from the periphery to cortex. The thalamic input results in relatively large postsynaptic potentials in the neurons of L4, resulting in a long-standing assumption of strong synaptic connections of the thalamic projections onto the dendrites of the L4 neurons. Indeed, there is evidence that sensory thalamocortical synapses are physically large compared to other cortical glutamatergic synapses, and presynaptic size correlates with synaptic efficacy in most regions of the brain (Holler et al., 2021). In vitro studies have suggested that single thalamocortical axons generate stronger monosynaptic responses in L4 cells than intracortical axons, on average, due to the release of more quanta (vesicles) per spike and, perhaps, also to higher release probability (Stratford et al., 1996). Despite these findings, however, in vivo measurements of these synaptic inputs suggest relatively weak synaptic input from individual thalamic afferents (Bruno and Sakmann, 2006). Specifically, in extracellular recordings of thalamic ventral posteromedial nucleus (VPm) neurons combined with intracellular recordings of cortical L4 neurons to which the VPm neurons are monosynaptically connected (Figure 62.3A), analysis of averaged unitary postsynaptic potentials reveals the relatively small magnitude of the thalamically driven cortical excitatory postsynaptic potentials (EPSPs) when the pathway is driven by a weak, sinusoidal whisker input. Taken together, the data suggest that instead of relying on strong individual synaptic inputs, the strong overall synaptic input to L4 may primarily rely on the synchronization of a relatively large number of weak thalamic inputs (Alonso et al., 1996).

Although the simultaneous recording of large numbers of neurons in deep structures such as the thalamus is very challenging experimentally, the pair-wise spiking activity can be quite illuminating of the overall degree of synchronization among the population. This is quantified by examining the relative timing of spiking across neuron pairs, as illustrated in Figure 62.3B. The spike cross-correlogram is formed by examining the spikes of neuron 2 relative to a particular spike of neuron 1 and creating a histogram of this across all spikes of neuron 1, as illustrated. Figure 62.3C exemplifies the high degree of synchronization of thalamic spiking activity in response to a punctate whisker stimulus in the anesthetized rat (Wang et al., 2010b).

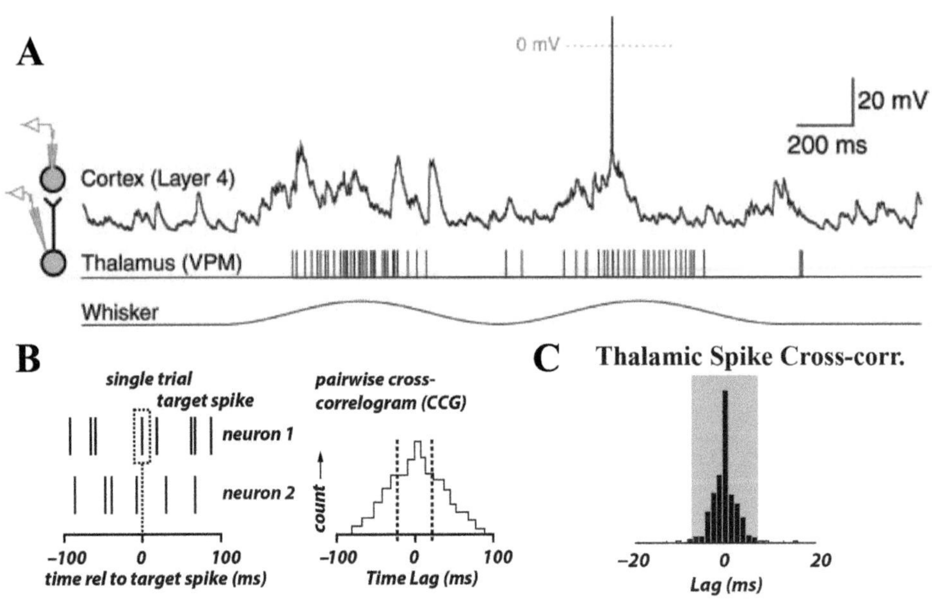

Figure 62.3. The synchrony of thalamic neurons provides a potent input to cortex. (A). Thalamic inputs to cortical L4 neurons are weak, but numerous. Panel shows simultaneous extracellular and intracellular recordings from connected VPm and cortical L4 S1 neurons in the anesthetized rat, respectively, illustrating the weak nature of the individual thalamocortical synaptic input. Adapted from Bruno and Sakmann (2006). (B). The degree of synchronization of simultaneously recorded pairs of neurons can be assessed through spike cross-correlation. The spike cross-correlogram is a histogram formed from the spiking of neuron 2 relative to each spike of neuron 1. Adapted from Borden et al. (2022). (C). Primary thalamic sensory neurons in VPm of the anesthetized rat exhibit strongly peaked spike cross-correlograms in response to punctate whisker stimulation, reflective of synchronization over a 10–15 ms timescale (highlighted with gray band). Adapted from Wang et al. (2010b).

Taken together, the cortical "window of opportunity" established by the direct feedforward thalamic excitation and the indirect, delayed feedforward inhibition, coupled with the potential for dynamic modulation of thalamic synchrony, sets the stage for an exquisite mechanism for the control of information flow in sensory pathways.

THE FUNCTIONAL IMPLICATIONS

STIMULUS SENSITIVITY

Cortical sensitivity to the synchronization of thalamic inputs means that any properties of sensory input that modulate thalamic synchrony could be reflected in cortical response properties. Again, turning to the rodent whisker/vibrissa system, neurons in the vibrissa pathway are sensitive to the velocity of whisker deflections on the contralateral side of the face. Individual neurons in the S1 cortex exhibit a transient response to a punctate whisker deflection that increases in magnitude with increasing velocity of the whisker deflection. Recordings of extracellular spiking activity in the VPm thalamic neurons that serve as the input to S1 (Figure 62.4A) reveal that S1 response magnitudes are not well correlated with the VPm response magnitude, as illustrated in the left of Figure 62.4B (Pinto et al., 2003). Instead, examination of the concentration of the thalamic spiking at the initial onset of the sensory-evoked response shows that the cortical response magnitude is very well correlated with the timing precision of the recorded VPm spiking, as shown in the right of Figure 62.4B. Note that the authors of this work originally described the measure as "Thalamic Initial Synchrony." However, timing measurements were made by evaluating the concentration of spiking in time following the delivery of a sensory stimulus across trials on a *single-neuron* basis. The plot was thus relabeled here as "Thalamic Timing Precision" to more accurately reflect the measure as computed. However, subsequent studies have shown the assumed relation between timing precision at the single neuron level and population synchrony to be well-founded in thalamus (Wang et al., 2010b). Building on this idea, although neurons in the lateral geniculate nucleus (LGN) in the visual pathway are generally not tuned for orientation or direction selectivity of visual stimuli, a rich array of orientation and direction selectivity emerges from the synchronous activity across the thalamic population (Stanley et al., 2012), providing the substrate upon which the classical notions of cortical tuning properties may be built.

ADAPTATION

Sensory adaptation has long been theorized to be an important computational construct for signaling in the complex natural environment (Barlow, 1961). Although rapid sensory adaptation is ubiquitous across sensory pathways, the neural basis has not been conclusively identified. A large body of work has pointed to the early sensory pathway; adaptation effects cascade toward and culminate in primary sensory cortex (Ganmor et al., 2010; Maravall et al., 2013; Lampl and Katz, 2017), and the nature of cortical adaptation is consistent with perceptual effects. In the vibrissa pathway of rodents, for example, adaptation induced through whisker stimulation shapes the magnitude (Webber and Stanley, 2004; Boloori and Stanley, 2006; Webber and Stanley, 2006; Maravall et al., 2007; Heiss

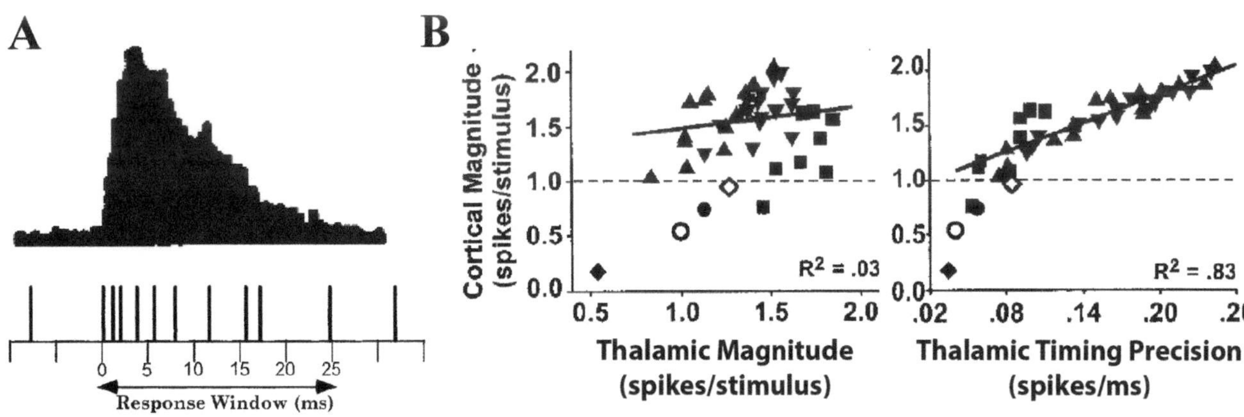

Figure 62.4. Cortical response magnitude is correlated with thalamic timing, but not thalamic magnitude. (A). Neurons in thalamic VPm respond to a punctate whisker stimulus with precise timing. More of the activity is concentrated earlier in the response. Adapted from Pinto et al. (2000, 2003). (B). Magnitude of the cortical response is not well correlated with thalamic response magnitude (left) but is well correlated with the timing precision of the thalamic response (right). Adapted from Pinto et al. (2003). Open Symbols – primary whisker across 8 deflection angles (Kyriazi et al., 1994); Closed Squares – sine wave stimulation of whisker (Hartings et al., 2000); Closed Circle – long plateau OFF response across 8 whisker deflection angles (Kyriazi et al., 1994); Closed Diamond – adjacent whisker in preferred direction (Simons and Carvell, 1989; Bruno and Simons, 2002); Closed Down Triangles – primary whisker in preferred direction (Pinto et al., 2000); Closed Up Triangles – primary whisker in caudal direction (Pinto et al., 2000).

et al., 2008; Boloori et al., 2010; Ganmor et al., 2010; Wang et al., 2010b; Cohen-Kashi Malina et al., 2013; Ollerenshaw et al., 2014; Kheradpezhouh et al., 2017) and spatial extent (Ollerenshaw et al., 2014; Zheng et al., 2015) of responses to subsequent stimuli in primary somatosensory cortex (S1). Yet significant mechanistic questions remain unanswered about rapid adaptation in the thalamocortical circuit (for a review, see Whitmire and Stanley, 2016). Previous work conducted in vitro or under anesthesia has implicated thalamocortical and/or intracortical synaptic depression (Castro-Alamancos and Oldford, 2002; Chung et al., 2002; Gabernet et al., 2005; Cruikshank et al., 2007, 2010; Cohen-Kashi Malina et al., 2013), as a means to explain why mean evoked rates are generally more adapted in the cortex than in the thalamus (Chung et al., 2002; Khatri et al., 2004; Gabernet et al., 2005). Yet adaptation also reduces thalamic synchrony (Wang et al., 2010b; Ollerenshaw et al., 2014). Given the sensitivity of the cortex to the timing of thalamic inputs (Usrey et al., 2000; Swadlow and Gusev, 2001; Bruno and Sakmann, 2006; Wang et al., 2010b; Ollerenshaw et al., 2014), this suggests a potential role for thalamic spike timing in cortical adaptation.

To determine the role of thalamic timing in rapid sensory adaptation, Wright et al. combined extracellular electrophysiological recordings across the thalamus and primary sensory cortex in the awake head-fixed mouse with computational network modeling (Wright et al., 2021). Specifically, multi-electrode silicon probe recordings were made in VPm thalamus and S1 cortex in the vibrissa pathway, while delivering computer-controlled stimulation of facial vibrissae in the presence or absence of a background adapting sensory white-noise stimulus, as illustrated in Figure 62.5A. Adaptation significantly reduces both putative excitatory (Figure 62.5B, left) and inhibitory (Figure 62.5B, right) neurons in S1, despite only a modest effect of the adapting stimulus on the evoked response in VPm thalamus (Figure 62.5C). The adaptation does, however, decrease thalamic synchrony, as reflected in the spike cross-correlogram (Figure 62.5D). In this work, several lines of evidence—including the recording of the thalamic inputs under a range of optogenetic controls and computational modeling—suggest that the observed adaptation in S1 primarily reflects reduced synchronous thalamic firing and robust thalamically driven feedforward inhibition in the adapted condition, with little contribution from thalamocortical and intracortical synaptic depression. Taken together, these results establish the role of the thalamocortical circuit in rapid adaptation during wakefulness and implicate a more critical role of thalamic input in controlling information flow than previously thought.

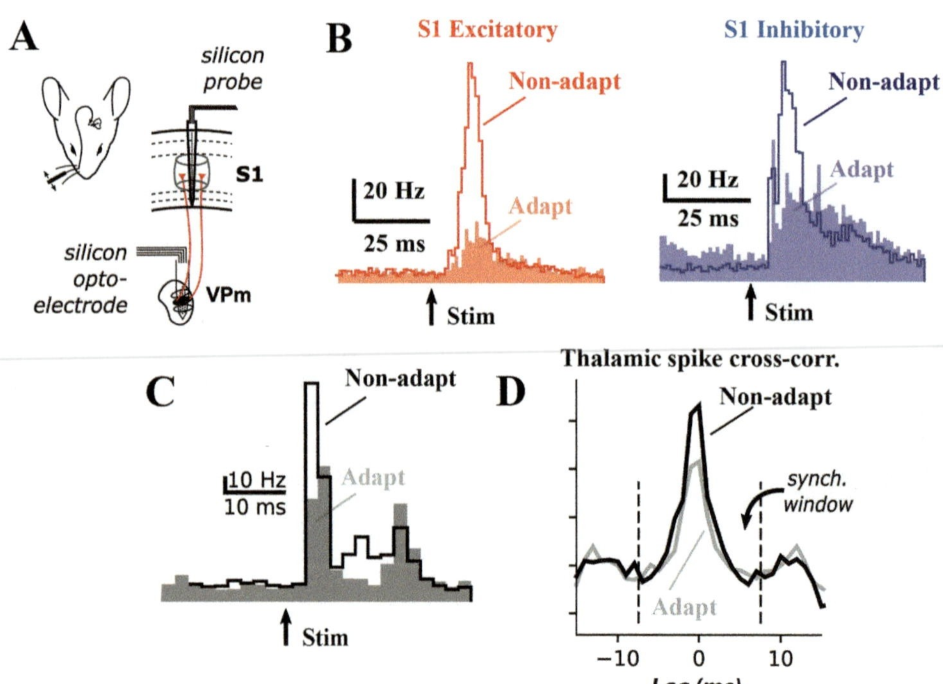

Figure 62.5. Rapid sensory adaptation desynchronizes thalamic population activity, and gates cortical sensory representations. (A). Simultaneous extracellular recordings in thalamic VPm and cortical S1 in awake, head-fixed mouse. (B). Grand PSTHs for excitatory (RS, n = 119) and inhibitory (FS, n = 95) neurons, for a 900 deg/sec punctate whisker deflection. For each sub-population, sensory evoked PSTHs are shown for the control (non-adapt., open) case, and after adaptation with a white noise whisker stimulation (adapt., shaded). (C). In contrast to the cortical neurons, neurons in VPm thalamus are only modestly adapted in overall firing. Shown are the control (non-adapt., open) and adapted (adapt., shaded) conditions. (D). Simultaneously recorded thalamic neurons pairs exhibit decrease in sensory-driven synchrony with adaptation. Adapted from Wright et al. (2021).

So, what is the functional implication of the rapid sensory adaptation? Wang et al. demonstrated the impact of rapid sensory adaptation on theoretical detectability and discriminability of sensory stimuli in a set of experiments in the anesthetized rat, in which they presented different velocities of whisker stimuli following adapting sensory inputs (Wang et al., 2010b). Shown in Figure 62.6A are the velocity sensitivity curves across putative excitatory cortical L4 neurons, in the non-adapted (dotted line) and adapted (solid line) states. The adaptation attenuates the response magnitude for all velocities, which degrades the detectability. However, the difficulty of the task of an ideal observer depends upon the overlap in distributions of the responses to each of the velocities. Shown is an example of the response distributions of a particular cortical neuron in response to 5 different stimuli (s1, s2, s3, s4, s5; Figure 62.6B), which are 5 different angular velocities of vibrissa deflection (50, 100, 300, 600, 1200 deg./sec, respectively), with an increase in separability, and thus discriminability with adaptation. How does this happen? The mean thalamic synchrony decreases with the adapting stimulus (Figure 62.6C), where synchrony is defined as the central area under the spike cross-correlogram (±7.5 ms). The synchrony across VPm neurons is also modulated by the vibrissa deflection velocity (Figure 62.6C,D). The different velocities (s1–s5) elicit greater variation in the synchrony across the pairs in the adapted than in the non-adapted states (Figure 62.6C). The adaptation acts to decrease the overall thalamic synchrony, while simultaneously increasing the sensitivity of thalamic synchrony to velocity (Figure 62.6D). This can be seen in the shape of the synchrony curve as a function of velocity in the adapted state, reminiscent of the cortical spike count curve shown in Figure 62.6A.

Previous studies of the thalamocortical circuit have demonstrated that adaptation modulates the width of the window of opportunity of the thalamic influence on the recipient L4 neurons (Gabernet et al., 2005). In the non-adapted case, the changes in the synchronization with stimulus velocity are small relative to the integration window, leading to a relative insensitivity of the cortical response to stimulus velocity. In contrast, in the adapted case, although the integration window is wider as compared to the non-adapted state, the overall synchrony is less. Importantly, the changes in synchrony with deflection velocity are pronounced. This leads to spiking of some of the VPm cells outside the window of integration and thus a less potent input to drive the downstream recipient cortical neurons. Despite the lack of VPm spike count sensitivity to velocity, Wang et al. showed that the spike count of a biophysically inspired model of the thalamically driven cortical circuit displays increased differential sensitivity to stimulus velocity with adaptation, and as a result, the discrimination performance is enhanced in the adapted state, as shown in Figure 62.6F. Thus, the feedforward inhibition at the thalamocortical junction, combined with the adaptive modulation of thalamic synchrony, mediates a trade-off between detectability and discriminability in the pathway, further demonstrated in subsequent behavioral work (Ollerenshaw et al., 2014).

BURST/TONIC GATING

One prominent and important characteristic of neurons in the thalamus is the presence of T-type calcium channels that are normally inactivated, but become deinactivated through prolonged hyperpolarization (Jahnsen and Llinás, 1984a, 1984b; Sherman and Koch, 1986; Suzuki and Rogawski, 1989; see also **Chapter 3 in Section II** and **Chapter 63 in Section XI** of this volume). Subsequent depolarizing inputs lead to calcium-mediated "bursts" of action potentials characterized by transient, high-frequency spiking. This is distinct from "tonic" spiking mediated through the usual Na^+/K^+ channel dynamics. This mechanism is thought to serve an important role in top-down modulation of sensory signaling in complex environments (Sherman and Guillery, 2002). However, the precise role of this mechanism in sensory signaling remains unclear, despite decades of speculation. At the level of the synapse, high-frequency thalamic burst events have a significant impact on downstream cortical activation (Swadlow and Gusev, 2001), where thalamic bursts are associated with an increased postsynaptic response in recipient cortical neurons, arising primarily from properties of the thalamocortical synapse. Coupled with sensitivity of the cortex to the timing of thalamic inputs through the window of opportunity established by the disynaptic feedforward cortical inhibition (Pinto et al., 2000; Pinto et al., 2003; Wehr and Zador, 2003; Wilent and Contreras, 2004, 2005; Gabernet et al., 2005; Isaacson and Scanziani, 2011), the gating of thalamic signaling through this array of mechanisms is hypothesized to serve a critical role in controlling sensory-information flow (Crick, 1984; Pinto et al., 2000; Sherman and Guillery, 2002; Wilent and Contreras, 2004; Sherman, 2005). However, early investigation and speculation about the functional consequences of this mechanism (Reinagel et al., 1999; Lesica and Stanley, 2004; Alitto et al., 2005; Denning and Reinagel, 2005; Lesica et al., 2006) were hampered by lack of methodological approaches to precisely measure and manipulate thalamic properties that would enable a full mechanistic exploration in the intact brain.

Using a combination of multi-site electrophysiology and optogenetic manipulation, along with computational network modeling, Borden et al. investigated how thalamic gating properties control sensory-evoked thalamic and cortical responses in the vibrissa pathway of the awake, head-fixed mouse (Borden et al., 2022). Specifically, optogenetic manipulation with the inhibitory opsin halorhodopsin was used to modulate the membrane potential of the thalamus while recording extracellular thalamic and cortical activity, as illustrated in Figure 62.7A. Despite enhanced bursting in

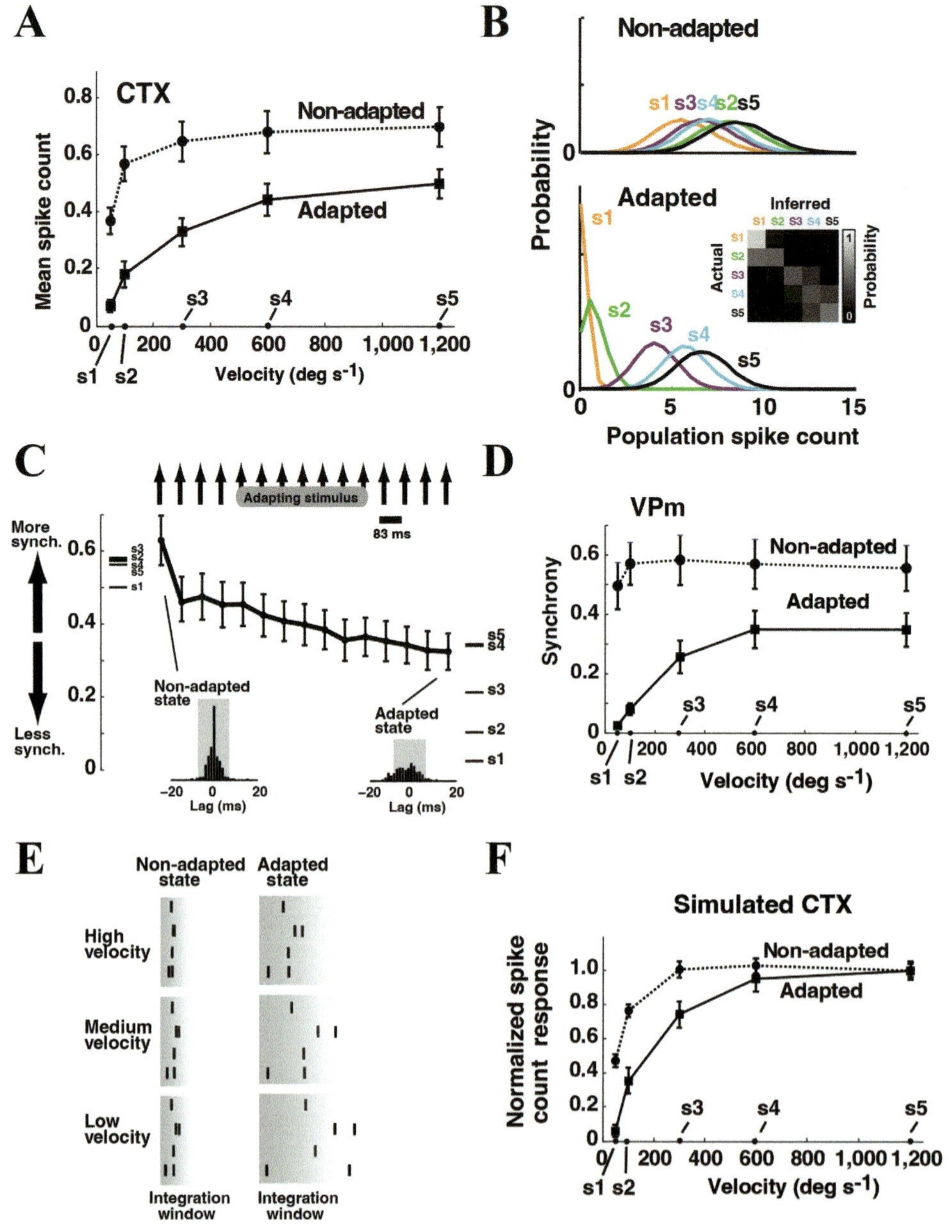

Figure 62.6. Rapid sensory adaptation mediates a trade-off between detection and discrimination, through cortical sensitivity to thalamic synchrony. (A). Neurons in barrel cortex are sensitive to the velocity of whisker stimulation, and this sensitivity changes with rapid sensory adaptation. Punctate whisker stimuli were presented to the anesthetized rat for a range of velocities (labeled s1 through s5), in the presence (adapted) or absence (non-adapted) of 12 Hz adapting whisker stimulation. (B). From the perspective of an ideal observer of cortical spike counts, the responses to the different whisker stimuli are stronger but more similar in the non-adapted condition (thus more detectable), in contrast to weaker but more separable (thus more discriminable) in the adapted condition. The inset shows the "confusion matrix," describing the probability of inferring each of the possible velocities, given the actual presentation of a specific velocity. (C). Thalamic synchrony is monotonically desynchronized with rapid sensory adaptation. (D). Adaptive changes in sensitivity to whisker velocity emerge in thalamic synchrony, as opposed to the magnitude of thalamic firing. (E). The high degree of thalamic synchronization for all velocities in the non-adapted condition results in all thalamic activity making it through the "window of opportunity" established by the feedforward inhibition (left). The desynchronization of thalamic activity with adaptation results in thalamic responses to high velocity being passed to cortex, while the lower synchrony for low velocities results in a weaker activation of cortex. (F). A thalamocortical network model simulation using experimentally recorded thalamic population spiking replicates the experimentally observed changes in cortical sensitivity to whisker velocity with adaptation. Adapted from Wang et al. (2010b).

the sensory-evoked thalamic response (Figure 62.7C), the corresponding magnitude of the sensory-evoked response of putative excitatory and inhibitory neurons in S1 cortex is remarkably invariant to the thalamic firing mode (Figure 62.7B). Investigation of the timing and synchronization of the thalamic population across these modes of firing revealed that the early phase of the sensory-evoked response is desynchronized across thalamic neurons (Figure 62.7D). A

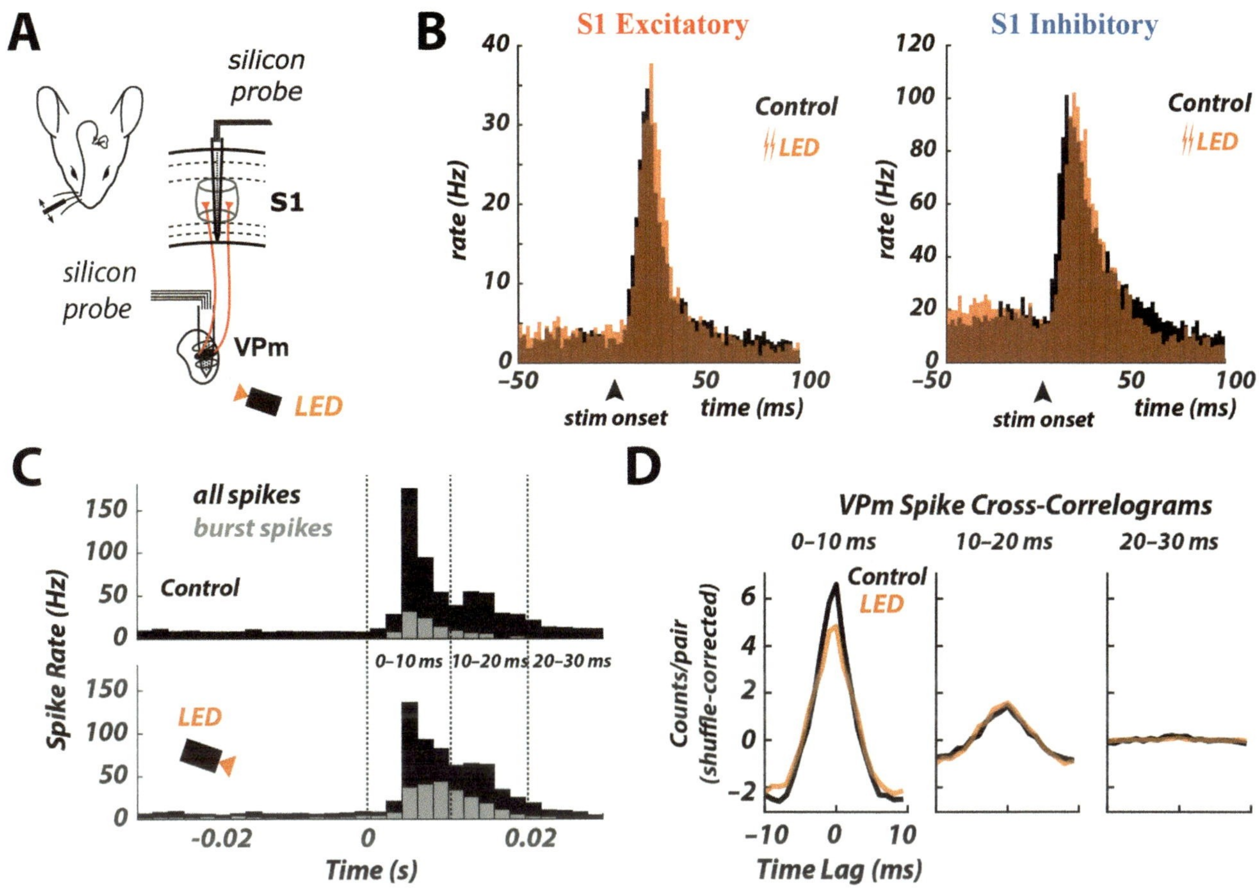

Figure 62.7. Thalamic burst effects on cortex are gated by cortical sensitivity to thalamic synchrony. (A). Simultaneous recording of VPm thalamic and cortical S1 activity in the awake, head-fixed mouse, during manipulation of thalamic burst mode firing with modulatory, inhibitory optogenetic input. Extracellular electrodes record whisker-evoked spiking activity in VPm thalamus in the presence (light-emitting diode, LED) and absence (control) of a hyperpolarizing optogenetic modulation of excitatory thalamic relay neurons expressing halorhodopsin (eNphR3.0), through a small fiber optic cable attached to the electrode. (B). Grand PSTHs for excitatory (RS, n = 86) and inhibitory (FS, n = 32) neurons, for a punctate whisker deflection. For each sub-population, sensory evoked PSTHs are shown for the control (black) case, and during hyperpolarization with an inhibitory opsin (LED, orange). (C). Grand single-unit PSTH for thalamic spiking in response to a single, punctate whisker stimulus (t = 0) for the control (no light, top panel, n = 54 units) and LED (bottom panel, n = 51 units) conditions. Instantaneous firing rates (bin size 2 ms) shown for all spikes (black) and putative burst spikes (gray). (D.) Grand mean spike cross-correlograms of the sensory-evoked response for the control (black) and LED (orange) conditions, when restricting spiking data to the (0–10 ms), (10–20 ms), and (20–30 ms) post-stimulus time windows (n = 187) simultaneously recorded pairs from 46 VPm single units, across 6 recording sessions). Adapted from Borden et al. (2022).

thalamocortical network model replicates the experimental observations of the invariant cortical response in the face of enhanced thalamic bursting and implicates changes in the precise timing and synchronization across the thalamic population as a likely mechanism underlying the experimental observations. In other words, whatever gains are made by the short inter-spike intervals of the thalamic burst at the individual thalamocortical synapse are compensated for by a disruption in the synchrony in the thalamic population when hyperpolarized. Importantly, however, while the magnitude of the cortical response was invariant to the enhanced thalamic bursting, cortical neurons exhibit an enhanced synchronization, which could have an even more potent impact on cortical signaling than magnitude alone. Thus, taken together, the evidence points to timing rather than response magnitude as a fundamental feature of the thalamocortical circuit, presenting a dynamic, timing-based control of information flow.

MODELING AND COMPUTATION

Given that the sensitivity of cortex to the timing and synchrony of thalamic input is strongly influenced by the interactions between the direct excitation and the delayed feedforward inhibition, it seems natural that the class of widely utilized network models that capture the general dynamics of the interplay between cortical excitation and inhibition would be applicable (Wilson and Cowan, 1972). Indeed, in a series of studies, Pinto and colleagues developed a framework for the circuit properties of the thalamocortical network that captures many aspects of the transformation from thalamus to cortex (Pinto et al., 1996; Pinto et al., 2000; Pinto et al., 2003). This modeling framework specifically relies on coupled nonlinear differential equations to represent the excitatory and inhibitory dynamics of the cortical circuit. When this network is driven with experimentally observed thalamic activity, the model output replicates several experimentally observed functional behaviors of the circuit, such as paired-pulse suppression and velocity tuning, resulting from the inherent sensitivity to input timing.

What are the computational consequences for the timing-based control of information flow at the thalamocortical junction? There is a significant cost to the elaborate anatomical wiring that is necessary to produce the extreme degree of timing sensitivity at the level of the input to sensory cortex. Large-scale, biophysically inspired network models of the thalamocortical circuit reveal that thalamic synchrony indeed enhances the reliability of cortical sensory-evoked responses, enhancing information rates for thalamic synchrony within a time window of ~5 ms (Wang et al., 2010a). Importantly, this also coincides with a range of time windows over which the cortical reliability is most energy efficient. The feedforward inhibition has the effect of stabilizing the cortical network, while maintaining stimulus selectivity (Kremkow et al., 2010). The optimality of the signaling is sensitive to the background levels of cortical excitation and inhibition, setting the stage for dynamic coordination of thalamocortical gating that is controlled through both regulation of thalamic synchrony and the cortical sensitivity to the synchrony of the thalamic input. Any and all aspects of function that impact cortical E/I balance thus engage this mechanism of signaling, such as bottom-up sensory adaptation, and states of arousal and attention. The consequence of this is the control of not only how much information is conveyed across the thalamocortical circuit, but also what information is conveyed. The dynamic regulation of thalamic synchrony is a key element in the trade-off between signaling that favors detection versus signaling that favors discrimination that has been demonstrated in the context of sensory adaptation (Wang et al., 2010b; Ollerenshaw et al., 2014).

CONCLUSIONS

A growing body of experimental work has revealed the thalamus as a dynamic gate of sensory signaling that plays a prominent role in controlling information flow from the sensory periphery to the cortical structures that ultimately lead to perception. Underlying this is a ubiquitous circuit motif of disynaptic inhibition that plays a major role in regulating signaling through sensitivity to timing and synchrony of thalamic inputs to cortex, establishing a "window of opportunity" for sensory signaling. This mechanism is precise, reliable, and energy efficient, and the dynamic regulation of the signaling through modulation of input synchrony and cortical timing sensitivity provides the signal-processing framework that enables the adaptive, context-dependent sensing of the complex, natural sensory world.

ACKNOWLEDGMENTS

The author would like to thank Elaida Dimwamwa for helpful discussions. This work was supported by NIH/National Institute of Neurological Disorders and Stroke BRAIN Initiative Grants RF1-NS128896 and R01-NS104928.

REFERENCES

Alitto HJ, Weyand TG, Usrey WM (2005). Distinct properties of stimulus-evoked bursts in the lateral geniculate nucleus. *J Neurosci.* 25:514–523.

Alonso JM, Usrey WM, Reid RC (1996). Precisely correlated firing in cells of the lateral geniculate nucleus. *Nature.* 383:815–819.

Barlow HB (1961). Possible principles underlying the transformations of sensory messages. In: Rosenblith WA, ed. *Sensory Communication.* MIT Press:217–234.

Boloori A-R, Jenks RA, Desbordes G, Stanley GB (2010). Encoding and decoding cortical representations of tactile features in the vibrissa system. *J Neurosci.* 30:9990–10005.

Boloori A-R, Stanley GB (2006). The dynamics of spatiotemporal response integration in the somatosensory cortex of the vibrissa system. *J Neurosci.* 26:3767–3782.

Borden PY, Wright NC, Morrissette AE, Jaeger D, Haider B, Stanley GB (2022). Thalamic bursting and the role of timing and synchrony in thalamocortical signaling in the awake mouse. *Neuron.* 110:2836–2853.

Bruno RM, Sakmann B (2006). Cortex is driven by weak but synchronously active thalamocortical synapses. *Science.* 312:1622–1627.

Bruno RM, Simons DJ (2002). Feedforward mechanisms of excitatory and inhibitory cortical receptive fields. *J Neurosci.* 22:10966–10975.

Castro-Alamancos MA, Oldford E (2002). Cortical sensory suppression during arousal is due to the activity-dependent depression of thalamocortical synapses. *J Physiol.* 541:319–331.

Chung S, Li X, Nelson SB (2002) Short-term depression at thalamocortical synapses contributes to rapid adaptation of cortical sensory responses in vivo. *Neuron.* 34:437–446.

Cohen-Kashi Malina K, Jubran M, Katz Y, Lampl I (2013). Imbalance between excitation and inhibition in the somatosensory cortex produces postadaptation facilitation. *J Neurosci.* 33:8463–8471.

Crick F (1984). Function of the thalamic reticular complex: The searchlight hypothesis. *Proc Natl Acad Sci U S A.* 81:4586–4590.

Cruikshank SJ, Lewis TJ, Connors BW (2007). Synaptic basis for intense thalamocortical activation of feedforward inhibitory cells in neocortex. *Nat Neurosci.* 10:462–468.

Cruikshank SJ, Urabe H, Nurmikko AV, Connors BW (2010). Pathway-specific feedforward circuits between thalamus and neocortex revealed by selective optical stimulation of axons. *Neuron.* 65:230–245.

Denning KS, Reinagel P (2005). Visual control of burst priming in the anesthetized lateral geniculate nucleus. *J Neurosci.* 25:3531–3538.

Gabernet L, Jadhav SP, Feldman DE, Carandini M, Scanziani M (2005). Somatosensory integration controlled by dynamic thalamocortical feed-forward inhibition. *Neuron*. 48:315–327.

Ganmor E, Katz Y, Lampl I (2010). Intensity-dependent adaptation of cortical and thalamic neurons is controlled by brainstem circuits of the sensory pathway. *Neuron*. 66:273–286.

Hartings JA (2000). *Somatosensory processing in inhibitory feedback circuits of the thalamus*. Ph.D. thesis. University of Pittsburgh.

Heiss JE, Katz Y, Ganmor E, Lampl I (2008). Shift in the balance between excitation and inhibition during sensory adaptation of S1 neurons. *J Neurosci*. 28:13320–13330.

Holler S, Köstinger G, Martin KAC, Schuhknecht GFP, Stratford KJ (2021). Structure and function of a neocortical synapse. *Nature*. 591:111–116.

Isaacson JS, Scanziani M (2011). How inhibition shapes cortical activity. *Neuron*. 72:231–243.

Jahnsen H, Llinás R (1984a). Voltage-dependent burst-to-tonic switching of thalamic cell activity: An in vitro study. *Arch Ital Biol*. 122:73–82.

Jahnsen H, Llinás R (1984b). Electrophysiological properties of guinea-pig thalamic neurones: an in vitro study. *J Physiol*. 349:205–226.

Khatri V, Hartings JA, Simons DJ (2004). Adaptation in thalamic barreloid and cortical barrel neurons to periodic whisker deflections varying in frequency and velocity. *J Neurophysiol*. 92:3244–3254.

Kheradpezhouh E, Adibi M, Arabzadeh E (2017). Response dynamics of rat barrel cortex neurons to repeated sensory stimulation. *Sci Rep*. 7:1–10.

Kremkow J, Perrinet LU, Masson GS, Aertsen A (2010). Functional consequences of correlated excitatory and inhibitory conductances in cortical networks. *J Comput Neurosci*. 28:579–594.

Kyriazi HT, Carvell GE, Simons DJ (1994). OFF response transformations in the whisker/barrel system. *J Neurophysiol*. 72:392–401.

Lampl I, Katz Y (2017). Neuronal adaptation in the somatosensory system of rodents. *Neuroscience*. 343:66–76.

Lesica NA, Stanley GB (2004). Encoding of natural scene movies by tonic and burst spikes in the lateral geniculate nucleus. *J Neurosci*. 24:10731–10740.

Lesica NA, Weng C, Jin J, Yeh C-I, Alonso J-M, Stanley GB (2006). Dynamic encoding of natural luminance sequences by LGN bursts. *PLoS Biol*. 4:e209.

Maravall M, Alenda A, Bale MR, Petersen RS (2013). Transformation of adaptation and gain rescaling along the whisker sensory pathway. *PLoS ONE*. 8:e82418.

Maravall M, Petersen RS, Fairhall AL, Arabzadeh E, Diamond ME (2007). Shifts in coding properties and maintenance of information transmission during adaptation in barrel cortex. *PLoS Biol*. 5(2):e19.

Ollerenshaw DR, Zheng HJV, Millard DC, Wang Q, Stanley GB (2014). The adaptive trade-off between detection and discrimination in cortical representations and behavior. *Neuron*. 81:1152–1164.

Pinto DJ, Brumberg JC, Simons DJ (2000). Circuit dynamics and coding strategies in rodent somatosensory cortex. *J Neurophysiol*. 83:1158–1166.

Pinto DJ, Brumberg JC, Simons DJ, Ermentrout GB (1996). A quantitative population model of whisker barrels: re-examining the Wilson-Cowan equations. *J Comput Neurosci*. 3:247–264.

Pinto DJ, Hartings JA, Brumberg JC, Simons DJ (2003). Cortical damping: analysis of thalamocortical response transformations in rodent barrel cortex. *Cereb Cortex N Y N 1991*. 13:33–44.

Reinagel P, Godwin D, Sherman SM, Koch C (1999). Encoding of visual information by LGN bursts. *J Neurophysiol*. 81:2558–2569.

Sherman SM (2005). Thalamic relays and cortical functioning. *Prog Brain Res*. 149:107–126.

Sherman SM, Guillery RW (2002). The role of the thalamus in the flow of information to the cortex. *Philos Trans R Soc Lond B Biol Sci*. 357:1695–1708.

Sherman SM, Koch C (1986). The control of retinogeniculate transmission in the mammalian lateral geniculate nucleus. *Exp Brain Res*. 63(1):1–20.

Simons DJ, Carvell GE (1989). Thalamocortical response transformation in the rat vibrissa/barrel system. *J Neurophysiol*. 61:311–330.

Stanley GB, Jin J, Wang Y, Desbordes G, Wang Q, Black MJ, Alonso J-M (2012). Visual orientation and directional selectivity through thalamic synchrony. *J Neurosci*. 32:9073–9088.

Stratford KJ, Tarczy-Hornoch K, Martin KA, Bannister NJ, Jack JJ (1996). Excitatory synaptic inputs to spiny stellate cells in cat visual cortex. *Nature*. 382:258–261.

Suzuki S, Rogawski MA (1989). T-type calcium channels mediate the transition between tonic and phasic firing in thalamic neurons. *Proc Natl Acad Sci U S A*. 86:7228–7232.

Swadlow HA, Gusev AG (2001). The impact of "bursting" thalamic impulses at a neocortical synapse. *Nat Neurosci*. 4:402–408.

Usrey WM, Alonso J-M, Reid RC (2000). Synaptic interactions between thalamic inputs to simple cells in cat visual cortex. *J Neurosci*. 20:5461–5467.

Wang H-P, Spencer D, Fellous J-M, Sejnowski TJ (2010a). Synchrony of thalamocortical inputs maximizes cortical reliability. *Science*. 328:106–109.

Wang Q, Webber RM, Stanley GB (2010b). Thalamic synchrony and the adaptive gating of information flow to cortex. *Nat Neurosci*. 13:1534–1541.

Webber RM, Stanley GB (2004). Nonlinear encoding of tactile patterns in the barrel cortex. *J Neurophysiol*. 91:2010–2022.

Webber RM, Stanley GB (2006). Transient and steady-state dynamics of cortical adaptation. *J Neurophysiol*. 95:2923–2932.

Wehr M, Zador AM (2005). Synaptic mechanisms of forward suppression in rat auditory cortex. *Neuron*. 47:437–445.

Wehr MS, Zador AM (2003). Balanced inhibition underlies tuning and sharpens spike timing in auditory cortex. *Nature*. 426:442–446.

Whitmire CJ, Stanley GB (2016). Rapid sensory adaptation redux: A circuit perspective. *Neuron*. 92:298–315.

Wilent WB, Contreras D (2004). Synaptic responses to whisker deflections in rat barrel cortex as a function of cortical layer and stimulus intensity. *J Neurosci*. 24:3985–3998.

Wilent WB, Contreras D (2005). Dynamics of excitation and inhibition underlying stimulus selectivity in rat somatosensory cortex. *Nat Neurosci*. 8:1364–1370.

Wilson HR, Cowan JD (1972). Excitatory and inhibitory interactions in localized populations of model neurons. *Biophys J*. 12:1–24.

Wright NC, Borden PY, Liew YJ, Bolus MF, Stoy WM, Forest CR, Stanley GB (2021). Rapid cortical adaptation and the role of thalamic synchrony during wakefulness. *J Neurosci*. 41:5421–5439.

Yu J, Gutnisky DA, Hires SA, Svoboda K (2016). Layer 4 fast-spiking interneurons filter thalamocortical signals during active somatosensation. *Nat Neurosci*. 19:1647–1657.

Zheng HJ V., Wang Q, Stanley GB (2015). Adaptive shaping of cortical response selectivity in the vibrissa pathway. *J Neurophysiol*. 113:3850–3865.

63.

HOW THE LGN FORWARDS RETINAL INFORMATION TO THE CORTEX

Friedrich T. Sommer and Judith A. Hirsch

INTRODUCTION

Retinal circuits translate the visual image into patterned trains of action potentials that are processed by the dorsal lateral geniculate nucleus (LGN) of the thalamus before reaching cortex. The first recordings from cat LGN revealed receptive fields that, just like those in retina, comprised concentrically arranged center and surround subregions with the opposite preference for luminance contrast, bright (ON) or dark (OFF)—ON center cells have OFF surrounds, and vice versa (Kuffler, 1953; Hubel, 1960; Hubel and Wiesel, 1961). This repetition in visual response profile across the retinogeniculate synapse contrasted sharply with the emergence of novel receptive field types and response properties in primary visual cortex (Hubel and Wiesel, 1962). Thus, the idea of LGN as a relay took root, despite the fact that thalamic circuits and neural membrane properties are complex.

Here we provide examples of combined computational and empirical approaches that explore how LGN contributes to visual processing, focusing on a seemingly simple question. By asking how the rate and timing of retinal spike trains change across the first synapse in thalamus, we address the broader issue of how LGN influences the coding and content of information that cortex receives. While there are many experimental approaches to address this topic, we confine ourselves to those that directly involve simultaneous recording from synaptically connected ganglion and relay cell pairs, as follows. Early investigators noticed that spikes recorded extracellularly in LGN were often preceded by smaller and slower waveforms that they rightly hypothesized derived from retinogeniculate excitatory postsynaptic potentials (EPSPs) and called S potentials (Bishop et al., 1962; Kaplan and Shapley, 1984). Only larger EPSPs generate detectable S potentials, however; thus, smaller retinal inputs are missed. Hence this technique yields an incomplete account of retinal input because relay cells in rodent (Morgan et al., 2016; Rompani et al., 2017), carnivore (Cleland et al., 1971; Usrey et al., 1998), and perhaps primate (Sincich et al., 2009), receive input from more than one ganglion cell (see **Chapter 17 in Section V**). Techniques with greater sensitivity employ two or more electrodes to record from retinal and thalamic cells at once, and infer connectivity between cell pairs using cross-correlation analysis (Usrey et al., 1998). It is also possible to leverage whole-cell patch techniques to resolve both retinogeniculate EPSPs and thalamic spikes in the same cell (Usrey and Reid, 1999; Wang et al., 2007; Koepsell et al., 2009; Wang et al., 2011b).

Results from all types of simultaneous recordings of retinal inputs and thalamic outputs highlight two fundamental features of retinogeniculate transmission. The first is that many retinal spikes fail to trigger the postsynaptic relay cell; retinogeniculate EPSPs often remain subthreshold (Wang et al., 2007; Koepsell et al., 2009; Wang et al., 2011b; Rathbun et al., 2016). The second is that there is a tight temporal coupling between retina and LGN—afferent inputs drive thalamic spikes at short latency and with millisecond precision (Usrey et al., 1998; Usrey and Reid, 1999; Koepsell et al., 2009). Thus, for any given retinal input, a relay cell either produces an impulse almost instantaneously, or not at all (relay cells fire in two modes, tonic and burst; we consider only tonic mode here; see **Chapter 3 in Section II**). Hence, transmission can be described as a binary editing process of the presynaptic spike train (Cleland et al., 1971; Kaplan et al., 1987; Carandini et al., 2007; Sincich et al., 2007). This concept of editing the spike train is akin to "thinning" in statistics, in which transformation of one point process (e.g., a spike train) to another is achieved by removing single points (Stoyan et al., 1995; Victor, 2005). In the following we describe different forms of the editing process and associated computations in LGN, paring down our discussion to connections between one retinal ganglion cell and one relay cell.

UNIFORM VERSUS NONUNIFORM EDITING PROCESSES

Editing (thinning) processes are governed by the chance that a point will be removed (deleted) rather than kept (relayed, transmitted). In the simplest case, the editing process is uniform; the probability (p) that any point is deleted is the same over time (in statistics, this is called p-thinning). Alternatively, the editing process is nonuniform; the chance that any point is deleted depends on its position with respect to other points (position-dependent thinning) (Shannon and Weaver, 1949; Stoyan et al., 1995). Thus, if the editing process across the retinogeniculate synapse were uniform, the probability that any given retinal impulse will be transmitted (i.e., that it will cause a relay cell to fire) is constant. By contrast, for a nonuniform editing process, the probability that a retinal spike will be relayed depends on its history—on the temporal pattern of preceding spikes (Wang et al., 2010; Alexander et al., 2022). These two forms of the editing process are illustrated in Figure 63.1. Although both uniform and nonuniform editing processes decrease the ratio of thalamic to retinal spikes and thereby reduce average firing rate, they differ in the way they change how and what information is encoded.

STANDARD MEASURES OF INFORMATION ENCODED IN SPIKE TRAINS

Shannon's information theory provides a means to quantify information (Shannon and Weaver, 1949). In essence, the amount of information contained in a message (for example, the outcome of a coin toss) can be described as the average number of yes/no questions one would have to ask to guess the message's content. If the probability of heads or tails is the same, the amount of information conveyed by reporting the outcome of a coin toss, heads or tails, is one "bit." Shannon's method can be used to measure the amount of information carried by each action potential (bits per spike), or each second of the spike train (bits per second); this latter quantity is called the information rate. In general, if the timing of the neural response to the stimulus is reliable, similar across repeated trials, then the amount of information that each spike encodes is higher than for instances in which there is substantial cross-trial variability (Rieke et al., 1999).

THE EDITING PROCESS ACROSS THE RETINOGENICULATE SYNAPSE IS NONUNIFORM AND TRANSFORMS THE REPRESENTATION OF THE STIMULUS

Recordings from synaptically coupled ganglion and relay cells have shown that the chance that a retinal action potential will be relayed depends on spike history; thus, the editing process is nonuniform. An early study reported that the probability of transmission was highest for retinal spikes that occurred <30 ms after another (i.e., the interspike interval, ISI, is brief) (Usrey et al., 1998). This facilitatory effect is called paired-spike enhancement (Usrey et al., 1998) (Figure 63.1). Later work revealed that the fraction of retinal spikes relayed by thalamus were more reliable (there was less jitter across trials) than those that were not transmitted (Rathbun et al., 2016). Accordingly, an information theoretic analysis of these results showed that single thalamic spikes carried more information about the stimulus than individual retinal impulses (Rathbun et al., 2016). Although the information rate of the postsynaptic spike train could not exceed that in the presynaptic spike train (the data-processing inequality holds that the amount of information transmitted along a channel can decrease but not increase (Cover and Thomas, 1991)), information was recoded into fewer, more informative, spikes across the retinogeniculate synapse.

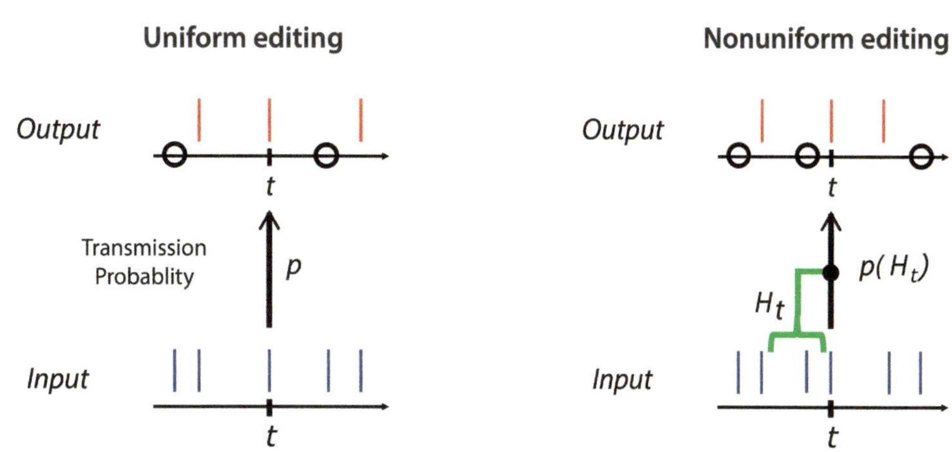

Figure 63.1. Editing processes (thinning). Left: Uniform editing, or p-thinning: each presynaptic spike is transmitted with same probability p. Right: Nonuniform editing, or position-dependent thinning: a presynaptic spike is deleted with a probability $p(H_t)$ that is a function of the input history H_t.

What stimulus features do these "more informative" thalamic spikes encode? Can one identify features that are inherited, are lost, or that emerge across the synapse? This question has been addressed by developing an information theoretic framework that subsumes stimulus features encoded by the presynaptic and postsynaptic neurons (Wang et al., 2010). The novel framework was used with a unique dataset obtained from whole-cell recording of the membrane voltage *in vivo*. This physiological method allows one to visualize synaptic potentials of various sizes and shapes, making it possible to determine if a relay cell receives only one versus multiple retinal inputs and, thus, to select single input relay cells for further analysis. The analysis yielded a surprising finding. Individual spikes in LGN encoded temporal features of the stimulus that single impulses in retina did not. An observation of novel temporal features was also reported in a study using the S potential technique in monkey (Sincich et al., 2009). (As an aside, these studies also highlight the importance of looking beyond encoding of spatial features to understand how different stages of the visual pathway contribute to image processing.)

How can a relay cell encode information about a feature that single action potentials in the presynaptic spike train do not represent? To address this question, it is important to recognize that information can be conveyed not only in single impulses, but also in stereotyped spike patterns (Brenner et al., 2000; Brenner et al., 2002). For example, information can be represented by a mixture of single spikes, and motifs such as spike pairs with a fixed ISI, and/or longer patterns of impulses (Rieke et al., 1999). One can determine if spike pairs are involved in the coding process by using a measure called synergy, defined as the information in spike pairs that remains after subtracting twice the information in single spikes. If synergy is positive, the pairs carry information beyond that encoded in single impulses. Conversely, in the case of negative synergy, spike pairs introduce redundancy (for example, replacing each spike in an independent spike code with a spike pair does not contribute new information). If synergy is zero, no additional information is encoded by spike pairs.

The application of this framework to LGN revealed positive synergy for retinal spike trains, whereas the synergy in thalamic impulse trains was negligible. Specifically, in retina, spike pairs conveyed information about the novel temporal feature that was not carried by independent spikes. In other words, as a result of a nonuniform editing process, information about the novel feature was transferred to single thalamic impulses. Fitting the data (spike trains of synaptically coupled retinal and relay cell pairs) with a nonuniform editing process in which the deletion probability was smaller after shorter (<20 ms) presynaptic ISIs was able to reproduce the emergence of the novel temporal feature (using the vocabulary of statistics, this is an example of position-dependent-thinning). By contrast, a model based on a uniform editing process failed to generate the emergent feature. These results are illustrated in Figure 63.2A and B. All told, this work provided the first direct biological example of the transformation of a synergistic code into an independent code across a synapse, and recalls early work that demonstrated the importance of paired-spike facilitation in retinogeniculate transmission (Usrey et al., 1998).

RETINOGENICULATE EDITING OF OSCILLATION-BASED ACTIVITY

So far, we have discussed how a nonuniform editing process can fundamentally transform the neural code by removing synergistic structure in the presynaptic spike train and representing novel features postsynaptically using trains of independent spikes. Here we focus on an aspect common to both uniform and nonuniform editing processes in LGN—that spikes are transmitted across the synapse with minimal jitter and only a small difference in latency (Usrey et al., 1998; Koepsell et al., 2009). This tight temporal coupling across stages means that relay cells can preserve information encoded in the fine (millisecond) timing of the input spike train.

Visual stimuli evoke changes in retinal spike rate with relatively slow dynamics (<20 Hz) (Butts et al., 2007). However, temporal structure in the spike train is also influenced by intrinsic retinal dynamics that often occupy a high frequency regime. For example, subsets of retinal ganglion cells fire rhythmically, with frequencies in the gamma range, 40–80 Hz. These oscillations have been observed in cat (Heiss and Bornschein, 1965; Neuenschwander et al., 1999), rodent (Saleem et al., 2017; Storchi et al., 2017), and other species, including frog (Ishikane et al., 1999; Ishikane et al., 2005). The phase of retinal oscillations is not entrained by the stimulus (Koepsell et al., 2009; Storchi et al., 2017), unlike the case in other visual areas such as superior colliculus (Brecht et al., 2001) or ventral LGN (Ciftcioglu et al., 2020). Still, the high temporal fidelity of the retinogeniculate synapse means that the periodicity and phase of retinal spike trains are conserved by thalamus and relayed to cortex. This preservation of oscillation frequency across the synapse can, in principle, be produced by either a uniform or nonuniform editing process.

Before discussing the functional significance of retinal oscillations (Sommer, 2014), it is important to analyze their capacity to carry information (Koepsell and Sommer, 2008; Agarwal and Sommer, 2013). Standard methods of calculating Shannon information involve the peristimulus time histogram (PSTH), which is computed by summing the number of spikes fired within a given time bin across stimulus trials. Because the phase of retinal oscillations

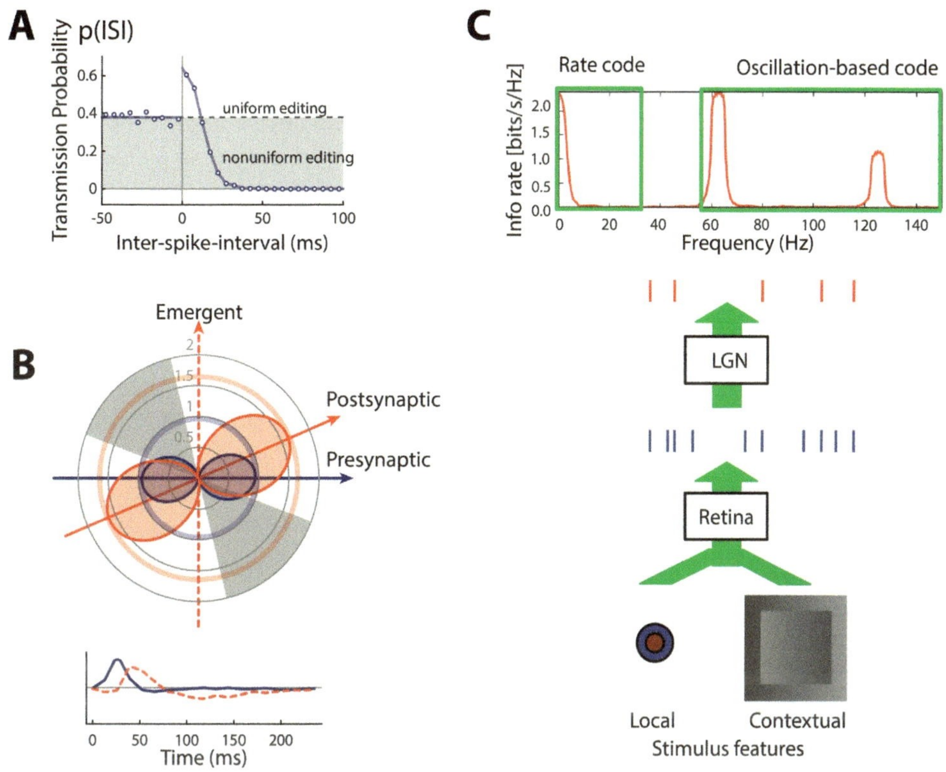

Figure 63.2. (A). Encoding and recoding of information across the retinogeniculate synapse. Editing processes fitted to simultaneously recorded retinal and thalamic spike trains. Fits using uniform editing (dashed line) reproduce mean spike rate, whereas the nonuniform editing process that depends on the history of presynaptic spikes (solid blue line) shows that spike pairs with ISIs <20 ms increase the probability of transmission. (B). Information content about the joint pre- and postsynaptic feature space illustrated as polar plots. The angular coordinate corresponds to different features and the radial coordinate represents information in single spikes; the solid arrows mark the presynaptic feature conveyed by single retinal spikes and the postsynaptic feature conveyed by single thalamic spikes (the linear combination of the presynaptic feature and the emergent feature [dashed line]). The information about particular features falls on the filled, dumbbell-shaped curves and is limited by the pale pink and blue circles, which mark the maximum amount of information in single spikes. Last, the gray sectors indicate the subset of features about which a thalamic spike conveyed less information that a retinal one. The plot at bottom depicts the normalized amplitude of the presynaptic feature (solid blue line) and the emergent feature (dashed red line). (C) The retinogeniculate editing process preserves both slow changes in rate and the fine timing of retinal spikes. Thus, the LGN conveys a multiplexed signal to the cortex in which rate-coded information about local features is carried in the low frequency band and oscillation-based information about contextual features is transmitted in the higher frequency band (and its harmonics).

varies randomly with respect to the stimulus, the PSTH averages the oscillations away. Thus, it is necessary to preprocess oscillating spike trains so that they can be analyzed with standard tools. Such preprocessing can be accomplished by extracting an oscillatory reference signal from the retinal inputs (Koepsell and Sommer, 2008; Koepsell et al., 2009). The reference signal is used to align, or de-jitter, oscillation phase differences across repeats of the stimulus. This alignment is done by shifting spikes in time, but with displacements so small that the structure of the spike train at low frequencies is unaltered. Thus, the PSTH that is constructed after de-jittering reveals both the slow changes in spike rate that are driven by the stimulus (rate-coding) and the rapid changes that are locked to oscillations (oscillation-based coding). Comparing the difference between the information computed from this PSTH to the one prepared before the de-jittering process yields the amount of information that single spikes convey about (the phase of) the retinal oscillation (Koepsell et al., 2009).

How might these oscillations serve sensory function? Recent work shows that retinal oscillation strength and frequency carry information about background intensity (irradiance) (Storchi et al., 2017). But the potential sensory functions of the oscillation-based channel extend far beyond signaling this single parameter, as can be appreciated by considering that retinal oscillations are formed by the joint activity of spatially distributed networks. Hence, oscillations stand to encode contextual features of the image that complement neural responses to local features (Ishikane et al., 2005; Gollisch and Meister, 2008; Koepsell et al., 2009; Koepsell et al., 2010). This idea was tested with a model of coupled oscillators whose interactions were influenced by local changes in luminance contrast within a larger scene. In essence, the strength of coupling between pairs of oscillators increased when both received input from regions, or segments, with similar luminance values. Thus, coupled oscillators synchronized and "labeled" areas of homogeneous contrast within the scene. When tested on a dataset used as a benchmark for image segmentation, this network outperformed a standard method (histogram-based thresholding of local contrast values) (Warner and Sommer, 2020). Although this model of coupled oscillators is abstract and does not map to neural circuits directly, the results it generates suggest the oscillation-based retinogeniculate channel is able to contribute unique information about extended features in the retinal image.

The same cells that transmit oscillation-based information also carry rate-coded information. This property of sending two messages in separate frequency bands of the same spike train is a form of multiplexing (conveying multiple messages along a single channel) and has been observed in other sensory systems (Ruderman and Bialek, 1994; Ahissar et al., 2000; O'Connor et al., 2002; Nadasdy, 2010). In the case of the retinogeniculate synapse, the low-frequency channel carries rate-coded information (estimation window 50–100ms) about the visual stimulus. The position of this band at the low end of the frequency spectrum reflects the temporal structure of the stimulus; the spectral power of natural signals decays as a power law with increasing temporal frequency (Ruderman and Bialek, 1994). Conversely, the high-frequency channel conveys information extracted from the phase of single spikes with respect to the oscillation. Because the oscillation frequency is sufficiently high, the variations in spike phase remain in a band that does not overlap with that carrying rate-coded information (Koepsell et al., 2010). This scenario recalls the auditory system, in which neurons phase-lock to stimuli in a range higher than the acoustic modulation spectrum (Langner, 1992). These results are summarized in Figure 63.2C.

In summary, the editing process across the retinogeniculate synapse serves multiple purposes. For example, a nonuniform editing process transforms visual features encoded as spike pairs by retina into an independent spike code in thalamus. Regardless of whether the editing process is uniform or nonuniform, the tight temporal coupling between retina and LGN allows the transmission of information in the high-frequency band, such as that encoded by fast oscillations.

MODELING BIOLOGICAL MECHANISMS THAT UNDERLIE THE EDITING PROCESS

Simulations using a simple leaky integrate-and-fire-model (essentially an isopotential neuron that fires a spike when the membrane voltage depolarizes to reach threshold, and then resets) suggest that the basis of paired-spike facilitation observed in LGN (Usrey et al., 1998) can be largely explained by a postsynaptic mechanism rather than presynaptic processes such as synaptic facilitation or depression (Carandini et al., 2007; Casti et al., 2008; Heiberg et al., 2013). Here visually evoked retinal spike trains are converted into a series of EPSPs that have a stereotyped tall and sharply peaked profile. These are then introduced to a (slightly tweaked) leaky-integrate-and-fire model of a relay cell. The correspondence between the simulated and biological spike trains is remarkably close (Carandini et al., 2007; Casti et al., 2008). Exploration of the modeled membrane voltage reveals that spikes are produced by EPSPs that arrive shortly after another such that successive events sum to reach spike threshold. Thus, this temporal summation of successive retinal inputs can explain a nonuniform editing process such as that based on paired-spike enhancement.

That the leaky integrate-and-fire models perform as well as they do in recapitulating the editing process across the retinogeniculate synapse might seem improbable given that relay cells also receive feedback excitation from cortex, and inhibition from local interneurons and the thalamic reticular nucleus (Babadi et al., 2010; Norheim et al., 2012; Butts et al., 2016). Remarkably, however, whole-cell recordings *in vivo* show that the biological membrane voltage is much like that generated by the models—dominated by retinogeniculate EPSPs that rise rapidly to reach tall peaks; presumed cortical inputs are so small that they are rarely visible at the soma. Moreover, inhibition does not interfere with the shapes of the retinogeniculate inputs because it is smooth and graded. Further, because relay cells have push-pull responses (e.g., a bright stimulus placed in an ON subregion excites, whereas a dark stimulus inhibits), periods of strongest excitation and inhibition do not overlap (Wang et al., 2007; Wang et al., 2011b). By contrast, whole-cell recordings from interneurons have a different, almost inverse, profile; excitation is graded and smooth, whereas inhibition takes the form of individual inhibitory postsynaptic potentials (IPSPs) that bite into the membrane voltage to give it a jagged appearance (Wang et al., 2011b), as illustrated in Figure 63.3. (The biological mechanisms that might underlie the shapes of different types of synaptic input to relay cells and interneurons are described elsewhere in the literature; Wang et al., 2011a; Wang et al., 2011b.)

How do these physiological observations relate to the preservation of retinal spike timing across the synapse? This question was addressed by developing a framework based on initial studies with leaky integrate-and-fire models (Wang et al., 2011b). A stimulus with the statistics of natural images was passed through a push-pull filter and Poisson process to generate trains of excitatory and inhibitory inputs that were introduced to a modeled relay cell. When the relay cell was provided with realistic afferent inputs—peaky EPSPs and graded inhibition—the resulting thalamic spike train was able to reproduce the

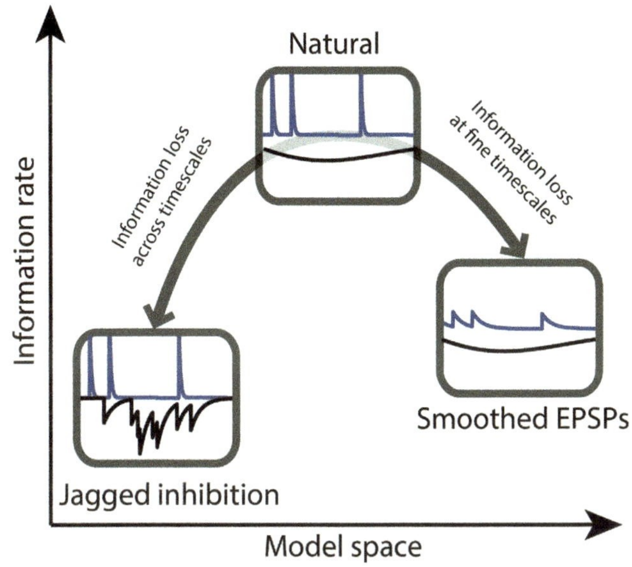

Figure 63.3. Shapes of synaptic potentials and relay of information. The amount of information relayed to cortex is highest for naturalistic, peaky EPSPS (blue) and graded inhibition (black). If excitation is smoothed or slowed, information at fine temporal resolution is lost; if inhibition is made jagged (as it is in local interneurons), information rates decline at all temporal resolutions.

temporal fidelity of the biological editing process and preserve retinal information. If, however, the sharp peaks of the EPSPs were smoothed, the tight temporal coupling at the synapse was degraded and information rates were dramatically reduced, especially in the high-frequency band. Substituting jagged (like that recorded from interneurons) for graded inhibition had an even more drastic impact and reduced information rates across frequency bands (Figure 63.3). While responses of interneurons have not yet been modeled, their synaptic physiology suggests that they provide means to rapidly inhibit and disinhibit relay cells, perhaps enhancing the ability of the system to respond with high temporal precision (Wang et al., 2007; Butts et al., 2011; Wang et al., 2011b).

CONCLUSION

Here we focused on how the statistical framework of the editing, or thinning, process can explain vital computations in LGN such as how information represented in spike patterns is transformed into single spikes that encode emergent stimulus features. This perspective complements other computational approaches that provide novel insight into how thalamic circuits and cells contribute to visual processing. Examples include studies of the role of burst firing in conveying visual information (Reinagel and Reid, 2000; Lesica and Stanley, 2004; Lesica et al., 2006; Wang et al., 2007); of how retinothalamic convergence and divergence, coupled with the spatial arrangement of excitation and inhibition in the relay cell's receptive field, not only improve the information capacity of signals sent to cortex, but enhance the visual image itself (Martinez et al., 2014; Hirsch et al., 2015; Alexander et al., 2022); and of contributions of cortical feedback (Murphy et al., 2021).

REFERENCES

Agarwal G, Sommer FT (2013). Measuring information in spike trains about intrinsic brain signals. In: DiLorenzo PM, Victor JD, eds. *Spike Timing: Mechanisms and Function*. CRC Press:37–152.

Ahissar E, Sosnik R, Haidarliu S (2000). Transformation from temporal to rate coding in a somatosensory thalamocortical pathway. *Nature*. 406:302–306.

Alexander PC, Alitto HJ, Fisher TG, Rathbun DL, Weyand TG, Usrey WM (2022). Dynamics of temporal integration in the lateral geniculate nucleus. *eNeuro*. 9:1–22.

Babadi B, Casti A, Xiao Y, Kaplan E, Paninski L (2010). A generalized linear model of the impact of direct and indirect inputs to the lateral geniculate nucleus. *J Vis*. 10:22.1–14.

Bishop PO, Burke W, Davis R (1962). The interpretation of the extracellular response of single lateral geniculate cells. *J Physiol*. 162:451–472.

Brecht M, Goebel R, Singer W, Engel AK (2001). Synchronization of visual responses in the superior colliculus of awake cats. *Neuroreport*. 12:43–47.

Brenner N, Agam O, Bialek W, de Ruter van Steveninck R (2002) Statistical properties of spike trains: Universal and stimulus dependent aspects. *Phys Rev E*. 66:031907.1–14.

Brenner N, Strong SP, Koberle R, Bialek W, de Ruyter van Steveninck RR (2000). Synergy in a neural code. *Neural Comput*. 12:1531–1552.

Butts DA, Cui Y, Casti AR (2016). Nonlinear computations shaping temporal processing of precortical vision. *J Neurophysiol*. 116:1344–1357.

Butts DA, Weng C, Jin J, Alonso JM, Paninski L (2011). Temporal precision in the visual pathway through the interplay of excitation and stimulus-driven suppression. *J Neurosci*. 31:11313–11327.

Butts DA, Weng C, Jin J, Yeh CI, Lesica NA, Alonso JM, Stanley GB (2007). Temporal precision in the neural code and the timescales of natural vision. *Nature*. 449:92–95.

Carandini M, Horton JC, Sincich LC (2007). Thalamic filtering of retinal spike trains by postsynaptic summation. *J Vision*. 7:20.1–11.

Casti A, Hayot F, Xiao Y, Kaplan E (2008). A simple model of retina-LGN transmission. *J Comput Neurosci*. 24:235–252.

Ciftcioglu UM, Suresh V, Ding KR, Sommer FT, Hirsch JA (2020). Visual information processing in the ventral division of the mouse lateral geniculate nucleus of the thalamus. *J Neurosci*. 40:5019–5032.

Cleland BG, Dubin MW, Levick WR (1971). Simultaneous recording of input and output of lateral geniculate neurones. *Nat New Biol*. 231:191–192.

Cover TM, Thomas JA (1991). *Elements of Information Theory*. Wiley.

Gollisch T, Meister M (2008). Rapid neural coding in the retina with relative spike latencies. *Science*. 319:1108–1111.

Heiberg T, Kriener B, Tetzlaff T, Casti A, Einevoll GT, Plesser HE (2013). Firing-rate models capture essential response dynamics of LGN relay cells. *J Comput Neurosci*. 35:359–375.

Heiss WD, Bornschein H (1965). Distribution of impulse of continuous activity of single optic nerve fibers: Effects of light, ischemia, strychnine and barbiturate. *Pflugers Arch Gesamte Physiol Menschen Tiere*. 286:1–18.

Hirsch JA, Wang X, Sommer FT, Martinez LM (2015). How inhibitory circuits in the thalamus serve vision. *Ann Rev Neurosci*. 38:309–329.

Hubel DH (1960). Single unit activity in lateral geniculate body and optic tract of unrestrained cats. *J Physiol.* 150:91–104.

Hubel DH, Wiesel TN (1961). Integrative action in the cat's lateral geniculate body. *J Physiol.* 155:385–398.

Hubel DH, Wiesel TN (1962). Receptive fields, binocular interaction and functional architecture in the cat's visual cortex. *J Physiol.* 160:106–154.

Ishikane H, Gangi M, Honda S, Tachibana M (2005). Synchronized retinal oscillations encode essential information for escape behavior in frogs. *Nat Neurosci.* 8:1087–1095.

Ishikane H, Kawana A, Tachibana M (1999). Short- and long-range synchronous activities in dimming detectors of the frog retina. *Vis Neurosci.* 16:1001–1014.

Kaplan E, Purpura KP, Shapley RM (1987). Contrast affects the transmission of visual information through the mammalian lateral geniculate nucleus. J Physiol. 391:267–288.

Kaplan E, Shapley R (1984). The origin of the S (slow) potential in the mammalian lateral geniculate nucleus. *Exp Brain Res.* 55:111–116.

Koepsell K, Sommer FT (2008). Information transmission in oscillatory neural activity. *Biol Cybern.* 99:403–416.

Koepsell K, Wang X, Hirsch JA, Sommer FT (2010), Exploring the function of neural oscillations in early sensory systems. *Front Neurosci.* 4:53–61.

Koepsell K, Wang X, Vaingankar V, Wei Y, Wang Q, Rathbun DL, Usrey WM, Hirsch J, Sommer FT (2009). Retinal oscillations carry visual information to cortex. *Front Syst Neurosci.* 3:1–18.

Kuffler SW (1953). Discharge patterns and functional organization of the mammalian retina. *J Neurophysiol.* 16:37–68.

Langner G (1992). Periodicity coding in the auditory system. *Hearing Res.* 60:115–142.

Lesica NA, Stanley GB (2004). Encoding of natural scene movies by tonic and burst spikes in the lateral geniculate nucleus. *J Neurosci.* 24:10731–10740.

Lesica NA, Weng C, Jin J, Yeh CI, Alonso JM, Stanley GB (2006). Dynamic encoding of natural luminance sequences by LGN bursts. *PLoS Biol.* 4:e209.

Martinez LM, Molano-Mazon M, Wang X, Sommer FT, Hirsch JA (2014). Statistical wiring of thalamic receptive fields optimizes spatial sampling of the retinal image. *Neuron.* 81:943–956.

Morgan JL, Berger DR, Wetzel AW, Lichtman JW (2016). The fuzzy logic of network connectivity in mouse visual thalamus. *Cell.* 165:192–206.

Murphy AJ, Shaw L, Hasse JM, Goris RLT, Briggs F (2021). Optogenetic activation of corticogeniculate feedback stabilizes response gain and increases information coding in LGN neurons. *J Comput Neurosci.* 49:259–271.

Nadasdy Z (2010). Binding by asynchrony: The neuronal phase code. *Front Neurosci.* 4:1–10.

Neuenschwander S, Castelo-Branco M, Singer W (1999). Synchronous oscillations in the cat retina. *Vision Res.* 39:2485–2497.

Norheim ES, Wyller J, Nordlie E, Einevoll GT (2012). A minimal mechanistic model for temporal signal processing in the lateral geniculate nucleus. *Cogn Neurodyn.* 6:259–281.

O'Connor SM, Berg RW, Kleinfeld D (2002). Coherent electrical activity between vibrissa sensory areas of cerebellum and neocortex is enhanced during free whisking. *J Neurophysiol.* 87:2137–2148.

Rathbun DL, Alitto HJ, Warland DK, Usrey WM (2016). Stimulus contrast and retinogeniculate signal processing. *Front Neural Circuits.* 10:8.1–10.

Reinagel P, Reid RC (2000). Temporal coding of visual information in the thalamus. *J Neurosci.* 20:5392–5400.

Rieke F, Warland D, de Ruter van Steveninck R, Bialek W (1999). *Spikes: Exploring the Neural Code.* MIT Press.

Rompani SB, Mullner FE, Wanner A, Zhang C, Roth CN, Yonehara K, Roska B (2017). Different modes of visual integration in the lateral geniculate nucleus revealed by single-cell-initiated transsynaptic tracing. *Neuron.* 93:767–776.

Ruderman DL, Bialek W (1994). Statistics of natural images: Scaling in the woods. *Phys Rev Lett.* 73:814–817.

Saleem AB, Lien AD, Krumin M, Haider B, Rosón MR, Ayaz A, Reinhold K, Busse L, Carandini M, Harris KD (2017). Subcortical source and modulation of the narrowband gamma oscillation in mouse visual cortex. *Neuron.* 93:315–322.

Shannon CE, Weaver W (1949). *The Mathematical Theory of Communication.* University of Illinois Press.

Sincich LC, Adams DL, Economides JR, Horton JC (2007). Transmission of spike trains at the retinogeniculate synapse. *J Neurosci.* 27:2683–2692.

Sincich LC, Horton JC, Sharpee TO (2009). Preserving information in neural transmission. *J Neurosci.* 29:6207–6216.

Sommer FT (2014). Neural oscillations and synchrony as mechanisms for coding, communication, and computation in the visual system. In: Werner JS, ed. *The New Visual Neurosciences.* MIT Press:1279–1288.

Storchi R, Bedford RA, Martial FP, Allen AE, Wynne J, Montemurro MA, Petersen RS, Lucas RJ (2017). Modulation of fast narrowband oscillations in the mouse retina and dlgn according to background light intensity. *Neuron.* 93:299–307.

Stoyan D, Kendall WS, Mecke J (1995). *Stochastic Geometry and Its Applications.* 2nd ed. Wiley.

Usrey WM, Reid RC (1999). Synchronous activity in the visual system. *Annu Rev Physiol.* 61:435–456.

Usrey WM, Reppas JB, Reid RC (1998). Paired-spike interactions and synaptic efficacy of retinal inputs to the thalamus. *Nature.* 395:384–387.

Victor JD (2005). Spike train metrics. *Curr Opin Neurobiol.* 15:585–592.

Wang X, Hirsch JA, Sommer FT (2010). Recoding of sensory information across the retinothalamic synapse. *J Neurosci.* 30:13567–13577.

Wang X, Sommer FT, Hirsch JA (2011a). Inhibitory circuits for visual processing in thalamus *Curr Opin Neurobiol.* 21:726–733.

Wang X, Vaingankar V, Sanchez CS, Sommer FT, Hirsch JA (2011b). Thalamic interneurons and relay cells use complementary synaptic mechanisms for visual processing. *Nat Neurosci.* 14:224–231.

Wang X, Wei Y, Vaingankar V, Wang Q, Koepsell K, Sommer FT, Hirsch JA (2007). Feedforward excitation and inhibition evoke dual modes of firing in the cat's visual thalamus during naturalistic viewing. *Neuron.* 55:465–478.

Warner C, Sommer F (2020). *A Model for Image Segmentation in Retina.*

64.

PREDICTIVE CODING IN CORTICAL AND CORTICOTHALAMIC HIERARCHICAL NETWORKS

Rajesh P. N. Rao, Linxing Preston Jiang, and Ares Fisher

INTRODUCTION

How do neural circuits in the brain learn internal models of the world, and how do such circuits combine prior beliefs with sensory evidence for Bayesian inference? The theory of predictive coding suggests possible answers for these questions. For example, the classical predictive coding model of Rao and Ballard (1999) assumes that the areas comprising the neocortical hierarchy (Hubel and Wiesel, 1959; Felleman and Van Essen, 1991) implement a hierarchical generative model of the world. The neural activities at each level of the hierarchy are assumed to represent the brain's internal belief of the hidden causes of the stimuli at a particular abstraction level (e.g., edges, object parts, objects). The model postulates that the "top-down" feedback connections from higher to lower order cortical areas convey predictions of lower-level activities while the "bottom-up" feedforward connections convey prediction errors, both of which jointly influence neural activities at each level (see **Chapter 41 in Section VII**). The goal of the cortex then is to minimize prediction errors across all levels. Within a Bayesian framework, the top-down predictions convey prior beliefs based on learned expectations, while the bottom-up prediction errors carry evidence from the current input. Predictive coding combines these 2 sources of information, weighted according to their reliability (inverse variances or "precisions"), to perform Bayesian inference by computing the posterior beliefs over hidden causes at each level. Minimizing prediction errors across all levels is therefore equivalent to finding the *maximum* a posteriori (MAP) estimates of the hidden causes. We start by reviewing this classical predictive coding model, extend the model to robust predictive coding and attention through corticothalamic filtering, and conclude by reviewing more recent models such as dynamic and active predictive coding, and experimental evidence for predictive coding.

THEORY OF PREDICTIVE CODING

Predictive coding was first used as a model of efficient coding to explain the center-surround receptive fields and biphasic temporal antagonism in cells in the retina and lateral geniculate nucleus (LGN): rather than conveying the local intensity directly, retinal and LGN cells can be interpreted as sending the differences (errors) between the local intensity and a prediction of that intensity, computed as a linear weighted sum of nearby values in space and preceding input values in time (Srinivasan et al., 1982; Dong and Atick, 1995a, 1995b; Huang and Rao, 2011). These models were derived from the information-theoretic principles of efficient coding (Attneave, 1954; Barlow, 1961; Simoncelli and Olshausen, 2001), rather than hierarchical generative models like the Rao-Ballard model. Under the efficient coding hypothesis, the goal of the visual system is to efficiently represent visual information by reducing redundancy arising from natural scene statistics (Field, 1987; Ruderman and Bialek, 1994; Dong and Atick, 1995b). A simple example of redundancy reduction is to remove aspects of an input that are predictable from nearby inputs. Neural activities then only need to represent information that deviates from the prediction.

HIERARCHICAL PREDICTIVE CODING

A Bayesian form of predictive coding was proposed for the visual cortex by Rao and Ballard (1999). The model assumes that the visual cortex learns a generative model of the visual world: sensory inputs are assumed to be generated by hidden states or "causes" in the external world via this generative mode, which the cortex attempts to learn over many inputs. Perception is equated with Bayesian inference (i.e., estimating the hidden state $\mathbf{r}$ that caused the input $\mathbf{I}$ given a learned generative model). Neural activities are assumed to represent estimates of **r**. The prior distribution of hidden states is assumed to be $p(\mathbf{r})$, which imposes a constraint on neural activities such as sparse activation, while the observation model $p(\mathbf{I}|\mathbf{r})$ is defined by the following equation:

$$\mathbf{I} = f(\mathbf{Ur}) + \mathbf{n} \tag{1}$$

where f is a linear or nonlinear function (e.g., rectification or sigmoidal function), **U** is a matrix assumed to be encoded in the "top-down" synaptic weights of the predictive coding network, and **n** is assumed to be zero mean Gaussian noise with covariance $\sigma^2\mathbb{1}$ ($\mathbb{1}$ is the identity matrix). The columns of **U** can be regarded as the "basis" vectors (e.g., edges or "parts" of an image or scene) that can be used to compose an input as a weighted sum of basis vectors with the weights given by the values in the vector **r**.

The generative model can be extended to multiple hierarchical levels by assuming that the hidden state **r** can be generated by a higher-level representation $\mathbf{r}^h$, corresponding to more abstract image properties than the lower-level representation:

$$\mathbf{r} = \mathbf{r}^{td} + \mathbf{n}^{td} = f(\mathbf{U}^h\mathbf{r}^h) + \mathbf{n}^{td} \tag{2}$$

where $\mathbf{r}^{td} = f(\mathbf{U}^h\mathbf{r}^h)$ is the top-down prediction of **r** and $\mathbf{n}^{td}$ is zero mean Gaussian noise with covariance $\sigma_{td}^2\mathbb{1}$. The lower-level neurons have smaller receptive fields and represent a local image region by estimating the hidden state **r**. The higher-level neurons estimate their state $\mathbf{r}^h$ based on *several* lower-level hidden states **r** associated with local image patches. This arrangement results in a progressive convergence of inputs from lower to higher levels and an increase in receptive field size as one ascends the hierarchical network until the receptive fields of the highest-level neurons span the entire input image.

Inference ("perception") corresponds to estimating the hidden state vectors **r** and $\mathbf{r}^h$, while *learning* corresponds to learning an estimate for **U** ("synaptic learning"), both with the goal of maximizing the joint probability $p(\mathbf{I},\mathbf{r})$. Since $p(\mathbf{I})$ is constant, this is equivalent to maximizing the posterior probability $p(\mathbf{r}|\mathbf{I})$, also known as MAP inference.

Given Equations (1) and (2), and the fact that **n** and $\mathbf{n}^{td}$ are zero mean Gaussians, the negative logarithm of the posterior probability $p(\mathbf{r}|\mathbf{I})$ can be shown to be proportional to:

$$H = \frac{1}{\sigma^2}(\mathbf{I} - f(\mathbf{Ur}))^{\mathrm{T}}(\mathbf{I} - f(\mathbf{Ur})) + \frac{1}{\sigma_{td}^2}(\mathbf{r} - \mathbf{r}^{td})^{\mathrm{T}}(\mathbf{r} - \mathbf{r}^{td}) + g(\mathbf{r}) + g(\mathbf{r}^h) + h(\mathbf{U}) + h(\mathbf{U}^h)$$

where the first term is the sum of squared errors between the image **I** and its reconstruction (or "prediction") $f(\mathbf{Ur})$ across all pixels, weighted by the inverse noise variance (or precision) $\frac{1}{\sigma^2}$; the second term is the sum of squared top-down predictions errors; the g and h terms are proportional to the negative logarithm of the priors (zero mean Gaussian or sparse priors) for the corresponding variables.

As mentioned above, the goal of both inference and learning is to maximize the posterior probability $p(\mathbf{r}|\mathbf{I})$, which can be seen to be equivalent to minimizing the negative logarithm of the posterior (i.e., minimizing H). In predictive coding, this is done by performing gradient descent on H. Intuitively, minimizing H involves minimizing prediction errors for predictions at all hierarchical levels (note the first 2 terms of H involving prediction error for **I** and prediction error for **r**), subject to constraints given by g and h on the neural responses and synaptic weights, respectively (constraints may include sparsity of responses, non-negativity, or keeping weight magnitudes small).

Network Dynamics and Synaptic Learning

Perceptual inference involves estimating hidden state **r** using gradient descent on H:

$$\check{\mathbf{r}}_t = \check{\mathbf{r}}_{t-1} + \frac{k_1}{\sigma^2}\mathbf{U}^{\mathrm{T}}\frac{\partial f^{\mathrm{T}}}{\partial \mathbf{x}_{t-1}}(\mathbf{I} - f(\mathbf{U}\check{\mathbf{r}}_{t-1})) + \frac{k_1}{\sigma_{td}^2}(\mathbf{r}_{t-1}^{td} - \check{\mathbf{r}}_{t-1}) - \frac{k_1}{2}g'(\check{\mathbf{r}}_{t-1}) \tag{3}$$

where t is the current time step, k_1 is a positive constant governing the rate of descent toward a minimum for H, $\mathbf{x} = \mathbf{Ur}$, and g' is the derivative of g with respect to **r**. This equation, derived from first principles, specifies recurrent network dynamics for hierarchical predictive coding in terms of how the firing rate (or neural response) vector **r** at a given level should be updated over time. At each time step, **r** is multiplied by the feedback matrix **U** and a new prediction is generated for the lower level (Figure 64.1A,B). This prediction is then subtracted from input **I** to generate the bottom-up error $(\mathbf{I} - f(\mathbf{Ur}))$ which is filtered by the feedforward weights $\mathbf{U}^{\mathrm{T}}$ and the gradient of the function f. The neural response **r** is updated based on a weighted combination of the bottom-up prediction error $(\mathbf{I} - f(\mathbf{Ur}))$ and the top-down prediction error $(\mathbf{r}^{td} - \mathbf{r})$ (Figure 64.1B). Each error is weighted by the inverse of the corresponding noise variance: the larger the noise variance, the smaller the weight given to that error term.

The learning rule for the feedback synaptic weights **U** (and feedforward weights $\mathbf{U}^{\mathrm{T}}$) is obtained by using gradient

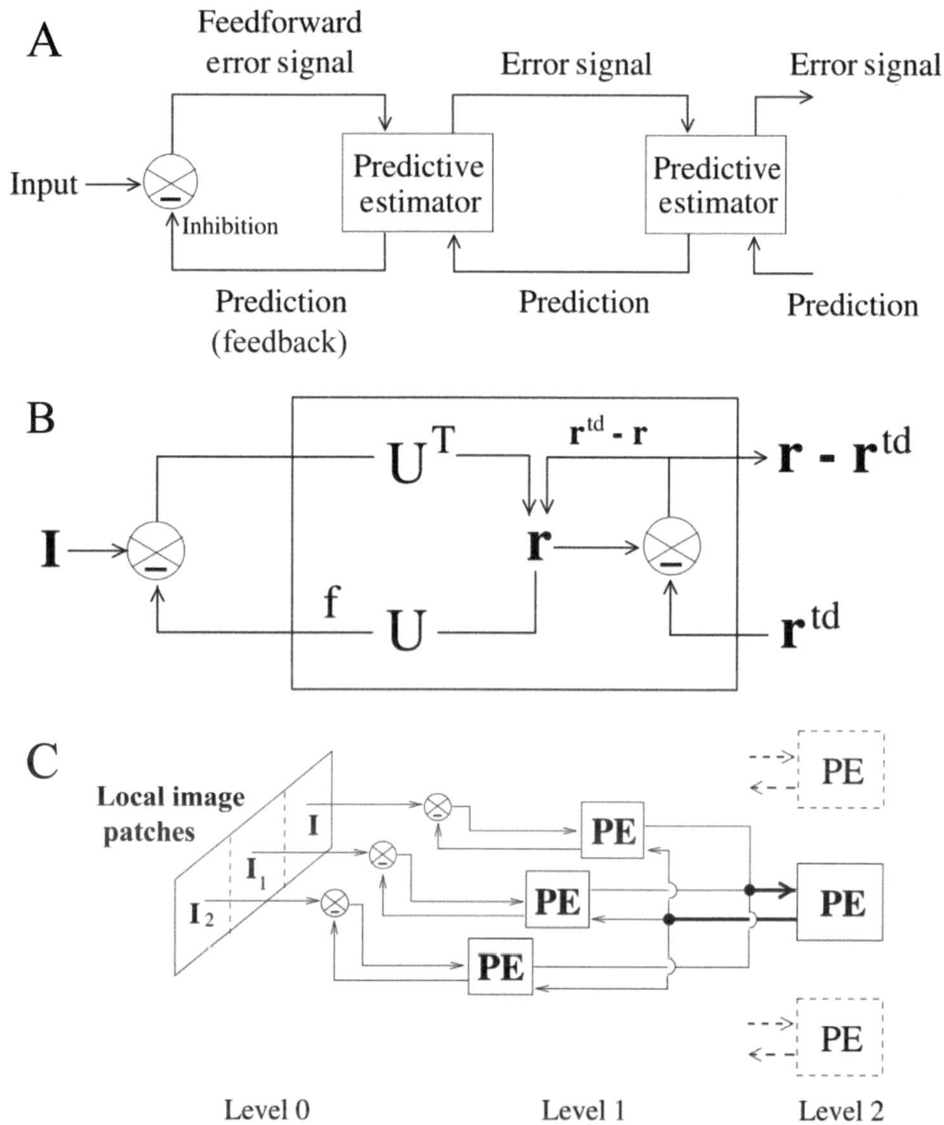

Figure 64.1. Hierarchical predictive coding model. (A). The general architecture of the hierarchical predictive coding model. Each predictive estimator (PE) module maintains an internal representation, generates top-down prediction (lower arrows, feedback), and receives bottom-up prediction errors (upper arrows, feedforward). (B). Components of a PE module. Feedforward connections carry bottom-up prediction errors from the lower level. Feedback connections deliver top-down predictions to the lower level. The internal representation neurons correct their current estimate r using both the bottom-up prediction error and the top-down prediction error. A separate class of error-detecting neurons compute the discrepancy between the current estimate and its top-down prediction, and send the error to the higher level. (C). An example of a 2-level hierarchical network. Three image patches at Level 0 are processed separately by 3 Level 1 PE mofidules. These 3 Level 1 modules converge to provide input (prediction errors) to a single Level 2 module, which attempts to predict the states r in all 3 of these Level 1 PE modules. This convergence effectively increases the receptive field size of neurons as one ascends the hierarchy.

descent on H with respect to $\mathbf{U}$ (written here as a differential equation, but implemented in discrete form, as in Equation 3):

$$\frac{d\mathbf{U}}{dt} = -\frac{k_2}{2}\frac{\partial H}{\partial \mathbf{U}} = \frac{k_2}{\sigma^2}\frac{\partial f^{\mathrm{T}}}{\partial \mathbf{x}}(\mathbf{I} - f(\mathbf{Ur}))\mathbf{r}^{\mathrm{T}} - k_2 \lambda \mathbf{U} \quad (4)$$

where k_2 is a positive parameter determining the learning rate of the network and $\mathbf{x} = \mathbf{Ur}$. Note that this learning rule is local and is a form of Hebbian plasticity: for the feedforward weights $\mathbf{U}^{\mathrm{T}}$, the input presynaptic activity is the residual error $(\mathbf{I} - f(\mathbf{Ur}))$ (weighted by $\frac{df^{\mathrm{T}}}{d\mathbf{x}}$) and the output postsynaptic activity is $\mathbf{r}$. To ensure stability, learning of synaptic weights operates on a slower timescale than the dynamics of $\mathbf{r}$: the learning rate k_2 is a much smaller value than the rate k_1. An example of a 2-level hierarchical network is depicted in Figure 64.1C.

ROBUST PREDICTIVE CODING AND ATTENTION

The predictive coding model can be extended to model top-down attention using a *robust optimization function* as first proposed in Rao (1998). Specifically, instead of using the squared error loss function:

$$H_1 = \frac{1}{\sigma^2}\|\mathbf{I} - f(\mathbf{Ur})\|_2^2 = \frac{1}{\sigma^2}(\mathbf{I} - f(\mathbf{Ur}))^{\mathrm{T}}(\mathbf{I} - f(\mathbf{Ur}))$$

the robust predictive coding model uses a robust optimization function (Huber, 1981):

$$H_1^R = \rho(\mathbf{I} - f(\mathbf{Ur}))$$

where ρ is a function that reduces the influence of outliers (large prediction errors) in the estimation of **r**.

As an example, ρ could be defined in terms of a diagonal matrix **S** as follows (Rao, 1998):

$$H^R = (\mathbf{I} - f(\mathbf{Ur}))^{\mathrm{T}}\mathbf{S}(\mathbf{I} - f(\mathbf{Ur}))$$

which is now a *weighted* sum of prediction errors across all input locations, with the diagonal entries $S_{i,i}$ determining the weight accorded to the prediction error $(I_i - f(\mathbf{u}_i r))^2$ at location i ($\mathbf{u}_i$ here is a row vector denoting the ith row of **U**).

The matrix **S** is often called the precision matrix and is traditionally defined as the inverse of the covariance matrix of the prediction error. Robust predictive coding generalizes this concept to allow robust estimation by decreasing the influence of outliers in the inference process, thereby allowing us to model visual attention as inference of an object or attribute in the presence of distractors and occluders (see examples below). Specifically, we use a simple nonlinear function given by:

$$S_{i,i} = \min\left\{1, \frac{c}{(I_i - f(\mathbf{u}_i\mathbf{r}))^2}\right\}$$

where c is a threshold parameter. This function has the following effect useful for ignoring outliers when estimating the state **r**: **S** clips the squared prediction error for the ith input location to a constant value c whenever $(I_i - f(\mathbf{u}_i r))^2$ exceeds the threshold c. Minimizing H^R leads to the following equation for robust predictive coding:

$$\check{\mathbf{r}}_t = \check{\mathbf{r}}_{t-1} + k_1\mathbf{U}^{\mathrm{T}}\mathbf{G}_t\frac{df^{\mathrm{T}}}{d\mathbf{x}_{t-1}}(\mathbf{I} - f(\mathbf{U}\check{\mathbf{r}}_{t-1})) \quad (5)$$

where $\boldsymbol{G}_t$ is a diagonal matrix whose diagonal entries at time instant t are given by: $\boldsymbol{G}_{i,i} = 0$ if $(I_i - f(\mathbf{u}_i\check{\mathbf{r}}_{t-1}))^2 > c_t$ and 1 otherwise, and c_t is a (potentially time-varying) threshold on the squared prediction error. Although we use a fixed threshold in our examples below, the threshold could potentially be changed by another network monitoring the current task or other context (e.g., in the prefrontal cortex).

The gain $\boldsymbol{G}_t$ acts as an "attentional filter" for outlier detection and filtering (potentially implemented via corticothalamic interaction through the thalamic reticular nucleus—see the section "Neural Implementation of Predictive Coding Models" below), allowing the predictive coding network estimating **r** (Figure 64.2, left) to suppress large prediction errors in parts of the input containing outliers. This enables the network to focus on verifying the feasibility of its current best hypothesis by trying to minimize prediction errors while ignoring outliers. Robust predictive coding thus allows the network to "focus its attention" on one object while ignoring occluders and background objects, and even "switch attention" from one object to another (Figure 64.2, right panels) (see Rao, 1998, 1999).

DYNAMIC PREDICTIVE CODING: LEARNING TEMPORAL HIERARCHIES

The hierarchical predictive coding model of Rao and Ballard described above learns a spatial hierarchy of representations but does not address how temporal hierarchies can be learned. In a temporal hierarchy, representations at higher levels encode stimuli and events that span longer timescales. There is growing evidence that the cortex exhibits such a temporal hierarchy (JD Murray et al., 2014; Runyan et al., 2017; Siegle et al., 2021).

Hierarchical temporal representations can be learned using dynamic predictive coding (DPC), a recently proposed model for predictive coding of sequences (Jiang et al., 2021; Jiang and Rao, 2022). The DPC model proposes that higher-level model neurons $\mathbf{r}^h$ modulate the *transition dynamics* of lower-level networks representing **r**. The top-down modulation is formulated as a set of mixture weights $\boldsymbol{w}$ for a group of learned transition matrices $\boldsymbol{V}^{(1)}, \ldots, \boldsymbol{V}^{(K)}$ (Figure 64.3A). These weights, which can be regarded as gain modulation on lower-level synapses, are predicted by the higher level. The DPC model postulates that the different timescales of neural responses observed in the cortex could be an emergent property of the cortex learning a hierarchical spatiotemporal generative model.

Figure 64.3B illustrates perceptual inference in the DPC model. The network generates top-down and lateral predictions (green) using the current higher-level

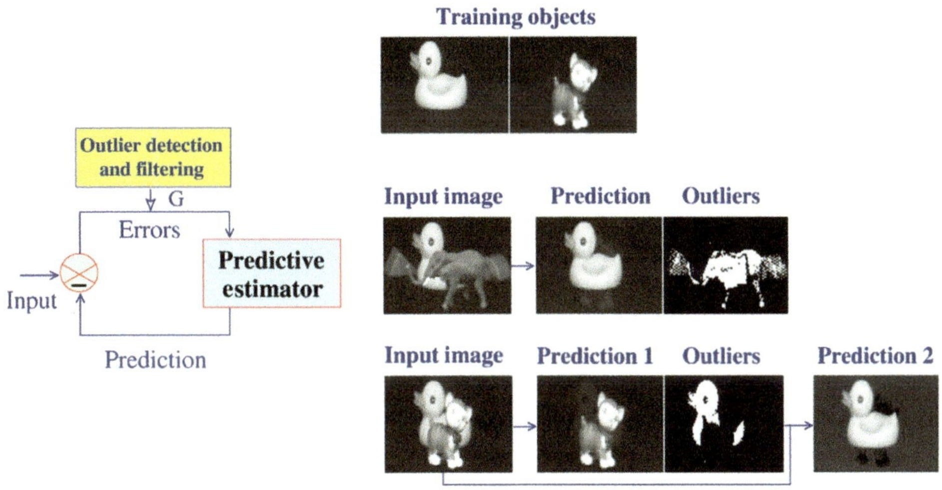

Figure 64.2. Robust predictive coding and attention. Left: The robust predictive coding model utilizes a gain or gating term **G** to modulate the prediction errors before they are fed to the predictive estimator network that estimates the state vector **r**. This allows the network to filter out any outliers dynamically as a function of the current top-down hypothesis, allowing the network to "focus attention" and test its hypothesis. Right: The top row shows images of two objects the network was trained on. The middle row illustrates how the network can filter out occluders and background objects as outliers, recovering an estimate of a training object (the duck). The bottom row shows how, when presented with an image containing both training objects, the network can sequentially focus and recognize each object.

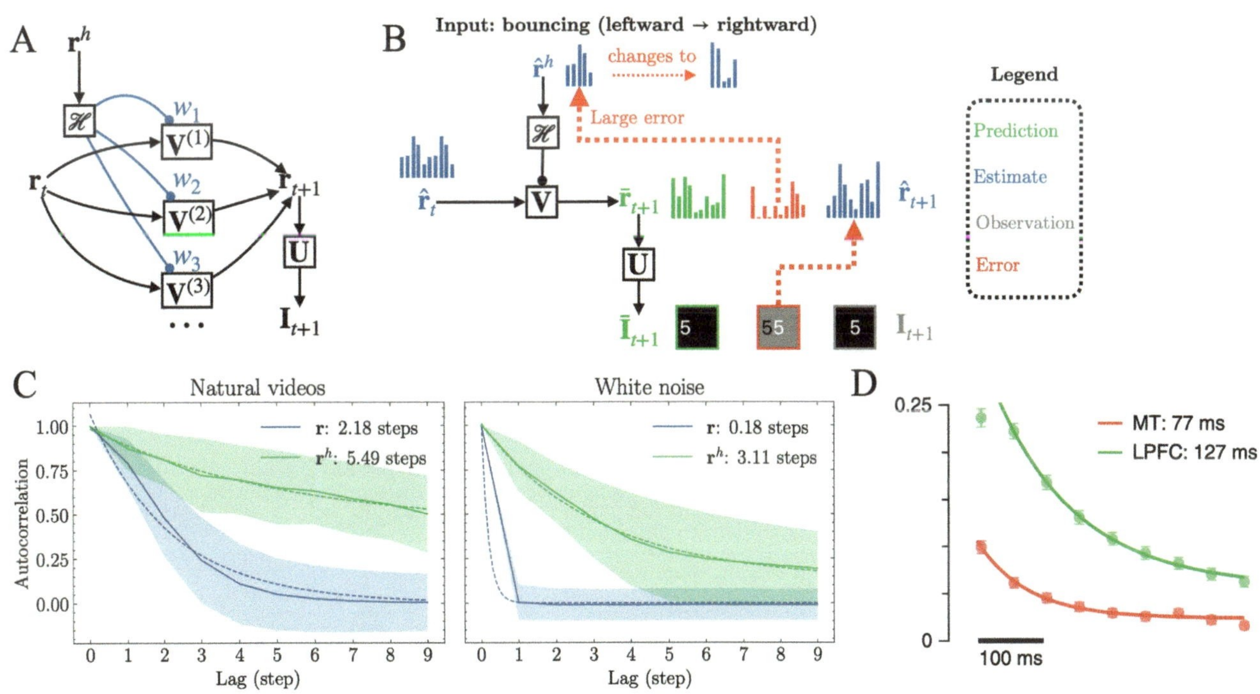

Figure 64.3. Dynamic predictive coding. (A). Parameterization of the DPC generative model. The higher-level state modulates the lower-level transition matrices through a top-down network $\mathscr{H}$. (B). Depiction of an inference step when the lower-level dynamics changes. The resulting large prediction errors drive updates to the higher-level state to account for the new lower-level dynamics. (C). Autocorrelation of the lower- and higher-level responses in the trained network. Shaded area denotes ±1 standard deviation. Dotted lines show fitted exponential decay functions. Left: responses recorded during natural video stimuli; right: white noise stimuli. (D) Autocorrelation of the neural responses recorded from MT and LPFC of monkeys. Panels A–C adapted from Jiang and Rao (2022); D adapted from Murray et al. (2014).

state estimates (blue). If the input sequence is predicted well by the top-down-modulated transition matrix V, the higher-level response $\mathbf{r}^h$ remains stable due to small prediction errors. When a non-smooth transition occurs in the input sequence (Figure 64.3B), the resulting large prediction errors are sent to the higher level via feedforward connections (red arrows), driving changes in $\mathbf{r}^h$ to predict new dynamics for the lower level.

Autocorrelation analysis on the lower- and higher-level responses of a trained 2-level DPC network revealed different levels of response timescales (Figure 64.3C). When tested with natural videos, the autocorrelation of the higher-level responses $\mathbf{r}^b$ is greater than that of the lower-level response $\mathbf{r}$ and has a slower decay rate (exponential time constant for $\mathbf{r}^b$: 5.49 steps; for $\mathbf{r}$: 2.18 steps). This distinction in response timescales still exists when tested with Gaussian noise (exponential time constant for $\mathbf{r}^b$: 3.11 steps; for $\mathbf{r}$: 0.18 steps). Such a difference in response timescales is qualitatively similar to that observed in the cortex of nonhuman primates (Figure 64.3D). Note that no explicit constraints such as larger time constants or smoothness were enforced on $\mathbf{r}^b$ during inference—the longer timescale and more stable responses are entirely a result of the higher-level neurons learning to predict the lower-level dynamics under the proposed generative model.

ACTIVE PREDICTIVE CODING: A UNIFYING SENSORY-MOTOR THEORY

Active predictive coding (APC) (Rao, 2022; Rao et al., 2022) is a recently introduced extension of predictive coding that acknowledges the central role of motor commands, goals, and actions (e.g., eye, head, body movements) in perception, cognition, and behavior. The model is motivated by the fact that each cortical area, even primary sensory areas such as V1, not only receives sensory or internal state information from the thalamus (most prominently in layer [L]4) but also sends outputs from L5 to sub-cortical motor centers (with axon collaterals to higher-order thalamus) (Sherman and Guillery, 2013) (Figure 64.4A; see **Chapter 6 in Section III**). The APC model assumes that each cortical area computes a "state-prediction" function f_s in the superficial layers that predicts the next sensory (or internal) state given a current state and action (Figure 64.4B, upper part of the figure). The deeper layers are hypothesized to

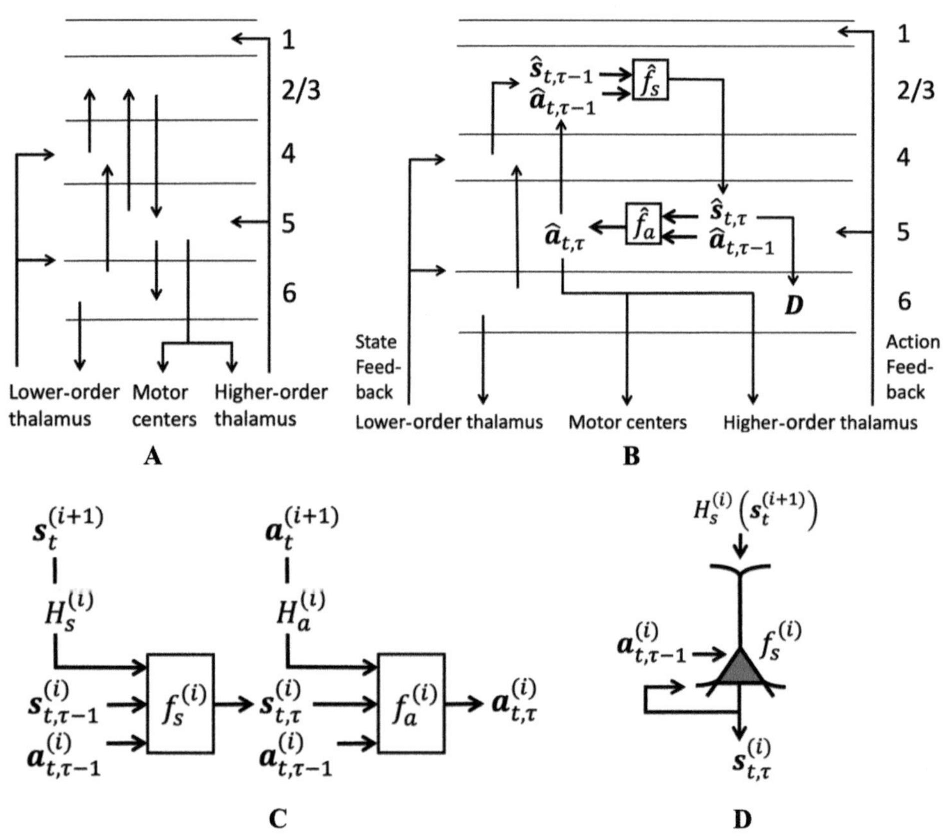

Figure 64.4. Active predictive coding. (A). Six-layered laminar structure of a cortical column, showing some of the major connections between layers and with the thalamus (not all connections are shown) (based on Douglas and Martin, 2004; Sherman and Guillery, 2013; EJ Kim et al., 2015). (B). Computation in a canonical APC cortical module with a state network (upper cortical layers) that learns the dynamics of the world at a particular level of abstraction, and an action network (deeper cortical layers) that specifies that level's actions/goals/coordinates within a reference frame (e.g., where a part of an object is located). (C). Two levels (levels i and i+1) of a hierarchical APC generative model. The higher-level state and action neurons maintain estimates of state and action vectors (s_t^{i+1} and a_t^{i+1}) and modulate the lower-level state and action networks via top-down inputs $H_s^i(s_t^{i+1})$ and $H_a^i(a_t^{i+1})$, respectively, allowing the higher level to compose a sequence of states and actions at the lower level to accomplish a goal or sub-goal. (D). As depicted for a single pyramidal neuron here, we hypothesize that top-down inputs $H_s^i(s_t^{i+1})$ (or $H_a^i(a_t^{i+1})$ from a higher cortical area to the apical dendrites of a pyramidal neuron in a lower area modulate the dynamics of a network of such neurons (cf. gain modulation; Larkum et al., 2004; Ferguson and Cardin, 2020), allowing the higher area to modulate the function f_s (or f_a) being computed by these neurons. Adapted from Rao (2022).

compute an action or "policy" function f_a that maps the current state and previous action to the current action (Figure 64.4B, lower part of the figure). The recurrent neuronal networks implementing f_s and f_a feed their outputs to each other, as shown in Figure 64.4B, leading naturally to a sequence of predictions of states and actions. Such a sequence can track environmental states during behavior or alternately, can be used for internal simulation and planning. During behavior, state and action feedback from the sensory periphery and sub-cortical motor centers are conveyed back to these networks via the thalamus (see **chapters in Sections V and VI**).

Hierarchical active predictive coding involves connecting a higher-level APC sensory-motor module to a lower-level module, as shown in Figure 64.4C. The higher-level state controls the function being computed by the lower-level state network by modulating it using top-down inputs: the top-down input is maintained as an additional contextual input to the lower-level state network, modulating its recurrent dynamics to change the input-output function. We hypothesize that such modulation can be achieved via the top-down inputs from higher cortical areas that are known to target the apical dendrites of pyramidal neurons in lower cortical areas. (Additional details can be found in Rao, 2022; Rao et al., 2022.)

The hierarchical APC architecture in Figure 64.4C is motivated by the theory, proposed by Mountcastle and others, that a common computational principle may underlie cortical processing across areas (Mountcastle, 1978; Sherman and Guillery, 2013; Hawkins, 2021). In keeping with this hypothesis, we provide 2 examples to illustrate the APC architecture's flexibility and versatility. Specifically, the model lends itself naturally to solving a variety of complex problems using a compositional divide-and-conquer approach, demonstrated here using the examples of active vision and navigation.

Active Vision

In this case, the sensory inputs are image patches ("glimpses") sampled by actions (eye movements) from a larger image containing an object. We trained a 2-level APC network to reconstruct images of objects based on a sequence of glimpses. As illustrated in Figure 64.5A, the lower-level network estimated visual states represented parts of a larger object using a sequence of lower-level actions (simulated eye movements within a reference frame defined by the higher level) while the higher level integrated the information gathered by a lower-level action at each "micro-step." The information integrated over a sequence of such "macro-steps" by the (recurrent) higher-level state network allows the network to perceive the object in an input image. By parsing images in this hierarchical manner, the 2-level APC network learned to represent objects in terms of part-whole hierarchies, addressing a major open problem in vision (see Rao et al., 2022, for details).

Navigation and Planning

The same APC architecture used above for the active vision example can also be used for solving a very different task: navigation problems using a compositional divide-and-conquer strategy: the APC network in this case learns a hierarchical model that captures the state-prediction function (the dynamics or "physics") of a potentially complex environment as a sequence of simpler state-prediction functions. Consider the problem of navigating from any starting location to any goal location in a large "multi-room" building environment such as the one in Figure 64.5B (top left panel; gray: walls, blue circle: current location, green square: current goal location). In this example, the APC model's states at the lowest level are discrete locations in the grid, and its actions at that level are going "one step north" (N), "one step east" (E), "one step south" (S), and "one step west" (W).

Just as an object consists of the same parts (e.g., strokes, curves) occurring at different locations, the multi-room environment is also made up of the same 2 components ("room types" S1 and S2), shown in Figure 64.5B, occurring at several different locations (some example locations are highlighted by yellow and red boxes). These components form part of the higher-level states in the APC defined by state vectors S1 and S2, which can be trained to generate, via the top-down network H_s (Figure 64.4C), the lower-level state-prediction functions f_s for rooms S1 and S2, respectively. Similar to how the APC model can be trained to reconstruct an image using higher-level action embedding vectors to generate policies and actions (locations) to compose parts from sub-parts, the APC model can compute higher-level action vectors a^{i+1} for the multi-room environment that generate, via top-down network H_a (Figure 64.4C), lower-level policies (or "options") f_a which produce primitive actions (N, E, S, W) to reach the goal encoded by a^{i+1}. At each time point, the model samples sequences of higher-level actions and the resulting states using its higher-level state network, selects the sequence with the largest cumulative reward (Figure 64.5B, lower right panel), executes the first higher-level action from that sequence, and repeats until the goal is reached. Figure 64.5B (right panels) illustrates how this hierarchical planning process allows the APC model to efficiently reach a goal location (green) from a start location (blue) (for additional details, see Rao et al., 2022).

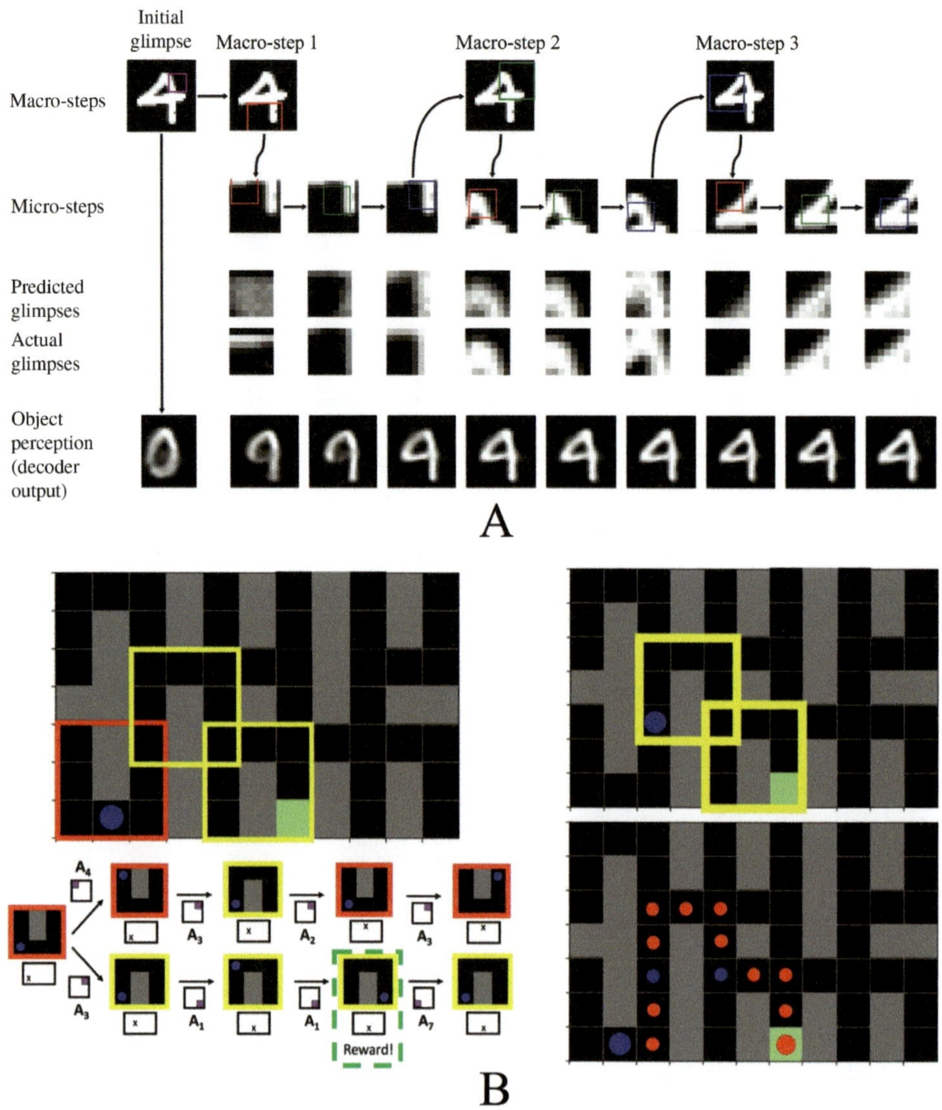

Figure 64.5. Examples of active predictive coding. (A). Hierarchical parsing of an image by the APC model. 1st row: initial glimpse (purple box) and higher-level parts selected by the model for this image (red, green, blue boxes); 2nd row: lower-level image regions visited by the model within each top-level frame; 3rd and 4th rows: predicted versus actual parts/"glimpses" of the image; and 5th row: "perception" of the model (object reconstructed by a decoder network from current network state) over time. Note the "perceptual" stability in the model despite jumps in the sampled glimpses (see row of "Actual glimpses") as the object hypothesis is gradually refined based on accumulated evidence. (B). Solving a navigation task using a state action hierarchy. The problem of navigating in a large building can be reduced to planning using high-level states (red and yellow outlined "rooms") and higher-level actions (navigating to a given corner of a room). Blue = current location; gray = walls; green = current goal location. See text for details. Adapted from Rao (2022).

NEURAL IMPLEMENTATION OF PREDICTIVE CODING MODELS

MAPPING THE COMPONENTS OF THE RAO-BALLARD PREDICTIVE CODING MODEL TO CORTICAL LAYERS

Rao and Ballard (1999) postulated 2 groups of neurons at each hierarchical level (Figure 64.6): (1) a group of internal representation neurons that maintain the current state estimate and generate top-down predictions of lower-level activities (hypothesized to be in the deep layers 5/6 (L5/6) of the cortex and expected to show sustained activity to maintain the predictions to lower levels); and (2) a different group of neurons (hypothesized to be L2/3, exhibiting transient activity and sending connections to "higher-order" cortical areas) that compute and convey prediction errors to the next higher level (see **Chapter 6 in Section III** and **chapters in Section VII**). Rao and Ballard proposed 2 sub-classes of error-detecting neurons, one subclass representing positive errors and another representing negative errors, similar to ON-center OFF-surround and

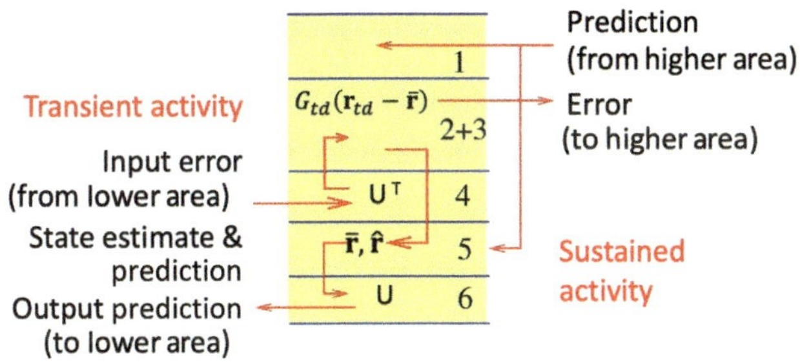

Figure 64.6. Suggested laminar implementation of the Rao-Ballard predictive coding model. A mapping of the hierarchical predictive coding model components (see Figure 64.1) to cortical layers.

OFF-center ON-surround neurons in the retina and LGN. Some evidence for such a representation has been found in the visual cortex (Jordan and Keller, 2020).

MODEL PREDICTIONS

The Rao-Ballard model predicts that L2/3 neurons are suppressed when the stimuli are predictable (i.e., consistent with natural image statistics), while deeper-layer neurons remain active. Stimuli that deviate from natural image statistics ("novel" stimuli), on the other hand, are predicted to elicit large responses in L2/3 neurons. This was illustrated in the original simulations of Rao and Ballard using the example of end-stopping, a property of complex cells in L2/3 of V1: cells that respond robustly to a small bar in their classical receptive field stop responding when the bar extends to the surrounding region. Similar responses are also elicited when an oriented grating in the classical receptive field is surrounded by a grating with the opposite orientation. Such phenomena are consistent with the predictive coding model, which learns that the statistics of adjacent natural image patches are locally smooth. Thus, the prediction error (L2/3 response) is small for stimuli in which the center orientation matches the surround, and large for orientation mismatches. The model also predicts that prediction error signals are used for unsupervised learning of the synaptic connections in the predictive coding network, driving the synaptic weights to better reflect the structure of the environment.

EVIDENCE FOR PREDICTION ERROR NEURONS AND INTERNAL REPRESENTATION NEURONS IN THE CORTEX

Keller et al. (2012) recorded neural activities from L2/3 cells in the monocular visual cortex of behaving mice that were head-fixed and running on a spherical treadmill. In normal "feedback" trials, full-field vertical gratings were coupled to the mouse's locomotion on the treadmill, producing visual flow. In "mismatch" trials, visual-locomotion mismatches were delivered randomly as brief visual flow halts (1 second) during a "feedback" session. The authors found that ~13.0% of the visual cortical neurons recorded responded predominately to feedback mismatches. The mismatch signals also encoded the degree of mismatch—a visual flow halt during faster locomotion resulted in a stronger response than during slow locomotion. V1 neurons have also been found to be predictive of spatial locations after adapting to a new environment. In an experiment by Fiser et al. (2016), mice went through a virtual tunnel with blocks of 2 different grating patterns (A or B) separated by distinct landmarks. The 5 trial conditions differed in the fifth block, where the patterns A and B occurred with different probabilities, or were omitted with no visual stimulus presented. After adaptation, some neurons developed predictive responses to specific visual stimuli based on spatial location. The authors also discovered prediction error responses similar to those reported by Keller et al. (2012).

Prediction and prediction error-like signals have also been found in cortical areas in the human visual cortex (e.g., Murray et al., 2002) and the hierarchical face-processing region of the monkey inferior temporal cortex (IT) (Tsao et al., 2006; Freiwald and Tsao, 2010). Schwiedrzik and Freiwald (2017) exposed macaque monkeys to fixed pairs of face images with different head orientations and identities such that the successor face image can be predicted from the preceding face image. Neurons in the lower-level face area ML (middle lateral section of the superior temporal sulcus) displayed large responses when the pair association was violated (either in identity, or head orientation, or both). Issa and colleagues (2018) used different face-part configuration stimuli (typical versus atypical) and found that the lower level areas of the hierarchy (posterior, central IT: pIT and cIT) signal deviations of their preferred features from the expected configurations, whereas the top level (anterior IT: aIT) maintained a preference for natural, frontal face-part configuration. The authors further discovered that the early responses in cIT and aIT are correlated with late responses in pIT: images that produced large responses in higher-level areas early are followed by reduced activities in lower-level areas, consistent with top-down predictions signal subduing lower-level responses.

Predictive coding-related effects have also been found in the primary auditory cortex (Schneider et al., 2018). Rats were given artificial auditory feedback coupled to their locomotion: the pitch of the sound was proportional to the rat's running speed. They found that a group of neurons in the rat's primary auditory cortex initially responded strongly to the artificial auditory feedback ("reafferent sound"), but over the course of several days, the neuronal circuits learned to suppress this activity. The gradual suppression of responses is consistent with how the predictive coding model learns an internal model of the environment based on prediction errors: as the network learns to predict the artificial sound coupled to the rat's locomotion, the predictions get better, resulting in decreasing prediction errors which manifest as suppression of the auditory neurons' activities.

The Rao-Ballard model additionally postulates that neurons in the deeper layers 5/6 maintain the state estimate. Recent experiments have attempted to test these predictions. While it is hard to distinguish the state estimating "internal representation" neurons from those driven by bottom-up sensory stimuli (see review by Keller and Mrsic-Flogel, 2018, for further discussions), there is a growing body of evidence suggesting that L2/3 neurons may indeed play a role in comparing bottom-up information and top-down predictions (Jordan and Keller, 2020).

PREDICTIVE CODING AND PROPERTIES OF L2/3 AND L5/6 CORTICAL NEURONS

For biological networks to correct state estimates using prediction errors, both positive and negative errors need to be represented, as originally suggested by Rao and Ballard (1999) for neurons in L2/3 in the cortex. To test this theory, Jordan and Keller (2020) used an experimental setup similar to the one used in Keller et al. (2012): mice ran on a treadmill with locomotion-coupled visual flow feedback. Whole-cell recordings were obtained from both L2/3 and L5/6 neurons in V1. Visual feedback was interrupted with a brief flow halt (1 second) at random times to generate visual-locomotion mismatch events. Out of 32 neurons recorded in L2/3, 17 neurons showed depolarizing activities and 6 neurons showed hyperpolarizing activities during mismatch trials, suggesting that these neurons may subserve the function of encoding positive and negative prediction errors. In addition, 30% of the neurons exhibited significant correlations between the mismatch responses and the speed of locomotion (visual halts that occurred during faster locomotion generated "stronger" mismatch signals). These results are consistent with Keller et al.'s (2012) calcium imaging study previously discussed, showing that the responses of L2/3 neurons could potentially signal the amount of prediction error.

Compared to L2/3 neurons, a much lower ratio of neurons in L5/6 (5 out of 14) responded predominantly to mismatch trials (Jordan and Keller, 2020). The difference in responses between superficial and deep-layer neurons was significant in mismatch trials but not in normal visual flow trials. Furthermore, the distribution of correlations between the activities of neurons in L2/3 and locomotion speed or visual flow speed was bimodal: activities of most dMM neurons were positively correlated with locomotion speed and negatively correlated with visual flow speed (and vice versa for hMM neurons); activities of L5/6 neurons were mostly positively correlated with both locomotion speed and visual flow speed. These results suggest that at least a subset of L2/3 neurons may compute the error between the locomotion-generated predictions of visual inputs and the actual visual input, whereas deeper L5/6 neurons may integrate top-down predictions (here, from motor areas) and bottom-up input to compute an estimate of the state r at the current hierarchical level.

Gillon et al. (2021) also investigated the difference in neural responses to expected versus unexpected visual flows in L2/3 versus L5 using an open-loop experiment (no sensorimotor coupling to locomotion) with stimuli consisting of moving squares. Expectation violations were created in some trials by making 25% of the visual squares move in the opposite direction from the other 75%. Somatic and distal apical dendritic populations in L5 did not exhibit significantly different responses to expected versus unexpected visual flow, whereas both L2/3 somatic and distal apical dendritic populations showed a significant difference in responses. Additionally, this difference increased over days of exposure. The researchers also found learning effects when mice were exposed to Gabor sequence stimuli for several consecutive days. The responses to unexpected stimuli (in this case, novel Gabor stimuli replacing an expected stimulus in a sequence) were predictive of how these responses evolve in subsequent sessions on a cell-by-cell basis. Besides implicating L2/3 neurons in computing prediction errors, these results further confirm that prediction errors (responses to unexpected stimuli) can drive learning in neural circuits, an important prediction of the predictive coding model (Rao and Ballard, 1999) (see Equation (4) above).

A larger fraction of error-detecting neurons in superficial layers than in deep layers was also found by Hamm et al. (2021) in awake mice with visual oddball paradigms. Additionally, optogenetic suppression of prefrontal inputs to V1 reduced the contextual selectivity of the error-detecting neurons, consistent with the effect of top-down signals in the predictive coding model. Finally, through laminar local field potential (LFP) recordings in monkeys, Bastos et al. (2020) showed that the predictability of visual stimuli affects neural activities in the superficial and deep layers differently—during predictable trials, there is an enhancement of alpha and beta band spectral power in the deep layers of the cortex, while an increase in spiking and gamma power was observed in the superficial layers during unpredictable trials.

ROBUST PREDICTIVE CODING MODEL AND CORTICOTHALAMIC LOOPS

In robust predictive coding (Figure 64.2), a gating network regulates the flow of feedforward prediction errors to a cortical area from a lower cortical area via the thalamus.

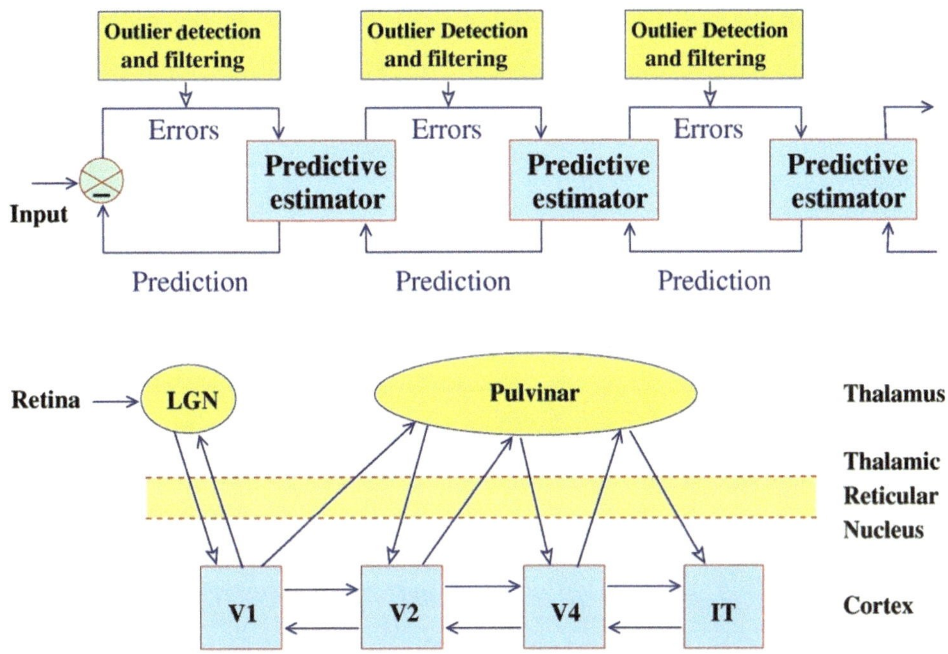

Figure 64.7. Suggested implementation of the robust predictive coding model. The figure illustrates how a hierarchical robust predictive coding model of the ventral visual cortical pathway can be implemented using corticothalamic interactions modulated by the inhibitory influence of the thalamic reticular nucleus (TRN) on thalamic relay neurons.

As shown in Figure 64.7, we suggest that a hierarchical form of robust predictive coding can be implemented in corticothalamic loops. Specifically, we postulate that the thalamic reticular nucleus (TRN), a network of inhibitory neurons between the thalamus and cortex, is well-positioned to function as a gating network. Cortical neurons in L6 send axon collaterals to reticular neurons and can therefore control the inhibitory influence of the thalamic reticular network on thalamic relay neurons. "Feedforward" axons going to the cortex from the thalamus also send collaterals to the TRN (see **Chapter 16 in Section IV**). If these feedforward axons carry prediction-error signals, as hypothesized by predictive coding, the TRN can help implement the "attentional" gain G_t in robust predictive coding for outlier detection and filtering (see Equation (5) above). In other words, the TRN and the thalamus can together enable the cortical predictive coding network estimating r to suppress large prediction errors in parts of the input containing outliers. This would enable the cortical network to test the feasibility of its current best hypothesis by trying to minimize prediction errors while ignoring outliers, as illustrated in the examples in Figure 64.2.

ACTIVE PREDICTIVE CODING AND CORTICOTHALAMIC-SUB-CORTICAL INTERACTIONS

As described above, the active predictive coding model is motivated by the fact that almost all cortical areas send outputs from L5 to sub-cortical motor centers, with axon collaterals targeting higher-order thalamus, which receives feedback from these motor centers (see Chapter 1 in Section I and **Chapter 15 in Section IV**). The thalamus in turn conveys this feedback, potentially as prediction errors, back to the superficial and deep layers of the cortical area that generated the output to the motor center. This motivates the sensory-motor computational module and hierarchical architecture proposed in Figure 64.4, which ascribes concrete computational roles to intracortical, cortico-cortical, cortico-subcortical, and corticothalamic interactions.

DISCUSSION

Predictive coding is a general computational framework that assumes that the brain maintains an internal model of the world and performs Bayesian inference and learning by minimizing prediction errors. The classical hierarchical predictive coding model (Rao and Ballard, 1999) postulates that feedback connections between cortical areas convey predictions of expected neural activity from higher to lower levels, while the feedforward connections convey the prediction errors, differentiating the model from other cortical models (Lee and Mumford, 2003; Heeger, 2017). Robust predictive coding extends the predictive coding model to robust optimization by allowing the model to ignore statistical outliers, linking the model to attention and a possible implementation within cortico-TRN-thalamic loops.

Dynamic predictive coding proposes that the cortical hierarchy encodes temporal sequences using representations that store information over increasingly longer timescales as one ascends the hierarchy. This allows the model to explain cortical space-time receptive fields, post-diction in perception, and episodic memory and cue-triggered recall.

Active predictive coding (APC) is a sensory-motor theory of the cortex, providing a unifying computational principle for perception, action, and cognition that is inspired by Mountcastle's prescient observation that the architecture of the cortical column is surprisingly uniform across cortical areas. The flexibility of the APC model and its ability for compositional problem-solving was illustrated using 2 different tasks: active vision for learning part-whole decompositions of objects, and navigation using hierarchical planning.

Experimental studies have discovered neurons in the visual and auditory cortex that encode predictions or prediction errors in sensory-motor tasks (Keller et al., 2012; Fiser et al., 2016; Schneider et al., 2018). Following up on the suggestion of Rao and Ballard (1999), some studies have tested the role of cortical L2/3 neurons in encoding prediction errors (Jordan and Keller, 2020) and driving learning in synaptic connections. Other aspects of predictive coding await further experimental testing, such as the role of L5 neurons in maintaining hidden state, the computation of inverse variances in Equation (3) ("precisions" terms in the free energy principle; Friston, 2010) and the contribution of this "precision"-based weighting to attention as in the robust predictive coding model. Dynamic predictive coding is supported by experimental evidence that cortical representations exhibit a hierarchy of timescales from lower-order to higher-order areas across both sensory and cognitive regions (Murray et al., 2014; Runyan et al., 2017; Siegle et al., 2021).

Although the predictive coding model includes variances for the top-down and bottom-up errors, as in Equation (3), it does not explicitly represent uncertainty via full probability distributions. The Kalman filter version of predictive coding (Rao, 1999) does represent uncertainty using a Gaussian posterior distribution, but it is unclear whether the cortex can compute covariance matrices (or even variances). There are alternate theories of Bayesian inference in the brain based on population coding and sampling (Rao, 2004, 2005; Ma et al., 2006; Huang and Rao, 2016; Orbán et al., 2016; Echeveste et al., 2020), and it may be worthwhile to connect predictive coding to some of these theories to leverage their ability to explicitly code for uncertainty.

Other directions worthy of exploration include relating predictive coding to cognition and memory. For example, recent results have shown that prediction errors (or "surprise"-related signals) can drive memory reactivation and reconsolidation (Kim et al., 2014; Sinclair and Barense, 2019; Bein et al., 2020; Rust and Palmer, 2021). Friston and colleagues have made important contributions in establishing the connections of predictive coding to a range of cognitive and neural phenomena through the free energy principle and active inference (Friston, 2010; Friston et al., 2017). With the advent of technologies for large-scale neural recording and stimulation (Jun et al., 2017; Steinmetz et al., 2021), the time may be ripe for in-depth experimental testing of the neuroanatomical and neurophysiological predictions of the various predictive coding theories reviewed in this chapter.

ACKNOWLEDGMENTS

This work is supported by a National Institutes of Health (NIH) eTeamBCP U01 award, a National Science Foundation (NSF) EFRI Program grant (no. 2223495), a Weill Neurohub Investigators Program grant, a grant from the Templeton World Charity Foundation, and a CJ and Elizabeth Hwang Professorship to RPNR. The opinions expressed in this publication are those of the authors and do not necessarily reflect the views of the funders.

REFERENCES

Attneave F (1954). Some informational aspects of visual perception. *Psychol Rev.* 61(3):183–193.

Barlow HB (1961). Possible principles underlying the transformation of sensory messages. In: Rosenblith W, ed. *Sensory communication* (pp. 217–234). MIT Press.

Bastos AM, Lundqvist M, Waite AS, Kopell N, Miller EK (2020). Layer and rhythm specificity for predictive routing. *Proc Natl Acad Sci.* 117(49):31459–31469.

Bein O, Duncan K, Davachi L (2020). Mnemonic prediction errors bias hippocampal states. *Nat Commun.* 11(1):3451.

Dong DW, Atick JJ (1995a). Temporal decorrelation: A theory of lagged and nonlagged responses in the lateral geniculate nucleus. *Network Comput Neural Syst.* 6(2):159–178.

Dong DW, Atick JJ (1995b). Statistics of natural time-varying images. *Network Comput Neural Syst.* 6(3):345–358.

Douglas RJ, Martin KAC (2004). Neuronal circuits of the neocortex. *Ann Rev Neurosci.* 27(1):419–451.

Echeveste R, Aitchison L, Hennequin G, Lengyel M (2020). Cortical-like dynamics in recurrent circuits optimized for sampling-based probabilistic inference. *Nat Neurosci.* 23(9):1138–1149.

Felleman D.J, Van Essen DC (1991). Distributed hierarchical processing in the primate cerebral cortex. *Cereb Cortex.* 1(1):1–47.

Ferguson KA, Cardin JA (2020). Mechanisms underlying gain modulation in the cortex. *Nat Rev Neurosci.* 21(2):80–92.

Field DJ (1987). Relations between the statistics of natural images and the response properties of cortical cells. *JOSA A.* 4(12):2379–2394.

Fiser A, Mahringer D, Oyibo HK, Petersen AV, Leinweber M, Keller GB (2016). Experience-dependent spatial expectations in mouse visual cortex. *Nat Neurosci.* 19(12):1658–1664.

Freiwald WA, Tsao DY (2010). Functional compartmentalization and viewpoint generalization within the macaque face-processing system. *Science*. 330(6005):845–851.

Friston K (2010). The free-energy principle: A unified brain theory? *Nat Rev Neurosci*. 11(2):127–138.

Friston K, FitzGerald T, Rigoli F, Schwartenbeck P, Pezzulo G (2017). Active inference: A process theory. *Neural Comput*. *29*(1):1–49.

Gillon CJ, Pina JE, Lecoq JA, Ahmed, R, Billeh Y, Caldejon S, Groblewski P, Henley TM, Kato I, Lee E, Luviano J, Mace K, Nayan C, Nguyen T, North K, Perkins J, Seid S, Valley MT, Williford A, Bengio Y, Lillicrap TP, Richards BA, Zylberberg J (2021). Learning from unexpected events in the neocortical microcircuit. *bioRxiv*. https://doi.org/10.1101/2021.01.15.426915

Hamm JP, Shymkiv Y, Han S, Yang W, Yuste R (2021). Cortical ensembles selective for context. *Proc Natl Acad Sci USA*. 118(14):e2026179118.

Hawkins J (2021). *A Thousand Brains: A New Theory of Intelligence*. Basic Books.

Heeger DJ (2017). Theory of cortical function. *Proc Natl Acad Sci*. 114(8):1773–1782.

Huang Y, Rao RPN (2011). Predictive coding. *WIREs Cogn Sci*. 2(5):580–593.

Huang Y, Rao RPN (2016). Bayesian inference and online learning in Poisson neuronal networks. *Neural Comput*. 28(8):1503–1526.

Hubel DH, Wiesel TN (1959). Receptive fields of single neurones in the cat's striate cortex. *J Physiol*. 148(3):574–591.

Huber PJ (1981). *Robust Statistics*. John Wiley & Sons.

Issa EB, Cadieu CF, DiCarlo JJ (2018). Neural dynamics at successive stages of the ventral visual stream are consistent with hierarchical error signals. *ELife*, *7*, e42870. https://doi.org/10.7554/eLife.42870

Jiang LP, Gklezakos DC, Rao RPN (2021). Dynamic Predictive Coding with Hypernetworks. *BioRxiv*, 2021.02.22.432194. https://doi.org/10.1101/2021.02.22.432194

Jiang LP, Rao RPN (2022). Dynamic predictive coding: A new model of hierarchical sequence learning and prediction in the cortex. *bioRxiv*. https://doi.org/10.1101/2022.06.23.497415

Jordan R, Keller GB (2020). Opposing influence of top-down and bottom-up input on excitatory layer 2/3 neurons in mouse primary visual cortex. *Neuron*. 108(6):1194–1206.

Jun JJ, Steinmetz NA, Siegle JH, Denman DJ, Bauza M, Barbarits B, Lee AK, Anastassiou CA, Andrei A, Aydın Ç, Barbic M, Blanche TJ, Bonin V, Couto J, Dutta B, Gratiy SL, Gutnisky DA, Häusser M, Karsh B, Ledochowitsch P, Lopez CM, Mitelut C, Musa S, Okun M, Pachitariu M, Putzeys J, Rich PD, Rossant C, Sun WL, Svoboda K, Carandini M, Harris KD, Koch C, O'Keefe J, Harris TD (2017). Fully integrated silicon probes for high-density recording of neural activity. *Nature*. 551(7679):232–236.

Keller GB, Bonhoeffer T, Hübener M (2012). Sensorimotor mismatch signals in primary visual cortex of the behaving mouse. *Neuron*. 74(5):809–815.

Keller GB, Mrsic-Flogel TD (2018). Predictive processing: A canonical cortical computation. *Neuron*. 100(2):424–435.

Kim EJ, Juavinett AL, Kyubwa EM, Jacobs MW, Callaway EM (2015). Three types of cortical layer 5 neurons that differ in brain-wide connectivity and function. *Neuron*. 88(6):1253–1267.

Kim G, Lewis-Peacock JA, Norman KA, Turk-Browne NB (2014). Pruning of memories by context-based prediction error. *Proc Natl Acad Sci*. 111(24):8997–9002.

Larkum ME, Senn W, Lüscher H-R (2004). Top-down dendritic input increases the gain of layer 5 pyramidal neurons. *Cereb Cortex*. 14(10):1059–1070.

Lee TS, Mumford D (2003). Hierarchical Bayesian inference in the visual cortex. *JOSA A*. 20(7):1434–1448.

Ma WJ, Beck JM, Latham PE, Pouget A (2006). Bayesian inference with probabilistic population codes. *Nat Neurosci*. 9(11):1432–1438.

Mountcastle VB (1978). An organizing principle for cerebral function: The unit model and the distributed system. In: Edelman GM, Mountcastle VM, eds. *The Mindful Brain: Cortical Organization and the Group-Selective Theory of Higher Brain Function* (pp. 7–50). MIT Press.

Murray JD, Bernacchia A, Freedman DJ, Romo R, Wallis JD, Cai X, Padoa-Schioppa C, Pasternak T, Seo H, Lee D, Wang X-J (2014). A hierarchy of intrinsic timescales across primate cortex. *Nat Neurosci*. 17(12):1661–1663.

Murray SO, Kersten D, Olshausen BA, Schrater P, Woods DL (2002). Shape perception reduces activity in human primary visual cortex. *Proc Natl Acad Sci*. 99(23):15164–15169.

Orbán G, Berkes P, Fiser J, Lengyel M (2016). Neural variability and sampling-based probabilistic representations in the visual cortex. *Neuron*. 92(2):530–543.

Rao RPN (1998). Correlates of attention in a model of dynamic visual recognition. *Adv Neural Inf Process Syst*. 10:80–86.

Rao RPN (1999). An optimal estimation approach to visual perception and learning. *Vision Res*. 39(11):1963–1989.

Rao RPN (2004). Bayesian computation in recurrent neural circuits. *Neural Comput*. 16(1):1–38.

Rao RPN (2005). Bayesian inference and attentional modulation in the visual cortex. *NeuroReport*. 16(16):1843–1848.

Rao RPN (2022). A sensory-motor theory of the neocortex based on active predictive coding. *bioRxiv*. https://doi.org/10.1101/2022.12.30.522267

Rao RPN, Ballard DH (1999). Predictive coding in the visual cortex: A functional interpretation of some extra-classical receptive-field effects. *Nat Neurosci*. *2*(1):79–87.

Rao RPN, Gklezakos DC, Sathish V (2022). Active predictive coding: A unified neural framework for learning hierarchical world models for perception and planning (arXiv:2210.13461). *arXiv*. https://doi.org/10.48550/arXiv.2210.13461

Ruderman DL, Bialek W (1994). Statistics of natural images: Scaling in the woods. *Phys Rev Lett*. 73(6):814–817.

Runyan CA, Piasini E, Panzeri S, Harvey CD (2017). Distinct timescales of population coding across cortex. *Nature*. 548(7665):92–96.

Rust NC, Palmer SE (2021). Remembering the past to see the future. *Ann Rev Vision Sci*. 7(1):349–365.

Schneider DM, Sundararajan J, Mooney R (2018). A cortical filter that learns to suppress the acoustic consequences of movement. *Nature*. 561(7723):391–395.

Schwiedrzik CM, Freiwald WA (2017). High-level prediction signals in a low-level area of the macaque face-processing hierarchy. *Neuron*. 96(1):89–97.

Sherman SM, Guillery RW (2013). *Functional Connections of Cortical Areas: A New View from the Thalamus*. MIT Press.

Siegle JH, Jia X, Durand S, Gale S, Bennett C, Graddis N, Heller G, Ramirez TK, Choi H, Luviano JA, Groblewski PA, Ahmed R, Arkhipov A, Bernard A, Billeh YN, Brown D, Buice MA, Cain N, Caldejon S, Casal L, Cho A, Chvilicek M, Cox TC, Dai K, Denman DJ, de Vries SEJ, Dietzman R, Esposito L, Farrell C, Feng D, Galbraith J, Garrett M, Gelfand EC, Hancock N, Harris JA, Howard R, Hu B, Hytnen R, Iyer R, Jessett E, Johnson K, Kato I, Kiggins J, Lambert S, Lecoq J, Ledochowitsch P, Lee JH, Leon A, Li Y, Liang E, Long F, Mace K, Melchior J, Millman D, Mollenkopf T, Nayan C, Ng L, Ngo K, Nguyen T, Nicovich PR, North K, Ocker GK, Ollerenshaw D, Oliver M, Pachitariu M, Perkins J, Reding M, Reid D, Robertson M, Ronellenfitch K, Seid S, Slaughterbeck C, Stoecklin M, Sullivan D, Sutton B, Swapp J, Thompson C, Turner K, Wakeman W, Whitesell JD, Williams D, Williford A, Young R, Zeng H, Naylor S, Phillips JW, Reid RC, Mihalas S, Olsen SR, Koch C (2021). Survey of spiking in the mouse visual system reveals functional hierarchy. *Nature*. 592(7852):86–92.

Simoncelli EP, Olshausen BA (2001). Natural image statistics and neural representation. *Ann Rev Neurosci*. 24:1193–1216.

Sinclair AH, Barense MD (2019). Prediction error and memory reactivation: How incomplete reminders drive reconsolidation. *Trends Neurosci.* 42(10):727–739.

Srinivasan MV, Laughlin SB, Dubs A (1982). Predictive coding: A fresh view of inhibition in the retina. *Proc R Soc Lond B Biol Sci.* 216(1205):427–459.

Steinmetz NA, Aydin C, Lebedeva A, Okun M, Pachitariu M, Bauza M, Beau M, Bhagat J, Böhm C, Broux M, Chen S, Colonell J, Gardner RJ, Karsh B, Kloosterman F, Kostadinov D, Mora-Lopez C, O'Callaghan J, Park J, Putzeys J, Sauerbrei B, van Daal RJJ, Vollan AZ, Wang S, Welkenhuysen M, Ye Z, Dudman JT, Dutta B, Hantman AW, Harris KD, Lee AK, Moser EI, O'Keefe J, Renart A, Svoboda K, Häusser M, Haesler S, Carandini M, Harris TD (2021). Neuropixels 2.0: A miniaturized high-density probe for stable, long-term brain recordings. *Science.* 372(6539):eabf4588.

Tsao DY, Freiwald WA, Tootell RBH, Livingstone MS (2006). A cortical region consisting entirely of face-selective cells. *Science.* 311(5761):670–674.

SECTION XII

DISEASE AND PATHOPHYSIOLOGY

Anna S. Mitchell

For over 120 years, clinical evidence and neuropathology reports indicated that damage to, or dysfunction of the thalamus, in addition to damage within the cerebral cortex, is associated with many neurological diseases and disorders. This clinical evidence prompted neuroscientific studies involving animal models that reaffirmed the critical importance of the cerebral cortex and thalamic nuclei in the physiological, cognitive, and behavioral symptoms associated with various neurological diseases and disorders. These studies and other work in animal models, together with in vivo neuroimaging studies in patient populations, have helped clinicians and neuroscience researchers to understand that widespread brain changes, from molecular to macroscopic levels in the thalamus, the cortex, and beyond, are critical for the underlying causes and manifestations of these many neurological, neurodevelopmental, and neurodegenerative diseases and disorders.

Consequently, it is now more widely recognized that cortico-thalamocortical neural networks with essential modulation from interconnected subcortical and brainstem structures govern normal brain functioning and dysfunction. Many neurological diseases and disorders now identify changes in the thalamus and cortico-thalamocortical neural networks, some of which are discussed in this Disease and Pathophysiology section, including absence epilepsy; affective disorders (e.g., schizophrenia); stroke injuries; neurodegeneration related to alcohol use disorder and Korsakoff's syndrome; and neurodegenerative diseases (e.g., Alzheimer's disease, Parkinson's disease, and frontotemporal dementia).

As indicated in other sections of this book, thalamic motifs are emerging as neuroscience evidence, using different methodologies, is converging to highlight similarities in the ways different thalamic nuclei are interconnected with cortical layers. Additionally, evidence is emerging about the heterogeneity within and between thalamic nuclei as we characterize their distinct subdivisions, investigate their genetics, and identify their unique cortico-thalamocortical connectivity and subcortical and brainstem inputs. This heterogeneity is also apparent when we consider changes to thalamic nuclei and cortico-thalamocortical connectivity in the many varied neurological diseases and disorders.

This heterogeneity must be embraced, as it will likely help us understand why distinct subdivisions of thalamic nuclei are changed in these many varied neurological diseases and disorders. Additionally, this heterogeneity may allow researchers using in vivo imaging and state-of-the-art genetic research techniques to help determine if specific genetic, structural, and/or functional connectivity changes in thalamic nuclei and cortico-thalamocortical pathways can clarify unique biomarkers to further characterize differences between diseases and help with the diagnosis of many neurological diseases within a patient's lifetime.

Furthermore, many different animal species (including rodents and monkeys) are used to model aspects of the underlying brain changes, or physiological, behavioral, and cognitive symptoms and phenotypes that occur with altered genetics or damage to individual neural structures that are linked to neurological diseases and disorders. However, differences in genetics, anatomy, physiology, and behavior exist in different animal species. Thus, it is vital that researchers, clinicians, and neuroscientists have the continued ability to conduct research using the most appropriate animal model to investigate specific disease mechanisms and pathophysiology, and determine effective treatment options.

This Disease and Pathophysiology section highlights our current understanding about the involvement of thalamic nuclei and cortico-thalamocortical neural networks from historical to modern-day underpinnings. The chapters cover technological advances in clinical evidence obtained from patient populations and advances gained from the use of different animal species to model underlying mechanisms and some of the physiological, cognitive, and behavioral symptoms of the varied neurological diseases and disorders. In Chapter 65, Leresche and Crunelli document advances in knowledge about the involvement of thalamocortical networks in absence epilepsy from studies involving patient populations and genetically modified rodents. In Chapter 66, Abram, Spiro, and Ford review evidence that characterizes thalamocortical neural networks associated with symptoms of schizophrenia, a neurodevelopmental affective disorder that remains elusive to effective treatment strategies for the varied cognitive symptoms of the disorder. In Chapter 67, Dalrymple-Alford, Anderson, and Melzer highlight how changes to the thalamus and cortico-thalamocortical networks occur earlier on in the progression of Parkinson's disease and consider how these changes may contribute to the non-motor symptoms. In Chapter 68, Forno and Hornberger identify thalamocortical neural network changes that are now documented in healthy aging, comparing them with Alzheimer's disease patients and considering how these thalamic and thalamocortical network changes may contribute to dementia. In Chapter 69, Pitel and Segobin review evidence from advanced neuroimaging analysis methods that details the degeneration that occurs within thalamocortical networks and beyond in patients diagnosed with alcohol use disorder, including Korsakoff's syndrome and consider how these changes impact the symptoms and continuation of the disorder. In Chapter 70, Bocchetta explains the genetic phenotyping that is now utilized to identify differences in thalamic nuclei affected in the different subtypes of frontotemporal dementia. This genotyping is helping to align thalamic nuclei with interconnected frontal and temporal structures that have traditionally been associated with the varied symptomology in this disease. Studies in humans and animals are characterizing the genetic uniqueness of individual thalamic nuclei, which in time will hopefully provide effective gene therapy treatment targets in many neurological diseases and disorders. In Chapter 71, Wilke and Kagan review cortico-thalamocortical networks linked to the pulvinar involved in visuospatial and motor processing. The review discusses the clinical symptoms observed in patients after stroke damage to the pulvinar which was used to inform empirical experiments in macaque monkeys that have documented specific motor and visuospatial deficits after selective pulvinar manipulations. Finally, Chapter 72 offers concluding remarks.

The authors in each of these chapters have reviewed different research methodologies that provide effective ways to capture the heterogeneity of thalamic nuclei and cortico-thalamocortical networks involved in these many varied neurological diseases and disorders. Studies that continue to identify the similarities as well as the heterogeneity of individual thalamic nuclei will help to better diagnose these diseases and disorders, and may provide a means to develop effective and safe treatment strategies that can improve the quality of life for patient populations, and in time, may delay the onset or reduce the many symptoms experienced in these debilitating neurological diseases and disorders.

65.

CORTICOTHALAMIC NETWORK ABNORMALITIES UNDERLYING ABSENCE SEIZURES

Nathalie Leresche and Vincenzo Crunelli

INTRODUCTION

As early as the first half of the 20th century, the German neurologist Hans Berger hypothesized that the 3Hz oscillations that he observed in the electroencephalograph (EEG) of a patient during an absence seizure (AS) resulted from an abrupt withdrawal of tonic inhibitory influences exerted by the thalamus on the cortex (see Avoli, 2012). Thus, although at that time the mechanism of these EEG oscillations (now known as spike-and-wave discharges, SWDs) was completely unknown, the cortex and the thalamus were already considered major players in the generation of this paroxysmal activity. Nowadays, there is a general consensus that ASs are a disease of corticothalamic (CT) networks with a critical modulation by the basal ganglia and brainstem (Crunelli et al., 2020).

Following a short summary of the clinical presentation, therapy, and comorbidities of ASs, we first discuss their genetics and the mechanisms involved in their initiation and maintenance in humans and in models of ASs. The main body of this chapter is a focused review of the current state of knowledge of the molecular and cellular abnormalities in different neuronal populations of CT networks that can bring about ASs, using evidence both from investigations that have used selective genetic manipulations in normal non-epileptic animals and from studies in 5 genetic models with spontaneous ASs that have been validated for their strong face-, construct-, and content-validity: the polygenic GAERS and WAG/Rij rats, and the monogenic stargazer (STG), tottering (TG), and lethargic (LH) mice (Pitkänen et al., 2017). Due to space limitations, studies using only pharmacological manipulations are not discussed.

CLINICAL PRESENTATION OF ASs

An AS is a sudden and relatively brief lapse of consciousness associated with lack of voluntary movements and distinctive SWDs at 2.5–4Hz in the EEG (Crunelli et al., 2020). ASs are present, together with other seizure types, in many genetic generalized epilepsies (GGEs); here we focus on ASs observed in childhood absence epilepsy (CAE) since in this epilepsy type ASs are the only clinical symptom present and are not compromised by other epileptic phenotypes. In CAE, ASs begin between 3 and 8 years of age and can occur up to ~200 per day. Approximately 70% of CAE patients go into spontaneous remission around adolescence (Crunelli and Leresche, 2002), though the mechanisms underlying remission have not been elucidated, since currently researched AS models lack spontaneous remission. Pharmacological treatment of ASs relies primarily on ethosuximide (ETX), valproic acid, lamotrigine, or their combinations, but ~30% of children are pharmaco-resistant and ~60% show neuropsychiatric comorbidities (see Crunelli et al., 2020).

GENETICS OF ASs

Strong evidence exists for a genetic component in CAE, possibly of a polygenic nature (Perucca et al., 2020) and concordances of 70%–85% and 33% have been reported in monozygotic twins and first-degree relatives, respectively (Gruenbaum et al., 2021). However, most CAE patients do not have a family history of epilepsy, and siblings of affected individuals have an 8% risk of ASs, which is lower than the expected risk for a recessive (25%) or dominant (50%) inherited trait (Perucca et al., 2020).

Most AS genetic studies have used cohorts with a complex phenotype that included various epilepsies and other neurological deficits, rendering it difficult to demonstrate an association between gene abnormalities and ASs. Further, these studies focused on a few families and reported diverse mutations mainly in γ-aminobutyric acid receptor ($GABA_AR$) sub-units and voltage-gated Ca^{2+} channels (reviewed in Crunelli and Leresche, 2002). More recently, in a large study of GGEs (that included CAE as well as juvenile absence epilepsy), no single gene reached exome-wide significance, though lead associations were found for *CACNA1G*, *EEF1A2*, and *GABRG2*, and, notably, $GABA_AR$ genes, but not excitatory receptor genes, were enriched for missense variants (Feng et al., 2019).

The first genome-wide association study of a large, pure CAE cohort (778 Caucasian, 15 African) has identified 2 significant loci: 2p16.1 and 2q22.3 (International League Against Epilepsy Consortium on Complex Epilepsies, 2018). Though the precise genes in these loci that contribute to the CAE phenotype are still unknown, prioritized genes include *FANCL*, *BCL11A*, and *ZEB2* (i.e., genes not linked to membrane channels or transmitter pathways that are generally considered key proteins for AS expression). Notwithstanding, mutations in *SCN1A*, *SCN8A*, *CACNA1H*, and *CLCN2* have been identified in different Chinese CAE cohorts (Xie et al., 2019).

A large number of spontaneous single gene mutations bring about spontaneous ASs in mice (Frankel, 2009), some of which have been validated as ASs models, including *CACNG2* (in STG mice), *CACNA1A* (in TG mice), and *CACNB4* (in LH mice). Whole-genome sequencing of the polygenic GAERS rats has identified *CACNA1H* (R1584P) as the only gene-variant that selectively cosegregates with high-seizuring F2 progenies (Casillas-Espinosa et al., 2017). Notably, a combined proteomic and metabolomic analysis of GAERS rats has found an increase in pathways involved in synaptic transmission and metabolism of neurotransmitters and the down-regulation of the lysine degradation pathway in both cortex and thalamus (Harutyunyan et al., 2022). However, adult rats were used in this study, raising the issue of whether the observed changes in proteins and metabolites are the cause or a consequence of ASs.

In summary, single gene variants are causally linked to the expression of spontaneous ASs in mice models. However, the human homologs of these genes, as well as the mutant genes found in the Chinese CAE cohorts, are not located in the significant loci (2p16.1 and 2q22) identified in a large CAE cohort.

CORTICAL INITIATION NETWORKS OF ASs

Although ASs are classified as generalized seizures, many studies in humans report that SWDs originate from cortical regions before spreading to the entire cortical surface and other brain regions, though the cohorts of these studies had a mixed phenotype including different GGEs (reviewed in Crunelli et al., 2020). The first study on a pure CAE cohort reported cortical-selective increases in the amplitude of blood-oxygen-level-dependent (BOLD) signals occurring as early as 14 s before the clinical and electrographic signatures of an AS could be observed (Bai et al., 2010). The cortical area(s) showing these pre-ictal increases in BOLD amplitude are mostly in frontal regions and can differ from child to child, but are generally consistent across seizures for a given child (Bai et al., 2010). Importantly, the same cortical areas show a post-ictal decrease in BOLD amplitude that can persist for up to 10 s after seizure offset (Bai et al., 2010).

Changes in BOLD signals also occur in the thalamus during ASs (particularly, in the anterior, centromedian, parafasicular, and reticular nuclei), supporting its critical involvement in the expression of these seizures (Bai et al., 2010; Guo et al., 2016).

The lack of generalization at the onset of an AS (i.e., the existence of a cortical initiation network, CIN) had been originally reported in 2 rat models of ASs. It was found that the peri-oral region of the primary somatosensory cortex (S1) was the cortical initiation network in GAERS and WAG/Rij rats (Figure 65.1A) and that S1 layer (L)5/6 excitatory pyramidal neurons were the first to fire at the start of SWDs (Figure 65.1B) (Meeren et al., 2002; Polack et al., 2007). Moreover, BOLD signals in awake GAERS rats recapitulate human BOLD results with an AS-linked decrease and increase in signal in S1 and ventrobasal (VB) thalamus, respectively (McCafferty et al., 2023).

All these pre-seizure changes observed with noninvasive imaging and EEG in humans and animal models may be the substrate for impaired consciousness, including loss of responses to external stimuli and a pause in ongoing interactive behaviors, which occur not only during, but even before the development of SWDs (Guo et al., 2016). Notably, the magnitude of BOLD and EEG changes in the cortex and thalamus that precede an AS are positively linked to the severity of altered behavioral responses and consciousness during a seizure and may thus be predetermined by the state of CT networks before seizure onset (Bai et al., 2010; Guo et al., 2016). Notably, in GAERS rats, the expression of the AS-related behavioral phenotype requires the mutual interactions of cortical and thalamic networks, as shown by the lack of any overt behavioral sign when short (1–2 sec) small-amplitude SWDs are present in the cortex but not in the thalamus (Figure 65.1C) (Polack et al., 2007), indicating that the recruitment of the thalamus is a necessary step for the expression and maintenance of ASs (see below for intralaminar thalamic nuclei).

In summary, it is now well established that (1) SWDs are not generalized from the start but originate from selective CINs, with S1 L5/6 pyramidal neurons being the first to fire in animal models; (2) the cortex and thalamus are critical elements for the full expression of ASs; and (3) the severity of the clinical symptoms of ASs may be predicted

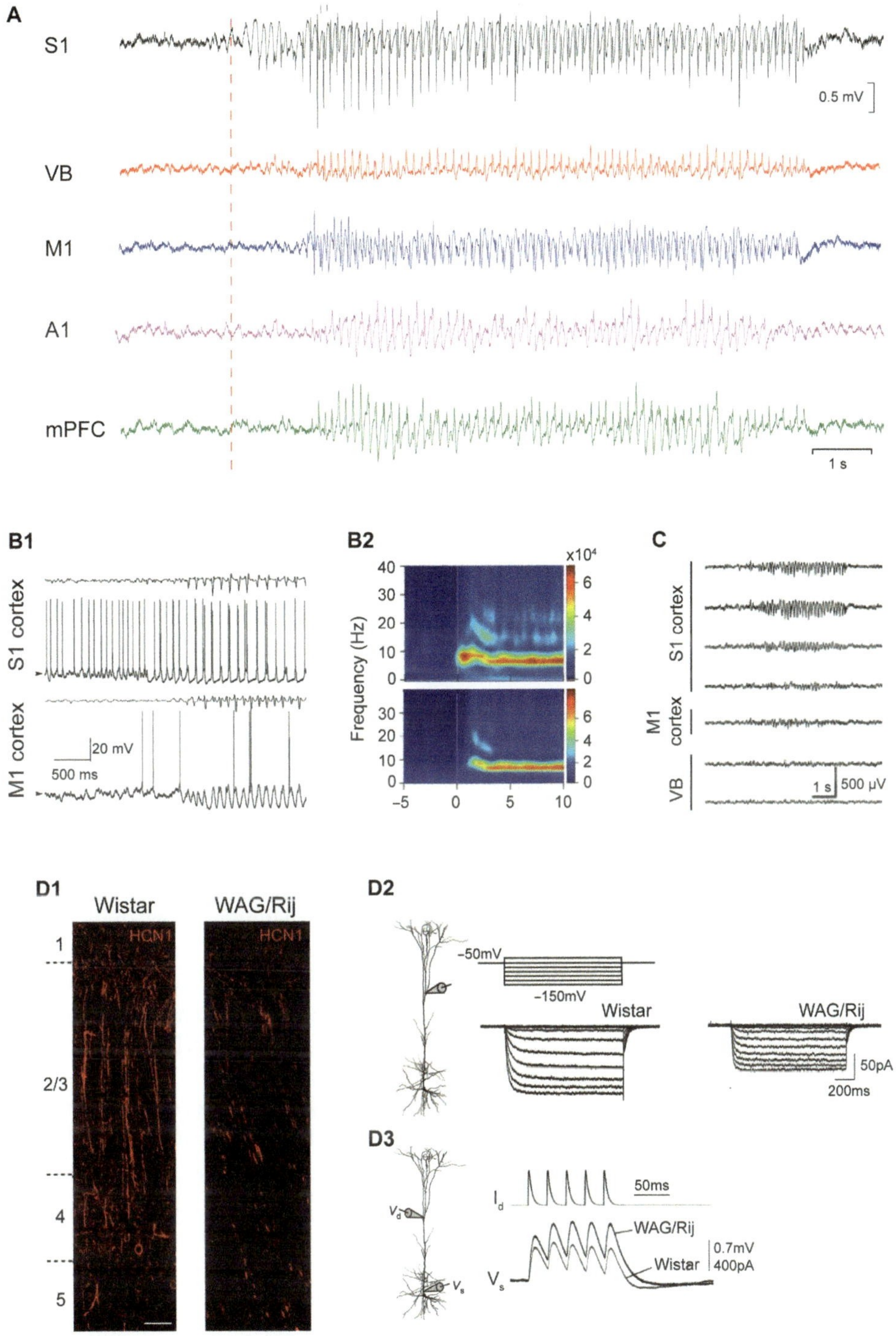

Figure 65.1. Cortical initiation of absence seizures (ASs) and cellular abnormalities in L5 neurons of AS models. (A). EEG traces from the indicated cortical and thalamic areas in a freely moving GAERS rat show the initiation of SWDs in the primary somatosensory cortex (S1); VB = ventrobasal thalamic nucleus; M1 = primary motor cortex; A1 = primary auditory cortex; mPFC = medial prefrontal cortex. (B1). Interictal and ictal intracellular activity (bottom trace in each pair) of L5 neurons, recorded in the same GAERS rat simultaneously with the corresponding EEG (top trace in each pair). Note the hyperactivity of the neuron located in the primary somatosensory cortex (S1 cortex), the cortical initiation network in this animal model, compared to the neuron in the primary motor cortex (M1 cortex). (B2). Averaged time-frequency maps showing the occurrence of 6–8 Hz SWDs in S1 before M1. (C). EEG traces from different sites in the CT network of a GAERS rat show large-amplitude SWDs in S1 (and very small SWDs in M1) but not in the VB, the somatotopic thalamic nucleus of S1, which were not accompanied by any behavioral sign. (D1). HCN1 immunofluorescence in S1 shows weaker labeling in a WAG/Rij rat compared to a control Wistar rat (scale bar: 100 µm). (D2). Dendritic cell-attached patch-clamp recordings in the apical dendrites of an S1 L5 pyramidal neuron show significantly lower amplitude of I_h in a WAG/Rij rat compared to a control Wistar rat. (D3). Simultaneous dendritic and somatic whole-cell recordings in an S1 L5 pyramidal neuron. EPSPs were simulated by current injection in the dendrite (I_d, top trace) and their summation at the soma (Vs, bottom traces) was increased in a WAG/Rij rat neuron compared to a control Wistar rat neuron. Panel A adapted from McCafferty et al. (2018); B and C: Polack et al. (2007); D1–D3: Kole et al. (2007).

by the degree of pre-seizure EEG and BOLD changes and predetermined by persistently abnormal CT networks.

MECHANISMS OF ALTERED EXCITABILITY IN DEEP LAYERS OF THE CIN

The presence of a CIN in animal models has allowed study of the mechanism(s) underlying the AS initiation. Strong evidence obtained in vivo and in vitro in 4 AS models (GAERS and WAG/Rij rats, STG and TG mice) has shown enhanced excitability of L5/6 pyramidal neurons (Kole et al., 2007; Polack et al., 2007) that is evident from the presence of spontaneous giant depolarizing potentials, an increased number of neurons with intrinsic strong bursting and N-methyl-D-aspartate (NMDA)-mediated excitatory postsynaptic potentials (EPSPs), as well as an enhanced firing rate, response to weak stimuli, and gain of input-output function of L5/6 neurons (D'Antuono et al., 2005; Polack et al., 2007). Moreover, in WAG/Rij rats, it has elegantly been shown that a selective decrease in HCN1, but not HCN2 channels in the distal apical dendrites of L5 pyramidal neurons leads to a decreased dendritic I_h current, larger and faster backpropagating action potentials, and increased somato-dendritic coupling, so that the summation of synaptic potentials is enhanced (Figure 65.1D) (Kole et al., 2007). These intrinsic HCN channel-dependent changes in excitability are strengthened by a decreased peak conductance and paired-pulse depression (PPD) of evoked $GABA_A$ inhibitory postsynaptic potentials (eIPSPs) in S1 L5 (and L2/3) pyramidal neurons and evoked excitatory postsynaptic potentials (eEPSPs) with a reduced PPD and sensitivity to baclofen, a $GABA_BR$ agonist. The net result of all these changes in synaptic activity in S1 L5/6 pyramidal neurons of WAG/Rij rats is an increased EPSP/inhibitory postsynaptic potential (IPSP) ratio since (1) the PPD of both EPSPs and IPSP is decreased, (2) the number of NMDA EPSPs is increased, and (3) the peak conductance of IPSCs is decreased.

Notably, a similar decrease in cortical inhibition is present in S1 of the TG mice, where a comprehensive study of EPSPs and IPSPs, evoked either by thalamic stimulation or locally (either in L4 or L5), has compellingly shown that the feedforward inhibition from the thalamus to L4 is markedly compromised and correlated to the developmental onset of ASs (Sasaki et al., 2006).

Notably, there is a lower expression of PV+ interneurons in all S1 layers of WAG/Rij rats (Papp et al., 2018). Functionally, chemogenetic silencing of S1 PV+ interneurons in normal mice is sufficient to generate ETX-sensitive ASs (Panthi and Leitch, 2019), whereas DREADD-mediated activation of S1 PV+ interneurons (i.e., activation of local feedforward inhibition) protects against pentylenetetrazol-induced SWDs (Panthi and Leitch, 2019) (see below).

In summary, in WAG/Rij rats, we have a clear mechanistic explanation of the aberrant behavior of S1 L5/6 pyramidal neurons where SWDs originate: intrinsic HCN1 loss-of-function (leading to enhanced synaptic potential summation) combines with an increased EPSPs/IPSPs ratio to bring about the enhanced excitability observed in this neuronal population. A decreased cortical inhibition is also present in TG mice, but similar detailed mechanistic studies of S1 are not available for GAERS rats or STG and LH mice. PV+ interneurons have an anti-absence role since their selective silencing in normal non-epileptic animals leads to AS expression.

CELLULAR ACTIVITY IN THE NEOCORTEX AND THALAMUS DURING ASs

Using electrophysiological and imaging approaches in non-anesthetized AS models, the ictal neuronal activities that occur in the cortex and thalamus have recently been studied with single-cell resolution. Using Ca^{2+}-sensitive dyes in the primary visual cortex in STG mice, Meyer et al. (2018) reported that ~80% of neurons in all cortical layers reduce their activity before an SWD is detected in the EEG, while smaller neuronal pools show an increased, or no change in, activity (Figure 65.2A). Interestingly, the ictal involvement of each neuron is not fixed, but fluctuates from seizure to seizure. An overall decrease in neuronal activity during ASs has been confirmed in S1 cortical neurons of GAERS rats (McCafferty et al., 2023). However, on closer examination, 4 patterns of ictal activity could be identified. While 37% of neurons experienced a sustained ictal decrease in firing, 28% showed a transient firing peak at seizure onset. Importantly, regardless of the type of firing, all cortical neurons show a marked increase in rhythmicity during SWDs compared to inter-ictal periods and all changes precede seizure onset by up 80 seconds.

Simultaneous ensemble recordings in cortical neurons and VB thalamocortical (TC) and nucleus reticularis thalami (NRT) neurons during spontaneous ASs in GAERS rats and γ-hydroxybutyric acid (GHB)-induced ASs in Wistar rats under freely moving conditions show that cortical afferent activity drives thalamic firing and temporally frames thalamic output via feedforward inhibition of TC neurons by NRT neurons (Figure 65.2D) (McCafferty et al., 2018). This results in a marked decrease in the overall firing of TC neurons (with only rare T-type Ca^{2+} channel-dependent burst firing being

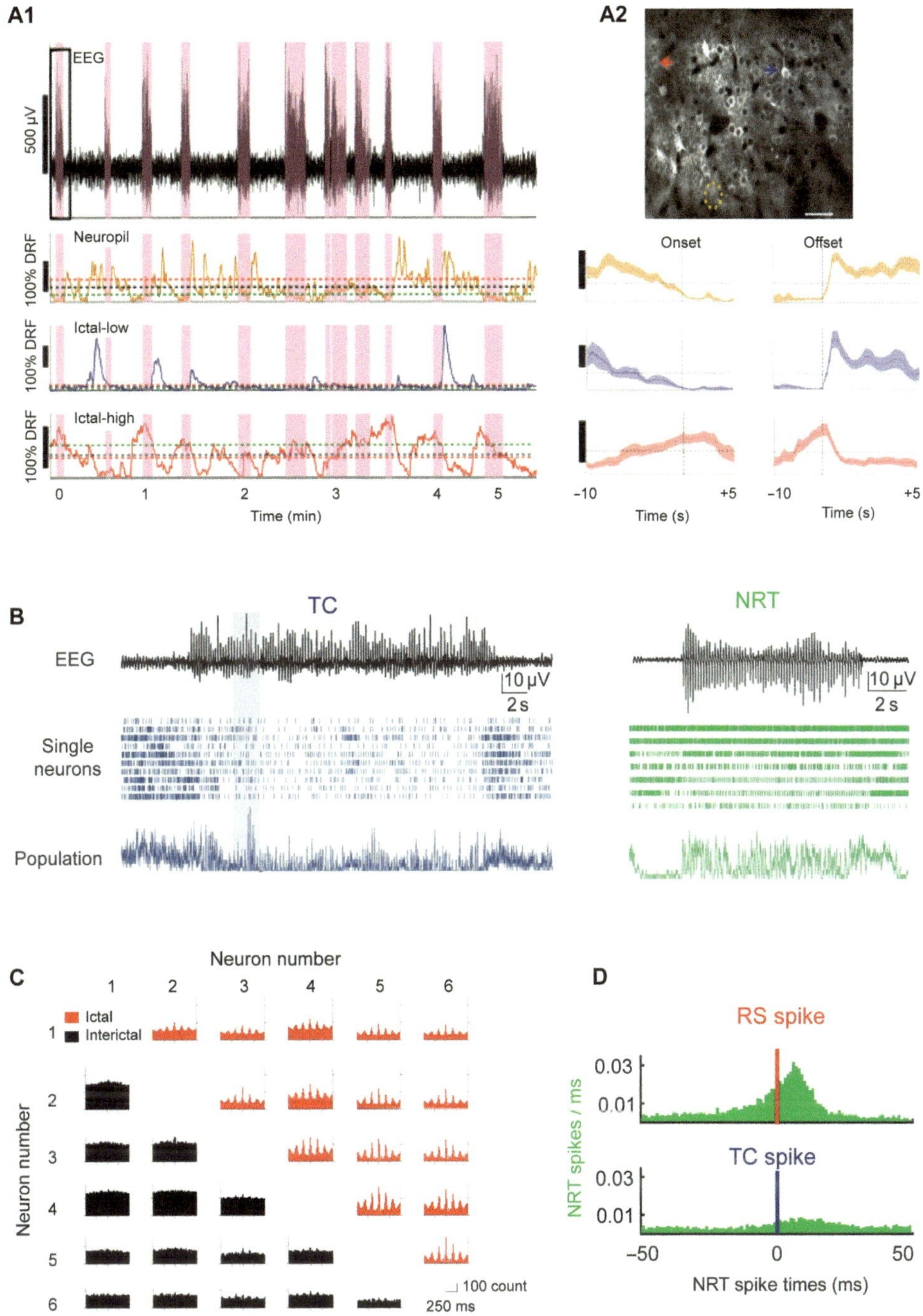

Figure 65.2. Activity of cortical and thalamic neurons during absence seizures (ASs). (A1). Concomitant EEG (top trace) and intracellular Ca^{2+} changes from L2/3 GCaMP6-filled neurons in the visual cortex of a head-restrained STG mouse. The Ca^{2+} changes of each neuron/neuropil patch over time are shown with the seizure epochs highlighted in pink. Mean ictal activity (green dashed line), mean interictal activity (red dashed line), and overall mean activity (black dashed line). (A2). The average activity (vertical bar = 20% $\Delta F/F$) time-locked to the seizure onset and offset is shown for exemplary neuropil (top), a neuron with low activity during ASs (ictal-low, middle) and a neuron with high activity during ASs (ictal-high, bottom). The ictal-low and ictal-high neurons are marked by a blue and red arrow, respectively, in the 2-photon microscopy image shown at the top, with the neuropil area highlighted by a yellow dotted circle. Scale bar = 50 μm. (B). EEG (top trace) and spike-time raster plots from 10 thalamocortical (TC) neurons in the ventrobasal thalamic nucleus and 7 nucleus reticularis thalami (NRT) neurons simultaneously recorded with a silicon probe in a freely moving GAERS rats. Most TC neurons show a decrease in firing rate during the ASs, which is also evident from the population activity (bottom trace). NRT neurons show 2 different types of ictal firing: one group increases their firing ictally (top 3 cells in raster-plot) and another group shows a decreased activity (bottom 4 cells). (C). Cross-correlograms of the total firing of 6 simultaneously recorded GAERS TC neurons during inter-ictal periods (black plots) and ASs (red plots). Despite the overall decrease in firing during ASs (red plots), there is a marked increase in neuronal synchrony. (D). Histograms of the firing of an NRT neuron recorded simultaneously with a cortical regular spiking (RS) neuron (top) or a TC neuron (bottom) in a freely moving GAERS rat. Note that the ictal NRT neuron firing (green bars) is mainly driven by the cortical RS neuron spikes rather than the TC neuron spikes. Time zero marks the firing of the RS and TC neurons (red and blue lines, respectively). Panel A adapted from Meyer et al. (2018); B–D: McCafferty et al. (2018).

present ictally) (Figure 65.2B), which is causally linked to the constitutively enhanced tonic $GABA_A$ current of TC neurons that is present in this AS model, as well as in STG and LH mice (Cope et al., 2009). Nevertheless, the total firing of TC (and NRT) neurons is strictly in phase with each SWD cycle (Figure 65.2C), indicating that the synchrony and strength of the ictal output of TC neurons to the neocortex is not a feature of single, rhythmically bursting TC neurons, as previously suggested (Sorokin et al., 2017), but an emergent property of this neuronal population. Notably, an altered firing that precedes the start of an AS is also present in WAG/Rij rats and STG mice (Sorokin et al., 2017).

A clear limitation in our understanding of thalamic activity during ASs is that the vast majority of available data have been obtained in first-order somatosensory thalamic nuclei (see below), whereas higher-order and intralaminar thalamic nuclei have been relatively neglected. Notwithstanding, the rodent thalamic posterior nucleus may be one of the key thalamic hubs in view of its critical involvement in pre- and post-ictal activity (Lüttjohann and Pape, 2019). Notably, TC neurons in intralaminar thalamic nuclei show a similar decrease in total firing as that of first-order thalamic nuclei during ASs, but present two patterns of ictal burst firing: one with a transient increase in bursts at seizure onset, and one with a sustained increase in burst firing that lasts for the duration of the seizure (Atherton et al., 2023). Moreover, it is interesting to note that extracellular alkalosis elicits ASs by activating intralaminar thalamic, but not cortical and first-order thalamic neurons of WAG/Rij rats (Salvati et al., 2022), a finding of key clinical significance since hyperventilation is a trigger of ASs in many CAE patients.

Finally, although ASs are generated by CT network, both the direct and indirect basal ganglia pathways provide a fine control of the expression of these seizures (Deransart and Depaulis, 2002). Ictal changes in MRI BOLD signals in the basal ganglia of CAE cohorts have been consistently reported (Bai et al., 2010). Moreover, genetic deletion of the Munc-18-1 protein in cortico-striatal L5 neurons elicits ETX-sensitive ASs, and activation of DREADD-transfected striatal fast-spiking interneurons blocks the spontaneous ASs of mice with whole-brain knock-out (KO) of Munc-18-1 (Miyamoto et al., 2019).

In summary, most cortical neurons decrease their firing during ASs, but there are significant neuronal pools that either increase or do not change their firing rate. In the thalamus, the total and burst firing rates of TC neurons are reduced, but that of the GABAergic NRT neurons is increased. Each cortical and thalamic neuron may show different activity changes from one seizure to the next. Intrathalamic thalamic nuclei are critical for AS synchronization and maintenance.

INDUCED GENETIC ALTERATIONS IN CT NETWORKS THAT LEAD TO ASs

To understand how alterations in different neuronal populations of CT networks lead to the expression of ASs (i.e., to an abnormal function of these networks), many studies have used genetic tools to manipulate voltage- and ligand-gated membrane channels in normal animals. In the following, we provide a detailed summary of all the studies that genetically manipulated voltage- and ligand-gated membrane channels in normal non-epileptic animals and compare these findings with studies that have investigated abnormalities in voltage- and ligand-gated membrane channels of 5 well-validated AS models: GAERS and WAG/Rij rats, and STG, TG, and LH mice. Due to space limitations, we report changes in the expression levels of intrinsic and synaptic membrane channels only if their functional consequences have been investigated. Moreover, the abnormalities are described in comparison with the respective non-epileptic control animals of each model.

LIGAND-GATED CHANNELS

$GABA_AR$ $\alpha3$ sub-unit KO mice do not express spontaneous ASs, but show a decrease in GHB-induced SWDs (Winsky-Sommerer et al., 2008), whereas $GABA_AR\beta3$ sub-unit KO mice show a complex epileptic phenotype that only at times includes ASs (DeLorey et al., 1998). Spontaneous ASs are observed in mice knock-in (KI) for the human $\gamma2$ sub-unit R43Q mutation (Tan et al., 2007), which had been identified in probands with febrile seizures and ASs (see Crunelli and Leresche, 2002), whereas whole-brain KO of $GABA_BR1$ or $GABA_BR2$ lead only to rare or no spontaneous SWDs, respectively (Gassmann et al., 2004; Schuler et al., 2001).

Paradoxically, whole brain ablation of one of the sub-units of the α-amino-3-hydroxy-5-methyl-4-isoxazolepropionic acid (AMPA) receptors (*GRIA4* KO mice) results in spontaneous ASs with decreased excitatory postsynaptic currents (EPSCs) in NRT but not TC neurons (Paz et al., 2011). Notably, decreased evoked EPSCs (eEPSCs) are present in TC neurons of TG mice and reduced NMDA and non-NMDA eEPSCs in LH mice (Caddick et al., 1999).

A common finding in GAERS rats and STG and LH mice is the constitutively enhanced tonic $GABA_A$ current of TC neurons that results from a non-genetic loss-of-function of GAT1 (one of the GABA transporters) that is necessary and sufficient for AS expression (Cope et al., 2009). In TC neurons of LH mice, $GABA_B$ evoked IPSCs (eIPSCs) are decreased but there is no change in $GABA_A$ eIPSCs and baclofen-evoked currents (Caddick et al., 1999). The lack of changes in $GABA_A$ IPSCs is shared by other models since no difference was found in $GABA_A$ miniature IPSCs (mIPSCs) and eIPSCs in TC neurons of GAERS rats (Bessaïh et al., 2006; Cope et al., 2009) and TG mice (Caddick et al., 1999), respectively. An increased amplitude and decreased frequency and duration of mIPSCs, but no change in PPD, were reported in GAERS NRT neurons (Bessaïh et al., 2006).

As far as excitatory transmitter function is concerned, a detailed analysis of synaptic potentials in NRT neurons has only been carried out in STG mice. Using a combination of electrical stimulation and photolysis of caged-glutamate, it was elegantly shown that a trafficking defect of AMPA receptors (i.e., decreased synaptic and increased perisynaptic receptors) leads to a reduction of AMPA EPSCs and a compensatory increase of NMDA EPSCs efficacy, leading to an increased NMDA/AMPA charge of EPSCs of both thalamic and cortical origin (Lacey et al., 2012).

VOLTAGE-GATED CA^{+2} CHANNELS

Selective knock-down (KD) of $Ca_V2.1$ channels in L6 pyramidal neurons of normal mice leads to spontaneous ETX-sensitive ASs accompanied by increased T-type Ca^{2+} currents in the TC and NRT neurons and increased burst firing in the latter but not the former thalamic neural population (Bomben et al., 2016) (Figure 65.3). Since no changes in cortical $Ca_V2.1$ channels expression were observed with light microscopy, T-type Ca^{2+} currents and associated burst firing were not investigated in cortical excitatory and inhibitory neurons in these L6 pyramidal neuron-selective KD mice, making it difficult to accept the authors' conclusion that changes in thalamic neuron excitability were the only ones causally linked to AS expression. Indeed, ablation of $Ca_V2.1$ channels in adult mice brings about ASs that are not associated with changes in T-type Ca^{2+} currents or burst firing in TC neurons (Miao et al., 2019). Similarly, selective KO of $Ca_V2.1$ channels in PV+ and somatostatin-positive (SST+) cortical interneurons leads to ASs with no change in T-type Ca^{2+} currents or burst firing of TC neurons (Figure 65.3) (Rossignol et al., 2013), demonstrating that a constitutive or induced increase in these TC neuron parameters, as observed in TG mice and mice with selective L6 pyramidal neurons KD of $Ca_V2.1$ channels, is not necessary for the expression of ASs.

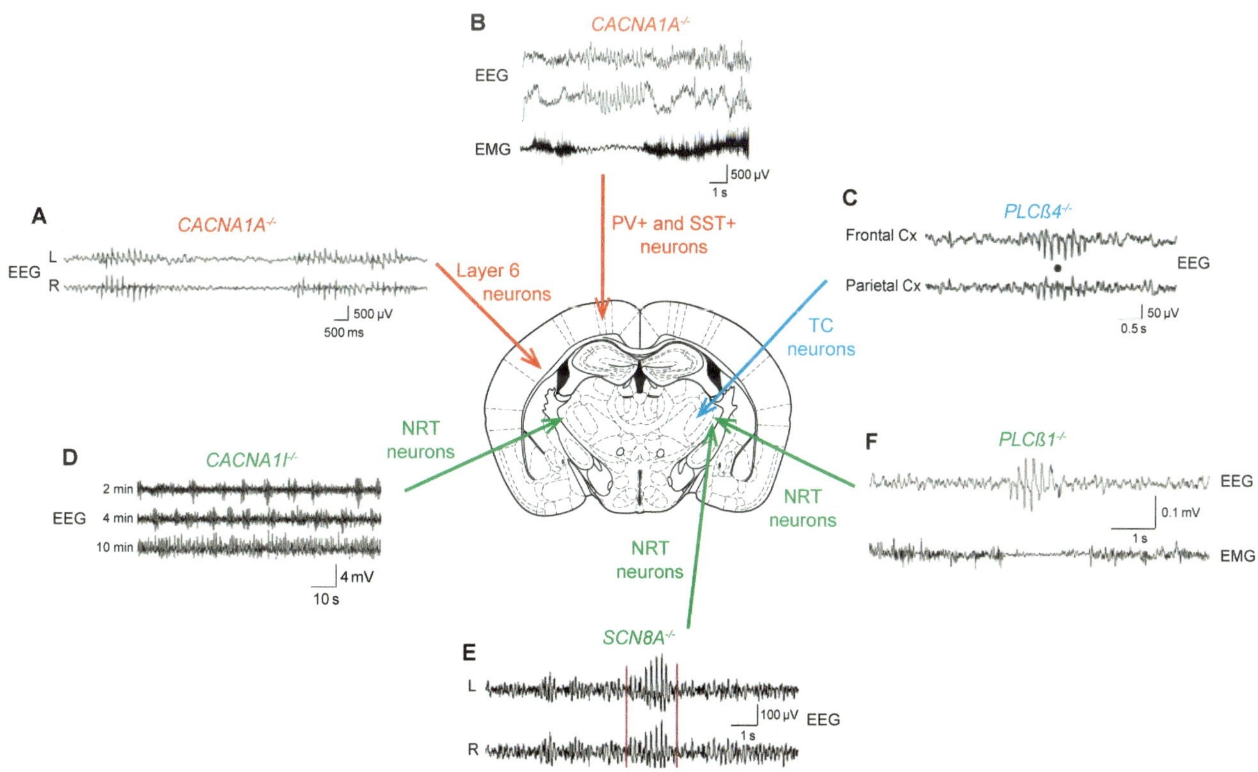

Figure 65.3. Single genetic ablations that lead to absence seizures (ASs). Schematic brain drawing (center panel) showing the location of the different single mutations induced in either L6 neurons, cortical parvalbumin- and somatostatin-positive (PV+ and SST+, respectively) interneurons, thalamocortical (TC) and nucleus reticularis thalami (NRT) neurons that have been shown to elicit spontaneous ASs (A, B, C, E, and F) or increase γ-hydroxybutyric acid-induced ASs (D). Panels show electroencephalographic (EEG) and electromyographic (EMG) activity in mice with the illustrated genetic ablation. Minutes on the left-hand side of the traces in D indicate the time from the γ-hydroxybutyric acid injection (Cx = cortex; L = left; R = right). Panel A adapted from Bomben et al. (2016); B: Rossignol et al. (2013); C: Cheong et al. (2009); D: Lee et al. (2014); E: Makinson et al. (2017); F: Chang et al. (2022).

$Ca_V2.3$ channels are expressed in cortical and NRT, but not TC neurons, and their whole-brain KO leads to a decrease in GHB-induced ASs (Zaman et al., 2011). However, whether this pathological phenotype is due to NRT or cortical $Ca_V2.3$ channels remains to be investigated.

Whole-brain overexpression of $Ca_V3.1$ channels lead to spontaneous ETX-sensitive ASs (Ernst et al., 2009), whereas mice with KO of these channels show a decrease in baclofen and GHB-induced SWDs (Kim et al., 2001). Notably, the spontaneous ASs observed in $Ca_V2.1$ KO mice (Bomben et al., 2016; Miao et al., 2019; Song et al., 2004) are abolished by the simultaneous KO of $Ca_V3.1$ channels. However, despite the substantial expression of $Ca_V3.1$ channels in cortical territories (Talley et al., 1999), the analysis of the cellular/network abnormalities in these single and double KO mice was only carried out in the thalamus. So, though these studies strongly point to a key role of $Ca_V3.1$ channels in the genesis of ASs, they do not clarify the brain area(s) and neuronal population(s) where these channels are critical for AS expression (see below).

Notably, elevated T-type Ca^{2+} currents (mainly due to $Ca_V3.1$ channels) are present in TC neurons of different thalamic nuclei of 4 AS models (WAG/Rij rat, and STG, TG. and LG mice) (Broicher et al., 2008; Zhang et al., 2002). The block of ASs by whole-brain genetic ablation of $Ca_V3.1$ channels in STG, TG, and LH mice has been causally linked to the loss of function of these channels in TC neurons (Song et al., 2004). However, such conclusion is flawed since the block of ASs in these mice could also be due to the simultaneous ablation of $Ca_V3.1$ channels in the cortex where they are highly expressed (Talley et al., 1999). Indeed, ASs in GAERS rats are not affected by selectively blocking $Ca_V3.1$ channels and related burst firing of TC neurons with a potent T-type Ca^{2+} channels antagonist (McCafferty et al., 2018). Notwithstanding, TC neuron firing is necessary for AS expression since brief optogenetic manipulation of TC neurons abort ASs in WAG/Rij rats and STG mice (Sorokin et al., 2017).

Mice with whole-brain KO or selective KD of $Ca_V3.3$ channels in the NRT do not show spontaneous ASs, but an unexpected increase in GHB-induced ASs (Lee et al., 2014) (Figure 65.3). The same phenotype is observed in double KO mice of $Ca_V3.3$ and $Ca_V3.2$ channels (Lee et al., 2014) where the T-type Ca^{2+} current of NRT neurons is completely abolished. Thus, notwithstanding the key role of the NRT neurons T-type Ca^{2+} current (and the burst firing that it generates) in physiological activities expressed by CT networks (Crunelli et al., 2018), full ablation of these NRT channels leads to potentiation of pharmacologically induced ASs. Whether this is due to the increased tonic firing of NRT neurons that is present in these single and double KO mice (Lee et al., 2014) or results from alterations in the activity of cortical neuronal populations remains to be elucidated since the analysis of the cellular and network abnormalities of these mice was only carried out in the thalamus.

Elevated T-type Ca^{2+} currents are also present in the NRT neurons of GAERS and WAG rats (Broicher et al., 2008; Cain et al., 2018), but have not been studied in the NRT of STG, TG, and LH mice. In GAERS rats, this increased current is due to the R158P missense variant of *CACNA1H* (the $Ca_V3.2$ channel gene) and selective KD of these channels in NRT neurons markedly reduces ASs and normalizes their elevated burst firing (Cain et al., 2018). Moreover, KI of the human *CACNA1H* variant C456S in S1 of normal non-epileptic rats elicits spontaneous ASs in >60% of the injected animals (Wang et al., 2015).

VOLTAGE-GATED NA^+ AND NA^+/K^+ CHANNELS

Ablation of $Na_V1.2$ channels in the whole brain or selectively in telencephalic and diencephalic neurons leads to spontaneous ETX-sensitive ASs, with smaller SWDs being present in the thalamus compared to those in S1, medial prefrontal cortex, and hippocampus (Ogiwara et al., 2018). Ablation of $Na_V1.6$ channels either in the whole brain, only in inhibitory neurons, or in the NRT (Figure 65.3) brings about ETX-sensitive ASs, whereas no seizures are observed when these channels are ablated in either all cortical neurons, in glutamatergic excitatory neurons, or only in inhibitory neurons of the cortex, hippocampus, and striatum (Makinson et al., 2017).

Mice with whole-brain KO of hyperpolarization-activated, cyclic nucleotide-gated Na^+/K^+ channels subtype 2 (HCN2) show spontaneous ASs (Ludwig et al., 2003) but HCN4 channel KO mice do not (Zobeiri et al., 2019). However, extensive analysis of thalamic neuron abnormalities but no data on cortical neurons were provided in these two studies, leaving a knowledge gap on the relative contribution of HCN2 and HCN4 channels in the cortical and thalamic sectors of the CT network. In TC neurons of different first-order thalamic nuclei of both GAERS and WAG/Rij rats, there is an increased I_h (Budde et al., 2005; Cain et al., 2015; Kanyshkova et al., 2012), that most likely results from an increased HCN1 channel density with no change in HCN2 and HCN4 expression (Cain et al., 2015). Notably, the increased density of HCN1 channels in thalamic nuclei of WAG/Rij rats (Kuisle et al., 2006) is in contrast to their decreased expression in S1 (Kole et al., 2007) (see above), indicating that in a given AS model, and potentially in humans with ASs, the same channels can be differentially expressed in diverse sectors of the CT network. Notably, pharmacological block of all HCN subtypes systemically, or locally in S1 or VB, blocks ASs (Iacone et al., 2021).

In summary, the following significant conclusions arise from these studies:

(1) Genetic ablations of diverse *single*-channel abnormalities in *single* neuronal populations of the CT network lead to spontaneous ETX-sensitive ASs (Figure 65.3); $Ca_V2.1$ channels in cortical L6 neurons (Bomben et al., 2016), $Ca_V2.1$ channels in cortical PV+ and SST+ interneurons (Rossignol et al., 2013), $Na_V1.6$ channels in NRT neurons (Makinson et al., 2017), phospholipase C β1sub-unit in NRT neurons (Chang et al., 2022), and phospholipase C β4 sub-unit in TC neurons (Cheong et al., 2009), whereas selective KD of $Ca_V3.3$ channels in NRT neurons does not lead to spontaneous ASs but increases GHB-induced SWDs (Lee et al., 2014).

(2) Secondary (compensatory) changes in a CT sector other than the one where the primary genetic manipulation had been made may be the critical element for ASs expression, indicated by the increased T-type Ca^{2+} currents in TC neurons of mice with selective KD of $Ca_V2.1$ channels in cortical L6 neurons (Bomben et al., 2016).

(3) Different changes of membrane proteins expression may occur in diverse areas of CT network, as indicated by the decreased and increased expression of HCN1 channels in the cortex and thalamus, respectively, of WAG/Rij rats (Cain et al., 2015; Kole et al., 2007).

(4) In view of the cortical location of SWD initiation in humans and animal models (see above), it is unfortunate that almost all investigations that used whole-brain genetic manipulations have studied abnormalities only in thalamic but not cortical neurons, even when the manipulated channels are (highly) expressed in both the cortical and thalamic sectors of CT networks.

CONCLUSIONS

Abnormal genes found either in small Chinese CAE cohorts or in mice AS models are not located in the significant loci (2p16.1 and 2q22) identified in the only large genome-wide CAE study, though *CACNA1H* variants are present in Chinese probands and the GAERS rat model of ASs. Selective ablation of $Ca_V2.1$, $Ca_V3.3$, and $Na_V1.6$ channels, as well as phospholipase C β1 and β4 sub-units in either L6 pyramidal neurons, cortical PV+, and SST+ interneurons, TC or NRT neurons of normal non-epileptic animals, leads to ASs. In contrast to focal seizures and other GGEs, the ictal activity of cortical and thalamic neurons is predominantly characterized by a decreased, but highly synchronized, firing with all component members of a neuronal population often changing firing pattern from one seizure to the next. Selective chemogenetic and optogenetic manipulation of S1 PV+ interneurons and TC neurons, respectively, conclusively demonstrates that these neuronal populations are necessary for AS expression. The evidence in normal non-epileptic animals that a single manipulation of different voltage- and ligand-gated membrane channels does lead to ASs indicates the liability of CT networks toward the generation of these seizures. On the other hand, the high remittance rate of children with CAE indicates that CT networks, and/or their modulatory inputs from the basal ganglia and brainstem, have the inherent capacity to compensate for the developmental abnormalities that lead to ASs. However, the mechanism of remittance has not been investigated since all available AS models lack this feature of the human disease.

ACKNOWLEDGMENTS

Work in our labs is supported by the CNRS, Sorbonne Université, and Inserm (grants to NL), and the Ester Floridia Neuroscience Research Foundation (grant 1502 to VC). The authors declare no conflict of interest.

REFERENCES

Atherton Z, Nagy O, Barcsai L, Sere P, Zsigri N, Földi T, Berényi A, Crunelli V, Lőrincz ML (2023). Higher-order thalamic nuclei facilitate the generalization and maintenance of spike-and-wave discharges of absence seizures. *Neurobiol Dis.* 178:106025.

Avoli M (2012). A brief history on the oscillating roles of thalamus and cortex in absence seizures. *Epilepsia.* 53:779–789.

Bai X, Vestal M, Berman R, Negishi M, Spann M, Vega C, Desalvo M, Novotny EJ, Constable RT, Blumenfeld H (2010). Dynamic time course of typical childhood absence seizures: EEG, behavior, and functional magnetic resonance imaging. *J Neurosci.* 30:5884–5893.

Bessaïh T, Bourgeais L, Badiu CI, Carter DA, Toth TI, Ruano D, Lambolez B, Crunelli V, Leresche N (2006). Nucleus-specific abnormalities of GABAergic synaptic transmission in a genetic model of absence seizures. *J Neurophysiol.* 96:3074–3081.

Bomben VC, Aiba I, Qian J, Mark MD, Herlitze S, Noebels JL (2016). Isolated P/Q calcium channel deletion in layer VI corticothalamic neurons generates absence epilepsy. *J Neurosci.* 36:405–418.

Broicher T, Kanyshkova T, Meuth P, Pape H-C, Budde T (2008). Correlation of T-channel coding gene expression, IT, and the low threshold Ca2+ spike in the thalamus of a rat model of absence epilepsy. *Mol Cell Neurosci.* 39:384–399.

Budde T, Caputi L, Kanyshkova T, Staak R, Abrahamczik C, Munsch T, Pape H-C (2005). Impaired regulation of thalamic pacemaker channels through an imbalance of subunit expression in absence epilepsy. *J Neurosci.* 25:9871–9882.

Caddick SJ, Wang C, Fletcher CF, Jenkins NA, Copeland NG, Hosford DA (1999). Excitatory but not inhibitory synaptic transmission is reduced in lethargic (Cacnb4 lh) and tottering (Cacna1a tg) mouse thalami. *J Neurophysiol.* 81:2066–2074.

Cain SM, Tyson JR, Choi H, Ko R, Lin PJC, LeDue JM, Powell KL, Bernier L, Rungta RL, Yang Y, Cullis PR, O'Brien TJ, MacVicar BA, Snutch TP (2018). CaV3.2 drives sustained burst–firing, which is critical for absence seizure propagation in reticular thalamic neurons. *Epilepsia.* 59:778–791.

Cain SM, Tyson JR, Jones KL, Snutch TP (2015). Thalamocortical neurons display suppressed burst-firing due to an enhanced Ih current in a genetic model of absence epilepsy. *Pflugers Arch - Eur J Physiol.* 467:1367–1382.

Casillas-Espinosa PM, Powell KL, Zhu M, Campbell CR, Maia JM, Ren Z, Jones NC, O'Brien TJ, Petrovski S (2017). Evaluating whole genome sequence data from the genetic absence epilepsy rat from Strasbourg and its related non-epileptic strain. *PLoS ONE.* 12:e0179924.

Chang B, Byun J, Kim KK, Lee SE, Lee B, Kim K-S, Ryu H, Shin H-S, Cheong E (2022). Deletion of phospholipase C β1 in the thalamic reticular nucleus induces absence seizures. *Exp Neurobiol.* 31:116–130.

Cheong E, Zheng Y, Lee K, Lee J, Kim S, Sanati M, Lee S, Kim Y-S, Shin H-S (2009). Deletion of phospholipase C β4 in thalamocortical relay nucleus leads to absence seizures. *Proc Natl Acad Sci U S A.* 106:21912–21917.

Cope DW, Di Giovanni G, Fyson SJ, Orbán G, Errington AC, Lorincz ML, Gould TM, Carter DA, Crunelli V (2009). Enhanced tonic GABAA inhibition in typical absence epilepsy. *Nat Med.* 15:1392–1398.

Crunelli V, Leresche N (2002). Childhood absence epilepsy: Genes, channels, neurons and networks. *Nat Rev Neurosci.* 3:371–382.

Crunelli V, Lőrincz ML, Connelly WM, David F, Hughes SW, Lambert RC, Leresche N, Errington AC (2018). Dual function of thalamic low-vigilance state oscillations: rhythm-regulation and plasticity. *Nat Rev Neurosci.* 19:107–118.

Crunelli V, Lőrincz ML, McCafferty C, Lambert RC, Leresche N, Di Giovanni G, David F (2020). Clinical and experimental insight into pathophysiology, comorbidity and therapy of absence seizures. *Brain.* 143:2341–2368.

D'Antuono M, Inaba Y, Biagini G, D'Arcangelo G, Tancredi V, Avoli M (2005). Synaptic hyperexcitability of deep layer neocortical cells in a genetic model of absence seizures: Neocortical excitability and absence epilepsy. *Genes Brain Behav.* 5:73–84.

DeLorey TM, Handforth A, Anagnostaras SG, Homanics GE, Minassian BA, Asatourian A, Fanselow MS, Delgado-Escueta A, Ellison GD, Olsen RW (1998). Mice lacking the β3 subunit of the GABAA receptor have the epilepsy phenotype and many of the behavioral characteristics of Angelman syndrome. *J Neurosci.* 18:8505–8514.

Deransart C, Depaulis A (2002). The control of seizures by the basal ganglia? A review of experimental data. *Epileptic Disord.* 4(Suppl 3):S61–S72.

Ernst WL, Zhang Y, Yoo JW, Ernst SJ, Noebels JL (2009). Genetic enhancement of thalamocortical network activity by elevating alpha 1g-mediated low-voltage-activated calcium current induces pure absence epilepsy. *J Neurosci.* 29:1615–1625.

Feng Y-CA, Howrigan DP, Abbott LE, Tashman K, Cerrato F, Singh T, Heyne H, Byrnes A, Churchhouse C, Watts N, Solomonson M, Lal D, Heinzen EL, Dhindsa RS, Stanley KE, Cavalleri GL, Hakonarson H, Helbig I, Krause R, May P, Weckhuysen S, Petrovski S, Kamalakaran S, Sisodiya SM, Cossette P, Cotsapas C, De Jonghe P, Dixon-Salazar T, Guerrini R, Kwan P, Marson AG, Stewart R, Depondt C, Dlugos DJ, Scheffer IE, Striano P, Freyer C, McKenna K, Regan BM, Bellows ST, Leu C, Bennett CA, Johns EMC, Macdonald A, Shilling H, Burgess R, Weckhuysen D, Bahlo M, O'Brien TJ, Todaro M, Stamberger H, Andrade DM, Sadoway TR, Mo K, Krestel H, Gallati S, Papacostas SS, Kousiappa I, Tanteles GA, Štěrbová K, Vlčková M, Sedláčková L, Laššuthová P, Klein KM, Rosenow F, Reif PS, Knake S, Kunz WS, Zsurka G, Elger CE, Bauer J, Rademacher M, Pendziwiat M, Muhle H, Rademacher A, van Baalen A, von Spiczak S, Stephani U, Afawi Z, Korczyn AD, Kanaan M, Canavati C, Kurlemann G, Müller-Schlüter K, Kluger G, Häusler M, Blatt I, Lemke JR, Krey I, Weber YG, Wolking S, Becker F, Hengsbach C, Rau S, Maisch AF, Steinhoff BJ, Schulze-Bonhage A, Schubert-Bast S, Schreiber H, Borggräfe I, Schankin CJ, Mayer T, Korinthenberg R, Brockmann K, Kurlemann G, Dennig D, Madeleyn R, Kälviäinen R, Auvinen P, Saarela A, Linnankivi T, Lehesjoki A-E, Rees MI, Chung S-K, Pickrell WO, Powell R, Schneider N, Balestrini S, Zagaglia S, Braatz V, Johnson MR, Auce P, Sills GJ, Baum LW, Sham PC, Cherny SS, Lui CHT, Barišić N, Delanty N, Doherty CP, Shukralla A, McCormack M, El-Naggar H, Canafoglia L, Franceschetti S, Castellotti B, Granata T, Zara F, Iacomino M, Madia F, Vari MS, Mancardi MM, Salpietro V, Bisulli F, Tinuper P, Licchetta L, Pippucci T, Stipa C, Minardi R, Gambardella A, Labate A, Annesi G, Manna L, Gagliardi M, Parrini E, Mei D, Vetro A, Bianchini C, Montomoli M, Doccini V, Marini C, Suzuki T, Inoue Y, Yamakawa K, Tumiene B, Sadleir LG, King C, Mountier E, Caglayan SH, Arslan M, Yapıcı Z, Yis U, Topaloglu P, Kara B, Turkdogan D, Gundogdu-Eken A, Bebek N, Uğur-İşeri S, Baykan B, Salman B, Haryanyan G, Yücesan E, Kesim Y, Özkara Ç, Poduri A, Shiedley BR, Shain C, Buono RJ, Ferraro TN, Sperling MR, Lo W, Privitera M, French JA, Schachter S, Kuzniecky RI, Devinsky O, Hegde M, Khankhanian P, Helbig KL, Ellis CA, Spalletta G, Piras Fabrizio, Piras Federica, Gili T, Ciullo V, Reif A, McQuillin A, Bass N, McIntosh A, Blackwood D, Johnstone M, Palotie A, Pato MT, Pato CN, Bromet EJ, Carvalho CB, Achtyes ED, Azevedo MH, Kotov R, Lehrer DS, Malaspina D, Marder SR, Medeiros H, Morley CP, Perkins DO, Sobell JL, Buckley PF, Macciardi F, Rapaport MH, Knowles JA, Fanous AH, McCarroll SA, Gupta N, Gabriel SB, Daly MJ, Lander ES, Lowenstein DH, Goldstein DB, Lerche H, Berkovic SF, Neale BM (2019). Ultra-rare genetic variation in the epilepsies: A whole-exome sequencing study of 17,606 individuals. *Am J Human Genet.* 105:267–282.

Frankel WN (2009). Genetics of complex neurological disease: challenges and opportunities for modeling epilepsy in mice and rats. *Trends Genet.* **25**:361–367.

Gassmann M, Shaban H, Vigot R, Sansig G, Haller C, Barbieri S, Humeau Y, Schuler V, Müller M, Kinzel B, Klebs K, Schmutz M, Froestl W, Heid J, Kelly PH, Gentry C, Jaton A-L, Van der Putten H, Mombereau C, Lecourtier L, Mosbacher J, Cryan JF, Fritschy J-M, Lüthi A, Kaupmann K, Bettler B (2004). Redistribution of GABAB(1) protein and atypical GABAB responses in GABAB(2)-deficient mice. *J Neurosci.* 24:6086–6097.

Gruenbaum BF, Sandhu MRS, Bertasi RAO, Bertasi TGO, Schonwald A, Kurup A, Gruenbaum SE, Freedman IG, Funaro MC, Blumenfeld H, Sanacora G (2021). Absence seizures and their relationship to depression and anxiety: Evidence for bidirectionality. *Epilepsia.* 62:1041–1056.

Guo J, Kim R, Chen Y, Negishi M, Jhun S, Weiss S, Ryu JH, Bai X, Xiao W, Feeney E, Rodriguez-Fernandez J, Mistry H, Crunelli V, Crowley MJ, Mayes LC, Todd Constable R, Blumenfeld H (2016). Mechanism of impaired consciousness in absence seizures: A cross-sectional study. *Lancet Neurol.* 15:1336–1345.

Harutyunyan A, Chong D, Li R, Shah AD, Ali Z, Huang C, Barlow CK, Perucca P, O'Brien TJ, Jones NC, Schittenhelm RB, Anderson A, Casillas-Espinosa PM (2022). An integrated multi-omic network analysis identifies seizure-associated dysregulated pathways in the GAERS model of absence epilepsy. *Int J Mol Sci.* 23:6063.

Iacone Y, Morais TP, David F, Delicata F, Sandle J, Raffai T, Parri HR, Weisser JJ, Bundgaard C, Klewe IV, Tamás G, Thomsen MS, Crunelli V, Lőrincz ML (2021). Systemic administration of ivabradine, a hyperpolarization-activated cyclic nucleotide-gated channel inhibitor, blocks spontaneous absence seizures. *Epilepsia.* 62:1729–1743.

International League Against Epilepsy Consortium on Complex Epilepsies (2018). Genome-wide mega-analysis identifies 16 loci and highlights diverse biological mechanisms in the common epilepsies. *Nat Commun.* 9:5269.

Kanyshkova T, Meuth P, Bista P, Liu Z, Ehling P, Caputi L, Doengi M, Chetkovich DM, Pape H-C, Budde T (2012). Differential regulation of HCN channel isoform expression in thalamic neurons of epileptic and non-epileptic rat strains. *Neurobiol Dis.* 45:450–461.

Kim D, Song I, Keum S, Lee T, Jeong M-J, Kim S-S, McEnery MW, Shin H-S (2001). Lack of the burst firing of thalamocortical relay neurons and resistance to absence seizures in mice lacking α1G T-Type Ca2+ channels. *Neuron.* 31:35–45.

Kole MH, Bräuer AU, Stuart GJ (2007). Inherited cortical HCN1 channel loss amplifies dendritic calcium electrogenesis and burst firing in a rat absence epilepsy model. *J Physiol.* 578:507–525.

Kuisle M, Wanaverbecq N, Brewster AL, Frère SGA, Pinault D, Baram TZ, Lüthi A (2006). Functional stabilization of weakened thalamic pacemaker channel regulation in rat absence epilepsy. *J Physiol.* 575:83–100.

Lacey CJ, Bryant A, Brill J, Huguenard JR (2012). Enhanced NMDA receptor-dependent thalamic excitation and network oscillations in stargazer mice. *J Neurosci.* 32:11067–11081.

Lee SE, Lee J, Latchoumane C, Lee B, Oh SJ, Saud ZA, Park C, Sun N, Cheong E, Chen CC, Choi EJ, Lee CJ, Shin H-S (2014). Rebound burst firing in the reticular thalamus is not essential for pharmacological absence seizures in mice. *Proc Natl Acad Sci U S A.* 111:11828–11833.

Ludwig A, Budde T, Stieber J, Moosmang S, Wahl C, Holthoff K, Langebartels A, Wotjak C, Munsch T, Zong X, Feil S, Feil R, Lancel M, Chien KR, Konnerth A, Pape H-C, Biel M, Hofmann F (2003). Absence epilepsy and sinus dysrhythmia in mice lacking the pacemaker channel HCN2. *EMBO J.* 22:216–224.

Lüttjohann A, Pape H-C (2019). Regional specificity of corticothalamic coupling strength and directionality during waxing and waning of spike and wave discharges. *Sci Rep.* 9:2100.

Makinson CD, Tanaka BS, Sorokin JM, Wong JC, Christian CA, Goldin AL, Escayg A, Huguenard JR (2017). Regulation of thalamic and cortical network synchrony by Scn8a. *Neuron.* 93(5):1–22.

McCafferty C, David F, Venzi M, Lőrincz ML, Delicata F, Atherton Z, Recchia G, Orban G, Lambert RC, Di Giovanni G, Leresche N, Crunelli V (2018). Cortical drive and thalamic feed-forward inhibition control thalamic output synchrony during absence seizures. *Nat Neurosci.* 21:744–756.

McCafferty C, Gruenbaum B, Tung R, Li J-J, Zheng X, Salvino P, Vincent P, Kratochvil Z, Ryu JH, Khalaf A, Swift K, Akbar R, Islam W, Antwi P, Johnson EA, Vitkovskiy P, Sampognaro J, Freedman I, Kundishora A, Depaulis A, David F, Crunelli V, Sanganahalli BG, Herman P, Hyder F, Blumenfeld H (2023). The neuronal basis of impaired consciousness in absence seizures. *Nat Commun.* 14(1):117.

Meeren HKM, Pijn JPM, Luijtelaar ELJMV, Coenen AML, Silva FHL da (2002). Cortical focus drives widespread corticothalamic networks during spontaneous absence seizures in rats. *J Neurosci.* 22:1480–1495.

Meyer J, Maheshwari A, Noebels J, Smirnakis S (2018). Asynchronous suppression of visual cortex during absence seizures in stargazer mice. *Nat Commun.* 9:1938.

Miao Q-L, Herlitze S, Mark MD, Noebels JL (2019). Adult loss of Cacna1a in mice recapitulates childhood absence epilepsy by distinct thalamic bursting mechanisms. *Brain.* 143:161–174.

Miyamoto H, Tatsukawa T, Shimohata A, Yamagata T, Suzuki T, Amano K, Mazaki E, Raveau M, Ogiwara I, Oba-Asaka A, Hensch TK, Itohara S, Sakimura K, Kobayashi Kenta, Kobayashi Kazuto, Yamakawa K (2019). Impaired cortico-striatal excitatory transmission triggers epilepsy. *Nat Commun.* 10:1917.

Ogiwara I, Miyamoto H, Tatsukawa T, Yamagata T, Nakayama T, Atapour N, Miura E, Mazaki E, Ernst SJ, Cao D, Ohtani H, Itohara S, Yanagawa Y, Montal M, Yuzaki M, Inoue Y, Hensch TK, Noebels JL, Yamakawa K (2018). Nav1.2 haplodeficiency in excitatory neurons causes absence-like seizures in mice. *Commun Biol.* 1:96.

Panthi S, Leitch B (2019). The impact of silencing feed-forward parvalbumin-expressing inhibitory interneurons in the corticothalamocortical network on seizure generation and behaviour. *Neurobiol Dis.* 132:104610.

Papp P, Kovács Z, Szocsics P, Juhász G, Maglóczky Z. 2018. Alterations in hippocampal and cortical densities of functionally different interneurons in rat models of absence epilepsy. *Epilepsy Res.* 145:40–50.

Paz JT, Bryant AS, Peng K, Fenno L, Yizhar O, Frankel WN, Deisseroth K, Huguenard JR. 2011. A new mode of corticothalamic transmission revealed in the Gria4(-/-) model of absence epilepsy. *Nat Neurosci.* 14:1167–1173.

Perucca P, Bahlo M, Berkovic SF (2020). The genetics of epilepsy. *Ann Rev Genom Hum Genet.* 21:205–230.

Pitkänen A, Buckmaster P, Galanopoulou AS, Moshé SL, eds. (2017). *Models of Seizures and Epilepsy.* 2nd ed. Elsevier.

Polack PO, Guillemain I, Hu E, Deransart C, Depaulis A, Charpier S (2007). Deep layer somatosensory cortical neurons initiate spike-and-wave discharges in a genetic model of absence seizures. *J Neurosci.* 27:6590–6599.

Rossignol E, Kruglikov I, van den Maagdenberg AMJM, Rudy B, Fishell G (2013). CaV2.1 ablation in cortical interneurons selectively impairs fast-spiking basket cells and causes generalized seizures. *Ann Neurol.* 74:209–222.

Salvati KA, Souza GM, Lu AC, Ritger ML, Guyenet P, Abbott SB, Beenhakker MP (2022). Respiratory alkalosis provokes spike-wave discharges in seizure-prone rats. *eLife.* 11:e72898.

Sasaki S, Huda K, Inoue T, Miyata M, Imoto K (2006). Impaired feed-forward inhibition of the thalamocortical projection in epileptic Ca2+ channel mutant mice, tottering. *J Neurosci.* 26:3056–3065.

Schuler V, Lüscher C, Blanchet C, Klix N, Sansig G, Klebs K, Schmutz M, Heid J, Gentry C, Urban L, Fox A, Spooren W, Jaton A-L, Vigouret J-M, Pozza M, Kelly PH, Mosbacher J, Froestl W, Käslin E, Korn R, Bischoff S, Kaupmann K, van der Putten H, Bettler B (2001). Epilepsy, hyperalgesia, impaired memory, and loss of pre- and postsynaptic GABAB responses in mice lacking GABAB(1). *Neuron.* 31:47–58.

Song I, Kim D, Choi S, Sun M, Kim Y, Shin H-S (2004). Role of the α1G T-type calcium channel in spontaneous absence seizures in mutant mice. *J Neurosci.* 24:5249–5257.

Sorokin JM, Davidson TJ, Frechette E, Abramian AM, Deisseroth K, Huguenard JR, Paz JT (2017). Bidirectional control of generalized epilepsy networks via rapid real-time switching of firing mode. *Neuron.* 93:194–210.

Talley EM, Cribbs LL, Lee JH, Daud A, Perez-Reyes E, Bayliss DA (1999). Differential distribution of three members of a gene family encoding low voltage-activated (T-type) calcium channels. *J Neurosci.* 19:1895–1911.

Tan HO, Reid CA, Single FN, Davies PJ, Chiu C, Murphy S, Clarke AL, Dibbens L, Krestel H, Mulley JC, Jones MV, Seeburg PH, Sakmann B, Berkovic SF, Sprengel R, Petrou S (2007). Reduced cortical inhibition in a mouse model of familial childhood absence epilepsy. *Proc Natl Acad Sci U S A.* 104:17536–17541.

Wang G, Bochorishvili G, Chen Y, Salvati KA, Zhang P, Dubel SJ, Perez-Reyes E, Snutch TP, Stornetta RL, Deisseroth K, Erisir A, Todorovic SM, Luo J-H, Kapur J, Beenhakker MP, Zhu JJ (2015). Ca V3.2 calcium channels control NMDA receptor-mediated transmission: a new mechanism for absence epilepsy. *Genes Dev.* 29:1535–1551.

Winsky-Sommerer R, Knapman A, Fedele DE, Schofield CM, Vyazovskiy VV, Rudolph U, Huguenard JR, Fritschy JM, Tobler I (2008). Normal sleep homeostasis and lack of epilepsy phenotype in GABAA receptor Î±3 subunit-knockout mice. *Neuroscience.* 154:595–605.

Xie H, Su W, Pei J, Zhang Yujia, Gao K, Li J, Ma X, Zhang Yuehua, Wu X, Jiang Y (2019). De novo SCN1A, SCN8A, and CLCN2 mutations in childhood absence epilepsy. *Epilepsy Res.* 154:55–61.

Zaman T, Lee K, Park C, Paydar A, Choi JH, Cheong E, Lee CJ, Shin H-S (2011). CaV2.3 channels are critical for oscillatory burst discharges in the reticular thalamus and absence epilepsy. *Neuron.* 70:95–108.

Zhang Y, Mori M, Burgess DL, Noebels JL (2002). Mutations in high-voltage-activated calcium channel genes stimulate low-voltage-activated currents in mouse thalamic relay neurons. *J Neurosci.* 22:6362–6371.

Zobeiri M, Chaudhary R, Blaich A, Rottmann M, Herrmann S, Meuth P, Bista P, Kanyshkova T, Lüttjohann A, Narayanan V, Hundehege P, Meuth SG, Romanelli MN, Urbano FJ, Pape H-C, Budde T, Ludwig A (2019). The hyperpolarization-activated HCN4 channel is important for proper maintenance of oscillatory activity in the thalamocortical system. *Cereb Cortex.* 29:2291–2304.

66.

THALAMIC DYSCONNECTIVITY ACROSS THE SCHIZOPHRENIA ILLNESS COURSE AND PSYCHOSIS SPECTRUM

Samantha V. Abram, Caleb M. Spiro, and Judith M. Ford

INTRODUCTION

Schizophrenia is a complex and heterogenous mental illness that has major impacts on a person's ability to think, behave, perceive reality, and manage emotion. The primary symptoms are organized in positive, negative, and disorganized domains. Positive symptoms include hallucinations (often auditory) and delusions (fixed, false beliefs). Negative symptoms reflect a decrease in speech, affective expression, motivation, and the ability to experience pleasure. Disorganization can be evident in both thinking (e.g., incoherence, frequent derailments or tangents) and behavior (e.g., dressing oddly). Schizophrenia is also associated with notable deficits in cognitive and social abilities that have major impacts on real-world functioning.

There is a long history implicating the thalamus in the pathophysiology of schizophrenia, as well as other psychiatric and neurologic diseases (see **Chapters 65, 67, 68, 69, 70, and 71 in Section XII**). Narrative accounts from people with schizophrenia highlighted difficulties filtering information that might in turn undermine one's ability to distinguish the self from the external world (McGhie and Chapman, 1961). Failures to filter information were hypothesized to produce the positive symptoms of schizophrenia, such as auditory hallucinations or persecutory delusions (Andreasen, 1997). The thalamus is a strong candidate for explaining these illness features given its role in gating input from the cortex and brainstem (Wang et al., 2010). In support of these early notions, empirical studies of the thalamus in schizophrenia reveal impaired neuronal integrity, reduced volume, and altered activity during cognitive tasks (Huang et al., 2021).

Several decades ago, it was proposed that schizophrenia is a disorder of connectivity (Friston and Frith, 1995); that is, symptoms of schizophrenia could relate to abnormal or inefficient communication across brain regions and/or disruptions in white matter integrity. This hypothesis grew, in part, from the pivotal observation of a functional dissociation between recruitment of frontal and temporal cortex in people with schizophrenia during a verbal fluency task (Frith et al., 1995). Since then, advances in magnetic resonance imaging (MRI) and analytic techniques have led to a proliferation of studies on this topic. Altered thalamocortical connectivity is now considered a central feature of psychosis (Barch, 2021) and is present among those at clinical high risk for psychosis (Anticevic et al., 2015; Fryer et al., 2021), and those with psychotic-spectrum disorders beyond schizophrenia, such as bipolar disorder (Anticevic et al., 2014a; Sheffield et al., 2020). Because the thalamus is a crucial node in most neural computations via extensive thalamocortical loops (Alexander et al., 1986), thalamic connectivity remains an important target for efforts seeking to characterize the diverse symptoms of schizophrenia.

The current chapter reviews evidence for thalamic connectivity abnormalities in schizophrenia and the implications with respect to cognitive and clinical illness features. We first cover common neuroimaging approaches for studying brain connectivity in humans, as well as the major thalamic connectivity deviations observed in schizophrenia, including both hypo- and hyperconnectivity patterns. We also discuss the extent to which thalamic connectivity abnormalities are seen at different points in the illness phase, including those with high-risk illness features who eventually convert to a full psychotic disorder, providing evidence to support notions of schizophrenia as a neurodevelopmental disorder. And while this literature provides compelling support for thalamic connectivity abnormalities as biomarkers for psychosis, such thalamic pathology may not be specific to schizophrenia. Accordingly, we close this review by covering recent and growing evidence of thalamic connectivity alterations observed transdiagnostically (i.e., across mental illness diagnostic categories).

MRI-BASED APPROACHES FOR EXAMINING THALAMIC CONNECTIVITY IN THE HUMAN BRAIN

Advances in neuroimaging methods make it possible to evaluate thalamocortical networks in vivo. While diffusion tensor imaging (DTI) was historically used to define large white-matter tracts in the human brain, novel work by Behrens and colleagues (2003) used DTI to map gray matter connections between the thalamus and cortex. The authors capitalized on the knowledge that the thalamus is not a homogenous region; rather, it is topologically organized such that parallel pathways connect specific thalamic nuclei with various cortical regions. Accordingly, Behrens et al. have partitioned the cortex into large, anatomically defined regions of interest (ROIs), such as prefrontal, motor, and somatosensory zones, based on known connections with major thalamic nuclei. (For a review of the nucleus-specific organization that underlies thalamocortical pathways, see **Chapter 11 in Section IV**.) These authors used probabilistic tractography to classify thalamic gray matter based on cortical connectivity patterns; this revealed separable thalamic subregions that corresponded to nuclei defined in prior histological studies. It was later shown that these connectivity-based parcellations strongly overlap with DTI segmentation of the thalamus (Johansen-Berg et al., 2005).

The application of functional connectivity methods to resting-state functional magnetic resonance imaging (rsfMRI) enables researchers to characterize temporal relationships between brain regions in the absence of extrinsic task demands, thus providing a unique window into the brain's intrinsic functional architecture. rsfMRI measures spontaneous fluctuations of the blood-oxygen-level-dependent (BOLD) signal. If two brain areas have highly correlated BOLD activity, it is likely these areas are often co-activated, and are thus more likely to be communicating. rsfMRI methods are particularly beneficial for studying clinical populations with known cognitive impairments, like schizophrenia, as the data are less likely to be confounded by task performance. Like Behrens et al. (2003), Zhang and colleagues (2008) characterized functional relationships between the thalamus and distinct cortical areas using rsfMRI. Using the cortical ROIs as seeds (i.e., prefrontal, parietal/occipital, motor/premotor, somatosensory, and temporal partitions), the authors discovered that distinct, and largely non-overlapping areas of the thalamus co-fluctuated with these different cortical regions. Other rsfMRI methods commonly used to study thalamic connectivity include seed-based analyses that define a specific a priori region for evaluating connectivity; studies can then vary in terms of their specific thalamic seed, for example, a whole bilateral thalamus seed, versus separate left/right hemispheres, or even specific thalamic nuclei (Ramsay, 2019). Alternatively, data-driven approaches, like independent component analysis (ICA), decompose rsfMRI data to find spatially and/or temporally segregated and statistically independent components. While many of the studies included in this chapter used seed-based approaches, we note that ICA methods are particularly powerful for identifying network-based biomarkers of psychiatric disorders, including schizophrenia (Du et al., 2018).

Neuroimaging research has converged on two general patterns of thalamic *dysconnectivity* in schizophrenia: (1) hypoconnectivity between thalamus with prefrontal, cerebellar, and thalamic areas; and (2) hyperconnectivity with motor and sensory cortices (Figure 66.1). The term *dysconnectivity* is appropriate here given the presence of both hypo- and hyperconnectivity. Consistent findings across neuroimaging modalities give support to the notion that thalamic dysconnectivity is a core neurobiological feature of schizophrenia (Giraldo-Chica and Woodward, 2017). Further, there is compelling evidence for a strong anticorrelation between the (1) prefrontal and cerebellum

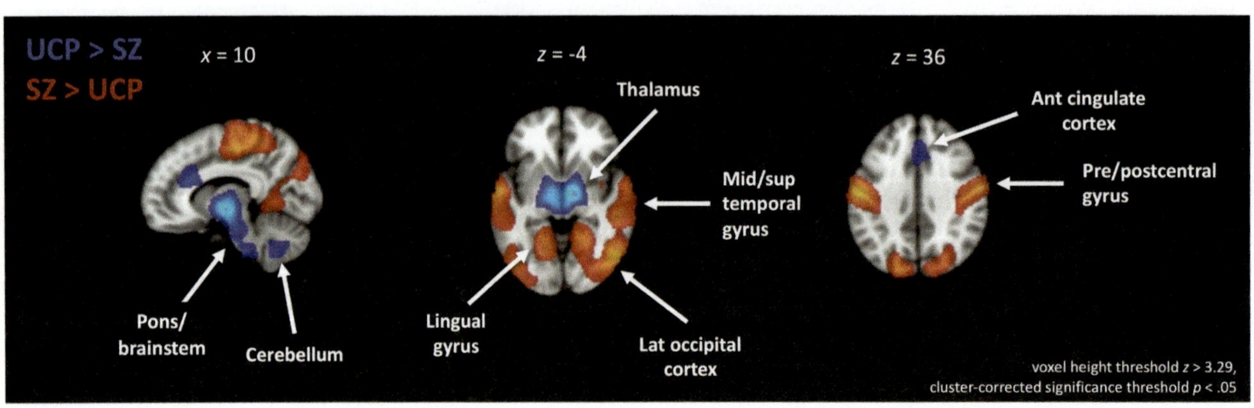

Figure 66.1. Example of concurrent hypo and hyperconnectivity patterns of thalamic rsfMRI dysconnectivity from a large, multi-site schizophrenia dataset. Abbreviations: ant = anterior; mid = middle; sup = superior; SZ = individuals with schizophrenia; UCP = unaffected comparison participants. * This is a novel figure including data that was previously published in: Ferri J, Ford JM, Roach BJ, et al. Resting-state thalamic dysconnectivity in schizophrenia and relationships with symptoms. *Psychol Med.* 2018;48(15):2492–2499.

hypoconnectivity and (2) sensorimotor hyperconnectivity patterns at rest (Anticevic et al., 2014a; Wu et al., 2022) and during task performance (Wu et al., 2022) among individuals with schizophrenia and unaffected comparison participants. In the next sections, we unpack these different connectivity patterns and their clinical correlates.

THALAMIC HYPOCONNECTIVITY

PREFRONTAL CORTEX

Numerous studies point to disruptions in thalamocortical connections with prefrontal cortex (PFC) in schizophrenia based on structural (e.g., Marenco et al., 2012; Giraldo-Chica et al., 2018; Sheffield et al., 2020), and functional (e.g., Woodward et al., 2012; Ferri et al., 2018; Chen et al., 2019) connectivity methods (although the exact location of the prefrontal areas showing reduced connectivity sometimes differ; Giraldo-Chica et al., 2018). Executive functions are notably impaired in schizophrenia and may stem from underlying abnormalities in fronto-thalamic circuits. A clever study (Marenco et al., 2012) used DTI and functional magnetic resonance imaging (fMRI) to test whether connectivity strength between the thalamus and PFC correlated with working memory performance in a small schizophrenia sample. Their seminal findings showed that enhanced connectivity between the thalamus and lateral prefrontal cortex equated to better working memory for people with schizophrenia and unaffected comparison participants. Later studies similarly showed that thalamo-frontal functional connectivity positively correlated with global cognition across individuals with schizophrenia and unaffected comparison participants (when controlling for group differences in these metrics) (Woodward and Heckers, 2016), and with working memory performance among people with schizophrenia (Wu et al., 2022). Although others did not detect significant relationships between thalamo-frontal structural hypoconnectivity and executive functions in schizophrenia; Sheffield et al., 2020. It may be possible to modulate this circuit to influence cognition (Parnaudeau et al., 2013). For instance, people with chronic schizophrenia who received a working memory–focused training exhibited enhanced thalamo-frontal connectivity that scaled with gains in global cognition scores (Ramsay et al., 2017).

Certain thalamic nuclei are particularly relevant to thalamo-frontal connectivity abnormalities in schizophrenia. The mediodorsal nucleus (MD) of the thalamus is a key player in this context, given its reciprocal connections with the PFC (Hwang et al., 2021). As discussed further in **Chapters 46, 47, and 48 in Section VIII**, MD-PFC relationships are central to executive functions, like working memory, learning, and decision-making. The involvement of the MD-PFC circuit in cognition is evident across species. For instance, in humans, lesions to the MD mirrored executive functioning deficits seen after PFC damage (De Witte et al., 2011), while inhibition of the MD in mice disrupted MD-PFC synchrony and produced working-memory impairments akin to schizophrenia (Parnaudeau et al., 2013).

The MD and pulvinar nuclei are heavily emphasized in thalamocortical dysfunction models of schizophrenia given the significant cognitive impairments in schizophrenia and the role of these higher-order nuclei in those functions (Giraldo-Chica and Woodward, 2017; Hwang et al., 2021). Furthermore, neuroimaging and post-mortem evidence point to structural abnormalities in these higher-order nuclei that may progressively deteriorate over the illness course (Hwang et al., 2021). Reduced anatomical connectivity between the thalamus and PFC in schizophrenia was localized to the MD (Giraldo-Chica et al., 2018); further MD-PFC hypoconnectivity correlated specifically with working-memory impairments among people with schizophrenia. Functional connectivity between the MD and PFC is also reduced in schizophrenia (see review by Pergola et al., 2015), and cognitive deficits have been linked to abnormal functional connectivity with a PFC thalamic subregion that corresponded with the MD (Chen et al., 2019).

CEREBELLUM

Hypoconnectivity between the thalamus and cerebellum during resting-state fMRI is evident in schizophrenia (see meta-analysis by Ramsay, 2019), and may relate to clinical and cognitive symptoms. One study observed thalamo-cerebellar functional hypoconnectivity among people with schizophrenia and unaffected comparison participants during a working-memory task (Wu et al., 2022). Across all participants, the combination of thalamic hypoconnectivity (including cerebellum and PFC regions) and thalamic hyperconnectivity (with sensorimotor cortices) corresponded with poorer accuracy and difficulty discriminating between targets and non-targets (i.e., *d'* score). This suggests that the balance across these thalamic systems is relevant to cognitive function. Reduced connectivity between the thalamus and cerebellum has also been linked with deficiencies in fluidly processing information in one's environment. For example, reduced rsfMRI connectivity between a prefrontal thalamic subregion and the right cerebellum corresponded with slower processing speed (Chen et al., 2019). In a different study, rsfMRI thalamo-cerebellar connectivity negatively correlated with positive symptoms, with evidence that greater hypoconnectivity equated to more severe delusions and bizarre behavior (Ferri et al., 2018).

Compelling theories regarding the role of the cerebellum in sensory prediction may help to explain the observed relations between thalamo-cerebellar connectivity abnormalities and schizophrenia symptoms. The cerebellum is highly connected to the cerebral cortex by way of a cortico-ponto-cerebellar-thalamocortical loop. Andreasen and colleagues (1998) first proposed that schizophrenia symptoms may result from disruptions within this circuit, a hypothesis termed *cognitive dysmetria*. The cerebellum is thought to influence motor and higher cognitive functions via feedback and feedforward loops that transmit information via the thalamus and pons (Ramnani, 2006). This circuit is also fundamental to sensory prediction. As described in Abram et al. (2022b), sensory prediction helps the brain to anticipate and distinguish incoming sensory information as being generated from the self, as opposed to the external environment. in a manner that is rapid and occurs outside of conscious awareness. The cerebellum has been called the *comparator* given the notion that it compares anticipated sensory feedback following self-generated action with incoming sensory experiences, producing an error when the anticipated and actual sensory experiences disagree. The error signal is then transmitted from the cerebellum to the thalamus, and the thalamus consequently updates the cerebral cortex to facilitate smooth, sensory-guided movements (Figure 66.2).

People with schizophrenia show deficits in sensory-prediction processes that may relate to clinical features of the disorder. For instance, failures to update motion direction during a random dot motion perceptual task correlated with delusion and hallucination severity (Bansal et al., 2022); that is, psychosis severity equated to an overweighting of initial information above updated sensory evidence. Sensory-prediction deficits may originate from abnormalities in the underlying cortico-ponto-cerebellar-thalamocortical loop (Abram et al., 2022b). People with schizophrenia show notable deviations at many points along this circuit, in addition to the thalamic connectivity abnormalities discussed in this chapter. Such deviations include evidence for impaired white matter integrity of corticopontine pathways (through which expected motor plans are transmitted to the cerebellum; Figure 66.2), and reduced functional connectivity of the pons within a midbrain/cerebellar network (see Abram et al., 2022b, for further discussion).

Taken together, one explanation for these thalamo-cerebellar symptom correlations is that poor synchrony between the cerebellum and thalamus disrupts the generation

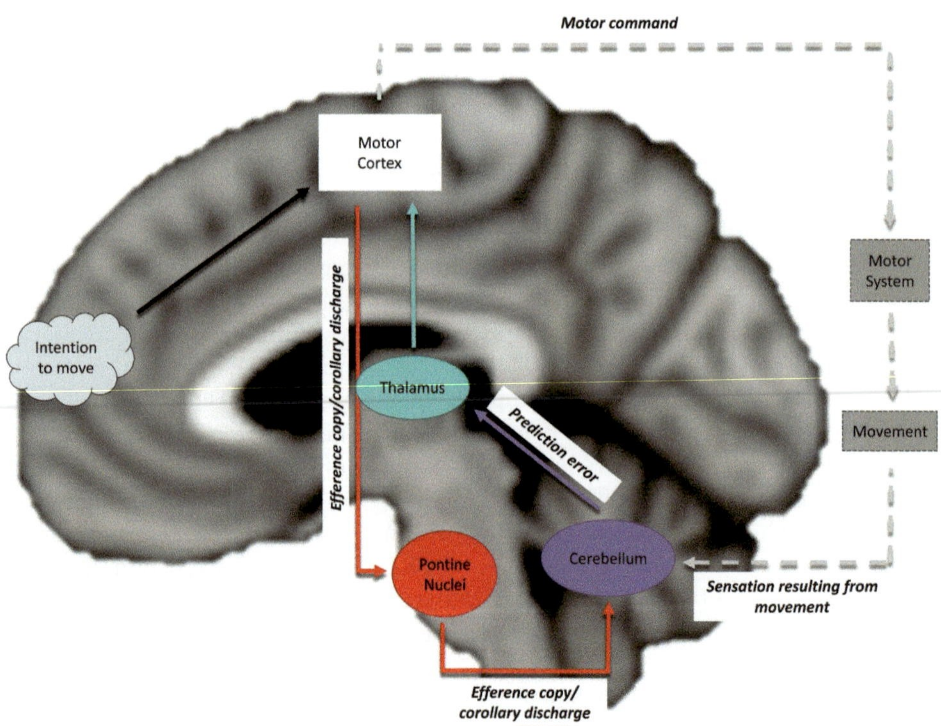

Figure 66.2. Representation of the sensory prediction system with the example of planned movement. The circuit begins with an intention to act/move that initiates transmission of the motor command (i.e., plan). The motor plan and anticipated sensory consequences (i.e., efference copy/corollary discharge) resulting from movement are transmitted from the cortex to cerebellum via the pontine nuclei. In parallel, the cerebellum receives sensory feedback following the movement, which it then compares to the expected sensory consequences based on the efference copy/corollary discharge. Discrepancies between anticipated and actual sensory feedback (i.e., prediction errors) are then sent to the thalamus, which updates several cortical targets to help coordinate smooth, sensory-guided actions.

or transmission of the prediction error (see **Chapter 64 in Section XI** for a discussion of inference and prediction in thalamocortical networks). As noted above, when a person executes movement, the cerebellum compares a copy of the expected motor plan/sensory consequences (sent from the pons) with the actual sensory feedback (received from other cortical areas) (Abram et al., 2022b). The cerebellum produces a prediction error when the expected and actual sensory feedback do not align. Following from this theoretical framework, breakdowns in the cortico-ponto-cerebellar-thalamocortical sensory-processing circuit may lead to abnormal perceptual experiences (hallucinations), overreliance on prior beliefs/failures to update beliefs with new information (delusions), and slower processing of sensory information (processing speed deficits).

THALAMUS

Several studies denote reduced intra-thalamic connectivity in schizophrenia (Anticevic et al., 2014a; Ferri et al., 2018; Gong et al., 2019; Fryer et al., 2021). In a meta-analysis using activation likelihood estimate (ALE) analysis, the largest cluster of thalamic hypoconnectivity in schizophrenia was with the right thalamus itself (Ramsay, 2019); this finding could reflect disruptions in local thalamic connectivity as found previously in schizophrenia (Wei et al., 2018). Reduced within-thalamus connectivity in schizophrenia may relate to longer illness durations (Gong et al., 2019). Both individuals early in the schizophrenia illness course and those at clinical high risk for psychosis (who have not yet crossed the threshold to meet criteria for a full-blown psychotic disorder) also exhibited decreased thalamus-to-thalamus connectivity relative to unaffected comparison participants (Fryer et al., 2021), indicating that these disruptions may begin even before the illness onset.

BASAL GANGLIA CIRCUIT

The thalamus is also pertinent to cortico-striatal-thalamic circuits that support a plethora of emotional drive and goal-directed behaviors (Haber, 2016), motivational processes that are notably impaired in schizophrenia (Barch and Dowd, 2010). There is compelling evidence for parallel, but segregated circuits linking the basal ganglia and thalamus based on their connectivity patterns in humans (Draganski et al., 2008). These circuits begin in the cortex, where they project to the striatum, pallidum, and thalamus, and then back to the original cortical projection area (Alexander et al., 1986). These circuits are particularly relevant to schizophrenia as they both regulate and are modulated by dopaminergic activity in the midbrain (Sabaroedin et al., 2023), and dysregulated dopamine is implicated in the pathophysiology of schizophrenia (Grace, 2016). An interesting question is whether deficient connections with the striatum have cascading effects to the broader thalamocortical dysconnectivity seen in schizophrenia. For instance, in a sample of medication-naïve individuals with schizophrenia, researchers observed reduced functional connectivity between the striatum and thalamus, as well as increased connectivity between the thalamus and sensorimotor cortex, within specific slow frequency bands thought to facilitate subcortical-cortical communication loops (Martino et al., 2018). Thalamocortical dysconnectivity has also been found to correlate with reduced striatal dopamine synthesis capacity, as evidenced by a small group of individuals with schizophrenia who were in remission from psychotic symptoms (Avram et al., 2020). Further research is needed to clarify the causal link between striatal dopamine abnormalities and thalamocortical dysconnectivity in schizophrenia, as well as the functional impacts of these relationships on dopamine-mediated behaviors.

THALAMIC HYPERCONNECTIVITY

In addition to the hypoconnectivity observed with prefrontal, cerebellar, and subcortical structures, widespread hyperconnectivity between the thalamus and sensory regions, including auditory, visual, motor, and association cortices, is a common feature of schizophrenia that is well documented in the rsfMRI literature (Giraldo-Chica and Woodward, 2017; Ramsay, 2019). Early reports of this hyperconnectivity pattern were especially intriguing in light of expectations that schizophrenia would generally be associated with patterns of decreased connectivity (Woodward et al., 2012). And while this pattern is now well established, we are only beginning to characterize its clinical significance.

Greater thalamo-temporal hyperconnectivity has been linked with positive symptom severity, including specific relationships with delusions and hallucinations (Ferri et al., 2018). The temporal lobe has been repeatedly implicated in hallucinatory symptoms of schizophrenia, consistent with its involvement in auditory and speech processing (and corresponding abnormalities among voice-hearers when processing external speech) (Richards et al., 2021). N-methyl-D-aspartate receptor (NMDAR) hypofunction is a possible mechanism to explain heightened relations between the thalamus and temporal cortex in schizophrenia. The NMDAR hypofunction hypothesis posits that abnormalities in glutamate signaling may underlie the symptoms of schizophrenia. It follows from the fact that acute administration of NMDAR antagonists, like ketamine, produce a robust, transient schizophrenia-like presentation (including positive, negative, and cognitive symptoms) (Abi-Saab et al., 1998). Ketamine infusions

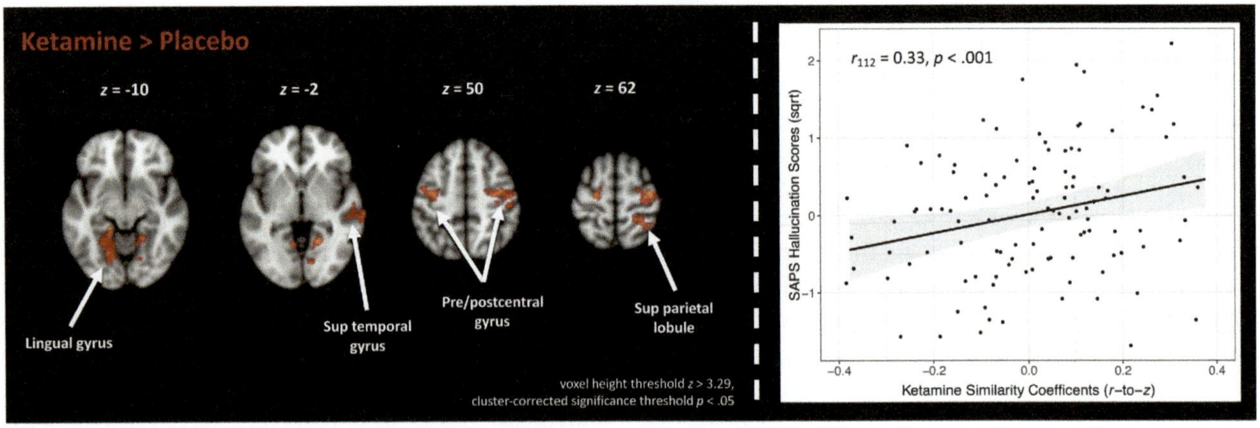

Figure 66.3. The effects of intravenous ketamine administered to healthy volunteers mimic thalamic hyperconnectivity as seen in schizophrenia when compared to placebo saline infusion (left). Among individuals with schizophrenia, similarity to the ketamine-induced thalamic dysconnectivity pattern (i.e., ketamine similarity coefficients) was positively correlated with hallucination severity (right). Sup = superior; SAPS = Scale for the Assessment of Positive Symptoms; sqrt = square root transform to normalize variable. * This is a novel figure including data that was previously published in: Abram SV, Roach BJ, Fryer SL, et al. Validation of ketamine as a pharmacological model of thalamic dysconnectivity across the illness course of schizophrenia. *Mol Psychiatry*. 2022;27(5):2448–2456.

administered to healthy volunteers during rsfMRI reproduce the thalamo-sensorimotor hyperconnectivity characteristic of schizophrenia, including enhanced connectivity with temporal gyri (Abram et al., 2022a) (Figure 66.3). A novel feature of this study is that the authors showed direct correspondence between thalamic dysconnectivity observed among individuals with schizophrenia and healthy volunteers under ketamine. To this end, the authors derived a template connectivity map of the ketamine effect to directly test the overlap between ketamine-induced thalamic dysconnectivity and the deviations seen among people with schizophrenia. rsfMRI thalamic connectivity maps from a large, multi-site sample of individuals with schizophrenia were correlated with the ketamine template map, producing a single "ketamine similarity coefficient" for each participant. Among people with schizophrenia, greater similarity to the ketamine-induced thalamic hyperconnectivity pattern (i.e., more positive ketamine similarity coefficients) correlated with greater hallucinatory symptoms (Figure 66.3). Interestingly, ketamine did not induce the typical thalamo-prefrontal and -cerebellar hypoconnectivity patterns common in schizophrenia. It is possible that this null effect is the result of an acute versus chronic dosing protocol. Ethical constraints prevent sub-chronic dosing protocols in humans in experimental studies, making naturalistic studies of drug-using populations helpful for deriving mechanistic hypotheses. For instance, a study of chronic ketamine users reported reduced connectivity between the thalamus and prefrontal cortex compared to drug-free comparison participants (Liao et al., 2016).

With respect to cognitive implications, hyperconnectivity between the thalamus with sensorimotor areas also correlated with poorer working-memory performance in people with schizophrenia (Wu et al., 2022). The combination of fronto-thalamic hypoconnectivity with thalamic-sensorimotor hyperconnectivity was also found to underlie working-memory deficits, raising the possibility that an imbalance between these systems may contribute to cognitive deficits (Wu et al., 2022).

RELEVANCE OF THALAMIC DYSCONNECTIVITY AS A SCHIZOPHRENIA BIOMARKER AND PREDICTOR OF CONVERSION

There is increasing appreciation of schizophrenia as a neurodevelopmental disorder, or rather a collection of neurodevelopment disorders, which follows from the facts that psychosis typically emerges in late adolescence or early adulthood, and associated brain abnormalities are likely present before illness has fully manifested (Insel, 2010). Thalamocortical connections undergo major developments during the translation from childhood to adolescence and early adulthood (Huang et al., 2021; Avery et al., 2022). Several research groups have thus postulated that altered thalamocortical connectivity in schizophrenia is the result of atypical development during this key neurodevelopmental phase (Woodward et al., 2012; Giraldo-Chica and Woodward, 2017). This notion is supported by observations of thalamocortical functional and structural dysconnectivity among individuals for whom the illness has not fully manifested, for example, those at clinical high risk for psychosis (Anticevic et al., 2015; Fryer et al., 2021) and youth on the psychosis spectrum (Huang et al., 2021), and among those early in the schizophrenia illness course (Woodward and Heckers, 2016; Fryer et al., 2021). Atypical thalamic development in adolescence may have lasting impacts on the PFC that, in turn, lead to persistent cognitive impairments. For instance, the integrity of white matter microstructure in fronto-thalamic pathways corresponded with better cognitive functioning in typically developing youth and those on the psychosis spectrum (Avery et al., 2022).

Individuals at clinical high risk show an intermediate level of dysconnectivity falling between those early in the schizophrenia illness course and unaffected comparison participants (Fryer et al., 2021), although at least some aspects of thalamic functional connectivity may follow a nonlinear trajectory during the early phase of the illness (Chan et al., 2022). The intermediate level could be due to the fact that some clinical high-risk individuals may develop schizophrenia or another psychotic illness, while others will not. In fact, clinical high-risk individuals who eventually converted to a full-blown psychotic illness showed greater thalamic dysconnectivity at baseline than those who did not (Anticevic et al., 2015). One of these samples observed thalamic dysconnectivity among medication-naïve participants, stressing that thalamic dysconnectivity is likely not an artifact of chronic medication exposure (Fryer et al., 2021).

Thalamocortical pathways among psychosis-risk youth may be sexually dimorphic. While the percent of total thalamic white matter connectivity was comparable across sexes, thalamocortical white matter microstructure values were generally higher for males than females and lowest among youths on the psychosis spectrum (Avery et al., 2022). One study observed decreased rsfMRI thalamocortical connectivity among younger males with psychotic-spectrum symptoms (PSS) compared to younger males without PSS, while older males with PSS had greater thalamocortical connectivity relative to non-PSS older males (Jacobs et al., 2019). Importantly, these age-related effects were specific to males, raising the possibility of protection against these neurodevelopmental abnormalities in females. The authors further speculated that these male-driven abnormalities could explain the lower prevalence and later onset of schizophrenia in females. Because this study was based on cross-sectional data, longitudinal analysis is needed to clarify whether sex-specific thalamocortical connectivity differences in PSS youth are the result of differences in symptom onset ages or are due to the progression of symptoms across development. As noted by the National Institutes of Health (NIH), elucidating biological sex differences is necessary for aiding the interpretation, validity, and generalizability of research findings (and sex as a biological variable is often overlooked in basic science). Future research that replicates and extends the thalamocortical sex differences described in Jacobs et al. (2019) and Avery et al. (2022) is thus an important initiative.

SPECIFICITY TO SCHIZOPHRENIA: THALAMIC DYSCONNECTIVITY AS A TRANSDIAGNOSTIC FEATURE

Growing evidence emphasizes the transdiagnostic utility of thalamic dysconnectivity (Hwang et al., 2021), as a clinical feature present beyond schizophrenia. While mental disorders were traditionally conceptualized as distinct, episodic conditions that fit into discrete categories, it has become increasingly clear that many psychopathology features are shared across categorically defined disorders. A general liability factor for psychopathology, called the *p factor*, represents the shared risk across mental disorders. The *p factor* is a latent dimension estimated from the shared variance of psychopathology symptoms in a dataset. Thalamic dysconnectivity has recently been appreciated as a transdiagnostic feature that may relate to the *p factor* (Hwang et al., 2021). This stems, in part, from evidence that a higher *p factor* score has been linked with structural disturbances in the cerebello-thalamocortical pathway (Romer et al., 2018).

Thalamic dysconnectivity is observed across the psychosis spectrum, including bipolar disorder and psychotic bipolar disorder (Anticevic et al., 2014a, 2014b; Sheffield et al., 2020; Ramsay et al., 2022), raising the possibility for a shared underlying mechanism of thalamic dysconnectivity (Hwang et al., 2021). Bipolar disorder is characterized by extreme changes in mood, energy, and activity levels that are reflected in periods of mania and depression. Sometimes people with bipolar disorder also experience psychotic symptoms, like hallucinations and delusions. Like those with schizophrenia, the combination of weaker coupling between the thalamus and frontal and cerebellar regions, with enhanced coupling between the thalamus and sensory and motor regions, has been documented in bipolar disorder (based on structural and functional connectivity studies; Anticevic et al., 2014a; Sheffield et al., 2020); albeit individuals with bipolar disorder may show an intermediate level of thalamic dysconnectivity (Anticevic et al., 2014a). Symptom severity and cognitive function also correlated with thalamocortical connectivity strength among individuals with schizophrenia, bipolar disorder, first-degree relatives of psychosis probands, and unaffected comparison participants (Ramsay et al., 2022). Disorder-specific differences may emerge when examining specific thalamic nuclei. While the MD and lateral geniculate nucleus (LGN) showed highly similar dysconnectivity patterns in schizophrenia, their overlap was less pronounced in bipolar disorder; rather, thalamic dysconnectivity was more widespread when seeding from the MD than LGN (Anticevic et al., 2014b). The breadth of reduced MD connectivity was also more restricted in bipolar disorder compared to schizophrenia, although bipolar participants with a psychosis history showed more similarity to schizophrenia. It has been theorized that thalamic disturbances in schizophrenia have more of a neurodevelopmental basis than in bipolar disorder. Rather, bipolar disorder may be a neuroprogressive illness for which there is relatively normal premorbid functioning and decline begins following illness onset, versus declines beginning in the prodromal phase of schizophrenia (Lewandowski et al., 2011). In support of this theory, abnormalities in the white matter integrity of the anterior thalamic radiation (which links the PFC with thalamus) were present among those with a bipolar diagnosis, but not those at risk for bipolar disorder (Hu et al., 2020).

Findings from major depressive disorder (MDD) provide further support for thalamocortical dysconnectivity as a transdiagnostic marker. MDD is associated with emotional impairments as well as cognitive deficits. Similar to schizophrenia, there is evidence for both thalamo-frontal hypoconnectivity and thalamic hyperconnectivity with temporal and sensorimotor regions in MDD (examples from rsfMRI literature; Brown et al., 2017; Tu et al., 2019), although when compared directly, thalamic dysconnectivity was greater among individuals with schizophrenia than those with MDD (Tu et al., 2019). rsfMRI thalamic-temporal hyperconnectivity correlated with more severe depressive symptoms (Brown et al., 2017). The temporal ROI used in this analysis encompassed areas critical to affective processing (i.e., hippocampus, amygdala), in addition to more lateral temporal structures (like the superior, middle, and inferior temporal gyri). Because the thalamocortical system helps to modulate the amygdala's response to negative affect, the authors (Brown et al., 2017) postulated that increased connectivity between the thalamus and temporal cortex could reflect an overreaction of the limbic system and may thus relate to abnormalities in the experience and processing of negative affect.

The presence of sensorimotor and cortical excitability lend support to thalamic dysfunction as a possible underlying mechanism in autism spectrum disorder (ASD), as well (Woodward et al., 2017). ASD is another set of neurodevelopmental disorders associated with marked deficits in communication and social skills. While distinct, ASD and schizophrenia may overlap in some aspects of their underlying neurobiology (e.g., imbalance of glutamate/GABA circuit dysfunctions) and clinical presentations (e.g., social deficits, sensory abnormalities) (Foss-Feig et al., 2017). Growing evidence points to thalamocortical hyperconnectivity in ASD (Woodward et al., 2017; Tomasi and Volkow, 2019), including evidence for functional and structural thalamocortical connectivity abnormalities among infants at familial high-risk for ASD (Nair et al., 2021), representing a possible early biomarker of ASD. The combination of reduced local connectivity and disrupted long-range connectivity described by Tomasi and colleagues (2019) parallels their earlier dysconnectivity findings in schizophrenia, indicating a possible neurobiological convergence across these disorders.

Obsessive-compulsive disorder (OCD) is a disabling disorder characterized by repeated, unwanted thoughts, urges, or images (obsessions) and/or repetitive behaviors (compulsions). OCD symptoms are prominent among individuals with schizophrenia compared to the general population (Grover et al., 2019). A recent systemic review and meta-analysis of rsfMRI connectivity studies found that relative to unaffected comparison participants, individuals with OCD showed reduced connectivity between the thalamus and the superior frontal gyrus, caudate, and cerebellum, as well as increased connectivity with the middle temporal gyrus, extending into the amygdala and parahippocampus (Liu et al., 2022). Cognitive inflexibility is a cause of impairment in both OCD (obsessions, compulsions) and schizophrenia (delusions) (Chamberlain et al., 2021). Reduced connectivity between the thalamus with the PFC and striatum may represent a shared mechanism underlying cognitive rigidity. In support of this hypothesis, deep brain stimulation to the anteromedial subthalamic nucleus led to increased cognitive flexibility among a small sample of men with treatment-refractory OCD (Tyagi et al., 2019). This subthalamic nucleus is an integral node within the basal ganglia circuitry that receives uninterrupted projections from the frontal cortex via a "hyper-direct" pathway; further, this hyper-direct pathway plays a role in pausing or suppressing programmed behaviors. Stimulation of the subthalamic nucleus may alleviate cognitive inflexibility in OCD by remediating a dysfunctional circuit and thus enabling individuals to interrupt programmed behaviors that are no longer appropriate (such as compulsive and repetitive acts) (Chamberlain et al., 2021).

CONCLUSIONS

Abnormal thalamic connectivity is a core feature of schizophrenia that is seen throughout the illness trajectory and is present across the psychosis spectrum. While there are certainly well-established aspects of thalamic dysconnectivity in schizophrenia, such as the presence of both hypo- and hyperconnections, our understanding of both the specificity and scope of the impact of thalamic dysconnectivity on clinical and behavioral symptoms is still in its nascent stages.

This chapter aimed to highlight some of the more consistent clinical correlates of thalamic dysconnectivity; however, it is important to note that the broader literature is more mixed. Meta-analytic findings suggest that thalamic dysconnectivity has small, variable, and generally nonspecific relationships with symptom domains (i.e., positive, negative, disorganized, or general psychopathology symptoms) (Ramsay, 2019). One approach for better clarifying these relationships is to evaluate clinical subdomains (e.g., delusions, hallucinations) versus solely examining total positive symptoms (Ferri et al., 2018; Abram et al., 2022b). This approach could similarly be applied to negative symptoms, which nicely parse into expressive (e.g., blunted affect, alogia) and experiential (e.g., low motivation, anhedonia) deficits. Utilizing behavioral paradigms that tap more specific processes may also help to better pinpoint the relevance of thalamic connectivity abnormalities in schizophrenia. For example, behavioral paradigms can capture the thalamocortical signals associated with successful sensory attenuation (i.e., where sensory feedback from self-generated actions is reduced in intensity relative to externally generated feedback) among healthy individuals (Hua et al., 2020).

Aligned with dimensional models of psychopathology, there is increasing evidence that thalamic dysconnectivity is observed in several major mental disorders and thus may not be specific to schizophrenia. Here we have discussed evidence from other psychotic-spectrum disorders (like bipolar disorder), as well as findings from MDD, ASD, and OCD (which each share some overlapping features with schizophrenia, such as anhedonia, social functioning impairments, and rigid thinking). Of course, the disorders included in this chapter are not exhaustive, and other characteristics associated with multiple psychopathologies, like trauma and early life stress, may also contribute to or be reflected in underlying abnormalities in thalamic connections. The convergence of thalamic connectivity abnormalities across diagnostic categories is highly relevant to the Research Domain Criteria (RDoC), which is a research framework for studying psychopathology developed by the National Institute of Mental Health (NIMH). RDoC aims to elucidate the systems and psychological constructs that span diagnostic boundaries (e.g., positive valence, negative valence, cognition, sensorimotor domains) using data from multiple units of analysis (e.g., genetics, circuits, behavior, physiology). Relatedly, an interesting question is whether the thalamic dysconnectivity present across these disorders shares similar underlying mechanisms, such as the implication of NMDAR hypofunction in thalamic hyperconnectivity in schizophrenia (Abram et al., 2022a), and whether thalamic connectivity abnormalities equate to similar clinical phenomena across disorders (such as the relationship between thalamocortical connectivity with cognition among individuals with schizophrenia and bipolar disorder; Ramsay et al., 2022). Studies that directly compare multiple diagnostic groups are especially valuable for assessing the specificity and relevance of thalamic abnormalities, and for identifying the underlying mechanisms, though such study designs can be costly and challenging to execute.

This chapter has covered the typical thalamic connectivity deviations seen in schizophrenia, as evidenced from structural and functional neuroimaging research. This includes a combination of reduced connectivity with prefrontal, cerebellar, and thalamic areas, in conjunction with excessive connectivity with sensory and motor regions. Continued efforts to clarify how global or specific thalamic connectivity deficits contribute to clinical phenomena have implications for not only schizophrenia, but other disorders with similar network abnormalities. An important next step is to determine whether modulating these circuits can alleviate symptoms both within and beyond schizophrenia, and accordingly, if there are critical windows for such interventions that depend on the diagnosis-specific developmental course of thalamic dysconnectivity.

ACKNOWLEDGMENTS

SVA. was supported by a Career Development Award from the Department of Veterans Affairs (IK2 CX002355). JMF was supported by funding from the National Institute of Mental Health (R03MH121900) and the Department of Veterans Affairs (IK6 CX002519).

REFERENCES

Abi-Saab WM, D'Souza DC, Moghaddam B, Krystal JH (1998). The NMDA antagonist model for schizophrenia: Promise and pitfalls. *Pharmacopsychiatry*. 31:104–109.

Abram SV, Roach BJ, Fryer SL, Calhoun VD, Preda A, Van Erp TGM, Bustillo JR, Lim KO, Loewy RL, Stuart BK, Krystal JH, Ford JM, Mathalon DH (2022a). Validation of ketamine as a pharmacological model of thalamic dysconnectivity across the illness course of schizophrenia. *Mol Psychiatry*. 27:2448–2456.

Abram SV, Hua JPY, Ford JM (2022b). Consider the pons: Bridging the gap on sensory prediction abnormalities in schizophrenia. *Trends Neurosci*. 45:798–808.

Alexander GE, DeLong MR, Strick PL (1986). Parallel organization of functionally segregated circuits linking basal ganglia and cortex. *Annu Rev Neurosci*. 9:357–381.

Andreasen NC (1997). The role of the thalamus in schizophrenia. *Can J Psychiatry*. 42:27–33.

Andreasen NC, Paradiso S, O'Leary DS (1998). "Cognitive dysmetria" as an integrative theory of schizophrenia: A dysfunction in cortical-subcortical-cerebellar circuitry? *Schizophr Bull*. 24:203–218.

Anticevic A, Haut K, Murray JD, Repovs G, Yang GJ, Diehl C, McEwen SC, Bearden CE, Addington J, Goodyear B, Cadenhead KS, Mirzakhanian H, Cornblatt BA, Olvet D, Mathalon DH, McGlashan TH, Perkins DO, Belger A, Seidman LJ, Tsuang MT, van Erp TGM, Walker EF, Hamann S, Woods SW, Qiu M, Cannon TD (2015). Association of thalamic dysconnectivity and conversion to psychosis in youth and young adults at elevated clinical risk. *JAMA Psychiatry*. 72:882–891.

Anticevic A, Cole MW, Repovs G, Murray JD, Brumbaugh MS, Winkler AM, Savic A, Krystal JH, Pearlson GD, Glahn DC (2014a). Characterizing thalamo-cortical disturbances in Schizophrenia and bipolar illness. *Cereb Cortex.* 24:3116–3130.

Anticevic A, Yang G, Savic A, Murray JD, Cole MW, Repovs G, Pearlson GD, Glahn DC (2014b). Mediodorsal and visual thalamic connectivity differ in schizophrenia and bipolar disorder with and without psychosis history. *Schizophr Bull.* 40:1227–1243.

Avery SN, Huang AS, Sheffield JM, Rogers BP, Vandekar S, Anticevic A, Woodward ND (2022). Development of thalamocortical structural connectivity in typically developing and psychosis spectrum youths. *Biol Psychiatry Cogn Neurosci Neuroimaging.* 7(8):782–792.

Avram M, Brandl F, Knolle F, Cabello J, Leucht C, Scherr M, Mustafa M, Koutsouleris N, Leucht S, Ziegler S, Sorg C (2020). Aberrant striatal dopamine links topographically with cortico-thalamic dysconnectivity in schizophrenia. *Brain.* 143:3495–3505.

Bansal S, Bae GY, Robinson BM, Hahn B, Waltz J, Erickson M, Leptourgos P, Corlett P, Luck SJ, Gold JM (2022). Association between failures in perceptual updating and the severity of psychosis in schizophrenia. *JAMA Psychiatry.* 79:169–177.

Barch DM (2021). Editorial: The centrality of both hyper- and hypothalamocortical connectivity in psychosis. *J Am Acad Child Adolesc Psychiatry.* 60:438–440.

Barch DM, Dowd EC (2010). Goal representations and motivational drive in schizophrenia: the role of prefrontal-striatal interactions. *Schizophr Bull.* 36:919–934. Accessed June 16, 2015. http://www.pubmedcentral.nih.gov/articlerender.fcgi?artid=2930335&tool=pmcentrez&rendertype=abstract.

Behrens TEJ, Johansen-Berg H, Woolrich MW, Smith SM, Wheeler-Kingshott CAM, Boulby PA, Barker GJ, Sillery EL, Sheehan K, Ciccarelli O, Thompson AJ, Brady JM, Matthews PM (2003). Non-invasive mapping of connections between human thalamus and cortex using diffusion imaging. *Nat Neurosci.* 6:750–757.

Brown EC, Clark DL, Hassel S, MacQueen G, Ramasubbu R (2017). Thalamocortical connectivity in major depressive disorder. *J Affect Disord.* 217:125–131.

Chamberlain SR, Solly JE, Hook RW, Vaghi MM, Robbins TW (2021). Cognitive inflexibility in OCD and related disorders. In: Fineberg NA, Robbins TW, eds. *The Neurobiology and Treatment of OCD: Accelerating Progress.* :Springer:125–145.

Chan SY, Brady RO, Lewandowski KE, Higgins A, Öngür D, Hall M (2022). Connectivity over the first five years of psychosis. *Mol Psychiatry.* 27:1177–1183.

Chen P, Ye E, Jin X, Zhu Y, Wang L (2019). Association between thalamocortical functional connectivity abnormalities and cognitive deficits in schizophrenia. *Sci Rep.* 9:1–10.

De Witte L, Brouns R, Kavadias D, Engelborghs S, De Deyn PP, Mariën P (2011). Cognitive, affective and behavioural disturbances following vascular thalamic lesions: A review. *Cortex.* 47:273–319.

Draganski B, Kherif F, Klöppel S, Cook PA, Alexander DC, Parker GJM, Deichmann R, Ashburner J, Frackowiak RSJ (2008). Evidence for segregated and integrative connectivity patterns in the human basal ganglia. *J Neurosci.* 28:7143–7152.

Du Y, Fryer SL, Lin D, Sui J, Yu Q, Chen J, Stuart B, Loewy RL, Calhoun VD, Mathalon DH (2018). Identifying functional network changing patterns in individuals at clinical high-risk for psychosis and patients with early illness schizophrenia: A group ICA study. *NeuroImage Clin.* 17:335–346.

Ferri J, Ford JM, Roach BJ, Turner JA, van Erp TG, Voyvodic J, Preda A, Belger A, Bustillo J, O'Leary D, Mueller BA, Lim KO, McEwen SC, Calhoun VD, Diaz M, Glover G, Greve D, Wible CG, Vaidya JG, Potkin SG, Mathalon DH (2018). Resting-state thalamic dysconnectivity in schizophrenia and relationships with symptoms. *Psychol Med.* 48:2492–2499.

Foss-Feig JH, Adkinson BD, Ji JL, Yang G, Srihari VH, McPartland JC, Krystal JH, Murray JD, Anticevic A (2017). Searching for cross-diagnostic convergence: Neural mechanisms governing excitation and inhibition balance in schizophrenia and autism spectrum disorders. *Biol Psychiatry.* 81:848–861.

Friston KJ, Frith CD (1995). Schizophrenia: A disconnection syndrome? *Clin Neurosci.* 3:89–97.

Frith CD, Friston KJ, Herold S, Silbersweig D, Fletcher P, Cahill C, Dolan RJ, Frackowiak RSJ, Liddle PF (1995). Regional brain activity in chronic schizophrenic patients during the performance of a verbal fluency task. *Br J Psychiatry.* 167:343–349.

Fryer SL, Ferri JM, Roach BJ, Loewy RL, Stuart BK, Anticevic A, Ford JM, Mathalon DH (2021). Thalamic dysconnectivity in the psychosis risk syndrome and early illness schizophrenia. *Psychol Med.* 52(13):2767–2775.

Giraldo-Chica M, Rogers BP, Damon SM, Landman BA, Woodward ND (2018). Prefrontal-thalamic anatomical connectivity and executive cognitive function in schizophrenia. *Biol Psychiatry.* 83:509–517.

Giraldo-Chica M, Woodward ND (2017). Review of thalamocortical resting-state fMRI studies in schizophrenia. *Schizophr Res.* 180:58–63.

Gong J, Luo C, Li X, Jiang S, Khundrakpam BS, Duan M, Chen X, Yao D (2019). Evaluation of functional connectivity in subdivisions of the thalamus in schizophrenia. *Br J Psychiatry.* 214:288–296.

Grace AA (2016). Dysregulation of the dopamine system in the pathophysiology of schizophrenia and depression. *Nat Rev Neurosci.* 17:524–532.

Grover S, Sahoo S, Surendran I (2019). Obsessive-compulsive symptoms in schizophrenia: A review. *Acta Neuropsychiatr.* 31:63–73.

Haber SN (2016). Corticostriatal circuitry. *Dialogues Clin Neurosci.* 18:7–21.

Hu R, Stavish C, Leibenluft E, Linke JO (2020). White matter microstructure in individuals with and at risk for bipolar disorder: evidence for an endophenotype from a voxel-based meta-analysis. *Biol Psychiatry Cogn Neurosci Neuroimaging.* 5:1104–1113.

Hua L, Recasens M, Grent-'T-Jong T, Adams RA, Gross J, Uhlhaas PJ (2020). Investigating cortico-subcortical circuits during auditory sensory attenuation: A combined magnetoencephalographic and dynamic causal modeling study. *Hum Brain Mapp.* 41:4419–4430.

Huang AS, Rogers BP, Sheffield JM, Vandekar S, Anticevic A, Woodward ND (2021). Characterizing effects of age, sex and psychosis symptoms on thalamocortical functional connectivity in youth. *NeuroImage.* 243:118562.

Hwang WJ, Kwak Y Bin, Cho KIK, Lee TY, Oh H, Ha M, Kim M, Kwon JS (2021). Thalamic connectivity system across psychiatric disorders: Current status and clinical implications. *Biol Psychiatry Glob Open Sci.* 2(4):332–340.

Insel TR (2010). Rethinking schizophrenia. *Nature.* 468:187–193.

Jacobs GR, Ameis SH, Ji JL, Viviano JD, Dickie EW, Wheeler AL, Stojanovski S, Anticevic A, Voineskos AN (2019). Developmentally divergent sexual dimorphism in the cortico-striatal-thalamic-cortical psychosis risk pathway. *Neuropsychopharmacology.* 44:1649–1658.

Johansen-Berg H, Behrens TEJ, Sillery E, Ciccarelli O, Thompson AJ, Smith SM, Matthews PM (2005). Functional-anatomical validation and individual variation of diffusion tractography-based segmentation of the human thalamus. *Cereb Cortex.* 15:31–39.

Lewandowski KE, Cohen BM, Öngur D (2011). Evolution of neuropsychological med dysfunction during the course of schizophrenia and bipolar disorder. *Psychol Med.* 41:225–241.

Liao Y, Tang J, Liu J, Xie A, Yang M, Johnson M, Wang X, Deng Q, Chen H, Xiang X, Liu T, Chen X, Song M, Hao W (2016). Decreased thalamocortical connectivity in chronic ketamine users. *PLoS ONE.* 11:e0167381.

Liu J, Cao L, Li H, Gao Y, Bu X, Liang K, Bao W, Zhang S, Qiu H, Li X, Hu X, Lu L, Zhang L, Hu X, Huang X, Gong Q (2022). Abnormal resting-state functional connectivity in patients with obsessive-compulsive disorder: A systematic review and meta-analysis. *Neurosci Biobehav Rev.* 135:104574.

Marenco S, Stein JL, Savostyanova AA, Sambataro F, Tan HY, Goldman AL, Verchinski BA, Barnett AS, Dickinson D, Apud JA, Callicott JH, Meyer-Lindenberg A, Weinberger DR (2012). Investigation of anatomical thalamo-cortical connectivity and fMRI activation in schizophrenia. *Neuropsychopharmacology*. 37:499–507.

Martino M, Magioncalda P, Yu H, Li X, Wang Q, Meng Y, Deng W, Li Y, Li M, Ma X, Lane T, Duncan NW, Northoff G, Li T (2018). Abnormal resting-state connectivity in a substantia nigra-related striato-thalamo-cortical network in a large sample of first-episode drug-naïve patients with schizophrenia. *Schizophr Bull.* 44: 419–443.

McGhie A, Chapman J (1961). Disorders of attention and perception in early schizophrenia. *Br J Med Psychol.* 34:103–116.

Nair A, Jalal R, Liu J, Tsang T, Mcdonald NM, Jackson L, Ponting C, Jeste SS, Bookheimer SY, Dapretto M (2021). Altered thalamocortical connectivity in 6-week-old infants at high familial risk for autism spectrum disorder. *Cereb Cortex*. 31:4191–4205.

Parnaudeau S, O'Neill PK, Bolkan SS, Ward RD, Abbas AI, Roth BL, Balsam PD, Gordon JA, Kellendonk C (2013). Inhibition of mediodorsal thalamus disrupts thalamofrontal connectivity and cognition. *Neuron*. 77:1151–1162.

Pergola G, Selvaggi P, Trizio S, Bertolino A, Blasi G (2015). The role of the thalamus in schizophrenia from a neuroimaging perspective. *Neurosci Biobehav Rev*. 54:57–75.

Ramnani N (2006). The primate cortico-cerebellar system: Anatomy and function. *Nat Rev Neurosci*. 7:511–522.

Ramsay IS (2019). An activation likelihood estimate meta-analysis of thalamocortical dysconnectivity in psychosis. *Biol Psychiatry Cogn Neurosci Neuroimaging*. 4:859–869.

Ramsay IS, Mueller B, Ma Y, Shen C, Sponheim SR (2022). Thalamocortical connectivity and its relationship with symptoms and cognition across the psychosis continuum. *Psychol Med*. 1–10.

Ramsay IS, Nienow TM, MacDonald AW (2017). Increases in intrinsic thalamocortical connectivity and overall cognition following cognitive remediation in chronic schizophrenia. *Biol Psychiatry Cogn Neurosci Neuroimaging*. 2:355–362.

Richards SE, Hughes ME, Woodward TS, Rossell SL, Carruthers SP (2021). External speech processing and auditory verbal hallucinations: A systematic review of functional neuroimaging studies. *Neurosci Biobehav Rev*. 131:663–687.

Romer AL, Knodt AR, Houts R, Brigidi BD, Moffitt TE, Caspi A, Hariri AR (2018). Structural alterations within cerebellar circuitry are associated with general liability for common mental disorders. *Mol Psychiatry*. 23:1084–1090.

Sabaroedin K, Tiego J, Fornito A (2023). Circuit-based approaches to understanding corticostriatothalamic dysfunction across the psychosis continuum. *Biol Psychiatry*. 93(2):113–124.

Sheffield JM, Huang AS, Rogers BP, Giraldo-Chica M, Landman BA, Blackford JU, Heckers S, Woodward ND (2020). Thalamocortical anatomical connectivity in schizophrenia and psychotic bipolar disorder. *Schizophr Bull*. 46:1062–1071.

Tomasi D, Volkow ND (2019). Reduced local and increased long-range functional connectivity of the thalamus in autism spectrum disorder. *Cereb Cortex*. 29:573–585.

Tu PC, Bai YM, Li CT, Chen MH, Lin WC, Chang WC, Su TP (2019). Identification of common thalamocortical dysconnectivity in four major psychiatric disorders. *Schizophr Bull*. 45:1143–1151.

Tyagi H, Apergis-Schoute AM, Akram H, Foltynie T, Limousin P, Drummond LM, Fineberg NA, Matthews K, Jahanshahi M, Robbins TW, Sahakian BJ, Zrinzo L, Hariz M, Joyce EM (2019). A randomized trial directly comparing ventral capsule and anteromedial subthalamic nucleus stimulation in obsessive-compulsive disorder: Clinical and imaging evidence for dissociable effects. *Biol Psychiatry*. 85:726–734.

Wang Q, Webber RM, Stanley GB (2010). Thalamic synchrony and the adaptive gating of information flow to cortex. *Nat Neurosci*. 13:1534–1543.

Wei Y, Chang M, Womer FY, Zhou Q, Yin Z, Wei S, Zhou Y, Jiang X, Yao X, Duan J, Xu K, Zuo XN, Tang Y, Wang F (2018). Local functional connectivity alterations in schizophrenia, bipolar disorder, and major depressive disorder. *J Affect Disord*. 236:266–273.

Woodward ND, Giraldo-Chica M, Rogers B, Cascio CJ (2017). Thalamocortical dysconnectivity in autism spectrum disorder: An analysis of the Autism Brain Imaging Data Exchange. *Biol Psychiatry Cogn Neurosci Neuroimaging*. 2:76–84.

Woodward ND, Heckers S (2016). Mapping thalamocortical functional connectivity in chronic and early stages of psychotic disorders. *Biol Psychiatry*. 79:1016–1025.

Woodward ND, Karbasforoushan H, Heckers S (2012). Thalamocortical dysconnectivity in schizophrenia. *Am J Psychiatry*. 169:1092–1099.

Wu G, Palaniyappan L, Zhang M, Yang J, Xi C, Liu Z, Xue Z, Ouyang X, Tao H, Zhang J, Luo Q, Pu W (2022). Imbalance between prefronto-thalamic and sensorimotor-thalamic circuitries associated with working memory deficit in schizophrenia. *Schizophr Bull*. 48:251–261.

Zhang D, Snyder AZ, Fox MD, Sansbury MW, Shimony JS, Raichle ME (2008). Intrinsic functional relations between human cerebral cortex and thalamus. *J Neurophysiol*. 100:1740–1748.

67.

THALAMUS, THALAMOCORTICAL NETWORKS, AND NON-MOTOR SYMPTOMS IN PARKINSON'S DISEASE

John C. Dalrymple-Alford, Tim J. Anderson, and Tracy R. Melzer

PARKINSON'S DISEASE: NON-MOTOR SYMPTOMS

The core features for the diagnosis of Parkinson's disease (PD) are marked slowness in making voluntary movements and their decrement in amplitude or speed over time (i.e., bradykinesia), combined with either rigidity ("lead pipe" resistance to passive movement) and/or tremor in a fully resting body part (Tolosa et al., 2021). There is also a remarkably diverse range of non-motor features that accompany PD (Kalia and Lang, 2015). One of the most common non-motor changes in PD concerns declining cognition. The impairment can meet a formal threshold for mild cognitive impairment (PD-MCI) at any time during the disease course. Cognitive decline often becomes highly debilitating, because it evolves into frank dementia (PDD) in about half of older patients by 10 years post-diagnosis and in the majority (~80%) when viewed from a long-term perspective (Aarsland et al., 2021). The array of other non-motor features includes autonomic nervous system (sympathetic/parasympathetic/enteric, especially gastrointestinal, cardiovascular, and urogenital function), sensory (olfaction, vision, central pain, and hearing), sleep (especially rapid eye movement behavior disorder [RBD] and daytime somnolence), central fatigue, and neuropsychiatric symptoms (visual hallucinations, apathy, anxiety, and depression) (Heinzel et al., 2019; Tolosa et al., 2021). These non-motor features are often associated with declining cognition and poor prognosis (Camacho et al., 2021).

Some non-motor signs can predate diagnosis by many years or emerge and worsen with disease progression (Figure 67.1). This evidence has shaped relatively new concepts of prodromal PD (based especially on RBD, anosmia, and gastrointestinal symptoms), and PD subtypes, with potentials to identify different phenotypes for risk management, early intervention, and personalized therapeutics (Mahlknecht et al., 2022). Nonetheless, a PD diagnosis is rare before age 50 and affects 2%–3% of those aged 65 or older (with a 3M:2F ratio), so its prevalence is growing in step with the world's rapidly aging population (Tolosa et al., 2021). This means that is it increasingly important to distinguish PD and its non-motor symptoms in particular from the subtle effects of aging and other movement disorders.

Our review highlights important changes in the thalamus, and thalamocortical networks, associated with non-motor features in PD. Thalamic nuclei provide a window on most aspects of brain function and are active partners in the regulation of functional connectivity across multiple cortical regions (Nakajima and Halassa, 2017). Studies on normal aging show that changes to the thalamus are among the most pronounced in the brain after age 60, particularly in thalamo-frontal, thalamo-parietal, and thalamo-limbic networks, rather than in thalamic sensorimotor networks (Choi et al., 2022). PD might be expected to impact these age-related effects, but the thalamic changes in PD are different from exaggerated aging. The context for these changes includes the key neuropathological substrates of PD, which are addressed in the next section.

CELLULAR AND SYNAPTIC CHANGES IN PARKINSON'S DISEASE: CURRENT AND FUTURE PERSPECTIVES

The neuropathology of PD is heterogenous, but there is no substantial brain volume loss. The primary hallmark is the presence of intracellular Lewy bodies and Lewy neurites in both the peripheral autonomic nervous system (including the enteric nervous system) and especially the central nervous system, in which misfolded alpha-synuclein protein is the salient molecular signature (Braak and Del Tredici, 2017; Del Tredici and Braak, 2020; Jellinger, 2019). In contrast to a PD diagnosis, the rapid onset of dementia

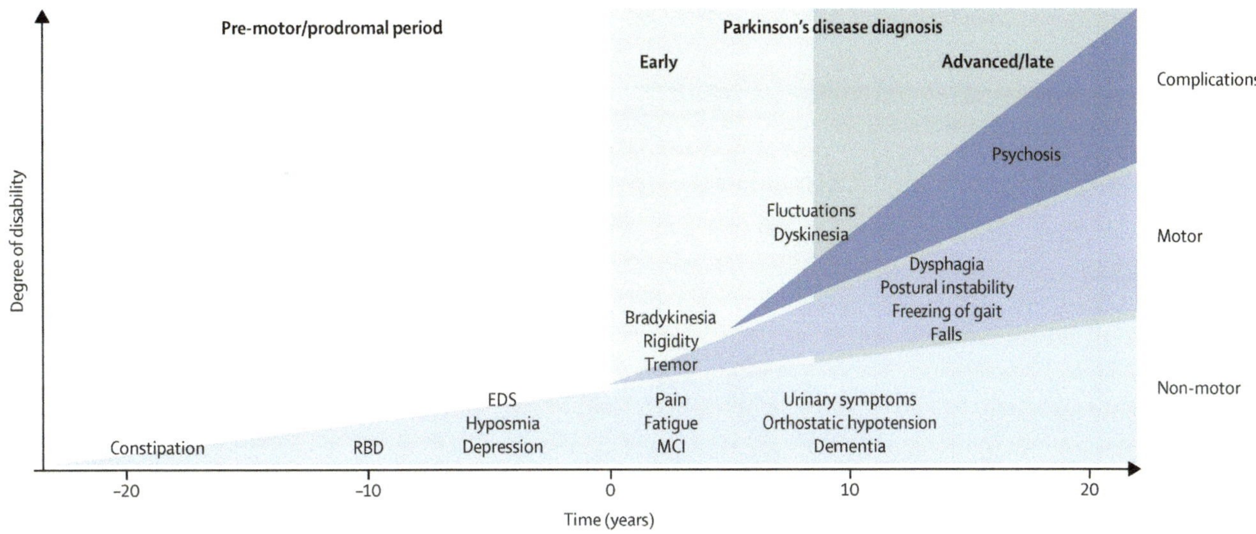

Figure 67.1. A diverse range of non-motor signs are associated with Parkinson's disease (PD). The timecourse shows that some non-motor signs often predate diagnosis (time 0 years), reach clinical significance at different time points, and/or worsen with progression and duration of disease. The precise order of clinical events and their presence or prominence varies across patients and may not follow the general scheme shown here. For example, subtle motor signs and early cognitive impairment may be evident before the advent of a clinical motor diagnosis; in others, the change to PD with mild cognitive impairment (MCI) may be steep or not emerge for a significant period of time; and, progression to PD with dementia can vary from a few years to over 20 years, even in older patients. REM sleep behavior disorder (RBD) is the highest risk factor by far for PD and most but not everyone with confirmed idiopathic RBD converts to either PD or another parkinsonian disorder; PD with RBD shows a more malignant course; but 25%–40% of PD patients do not exhibit RBD. EDS = excessive daytime sleepiness. Figure reproduced with permission from Kalia and Lang (2015).

either before or within ~1 year of parkinsonian motor signs is a separate but controversial clinical entity of "dementia with Lewy bodies" (DLB) (Boeve et al., 2016; Jellinger and Korczyn, 2018). Neither DLB nor its seldom-used research label of "PD-DLB subtype" is a focus here. About 10% of PD has a familial genetic basis (Surmeier et al., 2017). In idiopathic PD, it was evidence of the neural distribution of Lewy pathology and its broad association with clinical disease course that established the current perspective of PD as a multisystem neurodegenerative disorder. This complexity explains why the clinical evolution of PD extends far beyond the classic emphasis on the substantia nigra and impaired motor function (Surmeier et al., 2017; Jellinger and Korczyn, 2018; Del Tredici and Braak, 2020).

Lewy pathology in the thalamic neurons is believed to be emerging around the time of a PD diagnosis (Del Tredici and Braak, 2020); that is, evidence of thalamic pathology occurs during Stage 3 and especially Stage 4 within a generally accepted theory of a 6-stage scheme of progression based on Lewy neuropathology (Figure 67.2). This model depicts a variable continuum rather than a rigid step-wise process (Braak and del Tredici, 2017; Del Tredici and Braak, 2020). Nonetheless, this scheme proposes a relatively predictable caudal-to-rostral propagation of Lewy pathology in selectively vulnerable neurons that enables classification of the progressive 6 stages. The critical pathology is believed to spread most often from the olfactory bulb and/or the periphery to the lower brainstem (especially the dorsal motor nucleus of the vagus in the caudal medulla) and thence to other brain regions and ultimately the prosencephalon and its neocortical mantle. Changes to the thalamus worsen as cortical neuropathology becomes pronounced, but cortical neuropathology should not be viewed in isolation from changes to many vulnerable subcortical regions and vice versa (Gratwicke et al., 2015; Foffani and Obeso, 2018). The frequent description of PDD as a primarily cortical alpha-synuclein pathology is an oversimplification that obscures the importance of subcortical regions, such as the thalamus, and the impact of non-synuclein mechanisms on functional integrity in PD (Braak et al., 2005; Jellinger and Korczyn, 2018; Jellinger, 2022a; Martin et al., 2023). Also, alpha-synuclein pathology, cell loss, and impaired functional integrity are not always synonymous, which is also relevant when considering the thalamus in PD.

Many recent developments in the neuropathology of PD provide unchartered territory to examine thalamocortical non-motor function. First, there is inconsistency across patients in the pattern and regional presence of Lewy pathology that questions the commonality of the standard staging scheme, which has yet to be addressed with respect to the thalamus. Exceptions to Braak and Del Tredici's scheme exist in ~10% to 50% of some post-mortem samples, especially regarding the vagal pathway. As a consequence, there is a growing number of alternative proposals on the spread and significance of the hallmark Lewy pathology in PD (Surmeier et al., 2017; Foffani and Obeso, 2018; Jellinger, 2019; Blesa et al., 2022). For example, evidence supports the existence of either "brain-first" or "body-first" phenotypes,

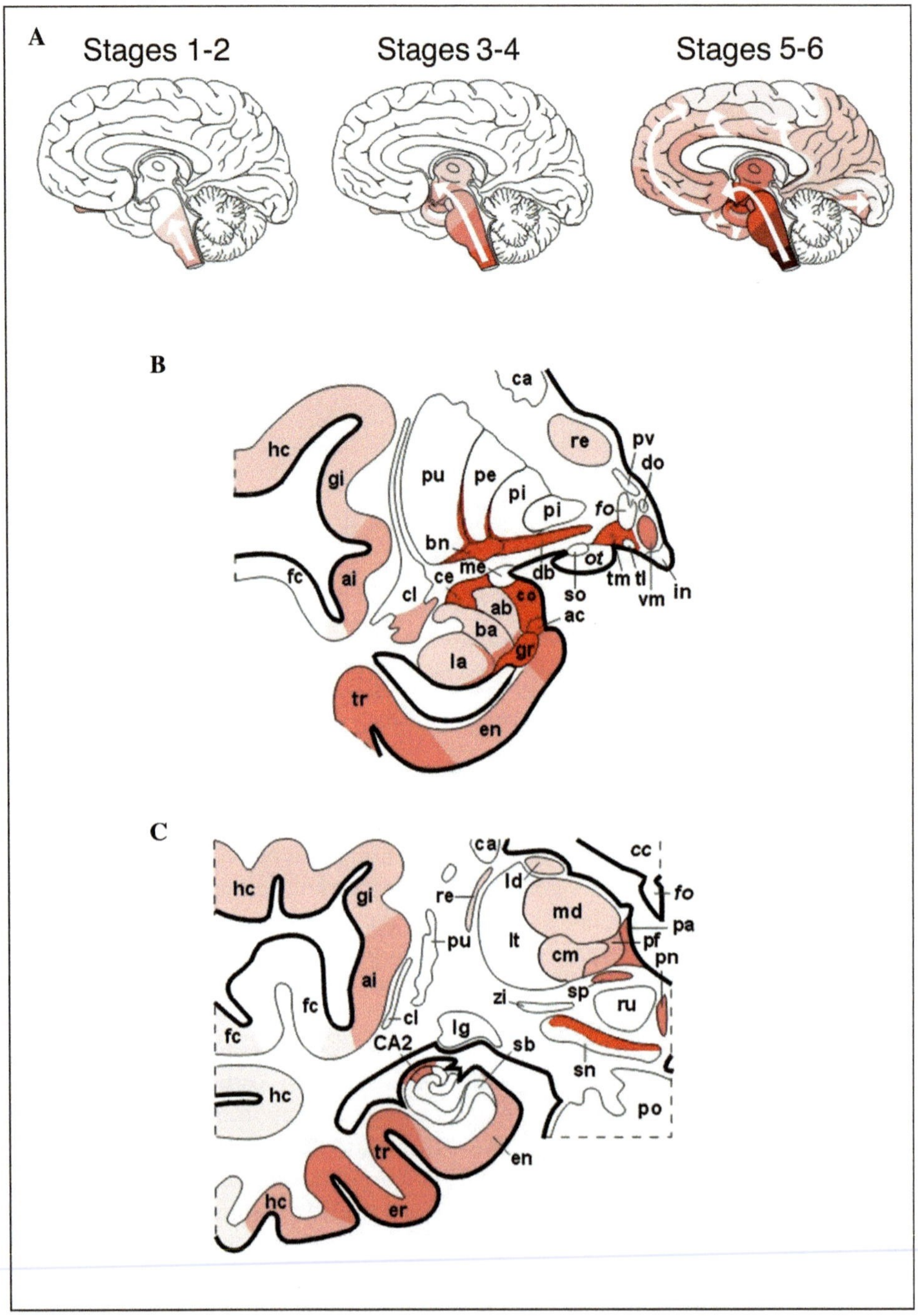

Figure 67.2. Braak and Del Tredici's 6 stages of idiopathic Parkinson's disease (PD). A. The expected regional distribution of Lewy neuropathology (Lewy neurites and Lewy bodies) is depicted by the density of shading for each stage. The expansion of this neuropathology from the brainstem to thalamic nuclei is evident by the mid-point of this scheme (Stages 3–4: typical clinical diagnosis) and worsens with disease progression. B and C. The insular cortex (ai = agranular and dysgranular insular cortex; gi = granular insular cortex), the parahippocampal gyrus (en = entorhinal cortex; er = ectorhinal cortex; tr = transentorhinal cortex) and hypothalamic nuclei (tm = tuberomammillary; vm = ventromedial nuclei) are also a focus for Lewy pathology. In C, the thalamic paraventricular (pa) and subparafascicular (sp) nuclei are often affected earliest as well as other anterior and midline thalamic nuclei; additional thalamic nuclei affected include the reticular (re), parafascicular (pf), centromedian (cm), mediodorsal (md), and laterodorsal (ld) nuclei (see Figure 67.3 for more detail on the thalamus). Abbreviations: ab = amygdala, accessory basal nucleus; ac = amygdala, accessory cortical nucleus; ai = agranular and dysgranular insular cortex; ba = amygdala, basal nucleus; bn = basal nucleus of Meynert; ca = caudate nucleus; CA2 = Ammon's horn, second sector of the Ammon's horn; cc = corpus callosum; ce = central nucleus of the amygdala; cl = claustrum; cm = centromedian nucleus of the thalamus; co = amygdala, cortical nucleus; db = interstitial nucleus of the diagonal band; do = hypothalamus, dorsomedial nucleus; en = entorhinal region; er = ectorhinal region (mesocortex); fo = fornix; fc = first order sensory association areas, premotor areas, primary sensory and motor fields; gi = granular insular cortex; gr = amygdala, granular nucleus; hc = high order sensory association areas and prefrontal fields; in = hypothalamus, infundibular nucleus; la = amygdala, lateral nucleus; ld = laterodorsal nucleus of the thalamus; lg = lateral geniculate body of the thalamus; lt = lateral nuclei of the thalamus; md = mediodorsal nucleus; me = amygdala, medial nucleus; ml = medial lemniscus; ot = optic tract; pa = thalamus, paraventricular nucleus; pe = pallidum, external segment; pf = thalamus, parafascicular nucleus; pi = pallidum, internal segment; pn = paranigral nucleus; pu = putamen; pv = hypothalamus, paraventricular nucleus; re = thalamus, reticular nucleus; sb = subiculum; sn = substantia nigra, pars compacta; so = supraoptic nucleus; sp = thalamus, subparafascicular nucleus; tl = hypothalamus, lateral tuberal nucleus; tm = hypothalamus, tuberomammillary nucleus; tr = transentorhinal region (mesocortex); vm = hypothalamus, ventromedial nucleus; zi = zona incerta. Figure and abbreviations reproduced with permission from Del Tredici and Braak (2020).

with the suggestion of more asymmetric spread from the olfactory bulb and/or amygdala in the "brain-first" subtype (Berg et al., 2021; Borghammer et al., 2022; but, see Banwinkler et al., 2022a, for cautionary notes). Second, a "triple hit" pathogenic route is possible, by adding blood-brain barrier integrity and neuroinflammation to the previously proposed dual-hit theory of primary brain (olfactory/amygdala) and body (vagal) routes of propagation (Dolgacheva et al., 2022). Other recent issues include top-down and/or compensatory mechanisms, debate on the neuropathological relevance of mature alpha-synuclein aggregations (toxicity versus a relative bystander or temporary neuroprotection), as well as the likely neurotoxicity and "seeding" (self-propagating) properties of misfolded oligomeric (small aggregate) and protofibril alpha-synuclein (Blesa et al., 2022). Moreover, the trans-synaptic (or other) propagation of Lewy pathology is dependent on a spectrum of cell-specific factors that define cell vulnerability. These include metabolic demands, exosomal or cell-to-cell transfer, locally activated microglia and astrocytes, and the level of alpha-synuclein expression (Surmeier et al., 2017; Henderson et al., 2022). Relatively weak myelination in hyper-branched thin axons and elaborate terminal arbors in non-relay thalamic projection neurons is another factor that is particularly relevant for critical thalamocortical circuits (Del Tredici and Braak, 2020). These neurons may be under stress from the combination of disproportionately expanded cortex and longevity in humans (Diederich et al. 2020).

Adding weight to this last factor is the significant axonal and somatodentritic disruption from both intracellular Lewy neurites and, perhaps especially, less-studied alpha-synuclein "micro-aggregates." These micro-aggregates are primarily located in presynaptic terminals. They are associated with impaired pre- and postsynaptic processes, dendritic spine degeneration, and dysregulated neurotransmitter release (Schulz-Schaeffer, 2014; Bridi and Hirth, 2018). Long before neuronal death, and separate to more classic Lewy pathology, this synaptic pathology would produce aberrant neural activity among spatially distributed neural systems (Bridi and Hirth, 2018; Kulkarni et al., 2022). Of course, these neural systems depend on thalamocortical integrity, but there is little direct evidence on the synaptic neuropathology associated with different thalamic connections.

Our understanding of the issues summarized above would be enhanced by examining them from the perspective of thalamocortical interactions. The reason to propose this is that there is sufficient post-mortem and in vivo neuroimaging evidence to implicate several excitatory thalamic nuclei as core features of many non-motor signs in PD. This evidence is addressed in the next 2 sections.

NEURONAL CHANGES IN THE THALAMUS IN PARKINSON'S DISEASE: THE POST-MORTEM PERSPECTIVE

Post-mortem studies provide insight to the involvement of the thalamus in PD and its central role in many motor and non-motor facets of this condition. Lewy bodies and Lewy neurites concentrate in selective thalamic regions, but there may also be some functional changes or cell loss that do not correlate with the presence of classic Lewy pathology. Thalamic neuropathology has been described beyond the structural and functional changes to sensorimotor ventral and ventrolateral thalamus that are traditionally associated with motor impairment (Halliday, 2009; Tressitore et al., 2019). Lewy had originally described some inclusions in lateral and paraventricular thalamic nuclei, but the observation that significant alpha-synuclein neuropathology occurs in selective thalamic nuclei in PD was established only 20 years ago (Rüb et al., 2002). Unfortunately, post-mortem studies often provide little if any specific information on thalamic neuropathology, at least in the primary text. For example, a large post-mortem study of 88 idiopathic PD patients gave only supplementary mention of Lewy pathology in the intralaminar thalamic nuclei as an example of established Stage 4 changes in PD (Braak et al., 2005). Remarkably, detailed neuropathology of the thalamus remains one of the most under-reported features of PD.

The first comprehensive assessment on the thalamus was conducted by Braak and colleagues (Rüb et al., 2002) in 12 older idiopathic PD cases (mean age = 75.7 yr). They were confirmed as mid-to-late Hoehn and Yahr motor-severity stages with Lewy pathology in the substantia nigra and amygdala. It was found that parts of the thalamus associated with "limbic system loops" suffer the greatest burden of Lewy pathology, which was rated as particularly severe in midline and anterior intralaminar thalamic regions (Figure 67.3). Pathology was completely absent in their age-matched controls. The list of severely affected thalamic nuclei in the patients is impressive: central lateral; central medial; fasciculosus; limitans-suprageniculate; paracentral; paraventricular; reuniens; and subparafascicular. The cucullar nucleus and the parafascicular nucleus were moderately affected also. There was mild Lewy pathology, and generally Lewy bodies only, in the more "conventional" limbic sites; that is, the anterior thalamic nuclei (anterior principal nucleus; and anterodorsal nucleus) and the mediodorsal nuclei. Thalamic nuclei associated with striatal putamen (centromedian [aka centremedian], ventral anterior) and cerebellar (ventral lateral) loops were also only mildly affected by Lewy pathology; the primary sensory nuclei were generally spared. It is, however, also important to recognize that any neural changes in "limbic

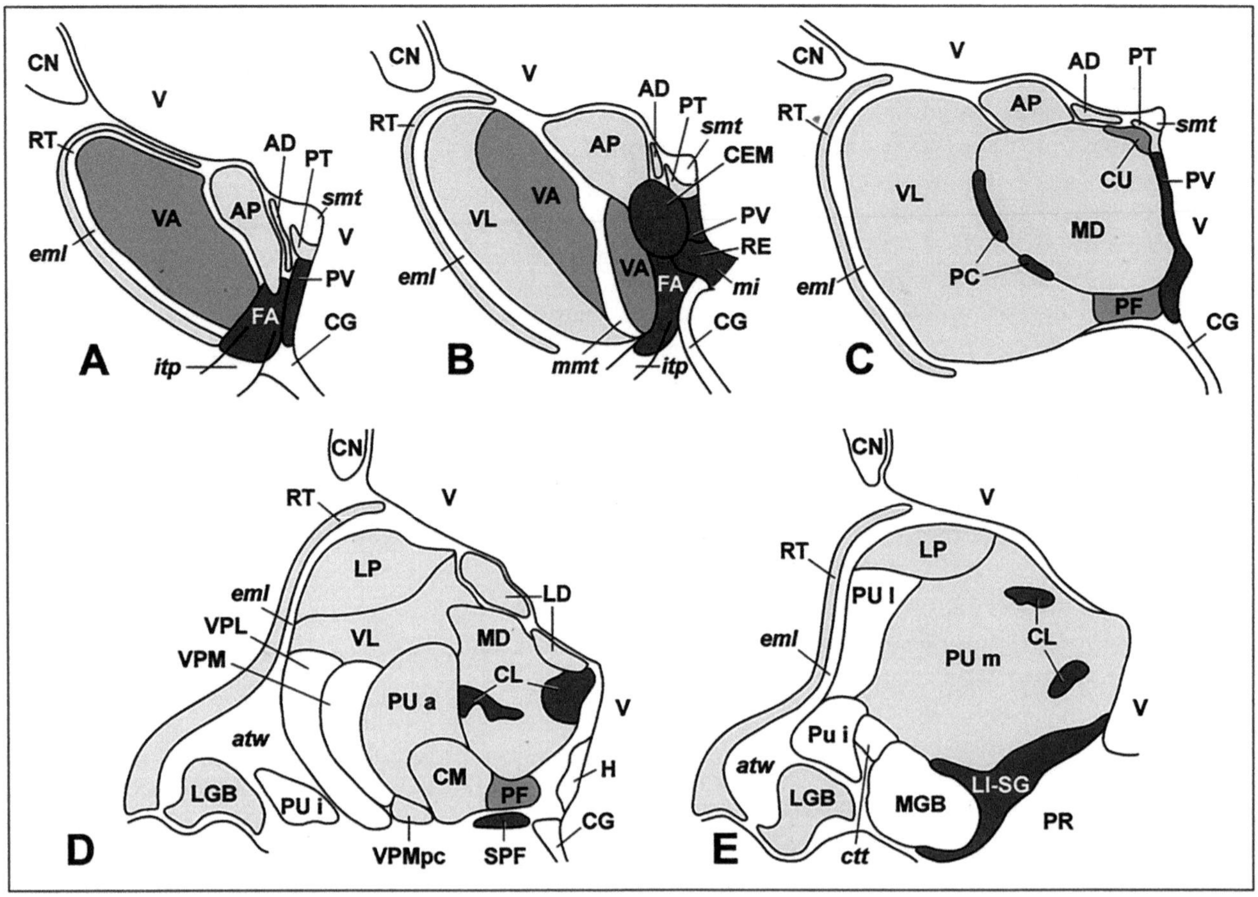

Figure 67.3. Lewy neuropathology in idiopathic Parkinson's disease is evident in many thalamic nuclei. Increasing levels of pathology are reflected by increasing density of shading (mild: light gray; moderate: darker gray; severe: black). The rostral (A) to caudal (E) human thalamus is shown. Abbreviations: AD = anterodorsal nucleus; AP = anterior principal nucleus; *atw* = triangular area of Wernicke; CEM = central medial nucleus; CG = central gray; CL = central lateral nucleus; CM = centromedian nucleus of the thalamic central complex; CN = caudate nucleus; *ctt* = corticotectal tract; CU = cucullar nucleus; *eml* = external medullary lamina; FA = fasciculosus nucleus; H = habenular nuclei; *itp* = inferior thalamic peduncle; LD = laterodorsal nucleus; LGB = lateral geniculate body; LI-SG = limitans-suprageniculate-complex; LP = lateral posterior nucleus; MD = mediodorsal nuclei; MGB = medial geniculate body; *mi* = massa intermedia; *mmt* = mammillothalamic tract (Vicq d'Azyr); PC = paracentral nucleus; PF = parafascicular nucleus of the thalamic central complex; PR = pretectum; PT = parataenial nucleus; PU a = pulvinar, anterior nucleus; PU i = Pulvinar, inferior nucleus; PU l = pulvinar, lateral nucleus; PU m = pulvinar, medial nucleus; PV = paraventricular nuclei; RE = reuniens nucleus; *rf* = tractus retroflexus; RT = reticular nucleus; *smt* = stria medullaris thalami; SPF = subparafascicular nucleus of the thalamic central complex; V = ventricle; VA = ventral anterior nucleus; VL = ventral lateral nucleus; VPL = ventral posterior lateral nucleus; VPM = ventral posterior medial nucleus; VPMpc = ventral posterior medial nucleus, parvocellular part. Figure and abbreviations reproduced from Rub et al. (2002) with permission.

thalamic loops," including the anterior thalamic nuclei and mediodorsal nuclei, exist in the context of significantly compromised cholinergic, dopaminergic, serotonergic, and, perhaps especially, noradrenergic inputs from the brainstem, which would have a major impact on thalamic functional integrity (Pifl et al., 2012; Gratwicke et al., 2015; Surmeier et al., 2017). The precise influence of these neuromodulatory deficits for thalamic function has yet to be determined.

Other detailed post-mortem studies have focused on the explicit loss of neurons in the thalamus, which does not always correspond with the level of observable Lewy pathology. Halliday and colleagues (Henderson et al., 2000a) used stereological (unbiased) cell counting techniques to examine this. They focused on the caudal intralaminar nuclei, which are primarily regarded as regulators of striatal outputs and sensorimotor processing. They found a substantial loss of neurons in the centromedian/parafascicular complex (caudal intralaminar; ~35% loss) in 9 non-dementia PD patients and in 6 slightly younger patients with progressive supranuclear palsy (~50% loss) compared to 10 controls. The PD samples expressed Lewy pathology in this thalamic region, whereas only tau pathology occurred in progressive supranuclear palsy. Their evidence was among the first to suggest a more direct role for the thalamus in PD motor deficits than was previously envisaged. As suggested below, however, the parafascicular nuclei in particular may have a broader set of functional characteristics. Henderson's study reported no cell loss in the anterior principal nucleus (one of the key anterior thalamic nuclei) or the mediodorsal nuclei. They reported nonsignificant smaller volumes in these nuclei PD when compared to controls; but this may reflect the sample size, as the standardized effect size differences were d ~1.5–2.0, and significant reductions were reported in a subsequent study (Matar et al., 2022).

In a second study, Henderson et al. (2000b) reported similar neuronal loss in the caudal intralaminar region in another group of 9 older PD patients. They showed 40% and 30% cell loss, respectively, in the centromedian and parafascicular nuclei, compared to 8 (mostly younger) controls. In this instance, there was generally little thalamic Lewy pathology in this caudal intralaminar complex, and cell loss was found in 2 cases with early Hoehn and Yahr stage ratings, suggesting that these thalamic changes need not be a disease end-stage phenomenon or necessarily linked with conventional Lewy pathology. This study also reported that parafascicular cell loss in PD was focused on parvalbumin calcium-binding glutamatergic neurons (which are 70% of all neurons at this site), but cell loss was evident only in the proportion non-parvalbumin calcium-binding glutamatergic cells of the centromedian nucleus. This suggests that functionally distinct neurons in the thalamus are susceptible to PD. Again, there was no cell loss in the anterior principal nucleus or mediodorsal nuclei.

Halliday (2009) summarized 4 of her group's studies on the thalamus, spanning 37 levodopa-responsive PD patients and 24 controls, which included those in Henderson et al. (2000b). This confirmed evidence for loss of neurons in the posterior intralaminar centromedian/parafascicular complex (52% loss), but one of these studies also found a relative preservation of neurons in the centromedian nucleus in PD patients who displayed early dystonia (abnormal postures and movements). It is possible that the lack of Lewy neuropathology in the centromedian nucleus reported by Rüb et al. (2002) depends on the motor characteristics of individual patients. Cell loss in the posterior intralaminar complex correlated with Lewy neurite pathology. Halliday (2009) reported that neuronal loss was, however, also clearly evident in the anterior intralaminar and midline thalamic nuclei (54% in the parataenial; 52% in the cucullar; and 31% in the centrolateral nuclei), where alpha-synuclein pathology was more severe than in the posterior intralaminar complex. Paratenial cell loss correlated with disease duration, cucullar degeneration associated with visual hallucinations, and the severity of alpha-synuclein pathology in the central lateral (anterior intralaminar) nucleus were more variable, but higher in patients with dementia. Other regions examined did not show cell loss, including some anterior midline nuclei (central medial; and paraventricular, although Lewy pathology here correlated with disease stage), anteromedial limbic nuclei (anterior thalamic nuclei and mediodorsal nuclei), and ventral sensorimotor regions (ventral anterior nuclei; ventrolateral nuclei).

A recent study highlights the need for additional clinicopathological post-mortem assessments in PD (Matar et al., 2022). Five non-demented PD cases, plus 9 with PDD and 10 with DLB, were compared with 10 controls. Here, the focus was the association between visual hallucinations and the integrity of the more conventional limbic thalamus (specifically, the anterior principal nucleus of the anterior thalamic nuclei, and the mediodorsal nuclei). Visual hallucinations are a neuropsychiatric feature in PD that can be more common as cognition worsens, although their presence is not an independent predictor for PDD (Horne et al., 2021). The level and distribution of Lewy pathology across thalamic nuclei in the patient groups were similar to that of Rüb et al. (2002). The anterior principal nucleus and mediodorsal nuclei showed little Lewy pathology and no loss of neurons compared to controls. However, the anterior principal nucleus showed reduced volume (global atrophy) to a similar degree in all 3 patient groups and reduced neuronal (somal) size in the PDD and DLB groups. Compared to controls, there was evidence of a graded effect of reduced volume in the mediodorsal nuclei across the patient groups; this volume in the PD group was intermediate between controls and PDD, but worse again in DLB. However, all 3 patient groups showed similarly reduced neuron size in the mediodorsal nuclei. Global atrophy of the mediodorsal nuclei, but not anterior principal nucleus volume, was associated with the presence of visual hallucinations. Unfortunately, the relationship between structural integrity in these thalamic regions and other clinical symptoms was not examined.

NEUROIMAGING: IN VIVO STRUCTURAL AND FUNCTIONAL CHANGES IN PARKINSON'S DISEASE

Neuroimaging studies have predominantly used a hypothesis-free whole brain approach to determine structural and functional changes in PD. The majority of studies, but not all, describe voxel-based changes where statistically robust effects are reported when adjustment is made for the large number of multiple comparisons being made. The disproportionately greater and heterogenous decline of the thalamus in normal aging (Choi et al., 2022) may reduce the chances of finding thalamic effects. Evidence on the thalamus has seldom been a focus in MRI studies on PD, but there is a growing number that demonstrate that changes to the thalamus and thalamocortical connections provide a highly relevant in vivo perspective on PD. Due to the relatively poor and variable resolution of brain imaging, findings are often based on broad regional terms such as "anterior," "medial," or "posterior" thalamus, although this is often used to infer specific nuclear aggregations or functional classification. Uncertainty with respect to thalamic boundaries and parcellations is an issue in thalamus neuroimaging especially (Mai and Majtanik, 2019) and sometimes complex regions spanning many nuclei are summarized by reference to one major nucleus, which may be misleading.

One recent meta-analysis of structural (T1, voxel-based morphometry and cortical thickness) and functional (fMRI; FDG-PET) neuroimaging summarized changes associated with cognitive status in about 1,400 PD patients, including those with normal cognition (PD-N), those classified as PD-MCI, and those with PDD. Their summary did not address changes in the thalamus (Mihaescu et al., 2019). They reported that the most consistent evidence for the structural gray matter loss in PD-MCI compared to PD-N was focused on the left (paralimbic) insular cortex, as well as the right supramarginal gyrus (sensorimotor cortex) and posterior mid-cingulate cortex. For PDD, compared to those without dementia, differences in the insula were more extensive and bilateral, with additional right hippocampal atrophy. Functional analyses emphasized changes in the bilateral dorsolateral prefrontal cortex and left angular gyrus in PD-MCI. Other recent summaries of neuroimaging and non-motor function in PD can be found in Banwinkler et al. (2022b), and Jellinger (2022a).

Changes to the insular cortex are particularly interesting. This is one of the cortical regions with a propensity for Lewy pathology by the mid-stages of PD (Del Tredici and Braak, 2020). A common focus in PD concerns the posterior insula and its association with sensorimotor processing and bodily awareness; it is extensively connected with the ventroposterior inferior, medial pulvinar, and suprageniculate thalamic nuclei. The anterior insula is perhaps more interesting from the perspective of non-motor function. It is associated with domain-general attentional processing, emotion and anxiety, self-awareness, cognitive control, and decision-making. The anterior insula has extensive neural connections with areas of the thalamus that show degeneration in PD: the reuniens nucleus; the centromedian-parafasicular complex; and, perhaps in a subtler way given the evidence presented above, the mediodorsal nuclei. Together with the anterior cingulate cortex, the anterior insula and its ventral region in particular have been described as a key hub within a large-scale, spatially distributed "salience network." The salience network is activated when responding to behaviorally relevant stimuli and is believed to play an important role in interoceptive-autonomic processing and anxiety, as well as switching attention between internal and external modes of cognition (stimuli or thought about oneself and responding to environmental stimuli, respectively). The dorsal anterior insula is associated with arousal, vigilance, and autonomic system function. Interestingly, a dorsomedial thalamus region (the spatial resolution cannot specify nuclei) provides one of the subcortical nodes in this salience network; whereas an anterior thalamus region (nuclei unspecified) is associated with the central executive network, which is concerned with cognitive control and directed attention, response flexibility, and working memory (Seeley et al., 2007) (Figure 67.4). However, both of these subcortical thalamic regions, in concert with the basal forebrain, appear to regulate the default mode network (DMN), which supports internal modes of cognitive processing such as self-reflection, moral decision-making, theory of mind, and autobiographical/episodic memory (Aguilar and McNally, 2022). So, there seems to be engagement with both separate and overlapping features of large-scale human neurocognitive networks that probably involve the mediodorsal nuclei and anterior thalamic nuclei, which remain to be specified.

Precisely how PD affects different intrinsic fMRI-derived networks for non-motor function is not yet clear. This approach is fraught with issues concerning specific fMRI methodology and analyses, the patients under study (disease stage; symptom profile, including non-motor status and whether they have "tremor-dominant" or "akinetic-rigid" motor symptoms), together with the potential impact of medication. In PD, functional networks may show either increases or decreases, which may (or may not) reflect compensatory changes.

One meta-analysis on resting state fMRI studies (rsfMRI), when the participant is awake but not engaged in any explicit task, concluded that changes in intrinsic functional brain activity in PD is most strongly evident in the bilateral inferior parietal lobule and supramarginal gyrus, which may be associated with changes in the DMN (Tahmasian et al., 2017). They concluded that functional connectivity may decrease in medicated patients, but increase in those off medication, including increased connectivity with unspecified central regions of the thalamus. There was suggestion of ineffective deactivation (uncoupling) of the DMN that is associated with dopamine depletion. A second meta-analysis on rsfMRI evidence (Wolters et al., 2019) focused on studies looking at cognitively impaired PD patients. They also concluded that cognitive impairment predominantly reduced functional connectivity in regions that are part of the DMN. A descriptive review (Filippi et al., 2019) again concluded that cognitive impairment in PD was associated with a disrupted DMN, perhaps especially in posterior brain hubs, but also reduction in the functional connectivity of the (frontoparietal) central executive network. Related studies have reinforced the suggestion that a key problem in PD may be the functional dynamics and coupling between networks. For example,

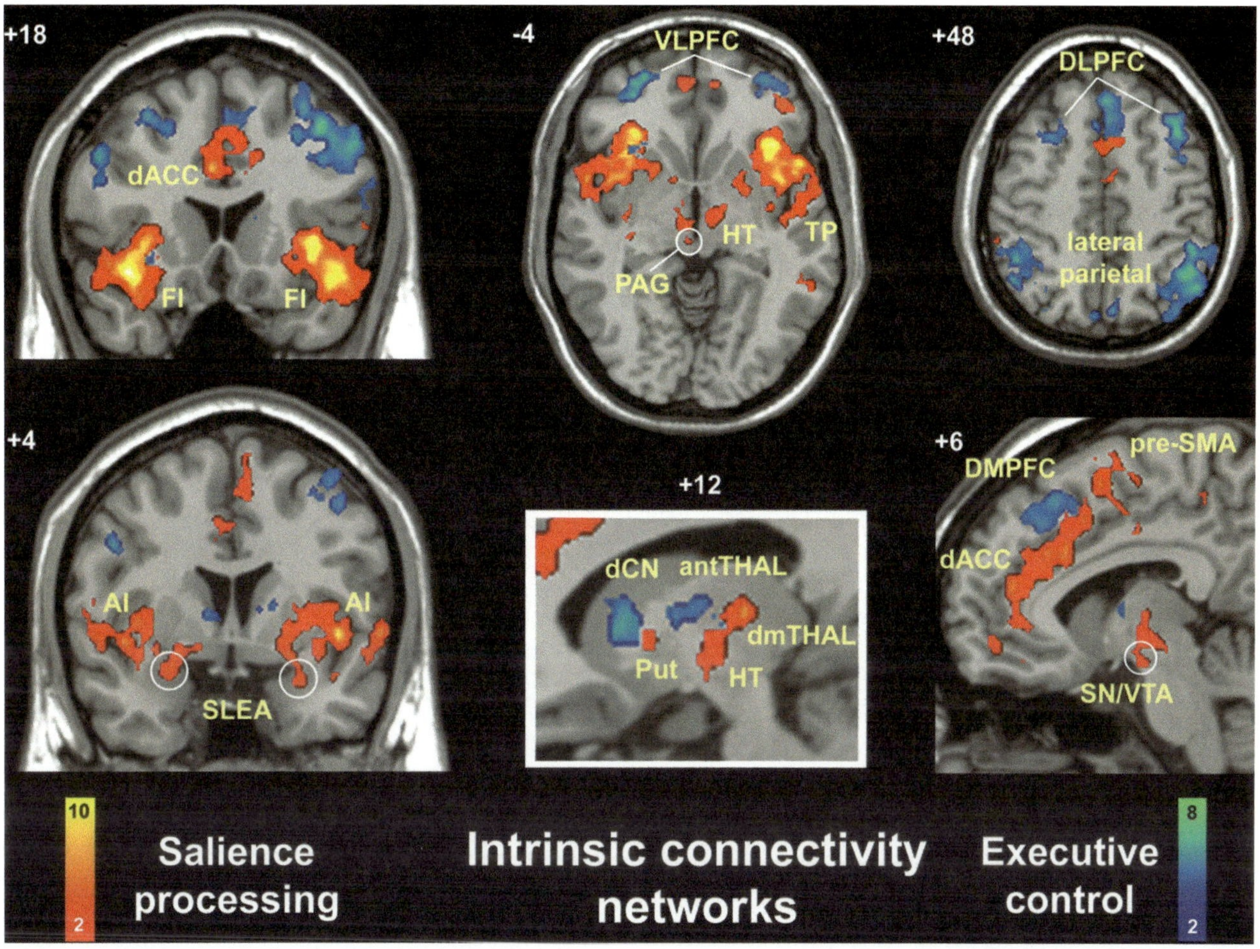

Figure 67.4. Independent component analysis of blood oxygenation dependent fMRI activity in the brain reveals that the intrinsic functional connectivity of the "salience network" (left; red-orange colorbar) includes the key hubs of dorsal anterior cingulate cortex (dACC), orbital fronto-insular cortex and anterior insular cortex (AI), together with a dorsomedial thalamic region (dmTHAL; nuclei unspecified given the spatial resolution). The intrinsic connectivity of the "executive control network" (right; blue-green colorbar) includes the dorsomedial prefrontal cortex (DMPFC) and the dorsal lateral parietal cortex, together with the anterior thalamic region (antTHAL; nuclei unspecified). Functional images are displayed on the MNI standard brain template. Abbreviations: AI = anterior insula; antTHAL = anterior thalamic region; dACC = dorsal anterior cingulate cortex; dCN = dorsal caudate nucleus; dmTHAL = dorsomedial thalamic region; DMPFC = dorsomedial prefrontal cortex; FI = orbital frontoinsular cortex; HT = hypothalamus; PAG = periaqueductal gray; pre-SMA = pre-supplementary motor area; Put = putamen; SLEA = sublenticular extended amygdala; SN/VTA = substantia nigra/ventral tegmental area; TP = temporal pole; VLPFC = ventrolateral prefrontal cortex. Figure and abbreviations reproduced with permission from Seeley et al. (2007) (Copyright 2007 Society for Neuroscience).

during rsfMRI, medicated PD patients with normal Mini-Mental State Examination scores showed less coupling between the salience network and the central executive network ("cognitive control"), but greater coupling between the DMN and the central executive network (Putcha et al., 2015). These differences might be expected to lead to faulty recruitment or suppression of different neurocognitive networks when confronted with external stimuli during challenging task demands. A fourth summary (Tressitore et al., 2019) reported that PD patients with hyposmia, compared to normosmia, showed decreased functional connectivity in both olfactory (limbic/paralimbic) and non-olfactory (parietal/occipital) cortical areas, but increased functional connectivity in the left cingulate cortex. This summary also reported that depression in PD was related to dysfunction in the dorsolateral prefrontal cortex, although clearly many networks and neurotransmitter systems contribute to the relationship between depression and PD (Jellinger, 2022b). Apathy was associated with disrupted limbic frontostriatal pathways. Fatigue revealed a divergent pattern of functional connectivity within the sensorimotor network and the DMN in drug-naïve PD patients. In RBD patients with mild memory impairment there is increased left thalamo-occipital functional connectivity and, in RBD with motor prodromal features, decreased functional connectivity in a cortico-striatal network but increased functional connectivity in cortico-cerebellar networks (Valli et al., 2022). With respect to cognition, there is evidence of increased functional connectivity between a composite "anterior/mediodorsal" thalamic seed-based region and frontal cortex when comparing PD-N with controls (perhaps to maintain cognitive performance in PD-N patients), but reduced thalamofrontal connectivity in PD-MCI versus PD-N, and decreased thalamic functional connectivity with the posterior cingulate cortex of the DMN in PDD versus PD-MCI (Owens-Walton et al., 2021).

With respect to global thalamic volume, MRI studies have reported increases and decreases, respectively reflecting an early motor stage or later stage and/or cognitively impaired characteristics of the sample (Laansma et al., 2021; Owens-Walton et al., 2021). Cognitively impaired patients generally show some gray matter cortical atrophy, but this worsens or becomes apparent with progression to PDD (Jellinger, 2022a). One recent study focused on the relationship between the overall clinical burden of non-motor symptoms and cortical and subcortical T1-weighted (i.e., mostly gray matter) morphological changes in non-demented PD patients who were divided on the basis of experiencing either severe or mild-to-moderate non-motor symptoms (Niccolini et al., 2019). In that study, severe non-motor symptoms, particularly sleep/fatigue and gastrointestinal dysfunction, were associated with reduced thalamic volume; there was no association with other subcortical volumes or cortical thickness. Using coarse segmentation of the thalamus, Li et al. (2020) found smaller volumes in all their 7 thalamic sub-regions in patients with PD-MCI compared to PD-N patients and controls. This evidence suggests that a general macroscopic thalamic structural decline accompanies a threshold of cognitive decline. Using whole left and right thalamus as seeds, Li et al. (2020) also found that PD-MCI was associated with decreased functional connectivity between the thalamus and both visual cortex and precuneus (a posterior hub in the DMN), but increased functional connectivity between the thalamus and mid-cingulate and posterior cingulate cortex. These observations reinforce an important role for the thalamus for non-motor symptoms in PD.

Recently, Park et al. (2020) provided the first neuroimaging study using T1-weighted imaging to segment individual thalamic nuclei, although they were focused on motor symptoms in de novo (drug-naïve) PD patients. Relative to controls, both PD patients meeting criteria for tremor-dominant (TD) motor subtype and those classified as postural instability and gait disturbance (PIGD) subtype showed increased volumes of the suprageniculate nuclei, but decreased volumes of the parafascicular nuclei. However, the intrinsic thalamic network (across 50 thalamic nuclei; 25 per hemisphere) showed greater global inefficiencies in the PIGD subtype, together with localized thalamic network differences between these motor-defined subtypes.

Diffusion-weighted (MR) imaging, such as diffusion tensor imaging (DTI; single within-voxel orientation) and—less frequently—high angular resolution diffusion imaging (HARDI; multiple within-voxel orientations), has been conducted to examine the microstructural integrity of brain regions for both gray tissue and white matter fiber pathways (tractography). These imaging sequences are sensitive to the movement of water molecules. They provide indirect measures of integrity based on different metrics that reflect how water diffusion is affected by natural barriers and local physiology in brain tissue (e.g., mean diffusivity and fractional anisotropy in DTI). Zhang and Burock (2020) describe a complex set of findings based on diffusion imaging, which may reflect the heterogeneity of the PD population used and/or methodological differences in imaging and analysis. Various DTI changes have been reported in the thalamus in PD. The thalamus changes are associated with the general progression of clinical signs in PD, especially decline in global cognition and depression. There is also reduced integrity in the fornix, which includes hippocampal-diencephalic fibers, that is more severe in patients with excessive daytime sleepiness, a non-motor feature that is strongly linked with impaired cognition (Jester et al., 2020). Some studies suggest that loss of gray matter and poor micro-structural integrity of the thalamus is among one of the strongest differences between PD-N and PD-MCI groups at baseline imaging, while reduced gray matter and microstructural integrity of the nucleus basalis of Meynert (nbM) is a strong predictor of cognitive decline (Jellinger, 2022a; Schulz et al., 2018). By contrast, Guttuso et al. (2022) reported evidence of equal or stronger influence of thalamic change for cognitive decline. It is noteworthy that they used a diffusion metric called "free-water" DTI, which is the fractional volume of free water within a voxel (higher values reflect less integrity) as it may be more robust than conventional DTI measures. Free-water DTI had previously been found to be a more sensitive measure for substantia nigra integrity compared to more standard diffusion metrics. Guttuso et al. (2022) found that the increase over time for free-water values in a dorsomedial thalamic segmentation (this was a composite of paratenial, reuniens, and mediodorsal nuclei), but not in an anterior thalamic nucleus segmentation or the hippocampus, was sensitive to declining cognition at both 1 year and 3 years after baseline assessment. Changes in the free-water values for the nbM were sensitive to cognitive change after 3 years, but not over 1 year. This evidence suggests that a combination of microstructural changes to the medial aspects of the thalamus and the cholinergic basal forebrain have significant influence on cognitive health in PD.

FUNCTIONAL SIGNIFICANCE OF THALAMIC CHANGES IN PARKINSON'S DISEASE

Neuropathology and functional connectivity changes in thalamic corticolimbic system loops would contribute

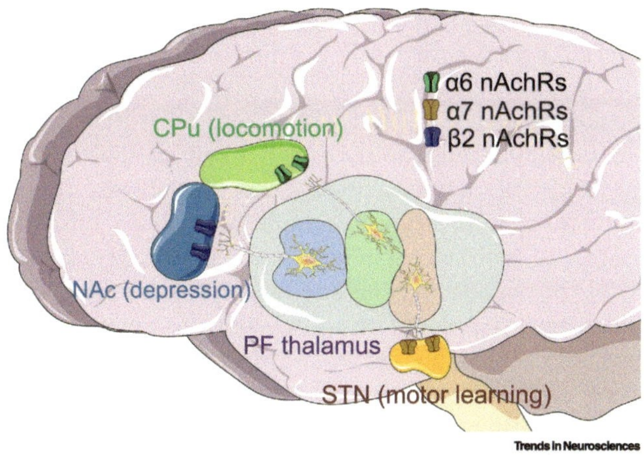

Figure 67.5. Separate subpopulations of neurons of the parafascicular (PF) thalamus (exaggerated in the cartoon) project to the caudate putamen (CPu), subthalamic nucleus (STN), and nucleus accumbens (NAc). A study using a PD mouse model (Zhang et al., 2022) provided evidence that each pathway is linked to separate populations of nicotinic acetylcholine receptors (α6, α7, and β2), which are associated with locomotion, motor learning, and depression, respectively, and with potentials for differential modulation. Figure reproduced with permission from Xiao and Tan (2023).

to many non-motor symptoms in PD. These include: (1) fluctuating arousal, vigilance, and attention, associated in particular with the intralaminar nuclei and their widespread thalamocortical connections, especially with posterior cortical regions (anterior intralaminar nuclei: paracentral, central lateral, and cucullar nuclei; posterior intralaminar nuclei: centromedian/parafasciular complex; Saalmann, 2014); (2) emotional memory, anxiety, stress, motivation, and sleep/wake transitions, related to the dorsal nuclei (paratenial; paraventricular nuclei) connected to ventromedial frontal regions (Colavito et al., 2015; Kirouac, 2015); (3) memory consolidation, executive function, and aspects of emotion, with respect to the ventral midline nuclei (fasciculosus nucleus; reuniens nucleus) and their prefrontal and hippocampal formation connections (Vertes et al., 2022); (4) many of these thalamic nuclei also contribute to a medial pain system (Rub et al., 2002); (5) the anterior thalamic nuclei and mediodorsal nuclei, often seen as key regions of the limbic thalamus, exhibit less explicit neuropathology but may have functional changes as a consequence, for example, of loss of volume and faltering neuromodulatory input from brainstem nuclei. These thalamic nuclei have strong interconnections with many frontal and limbic cortex regions and are consistently associated with episodic memory, executive function, attention, and distributed higher-order cognitive brain networks (Aggleton and O'Mara, 2022; Aguilar and McNally, 2022; Perry et al., 2023); (6) there is growing evidence that thalamic changes will influence many large-scale neurocognitive networks, which may in large part be due to disruption to their dynamic interactions. We speculate that such thalamic changes would contribute to de-differentiation of function during cognitive decline. These thalamic effects and their associated thalamocortical impact may be one reason that multiple cognitive domain impairments are often evident in PD once cognition declines and why PD-MCI so often heralds progression to dementia (Hoogland et al., 2017).

There is growing evidence of cognitive and noncognitive heterogeneity of function within thalamic nuclei, together with complementary function across thalamic nuclei (Halassa and Sherman, 2019); that is, there may be a broader impact of any thalamic changes beyond their traditional nuclear boundary and conventional function. An interesting example of this concerns the parafascicular nucleus, which is among the most prominent sites for thalamic neuropathology in PD. This region is usually associated with regulating motor function, but recent animal studies suggest that distinct neuronal subpopulations in the parafascicular region are differentially associated with movement, depression, and learning (Zhang et al., 2022) (Figure 67.5). The findings support the idea that parallel and independent cortical-thalamic-striatal neurocircuits associate with this posterior intralaminar nucleus to support limbic, associative, and sensorimotor function. Nonetheless, some degree of overlap in neural connections also exists in rodents: medial parafascicular axons project to infralimbic cortex, and ventral parts of the anterior cingulate and insular cortices; the central parafascicular projects to the same regions as the medial parafascicular, but also projects to motor and gustatory cortices; whereas the lateral parafascicular projects to somatosensory and gustatory cortices. Indeed, many individual nuclei in the thalamus may operate through sub-networks of multiple thalamocortical interactions that subserve thalamic control across a wide range of cortical functions (Roy et al., 2022). This perspective suggests that even relatively selective thalamic neuropathology in PD may have a wider impact than that expected from a classic view on the role of individual thalamic nuclei and certainly with regard to the outmoded view of the thalamus as a relay.

IMPORTANT AND UNANSWERED QUESTIONS

The preceding summary shows that the thalamus and thalamocortical interactions can be considered a "hot topic" to help explain many aspects of the varied non-motor symptoms in PD. A digest of some of the more important unanswered questions is provided below.

(1) There are many factors that may define selective cell vulnerability in PD, but we still need to establish this information from the perspective of the thalamus and its neural connections. This might particularly include close inspection of hyper-branching (non-relay) thalamocortical

neurons, especially their terminals. The same would be of interest in terms of brainstem inputs to vulnerable thalamic sites, including the various classic neuromodulators of thalamic function that are known to degenerate in PD (Pifl et al., 2012; Gratwicke et al., 2015). It is believed that some of the earliest changes in the brainstem concern noradrenergic cells in the locus coeruleus complex, less so in tremor-dominant PD, so this should be one focal point (Del Tredici and Braak, 2020). This could be examined in suitable rodent models in the first instance. The advent of alpha-synuclein propagation models offers special hope of understanding in a targeted way the progressive complexities of PD and the propagation of neuropathology (there is a long list of recent reviews on animal models for PD; e.g., Cenci and Bjorklund, 2020). This issue is of interest from the standpoint of dopaminergic input, in particular, because at least some antero-dorsal and midline thalamic nuclei that are associated with cognitive and emotional function have abundant dopaminergic input in humans and nonhuman primates, but this feature is lacking in rodents. In general, primate models in this regard would be especially informative; they represent a gyrencephalic brain and the benefit of cognitive testing that parallels some of that used for humans. Interleaving this work should be research to answer specifically the presence and impact of alpha-synuclein micro-aggregates in axon terminals in the thalamus and thalamocortical projections. Research that examines the impact of alpha-synuclein synaptic pathology on neural networks is growing (Kulkarni et al., 2022), but this has not yet been extended to the thalamus and non-motor function. The benefit of understanding and then seeking interventions for changes to key thalamic aggregates is that they may provide more tractable solutions as biomarkers and for translational intervention (e.g., Guttuso et al., 2022). In short, preventing or minimizing dysfunction in (relatively small) thalamic regions would have a greater impact on cortical function than attempting to do the same to a vast expanse of cortex (Barnett et al., 2021).

(2) The clinicopathological correlates of non-motor function in Parkinson's disease from the perspective of the thalamus remain surprisingly under-reported. Recent cohorts, such as the Parkinson's Progression Markers Initiative dataset, have an immense richness of data. This and similar cohorts and international consortia are also of course making the study of structure-function relationships more feasible and robust, and enable longitudinal designs. They will also help address detailed post-mortem analyses through international cooperation. A concerted effort, however, is needed to examine specific hypotheses concerning the thalamus, along the lines of the summary provided in the above sections. The literature is now sufficiently mature to examine targeted hypotheses to strengthen the interpretation of the data collected. Data-driven approaches are also valuable, of course, because they can provide unexpected insights and facilitate a flexible approach to model-building (e.g., Boeken et al., 2022). But we can anticipate that certain thalamic functions or the involvement of thalamic regions in large-scale distributed networks should now be testable, open to being disproved or modified, and thus refined to improve our understanding of the importance to PD of thalamic regions and their connections. Among the unanswered questions in PD concerns the role of the thalamus across different disease phenotypes, however defined. Looking at prodromal status and different disease progressions and their specific evolution should now be addressed using studies that focus on different regions in the thalamus and their thalamocortical function. This can include evidence on the complex genetics of idiopathic PD, as well as recognized familial genetic subtypes.

(3) The manner in which PD affects large-scale neurocognitive networks in the brain remains unanswered. Together with the anterior cingulate cortex, the anterior insula has been described as a key hub within a large-scale, spatially distributed "salience network." The insular cortex is clearly a major cortical factor in PD, but exactly how, as well as the role of the thalamus in this regard, remains a mystery. This cortex has extensive neural connections with the reuniens nucleus, the centromedian-parafasicular complex, and the mediodorsal nuclei, which show varying levels of neuropathology in PD. In addition, the potential separate and overlapping roles of the "anterior region" and "dorsomedial region" of the thalamus may hold one of the keys to understand how cognition begins to unravel in some people with PD. Understanding the thalamocortical functional gradients associated with limbic thalamic nuclei should be another focal point to learn more about thalamic regulation of key neurocognitive networks in PD.

ACKNOWLEDGMENTS

The authors gratefully acknowledge support from the Health Research Council of New Zealand, The Neurological Foundation of New Zealand, and Canterbury Medical Research Foundation. They report no conflicts of interest.

REFERENCES

Aarsland D, Batzu L, Halliday GM, Geurtsen GJ, Ballard C, Ray Chaudhuri K, Weintraub D (2021). Parkinson disease-associated cognitive impairment. *Nat Rev Dis Primers*. 7:1–21.

Aggleton JP, O'Mara SM (2022). The anterior thalamic nuclei: core components of a tripartite episodic memory system. *Nat Rev Neurosci*. 23:505–516.

Aguilar DD, McNally JM (2022). Subcortical control of the default mode network: Role of the basal forebrain and implications for neuropsychiatric disorders. *Brain Res Bull.* 185:129–139.

Banwinkler M, Dzialas V, Parkinson's Progression Markers Initiative, Hoenig MC, van Eimeren T (2022a). Gray matter volume loss in proposed brain–first and body–first Parkinson's disease subtypes. *Mov Disord.* 37:2066–2074.

Banwinkler M, Theis H, Prange S, van Eimeren T (2022b). Imaging the limbic system in Parkinson's disease - A review of limbic pathology and clinical symptoms. *Brain Sci.* 12:1248.

Barnett SC, Parr-Brownlie LC, Perry BAL, Young CK, Wicky HE, Hughes SM, McNaughton N, Dalrymple-Alford JC (2021). Anterior thalamic nuclei neurons sustain memory. *Curr Res Neurobiol.* 2:100022.

Berg D, Borghammer P, Fereshtehnejad SM, Heinzel S, Horsager J, Schaeffer E, Postuma RB (2021). Prodromal Parkinson disease subtypes: Key to understanding heterogeneity. *Nat Rev Neurol.* 17:349–361.

Blesa J, Foffani G, Dehay B, Bezard E, Obeso JA (2022). Motor and non-motor circuit disturbances in early Parkinson disease: Which happens first? *Nat Rev Neurosci.* 23:115–128.

Boeken OJ, Cieslik EC, Langner R, Markett S (2022). Characterizing functional modules in the human thalamus: Coactivation-based parcellation and systems-level functional decoding. *Brain Struct Funct.* 228:1–24.

Boeve BF, Dickson DW, Duda JE, Ferman TJ, Galasko DR, Galvin JE, Goldman JG, Growdon JH, Hurtig HI, Kaufer DI, Kantarci K, Leverenz JB, Lippa CF, Lopez OL, McKeith IG, Singleton AB, Taylor A, Tsuang D, Weintraub D, Zabetian CP (2016). Arguing against the proposed definition changes of PD. *Mov Disord.* 31:1619–1622.

Borghammer P, Just MK, Horsager J, Skjærbæk C, Raunio A, Kok EH, Savola S, Murayama S, Saito Y, Liisa Myllykangas L, Van Den Berge N (2022). A postmortem study suggests a revision of the dual-hit hypothesis of Parkinson's disease. *NJP Parkinsons Dis.* 8:1–11.

Braak H, Del Tredici K (2017). Neuropathological staging of brain pathology in sporadic Parkinson's disease: Separating the wheat from the chaff. *J Parkinsons Dis.* 7:S71–S85.

Braak H, Rüb U, Steur EJ, Del Tredici K, De Vos RA (2005). Cognitive status correlates with neuropathologic stage in Parkinson disease. *Neurology.* 64:1404–1410.

Bridi JC, Hirth F (2018). Mechanisms of α-synuclein induced synaptopathy in Parkinson's disease. *Front Neurosci.* 12:80.

Camacho M, Macleod AD, Maple-Grødem J, Evans JR, Breen DP, Cummins G, Wijeyekoon RS, Greenland JC, Alves G, Tysnes OB, Lawson RA, Barker RA, Williams-Gray CH (2021). Early constipation predicts faster dementia onset in Parkinson's disease. *NJP Parkinsons Dis.* 7:1–7.

Cenci MA, Björklund A (2020). Animal models for preclinical Parkinson's research: An update and critical appraisal. *Prog Brain Res.* 252:27–59.

Choi EY, Tian L, Su JH, Radovan MT, Tourdias T, Tran TT, Trelle AN, Mormino E, Wagner AD, Rutt BK (2022). Thalamic nuclei atrophy at high and heterogenous rates during cognitively unimpaired human aging. *NeuroImage.* 262:119584.

Colavito V, Tesoriero C, Wirtu AT, Grassi-Zucconi G, Bentivoglio M (2015). Limbic thalamus and state-dependent behavior: The paraventricular nucleus of the thalamic midline as a node in circadian timing and sleep/wake-regulatory networks. *Neurosci Biobehav Rev.* 54:3–17.

Del Tredici K, Braak H (2020). To stage, or not to stage. *Curr Opin Neurobiol.* 61:10–22.

Diederich NJ, Uchihara T, Grillner S, Goetz CG (2020). The evolution-driven signature of Parkinson's disease. *Trends Neurosci.* 43:475–492.

Dolgacheva LP, Zinchenko VP, Goncharov NV (2022). Molecular and cellular interactions in pathogenesis of sporadic Parkinson disease. *Int J Mol Sci.* 23:13043.

Filippi M, Sarasso E, Agosta F (2019). Resting–state functional MRI in Parkinsonian syndromes. *Mov Dis Clin Pract.* 6:104–117.

Foffani G, Obeso JA (2018). A cortical pathogenic theory of Parkinson's disease. *Neuron.* 99:1116–1128.

Gratwicke J, Jahanshahi M, Foltynie T (2015). Parkinson's disease dementia: A neural networks perspective. *Brain.* 138:1454–1476.

Guttuso Jr T, Sirica D, Tosun D, Zivadinov R, Pasternak O, Weintraub D, Baglio F, Bergsland N (2022). Thalamic dorsomedial nucleus free water correlates with cognitive decline in Parkinson's disease. *Mov Disord.* 37:490–501.

Halassa MM, Sherman SM (2019). Thalamocortical circuit motifs: A general framework. *Neuron.* 103:762–770.

Halliday GM (2009). Thalamic changes in Parkinson's disease. *Parkinsonism Relat Disord.* 15:S152–S155.

Heinzel S, Berg D, Gasser T, Chen H, Yao C, Postuma RB, MDS Task Force on the Definition of Parkinson's Disease (2019). Update of the MDS research criteria for prodromal Parkinson's disease. *Mov Disord.* 34:1464–1470.

Henderson JM, Carpenter K, Cartwright H, Halliday GM (2000a). Loss of thalamic intralaminar nuclei in progressive supranuclear palsy and Parkinson's disease: clinical and therapeutic implications. *Brain.* 123:1410–1421.

Henderson JM, Carpenter K, Cartwright H, Halliday GM (2000b). Degeneration of the centré median–parafascicular complex in Parkinson's disease. *Ann Neurol.* 47:345–352.

Henderson MX, Henrich MT, Geibl FF, Oertel WH, Brundin P, Surmeier DJ (2022). The roles of connectivity and neuronal phenotype in determining the pattern of α-synuclein pathology in Parkinson's disease. *Neurobiol Dis.* 168:105687.

Hoogland J, Boel JA, de Bie RM, Geskus RB, Schmand BA, Dalrymple–Alford JC, Marras C, Adler CH, Goldman JG, Tröster AI, Burn DJ, MDS Study Group Validation of Mild Cognitive Impairment in Parkinson Disease (2017). Mild cognitive impairment as a risk factor for Parkinson's disease dementia. *Mov Disord.* 32:1056–1065.

Horne KL, MacAskill MR, Myall DJ, Livingston L, Grenfell S, Pascoe MJ, Young B, Shoorangiz R, Melzer TR, Pitcher TL, Anderson TJ, Dalrymple-Alford JC (2021). Neuropsychiatric symptoms are associated with dementia in Parkinson's disease but not predictive of it. *Mov Disord Clin Pract.* 8:390–399.

Jellinger KA (2019). Is Braak staging valid for all types of Parkinson's disease? *J Neural Transm (Vienna).* 126:423–431.

Jellinger KA (2022a). Morphological basis of Parkinson disease-associated cognitive impairment: an update. *J Neural Transm (Vienna).* 129:977–999.

Jellinger KA (2022b). The pathobiological basis of depression in PD: Challenges and outlooks. *J Neural Transm.* 129:1397–1418.

Jellinger KS, Korczyn AO (2018). Are dementia with Lewy bodies and Parkinson disease-dementia the same disease? *BMC Med.* 16:34.

Jester DJ, Lee S, Molinari V, Volicer L (2020). Cognitive deficits in Parkinson's disease with excessive daytime sleepiness: a systematic review. *Aging Ment Health.* 24:1769–1780.

Kalia LV, Lang AE (2015). Parkinson's disease. *Lancet.* 386:896–912.

Kirouac GJ (2015). Placing the paraventricular nucleus of the thalamus within the brain circuits that control behavior. *Neurosci Biobehav Rev.* 56:315–329.

Kulkarni AS, Burns MR, Brundin P, Wesson DW (2022). Linking α-synuclein-induced synaptopathy and neural network dysfunction in early Parkinson's disease. *Brain Commun.* 4:fcac165.

Laansma MA, Bright JK, Al–Bachari S, Anderson TJ, Ard T, Assogna F, Baquero KA, Berendse HW, Blair J, Cendes F, Dalrymple–Alford JC, de Bie RMA, Debove I, Dirkx MF, Druzgal J, Emsley HCA, Garraux G, Guimaraes RP, Helmich RC, Klein JC, Mackay CE, McMillan CT, Melzer TR, Parkes LM, Piras F, Pitcher TL, Poston KL, Rango M, Ribeiro LF, Rocha CS, Rummel C, Santos LSR, Schmidt R, Schwingenschuh P, Spalletta G, Squarcina L, van den Heuvel OA, Vriend C, Wang J-J, Weintraub D, Wiest R, Yasuda CL, Jahanshad N, Thompson PM, van der Werf YD, and The

ENIGMA-Parkinson's Study (2021). International multicenter analysis of brain structure across clinical stages of Parkinson's disease. *Mov Disord.* 36:2583–2594.

Li MG, He JF, Liu XY, Wang ZF, Lou X, Ma L (2020). Structural and functional thalamic changes in Parkinson's disease with mild cognitive impairment. *J Magn Reson Imaging.* 52:1207–1215.

Mahlknecht P, Marini K, Werkmann M, Poewe W, Seppi K (2022). Prodromal Parkinson's disease: Hype or hope for disease-modification trials? *Transl Neurodegener.* 11:1–13.

Mai JK, Majtanil M (2019). Toward a common terminology for the thalamus. *Front Neuroanat.* 12:114.

Martin WW, Younce JR, Campbell MC, Racette BA, Norris SA, Ushe M, Criswell S, Davis AA, Alfradique-Dunham I, Maiti B, Cairns NJ, Perrin RJ, Kotzbauer PT, Perlmutter JS (2023). Neocortical Lewy body pathology parallels Parkinson's dementia, but not always. *Ann Neurol.* 93:184–195.

Matar E, Brooks D, Lewis SJ, Halliday GM (2022). Limbic thalamus atrophy is associated with visual hallucinations in Lewy body disorders. *Neurobiol Aging.* 112:122–128.

Mihaescu AS, Masellis M, Graff-Guerrero A, Kim J, Criaud M, Cho SS, Ghadery C, Valli M, Strafella AP (2019). Brain degeneration in Parkinson's disease patients with cognitive decline: A coordinate-based meta-analysis. *Brain Imaging Behav.* 13:1021–1034.

Nakajima M, Halassa MM (2017). Thalamic control of functional cortical connectivity. *Curr Opin Neurobiol.* 44:127–131.

Niccolini F, Wilson H, Giordano B, Diamantopoulos K, Pagano G, Chaudhuri KR, Politis M (2019). Sleep disturbances and gastrointestinal dysfunction are associated with thalamic atrophy in Parkinson's disease. *BMC Neurosci.* 20:1–9.

Owens-Walton C, Jakabek D, Power BD, Walterfang M, Hall S, van Westen D, Looi JC, Shaw M, Hansson O (2021). Structural and functional neuroimaging changes associated with cognitive impairment and dementia in Parkinson's disease. *Psychiatry Res Neuroimaging.* 312:111273.

Park J, Park KM, Jo G, Lee H, Choi BS, Shin KJ, Ha S, Park S, Lee H, Kim HY (2020). An investigation of thalamic nuclei volumes and the intrinsic thalamic structural network based on motor subtype in drug naïve patients with Parkinson's disease. *Parkinson Relat Disord.* 81:165–172.

Perry BA, Mendez JC, Mitchell AS (2023). Cortico–thalamocortical interactions for learning, memory and decision–making. *J Physiol.* 601.1:25–35.

Pifl C, Kish SJ, Hornykiewicz O (2012). Thalamic noradrenaline in Parkinson's disease: Deficits suggest role in motor and non–motor symptoms. *Mov Disord.* 27:1618–1624.

Putcha D, Ross RS, Cronin-Golomb A, Janes AC, Stern CE (2015). Altered intrinsic functional coupling between core neurocognitive networks in Parkinson's disease. *Neuroimage Clin.* 7:449–455.

Roy DS, Zhang Y, Halassa MM, Feng G (2022). Thalamic subnetworks as units of function. *Nat Neurosci.* 25:140–153.

Rüb U, Del Tredici K, Schultz C, Ghebremedhin E, De Vos RAI, Steur EJ, Braak H (2002). Parkinson's disease: The thalamic components of the limbic loop are severely impaired by α-synuclein immunopositive inclusion body pathology. *Neurobiol Aging.* 23:245–254.

Saalmann YB (2014). Intralaminar and medial thalamic influence on cortical synchrony, information transmission and cognition. *Front Syst Neurosci.* 8:83.

Schulz J, Pagano G, Fernández Bonfante JA, Wilson H, Politis M (2018). Nucleus basalis of Meynert degeneration precedes and predicts cognitive impairment in Parkinson's disease. *Brain.* 141:1501–1516.

Schulz-Schaeffer WJ (2014). Morbus Parkinson: A synaptic disorder? *Basal Ganglia.* 4:3–8.

Seeley WW, Menon V, Schatzberg AF, Keller J, Glover GH, Kenna H, Reiss AL, Greicius MD (2007). Dissociable intrinsic connectivity networks for salience processing and executive control. *J Neurosci.* 27:2349–2356.

Surmeier DJ, Obeso JA, Halliday GM (2017). Selective neuronal vulnerability in Parkinson disease. *Nat Rev Neurosci.* 18:101–113.

Tahmasian M, Eickhoff SB, Giehl K, Schwartz F, Herz DM, Drzezga A, van Eimeren T, Laird AR, Fox PT, Khazaie H, Zarei M, Eggers C, Eickhoff CR (2017). Resting-state functional reorganization in Parkinson's disease: An activation likelihood estimation meta-analysis. *Cortex.* 92:119–138.

Tessitore A, Cirillo M, De Micco R (2019). Functional connectivity signatures of Parkinson's disease. *J Parkinson Dis.* 9:637–652.

Tolosa E, Garrido A, Scholz SW, Poewe W (2021). Challenges in the diagnosis of Parkinson's disease. *Lancet Neurol.* 20:385–397.

Valli M, Uribe C, Mihaescu A, Strafella AP (2022). Neuroimaging of rapid eye movement sleep behavior disorder and its relation to Parkinson's disease. *J Neurosci Res.* 100:1815–1833.

Vertes RP, Linley SB, Viena TD (2022). Nucleus reuniens: Circuitry, function, and dysfunction. In: Vertes RP, Allen TA, eds. *Electrophysiological Recording Techniques.* 2nd ed. Humana:55–101.

Wolters AF, van de Weijer SC, Leentjens AF, Duits AA, Jacobs HI, Kuijf ML (2019). Resting-state fMRI in Parkinson's disease patients with cognitive impairment: A meta-analysis. *Parkinson Relat Disord.* 62:16–27.

Xiao B, Tan EK (2023). Thalamic pathways mediating motor and non-motor symptoms in a Parkinson's disease model. *Trends Neurosci.* 46:1–2.

Zhang Y, Burock MA (2020). Diffusion tensor imaging in Parkinson's disease and Parkinsonian syndrome: a systematic review. *Front Neurol.* 11:531993.

Zhang Y, Roy DS, Zhu Y, Chen Y, Aida T, Hou Y, Shen C, Lea NE, Schroeder ME, Skaggs KM, Sullivan HA, Fischer KB, Callaway EM, Wickersham IR, Dai J, Li XM, Lu Z, Feng G (2022). Targeting thalamic circuits rescues motor and mood deficits in PD mice. *Nature.* 607:321–329.

68.

THALAMIC CHANGES IN AGING AND ALZHEIMER'S DISEASE

Gonzalo Forno and Michael Hornberger

INTRODUCTION

Thalamic changes in Alzheimer's disease (AD) have been known since the 1980s but are still overlooked to this day. Like a midfield player in football, the thalamus's role is often underappreciated, despite it having a critical role in receiving input via corticostriatal connections and projecting back to several cortical structures through thalamocortical pathways (Behrens et al., 2003). It would be oversimplistic to assume the thalamus has simply a role of "distributing the ball" between cortical areas. Instead, the thalamus directs and modulates the flow of information between cortical-subcortical areas. This modulation has an important aspect for sensorimotor and cognitive processes and can also affect our aging and dementia processes (Tuch et al., 2005; Aggleton et al., 2010; Hornberger and Piguet, 2012; Hughes et al., 2012; Serbruyns et al., 2015). Indeed, it comes to many people as a surprise that the anterodorsal thalamus is, along with the entorhinal cortex, one of the first regions affected by AD pathophysiology (Braak and Braak, 1991) and is implicated with several other neurodegenerative diseases, such as frontotemporal dementia (Bocchetta et al., 2021) (see **Chapter 70 in Section XII**), dementia with Lewy bodies (DLB) (Ziabreva et al., 2006; Delli Pizzi et al., 2015; Watson et al., 2017), and Parkinson's disease (Laansma et al., 2021) (see **Chapter 67 in Section XII**). In this chapter, we will describe thalamic changes during aging and AD and how these impact cognitive functions and AD symptomology.

AGING

STRUCTURAL AND FUNCTIONAL CHANGES

It is well established that "healthy aging" leads to gradual structural and functional changes across several brain regions. Overall, increasing age has a global impact on the brain, but this is not a homogeneous process. It seems that morphological changes differ within different brain tissues, structures, and even sex. For instance, some brain regions seem to follow a more linear decline (i.e., cortex, amygdala, thalamus), whereas others show a more curvilinear relationship (i.e., cerebral white matter, hippocampus) with age (Walhovd et al., 2005). Regarding sex, some studies have shown a more gradual cortical gray matter volume decline in men compared with women (Sullivan et al., 2004), while others showed greater cortical thinning in women compared with men (Jiang et al., 2014). In general, morphological brain changes associated with age can be described as a heterogeneous process that differentially affects brain tissues, structures, and sex.

Similar aging changes have been reported for the thalamus, with evidence showing consistent volume and microstructural changes with increasing age (Cherubini et al., 2009; Hughes et al., 2012; Schippling et al., 2017; Goldstone et al., 2018). By contrast, thalamic differences across sex in aging are not as clear-cut, and no consensus has emerged to date. For example, a cross-sectional study did not find a significant effect of sex in thalamic changes (Sullivan et al., 2004), while evidence from a longitudinal study shows that men had a more rapid thalamic volume loss (Pfefferbaum et al., 2013). More importantly, thalamic volume loss in aging may begin before overall brain volume loss and can be steeper than total brain volume decreases (Van der Werf et al., 2001). Overall thalamic structural changes show an estimated average of 0.25% volume loss per year at the age of 35, which increases to 0.54% volume loss per year at the age of 70 (Schippling et al., 2017). Thalamus structural changes are steeper with increasing age, with accelerated volume loss particularly starting at the age of 60 (Pfefferbaum et al., 2013; Choi et al., 2022). Compared to other subcortical and cortical structures, the thalamus has been suggested to show some of the steepest volume declines with age (Walhovd et al., 2005). Within the thalamus, evidence shows differential volume loss in its subregions. For instance, shape analysis has shown increased vulnerability in its anterior and medial subregion, whereas no effect was found in the lateral portion of the thalamus with aging (Hughes et al., 2012). More recently, thalamic aging changes were explored across thalamic nuclei, and ranged from 0.06% to 1.18% per year (Choi et al., 2022). The anteroventral nucleus emerged as the most susceptible to aging effects (left: 1.18%; right: 1.08%), whereas the ventral lateral posterior nucleus (0.30%) and the ventral posterior lateral (0.28%) were the least affected with increasing age (Choi et al., 2022).

The thalamic volumetric changes are also mirrored by structural connectivity changes. Microstructural changes, as measured by diffusion tensor imaging (DTI) fractional anisotropy (FA), showed bilateral anterior medial thalamus with aging, whereas mean diffusivity (MD) changes affected more right anterior lateral, bilateral centromedian, and centrolateral thalamic nuclei (Ota et al., 2007). These connectivity changes are also somewhat reflected by volumetric changes between connected thalamic and cortical regions. For instance, greater thalamic-frontal projections volume decrease has been described compared to thalamocortical projections to the parietal, temporal, and occipital lobes with physiological aging (Hughes et al., 2012). This is further supported by the notion that a frontal-striatal-thalamic complex is the primary circuit accounting for functional age deterioration, potentially even more affected than the hippocampus and its connected regions (Bonifazi et al., 2018). There are, to our knowledge, much fewer functional connectivity investigations on the aging effect in the thalamus. But there have been reports of impaired functional connectivity between sensory thalamic subregions and the sensory cortex, as well as increased thalamic–right hippocampal connectivity in healthy older people (Goldstone et al., 2018), which need to be further investigated.

Overall, the evidence points to a significant association between thalamic macro- and microstructural changes and normal aging. Thalamic volume loss and increased mean diffusivity have been even proposed as one of the best predictors of physiological aging, compared with other subcortical structures (Cherubini et al., 2009). In the next section, we will explore how these structural aging changes relate to cognitive functioning in aging.

COGNITION

On a cognitive level, the thalamus has been mostly associated with episodic memory and executive functions, due to its dense reciprocal connections with the medial temporal and prefrontal cortex brain regions (Aggleton et al., 2016; Bonifazi et al., 2018; Aggleton et al., 2022). However, the thalamus plays an important role in a whole variety of cognitive processes, including spatial navigation (Coughlan et al., 2018), processing speed (Van der Werf et al., 2001), working memory (Menegaux et al., 2020), sensorimotor performance (Serbruyns et al., 2015) and neuropsychiatric changes (Carrera and Bogousslavsky, 2006). However, how aging processes are affected by thalamic changes has been much less explored.

There is agreement that aging-related thalamic changes are related to executive function and episodic memory changes, but there is also emerging evidence that complex attention tasks, such as selective and divided attention, processing speed, and visuospatial abilities, diminish with aging due to thalamic involvement (Harada et al., 2013). For instance, reduced speed processing and working memory have been associated with decreased thalamic volume due to the effect of aging (Van der Werf et al., 2001). This was further confirmed, as decreased attention and executive function were attributed to the diminishing volume of a thalamic-frontal projection and greater changes in the anterior and medial thalamic subregions (Hughes et al., 2012). Decreased visual short-term memory capacity associated with aberrant structural connectivity between the posterior thalamus and visual occipital cortices during aging has also been reported (Menegaux et al., 2020). On a functional level, thalamic functional connectivity changes in the dorsomedial thalamus-striatum network have been associated with age-related variability in an episodic memory test (Ystad et al., 2010), while decreased memory performance in visual-spatial memory has been associated with abnormal thalamic–right hippocampal functional connectivity (Goldstone et al., 2018). Regarding spatial navigation, a longitudinal study showed that decreased bilateral thalamic volume was, in part, responsible for poorer spatial learning and memory (Korthauer et al., 2016). Finally, atrophy in bilateral ventrolateral thalamic subregions has been described as partially responsible for decreased sensorimotor performance with increasing aging (Serbruyns et al., 2015).

In sum, anterior and medial thalamic subregions seem to be more vulnerable to physiological aging, leading to episodic memory, attention, executive function, and spatial memory changes during healthy aging. Further, frontal-thalamic and parietal-thalamic network changes may affect speed processing and sensorimotor function in older adults. Based on this, it is important to take into account the normal "function" and "cognitive profile" and associated changes for the thalamus during aging. This becomes particularly relevant when contrasting thalamic changes due to AD pathophysiology, which we will cover in the next sections.

ALZHEIMER'S DISEASE

STRUCTURAL AND FUNCTIONAL CHANGES

Alzheimer's disease is commonly associated with structural atrophy changes to the medial temporal lobes, in particular the hippocampus. This notion has led to AD being sometimes referred to as hippocampal dementia (Craig et al., 2011). However, this notion seems too narrow, since AD can also show significant medial parietal changes and subcortical changes from the earliest disease stages onward, affecting the so-called Papez circuit (Forno et al., 2021) (see Figure 68.1). Among these "extra-hippocampal" structures involved in AD, the thalamus emerges as one of the critical regions (Aggleton et al., 2016).

Indeed, from a neuropathological perspective, it is not surprising that the thalamus is involved in AD from the earliest disease stages. It was Braak and Braak who first described neurofibrillary tau deposition changes in the anterodorsal thalamic nucleus (starting in Stages I–II), before many other brain areas (Braak and Braak, 1991). Importantly, these thalamic tau changes occurred concurrently with transentorhinal cortex changes and before hippocampal changes in CA1, making it less likely that thalamic changes are simply due to Wallerian degeneration caused by the hippocampus, which is a common notion. This is further supported by the fact that the AD neurofibrillary tau changes do not homogeneously affect the thalamus, as some of the thalamic nuclei show significant neurofibrillary tau loads while others remain largely unaffected, including areas with strong hippocampal connectivity (Braak and Braak, 1991). Of the affected thalamic regions, the anterodorsal, anteroventral, laterodorsal, centromedian, paratenial, and reticular nuclei are those most affected by tau AD pathophysiology (Braak and Braak, 1991). These tau changes contrast starkly with amyloid changes, the other key protein deposition associated with AD pathophysiology since amyloid deposits are more sparsely distributed across most thalamic nuclei (Braak and Braak, 1991). The amyloid deposit starting point appears in the frontal, temporal, and occipital lobes, whereas subcortical regions (i.e., thalamus) are affected at the end stage of the disease (Braak and Braak, 1991). This has been confirmed in cohorts of older people at elevated risk for AD (Sakr et al., 2019) and in the Alzheimer's Disease Neuroimaging Initiative (ADNI) (Grothe et al., 2017). Most notably, thalamic tau deposits have been associated with worsened cognition in the very early stages of the disease (Bernstein et al., 2021), contrary to amyloid deposits which appear to affect cognition when "extra-neocortical" structures are affected (Grothe et al., 2017). Still, the limbic thalamus, mainly consisting of the anterior, laterodorsal, and mediodorsal thalamic nuclei (Taber et al., 2004), is more vulnerable to the neuropathological, structural, and functional impairment caused by the disease (Aggleton et al., 2016) and therefore the thalamic changes are clearly part of the cortical changes in AD.

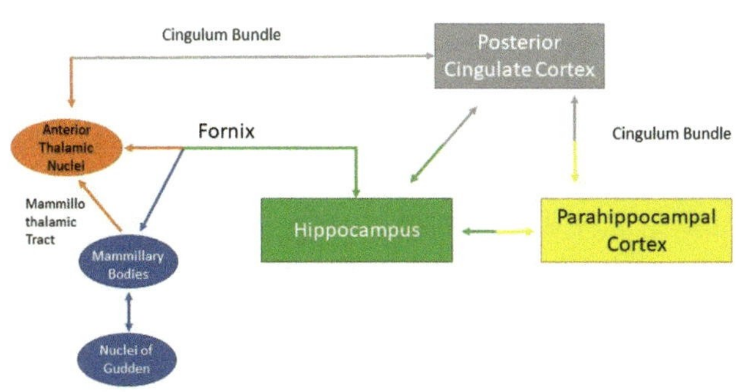

Figure 68.1. Diagram showing the Papez circuit and its subdivisions. Arrows represent afferent connections to the structure depicted in the same color.

Similar findings have been now reported via neuroimaging techniques, for instance, decreased anterior (Callen et al., 2001), dorsal, medial, and ventral (Zarei et al., 2010) but also whole thalamic volume (de Jong et al., 2008; Low et al., 2019) have been reported in people with clinical AD. Most studies have found bilateral thalamic changes in AD, but there are also a few studies which suggest more left hemispheric thalamic atrophy due to AD pathophysiology (Low et al., 2019). However, these findings were in smaller samples, and recent evidence using more sophisticated thalamic segmentation methods (e.g., Thalamus Optimized Multi Atlas Segmentation [THOMAS]) in the large ADNI dataset did not replicate the asymmetrical atrophy in any thalamic nuclei (Bernstein et al., 2021; Forno et al., 2023). Importantly, these more sophisticated segmentation methods have also revealed significant changes in clinical AD in smaller anteroventral, pulvinar, medial geniculate, centromedian, and mediodorsal nuclei (Bernstein et al., 2021), which needs to be confirmed in the future (Forno et al., 2023).

Besides the structural changes in clinical AD, there are also well-established functional and microstructural thalamic changes. For instance, impaired functional connectivity of the thalamus-default mode network, thalamus-cortical connectivity (Zhou et al., 2013), and decreased right thalamic fractional anisotropy (Rose et al., 2006) have been reported. It has been suggested that changes in the anterodorsal nuclei may account, at least in part, for decreased FA of the internal medullary lamina (IML) caused by IML degeneration (Zarei et al., 2010). From these imaging findings, it emerges that not the entire thalamus, but specific subregions and nucleus are affected due to AD pathophysiology in clinical AD.

Thalamic changes are seen not only in clinical AD (i.e., people who have a diagnosis of AD), but also in the prodromal stages of the disease, when a diagnosis of AD is yet to be confirmed and people often have a diagnosis of mild cognitive impairment (MCI). Thalamic changes in MCI have been known for quite some time (Nestor et al., 2003), but it is only recently that this has been confirmed in people with

MCI who have raised levels of amyloid and tau (Bernstein et al., 2021). Current evidence also shows that thalamic nuclei are differently affected according to the stage of disease severity in MCI. Intriguingly, decreased volume in the bilateral mediodorsal, left pulvinar, and left medial geniculate thalamic nucleus can be detected in the earliest stages of MCI, eventually affecting most of the thalamus in late MCI when compared to healthy controls (Bernstein et al., 2021). Note: It is to date not possible to image the anterodorsal nucleus specifically, which is clearly a shortcoming, as this would potentially improve the specificity of the findings further. Differences in the ventrolateral thalamus with its connectivity to the globus pallidus and the orbitofrontal cortex have also been reported in early MCI (van de Mortel et al., 2021). The same study showed over 2 years, that MCI patients who then convert to AD dementia had reduced right posterior thalamic volume compared to the ones who did not progress to dementia. This thalamic volume reduction was mirrored by the right amygdala and bilateral hippocampal volume changes (van de Mortel et al., 2021).

Prodromal Alzheimer's disease in the form of MCI is one way to look at the changes before a clinical diagnosis of AD. The other angle is to investigate people who are at genetic risk of AD. There are different types of genetic risk for AD with different actions on AD risk, which is beyond the scope of the chapter to discuss in more detail. Suffice to say that there are 2 main genetic risk approaches in AD: (1) investigations of people with familial AD, people who have inherited highly rare (<1% of people with AD) genes which *cause* AD, such as *Presenilin 1*, *Presenilin 2*, or the *Amyloid Precursor Protein* gene (Lee et al., 2013; Sala-Llonch et al., 2015; Pardilla-Delgado et al., 2021); (2) investigations of people who carry a common genetic risk gene for AD in the population, which *increases* the risk but does not necessarily cause AD, such as *APOE4* (Li et al., 2014; Novellino et al., 2019). In the case of *APOE*, depending on whether people are hetero- (e3:e4) or homozygous (e4:e4), their risk for AD is increased by 2–3-fold or 8–12-fold, respectively, compared to the general population, who are mostly e3:e3.

Interestingly, even when asymptomatic, thalamic changes have been described in familial AD (Lee et al., 2013; Ryan et al., 2013). Smaller whole (Lee et al., 2013) and left thalamic volume (Ryan et al., 2013), as well as bilateral thalamic increased FA, are described in such asymptomatic mutation carriers (AMC) and can be apparent up to 5 years before the expected age of symptom onset (Ryan et al., 2013). Although not significant, a longitudinal study also showed a trend toward significant right thalamic volume loss in AMC (Sala-Llonch et al., 2015). This same study also reported bilateral thalamic atrophy in symptomatic mutation carriers (SMC) and right thalamic volume loss in the longitudinal analysis (Sala-Llonch et al., 2015). A different study showed an association between thalamic volumetric loss and AD-related pathology in AMC. Importantly, the volumetric reduction did not affect the thalamus uniformly, but lower levels in the medial subregion were related to cortical amyloid-β burden (Pardilla-Delgado et al., 2021). Connectivity studies have also shown increased thalamic segregation associated with a higher cortical amyloid-β burden in AMC (Guzman-Velez et al., 2022). It seems that changes in the thalamus not only occur when symptomatic, but also are evident far from the estimated years to symptom onset in familial AD.

There also have been reports showing decreased volume and impaired functional connectivity in cognitively unimpaired people who carry the *APOE4* gene (Li et al., 2014; Novellino et al., 2019). Notably, hypoactivity between the hippocampus and the thalamus has been described in both cognitively unimpaired *APOE4* carriers and MCI carriers, but thalamic volume decrease is only described in MCI carriers with *APOE4* (Li et al., 2014; Novellino et al., 2019). It has been suggested that the e4 allele may contribute to the regional atrophic pattern which also affects the interaction between the hippocampus and the thalamus (Novellino et al., 2019). But clearly more thalamic studies are needed in these at-genetic-risk groups, which provide an important window into the preclinical stages of AD pathophysiology.

From the above, it should become clear that the thalamus is affected at least as early as the hippocampus in clinical AD on a structural and functional level (see Table 68.1). More importantly, there are significant thalamic changes in the prodromal stages of AD, including MCI and even in cognitively healthy at-genetic-risk people. It further underlines the point that the thalamic changes in AD might be due not only to "downstream effects," such as Wallerian degeneration from the hippocampus. Instead, thalamic AD changes should be regarded more as part of the well-known Papez circuit changes in AD, which make their unique contribution to the disease progression and symptomology (Aggleton et al., 2016; Forno et al., 2021). In the next section, we will explore how thalamic changes might impact the symptomology in AD.

SYMPTOMOLOGY

To date, the literature analyzing cognitive impairment due to thalamic dysfunction in AD has mainly focused on episodic memory. This makes sense since AD pathophysiology affects largely anterior and mediodorsal nuclei—at least at the beginning of the disease—which are known to be strongly involved in episodic memory processes. Although AD studies have most commonly focused on episodic memory, evidence has also shown associations between the thalamus and more generalized cognitive impairment (de Jong et al., 2008; Low et al., 2019; Forno et al., 2023; Bernstein et al., 2021; Wu et al., 2022; (Forno

TABLE 68.1 THALAMIC IMPAIRMENT ACROSS DIFFERENT DISEASE STAGES OF ALZHEIMER'S DISEASE

	Genetic at Risk (Healthy APOE4 carriers)	Familial AD (PSEN)	Prodromal AD (MCI)	AD
Hippocampus	↓↓	↓	↓↓	↓↓↓
Anterior thalamic nuclei	↓↓	↓↓	↓↓	↓↓↓
Mediodorsal thalamic nuclei	~	~	↓↓	↓↓↓
Pulvinar thalamic nuclei	~	~	↓↓	↓↓↓
Posterior cingulate cortex	↓↓	↓↓	↓↓	↓↓↓

Note: a higher number of arrows indicates greater impairment; ~: inconsistent or no significant evidence.

et al., 2023)), dysexecutive function (Zarei et al., 2010; Low et al., 2019), and decreased attention (van de Mortel et al., 2021). Further, thalamic changes have been associated with AD disease severity/progression (Low et al., 2019; Bernstein et al., 2021; van de Mortel et al., 2021), overall neuropsychiatric symptoms severity, and everyday functional impairment (Low et al., 2019).

Specifically, more recent studies have related AD symptomology with ventromedial, ventral posterior lateral, and ventral lateral posterior nuclei changes (Low et al., 2019), although these findings were not replicated in a more recent study which showed no ventral subthalamic nuclei associations with symptoms or cognition (Bernstein et al., 2021). Instead, Bernstein and colleagues reported impaired cognition, memory and disease severity associated with the anteroventral, mediodorsal, pulvinar, medial geniculate, and centromedian nucleus atrophy. It is currently unclear as to why the 2 studies showed such stark differences for cognitive correlates. One issue could be statistical power due to sample size differences, with the different segmentation methods employed (Freesurfer vs. THOMAS) also likely contributing to these contrasting findings. Future studies on thalamic nuclei are clearly needed to investigate how symptoms and cognitive changes in AD are related to thalamic changes. Similarly, the thalamocortical connectivity and its impact on AD symptomology remains largely unexplored and requires investigation.

Overall, there is consistent evidence showing the critical involvement of the thalamus in the clinical symptomatology of AD. But there is a need to understand better which role the thalamus plays in the symptom generation within the wider neural circuit vulnerable to AD pathophysiology (Aggleton et al., 2016; Forno et al., 2021). Future studies are also needed to analyze the role that specific thalamic nuclei play in AD symptomatology as, to date, the literature shows contradicting findings for thalamic nuclei symptomology correlations.

CONCLUSION AND OUTLOOK

This chapter shows that the thalamus plays an important role in aging and Alzheimer's disease. We have highlighted the structural and functional integrity of the thalamus in normal aging and Alzheimer's disease. We also have detailed how these thalamic changes impact normal cognition, as well as their impact on AD symptomatology.

Within normal aging, the thalamus is one of several brain structures affected by the aging process. Both macro- and microstructural changes have been described in detail, which are also mirrored by structural and functional changes in neuroimaging. Still, it remains unclear whether there are biological sex differences across aging since studies have produced mixed results in this regard. Despite this, it is clear that the thalamus is vulnerable to age, with an accelerated volume loss from the sixth decade onward (Pfefferbaum et al., 2013). This total volume loss seems one of the steepest compared to other cortical and subcortical structures (Walhovd et al., 2005) and should be taken into account when conducting age-related thalamic studies. Within the thalamus, the anterior and medial region seems more vulnerable to aging (Hughes et al., 2012), reinforcing the notion that the thalamus is not a homogenous structure, as different subregions are distinctly affected by these aging effects. Further, these aging changes contribute, at least in part, to the cognitive decline attributed to aging. As reviewed in this chapter, several cognitive domains are affected due to structural and functional thalamic changes. Importantly, these changes not only contribute to episodic memory changes, but also can explain attentional, executive, and spatial navigation age-related changes.

The thalamus is particularly vulnerable to AD pathology, which has been known for a long time on a neuropathological level (Braak and Braak, 1991). These changes have been also replicated now via neuroimaging techniques showing that different thalamic subregions and nuclei are affected by AD pathophysiology. It is, therefore, important to recognize that AD is not just a "hippocampal dementia," but that the concomitant thalamic and medial parietal changes contribute to the disease progression and symptomology. It begs the question as to whether it is time to move away from a conventional understanding of AD, with the medial temporal lobe as the central focus of the disease, and think more of a cortico-thalamic network which is central to the disease. Finally, there is increasing evidence that thalamic nuclei not only are affected by AD pathology, but also contribute to the common symptomology and cognitive changes in the disease. However, at this stage these investigations are in their infancy. In particular, symptomology changes would be important to establish as this would potentially allow better clinical management of thalamic changes in AD, which so far does not exist. Finally, it would be important to establish how AD-specific medications impact thalamus functioning.

Despite these exciting new directions for the thalamus in aging and AD, there remain several aspects which require further future work. In particular, there is a clear need to harmonize imaging and segmentation protocols for the thalamus, similar to what has been already occurring for hippocampal neuroimaging. Such better harmonization will allow the creation of not only larger datasets and better cross-validation of thalamic findings, but also better future neuroimaging atlases of thalamic nuclei. This becomes particularly relevant for aging and Alzheimer's disease studies, for which most atlases were never conceived. In this context, it would be also beneficial to have AD-specific regions of interest, such as the anterodorsal nucleus, which is already known from neuropathology to be critical for AD pathophysiology. Similarly, there is still a very limited comprehension of functional neuroimaging, including connectivity changes for aging and AD, which would improve understanding of the thalamocortical interactions during healthy and pathological aging.

Finally, the cognitive correlates of thalamic changes are still very poorly established. A lot of focus has been on the prominent episodic memory changes in aging and AD, which relate mostly to anterior and mediodorsal thalamic changes. However, it is at this stage unclear how one can delineate the hippocampal from the thalamic contributions to these memory deficits. This failure to delineate makes researchers too focused on the hippocampus as the singular contributor to episodic memory changes or frank amnesia, while subcortical changes and particularly the anterior thalamus likely contribute to these episodic memory changes (Aggleton et al., 2022). One interesting future solution might be to look more at accelerated forgetting processes which are more thalamic than hippocampal-dependent. Another interesting future avenue might be how the well-known spatial navigation/orientation deficits in AD might relate to thalamic changes. Specifically, head-direction cells have been known for a long time to be densely located in the anterior and mediodorsal thalamic nucleus while at the same time being much sparser in cortical regions (Coughlan et al., 2018). Such novel cognitive approaches might allow us to shed further light on thalamic-specific cognitive changes during aging and AD progression, which in turn will inform future cognitive measures and outcomes.

In summary, we hope that the reader can see that the investigations of the thalamus in aging and Alzheimer's disease are still in their infancy. This makes thalamic investigations in aging and Alzheimer's disease an exciting prospect since many thalamic aspects which are well-established for basic neuroscience or other health conditions have not yet been explored. We are, therefore, excited to share this chapter with you and hope many of you will join us in investigating the thalamus in aging and Alzheimer's disease, as it has significant potential to make a difference not only in advancing learning about the thalamus structure, but also in diagnosing and managing symptoms related to thalamic changes in this field.

REFERENCES

Aggleton JP, Nelson AJD, O'Mara SM (2022). Time to retire the serial Papez circuit: Implications for space, memory, and attention. *Neurosci Biobehav Rev*. 140:104813.

Aggleton JP, O'Mara SM, Vann SD, Wright NF, Tsanov M, Erichsen JT (2010). Hippocampal-anterior thalamic pathways for memory: Uncovering a network of direct and indirect actions *Eur J Neurosci*. 31:2292–2307.

Aggleton JP, Pralus A, Nelson AJ, Hornberger M (2016). Thalamic pathology and memory loss in early Alzheimer's disease: Moving the focus from the medial temporal lobe to Papez circuit. *Brain*. 139:1877–1890.

Behrens TE, Johansen-Berg H, Woolrich MW, Smith SM, Wheeler-Kingshott CA, Boulby PA, Barker GJ, Sillery EL, Sheehan K, Ciccarelli O, Thompson AJ, Brady JM, Matthews PM (2003). Non-invasive mapping of connections between human thalamus and cortex using diffusion imaging. *Nat Neurosci*. 6:750–757.

Bernstein AS, Rapcsak SZ, Hornberger M, Saranathan M, Alzheimer's Disease Neuroimaging Initiative (2021). Structural changes in thalamic nuclei across prodromal and clinical Alzheimer's disease. *J Alzheimers Dis*. 82:361–371.

Bocchetta M, Malpetti M, Todd EG, Rowe JB, Rohrer JD (2021). Looking beneath the surface: The importance of subcortical structures in frontotemporal dementia. *Brain Commun*. 3:fcab158.

Bonifazi P, Erramuzpe A, Diez I, Gabilondo I, Boisgontier MP, Pauwels L, Stramaglia S, Swinnen SP, Cortes JM (2018). Structure-function multi-scale connectomics reveals a major role of the fronto-striato-thalamic circuit in brain aging. *Hum Brain Mapp*. 39:4663–4677.

Braak H, Braak E (1991). Neuropathological stageing of Alzheimer-related changes. *Acta Neuropathol*. 82:239–259.

Callen DJ, Black SE, Gao F, Caldwell CB, Szalai JP (2001). Beyond the hippocampus: MRI volumetry confirms widespread limbic atrophy in AD. *Neurology*. 57:1669–1674.

Carrera E, Bogousslavsky J (2006). The thalamus and behavior: Effects of anatomically distinct strokes. *Neurology*. 66:1817–1823.

Cherubini A, Peran P, Caltagirone C, Sabatini U, Spalletta G (2009). Aging of subcortical nuclei: Microstructural, mineralization and atrophy modifications measured in vivo using MRI. *NeuroImage*. 48:29–36.

Choi EY, Tian L, Su JH, Radovan MT, Tourdias T, Tran TT, Trelle AN, Mormino E, Wagner AD, Rutt BK (2022). Thalamic nuclei atrophy at high and heterogenous rates during cognitively unimpaired human aging. *NeuroImage*. 262:119584.

Coughlan G, Laczo J, Hort J, Minihane AM, Hornberger M (2018). Spatial navigation deficits: Overlooked cognitive marker for preclinical Alzheimer disease? *Nat Rev Neurol*. 14:496–506.

Craig LA, Hong NS, McDonald RJ (2011). Revisiting the cholinergic hypothesis in the development of Alzheimer's disease. *Neurosci Biobehav Rev*. 35:1397–1409.

de Jong LW, van der Hiele K, Veer IM, Houwing JJ, Westendorp RG, Bollen EL, de Bruin PW, Middelkoop HA, van Buchem MA, van der Grond J (2008). Strongly reduced volumes of putamen and thalamus in Alzheimer's disease: An MRI study. *Brain.* 131:3277–3285.

Delli Pizzi S, Franciotti R, Taylor JP, Thomas A, Tartaro A, Onofrj M, Bonanni L (2015). Thalamic involvement in fluctuating cognition in dementia with Lewy Bodies: Magnetic resonance evidences. *Cereb Cortex.* 25:3682–3689.

Forno G, Llado A, Hornberger M (2021). Going round in circles: The Papez circuit in Alzheimer's disease. *Eur J Neurosci.* 54:7668–7687.

Forno G, Saranathan M, Contador J, Guillen N, Falgàs N, Tort-Merino A, Balasa M, Sanchez-Valle R, Hornberger M, Lladó A (2023). Thalamic nuclei changes in early and late onset Alzheimer's disease. *Curr Res Neurobiol.* 4:100084. doi: 10.1016/j.crneur.2023.100084. PMID: 37397807; PMCID: PMC10313877.

Goldstone A, Mayhew SD, Hale JR, Wilson RS, Bagshaw AP (2018). Thalamic functional connectivity and its association with behavioral performance in older age. *Brain Behav.* 8:e00943.

Grothe MJ, Barthel H, Sepulcre J, Dyrba M, Sabri O, Teipel SJ, Alzheimer's Disease Neuroimaging Initiative (2017). In vivo staging of regional amyloid deposition. *Neurology.* 89:2031–2038.

Guzman-Velez E, Diez I, Schoemaker D, Pardilla-Delgado E, Vila-Castelar C, Fox-Fuller JT, Baena A, Sperling RA, Johnson KA, Lopera F, Sepulcre J, Quiroz YT (2022). Amyloid-beta and tau pathologies relate to distinctive brain dysconnectomics in preclinical autosomal-dominant Alzheimer's disease. *Proc Natl Acad Sci U S A.* 119:e2113641119.

Harada CN, Natelson Love MC, Triebel KL (2013). Normal cognitive aging. *Clin Geriatr Med.* 29:737–752.

Hornberger M, Piguet O (2012). Episodic memory in frontotemporal dementia: A critical review. *Brain.* 135:678–692.

Hughes EJ, Bond J, Svrckova P, Makropoulos A, Ball G, Sharp DJ, Edwards AD, Hajnal JV, Counsell SJ (2012). Regional changes in thalamic shape and volume with increasing age. *NeuroImage.* 63:1134–1142.

Jiang J, Sachdev P, Lipnicki DM, Zhang H, Liu T, Zhu W, Suo C, Zhuang L, Crawford J, Reppermund S, Trollor J, Brodaty H, Wen W (2014). A longitudinal study of brain atrophy over two years in community-dwelling older individuals. *NeuroImage.* 86:203–211.

Korthauer LE, Nowak NT, Moffat SD, An Y, Rowland LM, Barker PB, Resnick SM, Driscoll I (2016). Correlates of virtual navigation performance in older adults. *Neurobiol Aging.* 39:118–127.

Laansma MA, Bright JK, Al–Bachari S, Anderson TJ, Ard T, Assogna F, Baquero KA, Berendse HW, Blair J, Cendes F, Dalrymple–Alford JC (2021) International multicenter analysis of brain structure across clinical stages of Parkinson's disease. *Mov Disord.* 36:2583–2594.

Lee GJ, Lu PH, Medina LD, Rodriguez-Agudelo Y, Melchor S, Coppola G, Braskie MN, Hua X, Apostolova LG, Leow AD, Thompson PM, Ringman JM (2013). Regional brain volume differences in symptomatic and presymptomatic carriers of familial Alzheimer's disease mutations. *J Neurol Neurosurg Psychiatry.* 84:154–162.

Li W, Antuono PG, Xie C, Chen G, Jones JL, Ward BD, Singh SP, Franczak MB, Goveas JS, Li SJ (2014). Aberrant functional connectivity in Papez circuit correlates with memory performance in cognitively intact middle-aged APOE4 carriers. *Cortex.* 57:167–176.

Low A, Mak E, Malpetti M, Chouliaras L, Nicastro N, Su L, Holland N, Rittman T, Rodriguez PV, Passamonti L, Bevan-Jones WR, Jones PS, Rowe JB, O'Brien JT (2019). Asymmetrical atrophy of thalamic subnuclei in Alzheimer's disease and amyloid-positive mild cognitive impairment is associated with key clinical features. *Alzheimers Dement (Amst).* 11:690–699.

Menegaux A, Bauerlein FJB, Vania A, Napiorkowski N, Neitzel J, Ruiz-Rizzo AL, Muller HJ, Sorg C, Finke K (2020). Linking the impact of aging on visual short-term memory capacity with changes in the structural connectivity of posterior thalamus to occipital cortices. *NeuroImage.* 208:1–13.

Nestor PJ, Fryer TD, Smielewski P, Hodges JR (2003). Limbic hypometabolism in Alzheimer's disease and mild cognitive impairment. *Ann Neurol.* 54:343–351.

Novellino F, Lopez ME, Vaccaro MG, Miguel Y, Delgado ML, Maestu F (2019). Association between hippocampus, thalamus, and caudate in mild cognitive impairment APOEepsilon4 carriers: A structural covariance MRI study. *Front Neurol.* 10:1–12.

Ota M, Obata T, Akine Y, Ito H, Matsumoto R, Ikehira H, Asada T, Suhara T (2007). Laterality and aging of thalamic subregions measured by diffusion tensor imaging. *Neuroreport.* 18:1071–1075.

Pardilla-Delgado E, Torrico-Teave H, Sanchez JS, Ramirez-Gomez LA, Baena A, Bocanegra Y, Vila-Castelar C, Fox-Fuller JT, Guzman-Velez E, Martinez J, Alvarez S, Ochoa-Escudero M, Lopera F, Quiroz YT (2021). Associations between subregional thalamic volume and brain pathology in autosomal dominant Alzheimer's disease. *Brain Commun.* 3:1–10.

Pfefferbaum A, Rohlfing T, Rosenbloom MJ, Chu W, Colrain IM, Sullivan EV (2013). Variation in longitudinal trajectories of regional brain volumes of healthy men and women (ages 10 to 85 years) measured with atlas-based parcellation of MRI. *NeuroImage.* 65:176–193.

Rose SE, McMahon KL, Janke AL, O'Dowd B, de Zubicaray G, Strudwick MW, Chalk JB (2006). Diffusion indices on magnetic resonance imaging and neuropsychological performance in amnestic mild cognitive impairment. *J Neurol Neurosurg Psychiatry.* 77:1122–1128.

Ryan NS, Keihaninejad S, Shakespeare TJ, Lehmann M, Crutch SJ, Malone IB, Thornton JS, Mancini L, Hyare H, Yousry T, Ridgway GR, Zhang H, Modat M, Alexander DC, Rossor MN, Ourselin S, Fox NC (2013). Magnetic resonance imaging evidence for presymptomatic change in thalamus and caudate in familial Alzheimer's disease. *Brain.* 136:1399–1414.

Sakr FA, Grothe MJ, Cavedo E, Jelistratova I, Habert MO, Dyrba M, Gonzalez-Escamilla G, Bertin H, Locatelli M, Lehericy S, Teipel S, Dubois B, Hampel H, for the I-psg, Alzheimer Precision Medicine Initiative (2019). Applicability of in vivo staging of regional amyloid burden in a cognitively normal cohort with subjective memory complaints: The INSIGHT-preAD study. *Alzheimers Res Ther.* 11:15.1–11.

Sala-Llonch R, Llado A, Fortea J, Bosch B, Antonell A, Balasa M, Bargallo N, Bartres-Faz D, Molinuevo JL, Sanchez-Valle R (2015). Evolving brain structural changes in PSEN1 mutation carriers. *Neurobiol Aging.* 36:1261–1270.

Schippling S, Ostwaldt AC, Suppa P, Spies L, Manogaran P, Gocke C, Huppertz HJ, Opfer R (2017). Global and regional annual brain volume loss rates in physiological aging. *J Neurol.* 264:520–528.

Serbruyns L, Leunissen I, Huysmans T, Cuypers K, Meesen RL, van Ruitenbeek P, Sijbers J, Swinnen SP (2015). Subcortical volumetric changes across the adult lifespan: Subregional thalamic atrophy accounts for age-related sensorimotor performance declines. *Cortex.* 65:128–138.

Sullivan EV, Rosenbloom M, Serventi KL, Pfefferbaum A (2004). Effects of age and sex on volumes of the thalamus, pons, and cortex. *Neurobiol Aging.* 25:185–192.

Taber KH, Wen C, Khan A, Hurley RA (2004). The limbic thalamus. *J Neuropsychiatry Clin Neurosci.* 16:127–132.

Tuch DS, Salat DH, Wisco JJ, Zaleta AK, Hevelone ND, Rosas HD (2005). Choice reaction time performance correlates with diffusion anisotropy in white matter pathways supporting visuospatial attention. *Proc Natl Acad Sci U S A.* 102:12212–12217.

van de Mortel LA, Thomas RM, van Wingen GA, Alzheimer's Disease Neuroimaging Initiative (2021). Grey matter loss at different stages of cognitive decline: A role for the thalamus in developing Alzheimer's disease. *J Alzheimers Dis.* 83:705–720.

Van der Werf YD, Tisserand DJ, Visser PJ, Hofman PAM, Vuurman E, Uyling HBM, Jolles J (2001). Thalamic volume predicts performance on tests of cognitive speed and decreases in healthy aging: A magnetic resonance imaging-based volumetric analysis. *Cogn Brain Res.* 11(3):377–385.

Walhovd KB, Fjell AM, Reinvang I, Lundervold A, Dale AM, Eilertsen DE, Quinn BT, Salat D, Makris N, Fischl B (2005). Effects of age on volumes of cortex, white matter and subcortical structures. *Neurobiol Aging.* 26:1261–1270.

Watson R, Colloby SJ, Blamire AM, Wesnes KA, Wood J, O'Brien JT (2017). Does attentional dysfunction and thalamic atrophy predict decline in dementia with Lewy bodies? *Parkinsonism Relat Disord.* 45:69–74.

Wu Y, Wu X, Gao L, Yan Y, Geng Z, Zhou S, Zhu W, Tian Y, Yu Y, Wei L, Wang K (2022). Abnormal functional connectivity of thalamic subdivisions in Alzheimer's disease: A functional magnetic resonance imaging study. *Neuroscience.* 496:73–82.

Ystad M, Eichele T, Lundervold AJ, Lundervold A (2010). Subcortical functional connectivity and verbal episodic memory in healthy elderly: A resting state fMRI study. *NeuroImage.* 52:379–388.

Zarei M, Patenaude B, Damoiseaux J, Morgese C, Smith S, Matthews PM, Barkhof F, Rombouts SA, Sanz-Arigita E, Jenkinson M (2010). Combining shape and connectivity analysis: An MRI study of thalamic degeneration in Alzheimer's disease. *NeuroImage.* 49:1–8.

Zhou B, Liu Y, Zhang Z, An N, Yao H, Wang P, Wang L, Zhang X, Jiang T (2013). Impaired functional connectivity of the thalamus in Alzheimer's disease and mild cognitive impairment: A resting-state fMRI study. *Curr Alzheimer Res.* 10:754–766.

Ziabreva I, Ballard CG, Aarsland D, Larsen JP, McKeith IG, Perry RH, Perry EK (2006). Lewy body disease: Thalamic cholinergic activity related to dementia and parkinsonism. Neurobiol Aging. 27:433–438.

69.

CONTRIBUTION OF THALAMOCORTICAL LOOPS TO THE DEVELOPMENT, PERSISTENCE, AND COMPLICATIONS OF ALCOHOL USE DISORDER

Anne Lise Pitel and Shailendra Segobin

Alcohol use disorder (AUD) is among the most prevalent mental disorders worldwide. It is a major risk factor for the global burden of mortality and disease, the leading cause of more than 60 diseases, and contributes to the development of 200 others (World Health Organization, 2018). Chronic and heavy alcohol consumption has dramatic consequences for the body and is associated with brain abnormalities and neurocognitive disorders even without clinically detectable evidence for alcohol-related neurological complications such as Korsakoff's syndrome (KS). KS is a major neurocognitive disorder (DSM-5, 2013) characterized by severe, persistent, and debilitating amnesia. It results from thiamine deficiency or altered thiamine metabolism and is most frequently observed in AUD patients who are at special risk for thiamine deficiency. Cognitive and behavioral alterations associated with AUD and KS include impairments in episodic memory, executive functions, motor skills, and craving. These alterations result from dysfunction of several brain networks involving different thalamocortical loops, suggesting that thalamic abnormalities may play a cardinal role in the pathophysiology of AUD with or without KS (Pitel et al., 2015). Three cerebral circuits are especially involved in the development, persistence, and complications of AUD (Figure 69.1): the Papez circuit, the frontocerebellar circuit, and the thalamo-insular circuit.

PAPEZ CIRCUIT AND ALCOHOL-RELATED EPISODIC MEMORY DEFICITS

EPISODIC MEMORY IN AUD AND KS

Episodic memory is the mnemonic system controlling the encoding, storage, and retrieval of personally experienced events, associated with a precise spatial and temporal context of encoding. Episodic memory allows the conscious recollection of personal happenings and events from one's personal past and the mental projection of anticipated events into one's subjective future. Recollection of episodic events includes autonoetic awareness: the impression of re-experiencing or reliving the past and mentally traveling back in subjective time (Tulving, 2001).

Severe AUD is associated, early after detoxification, with altered episodic memory encoding and retrieval abilities, as well as impaired contextual memory and recollection processes (Pitel et al., 2014). Both verbal memory and visual memory are affected by chronic and excessive alcohol drinking (Fama et al., 2021). While some episodic memory disorders observed in AUD may be related to poor effortful executive processes, they cannot solely be explained by executive dysfunction (Pitel et al., 2007). These findings suggest a genuine episodic memory deficit in AUD without KS.

In AUD, altered episodic memory disorders may lead to professional disabilities and may hamper treatment. During psychosocial treatment, AUD patients have to travel mentally through time in order to consider former and future high-risk situations. They have to picture the contexts in which they were in danger and might be again in the future. They also have to remember very precisely, with autonoetic consciousness, what they felt and how they experienced the craving for alcohol (Pitel et al., 2007). However, AUD patients with episodic memory impairments will have difficulties remembering their past experiences of drinking and projecting into their future life without alcohol, which may limit the benefit of treatment (Le Berre et al., 2017). Furthermore, psychoeducation, which is frequently included in AUD treatment, also requires efficient episodic memory abilities and may thus not be a relevant therapeutic choice for patients with episodic memory deficits. Globally, episodic memory disorders can be considered as a factor

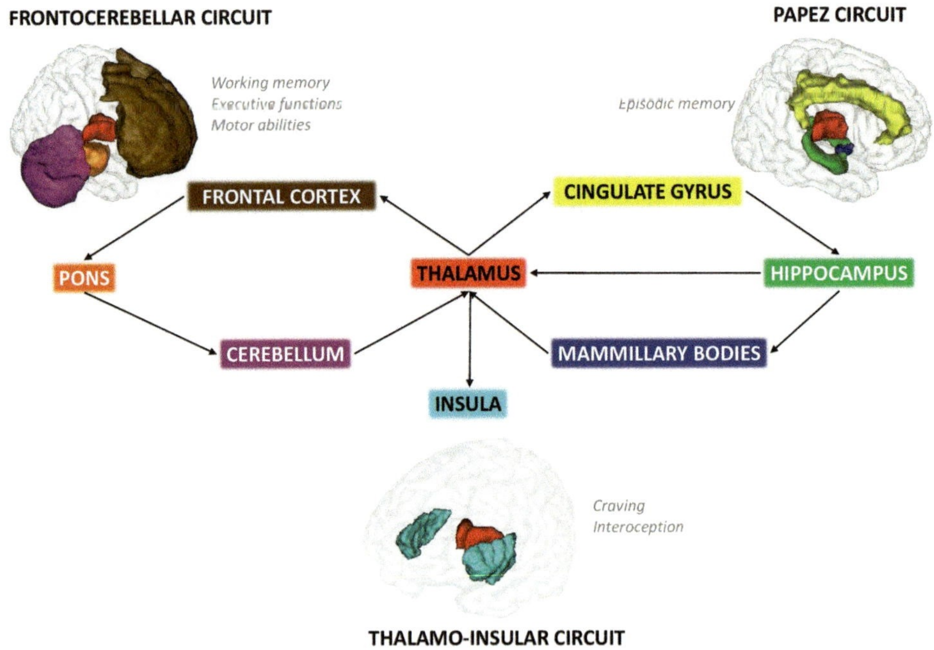

Figure 69.1. Three thalamocortical circuits involved in the development, persistence, and complications of alcohol use disorder. Orange: pons; brown: frontal cortex; pink: cerebellum; yellow: cingulate cortex; green: hippocampus; blue: mammillary bodies; turquoise: insular cortex; red: thalamus.

limiting the benefit of treatment and ultimately as a risk factor for relapse (Le Berre et al., 2017).

Severe anterograde amnesia and variable retrograde amnesia are the main features of KS. The few neuropsychological studies that made direct comparisons between AUD and KS patients confirmed the severity of episodic memory deficits in KS compared with those exhibited by AUD patients (Fama et al., 2012), in agreement with the diagnosis of major neurocognitive disorders (DSM-5, 2013) that requires an impact of cognitive impairments on autonomy in daily life. More qualitatively, some similarities could be found between AUD and KS in the profile of alterations of the different episodic memory components, except for the disproportionate impairment of encoding abilities compared with retrieval ones in KS (Pitel et al., 2014).

ANATOMY OF THE PAPEZ CIRCUIT

The Papez circuit was initially posed as a model of emotion, which later evolved to a model for long-term memory. Probably now in need of an update (Aggleton et al., 2022), it classically consists of the hippocampus, where the subiculum projects directly to the anterior thalamic nuclei through the fornix, or via the mammillary bodies through the mammillothalamic tract. The "loop" is closed from the thalamus to the cingulate cortex through the internal capsule and then back to the parahippocampal region and subiculum through the cingulum.

RELATIONSHIPS BETWEEN EPISODIC MEMORY AND PAPEZ CIRCUIT IN AUD AND KS

Historically, damage to the Papez circuit and its relationship to episodic memory has been studied through the prism of KS (Segobin and Pitel, 2021; Aggleton and O'Mara, 2022) (for further details on human memory and the anterior thalamic nuclei, see **Chapter 43 in Section VIII**). Following the first clinical description of patients with alcohol-related persistent and severe memory disorders, neuropathological studies described the nature of brain damage in what eventually became known as KS (Korsakoff, 1887, 1889). These studies (Victor et al., 1971; Brion and Mikol, 1978; Mair et al., 1979; Mayes et al., 1988; Harding et al., 2000) collectively reported lesions in the mammillary bodies and thalamic nuclei. The arrival of neuroimaging studies has allowed the in vivo study of KS, and brain imaging is still contributing to our understanding of the pathophysiological mechanisms that lead to KS.

Theoretically, the comparison between AUD and KS patients makes it possible to disentangle the effect of excessive and chronic alcohol drinking from that of thiamine and to highlight the specificity of KS amnesia compared with that of AUD. The first neuroimaging group comparisons between 7 AUD patients and 7 KS patients were performed using computed tomography (CT) imaging that showed lower tissue density in the thalamus of KS patients (Figure 69.2A). The tissue density in the thalamus of these KS patients correlated with their memory performance (Shimamura et al., 1988). CT scans also showed enlarged lateral and third ventricles in 25 KS patients compared with

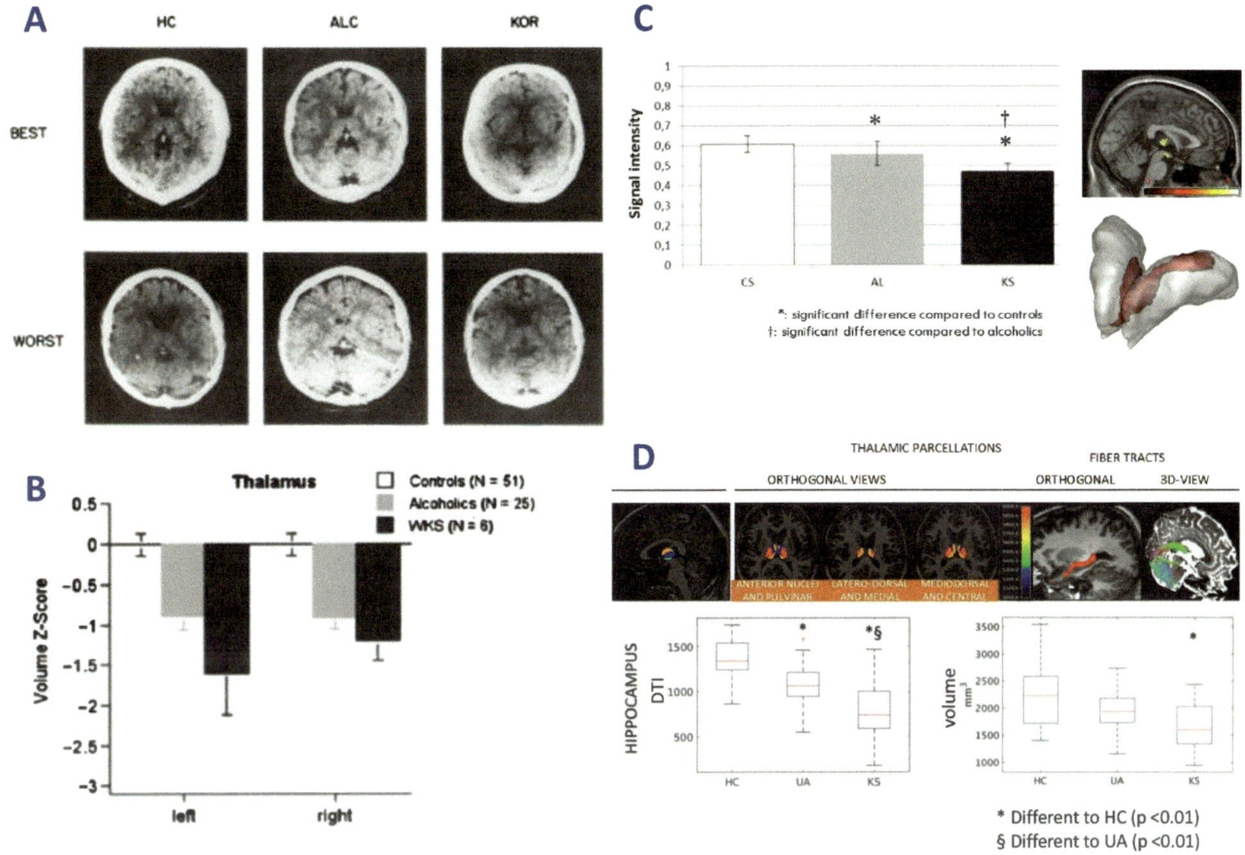

Figure 69.2. Neuroimaging data highlighting alterations to the thalamus within Papez circuit in alcohol use disorder (AUD) patients with and without Korsakoff's syndrome (KS). (A). Computed tomography scans (7 controls vs. 7 patients with AUD vs. 7 patients with KS) shows hypodensity of the thalamus (KS < AUD) (Shimamura et al., 1988). (B). Magnetic resonance imaging (MRI) volumetric measures (51 controls vs. 25 AUD vs. 6 KS) show more severe shrinkage in the thalamus (Sullivan and Pfefferbaum, 2009). (C). Voxel-based analysis of MRI data (25 controls vs. 34 AUD vs. 11 KS) shows more severe shrinkage in what appears to be anterior and midline thalamic nuclei (Pitel et al., 2012). (D). Parcellation of the thalamus using probabilistic tractography (20 controls vs. 26 AUD vs. 16 KS) shows that the hippocampus is principally connected to the anterior thalamic nuclei via the fornix. There is a graded effect in the number of connections from controls to AUD without KS to AUD with KS. Atrophy of the anterior thalamic nuclei is specific to the KS patients only (Segobin et al., 2019).

AUD patients and healthy controls (Jacobson et al., 1990). While no quantifications were performed on diencephalic regions, it can be argued that these ventricular enlargements would induce, or be the result of, shrinkage of the thalamus and/or mammillary bodies that are located in the neighborhood of the ventricles.

Magnetic resonance imaging (MRI) soon became the most favored for neuroimaging investigations as it could provide information on multiple modalities, including structure, connectivity, and functions, without having to administer any radioactive dose. From a volumetric point of view, gray matter comparisons between AUD and KS patients have corroborated previously observed alterations to the mammillary bodies (Shear et al., 1996; Sullivan et al., 1999; Pitel et al., 2012) and the thalamus (Figure 69.2B,C) (Pitel et al., 2009; Sullivan and Pfefferbaum, 2009; Pitel et al., 2012). Damage of the hippocampus was also reported (Sullivan and Pfefferbaum, 2009). Studies focusing on white matter in AUD without KS showed impaired microstructural integrity of the cingulum and fornix (Pfefferbaum et al., 2009; Segobin et al., 2015, 2019), which have direct connections to the thalamus. From a molecular point of view, positron emission tomography studies involving KS patients are rare. The few studies showed lower glucose uptake in the thalamus, hypothalamus, and mammillary bodies (Reed et al., 2003; Pitel et al., 2009). Future evaluation of molecular processes should go beyond synaptic ones to dissociate brain alterations as a result of ethanol toxicity versus those resulting from thiamine deficiency. So far, animal models, via both cross-sectional and longitudinal studies, have pointed toward separate markers for brain damage as a result of ethanol toxicity versus thiamine deficiency (Zahr and Pfefferbaum, 2017). For example, it has recently been shown that neuroinflammatory processes were more severe in the thalamus of adult rats, in which thiamine deficiency had been induced as opposed to ethanol exposure alone (Toledo Nunes et al., 2019).

Taken together, studies comparing AUD patients with and without KS have shown more severe deterioration of Papez circuit in patients with KS.

FRONTOCEREBELLAR CIRCUIT AND ALCOHOL-RELATED EXECUTIVE/MOTOR DYSFUNCTION

EXECUTIVE FUNCTIONS IN AUD AND KS

Executive functions are defined as the set of high-level cognitive processes whose main function is to facilitate the subject's adaptation to new situations, especially when action routines (i.e., over-learned cognitive skills) can no longer suffice. These functions are thus required for non-routine tasks, that is, tasks that are new, or that require the inhibition of a predominant action pattern, or that are complex (e.g., requiring rule deduction, problem-solving, sequencing of multiple operations). Depending on the cognitive models, executive functions refer to inhibition, flexibility, updating, or planning abilities.

Numerous studies examined executive functions in AUD, and they systematically showed altered executive functioning in recently detoxified patients with severe AUD. Response inhibition, problem-solving, deduction of rules, updating, cognitive flexibility, set-shifting, and decision-making are notably impaired (Le Berre et al., 2017).

AUD treatment and subsequent abstinence or alcohol reduction require executive abilities. Patients have to stop or inhibit their automatic drinking behavior to adopt new habits. In the face of temptation, they have to resist drinking alcohol, plan efficient avoidance strategies, and choose a behavior other than drinking alcohol. AUD patients with executive dysfunction are thus at increased risk for relapse.

In KS, amnesia can be associated with working memory and executive impairments. Studies that compared executive abilities in AUD and KS provided conflicting results. Some studies indicated more severe executive dysfunction in KS (Brokate et al., 2003; Maharasingam et al., 2013), while others found similar level of deficits (Krabbendam et al., 2000; Pitel et al., 2008). These different findings can be explained by the heterogeneity of the executive profile in AUD patients and/or may be related to the nature of the executive processes evaluated, which are highly multidimensional.

MOTOR ABILITIES IN AUD AND KS

Motor disorders are common in AUD patients. Impaired coordination and postural stability result in ataxia characterized by balance and gait deficits (Zhao et al., 2020; Müller-Oehring et al., 2022). Upper limb motor skills are generally less affected, but manual dexterity and grip strength can also be impaired (Zhao et al., 2020).

ANATOMY OF THE FRONTOCEREBELLAR CIRCUIT

Executive dysfunctions were initially considered to result from frontal abnormalities. It is now well known, however, that executive functions are linked to alterations within a larger network involving the cerebellum, pons, and thalamus, in addition to frontal regions. The frontocerebellar circuit (FCC) was identified first in nonhuman primates through retrograde and anterograde viral transneuronal tracing (Kelly and Strick, 2003). Two parallel closed-loops within the cortico-thalamo-cerebellar circuitry constitute the FCC (Chanraud et al., 2010; Segobin et al., 2015). In the first one (Figure 69.3A, red loop), which is involved in working memory and executive functions (Marvel and Desmond, 2010), cerebellar efferents mainly from the cerebellar crus II project to the thalamus and from there onto Brodmann areas 9 and 46 of the dorsolateral prefrontal cortex, before projecting back to the cerebellum via the pons. The second one (Figure 69.3A, blue loop) underlies motor functions like gait and balance, where the thalamus receives input from the cerebellar lobules IV–VI and projects these to the motor cortex before feeding back to the cerebellum via the pons (for further details, see **Chapter 31 in Section VI**).

RELATIONSHIPS BETWEEN EXECUTIVE/MOTOR FUNCTIONS AND THE FRONTOCEREBELLAR CIRCUIT IN AUD AND KS

Alcohol-related brain damage to the FCC, observed post-mortem (Harper, 1998; Harding et al., 2000) and in vivo using CT (Diener et al., 1986) and later MRI (Shear et al., 1996; Pfefferbaum et al., 1997) involves several nodes from the frontal cortex and the cerebellum. While the frontal cortex connects to the thalamus via the anterior corona radiata and anterior limb of the internal capsule, the cerebellum connects to the thalamus by the middle and superior cerebellar peduncles, among other systems (Pitel et al., 2015; Segobin et al., 2015). Volume deficits in white matter tissue were observed both in the midbrain and the cerebellum (Mechtcheriakov et al., 2007; Pitel et al., 2012). Studies using diffusion imaging have revealed lower values of fractional anisotropy, and hence decreased microstructural integrity, in the anterior limb of the internal capsule (Pfefferbaum et al., 2009) as well as the middle and superior cerebellar peduncles (Segobin et al., 2015), but no such apparent alterations in tracts from the mesencephalic region. However, fewer fiber tracts (up to 18%) were found in that region (Chanraud et al., 2009). From a molecular point of view, the pattern of glucose metabolism in the thalamus of AUD patients seems to be unaltered (Ritz et al., 2016).

Few imaging studies exist in the literature where there is a direct comparison of alterations within the FCC of AUD and KS patients (Sullivan and Pfefferbaum, 2009;

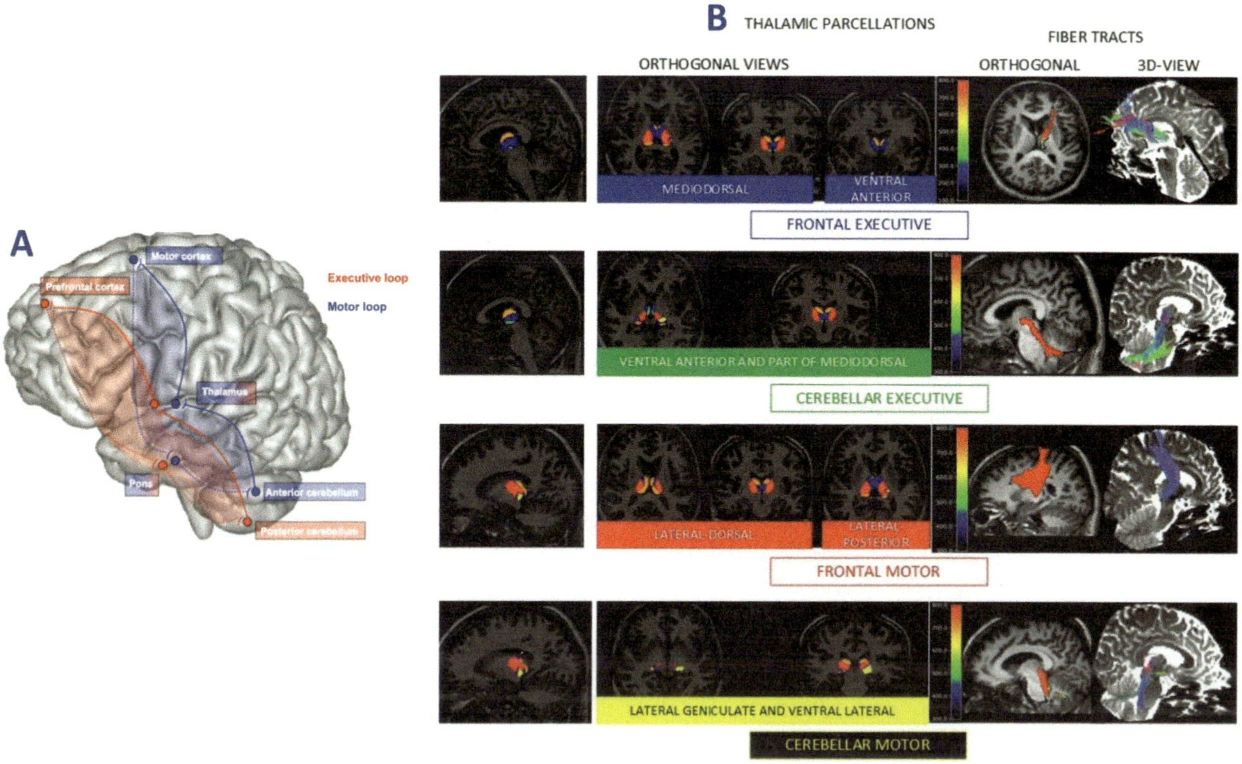

Figure 69.3. Involvement of thalamic nuclei within the frontocerebellar circuit. (A). Executive loop in red involving the cerebellar cruses, pons, frontal executive regions comprising BA9 and BA46 and motor loop in blue involving cerebellar lobes IV–VI, pons, frontal motor regions comprising the primary motor areas (precentral gyrus). The thalamus is shared between these 2 loops (Zhar et al., 2010). (B). Parcellation of the thalamus using probabilistic tractography shows loci of thalamic nuclei that are connected to the frontal and cerebellar executive and motor regions of the frontocerebellar circuit (Segobin et al., 2019).

Pitel et al., 2012; Segobin et al., 2015, 2019). They have shown similarities in the profiles of gray and white matter alterations within the FCC. The only subtle difference, identified using voxel-based morphometry, was located in the thalamus that is shared with the Papez circuit and is discussed below.

Regarding brain-behavior relationships, executive dysfunction and altered working memory capacities correlated with lower gray matter volumes of the thalamus and cerebellum (Chanraud et al., 2007). On the mesencephalic side, these cognitive measures correlated with white matter volume in the brainstem and fiber counts between midbrain and pons (Chanraud et al., 2009).

SPECIFICITY OF THALAMIC NUCLEI DAMAGE WITHIN PAPEZ AND FRONTOCEREBELLAR CIRCUITS

Thalamic damage is a leading candidate underlying pathology of AUD. As the thalamus is shared between the Papez circuit and FCC and consists of many nuclei, it paves the way to suggest that alterations to specific nuclei drive the pathophysiological mechanisms happening within each circuit. Thus, precision in location and quantification of thalamic alterations is essential.

POST-MORTEM INVESTIGATIONS

The post-mortem studies of specific thalamic nuclei report disparate results. In their 11 patients with KS, Brion and Mikol (1978) reported damage to the mediodorsal nuclei in 7 of them, while 6 had damage to the pulvinar, and 9 to the latero-dorsal nuclei. No damage to the anterior nuclei, now trusted to be one of the hallmarks of amnesia (Aggleton and O'Mara, 2022), was found. Another post-mortem study (Victor et al., 1971) investigated a larger sample. Eighty of the 82 KS patients were "alcoholics" and showed lesions in the mediodorsal and pulvinar nuclei. Five patients with Wernicke's encephalopathy but without KS (who thus did not have amnesia) were found to have preserved mediodorsal nuclei. This finding led to a proposition that the mediodorsal nuclei could be a critical site for the development of amnesia. By contrast, 2 case studies that each included 2 well-described KS patients showed gliosis along the midline nuclei of the thalamus and involved not only the most medial part of the mediodorsal nuclei, but also those nuclei adjacent to it, including the anterior thalamic nuclei (Mair et al., 1979; Mayes et al., 1988). Such contradictory findings can be explained by factors beyond sample size. For example, all patients studied in the large cohort of Victor and colleagues did not undergo the same neuropsychological tests to diagnose amnesia. Postmortem tissue measurements for all patients were not necessarily performed with the same equipment. With better access to improved technology, a more comprehensive and precise quantification was performed to compare 5 AUD patients without neurological complications, 5 AUD with Wernicke's encephalopathy but no amnesia, 8 KS patients having experienced Wernicke's encephalopathy, and 6 controls (Harding et al., 2000). The mammillary bodies and mediodorsal nuclei showed substantial neurodegeneration in both amnesic and non-amnesic AUD patients with Wernicke's encephalopathy. But the anterior thalamic nuclei were significantly and consistently damaged only in patients with KS, whereas these nuclei remained largely preserved in AUD patients with Wernicke's encephalopathy but no amnesia. The authors concluded that neuronal loss in the anterior thalamic nuclei is required for amnesia to develop.

IN VIVO NEUROIMAGING INVESTIGATIONS

In vivo neuroimaging of thalamic nuclei in AUD with and without KS is rare. One of the main reasons is that thalamic nuclei are inherently difficult to image and quantify. With no less than 35 nuclei bound within an ovoid structure that spans 20–30 millimeters in cross-sectional length, relaxation times and proton densities of the nuclei lie very close to each other. Contrast signal is poor, which makes it in turn challenging to separate these nuclei. Qualitatively, they look blurred as the signal from one nucleus spills into another. Subsequent quantitative measurements are likely to be artifactual and hence would not reflect actual alterations due to a particular pathology. As a result, labeling and quantification of thalamic nuclei using atlases become flawed, as it can be difficult to either identify or accurately delineate a nucleus of interest (Segobin and Pitel, 2021).

Only recently, a few studies have surfaced in the literature associating alterations to specific thalamic nuclei to AUD. The ventral posterior lateral nuclei of the thalamus have been observed to have greater functional connectivity with the cerebellum of AUD patients, while the connectivity between these nuclei and the medial temporal lobe were reversed (negative in controls and positive in AUD) compared to those between controls (Honnorat et al., 2021). These alterations to connectivity patterns, from controls to AUD, between the ventral posterior lateral nuclei and the cerebellum were linked to impaired balance. These findings are in line with initial exploration of the FCC, where the retrograde transneuronal transport of the rabies virus was found to pass through the ventrolateral thalamus, itself known to be a key component in motor functions (Percheron et al., 1996).

To our knowledge, only one study (Segobin et al., 2019) has explicitly compared the loci and nature of structural alterations in thalamic nuclei in AUD patients with and without KS. Instead of an atlas-based method, our research group explored the ability of probabilistic tractography algorithms to segment the thalamus into several nuclei according to their structural connectivity strength with nodes of the FCC and Papez circuit. We found that the thalamic parcellation that was principally connected to the hippocampus overlapped with 83.9% of the anterior nuclei (Figure 69.2D) according to the thalamic histological atlas (Krauth et al., 2010). The prefrontal cortex constituting BA9 and BA46 regions formed the key node for the frontal executive part of the FCC principally connected to a thalamic parcellation that covered 70.1% of the mediodorsal nuclei (Figure 69.3B). Similarly, the cerebellar cruses I and II forming the cerebellar-executive nodes of the FCC were connected to the ventral anterior nucleus and part of the mediodorsal nucleus. The frontal motor cortex was connected to the ventral lateral nucleus and laterodorsal nucleus of the thalamus, while the cerebellar-motor lobules IV–VI were connected to the ventral lateral thalamic nuclei (Figure 69.3B). The mean number of connections estimated within each parcellation and their volumes were compared across 20 healthy controls, 26 AUD patients, and 16 KS patients.

This study is unique in addressing 2 questions. The first one is how thalamic structural connectivity comports with the known literature regarding the loci of these nuclei and how they are involved in cognitive functions. The second is to observe how they are altered in AUD and KS, in terms of connectivity and volumetry, to propose a potential mechanism governing the pathophysiology of AUD and KS. The fact that the ventral lateral and laterodorsal nuclei of the thalamus play a major role in the motor loop has also been shown notably in healthy subjects (Johansen-Berg et al., 2005). That no group differences were found either in terms of connectivity, or in volumetry of the motor parcellations, potentially implies that these regions are largely preserved in AUD, as has been previously reported (Harper, 2009). That the cerebellum bears a contribution beyond motor abilities and extends to executive functions is relatively new (Buckner, 2013), which may explain the dearth of literature on connectivity between the cerebellar cruses and the ventral anterior nucleus and mediodorsal nucleus.

Most importantly, this study shows that the mediodorsal and anterior thalamic nucleus are both affected in AUD and KS. However, it is the nature of their alterations that differ. First, regarding the FCC, there were no apparent alterations in thalamic connectivity between the mediodorsal nucleus and the executive node of the FCC in any groups. However, within the Papez circuit, there was a graded effect in structural connectivity abnormalities between the anterior nuclei and the hippocampus from controls to AUD and KS (Figure 69.2D). Although atrophy of a thalamic parcellation that consisted mainly of the mediodorsal nucleus was observed to be statistically similar as compared to controls, atrophy of the anterior nucleus was significant in the KS group only, corroborating with post-mortem findings of Harding and colleagues (2000). Taken together, these findings would imply that while damage to the mediodorsal nucleus would likely be a direct result of the neurotoxic effect of chronic and excessive alcohol consumption and contribute to some cognitive deficits, the leading mechanism for amnesia in patients with KS would be a progressive disconnection between the anterior nuclei and hippocampus. This disconnection would finally cascade into atrophy of the anterior nuclei, triggering the observed persistent amnesia.

More recently, another thalamocortical loop has received attention in AUD and addiction more generally regarding its contribution to craving and more particularly to its interoceptive symptoms.

THALAMO-INSULAR CIRCUIT AND CRAVING

CRAVING IN ADDICTION

In the addiction field, *craving* is defined as an urge to consume a substance or to perform a behavior despite willingness or lack of desire to do so at that moment. In the fifth edition of the *Diagnostic and Statistical Manual of Mental Disorders* (American Psychiatric Association, 2013), craving has been included as one of the 11 criteria to define AUD. It is associated with severity of AUD, relapse to drinking following treatment, and in vivo alcohol consumption in laboratory studies. Level of craving is now often considered as a treatment outcome for clinical and pharmacological interventions.

A neurocognitive model of addiction proposed that craving would play a key role in the development and maintenance of addiction. This triadic model (Noel et al., 2013) proposes that addictive behaviors would result from the imbalance between 3 brain systems: (1) an impulsive system, supported by the amygdala and striatum, that would promote automatic and habitual behaviors; (2) a reflective system, sub-served by the prefrontal cortex, that would enable consideration of the future consequences of a behavior, and promote inhibitory control; and (3) a regulatory system, subtended by the insula, that would integrate interoceptive signals and promote emotional awareness. The insula regulatory system can potentiate the activity of the impulsive system, or weaken or hijack the resources of the reflective system. Thus, by responding to homeostatic perturbations, insula would translate interoceptive signals into a feeling of desire, anticipation, or urge that would disrupt decision-making processes or increase the drive and motivation for the addictive behavior.

ANATOMY OF THE THALAMO-INSULAR CIRCUIT

As described by Craig (2002), lamina I, the most superficial layer of the spinal dorsal horn, is composed of neurons that represent many aspects of the physiological condition of the body. These ascending projections from lamina I neurons are involved in homeostasis and contribute to half of the spinal information to the brainstem and thalamus through the spinothalamic tract. In the primate, lamina I spinothalamic neurons project to a specific thalamic nuclear group, the ventromedial nucleus, that receives all homeostatic information directly (both sympathetic and parasympathetic information for the posterior and basal parts of the ventromedial nucleus, respectively). In turn, neurons of the ventromedial nuclei project to the insula (Figure 69.4). The integration pattern of interoceptive processing is anatomically organized in the insula, with the posterior and middle parts of insula being involved in the interoceptive representation of the physiological states of the body (Craig, 2009), together with all sensory and sensorimotor information as well as with information from reward and salient networks, and the most anterior insula contributing to meta-representation and emotional awareness. Accordingly, a larger insula was found to be related to greater interoceptive and emotional awareness, empathy, and social behavior (Critchley et al., 2004; King-Casas et al., 2008).

RELATIONSHIPS BETWEEN CRAVING AND THALAMO-INSULAR CIRCUIT IN ADDICTION

Insula

The insula is thus considered as an interoception hub and, as proposed in the triadic neurocognitive model of addiction (Noel et al., 2013), may be the brain substrate of craving in addiction. On the one hand, a lesion of the insula was indeed associated with disruption of smoking addiction. Patients with insular lesion after a stroke were significantly more likely to quit smoking easily, immediately, without relapse and without a persistence of the urge to smoke (Naqvi et al., 2007). Functional imaging studies also showed insular cortex activation during food, cocaine, and cigarette cravings (Bonson et al., 2002; Pelchat et al., 2004; Wang et al., 2007). On the other hand, structural neuroimaging

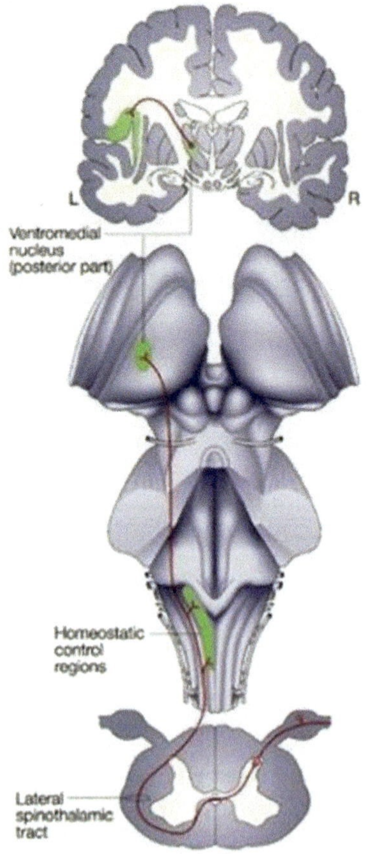

Figure 69.4. Thalamo-insular circuit involving the Lamina I spinothalamocortical pathway. Lamina I spinothalamic neurons project to the ventromedial thalamic nuclei, which project in turn to the insula with a functional mapping. From Craig et al. (2002).

revealed gray matter volume or density difference between patients with addiction (cocaine, heroin, cannabis, methamphetamine, tobacco) and control participants (Droutman et al., 2015).

In AUD more specifically, gray matter abnormalities were found in the insula: smaller volumes, thinner cortices, and lower cerebral blood flow than in controls (Campbell and Lawrence, 2021). A recent study investigated gray matter volume of the insular cortex, morphometry, white matter structural connectivity, and resting state functional connectivity in a large group of AUD patients (Manuweera et al., 2022). Results indicate structural abnormalities mostly in the anterior insula, whereas functional connectivity alterations affected both the anterior and posterior insula. Given the anatomical organization of the insula, these findings suggest a contribution of damage to the insula to cognitive deficits. A voxel-based comparison between AUD and KS revealed that the left insula was among the few brain regions to show more severe shrinkage in KS than in AUD (Pitel et al., 2012). Further studies are now required to specify whether these severe abnormalities in KS concern the entire insula or are more limited to certain parts of the insula associated with specific processes.

Activity of the insular cortex is increased in AUD during alcohol consumption or in response to alcohol-associated cues and contexts (Campbell and Lawrence, 2021). In a more mechanistic approach, a repetitive transcranial magnetic stimulation study did not show any effect of insular cortex stimulation on alcohol craving or consumption (Perini et al., 2020). These apparently conflicting results combining a "sensitized" (lesion and functional imaging studies) and a "desensitized" (structural imaging studies) insula in addiction could be explained by the lack of precise structural analyses of the different sub-regions that compose insula (Droutman et al., 2015) or by the fact that insula may not alone drive alcohol craving (Campbell and Lawrence, 2021), but instead may be a node of a more complex thalamo-insular network (Sullivan et al., 2013).

Thalamus

The contribution of the thalamus to craving and more generally to addiction is not well understood but is receiving increasing attention. In patients with addictive disorders, the thalamus has smaller gray matter volume and lower glucose metabolism, as well as altered white matter integrity and resting-state functional connectivity (Huang et al., 2018). In AUD, as previously mentioned, thalamic abnormalities are consistently described (Pitel et al., 2015). However, previous structural neuroimaging studies did not specifically investigate the ventromedial nuclei, which is supposed to be a key node of the thalamo-insular network involved in interoception and potentially in craving. The ventromedial thalamic nuclear group cannot be easily segmented in vivo with 3T MRI. For example, Freesurfer software includes this nucleus in its atlas, but THOMAS (Su et al., 2019) does not. In addition, the name of these nuclei is inconsistent, notably between the animal literature and in humans (basal part of the ventromedial nucleus or parvocellular part of the ventroposterior medial nucleus, for example).

Patients with addiction disorders usually show greater thalamic activation to drug cues compared with healthy individuals (Huang et al., 2018). However, most previous studies focused on cocaine-dependent users. A study combining skin-conductance responses and functional MRI investigated the whole brain for arousal-related regional activations to craving in non-dependent drinkers (Wang et al., 2019). Results indicated thalamic activation in relation to differences in skin-conductance responses to alcohol versus neutral cues. These findings suggest an involvement of the thalamus in cue-eliciting craving and a specific role in associating cue exposure to increases in physiological arousal and craving (Huang et al., 2018).

Thalamic activations were identified in the intralaminar nucleus and mediodorsal nucleus, which suggests that these nuclei (rather the ventromedial nuclei) would play a role in craving. Given that intralaminar nuclei are known to contribute to arousal and mediodorsal nuclei to high-level cognitive functions, the authors interpret their findings as underlying the cognitive and affective processing in relation to craving. However, given the limited spatial resolution of functional MRI and the difficulties previously mentioned related to labeling and parcellation of the thalamic nuclei with in vivo imaging, further investigations are necessary to specify which of the thalamic nuclear groups are especially involved in craving.

CONCLUSION AND PERSPECTIVES

Our understanding of the contribution of the thalamus to AUD has historically been driven by studies conducted in KS. Even in these early studies, the question related to the specificity of the thalamic nuclei involvement was addressed. Nowadays, the role of different thalamocortical loops is considered not only in alcohol-related neurological complication, but also in the development and persistence of AUD without KS. Recent technical and methodological improvements favor in vivo imaging of different thalamus nuclear groups and quantification of thalamic nuclei volume and connectivity. Applied to clinical populations such as patients with AUD or KS, they will provide novel insights regarding the specific role of the different thalamic nuclei in the pathophysiology of AUD and KS and may indicate new treatment avenues.

These new scientific advances based on anatomical and functional connectivity require us to continue our efforts in thalamus imaging by harmonizing the anatomical nomenclature not only in humans but also with animals in order to facilitate translational research. High-field anatomical imaging techniques have yielded promising results but do not seem easily accessible to clinical populations for the moment. The improvement of clinical imaging data-processing methods is therefore a priority and should soon allow reliable and reproducible segmentation of thalamic nuclei.

REFERENCES

Aggleton JP, Nelson AJD, O'Mara SM (2022). Time to retire the serial Papez circuit: Implications for space, memory, and attention. *Neurosci Biobehav Rev*. 140:104813.

Aggleton JP, O'Mara SM (2022). The anterior thalamic nuclei: Core components of a tripartite episodic memory system. *Nat Rev Neurosci*. 23:505–516.

American Psychiatric Association (2013). *Diagnostic and statistical manual of mental disorders* (5th ed.).

Bonson KR, Grant SJ, Contoreggi CS, Links JM, Metcalfe J, Weyl HL, Kurian V, Ernst M, London ED (2002). Neural systems and cue-induced cocaine craving. *Neuropsychopharmacology*. 26:376–386.

Brion S, Mikol J (1978). [Lesion of the lateral dorsal thalamic nucleus and alcoholic Korsakoff syndrome]. *J Neurol Sci*. 38:249–261.

Brokate B, Eling P, Hildebrandt H, Fichtner H, Runge K, Timm C (2003). Frontal lobe dysfunctions in Korsakoff's syndrome and chronic alcoholism: continuity or discontinuity? *Neuropsychology*. 17:420–428.

Buckner RL (2013). The cerebellum and cognitive function: 25 years of insight from anatomy and neuroimaging. *Neuron*. 80:807–815.

Campbell EJ, Lawrence AJ (2021). It's more than just interoception: The insular cortex involvement in alcohol use disorder. *J Neurochem*. 157:1644–1651.

Chanraud S, Martelli C, Delain F, Kostogianni N, Douaud G, Aubin H-J, Reynaud M, Martinot J-L (2007). Brain morphometry and cognitive performance in detoxified alcohol-dependents with preserved psychosocial functioning. *Neuropsychopharmacology*. 32:429–438.

Chanraud S, Pitel AL, Rohlfing T, Pfefferbaum A, Sullivan E V (2010). Dual tasking and working memory in alcoholism: relation to frontocerebellar circuitry. *Neuropsychopharmacology*. 35:1868–1878.

Chanraud S, Reynaud M, Wessa M, Penttilä J, Kostogianni N, Cachia A, Artiges E, Delain F, Perrin M, Aubin HJ, Cointepas Y, Martelli C, Martinot JL (2009). Diffusion tensor tractography in mesencephalic bundles: Relation to mental flexibility in detoxified alcohol-dependent subjects. *Neuropsychopharmacology*. 34:1223–1232.

Craig AD (2002). How do you feel? Interoception: The sense of the physiological condition of the body. *Nat Rev Neurosci*. 3:655–666.

Craig ADB (2009). How do you feel—now? The anterior insula and human awareness. *Nat Rev Neurosci*. 10:59–70.

Critchley HD, Wiens S, Rotshtein P, Öhman A, Dolan RJ (2004). Neural systems supporting interoceptive awareness. *Nat Neurosci*. 7:189–195.

Diener HC, Müller A, Thron A, Poremba M, Dichgans J, Rapp H (1986). Correlation of clinical signs with CT findings in patients with cerebellar disease. *J Neurol*. 233:5–12.

Droutman V, Read SJ, Bechara A (2015). Revisiting the role of the insula in addiction. *Trends Cogn Sci*. 19:414–420.

Fama R, Le Berre AP, Sassoon SA, Zahr NM, Pohl KM, Pfefferbaum A, Sullivan EV (2021). Memory impairment in alcohol use disorder is associated with regional frontal brain volumes. *Drug Alcohol Depend*. 228:109058.

Fama R, Pitel AL, Sullivan EV (2012). Anterograde episodic memory in Korsakoff syndrome. *Neuropsychol Rev*. 22:93–104.

Harding A, Halliday G, Caine D, Kril J (2000). Degeneration of anterior thalamic nuclei differentiates alcoholics with amnesia. *Brain*. 123 (Pt 1):141–154.

Harper C (1998). The neuropathology of alcohol-specific brain damage, or does alcohol damage the brain? *J Neuropathol Exp Neurol*. 57:101–110.

Harper C (2009). The neuropathology of alcohol-related brain damage. *Alcohol Alcohol*. 44:136–140.

Honnorat N, Saranathan M, Sullivan EV., Pfefferbaum A, Pohl KM, Zahr NM (2021). Performance ramifications of abnormal functional connectivity of ventral posterior lateral thalamus with cerebellum in abstinent individuals with alcohol use disorder. *Drug Alcohol Depend*. 220:108509.

Huang AS, Mitchell JA, Haber SN, Alia-Klein N, Goldstein RZ (2018). The thalamus in drug addiction: From rodents to humans. *Philos Trans R Soc Lond B Biol Sci*. 373:20170028.

Jacobson RR, Acker CF, Lishman WA (1990). Patterns of neuropsychological deficit in alcoholic Korsakoff's Syndrome. *Psychol Med*. 20:321–334.

Johansen-Berg H, Behrens TEJ, Sillery E, Ciccarelli O, Thompson AJ, Smith SM, Matthews PM (2005). Functional-anatomical validation and individual variation of diffusion tractography-based segmentation of the human thalamus. *Cereb Cortex*. 15:31–39.

Kelly RM, Strick PL (2003). Cerebellar loops with motor cortex and prefrontal cortex of a nonhuman primate. *J Neurosci*. 23:8432–8444.

King-Casas B, Sharp C, Lomax-Bream L, Lohrenz T, Fonagy P, Read Montague P (2008). The rupture and repair of cooperation in borderline personality disorder. *Science*. 321:806–810.

Korsakoff SS (1887). Disturbance of psychic function in alcoholic paralysis and its relation to the disturbance of the psychic sphere in multiple neuritis of non-alcoholic origin. *Vestn Psichiatr*. 4:1–102.

Korsakoff S (1889). Etude medico-psychologique sur une forme des maladies de la mémoire. *Rev Philos*. 20:501–530.

Krabbendam L, Visser PJ, Derix MMA, Verhey F, Hofman P, Verhoeven W, Tuinier S, Jolles J (2000). Normal cognitive performance in patients with chronic alcoholism in contrast to patients with Korsakoff's syndrome. *J Neuropsychiatry Clin Neurosci*. 12:44–50.

Krauth A, Blanc R, Poveda A, Jeanmonod D, Morel A, Székely G (2010). A mean three-dimensional atlas of the human thalamus: generation from multiple histological data. *NeuroImage*. 49:2053–2062.

Le Berre AP, Fama R, Sullivan EV (2017). Executive functions, memory, and social cognitive deficits and recovery in chronic alcoholism: A critical review to inform future research. *Alcohol Clin Exp Res*. 41:1432–1443.

Maharasingam M, Macniven JAB, Mason OJ (2013). Executive functioning in chronic alcoholism and Korsakoff syndrome. *J Clin Exp Neuropsychol*. 35:501–508.

Mair WGP, Warrington EK, Weiskrantz L (1979). Memory disorder in Korsakoff's psychosis: A neuropathological and neuropsychological investigation of two cases. *Brain*. 102:749–783.

Manuweera T, Kisner MA, Almira E, Momenan R (2022). Alcohol use disorder-associated structural and functional characteristics of the insula. *J Neurosci Res*. 100:2077–2089.

Marvel CL, Desmond JE (2010). Functional topography of the cerebellum in verbal working memory. *Neuropsychol Rev*. 20:271–279.

Mayes AR, Meudell PR, Mann D, Pickering A (1988). Location of lesions in Korsakoff's syndrome: Neuropsychological and neuropathological data on two patients. *Cortex*. 24:367–388.

Mechtcheriakov S, Brenneis C, Egger K, Koppelstaetter F, Schocke M, Marksteiner J (2007). A widespread distinct pattern of cerebral atrophy in patients with alcohol addiction revealed by voxel-based morphometry. *J Neurol Neurosurg Psychiatry*. 78:610–614.

Müller-Oehring EM, Schulte T, Pfefferbaum A, Sullivan E V. (2022). Disruption of cerebellar-cortical functional connectivity predicts balance instability in alcohol use disorder. *Drug Alcohol Depend*. 235:109435.

Naqvi NH, Rudrauf D, Damasio H, Bechara A (2007) Damage to the insula disrupts addiction to cigarette smoking. Science 315: 531–534.

Noel X, Brevers D, Bechara A (2013). A triadic neurocognitive approach to addiction for clinical interventions. *Front Psychiatry*. 4:1–14.

Pelchat ML, Johnson A, Chan R, Valdez J, Ragland JD (2004). Images of desire: food-craving activation during fMRI. *Neuroimage*. 23:1486–1493.

Percheron G, François C, Talbi B, Yelnik J, Fénelon G (1996). The primate motor thalamus. *Brain Res Rev*. 22:93–181.

Perini I, Kämpe R, Arlestig T, Karlsson H, Löfberg A, Pietrzak M, Zangen A, Heilig M (2020). Repetitive transcranial magnetic stimulation targeting the insular cortex for reduction of heavy drinking in treatment-seeking alcohol-dependent subjects: a randomized controlled trial. *Neuropsychopharmacology*. 45:842–850.

Pfefferbaum A, Rosenbloom M, Rohlfing T, Sullivan EV (2009). Degradation of association and projection white matter systems in alcoholism detected with quantitative fiber tracking. *Biol Psychiatry*. 65:680–690.

Pfefferbaum A, Sullivan EV, Mathalon DH, Lim KO (1997). Frontal lobe volume loss observed with magnetic resonance imaging in older chronic alcoholics. *Alcohol Clin Exp Res*. 21:521–529.

Pitel AL, Aupée AM, Chételat G, Mézenge F, Beaunieux H, de la Sayette V, Viader F, Baron JC, Eustache F, Desranges B (2009). Morphological and glucose metabolism abnormalities in alcoholic Korsakoff's syndrome: Group comparisons and individual analyses. *PLoS ONE*. 4:e7748.

Pitel AL, Beaunieux H, Witkowski T, Vabret F, De La Sayette V, Viader F, Desgranges B, Eustache F (2008). Episodic and working memory deficits in alcoholic korsakoff patients: The continuity theory revisited. *Alcohol Clin Exp Res*. 32:1229–1241.

Pitel AL, Beaunieux H, Witkowski T, Vabret F, Guillery-Girard B, Quinette P, Desgranges B, Eustache F (2007). Genuine episodic memory deficits and executive dysfunctions in alcoholic subjects early in abstinence. *Alcohol Clin Exp Res*. 31:1169–1178.

Pitel A-L, Chételat G, Le Berre AP, Desgranges B, Eustache F, Beaunieux H (2012). Macrostructural abnormalities in Korsakoff syndrome compared with uncomplicated alcoholism. *Neurology*. 78:1330–1333.

Pitel A-L, Eustache F, Beaunieux H (2014). Component processes of memory in alcoholism. In: Edith Sullivan and Adolf Pfefferbaum, eds. *Handbook of Clinical Neurology*. 1st ed. Elsevier: 211–225.

Pitel AL, Segobin SH, Ritz L, Eustache F, Beaunieux H (2015). Thalamic abnormalities are a cardinal feature of alcohol-related brain dysfunction. *Neurosci Biobehav Rev*. 54:38–45.

Reed LJ, Lasserson D, Marsden P, Stanhope N, Stevens T, Bello F, Kingsley D, Colchester A, Kopelman MD (2003). FDG-PET findings in the Wernicke-Korsakoff syndrome. *Cortex*. 39:1027–1045.

Ritz L, Segobin S, Lannuzel C, Boudehent C, Vabret F, Eustache F, Beaunieux H, Pitel AL (2016). Direct voxel-based comparisons between grey matter shrinkage and glucose hypometabolism in chronic alcoholism. *J Cereb Blood Flow Metab*. 36:1625–1640.

Segobin S, Laniepce A, Ritz L, Lannuzel C, Boudehent C, Cabé N, Urso L, Vabret F, Eustache F, Beaunieux H, Pitel AL (2019). Dissociating thalamic alterations in alcohol use disorder defines specificity of Korsakoff's syndrome. *Brain*. 142:1458–1470.

Segobin S, Pitel AL (2021). The specificity of thalamic alterations in Korsakoff's syndrome: Implications for the study of amnesia. *Neurosci Biobehav Rev*. 130:292–300.

Segobin S, Ritz L, Lannuzel C, Boudehent C, Vabret F, Eustache F, Beaunieux H, Pitel AL (2015). Integrity of white matter microstructure in alcoholics with and without Korsakoff's syndrome. *Hum Brain Mapp*. 36:2795–2808.

Shear PK, Sullivan EV., Lane B, Pfefferbaum A (1996). Mammillary body and cerebellar shrinkage in chronic alcoholics with and without amnesia. *Alcohol Clin Exp Res*. 20:1489–1495.

Shimamura AP, Jernigan TL, Squire LR (1988). Korsakoff's syndrome: radiological (CT) findings and neuropsychological correlates. *J Neurosci*. 8:4400–4410.

Su JH, Thomas FT, Kasoff WS, Tourdias T, Choi EY, Rutt BK, Saranathan M (2019). Thalamus Optimized Multi Atlas Segmentation (THOMAS): Fast, fully automated segmentation of thalamic nuclei from structural MRI. *NeuroImage*. 194:272–282.

Sullivan EV, Lane B, Deshmukh A, Rosenbloom MJ, Desmond JE, Lim KO, Pfefferbaum A (1999). In vivo mammillary body volume deficits in amnesic and nonamnesic alcoholics. *Alcohol Clin Exp Res*. 23:1629–1636.

Sullivan EV, Müller-Oehring E, Pitel AL, Chanraud S, Shankaranarayanan A, Alsop DC, Rohlfing T, Pfefferbaum A (2013). A selective insular perfusion deficit contributes to compromised salience network connectivity in recovering alcoholic men. *Biol Psychiatry*. 74:547–555.

Sullivan EV, Pfefferbaum A (2009). Neuroimaging of the Wernicke-Korsakoff syndrome. *Alcohol Alcohol*. 44:155–165.

Toledo Nunes P, Vedder LC, Deak T, Savage LM (2019). A pivotal role for thiamine deficiency in the expression of neuroinflammation markers in models of alcohol-related brain damage. *Alcohol Clin Exp Res*. 43:425–438.

Tulving E (2001). Episodic memory and common sense: how far apart? *Philos Trans R Soc B Biol Sci*. 356:1505–1515.

Victor M, Adams RD, Collins GH (1971). The Wernicke-Korsakoff syndrome. A clinical and pathological study of 245 patients, 82 with post-mortem examinations. *Contemp Neurol Ser.* 7:1–206.

Wang W, Zhornitsky S, Le TM, Dhingra I, Zhang S, Krystal JH, Li C, Shan R (2019). Cue-elicited craving, thalamic activity, and physiological arousal in adult non-dependent drinkers. *J Psychiatr Res.* 116:74–82.

Wang Z, Faith M, Patterson F, Tang K, Kerrin K, Wileyto EP, Detre JA, Lerman C (2007). Neural substrates of abstinence-induced cigarette cravings in chronic smokers. *J Neurosci.* 27:14035–14040.

World Health Organization, ed. (2018). *Global Status Report on Alcohol and Health.* Geneva.

Zahr NM, Pfefferbaum A (2017). Alcohol's effects on the brain: Neuroimaging results in humans and animal models. *Alcohol Res.* 38:183–206.

Zahr NM, Pitel A-L, Chanraud S, Sullivan EV (2010). Contributions of studies on alcohol use disorders to understanding cerebellar function. *Neuropsychol Rev.* 20:280–289.

Zhao Q, Pfefferbaum A, Podhajsky S, Pohl KM, Sullivan EV (2020). Accelerated aging and motor control deficits are related to regional deformation of central cerebellar white matter in alcohol use disorder. *Addict Biol.* 25:e12746.

70.

FRONTOTEMPORAL DEMENTIA

Martina Bocchetta

INTRODUCTION

Frontotemporal dementia (FTD) is a very heterogeneous group of neurodegenerative disorders with multiple genetic and pathological causes (Figure 70.1). Despite presenting clinically with different symptoms including behavioral, language, cognitive, and motor deficits (Woollacott and Rohrer, 2016), there are common anatomical features across cases, including atrophy and hypometabolism in the frontal and temporal lobes (Gordon et al., 2016).

Overall, FTD represents 5%–10% of all dementia cases, and it constitutes a more common cause of early onset dementia, approximately equal in frequency to Alzheimer's disease (AD) in people under the age of 65. The only known risk factors are genetic, with approximately 30%–50% of cases due to an autosomal dominant mutation, in the microtubule-associated protein tau (*MAPT*), progranulin (*GRN*), or chromosome 9 open reading frame 72 (*C9orf72*) genes. The age at onset is extremely variable within and between clinical and genetic forms, and therefore hard to predict (Moore et al., 2020).

Pathologically, 3 major groups are described according to the main abnormal protein seen in neuronal or glial inclusions: tau, transactive response DNA-binding protein-43 kDa (TDP-43), and fused-in-sarcoma (FUS) (Lashley et al., 2015).

Clinically, behavioral variant FTD (bvFTD) is the most common presentation, with impaired social conduct and personality changes (e.g., disinhibition, apathy, repetitive behaviors, loss of empathy), often misdiagnosed as psychiatric illness at the onset (Rascovsky et al., 2011). It could be caused by any type of pathology, and it is associated with extremely variable patterns of atrophy between patients.

Less frequently, patients present with progressive decline in speech and language domains (primary progressive aphasia, PPA), with 3 variants: semantic variant (svPPA), nonfluent variant (nfvPPA), and logopenic variant (lvPPA) (Gorno-Tempini et al., 2011). svPPA is characterized by progressive breakdown of semantic memory, preserved fluency, anomia, and impaired single-word comprehension, and it is associated with TDP-43 type C pathology (Rohrer et al., 2011). nfvPPA is characterized by agrammatism and apraxia of speech, and more frequently caused by tau rather than TDP-43 (type A or B) pathology. lvPPA tends to be associated with AD pathology and is characterized by word-finding difficulty, impaired repetition of sentences and phrases, and neologistic jargon (Gorno-Tempini et al., 2011). When language symptoms do not neatly fit into one of these 3 PPA variants, patients are classified as PPA not otherwise specified (PPA-NOS) (Harris et al., 2013).

Around 15% of people on the FTD spectrum can also develop motor features consistent with either amyotrophic lateral sclerosis (ALS) (or motor neurone disease, MND) or parkinsonism (including progressive supranuclear palsy, PSP, or corticobasal syndrome, CBS) (Woollacott and Rohrer, 2016).

THE CORTICAL AND THALAMIC INVOLVEMENT IN THE FTD SPECTRUM

For many years, FTD was mainly considered a cortical dementia, with atrophy predominantly in the frontal and temporal lobes, hence its name. However, recent studies have shown that brain involvement in FTD extends beyond the frontotemporal cortex (Figure 70.2), and that clear subcortical involvement is seen across the spectrum (Bocchetta et al., 2021a). In particular, thalamic atrophy is a common feature across all forms of FTD (Bocchetta et al., 2020; Hornberger et al., 2012; McKenna et al., 2022), and it is visible since the early clinical stages (Figure 70.3). The heterogeneity seen in FTD, with its different pathophysiological mechanisms and its genetic factors, could be potentially explained by damage in the interactions between the thalamus and the cortex, and the roles played by the specific regions in disrupting brain networks. The specific conjunctions between the localization of the initial

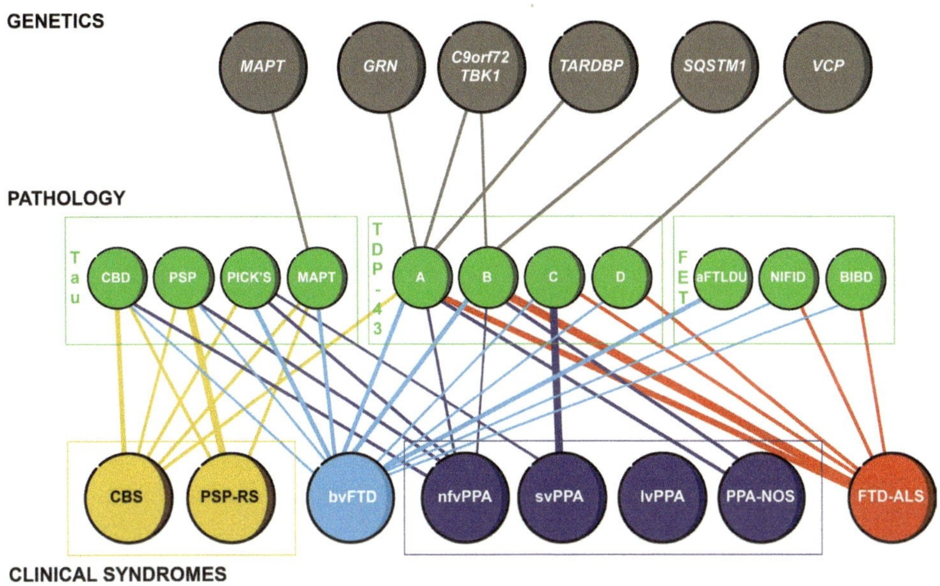

Figure 70.1. The complex relationship and the heterogeneity between the clinical syndromes of the frontotemporal dementia (FTD) spectrum (on the at the bottom), the underlying pathology (in the middle), and the known genetic causes (at the top). Adapted from FTDtalk.org. MAPT = microtubule-associated protein tau; GRN = progranulin; C9orf72 = chromosome 9 open reading frame 72; TBK1 = TANK-binding kinase 1; TARDBP = TAR DNA binding protein; SQSTM1 = sequestome1 p62; VCP = valosin-containing protein; TDP-43 TAR = DNA binding protein 43; FET = FUS (fused in liposarcoma), EWS (Ewing sarcoma) and TAF15 (TATA binding associated factor 15) proteins; CBD = corticobasal degeneration; aFTLDU = atypical frontotemporal lobal degeneration with ubiquitin-positive inclusions; NIFID = neurofilament inclusion body disease; BIBD = basophilic inclusion body disease; CBS = corticobasal syndrome; PSP-RS = progressive supranuclear palsy–Richardson syndrome; bvFTD = behavioral variant frontotemporal dementia; nfvPPA = nonfluent variant of primary progressive aphasia; svPPA = semantic variant of primary progressive aphasia; lvPPA = logopenic variant of primary progressive aphasia; PPA-NOS = primary progressive aphasia not otherwise specified; FTD-ALS = frontotemporal dementia and amyotrophic lateral sclerosis.

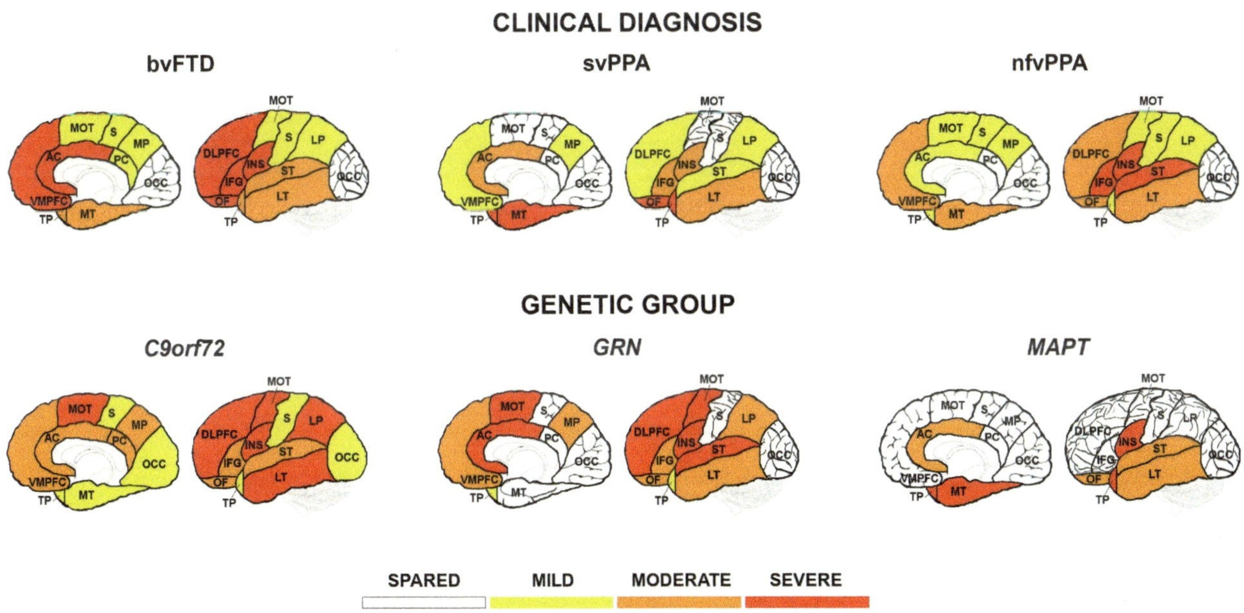

Figure 70.2. Overview of the involvement of the cerebral cortex in the most common clinical and genetic groups of frontotemporal dementia (FTD). VMPFC = ventromedial prefrontal; TP = temporal pole; MT = medial temporal; AC = anterior cingulate; PC = posterior cingulate; MOT = motor; S = sensory; MP = medial parietal; OCC = occipital; DLPFC = dorsolateral prefrontal; OF = orbitofrontal; IFG = inferofrontal gyrus; INS = insula; LT = lateral temporal; ST = supratemporal; LP = lateral parietal.

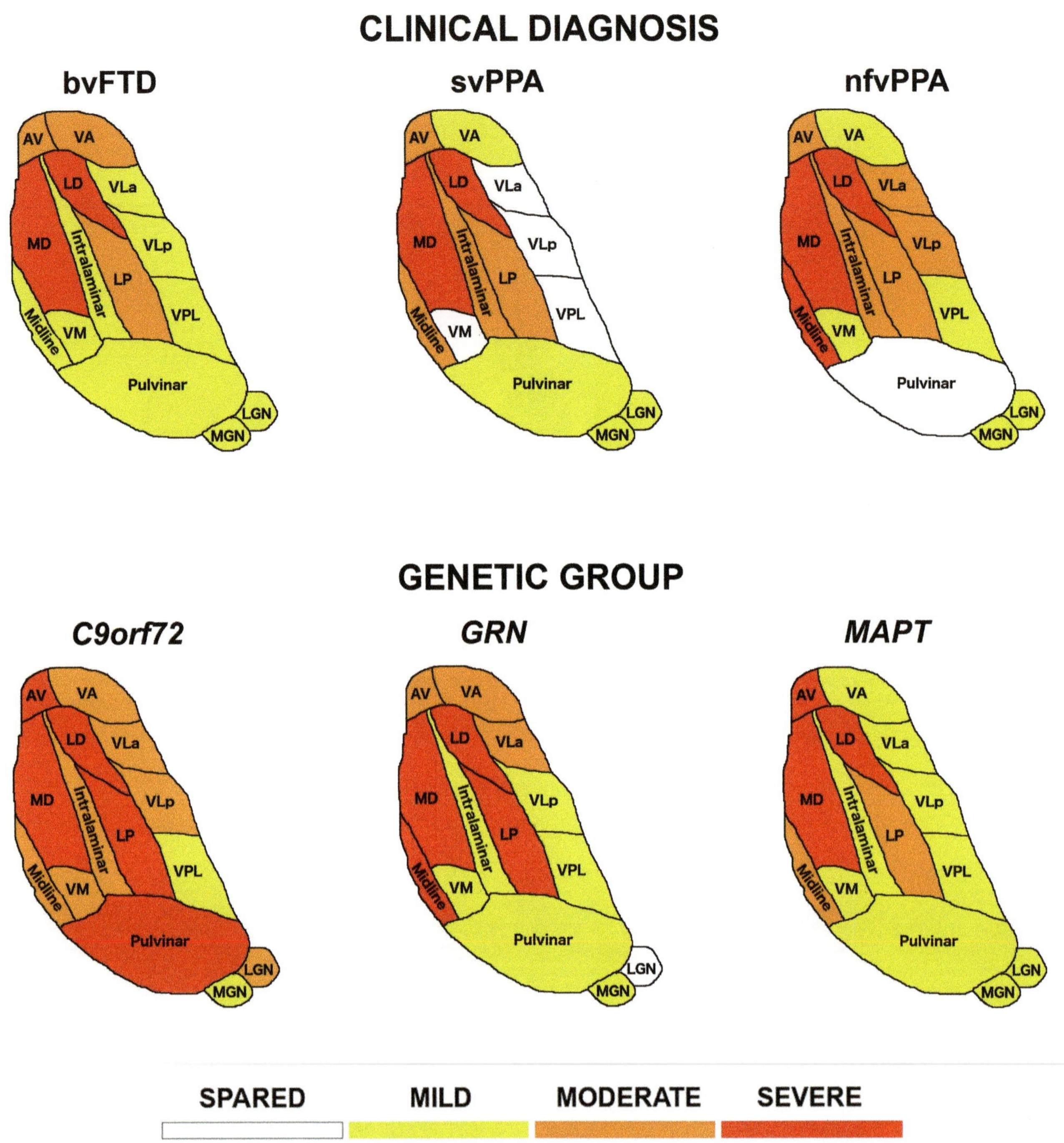

Figure 70.3. Overview of the involvement of the thalamic regions in the most common clinical and genetic groups of frontotemporal dementia (FTD). AV = anteroventral; VA = ventral anterior; LD = laterodorsal; VLa = ventral lateral anterior; MD = mediodorsal; LP = lateral posterior; VLp = ventral lateral posterior; VPL = ventral posterolateral; VM = ventromedial; LGN = lateral geniculate nucleus; MGN = medial geniculate.

damage, the spreading of the underlying pathology, and the rate of progression could lead to the heterogenous symptomatology among the FTD forms, and even within the same form. Indeed, according to the *molecular nexopathy* paradigm (Warren et al., 2013), within the thalamocortical networks there could be preferentially vulnerable hubs that can be differentially targeted by different abnormal proteins, leading to different elements of each network to be damaged, giving rise to diverse symptoms and signs in several domains.

BEHAVIOR, EMOTION, AND SOCIAL COGNITION

In bvFTD, the cortical regions that are commonly affected by atrophy and hypometabolism include the orbitofrontal cortex, medial and lateral prefrontal cortex, anteromedial temporal cortex, together with the anterior cingulate and insula, with relative sparing of the posterior regions (Gordon et al., 2016; Whitwell et al., 2009).

In the thalamus, the mediodorsal (MD) and laterodorsal (LD) nuclei are commonly affected among all clinical forms (Bocchetta et al., 2020; McKenna et al., 2022) including bvFTD, where thalamic involvement also spreads to the anterior and posterior thalamus (Bocchetta et al., 2020). These MD, LD, and anterior nuclei are considered limbic and associative structures (reviewed in Bocchetta et al., 2021a), with dense connections to the prefrontal cortex, anterior cingulate, insula, and amygdala (Vertes et al., 2015). These are all key regions affected in bvFTD and they are critical for social and emotional processing, as they are part of the salience network (Seeley et al., 2007). This network, activated by stimuli which are emotionally and personally significant, regulates behaviors, reward evaluation, cognitive control, homeostasis, and subjective feelings (Seeley et al., 2007; Schönecker et al., 2018). A damage in this network can lead to some of the core behavioral features and symptoms of FTD, and specifically lack of insight, and impaired personal and social conduct. In particular, the thalamic LD and the orbitofrontal cortex are affected in bvFTD and might lead to disinhibition (Hornberger et al., 2011), while atrophy in MD has been strongly associated with behavioral changes in bvFTD due to *C9orf72*.

The salience network also supports social cognition, which has been described as the cognitive process that allows us to perceive, understand, and interpret our "self" in relation to others and the surrounding environment, to generate an appropriate response to the intentions, behaviors, and feelings of others (Harciarek and Cosentino, 2013). Social cognition is typically affected in bvFTD, and this manifests as lack of socioemotional sensitivity, where patients struggle to understand social cues and adapt their behavior and maintain social attention, resulting in loss of empathy and disorganized attentional engagement (Rankin, 2020). In bvFTD the loss of empathy has been found to be directly linked to a selective loss of the Von Economo neurons in the core regions of the salience network (medial frontal, insula, anterior cingulate, and MD) (Pasquini et al., 2020). These neurons might be specialized for social cognition, and their selective damage in patients with bvFTD can explain their deficits in self-awareness, theory of mind, and moral reasoning, and specifically their impaired recognition of social and emotional signals, decreased attention to what might be warning signs of negative consequences, decreased semantic knowledge for social norms, and an inability to represent their own and others' perspectives and emotions (Harciarek and Cosentino, 2013; Shany-Ur and Rankin, 2011).

Among the language variants, both lvPPA and nfvPPA show selective difficulties in interpreting emotions from vocal prosody, while the deficit in svPPA is more widespread and related to social comprehension (Shany-Ur and Rankin, 2011). In svPPA, the limbic network, also known as the *anterior temporo-frontal semantic appraisal network* (Rankin, 2020), is typically impaired: due to these deficits in the emotion conceptual knowledge, patients with svPPA have difficulties in emotional processing, categorization, and recognition (mainly of negative emotions, as described in Kumfor et al., 2011), valence assignment, and in understanding socioemotional semantic context, eventually leading to lack of empathy and dysregulation in autonomic reactivity (Rankin, 2020; Fletcher et al., 2015). Looking at the neuroanatomical signatures in svPPA, this group shows initial atrophy and hypometabolism in the left antero-inferior temporal lobe (temporal pole, fusiform gyrus, the inferior-middle temporal gyri, entorhinal cortex) (Rohrer et al., 2011; Bejanin et al., 2020; Lombardi et al. 2021), and, as the disease progresses, in the left orbitofrontal cortex, anterior cingulate and insula, and antero-medial and lateral thalamus (mainly the MD, LD) (Bocchetta et al. 2020; McKenna et al., 2022; Tan et al., 2014), all connected within the limbic network.

Although more symmetrical and more mediotemporal than svPPA, *MAPT* mutation carriers also show similar involvement of the limbic structures, both as atrophy and or/hypometabolism in the cortical regions (anterior and medial temporal, orbitofrontal cortex, anterior cingulate), which is present before symptom onset (Rohrer et al., 2015; Whitwell et al., 2012), and as atrophy in the MD and lateral thalamic regions, which instead is present only after symptom onset (Bocchetta et al. 2020; Rohrer et al., 2015; Bocchetta et al., 2021b). The impairment in these structures might be linked with the *anterior cingulate circuit*, a crucial pathway for the control of motivation, and might result in the impaired social conduct, disinhibition, obsessions, and stereotyped repetitive behavior typically seen in *MAPT* mutation carriers (Woollacott and Rohrer, 2016).

In *GRN* mutation carriers, abnormal performance in empathy, social cognition, emotion processing, and theory of mind tasks correlated with atrophy in the insula, cingulate, frontal, inferomedial temporal cortex, and thalamus (Russell et al., 2020; Foster et al., 2022).

Another thalamic nucleus connected to insula and anterior cingulate within the salience network is the medial pulvinar. This association nucleus helps to guide visual attention by detecting salient stimuli, and it plays a crucial role in prosocial engagement (Sturm et al., 2017), consistent with an integrative role in social, affective, and reward processing. The medial pulvinar and its connections to the salience network regions are impaired in bvFTD, and especially in the *C9orf72* expansion carriers (Bocchetta et al., 2020; Lee et al., 2014), as discussed later in this chapter.

Limbic connectivity of the anterior thalamus with the insula and cingulate cortex is altered in bvFTD and svPPA patients and it is associated with greater levels of apathy and disinhibition (Farb et al., 2013): without proper communication between these regions, the emotional impulses cannot influence the behavioral-control processes in the prefrontal cortex (Farb et al., 2013). Apathy and impulsivity are strongly correlated and present in all clinical FTD syndromes, even when not specifically indicated in the diagnostic criteria, and even at the presymptomatic stage. The disruption of the fronto-striatal circuits involving the thalamus has been associated to impulsivity, while apathy has been linked to deficits in the regulation of motivational states, and specifically in orbitofrontal and ventromedial prefrontal connections to the ventral striatum and thalamus (MD and LD in particular) (Kumfor et al., 2018). Volume reduction in the ventral prefrontal cortex in bvFTD has been associated with *affective apathy*, defined as the "inability to use emotional context to guide behavior, which leads to emotional blunting and altered social interactions" (Kumfor et al., 2018, p. 351). The disruption of the ventral prefrontal cortex and MD and LD could indicate a deficit in evaluating the rewarding nature of social and emotional stimuli, which might lead to reduced social engagement and changes in goal-directed behaviors (Kumfor et al., 2018).

The connections between the limbic regions of the thalamus and the orbitofrontal-insular-striatal circuit and the alterations in reward processing and motivation (Johnen and Bertoux, 2019) might also lead to the aberrant eating behaviors seen mainly in bvFTD and svPPA (Ahmed et al., 2016).

The behavioral changes in FTD can also be explained by a deficit in interoception processing, the mechanism that allows us to perceive and predict our internal physiological state by integrating multimodal somatosensory signals (Johnen and Bertoux, 2019). In bvFTD there is a difficulty in the top-down processing of such internal signals, causing patients to misinterpret their bodily states (such as pain, temperature, satiety, and heartbeat perception; see Fletcher et al., 2015; Ahmed et al., 2016), and leading to an allostatic overload and behavioral symptoms, including underreacting to emotional stimuli, overreacting, and responding inappropriately to social situations (Migeot et al., 2022). The allostatic-interoceptive network that integrates visceromotor and interoceptive processes into a multimodal network connects a wide range of cognitive domains (e.g., memory, executive functions, emotions) and includes regions such as thalamus, insula, anterior cingulate, and somatosensory cortex (Schönecker et al., 2018; Johnen and Bertoux, 2019; Migeot et al., 2022).

EXECUTIVE FUNCTIONS

The disrupted connectivity between the thalamic MD and the prefrontal cortex might also result in an impairment in the cognitive set-shifting needed to meet changing behavioral contingencies and in regulating goal-directed actions (Vertes et al., 2015), leading to an exacerbation of the emotional and executive impairments seen in bvFTD. The thalamus and the prefrontal cortex both contribute to executive functioning by prioritizing and contextualizing data to formulate decisions and initiate actions (the cortex), and by filtering, integrating, and simplifying the most relevant information received from the sensory sources (the thalamus) (Vertes et al., 2015; Andreasen, 1997). Whether the dysexecutive symptoms seen in FTD (i.e., cognitive flexibility, planning, working memory, inhibitory control, abstract reasoning) (Harciarek and Cosentino, 2013; Johnen and Bertoux, 2019) are due to prefrontal or thalamic damage is still unknown, and it could vary across FTD forms. This is likely due to a combination of alterations to the cortex and thalamus, and within the thalamocortical interactions, leading to changes in the network. For instance, the frequently observed symptoms of imitation or utilization behavior in bvFTD (Johnen and Bertoux, 2019) could be explained as a problem in filtering the non-relevant information from the context, or as a problem in decision-making, or as a deficit in inhibiting the incorrect action.

An *executive control network* has been identified, linking the dorsolateral prefrontal and lateral parietal cortices to the anterior thalamus (Seeley et al., 2007). This network regulates working memory and cognitive control, by directing attention to the pertinent stimuli and choosing the most appropriate responses based on the context, the changing conditions, and the homeostatic demands. Interestingly, the anterior thalamus seems to be more affected in bvFTD than in PPA (Bocchetta et al., 2020; McKenna et al., 2022), reflecting the greater deficit in the executive functions with difficulties in regulating attention in this form of FTD (Aggleton et al., 2016).

Genetic mutation carriers all share severe MD, LP, and anterior thalamic involvement, together with anterior cingulate atrophy, and with additional dorsolateral prefrontal and parietal atrophy in presymptomatic *C9orf72* and *GRN* mutation carriers (Rohrer et al., 2015; Bocchetta et al., 2021b). Genetic mutation carriers also present with executive dysfunctions, particularly severe in *C9orf72* expansion carriers (Woollacott and Rohrer, 2016). It should be noted that LP and the anterior thalamic regions are more affected in *GRN* and *C9orf72* than in *MAPT* mutation carriers (Bocchetta et al., 2020): LP is connected to the parietal cortex, which is also found to be atrophic in *GRN* and *C9orf72*, but not in *MAPT* (Rohrer et al., 2015; Bocchetta et al., 2021b).

LANGUAGE

Language difficulties are a hallmark of FTD, specifically for the PPA variants. However, the role that the thalamus and its connections to the cortex play in relation to language is still unclear (Hebb and Ojemann 2013). The thalamic regions that are most frequently implicated in language deficits are the MD, LD, midline, ventrolateral and intralaminar nuclei in the left hemisphere (Bocchetta et al., 2020), and specifically the pulvinar in svPPA (associated with naming errors), and the ventrolateral in nfvPPA (associated with perseveration, naming, fluency, and articulation errors) (McKenna et al., 2022; Hebb and Ojemann 2013). Moreover, the interaction between MD and prefrontal cortex supports the ability of naming, word retrieval, and object representation for language processing (Zhang et al., 2022). With its connections to the frontal and superior temporal regions, the thalamus (and specifically the MD, pulvinar, and intralaminar) could help in regulating the downstream signals for the integration of lexico-syntactic with semantic information (Zhang et al., 2022); such signals would then be sent to the basal ganglia and other cortical regions to coordinate the release of the selected language motor plan into speech. Damage within these neural connections might lead to deficits in semantic retrieval (antero-inferior temporal lobe in svPPA), in fluency (inferior frontal and superior temporal gyrus in nfvPPA), or in word-finding difficulty and impaired repetition (temporo-parietal junction in lvPPA), and might contribute to the variability of language symptoms seen across the PPA variants.

MEMORY

Although it is often considered that episodic memory processing is intact across the FTD spectrum, recent findings have demonstrated that episodic memory is compromised in bvFTD and in svPPA, although some components are preserved in the latter (including visual memory and recall of recent autobiographical events) (Hornberger and Piguet, 2012).

Degeneration of the anterior thalamus has been associated with severity of amnesia in bvFTD (Hornberger et al., 2012) and svPPA (Tan et al., 2014): although the anterior cingulate cortex is severely atrophic only in bvFTD, both groups show loss of Von Economo neurons in this cortical area. Both the anterior cingulate cortex and the anterior thalamus play a role in episodic memory (Aggleton et al., 2016), receiving input from orbitofrontal and anterior temporal cortex, regions severely atrophic in bvFTD and svPPA.

The memory difficulties identified at the presymptomatic stages in *C9orf72* and *MAPT* mutation carriers (Poos et al., 2021) were related to a different neuroanatomical substrate. In particular, *MAPT* showed association between memory deficit and atrophy in the mediotemporal cortex, suggesting that this deficit is due to a "true" consolidation problem (Hornberger et al., 2012; Bertoux et al., 2016), while *C9orf72* showed an involvement of a more distributed network (Poos et al., 2021; Irish et al., 2013), including the prefrontal and parietal cortex and thalamus, suggesting that episodic memory impairments in this group might be a consequence of dysexecutive function, such as poor organization and lack of efficient learning and retrieval strategies.

THE CASE OF *C9ORF72*

Repeated expansions in *C9orf72* are the most frequent genetic cause for FTD (Moore et al., 2020), causing most frequently bvFTD, bvFTD-ALS, or ALS, with extremely variable disease duration and age at onset.

C9orf72 repeated expansion carriers showed widespread involvement of the cortex (particularly insula, dorsolateral prefrontal, motor, dorsolateral temporal, lateral parietal, and occipital cortex), the cerebellum, and the thalamus (Whitwell et al., 2012; Häkkinen et al., 2020), including changes at the early presymptomatic stages (Bocchetta et al., 2021b), over 25 years before estimated symptoms (Rohrer et al., 2015), and at an age younger than 40 years (Lee et al., 2016).

C9orf72 expansion carriers represent the most heterogenous group among the FTD spectrum, not only for clinical presentation, onset, and duration, but also for the brain regions which are most prominently targeted. For example, a study suggested that there are at least 2 distinct subtypes of atrophy patterns in *C9orf72*, one predominantly frontotemporal and one mainly subcortical (Young et al., 2018).

At group level, *C9orf72* expansion carriers show a peculiar profile of a distributed cortico-thalamo-cerebellar network degeneration, in line with the theory of *molecular nexopathy* (Warren et al., 2013; Mahoney et al., 2012). A damage in any point of this system could give rise to several cognitive, behavioral, and emotional symptoms, including neuropsychiatric symptoms such as abnormal body schema coding (Mahoney et al., 2012), altered processing of pain, and psychosis (Fletcher et al., 2015). In particular, the psychotic symptoms often seen in *C9orf72*, which include somatic delusions and hallucinations in all sensory modalities, might be caused by an impaired auditory and visual filtering at the level of the thalamus (pulvinar, LP, lateral geniculate nucleus, medial geniculate nucleus), or by dysexecutive functions at the level of the prefrontal cortex, or in the contextualization or coordination of the different information at the level of the cerebellum (Andreasen, 1997).

Is there a specific brain region where such degeneration starts in the network, and could this primary target of neurodegeneration vary among individuals, resulting in such heterogeneity?

While a more widespread cortical and thalamic involvement has been proven at late disease stages, it seems that the earliest changes are localized in the pulvinar (Bocchetta et al., 2021b; Lee et al., 2014; Lee et al., 2016; Vatsavayai et al., 2016).

Lee et al. (2014) have shown that the medial pulvinar could be the epicenter for the salience network disruption, as degeneration of this nucleus alone predicted more severe disruptions of the salience network connectivity than the typical foci (anterior insula and anterior cingulate) in bvFTD due to *C9orf72*. This finding was replicated in presymptomatic mutation carriers, where atrophy in the medial pulvinar explained dysconnectivity in several other brain networks (Lee et al., 2016). This suggests that the pulvinar could be the earliest and most vulnerable brain hub in *C9orf72*, and that the pathology often found abundant in this region (i.e., TDP-43 and dipeptide repeat proteins) might then spread within the thalamus and to other cortical and subcortical regions over the course of the disease. Conjunctions between pathogenic proteins and neural network characteristics might drive the specific disease manifestation of the clinico-anatomical phenotype: for instance, pathology might spread from the thalamus to the frontal cortex, leading to behavioral symptoms; or to the left supratemporal cortex, causing language difficulties; to parietal cortex, leading to atypical parkinsonism; or to the cerebellum, interfering with top-down control and producing cognitive, motor, affective and psychotic symptoms (Warren et al., 2013). The hypothesis of early and strategic damage to the thalamus could explain what happens in other genetic groups too. For instance, *GRN* mutation carriers can present with bvFTD, nfvPPA, or corticobasal syndrome. Presymptomatic *GRN* mutation carriers show an abnormal connectivity of the thalamus (particularly ventral) within multiple cerebral networks, including salience, somatosensory, and language (Lee et al., 2019), potentially giving rise to different clinical phenotypes.

The evidence of brain structure and function abnormalities at a very young age in *C9orf72* expansion carriers (Lee et al., 2016) has led to the hypothesis that *C9orf72* phenotype might not be a neurodegenerative disease, but rather a neurodevelopmental condition. The key role of thalamocortical connections in relation to the FTD symptoms still stands. From childhood to adulthood, there is a significant refinement of the connections between the thalamus and the cortex, particularly the prefrontal, primary sensory, and temporal cortex (Fair et al., 2010): pathological processes occurring during this window of time could interfere with the normal process of weakening and strengthening of the thalamocortical connections and result in the cognitive and behavioral symptoms seen at an older age.

CONCLUSIONS AND FUTURE DIRECTIONS

Given the central role of the thalamus in several cortico-striatal networks, thalamic atrophy is likely to contribute to the diverse clinical phenotypes seen in FTD. However, thalamic abnormalities alone could not explain all symptoms, but rather damage in the connections between the thalamus and the cortex might better explain the complex scenario of signs and symptoms across the FTD spectrum (Warren et al., 2013).

The specific vulnerability within a brain network can be due to a complex interplay between type of pathology, genetic cause, and/or environmental factors.

Damage in the salience network (mainly in the thalamus, and orbitofrontal, insular, anterior cingulate cortex) has been associated with disruption of social, reward, and emotional processing, and the core behavioral features of FTD. Dysexecutive symptoms are linked with disrupted connectivity between the thalamus and the prefrontal and parietal cortex, which could result in difficulty in regulating attention and choosing the most appropriate responses based on the situation, both internal and external. In fact, a disruption in the top-down processing from the prefrontal cortex or thalamus could lead to misinterpretation of the visceral inputs, causing behavioral and psychotic symptoms. Deficits in other cognitive domains are central to FTD and again could be linked to damage in specific thalamocortical connections: for instance, the anterior thalamus and mediotemporal cortex are crucial for memory, while language difficulties could emerge from involvement of the frontal and superior temporal cortex and specific subregions of the thalamus (mainly MD, pulvinar, and intralaminar).

There is a large heterogeneity in the anatomical and clinical phenotypes in FTD, not only between but also within the same type, where even individuals carrying the same genetic mutation can present with different neuroanatomical profiles, symptoms, and disease onset and duration.

The complexity and variability across FTD types make accurate diagnosis and prognosis extremely challenging, leaving patients and their families without clear information about what the future holds, and without the ability to properly support and care for their loved ones. There is currently no cure for FTD, but there are several promising clinical trials in the pipeline. Understanding what is happening in the brain of individuals with FTD, which type of pathology is causing the disease, where in the brain it accumulates, how fast it spreads, and how it correlates with cognitive and clinical symptoms are just some of the questions that future research has to answer, in order to provide evidence for the potential use of specific biomarkers (or combinations of biomarkers) as enrichment and stratification measures in future trials, and in clinical practice. Beside the large heterogeneity, a further challenge in FTD is its relative rarity compared to other forms of dementia (such as Alzheimer's disease; see **Chapter 68 in Section XII**). Joining forces in global initiatives such as the Frontotemporal Prevention Initiative (Rohrer and Boxer, 2021) is crucial to improving our understanding of

FTD, generating predictive modeling, and facilitating the development of therapies that might prevent the onset of symptoms or slow down their progression.

To conclude, being able to understand why certain neural networks become affected in an individual, which hub within the network is the first to become atrophic and likely to trigger the spread of neurodegeneration, and what is causing such heterogeneity across individuals will be fundamental for therapeutic treatments, personalized medicine, and accurate diagnosis and prognosis. I believe that the answers will be found researching the interactions between the thalamus and the cerebral cortex.

REFERENCES

Aggleton JP, Pralus A, Nelson AJ, Hornberger M (2016). Thalamic pathology and memory loss in early Alzheimer's disease: Moving the focus from the medial temporal lobe to Papez circuit. *Brain*. 139(Pt 7):1877–1890.

Ahmed RM, Irish M, Henning E, Dermody N, Bartley L, Kiernan MC, Piguet O, Farooqi S, Hodges JR (2016). Assessment of eating behavior disturbance and associated neural networks in frontotemporal dementia. *JAMA Neurol*. 73(3):282–290.

Andreasen NC (1997). The role of the thalamus in schizophrenia. *Can J Psychiatry*. 42(1):27–33.

Bejanin A, Tammewar G, Marx G, Cobigo Y, Iaccarino L, Kornak J, Staffaroni AM, Dickerson BC, Boeve BF, Knopman DS, Gorno-Tempini M, Miller BL, Jagust WJ, Boxer AL, Rosen HJ, Rabinovici GD (2020). Longitudinal structural and metabolic changes in frontotemporal dementia. *Neurology*. 95(2):e140–e154.

Bertoux M, Ramanan S, Slachevsky A, Wong S, Henriquez F, Musa G, Delgado C, Flanagan E, Bottlaender M, Sarazin M, Hornberger M, Dubois B (2016). So close yet so far: Executive contribution to memory processing in behavioral variant frontotemporal dementia. *J Alzheimers Dis*. 54(3):1005–1014.

Bocchetta M, Iglesias JE, Neason M, Cash DM, Warren JD, Rohrer JD (2020). Thalamic nuclei in frontotemporal dementia: Mediodorsal nucleus involvement is universal but pulvinar atrophy is unique to C9orf72. *Hum Brain Mapp*. 41(4):1006–1016.

Bocchetta M, Malpetti M, Todd EG, Rowe JB, Rohrer JD (2021a). Looking beneath the surface: the importance of subcortical structures in frontotemporal dementia. *Brain Communications*. 16;3(3):fcab158.

Bocchetta M, Todd EG, Peakman G, Cash DM, Convery RS, Russell LL, Thomas DL, Eugenio Iglesias J, van Swieten JC, Jiskoot LC, Seelaar H, Borroni B, Galimberti D, Sanchez-Valle R, Laforce R, Moreno F, Synofzik M, Graff C, Masellis M, Carmela Tartaglia M, Rowe JB, Vandenberghe R, Finger E, Tagliavini F, de Mendonça A, Santana I, Butler CR, Ducharme S, Gerhard A, Danek A, Levin J, Otto M, Sorbi S, Le Ber I, Pasquier F, Rohrer JD; Genetic Frontotemporal Dementia Initiative (GENFI) (2021b). Differential early subcortical involvement in genetic FTD within the GENFI cohort. *Neuroimage Clin*. 30:102646.

Fair DA, Bathula D, Mills KL, Dias TG, Blythe MS, Zhang D, Snyder AZ, Raichle ME, Stevens AA, Nigg JT, Nagel BJ (2010). Maturing thalamocortical functional connectivity across development. *Front Syst Neurosci*. 18(4):4–10.

Farb NA, Grady CL, Strother S, Tang-Wai DF, Masellis M, Black S, Freedman M, Pollock BG, Campbell KL, Hasher L, Chow TW (2013). Abnormal network connectivity in frontotemporal dementia: evidence for prefrontal isolation. *Cortex*. 49(7):1856–1873.

Fletcher PD, Downey LE, Golden HL, Clark CN, Slattery CF, Paterson RW, Rohrer JD, Schott JM, Rossor MN, Warren JD (2015). Pain and temperature processing in dementia: a clinical and neuroanatomical analysis. *Brain*. 138(Pt 11):3360–3372.

Foster PH, Russell LL, Peakman G, Convery RS, Bouzigues A, Greaves CV, Bocchetta M, Cash DM, van Swieten JC, Jiskoot LC, Moreno F, Sanchez-Valle R, Laforce R, Graff C, Masellis M, Tartaglia C, Rowe JB, Borroni B, Finger E, Synofzik M, Galimberti D, Vandenberghe R, de Mendonça A, Butler CR, Gerhard A, Ducharme S, Le Ber I, Tagliavini F, Santana I, Pasquier F, Levin J, Danek A, Otto M, Sorbi S, Rohrer JD; Genetic FTD Initiative (GENFI) (2022). Examining empathy deficits across familial forms of frontotemporal dementia within the GENFI cohort. *Cortex*. 150:12–28.

Gordon E, Rohrer JD, Fox NC (2016). Advances in neuroimaging in frontotemporal dementia. *J Neurochem*. 138(Suppl 1):193–210.

Gorno-Tempini ML, Hillis AE, Weintraub S, Kertesz A, Mendez M, Cappa SF, Ogar JM, Rohrer JD, Black S, Boeve BF, Manes F, Dronkers NF, Vandenberghe R, Rascovsky K, Patterson K, Miller BL, Knopman DS, Hodges JR, Mesulam MM, Grossman M (2011). Classification of primary progressive aphasia and its variants. *Neurology*. 76(11):1006–1014.

Häkkinen S, Chu SA, Lee SE (2020). Neuroimaging in genetic frontotemporal dementia and amyotrophic lateral sclerosis. *Neurobiol Dis*. 145:105063.

Harciarek M, Cosentino S (2013). Language, executive function and social cognition in the diagnosis of frontotemporal dementia syndromes. *Int Rev Psychiatry*. 25(2):178–196.

Harris JM, Gall C, Thompson JC, Richardson AM, Neary D, du Plessis D, Pal P, Mann DM, Snowden JS, Jones M (2013). Classification and pathology of primary progressive aphasia. *Neurology*. 81(21):1832–189.

Hebb AO, Ojemann GA (2013). The thalamus and language revisited. *Brain Lang*. 126(1):99–108.

Hornberger M, Geng J, Hodges JR (2011). Convergent grey and white matter evidence of orbitofrontal cortex changes related to disinhibition in behavioural variant frontotemporal dementia. *Brain*. 134(Pt 9):2502–2512.

Hornberger M, Piguet O (2012). Episodic memory in frontotemporal dementia: a critical review. *Brain*. 135(3):678–692.

Hornberger M, Wong S, Tan R, Irish M, Piguet O, Kril J, Hodges JR, Halliday G (2012). In vivo and post-mortem memory circuit integrity in frontotemporal dementia and Alzheimer's disease. *Brain*. 135(Pt 10):3015–3025.

Irish M, Devenney E, Wong S, Dobson-Stone C, Kwok JB, Piguet O, Hodges JR, Hornberger M (2013). Neural substrates of episodic memory dysfunction in behavioural variant frontotemporal dementia with and without C9ORF72 expansions. *Neuroimage Clin*. 2:836–843.

Johnen A, Bertoux M (2019). Psychological and cognitive markers of behavioral variant frontotemporal dementia: A clinical neuropsychologist's view on diagnostic criteria and beyond. *Front Neurol*. 10:594.

Kumfor F, Miller L, Lah S, Hsieh S, Savage S, Hodges JR, Piguet O (2011). Are you really angry? The effect of intensity on facial emotion recognition in frontotemporal dementia. *Soc Neurosci*. 6(5–6):502–514.

Kumfor F, Zhen A, Hodges JR, Piguet O, Irish M (2018). Apathy in Alzheimer's disease and frontotemporal dementia: distinct clinical profiles and neural correlates. *Cortex*. 103:350–359.

Lashley T, Rohrer JD, Mead S, Revesz T (2015). Review: An update on clinical, genetic and pathological aspects of frontotemporal lobar degenerations. *Neuropathol Appl Neurobiol*. 41(7):858–881.

Lee SE, Khazenzon AM, Trujillo AJ, Guo CC, Yokoyama JS, Sha SJ, Takada LT, Karydas AM, Block NR, Coppola G, Pribadi M,

Geschwind DH, Rademakers R, Fong JC, Weiner MW, Boxer AL, Kramer JH, Rosen HJ, Miller BL, Seeley WW (2014). Altered network connectivity in frontotemporal dementia with C9orf72 hexanucleotide repeat expansion. *Brain.* 137(Pt 11):3047–3060.

Lee SE, Sias AC, Kosik EL, Flagan TM, Deng J, Chu SA, Brown JA, Vidovszky AA, Ramos EM, Gorno-Tempini ML, Karydas AM, Coppola G, Geschwind DH, Rademakers R, Boeve BF, Boxer AL, Rosen HJ, Miller BL, Seeley WW (2019). Thalamo-cortical network hyperconnectivity in preclinical progranulin mutation carriers. *Neuroimage Clin.* 22:101751.

Lee SE, Sias AC, Mandelli ML, Brown JA, Brown AB, Khazenzon AM, Vidovszky AA, Zanto TP, Karydas AM, Pribadi M, Dokuru D, Coppola G, Geschwind DH, Rademakers R, Gorno-Tempini ML, Rosen HJ, Miller BL, Seeley WW (2016). Network degeneration and dysfunction in presymptomatic C9ORF72 expansion carriers. *Neuroimage Clin.* 14:286–297.

Lombardi J, Mayer B, Anderl-Straub S, Uttner I, Kassubek J, Diehl-Schmid J, Danek A, Levin J, Fassbender K, Fliessbach K, Schneider A, Huppertz HJ, Jahn H, Volk A, Kornhuber J, Landwehrmeyer B, Lauer M, Prudlo J, Wiltfang J, Schroeter ML, Ludolph A, Otto M; FTLD Consortium (2021). Quantifying progression in primary progressive aphasia with structural neuroimaging. *Alzheimers Dement.* 17(10):1595–1609.

Mahoney CJ, Beck J, Rohrer JD, Lashley T, Mok K, Shakespeare T, Yeatman T, Warrington EK, Schott JM, Fox NC, Rossor MN, Hardy J, Collinge J, Revesz T, Mead S, Warren JD (2012). Frontotemporal dementia with the C9ORF72 hexanucleotide repeat expansion: Clinical, neuroanatomical and neuropathological features. *Brain.* 135(Pt 3):736–750.

McKenna MC, Li Hi Shing S, Murad A, Lope J, Hardiman O, Hutchinson S, Bede P (2022). Focal thalamus pathology in frontotemporal dementia: Phenotype-associated thalamic profiles. *J Neurol Sci.* 436:120221.

Migeot JA, Duran-Aniotz CA, Signorelli CM, Piguet O, Ibáñez A (2022). A predictive coding framework of allostatic-interoceptive overload in frontotemporal dementia. *Trends Neurosci.* 45(11):838–853.

Moore KM, Nicholas J, Grossman M, McMillan CT, Irwin DJ, Massimo L, Van Deerlin VM, Warren JD, Fox NC, Rossor MN, Mead S, Bocchetta M, Boeve BF, Knopman DS, Graff-Radford NR, Forsberg LK, Rademakers R, Wszolek ZK, van Swieten JC, Jiskoot LC, Meeter LH, Dopper EG, Papma JM, Snowden JS, Saxon J, Jones M, Pickering-Brown S, Le Ber I, Camuzat A, Brice A, Caroppo P, Ghidoni R, Pievani M, Benussi L, Binetti G, Dickerson BC, Lucente D, Krivensky S, Graff C, Öijerstedt L, Fallström M, Thonberg H, Ghoshal N, Morris JC, Borroni B, Benussi A, Padovani A, Galimberti D, Scarpini E, Fumagalli GG, Mackenzie IR, Hsiung GR, Sengdy P, Boxer AL, Rosen H, Taylor JB, Synofzik M, Wilke C, Sulzer P, Hodges JR, Halliday G, Kwok J, Sanchez-Valle R, Lladó A, Borrego-Ecija S, Santana I, Almeida MR, Tábuas-Pereira M, Moreno F, Barandiaran M, Indakoetxea B, Levin J, Danek A, Rowe JB, Cope TE, Otto M, Anderl-Straub S, de Mendonça A, Maruta C, Masellis M, Black SE, Couratier P, Lautrette G, Huey ED, Sorbi S, Nacmias B, Laforce R Jr, Tremblay ML, Vandenberghe R, Damme PV, Rogalski EJ, Weintraub S, Gerhard A, Onyike CU, Ducharme S, Papageorgiou SG, Ng ASL, Brodtmann A, Finger E, Guerreiro R, Bras J, Rohrer JD; FTD Prevention Initiative (2020). Age at symptom onset and death and disease duration in genetic frontotemporal dementia: an international retrospective cohort study. *Lancet Neurol.* 19(2):145–156.

Pasquini L, Nana AL, Toller G, Brown JA, Deng J, Staffaroni A, Kim EJ, Hwang JL, Li L, Park Y, Gaus SE, Allen I, Sturm VE, Spina S, Grinberg LT, Rankin KP, Kramer JH, Rosen HJ, Miller BL, Seeley W (2020). Salience network atrophy links neuron type-specific pathobiology to loss of empathy in frontotemporal dementia. *Cereb Cortex.* 30(10):5387–5399.

Poos JM, Russell LL, Peakman G, Bocchetta M, Greaves CV, Jiskoot LC, van der Ende EL, Seelaar H, Papma JM, van den Berg E, Pijnenburg YAL, Borroni B, Sanchez-Valle R, Moreno F, Laforce R, Graff C, Synofzik M, Galimberti D, Rowe JB, Masellis M, Tartaglia C, Finger E, Vandenberghe R, de Medonça A, Tagliavini F, Butler CR, Santana I, Ber IL, Gerhard A, Ducharme S, Levin J, Danek A, Otto M, Sorbi S, Pasquier F, van Swieten JC, Rohrer JD; Genetic FTD Initiative, GENFI (2021). Impairment of episodic memory in genetic frontotemporal dementia: A GENFI study. *Alzheimers Dement (Amst).* 13(1):e12185.

Rankin KP (2020). Brain networks supporting social cognition in dementia. *Curr Behav Neurosci Rep.* 7:203–211.

Rascovsky K, Hodges JR, Knopman D, Mendez MF, Kramer JH, Neuhaus J, van Swieten JC, Seelaar H, Dopper EG, Onyike CU, Hillis AE, Josephs KA, Boeve BF, Kertesz A, Seeley WW, Rankin KP, Johnson JK, Gorno-Tempini ML, Rosen H, Prioleau-Latham CE, Lee A, Kipps CM, Lillo P, Piguet O, Rohrer JD, Rossor MN, Warren JD, Fox NC, Galasko D, Salmon DP, Black SE, Mesulam M, Weintraub S, Dickerson BC, Diehl-Schmid J, Pasquier F, Deramecourt V, Lebert F, Pijnenburg Y, Chow TW, Manes F, Grafman J, Cappa SF, Freedman M, Grossman M, Miller BL (2011). Sensitivity of revised diagnostic criteria for the behavioural variant of frontotemporal dementia. *Brain.* 134:2456–2477.

Rohrer JD, Boxer AL (2021). The Frontotemporal Dementia Prevention Initiative: Linking together genetic frontotemporal dementia cohort studies. *Adv Exp Med Biol.* 1281:113–121.

Rohrer JD, Lashley T, Schott JM, Warren JE, Mead S, Isaacs AM, Beck J, Hardy J, de Silva R, Warrington E, Troakes C, Al-Sarraj S, King A, Borroni B, Clarkson MJ, Ourselin S, Holton JL, Fox NC, Revesz T, Rossor MN, Warren JD (2011). Clinical and neuroanatomical signatures of tissue pathology in frontotemporal lobar degeneration. *Brain.* 134(Pt 9):2565–2581.

Rohrer JD, Nicholas JM, Cash DM, van Swieten J, Dopper E, Jiskoot L, van Minkelen R, Rombouts SA, Cardoso MJ, Clegg S, Espak M, Mead S, Thomas DL, De Vita E, Masellis M, Black SE, Freedman M, Keren R, MacIntosh BJ, Rogaeva E, Tang-Wai D, Tartaglia MC, Laforce R Jr, Tagliavini F, Tiraboschi P, Redaelli V, Prioni S, Grisoli M, Borroni B, Padovani A, Galimberti D, Scarpini E, Arighi A, Fumagalli G, Rowe JB, Coyle- Gilchrist I, Graff C, Fallström M, Jelic V, Ståhlbom AK, Andersson C, Thonberg H, Lilius L, Frisoni GB, Binetti G, Pievani M, Bocchetta M, Benussi L, Ghidoni R, Finger E, Sorbi S, Nacmias B, Lombardi G, Polito C, Warren JD, Ourselin S, Fox NC, Rossor MN (2015). Presymptomatic cognitive and neuroanatomical changes in genetic frontotemporal dementia in the Genetic Frontotemporal dementia Initiative (GENFI) study: A cross-sectional analysis. *Lancet Neurol.* 14(3):253–2562.

Russell LL, Greaves CV, Bocchetta M, Nicholas J, Convery RS, Moore K, Cash DM, van Swieten J, Jiskoot L, Moreno F, Sanchez-Valle R, Borroni B, Laforce R Jr, Masellis M, Tartaglia MC, Graff C, Rotondo E, Galimberti D, Rowe JB, Finger E, Synofzik M, Vandenberghe R, de Mendonça A, Tagliavini F, Santana I, Ducharme S, Butler C, Gerhard A, Levin J, Danek A, Otto M, Warren JD, Rohrer JD; Genetic FTD Initiative, GENFI (2020). Social cognition impairment in genetic frontotemporal dementia within the GENFI cohort. *Cortex.* 133:384–398.

Schönecker S, Neuhofer C, Otto M, Ludolph A, Kassubek J, Landwehrmeyer B, Anderl-Straub S, Semler E, Diehl-Schmid J, Prix C, Vollmar C, Fortea J; Deutsches FTLD-Konsortium, Huppertz HJ, Arzberger T, Edbauer D, Feddersen B, Dieterich M, Schroeter ML, Volk AE, Fließbach K, Schneider A, Kornhuber J, Maler M, Prudlo J, Jahn H, Boeckh-Behrens T, Danek A, Klopstock T, Levin J (2018). Atrophy in the thalamus but not cerebellum is specific for C9orf72 FTD and ALS patients: An atlas-based volumetric MRI study. *Front Aging Neurosci.* 10(45):1–11.

Seeley WW, Menon V, Schatzberg AF, Keller J, Glover GH, Kenna H, Reiss AL, Greicius MD (2007). Dissociable intrinsic connectivity networks for salience processing and executive control. *J Neurosci.* 27(9):2349–2356.

Shany-Ur T, Rankin KP (2011). Personality and social cognition in neurodegenerative disease. *Curr Opin Neurol.* 24(6):550–555.

Sturm VE, Perry DC, Wood K, Hua AY, Alcantar O, Datta S, Rankin KP, Rosen HJ, Miller BL, Kramer JH (2017). Prosocial deficits in behavioral variant frontotemporal dementia relate to reward network atrophy. *Brain Behav.* 7:1–14.

Tan RH, Wong S, Kril JJ, Piguet O, Hornberger M, Hodges JR, Halliday GM (2014). Beyond the temporal pole: Limbic memory circuit in the semantic variant of primary progressive aphasia. *Brain.* 137: 2065–2076.

Vatsavayai SC, Yoon SJ, Gardner RC, Gendron TF, Vargas JN, Trujillo A, Pribadi M, Phillips JJ, Gaus SE, Hixson JD, Garcia PA, Rabinovici GD, Coppola G, Geschwind DH, Petrucelli L, Miller BL, Seeley WW (2016). Timing and significance of pathological features in C9orf72 expansion-associated frontotemporal dementia. *Brain.* 139(Pt 12):3202–3216.

Vertes RP, Linley SB, Hoover WB (2015). Limbic circuitry of the midline thalamus. *Neurosci Biobehav Rev.* 54:89–107.

Warren JD, Rohrer JD, Schott JM, Fox NC, Hardy J, Rossor MN (2013). Molecular nexopathies: a new paradigm of neurodegenerative disease. *Trends Neurosci.* 36(10):561–569.

Whitwell JL, Przybelski SA, Weigand SD, Ivnik RJ, Vemuri P, Gunter JL, Senjem ML, Shiung MM, Boeve BF, Knopman DS, Parisi JE, Dickson DW, Petersen RC, Jack CR Jr, Josephs KA (2009). Distinct anatomical subtypes of the behavioural variant of frontotemporal dementia: A cluster analysis study. *Brain.* 132(Pt 11):2932–2946.

Whitwell JL, Weigand SD, Boeve BF, Senjem ML, Gunter JL, DeJesus-Hernandez M, Rutherford NJ, Baker M, Knopman DS, Wszolek ZK, Parisi JE, Dickson DW, Petersen RC, Rademakers R, Jack CR Jr, Josephs KA (2012). Neuroimaging signatures of frontotemporal dementia genetics: C9ORF72, tau, progranulin and sporadics. *Brain.* 135(Pt 3):794–806.

Woollacott IO, Rohrer JD (2016). The clinical spectrum of sporadic and familial forms of frontotemporal dementia. *J Neurochem.* 138(Suppl 1):6–31.

Young AL, Marinescu RV, Oxtoby NP, Bocchetta M, Yong K, Firth NC, Cash DM, Thomas DL, Dick KM, Cardoso J, van Swieten J, Borroni B, Galimberti D, Masellis M, Tartaglia MC, Rowe JB, Graff C, Tagliavini F, Frisoni GB, Laforce R Jr, Finger E, de Mendonça A, Sorbi S, Warren JD, Crutch S, Fox NC, Ourselin S, Schott JM, Rohrer JD, Alexander DC; Genetic FTD Initiative (GENFI); Alzheimer's Disease Neuroimaging Initiative (ADNI) (2018). Uncovering the heterogeneity and temporal complexity of neurodegenerative diseases with subtype and stage inference. *Nat Commun.* 9(1):4273.

Zhang J, Zhou Z, Li L, Ye J, Shang D, Zhong S, Yao B, Xu C, Yu Y, He F, Ye X, Luo B (2022). Cerebral perfusion mediated by thalamo-cortical functional connectivity in non-dominant thalamus affects naming ability in aphasia. *Hum Brain Mapp.* 43(3):940–954.

71.

VISUOSPATIAL AND MOTOR DEFICITS FOLLOWING PULVINAR LESIONS

Melanie Wilke and Igor Kagan

INTRODUCTION

EVOLUTIONARY PERSPECTIVE ON PRIMATE-SPECIFIC SPECIALIZATIONS OF COGNITION AND FLEXIBLE BEHAVIOR

The ability to explore the visual environment and to decide between alternative movement options underlies many everyday behaviors and is important for survival. In human and nonhuman primates, these functions are predominantly subserved by directing attention and gaze toward perceptually or motivationally salient spatial locations that are selected for detailed visual analysis and reach-grasp actions. The current view is that visually guided behaviors engage widely distributed brain regions in frontoparietal, superior temporal, and cingulate cortices, along with subcortical structures such as superior colliculus and basal ganglia. At the same time, how this distributed information is coordinated and integrated remains poorly understood. Apart from direct connections between cortical areas, the thalamic pulvinar is a promising candidate to perform this function, either by directly relaying stimulus or action-relevant signals to a given cortical region, or by modulating the signal transfer within and across cortical regions (Shipp, 2003; Halassa and Sherman, 2019; Kastner et al., 2020). Indeed, the dorsal portion of the pulvinar complex increased concurrently with frontoparietal neocortex during primate evolution, being largest in humans (Preuss, 2007; Homman-Ludiye and Bourne, 2019). It is therefore plausible that the functional significance of the dorsal pulvinar lies in supporting primate-specific specializations, such as eye-hand coordination and fine reach-grasp manipulations.

FUNCTIONAL NEUROANATOMY AND NEURONAL PROPERTIES OF THE PULVINAR

Pulvinar Parcellations and Connectivity

Pulvinar connectivity and its functional properties have been extensively studied in nonhuman primates, and more recently in humans. The pulvinar is a prototypic higher-order thalamic nucleus that receives its cortical inputs from layer (L) L5 and L6, and in turn projects back to the superficial layers of the same areas and L4 of "higher-order" areas (Shipp, 2003; de Souza et al., 2020). The primate pulvinar is a complex heterogeneous structure (see **Chapter 1 in Section I**; **Chapter 30 in Section V**). In both humans and nonhuman primates it has been traditionally subdivided cytoarchitectonically into anterior, medial, lateral, and inferior pulvinar (Benarroch, 2015; Bridge et al., 2016) (Figure 71.1A). The subnuclei of the inferior pulvinar (PI) and the ventral portion of the lateral pulvinar (PLvl) form together the ventral pulvinar (vPul) that is retinotopically organized, is primarily interconnected with visual cortices, and receives input from the superficial layers of the superior colliculus (Kaas and Lyon, 2007; Baldwin and Bourne, 2017) (see also **Chapter 20 in Section V**).

The dorsal pulvinar (dPul) roughly occupies the region above the brachium of the superior colliculus (bsc) and encompasses the medial/anterior (PM/PA) and the dorsal portion of the lateral pulvinar (PLdm, denoted Pdm in earlier studies), with a small portion of the inferior pulvinar (PI) also extending above the brachium (Gutierrez et al., 2000; Kaas and Lyon, 2007). Unlike the ventral pulvinar, the dorsal pulvinar does not have a clear retinotopic organization (Petersen et al., 1987; Benevento and Port, 1995; Arcaro et al., 2018). The dorsal pulvinar is heavily

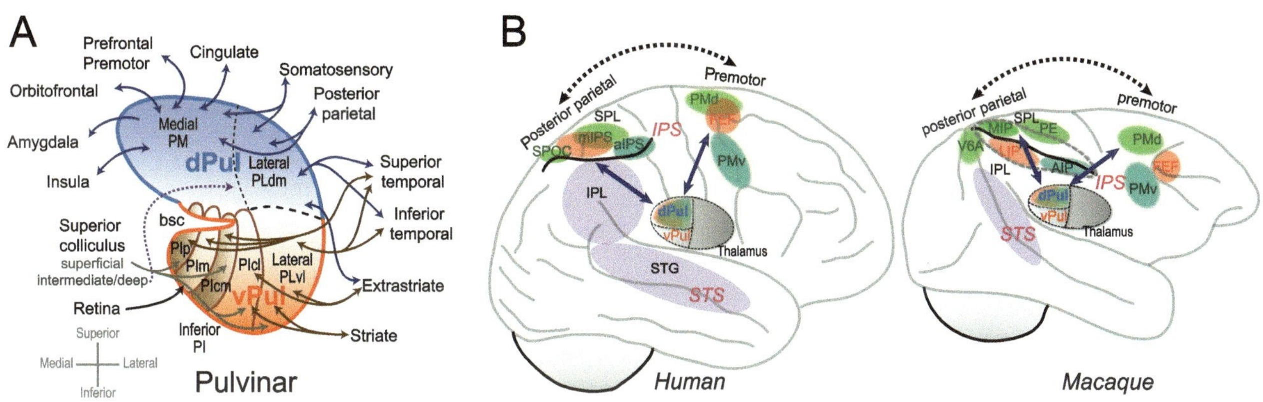

Figure 71.1. Pulvinar subdivisions and connectivity. (A). Pulvinar parcellation and anatomical connectivity (a schematic coronal section; anterior pulvinar (PA), part of the dorsal pulvinar, is not shown). Adapted from Kaas and Lyon (2007) and Grieve et al. (2000). dPul = dorsal pulvinar; vPul = ventral pulvinar; PM = medial pulvinar; PLdm = dorsal medial lateral pulvinar; PLvl = ventral lateral lateral pulvinar; Plcm = central medial inferior pulvinar; Plcl = central lateral inferior pulvinar; Plp = posterior inferior pulvinar; Plpl = posterior lateral inferior pulvinar; bsc = brachium of superior colliculus. (B). Schematic representation of frontoparietal and parieto-temporal regions important for spatial attention and action planning. Saccade (red), reach (green), and grasp (teal) frontoparietal circuitry in humans and macaques (not all regions are depicted). The effector specificity in human IPL/STG and potentially homologous macaque STS regions (purple) is not specified. SPL = superior parietal lobule; IPL = inferior parietal lobule; IPS = intraparietal sulcus; STS = superior temporal sulcus; STG = superior temporal gyrus; FEF = frontal eye field; PMd/v = dorsal/ventral premotor cortex; aIPS = anterior IPS; mIPS = mid IPS; SPOC = superior parieto-occipital cortex; dPul = dorsal pulvinar; vPul = ventral pulvinar; AIP = anterior intraparietal area; MIP = medial intraparietal area; LIP = lateral intraparietal area; PE = parietal Brodmann's area 5; V6A = medial parieto-occipital cortex; the caudal part of MIP and V6A are often denoted as PRR = the parietal reach region.

connected with cortical areas related to the guidance of goal-directed behavior such as prefrontal (FEF, dlPFC), premotor (PMv, PMd), posterior parietal (e.g., area 7, LIP, MIP, AIP), and superior temporal sulcus (STS) regions (Figure 71.1B). These connections were revealed by tracer studies in monkeys (Yeterian and Pandya, 1989; Gutierrez et al., 2000) and by effective and functional magnetic resonance imaging (fMRI) connectivity studies in monkeys and humans (Arcaro et al., 2018; Kagan et al., 2021) (see also **Chapter 20 in Section V**).

Pulvinar Neuronal Response Properties in Oculomotor Tasks

Neurons in both ventral and dorsal pulvinar respond to a variety of visual stimuli, and are modulated by their behavioral saliency, for example via spatial attention and relevance for upcoming actions, as well as subjective perception (Petersen et al., 1985; Robinson et al., 1986; Bender and Youakim, 2001; Wilke et al., 2009; Saalmann et al., 2012; Zhou et al., 2016). Beyond visual activity, pulvinar neurons often respond before, during, and especially after eye movements (Robinson et al., 1986; Benevento and Port, 1995; Berman and Wurtz, 2011; Schneider et al., 2023). Visual sensitivity of pulvinar neurons is furthermore influenced by the orbital position (Robinson et al., 1990). The ventral (inferior) pulvinar has been implicated in the pathway from the superior colliculus to cortex that carries saccadic suppression signals (Berman and Wurtz, 2011) (see also **Chapter 36 in Section VI**). A recent study that investigated dorsal pulvinar (at the border between PM and PLdm) in a memory-delay saccade task showed both space-specific visual and saccade-related enhancement, as well as predominately non-spatial pre-movement or movement-related suppression (Schneider et al., 2023). Pre-saccadic enhancement was rare, and only a few units exhibited classical visuomotor patterns such as continuous delay period activity up to the saccade onset, or pre-saccadic ramping. Furthermore, even in spatially tuned neurons, spiking activity in free-choice trials did not encode the upcoming spatial decision during the delay. Therefore, the dorsal pulvinar has only a limited role in prospective saccade processing and instead may participate in processing the consequences of oculomotor behavior such as saccadic updating. Indeed, many dorsal pulvinar neurons carry information about (static) eye position during fixation, and the spatial reference frame in the dorsal pulvinar is dynamic, transitioning from predominantly eye-centered visual goal encoding to final gaze representation in non-retinocentric coordinates (Schneider et al., 2020). These properties, together with the multisensory nature of many pulvinar responses (Froesel et al., 2021), suggest an involvement of the pulvinar in the integration of sensory inputs and consequences of oculomotor behavior.

Pulvinar Neuronal Response Properties in Reach-Grasp Tasks

Only a few studies have examined neural responses in higher-order thalamus such as the pulvinar to reach-grasp movements. Earlier electrophysiological work in macaque monkeys (Cudeiro et al., 1989; Acuna et al., 1990) investigated reach movements in a simple manual task, where monkeys responded to visual stimuli by pressing a switch, and found increased firing mainly during the arm-motion phase without a spatial preference. Furthermore, neurons responded stronger during task-related reaches as compared to spontaneous movements or external manipulation of the limb (Acuna et al., 1990). Many basic questions remain open, such as the extent to which pulvinar neurons exhibit specificity for hand (left, right) or effector (saccade vs. reach), interactions between effector and spatial selectivity, and the underlying spatial reference frames. The question whether there is a gradient from visual to motor or effector-specific signals as predicted by the gradient of anatomical connections along the posterior-anterior pulvinar axis (Arcaro et al., 2018; Froesel et al., 2021) also remains open. Although it has been hypothesized that the pulvinar integrates eye and hand signals (Grieve et al., 2000; Wilke et al., 2010), a systematic assessment of pulvinar activity during dissociated versus combined saccade and reach tasks and its role in eye-hand coordination is lacking.

VISUOSPATIAL DEFICITS IN HUMANS AND MONKEYS

CLINICAL NEGLECT AND EXTINCTION IN HUMANS

Visuospatial neglect is a frequent neuropsychological disorder with a prevalence of 38% after lesions in the right hemisphere and 18% in the left hemisphere, typically following stroke (Esposito et al., 2021). In short, spatial neglect is a heterogeneous syndrome that is characterized by an impaired or lost ability to perceive and explore space contralateral to the lesion and to react to stimuli presented on this side (Corbetta and Shulman, 2011). Clinically, it is diagnosed with cancellation, line bisection, and figure-copying tasks and by observing spontaneous spatial orienting and search behavior.

A hallmark of spatial neglect is that patients exhibit a strong bias of active behaviors towards the ipsilesional side (i.e., the same side of space as the lesion, e.g., the right visual hemifield after a lesion in the right hemisphere), which can be observed with eye and hand movements (Karnath, 2015). By definition, neglect cannot be explained by primary sensory or motor dysfunction, but occurs at higher processing levels where sensory and motor information are integrated. Accordingly, lesion sites associated with spatial neglect are widely distributed, involving frontal and parieto-temporal areas, as well as subcortical structures such as basal ganglia and pulvinar, and their white matter connections (Karnath et al., 2002; Karnath, 2015), leading to considerable variability of manifestations. Following isolated thalamic lesions, clinically relevant neglect is rare, but sometimes has been observed (Watson and Heilman, 1979; Sebastian et al., 2014). By contrasting lesion locations in thalamic stroke patients with and without spatial neglect, the pulvinar has been identified as the critical lesion site (Karnath et al., 2002).

VISUOSPATIAL DEFICITS FOLLOWING PULVINAR LESIONS IN HUMANS

While neglect in pulvinar patients is often not detected with standardized clinical tests, more sensitive experimental tasks have uncovered ipsilesional (i.e., toward the side of the lesion) spatial orienting bias, and contralesional (to the side opposite to the lesion) attentional allocation deficits (Rafal and Posner, 1987; Ward et al., 2002; Arend et al., 2008) (Figure 71.2A). For example, using a visual task that required shape and color identification, a case study in a patient with unilateral hemorrhage of the dorsal right pulvinar showed increased false conjunctions (shape and color) in the lower quadrant of the contralesional hemifield (Ward et al., 2002). A follow-up study that investigated whether the pulvinar lesion-induced deficits occur in a retinotopic or in an object-based spatial reference frame reported impairments in both (Ward and Arend, 2007). Another study that investigated unilateral pulvinar lesion patients with a temporal order judgment task found an ipsilesional saccade choice bias (Arend et al., 2008). It is important to note that the nature of the thalamic stroke lesions rarely allows systematic mapping to specific pulvinar subdivisions, necessitating perturbation studies in monkeys.

VISUOSPATIAL DEFICITS FOLLOWING PULVINAR PERTURBATIONS IN MONKEYS

In monkeys, similar to human patients, neglect-like symptoms such as spatial orienting bias and contralesional detection deficits under bilateral stimulation ("extinction") have been reported after permanent or reversible lesions in prefrontal, parietal and superior temporal sulcus cortex (as reviewed in Wardak et al., 2002). Pulvinar lesions in monkeys also impair spatial orienting and visual detection (Wilke et al., 2010; Zhou et al., 2016). Ventral lateral pulvinar inactivation during a cued change-detection task led to decreased detection performance in the contralesional hemifield when a target was presented among distractors (Zhou et al., 2016). A study that assessed the effect of dorsal pulvinar inactivation on saccade target selection found an ipsilesional choice bias that was alleviated with visual cues signaling high rewards in the contralesional field (Wilke et al., 2013) (Figure 71.2B). Visual salience manipulations were not effective, and a recent study that employed a

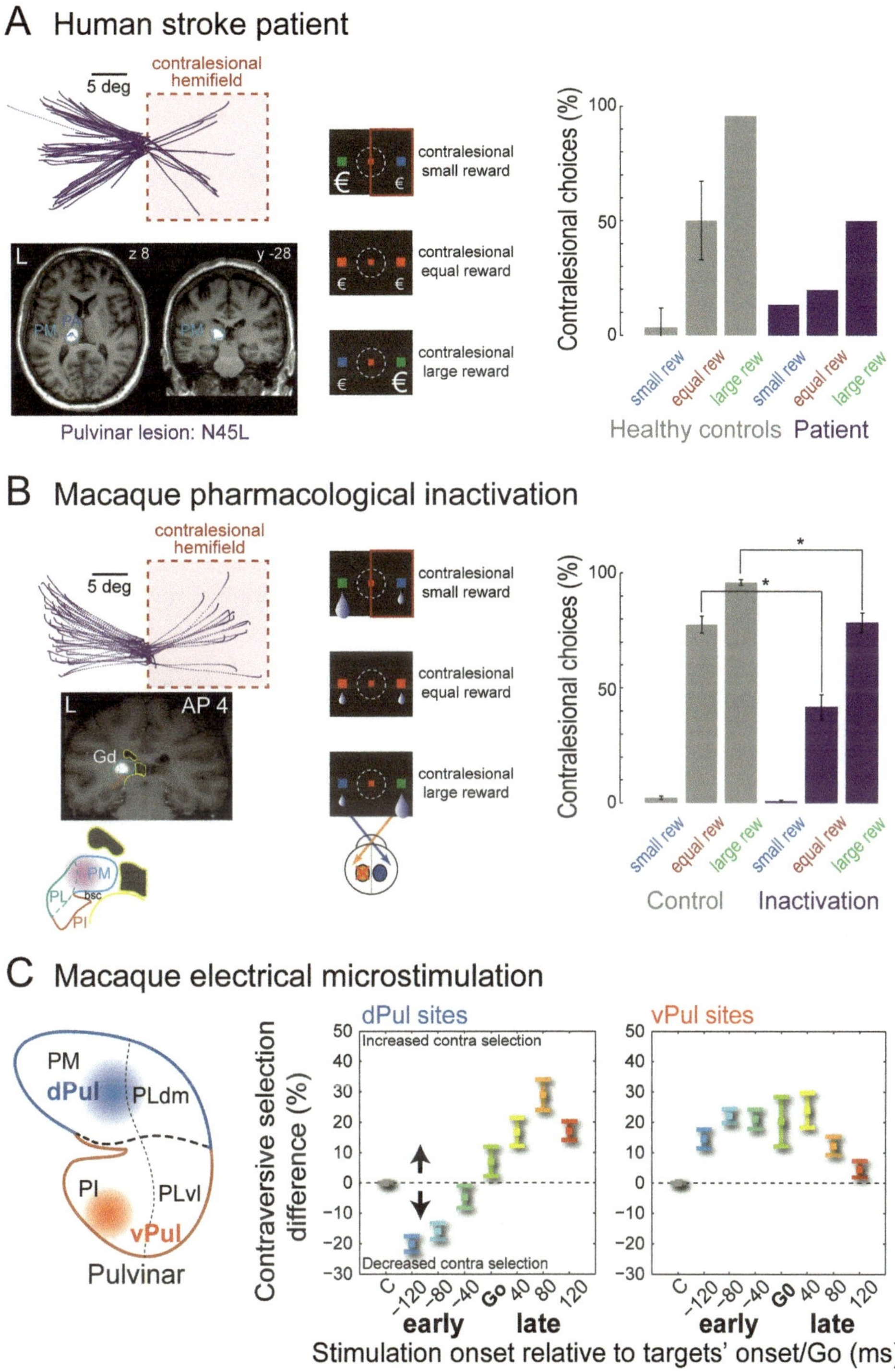

Figure 71.2. Pulvinar-related spatial saccade choice deficits. (A). Posterior thalamus (centered on the pulvinar) hemorrhagic stroke lesion in an example human patient, showing an ipsilesional bias (decrease of contralesional choices, as seen in eye movement traces and bar plots) compared to 20 age-matched healthy controls, and its partial compensation by color-indicated monetary reward. (B). Dorsal (PM and PLdm) pulvinar pharmacological GABA-A (γ-aminobutyric acid, the major inhibitory neurotransmitter in the central nervous system) agonist inactivation in macaques, showing a similar ipsilesional choice bias and reward compensation (Wilke et al., 2013). Gd = co-injected MRI contrast agent Gadolinium. (C) Dorsal (PM, at the atlas-aligned estimated border with PLdm) and ventral pulvinar (PI) microstimulation-induced choice bias in macaques (plotted as the difference to control trials without stimulation, "C"). Early stimulation was delivered before targets' onset; late stimulation was delivered after targets' onset ("Go") (Dominguez-Vargas et al., 2017).

perceptual discrimination saccade task showed that dorsal pulvinar inactivation leads to a shift in response criterion away from contralesional stimuli (i.e. reluctance to make saccades to contralesional targets or distractors) but not to a decrease in contralesional perceptual sensitivity (Kaduk et al., 2022). This is consistent with the absence of perceptual categorization impairments but diminished confidence after dorsal pulvinar inactivation (Komura et al., 2013). The distinct role of the dorsal pulvinar in spatial decision-making is also supported by a time-resolved electrical microstimulation study (Dominguez-Vargas et al., 2017). As predicted from inactivation results, in a visually-guided saccade task the stimulation of the dorsal medial pulvinar (PM) after the onset of choice targets led to an increased contraversive (i.e., opposite to the stimulated hemisphere) selection. Interestingly, pulvinar stimulation effects depended on the task context, as early stimulation during the fixation period before the targets' onset diminished contraversive choices (Figure 71.2C). Microstimulation effects of the ventral pulvinar, in particular in the inferior pulvinar (PI), had a different, unimodal temporal profile with an earlier contraversive bias (Figure 71.2C). These results, and the absence of a stimulation-induced bias in a memory-guided task (Dominguez-Vargas et al., 2017), suggest that transient perturbation of the dorsal pulvinar affects spatial choices only when the decision and action are close in time, while the ventral pulvinar is more perceptual/visual. To summarize, both ventral and dorsal pulvinar inactivation lead to specific visuospatial deficits that share features with stroke-induced spatial neglect, with the dorsal pulvinar primarily affecting saccadic target selection at the motor preparation stage. To further delineate distinct contributions of pulvinar subdivisions to visuomotor behavior, more studies using the same behavioral paradigms while manipulating different pulvinar sites are required.

VISUALLY GUIDED REACH-GRASP DEFICITS IN HUMANS AND MONKEYS

REACH-GRASP IMPAIRMENTS FOLLOWING PULVINAR LESIONS IN HUMANS

Reach-grasp movements are composed of multiple elements that include the ability to visually localize and identify the shape of an object, to move arm/hand in the correct direction with appropriate amplitude and shape of the hand, and to adjust grip strength (Castiello, 2005). Reach and grasp functions have been extensively studied in the cortex and are thought to rely on a distributed network of cortical regions such as motor, premotor, supplementary motor, superior parietal, and dorsal occipital cortex, as well as the cerebellum (Andersen et al., 2014). In addition, several thalamic nuclei with connections to these regions are important for motor behavior. The classical "motor thalamus" encompasses nuclei in the ventral anterior (VA), posterior (VP), and lateral (VL) thalamus that can be grouped into distinct sets based on whether they receive input from the cerebellum or the basal ganglia (Prevosto and Sommer, 2013) (see also **Chapter 31 in Section VI**). Clinically, lesions in the motor thalamus are associated with ataxia, dystonia, tremor, and delayed occurrence of involuntary movements (Gupta and Pandey, 2018). Although the pulvinar is not part of the classical "motor thalamus," lesion studies in humans and monkeys suggest that it is critical for limb movement behavior (Walker, 2015; Wilke et al., 2017, 2018). For example, a study in a patient with circumscribed bilateral medial pulvinar lesions showed severe standing/walking and reach-grasp deficits including impaired preshaping of the hand (Wilke et al., 2017; Wilke et al., 2018) (Figure 71.3A). Since reported pulvinar lesions in humans with limb movement impairments typically involved other thalamic nuclei as well (e.g., VP and CM/CL), it is necessary to derive more localized assessments using nonhuman primate models.

REACH-GRASP IMPAIRMENTS FOLLOWING PULVINAR LESIONS IN MONKEYS

Several monkey pulvinar lesion studies suggest that it is critical for reach-grasping behavior beyond visuospatial attention (Wilke et al., 2010; Mundinano et al., 2018; Homman-Ludiye and Bourne, 2019). For example, after reversible inactivation centered on the dorsal pulvinar, macaques exhibited a strong bias to use the ipsilesional hand in food-retrieval tasks independent of spatial location, and instructed reaches with the contralesional hand were imprecise and had longer durations as compared to the ipsilesional hand (Wilke et al., 2010). With the contralesional hand, monkeys also had problems to preshape their hand during the transport phase and to form an efficient grip to retrieve the objects (Figure 71.3B). In addition, studies in marmoset monkeys have underlined the importance of the pathway from the medial portion of the inferior pulvinar (PIm) to motion-sensitive area MT for the normal development of reach-grasping behavior and maturation of the visuomotor parietal cortices (Mundinano et al., 2018) (see also **Chapter 20 in Section V**). They employed excitotoxic lesions in marmoset monkeys shortly after birth and characterized the PIm lesion and its remote effects by a combination of MRI and retrograde tracing techniques. Lesioned animals had larger maximum grip apertures in the preshaping phase of the hand and showed a protracted deceleration period during the reach, which the authors interpreted as evidence for disturbed visual feedback. Analysis of the diffusion tensor imaging data across time after the lesion revealed

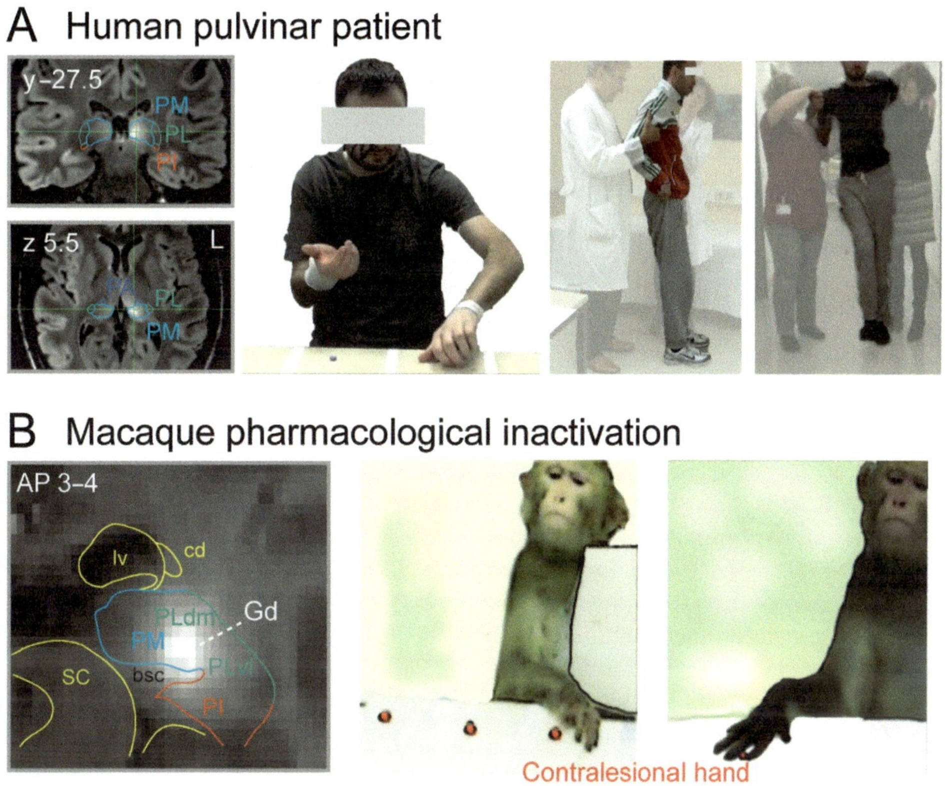

Figure 71.3. Reach-grasp and postural deficits following dorsal pulvinar lesions. (A). Bilateral pulvinar lesion patient (more extensive lesion in the left hemisphere), showing reach-grasp and standing/walking (astasia) deficits (Wilke et al., 2017; Wilke et al., 2018). (B). Similar (contralesional hand) reach-grasp deficits following unilateral reversible inactivation in the macaque that primarily affected the dorsal pulvinar at the border of PLdm and PM, above the brachium of the superior colliculus; Gd = co-injected MRI contrast agent Gadolinium; SC = superior colliculus; bsc = brachium of SC; lv = lateral ventricle (Wilke et al., 2010).

a progressive reduction of fractional anisotropy that was most pronounced in visuomotor reach/grasp-related cortical regions AIP, MIP, and PE in the parietal cortex—one of the main connectivity loci of the dorsal pulvinar in adult macaques. It remains to be tested if PIm lesions in neonatal macaques also lead to similar reach-grasp deficits, and more generally if ventral/inferior pulvinar lesions lead to visually guided manual deficits in adult monkeys.

OPTIC ATAXIA WITH THALAMIC LESIONS IN HUMANS AND MONKEYS

Optic ataxia (OA) is a higher-order type of reach-grasp deficit that consists of misreaching into the visual periphery, difficulties in preshaping the hand before object grasp (*apraxia*), and the inability to correct the hand online (e.g., when the object is moving) (Andersen et al., 2014). The defining feature of OA is the difficulty in executing reach-grasp actions to objects in the peripheral space (peripheral reach, i.e., when the gaze position is dissociated from the reach goal). In contrast, performance during foveal reach, when the patient is allowed to look at the object before reaching to it, remains largely unimpaired (Borchers et al., 2013). OA has been associated with unilateral and bilateral lesions in the superior parietal lobule, intraparietal sulcus, and the junction between inferior parietal lobule and superior occipital cortex (Perenin and Vighetto, 1988; Karnath and Perenin, 2005).

Thalamic lesions have rarely been associated with OA, although there is a strong prediction that dorsal pulvinar lesions would lead to OA based on parietal connectivity. Since the literature on visually guided reach-grasp behavior in thalamic patients is generally sparse, the dissociation between foveal and peripheral conditions was not tested in a large sample. Some single-case studies of thalamic patients are available, however. For instance, OA signified by misreaches toward contralesional, peripheral space was reported in a patient with left-sided lesions entailing midbrain, cerebellum, and thalamus (Frassinetti et al., 2007). In the case of a bilateral dorsal pulvinar lesion patient, the reach deficits differed from OA, as reach errors were not alleviated in the foveal condition (i.e. when the patient was allowed to look at the target; Wilke et al., 2018). Previous lesion/inactivation studies in macaque and marmoset monkeys reported "optic ataxia-like" symptoms following pulvinar lesions (Wilke et al., 2010; Mundinano et al., 2018). It is important to note, however, that these studies did not conduct the critical comparison between performance in foveal (gaze at the object) versus peripheral reach (gaze away from the object).

REMOTE EFFECTS OF THALAMIC/PULVINAR LESIONS

How do pulvinar lesions lead to spatial and reach-grasp deficits? Generally, there are several, non-mutually-exclusive mechanisms that could explain the symptoms. One possibility is the direct disruption of the functionally pertinent thalamic nucleus, while several other mechanisms are subsumed under the term *diaschisis* (Feeney and Baron, 1986; Carrera and Tononi, 2014; Sebastian et al., 2014). Diaschisis refers to effects of acute or chronic lesions that are measured as perfusion, metabolic, blood flow, and/or neural dysfunction at sites remote from the primary lesion. Diaschisis can be "focal" (i.e., consist of a remote general or task-specific reduction of metabolism or neural excitability) or be "non-focal" (i.e., affect functional coupling in distant neural networks due to the loss of afferent information with direct connections to the lesioned site; "connectional/connectomal diaschisis") (Carrera and Tononi, 2014).

REMOTE LESION EFFECTS IN HUMAN PATIENTS

From a clinical point of view, it is important to understand whether thalamic lesion-induced behavioral symptoms are at least partly related to cortical hypometabolism/interhemispheric asymmetry, as this could be used to develop pharmacological or stimulation approaches to promote functional recovery. Generally, stroke-induced lesions are determined by the vascular territory and typically affect multiple nuclei (Schmahmann, 2003). Earlier positron emission tomography (PET) and single-photon emission CT (SPECT) studies in patients with thalamic lesions reported a diffuse hypometabolism in the lesioned hemisphere that recovered over time and was in some studies linked to neuropsychological scores (Baron et al., 1986). An fMRI study that investigated somatosensory perception in patients with stroke-induced lesions involving VP and (anterior) pulvinar in 4 patients reported stimulus-dependent effects: lesions did not eliminate somatosensory responses when the affected (contralesional) hand was stimulated, but responses were drastically diminished when both hands were stimulated at the same time ("tactile extinction") (Staines et al., 2002). Post-stroke recovery of somatosensory detection thresholds with bilateral stimulation corresponded to fMRI blood-oxygenation-level-dependent (BOLD) activity increases in the somatosensory cortex in the stroke-affected hemisphere. Hwang et al. presented evidence for the occurrence of "connectomal diaschisis" following (chronic) stroke-induced thalamic lesions that also involved the pulvinar in 2 out of 4 patients (Hwang et al., 2017). They found increased (overall) cortical between-network connectivity in the thalamic lesion patients, which they interpreted as evidence for a (compensatory?) disinhibition of cortico-cortical connectivity (see **Chapter 47 in Section VIII**). Those studies are generally consistent with the idea that pulvinar lesions lead to spatial and reach-grasp deficits by reducing synaptic activity within and between sensorimotor cortices. An important next step is the direct experimental confirmation of these hypotheses in larger thalamic (pulvinar) patient cohorts during rest and task conditions.

REMOTE LESION EFFECTS IN MONKEYS

Several studies investigated the effects of pulvinar lesions on cortical activity in monkeys. Reversible inactivation of the pulvinar in prosimian primates abolished visual responses in supragranular layers of the primary visual cortex (V1) (Purushothaman et al., 2012). Inactivation of the *ventral* lateral pulvinar in macaques was shown to lead to pre-cued target detection deficits in the contralesional hemifield, reduction of visual responses, reduction of spike-local field potential (LFP) gamma coherence, and reduction of attentional modulation of spiking activity and coherence in V4 (Zhou et al., 2016). In line with these behavioral and neuronal deficits, an *increase* in low-frequency cortical oscillations (<20 Hz), often associated with low alertness or sleep, has been observed (Zhou et al., 2016). In contrast, inactivation of the *dorsal* lateral pulvinar during passive viewing (no active task) led to a *reduction* of LFP coherence between LIP and V4 in low frequencies (i.e., 4–15 Hz), without significant changes in firing rates or LFP power in either area (Eradath et al., 2021). Low frequency synchronization between pulvinar spikes and cortical LFP phase also decreased, while the synchronization increased in the opposite direction—between LIP spikes and pulvinar LFP (Eradath et al., 2021).

An imaging study that combined unilateral reversible medial pulvinar inactivation with event-related fMRI in monkeys performing memory saccade tasks toward instructed or chosen spatial locations found BOLD activity reduction in a wide range of task-related regions, including frontoparietal, superior temporal, and cingulate cortices in both hemispheres (Figure 71.4A). These BOLD activity changes were task-dependent, with reductions being strongest for the contralesional ("impaired") hemifield in instructed trials. Inactivation effects were mainly of the same sign in both hemispheres, suggesting that pulvinar modulates the balance between spatial representations rather than between hemispheres (Klink et al., 2021), similarly to parietal inactivation (Wilke et al., 2012). Ongoing

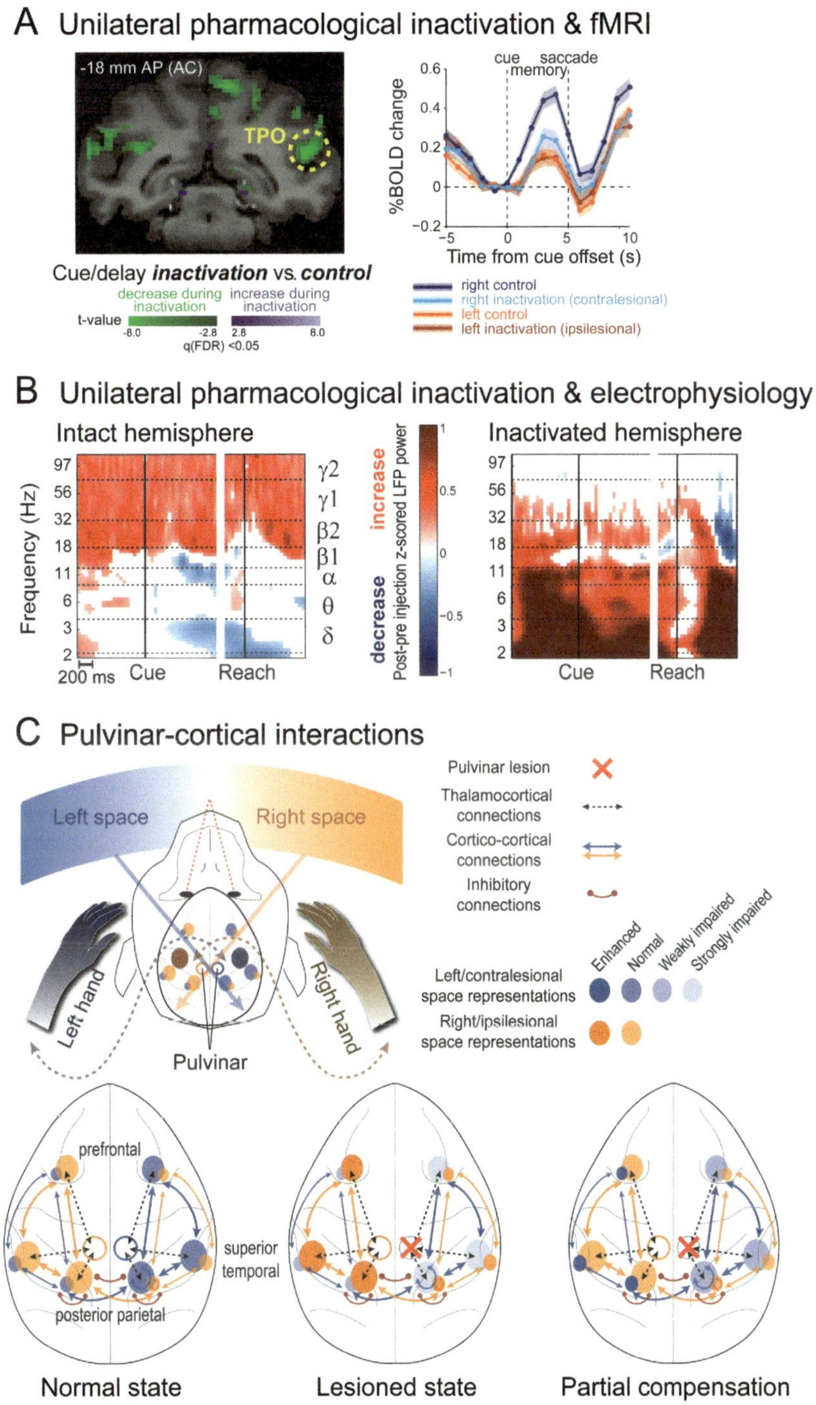

Figure 71.4. Remote effects of unilateral pulvinar perturbation on cortical activity in both hemispheres, and a conceptual model. (A). Pulvinar inactivation-induced a decrease of fMRI contralesional cue/delay activity in both hemispheres. Monkeys performed a memory saccade task. An example coronal section and the % BOLD signal timecourse derived from the outlined region of interest focusing on the temporal-parietal-occipital area TPO are shown (Klink et al., 2021). (B). Effect of dorsal pulvinar inactivation on the parietal reach-related area MIP during dissociated a delayed reach task: time-frequency plots showing a difference between post-injection and pre-injection blocks of local field potential (LFP) normalized power across sites, aligned to visual cue onset and reach onset (42 sites in the intact and 43 sites in the inactivated hemispheres). Only time-frequency bins with a significant difference are shown (t-test with False Discovery Rate [FDR] corrected critical *p*-value <0.05). (C). A conceptual model of hemifield- (and/or hand-) specific pulvinar-cortical and cortico-cortical interactions, before and after unilateral pulvinar lesion. In the normal state, the neural representations are mainly contralateral (with a weaker ipsilateral representation in each hemisphere) and balanced between hemifields. In the scheme, inhibitory intra- and inter-hemispheric connections (red) between representations of contralateral and ipsilateral space are shown only for the "posterior parietal" regions. After a unilateral lesion, mainly contralesional hemifield (blue) representations are suppressed, predominantly in the lesioned hemisphere, but also in the opposite hemisphere by means of interhemispheric connectivity. This imbalance is further amplified by the overactivation of ipsilesional hemifield representations due to disinhibition from contralesional hemifield representations. But it is also conceivable that due to rapid or long-term reorganization, the contralesional hemifield representations in the intact hemisphere are potentiated or recovered, leading to a partial compensation of the deficit.

work investigating the neuronal consequences of unilateral dorsal pulvinar inactivation in parietal areas LIP and MIP also shows a shift to more ipsilesional space (and hand) tuning of spiking activity. In addition, we observed inter-hemispheric push-pull effects (e.g., an increase in the low frequency LFP power in the inactivated hemisphere, and conversely a strong increase of high frequency power in the intact hemisphere) (Figure 71.4B). Overall, these results demonstrate that the dorsal pulvinar influences space and hand-specific tuning in parietal cortex, and helps maintaining an "alert cortical state". Additionally, the findings indicate that the removal of unilateral pulvinar drive leads to activity up-regulation in the opposite hemisphere, which might reflect maladaptive or compensatory processes (Figure 71.4C). Given diverse effects of pulvinar inactivation on the cortical electrophysiological and fMRI activity, future research is necessary to elucidate the causal mechanisms of task-dependent pulvinar-cortical interactions.

OPEN QUESTIONS AND OUTLOOK

The pulvinar studies reviewed in this chapter strongly suggest that pulvinar function goes well beyond visual attention. Lesions in several pulvinar subnuclei lead to reach-grasp and postural symptoms that point to a profound role in sensorimotor functions and primate-specific adaptations (Wilke et al., 2010; Wilke et al., 2017; Wilke et al., 2018; Mundinano et al., 2018; Homman-Ludiye and Bourne, 2019). Future combined electrophysiological and lesion studies should investigate the encoding of sensorimotor variables in the pulvinar and assess the causal influence of different subnuclei on activity in sensorimotor cortices. For example, does the pulvinar coordinate the relative timing of eye and reach signals in effector-specific frontoparietal cortices? Does the pulvinar contribute to motor learning in dynamic sensorimotor or flexible decisional environments similar to the medial thalamus? (See also **Chapter 46 in Section VIII; Chapter 59 in Section X.**)

For unilateral stroke-induced movement disorders and particularly for spatial neglect symptoms, the maladaptive versus compensatory role of regions in the intact hemisphere also remains an issue (Corbetta and Shulman, 2011). Systematic imaging and electrophysiological recordings in both hemispheres, in different task contexts (e.g., spatial competition vs. coordinated bimanual actions) should be conducted to study this fundamental question further (Figure 71.4C). Furthermore, since the main input to higher-order thalamic nuclei such as pulvinar originates from the cortex, future lesion-behavior-physiological/imaging studies should systematically test the impact of cortical lesions on pulvinar activity.

Finally, there is a growing body of literature investigating thalamocortical interactions in other species (e.g., cats and especially rodents) where advanced cell type- and pathway-specific manipulations such as optogenetics for circuit dissection and identification of driving and modulatory influences are readily available (Kirchgessner et al., 2021). For instance, rodent studies also show that the medial part of the lateral posterior nucleus, the presumed homolog of the primate pulvinar, is connected to amygdala and frontoparietal cortex and integrates multimodal sensorimotor information (Leow et al., 2022) (see also **Chapter 2 in Section II; Chapter 57 in Section X**). This provides an evolutionary perspective on the role of the pulvinar in coordination of perception and body movements. An important challenge will be to link between findings in the different species, with a clear delineation of homologies and distinct subnuclei, including a consistent nomenclature.

REFERENCES

Acuna C, Cudeiro J, Gonzalez F, Alonso JM, Perez R (1990). Lateral-posterior and pulvinar reaching cells: Comparison with parietal area 5a: A study in behaving *Macaca nemestrina* monkeys. *Exp Brain Res.* 82:158–166.

Andersen RA, Andersen KN, Hwang EJ, Hauschild M (2014). Optic ataxia: From Balint's syndrome to the parietal reach region. *Neuron.* 81:967–983.

Arcaro MJ, Pinsk MA, Chen J, Kastner S (2018). Organizing principles of pulvino-cortical functional coupling in humans. *Nat Commun.* 9:5382.

Arend I, Machado L, Ward R, McGrath M, Ro T, Rafal RD (2008). The role of the human pulvinar in visual attention and action: Evidence from temporal-order judgment, saccade decision, and antisaccade tasks. *Prog Brain Res.* 171:475–483.

Baldwin MKL, Bourne JA (2017). The evolution of subcortical pathways to the extrastriate cortex. In: Kaas JH, ed. *Evolution of Nervous Systems.* Elsevier:165–185. Accessed March 7, 2018. http://linkinghub.elsevier.com/retrieve/pii/B9780128040423000816.

Baron JC, D'Antona R, Serdaru M, Pantano P, Bousser MG, Samson Y (1986). [Cortical hypometabolism after a thalamic lesion in man: positron tomography study]. *Rev Neurol Paris.* 142:465–474.

Benarroch EE (2015). Pulvinar: Associative role in cortical function and clinical correlations. *Neurology.* 84:738–747.

Bender DB, Youakim M (2001). Effect of attentive fixation in macaque thalamus and cortex. *J Neurophysiol.* 85:219–234.

Benevento LA, Port JD (1995). Single neurons with both form/color differential responses and saccade-related responses in the nonretinotopic pulvinar of the behaving macaque monkey. *Vis Neurosci.* 12:523–544.

Berman RA, Wurtz RH (2011). Signals conveyed in the pulvinar pathway from superior colliculus to cortical area MT. *J Neurosci.* 31:373–384.

Borchers S, Synofzik M, Kiely E, Himmelbach M (2013). Routine clinical testing underestimates proprioceptive deficits in Friedreich's ataxia. *Cerebellum.* 12:916–922.

Bridge H, Leopold DA, Bourne JA (2016). Adaptive pulvinar circuitry supports visual cognition. *Trends Cogn Sci.* 20:146–157.

Carrera E, Tononi G (2014). Diaschisis: Past, present, future. *Brain.* 137:2408–2422.

Castiello U (2005). The neuroscience of grasping. *Nat Rev Neurosci.* 6:726–736.

Corbetta M, Shulman GL (2011) Spatial neglect and attention networks. *Annu Rev Neurosci.* 34:569–599.

Cudeiro J, Gonzalez F, Perez R, Alonso JM, Acuna C (1989). Does the pulvinar-LP complex contribute to motor programming? *Brain Res.* 484:367–370.

de Souza BOF, Cortes N, Casanova C (2020). Pulvinar modulates contrast responses in the visual cortex as a function of cortical hierarchy. *Cereb Cortex.* 30:1068–1086.

Dominguez-Vargas A-U, Schneider L, Wilke M, Kagan I (2017). Electrical microstimulation of the pulvinar biases saccade choices and reaction times in a time-dependent manner. *J Neurosci.* 37:2234–2257.

Eradath MK, Pinsk MA, Kastner S (2021). A causal role for the pulvinar in coordinating task-independent cortico-cortical interactions. *J Comp Neurol.* 529:3772–3784.

Esposito E, Shekhtman G, Chen P (2021). Prevalence of spatial neglect post-stroke: A systematic review. *Ann Phys Rehabil Med.* 64:101459.

Feeney DM, Baron JC (1986). Diaschisis. *Stroke.* 17:817–830.

Frassinetti F, Bonifazi S, Làdavas E (2007). The influence of spatial coordinates in a case of an optic ataxia-like syndrome following cerebellar and thalamic lesion. *Cogn Neuropsychol.* 24:324–337.

Froesel M, Cappe C, Ben Hamed S (2021). A multisensory perspective onto primate pulvinar functions. *Neurosci Biobehav Rev.* 125:231–243.

Grieve KL, Acuna C, Cudeiro J (2000). The primate pulvinar nuclei: vision and action. *Trends Neurosci.* 23:35–39.

Gupta N, Pandey S (2018). Post-thalamic stroke movement disorders: A systematic review. *Eur Neurol.* 79:303–314.

Gutierrez C, Cola MG, Seltzer B, Cusick C (2000). Neurochemical and connectional organization of the dorsal pulvinar complex in monkeys. *J Comp Neurol.* 419:61–86.

Halassa MM, Sherman SM (2019). Thalamocortical circuit motifs: A general framework. *Neuron.* 103:762–770.

Homman-Ludiye J, Bourne JA (2019). The medial pulvinar: Function, origin and association with neurodevelopmental disorders. *J Anat.* 235:507–520.

Hwang K, Bertolero MA, Liu WB, D'Esposito M (2017). The human thalamus is an integrative hub for functional brain networks. *J Neurosci.* 37:5594–5607.

Kaas JH, Lyon DC (2007). Pulvinar contributions to the dorsal and ventral streams of visual processing in primates. *Brain Res Rev.* 55:285–296.

Kaduk K, Wilke M, Kagan I (2022). Dorsal pulvinar inactivation leads to spatial selection bias without perceptual deficit. *bioRxiv*:2022.12.30.520322. Accessed January 6, 2023. https://www.biorxiv.org/content/10.1101/2022.12.30.520322v1.

Kagan I, Gibson L, Spanou E, Wilke M (2021). Effective connectivity and spatial selectivity-dependent fMRI changes elicited by microstimulation of pulvinar and LIP. *NeuroImage.* 240:118283.

Karnath HO (2015). Spatial attention systems in spatial neglect. *Neuropsychologia.* 75:61–73.

Karnath HO, Himmelbach M, Rorden C (2002). The subcortical anatomy of human spatial neglect: putamen, caudate nucleus and pulvinar. *Brain.* 125:350–360.

Karnath H-O, Perenin M-T (2005). Cortical control of visually guided reaching: Evidence from patients with optic ataxia. *Cereb Cortex.* 15:1561–1569.

Kastner S, Fiebelkorn IC, Eradath MK (2020). Dynamic pulvino-cortical interactions in the primate attention network. *Curr Opin Neurobiol.* 65:10–19.

Kirchgessner MA, Franklin AD, Callaway EM (2021). Distinct "driving" versus "modulatory" influences of different visual corticothalamic pathways. *Curr Biol.* 31:5121–5137.

Klink PC, Aubry JF, Ferrera VP, Fox AS, Froudist-Walsh S, Jarraya B, Konofagou EE, Krauzlis RJ, Messinger A, Mitchell AS, Ortiz-Rios M, Oya H, Roberts AC, Roe AW, Rushworth MFS, Sallet J, Schmid MC, Schroeder CE, Tasserie J, Tsao DY, Uhrig L, Vanduffel W, Wilke M, Kagan I, Petkov CI (2021). Combining brain perturbation and neuroimaging in non-human primates. *NeuroImage.* 235:118017.

Komura Y, Nikkuni A, Hirashima N, Uetake T, Miyamoto A (2013). Responses of pulvinar neurons reflect a subject's confidence in visual categorization. *Nat Neurosci.* 16:749–755.

Leow YN, Zhou B, Sullivan HA, Barlowe AR, Wickersham IR, Sur M (2022). Brain-wide mapping of inputs to the mouse lateral posterior (LP/Pulvinar) thalamus–anterior cingulate cortex network. *J Comp Neurol.* 530:1992–2013.

Mundinano I-C, Fox DM, Kwan WC, Vidaurre D, Teo L, Homman-Ludiye J, Goodale MA, Leopold DA, Bourne JA (2018). Transient visual pathway critical for normal development of primate grasping behavior. *Proc Natl Acad Sci.* 115:1364–1369.

Perenin MT, Vighetto A (1988). Optic ataxia: A specific disruption in visuomotor mechanisms. I. Different aspects of the deficit in reaching for objects. *Brain.* 111 (Pt 3):643–674.

Petersen SE, Robinson DL, Keys W (1985). Pulvinar nuclei of the behaving rhesus monkey: Visual responses and their modulation. *J Neurophysiol.* 54:867–886.

Petersen SE, Robinson DL, Morris JD (1987). Contributions of the pulvinar to visual spatial attention. *Neuropsychologia.* 25:97–105.

Preuss TM (2007). Evolutionary specializations of primate brain systems. In: Ravosa MJ, Dagosto M, eds. *Primate Origins: Adaptations and Evolution.* Springer:625–675.

Prevosto V, Sommer MA (2013). Cognitive control of movement via the cerebellar-recipient thalamus. *Front Syst Neurosci.* 7:56.

Purushothaman G, Marion R, Li K, Casagrande VA (2012). Gating and control of primary visual cortex by pulvinar. *Nat Neurosci.* 15:905–912.

Rafal RD, Posner MI (1987). Deficits in human visual spatial attention following thalamic lesions. *Proc Natl Acad Sci U S A.* 84:7349–7353.

Robinson DL, McClurkin JW, Kertzman C (1990). Orbital position and eye movement influences on visual responses in the pulvinar nuclei of the behaving macaque. *Exp Brain Res.* 82:235–246.

Robinson DL, Petersen SE, Keys W (1986). Saccade-related and visual activities in the pulvinar nuclei of the behaving rhesus monkey. *Exp Brain Res.* 62:625–634.

Saalmann YB, Pinsk MA, Wang L, Li X, Kastner S (2012). The pulvinar regulates information transmission between cortical areas based on attention demands. *Science.* 337:753–756.

Schmahmann JD (2003). Vascular syndromes of the thalamus. *Stroke.* 34:2264–2278.

Schneider L, Dominguez-Vargas A-U, Gibson L, Kagan I, Wilke M (2020). Eye position signals in the dorsal pulvinar during fixation and goal-directed saccades. *J Neurophysiol.* 123:367–391.

Schneider L, Dominguez- Vargas A- U, Gibson L, Wilke M, Kagan I (2023). Visual, delay and oculomotor timing and tuning in macaque dorsal pulvinar during instructed and free choice memory saccades. *Cereb Cortex.* bhad333.

Sebastian R, Schein MG, Davis C, Gomez Y, Newhart M, Oishi K, Hillis AE (2014). Aphasia or neglect after thalamic stroke: The various ways they may be related to cortical hypoperfusion. *Front Neurol.* 5:231.

Shipp S (2003). The functional logic of cortico-pulvinar connections. *Philos Trans R Soc Lond B Biol Sci.* 358:1605–1624.

Staines WR, Black SE, Graham SJ, McIlroy WE (2002). Somatosensory gating and recovery from stroke involving the thalamus. *Stroke.* 33:2642–2651.

Walker RH (2015). Jerky dystonic shoulder following infarction of the posterior thalamus. *J Clin Mov Disord.* 2:12.

Ward R, Arend I (2007). An object-based frame of reference within the human pulvinar. *Brain*. 130:2462–2469.

Ward R, Danziger S, Owen V, Rafal R (2002). Deficits in spatial coding and feature binding following damage to spatiotopic maps in the human pulvinar. *Nat Neurosci*. 5:99–100.

Wardak C, Olivier E, Duhamel JR (2002). Neglect in monkeys: Effect of permanent and reversible lesions. In: Karnath HO, Milner D, Vallar G, eds. *The Cognitive and Neural Bases of Spatial Neglect* (pp. 46–58). Oxford University Press.

Watson RT, Heilman KM (1979). Thalamic neglect. *Neurology*. 29:690–694.

Wilke M, Dechent P, Bähr M (2017). Sarcoidosis manifestion centered on the thalamic pulvinar leading to persistent astasia. *Mov Disord Clin Pract*. 4:898–900.

Wilke M, Kagan I, Andersen RA (2012). Functional imaging reveals rapid reorganization of cortical activity after parietal inactivation in monkeys. *Proc Natl Acad Sci U S A*. 109: 8274–8279.

Wilke M, Kagan I, Andersen RA (2013). Effects of pulvinar inactivation on spatial decision-making between equal and asymmetric reward options. *J Cogn Neurosci*. 25:1270–1283.

Wilke M, Mueller KM, Leopold DA (2009). Neural activity in the visual thalamus reflects perceptual suppression. *Proc Natl Acad Sci U S A*. 106:9465–9470.

Wilke M, Schneider L, Dominguez-Vargas A-U, Schmidt-Samoa C, Miloserdov K, Nazzal A, Dechent P, Cabral-Calderin Y, Scherberger H, Kagan I, Bähr M (2018). Reach and grasp deficits following damage to the dorsal pulvinar. *Cortex*. 99:135–149.

Wilke M, Turchi J, Smith K, Mishkin M, Leopold DA (2010). Pulvinar inactivation disrupts selection of movement plans. *J Neurosci*. 30:8650–8659.

Yeterian EH, Pandya DN (1989). Thalamic connections of the cortex of the superior temporal sulcus in the rhesus monkey. *J Comp Neurol*. 282:80–97.

Zhou H, Schafer RJ, Desimone R (2016). Pulvinar-cortex interactions in vision and attention. *Neuron*. 89:209–220.

72.

CONCLUDING THOUGHTS

W. Martin Usrey and S. Murray Sherman

A major purpose of this volume is to emphasize the point that attempts to study the functional organization of the cortex or thalamus in isolation represents a severely limited approach. This is because the cortex and thalamus are so intimately interrelated that any full understanding of either must take into account their functional relationships. Thus, this account has focused on these relationships.

To put this emphasis on interrelationships in perspective, it was not long ago that the view of thalamocortical interactions could be summarized by Figure 72.1A, which dominated thinking in the field and may still be seen as the standard textbook view of thalamocortical relationships. Textbook examples of this view include figure 4.51B of Luo (2020); figure 28-2 of Kandel et al. (2000); and figures 25.12 and 27.15 of Squire et al. (2008). Here, peripheral information is relayed to the cortex by certain thalamic nuclei (think of retinal input relayed by the LGN to visual cortex). That information is then processed via ascending hierarchical levels, such as higher visual areas, then sensorimotor, and finally motor areas until an executive level is reached, from which a motor command emanates. The only role of the thalamus is to relay peripheral information to the cortex, and all relevant additional processing is done strictly within the cortex, with no further role for the thalamus.

Figure 72.1B shows our current understanding, emphasized in earlier chapters in Sections IV and VII of this volume, and this transforms our appreciation of the key role played by the thalamus in ongoing cortical functioning (Sherman and Guillery, 2013; Usrey and Sherman, 2021). Most of thalamus by volume is higher order (HO), which is completely ignored in conventional accounts such as Figure 72.1A, and HO thalamus serves as a central element in transthalamic pathways. Available evidence indicates that these pathways are organized in parallel with direct corticocortical pathways and appear to play a crucial role in ongoing cortical functioning and corticocortical communication (Sherman and Guillery, 2013; Usrey and Sherman, 2021).

In this final chapter, rather than simply summarizing the content of preceding accounts, we focus on issues of thalamocortical relationships and functioning that we believe are of particular importance to the topic.

GENERAL PROPERTIES OF THALAMOCORTICAL RELATIONSHIPS

Several cellular, synaptic, and circuit properties appear ubiquitous across thalamic nuclei and thalamocortical projections. Among these are (1) driver and modulator glutamatergic inputs, (2) feedforward and feedback circuits, and (3) burst and tonic firing modes.

DRIVER AND MODULATOR GLUTAMATERGIC INPUTS

Glutamatergic inputs to neurons in the thalamus and cortex can be classified into driver and modulator types based on a series of criterion parameters (reviewed in Sections II and III) (Sherman and Guillery, 1998; Sherman and Guillery, 2013; Sherman, 2016; Usrey and Sherman, 2021). In particular, driver inputs have larger synaptic profiles, higher probabilities of neurotransmitter release, less convergence onto their target neurons, and larger initial excitatory postsynaptic potentials (EPSPs) compared with modulator inputs. Moreover, driver inputs seem much fewer in number compared to modulator afferents. Finally, driver inputs activate only ionotropic receptors, mainly AMPA and NMDA, whereas modulator inputs activate ionotropic and metabotropic receptors. The activation of metabotropic receptors by modulators is a key distinction from drivers, as metabotropic activation requires modulator inputs to fire at moderately high rates. Given these differences between driver and modulator glutamatergic inputs in the thalamus and cortex, it is tempting to speculate that drivers carry the main information to be communicated, and modulators

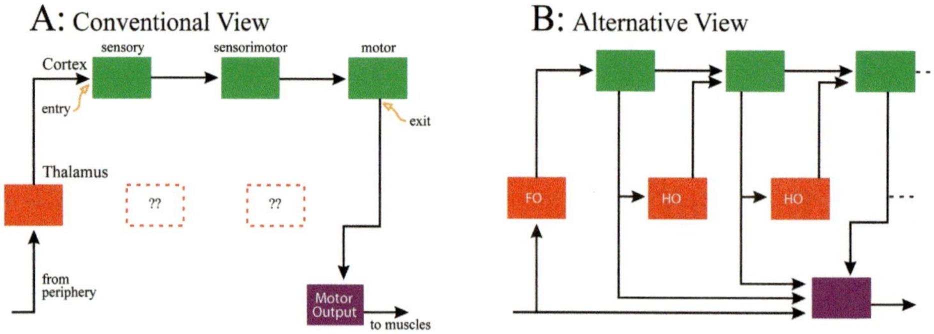

Figure 72.1. Different views of thalamocortical function. (A) Conventional view. Information initially reaches cortex via a thalamic relay, and it is then processed up a hierarchy of cortical areas, passing through various sensory, sensorimotor, and motor levels, finally reaching an executive level from which a motor command is initiated. Note that this involves a single entry and exit point for cortical processing. This view indicates no function for most of thalamus, indicated here by question marks (??). (B) Alternative view. Here, every cortical area has a projection from layer 5 that innervates motor subcortical regions, and, via branching axons, these also innervate the thalamic nuclei that are indicated by question marks in (A). The thalamic innervation initiates transthalamic cortico-cortical communication. Abbreviations: *FO*, first order; *HO*, higher order. Redrawn from (Sherman and Usrey, 2021).

affect how driver inputs are processed. From this perspective, the driver and modulator classification has proven useful for classifying thalamic nuclei into first order and higher order, with first-order nuclei receiving their driving input from pathways originating subcortically and higher-order nuclei receiving their driving input from the cortex, specifically from layer (L)5 corticothalamic neurons.

The need for driver inputs may seem self-evident given their proposed role in information transfer. However, given the presence of classical modulator inputs in the thalamus and cortex (e.g., ACh, NA, DA, etc.), one might wonder why glutamatergic modulator inputs are also needed. We suggest 2 reasons for this (Sherman, 2022). First, classical modulatory systems lack much topography in their projections, thus appearing to affect much of the neuraxis when active for such features as overall behavioral state, such as alertness, sleep, and so on. Only glutamatergic modulatory pathways possess significant topography, and such topographic modulation is needed for processes that require localized effects, such as focal or covert attention, adaptation, learning and memory, and so forth. Second, classic modulatory systems originate subcortically and thus do not have the benefit of thalamocortical processing, which is the source of most of the glutamatergic inputs discussed here, and such processing would obviously be important for modulation related to higher cognitive functioning.

FEEDFORWARD AND FEEDBACK CIRCUITS

Early studies of the neural pathways interconnecting the thalamus and cortex had a rather simple view of the hierarchical relationship between the 2 structures; projections from the thalamus to the cortex were classified as *feedforward*, and projections from the cortex to the thalamus were classified as *feedback*. With a better understanding of driver and modulator inputs, the classification of feedforward and feedback circuits requires some revision (see Section IV), depending on the type of thalamic nucleus involved. For first-order thalamic nuclei, including the LGN, VPm, and MGNv, the traditional view has not changed, with projections from the thalamus to the cortex being feedforward, and projections from cortex to first-order thalamus being feedback. Importantly, these feedback projections originate strictly from neurons in cortical L6, and they have modulator properties. For higher-order thalamus, including Pul, POm, and MGNd, there is also a feedforward projection to the cortex; however, higher-order thalamus receives 2 types of corticothalamic input, a feedback input from cortical L6 with modulator properties, and a feedforward input from cortical L5 with driver properties. By receiving feedforward drive from one cortical area and subsequently sending it to another cortical area, higher-order thalamus serves a unique role in cortico-thalamo-cortical (or trans-thalamic) communication. Although we lack a firm understanding of the functional significance of transthalamic communication, evidence indicates that it plays a role in general cortical processing of information as well as attention and cognitive control. (See **Chapter 42 in Section VII** for further details of transthalamic processing.) Moreover, various disorders and diseases may be associated with disruption of higher-order thalamus and transthalamic communication.

BURST AND TONIC FIRING MODES

Thalamic neurons have 2 distinct response modes for communicating with cortex: burst mode and tonic mode (see **Section I and Chapter 62 in Section XI**). Burst mode occurs when thalamic neurons have been sufficiently hyperpolarized to deinactivate the T-type ("T" for "transient") Ca^{2+} channels in the soma and dendritic membranes (Jahnsen and Llinás, 1984a, 1984b; Huguenard and McCormick, 1994; Zhan et al., 1999; Sherman, 2001; Llinás and Steriade, 2006), and tonic mode occurs when the T-type channels are in the inactivated state. (The mechanistic details of T-type channel states are described in greater depth in Chapter 62 in Section XI). Briefly, when the Ca^{2+} channels are deinactivated by sufficient hyperpolarization, an ensuing sufficient depolarization opens them, producing Ca^{2+} influx and a depolarization often known as a low threshold or Ca^{2+} spike that brings the neuron's membrane potential toward the threshold for activating voltage-gated Na^{+} channels (Jahnsen and Llinás, 1984a, 1984b; Suzuki and Rogawski, 1989). Because Ca^{2+} spikes last longer than the duration of an individual Na^{+} spike, a burst of Na^{+} spikes can ride on the Ca^{2+} spike (Wang et al., 1991; Huguenard and McCormick, 1992; McCormick and Huguenard, 1992; Huguenard, 1996; Perez-Reyes, 2003). Consistent with these properties, studies examining response mode in the LGN have shown that burst frequency is increased with natural scene stimuli (Lesica and Stanley, 2004), presumably due to the sustained inhibitory stimuli (which deinactivates the Ca^{2+} channels), followed by abrupt excitatory transitions (that activates them) that are common in natural scenes. More work is needed to understand the factors that underlie burst/tonic transitions in other thalamic nuclei, but given the greater efficacy of bursts in evoking cortical responses (Swadlow and Gusev, 2001), this knowledge is important for understanding the dynamics of thalamocortical communication.

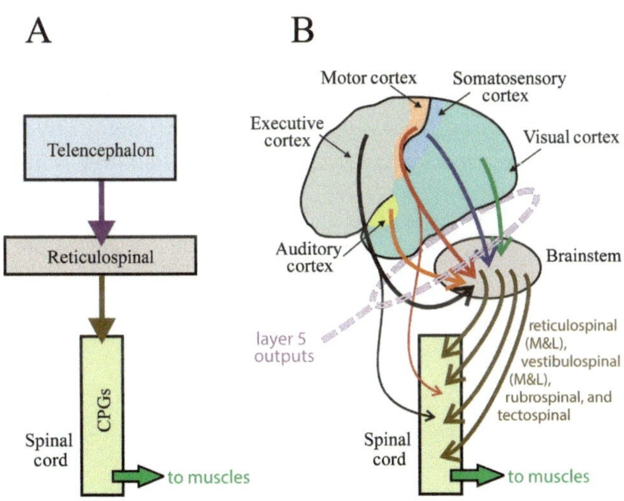

Figure 72.2. View of two evolutionary stages of the vertebrate central nervous system. (A) Lamprey central nervous system. Because the lamprey is the most primitive living vertebrate, it is likely representative of early vertebrate species. Here, the central nervous system is dominated largely by limited supraspinal control of central pattern generators (CPGs; for swimming behavior, etc.). Redrawn from (Grillner, 2003; Suryanarayana et al., 2017). (B) Mammalian central nervous system, showing human example. Evolution has led to cortical control of lower motor centers. Importantly, cortex must operate via layer 5 projections to centers evolved in (A) with additional subcortical brainstem control centers also appearing. Many of the circuits that evolved in (A) remain functional in (B). Redrawn from (Sherman and Usrey, 2021).

AN EVOLUTIONARY PERSPECTIVE ON THALAMOCORTICAL RELATIONSHIPS

It is important to understand that the thalamus and cortex and their interactions are the result of an extensive period of evolution. This point is emphasized throughout Section XII.

One outcome to understand about brain evolution is that it is a very messy process. Figure 72.2 helps to illustrate this point. We can imagine that the earliest vertebrates to evolve had little more than spinal circuitry to control movements, and simple behaviors, such as the rhythmic activity needed for swimming, could be accomplished by sensory activation of central pattern generators (CPGs) in the spinal cord. Next in evolution, we imagine, was the establishment of supraspinal mechanisms to control dynamically the collection of CPGs activated at any time. An example of such an early vertebrate is the lamprey, whose brain is schematically shown in Figure 72.2A.

Evolution then proceeded for eons, resulting in the mammalian brain, including the appearance of the thalamus and cortex, as shown schematically in Figure 72.2B. A comparison of the 2 panels of Figure 72.2 reveals several important results of this evolutionary process:

- When evolution produces a useful brain circuit, it tends to keep it, even after newly evolved circuits are added to improve responses to a dynamic environment. Little gets discarded. The result is a messy, complex assemblage of parts that must work together to control behavior. Thus most of the basic elements seen in the lamprey central nervous system (CNS) remain in that of the mammal, including spinal circuitry with its CPGs, and early evolved supraspinal control pathways. Evolution adds to this newer brainstem control circuitry and, eventually, thalamus and cortex. Note that this is in contrast to how an intelligent designer might construct things: keeping it neat by replacing old circuits with new ones.
- The evolution of cortex to provide the most flexible and effective behavior in response to a dynamic environment did not occur with the evolution of a motor plant to which the cortex has unique access. Instead, cortex must effect behavior by interacting with the older, conserved circuitry in the brainstem and spinal cord. This creates an apparent bottleneck as the cortex interacts and/or competes with older circuits to control behavior; as an example, the tectospinal tract can influence behavior,

and it may be activated by local circuits that evolved well before the cortex and/or by corticofugal projections, and so the cortex must play nice with older evolved neuronal elements. We have written previously how this interaction may be related to attentional mechanisms (Sherman and Usrey, 2021); see also **Chapter 45 in Section VIII**.

- The only feasible route by which the cortex may control behavior is via projections from L5 cells (Sherman and Guillery, 2013; Usrey and Sherman, 2021). The only other source of corticofugal projection comes from L6, but these almost all innervate thalamus only and do not noticeably innervate subcortical motor centers that can have fairly immediate effects on behavior. The L5 projections do provide such innervation of subcortical motor centers.
- All cortical areas so far studied, including primary sensory areas, have such L5 motor outputs (Prasad et al., 2020), and without these, the cortex would be pretty useless as an instrument to control behavior. The fact that primary sensory areas have such motor outputs belies their designation as "sensory." Indeed, an assertion about evolution is that any time a new sensory receptor or peripheral sensory process evolves, it will have no survival value if it lacks a fairly immediate motor output. This point is relevant to Figure 72.1: the assertion argues that Figure 72.1A is an unlikely product of evolution, because there is too much time and processing involved before an environmental event can possibly lead to a response; Figure 72.1B is more in accord with the above assertion.

AN EVOLUTIONARY PERSPECTIVE ON TRANSTHALAMIC PATHWAYS

Particularly relevant to this volume is the fact that these L5 projecting neurons also innervate the thalamus as initiators of transthalamic circuitry (Figure 72.1B). Indeed, individual L5 axons innervate the thalamus via branching axons that also innervate extrathalamic subcortical targets, often motor ones (Sherman and Guillery, 2013; Usrey and Sherman, 2019, 2021; Sherman, 2022). Because transthalamic circuitry begins with such branching axons, this means that the messages sent through the thalamic relay are exact copies of messages sent to the other subcortical targets (Cox et al., 2000; Raastad and Shepherd, 2003).

Given that the cortex can only influence behavior via motor messages sent through these L5 projections, given that these projecting axons generally branch to innervate the thalamus as well, and given that the branching means that the actual message projected is exactly the same along all branches of an axon (Cox et al., 2000; Raastad and Shepherd, 2003), the message sent to the thalamus for transthalamic processing is an exact copy of a motor message. Such a copy is a nice definition of an *efference copy* (also known as "corollary discharge").

Efference copies are ubiquitous features of brains of any animal that actively moves through its environment. A detailed description of efference copies is beyond the scope of this account, and readers are directed elsewhere for this (Crapse and Sommer, 2008; Sommer and Wurtz, 2008; Wolpert and Flanagan, 2010). These copies allow the animal to disambiguate sensory stimuli that are caused by the animal's own movements versus those that reflect actual changes in the environment. The role of efference copies in eye movements is a good example of this. We typically make saccades 3 to 5 times per second, and yet we do not normally perceive the world as spinning about when this occurs. This is because efference copies—copies of the command sent to the extraocular muscles to generate saccades—are set up to anticipate these eye movements and eliminate the sensory consequences of them from our perception.

Logic dictates that motor commands generated by the cortex must have efference copies associated with them. The fact that transthalamic pathways start with L5 projections that appear through branching to provide messages for thalamic relay that are copies of motor messages suggests the possibility that these pathways serve to provide efference copies. For instance, if V1 sends a command that may generate an eye movement to the superior colliculus via an L5 axon, the axon's branch to the pulvinar suggests that the extrastriate visual area targeted is kept apprised of any eye movements generated earlier in the visual hierarchy.

It is important to note that the output of any L5 neuron can be viewed as a singular message. Whereas the branching of its axon means that a copy of this message is sent to all postsynaptic targets, this does not mean that the effect of the message is the same everywhere. There are 2 related reasons for this. One is that synaptic processing may vary greatly at each target site. Perhaps more importantly, this message may have different meanings depending on the postsynaptic circuitry innervated.

A simple analogy may serve to elucidate this latter point. Think of an army commander who receives intelligence (i.e., a message) that the enemy is advancing toward the river and so a key bridge must be destroyed before the enemy can use it. The first part of the message can be regarded as sensory message, and the second, as motor. The commander must deal with both parts of the message. He sends copies of the message to 2 subordinates: one to the officer in charge of observing the progress of the enemy, and the other to the officer concerned with the control of river crossings. Each of the subordinates will read the whole message and react in accordance with their particular responsibilities.

SOME KEY QUESTIONS TO CONSIDER

Whereas the past decade or so has seen remarkable advances in our understanding of the functional relationships between the thalamus and the cortex, there is much that still needs to be addressed. Indeed, this is a typical result of successful research programs: the more one learns, the more new questions arise. We thought it would be useful to end this volume with a listing of such questions, and we emphasize that by no means do we regard this as a complete list. We encourage readers to consider the list and add to (or subtract from) it. Answers to these questions will inform the much larger question: Why do we have a thalamus?

CORTICAL AND THALAMIC CELL PROPERTIES

- What is the functional significance of bursting in the thalamus?
- What is the functional significance of bursting in L5 cells?

THALAMIC CIRCUITRY

- Are there parallel pathways through all thalamic nuclei, as is the case with the LGN?
- Is there a higher-order circuit through TRN to higher-order thalamus?
- What is the functional significance of triads and glomeruli?
- Higher-order thalamus seems to have special GABAergic inputs, such as the zona incerta, pretectum, and basal ganglia, that target exclusively higher-order relays; what is the significance of this?
- What is different in the functional organization of first- versus higher-order thalamus?
- What is the significance of why mice and rats lack interneurons in most of their thalamus?
- Are there distinct roles for local thalamic (interneuronal) inhibition versus inhibition from the TRN?
- What is the functional significance of interneurons having 2 different outputs, 1 from their dendrites and 1 from their axon?
- GABAergic cells in thalamus and cortex tend to also have neuroactive peptides like parvalbumin or somatostatin; what is the significance of that?

THALAMOCORTICAL (FEEDFORWARD)

- How many different unique classes of thalamocortical axons are?
- What is the role of presynaptic modulation of thalamic axon terminals in the cortex?
- Can we establish cell and pathway homologs in the mouse visual system with the W/X/Y cells and pathways in the cat or the parvocellular/magnocellular/koniocellular cells and pathways in the monkey?
- What is the functional significance of the W or koniocellular pathways, and how many classes are subsumed within them?
- Is the koniocellular pathway driver or modulator, or are some koniocellular neurons driver and others modulator?
- Do all retinal ganglion cells that innervate the LGN have branches that innervate the midbrain; that is, what about the parvo cells in the monkey?
- If our magnocellular/parvocellular/koniocellular pathways have very different axon conduction velocities, whereby their signals arrive to the cortex at very different times, how is it that we are able to stitch them together so that it seems as though they are happening simultaneously?

CORTICOTHALAMIC (FEEDBACK)

- What is the significance of branching axons from L5 neurons to the thalamus?
- Do L6 cells with axons that innervate the thalamus have local axons that target the same layers/neurons targeted by their thalamic targets?
- Do L6 axons that target both first-order and higher-order axons differ in their first-order targets compared to L6 axons that only target first-order thalamus?
- How many different unique classes of corticothlamic projection from L5 and L6?
- Do all L5 or L6 corticofugal cells innervate thalamic relay cells, interneurons, and TRN neurons?
- How do the L5 projections from cortical areas differ among cortical areas (e.g., primary sensory vs. motor vs. prefrontal)?

- Does every cortical area have L5 axons that innervate the thalamus and also innervate subcortical motor structures?
- If patchy horizontal connections in the cortex are ubiquitous across all cortical areas, how do L5 corticofugal neurons from the connected patches compare, and do they serve a common function?

CORTICOCORTICAL: DIRECT AND TRANS-THALAMIC

- When cortical cells are connected directly, are they always also connected by transthalamic pathways?
- What is different in the information content of direct corticocortical pathways versus transthalamic ones?
- Why are transthalamic pathways routed through the thalamus instead of sending the information directly?

EVOLUTION

- Is higher-order thalamus evolutionarily older than first-order thalamus?
- With evolution, did multiple streams to multiple cortical areas from the thalamus come about by adding new pathways or dividing existing ones?
- What are the homologous relationships between the thalamus and cortex in mammals and structures in non-mammalian vertebrates?

DISEASE

- Do disruptions of transthalamic processing have correlates with and/or consequences for mental health disorders/disease?

REFERENCES

Cox CL, Denk W, Tank DW, Svoboda K (2000). Action potentials reliably invade axonal arbors of rat neocortical neurons. *Proc Natl Acad Sci U S A*. 97:9724–9728.

Crapse TB, Sommer MA (2008). Corollary discharge across the animal kingdom. *Nat Rev Neurosci.* 9:587–600.

Grillner S (2003). The motor infrastructure: from ion channels to neuronal networks. *Nat Rev Neurosci.* 4:573–586.

Huguenard JR (1996). Low-threshold calcium currents in central nervous system neurons. *Ann Rev Physiol.* 58:329–348.

Huguenard JR, McCormick DA (1992). Simulation of the currents involved in rhythmic oscillations in thalamic relay neurons. *J Neurophysiol.* 68:1373–1383.

Huguenard JR, McCormick DA (1994). *Electrophysiology of the Neuron.* Oxford University Press.

Jahnsen H, Llinás R (1984a). Ionic basis for the electroresponsiveness and oscillatory properties of guinea-pig thalamic neurones in vitro. *J Physiol (Lond).* 349:227–247.

Jahnsen H, Llinás R (1984b). Electrophysiological properties of guinea-pig thalamic neurones: an in vitro study. *J Physiol (Lond).* 349:205–226.

Kandel ER, Schwartz JH, Jessell TM (2000). *Principles of Neural Science.* McGraw Hill.

Lesica NA, Stanley GB (2004). Encoding of natural scene movies by tonic and burst spikes in the lateral geniculate nucleus. *J Neurosci.* 24:10731–10740.

Llinás RR, Steriade M (2006). Bursting of thalamic neurons and states of vigilance. *J Neurophysiol.* 95:3297–3308.

Luo L (2020). *Principles of Neurobiology.* 2nd ed. CRC Press.

McCormick DA, Huguenard JR (1992). A model of the electrophysiological properties of thalamocortical relay neurons. *J Neurophysiol.* 68:1384–1400.

Perez-Reyes E (2003). Molecular physiology of low-voltage-activated t-type calcium channels. *Physiol Rev.* 83:117–161.

Prasad JA, Carroll BJ, Sherman SM (2020). Layer 5 corticofugal projections from diverse cortical areas: Variations on a pattern of thalamic and extrathalamic targets. *J Neurosci.* 40:5785–5796.

Raastad M, Shepherd GM (2003). Single-axon action potentials in the rat hippocampal cortex. *J Physiol.* 548:745–752.

Sherman SM (2001). Tonic and burst firing: Dual modes of thalamocortical relay. *Trends Neurosci.* 24:122–126.

Sherman SM (2016). Thalamus plays a central role in ongoing cortical functioning. *Nat Neurosci.* 19:533–541.

Sherman SM (2022). Thalamocortical circuitry matters. In: Halassa MM, ed. *The Thalamus.* Cambridge University Press:71–90.

Sherman SM, Guillery RW (1998). On the actions that one nerve cell can have on another: Distinguishing "drivers" from "modulators." *Proc Natl Acad Sci U S A.* 95:7121–7126.

Sherman SM, Guillery RW (2013). *Functional Connections of Cortical Areas: A New View from the Thalamus.* MIT Press.

Sherman SM, Usrey WM (2021). Cortical control of behavior and attention from an evolutionary perspective. *Neuron.* 109:3048–3064.

Sommer MA, Wurtz RH (2008). Brain circuits for the internal monitoring of movements. Ann Rev Neurosci. 31:317–338.

Squire LR, Berg D, Bloom FE, du Lac S, Ghosh A, Spitzer NC (2008). *Fundamental Neuroscience.* 3rd ed. Academic Press.

Suryanarayana SM, Robertson B, Wallen P, Grillner S (2017). The lamprey pallium provides a blueprint of the mammalian layered cortex. *Curr Biol.* 27(21):3264–3277.

Suzuki S, Rogawski MA (1989). T-type calcium channels mediate the transition between tonic and phasic firing in thalamic neurons. *Proc Natl Acad Sci U S A.* 86:7228–7232.

Swadlow HA, Gusev AG (2001). The impact of "bursting" thalamic impulses at a neocortical synapse. *Nat Neurosci.* 4:402–408.

Usrey WM, Sherman SM (2019). Corticofugal circuits: Communication lines from the cortex to the rest of the brain. *J Comp Neurol.* 527:640–650.

Usrey WM, Sherman SM (2021). *Exploring Thalamocortical Interactions: Circuitry for Sensation, Action, and Cognition.* Oxford University Press.

Wang X-J, Rinzel J, Rogawski MA (1991). A model of the T-Type calcium current and the low-threshold spike in thalamic neurons. *J Neurophysiol.* 66:839–850.

Wolpert DM, Flanagan JR (2010). Motor learning. *Curr Biol.* 20:R467–R472.

Zhan XJ, Cox CL, Rinzel J, Sherman SM (1999). Current clamp and modeling studies of low threshold calcium spikes in cells of the cat's lateral geniculate nucleus. *J Neurophysiol.* 81:2360–2373.

INDEX

For the benefit of digital users, indexed terms that span two pages (e.g., 52–53) may, on occasion, appear on only one of those pages.
Tables, figures, and boxes are indicated by an italic *t*, *f*, and *b* following the page/paragraph number.